pKₐ Values (Continued)

Acid	Base	pK_a
O_2N—⬡—OH	O_2N—⬡—O^-	7.2
⬡—SH	⬡—S^-	7.8
$CH_3CCH_2CCH_3$ (O O)	$CH_3CCHCCH_3$ (O O)	9.0
HCN	CN^-	9.1
NH_4^+	NH_3	9.4
$(CH_3)_3NH^+$	$(CH_3)_3N$	9.8
⬡—OH	⬡—O^-	10.0
HCO_3^-	CO_3^{2-}	10.2
CH_3NO_2	CH_2NO_2	10.2
CH_3CH_2SH	$CH_3CH_2S^-$	10.5
$CH_3NH_3^+$	CH_3NH_2	10.6
$CH_3CCH_2COCH_2CH_3$ (O O)	$CH_3CCHCOCH_2CH_3$ (O O)	11.0
cyclopentadiene	cyclopentadienyl anion	15.0
CH_3CNH_2 (O)	CH_3CNH (O)	15.0
CH_3OH	CH_3O^-	15.5
H_2O	OH^-	15.7
$CHCl_3$	CCl_3	16
CH_3CH_2OH	$CH_3CH_2O^-$	17
CH_3—C(CH_3)(CH_3)—OH	CH_3—C(CH_3)(CH_3)—O^-	19
CH_3CCH_3 (O)	CH_3CCH_2 (O)	20
$CH_3COCH_2CH_3$ (O)	$CH_2COCH_2CH_3$ (O)	23
$HC≡CH$	$HC≡C^-$	26
triphenylmethane (Ph)₃C–H	triphenylmethyl anion (Ph)₃C^-	31
NH_3	NH_2^-	36
$CH_2=CH_2$	$CH_2=CH^-$	36
CH_4	CH_3^-	47

Organic Chemistry

Seyhan N. Eğe

The University of Michigan

D. C. Heath and Company LEXINGTON, MASSACHUSETTS / TORONTO

ACKNOWLEDGMENTS

Cover photograph : ''Helioptix,'' designed by Henry Ries.

Mass spectra adapted from *Registry of Mass Spectral Data,* Vol. 1, by E. Stenhagen, S. Abrahamsson, and F. W. McLafferty. Copyright © 1974 by John Wiley & Sons, Inc. Reprinted by permission of John Wiley & Sons, Inc.

Carbon-13 NMR spectra adapted from *Carbon-13 NMR Spectra,* by LeRoy F. Johnson and William C. Jankowski. Copyright © 1972 by John Wiley & Sons, Inc. Reprinted by permission of John Wiley & Sons, Inc.

Ultraviolet spectra adapted from *UV Atlas of Organic Compounds,* Vols. I and II. Copyright © 1966 by Butterworth and Verlag Chemie. Reprinted by permission of Butterworth and Company, Ltd.

Infrared spectra from *The Aldrich Library of Infrared Spectra,* 3rd ed., by Charles J. Pouchert. Copyright © 1981 by Aldrich Chemical Company. Reprinted by permission of Aldrich Chemical Company.

NMR spectra from *The Aldrich Library of NMR Spectra,* 2nd ed., by Charles J. Pouchert. Copyright © 1983 by Aldrich Chemical Company. Reprinted by permission of Aldrich Chemical Company.

Published simultaneously in Canada.

Printed in the United States of America.

International Standard Book Number: 0-669-01944-5

Library of Congress Catalog Card Number: 83-80317

To E. Katherine Wright, teacher of chemistry, and Rachel L. Smith, teacher of geometry, who combined intellectual rigor with generosity of spirit to start a young scientist on her career.

Preface

My students are my teachers. Their responses have contributed greatly to the evolution of my current philosophy of teaching. Many years ago one student commented, ''You teach chemistry as process.'' Indeed, I try to. Many other students have said that they have learned an entirely new way of thinking, of analyzing problems, of sorting facts, of reasoning by analogy, of looking for patterns, and that as a consequence, their approach to all their other work has changed. I enjoy teaching organic chemistry precisely because the subject matter lends itself so well to the development of a discriminating eye and an intuitive but analytical mind.

My students have also taught me that the transformation of one's thinking is not easy. Only an exposition consistent in details and relentless in returning to fundamental principles again and again achieves this transformation. In this book, I try to present such an exposition, helping students to think as practicing chemists do in predicting reactivity from structure.

In the first two chapters of the book, the introduction to structures of organic compounds and the concept of functionality makes a connection between structure and reactivity right away. The theme is carried forward in Chapter 4, an examination of acid-base chemistry as an introduction to reactivity of compounds with functional groups containing oxygen, sulfur, and nitrogen. Students become accustomed to seeing and working with different elements and functional groups early and are able to recognize the similarity in their reactions right from the start. Once students master this way of looking at reactivity, the rest of the chemistry presented unfolds as a logical extension of a few fundamental concepts. Chapter 4 also introduces the symbolism used to write mechanisms of reactions. The ideas about acidity and basicity and detailed mechanisms for most of the reactions are used consistently in the rest of the book. When students are encouraged to think through the details of chemical transformations as presented in this book, they develop an independent understanding of the causes of chemical change.

The addition of hydrogen bromide to propene in the absence and in the presence of peroxides is examined in detail in Chapter 5 to introduce the concepts of reactive intermediates and of reaction pathways. Reaction mechanisms and kinetics are also explored here. Reactivity is introduced first, in Chapter 4 and then in Chapter 5, with ionic reactive intermediates. Free radical chemistry is emphasized where the reactions are important, in peroxide catalyzed additions to multiple bonds, in halogenations at allylic and benzylic positions, in reactions of polyenes with oxygen, and in oxidative coupling and polymerization reactions.

The reactions presented in the book have been carefully selected to keep from overwhelming the student with one reaction after another. Emphasis is on understanding fundamental reaction types such as acid-base reactions, oxidation-reduction reactions, and free radical reactions. The repeated and detailed discussion of these reaction types allows students time to learn the material thoroughly. Students achieve a sense of mastery and become confident that they can apply what they have learned to solve new problems. Each step they take in the direction of independent and correct thinking about reactivity reinforces their self-confidence and encourages them to explore more difficult problems. With this approach to organic chemistry, I have watched many students make the journey from trepidation to a joyful exercise of their minds.

All the examples in the text are taken from the chemical literature, many from the very recent literature, and represent real reactions. Very little use is made of general equations. Throughout the book I emphasize the importance of experimental observation and the difference between the observations made in the laboratory and the theories, models, and hypotheses created by scientists. The text emphasizes that the equations we write for mechanisms must conform to the actual conditions used in the laboratory. Students are encouraged throughout to make discriminating use of their eyes and their minds. To this end, for example, large numbers of spectra are provided in the text and in problems instead of distilling the spectral information into sets of numbers. Students learn to extract the necessary information themselves. With such a presentation, students gain a respect for experimental facts as the foundation of the science of organic chemistry.

Stereochemistry is introduced in Chapter 3 and used consistently throughout the book. It is not taught all in one chapter, but is introduced gradually as more and more complex ideas and conventions are needed. Every attempt is made to present stereochemical illustrations consistently to encourage the student to draw such structural formulas correctly. Three-dimensional space-filling drawings help the student appreciate the idea of steric hindrance.

Many students who study organic chemistry are interested in careers in biological sciences. Even those who are not are curious about components of the living world and about biological processes. As a result, chemical properties and syntheses of natural products and biochemical transformations are used as examples and in problems. This material is integrated into the book wherever it can be treated in depth from a chemical point of view. The use of such material teaches students to pick out sites of reactivity in complicated molecules and to recognize that biological processes are supported by organic reactions that are familiar to them.

The nomenclature presented in the book is consistent with the rules published in 1979 by IUPAC. Some of the changes introduced there, such as the new spelling of icosane, the dropping of ketal and hemiketal terminology, and the new suffix for aldehyde substituents on rings, have been adopted in this book. Every attempt has been made to be consistent in nomenclature, but the discriminating reader will find evidence of how difficult that is to achieve.

The book has a large number of problems varying in difficulty from routine exercises for drill to problems that require the student to apply the concepts just learned to new reactions. Regular review of earlier material has been incorporated into the problems at every level. With the exception of a few drill exercises, all the examples used in the problems are taken from the research literature, just as those in the text are.

The book is accompanied by a *Study Guide* containing detailed answers to all the problems, and, in some cases, explanations of the concepts used in arriving at the answers. The guide provides immediate reinforcement when the student has the correct answer and examples of how to approach a solution when the student is having difficulty. Because the problems are designed to reemphasize the concepts introduced in the text and to extend the application of these concepts to new systems, the *Study Guide* provides additional guidance in developing the thinking necessary for successful mastery of organic chemistry.

Many of my colleagues at this and other institutions have helped to shape the book. Odile Eisenstein, Masato Koreeda, Richard G. Lawton, Joseph P. Marino, Charles G. Overberger, George D. J. Phillies, Robert R. Sharp, Peter A. S. Smith, John R. Wiseman, and Peter G. M. Wuts guided me in their areas of expertise. Donald J. Carter kept my pharmacological facts straight.

Other reviewers including Sally Weersing, Muskegon Community College; D. M. S. Wheeler, University of Nebraska, Lincoln; James O. Schreck, University of Northern Colorado; Thomas T. Tidwell, University of Toronto; Jerrold P. Lokensgard, Lawrence University; Edwin M. Kaiser, University of Missouri, Columbia; Edwin R. Harris, California State University, Long Beach; Joseph Casanova, California State University, Los Angeles; and Philip W. LeQuesne, Northeastern University, reviewed the manuscript and made many valuable suggestions.

I owe special thanks to Paul J. Kropp, University of North Carolina at Chapel Hill, who has offered constructive criticism and encouragement at every stage of the development of the book.

Marjorie L. C. Carter and Roberta Kleinman deserve special mention. They have contributed to the book by supporting me in many ways: checking facts in the library, helping with manuscript preparation and indexing, reading and criticizing the manuscript, reading galley and page proofs, and taking off my hands a myriad practical details that go into the preparation of a book. Professor Kleinman helped me to clarify my ideas on ways to present equations and illustrations. Many of the drawings in the book originated with her.

Both Marjorie Carter and Roberta Kleinman have also contributed to the preparation of the *Study Guide* and are coauthors of that supplement to the text. Without their friendship and unstinting help the completion of this project would have been much more difficult.

Various versions of *Organic Chemistry* were made available to students here and at other institutions, and their suggestions and responses helped me develop the book. I am grateful to my own students, first in the special Integrated Premedical-Medical Program, then in the regular organic chemistry course, and finally in the Honors organic chemistry course, for their helpful suggestions and their interest in the progress of the book. My colleagues Roberta Kleinman and Peter Wuts used parts of the book as a supplement in their own organic chemistry courses at the University of Michigan at Ann Arbor and Dearborn, and at Lock Haven University of Pennsylvania. Their impressions of the book and the responses of their students were very valuable. I would also like to thank the students of Richard R. Doyle at Dennison University, Adriane G. Ludwick at Tuskegee Institute, and Lieutenant Thomas J. Haas at the U.S. Coast Guard Academy for helping me to class test the book.

Susan Seidman converted my original scribblings to typed copy in the first draft. Patricia Brennan has patiently and imaginatively followed the mazes of instructions that I have given her for the subsequent revisions of the manuscript. Both of them have worked extraordinary hours with a great sense of responsibility to meet deadlines. I owe them much thanks for making that part of the preparation of a book as easy as possible.

The book would not have been possible without Harvey Pantzis, my editor at D. C. Heath and Company, who saw possibilities when the manuscript was in a germinal state, and who encouraged me to go on writing. The most important person in making the final version of the book a reality has been Marret McCorkle, who has guided it through production with competence, humor, and great patience with the last-minute demands of the author. If I had not been able to trust that she was in complete command of the many details that go into putting a technical text together, the production of the book would have been very difficult indeed for me.

Finally I thank my family and friends for their forbearance with me when I disappeared to write and for their loving support of me when I resurfaced in need of renewal and encouragement.

Seyhan N. Eğe

Contents

3

Alkanes and Cycloalkanes
An Introduction to Nomenclature and Stereochemistry   74

4

Reactions of Organic Compounds
as Acids and Bases   124

5

Electrophilic Additions to Alkenes and Alkynes   158

6

Nucleophilic Substitution
and Elimination Reactions   221

7

Alcohols and Ethers   269

8

Hydrogenation and Oxidation Reactions of Alkenes and Alkynes 317

9

Aldehydes and Ketones 369

10
Carboxylic Acids and Their Derivatives   435

11
The Chemistry of Polyenes   513

12
The Chemistry of Aromatic Compounds   575

13
Nuclear Magnetic Resonance Spectroscopy
and Mass Spectrometry   660

14
Carbohydrates   740

15
The Chemistry of Carbanions;
Enols and Enolate Anions   802

17
Amino Acids, Peptides, Proteins   959

18
The Chemistry of Heterocyclic Compounds   1009

19
Concerted Reactions   1069

20
Macromolecular Chemistry   1120

Index   I1

An Introduction to Structure and Bonding in Organic Compounds

1

1.1

Introduction

Organic chemistry was born in 1828 when Friedrich Wöhler attempted to synthesize ammonium cyanate, NH_4CNO, and obtained urea

instead. Wöhler, who had studied to be a doctor of medicine before he decided to become a chemist, discovered that the compound he had made was identical with urea recovered from urine. Up to that time, scientists had thought that the compounds present in living plants and animals could not be synthesized in the laboratory from inorganic reagents. Wöhler recognized the importance of his experiment and wrote to a friend, "I must tell you that I can make urea without the use of kidneys, either man or dog. Ammonium cyanate is urea."

Wöhler's discovery was important because it gave impetus to a long series of experiments in which chemists probed the nature of the chemical substances that exist in living organisms and in petroleum and coal, which are formed from the remains of certain plants and animals that lived in the distant past. As early chemists struggled to isolate and purify the components of plants, animals, coal, and petroleum, they observed that many of the compounds that they isolated were composed of carbon and hydrogen. Many contained nitrogen, oxygen, sulfur, and phosphorus in addition.

Chemists quickly recognized that the chemistry of carbon was associated with life in a special way that distinguished that element from all others. Compounds of carbon were called **organic compounds** to reflect their origin in living systems and to distinguish them from the **inorganic compounds,** the acids, bases, and salts, derived from the other elements in the periodic table.

Organic chemistry is recognized today as an area of study central to many disciplines. Life processes are supported by the chemical reactions of complex organic compounds such as enzymes, hormones, proteins, carbohydrates, and lipids. The lipid cholesterol; hormones such as cortisone, progesterone, and testosterone; glucose, the most important source of energy in our bodies; and the pancreatic enzyme chymotrypsin, which is essential to our digestion, are among the compounds the structures and chemistry of which we will explore. Cells divide and grow in part according to signals carried and transmitted by giant organic molecules called deoxyribonucleic acids, DNA, and ribonucleic acids, RNA. An understanding of organic chemistry is essential to the study of the biological and medical sciences.

Chemists, in attempts to improve on nature, have created millions of organic compounds that did not exist in nature originally. The local anesthetic Novocain, for example, was developed to mimic the numbing effects of the natural alkaloid cocaine. The search continues for a synthetic painkiller that gives the merciful relief from pain afforded by natural opiates such as morphine without having the undesirable side effect of being addictive. The discovery in recent years that the brain manufactures endorphins, compounds that interact with the same sites in the brain as morphine does, is an exciting new development in our understanding of how the body copes with pain. Industrial chemists, on the other hand, have developed synthetic rubber, Neoprene, and synthetic silks, such as rayon and nylon, to improve on the properties of the natural substances and to meet shortages of natural supplies.

Crude petroleum is converted by organic reactions into fuels that supply energy for heat, transportation, and industry. Petroleum is also the chemical basis for giant molecules engineered to have properties that are useful. The names of these new materials, Teflon, Orlon, Acrilan, polystyrene, polypropylene, polyurethane, have become household words. In the areas of fuels, plastics, and polymers, organic chemistry touches chemical and materials engineering.

Food additives, dyes, artificial flavorings, artificial sweeteners, preservatives, and pesticides, most of them organic compounds, make the headlines in newspapers with regularity. The boxes our crackers and cereals come in carry the abbreviations BHT and BHA to designate organic chemicals that keep the food from becoming rancid. The carcinogenicity of saccharin and whether the use of it should be banned is a major political as well as scientific issue. Among the chemicals known as pheromones, used by insects to communicate with each other, the sex pheromone of the female Mediterranean fruit fly that lures the male to a mating rendezvous was an important tool as California fought an invasion of the pest. Organic chemists work to isolate and prove the structure of such compounds and synthesize analogues for research into the biological control of insects. Organic chemistry is thus relevant to agriculture, nutrition, and public health.

The early organic chemists were confronted with many puzzles. They determined molecular formulas for the compounds that they had recovered from natural sources. As early as 1824 they discovered that several compounds with very different properties might have the same molecular formula. In 1830, the Swedish chemist Jakob Berzelius named such compounds ''isomeric bodies'' from the Greek word *isos,* meaning "equal" and *meros,* meaning "part." This discovery intensified investigations into the nature of the forces that held atoms together and the ways in which chemical bonds could be symbolized in writing. A lively debate over these issues continued through the middle of the nineteenth century. Many different ways of representing the structures of molecules were proposed, disputed, and discarded.

Analyses of organic compounds provided empirical ratios of the weights of carbon, hydrogen, oxygen, and nitrogen. However, exact molecular formulas could not be assigned until a uniform and correct set of atomic weights was adopted by chemists. The Italian chemist, Stanislav Cannizzaro, proposed a system of experimentally accurate atomic weights at the first international congress of chemistry in Karlsruhe, Germany, in 1860. The gradual acceptance of his system made it possible for chemists to assign correct and universally recognized molecular formulas. This advance in turn allowed ideas about bonding and structure to develop.

In the early nineteenth century, chemists also experimented with the transformation of one chemical substance into another. As a result of these experiments, they observed that there were some structural units that were carried unchanged from one reaction to another. The recognition of these intact structural units, which they called **radicals,** enabled chemists to manipulate chemical substances with more precision. The art of organic synthesis was born. Chemists could use new pathways to construct compounds that already existed in nature, or they could create entirely new compounds.

By the end of the nineteenth century, the science of organic chemistry was flourishing. Serious study of the structural and chemical properties of the compounds of carbon had resulted in important discoveries about the chemistry of carbohydrates and proteins. Chemists had tackled the problems of representing the structures of organic compounds. They had wondered how molecules looked in three dimensions and had arrived at pictures remarkably similar to the ones we still use. Many organic compounds that did not exist in nature had been synthesized.

The next significant step forward in organic chemistry took place in the 1920s and 1930s. During these years, chemists explored the exact details of chemical transformations. They began to ask questions about the relative timing of the breaking and forming of bonds during chemical reactions. Chemists were also concerned with how the shapes of molecules, that is, their spatial properties, affected chemical reactivity. The question of three-dimensionality in reactions turned out to be a central one in biochemistry, the study of chemical transformations in living organisms.

While chemists were asking more and more subtle questions about chemical reactions, new instruments that extended their powers of observation were invented. These new instruments and techniques made it possible to probe more deeply into the processes of nature. The number of experimental observations that could be collected multiplied rapidly.

The progress of chemistry as a science depends on the experimental manipulations of substances in the laboratory leading to the observation of new phenomena. In thinking about these phenomena chemists arrive at ideas about the nature of the chemical substances under investigation. These ideas are tested by new experiments and new observations. The range of manipulations and observations available to the chemist has expanded enormously in recent years. The power that chemists have to transform chemical compounds into new ones, sometimes useful and sometimes harmful, has also increased.

Chemists have refined the way they visualize and think about the submicroscopic units called **atoms** and **molecules.** They have created models that help them to picture and understand experimental facts about chemical substances. Human beings are constantly creating and refining the models they use in dealing with the physical world. Some models are pictorial; others are more abstract and mathematical. Some models are widely adoped because they are useful; others are quickly modified on the basis of new observations.

The descriptions in this book of species such as atoms, ions, and molecules, and of phenomena, such as chemical bonding, present the models that organic chemists currently find most useful. The distinction between the facts, as determined experimentally in the laboratory, and the theories and models created by the human mind in thinking about them is important.

1.2

How To Study Organic Chemistry

Learning organic chemistry is like learning a new language, a language that is verbal and pictorial at the same time. Organic chemistry is a highly organized discipline, based on the premise that the structure of a compound determines its reactivity. Organic chemistry is, therefore, a study of the relationship between the structures of molecules and their reactions. As you develop an understanding of this relationship, you will be able to make predictions about molecules and reactions that are new to you. Your eyes will learn to dissect complex structures and to distinguish the pieces you recognize. You will be able to reason by analogy from systems and reactions that you have already learned to new systems and reactions that resemble the earlier ones in important ways.

Your pencil is an indispensable tool in your studies. While you are training your eye to look carefully, you should be training your hand to draw. Professional organic chemists cannot talk to each other without drawing structures. *You too,* if you expect to learn organic chemistry well, *must draw and redraw the structures of compounds and write out equations as you are studying.* The ways you will see structures illustrated in the book and drawn by your instructor on the blackboard represent the result of years of evolution in thinking about organic compounds. Different kinds of pictures represent different degrees of precision. You must train your hand to produce the correct pictures; precision in drawing leads to precision in thinking. The correct representation of a structure will often give you insight into the correct solution to a problem.

In organic chemistry, we are concerned about the shapes of molecules. You should acquire **molecular models*** and examine the structures of carbon compounds in three dimensions. Then you will learn to translate three-dimensional structures to the two dimensions of a page. Chemists have developed specific ways of representing three dimensionality on a flat surface in order to convey the maximum amount of information clearly and consistently.

You cannot have a conversation with another person if the two of you do not know the vocabulary and rules of grammar of a common language. Likewise, you will not be able to communicate with other chemists, and especially with your instructor, if you

---

*Framework Molecular Models available from Prentice-Hall, Inc., Englewood Cliffs, New Jersey 07632, and Fieser Molecular Models, available from Aldrich Chemical Co., Inc., Milwaukee, Wisconsin 53233, are recommended.

do not follow the correct rules for representing three-dimensional structures in two dimensions. If the rules are not followed exactly, the resulting pictures are meaningless.

Gradually, you will learn the names of a large number of organic compounds so you can communicate your thoughts about them. All compounds have formal names that can be assigned by the application of definite rules for naming agreed upon by chemists. Some compounds also have trivial names by which they have been known historically. These names still appear on labels of reagent bottles and in articles in scientific journals. To be literate in organic chemistry, you must be able to recognize both formal and trivial names.

The successful mastery of organic chemistry requires a lot of hard work and consistent studying. It is not a subject that can be crammed. Many students make the mistake of trying to memorize the text. An understanding of the basis of chemical transformations is more important than rote memorization. It is true that facts must be learned, but you will be overwhelmed by them unless you develop an ability to see relationships.

To study effectively, you should read the assignments before attending lectures. Spend most of your study time working with the problems. The reactions and ideas that are important to know will come up again and again in different problems throughout the book. If you find that you cannot do a problem, go back to the text and study the reactions and ideas that apply to it again, then try the problem once more. This method of studying will ensure that you spend the most time working with the ideas that you find difficult. As you go back and forth between the problems and the text, you will gradually learn the most important facts without making a specific effort to memorize them. You will also begin to learn a way of thinking, of looking for patterns, and of recognizing qualitative similarities between seemingly unrelated facts. The development of this skill will be one of the most important results of your study of organic chemistry because an ability to think this way can be applied to problems in every area of life.

There are two kinds of problems in the text. Some are drill problems designed to reinforce the concepts and reactions presented in the text. Others introduce new concepts that are logical extensions of the examples already studied. The information supplied in these problems is an important part of the book. As you practice with them, you will develop an ability to recognize patterns in the way molecules react. Consequently, you will learn to make predictions about related molecules and reactions. Work out all of the problems, no matter how simple they seem, in writing and in full detail. There is no other way to develop the skills you need to progress to the more and more complex structures and concepts you will encounter as you go through the book.

Although the study of organic chemistry requires a lot of concentrated work, many students find that learning organic chemistry is also fun. The thinking that goes into the study of organic chemistry is related to the thinking used in solving puzzles. For example, solving problems in organic chemistry requires you to recognize patterns and to fill in the missing pieces much the way you do when you put together a jigsaw puzzle. You also learn to be precise in thinking about qualitative concepts, just as you are already used to being precise in quantitative ways in mathematics. You will experience the power of your mind to analyze an unfamiliar problem and to arrive at a correct picture of the disparate facts that must be brought together for a solution. You will come to trust in your ability to think correctly to predict experimental outcomes. Such self-knowledge is exhilarating.

Ionic and Covalent Compounds

Compounds are divided broadly into two classes, ionic and covalent. **Ionic compounds** *are composed of ions, which are units of matter that may be single atoms or groups of atoms, bearing positive and negative charges. In* **covalent compounds,** *the structural units are molecules having no net charge.* Ionic compounds are usually crystalline solids with high melting points. These compounds often dissolve in water to form solutions that conduct electricity. Common table salt, NaCl, is a typical example of an ionic compound in which the ions are single atoms, Na^+ and Cl^-. Magnesium sulfate, $MgSO_4$, commonly known as Epsom salts, and sodium bicarbonate, $NaHCO_3$, which is baking soda, are other familiar ionic compounds. In these compounds, the negatively charged ions are composed of atoms held together by covalent bonds (the sulfate anion, SO_4^{2-}, and the bicarbonate anion, HCO_3^-). In other ionic compounds, such as ammonium chloride, NH_4Cl, the positively charged ion (the ammonium ion, NH_4^+) contains covalent bonds.

Covalent compounds may be gases, liquids, or solids. Methane, CH_4, is the principal component of natural gas. Carbon tetrachloride, CCl_4, is a typical covalent liquid that was once commonly used in dry cleaning. Naphthalene, $C_{10}H_8$, the constituent of old-fashioned mothballs, is a covalent solid. Methane, carbon tetrachloride, and naphthalene do not dissolve in water to any great extent. Such compounds are **nonpolar covalent compounds.**

Other covalent compounds, such as ethanol, CH_3CH_2OH, and glucose, $C_6H_{12}O_6$, are quite soluble in water and form solutions that do not conduct electricity. Such compounds ionize only slightly in aqueous solutions. Water, ethanol, and glucose are examples of **polar covalent compounds.** In the following sections of this chapter we shall look more closely at ideas about the nature of chemical bonding, and at the various factors that influence the different physical properties of compounds, such as boiling point, melting point, and solubility.

Ionic Bonding

Crystalline sodium chloride consists of an arrangement of positively charged sodium ions and negatively charged chloride ions arranged alternately in a three-dimensional array called the **crystal lattice.** Each sodium ion is surrounded by six chloride ions, and each chloride ion by six sodium ions. *The ions are held in place by strong electrostatic forces between the positively and negatively charged ions.* **Ionic bonding** *consists of electrostatic attractions between ions of opposite charge.* The individual ion is a sphere bearing symmetrical distribution of charge. For this reason, *there is no particular direction to bonding in ionic compounds.* In the solid state, there are no individual molecules of sodium chloride, composed of one Na^+ and one Cl^-.

Sodium chloride has a high melting point, 801 °C, and a very high boiling point, 1413 °C. These physical properties are an indication of the strength of the electrostatic forces holding the ions together. We must supply large amounts of energy to the sodium chloride crystal if we wish to overcome electrostatic forces and break down the crystal lattice by the process of melting. Even more energy is necessary to vaporize sodium chloride.

1.5

Covalent Bonding

A. Lewis Structures

Early in this century, Gilbert N. Lewis at the University of California at Berkeley proposed that the **covalent bond** be represented as *the sharing of a pair of electrons between two atoms.* He also proposed that with a few exceptions, *stable molecules or ions have eight electrons, or four pairs of electrons, in the outermost shell, the valence shell of each atom. This stable configuration of electrons is called an* **octet.** His suggestions for drawing structures of covalent compounds have proved enormously useful to organic chemists.

The **Lewis structure** of a covalent molecule shows all the electrons in the outer valence shell of each atom; the bonds between atoms are shown as shared pairs of electrons. The total number of electrons in the valence shell of each atom can be determined from its group number in the periodic table. *The shared electrons are called the* **bonding electrons** *and may also be represented by a line between the two atoms. The valence electrons that are not being shared are the* **nonbonding electrons;** they are shown by dots drawn in a square around the symbol of the atom. Lewis structures for a covalent compound, hydrogen chloride, are shown below.

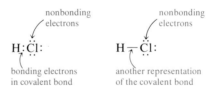

The hydrogen atom in hydrogen chloride shares two electrons, which is all its valence shell can hold. The chlorine atom has eight electrons around it in a stable octet.

Table 1.1 includes some more examples of Lewis structures of compounds and ions, first written in the abbreviated form known as the condensed formula, then written as the full Lewis structure showing all the valence electrons, and finally as the Lewis structure with the covalent bonds represented by lines and only the nonbonding electrons shown as dots.

In all the structures illustrated in Table 1.1, the valence electrons are arranged so that the atoms of all elements except hydrogen share eight electrons. The restriction that a hydrogen atom usually cannot share more than two electrons means that we cannot draw structures in which a hydrogen atom has more than one covalent bond. In constructing the water molecule, for example, a glance at the periodic table reminds us that the oxygen atom has six valence electrons, and the hydrogen atoms one each. A total of eight dots representing the electrons can be placed around the oxygen atom. The hydrogen atoms can each share two of these electrons with the oxygen atom, which then has two pairs of unshared, or nonbonding, electrons.

In drawing Lewis structures, we must keep carefully in mind the number of electrons available to form bonds and the location of the electrons. The hydronium ion, for

TABLE 1.1   Condensed Formulas and Lewis Structures of Some Ions and Compounds

Name	Condensed Formula	Lewis Structures	
water	H_2O	H:Ö:H	H—Ö—H
hydronium ion	H_3O^+	H:Ö:H (+) over O, H below	H—Ö—H (+) over O, H below
ammonia	NH_3	H:N̈:H, H below	H—N̈—H, H below
methane	CH_4	H:C̈:H, H above and below	H—C—H, H above and below
methanol	CH_3OH	H:C̈:Ö:H, H above and below C	H—C—Ö—H, H above and below C
methoxide anion	CH_3O^-	H:C̈:Ö:⁻, H above and below C	H—C—Ö:⁻, H above and below C
hydroxylamine	H_2NOH	H:N̈:Ö:H, H below N	H—N̈—Ö—H, H below N

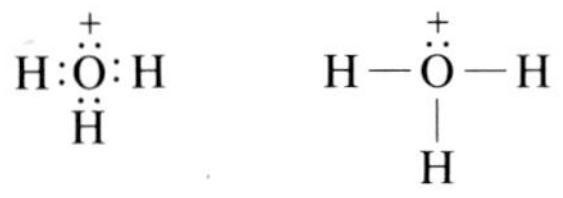

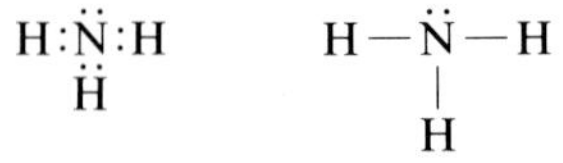

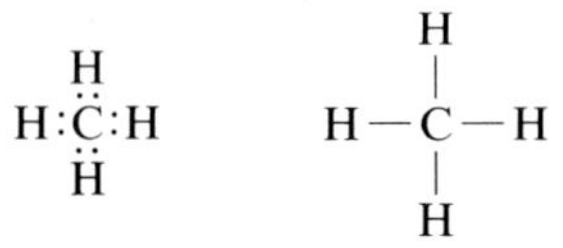

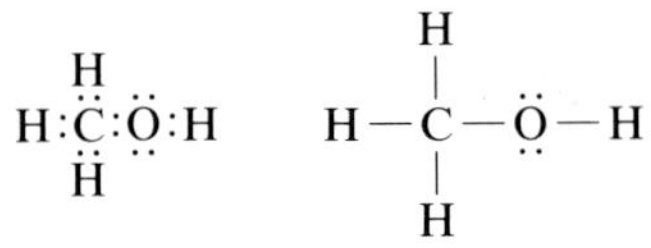

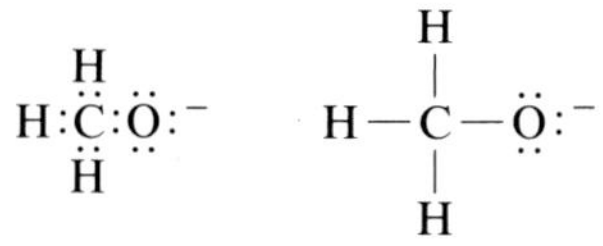

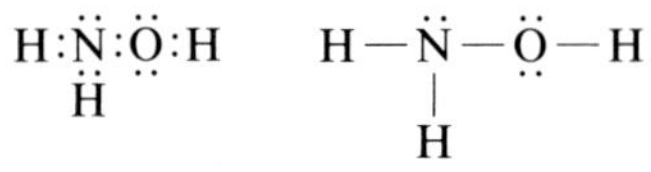

example, is formed when a water molecule accepts a proton (a hydrogen atom without its electron) from an acid.

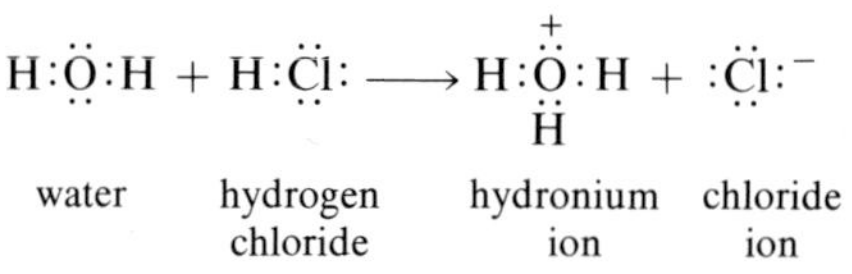

$$\text{H:}\overset{..}{\underset{..}{O}}\text{:H} + \text{H:}\overset{..}{\underset{..}{Cl}}\text{:} \longrightarrow \overset{+}{\underset{\overset{|}{H}}{\text{H:}\overset{..}{\underset{..}{O}}\text{:H}}} + \text{:}\overset{..}{\underset{..}{Cl}}\text{:}^-$$

water   hydrogen   hydronium   chloride
          chloride   ion   ion

Eight electrons are available to bond together one oxygen atom and three hydrogen atoms. The oxygen atom in the hydronium ion has two nonbonding electrons. It also has a half-share of the six electrons in the covalent bonds to hydrogen atoms, or three more electrons. Thus, there are effectively five electrons assigned to the oxygen atom, whereas oxygen in its neutral state has six electrons. The oxygen formally has one fewer electron than necessary for neutrality and, therefore, has a $+1$ **formal charge.**

The formal charge on an atom may be calculated using the formula

Formal charge = (number of valence electrons)

$-$ (number of nonbonding electrons)

$- \frac{1}{2}$ (number of bonded electrons)

In the case of the hydronium ion, the formal charge also represents the charge on the ion because no other atoms in the ion are charged. *The sum of the formal charges on the atoms in a neutral molecule is zero. For an ion, the sum of the formal charges on different atoms should add up to the charge on the ion.*

PROBLEM 1.1   Practice drawing Lewis structures for species represented by the following condensed formulas. When necessary, identify the atoms bearing formal charges.

(a) CCl_4 (b) CH_3Br (c) $CH_3OH_2{}^+$ (d) $NH_2{}^-$ (e) $CH_3NH_3{}^+$

(f) H_2NNH_2 (g) PH_3 (h) H_2S (i) CH_3CH_2OH

(j) $HOCH_2CH_2OH$ (k) $CH_3-O-CH_3{}^+$
                                    |
                                   CH_3

PROBLEM 1.2   Look at the following formulas and decide whether the central atom in each case is neutral, positively charged, or negatively charged. All nonbonding electrons are shown.

(a) $CH_3-\overset{\overset{\displaystyle CH_3}{|}}{\underset{\underset{\displaystyle CH_3}{|}}{N}}-CH_3$ (b) $:\!\ddot{B}r-\overset{\displaystyle |}{\underset{\displaystyle |}{\ddot{C}}}-\ddot{B}r\!:$ (c) $CH_3-\overset{\overset{\displaystyle H}{|}}{\underset{\underset{\displaystyle CH_3}{|}}{\ddot{O}}}-CH_3$
                                                                         $:\!\ddot{C}l\!:$

(d) $CH_3-\ddot{\ddot{N}}-H$ (e) $:\!\ddot{C}l-\overset{\displaystyle |}{\underset{\displaystyle |}{C}}-\ddot{C}l\!:$ (f) $CH_3-\overset{\overset{\displaystyle CH_3}{|}}{\underset{\underset{\displaystyle CH_3}{|}}{C}}-CH_3$

Sometimes there are not enough electrons in the system to provide a stable octet around the central atom. Boron trifluoride, BF_3, is such a molecule. Boron is in Group III of the periodic table and has only three electrons in its valence shell. These added to the twenty-one electrons contributed by three fluorine atoms make twenty-four electrons available for bonding. If we put octets around the fluorine atoms, the boron atom ends up with only six electrons.

$$:\!\ddot{F}\!:$$
$$:\!\ddot{F}\!:\!\ddot{B}\!\leftarrow \text{ six electrons around the boron atom in boron trifluoride; an \emph{open shell}}$$
$$:\!\ddot{F}\!:$$

$$:\!\ddot{F}\!:$$
$$:\!\ddot{F}-B$$
$$:\!\ddot{F}\!:$$

This leaves it with an *open shell,* meaning that the boron atom can accept another pair of electrons to complete an octet. As a result, boron trifluoride reacts with compounds such as ammonia that have nonbonding electrons.

$$:\!\ddot{F}\!:\quad H \qquad\qquad :\!\ddot{F}\!:\ H$$
$$:\!\ddot{F}-B\ +\ :\!N-H \rightleftharpoons\ :\!\ddot{F}-\overset{-}{B}-\overset{+}{N}-H$$
$$:\!\ddot{F}\!:\quad H \qquad\qquad :\!\ddot{F}\!:\ H$$

boron   ammonia   compound of boron
trifluoride   trifluoride and ammonia

PROBLEM 1.3   Formal charges are shown in the formula for the reaction product of boron trifluoride and ammonia in the equation above. Make sure that you understand how they were obtained.

PROBLEM 1.4   Look through the species shown in Table 1.1 and pick out the ones you would expect to react with BF_3. Write equations for the reactions. Be sure to include any formal charges that result.

Carbon, the central element in organic compounds, usually has a valence of four, meaning that it forms four covalent bonds to other atoms. In some compounds the carbon atoms are held together by double or triple covalent bonds. *Double or triple bonds are also called* **multiple bonds.** Lewis structures can be written to show the sharing of two or three pairs of electrons by two carbon atoms. For example, the carbon atoms in ethylene, C_2H_4, share two pairs of electrons with each other in a double bond. Acetylene, C_2H_2, is represented as having a triple bond formed by the sharing of three pairs of electrons so that each carbon atom has eight electrons around it and four bonds.

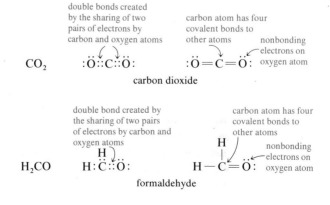

Many important compounds such as carbon dioxide, CO_2, and formaldehyde, CH_2O, have double bonds between carbon and oxygen. Triple bonds between carbon and nitrogen are found in cyanides, represented below by hydrogen cyanide.

double bonds created
by the sharing of two
pairs of electrons by
carbon and oxygen atoms

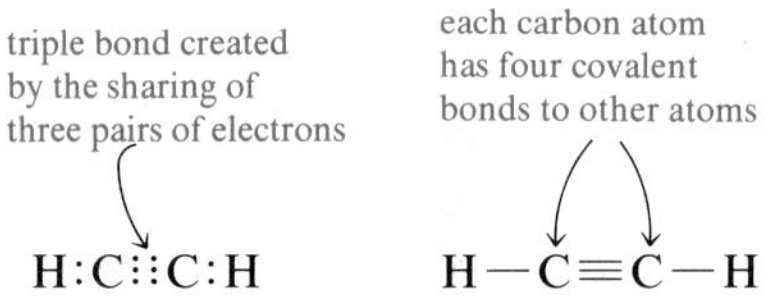

carbon atom has four
covalent bonds to
other atoms

nonbonding
electrons on
oxygen atom

CO_2 :Ö::C::Ö: :Ö=C=Ö:

carbon dioxide

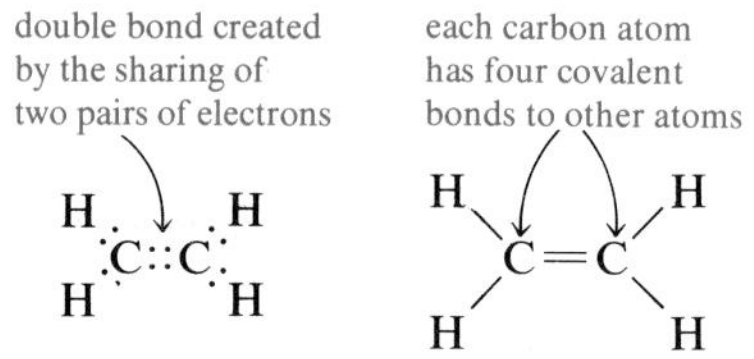

double bond created by
the sharing of two pairs
of electrons by carbon and
oxygen atoms

carbon atom has four
covalent bonds to
other atoms

nonbonding
electrons on
oxygen atom

H_2CO H:C::Ö: H—C=Ö:

formaldehyde

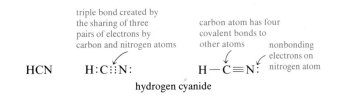

HCN   H:C::N: H—C≡N:

hydrogen cyanide

PROBLEM 1.5   Draw Lewis structures for the compounds represented by the following formulas:

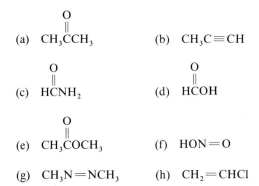

(a) $CH_3\overset{\displaystyle O}{\overset{\|}{C}}CH_3$

(b) $CH_3C\equiv CH$

(c) $H\overset{\displaystyle O}{\overset{\|}{C}}NH_2$

(d) $H\overset{\displaystyle O}{\overset{\|}{C}}OH$

(e) $CH_3\overset{\displaystyle O}{\overset{\|}{C}}OCH_3$

(f) $HON=O$

(g) $CH_3N=NCH_3$

(h) $CH_2=CHCl$

B. Isomerism. The Concept of the Functional Group

As we saw in the last section, carbon forms single, double, and triple bonds with other carbon atoms. Carbon also bonds to other elements, such as oxygen and nitrogen, with single or multiple bonds. Because of the different ways carbon can bond to carbon and to other elements, *a given molecular formula may correspond to more than one possible structure. This phenomenon is known as* **isomerism.** For example, two structural formulas may be written for the molecular formula C_4H_{10}.

$$\begin{array}{c} \underset{\displaystyle \overset{|}{H}}{\overset{\displaystyle \overset{|}{H}}{}} \end{array}$$

        H H H H   H
        | | | | |
     H—C—C—C—C—H   H—C—H
        | | | | H | H
        H H H H | | |
                               H—C——C——C—H
                                     | | |
             butane   H   H   H

           bp −0.6 °C   2-methylpropane
                                     bp − 10.2 °C

Butane and 2-methylpropane are **structural isomers** of each other, compounds that have the same molecular formula but different structural formulas. Because their structures differ, they have different physical properties, represented above by their different boiling points. Some of the ways that physical properties are influenced by structure are discussed in Section 1.8.

The molecular formula C_3H_6O gives rise to six structural isomers.

$$
\begin{array}{ccc}
\underset{\substack{|\\H}}{\overset{\substack{H\\|}}{H-C}}-\overset{:O:}{\overset{\|}{C}}-\underset{\substack{|\\H}}{\overset{\substack{H\\|}}{C}}-H
&
\underset{\substack{|\\H}}{\overset{\substack{H\\|}}{H-C}}-\underset{\substack{|\\H}}{\overset{\substack{H\\|}}{C}}-\overset{:O:}{\overset{\|}{C}}-H
&
\overset{H}{\underset{}{H-C}}=\underset{}{\overset{H}{C}}-\underset{\substack{|\\H}}{\overset{\substack{H\\|}}{C}}-\ddot{\text{O}}-H
\end{array}
$$

$$
\begin{array}{ccc}
\overset{O}{\overset{\|}{CH_3CCH_3}} & \overset{O}{\overset{\|}{CH_3CH_2CH}} & CH_2\!=\!CHCH_2OH \\
\text{acetone} & \text{propanal} & \text{2-propen-1-ol} \\
\text{bp 56.5 °C} & \text{bp 48.8 °C} & \text{allyl alcohol} \\
& & \text{bp 97 °C}
\end{array}
$$

$$
\begin{array}{ccc}
\overset{H}{\underset{}{H-C}}=\underset{\substack{|\\H}}{\overset{\substack{H\\|}}{C}}-\ddot{\text{O}}-\underset{\substack{|\\H}}{\overset{\substack{H\\|}}{C}}-H
&
\underset{\substack{|\\H}}{\overset{\substack{\cdot\ddot{O}\cdot\\}}{H-C}}-\underset{\substack{|\\H}}{\overset{\substack{\\|}}{C}}-\underset{\substack{|\\H}}{\overset{\substack{H\\|}}{C}}-H
&
\text{cyclopropanol structure}
\end{array}
$$

$$
\begin{array}{ccc}
CH_2\!=\!CHOCH_3 & \overset{O}{\overset{/\backslash}{CH_2-CHCH_3}} & \overset{CH_2}{\overset{/\backslash}{CH_2-CHOH}} \\
\text{methyl vinyl ether} & \text{2-methyloxirane} & \text{cyclopropanol} \\
\text{bp 8 °C} & \text{bp 35 °C} & \text{bp 100 °C}
\end{array}
$$

For the examples given above, the differences in structure and, as we will see later, in chemical and physical properties are large.

In organic chemistry, compounds are classified according to characteristic structural features that give rise to typical kinds of chemical reactivity. For example, *acetone and propanal both have a carbon atom doubly bonded to an oxygen atom. This structural unit is called a* **carbonyl group.** The carbonyl group is said to be the *functional group* in compounds such as acetone and propanal because it serves as a reactive site.

$$
\overset{:O:}{\overset{\|}{-C-}}
$$

the carbonyl group

*the functional group in
ketones and aldehydes*

$$
\begin{array}{cc}
\text{carbonyl group} & \text{carbonyl} \\
\overset{:O:}{\overset{\|}{CH_3-C-CH_3}} & \overset{:O:}{\overset{\|}{CH_3CH_2C-H}} \\
\text{acetone} & \text{propanal} \\
\textit{a ketone} & \textit{an aldehyde}
\end{array}
$$

Acetone is an example of a class of compounds known as **ketones,** in which the carbonyl group is bonded to two other carbon atoms. Propanal is an **aldehyde,** in which the carbonyl group is bonded to a carbon atom and a hydrogen atom.

A structural unit consisting of an atom or group of atoms that serves as a site of chemical reactivity in a molecule is called a **functional group.** Classification of organic compounds according to their functional groups allows us to recognize the compounds that are likely to have similar physical and chemical properties because they have similar structures. An important part of learning organic chemistry is to train yourself to understand the reactivity of functional groups in terms of their structure.

Allyl alcohol and methyl vinyl ether, both structural isomers with the molecular formula C_3H_6O, have one functional group in common. Both of them have a *carbon-carbon double bond,* and are, therefore, **alkenes.** In addition, allyl alcohol has a **hydroxyl group,** *an oxygen atom with a hydrogen atom bonded to it. The hydroxyl group is the functional group characteristic of* **alcohols.** In methyl vinyl ether, *the oxygen atom is bonded to two carbon atoms, a structural feature of the functional group class of* **ethers.**

$\diagdown C = C \diagup$	$-O-H$	$\diagdown C-O-C \diagup$
carbon-carbon double bond	hydroxyl group	ether group
functional group of alkenes	*functional group of alcohols*	*functional group of ethers*

alkene   alcohol   alkene   ether

$CH_2 = CHCH_2OH$ $CH_2 = CH - O - CH_3$

allyl alcohol   methyl vinyl ether

belongs to the functional group classes of alkenes and alcohols *belongs to the functional group classes of alkenes and ethers*

The structural features of different functional groups and some of the chemistry that results from their structures are explored in Chapter 2.

Butane and 2-methylpropane are examples of structural isomers that belong to the same functional group class, **alkanes** (Section 2.4 A). Because they are members of the same functional group class, they are more similar to each other than are the isomers of C_3H_6O. The structural isomers of C_3H_6O exhibit much greater structural variation and may be called **functional group isomers.** In Chapter 3, we shall learn about isomers that differ from each other in very subtle ways that can only be distinguished if we explore their structures in three dimensions.

PROBLEM 1.6   To which functional group class does 2-methyloxirane (p. 12) belong? How about cyclopropanol?

PROBLEM 1.7   Draw structural formulas for the isomers of C_3H_7Cl and C_3H_8O.

PROBLEM 1.8   Explore the different structures possible for a compound with the molecular formula C_5H_{12}.

C. Resonance

Some covalent molecules and ions cannot be represented satisfactorily by a single Lewis structure. The carbonate anion, CO_3^{2-}, is an example of such a species. The carbonate ion in calcite, $CaCO_3$, has been found by x-ray diffraction to be planar, with bond angles of 120° and three equivalent carbon-oxygen bonds each 1.29 Å (1.29 × 10^{-8} m) long.

O   2−

120°

O   O

1.29 Å

structure determined experimentally for
the carbonate anion, $CO_3{}^{2-}$

Yet, if there is to be an octet around each atom, the Lewis structure of $CO_3{}^{2-}$ must be drawn with a double bond to one of the oxygen atoms and single bonds to the others.

Lewis electron dot structures for
carbonate anion, $CO_3{}^{2-}$

Two of the oxygen atoms have formal charges of -1, giving a total charge of -2 for the anion. This picture of the carbonate anion suggests that one of the oxygen atoms should be different from the other two. The experimental evidence, however, indicates that all three oxygen atoms are equivalent. The Lewis structure does not adequately depict the experimental reality.

For many molecules and ions you cannot write a single Lewis structure that satisfactorily accounts for the chemical and physical properties of the species. The true state of the species is better represented by a combination of two or more Lewis structures. For example, for the carbonate anion the experimental observations would be better represented by a picture in which the electrons are equally distributed to all three oxygen atoms. Several structures may be drawn for the carbonate anion, differing only in the location of pairs of electrons. The individual structures are **resonance contributors** to the structure of the carbonate anion; the carbonate anion is pictured as a **resonance hybrid** of these contributors. These structures differ from each other only in the arrangement of electrons, and not in the position of the atoms.

resonance contributors to the
structure of the carbonate anion

The double-headed arrow used between the structures is a symbol of this relationship. It *does not* mean that these three forms are in equilibrium with each other. The symbol for equilibrium is two arrows, pointing in opposite directions, showing a reversible chemical reaction as, for example:

$$H_2CO_3 + H_2O \leftrightharpoons H_3O^+ + HCO_3{}^-$$

No reaction is implied by the double-headed arrow, $\leftrightarrow$. It signals that the real properties of the carbonate anion cannot be represented by any one of the three Lewis structures taken alone. Certain facts about the carbonate anion are reflected in all of the structures taken together. For example, each one of the oxygen atoms in the carbonate anion has negative charge and an equal probability of reacting with an acid to pick up a proton. Also, the distances between the carbon atom and each oxygen atom and the angles formed by the carbon atom and any two oxygen atoms are equal.

Resonance contributors are also significant for many other ions. The nitrate and acetate anions are examples of species for which a single Lewis structure is not satisfactory.

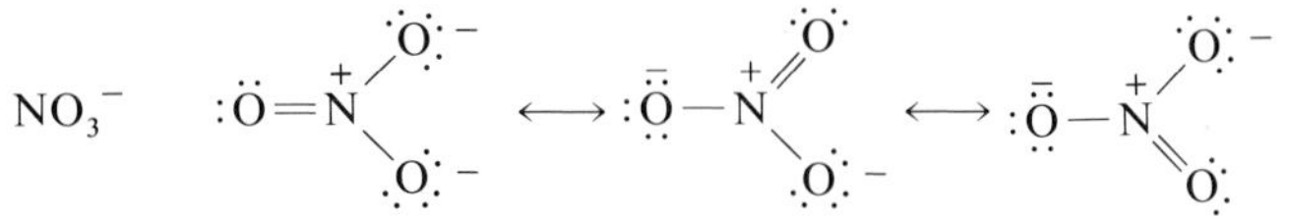

nitrate anion

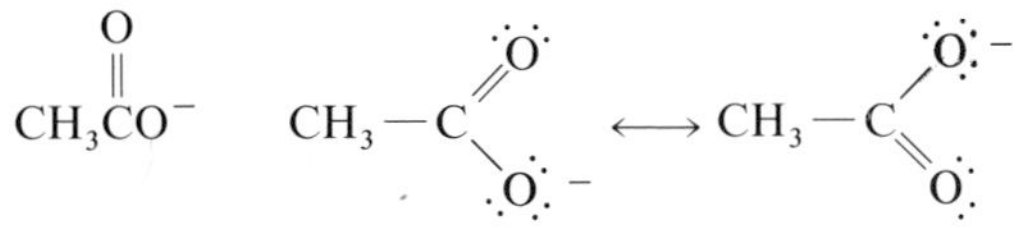

acetate anion

resonance contributors to the structures
of the nitrate and acetate anions

For each anion, the structures differ from each other only in the location of pairs of electrons. Note that in the nitrate anion two oxygen atoms bear -1 formal charges and the nitrogen atom has a $+1$ formal charge. The sum of these formal charges is the charge on the anion. Experimentally, the three oxygen atoms in the nitrate anion are equivalent to each other, as are the two oxgyen atoms of the acetate anion.

Structures for neutral molecules may also have resonance contributors. Nitromethane is an example of a molecule for which more than one Lewis structure is significant.

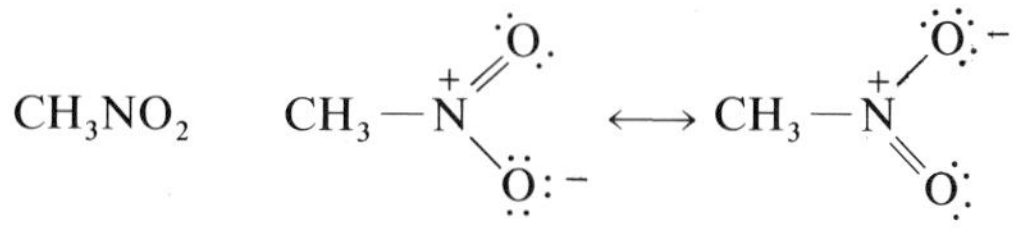

resonance contributors for nitromethane

In nitromethane, the $+1$ formal charge on the nitrogen atom and the -1 formal charge on the oxygen atom cancel each other so the molecule as a whole is not charged.

Resonance is an example of a model (Section 1.1) that was developed to deal with experimental observations that could not be explained in terms of a simpler model, such as a single Lewis structure for a molecular species. *It is important to remember that the individual representations of resonance contributors have no reality. The molecule,* such as nitromethane, *for which resonance contributors are written does not exist as a mixture of different forms. The actual species has properties suggested by all the resonance contributors taken together.* For example, in nitromethane, the nitrogen atom bears a positive charge. Each oxygen atom bears part of a negative charge. Both nitrogen-oxygen bonds are the same length.

Some representations of resonance contributors are more useful than others in predicting the observed physical and chemical properties of a compound. We classify some resonance contributors as "important," meaning that these representations are useful in making predictions about the system. Others we call "unimportant," meaning that the inclusion of such representations does not significantly improve the way our picture reflects experimental evidence.

The rules for writing resonance contributors are the following:

1. Only nonbonding electrons and electrons in multiple bonds change locations in different resonance contributors. The electrons are said to be delocalized to different atoms. The electrons in single covalent bonds are not involved.

2. The nuclei of atoms in different resonance contributors are in the same positions.
3. All resonance contributors must have the same number of paired or unpaired electrons.
4. Resonance contributors in which atoms of second period elements all have eight electrons around them are more likely to represent important contributors than those in which such atoms have fewer than eight electrons. Similarly, resonance contributors with a greater number of covalent bonds are more important than those with a smaller number. For atoms of elements beyond the second period, such as sulfur and phosphorus, it is possible to write structures with ten or more electrons around a central atom.
5. In writing resonance contributors, we sometimes create structures in which there is a separation of charge. One atom becomes positively charged and another negatively charged. In general, resonance contributors in which there is a large separation of charge are less important than those in which there is little or no separation of charge. When structures with a separation of charge are written, the more important resonance contributor has the negative charge on the more electronegative atom. (A review of the concept of electronegativity is found in Section 1.7A.)

The way these rules can be applied to the formate anion is shown below:

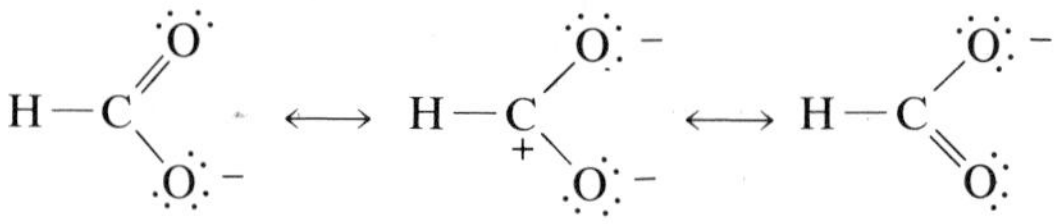

"important"   *"unimportant"*   *"important"*
resonance contributor *resonance contributor* *resonance contributor*
                         to the formate anion

Note that on paper one resonance contributor is converted into another one by a change in the location of a pair of electrons. The first transformation requires that one of the two pairs of electrons in the carbon-oxygen double bond (Rule 1) move to the more electronegative atom in the bond, the oxygen atom (Rule 5). The second resonance contributor is labeled "unimportant" for three reasons. It contains fewer bonds (Rule 4); the carbon atom has only six electrons around it (Rule 4) and there is a separation of charge between the carbon atom and the oxygen atom (Rule 5). This resonance contributor is converted into the third form by a shift of a pair of nonbonding electrons (Rule 1) from an oxygen atom towards the carbon atom. The first and third structures shown above are equivalent and could not be distinguished from each other if we had not kept track on paper of which oxygen atom is which. They are the "important" resonance contributors. In them, each atom (other than hydrogen) has eight electrons around it (Rule 4) and there is no separation of charge (Rule 5). Note that in going from one resonance contributor to another, only electrons in the double bond and nonbonding electrons on the oxygen atoms are moved. No electrons in single bonds are affected (Rule 1). In the structural formula of formate ion in which all of the electrons are paired, the electrons are also moved as pairs (Rule 3). The same atoms remain bonded to each other in all resonance contributors; the nuclei of the atoms do not move (Rule 2).

Sometimes compounds exhibit reactivity that is best explained by writing an "unimportant" resonance contributor. An example of such a compound is carbon dioxide, CO_2, which reacts readily with a reagent such as hydroxide ion, OH^-. The ease with which this reaction occurs can be rationalized by writing a resonance contributor in which the carbon atom has only six electrons around it and bears a positive charge.

$$:\ddot{O}\!=\!C\!=\!\ddot{O}: \longleftrightarrow :\ddot{O}\!=\!\overset{+}{C}\!-\!\ddot{O}:^{-} \longleftrightarrow {}^{-}:\ddot{O}\!-\!\overset{+}{C}\!=\!\ddot{O}:$$

resonance contributors for carbon dioxide, with a separation of charge and six electrons around carbon

$$:\ddot{O}\!=\!\overset{+}{C}\!-\!\ddot{O}:^{-} \longrightarrow :\ddot{O}\!=\!C\!-\!\ddot{O}:^{-}$$
$$^{-}:\ddot{O}\!-\!H \qquad\qquad :O\!-\!H$$

covalent bond formed by the sharing of a pair of electrons from the hydroxide ion with the positively charged carbon atom

bicarbonate anion

The concept of resonance was developed by Linus Pauling of the California Institute of Technology. In 1954, Pauling was awarded the Nobel Prize for his research into the nature of the chemical bond and his application of this knowledge to the determination of the structures of complex substances. Resonance is best understood in the context of different examples. The rationalization of the reactivity of carbon dioxide with the hydroxide ion illustrates the way the idea is most often used. Based on the experimental facts they know about a compound, chemists draw resonance contributors for the molecule to explain reactivity. In the case of carbon dioxide, resonance contributors are used to explain the reactivity of the carbon atom with reagents, such as hydroxide ion, that have pairs of electrons to share with atoms having open shells. In Section 2.10, another important application of the concept of resonance, the idea of stabilization of a species by resonance, is introduced. You will gradually develop an intuitive understanding of the concept as we apply it to many situations throughout the book.

PROBLEM 1.9   Write major resonance contributors for the following ions and molecules, including formal charges where applicable. Evaluate each resonance contributor you write in terms of the rules given in this section and decide which contributors are "important" and which "unimportant."

(a) HCO_3^- (b) H_2CO (c) NO_2^- (d) NCO^-

(e) HNO_3 (f) O_3 (g) NO_2^+

PROBLEM 1.10   The reaction between hydroxide ion and carbon dioxide is similar in some ways to the one shown between ammonia and boron trifluoride on p. 9.
What similarities do you see?

PROBLEM 1.11   On the basis of the resonance contributors that you have written for formaldehyde, H_2CO (Problem 1.9), and of your answer to Problem 1.10, predict whether ammonia would react with formaldehyde.

1.6

Shapes of Covalent Molecules

The covalent bond, in contrast to the ionic bond (p. 6), *has direction in space.* As soon as we have more than two atoms covalently bound to each other, we must decide how to arrange them in three dimensions. In 1916 when Lewis postulated that four pairs of electrons formed an octet around a central atom, he also suggested that the pairs of electrons were located at the corners of a tetrahedron, as far from each other as possible.

About twenty years after Lewis made his suggestion, experimental values for the distance and angles between atoms in simple covalent molecules were determined using a technique called electron diffraction. In this section, we will consider some of the experimental observations about the shapes of covalent compounds. Then in Chapter 2, we will study a theory of covalent bonding that explains the molecular geometries that have been observed.

Chemists are interested in two parameters, bond lengths and bond angles used to describe *the three-dimensional structure of a molecule. A* **bond length** *is the average distance between the nuclei of the atoms that are covalently bound together. A* **bond angle** *is the angle formed by the intersection of two covalent bonds at a third atom common to both.* Electron diffraction experiments have shown that the four hydrogens bonded to the carbon atom in methane, CH_4, lie at the corners of a regular tetrahedron, with the carbon atom itself at the center of the tetrahedron. A three-dimensional representation of methane, showing carbon-hydrogen bond lengths of 1.09 Å and H—C—H bond angles of 109.5°, the tetrahedral angle, is shown in Figure 1.1.

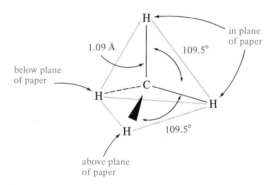

Figure 1.1 A three-dimensional representation of methane.

Figure 1.1 shows the molecule in perspective. The carbon atom and two of the hydrogen atoms are shown as lying in the plane of the paper by the use of ordinary lines in the drawing. The solid wedge used to attach one of the hydrogen atoms to the carbon atom indicates that this hydrogen atom is coming out of the plane of the paper towards the viewer. The dashed bond to the fourth hydrogen atom indicates that the atom is pointed away from the observer and is behind the plane of the paper. You should look at molecular models of tetrahedral carbon atoms so that you visualize these directional relationships clearly and can draw them.

Ethane, C_2H_6, has carbon-hydrogen bond lengths and H—C—H bond angles similar to those in methane. It also has a carbon-carbon bond that is 1.53 Å long. (See Figure 1.2.)

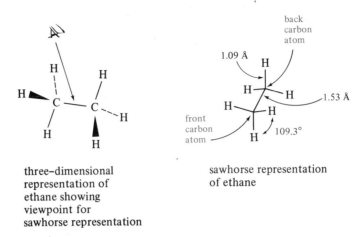

three–dimensional
representation of
ethane showing
viewpoint for
sawhorse representation

sawhorse representation
of ethane

Figure 1.2   Three-dimensional representations of ethane.

The ammonium ion, NH_4^+, has the same shape as methane, and ammonia, itself, is pyramidal with bond angles close to the tetrahedral value. Water is not a linear, but a triangular, molecule with a bond angle of 104.5°.

ammonium ion   ammonia   water

representations of the ammonium ion, ammonia, and water

In fact, whenever there are four pairs of electrons, either in single bonds or as non-bonding electrons, around a central atom from the second row of the periodic table, the bond angles for the species are close to tetrahedral, 109.5°, just as Lewis predicted. This generalization allows us to predict the shapes of many molecules and ions of interest in organic chemistry.

Bond lengths depend on the identity of the atoms being bonded. For example, a carbon-hydrogen bond is longer than a nitrogen-hydrogen bond, which is longer than an oxygen-hydrogen bond. The length of a carbon-carbon single bond as found in ethane, however, does not vary much from carbon-carbon single bonds in other organic molecules. Similarly, the carbon-hydrogen bond length in methane can be considered representative of carbon-hydrogen bond lengths in other organic compounds having carbon-hydrogen bonds at tetrahedral carbon atoms.

PROBLEM 1.12   Draw three-dimensional representations of the following compounds:

(a) CCl_4 (b) $CHCl_3$ (c) CH_3Br (d) CH_3CH_2Cl

(e) CH_3OH (f) CH_3NH_2 (g) CH_3OCH_3 (h) CH_3SCH_3

In the cases for which you have the necessary information, show the bond lengths and bond angles that you would expect to find in these compounds.

Not all carbon compounds have only tetrahedral carbon atoms in them. Organic molecules containing double or triple bonds have different shapes. Ethylene, $CH_2\!\!=\!\!CH_2$, is a flat molecule with bond angles of approximately 120° around each carbon atom. The carbon-carbon double bond length is 1.34 Å, considerably shorter than the carbon-carbon single bond in ethane. The carbon-hydrogen bonds are also a little shorter than the ones in methane and ethane. Two representations of ethylene, one in which the molecule is entirely in the plane of the paper, and the other in which we view the molecule from one edge, are shown below.

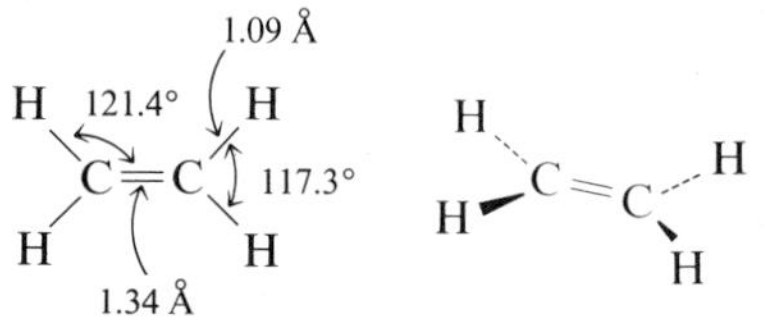

view from the top   *view from one edge*

ethylene

The triple bond in acetylene, $HC\!\!\equiv\!\!CH$, is shorter, 1.20 Å, than the carbon-carbon bond in ethane or ethylene. The carbon-hydrogen bond in acetylene is also shorter than it is in ethane and ethylene, only 1.06 Å. Acetylene is linear, with bond angles of 180°.

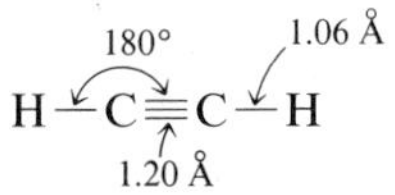

acetylene

The bond lengths in ethylene and acetylene can serve as models for the bond lengths of double and triple carbon-carbon bonds in other organic compounds. The bond angles are even more general. Whenever carbon is bonded by a double bond to one other atom, bond angles of approximately 120° are observed. If carbon is bonded by a triple bond to another atom, the portion of the molecule containing that functional group is linear.

PROBLEM 1.13   Draw three-dimensional representations of the following compounds. Give approximate bond lengths and bond angles whenever you can.

(a) $CH_2\!\!=\!\!CHCl$

(b) $CH_3\overset{\displaystyle CH_3}{\underset{|}{C}}\!\!=\!\!CH_2$

(c) $CHCl\!\!=\!\!CHCl$ (Two different arrangements of the atoms are possible.)

(d) $CH_3C\!\!\equiv\!\!CH$

(e) $H_2C\!\!=\!\!O$

1.7

The Polarity of Covalent Molecules

A. Polar Covalent Bonds

Electrons in covalent bonds may be shared equally when the two bonded atoms are the same, or unequally when different elements participate in bonding. For example, *in the hydrogen molecule, where electrons are shared equally by two hydrogen atoms, there is a* **nonpolar covalent bond.** In hydrogen chloride, the electrons of the covalent bond are drawn closer to the chlorine atom, which is the more electronegative of the two elements. *A covalent bond in which there is unequal sharing of electrons is a* **polar covalent bond.** *The presence of such a bond is shown by writing partial positive and partial negative charges,* $\delta+$ *and* $\delta-$, *over the bonded atoms, pointing to a permanent distortion of the electrons in the bond in one direction.* The polarized covalent bond may also be shown as a **bond dipole.** The dipole, with a negative pole and a positive pole, is represented by the symbol $\leftrightarrow$, with the point of the arrow drawn towards the more electronegative atom.

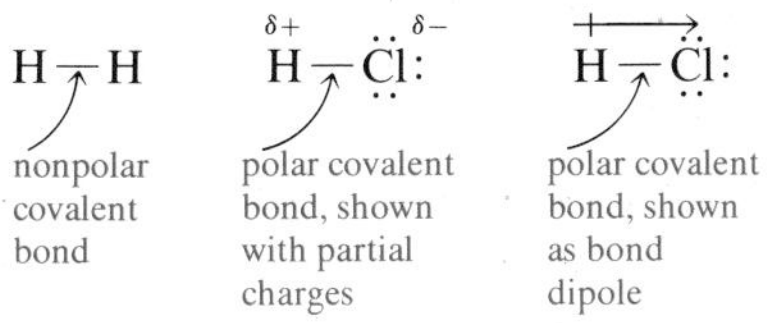

| nonpolar covalent bond | polar covalent bond, shown with partial charges | polar covalent bond, shown as bond dipole |

The polarity in a bond arises from the different electronegativities of the two atoms participating in the bond. The electronegativity of an element was defined by Pauling, who developed the concept, as "the power of an atom in a molecule to attract electrons to itself." The most electronegative elements are in the upper right hand corner of the periodic table, with electronegativity increasing as one moves up in a group and to the right in any period. The electronegativities of some elements of interest in organic chemistry are shown in Table 1.2, arranged according to the groups in the periodic table.

TABLE 1.2 Electronegativity Values for Some Elements

I	*II*		*III*	*IV*	*V*	*VI*	*VII*
H 2.1							
Li 1.0			B 2.0	C 2.5	N 3.0	O 3.5	F 4.0
Na 0.9	Mg 1.2		Al 1.5	Si 1.8	P 2.1	S 2.5	Cl 3.0
K 0.8							Br 2.8
							I 2.4

The greater the difference in electronegativity between the bonded atoms, the greater is the polarity of the bond. Compounds formed by the metals, at the left-hand side of the table, and nonmetals, at the extreme right, are ionic. Covalent bonds of all gradations of polarity form between nonmetals and between many metals and non-metals. For example, the electronegativities of carbon and hydrogen are close enough that carbon-hydrogen bonds do not have much polarity. Bonds of high polarity are those between hydrogen and the elements oxygen, fluorine, chlorine, and nitrogen. Carbon-halogen, carbon-oxygen, and carbon-nitrogen bonds are also polar. Bond polarities contribute significantly to the physical and chemical properties of molecules, and we will refer to them often as we consider chemical reactivity.

PROBLEM 1.14   Predict, by writing $\delta +$ and $\delta -$, the direction of polarization of the covalent bonds indicated by the arrows in the following compounds.

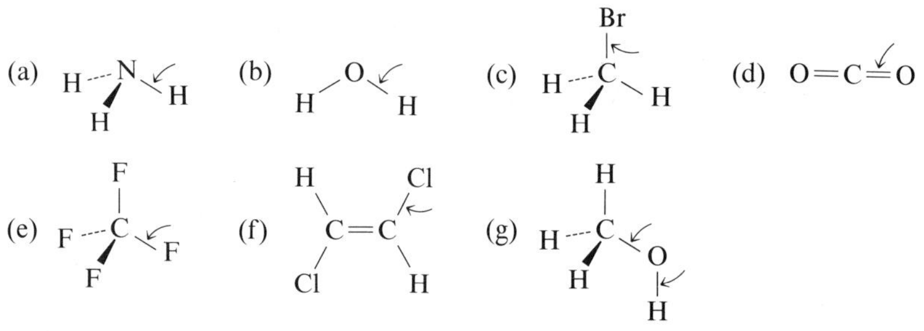

(a) [structure: H–N with H, H]
(b) [structure: O with H, H]
(c) [structure: H–C with Br, H, H]
(d) $O{=}C{=}O$

(e) [structure: F–C with F, F, F]
(f) [structure: C=C with H, Cl, Cl, H]
(g) [structure: H–C with H, H, O, H]

B. Dipole Moments of Covalent Molecules

For diatomic molecules, those containing two atoms, the bond dipole is also the dipole moment, μ, for the molecule. The dipole moment results from the separation of the centers of positive and negative charge, and is given a unit, D, the debye, that is derived from the magnitude of the charge and the distance of separation of the charges. Diatomic molecules in which both atoms are the same have no dipole moments. For other diatomic molecules, such as the hydrogen halides, the more electronegative the halogen, the larger is the dipole moment for the molecule. These facts are illustrated below.

$$:N{\equiv}N: \qquad :\ddot{Br}{-}\ddot{Br}:$$

no dipole moment for diatomic molecules in
which both atoms are the same

$$H{-}\ddot{F}: \qquad H{-}\ddot{Cl}: \qquad H{-}\ddot{Br}: \qquad H{-}\ddot{I}:$$
$$\mu, 1.98\ D \qquad \mu, 1.03\ D \qquad \mu, 0.78\ D \qquad \mu, 0.38\ D$$

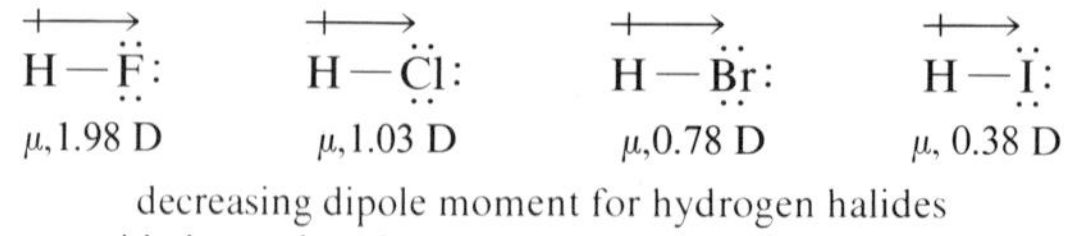

decreasing dipole moment for hydrogen halides
with decreasing electronegativity of the halogen atom

The overall dipole moment of a molecule containing more than two atoms is the vector sum of the individual bond dipole moments. A molecule may contain polar bonds, but have no overall dipole moment if the shape of the molecule is such that the individual bond moments cancel out. In such a case, the average position of the partial positive charges coincides with the average position of the partial negative charges.

This is so for carbon dioxide, CO_2, and carbon tetrachloride, CCl_4.

carbon dioxide   carbon tetrachloride
μ, 0 D μ, 0 D

cancellation of individual bond moments
in symmetrical polyatomic molecules

In carbon dioxide, bond moments of equal magnitude point in exactly opposite directions. The net result is that the molecule as a whole has no dipole moment even though the individual bond moments may be quite large. The absence of a measurable dipole moment is one of the reasons for assigning a linear structure to carbon dioxide.

The cancellation of the bond moments in carbon tetrachloride is not so easy to visualize. Because of the highly symmetrical nature of the tetrahedron, the vector sum of the bond moments of any three carbon-chlorine bonds is exactly equal to and opposite in direction to the bond moment for the fourth carbon-chlorine bond. The overall result is that the molecule as a whole has no permanent dipole, though the individual carbon-chlorine bonds remain polarized.

In most molecules, the vector sum of the individual bond moments is not zero, and the molecule has a dipole moment. Water, for example, has a dipole moment with the negative pole at the oxygen atom and the positive pole between the two hydrogen atoms. Similarly, the dipole moment in chloromethane, CH_3Cl, has the negative pole at the chlorine atom.

water   chloromethane
μ, 1.84 D μ, 1.86 D

molecules with dipole moments resulting from
vector sums of individual bond moments

Thus, the dipole moment of a molecule is related to the polarities of its individual bonds and to its molecular geometry. The dipole moments of most organic compounds have not been measured, yet we can predict the polarity of many compounds from a knowledge of the electronegativities of atoms, which allows us to predict bond polarities, and from a knowledge of molecular shapes, which enables us to decide whether the molecule will have a permanent dipole moment.

PROBLEM 1.15   Draw three-dimensional representations for the following compounds. For each compound, predict whether it will have a dipole moment, and if so, the direction of the moment.

(a) $CHCl_3$ (b) CH_3OH (c) ICl

(d) NH_3 (e) NH_4^+ (f) CH_2Br_2

(g) CH_3OCH_3 (h) $CH_2{=}CCl_2$

1.8

Nonbonding Interactions Between Molecules

Covalent compounds may be gases, liquids, or solids. Low molecular weight compounds with no dipole moments, such as methane and carbon dioxide, are gases. The forces that act between such molecules are very weak, so that condensation to the liquid or solid phases takes place only at low temperatures or at high pressures. *The forces that act between molecules are called* **intermolecular forces** *or* **intermolecular nonbonding interactions.** These interactions increase significantly as the molecular weight, and hence, the size of molecules, increases. They also increase with increasing polarity in molecules. *Three types of intermolecular forces are important: (1) dipole-dipole interactions, (2) hydrogen bonding, and (3) van der Waals forces.*

In the following sections of the chapter, a number of organic compounds with a variety of structures are introduced. These compounds appear many times in later chapters, and you will gradually become familiar with their structures, their names, and their reactions. For the moment, you should concentrate only on bond polarities and on the interactions that are possible between these kinds of molecules. As always, use the problems as a guide to what you have to learn in this chapter.

A. Dipole-Dipole Interactions

Molecules with dipole moments tend to orient themselves in the liquid and solid phases so that the negative end of one molecule is attracted to the positive end of another one. *The interactions of the permanent dipoles in different molecules are called* **dipole-dipole interactions.** Chloromethane has a dipole moment of 1.86 D, which lies along the carbon-chlorine bond. Chloromethane molecules orient themselves so that the positive end of one dipole is pointed towards the negative end of another dipole, a process that is depicted in a schematic representation in Figure 1.3.

An ordinary covalent bond has bond energy in the range of 30 to 100 kcal/mol (Section 5.4B). Dipole-dipole interactions are much weaker, approximately 1 to 3 kcal/mol.

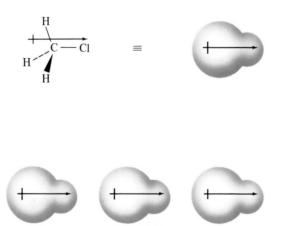

Figure 1.3   Schematic representation of dipole-dipole interactions in chloromethane.

B. Hydrogen Bonding

When a hydrogen atom is covalently bonded to a strongly electronegative atom such as oxygen, fluorine, or nitrogen, the bond is very polar. A hydrogen atom in this situation has a large affinity for nonbonding electrons on other oxygen, fluorine, or nitrogen atoms. The strong interaction that results is called a **hydrogen bond.** A hydrogen bond is a particularly strong dipole-dipole interaction. Significant hydrogen bond formation is seen only with hydrogen atoms covalently bonded to oxygen, fluorine, and nitrogen. Even though chlorine is as electronegative as nitrogen, neutral chlorine atoms are not significantly involved in hydrogen bonding. As an atom in the third period of the periodic table, chlorine is larger than fluorine, oxygen, or nitrogen. The partial negative charge on a chlorine atom in a covalent bond is, therefore, more diffuse and does not attract the hydrogen atom as strongly. Furthermore, *hydrogen atoms bonded to carbon are not usually involved in the formation of hydrogen bonds.*

The strength of hydrogen bonds is reflected in the physical properties of compounds in which such bonding occurs. Water boils at a much higher temperature than hydrogen sulfide, though it has a lower molecular weight. This difference in boiling points is attributed to the strong hydrogen bonds formed between oxygen and hydrogen atoms.

water
molecular weight 18
bp 100 °C

hydrogen sulfide
molecular weight 34
bp −62 °C

The attractive forces of hydrogen bonding are usually indicated by a dashed line rather than the solid line used for a covalent bond. The strength of a hydrogen bond involving oxygen, fluorine, or nitrogen atoms ranges from 3 to 10 kcal/mol, making hydrogen bonds the strongest type of intermolecular interaction known.

The effect of hydrogen bonding is demonstrated dramatically by the difference in the boiling points of ethanol, CH_3CH_2OH, and of its isomer dimethyl ether, CH_3OCH_3. Ethanol is an alcohol. *An alcohol resembles a water molecule in which one of the hydrogen atoms has been replaced by a nonpolar group containing carbon and hydrogen atoms. The polar part of the molecule is the hydroxyl group* (Section 1.5B).

ethanol
molecular weight 46
bp 78 °C

dimethyl ether
molecular weight 46
bp −24 °C

The hydrogen bonding that takes place between molecules of ethanol contributes to a much higher boiling point for this compound than for dimethyl ether, where the molecules are held together only by dipole-dipole interactions. *In an ether, both hydrogen atoms that are present in water are replaced by nonpolar groups containing carbon and hydrogen atoms.* As a result, an ether does not have a polar hydroxyl group.

PROBLEM 1.16   For each set of compounds, predict which one will have the highest boiling point. Indicate, with drawings, the reasoning behind your conclusions.

(a) $CH_3CH_2CH_2CH_2CH_3$, $CH_3CH_2CH_2CH_2OH$, $CH_3CH_2OCH_2CH_3$

(b) CH_3CH_3, CH_3F, CH_3OH

(c) $CH_3CH_2CH_3$, CH_3SH, CH_3OH

The hydrogen bond is important in determining the solubility of organic compounds in water. Molecules that can participate in the formation of hydrogen bonds with water will dissolve in water if the nonpolar part of the molecule (the part made up of carbon and hydrogen alone) is not too large. For example, the solubilities in water of three alcohols, ethanol, 1-butanol, and 1-hexanol vary with the size of the nonpolar portion of the molecule.

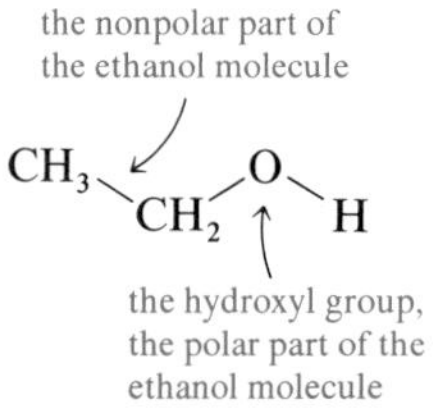

the nonpolar part of
the ethanol molecule

CH_3 — CH_2 — O — H

the hydroxyl group,
the polar part of the
ethanol molecule

*ethanol mixes with water
in all proportions*

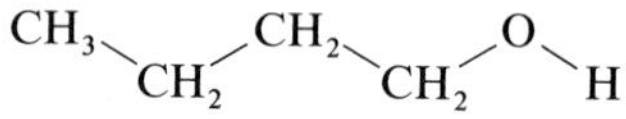

CH_3 — CH_2 — CH_2 — CH_2 — O — H
1-butanol

*7.9 g dissolves in
100 mL of water*

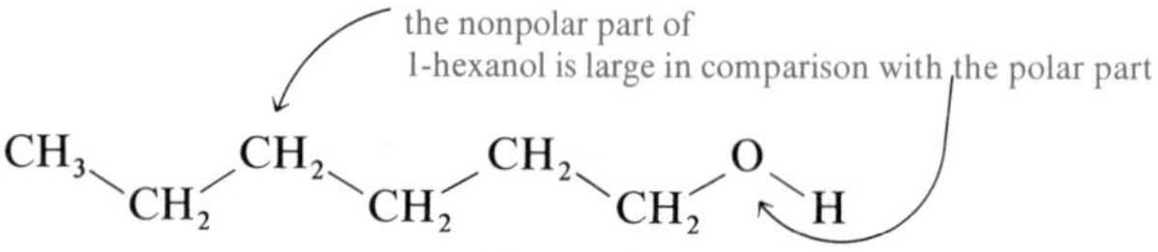

the nonpolar part of
1-hexanol is large in comparison with the polar part

CH_3 — CH_2 — CH_2 — CH_2 — CH_2 — O — H
1-hexanol

*0.59 g dissolves in
100 mL of water*

Ethanol, in which the polar part of the molecule, the hydroxyl group, is large in proportion to the whole molecule, is completely soluble in water. The hydroxyl group is a much less significant portion of the 1-hexanol molecule, so the compound has low water solubility. The structure of 1-butanol is intermediate between ethanol and 1-hexanol, and so is its solubility in water.

The interaction between a dissolved species and the molecules of a solvent is known as **solvation.** Hydrogen bonding is a particularly effective interaction between solute and solvent. The solvation of ethanol by water is shown below.

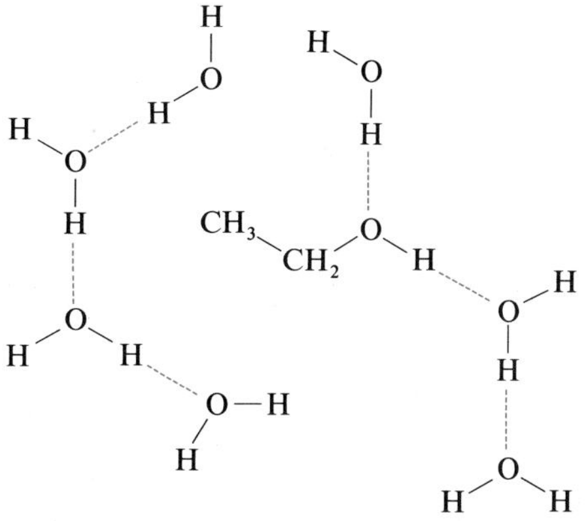

solvation of ethanol by water

Solvation is especially important in stabilizing ionic species. For example, when magnesium bromide, $MgBr_2$, is dissolved in water, the magnesium ions and the bromide ions are solvated by water molecules.

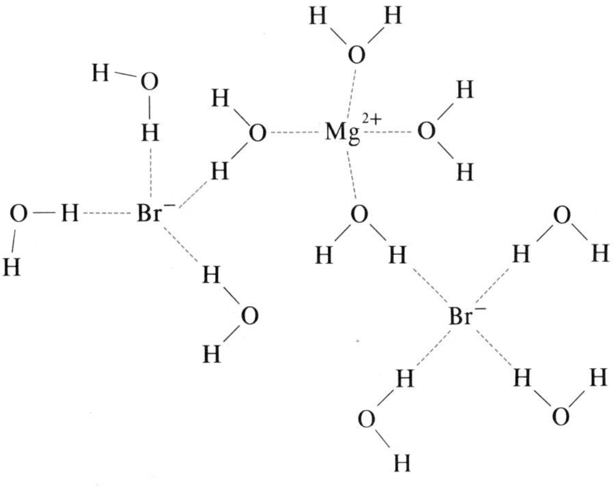

solvation of magnesium bromide by water

The interaction between water molecules and positively and negatively charged ions is a strong one known as **ion-dipole interaction.** The hydroxyl group in water and alcohols, with the large polarity of the oxygen-hydrogen bond and the high concentration of positive charge on the small hydrogen atom, is particularly effective at stabilizing anions.

Compounds that contain oxygen or nitrogen atoms, but lack hydrogen atoms bonded to these electronegative elements, may participate in hydrogen bonding as **hydrogen bond acceptors** even though they cannot function as **hydrogen bond donors.** The solubilities in water of diethyl ether, $CH_3CH_2OCH_2CH_3$, and pentane, $CH_3CH_2CH_2CH_2CH_3$, demonstrate this fact.

$$O{-}H$$
$$|$$
$$H$$
$$\vdots$$

CH₃—CH₂—O—CH₂—CH₃   CH₃—CH₂—CH₂—CH₂—CH₃

diethyl ether   pentane

7.5 g dissolves in   *0.036 g dissolves in*
100 mL of water   *100 mL of water*

Diethyl ether is almost as soluble in water as 1-butanol (p. 26) and much more so than pentane, which is a completely nonpolar compound containing only carbon and hydrogen atoms.

PROBLEM 1.17   For each pair of compounds, predict which one would be more soluble in water. Indicate with drawings the reasons for your conclusions.

(a) CH_3CH_2Cl or CH_3CH_2OH (b) $CH_3CH_2CH_2OH$ or $CH_3CH_2CH_2SH$

(c) $CH_3CH_2CH_2CH_2CH_2OH$ or $HOCH_2CH_2CH_2CH_2CH_2OH$

$$\qquad\qquad\quad O \qquad\qquad\qquad O$$
$$\qquad\qquad\quad \| \qquad\qquad\qquad \|$$
(d) CH_3CH_2COH or CH_3COCH_3 (e) $CH_3CH_2CH_2Cl$ or $CH_3CH_2CH_2NH_2$

The hydrogen bond plays an important role throughout chemistry and biochemistry. It is especially important in interactions between different parts of large molecules where many hydrogen bonds are possible. The shape of an enzyme that catalyzes chemical processes in our bodies, for example, is determined to a great extent by hydrogen bonding between distant parts of the large molecule. The two strands of the double helix of deoxyribonucleic acid are held together by a precise pattern of hydrogen bonding, believed to be responsible for the transmission of the genetic code. We will say much more about this when we look at the structures and the chemistry of proteins, carbohydrates, and other giant molecules of life.

C. Van der Waals Forces

Intermolecular forces act to attract even nonpolar molecules to each other, as demonstrated below by the physical properties of three nonpolar compounds: methane, CH_4; hexane, C_6H_{14}; and icosane, $C_{20}H_{42}$.

$$CH_4 \qquad\qquad\qquad\qquad CH_3CH_2CH_2CH_2CH_2CH_3$$

methane   hexane
molecular weight 16   molecular weight 86
bp -162 °C   bp 69 °C
gas at room temperature   liquid at room temperature

$$CH_3CH_2CH_2CH_2CH_2CH_2CH_2CH_2CH_2CH_2CH_2CH_2\ CH_2CH_2CH_2CH_2CH_2CH_2CH_2CH_3$$

icosane
molecular weight 282
mp 36 °C
solid at room temperature

The forces of attraction between the small molecules of methane are so weak that methane exists as a gas at room temperature. Molecules of hexane are larger than those of methane, and the attractive forces between them are increased enough that hexane is a liquid. The still larger molecules of icosane attract each other so strongly that the compound is a solid at ordinary temperatures.

The weak forces of attraction that exist between nonpolar molecules are called **van der Waals forces.** These forces are the result of the constant motion of electrons within bonds and molecules giving rise to effects known as London dispersion forces. The motion of electrons creates small distortions in the distribution of charge in nonpolar molecules. A small and momentary dipole results. This small dipole in one molecule can then create a dipole with the opposite orientation, an **induced dipole,** in a second molecule. Although the induced dipoles are constantly changing, the net result is a slight attraction between molecules. As the number of carbon and hydrogen atoms increases, the additive effect of these weak intermolecular forces becomes more significant, as evidenced by the increase in boiling and melting points from methane to hexane to icosane.

Van der Waals forces can act only through the parts of different molecules that are within a certain distance of each other. The three-dimensional shapes of molecules, therefore, determine to some extent the intermolecular interactions between molecules (see Figure 1.4). For example, the isomers butane and 2-methylpropane, both with the molecular formula C_4H_{10}, have different boiling points (Section 1.5B).

2-Methylpropane is a more compact molecule than butane. If you build models of the two compounds, you can see that 2-methylpropane is almost spherical, whereas

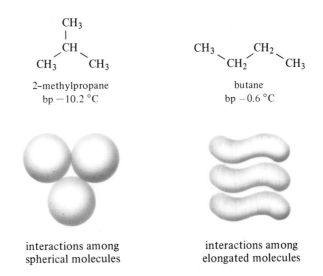

2–methylpropane
bp −10.2 °C

butane
bp −0.6 °C

interactions among
spherical molecules

interactions among
elongated molecules

Figure 1.4   A comparison of intermolecular interactions for 2-methylpropane and butane.

butane is elongated. The molecules of butane have a greater surface area for interaction with each other than do the molecules of 2-methylpropane. The stronger interactions that are possible for butane are reflected in its boiling point, which is higher than the boiling point of 2-methylpropane.

ADDITIONAL PROBLEMS

1.18 Which of the following compounds would you expect to be ionic, which covalent, and which to have both kinds of bonds? For the covalent bonds, show which ones have polarity.

(a) MgF_2 (b) SiF_4 (c) NaH   (d) ClF   (e) SCl_2 (f) OF_2

(g) SiH_4 (h) PH_3 (i) $NaOCH_3$ (j) CH_3Na (k) Na_2CO_3

1.19 Draw Lewis structures for the following species, using lines for covalent bonds, and showing any nonbonding electrons that are present. Show formal charges where relevant.

(a) NF_3 (b) $AlCl_3$ (c) CH_3SCH_3 (d) CH_3NH_2 (e) $CH_3CHClCH_3$

(f) OH^- (g) $CH_3CH_2CH_2OH$ (h) H_2O_2 (i) CH_3NHOH (j) SO_2

(k) CH_3SH (l) SiH_4 (m) H_2SO_4 (n) HNO_3

1.20 Draw the following compounds (or ions) in three dimensions, showing in each case the geometry that you would expect to find around the atom that is shaded with color.

(a) S Cl_2 (b) O F_2 (c) $CH_3 - \overset{\overset{\displaystyle H}{|}}{C} = CH_2$ (d) $CH_3 - C \equiv N$

(e) B F_4^- (f) C FCl_3 (g) $CH_3 - \overset{\overset{\displaystyle CH_3}{|}}{\underset{\underset{\displaystyle CH_3}{|}}{N}} - CH_3{}^+$ (h) P H_3

(i) CH_3 O $H_2{}^+$ (j) $CH_3 - \overset{}{\underset{\underset{\displaystyle CH_3}{|}}{N}} - CH_3$

1.21 Which of the following molecules would have a dipole moment? Show your reasoning by drawing a three-dimensional representation of each compound, and showing the direction of the dipole.

(a) $CH_2 = CHCl$ (b) CH_3SCH_3 (c) $CH_3C \equiv CCH_3$ (d) $FCBr_3$

(e) CH_3CH_2OH (f) $HC \equiv CCl$ (g) $Cl_2C = CCl_2$

1.22 For which of the following compounds will hydrogen bonding among molecules of the same kind be important?

(a) CH_2F_2 (b) $CH_3CH_2CH_2OH$ (c) $CH_3CH_2OCH_3$

(d) $HOCH_2CH_2CH_2CH_2OH$ (e) $CH_3CH_2CH_2NH_2$ (f) $CH_3CH_2\overset{\overset{\displaystyle O}{\|}}{C}NH_2$

(g) $CH_3CH_2CH_2SH$

1.23   Which of the following compounds will participate in hydrogen bonding to water? For each compound, indicate whether it will be a hydrogen bond donor, hydrogen bond acceptor, or both. Illustrate your reasoning with drawings.

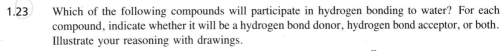

(a) $CH_3CH_2OCH_3$

(b) $CH_3CH_2NCH_2CH_3$
$\qquad\qquad\qquad\quad\; |$
$\qquad\qquad\qquad\quad\; CH_2CH_3$

(c) $\overset{\displaystyle O}{\overset{\displaystyle \|}{CH_3CNHCH_3}}$

(d) $CH_3C{\equiv}N$

(e) $\overset{\displaystyle O}{\overset{\displaystyle \|}{CH_3CH_2COH}}$

(f) $CH_3CH_2CH_2Cl$

(g) CH_3NHOH

(h) $\overset{\displaystyle CH_3}{\underset{\displaystyle CH_3}{}}C{=}C\overset{\displaystyle H}{\underset{\displaystyle Cl}{}}$

(i) $\overset{\displaystyle O}{\overset{\displaystyle \|}{CH_3SCH_3}}$

1.24   Which compound in each pair do you expect to be more soluble in water?

(a) CH_3Cl or $NaCl$

(b) $\overset{\displaystyle O}{\overset{\displaystyle \|}{HCOH}}$ or $\overset{\displaystyle O}{\overset{\displaystyle \|}{CH_3CH_2CH_2CH_2COH}}$

(c) $\overset{\displaystyle O}{\overset{\displaystyle \|}{HCOH}}$ or $\overset{\displaystyle O}{\overset{\displaystyle \|}{HCOCH_2CH_2CH_2CH_2CH_3}}$

(d) $CH_3CH_2SCH_2CH_3$ or $CH_3CH_2OCH_2CH_3$

(e) $\overset{\displaystyle O}{\overset{\displaystyle \|}{HOCCH_2CH_2CH_3}}$ or $\overset{\displaystyle O}{\overset{\displaystyle \|}{HOCCH_2CH_2}}\overset{\displaystyle O}{\overset{\displaystyle \|}{COH}}$

1.25   Draw a segment of the crystal lattice of sodium chloride (p. 6) and explain with diagrams why the compound dissolves easily in water even though a great deal of energy is required to melt or vaporize sodium chloride.

1.26   The boiling point of water (100 °C) is higher than the boiling points of hydrogen fluoride (19 °C) and ethanol (78 °C). Use drawings to explain this experimental observation.

1.27   Acetonitrile, $CH_3C{\equiv}N$, has a large dipole moment (greater than 3 D).

(a)   Draw a Lewis structure for acetonitrile.

(b)   Predict the direction of the dipole moment for acetonitrile.

(c)   The large dipole moment for acetonitrile has been interpreted to mean that the molecule has an important resonance contributor that reflects a separation of charge. Draw such a resonance contributor for acetonitrile.

1.28   Draw major resonance contributors for the species listed below. Be sure to show any formal charges. Analyze each resonance contributor on the basis of the rules in Section 1.5C as ''important'' or ''unimportant.''

(a)   Nitrous oxide, N_2O (a linear molecule in which the two nitrogen atoms are bonded to each other).

(b)   Azide anion, N_3^- (a linear ion).

(c)   Sulfate anion, SO_4^{2-} (an ion in which all of the oxygens are equivalent).

(d)   Acetic acid, $\overset{\displaystyle O}{\overset{\displaystyle \|}{CH_3COH}}$.

1.29   (a)   Carbon monoxide, CO, has a much smaller dipole moment than expected. This observation has puzzled chemists and led to much argument about the nature of the bonding in the molecule. The small dipole moment for CO can be rationalized by proposing that there are three important resonance contributors for the molecule. One of them has no formal charges, the other two are polarized in opposite directions. Write these resonance contributors and analyze each one according to whether: (1) each atom has an octet around it, and (2) the formal charges are in accord with the relative electronegativities of carbon and oxygen.

(b)   Carbon monoxide is highly toxic because it binds tightly to iron in hemoglobin and thus prevents that molecule from binding to and carrying oxygen in the blood. The carbon atom in carbon monoxide binds to iron(II), Fe^{2+}. What does this experimental fact indicate about the relative importance of the various resonance contributors to the structure of carbon monoxide?

1.30   Amides are compounds containing the following group of atoms.

This type of group is important in proteins. Amino acids are held together by amide bonds in building up a protein chain. The properties of an amide bond are best rationalized on the basis of resonance contributors for the amide. Write three resonance contributors for the amide group and discuss the relative importance of each.

1.31   For each set of structural formulas given below, decide which ones represent resonance contributors. Also, decide which resonance contributors for a given molecule are more important and which less so. Describe your reasoning.

(a) $CH_3 - \overset{..}{\underset{..}{O}} - \overset{+}{C}H_2$ $CH_3 - \overset{+}{\underset{..}{O}} = CH_2$

(b) $CH_2 = CH - \overset{\overset{\displaystyle \overset{..}{O}:}{\|}}{C} CH_3$ $CH_2 = CH\overset{:\overset{..}{O}H}{\overset{|}{C}} = CH_2$

(c) $CH_2 = CH - \overset{\overset{\displaystyle :O:}{\|}}{C} - CH_3$ $\overset{+}{C}H_2 - CH = \overset{\overset{\displaystyle :\overset{..}{O}:^-}{|}}{C} - CH_3$ $CH_2 = CH - \overset{\overset{\displaystyle :\overset{..}{O}:^-}{|}}{\underset{+}{C}} - CH_3$

(d) $CH_3\overset{\overset{\displaystyle \overset{..}{O}:}{\|}}{C} - \overset{..}{\underset{..}{O}}CH_3$ $CH_3\overset{\overset{\displaystyle :\overset{..}{O}:^-}{|}}{\underset{+}{C}} - \overset{..}{\underset{..}{O}}CH_3$ $CH_3\overset{\overset{\displaystyle :\overset{..}{O}:^-}{|}}{C} = \overset{+}{\underset{..}{O}}CH_3$

(e) $CH_2 = CH - \overset{+}{C}H_2$ $\overset{+}{C}H_2 - CH = CH_2$

(f) $CH_3 - \overset{..}{N} - \overset{..}{N} = \overset{..}{O}:$ $CH_3 - \overset{+}{\underset{\underset{\displaystyle CH_3}{|}}{N}} - \overset{..}{\underset{..}{N}} - \overset{..}{\underset{..}{O}}:^-$ $CH_3 - \overset{+}{N} = \overset{..}{N} - \overset{..}{\underset{..}{O}}:^-$

with CH_3 below each N as shown.

(g) $CH_3 - \overset{..}{N} = C = \overset{..}{O}:$ $CH_3 - \overset{..}{N} = \overset{+}{C} - \overset{..}{\underset{..}{O}}:^-$ $CH_3 - \overset{\overset{\displaystyle \bar{..}}{N}}{} - C = \overset{..}{O}:$

(h) $CH_2 = CH - \overset{+}{N} \overset{\nearrow \overset{..}{\underset{..}{O}:^-}}{\searrow \underset{..}{O}:}$ $\overset{+}{C}H_2 - CH = \overset{+}{N} \overset{\nearrow \overset{..}{\underset{..}{O}:^-}}{\searrow \underset{..}{O}:^-}$

(i)

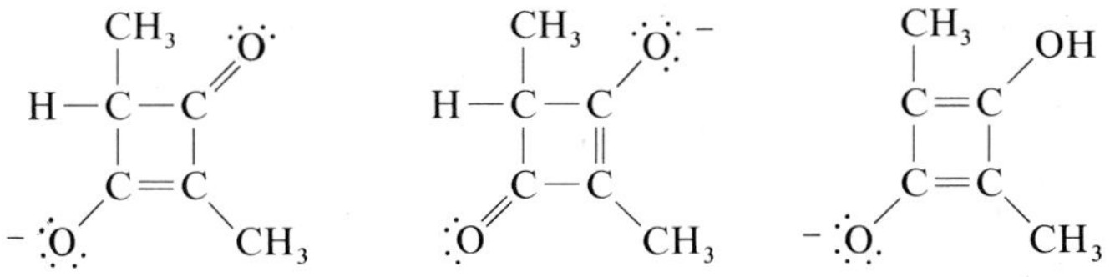

$$
\begin{array}{ccc}
\overset{\displaystyle CH_3}{\underset{\displaystyle}{|}} & \overset{\displaystyle \cdot \ddot{O}\cdot}{} & \\
H-C-C & & \\
\end{array}
$$

(i) Structure 1:

```
        CH₃   ·Ö·
         |    //
   H — C — C
         |    |
         C == C
       ⁻:Ö:      CH₃
```

Structure 2:

```
        CH₃   :Ö:⁻
         |    /
   H — C — C
         |    ‖
         C == C
      :O        CH₃
       ‖
```

Structure 3:

```
        CH₃    OH
         |    /
         C == C
         ‖    |
         C == C
     ⁻:Ö:       CH₃
```

(j)

Structure 1:
```
        :Ö       Ö:
         ‖        ‖
CH₃ — S — C̈H — S — CH₃
         ‖        ‖
         Ö:       Ö:
```

Structure 2:
```
       :Ö:⁻      Ö:
         |        ‖
CH₃ — S == CH — S — CH₃
         ‖        ‖
         O:       O:
```

Structure 3:
```
        Ö:       Ö:
         ‖        ‖
CH₃ — S == CH — S — CH₃
         |        ‖
       :O:⁻       O:
```

Structure 4:
```
        Ö:      :Ö:⁻
         ‖        |
CH₃ — S — CH == S — CH₃
         ‖        ‖
         O:       O:
```

Structure 5:
```
        Ö:       Ö:
         ‖        ‖
CH₃ — S — CH == S — CH₃
         ‖        |
         O:      :O:⁻
```

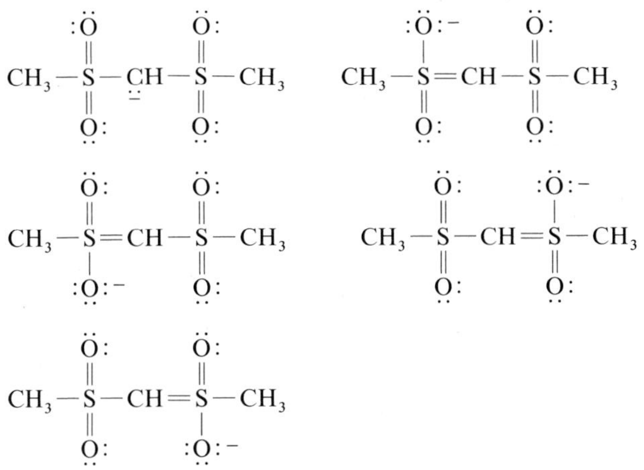

Covalent Bonding and Chemical Reactivity

Introduction

The structures of a number of covalent compounds were described in two ways in Chapter 1. On the one hand, Lewis electron dot structures were used to show the valence electrons of the atoms that participate in covalent bonding. On the other hand, the complete three-dimensional structures of some covalent compounds were described in terms of bond lengths and bond angles that are experimentally determined. For example, two carbon atoms are found bonded to each other and to three hydrogen atoms each in ethane, C_2H_6, a molecule in which the bond angles around the carbon atoms are tetrahedral. In ethylene, C_2H_4, the two carbon atoms are held together by a double bond. All of the atoms in ethylene lie in a plane with bond angles close to 120°. The two carbon atoms in acetylene, C_2H_2, share a triple bond, and each carbon is bonded to one hydrogen atom in a linear molecule. The experimental facts about these structures were given in Section 1.6.

How can we explain, on the basis of the electronic structure of carbon, the different kinds of bonding and the different molecular shapes that are observed for various organic compounds? Scientists have put forth many ideas about the nature of chemical bonding and the reasons for chemical reactivity. In this chapter, we examine one system of ideas, called molecular orbital theory, that is currently used by organic chemists to rationalize the facts known about the compounds of carbon.

Our ideas about electrons in molecules and how they participate in bonding are derived from quantum mechanics. In the mathematical equations of quantum mechanics, electrons are treated as if they have the properties of both waves and particles. The French physicist, Louis de Broglie, first introduced the idea that the motion of electrons can be described by equations similar to those associated with waves in 1923. It was further developed independently by Erwin Schrödinger of Austria and Werner Heisenberg of Germany. In the Schrödinger equation, the motion of the electron is related to a set of allowed energy values by mathematical expressions known as wave equations. Each energy level allowed for an electron corresponds to a particular solution of the wave equation called a wave function.

Nobel Prizes in Physics were awarded to de Broglie (1929), Heisenberg (1932), and Schrödinger (1933) in recognition of the importance of quantum mechanics to our understanding of the nature of chemical bonding. This way of looking at electrons and bonding is a highly sophisticated and mathematical model, in contrast with the Lewis structure for a covalent compound, which is a simple and pictorial one. Both systems, however, are our creations; we impose them on the facts of nature as explanations and aids to prediction.

2.2

Atomic Orbitals

A. The Atomic Orbitals of Hydrogen. A Review

According to the prevailing model of atomic structure, the exact location of an electron in an atom cannot be determined. Theories of atomic structure deal with the probability of finding an electron at a given distance and direction from the nucleus. *A region in space surrounding the nucleus of an atom in which there is a high probability of finding an electron is called an* **atomic orbital.** This probability is determined by the wave function for the electron. An orbital is usually represented by a picture that shows the boundary of the space within which there is some finite probability (such as 90%) of finding the electron.

The two atomic orbitals of lowest energy for the hydrogen atom, which has only a single proton and a single electron, are shown in Figure 2.1 (page 36).

The $1s$ and $2s$ atomic orbitals are spherically symmetrical. There is an equal probability of finding the electron at a given distance in all directions from the nucleus. The $2s$ orbital is larger than the $1s$ orbital. An electron in the $2s$ orbital is, on the average, farther away from the nucleus than one in the $1s$ orbital. The electron in the $2s$ orbital is less attracted by the nucleus, and is, thus, in a higher energy level.

The three $2p$ atomic orbitals of the hydrogen atom are oriented at right angles to each other and are usually given the designation of the coordinate axes $2p_x$, $2p_y$, and $2p_z$. Note that a p orbital does *not* have spherical symmetry, but does have symmetry about the axis along which it lies (Figure 2.2, page 36).

The probability of finding an electron in a $2p$ atomic orbital is greatest in the two lobes that are on opposite sides of the nucleus. The diagrams in Figure 2.2 illustrate that there is zero probability of finding an electron at the nucleus in a p orbital. *A region where the probability of finding an electron is zero is called a* **node.** A nodal plane passes through the nucleus for p orbitals. The $2s$ orbital also has a nodal region, the surface of a sphere buried within the boundary surface seen in Figure 2.1.

A more precise mathematical way of describing the p atomic orbital is to assign different mathematical signs to the two lobes of the orbital to indicate that the mathe-

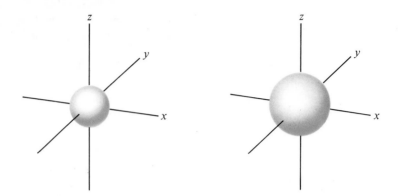

1s orbital in three
dimensions. It contains
one electron in the hydrogen
atom and is the lowest energy
orbital.

2s orbital in three
dimensions. The nodal
sphere is buried within.
This orbital is empty
in the hydrogen atom
and is the second lowest
energy orbital.

Figure 2.1   The two lowest-energy atomic orbitals of the
hydrogen atom.

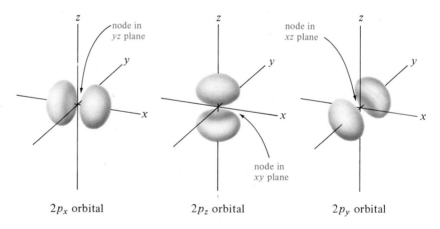

$2p_x$ orbital $2p_z$ orbital $2p_y$ orbital

The three $2p$ orbitals are empty in the hydrogen atom and are of equal energy. They
are higher in energy than the $2s$ orbital.

Figure 2.2   The three $2p$ orbitals of the hydrogen atom.

matical function, the wave function defining the p orbital, changes sign as it goes
through the nodal plane. Unfortunately, positive and negative signs also mean positive
and negative charges in the language of chemistry, so pictures with many such signs
in them can be confusing. Therefore, we will adopt the convention that the two lobes
of a p orbital be shown with different color shading to indicate the change in sign that
takes place at the nodal plane (Figure 2.3).

It is of no importance which lobe of a p orbital is seen as positive and which negative,
or which grey and which colored. Such an assignment is purely arbitrary. The change

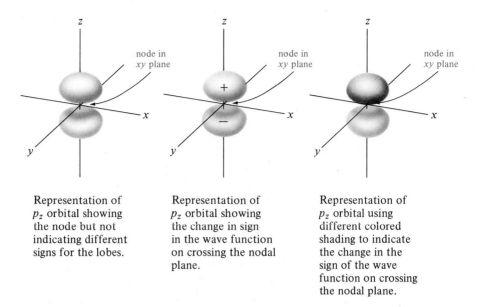

Representation of p_z orbital showing the node but not indicating different signs for the lobes.

Representation of p_z orbital showing the change in sign in the wave function on crossing the nodal plane.

Representation of p_z orbital using different colored shading to indicate the change in the sign of the wave function on crossing the nodal plane.

Figure 2.3   Different ways in which a *p* orbital may be represented.

in sign within an orbital is important because it determines how the orbital will interact with other orbitals.

The characteristics of the *s* and *p* atomic orbitals of the hydrogen atom are used as rough approximations of the properties of similar orbitals for elements such as carbon, oxygen, nitrogen, and the halogens. The relative energies and the exact sizes of atomic orbitals change with the charge on the nucleus of an atom and with the number of electrons it has. Nevertheless, chemists use the picture developed for the hydrogen atom in making predictions about bonding in more complex atoms.

B. Orbitals of Atoms in the Second Row of the Periodic Table

One of the complications that arises as soon as there is more than one electron in an atom is the question of electron spin. Electrons can have one of two possible spin orientations, symbolized by arrows pointing up or down, ↑ or ↓. Three rules govern the assignment of the most stable electronic configuration of an atom.

1. The Aufbau Principle states that electrons fill atomic orbitals in order of increasing energy. This means that the $1s$ orbital is filled first, then the $2s$, then the $2p$ orbitals.
2. The Pauli Exclusion Principle states that two electrons in the same orbital must have opposing spins.
3. Hund's Rule states that when orbitals of equal energy are available, one electron must be assigned to each one before any orbital receives two. This rule is based on the evidence that two paired electrons in the same orbital repel each other more than do two unpaired electrons in different orbitals.

Table 2.1 shows the assignment of electrons to the orbitals of the first ten elements in the periodic table. Note that the three $2p$ orbitals are of equal energy. They are said to be *degenerate*.

TABLE 2.1   Assignment of Electrons to Orbitals

Element	Electronic Configuration	Electrons in Orbitals				
		1s	2s	2p		
Hydrogen	$1s^1$	↑				
Helium	$1s^2$	↑↓				
Lithium	$1s^2 2s^1$	↑↓	↑			
Beryllium	$1s^2 2s^2$	↑↓	↑↓			
Boron	$1s^2 2s^2 2p^1$	↑↓	↑↓	↑		
Carbon	$1s^2 2s^2 2p^2$	↑↓	↑↓	↑	↑	
Nitrogen	$1s^2 2s^2 2p^3$	↑↓	↑↓	↑	↑	↑
Oxygen	$1s^2 2s^2 2p^4$	↑↓	↑↓	↑↓	↑	↑
Fluorine	$1s^2 2s^2 2p^5$	↑↓	↑↓	↑↓	↑↓	↑
Neon	$1s^2 2s^2 2p^6$	↑↓	↑↓	↑↓	↑↓	↑↓

PROBLEM 2.1   The orbitals that are in the next higher energy level after the $2p$ orbitals are the $3s$ and then the $3p$ orbitals. As a review exercise, assign electronic configurations to the elements: sodium, magnesium, aluminum, silicon, phosphorus, sulfur, chlorine, and argon.

2.3

Overlap of Atomic Orbitals. The Formation of Molecular Orbitals

A. The Hydrogen Molecule

The Lewis electron dot structure of a molecule shows a covalent bond arising from the sharing of a pair of electrons, one from each atom. The orbital picture of the covalent bond shows the formation of a molecular orbital from the overlap of two atomic orbitals each containing an electron. The directional properties of atomic orbitals are important in determining the extent of overlap possible and the geometry of the molecule that results when an atom forms more than one covalent bond.

The hydrogen molecule, arising from the combination of two hydrogen atoms, is the simplest molecule possible. An orbital picture of bonding requires that we imagine two hydrogen atoms, each with one electron in a $1s$ orbital, approaching each other. *If the orbitals have the same mathematical sign, they can interact to reinforce each other. They are said to be* **in phase.** The interaction of two orbitals of the same mathematical sign results in the formation of a **bonding molecular orbital,** shown schematically in

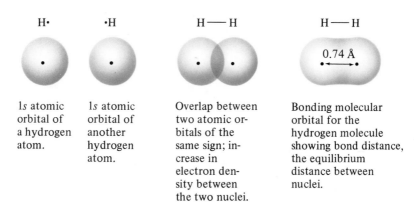

1s atomic orbital of a hydrogen atom.	1s atomic orbital of another hydrogen atom.	Overlap between two atomic orbitals of the same sign; increase in electron density between the two nuclei.	Bonding molecular orbital for the hydrogen molecule showing bond distance, the equilibrium distance between nuclei.

Figure 2.4   A schematic representation of the formation of a bonding molecular orbital in hydrogen by the overlap of two atomic 1s orbitals.

Figure 2.4. The two electrons, one from each hydrogen atom, occupy the bonding molecular orbital. There is an increase in electron density between the two nuclei.

The hydrogen molecule is more stable than the individual atoms because the electron of each hydrogen atom is attracted to the positively charged nucleus of the other atom as well as to its own nucleus. The interaction shown in Figure 2.4 results in a lowering of the overall energy of the system. *The bonding molecular orbital is lower in energy than the two separate atomic orbitals.* Note that the bonding molecular orbital of hydrogen is symmetrical about an axis connecting the two nuclei. *A bonding molecular orbital with cylindrical symmetry about an internuclear axis is called a σ (read "sigma")* **molecular orbital.**

The molecule is most stable when the nuclei of the hydrogen atoms are a certain distance apart. If the nuclei come any closer, they repel each other too strongly. If they are farther apart, the atomic orbitals do not overlap enough to form a good covalent bond. The equilibrium distance that allows for the most overlap without excessive nuclear repulsion is called the **bond distance,** which is 0.74 Å or 0.074 nm (a nanometer is 10^{-9} m) for the hydrogen molecule (Figure 2.4).

There is another possible combination of the two atomic 1s orbitals of hydrogen, one in which the orbitals are of opposite mathematical sign and are said to be **out of phase** with each other. In this combination, there is a node, a region of no electron density, between the two nuclei (Figure 2.5, page 40).

The lack of electron density between the two nuclei results in repulsion between them. This interaction gives rise to *an* **antibonding molecular orbital,** *a molecular orbital that is of higher energy than the two separate atomic orbitals.* The antibonding molecular orbital corresponding to the σ bonding orbital is called the σ* (read "sigma star") orbital.

When atomic orbitals are combined to give molecular orbitals, the number of molecular orbitals formed equals the number of atomic orbitals used. Thus, the combination of two atomic orbitals gives rise to two molecular orbitals, a bonding molecular orbital and an antibonding molecular orbital. The bonding and antibonding molecular orbitals of the hydrogen molecule are shown schematically in Figure 2.6 (page 40).

The two electrons of the hydrogen molecule are usually in the bonding molecular orbital (Figure 2.6). *Molecular orbitals, like atomic orbitals, can hold only two electrons of opposing spin.* When a bonding molecular orbital is occupied by two

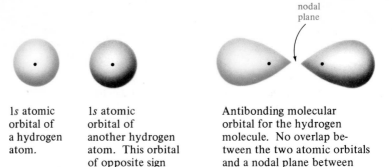

1s atomic orbital of a hydrogen atom.

1s atomic orbital of another hydrogen atom. This orbital of opposite sign to first one.

Antibonding molecular orbital for the hydrogen molecule. No overlap between the two atomic orbitals and a nodal plane between the two nuclei.

Figure 2.5 A schematic representation of the antibonding interaction between 1s atomic orbitals of two hydrogen atoms.

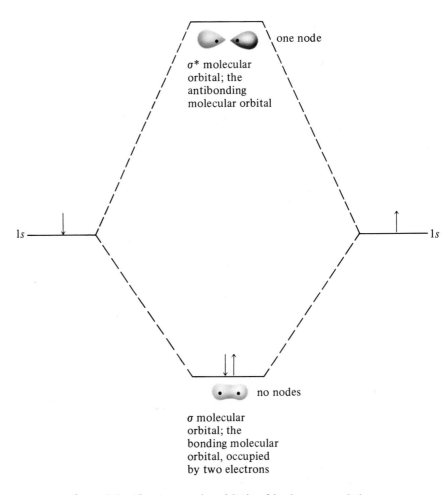

one node

σ* molecular orbital; the antibonding molecular orbital

1s

1s

no nodes

σ molecular orbital; the bonding molecular orbital, occupied by two electrons

Figure 2.6 The 1s atomic orbitals of hydrogen and the relative energies and shapes of the σ and σ* molecular orbitals for the hydrogen molecule.

electrons, a covalent bond results. The bond between the hydrogen atoms in a molecule of hydrogen is called a σ bond. Electrons in an antibonding molecular orbital do not contribute to bonding between atoms and, in fact, serve to destabilize a molecule.

A hydrogen molecule is more stable than two separate hydrogen atoms by about 104 kcal/mol. This is the amount of energy required to break the bond in the hydrogen molecule and separate the two hydrogen atoms. It is known as the **bond dissociation energy** (Section 5.4B) for hydrogen.

B. Sigma Bonds Involving *p* Orbitals

Different types of atomic orbitals may overlap to create σ bonds. For example, the covalent bond in a molecule of hydrogen chloride is thought to arise from the overlap of a 1*s* orbital of a hydrogen atom with a 3*p* orbital of a chlorine atom. All of the *p* orbitals are equivalent so any one of them may be used for the bond. The bonding molecular orbital that is formed when the spherically symmetrical 1*s* orbital of the hydrogen atom approaches the $3p_x$ orbital of the chlorine atom along the *x* axis is shown schematically in Figure 2.7.

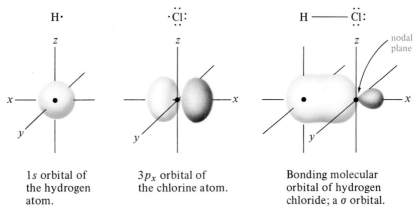

1*s* orbital of the hydrogen atom.

$3p_x$ orbital of the chlorine atom.

Bonding molecular orbital of hydrogen chloride; a σ orbital.

Figure 2.7   Schematic representation of the formation of a bonding molecular orbital in hydrogen chloride from the overlap of the 1*s* orbital of hydrogen and $3p_x$ orbital of chlorine.

The new molecular orbital in hydrogen chloride is a different shape from the atomic orbitals that go into its formation. In particular, the lobe of the $3p_x$ orbital that is not used for bonding is greatly contracted. The greatest electron density in the molecular orbital is observed between the two nuclei. The orbital is cylindrically symmetrical with respect to the axis between the two nuclei and is, therefore, a σ orbital. The presence of two electrons of opposing spin in the orbital gives rise to a single bond, also called a σ bond, between the hydrogen and chlorine atoms in hydrogen chloride.

The bond between two chlorine atoms in a molecule of chlorine is formed when two $3p_x$ orbitals, one on each chlorine atom, overlap with each other along the *x* axis (Figure 2.8, page 42). In this case, too, the molecular orbital that results is cylindrically symmetrical around the axis between the two nuclei. It is a σ orbital and, when occupied by two electrons, gives rise to a σ bond between the chlorine atoms.

The bonding in molecules of hydrogen, hydrogen chloride, and chlorine illustrates three different ways in which single bonds can be formed in covalent molecules. The examples that we have seen involve the overlap of *s* atomic orbitals, *s* and *p* atomic orbitals, and *p* atomic orbitals. *The bonding molecular orbitals that are formed by the*

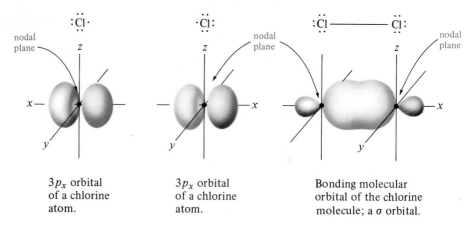

3p_x orbital
of a chlorine
atom.

3p_x orbital
of a chlorine
atom.

Bonding molecular
orbital of the chlorine
molecule; a σ orbital.

Figure 2.8   Schematic representation of the formation of a
bonding molecular orbital in chlorine from the
overlap of the 3p_x orbitals of two chlorine atoms.

*overlap of these atomic orbitals are called σ **orbitals** when they are symmetrical
around the axis joining the two atomic nuclei. All σ bonds have electron density
concentrated between the nuclei of the bonded atoms.*

In summary, the overlap of atomic orbitals gives rise to bonding (and antibonding)
molecular orbitals. Bonds are formed when bonding molecular orbitals are occupied
by pairs of electrons. For the sake of simplicity, in the future we will ignore the
antibonding molecular orbitals. In all the compounds we discuss, the bonding molecular orbitals are filled with electrons so distinctions between the formation of a bonding
orbital and the formation of a bond are not always made.

PROBLEM 2.2   Draw the atomic orbitals involved and show the bonding molecular orbitals of
hydrogen fluoride (HF) and the fluorine molecule (F$_2$).

PROBLEM 2.3   Look at the following drawings of atomic orbitals approaching each other along the
x axis for each atom. In each case, decide whether good overlap and bonding are going to occur
for those orientations of the orbitals.

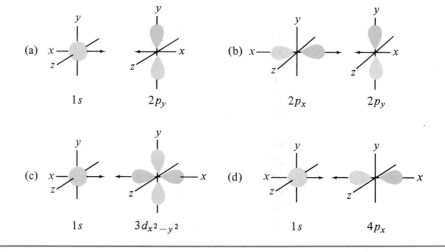

(a)   1s 2p_y (b)   2p_x 2p_y

(c)   1s 3$d_{x^2-y^2}$ (d)   1s 4p_x

2.4

Hybrid Orbitals

A. Tetrahedral Carbon Atoms

The simplest compound of carbon and hydrogen, methane, CH_4, is a symmetrical molecule with four carbon-hydrogen bonds of equal length directed towards the corners of a regular tetrahedron (Figure 1.1, Section 1.6). A similar tetrahedral arrangement of bonds is found experimentally whenever carbon is bound to four other atoms.

Methane has eight bonding electrons, four from the carbon atom and one from each of the four hydrogen atoms. In the language of molecular orbital theory, methane has four bonding molecular orbitals, each of which holds two of the bonding electrons. The hydrogen atoms (or the electron pairs) repel each other and, therefore, move as far as possible from each other in a tetrahedral arrangement. In this picture, no single molecular orbital represents a particular carbon-hydrogen bond, but instead more than one molecular orbital may contribute to each bond.

Chemists also explain the bonding at a carbon atom bound to four other atoms with tetrahedral geometry using the concept of **orbital hybridization.** According to this idea, overlap of hybrid atomic orbitals instead of the pure atomic orbitals of atoms forms molecular orbitals. **Hybrid orbitals** *are mathematical combinations of atomic orbitals. For a tetrahedral carbon atom, the wave functions for the four atomic orbitals, $2s$, $2p_x$, $2p_y$, and $2p_z$, are combined to create four new hybrid orbitals,* shown in Figure 2.9 (page 44).

The hybrid atomic orbitals are called sp^3 **hybrid orbitals,** meaning that they arise from a mathematical combination of one s orbital and three p orbitals, indicated by the superscript 3 over the p. *The number of hybrid orbitals generated is always equal to the number of atomic orbitals combined.*

Methane requires four identical σ bonds; therefore, four atomic orbitals are mixed. *The number of σ bonds present for a given atom dictates how many atomic orbitals are hybridized.* In other words, once chemists know what the structure of a molecule is, they decide which mathematical combination of atomic orbitals will result in new hybrid orbitals that have the directional properties necessary for that structure. The mathematical combination of one s and three p orbitals was chosen deliberately to create four hybrid orbitals directed to the corners of a tetrahedron. Like the pictures of atomic orbitals, the pictures of hybrid orbitals represent regions in space, or boundaries within which there is a finite probability of finding an electron.

Note that the general shape of a single sp^3 hybrid orbital resembles that of a p orbital in that it has a node at the nucleus of the carbon atom. An sp^3 orbital, however, no longer has the two equal lobes of a pure p orbital. The larger lobe of an sp^3 hybrid orbital extends farther into space from the nucleus of the carbon atom than any of the atomic orbitals. Thus, the new hybrid orbitals of carbon are better able to overlap with other orbitals and better able to form stronger covalent bonds than the pure atomic orbitals.

According to the orbital hybridization picture, each of the four carbon-hydrogen bonds in methane is formed by the overlap of a $1s$ orbital of a hydrogen atom and an sp^3 hybrid orbital of a carbon atom (Figure 2.10, page 45).

In this representation, the eight bonding electrons of methane are localized in four equivalent carbon-hydrogen bonds directed to the corners of a tetrahedron. Methane is pictured as having four equivalent σ bonds at tetrahedral angles to each other.

Organic chemists find the model of orbital hybridization to be useful in accounting

$$2s + 2p_x + 2p_y + 2p_z \xrightarrow[\text{combinations}]{\substack{\text{four} \\ \text{mathematical}}} \text{four } sp^3 \text{ hybrid orbitals}$$

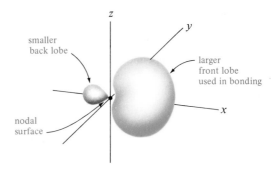

One sp^3 hybrid orbital shown
directed along the x axis.

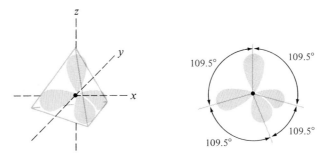

The four sp^3 hybrid orbitals shown
directed towards the corners of a regular
tetrahedron. The back lobes are not
shown, and front lobes are simplified.

The four sp^3 hybrid
orbitals and the
angles between them.

Figure 2.9 Hybridization of the $2s$, $2p_x$, $2p_y$, and $2p_z$ atomic
orbitals of carbon.

for many experimental observations and so use this way of thinking about bonding in organic compounds. We will use this language in describing the structure and bonding of a number of representative organic compounds in the remainder of this chapter.

Carbon tetrachloride, CCl_4, is another example of a compound containing a tetrahedral carbon atom. In carbon tetrachloride, four σ bonds are formed by the overlap of a p atomic orbital of each chlorine atom with an sp^3 hybrid orbital of the carbon atom (Figure 2.11).

A single bond between two carbon atoms is pictured as forming by the overlap of an sp^3 hybrid orbital from each one. For example, ethane, C_2H_6, has this type of single bond. Tetrahedral geometry is maintained around each carbon atom. In Figure 2.12 (page 46) the orbital picture of the bonding in ethane is shown, along with a three-dimensional structural formula showing the geometry of the final molecule.

The bonding in two simple members of the alkane family, methane and ethane, has been described. **Alkanes** *are a subclass of a much larger group of organic compounds called* **hydrocarbons.** *Hydrocarbons contain only the elements carbon and hydrogen.*

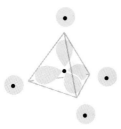

approach of four $1s$ orbitals of four hydrogen atoms to the four sp^3 hybrid orbitals of a single carbon atom

schematic drawing of the overlap between the $1s$ orbitals of the hydrogen atoms and the sp^3 hybrid orbitals of the carbon atom, forming four bonds directed towards the corners of a tetrahedron

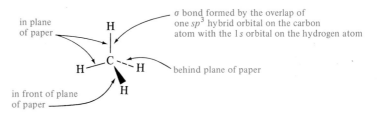

in plane of paper

H

σ bond formed by the overlap of one sp^3 hybrid orbital on the carbon atom with the $1s$ orbital on the hydrogen atom

H—C—H

behind plane of paper

in front of plane of paper

H

Figure 2.10   Formation of methane from overlap of hybrid sp^3 orbitals of a carbon atom with $1s$ orbitals of hydrogen atoms.

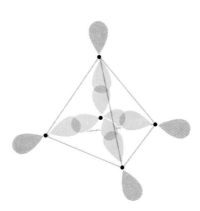

σ bond formed by overlap of sp^3 hybrid orbital on carbon atom with one of the $3p$ orbitals on the chlorine atom

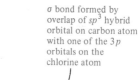

Cl

Cl—C—Cl

Cl

Picture showing overlap of atomic orbitals. Back lobes of sp^3 hybrid orbitals not shown, and orbitals not shown to scale.

Three–dimensional structural representation of carbon tetrachloride.

Figure 2.11   Formation of carbon tetrachloride by overlap of sp^3 orbitals of a carbon atom with p orbitals of chlorine atoms.

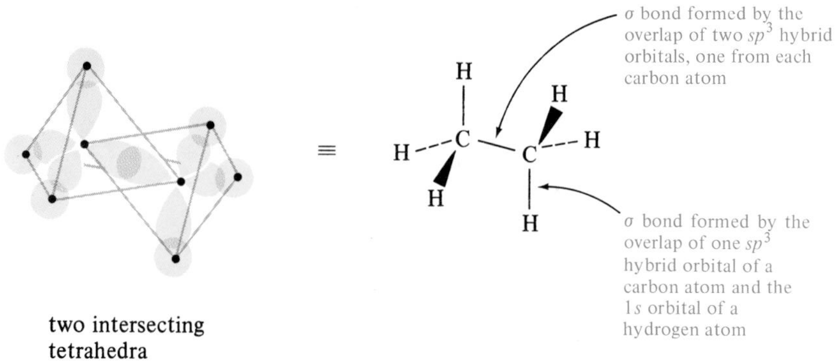

σ bond formed by the overlap of two sp^3 hybrid orbitals, one from each carbon atom

σ bond formed by the overlap of one sp^3 hybrid orbital of a carbon atom and the $1s$ orbital of a hydrogen atom

two intersecting tetrahedra

Figure 2.12   Orbital picture of ethane.

The alkane family of hydrocarbons is characterized by the presence of **tetrahedral carbon atoms** bonded to hydrogen atoms or to other tetrahedral carbon atoms. *The functional group (Section 1.5B) of an alkane is the tetrahedral carbon atom and the carbon-hydrogen bond.* As you learn the chemistry of various functional groups, it will become apparent that *alkanes have limited chemical reactivity.*

PROBLEM 2.4   Chloroform, $CHCl_3$, is another compound that contains a tetrahedral carbon atom. Sketch an orbital picture for it, showing how the bonding in the molecule arises. Also draw a three-dimensional representation of the molecule.

PROBLEM 2.5   Ethyl fluoride is C_2H_5F. It is like ethane, except that one of the hydrogen atoms has been replaced by a fluorine atom. Draw an orbital and a three-dimensional picture for it.

B. Some Functional Groups Containing sp^3-Hybridized Atoms Other than Carbon

The concept of sp^3 hybridization used in the description of bonding in tetrahedral carbon compounds can also be used to describe bonding in ammonia and related organic compounds. Like methane, ammonia has bond angles close to 109.5° (Section 1.6). Nitrogen, the central atom in ammonia, is bonded to three hydrogen atoms and has a pair of nonbonding electrons. Nitrogen has two electrons in the $2s$ orbital and one electron in each $2p$ orbital for a total of five electrons in its valence shell. Chemists explain the bonding in ammonia by postulating the hybridization of one $2s$ and three $2p$ orbitals to create the four sp^3 hybrid orbitals of nitrogen. One of the sp^3 hybrid orbitals of nitrogen is filled with a pair of electrons. The other three electrons of the nitrogen atom, along with three electrons from three hydrogen atoms, are in bonding molecular orbitals formed by the overlap of sp^3 hybrid orbitals of the nitrogen atom with $1s$ orbitals of the hydrogen atoms (Figure 2.13).

When ammonia reacts with an acid such as hydrogen chloride, the nonbonding electrons of nitrogen form a covalent bond with a proton, which has no electrons in the $1s$ orbital. The resulting ammonium ion, NH_4^+, contains four equal nitrogen-hydrogen bonds directed to the corners of a tetrahedron (Figure 2.13). The positively charged ion is held by an ionic bond to the chloride ion, which is also formed in the reaction.

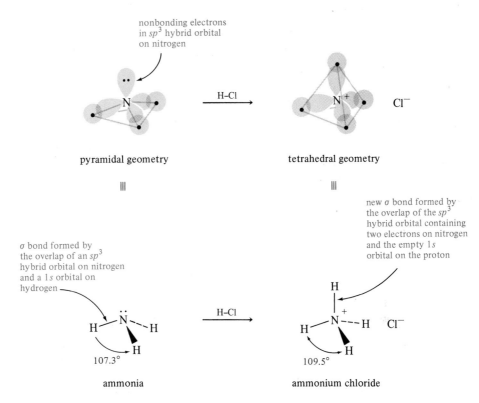

nonbonding electrons
in *sp*³ hybrid orbital
on nitrogen

pyramidal geometry

tetrahedral geometry

σ bond formed by
the overlap of an *sp*³
hybrid orbital on nitrogen
and a 1*s* orbital on
hydrogen

new σ bond formed by
the overlap of the *sp*³
hybrid orbital containing
two electrons on nitrogen
and the empty 1*s*
orbital on the proton

107.3°

109.5°

ammonia

ammonium chloride

Figure 2.13   Orbital and three-dimensional pictures of ammonia and ammonium chloride.

PROBLEM 2.6   What is wrong with a picture of the covalent bonding in ammonia that locates the nonbonding electrons of the molecule in the 2*s* orbital of nitrogen and postulates that the three nitrogen-hydrogen single bonds are formed by the overlap of 2*p* atomic orbitals of nitrogen with 1*s* atomic orbitals of hydrogen?

There are many organic compounds that contain nitrogen with tetrahedral geometry of the electron pairs. Most of them belong to a functional group class called amines. **Amines** *are organic derivatives (relatives) of ammonia in which one or more of the hydrogen atoms are replaced by a group that contains carbon.* Simple examples of amines are methylamine, CH_3NH_2, and trimethylamine, $(CH_3)_3N$. The carbon-nitrogen bonds in both of these amines are formed by the overlap of an *sp*³ hybrid orbital of a carbon atom with an *sp*³ hybrid orbital of a nitrogen atom. The C—N—H bond angle of methylamine, 107°, is close to that in ammonia, as is the C—N—C bond angle for trimethylamine, 108°. The carbon-nitrogen bond length for both amines is 1.47 Å, which is shorter than a carbon-carbon σ bond, 1.54 Å (Figure 2.14, page 48).

Another element that is often found in organic compounds is oxygen. The H—O—H bond angle in water is found experimentally to be 104.5°, somewhat smaller than the tetrahedral angle (Section 1.6). Oxygen has six valence electrons, two in the 2*s* orbital and four in three 2*p* orbitals. After hybridization, two of the *sp*³ hybrid orbitals are filled with pairs of nonbonding electrons. The σ bonds in water are formed when the remaining two *sp*³ hybrid orbitals of the oxygen atom overlap with 1*s* orbitals

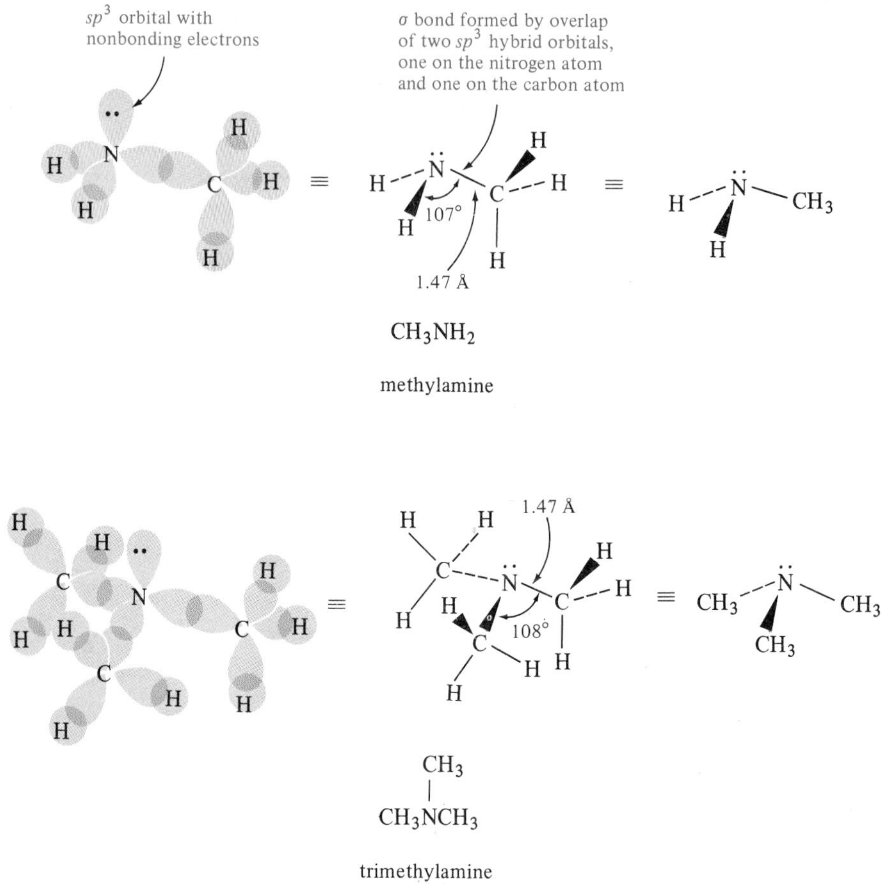

sp^3 orbital with nonbonding electrons

σ bond formed by overlap of two sp^3 hybrid orbitals, one on the nitrogen atom and one on the carbon atom

107°

1.47 Å

CH_3NH_2

methylamine

1.47 Å

108°

CH_3
|
CH_3NCH_3

trimethylamine

Figure 2.14 Orbital and three-dimensional pictures of methylamine and trimethylamine.

of two hydrogen atoms. Just as the nonbonding electrons in ammonia form a bond by overlapping with the empty $1s$ orbital of a proton, so does a water molecule react with an acid to form a hydronium ion. The molecular orbital pictures for water and for the hydronium ion are shown in Figure 2.15.

The deviation of the bond angle in the water molecule from the tetrahedral value is explained by postulating that the two pairs of nonbonding electrons on oxygen occupy more space and repel each other more strongly than do the two pairs of electrons in the oxygen-hydrogen bonds because bonding electrons are constrained in space by their interaction with two nuclei.

Alcohols and ethers, which are the organic relatives of water, were described in Sections 1.5B and 1.8B. Both of these classes of compounds have the same geometry around the oxygen atom as water does. In alcohols, one of the hydrogen atoms in water is replaced by a group containing a tetrahedral carbon atom. The structures of two typical alcohols, methanol, CH_3OH, and ethanol, CH_3CH_2OH, are shown in Figure 2.16 (page 50). The carbon-oxygen bond in each compound is formed by the overlap of an sp^3 hybrid orbital from each atom.

In ethers, both hydrogen atoms of water are substituted by groups containing carbon atoms. The structures of dimethyl ether and diethyl ether, which was once used as an

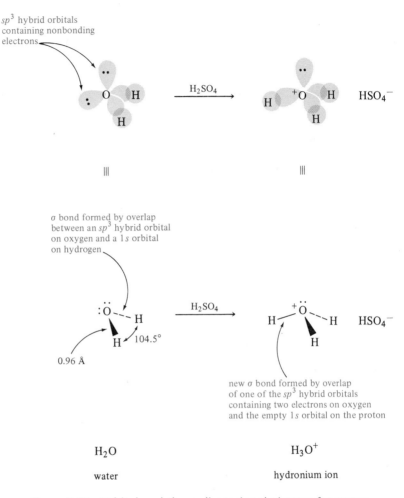

sp³ hybrid orbitals
containing nonbonding
electrons

σ bond formed by overlap
between an *sp³* hybrid orbital
on oxygen and a 1*s* orbital
on hydrogen

104.5°

0.96 Å

new σ bond formed by overlap
of one of the *sp³* hybrid orbitals
containing two electrons on oxygen
and the empty 1*s* orbital on the proton

H_2O

water

H_3O^+

hydronium ion

Figure 2.15   Orbital and three-dimensional pictures for water
and the hydronium ion.

anesthetic, are shown in Figure 2.17 (page 50). These compounds also contain sp^3-hybridized oxygen atoms and have two pairs of nonbonding electrons occupying two of the hybrid orbitals. The structures in Figure 2.17 are drawn in a simplified form to emphasize the importance of the oxygen atom and the nonbonding electrons in ethers.

In this section, we have been examining compounds that contain atoms of carbon, nitrogen, and oxygen that are covalently bonded to other atoms by single bonds with bond angles that are close to tetrahedral. The compounds fall into four functional group classes. *Alkanes* contain only tetrahedral carbon atoms bonded to hydrogen atoms. *Amines* are organic relatives of ammonia. *Alcohols* and *ethers* are related to water. The orbital and three-dimensional pictures of simple representatives of each of these functional group classes also apply to their more complex members.

The reactions of organic compounds are directly related to their structures. Thus, in many cases we can predict the chemistry of a complex molecule from what we know about the reactivity of a simple one in the same functional group class. The ability to predict reactivity from an inspection of the structure of a compound is important in the study of organic chemistry. In solving the following problems you can take the first steps in developing this skill.

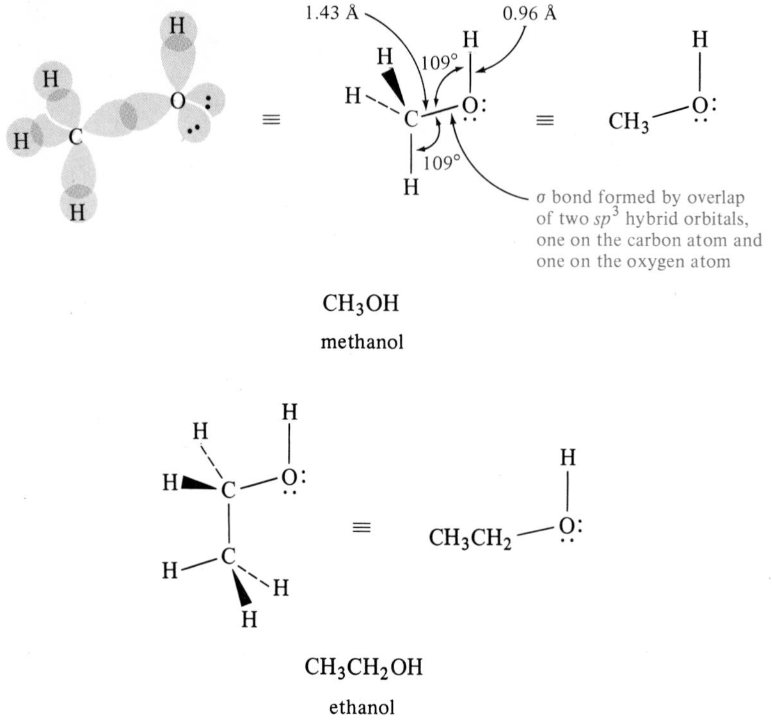

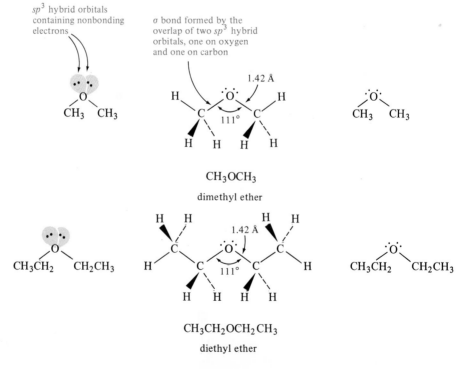

1.43 Å
0.96 Å
109°
109°
σ bond formed by overlap of two sp^3 hybrid orbitals, one on the carbon atom and one on the oxygen atom
CH_3OH
methanol
CH_3CH_2OH
ethanol

Figure 2.16 Orbital picture of methanol and three-dimensional pictures of methanol and ethanol.

sp^3 hybrid orbitals containing nonbonding electrons
σ bond formed by the overlap of two sp^3 hybrid orbitals, one on oxygen and one on carbon
1.42 Å
111°
CH_3OCH_3
dimethyl ether
1.42 Å
111°
$CH_3CH_2OCH_2CH_3$
diethyl ether

Figure 2.17 Structures of dimethyl ether and diethyl ether.

PROBLEM 2.7   Figure 2.13 illustrates the change in bonding that takes place when ammonia reacts with an acid. The nonbonding electrons on amines also interact with the proton from an acid in the same way. Draw structural formulas showing what happens when methylamine (Figure 2.14) reacts with hydrogen chloride.

PROBLEM 2.8   Alcohols and ethers react with acids the same way water does because of the presence of nonbonding electrons on the oxygen atom. Using Figure 2.15 as your guide, write structural formulas that show what happens when methanol (Figure 2.16) and dimethyl ether (Figure 2.17) are put into sulfuric acid.

PROBLEM 2.9   Write three-dimensional structural formulas for the following species. In each case, indicate which types of orbitals overlap to form each bond to the atom that is underlined. For some of the species, you may find it helpful to write Lewis structures first.

(a) $H_2\underline{N}-NH_2$ (b) $\underline{B}F_4^-$ (c) $CH_3-\overset{\displaystyle CH_3}{\overset{\displaystyle |+}{\underset{\displaystyle |}{\underset{\displaystyle CH_3}{N}}}}-CH_3$ (d) $CH_3-\overset{\displaystyle CH_3}{\overset{\displaystyle |+}{\underline{O}}}-CH_3$

PROBLEM 2.10   Identify each of the compounds shown below as belonging to one of the following functional group classes: alkanes, amines, alcohols, or ethers.

(a) $CH_3CH_2CH_2CH_2CH_2OH$ (b) $CH_3CH_2CH_2OCH_3$ (c) $CH_3CH_2CH_2CH_3$

(d) $CH_3CH_2NH_2$ (e) $CH_3\underset{\displaystyle |}{\underset{\displaystyle OH}{C}}HCH_2CH_3$ (f) $CH_3CH_2\overset{\displaystyle H}{\overset{\displaystyle |}{N}}CH_2CH_3$

PROBLEM 2.11   Draw structural formulas illustrating the important electronic features of the functional group in each of the compounds in Problem 2.10. Apply the knowledge you gained in doing the earlier exercises to predict which compounds will react with sulfuric acid.

2.5

The Orbital Picture for Compounds Containing Trigonal Carbon Atoms

A. Covalent Bonding in Alkenes

Not all compounds of carbon contain tetrahedral carbon atoms. Ethylene, $CH_2=CH_2$, for example, is a flat molecule in which all six atoms lie in the same plane (Section 1.6). Experiments show that the $H-C-H$ and $H-C-C$ bond angles are close to 120°, while the carbon-carbon bond length is 1.34 Å, which is shorter than the carbon-carbon bond in ethane. *Ethylene is the simplest member of the family of hydrocarbons called* **alkenes.** *The functional group of the alkenes is the carbon-carbon double bond,*

$$-\overset{\displaystyle |}{C}=\overset{\displaystyle |}{C}-$$

(Section 1.5B). In ethylene, each carbon atom is bonded to three other atoms, all lying in a plane. Such *a carbon atom with three other atoms around it in a plane is known as a* **trigonal carbon atom.** Other atoms, such as boron in boron trifluoride, BF_3, are also trigonal.

Each carbon atom in ethylene is bonded to only three other atoms. Therefore, to rationalize the bonding in ethylene we say that only two of the $2p$ orbitals of carbon are combined with the $2s$ orbital to create three new sp^2 **hybrid orbitals.** The shape of an sp^2 orbital is similar to that of an sp^3 hybrid orbital, but the spatial orientation is quite different. The three sp^2 hybrid orbitals lie in a plane and are directed to the corners of an equilateral triangle with 120° angles between them. Each carbon atom has a third $2p$ orbital, which is unhybridized. This orbital retains its shape as an atomic orbital and is perpendicular to the plane defined by the three sp^2 hybrid orbitals (Figure 2.18).

The skeleton of ethylene is made up of the σ bonds between all of the atoms. One carbon-carbon bond in ethylene is formed by the overlap of sp^2 hybrid orbitals from each carbon atom. The carbon-hydrogen bonds are created by the overlap of sp^2 hybrid

$$2s + 2p_x + 2p_y \xrightarrow[\text{combinations}]{\substack{\text{three} \\ \text{mathematical}}} \text{three } sp^2 \text{ hybrid orbitals}$$

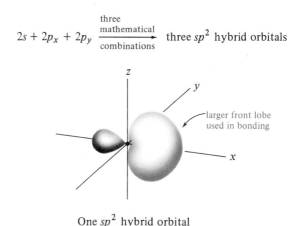

One sp^2 hybrid orbital

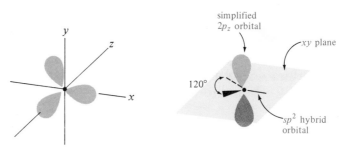

Three axes tipped 90°
to show sp^2 hybrid orbitals
lying in xy plane. Back lobes
not shown and front lobes
simplified.

p orbital perpendicular
to plane defined by
sp^2 hybrid orbitals.

Figure 2.18 sp^2 Hybrid orbitals of a trigonal carbon atom.

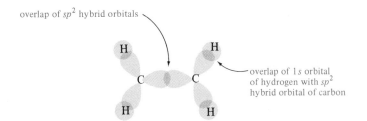

overlap of sp^2 hybrid orbitals

overlap of $1s$ orbital
of hydrogen with sp^2
hybrid orbital of carbon

Schematic drawing of the overlap between the
$1s$ orbitals of the hydrogen atoms and the
sp^2 hybrid orbitals of the carbon atoms forming
the C–H σ bonds of ethylene. The C–C σ bond is
formed by overlap of two sp^2 hybrid orbitals.

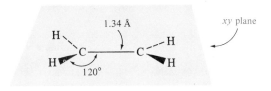

1.34 Å

xy plane

120°

σ bond skeleton of
ethylene lying in
xy plane.

Figure 2.19   Skeleton of σ bonds in ethylene.

orbitals of the carbon atoms with $1s$ orbitals of the hydrogen atoms. Each carbon atom
contributes three of its four valence electrons to these bonds (Figure 2.19).

The fourth valence electron of each carbon atom is in the unhybridized $2p$ orbital.
Two p atomic orbitals, if suitably placed on adjacent atoms, interact to create two new
molecular orbitals. They are the π (read ''pi'') **molecular orbitals.** Two p orbitals can
be placed next to each other so that lobes of the same sign, or color, are on the same
side of the nodal plane. In this case, the lobes that bear the same mathematical sign
interact to reinforce each other. The two orbitals are *in phase* and they combine to form
a bonding π orbital. If, on the other hand, the p orbitals are located so that lobes of
opposite sign (or differing colors) are next to each other, they are *out of phase* and no
bonding takes place. In this situation, there is a node (a region of no electron density)
between the nuclei of the two atoms and an antibonding π^* (read ''pi star'') orbital
results.

When the two carbon atoms in ethylene form a σ bond, the p atomic orbitals are close
enough that their parallel lobes interact as shown in Figure 2.20 (page 54). Two
electrons, one from each carbon atom, are in the bonding π molecular orbital, creating
a π bond between the two carbon atoms. *Thus, the double bond in ethylene consists
of a σ bond and a π bond.*

Each p orbital has two lobes for bonding and a node at the nucleus. The π molecular
orbital, because it is created by the side-to-side overlap of the p orbitals, also has two
lobes and a nodal plane. Thus, in a π bond there is a finite probability of finding
electrons in lobes above and below the plane of the molecule. Note (Figure 2.20) that
a π bond does not have the cylindrical symmetry of a σ bond.

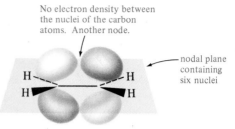

No electron density between
the nuclei of the carbon
atoms. Another node.

nodal plane
containing
six nuclei

Two $2p_z$ orbitals,
one on each carbon
atom, out of phase
with each other.

Antibonding π^* molecular orbital,
no interaction between two
adjacent p orbitals.

nodal plane
containing
six nuclei

Two $2p_z$ orbitals,
one on each carbon
atom, in phase with
each other.

Bonding π molecular orbital
showing electron density
between the two carbon atoms
in regions above and below the
plane containing the nuclei.

Figure 2.20   Molecular orbital picture of the π bond in
ethylene.

B. Covalent Bonding in Carbonyl Compounds. Aldehydes and Ketones

A number of organic compounds contain π bonds between carbon and oxygen atoms. *The* **carbonyl group,** *which contains a carbon-oxygen double bond,*

$$-\overset{|}{C}{=}O$$

is an important structural unit in such compounds. Compounds containing the carbonyl group are divided into different functional group classes depending on the other groups or atoms that are bonded to the carbon atom of the carbonyl group (Section 1.5B). *In* **aldehydes,** *for example, the carbon atom of the carbonyl group is bonded to one carbon atom and one hydrogen atom.* However, the simplest aldehyde, formaldehyde, $H_2C{=}O$, is an exception to this rule. *The carbonyl group of a* **ketone** *is bonded to two carbon atoms.* The bonds in acetaldehyde, a typical aldehyde, and in acetone, a typical ketone, are shown in Figure 2.21.

The carbon atom of the carbonyl group is bonded to three other atoms, all lying in a plane, and therefore fits the definition of a trigonal carbon atom. The bond angles around the carbonyl group are approximately 120°. The carbon atom of the carbonyl group is said to be sp^2 hybridized. The three hybrid orbitals form the backbone of σ bonds at the carbonyl group.

Just as with ethylene, we can identify a $2p$ orbital on the carbon atom of the carbonyl group that is perpendicular to the plane defined by the three sp^2 hybrid orbitals. To account for the double bond between the carbon and the oxygen atoms, chemists postulate that the oxygen atom is also sp^2 hybridized, and thus has a $2p$ orbital available for overlap with the $2p$ orbital of the carbon atom. A π bond is formed between the

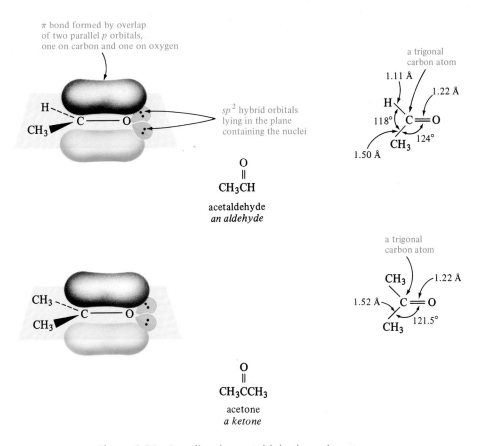

π bond formed by overlap of two parallel *p* orbitals, one on carbon and one on oxygen

sp² hybrid orbitals lying in the plane containing the nuclei

a trigonal carbon atom

1.11 Å

1.22 Å

H

118°   C═O

CH₃   124°

1.50 Å

O
‖
CH₃CH

acetaldehyde
an aldehyde

a trigonal carbon atom

CH₃   1.22 Å

1.52 Å   C═O

CH₃   121.5°

O
‖
CH₃CCH₃

acetone
a ketone

Figure 2.21   Bonding in acetaldehyde and acetone.

carbon and oxygen atoms. The π bond in a carbonyl compound has the same shape and symmetry as the π bond in ethylene.

The pictures shown in Figure 2.21 suggest, for the sake of consistency, that the two pairs of nonbonding electrons on the oxygen atom of the carbonyl group are in sp^2 hybrid orbitals.

PROBLEM 2.12   Using the drawings of ethylene in Figures 2.19 and 2.20 as a guide, construct the σ backbone for acetaldehyde (Figure 2.21). Show the different orbitals that must overlap to form the bonds. Also show how the π bond comes into being.

C. Carboxylic Acids and Esters

If another oxygen atom is attached by a single bond to the carbon atom of a carbonyl group, two new classes of organic compounds, carboxylic acids and esters, result. *In* **carboxylic acids,** *the carbonyl group is bonded to a hydroxyl group*. The oxygen atom of the hydroxyl portion of the carboxylic acid is represented as being sp^3 hybridized in the same way that the oxygen atom of an alcohol is (Section 2.4B). *If the hydrogen atom of the hydroxyl portion of a carboxylic acid is replaced by a group containing carbon, the new molecule that results is called an* **ester** (Figure 2.22, page 56).

Both functional groups, carboxylic acids and esters, contain a carbonyl group bonded to an sp^3-hybridized oxygen atom.

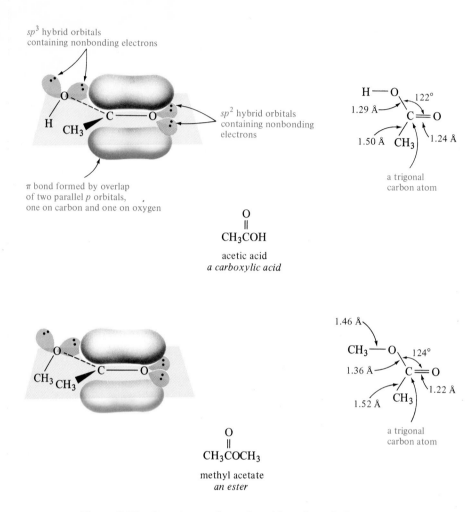

$$\underset{\substack{\text{acetic acid}\\ \textit{a carboxylic acid}}}{\overset{\displaystyle O \atop \displaystyle \|}{CH_3COH}}$$

$$\underset{\substack{\text{methyl acetate}\\ \textit{an ester}}}{\overset{\displaystyle O \atop \displaystyle \|}{CH_3COCH_3}}$$

Figure 2.22   Structures of acetic acid and methyl acetate.

PROBLEM 2.13   Practice identifying the hybridization of all the atoms and the origin of all the bonds in methyl acetate (Figure 2.22).

PROBLEM 2.14   The range of functional group classes that you have already learned to recognize includes alkanes, alkenes, amines, alcohols, ethers, aldehydes, ketones, carboxylic acids, and esters. To which functional group class does each of the following compounds belong?

(a) $\underset{\displaystyle OH}{CH_3CHCH_3}$

(b) $\underset{\displaystyle CH_3}{\overset{\displaystyle CH_3}{}}\underset{\displaystyle H}{\overset{\displaystyle CH_3}{C=C}}$

(c) $CH_3CH_2\overset{\displaystyle O \atop \displaystyle \|}{CH}$

(d) $CH_3CH_2\overset{\displaystyle O \atop \displaystyle \|}{C}CH_2CH_3$

(e) $CH_3CH_2CH_2NH_2$ (f) $CH_3CHCH_2CH_3$ (g) $CH_3CH_2CH_2\overset{\overset{\displaystyle O}{\|}}{C}OH$
$\qquad\qquad\qquad\qquad\qquad\;\;\; |$
$\qquad\qquad\qquad\qquad\quad\;\; CH_3$

(h) $CH_3CH_2OCHCH_3$ (i) $CH_3CH_2\overset{\overset{\displaystyle O}{\|}}{C}OCH_2CH_3$
$\qquad\qquad\quad\;\; |$
$\qquad\qquad\;\; CH_3$

PROBLEM 2.15 Some compounds contain more than one functional group. Identify the functional groups that are present in the following naturally occurring compounds.

(a) $CH_3\overset{\overset{\displaystyle O}{\|}}{C}HCOH$ (b) $CH_3\overset{\overset{\displaystyle O}{\|}}{C}HCOH$ (c) $CH_3\overset{\overset{\displaystyle OO}{\|\,\|}}{C}COH$ (d) $HOCH_2\overset{\overset{\displaystyle O}{\|}}{C}HCH$
$\quad\;\;\; |$ $\qquad\qquad\qquad\qquad |$ $\qquad\qquad\qquad\qquad\qquad\qquad\qquad\qquad\;\; |$
$\quad\; OH$ $\qquad\qquad\qquad\; NH_2$ $\qquad\qquad\qquad\qquad\qquad\qquad\qquad\quad\;\; OH$
     lactic acid   alanine   pyruvic acid   glyceraldehyde

(e) $CH_3CH_2CH_2CH_2CH_2CH_2CH_2CH_2CH=CHCH_2CH_2CH_2CH_2CH_2CH_2CH_2\overset{\overset{\displaystyle O}{\|}}{C}OH$
                                                oleic acid

PROBLEM 2.16 What is the hybridization of the atoms shown in color in each of the following compounds?

(a) $CH_3-\overset{\overset{\displaystyle :O:}{\|}}{C}-\overset{\displaystyle |}{\underset{\displaystyle |}{\ddot N}}-H$ (b) $CH_3CH_2\overset{\overset{\displaystyle :O:}{\|}}{C}-\ddot{\underset{\cdot\cdot}{O}}-\overset{\overset{\displaystyle H}{|}}{C}=\overset{\overset{\displaystyle H}{|}}{C}-H$ (c) $CH_3-\overset{\overset{\displaystyle :\ddot OCH_3}{|}}{\underset{\underset{\displaystyle :\ddot OCH_3}{|}}{C}}-CH_3$
$\qquad\qquad\;\; H$

(d) $H-\overset{\overset{\displaystyle H}{|}}{C}=\overset{\overset{\displaystyle H}{|}}{C}-\overset{\overset{\displaystyle H}{|}}{\underset{\underset{\displaystyle H}{|}}{C}}-H$ (e) $CH_3-\overset{\displaystyle \ddot N}{=}\overset{\overset{\displaystyle CH_3}{|}}{C}-CH_3$

2.6

The Orbital Picture for Linear Compounds

A. Bonding in Alkynes

Alkynes *are a class of hydrocarbons in which the functional group is a carbon-carbon triple bond,* $-C\equiv C-$. The simplest member of this class is acetylene, $HC\equiv CH$. Experimental measurements show that the four atoms in acetylene lie in a straight line with an $H-C-C$ bond angle of 180° (Section 1.6). The molecule is **linear.**

Each carbon atom in acetylene bonds with only two other atoms. Such a carbon atom is said to be *sp* hybridized. Two *sp* **hybrid orbitals** are formed by a combination of one 2*s* orbital and one 2*p* orbital of carbon. These hybrid orbitals of carbon point away from each other in a straight line. The two carbon atoms of acetylene are joined by the

$$2s + 2p_x \xrightarrow[\text{combinations}]{\begin{array}{c}\text{two}\\\text{mathematical}\end{array}} \text{two } sp \text{ hybrid orbitals}$$

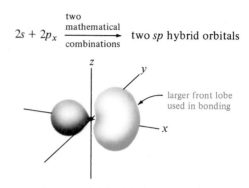

larger front lobe
used in bonding

One *sp* hybrid orbital.

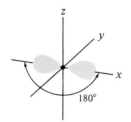

Two *sp* hybrid orbitals.
Back lobes not shown and
front lobes simplified.

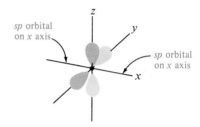

sp orbital
on *x* axis

sp orbital
on *x* axis

Two simplified *p* orbitals in *yz* plane
perpendicular to the two
sp hybrid orbitals.

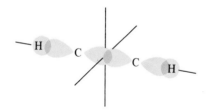

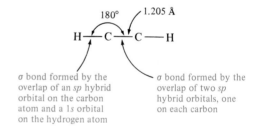

σ bond formed by the
overlap of an *sp* hybrid
orbital on the carbon
atom and a 1*s* orbital
on the hydrogen atom

σ bond formed by the
overlap of two *sp*
hybrid orbitals, one
on each carbon

Figure 2.23   Orbital hybridization and the σ bond skeleton of
acetylene.

overlap of one *sp* hybrid orbital from each atom. The carbon-hydrogen σ bond results
from the overlap of the *sp* hybrid orbital of carbon with the 1*s* orbital of hydrogen
(Figure 2.23).

Two of the valence electrons of each carbon atom are used to form the σ bonds. Each
carbon atom also has two unhybridized 2*p* orbitals at right angles to each other, with
one electron in each *p* orbital. As is the case in ethylene, the *p* orbitals of the carbon
atoms overlap to form bonding π molecular orbitals. Because there are two sets of *p*
orbitals, two sets of π molecular orbitals are formed. Each bonding π molecular orbital
has two electrons in it, giving rise to a π bond. Thus, the triple bond in acetylene
consists of a σ bond and two π bonds (Figure 2.24).

Two views of the π bonds are shown in Figure 2.24. A side view of the molecule
shows that there is electron density *above* and *below* the line defined by the nuclei, as
well as *in front* and *behind* it. A view from the end of the molecule shows that there
is actually electron density all around the axis of the molecule. There is a nodal axis
along the carbon-carbon bond.

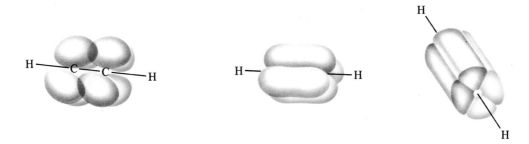

two perpendicular *p* orbitals on each
carbon atom, each one parallel to one
on the other carbon atom

side view of the two
π bonds perpendicular
to each other

view of the two π
bonds showing toroidal
symmetry

Figure 2.24 The π bonds in acetylene.

B. Bonding in Nitriles

Nitriles *comprise a class of compounds that contain a triple bond between a carbon and a nitrogen atom. The functional group in nitriles is the cyano group,* —C≡N. The carbon atom of the cyano group is bonded to nitrogen and to one other atom. A typical nitrile is acetonitrile, $CH_3C≡N$, shown in Figure 2.25.

The carbon atom and the nitrogen atom in the cyano group are *sp* hybridized and the carbon-nitrogen triple bond arises in exactly the same way as the carbon-carbon triple bond in acetylene, the simplest alkyne. The triple bond consists of a σ bond formed by the overlap of one *sp* hybrid orbital each from carbon and nitrogen, and two π bonds from the overlap of two *p* orbitals on each atom. One of the *sp* orbitals of carbon overlaps with the sp^3 orbital of the tetrahedral carbon atom to which it is bonded. The nonbonding electrons of nitrogen occupy the second *sp* orbital of that atom.

the orbital picture of
acetonitrile showing the
formation of the σ bonds

σ bond skeleton
of acetonitrile

sp hybrid orbital
containing the
nonbonding electrons

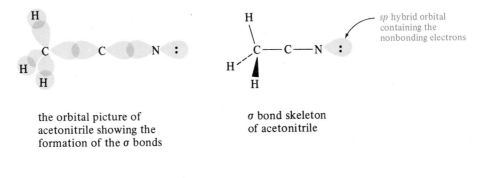

p orbitals used
to form π bonds

Figure 2.25 The orbital picture and structure of acetonitrile.

PROBLEM 2.17 Identify the functional group class to which each of the following compounds belongs.

(a) $CH_3CH_2C\equiv CH$ (b) $CH_3C=CHCH_2CH_2CH_3$ (c) $CH_3CH_2\overset{\displaystyle O}{\overset{\|}{C}}CH_2CH_3$

                                  CH_3

(d) $CH_3CH_2CH_2\overset{\displaystyle O}{\overset{\|}{C}}OCH_3$ (e) $CH_3CH_2CH_2C\equiv N$ (f) $CH_3CH_2CH_2\overset{\displaystyle O}{\overset{\|}{C}}H$

(g) $CH_3CH_2\overset{\displaystyle O}{\overset{\|}{C}}HCOH$ (h) $CH_3CH_2CHCH_2CH_3$

                  CH_3 NH_2

PROBLEM 2.18 How many σ and how many π bonds are present in each of the following structures?

(a) $H-C=O$ (b) CH_3CH_2OH (c) $H-C\equiv C-C\equiv C-H$

          H

(d) $H-C=C-C\equiv N$ (e) $CH_3-\overset{\displaystyle O}{\overset{\|}{C}}-OCH_3$ (f) $CH_3CH_2CH_3$

        H H

PROBLEM 2.19 Draw a detailed picture of the orbitals involved in the bonding in hydrogen cyanide, HCN.

PROBLEM 2.20 Draw a detailed picture of the orbitals involved in the bonding in propyne, $CH_3C\equiv CH$.

2.7

Covalent Bond Lengths and Their Relation to Orbital Hybridization

A. Bond Lengths in Hydrocarbons

Bond lengths and bond angles for a number of typical covalent compounds are given in Chapter 1 and in the earlier sections of this chapter. The values are remarkably constant for each particular kind of bond and are determined not only by the atoms involved but also by the hybridization of orbitals of those atoms.

Electrons in $2s$ atomic orbitals are closer to the nucleus than electrons in $2p$ atomic orbitals. Orbital hybridization affects bond length because a hybrid orbital with a greater percentage of s character is closer to the nucleus than a hybrid orbital with less s character. The percentage of s character of an orbital is the percentage of s orbital in the total number of orbitals used in hybridization. For example, in ethane, sp^3-hybridized carbon atoms have one-fourth or 25% s character and three-fourths or 75%

TABLE 2.2 Orbital Hybridization and Bond Lengths in Propane, Propene, and Propyne

Compound	*Bond*	*Bond Length, Å*	*Orbitals Used in Bonding*
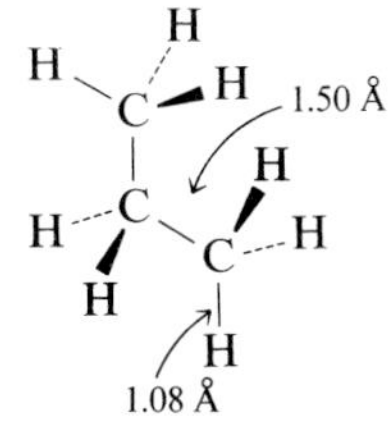 propane	C—C	1.50	sp^3-sp^3
	C—H	1.08	sp^3-s
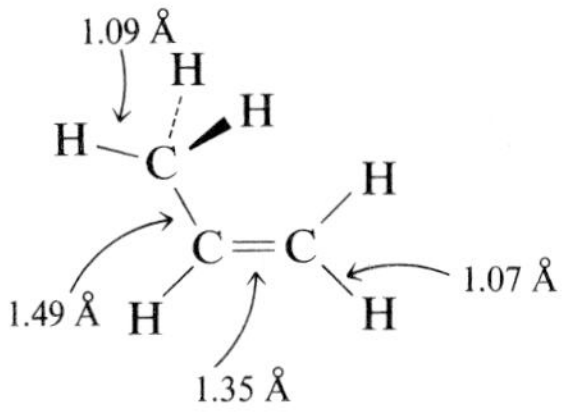 propene	C—C	1.49	sp^3-sp^2
	C=C	1.35	sp^2-sp^2
	C—H	1.09	sp^3-s
	C—H	1.07	sp^2-s
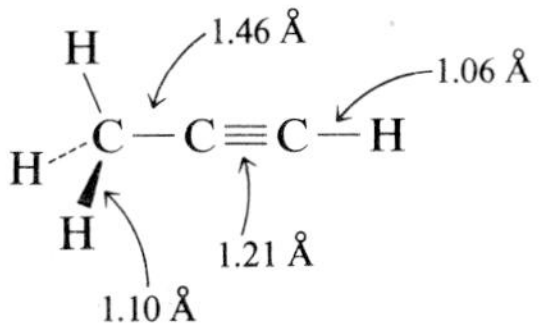 propyne	C—C	1.46	sp^3-sp
	C≡C	1.21	sp-sp
	C—H	1.10	sp^3-s
	C—H	1.06	sp-s

p character. For an *sp*-hybridized carbon atom, such as that found in acetylene, the hybrid orbitals have 50% *s* character.

The more *s* character there is in a hybrid orbital, the closer that orbital is to the nucleus and the more tightly held are its electrons. An orbital with a large percentage of *s* character, therefore, forms a shorter σ bond than one with a small percentage of *s* character. The lengths of carbon-carbon and carbon-hydrogen bonds in propane, propene, and propyne are compared in Table 2.2.

A carbon-carbon single bond between sp^3-hybridized carbon atoms, such as in propane, is longer (1.50 Å) than the carbon-carbon single bond between an sp^3-hybridized carbon atom and an sp^2-hybridized carbon atom (1.49 Å) in propene. The comparable bond between sp^3-hybridized and *sp*-hybridized carbon atoms in propyne is the shortest of all (1.46 Å). A similar trend is seen for carbon-hydrogen bonds as the hybridization of the carbon atom changes from sp^3 (1.10–1.08 Å) to sp^2 (1.07 Å) to *sp* (1.06 Å).

In this series we also see the effect of π bonding in addition to the σ bond between two carbon atoms. The carbon-carbon double bond in propene is shorter (1.35 Å) than the carbon-carbon single bond in propane (1.50 Å). Propyne, with two π bonds between two of the carbon atoms, has the shortest carbon-carbon bond length of all (1.21 Å). Carbon atoms are held closer together by multiple bonds than by single bonds. The shorter bond lengths also signify stronger bonds. This relationship is presented in detail in Section 5.4B.

B. Orbital Hybridization and the Electronegativities of Carbon Atoms

Chemists have rationalized many experimental observations of the properties and reactivity of compounds containing carbon atoms with various tetrahedral, trigonal, and linear geometries by suggesting that these carbon atoms have different electronegativities. Such a conclusion follows directly from the differing percentages of *s* character in the hybrid orbitals used in bonding by different types of carbon atoms. An electron in an *s* orbital is, on the average, closer to the nucleus than an electron in a *p* orbital. The carbon atoms in acetylene, with 50% *s* character, hold the electrons of the bond more tightly to the nucleus of the carbon atom, and are thus more electronegative than the carbon atoms in ethane, which have only 25% *s* character. This kind of reasoning is used many times in this book to explain the different chemical properties of compounds composed of different types of carbon atoms.

C. Bond Lengths in Organic Compounds Containing Oxygen and Nitrogen

Like carbon-carbon bonds, carbon-oxygen and carbon-nitrogen bond lengths are related to orbital hybridization and to the presence of multiple bonds. Information about some of the compounds containing such bonds is summarized in Table 2.3.

The trends for carbon-carbon bonds are also observed for carbon-nitrogen and carbon-oxygen bonds. The carbon-nitrogen single bond in methylamine is longer than the carbon-nitrogen triple bond in acetonitrile. Similarly, a carbon-oxygen single bond in methanol or dimethyl ether is longer than a carbon-oxygen double bond in acetaldehyde. Three different carbon-oxygen bond lengths are seen in methyl acetate. The shortest one (1.22 Å) is the double bond of the carbonyl group. The next (1.36 Å) is

TABLE 2.3   Orbital Hybridization and Bond Lengths in Organic Compounds Containing Nitrogen and Oxygen

Compound	Bond	Bond Length, Å	Orbitals Used for Bonding
CH_3NH_2 methylamine	$C-N$	1.47	C, sp^3; N, sp^3
$(CH_3)_2C=NOH$ acetoxime	$C=N$ $C-C$	1.29 1.49	C, sp^2; N, sp^2 C, sp^3; C, sp^2
$CH_3C\equiv N$ acetonitrile	$C\equiv N$ $C-C$	1.16 1.46	C, sp; N, sp C, sp^3; C, sp
CH_3OH methanol	$C-O$ $O-H$	1.43 0.96	C, sp^3; O, sp^3 O, sp^3; H, s
CH_3OCH_3 dimethyl ether	$C-O$	1.42	C, sp^3; O, sp^3
$\overset{\displaystyle O}{\overset{\displaystyle \|}{CH_3CH}}$ acetaldehyde	$C=O$ $C-C$	1.22 1.50	C, sp^2; O, sp^2 C, sp^3; C, sp^2
$\overset{\displaystyle O}{\overset{\displaystyle \|}{CH_3COCH_3}}$ methyl acetate	$C=O$ $C-O$ $O-C$	1.22 1.36 1.46	C, sp^2; O, sp^2 C, sp^2; O, sp^3 C, sp^3; O, sp^3

the bond between the sp^2-hybridized carbon atom of the carbonyl group and the sp^3-hybridized oxygen atom. The longest bond (1.46 Å) is between the sp^3-hybridized oxygen atom and a tetrahedral carbon atom.

We pay this much attention to the structural details of these few compounds because the types of bonding observed in them are representative of the bonding found in a wide range of organic compounds. The bond lengths for propane are typical of those in all compounds that have two or more tetrahedral carbon atoms bonded to each other. Whenever we see a double or a triple bond between carbon atoms, we will assume that they have properties similar to the bonds in propene or propyne. Similarly, the properties of the functional groups in the simple amines, alcohols, ethers, and carbonyl compounds analyzed here in detail closely resemble the properties of those functional groups in more complex compounds.

PROBLEM 2.21   Using the data in Tables 2.2 and 2.3 as a guide, assign approximate lengths to the bonds indicated by arrows in the following structural formulas.

(a) CH_3CH_2—CH_2—C—H (with H above and below C)

(b) CH_3—CH—CH_3, with O—H below

(c) CH_3CH_2—O—CH_3

(d) $CH_3CH_2C\equiv C$—H

(e) CH_3CH_2—C—CH_2CH_3, with O double-bonded above

(f) $CH_3CH_2C\equiv N$

(g) CH_3CH_2—C=C—H (with H H above)

(h) CH_3CH_2—NH_2

2.8

Effects of Bonding on Chemical Reactivity

Functional groups of organic compounds are the sites of chemical reactions. Among the hydrocarbons, ethane, as a representative of the alkane family, and ethylene, as a typical member of the alkene family, behave quite differently in the presence of bromine. When ethane, CH_3CH_3, is treated with bromine in carbon tetrachloride, a solvent, in the dark at room temperature, no reaction takes place. The solution remains red, the color of bromine.

$$H-C-C-H + Br_2 \xrightarrow[\substack{carbon \\ tetrachloride \\ dark}]{} \text{no reaction}$$

ethane   red   25 °C

an alkane

When ethylene, H_2C=CH_2, is treated with bromine under the same conditions, the red color of bromine disappears rapidly.

$$\underset{\substack{\text{ethylene} \\ \textit{an alkene}}}{\overset{\text{H}}{\underset{\text{H}}{\rangle}} \text{C}=\text{C} \overset{\text{H}}{\underset{\text{H}}{\langle}} {}^{sp^2}} + \underset{\substack{\text{red} \quad 25\,°\text{C}}}{\text{Br}_2} \xrightarrow[\substack{\text{carbon} \\ \text{tetrachloride} \\ \text{dark}}]{} \underset{\substack{\text{1,2-dibromoethane} \\ \text{colorless} \\ \textit{an alkyl halide}}}{\text{H}-\overset{\overset{\text{H}}{|}}{\underset{\underset{\text{Br}}{|}}{\text{C}}}-\overset{\overset{\text{H}}{|}}{\underset{\underset{\text{Br}}{|}}{\text{C}}} {}^{sp^3}-\text{H}}$$

an addition reaction

   The product of the reaction of bromine with ethylene, 1,2-dibromoethane, is colorless, so the progress of the reaction is easily seen. *The reaction of bromine with ethylene is an example of an* **addition reaction.** *In an addition reaction, the product contains all of the elements of the two reacting species.*

   In ethane, all of the bonding electrons are in σ bonds formed by sp^3-hybridized carbon atoms. Electrons in σ bonds are not readily involved in chemical reactions. The electrons in the π bond of ethylene, in contrast, react easily with bromine. The addition product, 1,2-dibromoethane, no longer has a π bond, and the sp^2 carbon atoms of ethylene are rehybridized and become sp^3 carbon atoms in the product. In general, electrons in π bonds (called π electrons) are involved more readily in chemical reactions than are electrons in σ bonds. The π molecular orbitals are at a higher energy level than σ molecular orbitals. We can visualize the π electrons as being more loosely held and, thus, more available to reagents that are seeking electrons than are electrons in a σ bond.

   Acetylene, $HC\equiv CH$, the simplest member of the alkyne family, also adds bromine in the dark. Bromine adds successively to each of the two π bonds of the alkyne.

$$\underset{\substack{\text{acetylene} \\ \textit{an alkyne}}}{\text{H}-\text{C}\overset{sp}{\equiv}\text{C}-\text{H}} \xrightarrow{\text{Br}_2} \underset{\substack{\text{1,2-dibromoethene} \\ \textit{an alkene}}}{\overset{\text{H}}{\underset{\text{Br}}{\rangle}}\text{C}\overset{sp^2}{=}\text{C}\overset{\text{Br}}{\underset{\text{H}}{\langle}}} \xrightarrow{\text{Br}_2} \underset{\substack{\text{1,1,2,2-tetrabromoethane} \\ \textit{an alkyl halide}}}{\text{H}-\overset{\overset{\text{Br}}{|}}{\underset{\underset{\text{Br}}{|}}{\text{C}}}-\overset{\overset{\text{Br}}{|}}{\underset{\underset{\text{Br}}{|}}{\text{C}}}{}^{sp^3}-\text{H}}$$

addition reactions

   At the end of the first stage of the reaction, the linear carbon atoms in acetylene have been converted to the trigonal carbon atoms of an alkene, 1,2-dibromoethene. The addition of another molecule of bromine to the remaining π bond converts the alkene to 1,1,2,2-tetrabromoethane, which has only tetrahedral carbon atoms.

   The addition of bromine is a reaction that is typical of many compounds having π electrons in carbon-carbon double or triple bonds. The final products formed by the addition of bromine to ethylene and acetylene are members of a class of organic compounds called alkyl halides. **Alkyl halides** *are compounds in which a halogen atom is bound to a tetrahedral carbon atom.* The details of the addition reactions of bromine to π bonds are presented in Section 5.8.

   Not all hydrocarbons that contain π bonds add bromine at room temperature in the dark. Benzene has the molecular formula C_6H_6, which indicates that the carbon atoms in benzene must be linked to each other by multiple bonds. *Benzene is the simplest member of a class of hydrocarbons known as* **aromatic hydrocarbons.** These hydrocarbons are pictured as having a six-membered ring structure with three double bonds in the ring. Yet benzene does not react with bromine under the same conditions as ethylene.

H
|
C
H   C   C   H
        ‖
C   C
H   C   C   H
|
H

+ Br$_2$ $\xrightarrow[\substack{\text{carbon} \\ \text{tetrachloride} \\ \text{dark} \\ 25\,°C}]{}$ no reaction

benzene

an aromatic hydrocarbon

How can this significant difference in reactivity between ethylene and benzene be explained? At first glance, we see that both molecules contain π bonds and π electrons, but the bonds of benzene must somehow be different from those of ethylene if an addition reaction wih bromine does not occur. The lack of reactivity of benzene and other aromatic hydrocarbons in addition reactions is explored in detail in Chapter 12. The molecular orbital picture of benzene, however, is related to the ideas already presented, so it can be examined now.

PROBLEM 2.22   Write equations showing the addition of bromine to the following compounds. How many moles of bromine will react with one mole of each compound before the reaction stops?

(a) $CH_3CH_2CH{=}CHCH_3$ (b) $CH_3C{\equiv}CCH_3$

(c) $HC{\equiv}CCH_2C{\equiv}CH$ (d) $CH_3CH{=}CHCH_2CH{=}CHCH_3$

PROBLEM 2.23   Different classes of compounds containing important functional groups have been introduced throughout this chapter. Review the chapter now, making a list of the different functional group classes and including the structural features that are typical of each one.

2.9

The Structure of Benzene

The experimental technique of x-ray crystallography provides convincing evidence that the structure of benzene is extraordinarily symmetrical. The six carbon atoms lie at the corners of a regular hexagon. Each carbon atom is bonded to two other carbon atoms and a hydrogen atom, all of which lie in a plane. Each carbon-carbon bond in benzene is 1.39 Å and each carbon-hydrogen bond, 1.09 Å long. All H—C—C and C—C—C bond angles are 120°, as would be expected in a regular hexagon. Benzene, in fact, is usually represented by a hexagon from which the symbols for the carbon and hydrogen atoms are omitted.

The carbon atoms in benzene are trigonal. In other words, they are bonded to three other atoms in a plane with bond angles of 120°. Such carbon atoms are postulated to be sp^2 hybridized. Thus the σ bonds in benzene result from the overlap of sp^2 hybrid orbitals between the carbon atoms, and of sp^2 orbitals with $1s$ orbitals of hydrogen (Figure 2.26).

Three valence electrons from each carbon atom are involved in the σ bonding in benzene. Each carbon atom also has a $2p$ orbital, which has a lobe above and below

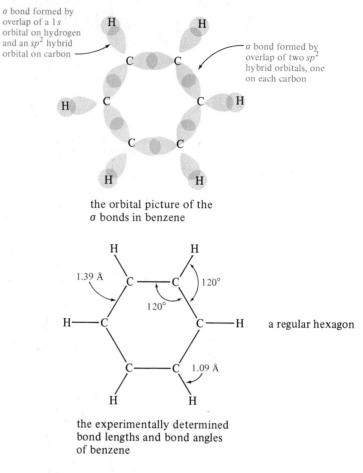

σ bond formed by overlap of a 1s orbital on hydrogen and an sp^2 hybrid orbital on carbon

σ bond formed by overlap of two sp^2 hybrid orbitals, one on each carbon

the orbital picture of the σ bonds in benzene

1.39 Å

120°

120°

1.09 Å

a regular hexagon

the experimentally determined bond lengths and bond angles of benzene

Figure 2.26 The σ bond skeleton of benzene.

the plane of the benzene ring, and an electron in that orbital. All of the six *p* orbitals are perpendicular to the plane of the ring in an orientation that allows for overlap of the orbitals to form π bonds between the carbon atoms. The overlap of six *p* orbitals in benzene gives rise to six molecular π orbitals. Three of them are bonding molecular orbitals and three are antibonding molecular orbitals. The six electrons from the *p* orbitals of the carbon atoms occupy the three bonding π orbitals of benzene (pictures and the relative energies of these orbitals are shown in Figure 12.1, Section 12.1). The lowest energy bonding molecular orbital in benzene is shown in Figure 2.27. This molecular orbital has two circular lobes, one above and one below the plane of the ring. *The π electrons in benzene are said to be completely delocalized over the entire ring.* Thus the molecule is highly symmetrical with all carbon-carbon bonds being of equal length (1.39 Å), longer than a double bond (1.34 Å) but shorter than a single bond (1.53 Å).

The structural formula for benzene is written with three double bonds in the ring. Such a representation is called the Kekulé formula for benzene, after the German chemist August Kekulé who first proposed that structure for benzene in 1865. Even that long ago, however, Kekulé recognized that a structure that depicts benzene with single

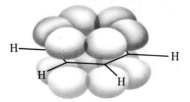

side view showing the
six parallel *p* orbitals
in phase

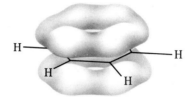

lowest energy bonding
π molecular orbital
showing delocalization

Figure 2.27   Molecular orbital picture of the π bonds in benzene.

and double bonds is not a satisfactory representation of the properties of the compound. He suggested that the structure of benzene should be represented by two formulas with different positions for single and double bonds.

Kekulé formulas for benzene

Kekulé thought that these structural formulas corresponded to different species that were in dynamic equilibrium with each other. We recognize today that *structures that differ from each other only in the location of electrons cannot be distinguished. Such structures are called resonance contributors* (Section 1.5C) of the true structure of the compound, which has properties best represented by all of the resonance contributors taken together. The correct structure for the compound is said to be a *resonance hybrid* of the various resonance contributors that can be written for it.

Each Kekulé formula for benzene suggests that benzene has three double bonds and three single bonds in the six-membered ring. Yet we saw in Section 2.8 that benzene does not react as if it has double bonds localized between two carbon atoms in the same way as compounds like ethylene. Thus, neither Kekulé formula is an accurate picture of the real benzene molecule. The *resonance hybrid,* the two Kekulé formulas taken together, suggests that the double bonds in benzene are not fixed in position and that the π electrons are delocalized over the entire ring. As a result, the reactivity of benzene is unlike that of compounds containing localized double bonds.

PROBLEM 2.24   Before the structure of benzene was finally determined experimentally, one of the many structures that had been proposed for the hydrocarbon with the molecular formula C_6H_6 was the following, called Dewar benzene:

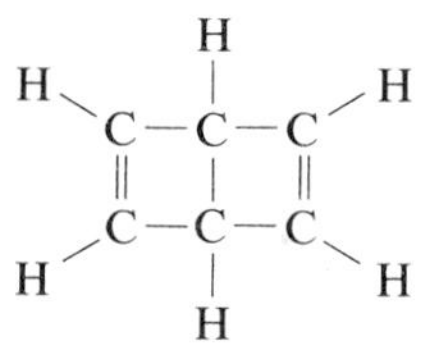

Dewar benzene

What would be the hybridization of each carbon atom in the structure shown? How would you expect such a compound to behave with bromine?

2.10

Resonance Stabilization of Benzene

The differing reactivity of ethylene and benzene towards bromine (Section 2.8) indicates that the π bonds in benzene are less reactive, somehow more stable, than the π bond of ethylene. Many experimental attempts have been made to measure the special stabilization of the benzene molecule. One experimental method is to add hydrogen to π bonds in compounds containing ordinary double bonds and to benzene and then compare the energy released in each reaction. *The addition of hydrogen to π bonds is called* **hydrogenation.** This reaction occurs when hydrogen and a compound containing π bonds are brought together in the presence of a metal catalyst. Hydrogenation reactions are explored in detail in Chapter 8. Now we are only interested in the overall reaction illustrated by the hydrogenation of ethylene:

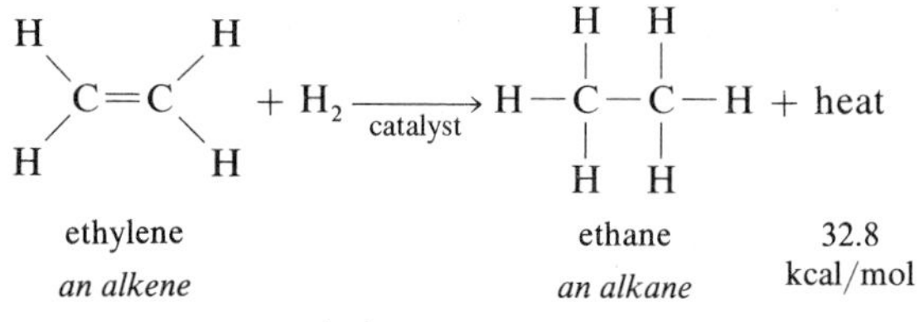

a hydrogenation reaction

One mole of ethylene reacts with one mole of hydrogen to give one mole of ethane. The reaction is exothermic; that is, heat is evolved. The **heat of the reaction** is also known as the **enthalpy** of the reaction, $\Delta H_r{}^\circ$. By convention, $\Delta H_r{}^\circ$ for an exothermic reaction is written as a negative quantity, emphasizing that the total internal energy of the system is less after the reaction than before. Some of the energy stored in the molecules on the left-hand side of the equation has been released to the environment as heat energy.

The energy of the system has many components, such as the kinetic energies of all of the molecules and the potential energy stored in the bonds of molecules and in the nonbonding interactions between molecules. After an exothermic reaction, the total of the components of energy for the products of the reaction is less than it was for the reactants. The system reaches a point of lower energy.

The enthalpy of the reaction of ethylene with hydrogen is also called the **heat of hydrogenation** for ethylene, and is -32.8 kcal/mol.

The heat of hydrogenation of an alkene varies with the structure of the alkene. The values for the heats of hydrogenation of a series of alkenes that give the same alkane are an experimental measure of the relative stabilities of the alkenes. Because we are interested in the heat of hydrogenation of benzene, we must compare it to the heats of hydrogenation for alkenes in other six-membered rings.

The cyclic alkene, cyclohexene, adds hydrogen to give a cyclic alkane, cyclohexane, with a heat of hydrogenation of -28.6 kcal/mol.

cyclohexene   cyclohexane
a cyclic alkene   *a cyclic alkane*

The structures for these cyclic hydrocarbons are usually represented by hexagons from which the symbols for all carbon and hydrogen atoms are omitted for the sake of clarity.

If a molecule contains several double bonds, the heat of hydrogenation for the compound is expected to be a multiple of that for the corresponding compound with one double bond. A six-membered ring compound with two double bonds in it is 1,3-cyclohexadiene. It reacts with two moles of hydrogen to give cyclohexane.

1,3-cyclohexadiene   cyclohexane
a cyclic alkene   *a cyclic alkane*
with two double bonds

The heat of hydrogenation expected for 1,3-cyclohexadiene is -57.2 kcal/mol, or twice -28.6 kcal/mol, the experimental value observed for cyclohexene, which has one double bond in a six-membered ring. The experimental heat of hydrogenation observed for 1,3-cyclohexadiene, -55.4 kcal/mol, is slightly less than the calculated value because the two double bonds influence each other.

The hypothetical compound with three localized double bonds in a six-membered ring would be "1,3,5-cyclohexatriene." There is, of course, no such compound because the compound corresponding to that structural formula is benzene, which has delocalized π bonds. Still, it is possible to calculate the heat of hydrogenation that would be expected of such a hypothetical "cyclohexatriene." The expected heat of hydrogenation would be three times -28.6 kcal/mol (the value for one localized double bond in a six-membered ring), or -85.8 kcal/mol. This value is the heat of hydrogenation we would expect to observe for the conversion of benzene to cyclohexane if benzene were a compound with three localized double bonds.

Benzene is much more resistant to the addition of hydrogen than cyclohexene is. Higher temperatures and higher pressures of hydrogen gas have to be used to convert benzene to cyclohexane.

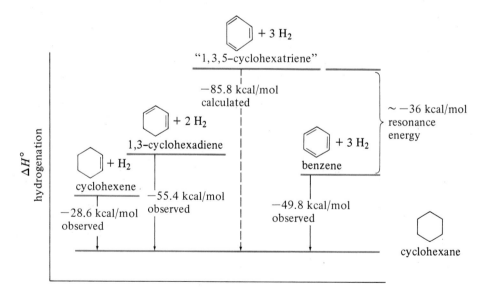

$$\Delta H° = -49.8 \text{ kcal/mole}$$

benzene   cyclohexane

The experimental heat of hydrogenation of benzene is only -49.8 kcal/mol instead of the -85.8 kcal/mol expected from the hydrogenation of three double bonds. Benzene is more stable than the hypothetical "cyclohexatriene" by the difference between these two figures, 36 kcal/mol (85.8 kcal/mol − 49.8 kcal/mol = 36 kcal/mol). *The value of 36 kcal/mol is* also *called the* **empirical resonance energy** *of benzene.*

The heats of hydrogenation of cyclohexene, 1,3-cyclohexadiene, and benzene are compared schematically in Figure 2.28. For example, cyclohexene and hydrogen lie 28.6 kcal/mol above cyclohexane in energy. In the hydrogenation reaction, the π bond of the alkene and the σ bond of hydrogen are broken and two new carbon-hydrogen bonds are formed. Some of the energy of the broken bonds is stored in the new carbon-hydrogen bonds, but some is released to the surroundings as heat, which is measured as the heat of hydrogenation, -28.6 kcal/mol. The negative sign given the heat of hydrogenation is an indication that the product of the reaction, cyclohexane, is more stable (lower in energy) than the reactants by that amount.

Cyclohexene, 1,3-cyclohexadiene, and benzene all give the same product, cyclohexane, in their reactions with hydrogen. The heats of hydrogenation, therefore, can be used to identify differences in energy in the starting compounds. The placement of cyclohexene, 1,3-cyclohexadiene, and benzene above the line representing cyclohexane shows the relative energies of the three compounds as determined by experimental heats of hydrogenation. The hypothetical "cyclohexatriene" with the calculated

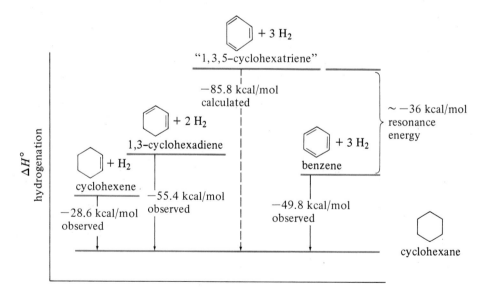

Figure 2.28   Relative stabilities and heats of hydrogenation for cyclohexane, 1,3-cyclohexadiene, and benzene.

value of a heat of hydrogenation is included to show how the empirical resonance energy of benzene is obtained.

The empirical resonance energy of benzene is a quantity that reflects the special stabilization of aromatic compounds that is attributed to the delocalization of π electrons. This **resonance stabilization** of benzene and other aromatic compounds is responsible for the lower reactivity of their π bonds in comparison to the π bonds of alkenes and alkynes.

Aromatic hydrocarbons represent a large class of organic compounds with many representatives and much interesting chemistry, often different from that of alkanes and alkenes. The chemistry of such compounds is described in detail, primarily in Chapter 12, but aromatic rings appear in many compounds in all parts of the book. Recognition of the special stability of this type of hydrocarbon is, therefore, important.

ADDITIONAL PROBLEMS

2.25 Draw Lewis electron dot structures for each of the following species and indicate the hybridization and geometry of the central atom.

(a) BeH_2 (b) CO_2 (c) CH_3^+ (d) CF_4 (e) Cl_2CO (f) CCl_3^-
(g) NF_3 (h) BF_4^-

2.26 How many σ and how many π bonds are present in each of the following?

(a) $CH_2{=}C{=}CHCH_3$ (b) $CH_2{=}C{=}O$ (c) $CH_3OCH{=}CHCH_3$
(d) $CH_2{=}NOH$

2.27 Indicate the hybridization of all of the carbon and oxygen atoms in each of the following.

(a) $CH_2{=}CHCH_2OH$ (b) $CH_3\overset{O}{\overset{\|}{C}}\overset{O}{\overset{\|}{OC}}CH$ ($\overset{\|}{CH_2}$) (c) $CH_3CH_2CH_3$

(d) $CH_3C{\equiv}CCH{=}CHCH_3$ (e) (ring)$-CH_3$

2.28 In each of the following groups of compounds, rank the bonds indicated in order of bond length.

(a) $CH_3C{\equiv}CH$, $CH_2{=}CHCH_3$, $CH_3CH_2{-}CH_3$, (ring)

(b) $CH_3{-}OH$, $CH_2{=}O$, $HC\overset{O}{\overset{\|}{{-}}}OCH_3$

(c) $CH_3C{\equiv}N$, $CH_3{-}NH_2$, $CH_2{=}NCH_3$

2.29 Circle and name all the functional groups in each of the following molecules.

(a) $CH_3C{\equiv}C\overset{CH_3}{\overset{|}{C}}{-}CH_3$ ($\overset{|}{OH}$) (b) $CH_3\overset{CH_3}{\overset{|}{C}}{=}CH\overset{O}{\overset{\|}{C}}CH_3$ (c) (ring)$-CH_2N\overset{CH_3}{\overset{|}{}}CH_3$

(d) $CH_3CH_2CHCH{=}CH_2$
       |
      Br

(e) $HOCCH_2CH_2COH$
    ‖   ‖
    O   O

(f) $CH_3COCH_2CH_2OCH_3$
      ‖
      O

(g) 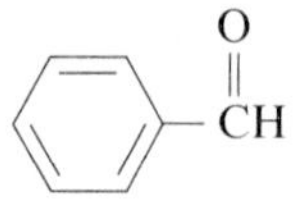 (benzene)—CH, with O double bond above CH

(h) CH_3CHCH_2CH, with O double bond above terminal CH
    |
    OH

(i) 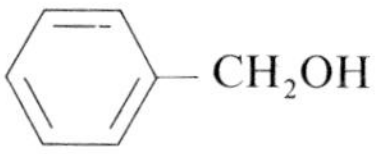 (benzene)—CH_2OH

(j) $CH_3CH_2CH_2C{\equiv}N$

2.30　For each of the following compounds, assign hybridization to each of the carbon, oxygen, or nitrogen atoms. Show which types of orbitals overlap to create the different kinds of bonds in the molecules.

(a) $CH_2{=}O$　(b) CH_3NHOH　(c) $CH_3CH{=}CHCH_3$　(d) $CH_3C{\equiv}CCH_3$

2.31　The formation of a bond between nonbonding electrons of nitrogen and oxygen atoms and the empty $1s$ orbital of a proton was shown in Figures 2.13 and 2.15. Keeping those figures in mind, predict, by writing an equation, what would happen if each of the following compounds were put into sulfuric acid.

(a) $CH_3C{\equiv}N{:}$
(b) $CH_3\overset{:O:}{\overset{‖}{C}}CH_3$
(c) $CH_3\overset{CH_3}{\overset{|}{C}}{=}NCH_3$
(d) $CH_3\overset{:O:}{\overset{‖}{C}}OCH_3$

2.32　Predict, by completing the equations, what would happen if the following reagents were mixed. (Review of Sections 2.8 and 2.10 might be helpful.)

(a) $CH_3CH{=}CHCH_3 + Br_2 \xrightarrow[\text{dark, 25 °C}]{\text{carbon tetrachloride}}$

(b) $CH_3CH_2CH{=}CH_2 + H_2 \xrightarrow{\text{catalyst}}$

(c) $CH_3CH_2CH_2CH_3 + Br_2 \xrightarrow[\text{dark, 25 °C}]{\text{carbon tetrachloride}}$

(d) 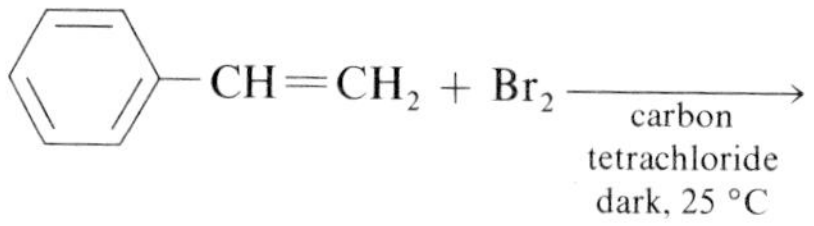 (benzene)—$CH{=}CH_2 + Br_2 \xrightarrow[\text{dark, 25 °C}]{\text{carbon tetrachloride}}$

(e) 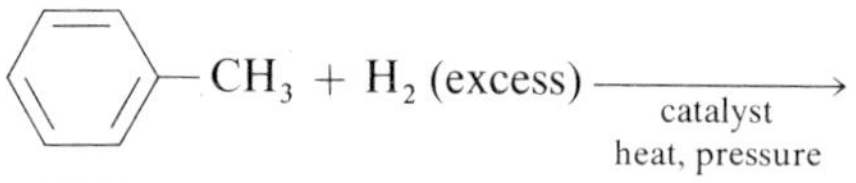 (benzene)—$CH_3 + H_2$ (excess) $\xrightarrow[\text{heat, pressure}]{\text{catalyst}}$

(f) 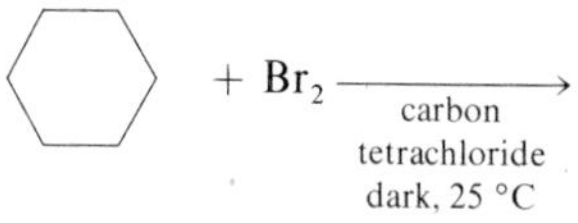 (cyclohexane) $+ Br_2 \xrightarrow[\text{dark, 25 °C}]{\text{carbon tetrachloride}}$

(g) $CH_3C{\equiv}CH + Br_2$ (excess) $\xrightarrow[\text{dark, 25 °C}]{\text{carbon tetrachloride}}$

(h) $CH_3C{\equiv}CH + H_2$ (excess) $\xrightarrow{\text{catalyst}}$

2.33   Allene has the following structural formula. What is the hybridization of each of the carbon atoms? How does the σ skeleton of allene arise? What are the orbitals involved in the formation of the π bonds in this molecule? Illustrate your answers with drawings. (Hint: Figures 2.20 and 2.24 may be helpful.)

$$CH_2 {=} C {=} CH_2$$
allene

2.34   Methylamine (Figure 2.14) forms a compound with boron trifluoride, just as ammonia does (Section 1.5A, p. 9). Write an equation for the reaction. Write a Lewis structure for the product and show any formal charges that are present. What are the hybridization and the geometry of the nitrogen atom and of the boron atom in the product? How would you show the bonding present in this system in a three-dimensional picture? (Hint: Another look at the representations of the structure of ethane in Figure 2.12 may be helpful.)

2.35   (a)   In Sections 2.5 and 2.6 we put forth the idea that a carbon atom that is trigonal planar is sp^2 hybridized and that a carbon atom that is bonded to two other groups in a linear molecule is sp hybridized. Similar reasoning can be applied to other elements. For example, water can also be viewed as a planar triangular molecule. What would the hybridization of the oxygen atom be in this model of water? Which orbitals would the two pairs of nonbonding electrons occupy?

        (b)   A recent experiment using photoelectron emission spectroscopy, a form of spectroscopy that measures how easily electrons can be removed from various energy levels in a molecule, indicates that the two pairs of nonbonding electrons in water occupy different energy levels. Which model for the water molecule better fits the experimental facts? The one in which the oxygen atom is sp^3 hybridized? The one in which it is sp^2 hybridized?

        (c)   Would the experimental facts described above be accommodated by any other model that you can think of for the bonding in water?

        (d)   Lithium hydroxide, LiOH, is a linear covalent molecule. Propose an orbital picture of the bonding in lithium hydroxide. What kind of hybridization does the oxygen atom have here?

Alkanes and Cycloalkanes

An Introduction to Nomenclature and Stereochemistry

3

Structural Isomerism and Physical Properties

Alkanes *are hydrocarbons with the general formula* C_nH_{2n+2}. *They have only σ bonds (Section 2.3A) and are described as saturated hydrocarbons, meaning that all of the valences of carbon are taken up in single bonds.* The structures of methane, CH_4, and ethane, C_2H_6, the first two members of the family of alkanes, were examined thoroughly in Section 2.4A. *Compounds that differ from each other in their molecular formulas by the unit* $-CH_2-$ *are called members of a* **homologous series.** Thus, methane and ethane belong to a homologous series that has many other members, all saturated hydrocarbons. These compounds have similar chemical properties, but their physical properties change with the molecular weight and the shape of the molecules. Table 3.1 shows the boiling and melting points of some representative alkanes.

The boiling points of the alkanes increase steadily with molecular weight from methane to butane, which are all gases at room temperature. In compounds with four or more carbon atoms, several different arrangements of the carbon chains are possible. The boiling points of different compounds having the same molecular formula and functional groups depend on the compactness of the molecular shape. Compounds with a long carbon chain tend to have higher boiling points than those with a more spherical shape (Section 1.8C).

On the other hand, compact, spherical molecules pack better in crystal lattices, so such compounds have higher melting points. These trends are illustrated by the different compounds having the molecular formula C_5H_{12}.

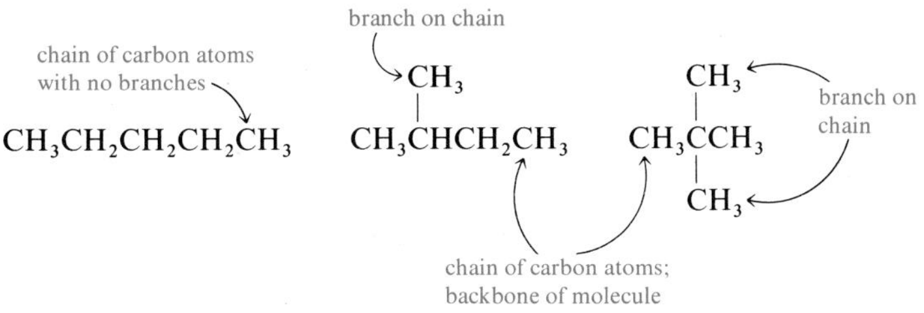

pentane	2-methylbutane	2,2-dimethylpropane
bp 36.1 °C	bp 27.9 °C	bp 9.5 °C
mp − 129.7 °C	mp − 159.9 °C	mp − 16.6 °C
straight chain alkane	*branched chain alkanes*	

Pentane has the most elongated molecular shape and the highest boiling point of the three compounds. 2,2-Dimethylpropane has the most compact and spherical shape, the lowest boiling point, and by far the highest melting point of the three.

Pentane, 2-methylbutane, and 2,2-dimethylpropane all have the molecular formula C_5H_{12} but exhibit different bonding arrangements of carbon atoms and are *structural isomers* (Section 1.5B). Pentane is a *straight* chain hydrocarbon; 2-methylbutane and 2,2-dimethylpropane are *branched* chain hydrocarbons. In pentane, all the carbon atoms are in one continuous chain. In the other two compounds, some of the carbon atoms may be defined as part of a chain but others branch off the backbone of the molecule. *Structural isomers have different physical properties,* as shown above for the isomers of C_5H_{12}.

As the numbers of carbon atoms in the alkanes, and their resulting molecular weights, increase, so do their boiling and melting points. The actual values are given in Table 3.1 for the straight chain hydrocarbons CH_4 to $C_{10}H_{22}$, and for $C_{20}H_{42}$. If a

TABLE 3.1 Boiling Points and Melting Points for Some Alkanes

Molecular Formula	Name	Molecular Weight	bp, °C	mp, °C
CH_4	methane	16	− 164	− 182.5
C_2H_6	ethane	30	− 88.6	− 183.3
C_3H_8	propane	44	− 42.1	− 189.7
C_4H_{10}	butane	58	− 0.6	− 138.4
C_4H_{10}	2-methylpropane	58	− 10.2	− 138.3
C_5H_{12}	pentane	72	36.1	− 129.7
C_5H_{12}	2-methylbutane	72	27.9	− 159.9
C_5H_{12}	2,2-dimethylpropane	72	9.5	− 16.6
C_6H_{14}	hexane	86	68.9	− 93.5
C_7H_{16}	heptane	100	98.4	− 90.6
C_8H_{18}	octane	114	125.7	− 56.8
C_9H_{20}	nonane	128	150.8	− 51.0
$C_{10}H_{22}$	decane	142	174.1	− 29.7
$C_{20}H_{42}$	icosane	282	343	36.8

linear alkane contains approximately eighteen carbon atoms, it is a solid at room temperature. Icosane, $C_{20}H_{42}$, has a melting point of 36.8 °C, which is very close to body temperature.

In Section 1.8B, we saw how the alkane-like portion of molecules bearing other functional groups such as alcohols influenced the solubility of the compounds in water. In later chapters we will concentrate on the chemistry of reactive functional groups such as alkenes, alkynes, alcohols, carbonyl compounds, and amines. Because most of these compounds have a hydrocarbon residue as part of the molecule, we shall examine alkanes in this chapter. The material presented here about alkanes provides an introduction to ideas about structure, isomerism, and nomenclature.

3.2

Methane

Methane, CH_4, is the simplest hydrocarbon. The details of the structure of methane were given in Sections 1.6 and 2.4A. The molecule is tetrahedral and has four equivalent carbon-hydrogen bonds. It may be represented in three dimensions, as a Lewis structure, or simply in a condensed formula (Figure 3.1).

Any one of the hydrogen atoms in methane may be substituted by another atom or group to give a new compound. Chloromethane, for example, is a compound in which one of the hydrogen atoms in methane has been substituted by a chlorine atom. Chloromethane is an alkyl halide and the hydrocarbon portion of the molecule is called a methyl group. Methanol, an alcohol, is another example of a compound containing a methyl group.

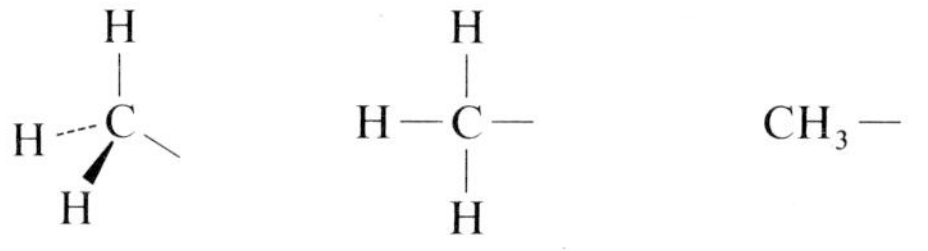

the methyl group
an alkyl group

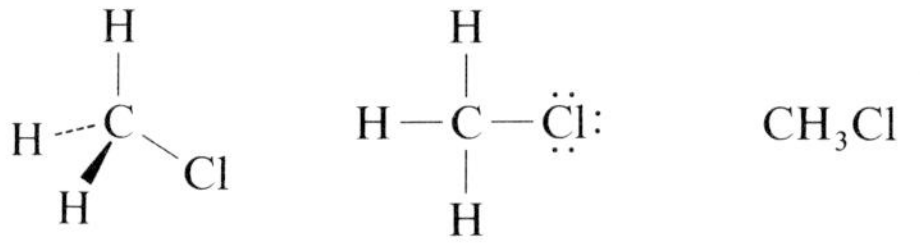

chloromethane
or methyl chloride
an alkyl halide

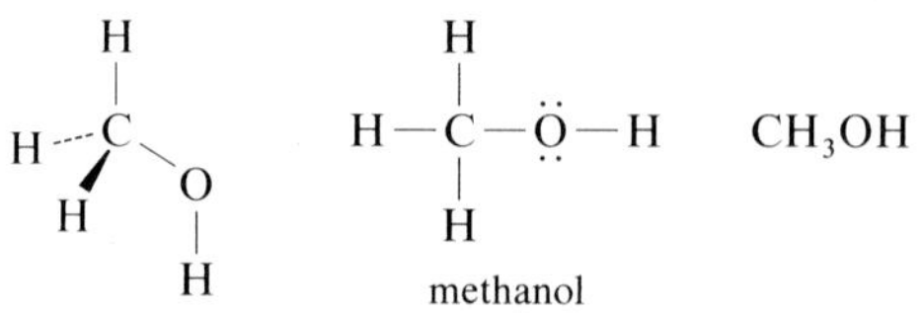

methanol
or methyl alcohol
an alcohol

In the illustration above, the first name given for each compound is its systematic name; the second is another acceptable and widely used name. The **methyl group** is

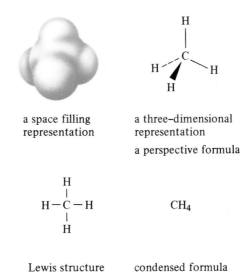

a space filling   a three–dimensional
representation   representation

a perspective formula

H
|
H—C—H   CH$_4$
|
H

Lewis structure   condensed formula

Figure 3.1 Different representations of methane.

the alkyl group derived from a particular alkane, methane. The name of an alkyl group is obtained by changing the **ane** ending of the name of an alkane to **yl.** Thus, the meth**yl** group is clearly related to meth**ane.** An alkane is sometimes represented in tables or equations by the symbol RH; the corresponding alkyl group is given the symbol R. An introduction to the rules for the systematic naming of organic compounds is given in Section 3.6. In this and the next three sections, you should concentrate on the structures and the names of the first four alkanes and the alkyl groups derived from them.

It is important to recognize that the chlorine atom in chloromethane or the hydroxyl group in methanol could have replaced any one of the hydrogen atoms in methane and the same compound would have been formed. The structural formulas below, for example, represent the tumbling of a single methanol molecule in space, not three different molecular species.

different orientations in space of a
molecule of methanol

This seems a trivial point to make, but the distinction is important in more complicated systems. *All of the hydrogen atoms in methane are equivalent.* Substitution of any one of them by a particular group results in the same compound.

Similarly, there can be only one compound in which two of the hydrogen atoms are substituted. For example, a molecule of dichloromethane may be represented in a number of ways.

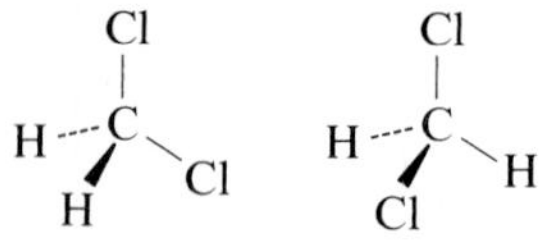

perspective formulas

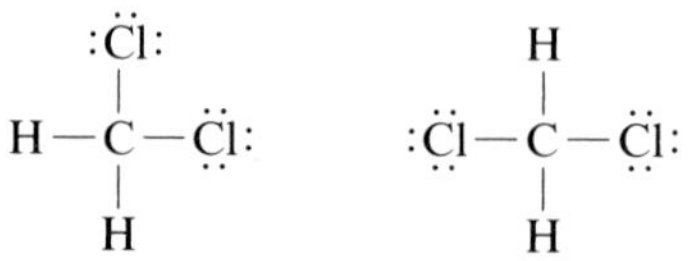

          Lewis structures   *condensed formula*
                     dichloromethane

The representation that we choose to use at any one time depends on the particular molecular features we are trying to emphasize.

PROBLEM 3.1   Write at least two different three-dimensional representations for each of the following compounds. (You may wish to review Sections 1.6 and 2.4 before attempting this.)

(a) CH_2ClBr (b) CH_3SH (c) CH_3OCH_3 (d) $CHCl_3$ (e) CH_3NH_2

3.3

Ethane. Conformation

A. The Ethyl Group

Ethane, C_2H_6, follows methane in the series of alkanes. The structure of ethane (considered in detail in Sections 1.6 and 2.4A) may also be represented by structural formulas showing the molecule in three dimensions, by a Lewis structure, or by a condensed formula. The second three-dimensional picture below is known as a *sawhorse representation* and will be the one that we use most frequently.

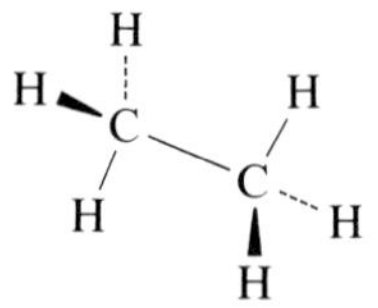

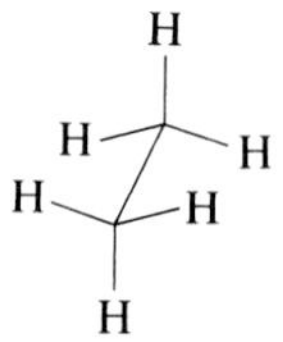

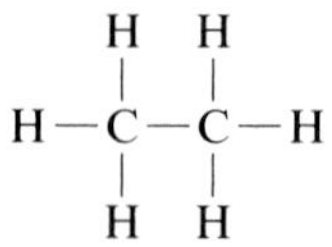

   perspective   *sawhorse*   Lewis structure   *condensed*
   formula   *representation*   *formula*
                              ethane

All of the hydrogen atoms in ethane are equivalent. Removal of any one of them gives the **ethyl group,** and substitution of any one of them by another atom or group results in a single compound.

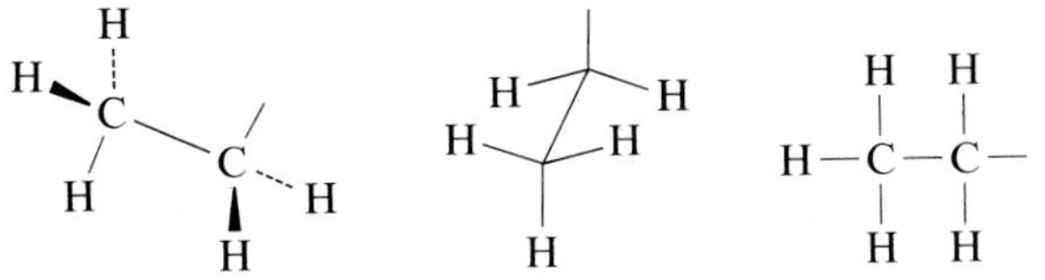

the ethyl group

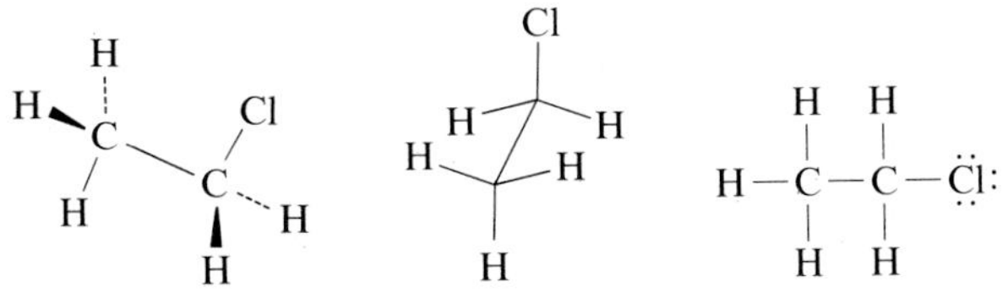

chloroethane
or ethyl chloride

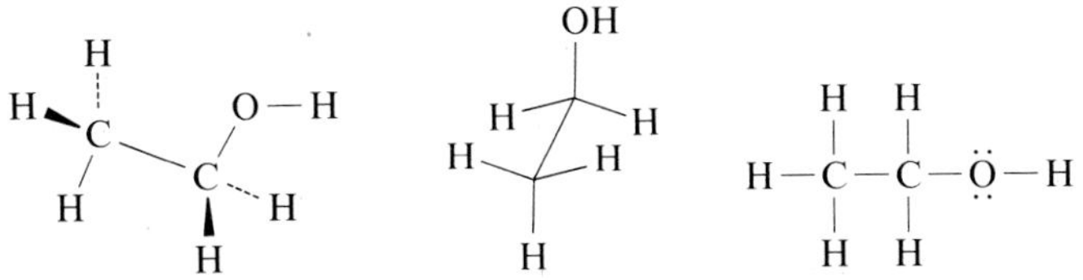

ethanol
or ethyl alcohol

The sawhorse representation of chloroethane is shown below, drawn in different ways to represent views of the molecule from different points in space.

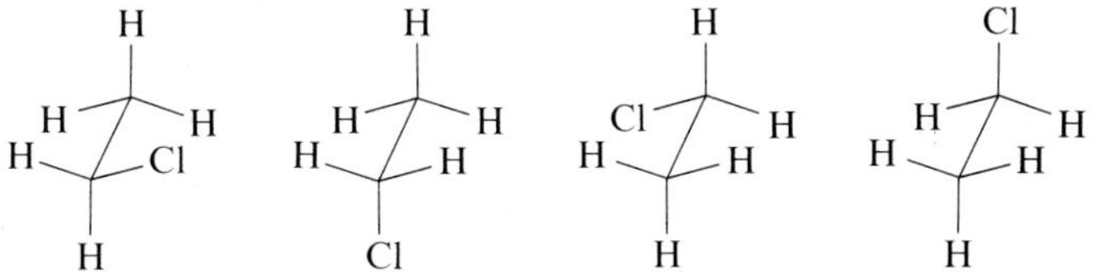

representations of chloroethane, viewed from
several different perspectives

All of the structural formulas represent the same molecule.

B. Conformation in Ethane

The different three-dimensional pictures of chloroethane shown in Section 3.3A are drawn so that the carbon-chlorine bond, for example, bisects the angle made by two of the hydrogen atoms on the adjacent carbon atom. The two carbon atoms in ethane, however, are not frozen in place. They are constantly in motion, rotating relative to each other. *The different orientations that the hydrogen atoms on one of the carbon atoms may have relative to the hydrogen atoms on the other carbon atom are called the* **conformations** *of the molecule.* In ethane, the conformation in which the bonding electrons of the carbon-hydrogen bonds on two adjacent carbon atoms are as far apart as possible is called the **staggered conformation.** This is a low-energy conformation. As the two carbon atoms rotate with respect to each other, at some point the carbon-hydrogen bonds on one carbon atom are directly across from similar bonds on the other carbon atom. The bonds are said to be **eclipsed,** and this conformation corresponds to a maximum in energy.

The relative positions of the hydrogen atoms on the two carbon atoms are best seen if *we view the molecule down the axis of the carbon-carbon bond. Such a picture is known as a* **Newman projection.** The front carbon atom is represented by a dot and the rear one by a circle. The atoms or groups on each carbon atom are shown as being bonded to the dot or the circle. Newman projections of the staggered and eclipsed conformations of ethane are shown below, alongside the corresponding perspective and sawhorse representations.

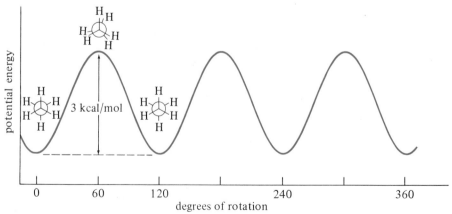

The difference in energy between the staggered and eclipsed conformations of ethane is approximately 3 kcal/mol. This energy difference is believed to arise from repulsion between the electrons in the carbon-hydrogen bonds when the bonds are closer together in the eclipsed conformation (Figure 3.2).

Figure 3.2   Energy diagram showing the energy difference between the staggered and eclipsed conformations of ethane.

The Newman projections of ethane in Figure 3.2 show the rear carbon atom rotating relative to the front one around the carbon-carbon bond. A rotation of 60° converts a staggered conformation into an eclipsed one. Another 60° rotation (or a total rotation of 120°) returns the molecule to a staggered conformation. At room temperature, molecules have enough kinetic energy to get over energy barriers up to approximately 20 kcal/mol. As Figure 3.2 illustrates, the energy required for rotation around the carbon-carbon bond is only 3 kcal/mol. As a result, the rotation around the carbon-carbon bond is constant. Yet the molecules spend most of their time in the energy valleys represented by the staggered conformations.

The different conformations of ethane are isomeric with each other. *The isomeric species that result from conformational changes are called* **conformers.** Usually conformers are being converted so rapidly into each other that they cannot be isolated as individual species.

PROBLEM 3.2   Draw sawhorse representations and Newman projections for the different conformations of chloroethane.

C. Isomerism in Disubstituted Ethanes

The replacement of two hydrogen atoms of methane by two other atoms gives a single compound (Section 3.2). An inspection of the structure of chloroethane, however, reveals that the five hydrogen atoms that remain are no longer equivalent to each other. There are two hydrogen atoms on the carbon atom already bearing a chlorine atom and three on the other carbon atom. If another chlorine is added, the new compound, $C_2H_4Cl_2$, could have one of two possible structures.

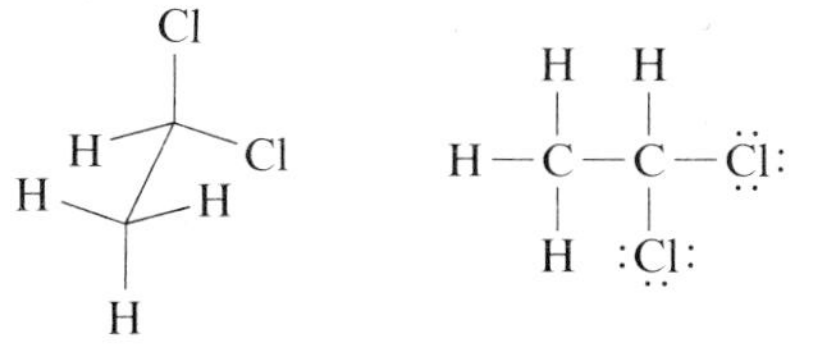

1,1-dichloroethane

bp 57 °C   mp −97 °C $\mu = 1.95$ D

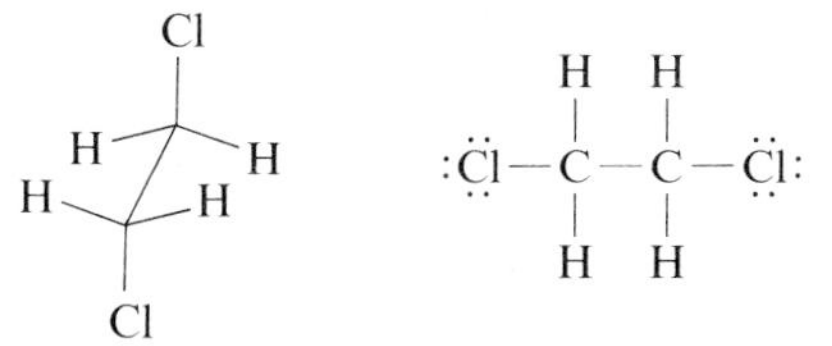

1,2-dichloroethane

bp 83 °C   mp −35 °C $\mu = 1.42$ D

isomers of $C_2H_4Cl_2$

1,1-Dichloroethane is derived from chloroethane by replacement of a hydrogen atom on the same carbon atom as the first chlorine atom. 1,2-Dichloroethane is obtained by replacement of one of the three hydrogen atoms on the originally unsubstituted carbon atom. The two compounds are another example of *structural isomers, compounds having the same molecular formula but different arrangements of the bonds.* They have different physical properties, including boiling and melting points and dipole moments. Their names reflect the difference in their structures. The name 1,1-dichloroethane indicates that both chlorine atoms are on the same carbon atom in ethane. 1,2-Dichloroethane identifies the positions of the chlorine atoms on adjacent carbon atoms.

PROBLEM 3.3   How many compounds with the molecular formula $C_2H_3Cl_3$ are possible? How about $C_2H_2Cl_4$? Draw condensed formulas, Lewis structures, and three-dimensional representations of these compounds.

3.4

Propane

The three carbon atoms and eight hydrogen atoms in propane, C_3H_8, can be combined in only one way. In propane, however, for the first time we encounter two different kinds of carbon atoms and hydrogen atoms.

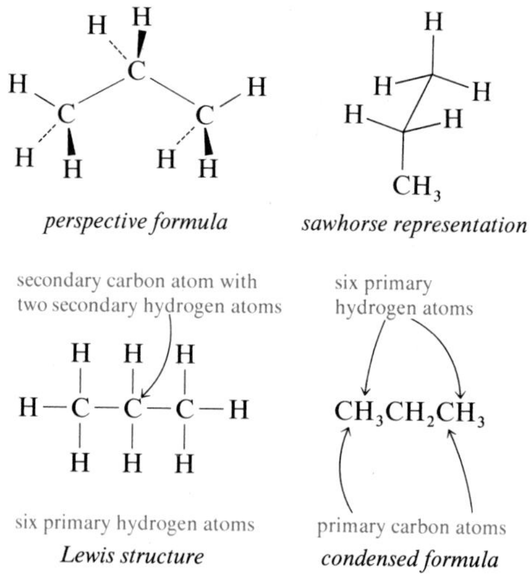

perspective formula   *sawhorse representation*

secondary carbon atom with   six primary
two secondary hydrogen atoms   hydrogen atoms

six primary hydrogen atoms   primary carbon atoms
Lewis structure   *condensed formula*

propane

The two end carbon atoms in propane that are each attached to only one other carbon atom are called **primary carbon atoms.** *The hydrogen atoms on these carbon atoms are designated as* **primary hydrogen atoms.** There are two equivalent primary carbon atoms and six equivalent primary hydrogen atoms in propane. *The middle carbon atom in propane is bonded to two carbon atoms and is defined as a* **secondary carbon atom,** *bearing two* **secondary hydrogen atoms.**

Two different propyl groups may be created from propane. Removal of a primary hydrogen atom gives the **propyl,** or *n***-propyl** (*normal* propyl) **group,** a primary alkyl group.

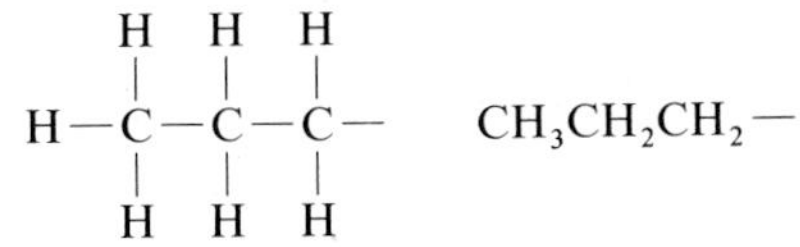

$$\begin{array}{ccc} & H & H & H \\ & | & | & | \\ H- &C- &C- &C- \\ & | & | & | \\ & H & H & H \end{array} \qquad CH_3CH_2CH_2-$$

propyl or *n*-propyl group

a primary alkyl group

$CH_3CH_2CH_2Cl$	$CH_3CH_2CH_2OH$
1-chloropropane	1-propanol
or *n*-propyl chloride	or *n*-propyl alcohol
bp 46.6 °C	bp 97.4 °C
a primary alkyl halide	*a primary alcohol*

If one of the secondary hydrogen atoms is removed, a secondary alkyl group known as the **1-methylethyl** or **isopropyl group** is formed.

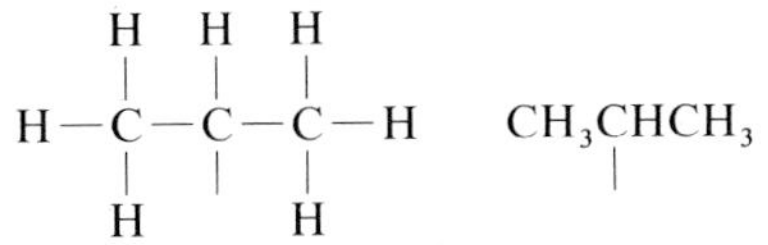

$$\begin{array}{ccc} & H & H & H \\ & | & | & | \\ H- &C- &C- &C-H \\ & | & & | \\ & H & & H \end{array} \qquad CH_3CHCH_3$$

1-methylethyl or isopropyl group

a secondary alkyl group

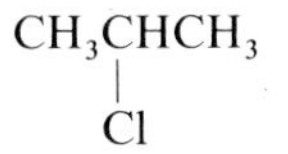

| CH_3CHCH_3 | CH_3CHCH_3 |
| | |
Cl	OH
2-chloropropane	2-propanol
or isopropyl chloride	or isopropyl alcohol
bp 35.7 °C	bp 82.4 °C
a secondary alkyl halide	*a secondary alcohol*

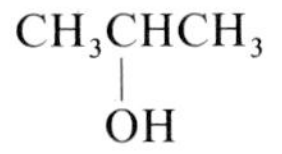

1-Chloropropane and 2-chloropropane are structural isomers of each other. They have different physical properties, as demonstrated by the boiling points given for them. One is classified as a **primary alkyl halide** because the halogen is bonded to a primary carbon atom; the other is a **secondary alkyl halide.** The two compounds have some chemical properties that are similar and some that are different. These are treated in more detail in Chapter 6. Likewise, 1-propanol and 2-propanol are isomeric alcohols. They share some chemical properties, but differ from each other because one is **a primary alcohol** and the other is a **secondary alcohol** (Chapter 7). The kind of carbon atom to which the functional group is bonded often makes a difference in reactivity. Therefore, it is important that you learn to distinguish isomers and to classify them according to their structural features.

PROBLEM 3.4   How many structural isomers of $C_3H_6Cl_2$ are possible? Draw them as condensed formulas, Lewis structures, and three-dimensional formulas.

3.5

Butanes

A. Structural Isomers of Butane

Alkanes with the molecular formula C_4H_{10} are butanes. If you think about it, you will realize that there are two ways that four carbon atoms and ten hydrogen atoms can be bonded together.

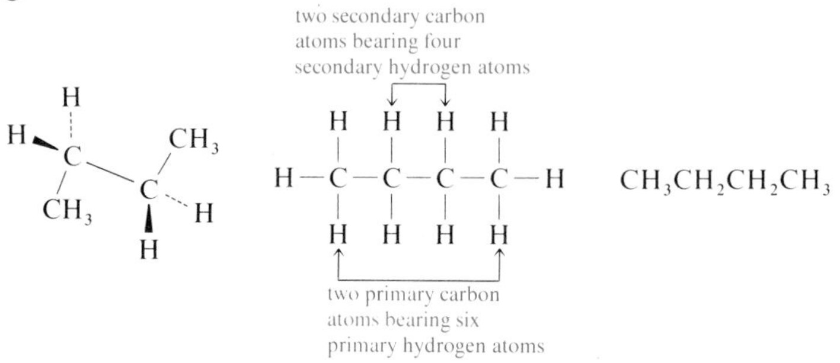

two secondary carbon
atoms bearing four
secondary hydrogen atoms

two primary carbon
atoms bearing six
primary hydrogen atoms

butane or *n*-butane
bp −0.6 °C

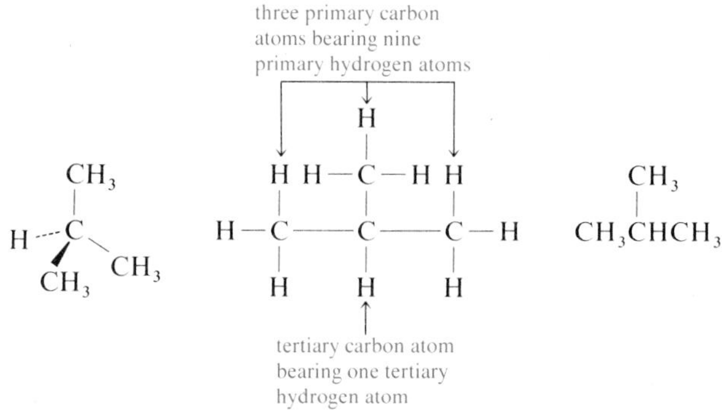

three primary carbon
atoms bearing nine
primary hydrogen atoms

tertiary carbon atom
bearing one tertiary
hydrogen atom

2-methylpropane or isobutane
bp −10 °C

One compound is the straight chain isomer with two equivalent primary and two equivalent secondary carbon atoms. The second isomer of C_4H_{10} is a branched chain compound. In this isomer, there are three carbon atoms that are attached to only one other carbon atom, and are, by definition, primary carbon atoms. All of the hydrogen atoms on these three carbon atoms are equivalent, and are also primary. The fourth carbon atom, which is bonded to three other carbon atoms, is called a **tertiary carbon atom** and bears a **tertiary hydrogen atom.**

The straight chain isomer of C_4H_{10} is called butane, or *normal* or *n*-butane. The systematic name for the branched chain isomer is 2-methylpropane, indicating that the compound has a three-carbon chain, propane, substituted on the second carbon atom by a methyl group. This compound is also called isobutane.

Four different alkyl groups with the molecular formula C_4H_9 can be created by removing different types of hydrogen atoms from butane and 2-methylpropane. Removal of a primary hydrogen atom from butane creates a primary alkyl group, the **butyl (or *n*-butyl) group.**

$$H-\overset{\overset{\displaystyle H}{|}}{\underset{\underset{\displaystyle H}{|}}{C}}-\overset{\overset{\displaystyle H}{|}}{\underset{\underset{\displaystyle H}{|}}{C}}-\overset{\overset{\displaystyle H}{|}}{\underset{\underset{\displaystyle H}{|}}{C}}-\overset{\overset{\displaystyle H}{|}}{\underset{\underset{\displaystyle H}{|}}{C}}- \qquad CH_3CH_2CH_2CH_2-$$

butyl or *n*-butyl group

a primary alkyl group

CH$_3$CH$_2$CH$_2$CH$_2$Cl   CH$_3$CH$_2$CH$_2$CH$_2$OH

1-chlorobutane   1-butanol
or *n*-butyl chloride   or *n*-butyl alcohol
bp 78.4 °C   bp 117 °C

a primary alkyl halide   *a primary alcohol*

Replacement of one of the secondary hydrogen atoms gives compounds containing the **1-methylpropyl (or *sec*-butyl) group.**

$$H-\overset{\overset{\displaystyle H}{|}}{\underset{\underset{\displaystyle H}{|}}{C}}-\overset{\overset{\displaystyle H}{|}}{\underset{\underset{\displaystyle H}{|}}{C}}-\overset{\overset{\displaystyle H}{|}}{C}-\overset{\overset{\displaystyle H}{|}}{\underset{\underset{\displaystyle H}{|}}{C}}-H \qquad CH_3CH_2\underset{\underset{\displaystyle |}{}}{C}HCH_3$$

1-methylpropyl or *sec*-butyl group

a secondary alkyl group

CH$_3$CH$_2$CHCH$_3$ CH$_3$CH$_2$CHCH$_3$
          | |
          Cl   OH

2-chlorobutane   2-butanol
or *sec*-butyl chloride   or *sec*-butyl alcohol
bp 68.3 °C   bp 99.5 °C

a secondary alkyl halide   *a secondary alcohol*

Two alkyl groups are also formed from 2-methylpropane. A primary alkyl group, the **2-methylpropyl or isobutyl group,** results from the removal of any one of the primary hydrogen atoms.

2-methylpropyl or isobutyl group

a primary alkyl group

CH$_3$ CH$_3$
| |
CH$_3$CHCH$_2$Cl   CH$_3$CHCH$_2$OH

1-chloro-2-methylpropane   2-methyl-1-propanol
or isobutyl chloride   or isobutyl alcohol
bp 68.9 °C   bp 108 °C

a primary alkyl halide   *a primary alcohol*

Removal of the tertiary hydrogen atom gives a tertiary alkyl group, usually called the **tertiary butyl (abbreviated *tert*-butyl) group.**

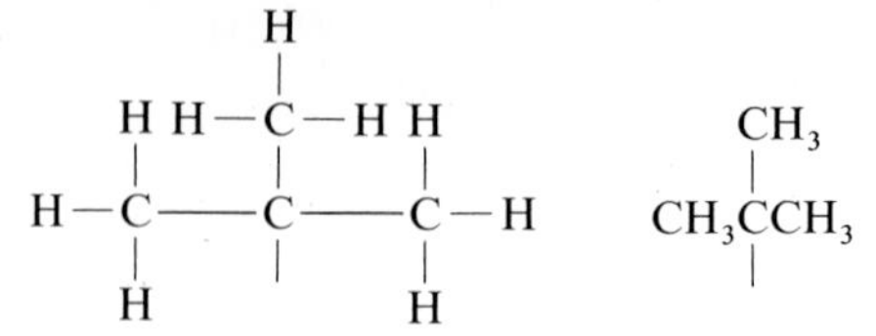

$$H-\overset{\overset{\displaystyle H}{|}}{\underset{\underset{\displaystyle H}{|}}{C}}-\overset{\overset{\displaystyle H-\overset{\overset{\displaystyle H}{|}}{\underset{\underset{\displaystyle H}{|}}{C}}-H}{|}}{\underset{\underset{\displaystyle H}{|}}{C}}-\overset{\overset{\displaystyle H}{|}}{\underset{\underset{\displaystyle H}{|}}{C}}-H$$

$$\overset{\overset{\displaystyle CH_3}{|}}{CH_3CCH_3}$$
$$|$$

1,1-dimethylethyl or *tert*-butyl group

a tertiary alkyl group

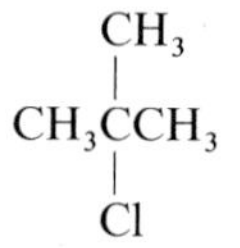

$$\overset{\overset{\displaystyle CH_3}{|}}{\underset{\underset{\displaystyle Cl}{|}}{CH_3CCH_3}}$$ $$\overset{\overset{\displaystyle CH_3}{|}}{\underset{\underset{\displaystyle OH}{|}}{CH_3CCH_3}}$$

2-chloro-2-methylpropane   2-methyl-2-propanol
tert-butyl chloride   *tert*-butyl alcohol
bp 52 °C   bp 82 °C

a tertiary alkyl halide   *a tertiary alcohol*

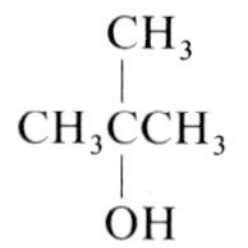

You should learn the structures and the names of the alkyl groups derived from methane, ethane, propane, and the two butanes because they are used in naming more complicated molecules.

The four alkyl halides with the molecular formula C_4H_9Cl shown above are structural isomers of each other. So are the four alcohols with the molecular formula $C_4H_{10}O$. These four alkyl halides (or the four alcohols) are examples of structural isomers that belong to the same functional group class (Section 1.5B). Each structural isomer has distinctive physical properties. For example, the illustrations in this section include the different boiling points for all of the isomeric compounds. Thus, the various isomers may, in principle, be separated from each other by methods such as fractional distillation of liquids and recrystallization of solids.

PROBLEM 3.5   Make a list of the alkyl groups derived from methane, ethane, propane, butane, and 2-methylpropane. Show the condensed formula for each alkyl group, as well as its common name. The common names (the names given second in the text) are the ones we will use most frequently in naming larger molecules.

PROBLEM 3.6   Many aromatic hydrocarbons have alkyl groups on benzene rings. Identify the alkyl group(s) on the benzene ring in each of the following compounds.

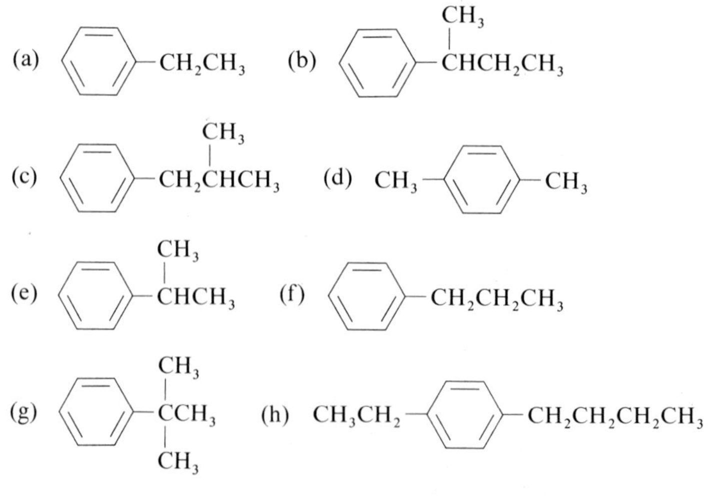

(a) ⟨benzene⟩—CH_2CH_3 (b) ⟨benzene⟩—$\overset{\overset{\displaystyle CH_3}{|}}{C}HCH_2CH_3$

(c) ⟨benzene⟩—$CH_2\overset{\overset{\displaystyle CH_3}{|}}{C}HCH_3$ (d) CH_3—⟨benzene⟩—CH_3

(e) ⟨benzene⟩—$\overset{\overset{\displaystyle CH_3}{|}}{C}HCH_3$ (f) ⟨benzene⟩—$CH_2CH_2CH_3$

(g) ⟨benzene⟩—$\overset{\overset{\displaystyle CH_3}{|}}{\underset{\underset{\displaystyle CH_3}{|}}{C}}CH_3$ (h) CH_3CH_2—⟨benzene⟩—$CH_2CH_2CH_2CH_3$

PROBLEM 3.7   Write structural formulas for all compounds having the molecular formula $C_4H_8Cl_2$.

PROBLEM 3.8   For each of the following sets of structural formulas decide whether they represent structural isomers or not. Of the ones that are structural isomers, which ones are functional group isomers (Section 1.5B)?

(a) $CH_3CH{=}CHCH_2OH$, $CH_3CH{=}CHOCH_3$

(b) $ClCH_2CH_2CH_2CH_3$, $CH_3CH_2CH_2CH_2Cl$

(c) $CH_3CH_2CH_2CH_2NH_2$, $CH_3CH_2NHCH_2CH_3$

(d) $CH_3CH_2CH_2\overset{\overset{\displaystyle O}{\|}}{C}H$, $CH_3CH_2\overset{\overset{\displaystyle O}{\|}}{C}CH_3$

(e) $ClCH_2CH_2CH_2Cl$, $CH_3\overset{\overset{\displaystyle Cl}{|}}{\underset{\underset{\displaystyle Cl}{|}}{C}}CH_3$

(f) $CH_2{=}CHCH_2CH_2OH$, $CH_3CH_2\overset{\overset{\displaystyle O}{\|}}{C}CH_3$

(g) $CH_3\underset{\underset{\displaystyle OH}{|}}{C}HCH_2CH_3$, $CH_3CH_2\underset{\underset{\displaystyle OH}{|}}{C}HCH_3$

B. Conformations of Butane

The three-dimensional representation of butane on p. 84 was drawn so that the first and last carbon atoms of the chain were as far apart from each other as possible, and the hydrogen atoms on the second and third carbon atoms were staggered. This arrangement, which is the most stable one, is called the **anti** conformation of butane. There are, however, other conformations, and one is converted into another by a 60° rotation around single bonds of the different carbon atoms with respect to each other. Below you can see sawhorse formulas and Newman projections (Section 3.3B) of the conformations of butane that result from rotation around the bond between the second and third carbon atoms in the chain.

sawhorse formulas

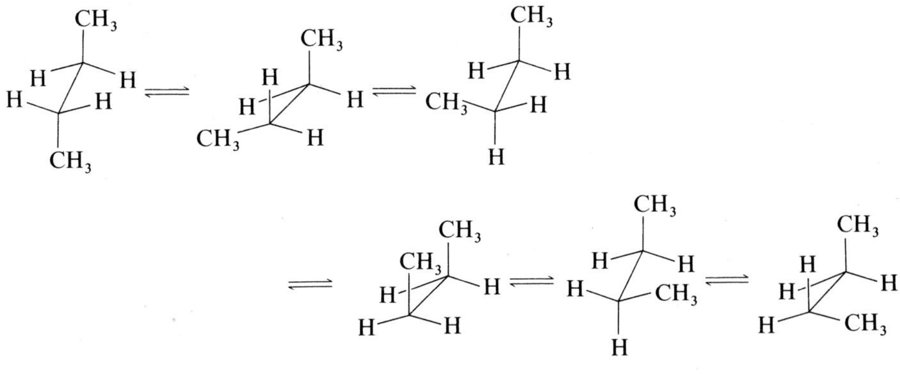

Newman projections

CH₃ ... anti A ⇌ (rotation of 60°) eclipsed B ⇌ (rotation of 60°) gauche C ⇌ (rotation of 60°)

eclipsed D ⇌ (rotation of 60°) gauche E ⇌ (rotation of 60°) eclipsed F

Starting with the anti conformer, rotation of the front carbon atom relative to the rear carbon atom gives different conformations of butane. Besides the anti conformer A, there are two other staggered conformers, C and E; these are called **gauche** conformers. In C and E the carbon 1–carbon 2 bond and the carbon 3–carbon 4 bond are at an angle of 60° to each other in the Newman projection. The angle between the two methyl groups in this representation of butane is called the **dihedral angle** (Figure 3.3).

When the methyl groups are that close, nonbonding interactions (Section 1.8C) known as van der Waals repulsion occur. The drawing on the right in Figure 3.3 shows the atoms of the methyl groups as spheres that interfere with each other and thus repel each other. In other words, atoms have space-filling properties and exert an influence that is not apparent just from looking at the symbols of the atoms in molecular formulas. The effective size of atoms in molecules is expressed in terms of van der Waals radii for those atoms. The van der Waals radius of a hydrogen atom is 1.2 Å. Nonbonding

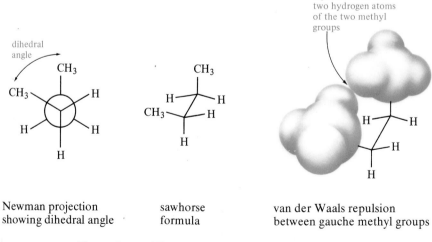

Newman projection   sawhorse   van der Waals repulsion
showing dihedral angle   formula   between gauche methyl groups

Figure 3.3 Different representations of a gauche conformation of butane:

interactions between molecules or between different parts of the same molecule result in van der Waals attraction (Section 1.8C) as long as the interacting atoms do not get too close to each other. At distances shorter than the van der Waals radii of the atoms, repulsion occurs. In the gauche conformation, the hydrogen atoms of the two methyl groups are close enough that van der Waals repulsions result. Consequently, these conformers are less stable than the anti form by approximately 0.9 kcal/mol.

At room temperature, butane exists about 25% in gauche conformations and 75% as the anti conformation. The energy barrier to rotation around the central bond in butane is small enough, approximately 3.8 kcal/mol, that the different conformers cannot be separated from each other at room temperature; they are rapidly interconverted. The physical properties of butane represent an average of all the properties that would be expected for all of the different forms present at any one time.

The interconversions of the anti and gauche conformers require rotation through three high-energy conformations, the eclipsed conformations B, D, and F. The highest-energy conformation is D, in which the two methyl groups are eclipsed. The energy relationships of the different conformations of butane are shown in Figure 3.4.

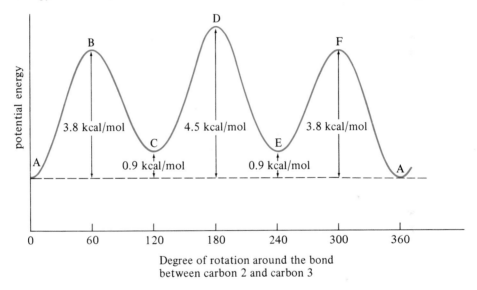

Figure 3.4   Energy diagram for the conformations of butane.

In larger alkanes, too, the most stable conformations are generally staggered, with the largest groups anti to each other. The gauche conformations are only slightly less stable than the anti ones, and are also present. Thus, while the structural formulas of alkanes are usually written in a straight line for convenience, it is important to keep in mind the zig-zag nature of hydrocarbon chains, as illustrated for octane, $CH_3CH_2CH_2CH_2CH_2CH_2CH_2CH_3$.

the most stable conformation of octane; largest
substituents at any one carbon-carbon bond anti
to each other

PROBLEM 3.9   Draw possible conformations of 1-chloropropane, $CH_3CH_2CH_2Cl$. Show them in the sawhorse representation and as Newman projections. Draw an energy diagram showing the relative energy of the different conformers.

PROBLEM 3.10   Show the low-energy conformers of 1,2-dichloroethane (Section 3.3C, p. 81) as sawhorse representations and in Newman projections. Predict, from looking at your pictures, whether you would expect each conformer to have a dipole moment or not. (You may wish to review Section 1.7B.) 1,2-Dichloroethane has a dipole moment of 1.42 D. What does this tell you about the conformational composition of 1,2-dichloroethane at room temperature?

3.6

Nomenclature

A. Introduction

Historically, compounds were given names that reflected their origin or their properties. For example, the pain killer morphine was named for Morpheus, the Greek god of dreams. Cholesterol, the chief component of gallstones, got its name from the Greek words for bile and solid. This way of naming compounds, although descriptive, does not lead to any systematic procedure for assigning names to new and related compounds.

By the end of the nineteenth century, the number of organic compounds that were being synthesized or isolated from natural sources was growing rapidly. Under these conditions, it eventually became impossible for chemists to learn the names randomly assigned to compounds by their discoverers, especially when there was no correlation between the names and the structures of the compounds. Since 1892, chemists from all over the world have met periodically to decide on systematic rules for naming organic compounds. These rules, which are constantly evolving, are called the International Union of Pure and Applied Chemistry (abbreviated as IUPAC) rules. The IUPAC system of nomenclature was developed so that each organic compound would have an unmistakable systematic name that would allow a unique structural formula to be written for it.

Ideally, one would have to learn only the systematic names of compounds. In the real world, however, there are three major complications. First, even the IUPAC rules allow for some variations, within limits, for the naming of compounds. Second, the correct IUPAC names for some common compounds are so complicated and cumbersome that everybody continues to use common unsystematic names for them. These names are usually derived from the origins of compounds in nature, or their shapes, or sometimes even the whimsy of their creators. The structural formulas and systematic names of morphine and cholesterol are shown below, not for you to memorize, but so that you will have some idea why the IUPAC names are not used in everyday language to refer to these compounds.

(5α,6α)-7,8-didehydro-4,5-
epoxy-17-methylmorphinan-
3,6-diol

morphine

(3β)-cholest-5-en-3-ol

cholesterol

Finally, many compounds had names before the IUPAC rules came into being, and these names are still used, especially by chemical supply houses and by industry. To be literate in the laboratory, you must be able to recognize many common names. Therefore, while the emphasis in this book is on the systematic names of compounds, the common names of important compounds are also given and are used in cases where they are the ones overwhelmingly used by chemists in their daily work.

B. Nomenclature of Alkanes

The names of the first four alkanes are methane, ethane, propane, and butane. The systematic nomenclature of the other members of the series is based on a prefix indicating the number of carbon atoms in the chain, followed by the suffix **ane**. The prefixes come from the Greek or Latin words for the numbers. The names of the first twelve straight chain alkanes, along with those of the higher alkanes that form the basis for names of biologically important compounds, are shown in Table 3.2. These names are the basis for the naming of all other types of organic compounds derived from alkanes, so you should learn them.

For a compound that is not a straight chain alkane, *the name is derived from the name of the alkane that corresponds to the longest continuous chain of carbon atoms in the molecule.* For example, the branched chain compounds with the molecular formula C_5H_{12} (Section 3.1, p. 75) are named 2-methyl*butane* and 2,2-dimethyl-*propane*. The longest chain in the first one is *four carbons* long, therefore the com-

TABLE 3.2   The Names of Some Straight Chain Alkanes

Molecular Formula	Name	Molecular Formula	Name
CH_4	methane	C_8H_{18}	octane
C_2H_6	ethane	C_9H_{20}	nonane
C_3H_8	propane	$C_{10}H_{22}$	decane
C_4H_{10}	butane	$C_{11}H_{24}$	undecane
C_5H_{12}	pentane	$C_{12}H_{26}$	dodecane
C_6H_{14}	hexane	$C_{16}H_{34}$	hexadecane
C_7H_{16}	heptane	$C_{18}H_{38}$	octadecane
		$C_{20}H_{42}$	icosane

pound is classified as a substituted *butane*. The longest chain in the second one is *three carbons* long, so the compound is a substituted *propane*.

Alkanes with the molecular formula C_6H_{14} exist in five isomeric forms. The straight chain isomer is called hexane. Two of the branched chain isomers have five carbon atoms in the longest continuous chain, and are named as substituted pentanes.

$$CH_3CH_2CH_2CH_2CH_2CH_3$$
$$1\quad2\quad\ 3\quad\ 4\quad\ 5\quad\ 6$$

hexane

straight chain isomer
of C_6H_{14}

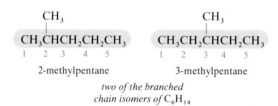

2-methylpentane   3-methylpentane

two of the branched
chain isomers of C_6H_{14}

The other two isomers of C_6H_{14} have only four carbon atoms in the longest continuous chain and are, therefore, substituted butanes.

$$\begin{array}{cc} CH_3 & CH_3\ \ CH_3 \\ | & |\ \ \ \ | \\ CH_3CCH_2CH_3 & CH_3CH-CHCH_3 \\ 1\ \ |2\ 3\ \ \ 4 & 1\ \ \ 2\ \ \ \ \ 3\ \ \ 4 \\ \ \ CH_3 & \end{array}$$

2,2-dimethylbutane   2,3-dimethylbutane

two more of the branched chain
isomers of C_6H_{14}

After the longest continuous chain of carbon atoms is identified, *the alkyl groups attached to that backbone are named. Each one is assigned a number indicating its position on the chain. The chain is numbered so that the substituents have the lowest possible numbers.* The name of the compound is created by listing the names of the substituents along with their positions on the chain that is the backbone of the molecule. For example, the substituents in the branched chain C_6H_{14} compounds are all methyl groups. The first isomer is called 2-methylpentane to indicate that the methyl group is on the second carbon atom from the end of the five-carbon chain. The alternate name, 4-methylpentane, which is arrived at by numbering the chain in the opposite direction, is incorrect because it violates the rule that substituents must be given the lowest possible numbers. The second isomer has the methyl group attached to the third carbon atom of the chain and is 3-methylpentane.

If there is more than one substituent of the same kind, the name of the alkyl group is given the prefix di (for two), tri (for three), or tetra (for four). As before, each substituent is also given a number indicating its position on the chain. These rules are illustrated by the names of the last two isomers of C_6H_{14}. The name of 2,2-dimethylbutane indicates that the compound has a chain of four carbon atoms with two methyl groups attached to the second carbon atom of the chain. Note that the number 2 is repeated for each methyl group so that there is no ambiguity about the placement of the groups. The name of the isomeric compound 2,3-dimethylbutane differs from 2,2-dimethylbutane in only one number. Note how the names are written. *The numbers are separated by commas and set apart from the name by a hyphen.*

An application of these rules to somewhat more complicated examples may be helpful. A $C_{12}H_{26}$ hydrocarbon is shown below.

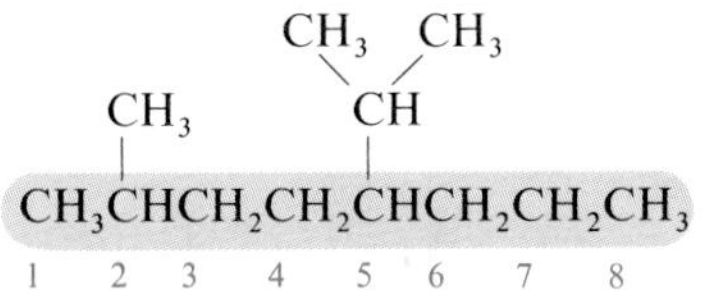

$$
\begin{array}{cc}
& \overset{\displaystyle CH_3 \;\; CH_3}{\underset{}{\diagdown \;\; \diagup}} \\
\overset{\displaystyle CH_3}{|} & \overset{\displaystyle CH}{|} \\
\underset{\;\;1\quad 2\quad 3\quad\;\; 4\quad\; 5\quad 6\quad\; 7\quad\;\; 8}{CH_3CHCH_2CH_2CHCH_2CH_2CH_3}
\end{array}
$$

5-isopropyl-2-methyloctane

The longest continuous chain contains *eight* carbon atoms. The compound is, therefore, named as a substituted *octane*. The alkyl substituents on the chain are identified as the methyl group (on carbon atom 2) and the isopropyl group (on carbon atom 5). The compound is given the name 5-isopropyl-2-methyloctane. *When there are several different kinds of substituents on the chain, they are listed in alphabetical order.* Prefixes such as di or tri are not considered when the substituents are alphabetized. Neither are italicized prefixes set off by hyphens such as, for example, *tert-*butyl, which is treated as a butyl group.

Another example further illustrates the rules.

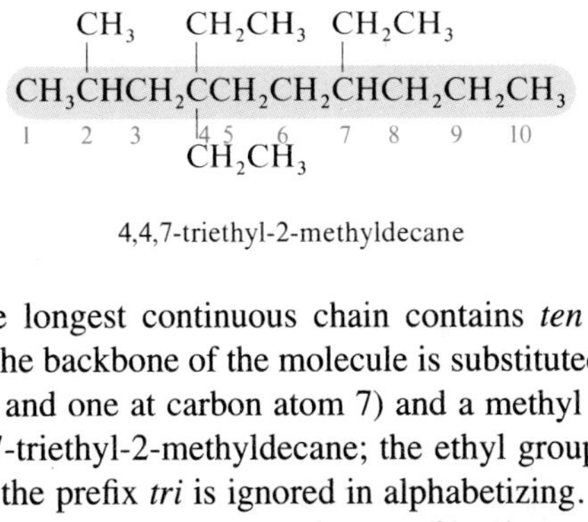

$$
\begin{array}{c}
\overset{\displaystyle CH_3 \quad\;\; CH_2CH_3 \;\; CH_2CH_3}{\underset{}{|\qquad\quad\; |\qquad\qquad |}} \\
\underset{\;\;1\quad 2\quad\; 3\quad\;\; 4\; 5\quad\; 6\quad\;\; 7\quad 8\quad 9\quad\; 10}{CH_3CHCH_2CCH_2CH_2CHCH_2CH_2CH_3} \\
\underset{}{|} \\
CH_2CH_3
\end{array}
$$

4,4,7-triethyl-2-methyldecane

In this compound, the longest continuous chain contains *ten* carbon atoms and is, therefore, a *decane*. The backbone of the molecule is substituted by three ethyl groups (two at carbon atom 4 and one at carbon atom 7) and a methyl group (at carbon atom 2). The name is 4,4,7-triethyl-2-methyldecane; the ethyl groups are listed before the methyl group because the prefix *tri* is ignored in alphabetizing. Note again the use of commas to separate a series of numbers and the use of hyphens to set the numbers apart from the substituents. The name of the last substituent to be listed is merged with that assigned to the backbone.

In summary, the rules for nomenclature of alkanes are:

1. Locate the longest continuous straight chain of carbon atoms in the molecule. The name of the straight chain alkane with the same number of carbon atoms becomes the last part of the name of the compound.
2. Find and name all of the alkyl groups that are branches on the backbone of the molecule. Assign each one a position on the chain, numbering the chain so that the substituents have the lowest possible numbers.
3. If there are several substituents of the same kind, indicate how many by using the prefixes di, tri, and tetra, and use a number to assign a position on the chain to each one.
4. Construct the name of the compounds by listing all the substituents in alphabetical order, ignoring the prefixes di, tri, tetra, and italicized prefixes such as *n-*, *sec-*, and *tert-* in alphabetizing.

5. Use commas to separate numbers that are grouped together. Separate numbers from names of groups by hyphens. Merge the name of the last named substituent with the name of the straight chain alkane that is the basis of the name of the compound.

One last example will provide a review of these points.

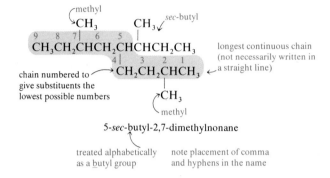

5-*sec*-butyl-2,7-dimethylnonane

PROBLEM 3.11   Name the following compounds.

(a)
$$
\underset{\begin{array}{c}|\\ CH_3\end{array}}{\overset{\begin{array}{c}CH_3\\|\end{array}}{CH_3CHCHCH_2CH_2CH_2}}\underset{\begin{array}{c}|\\ CH_3\end{array}}{\overset{\begin{array}{c}CH_3\\|\end{array}}{CCH_3}}
$$

(b)
$$
\overset{\begin{array}{c}CH_3\\|\end{array}}{CH_3CH_2CHCHCH_2CH_2}\underset{\begin{array}{c}|\\ CH_2CH_3\end{array}}{\overset{\begin{array}{c}CH_3\\|\end{array}}{CHCH_3}}
$$

(c)
$$
\overset{\begin{array}{c}CH_3\\|\end{array}}{CH_3CHCHCH_2CH_3}\underset{\begin{array}{c}|\\ CH_3\end{array}}{}
$$

(d)
$$
\overset{\begin{array}{c}CH_3\\|\\ CH_3CCH_3\\|\end{array}}{CH_3CH_2CH_2CHCH_2CH_2CH_3}
$$

(e)
$$
\overset{\begin{array}{c}CH_3\\|\end{array}}{CH_3CCH_2CH_2}\underset{\begin{array}{c}|\\ CH_3\end{array}}{}\overset{\begin{array}{c}CH_3\\|\end{array}}{CHCHCH_2CH_2CH_3}\underset{\begin{array}{c}|\\ CH_2CH_2CH_2CH_3\end{array}}{}
$$

(f)
$$
\overset{\begin{array}{c}CH_3\quad CH_3\\ \diagdown\;\diagup\\ CH_3\quad\; CH\\|\qquad\;|\end{array}}{CH_3CH_2CHCH_2CH_2CHCH_2CH_2CH_3}
$$

PROBLEM 3.12   Draw structural formulas for the following compounds.

   (a)   2,2-dimethylheptane
   (b)   6-isobutyl-2-methyldecane
   (c)   5,5-diisopropyl-2,8-dimethylnonane
   (d)   4-*tert*-butyl-3-ethyloctane
   (e)   6-ethyl-2,2,4-trimethyldodecane

PROBLEM 3.13   Draw structural formulas for the nine alkanes with the molecular formula C_7H_{16}. Identify primary, secondary, and tertiary carbon atoms in each one. Show which hydrogen atoms are equivalent to each other. Name each compound.

C. Nomenclature of Alkyl Halides and Alcohols

The systematic names of alkyl halides are assigned in the same way as the names of branched chain alkanes. The prefixes **fluoro, chloro, bromo,** and **iodo** are used to indicate the presence of halogens. For example:

$$\begin{array}{c} CH_3 \\ | \\ CH_3CHCHCH_2CH_3 \\ 1 \quad 2 \quad |3 \quad 4 \quad 5 \\ F \end{array} \qquad \begin{array}{c} Cl \\ | \\ CH_3CH_2CCH_3 \\ 4 \quad 3 \quad 2| \ 1 \\ Cl \end{array}$$

3-fluoro-2-methylpentane   2,2-dichlorobutane

$$\begin{array}{c} CH_3 \ \ CH_3 \\ | \quad \ \ | \\ CH_3CH_2C\!\!-\!\!CCH_2CH_3 \\ 1 \quad 2 \quad |3 \quad 4| \ 5 \quad 6 \\ Br \quad \ Br \end{array} \qquad \begin{array}{c} CH_2CH_3 \\ | \\ CH_3CH_2CCH_2CH_2CH_3 \\ 1 \quad 2 \quad 3| \ 4 \quad 5 \quad 6 \quad 7 \\ I \end{array}$$

3,4-dibromo-3,4-dimethylhexane   3-ethyl-3-iodoheptane

Some organic compounds are named by changing the *suffix* of the name of the hydrocarbon chain, instead of adding a prefix to it as we have done for alkyl and halogen substituents. Alcohols are usually named by changing the **e** ending of the name of the alkane to **ol** and using a number to indicate the position of the hydroxyl group. In this type of nomenclature, the hydroxyl group is not specifically named. An alcohol owes its characteristic properties and reactivity to the hydroxyl group, so in naming the alcohol, the carbon chain is numbered so that the carbon atom bearing the hydroxyl group has the lowest possible number.

$$\begin{array}{c} CH_3 \\ | \\ CH_3CHCH_2CH_2CH_2OH \end{array} \qquad \begin{array}{c} CH_3 \\ | \\ CH_3CCH_2OH \\ | \\ CH_3 \end{array}$$

4-methyl-1-pentanol   2,2-dimethyl-1-propanol

$$HOCH_2CH_2CH_2OH \qquad ClCH_2CH_2CH_2CH_2OH$$

1,3-propanediol   4-chloro-1-butanol

PROBLEM 3.14   Name the following compounds.

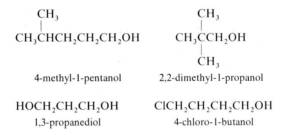

$$\begin{array}{c} CH_3 \ \ CH_3 \\ | \quad \ \ | \\ (a) \quad CH_3C\!\!-\!\!CHCH_2CH_3 \\ | \\ Br \end{array} \qquad (b) \quad \begin{array}{c} CH_3CH_2CHCH_2CH_3 \\ | \\ OH \end{array}$$

$$\begin{array}{c} Cl \\ | \\ (c) \quad CH_3CH_2CHCHCH_2CH_3 \\ | \\ OH \end{array} \qquad (d) \quad \begin{array}{c} Cl \ \ CH_3 \\ | \quad \ \ | \\ CH_3CH_2CCH_2CCH_3 \\ | \quad \ \ | \\ Cl \ \ CH_3 \end{array}$$

$$\begin{array}{c} CH_3 \ \ CH_3 \\ | \quad \ \ | \\ (e) \quad CH_3C\!\!-\!\!CCH_3 \\ | \quad \ \ | \\ Br \quad Br \end{array} \qquad (f) \quad CH_3CH_2CHCH_2CH_2CH_2CH_2CH_3 \\ | \\ I$$

$$\text{(g)} \quad \underset{\underset{\text{Cl}}{|}}{\overset{\overset{\text{Cl}}{|}}{\text{ClCCH}_2\text{CH}_2\text{CH}_3}} \qquad \text{(h)} \quad \text{CH}_3\text{CH}_2\text{CH}_2\text{CH}_2\text{CH}_2\underset{\underset{\text{OH}}{|}}{\text{CHCHCH}_2\text{CH}_3}$$

with the CH₃ group:

$$\overset{\overset{\text{CH}_3}{|}}{\underset{}{\text{CH}_3\text{CCH}_3}}$$

PROBLEM 3.15 Write structural formulas for the following compounds.

(a) 2-iodooctane   (b) 3-hexanol   (c) 2,2,2-trifluoroethanol
(d) 1-chloro-4,4-dimethyl-2-pentanol   (e) 1,2,3-propanetriol   (f) 1-pentanol
(g) 1,2-ethanediol   (h) 2,3-dichloro-3-ethylheptane
(i) 4-bromo-3,3-dimethylheptane   (j) 2-bromo-3-ethylhexane

PROBLEM 3.16 There are seventeen structural isomers with the molecular formula $C_6H_{13}Cl$. Write structural formulas for them. Name each one, and say whether it is a primary, secondary, or tertiary halide. Review of Section 3.5A may help to remind you of the structural features of alkyl halides.

PROBLEM 3.17 Name all the isomeric compounds having the molecular formula $C_4H_8Cl_2$ (Problem 3.7).

D. The Phenyl Group

There is one other group that you should learn at this stage. *The **phenyl group** is formed by the removal of one of the six equivalent hydrogen atoms on benzene. Groups such as the phenyl group, derived from aromatic hydrocarbons* (Section 2.8), *are known as **aryl groups**.* An aromatic hydrocarbon is sometimes represented by the symbol ArH, and the aryl group as Ar.

sometimes abbreviated C_6H_5—

benzene   phenyl group

The phenyl group is used in much the same way as the alkyl groups are used in naming compounds. When the hydrocarbon chain attached to the benzene ring is complex or contains more than five carbon atoms, the phenyl group is used in the name of the compound. However, benzene rings substituted by small alkyl groups are usually named as benzene derivatives.

—CH_2CH_3
ethylbenzene

—$\text{CH}_2\text{CH}_2\text{CH}_2\text{CH}_2\text{CH}_2\text{CH}_2\text{CH}_3$
1-phenylheptane

—$\underset{\underset{\text{Br}}{|}}{\overset{\overset{\text{CH}_3}{|}}{\text{CCH}_2\text{CH}_2\text{CH}_3}}$
2-bromo-2-phenylpentane

—$\underset{\underset{\text{OH}}{|}}{\overset{\overset{\text{CH}_3}{|}}{\text{CCH}_3}}$
2-phenyl-2-propanol

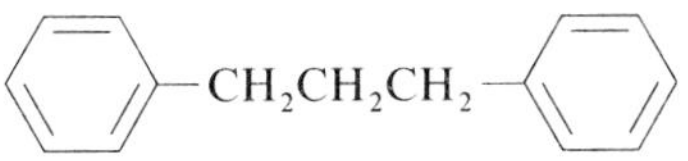

1,3-diphenylpropane 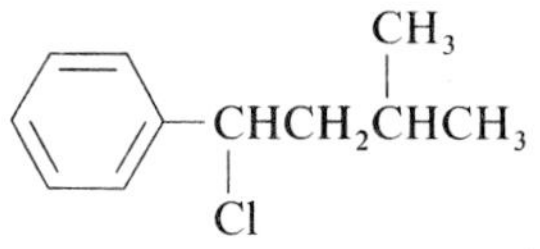   1-chloro-3-methyl-1-phenylbutane

---

PROBLEM 3.18   Name the compounds in Problem 3.6. (Hint: Positions on a benzene ring may be numbered too.)

PROBLEM 3.19   Name the following compounds.

(a) 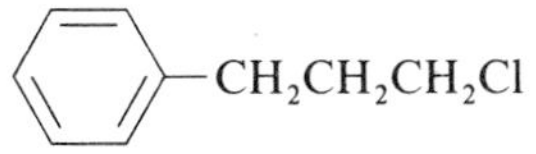 —$CH_2CH_2CH_2Cl$ (b)

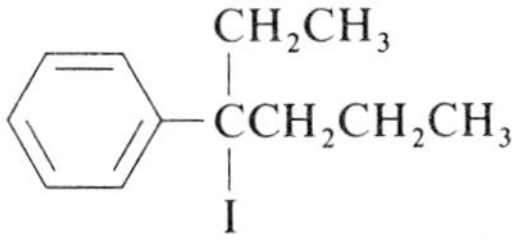

(c) 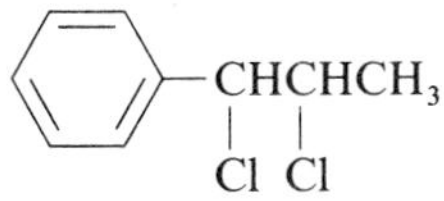   (d) 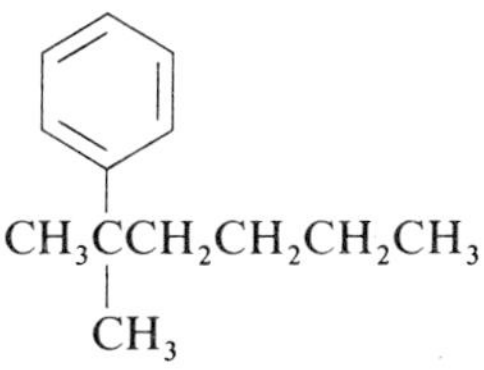 $CH_3CCH_2CH_2CH_2CH_3$

(e) 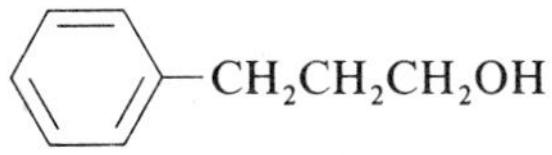 —$CH_2CH_2CH_2OH$ (f)

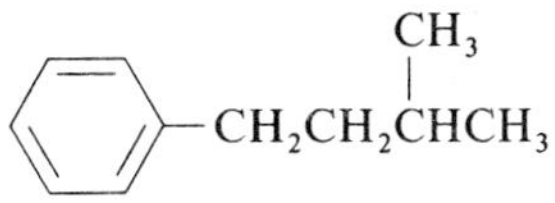

PROBLEM 3.20   Draw structural formulas for the following compounds.

(a)   1-phenyl-1-pentanol   (b)   3-bromo-3-methyl-1-phenylbutane
(c)   4-methyl-2,6-diphenylheptane   (d)   4-*tert*-butyl-2-phenyloctane
(e)   2-phenyl-1-ethanol

3.7

The Chemical Properties of Alkanes

Alkanes have structures in which all of the valence electrons of carbon are bonded to hydrogen atoms in σ bonds. Saturated hydrocarbons are inert towards most reagents. Their reactions generally take place at high temperatures or in the presence of catalysts that promote the cleavage of single bonds.

Alkanes are important constituents of fuels, and their combustion in air provides much of the energy consumed in the world today. Methane is the chief constituent of natural gas. Once ignited, it reacts with oxygen to evolve heat.

$$CH_4 + 2\,O_2 \longrightarrow CO_2 + 2\,H_2O \qquad \Delta H = -210.8 \text{ kcal/mol}$$
methane

The equation represents the complete combustion of methane and assumes that sufficient oxygen is available to convert the compound to carbon dioxide and water. Often, however, incomplete combustion takes place so that carbon monoxide, which is highly toxic, and carbon, which gives the flame of a Bunsen burner its yellow color when the air supply is limited, are also formed.

Petroleum is a mixture of a large variety of hydrocarbons, including alkanes and aromatic hydrocarbons (Section 2.8). The hydrocarbons are separated into fractions according to their boiling points. Propane and butane, for example, are gases that are liquefied under pressure and used as fuel that can be transported in small tanks such as in cigarette lighters. The hydrocarbons with boiling points up to 25 °C (Table 3.1) make up the fraction of petroleum known as gas and liquefied gas.

The next and largest fraction obtained from petroleum is gasoline, consisting of hydrocarbons with four to twelve carbon atoms, and a boiling point range of 20–200 °C. Hydrocarbons at the upper limit of this range are converted into more useful mixtures of lower molecular weight alkanes by heating in the presence of hydrogen gas and catalysts in a process known as **hydrocracking.** For example, undecane, $C_{11}H_{24}$, has a boiling point of 196 °C. It is *cracked* to a mixture of lower weight alkanes in which butanes, pentanes, hexanes, and heptanes are the major constituents.

$$C_{11}H_{24} \xrightarrow[\text{catalyst}]{H_2} C_4H_{10} + C_5H_{12} + C_6H_{14} + C_7H_{16}$$

undecane	275 °C	butanes	pentanes	hexanes	heptanes
bp 196 °C		bp ~ 0 °C	bp ~ 30 °C	bp ~ 68 °C	bp ~ 98 °C

The boiling points of these compounds range from approximately 0–100 °C; thus they are more easily volatilized than undecane and can be used more efficiently in internal combustion engines.

Kerosene, with hydrocarbons containing nine to sixteen carbon atoms and a boiling range of 175–275 °C, is a higher boiling fraction of petroleum. Gas oil and diesel oil come next, boiling at 200–400 °C and consisting of hydrocarbons with fifteen to twenty-five carbon atoms. Lubricating oil has even higher molecular weight alkanes, with twenty to seventy carbon atoms in the chain.

In comparison with other functional group classes, alkanes have limited chemical reactivity. Thus, the alkyl portions of molecules containing other functional groups usually remain unchanged in the reactions that transform the functional groups. This image of the unreactivity of alkyl groups is a useful one for you to carry into the study of the following chapters.

3.8

Cycloalkanes

A. Cyclopropane and Cyclobutane. Ring Strain

Cycloalkanes *are hydrocarbons with the general formula C_nH_{2n} in which some of the carbon atoms form a ring. The cycloalkanes are named for the alkane corresponding to the number of carbon atoms in the ring. Thus, alkanes are divided into two classes: the compounds containing a ring, the cycloalkanes, and the* **open chain alkanes,** *sometimes called* **acyclic** *compounds to distinguish them from the cyclic ones.*

The three carbon atoms in cyclopropane, C_3H_6, the smallest cycloalkane, define a plane. Cyclopropane has internal bond angles of 60° and external bond angles of 116° and 118°.

cyclopropane   cyclopropyl group   chlorocyclopropane cyclopropyl chloride

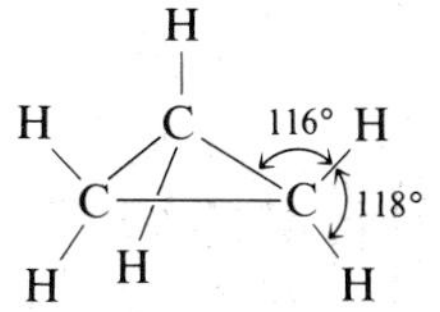

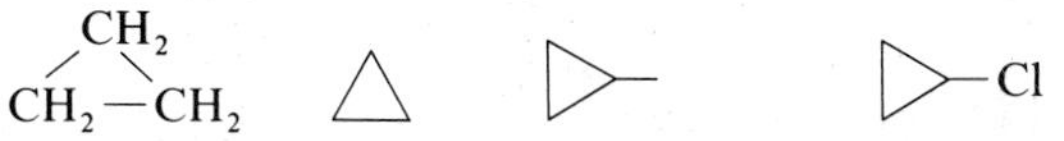

In cyclobutane, C_4H_8, the expected internal bond angles of 90° are reduced to 88° by puckering that minimizes the eclipsing of the hydrogen atoms on adjacent carbon atoms. One of the atoms in the ring is bent out of the plane of the other three by about 20°.

cyclobutane   cyclobutyl group   bromocyclobutane cyclobutyl bromide

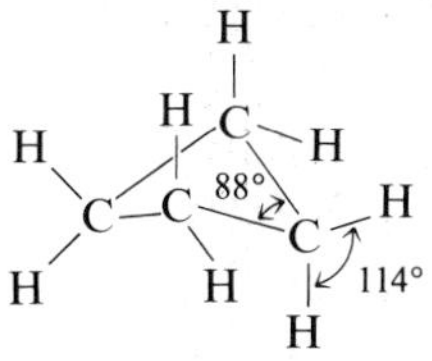

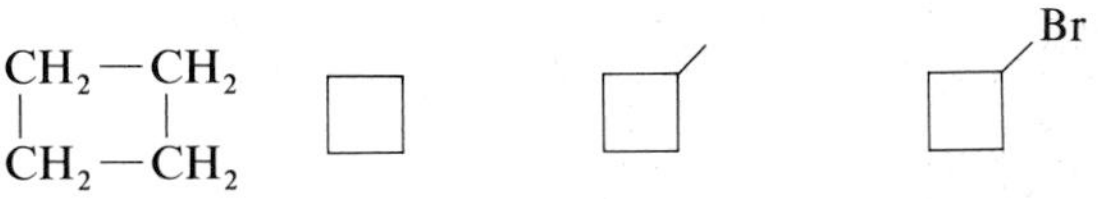

The cycloalkanes are usually symbolized in equations by regular polygons corresponding to the number of carbon atoms in the ring. Thus, cyclopropane is represented by a triangle and cyclobutane by a square. It is understood that each corner of the polygon represents a carbon atom and two hydrogen atoms unless another substituent is shown bonded to that position. Cycloalkyl groups are derived from cycloalkanes, just as alkyl groups are derived from alkanes. Thus cyclopropane gives a cyclopropyl group, and cyclobutane, a cyclobutyl group.

The bond angles in cyclopropane and in cyclobutane are quite different from normal tetrahedral bond angles of 109.5°. This distortion of the bond angles from those expected of sp^3-hybridized carbon atoms (Section 2.4A) results in instability that has been attributed to *ring strain*. Cyclopropane and cyclobutane have more energy stored in their bonds than the larger ring cycloalkanes such as cyclopentane and cyclohexane, which have tetrahedral bond angles. The *strained* state of the small ring compounds is demonstrated by the greater reactivity of those rings than is expected of alkanes. In some respects, the reactivity of the small ring compounds resembles that of alkenes. For example, compounds with double bonds, such as ethylene, react easily with hydrogen gas in the presence of metal catalysts (Section 2.10) to form alkanes.

$$\underset{\text{ethylene}}{CH_2{=}CH_2} \xrightarrow[\text{Ni, } 20\,^\circ\text{C}]{H_2} \underset{\text{ethane}}{CH_3CH_3}$$

Cyclopropane, even though it is not an alkene, also reacts with hydrogen. In the presence of a nickel catalyst, the ring opens and propane is formed.

$$\triangle \quad \xrightarrow[\text{Ni}]{\text{H}_2} \quad CH_3CH_2CH_3$$

cyclopropane   120 °C   propane

A higher temperature is required for this reaction than for the one with an alkene, and a still higher temperature is necessary to cleave a cyclobutane ring.

$$\square \quad \xrightarrow[\text{Ni}]{\text{H}_2} \quad CH_3CH_2CH_2CH_3$$

cyclobutane   200 °C   butane

The larger rings in cyclopentane and cyclohexane do not open upon treatment with hydrogen in the presence of a catalyst under these conditions

$$\text{cyclopentane} \quad \text{or} \quad \text{cyclohexane} \quad \xrightarrow[\text{Ni}]{\text{H}_2} \quad \text{no reaction}$$

200 °C

The reactivity of compounds containing a three-membered ring raises questions about the exact nature of the bonding in such compounds. The H—C—H bond angles in cyclopropane, for example, appear to fit better with sp^2-hybridized (Section 2.5A) rather than sp^3-hybridized (Section 2.4A) carbon atoms. The reactivity of the compound also suggests some unsaturated character to the ring. The small C—C—C bond angles in the molecule make overlap between orbitals on adjacent carbon atoms difficult, and certainly weaken the carbon-carbon bonds.

B. Cyclopentane

The bond angles in cyclopentane are found to be very close to the tetrahedral value, which is also close to the internal bond angle of 108° for a regular pentagon. We would expect cyclopentane to be stable in a planar form except that the hydrogen atoms on adjacent carbon atoms are eclipsed in this form. Cyclopentane is most stable when one of the carbon atoms is bent out of the plane of the other four. This shape is called the **envelope** form of cyclopentane. The carbon atom that is out of the plane moves around the ring in a phenomenon known as **pseudorotation,** so that the eclipsing of the hydrogen atoms at various points in the ring is relieved at least part of the time. Note, however, that the structure of the ring does not allow complete rotation around the carbon-carbon single bonds. The type of conformational isomerism seen for ethane or butane is not possible for a cyclic alkane.

eclipsed   envelope

cyclopentane   cyclopentyl   bromocyclopentane
planar representation   group   cyclopentyl bromide

Substitution of another group for a hydrogen atom on cyclopentane gives rise to cyclopentyl compounds.

PROBLEM 3.21 Name the following compounds.

(a) 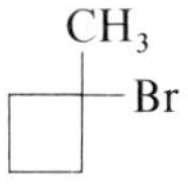 [structure: cyclobutane with CH₃ and Br]   (b) 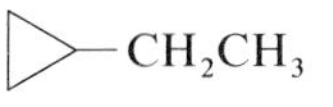 [structure: cyclopropane]—CH₂CH₃   (c) 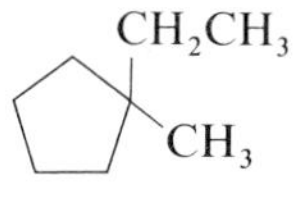 [structure: cyclopentane with CH₂CH₃ and CH₃]   (d)  [structure: cyclopentane]—I

(e) 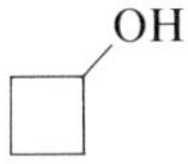 [structure: cyclobutane with OH]   (f) 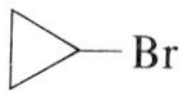 [structure: cyclopropane]—Br

---

PROBLEM 3.22 Draw structures for the following compounds.

(a) *tert*-butylcyclopentane   (b) cyclopropanol   (c) 1-ethyl-1-cyclopentanol
(d) isopropylcyclobutane   (e) chlorocyclopentane   (f) 2-cyclopentyloctane
(g) phenylcyclopentane (or cyclopentylbenzene)

C. Cyclohexane. Conformation in Cyclohexane and in Cyclohexanes with One Substituent

Cyclohexane, C_6H_{12}, was known to be an unstrained molecule in the late nineteenth century. This fact was puzzling because a planar regular hexagon with internal bond angles of 120°, which are larger than tetrahedral, would be expected to exhibit some ring strain. In 1890, the German chemist Ulrich Sachse pointed out that cyclohexane need not be a planar molecule. Indeed, puckering of the ring would allow for tetrahedral bond angles. He suggested that cyclohexane should exist in two forms, what we now call the **chair** and the **boat** forms. At that time, these forms could not be observed experimentally, so Sachse's ideas were not accepted until about 35 years later when experimental evidence was found for such puckered six-membered rings in some more complex compounds. In 1943 the Norwegian chemist Odd Hassel showed by electron diffraction that the chair form of cyclohexane was the predominant conformer in the gas phase. For this work Hassel shared the Nobel Prize in Chemistry in 1969 with Sir Derek Barton (p. 543).

In the chair form of cyclohexane, four of the carbon atoms in the ring are in a plane, a fifth carbon atom is above, and a sixth is below the plane.

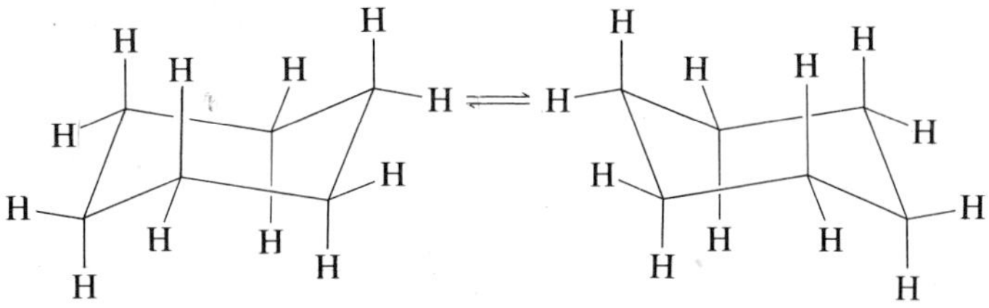

sawhorse representations of the
chair forms of cyclohexane

All of the carbon-hydrogen bonds on adjacent carbon atoms are staggered, as shown in the Newman projection for the chair form of cyclohexane.

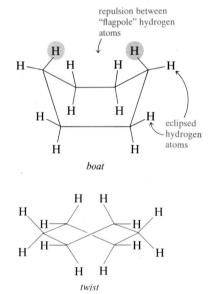

viewed along these bonds   *Newman projection*

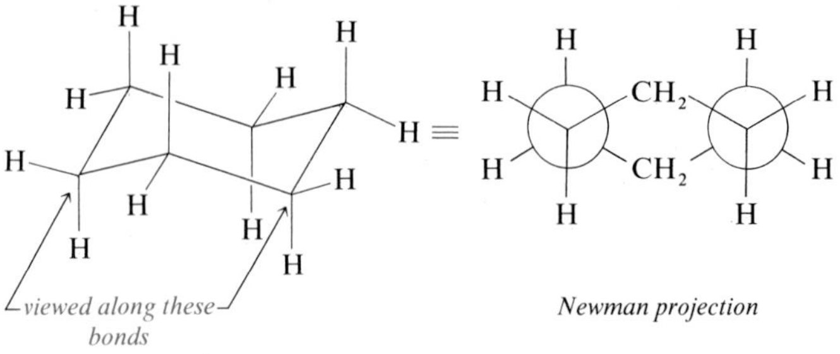

Partial rotation around the carbon-carbon bonds in cyclohexane results in the conversion of one chair form to another one. The energy barrier between the two forms is 10.8 kcal/mol, low enough that the ring is constantly flipping (approximately 10^5 times a second) from one conformation to the other at room temperature.

Besides the chair forms, cyclohexane has other higher energy, less stable conformations such as the boat form and the twist form.

repulsion between "flagpole" hydrogen atoms

eclipsed hydrogen atoms

boat

twist

high-energy conformations of cyclohexane

In the boat form, two of the carbon atoms lie above the plane defined by the other four carbon atoms. The hydrogen atoms on these two carbon atoms (called the "flagpole" positions) are brought close enough that they repel each other. In addition, the carbon-hydrogen bonds on the carbon atoms in the plane are all eclipsed. In the twist form, some of these interactions are relieved so that it is a little more stable than the boat form. At room temperature, most cyclohexane molecules exist in the most stable conformation, the chair form. The twist form is an intermediate in the conversion of one chair form to another.

There are two types of hydrogen atoms in the chair form of cyclohexane. Six of the hydrogen atoms lie in a ring roughly in the plane of the molecule. These hydrogen atoms are called the **equatorial** hydrogen atoms. The bonds to the other six hydrogen atoms are parallel to an axis that goes through the center of the cyclohexane ring. These

axial hydrogen atoms occur in an alternating arrangement; three are above the average plane of the ring and three are below.

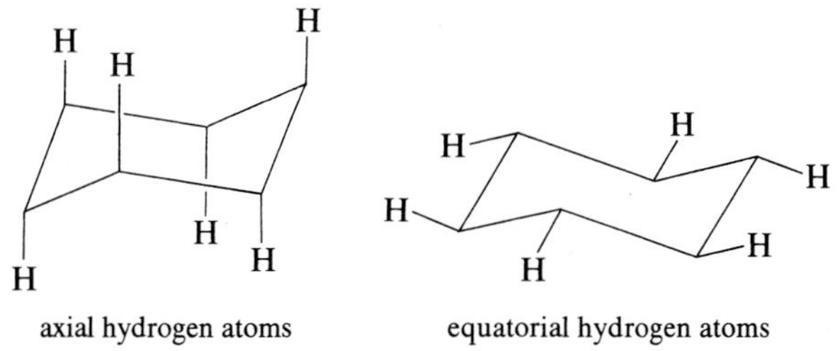

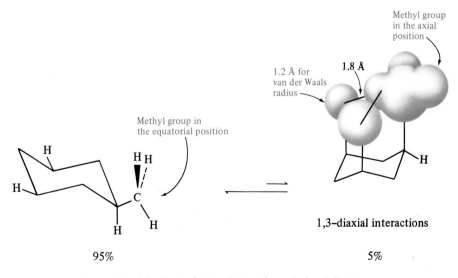

axial hydrogen atoms   equatorial hydrogen atoms

The rapid flipping of the ring at room temperature converts one type of hydrogen atom into another. You should demonstrate to yourself how this works by using molecular models. The easiest way to convert one chair form of a model of cyclohexane to another is to move the carbon atom that is below the plane of the other four up, and the one that is above the plane down, as demonstrated in Figure 3.5.

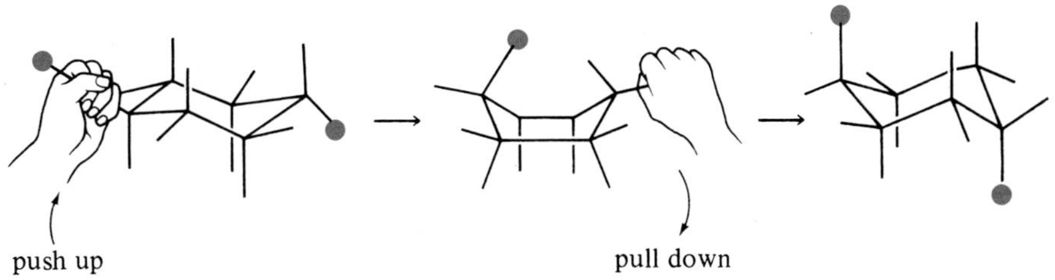

push up   pull down

Figure 3.5 The interconversions of axial and equatorial hydrogen atoms in a model of cyclohexane by the flipping of the ring.

When a substituent replaces one of the hydrogen atoms, the two chair conformations are no longer equivalent. For example, molecules of methylcyclohexane are mostly in the form in which the methyl group is in an equatorial position (Figure 3.6).

Methyl group in the axial position

1.2 Å for van der Waals radius

1.8 Å

Methyl group in the equatorial position

1,3–diaxial interactions

95%   5%

Figure 3.6 Chair conformations of methylcyclohexane.

This form is more stable by approximately 1.7 kcal/mol than the other chair form created by flipping the ring. When the methyl group is in an axial position, it is close to the two axial hydrogen atoms on the same side of the plane of the ring. The distance between a hydrogen atom of the methyl group and either one of the hydrogen atoms in the axial positions on the same side of the ring is 1.8 Å, shorter than 2.4 Å, the sum of the van der Waals radii of the two hydrogen atoms (p. 88). *The steric interaction, the repulsion between the methyl group and each of these two hydrogen atoms, is called a* **1,3-diaxial interaction.** The 1,3-diaxial interaction serves to make the system less stable than the conformer in which the methyl group is equatorial.

A methyl group in the axial position on a cyclohexane ring is gauche (Section 3.5B) to the largest substituent on the adjacent carbon atoms. These substituents happen to be other carbon atoms of the ring.

If a methyl group is equatorial on the ring, it is anti to the ring atoms that are the largest substituents on adjacent carbon atoms. This analysis also predicts a lower stability for the conformer of cyclohexane with a substituent in the axial position.

In general, the larger the substituent, the greater the repulsive interactions when it occupies an axial position. A *tert*-butyl group, for example, is so bulky that the conformation in which it is in the equatorial position is overwhelmingly favored (Figure 3.7). The difference in energy between the two chair forms of *tert*-butylcyclohexane is 5.6 kcal/mol. The presence of the large alkyl group effectively slows down the flipping of the ring because 1,3-diaxial interactions are so severe in one of the chair forms.

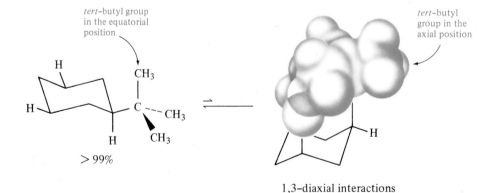

Figure 3.7 Chair conformations of *tert*-butylcyclohexane.

PROBLEM 3.23 Draw structures for chlorocyclohexane and cyclohexanol in the planar form and in the two chair conformations.

3.9

Stereoisomerism

A. Enantiomers

In the earlier sections of this chapter we explored the question of the equivalency of hydrogen atoms by removing different types of hydrogen atoms and replacing them with halogen atoms, to form structural isomers. If this exercise is carried a little farther, we discover a more subtle kind of isomerism called stereoisomerism. **Stereoisomers** *differ from each other only in the way that the different groups in the molecules are oriented in space.* They cannot be distinguished by looking at the condensed structural formulas, which are identical for stereoisomers. Molecular models will help you see this more subtle kind of isomerism. Use them as you follow the arguments in this section.

Bromochloromethane, CH_2BrCl, exists as a single isomer. At first glance, the two hydrogen atoms on the carbon atom may appear to be equivalent. First one then the other of the two hydrogen atoms may be replaced by a substituent different from bromine or chlorine. Iodine is used below for the sake of simplicity.

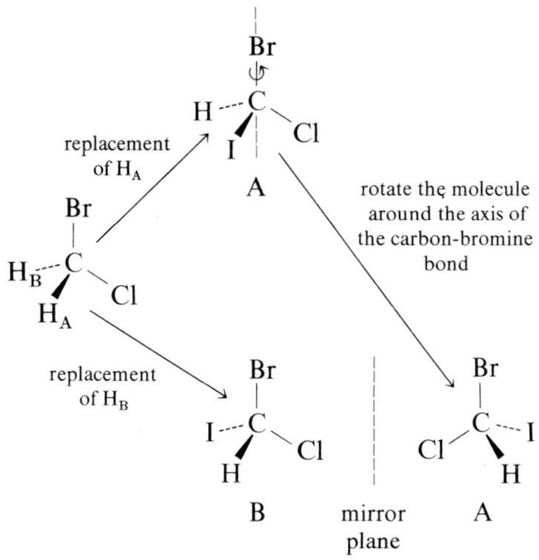

the creation of stereoisomers by the
replacement of one or the other of the
two hydrogen atoms in bromochloromethane
by a third substituent; enantiomers

The two molecular species that are formed in the exercise shown above are not identical. It should be clear to you that structure A cannot be made to coincide with structure B in all respects. If you were to try to pick A up out of the page and move it down and place it on the drawing of B, the bromine and chlorine atoms and the central carbon atom would fall in the same places, but the positions of the iodine and hydrogen atoms would be reversed. Thus, *structure A and structure B represent two different molecular species that differ from each other only in the orientation of the atoms in space. They are stereoisomers of bromochloroiodomethane.* If structure A is rotated around the axis of the carbon-bromine bond and placed next to structure B, the relationship between the two isomers becomes more clear. Structures A and B are related to each other as an object is related to its mirror image. If you imagine a mirror between them, you will see that *one structure looks like the reflection of the other one. Thus, the two molecular species represented by these structural formulas are mirror-image isomers, or* **enantiomers** *of each other.*

A mirror image may be drawn of any structure. The important criterion for enantiomerism is that the mirror images not be superimposable on each other. Structure A cannot be picked up and placed on structure B so that all points coincide no matter how the two structures are rotated in space. This is not always so. A molecule of bromochloromethane and its mirror image are shown below.

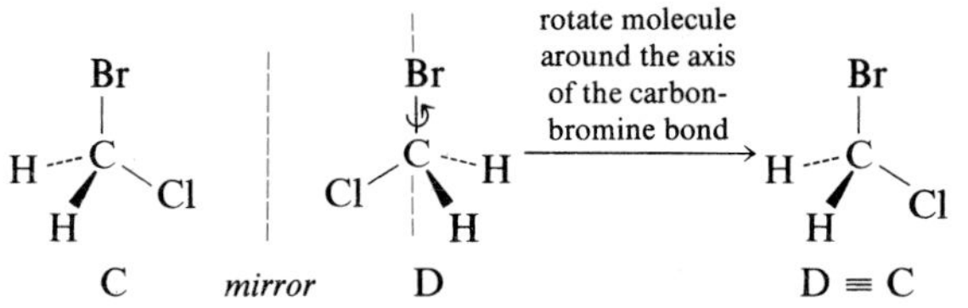

two representations of bromochloromethane
that are superimposable mirror images of
each other

Representation C is the mirror image of representation D, but these drawings do not represent different molecular species. If D is rotated around in space, it becomes identical with C and can be lifted out of the page, brought over, and fitted exactly on C. In other words, D is a superimposable mirror image of C.

We had replaced a primary hydrogen in butane to make 1-chlorobutane, and a secondary one to make 2-chlorobutane (p. 85). A closer examination of the secondary hydrogen atoms in butane reveals that they are not all equivalent. Enantiomeric 2-chlorobutanes are formed, depending on which hydrogen atom is replaced.

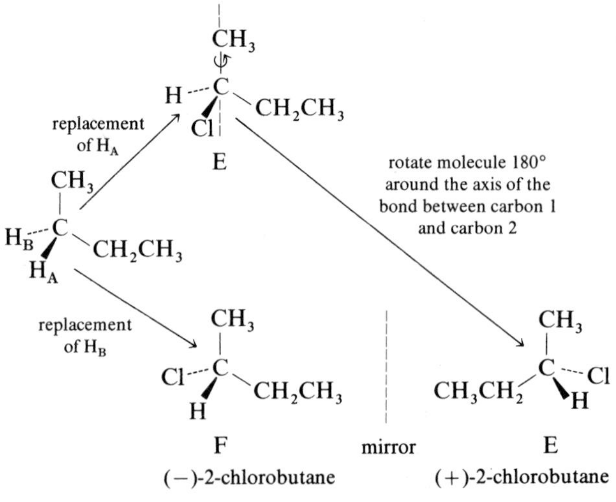

enantiomers of 2-chlorobutane

By maneuvers identical to those described earlier for bromochloroiodomethane (p. 105), it is possible to show that structural formula E represents a different molecular species than structural formula F. The relationship between the two forms is that of nonsuperimposable mirror images. Structural formulas E and F represent enantiomers of 2-chlorobutane.

The separate enantiomers have identical physical properties except in one respect. They rotate the plane of polarized light in opposite directions. This phenomenon known as the **optical activity** of such compounds is indicated by the plus and minus signs shown by the names of the two enantiomers of 2-chlorobutane. ($-$)-2-Chlorobutane rotates the plane of polarized light to the left, while ($+$)-2-chlorobutane rotates it by an equal amount to the right. In all other aspects, enantiomers have identical physical properties and, therefore, cannot be separated from each other by physical methods such as distillation or crystallization.

The phenomenon of optical activity and how it is measured will be taken up in more detail in Section 3.9D. In the next section, we examine why some compounds have nonsuperimposable mirror images, and therefore exist as enantiomers, while others do not.

B. Chirality

An object or a molecule that cannot be superimposed upon its mirror image is said to be **chiral.** Thus, bromochloroiodomethane (p. 105) and 2-chlorobutane (p. 106) are chiral compounds, whereas bromochloromethane (p. 106) is not. *An object that is superimposable upon its mirror image is* **achiral.**

The concept of chirality is important in the study of the chemistry of biological systems. Many of the compounds that occur in living organisms, such as carbohydrates and proteins, are chiral. Living organisms, including human beings, are chiral. Our internal organs are arranged asymmetrically, and we even have distinctive twists and whorls in the way our hair grows. Our left and right hands are nonsuperimposable mirror images of each other, as are our two feet (Figure 3.8, p. 108). Indeed the word chiral comes from the Greek word *cheir,* which means hand.

Many naturally occurring compounds exist as one of two possible enantiomers. Biological systems can distinguish between one enantiomer and its mirror image. For example, all amino acids that are present in proteins exhibit one particular spatial orientation of the groups around a central carbon atom. The natural enantiomer of the amino acid alanine and its mirror image are shown below.

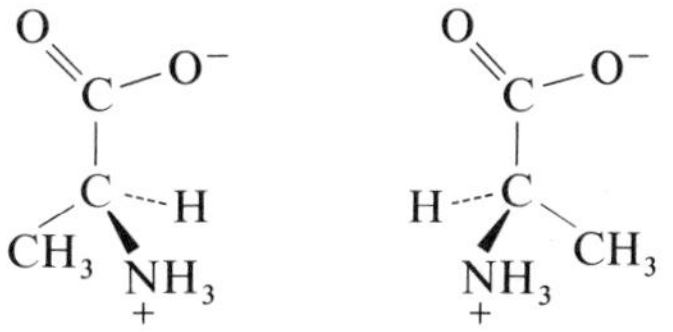

(+)-alanine   ($-$)-alanine

the natural enantiomer   *the mirror image*

The chirality of the human body, which is reflected at the most fundamental level in the stereochemistry of enzyme systems, requires that chemical reactions take place with a particular orientation in space. You will have some idea of what is involved if you try to put your left glove on your right hand, or your left foot into your right shoe. Thus, one enantiomer of a compound may be a hormone or may be active as a

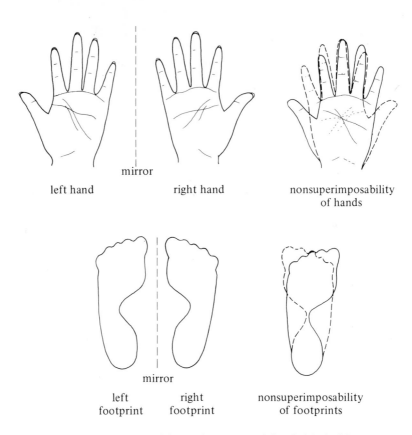

left hand   mirror   right hand   nonsuperimposability of hands

left footprint   mirror   right footprint   nonsuperimposability of footprints

Figure 3.8 Hands and footprints as models of chiral objects with nonsuperimposable mirror images.

medication; its mirror image may be biologically inactive. The sensitivity of biological systems to stereochemistry is one reason why we are always concerned with stereoisomerism of organic molecules, and with the way their stereochemistry is determined and changed by chemical reactions.

An easy way to test for chirality is to look for a **plane of symmetry.** If an object or a molecule can be divided by a plane into two equal halves that are mirror images of each other, a plane of symmetry exists and the object or molecule is achiral. For example, bromochloromethane has a plane of symmetry that bisects the bromine atom, the chlorine atom, the carbon atom, and the angle between the two hydrogen atoms. (Remember that individual atoms are spherically symmetrical even though the symbols we use to represent them are not.) Some achiral molecules and common objects are shown in Figure 3.9.

Acetone has a plane of symmetry that bisects the central carbon atom and the oxygen atom. The plane of symmetry of methylcyclobutane goes through the methyl group, the carbon atom to which it is bound and the hydrogen atom on the same carbon atom, as well as the carbon atom and two hydrogen atoms on the opposite side of the ring. A cube and a spool have many planes of symmetry; two are shown for each one. *An object or molecule that has a plane of symmetry is achiral.*

There are, however, complex molecules that do not have a plane of symmetry and yet are achiral because they possess other kinds of symmetry. For the molecules we are concerned with, the test of whether or not a plane of symmetry is present is sufficient to distinguish between chiral and achiral systems.

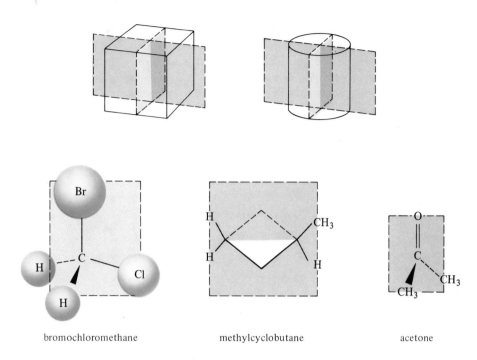

bromochloromethane   methylcyclobutane   acetone

Figure 3.9   Some achiral molecules and common objects shown with planes of symmetry.

PROBLEM 3.24

(a)   Suppose the cube in Figure 3.9 were a toy block with different letters of the alphabet written on its sides. Would it be chiral or achiral?

(b)   Would a spool with thread on it be chiral or achiral?

PROBLEM 3.25   Of the following common objects, which ones are chiral, which ones achiral?

(a)   a screw   (b)   a nail   (c)   a hammer   (d)   a spade   (e)   an eggbeater
(f)   socks   (g)   mittens   (h)   a stocking cap (with no decoration on it)
(i)   a shirt   (j)   a tee shirt with GO BLUE written on it

C. The Asymmetric Carbon Atom

Bromochloroiodomethane (p. 105), 2-chlorobutane (p. 106), and alanine (p. 107) are examples of chiral molecules and they have one thing in common. One carbon atom in each one has four different substituents on it. *A carbon atom that is bonded to four different substituents is called an* **asymmetric carbon atom.** *A molecule that contains a single asymmetric carbon atom has no plane of symmetry and is chiral.*

In Section 5.7 we will see that molecules containing more than one asymmetric carbon atom may or may not be chiral, depending upon whether they have a plane of symmetry and are superimposable on their mirror image. For the present time, we are only concerned with molecules containing a single asymmetric carbon atom. Such molecules are not superimposable upon their mirror images and are chiral.

Another way of looking at such molecules is to say that four different groups may be arranged around a carbon atom in two (and only two) different ways. Such arrangements may be considered to be left handed or right handed. *You should convince yourself by experimenting with models that any four different groups around a carbon atom can only have two different arrangements and that these two different forms are nonsuperimposable mirror images.*

The groups around the asymmetric carbon atom do not have to be very different in order to qualify as *different*. For example, the asymmetric carbon atom in 2-butanol has a hydrogen atom, a hydroxyl group, a methyl group, and an ethyl group as the four different substituents.

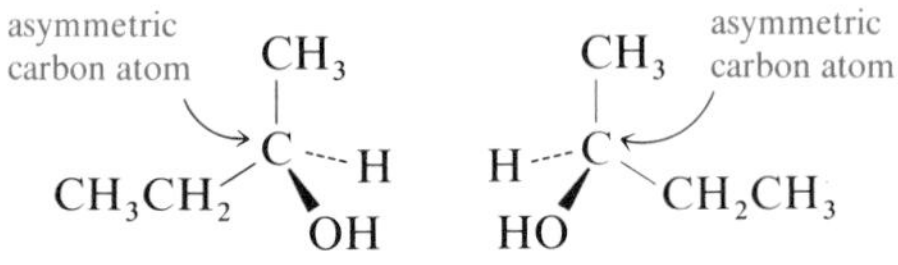

enantiomers of 2-butanol

Isotopes of the same element count as different, as do alkyl groups that differ from each other in subtle ways at a distance from the asymmetric carbon atom. To discover whether a compound has an asymmetric atom, one must search carefully in all directions, moving away from the central carbon atom until some difference is found or until it is established that at least two of the groups are the same.

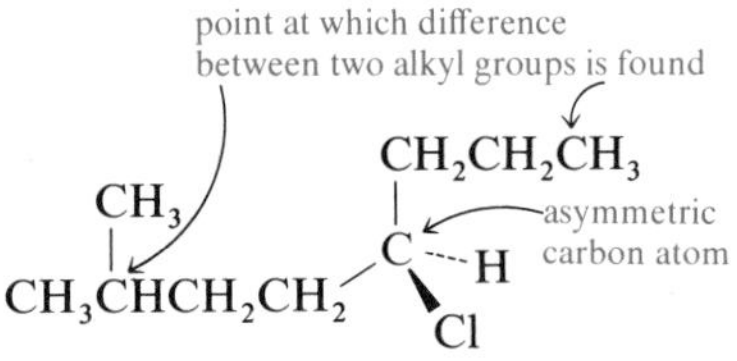

one enantiomer of 5-chloro-2-methyloctane

PROBLEM 3.26   For each of the following compounds, determine whether it contains an asymmetric carbon atom or not. If it does, draw the two enantiomers of the compound. If it does not, draw the compound in such a way that you can identify the plane of symmetry in the molecule.

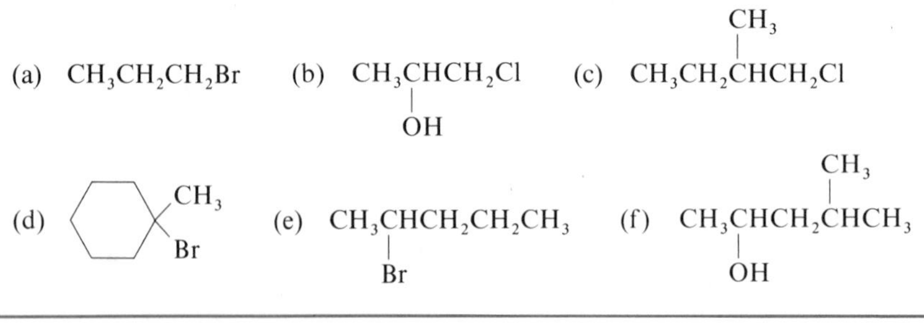

(a) $CH_3CH_2CH_2Br$

(b) CH_3CHCH_2Cl
       |
      OH

(c) $CH_3CH_2CHCH_2Cl$
          |
         CH_3

(d)

(e) $CH_3CHCH_2CH_2CH_3$
       |
      Br

(f) $CH_3CHCH_2CHCH_3$
       | |
      OH CH_3

D. Plane-Polarized Light and Optical Activity

We frequently characterize light as a wave with associated electric and magnetic fields. Thus, for a beam of light there is one vector describing the electric field strength, and one for the magnetic field strength. These vectors are in planes perpendicular to each other and to the direction of the wave motion (Figure 3.10). In ordinary light, there are

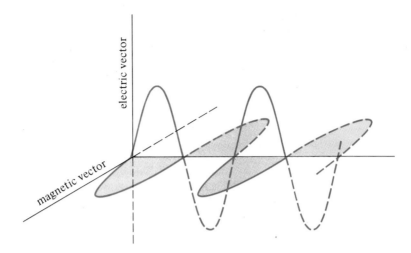

Figure 3.10   Electric and magnetic field strength vectors of a
light wave.

a great number of these electromagnetic waves; the planes of the electric and magnetic vectors of each individual wave are randomly oriented with respect to those of the other waves. Such light is said to be unpolarized.

Certain crystalline substances only allow the passage of light waves with their electric vectors in one single plane. Prisms made of such crystals—calcite is a good example—transmit light that is said to be plane polarized.

When plane-polarized light interacts with molecules, the direction of the plane of polarization is changed. If the molecules are achiral, the interactions that take place with molecules in many random orientations do not lead to an overall change in the plane of polarization. A twist given to the plane of polarization by a molecule in one orientation is canceled out by an equal twist in the opposite direction caused by an encounter with a molecule in the orientation that looks like the mirror image of the first one.

If the molecules are chiral, then no molecular orientation is the exact mirror image of any other one. In other words, such molecules are not superimposable upon their mirror images. Therefore, the changes that take place in the plane of polarized light as it interacts with the molecular species are not averaged out. Instead, they add up so that rotation of the plane of polarization is seen. The experimental observation is the result of the interaction of the plane-polarized light with a large number of molecules. The direction of the rotation may be counterclockwise, in which case the compound is said to be **levorotatory,** or it may be clockwise for a **dextrorotatory** species. *Compounds that rotate the plane of polarized light are said to have* **optical activity,** *or to be optically active.* Whether a compound rotates the plane of polarization clockwise or counterclockwise is indicated by the sign of rotation, (+) or (−), placed before the name of the compound.

The instrument that is used to measure optical activity is called a polarimeter. It has a source of light and two prisms, one that is used to make plane-polarized light and the other to detect any rotation in the plane of polarization. A tube containing a solution of the compound to be investigated is placed between the prisms. Finally, there is a viewing device with a scale on it, so that the angle by which the plane of polarized light has been rotated can be measured (Figure 3.11).

The light that goes through the polarizer prism emerges plane polarized. If the axis of the analyzer prism is aligned with that of the polarizer prism, and there is no optically

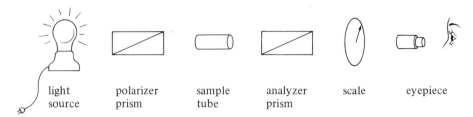

light   polarizer   sample   analyzer   scale   eyepiece
source   prism   tube   prism

Figure 3.11   Sketch of the main parts of a polarimeter.

active material in the sample tube, the light will go through unchanged. If the sample tube contains an optically active compound, however, the plane of polarization will be rotated, and not all of the light will be able to get through the analyzer prism. The analyzer is then rotated until its axis coincides with the new plane of polarization, and light is once more transmitted by the crystal. The angle by which the analyzer prism has to be rotated in order to allow the full transmission of light is the angle of rotation for the optically active compound (Figure 3.12).

How much the plane of polarization is rotated depends on several things. First, it depends on the particular optically active compound on which the measurement is being made. *The actual sign of rotation for a compound is an experimental measurement. There are some rules for predicting the direction and the extent of the rotation for some compounds, but ultimately the optical rotation for a compound must be determined experimentally. Once the optical rotation for one enantiomer is known, however, we can say with certainty that the other isomer will have the same degree of rotation, but with the opposite sign.*

Second, because the total rotation depends upon the number of interactions between the light beam and the molecules, the concentration of the solution and the length of the path of the beam of polarized light through the solution must be considered. The particular conformations of the molecules being examined and any interactions between the molecules and the solvent are important, so temperature and solvent are recorded too.

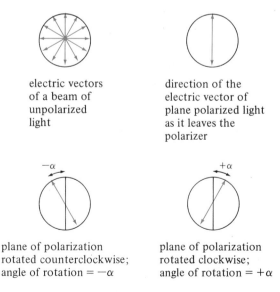

electric vectors
of a beam of
unpolarized
light

direction of the
electric vector of
plane polarized light
as it leaves the
polarizer

plane of polarization
rotated counterclockwise;
angle of rotation = $-\alpha$

plane of polarization
rotated clockwise;
angle of rotation = $+\alpha$

Figure 3.12   Schematic indication of the rotation of the plane of polarized light.

Finally, optical rotation changes with the wavelength of the light used in the measurement, so a complete report must also include that information. For purposes of comparing different compounds, the specific rotation of a compound is calculated using the formula,

$$[\alpha]_D^T = \frac{\alpha}{lc} \quad \text{(concentration; solvent)}$$

The symbol $[\alpha]_D^T$ means the specific rotation, calculated from data measured at temperature, t °C, using the light from the yellow line (so-called D line) in the visible spectrum of sodium, 589 nm. This wavelength corresponds to the yellow color seen when a sodium salt is heated in the flame of a Bunsen burner. In the nineteenth century, it was the only readily available source of light of a single wavelength. It is still used today, mainly by tradition, and so that contemporary results can be compared with those of earlier days. The measured angle of rotation is α. The length of the sample tube, l, is expressed in decimeters, and a standard tube is 10 cm or 1 dm long. The concentration of the sample, c, is given in g/mL of solution. In most cases, the concentration of the solution and the solvent used are also reported in parentheses after the value for the specific rotation.

For example, a solution of 16.5 g of the levorotatory form of camphor in 100 mL of ethanol solution has an optical rotation of $-7.29°$ when a 10-cm sample tube is used in the measurement. A sodium lamp is used and the temperature is 20°C. The specific rotation of the compound is calculated using the formula given above, and reported in the form below.

$$[\alpha]_D^{20} = \frac{\alpha}{1 \text{ dm} \times 16.5 \text{ g/100 mL}}$$

$$[\alpha]_D^{20} = \frac{\alpha \times 100}{1 \times 16.5} = \frac{-7.29° \times 100}{1 \times 16.5}$$

$$[\alpha]_D^{20} = -44.2° \quad (c \ 0.165; \text{ ethanol})$$

The optical rotation of a chiral compound is a specific physical property of the compound and is determined and reported just as other physical properties, such as boiling and melting points, are. Different chiral compounds have widely varying specific rotations. The specific rotations of some compounds isolated from natural sources illustrate this. In ethanol, cholesterol has $[\alpha]_D^{20}$ $-3.15°$, while for nicotine $[\alpha]_D^{20}$ is $+163.2°$. The specific rotation, $[\alpha]_D^{20}$, for cane sugar (sucrose) in water is $+66.4°$.

A mixture made up of equal numbers of molecules of two enantiomers has no optical rotation. Such a mixture is known as *a* **racemic mixture** and *contains equal concentrations of right-handed and left-handed molecules.* Plane-polarized light that is given a twist to the right by an encounter with a right-handed molecule is twisted back by an encounter with its mirror-image isomer. The plane of polarization remains unchanged. The presence of a racemic mixture is indicated by the use of the symbol ($\pm$) before the name of the compound. For example, ($+$)-2-butanol has $[\alpha]_D^{20}$ $+13.9°$, while for ($-$)-2-butanol $[\alpha]_D^{20}$ is $-13.9°$ when the determinations are done on the pure liquids. A mixture of equal quantities of the two enantiomers of 2-butanol is not optically active and is designated as ($\pm$)-2-butanol.

If there is any excess of one enantiomer over the other, the mixture will have optical rotation corresponding to the percentage of the species that is present in excess. Thus, if we had a sample that consisted of 75% ($+$)-2-butanol and 25% ($-$)-2-butanol, the sample would have $[\alpha]_D^{20} = +13.9 \times 0.5$ or $+6.9°$. The sample would be dextrorotatory because it contains an excess of the dextrorotatory form. The optical rotation of 25% of the (+)-2-butanol molecules is canceled out by the presence of 25% (−)-2-butanol. The optical rotation of the sample is only half (0.75 − 0.25 = 0.50) of what it would be for pure (+)-2-butanol. Such a sample is said to have an **enantiomeric excess** of 50%. The optical rotation observed for a mixture and the known optical rotation of a pure enantiomer are used to calculate the enantiomeric excess. For the example given above,

$$\frac{\text{measured specific rotation}}{\text{specific rotation for pure }(+)\text{-2-butanol}} = \frac{+6.9°}{+13.9°} \times 100 = 50\% \text{ enantiomeric excess}$$

A racemic mixture cannot be separated into its components by ordinary physical methods because the physical properties of the two components are identical except for optical activity. For example, (+)-2-butanol and (−)-2-butanol both boil at 99 °C. A separation of enantiomers must, therefore, always involve the use of chiral reagents that interact differently with right-handed and left-handed molecules. As we have already mentioned, enzymes are such reagents. One way of separating enantiomers is to use a living organism to metabolize one form and leave the other form untouched. Examples of such separations are given in the next section.

A more general method to separate enantiomers, known as the **resolution of a racemic mixture,** involves the formation of compounds between the enantiomers and a chiral reagent. Then the different properties that are created in this way are used to make a separation. To understand this method, we must first look at molecules containing more than one asymmetric carbon atom. For that reason, we shall postpone a discussion of this topic until Section 16.4B.

PROBLEM 3.27   The optical rotation of sugar is used in industry as a quick way to check on the concentration of sugar solutions. If 20 g of cane sugar dissolved to make up 100 mL of solution and placed in a tube 40 cm long rotates the plane of polarized light +53.2° at 20 °C, what is the concentration of a solution of sugar, measured at the same temperature and in the same polarimeter tube, if the optical rotation is +13.3°?

PROBLEM 3.28   Menthol has an optical rotation of +2.46° when the measurement is made with a sodium lamp at 20 °C on a solution containing 5 g of menthol in 100 mL of ethanol solution, using a sample tube 10 cm long. What is the specific rotation of menthol?

PROBLEM 3.29   When the analyzer prism is rotated 90° clockwise, it arrives at the same position as if it were rotated counterclockwise 270° (−270°). When the optical rotation of a compound is determined for the first time, how would it be possible to establish whether the rotation should be reported as +90° or −270°?

E. The Discovery of Molecular Dissymmetry

Optical activity was known early in the nineteenth century as a property of crystals, such as quartz, that are demonstrably dissymmetric (without symmetry) in appearance. In Paris in 1848, Louis Pasteur noticed that crystals of the sodium ammonium salt of (+)-tartaric acid, a by-product of winemaking, had dissymmetric crystals that could

Figure 3.13   Quartz crystals, demonstrating the dissymmetry in their forms. (From the collections of the Exhibit Museum of the University of Michigan)

not be superimposed upon their mirror images. He thought that crystalline dissymmetry might indicate a similar lack of symmetry in the molecules of (+)-tartaric acid. Another form of tartaric acid, known as paratartaric acid, did not rotate the plane of polarized light. Pasteur studied the sodium ammonium salt of paratartaric acid, expecting to find that its crystals were symmetrical. Instead, to his great surprise, he saw that some of the crystals had a right-handed appearance and others a left-handed one. The two crystalline forms were mirror images of each other.

Pasteur separated the two crystalline forms of sodium ammonium paratartrate with tweezers as he looked through a microscope. A solution of the right-handed crystals in water rotated the plane of polarized light to the right, exactly the way the (+)-tartaric acid salt did. The left-handed crystals gave a solution with an optical rotation of equal magnitude, but opposite sign. A mixture of equal weights of the two crystal forms, when dissolved in water, gave a solution with no optical rotation. With these experiments, Pasteur demonstrated that optical activity was the result of a molecular property that survived even when the crystal form was destroyed by solution. He saw molecular dissymmetry as the cause of the phenomenon of optical activity, and recognized that there were two dissimilar molecular forms of optically active tartaric acids.

In further experiments in 1854, Pasteur showed that the microorganism *Penicillum glaucum* consumed (+)-tartaric acid but not (−)-tartaric acid. His work with optically active compounds led Pasteur to say, ''Life is dominated by dissymmetrical actions. I can even foresee that all living species are primordially in their structure, in their external forms, functions of cosmic dissymmetry.''

While the phenomenon of molecular dissymmetry was recognized by the 1840s, there was no clear picture of how it came about until 1874. Up to that time, chemists were still struggling with ways to represent molecular structures, and had not yet made clear distinctions between structural isomers and stereoisomers. The idea that groups around a carbon atom are arranged in a tetrahedron was suggested in 1874 by the Dutch chemist, J. H. van't Hoff, who was twenty-two years old at the time. He recognized that it was necessary to think of structures in three dimensions in order to solve the problems of isomerism that were being discovered in the laboratory. A carbon atom with four different substituents arranged tetrahedrally around it would account for the two isomers observed experimentally for compounds with a single asymmetric carbon atom. The right-handed and left-handed arrangements that are possible for four groups at a tetrahedral carbon atom would explain the phenomenon of optical activity.

The French chemist J. A. Le Bel, starting with the work of Pasteur, also arrived at the idea that an asymmetric carbon atom with four different substituents around it is the basis for optical activity in molecules. He published his ideas in 1874, the same year as van't Hoff. He emphasized lack of symmetry, and in particular the lack of a plane of symmetry, at the molecular level as a necessary condition for optical activity. He came to the idea of a tetrahedral carbon atom by exploring the number of isomers that are formed as one, two, and then three different groups are substituted on a compound that originally had four identical groups on it. Our exercise in creating isomers by substituting the hydrogen atoms on methane followed very closely his way of thinking about stereoisomerism. He differed from van't Hoff in that van't Hoff took a tetrahedral carbon atom as his starting point.

Le Bel's ideas remained more abstract than van't Hoff's. For example, van't Hoff created molecular models of tetrahedral carbon atoms and sent them to the leading chemists of the time in an effort to gain acceptance of his ideas. He drew structural formulas in perspective to make his ideas clear. He made predictions about optical activity or the lack of it for compounds yet to be investigated, predictions that were found to be correct when the experimental work was done. In spite of this, with a few exceptions, his ideas were not widely accepted for a number of years. Part of the opposition was supported by conflicting experimental results. In those days, many organic compounds were isolated from natural sources and contained small amounts of optically active compounds as impurities. Because of this contamination, confusing data were obtained about optical activity for compounds that did not have asymmetric carbon atoms. Van't Hoff himself supervised much of the experimental work that proved that pure compounds were not optically active unless molecular dissymmetry was present. In 1901, he received the first Nobel Prize in Chemistry for his work in this and other areas.

By the end of the nineteenth century, the tetrahedral carbon atom was accepted as the basis of structural organic chemistry. Much of the research that was being done by that time on sugars, which contain several asymmetric carbon atoms, would not have been possible without such a basis. The experimental results obtained confirmed the correctness of van't Hoff's ideas. His predictions about the number of isomers possible and the kinds of compounds that should have optical activity were demonstrated to be true.

F. Configuration. Nomenclature of Stereoisomers

The orientation in space of the groups around an asymmetric carbon atom is called the **configuration** *of the compound.* The determination of configuration for a particular compound is not a trivial matter. In subsequent chapters, we discuss the reactions that take place with known stereochemical results. Experimental evidence from many such reactions has accumulated over the years so that the stereochemical relationships among many series of compounds are known. The ultimate determination of exactly how these molecules actually look, which was completed in 1951, depended upon the development of sophisticated x-ray diffraction determinations of crystal structures. The solution of this problem, known as the *determination of the absolute configuration of stereoisomers,* is described in Chapter 14. Meanwhile, it is sufficient for you to understand that *the sign of rotation given for a compound for which a three-dimensional representation is drawn is the result of an experimental determination.*

The configuration of a compound is unchanged unless bonds to the asymmetric carbon atom are broken. Thus, configuration must not be confused with conformation, which changes continuously by rotation about single bonds and by the flipping of rings in molecules. A stereoisomer has a single configuration, but may exist in a number of conformations depending upon the solvent and the temperature. For example, four representations of ($-$)-2-chlorobutane, three of them showing different conformations of the molecule and one of them a rotation of the whole molecule in space, are given below. In all of these, the configuration of the molecule remains unchanged.

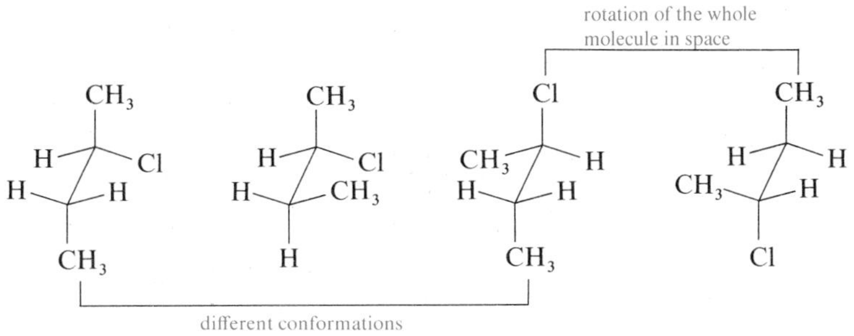

rotation of the whole
molecule in space

different conformations

($-$)-2-chlorobutane

Stereoisomers, such as enantiomers, are also known as **configurational isomers.** The rules of nomenclature described in Section 3.6 are not adequate for naming stereoisomers. For example, on p. 106, we had to resort to calling two isomers "stereoisomers of bromochloroiodomethane." Structural formulas for ($+$)-2-chloro-butane and ($-$)-2-chlorobutane were shown (p. 106) but unless you memorize the structures, there is no way for you to reproduce a unique structure for each isomer from the names alone. *As there is no simple correlation between the sign of rotation and structure, the sign of rotation by itself does not tell us what the configuration of the compound is.*

To solve this problem, another set of rules, the **Cahn-Ingold-Prelog rules,** were created. The configuration at an asymmetric carbon atom is described as being *R*, from the Latin *rectus* or *right* handed, or *S*, from the Latin *sinister* or *left* handed, depending upon the order in which the different substituents are arranged around that center. The rules that are applied in determining configuration are as follows:

1. The groups attached to the asymmetric carbon atom are assigned priorities. The higher the atomic number of the atom bonded directly to the asymmetric carbon atom, the higher the priority of the substituent. For example,

$$Cl > O > N > C > H$$

    Among isotopes, the isotope of higher atomic weight takes priority, thus, tritium, the isotope of hydrogen with atomic weight 3 has higher priority than deuterium, with atomic weight of 2. Hydrogen, with atomic weight of 1, has the lowest priority always.

$$T > D > H$$

2. If two atoms of the same atomic number are attached to the asymmetric carbon atom, the next atoms in the chains are investigated moving away from the asymmetric carbon atom until some difference is found. A decision is made at the *first*

point of difference. For example:

lowest
priority

asymmetric carbon atom with
three carbon atoms attached to it

$$
\begin{array}{c}
\text{H}\quad\text{H}\quad\text{H}\quad\quad\text{Cl}\quad\text{H}\\
|\quad\quad|\quad\quad|\quad\quad\quad|\quad\quad|\\
\text{Cl}-\text{C}-\text{C}-\text{C}-\!\!-\!\!-\text{C}-\!\!-\!\!-\text{C}-\text{H}\\
|\quad\quad|\quad\quad|\quad\quad\quad|\quad\quad|\\
\text{H}\quad\text{H}\quad\text{H}\quad\text{H}-\text{C}-\text{H}\quad\text{H}\\
\quad\quad\quad\quad\quad\quad\quad\quad|\\
\quad\quad\quad\quad\quad\text{H}-\text{C}-\text{H}\quad\text{H}\\
\quad\quad\quad\quad\quad\quad\quad\quad|\\
\quad\quad\quad\quad\quad\quad\quad\text{H}
\end{array}
$$

two hydrogen
atoms and one
carbon atom
on this carbon
atom; group
of second
priority

only hydrogen
atoms on this
carbon atom;
group of third
priority

one chlorine atom and two carbon
atoms on this carbon atom; the
chlorine atom has a higher
atomic number than any of
the substituents on the other
two carbon atoms, thus this
group has the highest priority

3. Double bonds are counted as two single bonds for both of the atoms involved in the multiple bond.

$$
\text{C}=\text{C}\ \equiv\ -\!\underset{\underset{\text{C}}{|}}{\overset{}{\text{C}}}-\!\underset{\underset{\text{C}}{|}}{\overset{}{\text{C}}}- \ ;\quad \text{C}=\text{O}\ \equiv\ -\!\underset{\underset{\text{O}}{|}}{\overset{}{\text{C}}}-\!\underset{\underset{\text{C}}{|}}{\overset{}{\text{O}}}
$$

The same principle is extended to triple bonds.

$$
-\text{C}\equiv\text{C}-\ \equiv\ -\!\underset{\underset{\text{C}}{|}}{\overset{\overset{\text{C}}{|}}{\text{C}}}-\!\underset{\underset{\text{C}}{|}}{\overset{\overset{\text{C}}{|}}{\text{C}}}- \ ;\quad -\text{C}\equiv\text{N}\ \equiv\ -\!\underset{\underset{\text{N}}{|}}{\overset{\overset{\text{N}}{|}}{\text{C}}}-\!\underset{\underset{\text{C}}{|}}{\overset{\overset{\text{C}}{|}}{\text{N}}}
$$

4. The molecule is then viewed with the substituent of lowest priority away from the viewer. If the eye travels in a clockwise direction from the group of highest priority to the one of second and then to the one of third priority, the asymmetric carbon atom is assigned the *R* configuration. If the arrangement of the groups in order of decreasing priority is counterclockwise, the configuration is *S*.

The rules will become clearer if we apply them to assign configuration to bromochloroiodomethane and to 2-chlorobutane.

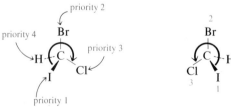

*eye must go clockwise
from priority 1 to 2 to 3,
therefore this is
(R)-bromochloroiodomethane*

*eye must go counterclockwise
from 1 to 2 to 3, therefore this
is (S)-bromochloroiodomethane*

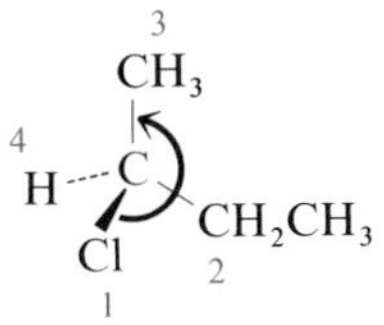

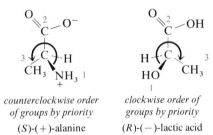

eye must go counterclockwise
from 1 to 2 to 3, therefore
this is (S)-2-chlorobutane

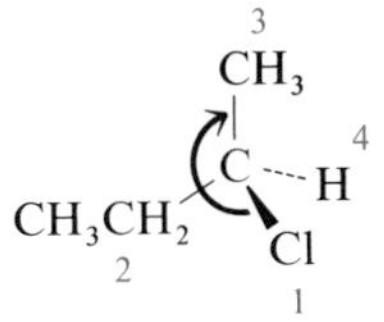

eye must go clockwise from
1 to 2 to 3, therefore this
is (R)-2-chlorobutane

Note that the molecules are drawn so that the group of lowest priority is projecting away from the viewer. The assignment of priorities in bromochloroiodomethane is straightforward; the priorities are determined by looking up the atomic numbers for the atoms attached to the asymmetric carbon atom. For 2-chlorobutane, two of the atoms bound to the asymmetric carbon atom are carbons. The ethyl group takes priority over the methyl group because the second carbon atom in the chain has priority over the three hydrogen atoms of the methyl group.

(+)-Alanine and (−)-lactic acid, two naturally occurring compounds, are assigned configuration below.

counterclockwise order
of groups by priority

(S)-(+)-alanine

clockwise order of
groups by priority

(R)-(−)-lactic acid

In alanine, the nitrogen atom takes priority over the methyl group and the carboxylic acid group. The carboxylic acid group with oxygen bonded to the carbon atom, is of higher priority than the methyl group in which only hydrogens are bonded to the carbon atom. Thus, the enantiomer of alanine that has a positive sign of rotation is assigned the S configuration. Similar reasoning is used to assign the R configuration to the form of lactic acid that is levorotatory.

Note that the designation of a compound as R or S has nothing to do with the sign of rotation. Any time we have a three-dimensional representation of a chiral compound, we can apply the Cahn-Ingold-Prelog rules to decide whether it is R or S. To find out whether a particular picture of a molecule corresponds to the dextrorotatory or levorotatory form requires much experimental work and cannot be determined just by looking at the structural formula. The only exception to this is the case in which we know the sign of rotation of the enantiomer. For example, knowing that (S)-(+)-alanine and (R)-(−)-lactic acid have the structures shown above, we can write the configurations of (R)-(−) alanine and (S)-(+)-lactic acid without hesitation.

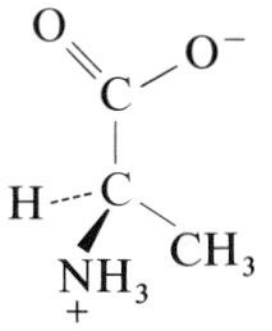

(R)-(−)-alanine
enantiomer of
(S)-(+)-alanine

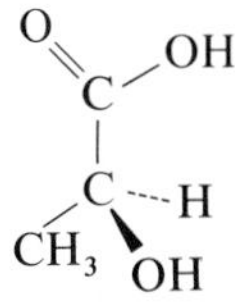

(S)-(+)-lactic acid
enantiomer of
(R)-(−)-lactic acid

The use of *R* and *S* in the name of a compound assigns a particular configuration, a particular orientation in space, to the atoms in the molecule. Thus, the Cahn-Ingold-Prelog rules enable us to describe the stereochemistry of a compound without having to draw a three-dimensional picture of the molecule. For example, if we are told that a certain compound is (*R*)-2-bromo-1-propanol, we are able to draw a structural formula that correctly depicts the orientation of the groups around the asymmetric carbon atom.

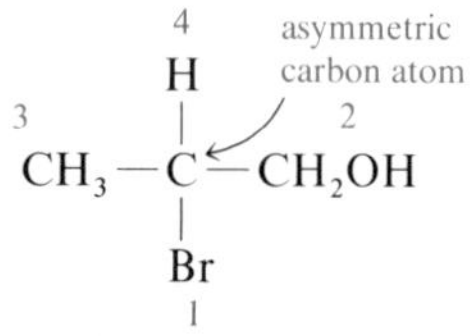

2-bromo-1-propanol

(no stereochemistry shown)
priorities assigned to the groups
on the asymmetric carbon atom

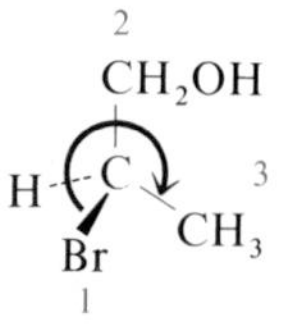

(*R*)-2-bromo-1-propanol

one possible three-dimensional
picture showing stereochemistry

The rules we have just learned fulfill the requirement for a good system of nomenclature: each name corresponds to a unique structure that can be reproduced from the name alone.

PROBLEM 3.30   Name the following compounds, including an assignment of configuration.

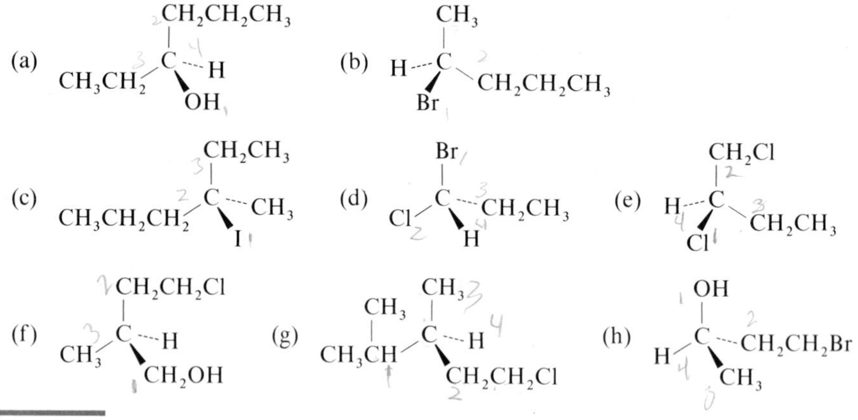

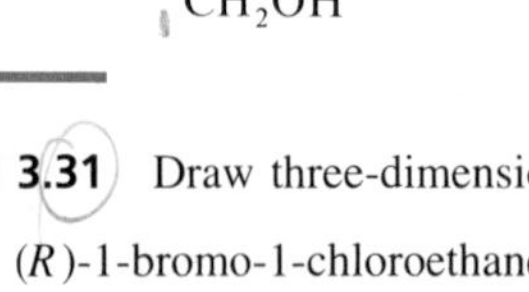

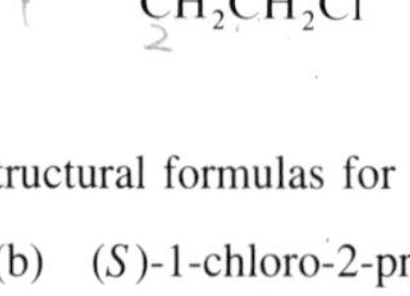

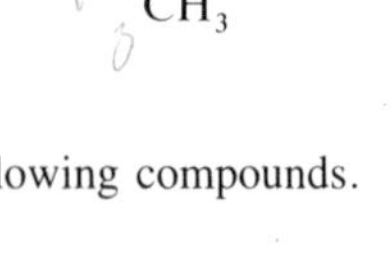

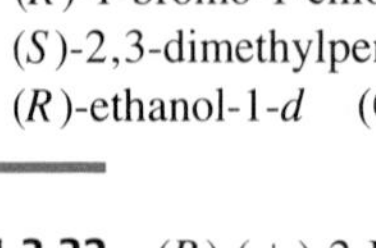

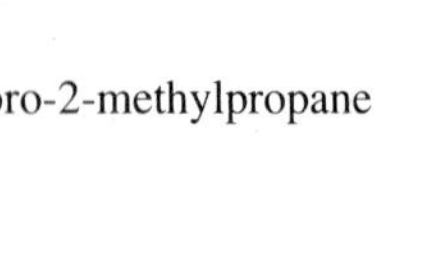

PROBLEM 3.31   Draw three-dimensional structural formulas for the following compounds.

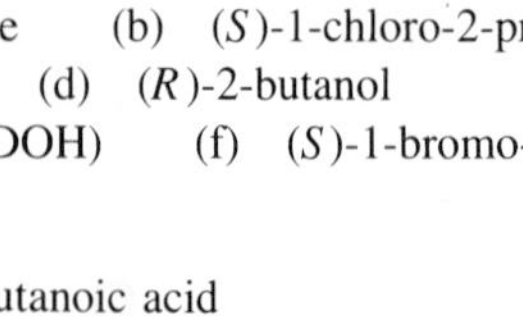

(a) (*R*)-1-bromo-1-chloroethane   (b) (*S*)-1-chloro-2-propanol
(c) (*S*)-2,3-dimethylpentane   (d) (*R*)-2-butanol
(e) (*R*)-ethanol-1-*d* (CH₃CHDOH)   (f) (*S*)-1-bromo-3-chloro-2-methylpropane

---

PROBLEM 3.32   (*R*)-(+)-2-Bromobutanoic acid

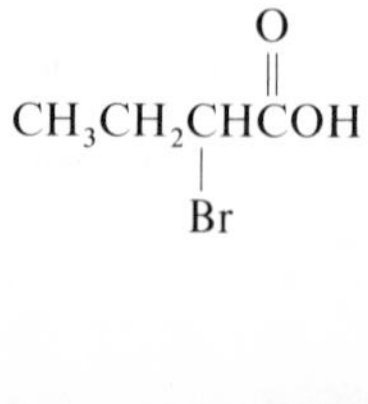

has $[\alpha]_D^{25} +39.5°$ (c 9.1; ether). A sample of 2-bromobutanoic acid having $[\alpha]_D^{25} -14.70°$ was recovered by resolution of a racemic mixture of the acid.

(a) What is the enantiomeric excess of the sample of acid recovered from the racemic mixture?
(b) Draw the correct stereochemical formulas for the isomers present in the sample and say what percent of the mixture each enantiomer is.

ADDITIONAL PROBLEMS

3.33 Name the following compounds.

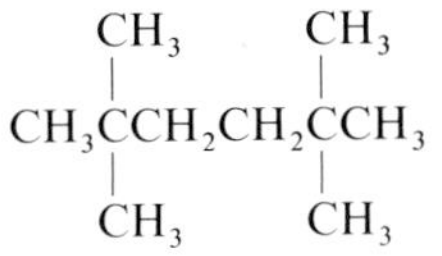

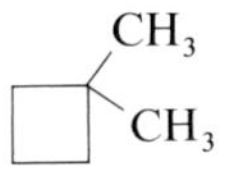

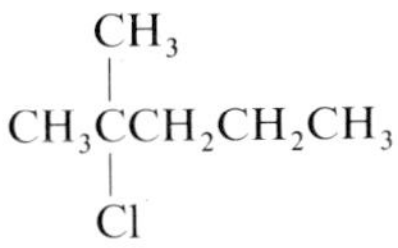

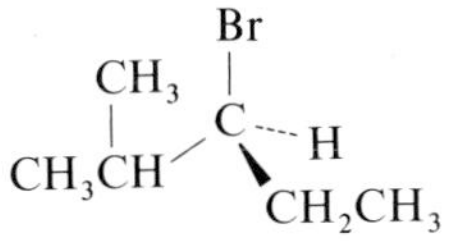

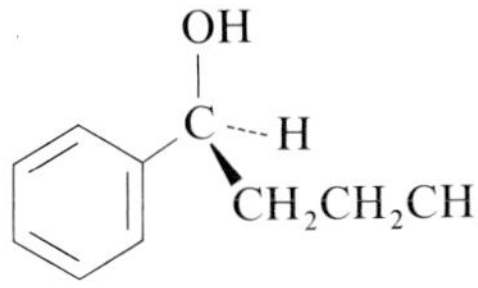

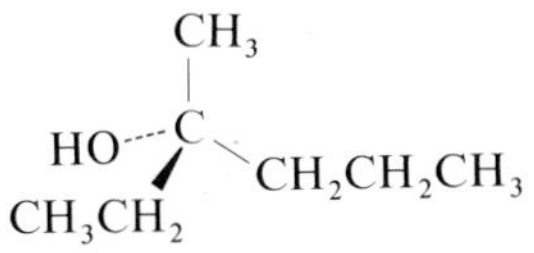

3.34 Draw structural formulas for the following compounds.

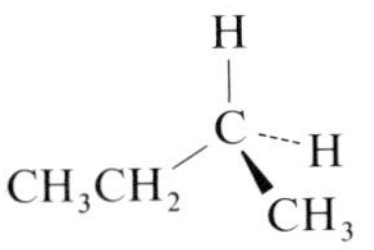

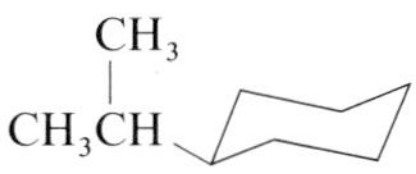

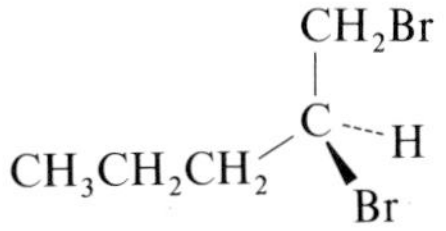

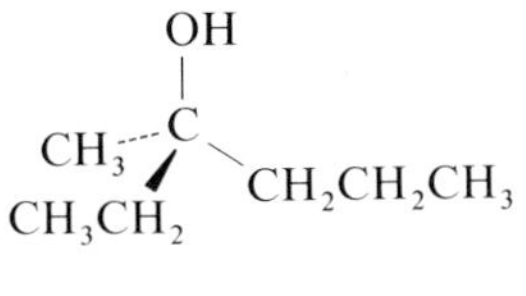

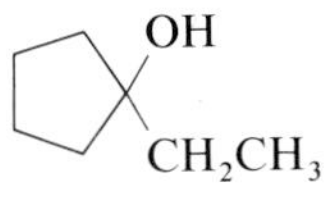

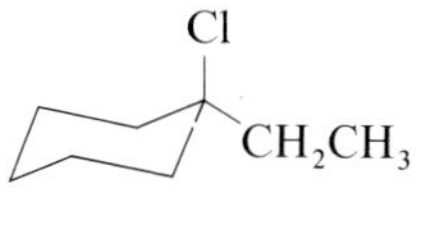

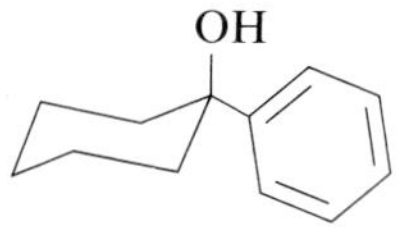

(a) 2-methyl-2-hexanol (b) (R)-4-bromo-4-methyloctane
(c) 1-methyl-1-cyclobutanol (d) (S)-2-bromo-1-phenylbutane
(e) iodocyclopropane (f) 4-*tert*-butylheptane (g) 5-cyclopropylnonane
(h) isobutylcyclopentane (i) 3-phenylheptane (j) (S)-2-methyl-4-heptanol
(k) 1-bromo-1-methylcyclohexane (l) 7-chloro-2,7-dimethyl-2-octanol

3.35 For each molecular formula below, draw structures of all the structural isomers and stereoisomers that are possible. Name each one.

(a) $C_5H_{12}O$ (Ethers may be named for the two alkyl groups they contain; for example, $CH_3OCH_2CH_3$ is methyl ethyl ether.)
(b) $C_3H_5Cl_3$

3.36 Each of the following pairs of structural formulas may represent identical molecular species, conformers of the same species, structural isomers, or enantiomers. Say which is the case for each pair.

(a)

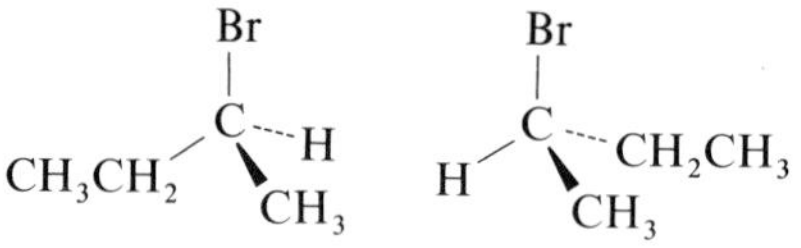

(b)

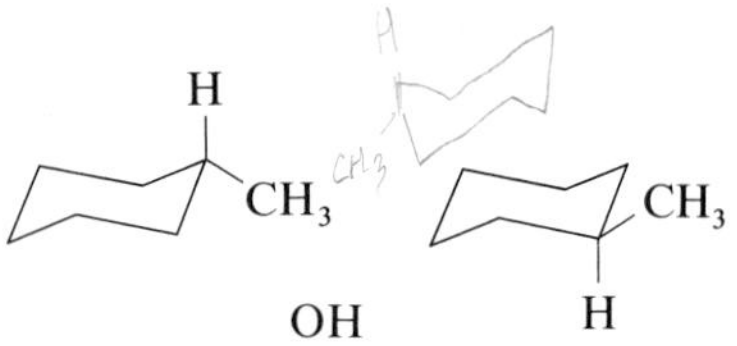

(c)

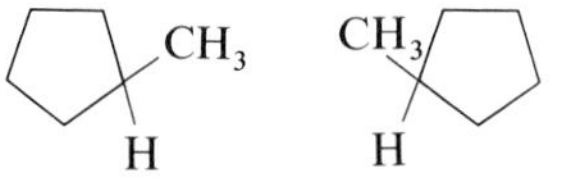

(d)

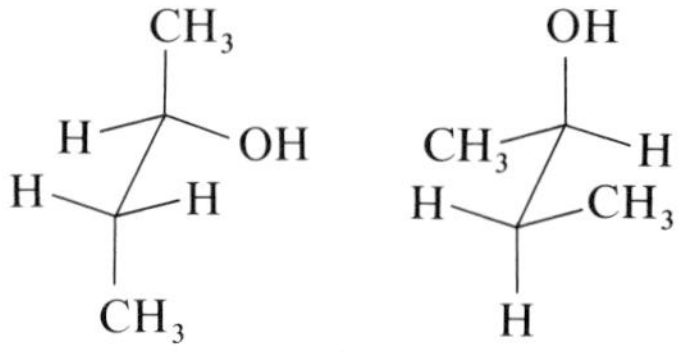

(e)

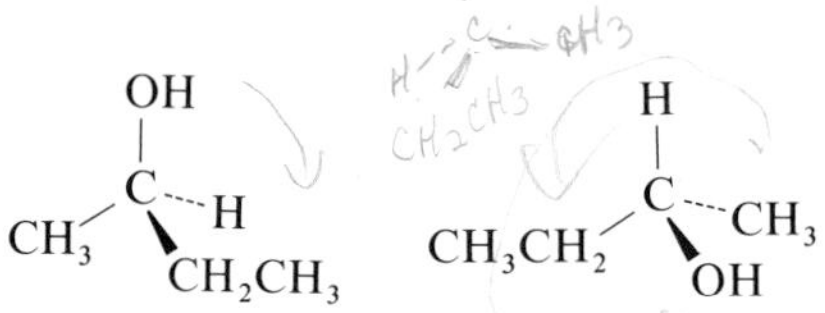

(f)

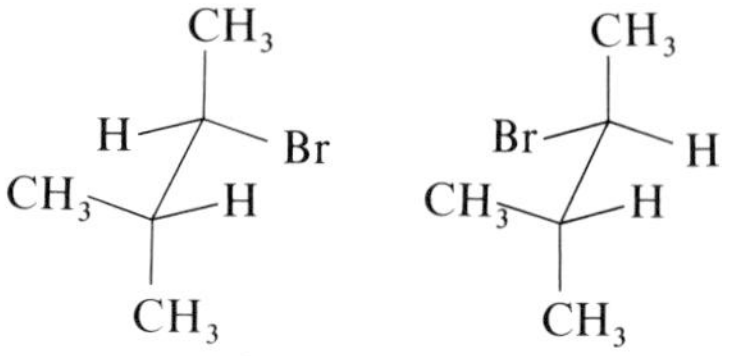

(g)

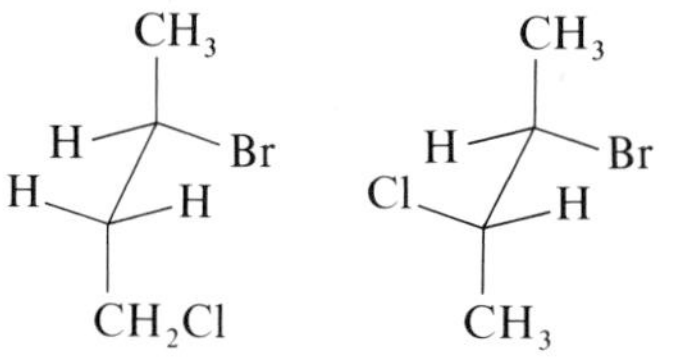

(h)

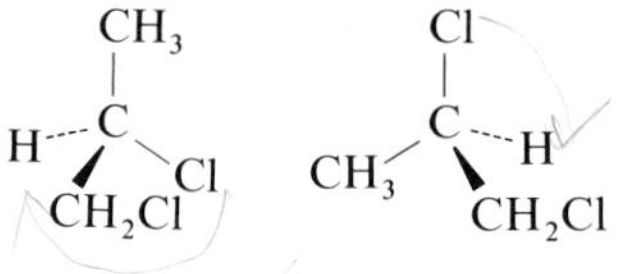

(i)

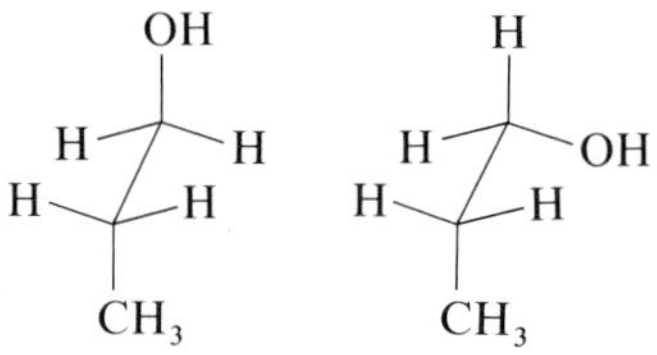

(j)

(k)

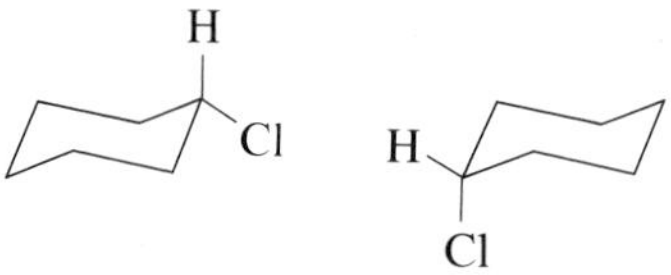

3.37 Which of the following compounds have asymmetric carbon atoms? Draw three-dimensional pictures of those that do, showing the enantiomers.

(a)

(b)  $ClCH_2CHCH_2CH_2Cl$ with Cl substituent

(c)

(d) 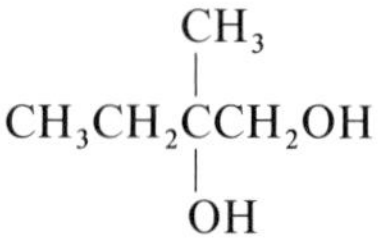 $CH_3CH_2CCH_2OH$ with CH_3 and OH substituents

(e) 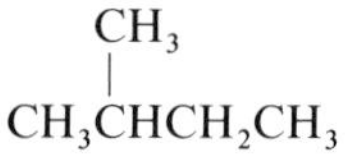 $CH_3CHCH_2CH_3$ with CH_3 substituent

(f)

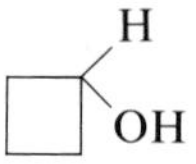

(g) 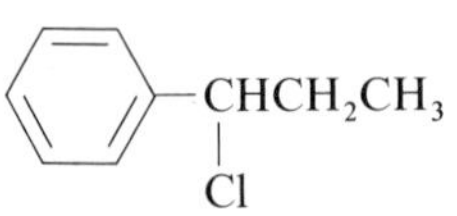 $-CHCH_2CH_3$ with Cl substituent

(h)  $CH_3CHCH_2CH_2Br$ with CH_3 substituent

3.38 Fruit sugar, fructose, has $[\alpha]_D^{20} -92°$ (c 2; water). Calculate the rotation that would be observed for a solution made up from 1 g of fructose in 100 mL of water and measured in a tube 5 cm long at 20 °C using a sodium lamp.

3.39 Chiral acetic acid in which two of the hydrogen atoms on the methyl group have been substituted by deuterium, D, and tritium, T (p. 117) has been synthesized for studies of the stereochemistry of biological reactions. Write three-dimensional formulas for (*R*)- and (*S*)-DHTCCOH.

$$\overset{\parallel}{\underset{O}{}}$$

3.40 Citronellol is a fragrant component of various plant oils such as geranium oil or rose oil. A synthetic sample of (−)-citronellol with an enantiomeric excess of 88% has $[\alpha]_D^{20}$ −4.1°. What is the optical rotation of the pure enantiomer? (−)-Citronellol has the *S* configuration. The structure of citronellol is

$$\begin{array}{c} CH_3 \qquad\qquad\qquad CH_3 \\ \diagdown \qquad\qquad\qquad\qquad | \\ C = CHCH_2CH_2CHCH_2CH_2OH \\ \diagup \\ CH_3 \end{array}$$

Draw a stereochemically correct formula for (−)-citronellol.

3.41 The enantiomers of 1-amino-2-propanol were separated and recovered as their hydrochlorides,

$$\begin{array}{c} CH_3CHCH_2NH_3{}^+Cl^- \\ | \\ OH \end{array}$$

The sample of (*R*)-(−)-1-amino-2-propanol hydrochloride had $[\alpha]_D^{25}$ −31.5° (*c* 1.0; methanol), while that of (*S*)-(+)-1-amino-2-propanol hydrochloride had $[\alpha]_D^{25}$ +35° (*c* 1.0; methanol).

(a) Write correct stereochemical formulas for the levo- and dextrorotatory isomers.

(b) Which enantiomer was recovered with the higher purity?

(c) If we assume that the optical rotation observed for the enantiomer of higher purity is the correct optical rotation for the compound, what is the enantiomeric excess in the sample of the other enantiomer?

3.42 For each of the following sets of compounds, identify the one you expect to have the lowest boiling point.

$$\begin{array}{c} \qquad\quad CH_3 \qquad\qquad\qquad\qquad\qquad\qquad\qquad\qquad\qquad\qquad CH_3 \\ \qquad\qquad | \qquad\qquad\qquad\qquad\qquad\qquad\qquad\qquad\qquad\qquad\qquad | \\ (a)\quad CH_3CH_2CCH_2CH_3,\quad CH_3CH_2CH_2CH_2CH_2CH_2CH_3,\quad CH_3CHCH_2CH_2CH_2CH_3 \\ \qquad\qquad | \\ \qquad\qquad CH_3 \end{array}$$

$$\begin{array}{c} \qquad\quad CH_3 \qquad\qquad\qquad\qquad\qquad\qquad\qquad\qquad\qquad CH_3 \\ \qquad\qquad | \qquad\qquad\qquad\qquad\qquad\qquad\qquad\qquad\qquad\quad | \\ (b)\quad CH_3CH_2CHCH_2OH,\quad CH_3CHCH_2CH_2CH_3,\quad CH_3CCH_2CH_3 \\ \qquad\qquad\qquad\qquad\qquad\qquad\quad | \qquad\qquad\qquad\qquad\quad | \\ \qquad\qquad\qquad\qquad\qquad\qquad\quad OH \qquad\qquad\qquad\qquad OH \end{array}$$

$$\begin{array}{c} \qquad\qquad\qquad\qquad\qquad\qquad\qquad\quad CH_3 \qquad\qquad\qquad CH_3 \\ \qquad\qquad\qquad\qquad\qquad\qquad\qquad\quad | \qquad\qquad\qquad\qquad | \\ (c)\quad CH_3CH_2CH_2CH_2CH_2OH,\quad CH_3CHCH_2CH_2OH,\quad CH_3CCH_2OH \\ \qquad\qquad\qquad\qquad\qquad\qquad\qquad\qquad\qquad\qquad\qquad\qquad\quad | \\ \qquad\qquad\qquad\qquad\qquad\qquad\qquad\qquad\qquad\qquad\qquad\qquad\quad CH_3 \end{array}$$

$$\begin{array}{c} \qquad\quad CH_3 \\ \qquad\qquad | \\ (d)\quad CH_3COCH_3,\quad CH_3CH_2CHCH_2CH_3,\quad CH_3CH_2CH_2CH_2CH_2OH \\ \qquad\qquad | \qquad\qquad\qquad\qquad\quad | \\ \qquad\qquad CH_3 \qquad\qquad\qquad\qquad OH \end{array}$$

$$\begin{array}{c} \qquad\qquad\qquad O \qquad\qquad\qquad\qquad\qquad OH \\ \qquad\qquad\qquad \parallel \\ (e)\quad CH_3CH_2CH_2CH, \end{array}$$

(Hint: You may want to review Sections 1.7 and 1.8 before answering the question.)

Reactions of Organic Compounds as Acids and Bases

4

Introduction

The first step in most chemical reactions is the interaction of a pair of nonbonding electrons in one molecule or ion with a center of electron deficiency (which may be an empty orbital or a partial positive charge) at an atom in another species. For example, the nonbonding electrons of a molecule of ammonia are donated to an empty orbital in boron trifluoride (p. 9) to give a new covalent bond.

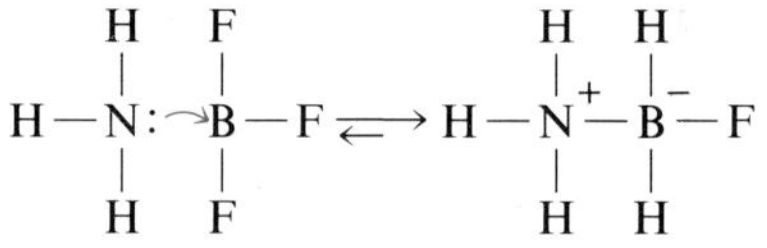

$$\begin{matrix} & H & F & & & H & H \\ & | & | & & & | & | \\ H-&N:&\curvearrowright B-F & \rightleftharpoons & H-&N\overset{+}{-}&\overset{-}{B}-F \\ & | & | & & & | & | \\ & H & F & & & H & H \end{matrix}$$

curved arrow symbolizing the donation of a pair of electrons from the nitrogen atom to the boron atom

An understanding of the generality of this way of looking at organic reactions is so fundamental to our study that we will explore it in this chapter using acid-base reactions as examples. In Chapter 1, you acquired the skills needed to begin predicting the reactivity of a wide variety of compounds. You developed these skills by locating the nonbonding electrons on Lewis structures, by practicing writing resonance contributors, and by making decisions about the polarities of covalent bonds. The relationship between the structure of a molecule and its chemical reactivity was further explored in Chapter 2. For example, oxygen and nitrogen atoms bearing nonbonding electrons were identified as reactive sites towards acids such as sulfuric acid. Your goal in this chapter should be to learn to see the sites of electron density and electron deficiency in the structures of different compounds. These sites make the molecules vulnerable to attack by a variety of chemical reagents. You should not attempt, at this point, to memorize structures or names of compounds; you will naturally become familiar with them as you draw structures and write reactions. The emphasis in this chapter is on learning to predict certain kinds of reactivity from the structural features that are apparent in the formulas.

4.2

Brønsted-Lowry and Lewis Acids and Bases

A. Brønsted-Lowry Acids and Bases

What are acids and bases? Two typical acid-base reactions are shown below.

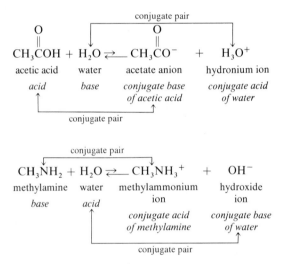

In each reaction, a proton is transferred from one of the molecules on the left-hand side of the equation to the other molecule. *In the* **Brønsted-Lowry theory** *of acids and bases, an* **acid** *is defined as a* **proton donor** *and a* **base** *as a* **proton acceptor.** In the first equation, therefore, acetic acid is the acid (the proton donor), and water is the base (the proton acceptor). Acetate ion, which is formed by the loss of a proton from acetic acid, is a base. It can accept a proton to become acetic acid again. *Two species that are related to each other by the loss and gain of a proton are called a* **conjugate acid** *and* **conjugate base** *pair.* Brønsted-Lowry acids are also called **protic acids,** or acids that react by the transfer of a proton. Note that water can be an acid or a base depending on what the other reagent is. It can gain a proton to become a hydronium ion, H_3O^+, its conjugate acid, or lose a proton to become the hydroxide ion, OH^-, its conjugate base.

PROBLEM 4.1   Identify the conjugate acid and conjugate base in each of the following pairs. Draw Lewis structures for each species and locate the formal charge.

(a) $(CH_3)_2O$, $(CH_3)_2OH^+$ (b) H_2SO_4, HSO_4^- (c) NH_2^-, NH_3

(d) $CH_3OH_2^+$, CH_3OH (e) H_2CO, H_2COH^+ (f) CH_3OH, CH_3O^-

---

It is not possible to define the acidity of a substance in absolute terms. Acidity and basicity are always described in terms of equilibria. A review of the quantitative treatment of acid-base equilibria is given in Section 4.6. Here we concentrate on achieving a qualitative understanding of the phenomena.

A useful way to picture acid-base reactions is to imagine two bases competing for the same proton. For example, in the reaction of acetic acid with water (p. 125), acetate anion and water are the bases in competition for the proton. Their relative strengths as bases determine where the equilibrium lies. Methylamine and hydroxide ion are the bases that are competing for the proton in an equilibrium process when methylamine reacts with water (p. 125). *An amine* (Section 2.4B) *has basic properties similar to those of ammonia and reacts with acids to give substituted ammonium ions.*

If two bases have quite different basicities, the transfer of a proton from one to the other may be virtually complete. The reaction of hydrogen chloride with dimethylamine, for example, is not written as an equilibrium reaction.

$$
\underset{\substack{\text{dimethylamine}\\ \textit{base}}}{(CH_3)_2NH} + \underset{\substack{\text{hydrogen}\\ \text{chloride}\\ \textit{acid}}}{HCl} \longrightarrow \underset{\substack{\text{dimethylammonium}\\ \text{ion}\\ \textit{conjugate acid of}\\ \textit{dimethylamine}}}{(CH_3)_2NH_2^+} + \underset{\substack{\text{chloride}\\ \text{ion}\\ \textit{conjugate base}\\ \textit{of hydrogen chloride}}}{Cl^-}
$$

conjugate pair

conjugate pair

Chloride ion, the conjugate base of the strong acid hydrogen chloride, is a weak base and does not compete effectively with dimethylamine for the proton. *Strong acids have weak conjugate bases, and vice versa.*

B. Lewis Acids and Bases

The proton acts as an acid because it has an empty $1s$ orbital that can accept a pair of nonbonding electrons from some species that is serving as a base. This view of an acid-base reaction is shown in Figures 2.13 and 2.15, and leads to another definition of acids and bases. *The* **Lewis theory** *of acids and bases defines an* **acid** *as an* **electron-pair acceptor** *and a* **base** *as an* **electron-pair donor.** Thus, a proton is only one of a large number of species that may function as an acid. *Any molecule or ion may be an acid if it has room in its orbitals to accept a pair of electrons. Any molecule or ion with an unshared pair of electrons to donate can be a base.* For example, dimethyl ether acts as a Lewis base towards boron trichloride, a Lewis acid.

dimethyl ether   boron   complex of dimethyl
Lewis base   trichloride   ether and boron
         Lewis acid   trichloride

Ethanol may donate a pair of electrons to a metal ion such as zinc(II).

ethanol   zinc cation   complex of zinc
Lewis base   *Lewis acid*   cation with ethanol

An alkyl halide may complex with aluminum chloride.

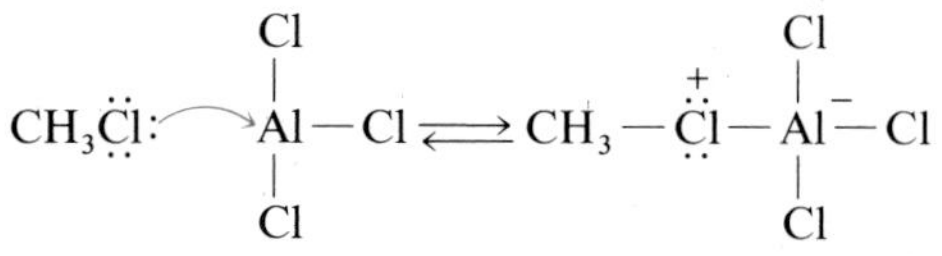

$$CH_3\ddot{Cl}: \quad \overset{Cl}{\underset{Cl}{\overset{|}{\underset{|}{Al}}}} - Cl \rightleftharpoons CH_3 - \overset{+}{\underset{\cdot\cdot}{Cl}} - \overset{Cl}{\underset{Cl}{\overset{|}{\underset{|}{Al}}}} \bar{-} Cl$$

chloromethane   aluminum   complex of aluminum
                  chloride   chloride with chloromethane
   Lewis base
               Lewis acid

In each of the cases shown above, a species with an empty orbital accepts a pair of electrons from a donor species. Note, too, that the Lewis acids do not necessarily contain hydrogen atoms. Nor do they owe their acidity to the ability to transfer a proton. They are known as **aprotic acids.**

PROBLEM 4.2   The following reactions take place. Write Lewis structures for the ions or molecules involved and point out which ones are Lewis acids, and which are Lewis bases.

(a) $Cu^{2+} + 4\,NH_3 \rightleftharpoons Cu(NH_3)_4{}^{2+}$

(b) $CH_3 - \overset{CH_3}{\underset{+}{\overset{|}{\underset{|}{C}}}} - CH_3 + H_2O \rightleftharpoons CH_3 - \overset{CH_3}{\underset{\underset{H}{\overset{|}{\underset{|}{O^+}}}}{\overset{|}{\underset{|}{C}}}} - CH_3$

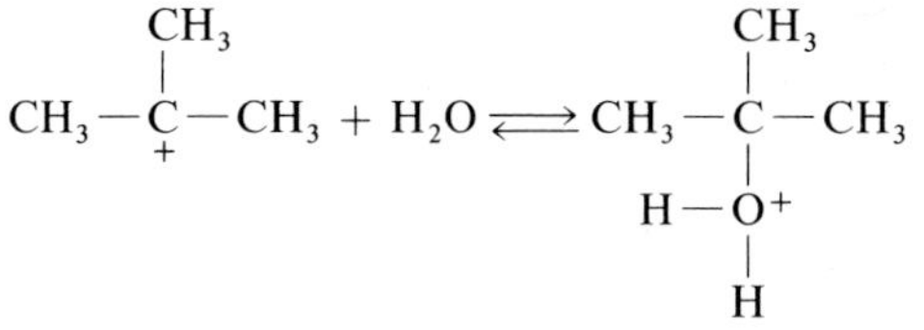

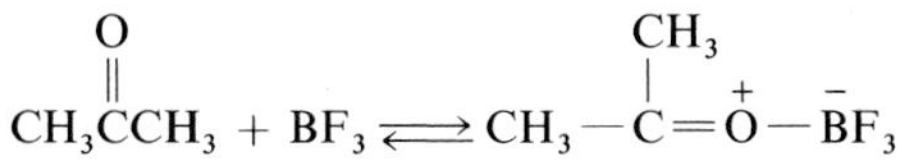

(c) $CH_3\overset{O}{\overset{\|}{C}}CH_3 + BF_3 \rightleftharpoons CH_3 - \overset{CH_3}{\underset{}{\overset{|}{C}}} = \overset{+}{O} - \bar{B}F_3$

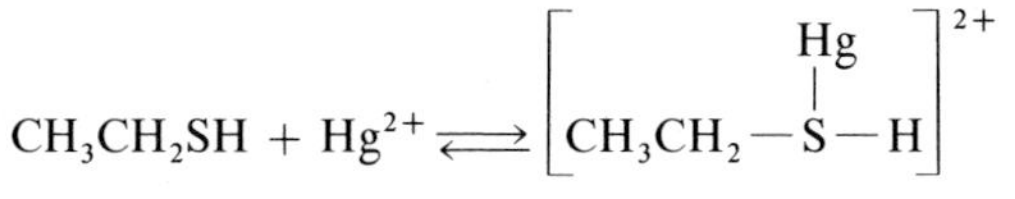

(d) $CH_3CH_2SH + Hg^{2+} \rightleftharpoons \left[CH_3CH_2 - \overset{Hg}{\underset{}{\overset{|}{S}}} - H \right]^{2+}$

The Reactions of Organic Compounds as Bases

Organic compounds with nonbonding electrons on atoms such as nitrogen, oxygen, or sulfur are bases. They react with protic acids or with Lewis acids. We considered the reactions of nitrogen bases or amines with acids in Sections 4.1 and 4.2A. The reaction of diethyl ether with concentrated hydriodic acid (57% HI in water) typifies that of an oxygen base with a protic acid.

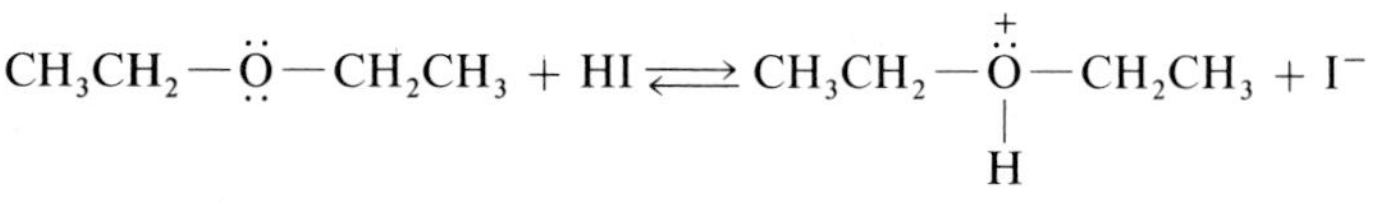

$$CH_3CH_2 - \ddot{O} - CH_2CH_3 + HI \rightleftharpoons CH_3CH_2 - \overset{+}{\underset{H}{\overset{|}{\underset{|}{O}}}} - CH_2CH_3 + I^-$$

diethyl ether   hydriodic   diethyloxonium ion   iodide
                      acid   ion
   base   *conjugate acid of*
                    acid   *diethyl ether*   *conjugate base*
                      acid   *of hydriodic*
                                                            acid

Alcohols such as ethanol also are protonated by acids.

$$CH_3CH_2-\overset{..}{\underset{..}{O}}-H + H_2SO_4 \rightleftharpoons CH_3CH_2-\overset{+}{\underset{\underset{H}{|}}{\overset{..}{O}}}-H + HSO_4^-$$

ethanol	sulfuric acid	ethyloxonium ion	hydrogen sulfate anion
base	*acid*	*conjugate acid of ethanol*	*conjugate base of sulfuric acid*

These reactions in which a proton is transferred from one species to another are usually rapid and reversible. *Just as water is protonated to give a hydronium ion, organic oxygen compounds are protonated to give* **oxonium ions.**

A ketone is another type of oxygen compound that can behave as a base (Sections 1.5B, 2.5B). For example, acetone, which is a ketone, donates electrons to boron trifluoride, a Lewis acid.

$$\overset{\overset{:O:}{\|}}{CH_3CCH_3} + BF_3 \rightleftharpoons \overset{\overset{CH_3}{|}}{CH_3C}=\overset{+}{O}-\overset{-}{BF_3}$$

acetone	boron trifluoride	a complex of boron trifluoride with acetone
base	*acid*	

A thiol is a typical organic sulfur compound, formed when an organic group replaces one of the hydrogen atoms in hydrogen sulfide, H_2S. Ethanethiol reacts with a strong protic acid, sulfuric acid.

$$CH_3CH_2-\overset{..}{\underset{..}{S}}-H + H_2SO_4 \rightleftharpoons CH_3CH_2-\overset{+}{\underset{\underset{H}{|}}{\overset{..}{S}}}-H + HSO_4^-$$

ethanethiol	sulfuric acid		hydrogen sulfate anion
base	*acid*	*conjugate acid of ethanethiol*	*conjugate base of sulfuric acid*

Compounds with π bonds, such as alkenes and alkynes, are also bases. They too react with strong acids.

$$CH_3CH=CHCH_3 + HBr \longrightarrow CH_3-\overset{\overset{H}{|}}{\underset{+}{C}}-\overset{\overset{H}{|}}{\underset{\underset{H}{|}}{C}}-CH_3 + Br^-$$

2-butene	hydrogen bromide	*sec*-butyl cation a carbocation	bromide ion
base	*acid*	*the conjugate acid of 2-butene*	*the conjugate base of hydrogen bromide*

When 2-butene reacts with the strong acid, hydrogen bromide, a *positively charged carbon atom, a* **carbocation,** is formed. The cationic carbon atom is sp^2 hybridized, with an empty *p* orbital (Problem 2.25c). In a subsequent reaction, the cation, as a Lewis acid, can combine with bromide anion, a Lewis base, to give 2-bromobutane.

$$:\overset{..}{\underset{..}{Br}}:^- + CH_3-\overset{\overset{H}{|}}{\underset{+}{C}}-\overset{\overset{H}{|}}{\underset{\underset{H}{|}}{C}}-CH_3 \longrightarrow CH_3-\overset{\overset{H}{|}}{\underset{\underset{Br}{|}}{C}}-\overset{\overset{H}{|}}{\underset{\underset{H}{|}}{C}}-CH_3$$

bromide ion	*sec*-butyl cation	2-bromobutane
base	*acid*	

Generally, nonbonding electrons on nitrogen, oxygen, and sulfur are protonated more easily than the electrons of π bonds.

PROBLEM 4.3 Write equations showing the reactions you would expect for the following compounds with $AlCl_3$ (a Lewis acid), and with H_2SO_4 (a strong protic acid).

(a) CH_3NHCH_3
dimethylamine

(b) $\overset{\displaystyle O}{\overset{\displaystyle \|}{HCH}}$
formaldehyde

(c) $HC\equiv CH$
acetylene

(d) $CH_3CH_2\overset{\displaystyle CH_2CH_3}{\overset{\displaystyle |}{P}}CH_2CH_3$
triethylphosphine

(e) CH_3CH_2OH
ethanol

(f) $CH_3CH_2SCH_2CH_3$
diethyl sulfide

Some organic compounds have more than one atom with nonbonding electrons, so that more than one site in the molecule can react with acids. Thus, acetamide has nonbonding electrons on an oxygen and a nitrogen atom, and either may be protonated.

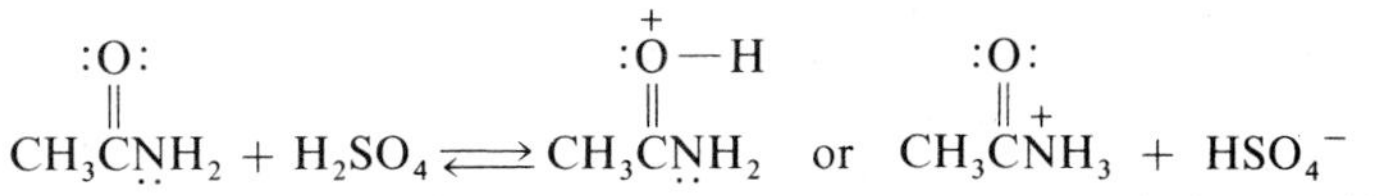

$$CH_3\overset{\displaystyle :O:}{\overset{\displaystyle \|}{C}}NH_2 + H_2SO_4 \rightleftharpoons CH_3\overset{\displaystyle \overset{+}{:O}-H}{\overset{\displaystyle \|}{C}}NH_2 \quad or \quad CH_3\overset{\displaystyle :O:}{\overset{\displaystyle \|}{C}}\overset{+}{NH_3} + HSO_4^-$$

acetamide   sulfuric acid   conjugate acid of acetamide   hydrogen sulfate anion

Acetic acid can react as a base at either of two different oxygen atoms.

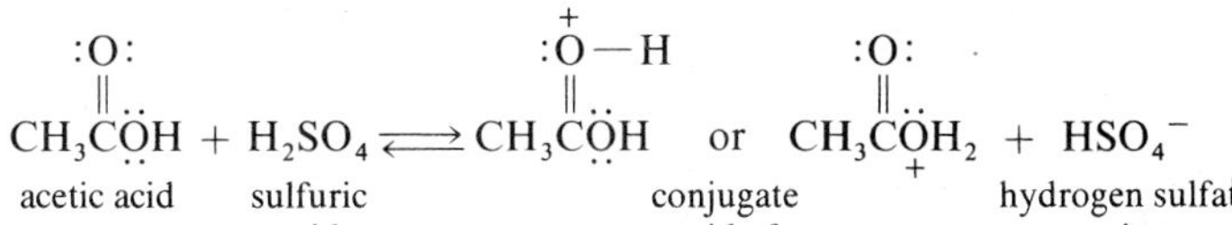

$$CH_3\overset{\displaystyle :O:}{\overset{\displaystyle \|}{C}}\ddot{O}H + H_2SO_4 \rightleftharpoons CH_3\overset{\displaystyle \overset{+}{:O}-H}{\overset{\displaystyle \|}{C}}\ddot{O}H \quad or \quad CH_3\overset{\displaystyle :O:}{\overset{\displaystyle \|}{C}}\overset{+}{\ddot{O}H_2} + HSO_4^-$$

acetic acid   sulfuric acid   conjugate acid of acetic acid   hydrogen sulfate anion

Note that we did not protonate either compound twice. Once a molecule is protonated and bears a positive charge, the availability of the other electrons in the system is greatly reduced. It is possible to protonate the compound again, but only under very strongly acidic conditions. Normally, the reaction stops when one proton is added to the molecule.

Acidity is always a relative quality. Acetic acid, a compound that we usually identify as an acid, is shown above reacting as a base with the much stronger sulfuric acid.

Acetamide and acetic acid are more readily protonated at the oxygen atom of the carbonyl group than at the other basic site in each molecule. This experimental observation can be explained by writing resonance contributors for each cation. The form of acetamide in which the oxygen atom is protonated, for example, has resonance contributors with the positive charge distributed among an oxygen, a carbon, and a nitrogen atom.

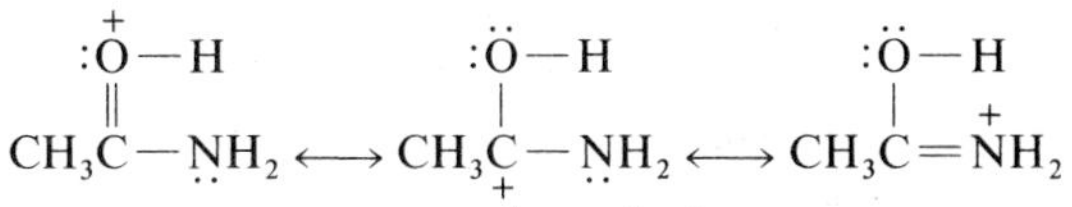

$$CH_3\overset{\displaystyle \overset{+}{:O}-H}{\overset{\displaystyle \|}{C}}-\ddot{N}H_2 \longleftrightarrow CH_3\overset{\displaystyle :\ddot{O}-H}{\overset{\displaystyle |}{\overset{+}{C}}}-\ddot{N}H_2 \longleftrightarrow CH_3\overset{\displaystyle :\ddot{O}-H}{\overset{\displaystyle |}{C}}=\overset{+}{N}H_2$$

resonance contributors for the cation
resulting from the protonation of acetamide
at the oxygen atom; the cation is
stabilized by delocalization of charge

The spreading of charge to two or more atoms is called **delocalization of charge.** *Cations (or anions) for which delocalization of charge is possible are more stable than species in which charge is concentrated in space.* We can compare the cation shown above with the cation that results if the nitrogen atom in acetamide is protonated.

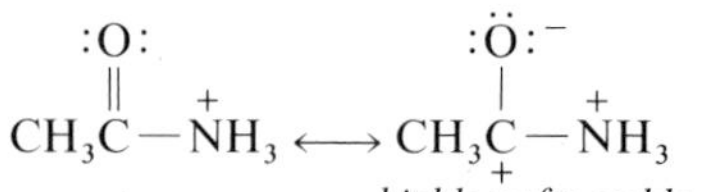

$$
\overset{\displaystyle :O:}{\underset{\displaystyle}{\overset{\displaystyle \|}{CH_3C}}} \overset{+}{-NH_3} \longleftrightarrow CH_3\overset{\displaystyle :\overset{..}{O}:^-}{\underset{\displaystyle |}{C}} \overset{+}{-\overset{+}{N}H_3}
$$

highly unfavorable resonance contributor; two positive charges adjacent to each other

no resonance contributor that delocalizes the positive charge at the nitrogen atom

This cation has a high concentration of positive charge at a single atom and is, therefore, less stable than the cation from acetamide in which the oxygen is protonated.

The protonation of the nonbonding electrons on the oxygen atom of a carbonyl or a hydroxyl group is an important first step in the reactions under acidic conditions of compounds such as acetamide, acetic acid, and alcohols. The conjugate acids of these compounds are usually more reactive towards Lewis bases than the unprotonated forms. Consequently, we will often use acids as catalysts to promote reactions of organic compounds.

PROBLEM 4.4   The following compounds all have more than one possible site where they can react with a strong protic acid such as sulfuric acid. Draw Lewis structures for each compound. Write equations showing the structures of the different products you would expect to obtain from the protonation reactions of each one.

(a)  $CH_3\overset{\displaystyle O}{\overset{\displaystyle \|}{C}}OCH_2CH_3$ (b) $HOCH_2CH_2\overset{\displaystyle CH_3}{\overset{\displaystyle |}{N}}CH_3$
      ethyl acetate   2-(*N,N*-dimethylamino) ethanol

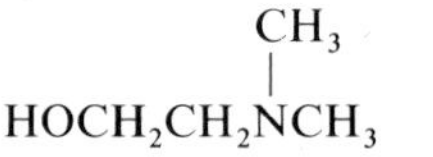

(c) $CH_2{=}CHCH_2OH$ (d) $CH_3N{=}C{=}O$
      2-propen-1-ol   methyl isocyanate

PROBLEM 4.5   Give the structural formula of the conjugate acid of each of the following species.

(a) 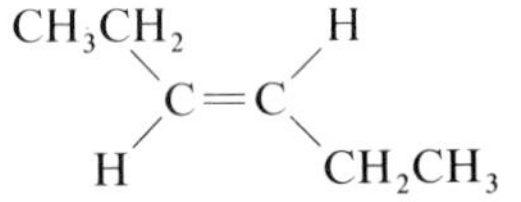 $\underset{\displaystyle H}{\overset{\displaystyle CH_3CH_2}{}}C{=}C\underset{\displaystyle CH_2CH_3}{\overset{\displaystyle H}{}}$ (b) $CH_3CH_2OCH_2CH_3$

(c) 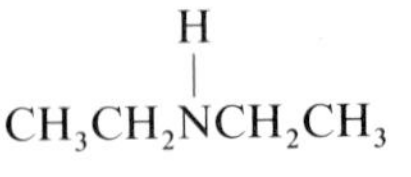 $CH_3CH_2\overset{\displaystyle H}{\overset{\displaystyle |}{N}}CH_2CH_3$ (d) $CH_3CH_2CH_2SCH_3$

(e) 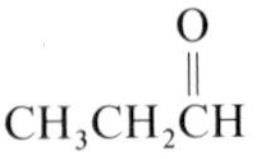 $CH_3CH_2\overset{\displaystyle O}{\overset{\displaystyle \|}{C}}H$ (f) $CH_3CH_2CH_2OH$

(g) 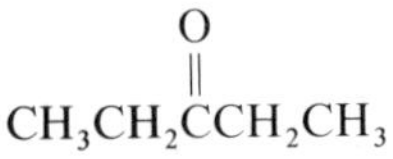 $CH_3CH_2\overset{\displaystyle O}{\overset{\displaystyle \|}{C}}CH_2CH_3$

4.4

An Introduction to Electron-Pushing

Chemists are not satisfied to write equations showing the reactants and the products of a reaction. They are also interested in the details of the transformations taking place as reactants are converted to products. In order to study these details, chemists examine the progress of reactions using many experimental techniques. For example, they determine the rate of a chemical reaction (Section 5.3E) and see how the rate and the products vary when experimental conditions are changed. They explore changes in stereochemistry that occur during the reaction. From observations such as these, they postulate the details of the pathway the reactants follow as they are converted into the products of a reaction. Such a detailed reaction pathway is called the **mechanism** of the reaction. Mechanisms serve as an aid to understanding the reactivity of different types of organic compounds. In subsequent sections of the book, we will often describe the mechanisms proposed for reactions. In this section we explore the symbolism widely used by organic chemists to trace the details of the changes that occur as reactants are converted to products.

For example, the equation that shows the reactants diethyl ether and hydriodic acid on one side of the equation and the products, diethyloxonium ion and iodide ion, on the other side (p. 127) can be rewritten to show greater detail.

$$
\begin{array}{cccc}
\text{CH}_3\text{CH}_2 & & \text{CH}_3\text{CH}_2 & \\
\diagdown & & \diagdown & \\
\text{O:} \quad \text{H} \frown \text{I:} \longrightarrow & & \text{O} \overset{+}{-} \text{H} + \text{:I:}^- & \\
\diagup & \overset{\delta+}{} \overset{\delta-}{} & \diagup & \\
\text{CH}_3\text{CH}_2 & & \text{CH}_3\text{CH}_2 & \\
\text{diethyl ether} & \text{hydriodic} & \text{diethyloxonium} & \text{iodide} \\
 & \text{acid} & \text{ion} & \text{ion}
\end{array}
$$

The curved arrows show the electronic changes that take place as covalent bonds are formed and broken during the reaction.

The curved arrows in the equation above depict a language known among organic chemists as *electron-pushing*. The important features of this kind of symbolism are the following. First, the structures of both reactants are written to emphasize the presence or absence of nonbonding electrons. Partial positive and negative charges are shown to indicate that *bonding is between an atom with unshared electrons in one reactant and an atom with a deficiency of electrons in the other reactant. A curved arrow is drawn from the site of electron density, such as nonbonding electrons, to the site of electron deficiency, such as an atom with a partial positive charge.* This arrow depicts the start of the bonding process. *It represents the flow of electrons, not the movement of atoms.* If the process results in too many bonds around an atom, such as two covalent bonds to a hydrogen atom, another curved arrow is drawn symbolizing the simultaneous release of bonding electrons to some atom or group of atoms that then detaches itself. Iodide ion is the species formed by this process in the equation above.

The reaction of ethanethiol with sulfuric acid provides another example.

$$
\begin{array}{cccc}
\text{CH}_3\text{CH}_2 & & \text{CH}_3\text{CH}_2 & \\
\diagdown & & \diagdown & \\
\text{S:} \quad \text{H} \frown \text{O} - \text{SO}_3\text{H} \longrightarrow & & \overset{+}{\text{S}} - \text{H} + \text{:O} - \text{SO}_3\text{H} & \\
\diagup & \overset{\delta+}{} \overset{\delta-}{} & \diagup & \\
\text{H} & & \text{H} & \\
\text{ethanethiol} & \text{sulfuric acid} & \text{conjugate acid} & \text{hydrogen sulfate} \\
 & & \text{of ethanethiol} & \text{ion}
\end{array}
$$

The nonbonding electrons on the sulfur atom are shown moving towards the positively charged hydrogen atom of sulfuric acid. As the proton is transferred to the sulfur atom of ethanethiol, the hydrogen sulfate ion is detached, taking with it the pair of electrons that bonded it to the proton.

The π electrons of a multiple bond may also serve as the site of electron density in a reaction.

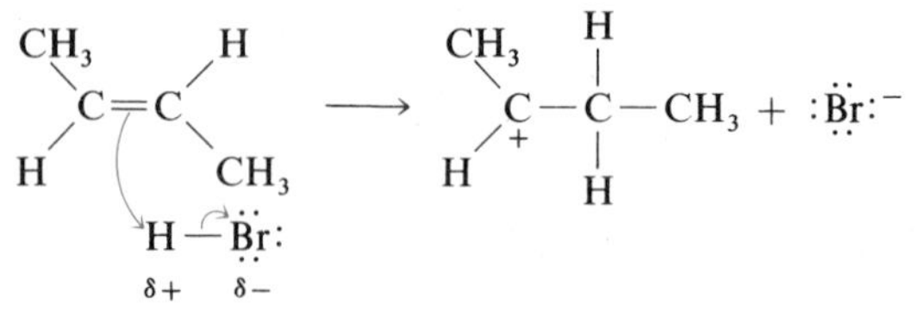

| 2-butene | hydrobromic acid | *sec*-butyl cation | bromide ion |

The reaction of 2-butene with hydrobromic acid starts with the flow of the π electrons towards the positively charged hydrogen atom of the hydrogen halide. A strict accounting for the electrons indicates that the carbon atom that lost its share of electrons in the π bond becomes a cation. Bromide ion is the other product of the reaction.

Whenever a mechanism is written for a reaction, it is important to make sure that the net charge is the same on both sides of the equation. Thus, in all the previous examples, two neutral reactants with a net zero charge give products with one positive and one negative charge, also a net zero charge.

Often the site of electron deficiency is an empty orbital or a positively charged atom. The reaction of the *sec*-butyl cation with bromide ion illustrates this kind of reaction.

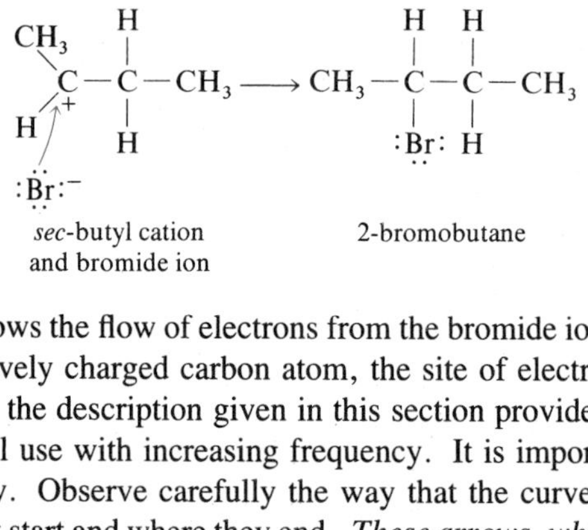

| *sec*-butyl cation and bromide ion | 2-bromobutane |

The curved arrow shows the flow of electrons from the bromide ion, the site of electron density, to the positively charged carbon atom, the site of electron deficiency.

The equations and the description given in this section provide an introduction to a language that we will use with increasing frequency. It is important that you start to practice it right away. Observe carefully the way that the curved arrows are drawn, especially where they start and where they end. *These arrows, which represent the flow of electrons as bonds are formed and broken, must be used with precision.*

PROBLEM 4.6 Go back to Problems 4.3 and 4.4 and rewrite the equations in your answers to show the flow of electrons in the reactions.

PROBLEM 4.7 In the following equations, reactants and arrows showing the flow of electrons are indicated. Supply the products of the reactions.

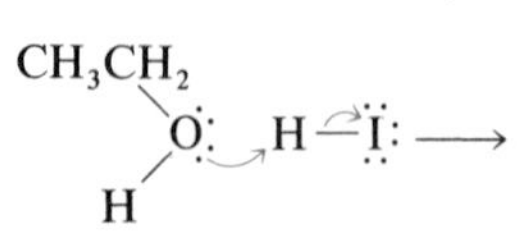

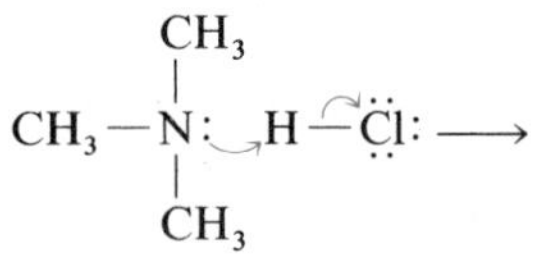

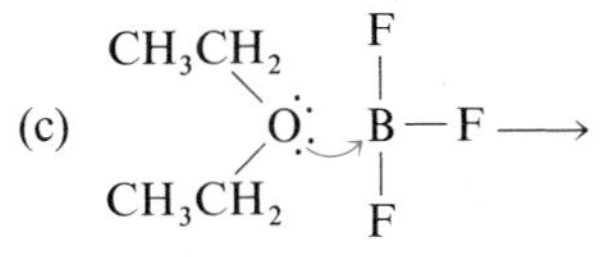

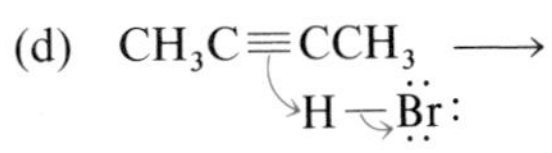

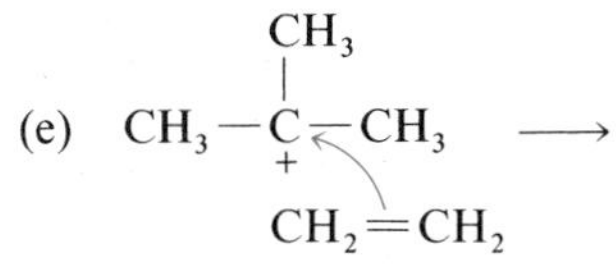

(c)

$$CH_3CH_2 \diagdown \ddot{O}: \diagup CH_3CH_2 \diagup$$

$$\begin{array}{c} F \\ | \\ B-F \\ | \\ F \end{array} \longrightarrow$$

(d) $CH_3C\equiv CCH_3 \longrightarrow$
   $H\text{—}\ddot{B}\ddot{r}:$

(e) $CH_3 - \underset{\underset{CH_2=CH_2}{+}}{\overset{\overset{CH_3}{|}}{C}} - CH_3 \longrightarrow$

4.5

Carbon, Nitrogen, Oxygen, Sulfur, and Halogen Acids

The reactions of organic compounds as acids depend upon the ease with which they can lose a proton to a base. Acetic acid has two kinds of bonds to hydrogen atoms: (1) carbon-hydrogen bonds, which are nonpolar, and (2) an oxygen-hydrogen bond, which is highly polar because of the difference in electronegativity between oxygen and hydrogen. When acetic acid behaves as an acid, it is the hydrogen atom on the oxygen atom that is lost as a proton.

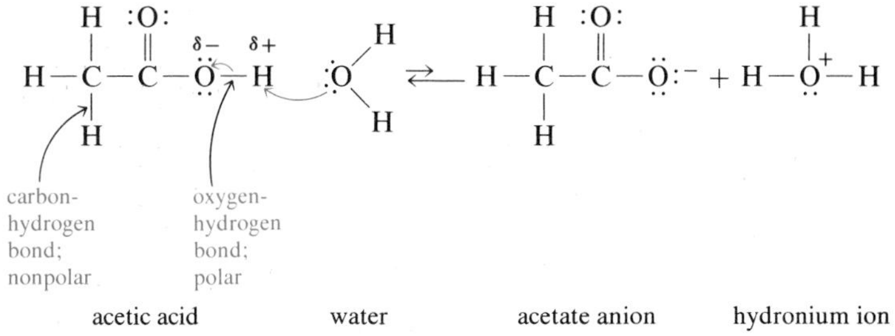

carbon-hydrogen bond; nonpolar	oxygen-hydrogen bond; polar		
acetic acid	water	acetate anion	hydronium ion

In the reaction of acetic acid with water, the equilibrium concentration of undissociated acetic acid is high. From this observation we can conclude that acetic acid is a weak acid.

Similar comparisons can be made for methanol.

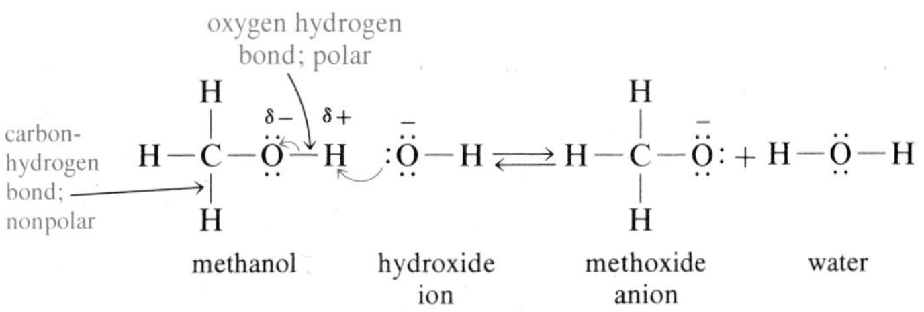

methanol   hydroxide ion   methoxide anion   water

The most acidic hydrogen atom in methanol is the one on the oxygen atom. The equilibrium is reached in the reaction of methanol with hydroxide ion when the concentrations of hydroxide ion and methoxide anion are roughly equal, because the acidities of methanol and water are similar. The substitution of a methyl group for a hydrogen atom in the water molecule does not affect the acidity of the hydroxyl group much.

In the reaction of methanethiol with hydroxide ion, the equilibrium lies much more to the right.

$$H-\overset{\overset{\displaystyle H}{|}}{\underset{\underset{\displaystyle H}{|}}{C}}-\ddot{\overset{..}{S}}-H \quad :\ddot{O}-H \rightleftharpoons H-\overset{\overset{\displaystyle H}{|}}{\underset{\underset{\displaystyle H}{|}}{C}}-\ddot{\overset{..}{S}}:^- + H-\ddot{O}-H$$

| methanethiol | hydroxide ion | methanethiolate anion | water |

A hydrogen atom on a sulfur atom is more acidic than one on oyxgen, and is transferred more completely to hydroxide ion. Another way of describing the same phenomenon is to say that the methanethiolate anion is a weaker base than hydroxide ion (or methoxide ion). Sulfur is below oxygen in the periodic table and is a larger atom. Therefore, a negative charge on a sulfur anion is more spread out, less concentrated at any one point in space, than one on oxygen. Consequently, a sulfur anion is less likely to attract a proton than an oxygen anion. Basicities of anions decrease as one goes down any group in the periodic table; the acidities of the corresponding conjugate acids increase in the same order.

$$R-\ddot{O}:^- \;>\; R-\ddot{S}:^-$$
stronger base   weaker base

$$R-\ddot{O}-H \;<\; R-\ddot{S}-H$$
weaker acid   stronger acid

The same trend is characteristic of the halogen acids, which become stronger as we go from hydrogen fluoride to hydrogen iodide.

$$HF < HCl < HBr < HI$$
weakest acid   strongest acid

$$F^- > Cl^- > Br^- > I^-$$
strongest base   weakest base

In general, then, protic acids become stronger as we go down any group in the periodic table.

PROBLEM 4.8   Which is the stronger acid, H_2O or H_2S?

PROBLEM 4.9   Write an equation showing the reaction you would expect between ethanethiol, CH_3CH_2SH, and methoxide anion, CH_3O^-.

In any period of the periodic table, the acidity of a hydrogen atom bound to a central atom increases as we go from left to right. Methane, in which the hydrogen atoms are bound to carbon, is a very weak acid. Methylamine does not give up a proton from its nitrogen atom to hydroxide ion, but methanol, with a hydrogen atom bound to oxygen, does (p. 133).

Group IV   *Group V*   *Group VI*

carbon   nitrogen   oxygen

$$H-\underset{\underset{\textstyle H}{|}}{\overset{\overset{\textstyle H}{|}}{C}}-H \qquad H-\underset{\underset{\textstyle H}{|}}{\overset{\overset{\textstyle H}{|}}{C}}-\underset{\underset{\textstyle H}{|}}{\overset{\overset{\textstyle H}{|}}{N}}-H \qquad H-\underset{\underset{\textstyle H}{|}}{\overset{\overset{\textstyle H}{|}}{C}}-\ddot{\text{O}}-H$$

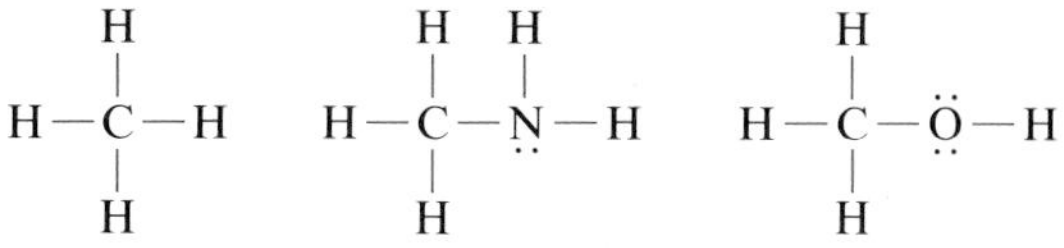

methane   methylamine   methanol

⟶

increasing acidity of hydrogen bonded to
carbon, nitrogen, and oxygen

The acidities of these compounds are closely related to the basicities of their conjugate bases. We can develop some idea of the basicity of the conjugate base of an amine by recalling what we know about the relative basicities of ammonia and water.

ammonia   water

stronger base   *weaker base*
nonbonding electrons   *nonbonding electrons*
more available for sharing *less available for sharing*

Nitrogen is less electronegative than oxygen, which means that it attracts and holds the electrons around it less firmly than oxygen. The nonbonding electrons on a nitrogen atom are more available for donation to a proton or a Lewis acid than the electrons on an oxygen atom. Therefore, nitrogen compounds are more basic than similar oxygen compounds. For example, if an amine and water compete for a proton, the proton is transferred to the amine.

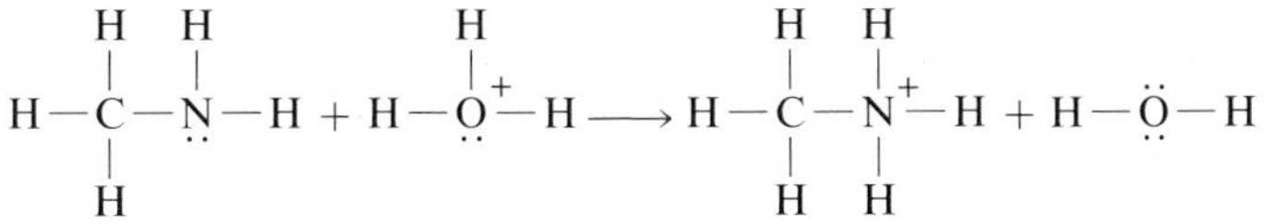

$$H-C-N-H + H-O^+-H \longrightarrow H-C-N^+-H + H-\ddot{O}-H$$

methylamine   hydronium ion   methylammonium   water
                                    ion
base   *acid*   *conjugate base*
                                    conjugate acid   *of hydronium*
                                    of methylamine   *ion*

In a similar way, the conjugate base of an amine, a nitrogen anion, is a much stronger base than an oxygen anion.

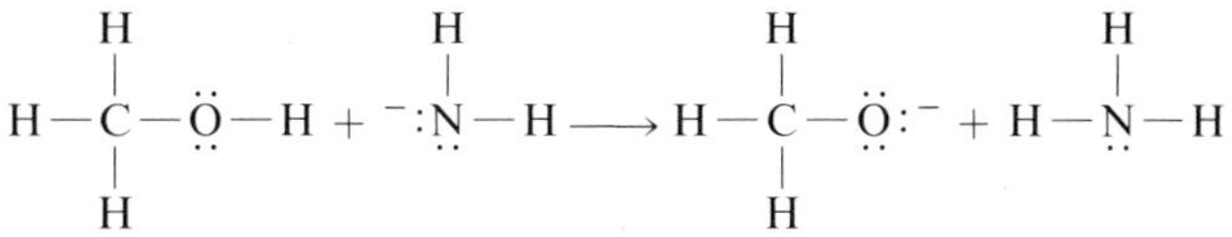

$$H-C-\ddot{O}-H + {}^-\!:\!N-H \longrightarrow H-C-\ddot{O}\!:^- + H-N-H$$

methanol   amide anion   methoxide anion   ammonia

acid   *base*   *conjugate base*   *conjugate acid*
                               of methanol   *of amide anion*

Amide anion, the conjugate base of ammonia, reacts with methanol to give methoxide anion and ammonia. As a base, amide anion is so much stronger than methoxide ion that the reaction shown above goes essentially to completion.

The conjugate base of methane is the methyl anion, a carbanion. *A* **carbanion** *has a carbon atom with a pair of nonbonding electrons and a negative charge on it, in contrast to a carbocation, which has only six electrons around a carbon atom and a positive charge* (p. 128). A carbanion is like ammonia in its shape (Problem 2.25f), but is much more basic than ammonia or even amide anion because carbon is less electronegative than nitrogen. The conjugate bases of methane, methylamine, and methanol (p. 135) are shown below.

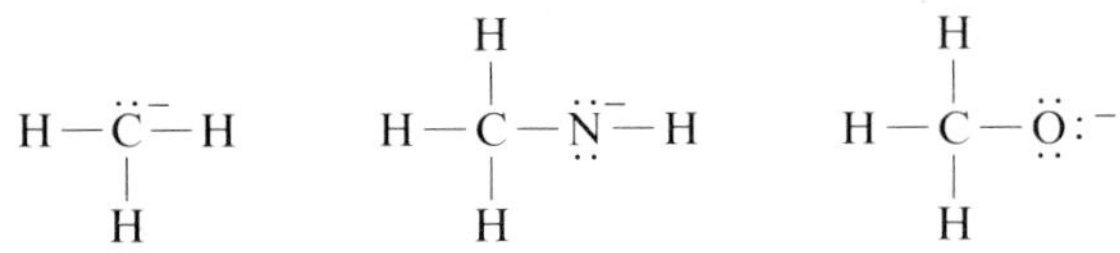

$$
\begin{array}{ccc}
\text{H} & \text{H} & \text{H} \\
| & | & | \\
\text{H}-\overset{..}{\text{C}}{}^{-}-\text{H} & \text{H}-\text{C}-\overset{..}{\underset{..}{\text{N}}}{}^{-}-\text{H} & \text{H}-\text{C}-\overset{..}{\underset{..}{\text{O}}}{:}^{-} \\
| & | & | \\
\text{H} & \text{H} & \text{H}
\end{array}
$$

methyl anion	methylamide	methoxide
a carbanion	anion	anion
conjugate base of methane	*conjugate base of methylamine*	*conjugate base of methanol*
strongest base		*weakest base*

The weakest acid, methane, has the strongest conjugate base, while the strongest acid in the series, methanol, has the weakest conjugate base.

Most carbanions are strong bases. In Section 4.8, some carbanions that are formed easily enough to be important intermediates in organic reactions are explored.

Arguments about acidity that depend upon the concept of electronegativity only work when we are comparing species in the same period of the periodic table. Comparisons among atoms in different periods based on electronegativity do not work. The effects of atomic size, mentioned earlier in comparing the acidity of alcohols and thiols (p. 134), are more important in such cases.

The protonated forms of organic compounds, which are organic cations, are also acids. The conjugate base of methylammonium ion is methylamine, a stable, neutral molecule, so methylammonium ion loses a proton relatively easily.

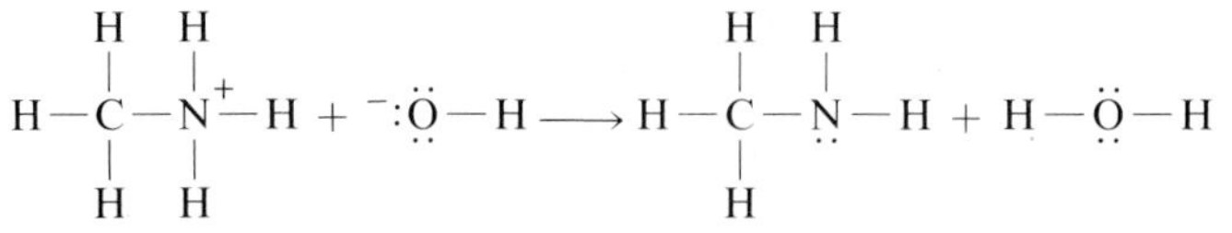

$$
\begin{array}{cccc}
\text{H}\ \ \text{H} & & \text{H}\ \ \text{H} & \\
|\ \ \ \ | & & |\ \ \ \ | & \\
\text{H}-\text{C}-\overset{+}{\text{N}}-\text{H} + {}^{-}\!:\!\overset{..}{\underset{..}{\text{O}}}-\text{H} \longrightarrow \text{H}-\text{C}-\overset{..}{\text{N}}-\text{H} + \text{H}-\overset{..}{\underset{..}{\text{O}}}-\text{H} \\
|\ \ \ \ | & & | & \\
\text{H}\ \ \text{H} & & \text{H} &
\end{array}
$$

methylammonium ion	hydroxide ion	methylamine	water
acid	base	*conjugate base of methylammonium ion*	*conjugate acid of hydroxide ion*

Protonation reactions of alcohols, ethers, and ketones yield oxonium ions, which are acidic. Methyloxonium ion, the conjugate acid of methanol, has an acidity similar to the hydronium ion.

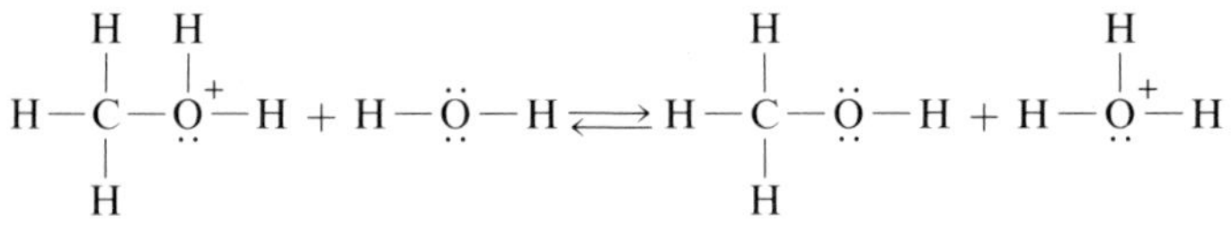

$$
\begin{array}{cccc}
\text{H}\ \ \text{H} & & \text{H} & \text{H} \\
|\ \ \ \ | & & | & | \\
\text{H}-\text{C}-\overset{+}{\underset{}{\text{O}}}-\text{H} + \text{H}-\overset{..}{\underset{..}{\text{O}}}-\text{H} \rightleftharpoons \text{H}-\text{C}-\overset{..}{\underset{..}{\text{O}}}-\text{H} + \text{H}-\overset{+}{\underset{}{\text{O}}}-\text{H} \\
| & & | & \\
\text{H} & & \text{H} &
\end{array}
$$

methyloxonium ion	water	methanol	hydronium ion
acid	base	*conjugate base of methyloxonium ion*	*conjugate acid of water*

PROBLEM 4.10 Arrange the species in each of the following sets in order of increasing acidity.

(a) CH_3CH_2OH, $CH_3CH_2CH_3$, $CH_3CH_2NH_2$, CH_3CH_2SH

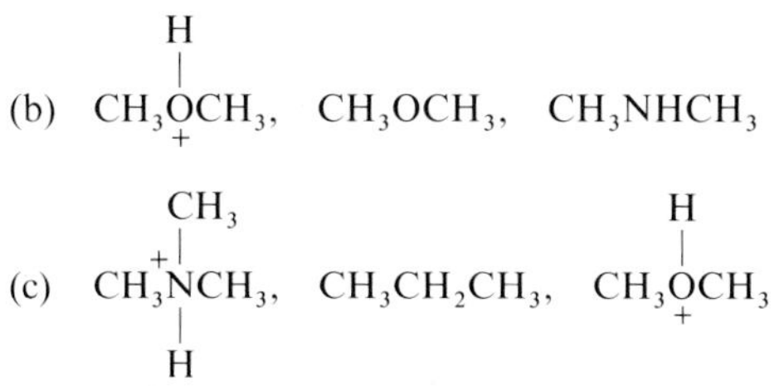

(b) $CH_3\overset{\overset{\displaystyle H}{|}}{\underset{+}{O}}CH_3$, CH_3OCH_3, CH_3NHCH_3

(c) $CH_3\overset{\overset{\displaystyle CH_3}{|}}{\underset{\underset{\displaystyle H}{|}}{\overset{+}{N}}}CH_3$, $CH_3CH_2CH_3$, $CH_3\overset{\overset{\displaystyle H}{|}}{\underset{+}{O}}CH_3$

PROBLEM 4.11 Arrange the species in each of the following sets in order of decreasing basicity.

(a) CH_3NHCH_3, $CH_3CH_2CH_3$, CH_3OCH_3 (b) NH_2^-, CH_3^-, F^-, OH^-

(c) $CH_3\bar{N}CH_3$, CH_3O^-

PROBLEM 4.12 Give the structural formula for the conjugate base of each of the following species.

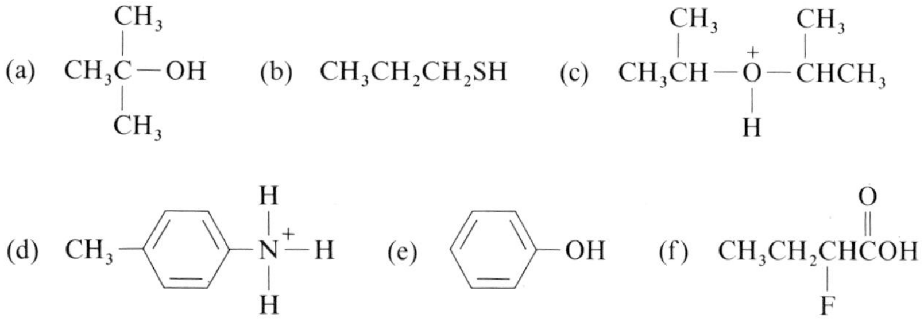

(a) $CH_3\overset{\overset{\displaystyle CH_3}{|}}{\underset{\underset{\displaystyle CH_3}{|}}{C}}-OH$ (b) $CH_3CH_2CH_2SH$ (c) $CH_3\overset{\overset{\displaystyle CH_3}{|}}{\underset{\underset{\displaystyle H}{|}}{CH}}-\overset{+}{O}-\overset{\overset{\displaystyle CH_3}{|}}{CH}CH_3$

(d) $CH_3-\!\!\!\left\langle\!\!\!\bigcirc\!\!\!\right\rangle\!\!\!-\overset{\overset{\displaystyle H}{|}}{\underset{\underset{\displaystyle H}{|}}{\overset{+}{N}}}-H$ (e) $\left\langle\!\!\!\bigcirc\!\!\!\right\rangle-OH$ (f) $CH_3CH_2\overset{\overset{\displaystyle O}{\|}}{\underset{\underset{\displaystyle F}{|}}{CH}}COH$

4.6

Equilibria in Acid-Base Reactions

A. Acidity Constants and pK_a

The measurement of the acidities of organic compounds involves the measurement of equilibrium constants. When acetic acid is dissolved in water, certain concentrations of acetic acid, water, acetate ions and hydronium ions are present at equilibrium. Acetate ion is more basic than water, so the equilibrium lies on the side of the undissociated acetic acid.

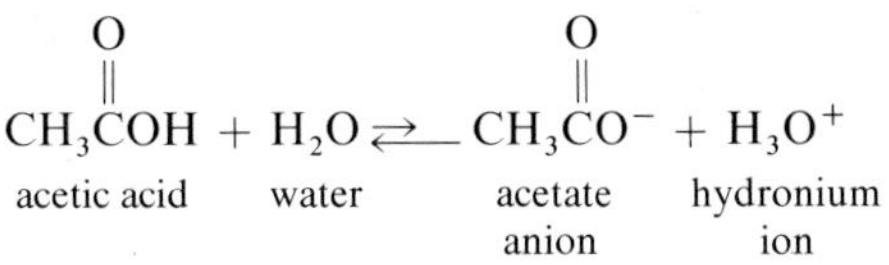

$$CH_3\overset{\overset{\displaystyle O}{\|}}{C}OH + H_2O \rightleftharpoons CH_3\overset{\overset{\displaystyle O}{\|}}{C}O^- + H_3O^+$$

acetic acid   water   acetate   hydronium
anion   ion

The **equilibrium constant** for the reaction, K_{eq}, is written as

$$K_{eq} = \frac{[CH_3\overset{\overset{O}{\|}}{C}O^-]\,[H_3O^+]}{[CH_3\underset{\underset{O}{\|}}{C}OH]\,[H_2O]}$$

For dilute solutions in which the concentration of water is large and almost constant, another expression, the **acidity constant, K_a**, is used. For acetic acid,

$$K_{eq}[H_2O] = K_a = \frac{[CH_3\overset{\overset{O}{\|}}{C}O^-]\,[H_3O^+]}{[CH_3\underset{\underset{O}{\|}}{C}OH]} = 1.75 \times 10^{-5}$$

The acidity constants for different acids have values ranging from 10^{14} to 10^{-50}. Such large and small numbers are most conveniently expressed as their logarithms. Just as pH is $-\log[H_3O^+]$, so pK_a is defined as $-\log K_a$. The pK_a for acetic acid is 4.76, a value that is typical of many organic acids. A strong inorganic acid such as sulfuric acid has $K_a \sim 10^9$ and $pK_a \sim -9$. The pK_a of hydronium ion is -1.7; that of ammonium ion is 9.4. Hydrocarbons, such as methane, are extraordinarily weak acids, with $pK_a \sim 50$.

Inside the front cover of this book there is a table of pK_a values for a variety of acids shown with their conjugate bases. Note that the *strong acids, with weak conjugate bases, are at the top of the table with negative or small positive pK_a values. The weaker the acid, the stronger its conjugate base, and the larger the pK_a value of the acid.* The pK_a values for the very weak acids at the bottom of the table can be measured only indirectly, and you may find different values for the same compound in different tables. These pK_a values are rough indications of relative acidities. Note that pK_a values are only given for protic acids. As you use the table, you will acquire some sense of which acids are strong and which are weak. Do not attempt to memorize pK_a values. You will be able to use the table if the solution of a problem requires it.

PROBLEM 4.13 Formic acid,

$$H\overset{\overset{O}{\|}}{C}OH$$

has $K_a = 1.99 \times 10^{-4}$ What is its pK_a? Is it a stronger or weaker acid than acetic acid, pK_a 4.76?

PROBLEM 4.14 The imidazole ring is present in the amino acid histidine. This amino acid plays an important role in proton transfer reactions in the human body. The protonated form of imidazole (imidazolinium ion) has pK_a 7.0. What is K_a for this acid?

imidazole

imidazolinium ion

the conjugate acid of imidazole

pK_a 7.0

B. The Use of the pK_a Table in Predicting Acid-Base Reactions

The table of pK_a values can be used to predict whether acid-base reactions will take place. In general, an acid will transfer a proton to the conjugate base of any acid that is below it in the table. If the pK_a values for the two acids are within one or two pK_a units, both acids will be present in substantial quantities at equilibrium. When large differences in acidity exist, the transfer of the proton will be nearly complete. For example, suppose we wanted to know whether we could use ammonia, NH_3, to prepare free trimethylamine, $(CH_3)_3N$, from trimethylammonium chloride, $(CH_3)_3NH^+Cl^-$. The proton donor in this reaction is the trimethylammonium ion, $(CH_3)_3NH^+$, and the proton acceptor is ammonia.

$$
\underset{\substack{\text{trimethylammonium} \\ \text{ion} \\ \text{p}K_a\ 9.8 \\ acid}}{CH_3-\overset{\overset{\displaystyle CH_3}{|}}{\underset{\underset{\displaystyle CH_3}{|}}{\overset{+}{N}}}-H} + \underset{\substack{\text{ammonia} \\ base}}{H-\overset{\overset{\displaystyle \cdot\cdot}{}}{\underset{\underset{\displaystyle H}{|}}{N}}-H} \rightleftharpoons \underset{\substack{\text{trimethylamine} \\ conjugate\ base\ of \\ trimethylammonium \\ ion}}{CH_3-\overset{\overset{\displaystyle CH_3}{|}}{\underset{\underset{\displaystyle CH_3}{|}}{N}}{:}} + \underset{\substack{\text{ammonium} \\ \text{ion} \\ \text{p}K_a\ 9.4 \\ conjugate\ acid \\ of\ ammonia}}{H-\overset{\overset{\displaystyle H}{|}}{\underset{\underset{\displaystyle H}{|}}{\overset{+}{N}}}-H}
$$

We look up the two acids, trimethylammonium ion and ammonium ion, in the pK_a table and find that their pK_a values are very close to each other, 9.4 for the ammonium ion, and 9.8 for the trimethylammonium ion. The larger pK_a value for the trimethylammonium ion indicates that trimethylamine is actually a stronger base than ammonia. In any competition for a proton, trimethylamine is more likely to retain the proton than to release it to ammonia. The species shown in the equation above are in equilibrium, with the arrow pointing to the left being larger than the one pointing to the right. These experimental conditions would not generate trimethylamine from trimethylammonium ion.

To remove the proton from trimethylammonium ion, we must find a stronger base than ammonia. A search of the table of pK_a values reveals that hydroxide ion, the conjugate base of water, pK_a 15.7, would be a good reagent to use.

$$
\underset{\substack{\text{trimethylammonium} \\ \text{ion} \\ \text{p}K_a\ 9.8 \\ acid}}{CH_3-\overset{\overset{\displaystyle CH_3}{|}}{\underset{\underset{\displaystyle CH_3}{|}}{\overset{+}{N}}}-H} + \underset{\substack{\text{hydroxide ion} \\ base}}{{}^-{:}\overset{\cdot\cdot}{\underset{\cdot\cdot}{O}}-H} \longrightarrow \underset{\substack{\text{trimethylamine} \\ conjugate\ base\ of \\ trimethylammonium \\ ion}}{CH_3-\overset{\overset{\displaystyle CH_3}{|}}{\underset{\underset{\displaystyle CH_3}{|}}{N}}{:}} + \underset{\substack{\text{water} \\ \text{p}K_a\ 15.7 \\ conjugate\ acid \\ of\ hydroxide \\ ion}}{H-\overset{\cdot\cdot}{\underset{\cdot\cdot}{O}}-H}
$$

The difference in pK_a values between the two acids, trimethylammonium ion and water, is large, indicating that hydroxide ion is a much stronger base than trimethylamine. In the competition between the two bases, the proton is transferred essentially completely from the trimethylammonium ion to the hydroxide ion, freeing the weaker base, trimethylamine.

The table of pK_a values cannot include all possible acids and bases. What happens when we need to make a prediction about a compound that does not appear in the table? For instance, we might wish to know whether propyne, $CH_3C{\equiv}CH$, will be deprotonated by amide anion, NH_2^-.

$$CH_3C\equiv C-H + \ ^-:\overset{H}{\underset{|}{N}}-H \longrightarrow CH_3C\equiv C:^- + H-\overset{H}{\underset{|}{N}}-H$$

propyne	amide anion	methylacetylide	ammonia
pK_a ~ 26	*base*	anion	pK_a ~ 36
acid		*conjugate base of propyne*	*conjugate acid of amide anion*

Propyne does not appear in the table of pK_a values. Acetylene, $HC\equiv CH$, which resembles propyne, is listed with pK_a 26. In assigning the same pK_a value to propyne, we are reasoning that the acidity of propyne, like that of acetylene, is derived from the hydrogen atom bonded to the *sp*-hybridized carbon atom (Section 2.6A) of the triple bond. (Because there are no other kinds of hydrogen atoms in acetylene, the acidity must come from that hydrogen.) We are also reasoning that the substitution of an alkyl group for a hydrogen atom in acetylene will not dramatically change the acidity of the hydrogen atom on the triple bond. We find, in general, that functional groups are affected only in subtle ways by the particular alkyl group attached to them. Therefore, we can generalize further that a series of alkynes such as

$$CH_3C\equiv CH \qquad CH_3CH_2C\equiv CH \qquad CH_3CH_2CH_2C\equiv CH$$

propyne	1-butyne	1-pentyne

will have similar pK_a values and similar reactivity. This process of reasoning by analogy to make judgments and estimates about unknown situations is one you should cultivate during your study of organic chemistry.

Once we conclude that the pK_a of propyne is approximately 26, we can complete the equation by showing that the alkyne will be deprotonated by amide anion, the conjugate base of ammonia, which has pK_a 36.

The table of pK_a values aids our predictions about a large number of reactions. Developing familiarity with the table and skill in using it will help you learn organic chemistry.

PROBLEM 4.15   From the table of pK_a values, select an oxonium ion, an ammonium ion, a carboxylic acid, an alcohol, a thiol, an amine, and an alkane, and arrange them in order of acidity. This small table summarizes the trends that were discussed in Section 4.5, and will be useful to you in answering the questions that follow.

PROBLEM 4.16   A number of acid-base reactions were given in Section 4.5. Rewrite each equation, leaving out the arrows between the reactants and the products. Use the pK_a table to assign pK_a values to the acids that appear in each equation. Make your own judgment in each case about whether the reaction goes essentially to completion as written or whether reactants and products are present together in significant quantities.

PROBLEM 4.17   Use the table of pK_a values to decide whether or not the following reactions will take place. Complete the equations for the reactions that will take place. Write ''no reaction'' for the others.

(a) $CH_3CH_2OH + KOH \longrightarrow$

(b) $CH_3CH_2\overset{+}{N}H_2 \ Cl^- + NaOH \longrightarrow$
$\qquad\quad \underset{|}{\ }$
$\qquad\quad CH_2CH_3$

(c) $CH_2F\overset{\overset{\displaystyle O}{\|}}{C}OH + NaHCO_3 \longrightarrow$

(d) $CH_3 -\!\!\left\langle\!\!\!\bigcirc\!\!\!\right\rangle\!\!- OH + NaHCO_3 \longrightarrow$

(e) $\left\langle\!\!\!\bigcirc\!\!\!\right\rangle\!\!- NH_2 + HCl \longrightarrow$

(f) $CH_3CH_2CH_2SH + CH_3CH_2ONa \longrightarrow$

(g) $CH_3 -\!\!\left\langle\!\!\!\bigcirc\!\!\!\right\rangle\!\!-\overset{\overset{\displaystyle O}{\|}}{C}OH + NaCN \longrightarrow$

(h) $CH_3\overset{\overset{\displaystyle O}{\|}}{C}CH_2\overset{\overset{\displaystyle O}{\|}}{C}CH_3 + CH_3ONa \longrightarrow$

C. Acidities in the Gas Phase and in Solution

The pK_a of an acid depends on the solvent it is dissolved in and the temperature at which the determination is made. The importance of the solvent has become much clearer since the development of mass spectrometers (Section 13.6), instruments that allow chemists to study reactions in the gas phase. For example, the proton transfer reaction of acetic acid with water in the gas phase has been examined.

$$CH_3\overset{\overset{\displaystyle O}{\|}}{C}OH(g) + H_2O(g) \rightleftharpoons CH_3\overset{\overset{\displaystyle O}{\|}}{C}\overset{-}{O}(g) + H_3\overset{+}{O}(g)$$

| acetic acid in the gas phase | water in the gas phase | acetate anion in the gas phase | hydronium ion in the gas phase |

pK_a 130

The pK_a of acetic acid under these conditions is 130, indicating that it is enormously difficult to separate the negatively charged acetate ion from the positively charged hydronium ion unless solvent molecules are present to stabilize the charged particles. Clearly, the interaction of the solvent with the undissociated acid and with anions and cations that form on dissociation contributes significantly to the stabilization of the various species and, therefore, to the position of the equilibrium.

Such interactions are particularly important in water, which has extensive hydrogen bonding (Section 1.8B). When an acid is dissolved in water, it disrupts the bonding between the solvent molecules, and new hydrogen bonds are formed between water and the solute particles. Careful experimentation now shows that such interactions, about which more is said in the next section, may be the most important factor in determining the relative acidities of closely related compounds.

In Section 4.7 we examine the relative acidities of a series of acids and offer rationalizations for the trends observed. The language that we are developing in talking about acidity is used by chemists to correlate facts and predict reactivity in all areas of organic chemistry. In order to understand the assumptions inherent in this language, we first (in Section 4.6D) review the relationship between equilibrium and the changes in energy that occur during a reaction.

D. Equilibrium, Free Energy, Enthalpy, and Entropy. A Review

The equilibrium constant for a reaction is related to the change in standard free energy by the following equation.

$$\Delta G^\circ = -2.303RT \log K_{eq}$$

ΔG° is the change in free energy for the reaction when the reactants and the products are in their standard states (1 M solution for solutions; 1 atm of pressure for gases). R is the gas constant, 1.987×10^{-3} kcal/deg·mol; T is the absolute temperature, and K_{eq} is the equilibrium constant for the reaction. If there is a decrease in free energy during the reaction, that is if ΔG° is negative, the equilibrium constant is greater than 1. In other words, the reaction proceeds so that more products than reactants are present at equilibrium. If ΔG° is positive, the reverse reaction is favored.

For example, chloroacetate anion and acetate anion compete for a proton in the following reaction.

$$
\underset{\substack{\text{chloroacetic}\\\text{acid}}}{\text{ClCH}_2\overset{\text{O}}{\overset{\|}{\text{C}}}\text{OH}} + \underset{\substack{\text{acetate}\\\text{anion}}}{\text{CH}_3\overset{\text{O}}{\overset{\|}{\text{C}}}\text{O}^-} \underset{\text{H}_2\text{O}}{\rightleftharpoons} \underset{\substack{\text{chloroacetate}\\\text{anion}\\\textit{conjugate base}\\\textit{of chloroacetic}\\\textit{acid}}}{\text{ClCH}_2\overset{\text{O}}{\overset{\|}{\text{C}}}\text{O}^-} + \underset{\substack{\text{acetic}\\\text{acid}\\\textit{conjugate acid}\\\textit{of acetate}\\\textit{anion}}}{\text{CH}_3\overset{\text{O}}{\overset{\|}{\text{C}}}\text{OH}}
$$

ΔG° for this reaction at 298 K has been determined to be -2.59 kcal/mol. There is a *decrease* in free energy during the reaction, therefore the equilibrium lies to the right. Acetate anion takes a proton away from chloroacetic acid; it is a stronger base than chloroacetate anion. Therefore, at equilibrium there is more chloroacetate, anion present than acetate anion.

The equilibrium constant for the reaction can be calculated from ΔG°, and is also related to the acidity constants for chloroacetic acid and acetic acid by the equation

$$K_{eq} = \frac{K_a \text{ for ClCH}_2\overset{\text{O}}{\overset{\|}{\text{C}}}\text{OH}}{K_a \text{ for CH}_3\underset{\underset{\text{O}}{\|}}{\text{C}}\text{OH}}$$

PROBLEM 4.18   Calculate K_{eq} for the reaction of chloroacetic acid and acetate anion and the pK_a for chloroacetic acid from the data given above. K_a for acetic acid is 1.75×10^{-5}.

The change in free energy, ΔG, of the system determines whether a reaction will take place spontaneously as written. ΔG has two components. The heat of the reaction, the change in enthalpy, ΔH, is one. The other component is ΔS, the change in entropy during the reaction. Entropy is a measure of disorder or randomness in the system. A restriction on the freedom of motion of molecules, or within molecules, leads to a decrease in entropy. For example, a reaction in which two molecules combine to give a single one would lead to a decrease in entropy. A reaction in which a molecule dissociates to give two fragments gives an increase in entropy because each piece is free to move in ways that were not possible when they were bound together.

The relationship between changes in free energy, enthalpy, and entropy during a reaction is summed up in the following equation.

$$\Delta G = \Delta H - T \Delta S$$

If ΔG is negative for a particular reaction—that is, there is a decrease in free energy during the reaction—the reaction will take place spontaneously. If ΔG is positive, the reaction will not proceed as written. Instead, the reverse reaction will be spontaneous. If ΔG is 0, the reaction must be at equilibrium with no net change being observed.

Negative values for ΔH and positive values for ΔS contribute to making ΔG negative. In other words, exothermic reactions, reactions in which heat energy is lost to the surroundings, will be spontaneous *if* there is not a large decrease in entropy. For many organic reactions, entropy changes tend to be small and predictions about the spontaneity of reactions can be based on the values of the heat of reaction alone. However, our intuitive feeling that exothermic reactions proceed easily while endothermic ones do not is not totally accurate.

The importance of entropy in determining the position of equilibrium becomes apparent when we compare the acidities of closely related compounds in the gas phase and in solution. For example, propanoic acid is a slightly weaker acid than acetic acid in water.

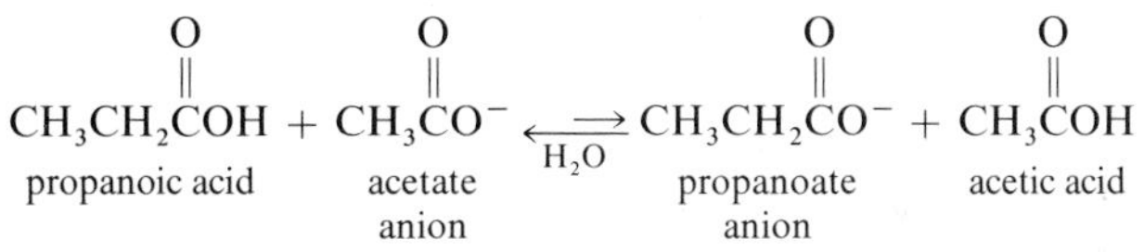

$$
\underset{\text{propanoic acid}}{CH_3CH_2\overset{\overset{\displaystyle O}{\|}}{C}OH} + \underset{\substack{\text{acetate}\\\text{anion}}}{CH_3\overset{\overset{\displaystyle O}{\|}}{C}O^-} \underset{H_2O}{\rightleftarrows} \underset{\substack{\text{propanoate}\\\text{anion}}}{CH_3CH_2\overset{\overset{\displaystyle O}{\|}}{C}O^-} + \underset{\text{acetic acid}}{CH_3\overset{\overset{\displaystyle O}{\|}}{C}OH}
$$

$\Delta G°$ for the reaction shown above is $+0.15$ kcal/mol, meaning that the equilibrium lies slightly to the left. In the gas phase, however, propanoic acid is stronger than acetic acid. $\Delta G°$ for the reaction in the gas phase is -1.2 kcal/mol, and the equilibrium is to the right. The difference between these two values is due to the difference in $\Delta S°$ values for the two reactions. In the gas phase there is a very small change in entropy during the reaction. The number of particles is the same on both sides of the equation.

In solution, however, we must consider the change not only in the reactants but in the degree of solvation of the various species. Propanoic acid and propanoate anion, having larger alkyl groups than acetic acid and acetate anion respectively, interact differently with the surrounding water molecules (Figure 4.1, page 144).

It has been determined experimentally that the reaction shown in the equation above is slightly exothermic, $\Delta H° = -0.12$ kcal/mol, but also goes with a small decrease in entropy, $\Delta S° = -0.0009$ kcal/deg·mol (or -0.9 e.u.). The decrease in entropy has been interpreted to mean that the more nonpolar solute, propanoate anion, disrupts the highly hydrogen-bonded structure of the surrounding water less than the more polar species, acetate anion. Thus, in going from acetate anions to propanoate anions, the hydrogen-bonded structure in the surrounding solvent increases and entropy decreases.

The overall effect on the free energy of the reaction can be calculated from the data given above. At 298 K, $T \Delta S° = 298 \times -0.0009$ kcal/deg·mol or -0.27 kcal/mol.

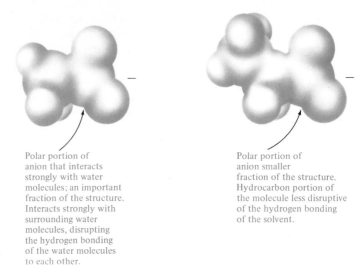

Polar portion of
anion that interacts
strongly with water
molecules; an important
fraction of the structure.
Interacts strongly with
surrounding water
molecules, disrupting
the hydrogen bonding
of the water molecules
to each other.

Polar portion of
anion smaller
fraction of the structure.
Hydrocarbon portion of
the molecule less disruptive
of the hydrogen bonding
of the solvent.

Figure 4.1   A comparison of acetate and propanoate anions.

$$\Delta G° = \Delta H° - T \, \Delta S°$$

$$\Delta G° = -0.12 - (-0.27) = +0.15 \text{ kcal/mol}$$

In water, the change in entropy has a larger effect than the change in enthalpy and determines the relative acidities of the two acids. In the gas phase, the change in enthalpy determines the change in free energy, and thus the relative acidities. We must be cautious, therefore, in interpreting experimental data obtained under any one set of conditions, especially when the structural differences between the compounds we are comparing are small.

4.7

The Effects of Structural Changes on Acidity

A. The Resonance Effect

Acetic acid is a typical member of the family of organic compounds known as the carboxylic acids (Section 2.5C). Carboxylic acids, in which the hydroxyl group is bound to a carbonyl group, are much stronger acids than are alcohols in which the hydroxyl group is on an sp^3-hybridized carbon atom. A simple experimental observation illustrates this difference in acidity. When a carboxylic acid is put into a solution containing bicarbonate ion, bubbles of carbon dioxide are seen.

$$\underset{\substack{\text{acetic acid} \\ \text{p}K_a \text{ 4.8} \\ acid}}{\overset{\overset{\displaystyle O \atop \|}{\text{CH}_3\text{COH}}}{}} + \underset{\substack{\text{bicarbonate} \\ \text{anion} \\ base}}{\text{HCO}_3{}^-} \rightleftharpoons \underset{\substack{\text{acetate} \\ \text{anion} \\ \textit{conjugate} \\ \textit{base of} \\ \textit{acetic acid}}}{\overset{\overset{\displaystyle O \atop \|}{\text{CH}_3\text{CO}^-}}{}} + \underset{\substack{\text{carbonic} \\ \text{acid} \\ \text{p}K_a \text{ 6.5} \\ \textit{conjugate acid} \\ \textit{of bicarbonate} \\ \textit{anion}}}{\text{H}_2\text{CO}_3} \rightleftharpoons \underset{\text{water}}{\text{H}_2\text{O}} + \underset{\substack{\text{carbon} \\ \text{dioxide} \\ \textit{bubbles}}}{\text{CO}_2 \uparrow}$$

carboxyl group · · · · · carboxylate anion

An alcohol placed in the same solution shows no evidence of reaction. Ethanol is not a strong enough acid to protonate bicarbonate ion.

$$CH_3CH_2OH + HCO_3^- \longrightarrow \text{no bubbles}$$

ethanol   bicarbonate
anion
pK_a 17

The carboxylic acid, pK_a 4.8, on the other hand, is a stronger acid than carbonic acid H_2CO_3, pK_a 6.5, and protonates bicarbonate anion, the conjugate base of carbonic acid. The stronger acid, acetic acid, transfers a proton to the conjugate base of the weaker acid, carbonic acid. Carbonic acid is in equilibrium with carbon dioxide dissolved in water; in the absence of external pressure (such as exists in bottles of soda pop), bubbles of carbon dioxide can be seen whenever carbonic acid is formed.

The conjugate base of acetic acid, acetate anion, is a weaker base than ethoxide anion, the conjugate base of ethanol. In acetate anion, the negative charge is distributed evenly to the two oxygen atoms (Section 1.5C). The delocalization of charge (p. 130) by resonance stabilizes the anion. An anion in which the charge is delocalized to two oxygen atoms is a weaker base than one with the charge concentrated on a single oxygen atom as it is in ethoxide anion.

$$CH_3CH_2-\overset{..}{\underset{..}{O}}:^- \qquad CH_3-\overset{\overset{:O:}{\|}}{C}-\overset{..}{\underset{..}{O}}:^- \longleftrightarrow CH_3-\overset{\overset{:\overset{..}{O}:^-}{|}}{C}=\overset{..}{\underset{..}{O}}$$

ethoxide anion   resonance contributors for the
acetate anion

charge localized on
one oxygen atom   *charge delocalized to two oxygen atoms*
strong base   *weak base*

In summary, one factor that affects the basicity of an anion and therefore the acidity of its conjugate acid is the likelihood of delocalization of charge over several atoms by resonance.

B. Inductive and Field Effects

Other changes in the structures of acids also affect their acidity. The acidities are known for a series of acids in which one of the hydrogen atoms on the carbon atom attached to the carboxyl group has been replaced by another group or atom. This carbon atom, adjacent to the carboxyl group, is called the **α-carbon** of the acid and the hydrogen atoms on it are the **α-hydrogen atoms.**

the carboxyl group

the α-hydrogen
atoms

$$H-\overset{\overset{\displaystyle H}{|}}{\underset{\underset{\displaystyle H}{|}}{C}}-\overset{\overset{\displaystyle O}{\|}}{C}-OH$$

the α-carbon
atom

In the series of compounds shown in Table 4.1, an α-hydrogen atom in acetic acid has been substituted in turn by a methyl group, a fluorine atom, a chlorine atom, a bromine atom, and an iodine atom. The pK_a values given indicate that acidity is decreased slightly by substitution of a methyl group, but increased by substitution of a halogen atom. Substitution by a fluorine atom at the α-carbon atom produces the

TABLE **The Acidity of Substituted Carboxylic Acids**
4.1   **Determined in Water at 25 °C**

Name	Structure	pK_a
acetic acid	$\begin{array}{c} O \\ \parallel \\ CH_2COH \\ \mid \\ H \end{array}$	4.76
propanoic acid	$\begin{array}{c} O \\ \parallel \\ CH_2COH \\ \mid \\ CH_3 \end{array}$	4.87
fluoroacetic acid	$\begin{array}{c} O \\ \parallel \\ CH_2COH \\ \mid \\ F \end{array}$	2.59
chloroacetic acid	$\begin{array}{c} O \\ \parallel \\ CH_2COH \\ \mid \\ Cl \end{array}$	2.86
bromoacetic acid	$\begin{array}{c} O \\ \parallel \\ CH_2COH \\ \mid \\ Br \end{array}$	2.90
iodoacetic acid	$\begin{array}{c} O \\ \parallel \\ CH_2COH \\ \mid \\ I \end{array}$	3.17

greatest increase in acidity. Acidities decrease as the substitution changes from fluorine to chlorine to bromine to iodine.

The pK_a values for another series of acids illustrate the effect of putting a halogen atom farther and farther away from the carboxyl group (Table 4.2). Butanoic acid,

$$\begin{array}{c} O \\ \parallel \\ CH_3CH_2CH_2COH \end{array}$$

has an acidity very close to that of propanoic acid (Table 4.1). Acidity does not change significantly with the size of the alkyl group attached to the carboxyl group. A chlorine atom on the α-carbon atom increases the acidity of butanoic acid, but the effect falls off rapidly as the chlorine atom is moved farther from the carboxyl group in 3-chlorobutanoic acid and still farther in 4-chlorobutanoic acid.

These data suggest that there is some process that increases the acidity of an acid when electronegative atoms are introduced into the molecule. The greater the electronegativity of the substituent, the greater is the increase in acidity (Table 4.1). This phenomenon also depends upon the distance between the substituent and the carboxyl group (Table 4.2).

Two effects have been proposed to account for these data. One is called the **inductive effect.** *It is postulated to operate through the σ bonds in the molecule.* An electronegative substituent such as a chlorine atom draws the electrons of the carbon-chlorine bond towards the chlorine atom, creating an electron deficiency in that carbon atom and in neighboring carbon atoms. *An electron-withdrawing group is said to have a negative inductive effect.* This effect results in a withdrawal of electron density from the carboxyl group. The effect is present in the carboxylic acid, but is especially important in the carboxylate anion. The withdrawal of electron density from the carboxylate anion stabilizes it by making the electrons less available for protonation. The carboxylate anion becomes a weaker base; consequently, the corresponding conjugate acid is stronger.

$$
\underset{\displaystyle \text{Cl}}{\overset{\displaystyle \overset{\textstyle O}{\overset{\|}{}}}{\text{CH}_2 \leftarrow \text{C} \leftarrow \text{O}^-}}
\qquad
\underset{\displaystyle \text{H}}{\overset{\displaystyle \overset{\textstyle O}{\overset{\|}{}}}{\text{CH}_2 - \text{C} - \text{O}^-}}
$$

chloroacetate anion   acetate anion

stabilized by the inductive
effect, a weaker base than
acetate anion

Because this effect is transmitted through σ bonds, it diminishes rapidly as the number of bonds between the substituent and the carboxyl group increases.

The other effect that has been suggested to explain the data reported in Tables 4.1 and 4.2 is called the *field effect.* The **field effect** *arises in the bond dipole moments of the molecule and is transmitted not through the bonds but through the environment around the molecule.* According to this model, for example, chloroacetate ion is stabilized relative to acetate ion by the bond dipole of the carbon-chlorine bond.

**TABLE
4.2**   The Effect on Acidity (Determined in Water at 25 °C) of the Distance Between the Substituent and the Carboxyl Group

Name	Structure	pK_a
butanoic acid	$\overset{\displaystyle \overset{\textstyle O}{\|}}{\text{CH}_3\text{CH}_2\text{CH}_2\text{COH}}$	4.82
2-chlorobutanoic acid	$\underset{\text{Cl}}{\overset{\displaystyle \overset{\textstyle O}{\|}}{\text{CH}_3\text{CH}_2\text{CHCOH}}}$	2.86
3-chlorobutanoic acid	$\underset{\text{Cl}}{\overset{\displaystyle \overset{\textstyle O}{\|}}{\text{CH}_3\text{CHCH}_2\text{COH}}}$	4.05
4-chlorobutanoic acid	$\underset{\text{Cl}}{\overset{\displaystyle \overset{\textstyle O}{\|}}{\text{CH}_2\text{CH}_2\text{CH}_2\text{COH}}}$	4.52

positive end of carbon-chlorine
bond dipole interacts through
space with negative charge
on carboxylate group

chloroacetate anion

*the field effect;
negative charge
on carboxylate group
is stabilized by the
positive end of the
carbon-chlorine
bond dipole*

acetate anion

*no additional stabilization
arises from the presence
of polar bonds in the
molecule; stronger base
than chloroacetate anion*

The positive end of the dipole of the carbon-chlorine bond in chloroacetate anion is postulated to interact directly through space with the negative charge on the carboxylate anion to reduce charge density at the carboxylate group. The effect of the carbon-chlorine bond dipole on the carboxylate anion decreases as the distance between the chlorine atom and the carboxylate group increases (Table 4.2).

The inductive effect and the field effect operate in the same direction. Many ingenious experiments have been devised to measure these effects separately. The most successful experiments indicate that the transmission of the effect through space by interaction of bond dipoles is more important than the transmission of the effect through bonds. Customarily, the term *inductive effect* is used to refer to the combination of both of these effects.

Propanoic acid is slightly weaker than acetic acid in water, but slightly stronger in the gas phase (Section 4.6D). Chemists have explained the acidity of propanoic acid in water by suggesting that there must be a slight release of electrons from the substituent methyl group towards the carboxyl group. Such a release of electrons destabilizes the anion, making it more basic and more easily protonated; hence, the conjugate acid is correspondingly weaker. This picture cannot be entirely correct because the methyl group apparently stabilizes the anion in the gas phase, making it less basic. Many experimental results, however, can be rationalized if we postulate that an alkyl group is a more electron-releasing entity than a hydrogen atom. This effect is especially important for alkyl groups bound to sp^2- and sp-hybridized carbon atoms. The theoretical explanations that are given for this effect vary, but the effect itself is real.

PROBLEM 4.19 Predict whether CF_3CH_2OH is more or less acidic than CH_3CH_2OH. Why?

Stronger acid

PROBLEM 4.20 Look up the pK_a values for chloroacetic acid, CH_2ClCOH (with O double bonded); dichloroacetic acid, $CHCl_2COH$ (with O double bonded) and trichloroacetic acid, CCl_3COH (with O double bonded); in the table of pK_a values. How do you explain the data?

TABLE   The Effects of Electron-Withdrawing Groups on
4.3   the Acidity of Carboxylic Acids

Name	Structure	pK_a
acetic acid	$\underset{\displaystyle H}{CH_2COH}$ (with $\overset{\displaystyle O}{\overset{\|}{}}$)	4.76
hydroxyacetic acid	$\underset{\displaystyle OH}{CH_2COH}$ (with $\overset{\displaystyle O}{\overset{\|}{}}$)	3.83
cyanoacetic acid	$\underset{\displaystyle CN}{CH_2COH}$ (with $\overset{\displaystyle O}{\overset{\|}{}}$)	2.46
nitroacetic acid	$\underset{\displaystyle NO_2}{CH_2COH}$ (with $\overset{\displaystyle O}{\overset{\|}{}}$)	1.68

Other substituents that increase the acidity of carboxylic acids are the hydroxyl, cyano, and nitro groups (Table 4.3). In hydroxyacetic acid, the electronegative oxygen atom substituted on the α-carbon atom reduces electron density at the carboxylate anion, making it a weaker base in comparison with the unsubstituted acetate ion. Consequently, hydroxyacetic acid is a stronger acid than acetic acid.

The cyano and nitro groups are strongly electron-withdrawing. Note that nitroacetic acid (Table 4.3) is a stronger acid than (and cyanoacetic acid almost as strong an acid as) fluoroacetic acid (Table 4.1). The powerful electron-withdrawing effects of these groups are rationalized by pointing to their electronic structures. In each case, the atom attached to the α-carbon atom of the acid is either positively charged or may be positively charged in an important resonance contributor. The presence of this positive charge in the substituent contributes to the stabilization of the carboxylate anion.

PROBLEM 4.21   Draw Lewis structures for cyanoacetic acid and nitroacetic acid (Table 4.3). Draw resonance contributors for them. Explain in each case why the substituent increases the acidity of the acid in comparison with acetic acid, and also why nitroacetic acid is stronger than cyanoacetic acid.

In summary, an electron-withdrawing atom or group substituted on the alkyl chain of a carboxylic acid stabilizes the corresponding carboxylate anion, thereby increasing the acidity of the carboxylic acid. Important electron-withdrawing groups are the halogens, hydroxyl and ether groups, the cyano group, and the nitro group. Alkyl groups behave as if they are weakly electron-releasing.

A word of caution is necessary. The relative acidities on which these generalizations are based were determined in water. In the gas phase, reversals in the order of related compounds are often seen. For example, in the gas phase bromoacetic acid is a stronger acid than fluoroacetic acid. The models presented above, which rationalize acidity only in terms of the electronegativity of the substituents and of the polarization of bonds within molecules, are too simple. Such models are retained, however, because they are enormously useful in rationalizing a wide range of experimental observations and in making predictions about reactivity in many different systems. We will invoke "resonance effects" and "inductive effects" frequently as we try to understand how the structures of organic compounds determine their reactivity.

4.8

Factors That Stabilize Carbon Anions

While methane is an extremely weak acid, and will not react with hydroxide ion, trichloromethane (chloroform) does react to give low concentrations of trichloromethyl anion.

$$H-\overset{\overset{\displaystyle H}{|}}{\underset{\underset{\displaystyle H}{|}}{C}}-H \;+\; :\overset{..}{\underset{..}{O}}{}^{-}-H \longrightarrow \text{no reaction}$$

methane
pK_a ~50   hydroxide ion

$$Cl-\overset{\overset{\displaystyle Cl}{|}}{\underset{\underset{\displaystyle Cl}{|}}{C}}-H \longleftarrow :\overset{..}{\underset{..}{O}}-H \rightleftarrows Cl-\overset{\overset{\displaystyle Cl}{|}}{\underset{\underset{\displaystyle Cl}{|}}{C}}:^{-} + H-\overset{..}{O}-H$$

trichloromethane   hydroxide   trichloromethyl   water
pK_a ~25   ion   anion   pK_a 15.7

With the substitution of electron-withdrawing chlorine atoms for three of the hydrogen atoms in methane, the remaining atom becomes acidic enough to react with hydroxide ion. Trichloromethane is a stronger acid than methane because the negative charge on the trichloromethyl anion can be stabilized by the negative inductive effect of the chlorine atoms.

methyl anion
very strong base

trichloromethyl anion
*weaker base than
methyl anion*

The carbon-chlorine bond dipoles act to disperse some of the negative charge at the anionic carbon atom. Thus, the carbanion from trichloromethane is a weaker base than the carbanion derived from methane.

PROBLEM 4.22 Would you expect triiodomethane, CHI_3 (iodoform), to be a stronger or a weaker acid than trichloromethane, $CHCl_3$? Explain.

Other groups are also effective in increasing the acidity of carbon-hydrogen bonds. The nitro group greatly increases the acidity of a carboxylic acid (Table 4.3). The electron-withdrawing effect of the nitro group is also seen in nitromethane, CH_3NO_2, which, with pK_a 10.2, is a much stronger acid than methane or even trichloromethane. The conjugate base of nitromethane is stabilized by the negative inductive effect of the nitro group and also by resonance.

nitromethane and   resonance contributors for the carbanion from   water
hydroxide ion   nitromethane

For the carbanion from nitromethane, we can write a resonance contributor that delocalizes the negative charge from the carbon atom to one of the oxygen atoms in the nitro group. The delocalization of charge from the carbon atom to the much more electronegative oxygen atom decreases the basicity of the anion in comparison with carbanions such as the trichloromethyl anion where the charge must remain on a carbon atom. *Whenever resonance contributors that delocalize charge can be written for an ion, stabilization of the ion is postulated. Stabilization by resonance is always much more important than stabilization by the inductive effect.* Thus, chemists use resonance to explain the low basicity of the carboxylate anion and the carbanion from nitromethane. In subsequent chapters there are many other examples in which we can rationalize the stability of ions by writing resonance contributors that delocalize the charge on the ion.

PROBLEM 4.23 For each of the following compounds, identify the most acidic proton(s) by writing an equation, using the symbolism of electron pushing, for the reaction with hydroxide ion. What factors are important in the stabilization of each organic anion that you get?

(a) $HOCH_2\overset{\displaystyle O}{\overset{\displaystyle \|}{C}}OH$ (b) $CH_3CH_2CH_2NO_2$

(c) $HSCH_2CH_2CH_2OH$ (d) $Br\overset{\displaystyle O}{\overset{\displaystyle \|}{C}}H\overset{\displaystyle }{C}CH_3$
                                              $\underset{\displaystyle Br}{|}$

4.9

A Brief Review

We can classify many organic reactions as acid-base reactions. Two ways of defining acids and bases are useful. In the Brønsted-Lowry definition, an acid is a proton donor, and a base, a proton acceptor. A Brønsted-Lowry acid is also known as a protic acid. In the Lewis definition, an acid is an electron-pair acceptor, and a base, an electron-pair donor. In this more general definition, any species, including a proton, that can accept a pair of electrons is classified as a Lewis acid. Important Lewis acids include compounds such as boron trifluoride, BF_3, and aluminum chloride, $AlCl_3$. Such acids are known as aprotic acids.

Atoms bearing nonbonding electrons, such as oxygen, nitrogen, and sulfur, are the basic sites in molecules or ions. Protic acids transfer a proton to such atoms and Lewis acids coordinate with them. The electrons in π bonds also react with protic acids or Lewis acids.

The basicity of a species depends upon the position in the periodic table of the atom bearing the nonbonding electrons. Basicity decreases in atoms going from left to right across any period and going down any group. An anion containing a given atom is always more basic than a neutral molecule containing the same atom (for example, OH^- is more basic than H_2O).

The acidities of protic acids are usually expressed as their pK_a values. The loss of a proton converts a protic acid into its conjugate base. Similarly, a base, when it gains a proton, is converted into its conjugate acid. A strong acid has a weak conjugate base and a weak acid has a strong conjugate base. Bases are stabilized by delocalization of electron density by resonance or by negative inductive effects. These stabilized bases are weaker and have stronger conjugate acids than bases in which such effects are not as important.

ADDITIONAL PROBLEMS

4.24   Arrange the following compounds in order of relative acidity of the hydrogen atoms.

AlH_3, H_2S, HCl, NaH

4.25   Arrange the following compounds in order of increasing acidity.

(a) $Cl^- \, H_3\overset{+}{N}CH_2\overset{O}{\overset{\|}{C}}OH$, $ClCH_2\overset{O}{\overset{\|}{C}}OH$, $CH_3CH_2\overset{O}{\overset{\|}{C}}OH$

(b) $CF_3\overset{O}{\overset{\|}{C}}OH$, $CCl_3\overset{O}{\overset{\|}{C}}OH$, $CH_3\overset{O}{\overset{\|}{C}}OH$

(c) CH_3CH_2OH, FCH_2CH_2OH, $ClCH_2CH_2OH$

(d) CH_3CH_2SH, CH_3CH_2OH, $CH_3CH_2NH_2$

(e) $CH_3\overset{O}{\overset{\|}{C}}OH$, $CH_3\overset{+OH}{\overset{\|}{C}}OH$, $CH_3\overset{O}{\overset{\|}{C}}OCH_3$

(f) $\underset{\text{O}}{\overset{\text{O}}{\parallel}}$ CH_3COH, CH_3CNH_2, CH_3COCH_3

4.26 Arrange the following compounds in order of increasing basicity.

(a) $Cl_3C:^-$, $CH_3\overset{\overset{\displaystyle CH_3}{|}}{N}CH_3$, $CH_3\ddot{O}CH_3$ (b) $\ddot{N}F_3$, $\ddot{N}H_3$, $\ddot{N}H_2\ddot{O}H$

(c) $\ddot{N}H_3$, $:\ddot{N}H_2^-$, NH_4^+ (d) $CH_3\ddot{S}CH_3$, $CH_3\overset{\overset{\displaystyle CH_3}{|}}{\ddot{P}}CH_3$, $CH_3\underset{\underset{\displaystyle CH_3}{|}}{\overset{\overset{\displaystyle CH_3}{|}}{Si}}CH_3$

4.27 Some acid-base reactions are given below. For each reaction, label acids, bases, conjugate acids and conjugate bases, and say whether Brønsted-Lowry acids or Lewis acids are present.

(a) $F^- + BF_3 \longrightarrow BF_4^-$ (b) $Ag^+ + 2\,NH_3 \longrightarrow Ag(NH_3)_2^+$

(c) $Al(H_2O)_6^{3+} + OH^- \longrightarrow Al(OH)(H_2O)_5^{2+} + H_2O$

(d) $CH_3CH_2\overset{\text{O}}{\overset{\parallel}{C}}OH + OH^- \longrightarrow CH_3CH_2\overset{\text{O}}{\overset{\parallel}{C}}O^- + H_2O$

(e) $CH_3\overset{\overset{\displaystyle CH_2CH_3}{|}}{C}H_2NCH_2CH_3 + CF_3\overset{\text{O}}{\overset{\parallel}{C}}OH \longrightarrow CH_3CH_2\overset{\overset{\displaystyle CH_2CH_3}{|}}{\underset{+}{N}}HCH_2CH_3 + CF_3\overset{\text{O}}{\overset{\parallel}{C}}O^-$

(f) $CH_3CH_2SCH_2CH_3 + BF_3 \longrightarrow CH_3CH_2\overset{\overset{\displaystyle ^-BF_3}{|}}{\underset{+}{S}}CH_2CH_3$

(g) $ClCH_2\overset{\text{O}}{\overset{\parallel}{C}}OH + HCO_3^- \longrightarrow ClCH_2\overset{\text{O}}{\overset{\parallel}{C}}O^- + H_2CO_3$

(h) $H_2PO_4^- + OH^- \longrightarrow HPO_4^{2-} + H_2O$

(i) $CH_3CH_2CH_2SH + CH_3CH_2O^- \longrightarrow CH_3CH_2CH_2S^- + CH_3CH_2OH$

4.28 Use the table of pK_a values to decide whether the following reactions will take place or not. Complete the equations for the reactions that will take place. Write ''no reaction'' for the others.

(a) 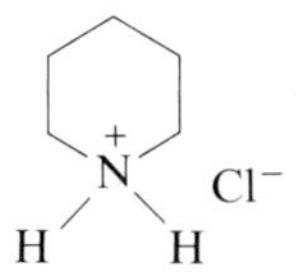 [piperidinium chloride structure: ring with $\overset{+}{N}$, H, H, Cl^-] $+ NaOH \longrightarrow$ (b) $CH_3CH_2\overset{\overset{\displaystyle H}{|}}{\underset{+}{O}}CH_2CH_3 + H_2O \longrightarrow$

(c) $CH_3\overset{\text{O}}{\overset{\parallel}{C}}OH + CH_3CH_2SNa \longrightarrow$ (d) $CCl_3\overset{\text{O}}{\overset{\parallel}{C}}OH + NaHCO_3 \longrightarrow$

(e) $CH_3CH_2\overset{\text{O}}{\overset{\parallel}{C}}OH + NH_3 \longrightarrow$ (f) $CH_3NO_2 + CH_3CH_2ONa \longrightarrow$

(g) $CH_3CH_2SH + NaNH_2 \longrightarrow$ (h) $HC\equiv CH + NaHCO_3 \longrightarrow$

4.29   For the following reactions, most of which you have not yet met, complete the Lewis structures for the pertinent parts of molecules and draw arrows to show the electronic changes taking place.

Example:

For $N\equiv C^- + H-\underset{\underset{H}{|}}{\overset{\overset{CH_3}{|}}{C}}-Cl \longrightarrow N\equiv C-\underset{\underset{H}{|}}{\overset{\overset{CH_3}{|}}{C}}-H + Cl^-$

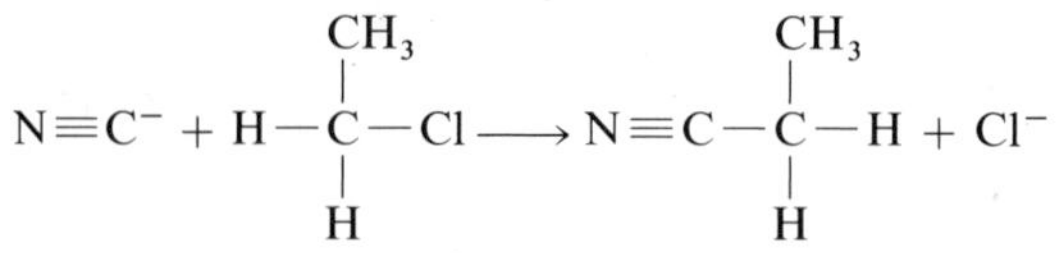

Write $:N\equiv C:^- \ H-\underset{\underset{H}{|}}{\overset{\overset{CH_3}{|}}{C}}\overset{\frown}{\ddot{C}l}: \longrightarrow :N\equiv C-\underset{\underset{H}{|}}{\overset{\overset{CH_3}{|}}{C}}-H + :\ddot{C}l:^-$

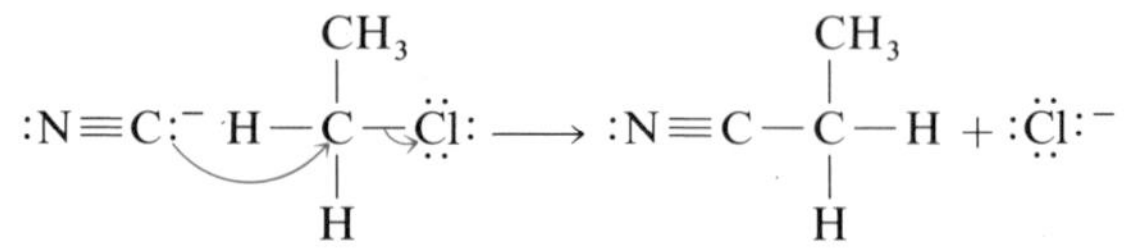

(a) $I^- + CH_3Cl \longrightarrow ICH_3 + Cl^-$

(b) 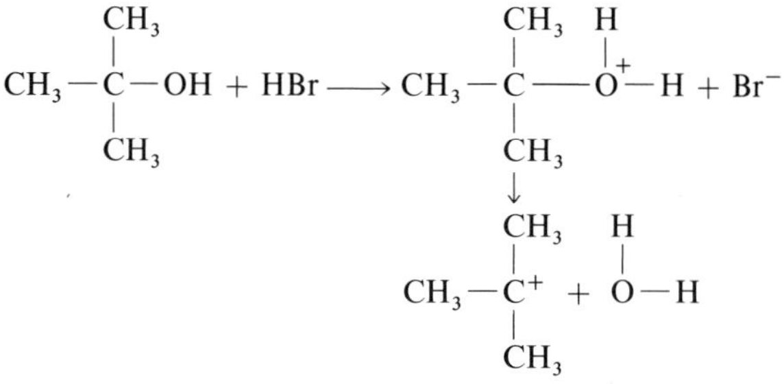 $CH_3-\underset{\underset{CH_3}{|}}{\overset{\overset{CH_3}{|}}{C}}-OH + HBr \longrightarrow CH_3-\underset{\underset{CH_3}{|}}{\overset{\overset{CH_3}{|}}{C}}-\overset{\overset{H}{|}}{\overset{+}{O}}-H + Br^-$

$\downarrow$

$CH_3-\underset{\underset{CH_3}{|}}{\overset{\overset{CH_3}{|}}{C^+}} + \overset{\overset{H}{|}}{O}-H$

(c) $\underset{\underset{H}{|}}{\overset{\overset{O}{\|}}{CH_2}}-C-CH_3 + OH^- \longrightarrow \left[\overset{\overset{O}{\|}}{\bar{C}H_2}-C-CH_3 \longleftrightarrow CH_2=\overset{\overset{O^-}{|}}{C}-CH_3 \right] + H_2O$

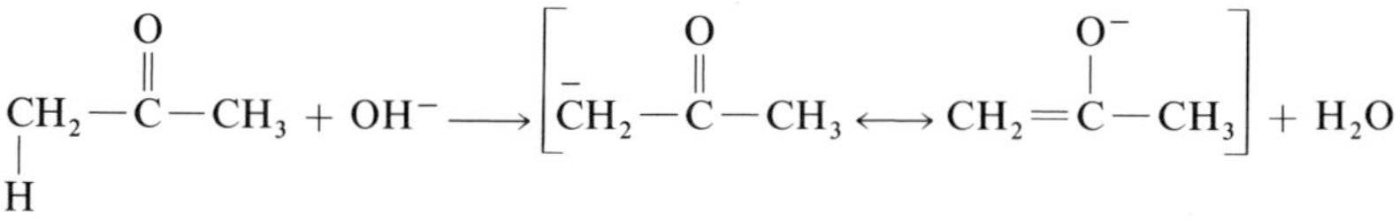

(d) 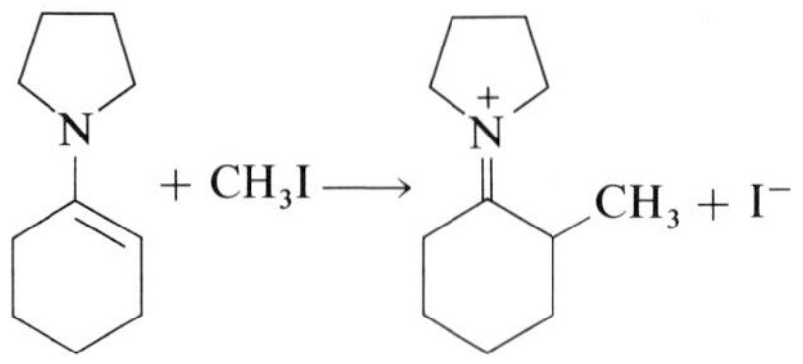   [pyrrolidine-cyclohexene structure] $+ CH_3I \longrightarrow$ [N-methylated iminium structure] $CH_3 + I^-$

(e) 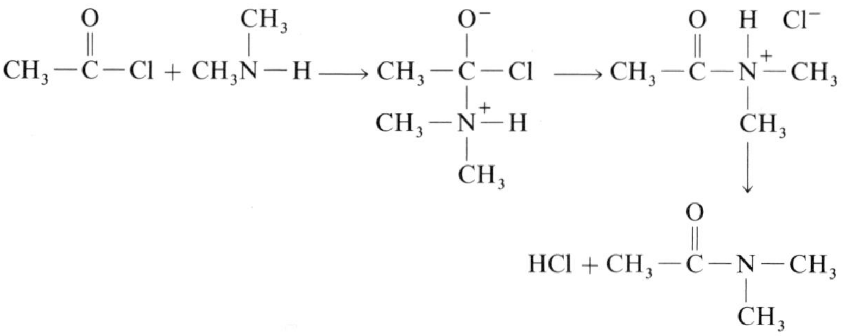 $\underset{}{\overset{\overset{O}{\|}}{CH_3}}-C-Cl + \overset{\overset{CH_3}{|}}{CH_3}-N-H \longrightarrow CH_3-\underset{\underset{CH_3-\overset{+}{N}-H}{}}{\overset{\overset{O^-}{|}}{C}}-Cl \longrightarrow CH_3-\overset{\overset{O}{\|}}{C}-\overset{\overset{H}{|}}{\underset{\underset{CH_3}{|}}{\overset{+}{N}}}-CH_3$

$CH_3-\overset{+}{N}-H$
$\underset{CH_3}{|}$

$\downarrow$

$HCl + CH_3-\overset{\overset{O}{\|}}{C}-\underset{\underset{CH_3}{|}}{N}-CH_3$

(f) 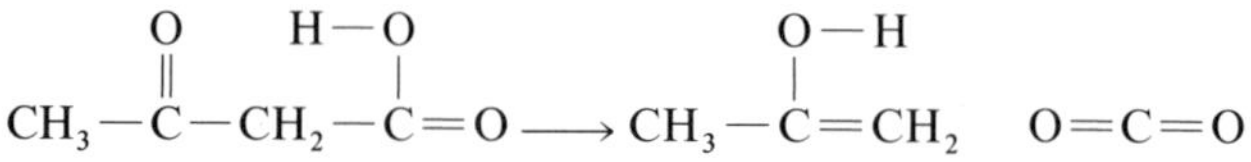 $CH_3-\overset{\overset{O}{\|}}{C}-CH_2-\overset{\overset{H-O}{|}}{C}=O \longrightarrow CH_3-\overset{\overset{O-H}{|}}{C}=CH_2 \quad O=C=O$

(g) 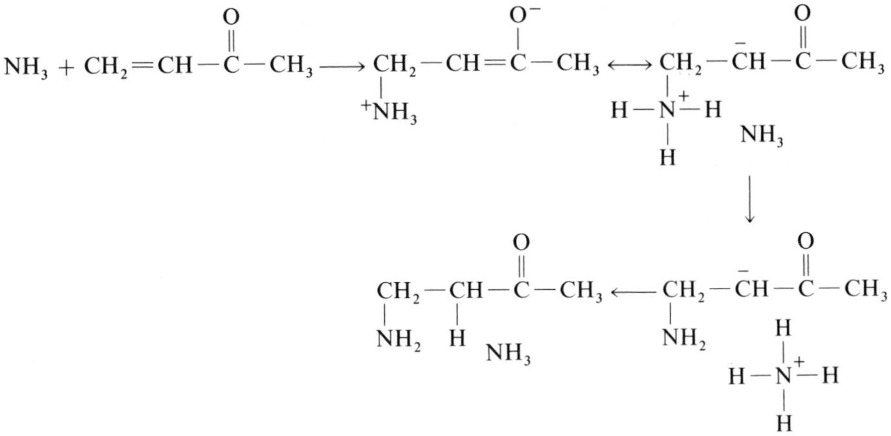 $NH_3 + CH_2{=}CH{-}\overset{\displaystyle O}{\overset{\|}{C}}{-}CH_3 \longrightarrow$ $\underset{\underset{+NH_3}{|}}{CH_2}{-}CH{=}\overset{\displaystyle O^-}{\overset{|}{C}}{-}CH_3 \longleftrightarrow CH_2{-}\overset{-}{C}H{-}\overset{\displaystyle O}{\overset{\|}{C}}{-}CH_3$

$$H{-}\overset{H}{\overset{|}{\underset{\underset{H}{|}}{\overset{+}{N}}}}{-}H \qquad NH_3$$

$$\downarrow$$

$$\underset{\underset{NH_2}{|}}{CH_2}{-}\underset{\underset{H}{|}}{CH}{-}\overset{\displaystyle O}{\overset{\|}{C}}{-}CH_3 \underset{NH_3}{\longleftarrow} CH_2{-}\underset{\underset{NH_2}{|}}{\overset{-}{C}H}{-}\overset{\displaystyle O}{\overset{\|}{C}}{-}CH_3$$

$$H{-}\overset{H}{\overset{|}{\underset{\underset{H}{|}}{\overset{+}{N}}}}{-}H$$

(h) $CH_3{-}\overset{\displaystyle O}{\overset{\|}{C}}{-}CH_3 + H{-}\overset{H}{\overset{+|}{O}}{\underset{\underset{H}{|}}{}} \longrightarrow CH_3{-}\overset{\displaystyle {}^+O{-}H}{\overset{\|}{C}}{-}CH_3 \longleftrightarrow CH_3{-}\overset{\displaystyle O{-}H}{\overset{|}{\underset{\underset{H}{|}}{C}}}{-}CH_2 \cdots$

$$\downarrow$$

$$\underset{CH_3{-}C{=}CH_2}{\overset{\displaystyle O{-}H}{\overset{|}{}}} \qquad H{\underset{\underset{H}{|}}{\overset{+}{O}}}H$$

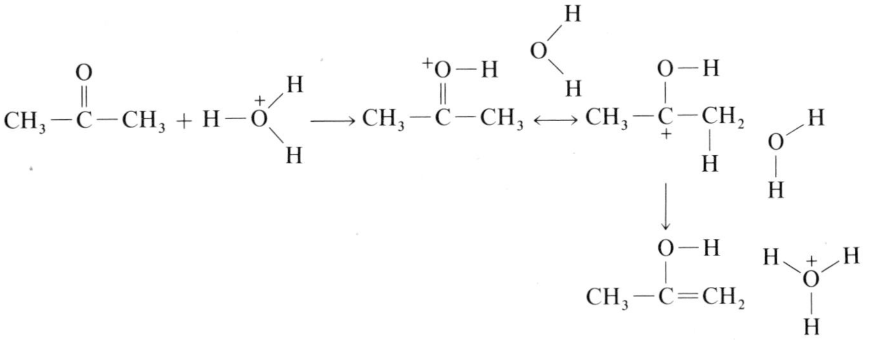

4.30 Complete the equations shown below.

(a) 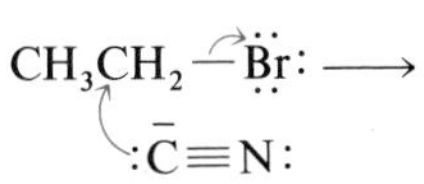 $CH_3CH_2{\overset{\frown}{\ddot{B}r}}: \longrightarrow$
$:\overset{-}{C}{\equiv}N:$

(b)
$$\overset{:\overset{\displaystyle O:}{\|}}{\underset{CH_3 \quad CH_3}{C}} \quad H{-}\overset{H}{\overset{\frown}{\ddot{O}}}{}^+ \atop H \longrightarrow$$
$$\underset{H \quad CH_3}{\ddot{O}}$$

(c) $CH_2{=}CH_2 \longrightarrow$
$H{-}\ddot{O}{-}SO_3H$

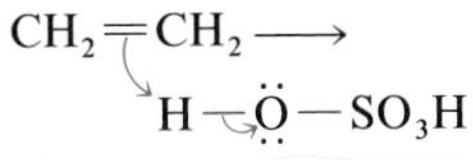

(d) 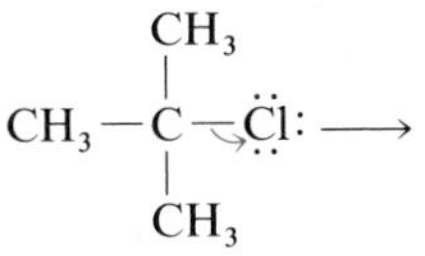 $CH_3{-}\underset{\underset{CH_3}{|}}{\overset{\overset{CH_3}{|}}{C}}{\overset{\frown}{\ddot{C}l}}: \longrightarrow$

(e) 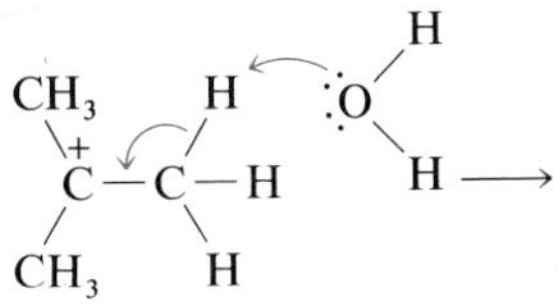 $\underset{\underset{CH_3}{|}}{\overset{\overset{CH_3}{|}}{\overset{+}{C}}}{-}\underset{\underset{H}{|}}{C}{-}H \quad :\overset{\overset{H}{\diagdown}}{\ddot{O}} \atop H \longrightarrow$

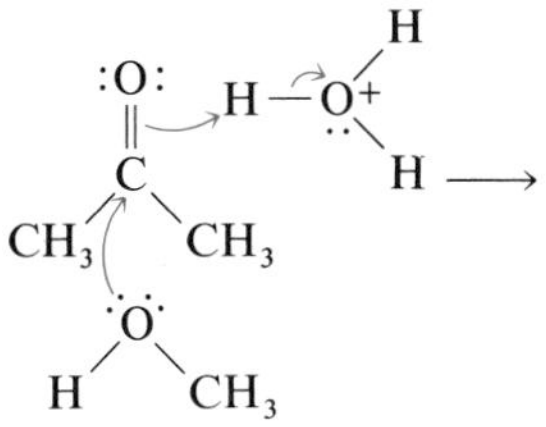

(f) 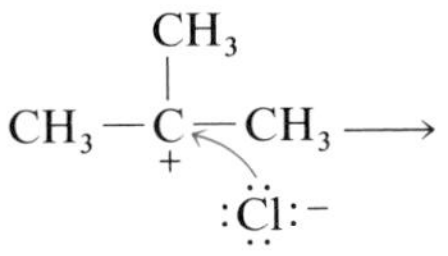 $CH_3{-}\underset{\underset{:\ddot{C}l:^-}{}}{\overset{\overset{CH_3}{|}}{\overset{+}{C}}}{\overset{\frown}{{}}}CH_3 \longrightarrow$

(g) 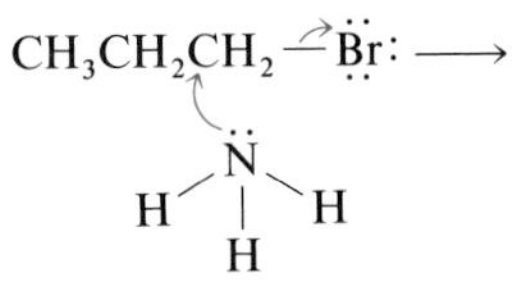 $CH_3CH_2CH_2{\overset{\frown}{\ddot{B}r}}: \longrightarrow$
$\underset{H \quad {\overset{|}{\underset{H}{}}} \quad H}{\overset{\ddot{N}}{}}$

4.31   For the following compounds for which pK_a values are given, calculate the acidity constants.

(a) $\overset{\displaystyle O}{\overset{\displaystyle \|}{CHCl_2C}}OH$, pK_a 1.3   (b) $CH_3NH_3{}^+$, pK_a 10.4   (c) CCl_3CH_2OH, pK_a 12.2

4.32   The following compounds have the acidity constants shown. What are their pK_a values?

(a) CH_3CH_2OH, $K_a = 10^{-17}$ (b) CH_3CH_2SH, $K_a = 3.16 \times 10^{-11}$

(c) $CH_3CH_2 - \overset{\displaystyle H}{\underset{\displaystyle +}{\overset{\displaystyle |}{O}}} - CH_2CH_3$, $K_a = 3.98 \times 10^3$

4.33   Proteins are built up from units called amino acids. The simplest amino acid is glycine, the structure of which can be written in two ways:

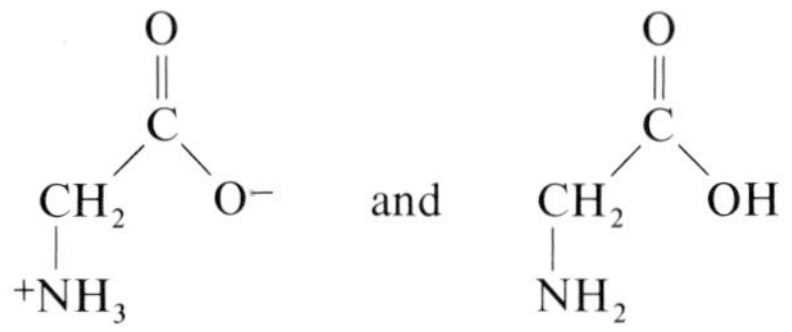

(a)   Consult the table of pK_a values and decide which way is the more nearly correct representation of the compound. (Hint: Which is more basic, an amino group or a carboxylate anion?)
(b)   Glycine is a crystalline solid that decomposes at 233 °C as it melts. It has a solubility of 26 g in 100 mL of water. Do these facts fit the structure that you have chosen for glycine? Explain.

4.34   Each of the following compounds has two sites for protonation. For each compound, write a reaction with hydronium ion showing the ionic species that will be formed. (Hint: You may wish to check the table of pK_a values to decide on the relative basicities of some of the sites.)

(a)  $CH_3CH_2OCH_2CH_2N\overset{\displaystyle CH_3}{\underset{\displaystyle CH_3}{<}}$ (b) $\overset{\displaystyle O}{\overset{\displaystyle \|}{CH_3\,C}}{\diagdown}CH_2CH_2CH_2CH_2CH_2OH$

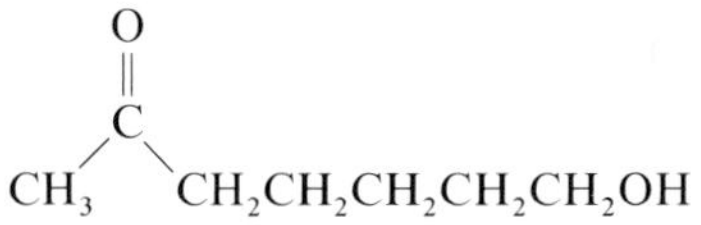

(c) $CH_3C{\equiv}CCH_2CH_2OCH_3$

4.35   The expression for the acidity constant for any acid, HA, is

$$K_a = \frac{[A^-][H_3O^+]}{[HA]}$$

The expression for K_a can be converted into another one showing the relationship between the pK_a of an acid and the pH of the solution:

$$pH = pK_a + \log\frac{[A^-]}{[HA]}$$

This expression, sometimes called the Henderson-Hasselbach equation, is useful in calculating the relative amounts of dissociated and undissociated acid present at any pH when the pK_a of an acid is known.

(a)   Derive the Henderson-Hasselbach equation from the expression for K_a.
(b)   What does the Henderson-Hasselbach expression tell you about the concentrations of HA and A^- when $pK_a = pH$?

(c) In Problem 4.14 you learned that the imidazole ring in the amino acid histidine is important in proton transfer reactions in the human body. The protonated form of imidazole has pK_a 7.0. The pH of body fluids is ~6.5. What can you say in a qualitative way about the relative quantities of protonated and unprotonated imidazole rings that are present in the body?

4.36 A research paper in the *Journal of the American Chemical Society* describes the preparation of trifluoromethanol, CF_3OH, and trifluoromethylamine, CF_3NH_2, two compounds that had not been prepared until recently because of their tendency to lose HF to give multiple bonds. The compounds were found to be reasonably stable at temperatures around 0 °C.

The authors of the paper report that "CF_3OH is certainly not a typical alcohol but rather an acid." They also say "CF_3OH is of surprisingly high volatility" (meaning that it has a much lower boiling point than expected). For CF_3NH_2, they report that "the basic character of CF_3NH_2 is lower than the basicity of normal organic bases like $(CH_3)_3N$."

(a) How would you rationalize each of the three observations quoted?
(b) On the basis of the information given above, complete the following equation by writing "no reaction" or writing in the products, if any would be formed.

$$CF_3NH_3^+ \ Cl^- + (CH_3)_3N \longrightarrow$$

4.37 When strong acids or strong bases are dissolved in a solvent such as water, a phenomenon known as the *leveling effect* is observed. Differences in acidity for strong acids such as sulfuric acid and hydrochloric acid cannot be measured in water. They appear to have the same acidity in that solvent. What is happening? What is the strongest acid that can exist in aqueous solution? What is the strongest base? Write equations illustrating your answers.

Electrophilic Additions to Alkenes and Alkynes

5

Structure and Isomerism in Alkenes and Alkynes. Nomenclature

A. Isomerism in Alkenes and Alkynes

Alkenes *are hydrocarbons that contain carbon-carbon double bonds.* The experimentally determined geometry of the double bond was described in Section 1.6. The carbon atoms of the double bond and the four other atoms attached to them lie in a plane with bond angles of approximately 120° around the two carbon atoms. Such trigonal carbon atoms are described as being sp^2 hybridized (Section 2.5A). The double bond in an alkene consists of a σ bond from the overlap of sp^2 hybrid orbitals on two carbon atoms and a π bond from the side-to-side overlap of the remaining p orbital on each carbon (Figure 2.20, Section 2.5A).

Alkenes, like alkanes, exist as a homologous series with the molecular formula C_nH_{2n}. The lower members of the series, ethylene (C_2H_4), propene (C_3H_6), and butenes (C_4H_8), are gases at room temperature. Higher alkenes to $C_{20}H_{40}$ are liquids. As with alkanes, the boiling points and melting points of alkenes increase with increasing molecular weight but show variations depending upon the shape of the molecule. Alkenes having the same molecular formula have isomerism because of the position and the stereochemistry of the double bond in the molecule. The origin of stereoisomerism in alkenes is examined next.

The bonding between the two carbon atoms of the double bond has been described (Section 2.5A) as consisting of a σ bond, which constitutes the backbone of the molecule, and the π bond, created by overlap of the p orbitals. *The necessity for the overlap of the p orbitals in the π bond imparts a certain rigidity to the double bond.* Rotation of the two halves of the double bond with respect to each other does not take place unless enough energy is supplied to break the π bond. A carbon-carbon single bond consisting of a σ bond with cylindrical symmetry (Section 2.3B), however, permits free rotation (Section 3.3) of the bonded atoms with respect to each other (Figure 5.1).

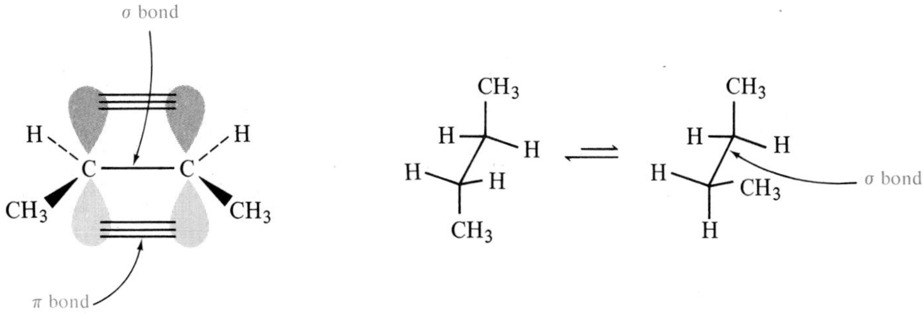

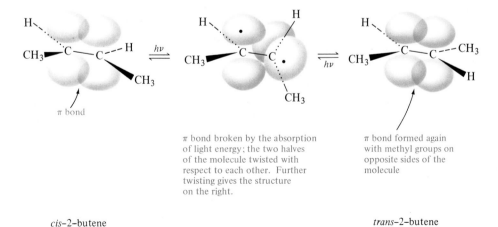

no free rotation around the carbon–carbon double bond in 2–butene

two conformers of butane showing rotation around the σ bond

Figure 5.1   The double bond in 2-butene compared with the single bond in butane.

The rigidity of the double bond gives rise to the possibility of stereoisomerism (Section 3.9A) *around the double bond.* For example, the structure of 2-butene was shown with two methyl groups on the same side of the molecule, or cis to each other. Another isomer of 2-butene in which the methyl groups are on opposite sides of the double bond, or trans to each other, also exists. Each isomer is converted to the other when enough energy is supplied, as is the case when they absorb ultraviolet radiation (or are heated to temperatures around 300 °C). The conversion is believed to take place because the π bond breaks when energy is absorbed, and the two halves of the molecule can then rotate with respect to each other before the π bond forms again (Figure 5.2).

π bond

π bond broken by the absorption of light energy; the two halves of the molecule twisted with respect to each other. Further twisting gives the structure on the right.

π bond formed again with methyl groups on opposite sides of the molecule

cis–2–butene

trans–2–butene

Figure 5.2   The transformation of *cis*-2-butene into *trans*-2-butene.

The interconversion of double bond stereoisomers requires much more energy (~65 kcal/mol) than the interconversion of conformational isomers by rotations around single bonds (<10 kcal/mol, Sections 3.3B, 3.5). Conformational isomers cannot be separated and isolated as individual molecular species at ordinary temperatures (where ~20 kcal/mol of kinetic energy is available to molecules), but double bond stereoisomers can. *The stereoisomers created by different spatial arrangements of groups around double bonds are configurational isomers* (Section 3.9F). *They are stereoisomers, but not mirror-image isomers or enantiomers. All stereoisomers that are not enantiomers are called* **diastereomers. Geometric isomerism** *is the term given also to the diastereomerism resulting from the presence of double bonds or rings that impose rigidity on the molecule preventing free rotation around certain bonds* (Section 3.8B).

Even with relatively simple structures, several kinds of isomerism can be seen with alkenes. For example, there are four different noncyclic structures that can be drawn for C_4H_8.

1-butene
bp −5 °C

identical substituents at one end of double bond; no stereoisomerism possible at a double bond

2-methylpropene
bp −6 °C

also identical in this case

different substituents at each end of the double bond; stereoisomerism possible

trans-2-butene
bp 2.5 °C

cis-2-butene
bp 1 °C

isomeric structures for alkenes with the molecular formula C_4H_8

1-Butene and 2-methylpropene are structural isomers of the 2-butenes. They have quite different arrangements of the carbon atoms and double bonds. They do not have stereoisomers because in each case at least one end of the double bond has identical substituents. *cis*-2-Butene and *trans*-2-butene are stereoisomers of each other. They differ from each other only in the spatial arrangement of the methyl groups and the hydrogen atoms. This similarity is reflected in their names, which are identical except for the designation for spatial orientation. *cis*-2-Butene and *trans*-2-butene have different physical properties, reflected by their different boiling points. *Diastereomers, stereoisomers that are not mirror-image isomers, have different physical properties such as melting points, boiling points, dipole moments, and solubilities.* This contrasts with enantiomers, which *are* mirror-image isomers (Section 3.9A), and have identical physical properties except for the direction they rotate plane-polarized light.

Alkynes are characterized by a carbon-carbon triple bond and linear geometry (Sections 1.6, 2.6A). The only isomerism possible for an alkyne is the position of the triple bond in the carbon chain.

$$CH_3CH_2CH_2C\equiv CH \qquad CH_3CH_2C\equiv CCH_3$$

1-pentyne
bp 40 °C

2-pentyne
bp 56 °C

a terminal alkyne

an internal alkyne

In a **terminal alkyne,** *the triple bond is at the end of the carbon chain. A terminal alkyne has a hydrogen atom bonded to an sp carbon atom,* a structural feature that is important in its chemistry (Section 9.4A). *In an **internal alkyne**, the triple bond is at least one carbon atom away from the end of the carbon chain.* There is, of course, no hydrogen atom bonded to the carbon atoms of the triple bond in an internal alkyne.

B. Nomenclature of Alkenes and Alkynes

The systematic name of an alkene or an alkyne is derived from the name of the alkane corresponding to the longest continuous chain of carbon atoms that contains the double or triple bond. The **-ane** ending of the name of the alkane is changed to **-ene** for an alkene and **-yne** for an alkyne. Thus, the IUPAC name for ethylene is ethene, and that for acetylene is ethyne.

$CH_2{=}CH_2$ $HC{\equiv}CH$

ethene   ethyne
ethylene   acetylene

An *alkene* with *five* carbon atoms in a continuous chain, for example, is named as a *pentene*. An alkyne with *six* carbon atoms in the longest continuous chain is a *hexyne*.

$\overset{5}{C}H_3\overset{4}{C}H_2\overset{3}{C}H_2\overset{2}{C}H{=}\overset{1}{C}H_2$ $\overset{6}{C}H_3\overset{5}{C}H_2\overset{4}{C}{\equiv}\overset{3}{C}\overset{2}{C}H_2\overset{1}{C}H_3$

1-pentene   3-hexyne

The chain is numbered so as to give the carbon atoms of the multiple bond the lowest possible number. The position of the multiple bond is indicated by the number of the *first* of the carbon atoms of the bond. Thus, the examples shown above are 1-pentene, meaning that the double bond is between carbon atoms 1 and 2 of a five-carbon chain, and 3-hexyne, with the triple bond between carbon atoms 3 and 4 of a six-carbon chain. If other substitutents are present, they too are named and their positions indicated by numbers.

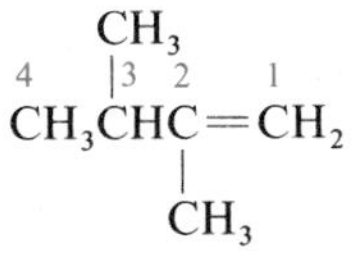

CH_3
$|$
$\overset{4}{C}H_3\overset{3}{C}H\overset{2}{C}{=}\overset{1}{C}H_2$
$|$
CH_3

2,3-dimethyl-1-butene

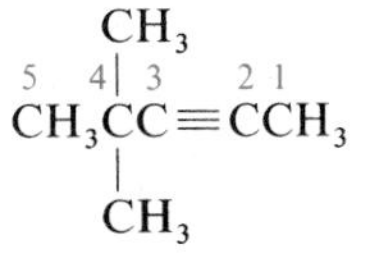

CH_3
$|$
$\overset{5}{C}H_3\overset{4}{C}\overset{3}{C}{\equiv}\overset{2}{C}\overset{1}{C}H_3$
$|$
CH_3

4,4-dimethyl-2-pentyne

CH_3 CH_3
$\diagdown$ $\diagup$
$C{=}C$
$\diagup\ \ \ \ \diagdown$
$ClCH_2CH_2CH_2$ CH_3
6   5   4   1

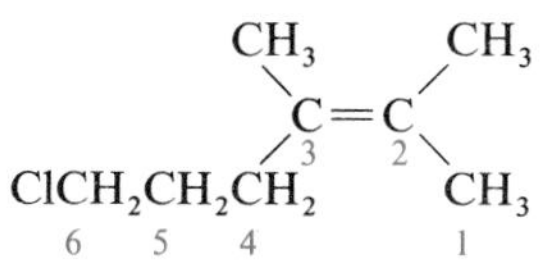

6-chloro-2,3-dimethyl-2-hexene

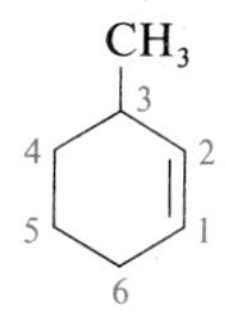

3-methylcyclohexene

*the first carbon of the
double bond is understood
to be carbon 1 in
the cyclic case*

Two unsaturated groups have common names that are used as substituents or to indicate structural features in molecules. *The **vinyl group** is formed when a hydrogen atom is removed from ethylene.*

vinylic hydrogen atoms,
attached to *sp²* carbon atom

ethylene

the vinyl group

$\equiv CH_2=CH-$

cyclohexylethene
vinylcyclohexane

chloroethene
vinyl chloride
a vinylic halide

Hydrogen atoms attached to the sp²-hybridized carbon atoms of a double bond are described as **vinylic hydrogen atoms.** Similarly, a halide with a halogen atom bonded directly to an *sp²*-hybridized carbon atom of a double bond is known as **vinylic halide.**

Removing a hydrogen atom from the sp³-hybridized carbon atom of propene gives rise to an **allyl group.**

vinylic hydrogen
atoms

allylic
hydrogen
atoms

the allylic
position

propene

$\equiv CH_2=CHCH_2-$

an allyl group

$CH_2=CHCH_2Cl$

3-chloropropene
allyl chloride
an allylic halide

3-phenylpropene
allylbenzene

In general, *sp³-hybridized carbon atoms next to a double bond are known as the* **allylic positions** in the molecule. *The hydrogen atoms on such a carbon atom are* **allylic hydrogen atoms.** Substitution of a halogen atom for one of the allylic hydrogen atoms gives rise to an **allylic halide.** Because hydrogen atoms (or halogen atoms) in a vinylic position are much less reactive than hydrogen atoms (or halogen atoms) in an allylic position (Sections 6.8, 11.2), we need to make the kind of structural distinctions described above.

When an alkene has stereoisomers, the complete name of the compound should include a definition of the stereochemistry, if it is known. In simple alkenes such as the 2-butenes, the designations cis and trans are used with no ambiguity. *In the* **cis** *isomer, similar groups are on the same side of the double bond; in the* **trans** *isomer, they are on opposite sides of the double bond.* In some simple cases such nomenclature is used with no confusion.

trans-1,2-dichloroethene

cis-1,2-dichloroethene

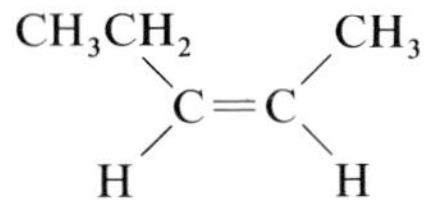

cis-2-pentene 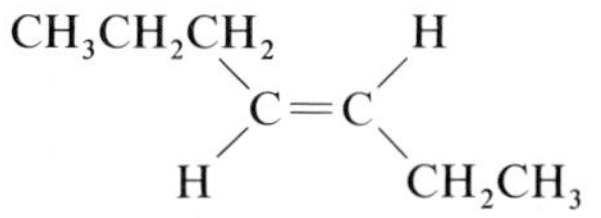   *trans*-3-heptene

The stereochemistry of more highly substituted alkenes is harder to define as cis or trans. In order to deal with these complications, chemists have developed a systematic way of indicating stereochemistry based on a system of assigning priorities to the substituents on the double bond. Priorities are assigned to the groups that are attached to the double bond in the order already described for the assignment of configuration at asymmetric carbon atoms (Section 3.9F). The rules are resummarized below.

1. The priority of a group depends upon the atomic number of the atom attached to the double bond. The higher the atomic number, the higher the priority of the group is.

2. If two atoms with the same priority are bound to the carbon atom of the double bond, a search is made along each group, moving away from the double bond until a difference is found, and priority is assigned on the basis of that difference.

3. Groups with multiple bonds are assigned priority as if each atom of a multiple bond were attached twice for a double bond (or three times for a triple bond) by a single bond to the other one.

4. The double bond is assigned a Z (for *zusammen*, German for *together*) configuration when the two groups with the higher priority at each end of the double bond are on the same side of the molecule. If the two groups of higher priority are on the opposite sides of the double bond, the molecule is given an E (for *entgegen*, German for *opposite*) configuration.

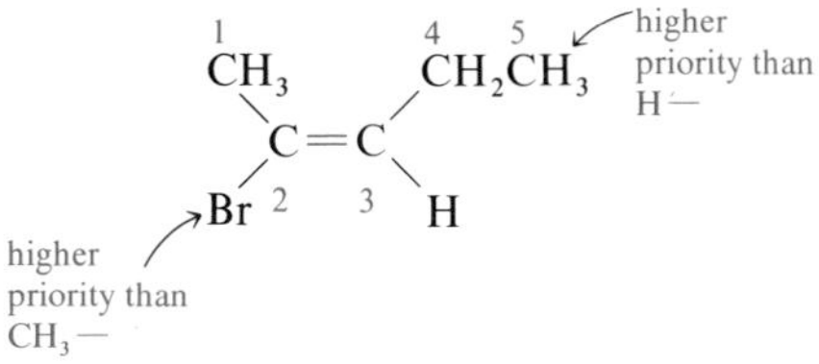

*groups of higher priority are on opposite
sides of the double bond; therefore this is
(E)-2-bromo-2-pentene*

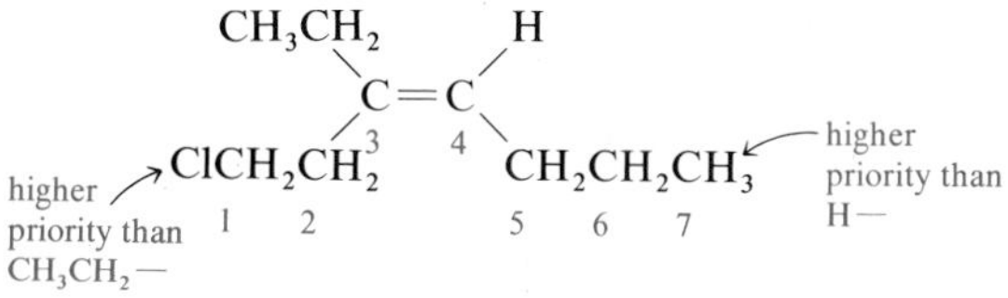

*groups of higher priority are on the same
side of the double bond; therefore this is
(Z)-1-chloro-3-ethyl-3-heptene*

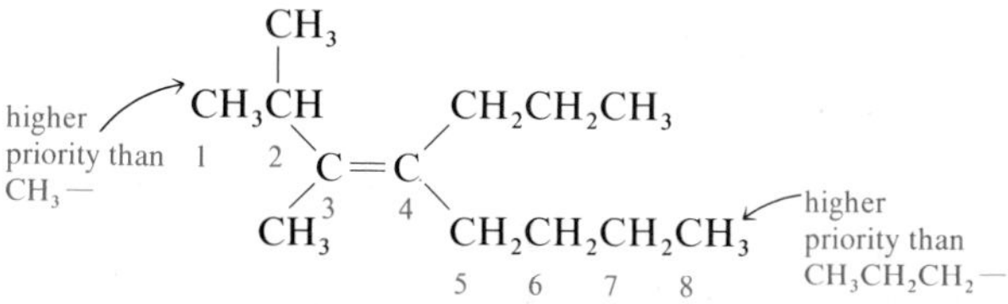

*groups of higher priority are on opposite
sides of the double bond; therefore this
is (E)-2,3-dimethyl-4-propyl-3-octene*

To avoid any possible confusion, this system of assigning configuration to stereo-isomers of alkenes is used in this book. The terms cis and trans are reserved for descriptions of the spatial relationship between groups. For example, in (*E*)-2-bromo-2-pentene shown above, the methyl and the ethyl group on the carbon atoms of the double bond are cis to each other. The molecule, however, has the *E* configuration.

PROBLEM 5.1   Name the following compounds, assigning the correct configuration to any alkenes for which stereochemistry is shown.

(a) 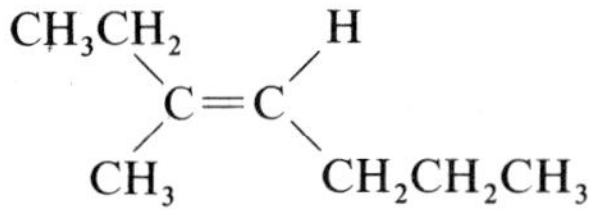
$$CH_3CH_2 \quad H$$
$$C=C$$
$$CH_3 \quad CH_2CH_2CH_3$$

(b)
$$CH_3CH_2 \quad CH_3$$
$$C=C$$
$$CH_3CHCH_2 \quad CH_3$$
$$\quad\quad CH_3$$

(c)
$$CH_3CH_2CH_2 \quad Cl$$
$$C=C$$
$$Cl \quad CH_2CH_3$$

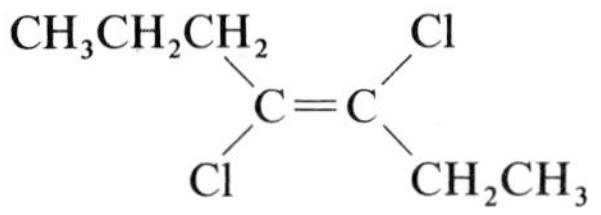

(d) 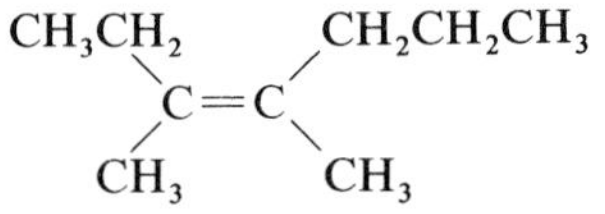
$$CH_3CH_2 \quad CH_2CH_2CH_3$$
$$C=C$$
$$CH_3 \quad CH_3$$

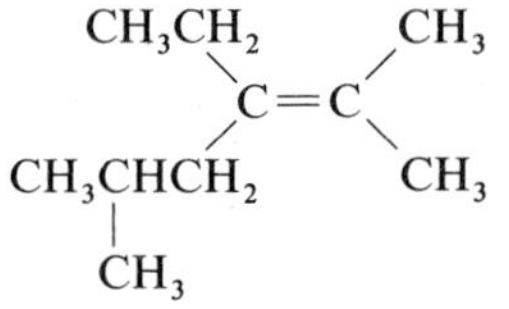

(e) 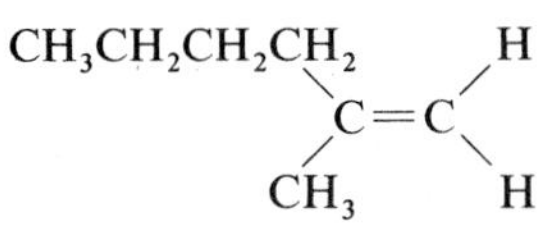
$$CH_3CH_2CH_2CH_2 \quad H$$
$$C=C$$
$$CH_3 \quad H$$

(f) 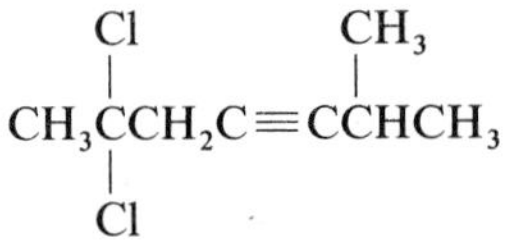
$$\quad Cl \quad\quad CH_3$$
$$CH_3CCH_2C\equiv CCHCH_3$$
$$\quad Cl$$

(g) 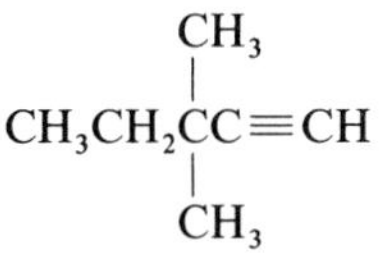
$$\quad\quad CH_3$$
$$CH_3CH_2CC\equiv CH$$
$$\quad\quad CH_3$$

(h)
$$\quad\quad Cl$$
$$CH_3CH \quad CH_2CH_3$$
$$C=C$$
$$CH_3CH_2 \quad H$$

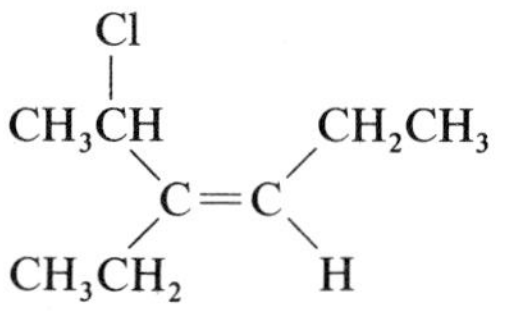

PROBLEM 5.2   Draw structural formulas for the following compounds, showing correct stereochemistry when it is designated.

(a)   1-phenylcyclohexene   (b)   3,3-dimethylcyclopentene
(c)   (*Z*)-2-phenyl-2-butene   (d)   2-chloro-2-methyl-3-heptyne
(e)   (*E*)-1,3-dichloro-2-methyl-2-pentene

5.2

The Relative Stabilities of Alkenes

In Section 2.10, we used heats of hydrogenation of cyclohexene, 1,3-cyclohexadiene, and the hypothetical 1,3,5-cyclohexatriene to reach conclusions about the stability of benzene. In the same way, experimental heats of hydrogenation of alkenes indicate whether the position of a double bond and its stereochemistry affect the stability of the molecule.

1-Butene and (*Z*)- and (*E*)-2-butene each add one mole of hydrogen to become butane.

$$CH_3CH_2CH{=}CH_2 \; + \; H_2 \; \xrightarrow[\text{catalyst}]{} \; CH_3CH_2CH_2CH_3 \qquad \Delta H^\circ = -30.3 \,\text{kcal/mol}$$

1-butene   butane

$$\underset{\text{(Z)-2-butene}}{\overset{\displaystyle CH_3 \quad\;\; CH_3}{\underset{\displaystyle H \qquad\;\; H}{C{=}C}}} \; + \; H_2 \; \xrightarrow[\text{catalyst}]{} \; CH_3CH_2CH_2CH_3 \qquad \Delta H^\circ = -28.6 \,\text{kcal/mol}$$

butane

$$\underset{\text{(E)-2-butene}}{\overset{\displaystyle CH_3 \qquad\;\; H}{\underset{\displaystyle H \qquad\;\; CH_3}{C{=}C}}} \; + \; H_2 \; \xrightarrow[\text{catalyst}]{} \; CH_3CH_2CH_2CH_3 \qquad \Delta H^\circ = -27.6 \,\text{kcal/mol}$$

butane

Each reaction has an identical product, butane, and one identical reagent, hydrogen. The differences in the heats of hydrogenation, the energy evolved as the alkene is converted to the alkane, must reflect the differences in energy in the alkenes. These differences are shown schematically in Figure 5.3. Less heat is evolved when (*E*)-2-butene is hydrogenated than when (*Z*)-2-butene undergoes the same reaction. Thus, the trans compound must be closer to butane in energy before the reaction. It is at a lower energy level than (*Z*)-2-butene or, in other words, it is more stable than (*Z*)-2-butene. The same argument demonstrates that 1-butene is less stable than either

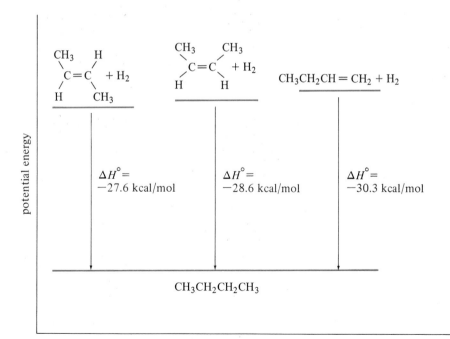

Figure 5.3 Differences in energy levels of butane, compared with hydrogen plus 1-butene, (*Z*)-, and (*E*)-2-butene, as reflected by heats of hydrogenation.

of the 2-butenes. More heat is evolved as 1-butene is transformed into butane than in the other two cases. 1-Butene starts at a higher energy level than either of the 2-butenes, and has farther to drop to reach the energy level of butane.

In other words, *a* **terminal alkene,** *in which the double bond is at the end of a chain, is less stable than an* **internal alkene** *with the double bond in the middle of the chain. An alkene, in which the larger substituents on the double bond are farther apart from each other, is more stable than the corresponding alkene in which the larger substituents are closer together.*

An important factor seems to be the number of substituents on the double bond. The experimental heats of hydrogenation for three more alkenes, 3-methyl-1-butene, 2-methyl-1-butene, and 2-methyl-2-butene, all of which are hydrogenated to the same alkane, illustrate this.

$$\underset{\text{3-methyl-1-butene}}{CH_3\overset{\overset{\displaystyle CH_3}{|}}{C}HCH=CH_2} + H_2 \xrightarrow[\text{catalyst}]{} \underset{\text{2-methylbutane}}{CH_3\overset{\overset{\displaystyle CH_3}{|}}{C}HCH_2CH_3} \qquad \Delta H^\circ = -30.3 \text{ kcal/mol}$$

$$\underset{\text{2-methyl-1-butene}}{CH_3CH_2\overset{\overset{\displaystyle CH_3}{|}}{C}=CH_2} + H_2 \xrightarrow[\text{catalyst}]{} \underset{\text{2-methylbutane}}{CH_3CH_2\overset{\overset{\displaystyle CH_3}{|}}{C}HCH_3} \qquad \Delta H^\circ = -28.5 \text{ kcal/mol}$$

$$\underset{\text{2-methyl-2-butene}}{CH_3CH=\overset{\overset{\displaystyle CH_3}{|}}{C}CH_3} + H_2 \xrightarrow[\text{catalyst}]{} \underset{\text{2-methylbutane}}{CH_3CH_2\overset{\overset{\displaystyle CH_3}{|}}{C}HCH_3} \qquad \Delta H^\circ = -26.9 \text{ kcal/mol}$$

The product in each case is the same compound, 2-methylbutane. One of the reactants, hydrogen, is the same for each reaction. The differences in the heats of reaction must arise from the differences in energy of the remaining reagents, the three different alkenes.

The first alkene, 3-methyl-1-butene, has the double bond at the end of the chain, and has only one alkyl substituent on it. The other positions on the double bond are occupied by hydrogen atoms. 2-Methyl-1-butene is a terminal alkene, but it has two alkyl substituents and two hydrogen atoms on the double bond. A comparison of the heats of hydrogenation of the two compounds indicates that the second alkene is more stable than the first one. Finally, in 2-methyl-2-butene, the double bond in the middle of the chain is an internal double bond, and has three alkyl sustituents and one hydrogen atom on it. 2-Methyl-2-butene has the lowest heat of hydrogenation of the three, and is, consequently, the most stable of the three alkenes. Note that the heat of hydrogenation of 3-methyl-1-butene is the same as that of 1-butene.

In summary, the position of the double bond and the number of alkyl substituents on it seem to be more important in predicting the relative stability of alkenes than the nature of the alkyl group on the double bond. Internal alkenes are more stable than terminal alkenes. The stability of an alkene increases with the number of alkyl groups that are substituted on carbon atoms of the double bond. Alkenes in which bulky substituents are trans to each other are more stable than the corresponding cis alkenes.

PROBLEM 5.3   Are Z and E isomers possible for any of the methylbutenes given above?

PROBLEM 5.4   Draw an energy diagram like Figure 5.3 for the methylbutenes to illustrate their relative stabilities in a schematic way.

5.3

The Addition of Hydrogen Bromide to Propene.   Mechanism, Energetics, and Kinetics

A. Selectivity in the Reaction of Hydrogen Bromide with Propene

The π electrons in alkenes are protonated by strong acids (Section 4.3). The vulnerability of the double bond to acids is illustrated by the addition of hydrogen bromide to propene.

$$CH_3CH=CH_2 + HBr \xrightarrow[\substack{18\ h \\ absence \\ of\ oxygen}]{25\ °C} CH_3\underset{\underset{Br}{|}}{C}HCH_3$$

| propene | hydrogen bromide | | 2-bromopropane isopropyl bromide 95% |

When the reaction is carried out in the gas phase in the absence of air and with carefully purified reagents for eighteen hours at room temperature, a 95% yield of isopropyl bromide is obtained.

Propene and hydrogen bromide are both unsymmetrical reagents. They could react to form n-propyl bromide, $CH_3CH_2CH_2Br$, as well as isopropyl bromide, but under these reaction conditions only the latter is produced. *Of two possible ways in which the same reagents can react, one reaction pathway and one product is favored. Such selectivity in the direction in which reagents bind to each other is called the* **regioselectivity of the reaction.** This selectivity is useful if we wish to synthesize the compound, in this case isopropyl bromide, that is favored in the reaction. On the other hand, it works against us if we plan to use this reaction under these conditions to make n-propyl bromide.

The selectivity in the addition of protic acids to alkenes has been recognized for a long time. The Russian chemist, Vladimir Markovnikov, in 1869 summarized the experimental observations in a rule that bears his name. Markovnikov said, "When a hydrocarbon of unsymmetrical structure combines with a halogen hydracid, the halogen adds itself to the less hydrogenated carbon atom, i.e., to the carbon that is more under the influence of other carbon atoms." Thus, the bromine atom from hydrogen bromide adds to the carbon atom of the double bond in propene that has fewer hydrogen atoms on it, giving rise to isopropyl bromide in accord with Markovnikov's Rule.

PROBLEM 5.5 Predict the major products(s) of the following reactions.

$$\text{CH}_3$$
$$|$$
(a) $CH_3C=CH_2$ + HBr $\longrightarrow$

(b) $CH_3CH_2CH=CH_2$ + HBr $\longrightarrow$

(c) $CH_3CH_2CH=CHCH_3$ + HBr $\longrightarrow$

B. The Mechanism for the Addition of Hydrogen Bromide to Propene. Carbocations as Reactive Intermediates

The experimental facts about the reaction of hydrogen bromide with propene can be rationalized by writing a mechanism for the reaction (Section 4.4). The mechanism postulated must take into account the electronic character of the reagents and the polarities of the bonds involved. It must give a step-by-step picture of the details of bond making and bond breaking that take place as the reagents are transformed into products.

The π bond is a source of electron density. The hydrogen bromide molecule is polarized so that the hydrogen atom has a partial positive charge while the more electronegative bromine atom has the greater share of the electrons. These facts are symbolized by writing $\delta+$ and $\delta-$ in the appropriate places. The approach of the polarized hydrogen bromide to the π bond of propene initiates the reaction. *The proton in hydrogen bromide is an electron-seeking reagent, an* **electrophile.** *The reaction is classified as an electrophilic attack on the π electrons of the double bond.*

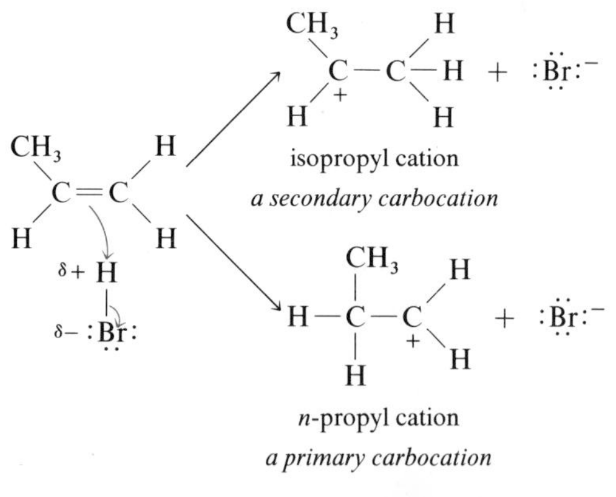

isopropyl cation
a secondary carbocation

n-propyl cation
a primary carbocation

In principle, a bond can form between the proton and the double bond in two ways. The proton can bond to the first carbon atom of propene, leaving the second carbon atom deficient in electrons. This creates a **secondary carbocation,** *a cation in which the electron-deficient carbon atom has two alkyl groups attached to it.* On the other hand, bonding between the proton and the second carbon atom in the chain would leave the end carbon atom with a positive charge, creating a **primary carbocation,** *with only one alkyl group attached to the electron-deficient carbon atom.*

Carbocations are Lewis acids. They are electron-seeking reagents or electrophiles and react with Lewis bases, electron-rich reagents that are present in the reaction mixture. Such reagents with electrons to share are also called **nucleophiles** *or* *nucleus-seeking reagents.* Bromide anion, which is one of the nucleophiles in this reaction mixture, combines with isopropyl cation to give the observed product.

$$CH_3\overset{+}{C}HCH_3 \longrightarrow CH_3CHCH_3$$

$$:\overset{..}{\underset{..}{Br}}:^- \qquad\qquad :\overset{..}{\underset{..}{Br}}:$$

secondary carbocation
and bromide ion

isopropyl bromide

No *n*-propyl bromide is obtained, suggesting that the primary carbocation is not formed in the reaction mixture to any large extent.

PROBLEM 5.6   What is another possible nucleophile in the reaction mixture formed by the reaction of propene with hydrogen bromide? Write an equation for the reaction of isopropyl cation with this nucleophile, showing electronic changes that must occur. (Hint: A review of Section 4.4 may be helpful.)

Carbocations are **reactive intermediates** *that are formed on the way from the reactants, propene and hydrogen bromide, to the product. They are unstable species relative to the reactants and the product of the reaction, and have only a very short lifetime in the reaction mixture.* Nevertheless, there is good experimental evidence for the existence of such intermediates, including physical measurements that can be made on some solutions. The cationic carbon atom, bonded to three other atoms, is trigonal and is pictured as being *sp*^{2} hybridized. An empty *p* orbital is perpendicular to the plane containing the carbon atom and the three atoms to which it is bonded (Figure 5.4).

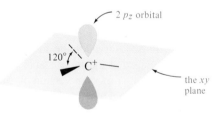

Figure 5.4   The geometry at a carbocation.

Why are secondary carbocations formed more easily in this reaction mixture than primary cations? The observations are explained by postulating that the methyl group is slightly electron releasing in its effect on neighboring atoms (Section 4.7B). We say that a methyl group is more **polarizable** than a hydrogen atom; it has a larger density of electrons that can be drawn towards a site of electron deficiency. This electron-releasing effect is particularly noticeable when the *sp*^{3}-hybridized carbon atom of an alkyl group is bonded to a more electronegative *sp*^{2}- (or *sp*-) hybridized carbon atom (Section 2.7B). The electrons around the carbon atom of the alkyl group are drawn towards the more electronegative carbon atom to which it is bonded. In the case of a carbocation, this effect results in the delocalization of some of the positive charge to the alkyl group and a resulting stabilization of the ion.

In a primary carbocation, one alkyl group and two hydrogen atoms are found on the carbon atom bearing the positive charge. In the secondary carbocation, there are two alkyl groups and one hydrogen atom on the electron-deficient carbon atom. A secondary carbocation is more stable than a primary carbocation because the electron deficiency in the secondary carbocation is stabilized by the electron-releasing effect of two methyl groups. Only one alkyl group stabilizes a primary carbocation.

$$
\begin{array}{cc}
\overset{\displaystyle H}{\underset{\displaystyle H}{CH_3\,\overset{\delta+}{CH_2}\!\rightarrow\!\overset{|\ \delta+}{C}}} & \overset{\displaystyle H}{CH_3\!\rightarrow\!\overset{|}{\underset{\delta+}{C}}\!\leftarrow\! CH_3}
\end{array}
$$

n-propyl cation	isopropyl cation
a primary carbocation	*a secondary carbocation*
delocalization of charge	*delocalization of*
to one alkyl group	*charge to two alkyl*
	groups

PROBLEM 5.7 Would you expect the methyl cation, CH_3^+, to be more or less stable than the *n*-propyl cation, $CH_3CH_2CH_2^+$?

PROBLEM 5.8 Would a tertiary carbocation, such as the *tert*-butyl cation, $(CH_3)_3C^+$, be more or less stable than the isopropyl cation, $(CH_3)_2CH^+$, a secondary carbocation? Make a list showing the four cations given in this problem and in Problem 5.7 in order of decreasing stability.

The regioselectivity observed in the addition of hydrogen bromide to propene can be rationalized by saying that the more stable of two possible cationic intermediates is formed in the reaction. A modern version of Markovnikov's Rule may be stated thus: In the addition of a hydrogen halide to a double bond, the major product is that isomer that results from the formation of the more stable cationic intermediate. Markovnikov himself recognized the importance of the "carbon that is under the influence of other carbon atoms" as the site at which the halogen would attach itself. The stability of a cationic intermediate increases with the number of carbon atoms bonded to the positively charged carbon atom. Thus, a tertiary carbocation is more stable than a secondary one, which in turn is more stable than a primary carbocation. The methyl cation is the least stable of all. For example:

$CH_3\overset{\displaystyle CH_3}{\underset{\displaystyle +}{-\overset{|}{C}-}}CH_3$	$CH_3\overset{\displaystyle CH_3}{\underset{\displaystyle +}{-\overset{|}{C}-}}H$	$CH_3CH_2\overset{\displaystyle H}{\underset{\displaystyle +}{-\overset{|}{C}-}}H$	$H\overset{\displaystyle H}{\underset{\displaystyle +}{-\overset{|}{C}-}}H$
tert-butyl cation	isopropyl cation	*n*-propyl cation	methyl cation
most stable			*least stable*

Experimental evidence for this order of stability is given in Section 5.6A, where we further examine reactions of carbocations.

C. A Comparison of the Thermodynamics of the Reactions Leading from Propene and Hydrogen Bromide to Isopropyl Bromide and *n*-Propyl Bromide

The reaction of propene with hydrogen bromide can, in principle, give two products, isopropyl bromide and *n*-propyl bromide (Section 5.3A). The two products do not differ much from each other in energy. It is possible to calculate the change in standard free energy (Section 4.6D) in going from the reactants to either one of the products. To do this we need to look up the standard free energy of formation, ΔG_f°, for each of the reactants and products. The standard free energy change for each reaction, ΔG_r°, is calculated using the following equation.

$$\Delta G_r^\circ = \Delta G_f^\circ(\text{products}) - \Delta G_f^\circ(\text{reactants})$$

$$CH_3CH{=}CH_2(g) \;+\; HBr(g) \longrightarrow CH_3CH_2CH_2Br(g)$$

ΔG_f° +14.99 kcal/mol   −12.73 kcal/mol   −5.37 kcal/mol

$$\Delta G_r^\circ = -5.37 - (14.99 - 12.73)$$
$$= -7.63 \text{ kcal/mol}$$

$$CH_3CH{=}CH_2(g) \;+\; HBr(g) \longrightarrow CH_3\overset{\displaystyle |}{\underset{\displaystyle Br}{C}}HCH_3$$

ΔG_f° +14.99 kcal/mol   −12.73 kcal/mol   −6.51 kcal/mol

$$\Delta G_r^\circ = -6.51 - (14.99 - 12.73)$$
$$= -8.77 \text{ kcal/mol}$$

The change in standard free energy for each reaction is negative. These values indicate that the formation of *n*-propyl bromide and isopropyl bromide are both spontaneous reactions (Section 4.6D). Note that isopropyl bromide is more stable than *n*-propyl bromide. It has a lower ΔG_f° by 1.14 kcal/mol.

Equilibrium constants for the two reactions can be calculated using the expression relating free energy changes to the equilibrium constant (Section 4.6D). The equilibrium constant for the formation of *n*-propyl bromide from propene and hydrogen bromide is 3.89×10^5. The equilibrium constant for the formation of isopropyl bromide is 2.63×10^6. The ratio of the two equilibrium constants is 6.76, indicating that the equilibrium mixture should contain 87% isopropyl bromide and 13% *n*-propyl bromide. Experimentally, isopropyl bromide is the only product. The composition of the product is *not* the equilibrium mixture that is expected from energy considerations. Therefore, the regioselectivity of the reaction is *not* determined by the relative stabilities of isopropyl bromide and *n*-propyl bromide.

PROBLEM 5.9   Carry out the actual calculations of the equilibrium constants for the reactions above.

D. The Transition State and the Energy of Activation

The reaction of propene with hydrogen bromide is postulated to go by means of a carbocation, a reactive intermediate that is higher in energy than the reactants or the product (Section 5.3B). These energy relationships can be represented on an **energy**

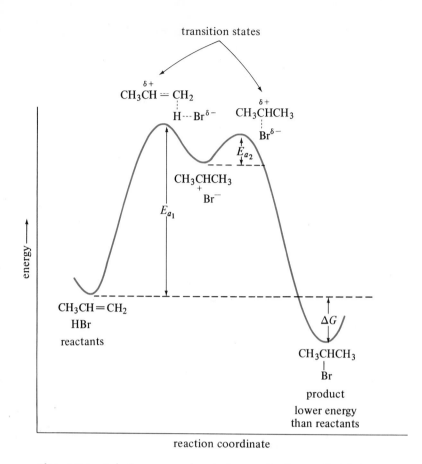

Figure 5.5   Relative energy levels for reactants, reactive intermediate, and product in the conversion of propene and hydrogen bromide to isopropyl bromide.

diagram (Figure 5.5). The ordinate represents energy, roughly the potential energy of the system. The abscissa is called the **reaction coordinate.** For a diatomic molecule, one possible reaction coordinate would be the distance between two nuclei. For molecules with more than two atoms in them, it is difficult to pick out one dimension that changes consistently and significantly during a reaction because many bonds and bond angles are involved. However, the reaction coordinate can still represent changes that must take place in the relationships between molecules and in bond lengths and bond angles within molecules as reactants are converted to products. Motion along that axis corresponds to the progress of the reaction from reactants to products.

In going from the reactants to the reactive intermediate, the system has to go over an energy barrier that is greater than the difference in energy between the two species. This energy barrier represents the most unstable position on the reaction pathway, the state in which some bonds are partially broken and some are partially formed. At this point the whole group of atoms is sticking together in a rather precarious way and may roll off the energy hill in either direction, going back to being starting materials, or going on to the next energy valley, the carbocation. *This high-energy state is the* **transition state.** *The energy that the molecules must have to get to the transition state is called the* **energy of activation** *for the reaction,* E_a. The energy of activation is a quantity that has to be determined experimentally for a particular reaction. It cannot be predicted from the data available to us. However, in most cases, the relative heights of energy barriers can be estimated even when the exact values for energies of activation are not known. Note that the peaks labeled transition state in Figure 5.5 represent the highest points from reactants to reactive intermediate and from the intermediate to product. The transition state cannot be observed experimentally. Nevertheless we identify some stage of the changes in bonding as the transition state. The molecular configuration postulated as the transition state for the formation of the secondary carbocation is shown in detail below.

starting materials   *transition state*   *intermediates;*
isopropyl cation
and bromide ion

The two reagents, propene and hydrogen bromide, approach each other with some critical orientation of the two molecules so that the positive end of the hydrogen bromide molecule points towards the π electron cloud of propene. A flow of electrons starts from the π bond to the hydrogen atom. As this bonding starts, the bond between the hydrogen and bromine atoms loosens with the bonding electrons moving closer to bromine, which becomes more like bromide ion. At the same time, one of the carbon atoms of the double bond develops a partial positive charge and begins to look like a carbocation. The hybridization at the carbon atom that is bonding to the hydrogen atom starts to change from sp^2 to sp^3. The transition state is a high-energy situation because bonds are being broken while all of the energy that will ultimately be released from the formation of new bonds is not yet fully available.

The energy diagram in Figure 5.5 indicates that the formation of the secondary carbocation from propene and hydrogen bromide is an endothermic reaction. The molecules must absorb energy from their surroundings in order to get to that state. The energy of activation, the energy necessary to get to the transition state, is actually even greater than the energy difference between the starting materials and the intermediate.

The isopropyl cation is converted to isopropyl bromide by a highly exothermic reaction. Even this reaction has a transition state and an energy of activation—a small one in this case.

The transition state for a reaction may resemble either the reactants or the products in structure. If a reaction has a low energy of activation, the transition state is like the reactants in structure. In other words, the breaking of bonds has not progressed far when the transition state is reached, therefore the process does not require a large input of energy. If the reaction has a high energy of activation, the transition state is closer in structure to the intermediates or products of the reaction. The breaking of bonds, which requires an input of energy, has progressed to a considerable degree by the time the transition state is reached.

E. Kinetics. The Rate of a Reaction

In Section 5.3C we dealt with differences in energy between reactants and products, with the **thermodynamics** of the reaction of propene with hydrogen bromide. The equilibrium between isopropyl bromide and *n*-propyl bromide, predicted by thermodynamic considerations, is not observed experimentally. *Thermodynamics deals only* *with the change in energy between the initial state of the system, the reactants, and the final state of the system when the products have been formed.* It defines where the equilibria between reactants and products, and between different possible products, will lie. It says nothing about the pathway that leads from starting materials to products. Another factor that determines the practicality of a reaction is how fast the reaction will go under a given set of conditions.

Such a question leads us to the realm of kinetics. **Kinetics** *deals with the rate of a chemical reaction and the factors that influence the rate.* The mechanism proposed for a reaction must be consistent with observations on how the rate of the reaction is affected by changes in conditions, such as concentration of reactants and polarity of the solvent. When a set of reactants can give rise to more than one product, often it is the relative rates of different competing reactions that determine which is the major product. Kinetics is concerned with the exact pathway between reactants and products.

In the last section, we postulated that the reaction of propene with hydrogen bromide to give the isopropyl cation and bromide anion occurs by way of a high-energy transition state. In the transition state, partial bonding between propene and hydrogen bromide has occurred. The transition state represents an unstable situation that can return to the starting materials or go on to the reactive intermediate. The rate of the reaction depends on several factors. First, because we are postulating that a molecule of propene and a molecule of hydrogen bromide must get within bonding distance of each other, the rate will depend on the concentrations of the two species. If the concentrations of propene and hydrogen bromide are low, the likelihood that the molecules will come within colliding distance is also low. Higher concentrations of the reactants will increase the probability that collision, and reaction, will occur. Thus,

$$\text{rate of the reaction} = k_r[\text{CH}_3\text{CH}=\text{CH}_2][\text{HBr}]$$

where k_r is the **rate constant** for the reaction at a given temperature.

The rate of a reaction and the dependence of rate on the concentrations of the reactants is determined experimentally. Because two concentration terms appear in the rate equation above, the reaction is said to be a **second-order reaction,** first order with respect to propene, and first order with respect to hydrogen bromide. The order of a reaction is always determined experimentally.

The rate constant k_r has many components. For example, not every collision between propene and hydrogen bromide leads to reaction. The molecules must be oriented properly towards each other at the moment of collision. If two molecules collide so that the hydrogen atom of hydrogen bromide is in a line with the methyl group of propene, reaction is not likely to take place. On the other hand, if the hydrogen atom is oriented towards the π electrons of the double bond, the collision is more likely to be fruitful.

Also, not all collisions result in reaction because the colliding molecules do not always have enough energy to react. Remember that the formation of isopropyl cation from propene and hydrogen bromide is an endothermic reaction, and that there is, besides, an additional amount of energy contributing to the total energy barrier that we defined as the experimental energy of activation (Figure 5.5, Section 5.3D).

At any given temperature, the molecules in a reaction mixture have a wide distribution of energies as shown in Figure 5.6. Most molecules have energies close to some average value, but some have much lower or higher energies. The molecules are moving rapidly and collide frequently with each other. As a result of these collisions, energy is transferred from one molecule to another. Therefore, the energy available to a given molecule is constantly changing.

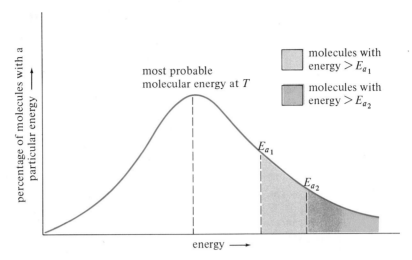

Figure 5.6   Distribution of energy among molecules at temperature, T.

If two reactions, one with a low energy of activation, E_{a_1}, and the other with a high energy of activation, E_{a_2}, are compared, a larger number of molecules that collide with each other during the reaction will have energy equal to or greater than E_{a_1}, the lower energy of activation (Figure 5.6). Therefore, during a given period of time, more molecules will take the pathway leading over the hill corresponding to the lower energy of activation. The observed rate of the reaction with the lower energy of activation will be greater than that of the reaction with the higher energy of activation.

An increase in temperature increases the kinetic energy of the molecules and increases the number of molecules that have enough energy to go over the energy barrier represented by the energy of activation. Thus, an increase in temperature increases the rate of a reaction. A more exact way of judging this effect is given in the next section.

In summary then, the observed rate of a reaction is affected by four factors: (1) concentration of reagents, (2) energy of activation, (3) temperature, and (4) steric effects, derived from the sizes and shapes of colliding molecules, that determine which orientations lead to reaction. The rate expression that appears on p. 174 includes all of these factors except concentration in the experimental rate constant, k_r. In the next section we look more closely at the relationship between k_r and these other factors.

PROBLEM 5.10   The reaction in the gas phase of propene with hydrogen iodide to give isopropyl iodide has been studied. At 238.5 °C when propene and hydrogen iodide are both present at an initial pressure of 45 mm, the second-order rate constant is determined to be 1.66×10^{-6}/atm · s. What is the overall rate of the reaction under these conditions? (760 mm = 1 atm)

F. The Absolute Rate Theory

A more exact relationship between the experimental rate constant and the energy of activation is based on the assumption that the molecular complex at the transition state is in equilibrium with the reactants. This molecular complex is also called the **activated complex.**

$$CH_3CH{=}CH_2 + HBr \rightleftharpoons \begin{bmatrix} CH_3\,\underset{\delta+}{CH} {-\!\!-} CH_2 \\ \quad\quad\quad\; | \\ \underset{\delta-}{Br} {-\!-\!-\!-} H \end{bmatrix}^{\ddagger}$$

$$\underset{reactants}{\phantom{CH_3CH{=}CH_2}}$$

the activated complex;
the molecular configuration
at the transition state

The activated complex is represented in equations by the symbol $\ddagger$, a double dagger used as a superscript. As is true for any equilibrium reaction, an equilibrium constant, $K^{\ddagger}$, can be written for the reaction above.

$$K^{\ddagger} = \frac{[\text{activated complex}]^{\ddagger}}{[\text{reactants}]}$$

The experimental rate constant for the reaction is proportional to the equilibrium constant, $K^{\ddagger}$.

$$k_r = k_0 K^{\ddagger}$$

In other words, the rate of the reaction is determined by the equilibrium concentration of the activated complex. Once the activated complex forms, it is converted to product at some absolute rate, k_0. At 25 °C, k_0 is 6.2×10^{12}/s. This number has the dimensions of frequency, and is of the same order of magnitude as the frequency of the vibration of a covalent bond in a molecule (see Section 7.7B for a description of some vibrational motions of covalent bonds). Thus, it may be said that once a molecule acquires enough energy to become an activated complex, bonds break and new bonds form during the time required for a molecular vibration.

The equilibrium constant, $K^{\ddagger}$, is also related to the change in standard free energy for the reaction in which propene and hydrogen bromide are converted to the activated complex. This change in free energy is the **free energy of activation, $\Delta G^{\ddagger}$**. $\Delta G^{\ddagger}$ has the **enthalpy of activation, $\Delta H^{\ddagger}$**, and the **entropy of activation, $\Delta S^{\ddagger}$**, as components.

$$\Delta G^{\ddagger} = \Delta H^{\ddagger} - T\Delta S^{\ddagger}$$

These two components may be related to two of the factors, other than concentration of reactants, mentioned on p. 175 as affecting the rate of the reaction. The enthalpy of activation determines whether the colliding molecules have enough energy to react and is similar to, but not identical with, the experimental energy of activation (Section 5.3D). The entropy of activation is related to questions of orientation between the colliding molecules and to the probability that an activated complex will be formed on any one collision.

The free energy of activation, $\Delta G^{\ddagger}$, is related to $K^{\ddagger}$ in the same way that any equilibrium constant is related to the standard free energy change for that reaction (Section 4.6D). Thus,

$$\Delta G^{\ddagger} = -RT \ln K^{\ddagger}$$

$$= -2.303\, RT \log K^{\ddagger}.$$

The relationship is an exponential one and may be rewritten

$$\ln K^{\ddagger} = -\Delta G^{\ddagger}/RT$$

$$K^{\ddagger} = e^{-(\Delta G^{\ddagger}/RT)}$$

in which e is 2.718, the base of natural logarithms. By substituting this expression for $K^{\ddagger}$ in the equation giving the relationship between the experimental rate constant and $K^{\ddagger}$, we get

$$k_r = k_0 e^{-(\Delta G^{\ddagger}/RT)}.$$

The negative exponential relationship between the rate constant and $\Delta G^{\ddagger}$ is very important. From the expression we can see that as $\Delta G^{\ddagger}$ gets larger, $e^{-(\Delta G^{\ddagger}/RT)}$ must get smaller, therefore the observed rate of the reaction decreases. Conversely a smaller $\Delta G^{\ddagger}$ will give an increased rate of reaction. This is the same conclusion that we reached by qualitative arguments about the range of molecular energies available to the reacting species (p. 175, Figure 5.6).

The exponential relationship between the rate constant and $\Delta G^{\ddagger}$ also ensures that small differences in energy of activation make large differences in rate. For example, at 25 °C a reaction with $\Delta G^{\ddagger}$ of 10 kcal/mol will have a rate constant,

$$k_r = (6.2 \times 10^{12}/s) \times \exp\left[-\frac{10 \text{ kcal/mol}}{(1.99 \times 10^{-3} \text{ kcal/mol} \cdot \text{K}) \times 298 \text{ K}}\right]$$

$$k_r = (6.2 \times 10^{12}/s) \times \exp(-16.9)$$

$$k_r = (6.2 \times 10^{12}/s) \times (4.6 \times 10^{-8})$$

$$k_r = 2.9 \times 10^5/s.$$

If $\Delta G^{\ddagger}$ is 20 kcal/mol instead, the rate constant will be

$$k_r = (6.2 \times 10^{12}/s) \times \exp\left[-\frac{20 \text{ kcal/mol}}{(1.99 \times 10^{-3} \text{ kcal/mol} \cdot \text{K}) \times 298 \text{ K}}\right]$$

$$k_r = (6.2 \times 10^{12}/s) \times \exp(-33.7)$$

$$k_r = (6.2 \times 10^{12}/s) \times (2.3 \times 10^{-15})$$

$$k_r = 1.4 \times 10^{-2}/s.$$

A doubling of the free energy of activation leads to a ten-million-fold decrease in the rate of reaction.

The same exponential relationship determines the effect of temperature on the rate of the reaction. Temperature, T, appears in the denominator of the exponent, therefore an increase in T increases the value of $e^{-(\Delta G^{\ddagger}/RT)}$, and also of k_r. We assume that $\Delta G^{\ddagger}$ does not vary much with temperature.

PROBLEM 5.11   Calculate the rate constant of a reaction with $\Delta G^{\ddagger}$ 10 kcal/mol at 30 °C and at 40 °C. Remember that T in the equation is an absolute temperature. What is the effect on the rate of reaction of a temperature increase of 10 degrees?

From experimental determinations of the rate of a reaction at different temperatures, a relationship between the rate constant for a reaction and the experimental energy of activation can be derived.

$$k_r = A e^{-(E_a/RT)}$$

This relationship, known as the Arrhenius equation, is similar in form to the expression for the relationship between the rate constant and $\Delta G^{\ddagger}$ (seen above). In this expression, A is known as the **frequency factor** and is also determined experimentally for each reaction studied. Note that A is similar to, but not identical with, the absolute rate constant, k_0 (p. 176).

PROBLEM 5.12   The reaction of 1-butene with hydrogen iodide at 211 °C has been determined to have E_a 20.9 kcal/mol and a frequency factor, A, of 5×10^7 L/mol·s. Using the Arrhenius equation, calculate the rate constant for the reaction.

The expression for the overall rate of a reaction has, of course, in addition to k_r, terms for the concentrations of the reactants (p. 174). Any calculations to determine the rate at which reactants are converted to products must include these concentrations (Problem 5.10).

G. Kinetic Versus Thermodynamic Control of a Reaction

While the possible products of the reaction of propene with hydrogen bromide, iso-propyl bromide and *n*-propyl bromide, differ from each other by about 1 kcal/mol in stability (Section 5.3C), the isopropyl cation is more stable than the *n*-propyl cation by about 16 kcal/mol (Section 5.6A). This large difference in stability between the secondary and primary cations is also reflected by a large difference in energy of activation for the formation of the two species, shown in the energy diagrams in Figure 5.7. The lower energy pathway for the reaction goes by way of the more stable secondary cation. The rate at which the secondary cation is formed is much greater than

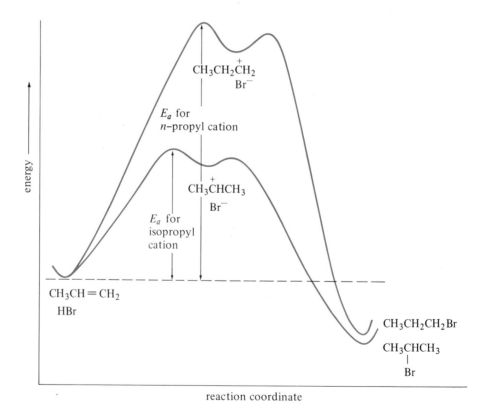

Figure 5.7   Comparison of the energies of activation for the formation of primary and secondary carbocations from propene and hydrogen bromide.

the rate at which the primary cation is formed (Section 5.3F). Therefore, the product of the reaction comes exclusively from the intermediate formed with the lower energy of activation, even though overall energy considerations (Section 5.3C) allow formation of the other product.

A reaction is said to be **kinetically controlled** *when its products are determined by the relative rates at which they are formed.* In other reactions, such as the acid-base reactions in Section 4.3, the products are determined by equilibria between the different species. *A reaction is* **thermodynamically controlled** *when the products are present in equilibrium in the relative amounts dictated by energy considerations.* The reaction of propene with hydrogen bromide is kinetically controlled. The regioselectivity of the reaction comes from the relative rates at which the two intermediate cations form. Later (Sections 11.1C, 15.1D, for example) we will see that *in some reactions kinetic factors become so important that the thermodynamically less stable product becomes the major product of the reaction.*

PROBLEM 5.13   The experimental energy of activation of the reaction of hydrogen iodide with propene (Problem 5.10) has been determined to be 22.4 kcal/mol. The observed second-order rate constant, k_r, for the reaction at 490 K is 2.61×10^{-3} L/mol·s. The relationship between k_r and the experimental energy of activation is given by the Arrhenius equation (p. 177). Calculate A, the frequency factor for the reaction.

PROBLEM 5.14   Assume that the energy of activation for the formation of n-propyl cation from propene and hydrogen iodide is at least 16 kcal/mol higher than that for the isopropyl cation, or approximately 38 kcal/mol. Using the Arrhenius equation and the value of A calculated in Problem 5.13, calculate k_r for the formation of the n-propyl cation at 490 K. How much faster is the reaction giving the isopropyl cation (Problem 5.13) than the one going to the n-propyl cation?

H. Kinetics of a Reaction with Two Steps. The Rate-Determining Step

In the discussions so far, we have focused on the conversion of propene to isopropyl cation or n-propyl cation, and have not considered the reaction leading from isopropyl cation to isopropyl bromide, the product obtained from the reaction. Isopropyl cation reacts with bromide ion in the second step of the reaction with a small energy of activation, goes through a second transition state (Figure 5.5, p. 172), and gives isopropyl bromide in an exothermic reaction. Because the energy of activation for this second reaction is so small in comparison to the energy of activation for the first step, the rate constant for the reaction of isopropyl cation with bromide ion is large. If the molecules have enough energy to get up the first hill, they have more than enough energy to make it over the second hill, too. Thus while there are actually two different rate constants involved in the overall reaction leading to the formation of isopropyl bromide, experimentally we measure only one of them, k_1, the rate constant for the slow step of the reaction.

$$CH_3CH{=}CH_2 + HBr \xrightarrow{k_1 \text{ (slow)}} CH_3\overset{+}{C}HCH_3 + Br^- \xrightarrow{k_2 \text{ (fast)}} CH_3\underset{Br}{C}HCH_3$$

The slow step of a reaction sequence, the step with the higher energy of activation, is known as the **rate-determining step** *of the reaction.* The overall rate of the reaction cannot be any faster than the rate-determining step. In this case, the formation of the high-energy reactive intermediate, isopropyl cation, is rate-determining. Once the cation is formed, it reacts rapidly with bromide ion to give the product.

5.4

Free Radicals as Reactive Intermediates

A. Free Radical Addition of Hydrogen Bromide to Propene. Free Radical Chain Reactions

In Section 5.3A, the reaction conditions under which isopropyl bromide is formed from propene and hydrogen bromide were carefully specified. A change in reaction conditions completely changes the product that is obtained in this reaction. When hydrogen bromide and propene are mixed at $-78\ °C$ in the presence of benzoyl peroxide and air, a very rapid reaction takes place, and the product is 96% *n*-propyl bromide and 4% isopropyl bromide.

$$CH_3CH=CH_2 + HBr \xrightarrow[\substack{benzoyl \\ peroxide \\ -78\ °C}]{} CH_3CH_2CH_2Br + CH_3CHCH_3$$

$$\overset{\displaystyle |}{\underset{\displaystyle Br}{}}$$

| propene | hydrogen bromide | *n*-propyl bromide 96% | isopropyl bromide 4% |

A different mechanism involving a different type of reactive intermediate is postulated for the reaction of propene with hydrogen bromide under these conditions. The reaction starts with the homolytic cleavage of a weak oxygen-oxygen single bond in benzoyl peroxide.

benzoyl peroxide   benzoyloxy radical
a homolytic cleavage of a covalent bond

In a **homolytic cleavage** *of a bond, one electron of the covalent bond being broken goes to each fragment of the molecule. Fishhooks,* ⌢, *are used to represent the motion of a* single *electron in homolytic transformations of bonds. By contrast, arrows,* ⌢, *are used to indicate the movement of* pairs *of electrons when bonds are broken in heterolytic cleavages. In the* **heterolytic cleavage** *of a bond, both electrons of the covalent bond remain with one of the fragments of the molecule. A heterolytic cleavage of a bond usually gives rise to ions. The homolytic cleavage of a bond, on the other hand, creates uncharged species called* **free radicals.** *A radical has an odd number of electrons, and hence one unpaired electron. It behaves as an electrophile.*

The radicals from benzoyl peroxide abstract hydrogen atoms from hydrogen bromide, giving rise to bromine atoms.

| benzoyloxy radical | hydrogen bromide | benzoic acid | bromine atom |

The bromine atom is an electrophile and is attacked by the π electrons of the double bond in propene. New reactive intermediates known as carbon radicals form.

$$
\begin{array}{c}
CH_3 \qquad H \\
\diagdown \qquad \diagup \\
C = C \\
\diagup \qquad \diagdown \\
H \qquad\qquad H \\
\qquad\qquad \cdot\ddot{B}r\!:\!
\end{array}
\longrightarrow
CH_3\overset{.}{C}HCH_2 + CH_3CHCH_2
$$

$$
\underset{\substack{a\ secondary \\ radical}}{\quad\quad\quad|\quad\quad} \underset{\substack{a\ primary \\ radical}}{\quad\quad\quad|\quad\quad}
$$
:Br: :Br:
a secondary   *a primary*
radical   *radical*

Whether a primary or secondary radical forms depends upon which carbon atom of the double bond becomes attached to the bromine atom.

The carbon radicals are the new electrophiles in the reaction mixture. The product, *n*-propyl bromide, is formed when the secondary carbon radical abstracts a hydrogen atom from hydrogen bromide.

$$
CH_3\overset{.}{C}HCH_2Br \longrightarrow CH_3CHCH_2Br + :\overset{.}{\ddot{B}r}\cdot
$$

$$
H\!-\!\ddot{B}r\!: \qquad\qquad \underset{\substack{n\text{-}propyl \\ bromide}}{\overset{|}{H}} \qquad \underset{\substack{bromine \\ atom}}{}
$$

Note that it is *not* a proton that is being transferred in hydrogen abstraction reactions but a hydrogen atom with no charge.

As the overall result of this sequence of steps, hydrogen bromide adds to the double bond of propene with an orientation opposite to that seen when air and peroxides are carefully excluded from the reaction mixture. This orientation has been called the **anti-Markovnikov addition** of hydrogen bromide to alkenes.

The observation that *n*-propyl bromide is the major product in this reaction leads us to conclude that the secondary carbon radical is a more stable reactive intermediate than the primary carbon radical. (The relative stabilities of radicals will be discussed more fully in the next section.) The change in orientation comes from the change in the mechanism of the reaction.

Historically, this reaction caused much confusion among chemists because alkenes form peroxides in the presence of oxygen in air. Reactions carried out with impure alkenes and solvents gave varying mixtures of products. This generated a lively controversy in the literature of organic chemistry until Morris S. Kharasch and his coworkers at the University of Chicago in the 1930s, working with carefully purified reagents under conditions that excluded air, discovered the source of the problem.

Once we understand how a change in mechanism and in the reactive intermediates, caused by a change in reaction conditions, alters the regioselectivity of a reaction, we can take advantage of this to create different products at will.

PROBLEM 5.15   Does the change in mechanism for the addition reaction and the fact that *n*-propyl bromide is now the major product change the relative energy levels for propene and hydrogen bromide, *n*-propyl bromide and isopropyl bromide (Figure 5.7)? Explain your answer.

PROBLEM 5.16   Predict what will happen when 2-methylpropene is treated with hydrogen bromide in the presence and in the absence of peroxides. Write detailed mechanisms for the two reactions. In accounting for the products, analyze the relative stabilities of the various intermediates that might be formed.

The steps described above for the free radical addition of hydrogen bromide to propene are typical of a free radical chain reaction. A **free radical chain reaction** has initiation, chain-propagation, and termination steps.

The **initiation step** is the one that gives rise to the first free radical intermediate that is part of the chain. Heat or light energy is often used to cleave a covalent bond homolytically to start a chain reaction. Sometimes a small amount of a compound with a weak covalent bond is used as an **initiator** to start the chain.

Benzoyl peroxide is the initiator in the free radical addition of hydrogen bromide to propene. Abstraction of a hydrogen atom from hydrogen bromide by the benzoyloxy radical to give a bromine atom is the initiation step.

The bromine atom reacts with the π bond of propene to create an intermediate carbon radical. The carbon radical, in turn, abstracts a hydrogen atom from hydrogen bromide to give a new bromine atom. Reaction of the bromine atom with a second molecule of propene creates a carbon radical once more. These two steps repeated many times carry the reaction forward, and are known as the **chain-propagation steps** of the reaction. *A chain reaction occurs when a product of one step of a reaction is a reactive intermediate for the next step.* Thus, the propagation step of a free radical chain reaction always creates a new radical along with a stable product. The reaction is self-perpetuating.

A chain reaction stops when it runs out of reagents or when reactive intermediates are destroyed in reactions known as **termination steps.** A possible termination step for the reaction described here would be the combination of two radicals, using up the reactive intermediates.

$$:\!\overset{..}{Br}\!\cdot \curvearrowright \cdot\overset{..}{Br}\!: \longrightarrow \;:\!\overset{..}{Br}\!-\!\overset{..}{Br}\!:$$

$$CH_3CHCH_2Br \longrightarrow CH_3CHCH_2Br$$

three possible termination steps

Because the reactive intermediate radicals are usually present in the reaction mixture in low concentrations and collisions between them are rare, the termination reactions are not likely. The importance of controlling the chain-propagating and chain-terminating steps in industrial chemical manufacturing processes is the subject of Section 20.2A.

We have been examining one chemical reaction in great detail. The particular reaction is not as important as the general ideas about chemical reactions that have been developed. The concepts of thermodynamics and kinetics will be used over and over again. The selectivity of reactions will be explained in terms of the relative stabilities of different transition states leading to products. We will speculate on how transition states and intermediates look, and what factors stabilize them. We will write mechanisms for reactions, postulating various reactive intermediates on the basis of the reaction conditions. Distinctions among primary, secondary, and tertiary positions on carbon chains will be important in describing reactive intermediates and structures of reactants and products. We have been engaged in an exercise in learning to understand and predict the reactivity of organic molecules by looking at their structures in a discriminating way.

B. Bond Dissociation Energies. The Relative Stabilities of Radicals

The regioselectivity of the addition of hydrogen bromide to propene was explained in Section 5.4A by postulating that a secondary carbon radical is more stable than a primary carbon radical. The relative stabilities of radicals may be determined by measurement of **standard bond dissociation energies, DH°**. *The standard bond dissociation energy is defined as the change in enthalpy for a reaction in which one specific covalent bond in a molecule is broken homolytically while the reactants and the products are in their standard state* (in the gas phase at 1 atm and 25 °C). For example, it takes 104 kcal/mol to break the bond in a hydrogen molecule to give two hydrogen atoms.

$$H_2(g) \longrightarrow 2\,H(g) \qquad \Delta H° = 104\text{ kcal/mol}$$

The enthalpy of this endothermic reaction is called the bond dissociation energy, $DH°$, for the hydrogen molecule.

Bond dissociation energies for carbon-hydrogen bonds in some hydrocarbons are shown in Table 5.1.

TABLE 5.1 Bond Dissociation Energies for the Reaction
$RH(g) \longrightarrow R\cdot(g) + H\cdot(g)$

R—H	Dissociation Energy (kcal/mol)
CH_3—H	104
CH_3CH_2—H	98
$CH_3CH_2CH_2$—H	98
$CH_3\overset{\displaystyle CH_3}{\underset{\displaystyle |}{C}H}$—H	95
$CH_3\overset{\displaystyle CH_3}{\underset{\displaystyle |}{C}}HCH_2$—H	98
$CH_3-\overset{\displaystyle CH_3}{\underset{\displaystyle CH_3}{C}}-H$	91
$CH_2{=}CH$—H	108
$HC{\equiv}C$—H	125
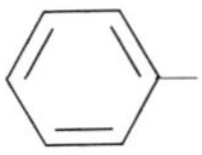 ⬡—H	110
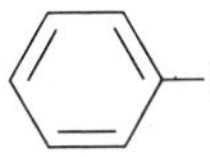 ⬡—CH_2—H	85
$CH_2{=}CHCH_2$—H	89

If we compare the bond dissociation energies for two different carbon-hydrogen bonds for the same alkane, we arrive at a value for the relative stabilities of two isomeric radicals.

$$CH_3CH_2CH_2-H \longrightarrow CH_3CH_2\dot{C}H_2 + H\cdot \qquad \Delta H° = +98\,kcal/mol$$

propane   *n*-propyl   hydrogen
               radical   atom

$$\underset{\underset{H}{|}}{CH_3CHCH_3} \longrightarrow CH_3\dot{C}HCH_3 + H\cdot \qquad \Delta H° = +95\,kcal/mol$$

propane   isopropyl   hydrogen
                 radical   atom

To break a carbon-hydrogen bond at a primary carbon atom in propane requires 3 kcal/mol more energy than breaking a carbon-hydrogen bond at the secondary carbon atom. The reactant, propane, and one of the products, the hydrogen atom, in each reaction are the same, so the difference in the $\Delta H°$ values for the two reactions reflects the different stabilities of the *n*-propyl radical and the isopropyl radical (Figure 5.8).

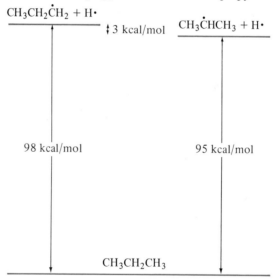

Figure 5.8   A comparison of the relative stabilities of *n*-propyl and isopropyl radicals, determined from bond dissociation energies for primary and secondary carbon-hydrogen bonds in propane.

A similar comparison can be made for the primary and tertiary radicals derived from 2-methylpropane.

$$\underset{\underset{}{|}}{\overset{\overset{CH_3}{|}}{CH_3CHCH_2}}-H \longrightarrow CH_3\overset{\overset{CH_3}{|}}{CH}\dot{C}H_2 + H\cdot \qquad \Delta H° = +98\,kcal/mol$$

2-methylpropane   isobutyl   hydrogen
                 radical   atom

$$\underset{\underset{H}{|}}{\overset{\overset{CH_3}{|}}{CH_3CCH_3}} \longrightarrow \overset{\overset{CH_3}{|}}{CH_3\dot{C}CH_3} + H\cdot \qquad \Delta H° = +91\,kcal/mol$$

2-methylpropane   *tert*-butyl   hydrogen
                  radical   atom

The tertiary carbon radical is more stable than the primary one by approximately 7 kcal/mol.

An inspection of Table 5.1 reveals that bond dissociation energies for all the carbon-hydrogen bonds giving rise to a primary carbon radical, RCH_2—H, whether R is CH_3, CH_3CH_2, or $CH_3CH_2CH_2$, are approximately 98 kcal/mol. This close agreement suggests that for compounds of similar structure, the stability of the radicals being formed is an important factor in the bond dissociation energy. Thus, we expect tertiary carbon radicals to be approximately 4 kcal/mol more stable than secondary carbon radicals, which are in turn about 3 kcal/mol more stable than primary radicals.

$$\underset{\substack{\text{most}\\\text{stable}}}{\overset{\displaystyle CH_3\\\displaystyle |}{\underset{\displaystyle |\\\displaystyle CH_3}{CH_3C\cdot}}} > \overset{\displaystyle CH_3\\\displaystyle |}{\underset{\displaystyle |\\\displaystyle H}{CH_3C\cdot}} > \overset{\displaystyle H\\\displaystyle |}{\underset{\displaystyle |\\\displaystyle H}{CH_3C\cdot}} > \underset{\substack{\text{least}\\\text{stable}}}{\overset{\displaystyle H\\\displaystyle |}{\underset{\displaystyle |\\\displaystyle H}{HC\cdot}}}$$

While a radical does not bear a positive charge, it is still an electron-deficient species in the sense that the carbon atom with the odd electron does not have a stable octet. Alkyl groups, therefore, stabilize radicals just as they stabilize carbocations, by delocalization of the electron deficiency.

The value for bond dissociation energies for carbon-hydrogen bonds in ethylene and acetylene show the extra strength of bonds at sp^2- and sp-hybridized carbon atoms (Table 5.1). The bond dissociation energies may be correlated with the decreasing bond lengths for carbon-hydrogen bonds going from propane to propene to propyne (Section 2.7A).

$CH_3CH_2CH_2$—H	$CH_3CH{=}CH$—H	$CH_3C{\equiv}C$—H
carbon-hydrogen bond at sp^3-hybridized carbon atom; 1.08 Å	carbon-hydrogen bond at sp^2-hybridized carbon atom; 1.07 Å	carbon-hydrogen bond at sp-hybridized carbon atom; 1.06 Å
$DH° = 98$ kcal/mol	$DH° = 108$ kcal/mol	$DH° = 125$ kcal/mol
longest carbon-hydrogen bond; weakest bond		*shortest carbon-hydrogen bond; strongest bond*

The lowest bond dissociation energies reported in Table 5.1 are for carbon-hydrogen bonds at tetrahedral carbon atoms adjacent to an aryl ring (85 kcal/mol) or a double bond (89 kcal/mol), suggesting special stability for radicals formed by the loss of those hydrogen atoms. The chemistry of such radicals is the subject of Sections 11.2 and 12.4A.

Bond dissociation energies for some other types of bonds are given in Table 5.2.

TABLE 5.2   Bond Dissociation Energies for Representative Single, Double, and Triple Bonds

Bond	Dissociation Energy (kcal/mol)	Bond	Dissociation Energy (kcal/mol)
CH_3 — CH_3	88	CH_2 = CH_2	145
R — Cl	~81	CH_2 — CH_2	59
R — Br	~68	$(CH_3)_2C$ = O	176
R — I	~53	$HC \equiv CH$	196

The triple bond in acetylene is shorter (1.21Å) and stronger than the double bond in ethylene (1.34 Å), which is in turn shorter and stronger than the single bond in ethane (1.54 Å). However, it takes only 59 kcal/mol to break the π bond in ethylene.

PROBLEM 5.17 The *peroxide effect* is seen only in the addition of hydrogen bromide to propene (Section 5.4A). Use the bond dissociation energies in Tables 5.1 and 5.2, as well as those of HCl (103 kcal/mol), HBr (87 kcal/mol), and HI (71 kcal/mol), to calculate the enthalpy of reaction, $\Delta H°$, for each of the following two steps of the general mechanism proposed for the free radical addition of hydrogen halide to propene when X is Cl, Br, or I.

(1) $CH_3CH{=}CH_2 + :\ddot{X}\cdot \longrightarrow CH_3\dot{C}HCH_2{-}\ddot{X}:$

(2) $CH_3\dot{C}HCH_2\ddot{X}: + H{-}\ddot{X}: \longrightarrow CH_3\underset{\underset{H}{|}}{C}HCH_2X + :\ddot{X}\cdot$

(Hint: Bond dissociation is an *endothermic* reaction. Energy must be put into the system to break bonds. What happens when a bond is formed by the combination of two species?) Explain why the peroxide effect is only seen with hydrogen bromide.

5.5

Other Additions of Acids to Alkenes and Alkynes

A. Addition of Hydrogen Halides to Alkynes and Vinyl Halides

The ionic addition of hydrogen bromide to an alkyne is similar to the addition reaction with alkenes, except that two moles of hydrogen bromide are added before a saturated compound is formed. This reaction, too, is regioselective and the product is one that would be predicted by Markovnikov's Rule. 1-Hexyne, for example, adds hydrogen bromide twice to give 2,2-dibromohexane.

$$CH_3CH_2CH_2CH_2C{\equiv}CH \xrightarrow[\substack{15\ °C \\ \text{no peroxides}}]{HBr} CH_3CH_2CH_2CH_2\underset{\underset{Br}{|}}{C}{=}CH_2 \xrightarrow{HBr}$$

1-hexyne   2-bromohexene

$$CH_3CH_2CH_2CH_2\underset{\underset{Br}{|}}{\overset{\overset{Br}{|}}{C}}CH_3$$

2,2-dibromohexane

The reaction goes by way of a vinyl cation formed by electrophilic attack on the triple bond.

$$CH_3CH_2CH_2CH_2C{\equiv}CH \longrightarrow CH_3CH_2CH_2CH_2\overset{+}{C}{=}C\overset{H}{\underset{H}{\diagup}} \longrightarrow \underset{:\ddot{Br}:}{\overset{CH_3CH_2CH_2CH_2}{C}}{=}C\overset{H}{\underset{H}{\diagup}}$$

1-hexyne and hydrogen   vinyl cation   2-bromo-1-hexene
bromide   *an unstable carbocation*

The electron-deficient carbon atom of the vinyl cation, bound to two other atoms, may be considered to be *sp* hybridized. As such, it is more electronegative (Section 2.7B) than the sp^2-hybridized carbon atom found in an alkyl cation, and thus less able to bear a positive charge. Vinyl cations are intermediate in energy between methyl and ethyl (or *n*-propyl) cations (p. 170, also Table 5.3, p. 194) and less stable than secondary or tertiary cations. However, an alkyne is also a higher-energy molecule than an alkene, so the energy of activation in going from an alkyne to a vinyl cation need not be much larger than the energy of activation for the formation of an alkyl cation from an alkene. The π bond in an alkyne is, thus, about as reactive as the π bond of an alkene towards acids.

2-Bromo-1-hexene is less reactive towards electrophilic addition than an alkene unsubstituted by halogen. The halogen atom on the double bond is electron withdrawing and decreases the availability of the π electrons for reaction with a second molecule of hydrogen bromide. The orientation of the second addition reaction is also interesting. Two intermediates are possible.

a primary carbocation	2-bromo-1-hexene and hydrogen bromide	*a secondary carbocation stabilized further by delocalization of charge to bromine atom*

$$\text{CH}_3\text{CH}_2\text{CH}_2\text{CH}_2\underset{\text{Br}}{\overset{\text{Br}}{\text{C}}}\text{CH}_3$$

2,2-dibromohexane

One is a primary carbocation, the other a secondary carbocation with an electron-withdrawing bromine atom at the cationic center. The halogen, while electron withdrawing in its inductive effect, does stabilize the carbocation by some delocalization of charge to the bromine atom by resonance. The product from the addition of hydrogen bromide to 2-bromo-1-hexene is the one derived from the reaction of bromide ion with the more stable of the two possible intermediates.

Resonance stabilization of a cation by a halogen atom is even more clearly seen in the addition of hydrogen chloride to acetylene.

acetylene	vinyl chloride	1,1-dichloroethane

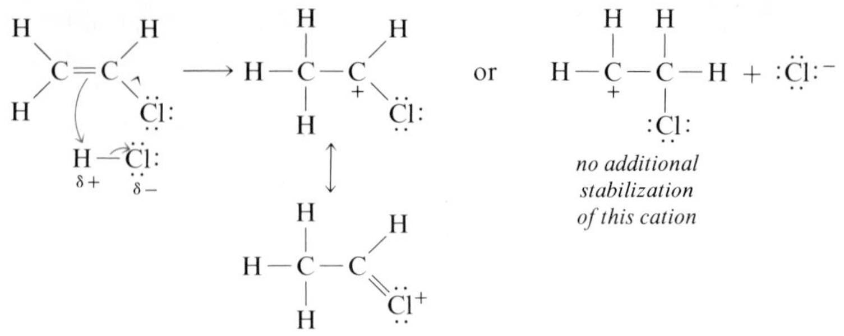

cation stabilized by delocalization
of charge to the chlorine atome

Both possible intermediates are primary cations, and in one case a chlorine atom with its electron-withdrawing effect would be expected to further destabilize the cation. The observed product can be rationalized by postulating resonance stabilization of one of the cations by the chlorine atom. Metal salts such as zinc(II) chloride and mercury(II) chloride, which act as Lewis acids, are used to catalyze the reactions of hydrogen chloride with alkynes and vinyl halides because of the lower reactivity of these compounds.

These equations and the ones in previous sections of this chapter are attempts to represent accurately the reagents and reaction conditions used in these organic reactions. The exact details are given because it is important that you appreciate the nature of experiments in organic chemistry. It is not necessary, however, for you to memorize all the details that appear in equations. The important reagents will be given many times in problems. If you use the problems as your guide, as you were advised to do in Section 1.2, your attention will be focused on the most important reactions and reagents. In other words, you need learn only the detail that is necessary to answer the problems.

In addition, to help you distinguish between the chief reagent and other reaction conditions, we will adopt the convention that reagents appear *over the arrow;* catalysts, solvents, and other conditions, such as temperature and reaction time, appear *below the arrow*. For example, in the reaction of acetylene with hydrogen chloride, shown above, HCl is the important reagent. It adds to the triple bond and appears over the arrow in the equations. The catalysts, the Lewis acids, zinc(II) chloride and mercury(II) chloride appear below the arrows. Temperatures are also given. The important reaction for you to recognize is the addition of HCl to the multiple bond. When you see zinc(II) chloride, for example, you should recognize it as a Lewis acid. It is not necessary, however, that you remember it as a reagent, nor that you try to memorize the temperature used for the reaction.

PROBLEM 5.18 Complete the following equations.

(a) 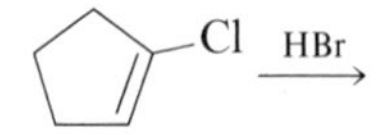 $\xrightarrow{\text{HBr}}$

(b) $CH_3C{\equiv}CCH_3 \xrightarrow{\text{HBr (excess)}}$

(c) $CH_3CH_2CH_2CH{=}CH_2 \xrightarrow[\text{peroxide}]{\text{HBr}}$

(d) 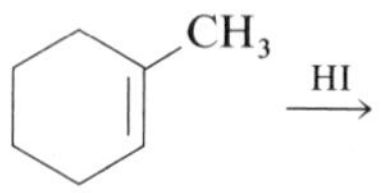 $\xrightarrow{\text{HI}}$

(e) $HC{\equiv}CH \xrightarrow{\text{HBr (excess)}}$

(f) 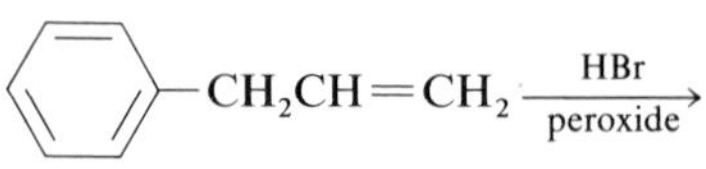 $-CH_2CH{=}CH_2 \xrightarrow[\text{peroxide}]{\text{HBr}}$

B. Addition of Water to Alkenes

Alkenes add water in the presence of acids to form alcohols. The reaction is of practical importance in the case of simple alkenes that are available from petroleum as by-products in the manufacture of gasoline. The *hydration* of 2-methylpropene is such an example.

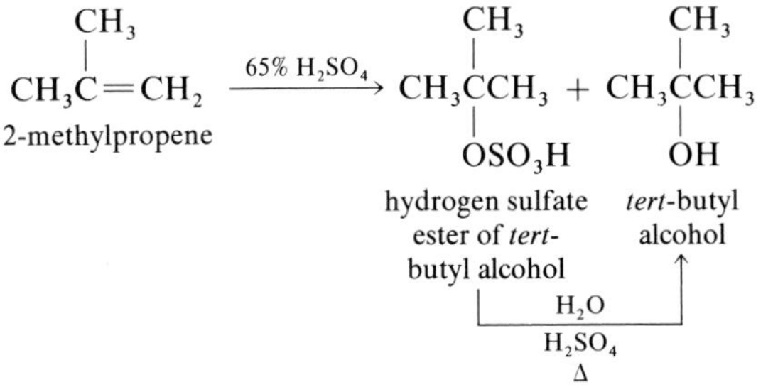

The alkene reacts with acid to give a tertiary carbocation. The nucleophiles that are present in the reaction mixture are water and hydrogen sulfate ion, and the carbocation combines with both. The products of these two reactions are *tert-*butyl alcohol and the hydrogen sulfate ester of the alcohol, respectively. In the industrial process, the sulfate ester is decomposed by heating with dilute aqueous acid, and the alcohol is isolated as the product. A detailed mechanism for the reaction is shown below.

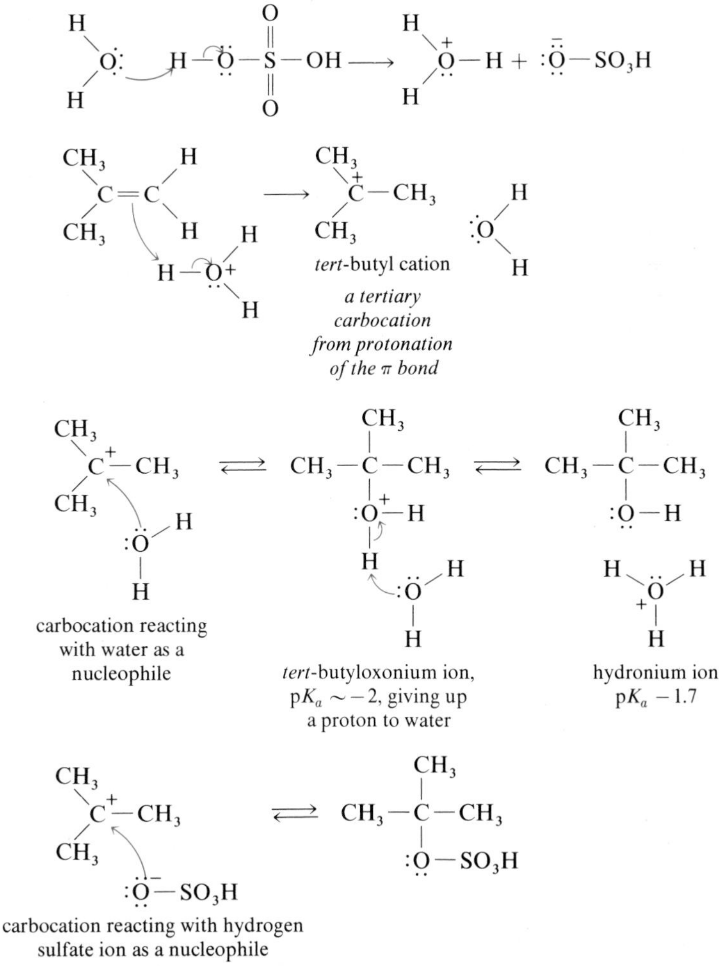

The reactivity of an alkene towards acids depends heavily upon the stability of the carbocation that forms on protonation. For example, for reactions in aqueous acid, 2,3-dimethyl-2-butene, 2-methyl-2-butene, and 2-methylpropene (all of which give tertiary carbocations on protonation) react about ten thousand times as fast as (*E*)-2-butene and propene (both of which give secondary carbocations) do.

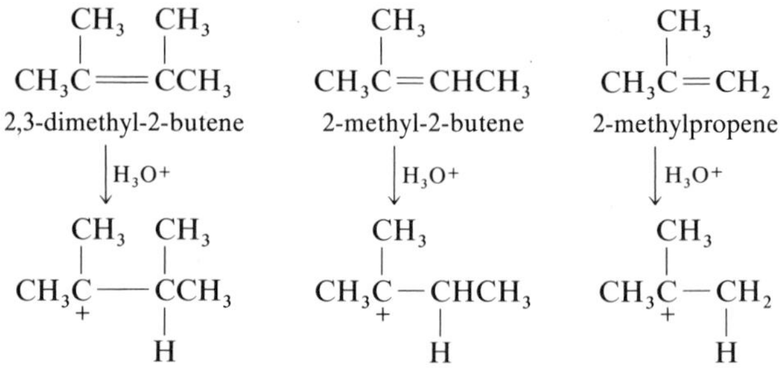

2,3-dimethyl-2-butene   2-methyl-2-butene   2-methylpropene

tertiary carbocations

These reactions are approximately 10,000 times as fast as

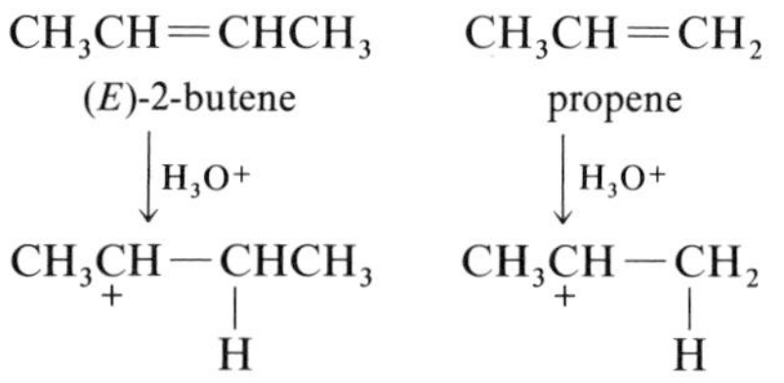

(*E*)-2-butene   propene

secondary carbocations

Such experimental evidence reinforces the idea that alkyl groups are electron releasing when bonded to sp^2-hybridized carbon atoms and stabilize carbocation intermediates (p. 169).

The hydration reaction is a reversible one. Alcohols, when heated in acid, are converted to alkenes, and the reaction is postulated to go through a series of steps that are the reverse of the ones shown for the hydration reaction. Cyclopentanol, for example, gives cyclopentene when heated with phosphoric acid.

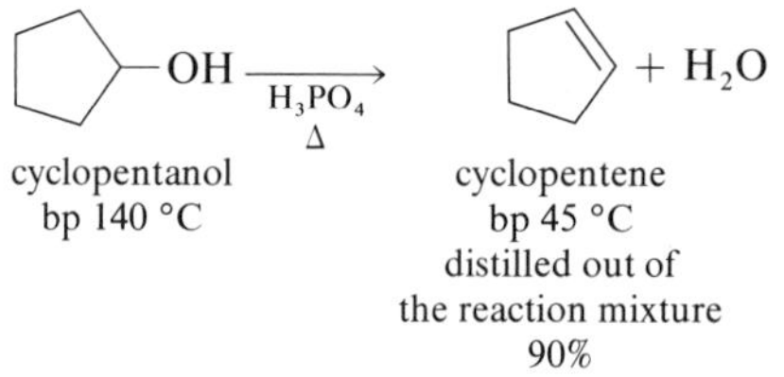

cyclopentanol
bp 140 °C

cyclopentene
bp 45 °C
distilled out of
the reaction mixture
90%

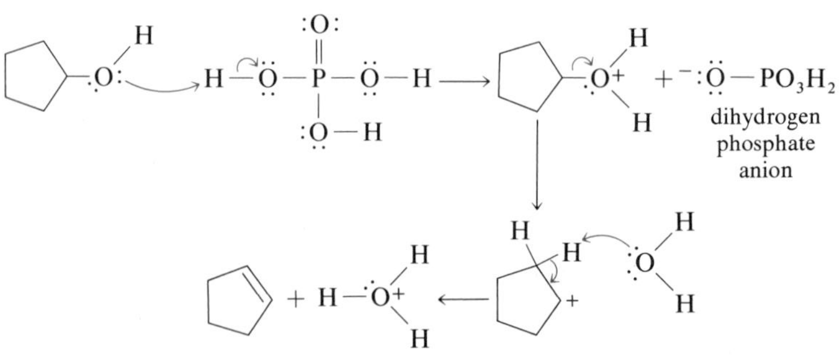

dihydrogen
phosphate
anion

The alcohol is protonated by the acid, giving an oxonium ion. The oxonium ion loses a molecule of water creating a carbocation, which loses a proton to water acting as a base. Thus, *when alkenes are heated with dilute acid at low temperatures, the hydration reaction takes place. If,* on the other hand, *alcohols are heated with more concentrated acid while the products are removed from the reaction mixture,* the reverse reaction occurs and *good yields of alkenes result.*

PROBLEM 5.19   Predict the products of the following reactions.

(a)  $\underset{\text{H}_2\text{SO}_4}{\xrightarrow{\text{H}_2\text{O}}}$

(b) $\underset{\text{H}_2\text{SO}_4}{\xrightarrow{\text{H}_2\text{O}}}$

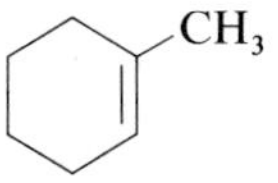

(c) $\text{CH}_3\text{CH}_2\text{CH}_2\text{CH}_2\text{CH}=\text{CH}_2 \underset{\text{H}_2\text{SO}_4}{\xrightarrow{\text{H}_2\text{O}}}$

(d) $\underset{\underset{\text{OH}}{|}}{\text{CH}_3\text{CCH}_3} \overset{\text{CH}_3}{\underset{\Delta}{\overset{|}{\xrightarrow{\text{H}_3\text{PO}_4}}}}$

(e) $\underset{\underset{\text{OH}}{|}}{\text{CH}_3\text{CHCH}_3} \underset{\Delta}{\xrightarrow{\text{H}_3\text{PO}_4}}$

(f) $\text{CH}_2=\text{CH}_2 \underset{\text{H}_2\text{SO}_4}{\xrightarrow{\text{H}_2\text{O}}}$

(g) $\underset{}{\text{CH}_3\overset{\overset{\text{CH}_3}{|}}{\text{C}}=\text{CHCH}_2\text{CH}_3} \underset{\text{H}_2\text{SO}_4}{\xrightarrow{\text{H}_2\text{O}}}$

(h) $\underset{\Delta}{\xrightarrow{\text{H}_3\text{PO}_4}}$ (cyclohexanol with OH)

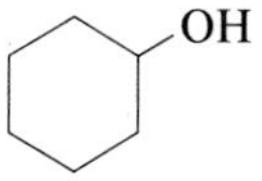

C. Addition of Water to Alkynes

Water adds to acetylene in the presence of acid and mercury(II) and iron(III) salts to give acetaldehyde in an important industrial process.

$$\underset{\text{acetylene}}{\text{HC}\equiv\text{CH}} \underset{\underset{\underset{\text{Fe}_2(\text{SO}_4)_3}{\text{HgSO}_4}}{\text{H}_2\text{SO}_4}}{\xrightarrow{\text{H}_2\text{O}}} \underset{\text{acetaldehyde}}{\overset{\overset{\text{O}}{\|}}{\text{CH}_3\text{CH}}}$$

All other alkynes add water to give ketones. Thus, 1-hexyne is converted into 2-hexanone.

$$\underset{\text{1-hexyne}}{\text{CH}_3\text{CH}_2\text{CH}_2\text{CH}_2\text{C}\equiv\text{CH}} \underset{\underset{\text{HgSO}_4}{\text{H}_2\text{SO}_4}}{\xrightarrow{\text{H}_2\text{O}}} \underset{\underset{80\%}{\text{2-hexanone}}}{\text{CH}_3\text{CH}_2\text{CH}_2\text{CH}_2\overset{\overset{\text{O}}{\|}}{\text{C}}\text{CH}_3}$$

The hydration of an alkyne proceeds by electrophilic attack on the π electrons of the triple bond. Mercury(II) ion, Hg^{2+}, is a Lewis acid and serves as a catalyst in the hydration reaction. Alkynes that are substituted on both sides of the triple bond are more reactive and can be hydrated without the use of a catalyst as shown for 2-pentyne.

$$\underset{\text{2-pentyne}}{\text{CH}_3\text{C}\equiv\text{CCH}_2\text{CH}_3} \underset{\underset{0\text{ °C, 10 min}}{\text{H}_2\text{SO}_4}}{\xrightarrow{\text{H}_2\text{O}}} \underset{\underset{\sim50\%}{\text{3-pentanone}}}{\text{CH}_3\text{CH}_2\overset{\overset{\text{O}}{\|}}{\text{C}}\text{CH}_2\text{CH}_3} + \underset{\underset{\sim50\%}{\text{2-pentanone}}}{\text{CH}_3\text{CH}_2\text{CH}_2\overset{\overset{\text{O}}{\|}}{\text{C}}\text{CH}_3}$$

A general mechanism for the hydration of an alkyne is given below, using 2-pentyne as an example.

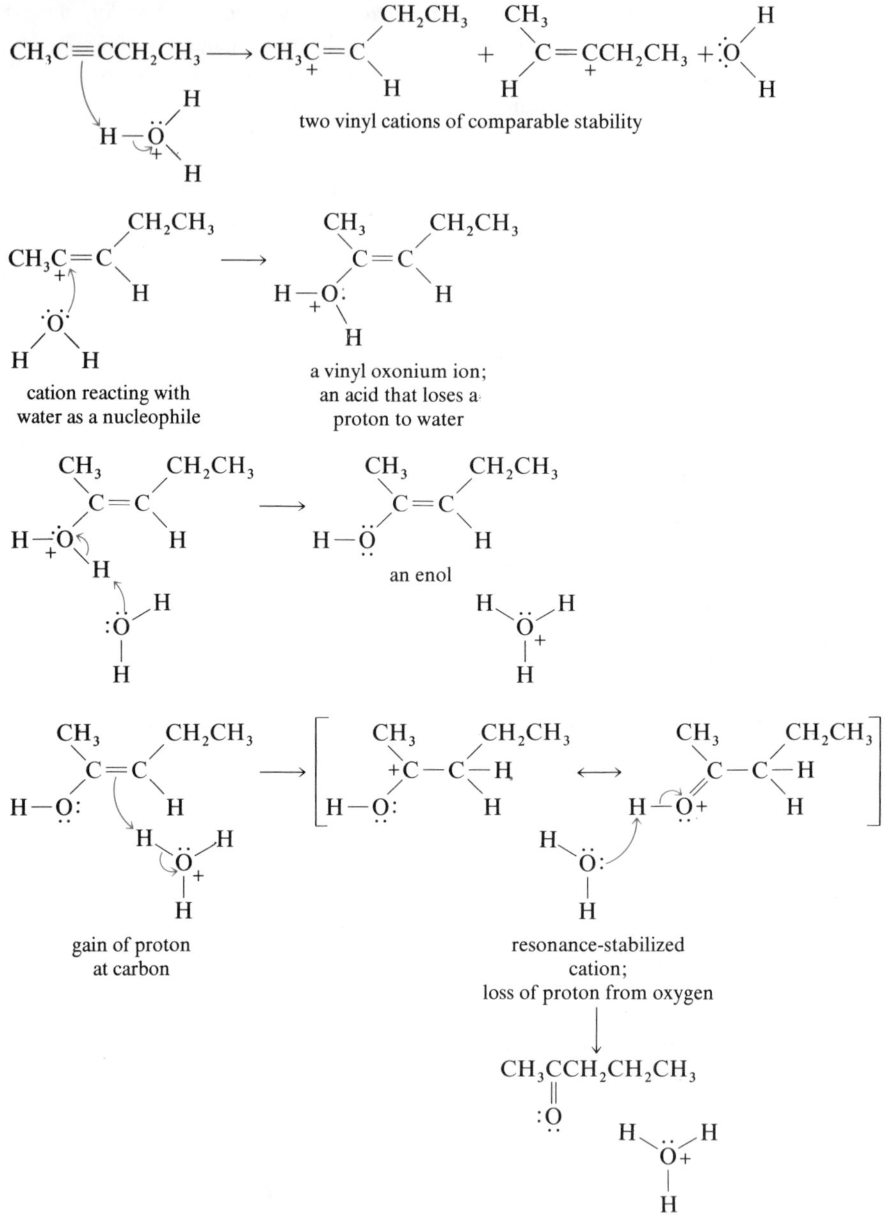

$$CH_3C\equiv CCH_2CH_3 \longrightarrow CH_3\overset{+}{C}=C\overset{CH_2CH_3}{\underset{H}{\Big\backslash}} \quad + \quad \overset{CH_3}{\underset{H}{\Big/}}C=\overset{+}{C}CH_2CH_3 \quad + \ddot{\underset{H}{\overset{H}{O}}}$$

two vinyl cations of comparable stability

cation reacting with water as a nucleophile

a vinyl oxonium ion; an acid that loses a proton to water

an enol

gain of proton at carbon

resonance-stabilized cation; loss of proton from oxygen

$$CH_3CCH_2CH_2CH_3$$
$$\overset{\|}{\underset{:O:}{}}$$

Similarly,

$$\overset{CH_3}{\underset{H}{\Big\backslash}}C=\overset{+}{C}CH_2CH_3 \longrightarrow \longrightarrow \longrightarrow CH_3CH_2CCH_2CH_3$$
$$\overset{}{\underset{\overset{\|}{O}}{}}$$

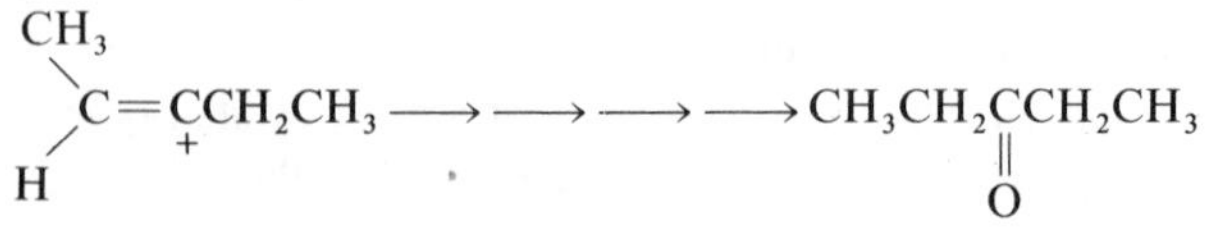

The protonation of an alkyne gives a vinyl cation as an intermediate. In the case of 1-alkynes, the secondary cationic intermediate is favored over the other one, but for 2-pentyne both intermediates are similar in structure and, therefore, in energy.

The cationic intermediates react with water, the nucleophile in the reaction mixture, to give unstable compounds, **enols.** The name enol indicates that such compounds contain a hydroxyl group, the *-ol,* on a double bond, the *en-.* Enols are important intermediates in many organic reactions and they are further discussed in Section 9.1A and Chapter 15.

The enol is protonated easily at the double bond because the carbocation formed adjacent to an oxygen atom is stabilized by resonance. In one of the resonance contributors of the cation, carbon and oxygen each have an octet of electrons around them and the positive charge is delocalized to the oxygen atom. Such cations stabilized by oxygen atoms are important intermediates (see Section 9.7B, for example), and we will encounter them many times.

A close inspection of the resonance-stabilized cation on p. 192 reveals that it is identical to a protonated ketone. Loss of the proton to water, acting as a base, gives 2-pentanone. Hydration of alkynes gives rise to ketones, except for acetylene, in which case acetaldehyde is the product.

PROBLEM 5.20   Fill in the steps for the detailed mechanism for the conversion of 2-pentyne to 3-pentanone.

PROBLEM 5.21   Give a detailed mechanism for the formation of acetaldehyde from acetylene.

PROBLEM 5.22   9-Undecynoic acid, shown below, is hydrated in 80% H_2SO_4. Write an equation for the reaction.

$$CH_3C \equiv C(CH_2)_7 \overset{\displaystyle O}{\overset{\displaystyle \|}{C}} OH$$

5.6

Reactions of Carbocations

A. Relative Stabilities of Carbocations

From experimental observations on the orientation of the addition of hydrogen bromide to propene (Section 5.3A) or the ease with which different alkenes are hydrated (Section 5.5B), we have already reached the qualitative conclusion that tertiary carbocations form more easily than secondary carbocations, which in turn are more accessible than primary carbocations. Quantitative measurements of the relative stabilities of carbocations support these conclusions. Most carbocations are so reactive that they have very short lifetimes in solutions that contain nucleophiles. However, they may be obtained in the gas phase. One reaction that has been observed is the formation of a cation by the loss of an electron from a radical.

$$R^{\cdot}(g) \longrightarrow R^{+}(g) + e^{-}(g)$$

This reaction requires the input of considerable energy, usually by collision with a beam of electrons in a mass spectrometer (Section 13.6). *The energy required to remove an electron from the radical is the* **ionization potential** *of the radical.* The ionization potentials of a number of radicals are given in Table 5.3.

TABLE 5.3   Ionization Potentials of a Series of Radicals

$R \cdot$	Ionization Potential (kcal/mol)
$\overset{\cdot}{C}H_3$	227
$CH_2 = \overset{\cdot}{C}H$	217
$CH_3\overset{\cdot}{C}H_2$	193
$CH_3CH_2\overset{\cdot}{C}H_2$	187
$CH_3CH_2CH_2\overset{\cdot}{C}H_2$	185
$\begin{array}{c} CH_3 \\ | \\ CH_3CH\overset{\cdot}{C}H_2 \end{array}$	185
$\begin{array}{c} CH_3 \\ | \\ CH_3\underset{\cdot}{C}H \end{array}$	174
$\begin{array}{c} CH_3 \\ | \\ CH_3CH_2\underset{\cdot}{C}H \end{array}$	171
$\begin{array}{c} CH_3 \\ | \\ CH_3\underset{|}{C}\cdot \\ CH_3 \end{array}$	160

An inspection of Table 5.3 reveals some trends. The most energy is required to remove an electron from the methyl radical to form a methyl cation. A *tert*-butyl cation, on the other hand, is formed most easily from the corresponding radical. Primary radicals, such as ethyl, *n*-propyl, and *n*-butyl, are converted to primary cations by the expenditure of approximately 190 kcal/mol. Secondary radicals, such as iso-propyl and *sec*-butyl, can be converted to secondary cations with the input of about 170 kcal/mol.

As indicated by bond dissociation energies for hydrocarbons (Section 5.4B), tertiary radicals are more stable than secondary radicals, which in turn are more stable than primary radicals. Therefore, the relative ease with which cations are formed from the corresponding radicals gives us the relative stabilities of the cations. Primary, second-ary, and tertiary carbocations, however, differ more from each other in energy than the corresponding radicals do.

If we wish to attach numbers to the relative stabilities of cations, we must compare isomeric species and take into account the relative stabilities of the starting radicals. For example, we know from comparing the bond dissociation energies for a primary carbon-hydrogen bond and a secondary carbon-hydrogen bond in propane that the isopropyl radical is about 3 kcal/mol more stable than the *n*-propyl radical (Section 5.4B). The ionization potentials for these two radicals differ by 13 kcal/mol (Table 5.3). When these facts are combined, you can see that the isopropyl cation is 16 kcal/mol more stable than the *n*-propyl cation (Figure 5.9).

A similar treatment for the isobutyl radical and the *tert*-butyl radical (Section 5.4B) leads to the conclusion that the *tert*-butyl cation is 32 kcal/mol (7 kcal/mol for the radicals + 25 kcal/mol difference in ionization potentials) more stable than the isobutyl cation, a primary cation.

The formation of ions in the gas phase is a highly endothermic process because it is

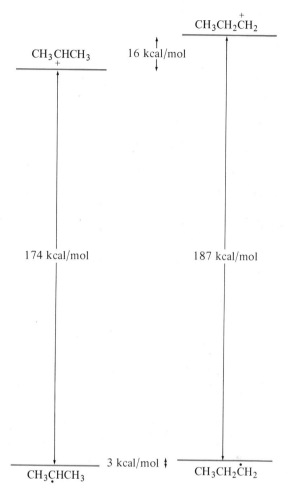

Figure 5.9 A comparison of the relative stabilities of the isopropyl and *n*-propyl radicals, and the corresponding cations.

difficult to separate particles of opposite charge from each other in the absence of a solvent (Section 4.6C). The ionization potential, for example, measures the energy required to remove a negatively charged electron from a positively charged carbocation. Ions are much more easily formed in solution where they are stabilized by ion-dipole interactions with the solvent (Section 1.8B). In recent years, carbocations have been generated by the ionization of alkyl halides in solvents that are highly polar but do not react with carbocations as nucleophiles. In these solvents, the formation of a carbocation can be an exothermic reaction. For example, in sulfuryl chlorofluoride, SO_2ClF, to which antimony pentafluoride, SbF_5, has been added, *tert*-butyl chloride ionizes at $-55\ °C$ with an enthalpy of ionization, ΔH_i, of -24.8 kcal/mol. The ionization of isopropyl chloride is less exothermic under the same conditions with ΔH_i of -15.3 kcal/mol. The ionization is helped by the interaction of the Lewis acid, SbF_5, with chloride ion, a Lewis base.

$$RCl + SbF_5 \xrightarrow[\substack{SO_2ClF \\ -55\ °C}]{} R^+\, SbF_5Cl^-$$

In solution, as in the gas phase, the tertiary carbocation is formed more easily than the secondary one.

B. Rearrangements of Carbocations

At the end of the nineteenth century, Georg Wagner, who was at the University of Warsaw, investigated the reactions of some naturally occurring alkenes and alcohols with acidic reagents and found that some of the products had carbon skeletons that differed from those in the starting materials. These reactions were further investigated by Hans Meerwein in Germany in the 1920s, and he confirmed that the bonding between carbon atoms was being rearranged.

Chemists also observed unexpected products in other reactions with simple alkenes. For example, 3-methyl-1-butene reacts with hydrogen chloride to give 2-chloro-2-methylbutane, in which the chlorine atom is bound to a tertiary carbon atom, as well as 2-chloro-3-methylbutane, which is the expected product.

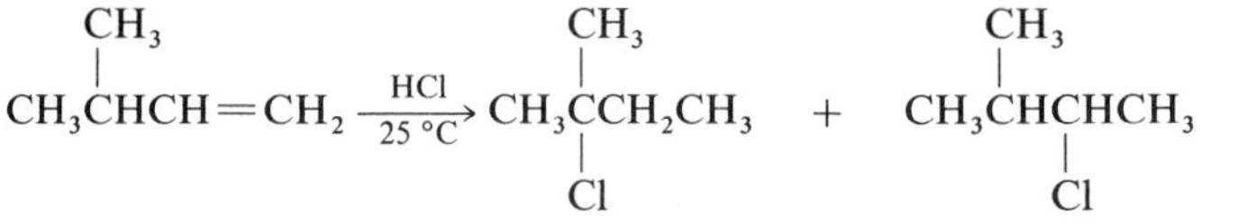

$$
\underset{\substack{\text{3-methyl-1-butene}}}{\overset{CH_3}{\underset{|}{CH_3CHCH=CH_2}}} \xrightarrow[\text{25 °C}]{HCl} \underset{\substack{\text{2-chloro-2-methylbutane} \\ \sim 50\%}}{\overset{CH_3}{\underset{|}{\underset{Cl}{\overset{|}{CH_3CCH_2CH_3}}}}} + \underset{\substack{\text{2-chloro-3-methylbutane} \\ \sim 50\%}}{\overset{CH_3}{\underset{|}{\underset{Cl}{\overset{|}{CH_3CHCHCH_3}}}}}
$$

Frank C. Whitmore, who did much of the early work on the chemistry of alkenes at Pennsylvania State University, postulated that *the secondary carbocation* that formed as an intermediate in the reaction *rearranged to a more stable tertiary carbocation.* The two products were derived from combination of chloride ion with the secondary and tertiary carbocation intermediates.

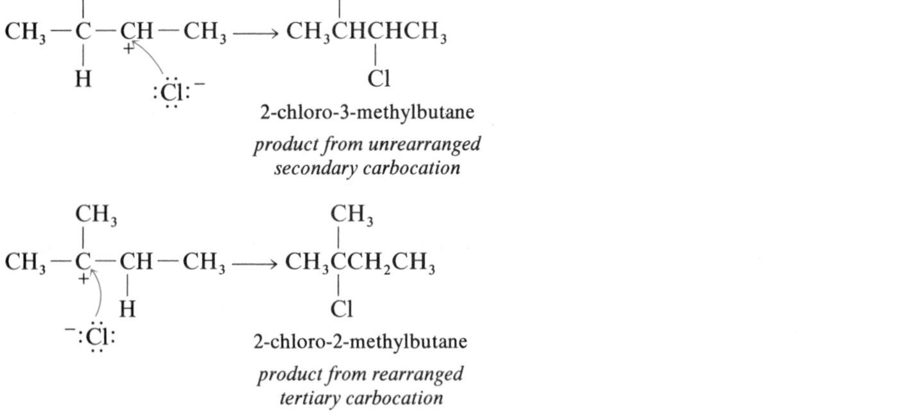

$$
\overset{CH_3}{\underset{H}{\overset{|}{\underset{|}{CH_3-C-CH=CH_2}}}} \longrightarrow \underset{\text{secondary carbocation}}{\overset{CH_3}{\underset{H}{\overset{|}{\underset{|}{CH_3-C-CH-CH_3}}}}} \xrightarrow{\text{1,2-hydride shift}} \underset{\substack{\text{tertiary carbocation} \\ \textit{more stable}}}{\overset{CH_3}{\underset{H}{\overset{|}{\underset{|}{CH_3-C-CH-CH_3}}}}}
$$

$$
\overset{CH_3}{\underset{H}{\overset{|}{\underset{|}{CH_3-C-CH-CH_3}}}} \longrightarrow \underset{\substack{\text{2-chloro-3-methylbutane} \\ \textit{product from unrearranged} \\ \textit{secondary carbocation}}}{\overset{CH_3}{\underset{Cl}{\overset{|}{\underset{|}{CH_3CHCHCH_3}}}}}
$$

$$
\overset{CH_3}{\underset{H}{\overset{|}{\underset{|}{CH_3-C-CH-CH_3}}}} \longrightarrow \underset{\substack{\text{2-chloro-2-methylbutane} \\ \textit{product from rearranged} \\ \textit{tertiary carbocation}}}{\overset{CH_3}{\underset{Cl}{\overset{|}{\underset{|}{CH_3CCH_2CH_3}}}}}
$$

The rearrangement shown above results when a hydrogen atom with the pair of electrons that bind it to a carbon atom moves to an adjacent carbon atom that has a deficiency of electrons. A hydrogen atom with a pair of electrons resembles a hydride ion in its properties, so the rearrangement is called a **hydride shift.** Because it takes place between adjacent carbon atoms, it is a *1,2-hydride shift. The loss of the hydrogen atom and its bonding electrons from the affected carbon atom leaves a new cationic center in the molecule.* In the example shown above, a secondary carbocation is transformed into a more stable tertiary carbocation. Some of the secondary carbocation intermediate reacts with chloride ions without rearrangement, so products derived from both secondary and tertiary carbocations are seen.

Carbon atoms as well as hydrogen atoms can migrate in carbocation rearrangements. Another reaction studied by Whitmore illustrates this.

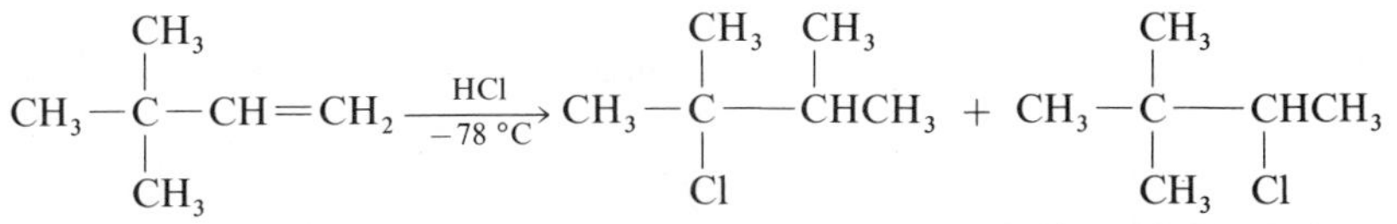

3,3-dimethyl-1-butene   2-chloro-2,3-dimethylbutane   3-chloro-2,2-dimethylbutane
                                         61%   37%

3,3-Dimethyl-1-butene reacts with hydrogen chloride to give a secondary halide and a tertiary halide. The following mechanism for the formation of these compounds is postulated.

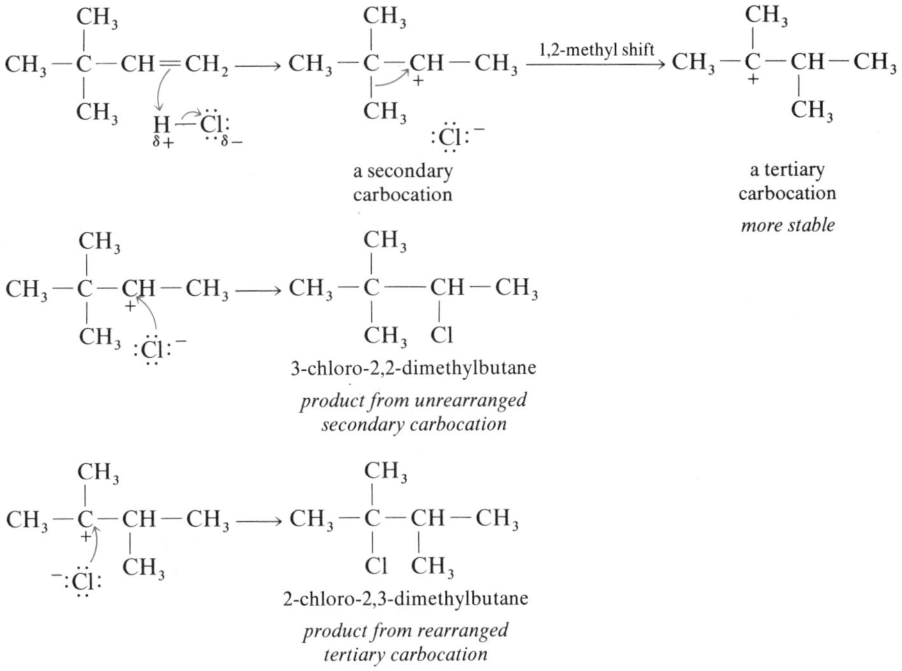

a secondary   a tertiary
carbocation   carbocation

more stable

3-chloro-2,2-dimethylbutane

*product from unrearranged
secondary carbocation*

2-chloro-2,3-dimethylbutane

*product from rearranged
tertiary carbocation*

A secondary carbocation forms when 3,3-dimethyl-1-butene reacts with hydrogen chloride. The shift of a methyl group, along with the pair of electrons that binds it, from a carbon atom to the cationic carbon atom adjacent to it creates a new tertiary carbocation. This rearrangement is called a **1,2-methyl shift.** More generally, alkyl or aryl groups of all sorts may participate in such rearrangements. Note that a 1,2-hydride shift does not change the carbon skeleton of the molecule (p. 196), but a 1,2-alkyl shift does.

Carbocation intermediates may be formed by the loss of water from an alcohol in acid, as well as by the addition of a proton to an alkene. For example, the dehydration of cyclopentanol by phosphoric acid was shown to occur by way of a carbocation on p. 190 (Section 5.5B). Rearrangements occur in the conversion of some alcohols to alkenes, pointing to carbocation intermediates. The dehydration of 3,3-dimethyl-2-butanol is an example. The unrearranged alkene, 3,3-dimethyl-1-butene, is formed in very small amounts. The major products, 2,3-dimethyl-2-butene and 2,3-dimethyl-1-butene, are formed by rearrangement.

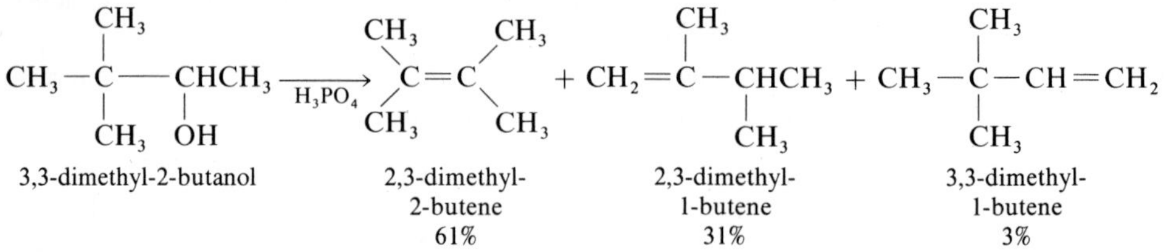

| 3,3-dimethyl-2-butanol | 2,3-dimethyl-2-butene 61% | 2,3-dimethyl-1-butene 31% | 3,3-dimethyl-1-butene 3% |

A mechanism for the dehydration of 3,3-dimethyl-2-butanol is shown below.

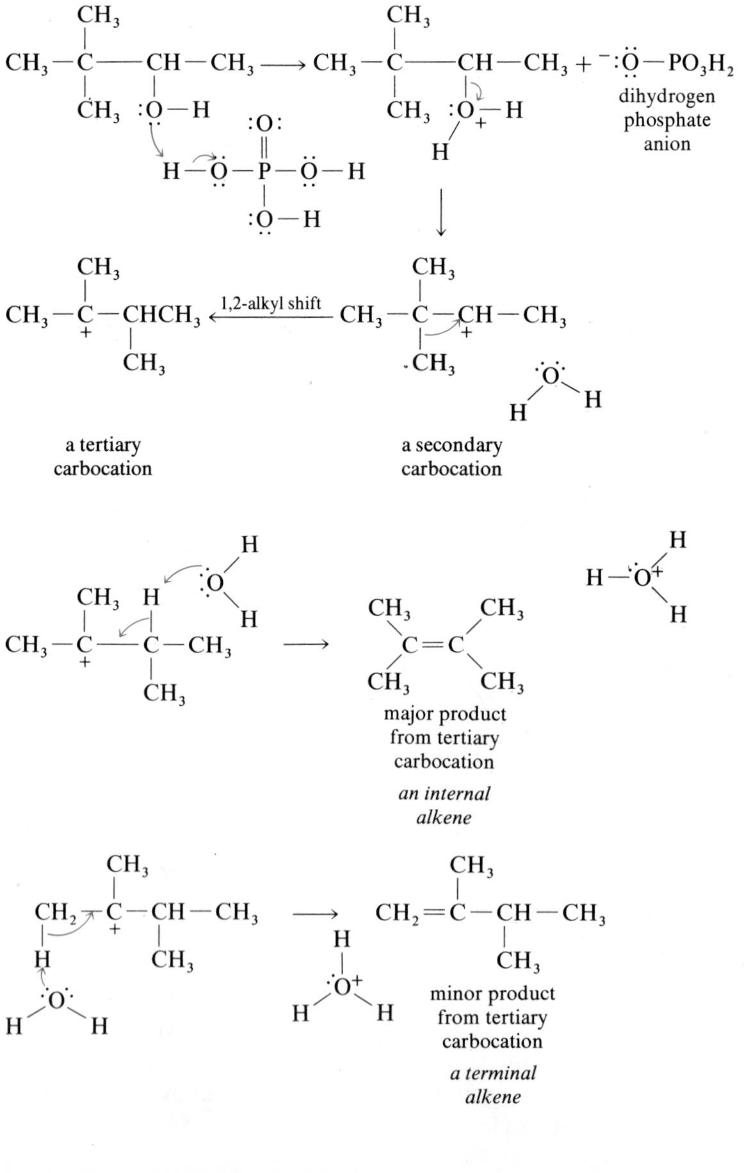

dihydrogen phosphate anion

a tertiary carbocation   a secondary carbocation

major product from tertiary carbocation

an internal alkene

minor product from tertiary carbocation

a terminal alkene

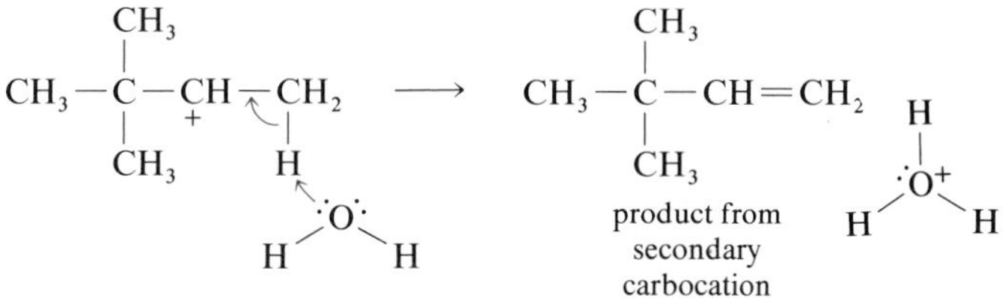

The alcohol is protonated by phosphoric acid and loses water to form a secondary carbocation. The shift of a methyl group creates a tertiary carbocation that has two different kinds of hydrogen atoms. Loss of the tertiary hydrogen atom as a proton to a base, such as water, dihydrogen phosphate anion, or another molecule of alcohol, gives the major product of the reaction, the alkene with the most highly substituted double bond (Section 5.2). Loss of any one of six primary hydrogen atoms from the same cation gives the less highly substituted terminal alkene as the minor product. The secondary carbocation has only one methyl group from which it can lose a proton, and a small amount of product results from that pathway.

The equation above reemphasizes a fundamental reaction of carbocations. *Carbocations are strong acids, and lose a proton easily, even to a weak base. Most reactions that proceed through a carbocation intermediate give rise to some alkene as product.*

Interconversions of carbocations are quite general phenomena. The alkene mixture seen when 3,3-dimethyl-2-butanol is dehydrated is formed whenever any one of the three product alkenes isolated in the pure state is subjected to the acid conditions of the dehydration reaction. Such a product composition represents, therefore, an equilibrium mixture of secondary and tertiary carbocations.

In this section, we have reviewed two ways of forming alkyl cations: by the protonation of an alkene and by loss of water from a protonated alcohol. We have also explored further the major reactions of carbocations. They react as Lewis acids with electron-rich species and as Brønsted-Lowry acids to lose a proton to a base. If their structure permits, they also rearrange to other carbocations in equilibrium reactions.

PROBLEM 5.23   The following was observed:

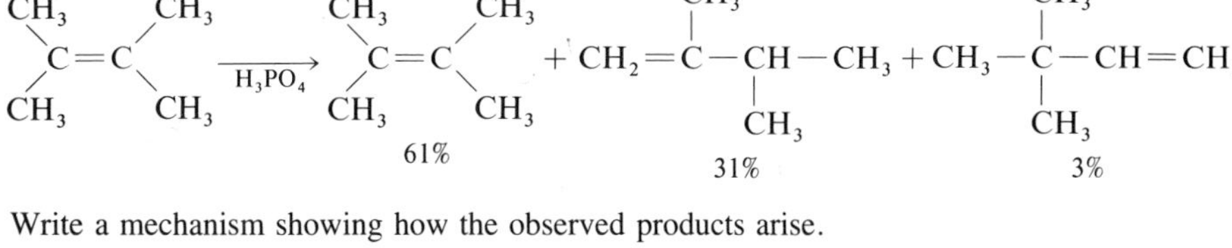

Write a mechanism showing how the observed products arise.

PROBLEM 5.24   The following reaction was observed. Write a mechanism rationalizing the product that was obtained.

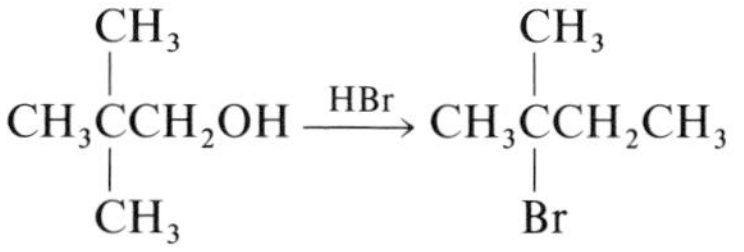

PROBLEM 5.25   A mixture of two primary alcohols obtained from natural sources, 3-methyl-1-butanol and 2-methyl-1-butanol, gives chiefly 2-methyl-2-butene when dehydrated with acid.

2-Methyl-2-butene is converted to 2-methyl-2-butanol in 90% yield by treatment with 50% aqueous sulfuric acid at 0 °C. Write equations showing the details of the conversion of the mixture of primary alcohols into a single tertiary alcohol.

C. Reaction of Carbocations with Alkenes

Alkenes are electron-rich reagents because the electrons of the π bond are available for reaction with electrophiles. Whenever carbocations are formed in reaction mixtures containing alkenes, some reaction between the electron-deficient intermediate and an alkene can be expected. Under some conditions, this reaction may be a major one, and useful too. An example is the dimerization of 2-methylpropene that takes place in sulfuric acid. The same product mixture is obtained whether the alkene or the corresponding alcohol, *tert*-butyl alcohol, is used as the starting material, pointing to the identity of the intermediates in both cases.

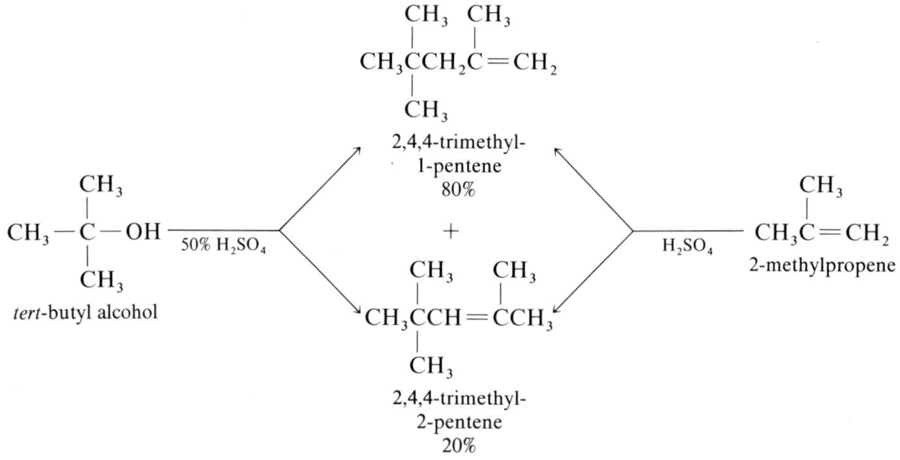

The reaction is postulated to go through the *tert*-butyl cation, which adds to the double bond of the alkene to give a new cation. The two alkenes are formed by the loss of a proton in two different ways from the intermediate.

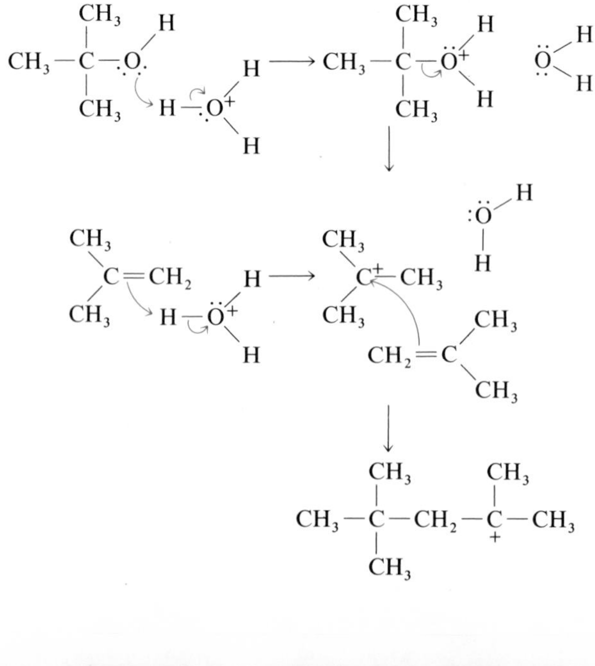

CH₃   CH₃   CH₃   CH₃
 | | | |
CH₃—C—CH₂—C⁺—CH₂   ⟶   CH₃—C—CH₂—C=CH₂
 | | | H
CH₃   H:Ö:H   CH₃   +|
              | 2,4,4-trimethyl-   :O:
              H   1-pentene   H   H

CH₃   CH₃   CH₃   CH₃
 | | | |
CH₃—C——CH—C⁺—CH₃   ⟶   CH₃—C—CH=C—CH₃
 | | |
CH₃ H   H   CH₃
                                          | 2,4,4-trimethyl-
        :O: +O: 2-pentene
      H   H   H   H

The two products have the same carbon skeleton and can be converted, by the addition
of hydrogen (Sections 2.8, 8.2B), to the same alkane, 2,2,4-trimethylpentane, which
is an important constituent of high octane gasoline.

CH₃ CH₃   CH₃ CH₃   CH₃ CH₃
 | | | | | |
CH₃CCH₂C=CH₂ + CH₃CCH=CCH₃ —H₂→   CH₃CCH₂CHCH₃
 | | catalyst |
CH₃   CH₃   CH₃

2,4,4-trimethyl-   2,4,4-trimethyl-   2,2,4-trimethylpentane
  1-pentene   2-pentene

The reaction does not have to stop with the combination of two alkene units and, in
fact, with longer heating, complex mixtures containing higher molecular weight alkenes are formed. Further additions to 2-methylpropene can occur.

CH₃   CH₃   CH₃   CH₃   CH₃
 | | | | |
CH₃—C—CH₂—C⁺—CH₃ ⟶ CH₃—C—CH₂—C—CH₂—C⁺—CH₃
 | | | |
CH₃   CH₃   CH₃   CH₃
dimeric   CH₃   trimeric   CH₃
cation   CH₂=C   cation   CH₂=C
                 CH₃   CH₃
                                                               ↓

                        CH₃   CH₃   CH₃   CH₃
                         | | | |
                CH₃—C—CH₂—C—CH₂—C—CH₂—C⁺—CH₃
                         | | |
                        CH₃   CH₃   CH₃

                                  tetrameric
                                    cation
                                      ↓
                                      ↓
                                      ↓
                             polymeric material

    In these reactions, 2-methylpropene is *the* **monomer,** *a low molecular weight unit
that adds to itself in a repetitious fashion to give the higher molecular weight material.
Two units of the monomer combine to give a* **dimer;** *three form a* **trimer,** *and so on.
A large molecule consisting of many units of monomer bonded together is called a*
polymer. The synthesis of polymeric materials with useful properties is an important
area of organic chemistry that we will consider in Chapter 20.

5.7

An Exploration of the Stereochemistry of Addition Reactions of Alkenes

A. The Addition of Hydrogen Bromide to 1-Butene

Addition of hydrogen bromide to 1-butene in the absence of peroxides gives 2-bromobutane.

(*R*)-2-bromobutane (*S*)-2-bromobutane

racemic mixture

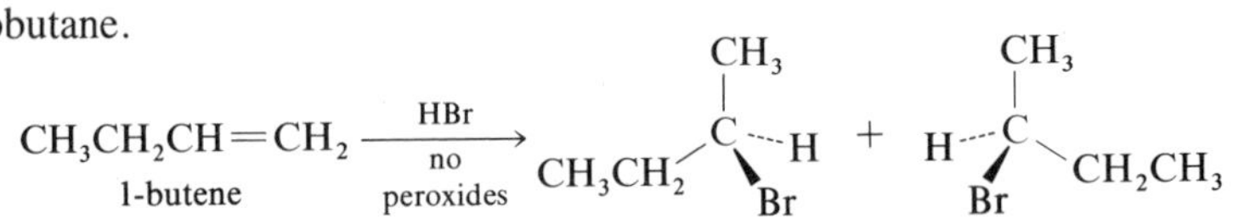

2-Bromobutane has an asymmetric carbon atom and is formed as a racemic mixture (Section 3.9D) containing equal amounts of the two enantiomers, (*R*)- and (*S*)-2-bromobutane (Section 3.9F).

The two stereoisomeric 2-bromobutanes are formed in equal amounts because the reaction goes by way of a symmetrical carbocation intermediate. The bromide ion reacts with equal probability at either face of the cation.

(*S*)-2-bromobutane (*R*)-2-bromobutane

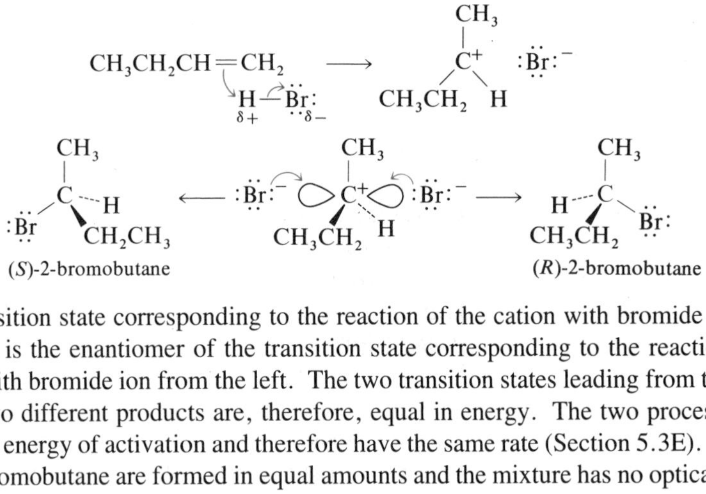

The transition state corresponding to the reaction of the cation with bromide ion from the right is the enantiomer of the transition state corresponding to the reaction of the cation with bromide ion from the left. The two transition states leading from the cation to the two different products are, therefore, equal in energy. The two processes have the same energy of activation and therefore have the same rate (Section 5.3E). (*R*)- and (*S*)-2-Bromobutane are formed in equal amounts and the mixture has no optical activity (Section 3.9D). The two enantiomers have the same physical properties, except for optical rotation, and cannot be separated from each other by ordinary physical methods, such as distillation.

The equations written in this section show the formation of both enantiomers of 2-bromobutane as an exercise to increase our awareness of what is happening stereochemically in such a reaction. Normally, when we use either an achiral or a racemic starting material, we show different diastereomers that are formed but do *not* show both enantiomers if chiral products are obtained.

B. Introduction of a Second Chiral Center into a Molecule Containing One Asymmetric Carbon Atom

Adding hydrogen bromide to 1-butene converted a compound with no chirality into one with one chiral center. What are the stereochemical consequences of introducing a second chiral center into a molecule that is already chiral? Suppose we were to add hydrogen bromide to (*R*)-3-methyl-1-pentene in the absence of peroxides. We would expect Markovnikov addition to the double bond to give two compounds. The second carbon atom in the chain is converted to a chiral center in this reaction and may have either the *R* or the *S* configuration because the bromide ion may add to the planar carbocation intermediate (Section 5.7A) at either face.

(2*S*, 3*R*)-2-bromo-3-methylpentane

(a) addition of bromide from above

(b) addition of bromide from below

(2*R*, 3*R*)-2-bromo-3-methylpentane

The products of this reaction are diastereomers of each other (Section 5.1A). They retain the chirality that was present in (*R*)-3-methyl-1-pentene because no bonds are broken at the asymmetric carbon atom. They have opposite chirality at the new chiral center created in the reaction. Thus, they are stereoisomers because they differ in the configuration of one asymmetric carbon atom, but they are not mirror-image isomers because they have the same configuration at the other asymmetric center.

   More interestingly, the transition states for the formation of the two products from the intermediate carbocation and bromide ion are also diastereomeric, and therefore differ in stability. The process that leads from the carbocation to one product differs in energy of activation from the process giving rise to the other product. Therefore, we expect the two products to be formed in different amounts (Section 5.3E).

   The mixture that is formed has optical activity (Section 3.9D). The optical rotations of the two products are different from each other, and do not (except by chance) cancel each other. The diastereomers also have different physical properties, such as boiling points. In principle, we can separate them from each other by ordinary methods of purification, such as fractional distillation. The separated isomers are, of course, optically active.

PROBLEM 5.26   Suppose that we used racemic 3-methyl-1-pentene instead of the (*R*)-isomer in the reaction with hydrogen bromide. What would the stereochemical consequences be? How many isomers would be formed? Into how many fractions could the product be separated by fractional distillation? Would the original mixture have optical activity? Would the separated fractions have optical activity?

C. Generation of Two Chiral Centers in the Addition of Bromine to Cyclopentene

The addition of bromine to the π bond in cyclopentene, a typical reaction of alkenes (Section 2.8), gives 1,2-dibromocyclopentane in which the two bromine atoms are on opposite sides of the ring.

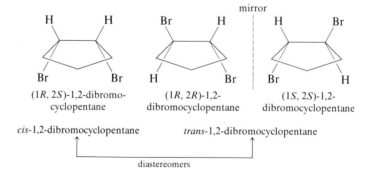

Two stereoisomers, enantiomers of each other, are formed in equal amounts in the reaction.

The rigidity of the ring and the lack of free rotation around the ring carbon-carbon bonds (Section 3.8B) is comparable to the rigidity of the π bond in alkenes (Section 5.1A). Therefore, the language used to describe stereoisomerism in alkenes applies to cyclic compounds, too. Two kinds of 1,2-dibromocyclopentanes are possible. The two substituents may be on opposite sides of the plane of the ring, or trans to each other. Alternatively, they may be on the same side of the ring, or cis to each other (Figure 5.10).

Figure 5.10 Different isomers of the 1,2-dibromocyclopentanes.

Each compound has two chiral centers. In the *trans*-1,2-dibromocyclopentanes, one isomer has the R configuration while the other one has the S configuration at each asymmetric carbon atom. Thus the two isomers are mirror-image isomers, enantiomers, of each other. The molecules of the isomers are not superimposable on each other and are, therefore, truly different species.

cis-1,2-Dibromocyclopentane is a stereoisomer of the two trans compounds, but is *not* a mirror-image isomer of either one. With the R configuration at one chiral center and the S configuration at the other one, it is a diastereomer (Section 5.1A) of the trans compounds. *cis*-1,2-Dibromocyclopentane is interesting in another way. It does not have an enantiomer. A mirror image of *cis*-1,2-dibromocyclopentane can be drawn, but it is superimposable on the original compound (Figure 5.11). The super-imposability of the mirror image of *cis*-1,2-dibromocyclopentane on itself points to the symmetry of the compound (Section 3.9B). A plane of symmetry bisects carbon atom

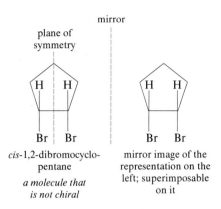

Figure 5.11   The symmetry of *cis*-1,2-dibromocyclopentane.

4, the two hydrogen atoms substituted on it, and the carbon-carbon bond between carbon atoms 1 and 2. *cis*-1,2-Dibromocyclopentane is not chiral and has no optical activity. It is known as a **meso** form. *A meso compound has asymmetric carbon atoms, but no overall chirality because of the existence of internal symmetry.* The *trans*-1,2-dibromocyclopentanes are chiral. Each one individually would be optically active. But in the reaction of bromine with cyclopentene, they form in equal amounts, resulting in a racemic mixture. Such a mixture, as we already know, does not exhibit optical activity (Section 3.9E).

PROBLEM 5.27   How many stereoisomers are possible for 1,3-dimethylcyclopentane? Draw the structural formulas, name the compounds, and discuss the stereochemical relationships among them.

PROBLEM 5.28   How many stereoisomers are possible for 1-bromo-2-methylcyclopentane? Explore the possibilities with structural formulas and names giving stereochemical designations at the chiral centers. You may want to use molecular models.

D. Configuration and Conformation of Disubstituted Cyclohexanes

The stereochemical relationships in substituted cyclopentanes, which have rings that are nearly planar (Section 3.8B), are easier to see than those in cyclohexane derivatives, in which the ring is puckered (Section 3.8C). Cyclohexene, too, adds bromine to give a mixture of enantiomeric *trans*-1,2-dibromocyclohexanes.

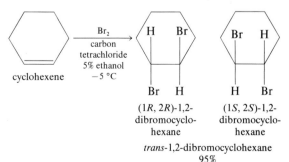

As is the case for the *trans*-1,2-dibromocyclopentanes, the two bromine atoms are on opposite sides of the ring. Because of the flexibility of the ring in cyclohexane, and the interconversion of the ring between two chair forms, each enantiomer of 1,2-dibromocyclohexane exists mainly in two conformations. In one, the two bromine atoms occupy equatorial positions on the ring; in the other, the two bromine atoms are both in axial positions (Section 3.8C). The conformers of (1*S*,2*S*)-1,2-dibromocyclohexane are shown below.

planar form

diaxial form of (1*S*, 2*S*)-1,2-dibromocyclohexane

diequatorial form of (1*S*, 2*S*)-1,2-dibromocyclohexane

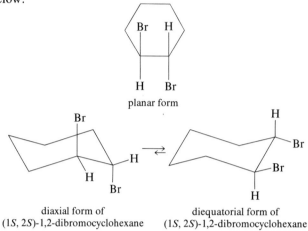

The diequatorial form of *trans*-1,2-dibromocyclohexane predominates in the equilibrium mixture at room temperature. In the following pages we will explore the configurational and conformational relationships of disubstituted cyclohexanes. We

(1*R*, 2*S*)-1,2-dimethylcyclohexane

(1*S*, 2*R*)-1,2-dimethylcyclohexane

cis-1,2-dimethylcyclohexane

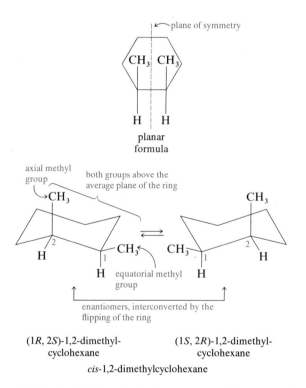

Figure 5.12 *cis*-1,2-Dimethylcyclohexane, with one axial and one equatorial methyl group.

will use the dimethylcyclohexanes, which have been well investigated, as examples, and will be especially aware of the relative stabilities of different conformers.

For a cyclohexane with a single substituent on the ring, the preferred conformation is the chair form in which the substituent is in the equatorial position (Section 3.8C). When there are two substituents on the ring, they may be either cis or trans to each other. 1,2-Dimethylcyclohexane has three different configurational isomers. In *cis*-1,2-dimethylcyclohexane, one methyl group is axial and the other one is equatorial (Figure 5.12).

When *cis*-1,2-dimethylcyclohexane flips from one chair form to the other, the methyl group that is in the equatorial position becomes axial, while the methyl group that is axial becomes equatorial. In both conformations, the number and type of axial and equatorial substituents remain the same, and the two conformers are equal in energy and are present in equal amounts. The second conformer is, in fact, the enantiomer of the first one. The two enantiomeric forms of *cis*-1,2-dimethylcyclohexane interconvert rapidly, with the practical consequence that the compound cannot be separated into optically active mirror-image isomers. In effect, a *cis*-1,2-disubstituted cyclohexane with identical substituents behaves like a meso compound with a plane of symmetry as suggested by the simplified planar formula above.

trans-1,2-Dimethylcyclohexane exists as two enantiomers that, in principle, can be separated and are optically active (Figure 5.13). Each enantiomer of *trans*-1,2-dimethylcyclohexane has two conformations, one in which both methyl groups are equatorial, and the other in which they are both axial. The trans stereochemistry for

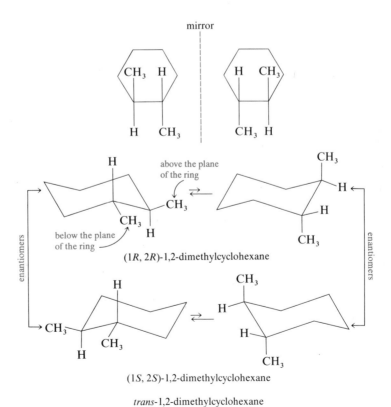

Figure 5.13 The conformations of the two enantiomers of *trans*-1,2-dimethylcyclohexane.

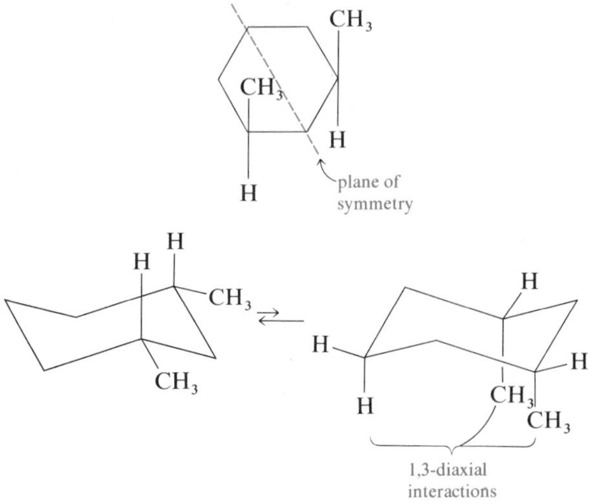

Figure 5.14   The conformations of *cis*-1,3-dimethylcyclohexane.

the two groups is most easily seen for the conformer in which the groups are diaxial, but it is nevertheless true that one of the equatorial methyl groups lies above the plane of the ring and the other one below it in the other conformer. The diaxial conformers are higher in energy by about 2.7 kcal/mol and more of the diequatorial form is present at equilibrium. *cis*-1,2-Dimethylcyclohexane is diastereomeric with either one of the *trans*-1,2-dimethylcyclohexanes.

   1,3-Dimethylcyclohexane also has three configurational isomers. In one conformer of *cis*-1,3-dimethylcyclohexane, both of the alkyl groups are equatorial; in the other conformer both are axial (Figure 5.14). The diequatorial conformer of *cis*-1,3-di-methylcyclohexane is more stable than the diaxial one by about 5.4 kcal/mol, largely because of 1,3-diaxial interactions (Section 3.8C) of the two methyl groups. *cis*-1,3-Dimethylcyclohexane has a plane of symmetry that is most easily seen in its planar formula, but exists in the two chair forms as well. If you build a model of *cis*-1,3-dimethylcyclohexane in the chair form and then build a model of the mirror image of the first form that you created, you will see that the two structures are superimposable upon each other. *cis*-1,3-Dimethylcyclohexane is a true meso form, and not optically active. It is an example of a compound in which there are asymmetric carbon atoms, but the molecule as a whole is achiral owing to the presence of a plane of symmetry (compare Section 5.7C).

   trans-1,3-Dimethylcyclohexane exists as two enantiomers that are not inter-convertible by flipping the cyclohexane ring. Each enantiomer has one axial and one equatorial methyl group, and flipping the cyclohexane ring gives the same structure back in each case (Figure 5.15). *trans*-1,3-Disubstituted cyclohexanes can, in prin-ciple, be separated into optically active components.

   Neither *cis*- nor *trans*-1,4-dimethylcyclohexane has asymmetric carbon atoms, and both are optically inactive. Starting with either carbon atom 1 or carbon atom 4, an investigation of the substituents on that carbon atom turns up a hydrogen atom, a methyl group, and then two branches of the ring that are identical going in either direction. The molecules as a whole have planes of symmetry bisecting carbon atoms 1 and 4 and the methyl group and hydrogen atom on each one.

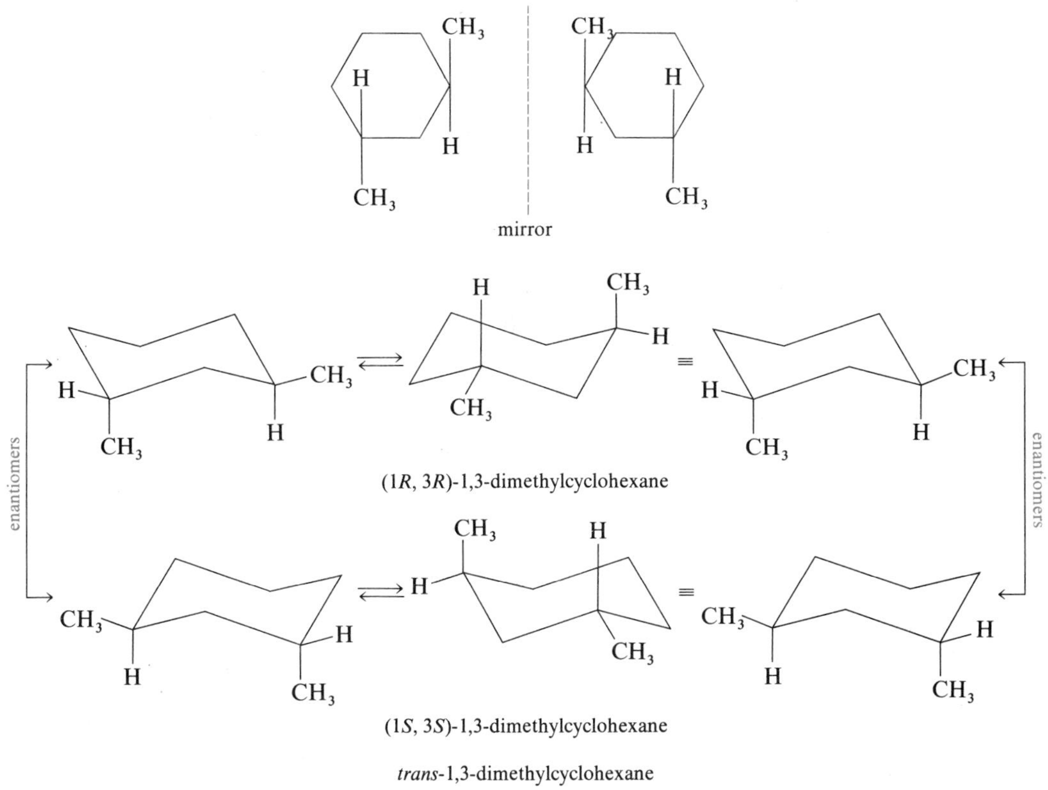

mirror

(1*R*, 3*R*)-1,3-dimethylcyclohexane

enantiomers

enantiomers

(1*S*, 3*S*)-1,3-dimethylcyclohexane

trans-1,3-dimethylcyclohexane

Figure 5.15   The enantiomeric *trans*-1,3-dimethylcyclohexanes.

In *cis*-1,4-dimethylcyclohexane one methyl group is axial and the other equatorial (Fig. 5.16). Flipping the ring gives a form that is identical with the original structure.

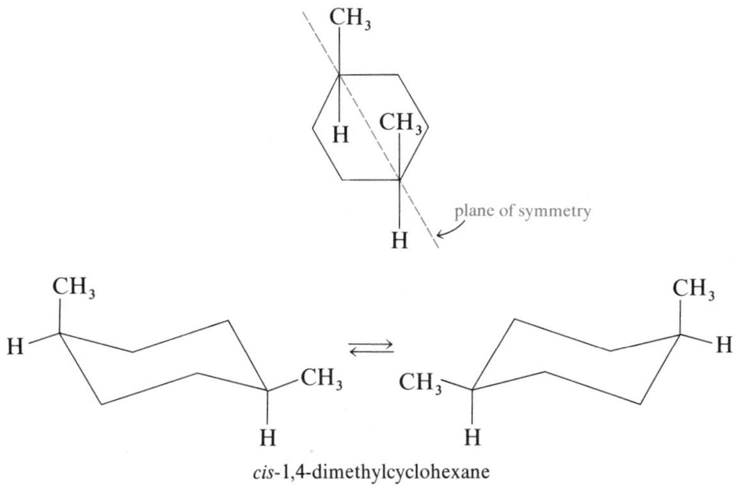

plane of symmetry

cis-1,4-dimethylcyclohexane

Figure 5.16   The symmetry of *cis*-1,4-dimethylcyclohexane.

msp. 137

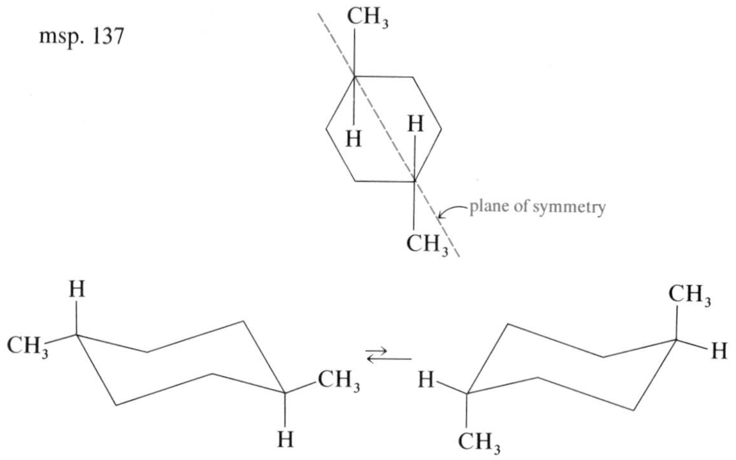

trans-1,2-dimethylcyclohexane

Figure 5.17 Conformers of *trans*-1,4-dimethylcyclohexane.

trans-1,4-Dimethylcyclohexane has two conformers, one in which the two methyl groups are equatorial, and the other in which they are axial (Fig. 5.17). The form with two equatorial substituents is more stable than the other by about 3.6 kcal/mol.

Although we used dimethylcyclohexanes as typical examples of configurational isomers and conformers of disubstituted cyclohexanes, you should keep in mind that similar relationships exist for other substituents whenever the two groups on the ring are the same. If the substituents are not identical, the *cis*-1,2- and *cis*-1,3- compounds will no longer have planes of symmetry, but will have isolable enantiomers and optical activity.

PROBLEM 5.29 How many different configurational isomers are possible for 1-bromo-2-methyl-cyclohexane? Illustrate by drawing planar formulas and chair conformations.

PROBLEM 5.30
(a) In a stereoisomer of 2-isopropyl-5-methylcyclohexanol, the methyl group is cis to the hydroxyl group while the isopropyl group is trans. Draw the two chair conformers of the compound and decide which conformer would be the more stable.
(b) Draw the chair conformations of a stereoisomer of the compound in (a). There are several possible stereoisomers.

5.8

The Addition of Bromine to Alkenes

A. Bromine as an Electrophile

A halogen is an electrophile in the sense that it can accept electrons to form halide anions. This property of halogens leads to their addition to carbon-carbon double bonds (Section 2.8, 5.7C, 5.7D). The reaction is most useful in the case of bromine. Even though a bromine molecule is symmetrical, the approach of the π electrons of the alkene changes the distribution of the electrons in the covalent bond. One of the bromine atoms becomes more positive, the other one more negative. This polarization of the bond in bromine enables bonding to take place between the alkene and the halogen.

induced polarization of the bromine molecule on approach of a π bond

A solution of bromine in carbon tetrachloride has the typical reddish brown color of elemental bromine. When such a solution is added to an alkene, the color of bromine rapidly disappears. This reaction serves as a test for the presence of carbon-carbon multiple bonds and distinguishes alkenes and alkynes with π bonds from alkanes, which have no π bonds. Aromatic compounds such as benzene, in which π bonds are present in an especially stable arrangement, do not react with bromine under these conditions (Section 2.8).

The addition of bromine to ethylene appears to proceed in two steps. Evidence is provided by the following experiments. If bromination is carried out in the presence of species that can act as nucleophiles, such as negatively charged ions other than bromide, or in a solvent such as water, mixtures of products are obtained. When bromide ion or chloride ion is the nucleophile, the products are alkyl halides. If water is the nucleophile, the product has a hydroxyl group and a halogen atom on adjacent carbon atoms. Such compounds are known as **halohydrins.**

$$CH_2\!\!=\!\!CH_2 + Br_2 \xrightarrow[H_2O]{NaCl\ (saturated)} BrCH_2CH_2Br + BrCH_2CH_2Cl$$

ethylene   bromine   1,2-dibromoethane   1-bromo-2-chloro-
                                                                                   54%   ethane
                                                                                                               46%

$$CH_2\!\!=\!\!CH_2 + Br_2 \xrightarrow[0\,°C]{H_2O} BrCH_2CH_2OH + BrCH_2CH_2Br$$

ethylene   bromine   2-bromoethanol   1,2-dibromoethane
                                                           54%   37%

These facts have led to the proposal that reaction of bromine with the double bond gives an intermediate cation that then reacts with any nucleophilic species present.

B. The Bromonium Ion and the Stereochemistry of the Addition of Bromine to Alkenes

A simple carbocation is a symmetrical planar intermediate, and reaction can take place from either side of the plane (Sections 5.7A, 5.7B). If the addition of bromine to cyclopentene (Section 5.7C) proceeds by nucleophilic attack of the double bond on bromine with the resulting formation of a carbocation as an intermediate, some *cis*-1,2-dibromocyclopentane would be expected as a product.

cis-1,2-dibromo-   *trans*-1,2-dibromo-
cyclopentane   cyclopentane

In fact, only the trans isomer is obtained from the reaction. This observation, combined with other experiments in which the reaction did not proceed as would be expected of a carbocation intermediate, led to the idea that the cation is a cyclic **bromonium ion.**

bromonium ion shown as three
resonance contributors

The bromonium ion is written as if the bromine atom were bonded equally to both carbon atoms of the double bond. It need not be so. A strong electrostatic interaction between the electron cloud of the bromine atom and the carbocation at one of the carbon atoms would also lead to the same results. In effect, the bromine blocks one side of the double bond. The incoming nucleophile can approach the molecule only from the other side, so that the overall addition to the double bond leads to a trans stereochemistry. *A reaction in which one of several possible stereochemical results is favored is called a* **stereoselective reaction.** The addition of bromine to the double bond in cyclopentene is 100% stereoselective, leading only to *trans*-1,2-dibromocyclopentane as a product. The bromonium ion is symmetrical in the case of cyclopentene, so nucleophilic attack at either one of the carbon atoms is equally likely. A racemic mixture of the two enantiomeric *trans*-1,2-dibromocyclopentanes is formed.

An addition reaction in which the two components that add to the double bond end up trans to each other is said to proceed with **anti** *stereochemistry.* The origin of this nomenclature is seen more clearly in the addition of bromine to an acyclic alkene. The reaction of (Z)-2-pentene with bromine gives (2S,3S)-2,3-dibromopentane and (2R,3R)-2,3-dibromopentane (Figure 5.18). The products from the reaction of (E)-2-pentene with bromine are (2S,3R)-2,3-dibromopentane and (2R,3S)-2,3-dibromopentane (Figure 5.19). In each case, a bromonium ion of specific stereochemistry is formed. The two possible ways that the bromonium ion can be opened give rise to mixtures of enantiomers from each alkene. The anti orientation (Section 3.5B) of the bromine atoms in the products is seen clearly for these acyclic compounds. (Z)-2-Pentene gives rise to one set of enantiomeric 2,3-dibromopentanes, the (E)-isomer, to another set. The addition of bromine to these two different alkenes proceeds with high stereoselectivity.

There are four different stereoisomers of 2,3-dibromopentane, which has two different asymmetric carbon atoms. They exist as pairs of enantiomers. (2S,3S)-2,3-Dibromopentane and (2R,3R)-2,3-dibromopentane are enantiomers. Note that the names of these compounds include the correct designation of configuration for each of the asymmetric carbon atoms. The configuration at both carbon atoms 2 and 3 for one compound is S, for the other compound is R. Thus, these two compounds have the

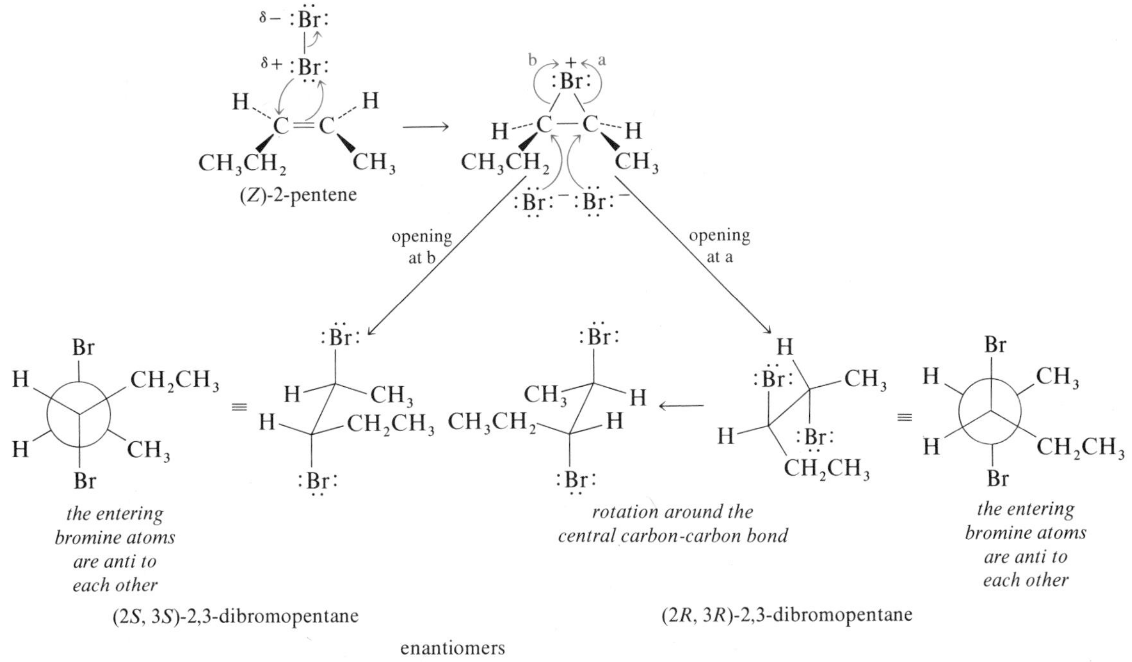

Figure 5.18   The reaction of (*Z*)-2-pentene with bromine to form (2*S*,3*S*)-2,3-dibromopentane and (2*R*,3*R*)-2,3-dibromopentane.

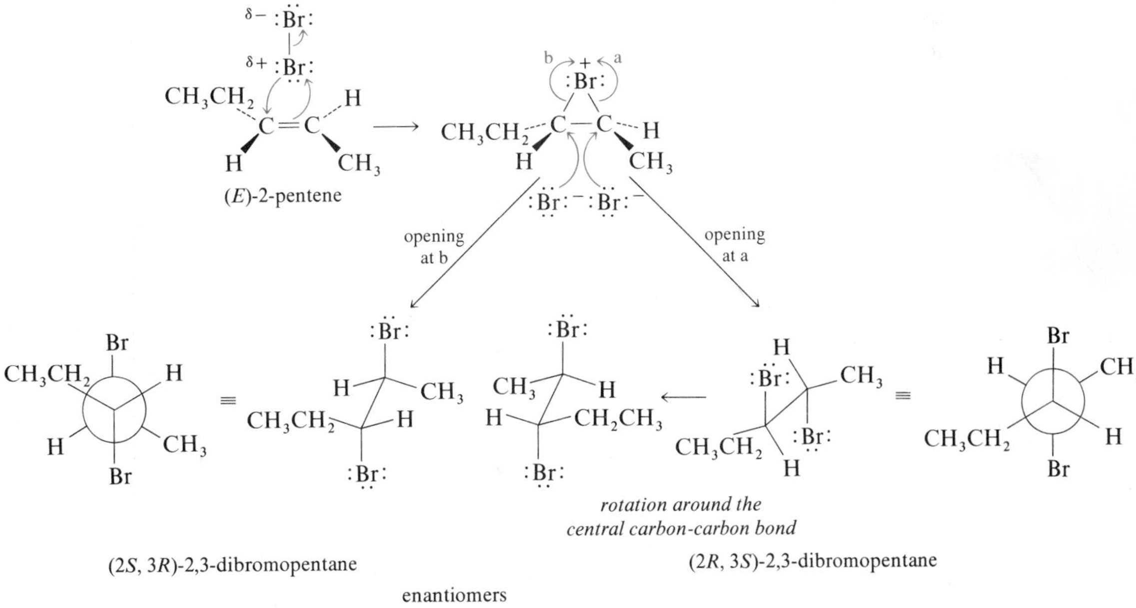

Figure 5.19   The reaction of (*E*)-2-pentene with bromine to form (2*S*,3*R*)-2,3-dibromopentane and (2*R*,3*S*)-2,3-dibromopentane.

opposite configuration at each carbon atom, and are mirror-image isomers (enantiomers) of each other. A similar analysis shows that $(2S,3R)$-2,3-dibromopentane and $(2R,3S)$-2,3-dibromopentane are also mirror-image isomers of each other. They have exactly the opposite configuration at each asymmetric carbon atom. $(2S,3S)$-2,3-Dibromopentane (or its enantiomer), however, is a diastereomer of $(2S,3R)$-2,3-dibromopentane (or its enantiomer). They are stereoisomers but not mirror-image isomers.

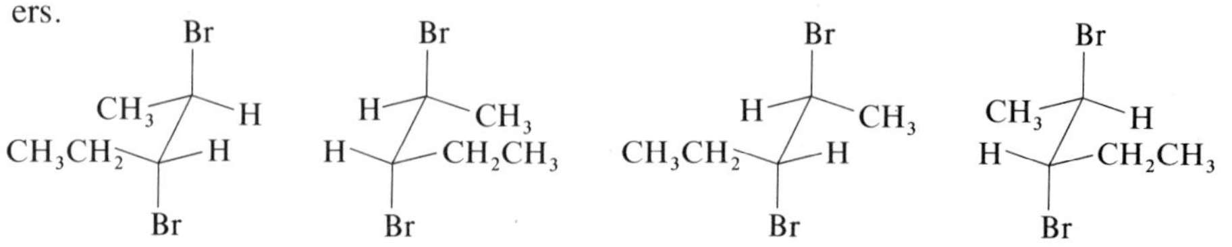

enantiomeric
2,3-dibromopentanes

$(2R, 3R)$ $(2S, 3S)$

either one of these   *is diastereomeric with*

enantiomeric
2,3-dibromopentanes

$(2S, 3R)$ $(2R, 3S)$

either one of these

The presence of two asymmetric carbon atoms gives rise to 2^2 or four different stereoisomers. *The maximum number of stereoisomers expected for a structure containing n asymmetric carbon atoms is 2^n.* This maximum number is not observed when molecules that have asymmetric carbon atoms also have some element of symmetry. For example, 2^2 or four stereoisomers might have been expected in the case of 1,2-dibromocyclopentanes, but the symmetry of *cis*-1,2-dibromocyclopentane reduces to three the number of isomers actually observed.

PROBLEM 5.31 Complete the following equations.

(a) (E)-$CH_3CH_2CH{=}CHCH_2CH_3$ $\xrightarrow[\text{carbon tetrachloride}]{Br_2}$

(b) (Z)-$CH_3CH_2CH{=}CHCH_2CH_3$ $\xrightarrow[\text{carbon tetrachloride}]{Br_2}$

(c) $CH_3CH_2C{\equiv}CCH_2CH_3$ $\xrightarrow[\text{carbon tetrachloride}]{Br_2 \text{ (excess)}}$

(d) 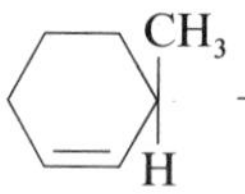 $\xrightarrow[\text{carbon tetrachloride}]{Br_2}$

(e) 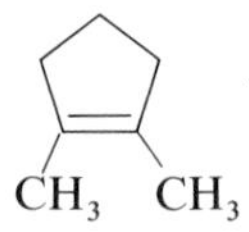 $\xrightarrow[\text{carbon tetrachloride}]{Br_2}$

PROBLEM 5.32 Write the structures for all possible stereoisomers for the following compounds. (Hint: Explore all conformations for an acyclic molecule in making decisions about the presence or absence of symmetry.)

(a)

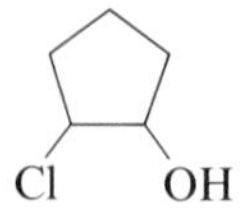

(b) $\underset{\underset{HO \;\; OH}{|\quad\;\; |}}{\overset{\overset{O \qquad O}{\|\quad\;\; \|}}{HOCCHCHCOH}}$

(c)

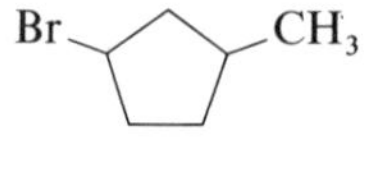

(d) $\underset{\underset{OH}{|}}{CH_3CH{=}CHCHCH_3}$

(e)

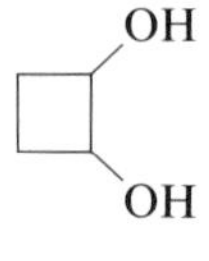

(f) $\underset{\underset{HO \;\; Cl}{|\quad |}}{CH_3CHCHCH_3}$

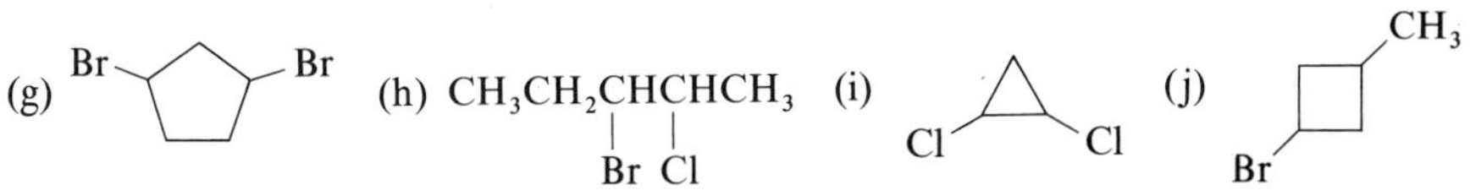

(g) Br–⬠–Br   (h) $CH_3CH_2CHCHCH_3$ (with Br and Cl substituents)   (i) ◁ with Cl and Cl   (j) ⬜ with CH_3 and Br

C. The Bromonium Ion and Nucleophiles

As described in Section 5.8A, the addition of bromine to ethylene in the presence of high concentrations of chloride ions gives 1-bromo-2-chloroethane, as well as 1,2-dibromoethane. Chloride ion does not add to the double bond unless bromine is also present, and no 1,2-dichloroethane is obtained. Similarly, the use of bromine in water instead of in an unreactive solvent, results in the formation of 2-bromoethanol (p. 211).

These facts provide additional experimental evidence for a bromonium ion intermediate. Mechanisms for the formation of 1,2-dibromoethane, 1-bromo-2-chloroethane, and 2-bromoethanol are shown below.

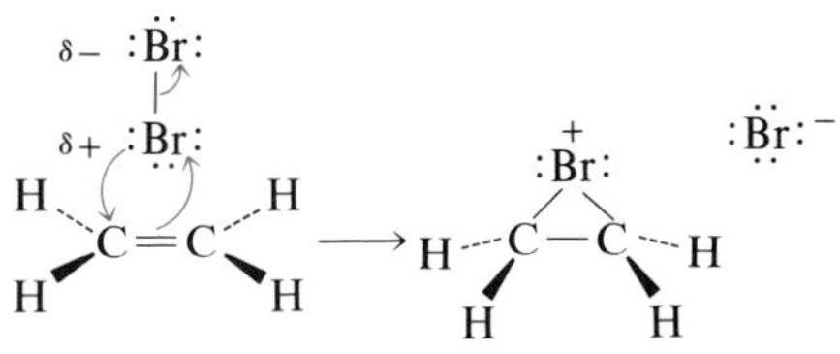

bromonium
ion

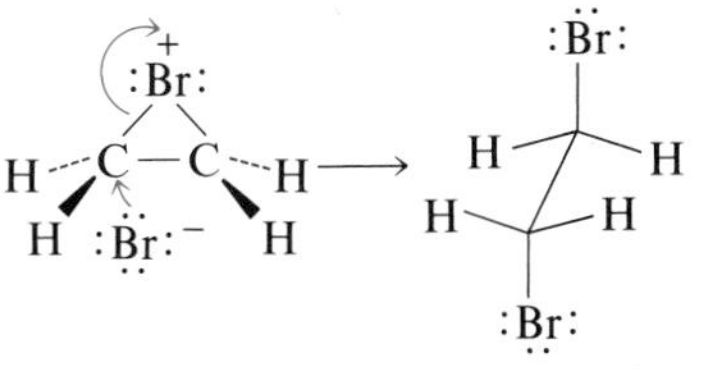

1,2-dibromoethane

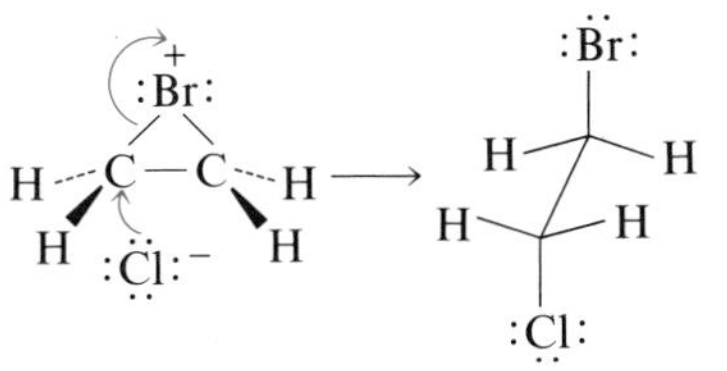

1-bromo-2-chloroethane

2-bromoethanol

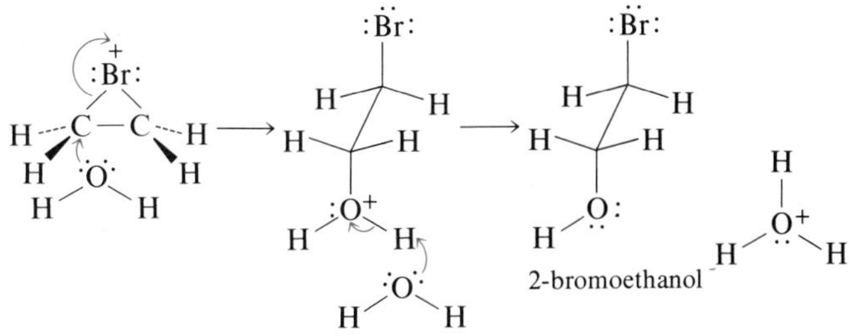

The intermediate bromonium ion is attacked by a nucleophile, bromide ion, chloride ion, or water, to give the final products observed.

The evidence for a bromonium ion intermediate is strengthened by the discovery that alkenes that easily undergo carbocation rearrangements react with bromine in a nucleophilic solvent with no rearrangement. For example, 3,3-dimethyl-1-butene, which reacts with hydrogen halides with rearrangement (p. 197, Section 5.6B), reacts with bromine in methanol to give, besides 1,2-dibromo-3,3-dimethylbutane, an unrearranged product, 2-bromo-1-methoxy-3,3-dimethylbutane. No rearranged products are found.

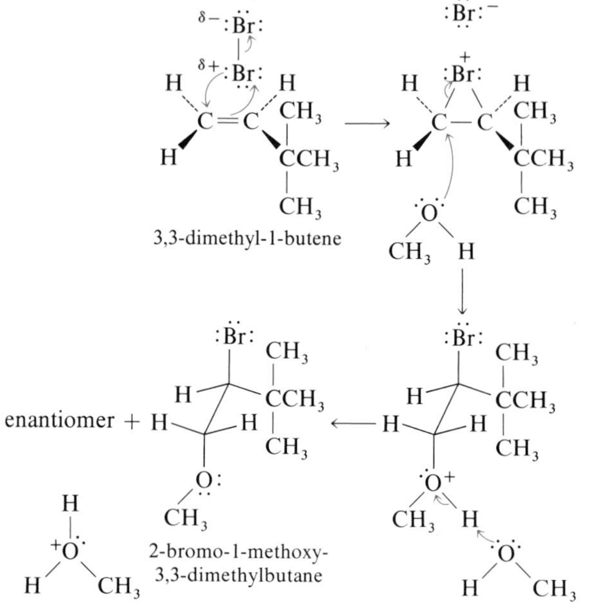

3,3-dimethyl-1-butene

2-bromo-1-methoxy-
3,3-dimethylbutane

enantiomer +

The bromonium ion from 3,3-dimethyl-1-butene is unsymmetrical, and methanol attacks the less hindered carbon atom preferentially. If a carbocation had been formed instead, not only rearrangement but the opposite orientation of addition of bromine and the methoxyl group would also have been expected.

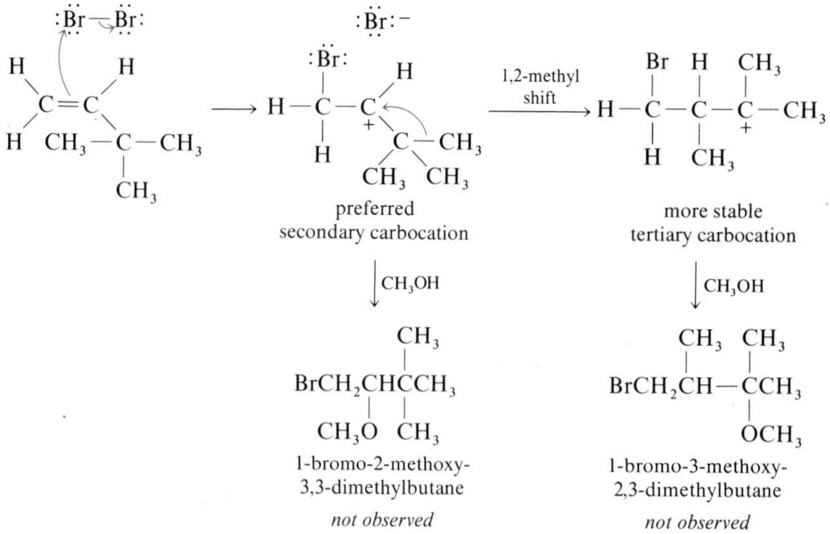

preferred
secondary carbocation

more stable
tertiary carbocation

1,2-methyl
shift

BrCH$_2$CHCCH$_3$
          | |
        CH$_3$O CH$_3$

1-bromo-2-methoxy-
3,3-dimethylbutane

not observed

BrCH$_2$CH—CCH$_3$
                |
              OCH$_3$

1-bromo-3-methoxy-
2,3-dimethylbutane

not observed

The absence of the products that would be predicted to arise from a carbocation points to another type of intermediate for this reaction.

Note that the proton, which is small in size and has no nonbonding electrons, does not usually form bridged cations the way bromine does. Carbocations formed when acids add to alkenes therefore give more rearranged products (Section 5.6B) and show less stereoselectivity (Sections 5.7A, 5.7B, 5.8B) than bromonium ions do.

PROBLEM 5.33 *trans-* 1-Bromo-2-chlorocyclopentane is synthesized in the laboratory by treating cyclopentene with bromine and dry hydrogen chloride in dichloromethane. Propose a mechanism for its formation. Be sure to show correct stereochemistry.

PROBLEM 5.34 The enantiomer of the 2-bromo-1-methoxy-3,3-dimethylbutane shown in the equation on p. 216 is also formed. Write a mechanism for its formation.

ADDITIONAL PROBLEMS

5.35 Name the following compounds.

(a) $CH_3CH_2CH_2C \equiv CCH_2CH_3$ (b) $ClCH_2C \equiv CH$ (c) 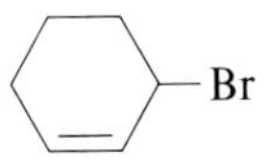 —Br

(d) 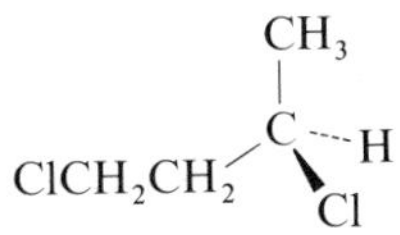   (e) 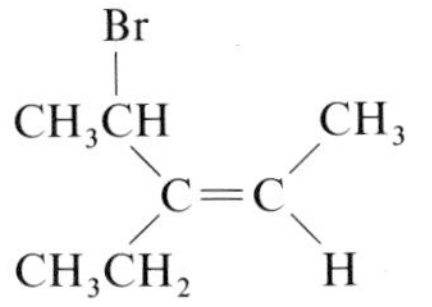   (f)

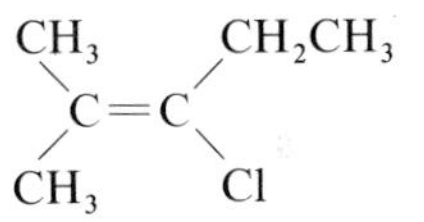

(g) 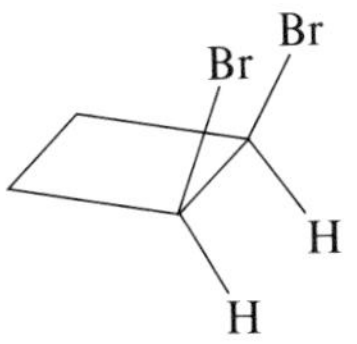   (h)

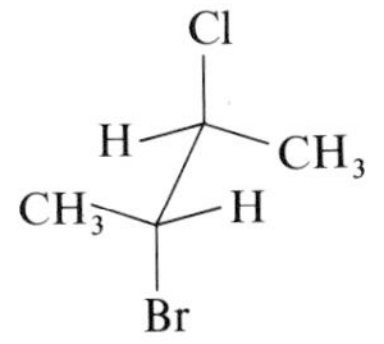

5.36 Draw structural formulas for the following compounds. Be sure to show stereochemistry when it is indicated.

(a) (*Z*)-5-chloro-2-pentene   (b) 2,2-dimethyl-3-hexyne
(c) *trans*-1,2-dimethylcyclopropane   (d) *meso*-2,3-dibromobutane
(e) 3-chloro-2-methyl-1-pentene   (f) *cis*-1,2-dichlorocyclopentane
(g) 2-bromo-2-methyl-4-octyne   (h) 1,2-dimethylcyclohexene

5.37 Using 1-butene as a typical alkene, write equations predicting its reaction with the following reagents. Show the stereochemistry of the resulting products if they have chirality.

(a) HBr, no peroxides   (b) HBr, peroxide   (c) H_2O, H_2SO_4
(d) Br_2, carbon tetrachloride   (e) Br_2, H_2O (f) Br_2, saturated NaCl solution
(g) Br_2, CH_3OH solution   (h) HI

5.38   Using 1-butyne as a typical alkyne, write equations predicting its reaction with the following reagents.

   (a) HBr, 1 equivalent, no peroxides   (b) HBr, excess, no peroxides
   (c) Br$_2$, 1 equivalent, carbon tetrachloride   (d) Br$_2$, excess, carbon tetrachloride
   (e) H$_2$O, H$_2$SO$_4$, HgSO$_4$

5.39   Complete the following equations. Show stereochemistry whenever it is known.

   (a) $\boxed{=}$ $\xrightarrow[\text{carbon tetrachloride}]{\text{Br}_2}$

   (b) $\underset{\underset{\displaystyle CH_3}{|}}{CH_3CHC}\!\equiv\!CH$ $\xrightarrow{\text{HBr (excess)}}$

   (c) $\underset{H}{\overset{CH_3}{C}}\!=\!\underset{H}{\overset{CH_3}{C}}$ $\xrightarrow[\text{carbon tetrachloride}]{\text{Br}_2}$

   (d) $\underset{H}{\overset{CH_3}{C}}\!=\!\underset{CH_3}{\overset{H}{C}}$ $\xrightarrow[\text{carbon tetrachloride}]{\text{Br}_2}$

   (e) $CH_3CH_2C\!\equiv\!CCH_2CH_3$ $\xrightarrow[\text{H}_2\text{SO}_4]{\text{H}_2\text{O}}$

   (f) ⬡ $\xrightarrow[\text{H}_2\text{SO}_4]{\text{H}_2\text{O}}$

   (g) $\underset{CH_3}{\overset{CH_3CH_2}{C}}\!=\!\underset{H}{\overset{CH_3}{C}}$ $\xrightarrow[\text{no peroxides}]{\text{HBr}}$

   (h) $CH_3CH_2C\!\equiv\!CCH_3$ $\xrightarrow{\text{Br}_2\text{ (excess)}}$

   (i) $CH_3CH_2\underset{\underset{\displaystyle OH}{|}}{CH}CH_2CH_3$ $\xrightarrow[\Delta]{\text{H}_3\text{PO}_4}$

   (j) ⬡$-CH\!=\!CH_2$ $\xrightarrow[\text{carbon tetrachloride}]{\text{Br}_2}$

   (k) $CH_3\underset{\underset{\displaystyle OH}{|}}{\overset{\overset{\displaystyle CH_3}{|}}{C}}CH_2CH_3$ $\xrightarrow[\Delta]{\text{H}_2\text{SO}_4}$

   (l) $CH_2\!=\!CHBr$ $\xrightarrow[\text{no peroxides}]{\text{HBr}}$

   (m) $CH_2\!=\!CHBr$ $\xrightarrow[\text{peroxides}]{\text{HBr}}$

   (n) $CH_2\!=\!CHCH_2Br$ $\xrightarrow[\substack{\text{carbon tetrachloride}\\-5\,°C}]{\text{Br}_2}$

5.40   Draw structural formulas for the stereoisomers of the following compounds.

   (a) 3-hexene   (b) 1,2-dimethylcyclopropane
   (c) 1-bromo-2-methylcyclopentane   (d) 2-chloropentane
   (e) 3,4-dimethylhexane   (f) 2-bromo-3-methylpentane
   (g) 5-bromo-2-hexene

5.41   Fill in structural formulas for the starting materials or reagents that would be necessary for the following transformations.

   (a) A $\xrightarrow[\text{H}_2\text{SO}_4]{\text{H}_2\text{O}}$ (pentagon with CH_3 and OH)

   (b) B $\xrightarrow{C}$ (cyclopentane with Br H, H; OH) + (cyclopentane with H Br, H; HO)

   (c) D $\xrightarrow[\substack{\text{carbon}\\\text{tetrachloride}}]{\text{Br}_2}$ $CH_3CHCH_2\underset{Br}{\overset{\overset{\displaystyle CH_2Br}{|}}{C}}\!-\!H$ + $H\!-\!\underset{Br}{\overset{\overset{\displaystyle CH_2Br}{|}}{C}}CH_2CHCH_3$

(d) E $\xrightarrow[\text{carbon tetrachloride}]{\text{Br}_2}$

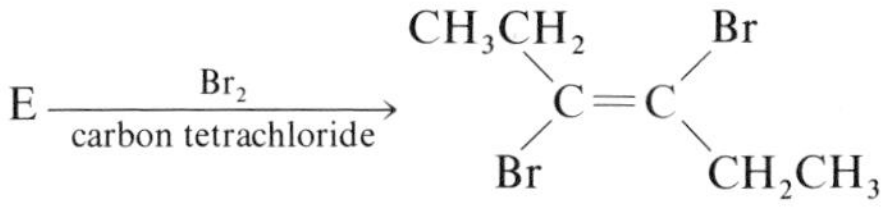

$$\underset{\text{Br}}{\overset{\text{CH}_3\text{CH}_2}{\diagdown}}\text{C}=\text{C}\underset{\text{CH}_2\text{CH}_3}{\overset{\text{Br}}{\diagup}}$$

(e) F $\xrightarrow[\substack{\text{H}_2\text{SO}_4 \\ \text{HgSO}_4}]{\text{H}_2\text{O}}$ CH$_3$CH$_2$$\overset{\overset{\text{O}}{\|}}{\text{C}}CH_2CH_2CH_3$

(f) G $\xrightarrow[\substack{\text{carbon} \\ \text{tetrachloride}}]{\text{Br}_2}$

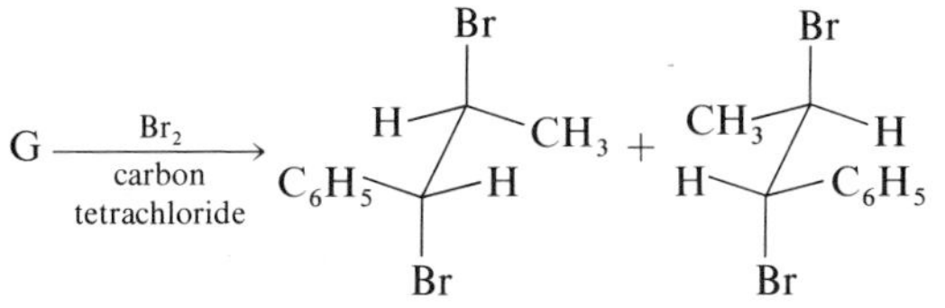

(g) CH$_3$(CH$_2$)$_6$CH=CH$_2$ $\xrightarrow{\text{H}}$ CH$_3$(CH$_2$)$_6$CH$_2$CH$_2$Br

(h) I $\xrightarrow[\Delta]{\text{H}_2\text{SO}_4}$

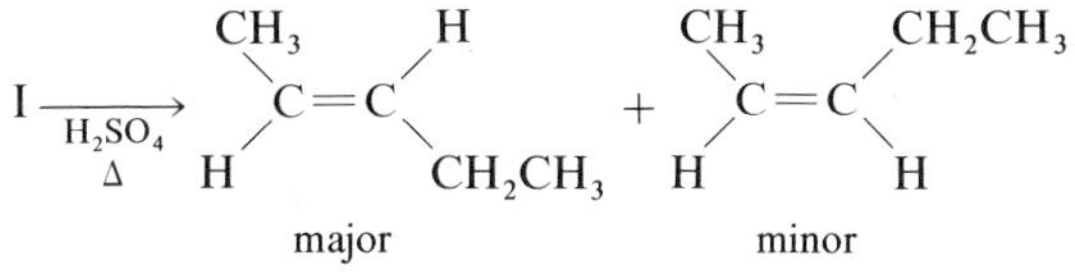

major   minor

5.42 When 2,2-dimethylcyclohexanol is treated with acid, 1,2-dimethylcyclohexene and isopropylidenecyclopentane are the products obtained. Draw a detailed mechanism that explains this observation.

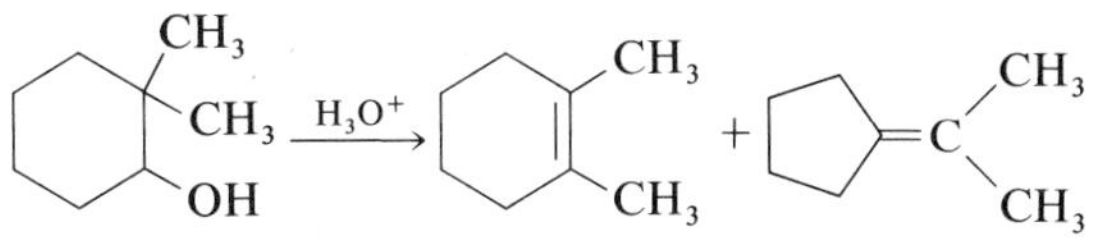

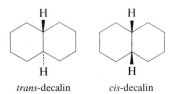

2,2-dimethyl-
cyclohexanol   1,2-dimethyl-
cyclohexene   isopropylidenecyclopentane

5.43 When 2,2,4-trimethyl-3-pentanol is heated over alumina (which is an acidic reagent), it loses water to give a mixture of alkenes. The alkenes that were identified are:

2,4,4-trimethyl-2-pentene	24%	2,3,4-trimethyl-2-pentene	18%
2,4,4-trimethyl-1-pentene	24%	2-isopropyl-3-methyl-1-butene	3%
2,3,4-trimethyl-1-pentene	29%	3,3,4-trimethyl-1-pentene	2%

(a) Write structural formulas for the alcohol and the product alkenes.
(b) Write a mechanism that accounts for the formation of the various products.

5.44 The compound decalin has two cyclohexane rings fused together. The hydrogen atoms at the ring junction are on the same side of the ring in *cis*-decalin and on opposite sides of the ring in *trans*-decalin, as shown below in the planar formulas. Decide which stereoisomer of decalin is more stable. (Hint: Think of the rest of one ring as substituents on the other ring. It may help you to construct the ring systems with models.)

trans-decalin   *cis*-decalin

5.45 Assume that you add hydrogen bromide to (*R*)-(+)-4-methylcyclohexene. Explore the reaction thoroughly, worrying about the orientation of the addition reaction and the stereochemistry of the products. Write planar formulas for all of the possible products. Decide how many of the compounds that form can be separated by ordinary physical methods. Which fractions will be optically active? Identify the types of isomerism that exist for the products of the reaction.

5.46 Explore the different chair conformations that are possible for the products of the addition of hydrogen bromide to (*R*)-(+)-4-methylcyclohexene.

5.47 Some of the evidence for a bromonium ion as a reaction intermediate comes from experiments in which the reactions expected of carbocations are not seen. In particular, one can see differences between the products observed when the reagent is chlorine in water compared with bromine in water. Below, two sets of experimental observations are given. Write out mechanisms that explain the difference in the products seen for the different reagents.

(a)

(b)

5.48 The following experimental observations are made. How would you explain these facts?

(a)

(b) $CF_3CH{=}CH_2 \xrightarrow{HCl} CF_3CH_2CH_2Cl$ $\underset{\underset{0\%}{Cl}}{CF_3CHCH_3}$

    100%

5.49 A vinyl ether reacts with alcohols in the presence of a trace of acid as a catalyst. Predict what the product of the reaction will be by making decisions about the relative stabilities of possible intermediates in the reaction.

$$CH_3OCH{=}CH_2 + CH_3OH \xrightarrow{HCl}$$

5.50 The kinetics of the addition of hydrogen chloride to 2-methylpropene and to ethyl vinyl ether, $CH_2{=}CHOCH_2CH_3$, in the gas phase have been studied. The rate constants for the two reactions may be calculated using the Arrhenius equation (p. 177). The reaction of 2-methylpropene with hydrogen chloride has a frequency factor, A, of 1×10^{11} mL/mol·s and E_a 28.8 kcal/mol. For the reaction of ethyl vinyl ether with hydrogen chloride, A is 5×10^8 mL/mol·s and E_a 14.7 kcal/mol. What is the second order rate constant for each reaction at 25 °C? How do you account for the large difference in E_a for the two reactions? Write equations and draw energy diagrams illustrating your answer.

Nucleophilic Substitution and Elimination Reactions

6.1

Comparison of Addition, Substitution, and Elimination Reactions

We explored the addition reactions of alkenes and alkynes in considerable detail in the last chapter. The reactions were initiated by the attack of the electrons of the π bond on some electron-deficient reagent, an electrophile. They went through a reactive intermediate that was also electron-deficient, usually a carbocation. Reactions were completed when this new electrophilic species became bonded to a pair of unshared electrons from a nucleophile.

Isopropyl bromide, an alkyl halide that is the product of the addition reaction of hydrogen bromide to propene, is also reactive. The presence of the halogen on the alkyl chain creates polarity in the molecule, and, consequently, introduces a point of vulnerability. Two reactions are common for alkyl halides such as isopropyl bromide. When it is treated with a strong base such as sodium ethoxide, in ethanol as a solvent, at 55°C, a mixture of products is formed.

$$\underset{\substack{\text{isopropyl bromide}}}{\underset{\substack{|\\ \text{Br}}}{\text{CH}_3\text{CHCH}_3}} + \underset{\substack{\text{sodium ethoxide}}}{\text{CH}_3\text{CH}_2\text{O}^-\text{Na}^+} \xrightarrow[\substack{55\,°\text{C}}]{\text{ethanol}} \underset{\substack{\text{propene}\\79\%}}{\text{CH}_3\text{CH}=\text{CH}_2} + \underset{\substack{\text{2-ethoxypropane}\\21\%}}{\underset{\substack{|\\ \text{OCH}_2\text{CH}_3}}{\text{CH}_3\text{CHCH}_3}} + \underset{\substack{\text{ethanol}}}{\text{CH}_3\text{CH}_2\text{OH}} + \text{Na}^+\text{Br}^-$$

The major product, propene, results from the removal of the elements of hydrogen bromide from isopropyl bromide.

221

A reaction in which a stable species is lost with the formation of a multiple bond is called an **elimination reaction.** The strong base, ethoxide ion, removes a proton from isopropyl bromide, giving rise to its conjugate acid, ethanol. A double bond is created, with the loss of bromide ion, a stable species, as the other fragment.

The minor product of the reaction results from ethoxide anion attaching itself to the carbon that held the bromine atom in isopropyl bromide, while bromide ion is lost from the molecule. *A reaction in which one atom or group replaces another one in the molecule is called a* **substitution reaction.** Elimination and substitution reactions usually occur in competition with each other.

Reaction of isopropyl bromide with an anion that is less basic than ethoxide ion gives chiefly a substitution reaction. Iodide ion in acetone as a solvent at room temperature substitutes for bromide ion to give isopropyl iodide.

$$CH_3CHCH_3 + Na^+ + I^- \xrightarrow[25\ °C]{acetone} CH_3CHCH_3 + Na^+Br^- \downarrow$$

isopropyl bromide (Br)   isopropyl iodide (I)   *insoluble in acetone*

The reaction is reversible. Treatment of isopropyl iodide with sodium bromide gives isopropyl bromide. However, sodium bromide, which is not as soluble in acetone as sodium iodide, precipitates out of the reaction mixture, drawing the reaction towards completion.

The reactivity of isopropyl bromide can be explained by an examination of its structure. In isopropyl bromide, the electronegativity of the halogen polarizes the carbon-bromine bond, creating a partial positive character at the carbon atom and at hydrogen atoms close to it. This is symbolized by writing $\delta +$ next to those atoms.

electrophilic carbon atom

$$CH_3 \underset{\delta+}{-}C \leftarrow C \leftarrow H^{\delta+}$$

$H^{\delta+}\quad H^{\delta+}$

$\delta^- :\ddot{B}r:\qquad H^{\delta+}$

*polarization of bonds
by the electronegative
bromine atom*

The electronegativity of the bromine atom thus creates electrophilic character at the carbon atom to which it is bound and, to a lesser extent, at the hydrogen atoms on that carbon atom and adjacent carbon atoms. The molecule is vulnerable to attack at these points by reagents that are rich in electrons.

A reagent that is a strong base, the ethoxide ion, attacks the hydrogen atoms of the methyl group and removes one as a proton in the major reaction. A reagent that is a weaker base (like ethanol) but still a nucleophile because it has unshared pairs of electrons that it is willing to share, reacts at the carbon atom bearing the halogen. The hydrogen atom on the carbon atom bound to the bromine can also be removed by a basic reagent, but this reaction does not usually give a stable product. In Section 6.10 we will see cases where such a reaction can be important.

The bromine atom in isopropyl bromide is lost as a bromide anion ($:\ddot{B}r:^-$). *A stable species that can be detached from a molecule during a reaction is called* **a leaving group.** Bromide anion is the leaving group in both the elimination and the substitution reactions.

In the elimination reaction, a new π bond is formed between two carbon atoms in the starting material.

$$CH_3-CH-CH_2-\overset{\delta+}{H} + :\overset{..}{O}CH_2CH_3 \longrightarrow CH_3CH=CH_2 + CH_3CH_2OH + :\overset{..}{Br}:^-$$

electrophilic hydrogen atom

new π bond

$\delta-:\overset{..}{Br}:$

leaving group

nucleophile that is also a strong base

conjugate acid of base

leaving group as stable anion in solution

In the substitition reaction, a σ bond is created between the substituent and the carbon atom that bore the leaving group.

electrophilic carbon atom

$$CH_3-\overset{\delta+}{C}H-CH_3 + :\overset{..}{I}:^- + Na^+ \xrightarrow{\text{acetone}} CH_3-CH-CH_3 + Na^+Br^-\downarrow$$

$\delta-:\overset{..}{Br}:$

leaving group

nucleophile that is a weak base

$:\overset{..}{I}:$

new σ bond

leaving group as stable anion in salt crystal

PROBLEM 6.1   Isopropyl bromide is converted into mixtures of isopropyl alcohol

$$CH_3CHCH_3$$
$$|$$
$$OH$$

and propene by reaction with water, or by treatment with sodium hydroxide dissolved in water. Do you expect to get the same ratio of isopropyl alcohol to propene under these two different conditions? (Hint: What is the nucleophile in each reaction? What are their relative basicities?)

These reactions have tremendous potential for creating transformations in organic molecules, and because of this, they have been much studied. Early work on substitution reactions of alkyl halides was done by Sir Christopher Ingold and Edward Hughes and their collaborators at University College, London, in the 1930s and 1940s. Their ideas have influenced the thinking of organic chemists about the mechanisms of many different reactions and have inspired much experimental work and vigorous debate that still continues.

In principle, the kinds of substituents that can be introduced into organic molecules depend upon the availability of starting materials with suitable leaving groups, and nucleophiles that will give substitution reactions with them. *Elimination and substitution reactions almost always appear in competition with each other.* Factors such as the nature of reagents and solvents play a part in determining which reaction is the major one. In the remainder of the chapter, we will explore the detailed mechanisms for these transformations so that we understand the limitations on them and how to control them.

6.2

Nucleophilicity. Solvent Effects

A. A Comparison of Basicity and Nucleophilicity

Ethoxide ion, a nucleophile that is a strong base as well, gives chiefly the elimination product rather than the substitution product from isopropyl bromide (Section 6.1). Iodide ion, a good nucleophile but a weak base, gives the substitution product. It appears that basicity and nucleophilicity are not necessarily the same thing. A better understanding of the difference is necessary before substitution and elimination reactions are further examined.

The basicity of a species as indicated by the pK_a of its conjugate acid is a measure of a thermodynamic property of the system (Section 4.6D). The pK_a is derived from an equilibrium constant, the acidity constant, which is determined by the relative energy levels for the base and its protonated form. When a nucleophile is removing a proton from an alkyl halide, it is behaving as a Brønsted-Lowry base. We can predict how effective a base would be at this task by looking it up in the table of pK_a values.

Nucleophilicity in substitution reactions is related to kinetics, to how fast a given nucleophile reacts at an electrophilic carbon atom. Here other factors besides the basicity of the Lewis base matter, as demonstrated in the following example.

A thiol is a stronger acid than an alcohol for the reasons discussed in Section 4.5. The thiolate anion is a weaker base towards a proton than the alkoxide ion is. Both will behave as nucleophiles in substitution reactions. In an experiment comparing an oxygen anion (phenolate ion), with a sulfur anion (thiophenolate ion), the reaction with the sulfur compound is over in a few seconds, while that with the oxygen compound requires several hours of heating at the boiling point of ethanol.

sodium phenolate	*n*-butyl bromide	phenyl *n*-butyl ether

sodium thiophenolate	*n*-butyl bromide	phenyl *n*-butyl thioether

Even though thiophenolate anion is a weaker base than phenolate anion, it is the better nucleophile.

The sulfur atom is larger than the oxygen atom; its outermost electrons are farther away from the nucleus and more loosely held. When sulfur approaches a center of partial positive charge, its electrons are more easily distorted in the direction of the charge, so the bonding process starts more easily. *The ease with which the electron cloud on an atom can be distorted is called the* **polarizability** *of the atom.* In general, as atoms become larger going down a group in the periodic table, they also become more polarizable. They become less basic but more nucleophilic. This trend is seen in the halogen family too.

$$:\ddot{I}:^- \qquad :\ddot{Br}:^- \qquad :\ddot{Cl}:^- \qquad :\ddot{F}:^-$$

least basic	*most basic*
most nucleophilic	*least nucleophilic*

By contrast, within any period in the periodic table, nucleophilicity increases with basicity.

$$\overset{\textstyle R}{\underset{\textstyle R}{R-\overset{|}{\underset{|}{C}}:^-}} \qquad \overset{\textstyle R}{R-\overset{|}{N}:^-} \qquad R-\ddot{O}:^- \qquad :\ddot{F}:^-$$

most basic	*least basic*
most nucleophilic	*least nucleophilic*

The arguments made about electronegativity and basicity in Section 4.5 apply to nucleophilicity too. For example, nitrogen is less electronegative than oxygen and holds the electrons around it less firmly than oxygen does. Consequently, ammonia, in which nonbonding electrons are on nitrogen, is more nucleophilic than water, in which the nonbonding electrons are on oxygen. The lower electronegativity of the nitrogen atom makes it more polarizable. Both the electronegativity and the size of the atom bearing the nonbonding electrons are important in determining nucleophilicity.

In general, the electrons on a nucleophile with a negative charge are more polarizable than those on the uncharged conjugate acid. As a result, anions are more powerful nucleophiles than the corresponding conjugate acids.

$$H-\ddot{O}{:}^- > \overset{H}{\underset{H}{\diagdown\!\diagup}}\ddot{O}{:} \; ; \quad R-\ddot{O}{:}^- > \overset{R}{\underset{H}{\diagdown\!\diagup}}\ddot{O}{:} \; ; \quad \overset{H}{\underset{H}{\diagdown\!\diagup}}\ddot{N}{:}^- > H-\overset{H}{\underset{H}{\diagdown\!\diagup}}N{:}$$

anions more nucleophilic than the
corresponding conjugate acids

PROBLEM 6.2   In each pair of reagents, decide which species would be more nucleophilic.

(a) SH^- or OH^- (b) OH^- or $CH_3\overset{\displaystyle O}{\overset{\displaystyle \|}{C}}O^-$ (c) OH^- or NO_3^-

(d) $(CH_3)_2NH$ or NH_3 (e) OH^- or CH_3O^-

B. Solvent Effects

The relative orders of basicity and nucleophilicity are never absolute because they depend on the other reagent in the reaction, on the other ions present in the solution, on the solvent, and on the degree of solvation of the ions. The relative order of nucleophilicity given above for the halide ions was determined in solvents such as ethanol or methanol. These solvents have hydroxyl groups and participate in hydrogen bonding. The smaller the anion, the more concentrated the negative charge, and the more strongly solvated it is by a solvent such as ethanol (Section 1.8B). Thus, chloride anion is more strongly hydrogen bonded than bromide anion. Hydrogen bonding diminishes the availability of the nonbonding electrons on the anion and decreases the nucleophilicity of the ion. In hydroxylic solvents, chloride ion is a weaker nucleophile than bromide ion.

A number of polar solvents do not have functional groups that serve as donors in hydrogen bonding. Examples of such polar but aprotic solvents are acetone, acetonitrile, dimethylformamide, dimethyl sulfoxide, and hexamethylphosphoric triamide.

$CH_3\overset{\displaystyle O}{\overset{\displaystyle \|}{C}}CH_3$	$CH_3C{\equiv}N$	$HC\overset{\displaystyle O}{\overset{\displaystyle \|}{}}NCH_3$, with CH_3 below N	$CH_3\overset{\displaystyle O}{\overset{\displaystyle \|}{S}}CH_3$	$(CH_3)_2N\!-\!\overset{\displaystyle O}{\overset{\displaystyle \|}{P}}\!-\!N(CH_3)_2$, $N(CH_3)_2$ below P
acetone	acetonitrile	dimethylformamide	dimethyl sulfoxide	hexamethylphosphoric triamide
bp 56.5 °C	bp 81.6 °C	bp 153 °C	bp 189 °C	bp 232 °C
μ 2.88	μ 3.92	μ 3.82	μ 3.96	μ 4.30

All of these solvents have high dipole moments because of the presence of the nitrile group (the carbon-nitrogen triple bond), or of polar groups such as carbon, sulfur, or phosphorus doubly bonded to oxygen. Because of their polarity, they can be used as solvents for organic reactions involving ionic reagents or reaction intermediates. When ionic compounds are dissolved in these solvents, the anions are not solvated as strongly as they are in hydroxylic solvents, and so are freer to participate in nucleophilic substitution reactions.

In polar but aprotic solvents, nucleophilicity more closely approximates basicity. For example, while bromide ion is a better nucleophile than chloride ion in methanol, the order is reversed in dimethylformamide where chloride ion is not involved in hydrogen bonding. The more basic chloride ion is, therefore, the better nucleophile in the aprotic solvent. While it is not possible to assign absolute nucleophilicities to Lewis bases, the trends discussed in the last two sections provide useful generalizations.

PROBLEM 6.3   The rate of the reaction of methyl iodide with chloride ion was measured in a series of solvents. The structures of the solvents and the relative rates are shown below. How would you explain these observations?

$$CH_3I \quad + \quad Cl^- \quad \longrightarrow \quad CH_3Cl \quad + \quad I^-$$

Solvent	CH_3OH	$\overset{\displaystyle O}{\overset{\displaystyle \|}{HCNH_2}}$	$\overset{\displaystyle O}{\overset{\displaystyle \|}{HCNHCH_3}}$	$\overset{\displaystyle O}{\overset{\displaystyle \|}{CH_3CN(CH_3)_2}}$
Relative rate	1	1.25×10^1	1.2×10^6	7.4×10^6

PROBLEM 6.4   (a) Hydroxylamine, H_2NOH, has two sites of potential basicity and nucleophilicity. Predict what the product of the reaction will be in each of the following reactions.

1. $H_2NOH \xrightarrow{\text{HCl}}$ 2. $H_2NOH \xrightarrow{\text{CH}_3\text{I (1 equivalent)}}$

(b) The conjugate acid of hydroxylamine has pK_a 5.97. What does this tell you about the structure of the conjugate acid? To which acid in the table of pK_a values would you compare it? How do you explain the difference in acidity between the two species?

6.3

Carbocation Intermediates in Substitution Reactions

A. Experimental Evidence for the Formation of Carbocations in Substitution Reactions

When *tert*-butyl chloride is put into a solution of silver nitrate in ethanol at room temperature, a white precipitate of silver chloride forms instantly. This indicates that *tert*-butyl chloride easily gives chloride ions in solution. The ionization of *tert*-butyl chloride to give chloride ion suggests that a carbocation must also be formed at the same time.

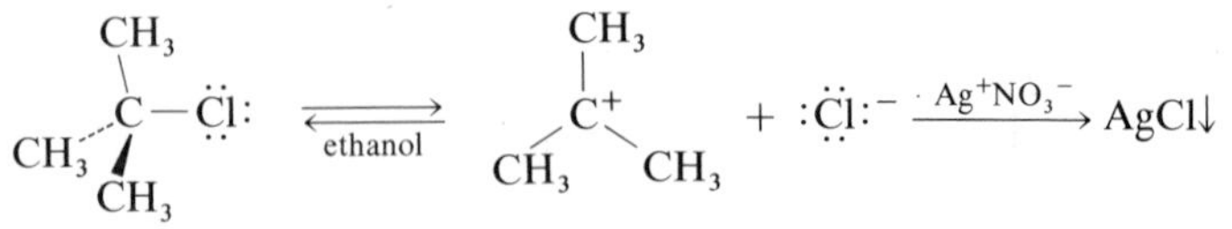

| 2-chloro-2-methylpropane | *tert*-butyl | silver |
| *tert*-butyl chloride | cation | chloride |

If *n*-butyl chloride is put into a solution of silver nitrate in ethanol, no precipitate forms.

$$CH_3CH_2CH_2$$
$$\overset{H}{\underset{H}{\diagdown}}C-Cl + Ag^+ + NO_3^- \xrightarrow[\text{ethanol}]{} \text{no precipitate}$$

n-butyl chloride

Chloride ions are not formed from *n*-butyl chloride, and *n*-butyl cations are not present either. A change in structure in the alkyl halide from *tert*-butyl chloride, where the chlorine atom is bound to a tertiary carbon atom, to *n*-butyl chloride, where the chlorine is on a primary carbon atom, results in a dramatic difference in reactivity.

Investigation of the reaction of *tert*-butyl chloride with ethanol shows that the reaction mixture gets progressively more acidic with time. The course of the reaction can be followed by titrating the reaction mixture with base. The products derived from *tert*-butyl chloride are *tert*-butyl ethyl ether and 2-methylpropene.

$$\underset{\underset{CH_3}{|}}{\overset{\overset{CH_3}{|}}{CH_3CCl}} \xrightarrow[25\,^\circ C]{CH_3CH_2OH} \underset{\underset{CH_3}{|}}{\overset{\overset{CH_3}{|}}{CH_3COCH_2CH_3}} + \overset{\overset{CH_3}{|}}{CH_3C}=CH_2 + \overset{\overset{H}{|}}{CH_3CH_2\overset{+}{O}H} + Cl^-$$

| *tert*-butyl chloride | *tert*-butyl ethyl ether ~80% | 2-methylpropene ~20% | conjugate acid of ethanol | chloride anion |

Thus, even in a system in which no strong base is present, some elimination reaction takes place for the tertiary halide.

B. The Formation and Reactions of the *tert*-Butyl Cation. A Review of the Relative Stabilities of Carbocations

In the ionization of *tert*-butyl chloride in ethanol, the bond that is already polarized because of the difference in electronegativity between carbon and chlorine is broken to give *tert*-butyl cation and chloride anion.

$$\underset{\underset{CH_3}{|}}{\overset{CH_3}{\diagdown}}\overset{\delta+}{C}-\overset{\delta-}{\overset{..}{\underset{..}{Cl}}}: \underset{\text{ethanol}}{\rightleftharpoons} \underset{CH_3}{\overset{\overset{CH_3}{|}}{C^+}}\diagup_{CH_3} + :\overset{..}{\underset{..}{Cl}}:^-$$

| *tert*-butyl chloride | *tert*-butyl cation | chloride ion |

In this heterolytic bond cleavage (Section 5.4A) both electrons that make up the covalent bond leave with the more electronegative atom, chlorine. The bond cleavage is represented by the arrow showing the transfer of the pair of electrons that constitutes the carbon-chlorine bond to the chlorine atom.

The experiments with silver nitrate described in the last section indicate that this kind of bond cleavage does not take place easily in *n*-butyl chloride. The structure of *tert*-butyl chloride is such that the departing chloride ion leaves behind a carbocation that is stabilized by three methyl groups. It is a tertiary carbocation, in contrast to the corresponding ion from *n*-butyl chloride, which is a primary carbocation.

$$\underset{CH_3 \quad CH_3}{\overset{\overset{\overset{CH_3}{\downarrow}}{C^+}}{}} \qquad \underset{CH_3CH_2CH_2 \quad H}{\overset{\overset{\overset{H}{|}}{C^+}}{}}$$

| *tert*-butyl cation | *n*-butyl cation |
| *a tertiary carbocation* | *a primary carbocation* |

The electron-releasing effect of the methyl groups sufficiently stabilizes the positive charge in the case of the tertiary carbocation that it becomes a realistic intermediate for the reaction under these conditions. The primary carbocation is too unstable to be formed (Section 5.6A). The chloride ion, as the conjugate base of a strong acid, hydrochloric acid, is a stable species and a good leaving group in this ionization.

Another important factor in this reaction is the solvent, ethanol, which can interact with and further stabilize by solvation the ionic species that are formed (Section 1.8B). In ethanol, the electrostatic interaction of the negative ends of the dipoles of the oxygen-hydrogen bonds with the carbocation, and the positive ends of the dipoles with the chloride ion releases energy.

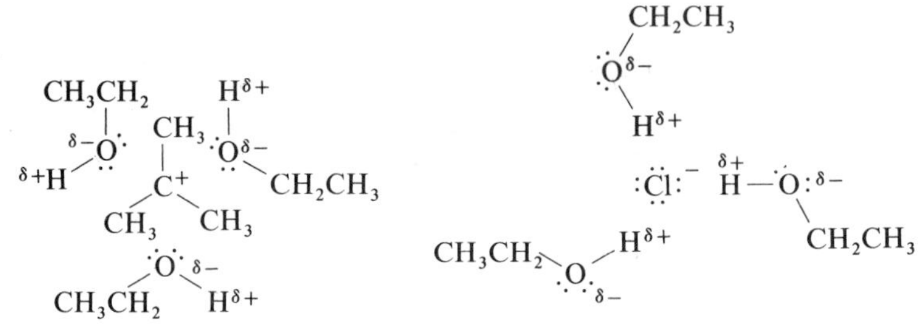

solvation of tert-*butyl cation and chloride ion by ethanol*

This solvation energy helps to compensate for the energy necessary to break the carbon-chlorine bond. *tert*-Butyl chloride ionizes much less readily in another solvent such as acetone. Ethanol, because of its polar oxygen-hydrogen bonds, stabilizes chloride anions, as well as carbocations, and is, thus, a better ionizing solvent than acetone. The importance of solvent in stabilizing ionic species has been emphasized in earlier comparisons of reactions in the gas phase and in solution (Sections 4.6C, 5.6A).

Once the carbocation is formed, it has two reaction pathways open to it: reaction with a nucleophile or loss of a proton. The carbocation behaves as a Lewis acid (Sections 4.3, 5.3B) because it lacks a pair of electrons and can accept them to acquire a stable octet. Such a cation is a good electrophile. When the *tert*-butyl cation is formed in ethanol, the cation and the alcohol, which is a Lewis base or nucleophile, react. The oxonium ion formed is a strong acid and loses a proton to the solvent.

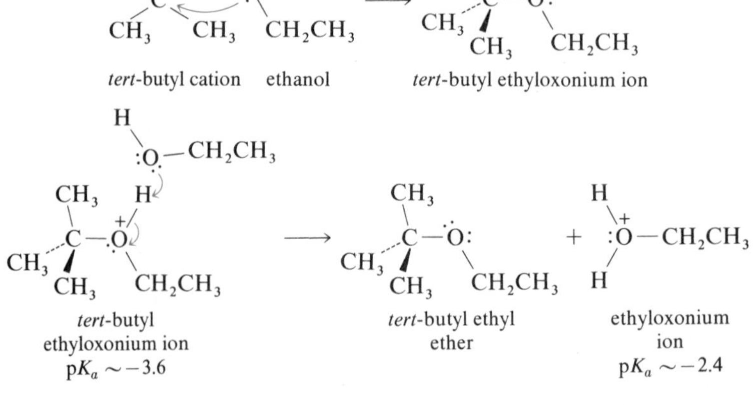

tert-butyl cation   ethanol   *tert*-butyl ethyloxonium ion

tert-butyl ethyloxonium ion $pK_a \sim -3.6$ *tert*-butyl ethyl ether   ethyloxonium ion $pK_a \sim -2.4$

In this reaction, the substitution product is derived from the solvent. *A substitution reaction in which the solvent acts as the nucleophile is called a* **solvolysis reaction.**

Very often when there are a number of potential bases in the reaction mixture, as there are in this case, the symbol :B, a general base, is used to represent any one of them. The last step of the solvolysis reaction can be rewritten to show this.

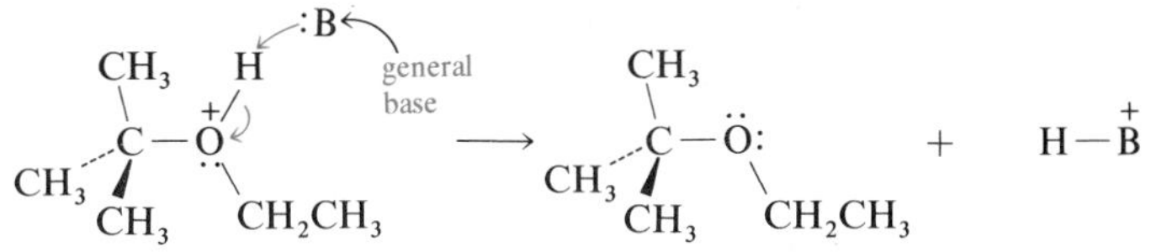

tert-butyl ethyloxonium ion   tert-butyl ethyl ether   conjugate acid of
                                                                   the general base

---

PROBLEM 6.5   Look at all the species present in the reaction mixture when *tert*-butyl chloride is placed in ethanol (Section 6.3A, p. 227) and make a list of those that can act as bases.

The carbocation is also a Brønsted-Lowry acid (Section 5.6B). The cationic carbon atom creates partial positive character on the hydrogen atoms on carbon atoms adjacent to it. These hydrogen atoms are acidic enough to be removed by any of the bases present in the reaction mixture to give the minor product, the product of the elimination reaction.

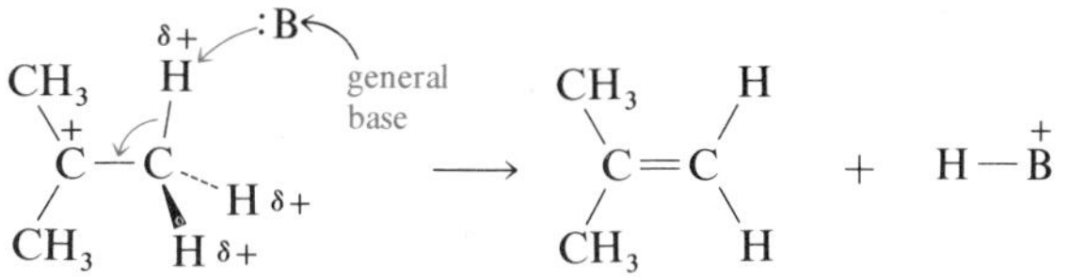

tert-butyl cation   2-methylpropene   conjugate acid of
                                                  the general base

A carbocation intermediate for a substitution reaction usually gives rise to some alkene because of the ease with which protons are lost from that intermediate.

6.4

Leaving Groups

A. Leaving Groups and Basicity

The experimental facts presented at the beginning of this chapter indicate that *tert*-butyl chloride ionizes easily in an ionizing solvent such as ethanol, and *n*-butyl chloride does not. We attributed this difference to the relative stabilities of the two carbocation intermediates that would be formed.

The stability of the potential carbocation is not the only factor that determines whether ionization takes place. Another experiment would be to put *tert*-butyl alcohol into water to see whether ionization into *tert*-butyl cations and hydroxide ions (detected by a change in the pH of the solution) would take place.

tert-butyl alcohol

TABLE
6.1 **A Comparison of the Acidities of Water and Hydrogen Chloride**

Acid	Conjugate Base	pK_a
HCl	Cl⁻	-7
H_2O	OH⁻	15.7

In fact, because the solution does not become more basic, we can conclude that there is no ionization to *tert*-butyl cations and hydroxide ions taking place.

The difference between *tert*-butyl chloride and *tert*-butyl alcohol is the difference in the leaving groups, chloride ion and hydroxide ion.

$$CH_3 - \overset{\overset{\displaystyle CH_3}{|}}{\underset{\underset{\displaystyle CH_3}{|}}{C}} - Cl \qquad CH_3 - \overset{\overset{\displaystyle CH_3}{|}}{\underset{\underset{\displaystyle CH_3}{|}}{C}} - OH$$

tert-butyl chloride   *tert*-butyl alcohol

Hydroxide ion is like chloride ion in being a stable species with an octet of electrons and a negative charge accommodated on an electronegative atom. It differs from chloride ion in being the conjugate base of water, a much weaker acid than the conjugate acid of chloride ion (Table 6.1).

In Chapter 4, acidity was discussed in terms of different conjugate bases competing with each other for the same proton. The more successful a base was in the competition, the stronger a base it was, and the weaker its conjugate acid. Strong bases with weak conjugate acids appeared in the lower part of the table of pK_a values. The table can be used to make predictions about how good a leaving group an ion or molecule will be. The same factors that make a species a weak base and a weak nucleophile—in other words its ability to accommodate a negative charge or its reluctance to share a pair of electrons—also make it a good leaving group. Generally, the conjugate bases near the top of the pK_a table, to pK_a 8, are reasonably good leaving groups. More basic species generally do not make very good leaving groups.

B. Making Better Leaving Groups

Once we know what kinds of species are good leaving groups, we can change reaction conditions in order to make poor leaving groups into better ones. A naive look at the transformation of *tert*-butyl alcohol into *tert*-butyl bromide suggests that a hydroxyl group has to be replaced by a bromine atom.

$$CH_3 - \overset{\overset{\displaystyle CH_3}{|}}{\underset{\underset{\displaystyle CH_3}{|}}{C}} - OH \qquad CH_3 - \overset{\overset{\displaystyle CH_3}{|}}{\underset{\underset{\displaystyle CH_3}{|}}{C}} - Br$$

tert-butyl alcohol   *tert*-butyl bromide

This reaction looks like a substitution to be achieved by addition of bromide ion to the reaction mixture to give displacement of hydroxide ion. In fact, nothing happens when sodium bromide is added to *tert*-butyl alcohol.

$$CH_3\overset{\overset{\displaystyle CH_3}{|}}{\underset{\underset{\displaystyle CH_3}{|}}{C}}OH + Na^+ + Br^- \xrightarrow[H_2O]{} \text{no reaction}$$

tert-butyl alcohol

TABLE   **A Comparison of the Acidities of**
  6.2   **Hydronium Ion and Water**

Acid	Conjugate Base	pK_a
H_3O^+	H_2O	-1.7
H_2O	OH^-	15.7

Hydroxide ion is not a good leaving group, so the substitution reaction cannot take place.

If the reaction conditions are changed by using hydrobromic acid instead of sodium bromide, *tert*-butyl bromide is formed.

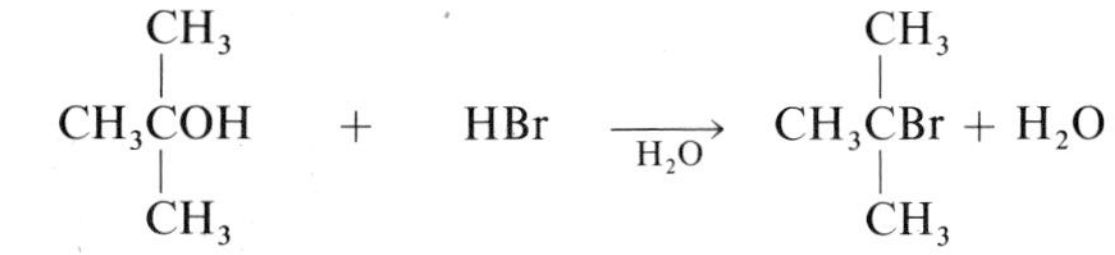

$$\underset{\substack{\text{CH}_3 \\ | \\ \text{CH}_3\text{COH} \\ | \\ \text{CH}_3}}{} + \text{HBr} \xrightarrow{\text{H}_2\text{O}} \underset{\substack{\text{CH}_3 \\ | \\ \text{CH}_3\text{CBr} + \text{H}_2\text{O} \\ | \\ \text{CH}_3}}{}$$

    tert-butyl alcohol   hydrobromic   *tert*-butyl bromide
                               acid

The acid protonates the unshared electrons on the oxygen atom of the alcohol to give a new leaving group, water, the conjugate base of the strong acid, hydronium ion (Table 6.2). Water is a good leaving group and ionization can take place.

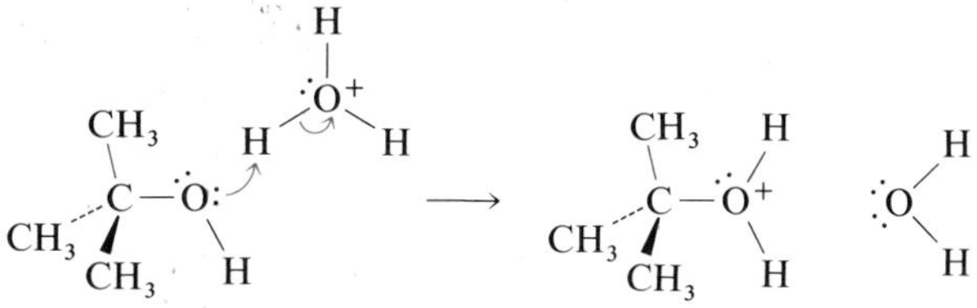

tert-butyl alcohol and   *tert*-butyloxonium ion
hydronium ion

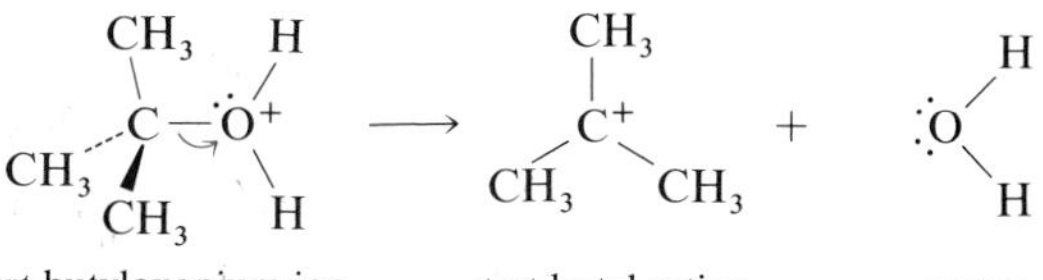

tert-butyloxonium ion   *tert*-butyl cation   water

                                                   the leaving group

The *tert*-butyl cation can now do all the things that carbocations do, including combining with the nucleophile, bromide ion.

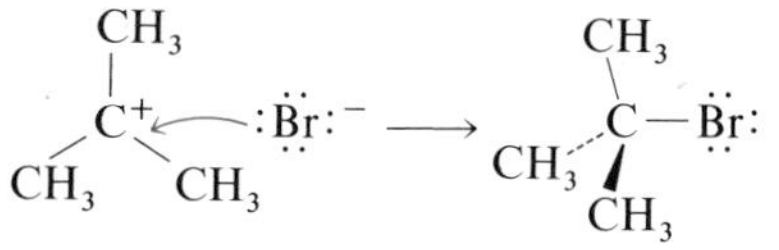

tert-butyl cation   *tert*-butyl bromide

PROBLEM 6.6   Where did the hydronium ion in the first step of the mechanism shown above come from?

PROBLEM 6.7   What other reactions would you predict for the carbocation in the system created when *tert*-butyl alcohol, hydrobromic acid, and water are mixed?

A hydroxyl group is frequently protonated in order to convert it into water, a good leaving group. Yet the protonated alcohol is an ionic intermediate with only a fleeting lifetime; it cannot be isolated and put in a bottle. Sometimes it is convenient to convert alcohols into stable, isolable compounds that have good leaving groups on them.

A successful strategy is to convert the alcohol into the ester of the relatively strong acid, p-toluenesulfonic acid, $pK_a \sim -0.6$. The sulfonic acid is an organic derivative of sulfuric acid. p-Toluenesulfonyl chloride, as its acid chloride, in which a chlorine atom, potentially a good leaving group, has been substituted for the hydroxyl group, reacts easily with the alcohol.

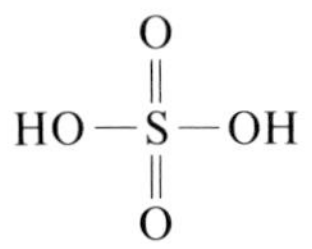

sulfuric acid

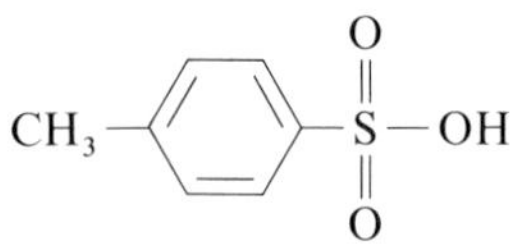

p-toluenesulfonic acid
TsOH

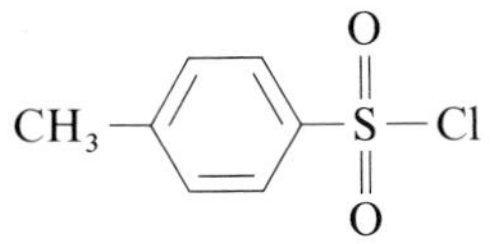

p-toluenesulfonyl chloride
tosyl chloride
TsCl

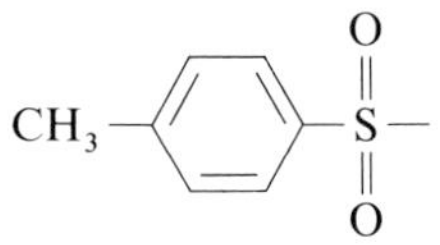

p-toluenesulfonyl group
tosyl group
Ts—

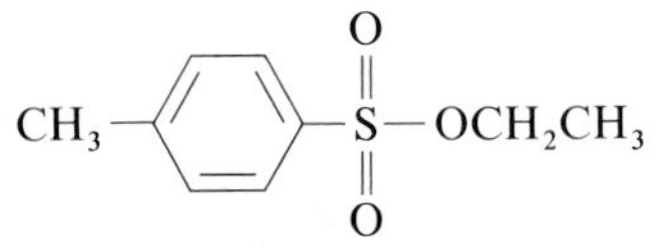

ethyl p-toluenesulfonate
ethyl tosylate
TsOCH$_2$CH$_3$

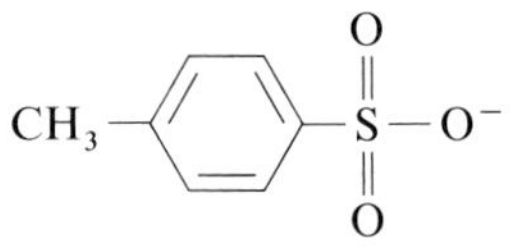

p-toluenesulfonate anion
tosylate anion
TsO$^-$

The p-toluenesulfonyl group has a large structure and a long name. It is used so often in organic reactions that it has been given an abbreviation for the sake of convenience. From the name p-toluenesulfonyl, the underlined sections have been condensed into **tosyl**, and further abbreviated in structures to the symbol Ts, as shown above.

An alcohol, cyclopentanol, reacts with tosyl chloride.

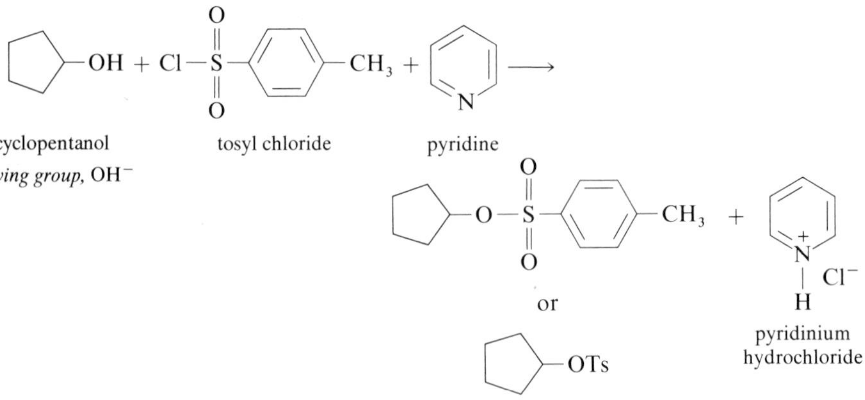

cyclopentanol
leaving group, OH$^-$

tosyl chloride

pyridine

or

cyclopentyl tosylate
leaving group, TsO$^-$

pyridinium
hydrochloride

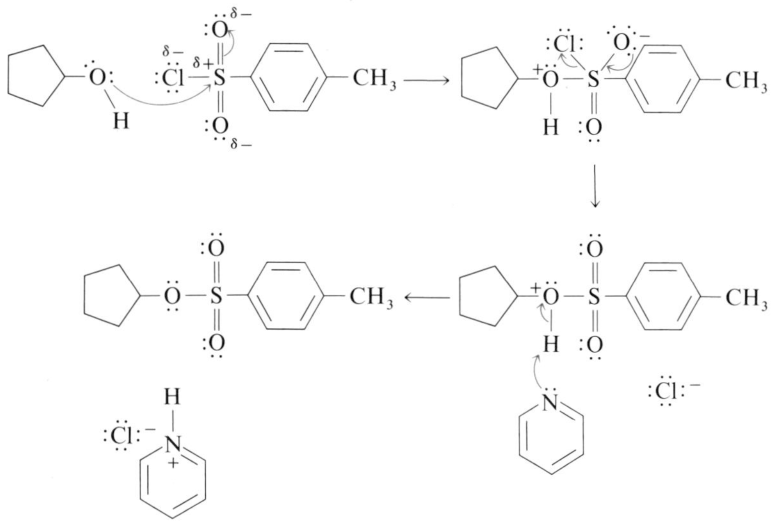

A basic solvent, pyridine, is used to neutralize the hydrogen chloride that is formed. In the reaction, the poor leaving group, the hydroxide ion, is converted into a good leaving group, the tosylate anion.

Note that alcohol is converted to tosylate without breaking the carbon-oxygen bond in the alcohol. Later in the chapter, we will be concerned with the stereochemistry of substitution reactions, and will notice carefully when bonds to an asymmetric carbon atom are broken. The tosylate of a chiral alcohol can be prepared without affecting stereochemistry because the bond to the asymmetric carbon atom is not broken until tosylate anion is displaced in a nucleophilic substitution reaction.

Cyclopentyl tosylate is converted mostly into ethoxycyclopentane in ethanol, with *p*-toluenesulfonic acid as the other product of the reaction.

cyclopentyl tosylate; ethoxycyclopentane 93%; cyclopentene 7%; *p*-toluenesulfonic acid

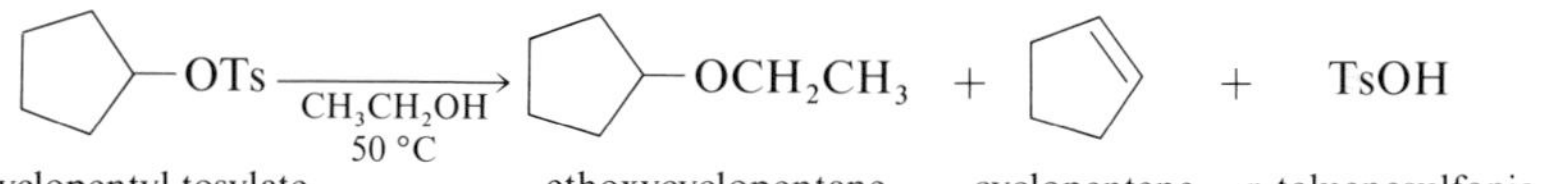

No ethoxycyclopentane is formed if cyclopentanol is put into ethanol. The substitution reaction takes place only when a poor leaving group has been converted into a good leaving group.

6.5 Substitution Reactions of Primary and Secondary Alkyl Halides

A. Experimental Evidence for Substitution Without Carbocation Intermediates

When *n*-butyl bromide is put into a solution of sodium iodide in acetone, a white precipitate of sodium bromide appears. Sodium bromide precipitates because it is not very soluble in acetone.

$$\underset{n\text{-butyl bromide}}{CH_3CH_2CH_2CH_2Br} + Na^+ + I^- \xrightarrow[25\ ^\circ C]{\text{acetone}} \underset{n\text{-butyl iodide}}{CH_3CH_2CH_2CH_2I} + Na^+Br^-\downarrow$$

fast reaction

If *tert*-butyl bromide is placed in the same solution, no precipitate is observed and no reaction takes place.

$$\underset{\text{\textit{tert}-butyl bromide}}{\overset{\displaystyle \text{CH}_3}{\underset{\displaystyle \text{Br}}{\text{CH}_3\text{CCH}_3}}} + \text{Na}^+ + \text{I}^- \xrightarrow[\substack{\text{acetone} \\ 25\,°\text{C}}]{} \text{no precipitate}$$

Isopropyl bromide reacts with sodium iodide in acetone, but more slowly than *n*-butyl bromide.

$$\underset{\substack{\text{isopropyl bromide}}}{\overset{}{\underset{\displaystyle \text{Br}}{\text{CH}_3\text{CHCH}_3}}} + \text{Na}^+ + \text{I}^- \xrightarrow[\substack{\text{acetone} \\ 25\,°\text{C} \\ \textit{slow reaction}}]{} \underset{\substack{\text{isopropyl iodide}}}{\overset{}{\underset{\displaystyle \text{I}}{\text{CH}_3\text{CHCH}_3}}} + \text{Na}^+\text{Br}^-\!\downarrow$$

The order of reactivity does not fit the order of stability of carbocations. *tert*-Butyl bromide, which would give the most stable cation in the series, does not react to a noticeable extent with this reagent. No sodium bromide precipitates, indicating that bromide ions are not formed from *tert*-butyl bromide under these conditions, and no substitution reaction takes place. Acetone is not a good ionizing solvent, so any reaction proceeding in this solvent must go by a pathway that does not involve the formation of carbocations.

PROBLEM 6.8   The order of reactivity that we have described above for the reactions of *n*-butyl bromide, isopropyl bromide, and *tert*-butyl bromide with sodium iodide in acetone comes from a laboratory experiment actually carried out by students in organic chemistry laboratory courses. Unfortunately, they very often do not see these results because they use *wet* test tubes for the experiment in spite of being given careful instructions that the test tubes must be *dry*. Why is this so important? What effect does water have?

B. The Kinetics of Substitution Reactions of Primary Alkyl Halides. The S$_N$2 Reaction

We often study the mechanism of a reaction by investigating how the rate of the reaction depends upon the concentrations of the reacting species (Section 5.3E). Measurements of the rate of the reaction of *n*-butyl bromide with iodide ion in acetone show that the rate of the reaction depends upon the concentrations of both reactants. An increase in the concentration of either reagent causes a proportional increase in the rate of the reaction. The rate expression for this second-order reaction is

$$R = k_r[n\text{-BuBr}][\text{I}^-] \qquad \text{Bu} \equiv \text{butyl}$$

where R is the rate of the reaction and k_r is the second-order rate constant for this particular reaction at a given temperature and for a given solvent.

PROBLEM 6.9   The rate constant, k, for the reaction

$$\text{CH}_3\text{CH}_2\text{Br} + \text{OH}^- \xrightarrow[\substack{80\% \text{ ethanol} \\ 55\,°\text{C}}]{} \text{CH}_3\text{CH}_2\text{OH} + \text{Br}^-$$

is 1.7×10^{-3} L/mol · s. What is the initial rate of the reaction when 0.05 M ethyl bromide is allowed to react with 0.07 M sodium hydroxide? What are the units in which the rate of the reaction, R, is expressed? What would the initial rate of the reaction be if 0.1 M ethyl bromide were used instead?

PROBLEM 6.10   *n*-Butyl bromide, 0.06 M, was allowed to react with potassium iodide, 0.02 M, in acetone at 25 °C. The rate constant, k_r, for the reaction at that temperature is 1.09×10^{-1} L/mol · min. What is the overall rate at which *n*-butyl bromide disappears from the reaction mixture under these conditions?

For a reaction involving breaking bonds and making new ones, the reacting molecules must meet with the proper orientation and with sufficient energy to get them to the transition state for the reaction. The amount of energy necessary for the reactants to achieve the transition state is the energy of activation (Section 5.3D).

Experimental evidence indicates that no reactive intermediate, such as a carbocation, is involved in the reaction of *n*-butyl bromide with iodide ion (Section 6.5A). The energy diagram for the reaction shows the reactants being converted to products by way of one transition state, with no energy minimum representing a reactive intermediate (Figure 6.1).

For the reaction of *n*-butyl bromide and iodide ion, the transition state is postulated to involve both species. The process is pictured below.

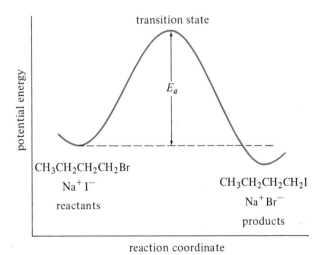

Figure 6.1   Energy relationship between the reactants and the transition state for a reaction involving one step.

Iodide ion, a nucleophile, approaches *n*-butyl bromide. The electrons on the iodide ion are drawn towards the electrophilic carbon atom, while the bond to the bromine atom loosens and lengthens. Notice that the lowest energy approach for the iodide ion, and the least crowded one for it, is from the side opposite the departing bromide ion. In this way, the negative charge on the iodide ion and the developing negative charge on the bromide ion can be kept as far apart as possible.

At the transition state, there is partial bonding between carbon and iodine. The carbon atom is also still partially bonded to the bromine atom, which is beginning to look like a bromide ion. The carbon atom develops a small partial positive charge in this transition state. *In the transition state, there are five groups around the carbon atom; it is pentacoordinate.* The transition state can revert to starting materials or proceed to products.

The energy diagram for a reaction occurring in one step (Figure 6.1) should be compared with that of a two-step reaction going by way of a reactive intermediate (Figure 5.5, p. 172). In the reaction that proceeds in two steps, the step leading to the transition state of highest energy on the path from reactants to products was defined as the rate-determining step (Section 5.3H). When a reaction has only one step, the overall rate is, of course, determined by the rate at which molecules pass over the hill represented by the transition state.

The rate of the reaction of *n*-butyl bromide with iodide ion is determined by the equilibrium concentration of the activated complex at the transition state and the rate at which the complex is converted to products (Section 5.3F). The equilibrium concentration of the activated complex, in turn, is dependent on the free energy of activation, $\Delta G^{\ddagger}$, of the reaction (Section 5.3F). The number of molecules having enough energy to reach the transition state is greater for low energies of activation than for high ones. More molecules will react in a given period of time in a reaction with a low energy of activation than one with a higher energy of activation (Section 5.3E).

PROBLEM 6.11   The energy of activation for the reaction of *n*-butyl chloride with iodide ion to give *n*-butyl iodide has been determined to be 22.2 kcal/mol. The frequency factor, *A* (p. 177), for the reaction is 2.24×10^{11} L/mol · s. What is the rate constant, k_r, for the reaction at 60 °C?

The reaction of *n*-butyl bromide with iodide ion belongs to a class of reactions known as **bimolecular nucleophilic substitution reactions.** *The reactions are classified as bimolecular reactions because two species undergo bonding changes in the transition state.* The reaction is initiated by the approach of a nucleophile, and hence is a nucleophilic substitution reaction. Chemists refer to these reactions as **S$_N$2 reactions,** the symbols standing for **substitution, nucleophilic,** and **bimolecular,** in sequence. The overall mechanism for the reaction is shown below.

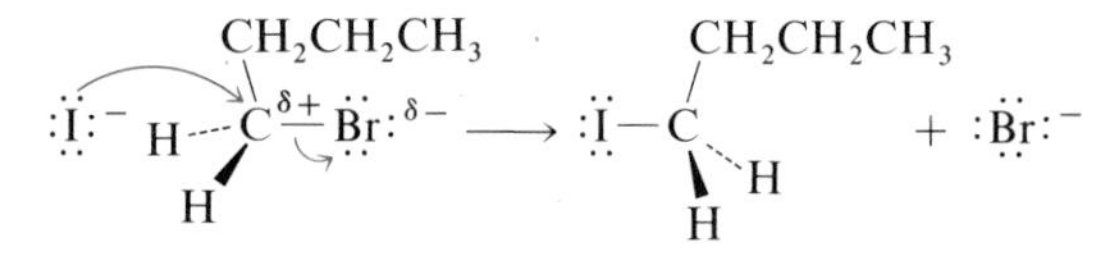

The reaction takes place in one step, unlike the reaction of *tert*-butyl chloride in ethanol (Section 6.3A), where an ionization step and a second step, the combination of the cation with a nucleophile, occur.

PROBLEM 6.12   When ethanol is heated with a small amount of sulfuric acid, diethyl ether is formed and can be distilled out of the reaction mixture. This is, in fact, how the ether once used for anesthesia was prepared.

$$2\ CH_3CH_2OH \xrightarrow[\Delta]{H_2SO_4} CH_3CH_2OCH_2CH_3 + H_2O$$

Write a detailed mechanism for this reaction, using all the ideas that we have developed so far about acids, bases, conversion of poor leaving groups into good leaving groups, and nucleophiles displacing leaving groups.

PROBLEM 6.13   When diethyl ether is heated with hydrogen iodide, iodoethane is the product.

$$CH_3CH_2OCH_2CH_3 + 2\ HI \xrightarrow{\Delta} 2\ CH_3\,CH_2I + H_2O$$

Write a detailed mechanism for the reaction, using the ideas in the preceding sections.

C. Steric Effects in S_N2 Reactions

In an experiment exploring the effect of structure on reactivity in S_N2 reactions, radioactive bromide ion was used to see how fast it was incorporated into different alkyl bromides in acetone at 25 °C. The relative rates of reaction for different alkyl groups, R, are given in Table 6.3.

Methyl bromide reacts the fastest. Even ethyl bromide, in which a methyl group has replaced one of the hydrogen atoms in methyl bromide, reacts much more slowly than methyl bromide. The reaction is very slow for the tertiary halide, *tert*-butyl bromide, and what reaction does occur probably goes by way of the *tert*-butyl cation.

Why is the bimolecular substitution reaction slower for a secondary halide than for a primary one and why does the reaction hardly go at all with a tertiary halide? If we look at the transition states for these reactions, we see that they become progressively more crowded as we increase the number of alkyl groups on the carbon atom undergoing substitution.

The transition state for the reaction of iodide ion with *tert*-butyl bromide is shown in Figure 6.2 (p. 238). A methyl group is considerably larger than a hydrogen atom

TABLE 6.3   Relative Rates for the Reaction of Alkyl Bromides, RBr, with Radioactive Bromide Ion in Acetone at 25 °C

$$Br^{*-} + R-Br \underset{\substack{\text{acetone}\\25\ °C}}{\rightleftharpoons} R-Br^{*} + Br^{-}$$

R	CH_3-	CH_3CH_2-	$CH_3CH_2CH_2-$	$CH_3\overset{\displaystyle CH_3}{\overset{\displaystyle |}{CH}}-$	$CH_3\overset{\displaystyle CH_3}{\overset{\displaystyle |}{\underset{\displaystyle |}{\underset{\displaystyle CH_3}{C}}}}-$
	methyl	ethyl	*n*-propyl	isopropyl	*tert*-butyl
Relative rates	100	1.31	0.81	0.015	0.004
Nature of R		primary	primary	secondary	tertiary

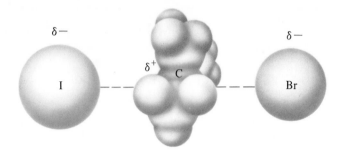

Figure 6.2   Transition state for the reaction of iodide ion with *tert*-butyl bromide.

in its effective radius (Figure 3.6, p. 103). A transition state with five groups crowded around the central carbon atom is a high-energy transition state. Only a few molecules have enough energy to react by that pathway. When the solvent does not help the process of ionization, there is no lower energy pathway available for substitution at the tertiary carbon atom, and reaction is very slow.

The structure of the alkyl halide, especially how many alkyl groups are at the carbon atom undergoing substitution, determines how high the energy of the transition state for the reaction will be. Reactions that can go through transition states of relatively low energy will proceed faster than those that must reach high-energy states.

Whenever the structure of the alkyl group bonded to the halogen is branched in such a way as to hinder the approach of the nucleophile to the back side of the carbon atom that will undergo substitution, the rate of the reaction decreases. This effect is seen in the relative rates for two series of compounds reacting with the strong nucleophile, ethoxide anion (Table 6.4). In the first series, the alkyl group becomes larger as we go from methyl to n-pentyl. The big difference in rate comes between the methyl and ethyl groups. Once the alkyl group is as large as propyl, the rates do not differ much. In the second series, branching at the carbon next to the reaction site (the β-position) increases, apparently hindering the bimolecular reaction severely. The rate of the S_N2

TABLE 6.4   Relative Rates for the Reaction of Alkyl Bromides, RBr, with Ethoxide Anion in Ethanol at 55 °C

$$RBr + CH_3CH_2O^- \xrightarrow[\text{55 °C}]{\text{ethanol}} ROCH_2CH_3 + Br^-$$

R	CH_3- methyl	CH_3CH_2- ethyl	$CH_3CH_2CH_2-$ n-propyl	$CH_3CH_2CH_2CH_2-$ n-butyl	$CH_3CH_2CH_2CH_2CH_2-$ n-pentyl
Relative rates	34.3	1.95	0.60	0.44	0.41

R	CH_3- methyl	CH_3CH_2- ethyl	$CH_3CH_2CH_2-$ n-propyl	CH_3CHCH_2- isobutyl	CH_3CCH_2- neopentyl
Relative rates	34.3	1.95	0.60	0.058	0.0000083

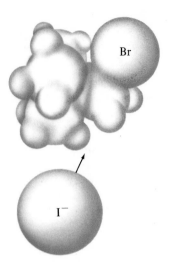

Figure 6.3   Methyl groups on the β-carbon atom in neopentyl bromide hindering the approach of the nucleophile.

reaction falls off rapidly from the propyl group (with one methyl group at the β-carbon) to the isobutyl group (with two methyl groups at the β-carbon). In neopentyl bromide, with three methyl groups at the β-carbon, the bimolecular reaction is negligibly slow (Figure 6.3). In fact, neopentyl halides react in substitution reactions only if they are placed in good ionizing solvents so that a carbocation intermediate can form.

PROBLEM 6.14   Neopentyl bromide undergoes reaction in 50% aqueous ethanol at high temperature (125 °C) at a rate that is independent of the concentration of added sodium hydroxide. Under these conditions, the major products of the reaction are 2-ethoxy-2-methylbutane and 2-methyl-2-butene.

$$
\underset{\substack{\text{neopentyl bromide}}}{\overset{\substack{\text{CH}_3 \\ |}}{\underset{\substack{| \\ \text{CH}_3}}{\text{CH}_3\text{CCH}_2\text{Br}}}}
\xrightarrow[\substack{\text{in H}_2\text{O} \\ 125\,°\text{C}}]{\text{50\% CH}_3\text{CH}_2\text{OH}}
\underset{\substack{\text{2-ethoxy-2-methylbutane}}}{\overset{\substack{\text{CH}_3 \\ |}}{\underset{\substack{| \\ \text{OCH}_2\text{CH}_3}}{\text{CH}_3\text{CCH}_2\text{CH}_3}}}
+ \underset{\substack{\text{2-methyl-2-} \\ \text{butene}}}{\overset{\substack{\text{CH}_3 \\ |}}{\text{CH}_3\text{C}=\text{CHCH}_3}}
$$

It is postulated that a carbocation is the intermediate for this reaction. Write a mechanism for the reaction that explains the products that are observed. (Hint: It may help you to review Section 5.6B.)

D. Stereochemical Evidence for S$_N$2 Reactions

When (S)-(+)-2-bromobutane is allowed to react with iodide ion in acetone, the 2-iodobutane that is formed has the R configuration. As would be expected from the transition state postulated for the S$_N$2 reaction, the iodine atom has become attached to the asymmetric carbon atom on the side opposite to the position occupied by the bromine atom.

(S)-$(+)$-2-bromobutane and iodide ion

reactants

transition state

(R)-$(-)$-2-iodobutane and bromide ion

products

An inversion of configuration at the asymmetric carbon atom has taken place. Another example may make this a little clearer. If bromide ion is used to displace the bromine in (S)-$(+)$-2-bromobutane, the original chiral compound is converted into its enantiomer, (R)-$(-)$-2-bromobutane.

(S)-$(+)$-2-bromobutane and bromide ion

reactants

transition state

(R)-$(-)$-2-bromobutane and bromide ion

products

enantiomers

Remember that the assignment of S and R designations can be made just by looking at the three-dimensional pictures. The use of $(+)$ and $(-)$ to indicate whether the compounds actually rotate the plane of polarized light to the right or left is an experimental determination that must be made and cannot be assigned by looking at the structure. Once it is determined for one enantiomer, the other must have optical rotation of the opposite sign (Section 3.9D).

PROBLEM 6.15 What will happen to the optical rotation observed for pure (S)-$(+)$-2-bromobutane after it has been exposed to bromide ion in acetone over a period of time?

Inversion of configuration, *the conversion of an asymmetric center from one configuration to the opposite one* was discovered in 1893 by Paul Walden at the University of Rostock. This stereochemical transformation is known as the Walden inversion. *It can take place only if bonds are broken and reformed at the asymmetric carbon atom.* If the compound that has undergone inversion has the same groups on it as before inversion, the enantiomer is formed. If the groups are different, it is still an inversion if the point of attachment of the new group is on the opposite side of the asymmetric carbon from that of the original leaving group. In the iodide displacement reaction, iodide ion has displaced bromide ion with inversion of configuration because its point of attachment to the asymmetric carbon atom is clearly on the opposite side of the molecule from where the bromine atom originally was.

(S)-$(+)$-2-bromobutane

(R)-$(-)$-2-iodobutane

inversion of configuration

Not all reactions proceed with inversion of configuration. If (R)-$(-)$-2-butanol is converted into its tosylate with tosyl chloride, the tosylate also has the R configuration because no bonds to the asymmetric carbon atom are broken in this reaction (and we have not done anything to any group that changes its priority in the rules for nomenclature of chiral compounds). The configuration of the molecule is untouched; the reaction goes with **retention of configuration.** The tosylate anion is a good leaving group, and a nucleophilic substitution reaction with iodide ion gives 2-iodobutane with the S configuration. This reaction goes by the S_N2 mechanism with inversion of configuration at the chiral center.

(R)-$(-)$-2-butanol (R)-2-butyl tosylate (S)-$(+)$-2-iodobutane

retention of configuration   inversion of configuration

All experimental evidence shows that when a bimolecular nucleophilic substitution reaction takes place, there is complete inversion of configuration at the carbon atom undergoing the substitution.

PROBLEM 6.16   You can obtain experimental evidence for inversion of configuration by carefully working out configurational relationships between different optically active compounds. One such cycle is outlined below. Look at all of the reactions, decide when bonds are being broken at an asymmetric center and when not, and predict for each reaction whether it goes with inversion or retention of configuration.

1. $CH_3(CH_2)_5CHCH_3 \xrightarrow{TsCl} CH_3(CH_2)_5CHCH_3$
         | |
         OH   OTs
   $(+)$-2-octanol $(+)$-2-octyl tosylate

2. $CH_3(CH_2)_5CHCH_3 \xrightarrow{CH_3COCCH_3} CH_3(CH_2)_5CHCH_3 + CH_3COH$
         | |
         OH   OCCH_3
                                             ||
                                             O
   $(-)$-2-octanol $(-)$-2-octyl acetate

3. $CH_3(CH_2)_5CHCH_3 \xrightarrow{CH_3CO^-Na^+} CH_3(CH_2)_5CHCH_3 + TsO^-$
         | |
         OTs   OCCH_3
                                             ||
                                             O
   $(+)$-2-octyl tosylate $(-)$-2-octyl acetate

PROBLEM 6.17   One of the classical experiments used to explore the stereochemistry of S_N2 reactions is outlined below. Two different rates were measured for this reaction. Radioactive iodide ion was used to measure the rate at which radioactive iodine was substituted in

(S)-(+)-2-iodooctane. The rate at which the starting alkyl iodide lost its optical activity, the **rate of racemization,** was also measured. The rate of racemization is twice the rate of the substitution reaction. Explain these observations.

$$
\begin{array}{c}
CH_3 \\
| \\
I \cdots C \cdots H \\
\diagdown \\
C_6H_{13}
\end{array}
\quad + \quad I^{*-} \xrightarrow{\text{acetone}} \text{racemic } (\pm)\text{-2-iodooctane}
$$

(S)-(+)-2-iodooctane   radioactive iodide ion   *containing radioactive iodine*

6.6

Kinetics and Stereochemistry for Substitution Reactions with Carbocationic Intermediates

A. The Kinetics of Substitution Reactions of *tert*-Butyl Chloride. The S_N1 Reaction

When *tert*-butyl chloride is allowed to react with potassium hydroxide in a solvent mixture of 80% ethanol and 20% water at 25°C, we find that the rate at which the chloride is converted to product is independent of whether any base is there or not. The reaction proceeds at the same rate whether hydroxide ions are present or whether, in fact, the solution is allowed to become acidic.

$$
\begin{array}{c}
CH_3 \\
| \\
CH_3CCl \\
| \\
CH_3
\end{array}
\quad + \quad K^+OH^- \xrightarrow[\substack{20\% \ H_2O \\ 25 \ °C}]{80\% \ CH_3CH_2OH}
$$

tert-butyl chloride

$$
\begin{array}{ccccc}
CH_3 & & CH_3 & & CH_3 \\
| & & | & & | \\
CH_3COH & + & CH_3COCH_2CH_3 & + & CH_3C{=}CH_2 \quad + \ K^+Cl^- \\
| & & | & & \\
CH_3 & & CH_3 & &
\end{array}
$$

tert-butyl alcohol 56%   *tert*-butyl ethyl ether 27%   2-methylpropene 17%

The rate of the reaction is proportional only to the concentration of the alkyl halide in the reaction mixture. The reaction is said to be first order.

$$R = k[\textit{tert-}BuCl]$$

Because only one species is postulated to undergo changes in bonding in the transition state of this reaction, this is classified as a **unimolecular reaction.**

As had been postulated earlier (Section 6.3A) on the basis of more qualitative experimental evidence, the rate-determining step is the ionization of the alkyl halide.

$$
\begin{array}{c}
CH_3 \\
| \\
CH_3 \cdots C \diagdown \\
\diagup \quad Cl \\
CH_3
\end{array}
\xrightarrow[\substack{20\% \ H_2O}]{80\% \ ethanol}
\begin{array}{c}
CH_3 \\
| \\
C^+ \\
CH_3 \diagdown CH_3
\end{array}
\quad + \ Cl^-
$$

tert-butyl chloride   *tert*-butyl cation

slow step of the reaction

The high-energy step for the reaction is the breaking of the carbon-chlorine bond, assisted, of course, by the solvation of the ions that are formed. The rate of the reaction increases 750 times if the solvent is changed from 90% ethanol/10% water, to 40% ethanol/60% water, indicating how important solvation is. Water, as a more polar molecule than ethanol, increases the ionizing capacity of the solvent and better solvates the ions that are formed. Because the transition state is already beginning to look like the product ions, a more polar solvent also helps to stabilize the transition state, and lowers the energy of activation for the reaction. This increases the rate of the reaction.

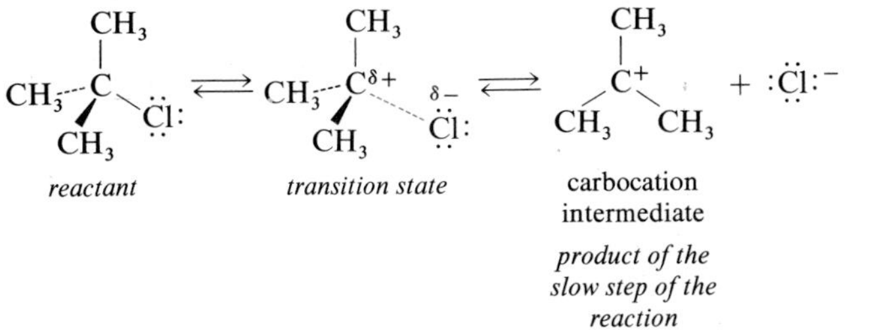

Once the intermediate tertiary carbocation is formed, it reacts rapidly with the nucleophiles (water, hydroxide ion, ethanol) that are present, or loses a proton to give the products observed.

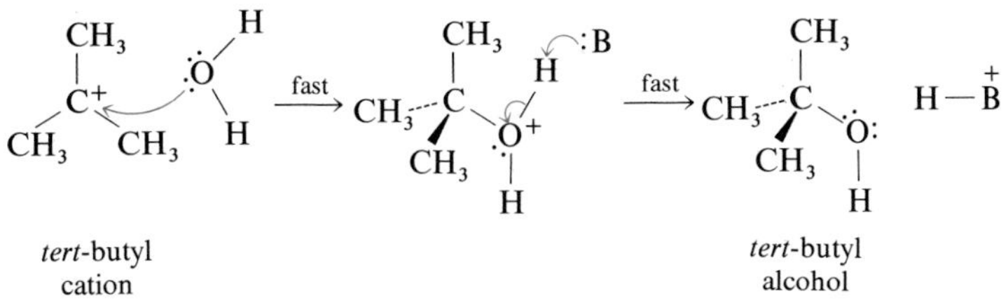

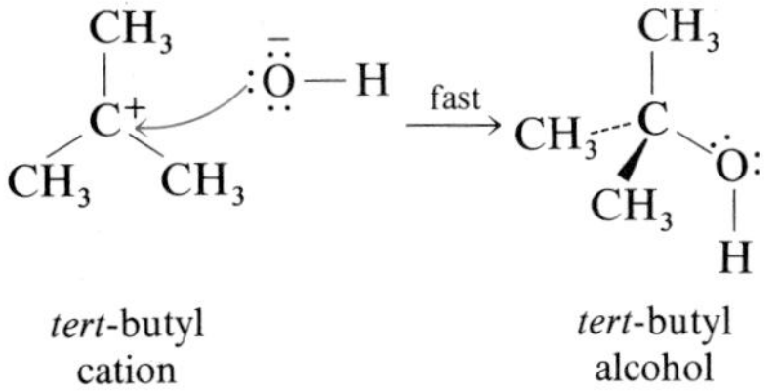

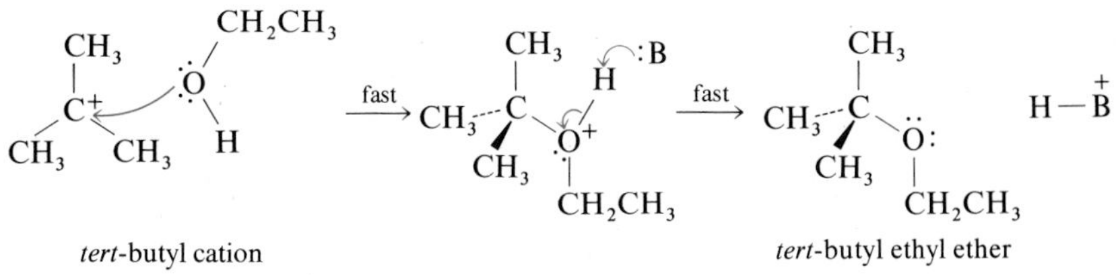

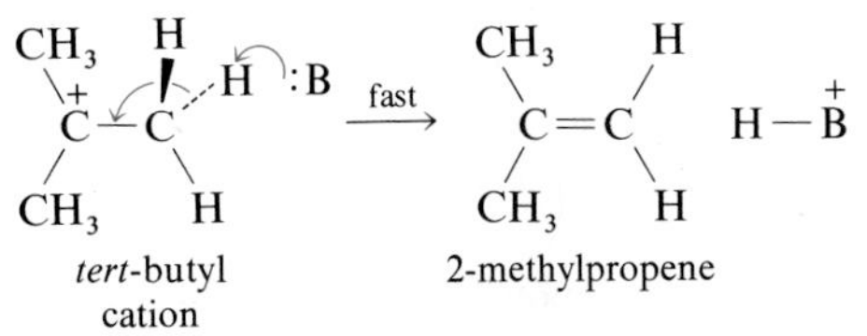

The nucleophiles present in the largest concentrations are the solvent molecules, so this is a solvolysis reaction.

The unimolecular nucleophilic substitution reaction is called the **S$_N$1** reaction for **substitution, nucleophilic, unimolecular,** just as the bimolecular nucleophilic reaction is called the S$_N$2 reaction.

PROBLEM 6.18 The rate constant, k, for the solvolysis of *tert*-butyl chloride in 70% aqueous ethanol at 25 °C was determined to be 0.145/h. If the initial concentration of *tert*-butyl chloride is 0.0824 M, what is the initial rate of the reaction? What units are used to express this rate? Will this rate remain the same as the reaction progresses? What will it be when half of the chloride that was initially there has reacted?

B. Stereochemistry of the S$_N$1 Reaction

What kind of stereochemistry should we expect from a reaction that goes through a carbocation intermediate? The simplest tertiary alkyl halide that would be chiral is 3-bromo-3-methylhexane. The rate-determining step in the hydrolysis of (S)-3-bromo-3-methylhexane would be expected to give a planar carbocation solvated on both sides by water molecules.

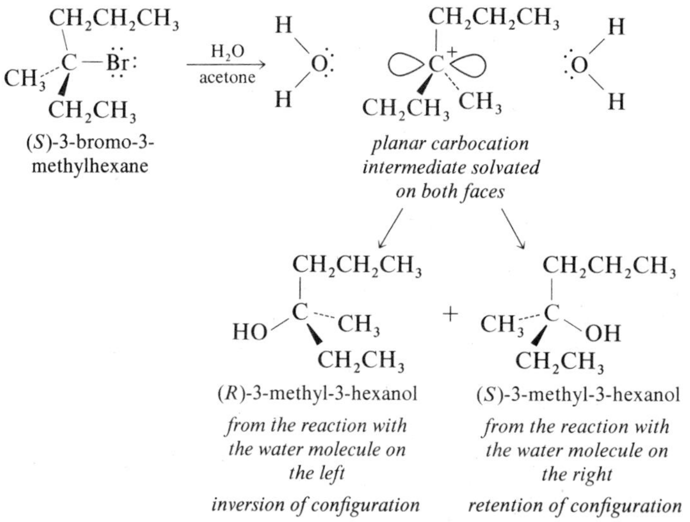

Two of the water molecules are in position to donate an electron pair to the empty *p* orbital. If reaction takes place with the water molecule on the right, the alcohol, (S)-3-methyl-3-hexanol, is formed with retention of configuration. If the other water molecule reacts, the alcohol with the inverted configuration, (R)-3-methyl-3-hexanol, is formed. *When a symmetrical intermediate, such as the planar carbocation, is formed in a reaction at the asymmetric carbon atom, the chirality of the starting material is lost.* In this case, the products are enantiomers. If they are formed in equal amounts, there will be no optical activity detected in the product, which will be a racemic mixture of alcohols.

Experiments with chiral compounds that can undergo S$_N$1 reactions show that a great deal of racemization does take place. The reactions of (S)-($-$)-1-chloro-1-phenylethane illustrate this best.

	(S)-(−)-1-chloro-1-phenylethane	(S)-(−)-1-phenyl-1-ethanol	(R)-(+)-1-phenyl-1-ethanol
H_2O		41%	59%
40% H_2O 60% acetone		47%	53%
20% H_2O 80% acetone		49%	51%

1-Chloro-1-phenylethane, though it is a secondary alkyl halide, ionizes easily because the carbocation that forms has special stability. The ease with which carbocations on a carbon atom adjacent to the aromatic ring are formed can be rationalized by invoking resonance. Resonance contributors in which the positive charge is delocalized to the aromatic ring contribute to the stabilization of the cation.

resonance contributors of the
1-phenylethyl cation

An examination of the experimental facts shows that 1-chloro-1-phenylethane undergoes the S_N1 reaction with extensive racemization, but with a slight excess of inversion. For example, when the reaction is carried out in water, there is approximately an 18% excess of the inverted product formed. As the percentage of water in the reaction mixture decreases to 40%, and then to 20%, and is replaced by an increasing amount of acetone, the product is more and more completely racemized.

We can interpret these facts to mean that the carbocation that is originally formed is not completely free to react with water on both sides. One side of the carbocation is shielded by the departing chloride ion for some time. In fact, the carbocation and the chloride ion are an **ion pair,** held together for a while by a **cage of solvent molecules.** At this stage, only the water molecules on the side of the cation opposite the departing chloride ion can react with the cation, giving rise to inversion (Figure 6.4).

If the cation is stable enough to survive this stage, the ions can diffuse farther apart, and water molecules can get around to the other side of the cation. Now *the ion pair is described as being separated by the solvent molecules* (Figure 6.5). Products with retention as well as inversion of configuration result.

If the solvent is pure water, the concentration of water molecules is high enough that some of the reaction takes place while the cation is still shielded by chloride ion in the front. As the concentration of water molecules in the solvent decreases from 100% to 40% and to 20%, more and more of the cations survive long enough without reacting with water that the ions can diffuse apart, and reaction with water can take place on both sides of the cation.

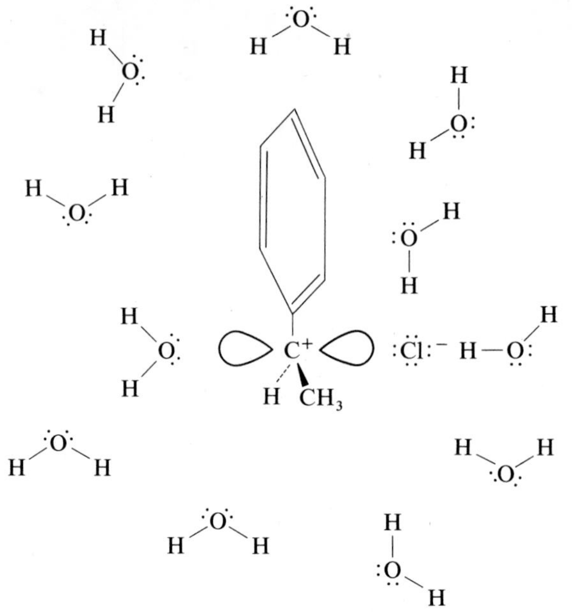

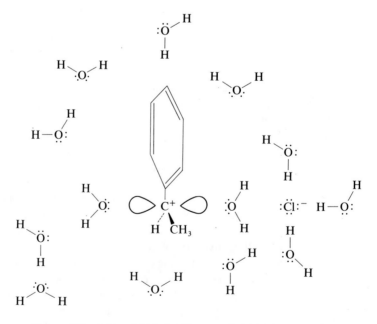

Figure 6.4 1-Phenylethyl cation and chloride ion as an ion pair in solvent cage; reaction of cation with water occurs with inversion.

Figure 6.5 1-Phenylethyl cation and chloride ion as a solvent-separated ion pair; reaction of cation with water occurs with racemization.

6.7

A Unified View of Nucleophilic Substitution Reactions

S_N1 and S_N2 reactions represent mechanistic extremes for nucleophilic substitution. Most systems undergo substitution reactions by some intermediate mechanism involving different degrees of ionization of the carbon-halogen bond and different degrees of bonding with nucleophiles in the transition state. For example, *tert*-butyl chloride was shown on p. 242 as undergoing ionization in a single step to give the *tert*-butyl cation and the chloride ion, which were represented as being completely separated from each other. Yet in the last section we saw stereochemical evidence that the ionization of an alkyl halide may actually occur in several stages. When ionization first occurs, the ions are held together by the surrounding solvent molecules in an ion pair, also called a **tight ion pair.** Then they are separated by one or more solvent molecules and become a **solvent-separated ion pair.** Finally they drift apart entirely and become free, totally solvated ions. These stages are shown schematically for an alkyl halide, RX, in an ionizing solvent such as water.

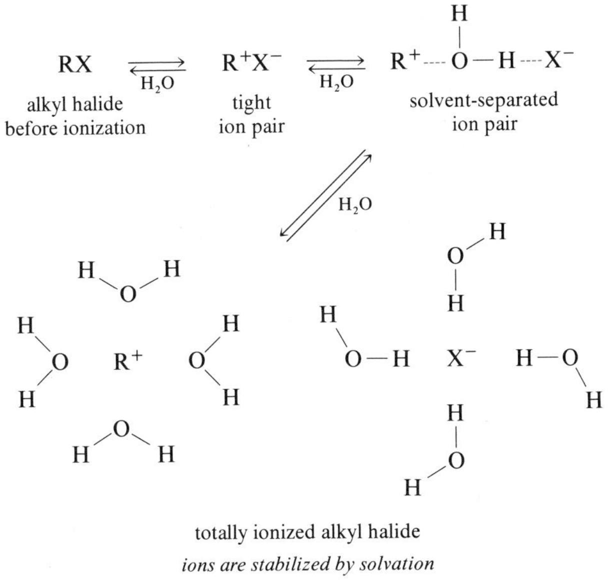

$$RX \underset{H_2O}{\rightleftharpoons} R^+X^- \underset{H_2O}{\rightleftharpoons} R^+ \cdots \overset{\overset{\textstyle H}{|}}{O}-H \cdots X^-$$

alkyl halide	tight	solvent-separated
before ionization	ion pair	ion pair

$\Big\updownarrow H_2O$

totally ionized alkyl halide

ions are stabilized by solvation

The extent of ionization before reaction with a nucleophile depends on the stability of the potential carbocation, the nature of the leaving group, and the ionizing power of the solvent.

The alkyl halide may, in principle, react with the nucleophile at any one of the stages shown above. For example, a primary alkyl halide, in most solvents, will interact with the nucleophile before there is any significant ionization. Such a reaction is the typical S_N2 reaction described earlier. Tertiary alkyl halides, or other compounds that give rise to stabilized carbocations, react with nucleophiles in a typical S_N1 reaction after the ion pairs are separated by solvent.

Note that reaction of a nucleophile with the alkyl halide before ionization or with a tight ion-pair, gives inversion of configuration. If the nucleophile is the solvent, reaction of the cation with solvent in solvent-separated ion pairs or as the free cation gives extensive racemization. With careful experiments, chemists have shown that secondary alkyl compounds undergo nucleophilic substitution reactions with almost complete inversion of configuration, even in solvents that encourage ionization.

Therefore, secondary alkyl compounds react by nucleophilic attack either on the neutral alkyl halide or on the tight ion pair.

This unified way of looking at nucleophilic substitution reactions was proposed by Saul Winstein of the University of California at Los Angeles after much careful experimentation. If we understand the factors that govern these reactions, we can manipulate the conditions such as the solvent and the type and concentration of nucleophile so that we obtain the products we want.

In summary, *more nearly pure S_N2 reactions are favored with primary alkyl halides in solvents, such as dry acetone, that do not promote ionization and with good nucleophiles such as iodide, bromide, hydroxide, and ethoxide ions.* Under these conditions, methyl compounds react the fastest. The rate of the reaction falls off as we go to bulky substituents and to secondary and tertiary halides.

For S_N2 conditions:

$$\text{CH}_3- \;>\; \text{CH}_3\text{CH}_2- \;>\; \underset{\text{CH}_3}{\text{CH}_3\overset{\displaystyle\text{CH}_3}{\text{CH}}-} \;>\; \underset{\displaystyle\text{CH}_3}{\overset{\displaystyle\text{CH}_3}{\text{CH}_3\text{C}-}}$$

       reacts fastest   *reacts most slowly*

More nearly pure S_N1 reactions take place with tertiary alkyl halides, or other compounds that give relatively stable carbocations, in ionizing solvents such as water, aqueous alcohol, or aqueous acetone. In ionizing solvents, especially with no added nucleophiles, tertiary alkyl halides react the fastest. The rate of the reaction decreases with secondary and primary halides.

For S_N1 conditions:

$$\underset{\displaystyle\text{CH}_3}{\overset{\displaystyle\text{CH}_3}{\text{CH}_3\text{C}-}} \;>\; \underset{\text{CH}_3}{\overset{\displaystyle\text{CH}_3}{\text{CH}_3\text{CH}-}} \;>\; \text{CH}_3\text{CH}_2- \;>\; \text{CH}_3-$$

       reacts fastest   *reacts most slowly*

Some of the various reagents and solvents that are actually used in syntheses involving nucleophilic substitution reactions are shown in Section 6.11.

6.8

Nucleophilic Substitution of Allylic and Benzylic Halides

S_N1 and S_N2 reactions usually take place at tetrahedral carbon atoms. Compounds in which the leaving group is bound to trigonal, sp^2-hybridized carbon atoms do not undergo nucleophilic substitution reactions easily. For example, chlorobenzene and vinyl chloride give no reaction with either silver nitrate in ethanol or sodium iodide in acetone.

$$\underset{\text{chlorobenzene}}{\boxed{}-\text{Cl}}\quad\begin{array}{l}\xrightarrow[\text{ethanol}]{\text{AgNO}_3}\text{ no reaction}\\[1em]\xrightarrow[\text{acetone}]{\text{NaI}}\text{ no reaction}\end{array}$$

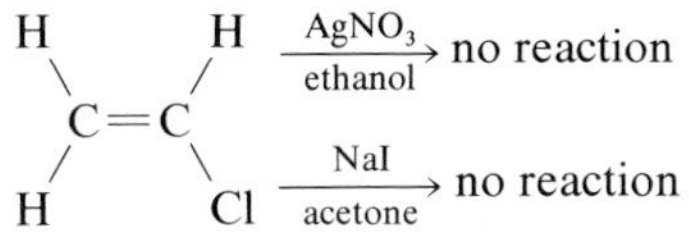

$$\underset{\text{vinyl chloride}}{\overset{\displaystyle \underset{H}{\overset{H}{\underset{|}{\overset{|}{C}}}}=\underset{Cl}{\overset{H}{\underset{|}{\overset{|}{C}}}}}{}} \quad \begin{array}{c} \xrightarrow[\text{ethanol}]{\text{AgNO}_3} \text{no reaction} \\[2mm] \xrightarrow[\text{acetone}]{\text{NaI}} \text{no reaction} \end{array}$$

There are several reasons for this unreactivity. The carbon-halogen bond in these compounds is stronger than it is in the corresponding alkyl halides. For example, the bond dissociation energy (Section 5.4B) of the carbon-chlorine bond in vinyl chloride is 89 kcal/mol; that in chlorobenzene is 95 kcal/mol. Ethyl chloride, by comparison, has a carbon-chlorine bond dissociation energy of 81 kcal/mol. In addition, reaction by either an S_N1 or an S_N2 mechanism is difficult. If the reaction were to proceed by the S_N2 mechanism, a nucleophile would have to approach the rear of a carbon atom that is involved in π bonding. Such an approach is both electronically and sterically unfavorable, as shown below for an aryl halide. The illustration shows a nucleophile blocked by π bonds from approaching the rear of the carbon atom bearing the leaving group.

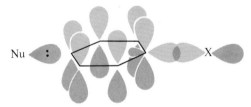

Nucleophilic substitution reactions do occur with especially reactive aryl halides, but by a different mechanism (Section 12.5).

S_N1 reactions would require that aryl or vinyl cations (Section 5.5A) be formed.

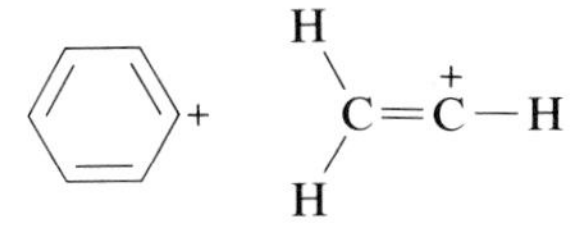

phenyl cation   vinyl cation

carbocations of higher energy than alkyl cations

These species are of higher energy than alkyl cations. Vinyl cations are formed as intermediates in substitution reactions only when an exceptionally good leaving group is present in a solvent, such as trifluoroacetic acid, that strongly favors ionization.

If the halogen is on a carbon atom next to an aromatic ring, the compound ionizes easily (Section 6.6B). *Such reactivity is typical of these compounds, which are known as* **benzylic halides.** Benzyl chloride itself reacts in 50% aqueous acetone in the presence of hydroxide ion by both the S_N1 and the S_N2 mechanisms.

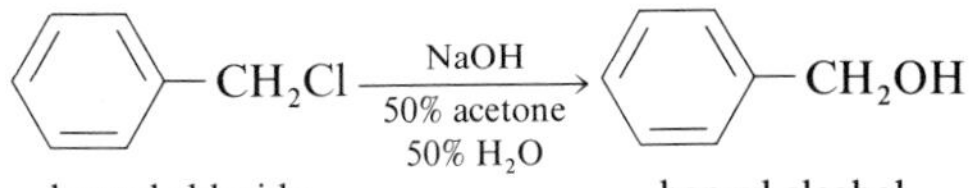

benzyl chloride $-\text{CH}_2\text{Cl} \xrightarrow[\substack{\text{50\% acetone}\\\text{50\% H}_2\text{O}}]{\text{NaOH}}$ benzyl alcohol $-\text{CH}_2\text{OH}$

It can undergo ionization, but it is also a primary halide, and the S_N2 reaction is not too difficult.

PROBLEM 6.19   Write an equation for the ionization of benzyl chloride and draw resonance contributors showing the delocalization of charge in the benzylic cation. (Hint: Before you do this, you may want to review Section 6.6B.)

PROBLEM 6.20   Write detailed mechanisms showing the conversion of benzyl chloride to benzyl alcohol by both the S_N1 and the S_N2 pathways.

The same kind of reactivity is seen when the halogen is on a tetrahedral carbon atom adjacent to a double bond, in the class of compounds known as **allylic halides** *(Section 5.1B). 3-Chloropropene (allyl chloride) gives an instant precipitate when it is put into a solution of silver nitrate in ethanol.*

$$CH_2{=}CHCH_2Cl \xrightarrow[\text{ethanol}]{\text{AgNO}_3} \text{white precipitate}$$

3-chloropropene
allyl chloride

On the other hand, when it reacts with potassium hydroxide in 50% aqueous ethanol, the rate of the reaction increases with an increase in the hydroxide ion concentration, though the alcohol is also formed in the absence of hydroxide ion.

$$CH_2{=}CHCH_2Cl \xrightarrow[\substack{\text{50\% ethanol}\\ \text{in water}}]{\text{KOH}} CH_2{=}CHCH_2OH$$

allyl chloride   2-propen-1-ol
                                                        allyl alcohol

Allyl chloride acts as a primary halide in participating in S_N2 reactions. It can also undergo S_N1 reactions because it forms a carbocation with delocalized charge.

$$CH_2{=}CHCH_2Cl \longrightarrow [CH_2{=}CH-\overset{+}{C}H_2 \longleftrightarrow \overset{+}{C}H_2-CH{=}CH_2] + Cl^-$$

*resonance contributors for the
allyl cation*

The two resonance contributors for the allyl cation are identical, giving good stabilization of the cation, which should react with nucleophiles at either one of the two carbon atoms shown with the positive charge. We can demonstrate that this does, indeed, happen by labeling one of the carbon atoms in allyl chloride with radioactive carbon, ^{14}C.

$$CH_2{=}CH^{14}CH_2Cl \xrightarrow{\text{H}_2\text{O}} CH_2{=}CH^{14}CH_2OH + {}^{14}CH_2{=}CHCH_2OH$$

allyl chloride   ~50%   ~50%

*labeled with
radioactive carbon*

The allyl alcohol that is formed has the label at the first carbon atom of the double bond, and at the carbon atom bearing the hydroxyl group in about equal amounts.

PROBLEM 6.21   Write a detailed mechanism for the conversion of the allyl chloride labeled with radioactive carbon to the alcohol by the S_N1 mechanism. Would you expect to see the same distribution of radioactivity if the reaction went by an S_N2 mechanism in the presence of hydroxide ion?

6.9

Elimination Reactions

A. The E_1 Reaction in Competition with the S_N1 Reaction. Orientation in the E_1 Reaction

Elimination reactions accompany substitution reactions to a lesser or greater degree. The intermediate carbocation that is formed in all S_N1 reactions loses a proton to a base in the reaction mixture to give an alkene in what is known as the **unimolecular elimination,** *the* E_1 *reaction. The rate-determining step for this elimination reaction is the unimolecular ionization of the alkyl halide.*

For the *tert*-butyl cation, only one alkene is possible as the product.

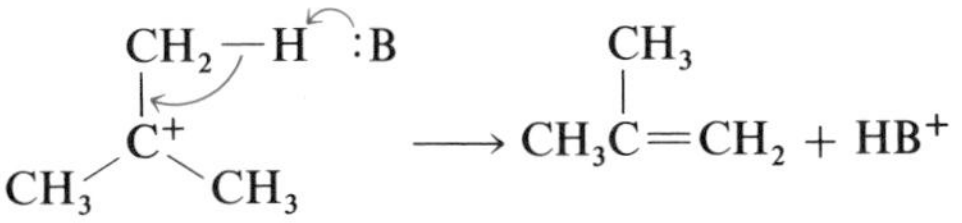

tert-butyl cation   2-methylpropene

If the structure of the carbocation is such that different kinds of protons can be lost to give different alkenes, the more highly substituted alkene is formed in larger amounts.

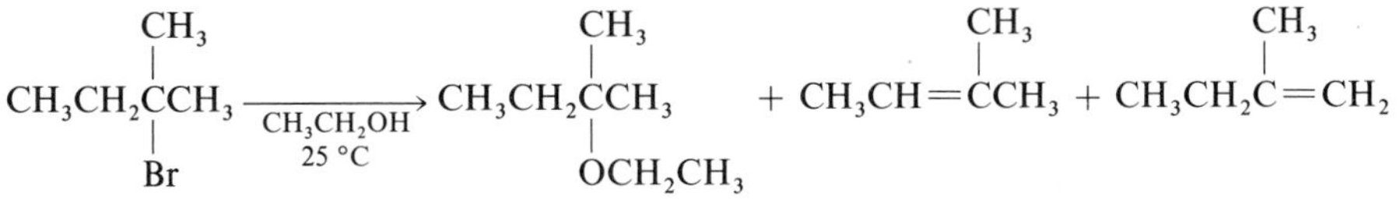

| 2-bromo-2-methylbutane | 2-ethoxy-2-methylbutane 64% | 2-methyl-2-butene 30% | 2-methyl-1-butene 6% |

S_N1 *product* E_1 *products*

The transition state for the formation of an alkene resembles the alkene in structure, and is stabilized by the same factors that stabilize the alkene. The reaction goes chiefly by the pathway leading through the transition state of lower energy to give the more highly substituted alkene, which is the more stable one (Section 5.2).

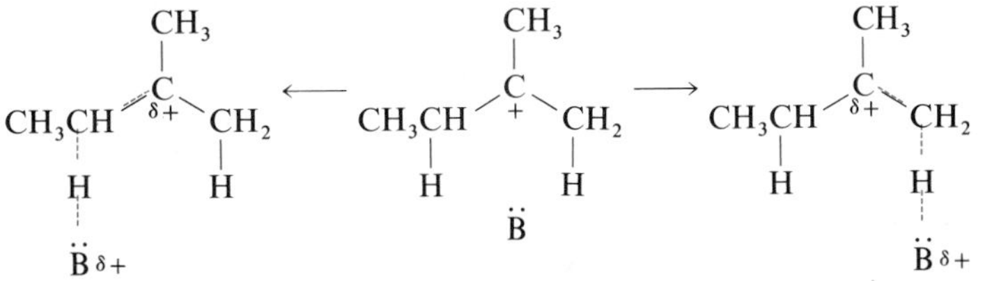

transition state for the formation of 2-methyl-2-butene; lower energy transition state resembling the more highly substituted alkene product

2-methyl-2-butyl cation and base

transition state for the formation of 2-methyl-1-butene; higher energy transition state resembling the less highly substituted alkene product

B. Competition Between S_N2 and E_2 Reactions

When reactions are carried out in the presence of strong bases, a **bimolecular elimination reaction,** the E_2 reaction, accompanies S_N2. How much elimination takes place depends on the structure of the alkyl halide. The experimental facts shown in the next three equations demonstrate that it is highly impractical to try to carry out nucleophilic substitution reactions on secondary and tertiary alkyl halides with a strongly basic reagent because the chief product of such a reaction is elimination.

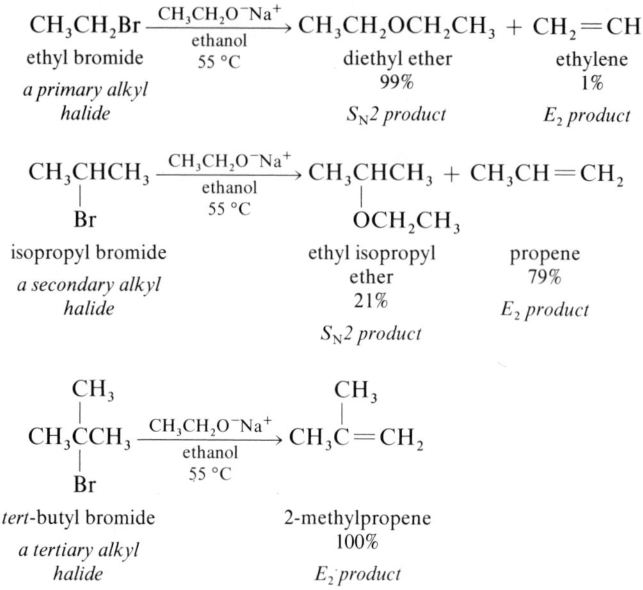

$$CH_3CH_2Br \xrightarrow[\substack{\text{ethanol} \\ 55\,°C}]{CH_3CH_2O^-Na^+} CH_3CH_2OCH_2CH_3 + CH_2{=}CH_2$$

ethyl bromide		diethyl ether	ethylene
a primary alkyl halide		99%	1%
		S_N2 *product*	E_2 *product*

$$\underset{\substack{|\\Br}}{CH_3CHCH_3} \xrightarrow[\substack{\text{ethanol} \\ 55\,°C}]{CH_3CH_2O^-Na^+} \underset{\substack{|\\OCH_2CH_3}}{CH_3CHCH_3} + CH_3CH{=}CH_2$$

isopropyl bromide	ethyl isopropyl ether	propene
a secondary alkyl halide	21%	79%
	S_N2 *product*	E_2 *product*

$$\underset{\substack{|\\Br}}{\overset{\substack{CH_3\\|}}{CH_3CCH_3}} \xrightarrow[\substack{\text{ethanol} \\ 55\,°C}]{CH_3CH_2O^-Na^+} \overset{\substack{CH_3\\|}}{CH_3C}{=}CH_2$$

tert-butyl bromide	2-methylpropene
a tertiary alkyl halide	100%
	E_2 *product*

PROBLEM 6.22   The **Williamson synthesis** *of an ether* involves nucleophilic substitution on an alkyl halide by an alkoxide ion. For example, ethyl bromide reacts with ethoxide ion to give diethyl ether.

$$CH_3CH_2Br + CH_3CH_2O^-Na^+ \xrightarrow{\text{ethanol}} CH_3CH_2OCH_2CH_3 + Na^+Br^-$$

| ethyl bromide | sodium ethoxide | diethyl ether |

On paper, there are two ways to synthesize *tert*-butyl ethyl ether by this method. The two equations are given below.

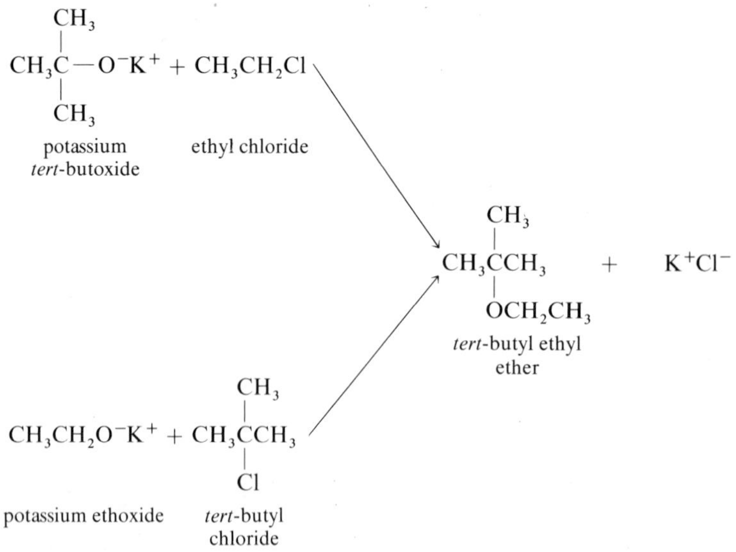

$$\underset{\substack{|\\CH_3}}{\overset{\substack{CH_3\\|}}{CH_3C}}{-}O^-K^+ + CH_3CH_2Cl$$

| potassium *tert*-butoxide | ethyl chloride |

$$\underset{\substack{|\\OCH_2CH_3}}{\overset{\substack{CH_3\\|}}{CH_3CCH_3}} + K^+Cl^-$$

tert-butyl ethyl ether

$$CH_3CH_2O^-K^+ + \underset{\substack{|\\Cl}}{\overset{\substack{CH_3\\|}}{CH_3CCH_3}}$$

| potassium ethoxide | *tert*-butyl chloride |

If you had to make this ether in the laboratory, which of these routes would you choose? Why?

If the formation of more than one alkene is possible, the more highly substituted alkene is favored. Thus, *sec*-butyl bromide in base gives chiefly 2-butene.

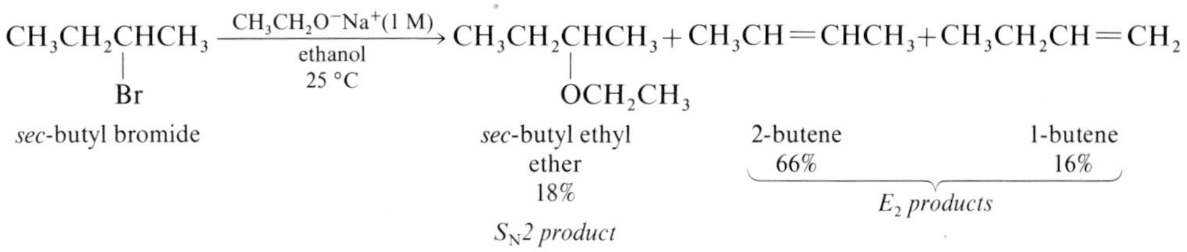

$$CH_3CH_2CHCH_3 \xrightarrow[\substack{\text{ethanol} \\ 25\,°C}]{CH_3CH_2O^-Na^+(1\,M)} CH_3CH_2CHCH_3 + CH_3CH=CHCH_3 + CH_3CH_2CH=CH_2$$

Br		OCH_2CH_3		
sec-butyl bromide		*sec*-butyl ethyl ether 18% S_N2 *product*	2-butene 66%	1-butene 16%
			E_2 *products*	

The formation of alkene is increased with increasing concentration of base.

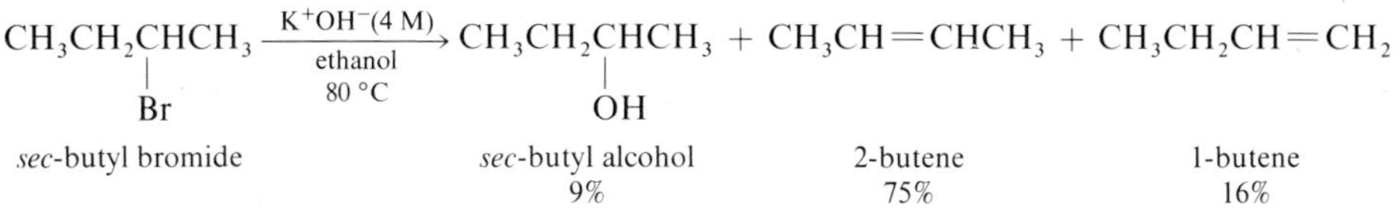

$$CH_3CH_2CHCH_3 \xrightarrow[\substack{\text{ethanol} \\ 80\,°C}]{K^+OH^-(4\,M)} CH_3CH_2CHCH_3 + CH_3CH=CHCH_3 + CH_3CH_2CH=CH_2$$

Br	OH		
sec-butyl bromide	*sec*-butyl alcohol 9%	2-butene 75%	1-butene 16%

The conditions shown above, alcoholic potassium hydroxide, are those classically used when organic chemists wish to convert alkyl halides into alkenes.

In summary, elimination reactions of alkyl halides become prevalent when the reagent is a strong base, such as an alkoxide ion, in high concentration. Tertiary halides are more likely to undergo elimination reactions than secondary halides, which, in turn, are more prone to elimination than primary halides. Because elimination reactions compete with substitution reactions, factors that hinder nucleophilic substitution, such as bulky substituents at the β-carbon atom of the alkyl halide (Section 6.5C), also increase the relative amount of elimination (see Problem 6.14, for example).

PROBLEM 6.23   Write the structure(s) of the product(s) of the elimination reactions that you would expect in the following cases. If more than one product can be formed, predict which one would be the major product.

(a) $CH_3CH_2CH_2CH_2CH_2Br \xrightarrow[\substack{\text{ethanol} \\ \Delta}]{KOH}$

(b) $CH_3CHCHCH_2CH_3 \xrightarrow[\substack{\text{ethanol} \\ \Delta}]{KOH}$ with CH_3 on upper carbon and Br on middle carbon

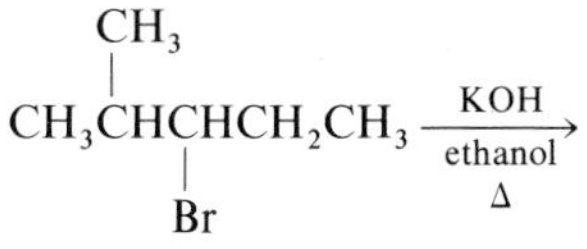

(c)   (cyclohexene ring with Br substituent) $\xrightarrow[\substack{\text{ethanol} \\ \Delta}]{KOH}$

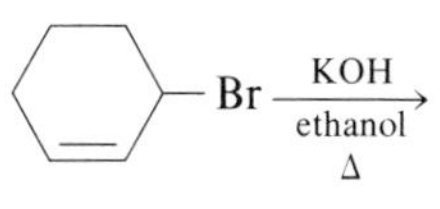

(d)   (bicyclic structure with CH_3, Br, H, CH_3) $\xrightarrow[\substack{\text{ethanol} \\ \Delta}]{KOH}$

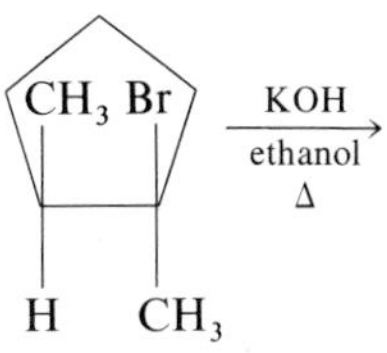

C. Mechanism and Stereochemistry for the E_2 Reaction

The rate of the E_2 reaction of *sec*-butyl bromide, like that of the S_N2 reaction, depends upon the concentrations of both the alkyl halide and the base. Any mechanism for the reaction must involve both species in the rate-determining step.

$$R = k[sec\text{-BuBr}]\,[OH^-]$$

The mechanism that has been proposed for E$_2$ reactions is shown for *sec*-butyl bromide.

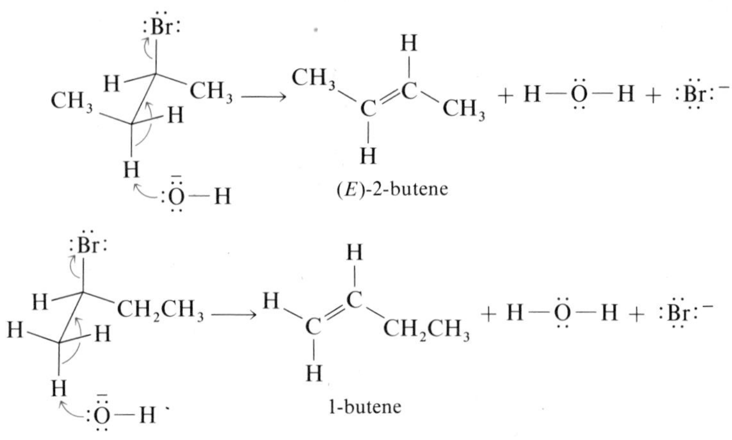

(E)-2-butene

1-butene

The lower energy transition state, and, therefore, the one that leads to most of the product, is the one giving rise to the more highly substituted alkene, 2-butene.

This generalization is usually true if the leaving group is a halide ion and the base is a hydroxide or ethoxide ion. Much experimental work has been done to show that the nature and size of the leaving group, the size of the base used to remove the proton, and, especially, the stereochemistry of the compound undergoing elimination can affect the exact proportions of the different alkenes formed. How the orientation of the elimination reaction can change with the nature of the leaving group is the subject of Section 16.7A.

PROBLEM 6.24   Is another stereochemistry possible for the 2-butene that is formed in the elimination reaction of *sec*-butyl bromide? Draw any other conformations of the halide that will give 2-butene as product.

A definite stereochemical relationship between the proton that is removed by the base and the leaving group, the bromide ion, has been suggested by the drawings above. There is evidence that the E$_2$ elimination proceeds best when the hydrogen to be removed and the leaving group are in the anti orientation with respect to each other. This is easiest to see in cyclohexane compounds. *cis*-2-Methylcyclohexyl tosylate, when treated with potassium *tert*-butoxide in dimethyl sulfoxide as solvent gives a mixture of two alkenes, 1-methylcyclohexene and 3-methylcyclohexene. The more highly substituted alkene is formed in the larger amount.

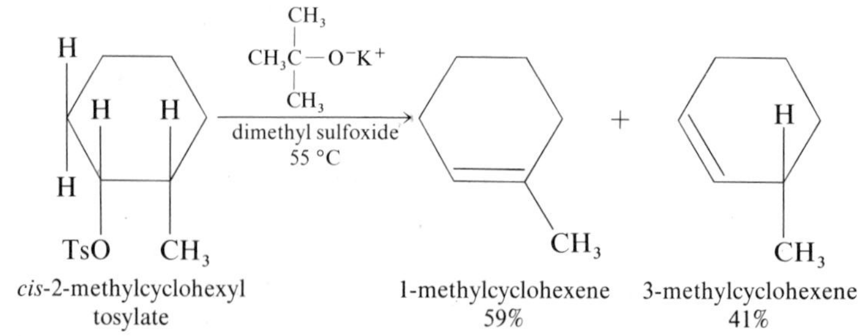

cis-2-methylcyclohexyl tosylate   1-methylcyclohexene 59%   3-methylcyclohexene 41%

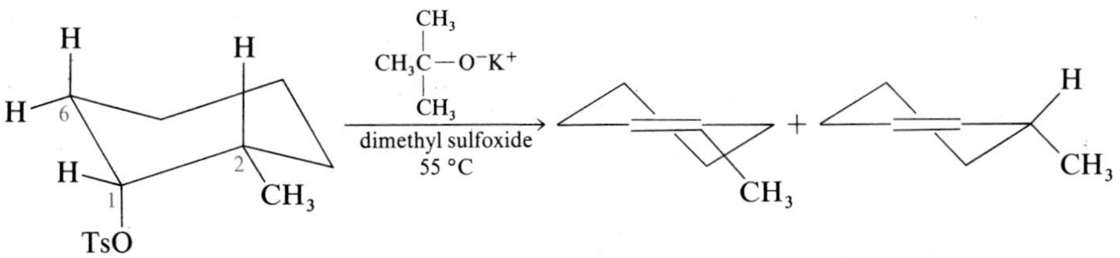

cis-2-methylcyclohexyl   1-methylcyclohexene   3-methylcyclohexene
tosylate

When *trans*-2-methylcyclohexyl tosylate is treated with the same reagent under the same conditions, only the less highly substituted alkene, 3-methylcyclohexene is formed.

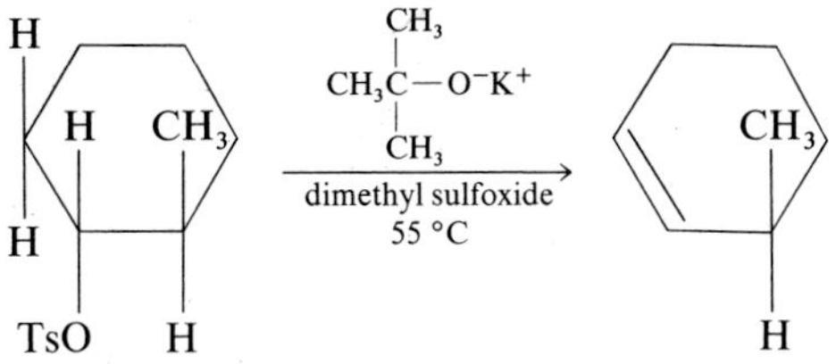

trans-2-methylcyclohexyl   3-methylcyclohexene
tosylate   100%

In the cis compound, two hydrogens are anti to the tosylate group in the most stable conformation of the compound.

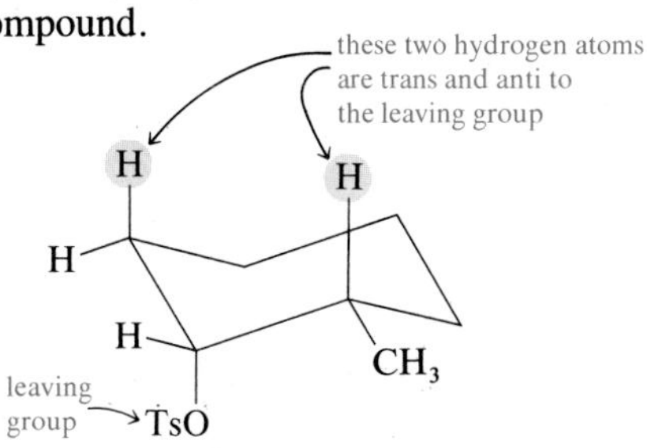

cis-2-methylcyclohexyl
tosylate

For the trans isomer, the most stable conformation of the compound has the leaving group gauche to three hydrogen atoms, one on the carbon bearing the methyl group, and two on carbon 6. This orientation between the leaving group and the proton that must be removed to give an alkene is called the **syn** orientation.

The tosylate group in *trans*-2-methylcyclohexyl tosylate can become anti to one of the hydrogen atoms on C-6 if the cyclohexane ring flips, and the compound assumes its less stable conformation.

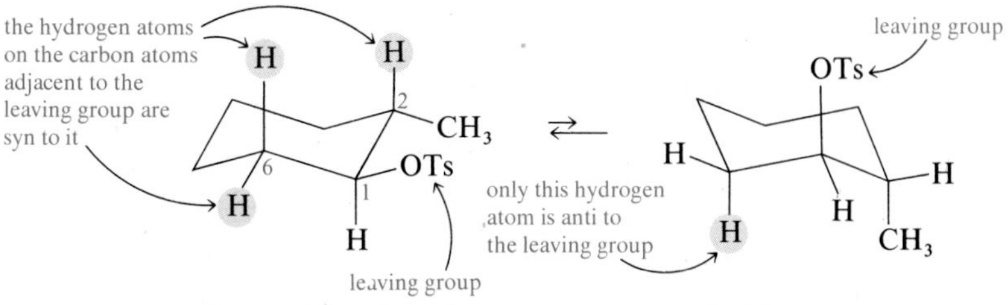

diequatorial conformation   diaxial conformation

trans-2-methylcyclohexyl tosylate

The only product is the alkene that would be formed by the removal of that proton, suggesting that the molecule reacts in this conformation. There is always a small amount of this conformation in equilibrium with the more stable conformation at room temperature.

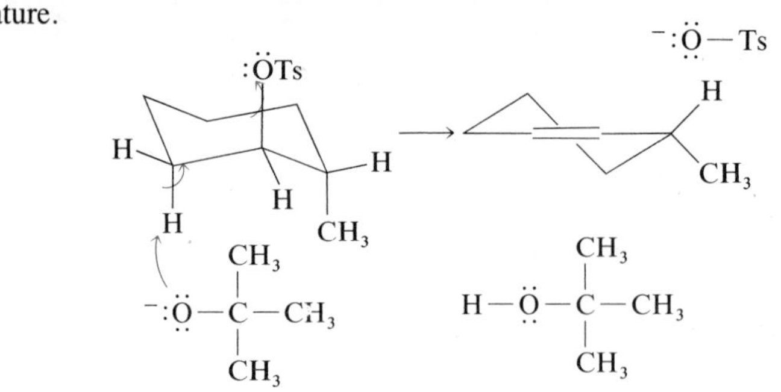

With *cis*-2-methylcyclohexyl tosylate, the reaction can proceed in two directions.

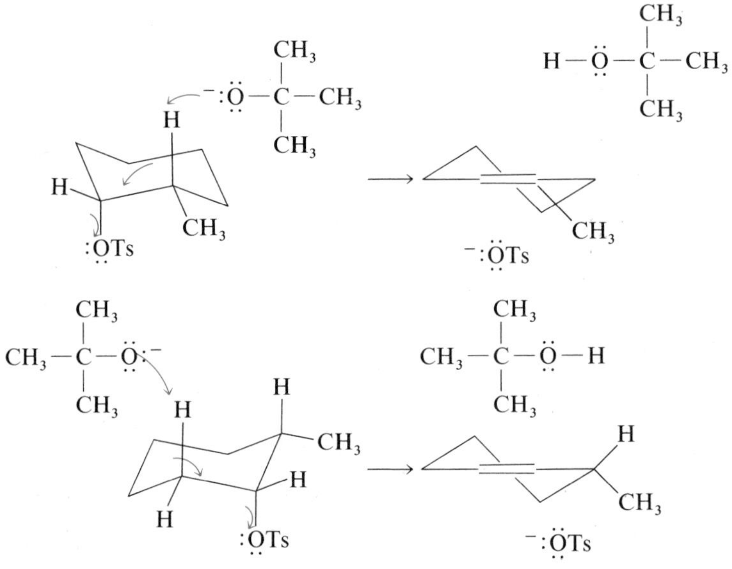

Similar stereochemical requirements are seen in compounds that are not cyclic. (2*R*,3*R*)-3-Phenyl-2-butyl tosylate gives exclusively (*E*)-2-phenyl-2-butene, while its diastereomer, (2*S*,3*R*)-3-phenyl-2-butyl tosylate gives (*Z*)-2-phenyl-2-butene.

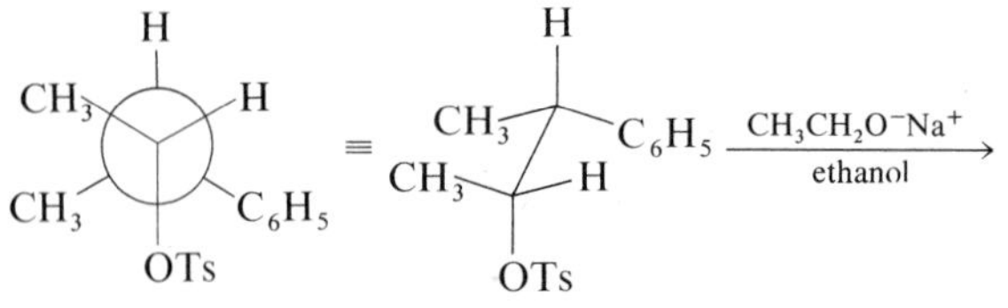

(2*R*,3*R*)-3-phenyl-2-butyl tosylate

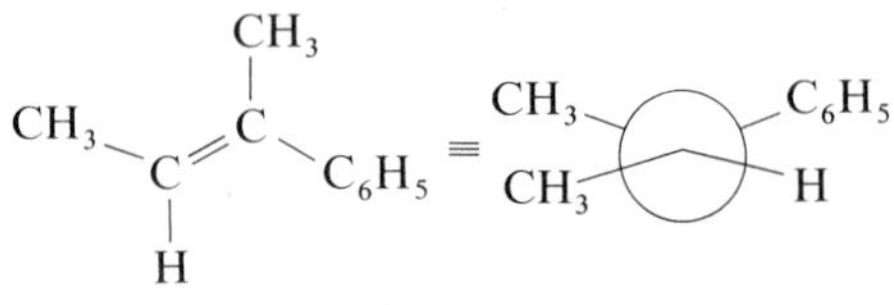

(*E*)-2-phenyl-2-butene

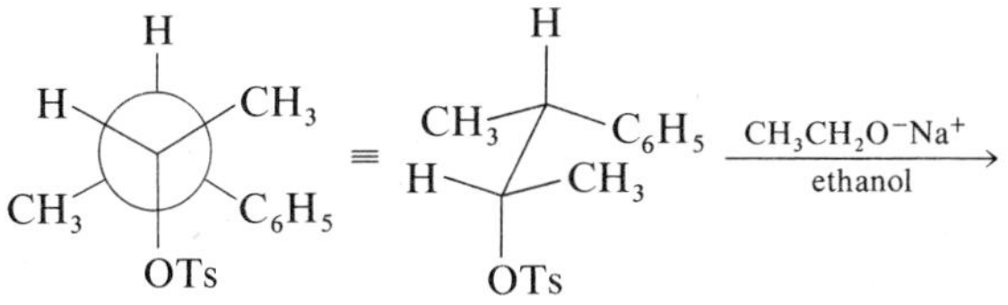

(2S,3R)-3-phenyl-2-butyl tosylate

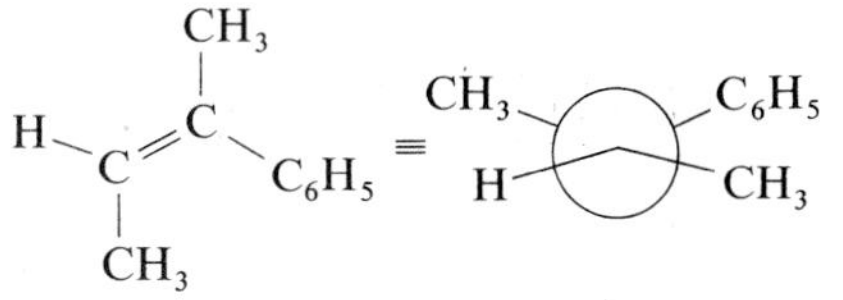

(Z)-2-phenyl-2-butene

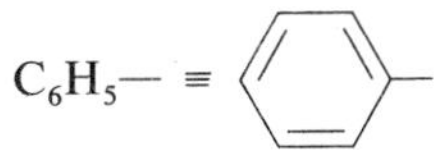

$$C_6H_5\!\!-\ \equiv\ \langle\!\langle\ \rangle\!\rangle\!\!-$$

The stereochemistry of the alkene that is formed is governed by the conformation that the starting materials must adopt if the leaving group and the proton being removed by the base are to be anti to each other.

PROBLEM 6.25   Two diastereomers of 1,2-dibromo-1,2-diphenylethane have been found to give two different bromoalkenes on undergoing E_2 reactions.

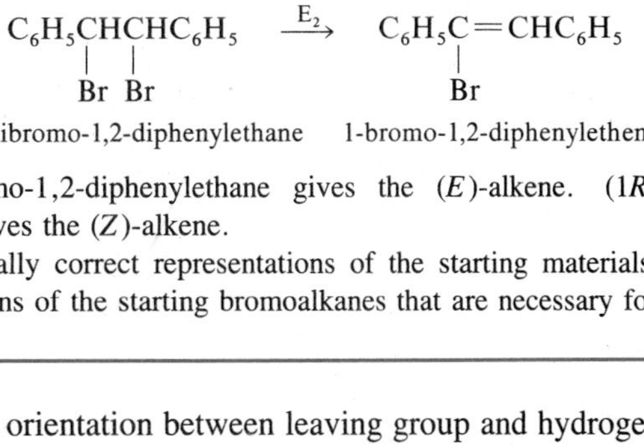

$$C_6H_5CHCHC_6H_5 \ \xrightarrow{\ E_2\ } \ C_6H_5C\!\!=\!\!CHC_6H_5$$

Br Br	Br

1,2-dibromo-1,2-diphenylethane   1-bromo-1,2-diphenylethene

(1R,2S)-1,2-Dibromo-1,2-diphenylethane gives the (E)-alkene. (1R,2R)-1,2-Dibromo-1,2-diphenylethane gives the (Z)-alkene.

Draw stereochemically correct representations of the starting materials and the products. Show the conformations of the starting bromoalkanes that are necessary for the E_2 reaction.

Although the anti orientation between leaving group and hydrogen atom appears to be the most favorable one for elimination reactions, it is not the only one possible. In cyclopentyl compounds especially, the elimination of a hydrogen and a leaving group that are trans to each other is easier, but the elimination of groups that are cis to each other also takes place. *cis*-2-Phenylcyclopentyl tosylate and *trans*-2-phenylcyclopentyl tosylate both give 1-phenylcyclopentene as the product. The first reaction, in which the groups that must be eliminated to give the more stable alkene are trans to each other, is nine times faster than the second reaction, in which the groups eliminated are cis to each other.

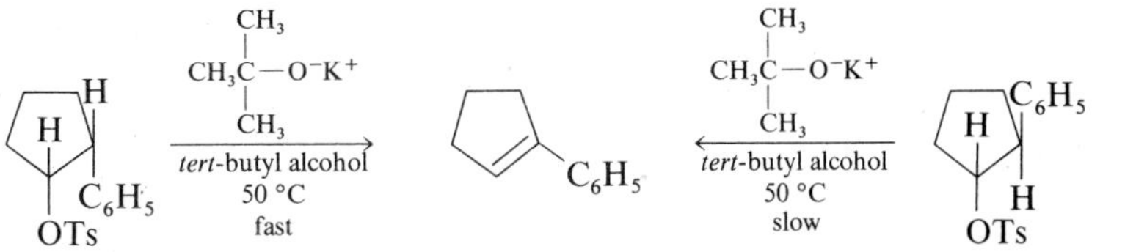

cis-2-phenylcyclopentyl   1-phenylcyclopentene   *trans*-2-phenylcyclopentyl
tosylate   tosylate

6.10

α-Elimination Reactions. Formation of a Carbene

Elimination of a hydrogen atom and a leaving group from adjacent carbon atoms gives a π bond. This kind of elimination is also called **β-elimination.**

$$H \overset{:\ddot{O}-H}{\curvearrowleft} \quad \longrightarrow \quad \overset{}{\underset{|}{C}} = \overset{}{\underset{|}{C}} + H-\ddot{O}-H + :\ddot{B}r:^-$$

If there is no possibility for β-elimination to take place, a proton and a leaving group can be lost from the same carbon atom. For example, the trichloromethyl anion, formed when trichloromethane loses a proton (Section 4.8), can lose a chloride ion, a good leaving group, to give a new, very reactive intermediate, a **carbene.** The overall reaction is an **α-elimination,** taking place in two steps.

$$Cl-\overset{Cl}{\underset{Cl}{\underset{|}{\overset{|}{C}}}}-H \curvearrowleft :\ddot{O}-H \rightleftharpoons Cl-\overset{Cl}{\underset{Cl}{\underset{|}{\overset{|}{C}}}}:^- + H-\ddot{O}-H$$

trichloromethane   trichloromethyl anion
chloroform   *a stabilized carbanion*

$$Cl-\overset{:\ddot{Cl}:}{\underset{Cl}{\underset{|}{\overset{|}{C}}}}:^- \longrightarrow \overset{Cl}{\underset{Cl}{\diagdown}}C: + :\ddot{Cl}:^-$$

trichloromethyl   dichlorocarbene   chloride
anion   anion

A carbene has no charge, but is an electron-deficient species having only six electrons around the carbon atom. A carbene is an esoteric but useful intermediate in organic reactions. Carbenes react, for example, with the π electrons of a double bond to give a three-membered ring, a cyclopropane.

$$CH_2{=}CH_2 + Cl-\ddot{C}-Cl \longrightarrow \overset{Cl \quad Cl}{\triangle}$$

ethylene   dichlorocarbene   1,1-dichlorocyclopropane

The reactions of carbenes are considered in more detail in Section 19.6, but you should be aware that they can be formed by α-elimination reactions and that they react as electron-deficient reagents.

6.11

Substitution Reactions in Syntheses

Nucleophilic substitution reactions permit the synthesis of a variety of compounds from alkyl halides. As shown in the previous sections, substitution reactions proceed with the fewest complications when primary alkyl halides react with nucleophiles. In this section, the reactions of *n*-propyl halides to give compounds containing other functional groups are shown. Some of the further transformations that are possible for the functional groups are already familiar to you. Other transformations are indicated and discussed more fully in later chapters.

The reactions of halides with alkoxide or hydroxide ions to give ethers or alcohols have been used as examples many times in previous sections of this chapter. Especially important is the Williamson synthesis of ethers (Problem 6.22).

$$CH_3CH_2CH_2Br + CH_3O^-Na^+ \xrightarrow[\text{methanol}]{} CH_3CH_2CH_2OCH_3 + Na^+Br^-$$

n-propyl bromide   sodium   methyl *n*-propyl
methoxide   ether

Most alkyl halides are prepared from the corresponding alcohols (Section 7.5); therefore the reverse reaction is not of great synthetic importance, but is useful sometimes.

$$CH_3CH_2CH_2Br + Na^+OH^- \xrightarrow[\text{H}_2\text{O}]{} CH_3CH_2CH_2OH + Na^+Br^-$$

n-propyl bromide   *n*-propyl alcohol

Aryl ethers and aryl thioethers are formed when the corresponding sodium salts of phenols (Section 12.6A) or thiophenols react with propyl halides.

$$CH_3CH_2CH_2I + \text{(phenyl ring)}{-}O^-Na^+ \xrightarrow[\text{acetone}]{} CH_3CH_2CH_2O{-}\text{(phenyl ring)} + Na^+I^-$$

n-propyl iodide   sodium phenolate   phenyl *n*-propyl ether

$$CH_3CH_2CH_2Br + \text{(phenyl ring)}{-}S^-Na^+ \xrightarrow[\text{methanol}]{} CH_3CH_2CH_2S{-}\text{(phenyl ring)} + Na^+Br^-$$

n-propyl bromide   sodium thiophenolate   phenyl *n*-propyl thioether

Ammonia reacts with alkyl halides to give amines.

$$CH_3CH_2CH_2Br + NH_3 \xrightarrow[\text{ethanol}]{} CH_3CH_2CH_2NH_2 + NH_4^+Br^-$$

n-propyl bromide ammonia   *n*-propylamine   ammonium
                 (excess)   bromide

If an amine is used as the nucleophilic reagent, a more highly substituted amine is the product.

$$CH_3CH_2CH_2Br \; + \; \text{(piperidine, N--H)} \xrightarrow[\text{solvent}]{\text{alkane}} \text{(N-propylpiperidine, N--CH}_2\text{CH}_2\text{CH}_3) \; + \; \text{(piperidine hydrobromide)} \; Br^-$$

n-propyl bromide   piperidine   *N*-propylpiperidine   piperidine
                     (excess)   hydrobromide

Another good nucleophile containing nitrogen is the azide ion.

$$CH_3CH_2CH_2I + Na^+N_3^- \xrightarrow[\substack{\text{carbitol}\\\text{H}_2\text{O}}]{} CH_3CH_2CH_2N_3 + Na^+I^-$$

n-propyl iodide   sodium   *n*-propyl azide
                  azide

$$\text{carbitol} \equiv CH_3CH_2OCH_2CH_2OCH_2CH_2OH$$

an ether and an alcohol

Some of the most useful nucleophilic substitution reactions occur with reagents in which carbon atoms are the nucleophilic centers. Such reagents are used to create carbon-carbon bonds. Cyanide ion is used, for example, to extend the chain by one carbon atom.

$$CH_3CH_2CH_2I + Na^+CN^- \xrightarrow[\substack{ethanol \\ H_2O}]{} CH_3CH_2CH_2CN + Na^+I^-$$

n-propyl iodide   sodium cyanide   butanenitrile
                                                        n-propyl cyanide

The product of the reaction can be reduced to an amine (Section 16.3D) or will react with water in the presence of acid to give a carboxylic acid (Section 10.5C).

$$\underset{\text{butanoic acid}}{CH_3CH_2CH_2\overset{\overset{\displaystyle O}{\|}}{C}OH} + NH_4^+Cl^- \xleftarrow[\substack{H_2O \\ \Delta}]{HCl} \underset{\text{butanenitrile}}{CH_3CH_2CH_2CN} \xrightarrow[\text{catalyst}]{H_2} \underset{n\text{-butylamine}}{CH_3CH_2CH_2CH_2NH_2}$$

Carbanions generated from terminal alkynes (Section 9.4A) react with alkyl halides to give internal alkynes.

$$\underset{n\text{-propyl bromide}}{CH_3CH_2CH_2Br} + \underset{\substack{\text{carbanion from} \\ \text{1-pentyne}}}{CH_3CH_2CH_2C\equiv C{:}^-Na^+} \xrightarrow[\substack{\text{ammonia} \\ \text{(liquid)}}]{}$$

$$\underset{\text{4-octyne}}{CH_3CH_2CH_2C\equiv CCH_2CH_2CH_3} + Na^+Br^-$$

The triple bond in alkynes is a reactive site (Chapters 5 and 8). In addition, alkynes are reduced to alkenes, which also undergo a variety of reactions (Chapters 5 and 8). Some possible reactions are indicated schematically in Figure 6.6.

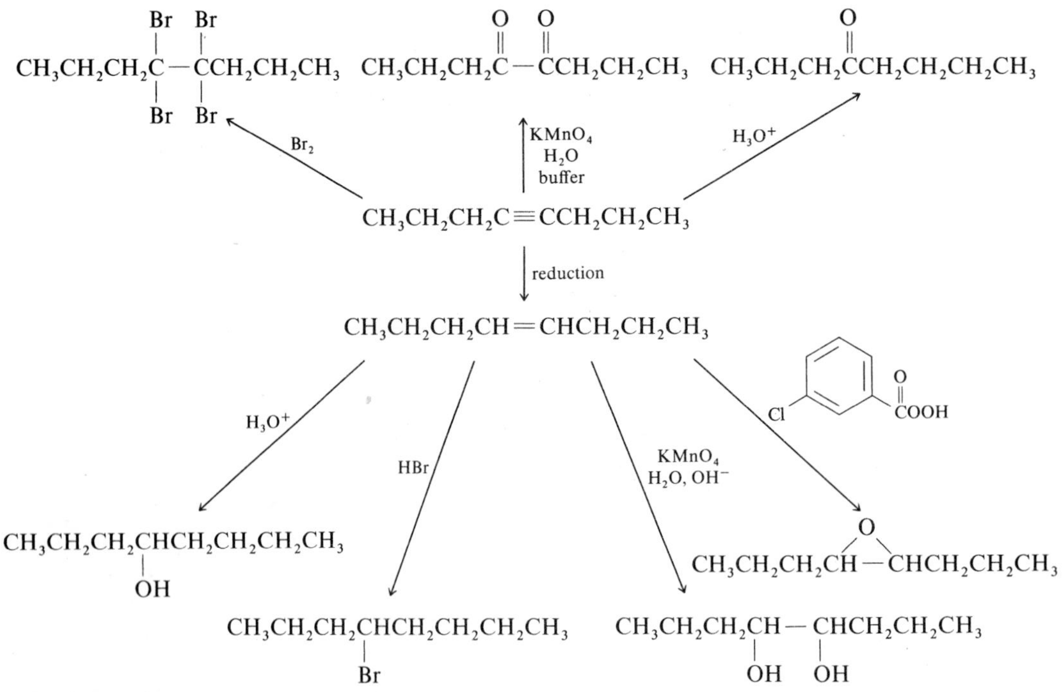

Figure 6.6 Some chemical transformations of alkenes and alkynes.

Other types of carbanions also react with alkyl halides. Among the most important of these are the carbanions prepared from the ester, diethyl malonate (Section 15.4B).

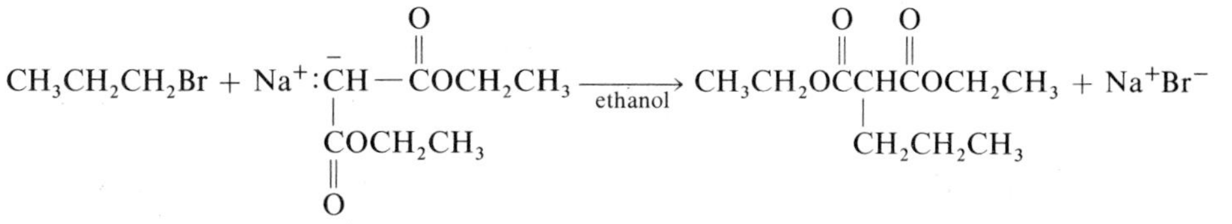

$$\underset{\substack{n\text{-propyl bromide}}}{CH_3CH_2CH_2Br} + Na^+:\overset{-}{\underset{\substack{\overset{|}{\underset{\parallel}{COCH_2CH_3}} \\ \parallel \\ O \\ \text{carbanion from} \\ \text{diethyl malonate}}}{\overset{O}{\overset{\parallel}{C}}H-\overset{O}{\overset{\parallel}{C}}OCH_2CH_3}} \xrightarrow[\text{ethanol}]{} \underset{\substack{\text{diethyl propylmalonate}}}{CH_3CH_2O\overset{O}{\overset{\parallel}{C}}\underset{\substack{| \\ CH_2CH_2CH_3}}{CH}\overset{O}{\overset{\parallel}{C}}OCH_2CH_3} + Na^+Br^-$$

All the reactions that create new carbon-carbon bonds are discussed in detail in later chapters. They are shown here so you can recognize the variety of nucleophilic substitution reactions and appreciate their usefulness.

ADDITIONAL PROBLEMS

6.26 Write equations for the reaction of *n*-butyl bromide as a typical primary alkyl halide, with the following reagents.

(a) alcoholic KOH, heat (b) NaOH, 0.1 M in 50% aqueous ethanol (c) NH_3

(d) NaN_3 (e) NaCN (f) CH_3CH_2SNa

(g) $CH_3\overset{O}{\overset{\parallel}{C}}ONa$ (h) $CH_3CH_2C\equiv CNa$ (i) $NaCH(\overset{O}{\overset{\parallel}{C}}OCH_2CH_3)_2$

6.27 The following equations represent some nucleophilic substitution reactions that are possible. Only primary alkyl groups are used so that the reactions will be S_N2, and competing elimination reactions will not be a problem. Complete the equations showing the products that will form. In each case, it will be helpful to you to label the leaving group and the incoming nucleophile.

(a) 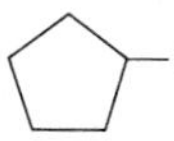 $\langle\text{cyclopentyl}\rangle-CH_2CH_2CH_2OTs + NaN_3 \longrightarrow$

(b)  $\langle\text{cyclohexenyl}\rangle-CH_2CH_2Cl + NaCN \longrightarrow$

(c) $CH_3CH_2CH_2CH_2Br + CH_3CH_2SNa \longrightarrow$

(d) $CH_3CH_2CH_2Cl + CH_3C\equiv CNa \longrightarrow$

(e) $\underset{\substack{\\(\text{excess})}}{CH_3\overset{\overset{\displaystyle CH_3}{|}}{C}HCH_2CH_2Br} + NH_3 \longrightarrow$

(f) $CH_3I + NaCH(\overset{O}{\overset{\parallel}{C}}OCH_2CH_3)_2 \longrightarrow$

(g) $CH_3CH_2I + (CH_3CH_2CH_2CH_2)_3P \longrightarrow$

(h) $HOCH_2CH_2Cl + CH_3SNa \longrightarrow$

6.28   For each of the following bromides, predict whether you expect to get a precipitate with silver nitrate in ethanol, or with sodium iodide in acetone. Some may react with both, some with neither.

(a) $CH_3CH_2CH_2Br$ (b) 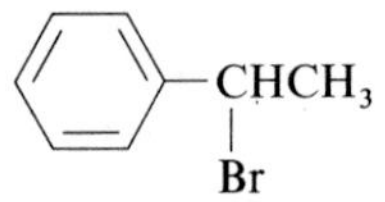 ⬡—$CHCH_3$ (c) $CH_3CH_2CHCH_3$
                                        | |
                                        Br   Br

(d) 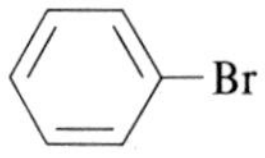 ⬡—Br   (e) $CH_3CH=CHCH_2Br$ (f) 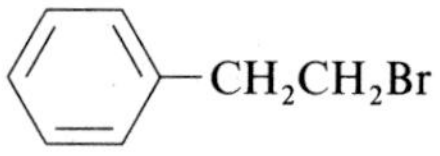 ⬡—CH_2CH_2Br

          CH_3
          |
(g) $CH_3CCH_2CH_3$ (h) 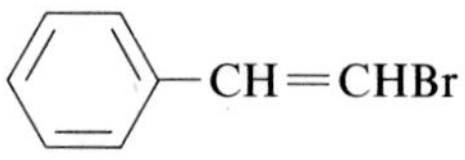 ⬡—$CH=CHBr$
          |
          Br

6.29   The following equations give reagents for elimination reactions that are possible. Complete the equations showing the products from the elimination reactions. Whenever you can, indicate which product would be a major product, which a minor one.

                                                 CH_3
                                             |
(a) 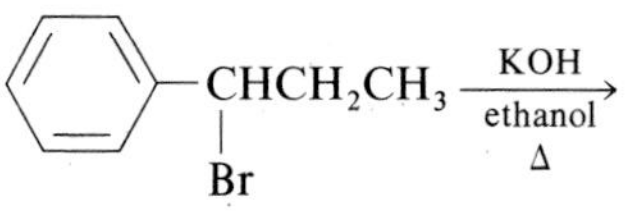 ⬡—$CHCH_2CH_3$ $\xrightarrow[\Delta]{\text{KOH} \atop \text{ethanol}}$ (b) $CH_3CH_2CCH_2CH_2CH_3$ $\xrightarrow[\Delta]{\text{KOH} \atop \text{ethanol}}$
        | |
        Br   Br

          CH_3 O
            | ‖
(c) 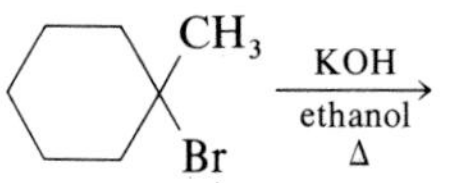 ⬡(CH_3, Br) $\xrightarrow[\Delta]{\text{KOH} \atop \text{ethanol}}$ (d) $CH_3CH_2CH_2CHCOH$ $\xrightarrow[\Delta]{\text{KOH} \atop \text{ethanol}}$
                                                             |
                                                             Br

(e) 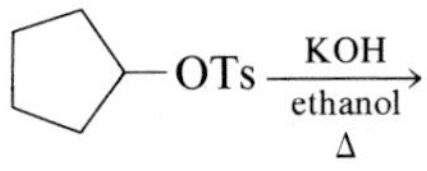 ⬠—OTs $\xrightarrow[\Delta]{\text{KOH} \atop \text{ethanol}}$

6.30   Complete the following reaction sequences, showing the major product you would expect at each stage. When you know what the stereochemistry of a product will be, include it in your answer by the use of some acceptable convention for indicating three dimensionality.

(a) $CH_3CH=CH_2$ $\xrightarrow[\text{H}_2\text{SO}_4]{\text{H}_2\text{O}}$ A $\xrightarrow[\text{pyridine}]{\text{TsCl}}$ B $\xrightarrow{\text{NaN}_3}$ C

(b) $CH_3C\equiv CCH_3$ $\xrightarrow[\text{HgSO}_4]{\text{H}_2\text{O} \atop \text{H}_2\text{SO}_4}$ D

       CH_3
       |
(c) $CH_3CHCH_2CH_2CH=CH_2$ $\xrightarrow[\text{peroxides}]{\text{HBr}}$ E $\xrightarrow[\text{ethanol}]{\text{NaCN}}$ F

(d) 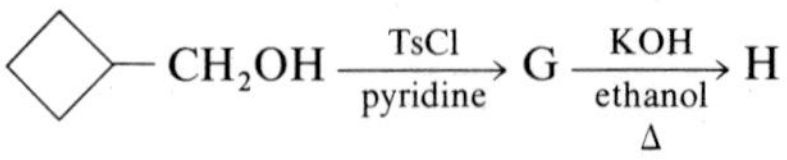 ⬜—CH_2OH $\xrightarrow[\text{pyridine}]{\text{TsCl}}$ G $\xrightarrow[\Delta]{\text{KOH} \atop \text{ethanol}}$ H

    (Why were sulfuric acid and heat not used as the reaction conditions for transforming the starting material to H?)

(e) 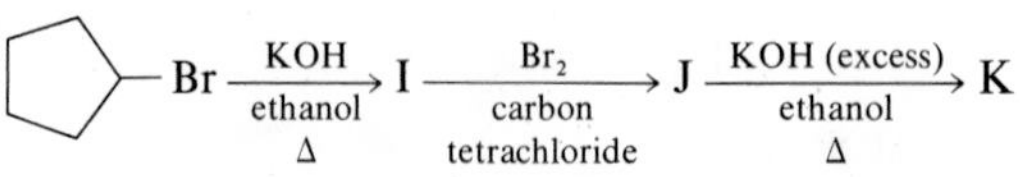 ⬠—Br $\xrightarrow[\Delta]{\text{KOH} \atop \text{ethanol}}$ I $\xrightarrow[\text{tetrachloride}]{\text{Br}_2 \atop \text{carbon}}$ J $\xrightarrow[\Delta]{\text{KOH (excess)} \atop \text{ethanol}}$ K

(f)

$$\overset{\overset{\displaystyle CH_3}{|}}{CH_3CH_2CH_2C} = CHCH_2CH_2OH \xrightarrow[\text{pyridine}]{\text{TsCl}} L \xrightarrow[\text{acetone}]{\text{NaI}} M$$

(g)

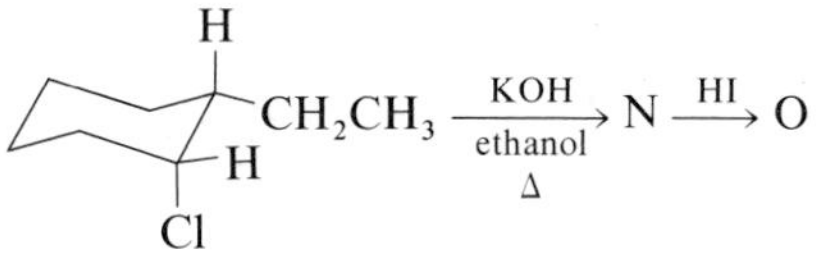

$CH_2CH_3 \xrightarrow[\text{ethanol}]{\text{KOH}} N \xrightarrow{\text{HI}} O$

6.31   In each of the following problems, a starting material and a product that can be prepared from it are shown. Show the reagents that would be necessary to carry out these transformations and any major products that would be formed in the intermediate stages. There may be more than one correct way to complete each sequence.

(a)

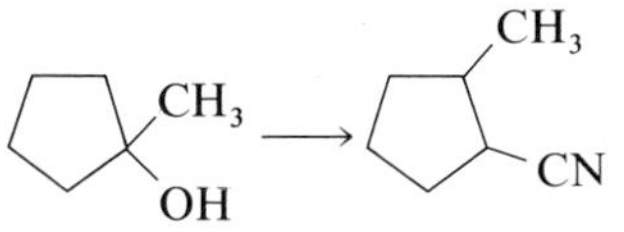

(b)

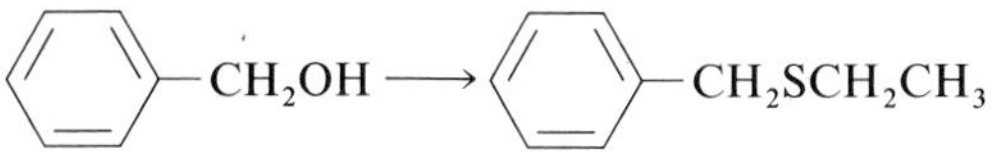

(c)

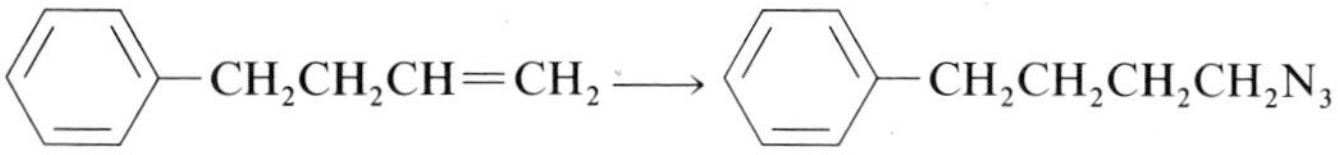

(d)

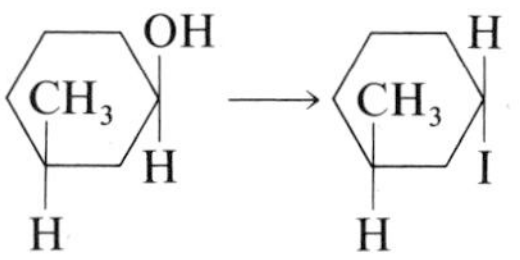

(e) $CH_3CH_2CH_2CH_2Br \longrightarrow CH_3CH_2CH_2CH_2C\equiv CH \longrightarrow CH_3CH_2CH_2CH_2\overset{\overset{\displaystyle O}{||}}{C}CH_3$

(f)

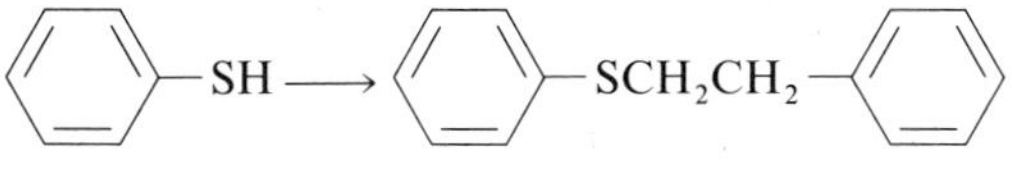

(g)

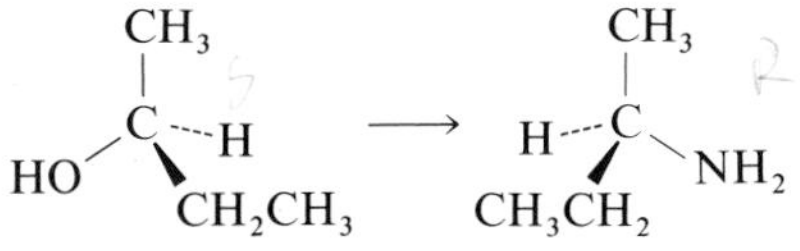

(h) $CH_3CH=CH_2 \longrightarrow CH_3CH_2CH_2C\equiv N$

(i) $CH_3CH_2CH_2CH_2CH_2Br \longrightarrow CH_3CH_2CH_2CH_2CH_2CH(\overset{\overset{\displaystyle O}{||}}{C}OCH_2CH_3)_2$

6.32   Decide for each pair below, which species would be the more nucleophilic.

(a) SH^- or Cl^- (b) $(CH_3)_3B$ or $(CH_3)_3P$ (c) CH_3NH^- or CH_3NH_2

(d) CH_3CH_2OH or CH_3CH_2Cl (e) CH_3SCH_3 or CH_3OCH_3

6.33   Which of the reaction conditions listed below will be likely to give the most inversion of configuration for the following reaction if we start with an optically active halide?

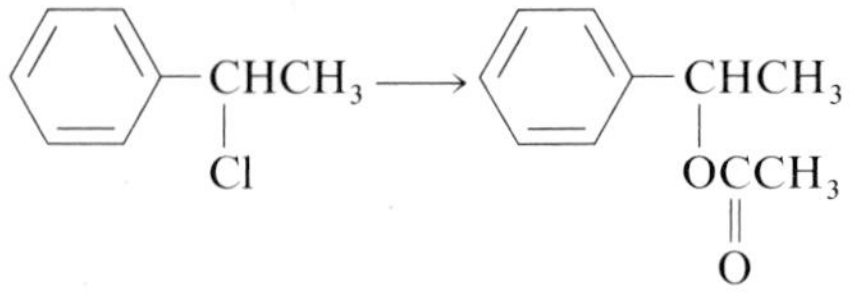

$$\text{C}_6\text{H}_5\!-\!\underset{\underset{\text{Cl}}{|}}{\text{CHCH}_3} \longrightarrow \text{C}_6\text{H}_5\!-\!\underset{\underset{\underset{\text{O}}{\|}}{\overset{|}{\text{OCCH}_3}}}{\text{CHCH}_3}$$

(a) $\text{CH}_3\overset{\overset{\text{O}}{\|}}{\text{C}}\text{O}^-\text{K}^+$ in acetic acid, 50 °C   (b) $(\text{CH}_3\text{CH}_2)_4\text{N}^+ {}^-\overset{\overset{\text{O}}{\|}}{\text{O}}\text{CCH}_3$ in acetone, 50 °C

(c) $\text{CH}_3\overset{\overset{\text{O}}{\|}}{\text{C}}\text{O}^-\text{Na}^+$ in 60% aqueous ethanol, 25 °C

6.34   For the S_N2 reaction,

$$\text{CH}_3\text{CH}_2\text{CH}_2\text{CH}_2\text{Br} + \text{NaOH} \xrightarrow[\text{ethanol}]{} \text{CH}_3\text{CH}_2\text{CH}_2\text{CH}_2\text{OH} + \text{NaBr}$$

answer the following questions:

(a)   What is the rate expression for the reaction?
(b)   Draw an energy diagram for the reaction. Label all parts. You may assume that the products are lower in energy than the starting reagents.
(c)   What will be the effect on the rate of reaction if the concentration of *n*-butyl bromide is doubled?
(d)   What will be the effect on the rate of reaction if the concentration of sodium hydroxide is halved?
(e)   Will the rate of the S_N2 reaction change significantly if the solvent is changed to 80% ethanol, 20% water?

6.35   For the S_N1 reaction,

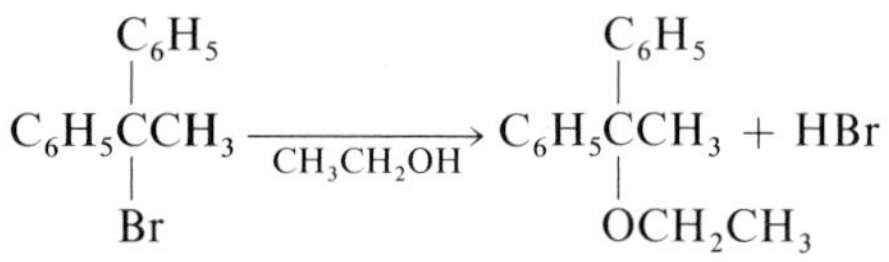

$$\text{C}_6\text{H}_5\underset{\underset{\text{Br}}{|}}{\overset{\overset{\text{C}_6\text{H}_5}{|}}{\text{C}}}\text{CH}_3 \xrightarrow[\text{CH}_3\text{CH}_2\text{OH}]{} \text{C}_6\text{H}_5\underset{\underset{\text{OCH}_2\text{CH}_3}{|}}{\overset{\overset{\text{C}_6\text{H}_5}{|}}{\text{C}}}\text{CH}_3 + \text{HBr}$$
1-bromo-1,1-diphenylethane

answer the following questions:

(a)   What is the rate expression for the reaction?
(b)   Draw an energy diagram for the reaction. Label all parts. You may assume that the products are lower in energy than the starting reagents.
(c)   What will be the effect on the rate of the reaction of doubling the initial concentration of 1-bromo-1,1-diphenylethane?
(d)   Will the rate of the S_N1 reaction change significantly if some water is added to the solvent ethanol?

6.36   When 1-chloro-2-butene is allowed to react in 50% aqueous acetone at 47 °C, the product is a mixture of two alcohols. Write a detailed mechanism that accounts for the experimental facts shown below.

$$CH_3CH=CHCH_2Cl \xrightarrow[\substack{50\% \; acetone \\ 47\,°C}]{50\% \; H_2O} CH_3CH=CHCH_2OH \;+\; CH_3CHCH=CH_2$$

$$\underset{OH}{|}$$

1-chloro-2-butene	2-buten-1-ol	3-buten-2-ol
	56%	44%

6.37   *cis*-2-Phenylcyclohexyl tosylate reacts with potassium *tert*-butoxide in *tert*-butyl alcohol at 50 °C to give exclusively 1-phenylcyclohexene. *trans*-2-Phenylcyclohexyl tosylate does not give any alkene under the same conditions. Draw structures for the two compounds and explain these observations.

6.38   One of the classical experiments in which the stereochemistry of E_2 reactions was explored was with menthyl chloride and neomenthyl chloride. The two compounds differ from each other only in the stereochemistry at the carbon atom to which the chlorine atom is attached. The following equations show the reactions that they undergo without implying anything about stereochemistry.

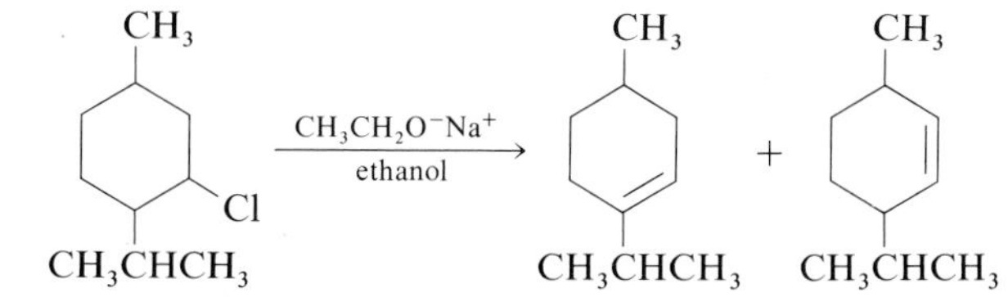

neomenthyl chloride $\xrightarrow[ethanol]{CH_3CH_2O^-Na^+}$ 78%   22%

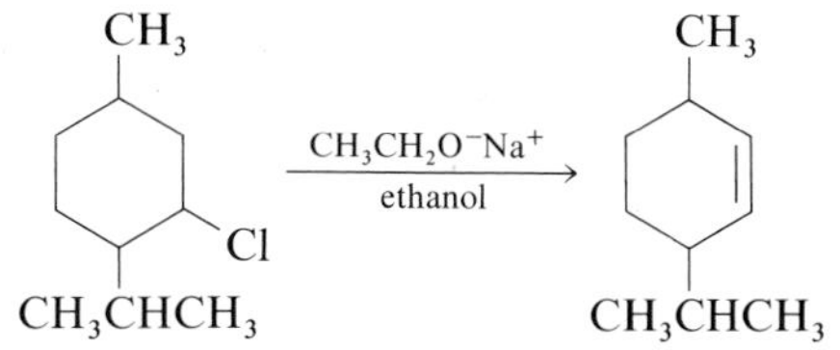

menthyl chloride $\xrightarrow[ethanol]{CH_3CH_2O^-Na^+}$ 100%

(a)   Using the information given, assign relative stereochemistry to menthyl chloride and neomenthyl chloride. Menthyl chloride is derived from natural sources, and the substituents on the cyclohexane ring are in the most stable configuration.

(b)   When the reaction with menthyl chloride is carried out in 80% aqueous ethanol with no added base, the following products from the elimination reaction are obtained. How do you explain these results?

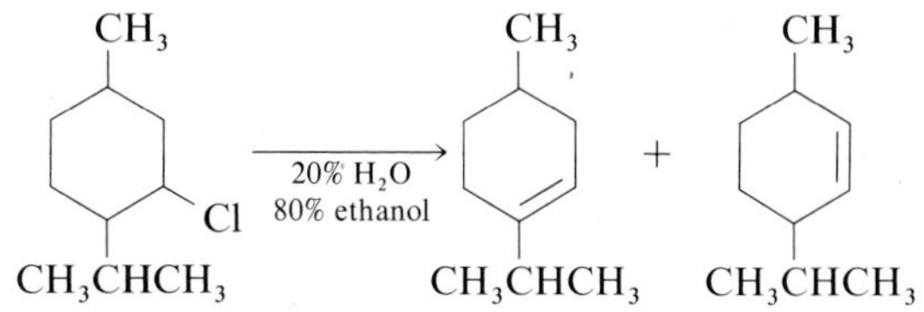

menthyl chloride $\xrightarrow[80\% \; ethanol]{20\% \; H_2O}$

6.39   The stereochemical requirements for the E_2 elimination reaction were explored using alkyl halides labeled with deuterium atoms. 2-Bromo-3-deuteriobutane was used in two isomeric forms. The products observed in each case are outlined as shown on the next page.

1. $(2S,3R)$CH$_3$CHCHCH$_3$

                                                    D   Br

                                                    CH$_3$CH$_2$O$^-$K$^+$
                                                    ethanol
                                                    70 °C

   (E)-2-butene   + (Z)-2-butene   +   1-butene
   with no deuterium   with one deuterium   with one deuterium
          30%   34%   35%

2. $(2S,3S)$CH$_3$CHCHCH$_3$

                                                    D   Br

                                                    CH$_3$CH$_2$O$^-$K$^+$
                                                    ethanol
                                                    70 °C

   (E)-2-butene   + (Z)-2-butene   +   1-butene
   with one deuterium   with no deuterium   with one deuterium
          58%   13%   29%

(a) Draw stereochemically accurate representations of $(2S,3R)$- and $(2S,3S)$-2-bromo-3-deuteriobutane.

(b) Show the transition states for the reactions leading to (E)-2-butene and (Z)-2-butene in each case.

6.40 A solution of triphenylchloromethane in liquid sulfur dioxide, SO$_2$, conducts an electric current. The conductivity of the solution *increases* as the concentration of triphenylchloromethane *decreases*. Write equations that account for these facts. Liquid sulfur dioxide is a polar solvent that is not nucleophilic. Triphenylchloromethane has the following structure:

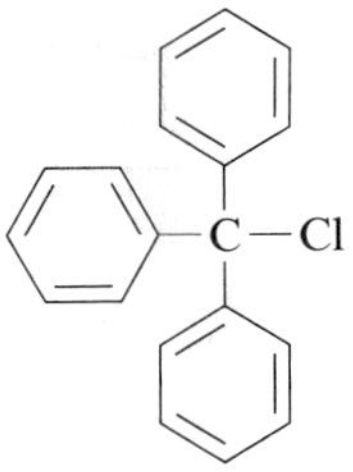

6.41 The following experimental observations were made. Even though the two reactions proceeded at very different rates, the relative amounts of the products obtained were practically identical.

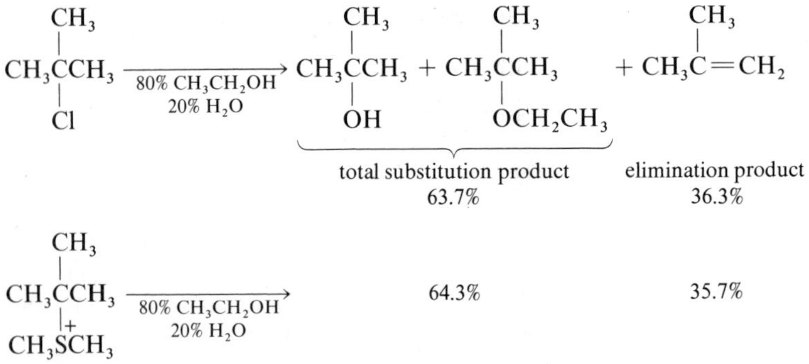

                                    total substitution product   elimination product
                                              63.7%   36.3%

                                              64.3%   35.7%

The first reaction is 7.5 times faster than the second reaction.

(a)   What do these data suggest about the mechanism of the reaction?

(b)   Why are the rates of the two reactions different?

6.42   Write detailed mechanisms to account for the experimental facts outlined below.

$$(S)\text{-2-octyl tosylate} \xrightarrow[\text{acetic acid}]{\text{CH}_3\text{CO}^-\text{Na}^+} (R)\text{-2-octyl acetate} + (S)\text{-2-octyl acetate}$$
$$\qquad\qquad\qquad\qquad\qquad\qquad\qquad 65\% \qquad\qquad\qquad 35\%$$

$$(S)\text{-2-octyl tosylate} \xrightarrow[\substack{50\% \text{ acetone} \\ 50\% \text{ water}}]{\text{CH}_3\text{CO}^-\overset{+}{\text{N}}(\text{CH}_2\text{CH}_3)_4} (R)\text{-2-octyl acetate}$$
$$\qquad\qquad\qquad\qquad\qquad\qquad\qquad\qquad 100\%$$

6.43   When an alkyl halide reacts with a thiocyanate ion, SCN⁻, reaction takes place at the sulfur atom while with the cyanate ion, OCN⁻, reaction takes place at the nitrogen.

$$\text{RX} + \text{SCN}^- \longrightarrow \text{RSCN} + \text{X}^-$$

thiocyanate   alkyl
ion   thiocyanate

$$\text{RX} + \text{OCN}^- \longrightarrow \text{RNCO} + \text{X}^-$$

cyanate   alkyl
ion   isocyanate

Write structures for the thiocyanate and cyanate ions, and their alkyl derivatives, and explain the difference in the reactivity of the ions. (Hint: You might find it useful to review Section 6.2.)

6.44   Optically active 3,7-dimethyl-3-octanol was prepared from naturally occurring optically active (−)-linalool by catalytic hydrogenation.

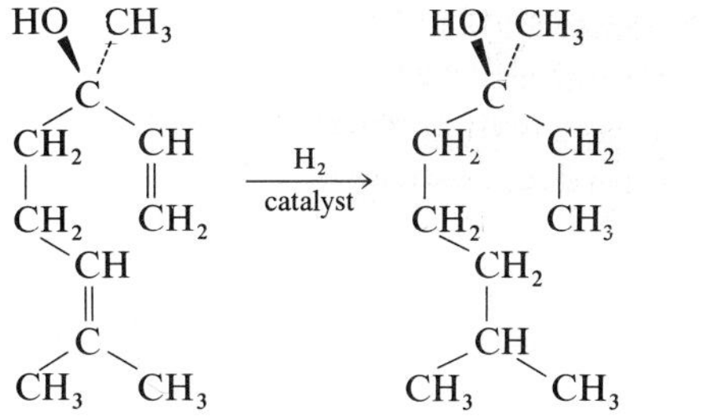

(−)-linalool   (−)-3,7-dimethyl-3-octanol

(a)   What is the configuration (*R* or *S*?) of (−)-linalool? What is the configuration of the octanol?

(b)   The octanol was converted into a tertiary alkyl halide so that the stereochemistry of S$_N$1 reactions at a tertiary center could be studied. The reaction that was used was reported to go mostly with inversion of configuration as shown below. Assign configuration to the chloro compound.

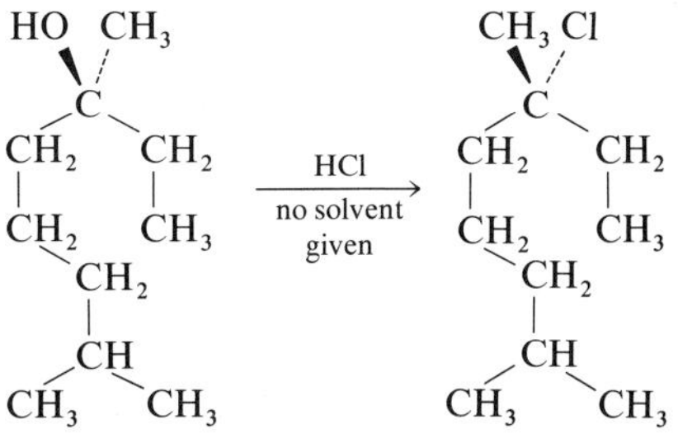

(c)   Discuss, with drawings, the mechanism for the conversion of the tertiary alcohol to the tertiary chloride, presumably in a nonpolar solvent. Would you expect to get chloro compound of the same, higher, or lower optical purity if the reaction were carried out in concentrated hydrochloric acid (37% hydrogen chloride by weight in water)?

6.45   When the following dibromo compound is dissolved in methanol, one of the bromine atoms is replaced by a methoxyl group. Predict what the structure of the product of this reaction is, and explain why this product is the one formed.

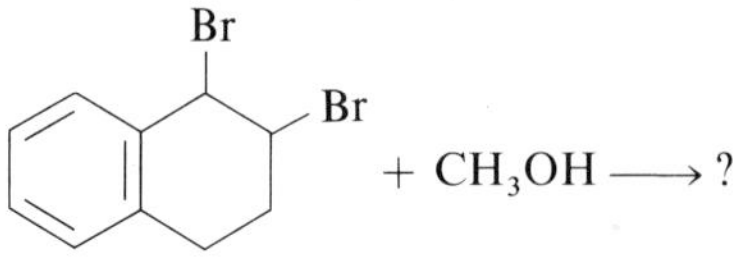

$+ \ CH_3OH \longrightarrow \ ?$

6.46   When 3,3-dimethyl-1-butyne is treated with 5 equivalents of hydrogen chloride at 25 °C, the following mixture of products is obtained. Write detailed mechanisms rationalizing the experimental facts. (Hint: A review of Sections 5.5A, 5.6B, and 6.8 will be helpful.)

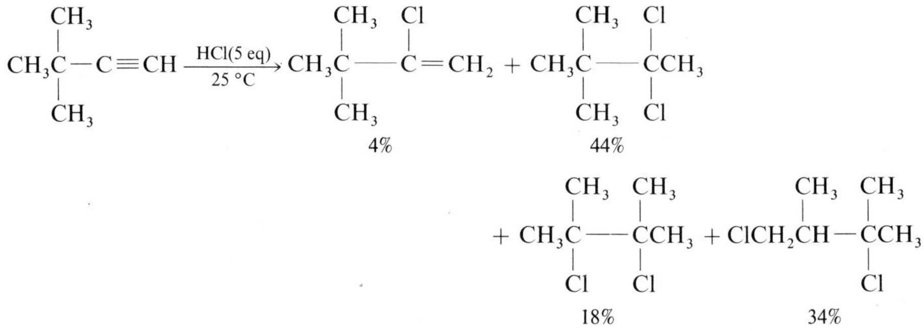

6.47   1-Chloro-4-*tert*-butylcyclohexene reacts with hydrogen bromide in pentane at -78 °C when exposed to light to give a single product. This product, when treated with sodium hydroxide in 80% ethanol at 25 °C gives the starting alkene back. What does this tell us about the regioselectivity and the stereoselectivity of the free radical addition of hydrogen bromide (Section 5.4A) to 1-chloro-4-*tert*-butylcyclohexene? Remember that the *tert*-butyl group prefers an equatorial orientation on the cyclohexane ring (Section 3.8C). Give a structural formula for the addition product showing stereochemistry and conformation.

Alcohols and Ethers

7

Review and Nomenclature

A. A Review of the Acid-Base Properties of Alcohols. Preparation of Ethers.

Alcohols *are compounds that contain a hydroxyl group attached to a tetrahedral carbon atom.* The presence of this hydroxyl group greatly influences the physical and chemical properties of alcohols. The polarity of the oxygen-hydrogen bond creates hydrogen bonding between alcohol molecules, which results in the high boiling points characteristic of alcohols. The water solubility of low molecular weight alcohols and of larger compounds with several hydroxyl groups can also be attributed to hydrogen bonding (Section 1.8B). The same properties make alcohols polar solvents for the reactions considered in Chapter 6.

Alcohols have basic properties because of the presence of nonbonding electrons on the oxygen atom; these can be donated to a proton or a Lewis acid.

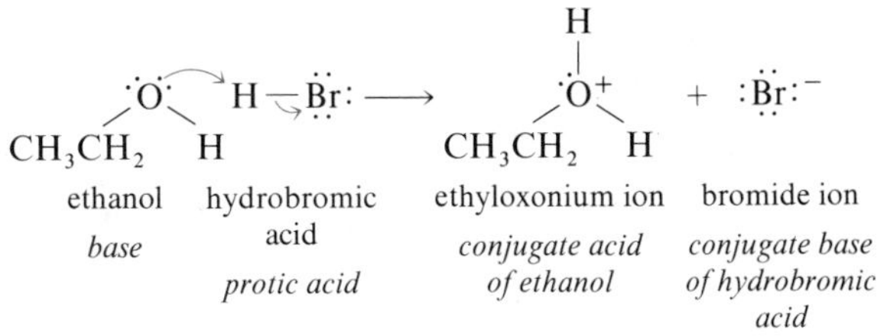

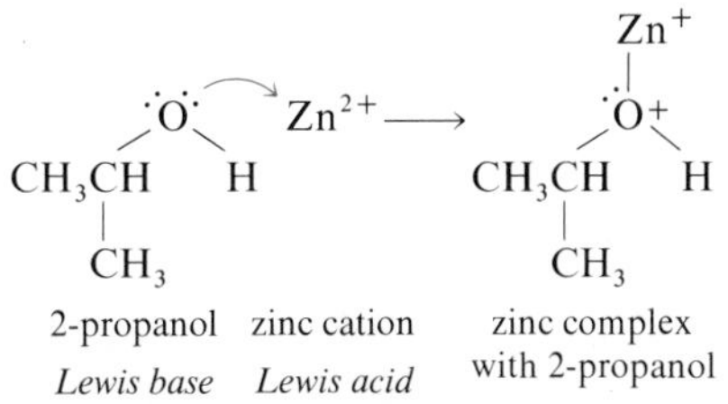

$$
\underset{\substack{\text{ethanol} \\ \textit{base}}}{\overset{\overset{\ddot{\text{O}}}{\underset{\text{CH}_3\text{CH}_2 \quad \text{H}}{}}}{}}
\quad
\underset{\substack{\text{hydrobromic} \\ \text{acid} \\ \textit{protic acid}}}{\text{H}-\ddot{\text{Br}}:}
\longrightarrow
\underset{\substack{\text{ethyloxonium ion} \\ \textit{conjugate acid} \\ \textit{of ethanol}}}{\overset{\text{H}}{\underset{\text{CH}_3\text{CH}_2 \quad \text{H}}{\overset{\ddot{\text{O}}^+}{}}}}
\quad + \quad
\underset{\substack{\text{bromide ion} \\ \textit{conjugate base} \\ \textit{of hydrobromic} \\ \textit{acid}}}{:\ddot{\text{Br}}:^-}
$$

$$
\underset{\substack{\text{2-propanol} \\ \textit{Lewis base}}}{\overset{\ddot{\text{O}}}{\underset{\underset{\text{CH}_3}{|}}{\text{CH}_3\text{CH} \quad \text{H}}}}
\quad
\underset{\substack{\text{zinc cation} \\ \textit{Lewis acid}}}{\text{Zn}^{2+}}
\longrightarrow
\underset{\substack{\text{zinc complex} \\ \text{with 2-propanol}}}{\overset{\text{Zn}^+}{\underset{\underset{\text{CH}_3}{|}}{\overset{\ddot{\text{O}}^+}{\text{CH}_3\text{CH} \quad \text{H}}}}}
$$

The nonbonding electrons on the oxygen atom also allow the alcohol to participate in nucleophilic substitution reactions. The solvolyses of tertiary alkyl halides that we studied in Chapter 6 are examples of this kind of reaction.

$$
\begin{array}{ccc}
\text{CH}_3 & \text{CH}_3 & \text{CH}_3 \\
| & | & | \\
\text{CH}_3\text{CBr} \underset{\text{ethanol}}{\rightleftarrows} & \overset{+}{\text{C}}\quad + \text{Br}^- \underset{\text{CH}_3\text{CH}_2\text{OH}}{\rightleftarrows} & \text{CH}_3\text{CCH}_3 \\
| & \diagup\quad\diagdown & | \\
\text{CH}_3 & \text{CH}_3\quad\text{CH}_3 & \text{OCH}_2\text{CH}_3 \\
\textit{tert-}\text{butyl} & \textit{tert-}\text{butyl} & \textit{tert-}\text{butyl ethyl} \\
\text{bromide} & \text{cation} & \text{ether}
\end{array}
$$

The alcohol, as a polar solvent, stabilizes the ions that are formed in the S_N1 reaction, and also serves as the nucleophile that reacts with the carbocation intermediate.

Alcohols also have acidic properties because of the presence of the hydrogen atom bonded to the oxygen atom of the hydroxyl group. This hydrogen atom is approximately as acidic as the hydrogen atom in water (Section 4.5). Thus, the concentrations of hydroxide ion and methoxide anion are roughly equal when sodium hydroxide is dissolved in methanol.

$$
\begin{array}{ccc}
\overset{\cdot\cdot}{\underset{\cdot\cdot}{\text{O}}} & \quad\rightleftharpoons\quad \text{CH}_3-\overset{\cdot\cdot}{\underset{\cdot\cdot}{\text{O}}}{:}^- \quad + & \overset{\cdot\cdot}{\underset{\cdot\cdot}{\text{O}}} \\
\diagup\quad\diagdown & & \diagup\quad\diagdown \\
\text{CH}_3\quad\text{H} & & \text{H}\quad\quad\text{H} \\
\quad\quad{}^-{:}\overset{\cdot\cdot}{\text{O}}-\text{H} & & \\
\text{methanol} & \text{methoxide} & \text{water} \\
& \text{anion} & \\
\textit{protic acid} & \textit{conjugate base} & \textit{conjugate acid} \\
p\text{K}_\text{a} \sim 15.5 & \textit{of methanol} & \textit{of hydroxide ion} \\
& & p\text{K}_\text{a}\ 15.7
\end{array}
$$

The comparable acidities of water and alcohols make it impossible to prepare high concentrations of alkoxide anions from alcohols by the use of hydroxide ions. When a reaction requires high concentrations of alkoxide anions (as reagents in nucleophilic substitution reactions, for example), they are prepared by dissolving alkali metals in an excess of dry alcohol. Potassium metal dissolved in *tert*-butyl alcohol gives potassium *tert*-butoxide in this way.

$$
\begin{array}{cccc}
\text{CH}_3 & & \text{CH}_3 & \\
| & & | & \\
2\ \text{CH}_3\text{COH} + 2\ \text{K} & \longrightarrow & 2\ \text{CH}_3\text{CO}^-\text{K}^+ + \text{H}_2\uparrow & \\
| & & | & \\
\text{CH}_3 & & \text{CH}_3 & \\
\textit{tert-}\text{butyl} & \text{potassium} & \text{potassium } \textit{tert-} & \text{hydrogen} \\
\text{alcohol} & \text{metal} & \text{butoxide} &
\end{array}
$$

While this reaction for the preparation of potassium *tert*-butoxide looks like an acid-base reaction, it is really an oxidation-reduction reaction, with potassium metal being oxidized and the hydrogen atom of the hydroxyl group being reduced to elemental hydrogen. The reaction is not reversible, so good yields of alkoxide anion can be generated. In a similar way, metallic sodium reacts with methanol to give sodium methoxide and hydrogen gas.

$$
\begin{array}{cccc}
2\ \text{CH}_3\text{OH} + 2\ \text{Na} \longrightarrow 2\ \text{CH}_3\text{O}^-\text{Na}^+ + \text{H}_2\uparrow \\
\text{methanol} \quad\quad \text{sodium} \quad\quad \text{sodium methoxide} \quad \text{hydrogen} \\
\text{metal}
\end{array}
$$

Potassium is more reactive than sodium in this reaction and primary alcohols are more reactive than secondary or tertiary alcohols. The reaction of potassium with methanol occurs so rapidly and is so exothermic that the hydrogen is ignited and an explosion occurs. Therefore, sodium is usually used with methanol. On the other hand, because sodium reacts only very slowly with *tert*-butyl alcohol, potassium metal is used instead.

Alkoxide ions, in turn, can be used to form ethers with primary alkyl halides in an S_N2 reaction, the Williamson synthesis (Problem 6.22 and Section 6.11). *n*-Butyl iodide in a substitution reaction with isopropoxide anion, gives an unsymmetrical ether, an ether with two different organic groups on the oxygen atom.

$$2\ \overset{\overset{\displaystyle CH_3}{|}}{CH_3CHOH} + 2\,Na \longrightarrow 2\ \overset{\overset{\displaystyle CH_3}{|}}{CH_3CHO^-Na^+} + H_2\uparrow$$

isopropyl   sodium   sodium isopropoxide
alcohol

$$CH_3CH_2CH_2CH_2 \overset{\curvearrowleft}{\ddot{I}}: \longrightarrow CH_3CH_2CH_2CH_2 - \overset{..}{\underset{..}{O}} - \overset{\overset{\displaystyle CH_3}{|}}{CHCH_3} + \ :\overset{..}{\underset{..}{I}}:^-$$

n-butyl iodide   *n*-butyl isopropyl   iodide
                          ether   ion
                          72%

$$:\overset{..}{\underset{..}{O}} - \overset{\overset{\displaystyle CH_3}{|}}{CHCH_3}$$

isopropoxide
anion

If 4-chloro-1-butanol is treated with sodium hydroxide, a cyclic ether, tetrahydrofuran, is formed in high yield.

$$ClCH_2CH_2CH_2CH_2OH \xrightarrow[\text{H}_2\text{O}]{\text{NaOH}}$$

4-chloro-1-butanol   tetrahydrofuran
                                             95%

The formation of the cyclic ether is *an* **intramolecular reaction,** *one in which the reaction occurs between two functional groups in the same molecule.* Note that the intramolecular reaction is favored over an **intermolecular reaction,** a reaction between two different molecules. For example, no 1,4-butanediol from the displacement of chloride ion by hydroxide ion is seen. The reacting centers in 4-chloro-1-butanol are held in close proximity by the chain of four carbon atoms so that collision and reaction between them is more probable than reaction between the same reactive groups on separate molecules.

It is particularly easy to form strainless rings of five and six carbon atoms (Sections 3.8B, 3.8C). We also observe enhanced reactivity when a three-membered ring is the product (Section 8.4E).

PROBLEM 7.1 Synthesize the following ethers by the Williamson method, choosing alkoxide anions and alkyl halides that will give the best yields. Show the preparation of the alkoxide anions you will need.

(a) $CH_3CHOCH_2CH_2CH_3$

(with CH_3 on the carbon)

(b) 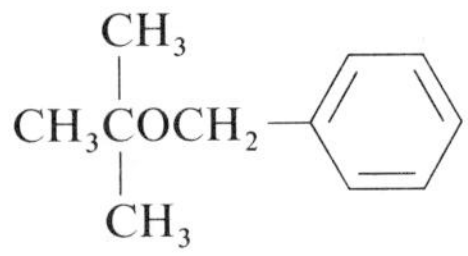 CH_3COCH_2— (benzene ring)

(with CH_3 above and CH_3 below the central carbon)

(c) $CH_3CH_2OCH_2CH=CH_2$

(d) $CH_3OCH_2CH_2CH_2CH_3$

(e) $CH_3CH_2OCH_2CH_2CH_2CH_2CH_2CH_3$

(f) 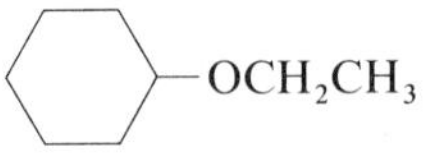 (cyclohexyl)—OCH_2CH_3

(g)  (six-membered ring with O)

PROBLEM 7.2 Tetrahydrofuran is a stable ether but the ring will open when the molecule is heated with acid,

$$\text{(tetrahydrofuran ring with O)} \xrightarrow[\Delta]{HCl} ClCH_2CH_2CH_2CH_2OH$$

Propose a mechanism for this reaction.

B. Some Typical Alcohols

The simplest alcohol is methanol, or methyl alcohol, a compound that is highly toxic, and when drunk, can cause blindness and death. Methanol can be generated by heating wood, and is sometimes called **wood alcohol.** Ethanol, or ethyl alcohol, is the familiar **grain alcohol.** Some alcohols have more than one hydroxyl group. Such is the case with 1,2-ethanediol, also known as ethylene glycol, which is used as antifreeze in cars. Other alcohols are large complicated molecules of which the hydroxyl group is a small but important part. Cholesterol, which has been implicated as a possible risk factor in arterial and heart disease, is such an alcohol.

CH_3OH	CH_3CH_2OH	$HOCH_2CH_2OH$
methanol	ethanol	1,2-ethanediol
methyl alcohol	ethyl alcohol	ethylene glycol
bp 64.7 °C	bp 78.5 °C	bp 197.2 °C

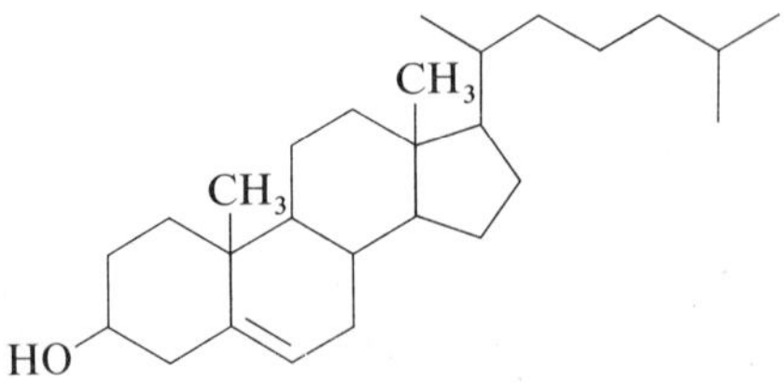

cholesterol
mp 148.5 °C

PROBLEM 7.3 The boiling point of 1,2-ethanediol is 197.2 °C. Why is it a good compound to use as an antifreeze?

PROBLEM 7.4 Would you expect cholesterol to be particularly soluble in water? Explain your answer.

PROBLEM 7.5 Cholesterol has more than one functional group. What other class of compounds does it belong to? Write equations using cholesterol to illustrate three reactions that are typical of this other functional group class.

C. Nomenclature of Alcohols

Simple alcohols can be named by using the name of the alkyl group they contain with the word **alcohol.** Methyl alcohol, ethyl alcohol, isopropyl alcohol, *tert*-butyl alcohol, allyl alcohol are compounds that are often named in this way. But the system becomes cumbersome when the alkyl groups are large and complicated, or when other substituents are also present in the molecule.

Alcohols are named systematically according to the IUPAC rules by substituting the ending **-ol** *for the final* **-e** *of the systematic name of the hydrocarbon portion of the molecule.* The location of the hydroxyl group on the hydrocarbon chain is indicated by a number, usually placed at the beginning of the name. Below are some examples of the IUPAC nomenclature of simple alcohols. Common names are also shown for some compounds.

CH_3OH	CH_3CH_2OH	$CH_3CH_2CH_2OH$
methanol	ethanol	1-propanol
methyl alcohol	ethyl alcohol	*n*-propyl alcohol
	a primary alcohol	*a primary alcohol*

2-propanol	2-butanol	2-methyl-1-propanol
isopropyl alcohol	*sec*-butyl alcohol	isobutyl alcohol
a secondary alcohol	*a secondary alcohol*	*a primary alcohol*

2-methyl-2-propanol	3-pentanol	3-methyl-3-hexanol
tert-butyl alcohol		
a tertiary alcohol	*a secondary alcohol*	*a tertiary alcohol*

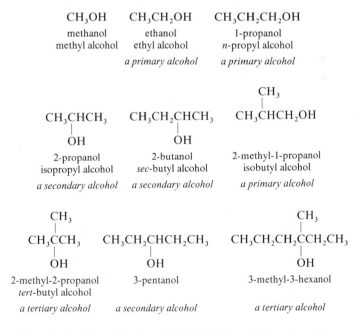

The reactivity of alcohols, like that of alkyl halides, differs depending on whether the functional group is on a primary, secondary, or tertiary carbon atom (Sections 3.4, 3.5). Therefore, alcohols are designated as primary, secondary, or tertiary alcohols, as shown above.

The alcohol function is considered to be more important than alkene, alkyne, or halo groups and is consequently given the lower number in naming compounds with several functional groups.

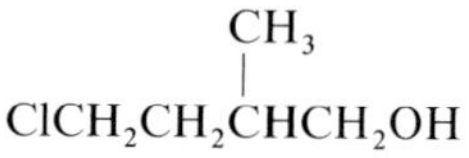

$$\underset{\text{CH}_3}{|}$$
$$\text{ClCH}_2\text{CH}_2\text{CHCH}_2\text{OH}$$

$$\text{CH}_3\text{CHCHCH}_2\text{CH}_2\text{CH}_2\text{OH}$$
$$\underset{\text{OH}}{|}\ \ \ \underset{\text{CH}_3}{|}$$

4-chloro-2-methyl-1-butanol

a primary alcohol

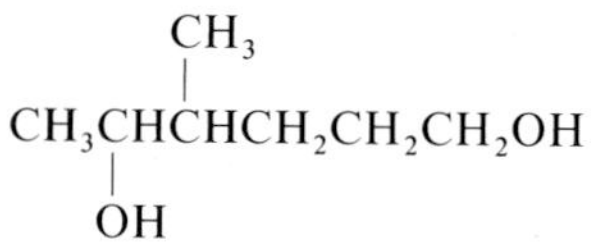

4-methyl-1,5-hexanediol

a primary and secondary alcohol

$$\text{HOCH}_2\text{C}\equiv\text{CCH}_2\text{OH}$$
2-butyne-1,4-diol

a primary alcohol

$$\text{CH}_2=\text{CHCH}_2\text{OH}$$
2-propen-1-ol
allyl alcohol

*a primary alcohol,
also an allylic alcohol*

*note location of numbers for the hydroxyl groups
in these two names; necessary for clarity*

D. Nomenclature of Ethers

Ethers are closely related to alcohols, and *have two organic groups substituted on an oxygen atom.* There are two ways of naming an ether. You can name simple ethers by naming each of the organic groups on the oxygen atom and adding the word *ether*. For example,

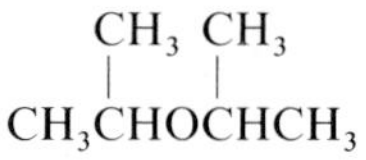

$$\underset{\text{CH}_3}{|}\ \ \underset{\text{CH}_3}{|}$$
$$\text{CH}_3\text{CHOCHCH}_3$$
diisopropyl ether

$$\text{CH}_3\text{OCH}_2\text{CH}_2\text{CH}_2\text{CH}_3$$
methyl *n*-butyl ether

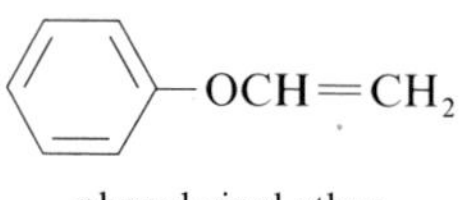

⬡—OCH=CH$_2$

phenyl vinyl ether

⬡—OCH$_2$CH$_2$CH$_2$CH$_2$CH$_3$

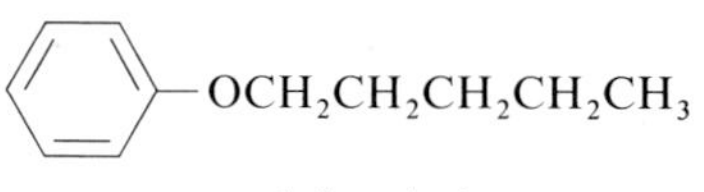

pentyl phenyl ether

For more complex molecules, the simplest organic group along with the oxygen atom is named as an **alkoxy group** and used as a substituent on a more complex chain. Important alkoxy or aryloxy groups are the following:

$$\text{CH}_3\text{O}—$$
methoxy

$$\text{CH}_3\text{CH}_2\text{O}—$$
ethoxy

$$\text{CH}_3\text{CH}_2\text{CH}_2\text{O}—$$
propoxy

⬡—O—

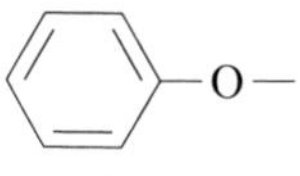

phenoxy

⬡—CH$_2$O—

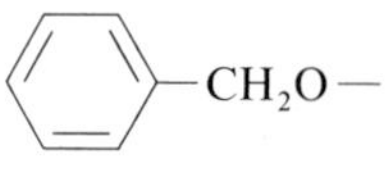

benzyloxy

Examples of their use in nomenclature are shown below.

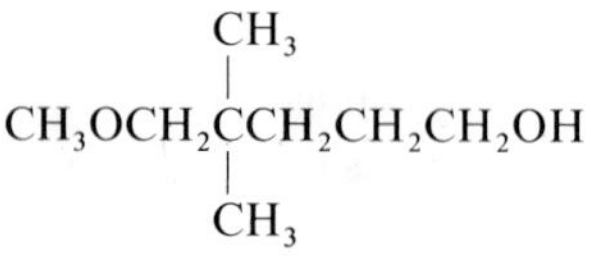

$$\underset{\text{CH}_3}{|}$$
$$\text{CH}_3\text{OCH}_2\text{CCH}_2\text{CH}_2\text{CH}_2\text{OH}$$
$$\underset{\text{CH}_3}{|}$$

5-methoxy-4,4-dimethyl-1-pentanol

$$\text{CH}_3\text{OCH}_2\text{CH}_2\text{OCH}_2\text{CH}_2\text{CH}_3$$

1-methoxy-2-propoxyethane

$$CH_3CH_2O-\!\!\left\langle\rule{0pt}{10pt}\right\rangle\!\!-OCH_2CH_3$$

1,4-diethoxybenzene

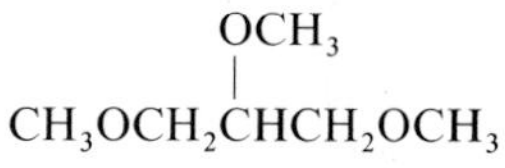

$$\overset{\displaystyle OCH_3}{\underset{\displaystyle |}{CH_3OCH_2CHCH_2OCH_3}}$$

1,2,3-trimethoxypropane

$$\left\langle\rule{0pt}{10pt}\right\rangle\!\!-CH_2OCH_2CH_2CH\!=\!CH_2$$

4-benzyloxy-1-butene

$$CH_3CH_2CH_2\overset{\displaystyle O}{\underset{\displaystyle |}{CHCH_2CH_2CH_3}}$$

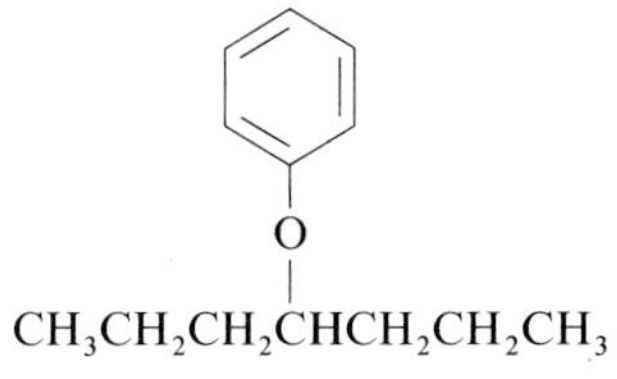

4-phenoxyheptane

PROBLEM 7.6 Name the following compounds, including stereochemistry if shown, according to IUPAC rules.

(a) 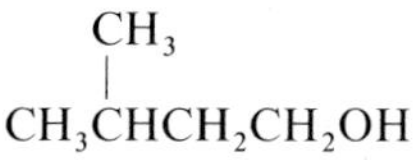 $\overset{\displaystyle CH_3}{\underset{\displaystyle |}{CH_3CHCH_2CH_2OH}}$

(b) 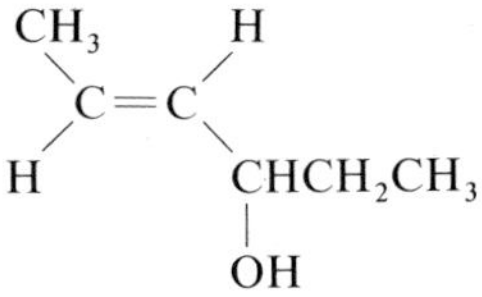 $\underset{H}{\overset{CH_3}{C}}\!\!=\!\!\underset{\underset{\displaystyle |}{\underset{\displaystyle OH}{CHCH_2CH_3}}}{\overset{H}{C}}$

(c) $CH_3CH_2C\!\equiv\!CCH_2OH$

(d)  $\underset{\quad Cl\ \ Cl}{CH_3CHCHCH_2CH_2CH_2OH}$

(e) $CH_3CH_2OCH_2CH_2CH_2Cl$

(f) 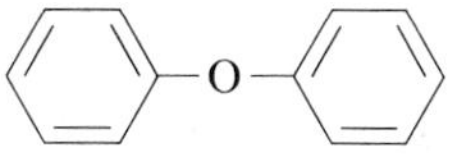 $\left\langle\rule{0pt}{8pt}\right\rangle\!\!-O\!\!-\!\!\left\langle\rule{0pt}{8pt}\right\rangle$

(g) 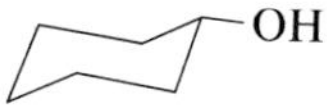 (cyclohexane with —OH)

(h) 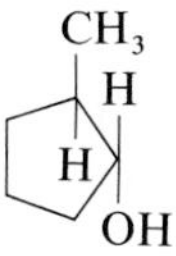 (cyclopentane structure with CH₃, H, H, OH)

(i) 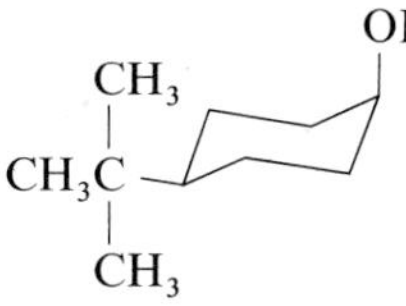 $CH_3\overset{CH_3}{\underset{CH_3}{C}}$ (cyclohexane with OH)

(j) 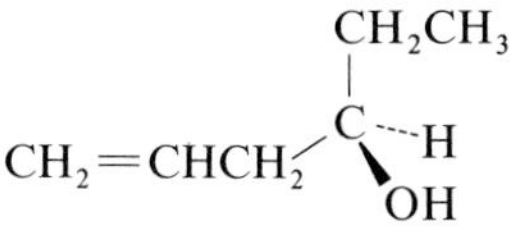 $CH_2\!=\!CHCH_2\overset{CH_2CH_3}{\underset{OH}{C}}\!\cdots\!H$

(k) $CH_3OCH_2CH_2CH_2OCH_3$

(l) 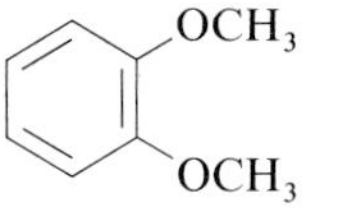 (benzene with two OCH₃ groups, ortho)

7.2

A Review of Some Reactions That Form Alcohols

Alcohols are formed when water adds to the double bond in alkenes. For example, propene is converted to isopropyl alcohol in dilute acid (Section 5.5B).

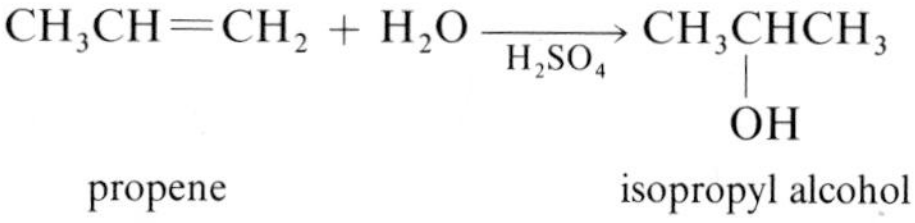

$$CH_3CH\!=\!CH_2 + H_2O \xrightarrow[H_2SO_4]{} \overset{\displaystyle CH_3CHCH_3}{\underset{\displaystyle OH}{}}$$

propene   isopropyl alcohol

The alcohol results from Markovnikov addition of water to the double bond in a reaction initiated by the reaction of the electrophile, hydronium ion, with the double bond. The reaction is the basis of an important industrial process for the production of simple alcohols, but is not useful for the preparation of more complex alcohols. The acidic reagent can attack other functional groups in the molecule and cause unwanted side reactions. Alkenes also isomerize in acids through carbocation intermediates (Section 5.6B). Attempts to add water to such alkenes with aqueous acid give rise to alcohols with rearranged carbon skeletons.

For example, the Markovnikov addition product is not obtained from the reaction of 3-methyl-1-butene with 60% aqueous sulfuric acid. The only alcohol recovered is 2-methyl-2-butanol.

$$
\underset{\text{3-methyl-1-butene}}{CH_3CHCH{=}CH_2} \xrightarrow[\text{60\% H}_2\text{SO}_4]{\text{H}_2\text{O}} \underset{\text{2-methyl-2-butanol}}{CH_3CCH_2CH_3}
$$

with CH_3 substituent above the first carbon and CH_3 / OH substituents on the product carbon.

PROBLEM 7.7   (a) What would be the structure of the Markovnikov addition product for the reaction of 3-methyl-1-butene with dilute acid? (b) Write a detailed mechanism that accounts for the observed product.

PROBLEM 7.8   3,3-Dimethyl-1-butene is isomerized by acid (Section 5.6B). Predict, by showing a mechanism for the reaction, what the structures of hydration products would be if a reaction with aqueous acid were attempted.

Alcohols are also formed when a leaving group is displaced from an alkyl halide or alkyl tosylate by hydroxide ion or by water in a nucleophilic substitution reaction. Elimination reactions often accompany these reactions, so they are not useful for the preparation of high-purity alcohols in good yield (Section 6.9). Besides, alkyl halides or tosylates are often prepared from the corresponding alcohol in the first place, and so cannot serve as an independent synthesis of an alcohol.

7.3

Hydroboration

A. The Addition of Diborane to Alkenes

The **hydroboration reaction,** discovered in 1956 by Herbert C. Brown at Purdue University, has made it possible to synthesize alcohols in high yields under mild conditions with high regioselectivity (Section 5.3A) and stereoselectivity (Section 5.8B). In this reaction, diborane, B_2H_6, adds to an alkene to give an organoborane. Oxidation of the carbon-boron bond in the organoborane with basic hydrogen peroxide gives an alcohol, as illustrated by the overall reaction for the hydroboration of 1-butene.

$$CH_3CH_2CH{=}CH_2 \xrightarrow{B_2H_6} (CH_3CH_2CH_2CH_2)_3B + (CH_3CH_2\overset{\overset{\displaystyle CH_3}{|}}{CH}{-})_3B$$

1-butene   tri(*n*-butyl)borane   tri(*sec*-butyl)borane

$$\Big\downarrow {\substack{H_2O_2 \\ NaOH \\ H_2O}}$$

$$CH_3CH_2CH_2CH_2OH + CH_3CH_2\overset{\overset{\displaystyle \,}{|}}{\underset{\underset{\displaystyle OH}{|}}{CH}}CH_3$$

n-butyl alcohol   *sec*-butyl alcohol
93%   7%

Notice that the chief product of the reaction sequence, n-butyl alcohol, arises from *anti-Markovnikov addition of the elements of water to the double bond.*

Diborane is an interesting compound that has only six pairs of bonding electrons. The question of what holds the two boron atoms together has been a controversial one that has fostered much lively discussion and research.

Diborane exists as B_2H_6, which may be regarded as a dimer of BH_3, in the absence of Lewis bases. It dissociates easily in the presence of ethers, which have nonbonding electrons, to give ether-borane complexes.

$$B_2H_6 \; + \; 2 \;\text{THF} \;\longrightarrow\; 2\; H{-}\overset{\overset{\displaystyle \overset{+}{\cdot \cdot}O}{|}}{\underset{\underset{\displaystyle H}{|}}{\overset{-}{B}}}{-}H$$

diborane   tetrahydrofuran   borane-tetrahydrofuran
a cyclic ether   complex

These complexes in the ethers used as solvents for the reaction are the reagents in hydroboration reactions, so the formula BH_3 is used to represent the reactive species.

PROBLEM 7.9   Write the Lewis structure for BH_3. What shape do you expect the molecule to have? What kind of hybridization does the boron atom have in BH_3? Draw an orbital picture of the bonding in the molecule.

BH_3 is a Lewis acid and, as such, reacts with the π electrons of an alkene with an orientation that gives rise to partial positive charge at the carbon atom that would develop into the more stable carbocation.

$$CH_3CH_2CH{=}CH_2 \longrightarrow CH_3CH_2\overset{\delta+}{CH}{=}CH_2$$

H—B—H $\overset{\delta-}{H}{-}\overset{\delta-}{B}{-}H$
| |
H   H

1-butene and borane   partial bonding between
1-butene and borane

At the same time, negative charge develops at the boron atom, and the hydrogen atoms in the boron hydride acquire partial negative character and can be transferred as hydride ion to the carbon atom with a partial positive charge. The reaction may be represented as a single step.

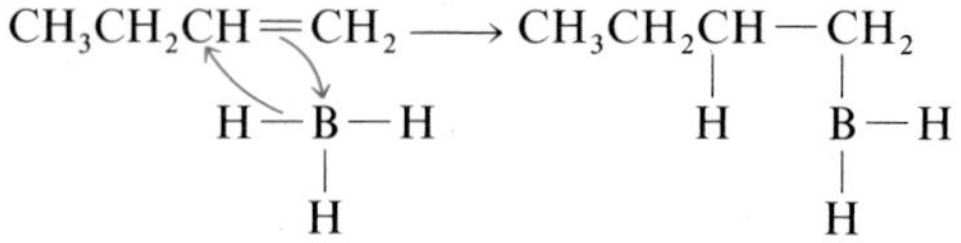

$$CH_3CH_2CH{=}CH_2 \longrightarrow CH_3CH_2CH{-}CH_2$$

$$H{-}B{-}H \qquad\qquad H \quad B{-}H$$

$$H \qquad\qquad\qquad H$$

1-butene and borane   n-butylborane

an organoborane

The reaction is said to have a **four-center transition state,** with four atoms (two carbon, one boron, and one hydrogen) undergoing changes in bonding at the same time.

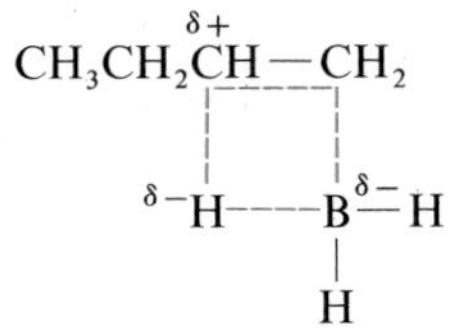

$$CH_3CH_2\overset{\delta+}{C}H{-}CH_2$$

$$\overset{\delta-}{H}{-}{-}{-}B\overset{\delta-}{=}H$$

$$H$$

four-center transition state for the
reaction of 1-butene with borane

The product from the addition of borane to an alkene is known as an organoborane, *n*-butylborane, for example. This compound is still a Lewis acid in which the boron atom can accept another pair of electrons from an alkene and can transfer a hydride ion to it. The reaction proceeds until all three hydrogen atoms on the boron hydride have been substituted by alkyl groups.

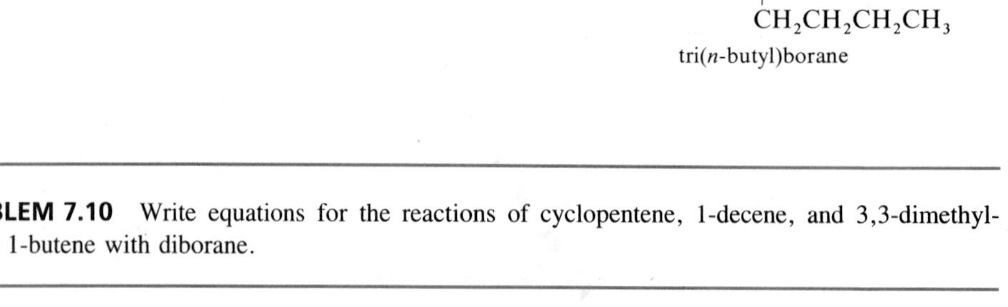

$$CH_3CH_2CH{=}CH_2 \longrightarrow CH_3CH_2CH_2CH_2{-}B{-}CH_2CH_2CH_2CH_3$$

$$H{-}B{-}CH_2CH_2CH_2CH_3 \qquad\qquad H$$

$$H$$

1-butene   n-butylborane   di(n-butyl)borane

$$\Big\downarrow CH_3CH_2CH{=}CH_2$$

$$CH_3CH_2CH_2CH_2{-}B{-}CH_2CH_2CH_2CH_3$$

$$CH_2CH_2CH_2CH_3$$

tri(n-butyl)borane

PROBLEM 7.10   Write equations for the reactions of cyclopentene, 1-decene, and 3,3-dimethyl-1-butene with diborane.

B. Regioselectivity in Hydroboration Reactions

In the unsymmetrical alkene, 1-butene, the boron atom adds chiefly to the less highly substituted carbon atom of the terminal alkene. In the case of internal alkenes with different degrees of substitution, the boron atom also attaches itself to the less substituted carbon atom. For example, 2-methyl-2-butene reacts with diborane to give chiefly the organoborane in which the boron atom is attached to the less highly substituted carbon atom.

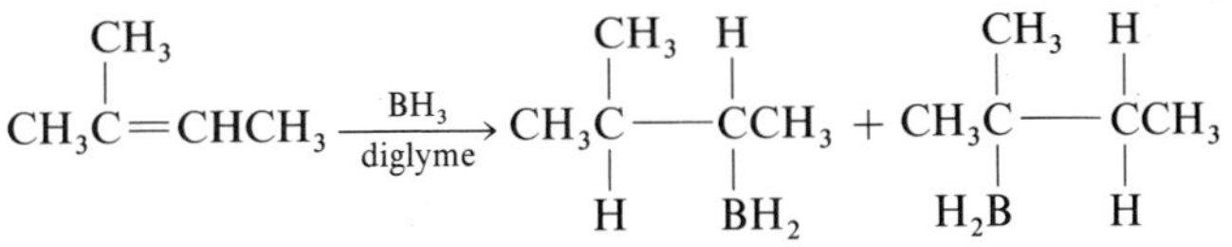

CH_3
$CH_3C{=}CHCH_3$
BH_3
diglyme
CH_3 H
$CH_3C{-}CCH_3$
H BH_2
+
CH_3 H
$CH_3C{-}CCH_3$
H_2B H

2-methyl-2-butene 98% 2%

diglyme = $CH_3OCH_2CH_2OCH_2CH_2OCH_3$

an ether

For more or less symmetrically disubstituted alkenes, however, two products are possible. They are formed in roughly equal amounts even when the bulkiness of one of the substituents would seem to favor addition of the boron atom to the other carbon atom, as illustrated by the reaction of (*E*)-4-methyl-2-pentene.

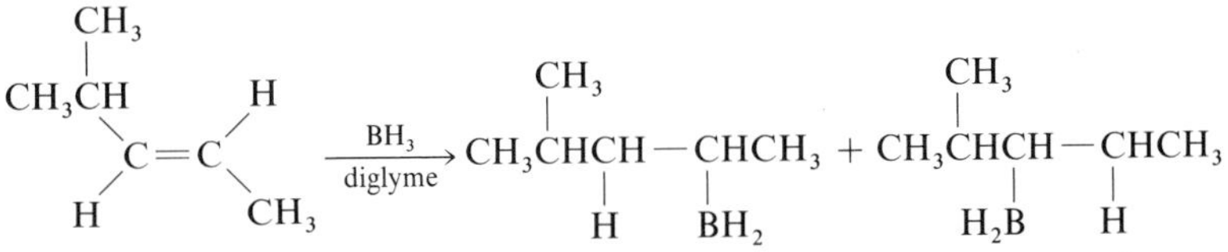

CH_3
CH_3CH
H
C=C
H
CH_3
BH_3
diglyme
CH_3
$CH_3CHCH{-}CHCH_3$
H BH_2
+
CH_3
$CH_3CHCH{-}CHCH_3$
H_2B H

(*E*)-4-methyl-2-pentene 1,3-dimethylbutylborane 57% 1-isopropylpropylborane 43%

Herbert C. Brown reasoned that a bulkier reagent than borane might show greater selectivity in the way that it added to a double bond, since steric interactions between substituents on the double bond and on the reagent would increase and become more important. One of the reagents that he developed for greater selectivity is the organoborane 9-borabicyclo[3.3.1]nonane (9-BBN), which is formed when borane adds twice to the cyclic diene, 1,5-cyclooctadiene, to give a bicyclic compound. In a bicyclic compound, two or more atoms are shared by two different rings (Section 11.5).

BH_3
tetrahydro-
furan
H
B
H
H
H
B
9
B
H
5
4
6
1
2
8
3
7

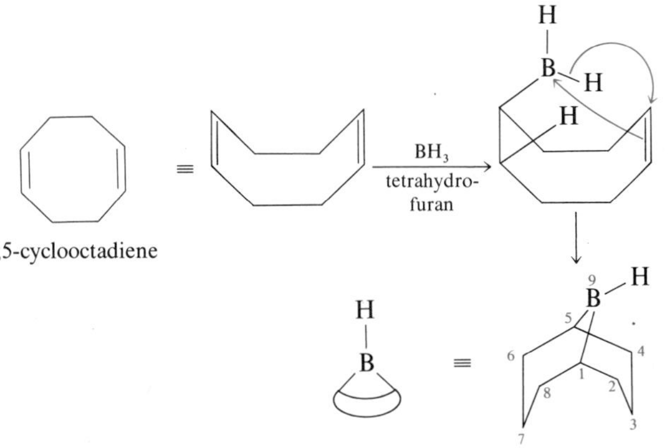

1,5-cyclooctadiene

a symbol frequently used for 9-BBN

9-borabicyclo[3.3.1]nonane or 9-BBN

9-Borabicyclo[3.3.1]nonane (9-BBN) is a reasonably stable compound that can be purchased in a bottle and handled with minor precautions in the air, in contrast to diborane, which is a gas and reacts violently with air. This makes the organoborane a convenient reagent for hydroboration reactions. The bulkiness of substituents on the boron atom gives the reagent high selectivity in its reaction with substituted alkenes.

With (*Z*)-4-methyl-2-pentene, for example, only the product in which the boron atom is bonded to the carbon atom with the less bulky of the two substituents is formed.

(*Z*)-4-methyl-2-pentene   →(9-BBN, tetrahydrofuran)→   99.8%

This contrasts with the equal distribution of products seen for the reaction of diborane with the (*E*)-isomer (p. 279).

PROBLEM 7.11 Write equations showing the chief product from the reaction of 9-borabicyclo-[3.3.1]nonane with: (a) 2-methyl-1-pentene; (b) styrene (phenylethene); (c) (*Z*)-4,4-dimethyl-2-pentene; (d) 2,3-dimethyl-2-butene; and (e) (*Z*)-3-hexene.

C. Stereochemistry of Hydroboration Reactions

The transition state postulated in Section 7.3A for the addition of borane to an alkene predicts that *the boron atom and the hydrogen atom that is transferred from it to the adjacent carbon atom end up bonded to the same side of the molecule. An addition reaction in which the incoming groups are added to the same side of a molecule is called a* **syn addition.** *A syn addition is in contrast to an anti addition, an example of which is the addition of bromine to a π bond (Section 5.8B). The words syn and anti refer to the mechanism of a reaction, while the words cis and trans are used to describe the stereochemistry of the products of the reaction.* Depending upon the structure of the starting reagent, syn addition may give products with cis (see Section 8.4A) or trans (see below) stereochemistry. In the hydroboration reaction, syn addition of boron and hydrogen to the double bond is expected.

There is no easy way to determine the stereochemistry of organoboranes. The alcohols formed from the oxidation of the organoborane intermediates show overall syn addition of the elements of water to the double bond in the cases where stereochemistry can be determined. For example 1-methylcyclohexene gives *trans*-2-methylcyclohexanol in the hydroboration reaction.

1-methylcyclohexene and 9-borabicyclo[3.3.1]nonane   organoborane from syn addition   *trans*-2-methylcyclohexanol 99.8%

Careful experiments with several different systems have led chemists to conclude that the overall syn addition of water is the result of syn addition of the boron hydride to the double bond followed by oxidation of the carbon-boron bond with retention of configuration. We will examine the details of the oxidation process in the next section.

D. Oxidation of Organoboranes

The oxidation of alkylboranes with alkaline hydrogen peroxide gives alcohols. Both the carbon atom and the boron atom in the organoborane are oxidized, while hydrogen peroxide is reduced to water. The reaction requires the addition of 6 M sodium hydroxide and 30% hydrogen peroxide directly to the hydroboration reaction mixture, so that the conversion of an alkene to an alcohol takes place in one reaction vessel without the isolation of any of the intermediates.

The reaction is believed to proceed by the formation of the hydroperoxide anion in base, and nucleophilic attack of the anion on the boron atom. The important step is the migration of an alkyl group to oxygen, a rearrangement that takes place with retention of configuration. The details of the mechanism of the oxidation reaction are shown for the organoborane from 1-methylcyclopentene.

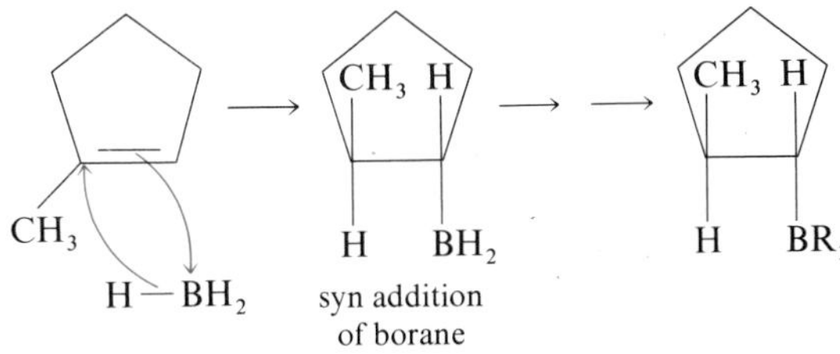

syn addition of borane

$$HOOH + Na^+OH^- \longrightarrow Na^+OOH^- + H_2O$$

hydrogen peroxide

more acidic than H_2O because of the inductive effect of the second oxygen atom

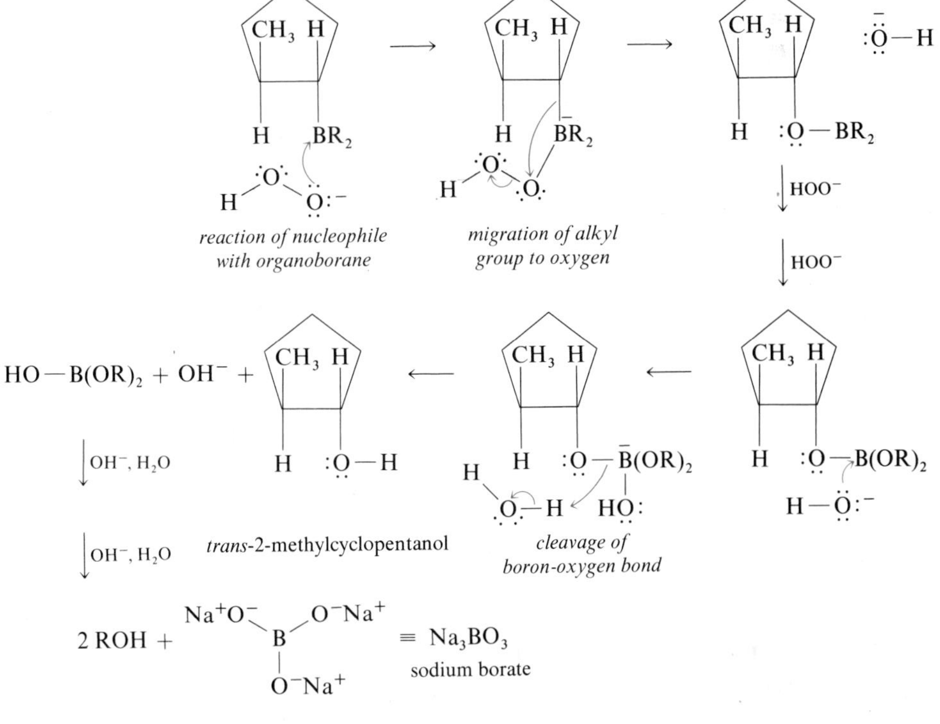

reaction of nucleophile with organoborane

migration of alkyl group to oxygen

$$HO-B(OR)_2 + OH^- +$$

trans-2-methylcyclopentanol

cleavage of boron-oxygen bond

$$2\,ROH + \quad \underset{\overset{|}{O^-Na^+}}{Na^+O^- \diagdown \underset{B}{} \diagup O^-Na^+} \quad \equiv Na_3BO_3$$

sodium borate

The hydroxyl group in *trans*-2-methylcyclopentanol ends up on the same side of the molecule where the boron atom was originally attached in the organoborane. Boron is oxidized to boric acid, present as the sodium salt, sodium borate.

The hydroboration-oxidation sequence thus provides a very mild and convenient way to prepare alcohols from alkenes with the elements of water adding to the double bond with high regioselectivity and stereoselectivity. The rearrangements and polymerization reactions typical of acid-catalyzed hydrations of alkenes do not take place under these conditions. The importance of this reaction was recognized in 1979 when Herbert C. Brown was honored with the Nobel Prize for his discovery of this and related transformations of organoboranes.

PROBLEM 7.12   Go back to Problems 7.10 and 7.11 and show the products you would expect to get from alkaline peroxide oxidation of the organoboranes you obtained there.

7.4

The Oxymercuration Reaction

Markovnikov addition of water to double bonds is achieved by the oxymercuration reaction under mild conditions that do not give rise to carbocation rearrangements. In this reaction, mercury(II) acetate

$$\underset{\displaystyle Hg(O\overset{\textstyle O}{\overset{\|}{C}}CH_3)_2}{}$$

adds to an alkene in aqueous tetrahydrofuran to give an intermediate organomercury compound. The carbon-mercury bond is converted with sodium borohydride, $NaBH_4$, in base to an alcohol and metallic mercury. The reaction is illustrated below with 1-hexene.

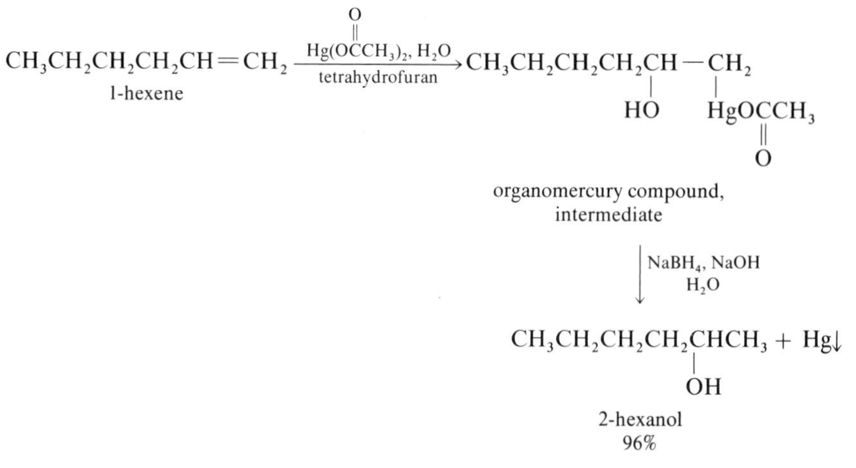

Mercury(II), a Lewis acid and electrophile, reacts with the double bond. A nucleophile from the reaction mixture, in this case the solvent, water, reacts with the developing cationic center.

$$CH_3CH_2CH_2CH_2CH = CH_2 \longrightarrow$$

$$\underset{\substack{\delta- \\ \|\\ :O:}}{CH_3CO} - \underset{\delta+}{Hg} - \underset{\substack{\delta- \\ \|\\ :O:}}{\overset{..}{O}CCH_3}$$

reaction of double bond with Lewis acid

$$CH_3CH_2CH_2CH_2\overset{\delta+}{CH} - CH_2$$

(with $\overset{H \quad H}{\underset{:O:}{}}$ above CH and $Hg^{\delta+}$ below CH_2)

$$\underset{\substack{\|\\ :O:}}{CH_3C} - \overset{..}{\underset{..}{O}}: \qquad :\overset{..}{\underset{..}{O}}\underset{\substack{\|\\ :O:}}{CCH_3}$$

reaction of cationic center with nucleophilic solvent

$$CH_3CH_2CH_2CH_2\overset{\overset{\displaystyle H}{\underset{\displaystyle |}{\overset{..}{O}:}}}{CHCH_2}$$

$$\underset{\substack{\|\\ :O:}}{Hg\overset{..}{O}CCH_3} \longleftarrow CH_3CH_2CH_2CH_2CH - CH_2$$

(with $\overset{H \quad H \frown :B}{\underset{:\overset{..}{O}^+}{}}$ above CH and $Hg - \overset{..}{\underset{..}{O}}CCH_3$ below, with $\overset{\|}{:O:}$)

deprotonation

organomercury compound

In most cases the stereochemistry of the addition reaction is <u>anti</u>. The complex between mercury(II) and the alkene is much like the bromonium ion (Section 5.8B) and directs the incoming nucleophile to the other side of the double bond. The carbon atom of the double bond that gives rise to the more stable carbocation on reaction with an electrophile is the one that forms a bond to the nucleophile.

The details of the replacement of mercury by hydrogen are not as clear. The hydrogen atom that replaces the mercury atom comes from sodium borohydride and not from the solvent. This step is not stereoselective, and is believed to go by way of a radical intermediate. Borohydride anion is a donor of hydride ions (Section 9.3). We picture the reaction as taking place by way of a four-center transition state in which borohydride ion donates a hydride ion to mercury in the organomercury compound. In the reaction, mercury(II) is ultimately reduced to mercury(0).

$$CH_3CH_2CH_2CH_2\overset{\overset{\displaystyle OH}{|}}{C}HCH_2 - \underset{\substack{H \\ \overset{\delta-}{|} \\ BH_3^- \\ \delta-}}{\overset{\delta+}{Hg}} \underset{\delta-}{\overset{:O:}{\underset{\|}{\overset{..}{O}CCH_3}}} \longrightarrow [CH_3CH_2CH_2CH_2\overset{\overset{\displaystyle OH}{|}}{C}HCH_2HgH]$$

unstable

$$H_3\bar{B} - \underset{\substack{\|\\ O}}{OCCH_3}$$

$$[CH_3CH_2CH_2CH_2\overset{\overset{\displaystyle OH}{|}}{C}HCH_2HgH] \longrightarrow CH_3CH_2CH_2CH_2\overset{\overset{\displaystyle OH}{|}}{C}HCH_2$$

(with $H \overset{\frown}{-} Hg\cdot$ below)

$$\downarrow$$

$$CH_3CH_2CH_2CH_2\overset{\overset{\displaystyle OH}{|}}{C}H\underset{\substack{|\\ H}}{CH_2} + Hg:\downarrow$$

The overall reaction is Markovnikov addition of water to the double bond under mild, basic conditions. Thus, it is possible to prepare alcohols from alkenes such as 3,3-dimethyl-1-butene that rearrange in acid (Problem 7.8).

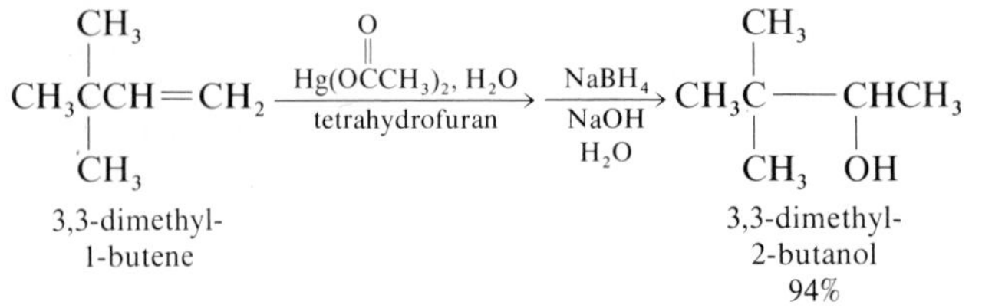

3,3-dimethyl-
1-butene

3,3-dimethyl-
2-butanol
94%

If the oxymercuration reaction is carried out in the presence of alcohols, alkoxy mercury intermediates are formed. The product of the reaction after reduction is an ether. The other group attached to the mercury atom may also vary depending upon the other ions present in the reaction mixture. For example, cyclohexene reacts with mercury(II) acetate in methanol to give an organomercury ether. The organomercury compound is isolated as the chloride if sodium chloride is added to the reaction mixture.

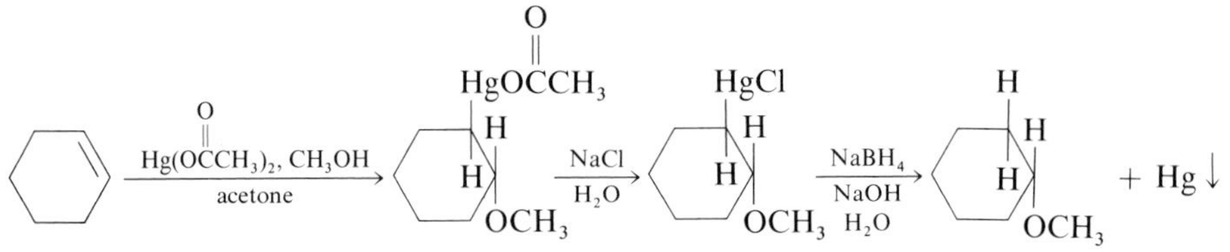

cyclohexene

Note the trans relationship between the methoxyl group and the mercury atom resulting from anti addition to the double bond.

The hydroboration and oxymercuration reactions complement each other. At this point, you will find it helpful to review both reactions to make sure that you understand why the first reaction gives anti-Markovnikov and the second one, Markovnikov addition of water to the double bond. Asking questions such as "What is the electrophile?, What is the nucleophile?, What is the most stable intermediate for each reaction?" will lead you to a better understanding of the reactions. Make sure you detect the pattern that lies beneath what would otherwise be a large number of unrelated facts to be memorized.

PROBLEM 7.13 Predict the products of the following reactions.

(a) 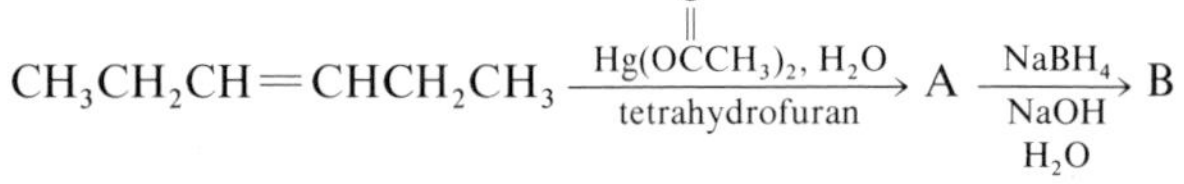 $CH_3CH_2CH=CHCH_2CH_3$ $\xrightarrow[\text{tetrahydrofuran}]{Hg(O\overset{O}{\overset{\|}{C}}CH_3)_2,\ H_2O}$ A $\xrightarrow[\substack{\text{NaOH} \\ H_2O}]{NaBH_4}$ B

(b) 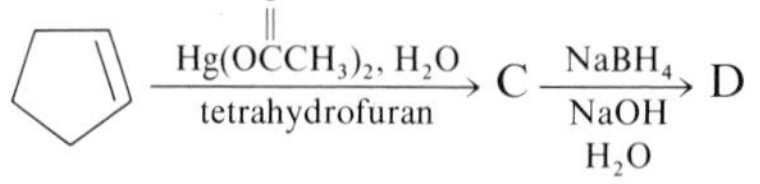 $\xrightarrow[\text{tetrahydrofuran}]{Hg(O\overset{O}{\overset{\|}{C}}CH_3)_2,\ H_2O}$ C $\xrightarrow[\substack{\text{NaOH} \\ H_2O}]{NaBH_4}$ D

(c) $CH_3\overset{\overset{\displaystyle CH_3}{|}}{C}=\overset{\overset{\displaystyle CH_3}{|}}{C}CH_3$ $\xrightarrow[\text{tetrahydrofuran}]{Hg(O\overset{O}{\overset{\|}{C}}CH_3)_2,\ H_2O}$ E $\xrightarrow[\substack{\text{NaOH} \\ H_2O}]{NaBH_4}$ F

(d) $=CH_2$ $\xrightarrow[\text{tetrahydrofuran}]{Hg(O\overset{O}{\overset{\|}{C}}CH_3)_2,\ H_2O}$ G $\xrightarrow[\substack{\text{NaOH} \\ H_2O}]{NaBH_4}$ H

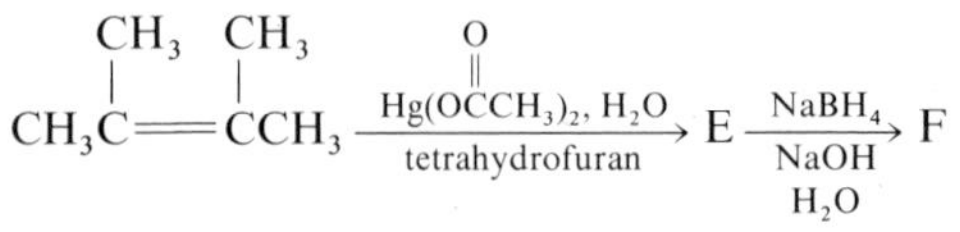

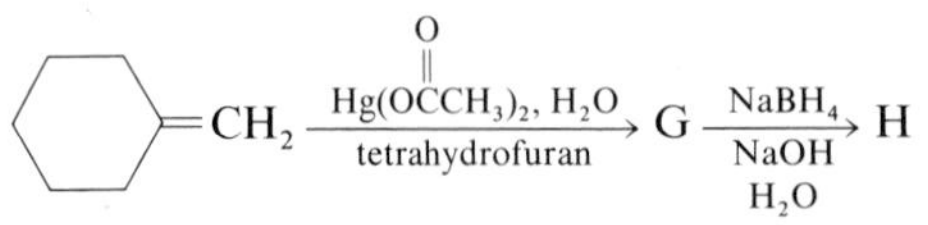

(e) [cyclobutane] $\xrightarrow[\text{acetone}]{\text{Hg(O}\overset{\text{O}}{\overset{\|}{\text{C}}}\text{CH}_3)_2,\ \text{CH}_3\text{OH}}$ I $\xrightarrow[\text{H}_2\text{O}]{\text{NaCl}}$ J $\xrightarrow[\substack{\text{NaOH}\\\text{H}_2\text{O}}]{\text{NaBH}_4}$ K

(f) [phenyl]—CH=CH$_2$ $\xrightarrow[\text{tetrahydrofuran}]{\text{Hg(O}\overset{\text{O}}{\overset{\|}{\text{C}}}\text{CH}_3)_2,\ \text{H}_2\text{O}}$ L $\xrightarrow[\substack{\text{NaOH}\\\text{H}_2\text{O}}]{\text{NaBH}_4}$ M

PROBLEM 7.14   The stereochemistry of oxymercuration was investigated using (*E*)- and (*Z*)-1,2-dideuterioethylene, the stereoisomers of DHC=CHD. Show in full stereochemical detail the reaction of these compounds with mercury(II) acetate in methanol, followed by sodium chloride in water.

7.5

Conversions of Alcohols to Alkyl Halides

A. Reactions with Hydrogen Halides

The hydroxyl group can undergo a wide variety of reactions. Alcohols are, therefore, important intermediates in chemical interconversions. Alkyl halides, which participate in many different substitution reactions (Section 6.11), are prepared from alcohols in several different ways. The hydroxyl group in an alcohol can be converted in two ways into a good leaving group (Section 6.4B) so that it can be displaced by an incoming halide ion. In the first method, an acid protonates the hydroxyl group, creating a new leaving group, water. If this reaction is carried out on a primary alcohol, the water is displaced by halide ion in an S_N2 reaction, giving the alkyl halide. Thus, dissolving gaseous hydrogen bromide in 1-heptanol converts it to 1-bromoheptane.

$$CH_3(CH_2)_5CH_2-\overset{H}{\overset{|}{\ddot{O}}} \qquad \longrightarrow \qquad CH_3(CH_2)_5CH_2-\overset{H}{\overset{|}{\overset{+}{\ddot{O}}}}\overset{|}{H}$$

1-heptanol   hydrogen bromide   :Br:⁻

$$\downarrow$$

$$\overset{\ddot{O}}{\underset{H \quad H}{}} \ + \ CH_3(CH_2)_5CH_2-\ddot{Br}:$$

1-bromoheptane
90%

In secondary alcohols, or other alcohols in which S_N2 reactions are hindered, the substitution must proceed through a carbocation intermediate, which can rearrange. For example, when 3-pentanol is treated with hydrogen bromide, 3-bromopentane and 2-bromopentane are obtained in mixtures of varying composition depending on whether aqueous or gaseous hydrogen bromide is used as the reagent.

$$CH_3CH_2\underset{\overset{|}{OH}}{C}HCH_2CH_3 \xrightarrow[\Delta]{\text{HBr}} CH_3CH_2\underset{\overset{|}{Br}}{C}HCH_2CH_3 + CH_3CH_2CH_2\underset{\overset{|}{Br}}{C}HCH_3$$

3-pentanol   3-bromopentane   2-bromopentane

PROBLEM 7.15 Write out the full mechanism for the conversion of 3-pentanol to the two bromopentanes in 48% aqueous hydrobromic acid.

PROBLEM 7.16 Whenever either 2-chloropentane or 3-chloropentane is allowed to stand in concentrated hydrochloric acid in which zinc chloride has been dissolved, a mixture of the two consisting of 66% 2-chloropentane and 34% 3-chloropentane is obtained. Write a mechanism for this interconversion starting with 3-chloropentane. What is the role of the zinc chloride? What does the fact that the equilibrium mixture is 66% 2-chloropentane tell you about the relative stabilities of 2-pentyl and 3-pentyl cations? (Hint: What are the relative numbers of 2-pentyl and 3-pentyl cations that can form?)

Another way to convert the hydroxyl group in an alcohol into a good leaving group is to make the tosylate ester, which can be displaced by halide ion in a nucleophilic substitution reaction (Section 6.4B). In this way, 3-pentanol can be converted to 3-bromopentane free of 2-bromopentane, suggesting that a carbocation intermediate is not formed under these conditions.

$$\underset{\substack{| \\ \text{OH} \\ \text{3-pentanol}}}{CH_3CH_2CHCH_2CH_3} + \underset{\substack{p\text{-toluenesulfonyl} \\ \text{chloride}}}{TsCl} \xrightarrow[\text{pyridine}]{} \underset{\substack{| \\ \text{OTs} \\ \text{3-pentyl tosylate}}}{CH_3CH_2CHCH_2CH_3} \xrightarrow[\text{dimethyl sulfoxide}]{Na^+Br^-} \underset{\substack{| \\ \text{Br} \\ \text{3-bromopentane} \\ 85\%}}{CH_3CH_2CHCH_2CH_3}$$

B. Reactions with Thionyl Chloride and Phosphorus Halides

Other reagents also avoid carbocation intermediates and minimize rearrangements in the conversion of alcohols to alkyl halides. One of these is thionyl chloride, $SOCl_2$. 3-Pentanol will react with thionyl chloride in pyridine to give 3-chloropentane, uncontaminated with 2-chloropentane.

$$\underset{\substack{| \\ \text{OH} \\ \text{3-pentanol}}}{CH_3CH_2CHCH_2CH_3} + \underset{\substack{\text{thionyl} \\ \text{chloride}}}{SOCl_2} + \underset{\substack{\\ \text{pyridine}}}{\underset{N}{\bigcirc}} \longrightarrow \underset{\substack{| \\ \text{Cl} \\ \text{3-chloropentane}}}{CH_3CH_2CHCH_2CH_3} + \underset{\substack{\\ \text{pyridine} \\ \text{hydrochloride}}}{\underset{\underset{H}{N^+}}{\bigcirc} Cl^-} + \underset{\substack{\text{sulfur} \\ \text{dioxide}}}{SO_2}\uparrow$$

However, this reaction is often accompanied by elimination reactions and does not give high yields.

The mechanism postulated for this reaction starts with the conversion of the alcohol to a compound with a good leaving group, an intermediate chlorosulfite ester, which decomposes to give the alkyl halide. In the presence of organic bases such as pyridine, inversion of configuration is seen in chiral compounds.

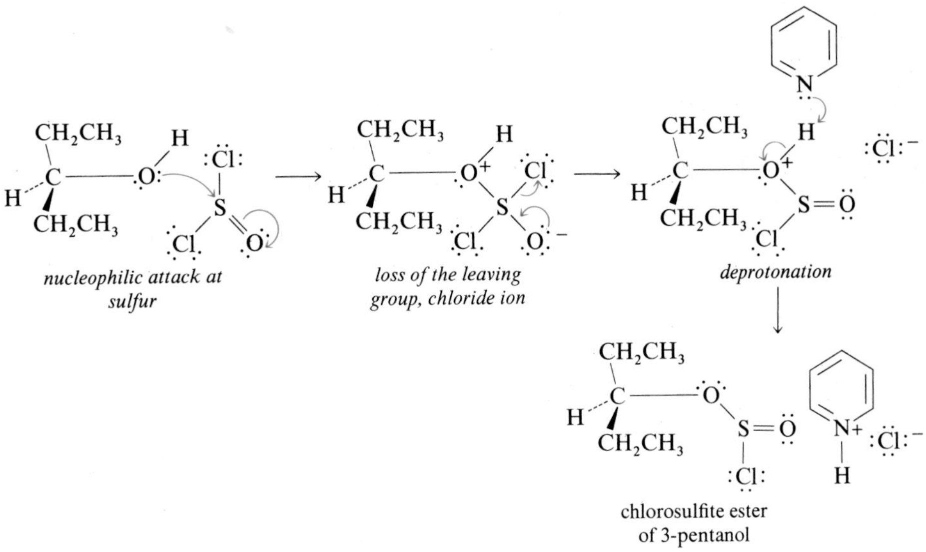

nucleophilic attack at
sulfur

loss of the leaving
group, chloride ion

deprotonation

chlorosulfite ester
of 3-pentanol

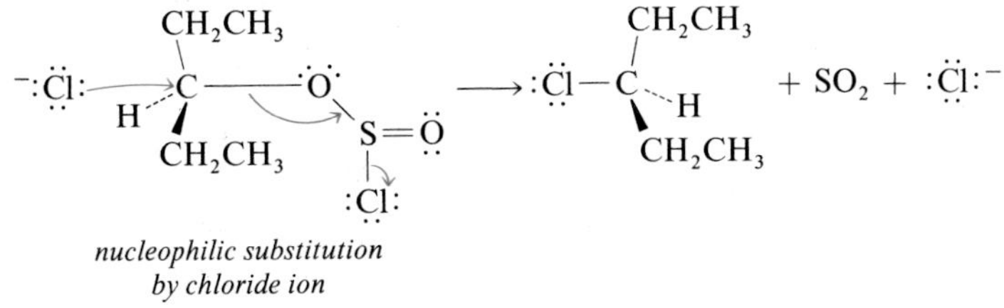

nucleophilic substitution
by chloride ion

PROBLEM 7.17 When 2-ethyl-1-butanol is treated with zinc chloride in concentrated hydrochloric acid, a mixture of chloro compounds forms. Chief among those are 2-ethyl-1-chlorobutane, 3-chlorohexane, 2-chlorohexane, and 3-chloro-3-methylpentane. When 2-ethyl-1-butanol is treated with thionyl chloride in pyridine, only 1-chloro-2-ethylbutane is formed. Write out detailed mechanisms that account for these observations. What is the function of the zinc chloride?

Another reagent that is used to substitute halogen for a hydroxyl group with a minimum of rearrangement is phosphorus tribromide. It is used to convert the primary alcohol, isobutyl alcohol, into isobutyl bromide.

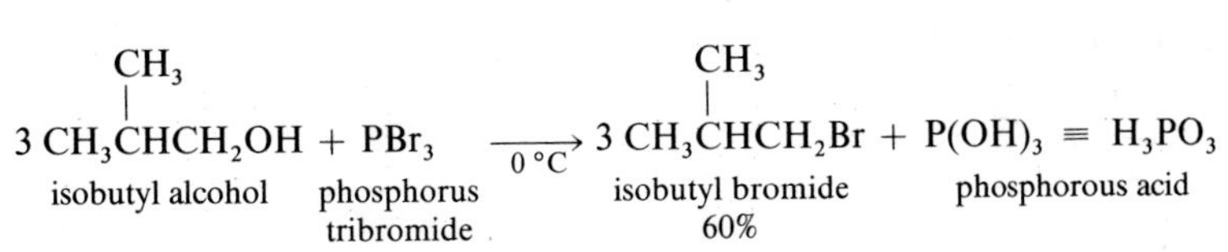

$$3 \ CH_3\underset{\overset{|}{CH_3}}{C}HCH_2OH + PBr_3 \xrightarrow{0\,°C} 3 \ CH_3\underset{\overset{|}{CH_3}}{C}HCH_2Br + P(OH)_3 \equiv H_3PO_3$$

isobutyl alcohol   phosphorus   isobutyl bromide   phosphorous acid
           tribromide   60%

The reagent is especially important for allylic alcohols, which are prone to give isomeric allylic halides. 2-Buten-1-ol can be converted into the corresponding bromide if phosphorus tribromide is used and the temperature is kept low.

$$CH_3CH=CHCH_2OH \xrightarrow[\substack{pyridine \\ -15\,°C}]{PBr_3} CH_3CH=CHCH_2Br + CH_3CHCH=CH_2$$

$$\underset{|}{}$$

$$Br$$

2-buten-1-ol	1-bromo-2-butene	3-bromo-1-butene
	94%	6%

The mechanism suggested for the conversion of an alcohol to an alkyl halide with phosphorus tribromide resembles the one proposed for the reaction with thionyl chloride. Its first step is the formation of a phosphorus-oxygen bond between the alcohol and phosphorus tribromide.

PROBLEM 7.18 When optically active 2-octanol is treated with phosphorus tribromide, 2-bromooctane is formed with almost complete inversion of configuration. The other product of the reaction is phosphorous acid, $P(OH)_3$. Write out the details of a mechanism for the reaction that would account for these facts.

PROBLEM 7.19 Complete the following equations.

(a) 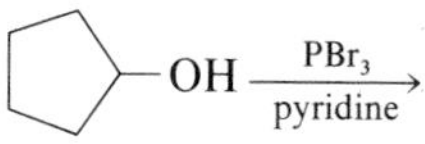 ⬠—OH $\xrightarrow[pyridine]{PBr_3}$

(b) 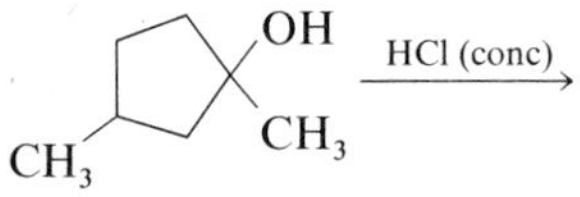 [cyclopentane structure with OH, CH_3, CH_3] $\xrightarrow{HCl\ (conc)}$

(c) $CH_3(CH_2)_9CH_2OH \xrightarrow{SOCl_2}_{\Delta}$

(d) $CH_3(CH_2)_8CH_2OH \xrightarrow{HBr}_{\Delta}$

(e) $CH_3SCH_2CH_2OH \xrightarrow[chloroform]{SOCl_2}$

(f) $CH_3CH_2OCH_2CH_2OH \xrightarrow[pyridine]{PBr_3}$

(g)

$$\underset{}{SCH_2\overset{\displaystyle O}{\overset{\displaystyle \|}{C}}OCH_3}$$

[β-lactam ring structure with N, O, $CHCOCH_2$—, OH] —⬡—NO_2 $\xrightarrow[\substack{CH_3\ N\ CH_3 \\ 0\,°C}]{SOCl_2}$

7.6

Oxidation Reactions of Alcohols

A. Oxidation-Reduction in Organic Compounds

In Chapter 4, we explored the properties of molecules from the two opposite poles of acidity and basicity. We examined various molecular structures for features that enabled them to lose or gain protons, or to share nonbonding electrons. In that chapter, we developed the concepts and language that we have since used to describe relationships between reactions that seem to be quite different from each other at first glance.

Oxidation and reduction reactions are another pair of polar opposites. For many metals and their ions, the processes of oxidation and reduction are easily seen. When copper metal is put into a solution of silver nitrate, crystals of silver metal grow on the surface of the copper and the solution turns the blue color that is typical of water solutions of copper (II) ion. Copper metal loses electrons to silver ions and is oxidized, while silver ions are reduced to silver metal by a gain of electrons.

$$Cu + 2\ Ag^+ \xrightarrow[\text{water}]{} Cu^{2+} + 2\ Ag$$

shiny	colorless		blue	silver-
red				grey
metal				crystals

For organic compounds, however, oxidation states are not so immediately obvious. Yet, it is of great importance in organizing reaction types to be able to decide whether a given compound has been oxidized or reduced. You can look for two types of structural changes that occur in going from a reactant to a product in order to decide whether an organic compound is being oxidized or reduced. The first thing to look for is whether any hydrogen atoms or oxygen atoms are lost or gained in the reaction. Loss of hydrogen atoms or gain of oxygen is an oxidation. For example,

$$CH_3CH_2OH \xrightarrow[\substack{\text{loss of two}\\\text{hydrogen atoms,}\\\text{therefore oxidation}}]{} CH_3\overset{\displaystyle O}{\overset{\|}{C}}H \xrightarrow[\substack{\text{gain of an}\\\text{oxygen atom,}\\\text{therefore oxidation}}]{} CH_3\overset{\displaystyle O}{\overset{\|}{C}}OH$$

ethanol   acetaldehyde   acetic acid

The changes that ethanol must undergo to be converted first to acetaldehyde and then to acetic acid are oxidation reactions and require the use of an oxidizing agent. The reverse reactions must be reduction reactions, carried out by a reducing agent. Formally, they involve the loss of oxygen or the gain of hydrogen atoms.

$$CH_3\overset{\displaystyle O}{\overset{\|}{C}}OH \xrightarrow[\substack{\text{loss of an}\\\text{oxygen atom,}\\\text{therefore reduction}}]{} CH_3\overset{\displaystyle O}{\overset{\|}{C}}H \xrightarrow[\substack{\text{gain of two}\\\text{hydrogen atoms,}\\\text{therefore reduction}}]{} CH_3CH_2OH$$

acetic acid   acetaldehyde   ethanol

Note that it is only one of the carbon atoms in this series of compounds that is undergoing oxidation or reduction. The methyl group is unchanged in the reactions.

By a similar reasoning process we can decide that the carbon atoms in an alkyne are more oxidized than the carbon atoms in an alkene, which in turn are more oxidized than those in an alkane.

$$HC\equiv CH \xrightarrow[\substack{\text{gain of two}\\\text{hydrogen atoms,}\\\text{therefore reduction}}]{} CH_2=CH_2 \xrightarrow[\substack{\text{gain of two}\\\text{hydrogen atoms,}\\\text{therefore reduction}}]{} CH_3CH_3$$

acetylene   ethylene   ethane

A reducing agent is necessary to convert an alkyne to an alkene and then to an alkane. In this particular case, you know from Section 2.10 that this conversion is carried out by adding hydrogen to the multiple bonds. For the purposes of this section, however, it is not necessary to know all the reagents that will be used for the transformations that are shown. Your goal here is to recognize which reactions involve oxidation and which, reduction. The ability to distinguish between these two reactions is important in itself.

If we attach oxidation numbers to various substituents, we can be quantitative about the oxidation states of carbon. A carbon-carbon bond contributes nothing to the oxidation state of a carbon atom. Each hydrogen atom is in a $+1$ oxidation state, a halogen or hydroxyl group is in a -1 oxidation state, and a double-bonded oxygen atom is in a -2 oxidation state. Using these numbers, you can assign oxidation states to carbon atoms in the compounds shown earlier in this section. For example,

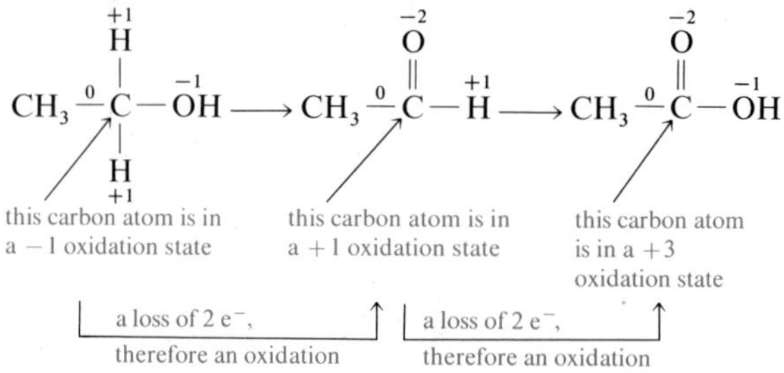

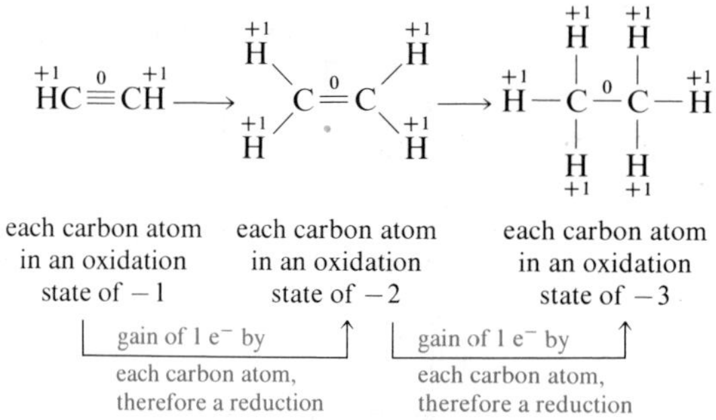

this carbon atom is in   this carbon atom is in   this carbon atom
a -1 oxidation state   a $+1$ oxidation state   is in a $+3$
                                                           oxidation state

a loss of 2 e⁻,   a loss of 2 e⁻,
therefore an oxidation   therefore an oxidation

each carbon atom   each carbon atom   each carbon atom
in an oxidation   in an oxidation   in an oxidation
state of -1 state of -2 state of -3

gain of 1 e⁻ by   gain of 1 e⁻ by
each carbon atom,   each carbon atom,
therefore a reduction therefore a reduction

Both the qualitative approach presented earlier and this quantitative one lead us to the same conclusions.

You should work to develop reasonable skill in deciding when processes involve oxidation-reduction. If you are able to recognize the different oxidation states of carbon, you will be able to make logical predictions about the kinds of reagents needed for the transformations you wish to achieve.

PROBLEM 7.20   Look at the following transformations and decide in each case whether the starting material has undergone oxidation or reduction. Specify whether you would need an oxidizing or reducing agent to carry out the change that is shown.

(a)

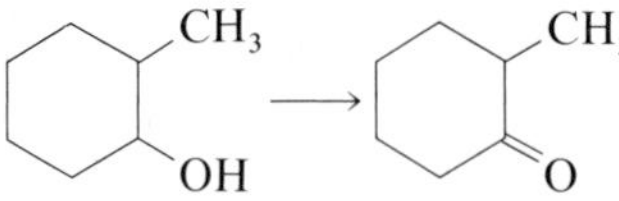

(b)

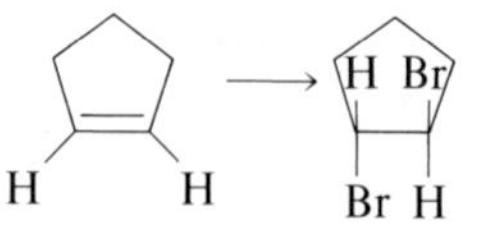

(c) $CH_3CH_2Br \longrightarrow CH_3CH_2NH_2$ (d)

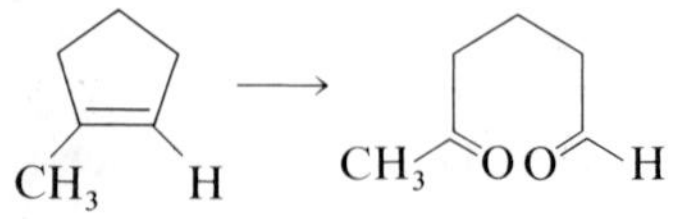

(e) 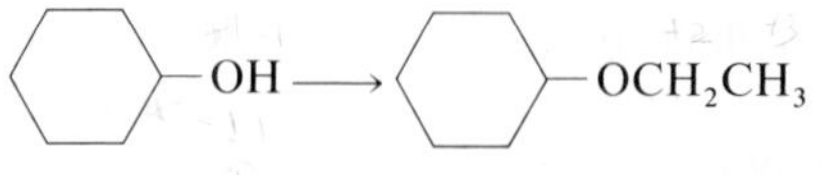 $\longrightarrow$—OH $\longrightarrow$ $\longrightarrow$—OCH$_2$CH$_3$ (f) $CH_3CH_2\overset{O}{\overset{\|}{C}}H \longrightarrow CH_3CH_2CH_2OH$

(g) $CH_3CH=CHCH_3 \longrightarrow CH_3CH-CHCH_3$
                                          $\quad\quad\;\;$ $|\quad\;\;|$
                                          $\quad\quad\;\;$ OH   OH

(h) $CH_3C\equiv CCH_2CH_3 \longrightarrow CH_3CH=CHCH_2CH_3$

(i) $CH_3C\equiv CCH_3 \longrightarrow CH_3\overset{\overset{O}{\|}}{C}-\overset{\overset{O}{\|}}{C}CH_3$ (j) $CH_3CH_2CH_2OH \longrightarrow CH_3CH_2\overset{\overset{O}{\|}}{C}OH$

(k) $CH_3\overset{\overset{O}{\|}}{C}H \longrightarrow CH_3\overset{\overset{OCH_3}{|}}{C}HOCH_3$

PROBLEM 7.21   If you have not already done so in answering Problem 7.20, assign oxidation states to the carbon atoms that undergo transformations in the compounds shown. Do your quantitative assignments support the conclusions you reached qualitatively?

PROBLEM 7.22   Methane is converted to carbon tetrachloride by successive replacement of hydrogen atoms by chlorine atoms. Methane represents the most highly reduced state of carbon, and carbon tetrachloride, the most highly oxidized. Calculate oxidation numbers for the carbon atom in methane, in carbon tetrachloride, and in the intermediate alkyl halides.

B. Oxidation with Sodium Dichromate

Primary and secondary alcohols are easily oxidized by reagents containing chromium(VI). For low molecular weight alcohols that have water solubility, water solutions of sodium dichromate in the presence of acid are used. A primary alcohol such as *n*-butyl alcohol is oxidized first to an aldehyde, which is oxidized further by the same reagent to a carboxylic acid.

$$3\ CH_3CH_2CH_2CH_2OH + Cr_2O_7^{2-} + 4\ H_2SO_4 \longrightarrow 3\ CH_3CH_2CH_2\overset{\overset{O}{\|}}{C}H + 7\ H_2O + 2\ Cr^{3+} + 4\ SO_4^{2-}$$

<table>
<tr><td>n-butyl alcohol
bp 118 °C
a primary alcohol</td><td>dichromate ion
orange</td><td>butanal
bp 75 °C
distilled out of the
reaction mixture</td><td>blue-green</td></tr>
</table>

$$3\ CH_3CH_2CH_2\overset{\overset{O}{\|}}{C}H + Cr_2O_7^{2-} + 4\ H_2SO_4 \longrightarrow 3\ CH_3CH_2CH_2\overset{\overset{O}{\|}}{C}OH + 2\ Cr^{3+} + 4\ SO_4^{2-} + 4\ H_2O$$

<table>
<tr><td>butanal</td><td>dichromate ion
orange</td><td>butanoic acid</td><td>blue-green</td></tr>
</table>

This process is not useful in preparing aldehydes from primary alcohols except for a few special cases where the aldehyde that is formed has a boiling point low enough that it can be distilled out of the reaction mixture before further reaction takes place. Butanal is such an aldehyde.

A secondary alcohol, such as *sec*-butyl alcohol, is oxidized to a ketone, which is usually stable to further oxidation. For a ketone to be oxidized, carbon-carbon bonds would have to be broken because there are no more carbon-hydrogen bonds available on the carbon atom undergoing oxidation. Ketones, therefore, survive oxidation conditions that readily oxidize aldehydes to carboxylic acids.

$$3\ CH_3CH_2\overset{\overset{}{\underset{\underset{OH}{|}}{C}}}HCH_3 + Cr_2O_7^{2-} + 4\ H_2SO_4 \longrightarrow 3\ CH_3CH_2\overset{\overset{O}{\|}}{C}CH_3 + 2\ Cr^{3+} + 4\ SO_4^{2-} + 7\ H_2O$$

<table>
<tr><td>sec-butyl alcohol
a secondary alcohol</td><td>dichromate ion
orange</td><td>2-butanone</td><td>blue-
green</td></tr>
</table>

The tertiary alcohol, *tert*-butyl alcohol, is not oxidized by the acidic dichromate solution. The carbon atom bearing the hydroxyl group in this alcohol does not have a hydrogen atom on it. Here, too, carbon-carbon bonds would have to be broken to allow oxidation to take place.

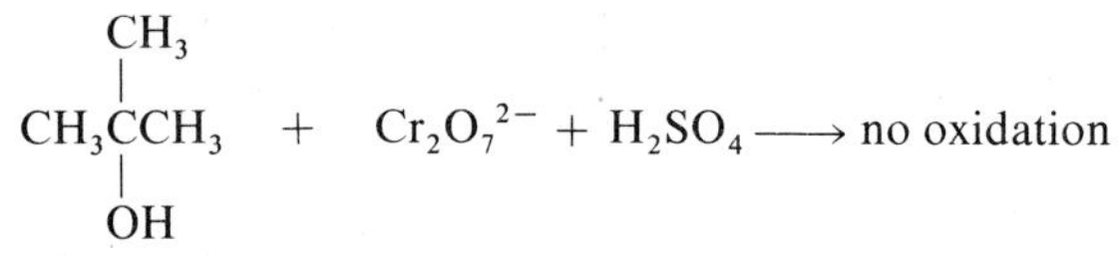

$$\underset{\substack{\text{OH}}}{\overset{\substack{\text{CH}_3}}{\text{CH}_3\text{CCH}_3}} \quad + \quad \text{Cr}_2\text{O}_7{}^{2-} + \text{H}_2\text{SO}_4 \longrightarrow \text{no oxidation}$$

tert-butyl alcohol	dichromate ion	*no change*
a tertiary alcohol	*orange*	*in color*

The oxidizing solution in these cases, an acidic solution of sodium dichromate, contains a variety of species in equilibrium. Most important of these is the acid chromate ion, $\text{HCrO}_4{}^-$.

$$\text{H}_2\text{O} + \text{Cr}_2\text{O}_7{}^{2-} \rightleftharpoons 2\ \text{HCrO}_4{}^-$$

dichromate ion	acid chromate ion

Chromium(VI), present in both the dichromate and the acid chromate ions, is the oxidizing agent. It is reduced to chromium(III) in several steps with chromium(IV) and chromium(V) present at some stages of the reaction. The reaction is postulated to proceed through the formation of a complex between the alcohol and the chromium atom with a series of deprotonation and protonation steps leading to the loss of water from the complex and the formation of the chromate ester of the alcohol. The mechanism for this part of the reaction is shown below.

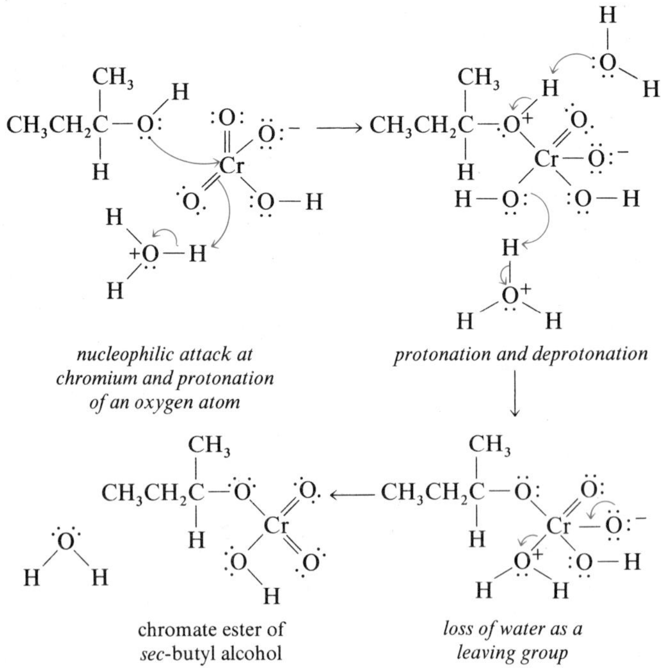

nucleophilic attack at chromium and protonation of an oxygen atom	protonation and deprotonation

chromate ester of *sec*-butyl alcohol	loss of water as a leaving group

If you examine the flow of electrons indicated by the curved arrows, you will recognize the individual processes as acid-base reactions even though the reaction looks complicated at first glance. A hydroxyl group (on the chromium atom this time) is converted to a good leaving group and the chromate ester of the alcohol results. The oxidation states of carbon and of chromium do not change during the formation of the chromate ester. The removal of the hydrogen atom from the carbon atom that originally bore the hydroxyl group is the oxidation-reduction step of the reaction. Chromium(IV) is formed in this step and is ultimately reduced to chromium(III). The exact details of all the processes involved are quite complex, but it is sufficient that you understand that *a hydroxyl group and a hydrogen atom on the same carbon atom are necessary for the oxidation reaction that proceeds through the formation of the chromate ester of the alcohol.*

carbon in the 0 oxidation state

chromium in the +6 oxidation state

carbon in the +2 oxidation state; has been oxidized

chromium in the +4 oxidation state; has been reduced

PROBLEM 7.23   The orange color of chromium(VI) and the blue-green color of chromium(III) have been used as the basis of a qualitative analysis test for primary and secondary alcohols. The alcohol is dissolved in acetone, and a solution of chromium trioxide, CrO_3, dissolved in water and sulfuric acid is added dropwise. Primary and secondary alcohols give an instant blue-green cloudiness. Write out the structures of the following alcohols and predict in each case whether they will give a positive test with chromium trioxide reagent. For the alcohols that you expect to be positive, write an equation for the reaction.

(a) 4-methyl-2-pentanol   (b) 2-*tert*-butylcyclohexanol   (c) cholesterol (p. 272)
(d) 1-ethylcyclopentanol   (e) 1-pentanol   (f) 2,3,3-trimethyl-2-butanol

---

C. Selective Oxidations. Pyridinium Chlorochromate

Aqueous acidic dichromate solutions are not suitable for all oxidations of alcohols to carbonyl compounds. Some alcohols are not soluble in water, so other solvents have to be used. In complex molecules with multiple functional groups, use of strong acids may lead to undesirable side reactions. For such systems, milder and more selective reagents have been developed.

A highly selective reagent is made by dissolving chromium trioxide in hydrochloric acid and adding the basic solvent pyridine to the solution.

chromium   hydrochloric   pyridine   pyridinium
trioxide   acid   chlorochromate
                                                              84%

yellow-orange crystals

The resulting reagent is called pyridinium chlorochromate, and is used in organic solvents such as dichloromethane to oxidize primary and secondary alcohols. For example, 1-decanol is oxidized to decanal in high yield.

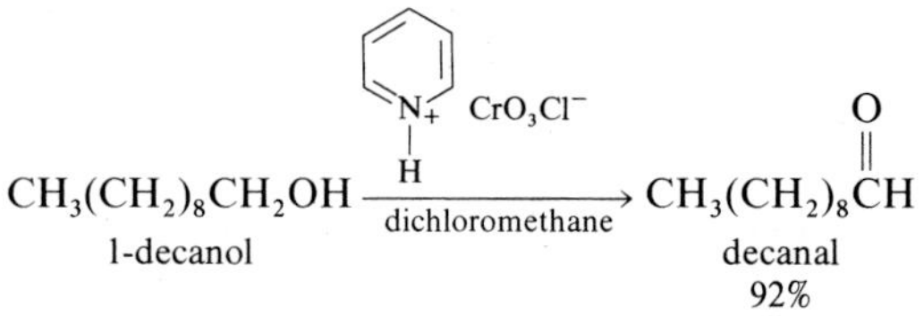

$$CH_3(CH_2)_8CH_2OH \xrightarrow[\text{dichloromethane}]{\overset{\displaystyle\bigcirc}{\underset{H}{N_+}}\ CrO_3Cl^-} CH_3(CH_2)_8\overset{\displaystyle O}{\overset{\|}{C}}H$$

1-decanol

decanal
92%

The oxidation stops at the aldehyde stage because no water is present to form the hydrate of the aldehyde, which is the intermediate necessary for further oxidation to the carboxylic acid by chromium(VI).

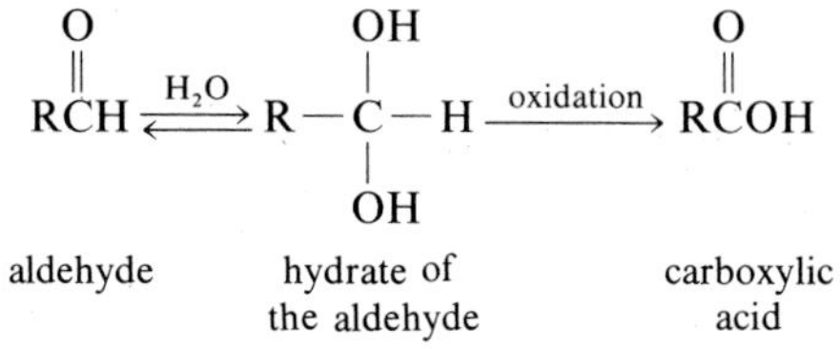

$$R\overset{\displaystyle O}{\overset{\|}{C}}H \underset{\longleftarrow}{\overset{H_2O}{\rightleftharpoons}} R-\underset{\underset{OH}{|}}{\overset{\overset{OH}{|}}{C}}-H \xrightarrow{\text{oxidation}} R\overset{\displaystyle O}{\overset{\|}{C}}OH$$

aldehyde

hydrate of
the aldehyde

carboxylic
acid

Similarly, 4-*tert*-butylcyclohexanol is oxidized to 4-*tert*-butylcyclohexanone.

4-*tert*-butylcyclohexanol

4-*tert*-butylcyclohexanone

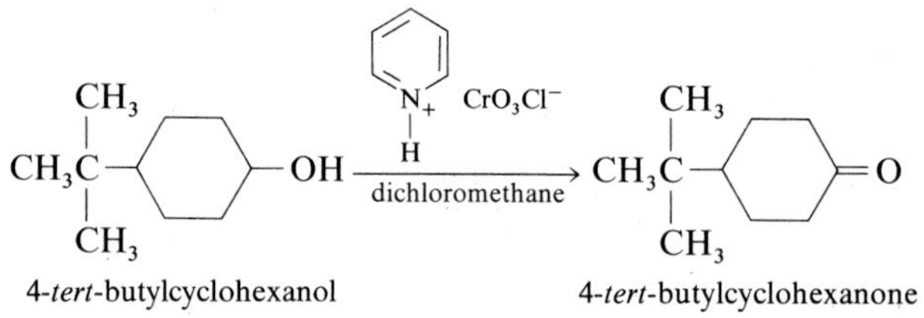

Pyridinium chlorochromate is also useful in the oxidation of alcohols containing other functional groups that may react with aqueous acid, such as double bonds. For example, the unsaturated alcohol citronellol, which is one of the fragrant compounds in rose oil and geranium oil, is oxidized by this reagent to the corresponding aldehyde in good yield.

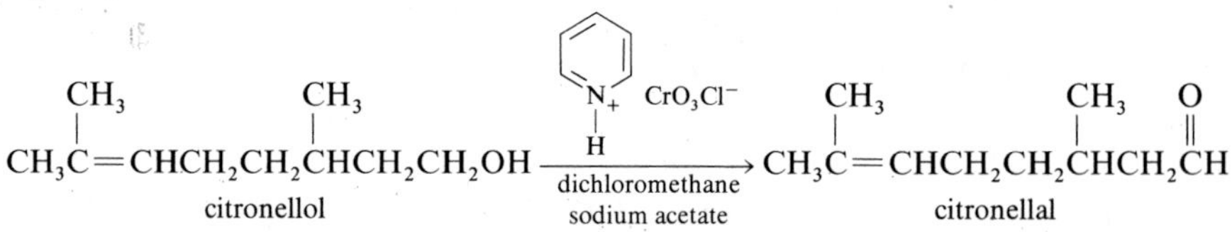

citronellol

citronellal

The double bond is untouched in this reaction.

PROBLEM 7.24   Write an equation showing the product that you would have expected if the oxidation of citronellol had been carried out with chromic acid in the presence of water. Can you foresee any other complications from having a strong acid such as sulfuric acid in the reaction mixture?

PROBLEM 7.25   Complete the following equations.

(a) $CH_3\underset{\underset{OH}{|}}{\overset{\overset{CH_3}{|}}{C}}H\overset{\overset{CH_3}{|}}{C}HCH_2CH_3 \xrightarrow[\substack{H_2SO_4 \\ H_2O \\ 40\,°C}]{Na_2Cr_2O_7}$

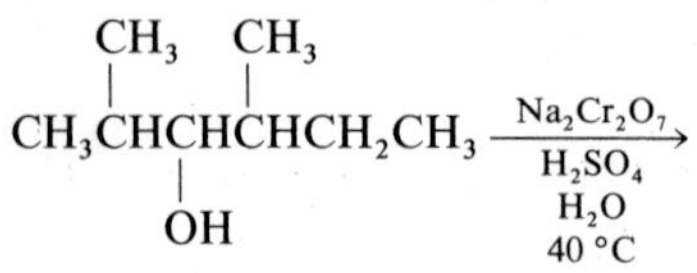

(b) 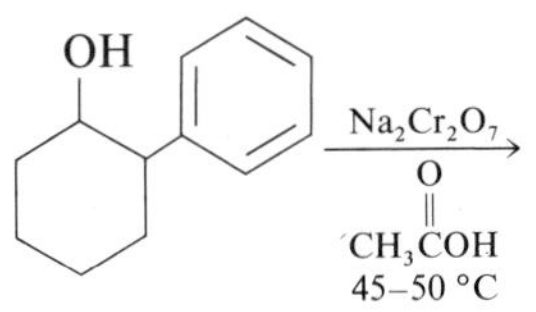

(c) 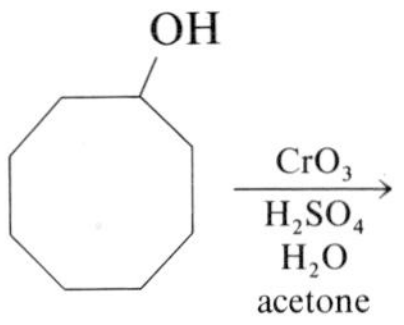

(d) 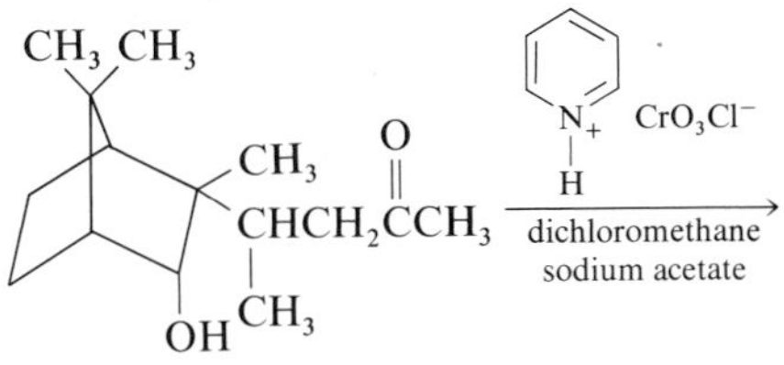

(e) 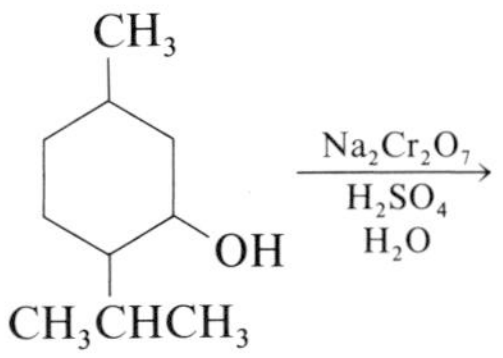

7.7

Infrared Spectroscopy

A. The Electromagnetic Spectrum and Absorption Spectroscopy

Chemists generally use two methods to obtain information about the structures of molecules. One is to carry out chemical transformations on molecules and observe the results. The reagents that are used in organic qualitative analysis tests are those that give immediate information about the presence or absence of some functional groups. Examples of such tests are the bromine test for unsaturation (Section 5.8) and the chromic acid test for primary and secondary alcohols (Problem 7.23).

Another way that chemists gather information about the structure of molecules is by making physical measurements on compounds. In Chapter 1, we used the dipole moment of a compound to deduce the shape and the symmetry of its molecules. Electron diffraction was also mentioned as a method of obtaining information about bond angles and bond lengths.

For organic chemists, spectroscopy is the most widely used physical method of investigation. A molecule interacts with electromagnetic radiation to absorb energy that corresponds to transitions between fixed energy levels for the molecular species. The idea of electronic energy levels, which are the orbitals of an atom, is a familar one. When an atom absorbs light of a given frequency, an electron moves from a lower energy level to a higher one. The difference in energy betwen the two energy levels is proportional to the frequency of the light absorbed, and is expressed as

$$E = h\nu$$

where h is the universal Planck's constant, 6.624×10^{-27} erg · s, and ν is the frequency of the light in Hz (Hertz), cycles/s. The frequency of the light is related to the wave length, λ, of the light by the expression

$$\nu = \frac{c}{\lambda}$$

where c is the velocity of light in a vacuum, 2.998×10^{10} cm/s.

An atom that has absorbed light energy is said to be in an **excited state,** above its normal **ground state.** It returns to the ground state by the loss of energy, usually by emitting heat or, less frequently, light.

Molecules have several different kinds of energy levels. For example, a molecule as a whole rotates in space. The various types of rotational motion possible for a given molecule correspond to **rotational energy levels.** The atoms within a molecule are in constant motion relative to each other. These motions are related to **vibrational energy levels.** Finally, just as atoms have energy levels among which electronic transitions take place, so do molecules. The different types of energy levels possible for a molecule are shown schematically in Figure 7.1. The gaps between **electronic energy levels** are much larger than those between vibrational energy levels. Energy differences between rotational energy levels are the smallest. The differences in electronic, vibrational, and rotational energy levels correspond to energy in different parts of the electromagnetic spectrum. The absorption of energy of definite frequencies in these portions of the electromagnetic spectrum can be monitored by instruments known as **spectrophotometers,** which record the changing absorption of energy as a function of the wavelength or the frequency of the light being used. The tracings obtained from such instruments are called **spectra,** and are used to make diagnoses about the structure of the molecule that gave rise to them. As our knowledge of chemical structures grows, we will learn to analyze different types of spectra for specific structural information.

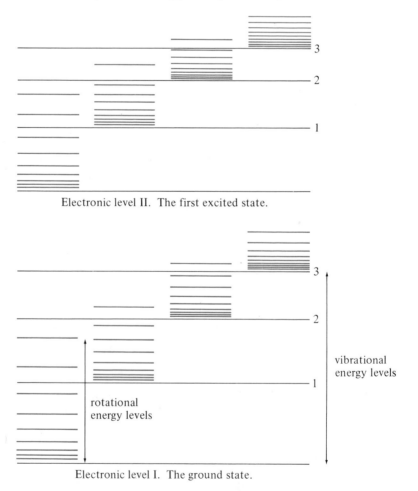

Electronic level II. The first excited state.

Electronic level I. The ground state.

rotational energy levels

vibrational energy levels

Figure 7.1   Energy-level diagram for a molecule. Two electronic levels, with the associated vibrational and rotational energy levels shown.

TABLE 7.1   Regions of the Electromagnetic Spectrum Used in Spectroscopy

Region of Spectrum	Molecular Change	Energy, kcal/mol	Frequency, Hz	Wavelength, cm
ultraviolet	electronic transitions (Chapter 11)	100	10^{15}	2×10^{-5}–4×10^{-5}
visible	electronic transitions (Chapter 11)	50	5×10^{14}	4×10^{-5}–8×10^{-5}
infrared	molecular vibrations (Chapter 7)	5	10^{13}–10^{14}	10^{-4}–10^{-2}
microwave	molecular rotations	3×10^{-3}	3×10^{10}	1
radio-frequency	orientation of spin of nucleus in magnetic field (Chapter 13)	10^{-7}	10^{6}	5×10^{5}

The electromagnetic spectrum has a tremendous range of energy. At one end of the spectrum are cosmic rays, which have energies of approximately 10^{9} kcal/mol (1 erg/molecule $= 1.439 \times 10^{13}$ kcal/mol). They are very short wavelength (10^{-12} cm), high frequency (10^{22} Hz) radiation. Medically useful x-rays have wavelengths of 10^{-8} to 10^{-6} cm, and energies in the range of 10^{3} kcal/mol. On the low energy side are radiowaves with long wavelengths (10^{3}–10^{6} cm) and low frequencies (10^{7}–10^{4} Hz).

The regions of the electromagnetic spectrum that are appropriate for looking at molecular structures are summarized in Table 7.1. In the following sections, we will focus on one of these regions, the infrared, to investigate how molecular interactions with radiation in that energy range give us information about molecular structure.

B. An Introduction to Infrared Spectroscopy

The atoms within a molecule are constantly in motion, distorting the chemical bonds. These motions are called **molecular vibrations.** One possible type of vibration for a molecule causes changes in bond length. Such vibrations are **stretching vibrations** (Figure 7.2). Others cause changes in bond angles and are called **bending vibrations**

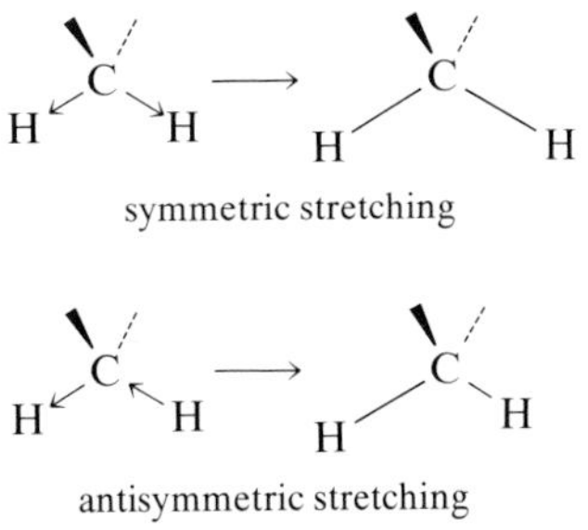

symmetric stretching

antisymmetric stretching

Figure 7.2 Typical stretching vibrations of a methylene group.

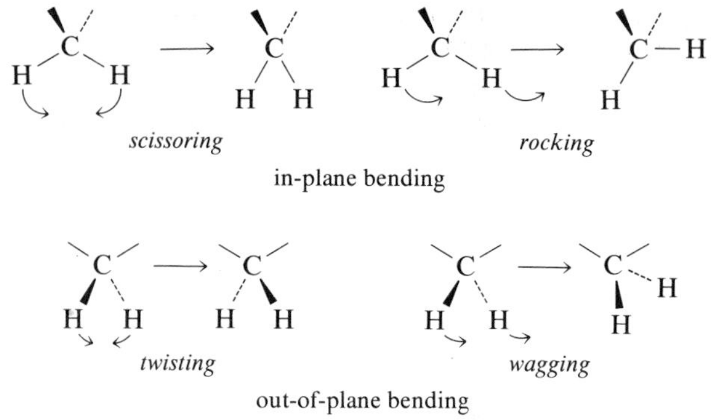

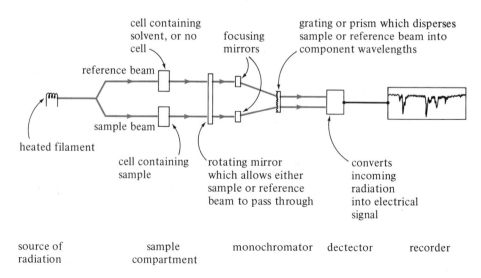

scissoring   *rocking*

in-plane bending

twisting   *wagging*

out-of-plane bending

Figure 7.3   Different kinds of bending vibrations for a methylene group.

(Figure 7.3). For any given molecule there are a number of energy levels correspond-ing to the different vibrational states possible for that molecule. The spacing between these energy levels corresponds to energy in the infrared region of the electromagnetic spectrum. Energy can be given units of wavenumbers, the number of waves per centimeter. The differences in the energy levels in the region of the infrared that is most useful to chemists correspond to radiation having frequencies of 4000 to 666 cm^{-1}, or wavelengths of 2.5 to 15.0 μm (a micrometer, μm, is 10^{-4} cm, 10^{-6} m).

When radiation having energy that is the same as the difference in energy between molecular vibrational energy levels impinges upon a molecule, the radiation is absorbed and the amplitude of the molecular vibration increases. The molecule moves from one vibrational energy level to a higher one. The energy that is absorbed is ultimately returned to the environment as heat energy. This energy is roughly 1 to 10 kcal/mol, not enough to break chemical bonds or cause chemical reactions. The fact that the radiation is absorbed is recorded as a tracing by an instrument called an infrared spectrophotometer (Figure 7.4).

Figure 7.4   Schematic diagram of a typical infrared spectrophotometer.

Even a simple molecule can have a large number of possible molecular vibrations, and an infrared spectrum usually has many absorption bands. The spectrum is useful to a chemist because the different functional groups absorb at different frequencies, corresponding to certain vibrations typical of that portion of the molecule. Tables that list the infrared absorption frequencies for different types of functional groups are available. Table 7.2 is an abbreviated listing of such group frequencies. It is reproduced inside the back cover of the book for easy reference.

The frequencies at which functional groups absorb energy are related to the types of bonding present in them. Table 7.3 (p. 300) makes this clearer by summarizing some trends that can be seen upon careful examination of Table 7.2.

The major frequencies that are typical of functional groups usually appear between 4000 and 1400 cm^{-1}. The portion of the spectrum between 1400 cm^{-1} and 200 cm^{-1} is called the **fingerprint region** because it is more difficult to make specific assignments of bands in that region as they are more dependent upon the structure of the molecule as a whole. This is immensely useful in making a positive identification of a compound. The presence of all the bands with the same relative intensities, in this region

TABLE 7.2   Characteristic Infrared Absorption Frequencies

Bond Type	Stretching, cm^{-1}	Bending, cm^{-1}
C—H alkanes	2960–2850 (*s*)	1470–1350 (*s*)
C—H alkenes	3080–3020 (*m*)	1000–675 (*s*)
C—H aromatic	3100–3000 (*v*)	870–675 (*v*)
C—H aldehyde	2900, 2700 (*m*, 2 bands)	
C—H alkyne	3300 (*s*)	
C≡C alkyne	2260–2100 (*v*)	
C≡N nitrile	2260–2220 (*v*)	
C=C alkene	1680–1620 (*v*)	
C=C aromatic	1600–1450 (*v*)	
C=O ketone	1725–1705 (*s*)	
C=O aldehyde	1740–1720 (*s*)	
C=O α, β-unsaturated ketone	1685–1665 (*s*)	
C=O aryl ketone	1700–1680 (*s*)	
C=O ester	1750–1735 (*s*)	
C=O acid	1725–1700 (*s*)	
C=O amide	1690–1650 (*s*)	
O—H alcohols (not hydrogen bonded)	3650–3590 (*v*)	
O—H hydrogen bonded	3600–3200 (*s*, broad)	
O—H acids	3000–2500 (*s*, broad)	
N—H amines	3500–3300 (*m*)	1620–1590 (*v*)
N—H amides	3500–3350 (*m*)	1655–1510 (*s*)
C—O alcohols, ethers, esters	1300–1000 (*s*)	
C—N amines, aliphatic	1220–1020 (*w*)	
C—N amines, aromatic	1360–1250 (*s*)	
NO_2 nitro	1560–1515 (*s*)	
	1385–1345 (*s*)	

s = strong absorption
m = medium absorption
w = weak absorption
v = variable absorption

TABLE 7.3   **Infrared Regions for Different Bond Types**

Bond Type	cm^{-1}
O—H , N—H, C—H, stretching	3650–2500
C≡N, C≡C, stretching	2260–2100
C=O, C=N, C=C, N=O, stretching	1800–1390
C—O, C—N, stretching	1360–1030
N—H, bending	1620–1590
C—H, bending	1470–1350
C=C—H, bending	1000–675

as well as in the higher frequency region of the spectrum, for two samples is considered to be proof of the identity of the two species.

Not all bands in the infrared spectrum have the same intensity, and this too is useful in identifying functional groups. In general, during a vibration that corresponds to a change in the dipole of the molecule, the molecule absorbs radiation strongly, while for a vibration in which there is a small or no change in the dipole of the molecule, an absorption band may not be seen. For example, the absorption corresponding to the stretching of a carbonyl group is a strong one, that corresponding to the stretching of a carbon-carbon double bond is a weak one.

$$\overset{\longrightarrow}{\underset{/}{\backslash}\!\!\overset{\delta+}{C}\!\!=\!\!\overset{\delta-}{O}} \qquad\qquad \overset{\longrightarrow}{\underset{/}{\backslash}\!\!\overset{\delta+}{C}\!\!=\!\!=\!\!\overset{\delta-}{O}}$$

polar bond
in
carbonyl group

*increase in dipole moment
on stretching of bond in
carbonyl group*

intense absorption

$$\underset{/}{\backslash}C\!\!=\!\!C\underset{\backslash}{/} \qquad\qquad \underset{/}{\backslash}C\!\!=\!\!=\!\!C\underset{\backslash}{/}$$

nonpolar
carbon-carbon
double bond

*small or no change in dipole
moment on stretching
of carbon-carbon
double bond*

*absorption weak
or absent*

Variations such as these are the reason that the interpretation of infrared spectra is an art requiring careful observation of relative intensities and appearances of bands rather than a slavish application of numbers from tables. Some examples of spectra for compounds with different functional groups presented in the next section will make the important points of this discussion clearer.

C. The Infrared Spectra of Alcohols

The most important absorption band in the infrared spectra of alcohols is the hydroxyl stretching frequency. In dilute solutions where hydrogen bonding is minimized, it appears at 3640–3610 cm^{-1}. Most of the time, however, it is present as a strong broad band at 3600–3200 cm^{-1} indicating the presence of a hydrogen-bonded hydroxyl

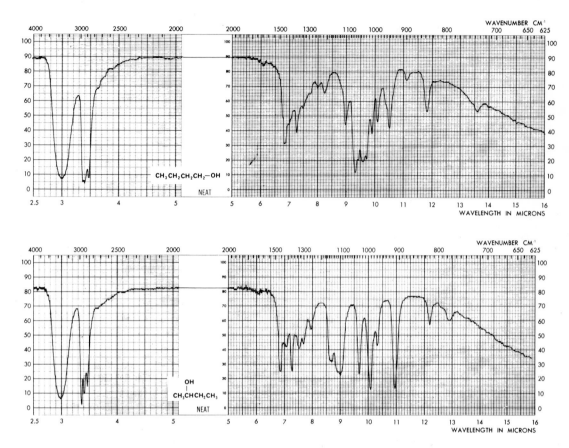

Figure 7.5   Infrared spectra of *n*-butyl alcohol, and *sec*-butyl alcohol.
(From *The Aldrich Library of Infrared Spectra*)

group. The infrared spectra of *n*-butyl alcohol and *sec*-butyl alcohol are shown in Figure 7.5. In each one, the broad strong band appearing at highest frequency is the hydroxyl stretching band. Other bands that can be identified are the carbon-hydrogen stretching bands appearing in the region of 2900 cm^{-1}, characteristic of alkane type of carbon-hydrogen bonds. The presence of such bonds is confirmed by absorption around 1400 cm^{-1}, where the bending vibrations for such bonds absorb. There are also strong bands in the region of 1080 to 1300 cm^{-1}, which are attributable to the carbon-oxygen single bond stretching vibrations.

The bands that have been singled out above are the most important in the spectra. There are many other bands that cannot be assigned to any one simple vibration of the molecule. They are the result of combinations of vibrations, or overtones of bands. They are, however, characteristic of each molecule, and can be used for its identification. Note that while the spectra of the two alcohols look fairly similar from 4000–1400 cm^{-1}, below that frequency, in the fingerprint region, they are quite individual in appearance. If we had a spectrum of an unknown alcohol to identify as one of these compounds, we would have no difficulty deciding which one it was.

Figure 7.6 shows the infrared spectrum of a much higher molecular weight alcohol, cholesterol, to demonstrate that the characteristic features of an alcohol show up even when the hydroxyl group is a much smaller portion of the molecule. For example, the

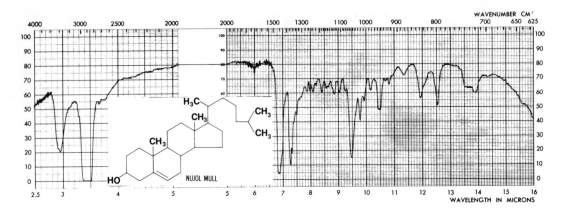

Figure 7.6 Infrared spectrum of cholesterol. (From *The Aldrich Library of Infrared Spectra*)

hydroxyl absorption at 3400 cm^{-1} and the carbon-oxygen single bond stretching frequency at 1060 cm^{-1} are clearly visible. The spectrum of cholesterol was taken as a mull in Nujol, a mixture of high molecular weight alkanes, so carbon-hydrogen stretching (2900 cm^{-1}) and bending (1470–1375 cm^{-1}) frequencies are prominent in the spectrum.

The spectrum also illustrates a phenomenon mentioned earlier. The carbon-carbon double bond that is present in the molecule is not easily detectible. The double bond is buried in a complex molecule, and does not undergo any large change in its dipole moment on stretching, therefore the intensity of its absorption band is weak. Its presence cannot be determined from the spectrum.

D. The Infrared Spectra of Aldehydes and Ketones

In contrast to the weak absorption of carbon-carbon double bonds, the band for the stretching frequency of a carbonyl group is usually one of the strongest in the spectrum. Figure 7.7 shows the spectra of butanal and 2-butanone, the oxidation products of *n*-butyl alcohol and *sec*-butyl alcohol, respectively.

Butanal has a strong carbonyl absorption at 1725 cm^{-1}. Another band that is typical of aldehydes is the carbon-hydrogen stretching frequency for the carbon-hydrogen bond on the carbonyl group. Two bands appear for that, one around 2700 cm^{-1}, the other buried in the alkane carbon-hydrogen absorption band around 2900 cm^{-1}. The remainder of the spectrum in the region down to 1400 cm^{-1} is derived from the stretching and bending frequencies for the alkane portion of the molecule. The region of the spectrum around 3420 cm^{-1} has a small band that might be erroneously assigned to a hydroxyl group if we were to apply numbers with no regard for intensities and appearances of bands. If we turn back to Figure 7.5 to remind ourselves of the appearance of the hydroxyl absorption bands, we see that the band in this spectrum is too weak to support such a diagnosis.

The phenomenon of a weak band in the hydroxyl stretching region also appears in the spectrum of 2-butanone (Figure 7.7). Such bands are often seen for compounds that have strong carbonyl bands and are the result of overtones (appearing at some multiple

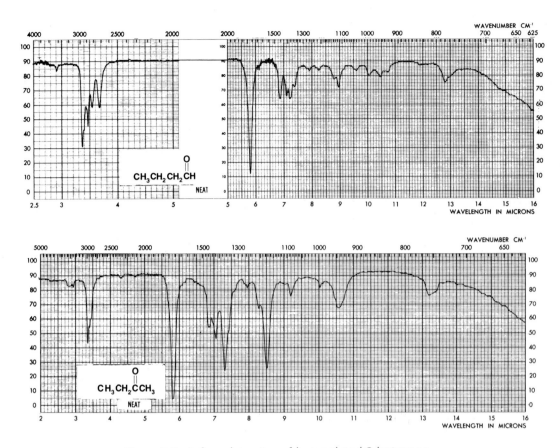

Figure 7.7 Infrared spectra of butanal and 2-butanone.
(From *The Aldrich Library of Infrared Spectra*)

of the frequency) of the carbonyl band. The carbonyl band in 2-butanone appears at 1710 cm^{-1}. Note that the carbon-hydrogen stretching frequency around 2700 cm^{-1}, seen in the spectrum of the aldehyde, is missing from the spectrum of the ketone.

A comparison of the infrared spectra of *n*-butyl alcohol and *sec*-butyl alcohol (Figure 7.5) with those of butanal and 2-butanone (Figure 7.7) illustrates how useful infrared spectroscopy is in following the course of a chemical reaction. It is possible to actually see the disappearance of the absorption band for a hydroxyl group, and the appearance of one for a carbonyl group as the oxidation of an alcohol progresses.

In contrast to compounds containing hydroxyl and carbonyl groups, which have distinctive absorption bands in their infrared spectra, the spectra of compounds belonging to many other functional group classes lack such bands in the region between 4000 and 1400 cm^{-1}. For example, the most distinctive band in the spectrum of dipropyl ether, shown in Figure 7.8, is the carbon-oxygen single bond stretching frequency near 1150 cm^{-1}. If a compound that contains oxygen does not show hydroxyl or carbonyl absorption, yet has strong absorption bands in the region of 1300 to 1060 cm^{-1}, it is probably an ether.

Features that are important in the infrared spectra of the other functional group classes are introduced in later chapters (Sections 8.8, 9.10, 10.11, 12.8B, and 16.8).

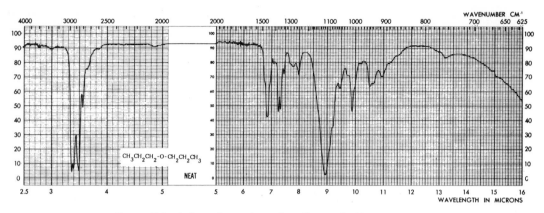

Figure 7.8 Infrared spectrum for dipropyl ether.
(From *The Aldrich Library of Infrared Spectra*)

PROBLEM 7.26 Compound A has the molecular formula $C_7H_{16}O$. On oxidation with pyridinium chlorochromate in dichloromethane it gives Compound B, $C_7H_{14}O$. The infrared spectra of Compounds A and B are given in Figure 7.9. Assign structures to the compounds that are compatible with their spectra. Point out the features in the spectra of Compounds A and B that are useful in making structural assignments.

Compound A

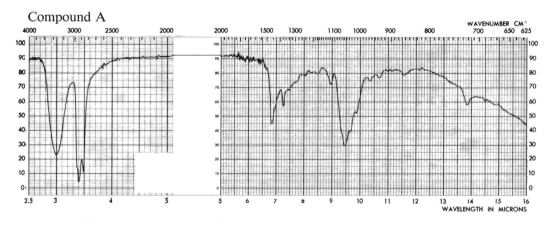

Compound B

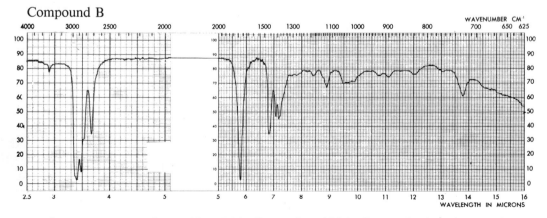

Figure 7.9 Spectra for Problem 7.26. (From *The Aldrich Library of Infrared Spectra*)

PROBLEM 7.27   Compound C, $C_6H_{14}O$, is oxidized by chromic acid to Compound D, $C_6H_{12}O$. The infrared spectra for the compounds are given in Figure 7.10. Assign possible structures to the compounds.

Compound C

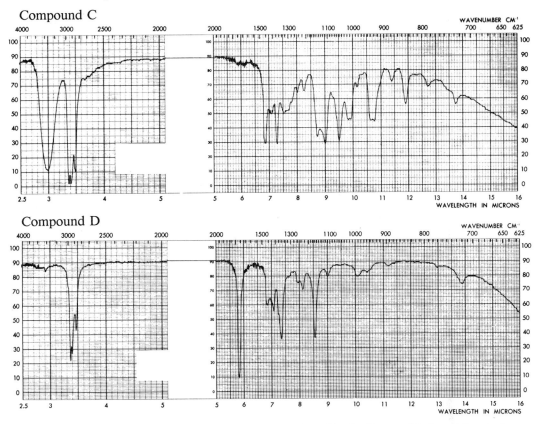

Compound D

Figure 7.10   Spectra for Problem 7.27. (From *The Aldrich Library of Infrared Spectra*)

PROBLEM 7.28   The infrared spectra of a series of compounds are given in Figure 7.11 with their molecular formulas. Decide whether each compound is an alcohol, an aldehyde, a ketone, or an ether.

Compound E, $C_4H_{10}O$

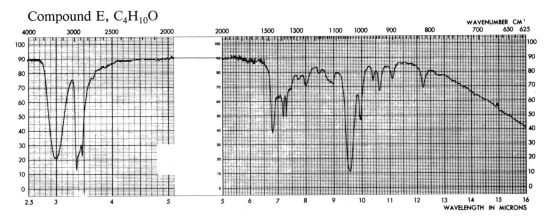

Figure 7.11   Spectra for Problem 7.28. (From *The Aldrich Library of Infrared Spectra*)

Compound F, C$_4$H$_{10}$O

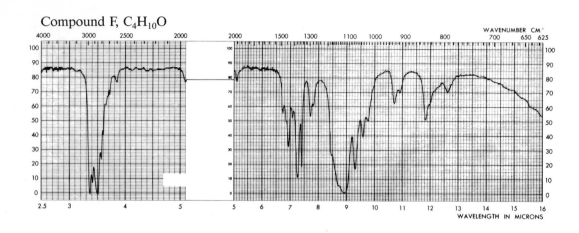

Compound G, C$_6$H$_{12}$O

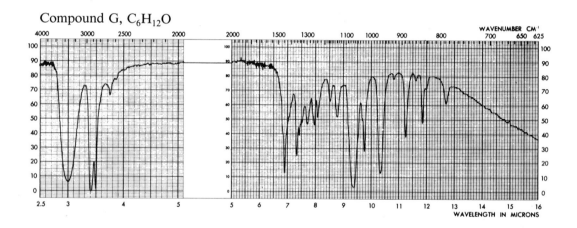

Compound H, C$_6$H$_{12}$O

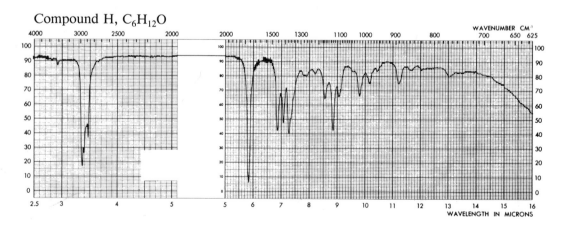

Figure 7.11 Spectra for Problem 7.28 (*Continued*)

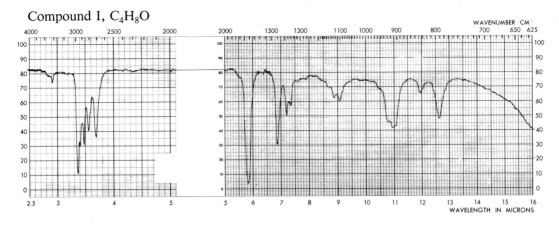

Compound I, C₄H₈O

Compound J, C₈H₁₈O

Figure 7.11 Spectra for Problem 7.28 (*Continued*)

ADDITIONAL PROBLEMS

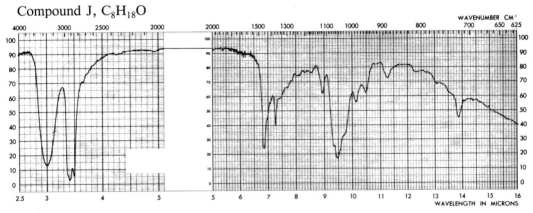

7.29 Name the following compounds, including stereochemistry, according to the IUPAC rules.

(a) HOCH₂CH₂CH₂OH

(b) CH₃OCH₂CH₂OCH₃

(c)
$$CH_3CH_2OCH_2\overset{\overset{\displaystyle CH_3}{|}}{\underset{\underset{\displaystyle CH_3}{|}}{C}}CH_2OH$$

(d) CH₃OCH₂CH₂Br

(e) ⬡—OCH₂CH₃

(f) CH₃CH₂OCH₂CH₂OH

(g) ▢—OH

(h)
$$H\overset{\overset{\displaystyle CH_3}{|}}{\underset{\underset{\displaystyle HO}{}}{C}}{-}CH_2\,\,\,\,C{=}C\,\,\,\,\begin{smallmatrix}CH_3\\CH_2CH_3\end{smallmatrix}$$

(i)
Br
H
H
OH

(j) ⬡—CH₂OCH₂—⬡

(k) OCH_3 / OH
(l) $\text{CH}_3\text{CH}_2\text{OCOCH}_2\text{CH}_3$ with $\overset{\text{CH}_3}{|}$ above the first C and $\overset{|}{\text{OCH}_2\text{CH}_3}$ below

7.30 Draw structures for the following compounds. When necessary, show stereochemistry by the use of appropriate conventions.

(a) 1-chloro-1-ethoxyethene   (b) (*E*)-1-methoxy-2-propoxyethene
(c) 1,3-dichloro-2-propanol   (d) cyclobutylmethanol
(e) (*R*)-3-methyl-5-hexen-3-ol   (f) (*S*)-2-chloro-1-propanol
(g) (*S*)-3-methyl-1-pentyn-3-ol   (h) 2-nitroethanol
(i) (*R*)-5,5-dimethyl-3-heptanol   (j) (3*S*,4*R*)-4-methyl-3-hexanol

7.31 Using 1-propanol as a typical primary alcohol, write equations for its reactions with the following reagents under the conditions shown.

(a) HBr, Δ   (b) PBr$_3$ (c) SOCl$_2$, pyridine   (d) Na metal
(e) H$_3$O$^+$, cold   (f) H$_2$SO$_4$, Δ   (g) Na metal, then CH$_3$CH$_2$CH$_2$CH$_2$Br
(h) Na$_2$Cr$_2$O$_7$, H$_2$SO$_4$, Δ   (i) pyridinium chlorochromate, dichloromethane

7.32 Using 2-hexanol as a typical secondary alcohol, write equations for its reactions with the following reagents under the conditions shown.

(a) HBr, Δ   (b) PBr$_3$ (c) SOCl$_2$, pyridine   (d) Na metal
(e) ZnCl$_2$, HCl   (f) H$_2$SO$_4$, Δ   (g) Na$_2$Cr$_2$O$_7$, H$_2$SO$_4$, Δ
(h) pyridinium chlorochromate, dichloromethane
(i) CrO$_3$, H$_2$O, H$_2$SO$_4$, acetone as solvent (see Problem 7.23)

7.33 The two columns on the left contain structural formulas for some organic compounds. The right-hand column is a list of reagents that may react with some of them. For each compound on the left, list all the reagents from the right that would result in a reaction detectable by the human senses when the compound and reagent are mixed in a test tube. (The human senses, for example, can detect a change in color, the formation of a solution, the evolution of a gas, the evolution of heat, or a change in odor.)

(a) —OH

(b) $\overset{\text{CH}_3}{\underset{\text{H}}{}}\text{C}=\text{C}\overset{\text{H}}{\underset{\text{CH}_2\text{CH}_3}{}}$

1. H$_2$O, cold
2. H$_2$SO$_4$, cold, conc
3. Br$_2$, carbon tetrachloride
4. CrO$_3$, H$_2$O, H$_2$SO$_4$
5. AgNO$_3$, CH$_3$CH$_2$OH
6. NaI, acetone

(c) $\overset{\text{CH}_3}{\underset{\text{H}}{}}\text{C}=\text{C}\overset{\text{H}}{\underset{\text{CH}_2\text{Cl}}{}}$

(d) CH$_3$ / OH

(e) —CH$_2$Cl

(f) CH$_3$CH$_2$CH$_2$CH$_2$CH$_2$OH

(g) CH$_3$C≡CCH$_2$CH$_3$

(h)

(i) HOCH$_2$CH$_2$OH

(j) $\text{CH}_3\text{CH}_2\overset{\text{CH}_3}{\underset{\text{Br}}{\text{C}}}\text{CH}_3$

(k) —OCH$_2$CH$_3$

(l) CH$_3$CH$_2$CH$_2$CH$_2$CH$_2$CH$_2$Cl

7.34 For each of the following pairs of compounds, predict which one would have the higher boiling point.

(a) $HOCH_2CH_2OH$ or CH_3CH_2OH

(b) $ClCH_2CH_2OH$ or FCH_2CH_2OH

(c) CH_3CHCH_2OH or CH_3CHCH_2OH
      | |
     OCH_3 OH

(d) $ClCH_2CH_2OH$ or $H_2NCH_2CH_2OH$

(e) $BrCH_2CH_2CH_2OH$ or $H_2NCH_2CH_2CH_2OH$

(f) $CH_3CH_2\overset{\displaystyle CH_2CH_3}{\overset{\displaystyle |}{N}}CH_2CH_2OH$ or $H_2NCH_2CH_2OH$

7.35 For each set of experimental facts presented below, give an explanation for the observed trends, using the concepts about the relationships between structure and physical properties that we have developed so far.

(a) $CH_3CH_2CH_2CH_2OH$

     solubility 7.9 g
     in 100 mL H_2O

$CH_3CH_2\overset{\displaystyle |}{\underset{\displaystyle OH}{C}}HCH_3$

     solubility 12.5 g
     in 100 mL H_2O

$CH_3\overset{\displaystyle CH_3}{\overset{\displaystyle |}{\underset{\displaystyle |}{\underset{\displaystyle OH}{C}}}}CH_3$

     infinitely
     soluble in H_2O

(b) $CH_3CH_2\overset{\displaystyle |}{\underset{\displaystyle OH}{C}}HCH_3$

     solubility 12.5 g
     in 100 mL H_2O

$CH_3CH_2\overset{\displaystyle |}{\underset{\displaystyle Cl}{C}}HCH_3$

     very slightly
     soluble in H_2O

$CH_3CH_2\overset{\displaystyle |}{\underset{\displaystyle Br}{C}}HCH_3$

     insoluble in
     H_2O

(c) $CH_3CH_2\overset{\displaystyle |}{\underset{\displaystyle CH_2CH_3}{C}}HCH_2OH$

     solubility 0.63 g
     in 100 mL H_2O

$CH_3CH_2\overset{\displaystyle |}{\underset{\displaystyle CH_2CH_3}{C}}H\overset{\displaystyle O}{\overset{\displaystyle \|}{O C}}CH_3$

     solubility 0.06 g
     in 100 mL H_2O

$CH_3CH_2\overset{\displaystyle |}{\underset{\displaystyle NH_2}{C}}HCH_2OH$

     infinitely soluble
     in H_2O

(d) CH_3CH_2OH

     pK_a 15.9

$CH_3\overset{\displaystyle CH_3}{\overset{\displaystyle |}{\underset{\displaystyle |}{\underset{\displaystyle OH}{C}}}}CH_3$

     pK_a ~ 18

CF_3CH_2OH

     pK_a 12.4

(e) $CH_3CH_2CH_2CH_2OH$

     bp 118 °C
     solubility 7.9 g
     in 100 mL H_2O

$CH_3CH_2CH_2CH_2CH_2CH_2OH$

     bp 157 °C
     solubility 0.59 g
     in 100 mL H_2O

$CH_3\overset{\displaystyle |}{\underset{\displaystyle HO}{C}}H\overset{\displaystyle |}{\underset{\displaystyle OH}{C}}HCH_2CH_2CH_3$

     bp 207°C
     infinitely soluble
     in H_2O

(f) $CH_3CH_2CH_2CH_2OH$

     solubility 7.9 g
     in 100 mL H_2O

$CH_3CH_2CH_2CH_2SH$

     slightly soluble
     in H_2O

7.36   Complete the following equations.

(a) $\underset{\displaystyle CH_3}{CH_3CHCH_2OH} \xrightarrow[\text{pyridine}]{SOCl_2}$

(b) $\underset{\underset{\displaystyle OH}{|}}{\overset{\overset{\displaystyle CH_3}{|}}{CH_3CCH_2CH_3}} \xrightarrow{\text{HCl (conc)}}$

(c) $\underset{\displaystyle CH_3}{CH_3CH_2C}{=}CH_2 \xrightarrow[\text{tetrahydrofuran}]{Hg(O\overset{\displaystyle O}{\overset{\|}{C}}CH_3)_2,\ H_2O} \xrightarrow[\substack{NaOH \\ H_2O}]{NaBH_4}$

(d) $\underset{\displaystyle OH}{CH_3CH_2CHCH_3} \xrightarrow{PBr_3}$

(e) 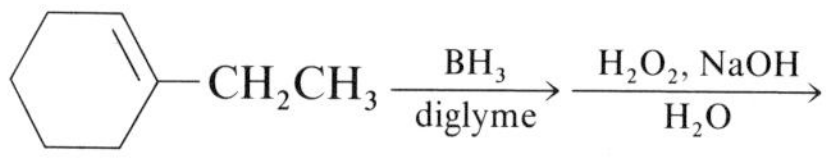 $\text{(cyclohexene)}{-}CH_2CH_3 \xrightarrow[\text{diglyme}]{BH_3} \xrightarrow[H_2O]{H_2O_2,\ NaOH}$

(f) $HOCH_2(CH_2)_4CH_2OH \xrightarrow{HBr(g),\ \text{excess}}$

(g) 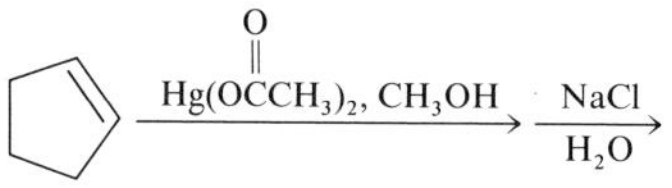 $\text{(cyclopentene)} \xrightarrow[H_2O]{Hg(O\overset{\displaystyle O}{\overset{\|}{C}}CH_3)_2,\ CH_3OH} \xrightarrow{NaCl}$

(h) $\underset{\displaystyle CH_3}{CH_3C}{=}CHCH_3 \xrightarrow[\text{tetrahydrofuran}]{\text{9-BBN}} \xrightarrow[H_2O]{H_2O_2,\ NaOH}$

(i) 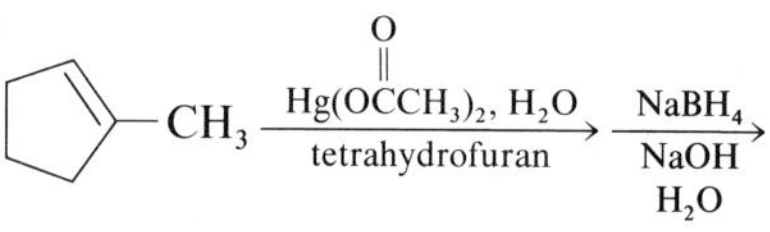 $\text{(cyclopentene)}{-}CH_3 \xrightarrow[\text{tetrahydrofuran}]{Hg(O\overset{\displaystyle O}{\overset{\|}{C}}CH_3)_2,\ H_2O} \xrightarrow[\substack{NaOH \\ H_2O}]{NaBH_4}$

(j) 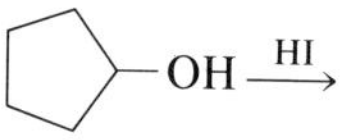 $\text{(cyclopentane)}{-}OH \xrightarrow{HI}$

(k) 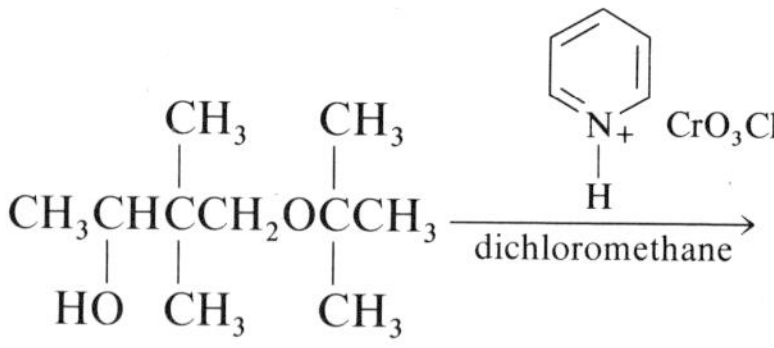 $\underset{\underset{\displaystyle HO\ \ \ CH_3\ \ \ \ CH_3}{|\ \ \ \ \ \ |\ \ \ \ \ \ \ |}}{CH_3CHCCH_2O\overset{\overset{\displaystyle CH_3}{|}}{C}CH_3} \xrightarrow[\text{dichloromethane}]{\underset{\underset{\displaystyle H}{|}}{N^+}\ CrO_3Cl^-}$

(l) 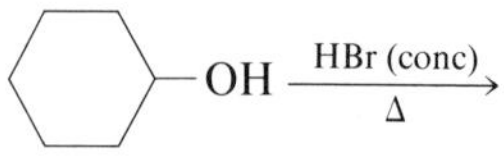 $\text{(cyclohexane)}{-}OH \xrightarrow[\Delta]{HBr\ (conc)}$

(m) 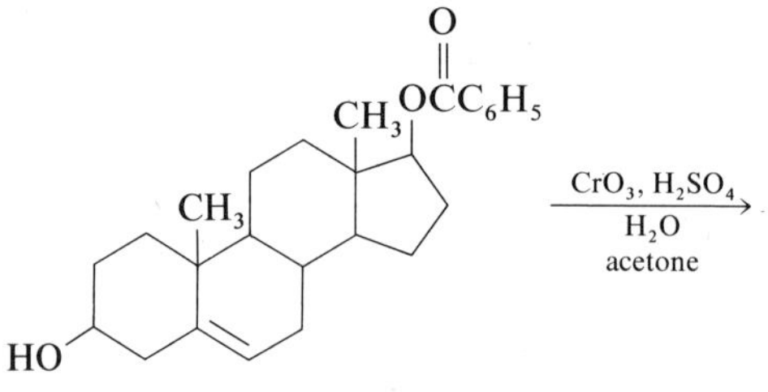 $\xrightarrow[\substack{H_2O \\ \text{acetone}}]{CrO_3,\ H_2SO_4}$

(n) 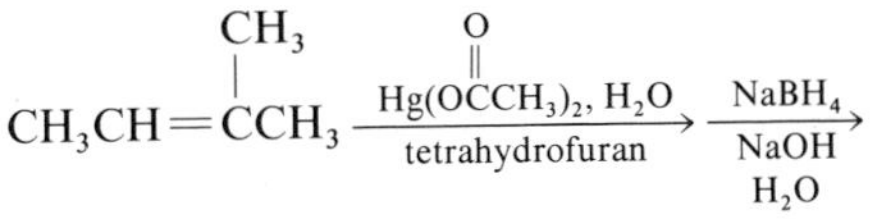 $CH_3CH = \overset{\overset{\displaystyle CH_3}{|}}{C}CH_3 \xrightarrow[\substack{tetrahydrofuran}]{Hg(OCCH_3)_2, H_2O} \xrightarrow[\substack{NaOH \\ H_2O}]{NaBH_4}$

(o) 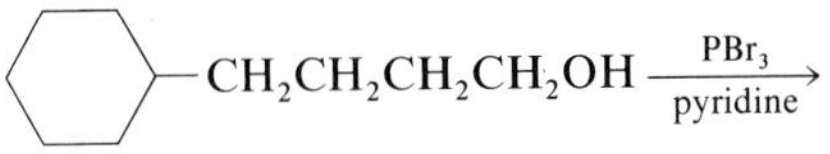 [cyclohexyl]$-CH_2CH_2CH_2CH_2OH \xrightarrow[\substack{pyridine}]{PBr_3}$

(p) 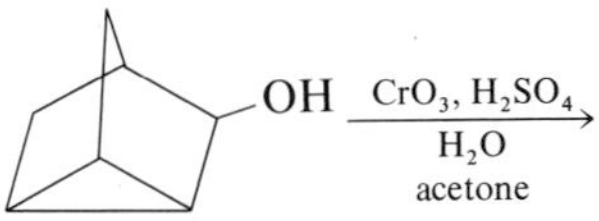 [norbornyl]$-OH \xrightarrow[\substack{H_2O \\ acetone}]{CrO_3, H_2SO_4}$

(q) 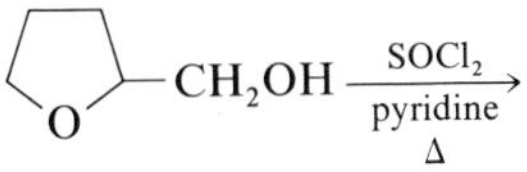 [tetrahydrofuranyl]$-CH_2OH \xrightarrow[\substack{pyridine \\ \Delta}]{SOCl_2}$

7.37 Predict the products corresponding to each of the capital letters. Show stereochemistry by appropriate conventions. When a racemic mixture forms, show the structure for one enantiomer and write "and enantiomer" below it.

(a) 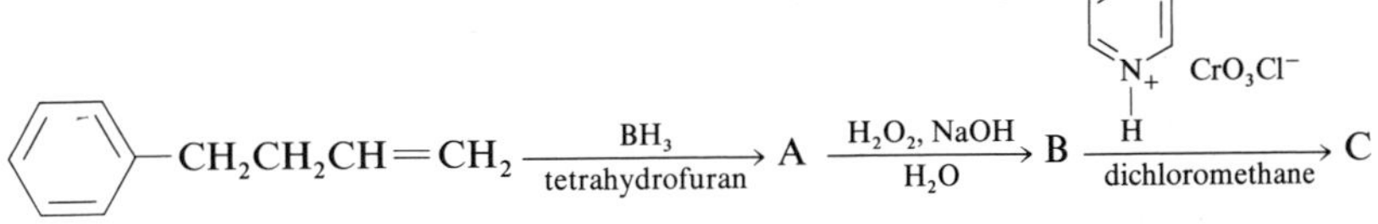 [phenyl]$-CH_2CH_2CH=CH_2 \xrightarrow[\substack{tetrahydrofuran}]{BH_3}$ A $\xrightarrow[\substack{H_2O}]{H_2O_2, NaOH}$ B $\xrightarrow[\substack{dichloromethane}]{[pyridinium\ chlorochromate]\ CrO_3Cl^-}$ C

(b) [methylcyclopentene] $\xrightarrow[\substack{major\ minor}]{9-BBN}$ D + E; D $\xrightarrow[\substack{H_2O}]{H_2O_2, NaOH}$ F; E $\xrightarrow[\substack{H_2O}]{H_2O_2, NaOH}$ G

(c) [cyclohexenyl-phenyl] $\xrightarrow{9-BBN}$ H $\xrightarrow[\substack{H_2O}]{H_2O_2, NaOH}$ I $\xrightarrow[\substack{H_2O}]{CrO_3, H_2SO_4}$ K

$\downarrow \substack{H_2O \\ H_2SO_4}$ (to L) $\downarrow \substack{PBr_3 \\ pyridine}$ (to J)

L J

(d) [cyclopentyl]$-CH_2CH_2Br \xrightarrow[\substack{NH_3(l)}]{HC \equiv \bar{C}: Na^+}$ M $\xrightarrow[\substack{HgSO_4}]{H_2SO_4, H_2O}$ N

$\downarrow \substack{KOH\ (4\ M) \\ ethanol \\ \Delta}$

O $\xrightarrow[\substack{carbon \\ tetrachloride}]{Br_2}$ P; O $\xrightarrow[\substack{no \\ peroxides}]{HBr}$ Q $\xrightarrow[\substack{ethanol}]{NaCN}$ R

(e) $HOCH_2(CH_2)_4CH_2OH \xrightarrow[\substack{dichloromethane}]{[pyridinium\ chlorochromate]\ CrO_3Cl^-}$ S

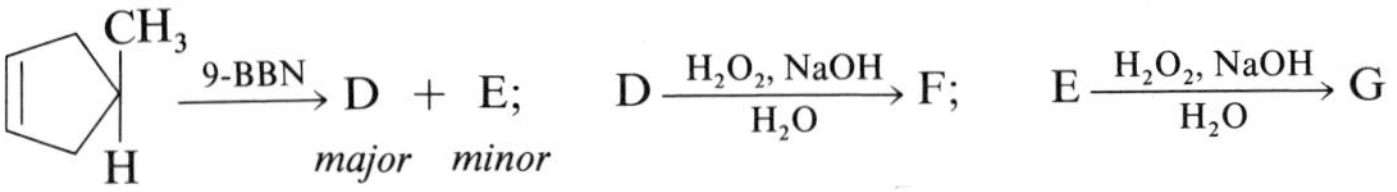

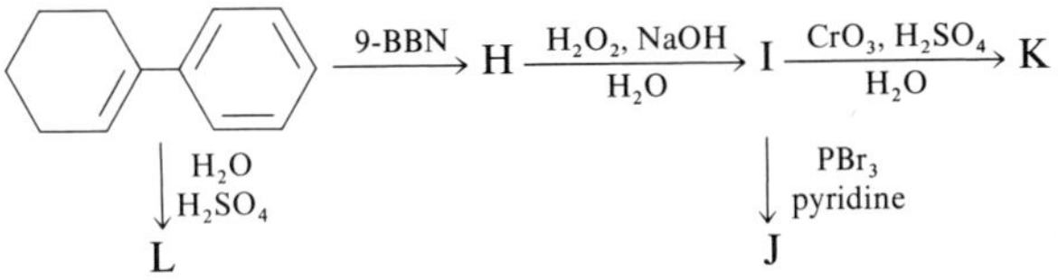

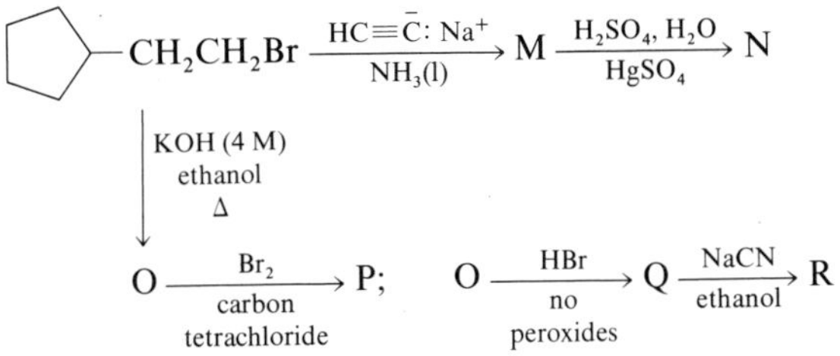

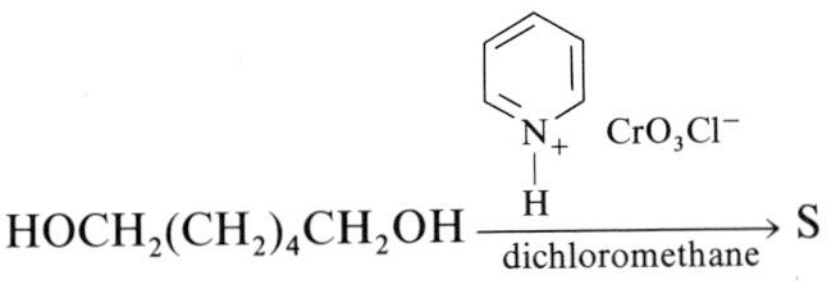

(f)

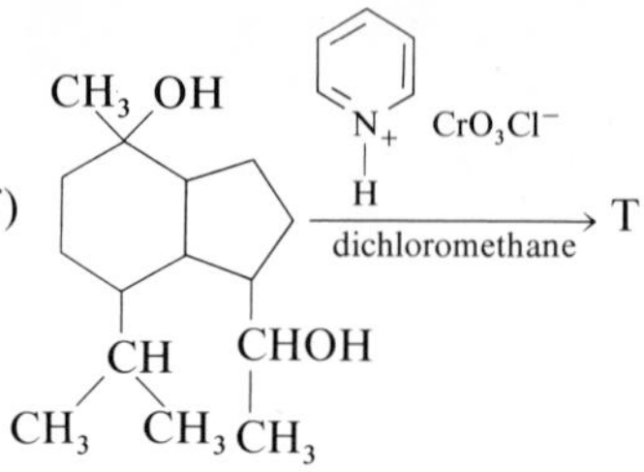

$$\xrightarrow[\text{dichloromethane}]{} \text{T}$$

(g)

$$\xrightarrow[\text{tetrahydrofuran}]{\text{Hg(OCCH}_3)_2, \text{H}_2\text{O}} \text{U} \xrightarrow[\substack{\text{NaOH} \\ \text{H}_2\text{O}}]{\text{NaBH}_4} \text{V}$$

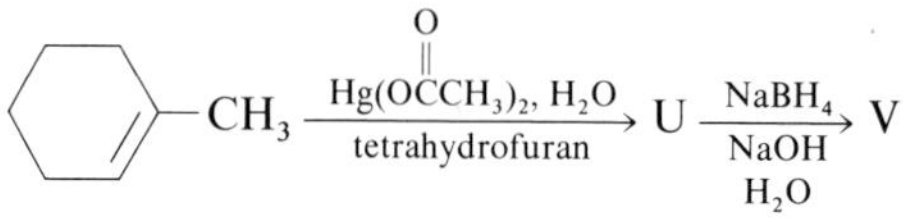

7.38 Supply the reagents that are necessary and the intermediate compounds that will form in the following transformations. There may be more than one good way to carry out each synthesis.

(a) $\underset{\underset{\text{CH}_3}{|}}{\text{CH}_3\text{CHCH}_2\text{CH}=\text{CH}_2} \longrightarrow \underset{\underset{\text{CH}_3}{|}}{\text{CH}_3\text{CHCH}_2}\overset{\overset{\text{O}}{\|}}{\text{C}}\text{CH}_3$

(b) $\text{CH}_3(\text{CH}_2)_4\text{CH}=\text{CH}_2 \longrightarrow \text{CH}_3(\text{CH}_2)_4\text{CH}_2\overset{\overset{\text{O}}{\|}}{\text{C}}\text{H}$

(c)

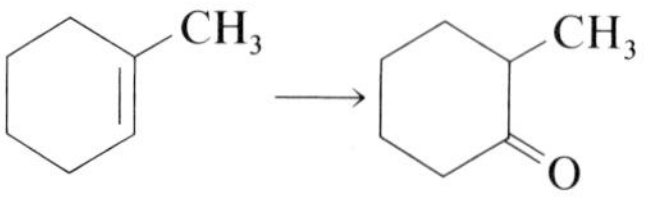

(d)

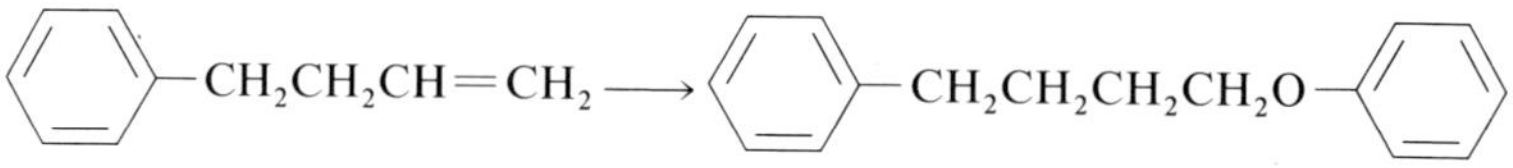

(e) $\underset{\underset{\text{CH}_3}{|}}{\text{CH}_3\text{CHCH}_2\text{CH}=\text{CH}_2} \longrightarrow (\underset{\underset{\text{CH}_3}{|}}{\text{CH}_3\text{CHCH}_2\text{CH}_2\text{CH}_2})_2\text{O}$

(f) $\text{CH}_3\text{CH}_2\text{CH}=\text{CH}_2 \longrightarrow \underset{\underset{\text{NH}_2}{|}}{\text{CH}_3\text{CH}_2\text{CHCH}_3}$

(g)

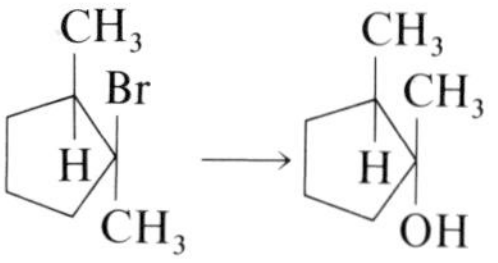

(h)

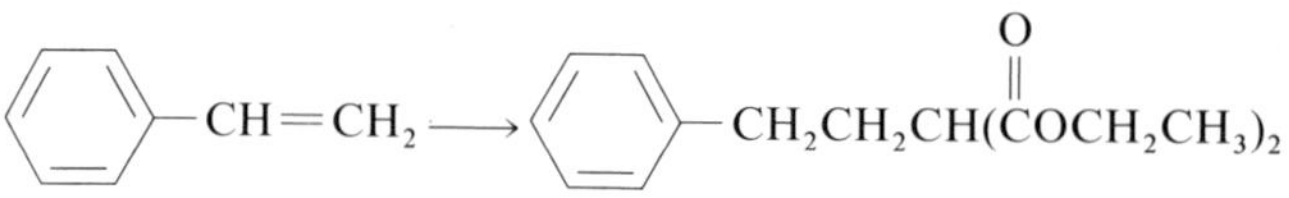

(i) $\text{CH}_3(\text{CH}_2)_3\text{CH}=\text{CH}_2 \longrightarrow \text{CH}_3(\text{CH}_2)_4\text{CH}_2\text{OCH}_3$

(j)

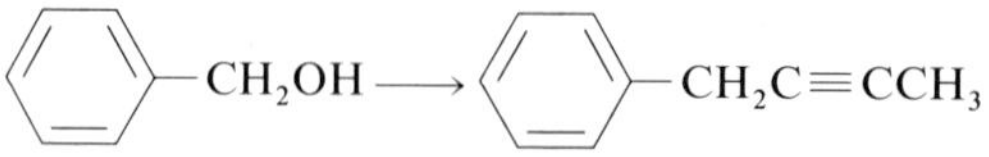

(k)

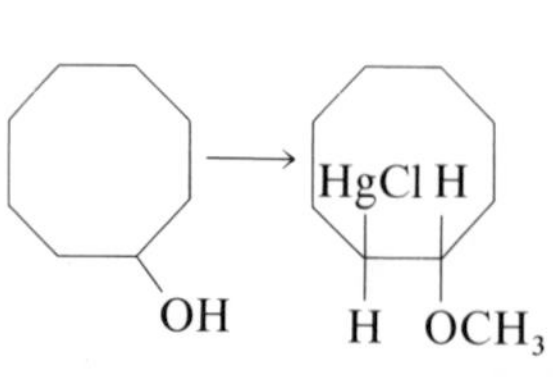

7.39 The following sequence of reactions is a good way to convert a 1-alkylcyclohexene cleanly to a 3-alkylcyclohexene. Draw full stereochemical structural formulas for each step of the reaction sequence and show why it works to give the desired product.

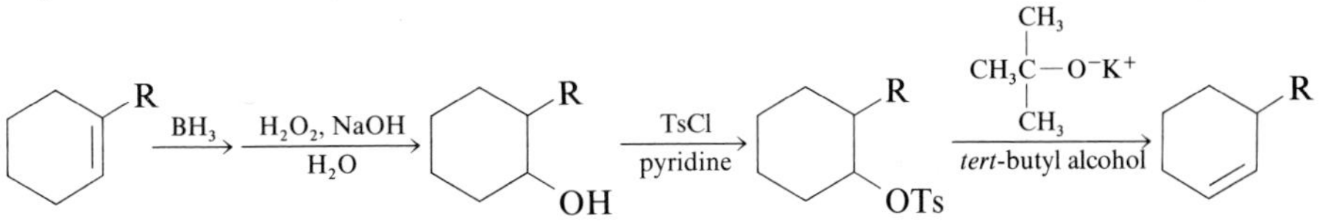

7.40 An important component of the visual pigment in the eye is retinal, the aldehyde structurally related to Vitamin A. Retinal, as it exists in the retina before absorption of light, has the *Z* configuration at the double bond at carbon 11. Absorption of light converts it to the *E* configuration at that double bond.

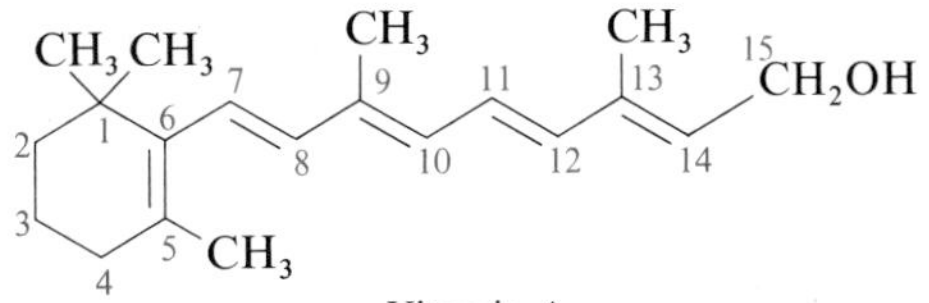

Vitamin A

(a) Draw the structure of (11*E*)-retinal.
(b) Draw the structure of (11*Z*)-retinal.
(c) What reagent would you use to oxidize Vitamin A to retinal?

7.41 Cholesterol gives primarily one product when it is treated first with diborane and then with alkaline peroxide. Predict the structure of the product if you assume that the side of the molecule away from the angular methyl groups (carbons 18 and 19), is less hindered.

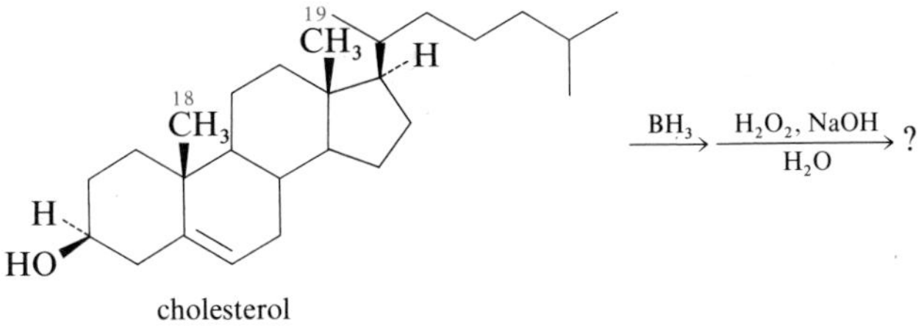

cholesterol

7.42 The following transformation has been observed. Assign a structure to Compound A and suggest a mechanism for the conversion of A to the final product.

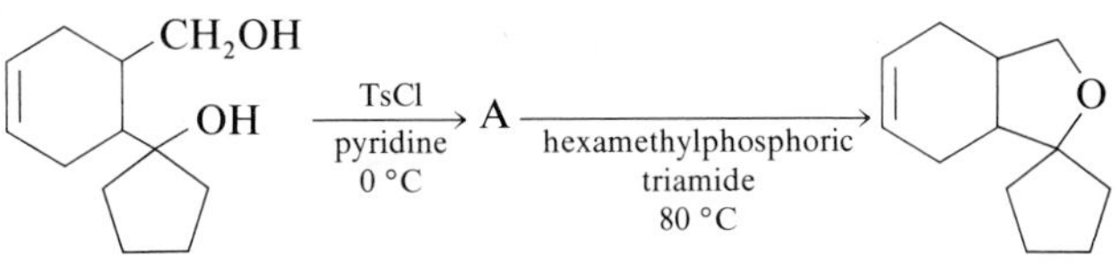

84%

7.43 The following reaction was observed.

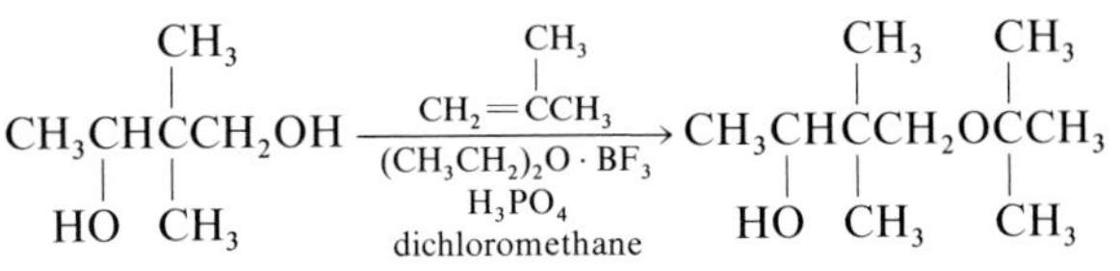

Propose a mechanism for the reaction.

7.44 Predict what the product of the reaction of cyclohexene with mercury(II) acetate in acetic acid will be. Include stereochemistry in your answer.

7.45 Dextrorotatory (+)-2-propyl-1,4-butanediol having the following structure was treated with excess PBr$_3$ in pyridine. The product that was formed was levorotatory.

$$
\begin{array}{c}
\text{CH}_2\text{CH}_2\text{CH}_3 \\
| \\
\text{C} \cdots \text{H} \\
\text{HOCH}_2\text{CH}_2 \diagup \ \ \diagdown \\
\text{CH}_2\text{OH}
\end{array}
\xrightarrow[\text{pyridine}]{\text{PBr}_3 \text{ (excess)}}
$$

(+)-2-propyl-1,4-butanediol

(−)-1-bromo-3-(bromomethyl)-
hexane

(a) Assign configuration to (+)-2-propyl-1,4-butanediol.
(b) Draw the structure for (−)-1-bromo-3-(bromomethyl)hexane showing stereochemistry. Assign configuration to the bromo compound.

7.46 Compound K, C$_5$H$_{12}$O, has the infrared spectrum shown in Figure 7.12. When it is dissolved in acetone, and a solution of chromium trioxide in sulfuric acid and water is added, the solution remains orange. On the other hand, Compound L, also C$_5$H$_{12}$O, turns the solution of chromium trioxide blue-green under the same conditions. Its infrared spectrum is also shown in Figure 7.12. Assign structures to Compounds K and L that are compatible with all the information given.

Compound K, C$_5$H$_{12}$O

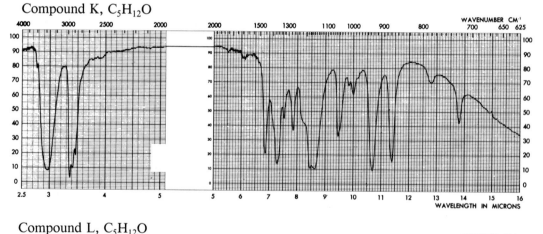

Compound L, C$_5$H$_{12}$O

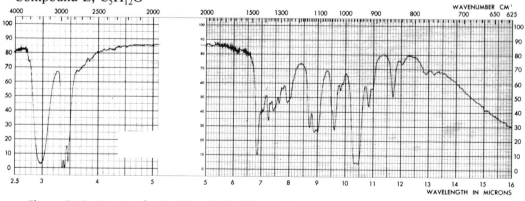

Figure 7.12 Spectra for Problem 7.46. (From *The Aldrich Library of Infrared Spectra*)

7.47   3β-Hydroxyandrost-5-en-17-one is a steroid related to the male sex hormones. Its infrared spectrum and structural formula appear in Figure 7.13. Assign as many of the absorption bands as you can to specific vibrational transitions in the molecule.

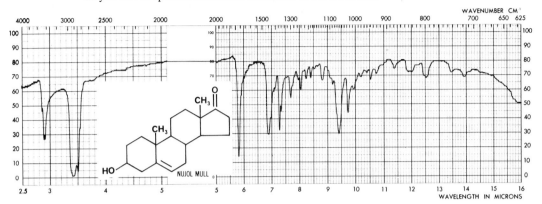

Figure 7.13   Spectrum for Problem 7.47. (From *The Aldrich Library of Infrared Spectra*)

7.48   In Figure 7.14, a series of infrared spectra are given for unknown compounds. They may be alcohols, aldehydes, ketones, or ethers. Assign each one to a functional group class.

Compound M

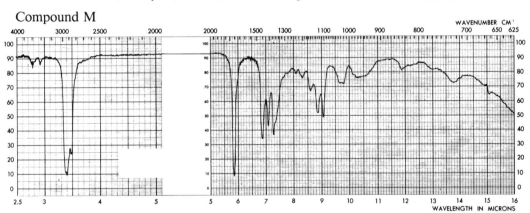

Compound N

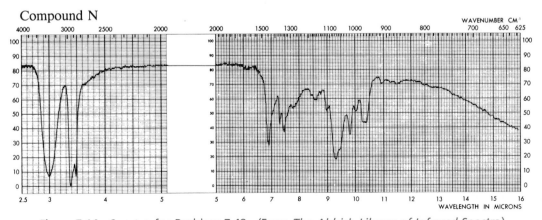

Figure 7.14   Spectra for Problem 7.48. (From *The Aldrich Library of Infrared Spectra*)

Compound O

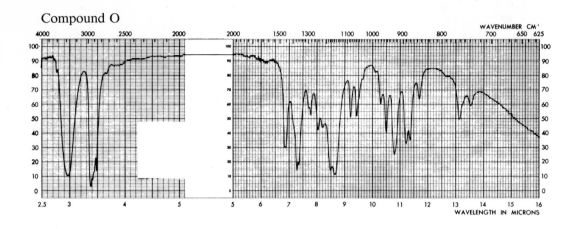

Compound P

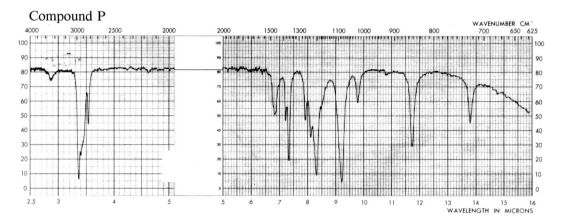

Figure 7.14   Spectra for Problem 7.48 (*Continued*)

Hydrogenation and Oxidation Reactions of Alkenes and Alkynes

8

8.1

Introduction

In Chapter 5, we explored the reactions of alkenes and alkynes with some electrophilic reagents that are attacked by the π electrons of the multiple bond. Reagents such as hydrogen halides and halogens convert alkenes to alkyl halides, and alkynes to vinyl halides and then to alkyl halides. Water, in the presence of acid catalysts, adds to alkenes to produce alcohols, and to alkynes to give carbonyl compounds.

Alkenes and alkynes also react with oxidizing agents to give alcohols and carbonyl compounds. Alkenes are converted by reducing agents, such as hydrogen, to alkanes (Sections 2.10, 5.2), while alkynes undergo reduction in two stages, first to alkenes, then to alkanes. In this chapter, as we investigate the oxidation and reduction reactions of alkenes and alkynes, we will be concerned with questions of the selectivity and the stereochemistry of the reactions in cases where such distinctions can be made.

8.2

Catalytic Hydrogenation of Alkenes

A. Heterogeneous Catalysis

Hydrogen adds to a multiple bond in the presence of metallic catalysts in a **hydrogenation reaction.** A hydrogenation reaction is usually carried out in some inert solvent. The alkene or alkyne is stirred in the presence of the solid catalyst while hydrogen gas is introduced into the reaction mixture. The amount of hydrogen used up in the reaction can be monitored by watching the drop in pressure or the decrease in the volume of the gas in the system. *Hydrogen gas does not add to multiple bonds unless a specially prepared metal surface is present.*

The presence of the catalyst as a separate solid phase makes the reaction mixture a **heterogeneous** *one, in contrast to reactions in which all reagents are present in a single* **homogeneous** *phase.* Metals that are commonly used as **hydrogenation catalysts** are palladium, platinum, and nickel. Palladium and platinum catalysts are usually prepared by the reduction of a salt of the metal by hydrogen, very often in the presence of a larger amount of an inert material that serves as a diluent and support for the catalyst. An example is palladium chloride on carbon, which when reduced by hydrogen gives a solid that consists of palladium metal in a finely divided state, supported on powdered carbon. Reduction of platinum oxide gives a platinum catalyst. A form of nickel called Raney nickel is made by using sodium hydroxide to dissolve out the aluminum in a nickel-aluminum alloy. The reaction of aluminum with sodium hydroxide gives hydrogen gas, which is adsorbed on finely divided nickel from the alloy.

Exactly how a catalyst functions is a matter of debate, and varies with the particular catalyst, the relative amounts of alkene and hydrogen gas used, and the temperature. The important factor in catalysis is the nature and extent of the available metal surface. The alkene or alkyne is believed to be adsorbed onto the surface of the catalyst, forming bonds to the metal atoms. The π bond is broken at this stage. This reaction may be viewed as a Lewis acid-base interaction between the metal atoms, which have room in their orbitals for electrons, and the π electrons of the multiple bond (Figure 8.1).

The carbon atoms that are bonded to the surface of the metal resemble radicals in character, and in this state they react with hydrogen, forming a bond to a hydrogen atom at each of the carbon atoms. In some cases, the hydrogen molecule may also be

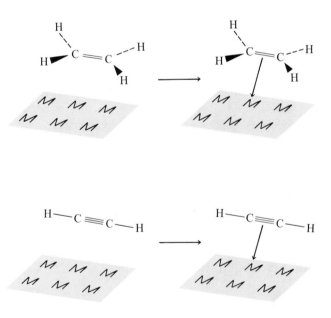

Figure 8.1 A schematic representation of the interaction between an alkene or alkyne and the metal atoms at the surface of a metallic catalyst.

adsorbed to the surface of the metal near the organic molecule with some loosening of the hydrogen-hydrogen bond.

The addition of hydrogen to an alkene to give an alkane is an exothermic process (Section 5.2). In the absence of a catalyst, however, the reaction has a high energy of activation and takes place at a negligible rate. The catalyst changes the nature of the transition state for the reaction and thereby lowers the energy of activation. The reaction can then proceed at a reasonable rate under practical conditions of temperature and pressure. Figure 8.2 is an energy diagram showing how the catalyzed reaction compares with the uncatalyzed reaction.

Because the uncatalyzed reaction has a very high energy of activation, no reaction takes place unless a catalyst is used. By showing the formation of an intermediate on the pathway from reagents to products, Figure 8.2 indicates that the mechanism of the catalyzed reaction is different from that of the uncatalyzed reaction. The interaction between the alkene and the surface of the catalyst, depicted in Figure 8.1, changes the nature of the bonding in the alkene and creates a species with greater reactivity towards hydrogen. The change in the energy of activation that accompanies catalysis occurs because the nature of the transition state is changed by the presence of the catalyst. You should notice that the catalyst also lowers the energy of activation for the reverse reaction, the dehydrogenation of an alkane to an alkene. Such reactions are known and do take place in the presence of hydrogenation catalysts if no external hydrogen gas is added to the reaction mixture.

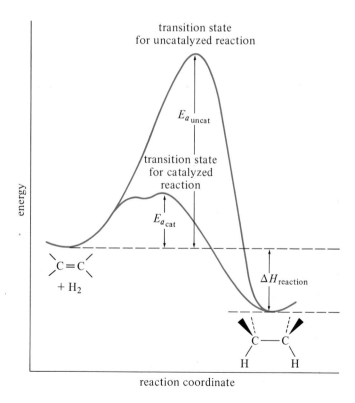

Figure 8.2   Energy diagram comparing the energy of activation for a catalyzed hydrogenation reaction with that of an uncatalyzed reaction.

B. Some Hydrogenation Reactions of Alkenes

Alkenes are converted to alkanes by catalytic hydrogenation. Some typical examples are the conversion of 2-methyl-2-pentene to 2-methylpentane and 3-methylcyclopentene to methylcyclopentane.

$$\underset{\substack{\text{2-methyl-2-pentene}}}{\overset{\overset{\displaystyle CH_3}{|}}{CH_3C}=CHCH_2CH_3} \xrightarrow[\substack{25\,°C}]{\substack{H_2 \\ Pt}} \underset{\text{2-methylpentane}}{\overset{\overset{\displaystyle CH_3}{|}}{CH_3CHCH_2CH_2CH_3}}$$

3-methylcyclopentene $\xrightarrow[\substack{25\,°C}]{\substack{H_2 \\ Pt}}$ methylcyclopentane

Hydrogenation reactions are generally efficient reactions giving the products in high yield and are important in industrial processes. For example, hydrogenation of the mixture of trimethylpentenes obtained from the acid-catalyzed dimerization of 2-methylpropene (Section 5.6C) gives 2,2,4-trimethylpentane, known commercially as isooctane and used as a component of high octane gasoline.

$$\underset{\substack{\text{2,4,4-trimethyl-1-pentene}}}{\overset{\overset{\displaystyle CH_3\;CH_3}{|\quad\;\;|}}{\underset{\underset{\displaystyle CH_3}{|}}{CH_3CCH_2C}=CH_2}} + \underset{\substack{\text{2,4,4-trimethyl-2-pentene}}}{\overset{\overset{\displaystyle CH_3\;\;CH_3}{|\quad\;\;\;|}}{\underset{\underset{\displaystyle CH_3}{|}}{CH_3CCH}=CCH_3}} \xrightarrow[\substack{\text{acetic acid}}]{\substack{H_2 \\ Pt}} \underset{\substack{\text{2,2,4-trimethylpentane} \\ \text{isooctane}}}{\overset{\overset{\displaystyle CH_3\;CH_3}{|\quad\;\;|}}{\underset{\underset{\displaystyle CH_3}{|}}{CH_3CCH_2CHCH_3}}}$$

Another important industrial application of hydrogenation is the conversion of vegetable oils, which are mixtures of esters of unsaturated acids, into solid fats, which are esters of saturated acids. Oleic acid, the acid found in olive oil, is (Z)-9-octadecenoic acid. Hydrogenation of oleic acid converts it to stearic acid, also called octadecanoic acid, which is found in butter and beef fat.

$$CH_3(CH_2)_7 \diagdown \qquad \diagup (CH_2)_7\overset{\overset{\displaystyle O}{\|}}{C}OH$$
$$C=C$$
$$H \diagup \qquad \diagdown H$$

$$\xrightarrow[Ni]{H_2} CH_3(CH_2)_{16}\overset{\overset{\displaystyle O}{\|}}{C}OH$$

(Z)-9-octadecenoic acid
oleic acid
mp 14 °C

liquid at room temperature,
a component of vegetable oil

octadecanoic acid
stearic acid
mp 69 °C

solid at room temperature,
a component of fats

Oils and fats are similar in structure except for the presence of double bonds in the acid components of oils, which lowers their melting points and makes them liquids at room temperature. Because fats are more stable towards oxidation by air and are more convenient to handle and store, it is practical to use hydrogenation to convert oils to fats such as margarine or vegetable shortening. The chemistry of fats and oils is further explored in Sections 10.7C and 11.7A.

PROBLEM 8.1   Complete the following equations.

(a) $\underset{\displaystyle CH_3}{\overset{\displaystyle CH_3}{CH_3\overset{|}{\underset{|}{C}}CH=CH_2}} \xrightarrow[\text{Ni}]{H_2} A$ (b) $CH_3\overset{\displaystyle CH_3}{\overset{|}{C}}=\overset{\displaystyle CH_3}{\overset{|}{C}}CH_3 \xrightarrow[\text{Ni}]{H_2} B$

(c) $CH_3CH_2\overset{\displaystyle CH_3}{\underset{\displaystyle OH}{\overset{|}{\underset{|}{C}}}}CH_2CH_3 \xrightarrow[\underset{\Delta}{H_2SO_4}]{} C + D \xrightarrow{H_2}{\text{Pt}} E$ (d) $CH_3\overset{\displaystyle CH_3}{\overset{|}{C}}H-\overset{\displaystyle CH_3}{\overset{|}{C}}=CH_2 \xrightarrow[\text{Ni}]{H_2} F$

(e) $CH_3\overset{\displaystyle CH_3}{\underset{\displaystyle OH}{\overset{|}{\underset{|}{C}}}}CH_2CH_2CH_3 \xrightarrow[\underset{\Delta}{H_3PO_4}]{} G + H \xrightarrow{H_2}{\text{Pt}} I$

C. Homogeneous Catalysis. Organometallic Compounds as Hydrogenation Catalysts

Some hydrogenation catalysts contain transition metal atoms bonded to organic groups. Such *compounds containing bonds between metal atoms and carbon atoms are known as* **organometallic compounds.** The transition metals, appearing in the central portion of the periodic table, with up to 12 electrons in their *s, p,* and *d* orbitals, have room for a total of 18 electrons in their valence shells. Consequently, transition metals function as Lewis acids and form covalent bonds with Lewis bases, such as anions or organic compounds with nonbonding electrons to donate. The Lewis bases that bind to the central metal atom are called **ligands.** The presence of organic ligands on the metal makes the organometallic compound soluble in organic solvents.

Some soluble transition metal compounds containing organic ligands function as hydrogenation catalysts. Because the organometallic compound is soluble in the solvent in which the reaction is taking place, the process is known as **homogeneous catalysis** (p. 318). Catalysis occurs not at the surface of a metal layer, but throughout the solution. The catalytic properties of the transition metal compounds that serve as catalysts for hydrogenation reactions are derived from the ease with which these metals accept and release electrons.

An organometallic compound that functions as a catalyst for hydrogenation reactions is chlorotris(triphenylphosphine)rhodium(I), $RhCl[P(C_6H_5)_3]_3$. A look at the periodic table tells us that rhodium has nine electrons in its valence shell, and as rhodium(I) (rhodium in the oxidation state of +1), retains eight of them. In the organometallic compound, the metal has accepted four pairs (a total of eight) electrons. One pair comes from a chloride ion and three pairs from three triphenylphosphine molecules. The rhodium metal thus has 16 electrons around it in this organometallic compound.

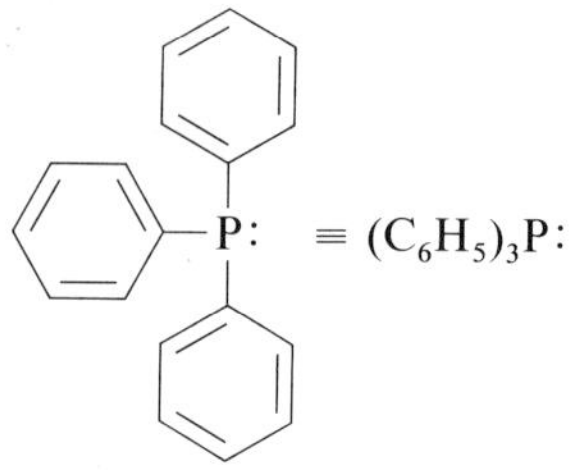

triphenylphosphine

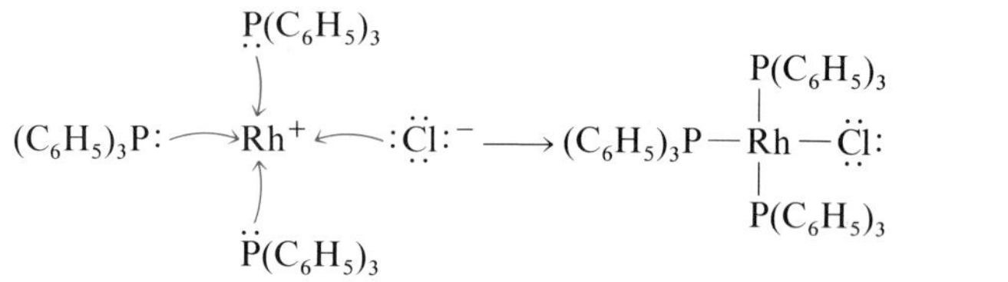

rhodium ion with 8 electrons in its valence shell, gaining 8 electrons by coordination with 4 ligands, which each donate 2 electrons

chlorotris(triphenylphosphine)rhodium(I)

rhodium now has 16 electrons around it

Chlorotris(triphenylphosphine)rhodium(I) dissolves in benzene in the presence of hydrogen gas to give a new organometallic compound that contains metal-hydrogen bonds. The number of electrons around the central metal atom is now 18, the maximum number that its orbitals can hold. The geometry of the new complex is octahedral (Figure 8.3).

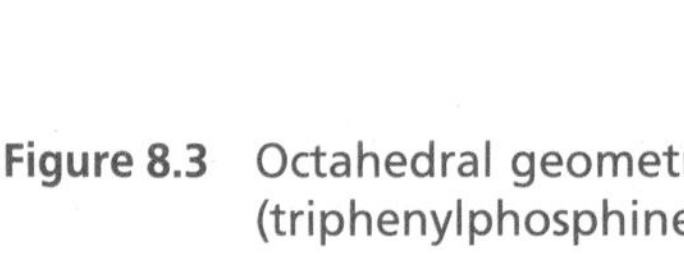

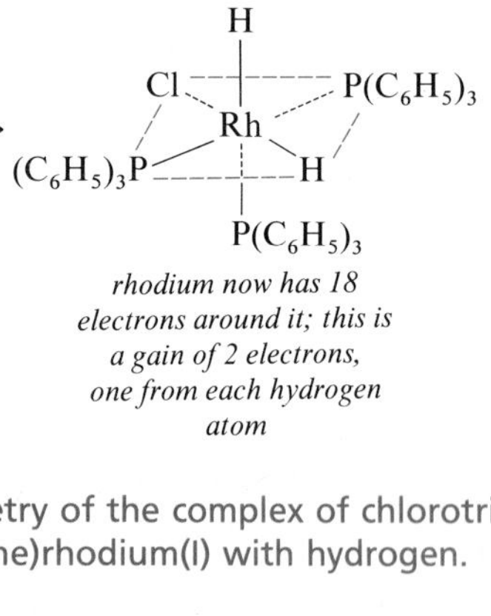

rhodium now has 18 electrons around it; this is a gain of 2 electrons, one from each hydrogen atom

Figure 8.3 Octahedral geometry of the complex of chlorotris-(triphenylphosphine)rhodium(I) with hydrogen.

When an alkene is added to the solution, it donates its π electrons to the rhodium, displacing one of the triphenylphosphine ligands. A reorganization of bonding occurs with the hydrogen atoms being transferred from the metal to the carbon atoms of the double bond. For the sake of clarity, we show the process on the next page in two steps, but it probably occurs in a single step (Figure 8.4).

The rhodium metal undergoes successive changes in the number of electrons in its valence shell. As the alkene is converted to an alkane in a reduction reaction, a corresponding loss of ligands with their bonding electrons occurs at rhodium. The organometallic rhodium compound serves as a catalyst because it picks up triphenyl-phosphine and hydrogen from the solution to regenerate the 18-electron species that reacts with the alkene to start the cycle again.

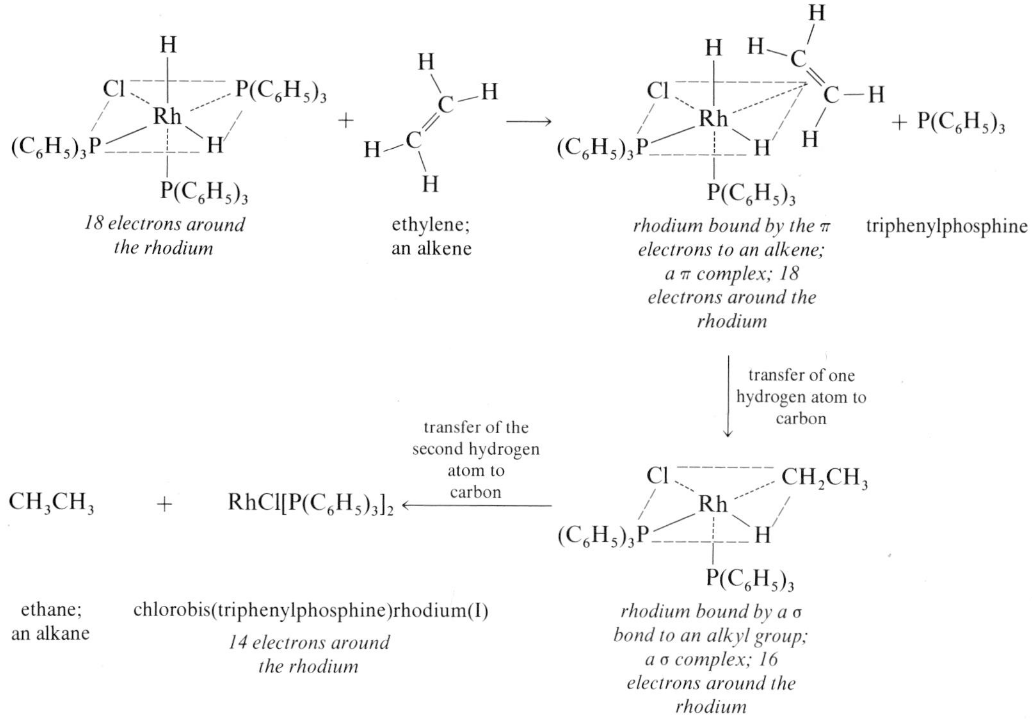

Figure 8.4   A hydrogenation reaction catalyzed by chlorotris(triphenylphosphine)rhodium(I).

Chlorotris(triphenylphosphine)rhodium(I) catalyzes the hydrogenation of alkenes and alkynes to alkanes and usually does not affect other functional groups.

A typical hydrogenation reaction with this catalyst is the conversion of vinylcyclopropane to ethylcyclopropane.

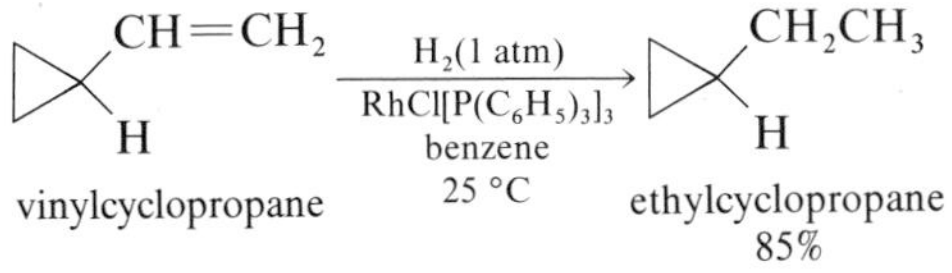

The rhodium catalyst differs from the heterogeneous catalysts discussed in Sections 8.2A and 8.2B by being more selective in its action. Terminal double bonds are hydrogenated much more rapidly by the rhodium catalyst than internal double bonds. In general, the higher the substitution on the double bond, the slower the hydrogenation reaction when this homogeneous catalyst is used. Thus it is possible to carry out selective reductions of one double bond even in the presence of another one in the same molecule. The selectivity of this catalyst is illustrated by the reduction of 3,7-dimethyl-1,6-octadien-3-ol to 3,7-dimethyl-6-octen-3-ol in high yield.

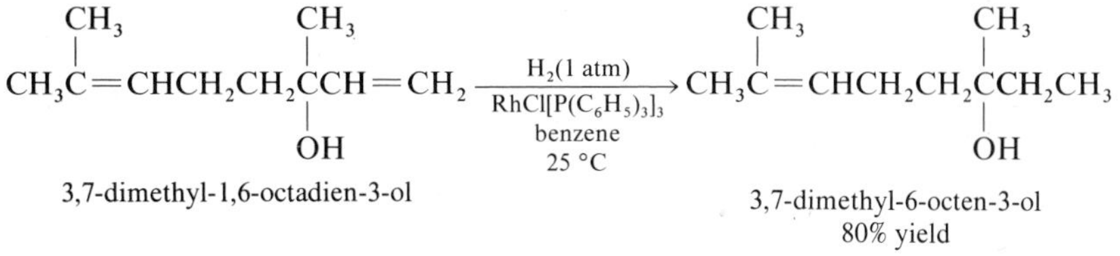

The rhodium compound catalyzes the hydrogenation of the terminal double bond, leaving the internal double bond untouched.

PROBLEM 8.2   Complete the following equations.

(a) $CH_3CH_2CH{=}CHCH_2CH_3 \xrightarrow{\text{H}_2}{\text{Pt}}$

(b) $CH_3CH_2CH_2\overset{\displaystyle CH_3}{\overset{|}{C}}{=}CH_2 \xrightarrow[\substack{\text{RhCl[P(C}_6\text{H}_5)_3]_3 \\ \text{benzene}}]{\text{H}_2}$

(c) 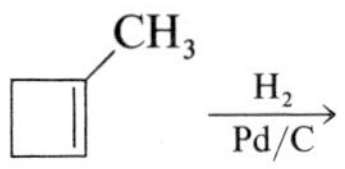
$\xrightarrow[\text{Pd/C}]{\text{H}_2}$

(d) 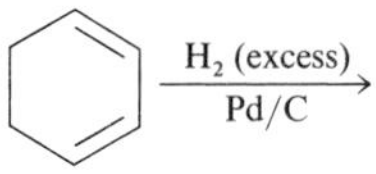
$\xrightarrow[\text{Pd/C}]{\text{H}_2 \text{ (excess)}}$

(e) $CH_3\overset{\displaystyle CH_3}{\underset{\displaystyle CH_3}{\overset{|}{\underset{|}{C}}}}CH{=}CH_2 \xrightarrow[\substack{\text{RhCl[P(C}_6\text{H}_5)_3]_3 \\ \text{benzene}}]{\text{H}_2}$

(f) 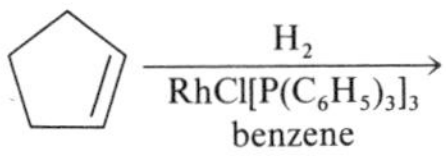
$\xrightarrow[\substack{\text{RhCl[P(C}_6\text{H}_5)_3]_3 \\ \text{benzene}}]{\text{H}_2}$

(g) $CH_3CH_2CH_2CH_2CH_2CH{=}CH_2 \xrightarrow[\substack{\text{RhCl[P(C}_6\text{H}_5)_3]_3 \\ \text{benzene}}]{\text{H}_2}$

8.3

Reduction of Alkynes

A. Catalytic Hydrogenation of Alkynes

Adding one mole of hydrogen to an alkyne is a useful way to synthesize an alkene. The catalysts that are used in such hydrogenation reactions are specially prepared to be active towards alkynes but not alkenes, so the reaction stops when one mole of hydrogen has reacted with the triple bond. One such catalyst, known as a **poisoned catalyst,** is palladium on barium sulfate or calcium carbonate to which a small amount of lead or an organic base, quinoline, has been added to make the catalyst less reactive. For example, phenylethyne can be converted to phenylethene (styrene) by the use of such a catalyst.

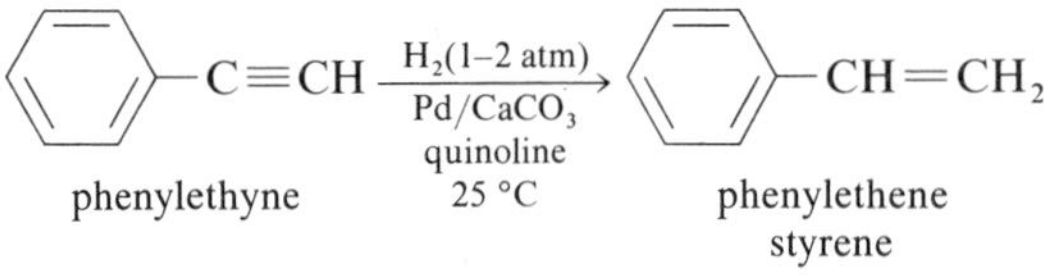

phenylethyne   25 °C   phenylethene
styrene

quinoline $\equiv$

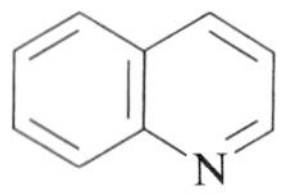

The reaction stops when one mole of hydrogen has been absorbed. The alkene is not further reduced to an alkane on this catalyst even though this is a reaction that such a double bond normally undergoes (p. 320, for example). Note that the aromatic ring is untouched under these conditions. In fact, aromatic rings are not usually hydrogenated under the conditions that add hydrogen to double and triple bonds.

Various experiments with different metallic catalysts have shown that internal alkynes give predominantly cis alkenes on catalytic hydrogenation. For example, the symmetrical alkyne, dimethyl 5-decynedioate, which has an internal triple bond and two ester functions, is reduced to the corresponding cis diester.

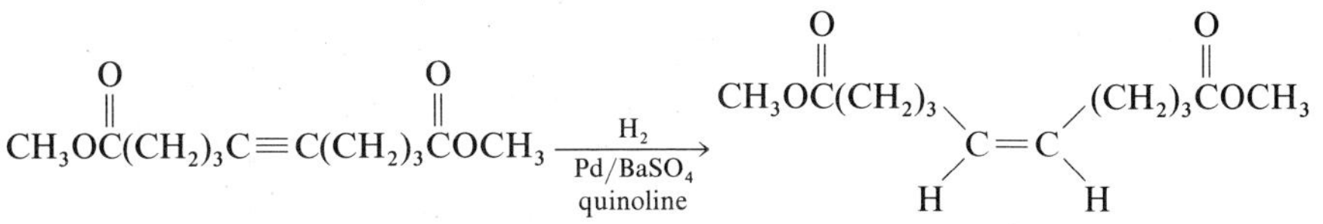

dimethyl 5-decynedioate   dimethyl (Z)-5-decenedioate
                                                        97%

The reaction stops cleanly when exactly one mole of hydrogen has been absorbed. The stereochemistry of catalytic reduction of alkynes shows that hydrogen atoms are added at the surface of the catalyst to the same side of the alkyne bond.

B. Stereoselective Reduction of Alkynes by Hydroboration

Alkynes add diborane just as alkenes do (Section 7.3) and the vinylboranes that result react with carboxylic acids to give cis alkenes. 3-Hexyne, for example, forms a vinylborane, which reacts with acetic acid to give (Z)-3-hexene with high stereochemical purity.

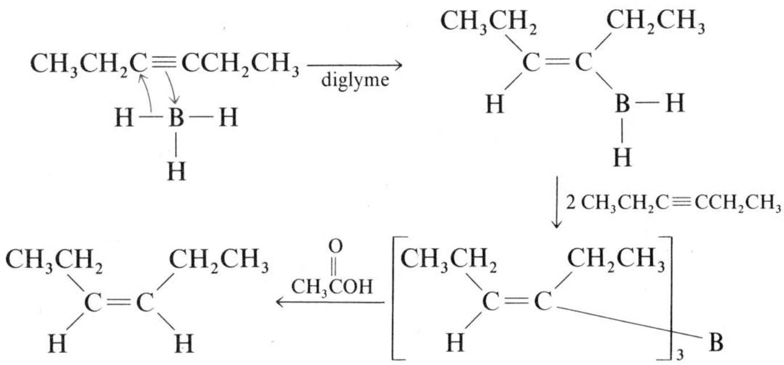

(Z)-3-hexene   a vinylborane
    83%
> 98% stereochemical
      purity

The addition of borane to a triple bond occurs with the same stereoselectivity as seen for the addition of borane to alkenes. The boron atom and the hydrogen atom that is transferred to the other carbon atom of the triple bond are on the same side of the molecule. The proton from the carboxylic acid replaces the boron atom with retention of configuration.

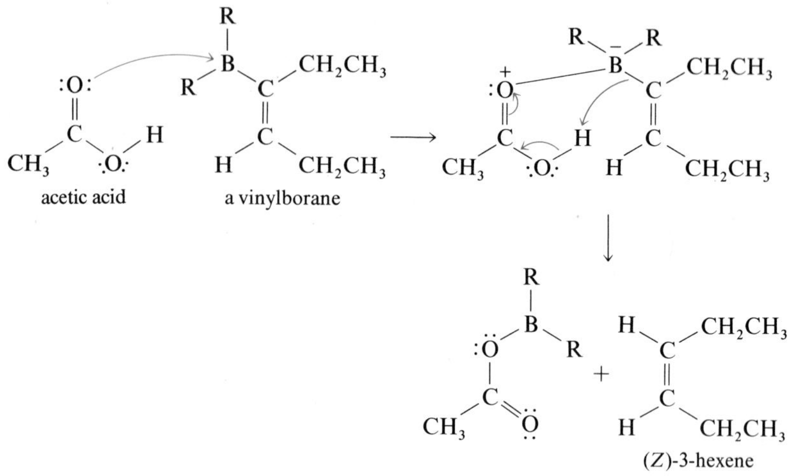

acetic acid   a vinylborane

(Z)-3-hexene

The stereoselectivity of the catalytic hydrogenation of internal alkynes (Section 8.3A) depends on the care with which the catalyst is prepared and seldom exceeds 95%. Acid cleavage of vinyl boranes is a better method for preparing cis alkenes of high stereochemical purity.

PROBLEM 8.3   What would be the structure of the alkene formed if the vinylborane from 3-hexyne were treated with deuterioacetic acid, CH_3CO_2D?

PROBLEM 8.4   The carbon-boron bond in alkylboranes is also cleaved by carboxylic acids in a way that is exactly analogous to the reaction shown above. Draw a mechanism for the reaction using BR_3 to represent a trialkylborane.

PROBLEM 8.5   Write equations showing what would happen if 1-methylcyclopentene were treated with diborane and then with deuterioacetic acid, CH_3CO_2D.

C. Reduction of Alkynes by Dissolving Metals

Alkynes, but not simple internal alkenes, are also reduced by sodium or lithium metal in liquid ammonia. In contrast to catalytic hydrogenation, this reaction gives rise to trans alkenes. For example, 3-octyne is reduced to (E)-3-octene with this reagent.

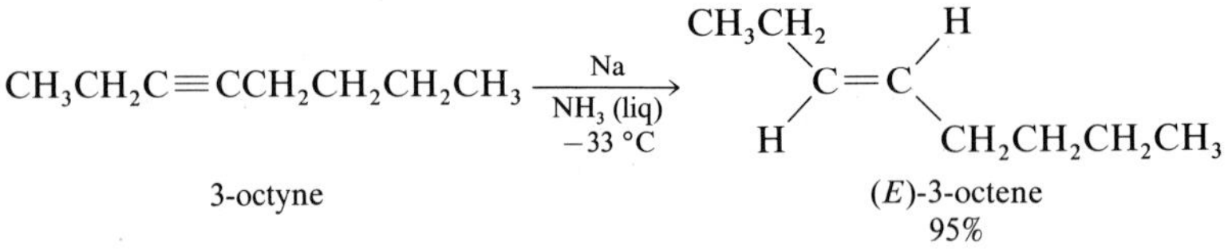

$$CH_3CH_2C\equiv CCH_2CH_2CH_2CH_3 \xrightarrow[\substack{NH_3\ (liq) \\ -33\ °C}]{Na}$$

3-octyne

(E)-3-octene
95%

This reaction is postulated to occur by a transfer of an electron from sodium metal to the alkyne. The intermediate formed is a **radical anion,** a species that bears a negative charge and has an unpaired electron. Such a strongly basic carbanionic species is protonated by ammonia. The radical that is formed is reduced again by sodium metal, and then protonated to give the most stable alkene (Section 5.2) having the E configuration.

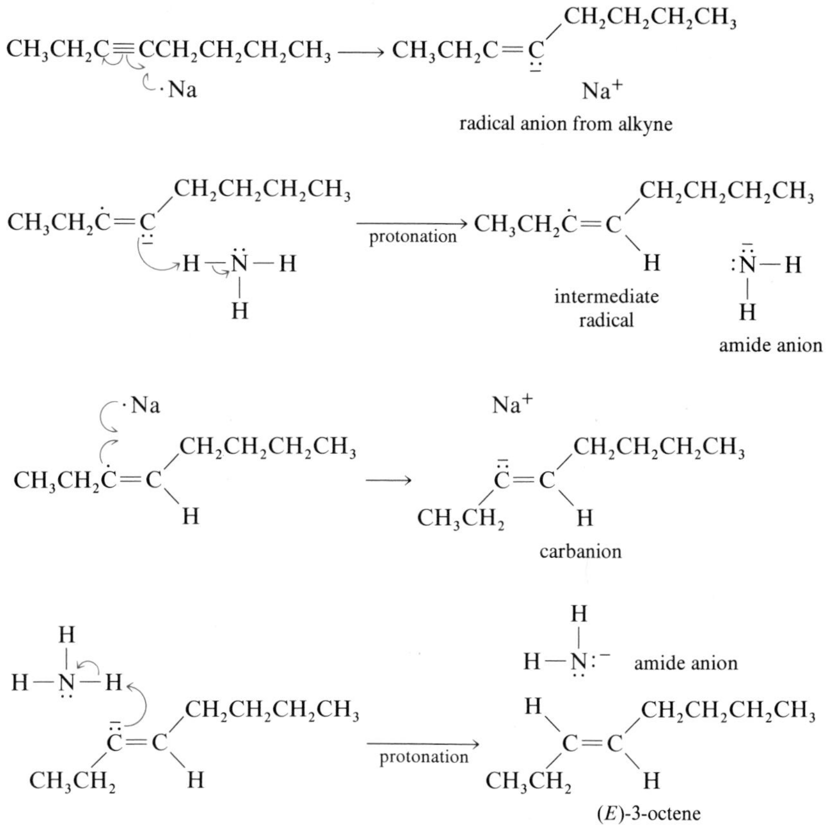

$$CH_3CH_2C \equiv CCH_2CH_2CH_2CH_3 \longrightarrow CH_3CH_2C = C \Big\langle \begin{array}{c} CH_2CH_2CH_2CH_3 \\ \end{array}$$

$$Na^+$$

radical anion from alkyne

$$CH_3CH_2\dot{C} = C \Big\langle \begin{array}{c} CH_2CH_2CH_2CH_3 \\ H - \ddot{N} - H \\ | \\ H \end{array} \xrightarrow[\text{protonation}]{} CH_3CH_2\dot{C} = C \Big\langle \begin{array}{c} CH_2CH_2CH_2CH_3 \\ H \qquad :\ddot{N} - H \\ | \\ H \end{array}$$

intermediate radical

amide anion

$$CH_3CH_2\dot{C} = C \Big\langle \begin{array}{c} \cdot Na \\ CH_2CH_2CH_2CH_3 \\ H \end{array} \longrightarrow \begin{array}{c} Na^+ \\ \ddot{C} = C \Big\langle \begin{array}{c} CH_2CH_2CH_2CH_3 \\ CH_3CH_2 \qquad H \end{array}$$

carbanion

$$H - N - H \begin{array}{c} H \\ | \\ \ddot{C} = C \Big\langle \begin{array}{c} CH_2CH_2CH_2CH_3 \\ CH_3CH_2 \qquad H \end{array} \xrightarrow[\text{protonation}]{} \begin{array}{c} H - N:^- \quad \text{amide anion} \\ | \\ H \\ H \Big\rangle C = C \Big\langle \begin{array}{c} CH_2CH_2CH_2CH_3 \\ CH_3CH_2 \qquad H \end{array}$$

(*E*)-3-octene

In contrast to the previous reaction, when 3-octyne is hydrogenated on a metal surface the product is predominantly (*Z*)-3-octene.

$$CH_3CH_2C \equiv CCH_2CH_2CH_2CH_3 \xrightarrow[\text{Ni}]{H_2 \text{ (1 molar equivalent)}} \begin{array}{c} CH_3CH_2 \\ \Big\rangle C = C \Big\langle \begin{array}{c} CH_2CH_2CH_2CH_3 \\ H \qquad H \end{array}$$

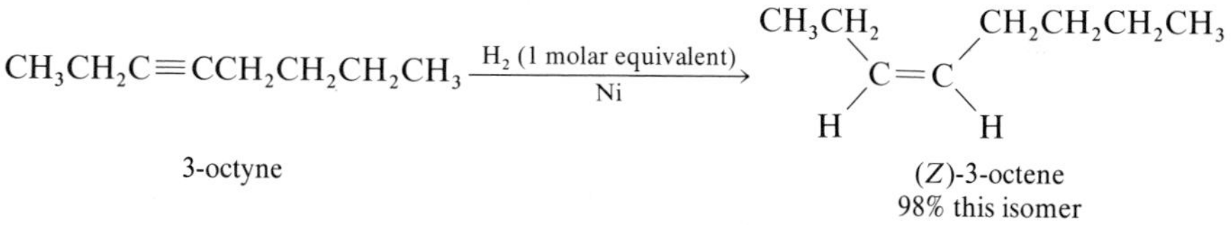

3-octyne

(*Z*)-3-octene
98% this isomer

The equations above demonstrate that alkynes can be converted with stereoselectivity (p. 212) either to cis or trans alkenes depending on the reagents used.

PROBLEM 8.6 Complete the following equations. Be sure to show the stereochemistry that you expect to get in the cases where it is pertinent.

(a) $CH_3CH_2CH_2CH_2C \equiv CH \xrightarrow[\substack{\text{Pd/BaSO}_4 \\ \text{quinoline}}]{H_2}$

(b) $CH_3CH_2CH_2CH_2C \equiv CCH_3 \xrightarrow[\substack{\text{Pd/BaSO}_4 \\ \text{quinoline}}]{H_2}$

(c) $CH_3CH_2CH_2CH_2C \equiv CCH_3 \xrightarrow[\text{NH}_3 \text{ (liq)}]{Na}$

(d)

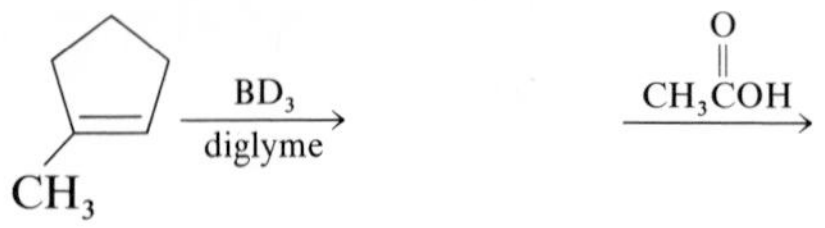

(e) $CH_3CH_2CH_2CH_2C\equiv CCH_3 \xrightarrow[\substack{Pt \\ acetic\ acid}]{H_2\ (excess)}$

(f) $CH_3C\equiv CCH_2CH_2CH_3 \xrightarrow[diglyme]{BH_3}$

PROBLEM 8.7 When 2-butyne in the vapor phase is treated with deuterium gas (D_2), with palladium on an alumina support as the catalyst, one alkene is obtained in 99% yield. Predict the structure of the alkene.

8.4

Oxidation of Alkenes and Alkynes

A. Oxidation of Alkenes with Permanganate and Osmium Tetroxide

An aqueous solution of potassium permanganate oxidizes multiple bonds. In this reaction, the purple permanganate ion is reduced to manganese dioxide, a brown solid. The change in color and appearance of the reaction mixture is the basis of a test for the presence of double and triple bonds known as the **Baeyer test for unsaturation.**

The first stage of the oxidation of an alkene leads to the formation of a compound with hydroxyl groups on the carbon atoms that were part of the double bond, a 1,2-diol. For example, cyclopentene is oxidized to *cis*-1,2-cyclopentanediol by potassium permanganate.

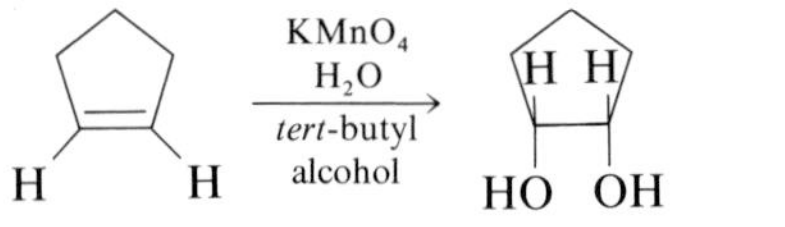

cyclopentene   *cis*-1,2-cyclopentanediol

Manganese(VII) in permanganate ion is ultimately reduced to manganese(IV) in manganese dioxide. The carbon atoms of the double bond are oxidized. Even if no base is added at first, the solution becomes progressively more basic as the reaction proceeds. Reactions that are being used to synthesize diols go best with some added base.

In this oxidation reaction, the two hydroxyl groups become attached to the same side of the double bond. The permanganate ion is believed to add to the double bond to give a cyclic intermediate, a manganate ester. The intermediate breaks down in water and base to give the diol.

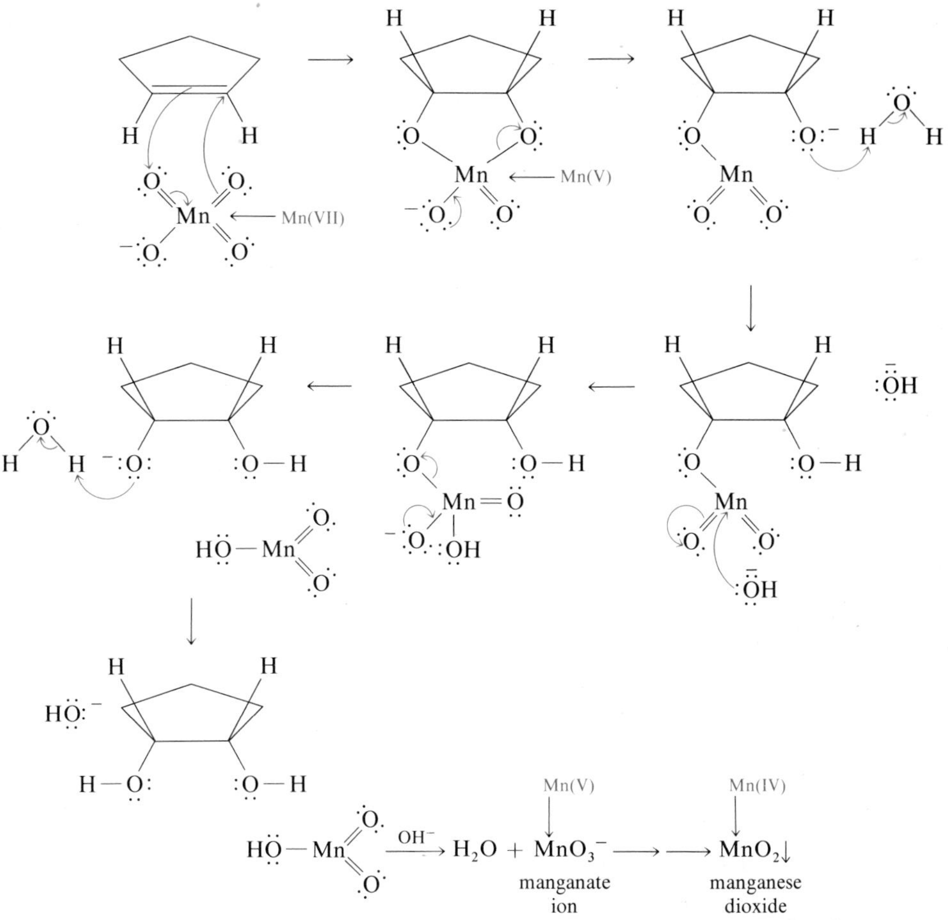

The first stage of the reduction of permanganate ion gives manganate ion with Mn(V). Manganate ion is eventually converted to manganese dioxide, which precipitates. The oxygen atoms of the diol come from the permanganate ion, but water and hydroxide ion participate in breaking up the ester. The cis stereochemistry of the product is derived from the stereochemistry of the cyclic intermediate. Syn addition (Section 7.3C) of permanganate ion to the double bond occurs.

Another example of the oxidation of an alkene by potassium permanganate is the conversion of oleic acid to 9,10-dihydroxystearic acid.

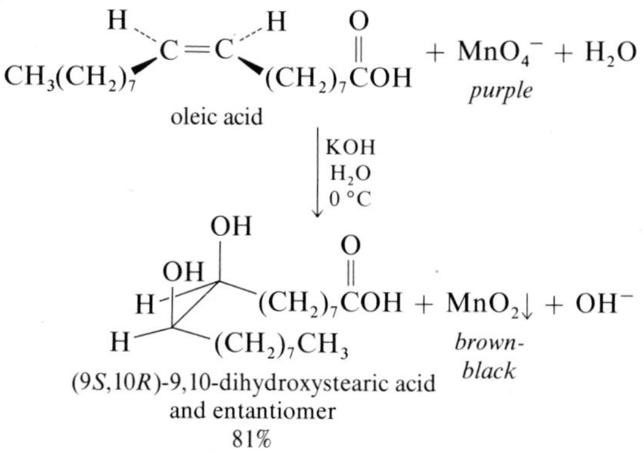

The product of the reaction is written in a conformation that emphasizes the syn addition of the hydroxyl groups to the double bond.

Another reagent that forms cis 1,2-diols by way of a similar intermediate is osmium tetroxide, an expensive and highly toxic reagent. While the cyclic manganate ester breaks down very rapidly in water and cannot be isolated, the corresponding osmate esters, which can be prepared in organic solvents, have been isolated in some cases. The oxidation of 1,2-dimethylcyclopentene to *cis*-1,2-dimethyl-1,2-cyclopentanediol is a typical example of the use of osmium tetroxide.

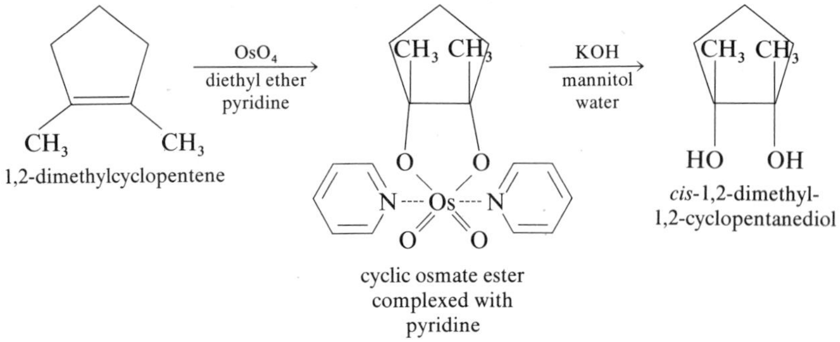

1,2-dimethylcyclopentene

cyclic osmate ester complexed with pyridine

cis-1,2-dimethyl-1,2-cyclopentanediol

$$\text{mannitol} \equiv HOCH_2(CH)_4CH_2OH$$
with an OH on the indicated carbon.

The reaction is carried out in ether. Pyridine catalyzes the addition of osmium tetroxide to the double bond and complexes with the osmium in the ester. Various methods are used to decompose the esters. In this case, base and mannitol, a poly-hydroxy compound that complexes with osmium, are used. Because osmium tetroxide is such an expensive and toxic reagent, it is often used in catalytic amounts with another oxidizing agent present to keep recycling the osmium back to the tetroxide stage. (See Problem 8.43, for example.)

PROBLEM 8.8   Complete the following equations, predicting what products you would expect. Be sure to show stereochemistry when you know it.

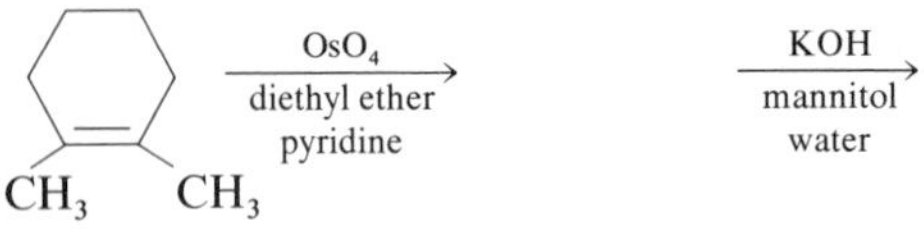

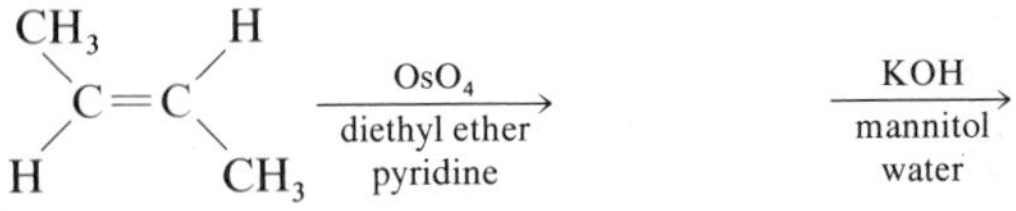

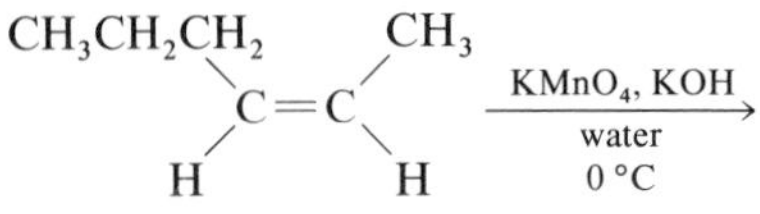

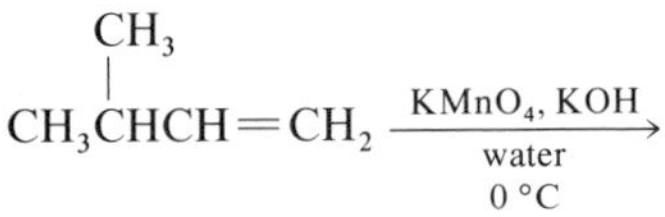

PROBLEM 8.9   Permanganate ion in which the oxygen atoms were labeled with ^{18}O was used to oxidize an alkene. Using

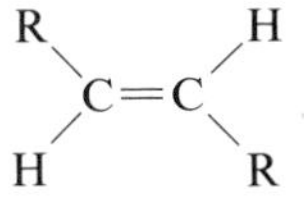

as the alkene, write out the mechanism for the oxidation reaction showing where the labeled oxygen atoms will be found in the product, and what the stereochemistry of the products will be.

PROBLEM 8.10   The cyclic osmate ester is broken down into the diol in the same way as the cyclic manganate ester is. Write a detailed mechanism for the reaction.

B. Oxidation of Alkynes with Potassium Permanganate

Aqueous potassium permanganate reacts with alkynes, as well as alkenes. The reaction produces a diketone as long as the solution is kept from getting too alkaline. An example is the oxidation of 1,2-diphenylethyne to 1,2-diphenyl-1,2-ethanedione.

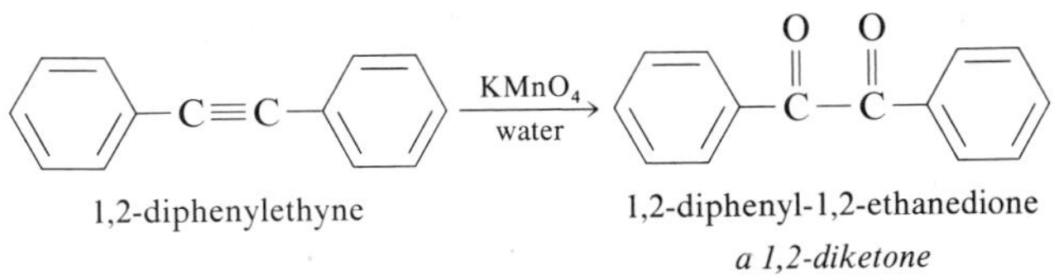

You can visualize the reaction as going through the same kind of intermediate as the oxidation of an alkene (p. 329) except that the 1,2-diol formed from an alkyne is an enol (an ene*diol*, to be accurate). Enols isomerize to ketones (Section 5.5C) so the initial product of the reaction would be a 2-hydroxyketone, which is very easily oxidized further to a 1,2-diketone. Note that phenyl rings are stable to oxidation under the conditions that affect multiple bonds; this is another example of the stability of aromatic rings (Sections 2.8, 2.10, 8.3A).

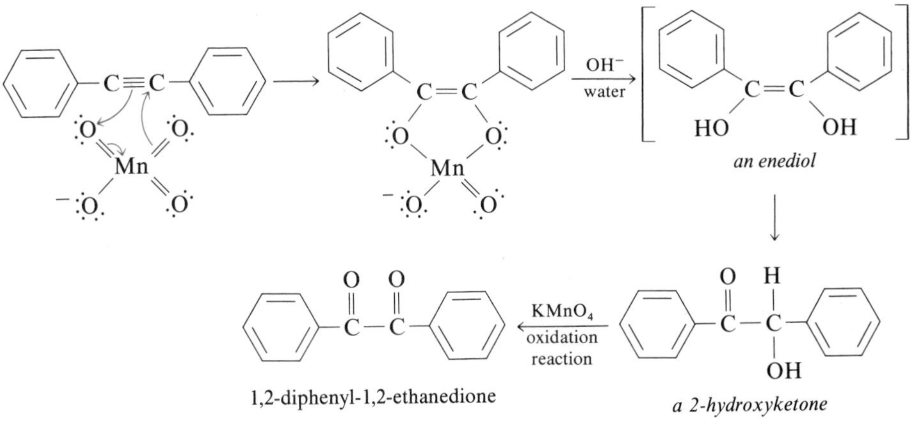

C. Cleavage of Carbon-Carbon Bonds by Oxidation with Permanganate

If permanganate oxidation reactions on alkenes are carried out under vigorous conditions, the molecule is cleaved at the double bond. Cleavage is especially likely to occur when the reaction conditions are not strongly basic. The products that are formed depend upon the structure of the alkene. The reaction is used to make carboxylic acids from alkenes with longer chains, and is most useful in determining the position of the double bond in unsaturated acids. For example, the position of the double bond in oleic acid is confirmed by oxidizing its methyl ester with potassium permanganate in acetic acid.

$$CH_3(CH_2)_7CH=CH(CH_2)_7\overset{\overset{\displaystyle O}{\displaystyle \|}}{C}OCH_3 \xrightarrow[\substack{\text{acetic acid}\\ 50\,°C}]{KMnO_4} CH_3(CH_2)_7\overset{\overset{\displaystyle O}{\displaystyle \|}}{C}OH + HO\overset{\overset{\displaystyle O}{\displaystyle \|}}{C}(CH_2)_7\overset{\overset{\displaystyle O}{\displaystyle \|}}{C}OCH_3$$

methyl oleate   nonanoic acid   monomethyl ester
of
nonanedioic acid
95%

The chain length of each acid isolated indicates that the double bond is between carbon atoms 9 and 10 in the 18-carbon chain of oleic acid.

The triple bond in alkynes is also vulnerable to vigorous oxidation. For example, potassium 9-octadecynoate is oxidized to a diketone by potassium permanganate in a buffered solution, but is cleaved at the triple bond in a strongly basic solution.

$$CH_3(CH_2)_7C{\equiv}C(CH_2)_7\overset{\overset{\displaystyle O}{\displaystyle \|}}{C}O^-K^+ \xrightarrow[\substack{\text{buffer}\\ pH\,7}]{KMnO_4} CH_3(CH_2)_7\overset{\overset{\displaystyle O}{\displaystyle \|}}{C}-\overset{\overset{\displaystyle O}{\displaystyle \|}}{C}(CH_2)_7\overset{\overset{\displaystyle O}{\displaystyle \|}}{C}O^-K^+$$

potassium 9-octadecynoate   potassium 9,10-dioxooctadecanoate

$$\Big\downarrow \substack{KMnO_4\\ pH\,12}$$

$$\Big\downarrow H_3O^+$$

$$CH_3(CH_2)_7\overset{\overset{\displaystyle O}{\displaystyle \|}}{C}OH + HO\overset{\overset{\displaystyle O}{\displaystyle \|}}{C}(CH_2)_7\overset{\overset{\displaystyle O}{\displaystyle \|}}{C}OH$$

nonanoic acid   nonanedioic acid
80%

PROBLEM 8.11   Complete the following equations, predicting what the products would be if these reactions were carried out. Show stereochemistry when it is pertinent.

(a) $CH_3CH_2C{\equiv}CCH_3 \xrightarrow[\substack{\text{water}\\ pH\,7}]{KMnO_4}$

(b) $-C{\equiv}CCH_3 \xrightarrow[\substack{\text{water}\\ pH\,12}]{KMnO_4}$

(c) $\xrightarrow[\substack{\text{ethanol}\\ \text{water}\\ -40\,°C}]{KMnO_4}$

(d) 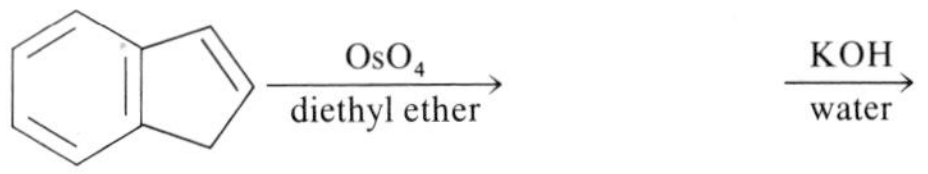 [structure: indene with OsO₄ / diethyl ether arrow, then KOH / water arrow]

(e) 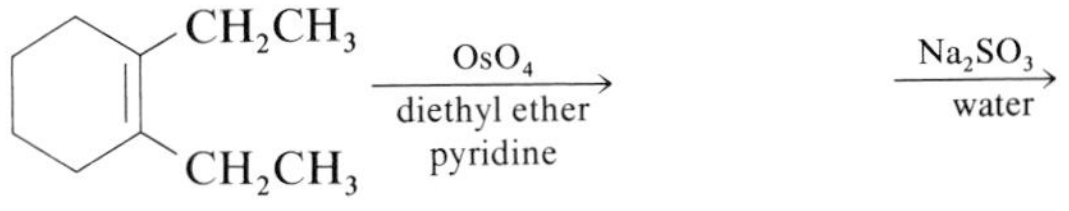 [structure: cyclohexene bearing two CH₂CH₃ groups, with OsO₄ / diethyl ether / pyridine arrow, then Na₂SO₃ / water arrow]

D. Peroxyacid Oxidations of Alkenes

Carboxylic acids have the general formula

$$\underset{\text{RCOH}}{\overset{\displaystyle\overset{O}{\|}}{}}$$

and may be regarded as being derived from water, HOH, by the replacement of one of the hydrogen atoms by an acyl group,

$$\overset{\displaystyle\overset{O}{\|}}{R-C-}$$

Carboxylic acids have an acidic hydrogen atom and act as Brønsted-Lowry acids. They are *not* oxidizing agents. **Peroxycarboxylic acids,**

$$\overset{\displaystyle\overset{O}{\|}}{R-C-O-O-H}$$

may be viewed as being related to the oxidizing agent hydrogen peroxide, HOOH, in the same way as carboxylic acids are related to water. Replacement of one of the hydrogen atoms in hydrogen peroxide by an acyl group gives the formula for a peroxycarboxylic acid. Peroxyformic acid, for example, is an unstable compound that can be made by mixing formic acid with hydrogen peroxide.

$$\underset{\substack{\text{formic acid} \\ }}{\overset{\displaystyle\overset{O}{\|}}{HCOH}} + \underset{\substack{\text{hydrogen} \\ \text{peroxide}}}{HOOH} \rightleftharpoons \underset{\substack{\text{peroxyformic} \\ \text{acid}}}{\overset{\displaystyle\overset{O}{\|}}{HCOOH}} + \underset{\substack{\text{water} \\ }}{H_2O}$$

Peroxyacids are oxidizing agents and react with alkenes to give three-membered cyclic ethers known as **epoxides** or **oxiranes.** These small ring compounds are quite reactive, and the ring opens easily, especially in acidic solutions. These reactions are considered in more detail in Section 8.5. In this section, we will examine the formation of oxiranes from alkenes using *m*-chloroperoxybenzoic acid as the reagent.

m-Chloroperoxybenzoic acid is sold commercially and is stable enough that it can be stored and used over a period of time. Reactions are carried out in organic solvents such as ethers or halogenated hydrocarbons, so that the solubility of starting materials and products is not a problem. The oxiranes that are formed are stable in the reaction mixture and are isolated in reasonably good yields.

A typical peroxyacid oxidation of an alkene, 1-hexene, gives 2-butyloxirane.

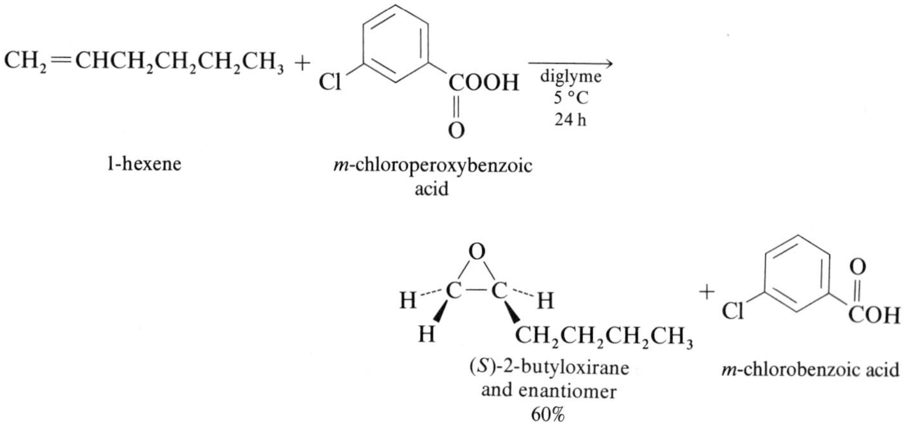

1-hexene   *m*-chloroperoxybenzoic acid

(*S*)-2-butyloxirane and enantiomer 60%   *m*-chlorobenzoic acid

An oxygen atom is transferred from the peroxyacid to the double bond giving the cyclic ether. The other product of the reaction is *m*-chlorobenzoic acid, no longer a peroxyacid. The mechanism of the reaction is traditionally believed to involve a single-step transfer of the peroxy oxygen atom to the double bond.

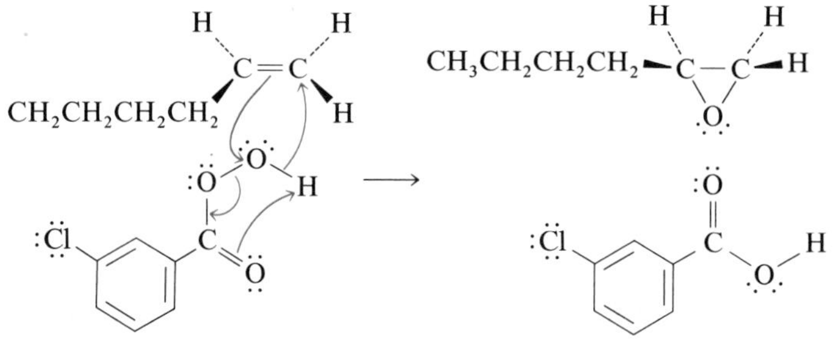

If this mechanism is correct, we would expect the reaction to be stereoselective. Stereoisomeric alkenes should give the corresponding oxiranes with no loss of stereochemistry. This has been found to be so. (*Z*)-2-Butene is oxidized to *cis*-2,3-dimethyloxirane, while (*E*)-2-butene gives *trans*-2,3-dimethyloxirane with over 99% stereochemical purity.

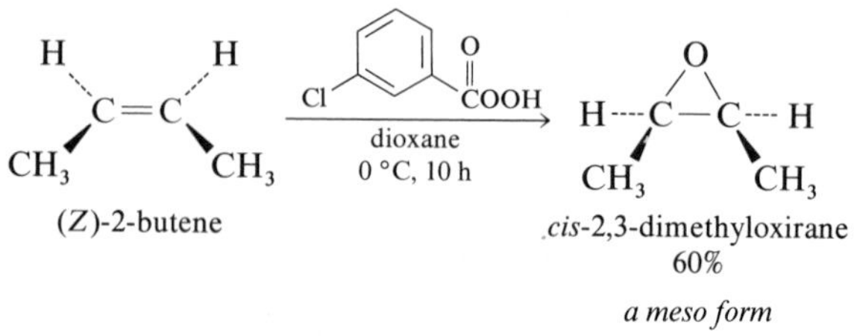

(*Z*)-2-butene   *cis*-2,3-dimethyloxirane 60%

a meso form

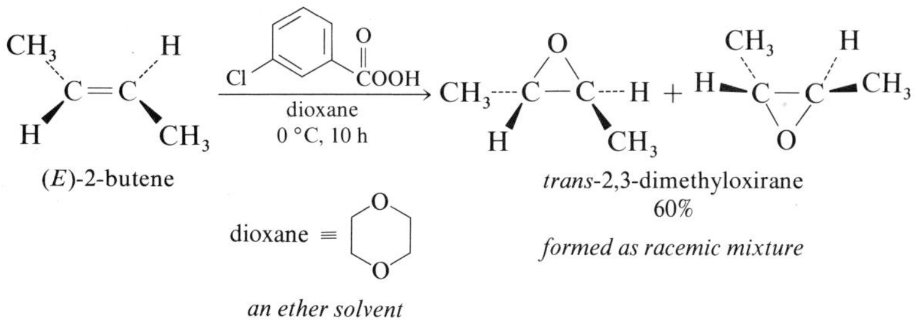

(E)-2-butene

dioxane ≡

an ether solvent

trans-2,3-dimethyloxirane
60%

formed as racemic mixture

cis-2,3-Dimethyloxirane is not optically active because it is a meso form, with a plane of symmetry bisecting the oxygen atom and the carbon-carbon bond of the oxirane ring. *trans*-2,3-Dimethyloxirane is formed as a racemic mixture, the two enantiomers resulting from the equal probability of attack from above and below the plane of the double bond.

The peroxyacid is behaving as an electrophile in this reaction. Its electrophilic character is seen in the selectivity with which it reacts with double bonds having different degrees of substitution. The more alkyl substituents on a double bond, the more readily it is attacked by the peroxyacid. This fact suggests that higher electron density in the π bond favors the reaction with peroxyacids. 1,2-Dimethyl-1,4-cyclohexadiene, for example, reacts selectively with one molar equivalent of *m*-chloroperoxybenzoic acid at the more highly substituted of the two double bonds.

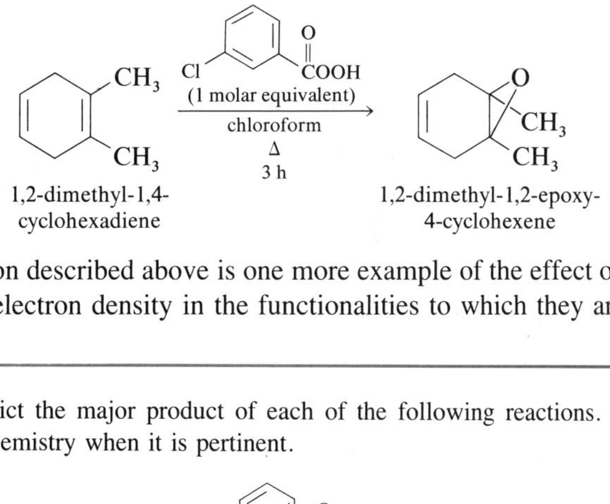

1,2-dimethyl-1,4-
cyclohexadiene

1,2-dimethyl-1,2-epoxy-
4-cyclohexene

The phenomenon described above is one more example of the effect of alkyl groups in increasing the electron density in the functionalities to which they are attached.

PROBLEM 8.12 Predict the major product of each of the following reactions. Write a structure showing stereochemistry when it is pertinent.

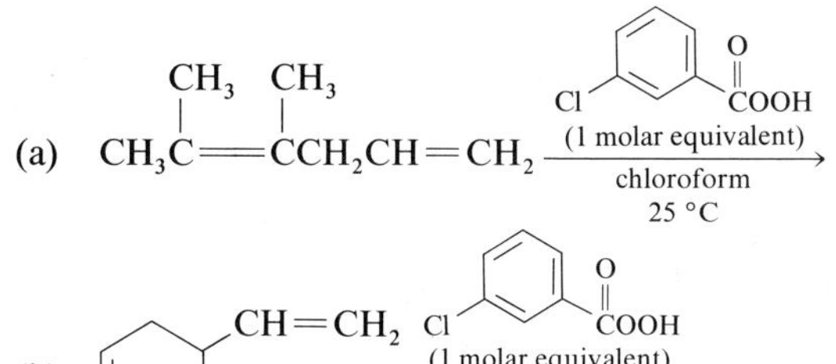

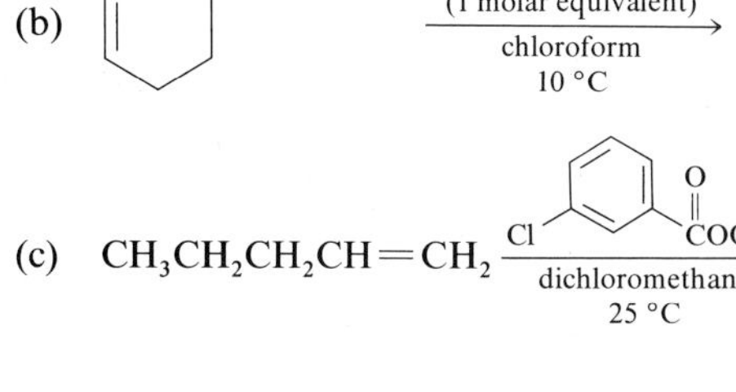

(d)

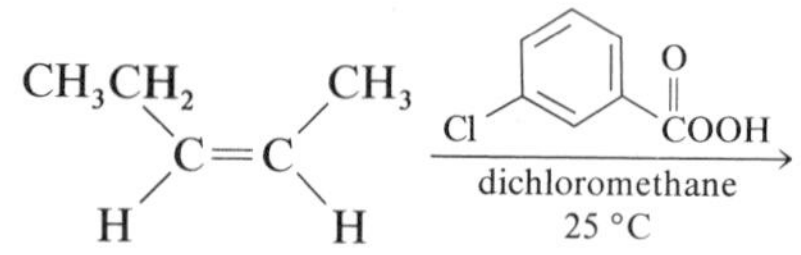

(e)

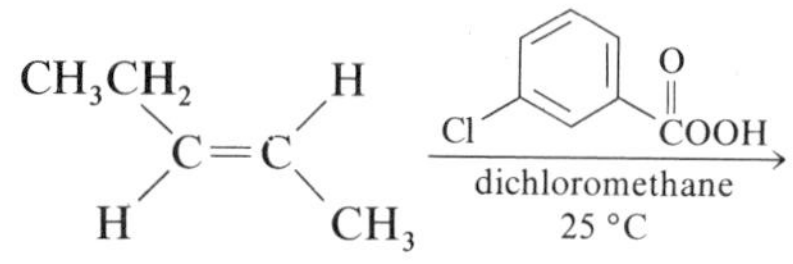

E. Formation of Oxiranes from Halohydrins. The Neighboring Group Effect

The reaction of halogens with alkenes in the presence of water gives rise to halohydrins, compounds containing a halogen atom and a hydroxyl group on adjacent carbon atoms (Section 5.8A). The same reaction is observed when hypochlorous acid or hypobromous acid, prepared by the acidification of solutions of sodium or calcium hypohalites, is used. Sodium hypochlorite is the chief constituent of laundry bleach and gives a solution of hypochlorous acid on treatment with cold dilute nitric acid.

$$\underset{\substack{\text{sodium} \\ \text{hypochlorite}}}{NaOCl} + HNO_3 \xrightarrow[\substack{H_2O \\ \text{cold}}]{} \underset{\substack{\text{hypochlorous} \\ \text{acid}}}{HOCl} + NaNO_3$$

Hypochlorous acid adds to an alkene, such as cyclohexene, to give a trans chlorohydrin.

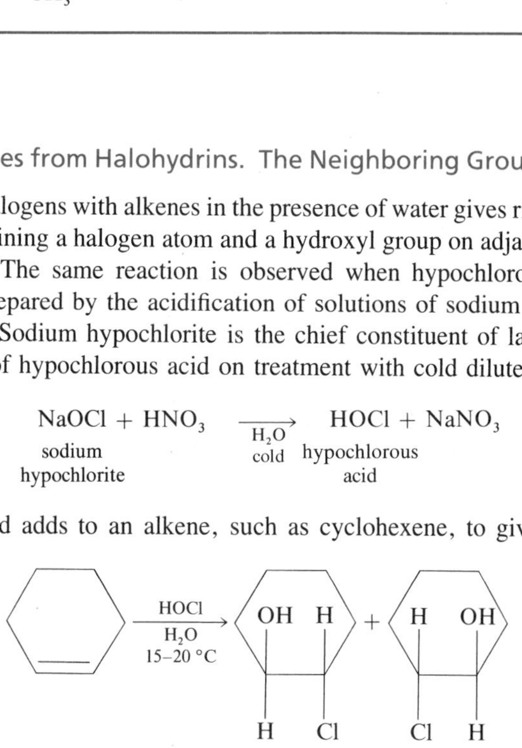

cyclohexene   *trans*-2-chlorocyclohexanol
chlorohydrin of cyclohexene
70%

racemic mixture

The reaction is postulated to go by way of a chloronium ion intermediate (Section 5.8B) resulting from attack of the double bond on chlorine, which is present in solutions of hypochlorous acid.

The trans halohydrin from cyclohexene, when treated with base, gives an oxirane.

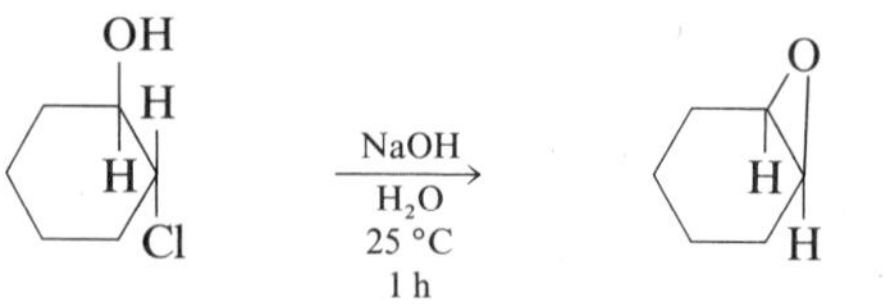

trans-2-chlorocyclohexanol   cyclohexene oxide
70%

The reaction occurs easily because in one conformation of the molecule a nucleophilic alkoxide ion is in position to displace the halide ion in an internal S_N2 reaction.

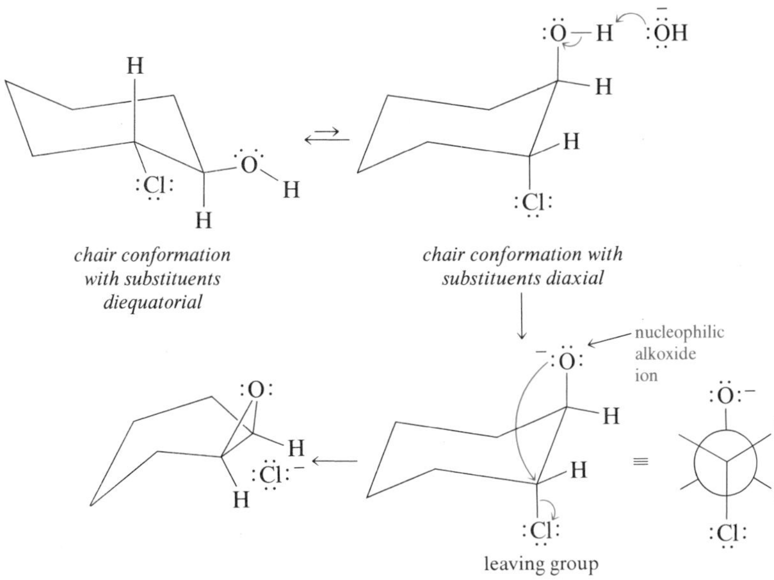

*chair conformation
with substituents
diequatorial*

*chair conformation with
substituents diaxial*

nucleophilic
alkoxide
ion

leaving group

In this conformation the alkoxide ion and the leaving group are anti to each other as shown in the Newman projection for that portion of the cyclohexane ring. Halohydrins that cannot achieve an anti orientation between the nucleophile and the leaving group do not give oxiranes.

The intramolecular reaction, in which the alkoxide ion is held close to the carbon atom bearing the leaving group, is favored over an intermolecular reaction in which the alkoxide ion collides and reacts with an alkyl halide functionality in another molecule. The formation of tetrahydrofuran from 4-chloro-1-butanol is another example of an intramolecular S_N2 reaction (Section 7.1A).

Often the rates of intramolecular reactions are higher than the rates of comparable intermolecular reactions. This increase in reactivity when nucleophiles react with other functional groups within the same molecule is known as the **neighboring group effect.**

PROBLEM 8.13 Write a detailed mechanism for the following reaction, which has been observed. Be sure to show the conformation of the molecule in the transition state.

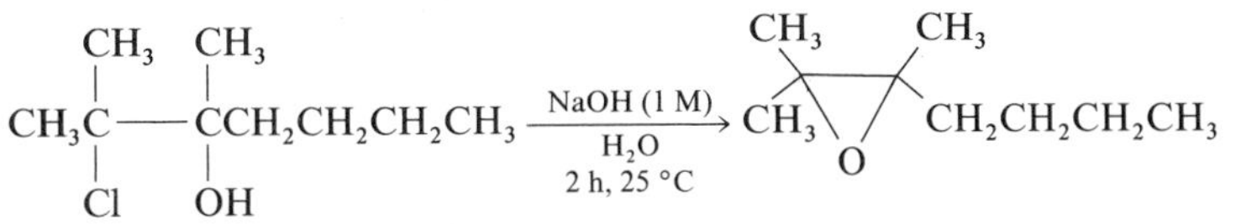

2-chloro-2,3-dimethyl-
3-heptanol

2,2,3-trimethyl-3-butyloxirane

PROBLEM 8.14 Supply the products in the following reactions. Show stereochemistry when it is known.

(a) 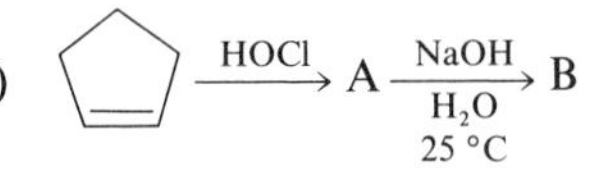 $\xrightarrow{\text{HOCl}}$ A $\xrightarrow[\text{H}_2\text{O}]{\text{NaOH}}$ B
                                                                25 °C

(b) $\text{CH}_3\text{CH}_2\text{CH}_2\text{CHCHCH}_2\text{CH}_2\text{CH}_3$ $\xrightarrow[\text{H}_2\text{O}]{\text{NaOH}}$ C
                        | | 25 °C
                        Cl   OH

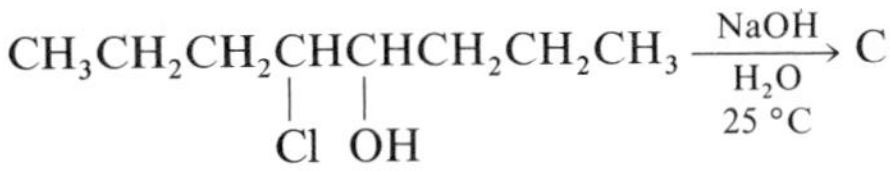

(c) 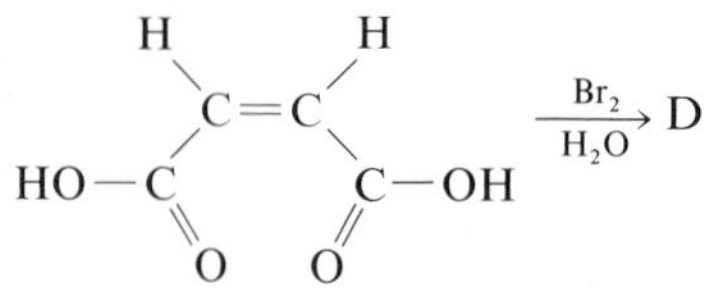 $\xrightarrow[\text{H}_2\text{O}]{\text{Br}_2}$ D

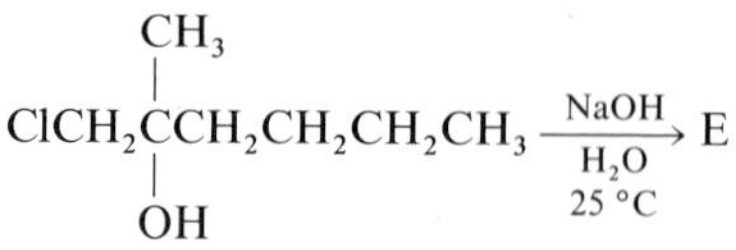

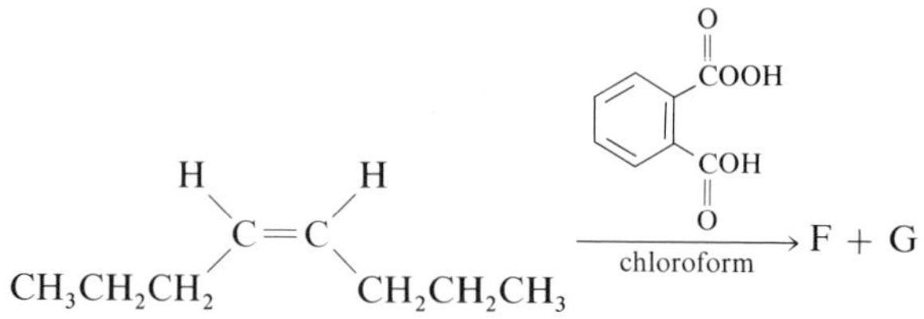

(d) $\text{ClCH}_2\overset{\overset{\displaystyle \text{CH}_3}{|}}{\underset{\underset{\displaystyle \text{OH}}{|}}{\text{C}}}\text{CH}_2\text{CH}_2\text{CH}_2\text{CH}_3 \xrightarrow[\substack{\text{H}_2\text{O}\\25\,°\text{C}}]{\text{NaOH}}$ E

(e) $\xrightarrow{\text{chloroform}}$ F + G

(f) $\xrightarrow{\text{Ca(OBr)}_2,\ \text{CH}_3\overset{\overset{\displaystyle \text{O}}{\|}}{\text{C}}\text{OH}}$ H + I

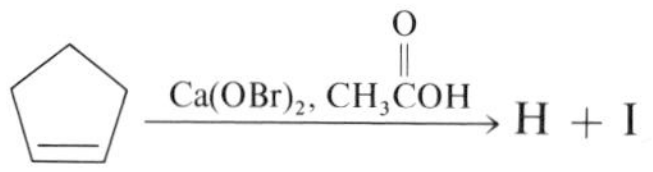

PROBLEM 8.15 Consider the kinetics of the formation of oxirane from 2-chloroethanol and of tetrahydrofuran from 4-chlorobutanol. The reactions were found to have the following heats and entropies of activation at 30 °C. Which reaction do you expect to be faster? Why? (You may want to review Section 5.3F.)

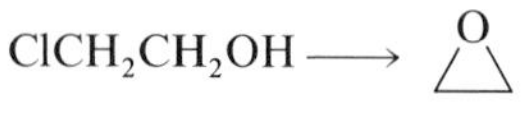

at 30 °C

$\text{ClCH}_2\text{CH}_2\text{OH} \longrightarrow$ $\Delta H^{\ddagger} = 23.2\ \text{kcal/mol}$
$\Delta S^{\ddagger} = 9.9 \times 10^{-3}\ \text{kcal/mol} \cdot \text{deg}$

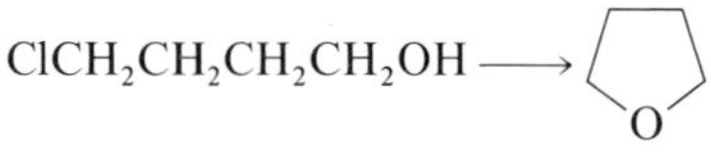

$\text{ClCH}_2\text{CH}_2\text{CH}_2\text{CH}_2\text{OH} \longrightarrow$ $\Delta H^{\ddagger} = 19.8\ \text{kcal/mol}$
$\Delta S^{\ddagger} = -5.0 \times 10^{-3}\ \text{kcal/mol} \cdot \text{deg}$

8.5

Ring-Opening Reactions of Oxiranes

A. Reactions of Oxiranes with Aqueous Acid. The Stereoselective Formation of trans 1,2-Diols

The ring in an oxirane opens readily with acidic reagents. For example, the oxirane from cyclopentene gives a racemic mixture of *trans*-1,2-cyclopentanediols with aqueous acid.

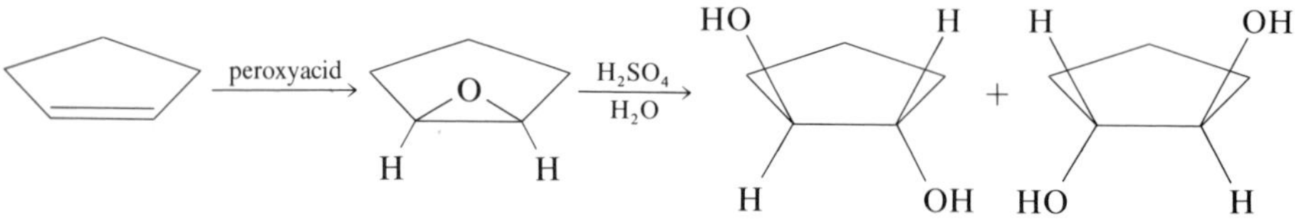

cyclopentene

cyclopentene oxide
1,2-epoxycyclopentane

trans-1,2-cyclopentanediol

a racemic mixture

The reaction starts with the protonation of the oxygen atom in the oxirane ring. The protonated oxirane, an oxonium ion, is electronically similar to the bromonium ion suggested as an intermediate in the addition of bromine to a double bond (Section

5.8B). Nucleophilic attack by water on either one of the carbon atoms of the oxirane ring opens the ring to give the trans orientation of the two hydroxyl groups.

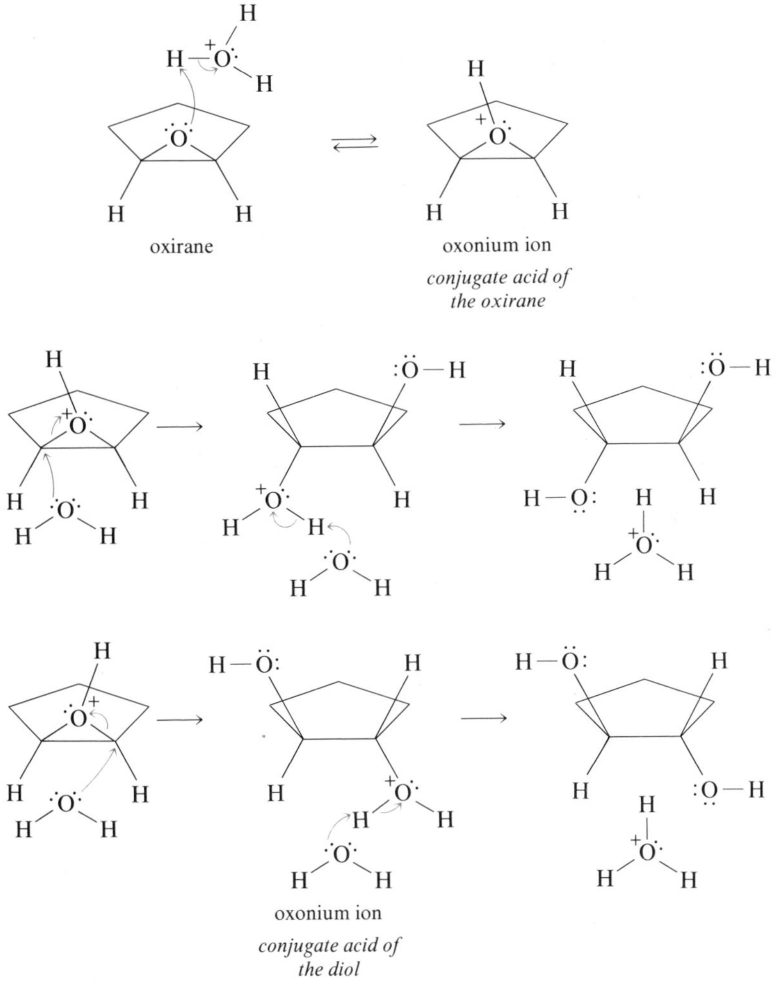

oxirane

oxonium ion

conjugate acid of
the oxirane

oxonium ion

conjugate acid of
the diol

The intermediate formed when the oxirane ring is first opened by the attack of water is an oxonium ion, the conjugate acid of the diol. The oxonium ion loses a proton to the solvent in the final step of the reaction. The reaction is stereoselective, giving only the trans diol. You can think of this sequence of reactions as complementary to the oxidation of cyclopentenes with potassium permanganate or osmium tetroxide, which give *cis*-1,2-cyclopentanediols (pp. 328 and 330).

Cyclohexene, like cyclopentene (Section 8.4A), reacts with aqueous potassium permanganate or osmium tetroxide to give a cis 1,2-diol.

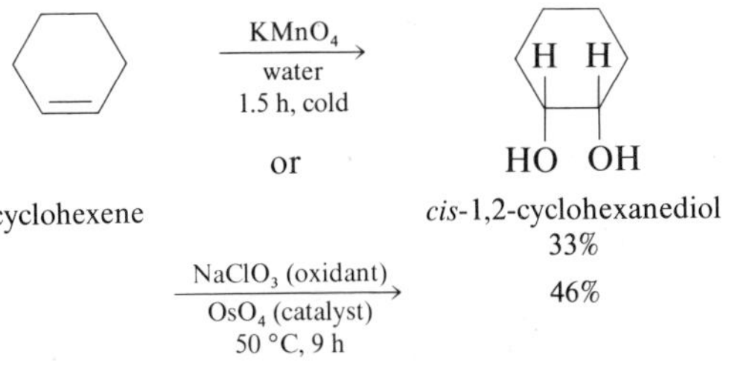

cyclohexene

$\xrightarrow[\substack{\text{water} \\ \text{1.5 h, cold}}]{\text{KMnO}_4}$

or

$\xrightarrow[\substack{\text{OsO}_4 \text{ (catalyst)} \\ 50\,^\circ\text{C, 9 h}}]{\text{NaClO}_3 \text{ (oxidant)}}$

cis-1,2-cyclohexanediol
33%

46%

The stereochemistry of the diol is determined by the cyclic intermediate formed when either potassium permanganate or osmium tetroxide reacts with the alkene.

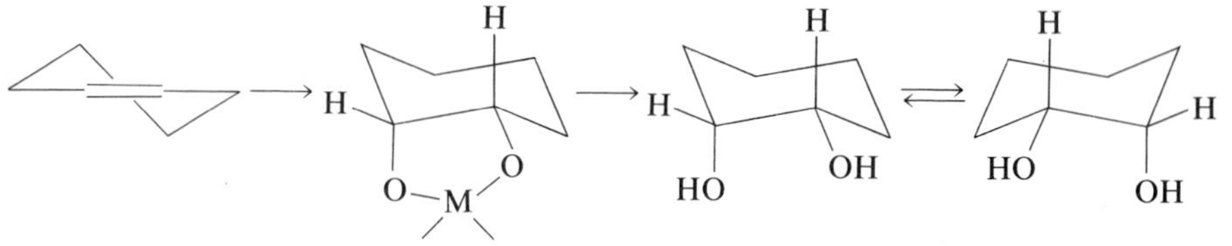

If, on the other hand, cyclohexene is oxidized by a peroxyacid, and the resulting epoxide opened in aqueous acid, *trans*-1,2-cyclohexanediol is formed as a racemic mixture.

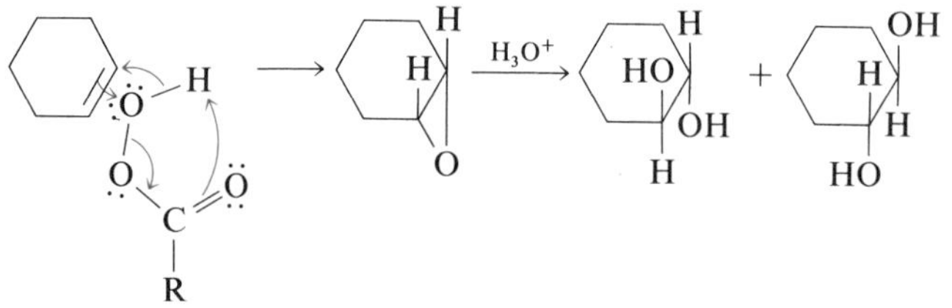

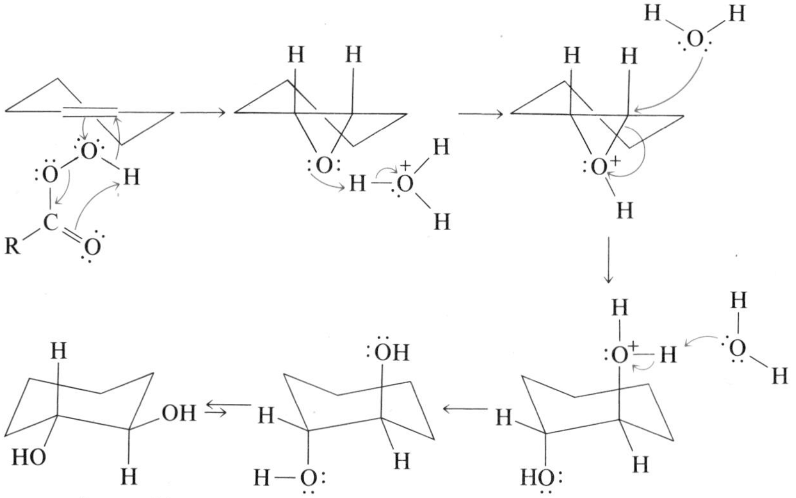

trans diequatorial   trans diaxial

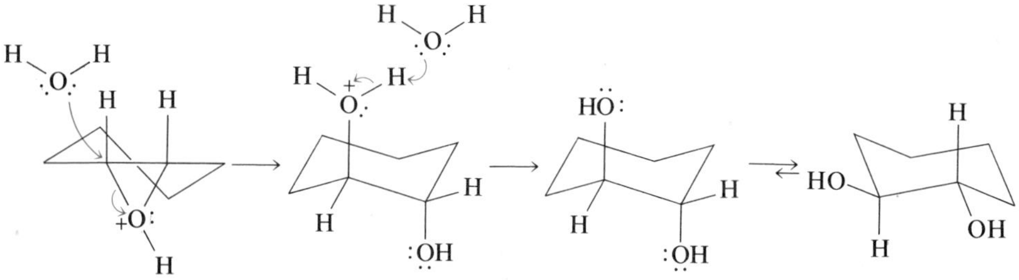

As the conformational formulas demonstrate, the opening of the three-membered oxirane ring first gives the two hydroxyl groups in a trans diaxial orientation. The more stable conformation for the diol is the diequatorial one, which predominates at equilibrium. Attack by the nucleophile, water, is equally probable at either carbon atom of the oxirane ring, so equal amounts of the two enantiomeric *trans*-1,2-cyclohexanediols are formed.

B. Reactions with Hydrogen Halides

The nucleophilic reagent that opens the protonated oxirane ring does not have to be a water molecule. For example, cyclopentene oxide reacts with hydrogen chloride in ether to give chlorocyclopentanols.

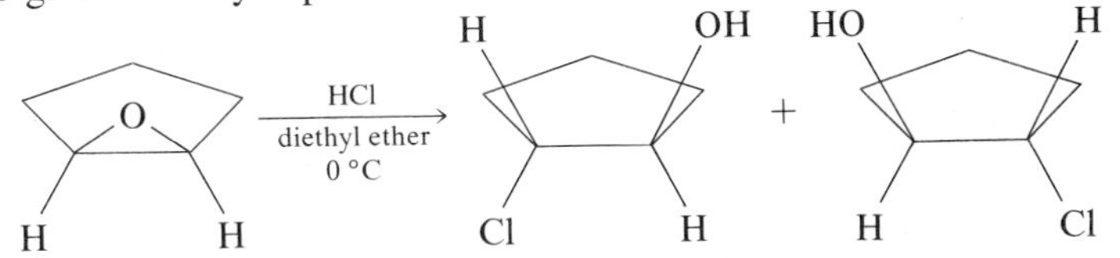

cyclopentene oxide → *trans*-2-chlorocyclopentanol
a racemic mixture

Protonation of the oxirane gives an oxonium ion intermediate that is attacked at either carbon atom by the nucleophilic halide ion.

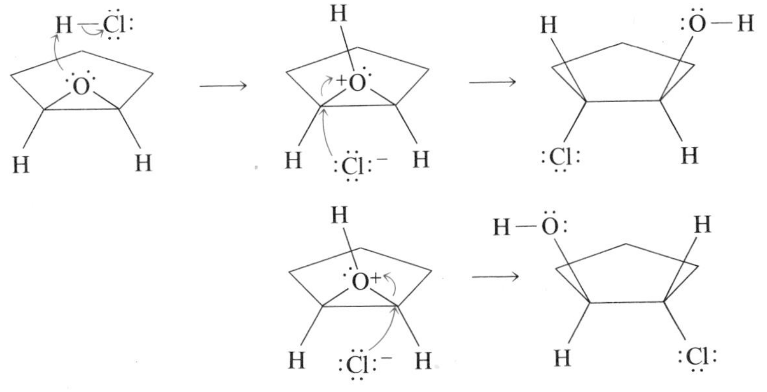

The two possible enantiomers of the trans isomer are formed in equal amounts to give the racemic mixture.

Cyclopentene oxide can have only one stereochemistry because of the constraint of the five-membered ring. Two substituents on a cyclopentane ring may be cis or trans to each other. Only one of these two possible outcomes is seen when the oxirane ring in cyclopentene oxide is opened with acidic reagents. Thus, a starting material of one particular stereochemistry is converted with high stereoselectivity to the product.

Oxiranes from acyclic alkenes may exist as cis or trans isomers. For example, (Z)- and (E)-2-butene are converted into the corresponding *cis*- and *trans*-2,3-dimethyloxiranes (p. 334). Each of these oxiranes gives a pair of enantiomeric 3-bromo-2-butanols with 48% aqueous hydrobromic acid. *cis*-2,3-Dimethyloxirane gives one racemic mixture.

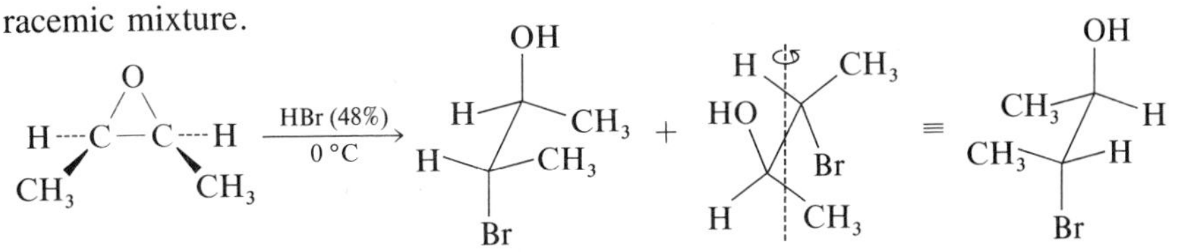

cis-2,3-dimethyloxirane

(2S,3S)-3-bromo-2-butanol + (2R,3R)-3-bromo-2-butanol

racemic mixture
85%

trans-2,3-Dimethyloxirane, which is a racemic mixture, gives another pair of enantiomeric 3-bromo-2-butanols.

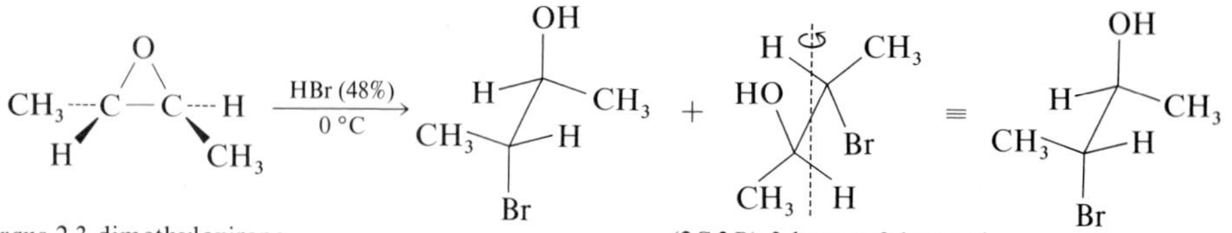

trans-2,3-dimethyloxirane
one enantiomer

(2*S*,3*R*)-3-bromo-2-butanol

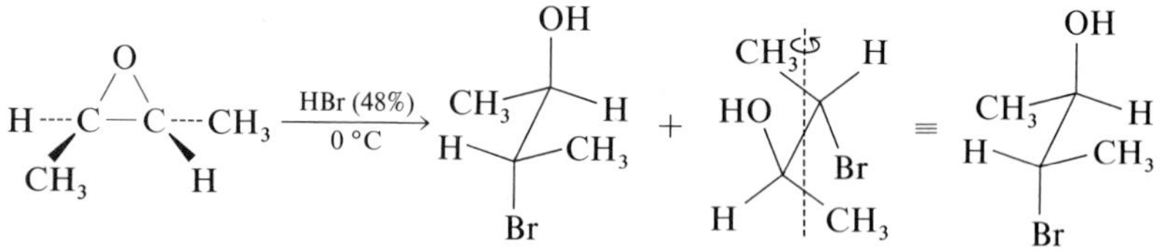

trans-2,3-dimethyloxirane
the other enantiomer

(2*R*,3*S*)-3-bromo-2-butanol

In each case, the products shown are derived from nucleophilic attack at the carbon atoms of a protonated oxirane ring, as illustrated for *cis*-2,3-dimethyloxirane.

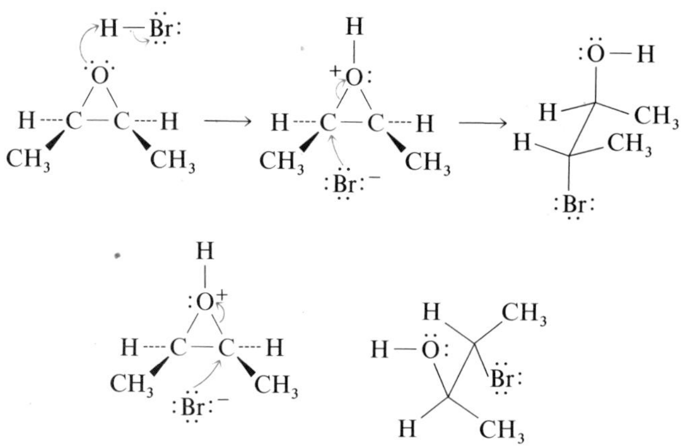

The 3-bromo-2-butanols obtained from *cis*-2,3-dimethyloxirane are diastereomers of those obtained from the trans oxirane.

These reactions are also examples of stereoselectivity in chemical transformations. Two starting materials with differing stereochemistry are converted into products that are also stereochemically different from each other.

PROBLEM 8.16 When *cis*-2,3-dimethyloxirane is treated with water containing a trace of perchloric acid, $HClO_4$, a racemic mixture of 2,3-butanediols is formed. *trans*-2,3-Dimethyloxirane, under the same conditions, gives *meso*-2,3-butanediol. Write equations for these reactions, including the mechanism and all the stereochemistry.

PROBLEM 8.17 If one of the two enantiomeric *trans*-2,3-dimethyloxiranes, (2*R*,3*R*)-(+)-2,3-dimethyloxirane, is treated with methanol containing a trace of sulfuric acid, a 57% yield of a single, optically active, 3-methoxy-2-butanol,

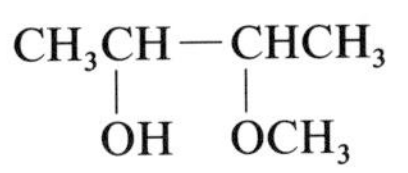

$$CH_3CH-CHCH_3$$
$$\quad\; | \qquad\quad |$$
$$\quad\; OH \quad\; OCH_3$$

is formed. Draw the correct structure for the oxirane and write a stereochemically correct mechanism showing how the 3-methoxy-2-butanol is formed. Assign configuration to the asymmetric carbon atoms in the product and give the correct name for the compound.

The oxiranes shown above are all symmetrical in structure. The question of regioselectivity only arises in ring-opening reactions of unsymmetrical oxiranes. For example, the reaction of 2-methyloxirane with hydrobromic acid gives a mixture of two halohydrins.

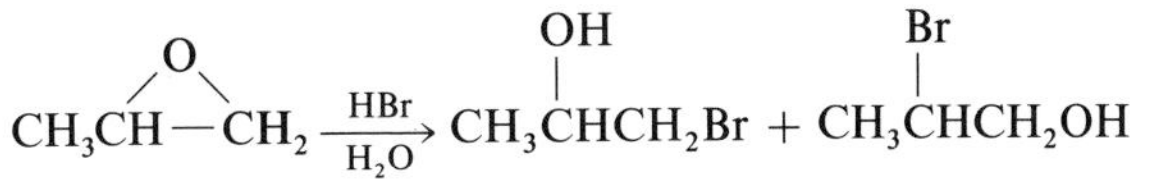

Attack by the nucleophile, bromide ion, on the less hindered carbon atom of the protonated oxirane in an S_N2 reaction is the favored reaction pathway.

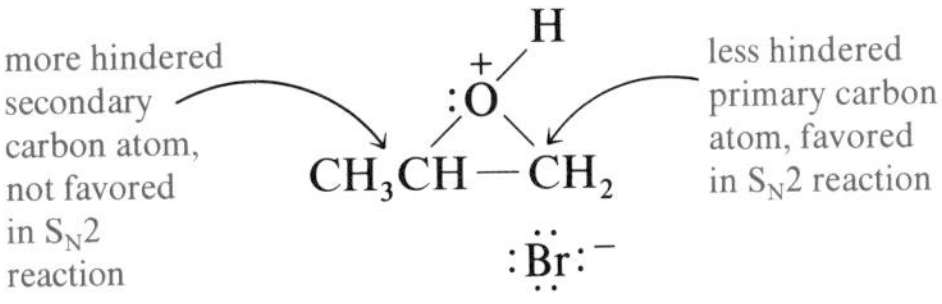

The protonated oxygen atom of the oxirane ring, however, is a good leaving group, and the carbon-oxygen bonds in the strained ring may start to ionize before the approach of the nucleophile. In an unsymmetrically substituted oxirane, one carbocationic intermediate will be favored over the other. For 2-methyloxirane, the favored intermediate will have cationic character at the secondary carbon atom instead of at the primary carbon atom.

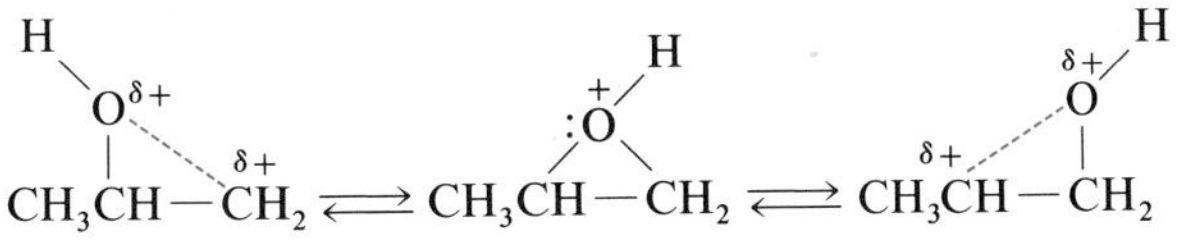

Consequently, 2-bromo-1-propanol, the minor product from the reaction of 2-methyloxirane with hydrobromic acid, may be postulated to arise by a substitution reaction on the protonated oxirane in which some positive charge develops at the secondary carbon atom.

The regioselectivity of the reaction is further changed when a tertiary carbocationic intermediate is possible. 2,2,3-Trimethyloxirane reacts with methanol in the presence of sulfuric acid to give 2-methoxy-2-methyl-3-butanol as the major product.

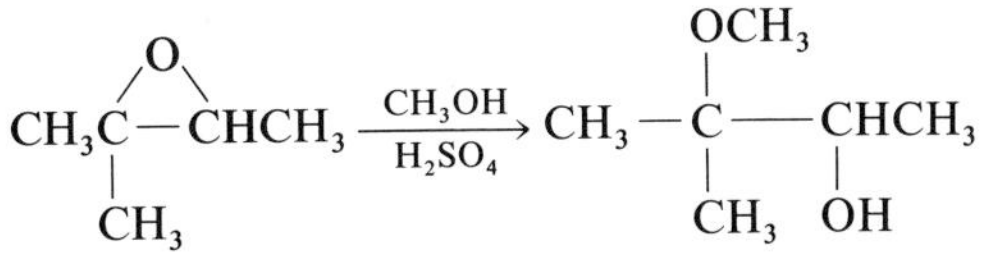

PROBLEM 8.18   For the reaction of 2,2,3-trimethyloxirane with methanol and sulfuric acid, write a detailed mechanism that explains the regioselectivity observed.

C. Reactions of Oxiranes with Nucleophiles

The strained ring in an oxirane is opened easily by nucleophilic reagents even when the oxygen atom is not protonated. These reactions are important in syntheses, especially when the attacking reagent is a nucleophilic carbon atom, as will be seen in Section 9.4B. However, other nucleophiles such as alkoxide ions or amines also react with oxiranes. For example, 2,2,3-trimethyloxirane reacts with sodium methoxide to give 3-methoxy-2-methyl-2-butanol.

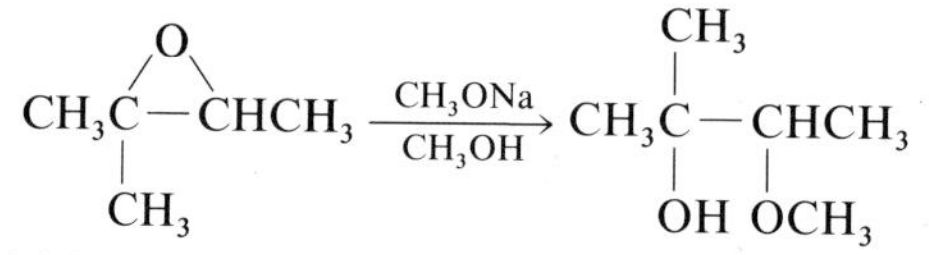

2,2,3-trimethyloxirane   3-methoxy-2-methyl-
2-butanol
53%

In the absence of a strong acid, there is no question of a cationic intermediate for the reaction. An S_N2 reaction occurs at the less highly substituted carbon atom. The new alkoxide ion that forms is protonated by the solvent.

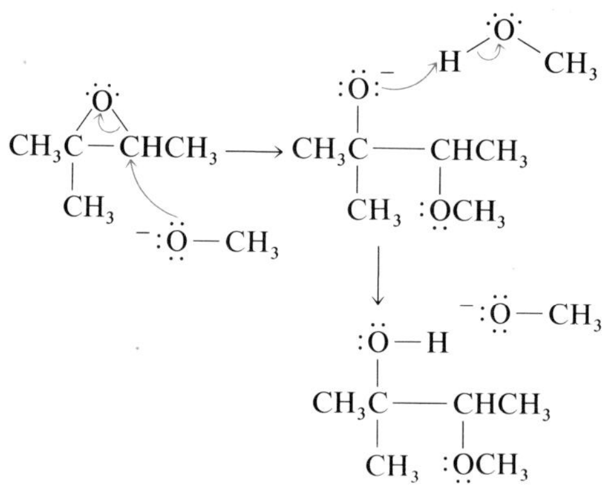

PROBLEM 8.19   Suggest reagents for the following transformations.

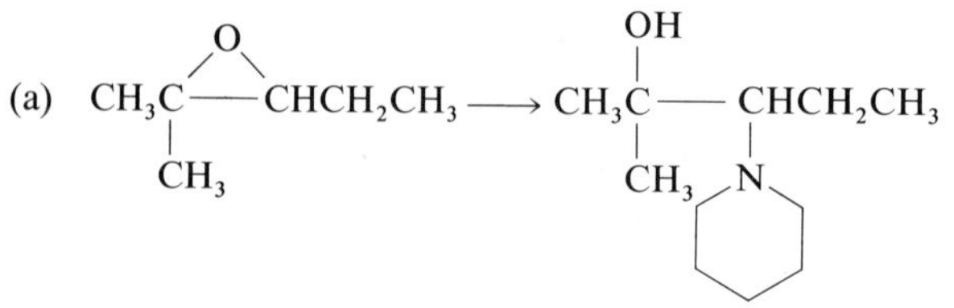

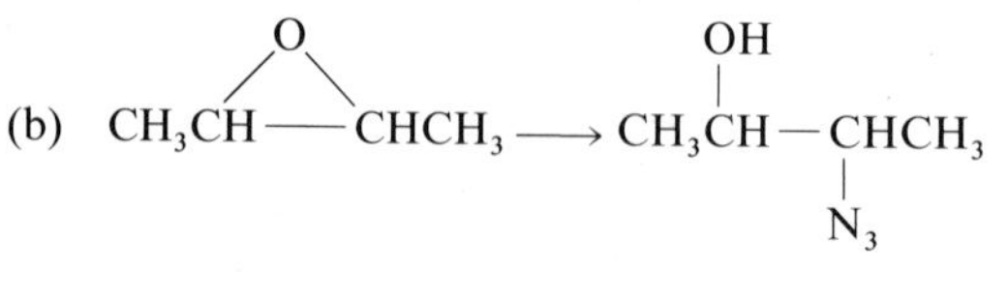

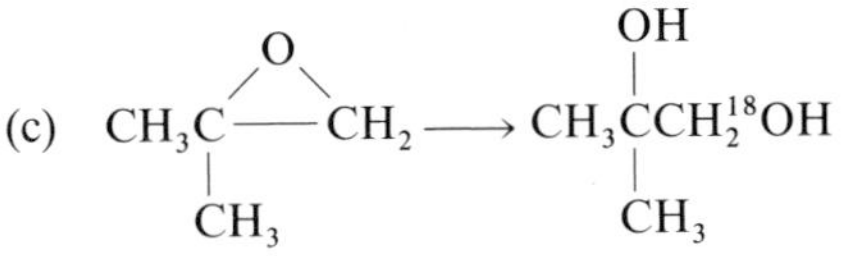

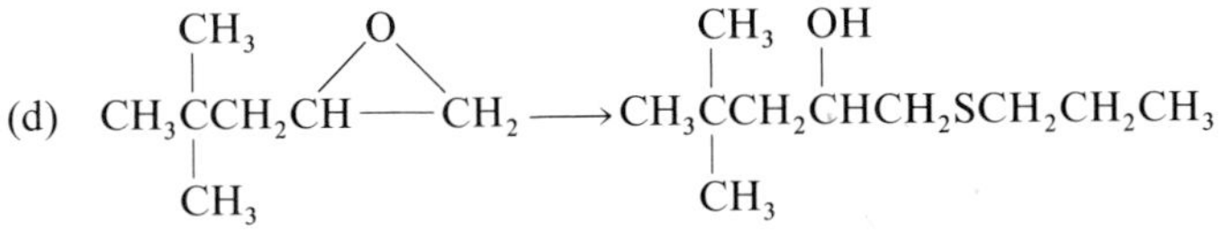

(c) CH_3C (epoxide) $CH_2 \longrightarrow CH_3CCH_2^{18}OH$ with OH and CH_3 substituents

(d) CH_3CCH_2CH (epoxide) $CH_2 \longrightarrow CH_3CCH_2CHCH_2SCH_2CH_2CH_3$ with CH_3, OH, and CH_3 substituents

PROBLEM 8.20   Because of the bulk of the *tert*-butyl group, *tert*-butylcyclohexane exists mainly in one conformation (Section 3.8C). *trans*-4-*tert*-Butylcyclohexene oxide, when treated with methanethiol, CH_3SH, in ethanol in the presence of sodium ethoxide, gives *trans*-2-methylthio-*trans*-5-*tert*-butylcyclohexanol. Write a mechanism for the reaction showing the conformation you expect for the product. A review of Section 8.5A may help you.

8.6

Reactions of Alkenes and Alkynes with Ozone

A. Ozonolysis of Alkenes

Ozone, O_3, is formed when an electric discharge is passed through oxygen gas. The ozone molecule, which cannot be symbolized satisfactorily by a single Lewis structure, is represented by a set of resonance contributors.

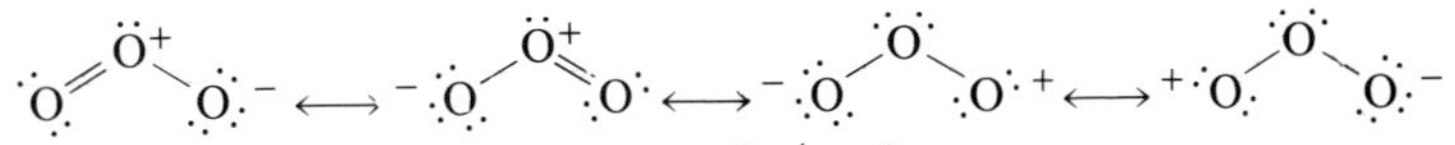

resonance contributors of ozone

Ozone is an electrophile that adds to an alkene to give an unstable cyclic compound called a **molozonide.** The molozonide decomposes to a charged intermediate and a carbonyl compound held close together in a solvent cage. If the carbonyl fragment is an aldehyde, the two pieces recombine to form an **ozonide.** These reactions are illustrated for 2-methylpropene.

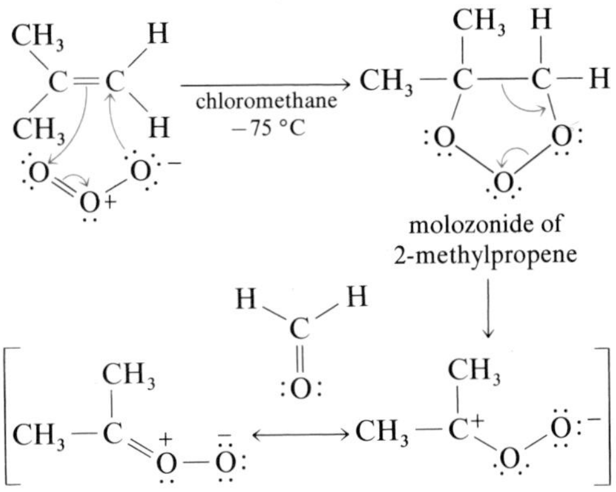

molozonide of
2-methylpropene

formaldehyde and the charged
intermediate that is stabilized
by resonance

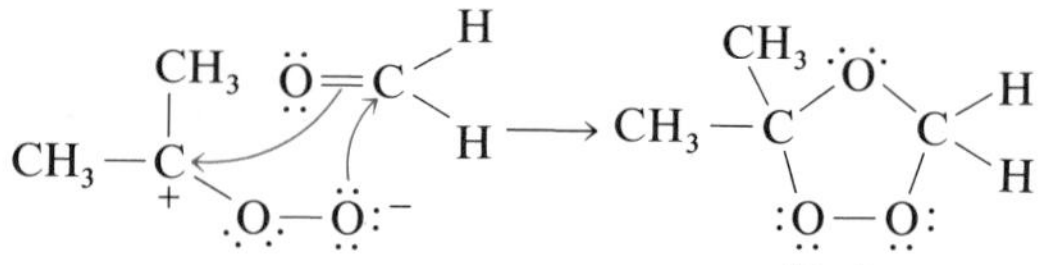

ozonide from
2-methylpropene

We would expect an ionic intermediate with a positively charged carbon atom, such as the one proposed for this reaction, to react with nucleophilic solvents—alcohols, for example. 2,3-Dimethyl-2-butene reacts with ozone in methanol to give a compound containing the methoxyl group.

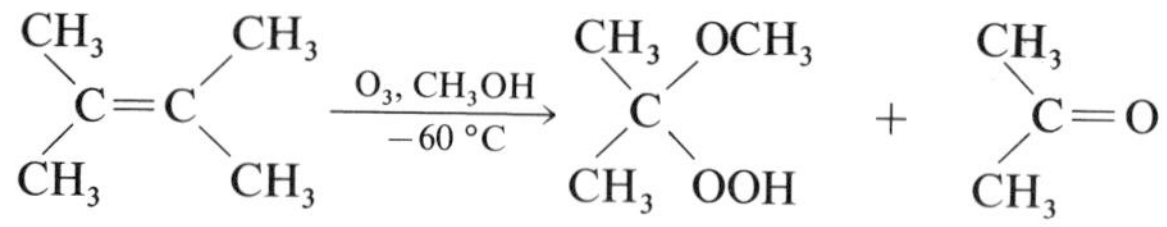

2,3-dimethyl-2-butene   2-methoxy-2-propanehydroperoxide   acetone

The reaction of methanol with the ionic intermediate resulting from fragmentation of the molozonide would give the observed product.

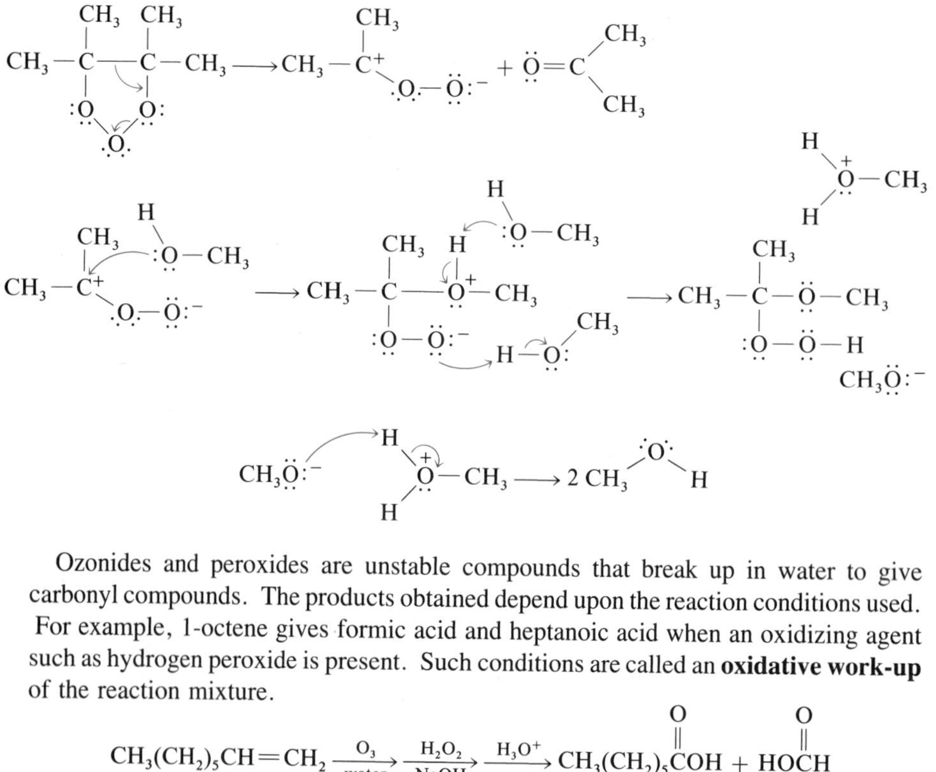

Ozonides and peroxides are unstable compounds that break up in water to give carbonyl compounds. The products obtained depend upon the reaction conditions used. For example, 1-octene gives formic acid and heptanoic acid when an oxidizing agent such as hydrogen peroxide is present. Such conditions are called an **oxidative work-up** of the reaction mixture.

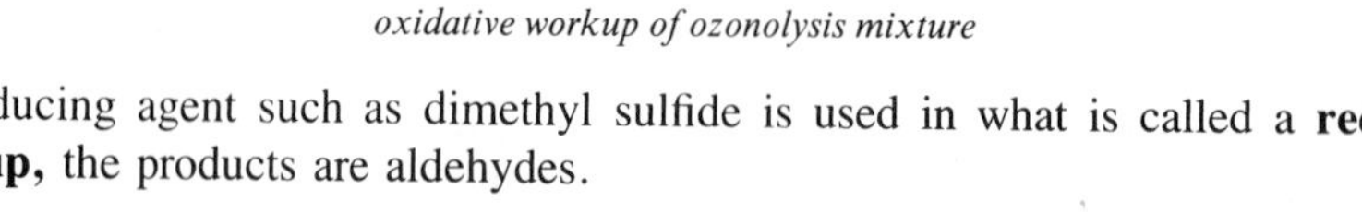

$$CH_3(CH_2)_5CH{=}CH_2 \xrightarrow[\text{water} \atop 10\,°C]{O_3} \xrightarrow[\text{NaOH}]{H_2O_2} \xrightarrow{H_3O^+} CH_3(CH_2)_5\overset{\displaystyle O}{\overset{\|}{C}}OH + HO\overset{\displaystyle O}{\overset{\|}{C}}H$$

1-octene   heptanoic   formic
                                        acid   acid

oxidative workup of ozonolysis mixture

If a reducing agent such as dimethyl sulfide is used in what is called a **reductive work-up,** the products are aldehydes.

$$CH_3(CH_2)_5CH{=}CH_2 \xrightarrow[-60\,^{\circ}C]{O_3,\,CH_3OH} \xrightarrow[-60\,^{\circ}C]{(CH_3)_2S} \underset{\text{heptanal}}{CH_3(CH_2)_5\overset{\overset{\displaystyle O}{\|}}{C}H} + \underset{\text{formaldehyde}}{H\overset{\overset{\displaystyle O}{\|}}{C}H} + \underset{\substack{\text{dimethyl}\\ \text{sulfoxide}}}{CH_3\overset{\overset{\displaystyle O}{\|}}{S}CH_3}$$

1-octene

oxidation product of dimethyl sulfide

reductive work-up of ozonolysis mixture

Dimethyl sulfide acts as a reducing agent because it reacts with oxygen to form a stable compound, dimethyl sulfoxide and thereby prevents oxidation of the aldehyde products to carboxylic acids.

The reaction of ozone with cyclohexene also demonstrates the different products obtained with changing reaction conditions. Cyclohexene is a cyclic alkene so reaction with ozone gives a compound with two aldehyde groups still attached to each other by the remainder of the chain when dimethyl sulfide is used in the decomposition of the ozonide.

cyclohexene

reductive work-up

1,6-hexanedial

dialdehyde from cleavage of the double bond in cyclohexene by ozone

62%

+   CH_3SCH_3 (with O above S)

dimethyl sulfoxide

If hydrogen peroxide is used after the addition of ozone, the corresponding dicarboxylic acid is formed.

cyclohexene

oxidative work-up

1,6-hexanedioic acid
adipic acid
85%

Thus, if a reducing agent is not used in the decomposition of an ozonide, aldehyde products are oxidized to carboxylic acids. In fact, ozone is used to synthesize carboxylic acids by cleavage of multiple bonds (Section 10.5A).

If the alkene has two organic groups on one or both sides of the double bond, a product of the reaction with ozone is a ketone. 2,3-Dimethyl-2-butene, for example, gives acetone on treatment with ozone (p. 346). A ketone, in contrast to an aldehyde, cannot be easily oxidized. The addition of any more oxygen atoms to the carbonyl group would require that a carbon-carbon single bond be broken. 2-Methyl-1-hexene is another example of an alkene that gives a ketone with ozone. Zinc metal is used as the reducing agent in the decomposition of the ozonide to prevent the buildup of explosive peroxides in the reaction mixture. Formaldehyde would also be expected as a product of this reaction, but is not isolated from the reaction mixture.

$$CH_3CH_2CH_2CH_2\overset{\overset{\displaystyle CH_3}{|}}{C}=CH_2 \xrightarrow[\substack{\text{dichloromethane} \\ -78\,°C}]{O_3} \xrightarrow[\substack{\text{acetic acid} \\ \text{water}}]{Zn} CH_3CH_2CH_2CH_2\overset{\overset{\displaystyle O}{\|}}{C}CH_3 + [H\overset{\overset{\displaystyle O}{\|}}{C}H]$$

2-methyl-1-hexene		2-hexanone	*not*
	reductive work-up	60%	*isolated*

B. Structure Determination by Ozonolysis

The final result of the oxidation of an alkene with ozone is the cleavage of the carbon-carbon double bond with the formation of a carbon-oxygen double bond on each fragment of the molecule. The overall reaction is called an **ozonolysis reaction,** meaning a cleavage of bonds by the use of ozone. The compounds formed as a result of ozonolysis—aldehydes, ketones, and carboxylic acids—are easily identified. A knowledge of the fragments that form allows us to reconstruct the original molecule. For example, if we identify heptanal and formaldehyde as ozonolysis products of an alkene of unknown structure, we would be able to conclude that the original alkene was 1-octene.

heptanal	formaldehyde	1-octene

We do this by mentally removing the oxygen atoms and joining the carbon atoms of the two carbonyl groups with a double bond.

Ozonolysis was used to determine the structures of the two alkenes formed from the acid-catalyzed dimerization of 2-methylpropene (Section 5.6C). One of the products gave formaldehyde and 4,4-dimethyl-2-pentanone when treated first with ozone, then with zinc and water.

$$CH_3\overset{\overset{\displaystyle CH_3}{|}}{\underset{\underset{\displaystyle CH_3}{|}}{C}}CH_2\overset{\overset{\displaystyle CH_3}{|}}{C}=CH_2 \xrightarrow{O_3} \xrightarrow[\text{water}]{Zn} CH_3\overset{\overset{\displaystyle CH_3}{|}}{\underset{\underset{\displaystyle CH_3}{|}}{C}}CH_2\overset{\overset{\displaystyle CH_3}{|}}{C}=O + O=CH_2$$

2,4,4-trimethyl-1-pentene	4,4-dimethyl-2-pentanone	formaldehyde

This alkene was assigned the structure 2,4,4-trimethyl-1-pentene. The other alkene gave acetone and 2,2-dimethylpropanal as ozonolysis products, and hence was 2,4,4-trimethyl-2-pentene.

$$CH_3\overset{\overset{\displaystyle CH_3}{|}}{\underset{\underset{\displaystyle CH_3}{|}}{C}}CH=\overset{\overset{\displaystyle CH_3}{|}}{C}CH_3 \xrightarrow{O_3} \xrightarrow[\text{water}]{Zn} CH_3\overset{\overset{\displaystyle CH_3}{|}}{\underset{\underset{\displaystyle CH_3}{|}}{C}}-\overset{\overset{\displaystyle H}{|}}{C}=O + O=\overset{\overset{\displaystyle CH_3}{|}}{C}CH_3$$

2,4,4-trimethyl-2-pentene	2,2-dimethylpropanal	acetone

C. Ozonolysis of Alkynes

Alkynes react with ozone to give carboxylic acids. 1-Hexyne, for example, is converted into formic acid and pentanoic acid upon ozonolysis.

$$CH_3CH_2CH_2CH_2C\equiv CH \xrightarrow[\substack{\text{carbon} \\ \text{tetrachloride} \\ 0\,°C}]{O_3} \xrightarrow{H_2O} CH_3CH_2CH_2CH_2\overset{\overset{\displaystyle O}{\|}}{C}OH + H\overset{\overset{\displaystyle O}{\|}}{C}OH$$

1-hexyne   pentanoic acid   formic acid

In another example, 9-octadecynedioic acid is converted to two moles of nonanedioic acid by ozonolysis.

$$HO\overset{\overset{\displaystyle O}{\|}}{C}(CH_2)_7C\equiv C(CH_2)_7\overset{\overset{\displaystyle O}{\|}}{C}OH \xrightarrow[\substack{\text{acetic} \\ \text{acid} \\ 25\,°C}]{O_3} \xrightarrow{H_2O} 2\ HO\overset{\overset{\displaystyle O}{\|}}{C}(CH_2)_7\overset{\overset{\displaystyle O}{\|}}{C}OH$$

9-octadecynedioic acid   nonanedioic acid 81%

Thus, ozonolysis breaks the multiple bond in alkynes to give fragments that have carboxylic acid groups at the carbon atoms that were part of the original triple bond. It is possible to reconstruct the structure of an alkyne if we know the fragments that result from its ozonolysis. For example, ozonolysis of an alkyne with the molecular formula $C_{15}H_{24}$ gives two moles of hexanoic acid and one of malonic acid.

$$\text{alkyne} \xrightarrow[\substack{\text{carbon} \\ \text{tetrachloride}}]{O_3} \xrightarrow{H_2O} 2\ CH_3(CH_2)_4\overset{\overset{\displaystyle O}{\|}}{C}OH + HO\overset{\overset{\displaystyle O}{\|}}{C}CH_2\overset{\overset{\displaystyle O}{\|}}{C}OH$$

$C_{15}H_{24}$ hexanoic acid   malonic acid

The only way these fragments can be put back together to give one molecule is shown below.

$$CH_3(CH_2)_4\overset{\overset{\displaystyle OH}{|}}{C}=O \qquad O=\overset{\overset{\displaystyle OH}{|}}{C}CH_2\overset{\overset{\displaystyle OH}{|}}{C}=O \qquad O=\overset{\overset{\displaystyle OH}{|}}{C}(CH_2)_4CH_3$$

hexanoic acid   malonic acid   hexanoic acid

↑ ozonolysis

$$CH_3(CH_2)_4C\equiv CCH_2C\equiv C(CH_2)_4CH_3$$

6,9-pentadecadiyne
$C_{15}H_{24}$

In the next section, we will explore the use of the molecular formula of a compound to extract information about the kinds of functional groups it may contain. Combining this analysis with the results from ozonolysis allows us to assign structure to a variety of compounds.

PROBLEM 8.21 Complete the following equations.

(a) $CH_3CH_2C\equiv CCH_3 \xrightarrow{O_3} \xrightarrow{H_2O}$

(b) $CH_3\overset{\overset{\displaystyle CH_3}{|}}{C}=CHCH_3 \xrightarrow[\text{water}]{O_3 \quad Zn}$

(c) $\underset{\underset{CH_3}{|}}{CH_3CHCH}=\underset{\underset{CH_2CH_3}{|}}{CCH_2CH_3}$ $\xrightarrow{O_3,\ CH_3OH}$ $\xrightarrow{(CH_3)_2S}$

(d) $\underset{\underset{CH_3}{|}}{CH_3CHCH}=\underset{\underset{CH_2CH_3}{|}}{CCH_2CH_3}$ $\xrightarrow[\text{water}]{O_3}$ $\xrightarrow{H_2O_2,\ NaOH}$ $\xrightarrow{H_3O^+}$

(e) [cyclohexene ring with CH_3] $\xrightarrow{O_3,\ CH_3OH}$ $\xrightarrow{(CH_3)_2S}$

(f) [large ring with two double bonds] $\xrightarrow[\substack{\text{dichloromethane}\\0\ °C}]{O_3}$ $\xrightarrow{H_2O_2,\ NaOH}$

(g) $CH_3CH_2\underset{\underset{CH_3}{|}}{C}=CH_2$ $\xrightarrow[\text{water}]{O_3}$ $\xrightarrow{Zn}$

(h) $CH_3(CH_2)_{13}CH=CH_2$ $\xrightarrow[\text{water}]{O_3}$ $\xrightarrow{H_2O_2,\ NaOH}$ $\xrightarrow{H_3O^+}$

PROBLEM 8.22   Predict the structure of the starting alkene or alkyne that would be expected to give the ozonolysis products shown in each case.

(a) $CH_3\overset{O}{\overset{||}{C}}CH_3 + CH_3CH_2CH_2\overset{O}{\overset{||}{C}}H$ (b) $H\overset{O}{\overset{||}{C}}CH_2CH_2CH_2\overset{O}{\overset{||}{C}}H$

(c) $CH_3\overset{O}{\overset{||}{C}}OH + CH_3CH_2\underset{\underset{CH_3}{|}}{CH}-\overset{O}{\overset{||}{C}}OH$ (from an alkyne)

(d) $CH_3\overset{O}{\overset{||}{C}}CH_2CH_2CH_2CH_2\overset{O}{\overset{||}{C}}CH_3$ (e) $CH_3\overset{O}{\overset{||}{C}}CH_2CH_2CH_3 + CH_3\overset{O}{\overset{||}{C}}H$

(f) [benzene ring with $\overset{O}{\overset{||}{C}}OH$ and $CH_2\overset{O}{\overset{||}{C}}OH$]   (g) $CH_3\overset{O}{\overset{||}{C}}CH_2CH_2CH_2CH_3 + H\overset{O}{\overset{||}{C}}H$

(h) $CH_3CH_2\overset{O}{\overset{||}{C}}H + CH_3\overset{O}{\overset{||}{C}}CH_3$

PROBLEM 8.23   In early research into the structure of branched hydrocarbons, chemists determined the structure of a number of alkenes by ozonolysis. What products would you expect to get from the ozonolysis of a mixture of alkenes consisting of 90% 3-ethyl-5,5-dimethyl-2-hexene and 10% 4-ethyl-2,2-dimethyl-3-hexene? The reaction mixture was treated with zinc and water.

PROBLEM 8.24   A tertiary alcohol was dehydrated to give an alkene the structure of which was determined by ozonolysis. Treatment of the alkene with ozone, then with water and zinc gave equal amounts of 3-pentanone.

$$\overset{\text{O}}{\overset{\|}{\text{CH}_3\text{CH}_2\text{CCH}_2\text{CH}_3}}$$

3-pentanone

$$\overset{\text{O}}{\overset{\|}{\text{CH}_3\text{CH}}}$$

acetaldehyde

What is the structure of the alcohol?

PROBLEM 8.25   In Section 6.8 (p. 250), you learned about the use of radioactive carbon, ^{14}C, to show that the allyl cation had delocalization of charge. Describe how you would do the chemical degradation necessary to obtain experimental evidence to prove where the radioactivity appears in the product molecules.

8.7

Structural Information from Molecular Formulas

A molecular formula tells how many atoms of each of the different elements are present in the molecules of a given compound. One of the pieces of information that can be extracted from the molecular formula is whether an organic compound is **saturated,** meaning that there are $2n + 2$ hydrogen atoms present for n carbon atoms.

If the molecular formula for the compound fits the formula $C_nH_{2n + 2}$, it cannot contain multiple bonds or rings. Every time the molecular formula has fewer hydrogen atoms per carbon atom than this formula requires, the compound is said to have **units of unsaturation.** The compound is assigned *one unit of unsaturation for each two hydrogen atoms that are missing* when the molecular formula of the compound is compared with the formula corresponding to $C_nH_{2n + 2}$. A unit of unsaturation may be a double bond between two carbon atoms or between a carbon and an oxygen atom, or it may be a ring. A triple bond introduces two units of unsaturation. These concepts are easier to grasp in concrete examples.

Hexane has the molecular formula C_6H_{14}, where $n = 6$, $2n + 2 = 14$. Hexane is a saturated compound, containing no rings or double bonds. A compound having the molecular formula C_6H_{12} has one unit of unsaturation. The reasoning is as follows: A compound containing six carbon atoms is saturated when it has the molecular formula C_6H_{14} ($C_nH_{2n + 2}$ when $n = 6$). The compound C_6H_{12} is missing two hydrogen atoms, and therefore has *one* unit of unsaturation. The unit of unsaturation may be a ring, so C_6H_{12} may be cyclohexane, or methylcyclopentane, or ethylcyclobutane, or a dimethylcyclobutane, among other compounds. The unit of unsaturation may also be a double bond, so C_6H_{12} may be any one of a number of open-chain alkenes such as 1-hexene, 3-hexene, 2-methyl-1-pentene, and so on.

cyclohexane   methylcyclopentane   ethylcyclobutane   1,2-dimethylcyclobutane

some C_6H_{12} compounds in which the unit of unsaturation is a ring

$$CH_2CH=CHCH_2CH_3 \qquad CH_3CH_2CH_2CH_2CH=CH_2 \qquad CH_3CH_2CH_2\overset{\displaystyle CH_3}{\underset{|}{C}}=CH_2$$

$$\text{3-hexene} \qquad\qquad\qquad \text{1-hexene} \qquad\qquad\qquad \text{2-methyl-1-pentene}$$

some C_6H_{12} compounds in which the unit of unsaturation is
a double bond

The molecular formula alone does not allow us to say whether the compound is an alkene or a cycloalkane, or tell us which alkene or cycloalkane it might be. For such a determination, we need further experimental evidence. For example, we might go to the laboratory and find that the compound with the molecular formula C_6H_{12} rapidly decolorizes a solution of bromine in carbon tetrachloride, and gives a brown precipitate with cold aqueous potassium permanganate. These experimental data make it highly probable that the compound is an alkene. For further evidence we might investigate the products of hydrogenation and ozonolysis of the compound. If the compound reacts in the presence of a catalyst with one molar equivalent of hydrogen to give 2-methyl-pentane, and yields 2-pentanone and formaldehyde upon ozonolysis, we could state with some certainty that it is 2-methyl-1-pentene.

$$CH_3CH_2CH_2\overset{\displaystyle CH_3}{\underset{\displaystyle OH}{\underset{|}{\overset{|}{C}}}}CH_2OH \xleftarrow[\text{water}]{KMnO_4} CH_3CH_2CH_2\overset{\displaystyle CH_3}{\underset{|}{C}}=CH_2 \xrightarrow[\substack{\text{carbon} \\ \text{tetrachloride}}]{Br_2} CH_3CH_2CH_2\overset{\displaystyle CH_3}{\underset{\displaystyle Br}{\underset{|}{\overset{|}{C}}}}CH_2Br$$

$$\text{2-methyl-1-pentene}$$
$$C_6H_{12}$$

$$H_2 \diagup\qquad\qquad \diagdown O_3,\ CH_3OH$$
$$Pt \qquad\qquad\qquad (CH_3)_2S$$

$$CH_3CH_2CH_2\overset{\displaystyle CH_3}{\underset{|}{CH}}CH_3 \qquad\qquad CH_3CH_2CH_2\overset{O}{\overset{\|}{C}}CH_3 + H\overset{O}{\overset{\|}{C}}H$$

$$\text{2-methylpentane} \qquad\qquad\qquad \text{2-pentanone} \qquad \text{formaldehyde}$$
$$C_6H_{14}$$

The experiments carried out on 2-methyl-1-pentene demonstrate the ways in which different kinds of unsaturation can be detected. Especially important is the use of hydrogenation. Units of unsaturation that resist catalytic hydrogenation are identified as rings. For example, benzene has the molecular formula C_6H_6, and thus has four units of unsaturation (C_6H_6 is eight hydrogen atoms short of being C_6H_{14}). Benzene can be catalytically hydrogenated, usually at higher temperatures or pressures or with special catalysts, to C_6H_{12} with one unit of unsaturation. Thus, three of the units of un-saturation in benzene are multiple bonds that add hydrogen; the fourth one is a ring that is stable to hydrogenation.

$$\text{benzene} \xrightarrow[\substack{\text{catalyst} \\ \Delta,\ P}]{H_2} \text{cyclohexane}$$

benzene
C_6H_6

four units of
unsaturation;
one ring, three
double bonds

cyclohexane
C_6H_{12}

one unit of
unsaturation;
a ring

In general, compounds with very large numbers of units of unsaturation are likely to be aromatic hydrocarbons (Section 2.8). Cyclopropane rings may be opened on hydrogenation (Section 3.8A). For the purposes of our exercises we will assume, therefore, that rings have four or more carbon atoms in them.

No modification is necessary in the calculation of units of unsaturation if oxygen is present in the molecule. For example, $C_4H_{10}O$ is a saturated compound; its molecular formula corresponds to C_nH_{2n+2}. The oxygen atom can be disregarded in the calculation. A compound with the molecular formula $C_4H_{10}O$ cannot contain any rings or double bonds. It might be one of the butyl alcohols or an ether, such as diethyl ether.

A compound with molecular formula C_4H_8O, on the other hand, has one unit of unsaturation. It might contain a carbon-carbon double bond, a carbon-oxygen double bond, or a ring containing just carbon atoms, or one including an oxygen atom. Some of the stable compounds representing these different categories of functionality are shown below.

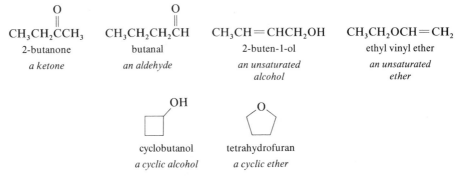

O ‖ $CH_3CH_2CCH_3$	O ‖ $CH_3CH_2CH_2CH$	$CH_3CH{=}CHCH_2OH$	$CH_3CH_2OCH{=}CH_2$
2-butanone	butanal	2-buten-1-ol	ethyl vinyl ether
a ketone	*an aldehyde*	*an unsaturated alcohol*	*an unsaturated ether*

cyclobutanol
a cyclic alcohol

tetrahydrofuran
a cyclic ether

some compounds having the molecular formula C_4H_8O

Other experimental evidence such as the origin of the compound or the results of qualitative analysis tests would be necessary to assign a definite structure to such a compound.

PROBLEM 8.26   Suppose a compound with the molecular formula C_4H_8O rapidly decolorized bromine in carbon tetrachloride and gave acetaldehyde, CH_3CHO, and 2-hydroxyethanal, $HOCH_2CHO$, upon ozonolysis. What structure would you assign to it?

The formula can be applied to alkyl halides if hydrogen is substituted for halogen before the calculations are done. A compound with the molecular formula C_5H_9Br has one unit of unsaturation (C_5H_9Br corresponds to C_5H_{10}, which is two hydrogen atoms short of saturation, C_5H_{12}). This compound does not react with bromine in carbon tetrachloride, nor with cold aqueous potassium permanganate; therefore, it does not have a double bond. The unit of unsaturation must be present as a ring.

PROBLEM 8.27   Write structural formulas, including stereochemistry, for all possible cyclic compounds having a four-membered ring or larger, with the molecular formula C_5H_9Br.

A triple bond corresponds to two units of unsaturation. For example, a compound with the molecular formula C_7H_{12} has two units of unsaturation (C_7H_{12} is four hydrogen atoms short of being C_7H_{16}). Such a compound may contain one triple bond, two double bonds, a ring and a double bond, or two rings.

PROBLEM 8.28   Write one structure corresponding to the molecular formula C_7H_{12} for each of the categories of compounds described as possible on the previous page.

If the compound C_7H_{12} decolorizes bromine in carbon tetrachloride and adds hydrogen to give a new compound with the molecular formula C_7H_{16}, structures containing rings can be eliminated as a possibility. If we have further information that ozonolysis of the compound gives butanoic acid and propanoic acid, the compound must be 3-heptyne.

$$CH_3CH_2CH_2C\equiv CCH_2CH_3 \xrightarrow[\text{carbon tetrachloride}]{Br_2} CH_3CH_2CH_2\overset{\displaystyle Br}{\underset{\displaystyle Br}{\overset{|}{\underset{|}{C}}}}-\overset{\displaystyle Br}{\underset{\displaystyle Br}{\overset{|}{\underset{|}{C}}}}CH_2CH_3$$

$$\underset{\substack{C_7H_{12} \\ \text{3-heptyne}}}{}$$

with $\xrightarrow[\text{Pt}]{H_2}$ giving

$$CH_3CH_2CH_2CH_2CH_2CH_2CH_3$$
$$\underset{\substack{\text{heptane} \\ C_7H_{16}}}{}$$

and with O_3 then H_2O_2 giving

$$\underset{\text{butanoic acid}}{CH_3CH_2CH_2\overset{\displaystyle O}{\overset{\|}{C}}OH} \qquad \underset{\text{propanoic acid}}{HO\overset{\displaystyle O}{\overset{\|}{C}}CH_2CH_3}$$

PROBLEM 8.29   For each of the following cases, identify the number of units of unsaturation in the original compound and say how many of them are multiple bonds and how many are rings.

(a) $C_5H_6 \xrightarrow[\text{catalyst}]{H_2} C_5H_{10}$ (b) $C_7H_8 \xrightarrow[\text{catalyst}]{H_2} C_7H_{14}$

(c) $C_6H_{10} \xrightarrow[\text{catalyst}]{H_2} C_6H_{14}$ (d) $C_{10}H_{16} \xrightarrow[\text{catalyst}]{H_2} C_{10}H_{18}$

PROBLEM 8.30   An alcohol, $C_9H_{20}O$, was dehydrated to give a mixture of alkenes. Treatment of the mixture of alkenes with ozone in an alkane solvent and then with zinc and water gave the following compounds:

$$\underset{\text{acetaldehyde}}{CH_3\overset{\displaystyle O}{\overset{\|}{C}}H}$$

$$\underset{\text{propanal}}{CH_3CH_2\overset{\displaystyle O}{\overset{\|}{C}}H}$$

$$\underset{\text{3-hexanone}}{CH_3CH_2\overset{\displaystyle O}{\overset{\|}{C}}CH_2CH_2CH_3}$$

$$\underset{\text{4-heptanone}}{CH_3CH_2CH_2\overset{\displaystyle O}{\overset{\|}{C}}CH_2CH_2CH_3}$$

What are the structures of the alkenes formed in the dehydration reaction? What is the structure of the alcohol?

8.8

Infrared Spectra of Alkenes and Alkynes

A. Infrared Spectra of Alkenes

In Section 7.7B, the stretching frequency for the carbon-oxygen double bond was contrasted with that of the carbon-carbon double bond. The absorption band arising from the stretching vibration of a carbonyl group is a strong one and is usually unmistakable. The carbon-carbon double bond stretching vibration absorbs at $1680-1640$ cm^{-1}, and is usually much weaker than the carbonyl band. It is seen in the spectra of terminal alkenes, but may be completely missing in the spectra of internal alkenes. The spectra of 3-methyl-1-pentene and (E)-3-hexene illustrate this (Figure 8.5). In the spectrum of 3-methyl-1-pentene, the carbon-carbon double bond stretching frequency is seen at 1645 cm^{-1}. In addition, the carbon-hydrogen stretching frequency for the terminal vinyl hydrogens appears near 3100 cm^{-1}. The bending vibration for the same carbon-hydrogen bonds is at 1000 cm^{-1}. The spectrum of (E)-3-hexene

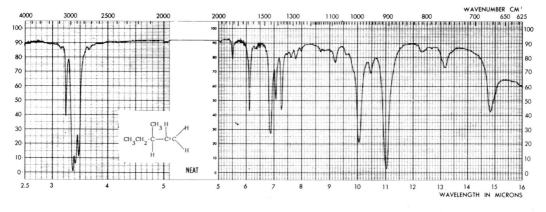

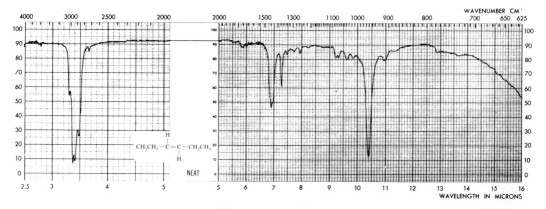

Figure 8.5   Infrared specta of 3-methyl-1-pentene and (E)-3-hexene. (From *The Aldrich Library of Infrared Spectra*)

shows very little absorption where the carbon-carbon double bond stretching frequency comes. The symmetry of the molecule results in little or no change in dipole moment during the stretching vibration and hence, weak absorption of radiation. The stretching frequencies for the vinylic carbon-hydrogen bonds are also hardly distinguishable from the alkane carbon-hydrogen stretching frequencies, appearing only as shoulders on the larger bands. The bending frequency for the vinylic carbon-hydrogen bonds that are trans to each other appears at 960 cm^{-1}.

B. Infrared Spectra of Alkynes

The carbon-carbon triple bond has its stretching frequency in the region of 2260–2100 cm^{-1}. A terminal triple bond is easier to see in the infrared than an internal triple bond as the symmetry of a molecule containing an internal triple bond results in little or no change in dipole moment during the stretching vibration. Figure 8.6 demonstrates this with the spectra of 5-phenyl-1-pentyne and 6-phenyl-2-hexyne.

The carbon-carbon triple bond stretching frequency in 5-phenyl-1-pentyne appears slightly above 2100 cm^{-1}. In addition, the carbon-hydrogen bond stretching frequency for the terminal alkyne appears as a strong, sharp band at 3300 cm^{-1}. These two features taken together identify a terminal alkyne. The spectrum of 6-phenyl-2-hexyne has neither of these features. As you can see, an internal triple bond is not easily identified by infrared spectroscopy.

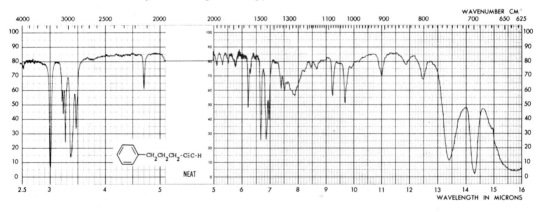

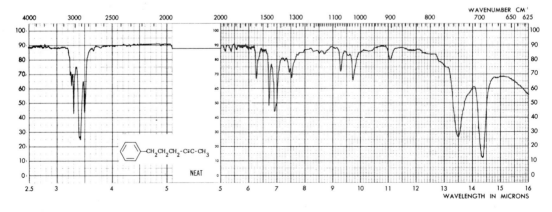

Figure 8.6   Infrared specta for 5-phenyl-1-pentyne and 6-phenyl-2-hexyne. (From *The Aldrich Library of Infrared Spectra*)

PROBLEM 8.31   The infrared spectra of a series of unknown compounds appear in Figure 8.7. The compounds may be alcohols, aldehydes, ketones, alkenes, or alkynes. Decide to which functional group class each compound belongs. Point out the bands that you used in making the assignments.

Compound A

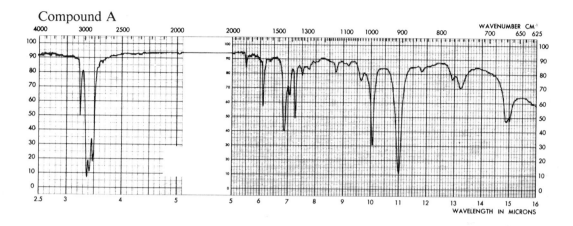

Compound B

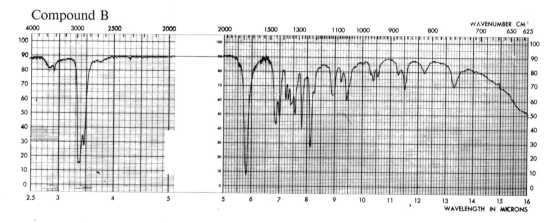

Compound C

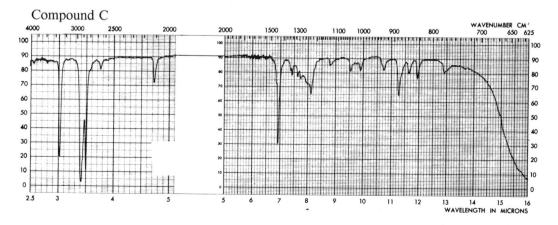

Figure 8.7   Spectra for Problem 8.31. (From *The Aldrich Library of Infrared Spectra*)

Compound D

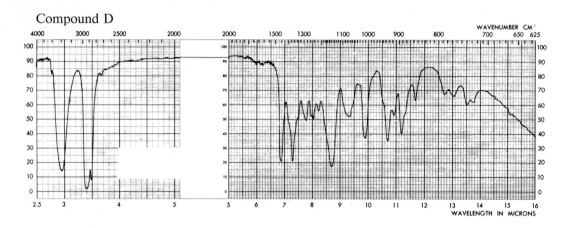

Compound E

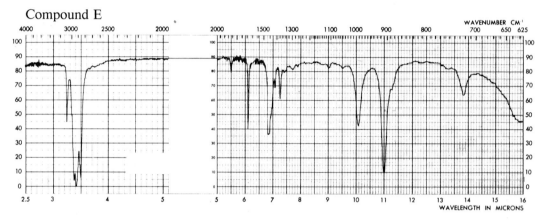

Figure 8.7   *(Continued)*

ADDITIONAL PROBLEMS

8.32   Using (*E*)-3-methyl-2-pentene as a typical alkene, write equations predicting its reactions with each of the following reagents. In cases where you know what the stereochemistry of the product(s) should be, illustrate it by appropriate three-dimensional representations.

(a) H_2, Pt   (b) H_2O, H_2SO_4 (c) HBr, no peroxides

(d) ![structure of m-chloroperbenzoic acid: benzene ring with Cl and C(=O)OOH substituents], chloroform   (e) $KMnO_4$, OH^-, water

(f) (1) O_3, CH_3OH; (2) $(CH_3)_2S$ (g) (1) OsO_4, diethyl ether; (2) KOH, water

(h) (1) O_3, water; (2) H_2O_2, NaOH; (3) H_3O^+ (i) product of (d) + HBr in methanol

(j) Br_2, carbon tetrachloride   (k) Br_2, H_2O (l) H_2, $RhCl[P(C_6H_5)_3]_3$

8.33   Using 2-pentyne as a typical alkyne, write equations predicting its reactions with each of the following reagents. In cases where you know what the stereochemistry of the product(s) should be, use appropriate conventions to illustrate it.

(a) H_2, $Pd/CaCO_3$, quinoline   (b) H_2 (excess), Pt

(c)   Na, NH_3(liq)   (d)   product of (a) +

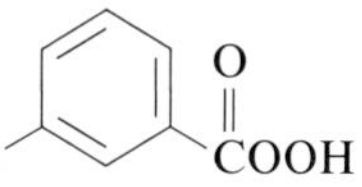

(e)   product of (c) +

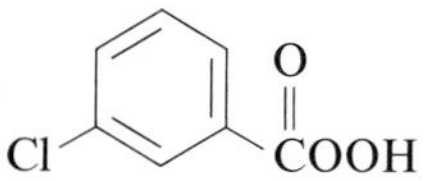

(f)   HBr (1 molar equivalent)

(g)   HBr (2 molar equivalents)   (h) Br_2 (1 molar equivalent)

(i) Br_2 (2 molar equivalents)   (j) $KMnO_4$, H_2O, buffer at pH 7

(k)   (1) O_3, carbon tetrachloride; (2) H_2O (l) H_2O, H_2SO_4, $HgSO_4$

(m) H_2, $RhCl[P(C_6H_5)_3]_3$ (n)   (1) BH_3, tetrahydrofuran; (2) CH_3CO_2H

8.34   Supply structural formulas for the intermediates and products designated by capital letters. Show the stereochemistry of the product(s) when known.

(a)

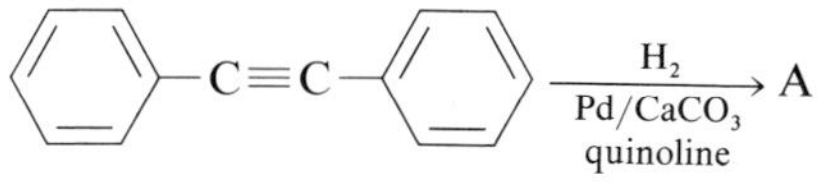

(b)

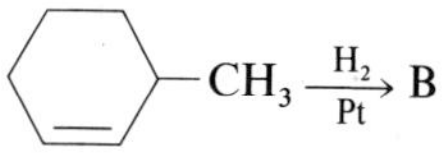

(c)

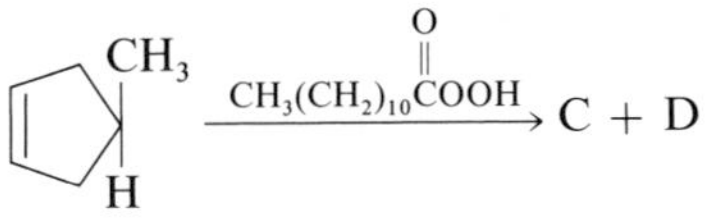

(d)

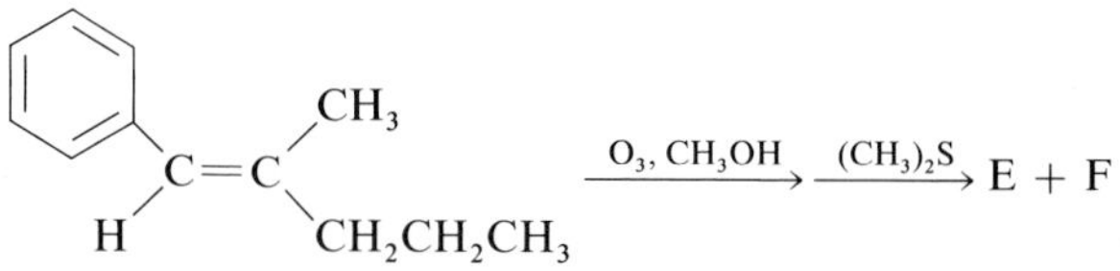

(e) $CH_3\overset{\displaystyle CH_3}{\underset{}{CH}}C\equiv CCH_2CH_3 \xrightarrow[\text{carbon tetrachloride}]{O_3} \xrightarrow{H_2O} G + H$

(f)

(g)

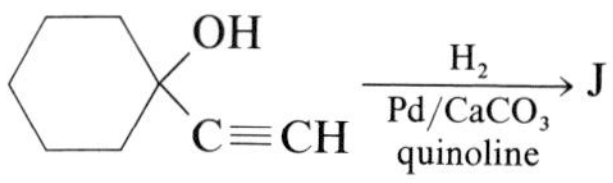

(h)

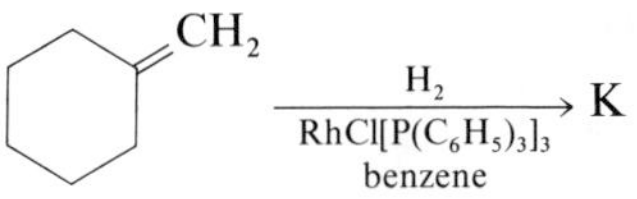

(i) $CH_3(CH_2)_7C\equiv C(CH_2)_7CO_2H \xrightarrow[\text{carbon tetrachloride}]{O_3} \xrightarrow{H_2O} L + M$

(j)

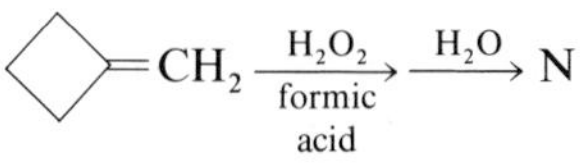

(k) $CH_3\overset{\displaystyle CH_3}{\underset{}{C}}=\overset{\displaystyle CH_3}{\underset{}{C}}CH_3 \xrightarrow[\text{diethyl ether}]{OsO_4} \xrightarrow[\text{water}]{KOH} O$

(l)

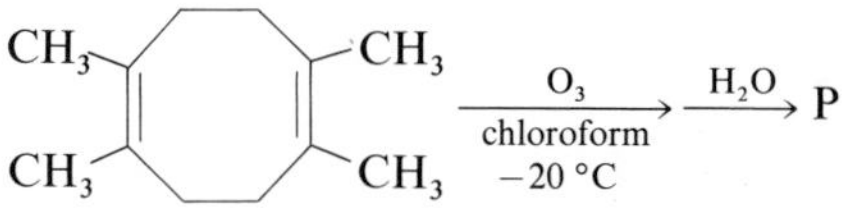

(m)

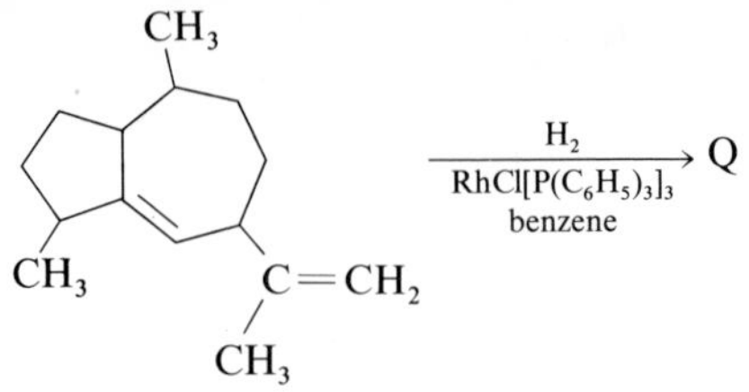

$$\xrightarrow[\substack{\text{RhCl[P(C}_6\text{H}_5)_3]_3 \\ \text{benzene}}]{\text{H}_2} \text{Q}$$

(n) 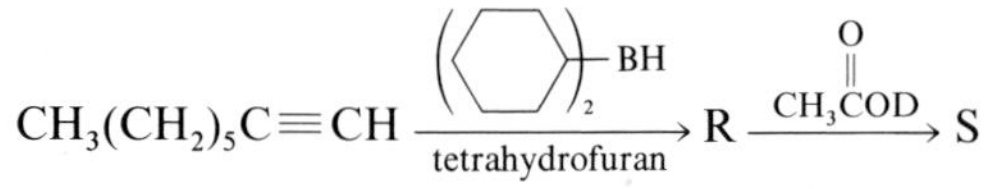 $CH_3(CH_2)_5C\equiv CH \xrightarrow[\text{tetrahydrofuran}]{\left(\begin{array}{c}\end{array}\right)_2 BH} R \xrightarrow{CH_3\overset{O}{\overset{\|}{C}}OD} S$

(o) 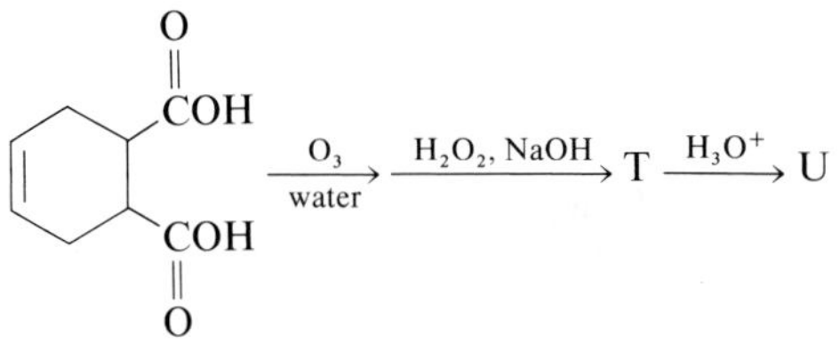 $\xrightarrow[\text{water}]{O_3} \xrightarrow{H_2O_2,\ NaOH} T \xrightarrow{H_3O^+} U$

(p) $\xrightarrow{OsO_4} \xrightarrow[\text{water}]{HClO_3} V$

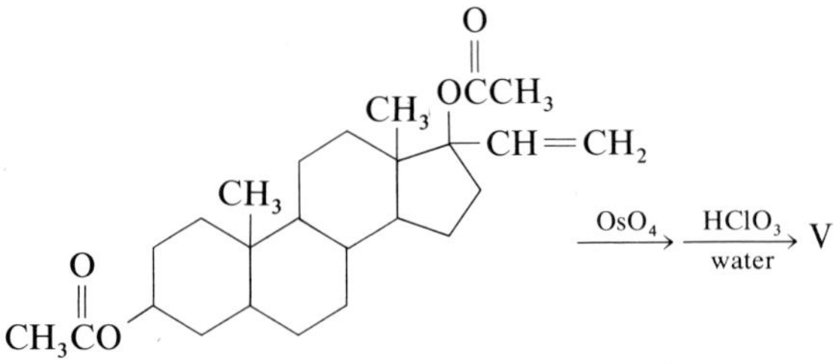

(q) $CH_3\overset{CH_3}{\overset{|}{C}}=CHCH_2CH_2\overset{O}{\overset{\|}{C}}OCH_3 \xrightarrow[\substack{\text{dichloromethane} \\ -78\ °C}]{O_3} \xrightarrow{(CH_3)_2S} W + X$

(r) 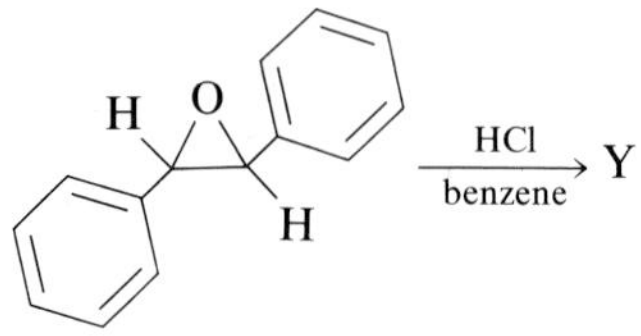 $\xrightarrow[\text{benzene}]{HCl} Y$

8.35   Show the reagents that would be necessary to carry out the following transformations. For some, more than one step may be necessary, in which case you should also show the compounds that will be formed as intermediates in the syntheses.

(a)

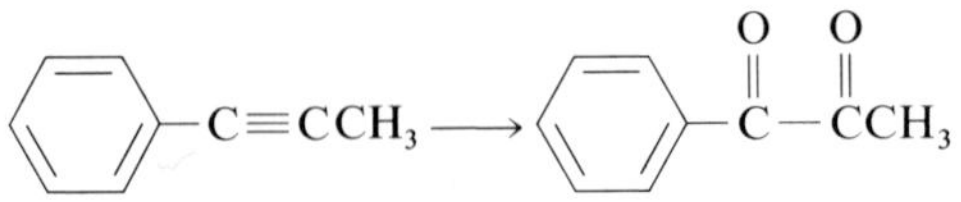

(b)

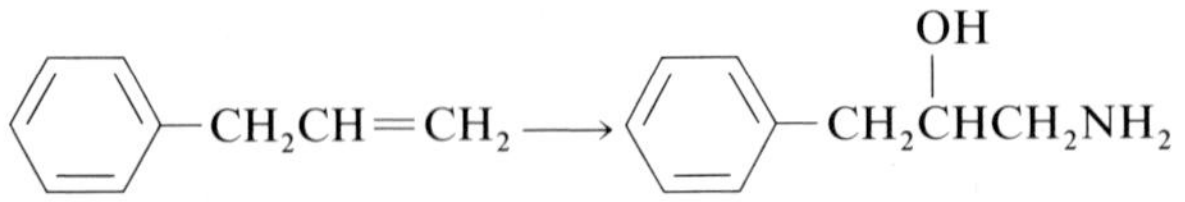

(c) 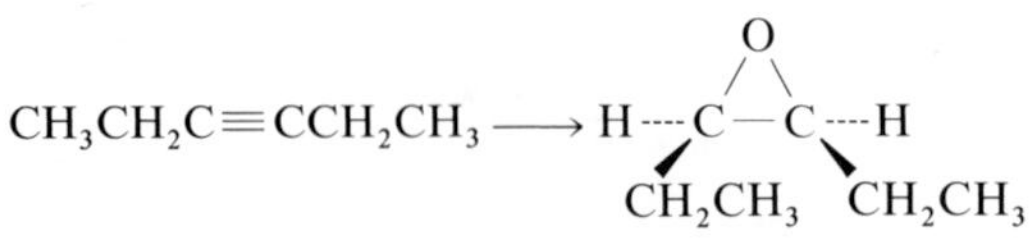 $CH_3CH_2C\equiv CCH_2CH_3 \longrightarrow$

(d) 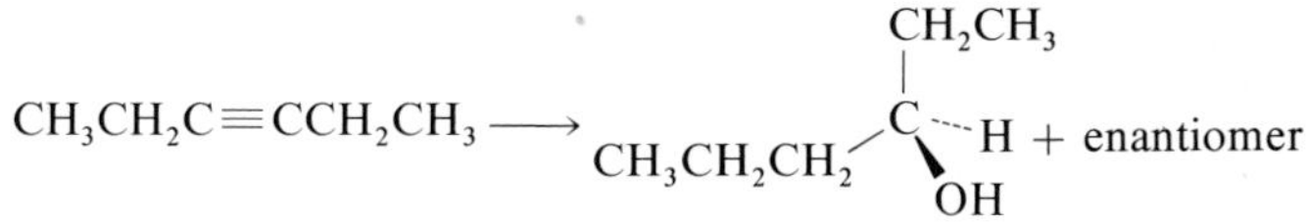 $CH_3CH_2C\equiv CCH_2CH_3 \longrightarrow$ + enantiomer

(e) 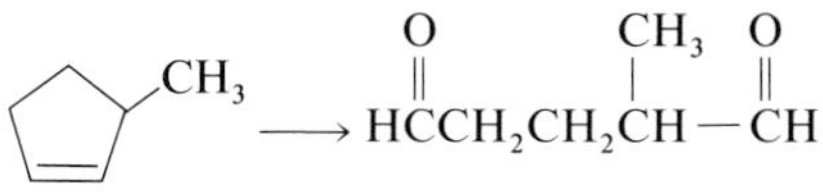 $\longrightarrow$

(f) 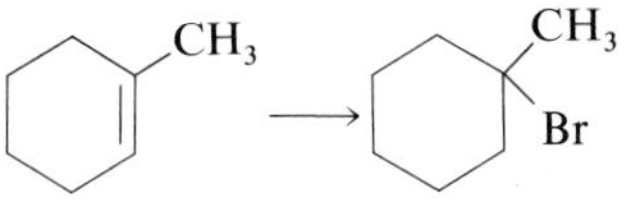 $\longrightarrow$

(g) 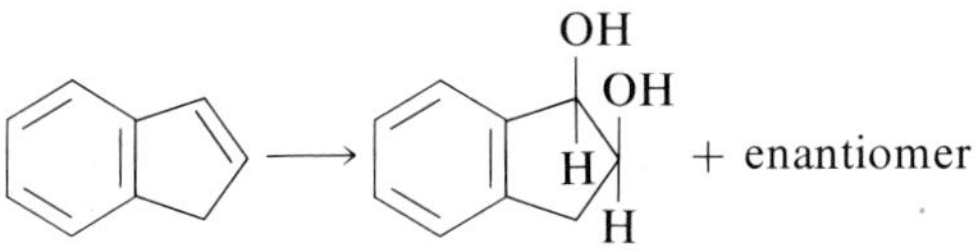 $\longrightarrow$ + enantiomer

(h) 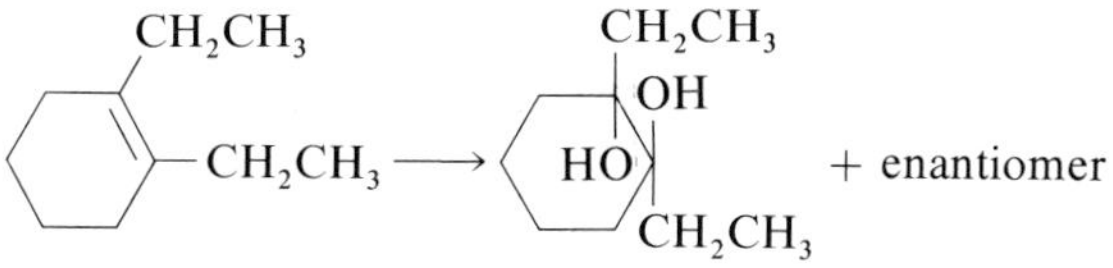 $\longrightarrow$ + enantiomer

(i) 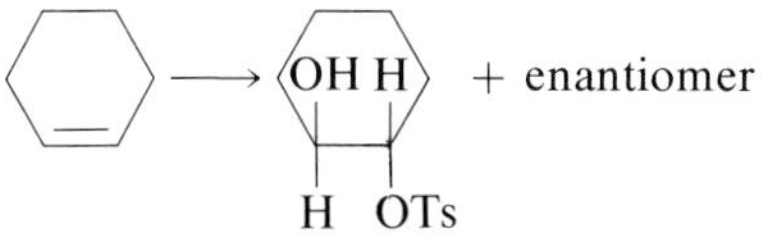 $\longrightarrow$ + enantiomer

(j) 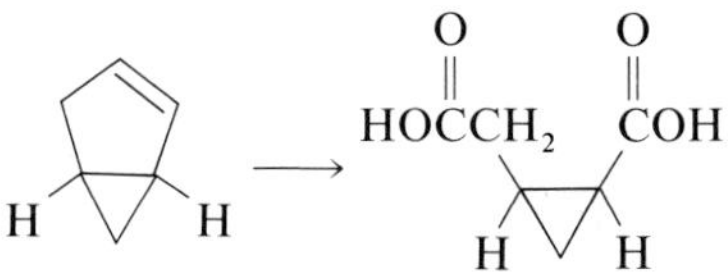 $\longrightarrow$

(k) 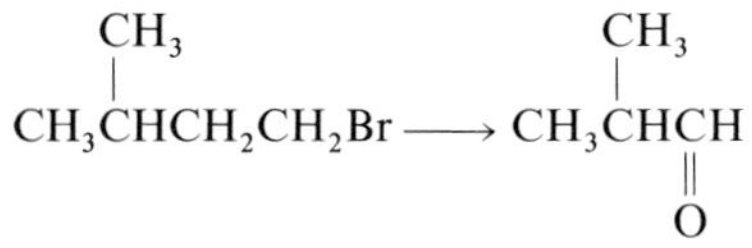 $\longrightarrow$

(l) 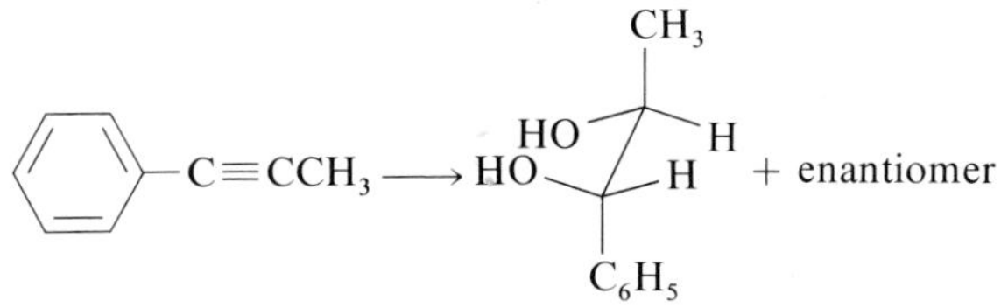 $-C\equiv CCH_3 \longrightarrow$ + enantiomer

(m) 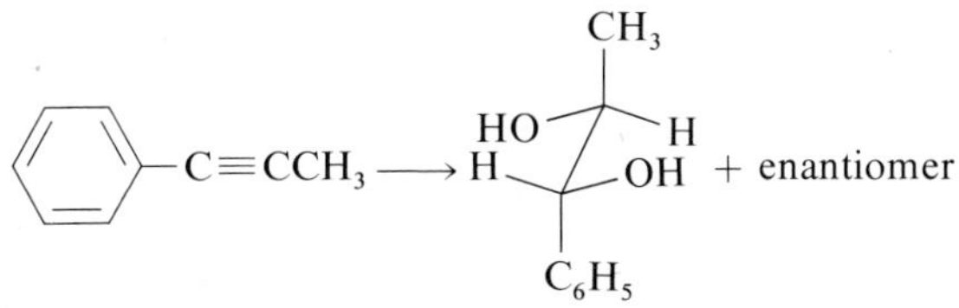 $-C\equiv CCH_3 \longrightarrow$ + enantiomer

(n)

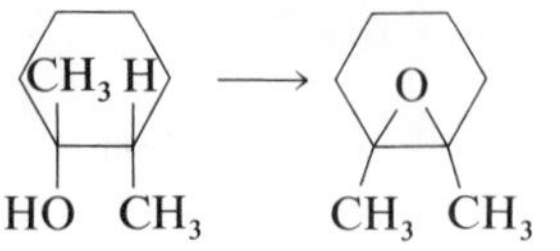

(o) 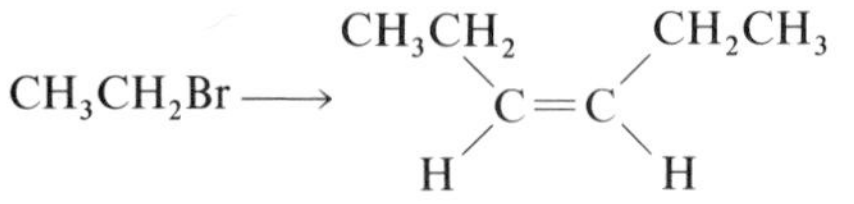 $CH_3CH_2Br \longrightarrow$

(p) 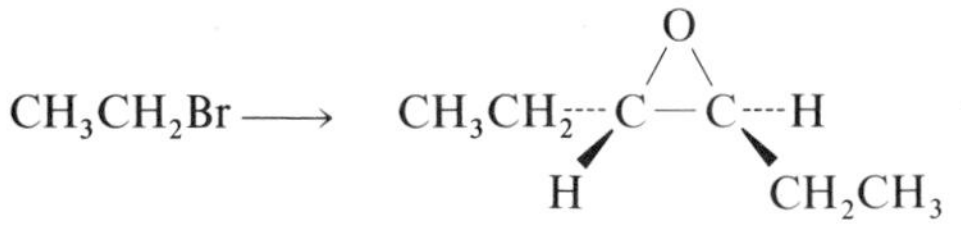 $CH_3CH_2Br \longrightarrow$ + enantiomer

8.36   For each of the following compounds, indicate how many units of unsaturation are present, and how many of those are rings.

(a) $C_{10}H_8 \xrightarrow[\text{catalyst}]{H_2} C_{10}H_{18}$ (b) $C_7H_{10} \xrightarrow[\text{catalyst}]{H_2} C_7H_{12}$

(c) $C_6H_{12}O \xrightarrow[\text{catalyst}]{H_2} C_6H_{14}O$ (d) $C_6H_8 \xrightarrow[\text{catalyst}]{H_2} C_6H_{12}$

(e) $C_5H_9Br \xrightarrow[\text{catalyst}]{H_2} C_5H_{11}Br$

8.37   The orientation of the dehydration reaction in unsymmetrical tertiary alcohols was the subject of considerable interest in the early days of research into the chemistry of hydrocarbons. An alcohol was dehydrated to give a mixture of alkenes, which on hydrogenation on platinum in acetic acid gave 2,2,4-trimethyloctane. When the mixture of alkenes was subjected to ozonolysis, and the reaction products decomposed in water in the presence of zinc, the major products were butanal and 4,4-dimethyl-2-pentanone. Small amounts of formaldehyde and 2,2-dimethyl-4-octanone were also formed. Traces of other carbonyl compounds were also observed.

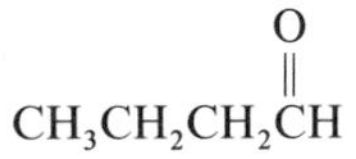

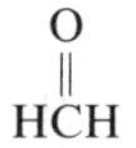

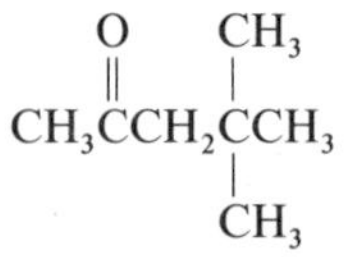

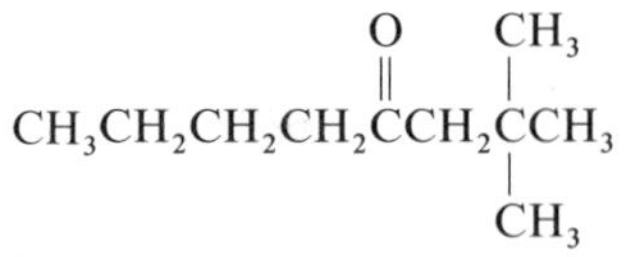

(a)   What are the structures of the alkenes that are formed in the dehydration reaction?
(b)   What is the structure of the alcohol?
(c)   Are other alkenes possible as dehydration products of the alcohol? What would their ozonolysis products be?
(d)   Can you suggest reasons(s) for the orientation of the major dehydration reaction that was observed?

8.38   (a)   The structure of the following bicyclic compound was investigated by hydrogenation and ozonolysis reactions. Give structural formulas for the products of the reactions shown.

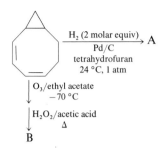

(b) Another bicyclic compound, Compound C, C_7H_{10}, also gives Compound B upon ozonolysis under the same conditions. What is the structure of Compound C?

8.39 (a) Compound A has the molecular formula C_7H_{12}. Reaction of Compound A with aqueous potassium permanganate in acetone gives Compound B, $C_7H_{14}O_2$. When Compound A is treated with peroxybenzoic acid in chloroform, Compound C, $C_7H_{12}O$, is formed. Compound C reacts with dilute aqueous sulfuric acid to give Compound D, $C_7H_{14}O_2$, which is isomeric with Compound B. Ozonolysis of Compound A gives 2,6-heptanedione. Assign structures to Compounds A, B, C, and D.

$$\underset{\text{2,6-heptanedione}}{CH_3\overset{\overset{\displaystyle O}{\|}}{C}CH_2CH_2CH_2\overset{\overset{\displaystyle O}{\|}}{C}CH_3}$$

(b) Compound E has the molecular formula C_7H_{10}. Hydrogenation of E gives Compound F, C_7H_{12}. Compound E decolorizes aqueous potassium permanganate and reacts with bromine in chloroform to give Compound G, $C_7H_{10}Br_2$. Compound E reacts with ozone in the presence of sodium hydroxide and hydrogen peroxide to give, after acidification, *cis*-1,3-cyclopentanedicarboxylic acid. What are the structures of Compounds E, F, and G?

$$\underset{\substack{\textit{cis-}\text{1,3-cyclopentanedicarboxylic} \\ \text{acid}}}{\underset{O}{\overset{H\qquad H}{HOC\qquad\qquad COH}}}$$

(c) Compound H, C_8H_{14}, is hydrogenated to Compound I, C_8H_{16}. Compound H also reacts with bromine in chloroform to give Compound J, $C_8H_{14}Br_2$. Ozonolysis of Compound H in an aqueous solution containing sodium hydroxide and hydrogen peroxide, followed by acidification of the reaction mixture gives octanedioic acid. What are the structures of Compounds H, I, and J?

$$\underset{\text{octanedioic acid}}{HO\overset{\overset{\displaystyle O}{\|}}{C}(CH_2)_6\overset{\overset{\displaystyle O}{\|}}{C}OH}$$

(d) Compound K, $C_{19}H_{36}O_2$, reacts with bromine in carbon tetrachloride to give Compound L, $C_{19}H_{36}Br_2O_2$. Hydrogenation of Compound K over a nickel catalyst gives Compound M, $C_{19}H_{38}O_2$. When Compound K is treated with ozone in methanol, and then with zinc in acetic acid, nonanal and methyl 9-oxononanoate are formed. Assign structures to Compounds K, L, and M.

$$\underset{\text{nonanal}}{CH_3(CH_2)_7\overset{\overset{\displaystyle O}{\|}}{C}H} \qquad \underset{\text{methyl 9-oxononanoate}}{H\overset{\overset{\displaystyle O}{\|}}{C}(CH_2)_7\overset{\overset{\displaystyle O}{\|}}{C}OCH_3}$$

(e) Compound N has the molecular formula C_8H_{12}. It reacts with hydrogen in the presence of palladium on barium sulfate to give Compound I, C_8H_{16} [part (c) of this problem]. When Compound N is treated with hydrogen bromide in acetic acid, Compound O, $C_8H_{13}Br$, is isolated. Ozonolysis of Compound N in acetic acid, and cleavage of the ozonide by heating the reaction mixture gives an 84% yield of succinic acid. What are the structures of Compounds N and O?

$$\underset{\text{succinic acid}}{\overset{\displaystyle O \qquad\quad O}{\overset{\displaystyle \| \qquad\quad \|}{HOCCH_2CH_2COH}}}$$

(f) Compound P, $C_{10}H_8$, is reduced by catalytic hydrogenation to Compound Q, $C_{10}H_{18}$. Reduction of Compound P with sodium metal in liquid ammonia gives Compound R, $C_{10}H_{10}$. Compound R reacts with bromine to give Compound S, $C_{10}H_{10}Br_2$. Treatment of Compound R with aqueous potassium permanganate gives mostly Compound T, $C_{10}H_{12}O_2$, and a small amount of 3-(2-carboxyphenyl)propanoic acid. Assign structures to Compounds P, Q, R, S, and T.

3-(2-carboxyphenyl)propanoic acid

8.40 Reactions of *cis*-4,5-dimethylcyclohexene and 3-*tert*-butylcyclohexene with *m*-chloroperoxybenzoic acid give predominantly one product in each case. Show the conformation of the starting alkene and predict, by showing the reaction with the peroxyacid, what the structure of the expected product should be for each starting material. (Hint: Using molecular models will help you.)

8.41 Compound A, $C_{13}H_{10}O$, isolated from the roots of *Carlina acaulis* and named Carlina-oxide, is hydrogenated to Compound B, $C_{13}H_{14}O$, which has the following structural formula.

Compound B

Compound B has a three-carbon chain with a phenyl ring and a furan ring, which resembles the phenyl ring in its resistance to hydrogenation at low pressures and temperatures.

(a) How many units of unsaturation does Carlina-oxide contain? How many of them were saturated by reaction with hydrogen? What structures are possible for Carlina-oxide?
(b) Carlina-oxide was treated with ozone in acetic acid. After removal of the acetic acid, the ozonide was decomposed with water. No reducing agent was used. Phenylacetic acid was isolated. What must the structure of Carlina-oxide be?

phenylacetic acid

8.42 In Section 8.5A, the analogy between a bromonium ion and a protonated oxirane is shown. We also considered the stereochemistry of the ring-opening reaction in epoxycyclohexane in the same section. (*R*)-(+)-4-Methylcyclohexene reacts with bromine to give two stereoisomeric 1,2-dibromo-4-methylcyclohexanes. One stereoisomer is by far the major product. Write the mechanism for the bromination reaction of (*R*)-(+)-4-methylcyclohexene and predict the stereochemistry of the products. Identify the major product. Do you expect the products to be optically active?

8.43 A number of different oxidizing agents have been used to recycle osmium tetroxide in reactions that convert alkenes to diols (Section 8.4A). One of the most successful was developed recently by Barry Sharpless at the Massachusetts Institute of Technology. In this reaction, *tert*-butyl hydroperoxide, $(CH_3)_3COOH$, in the presence of the base tetraethylammonium hydroxide, $(CH_3CH_2)_4N^+OH^-$, is used as the oxidizing agent. The osmate esters are cleaved by sodium bisulfite, $NaHSO_3$, and water. For the following alkenes, predict what the major products of these reactions will be: 2,3-dimethyl-2-butene, 1-decene, (*E*)-4-octene, and (*Z*)-4-octene. Include stereochemistry when it is pertinent.

8.44 Compound A, which is a degradation product of the antibiotic vermiculine (p. 481), has the following structure.

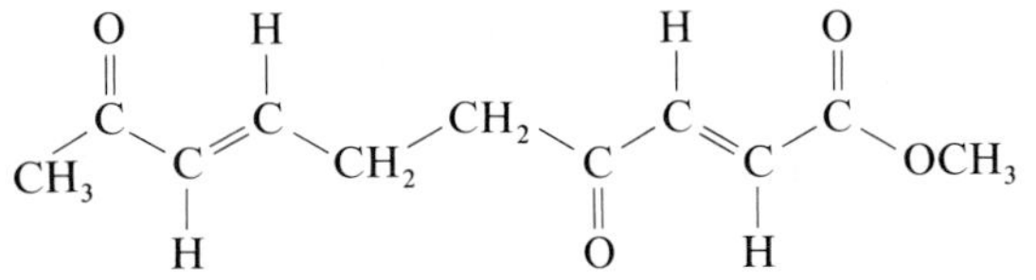

Compound A

The structure was confirmed by conversion to Compound B, $C_{11}H_{18}O_4$, which was also prepared by ozonolysis of Compound C, $C_{11}H_{18}O_2$.

$$\underset{C_{11}H_{14}O_4}{A} \xrightarrow[\substack{Pd/C \\ ethanol}]{H_2} \underset{C_{11}H_{18}O_4}{B} \xleftarrow{(CH_3)_2S} \xleftarrow[\substack{dichloromethane \\ -78\,°C}]{O_3} \underset{C_{11}H_{18}O_2}{C}$$

Assign structures to Compounds B and C.

8.45 4,4-Dimethyl-2-cyclohexen-1-ol reacts with *m*-chloroperoxybenzoic acid in dichloromethane. The product of this reaction is treated with pyridinium chlorochromate in dichloromethane in the presence of sodium acetate. Write equations for these reactions showing the products you would expect. What complications would have resulted if aqueous sodium dichromate had been used as the oxidizing agent in the last step?

8.46 An intermediate in the synthesis of cerulein, an antibiotic, was prepared in the following way. Fill in structural formulas for the missing compounds.

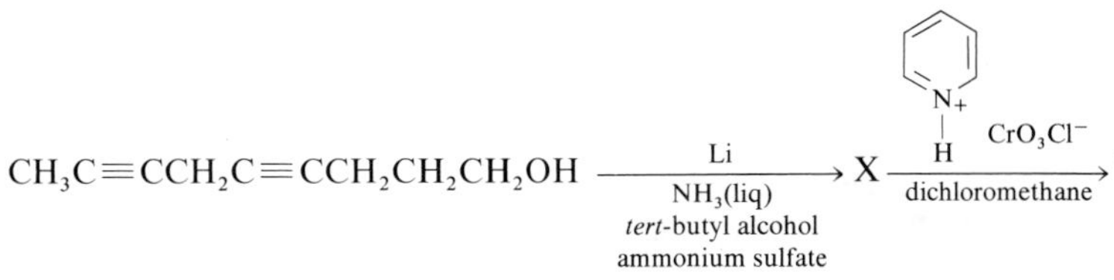

$$CH_3C{\equiv}CCH_2C{\equiv}CCH_2CH_2CH_2OH \xrightarrow[\substack{NH_3(liq) \\ tert\text{-butyl alcohol} \\ ammonium\ sulfate}]{Li} X \xrightarrow[dichloromethane]{} Y$$

8.47 2-Phenyloxirane has been treated with sulfuric acid in methanol, and in a separate reaction with sodium methoxide in methanol. Predict what the major product in each reaction will be and write detailed mechanisms to rationalize your predictions.

8.48 Cholesterol is converted into a saturated ketone, cholestanone, in several steps. Supply the reagents necessary to make this synthetic transformation.

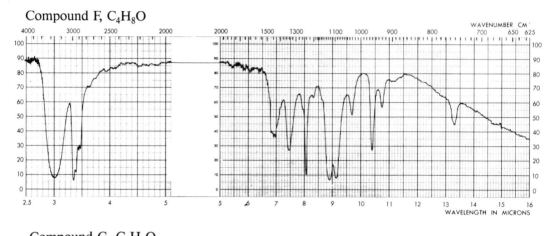

cholesterol   cholestanone

8.49 Two compounds, F and G, have the same molecular formula, C_4H_8O. Their infrared spectra are given in Figure 8.8.

Compound F, C_4H_8O

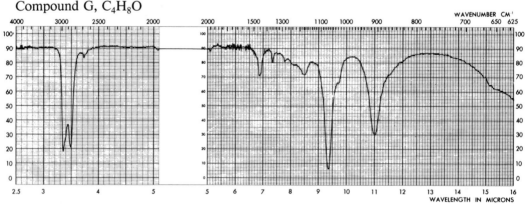

Compound G, C_4H_8O

Figure 8.8 Spectra for Problem 8.49. (From *The Aldrich Library of Infrared Spectra*)

Compound G is frequently used as a solvent in hydroboration reactions. Both compounds are unaffected by cold dilute potassium permanganate and bromine in carbon tetrachloride. Can you make firm structural assignments based on this information?

8.50 Compound H, $C_5H_{10}O$, is oxidized to Compound I, C_5H_8O, with chromic acid. Their infrared spectra are given in Figure 8.9. Neither H nor I reacts with bromine in carbon tetrachloride or with dilute aqueous potassium permanganate solution. Assign possible structures to the compounds.

Compound H, $C_5H_{10}O$

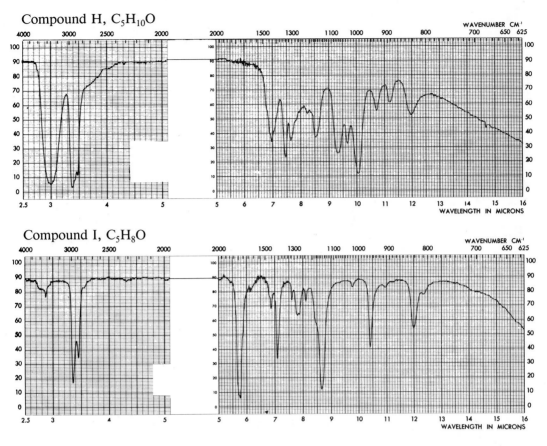

Compound I, C_5H_8O

Figure 8.9 Spectra for Problem 8.50. (From *The Aldrich Library of Infrared Spectra*)

8.51 The spectra numbered 1 through 5 in Figure 8.10 belong to the following compounds. Match each compound with the correct spectrum and identify the major absorption bands that allowed you to make the assignment.

(a) 2,3-dimethyl-1-pentanol   (b) 3-methyl-4-nonanone   (c) 1-heptene
(d) 3-chloro-1-propyne   (e) cyclopentene

Spectrum 1

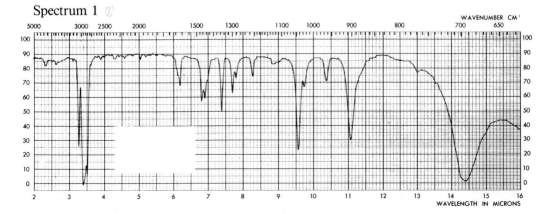

Figure 8.10 Spectra for Problem 8.51. (From *The Aldrich Library of Infrared Spectra*)

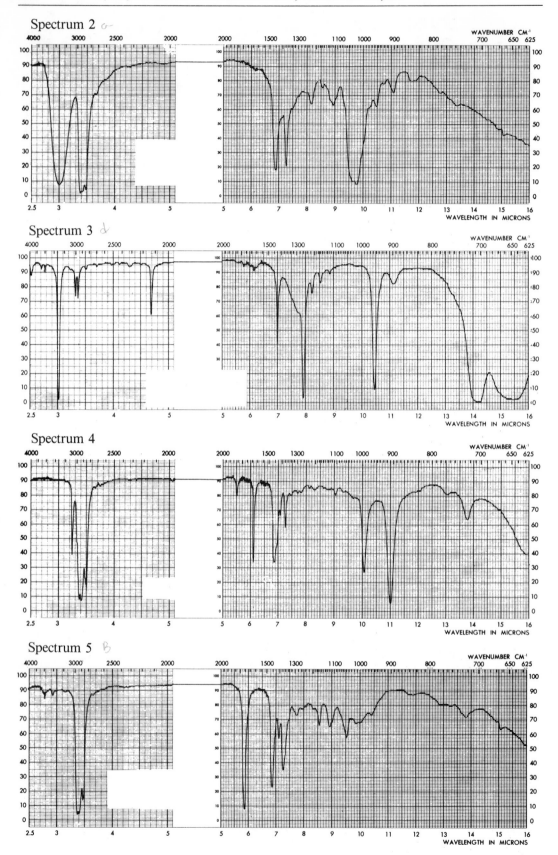

Figure 8.10 *(Continued)*

Aldehydes and Ketones

9

9.1

Introduction and Nomenclature

A. The Properties and Occurrence of Aldehydes and Ketones

The **carbonyl group,** *a carbon atom doubly bonded to an oxygen atom, dominates the physical and chemical properties of aldehydes and ketones.* In aldehydes, the carbonyl group is bonded to an organic group on one side and a hydrogen atom on the other. In ketones, it is between two organic groups. Oxygen is more electronegative than carbon, so the carbonyl group has a permanent dipole moment with the negative end of the dipole at the oxygen atom. For example, acetone, a typical ketone, has a dipole moment of 2.88 D. The high polarity of the carbonyl group is rationalized by pointing to a resonance contributor in which the carbon atom bears a positive formal charge and the oxygen atom, a negative one.

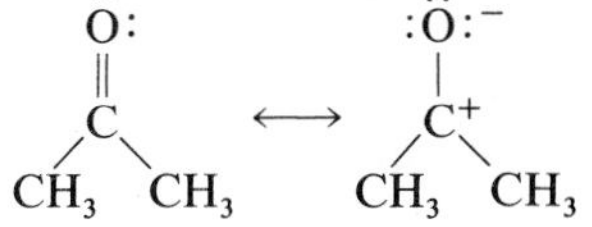

resonance contributors for acetone

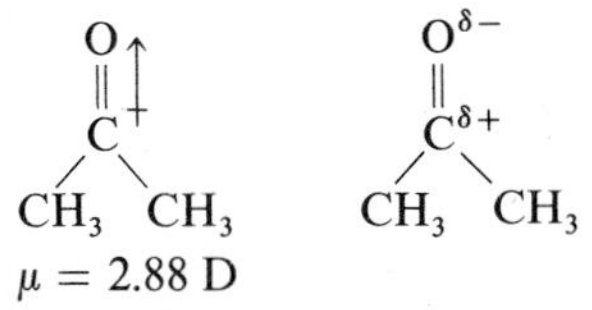

$\mu = 2.88$ D

different representations of the polarity of the carbonyl group

The polarity of the carbonyl group makes low molecular weight aldehydes and ketones soluble in water. The oxygen atom acts as an acceptor of hydrogen bonds, and the carbonyl compound dissolves in water as long as the hydrocarbon portion of the molecule does not contain more than four or five carbon atoms. The kinds of interactions that occur between water molecules and acetone representing a low molecular weight carbonyl compound, are shown below.

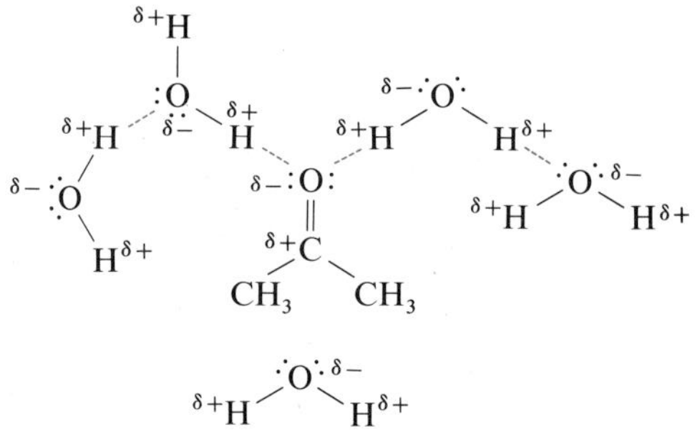

In Chapter 4, we explored the basicity of the carbonyl group and saw how the nonbonding electrons on the oxygen atom could accept a proton from a protic acid or be shared with a Lewis acid.

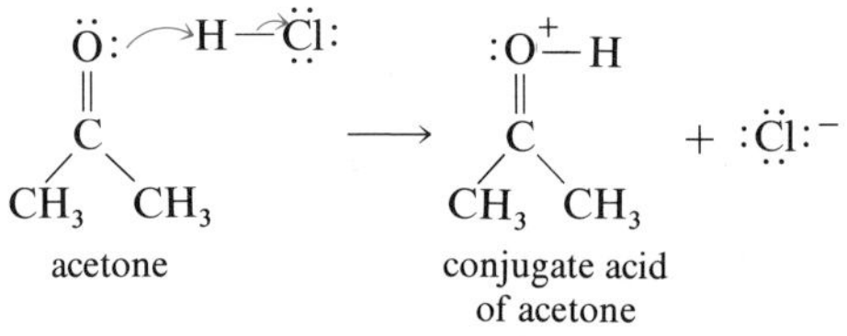

acetone

conjugate acid
of acetone

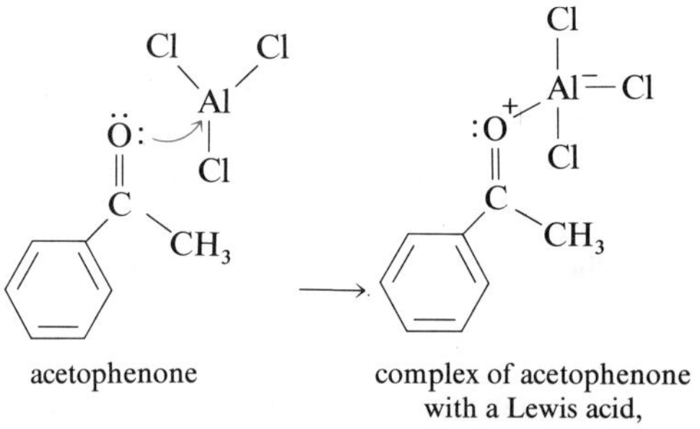

acetophenone

complex of acetophenone
with a Lewis acid,
aluminum chloride

Aldehydes and ketones are also weak acids. Acetone, for example, has a pK_a of 20. A hydrogen atom from the carbon atom adjacent to the carbonyl group can be removed by a base to give a carbanion that is stabilized by delocalization of the electrons to the oxygen atom of the carbonyl group.

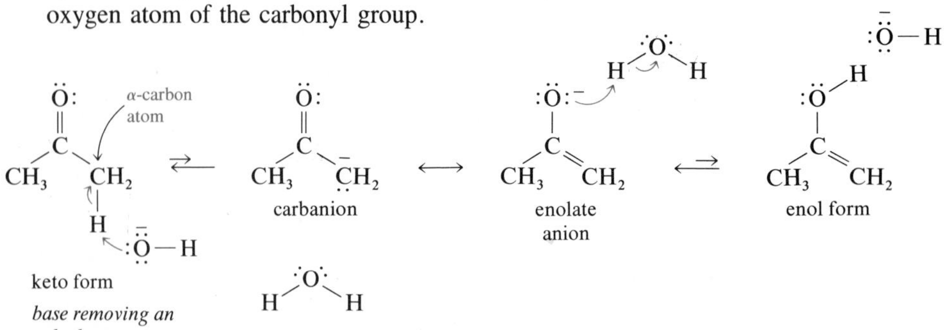

carbanion

enolate
anion

enol form

keto form

base removing an
α-hydrogen atom

The enolate anion, in which the negative charge is on the oxygen atom, is protonated to give an enol (Section 5.5C). The enol and keto forms of the carbonyl compound are isomers in equilibrium with each other, each step in the transformation being reversible. The two forms are *not* resonance contributors because to go from one to the other requires the movement of a hydrogen atom as well as a change in the position of electrons. *Isomers that differ from each other only in the location of a double bond and a hydrogen atom are called* **tautomers.** The term is used to refer to isomers in which the differing locations of the hydrogen atom and the double bond involve **heteroatoms,** atoms other than carbon, such as oxygen and nitrogen. This type of isomerism is known as **tautomerism.** For acetone, the form in which the hydrogen atom is on the carbon atom is the more stable one, and predominates in the equilibrium mixture. In some compounds in which a hydrogen atom may be removed from a carbon atom that is between two carbonyl groups, the enol form becomes important. The reactions that carbonyl compounds undergo because of the acidity of the hydrogen atoms on the carbon atom adjacent to the carbonyl group are explored extensively in Chapter 15.

PROBLEM 9.1   Decide whether each set of structural formulas shown below represents resonance contributors or tautomers.

(a)

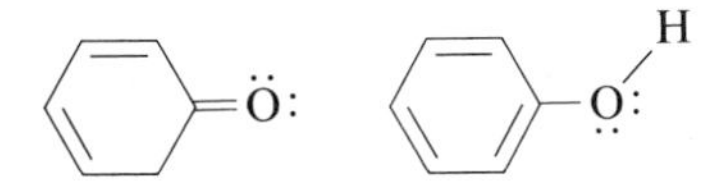

(b)

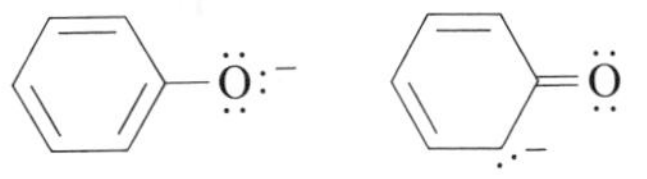

(c)

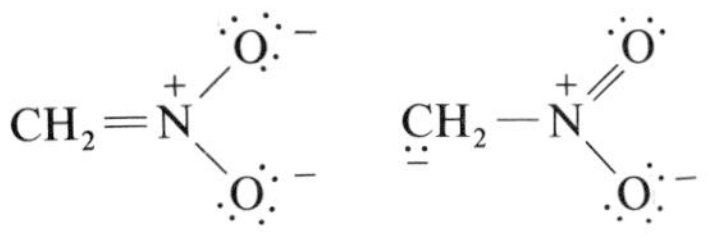

(d)

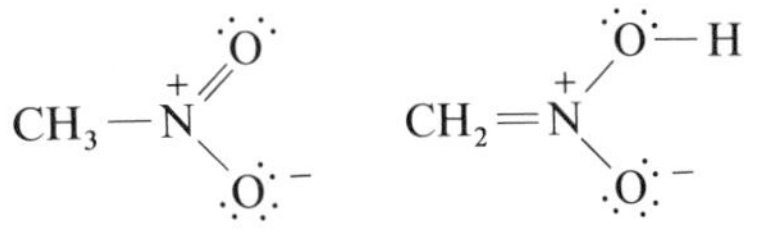

PROBLEM 9.2   Write equations for each of the following compounds showing the reactions you would expect with (a) concentrated sulfuric acid, and (b) sodium ethoxide. (You may want to review Sections 4.3, 4.5, and 4.8.) Show resonance contributors that explain the stabilization of any ions formed.

(a) 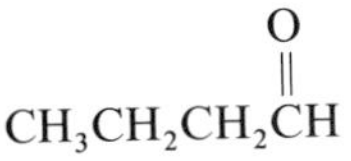 $CH_3CH_2CH_2\overset{\displaystyle O}{\overset{\displaystyle \|}{C}}H$ (b) $CH_3\overset{\displaystyle O}{\overset{\displaystyle \|}{C}}$—

(c) $CH_3\overset{\displaystyle CH_3}{\overset{\displaystyle |}{C}}=CHCH_2CH_2\overset{\displaystyle O}{\overset{\displaystyle \|}{C}}CH_3$ (d)

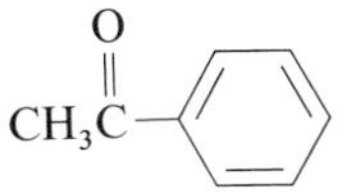

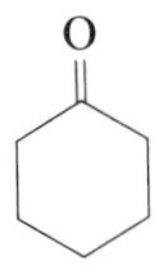

Reactions with carbon-carbon multiple bonds are initiated by reagents deficient in electrons, in other words, by electrophiles. In Chapters 5 and 8 we considered a number of cases where an acid or an oxidizing agent, seeking electrons, reacted with a carbon-carbon double or triple bond. A carbon-carbon multiple bond does not have a pronounced polarity, and the electrons in the π bond are nucleophilic, so alkenes and alkynes usually react with electrophiles.

Most of the important reactions of aldehydes and ketones are initiated by attack of an electron-rich species, a nucleophile, at the electrophilic carbon atom of the carbonyl group. In some cases, the reaction is catalyzed by a protic acid that protonates the oxygen atom of the carbonyl group, thus increasing the electron deficiency in the functional group.

B. The Occurrence of Aldehydes and Ketones in Nature

Many compounds of biological interest are aldehydes and ketones. The most important natural aldehyde is glucose, the carbohydrate that is metabolized in our bodies to produce energy. Many of the important steroid hormones are ketones, including testosterone, the hormone that controls the development of male sex characteristics; progesterone, the hormone secreted at the time of ovulation in females; and cortisone, a hormone of the adrenal glands that is used medicinally to relieve inflammation. Another hormone from the adrenal glands, aldosterone, is an aldehyde as well as a ketone. Aldosterone is important in regulating the presence of sodium ions, and hence, water retention in the body. Prednisone is an example of a synthetic drug; it is designed to substitute for cortisone to relieve the symptoms of rheumatoid arthritis with fewer undesirable side effects.

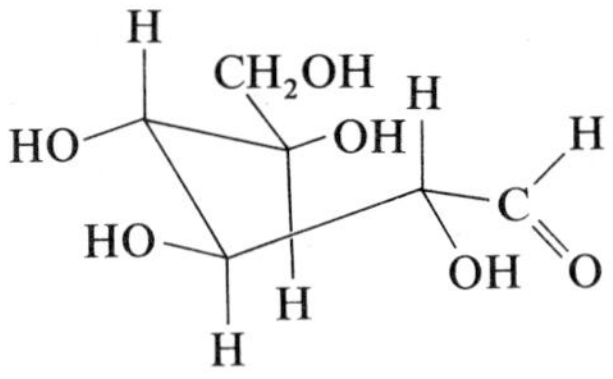

glucose 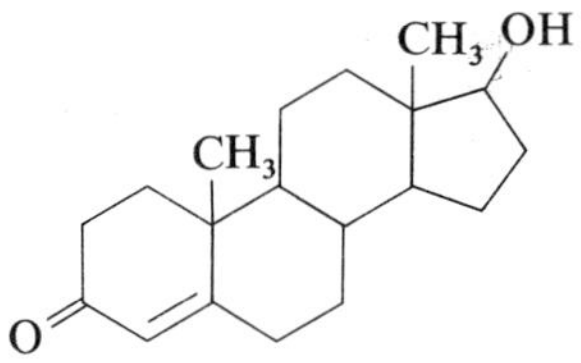   testosterone

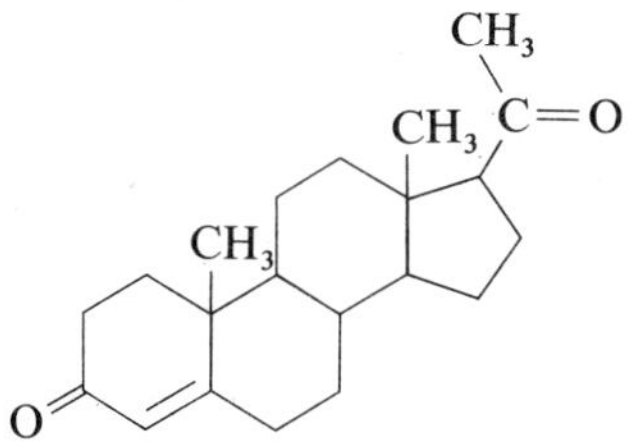

progesterone 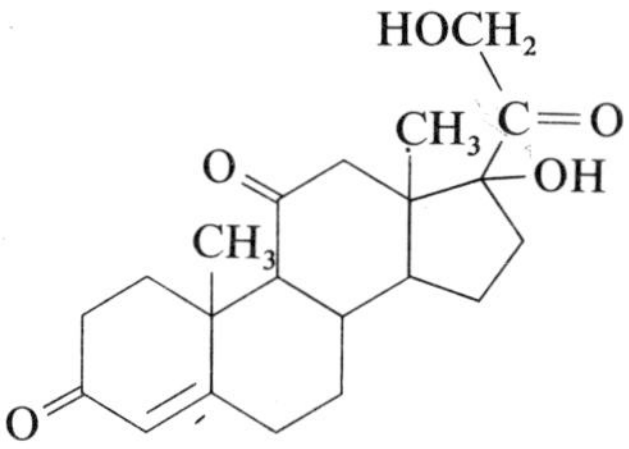   cortisone

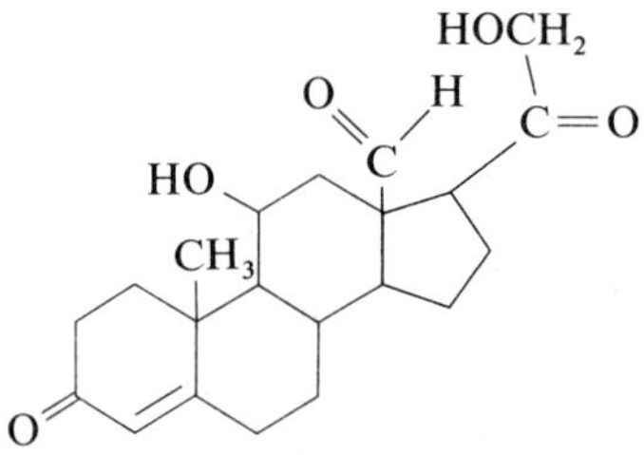

aldosterone 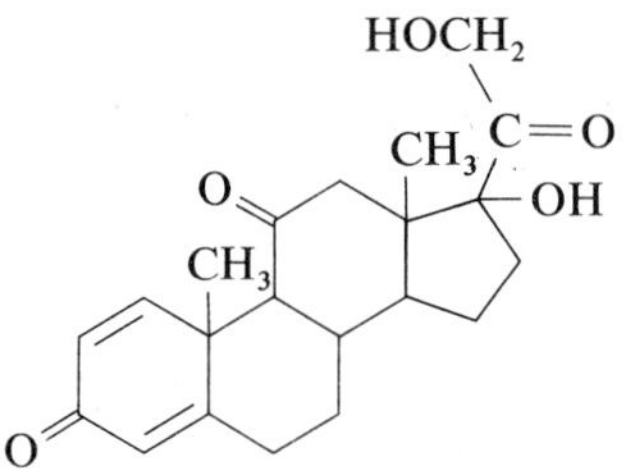   prednisone

If you carefully inspect the structural formulas of the steroids shown above, you may wonder how molecules that are so similar to each other in structure can perform functions so different from each other in the body. What makes testosterone a male hormone and progesterone a female one, for example, while the function of cortisone is something else entirely? Contemplating such questions will give you an appreciation of the finely discriminating systems that exist in living organisms. In no case do we fully comprehend how the structures of the hormones enable them to perform their physiological functions. Organic chemists and biochemists constantly refine their ways of looking at such compounds in order to understand their interactions with large molecules, such as the proteins known as hormone receptors that occur at the surface of cells or inside them. The binding of the hormone to the receptor triggers a series of reactions that ultimately result in the physiological changes associated with that hormone. Interactions between hormones and receptors are described using the same concepts that we use in describing the reactivity of simpler compounds. Polarity of bonds, hydrogen bonding, ionic interactions, van der Waals forces between nonpolar portions of molecules, transfer of protons from one basic site to another one, and nucleophilic centers reacting with electrophilic sites are among the ideas that chemists use as they struggle to make sense of the multiple reactions that proceed in our bodies.

C. Nomenclature of Aldehydes and Ketones

Simple aldehydes are named systematically by changing the **-e** *ending of the name of the hydrocarbon with the same number of carbon atoms to* **-al.** The aldehyde function is assumed to be at the first carbon atom of the chain. Other substituents are named as prefixes and numbered to indicate their positions relative to the aldehyde group. The common names of the simple aldehydes, formaldehyde and acetaldehyde, are almost always used.

methanal
formaldehyde

ethanal
acetaldehyde

propanal

pentanal

2-chloro-5-methylhexanal

If the aldehyde group is a substituent on a ring, the suffix **-carbaldehyde** is used.

cyclohexanecarbaldehyde

trans-2-methylcyclo-
pentanecarbaldehyde

Note that the suffix *-carbaldehyde* is used as a substituent. The carbon atom of the carbonyl group is *not* counted in deciding upon a name for the hydrocarbon portion of the molecule.

Benzene substituted by an aldehyde group is usually called benzaldehyde. Another substituent on the aromatic ring can have one of three positions in relation to the aldehyde group. It can be on the carbon atom adjacent to the one to which the carbonyl group is bonded. This position is called the **ortho** position, abbreviated as *o*- in the name of the compound. The next position away from the carbonyl group is known as the **meta** or *m*- position. The position on the ring opposite the carbonyl group is the **para** or *p*- position.

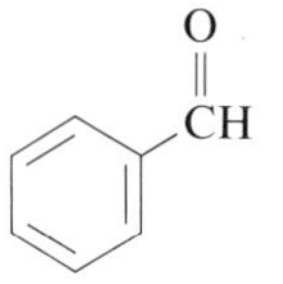

benzaldehyde 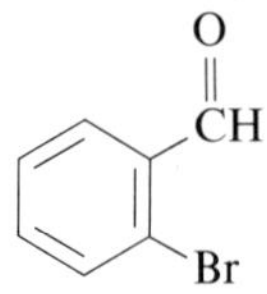   *o*-bromobenzaldehyde

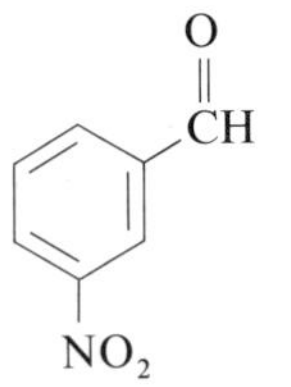

m-nitrobenzaldehyde 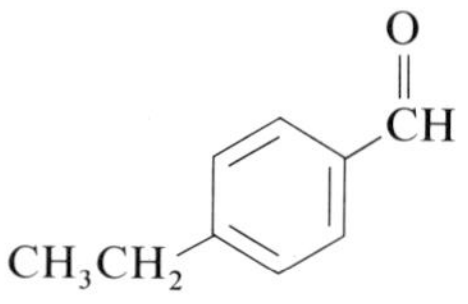   *p*-ethylbenzaldehyde

Ketones are named according to the IUPAC rules by substituting the ending **-one** *for the final* **-e** *of the name of the hydrocarbon with the same number of carbon atoms and using a number to indicate the position of the carbonyl group.* Other substituents are named as prefixes and their positions indicated by numbers. In a few cases, it is useful to name the organic groups that are present as substituents on the carbonyl group, which is then designated as **ketone.** A few ketones, such as acetone, have common names that you should learn.

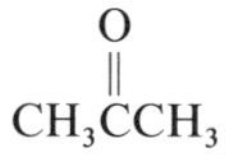

propanone   3-bromo-4-heptanone 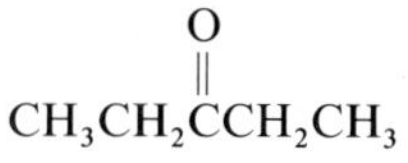   3-pentanone
acetone

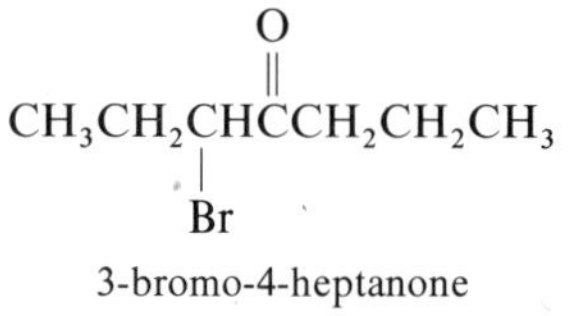

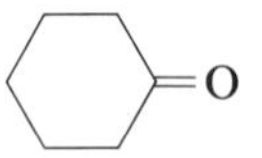

cyclohexanone 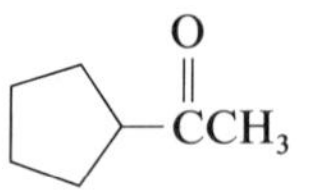   cyclopentyl methyl   2-chlorocyclopentanone
                            ketone

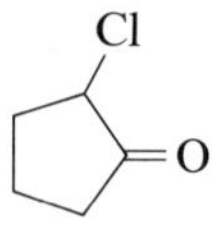

For compounds that contain more than one major functional group, the suffix for only one of them can be used as the ending of the name. The IUPAC rules establish priorities that specify which one will serve as the suffix to the name. The priorities as they apply to the common functional groups are given in Table 9.1.

The presence of double or triple bonds in a molecule is indicated by naming the chain (or ring) as an alkene or alkyne. Halogen substituents are always named as prefixes, as are alkoxyl groups. Thus, an aldehyde or a ketone function takes precedence over

TABLE 9.1 Decreasing Order of Precedence of Principal Functional Groups Used as Suffixes in Nomenclature

Chemical Structure	Functional Group
$\overset{\displaystyle O}{\overset{\displaystyle \|}{-COH}}$	carboxylic acid
$\overset{\displaystyle O}{\underset{\displaystyle O}{\overset{\displaystyle \|}{\underset{\displaystyle \|}{-SOH}}}}$	sulfonic acid
$\overset{\displaystyle O}{\overset{\displaystyle \|}{-COR}}$	ester
$\overset{\displaystyle O}{\overset{\displaystyle \|}{-CNH_2}}$	amide
$-C\equiv N$	nitrile
$\overset{\displaystyle O}{\overset{\displaystyle \|}{-CH}}$	aldehyde
$\overset{\displaystyle O}{\overset{\displaystyle \|}{-C-}}$	ketone
$-OH$	alcohol
$-NH_2$	amine

an alkene or an alcohol function as the suffix used in naming the compound. Here are some examples to illustrate this kind of nomenclature.

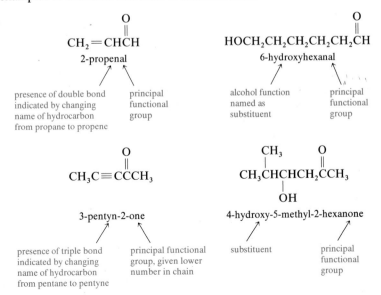

$$CH_2=CH\overset{\displaystyle O}{\overset{\displaystyle \|}{CH}}$$

2-propenal

presence of double bond indicated by changing name of hydrocarbon from propane to propene / principal functional group

$$HOCH_2CH_2CH_2CH_2CH_2\overset{\displaystyle O}{\overset{\displaystyle \|}{CH}}$$

6-hydroxyhexanal

alcohol function named as substituent / principal functional group

$$CH_3C\equiv C\overset{\displaystyle O}{\overset{\displaystyle \|}{C}}CH_3$$

3-pentyn-2-one

presence of triple bond indicated by changing name of hydrocarbon from pentane to pentyne / principal functional group, given lower number in chain

$$CH_3\overset{\displaystyle CH_3}{\underset{\displaystyle OH}{\overset{\displaystyle |}{\underset{\displaystyle |}{CH}}}}CHCH_2\overset{\displaystyle O}{\overset{\displaystyle \|}{C}}CH_3$$

4-hydroxy-5-methyl-2-hexanone

substituent / principal functional group

When it is necessary to name the carbonyl group as a substituent, the prefix **oxo-** and a number to indicate its position are used. This type of nomenclature is needed when another functional group of higher priority than an aldehyde or ketone, such as a carboxylic acid or ester, is present. Oxo- may be used to indicate either an aldehyde or a ketone, as illustrated below.

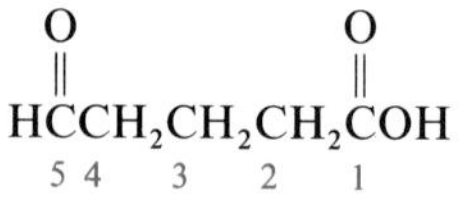

$$
\underset{\substack{5\ 4\quad\ 3\quad\ 2\quad\ 1}}{\overset{\overset{O}{\|}\qquad\qquad\overset{O}{\|}}{HCCH_2CH_2CH_2COH}}
\qquad
\underset{\substack{6\quad 5\quad 4\ 3\quad 2\quad 1}}{\overset{\overset{O}{\|}\qquad\quad\overset{O}{\|}}{CH_3CH_2CCH_2CH_2COCH_3}}
$$

5-oxopentanoic acid 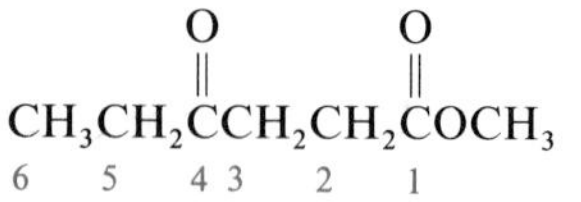   methyl 4-oxohexanoate

Ketones having a phenyl group on the carbonyl group have common names ending in **-phenone.** You should be able to recognize benzophenone and acetophenone. Other ketones containing aromatic rings are best named systematically, or as substituted ketones.

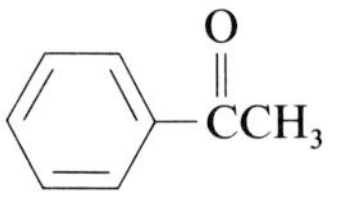

acetophenone 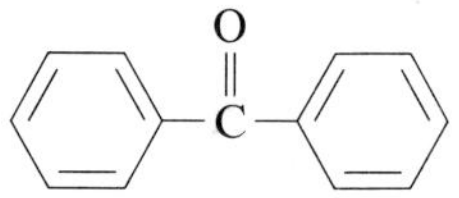   benzophenone

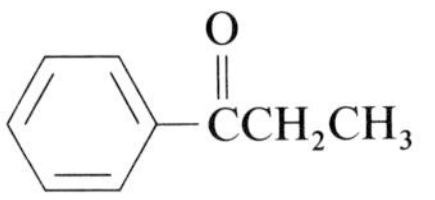

1-phenyl-1-propanone   1-phenyl-2-propanone
ethyl phenyl ketone   benzyl methyl ketone

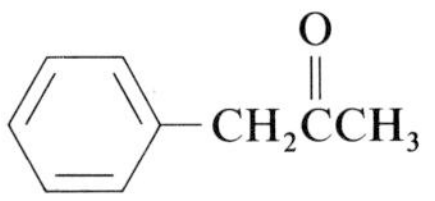

Greek letters are used to refer to positions on the chain in relation to a carbonyl group in describing positions of special reactivity or relationships between functional groups. For example, the carbon atoms adjacent to the carbonyl group are called the α-carbon atoms and the hydrogen atoms on them, the α-hydrogen atoms.

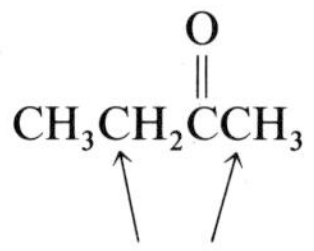

$$
\overset{\overset{O}{\|}}{CH_3CH_2CCH_3}
\qquad\qquad
\overset{\overset{O}{\|}}{CH_3CH_2CH_2CH}
$$

α-carbon atoms α-carbon atom
five α-hydrogen atoms   two α-hydrogen atoms

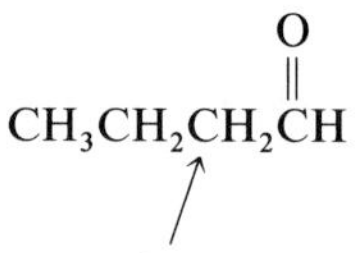

We refer many times to positions alpha to the carbonyl group because of the special reactivity of the α-hydrogen atoms (Section 9.1A). This type of nomenclature was also introduced in Section 4.7B in connection with substituted carboxylic acids. The Greek letters, besides alpha, most often used for positions on the chain are β, beta; γ, gamma; δ, delta; and ϵ, epsilon. These are shown below, with some examples of their use.

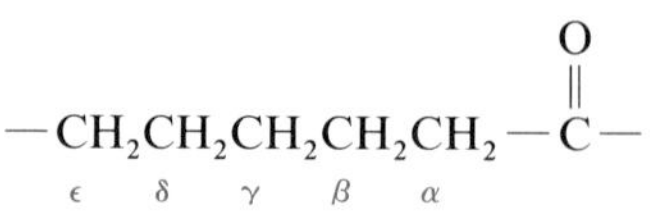

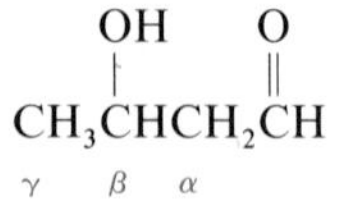

$$
\underset{\substack{\ \ \epsilon\quad\ \delta\quad\ \gamma\quad\ \beta\quad\ \alpha}}{-CH_2CH_2CH_2CH_2CH_2-\overset{\overset{O}{\|}}{C}-}
$$

$$
\underset{\substack{\gamma\quad\ \beta\quad\ \alpha}}{\overset{\overset{OH}{|}\quad\ \overset{O}{\|}}{CH_3CHCH_2CH}}
\qquad
\underset{\substack{\alpha\quad\ \alpha}}{\overset{\overset{O}{\|}}{CH_3CCH_2Br}}
\qquad
\underset{\substack{\delta\quad\ \gamma\ \beta\quad\ \alpha}}{\overset{\overset{O}{\|}\qquad\ \overset{O}{\|}}{CH_3CCH_2CH_2COH}}
$$

3-hydroxybutanal   1-bromo-2-propanone   4-oxopentanoic acid

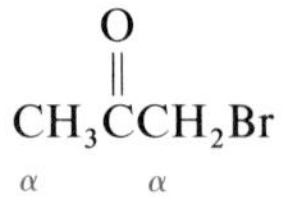

a β-hydroxyaldehyde   *an α-bromoketone*   *a γ-ketoacid*

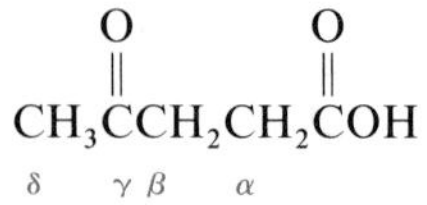

PROBLEM 9.3 Name the following compounds according to the IUPAC rules. Include stereo-chemistry when pertinent.

(a) 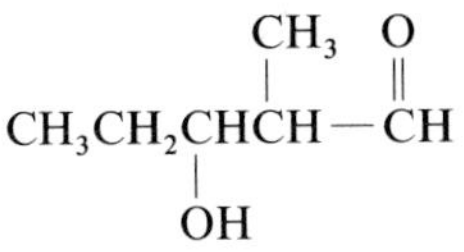 $\underset{\underset{\displaystyle OH}{|}}{CH_3CH_2CHCH}\overset{\overset{\displaystyle CH_3}{|}\ \overset{\displaystyle O}{\|}}{-CH}$

(b)  $CH_3\overset{\overset{\displaystyle O}{\|}}{C}CH_2CH_2CH_3$

(c) 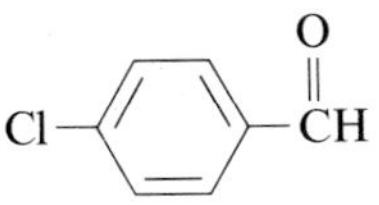   Cl—⟨benzene ring⟩—$\overset{\overset{\displaystyle O}{\|}}{CH}$

(d) 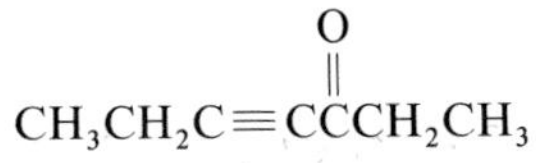 $CH_3CH_2C\equiv C\overset{\overset{\displaystyle O}{\|}}{C}CH_2CH_3$

(e) 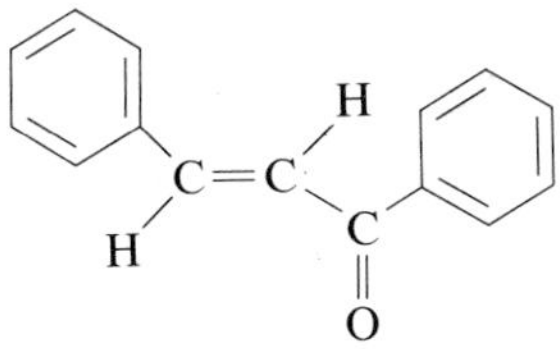   phenyl$\underset{\underset{\displaystyle H}{}}{C}$=$\underset{\underset{\displaystyle \overset{C}{\underset{O}{\|}}}{}}{C}$phenyl   with H

(f) 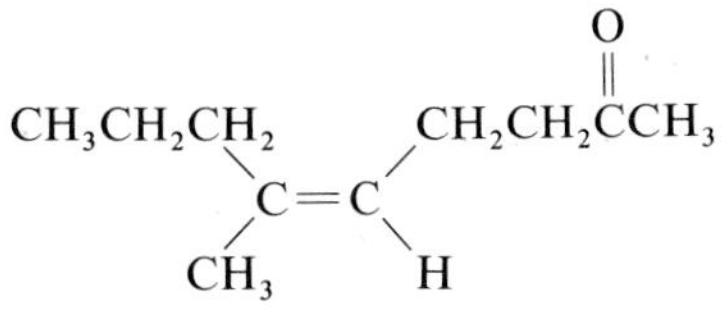 $CH_3CH_2CH_2$, CH_3 C=C $CH_2CH_2\overset{\overset{\displaystyle O}{\|}}{C}CH_3$, H

(g) 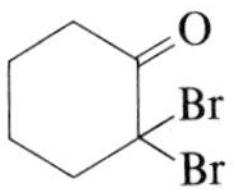   ⟨cyclohexanone with Br, Br at C2⟩

(h) $Cl_3C\overset{\overset{\displaystyle O}{\|}}{C}H$

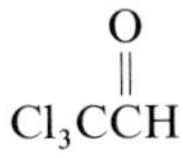

(i) 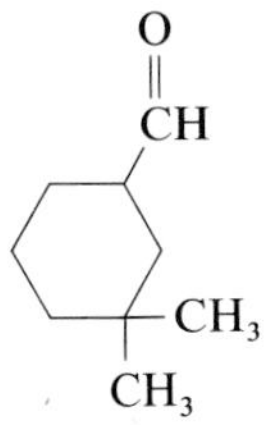   ⟨cyclohexane with CHO and two CH₃ groups⟩

(j) 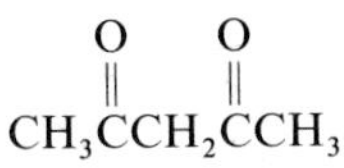 $CH_3\overset{\overset{\displaystyle O}{\|}}{C}CH_2\overset{\overset{\displaystyle O}{\|}}{C}CH_3$

(k) 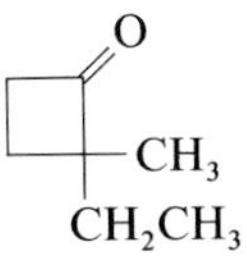   ⟨cyclobutanone with CH₃ and CH₂CH₃⟩

(l) 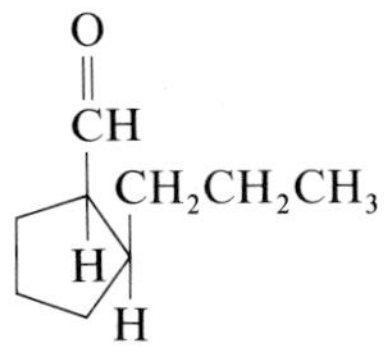   ⟨bicyclic ring with CHO, CH₂CH₂CH₃, H, H⟩

PROBLEM 9.4 Draw structural formulas for the following compounds:

(a)   3-methylcyclobutanone   (b)   4,4-dimethyl-2,5-cyclohexadienone
(c)   phenyl cyclopentyl ketone   (d)   (*E*)-4-heptenal   (e)   *p*-nitrobenzaldehyde
(f)   1-hydroxy-3-pentanone   (g)   hexanedial   (h)   2-hydroxy-1,2-diphenylethanone
(i)   1,4-cyclohexanedione

<hr/>

9.2

Syntheses of Aldehydes and Ketones

A. Oxidation of Alcohols: A Review

Aldehydes and ketones are often prepared by oxidation of alcohols. In Sections 7.6B and 7.6C, we saw how the conditions for the reaction of alcohols with compounds of chromium(VI) could be controlled to allow the oxidation of different types of alcohols to carbonyl compounds.

For example, cyclohexanol is a secondary alcohol with no other functional groups in the molecule, and is low enough in molecular weight to be soluble in water. It is oxidized to cyclohexanone by an aqueous, acidic solution of sodium dichromate.

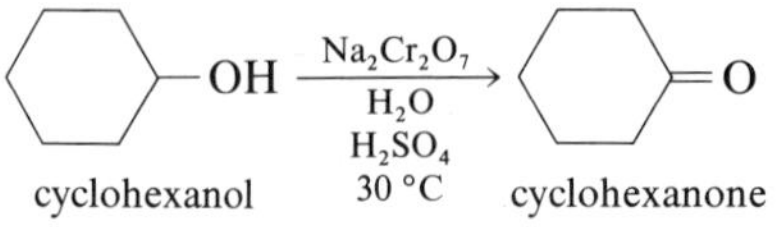

cyclohexanol   cyclohexanone

5-Methyl-4-hexen-1-ol is a primary alcohol. Oxidation of a primary alcohol in the presence of water leads to the formation of the corresponding carboxylic acid (Section 7.6B). Pyridinium chlorochromate is used to oxidize this alcohol so that the reaction stops at the aldehyde stage (Section 7.6C).

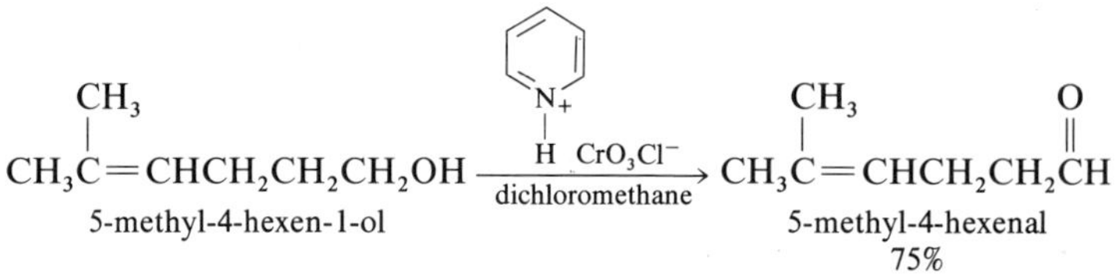

5-methyl-4-hexen-1-ol   5-methyl-4-hexenal
                                        75%

PROBLEM 9.5   Write an equation showing side reactions that would have taken place if 5-methyl-4-hexen-1-ol had been oxidized with sodium dichromate in aqueous acid.

B. Hydroboration of Alkynes

Alkynes add diborane just as alkenes do (Sections 7.3, 8.3B), *and the vinylboranes that result are oxidized to aldehydes and ketones.* When the triple bond is internal, the reaction is uncomplicated. 3-Hexyne, for example, forms a vinylborane that is oxidized by hydrogen peroxide and sodium hydroxide to an enol (Section 9.1A), which tautomerizes to the more stable keto form.

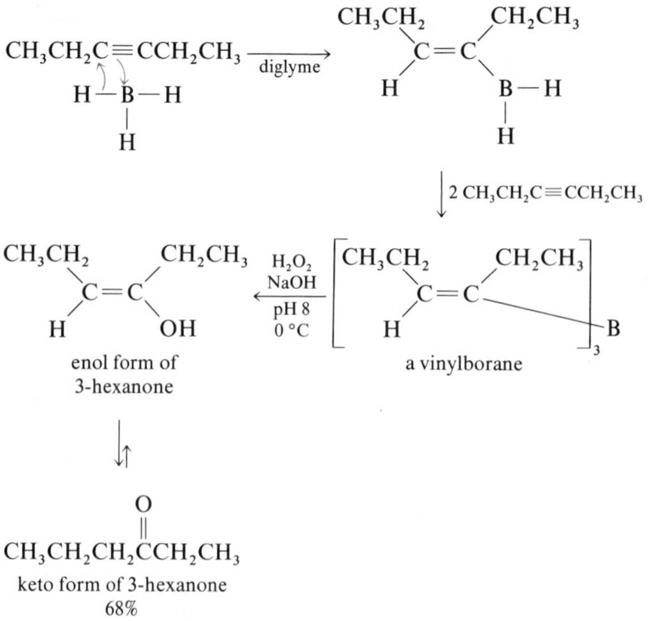

enol form of
3-hexanone

a vinylborane

keto form of 3-hexanone
68%

The ketones formed from internal alkynes by hydroboration of the triple bond and oxidation of the resulting vinylborane are the same products obtained by addition of water to the triple bond catalyzed by acid and mercury(II) ion (Section 5.5C). Hydroboration of a terminal alkyne, and oxidation of that vinylborane, however, give aldehydes, whereas the Markovnikov addition of water to terminal alkynes gives ketones from all alkynes except acetylene. The anti-Markovnikov orientation of the addition of borane can thus be used to convert 1-alkynes to aldehydes. The borane used for the reaction must be an especially bulky one, or side reactions in which borane adds twice to the terminal carbon atom become important. Even 9-borabicyclo[3.3.1]-nonane (Section 7.3B) adds twice to 1-alkynes unless a large excess of the alkyne is used. Dicyclohexylborane, however, is one of the reagents hindered enough that it adds only once to a 1-alkyne. The resulting vinylborane can be oxidized to an aldehyde through an enol intermediate.

$$BH_3 + 2 \text{ (cyclohexene)} \longrightarrow \text{(dicyclohexylborane)} \equiv R - B - R$$

borane   cyclohexene   dicyclohexylborane

a borane in which R is a bulky alkyl group

$$CH_3CH_2CH_2CH_2C\equiv CH \xrightarrow[\substack{H-B-R \\ R}]{\substack{\text{tetrahydrofuran} \\ 0\,°C}} CH_3CH_2CH_2CH_2 \bigg/ H \atop C=C \atop H \big/ \ \ B-R \atop R$$

pH 8, 0 °C | H₂O₂, NaOH

$$CH_3CH_2CH_2CH_2CH_2\overset{\overset{\displaystyle O}{\|}}{C}H \ \rightleftharpoons \ \left[CH_3CH_2CH_2CH_2 \atop H \bigg/ C=C \atop H \big/ \ \ OH \right]$$

keto form of hexanal
88%

enol form of hexanal

PROBLEM 9.6   When 1-cyclohexyl-1-propyne is treated with diborane, boron binds to carbon 2 of the alkyne in 74% of the product. When dicyclohexylborane is used as the reagent, 92% of the product has boron bonded to that position. Write mechanisms for the reactions showing the full structure of the alkyne and the boranes. Explain why dicyclohexylborane is more regioselective in its reaction with the alkyne.

9.3

Reductions of Aldehydes and Ketones to Alcohols

Aldehydes are reduced by metal hydrides to primary alcohols, while ketones are reduced to secondary alcohols. These reductions are usually carried out by the use of reagents such as sodium borohydride and lithium aluminum hydride. In both reagents, the anion is a source of hydride ion, which is a strongly basic and nucleophilic species.

For example, lithium aluminum hydride reacts violently with water and alcohols to generate hydrogen gas.

$$\text{Li}^+ \ \text{H}-\overset{\overset{\displaystyle H}{|}}{\underset{\underset{\displaystyle \overset{..}{O}}{|}}{\underset{\displaystyle H}{\text{Al}}}}{}^{\!-}\!\!\!\!\diagdown\text{H} \longrightarrow \text{Li}^+ \ \text{H}-\overset{\overset{\displaystyle H}{|}}{\underset{\underset{\displaystyle \overset{..}{O}-H}{|}}{\text{Al}}}{}^{\!-}\!-\text{H} + \text{H}_2\!\uparrow \xrightarrow{3\,\text{H}_2\text{O}} \text{Li}^+\text{Al(OH)}_4^- + 3\,\text{H}_2\!\uparrow$$

Sodium borohydride reacts in the same way, though much more slowly than lithium aluminum hydride.

$$\text{Na}^+ \ \text{H}-\overset{\overset{\displaystyle H}{|}}{\underset{\underset{\displaystyle CH_3}{|}}{\underset{\displaystyle H\overset{..}{\underset{..}{O}}}{\text{B}}}}\!\diagdown\text{H} \longrightarrow \text{Na}^+ \ \text{H}-\overset{\overset{\displaystyle H}{|}}{\underset{\underset{\displaystyle \overset{..}{\underset{..}{O}}-CH_3}{|}}{\text{B}}}-\text{H} + \text{H}_2\!\uparrow \xrightarrow{3\,\text{CH}_3\text{OH}} \text{Na}^+\text{B(OCH}_3)_4^- + 3\,\text{H}_2\!\uparrow$$

The reduction of an aldehyde to a primary alcohol by lithium aluminum hydride proceeds in two steps. The reduction step gives an alkoxide anion, which complexes with the metal ions present. To recover the alcohol, the complex is treated with water and dilute acid. Thus heptanal is converted into 1-heptanol.

$$4\ \text{CH}_3(\text{CH}_2)_5\overset{\overset{\displaystyle O}{\|}}{\text{CH}} \xrightarrow[\text{diethyl ether}]{\text{LiAlH}_4} \left[\text{CH}_3(\text{CH}_2)_5\underset{\underset{\displaystyle H}{|}}{\text{CH}}-\text{O}-\right]_4 \text{Al}^-\text{Li}^+ \xrightarrow{\text{H}_3\text{O}^+} 4\ \text{CH}_3(\text{CH}_2)_5\text{CH}_2\text{OH}$$

heptanal   1-heptanol
                                                                                              86%

A similar sequence of reactions gives a secondary alcohol from 2-butanone.

$$4\ \text{CH}_3\overset{\overset{\displaystyle O}{\|}}{\text{C}}\text{CH}_2\text{CH}_3 \xrightarrow[\text{diethyl ether}]{\text{LiAlH}_4} (\text{CH}_3\text{CH}_2\underset{\underset{\displaystyle H}{|}}{\overset{\overset{\displaystyle CH_3}{|}}{\text{C}}}-\text{O}-)_4 \ \text{Al}^-\text{Li}^+ \xrightarrow{\text{H}_3\text{O}^+} 4\ \text{CH}_3\text{CH}_2\underset{\underset{\displaystyle OH}{|}}{\text{CH}}\text{CH}_3$$

2-butanone   *sec*-butyl alcohol
                                                                                              80%

Lithium aluminum hydride is used in dry ether solvents because of its reactivity with water and alcohols. It is a much more reactive donor of hydride ion than sodium borohydride, and it reacts with carboxylic acids and esters (Section 10.9) as well as aldehydes and ketones. Sodium borohydride is easier to handle than lithium aluminum hydride, and may be used in water or alcohol solvents because its reactions with the carbonyl compounds are fast enough to compete with the decomposition of the reagent. It is a more selective reagent than lithium aluminum hydride and is commonly used to reduce aldehydes and ketones, even in the presence of carboxylic acids and esters. The reduction of butanal to *n*-butyl alcohol and of cyclopentanone to cyclopentanol by sodium borohydride in water are typical reactions of this reagent.

$$\text{CH}_3\text{CH}_2\text{CH}_2\overset{\overset{\displaystyle O}{\|}}{\text{CH}} \xrightarrow[\text{H}_2\text{O}]{\text{NaBH}_4} \text{CH}_3\text{CH}_2\text{CH}_2\text{CH}_2\text{OH}$$

butanal   *n*-butyl alcohol
                                                          85%

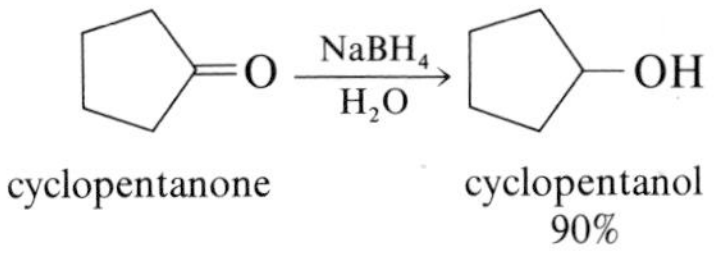

cyclopentanone   cyclopentanol
                                  90%

The mechanism for the reduction of 2-butanone with lithium aluminum hydride is shown below. A hydride ion, a nucleophile, is transferred to the electrophilic carbon atom of the carbonyl group. The alkoxide ion that forms is stabilized by complexation with a metal ion, which may be aluminum or lithium. The carbonyl group, with a trigonal carbon atom, is planar and thus symmetrical. There is equal probability that hydride ion will be donated to the carbonyl group from either side of the molecule, giving rise to equal numbers of molecules of the two enantiomers of 2-butanol. The alcohol recovered when the alkoxide ion is protonated is a racemic mixture and not optically active.

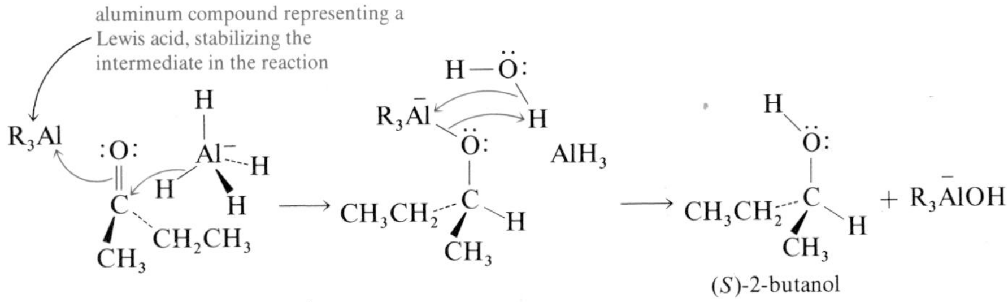

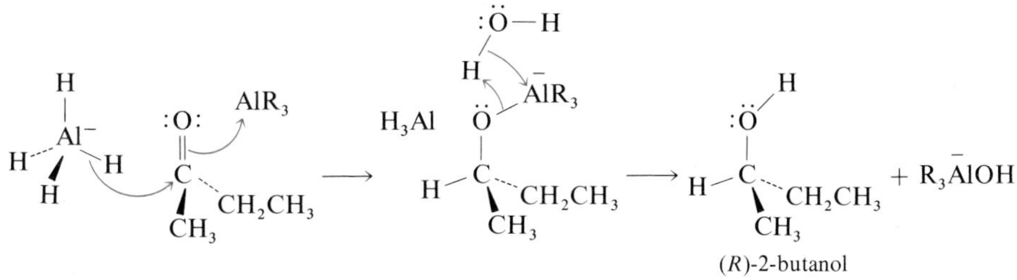

For the sake of simplicity, the transfer of only one of the four possible hydride ions is shown in the mechanism. Actually, as seen in the equations on p. 380, each molecule of lithium aluminum hydride (or sodium borohydride) can reduce four carbonyl groups.

The reductions with sodium borohydride usually take place in the presence of protic solvents, so that the alkoxide ion formed when a hydride ion is transferred to the carbonyl group is protonated by the solvent.

PROBLEM 9.7   Write out the mechanism for the reduction of cyclopentanone with sodium borohydride in water.

Sodium borohydride and lithium aluminum hydride do not reduce isolated carbon-carbon double bonds. Thus, selective reduction of a carbonyl group in a molecule containing a double bond is possible. An example of such a selective reduction is the synthesis of 6-methyl-5-hepten-2-ol, a compound used by the ambrosia beetle to communicate with others of its species. (Chemical substances used by insects and animals to communicate with each other are known as **pheromones.** They are discussed further in Section 11.7C.)

$$
\underset{\text{6-methyl-5-hepten-2-one}}{CH_3\overset{\overset{\displaystyle CH_3}{|}}{C}=CHCH_2CH_2\overset{\overset{\displaystyle O}{\|}}{C}CH_3} \xrightarrow[\text{ethanol}]{NaBH_4} \underset{\substack{\text{6-methyl-5-hepten-2-ol} \\ 95\%}}{CH_3\overset{\overset{\displaystyle CH_3}{|}}{C}=CHCH_2CH_2\overset{\overset{\displaystyle OH}{|}}{C}HCH_3}
$$

Reductions of double bonds adjacent to carbonyl groups do sometimes occur with metal hydride reagents, but these reactions depend on conditions such as solvent, temperature, and the presence of excess hydride. Even these compounds can be reduced selectively at the carbonyl group if mild reaction conditions are used. Thus, 2-butenal is reduced to 2-buten-1-ol by either sodium borohydride or lithium aluminum hydride.

$$
\underset{\text{2-butenal}}{CH_3CH=CH\overset{\overset{\displaystyle O}{\|}}{C}H} \quad \begin{matrix} \xrightarrow[\text{H}_2\text{O}]{NaBH_4} \\ \text{or} \\ \xrightarrow[\text{diethyl ether}]{LiAlH_4} \xrightarrow{H_3O^+} \end{matrix} \quad \underset{\text{2-buten-1-ol}}{CH_3CH=CHCH_2OH}
$$

PROBLEM 9.8   Complete the following equations.

(a) CH_3O—⬡—$\overset{\overset{\displaystyle O}{\|}}{C}H \xrightarrow[\text{methanol}]{NaBH_4}$

(b) $CH_3(CH_2)_5\overset{\overset{\displaystyle CH_3}{|}}{C}HCH_2\overset{\overset{\displaystyle O}{\|}}{C}H \xrightarrow[\text{diethyl ether}]{LiAlH_4} \xrightarrow{H_3O^+}$

(c) ⬡—$CH_2\overset{\overset{\displaystyle O}{\|}}{C}CH_3 \xrightarrow[\text{diethyl ether}]{LiAlH_4} \xrightarrow{H_3O^+}$

(d) ⬡—$\overset{\overset{\displaystyle O}{\|}}{C}H \xrightarrow[\text{diethyl ether}]{LiAlH_4} \xrightarrow{H_3O^+}$

(e) $CH_3\overset{\overset{\displaystyle O}{\|}}{C}CH_2\overset{\overset{\displaystyle CH_3}{|}}{C}HCH_3 \xrightarrow[\text{diethyl ether}]{LiAlH_4} \xrightarrow{H_3O^+}$

(f) $CH_3\overset{\overset{\displaystyle CH_3}{|}}{C}=CH\overset{\overset{\displaystyle O}{\|}}{C}CH_3 \xrightarrow[\text{H}_2\text{O}]{NaBH_4}$

(g) ⬡$=O \xrightarrow[\text{isopropyl alcohol}]{NaBH_4}$

(h) ◇$=O \xrightarrow[\text{diethyl ether}]{LiAlH_4} \xrightarrow{H_3O^+}$

9.4

Additions of Organometallic Reagents to Aldehydes and Ketones

A. Preparation of Organometallic Reagents

Organometallic compounds, compounds containing carbon-metal bonds, were introduced in Section 8.2C as catalysts for homogeneous hydrogenation reactions. In such catalysts, the central atom is a transition metal, which has room in its orbitals for up to 18 electrons. The transition metal atom serves as a Lewis acid and accepts pairs of electrons from species that function as Lewis bases.

     The alkali metals, such as lithium, sodium, and potassium, as well as alkaline earth metals such as magnesium, also form organometallic compounds. These types of organometallic compounds are usually prepared in two general ways: reaction of organic halides with metals, and reaction of organic compounds containing acidic carbon-hydrogen bonds with strong bases.

Alkyl, aryl, and vinyl halides will react with metals to give organometallic reagents. The reagents prepared in this way with magnesium are named **Grignard reagents,** after Victor Grignard, the French chemist who developed them at the beginning of this century. When he did this work, for which he received the Nobel Prize in 1912, he was a graduate student in the laboratory of François-Antoine-Philippe Barbier.

The reaction of methyl iodide with magnesium metal in ether gives methyl-magnesium iodide, a methyl Grignard reagent.

$$CH_3I \ + \ Mg \ \xrightarrow{\text{diethyl ether}} \ CH_3MgI$$

methyl   magnesium   methylmagnesium
iodide   metal   iodide

*a methyl Grignard
reagent*

The first step in the formation of the organomagnesium compound is the transfer of an electron from the metal to the alkyl halide.

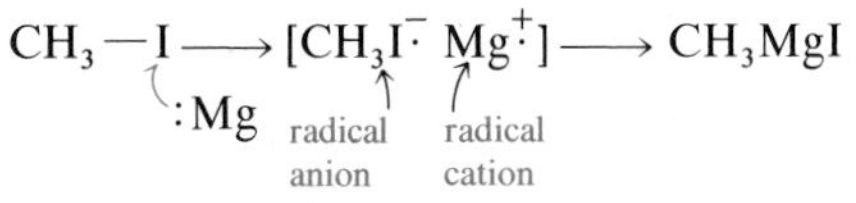

$$CH_3-I \longrightarrow [CH_3I^{\cdot-} \ Mg^{\cdot+}] \longrightarrow CH_3MgI$$

:Mg   radical   radical
       anion   cation

*held together by solvent
as a radical ion pair*

Transfer of another electron from the metal to the radical anion from the alkyl halide occurs with breakage of the carbon-halogen bond and complexation of the carbon atom and the halide ion with the magnesium ion. Chemists have not worked out the exact details and timing of these steps. The reaction is thus an overall reduction of the carbon atom and an oxidation of the magnesium atom. Note that it is *not* a nucleophilic substitution reaction. Grignard reagents of aryl and vinyl halides are also formed. Phenylmagnesium bromide, for example, is a useful reagent.

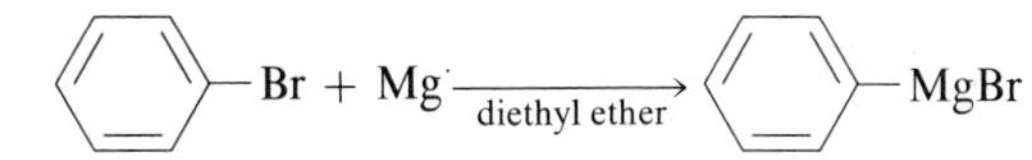

$$\text{C}_6\text{H}_5-Br + Mg \xrightarrow{\text{diethyl ether}} \text{C}_6\text{H}_5-MgBr$$

bromobenzene   phenylmagnesium
                                         bromide

a phenyl Grignard reagent

The exact structure of a Grignard reagent is not known. The solution contains a number of species with different numbers of organic groups or halide ions complexed with the metal. The solvent, usually some kind of ether, plays an important part in stabilizing the reagent.

When lithium metal reacts with organic halides, **organolithium reagents** are formed. The reaction is illustrated below for isopropyl bromide and bromobenzene.

$$\underset{\text{isopropyl bromide}}{CH_3CHBr} \ + \ \underset{\text{lithium metal}}{2\,Li} \ \xrightarrow{\text{pentane}} \ \underset{\text{isopropyllithium}}{CH_3CHLi} + Li^+Br^-$$

with CH_3 groups above CH_3CHBr and CH_3CHLi

$$\text{C}_6\text{H}_5-Br \ + \ 2\,Li \ \xrightarrow{\text{diethyl ether}} \ \text{C}_6\text{H}_5-Li + Li^+Br^-$$

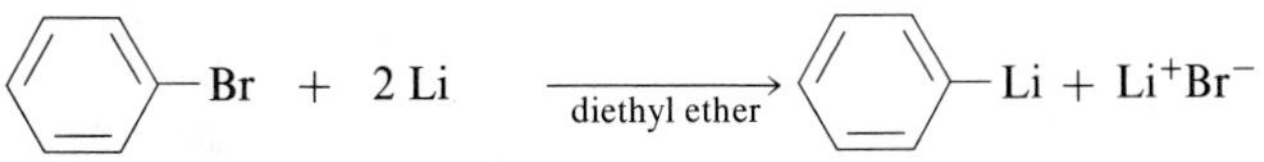

bromobenzene   lithium metal   phenyllithium

In organometallic compounds, the carbon atom bonded to the metal is carbanionlike in character, and is a strong nucleophile as well as a strong base. All of these reagents must be prepared in the absence of water and alcohols, which are acidic enough to protonate carbanions.

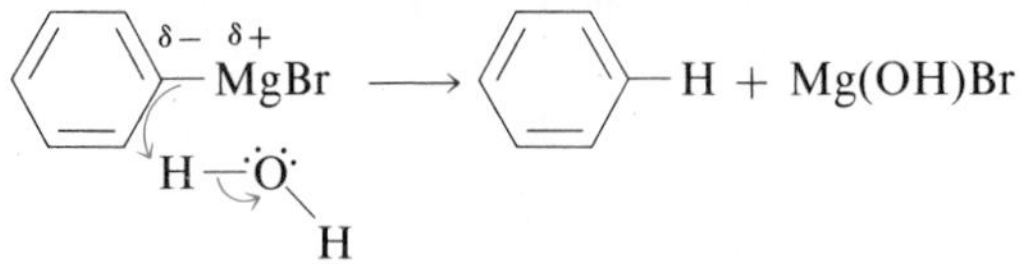

phenylmagnesium
bromide and water

benzene
*the conjugate acid
of phenyl anion*

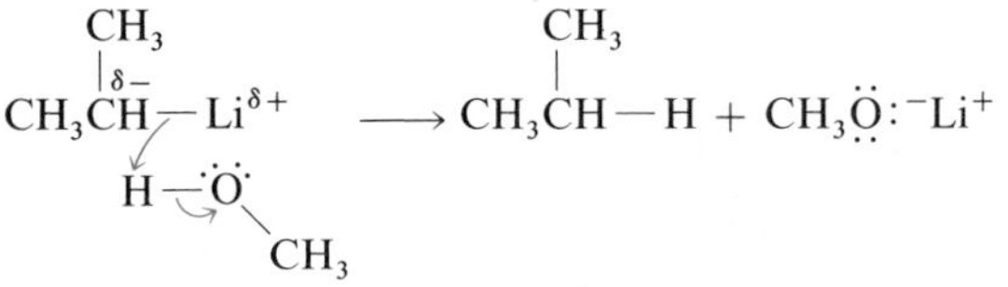

isopropyllithium
and methanol

propane
*the conjugate acid
of isopropyl anion*

Therefore, *Grignard reagents cannot be prepared from compounds that contain acidic hydrogen atoms.* Organic halides that have hydroxyl, carboxyl, thiol, or amino groups elsewhere in the molecule cannot be used to make Grignard reagents.

The basicity of the carbon atom bound to magnesium in the Grignard reagent can be utilized to make Grignard reagents from terminal alkynes in the second way of generating carbon-metal bonds. The relatively acidic hydrogen atom of a 1-alkyne is removed by an alkyl Grignard reagent.

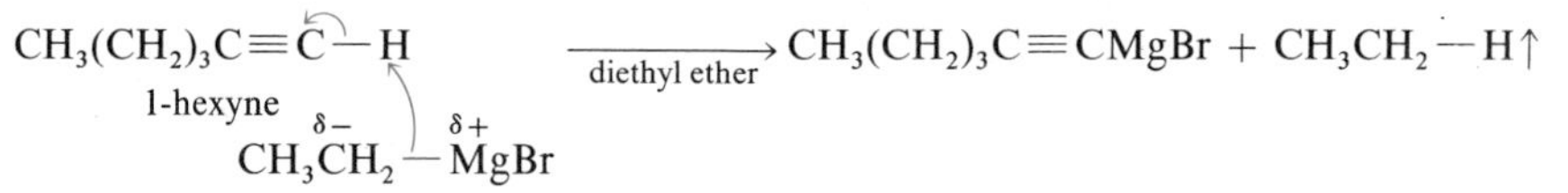

1-hexyne

ethylmagnesium
bromide

1-hexynylmagnesium
bromide

*from the conjugate
base of
1-hexyne*

ethane

*the conjugate
acid of ethyl
anion*

Ethane, in which all of the hydrogen atoms are bound to sp^3-hybridized carbon atoms, is a weaker acid than 1-hexyne, in which one of the hydrogen atoms is bound to the more electronegative sp-hybridized carbon atom (Section 2.7B). In a competition between the two bases, the proton is transferred to the ethyl group, and ethane gas evolves from the reaction mixture, pulling the reaction to completion.

Terminal alkynes also react with strong bases to give organometallic compounds. Acetylene, for example, reacts with sodium amide, prepared from sodium in liquid ammonia, to form sodium acetylide.

$$2\,Na + 2\,NH_3 \longrightarrow 2\,Na^+NH_2^- + H_2\uparrow$$

sodium ammonia
metal

sodium hydrogen
amide

$$HC\equiv CH + Na^+:\ddot{N}H_2^- \xrightarrow[\substack{NH_3(liq) \\ -33\,°C}]{} HC\equiv CNa + NH_3$$

acetylene   sodium   sodium
                   amide   acetylide

The anions from alkynes are also very strong nucleophiles and bases, and must be protected from water and other acids.

PROBLEM 9.9   Complete the following equations showing the products that you would expect.

(a) $CH_3CH_2CH_2CH_2Br \xrightarrow[\text{tetrahydrofuran}]{Li}$

(b) 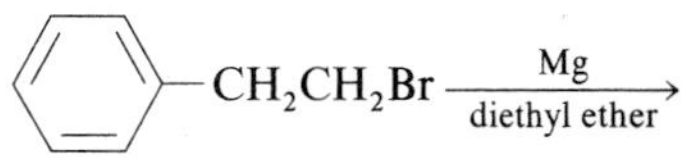   ⬡—$CH_2CH_2Br \xrightarrow[\text{diethyl ether}]{Mg}$

(c) $CH_2\!=\!CHCl \xrightarrow[\substack{\text{tetrahydrofuran} \\ \Delta}]{Mg}$

(d) 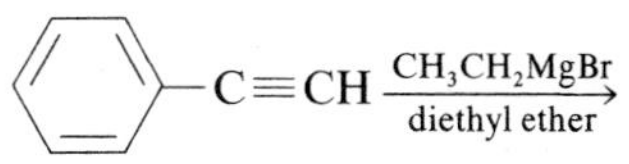   ⬡—$C\equiv CH \xrightarrow[\text{diethyl ether}]{CH_3CH_2MgBr}$

(e) 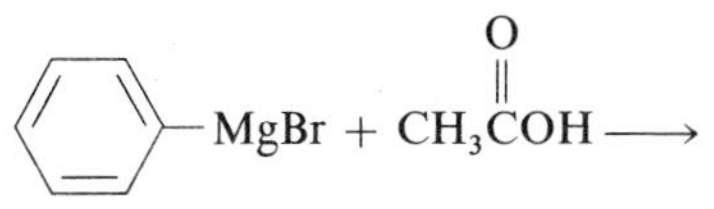   ⬡—$MgBr + CH_3\overset{\overset{\displaystyle O}{\|}}{C}OH \longrightarrow$

(f) $CH_3CH_2C\equiv CNa + H_2O \longrightarrow$

(g) 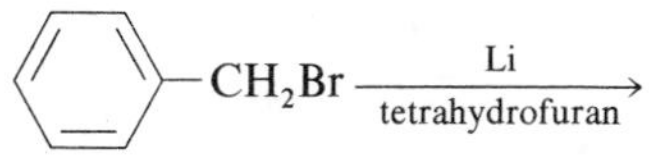   ⬡—$CH_2Br \xrightarrow[\text{tetrahydrofuran}]{Li}$

(h) $CH_3CH_2CH_2Li + NH_3 \longrightarrow$

(i) $CH_3CH_2MgI + CH_3OH \longrightarrow$

(j) 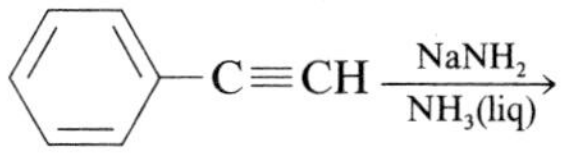   ⬡—$C\equiv CH \xrightarrow[\substack{NH_3(liq)}]{NaNH_2}$

B. Addition of Organometallic Compounds to Aldehydes and Ketones

All of the organometallic compounds prepared in Section 9.4A will react with aldehydes and ketones to produce alcohols. The reactions take place in two stages. The first involves attack by the nucleophilic carbon atom of the organometallic reagent on the electrophilic carbon atom of the carbonyl group to give an alkoxide anion, complexed with a metal ion. The second stage requires that this complex be broken up with water, and usually some dilute acid, to produce the alcohol.

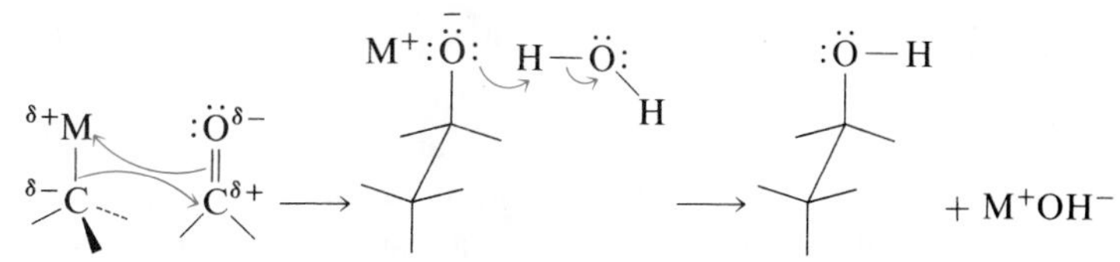

nucleophilic attack   *complex of*   *product*
by organometallic   *alkoxide ion*   *alcohol*
reagent on carbonyl   *with metal,*
compound   *reacting with*
                           water

Because it is possible to vary the structure of the aldehyde or ketone as well as that of the organometallic reagent, many different alcohols may be prepared in this way. Historically, the Grignard reagent has been used extensively to synthesize alcohols by means of the **Grignard reaction.** Three examples are shown as follows.

$$\underset{\text{isopropyl bromide}}{\overset{\overset{\displaystyle CH_3}{|}}{CH_3CHBr}} + Mg \xrightarrow{\text{diethyl ether}} \underset{\substack{\text{isopropylmagnesium} \\ \text{bromide}}}{\overset{\overset{\displaystyle CH_3}{|}}{CH_3CHMgBr}} \xrightarrow{\overset{\overset{\displaystyle O}{\|}}{CH_3CH \text{ (acetaldehyde)}}}$$

$$\underset{\substack{|\\OMgBr}}{\overset{\overset{\displaystyle CH_3}{|}}{CH_3CHCHCH_3}} \xrightarrow{H_3O^+} \underset{\substack{\\OH \\ \text{3-methyl-2-butanol} \\ 50\%}}{\overset{\overset{\displaystyle CH_3}{|}}{CH_3CHCHCH_3}}$$

$$\underset{\text{1-hexyne}}{CH_3(CH_2)_3C\equiv CH} + CH_3CH_2MgBr \xrightarrow{\text{diethyl ether}} \underset{\substack{\text{1-hexynylmagnesium} \\ \text{bromide}}}{CH_3(CH_2)_3C\equiv CMgBr} + CH_3CH_3\uparrow \xrightarrow{\overset{\overset{\displaystyle O}{\|}}{HCH \text{ (formaldehyde)}}}$$

$$CH_3(CH_2)_3C\equiv CCH_2OMgBr \xrightarrow{H_3O^+} \underset{\substack{\text{2-heptyn-1-ol} \\ 82\%}}{CH_3(CH_2)_3C\equiv CCH_2OH}$$

$$\underset{\text{methyl chloride}}{CH_3Cl} + Mg \xrightarrow{\text{diethyl ether}} \underset{\substack{\text{methylmagnesium} \\ \text{chloride}}}{CH_3MgCl} \xrightarrow{\overset{\overset{\displaystyle O}{\|}}{\triangleright-CCH_3 \text{ (cyclopropyl methyl ketone)}}}$$

$$\underset{\substack{|\\OMgCl}}{\overset{\overset{\displaystyle CH_3}{|}}{\triangleright-CCH_3}} \xrightarrow[H_2O]{NH_4Cl} \underset{\substack{\\OH \\ \text{2-cyclopropyl-} \\ \text{2-propanol} \\ 68\%}}{\overset{\overset{\displaystyle CH_3}{|}}{\triangleright-CCH_3}}$$

In each of the three reactions shown, the Grignard reagent is prepared, an aldehyde or a ketone is added to it to give an alkoxide ion complexed with magnesium ion, and the alcohol is formed by protonation of the alkoxide ion. In the last equation, a very weak acid, ammonium chloride, is used to protect the tertiary alcohol from dehydration.

The type of alcohol formed depends upon the carbonyl compound used. An aldehyde such as acetaldehyde produces a secondary alcohol. Formaldehyde produces a primary alcohol that has one more carbon atom in it than the alkyl halide used to make the Grignard reagent, and cyclopropyl methyl ketone produces a tertiary alcohol. These facts about the Grignard reaction are summarized in the general equations below.

formaldehyde   alkoxide anion complexed   primary
                    with magnesium ion   alcohol

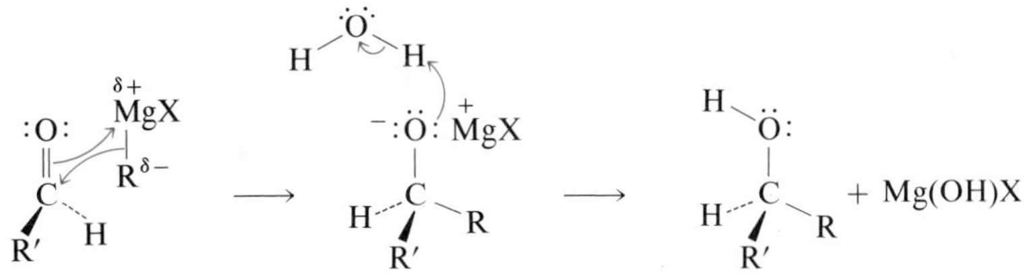

aldehyde other than formaldehyde → alkoxide anion complexed with magnesium ion → secondary alcohol + Mg(OH)X

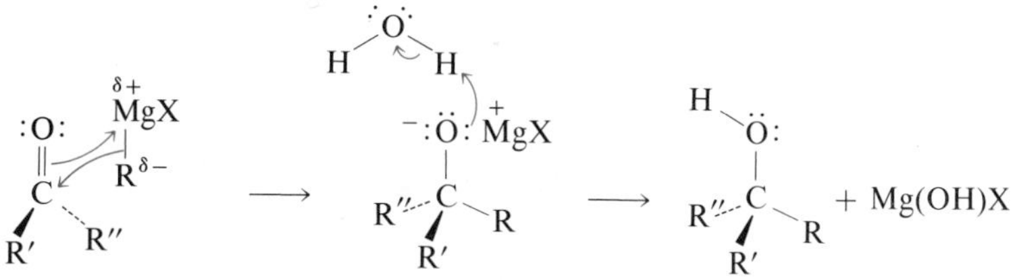

ketone → alkoxide anion complexed with magnesium ion → tertiary alcohol + Mg(OH)X

The secondary and tertiary alcohols formed in the Grignard reaction may contain asymmetric carbon atoms, depending upon the nature of the organic groups present. The stereochemical arguments that were made for the reduction of a carbonyl group by lithium aluminum hydride (Section 9.3) apply here too. The organometallic reagent reacts with equal probability at either face of the carbonyl group to give a racemic alcohol unless, of course, there is chirality already present in the Grignard reagent or elsewhere in the carbonyl compound.

Grignard reagents are reactive towards oxygen and carbon dioxide as well as towards carbonyl groups. The reaction with carbon dioxide is used to prepare carboxylic acids (Section 10.5B). The reaction with oxygen gives salts of hydroperoxides, compounds related to hydrogen peroxide. Such reactions diminish the yield of the desired product when Grignard reagents are used to synthesize alcohols. In the laboratory, the reaction is run under an atmosphere of dry nitrogen to ensure the absence of water vapor, oxygen, and carbon dioxide in the reaction mixture.

Organolithium reagents also add to aldehydes and ketones to form alcohols. The reactions parallel those of the Grignard reagents except that in many cases the organolithium reagents give better yields and, in fact, form alcohols with some sterically hindered ketones that do not react with Grignard reagents. Some representative reactions are shown below.

$$CH_3CH_2CH_2CH_2CH_2CH_2Br + 2\ Li \xrightarrow[0\ °C]{\text{tetrahydrofuran}} CH_3CH_2CH_2CH_2CH_2CH_2Li + Li^+Br^-$$

1-bromohexane   1-hexyllithium

$$\downarrow \overset{O}{\overset{\|}{\text{HCH}}}\ \text{(formaldehyde)}$$

$$CH_3CH_2CH_2CH_2CH_2CH_2CH_2OH \xleftarrow{H_3O^+} CH_3CH_2CH_2CH_2CH_2CH_2CH_2O^-Li^+$$

1-heptanol
72%

a primary alcohol

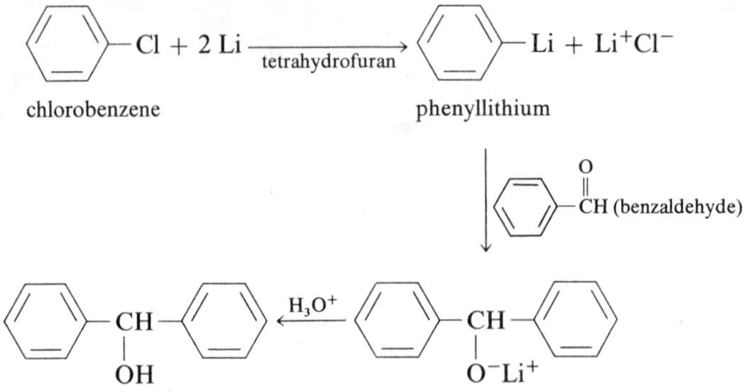

chlorobenzene → phenyllithium

diphenylmethanol
100%
a secondary alcohol

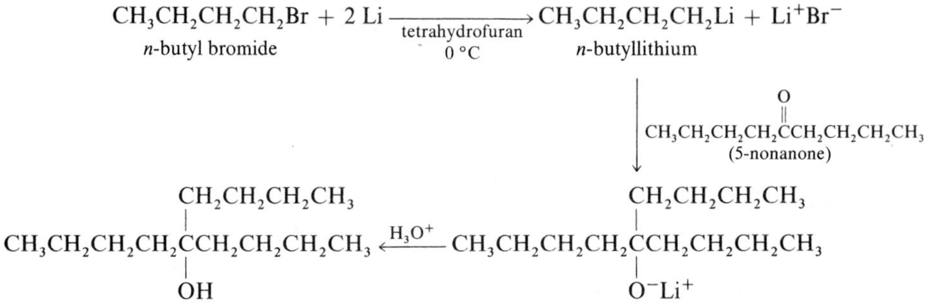

$CH_3CH_2CH_2CH_2Br + 2\ Li \xrightarrow[\substack{tetrahydrofuran \\ 0\ °C}]{} CH_3CH_2CH_2CH_2Li + Li^+Br^-$

n-butyl bromide *n*-butyllithium

(5-nonanone)

5-butyl-5-nonanol
91%
a tertiary alcohol

In each case, the organometallic reagent adds to the carbonyl group, and the resulting alkoxide anion is protonated by dilute acid to recover the alcohol. Formaldehyde yields a primary alcohol, other aldehydes give secondary alcohols, and ketones give tertiary alcohols.

Sodium acetylide also adds to carbonyl groups. Sodium acetylide, prepared from sodium amide and acetylene in liquid ammonia, reacts with benzaldehyde to give an acetylenic secondary alcohol, and with cyclohexanone to form a tertiary alcohol.

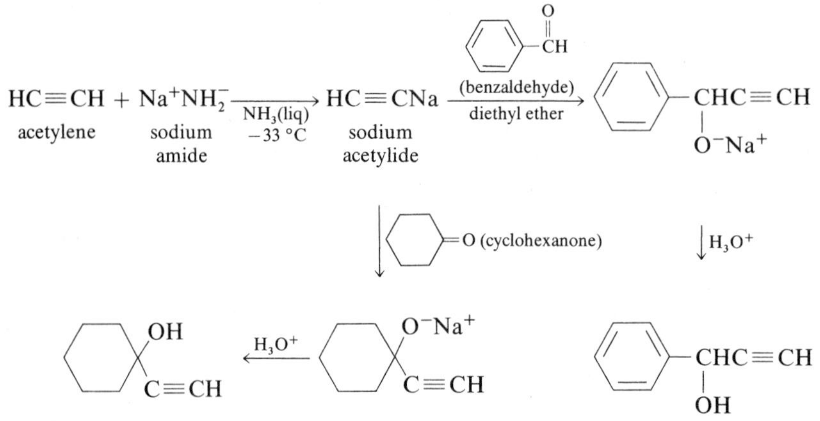

$HC\equiv CH + Na^+NH_2^- \xrightarrow[\substack{NH_3(liq) \\ -33\ °C}]{} HC\equiv CNa \xrightarrow[diethyl\ ether]{(benzaldehyde)}$

acetylene sodium amide sodium acetylide

1-ethynylcyclohexanol
70%
a tertiary alcohol

1-phenyl-2-propyn-1-ol
65%
a secondary alcohol

The three-membered ring of an oxirane is strained enough to open under nucleophilic attack (Section 8.5C) by organometallic reagents. In this respect, the oxirane behaves as if it were a carbonyl group with the nucleophilic carbon atom of the organometallic reagent bonding to one of the carbon atoms of the oxirane ring, and the metal complexing with the alkoxide ion that forms when the ring opens. *n*-Butylmagnesium bromide, for example, reacts with oxirane to give 1-hexanol, a primary alcohol in which the carbon chain has been lengthened by two carbon atoms.

$$CH_3CH_2CH_2\overset{\delta-}{C}H_2\overset{\delta+}{M}gBr + \overset{\delta-}{\underset{\delta+}{C}H_2}\overset{O}{\underset{}{C}H_2} \xrightarrow[\text{diethyl ether}]{} CH_3CH_2CH_2CH_2CH_2CH_2\overset{-}{O}\overset{+}{M}gBr \xrightarrow{H_3O^+}$$

n-butylmagnesium   oxirane
bromide

$$CH_3CH_2CH_2CH_2CH_2CH_2OH$$
1-hexanol
60%

In Chapter 6 we emphasized the use of a nucleophile to create a new σ bond at an electrophilic carbon atom bearing a good leaving group. In Section 6.11, especially, reagents containing nucleophilic carbon atoms were used to create new carbon-carbon bonds in nucleophilic substitution reactions. The idea that reaction between a carbon atom that is nucleophilic and one that is electrophilic is necessary in forming carbon-carbon bonds is an important one. In this section, we have seen that the electrophilic carbon atom of a carbonyl group reacts with the nucleophilic carbon atom of an organometallic reagent to give a product with a modified carbon chain. In the next section, we will apply these ideas to devise syntheses of a variety of compounds.

PROBLEM 9.10   Complete the following equations, showing the structures of all the intermediates and final products in the reactions.

(a) $\bigcirc$=O + CH$_3$CH$_2$CH$_2$CH$_2$Li $\longrightarrow$ A $\xrightarrow{H_3O^+}$ B

(b) $\bigcirc$—Br $\xrightarrow[\text{diethyl ether}]{\text{Mg}}$ C $\xrightarrow{CH_2CH_2 \text{(O)}}$ D $\xrightarrow{H_3O^+}$ E

(c)   CH$_3$CH$_2$CH$_2$CH (=O) $\xrightarrow[\text{tetrahydrofuran}]{\bigcirc-Br, \text{ Li}}$ F $\xrightarrow{H_3O^+}$ G

(d)   CH$_3$CH$_2$C≡CH $\xrightarrow[\text{NH}_3\text{(liq)}]{\text{NaNH}_2}$ H $\xrightarrow[\text{diethyl ether}]{\bigcirc=O}$ I $\xrightarrow{H_3O^+}$ J

(e)   [structure: cyclohexenone with gem-dimethyl (CH$_3$CH$_3$) group] $\xrightarrow[\text{diethyl ether}]{\text{CH}_3\text{Li}}$ K $\xrightarrow{H_3O^+}$ L

(f) $\underset{\text{CH}_3}{\overset{\text{CH}_3}{\underset{|}{\text{CH}_3\text{C}}}}{=}\text{CHCCH}_3 \xrightarrow[\text{diethyl ether}]{\text{CH}_3\text{CH}_2\text{CH}_2\text{CH}_2\text{Li}} \text{M} \xrightarrow{\text{H}_3\text{O}^+} \text{N}$

(g) $\text{CH}_3\overset{\text{CH}_3}{\underset{|}{\text{C}}}{=}\text{CHCH}_2\text{CH}_2\overset{\text{CH}_3}{\underset{|}{\text{C}}}{=}\text{CHCH}_2\text{CH}_2\overset{\text{O}}{\overset{\|}{\text{CH}}} \xrightarrow[\text{dichloromethane}]{\text{HC}\equiv\text{CMgBr}} \text{O} \xrightarrow{\text{H}_3\text{O}^+} \text{P}$

C. Syntheses Using Organometallic Reagents

The addition of organometallic compounds to carbonyl groups gives chemists a tool with which to create carbon-carbon bonds using a wide variety of reagents. The alcohols formed in these reactions are easily converted into alkenes, alkyl halides, and, in some cases, new carbonyl compounds. Each of these functional group classes has further reactions so that a wide range of compounds can be synthesized. In this section, we explore the kind of reasoning that goes into designing organic syntheses. One of the realistic constraints always present in planning a synthesis is the availability of reagents. For the sake of this exercise, we will assume that we have a good supply of bromobenzene, any other organic chemicals containing no more than three carbon atoms, and any inorganic reagents that we wish to use.

Synthesis of 2-phenyl-2-pentanol 2-Phenyl-2-pentanol is a tertiary alcohol that can be dissected at several points to give fragments that could be used to synthesize the molecule from smaller molecules. Tertiary alcohols can be synthesized from ketones and appropriate Grignard or organolithium reagents.

2-phenyl-2-pentanol   phenyllithium   2-pentanone

acetophenone   *n*-propyllithium   phenyl propyl ketone   methyllithium

This reaction scheme illustrates three important points about the design of a synthesis. First, the molecule is analyzed for the functional group or groups that are present. Second, the molecule is dissected at points next to functionalities to examine what the smaller fragments for the synthesis might be. Third, all the starting materials are listed that would be needed to create those functional groups. In this case, because there are restrictions on the starting materials available, only syntheses that start from smaller fragments are considered. Grignard reagents could be substituted for organolithium reagents without changing the logic of the analysis.

Whichever pathway we choose, one of the fragments will have to be prepared by further synthesis. Path c, for example, requires *n*-propyl bromide. The ketone acetophenone, however, must be synthesized from smaller pieces, so it becomes the target of a second synthetic analysis. Once again, a process of working backward is necessary. Acetophenone is a ketone and can be prepared by oxidation of the secondary alcohol that corresponds to its structure. The secondary alcohol is produced from the reaction of an organometallic reagent with an aldehyde.

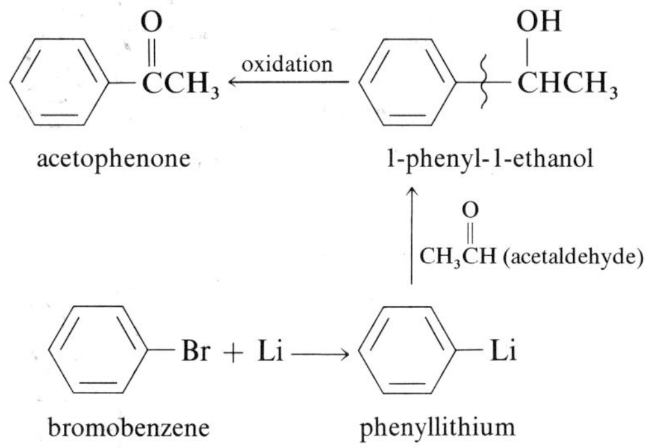

The total synthesis of 2-phenyl-2-pentanol from the available reagents is outlined below.

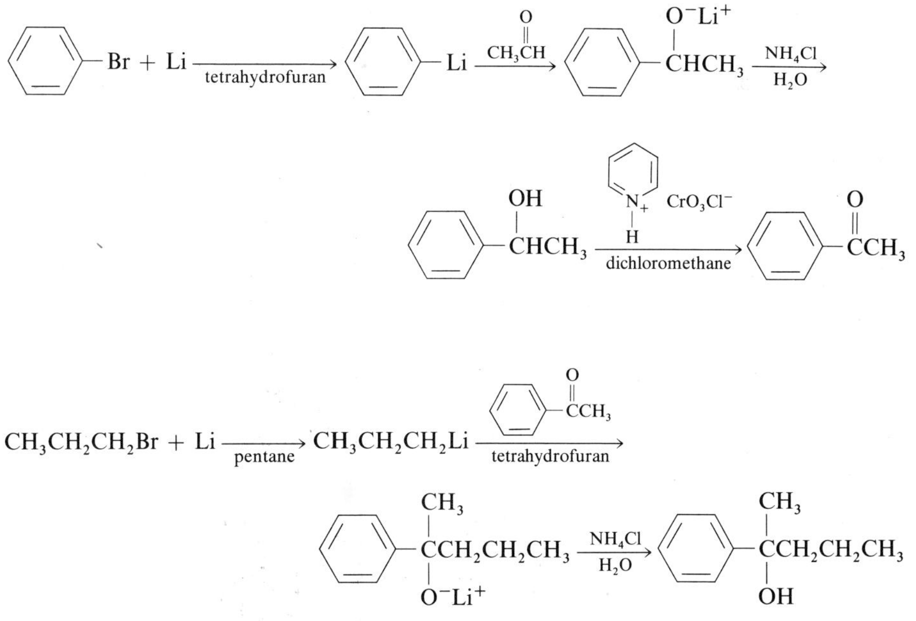

Ammonium chloride is used instead of a stronger acid in the last step of the synthesis to avoid dehydration of the tertiary alcohol. Dehydration would take place easily in this case because the *double bond* would be formed *adjacent to the aromatic ring*. Such a double bond *is said to be in* **conjugation** *with the ring and has special reactivity*. The properties of systems with conjugated multiple bonds are discussed in Chapter 11.

PROBLEM 9.11   Show complete syntheses of 2-phenyl-2-pentanol by the other possible pathways (p. 390).

Synthesis of (Z)-3-hexen-2-ol   (Z)-3-Hexen-2-ol has two functional groups: it is an alkene and a secondary alcohol. The double bond has the less stable cis configuration; therefore, it must be formed late in the synthesis so that it will not isomerize to the more stable trans form during other manipulations of the compound. A dissection of the structure suggests a synthetic route for its preparation from the available starting materials.

$$
\begin{array}{cc}
\text{OH} & \\
\text{CH}_3\text{CH} & \quad\text{CH}_2\text{CH}_3 \\
\qquad \text{C}=\text{C} & \\
\quad\text{H} \qquad\quad \text{H} &
\end{array}
$$

↑ cis double bond from the catalytic hydrogenation or hydroboration of an alkyne

$$
\begin{array}{c}
\text{OH} \\
\text{CH}_3\{\text{CH}\}\text{C}\equiv\text{CCH}_2\text{CH}_3
\end{array}
$$

↑ secondary alcohol from the reaction of a Grignard reagent with an aldehyde

$$
\underset{\text{2-pentynal}}{\text{CH}_3\text{MgI} + \overset{\text{O}}{\overset{\|}{\text{HCC}}}\equiv\text{CCH}_2\text{CH}_3} \quad \text{or} \quad \text{CH}_3\overset{\text{O}}{\overset{\|}{\text{CH}}} + \text{BrMgC}\equiv\text{CCH}_2\text{CH}_3
$$

↑ this Grignard reagent easily available from 1-butyne

$$
\text{HC}\equiv\text{C}\{\text{CH}_2\text{CH}_3\}
$$

↑ four-carbon alkyne can be prepared by S_N2 reaction

$$
\text{HC}\equiv\text{CNa} + \text{BrCH}_2\text{CH}_3
$$

The cis double bond is formed in a stereoselective catalytic hydrogenation (Section 8.3A) or hydroboration (Section 8.3B) reaction of an alkyne as the last step of the synthesis.

The secondary alcohol function is introduced by the reaction of an acetylenic Grignard reagent with an aldehyde. 1-Butyne is synthesized by a nucleophilic substitution reaction of sodium acetylide on bromoethane. *Note that the points at which the molecular structure is dissected in the synthetic analysis are always at or next to a functional group. These are the sites of reactivity and change in the molecule.*

A synthesis of (Z)-3-hexen-2-ol, showing all the necessary reagents, is outlined below.

$$
\text{HC}\equiv\text{CH} \xrightarrow[\text{NH}_3\text{(liq)}]{\text{NaNH}_2} \text{HC}\equiv\text{CNa} \xrightarrow{\text{CH}_3\text{CH}_2\text{Br}} \text{HC}\equiv\text{CCH}_2\text{CH}_3 \xrightarrow[\text{diethyl ether}]{\text{CH}_3\text{CH}_2\text{MgBr}}
$$

$$
\text{BrMgC}\equiv\text{CCH}_2\text{CH}_3 \xrightarrow{\overset{\text{O}}{\overset{\|}{\text{CH}_3\text{CH}}}} \text{CH}_3\overset{\overset{-\ +}{\text{OMgBr}}}{\text{CHC}}\equiv\text{CCH}_2\text{CH}_3 \xrightarrow[\text{H}_2\text{O}]{\text{NH}_4\text{Cl}}
$$

$$
\underset{}{\text{CH}_3\overset{\text{OH}}{\text{CHC}}\equiv\text{CCH}_2\text{CH}_3} \xrightarrow[\substack{\text{Pd}\\\text{quinoline}}]{\text{H}_2} \begin{array}{c}\text{OH}\\ \text{CH}_3\text{CH} \qquad \text{CH}_2\text{CH}_3 \\ \qquad\text{C}=\text{C}\\ \text{H}' \qquad\quad \text{H}\end{array}
$$

(Z)-3-hexen-2-ol

PROBLEM 9.12   Show the steps that would be necessary to synthesize 2-pentynal, the acetylenic precursor for the other synthetic pathway for (Z)-3-hexen-2-ol.

PROBLEM 9.13   Work out syntheses for the following compounds. You have bromobenzene, any organic reagents containing three or fewer carbon atoms, and any inorganic reagents you need. Be as complete as possible about all necessary conditions, including solvents.

(a) [benzene ring]—$CH_2CH_2\overset{\underset{\displaystyle |}{Cl}}{C}HCH_3$ (b) $CH_3\overset{\underset{\displaystyle |}{CH_3}}{C}HCH_2\overset{\overset{\displaystyle O}{\|}}{C}OH$

(c) [benzene ring]—$CH_2CH_2CH_2\overset{\overset{\displaystyle O}{\|}}{C}H$ (d) $\underset{H}{\overset{CH_3CH_2}{}}C=C\underset{CH_3}{\overset{H}{}}$

PROBLEM 9.14   The pheromone of the ambrosia beetle, 6-methyl-5-hepten-2-ol, prepared by sodium borohydride reduction of 6-methyl-5-hepten-2-one in Section 9.3, has also been synthesized by another route.

$$CH_3\overset{\underset{\displaystyle |}{CH_3}}{C}=CHCH_2CH=CH_2 \longrightarrow A \longrightarrow B \longrightarrow CH_3\overset{\underset{\displaystyle |}{CH_3}}{C}=CHCH_2CH_2\overset{\underset{\displaystyle |}{OH}}{C}HCH_3$$

Supply the structures of Compounds A and B and the reagents necessary to transform 5-methyl-1,4-hexadiene into 6-methyl-5-hepten-2-ol.

9.5

Cyanohydrin Formation

Another nucleophilic addition reaction of aldehydes and ketones is the addition of the elements of hydrogen cyanide, illustrated with acetone.

$$CH_3\overset{\overset{\displaystyle O}{\|}}{C}CH_3 \xrightarrow[H_2O]{NaCN} \xrightarrow[H_2O]{H_2SO_4} CH_3\overset{\underset{\displaystyle CN}{\underset{\displaystyle |}{|}}}{\overset{\displaystyle OH}{C}}CH_3$$

acetone   acetone
                                    cyanohydrin

The usual reagents are sodium or potassium cyanide followed by acid. The reaction is initiated when the cyanide ion attacks the carbonyl group. Simultaneous protonation of the alkoxide ion that is generated occurs.

$$CH_3-\overset{\overset{\displaystyle :O:}{\|}}{\underset{\underset{\displaystyle :C\equiv N:}{|}}{C}}-CH_3 \quad \xrightleftharpoons \quad CH_3-\overset{\overset{\displaystyle :\ddot{O}-H}{|}}{\underset{\underset{\displaystyle C\equiv N:}{|}}{C}}-CH_3$$

$H-B^+$ $:B$

The products of this reaction are called **cyanohydrins** because they include a hydroxyl and a cyano group in the same molecule.

The addition of hydrogen cyanide to aldehydes and ketones increases the number of carbon atoms in the molecule by one, while introducing two functional groups that can undergo further reactions. An important reaction of the cyano group is the addition of water to the carbon-nitrogen triple bond, a reaction that is presented in detail in Sections 10.5C and 10.8B. Conversion of acetaldehyde to lactic acid, which is one of the products of the metabolism of glucose in muscle and is also the acid that gives sour milk its taste, is an example of the formation and further transformation of a cyanohydrin.

$$
\underset{\text{acetaldehyde}}{\overset{\overset{\displaystyle O}{\|}}{CH_3CH}} \xrightarrow[\text{H}_2\text{O}]{\text{NaCN}} \xrightarrow[\text{H}_2\text{O}]{\text{H}_2\text{SO}_4} \underset{\substack{\text{acetaldehyde}\\\text{cyanohydrin}}}{\overset{\overset{\displaystyle OH}{|}}{CH_3CHCN}} \xrightarrow[\Delta]{\text{H}_3\text{O}^+}
$$

$$
\begin{bmatrix} \overset{\overset{\displaystyle O}{\|}}{\underset{\underset{\displaystyle OH}{|}}{CH_3CHCNH_2}} \end{bmatrix} \xrightarrow[\Delta]{\text{H}_3\text{O}^+} \underset{\underset{\displaystyle OH}{|}}{\overset{\overset{\displaystyle O}{\|}}{CH_3CHCOH}} + NH_4^+
$$

2-hydroxypropanamide   2-hydroxypropanoic
amide of lactic acid   acid

lactic acid

PROBLEM 9.15   The lactic acid present in the muscle is (R)-$(-)$-lactic acid, while the lactic acid found in sour milk is (S)-$(+)$-lactic acid. What is the stereochemistry of the addition of hydrogen cyanide to acetaldehyde? What is the stereochemistry of the lactic acid formed by the hydrolysis of the cyano group?

Acid-catalyzed dehydration is typical of an alcohol. The cyanohydrin of acetaldehyde undergoes this reaction to form acrylonitrile, the starting material for the polymer polyacrylonitrile, Orlon, which is used in many synthetic fabrics.

$$
\underset{\substack{\text{acetaldehyde}\\\text{cyanohydrin}}}{\underset{\underset{\displaystyle OH}{|}}{CH_3CHCN}} \xrightarrow[\Delta]{\text{H}_3\text{PO}_4} \underset{\substack{\text{propenenitrile}\\\text{acrylonitrile}\\90\%}}{CH_2{=}CHCN} \xrightarrow[\text{catalyst}]{} \underset{\substack{\text{polyacrylonitrile}\\\text{Orlon}}}{(CH_2CHCH_2CH)_{\overline{n}}}
$$
$$
\underset{CN \quad CN}{}
$$

where n = 200–1000

9.6

Addition Reactions of Compounds Related to Ammonia

A. Compounds Related to Ammonia and Their Use in the Characterization of Aldehydes and Ketones

Amines and other compounds related to ammonia are an important class of reagents that behave as nucleophiles towards the carbonyl group. Some typical compounds with this kind of reactivity are shown as follows.

$$RNH_2$$

alkylamine

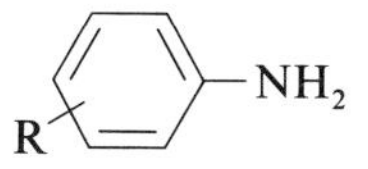

$$R\text{—}\langle\text{ring}\rangle\text{—}NH_2$$

arylamine

may have other substituents on the ring

$$HONH_2$$

hydroxylamine

$$H_2NNH_2$$

hydrazine

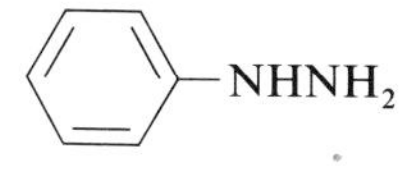

$$\langle\text{ring}\rangle\text{—}NHNH_2$$

phenylhydrazine

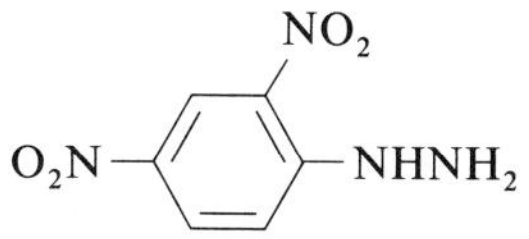

$$O_2N\text{—}\langle\text{ring with }NO_2\rangle\text{—}NHNH_2$$

2,4-dinitrophenylhydrazine

Some of these reagents convert low molecular weight liquid aldehydes and ketones into solid compounds. *A compound prepared from another compound by some standard reaction typical of the functional group is called a* **derivative.** Derivatives are used to check the identity of the original compound by comparing the physical properties of the derivative, such as the melting point, with those reported for derivatives of similar known compounds. Solid derivatives are easier to handle in small quantities than liquids are, and it is easier to determine an accurate melting point on a small amount of material than it is to determine an accurate boiling point. Solid derivatives are, therefore, useful in characterizing compounds.

Cyclohexanone, for example, is a liquid that reacts with hydroxylamine (available in the laboratory as its hydrochloride salt) to form a solid derivative called an **oxime.**

$$HONH_3^+\ Cl^- + CH_3\overset{\displaystyle O}{\overset{\|}{C}}O^-Na^+ \longrightarrow HONH_2 + CH_3\overset{\displaystyle O}{\overset{\|}{C}}OH + Na^+Cl^-$$

hydroxylamine hydrochloride   sodium acetate   hydroxylamine

$$\langle\text{cyclohexanone}\rangle{=}O + H_2NOH \xrightarrow[\substack{CH_3\overset{O}{\overset{\|}{C}}O^-Na^+ \\ \text{acetic acid} \\ \Delta}]{} \langle\text{ring}\rangle{=}NOH + H_2O$$

cyclohexanone
bp 156 °C

cyclohexanone oxime
mp 91 °C

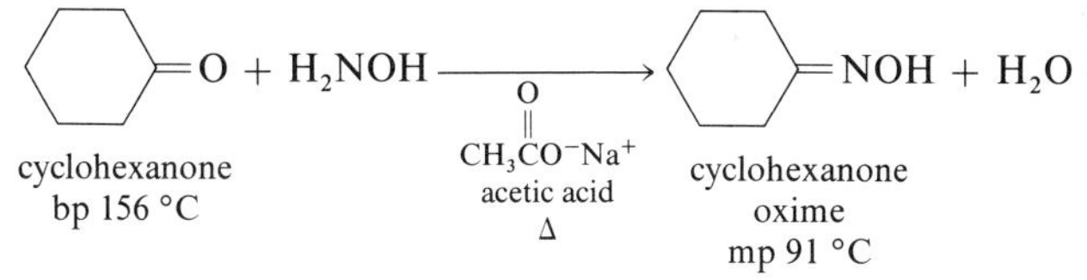

The reaction shown above is typical of the reactions of amines and similar compounds with aldehydes and ketones. A nitrogen atom is bound to the carbon atom of the carbonyl group by a double bond, and water is eliminated. The reaction usually goes best with mild acid catalysis at a pH of about 5. A mechanism consistent with these experimental facts is shown in the next section.

B. A Mechanism for the Reaction of Ammonialike Compounds with Aldehydes and Ketones

The reaction of cyclohexanone with hydroxylamine, shown in the last section, will be used as a model for the general reactions of such nitrogen compounds with aldehydes and ketones. Sodium acetate is used as a base to remove a proton from the nitrogen atom in hydroxylamine hydrochloride to form the weak acid, acetic acid (above). Sodium acetate and acetic acid together form a buffer system so that the acidity of the reaction mixture remains close to pH 5. It is important that the solution not be too acidic because the unprotonated nitrogen atom in hydroxylamine is the nucleophile that reacts with the electrophilic carbon atom of the carbonyl group. On the other hand, some protonation of the oxygen atom of the carbonyl group is useful in enhancing the electrophilicity of the carbon atom that will undergo reaction. The most important function of the acid is in assisting in the loss of water in the final stages of the reaction.

The reaction of hydroxylamine with cyclohexanone takes place in two steps. In the first step, the nucleophile attacks the carbonyl group, shown below as being protonated first.

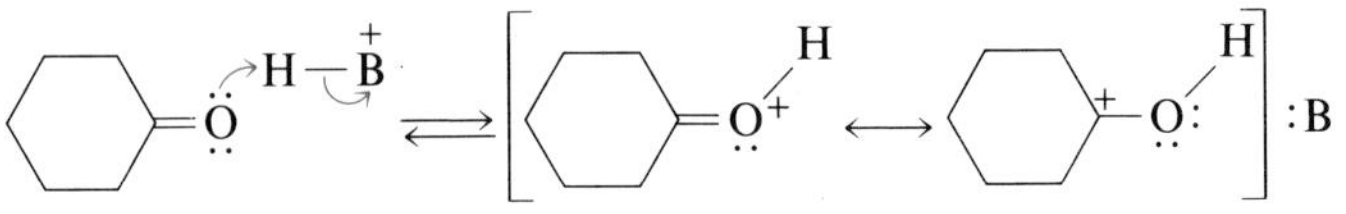

*protonation of
the carbonyl group*

*resonance contributors showing
enhanced electrophilic
character of carbon atom in
protonated carbonyl group*

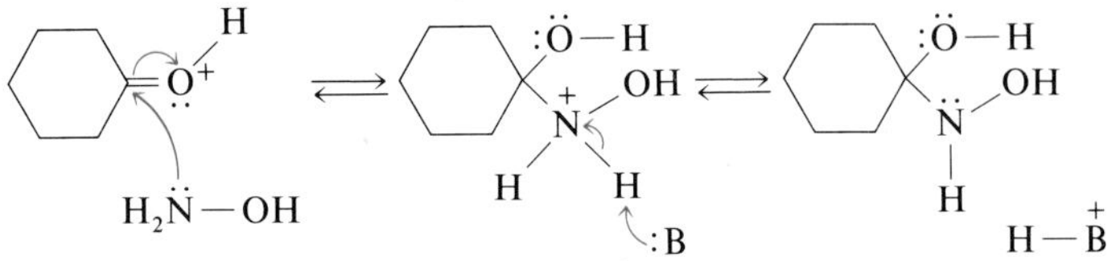

*nucleophilic attack
at the carbonyl group*

*deprotonation of
nitrogen atom*

*a carbinolamine
intermediate*

The intermediate that forms loses a proton from the positively charged nitrogen atom. The species with a hydroxyl group and an amino group on the same carbon atom loses water easily to give a carbon-nitrogen double bond. The loss of water is catalyzed by acid and is the rate-determining step for the reaction at moderate acidity.

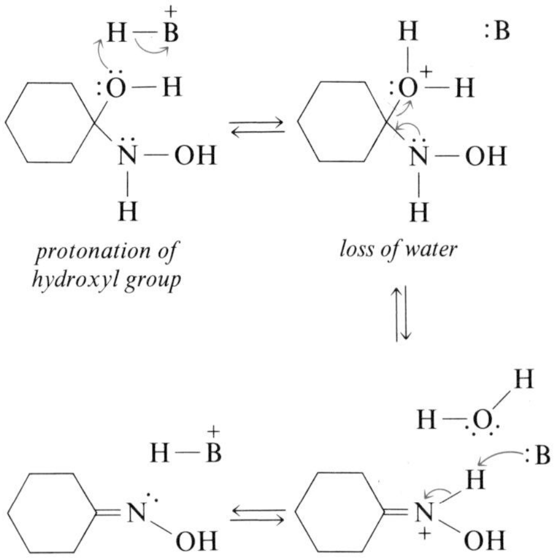

*protonation of
hydroxyl group*

loss of water

*deprotonation of
the oxime*

The hydroxyl group is converted into a good leaving group by protonation, and the water molecule is displaced by the nonbonding electrons on the nitrogen atom. Removal of a proton from the nitrogen atom gives the oxime. All of the processes shown in the equations above for the formation of cyclohexanone oxime may be applied to reactions with aldehydes and ketones of the other compounds that are related to ammonia.

C. Reactions with Amines

Aldehydes react readily with both alkyl and arylamines to give compounds known as **imines** or **Schiff bases.** Usually heating the aldehyde and the amine together while distilling off the water that forms is all that is necessary, as illustrated below by the formation of *N*-ethylpropanalimine and *N*-methylbenzaldimine.

$$
\underset{\substack{\text{propanal}}}{CH_3CH_2\overset{\displaystyle O}{\overset{\|}{C}}H} + \underset{\substack{\text{ethylamine}}}{H_2NCH_2CH_3} \xrightarrow{\Delta} \underset{\substack{N\text{-ethylpropanalimine}\\81\%}}{CH_3CH_2CH{=}NCH_2CH_3} + H_2O\uparrow
$$

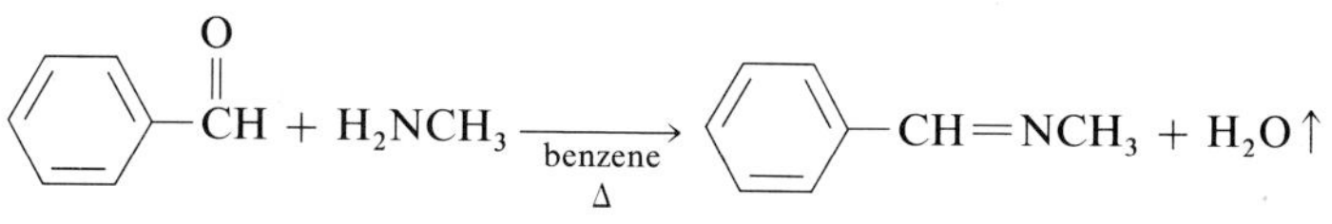

$$
\underset{\substack{\text{benzaldehyde}}}{\text{C}_6\text{H}_5{-}\overset{\displaystyle O}{\overset{\|}{C}}H} + \underset{\substack{\text{methylamine}}}{H_2NCH_3} \xrightarrow[\Delta]{\text{benzene}} \underset{\substack{N\text{-methylbenzaldimine}\\90\%}}{\text{C}_6\text{H}_5{-}CH{=}NCH_3} + H_2O\uparrow
$$

The reactions are reversible, the imines being converted back to aldehydes and amines by reaction with water. The imines derived from aromatic aldehydes and amines are more stable than the ones derived from alkyl components.

Ketones form imines, too, but generally with greater difficulty than aldehydes. A typical reaction is that of acetone with propylamine. The reaction is catalyzed by hydrochloric acid, and the acid is then neutralized by sodium hydroxide.

$$
\underset{\substack{\text{acetone}}}{CH_3\overset{\displaystyle O}{\overset{\|}{C}}CH_3} + \underset{\substack{\text{propylamine}}}{CH_3CH_2CH_2NH_2} \xrightarrow[\text{HCl}]{\text{NaOH}} \underset{\substack{\text{imine of acetone}\\\text{and propylamine}\\67\%}}{CH_3\overset{\displaystyle CH_3}{\overset{|}{C}}{=}NCH_2CH_2CH_3} + H_2O
$$

PROBLEM 9.16   Suggest a mechanism for the formation of *N*-methylbenzaldimine.

PROBLEM 9.17   The individual steps shown for the mechanism for the formation of cyclohexanone oxime in Section 9.6B are all reversible. Using them as a guide, propose a mechanism for the reaction of *N*-methylbenzaldimine with water to give benzaldehyde and methylamine.

PROBLEM 9.18   Suggest a mechanism for the formation of the imine of acetone and propylamine. What role does the catalytic amount of hydrochloric acid play? Why is sodium hydroxide used in the next step of the reaction?

D. Reactions with Hydrazines

Phenylhydrazine and 2,4-dinitrophenylhydrazine are used to prepare the derivatives of aldehydes and ketones known as **phenylhydrazones.** Not all phenylhydrazones are solids. For example, acetone phenylhydrazone is a low-melting solid that is often isolated as a liquid.

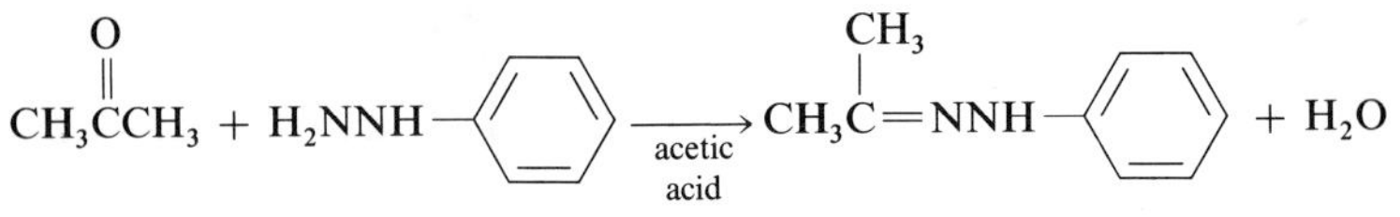

$$
\underset{\substack{\text{acetone} \\ \text{bp 56 °C}}}{CH_3\overset{O}{\overset{\|}{C}}CH_3} + H_2NNH\!-\!\!\underset{\text{phenylhydrazine}}{\bigcirc} \xrightarrow[\substack{\text{acetic} \\ \text{acid}}]{} \underset{\substack{\text{acetone phenylhydrazone} \\ \text{mp 42 °C} \\ \text{bp 90 °C at 0.5 mm}}}{CH_3\overset{CH_3}{\overset{\|}{C}}\!=\!NNH\!-\!\!\bigcirc} + H_2O
$$

For this reason, 2,4-dinitrophenylhydrazine is more useful in the preparation of derivatives. The two nitro groups add to the molecular weight and the polarity of the compound so that the derivatives are more crystalline and have a higher melting point than the corresponding unsubstituted phenylhydrazones, as illustrated for 2-methylpropanal.

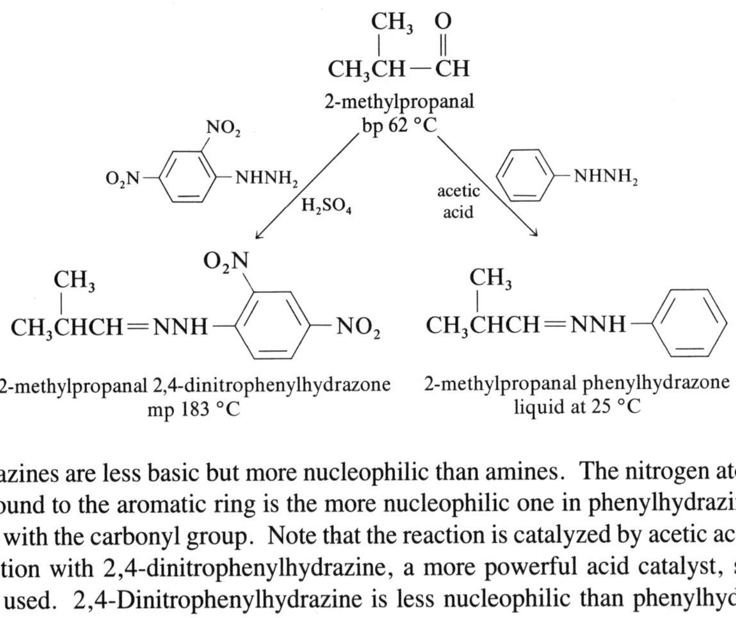

$$
\underset{\substack{\text{2-methylpropanal} \\ \text{bp 62 °C}}}{CH_3\overset{CH_3}{\overset{|}{C}}H\!-\!\overset{O}{\overset{\|}{C}}H}
$$

2-methylpropanal 2,4-dinitrophenylhydrazone
mp 183 °C   2-methylpropanal phenylhydrazone
liquid at 25 °C

Hydrazines are less basic but more nucleophilic than amines. The nitrogen atom that is not bound to the aromatic ring is the more nucleophilic one in phenylhydrazine, and it reacts with the carbonyl group. Note that the reaction is catalyzed by acetic acid. For the reaction with 2,4-dinitrophenylhydrazine, a more powerful acid catalyst, sulfuric acid, is used. 2,4-Dinitrophenylhydrazine is less nucleophilic than phenylhydrazine, and the conjugate acid of the carbonyl compound is necessary for a reaction to take place.

The difference in nucleophilicity in the two nitrogen atoms in phenylhydrazine and the still weaker nucleophilicity of the nitrogen atoms in 2,4-dinitrophenylhydrazine can be rationalized by drawing resonance contributors for the molecules.

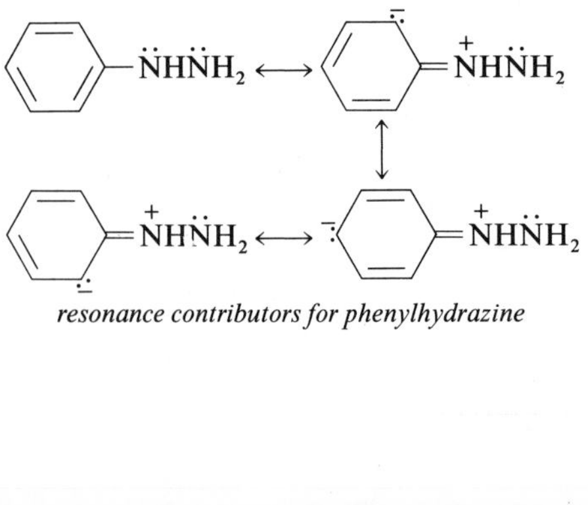

resonance contributors for phenylhydrazine

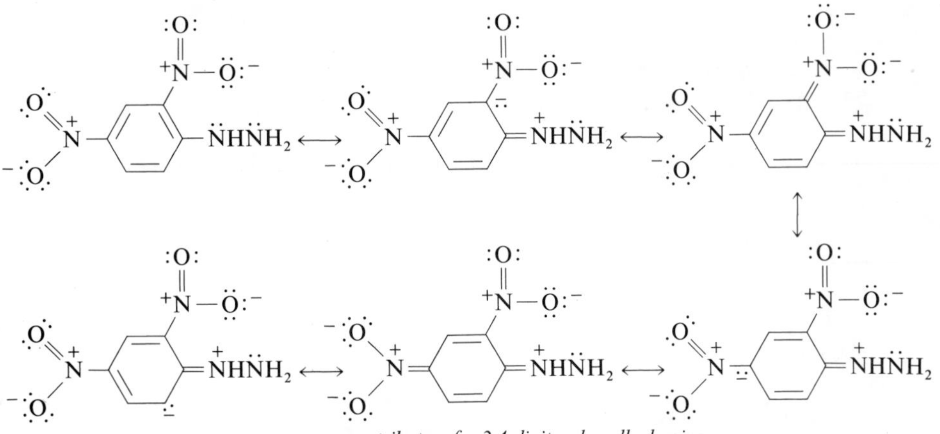

resonance contributors for 2,4-dinitrophenylhydrazine

The nonbonding electrons on the nitrogen atom adjacent to the aromatic ring can be delocalized to the ortho and para positions of the ring by resonance. They are, therefore, less available for reaction. The delocalization possible for the nonbonding electrons from the analogous nitrogen atom in 2,4-dinitrophenylhydrazine is even more extensive. Resonance contributors can be written delocalizing the electrons to oxygen atoms of the nitro groups. The electron density at the nitrogen atom adjacent to the aromatic ring is less in 2,4-dinitrophenylhydrazine than it is in phenylhydrazine. The partial positive character at the nitrogen atom next to the ring in 2,4-dinitrophenyl-hydrazine also reduces electron density to some extent at the other nitrogen atom. For this reason, reactions with 2,4-dinitrophenylhydrazine are carried out under strongly acidic conditions so that the carbonyl group, by protonation, is made more reactive towards the weaker nucleophile.

PROBLEM 9.19 Complete the following equations by writing structures for the expected products.

(a) $\underset{\underset{CH_3}{|}}{CH_3CHCH_2CH_2} \overset{O}{\overset{||}{CH}} + HO\overset{+}{N}H_3Cl^- \xrightarrow{CH_3\overset{O}{\overset{||}{C}}O^-Na^+}$

(b) 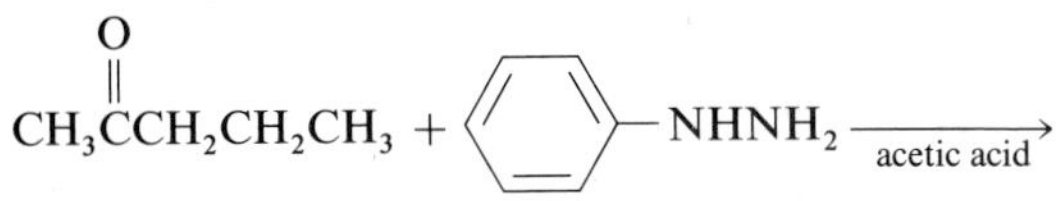 $CH_3\overset{O}{\overset{||}{C}}CH_2CH_2CH_3 + \langle\!\!\!\bigcirc\!\!\!\rangle\!-\!NHNH_2 \xrightarrow{\text{acetic acid}}$

(c) $\langle\!\!\!\bigcirc\!\!\!\rangle\!=\!O + CH_3\!-\!\langle\!\!\!\bigcirc\!\!\!\rangle\!-\!NH_2 \xrightarrow{\Delta}$

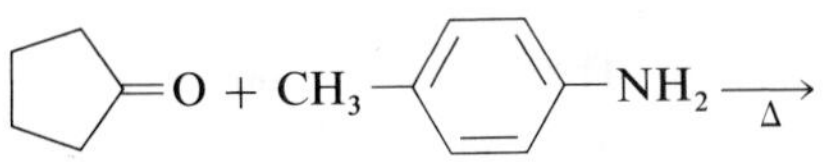

(d) 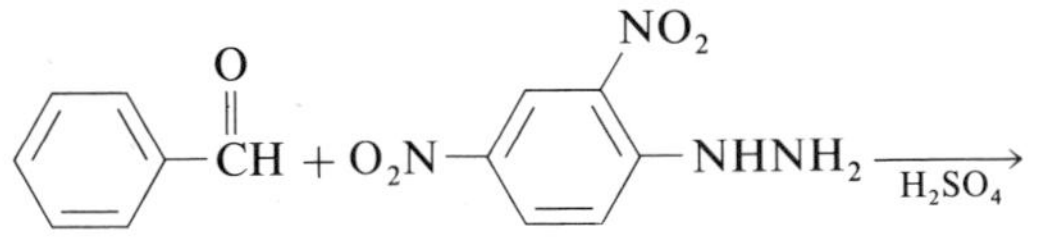 $\langle\!\!\!\bigcirc\!\!\!\rangle\!-\!\overset{O}{\overset{||}{CH}} + O_2N\!-\!\underset{}{\overset{NO_2}{\langle\!\!\!\bigcirc\!\!\!\rangle}}\!-\!NHNH_2 \xrightarrow{H_2SO_4}$

(e) 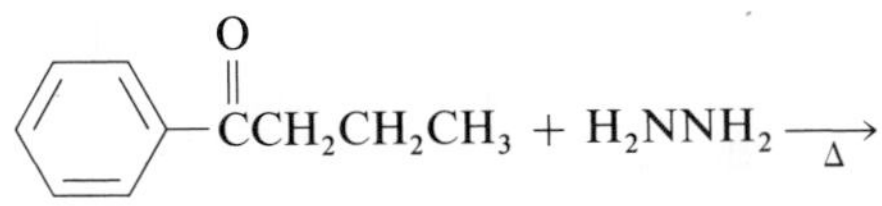 $\langle\!\!\!\bigcirc\!\!\!\rangle\!-\!\overset{O}{\overset{||}{C}}CH_2CH_2CH_3 + H_2NNH_2 \xrightarrow{\Delta}$

(f) 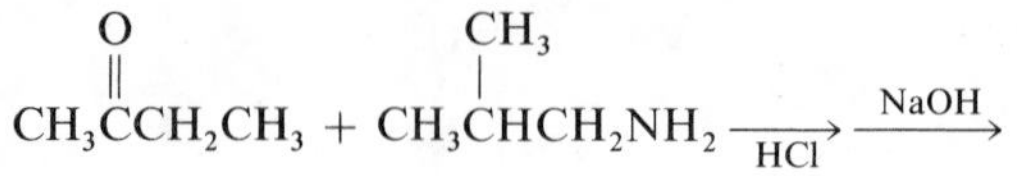　$\underset{\text{O}}{\overset{\text{O}}{\|}}$ CH$_3$CCH$_2$CH$_3$ + $\underset{\overset{|}{\text{CH}_3}}{}$ CH$_3$CHCH$_2$NH$_2$ $\xrightarrow[\text{HCl}]{\text{NaOH}}$

(g) 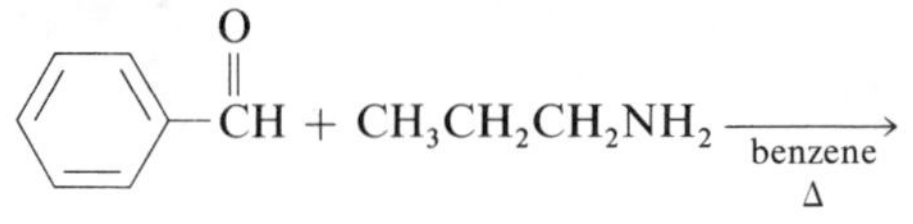 ⬡—$\overset{\overset{\text{O}}{\|}}{\text{CH}}$ + CH$_3$CH$_2$CH$_2$NH$_2$ $\xrightarrow[\substack{\text{benzene} \\ \Delta}]{}$

9.7

Acetal Formation

A. Hydrates, Hemiacetals, Acetals

Water and alcohols are also nucleophiles that react with carbonyl compounds. The oxidation of an aldehyde to a carboxylic acid by aqueous chromic acid (Section 7.6B) is postulated to go by way of the **hydrate** of the aldehyde, as shown for butanal.

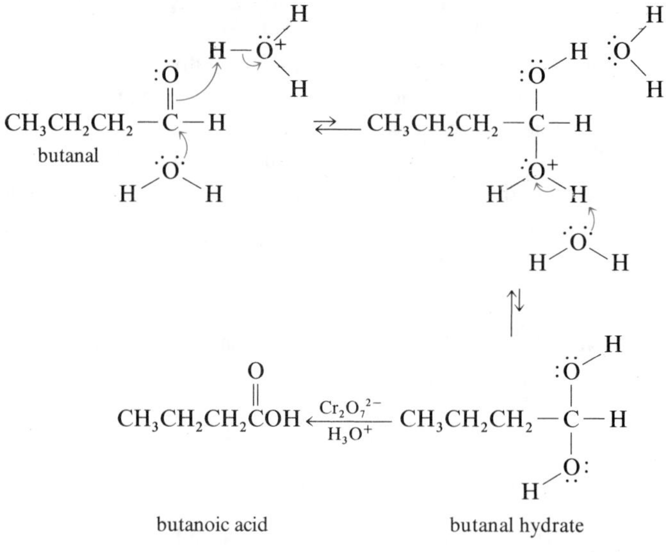

butanoic acid   butanal hydrate

The interaction of water with the carbonyl group is reversible, and the equilibrium lies with the free carbonyl group in most cases. 2,2,2-Trichloroethanal, commonly known as chloral, is an exception. The electron-withdrawing effect of the three halogen atoms makes the carbonyl group more electrophilic and stabilizes the hydrate of the aldehyde so that it is an isolable compound.

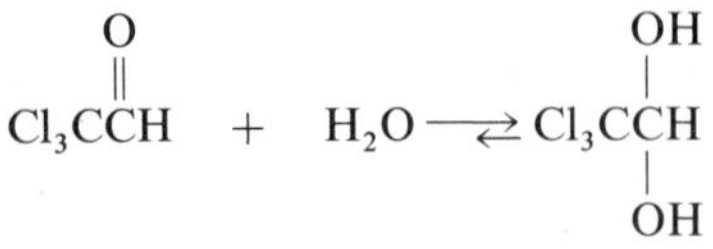

$$\underset{\substack{\text{2,2,2-trichloroacetaldehyde} \\ \text{chloral}}}{\text{Cl}_3\text{C}\overset{\overset{\text{O}}{\|}}{\text{C}}\text{H}} + \text{H}_2\text{O} \;\rightleftharpoons\; \underset{\text{chloral hydrate}}{\text{Cl}_3\text{C}\underset{\overset{|}{\text{OH}}}{\overset{\overset{\text{OH}}{|}}{\text{C}}}\text{H}}$$

Chloral hydrate was used at one time in hospitals and mental hospitals as a hypnotic, a medication that brings on sleep.

Alcohols will interact with carbonyl compounds in an equilibrium process, just as water does, to yield compounds known as **hemiacetals.** In a hemiacetal, a hydroxyl group and an alkoxyl group are bonded to the same carbon atom. Hemiacetals, after loss of water, react with yet another molecule of alcohol to give **acetals.** In an acetal, two alkoxyl groups are bonded to the same carbon atom. When acetaldehyde is placed in an excess of methanol, the aldehyde, its hemiacetal, and its acetal are all present at equilibrium.

$$
\underset{\substack{\text{acetaldehyde} \quad \text{methanol}}}{CH_3\overset{\overset{\displaystyle O}{\|}}{C}H + CH_3OH} \rightleftharpoons \underset{\substack{\text{1-methoxy-} \\ \text{1-ethanol} \\ \textit{a hemiacetal}}}{CH_3\overset{\overset{\displaystyle OH}{|}}{\underset{\underset{\displaystyle OCH_3}{|}}{C}}H} \quad + \quad \underset{\substack{\text{1,1-dimethoxyethane} \\ \textit{an acetal}}}{CH_3\overset{\overset{\displaystyle OCH_3}{|}}{\underset{\underset{\displaystyle OCH_3}{|}}{C}}H} + H_2O
$$

The hemiacetals and acetals of ketones do not form as spontaneously as those of aldehydes do. The amount of hemiacetal in equilibrium with acetone dissolved in methanol, for example, is very small.

$$
\underset{\substack{\text{acetone} \qquad \text{methanol}}}{CH_3\overset{\overset{\displaystyle O}{\|}}{C}CH_3 + CH_3OH} \rightleftharpoons \underset{\substack{\text{2-methoxy-2-propanol} \\ \text{0.3\% at equilibrium} \\ \textit{a hemiacetal}}}{CH_3\overset{\overset{\displaystyle OH}{|}}{\underset{\underset{\displaystyle OCH_3}{|}}{C}}CH_3}
$$

Hemiacetals are generally unstable compounds unless the alcohol and carbonyl groups are part of the same molecule, in which case stable cyclic hemiacetals form. The most important examples of these structures are found in carbohydrates. For example, glucose exists as a cyclic hemiacetal with a six-membered ring.

glucose, shown as   glucose, in the stable
a hydroxyaldehyde   hemiacetal form

Other simpler hydroxyaldehydes also form stable hemiacetals. 4-Hydroxybutanal, for example, forms a five-membered ring hemiacetal.

4-hydroxybutanal   2-hydroxytetrahydrofuran
11% at equilibrium   89% at equilibrium

PROBLEM 9.20   5-Hydroxypentanal gives a six-membered ring hemiacetal. Write an equation for the formation of the hemiacetal, showing the product in the chair form. How many stereoisomers can the hemiacetal have?

If an aldehyde or ketone is placed in an excess of an alcohol in the presence of an acid, an acetal is formed. Acetone and methanol, in a reaction that is catalyzed by an arylsulfonic acid, are converted into the dimethyl acetal of acetone, 2,2-dimethoxypropane.

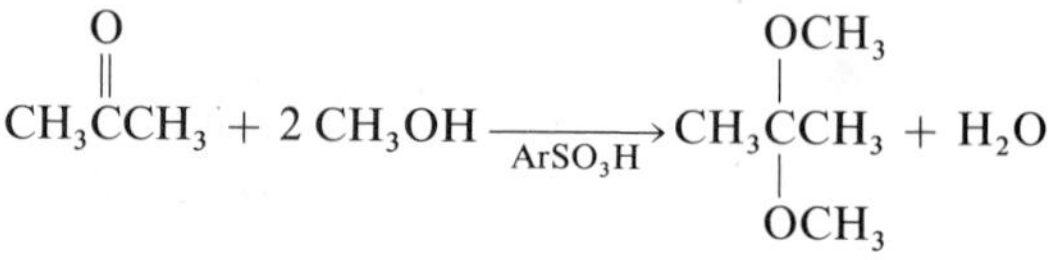

$$
\underset{\substack{\text{acetone}}}{\overset{\overset{\displaystyle O}{\|}}{CH_3CCH_3}} + 2\ \underset{\substack{\text{methanol}}}{CH_3OH} \xrightarrow[\text{ArSO}_3\text{H}]{} \underset{\substack{\text{2,2-dimethoxypropane}\\ \textit{an acetal}}}{\overset{\displaystyle OCH_3}{\underset{\displaystyle OCH_3}{CH_3CCH_3}}} + H_2O
$$

An acetal is structurally like a diether and as such is stable to bases and nucleophilic reagents. Acetals, however, are sensitive to acid and are hydrolyzed easily back to the original carbonyl compound and alcohol, as shown for 2,2-dimethoxypropane.

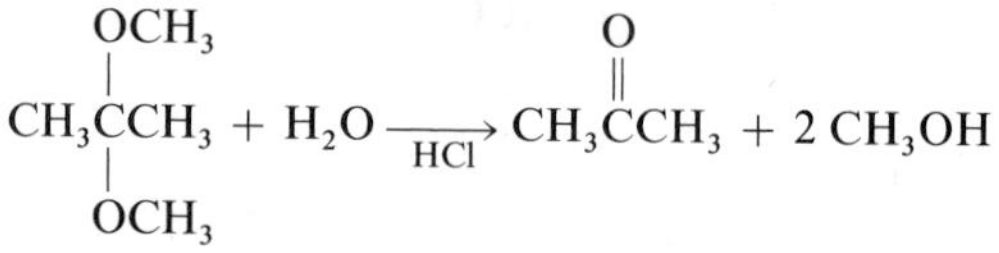

$$
\underset{\substack{\text{2,2-dimethoxypropane}}}{\overset{\displaystyle OCH_3}{\underset{\displaystyle OCH_3}{CH_3CCH_3}}} + H_2O \xrightarrow[\text{HCl}]{} \underset{\substack{\text{acetone}}}{\overset{\overset{\displaystyle O}{\|}}{CH_3CCH_3}} + 2\ \underset{\substack{\text{methanol}}}{CH_3OH}
$$

In order to get good yields of acetals, the water formed in the reaction must be removed; otherwise, the hydrolysis reaction takes place.

B. Mechanism of Acetal Formation and Hydrolysis

Any mechanism proposed for the formation of an acetal in acid must also explain the ease with which it is hydrolyzed by aqueous acid. The details of the reaction of acetone and methanol are illustrated below.

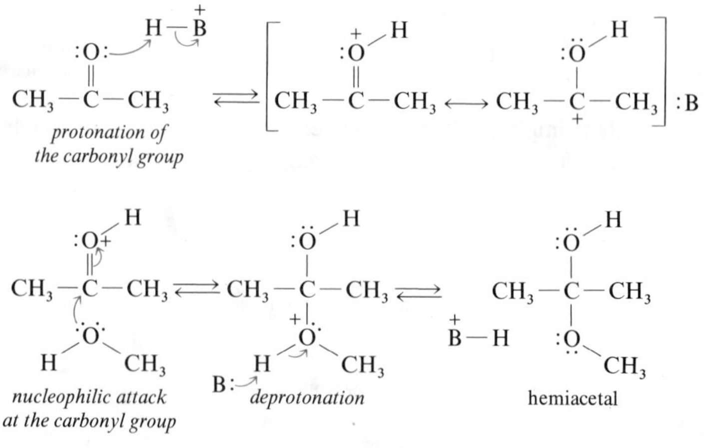

protonation of the carbonyl group

nucleophilic attack at the carbonyl group   *deprotonation*   hemiacetal *protonation of the hydroxyl group*   *loss of water*   *carbocation stabilized by delocalization of charge to oxygen*

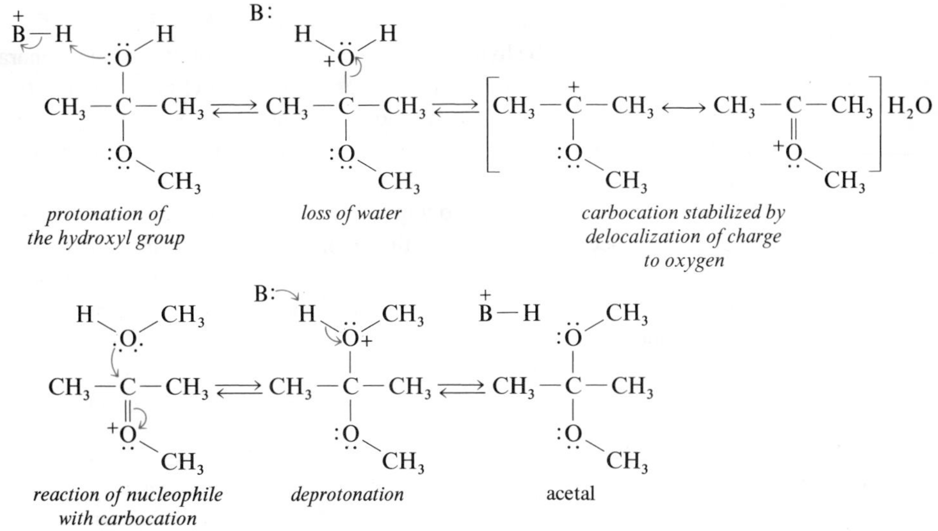

reaction of nucleophile with carbocation   *deprotonation*   *acetal*

The reaction starts with a protonation of the carbonyl group, making its carbon atom more electrophilic, and thus more susceptible to attack by the weakly nucleophilic alcohol. Deprotonation of the first intermediate that forms gives the hemiacetal. Each of the steps is reversible. For example, protonation of the hemiacetal on the ether oxygen atom and loss of methanol leads back to the conjugate acid of acetone.

Protonation of the hydroxyl group in the hemiacetal and loss of water forms a carbocation that is particularly stable because the charge can be delocalized to an oxygen atom in a resonance contributor in which carbon and oxygen atoms both have octets of electrons. The nonbonding electrons of methanol are donated to the carbocation, and deprotonation of the new intermediate gives the dimethyl acetal of acetone. The reaction is carried out by continuous removal from the reaction mixture of the water that is formed so that the equilibrium is drawn towards product. In practice, these reactions are often carried out in the presence of solvents such as benzene or toluene, that co-distill with water carrying it out of the reaction mixture.

The hydrolysis of the acetal is a reversal of all of the steps in its formation. If the acetal is put in excess of water with a trace of acid, an ether oxygen atom is protonated and methanol is lost. Water adds to the carbocation intermediate, the hemiacetal is obtained, and the reverse process continues until ketone and alcohol are regenerated.

PROBLEM 9.21   When acetals are hydrolyzed in water containing ^{18}O, the resulting alcohols contain very little ^{18}O. Write the mechanism for the hydrolysis of 2,2-dimethoxypropane in $H_2{}^{18}O$ with a trace of hydrochloric acid added, and show where the ^{18}O will turn up in the products.

PROBLEM 9.22   Acetals, though they resemble ethers in structure, are hydrolyzed much more easily than ethers are (see Problem 6.13 for the conditions necessary to cleave an ether linkage). Vinyl ethers, also called **enol ethers** are hydrolyzed easily with dilute acid, just as acetals are, to give aldehydes or ketones as products. Propose a mechanism for the following reaction.

$$CH_2{=}CH{-}OCH_2CH_3 \xrightarrow[\text{HClO}_4\ (0.1\ M)]{H_2O} CH_3\overset{\displaystyle O}{\overset{\displaystyle \|}{C}}H + CH_3CH_2OH$$

(A review of Problem 5.49 may be helpful.) Why does this hydrolysis reaction occur so easily?

C. Cyclic Acetals

Because acetals can be hydrolyzed under very mild conditions to regenerate the original carbonyl compound, they are used to **protect** carbonyl groups in multifunctional compounds while reactions are carried out elsewhere in the molecule. The acetal function is stable to bases, nucleophiles, and reducing agents, so it can be used with a wide variety of reagents as long as acidic conditions are avoided. While the acetals of simple aldehydes and ketones with low molecular weight alcohols are relatively easy to form, the reaction becomes difficult if the ketone or alcohol is large and sterically hindered.

Ethylene glycol is used to make cyclic acetals in which both alcohol groups that react with the carbonyl group come from the same molecule. The reaction is shown below for cyclohexanone.

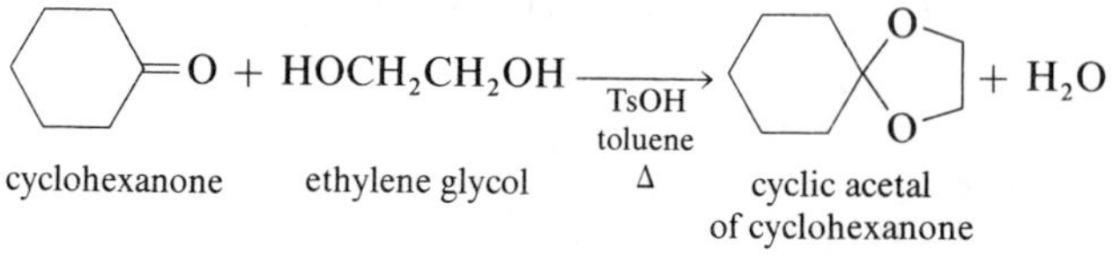

cyclohexanone   ethylene glycol   cyclic acetal of cyclohexanone

The second alcohol function is held in close proximity to the intermediate carbocation by a two-carbon chain, and an acetal is produced easily because of the ease with which five-membered rings form. An example illustrates the usefulness of the reaction.

Methyl 3-oxocyclohexanecarboxylate contains two carbonyl groups. One is a ketone function, the other is in an ester group. The two kinds of carbonyls differ in reactivity.

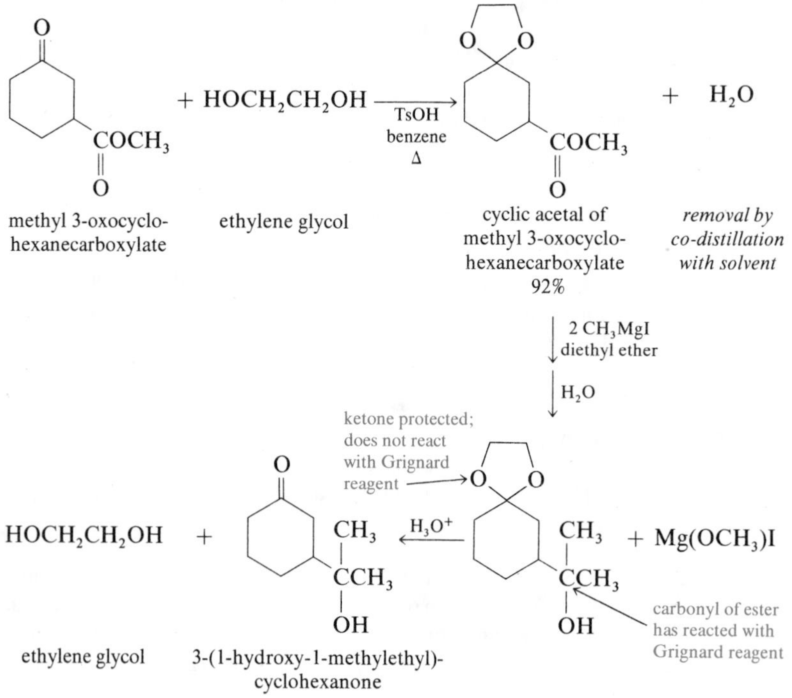

methyl 3-oxocyclo-   ethylene glycol   cyclic acetal of methyl 3-oxocyclo-hexanecarboxylate 92%   *removal by co-distillation with solvent*
hexanecarboxylate

2 CH$_3$MgI
diethyl ether

H$_2$O

ketone protected; does not react with Grignard reagent

HOCH$_2$CH$_2$OH   +   3-(1-hydroxy-1-methylethyl)-cyclohexanone   carbonyl of ester has reacted with Grignard reagent

ethylene glycol   3-(1-hydroxy-1-methylethyl)-cyclohexanone

+ Mg(OCH$_3$)I

The ketone is converted into a cyclic acetal in high yield, while the ester group is untouched. The carbonyl group in the ester function reacts with two moles of a Grignard reagent to give a tertiary alcohol while the ketone carbonyl is protected.

Dilute acid removes ethylene glycol and regenerates the ketone. The details of the addition of a Grignard reagent to an ester are discussed in Section 10.10A. The important point in this example is that by protecting the ketone as an acetal, it is possible to do a Grignard reaction selectively on an ester group in a molecule that also contains a ketone carbonyl.

D. Thioacetals

Thiols are more nucleophilic than alcohols (Section 6.2) so 1,2-ethanedithiol can be used also to make cyclic acetals. The reaction is catalyzed by a Lewis acid, boron trifluoride. Acetaldehyde is converted into its cyclic dithioacetal in this way.

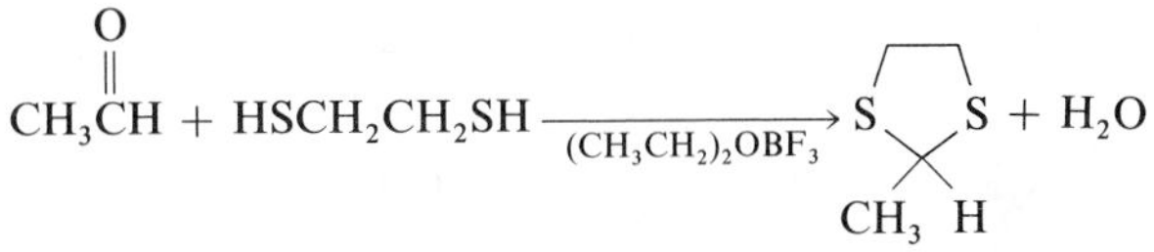

acetaldehyde   1,2-ethanedithiol   cyclic dithioacetal
                                                     of acetaldehyde

The sulfur atoms in thioacetals are replaced by hydrogen atoms when the compounds are treated with Raney nickel (Section 8.2A). This reaction can be used to remove carbonyl groups from complex molecules in a mild way. Thus, the steroid diketone cholestan-3,6-dione is reduced to cholestane by this method.

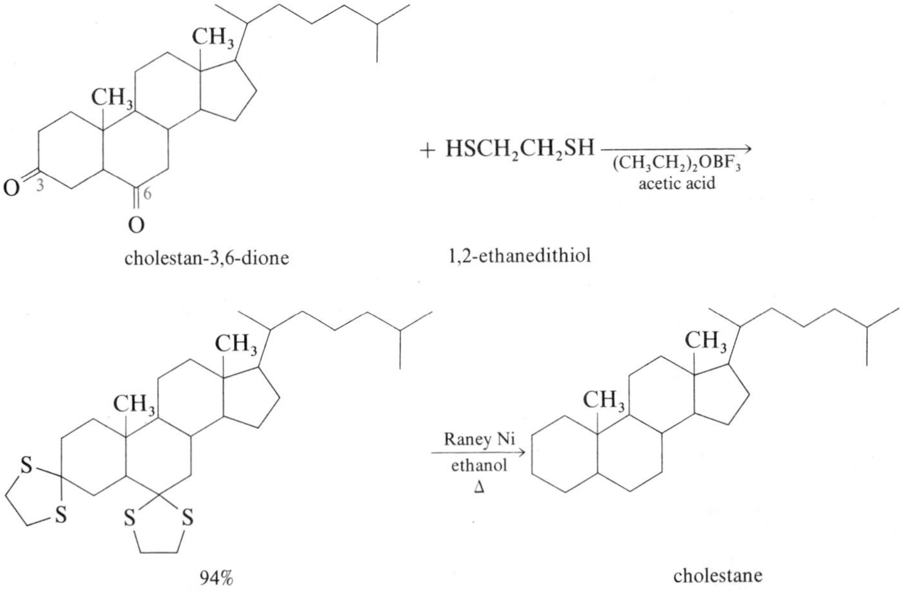

cholestan-3,6-dione   1,2-ethanedithiol

94%   cholestane

Cholestan-3,6-dione is converted into cyclic dithioacetals at both ketone groups, and the sulfur atoms are replaced by hydrogen atoms to yield the hydrocarbon cholestane. This way of simplifying the structure of a steroid can be used to prove that a series of reactions did not change the carbon skeleton of the molecule, or to establish the stereochemistry at various ring junctions of a steroid.

PROBLEM 9.23 Predict the products of the following reactions.

(a) 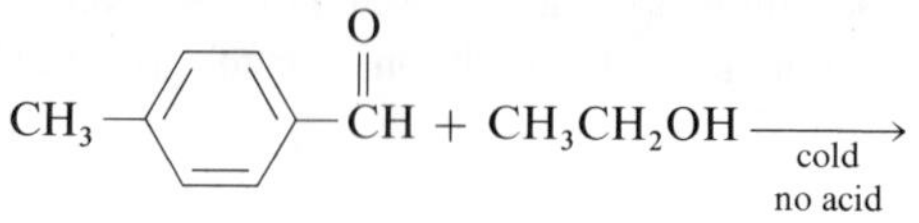 CH_3-⟨benzene ring⟩$-\overset{\overset{\displaystyle O}{\|}}{C}H + CH_3CH_2OH \xrightarrow[\text{no acid}]{\text{cold}}$

(b) 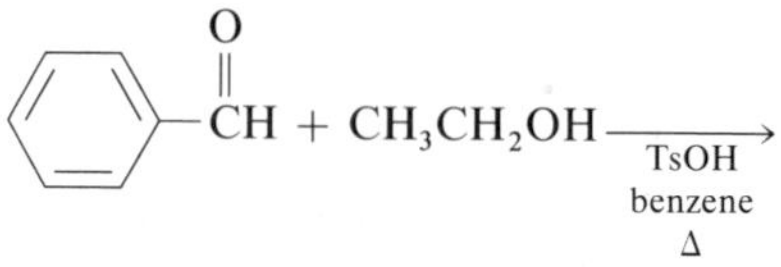 ⟨benzene ring⟩$-\overset{\overset{\displaystyle O}{\|}}{C}H + CH_3CH_2OH \xrightarrow[\substack{\text{benzene}\\ \Delta}]{\text{TsOH}}$

(c) $CH_3CH_2CH_2\overset{\overset{\displaystyle O}{\|}}{C}H + CH_3OH \xrightarrow[\substack{\text{benzene}\\ \Delta}]{\text{TsOH}}$

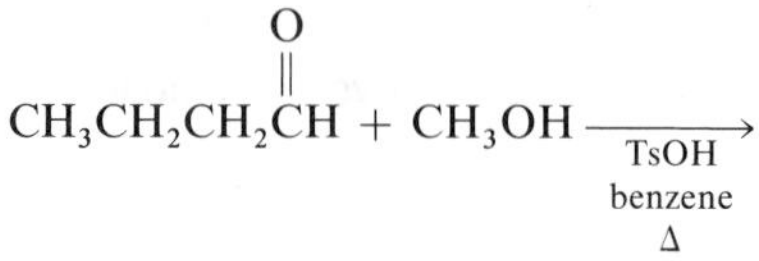

(d)  $CH_3\overset{\overset{\displaystyle O}{\|}}{C}CH_2CH_2CH_3 + HSCH_2CH_2SH \xrightarrow[(CH_3CH_2)_2OBF_3]{}$

(e) 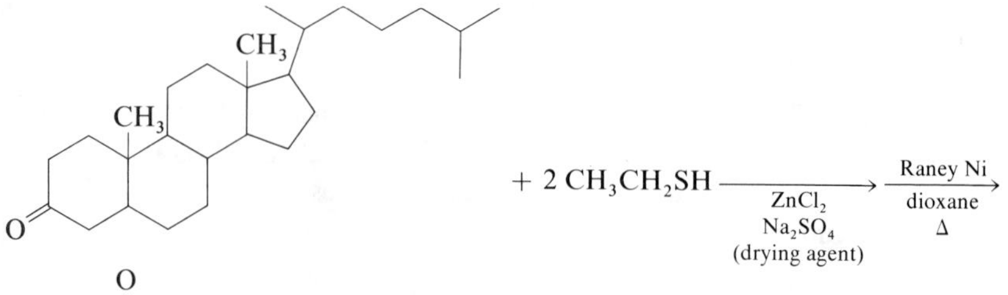 ⟨steroid structure with CH₃ groups and O⟩ $+ 2\ CH_3CH_2SH \xrightarrow[\substack{ZnCl_2\\ Na_2SO_4\\ \text{(drying agent)}}]{} \xrightarrow[\substack{\text{dioxane}\\ \Delta}]{\text{Raney Ni}}$

(f) 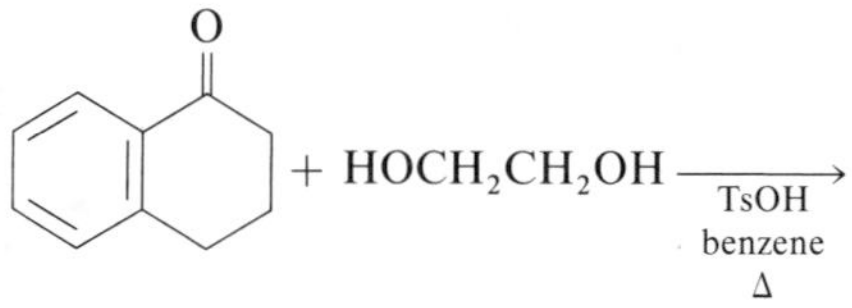 ⟨bicyclic ketone structure⟩ $+ HOCH_2CH_2OH \xrightarrow[\substack{\text{benzene}\\ \Delta}]{\text{TsOH}}$

PROBLEM 9.24 Acetone reacts with glycerol (1,2,3-propanetriol) in the presence of *p*-toluenesulfonic acid to give a product that has the molecular formula $C_6H_{12}O_3$. Propose a structure for the compound.

9.8

Reduction of Carbonyl Groups to Methylene Groups

A. The Clemmensen Reduction

A general reaction of aldehydes and ketones is the reduction of the carbonyl group all the way to a **methylene** ($-CH_2-$) group. For example, heptanal is reduced to the hydrocarbon, heptane.

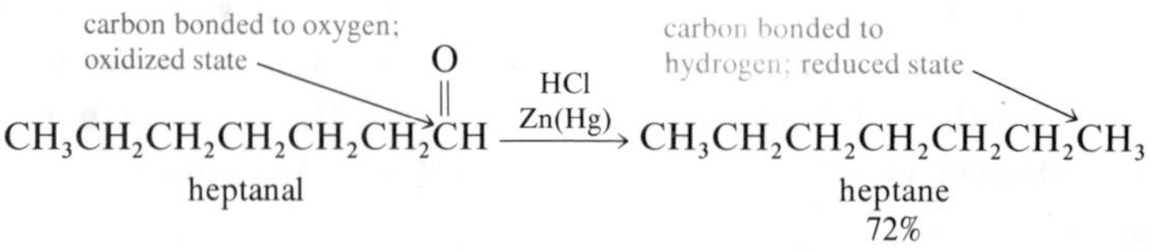

carbon bonded to oxygen;
oxidized state ⟶ $\overset{\overset{\displaystyle O}{\|}}{}$

carbon bonded to
hydrogen; reduced state ⟶

$$CH_3CH_2CH_2CH_2CH_2CH_2CH \xrightarrow[\text{Zn(Hg)}]{\text{HCl}} CH_3CH_2CH_2CH_2CH_2CH_2CH_3$$

heptanal   heptane
72%

This reaction, known as the **Clemmensen reduction,** is carried out with zinc amalgam (zinc dissolved in mercury) and concentrated hydrochloric acid. The reaction is selective for an aldehyde (or ketone) carbonyl group and does not affect the carbonyl group of a carboxylic acid. Thus, 4-oxo-4-phenylbutanoic acid is reduced to 4-phenylbutanoic acid.

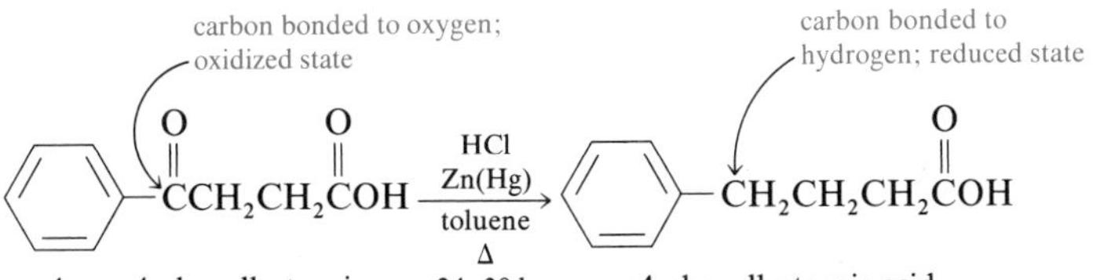

The Clemmensen reduction proceeds by transfer of electrons from the metal and protonation of intermediate species by the strongly acidic reaction mixture. The mechanism is complex and not completely understood. It is sufficient that you recognize that the conversion of a carbonyl group to a methylene group is a reduction reaction and requires a reducing agent, zinc metal in an acidic solution.

B. The Wolff-Kishner Reduction

The conditions for the Clemmensen reduction are harshly acidic and may give rise to side reactions. An alternative way of reducing carbonyl groups to methylene groups is the **Wolff-Kishner reduction,** which uses a basic reagent. In the Wolff-Kishner reduction, the hydrazone of an aldehyde or ketone is prepared and decomposed under basic conditions. An example is the reduction of 2-octanone to octane.

The reaction starts with the formation of a hydrazone. Under the strongly basic conditions, the hydrazone is deprotonated at the nitrogen atom.

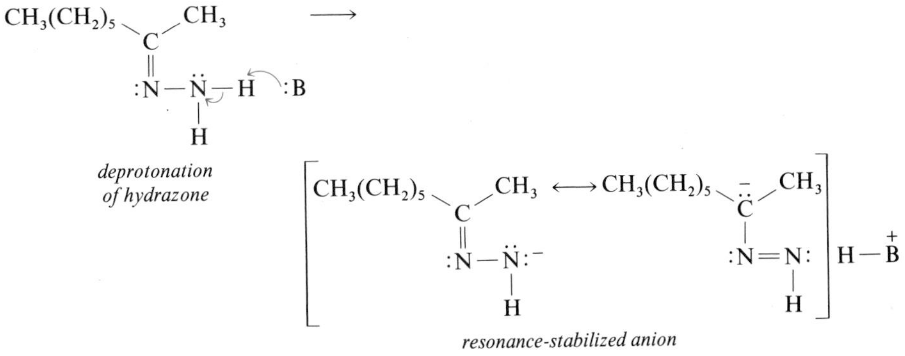

resonance-stabilized anion

Such a deprotonation is possible because there is delocalization of the charge to give anionic character to the carbon atom that was part of the original carbonyl group. A negatively charged carbon atom is more basic than a negatively charged nitrogen atom because carbon is less electronegative than nitrogen (Section 4.5). The anion receives a proton at the carbon atom and loses another one from the nitrogen atom.

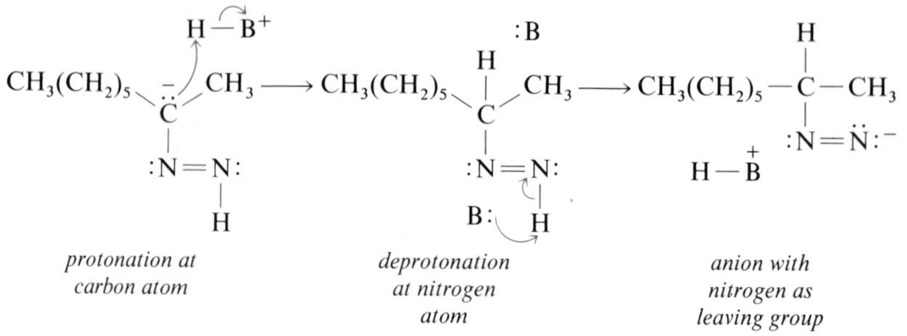

protonation at carbon atom	deprotonation at nitrogen atom	anion with nitrogen as leaving group

This new anion is set up to lose the very stable nitrogen molecule as a leaving group. Usually the loss of a stable molecule as a leaving group creates a cation. In this case, the species as a whole is negatively charged so the loss of the neutral molecule, nitrogen, leaves behind a carbanion. Protonation of the carbanion completes the reduction.

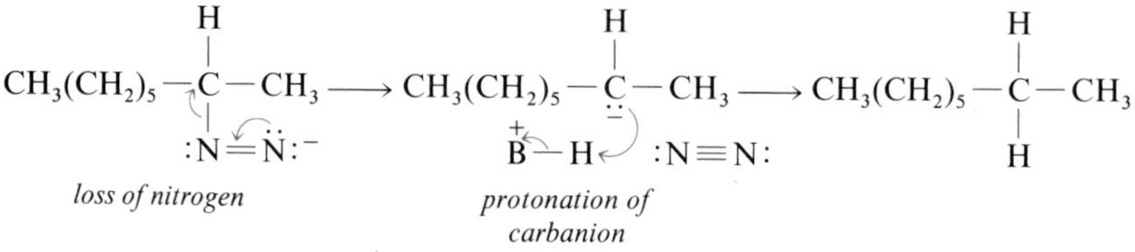

loss of nitrogen	protonation of carbanion

An example of a system in which reduction with zinc amalgam fails but the Wolff-Kishner reaction succeeds belongs to a class of compounds known as *sapogenins*. The sapogenins are toxic compounds, found in a variety of plants, especially desert plants. Their name derives from their ability to form soapy solutions in water. Chemists are interested in sapogenins because they contain a functionalized steroid structure that can be chemically modified to give medicinal steroids with hormonal activity. The sapogenin, hecogenin, found in a number of agave plants, has the structure shown below. The carbonyl group on the ring designated by the letter C (Section 11.6C) is reduced to a methylene group by the Wolff-Kishner reaction to give another naturally occurring sapogenin, tigogenin.

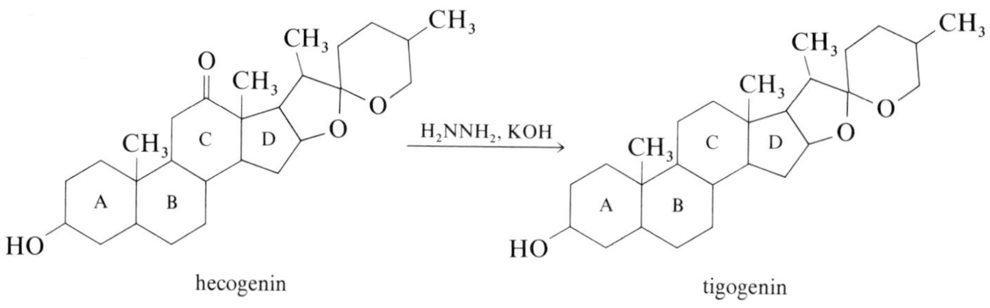

hecogenin   tigogenin

PROBLEM 9.25   What side reaction(s) would you anticipate if hecogenin were put in hydrochloric acid?

The Clemmensen and Wolff-Kishner reactions, as well as conversion of a carbonyl group to a dithioacetal and reduction of the carbon-sulfur bonds to carbon-hydrogen bonds with Raney nickel (Section 9.7D), all lead to the transformation of a carbonyl group to a methylene group.

PROBLEM 9.26   Predict what the products of the following reactions will be.

(a) 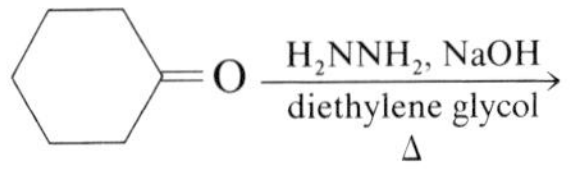 $\xrightarrow[\substack{\text{diethylene glycol} \\ \Delta}]{\text{H}_2\text{NNH}_2,\ \text{NaOH}}$

(b) 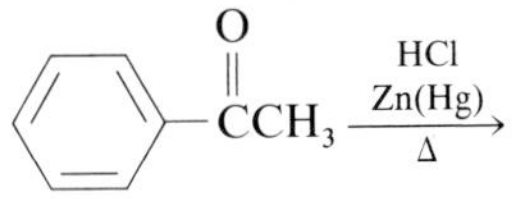 $\xrightarrow[\Delta]{\substack{\text{HCl} \\ \text{Zn(Hg)}}}$

(c) 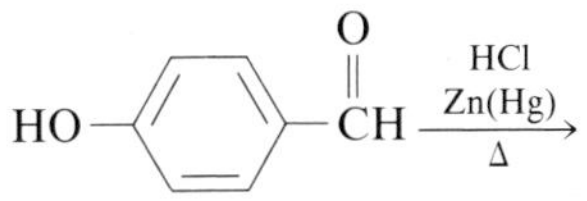 HO— $\xrightarrow[\Delta]{\substack{\text{HCl} \\ \text{Zn(Hg)}}}$

(d) 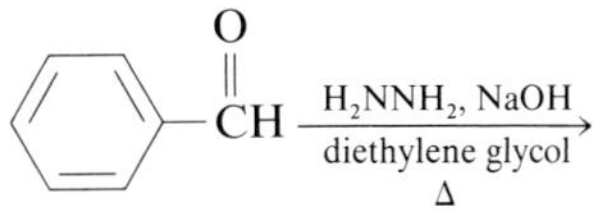 $\xrightarrow[\substack{\text{diethylene glycol} \\ \Delta}]{\text{H}_2\text{NNH}_2,\ \text{NaOH}}$

(e) 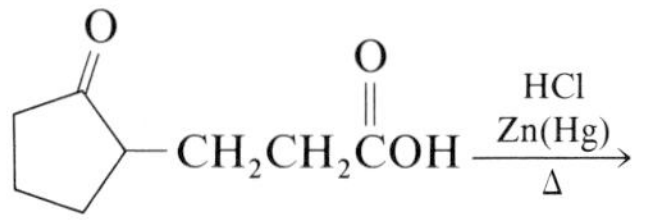 $\xrightarrow[\Delta]{\substack{\text{HCl} \\ \text{Zn(Hg)}}}$

(f) 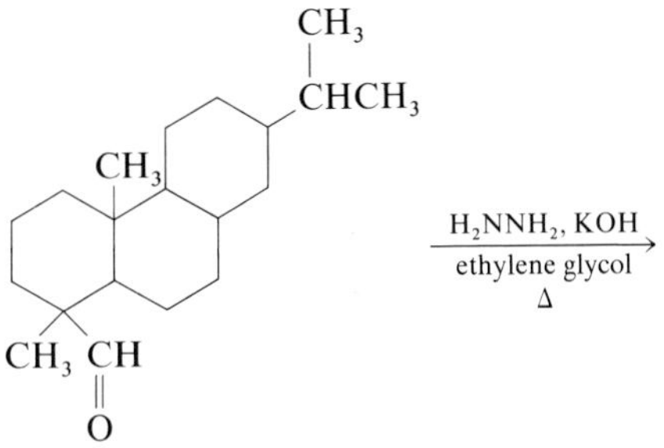 $\xrightarrow[\substack{\text{ethylene glycol} \\ \Delta}]{\text{H}_2\text{NNH}_2,\ \text{KOH}}$

(g) 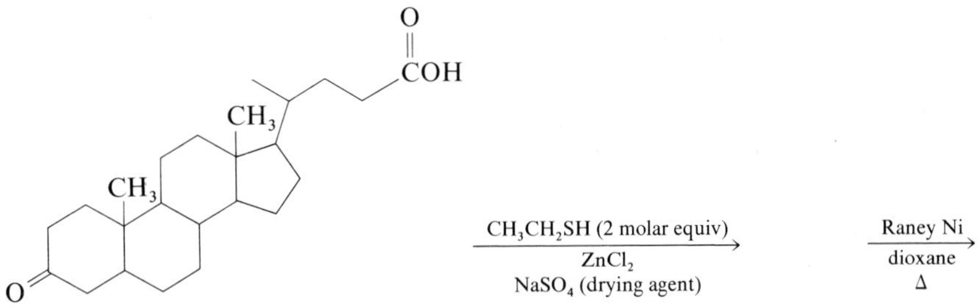 $\xrightarrow[\substack{\text{ZnCl}_2 \\ \text{NaSO}_4\ (\text{drying agent})}]{\text{CH}_3\text{CH}_2\text{SH (2 molar equiv)}}$ $\xrightarrow[\substack{\text{dioxane} \\ \Delta}]{\text{Raney Ni}}$

9.9

Qualitative Analysis of Hydrocarbons, Halides, Alkenes, Alkynes, Alcohols, Aldehydes, Ketones, and Ethers

In earlier sections of the book, we have identified certain reactions as being useful in detecting the presence of specific functional groups. Each such reaction is typical of the particular reactivity of that functional group, it can be carried out in a few minutes in a test tube, and it involves some change that is easily detected by the human senses. At this point, we will review the reactions that meet these requirements and can be used to distinguish compounds that fall into the major functional group classes of alkanes, alkenes, alkynes, alkyl and aryl halides, alcohols, aldehydes, ketones, and ethers. These reactions reflect important properties of the different functional groups. Doing organic qualitative analysis is an excellent way to review organic chemistry and to develop logical thinking about the relationship between structure and reactivity.

Any systematic attempt to identify organic compounds always starts with a determination of the solubility of the compound in water. Water-insoluble compounds in these functional group classes are then tested with concentrated sulfuric acid for solubility or for any other sign of reaction, such as evolution of heat or change in color. Determining water solubility enables us to separate out low molecular weight or polyfunctional alcohols, aldehydes, ketones, and ethers (Sections 1.8B, 7.1A, 9.1A) that are polar enough to participate in hydrogen bonding and are, therefore, soluble in the polar and hydrogen-bonding solvent, water. Alcohols, aldehydes, ketones, and ethers with more than four to five carbon atoms for each polar functional group are not significantly soluble in water.

Concentrated sulfuric acid can be used to detect the presence of nonbonding electrons on oxygen, nitrogen, and sulfur, and of electrons in the π bonds of alkenes and alkynes (Section 4.3). A compound with such structural features acts as a Brønsted-Lowry base towards the strong acid and is protonated. The conjugate acid of the compound is ionic and polar. It may react further to give polymers and colored decomposition products, as is the case with alkenes. Alternatively, it may go into solution in the polar solvent, sulfuric acid. The nonbonding electrons on halogens and the π electrons of simple aromatic compounds such as benzene are not protonated by cold concentrated sulfuric acid. The compounds that are soluble in water are not tested in sulfuric acid, because any compound polar enough to be water soluble is certain to contain the type of functionality that will be protonated by sulfuric acid.

Using water and sulfuric acid, we can make a start at separating compounds in the functional group classes into three categories. The distinctions that can be made are shown schematically in Figure 9.1 using general formulas to indicate the types of functional groups present.

The compounds soluble in sulfuric acid, as well as the ones soluble in water, are separated into different functional groups by the application of further tests. 2,4-Di-

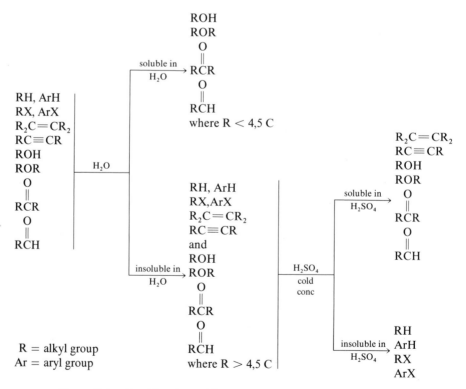

Figure 9.1 Classification of hydrocarbons, halides, alcohols, ethers, and carbonyl compounds according to their solubility in water and concentrated sulfuric acid.

nitrophenylhydrazine reagent, made by dissolving the hydrazine in ethanol with some sulfuric acid, gives a precipitate when an aldehyde or a ketone is added to it (Section 9.6D). Aldehydes are easily distinguished from ketones by the ease with which they are oxidized to carboxylic acids. An acidic solution of sodium dichromate will turn from orange to green, or a solution of potassium permanganate will give a brown precipitate in the presence of an aldehyde.

One other test, the **Tollens test,** is useful in distinguishing aldehydes from ketones by the ease with which they are oxidized. Tollens reagent is made by adding sodium hydroxide to silver nitrate, then dissolving the resulting precipitate in dilute ammonium hydroxide.

$$2\,Ag^+ \xrightarrow[H_2O]{NaOH} Ag_2O\downarrow \xrightarrow[H_2O]{NH_3} 2\,Ag(NH_3)_2{}^+OH^-$$

$$\overset{O}{\overset{\|}{R C H}} + 2\,Ag(NH_3)_2{}^+OH^- \xrightarrow{H_2O} \overset{O}{\overset{\|}{R C}}O^-NH_4{}^+ + 2\,Ag\downarrow + H_2O + 3\,NH_3$$

The aldehyde is oxidized to a carboxylic acid, which is present in the basic solution as its salt. Silver(I) is reduced to metallic silver. If the reaction is done in a clean vessel, a silver mirror deposits on the glass, so the reaction is also known as the silver mirror test. If the glass surface is not clean, silver precipitates as a finely divided black powder. These distinctions among the compounds that react with sulfuric acid are presented in Figure 9.2.

The compounds containing π bonds, alkenes and alkynes, can be distinguished from alcohols and ethers by their reactions with electrophiles and oxidizing agents. Bromine adds readily to π bonds to give colorless alkyl halides, so the loss of the color of bromine is an indication that a reaction has taken place (Section 5.8). Cold, aqueous potassium permanganate oxidizes compounds containing π bonds and is reduced to brown manganese dioxide in the process (Sections 8.4A and 8.4B). Alcohols and ethers are untouched by these reagents under these conditions.

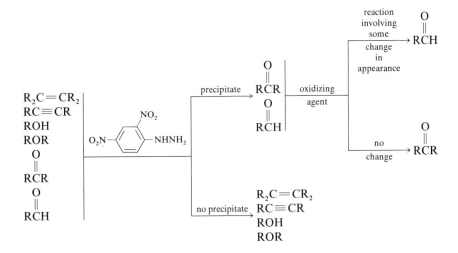

Figure 9.2 Classification of some compounds that are soluble in concentrated sulfuric acid.

$$
\begin{array}{l}
R_2C{=}CR_2 \\
RC{\equiv}CR \\
ROH \\
ROR
\end{array}
\quad
\underset{\text{brown solution}}{\overset{\underset{\text{carbon tetrachloride}}{\text{Br}_2 \text{ in}}}{\big|}}
$$

$\xrightarrow[\text{color}]{\text{loss of}}$ $\begin{array}{l} R_2C{=}CR_2 \\ RC{\equiv}CR \end{array}$

$\begin{array}{l} R_2C{=}CR_2 \\ RC{\equiv}CR \\ ROH \\ ROR \end{array}$ $\underset{\substack{\text{cold}\\ \text{purple color}}}{\overset{\underset{\text{in H}_2\text{O}}{\text{KMnO}_4}}{\big|}}$

$\xrightarrow[\text{brown precipitate}]{\text{MnO}_2\downarrow}$ $\begin{array}{l} R_2C{=}CR_2 \\ RC{\equiv}CR \end{array}$

$\xrightarrow[\text{in color}]{\text{no change}}$ $\begin{array}{l} ROH \\ ROR \end{array}$

$\xrightarrow[\text{in color}]{\text{no change}}$ $\begin{array}{l} ROH \\ ROR \end{array}$

There is no simple way to distinguish an alkene from an internal alkyne by test-tube reactions.

Primary and secondary alcohols are instantly oxidized by acidic chromium trioxide in acetone solution (Problem 7.23, Section 7.6B). The orange color of the reagent changes to the typical blue-green of chromium(III). Thus, primary and secondary alcohols may be distinguished from tertiary alcohols and ethers by this reaction.

Low molecular weight primary, secondary, and tertiary alcohols react at different rates with a solution of zinc chloride in concentrated hydrochloric acid (**Lucas reagent**) to give alkyl halides, which are not soluble in the reaction mixture and can be seen as a separate oily layer. The ease with which the alkyl halide forms depends upon the ease with which the alcohol is converted into a carbocation. Tertiary alcohols react instantly, secondary alcohols within two or three minutes, primary alcohols only very slowly.

$$
\underset{\substack{\text{tertiary}\\ \text{alcohol}}}{R-\overset{\displaystyle R}{\underset{\displaystyle OH}{C}}-R}
\xrightarrow[\substack{\text{ZnCl}_2\\ \text{fast}}]{\text{HCl}}
R-\overset{\displaystyle R}{\underset{+}{C}}-R \;\; Cl^-
\xrightarrow{\text{fast}}
\underset{\substack{\text{tertiary halide}\\ \textit{appears instantly}}}{R-\overset{\displaystyle R}{\underset{\displaystyle Cl}{C}}-R}
$$

$$
\underset{\substack{\text{secondary}\\ \text{alcohol}}}{R-\overset{\displaystyle R}{\underset{\displaystyle OH}{C}}-H}
\xrightarrow[\substack{\text{ZnCl}_2\\ \text{slow}}]{\text{HCl}}
R-\overset{\displaystyle R}{\underset{+}{C}}-H \;\; Cl^-
\xrightarrow{\text{fast}}
\underset{\substack{\text{secondary halide}\\ \textit{appears in 2–3}\\ \textit{minutes}}}{R-\overset{\displaystyle R}{\underset{\displaystyle Cl}{C}}-H}
$$

$$
\underset{\substack{\text{primary}\\ \text{alcohol}}}{R-\overset{\displaystyle H}{\underset{\displaystyle OH}{C}}-H}
\begin{cases}
\xrightarrow[\substack{\text{ZnCl}_2\\ \text{very slow}}]{\text{HCl}} R-\overset{\displaystyle H}{\underset{+}{C}}-H \;\; Cl^- \\[1em]
\xrightarrow[\substack{\text{ZnCl}_2\\ \text{very fast}}]{\text{HCl}} R-\overset{\displaystyle H}{\underset{\underset{Cl^-}{\overset{+}{H-O-H}}}{C}}-H
\end{cases}
\xrightarrow{\text{slow}}
\underset{\substack{\text{primary halide}\\ \textit{takes a long time}\\ \textit{to form}}}{R-\overset{\displaystyle H}{\underset{\displaystyle Cl}{C}}-H}
$$

alcohols are
soluble in the
reagent if they
contain < 7 carbons

alkyl halides
are insoluble
in the reagent

Tertiary and secondary alcohols react by S_N1 mechanisms. Reaction by an S_N1 mechanism is very slow for primary alcohols, and the alkyl halide that forms may be the product of an S_N2 reaction on the protonated alcohol. Because chloride ion is not a good nucleophile, the S_N2 reaction is also slow. Ethers do not react with any of these reagents under the conditions and within the time in which alcohols react. The reactions that allow distinctions among alcohols and ethers are outlined below.

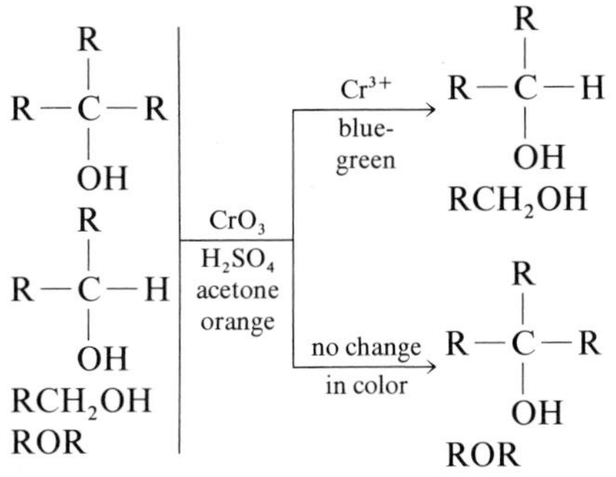

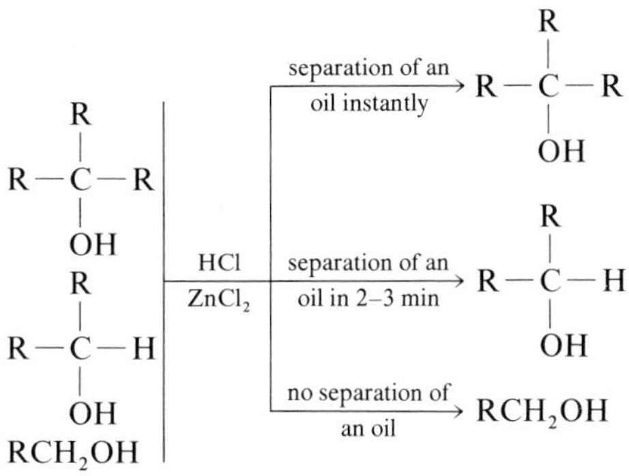

The functional group classes that are insoluble in sulfuric acid are the alkanes and alkyl and aryl halides. Some alkyl halides may be detected by their reactivity in S_N1 and S_N2 reactions (Sections 6.5, 6.6). Tertiary alkyl halides and benzylic halides react instantly with silver nitrate in ethanol to give precipitates of silver halides. Primary alkyl halides, as well as benzylic halides, react with sodium iodide in acetone. Secondary alkyl halides react slowly with both reagents. Allylic halides can also react with both of these reagents, but would have been classified earlier as compounds soluble in sulfuric acid.

Aryl halides (and vinylic halides) do not react by S_N1 or S_N2 reactions (Section 6.8). They can be distinguished from hydrocarbons containing no halogen by the **Beilstein test** in which a clean copper wire is heated in a flame, dipped into the compound to be tested, and held in the flame again. If the organic compound contains halogen, some copper halide is formed on the hot surface of the wire, and the copper salt gives a brilliant blue-green color to the flame. The test may be used to detect halogen in compounds belonging to any functional group class.

Reactions that can be used to distinguish among compounds insoluble in sulfuric acid are outlined as follows.

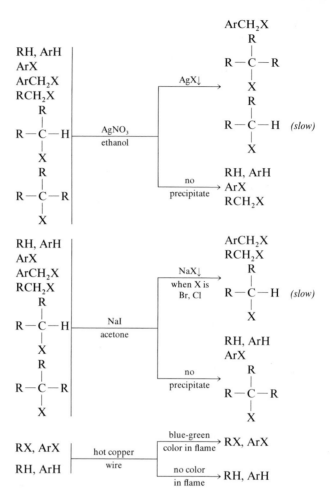

The order in which tests are carried out is not arbitrary. For example, aldehydes react with potassium permanganate to give a result that closely resembles the one observed when potassium permanganate reacts with alkenes and alkynes. Aldehydes also reduce chromic acid solutions in a reaction that looks like the reaction of a primary or secondary alcohol. For these reasons, 2,4-dinitrophenylhydrazine is used to identify compounds that may be aldehydes before the other reagents are used to test for alkenes, alkynes, and alcohols. As other functional group classes are added to the list of compounds being tested, the possibility of such conflicts increases. The successful application of tests for qualitative analysis will always require the ability to think logically and to understand the reactivity responsible for the observed results.

PROBLEM 9.27   How would you distinguish the compounds in each of the following sets from each other? Give the reagents you would use and the experimental observations you would expect to make in each case.

(a) CH_3CH_2OH, $CH_3CH_2CH_2CH_2CH_2CH_2OH$, $CH_3CH_2CH_2\overset{\displaystyle O}{\overset{\displaystyle \|}{C}}CH_2CH_3$

(b) 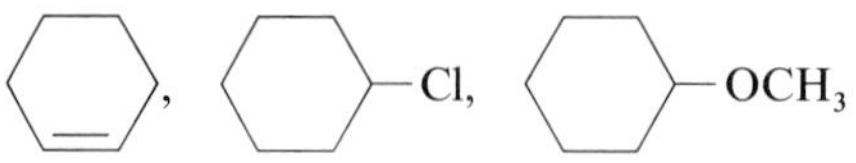   ⬡̶,   ⬡—Cl,   ⬡—OCH_3

(c) ⬡=O, ⬡—OH, ⬡—Br, ⬡

(d) ⬡—CH₂Cl, ⬡—Cl,   CH₂=CHCH₂Cl

O
‖

(e) CH₃CH₂CH₂CH₂CCH₃,   CH₃CH₂CH₂CH₂CH₂CH,   CH₃CH₂CH₂CH₂CH₂OCH₃

9.10

Infrared Spectra of Aldehydes and Ketones. Conjugated Carbonyl Groups

The carbonyl stretching frequency of alkyl aldehydes appears at 1740–1720 cm^{-1}; that of alkyl ketones at 1725–1705 cm^{-1}. The band is usually strong and is important in diagnosing the presence of a carbonyl group (Section 7.7D).

A ketone or aldehyde in which the carbonyl group is conjugated with a double bond (or with an aromatic ring) has a lower carbonyl stretching frequency than an alkyl ketone or aldehyde. For example, if we compare the infrared spectrum of 5-hexen-2-one, in which the carbonyl group is not conjugated with the double bond, with that of 2-cyclohexen-1-one, in which it is, the shift to lower frequency of the carbonyl band is clearly seen (Figure 9.3). In 5-hexen-2-one, the carbonyl stretching frequency is

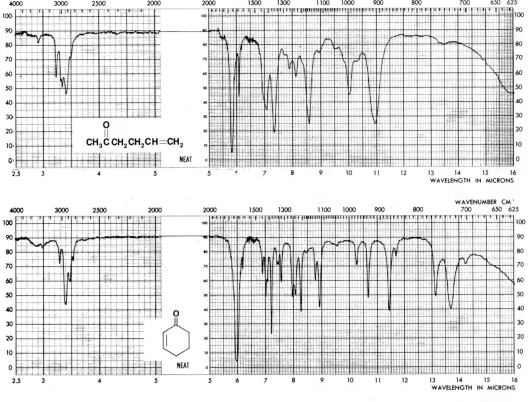

Figure 9.3   The infrared spectra of 5-hexen-2-one and 2-cyclohexen-1-one. (From *The Aldrich Library of Infrared Spectra*)

1710 cm^{-1}; in 2-cyclohexen-1-one, it is 1675 cm^{-1}. Note, too, the carbon-carbon double bond stretching frequency at 1630 cm^{-1} and the stretching frequency at 3080 cm^{-1} for the hydrogen atoms of the terminal double bond (Section 8.8A) in the spectrum of 5-hexen-2-one.

A similar comparison may be made between the infrared spectra of 1-phenyl-2-butanone and 1-phenyl-1-butanone (Figure 9.4). The carbonyl group in 1-phenyl-2-butanone is separated from the aromatic ring by an sp^3-hybridized carbon atom. Its carbonyl absorption band appears at 1715 cm^{-1}, a typical stretching frequency for an alkyl ketone. The carbonyl stretching frequency for 1-phenyl-1-butanone is lower, at 1695 cm^{-1}, indicating that in this compound the carbonyl group is conjugated with an unsaturated system. Note the strong, sharp bands between 1600–1450 cm^{-1} and between 850–700 cm^{-1} in the spectra in Figure 9.4. These bands indicate the presence of aromatic rings in the compounds (Section 12.8B).

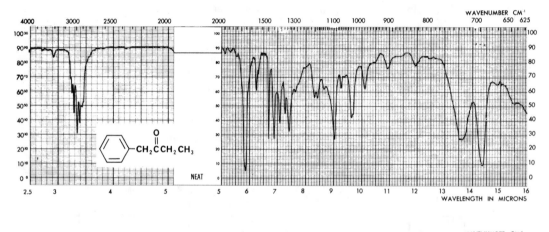

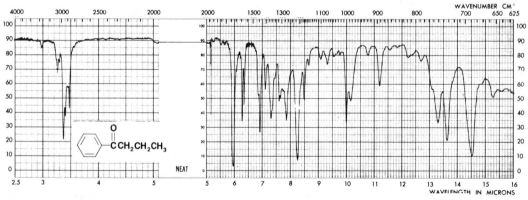

Figure 9.4   The infrared spectra of 1-phenyl-2-butanone and 1-phenyl-1-butanone. (From *The Aldrich Library of Infrared Spectra*)

PROBLEM 9.28   The following spectra (Figure 9.5) belong to 10-undecenal and 2-butenal. Draw structural formulas for the compounds and decide which spectrum belongs to which compound.

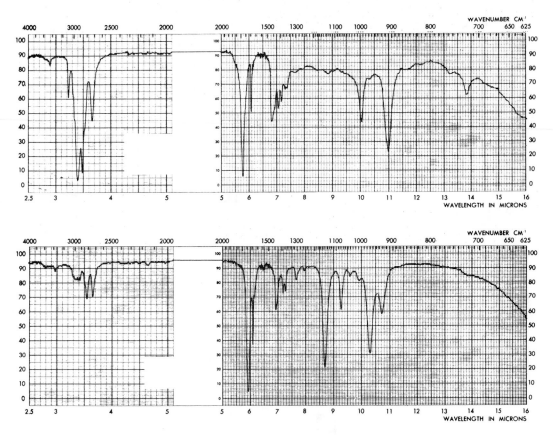

Figure 9.5   Spectra for Problem 9.28. (From *The Aldrich Library of Infrared Spectra*)

PROBLEM 9.29   The following compounds, A to E, have the infrared spectra (1–5) shown in Figure 9.6. Match each spectrum to the proper compound. For each spectrum, point out the important absorption bands that helped you to make the assignment.

A.   2,4-dimethyl-3-pentanol   B.   2-methylcyclohexanone   C.   1,8-nonadiyne
D.   1-phenyl-1-heptanone   E.   2-methyl-1-hexene

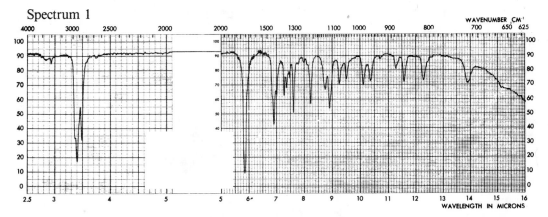

Figure 9.6   Spectra for Problem 9.29. (From *The Aldrich Library of Infrared Spectra*)

Spectrum 2

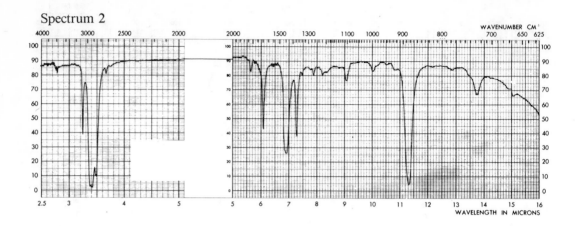

Spectrum 3

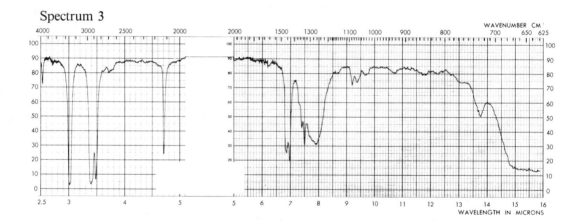

Spectrum 4

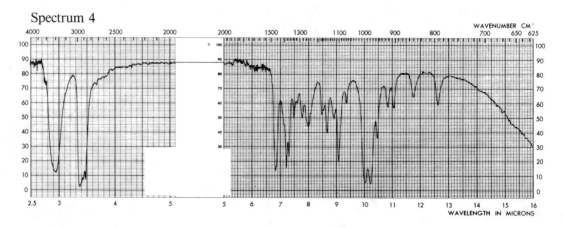

Figure 9.6 (*Continued*)

Spectrum 5

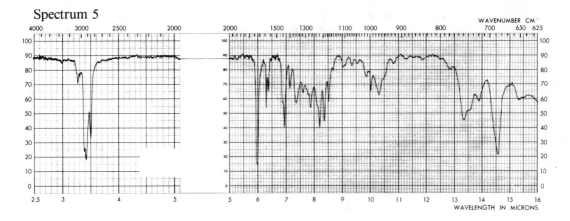

Figure 9.6 (*Continued*)

ADDITIONAL PROBLEMS

9.30 Name the following compounds according to the IUPAC rules, including stereochemistry when pertinent.

(a) cyclopentyl–CH=O   (b) cyclobutanone   (c)
$$CH_3CH_2CH_2 \quad H \\ C=C \\ H \quad CH=O$$

(d)
$$\begin{array}{c} O \\ \parallel \\ CH \\ \mid \\ C\text{---}H \\ CH_3 \quad OH \end{array}$$
(e) cyclohexanone with =O and CH₃   (f)
$$\begin{array}{c} CH_3 \quad O \\ \mid \qquad \parallel \\ CH_3CHCH_2CH_2CCH_3 \end{array}$$

(g)
$$CH_3O-\!\!\!\!\!\bigcirc\!\!\!\!\!-CH=O$$
(h) cyclohexenone with CH₃   (i)
$$\begin{array}{c} Cl \\ \mid \\ H \quad CH_2CCH_3 \\ \diagup \qquad \parallel \\ CH_3 \quad H \quad O \\ \mid \\ CH_2CH_3 \end{array}$$

(j)
$$\bigcirc\!\!\!\!\!-CH_2CH_2CH_2CH=O$$

9.31 Draw structural formulas for the following compounds.

(a) 3,3-dimethylcyclopentanone   (b) 1-bromo-2-hexanone
(c) *m*-chlorobenzaldehyde   (d) *p*-methylacetophenone   (e) 3,5-hexadien-2-one
(f) 2-methylpentanal   (g) (*R*)-2-chlorocyclobutanone   (h) (*S*)-3-bromobutanal

9.32   Using butanal as a typical aldehyde, write equations showing the reactions you would predict if the following reagents were used. Write "no reaction" if you expect none.

(a) H_2SO_4, cold, conc   (b) $NaBH_4$, H_2O

(c) 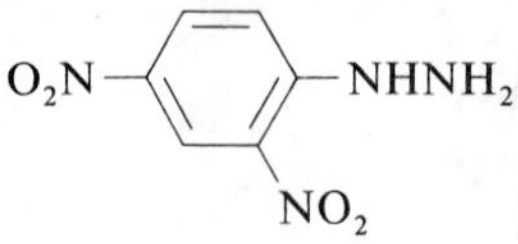 O_2N—⬡—$NHNH_2$, H_2SO_4 (d)   1. $CH_3CH_2CH_2CH_2Li$; 2. H_3O^+
          |
          NO_2

(e) 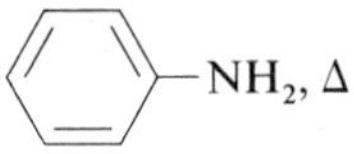   ⬡—NH_2, Δ (f) $NaCN$, H_2SO_4(aq)   (g) $Na_2Cr_2O_7$, H_2O, H_3O^+

$$\overset{O}{\overset{\|}{}}$$

(h) NaI, acetone   (i) $HONH_3{}^+Cl^-$, $CH_3\overset{O}{\overset{\|}{C}}O^-Na^+$ (j) $KMnO_4$, H_2O, cold

(k)   1. CH_3MgI, diethyl ether; 2. H_3O^+ (l)   5% $NaHCO_3$, H_2O

(m)   1. $LiAlH_4$, diethyl ether; 2. H_3O^+ (n) $Ag(NH_3)_2{}^+OH^-$, H_2O

(o) H_2NNH_2, KOH, diethylene glycol, Δ (p) $Zn(Hg)$, HCl, Δ

9.33   Using 2-pentanone as a typical ketone, write equations showing the reactions you would predict if the following reagents were used. Write "no reaction" if you expect none.

(a) H_2SO_4, cold, conc   (b) $NaBH_4$, H_2O

(c) 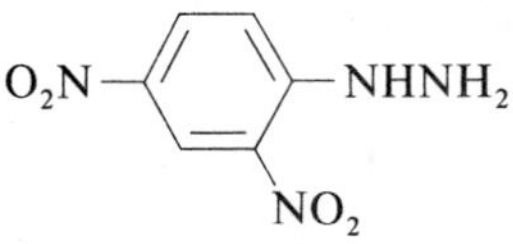 O_2N—⬡—$NHNH_2$, H_2SO_4 (d) 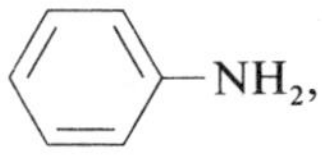   ⬡—NH_2, Δ
          |
          NO_2

(e)   1. 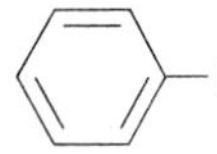 ⬡—Li, tetrahydrofuran; 2. NH_4Cl, H_2O (f) $NaCN$, H_2SO_4(aq)

(g) $Na_2Cr_2O_7$, H_2O, H_3O^+ (h) $AgNO_3$, ethanol

(i) $HONH_3{}^+Cl^-$, $CH_3\overset{O}{\overset{\|}{C}}O^-Na^+$ (j) $KMnO_4$, H_2O, cold

(k)   1. CH_3CH_2MgBr, diethyl ether; 2. H_3O^+ (l)   5% $NaHCO_3$, H_2O

(m)   1. $CH_3C{\equiv}CNa$; 2. NH_4Cl, H_2O (n) $Ag(NH_3)_2{}^+OH^-$, H_2O

(o) H_2NNH_2, KOH, diethylene glycol, Δ (p) $Zn(Hg)$, HCl, Δ

9.34   Complete each of the following equations, showing the structures of the products you predict will be formed.

(a) 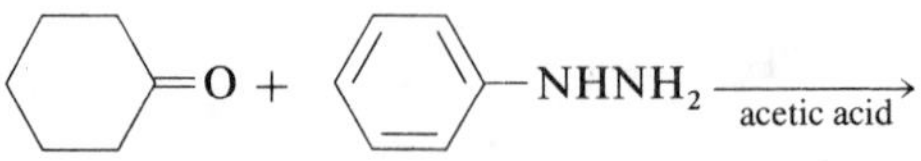   ⬡=O + ⬡—$NHNH_2$ $\xrightarrow[\text{acetic acid}]{}$

(b) $CH_3CH_2CH_2C{\equiv}CCH_3$ $\xrightarrow[\text{tetrahydrofuran}]{BH_3}$ $\xrightarrow{H_2O_2,\ NaOH}$

(c) 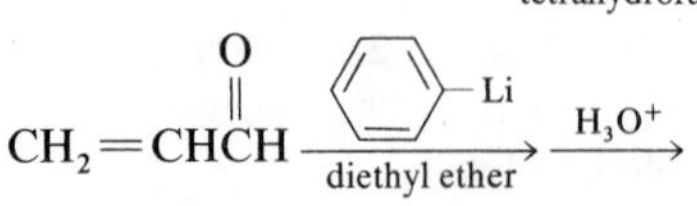 $CH_2{=}CHCH\overset{O}{\overset{\|}{C}}$ $\xrightarrow[\text{diethyl ether}]{⬡—Li}$ $\xrightarrow{H_3O^+}$

(d) 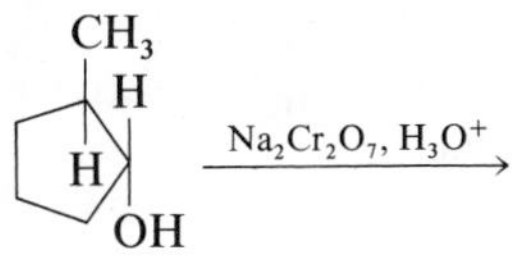 [structure: bicyclic compound with CH₃, H, H, OH] $\xrightarrow{\text{Na}_2\text{Cr}_2\text{O}_7, \text{H}_3\text{O}^+}$

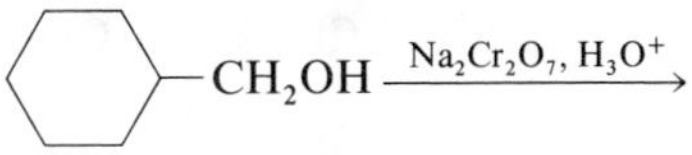

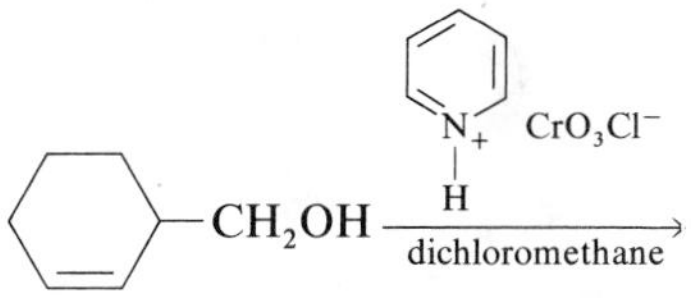

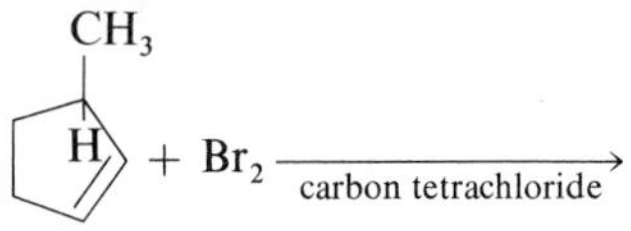

(e) [cyclohexyl]—$CH_2OH \xrightarrow{\text{Na}_2\text{Cr}_2\text{O}_7, \text{H}_3\text{O}^+}$

(f) [cyclohexenyl]—$CH_2OH \xrightarrow[\text{dichloromethane}]{\text{pyridinium chlorochromate } (C_5H_5NH^+ \ CrO_3Cl^-)}$

(g) [bicyclic compound with CH₃, H] $+ \ Br_2 \xrightarrow{\text{carbon tetrachloride}}$

(h) [bicyclic compound with CH₃, H] $+ \ KMnO_4 \xrightarrow[\text{H}_2\text{O}]{\text{NaOH}}$

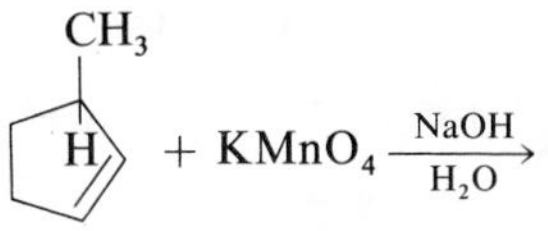

(i) 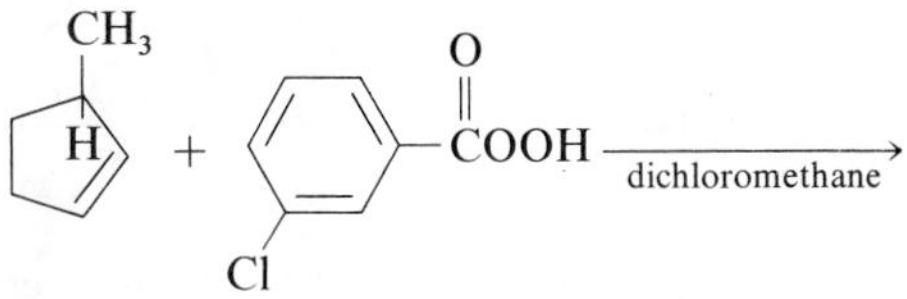 [bicyclic compound with CH₃, H] $+$ [3-chlorobenzoic acid structure with $\overset{O}{\underset{\|}{C}}OOH$, Cl] $\xrightarrow{\text{dichloromethane}}$

(j) 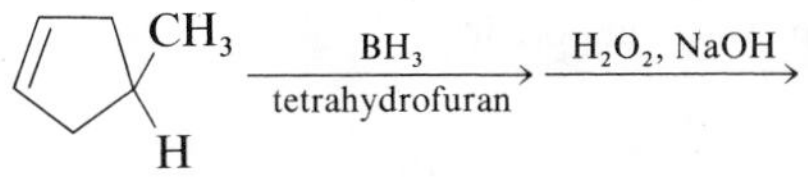 [cyclopentene with CH₃, H] $\xrightarrow[\text{tetrahydrofuran}]{\text{BH}_3} \xrightarrow{\text{H}_2\text{O}_2, \text{NaOH}}$

(k) 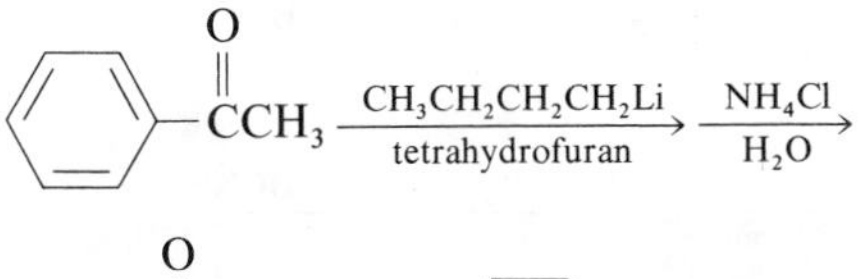 [phenyl]—$\overset{O}{\underset{\|}{C}}CH_3 \xrightarrow[\text{tetrahydrofuran}]{\text{CH}_3\text{CH}_2\text{CH}_2\text{CH}_2\text{Li}} \xrightarrow[\text{H}_2\text{O}]{\text{NH}_4\text{Cl}}$

(l) 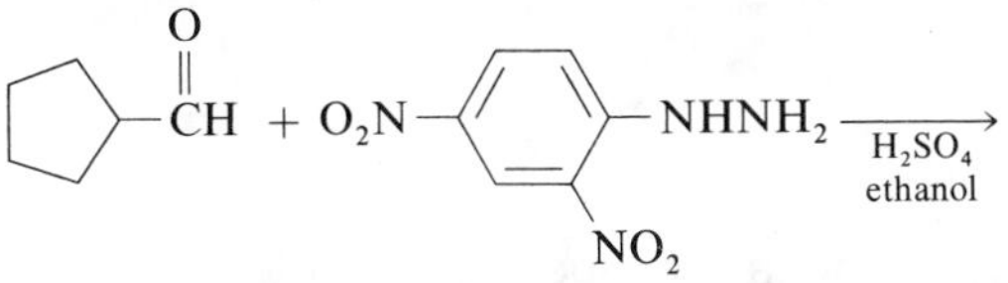 [cyclopentyl]—$\overset{O}{\underset{\|}{C}}H \ + \ O_2N$—[benzene ring with NO₂]—$NHNH_2 \xrightarrow[\text{ethanol}]{\text{H}_2\text{SO}_4}$

(m) 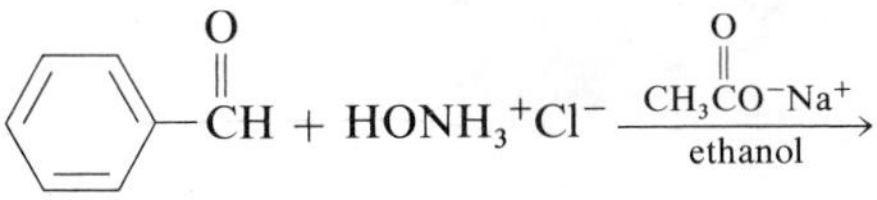 [phenyl]—$\overset{O}{\underset{\|}{C}}H \ + \ HONH_3^+Cl^- \xrightarrow[\text{ethanol}]{\text{CH}_3\overset{O}{\underset{\|}{C}}O^-\text{Na}^+}$

(n) 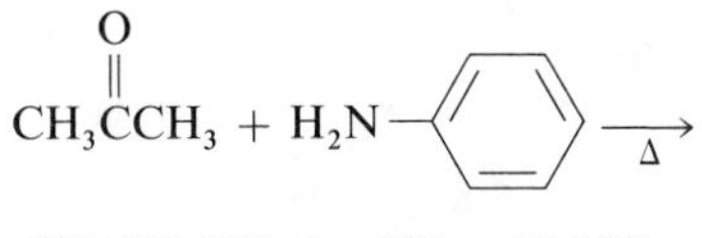 $CH_3\overset{O}{\underset{\|}{C}}CH_3 \ + \ H_2N$—[phenyl] $\xrightarrow{\Delta}$

(o) $CH_3CH_2CH_2C{\equiv}CH \ + \ NaNH_2 \xrightarrow{\text{NH}_3\text{(liq)}}$

(p) $CH_3CH_2CH_2C{\equiv}CH \ + \ CH_3CH_2MgBr \xrightarrow{\text{diethyl ether}}$

(q) 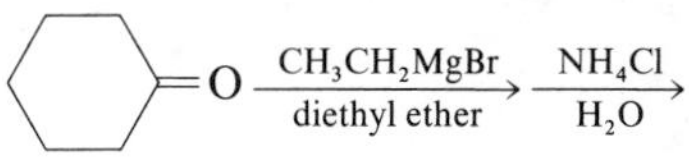 [cyclohexanone]${=}O \xrightarrow[\text{diethyl ether}]{\text{CH}_3\text{CH}_2\text{MgBr}} \xrightarrow[\text{H}_2\text{O}]{\text{NH}_4\text{Cl}}$

(r) 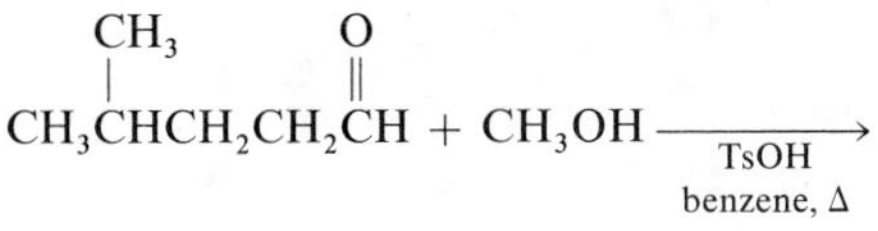 $CH_3\overset{CH_3}{\underset{|}{C}}HCH_2CH_2\overset{O}{\underset{\|}{C}}H \ + \ CH_3OH \xrightarrow[\text{benzene, }\Delta]{\text{TsOH}}$

(s) 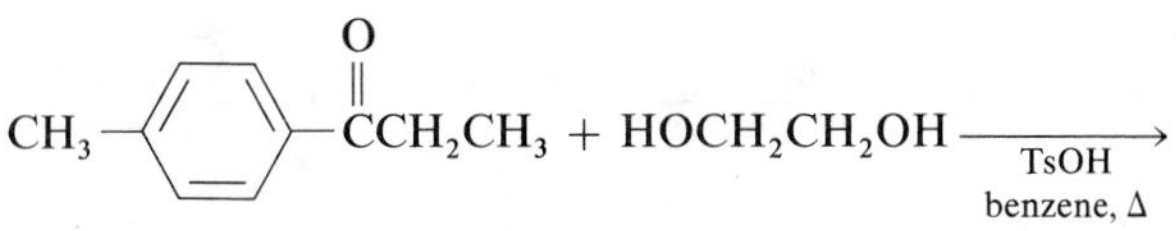 CH_3—[benzene ring]—$\overset{O}{\underset{\|}{C}}CH_2CH_3 \ + \ HOCH_2CH_2OH \xrightarrow[\text{benzene, }\Delta]{\text{TsOH}}$

(t) 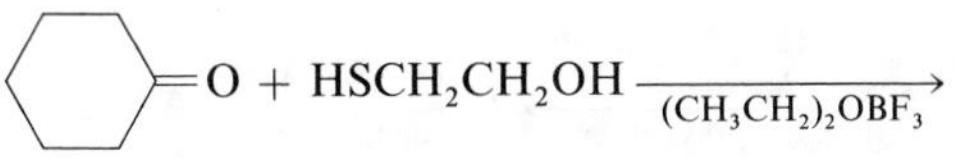 [cyclohexanone]${=}O \ + \ HSCH_2CH_2OH \xrightarrow{(\text{CH}_3\text{CH}_2)_2\text{OBF}_3}$

(u) $\overset{O}{\overset{\|}{CH_3CCH_2}}\overset{CH_3}{\overset{|}{CHCH_3}}$ + $\overset{CH_3}{\overset{|}{CH_3CHCH_2NH_2}}$ $\xrightarrow[HCl]{}$ $\xrightarrow{NaOH}$

(v) $\overset{CH_3}{\overset{|}{HOCH_2CH_2CCH_2}}\overset{O}{\overset{\|}{CH=CHCOCH_3}}$ $\xrightarrow[\text{dichloromethane}]{\text{pyridinium } CrO_3Cl^-}$
              $\overset{|}{CH_3}$

(w) $\overset{O}{\overset{\|}{CH_3C(CH_2)_8CH_3}}$ $\xrightarrow[\Delta]{Zn(Hg),\ HCl}$
   (x) $\overset{CH_3}{\overset{|}{CH_3CHCH_2}}\overset{O}{\overset{\|}{CCH_2}}\overset{CH_3}{\overset{|}{CHCH_3}}$ $\xrightarrow[\text{diethylene glycol, }\Delta]{H_2NNH_2,\ NaOH}$

9.35 Predict structures of the products or intermediates designated by letters in the following equations. Show stereochemistry when it is important.

(a) cyclopentyl—OH $\xrightarrow{Na_2Cr_2O_7,\ H_3O^+}$ A $\xrightarrow[\Delta]{\text{phenyl—}NH_2}$ B

(b) $CH_3C{\equiv}CCH_3$ $\xrightarrow[\text{tetrahydrofuran}]{BH_3}$ C $\xrightarrow{H_2O_2,\ NaOH}$ D $\xrightarrow[\underset{\|}{CH_3CO^-Na^+}\ O]{HONH_3^+Cl^-}$ E

(c) $CH_3CH_2CH_2CH_2\overset{O}{\overset{\|}{CH}}$ $\xrightarrow[H_2O]{NaBH_4}$ F $\xrightarrow[\text{pyridine}]{SOCl_2}$ G $\xrightarrow{NH_3\ (\text{excess})}$ H

(d) $CH_3CH_2CH_2\overset{O}{\overset{\|}{CCH_3}}$ $\xrightarrow[H_2O]{NaBH_4}$ I $\xrightarrow[\text{pyridine}]{TsCl}$ J $\xrightarrow[\text{ethanol, }\Delta]{KOH\ (4\ M)}$ K $\xrightarrow[\text{dichloromethane}]{\text{m-chloroperoxybenzoic acid}}$ L
                                                       (major product)

(e) $\overset{CH_3}{\overset{|}{CH_3CHCH_2CH_2}}\overset{O}{\overset{\|}{C}}$—phenyl $\xrightarrow[\text{diethyl ether}]{CH_3CH_2MgBr}$ M $\xrightarrow[H_2O]{NH_4Cl}$ N $\xrightarrow[\Delta]{H_3PO_4}$ O

(f) cyclopentyl—$\overset{O}{\overset{\|}{CH}}$ $\xrightarrow[\text{tetrahydrofuran}]{\text{phenyl—Li}}$ P $\xrightarrow{H_3O^+}$ Q $\xrightarrow[\text{acetone}]{CrO_3,\ H_3O^+}$ R $\xrightarrow[\text{acetic acid}]{\text{phenyl—}NHNH_2}$ S

(g) $\overset{CH_3}{\overset{|}{CH_3C}}{=}CHCH_3$ $\xrightarrow[\text{tetrahydrofuran}]{BH_3}$ T $\xrightarrow{H_2O_2,\ NaOH}$ U $\xrightarrow[H_2O]{Na_2Cr_2O_7,\ H_2SO_4}$ V $\xrightarrow{NaCN}$ W $\xrightarrow{H_3O^+}$ X

(h) $HC{\equiv}CH$ $\xrightarrow[NH_3(\text{liq})]{NaNH_2}$ Y $\xrightarrow[\text{diethyl ether}]{\text{phenyl—}CH_2Br}$ Z $\xrightarrow[\text{diethyl ether}]{CH_3CH_2MgBr}$ AA $\xrightarrow{\text{cyclopentanone}}$ BB $\xrightarrow[H_2O]{NH_4Cl}$ CC

(i)

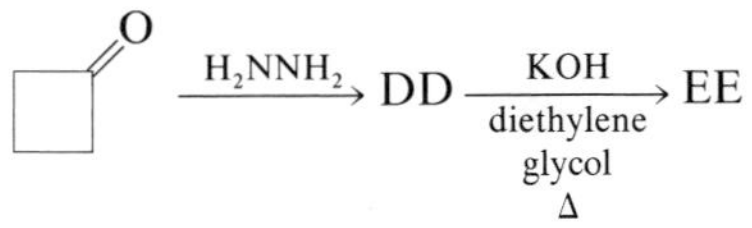

$$\text{(cyclobutanone)} \xrightarrow{\text{H}_2\text{NNH}_2} \text{DD} \xrightarrow[\substack{\text{diethylene} \\ \text{glycol} \\ \Delta}]{\text{KOH}} \text{EE}$$

9.36 Here are some more exercises in recognizing reactions. Show the structures of the products you expect.

(a) 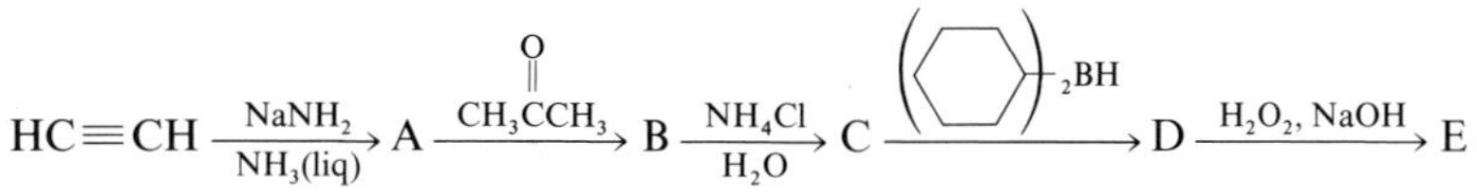 $\text{HC}\equiv\text{CH} \xrightarrow[\text{NH}_3(\text{liq})]{\text{NaNH}_2} \text{A} \xrightarrow{\text{CH}_3\overset{\text{O}}{\overset{\|}{\text{C}}}\text{CH}_3} \text{B} \xrightarrow[\text{H}_2\text{O}]{\text{NH}_4\text{Cl}} \text{C} \xrightarrow{\left(\text{cyclohexyl}\right)_2\text{BH}} \text{D} \xrightarrow{\text{H}_2\text{O}_2, \text{NaOH}} \text{E}$

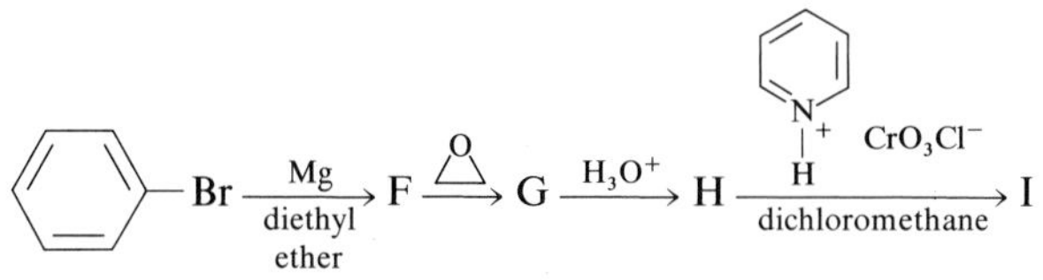

(b)

(c)

$$\xrightarrow{\text{HOCH}_2\text{CH}_2\text{OH}}_{\substack{\text{TsOH} \\ \text{benzene, } \Delta}} \text{J} \xrightarrow[\text{tetrahydrofuran}]{\text{BH}_3} \text{K} \xrightarrow{\text{H}_2\text{O}_2, \text{NaOH}}$$

$$\text{L} \xrightarrow{\text{H}_3\text{O}^+} \text{M}$$

(d)

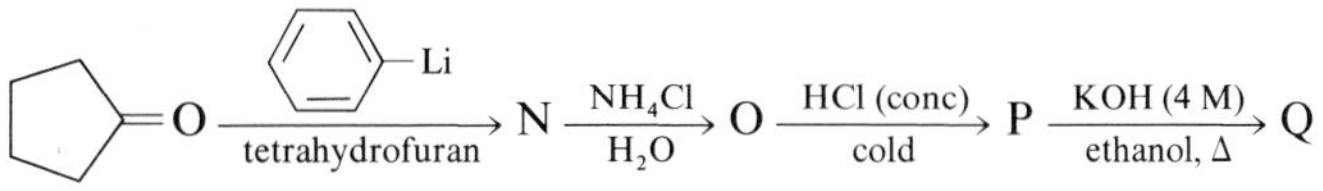

(e)

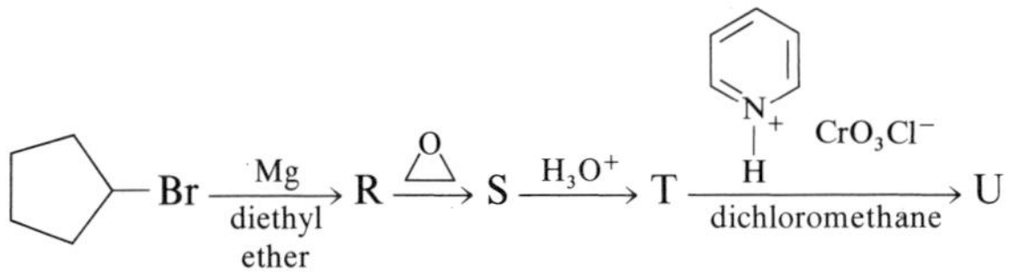

(f)

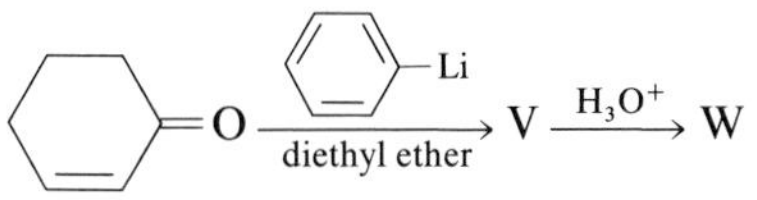

(g)

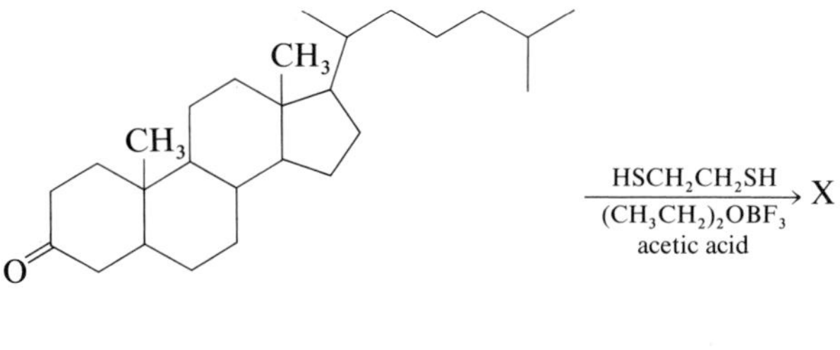

$$\xrightarrow[\substack{(\text{CH}_3\text{CH}_2)_2\text{OBF}_3 \\ \text{acetic acid}}]{\text{HSCH}_2\text{CH}_2\text{SH}} \text{X}$$

(h) 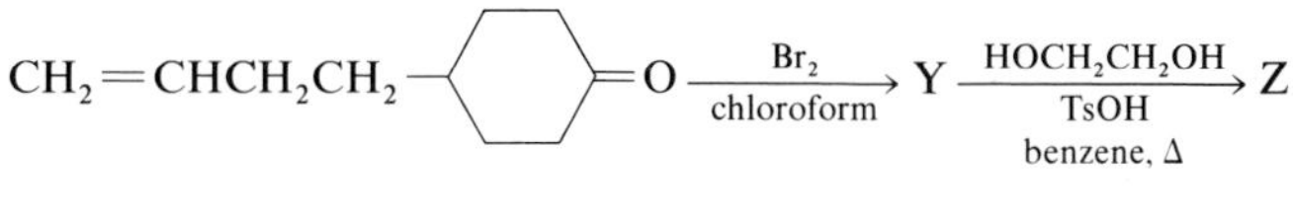 $\text{CH}_2=\text{CHCH}_2\text{CH}_2$—

$$\xrightarrow[\text{chloroform}]{\text{Br}_2} \text{Y} \xrightarrow[\substack{\text{TsOH} \\ \text{benzene, } \Delta}]{\text{HOCH}_2\text{CH}_2\text{OH}} \text{Z}$$

(i) $\underset{\text{CH}_3}{\text{CH}_3\text{CHCH}_2\text{CH}_2\overset{\text{O}}{\overset{\|}{\text{C}}}\text{CH}_2\text{CH}_2\overset{\text{O}}{\overset{\|}{\text{C}}}\text{OH}}$ $\xrightarrow[\substack{\text{diethylene}\\\text{glycol}\\\Delta\Delta}]{\text{H}_2\text{NNH}_2,\text{ NaOH}}$ AA $\xrightarrow{\text{H}_3\text{O}^+}$ BB

(j) $\text{CH}_3\text{CH}_2\text{CH}_2\text{CH}_2\overset{\text{O}}{\overset{\|}{\text{C}}}$—⟨benzene ring⟩ $\xrightarrow[\Delta]{\substack{\text{HCl}\\\text{Zn(Hg)}}}$ CC

(k)   ⟨bicyclic structure with CH$_2$ and CH$_3$CHCH$_3$, CH groups⟩ $\xrightarrow[\substack{\text{diethyl}\\\text{ether}}]{\text{CH}_3\text{Li}}$ DD $\xrightarrow{\text{H}_3\text{O}^+}$ EE $\xrightarrow{\text{Hg(O}\overset{\text{O}}{\overset{\|}{\text{C}}}\text{CH}_3)_2,\text{ H}_2\text{O}}$ FF $\xrightarrow[\substack{\text{NaOH}\\\text{H}_2\text{O}}]{\text{NaBH}_4}$ GG

9.37 Show syntheses for the following compounds. You have bromobenzene, any organic reagents containing three or fewer carbon atoms, and any inorganic reagents you need. There may be more than one acceptable answer to each problem. Try to find the shortest route to the desired product.

(a) ⟨benzene ring⟩—$\text{CH}_2\overset{\text{O}}{\overset{\|}{\text{C}}}\text{CH}_3$

(b) $\underset{\text{H}}{\overset{\text{CH}_3\text{CH}_2}{}}\text{C}=\text{C}\underset{\text{H}}{\overset{\text{CH}_2\text{CH}_2\text{OH}}{}}$

(c) $\text{CH}_3\text{CH}_2\overset{\text{O}}{\overset{\|}{\text{C}}}\text{CH}_2\text{CH}_2\text{CH}_3$

(d) $\text{CH}_3\overset{\text{CH}_3}{\overset{|}{\text{C}}}=\text{CHCH}_2\text{CH}_2\text{CH}_3$

(e) ⟨benzene ring⟩—$\text{CH}_2\text{CH}_2\text{Br}$

(f) $\text{CH}_3\text{CH}_2\text{CH}_2$—$\underset{\text{Br}}{\overset{\text{Br}}{}}$—H, CH$_3$ + enantiomer

(g) $\text{CH}_3(\text{CH}_2)_8\overset{\text{O}}{\overset{\|}{\text{C}}}\text{CH}=\text{CH}_2$

(h) ⟨benzene ring⟩—$\overset{\text{O}}{\overset{\|}{\text{C}}}\text{CH}_2\text{CH}_3$

(i) $\underset{\text{H}}{\overset{\text{H}}{}}$—$\underset{\text{CH}_3}{\overset{\text{OH}}{\overset{|}{\text{C}}}}$—CH$_3$, OH (structure with OH, CH$_3$)

(j) $\text{CH}_3\text{CHCH}_2\text{CH}_2\overset{\text{CH}_3}{\overset{|}{\text{C}}}\text{HCH}_3$ with OH on second carbon

(k) ⟨cyclohexadiene ring⟩—$\text{CH}_2\text{CH}_2\text{CH}_3$

(l) ⟨epoxide with two phenyl groups and H's⟩ + enantiomer

9.38 Provide the reagents that would be necessary to carry out the following transformations.

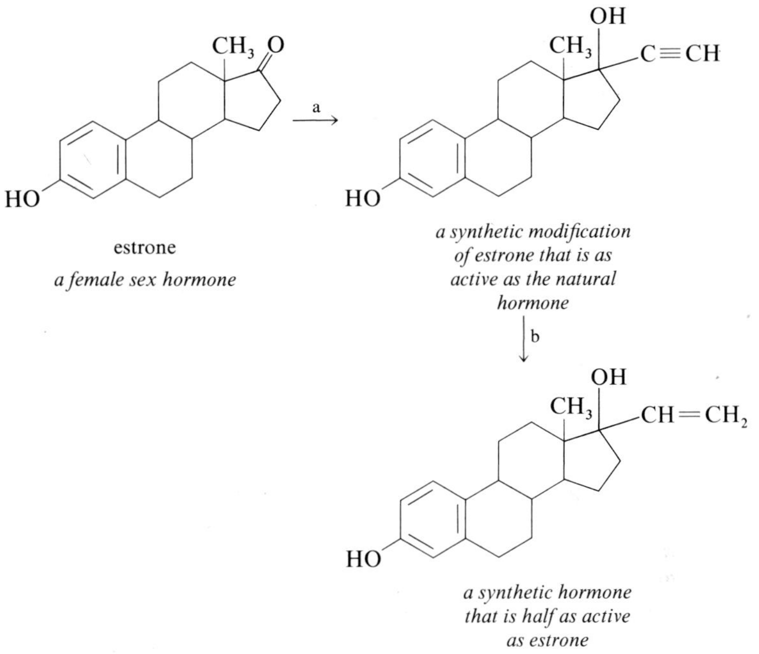

CH3 O
HO
a
OH
CH3
C≡CH
HO
estrone
a female sex hormone
a synthetic modification of estrone that is as active as the natural hormone
b
OH
CH3
CH=CH2
HO
a synthetic hormone that is half as active as estrone

9.39 Provide the reagents that are necessary to carry out the following transformations in the steroid series.

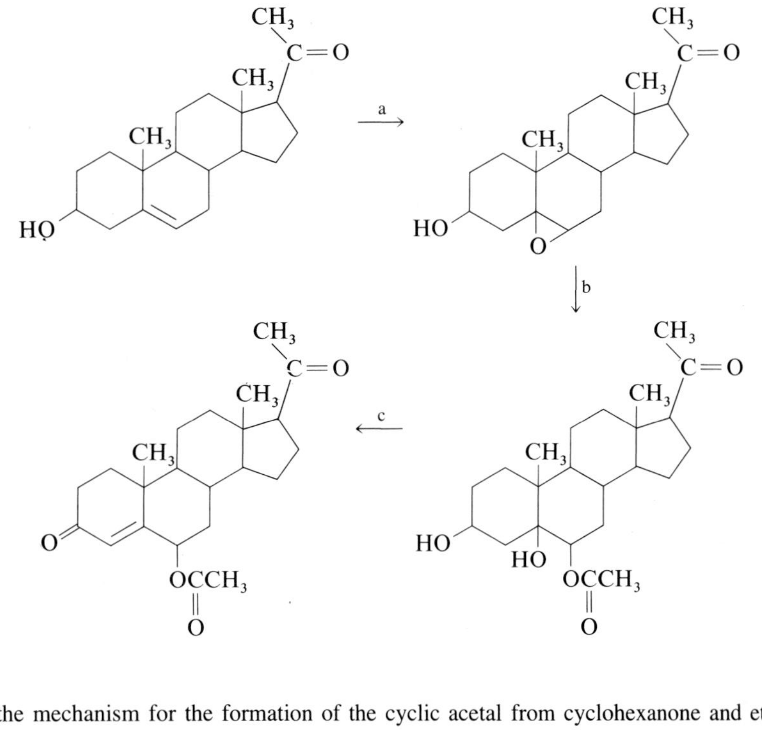

CH3
C=O
CH3
CH3
HO
a
O
b
c
HO
OCCH3
O

9.40 Show the mechanism for the formation of the cyclic acetal from cyclohexanone and ethylene glycol in the presence of toluenesulfonic acid.

9.41   A laboratory synthesis of chrysomelidial, a compound secreted by the larvae of some beetles to defend themselves from attack, starts in the following way.

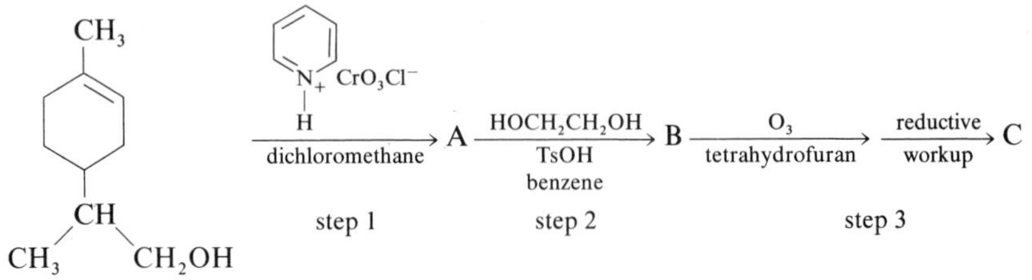

$$\text{A} \xrightarrow[\text{TsOH}]{\text{HOCH}_2\text{CH}_2\text{OH}} \text{B} \xrightarrow[\text{tetrahydrofuran}]{\text{O}_3} \xrightarrow[\text{workup}]{\text{reductive}} \text{C}$$

step 1   step 2   step 3

Fill in the structures of Compounds A, B, and C. Why was the second step of this synthesis necessary?

9.42   2,2-Dimethoxypropane is converted into 2,2-dibutoxypropane when it is heated with 1-butanol and a trace of acid. Some facts are summarized in the equation below.

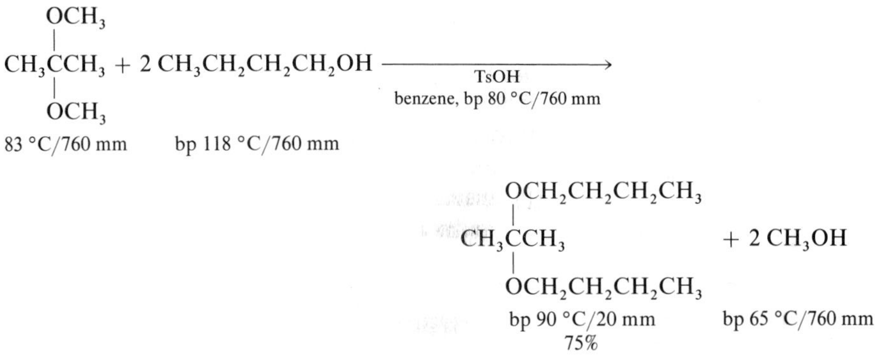

$$\begin{array}{c} \text{OCH}_3 \\ | \\ \text{CH}_3\text{CCH}_3 + 2\ \text{CH}_3\text{CH}_2\text{CH}_2\text{CH}_2\text{OH} \xrightarrow[\text{benzene, bp 80 °C/760 mm}]{\text{TsOH}} \\ | \\ \text{OCH}_3 \end{array}$$

bp 83 °C/760 mm   bp 118 °C/760 mm

$$\begin{array}{c} \text{OCH}_2\text{CH}_2\text{CH}_2\text{CH}_3 \\ | \\ \text{CH}_3\text{CCH}_3 \\ | \\ \text{OCH}_2\text{CH}_2\text{CH}_2\text{CH}_3 \end{array} \qquad + 2\ \text{CH}_3\text{OH}$$

bp 90 °C/20 mm   bp 65 °C/760 mm
75%

Write a mechanism for the reaction and suggest what practical steps could be taken to ensure that a good yield of 2,2-dibutoxypropane is formed.

9.43   Allyl alcohol and benzyl alcohol, even though they are primary alcohols, react with concentrated hydrochloric acid in the presence of zinc chloride to give an instant appearance of an oily layer. Explain why this happens.

9.44   How would you distinguish the compounds in each of the following sets from each other? Give the reagents you would use and the experimental observations you would expect to make.

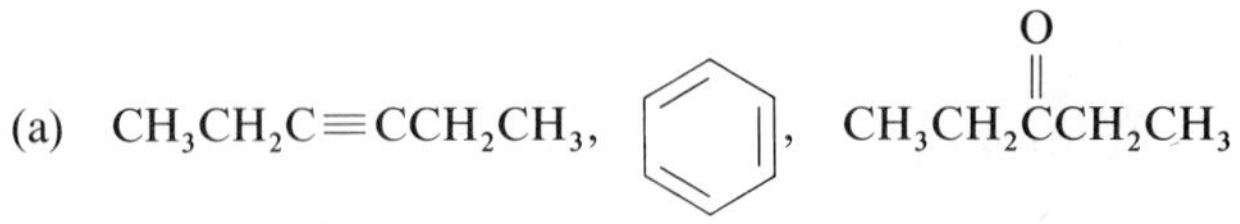

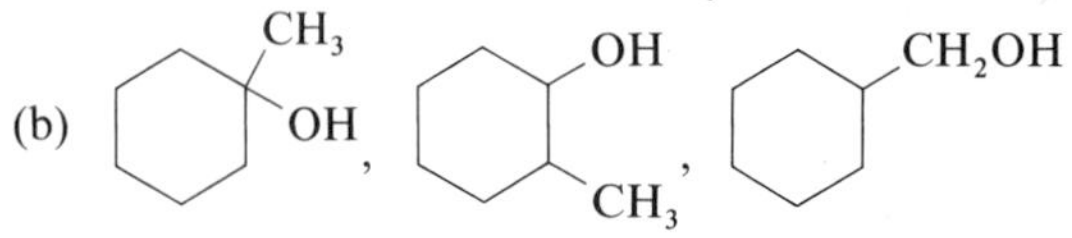

(a) $CH_3CH_2C\equiv CCH_2CH_3$,   ⬡ , $CH_3CH_2\overset{\displaystyle O}{\overset{\displaystyle \|}{C}}CH_2CH_3$

(b)

(c) $CH_3CH_2CH_2CH_2CH_2CH_3$, $CH_3CH_2CH_2OCH_2CH_3$,

$CH_3CH_2CH_2CH_2CH_2CH_2OH$, $CH_3CH_2CH_2CH_2CH_2Br$

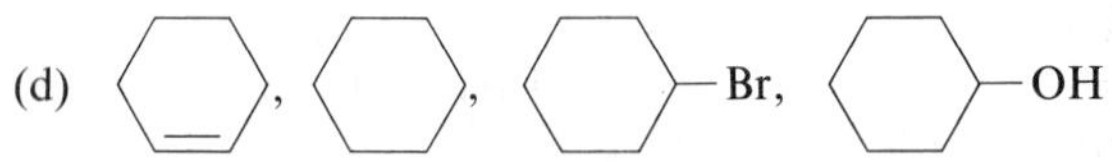

(d)

(e)

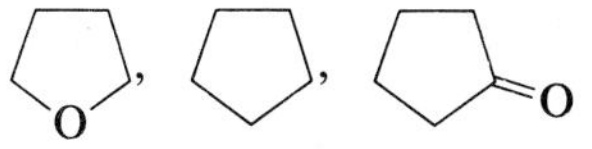

(f) $CH_2{=}CHCH_2Br,$ $CH_3CH_2CH_2Br,$ $CH_3\overset{\overset{\displaystyle CH_3}{|}}{\underset{\underset{\displaystyle CH_3}{|}}{C}}Br$

(g) 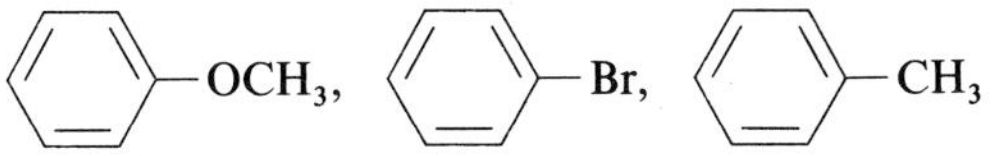 $-OCH_3,$ $-Br,$ $-CH_3$

(h) $CH_3CH_2CH_2CH_2\overset{\overset{\displaystyle O}{||}}{C}H,$ $CH_3CH_2CH_2\overset{\overset{\displaystyle O}{||}}{C}CH_3,$ $CH_3CH_2CH_2OCH_2CH_3$

9.45   The right-hand column below contains a list of some common laboratory reagents. For each compound in the left-hand columns, list all the reagents with which, when the two are mixed together in a test tube, you would expect to get a change that can be detected by the human senses.

Compounds

1. $CH_3CH_2CH_2OH$

2. 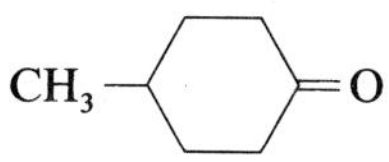 $CH_3-$$=O$

3.

4. $-\overset{\overset{\displaystyle O}{||}}{C}H$

5.

6. $CH_3CH_2CH_2\overset{\overset{\displaystyle CH_3}{|}}{\underset{\underset{\displaystyle Br}{|}}{C}}CH_3$

7. $-Br$

8. $HOCH_2CH_2OH$

9. $CH_3C{\equiv}CCH_2CH_2OH$

10.

11.

12. $-OCH_3$

13. $CH_2{=}CHCHCH_3$
       $\quad\quad\quad\;\;|$
       $\quad\quad\quad\;\;OH$

14. $CH_3CH_2CH_2Br$

15. $CH_3CH_2\overset{\overset{\displaystyle O}{||}}{C}CH_3$

Reagents

(a) $H_2SO_4,$ cold, conc

(b) $KMnO_4,$ $H_2O,$ cold

(c) $CrO_3,$ $H_2SO_4,$ acetone

(d) $Br_2,$ carbon tetrachloride

(e) $AgNO_3,$ ethanol

(f) $H_2O,$ cold

(g) $NaI,$ acetone

(h) $Ag(NH_3)_2{}^+OH^-$

(i) $O_2N-$$-NHNH_2,$ NO_2

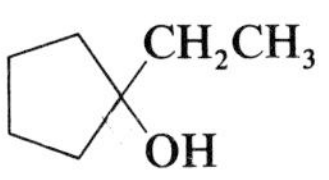

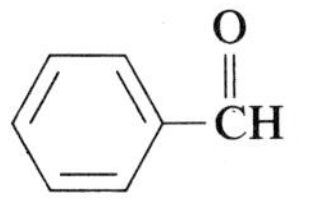

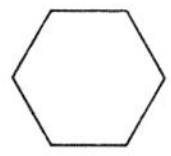

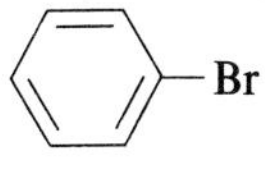

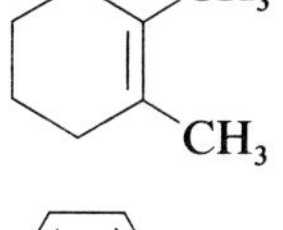

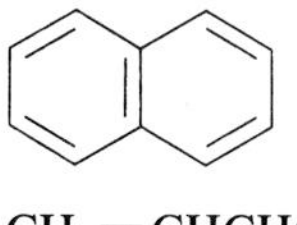

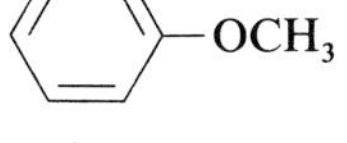

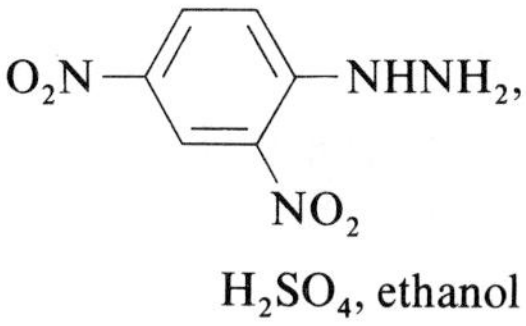

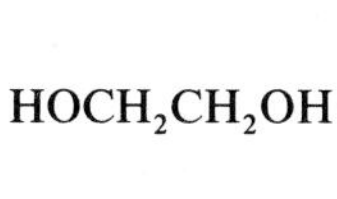

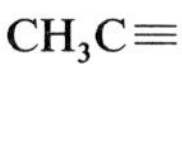

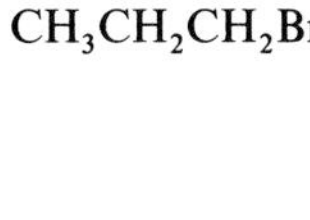

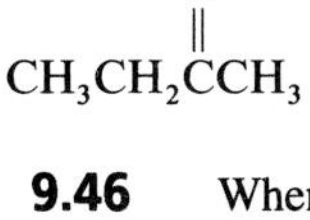

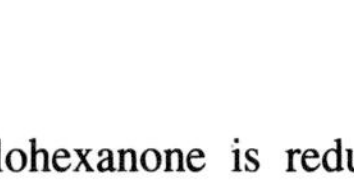

$\quad\quad$ $H_2SO_4,$ ethanol

9.46   When 4-*tert*-butylcyclohexanone is reduced with sodium borohydride, the product is 86% *trans*-4-*tert*-butylcyclohexanol and 14% *cis*-4-*tert*-butylcyclohexanol. When 3,3,5-trimethyl-cyclohexanone is reduced with sodium borohydride, the product is a mixture with 48% having the hydroxyl group cis to the methyl group at carbon 5 and 52% having the hydroxyl group trans to that methyl group.

Draw conformationally correct structures for the starting materials and the products of these reactions. Offer an explanation for the facts that we observe experimentally.

9.47   When 4-*tert*-butylcyclohexanone is reduced with lithium aluminum hydride, the product is almost exclusively *trans*-4-*tert*-butylcyclohexanol. If lithium tri(*sec*-butyl)borohydride is used as the reducing agent, the product is *cis*-4-*tert*-butylcyclohexanol. How do you account for this observation?

Lithium tri(*sec*-butyl)borohydride is

$$\underset{\displaystyle (CH_3CH_2\overset{\displaystyle CH_3}{\overset{|}{CH}}-)_3BH^-Li^+}{}$$

9.48   The following reaction was carried out.

$$CH_3\overset{\displaystyle CH_3}{\overset{|}{CH}}OCH{=}CH_2 \xrightarrow[\substack{H_2SO_4 \\ 145\,°C}]{H_2{}^{18}O}$$

Predict what the products will be, paying special attention to where the labeled oxygen atom(s) will be.

9.49   An intermediate in the synthesis of a natural product with antitumor activity is prepared by the following sequence of steps.

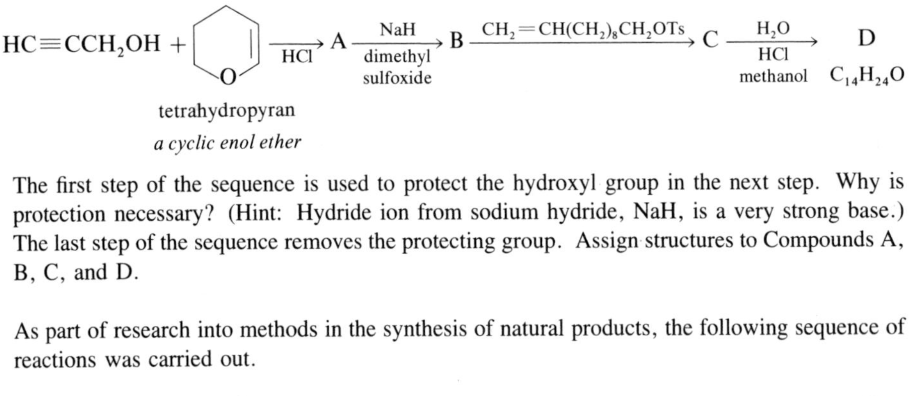

$$HC{\equiv}CCH_2OH + \underset{\substack{\text{tetrahydropyran} \\ \textit{a cyclic enol ether}}}{\boxed{}} \xrightarrow[HCl]{} A \xrightarrow[\substack{\text{dimethyl} \\ \text{sulfoxide}}]{NaH} B \xrightarrow{CH_2{=}CH(CH_2)_8CH_2OTs} C \xrightarrow[\substack{HCl \\ \text{methanol}}]{H_2O} \underset{C_{14}H_{24}O}{D}$$

The first step of the sequence is used to protect the hydroxyl group in the next step. Why is protection necessary? (Hint: Hydride ion from sodium hydride, NaH, is a very strong base.) The last step of the sequence removes the protecting group. Assign structures to Compounds A, B, C, and D.

9.50   As part of research into methods in the synthesis of natural products, the following sequence of reactions was carried out.

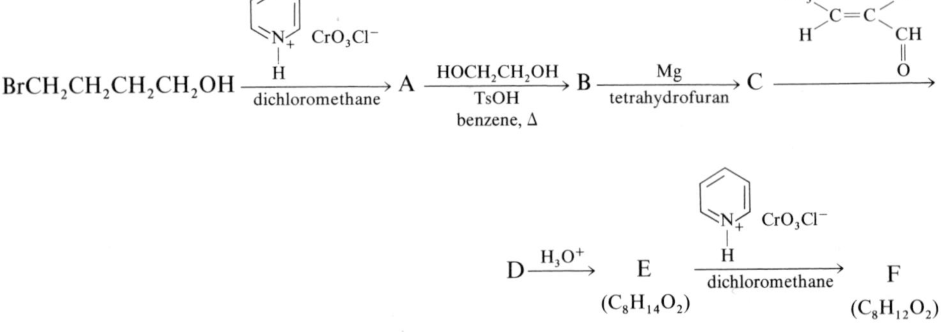

$$BrCH_2CH_2CH_2CH_2OH \xrightarrow[\text{dichloromethane}]{\overset{\text{CrO}_3\text{Cl}^-}{}} A \xrightarrow[\substack{\text{TsOH} \\ \text{benzene, }\Delta}]{HOCH_2CH_2OH} B \xrightarrow[\text{tetrahydrofuran}]{Mg} C \xrightarrow{}$$

$$D \xrightarrow{H_3O^+} \underset{(C_8H_{14}O_2)}{E} \xrightarrow[\text{dichloromethane}]{\overset{\text{CrO}_3\text{Cl}^-}{}} \underset{(C_8H_{12}O_2)}{F}$$

Assign structures to Compounds A, B, C, D, E, and F. (Hint: A review of Section 9.7A may be helpful in assigning a structure to E.) How many units of unsaturation does F have? How do you account for all of them?

9.51 When 1-methylcyclopentene is treated with ozone in the presence of methanol and then with dimethyl sulfide, the product is 6,6-dimethoxy-2-hexanone.

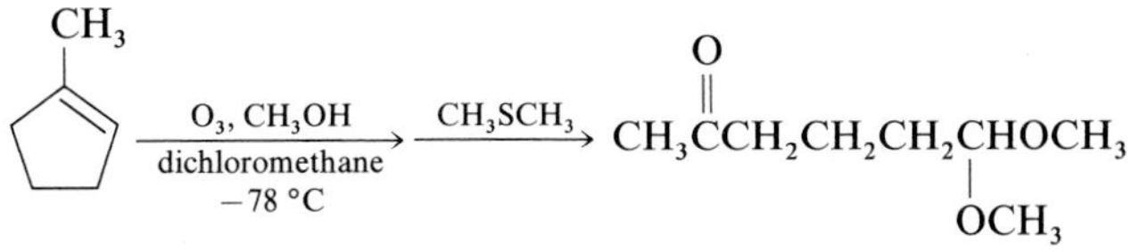

How do you account for this observation? The product from the reaction above is treated with sodium borohydride in ethanol. What is the structure of the product of this reaction?

9.52 Supply reagents for the following transformations. More than one step may be necessary for some.

(a)

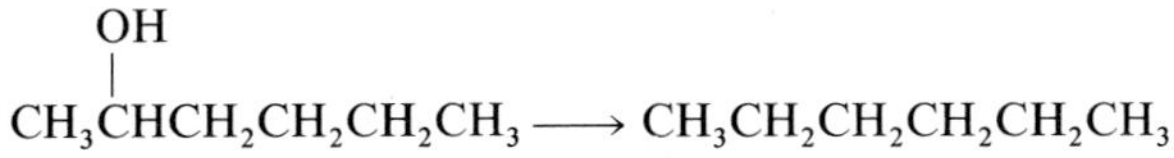

(b)

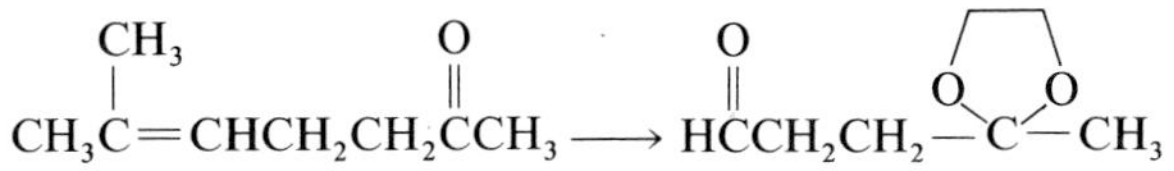

(c)

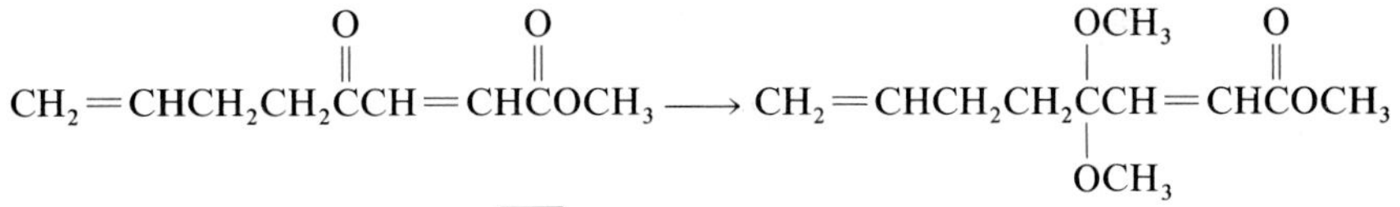

(d)

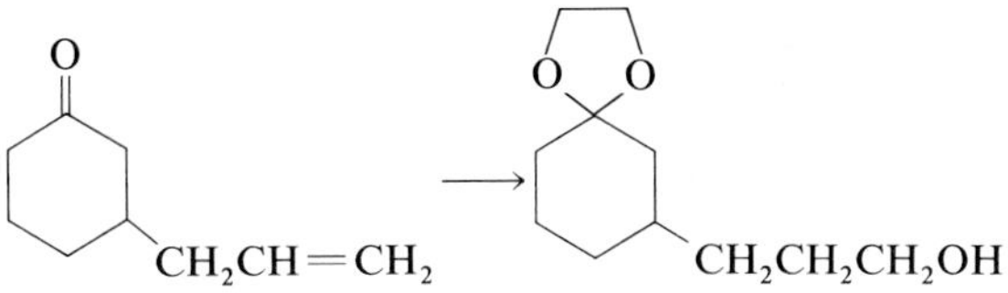

(e)

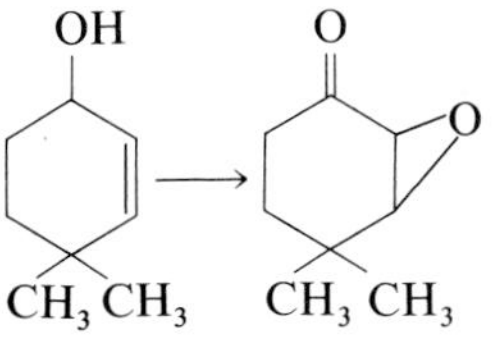

(f)

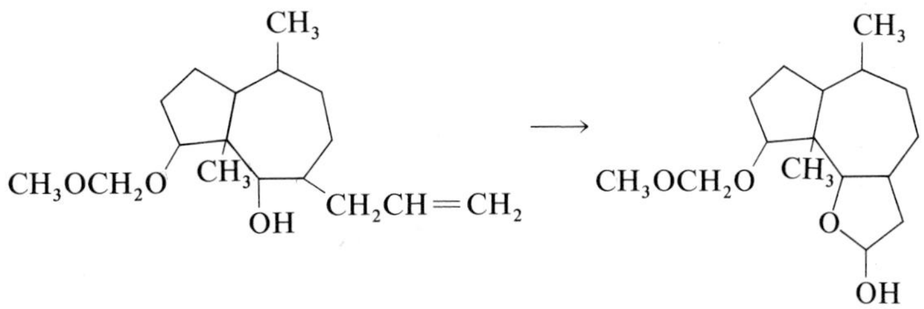

(g)

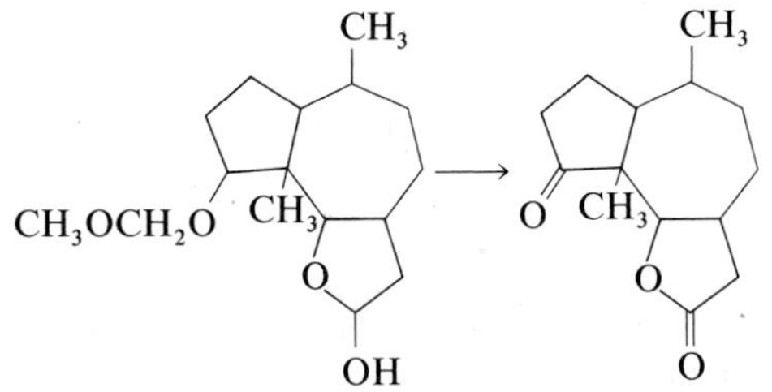

(h)

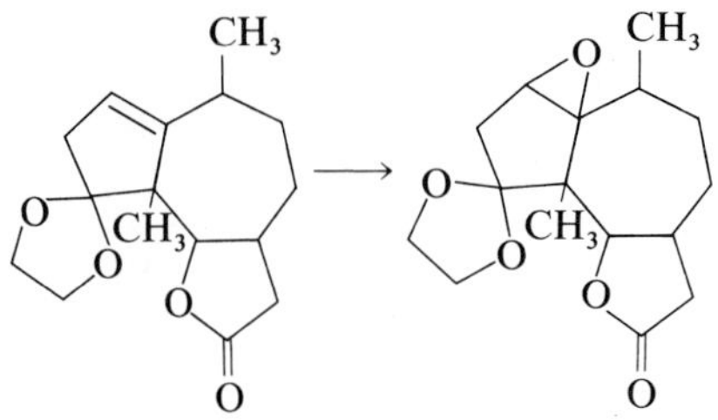

(i)

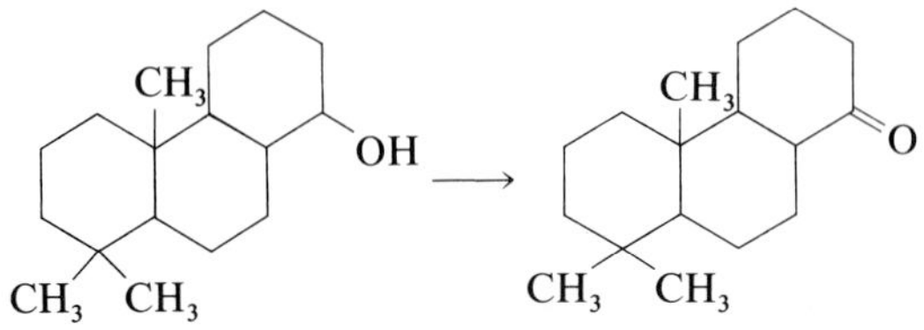

(j)

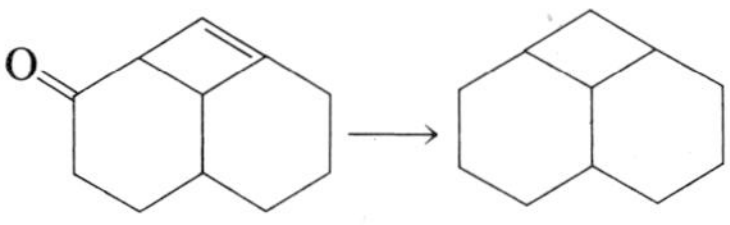

9.53   The Grignard reaction has been used extensively in the synthesis of hydrocarbons. The following are some of the hydrocarbons that have been synthesized by this method. How would you prepare these hydrocarbons from the starting material indicated in each case?

(a)

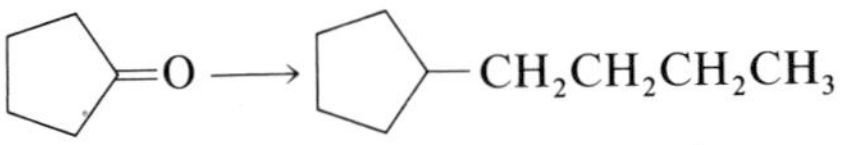

(b) $CH_3\overset{\overset{\displaystyle CH_3}{|}}{C}HCH_2Cl \longrightarrow CH_3\overset{\overset{\displaystyle CH_3}{|}}{C}HCH_2\overset{\overset{\displaystyle CH_3}{|}}{C}HCH_2CH_3$

(c)

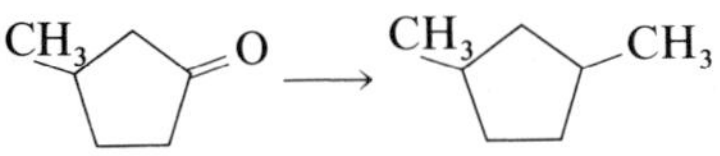

(d) $CH_3CH_2CH_2Cl \longrightarrow CH_3\overset{\overset{\displaystyle CH_3}{|}}{\underset{\underset{\displaystyle CH_3}{|}}{C}}HCHCH_2CH_2CH_3$

(e) $CH_3\overset{\overset{\displaystyle CH_3}{|}}{C}H-\overset{\overset{\displaystyle O}{\|}}{C}H \longrightarrow CH_3\overset{\overset{\displaystyle CH_3}{|}}{C}HCH_2CH_2CH_2CH_2CH_3$

9.54   The following sequences of reactions have been carried out in a synthesis of pentalenene, a natural product related to an antibacterial and antifungal agent. Supply the reagents you would use for the transformations shown.

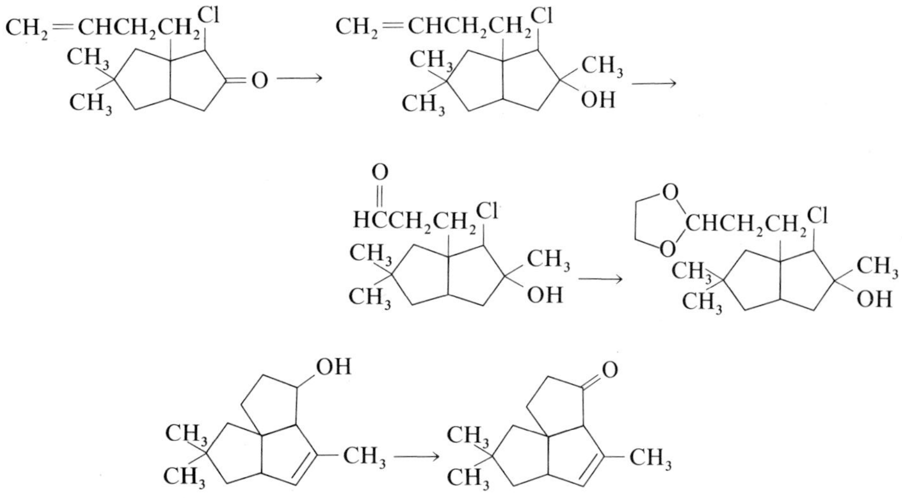

9.55 The following transformation is observed. Write detailed mechanisms that account for the products observed.

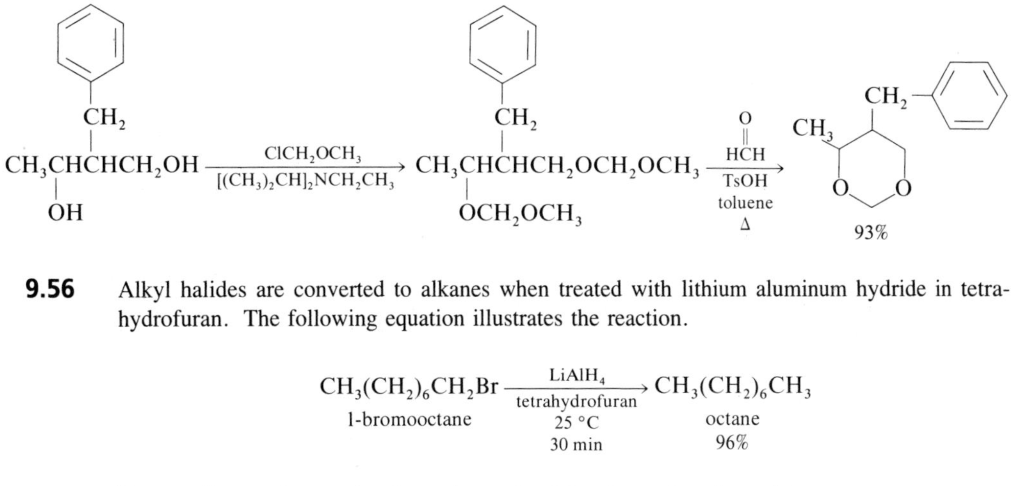

9.56 Alkyl halides are converted to alkanes when treated with lithium aluminum hydride in tetrahydrofuran. The following equation illustrates the reaction.

$$CH_3(CH_2)_6CH_2Br \xrightarrow[\substack{\text{tetrahydrofuran} \\ 25\ °C \\ 30\ \text{min}}]{\text{LiAlH}_4} CH_3(CH_2)_6CH_3$$

1-bromooctane   octane
                                                  96%

Rate studies on the reaction have shown the following order of reactivity.

$$CH_3CH_2CH_2CH_2Br > CH_3\underset{\underset{H}{|}}{\overset{\overset{CH_3}{|}}{C}}CH_2Br \ggg CH_3\underset{\underset{CH_3}{|}}{\overset{\overset{CH_3}{|}}{C}}CH_2Br$$

$$CH_3CH_2CH_2CH_2I > CH_3CH_2CH_2CH_2Br > CH_3CH_2CH_2CH_2Cl$$

Propose a mechanism for the reaction and discuss how the rate data support your mechanism.

9.57 The following sequences of reactions were carried out in order to synthesize isomeric deuterated methoxycyclohexanes for a study of the stereochemistry of the oxymercuration reaction (Section 7.4). Give structural formulas including conformation and stereochemistry for each of the species represented below by the capital letters. (Hint: A review of Sections 7.3, 8.4D, and 8.5C will be helpful.)

(a) ⟨triangle-cyclohexene⟩ $\xrightarrow[\text{tetrahydrofuran}]{\text{BD}_3}$ A $\xrightarrow[\text{H}_2\text{O}]{\text{H}_2\text{O}_2,\ \text{NaOH}}$ B $\xrightarrow[\text{pyridine}]{\text{TsCl}}$ C $\xrightarrow[\text{methanol}]{\text{CH}_3\text{ONa}}$ D

(b) ⟨triangle-cyclohexene⟩ $\xrightarrow[\text{dichloromethane}]{\text{Cl–C}_6\text{H}_4\text{COOH}}$ E $\xrightarrow[\text{diethyl ether}]{\text{LiAlD}_4}$ $\xrightarrow{\text{H}_2\text{O}}$ F $\xrightarrow[\text{pyridine}]{\text{TsCl}}$ G $\xrightarrow[\text{methanol}]{\text{CH}_3\text{ONa}}$ H

(Hint: LiAlD$_4$ is a source of which nucleophilic species?)

9.58   The infrared spectra presented in Figure 9.7 were taken of Compounds F to I, each of which has two major functional groups. Use the spectra to identify the functional groups that are present.

Compound F

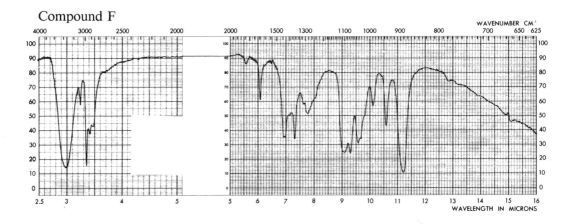

Compound G

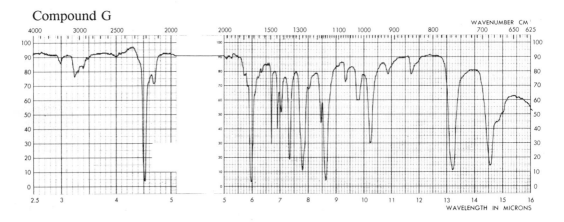

Figure 9.7   Spectra for Problem 9.58. (From *The Aldrich Library of Infrared Spectra*)

Compound H

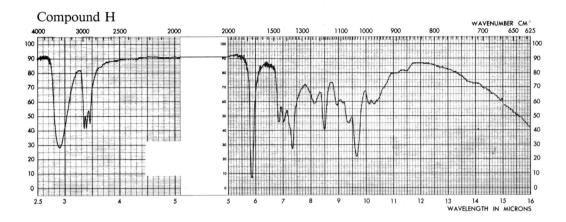

Compound I

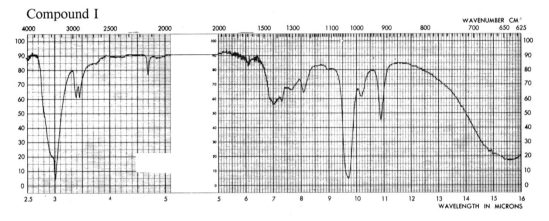

Figure 9.7 (*Continued*)

9.59 Compound J, $C_8H_{12}O$, is prepared by the reaction of Compound K, $C_6H_{10}O$, with an organo-metallic reagent. Assign structures to Compounds J and K using the infrared spectra given in Figure 9.8 and write an equation for the conversion of the one to the other.

Compound J, $C_8H_{12}O$

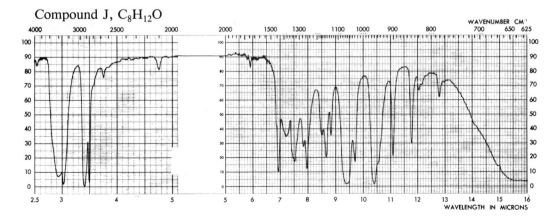

Figure 9.8 Spectra for Problem 9.59. (From *The Aldrich Library of Infrared Spectra*)

Compound K, $C_6H_{10}O$

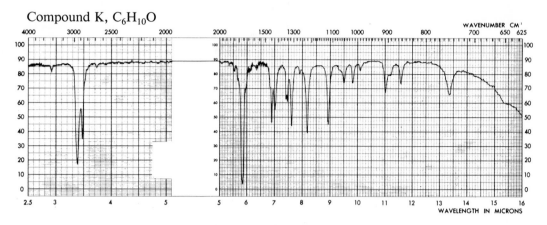

Figure 9.8 (*Continued*)

9.60 Compound L, $C_9H_{10}O$, is prepared from Compound M, $C_9H_{12}O$. The infrared spectra for the compounds appear in Figure 9.9. Assign structures to Compounds L and M and write an equation showing how you would carry out the conversion.

Compound L, $C_9H_{10}O$

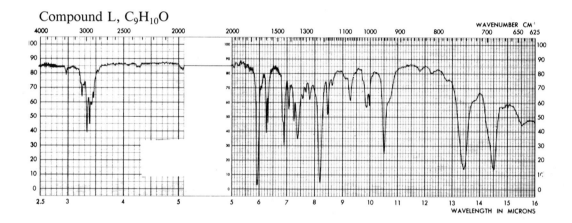

Compound M, $C_9H_{12}O$

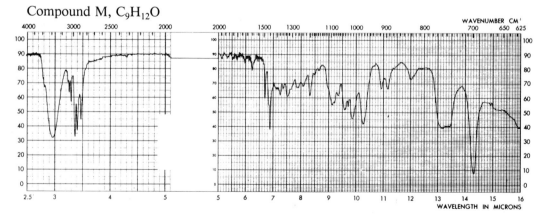

Figure 9.9 Spectra for Problem 9.60. (From *The Aldrich Library of Infrared Spectra*)

Carboxylic Acids and Their Derivatives

10

Properties of the Functional Groups in Carboxylic Acids and Their Derivatives

A. The Functional Group in Carboxylic Acids. Acidity

Carboxylic acids *are organic compounds that contain the* **carboxyl group**

$$\overset{\displaystyle O}{\overset{\displaystyle \|}{-C}}-OH$$

a functional group in which a hydroxyl group is directly bonded to a carbonyl group. The carbonyl group affects the properties of the hydroxyl group so profoundly that the hydroxyl group in a carboxylic acid is unlike that in an alcohol. For example, the hydrogen atom on the carboxyl group is much more easily lost as a proton than the hydrogen atom on the hydroxyl group of an alcohol is. Loss of a proton from the carboxyl group gives a carboxylate anion, which is stabilized by resonance. The anion of an alcohol has no such stabilization (Section 4.7A). The strength of a particular carboxylic acid varies depending on the other groups that are substituted on the hydrocarbon portion of the molecule (Tables 4.1, 4.2, 4.3). Inductive and field effects are used to explain the experimental observation of such differences in acidity (Section 4.7B).

In some carboxylic acids, the carboxyl group is bonded directly to an aromatic ring. Benzoic acid, for example, has a pK_a of 4.2 and is a slightly stronger acid than acetic acid, pK_a 4.8. In benzoic acid, the carboyxl group is bonded to an sp^2-hybridized carbon atom, which is more electronegative than an sp^3-hybridized carbon atom (Section 2.7B).

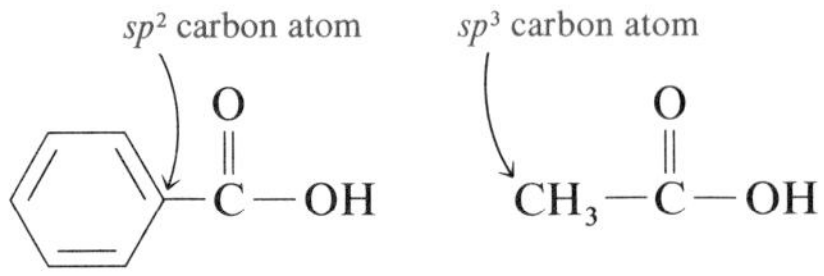

benzoic acid
pK_a 4.2
stronger acid than acetic acid, carboxyl group bonded to a more electronegative carbon atom

acetic acid
pK_a 4.8

The greater electronegativity of an sp^2-hybridized carbon atom is used to explain the apparent electron-withdrawing inductive effect that the phenyl group demonstrates in stabilizing the carboxylate anion.

Just as substitution on acetic acid (Table 4.1) changes its acidity, so does substitution on the aromatic ring in benzoic acid. For example, the fluorobenzoic acids are all more acidic than benzoic acid itself.

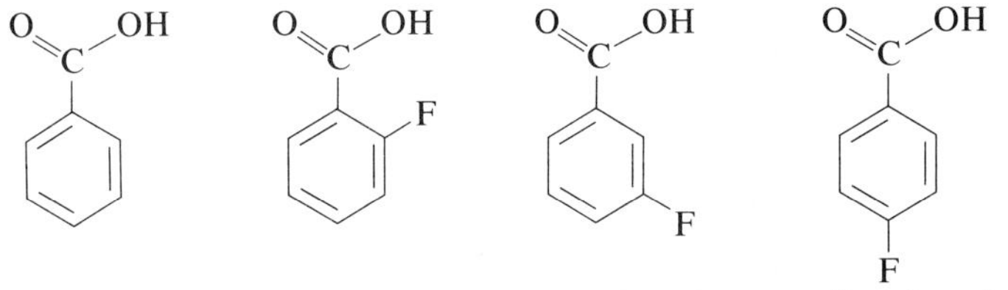

benzoic acid

pK_a 4.2

o-fluorobenzoic acid

pK_a 3.3

m-fluorobenzoic acid

pK_a 3.9

p-fluorobenzoic acid

pK_a 4.1

o-Fluorobenzoic acid, in which the fluorine atom is closest to the carboxyl group, is the strongest acid in the series, while *p*-fluorobenzoic acid, in which the fluorine atom is removed from the carboxyl group, is only slightly more acidic than benzoic acid. The electronegative fluorine atom operates through bonds and through space to withdraw electronic charge from the vicinity of the carboxyl group, thus stabilizing the conjugate base.

For some substituents, the effect varies with the position of the substituent. *m*-Methoxybenzoic acid is a stronger acid than benzoic acid while *p*-methoxybenzoic acid is weaker.

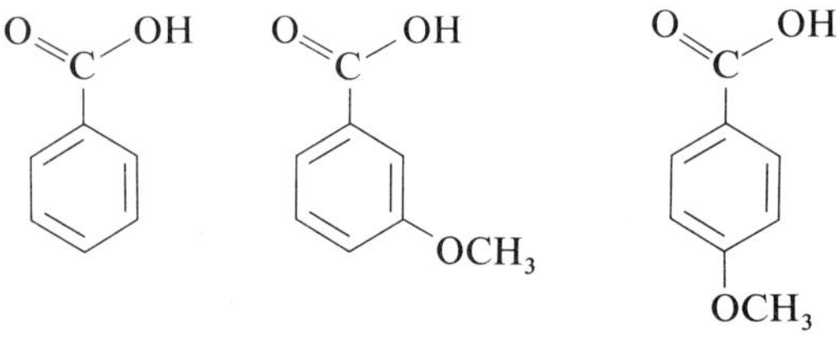

benzoic acid
pK_a 4.2

m-methoxybenzoic acid
pK_a 4.1

p-methoxybenzoic acid
pK_a 4.5

We can account for the acidity of *m*-methoxybenzoic acid by pointing to the electronegativity of the oxygen atom and postulating a negative inductive effect that decreases electron density at the carboxyl group. In *p*-methoxybenzoic acid, an effect seems to be operating in the opposite direction. The lower acidity of *p*-methoxybenzoic acid in comparison with benzoic acid indicates increased electron density at the carboxyl group.

Chemists attribute the increased basicity of the *p*-methoxybenzoate anion to resonance.

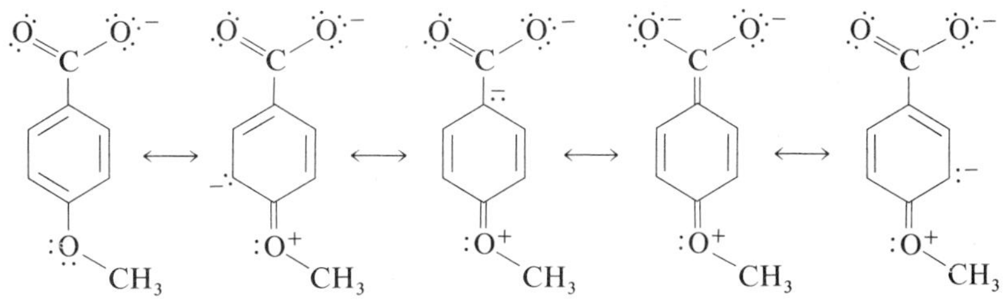

*resonance contributors for the anion
of p-methoxybenzoic acid*

The oxygen atom in the methoxyl group already has eight electrons around it. Resonance contributors that draw still more electrons towards the oxygen are not possible. The only way that the methoxyl group can participate in resonance is by donating electrons towards the aromatic ring. In the anion of *p*-methoxybenzoic acid, the donation of electrons results in an increased negative charge density on the carbon atom bearing the carboxylate group and on the oxygen atoms of the anion. The greater basicity of the anion of *p*-methoxybenzoic acid results in a higher pK_a for the acid.

None of the resonance contributors for the anion of *m*-methoxybenzoic acid has negative charge on the carbon atom to which the carboxylate group is bound.

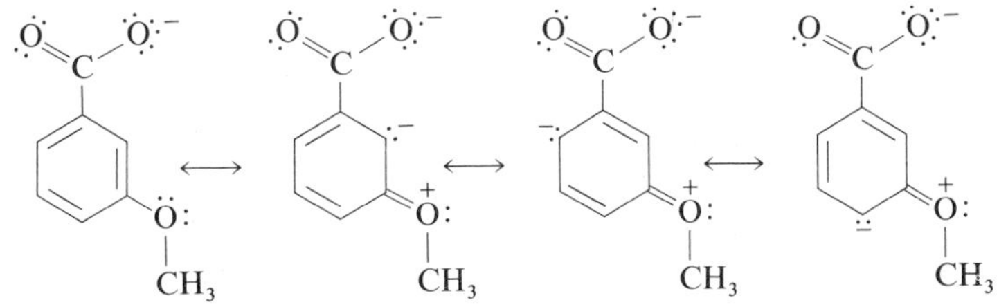

*resonance contributors for the anion
of m-methoxybenzoic acid*

Thus, a methoxyl group in the meta position does not increase electron density at the carboxylate anion as a methoxyl group in the para position does. Note that a resonance effect is generally larger and more important than an inductive effect.

p-Nitrobenzoic acid is a stronger acid than *m*-nitrobenzoic acid. This observation differs from the case of the fluorobenzoic acids, in which acid strength decreases as the electron-withdrawing group is moved away from the carboxyl group (p. 436). Resonance contributors that have a positive charge at the carbon atom bearing the carboxylate anion can be written for the ortho and para isomers, whereas the nitro group in the meta position exerts primarily an inductive effect.

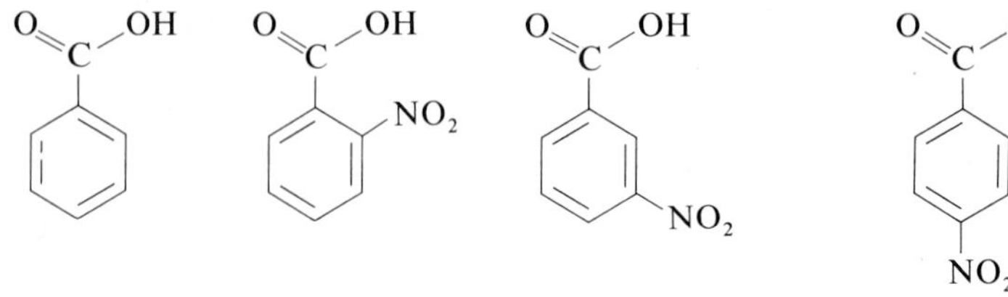

benzoic acid	o-nitrobenzoic acid	m-nitrobenzoic acid	p-nitrobenzoic acid
pK_a 4.2	pK_a 2.2	pK_a 3.5	pK_a 3.4
	strong inductive effect and resonance effect	*inductive effect*	*weak inductive effect and strong resonance effect*

The carboxylate anion in each case is stabilized by combinations of electron-withdrawing effects. All of the nitrobenzoic acids are stronger than benzoic acid. In o-nitrobenzoic acid, the strongest acid, the inductive effect is strong and a resonance effect also is present. In p-nitrobenzoic acid, a resonance effect is seen, but we would expect the inductive effect to be weak.

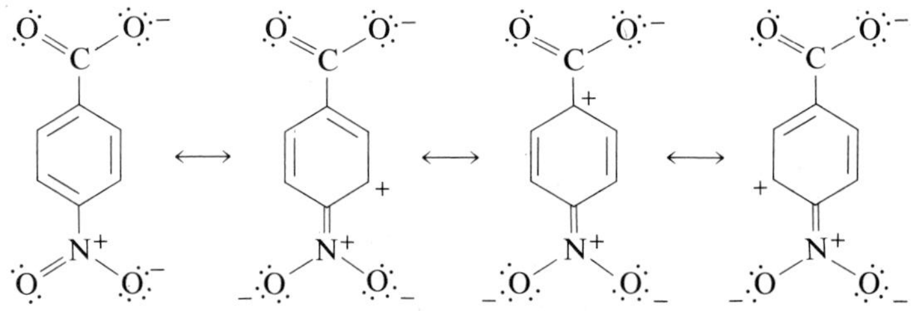

this resonance contributor puts a positive charge on the carbon atom to which the carboxylate group is bound

The fact that m-nitrobenzoic acid is a weaker acid than p-nitrobenzoic acid reconfirms that *the resonance effect is more important than the inductive effect when it operates.*

PROBLEM 10.1   Draw resonance contributors for the anions of o- and m-nitrobenzoic acids. Point out the contributor that is important in explaining the acidity of the ortho compound, and prove to yourself that no such effect is present in the meta isomer.

B. The Electronic Character of the Carbonyl Group in Carboxylic Acids and Acid Derivatives

In the previous section, we explained the influence of the carbonyl group on the hydroxyl group in a carboxylic acid. It is also true that the hydroxyl group changes the properties of the carbonyl portion of the carboxylic acid function. A comparison of the resonance contributors possible for acetone and acetic acid will indicate why this is so.

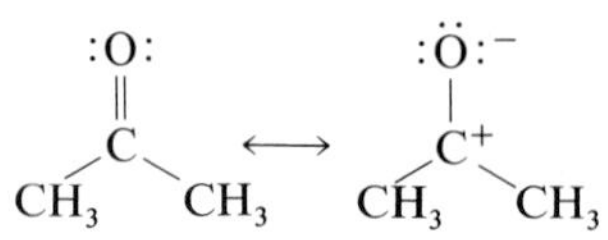

acetone

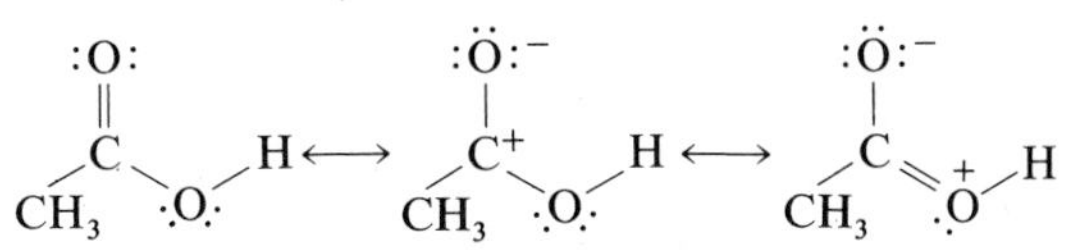

acetic acid

In acetone, the carbonyl group is strongly polarized, with the oxygen atom being the negative end of the dipole and the carbon atom the positive end. The carbon atom of the carbonyl group in an aldehyde or ketone is a strongly electrophilic center that reacts with ease with a variety of nucleophiles, such as alcohols (Section 9.7), amine derivatives (Section 9.6), and organometallic reagents (Section 9.4B, 9.4C).

While the carbon atom of the carbonyl group in acetic acid is also electrophilic, its electrophilicity is modified by the presence of nonbonding electrons on the oxygen atom of the hydroxyl group. Donation of those electrons towards the carbonyl group transfers some of the positive character of the carbon atom to that oxygen atom. The polarization of the carbonyl group in a carboxylic acid is not so strong as it is in an aldehyde or a ketone. Many of the reagents that react easily with the carbonyl group of aldehydes or ketones do so more slowly, or need more powerful catalysts when attacking the carbonyl group of a carboxylic acid or some of its derivatives. This difference in reactivity is considered more fully in Section 10.4.

The main types of carboxylic acid derivatives are shown below, using acetic acid as an example.

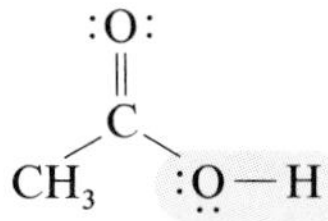

acetic acid

a carboxylic acid

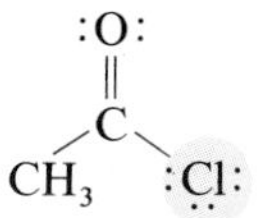

acetyl chloride

an acid chloride

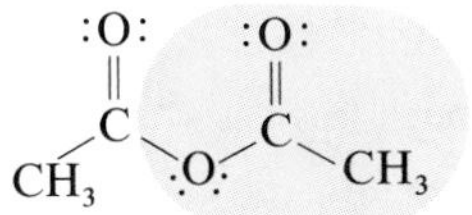

acetic anhydride

an acid anhydride

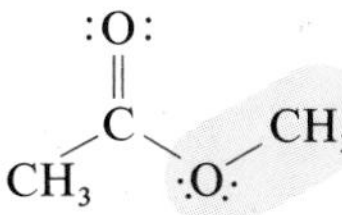

methyl acetate

an ester

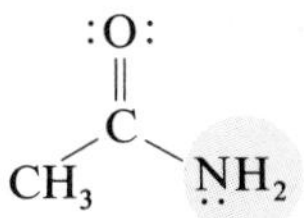

acetamide

an amide

The acid derivatives may be considered formally to have been created from the parent carboxylic acid by the replacement of the hydroxyl portion of the carboxyl group by another atom or group. Thus *in an* **acid chloride** *the hydroxyl group of a carboxylic acid is replaced by a chlorine atom. An* **acid anhydride** *has the anion corresponding to a carboxylic acid in place of the original hydroxyl group. In* **esters,** *an alkoxyl group replaces the hydroxyl group, while in* **amides** *an amino group is the replacement.* In each derivative, the atom bonded directly to the carbonyl group has at least one pair of nonbonding electrons on it. Each derivative can be converted back to the carboxylic acid by reaction with water (hydrolysis reactions are explored in Sections 10.4, 10.7D, 10.8B).

A close inspection of the structures of acetic acid and its derivatives reveals another important fact in their chemistry. One of the groups bonded to the carbonyl group (indicated by the area shaded in color in the structural formulas above) is either a good leaving group, or may be converted to a good leaving group by protonation. For example, chloride ion from acetyl chloride and acetate anion from acetic anhydride are good leaving groups. The hydroxyl group of the acid, the alkoxyl group of the ester, and the amino group of the amide are converted into good leaving groups on protonation (Section 6.4B). Acid derivatives differ in this way from aldehydes and ketones, which do not have good leaving groups bonded to the carbonyl group, nor the possibility of creating good leaving groups by protonation. Thus, many of the chemical reactions of acid derivatives differ from those of aldehydes and ketones. In the remainder of the chapter, we will explore in detail how the structures of acids and their derivatives determine their chemistry.

PROBLEM 10.2

(a)   Write structural formulas for propanoic acid, $CH_3CH_2CO_2H$, and its acid chloride, anhydride, ethyl ester, and amide.

(b)   Write equations for the reactions that you would expect between propanoic acid and concentrated sulfuric acid. Repeat the process for ethyl propanoate and propanamide. (Reviewing Section 4.3 may be helpful.)

(c)   Encircle any good leaving groups that you see in the structural formulas you have written in (a) and (b).

PROBLEM 10.3   Write resonance contributors for propanoic acid, its acid chloride, anhydride, ethyl ester, and amide, showing in each case how the polarity of the carbonyl group is affected by the presence of an adjacent atom with a pair of nonbonding electrons.

PROBLEM 10.4   Propanamide is much less basic than propylamine.

$$\overset{\displaystyle O}{\overset{\displaystyle \|}{CH_3CH_2CNH_2}} \qquad\qquad CH_3CH_2CH_2NH_2$$

How would you rationalize this observation in light of the resonance contributors that you wrote for the amide in Problem 10.3? (You may want to review the factors affecting basicity, covered in Sections 4.3 and 4.5.)

10.2

Nomenclature

A. Carboxylic Acids

The systematic name of an aliphatic carboxylic acid is derived from the name of the hydrocarbon containing the same number of carbon atoms in the chain with the ending **-oic acid** replacing the **-e** at the end of the name of the hydrocarbon. The carboxyl function is always assumed to be the first carbon atom of the chain. The presence of other substituents is indicated by assigning a name and a position number to each one.

The first two acids in the series, formic acid (from *formica,* Latin for ant) and acetic acid (from *acetum,* Latin for vinegar) are usually known by their common names.

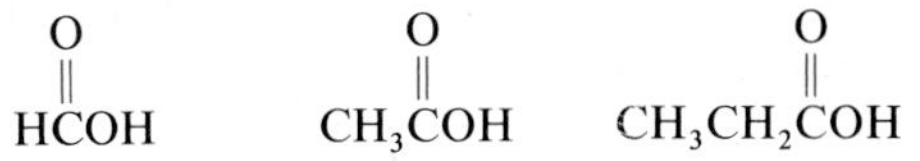

$$\underset{\substack{\text{methanoic acid}\\\text{formic acid}}}{\overset{\overset{\displaystyle O}{\|}}{HCOH}} \qquad \underset{\substack{\text{ethanoic acid}\\\text{acetic acid}}}{\overset{\overset{\displaystyle O}{\|}}{CH_3COH}} \qquad \underset{\substack{\text{propanoic acid}\\\text{propionic acid}}}{\overset{\overset{\displaystyle O}{\|}}{CH_3CH_2COH}}$$

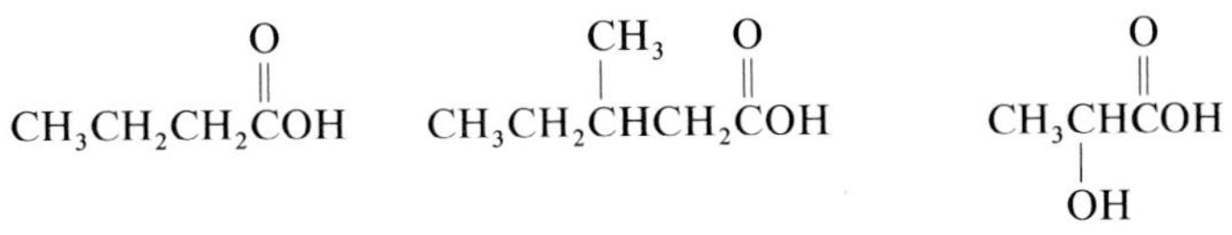

$$\underset{\substack{\text{butanoic acid}\\\text{butyric acid}}}{\overset{\overset{\displaystyle O}{\|}}{CH_3CH_2CH_2COH}} \qquad \underset{\text{3-methylpentanoic acid}}{CH_3CH_2\overset{\overset{\displaystyle CH_3}{|}}{CH}CH_2\overset{\overset{\displaystyle O}{\|}}{C}OH} \qquad \underset{\substack{\text{2-hydroxypropanoic acid}\\\text{lactic acid}}}{CH_3\overset{\overset{\displaystyle O}{\|}}{\underset{\underset{\displaystyle OH}{|}}{C}H}COH}$$

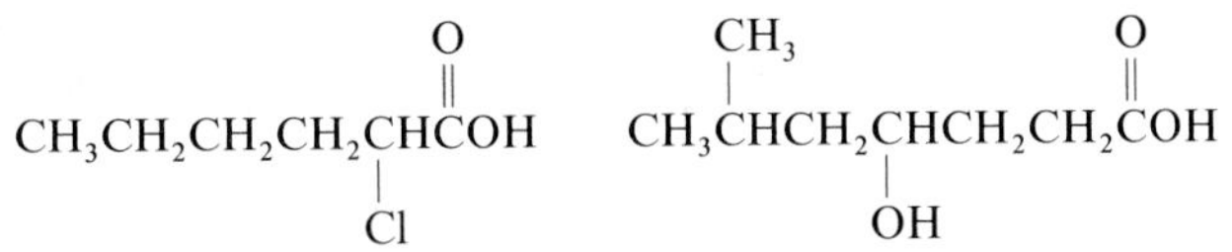

$$\underset{\text{2-chlorohexanoic acid}}{CH_3CH_2CH_2CH_2\overset{\overset{\displaystyle O}{\|}}{\underset{\underset{\displaystyle Cl}{|}}{C}H}COH} \qquad \underset{\text{4-hydroxy-6-methylheptanoic acid}}{CH_3\overset{\overset{\displaystyle CH_3}{|}}{C}HCH_2\overset{}{\underset{\underset{\displaystyle OH}{|}}{C}H}CH_2CH_2\overset{\overset{\displaystyle O}{\|}}{C}OH}$$

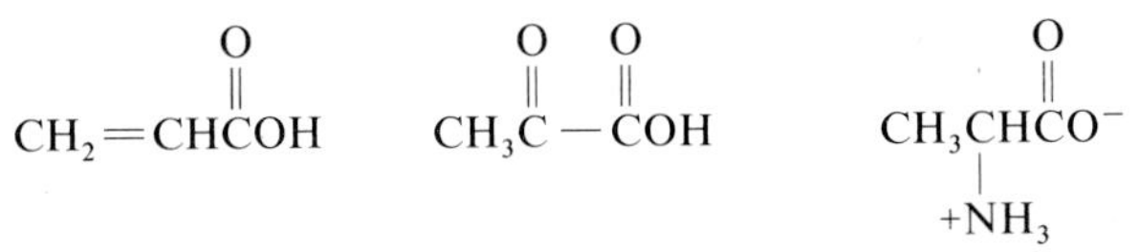

$$\underset{\substack{\text{propenoic acid}\\\text{acrylic acid}}}{\overset{\overset{\displaystyle O}{\|}}{CH_2=CHCOH}} \qquad \underset{\substack{\text{2-oxopropanoic acid}\\\text{pyruvic acid}}}{\overset{\overset{\displaystyle O\quad O}{\| \quad\|}}{CH_3C-COH}} \qquad \underset{\substack{\text{2-aminopropanoic acid}\\\text{alanine}}}{CH_3\overset{\overset{\displaystyle O}{\|}}{\underset{\underset{\displaystyle +NH_3}{|}}{C}H}CO^-}$$

Among biologically important carboxylic acids, also usually known by their common names, are the hydroxyacid lactic acid, the α-ketoacid pyruvic acid, and the amino acid alanine, also shown above.

The presence of two carboxyl groups is indicated by using the ending **-dioic acid** after the full name (including the **-e**) of the hydrocarbon having the same number of carbon atoms in the chain. The dicarboxylic acids shown below are also best known by their common names.

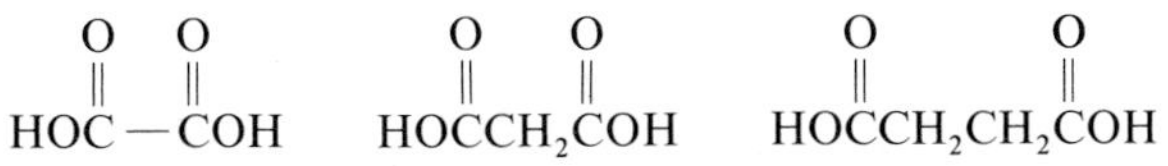

$$\underset{\substack{\text{ethanedioic acid}\\\text{oxalic acid}}}{\overset{\overset{\displaystyle O\quad O}{\| \quad\|}}{HOC-COH}} \qquad \underset{\substack{\text{propanedioic acid}\\\text{malonic acid}}}{\overset{\overset{\displaystyle O\quad O}{\| \quad\|}}{HOCCH_2COH}} \qquad \underset{\substack{\text{butanedioic acid}\\\text{succinic acid}}}{\overset{\overset{\displaystyle O\quad\quad O}{\| \quad\quad\|}}{HOCCH_2CH_2COH}}$$

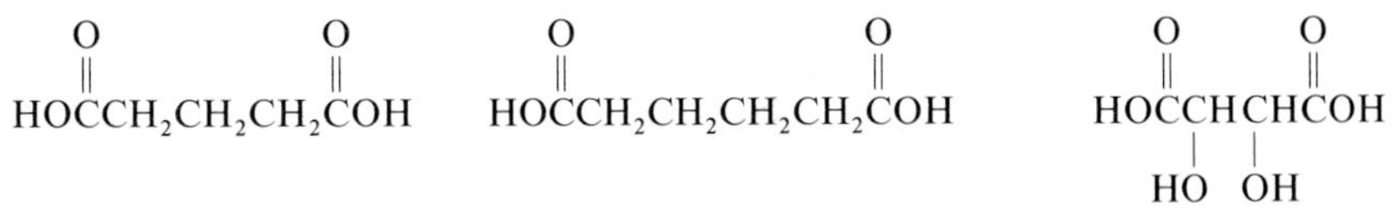

$$\underset{\substack{\text{pentanedioic acid}\\\text{glutaric acid}}}{\overset{\overset{\displaystyle O\quad\quad\quad O}{\| \quad\quad\quad\|}}{HOCCH_2CH_2CH_2COH}} \qquad \underset{\substack{\text{hexanedioic acid}\\\text{adipic acid}}}{\overset{\overset{\displaystyle O\quad\quad\quad\quad O}{\| \quad\quad\quad\quad\|}}{HOCCH_2CH_2CH_2CH_2COH}} \qquad \underset{\substack{\text{2,3-dihydroxybutanedioic acid}\\\text{tartaric acid}}}{\overset{\overset{\displaystyle O\quad\quad O}{\| \quad\quad\|}}{HOCC\underset{\underset{\displaystyle HO\;\;OH}{|\;\;\;|}}{H}CHCOH}}$$

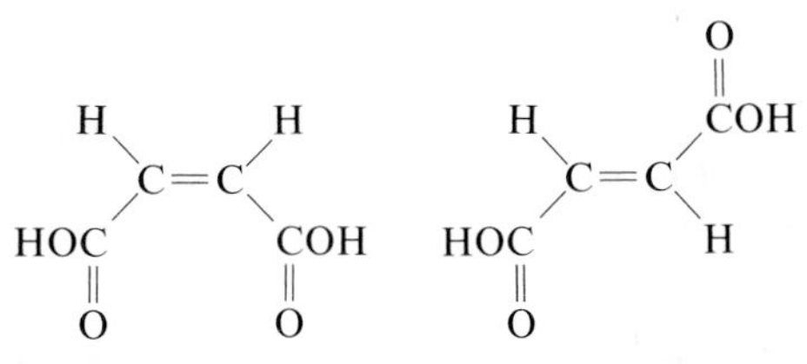

$$\underset{\substack{\text{(Z)-butenedioic acid}\\\text{maleic acid}}}{\underset{\underset{\displaystyle O}{\|}}{HOC}\diagdown \overset{\displaystyle H \quad\quad H}{\diagup} C=C \overset{\diagup}{\diagdown \underset{\underset{\displaystyle O}{\|}}{COH}}} \qquad \underset{\substack{\text{(E)-butenedioic acid}\\\text{fumaric acid}}}{\underset{\underset{\displaystyle O}{\|}}{HOC}\diagdown \overset{\displaystyle H \quad\quad \overset{\overset{\displaystyle O}{\|}}{COH}}{\diagup} C=C \overset{\diagup}{\diagdown H}}$$

Not all carboxylic acids have the carboxyl function as part of a chain. The aromatic carboxylic acids, for example, have the carboxyl group on an aromatic ring. Benzoic acid, which is the simplest unsubstituted aromatic acid, and some other examples are shown below.

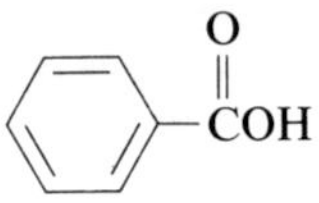

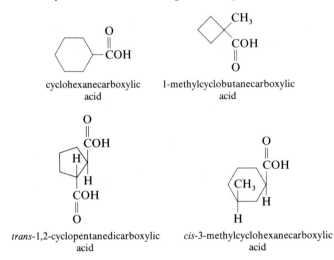

benzoic acid 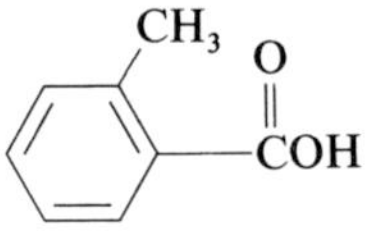   o-methylbenzoic acid
o-toluic acid

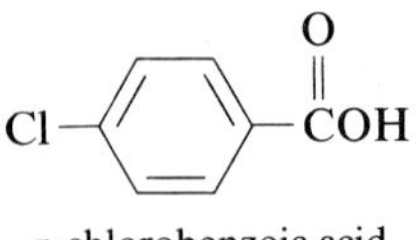

p-chlorobenzoic acid 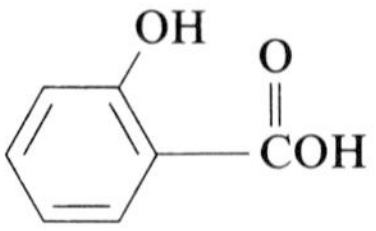   o-hydroxybenzoic acid
salicylic acid

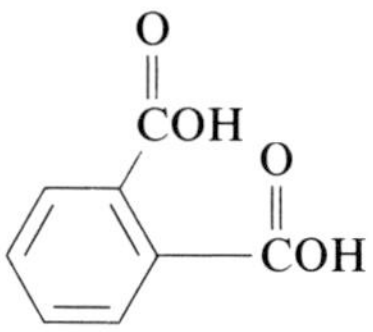

phthalic acid 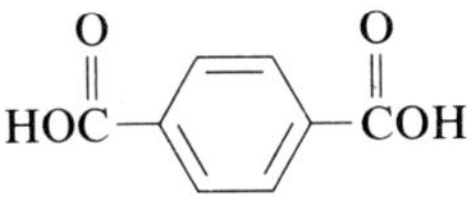   terephthalic acid

When the carboxyl group is on a cycloalkane, the suffix **carboxylic acid** is added to the name of the *hydrocarbon constituting the rest of* the molecule.

cyclohexanecarboxylic acid   1-methylcyclobutanecarboxylic acid

trans-1,2-cyclopentanedicarboxylic acid   *cis*-3-methylcyclohexanecarboxylic acid

---

PROBLEM 10.5   Some of the carboxylic acids shown in this section have more than one stereoisomer. Identify the compounds for which this is possible and draw all possible stereoisomers, naming each one correctly.

PROBLEM 10.6   Name the following compounds, indicating stereochemistry when known.

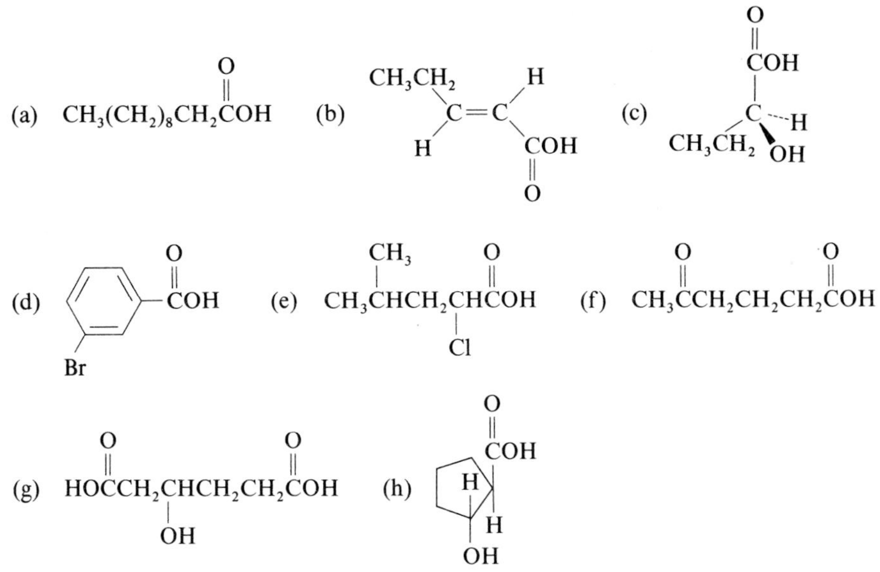

(a) $CH_3(CH_2)_8CH_2\overset{\overset{\displaystyle O}{\|}}{C}OH$

(b)
$$\begin{array}{c} CH_3CH_2 \qquad\qquad H \\ \diagdown \;\; / \\ C=C \\ / \qquad\qquad \diagdown \\ H \qquad\qquad \underset{\underset{\displaystyle O}{\|}}{C}OH \end{array}$$

(c)
$$\overset{\overset{\displaystyle O}{\|}}{C}OH$$
$$CH_3CH_2 \;\overset{|}{\underset{OH}{C}}\cdots H$$

(d)
$-\overset{\overset{\displaystyle O}{\|}}{C}OH$ with Br substituent

(e) $CH_3\overset{\overset{\displaystyle CH_3}{|}}{C}HCH_2\overset{\overset{\displaystyle O}{\|}}{C}\overset{\underset{\displaystyle Cl}{|}}{H}COH$

(f) $CH_3\overset{\overset{\displaystyle O}{\|}}{C}CH_2CH_2CH_2\overset{\overset{\displaystyle O}{\|}}{C}OH$

(g) $HO\overset{\overset{\displaystyle O}{\|}}{C}CH_2\overset{\overset{\displaystyle }{\underset{\displaystyle OH}{|}}}{C}HCH_2CH_2\overset{\overset{\displaystyle O}{\|}}{C}OH$

(h)
$$\overset{\overset{\displaystyle O}{\|}}{C}OH$$
cyclopentane ring with H, H and OH substituents

B. Acyl Groups, Acid Chlorides, and Anhydrides

The group obtained from a carboxylic acid by the removal of the hydroxyl portion is known as an **acyl group.** The name of a particular acyl group is created by changing the **-ic** ending of the name of the carboxylic acid to **-yl.** This applies to the common as well as the systematic names of acids. If, on the other hand, the **carboxylic acid** ending is used for a compound, this ending is changed to **carbonyl** for the corresponding acyl group. Some important acyl groups are shown below with their related acids.

$$\underset{\text{formic acid}}{HC\!-\!OH} \qquad \underset{\text{formyl group}}{HC\!-\!} \qquad \underset{\text{acetic acid}}{CH_3C\!-\!OH} \qquad \underset{\text{acetyl group}}{CH_3C\!-\!}$$
(each with O double bond above C)

$$\underset{\text{butanoic acid}}{CH_3CH_2CH_2C\!-\!OH} \qquad \underset{\text{butanoyl group}}{CH_3CH_2CH_2C\!-\!}$$
(each with O double bond above C)

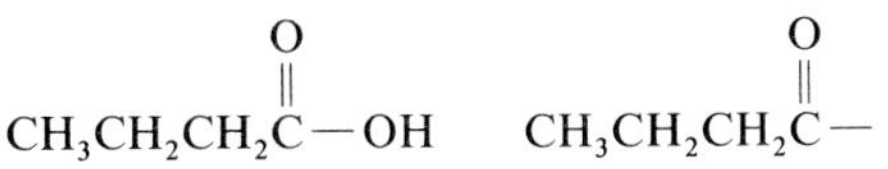

$$\underset{\text{benzoic acid}}{}\!-\!C\!-\!OH \qquad \underset{\text{benzoyl group}}{}\!-\!C\!-\!$$
(benzene ring with C=O)

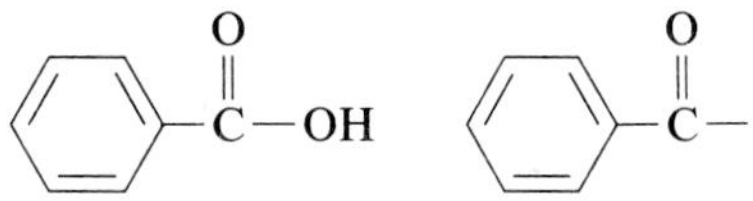

$$\underset{\substack{\text{cyclopentanecarboxylic}\\\text{acid}}}{}\!-\!C\!-\!OH \qquad \underset{\substack{\text{cyclopentanecarbonyl}\\\text{group}}}{}\!-\!C\!-\!$$
(cyclopentane ring with C=O)

Acid chlorides are named systematically as acyl chlorides.

$$\underset{\text{acetyl chloride}}{CH_3\overset{\displaystyle O}{\overset{\|}{C}}Cl}\qquad\qquad\underset{\text{pentanoyl chloride}}{CH_3CH_2CH_2CH_2\overset{\displaystyle O}{\overset{\|}{C}}Cl}$$

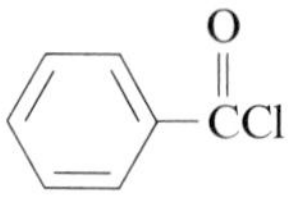

benzoyl chloride 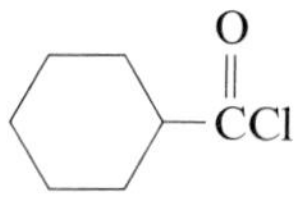   cyclohexanecarbonyl chloride

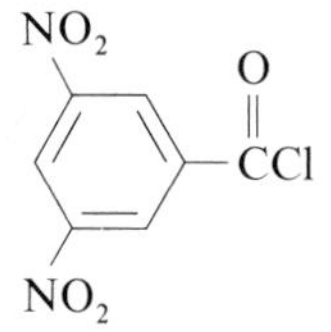

3,5-dinitrobenzoyl chloride

$$\underset{\substack{\text{butanedioyl dichloride}\\\text{succinyl dichloride}}}{Cl\overset{\displaystyle O}{\overset{\|}{C}}CH_2CH_2\overset{\displaystyle O}{\overset{\|}{C}}Cl}$$

Of the acid anhydrides, the anhydride of acetic acid and some cyclic anhydrides formed from dicarboxylic acids are important. They are named by substituting **anhydride** for **acid** in the name of the acid from which they are derived. Cyclic anhydrides containing five- or six-membered rings are stable and easily formed. Of the aromatic dicarboxylic acids, only the ones with the carboxylic groups ortho to each other form cyclic anhydrides.

$$\underset{\text{acetic anhydride}}{CH_3\overset{\displaystyle O}{\overset{\|}{C}}O\overset{\displaystyle O}{\overset{\|}{C}}CH_3}$$

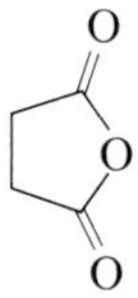

succinic anhydride

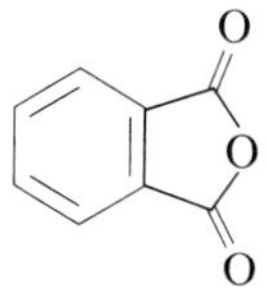

phthalic anhydride 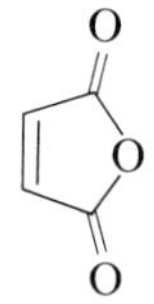   maleic anhydride

C. Salts and Esters

Esters of carboxylic acids are named in the same way as their salts. The name of the cation (in the case of a salt) or of the organic group attached to the oxygen of the carboxyl group (in the case of an ester) precedes the name of the acid, in which the **-ic** ending is converted to **-ate.**

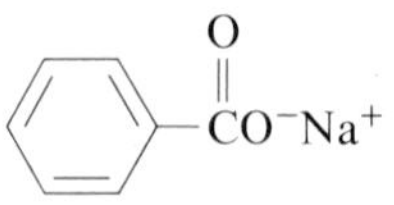

sodium benzoate $(CH_3CH_2\overset{\displaystyle O}{\overset{\|}{C}}O^-)_2Ca^{2+}$

calcium propanoate
calcium propionate

both salts used to retard spoilage in foods

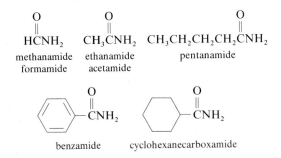

$$O_2N\!-\!\langle\ \rangle\!-\!\overset{\overset{\displaystyle O}{\|}}{C}OCH_2CH_3 \qquad CH_3O\overset{\overset{\displaystyle O}{\|}}{C}CH_2CH_2\overset{\overset{\displaystyle O}{\|}}{C}OCH_3$$

ethyl *p*-nitrobenzoate   dimethyl succinate

$$H_2N\!-\!\langle\ \rangle\!-\!\overset{\overset{\displaystyle O}{\|}}{C}OCH_2CH_3 \qquad \langle\ \rangle\!-\!\overset{\overset{\displaystyle O}{\|}}{C}O\overset{\overset{\displaystyle CH_3}{|}}{C}HCH_3$$

ethyl *p*-aminobenzoate   isopropyl cyclohexanecarboxylate
a local anesthetic

PROBLEM 10.7 Name the following compounds, indicating stereochemistry when known.

(a) $CH_3CH_2CH_2CH_2CH_2\overset{\overset{\displaystyle O}{\|}}{C}Cl$ (b) $CH_3CH_2CH_2\overset{\overset{\displaystyle O}{\|}}{C}O\overset{\overset{\displaystyle O}{\|}}{C}CH_2CH_2CH_3$

(c) $\underset{H}{\overset{CH_3}{}}C\!=\!C\underset{H}{\overset{CH_2CH_2\overset{\overset{\displaystyle O}{\|}}{C}OCH_3}{}}$ (d) $Cl\!-\!\langle\ \rangle\!-\!\overset{\overset{\displaystyle O}{\|}}{C}OCH_2\overset{\overset{\displaystyle CH_3}{|}}{C}HCH_3$

(e) $\langle\ \rangle\!-\!\overset{\overset{\displaystyle O}{\|}}{C}Cl$ (f) $Cl\!-\!\langle\ \rangle\!-\!\overset{\overset{\displaystyle O}{\|}}{C}O\overset{\overset{\displaystyle O}{\|}}{C}\!-\!\langle\ \rangle\!-\!Cl$

D. Amides, Imides, Nitriles

The names of acid amides are derived by replacing **-oic acid** (or **-ic acid** for common names) by **-amide,** or **carboxylic acid** by **-carboxamide.**

$$HC\overset{\overset{\displaystyle O}{\|}}{}NH_2 \qquad CH_3\overset{\overset{\displaystyle O}{\|}}{C}NH_2 \qquad CH_3CH_2CH_2CH_2\overset{\overset{\displaystyle O}{\|}}{C}NH_2$$

methanamide   ethanamide   pentanamide
formamide   acetamide

$$\langle\ \rangle\!-\!\overset{\overset{\displaystyle O}{\|}}{C}NH_2 \qquad \langle\ \rangle\!-\!\overset{\overset{\displaystyle O}{\|}}{C}NH_2$$

benzamide   cyclohexanecarboxamide

If the nitrogen atom of the amide is substituted by alkyl groups, the name of the amide is prefixed by the capital letter *N*-, to indicate substitution on nitrogen, followed by the name of the alkyl group(s).

$$HC\!-\!N\underset{\overset{\overset{\displaystyle O}{\|}}{}}{}\overset{CH_3}{\underset{CH_3}{}} \qquad O_2N\!-\!\langle\ \rangle\!-\!\overset{\overset{\displaystyle O}{\|}}{C}\!-\!N\overset{CH_3}{\underset{CH_2CH_3}{}}$$

N,N-dimethylformamide   *N*-ethyl-*N*-methyl-*p*-nitrobenzamide
abbreviated DMF

If the substituent on the nitrogen atom of an amide is a phenyl group, the ending for the name of the carboxylic acid is changed to **-anilide.**

acetanilide   benzanilide

Some dicarboxylic acids form cyclic amides in which two acyl groups are bonded to the nitrogen atom. The suffix **-imide** is given to such compounds.

phthalimide   succinimide   *N*-bromosuccinimide
                                                    abbreviated NBS

Nitriles, compounds containing the cyano, —C≡N, group, are considered to be acid derivatives because they are hydrolyzed to form amides and carboxylic acids (Sections 10.5C, 10.8A). In the systematic nomenclature of these compounds, the suffix **-nitrile** is added to the name of the hydrocarbon containing the same number of carbon atoms, *including* the carbon atom of the cyano group.

$$\overset{5}{C}H_3\overset{4}{C}H_2\overset{3}{C}H_2\overset{2}{C}H_2\overset{1}{C}\equiv N \qquad \overset{4}{C}H_3\overset{3}{C}=\overset{2}{C}H\overset{1}{C}\equiv N$$

pentanenitrile   3-methyl-2-butenenitrile

The nitriles related to acetic acid and benzoic acid are called acetonitrile and benzonitrile.

acetonitrile   benzonitrile

When the cyano group is on a cycloalkane, the suffix **-carbonitrile** is used with the name of the hydrocarbon.

cyclohexanecarbonitrile   *trans*-2-methylcyclo-
                            propanecarbonitrile

If other functional groups are present, the —C≡N group is indicated by the prefix **cyano-** when it is treated as a substituent. Note that when the *-nitrile* nomenclature is used, the carbon atom bonded to the nitrogen is included in the count of carbon atoms that determines the name. When *cyano-* (or *-carbonitrile*) is used, that carbon atom is part of a substituent and is not included in the numbering of the rest of the chain (or ring).

$$\underset{\text{4-cyanobutanoic acid}}{HO\overset{\displaystyle O}{\overset{\|}{C}}CH_2CH_2CH_2C\equiv N}$$

$$\underset{\text{ethyl }p\text{-cyanobenzoate}}{N\equiv C\!-\!\!\bigcirc\!\!-\!\overset{\displaystyle O}{\overset{\|}{C}}OCH_2CH_3}$$

PROBLEM 10.8 Write structural formulas for the following compounds.

(a) (Z)-4-heptenoic acid   (b)   *trans*-2-methylcyclobutanecarboxylic acid
(c) (R)-2-bromopentanoic acid   (d)   octanedioic acid
(e) dimethyl propanedioate (dimethyl malonate)   (f)   benzoic anhydride
(g) *N,N*-dimethylbenzamide   (h)   pentanedioyl dichloride (glutaryl dichloride)
(i) disodium ethanedioate (sodium oxalate)   (j)   methyl 3-nitrobenzoate

PROBLEM 10.9 Name the following compounds, including the stereochemistry when known.

(a) $\bigcirc\!\!-\!\overset{\displaystyle O}{\overset{\|}{C}}OCH_2CH_2CH_3$ (b) $CH_3CH_2\overset{\displaystyle O}{\overset{\|}{C}}CH_2CH_2CH_2CH_2\overset{\displaystyle O}{\overset{\|}{C}}OCH_2CH_3$

(c) $CH_3(CH_2)_8\overset{\displaystyle O}{\overset{\|}{C}}OH$ (d)   Br–$\bigcirc$–$\overset{\displaystyle O}{\overset{\|}{C}}Cl$

(e) $CH_3(CH_2)_{16}\overset{\displaystyle O}{\overset{\|}{C}}O^-Na^+$ (*a soap*)   (f) $\bigcirc\!\!-\!CH_2\overset{\displaystyle O}{\overset{\|}{C}}OH$

(g) $CH_3\overset{\displaystyle O}{\overset{\|}{C}}NHCH_3$ (h) $CH_3\!-\!\bigcirc\!\!-\!\overset{\displaystyle O}{\overset{\|}{C}}NH_2$

(i) cyclopentane ring with $\overset{\displaystyle O}{\overset{\|}{C}}OH$, H, H, Br substituents   (j) $\underset{H}{\overset{CH_3}{>}}C=C\underset{\overset{\displaystyle O}{\overset{\|}{C}}OH}{\overset{H}{<}}$

(k) $CH_3CH_2\overset{\displaystyle O}{\overset{\|}{C}}CH_2\overset{\displaystyle O}{\overset{\|}{C}}OCH_2CH_3$ (l) $H\cdots\overset{\overset{\displaystyle CH_2\overset{\displaystyle O}{\overset{\|}{C}}OH}{|}}{\underset{OH}{C}}CH_3$

(m) $CH_3CH_2CH_2CH_2CH_2CH_2C\equiv N$ (n) $CH_3\overset{\overset{CH_3}{|}}{C}HCH_2\overset{\displaystyle O}{\overset{\|}{C}}O\overset{\displaystyle O}{\overset{\|}{C}}CH_2\overset{\overset{CH_3}{|}}{C}HCH_3$

(o) $CH_3CH_2CH_2\overset{\displaystyle O}{\overset{\|}{C}}NH\!-\!\bigcirc$

Physical Properties of Carboxylic Acids and Their Derivatives

A. Properties of Low Molecular Weight Acids and Acid Derivatives

Low molecular weight carboxylic acids have boiling points that are high in relation to their molecular weights, and are very soluble in water. Molecular weight determinations on carboxylic acids indicate that they exist as dimers even in the vapor state. All of these data suggest the participation of the carboxyl group both as a donor and an acceptor in extensive hydrogen bonding, illustrated below for acetic acid in the vapor state, in the liquid state, and in aqueous solution.

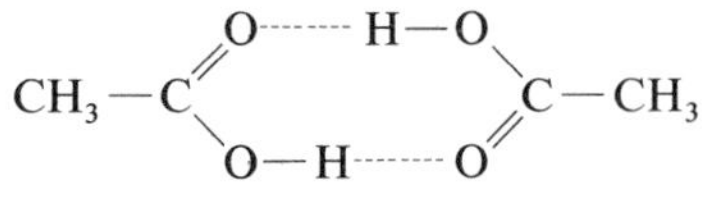

dimer of acetic acid
held together by hydrogen
bonding in the vapor state

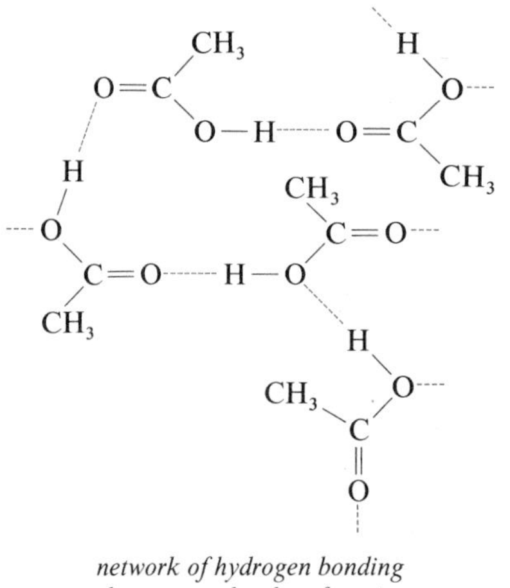

network of hydrogen bonding
between molecules of acetic
acid in the liquid state

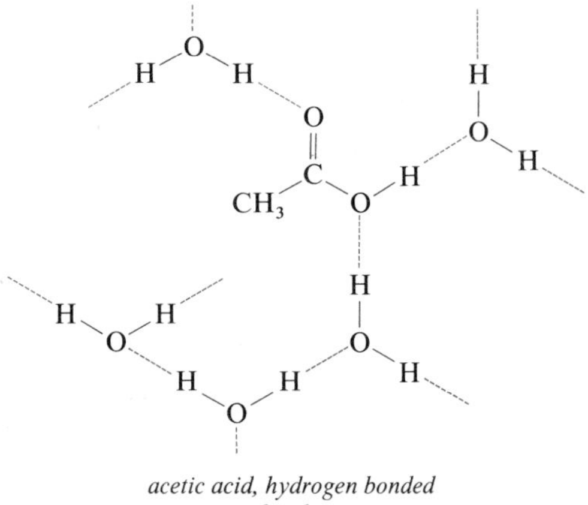

acetic acid, hydrogen bonded
to water molecules in aqueous
solution

The carboxylic acids containing a single functional group are liquids when they have fewer than ten carbon atoms in the chain. Acetic acid has a particularly high melting point, 16.7 °C, for such a low molecular weight compound, and is, in fact, known as **glacial acetic acid** in its pure state. It is a liquid at room temperature but freezes easily in an ice bath, a phenomenon of practical importance in the laboratory.

Dicarboxylic acids, from oxalic acid up, as well as the aromatic carboxylic acids, are all solids at room temperature.

$\overset{\text{O}}{\overset{\|}{\text{HCOH}}}$	$\overset{\text{O}}{\overset{\|}{\text{CH}_3\text{COH}}}$	$\overset{\text{O}}{\overset{\|}{\text{CH}_3\text{CH}_2\text{CH}_2\text{CH}_2\text{COH}}}$
formic acid	acetic acid	pentanoic acid
bp 100.5 °C	bp 118.2 °C	bp 186.4 °C
mp 8.4 °C	mp 16.7 °C	mp −34.5 °C
completely soluble in water	completely soluble in water	solubility 3.7 g in 100 g of water

$$\underset{\substack{\text{decanoic acid} \\ \text{bp 270.0 °C} \\ \text{mp 31.3 °C} \\ \text{solubility 0.015 g} \\ \text{in 100 g of water}}}{\text{CH}_3(\text{CH}_2)_8\overset{\displaystyle O}{\overset{\displaystyle \|}{\text{C}}}\text{OH}} \qquad \underset{\substack{\text{oxalic acid} \\ \\ \text{mp 187 °C} \\ \text{solubility 9.0 g} \\ \text{in 100 g of water}}}{\text{HO}\overset{\displaystyle O}{\overset{\displaystyle \|}{\text{C}}}-\overset{\displaystyle O}{\overset{\displaystyle \|}{\text{C}}}\text{OH}} \qquad \underset{\substack{\text{benzoic acid} \\ \\ \text{mp 122 °C} \\ \text{solubility 0.29 g} \\ \text{in 100 g of water}}}{\text{C}_6\text{H}_5-\overset{\displaystyle O}{\overset{\displaystyle \|}{\text{C}}}\text{OH}}$$

The low molecular weight monocarboxylic acids and dicarboxylic acids are soluble in water. Once the hydrocarbon portion of the molecule exceeds approximately five carbon atoms for each carboxyl group, solubility decreases. The high molecular weight carboxylic acids are almost insoluble in water.

The importance of hydrogen bonding to the physical properties and solubility in water of carboxylic acids becomes evident when we compare its properties with those of two acid derivatives, esters and amides.

$$\underset{\substack{\text{acetic acid} \\ \text{bp 118 °C} \\ \text{completely soluble} \\ \text{in water}}}{\text{CH}_3\overset{\displaystyle O}{\overset{\displaystyle \|}{\text{C}}}\text{OH}} \qquad \underset{\substack{\text{ethyl acetate} \\ \text{bp 77 °C} \\ \text{solubility 8.6 g} \\ \text{in 100 g of water}}}{\text{CH}_3\overset{\displaystyle O}{\overset{\displaystyle \|}{\text{C}}}\text{OCH}_2\text{CH}_3} \qquad \underset{\substack{\text{acetamide} \\ \text{mp 82 °C} \\ \text{solubility 97.5 g} \\ \text{in 100 g of water}}}{\text{CH}_3\overset{\displaystyle O}{\overset{\displaystyle \|}{\text{C}}}\text{NH}_2}$$

Acetic acid boils at 118 °C and is fully miscible with water, but its ethyl ester has a boiling point of 77 °C and a solubility of 8.6 g in 100 g of water. Ethyl acetate cannot hydrogen bond to itself in the liquid state. In water it can serve only as a hydrogen bond acceptor at its oxygen atoms. Hence, it has a low boiling point, and relatively low solubility in water. Acetamide, on the other hand, is a solid, mp 82 °C, with a very high solubility in water. The hydrogen atoms on the amide nitrogen atom participate strongly in hydrogen bonding, a fact of utmost importance to the structure of proteins (Section 17.5).

A carboxylic acid that is not soluble in water can be converted into a water-soluble salt. The salt, an ionic compound, is more polar than the original carboxylic acid, a covalent compound, and is, as a result, usually more soluble in the polar solvent water. Thus, adding base to an aqueous suspension of a carboxylic acid dissolves the acid. Addition of a strong mineral acid to a solution of an acid salt causes the acid to precipitate out of solution. These reactions are illustrated for benzoic acid.

$$\underset{\substack{\text{benzoic acid} \\ \textit{covalent compound} \\ \textit{low solubility in} \\ \textit{the polar solvent,} \\ \textit{water}}}{\text{C}_6\text{H}_5-\overset{\displaystyle O}{\overset{\displaystyle \|}{\text{C}}}\text{OH}} \quad \overset{\substack{\text{NaHCO}_3 \\ \text{H}_2\text{O}}}{\underset{\text{HCl}}{\rightleftharpoons}} \quad \underset{\substack{\text{sodium benzoate} \\ \textit{ionic compound} \\ \textit{soluble in the} \\ \textit{polar solvent,} \\ \textit{water}}}{\text{C}_6\text{H}_5-\overset{\displaystyle O}{\overset{\displaystyle \|}{\text{C}}}\text{O}^-\text{Na}^+} + \text{CO}_2\uparrow + \text{H}_2\text{O}$$

Many separations and purifications of organic compounds are based on these phenomena.

PROBLEM 10.10   For each of the following series of compounds, predict which one will have the highest and which the lowest solubility in water.

(a) $CH_3CH_2CH_2CH_2\overset{\overset{\displaystyle O}{\|}}{C}OH$, $CH_3CH_2\overset{\overset{\displaystyle O}{\|}}{C}OCH_2CH_3$, $CH_3CH_2CH_2CH_2\overset{\overset{\displaystyle O}{\|}}{C}O^-\,Na^+$

(b) $CH_3CH_2CH_2\overset{\overset{\displaystyle O}{\|}}{C}OH$, $CH_3CH_2CH_2CH_2OH$, $CH_3CH_2\overset{\overset{\displaystyle O}{\|}}{C}OCH_2CH_3$

(c) $CH_3CH_2CH_2\overset{\overset{\displaystyle O}{\|}}{C}OCH_2CH_3$, $CH_3CH_2CH_2\overset{\overset{\displaystyle O}{\|}}{C}NH_2$, $CH_3CH_2CH_2\overset{\overset{\displaystyle O}{\|}}{C}O(CH_2)_4CH_3$

B. Surface-Active Compounds. Soaps

While the salt of a carboxylic acid is usually more soluble in water than the parent acid itself, an interesting phenomenon arises when the hydrocarbon portion of the acid is very large in comparison with the carboxyl group. The ionic portion of the molecule interacts well with water and dissolves in it but is not able to bring the rest of the long molecule into solution. The hydrocarbon portions of adjacent molecules are more attracted to each other by van der Waals forces than they are to the polar water molecules. They are in fact hydrophobic, water-repelling, in their behavior. The salt then has two domains, a **hydrophilic** head, the ionic carboxylate group that is soluble in water, and a **hydrophobic** tail, the long hydrocarbon portion that is repelled by water molecules and attracted to other hydrocarbon residues.

The structure of such a compound allows for a particular orientation of its molecules at the surface of water; the heads are in water and the hydrocarbon tails stick up into the air. The concentration of molecules at the surface of the water has the effect of lowering the surface tension of water. Compounds that behave in this way are said to be **surface-active** or **surfactants.** All good surfactants have structures with a hydrophilic head and a long hydrophobic tail. At a certain critical concentration of the surfactant, the surface layer is broken up into smaller units, clusters of ions called micelles.

Micelles are particles in which the long hydrocarbon tails, repelled by the water molecules and attracted to each other, make up the interior of the particle while the negatively charged heads coat the surface and interact with the surrounding water molecules and positive ions. Figure 10.1 depicts a spherical micelle. Such micelles are formed when sodium octadecanoate (sodium stearate), which is a common ingredient of soap, is put into water.

$$CH_3CH_2CH_2CH_2CH_2CH_2CH_2CH_2CH_2CH_2CH_2CH_2CH_2CH_2CH_2CH_2CH_2\overset{\overset{\displaystyle O}{\|}}{C}O^-Na^+$$

$\longleftarrow$————————— hydrophobic tail —————————$\longrightarrow$ ionic head

sodium octadecanoate
sodium stearate

a soap

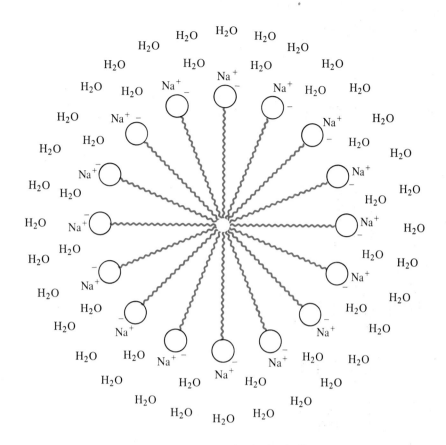

Figure 10.1   A spherical micelle.

That the particles in a soap solution, which is colloidal, are larger than individual molecules is evident from the opalescent appearance of the solution. The particles in a colloidal solution are large enough that light is scattered instead of being transmitted through the solution, as is the case with a true solution, which appears clear to the eye.

A soap emulsifies a grease because grease, in its chemical composition, resembles the hydrocarbon tails in the micelle. Rubbing a greasy spot with a soap solution causes the grease to break up into small enough particles to be trapped inside the micelles. The particles are held in solution by the hydrocarbon portion of the soap, but are kept in colloidal suspension by the interaction of the ionic surface of the micelle with the surrounding water. The grease is said to be **emulsified,** held in suspension in a medium in which it would not normally be soluble.

Soaps are usually the sodium salts of long-chain carboxylic acids. Ordinary soaps have disadvantages in hard water. Hard water has dissolved calcium and magnesium ions in it, so when soap is used in such water, the calcium and magnesium salts of the carboxylic acids in the soap precipitate out. This precipitate is the scum you see in hard water and causes the ring around the bathtub. A great variety of surfactants with more soluble calcium and magnesium salts have been developed. An example of a simple one prepared from nut oils is sodium dodecanyl sulfate (sodium lauryl sulfate), the sodium salt of the ester of 1-dodecanol with sulfuric acid. The property of surfactancy starts with about 12 carbon atoms in the hydrocarbon portion of the molecule.

$$CH_3CH_2CH_2CH_2CH_2CH_2CH_2CH_2CH_2CH_2CH_2CH_2O\overset{\displaystyle O}{\underset{\displaystyle O}{\overset{\displaystyle \|}{\underset{\displaystyle \|}{S}}}}O^-Na^+$$

←————————hydrophobic tail————————→

ionic head

sodium dodecanyl sulfate
sodium lauryl sulfate

*a surfactant used in commercial
detergents*

In sodium dodecanyl sulfate, sulfuric acid substitutes for the carboxylic acid group in providing the ionic portion of the molecule. Soaps, the salts of long-chain fatty acids, are only one category of a large variety of surfactant compounds.

Molecules that have hydrophilic heads and long-chain hydrophobic tails made up of the hydrocarbon portions of saturated and unsaturated fatty acids also play an important part in the structure of cell membranes, which provide the boundaries between the outer and inner worlds of the cell. More will be said about this in Section 16.7B, after we study the nature of the hydrophilic groups in such compounds.

PROBLEM 10.11   Pick out the compounds among the following that you would expect to be surfactants.

(a) $\text{C}_6\text{H}_5-\overset{\displaystyle O}{\overset{\displaystyle \|}{C}}O^-Na^+$ (b) $CH_3(CH_2)_{12}CH_2O\overset{\displaystyle O}{\underset{\displaystyle O}{\overset{\displaystyle \|}{\underset{\displaystyle \|}{S}}}}O^-Na^+$

(c) $CH_3(CH_2)_{10}CH_2\overset{+}{N}(CH_3)_3\ Cl^-$ (d) $\text{C}_6\text{H}_5-\overset{\displaystyle O}{\underset{\displaystyle O}{\overset{\displaystyle \|}{\underset{\displaystyle \|}{S}}}}O^-Na^+$

(e) $CH_3(CH_2)_{10}CH_2-\text{C}_6\text{H}_4-\overset{\displaystyle O}{\underset{\displaystyle O}{\overset{\displaystyle \|}{\underset{\displaystyle \|}{S}}}}O^-Na^+$ (f) $CH_3(CH_2)_{15}CH_2\overset{\displaystyle O}{\overset{\displaystyle \|}{C}}O^-\ \overset{+}{H}N(CH_2CH_2OH)_3$

(g) $Na^+\ {}^-O\overset{\displaystyle O}{\overset{\displaystyle \|}{C}}(CH_2)_{12}\overset{\displaystyle O}{\overset{\displaystyle \|}{C}}O^-\ Na^+$

10.4

Reactivity in Acid Derivatives

In Section 10.1B, the functional groups in acids and acid derivatives were characterized as having an electrophilic carbon atom in the carbonyl group, and some atom or group attached to it that is either a good leaving group or can be converted to one by protonation. *Acids and acid derivatives will react with nucleophilic reagents that attack the carbonyl group in a substitution reaction with loss of the portion of the acid function that is a leaving group.*

A comparison of the reactions of benzaldehyde and of benzoyl chloride with aniline makes these points clear.

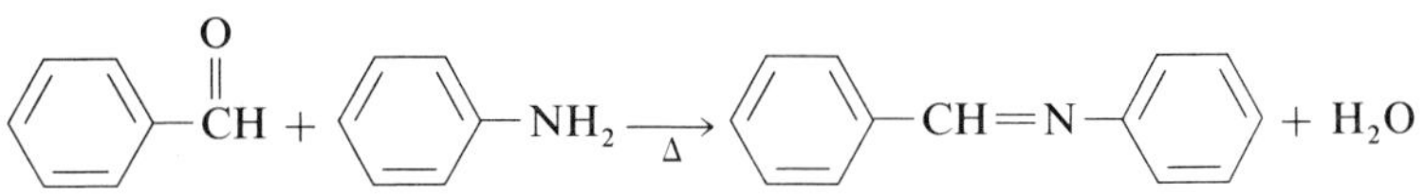

benzaldehyde   aniline   phenylimine of
                                               benzaldehyde

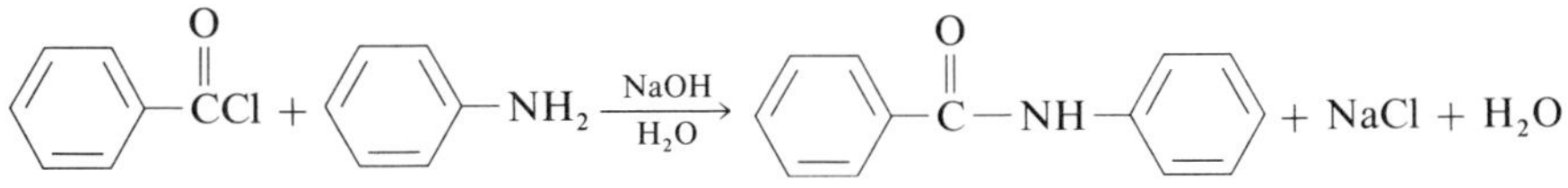

benzoyl chloride   aniline   benzanilide

Both of these reactions start with the attack of the nucleophilic nitrogen atom on the carbon atom of a carbonyl group. In both cases, an intermediate with four groups around the carbonyl carbon atom is formed. In the case of benzaldehyde, there is no good leaving group until a series of protonation and deprotonation steps takes place and water can finally be lost (Section 9.6B).

Benzoyl chloride is highly reactive towards aniline. The electronegative chlorine atom increases the electrophilicity of the carbon atom by its inductive effect while diminishing it somewhat by its resonance effect (Problem 10.3). The ease with which a chloride ion is lost as a leaving group also contributes to the reactivity of benzoyl chloride. The mechanism for the reaction of benzoyl chloride with aniline is shown below.

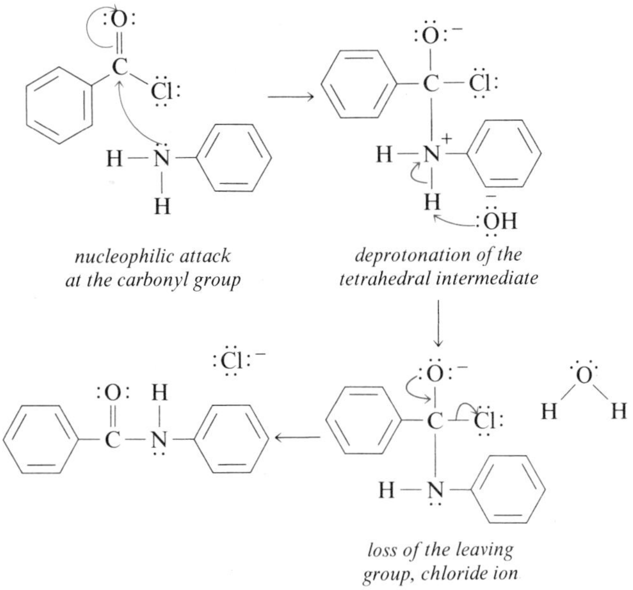

nucleophilic attack   deprotonation of the
at the carbonyl group   tetrahedral intermediate

loss of the leaving
group, chloride ion

The reaction of benzoyl chloride with aniline is carried out in the presence of aqueous sodium hydroxide. Benzoyl chloride is not very soluble in water and reacts faster with the organic amine than with the aqueous base. The hydroxide ion removes a proton from the tetrahedral intermediate, and chloride ion is lost as a leaving group. Reaction of an acid chloride with an amine (or an alcohol) in the presence of aqueous base is known as the **Schotten-Baumann reaction.**

The essential features of reactions of acid derivatives are demonstrated by the reaction of benzoyl chloride with aniline. *The carbon atom of the carbonyl group is attacked by some nucleophile, a tetrahedral intermediate is formed, and a leaving group is lost with the formation of the carbonyl group again. The overall reaction is a nucleophilic substitution at the carbon atom of the carbonyl group.*

The order of reactivity of acid derivatives depends upon the electrophilicity of the carbon atom of the carbonyl group and on how good a leaving group the compounds contain. For example, low molecular weight acid chlorides and acid anhydrides such as acetyl chloride and acetic anhydride are soluble in water and react very rapidly with it.

$$\underset{\text{acetyl chloride}}{CH_3\overset{\overset{\displaystyle O}{\|}}{C}Cl} + H_2O \longrightarrow \underset{\text{acetic acid}}{CH_3\overset{\overset{\displaystyle O}{\|}}{C}OH} + HCl$$

$$\underset{\text{acetic anhydride}}{CH_3\overset{\overset{\displaystyle O}{\|}}{C}O\overset{\overset{\displaystyle O}{\|}}{C}CH_3} + H_2O \longrightarrow 2\ \underset{\text{acetic acid}}{CH_3\overset{\overset{\displaystyle O}{\|}}{C}OH}$$

Both compounds have a good leaving group, chloride ion for the acid chloride and acetate ion for the anhydride. Water is the nucleophile in these reactions. The details of the reaction are illustrated for acetic anhydride.

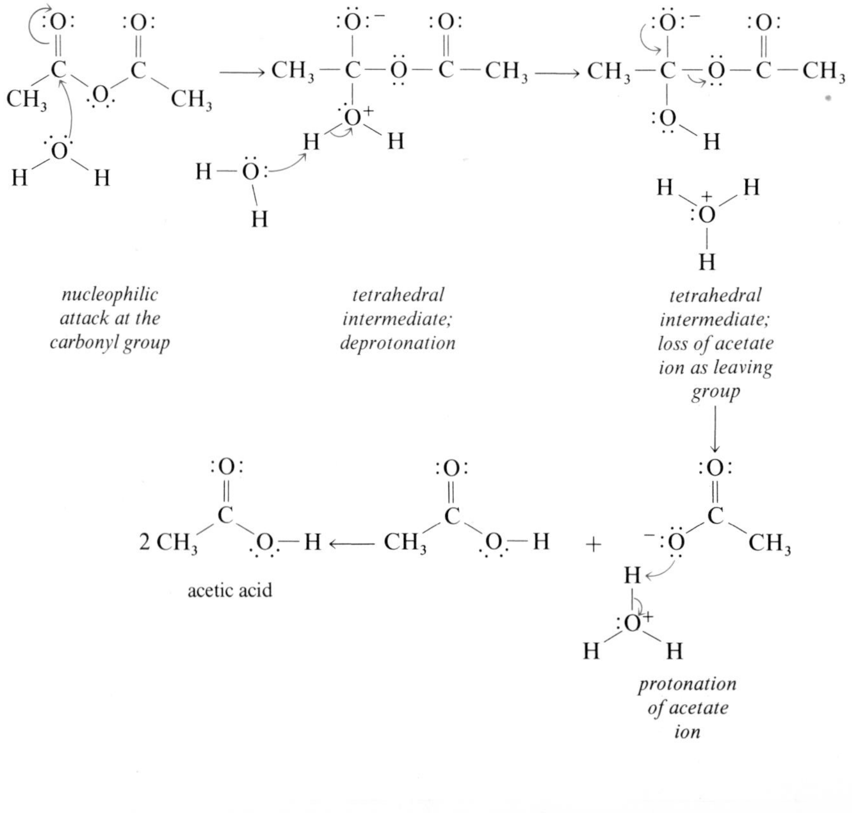

nucleophilic attack at the carbonyl group   *tetrahedral intermediate; deprotonation*   *tetrahedral intermediate; loss of acetate ion as leaving group*

acetic acid   *protonation of acetate ion*

PROBLEM 10.12   Write out the detailed mechanism for the reaction of acetyl chloride with water.

Acid anhydrides are generally less reactive than acid chlorides. The carbon atom of the carbonyl group is not as electrophilic in an anhydride as it is in an acid chloride, and acetate ion is not as good a leaving group as chloride ion. Anhydrides will, however, react readily with alcohols or with amines as well as with water.

$$\underset{\text{acetic anhydride}}{CH_3\overset{O}{\overset{||}{C}}O\overset{O}{\overset{||}{C}}CH_3} + \underset{\substack{\text{cholesterol}\\ \textit{an alcohol}}}{HO-}$$

$$\downarrow$$

$$\underset{\text{acetic acid}}{CH_3\overset{O}{\overset{||}{C}}OH} + \underset{\substack{\text{cholesteryl acetate}\\ \textit{an ester}}}{CH_3\overset{O}{\overset{||}{C}}O-}$$

$$\underset{\text{acetic anhydride}}{CH_3\overset{O}{\overset{||}{C}}O\overset{O}{\overset{||}{C}}CH_3} + \underset{\text{\textit{p}-methoxyaniline}}{H_2N-\!\!\!\bigcirc\!\!\!-OCH_3} \xrightarrow[\substack{H_2O\\0-5\,°C}]{\text{acetic acid}} \underset{\text{acet-\textit{p}-methoxyanilide}}{CH_3\overset{O}{\overset{||}{C}}NH-\!\!\!\bigcirc\!\!\!-OCH_3} + \underset{\text{acetic acid}}{CH_3\overset{O}{\overset{||}{C}}OH}$$

Acids, esters, and amides are much less reactive towards nucleophilic substitution reactions than either acid chlorides or anhydrides. The carbon atom of the carbonyl group is less electrophilic in these compounds, and the groups attached to the carbonyl group are not good leaving groups unless protonated first. For example, acetic acid does not react with an alcohol at an appreciable rate unless the reaction is catalyzed by a strong acid such as sulfuric acid or hydrochloric acid. Reactions that give esters as products are called **esterification reactions.** The equation for a typical esterification reaction is given below.

$$\underset{\text{acetic acid}}{CH_3\overset{O}{\overset{||}{C}}OH} + \underset{\text{methanol}}{CH_3OH} \xrightarrow{\text{HCl (g)}} \underset{\text{methyl acetate}}{CH_3\overset{O}{\overset{||}{C}}OCH_3} + H_2O$$

The reactants, acetic acid and methanol, contain several sites at which protonation takes place. A strong acid, dry gaseous hydrogen chloride, is used in catalytic amounts. A series of protonation and deprotonation reactions are in equilibrium with each other. Acetic acid is protonated at the carbonyl group, which makes the carbon atom of the carbonyl group more electrophilic. The unprotonated alcohol is the nucleophile; the protonated form of the alcohol would, of course, not be nucleophilic.

*protonation
of the carbonyl
group*

*protonated acetic acid has
greater electrophilicity at the
carbon atom of the carbonyl
group*

*nucleophilic
attack at the
carbonyl group*

*deprotonation of the
intermediate*

*tetrahedral intermediate
being protonated at
another site*

*deprotonation of
the carbonyl
group*

*loss of the leaving
group, water*

Note that the steps of the reaction fall into the main categories described earlier in this section. The carbon atom of the carbonyl group is attacked by a nucleophile, a tetrahedral intermediate is formed, and the carbonyl group is once more created by the loss of a leaving group. Because this reaction is carried out in strongly acidic media, protonation and deprotonation steps start and finish the reaction. Note also that all of the steps are written as equilibria. This suggests that if an ester were put into an excess of water in the presence of a strong acid, it would be converted into an acid and an alcohol. This is exactly what happens, when the equilibrium constant allows, in a reaction known as the *hydrolysis of an ester*.

A **hydrolysis reaction** *is one in which a σ bond is cleaved by the addition of the elements of water to the fragments formed in the cleavage.* Note that *hydrolysis* in which *lysis* (meaning *to loosen*) takes place contrasts with *hydration* in which water is added to a multiple bond, but no fragmentation of the molecule occurs (Sections 5.5B, 5.5C). The steps of esterification and hydrolysis reactions are very similar to the ones that were described for the formation and hydrolysis of acetals (Section 9.7B).

Chemists recognize the importance of esterification and hydrolysis reactions and have studied them in detail. In Sections 10.7B, 10.7D, and 10.7E, further examples and evidence for their mechanisms are given. For the moment, you should recognize that these reactions belong to the general pattern set down in this section for reactions of acids and their derivatives.

In summary, acid derivatives undergo nucleophilic substitution reactions at the carbonyl group. The order of reactivity for these compounds is

$$
\underset{\text{RCCl}}{\overset{\text{O}}{\overset{\|}{}}} > \underset{\text{RCOCR}}{\overset{\text{O}\quad\text{O}}{\overset{\|\ \ \|}{}}} > \underset{\text{RCOR}'}{\overset{\text{O}}{\overset{\|}{}}} \geqslant \underset{\text{RCOH}}{\overset{\text{O}}{\overset{\|}{}}} > \underset{\text{RCNH}_2}{\overset{\text{O}}{\overset{\|}{}}} > \underset{\text{RCO}^-}{\overset{\text{O}}{\overset{\|}{}}}
$$

The species that is least reactive towards nucleophilic substitution reactions is the carboxylate anion. Of all the acid derivatives, the anion is the most stabilized by resonance, and is formed whenever any of the other compounds react with water in the presence of base. An acid chloride is least stabilized by resonance, and an amide is the most stabilized among the other acid derivatives. Resonance stabilization is lost when the carbonyl group is converted to the tetrahedral intermediate. The energy of activation (Section 5.3D) for the conversion of an acid chloride, with little resonance stabilization, to the tetrahedral intermediate is lower than it is for the conversion of the more stable amide to a similar intermediate. Thus the ease with which an acid derivative is converted to the other acid derivatives by nucleophilic substitution reactions decreases from left to right in the reactivity series above. For example, an acid chloride is converted easily to any of the other compounds in the series. An amide, on the other hand, can be hydrolyzed to a carboxylic acid or carboxylate anion, but is not easily transformed into an ester, acid anhydride, or acid chloride.

PROBLEM 10.13   Draw an energy diagram for the reaction of an acid chloride with water (Problem 10.12). Ignore the different intermediates corresponding to proton transfer steps and show only reactants, the tetrahedral intermediate, and products. How does this energy diagram compare with the one for the reaction with water of an amide derived from the same acid? How do the energy diagrams you have drawn relate to the rates of the two reactions? You may want to review Sections 5.3D, 5.3E, and 5.3F.

It is interesting to compare the reactivity of an acid chloride towards nucleophilic substitution with that of an alkyl halide (Chapter 6). *An acid chloride reacts much faster in a bimolecular nucleophilic substitution reaction than even a primary alkyl halide does.* There are two reasons for this. *First, the carbon atom of the carbonyl group is much more electrophilic than the carbon atom bearing the halogen in an alkyl halide. Second, in an acid chloride the attack of the nucleophile takes place at an sp^2-hybridized carbon atom, a planar center with only three groups around it.* In the transition state, the carbon atom undergoing the substitution reaction becomes tetrahedral. In an alkyl halide, nucleophilic attack must take place at a carbon atom that already has four groups around it, and the transition state is crowded with a pentacoordinate carbon atom as its center (Section 6.5C). *So, bimolecular nucleophilic substitution at the carbonyl group of an acid chloride proceeds much faster than the same reaction at the tetrahedral carbon atom of an alkyl halide for both electronic and steric reasons.*

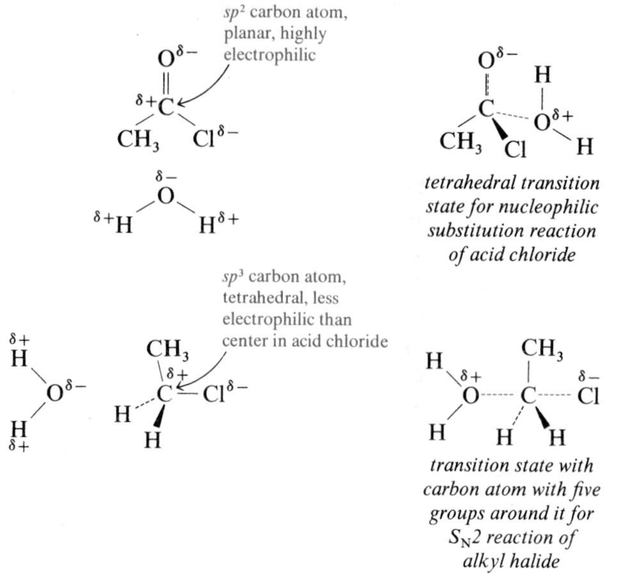

sp² carbon atom, planar, highly electrophilic

tetrahedral transition state for nucleophilic substitution reaction of acid chloride

sp³ carbon atom, tetrahedral, less electrophilic than center in acid chloride

transition state with carbon atom with five groups around it for S_N2 reaction of alkyl halide

---

PROBLEM 10.14   The overall reaction for the hydrolysis of an ester is given below.

$$CH_3\overset{\displaystyle O}{\overset{\|}{C}}OCH_3 + H_2O \xrightarrow[\Delta]{HCl} CH_3\overset{\displaystyle O}{\overset{\|}{C}}OH + CH_3OH$$

Using the mechanism shown on p. 456 as a guide, write a mechanism for the hydrolysis reaction.

PROBLEM 10.15   Complete the following equations.

(a) 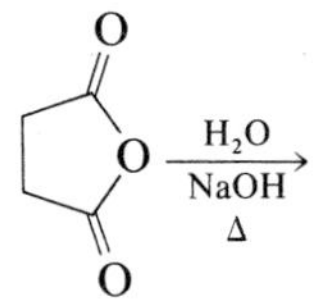 [cyclic anhydride structure] $\xrightarrow[\Delta]{\substack{H_2O \\ NaOH}}$

(b) 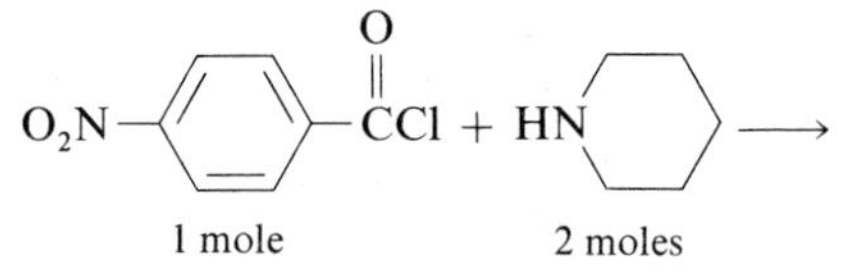 $O_2N-\langle\text{benzene ring}\rangle-\overset{\displaystyle O}{\overset{\|}{C}}Cl + HN\langle\text{ring}\rangle \longrightarrow$

                           1 mole   2 moles

(c) [benzene ring]$-\overset{\displaystyle O}{\overset{\|}{C}}OH + CH_3CH_2OH \xrightarrow[\Delta]{H_2SO_4}$

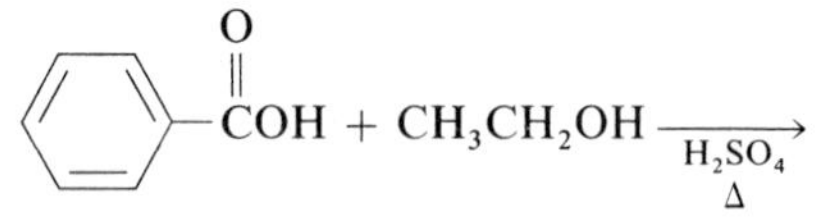

(d) $CH_3\overset{\displaystyle O}{\overset{\|}{C}}Cl + NH_3 \text{ (excess)} \longrightarrow$

(e) 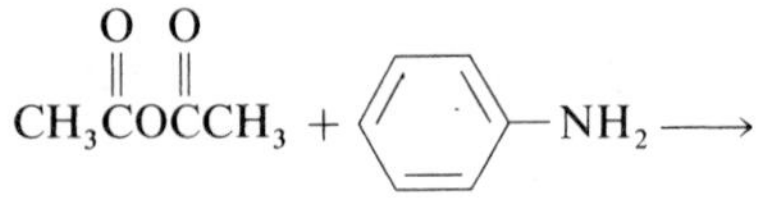 $CH_3\overset{\displaystyle O}{\overset{\|}{C}}O\overset{\displaystyle O}{\overset{\|}{C}}CH_3 + \langle\text{benzene ring}\rangle-NH_2 \longrightarrow$

(f) $CH_3CH_2CH_2\overset{\displaystyle O}{\overset{\|}{C}}OH + CH_3OH \xrightarrow[\Delta]{HCl}$

10.5

Preparation of Carboxylic Acids

A. Carboxylic Acids as Products of Oxidation Reactions

Carboxylic acids are formed as products of oxidation reactions. We have discussed many of these reactions in earlier chapters. Let's briefly review these methods.

Primary alcohols and aldehydes are oxidized to carboxylic acids containing the same number of carbon atoms (Section 7.6B). Oxidations of alcohols with acidic chromic acid solutions often give esters, the product of the reaction of the starting material alcohol with the product carboxylic acid (Section 10.4). For this reason, base-catalyzed oxidations with potassium permanganate are preferred when the acid is the desired product. 2-Ethyl-1-hexanol and the corresponding aldehyde have both been converted to 2-ethylhexanoic acid by this method.

$$
\begin{array}{c}
\text{CH}_3\text{CH}_2 \\
\mid \\
\text{CH}_3\text{CH}_2\text{CH}_2\text{CH}_2\text{CHCH}_2\text{OH}
\end{array}
$$

2-ethyl-1-hexanol

or

$$
\begin{array}{c}
\quad\quad\quad\quad\quad\quad\quad\quad\text{O} \\
\quad\quad\quad\quad\quad\quad\quad\quad\parallel \\
\text{CH}_3\text{CH}_2\text{CH}_2\text{CH}_2\text{CHCH}
\end{array}
$$
$$
\begin{array}{c}
\mid \\
\text{CH}_3\text{CH}_2
\end{array}
$$

2-ethylhexanal

$$
\xrightarrow[\text{H}_2\text{O}]{\overset{\text{KMnO}_4}{\text{NaOH}}}
$$

$$
\begin{array}{c}
\quad\quad\quad\quad\quad\quad\quad\quad\quad\text{O} \\
\quad\quad\quad\quad\quad\quad\quad\quad\quad\parallel \\
\text{CH}_3\text{CH}_2\text{CH}_2\text{CH}_2\text{CHCO}^-\text{Na}^+ \\
\mid \\
\text{CH}_3\text{CH}_2
\end{array}
$$

$$
\xrightarrow[\substack{\text{SO}_2 \\ \text{reducing} \\ \text{agent for} \\ \text{excess MnO}_4^-}]{\text{H}_3\text{O}^+}
$$

$$
\begin{array}{c}
\quad\quad\quad\quad\quad\quad\quad\quad\quad\text{O} \\
\quad\quad\quad\quad\quad\quad\quad\quad\quad\parallel \\
\text{CH}_3\text{CH}_2\text{CH}_2\text{CH}_2\text{CHCOH} \\
\mid \\
\text{CH}_3\text{CH}_2
\end{array}
$$

2-ethylhexanoic
acid
~75%

In basic solution, the acid is formed as its salt and hence is not reactive towards unused alcohol in the reaction mixture. The free carboxylic acid is generated with sulfuric acid after the oxidation is completed (Section 10.3A).

A very mild oxidizing agent for aldehydes is moist silver oxide, the same reagent used in the Tollens test (Section 9.9). On a larger scale, it serves as a preparation for carboxylic acids. For example, heptanal is oxidized in very high yield to heptanoic acid with this reagent.

$$
\begin{array}{c}
\quad\quad\text{O} \\
\quad\quad\parallel \\
\text{CH}_3(\text{CH}_2)_5\text{CH}
\end{array}
\xrightarrow[\substack{\text{H}_2\text{O} \\ 95\,°\text{C}}]{\overset{\text{Ag}_2\text{O}}{\text{NaOH}}}
\text{Ag}\downarrow +
\begin{array}{c}
\text{O} \\
\parallel \\
\text{CH}_3(\text{CH}_2)_5\text{CO}^-\text{Na}^+
\end{array}
\xrightarrow{\text{H}_3\text{O}^+}
\begin{array}{c}
\text{O} \\
\parallel \\
\text{CH}_3(\text{CH}_2)_5\text{COH}
\end{array}
$$

heptanal   heptanoic acid
                                                                                        97%

In compounds that would be sensitive to stronger oxidizing agents, such as potassium permanganate (Section 8.4), an aldehyde function can be successfully oxidized to the acid by means of silver oxide. For example, the unsaturated aldehyde, 9,12-octadeca-diynal is converted to the corresponding acid.

$$CH_3(CH_2)_4C\equiv CCH_2C\equiv C(CH_2)_7\overset{\overset{\displaystyle O}{\parallel}}{C}H \xrightarrow[\substack{\text{ethanol} \\ N_2 \text{ atmosphere}}]{\substack{Ag_2O \\ NaOH}} CH_3(CH_2)_4C\equiv CCH_2C\equiv C(CH_2)_7\overset{\overset{\displaystyle O}{\parallel}}{C}O^-Na^+$$

9,12-octadecadiynal

$$\downarrow H_3O^+$$

$$CH_3(CH_2)_4C\equiv CCH_2C\equiv C(CH_2)_7\overset{\overset{\displaystyle O}{\parallel}}{C}OH$$

9,12-octadecadiynoic acid
78%

The reaction is carried out in an atmosphere of nitrogen gas, because these unsaturated compounds are sensitive to oxidation, even by the oxygen in air. (The reasons for the great sensitivity of such polyunsaturated compounds to oxygen are discussed in Section 11.7A.) The oxidation reaction with silver oxide is highly selective. The aldehyde is oxidized and the multiple bonds are untouched.

Cleavage of the multiple bond in alkenes or alkynes by oxidative reactions leads to the production of carboxylic acids if the conditions are vigorous enough (Section 8.4C). For example, 3,7-dimethyl-1-octene gives 2,6-dimethylheptanoic acid when heated with potassium permanganate in acetone in the presence of base.

$$\overset{\overset{\displaystyle CH_3}{|}}{CH_3CH}CH_2CH_2CH_2\overset{\overset{\displaystyle CH_3}{|}}{CH}CH=CH_2 \xrightarrow[\substack{NaHCO_3 \\ \text{acetone}}]{KMnO_4} \xrightarrow{H_3O^+} \overset{\overset{\displaystyle CH_3}{|}}{CH_3CH}CH_2CH_2CH_2\overset{\overset{\displaystyle CH_3}{|}}{CH}\overset{\overset{\displaystyle }{}}{C}OH + CO_2 + H_2O$$

3,7-dimethyl-1-octene

2,6-dimethylheptanoic acid
45%

Ozonolysis is one of the most effective ways to cleave a multiple bond to produce acids by way of aldehyde intermediates. You studied many examples of this reaction in Section 8.6. The conversion of 1-tridecene to dodecanoic acid serves as a reminder. In this reaction, silver oxide oxidizes the aldehyde that is the product of the breakdown of 1-tridecene ozonide to the acid.

$$CH_3(CH_2)_{10}CH=CH_2 \xrightarrow{O_3} CH_3(CH_2)_{10}\overset{}{C}H\underset{O}{\overset{O-O}{\diagup \quad \diagdown}}CH_2 \xrightarrow[NaOH]{Ag_2O}$$

1-tridecene

1-tridecene ozonide

$$H\overset{\overset{\displaystyle O}{\parallel}}{C}O^-Na^+ + CH_3(CH_2)_{10}\overset{\overset{\displaystyle O}{\parallel}}{C}O^-Na^+ \xrightarrow{H_3O^+} H\overset{\overset{\displaystyle O}{\parallel}}{C}OH + CH_3(CH_2)_{10}\overset{\overset{\displaystyle O}{\parallel}}{C}OH$$

sodium formate · sodium dodecanoate · formic acid · dodecanoic acid 94%

These oxidative cleavage reactions give rise to carboxylic acids with fewer carbon atoms than the starting reagents.

B. The Grignard Reaction

The Grignard reaction is used to prepare acids that have one more carbon atom than the alkyl or aryl halide used to make the organomagnesium reagent. The Grignard reagent adds to carbonyl groups. In this reaction the nucleophilic carbon atom of the organo-metallic reagent attacks the carbon atom of the carbonyl group, while the magnesium atom complexes with the oxygen atom (Section 9.4B). Carbon dioxide may be regarded as a carbonyl compound and, as such, reacts with Grignard reagents just as aldehydes and ketones do. The resulting compound is the magnesium salt of a carboxylic acid. The free carboxylic acid is formed when acid is added to the reaction mixture. The conversion of 2-chlorobutane to 2-methylbutanoic acid illustrates this synthesis.

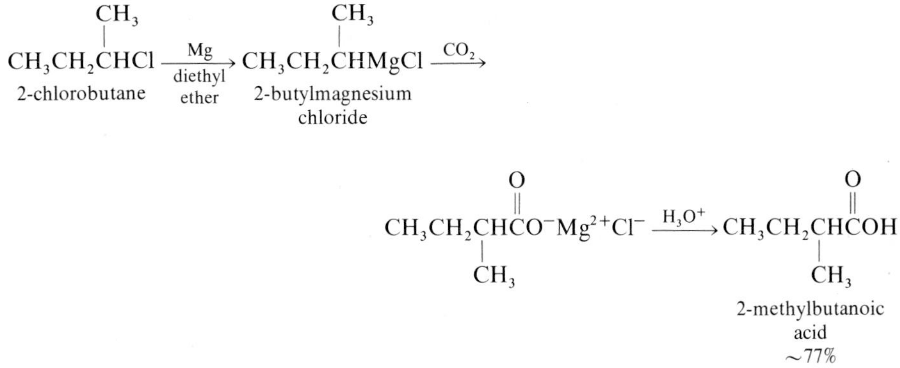

$$
\underset{\text{2-chlorobutane}}{\overset{\overset{\displaystyle CH_3}{|}}{CH_3CH_2CHCl}} \xrightarrow[\text{diethyl ether}]{Mg} \underset{\substack{\text{2-butylmagnesium} \\ \text{chloride}}}{\overset{\overset{\displaystyle CH_3}{|}}{CH_3CH_2CHMgCl}} \xrightarrow{CO_2}
$$

$$
\underset{\overset{\displaystyle |}{CH_3}}{CH_3CH_2CH\overset{\overset{\displaystyle O}{\|}}{C}O^-Mg^{2+}Cl^-} \xrightarrow{H_3O^+} \underset{\substack{\overset{\displaystyle |}{CH_3} \\ \text{2-methylbutanoic} \\ \text{acid} \\ \sim 77\%}}{CH_3CH_2CH\overset{\overset{\displaystyle O}{\|}}{C}OH}
$$

Aromatic as well as alkyl Grignard reagents undergo these reactions. For example, 1-bromonaphthalene gives 1-naphthoic acid by the Grignard reaction.

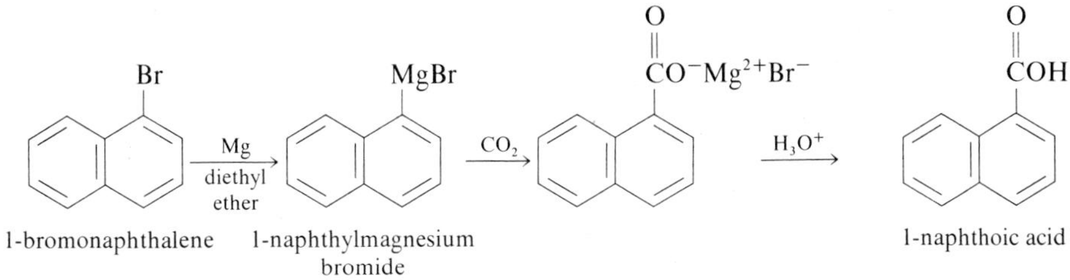

1-bromonaphthalene   1-naphthylmagnesium bromide   1-naphthoic acid

---

PROBLEM 10.16 Write out the details of the mechanism for the reaction of carbon dioxide with a Grignard reagent of your choice.

PROBLEM 10.17 Complete the following equations.

(a) 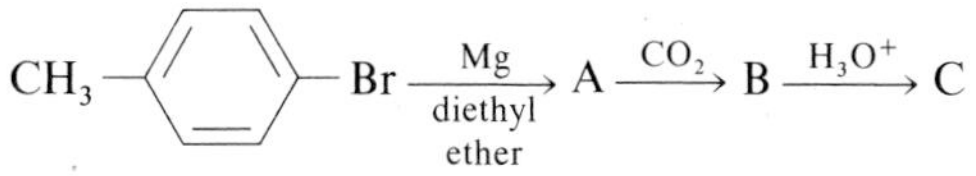 $CH_3\!\!-\!\!\underset{}{\bigcirc}\!\!-\!\!Br \xrightarrow[\text{diethyl ether}]{Mg} A \xrightarrow{CO_2} B \xrightarrow{H_3O^+} C$

(b) $\underset{}{\overset{\overset{\displaystyle CH_3}{|}}{CH_3CH_2CH_2CHCH_2CH_2OH}} \xrightarrow[\substack{\text{NaOH} \\ H_2O \\ \Delta}]{KMnO_4} D \xrightarrow{H_3O^+} E$

(c) $\underset{}{\bigcirc} \xrightarrow[\text{chloroform}]{O_3} F \xrightarrow[\substack{\text{NaOH} \\ H_2O \\ \Delta}]{Ag_2O} G \xrightarrow{H_3O^+} H$

(d) $\underset{}{\bigcirc}\!\!-\!\!CH_2CH_2\overset{\overset{\displaystyle O}{\|}}{CH} \xrightarrow[\substack{\text{NaOH} \\ H_2O \\ \Delta}]{Ag_2O} I \xrightarrow{H_3O^+} J$

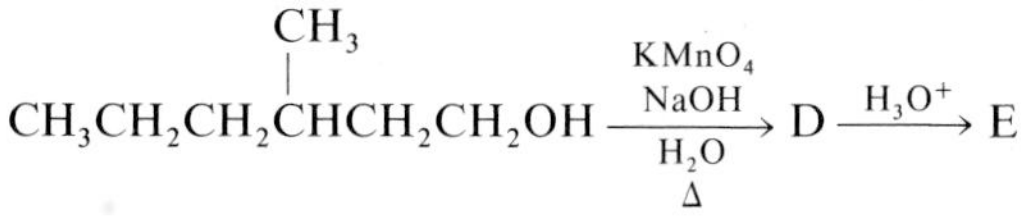

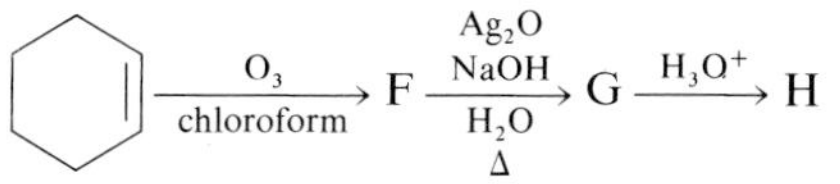

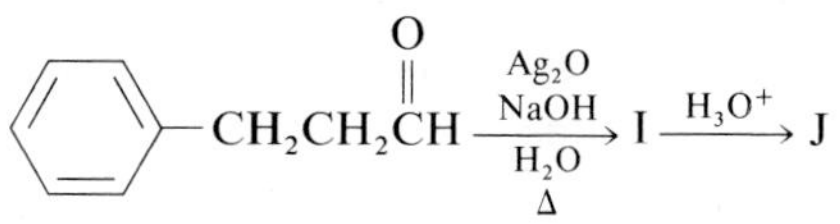

(e) $\text{C}_6\text{H}_5\text{—CH}_2\text{OH} \xrightarrow[\substack{\text{H}_2\text{O} \\ \Delta}]{\substack{\text{CrO}_3 \\ \text{H}_2\text{SO}_4}} \text{K}$

(f) $\underset{\text{CH}_3}{\text{CH}_3\overset{|}{\text{C}}\text{HCH}_2\text{CH}=\text{CHCH}_2\text{CH}_3} \xrightarrow[\substack{\text{H}_2\text{O} \\ \Delta}]{\substack{\text{KMnO}_4 \\ \text{NaOH}}} \text{L} + \text{M} \xrightarrow{\text{H}_3\text{O}^+} \text{N} + \text{O}$

(g) $\text{CH}_3(\text{CH}_2)_5\overset{\overset{\text{O}}{\|}}{\text{C}}\text{H} \xrightarrow[\text{H}_2\text{O}]{\substack{\text{CrO}_3 \\ \text{H}_2\text{SO}_4}} \text{P}$

(h) $\text{CH}_3(\text{CH}_2)_9\text{CH}=\text{CH}_2 \xrightarrow[\text{chloroform}]{\text{O}_3} \xrightarrow{\substack{\text{H}_2\text{O}_2 \\ \text{NaOH}}} \xrightarrow{\text{H}_3\text{O}^+} \text{Q} + \text{R}$

C. Nitriles

Nitriles, prepared by reactions of cyanide ion with alkyl halides (Section 6.11) or with aldehydes and ketones in the presence of acid (Section 9.5), allow for the preparation of acids having one more carbon atom than the starting material. Vigorous hydrolysis of a nitrile gives a carboxylic acid. For example, glutaric acid is synthesized in two steps from 1,3-dibromopropane, by nucleophilic displacement of halide ion by cyanide ion and then hydrolysis of the dinitrile to the diacid.

$\underset{\text{1,3-dibromopropane}}{\text{BrCH}_2\text{CH}_2\text{CH}_2\text{Br}} + 2\,\text{Na}^+\text{CN}^- \xrightarrow[\substack{\text{H}_2\text{O} \\ \text{ethanol}}]{} \underset{\text{pentanedinitrile}}{\text{N}\equiv\text{CCH}_2\text{CH}_2\text{CH}_2\text{C}\equiv\text{N}} \xrightarrow[\text{100 °C, 4 h}]{\text{H}_3\text{O}^+}$

$$\underset{\substack{\text{glutaric acid} \\ \sim 85\%}}{\text{HO}\overset{\overset{\text{O}}{\|}}{\text{C}}\text{CH}_2\text{CH}_2\text{CH}_2\overset{\overset{\text{O}}{\|}}{\text{C}}\text{OH}} + \underset{\substack{\text{ammonium} \\ \text{chloride}}}{2\,\text{NH}_4^+\text{Cl}^-}$$

Cyanohydrins can also be hydrolyzed to carboxylic acids. The cyanohydrin, 2-hydroxy-2-phenylacetonitrile, resulting from the addition of the elements of hydrogen cyanide to benzaldehyde (Section 9.5) is hydrolyzed to 2-hydroxy-2-phenylacetic acid with cold hydrochloric acid.

$\underset{\text{benzaldehyde}}{\text{C}_6\text{H}_5\text{—}\overset{\overset{\text{O}}{\|}}{\text{C}}\text{H}} \xrightarrow{\text{[HCN]}} \underset{\substack{\text{2-hydroxy-2-phenylacetonitrile} \\ \textit{cyanohydrin of} \\ \textit{benzaldehyde}}}{\text{C}_6\text{H}_5\text{—}\underset{\text{OH}}{\overset{|}{\text{C}}}\text{HC}\equiv\text{N}} \xrightarrow[\substack{\text{cold} \\ \text{12 h}}]{\text{H}_3\text{O}^+} \underset{\substack{\text{2-hydroxy-2-phenylacetic} \\ \text{acid}}}{\text{C}_6\text{H}_5\text{—}\underset{\text{OH}}{\overset{|}{\text{C}}}\text{H}\overset{\overset{\text{O}}{\|}}{\text{C}}\text{OH}} + \text{NH}_4^+\text{Cl}^-$

The hydrolysis of nitriles is catalyzed by base as well as by acid. The following equations show the conversion of chloroacetic acid to malonic acid. Chloroacetic acid, in the form of its sodium salt, is converted to sodium cyanoacetate. The nitrile function is hydrolyzed in base, and malonic acid is recovered from its sodium salt by careful acidification.

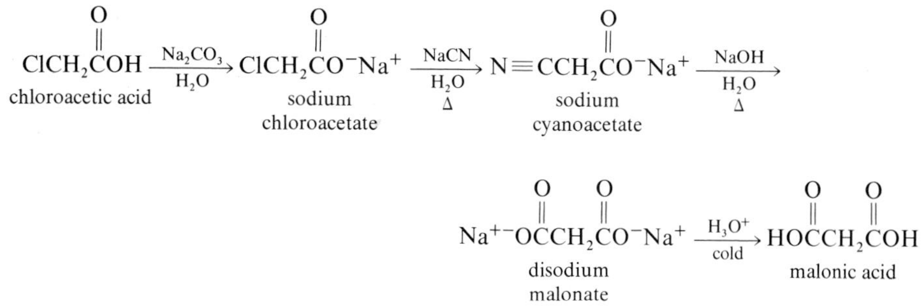

$$\underset{\text{chloroacetic acid}}{\text{ClCH}_2\overset{\overset{\displaystyle O}{\|}}{\text{C}}\text{OH}} \xrightarrow[\text{H}_2\text{O}]{\text{Na}_2\text{CO}_3} \underset{\substack{\text{sodium} \\ \text{chloroacetate}}}{\text{ClCH}_2\overset{\overset{\displaystyle O}{\|}}{\text{C}}\text{O}^-\text{Na}^+} \xrightarrow[\text{H}_2\text{O}]{\text{NaCN}} \underset{\substack{\text{sodium} \\ \text{cyanoacetate}}}{\text{N}\equiv\text{CCH}_2\overset{\overset{\displaystyle O}{\|}}{\text{C}}\text{O}^-\text{Na}^+} \xrightarrow[\substack{\text{H}_2\text{O} \\ \Delta}]{\text{NaOH}}$$

$$\underset{\substack{\text{disodium} \\ \text{malonate}}}{\text{Na}^+{}^-\text{OC}\text{CH}_2\text{C}\text{O}^-\text{Na}^+} \xrightarrow[\text{cold}]{\text{H}_3\text{O}^+} \underset{\text{malonic acid}}{\text{HOCCH}_2\text{COH}}$$

The mechanism for the hydrolysis of nitriles is discussed in Section 10.8B, when the hydrolysis of amides is considered.

PROBLEM 10.18   Why is chloroacetic acid converted to its salt before sodium cyanide is added to the reaction mixture? (Hint: The table of pK_a values may be helpful.)

PROBLEM 10.19   Write equations for the following conversions.   More than one step may be necessary.

(a)

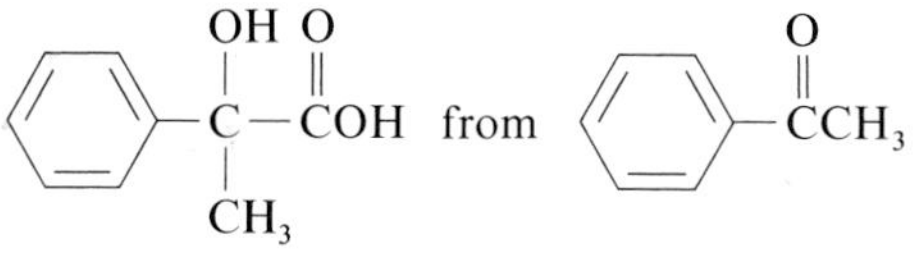

(b) $\text{HO}\overset{\overset{\displaystyle O}{\|}}{\text{C}}(\text{CH}_2)_4\overset{\overset{\displaystyle O}{\|}}{\text{C}}\text{OH}$ from $\text{Br}(\text{CH}_2)_4\text{Br}$

(c)

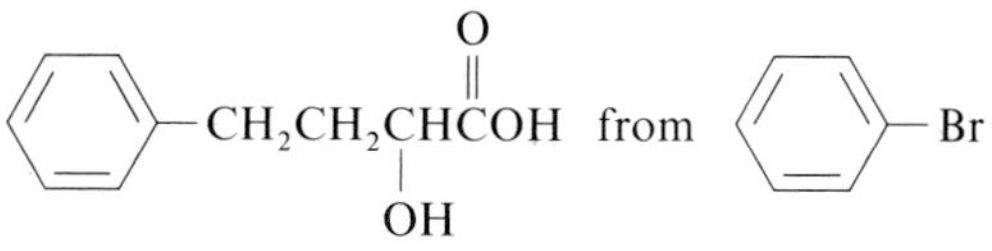

(d) $\underset{\substack{| \\ \text{CH}_3}}{\text{CH}_3\text{CHCH}_2\text{CH}_2\text{CH}_2\overset{\overset{\displaystyle O}{\|}}{\text{C}}\text{OH}}$ from $\underset{\substack{| \\ \text{CH}_3}}{\text{CH}_3\text{CHCH}_2\text{CH}_2\text{CH}_2\text{CH}=\text{CH}_2}$

10.6

Preparation of Acid Chlorides and Anhydrides

A. Preparation of Acid Chlorides

The same reagents, phosphorus halides and thionyl chloride, that convert alcohols into alkyl halides (Section 7.5B) change carboxylic acids into acid halides. For example, glacial acetic acid, when heated with phosphorus trichloride, is converted to acetyl chloride.

$$3 \underset{\substack{\text{acetic acid}}}{CH_3\overset{\displaystyle O}{\overset{\|}{C}}OH} + \underset{\substack{\text{phosphorus} \\ \text{trichloride}}}{PCl_3} \xrightarrow{\Delta} 3 \underset{\substack{\text{acetyl} \\ \text{chloride} \\ 67\%}}{CH_3\overset{\displaystyle O}{\overset{\|}{C}}Cl} + \underset{\substack{\text{phosphorous} \\ \text{acid}}}{P(OH)_3}$$

*distilled out
of reaction mixture*

Thionyl chloride is particularly useful in the preparation of acid halides because it has a low boiling point, 79 °C, and the other products of the reaction, sulfur dioxide and hydrogen chloride, can be easily removed from the reaction mixture because they are gases. The higher acid chlorides can be purified sufficiently for most uses by heating to expel the gases and then distilling out excess thionyl chloride. Butanoyl chloride, which boils at 100 °C, is prepared in this way.

$$\underset{\substack{\text{butanoic acid} \\ \text{bp 186 °C}}}{CH_3CH_2CH_2\overset{\displaystyle O}{\overset{\|}{C}}OH} + \underset{\substack{\text{thionyl} \\ \text{chloride} \\ \text{bp 79 °C}}}{SOCl_2} \longrightarrow \underset{\substack{\text{butanoyl chloride} \\ \text{bp 100 °C} \\ 85\%}}{CH_3CH_2CH_2\overset{\displaystyle O}{\overset{\|}{C}}Cl} + SO_2\uparrow + HCl\uparrow$$

Similarly, benzoyl chloride can be prepared by treating benzoic acid with thionyl chloride.

$$\underset{\substack{\text{benzoic acid} \\ \text{mp 122 °C}}}{\text{(benzene ring)}-\overset{\displaystyle O}{\overset{\|}{C}}OH} + \underset{\substack{\text{thionyl} \\ \text{chloride}}}{SOCl_2} \longrightarrow \underset{\substack{\text{benzoyl chloride} \\ \text{bp 198 °C} \\ 91\%}}{\text{(benzene ring)}-\overset{\displaystyle O}{\overset{\|}{C}}Cl} + SO_2\uparrow + HCl\uparrow$$

Acid chlorides are sensitive to moisture and nucleophiles (Section 10.4), so in most cases they are prepared just before they are used in reactions with alcohols or amines, though a few common ones can be bought commercially.

B. Preparation of Acid Anhydrides

As their names indicate, acid anhydrides are compounds that are formally related to carboxylic acids by the loss of water. In some cases, especially for cyclic anhydrides formed from dicarboxylic acids, they can actually be prepared by heating the acid and driving off the water. Acetic anhydride, a compound of great commercial importance, is prepared in industry by heating acetic acid to 800 °C.

$$2 \underset{\substack{\text{acetic acid}}}{CH_3\overset{\displaystyle O}{\overset{\|}{C}}OH} \xrightarrow[\substack{\text{quartz tube} \\ \text{porcelain chips} \\ \Delta \\ 800\,°C}]{} \underset{\substack{\text{acetic anhydride}}}{CH_3\overset{\displaystyle O}{\overset{\|}{C}}O\overset{\displaystyle O}{\overset{\|}{C}}CH_3} + H_2O\uparrow$$

Acetic anhydride itself serves as a dehydrating agent, and anhydrides that have boiling points higher than acetic acid can be prepared by heating the acid with acetic anhydride and distilling out acetic acid as it forms. Thus, dodecanoic anhydride is prepared by heating the acid with an excess of acetic anhydride.

$$2\ CH_3(CH_2)_{10}\overset{\overset{\displaystyle O}{\|}}{C}OH\ +\ CH_3\overset{\overset{\displaystyle O}{\|}}{C}O\overset{\overset{\displaystyle O}{\|}}{C}CH_3 \xrightarrow[\substack{6\text{--}8\ h}]{\Delta} \begin{matrix} CH_3(CH_2)_{10}\overset{\overset{\displaystyle O}{\|}}{C} \\ \diagdown \\ \diagup \\ CH_3(CH_2)_{10}\overset{\underset{\displaystyle O}{\|}}{C} \end{matrix} O\ +\ 2\ CH_3\overset{\overset{\displaystyle O}{\|}}{C}OH$$

dodecanoic acid	acetic anhydride	dodecanoic	acetic acid
bp 131 °C	bp 138 °C	anhydride	bp 118 °C
mp 44 °C		mp 42 °C	

As mentioned earlier, cyclic anhydrides, especially those with a five- or six-membered anhydride ring, form easily. Consider, for example, the preparation of maleic anhydride from maleic acid.

maleic acid	maleic *distilled out*
	anhydride *of the reaction*
	89% *mixture with*
	the solvent

In all of the preparations shown, some method to remove water from the reaction mixture must be used in order to produce a high yield of acid anhydride. Otherwise, the reverse reaction, the hydrolysis of an anhydride by water (Section 10.4), will take place. In the preparation of acetic anhydride, the water is vaporized by the high temperatures and distills out of the reaction mixture. The water formed when dodecanoic acid is converted into its anhydride reacts with the inexpensive, readily available acetic anhydride present in excess, and is thus prevented from hydrolyzing the desired dodecanoic anhydride. In the preparation of maleic anhydride, a solvent, 1,1,2,2-tetrachloroethane, that co-distills with water is used to remove it from the reaction mixture at relatively low temperatures.

One other method of making an anhydride is to carry out a nucleophilic substitution reaction on an acid chloride with the salt of the same acid. This reaction is illustrated by the preparation of 2-methylpropanoic anhydride.

$$CH_3\overset{\overset{\displaystyle O}{\|}}{C}HCCl\ +\ Na^{+-}O\overset{\overset{\displaystyle O}{\|}}{C}CHCH_3 \longrightarrow CH_3\overset{\overset{\displaystyle O}{\|}}{C}HCO\overset{\overset{\displaystyle O}{\|}}{C}CHCH_3\ +\ Na^{+}Cl^{-}$$

CH$_3$	CH$_3$	CH$_3$ CH$_3$
2-methylpropanoyl chloride	sodium 2-methyl-propanoate	2-methylpropanoic anhydride

PROBLEM 10.20   Complete the following equations.

(a) $\langle\!\!\!\bigcirc\!\!\!\rangle\!-\!CH_2CH_2\overset{\overset{\displaystyle O}{\|}}{C}OH \xrightarrow[\Delta]{SOCl_2}$

(b) $CH_3CH_2CH_2\overset{\overset{\displaystyle O}{\|}}{C}OH \xrightarrow{PCl_3}$

(c)
$$CH_3CH_2CH_2\overset{\overset{\displaystyle O}{\|}}{C}Cl + Na^{+-}\overset{\overset{\displaystyle O}{\|}}{O}CCH_2CH_2CH_3 \longrightarrow$$

(d)
$$CH_3(CH_2)_{10}\overset{\overset{\displaystyle O}{\|}}{C}OH \xrightarrow[\Delta]{SOCl_2}$$

(e)
$$HO\overset{\overset{\displaystyle O}{\|}}{C}CH_2CH_2\overset{\overset{\displaystyle O}{\|}}{C}OH \xrightarrow{CH_3\overset{\overset{\displaystyle O}{\|}}{C}O\overset{\overset{\displaystyle O}{\|}}{C}CH_3}$$

(f)
$$CH_3CH_2\overset{\overset{\displaystyle O}{\|}}{C}OH \xrightarrow[\substack{\Delta \\ 650\,°C}]{clay}$$

(g)
$$HO\overset{\overset{\displaystyle O}{\|}}{C}(CH_2)_5\overset{\overset{\displaystyle O}{\|}}{C}OH \xrightarrow[\Delta]{excess\ SOCl_2}$$

10.7

Esters

A. Formation of Esters by Reactions of Alcohols with Acid Chlorides and Anhydrides

Nucleophilic substitution reactions of acid chlorides and acid anhydrides with alcohols were shown to give esters (Section 10.4). The reaction with acid chlorides is used in particular when a derivative (Section 9.6A) of a small amount of a valuable alcohol is desired. For example, cyclohexanol and 2-heptanol are both secondary alcohols having a boiling point of 160 °C. Cyclohexanol can be converted to a 3,5-dinitrobenzoic ester that has a melting point of 112 °C, whereas the melting point of the corresponding ester of 2-heptanol is 49 °C.

cyclohexanol   3,5-dinitrobenzoyl   cyclohexyl 3,5-dinitrobenzoate
bp 160 °C   chloride   mp 112 °C

2-heptanol   2-heptyl 3,5-dinitrobenzoate
bp 160 °C   mp 49 °C

We cannot make a distinction between cyclohexanol and 2-heptanol on the basis of their boiling points or qualitative analysis tests. Their 3,5-dinitrobenzoic esters are prepared by heating a small amount of the alcohol with 3,5-dinitrobenzoyl chloride. The solid esters are purified by washing away the hydrochloric acid as well as some 3,5-dinitrobenzoic acid, which are the other products of the reaction, with sodium carbonate solution. Identification of the alcohols is then easy because of the distinctive melting points of the esters.

PROBLEM 10.21   Schotten-Baumann conditions for converting an alcohol into its benzoic ester with benzoyl chloride call for mixing the alcohol and the acid chloride in 10% aqueous sodium hydroxide. Write an equation for the reaction using *n*-butyl alcohol. Show any other products that will be formed under these reaction conditions.

PROBLEM 10.22   In a well-equipped laboratory, another method of distinguishing cyclohexanol from 2-heptanol should also be available. What is it?

Acetic anhydride is used most often in converting polyhydroxy alcohols into the corresponding acetic esters. Such reactions are especially important in the chemistry of carbohydrates, which are polyhydroxy aldehydes or ketones, usually with high melting points because of very strong hydrogen bonding between molecules. Carbohydrates are notorious for not crystallizing easily, again because they hydrogen bond so strongly to water that they form syrups instead. The formation of acetic esters of such compounds often lowers the melting point and makes them easier to crystallize and thus to purify. The application of these reactions to the chemistry of carbohydrates is discussed in Sections 14.5A and 14.9B.

A simpler example of the reaction of acetic anhydride with a polyhydroxy compound is that of 2,2-bis(hydroxymethyl)-1,3-propanediol (pentaerythritol).

$$
\begin{array}{c}
CH_2OH \\
| \\
HOCH_2CCH_2OH \\
| \\
CH_2OH
\end{array}
+ \; 4\; CH_3\overset{O}{\overset{||}{C}}O\overset{O}{\overset{||}{C}}CH_3
\xrightarrow[\Delta]{CH_3CO^-Na^+}
\begin{array}{c}
O \\
|| \\
CH_2OCCH_3 \\
| \\
CH_3\overset{O}{\overset{||}{C}}OCH_2CCH_2OCCH_3 \\
| \\
CH_3COCH_2 \quad O \\
|| \\
O
\end{array}
+ \; 4\; CH_3\overset{O}{\overset{||}{C}}OH
$$

| 2,2-bis(hydroxymethyl)-1,3-propanediol pentaerythritol mp 253 °C | acetic anhydride | | tetraacetate of 2,2-bis(hydroxymethyl)-1,3-propanediol mp 84 °C | acetic acid |

2,2-Bis(hydroxymethyl)-1,3-propanediol, though not very high in molecular weight, has a high melting point. It is a symmetrical molecule that packs well in a crystal lattice and can participate in much intermolecular hydrogen bonding. The tetraacetate has a much lower melting point (84 °C) than the parent tetraol (253 °C). The loss of hydrogen bonding is clearly reflected in this change of physical properties.

PROBLEM 10.23   Complete the following equations.

(a) $\begin{array}{c} CH_3 \\ | \\ CH_3COH \\ | \\ CH_3 \end{array} + \; CH_3\overset{O}{\overset{||}{C}}O\overset{O}{\overset{||}{C}}CH_3 \longrightarrow$

(b) $CH_3CH_2CH_2CH_2OH + CH_3CH_2CH_2\overset{O}{\overset{||}{C}}Cl \longrightarrow$

(c)   (d)

B. Esterification

Esters of methanol and ethanol are usually prepared by heating the carboxylic acid in an excess of the alcohol in the presence of a strong acid such as dry hydrogen chloride, concentrated sulfuric acid, or p-toluenesulfonic acid. Such a reaction of acetic acid with methanol was used as an example in Section 10.4, and details of its mechanism were given.

The reaction between a carboxylic acid and an alcohol is an equilibrium reaction. For example, when one mole each of acetic acid and ethanol are mixed, 0.667 mole each of ester and water are produced and 0.333 mole each of acid and alcohol remain at equilibrium.

$$CH_3COH + CH_3CH_2OH \rightleftharpoons CH_3COCH_2CH_3 + H_2O$$

At start	1 mole	1 mole	0 mole	0 mole
At equilibrium	0.333 mole	0.333 mole	0.667 mole	0.667 mole

The yield of ester can be increased either by removing one of the products of the reaction as it is formed or by increasing the concentration of one of the reactants. In practice, either water is distilled out of the reaction mixture or a greatly increased amount of alcohol, which is relatively inexpensive and abundant in the case of methanol or ethanol, is used. In some particularly tough cases, both precautions are taken.

2-Hydroxy-2-phenylacetic acid is esterified using an excess of ethanol with dry hydrogen chloride as the catalyst.

2-hydroxy-2-phenylacetic acid   ethanol   ethyl 2-hydroxy-2-phenylacetate
α-hydroxyphenylacetic acid   large   85%
                      excess

Under these conditions, the reaction is pushed towards completion by the presence of a large excess of the cheaper of the two reagents. The acid is a solid, so the excess of alcohol serves as a solvent as well as a reagent for the reaction.

In another example of esterification, some excess of alcohol combined with provisions to remove water as it is formed results in a high yield of ester. Adipic acid is converted into its diester by heating with ethanol in the presence of sulfuric acid and toluene. Toluene, ethanol, and water form a constant boiling mixture that allows for the removal of water from the reaction mixture as it is formed.

$$\underset{\substack{\text{adipic acid}\\ \text{1 mole}}}{HO\overset{O}{\overset{\|}{C}}(CH_2)_4\overset{O}{\overset{\|}{C}}OH} + \underset{\substack{\text{ethanol}\\ \text{3 moles}}}{2\,CH_3CH_2OH} \xrightarrow[\substack{\text{toluene}\\ \Delta}]{H_2SO_4} \underset{\substack{\text{diethyl adipate}\\ 96\%}}{CH_3CH_2O\overset{O}{\overset{\|}{C}}(CH_2)_4\overset{O}{\overset{\|}{C}}OCH_2CH_3} + 2\,H_2O$$

PROBLEM 10.24 Complete the following equations.

(a) $HO\overset{O}{\overset{\|}{C}}(CH_2)_9\overset{O}{\overset{\|}{C}}OH \xrightarrow[\substack{H_2SO_4\\ \Delta}]{CH_3OH\ (\text{excess})} A$

(b) 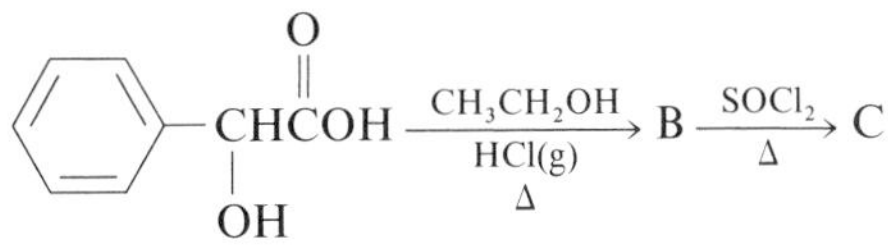

(c) 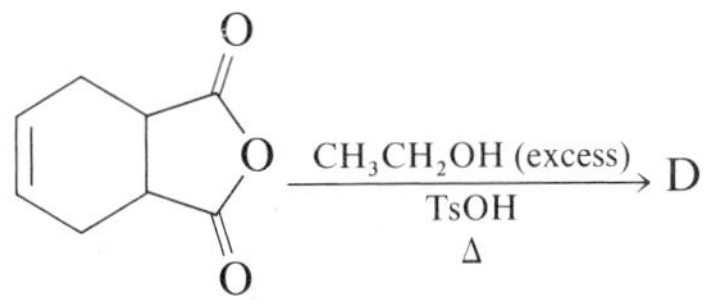

(d) $HO\overset{O}{\overset{\|}{C}}C{\equiv}C\overset{O}{\overset{\|}{C}}OH \xrightarrow[\substack{H_2SO_4\\ \Delta}]{CH_3OH\ (\text{excess})} E$

(e) $BrCH_2\overset{O}{\overset{\|}{C}}OH \xrightarrow[\substack{H_2SO_4\\ \Delta}]{CH_3CH_2OH} F$

(f) $CH_3CH{=}CH\overset{O}{\overset{\|}{C}}OH \xrightarrow[\substack{H_2SO_4\\ \text{benzene}\\ \Delta}]{\substack{CH_3CH_2CHCH_3\\ |\\ OH}} G$

PROBLEM 10.25 Methyl benzoate is prepared by heating 10 g of benzoic acid with 25 mL of methanol and 3 mL of concentrated sulfuric acid in an equilibrium reaction. (a) Write an equation for the reaction. (b) How would you remove excess methanol, sulfuric acid, and unreacted benzoic acid from methyl benzoate so that you were left with pure ester? This is an exercise in thinking about the physical and chemical properties of each reagent. Reviewing Sections 1.8B, 7.1A, and 10.3A will be useful to you.

C. Lipids: Fats, Oils, and Waxes

Esters are the chief constituents of a class of naturally occurring compounds known as **lipids** that are insoluble in water but soluble in hydrocarbon solvents. Some large, high molecular weight alcohols such as cholesterol (p. 272) and its esters, and other similar steroids (p. 543), are classified as lipids. Most lipids, however, are the fats in body tissues and in the blood stream; the oils isolated from plant sources such as sunflower seed oil, safflower seed oil, peanut oil, and olive oil; and waxes, which can have both animal and plant sources. All of these compounds are esters.

Fats and **oils** are esters of glycerol (1,2,3-propanetriol) with long-chain carboxylic acids, usually having 12 or more carbon atoms. The early isolation of carboxylic acids from fats has led to their being called **fatty acids.** The acids found in fats are predominantly saturated, while those in oils tend to be unsaturated, with one or more double bonds in the chain. The important fatty acids are shown as follows.

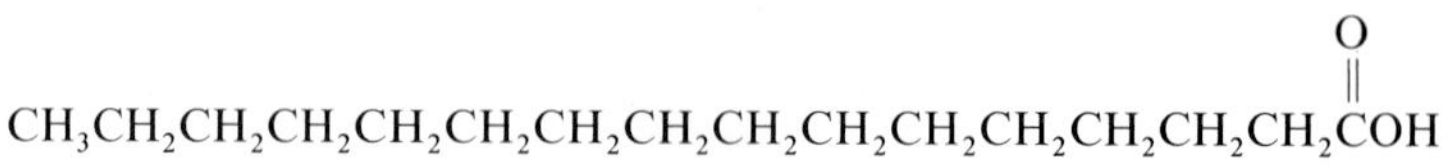

$$CH_3CH_2CH_2CH_2CH_2CH_2CH_2CH_2CH_2CH_2CH_2CH_2CH_2CH_2CH_2\overset{\displaystyle O}{\overset{\|}{C}}OH$$

hexadecanoic acid
palmitic acid
mp 62.9 °C

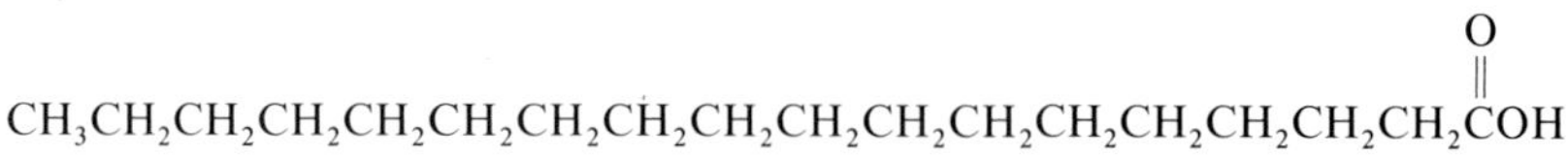

$$CH_3CH_2CH_2CH_2CH_2CH_2CH_2CH_2CH_2CH_2CH_2CH_2CH_2CH_2CH_2CH_2CH_2\overset{\displaystyle O}{\overset{\|}{C}}OH$$

octadecanoic acid
stearic acid
mp 69.6 °C

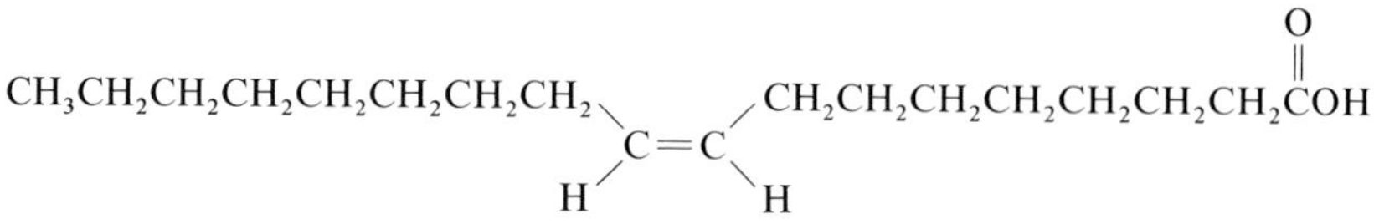

(Z)-9-octadecenoic acid
oleic acid
mp 13 °C

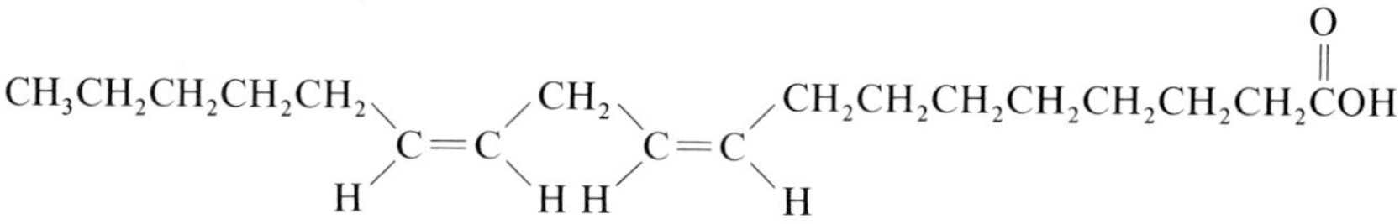

(9Z,12Z)-9,12-octadecadienoic acid
linoleic acid
mp −5 °C

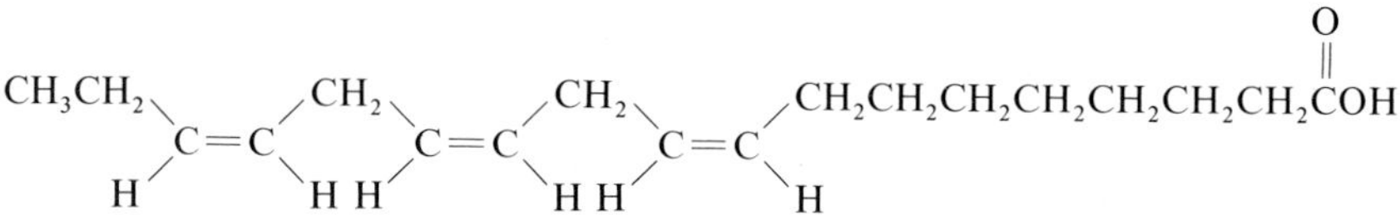

(9Z,12Z,15Z)-9,12,15-octadecatrienoic acid
linolenic acid
mp −16 °C

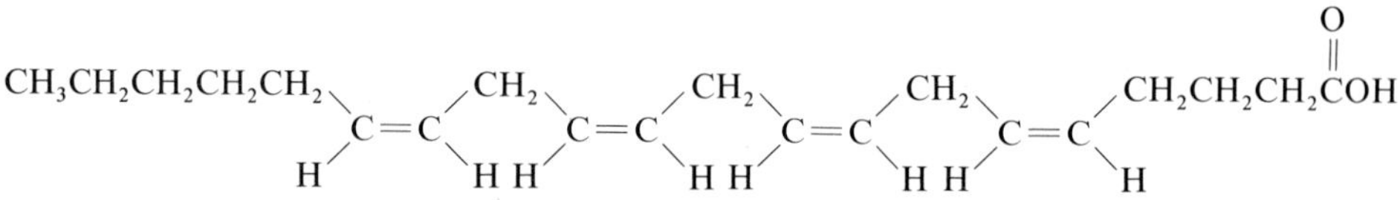

(5Z,8Z,11Z,14Z)-5,8,11,14-icosatetraenoic acid
arachidonic acid

The long-chain acids, palmitic acid and stearic acid, are found extensively in nature as the **saturated fatty acids** in solid fats. Note that they are solids at room temperature and at body temperature (37 °C). The melting point of a long-chain fatty acid decreases with the introduction of double bonds into the chain. Oleic acid melts below body temperature, but will solidify in a refrigerator. It is the chief fatty acid in olive oil and also in human body fat. As unsaturation increases, the **polyunsaturated fatty acids,** linoleic acid, linolenic acid, and arachidonic acid, have lower and lower melting points. The melting point of arachidonic acid is so low that it has not been determined accurately. These fatty acids are found as constituents of plant oils such as safflower seed oil, sesame oil, and peanut oil. Hydrogenation of the double bonds in oils (Section 8.2B) converts the liquid oils into fats.

The properties of the fatty acids are reflected in the properties of their esters. There is particular interest in the esters of cholesterol with fatty acids. Esters of cholesterol with saturated fatty acids are solids and have been implicated in the formation of solid deposits on the walls of blood vessels, causing diseases of the heart and the arteries. Physicians have recommended that we substitute vegetable oils containing poly-unsaturated fatty acids for animal fats, cooking oils, and margarines, which are rich in saturated fatty acids.

The glycerol esters of fatty acids are known as **triglycerides.** Two triglycerides, one containing saturated fatty acids and the other unsaturated ones, are shown below.

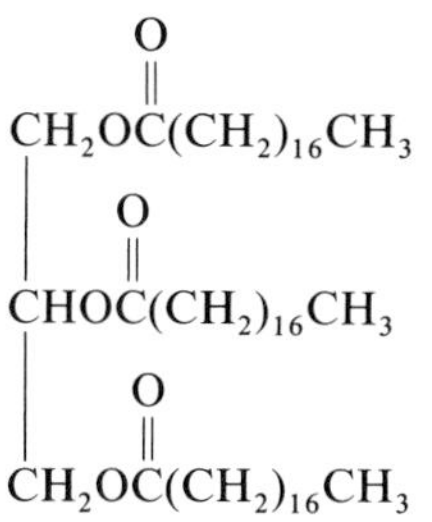

$$
\begin{array}{c}
\quad\quad\quad \overset{O}{\underset{\|}{}} \\
CH_2OC(CH_2)_{16}CH_3 \\
\mid \quad\quad \overset{O}{\underset{\|}{}} \\
CHOC(CH_2)_{16}CH_3 \\
\mid \quad\quad \overset{O}{\underset{\|}{}} \\
CH_2OC(CH_2)_{16}CH_3
\end{array}
$$

the ester of glycerol with three molecules of stearic acid
tristearin

a saturated triglyceride
a component of beef fat

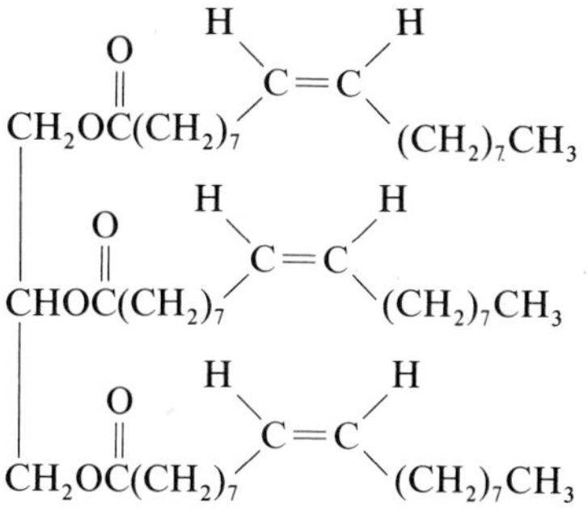

$$
\begin{array}{c}
\overset{O}{\underset{\|}{}} \quad\quad H \quad\quad\quad H \\
CH_2OC(CH_2)_7 \quad C=C \quad (CH_2)_7CH_3 \\
\mid \quad\quad \overset{O}{\underset{\|}{}} \quad H \quad\quad H \\
CHOC(CH_2)_7 \quad C=C \quad (CH_2)_7CH_3 \\
\mid \quad\quad \overset{O}{\underset{\|}{}} \quad H \quad\quad H \\
CH_2OC(CH_2)_7 \quad C=C \quad (CH_2)_7CH_3
\end{array}
$$

the ester of glycerol with three molecules of oleic acid
triolein

an unsaturated triglyceride
a component of olive oil

Because they were chosen for the sake of simplicity, the triglycerides shown have the same fatty acid esterifying each hydroxyl group in glycerol. In nature, two or three different fatty acids may be found in a single triglyceride.

As a result of a growing consciousness of the connections between nutrition and heart disease, many people monitor the levels of triglycerides and cholesterol in their blood. High levels of these two components of the blood have been correlated with higher risks of heart and artery disease. Researchers have recently discovered, however, that some of the cholesterol and blood fat can be tied up with blood proteins in *high-density lipoproteins,* or HDL. High-density lipoproteins seem to have some protective value against heart disease, as opposed to *low-density lipoproteins,* or LDL, which do not have such properties. Regular exercise is especially helpful in raising the ratio of high-density lipoproteins to low-density lipoproteins in the blood.

Waxes are esters of long-chain fatty acids, usually containing 24 to 28 carbon atoms, with long-chain (16 to 36 carbon atoms) primary alcohols, or with alcohols of the steroid series. They are solids used for lubricating and protecting surfaces. In plants, for example, waxes coat leaves and fruits with a waterproof layer.

$$CH_3(CH_2)_{15}O\overset{O}{\overset{\|}{C}}(CH_2)_{16}CH_3 \qquad CH_3(CH_2)_{15}O\overset{O}{\overset{\|}{C}}(CH_2)_{14}CH_3 \qquad CH_3(CH_2)_{29}O\overset{O}{\overset{\|}{C}}(CH_2)_{14}CH_3$$

cetyl stearate   cetyl palmitate   triacontyl palmitate

common in bacteria   *chief wax ester of spermaceti*   *a wax ester typical of those*
                                          from the sperm whale   *found in beeswax*

PROBLEM 10.26 9,12-Octadecadiynoic acid (p. 460) is a synthetic precursor to which of the poly-unsaturated fatty acids? How can 9,12-octadecadiynoic acid be transformed to the natural product?

D. Hydrolysis of Esters

The formation of an ester is a reversible process, and heating an ester with water and an acid or base as catalyst results in the formation of a carboxylic acid and an alcohol (Sections 10.4, 10.7B, and Problem 10.13). A number of studies have provided evidence for the mechanism that usually operates. The carbon atom of the carbonyl group of the ester is attacked by a nucleophile, a tetrahedral intermediate is formed, and the carbonyl group reappears with the loss of the alkoxyl group as a leaving group. The details of the reaction depend on whether conditions are acidic or basic.

In aqueous acid, the strongest and most abundant nucleophile is water itself. Water does not react very quickly with the unprotonated carbonyl group of an ester. In other words, the uncatalyzed hydrolysis of an ester in pure water is slow. Thus, the reversible protonation of the carbonyl group of an ester is postulated to be the first step in the mechanism for hydrolyses in acidic solutions.

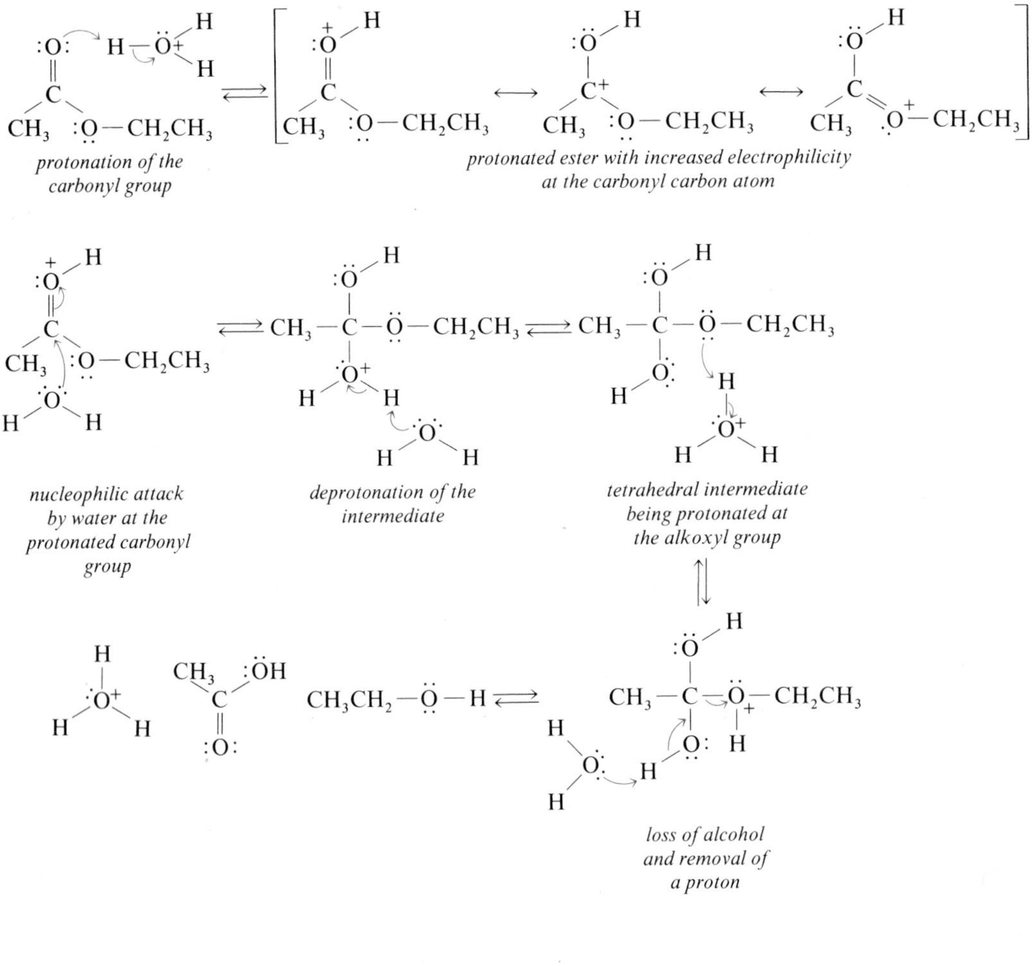

protonation of the carbonyl group

protonated ester with increased electrophilicity at the carbonyl carbon atom

nucleophilic attack by water at the protonated carbonyl group

deprotonation of the intermediate

tetrahedral intermediate being protonated at the alkoxyl group

loss of alcohol and removal of a proton

The carbonyl group, its electrophilicity increased by protonation, is attacked by water. Reversible loss of a proton from the resulting intermediate gives the tetrahedral intermediate. Protonation of the intermediate on either one of the hydroxyl groups is merely a reversal of the previous step and could lead, by loss of water, back to the protonated ester. Only when the oxygen atom of the alkoxyl group is protonated does a new reaction become possible. The exact timing of what happens next is not easy to determine. Simultaneous removal of a proton, with formation of the carbonyl group assisting in the departure of the alcohol, is shown. Each step of the acid-catalyzed hydrolysis reaction should be compared with the steps of the esterification reaction in Section 10.4, p. 456.

If the hydrolysis reaction is carried out in base, it is called a **saponification reaction.** The process is often described in stories of the Old West, where pioneer women made soap by heating animal fats saved from their household cooking with base, essentially potassium hydroxide derived from wood ashes. Animal fats are a mixture of the esters of long-chain fatty acids with glycerol (Section 10.7C). Heating with base gives glycerol and the salts of the fatty acids, which form micelles, and hence have the property of emulsifying and solubilizing grease (Section 10.3). A typical fat and its conversion to glycerol and soap is shown below.

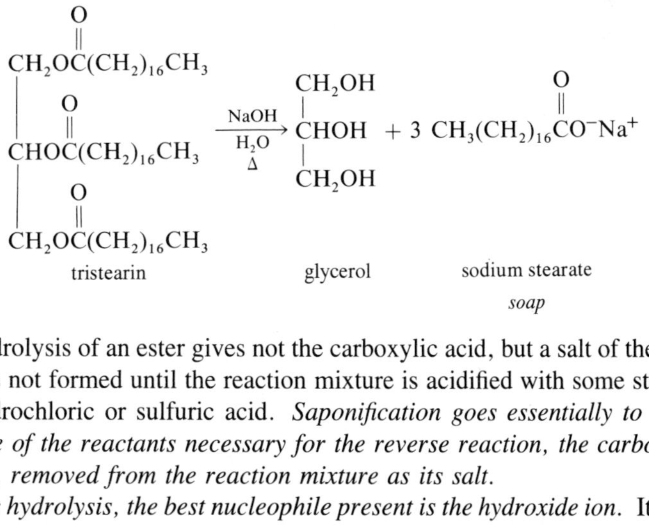

$$CH_2OC(CH_2)_{16}CH_3 \ \ | \ \ CHOC(CH_2)_{16}CH_3 \ \ | \ \ CH_2OC(CH_2)_{16}CH_3 \quad \xrightarrow[\Delta]{\underset{H_2O}{NaOH}} \quad CH_2OH \ | \ CHOH \ | \ CH_2OH \quad + \ 3 \ CH_3(CH_2)_{16}CO^-Na^+$$

tristearin   glycerol   sodium stearate

soap

Basic hydrolysis of an ester gives not the carboxylic acid, but a salt of the acid. The acid itself is not formed until the reaction mixture is acidified with some stronger acid such as hydrochloric or sulfuric acid. *Saponification goes essentially to completion because one of the reactants necessary for the reverse reaction, the carboxylic acid, is, in effect, removed from the reaction mixture as its salt.*

In a basic hydrolysis, the best nucleophile present is the hydroxide ion. It attacks the carbonyl group directly to give a tetrahedral intermediate, which falls apart to the acid and alkoxide ion.

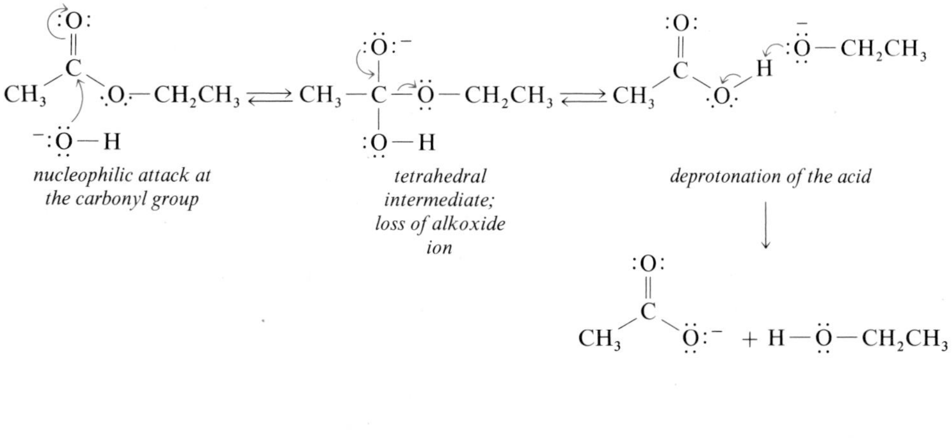

nucleophilic attack at the carbonyl group   tetrahedral intermediate; loss of alkoxide ion   deprotonation of the acid

Transfer of the proton from the acid to the strongly basic ethoxide ion completes the reaction.

In both mechanisms, the bond that is broken is the one between the carbonyl group and the alkoxyl group. A great deal of research has been carried out to prove that this is indeed the case. One kind of evidence comes from isotopic tracer studies. When pentyl acetate is allowed to stand in water enriched with ^{18}O in the presence of sodium hydroxide, no ^{18}O appears in the alcohol recovered from the reaction mixture.

$$2 H_2{}^{18}O + 2 Na \longrightarrow 2 Na^+ \, {}^{18}OH^- + H_2\uparrow$$

$$\underset{\text{pentyl acetate}}{CH_3\overset{O}{\overset{\|}{C}}OCH_2CH_2CH_2CH_2CH_3} + Na^+ \, {}^{18}OH^- \xrightarrow[H_2{}^{18}O]{}$$

$$\underset{\substack{\text{1-pentanol}\\\text{containing no }{}^{18}O}}{CH_3CH_2CH_2CH_2CH_2OH} + \underset{\substack{\text{sodium acetate}\\\text{containing }{}^{18}O\\\text{distributed between}\\\text{the two oxygen atoms}}}{CH_3\overset{{}^{18}O}{\overset{\|}{C}}-{}^{18}O^- \, Na^+}$$

This experimental determination excludes a mechanism such as one involving an S_N2 reaction by hydroxide ion on the pentyl group.

$$CH_3-\overset{:O:}{\overset{\|}{C}}-\ddot{O}-CH_2CH_2CH_2CH_2CH_3 \xrightarrow{\;/\!/\;}$$
$$^{18}:\ddot{O}-H$$

$$CH_3-\overset{:O:}{\overset{\|}{C}}-\ddot{O}:^- + \underset{\substack{\text{1-pentanol containing }{}^{18}O\\\textit{not} \text{ found experimentally}}}{CH_3CH_2CH_2CH_2CH_2-{}^{18}\ddot{O}H}$$

PROBLEM 10.27 ^{18}O is shown as being present as either one of the two oxygen atoms in acetate ion in the equation for the hydrolysis of pentyl acetate in $H_2{}^{18}O$. Why is this so?

Another probe for the mechanism of a reaction is to determine the stereochemistry of the reaction. Especially if you are considering an S_N2 reaction at an alkyl group, using a chiral alkyl group makes it possible to discover whether the reaction goes by inversion at the chiral center. Racemization of a chiral alkyl group, on the other hand, would suggest that a carbocation intermediate had formed. In an actual experiment, optically active (S)-$(-)$-2-hydroxybutanedioic acid (malic acid), isolated from natural sources, was converted into its acetic ester by a reaction that does not break the carbon-oxygen bond at the chiral center (Section 10.4). The ester was hydrolyzed with aqueous potassium hydroxide.

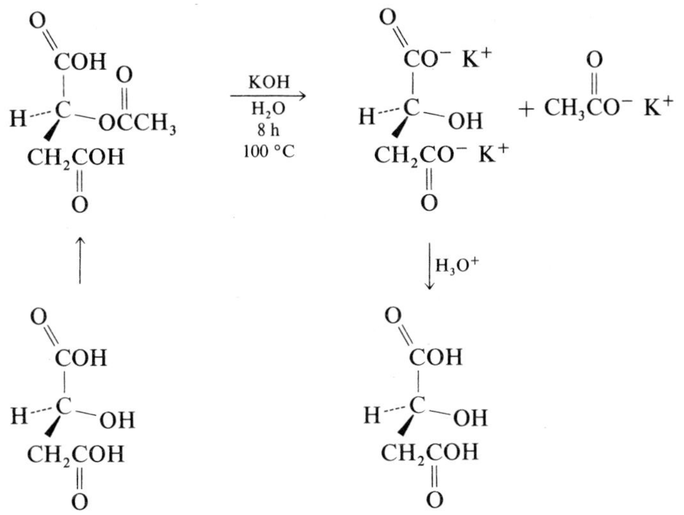

(S)-(−)-2-hydroxybutanedioic acid,   (S)-(−)-2-hydroxybutanedioic acid,
used as starting material   recovered

The optical rotation of the recovered 2-hydroxybutanedioic acid was found to be essentially identical with that of the sample of 2-hydroxybutanedioic acid used as a starting material for the sequence of reactions. This experiment confirms that nucleophilic attack is at the carbon atom of the carbonyl group and not at the alkyl carbon atom. Cleavage of the ester bond takes place between the acyl group and the alkoxyl group.

PROBLEM 10.28 Write a detailed mechanism for the basic hydrolysis of the acetate ester of (S)-(−)-2-hydroxybutanedioic acid showing first the stereochemistry that would result from attack at the carbonyl carbon atom, and then from attack at the chiral alkyl carbon atom. Prove to yourself that the experimental result obtained above does indeed show that nucleophilic attack takes place at the carbonyl group.

E. Steric and Electronic Effects in the Formation and Hydrolysis of Esters

If the rate-determining step in esterification and hydrolysis reactions is the formation of the tetrahedral intermediate, the rate of these reactions will be affected by the ease with which a nucleophile approaches the carbonyl group. Sterically, hindrance at the carbonyl group would be expected to decrease the rate of hydrolysis, and of esterification. The presence of electron-withdrawing groups next to the carbonyl group of an ester would be expected to increase the rate of hydrolysis of the ester by increasing the electrophilicity of the carbonyl group and, hence, its reactivity towards a nucleophile.

Both of these expectations are borne out by experiments. For example, 2,2-dimethylpropanoic acid reacts with methanol at a rate 27 times slower than acetic acid itself. For 2,2-diethylbutanoic acid, the rate falls even more dramatically.

O   CH₃ O   CH₃CH₂ O

$$\underset{\text{acetic acid}}{\text{CH}_3\text{COH}} \qquad \underset{\substack{| \\ \text{CH}_3 \\ \text{2,2-dimethylpropanoic acid}}}{\text{CH}_3\text{C}-\text{COH}} \qquad \underset{\substack{| \\ \text{CH}_3\text{CH}_2 \\ \text{2,2-diethylbutanoic acid}}}{\text{CH}_3\text{CH}_2\text{C}-\text{COH}}$$

Relative rates   1.0   0.037   0.00016

the effect of steric hindrance on the esterification with methanol of a series of carboxylic acids at 40 °C

Clearly, crowding around the carbonyl group affects the rate of esterification.

Hydrolysis reactions are as hindered by steric factors as esterification reactions are. Rates of hydrolysis for a series of ethyl esters are given below.

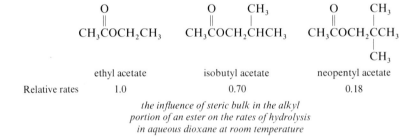

ethyl acetate   ethyl 2-methylpropanoate   ethyl 2,2-dimethylpropanoate

Relative rates   1.0   0.10   0.011

the effect of steric hindrance on basic hydrolysis rates for a series of esters, in 88% ethanol at 30 °C

As the ester becomes more crowded by the substitution of alkyl groups on the α-carbon atom, the rate of hydrolysis declines.

Increased bulkiness on the alkyl portion of an ester also decreases the rate of hydrolysis, as shown by the relative rates of hydrolysis for a series of esters.

$$\underset{\text{ethyl acetate}}{\text{CH}_3\text{COCH}_2\text{CH}_3} \qquad \underset{\text{isobutyl acetate}}{\text{CH}_3\text{COCH}_2\text{CHCH}_3} \qquad \underset{\text{neopentyl acetate}}{\text{CH}_3\text{COCH}_2\text{CCH}_3}$$

Relative rates   1.0   0.70   0.18

the influence of steric bulk in the alkyl portion of an ester on the rates of hydrolysis in aqueous dioxane at room temperature

The figures show that an increase in bulk, even when one carbon atom removed from the ester function, hinders the hydrolysis reaction.

We can see the electronic effects on rates of hydrolysis by comparing the rates of reaction for methyl chloroacetate and methyl dichloroacetate with that for methyl acetate.

methyl acetate   methyl chloroacetate   methyl dichloroacetate

Relative rates   1.0   761   16,000

the effect of substitution of electron-withdrawing groups on the carbon atom adjacent to the carbonyl group on the rate of basic hydrolysis of esters, in water at 25 °C

The rate of hydrolysis increases greatly with the substitution of an electron-withdrawing chlorine atom on the carbon atom adjacent to the carbonyl group. This result is believed to reflect increased positive character at the carbonyl group, and therefore increased ease of reaction with an incoming nucleophilic reagent.

The data presented in this section indicate that the rate-determining step for the reaction of carboxylic acids with alcohols and for the reverse reaction, the hydrolysis of esters, is the attack by a nucleophile on the carbonyl group. Once the tetrahedral intermediate is formed, its decomposition is rapid.

PROBLEM 10.29   In each of the following series, predict which reaction would be expected to go the fastest and which the slowest.

1.   For the reaction

$$\underset{\displaystyle \text{RCOR}'}{\overset{\displaystyle O}{\overset{\displaystyle \|}{}}} + \text{NaOH} \xrightarrow[\Delta]{H_2O} \underset{\displaystyle \text{RCO}^-\text{Na}^+}{\overset{\displaystyle O}{\overset{\displaystyle \|}{}}} + \text{R'OH}$$

(a) $CH_3CH_2\overset{O}{\overset{\|}{C}}OCH_2CH_3, \quad CH_3\overset{O}{\overset{\|}{C}}OCH_2CH_3, \quad CH_3\overset{O}{\overset{\|}{C}}-\overset{O}{\overset{\|}{C}}OCH_2CH_3$

(b) $-\overset{O}{\overset{\|}{C}}OCH_3, \quad O_2N-$ $-\overset{O}{\overset{\|}{C}}OCH_3, \quad CH_3O-$ $-\overset{O}{\overset{\|}{C}}OCH_3$

2. For the reaction

$$\underset{\displaystyle \text{RCOH}}{\overset{\displaystyle O}{\overset{\displaystyle \|}{}}} + CH_3OH \xrightarrow[\text{acid}]{} \underset{\displaystyle \text{RCOCH}_3}{\overset{\displaystyle O}{\overset{\displaystyle \|}{}}} + H_2O$$

$$CH_3\overset{O}{\overset{\|}{C}}OH, \quad CH_3\underset{\underset{\displaystyle CH_3}{|}}{\overset{\overset{\displaystyle CH_3}{|}}{C}}CH_2\overset{O}{\overset{\|}{C}}OH, \quad CH_3\underset{\underset{\displaystyle CH_3}{|}}{CH}-\underset{\underset{\displaystyle CH}{|}\underset{\diagdown \,\diagup}{}}{CH}-\overset{O}{\overset{\|}{C}}OH$$

F.   Transesterification Reactions

Transesterification *is the conversion of one ester into another by heating it with an excess of either a new alcohol or a new carboxylic acid in the presence of an acidic or a basic catalyst.* In an equilibrium reaction, either the alcohol or the acid portion of the original ester is freed. Such a reaction is carried out when there are practical reasons why an ordinary hydrolysis reaction would not work. For example, 2-chloro-2-phenylacetic acid is freed from its ethyl ester by a transesterification reaction.

ethyl 2-chloro-2-phenylacetate	acetic acid	2-chloro-2-phenylacetic	ethyl acetate
1 mole	7 moles	acid	
	bp 118 °C	mp 78 °C	bp 77 °C

The presence of the chlorine substituent on the carboxylic acid makes prolonged heating in aqueous base unwise; a nucleophilic substitution at that carbon atom as well as the saponification reaction may take place. Even long heating in dilute acid may lead to some substitution reaction at the benzylic position. Concentrated hydrochloric acid is used to catalyze the transesterification reaction in which the ethoxyl group is transferred from 2-chloro-2-phenylacetic acid to acetic acid. The other product of the reaction, ethyl acetate, and the excess of acetic acid, are easily separated from the solid 2-chloro-2-phenylacetic acid by distillation.

In another example, methyl acrylate is converted into *n*-butyl acrylate by heating with *n*-butyl alcohol in the presence of *p*-toluenesulfonic acid as the catalyst.

$$CH_2{=}CHCOCH_3 + CH_3CH_2CH_2CH_2OH \xrightarrow[\Delta]{TsOH} CH_2{=}CHCOCH_2CH_2CH_2CH_3 + CH_3OH$$

methyl acrylate   *n*-butyl alcohol   *n*-butyl acrylate   methanol
bp 81 °C   bp 117 °C   bp 145 °C   bp 65 °C

The difference in boiling points between the alcohols allows the equilibrium to be shifted towards the higher molecular weight ester by distillation of the methanol out of the reaction mixture.

PROBLEM 10.30   Write out a detailed mechanism for the following transesterification reaction.

$$CH_3COCH_2CH_3 + CH_3OH \underset{H_2SO_4}{\rightleftharpoons} CH_3COCH_3 + CH_3CH_2OH$$

PROBLEM 10.31   Coconut oil consists mainly of triglycerides of octanoic, dodecanoic, and tetradecanoic acids. It is used to synthesize the ethyl esters of those acids. Write an equation showing how this could be done.

G. Biological Transesterification Reactions

The process of transesterification goes on all the time in the body, usually with thioesters. **Thioesters** are compounds in which the oxygen atom of the alkoxyl group in an ester has been replaced by a sulfur atom. An important thioester that participates in a myriad of physiological processes is acetyl coenzyme A.

$$ROPOCH_2C\!-\!CH\!-\!CNHCH_2CH_2CNHCH_2CH_2SCCH_3$$

acetyl coenzyme A

thioester

usually abbreviated $CH_3CS\!-\!CoA$ *or* acetyl CoA

A relatively small molecule, usually incorporating a vitamin, that cooperates with enzymes in the catalysis of chemical reactions under the conditions that prevail in living organisms is called a **coenzyme.** The vitamin present in coenzyme A is pantothenic acid, one of the B vitamins. It is linked by an amide bond at one end to

2-aminoethanethiol, and at the other end through a phosphate group to a nucleotide, adenosine 3′,5′-diphosphate. The structures of nucleotides are discussed in Sections 14.5B and 18.6B. At present, we are interested in the thiol group at the end of coenzyme A.

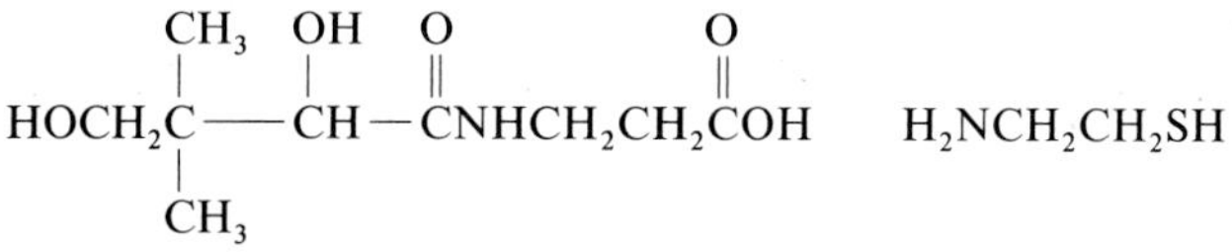

$$
\underset{\substack{\text{pantothenic acid} \\ \textit{a B vitamin}}}{\text{HOCH}_2\overset{\overset{\displaystyle CH_3}{|}}{\underset{\underset{\displaystyle CH_3}{|}}{C}}\!-\!\overset{\overset{\displaystyle OH}{|}}{CH}\!-\!\overset{\overset{\displaystyle O}{\|}}{C}NHCH_2CH_2\overset{\overset{\displaystyle O}{\|}}{C}OH} \qquad \underset{\text{2-aminoethanethiol}}{H_2NCH_2CH_2SH}
$$

$$
\underset{\underset{\text{nucleotide}}{\underbrace{}}}{RO\overset{\overset{\displaystyle O}{\|}}{P}\underset{\underset{\displaystyle O_-}{|}}{O}CH_2}\overset{\overset{\displaystyle CH_3}{|}}{\underset{\underset{\displaystyle CH_3}{|}}{C}}\!-\!\overset{\overset{\displaystyle OH}{|}}{CH}\!-\!\overset{\overset{\displaystyle O}{\|}}{C}NHCH_2CH_2\overset{\overset{\displaystyle O}{\|}}{C}NHCH_2CH_2SH \underset{\text{thiol group}}{\nwarrow}
$$

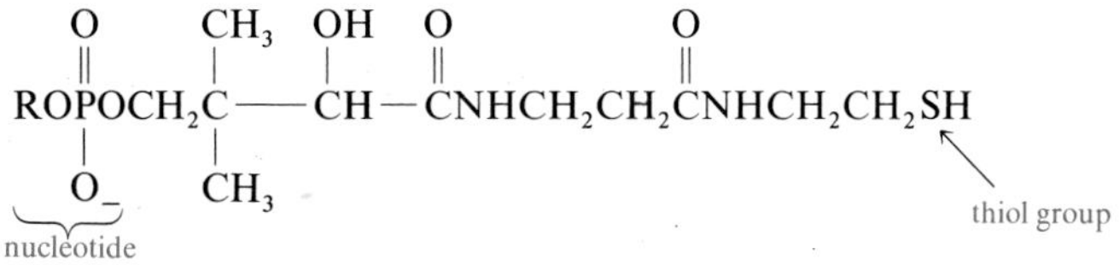

coenzyme A

abbreviated CoA—SH

A thioester is more reactive than the corresponding oxyester. There is less resonance interaction between the larger sulfur atom and the adjacent carbonyl group, so the acyl group is more vulnerable to nucleophilic attack. Thus, the acyl group of a thioester is easily transferred to another sulfur, oxygen, or nitrogen atom. Compounds that lose their acyl group easily in such reactions are called **acyl transfer agents.**

Acetyl coenzyme A is a good acyl transfer agent. For example, an important reaction in the smooth functioning of the nervous system is the synthesis of acetylcholine at the nerve synapses as required (Section 16.7B). Acetyl coenzyme A participates in what is essentially a transesterification reaction.

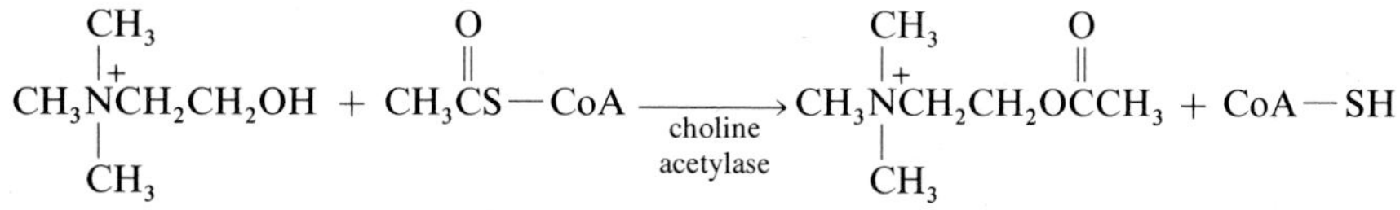

$$
\underset{\substack{\text{2-(trimethylammonium)ethanol} \\ \text{choline}}}{\overset{\overset{\displaystyle CH_3}{|}}{CH_3\overset{+}{N}CH_2CH_2OH}} + \underset{\text{acetyl CoA}}{CH_3\overset{\overset{\displaystyle O}{\|}}{C}S\!-\!CoA} \xrightarrow[\substack{\text{choline} \\ \text{acetylase}}]{} \underset{\substack{\text{acetylcholine} \\ \textit{essential in the} \\ \textit{transmission of} \\ \textit{nerve impulses}}}{\overset{\overset{\displaystyle CH_3}{|}}{CH_3\overset{+}{N}CH_2CH_2O}\overset{\overset{\displaystyle O}{\|}}{C}CH_3} + \underset{\text{coenzyme A}}{CoA\!-\!SH}
$$

The thioester, acetyl coenzyme A, is converted into an oxyester by the alcohol function in choline. Coenzyme A is the leaving group. In the body, the process is catalyzed by an enzyme called appropriately choline acetylase.

PROBLEM 10.32   Write a complete mechanism for the acyl transfer reaction shown above. You may assume that the enzyme is a good source of protons, HB^+, and bases to remove protons, $B:$, as necessary.

PROBLEM 10.33   The vitamin pantothenic acid is sold over the counter as calcium pantothenate. Write the structural formula for calcium pantothenate.

H. Lactones

Lactones *are cyclic esters formed when a carboxyl function and a hydroxyl group are present within the same molecule.* In fact, the formation of a five- or six-membered ring is so highly favored that reactions giving compounds with a carboxyl group and a hydroxyl group separated by three or four carbon atoms usually give a lactone directly as the product.

An example of such a reaction is the reduction of 4-oxopentanoic acid with sodium borohydride. Sodium borohydride reduces the ketone function (Section 9.3) but not the carboxylic acid. The product of the reaction after acidification is the lactone of 4-hydroxypentanoic acid.

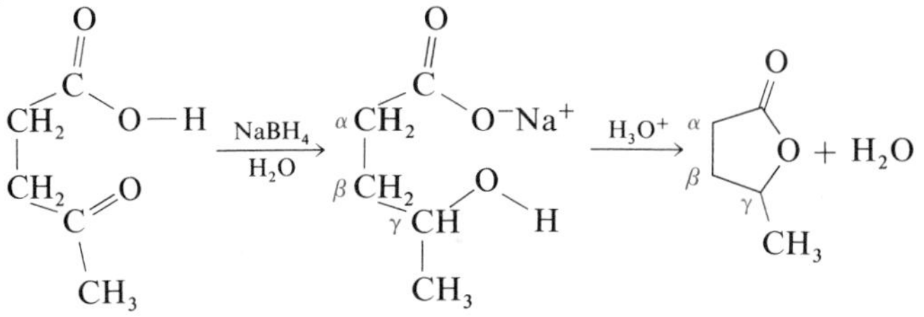

4-oxopentanoic acid	sodium salt of 4-hydroxypentanoic acid	lactone of 4-hydroxypentanoic acid
		81%
		a γ-lactone

A five-membered ring lactone is called a γ-lactone because it is formed when an intramolecular reaction takes place between the carboxyl group and the hydroxyl group on the γ-carbon atom of the acid (Section 9.1C).

A six-membered ring lactone, or δ-lactone, is formed when ethyl 5-acetoxypentanoate, synthesized by a nucleophilic substitution reaction on ethyl 5-chloropentanoate, is hydrolyzed.

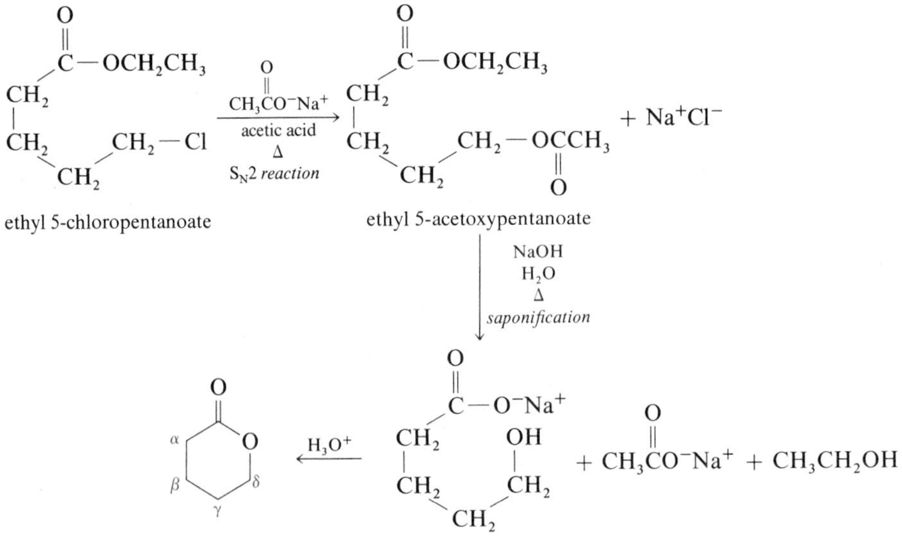

ethyl 5-chloropentanoate   ethyl 5-acetoxypentanoate

NaOH
H_2O
Δ
saponification

lactone of 5-hydroxypentanoic acid
50%

sodium 5-hydroxypentanoate
both ester functions hydrolyzed

a δ-lactone

Note that for both examples given, the lactone does not form as long as the acid exists as a carboxylate anion. Only upon acidification of the solution does ring closure take place. Lactones, like all esters, can be hydrolyzed into an acid and an alcohol component, which happen to be parts of the same molecule. The ring will remain open as long as the acid is present as its salt. Acidification usually leads to the isolation of the lactone instead of the hydroxyacid.

While it is harder to form lactones with more than five or six atoms in the ring in the laboratory, some large-ring lactones are found in nature. Some of them, along with the large-ring ketones, muscone, isolated from the scent glands of the musk deer, and civetone, from the civet cat, are highly valued by the perfume industry. In concentrated form, all of these compounds have rather unpleasant odors. The odors become attractive in high dilution and are particularly useful for perfumes because they enhance and fix scents derived from other sources. The structures of four of the large-ring compounds found in nature, two ketones and two lactones, are shown below.

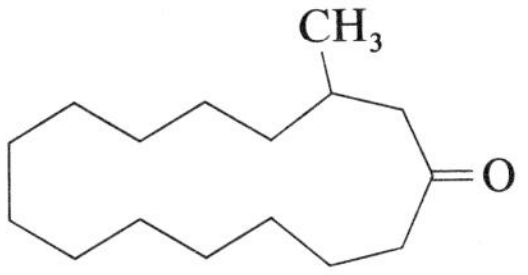

muscone
3-methylcyclopentadecanone
from the musk deer

15-membered ring

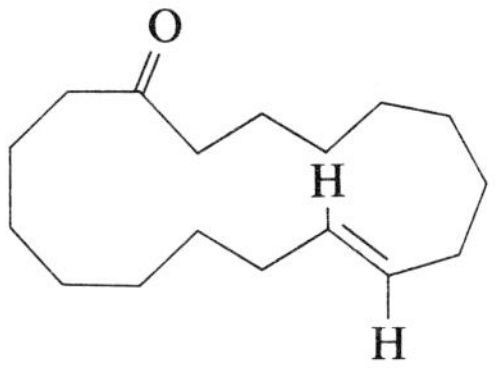

civetone
(*E*)-9-cycloheptadecenone
from the civet cat

17-membered ring

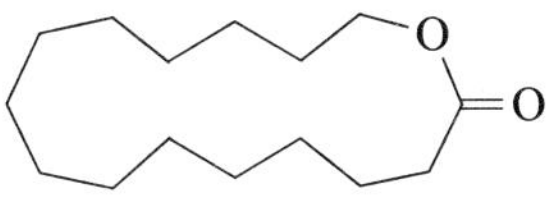

lactone of 15-hydroxypentadecanoic
acid from angelica oil, from the
roots of *Angelica archangelica*

16-membered ring

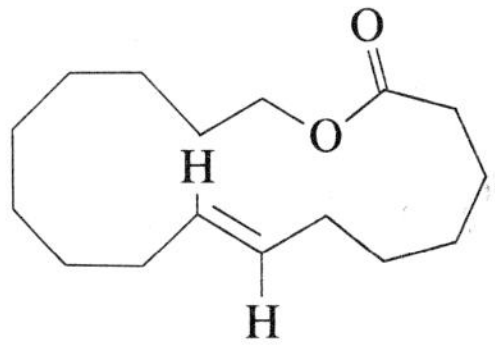

lactone of (*E*)-16-hydroxy-7-hexadecenoic acid
musk ambrette from the oil from the seeds of
Hibiscus abelmoschus

17-membered ring

The rings in these compounds contain 15 to 17 atoms, whether they are lactones or ketones. The odor of cyclic ketones has been found to correlate with ring size; the musky odors of interest to the perfume industry are found in compounds having 14 to 17 members to the ring. Larger ring compouunds have much less odor—at least, to the human nose.

Many other large-ring lactones that have recently been isolated from natural sources are interesting because they have antibiotic or antitumor activity. These lactones are known as **macrolides.** Two macrolides are shown below.

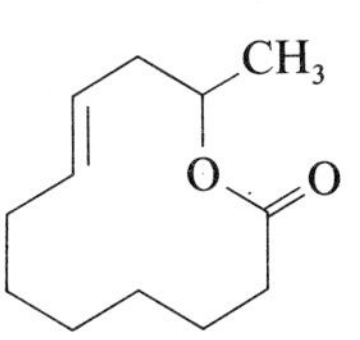

recifeiolide
from the fungus
Cephalosporium recifei

*a 12-membered ring
lactone natural product*

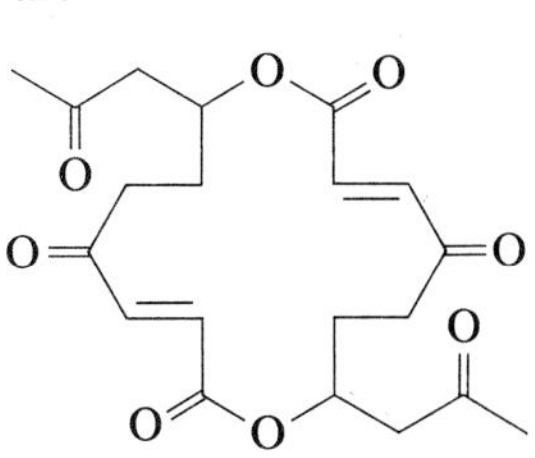

vermiculine
from the microorganism
Penicillium vermiculatum

*a 16-membered ring
dilactone*

PROBLEM 10.34   Complete the following equations.

(a) 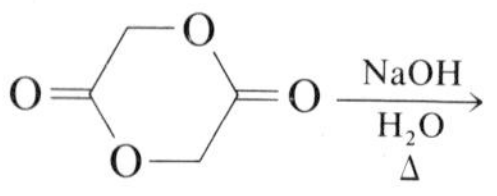 $\xrightarrow[\substack{H_2O \\ \Delta}]{NaOH}$

(b) 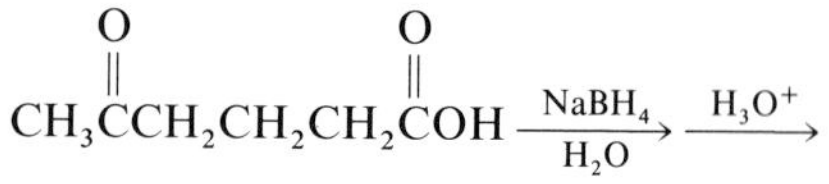 $CH_3\overset{O}{\overset{\|}{C}}CH_2CH_2CH_2\overset{O}{\overset{\|}{C}}OH \xrightarrow[H_2O]{NaBH_4} \xrightarrow{H_3O^+}$

(c) $\xrightarrow[\substack{H_2O \\ \Delta}]{NaOH}$

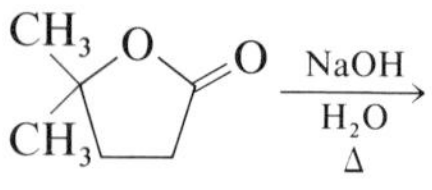

(d) 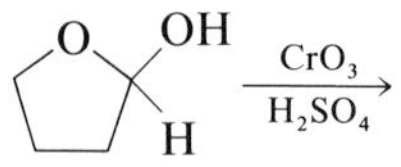 $\xrightarrow{\substack{CrO_3 \\ H_2SO_4}}$

PROBLEM 10.35   An intermediate in the synthesis of the vitamin biotin is prepared by the following sequence of reactions.

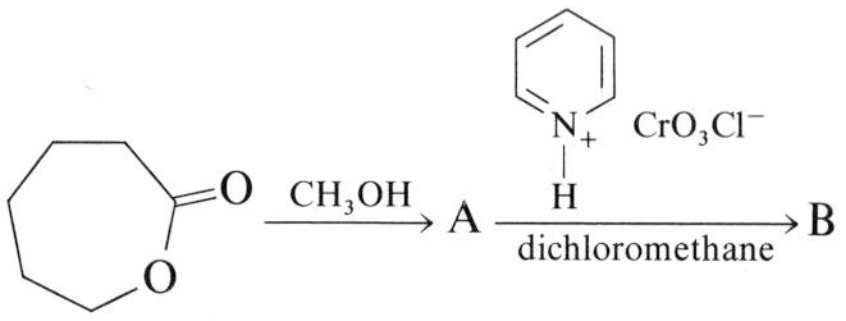

Assign structures to Compounds A and B. Biotin has the following structure.

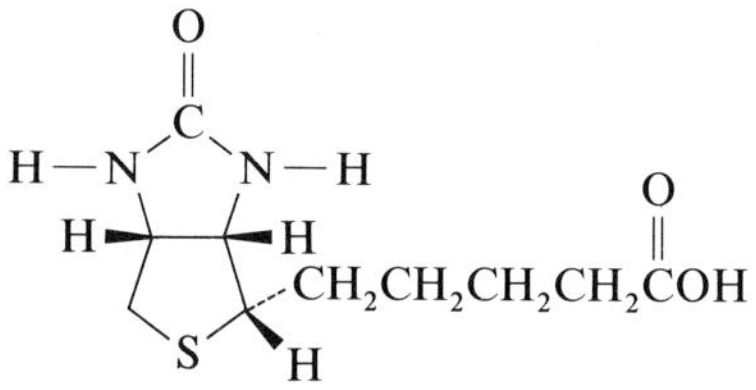

10.8

Amides

A. Preparation of Amides

Amides are prepared by nucleophilic substitution reaction of ammonia derivatives on acid chlorides, anhydrides, and esters. Most amides are solids and as such are useful derivatives of amines. For example, *p*-toluidine is converted to *N*-acet-*p*-toluidide by acetic anhydride.

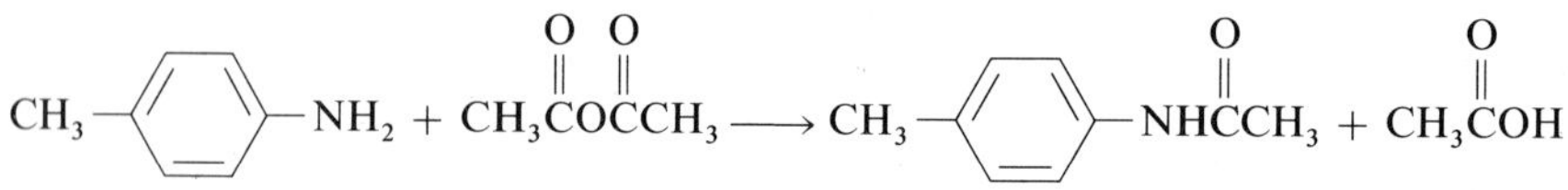

N,N-Disubstituted amides are formed when acid chlorides or anhydrides react with amines with two substituents on them. An example is the conversion of cyclohexane-carboxylic acid to *N,N*-dimethylcyclohexanecarboxamide.

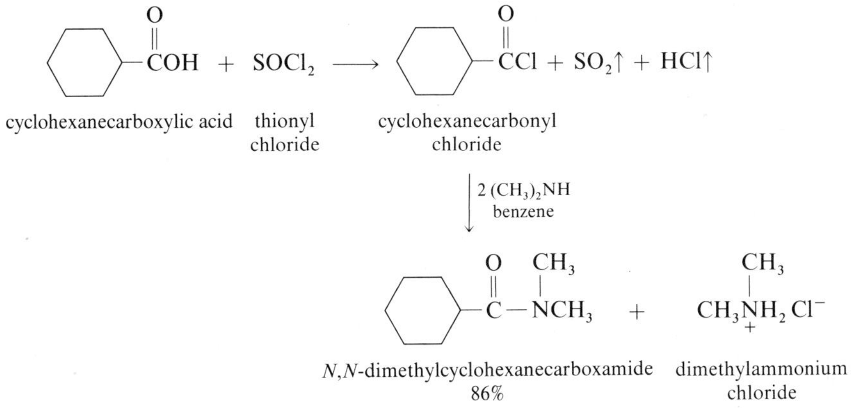

cyclohexanecarboxylic acid   thionyl
chloride

cyclohexanecarbonyl
chloride

$2\,(CH_3)_2NH$
benzene

N,N-dimethylcyclohexanecarboxamide   dimethylammonium
86%   chloride

If the amine has no hydrogen atom on the nitrogen atom, an isolable amide is not formed when it reacts with an acid chloride or an acid anhydride. Such amines do, however, form unstable reactive intermediates with acid chlorides, as illustrated for the reaction of benzoyl chloride and pyridine.

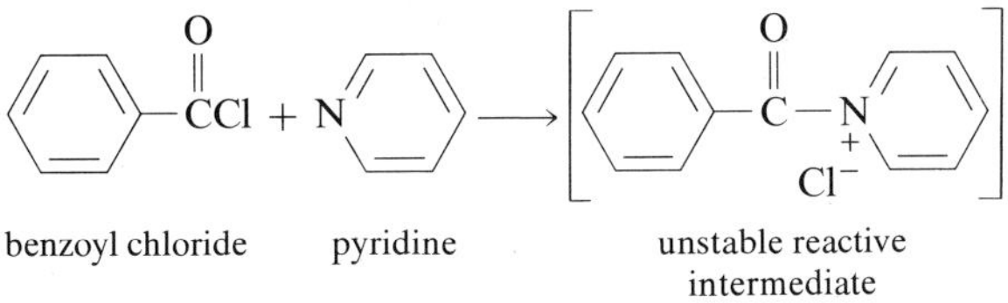

benzoyl chloride   pyridine   unstable reactive
intermediate

Just as esters are subject to hydrolysis reactions in which the ester bond is cleaved by water, they also react in **aminolysis reactions,** in which ammonia (or an amine) cleaves the ester. The products of such reactions are amides and alcohols. For example, ethyl lactate is converted into lactamide by treatment with liquid ammonia.

ethyl lactate   ammonia   lactamide   ethanol
70%

Another example demonstrates clearly the difference in reactivity towards nucleophilic substitution reactions, contrasted in Section 10.4, of a carbonyl group and a carbon-halogen bond. Ethyl chloroacetate, when treated with aqueous ammonium hydroxide at low temperatures and for short periods of time, undergoes substitution quite selectively at the carbonyl group.

ethyl chloroacetate   ammonia   chloroacetamide   ethanol

All of the methods shown above start with an acid derivative that already has the same carbon skeleton as the amide being prepared. *Nitriles* (Section 10.5C) *are a route to amides containing one more carbon atom than the starting material,* an alkyl halide. While vigorous hydrolysis of a nitrile group gives an acid, an amide is formed as an intermediate and may be isolated if the reaction is conducted with care. For example, phenylacetonitrile gives phenylacetamide if treated with hydrochloric acid for one hour at 40 °C, but gives phenylacetic acid if heated to boiling with aqueous sulfuric acid for three hours.

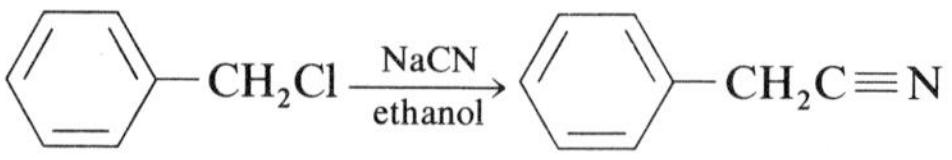

$$\text{benzyl chloride} \quad \xrightarrow[\text{ethanol}]{\text{NaCN}} \quad \text{phenylacetonitrile}$$

benzyl chloride — CH_2Cl

phenylacetonitrile — $CH_2C{\equiv}N$

contains one more carbon atom than starting material

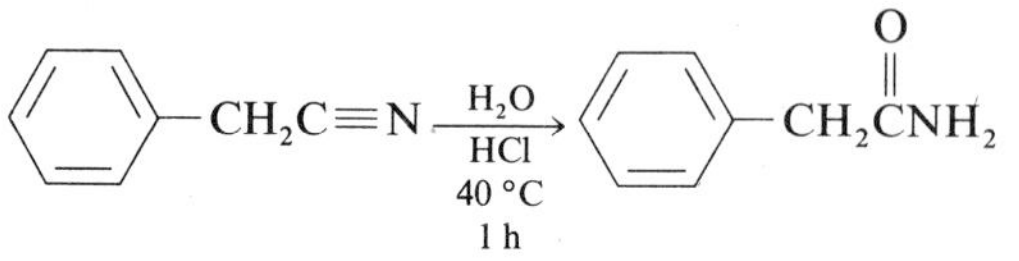

$$-CH_2C{\equiv}N \xrightarrow[\substack{HCl \\ 40\,°C \\ 1\,h}]{H_2O} -CH_2\overset{O}{\overset{\|}{C}}NH_2$$

phenylacetonitrile

phenylacetamide
80%

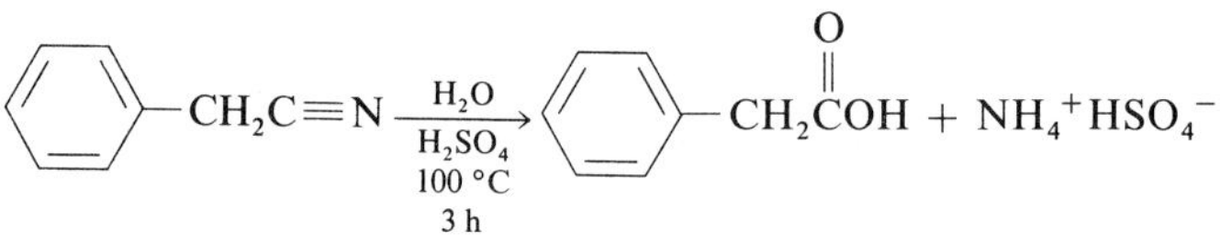

$$-CH_2C{\equiv}N \xrightarrow[\substack{H_2SO_4 \\ 100\,°C \\ 3\,h}]{H_2O} -CH_2\overset{O}{\overset{\|}{C}}OH + NH_4{}^+HSO_4{}^-$$

phenylacetonitrile

phenylacetic acid

ammonium hydrogen sulfate

PROBLEM 10.36   Complete the following equations.

(a) $CH_3CH_2O\overset{O}{\overset{\|}{C}}CH{=}CH\overset{O}{\overset{\|}{C}}OCH_2CH_3 \xrightarrow[\substack{NH_4Cl \\ H_2O}]{NH_3 \text{ (excess)}} A$

(b) 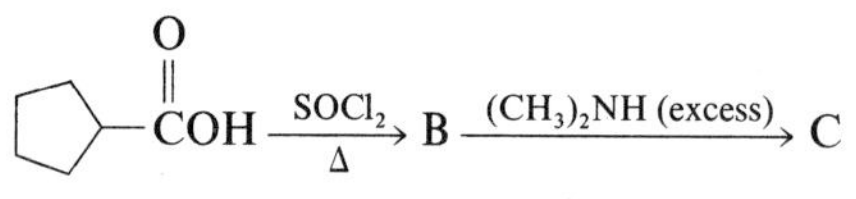 $\langle\text{cyclopentyl}\rangle-\overset{O}{\overset{\|}{C}}OH \xrightarrow[\Delta]{SOCl_2} B \xrightarrow{(CH_3)_2NH \text{ (excess)}} C$

(c) $CH_3CH_2O\overset{O}{\overset{\|}{C}}CH_2CH_2CH_2CH_2\overset{O}{\overset{\|}{C}}OCH_2CH_3 \xrightarrow{H_2NNH_2 \text{ (excess)}} D$

(d) $CH_3(CH_2)_4\underset{\underset{CH_2CH_3}{|}}{CH}\overset{O}{\overset{\|}{C}}OH \xrightarrow[\Delta]{SOCl_2} E \xrightarrow[H_2O]{NH_3} F$

(e) 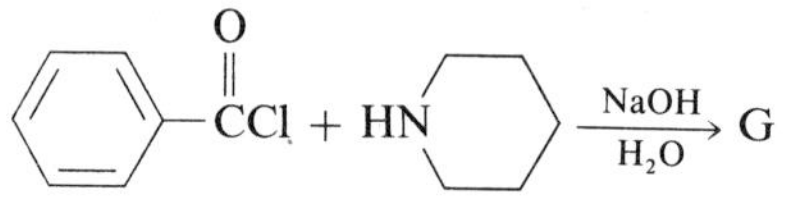 $\langle\text{phenyl}\rangle-\overset{O}{\overset{\|}{C}}Cl + HN\langle\text{piperidine}\rangle \xrightarrow[H_2O]{NaOH} G$

(f) $N{\equiv}CCH_2\overset{O}{\overset{\|}{C}}OCH_2CH_3 \xrightarrow[\substack{H_2O \\ 0\,°C}]{NH_3} H$

B. Mechanisms for the Hydrolysis of Nitriles and Amides

The cyano group in a nitrile is similar to a carbonyl group in its polarization. The carbon atom is the positive end of a dipole, while the more electronegative nitrogen atom is the negative end. The nitrogen atom also has a pair of nonbonding electrons available for protonation or complexing with Lewis acids. Thus, the mechanism for the hydrolysis of a nitrile should be very similar to that for an ester (or an amide). Such a mechanism is outlined for the hydrolysis of a nitrile to an amide in aqueous acid:

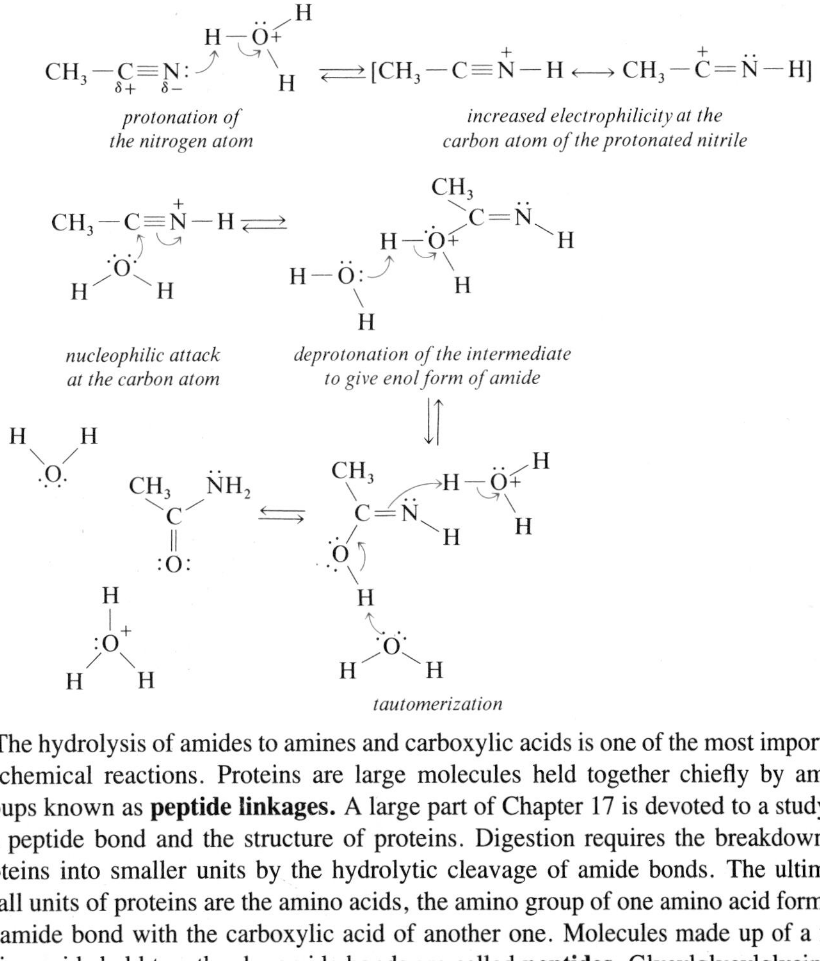

protonation of
the nitrogen atom

increased electrophilicity at the
carbon atom of the protonated nitrile

nucleophilic attack
at the carbon atom

deprotonation of the intermediate
to give enol form of amide

tautomerization

The hydrolysis of amides to amines and carboxylic acids is one of the most important of chemical reactions. Proteins are large molecules held together chiefly by amide groups known as **peptide linkages.** A large part of Chapter 17 is devoted to a study of the peptide bond and the structure of proteins. Digestion requires the breakdown of proteins into smaller units by the hydrolytic cleavage of amide bonds. The ultimate small units of proteins are the amino acids, the amino group of one amino acid forming an amide bond with the carboxylic acid of another one. Molecules made up of a few amino acids held together by amide bonds are called **peptides.** Glycylglycylglycine is an example of a simple peptide, made up of three units of the amino acid glycine. Hydrolysis of the peptide with aqueous acid gives the free glycine amino acid units.

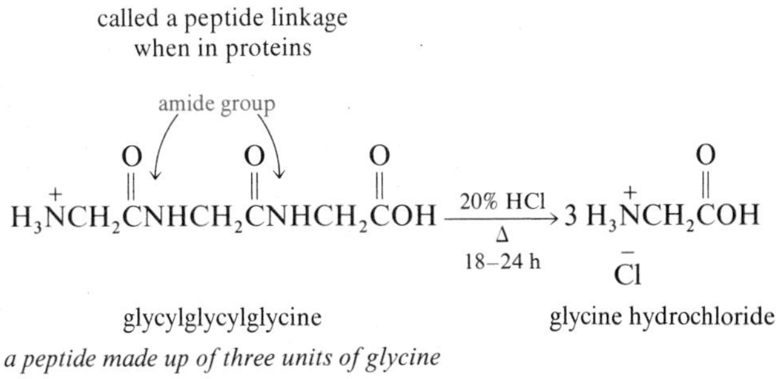

called a peptide linkage
when in proteins

amide group

$$H_3\overset{+}{N}CH_2\overset{O}{\overset{\|}{C}}NHCH_2\overset{O}{\overset{\|}{C}}NHCH_2\overset{O}{\overset{\|}{C}}OH \xrightarrow[\substack{\Delta \\ 18-24\ h}]{20\%\ HCl} 3\ H_3\overset{+}{N}CH_2\overset{O}{\overset{\|}{C}}OH \quad \overset{-}{Cl}$$

glycylglycylglycine   glycine hydrochloride
a peptide made up of three units of glycine

In the body, the cleavage of the peptide linkages occurs under very mild conditions and is catalyzed by enzymes (Sections 17.3C, 17.6). The essential steps of the process, as far as researchers have been able to determine, are the same as those postulated in Section 10.7D for the hydrolysis of an ester. The carbonyl group of the amide is attacked by a nucleophile before or after the carbonyl group is activated by protonation, a tetrahedral intermediate is formed, and the intermediate breaks down with the release of an amine and the formation of a carboxylic acid. The process in an acidic solution is shown below.

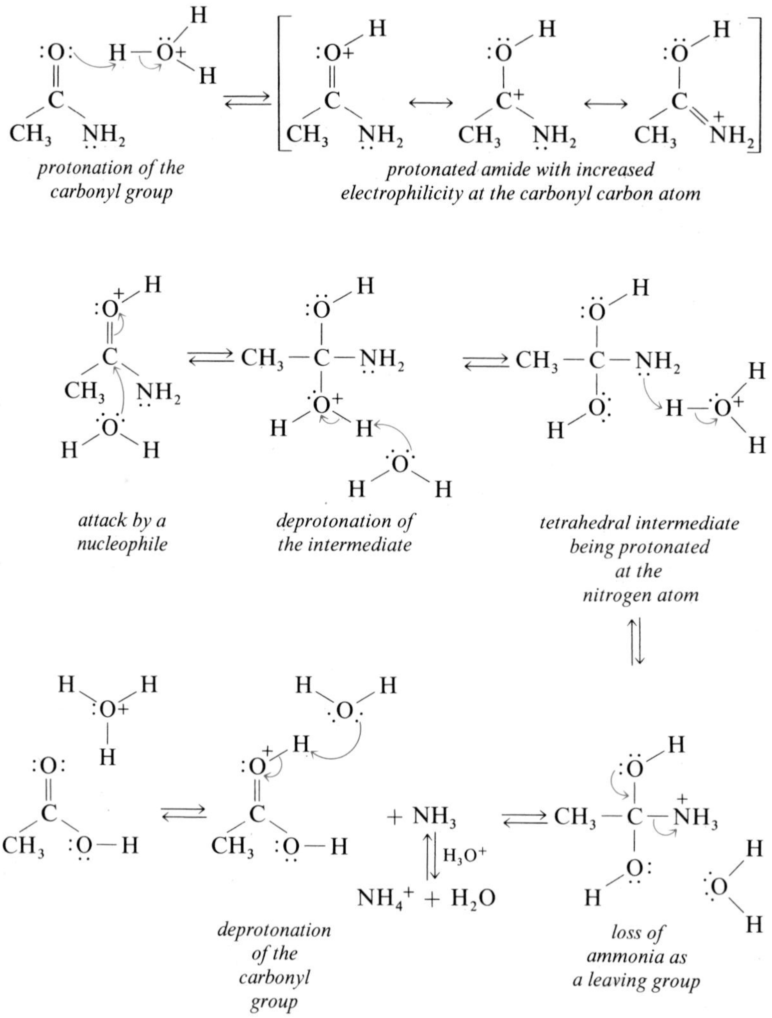

protonation of the carbonyl group

protonated amide with increased electrophilicity at the carbonyl carbon atom

attack by a nucleophile

deprotonation of the intermediate

tetrahedral intermediate being protonated at the nitrogen atom

deprotonation of the carbonyl group

loss of ammonia as a leaving group

Thus, the products of the hydrolysis of an amide in acidic solution are the free carboxylic acid and the salt of the amine portion of the molecule.

In basic solutions, the strong nucleophile hydroxide ion is available to attack the carbonyl group of the amide. Again a tetrahedral intermediate forms and breaks apart to give a carboxylate anion and the free amine as the products.

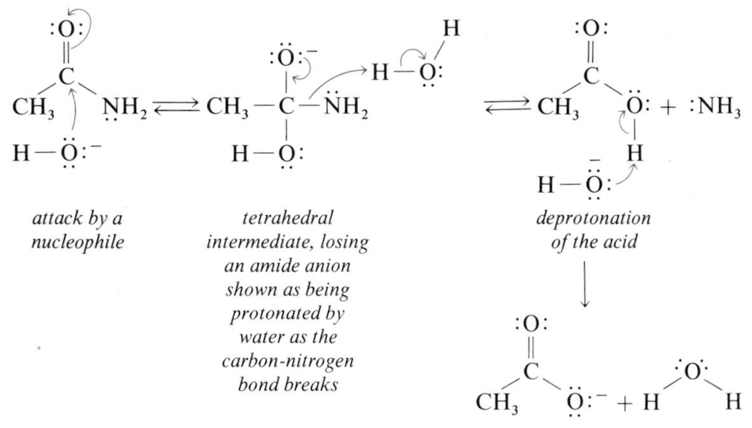

attack by a
nucleophile

tetrahedral
intermediate, losing
an amide anion
shown as being
protonated by
water as the
carbon-nitrogen
bond breaks

deprotonation
of the acid

For the carbonyl-nitrogen bond to cleave in an amide, the leaving group must be an amine or the conjugate base of an amine, an amide anion. In acidic solutions, the nitrogen is protonated before the cleavage takes place so that the leaving group is an amine. In a basic solution, it is unlikely that protonation will take place before the carbonyl-nitrogen bond begins to break. On the other hand, an amide anion is such a strong base that it is improbable that it has any real existence in water. The mechanism shown above accounts for this by showing the amide ion taking a proton from water as the tetrahedral intermediate starts to break up.

In all these mechanisms, the rate-determining step is the attack on the carbonyl group by a nucleophile, as it is for esterification of acids and hydrolysis of esters. Protonation and deprotonation steps are assumed to be fast equilibrium reactions, and the tetrahedral intermediate, once formed, also breaks apart rapidly.

PROBLEM 10.37   Benzanilide, the amide of benzoic acid and aniline, is a white solid that is insoluble in water. Both benzoic acid, a white solid, and aniline, a pale yellow oil, are insoluble in cold water. Their salts are soluble in cold water. For each of the following sets of conditions, complete the equations shown below and predict what you will see at the end of each process. You may assume complete reaction in each case.

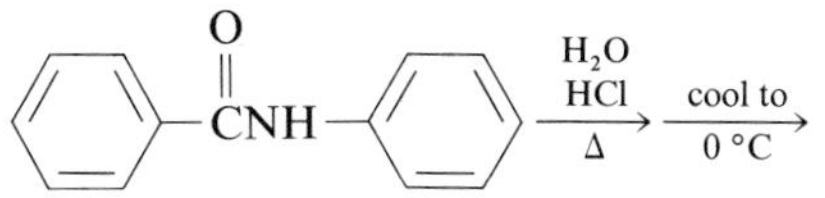

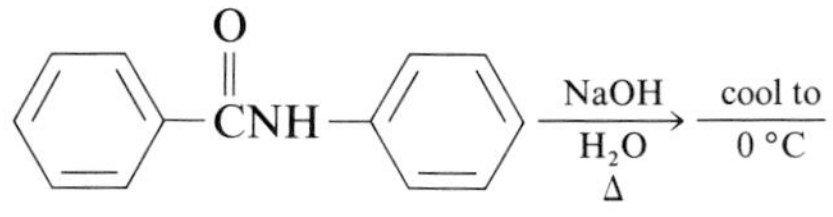

PROBLEM 10.38   Using acetonitrile, CH_3CN, as an example, write a mechanism for the basic hydrolysis of a nitrile.

Reduction of Carboxylic Acids and Their Derivatives by Hydrides

Carboxylic acids and their derivatives are reduced by metal hydride reagents just as aldehydes and ketones are. Lithium aluminum hydride is generally useful in reducing acids and acid derivatives. Sodium borohydride, while it reduces aldehydes and ketones with ease (Section 9.3), usually reacts with acid derivatives only with difficulty. Thus, it is possible to reduce aldehydes or ketones selectively in the presence of an acid or ester with sodium borohydride if reaction times are short and temperatures are kept low. Such a selective reduction of a cyclic ketone in the presence of an ester group is shown below.

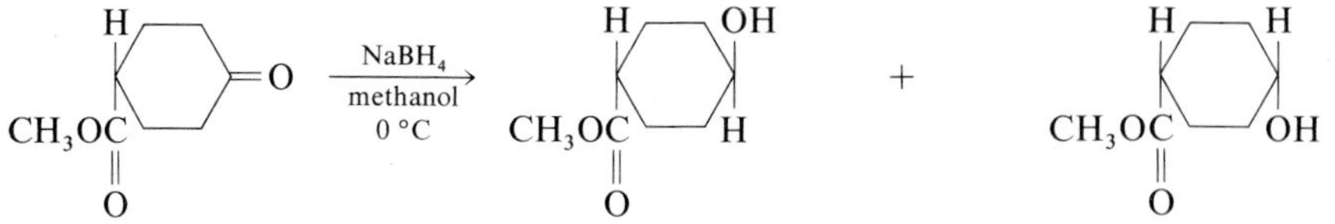

methyl
4-oxocyclohexanecarboxylate   *trans*-4-hydroxycyclohexanecarboxylate   *cis*-4-hydroxycyclohexanecarboxylate
                                        major product   *minor product*

Lithium aluminum hydride, however, reduces acids, esters, amides, and nitriles.

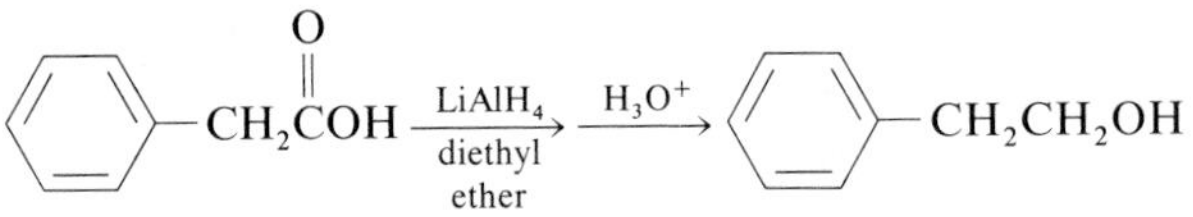

phenylacetic acid   2-phenyl-1-ethanol
                                                      92%

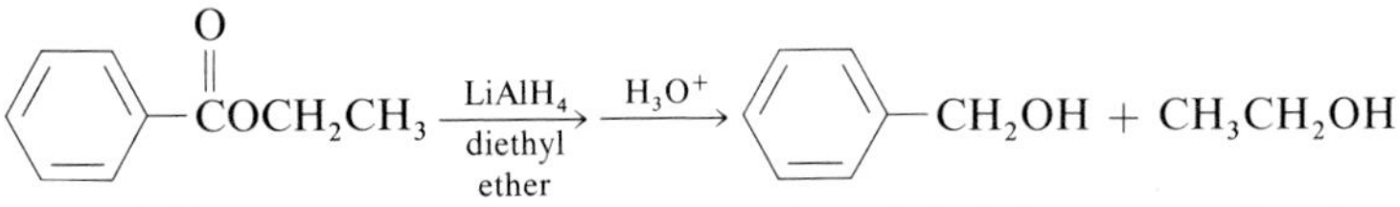

ethyl benzoate   benzyl alcohol   ethanol
                                             90%

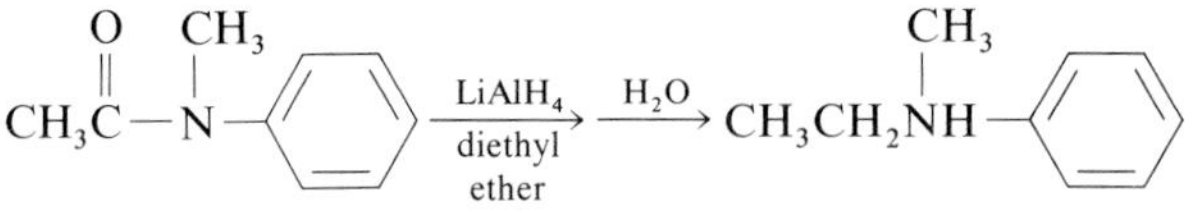

N-methylacetanilide   *N*-ethyl-*N*-methylaniline
                                                    91%

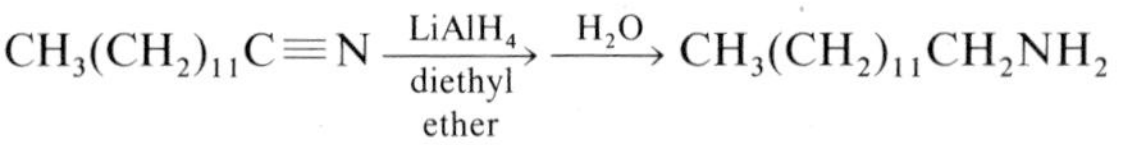

$$CH_3(CH_2)_{11}C \equiv N \xrightarrow[\text{diethyl ether}]{\text{LiAlH}_4} \xrightarrow{\text{H}_2\text{O}} CH_3(CH_2)_{11}CH_2NH_2$$

tridecanenitrile   tridecylamine
                                                  90%

A close examination of these chemical transformations reveals that in each case the carbonyl group (or the carbon atom of the nitrile) has been reduced to a methylene, —CH$_2$—, group. A carboxylic acid is reduced to a primary alcohol. In the case of an ethyl ester, we can clearly see the loss of ethanol as a leaving group. An ester is reduced to two alcohols: a primary alcohol corresponding to the carboxylic acid portion of the molecule, and the alcohol that corresponds to the alkyl group in the name of the ester. Amides and nitriles are reduced to amines.

With these facts in mind, we can propose a mechanism for the reaction of acid derivatives with lithium aluminum hydride. The original attack will be on the electrophilic carbon atom of the carbonyl group by the nucleophilic metal hydride ion in a reaction analogous to that seen for aldehydes and ketones (Section 9.3). The details of the proposed mechanism are shown below.

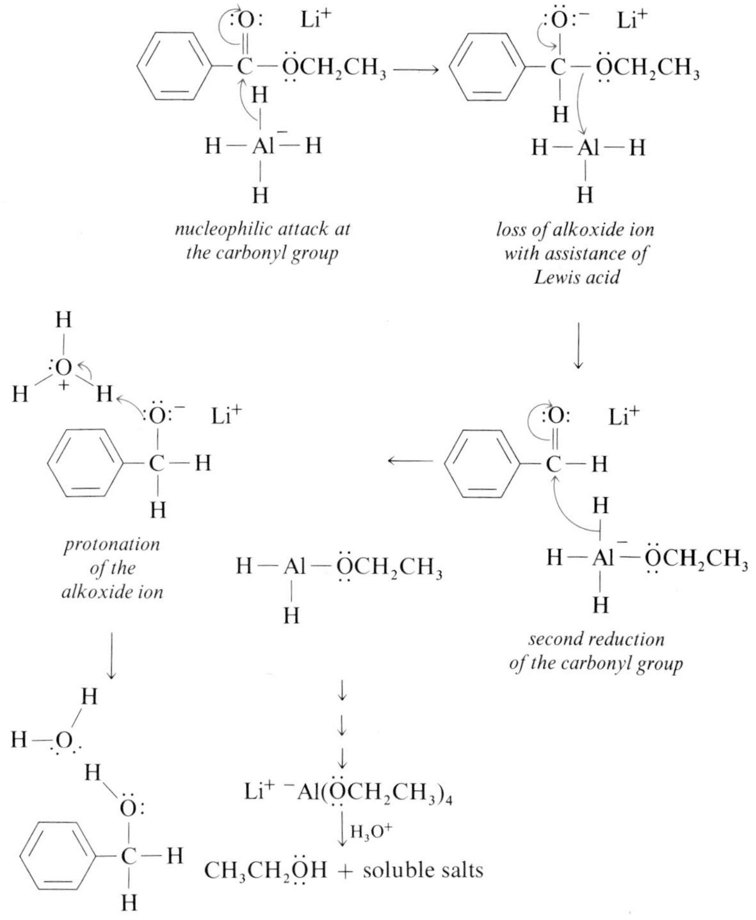

nucleophilic attack at the carbonyl group

loss of alkoxide ion with assistance of Lewis acid

protonation of the alkoxide ion

second reduction of the carbonyl group

$$Li^+ \ ^-Al(\ddot{O}CH_2CH_3)_4$$

$$\downarrow H_3O^+$$

$$CH_3CH_2\ddot{O}H + \text{soluble salts}$$

The addition of hydride ion to the carbonyl group of the ester gives a tetrahedral intermediate that resembles an acetal in structure and is therefore at the oxidation state of an aldehyde. It loses ethoxide ion with the assistance of a Lewis acid such as aluminum hydride. The carbon atom of the carbonyl group accepts another hydride ion and is reduced to a primary alcohol.

The complexes between aluminum hydride and the alkoxide ions from the ester are still sources of hydride ion and will continue to reduce carbonyl groups until all the hydride ions are used up. Cautious addition of a dilute acid protonates the alkoxide ions.

The reductions of carboxylic acids and unsubstituted amides with lithium aluminum hydride do not proceed as smoothly as the reactions of esters. The hydride ion is a powerful base as well as being a good nucleophile, and it deprotonates the acid or the amide, giving salts that are often insoluble in the reaction mixture and therefore react slowly.

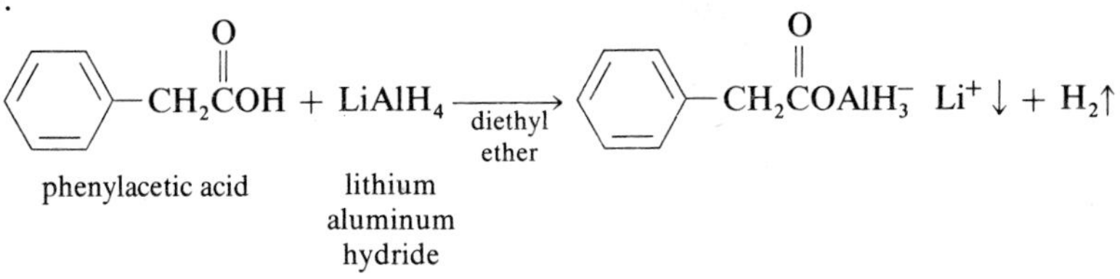

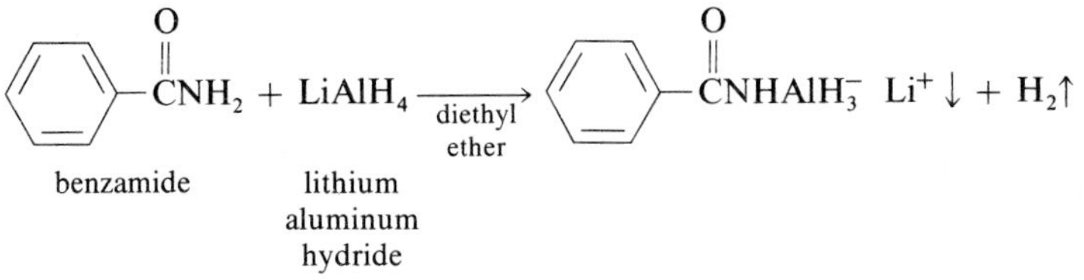

phenylacetic acid   lithium aluminum hydride

benzamide   lithium aluminum hydride

An imine or an iminium ion intermediate with a carbon-nitrogen double bond is postulated for the reduction of an amide or a nitrile by lithium aluminum hydride.

nucleophilic attack at the carbonyl group

alkoxide ion reacting with Lewis acid

iminium ion intermediate being reduced

tetrahedral intermediate

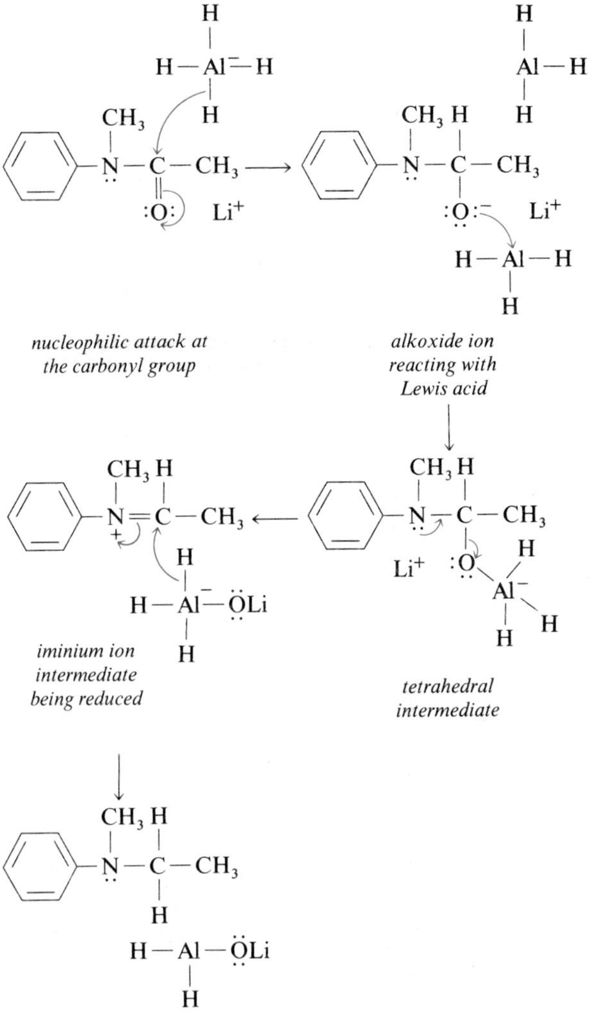

The loss of the oxygen atom of the carbonyl group is thought to take place through the donation of nonbonding electrons from the nitrogen atom of the amide, accompanied by the formation of an iminium ion that is reduced once more in another step.

PROBLEM 10.39 Write out the complete mechanism for the reduction of benzonitrile with lithium aluminum hydride.

PROBLEM 10.40 Complete the following equations.

(a) [structure: benzene ring with NH₂ and COH (C=O) substituents] $\xrightarrow[\text{diethyl ether}]{\text{LiAlH}_4}$ $\xrightarrow{\text{H}_2\text{O}}$

(b) [structure: cyclic imide N-phenyl] $\xrightarrow[\text{diethyl ether}]{\text{LiAlH}_4 \text{ (excess)}}$ $\xrightarrow{\text{H}_2\text{O}}$

(c) $\text{CH}_3\text{CH}{=}\text{CHCH}_2\text{CH}_2\overset{\overset{\text{O}}{\|}}{\text{C}}\text{OCH}_3$ $\xrightarrow[\text{diethyl ether}]{\text{LiAlH}_4}$ $\xrightarrow{\text{H}_3\text{O}^+}$

(d) $\text{CH}_3\text{CH}_2\text{O}\overset{\overset{\text{O}}{\|}}{\text{C}}(\text{CH}_2)_4\overset{\overset{\text{O}}{\|}}{\text{C}}\text{OCH}_2\text{CH}_3$ $\xrightarrow[\text{diethyl ether}]{\text{LiAlH}_4}$ $\xrightarrow{\text{H}_3\text{O}^+}$

(e) $\text{CH}_3\text{CH}_2\overset{\overset{\text{O}}{\|}}{\text{C}}\text{NHCH}_2\text{CH}_3$ $\xrightarrow[\text{diethyl ether}]{\text{LiAlH}_4}$ $\xrightarrow{\text{H}_2\text{O}}$

(f) [structure: cyclic lactone with =O] $\xrightarrow[\text{diethyl ether}]{\text{LiAlH}_4}$ $\xrightarrow{\text{H}_3\text{O}^+}$

(g) [benzene ring]$-\text{C}{\equiv}\text{N}$ $\xrightarrow[\text{diethyl ether}]{\text{LiAlH}_4}$ $\xrightarrow{\text{H}_2\text{O}}$

PROBLEM 10.41 Write equations showing how you would synthesize the following compounds. Bromobenzene and any other organic compounds containing three or fewer carbon atoms are available as starting materials.

(a) [benzene ring]$-\text{CH}_2\text{CH}_2\text{CH}_2\overset{\overset{\text{O}}{\|}}{\text{C}}\text{NH}_2$

(b) $\text{CH}_3\overset{\overset{\text{CH}_3}{|}}{\text{CH}}\text{CH}_2\text{CH}_2\overset{\underset{\underset{\text{CH}_3}{|}}{}}{\text{CH}}\overset{\overset{\text{O}}{\|}}{\text{C}}\text{OH}$

(c) $\text{HOCH}_2\text{CH}_2\text{CH}_2\text{CH}_2\text{CH}_2\text{OH}$

(d) $\text{CH}_3\text{CH}_2\overset{\underset{\text{[benzene ring]}}{|}}{\text{CH}}\text{CH}_2\text{CH}_2\text{CH}_2\text{NH}_2$

PROBLEM 10.42 The following sequence of reactions is part of a synthesis of compounds that are analogs of vitamin D (p. 1073).

[steroid structure with $\text{CH}_3\overset{\overset{\text{O}}{\|}}{\text{C}}\text{OCH}_2$ and CH₃O substituents] $\xrightarrow[\text{diethyl ether}]{\text{LiAlH}_4}$ $\xrightarrow{\text{H}_3\text{O}^+}$ A $\xrightarrow[\text{dichloromethane}]{\underset{\text{CrO}_3\text{Cl}^-}{\text{[pyridinium] (excess)}}}$ B

What are the structures of Compounds A and B?

10.10

Reactions of Acids and Acid Derivatives with Organometallic Reagents

A. The Grignard Reagent

Esters are the only acid derivatives that react with Grignard reagents in a synthetically useful way. A Grignard reagent (Section 9.4A) is a strong enough base that it reacts with carboxylic acids and unsubstituted amides (which have $pK_a \sim 15$, and are thus as acidic as water) to deprotonate them much as hydride ion does (p. 490). The anions that result are stabilized by resonance, which leads to greatly reduced electrophilicity at the carbon atom of the carbonyl group. The Grignard reagent, even if a molar excess were used, is not sufficiently nucleophilic to react with these ions.

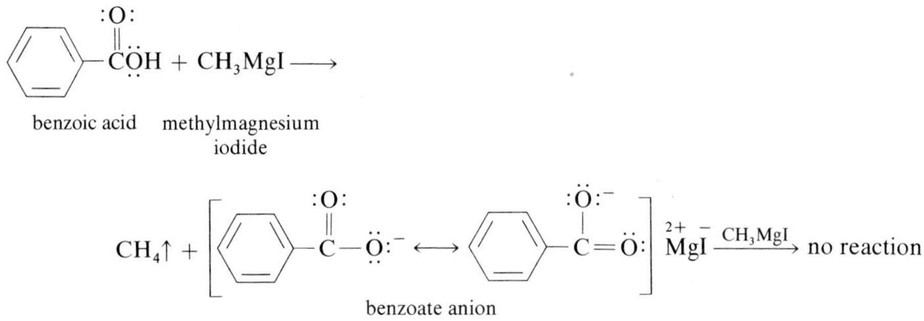

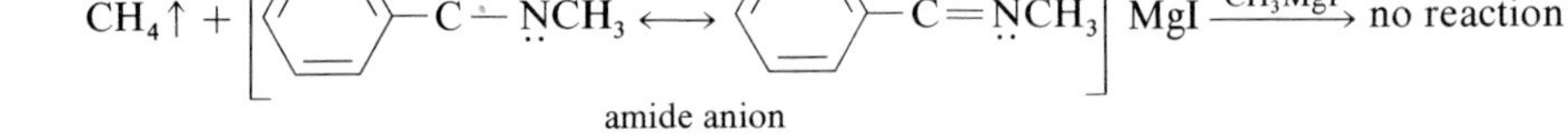

benzoic acid   methylmagnesium iodide

benzoate anion

N-methylbenzamide   methylmagnesium iodide

amide anion

An ester, however, has no acidic proton to destroy the Grignard reagent. Two moles of Grignard reagent react with one mole of ester to give an alcohol. A classical application of this reaction is the preparation of triphenylmethanol from bromobenzene and methyl benzoate.

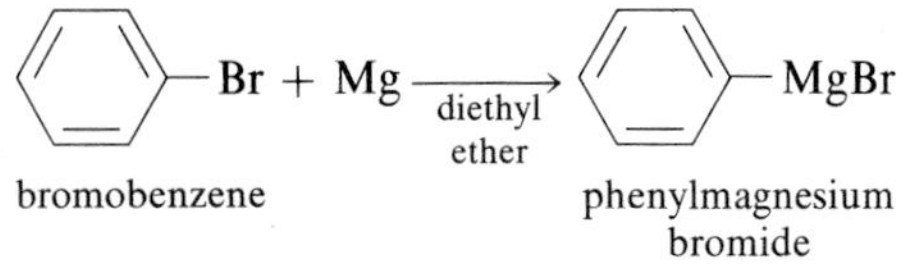

bromobenzene   phenylmagnesium bromide

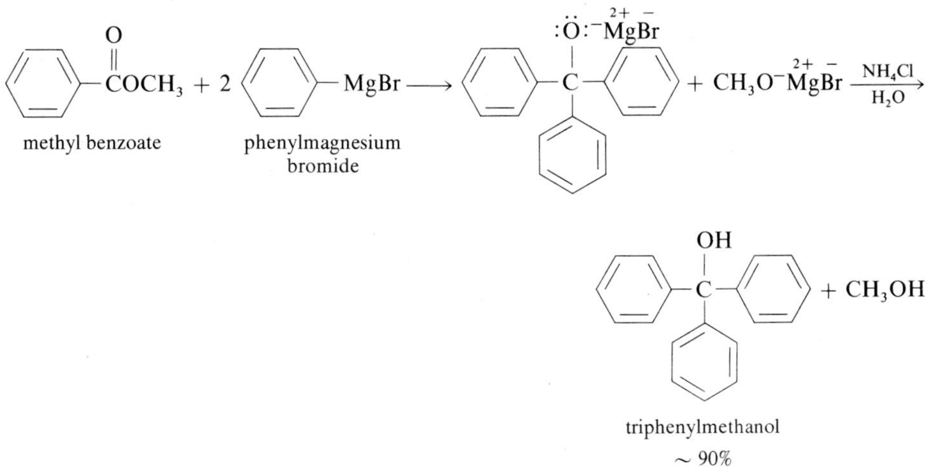

methyl benzoate   phenylmagnesium
bromide

triphenylmethanol
~ 90%

The addition of one mole of the Grignard reagent to the carbonyl group of the ester leads to an intermediate that loses methoxide ion, regenerating a carbonyl group, which reacts with a second mole of Grignard reagent.

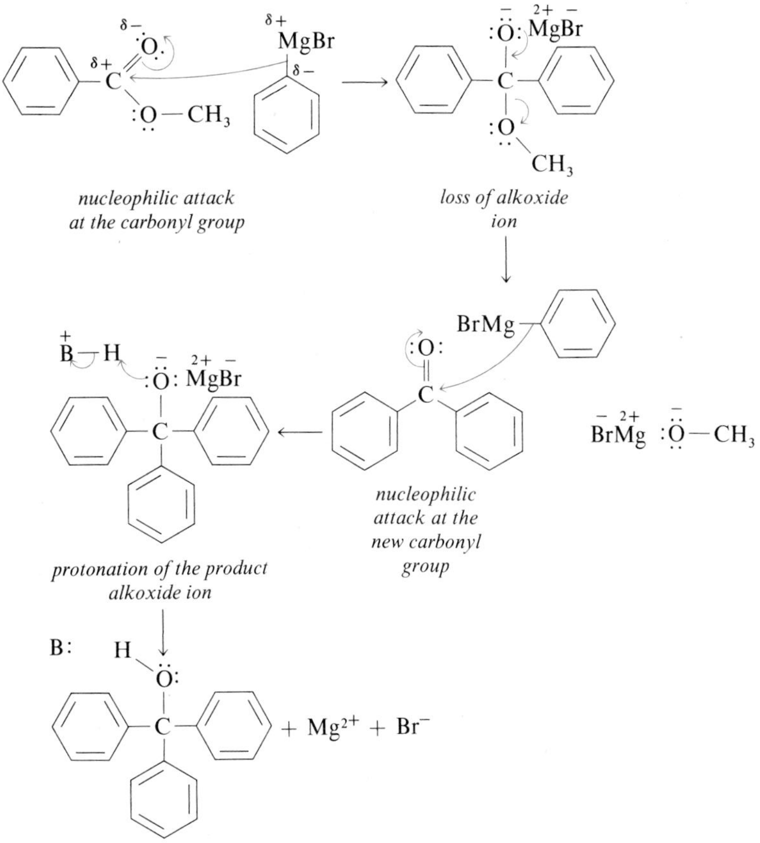

*nucleophilic attack
at the carbonyl group*

*loss of alkoxide
ion*

*protonation of the product
alkoxide ion*

*nucleophilic
attack at the
new carbonyl
group*

The resulting complex contains an alkoxide ion, which is protonated by a weakly acidic aqueous solution to give the free alcohol. You should be familiar with all of these steps from the reactions of aldehydes and ketones with Grignard reagents (Section 9.4B).

The only difference between the reactions of esters and those of aldehydes and ketones derives from the presence in the ester of the alkoxyl group that can act as a leaving group.

Most esters give a tertiary alcohol with at least two identical groups bonded to the tertiary carbon atom. In triphenylmethanol, all three groups are the same because an ester with a phenyl group was used as a starting material. If a formate ester is used, a secondary alcohol with two identical groups bound to the secondary carbon atom can be made, as in the synthesis of 5-nonanol.

$$CH_3CH_2CH_2CH_2Br + Mg \xrightarrow[\text{ether}]{\text{diethyl}} CH_3CH_2CH_2CH_2MgBr$$

n-butyl bromide   n-butylmagnesium bromide

O
‖
$2\ CH_3CH_2CH_2CH_2MgBr + HCOCH_2CH_3 \xrightarrow[\text{ether}]{\text{diethyl}}$

n-butylmagnesium bromide   ethyl formate

$$\overset{O^-\overset{2+}{M}gBr^-}{|}$$
$$CH_3CH_2CH_2CH_2CHCH_2CH_2CH_2CH_3 + CH_3CH_2O^-\overset{2+}{M}gBr^-$$

$\downarrow H_3O^+$

$$CH_3CH_2CH_2CH_2\underset{\underset{OH}{|}}{C}HCH_2CH_2CH_2CH_3$$

5-nonanol
84%

PROBLEM 10.43   Write out the mechanism for the formation of 5-nonanol from butylmagnesium bromide and ethyl formate.

PROBLEM 10.44   Suppose we wished to synthesize 3-ethyl-3-pentanol with a radioactive carbon atom as the tertiary carbon atom. Carbon dioxide labeled with radioactive carbon-14, $^{14}CO_2$, is readily available. Devise a synthesis for 3-ethyl-3-pentanol-3-^{14}C.

PROBLEM 10.45   It would be possible to make triphenylmethanol using a Grignard reagent by a different route from the one shown on p. 493. Work out another synthesis for it.

B. Organolithium Reagents

Organolithium reagents are more powerfully nucleophilic than Grignard reagents and thus react with carboxylate anions as well as with other carbonyl compounds (Section 9.4B). This reaction provides a useful way to synthesize ketones. In a typical reaction, hexanoic acid is converted to 2-heptanone by methyllithium.

O   O
‖   ‖
$CH_3CH_2CH_2CH_2CH_2COH \xrightarrow[\text{diethyl ether}]{CH_3Li\ (2\ \text{molar equivalents})} \xrightarrow{H_3O^+} CH_3CH_2CH_2CH_2CH_2CCH_3$

hexanoic acid   2-heptanone
                                                                         83%

Methyllithium is, of course, a strong base as well as a good nucleophile. The first stage of the reaction is the conversion of the carboxylic acid to its anion, which then reacts with a second mole of the strongly nucleophilic methyllithium.

$$CH_3CH_2CH_2CH_2CH_2 - C \overset{\overset{\cdot\cdot O\cdot}{\diagup}}{\underset{\cdot\cdot O - H}{\diagdown}} \longrightarrow CH_3CH_2CH_2CH_2CH_2 C \overset{\overset{\cdot\cdot O\cdot}{\diagup}}{\underset{\cdot\cdot O\cdot^- Li^+}{\diagdown}} \quad + CH_4\uparrow$$

$$\underset{\delta-\quad\delta+}{CH_3 - Li}$$

deprotonation
of the acid by
methyllithium

lithium hexanoate   methane

$$CH_3CH_2CH_2CH_2CH_2 - C \overset{\overset{Li^{\delta+}}{\underset{\cdot\cdot O \quad CH_3{}^{\delta-}}{\diagup}}}{\underset{\cdot\cdot O\cdot^- Li^+}{\diagdown}} \longrightarrow CH_3CH_2CH_2CH_2CH_2 - \overset{\overset{H}{\overset{:\overset{+}{O} - H}{|}}}{\underset{\underset{H}{\underset{:\overset{+}{O} - H}{|}}}{\underset{|}{C}}} - CH_3$$

$$:\ddot{O}:^- Li^+$$

nucleophilic addition   *dialkoxide anion being protonated*
to carboxylate anion

↓

$$H_2O + \underset{2\text{-heptanone}}{CH_3CH_2CH_2CH_2CH_2\overset{\overset{O}{||}}{C}CH_3} \rightleftharpoons \left[CH_3CH_2CH_2CH_2CH_2 - \overset{\overset{:\ddot{O}-H}{|}}{\underset{:\ddot{O}-H}{\underset{|}{C}}} - CH_3 \right]$$

hydrate of
2-heptanone

Once methyllithium adds to the carbonyl group of the acid salt, the intermediate is a dialkoxide ion. Protonation of the oxygen atoms with dilute acid gives the hydrate of the ketone, which is in equilibrium with the ketone itself (Section 9.7A).

It is the special properties of the lithium intermediates that allow this reaction to work well when the corresponding reaction with Grignard reagents is so poor. For example, lithium carboxylate salts are less highly ionic and therefore more soluble in ether than their magnesium counterparts. Because the salt is less ionic, the carbonyl group retains more of its electrophilic character. The dialkoxide anion is more stable with lithium as a counter ion; the lithium-oxyanion, LiO$^-$, seems to be a poorer leaving group than the corresponding magnesium ion. The dialkoxide anion survives until it is protonated when water is added to the reaction mixture.

The organolithium reagent used does not have to be methyllithium. For example, phenyllithium reacts with lithium butanoate, prepared from butanoic acid.

$$\underset{\text{lithium butanoate}}{CH_3CH_2CH_2\overset{\overset{O}{||}}{C}O^-Li^+} \quad \xrightarrow[\substack{\text{diethyl ether} \\ 24\,h}]{\bigcirc-Li} \xrightarrow{H_3O^+} \underset{\substack{\text{1-phenyl-1-butanone} \\ 85\%}}{CH_3CH_2CH_2\overset{\overset{O}{||}}{C}-\bigcirc}$$

PROBLEM 10.46 Complete the following equations, writing structural formulas for the intermediates and products indicated by the letters.

(a)

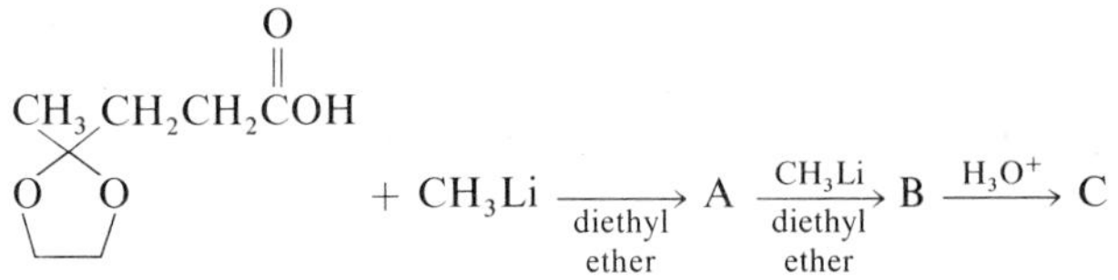

$$CH_3CH_2CH_2\overset{\overset{\displaystyle O}{\|}}{C}OH$$

$$\underset{\text{(epoxide)}}{\triangle O\quad O} \quad + CH_3Li \xrightarrow[\text{diethyl ether}]{} A \xrightarrow[\text{diethyl ether}]{CH_3Li} B \xrightarrow{H_3O^+} C$$

(b) $CF_3\overset{\overset{\displaystyle O}{\|}}{C}O^-Li^+ + CH_3CH_2CH_2CH_2Li \xrightarrow[\text{diethyl ether}]{} D \xrightarrow{H_3O^+} E$

(c)

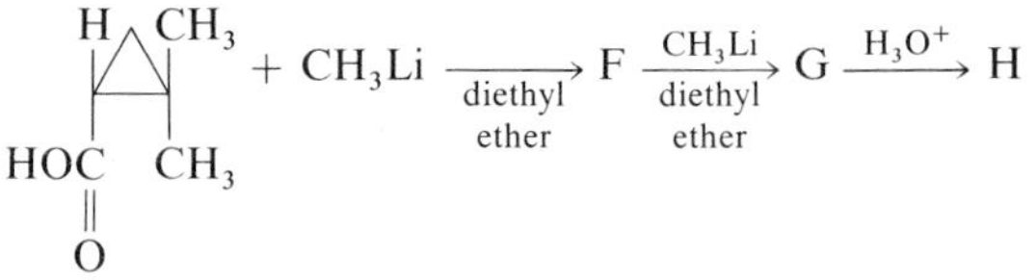

$$\begin{array}{c} H\quad CH_3 \\ \triangle \\ HO\overset{\underset{\displaystyle \|}{C}}{}\quad CH_3 \\ O \end{array} + CH_3Li \xrightarrow[\text{diethyl ether}]{} F \xrightarrow[\text{diethyl ether}]{CH_3Li} G \xrightarrow{H_3O^+} H$$

(d) 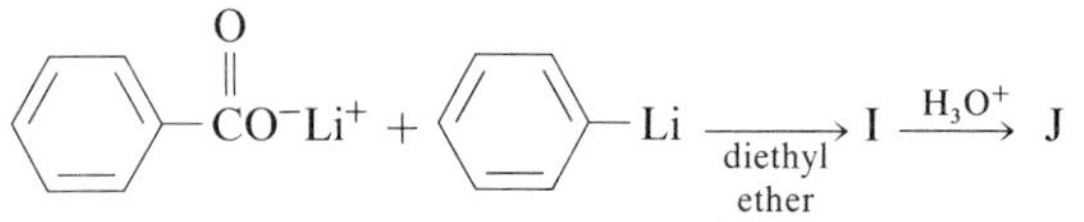 $\overset{\overset{\displaystyle O}{\|}}{C}O^-Li^+ + \quad\text{(Ar)}-Li \xrightarrow[\text{diethyl ether}]{} I \xrightarrow{H_3O^+} J$

PROBLEM 10.47 Bromobenzene and any other organic compound containing not more than three carbon atoms are available to you. Devise syntheses for the following compounds. There may be more than one synthetic pathway possible, so try to find the one with the fewest steps.

(a) 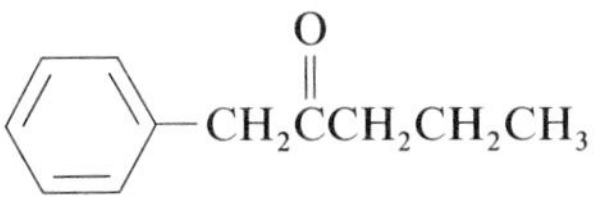 $\text{(Ar)}-CH_2\overset{\overset{\displaystyle O}{\|}}{C}CH_2CH_2CH_3$ (b) $CH_3CH_2CH_2CH_2NH\overset{\overset{\displaystyle O}{\|}}{C}CH_3$

(c) $CH_3CH_2\underset{\underset{\displaystyle NOH}{\|}}{C}CH_2CH_2CH_3$ (d) $CH_3CH_2CH_2\overset{\overset{\displaystyle O}{\|}}{C}OCH_2CH_3$

10.11

Infrared Spectroscopy of Acids and Acid Derivatives

A. Infrared Spectra of Carboxylic Acids and Esters

The infrared spectra of carboxylic acids display two important features illustrated in Figure 10.2 by the spectra of 10-undecenoic acid and benzoic acid. First, because of the very strong hydrogen bonding between the carboxyl groups of the acid, a strong and broad absorption band appears from 3300 to as low as 2500 cm^{-1} in the region of the spectrum for stretching frequencies for the oxygen-hydrogen single bond. The stretching frequencies for the carbon-hydrogen bond for carboxylic acids are usually buried

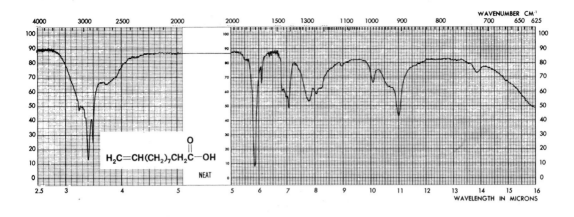

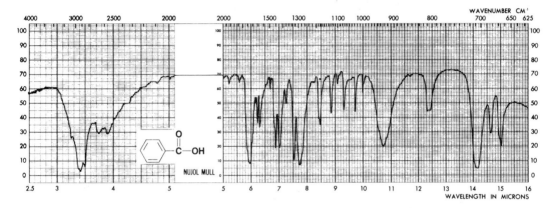

Figure 10.2   Infrared spectra of 10-undecenoic and benzoic acids. (From *The Aldrich Library of Infrared Spectra*)

in this band. Second, the stretching frequency for the carbonyl group of a carboxylic acid appears in one of two regions. Acids in which the carbonyl group is bonded to a tetrahedral carbon atom absorb around $1720–1706$ cm^{-1}. For acids in which the carbonyl group is conjugated with a double bond or with an aromatic ring, of which benzoic acid is an example, the carbonyl stretching frequency is around $1710–1680$ cm^{-1}.

The presence of other functional groups in acids also shows up in spectra. For example, the presence of a double bond is easily detected from the spectrum of 10-undecenoic acid (1645 cm^{-1}). In 10-undecenoic acid, the double bond is at the end of the chain, and a stretching vibration leads to a change in dipole moment, and thus to the absorption of energy (Section 7.7B).

The important absorption bands for esters lie in the carbonyl stretching and the carbon-oxygen single bond stretching regions of the spectrum. The carbonyl stretching frequency in alkyl esters is in the region of $1750–1735$ cm^{-1}. If the carbonyl group is conjugated, absorption occurs at $1730–1715$ cm^{-1}. The carbon-oxygen single bond stretching frequency varies with the exact nature of the group attached to the oxygen atom. In all esters, two bands are seen in the region between $1300–1000$ cm^{-1}, the same region in which a similar absorption is seen for alcohols and ethers (Section 7.7C). Typical spectra of esters are those of ethyl acetate and ethyl benzoate (Figure 10.3).

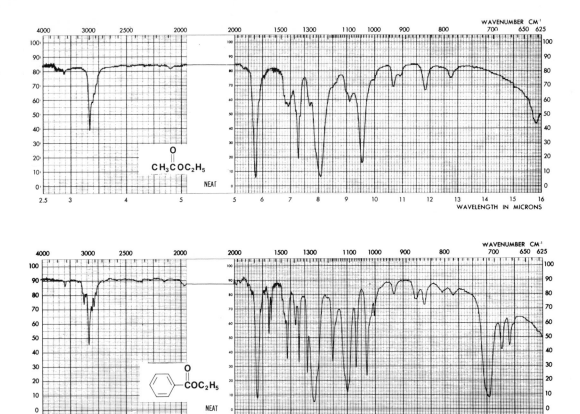

Figure 10.3 Infrared spectra of ethyl acetate and ethyl benzoate. (From *The Aldrich Library of Infrared Spectra*)

B. The Infrared Spectra of Amides

All amides have a strong band in the carbonyl stretching region of the spectrum, usually between 1690 and 1650 cm^{-1}. This is known as the amide I band. Another strong band in the same region, known as the amide II band, appears for amides that have at least one hydrogen atom on the amide nitrogen atom. This band arises from bending vibrations of the nitrogen-hydrogen single bond and usually falls between 1655 and 1590 cm^{-1}. The positions of both of these bands depend strongly on hydrogen bonding and therefore on the concentration of the amide in the media in which the spectra are taken. Such amides also show nitrogen-hydrogen stretching bands in the region between 3500 and 3200 cm^{-1}. For a primary amide with two hydrogen atoms on its nitrogen atom, two bands, corresponding to symmetrical and antisymmetrical stretching vibrations for the two bonds, are often seen. Secondary amides, with only one hydrogen atom on the nitrogen atom, show one band in this region.

The spectra of acetamide, *N*-methylacetamide, and *N*,*N*-dimethylacetamide, shown in Figure 10.4, demonstrate these features. The amide II bands for acetamide and *N*-methylacetamide in these spectra are almost coincident with the carbonyl stretching frequencies and appear only as a shoulder or a thickening of the carbonyl band.

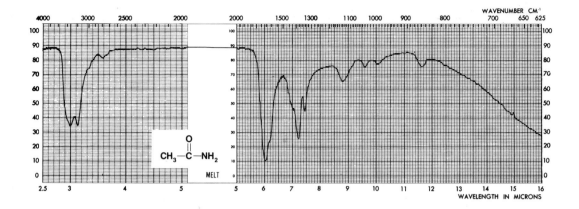

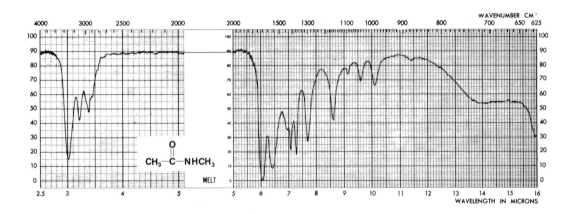

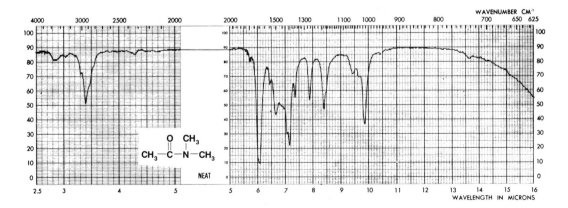

Figure 10.4 Infrared spectra of acetamide, *N*-methylacetamide, and *N,N*-dimethylacetamide. (From *The Aldrich Library of Infrared Spectra*)

PROBLEM 10.48   The spectra in Figure 10.5 belong to the compounds shown below. Assign each spectrum to the correct compound.

(a)
$$\text{C}_6\text{H}_5\text{—CNH—C}_6\text{H}_5$$
(benzanilide) *amide*

(b) $CH_3CHCH_2COCH_2CH_3$ (ethyl 3-hydroxybutanoate) *ester alch*
$|$
OH

(c) $CH_2{=}CHCH_2CH_2CH_2OCCH_3$ (4-penten-1-yl acetate) *ester*

(d)
$$\text{C}_6\text{H}_5\text{—CH}_2\text{OCCH}_3$$
(benzyl acetate) *ester*

(e) $HCNHCH_3$ (*N*-methylformamide) *amide*

Spectrum 1

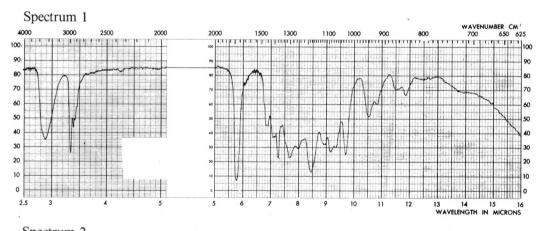

Spectrum 2

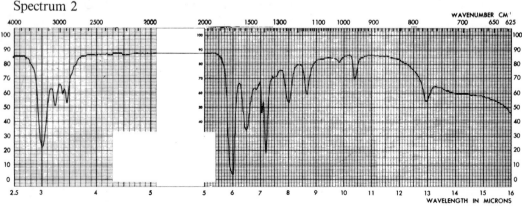

Figure 10.5 Spectra for Problem 10.48. (From *The Aldrich Library of Infrared Spectra*)

Spectrum 3

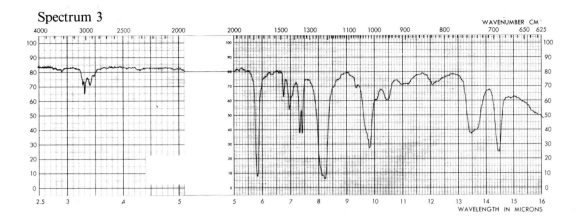

Spectrum 4

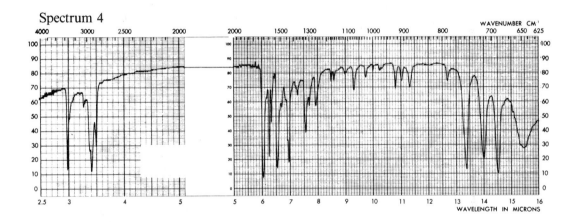

Spectrum 5

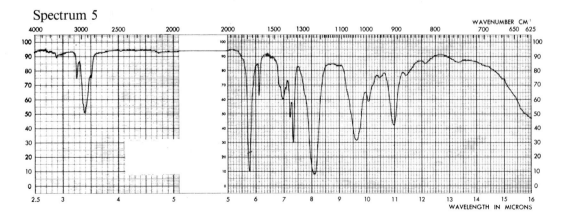

Figure 10.5 *(Continued)*

PROBLEM 10.49   The spectra in Figure 10.6 belong to compounds in the following functional group classes. Decide which major functional group class is represented by each spectrum. Point out the important bands in the spectrum of each compound.

(a) an alcohol   (b) an alkyl carboxylic acid   (c) an ester of an alkyl carboxylic acid
(d) a ketone   (e) an alkene

Spectrum 1

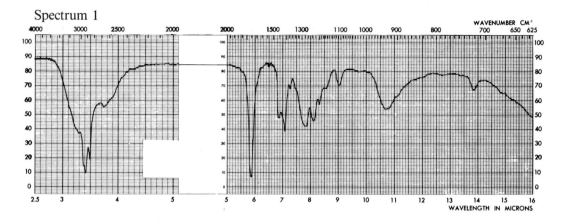

Spectrum 2

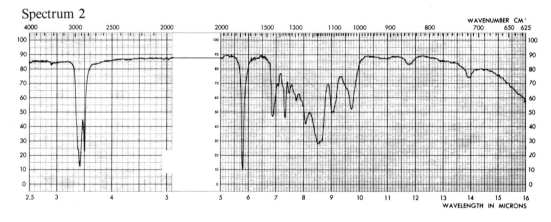

Spectrum 3

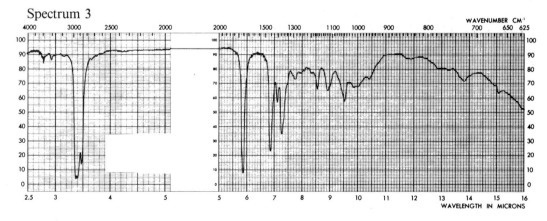

Figure 10.6   Spectra for Problem 10.49. (From *The Aldrich Library of Infrared Spectra*)

Spectrum 4

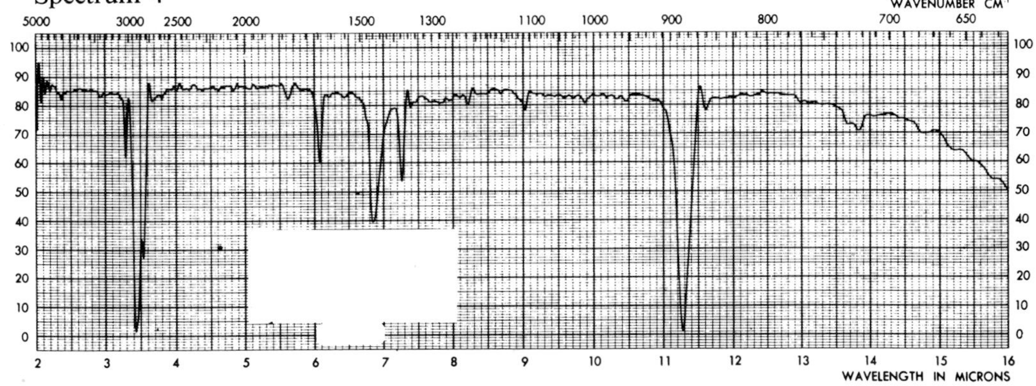

Spectrum 5

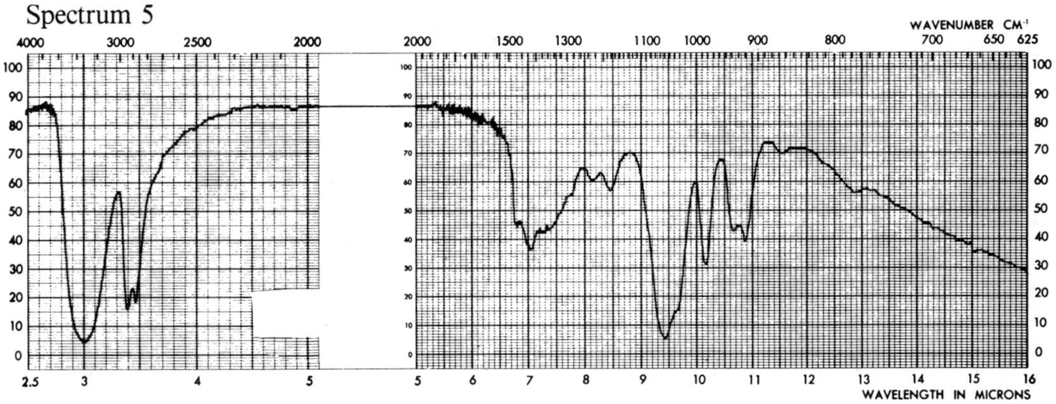

Figure 10.6 (*Continued*)

ADDITIONAL PROBLEMS

10.50 Name the following compounds, including stereochemistry.

(a) 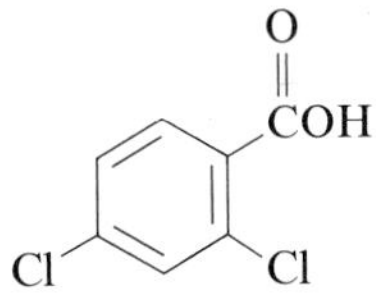

(b) 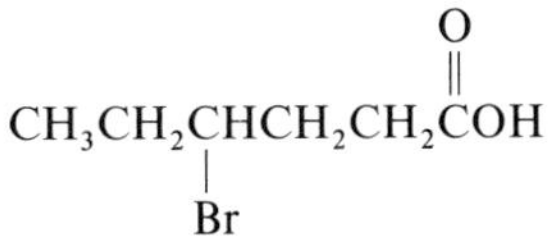

(c) 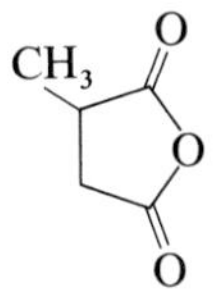

(d) 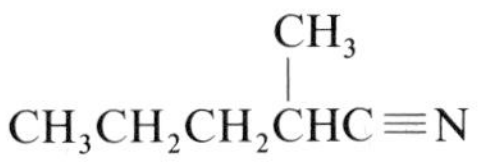

(e) 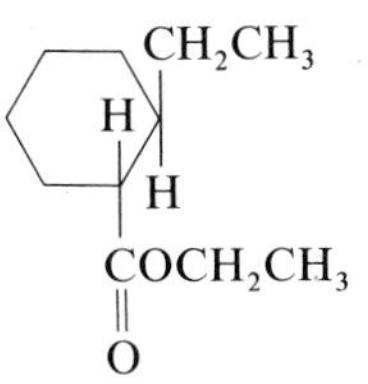

(f) 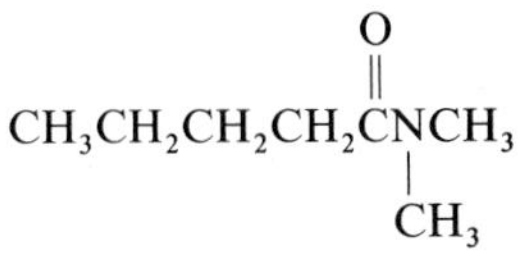

(g) 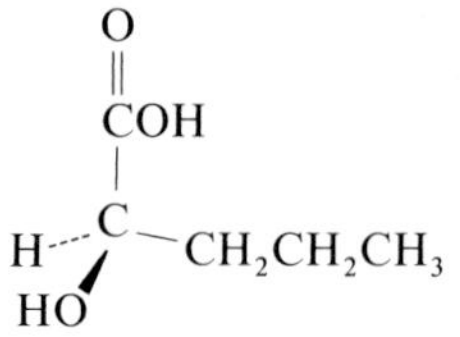
$$
\begin{array}{c}
\text{COH} \ (=\text{O}) \\
| \\
\text{H}\cdots\text{C}-\text{CH}_2\text{CH}_2\text{CH}_3 \\
| \\
\text{HO}
\end{array}
$$

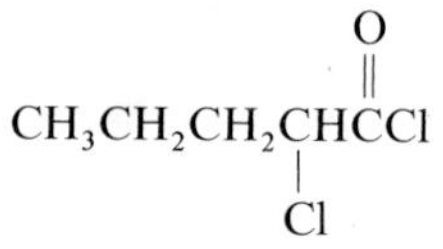

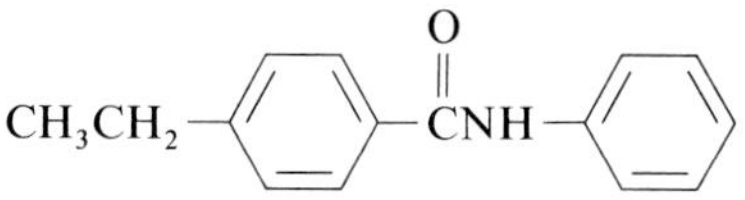

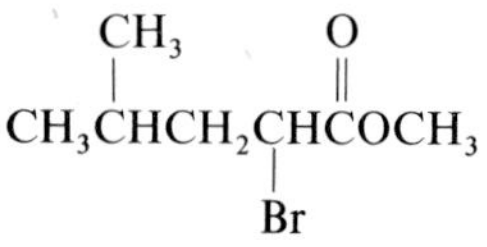

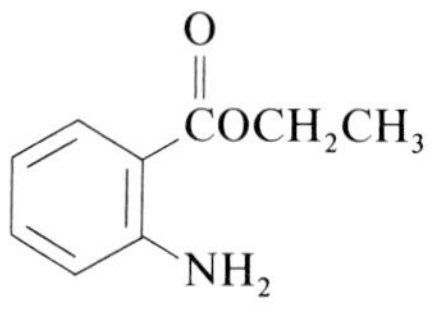

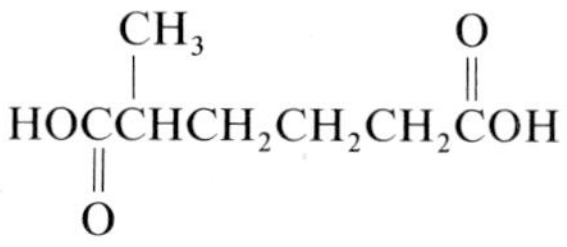

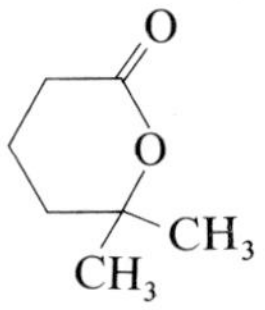

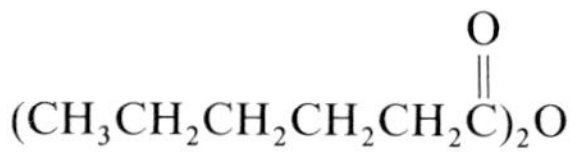

(h) $CH_3CH_2CH_2CHClCCl(=O)$

(i) $CH_3CH_2-\text{C}_6H_4-C(=O)NH-\text{C}_6H_5$

(j) $CH_3CH(CH_3)CH_2CH(Br)C(=O)OCH_3$

(k) $\text{2-aminophenyl } C(=O)OCH_2CH_3$

(l) $HOC(=O)CH(CH_3)CH_2CH_2CH_2COH(=O)$

(m) ring lactone with two CH_3 groups

(n) $(CH_3CH_2CH_2CH_2CH_2C(=O))_2O$

(o)
$$
\begin{array}{c}
CH_3CH_2CH_2 \qquad\quad H \\
\text{C}=\text{C} \\
H \qquad\quad CH_2CH_2COH(=O)
\end{array}
$$

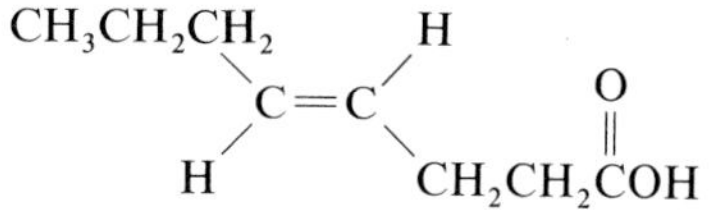

(p) 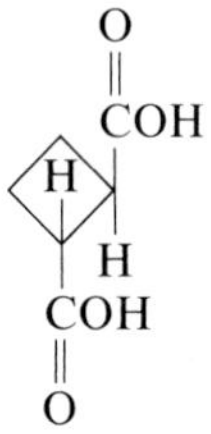 bicyclic structure with two $COH(=O)$ groups and two H

10.51   Draw structural formulas for the following compounds.

(a) methyl (*S*)-3-bromopentanoate   (b) *cis*-3-hydroxycyclopentanecarboxylic acid
(c) 2,4-dibromobenzamide   (d) ethyl (*Z*)-3-hexenoate   (e) heptanenitrile
(f) *N,N*-diethylhexanamide   (g) diethyl 2,2-dimethylpentanedioate
(h) methyl *p*-nitrobenzoate   (i) 3-hydroxy-5,5-dimethyldecanoic acid
(j) *p*-bromobenzoic anhydride   (k) *p*-methoxybenzoyl chloride

10.52   Complete the following equations, giving structural formulas for all the major organic intermediates and products indicated by letters.

(a) 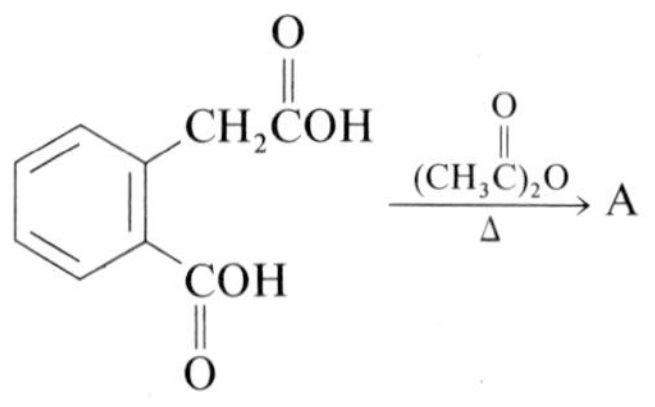
$$
\text{1,2-benzene with } CH_2COH(=O) \text{ and } COH(=O) \xrightarrow[\Delta]{(CH_3C(=O))_2O} A
$$

(b) 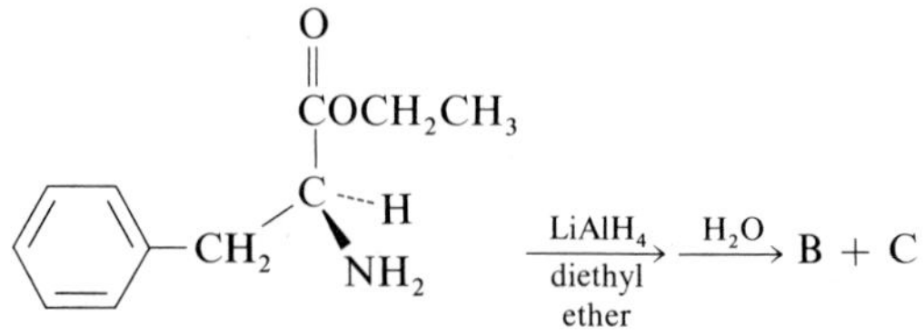
$$
\text{C}_6H_5-CH_2-C(H)(NH_2)-COCH_2CH_3(=O) \xrightarrow[\text{diethyl ether}]{LiAlH_4 \quad H_2O} B + C
$$

(c) $CH_3(CH_2)_{10}\overset{\overset{\displaystyle O}{\|}}{C}OCH_2(CH_2)_{10}CH_3 + CH_3OH \xrightarrow[\Delta]{H_2SO_4} D + E$

(d)

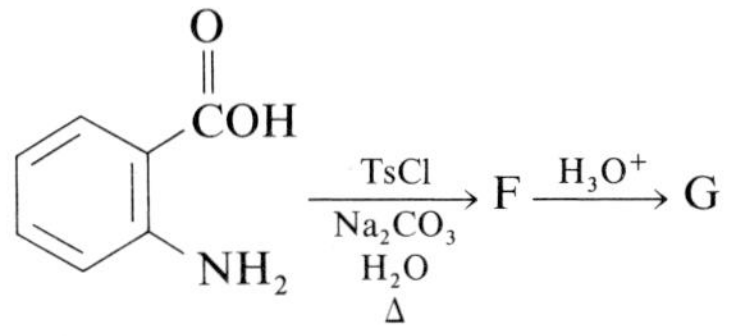

$\xrightarrow[\substack{Na_2CO_3 \\ H_2O \\ \Delta}]{TsCl} F \xrightarrow{H_3O^+} G$

(e) $HO\overset{\overset{\displaystyle O}{\|}}{C}(CH_2)_4\overset{\overset{\displaystyle O}{\|}}{C}OH \xrightarrow{SOCl_2 \text{ (excess)}} H$

(f) 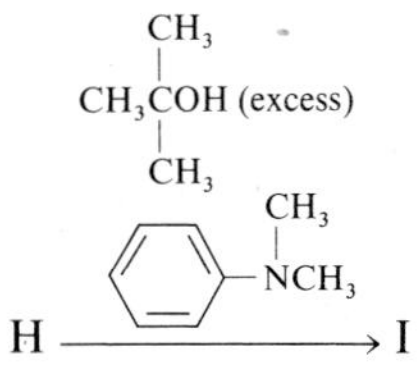 $H \xrightarrow{\quad\quad} I$

over the arrow:
$\underset{\overset{|}{CH_3}}{CH_3\overset{\overset{\displaystyle CH_3}{|}}{C}OH}$ (excess)

$\overset{\overset{\displaystyle CH_3}{|}}{\underset{}{}}$ N(CH$_3$) group on benzene

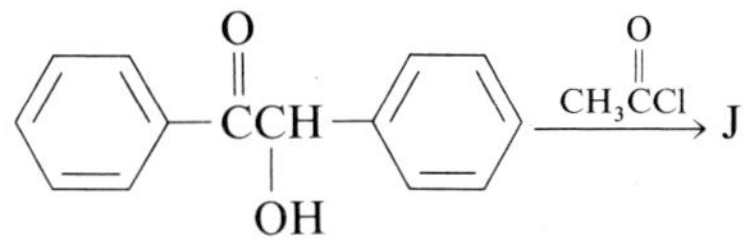

(g)

$\xrightarrow{CH_3\overset{\overset{\displaystyle O}{\|}}{C}Cl} J$

(h)

$\xrightarrow[\substack{H_2O \\ \Delta}]{KMnO_4} K \xrightarrow{H_3O^+} L$

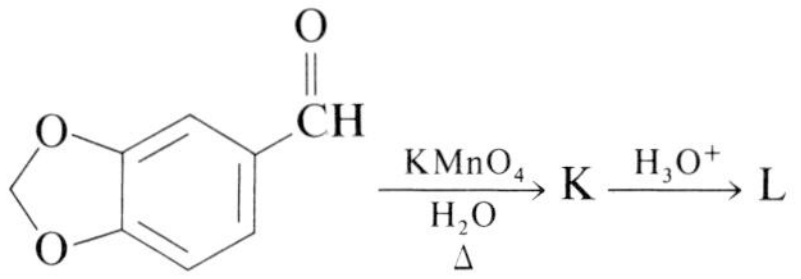

(i) $CH_3(CH_2)_{14}CH_2OH \xrightarrow[pyridine]{TsCl} M$

(j)

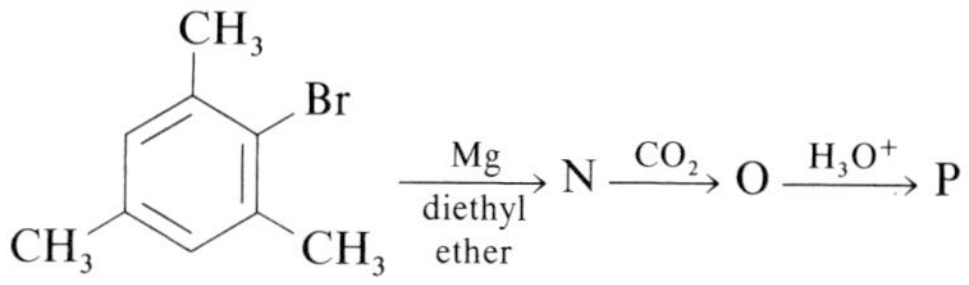

$\xrightarrow[\substack{diethyl \\ ether}]{Mg} N \xrightarrow{CO_2} O \xrightarrow{H_3O^+} P$

(k)

$\xrightarrow[H_2O]{Ag_2O} Q \xrightarrow{H_3O^+} R$

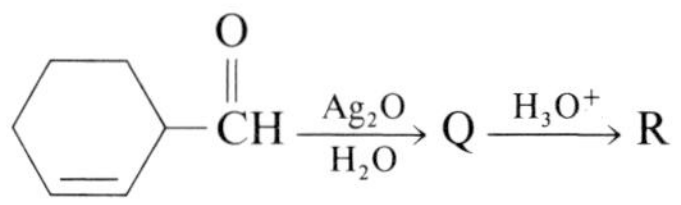

(l) $CCl_3CH_2CH_2CH_2CH_2OH \xrightarrow[\substack{H_2O \\ \Delta}]{KMnO_4} S \xrightarrow{H_3O^+} T$

(m) $BrCH_2(CH_2)_8CH_2OH \xrightarrow[\substack{H_2O \\ acetic\ acid}]{CrO_3} U$

(n) $CH_3\underset{\overset{|}{OH}}{CH}\overset{\overset{\displaystyle O}{\|}}{C}OH + CH_3\overset{\overset{\displaystyle CH_3}{|}}{CH}OH \xrightarrow[\substack{benzene \\ \Delta}]{H_2SO_4} V$

(o)

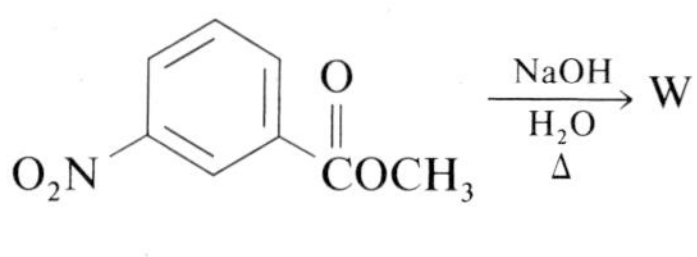

$\xrightarrow[\substack{H_2O \\ \Delta}]{NaOH} W$

(p)

$\xrightarrow[diethyl\ ether]{2\ CH_3CH_2CH_2CH_2Li} X \xrightarrow{H_3O^+} Y$

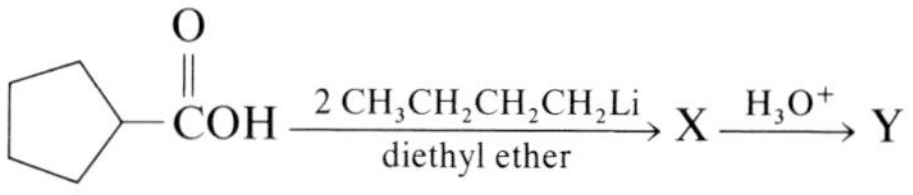

10.53 Additional exercises on the reactions of carboxylic acids and acid derivatives follow. Give structural formulas for the intermediates and products indicated by the letters.

(a) 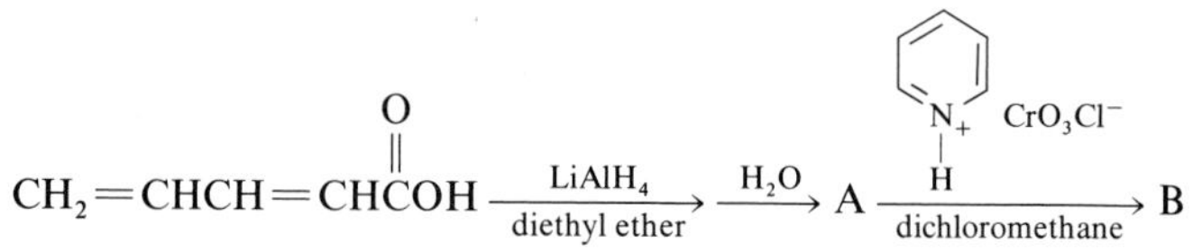 $CH_2=CHCH=CHCOH$ $\xrightarrow[\text{diethyl ether}]{LiAlH_4}$ $\xrightarrow{H_2O}$ A $\xrightarrow[\text{dichloromethane}]{\text{pyridinium } CrO_3Cl^-}$ B

(b) $CH_3CH_2CH_2CH_2COCH_2CH_3$ $\xrightarrow[\text{diethyl ether}]{2\ CH_3CH_2MgBr}$ C $\xrightarrow[\text{H}_2\text{O}]{NH_4Cl}$ D

(c) $CH_3CH_2CNHCH_2CH_3$ $\xrightarrow[\text{diethyl ether}]{LiAlH_4}$ $\xrightarrow{H_2O}$ E

(d) $CH_3CH=CHCH_2C\equiv CCH_2CH_2COH$ $\xrightarrow[\substack{NH_3(\text{liq}) \\ \text{ethanol}}]{Li}$ $\xrightarrow{H_3O^+}$ F

(e) 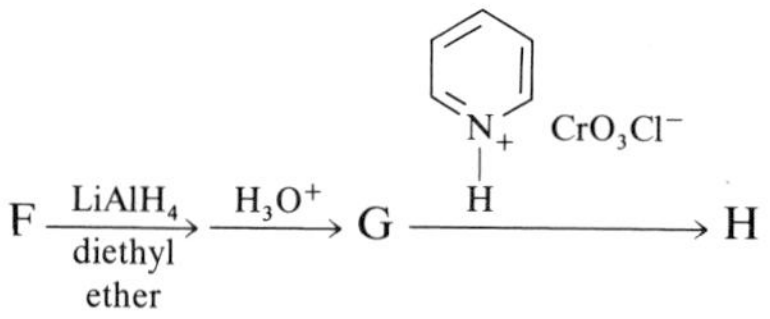 F 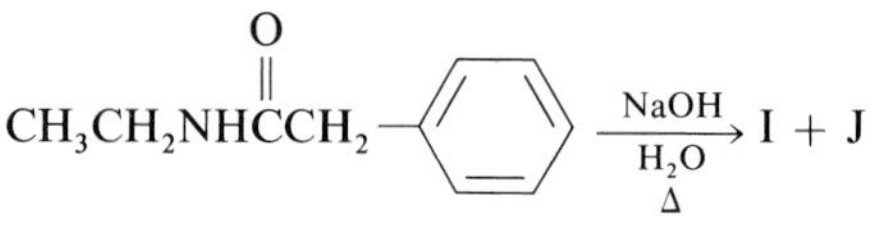 $\xrightarrow[\text{diethyl ether}]{LiAlH_4}$ $\xrightarrow{H_3O^+}$ G $\xrightarrow{\text{pyridinium } CrO_3Cl^-}$ H

(f) $CH_3CH_2NHCCH_2$—⟨benzene⟩ $\xrightarrow[\substack{H_2O \\ \Delta}]{NaOH}$ I + J

(g) $CH_3CCH_2OCH + CH_3OH$ $\xrightarrow[\Delta]{KOH}$ K + L

(h) $CH_3CHCH_2CO^-Na^+ + CH_3CH_2CH_2CCl$ $\longrightarrow$ M
(with CH_3 and O substituents)

(i) $CH_3C(CH_3)(CH_3)CO^-Li^+$ + ⟨benzene⟩—Li $\xrightarrow[\text{diethyl ether}]{}$ N $\xrightarrow{H_3O^+}$ O

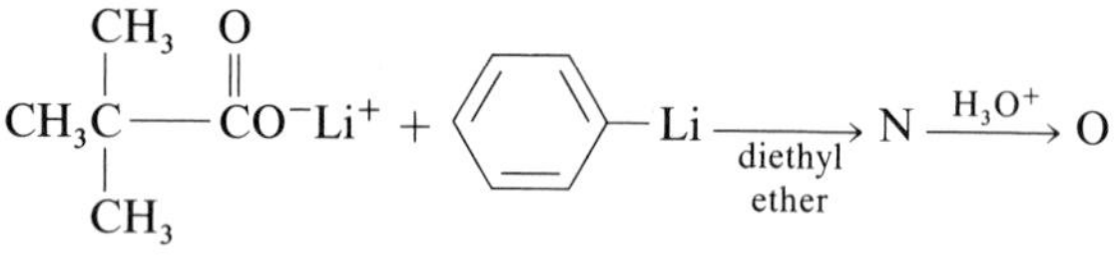

(j) ⟨benzene⟩—CCl + $H_2NCH_2CH_2CH_2CH_2CH_2COH$ $\xrightarrow[\text{H}_2\text{O}]{NaOH}$ P $\xrightarrow{H_3O^+}$ Q

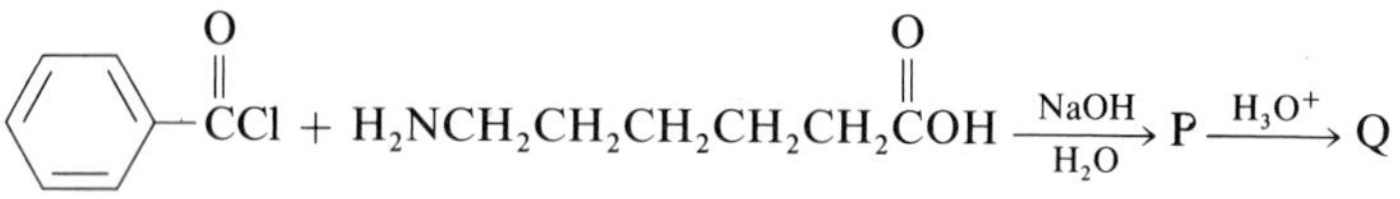

10.54 Propose syntheses for the following compounds. Assume that you have bromobenzene and any other organic compound containing up to three carbon atoms as starting materials. Several synthetic pathways are possible for some of the compounds. Try to find the one with the fewest steps.

(a) 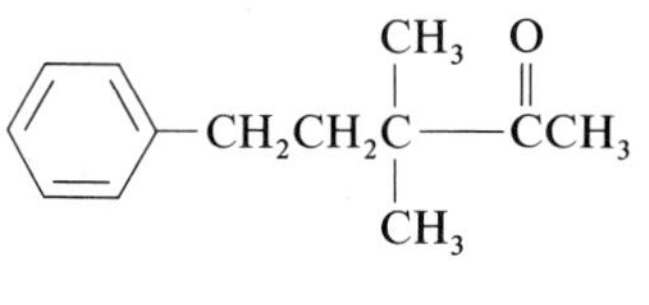 ⟨benzene⟩—$CH_2CH_2C(CH_3)(CH_3)CCH_3$ (with O)

(b) 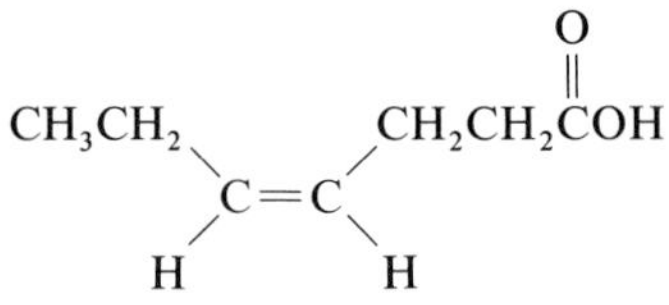 CH_3CH_2 and CH_2CH_2COH (with O) on $C=C$ with H and H

(c)

(d)

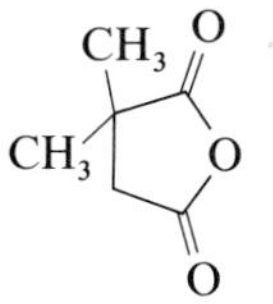

(e) 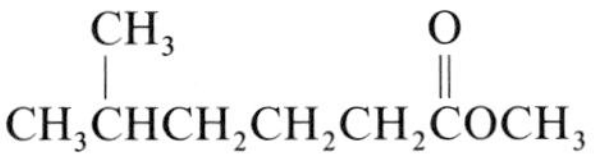 $CH_3CHCH_2CH_2CH_2COCH_3$
   with CH_3 above first CH and O above the CO

(f) 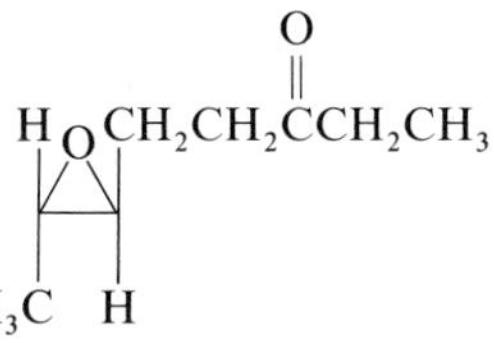 $HOCH_2CH_2CCH_2CH_3$ + enantiomer
   with epoxide structure H_3C H below and O above

(g)

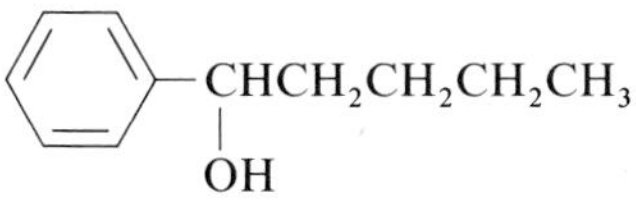

10.55 Polycyclic hydrocarbon derivatives are used as medication for viral infections. A new compound that inactivates viruses was synthesized by the following sequence of reactions. What is its structure?

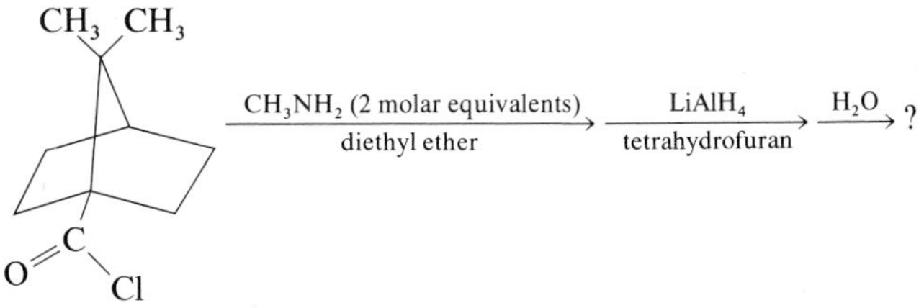

$$\xrightarrow[\text{diethyl ether}]{\text{CH}_3\text{NH}_2 \text{ (2 molar equivalents)}} \xrightarrow[\text{tetrahydrofuran}]{\text{LiAlH}_4} \xrightarrow{\text{H}_2\text{O}} ?$$

10.56 The following transformations have been carried out. Suggest synthetic routes for them, showing all reagents and intermediate products.

(a)

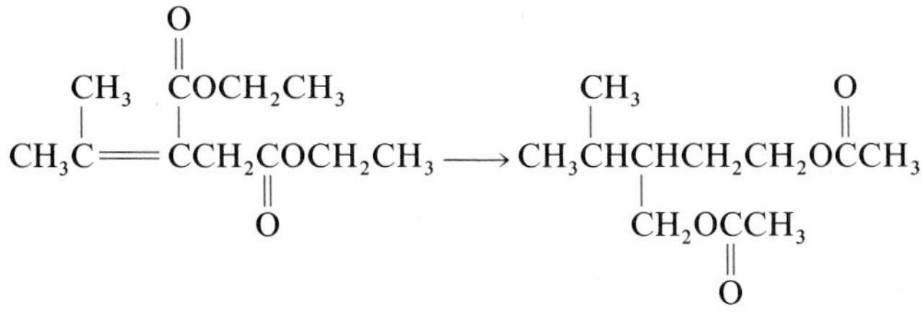

(b)

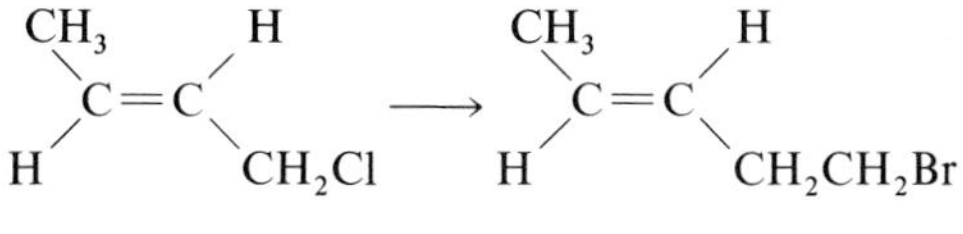

(c)

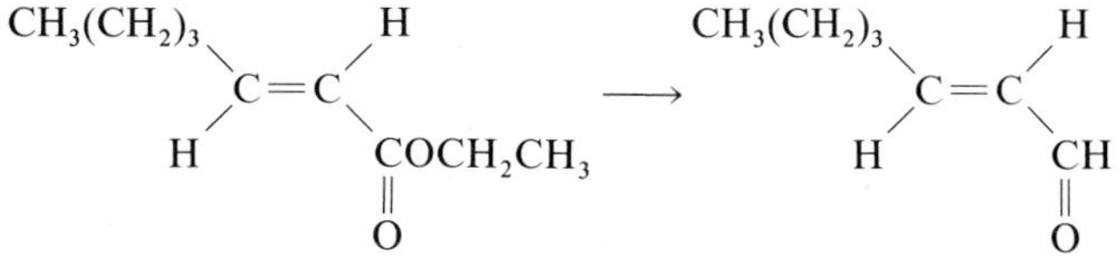

(d) 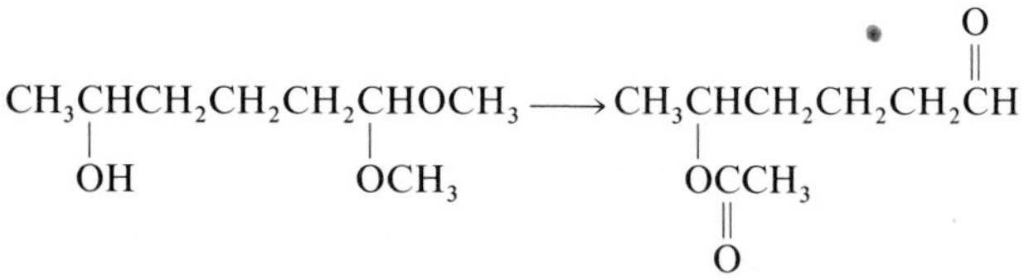 $CH_3CHCH_2CH_2CH_2CHOCH_3 \longrightarrow CH_3CHCH_2CH_2CH_2CH$
   with OH and OCH_3 below left; $OCCH_3$ and O below right, O above right

(e)

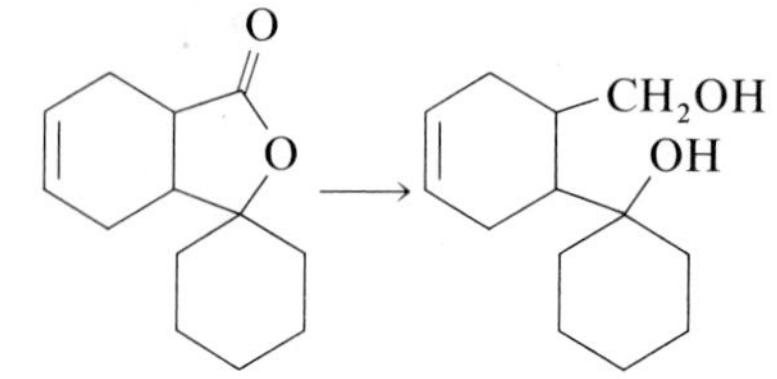

(f)

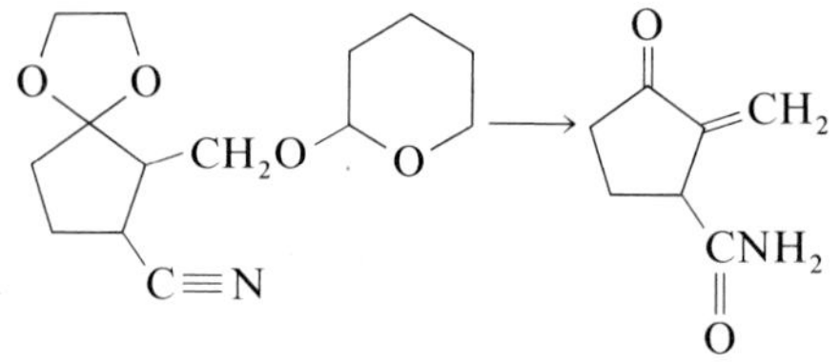

(g) $HC \equiv CCOCH_2CH_3 \longrightarrow CH_3(CH_2)_7\overset{OH}{\underset{|}{C}}HC \equiv C\overset{O}{\underset{\parallel}{C}}OCH_2CH_3$

(h) $CH_3(CH_2)_7\overset{OH}{\underset{|}{C}}HC \equiv C\overset{O}{\underset{\parallel}{C}}OCH_2CH_3 \longrightarrow CH_3(CH_2)_7$

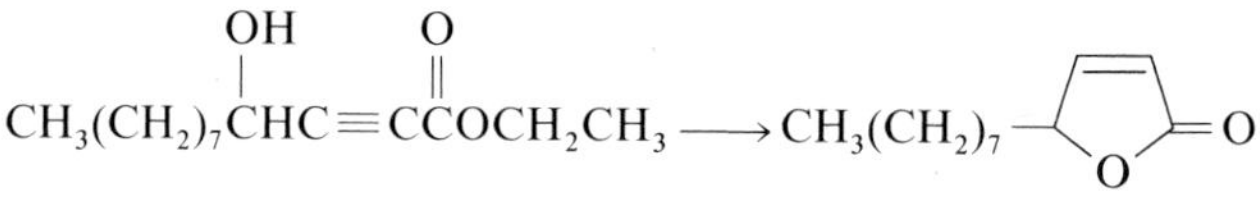

(i)

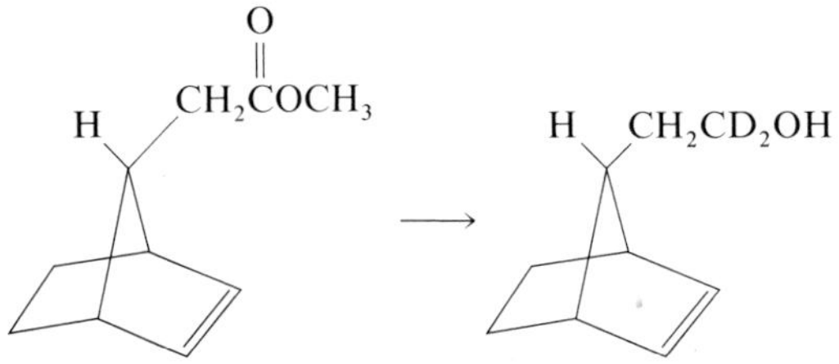

(j)

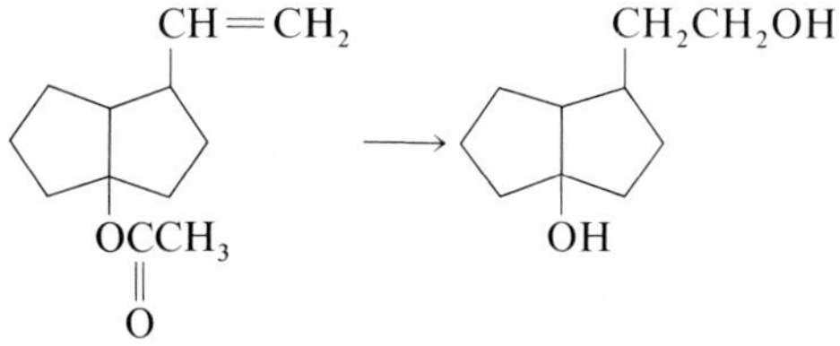

(k) $CH_2=CH(CH_2)_8\overset{O}{\underset{\parallel}{C}}OH \longrightarrow$

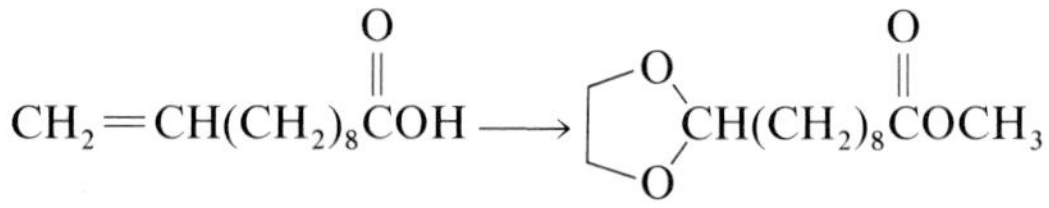

(l) $\longrightarrow$

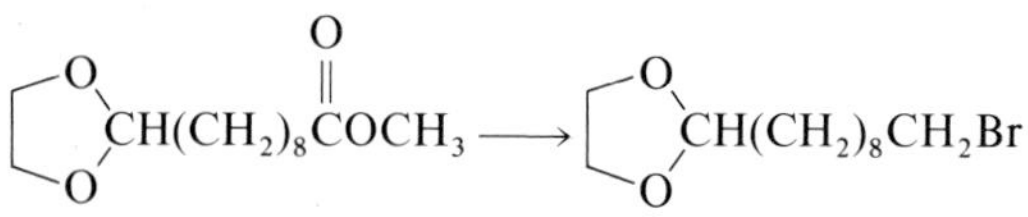

(m) $\longrightarrow HC(CH_2)_9C \equiv CH$

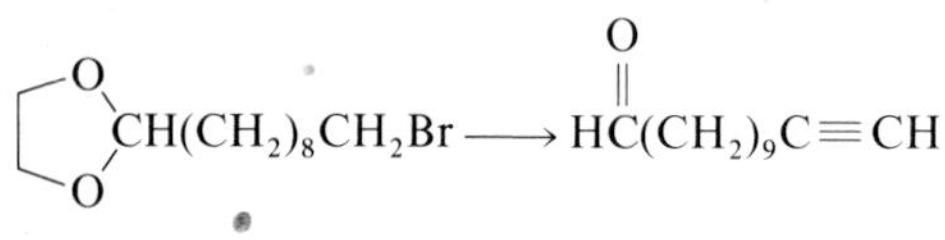

10.57   The following series of reactions was carried out during research into methods for the synthesis of large-ring lactones, macrolides such as recifeiolide (p. 481).

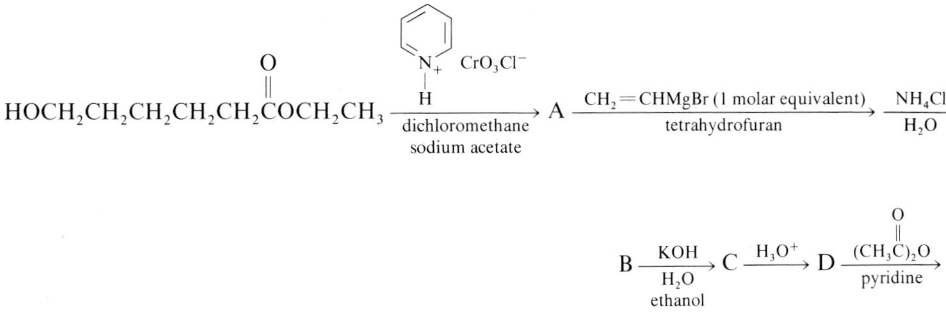

$$\underset{\substack{\\ \text{dichloromethane}\\ \text{sodium acetate}}}{\xrightarrow{\text{CrO}_3\text{Cl}^-,\ \text{N}^+_H\text{-pyridinium}}}$$

HOCH$_2$CH$_2$CH$_2$CH$_2$CH$_2$COCH$_2$CH$_3$ $\xrightarrow[\substack{\text{dichloromethane}\\\text{sodium acetate}}]{}$ A $\xrightarrow[\text{tetrahydrofuran}]{\text{CH}_2\!=\!\text{CHMgBr (1 molar equivalent)}}$ $\xrightarrow[\text{H}_2\text{O}]{\text{NH}_4\text{Cl}}$

B $\xrightarrow[\substack{\text{H}_2\text{O}\\\text{ethanol}}]{\text{KOH}}$ C $\xrightarrow{\text{H}_3\text{O}^+}$ D $\xrightarrow[\text{pyridine}]{(\text{CH}_3\text{C})_2\text{O}}$ E

What are the structures of Compounds A, B, C, D, and E? How do you account for the selectivity of the Grignard reaction?

10.58 The following sequence of reactions takes place. Suggest a mechanism for it.

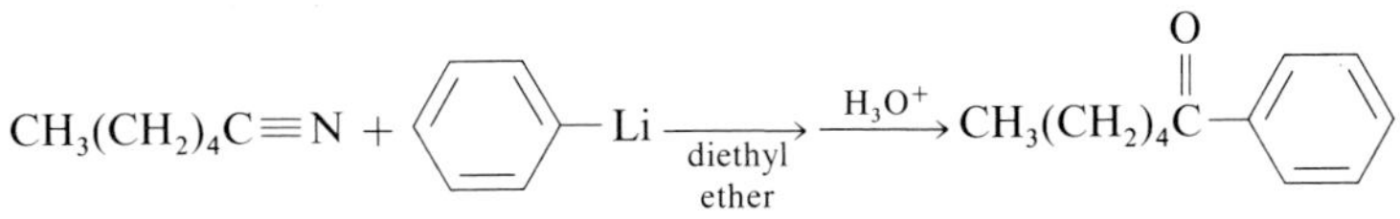

CH$_3$(CH$_2$)$_4$C$\equiv$N + ⬡—Li $\xrightarrow[\text{diethyl ether}]{}$ $\xrightarrow{\text{H}_3\text{O}^+}$ CH$_3$(CH$_2$)$_4$C(=O)—⬡

10.59 The following reaction is observed. Offer an explanation for the selectivity of the reduction.

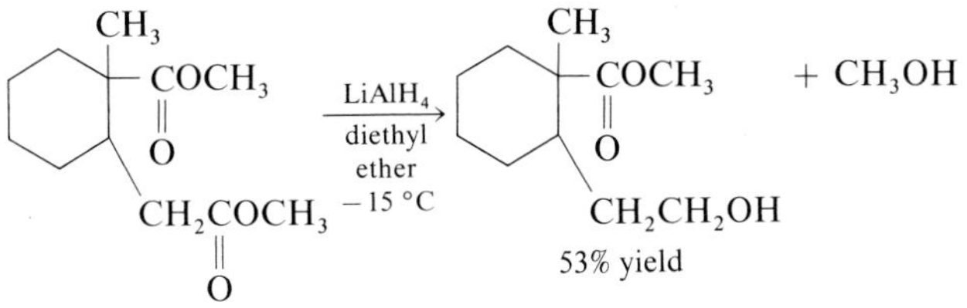

[structure: cyclohexane with CH$_3$, COCH$_3$, CH$_2$COCH$_3$ substituents] $\xrightarrow[\substack{\text{diethyl ether}\\-15\,°\text{C}}]{\text{LiAlH}_4}$ [structure: cyclohexane with CH$_3$, COCH$_3$, CH$_2$CH$_2$OH substituents] + CH$_3$OH

53% yield

10.60 A general reaction (the **Reformatsky reaction**) for the preparation of 3-hydroxycarboxylic acid esters involves the reaction of an aldehyde or ketone with a 2-bromoester in the presence of zinc. A typical reaction is shown below.

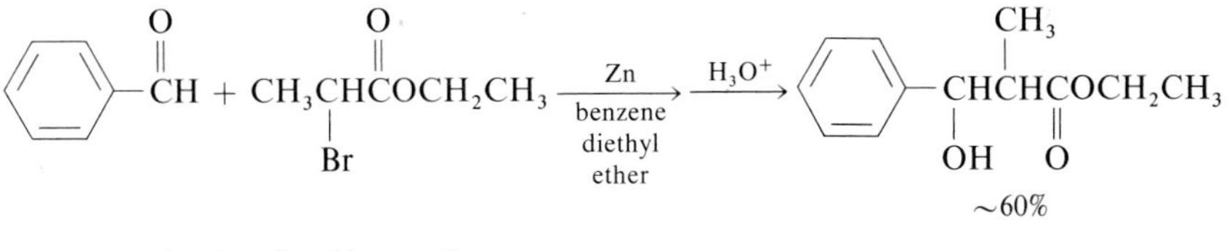

⬡—CHO + CH$_3$CHCOCH$_2$CH$_3$ (Br) $\xrightarrow[\substack{\text{benzene}\\\text{diethyl ether}}]{\text{Zn}}$ $\xrightarrow{\text{H}_3\text{O}^+}$ ⬡—CHCHCOCH$_2$CH$_3$ (OH) (CH$_3$) (O)

~60%

Suggest a mechanism for this reaction.

10.61 Ethyl benzoate, labeled with ^{18}O at the carbonyl group, was hydrolyzed in ordinary water, H$_2$^{16}O, with acidic or basic catalysis. The reaction was interrupted from time to time and the ^{18}O content of the *unhydrolyzed* ester was examined. It was found that ^{18}O was lost from the carbonyl groups of ester molecules and replaced by ^{16}O. How do you account for this observation?

10.62 The chemistry used to determine the structure of musk ambrette (p. 481) is a review of the chemistry of alkenes, alcohols, aldehydes, and acid derivatives. Some of the reactions are given in the form of the incomplete equations below. Complete the equations, giving structures for all the compounds indicated by Roman numerals.

(a) musk ambrette $\xrightarrow[\substack{\text{H}_2\text{O}\\\Delta}]{\text{NaOH}}$ $\xrightarrow{\text{H}_3\text{O}^+}$ I   (b) I $\xrightarrow[\text{ethanol}]{\text{H}_2,\ \text{Pt}}$ II $\xrightarrow{(\text{CH}_3\text{C})_2\text{O}}$ III (C$_{18}$H$_{34}$O$_4$)

(c) II $\xrightarrow[\substack{H_2SO_4 \\ \text{acetic acid}}]{CrO_3}$ IV ($C_{16}H_{30}O_4$)   (d) I $\xrightarrow[\text{}]{O_3 \quad \text{reductive work up}}$ V + VI

(e) VI $\xrightarrow[H_2O]{KMnO_4} \xrightarrow{H_3O^+}$ VII ($C_9H_{18}O_3$)   (f) VII $\xrightarrow[H_2SO_4]{CrO_3}$ VIII ($C_9H_{16}O_4$)

(g) V $\xrightarrow[\substack{HCl \\ Na_2CO_3}]{NH_2OH}$ IX ($C_7H_{13}O_3N$)   (h) IX $\xrightarrow[\Delta]{\overset{\displaystyle O \atop \displaystyle \|}{(CH_3C)_2O}}$ X ($C_7H_{11}O_2N$)

(i) X $\xrightarrow[\substack{H_2O \\ \Delta}]{NaOH} \xrightarrow{H_3O^+}$ heptanedioic acid

10.63 (a) When pyridine is used as a solvent for the reactions of acid chlorides, a reactive intermediate is formed (p. 483, Section 10.8A). This intermediate is more reactive towards
nucleophilic substitution at the carbon atom of the carbonyl group than benzoyl chloride
itself. Offer an explanation.

(b) If water were added to the system resulting from mixing an equimolar amount of benzoyl
chloride and pyridine, what would happen? Illustrate by writing a mechanism.

(c) When 1 mole of heptanoic acid, 0.5 mole of thionyl chloride, and 1 mole of pyridine are
mixed at $-10°$ C in ether solution, a 97% yield of heptanoic anhydride is formed. Propose
a reaction scheme that accounts for this observation.

10.64 The following sequence of reactions was used to synthesize an optically active cyclic ether,
Compound E. Fill in the structural formula, showing stereochemistry, for the major organic
product designated by a letter at each stage of the transformation.

$$H \cdots \overset{\overset{\displaystyle CH_3}{\displaystyle |}}{\underset{\displaystyle HO}{C}} \diagdown \overset{\overset{\displaystyle O}{\displaystyle \|}}{CH_2COCH_2CH_3} \xrightarrow[\underset{\displaystyle CF_3COH}{\overset{\displaystyle O \atop \displaystyle \|}{}}]{CH_3CH_2OCH=CH_2} A \xrightarrow[\text{diethyl ether}]{LiAlH_4} \xrightarrow{H_2O}$$

$$B \xrightarrow[\substack{\text{pyridine} \\ \text{dichloromethane}}]{TsCl} C \xrightarrow[\substack{\text{tetrahydrofuran} \\ \text{cold}}]{H_3O^+} D \xrightarrow{KOH} E \quad (C_4H_8O)$$

10.65 Infrared spectra are given in Figure 10.7 for the following compounds. Match each compound
with its spectrum. Point out which bands were important in making the assignment.

(a) 3-methyl-1-pentanol   (b) cycloheptanone   (c) (*E*)-2-hexenal
(d) cyclobutanecarboxylic acid   (e) *N*-cyclohexylformamide   (f) 3-butenoic acid
(g) ethyl decanoate

Spectrum 1

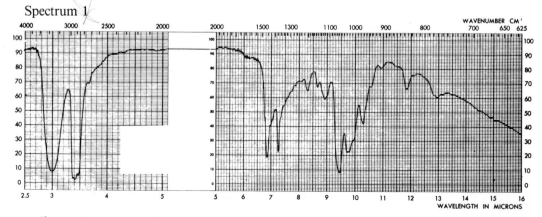

Figure 10.7 Spectra for Problem 10.65. (From *The Aldrich Library of Infrared Spectra*)

Spectrum 2

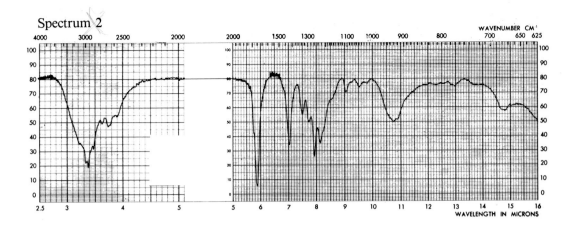

Spectrum 3

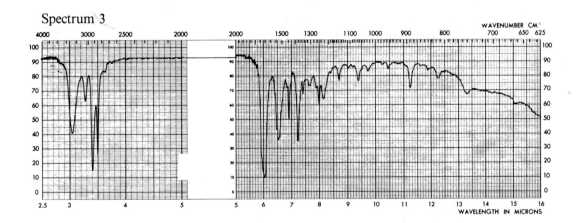

Spectrum 4

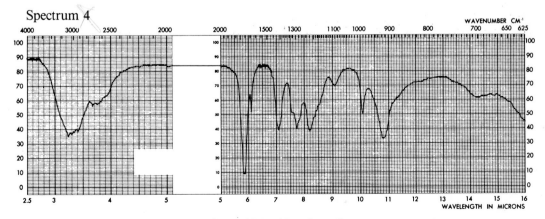

Figure 10.7 *(Continued)*

Spectrum 5

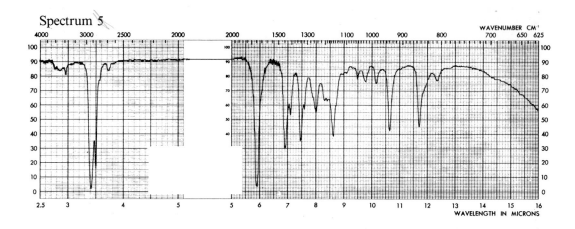

Spectrum 6

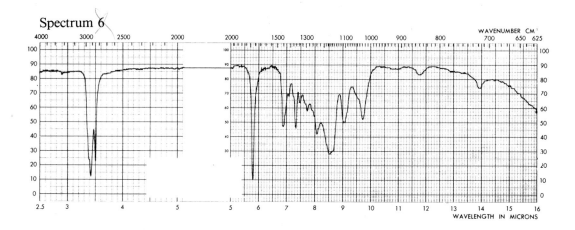

Spectrum 7

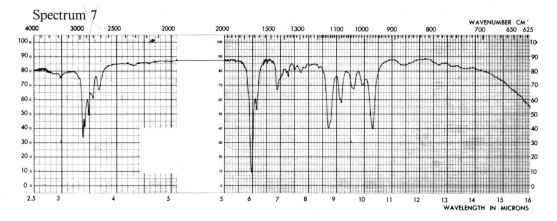

Figure 10.7   (*Continued*)

The Chemistry of Polyenes

11

Introduction. Conjugation

A. Isolated, Skipped, and Conjugated Multiple Bonds

By this point in your studies, you are familiar with the chemistry of a carbon-carbon multiple bond. Compounds containing double or triple bonds add halogens and acids. They are reduced by hydrogen and oxidized by reagents such as ozone, potassium permanganate, and peroxyacids. They react with the Lewis acids such as diborane or mercuric acetate to give intermediates that are converted to alcohols. This chemistry also applies to compounds that contain more than one multiple bond, as we have seen in many examples in preceding chapters. In all of those cases, however, the multiple bonds were separated from each other by one or more tetrahedral carbon atoms.

Multiple bonds that are separated from each other by two or more tetrahedral carbon atoms are said to be **isolated multiple bonds.** Some examples of compounds containing isolated multiple bonds and their nomenclature are given below.

$$CH_2 = CHCH_2CH_2CH = CH_2 \qquad\qquad HC \equiv CCH_2CH_2CH_2C \equiv CH$$
1,5-hexadiene   1,6-heptadiyne

$$CH_2 = CHCH_2CH_2CH = CHCH_2CH_2CH = CH_2 \qquad\qquad HC \equiv CCH_2CH_2CH = CH_2$$
1,5,9-decatriene   1-hexen-5-yne

Such compounds are named by indicating the number of multiple bonds in the chain by the suffixes **-adiene** and **-atriene,** or **-adiyne** and **-atriyne,** and also indicating the position of the multiple bond by giving the number in the chain of the first carbon atom of each multiple bond. If a compound contains both a double bond and a triple bond, the triple bond is named as the suffix.

Multiple bonds that are separated from each other by only one tetrahedral carbon atom are called **skipped multiple bonds.** The polyunsaturated fatty acids (Section 10.7C), for example, contain skipped double bonds. This structural feature makes such compounds especially reactive in ways that are explored in Section 11.7A.

If the multiple bonds are separated from each other by one single bond, the p orbitals on adjacent carbon atoms can interact. The prime example of this kind of interaction occurs in benzene. Chemists postulate sideways overlap of a p orbital from each of the six carbon atoms on the six-membered ring, resulting in delocalization of six electrons over the entire benzene ring (Section 2.10).

Multiple bonds that are separated from each other by one single bond are said to be **conjugated.** The concept of conjugation was first introduced in connection with a double bond (Section 9.4C) or a carbonyl group separated by a single bond from a double bond or an aromatic ring (Section 9.10). Some other examples of conjugated systems are shown below.

$$CH_2=CH-CH=CH_2 \qquad\qquad CH_2=CH-CH=CH-CH=CH_2$$
1,3-butadiene $\qquad\qquad\qquad\qquad\qquad$ 1,3,5-hexatriene

$$HC\equiv C-C\equiv CCH_2CH_3 \qquad\qquad HC\equiv C-CH=CHCH_3$$
1,3-hexadiyne $\qquad\qquad\qquad\qquad\qquad$ 2-penten-4-yne

$$\overset{\displaystyle CH_3}{\underset{\displaystyle }{CH_2=CH-C=O}}$$
3-buten-2-one

PROBLEM 11.1   Name the following compounds.

(a) $CH_3C\equiv CCH_2C\equiv CH$ (b) $CH_3CH=CHC\equiv CCH_3$ (c)

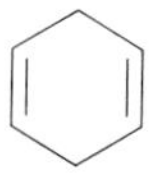

(d) 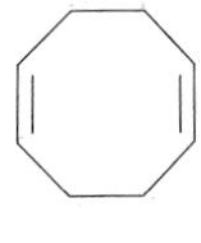   (e) 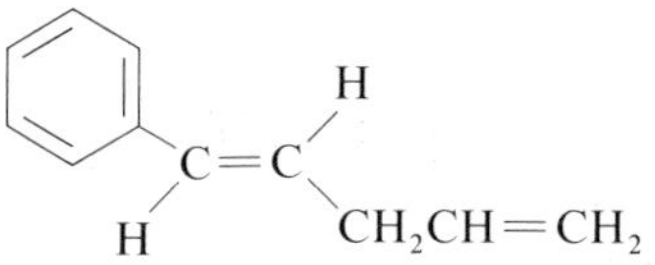   H   (f) $ClCH_2C\equiv CC\equiv CCH_2CH_3$

$$C=C$$
H $CH_2CH=CH_2$

PROBLEM 11.2   Write structural formulas for the following compounds.

(a)   5-hexen-2-one   (b)   1,5-hexadiyne   c)   1,4-pentadiene   (d)   2,5-heptadiyne
(e)   (*E*)-4,4-dimethyl-1,6-octadiene   (f)   1,3-hexadien-5-yne

---

B. 1,3-Butadiene

1,3-Butadiene is the simplest compound that contains two double bonds separated from each other by a single bond. 1,3-Butadiene has been determined by electron diffraction methods to be a planar molecule. The double bonds in 1,3-butadiene are 1.34 Å long, just about the length of the double bond in ethylene, while the single bond length is

1.48 Å, slightly shorter than a single bond (1.49 Å) adjacent to the double bond in propene and significantly shorter than the single bond in ethane (1.54 Å).

structure of 1,3-butadiene

The resonance energy (Section 2.10) obtained by comparing the heat of hydrogenation of 1,3-butadiene with that molecule in which there is no interaction between two double bonds is on the order of 4 kcal/mol, as compared with 36 kcal/mol for benzene.

A simple orbital picture of the bonding in 1,3-butadiene is shown in Figure 11.1.

1,3-Butadiene has a skeleton of four carbon atoms bonded to each other and to six hydrogen atoms by σ bonds. The carbon atoms are sp^2 hybridized, and a p orbital containing one electron is available to each carbon atom in the chain (Section 2.5A). We can picture the double bonds in 1,3-butadiene as arising from the overlap of pairs of p orbitals on adjacent carbon atoms. But this model for 1,3-butadiene, with double bonds localized between carbons 1 and 2 and carbons 3 and 4, does not explain the planarity of the molecule, the shorter than usual single bond, and the resonance energy determined experimentally for the compound.

We can develop a better model for the bonding in 1,3-butadiene by using molecular orbital theory (Sections 2.3 and 2.5). In this model, the four atomic p orbitals on the

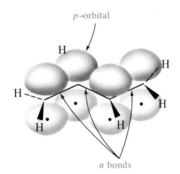

1,3–butadiene showing σ bonds
and p orbitals, each orbital
containing one electron

1,3–butadiene shown with
localized double bonds

Figure 11.1   An orbital picture of the structure of
1,3-butadiene.

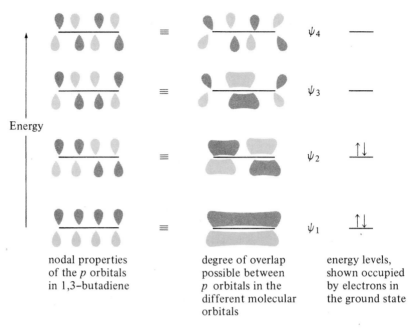

nodal properties
of the *p* orbitals
in 1,3–butadiene

degree of overlap
possible between
p orbitals in the
different molecular
orbitals

energy levels,
shown occupied
by electrons in
the ground state

Figure 11.2   The four molecular orbitals of 1,3-butadiene,
created from four *p* atomic orbitals.

carbon atoms are combined to give four π molecular orbitals, designated as ψ_1, ψ_2, ψ_3, and ψ_4 (Figure 11.2).

Of the molecular orbitals, ψ_1 and ψ_2 are bonding molecular orbitals and each has two electrons in it. The other two molecular orbitals are antibonding and do not contain any electrons. According to this picture of 1,3-butadiene, the molecule must be planar to allow interaction between the *p* orbitals on carbon atoms 2 and 3. There is some double bond character between those carbon atoms from the contribution of the lowest energy bonding molecular orbital, ψ_1, and, therefore, a shortening of the carbon-carbon single bond. 1,3-Butadiene is more stable by 4 kcal/mol than a molecule with two isolated double bonds because of the delocalization of electrons over the whole chain represented by ψ_1.

1,3-Butadiene was shown in Figure 11.1 in an extended conformation in which the two double bonds are trans to each other around the central single bond. This conformation is known as the **s-trans** conformation, the *s* indicating that the stereochemistry refers to the single bond. A higher energy conformation, but one that is important in the reactions of butadiene, is the **s-cis** conformation. Both of these conformations are in equilibrium with each other.

s-trans conformation
of 1,3-butadiene
~95%

s-cis conformation
of 1,3-butadiene
5%

In both of these conformations, all four of the carbon atoms and all the hydrogen atoms lie in the same plane.

C. Addition Reactions in Conjugated Systems

1,3-Butadiene undergoes addition reactions with electrophilic reagents such as chlorine and hydrogen chloride. If one molar equivalent of the reagent is used, a mixture of two products is obtained. The experimental observations about the addition of chlorine to butadiene are given below.

$$CH_2=CHCH=CH_2 + Cl_2$$

chloroform $\Big\downarrow -15\ °C$

$$ClCH_2CHCH=CH_2 + ClCH_2CH=CHCH_2Cl + ClCH_2CHCHCH_2Cl$$
$$\quad\quad\quad |$$
$$\quad\quad\quad Cl \quad\quad\quad\quad\quad\quad\quad\quad\quad\quad\quad\quad\quad\quad\quad Cl\ Cl$$

3,4-dichloro-1-butene   1,4-dichloro-2-butene   1,2,3,4-tetrachlorobutane

~60% of   *~40% of*
dichloro products   *dichloro products*

mixture formed
at −15 °C

$$\xrightarrow[200\ °C]{\Delta}$$

or

$$\xrightarrow{ZnCl_2}$$

$$\left\{ \begin{array}{l} \text{3,4-dichloro-1-butene} \\ \quad\quad 30\% \\ \\ \quad\quad + \\ \\ \text{1,4-dichloro-2-butene} \\ \quad\quad 70\% \end{array} \right.$$

3,4-Dichloro-1-butene is the product from the normal addition of chlorine to one of the double bonds in 1,3-butadiene. It is called the **1,2-addition product** of the reaction. In 1,4-dichloro-2-butene, the chlorine atoms have been attached to the first and fourth carbon atoms of the conjugated system, and the double bond that remains has shifted position in the molecule. Such a reaction is called a **1,4-addition** to the conjugated double bonds.

Experimentally, 3,4-dichloro-1-butene is formed in larger amounts at low temperatures. The mixture formed at low temperatures can be equilibrated at high temperatures or in the presence of a Lewis acid, $ZnCl_2$, which aids in the removal of chloride ion (Problem 7.17, Section 9.9). At equilibrium, the mixture is 70% 1,4-dichloro-2-butene.

1,4-Dichloro-2-butene is thermodynamically more stable than 3,4-dichloro-1-butene, as indicated by its predominance under equilibrium conditions. Yet, at low temperatures, under reaction conditions that do not allow for equilibrium to be established, 3,4-dichloro-1-butene is formed to a larger extent. In other words, 3,4-dichloro-1-butene is the *kinetic product* of the reaction, *the one that is formed faster at low temperatures* (Section 5.3G).

A conjugated diene system is always attacked at one end by an electrophile, because this gives the most highly stabilized cationic intermediate.

$$CH_2=CH-CH=CH_2$$

$$:\ddot{C}l\rightleftharpoons\ddot{C}l:$$
$$\delta+ \quad \delta-$$

$$CH_2-CH-CH=CH_2$$
$$^+ \quad |$$
$$\quad :\ddot{C}l: \quad\quad :\ddot{C}l:^-$$

$$:\ddot{C}l:^- \left[:\ddot{C}l-CH_2-CH-CH=CH_2 \longleftrightarrow :\ddot{C}l-CH_2-CH=CH-CH_2 \right]$$
$$\quad\quad\quad\quad\quad\quad\quad\quad + \quad\quad\quad\quad\quad\quad\quad\quad\quad\quad\quad\quad\quad\quad +$$

$$\text{|||}$$

$$:\ddot{C}l-CH_2-CH-CH-CH_2$$
$$\quad\quad\quad\quad\quad\quad +$$

primary carbocation, with
no resonance stabilization;
would be formed if attack
took place at the central
carbon atom

allylic carbocation, stabilized by
delocalization of charge; formed
by attack at terminal carbon
atom

The intermediate formed by addition of a chlorine atom to the end of the conjugated system is an allylic carbocation. An allylic cation is stabilized by delocalization of the positive charge (Section 6.8). Addition of a chlorine atom to one of the central carbon atoms in 1,3-butadiene would result in an unstable primary carbocation with no possibility of resonance stabilization.

An examination of the two contributing resonance structures to the allylic cation shows why 1,2- and 1,4-addition products are formed. The allylic cation has positive charge distribution at carbon atoms 2 and 4 in the chain. The negatively charged chloride ion can react with the cation at either one of these sites.

$$:\!\ddot{C}l\!-\!CH_2\!-\!\overset{+}{C}H\!-\!CH\!=\!CH_2 \longleftrightarrow :\!\ddot{C}l\!-\!CH_2\!-\!CH\!=\!CH\!-\!\overset{+}{C}H_2$$

$$:\!\ddot{C}l\!:^- \qquad\qquad\qquad\qquad\qquad\qquad :\!\ddot{C}l\!:^-$$

$$\downarrow \qquad\qquad\qquad\qquad\qquad\qquad\qquad \downarrow$$

$$:\!\ddot{C}l\!-\!CH_2\!-\!\underset{\underset{:\!\ddot{C}l\!:}{|}}{C}H\!-\!CH\!=\!CH_2 \qquad\qquad :\!\ddot{C}l\!-\!CH_2\!-\!CH\!=\!CH\!-\!CH_2\!-\!\ddot{C}l\!:$$

1,2-addition product   1,4-addition product

The 1,4-addition product, with the internal double bond, is thermodynamically more stable than the 1,2-addition product, with the terminal double bond (Section 5.2).

It is not so obvious why the 1,2-addition product should be the one favored kinetically. In other words, the energy of activation (Section 5.3D) for the formation of the 1,2-addition product from the intermediate is lower than the energy of activation leading to the 1,4-addition product from the same intermediate. These energy relationships are shown in Figure 11.3.

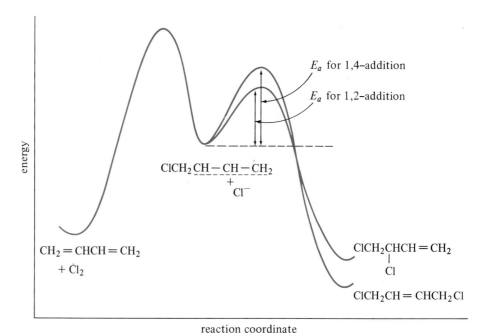

Figure 11.3   Energy diagram illustrating the relative energies of activation for 1,2- and 1,4-addition of chlorine to 1,3-butadiene.

1,4-Dichloro-2-butene is shown to be more stable and of lower energy than 3,4-dichloro-1-butene on the energy diagram. However, the energy of activation, the hill that must be climbed to get to the transition state from the common cationic intermediate, is higher for the 1,4-addition product. The highest energy of activation of all is, of course, the one leading to the formation of the reactive intermediate, the cation, from the starting reagents butadiene and chlorine. That is the rate-determining step (Section 5.3H) for the overall reaction.

Why should the energy of activation for 1,4-addition be higher than it is for 1,2-addition? Chemists offer two explanations. One is that the intermediate has some cyclic chloronium ion character (Section 5.8B), in which case attack at carbon atom 2 to open the three-membered ring is favored initially.

$$\overset{+}{\underset{\underset{-\;:\ddot{C}l:}{}}{\overset{\overset{..}{\ddot{C}l}}{CH_2-CH}}-CH=CH_2 \longrightarrow :\ddot{C}l-CH_2-\underset{\underset{:\ddot{C}l:}{|}}{CH}-CH=CH_2$$

chloronium ion
undergoing attack
by chloride ion
at carbon atom 2

The other explanation is that the two resonance contributors to the allylic cation are not equivalent. Positive charge density is greater at the secondary site, and the chloride ion reacts faster at that carbon atom.

$$\overset{\overset{..}{\ddot{C}l:}}{CH_2}-\underset{+}{CH}-CH=CH_2 \longleftrightarrow \overset{\overset{..}{\ddot{C}l:}}{CH_2}-CH=CH-\underset{+}{CH_2}$$

Resonance contributor   *Resonance contributor*
in which the positive   *in which the positive*
charge is localized   *charge is localized*
on a secondary as   *on a primary and*
well as allylic, carbon   *allylic carbon atom.*
atom. This resonance
contributor is favored,
hence there is a larger
overall density of
positive charge at
carbon atom 2.

An allylic halide ionizes easily (Section 6.8), so either one of the products can lose a halide ion and return to the allylic cation, allowing for equilibrium to be set up between the two products.

$$\underset{\underset{Cl}{|}}{ClCH_2CHCH}=CH_2 \quad \text{or} \quad ClCH_2CH=CHCH_2Cl$$

$$\Big\Updownarrow \Delta \text{ or ZnCl}_2$$

$$[ClCH_2\underset{+}{CHCH}=CH_2 \longleftrightarrow ClCH_2CH=CH\underset{+}{CH_2}]\;Cl^-$$

The competition between 1,2- and 1,4-addition reactions described in detail above applies to additions of halogens and hydrogen halides to other conjugated dienes. In general, the 1,4-addition products are favored under equilibrium conditions, while the 1,2-addition products are the major ones at low temperatures.

D. Reduction Reactions of Conjugated Dienes

Dienes add hydrogen catalytically (Section 8.3A) as well as in dissolving metal reactions (Section 8.3C). Catalytic hydrogenation reactions that are stopped before two molar equivalents of hydrogen are added to the diene give mixtures of alkenes and very little alkane, indicating that the diene reacts more readily with hydrogen than isolated double bonds do. For example, when butadiene is hydrogenated to the point where no diene is left, it has absorbed 54% of the hydrogen that would be required to hydrogenate it completely and only 6% of the product mixture is butane. The rest is a mixture of 1-butene and the (*E*)- and (*Z*)-2-butenes.

$$CH_2{=}CHCH{=}CH_2 + H_2$$

1,3-butadiene

$$\xrightarrow[\substack{\text{ethanol} \\ -12\,°C}]{\text{Pd}}$$

$$CH_3CH_2CH_2CH_3 + CH_3CH_2CH{=}CH_2 + \underset{\substack{(E)\text{-2-butene} \\ 38\%}}{\overset{CH_3\quad H}{\underset{H\quad CH_3}{C{=}C}}} + \underset{\substack{(Z)\text{-2-butene} \\ 10\%}}{\overset{CH_3\quad CH_3}{\underset{H\quad H}{C{=}C}}}$$

butane	1-butene
6%	45%

Dienes, in contrast to isolated internal double bonds, are easily reduced by metals dissolving in liquid ammonia.

$$CH_2{=}CHCH{=}CH_2 \xrightarrow[\substack{NH_3(\text{liq}) \\ -33\,°C}]{Na} \underset{\substack{(E)\text{-2-butene} \\ \textit{major product}}}{\overset{CH_3\quad H}{\underset{H\quad CH_3}{C{=}C}}} + \underset{\substack{(Z)\text{-2-butene} \\ \textit{minor product}}}{\overset{CH_3\quad CH_3}{\underset{H\quad H}{C{=}C}}}$$

1,3-butadiene

As is the case for the dissolving metal reductions of alkynes, the reaction proceeds by the formation of radical anion intermediates, which are protonated by the solvent, ammonia. Under these conditions, 1,4-addition of hydrogen to the conjugated system occurs.

PROBLEM 11.3   Complete the following equations, showing the product(s) you would expect.

(a) [cyclohexadiene structure] $\xrightarrow[\substack{\text{carbon tetrachloride} \\ 0\,°C}]{\text{Br}_2\ (1\ \text{molar equivalent})}$

(b) $CH_2{=}CHCH{=}CH_2 \xrightarrow{\text{HCl (1 molar equivalent)}}$

(c) $CH_2{=}\overset{\overset{\displaystyle CH_3}{|}}{C}{-}\overset{\overset{\displaystyle CH_3}{|}}{C}{=}CH_2 \xrightarrow[\substack{\text{carbon tetrachloride} \\ 0\,°C}]{\text{Br}_2\ (1\ \text{molar equivalent})}$

(d) [cyclopentadiene with CH_3 groups] $\xrightarrow[\text{Pd}]{\text{H}_2\ (1\ \text{molar equivalent})}$

(e) [cyclohexadiene with CH_3 groups] $\xrightarrow[\text{NH}_3(\text{liq})]{\text{Na}}$

11.2

Free Radical Substitution Reactions of Alkenes

The properties and reactions of radicals, reactive intermediates that behave as electrophiles, were discussed in Section 5.4. Whether a reaction proceeds with ionic or radical intermediates depends on the reaction conditions. For example, hydrogen bromide adds to an alkene by an ionic mechanism in the absence of peroxides and by a free radical mechanism in the presence of peroxides. Similarly, an alkene reacts with bromine in carbon tetrachloride solution at room temperature by addition, postulated to go through a bromonium ion intermediate (Section 5.8). If a reagent that generates bromine in low concentrations is used, another type of reaction takes place with alkenes that have allylic hydrogen atoms. A substitution of one of the allylic hydrogen atoms by a bromine atom occurs. For example, 2-heptene reacts with N-bromosuccinimide in boiling carbon tetrachloride in the presence of benzoyl peroxide to give 4-bromo-2-heptene.

$$CH_3CH_2CH_2CH_2CH{=}CHCH_3 \;+\; \underset{\substack{\text{N-bromosuccinimide}\\\text{NBS}}}{\overset{\text{O}}{\underset{\text{O}}{\bigvee}}\!\!N{-}Br} \xrightarrow[\substack{\text{benzoyl peroxide}\\\text{carbon tetrachloride}}]{}$$

2-heptene N-bromosuccinimide
                      NBS

$$CH_3CH_2CH_2\underset{\underset{\text{Br}}{|}}{C}HCH{=}CHCH_3 \;+\; \overset{\text{O}}{\underset{\text{O}}{\bigvee}}\!\!N{-}H$$

4-bromo-2-heptene   succinimide
    $\sim60\%$

The reaction is quite general. *N-Bromosuccinimide reacts with a variety of alkenes by substituting bromine for an allylic hydrogen atom.* The reaction is catalyzed by the presence of peroxides or by light. It is postulated that the reaction starts with the formation of a bromine atom, which abstracts an allylic hydrogen atom from the alkene. This reaction is shown by the use of a *fish-hook* representing the transfer of a hydrogen atom with *one* of the electrons of the covalent bond to the bromine atom.

$$\overset{\text{O}}{\underset{\text{O}}{\bigvee}}\!\!:N{-}\ddot{Br}: \xrightarrow[\Delta]{h\nu\text{ or}} \overset{\text{O}}{\underset{\text{O}}{\bigvee}}\!\!:N\cdot \;+\; \cdot\ddot{Br}:$$

$$CH_3CH_2CH_2\underset{\underset{\cdot\,\ddot{Br}:}{\overset{|}{\underset{}{H}}}}{C}HCH{=}CHCH_3 \longrightarrow CH_3CH_2CH_2\dot{C}HCH{=}CHCH_3$$

$$H{-}\ddot{Br}:$$

abstraction of an   secondary and allylic
allylic hydrogen atom   radical

most stable intermediate

$$CH_3CH_2CH_2CH_2CH=CHCH_2 \longrightarrow CH_3CH_2CH_2CH_2CH=CHCH_2$$
$$\overset{|}{\underset{\overset{}{\text{H}}}{}}$$

:Br:

H—Br:

abstraction of an   *primary and allylic*
allylic hydrogen atom   *radical*

The stabilities of various carbon radical species, as electron-deficient species, parallel those of carbocations (Section 5.4B). *A radical at a tertiary carbon atom is more stable than one at a secondary carbon atom, which in turn is more stable than a primary radical. Radicals stabilized by resonance delocalization of the radical character are the most stable of all* and, therefore, the most easily formed. A radical adjacent to a double bond, at an allylic position, is one for which resonance stabilization is possible.

$$\overset{|}{\underset{}{C}}=\overset{}{\underset{}{C}}-\overset{\cdot}{\underset{}{C}} \longleftrightarrow \overset{}{\underset{}{C}}-\overset{|}{\underset{}{C}}=\overset{}{\underset{}{C}}$$

resonance contributors for an allylic radical

Besides abstraction of a hydrogen atom, the bromine atom can also react with the double bond to give a secondary alkyl radical with no additional stabilization.

$$CH_3CH_2CH_2CH_2CH=CHCH_3 \longrightarrow CH_3CH_2CH_2CH_2\overset{\cdot}{C}H-CHCH_3$$

·Br:

:Br:

electrophilic attack at   *secondary radical*
the double bond

least stable intermediate

The most stable intermediate for reaction of a bromine atom with 2-heptene is the radical that is secondary and allylic (p. 521). The major product of the reaction is derived from that intermediate, formed by the abstraction of a hydrogen atom by a bromine atom. The other product of this abstraction reaction is hydrogen bromide, which reacts with *N*-bromosuccinimide to give bromine and succinimide. The bromine atom in *N*-bromosuccinimide is somewhat positive in character because of the electronegativity of the nitrogen atom and the electron-withdrawing effects of the adjacent carbonyl groups. It can be attacked by bromide ion to give bromine and a stabilized nitrogen anion, which is quickly protonated by hydrogen bromide.

N-bromosuccinimide
and
bromide ion from
hydrogen bromide

stabilized anion

protonation of anion   succinimide

As a result of this sequence of reactions, hydrogen bromide, which is a product of the hydrogen abstraction reaction, is converted to bromine. The amount of bromine formed in the reaction mixture is limited by the amount of hydrogen bromide formed in the previous step of the reaction. The allylic radical formed in the first step of the reaction abstracts a bromine atom from bromine, to give 4-bromo-2-heptene and a bromine atom, which starts the cycle all over again.

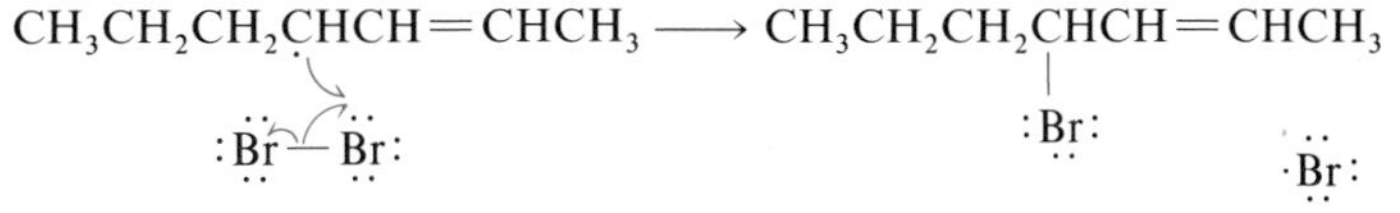

$$CH_3CH_2CH_2CHCH=CHCH_3 \longrightarrow CH_3CH_2CH_2CHCH=CHCH_3$$

$$:Br\frown Br: \qquad\qquad\qquad :Br: \qquad \cdot Br:$$

You may wonder why the bromine formed in this reaction does not add to the double bond in heptene. The answer is related to the very low concentrations of bromine present at any one time. The initial addition of bromine to the double bond involves only one of the two bromine atoms in a cyclic bromonium ion intermediate (Section 5.8B). The formation of the intermediate is reversible. If there is not a high concentration of bromide ion nearby, the formation of the dibromide cannot be completed. *N*-Bromosuccinimide competes with the bromonium ion for bromide ion. Under these conditions, the allylic substitution reaction is favored.

The reaction illustrated for 2-heptene has been widely applied to other systems. The selectivity between primary and secondary allylic positions shown for 2-heptene is quite general. For example, 1,1-diphenyl-1-propene, which has only primary allylic hydrogen atoms, reacts with *N*-bromosuccinimide (abbreviated in reaction schemes as NBS) much more slowly than 1,1-diphenyl-1-butene, where secondary allylic hydrogen atoms are available.

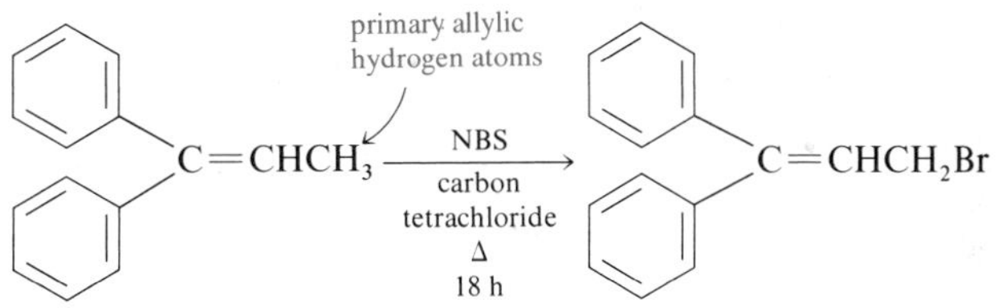

primary allylic
hydrogen atoms

$$C=CHCH_3 \xrightarrow[\substack{\text{carbon} \\ \text{tetrachloride} \\ \Delta \\ \text{18 h}}]{\text{NBS}} C=CHCH_2Br$$

1,1-diphenyl-1-propene

3-bromo-1,1-diphenyl-
1-propene
86%

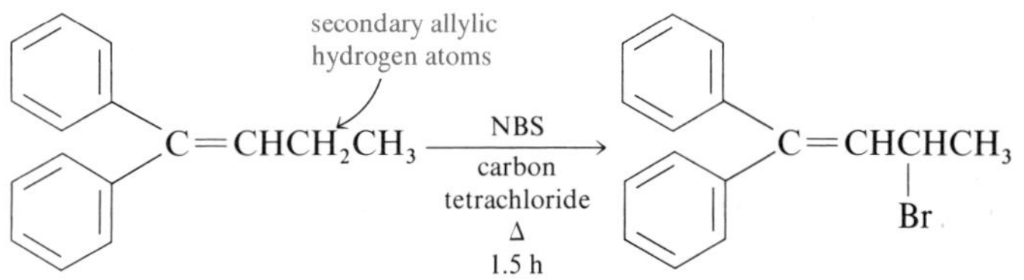

secondary allylic
hydrogen atoms

$$C=CHCH_2CH_3 \xrightarrow[\substack{\text{carbon} \\ \text{tetrachloride} \\ \Delta \\ \text{1.5 h}}]{\text{NBS}} C=CHCHCH_3$$
$$\qquad\qquad\qquad\qquad\qquad\qquad\qquad\qquad Br$$

1,1-diphenyl-1-butene

3-bromo-1,1-diphenyl-
1-butene
yield not known

PROBLEM 11.4 Predict the product(s) of the following reactions.

(a) 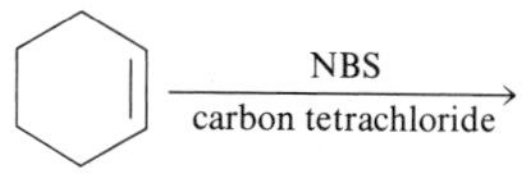 $\xrightarrow{\text{NBS}}$
      carbon tetrachloride

(b) 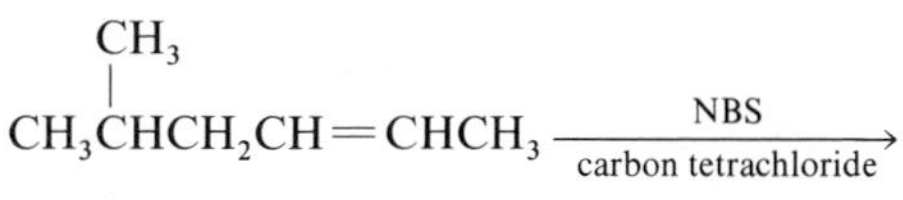 $CH_3CHCH_2CH=CHCH_3 \xrightarrow{\text{NBS}}$
     $|$ carbon tetrachloride
     CH_3

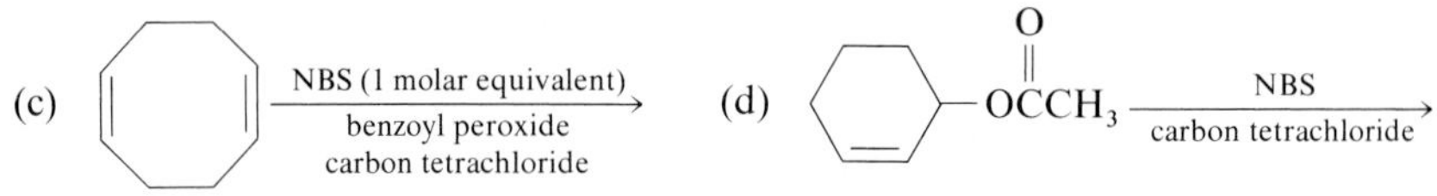

(c) ——NBS (1 molar equivalent)——→
       benzoyl peroxide
       carbon tetrachloride

(d) ——NBS——→
       carbon tetrachloride

PROBLEM 11.5 When 3-phenyl-1-propene is treated with *N*-bromosuccinimide in the presence of benzoyl peroxide, 3-bromo-1-phenyl-1-propene (50%) and 3-bromo-3-phenyl-1-propene (10%) are formed. Write a mechanism for the reaction and explain why the major product of the reaction is 3-bromo-1-phenyl-1-propene.

PROBLEM 11.6 When 2-heptene reacts with *N*-bromosuccinimide (p. 521), 4-bromo-2-heptene is isolated in only 60% yield. What are other possible substitution products of the reaction? Write equations supporting your answer.

11.3

Elimination Reactions Leading to Dienes

Dehydrohalogenation of alkyl halides by base (Sections 6.1, 6.9A and 6.9B) and dehydration of alcohols under acidic conditions (Section 5.5B) lead to the formation of alkenes. These reactions are used to synthesize dienes from suitably substituted starting materials. If the product dienes are conjugated, they tend to be highly reactive and their isolation is not always easy, often resulting in low yields.

Cyclohexene, for example, is converted to 1,3-cyclohexadiene by allylic bromination and then elimination of hydrogen bromide by the use of the organic base, quinoline.

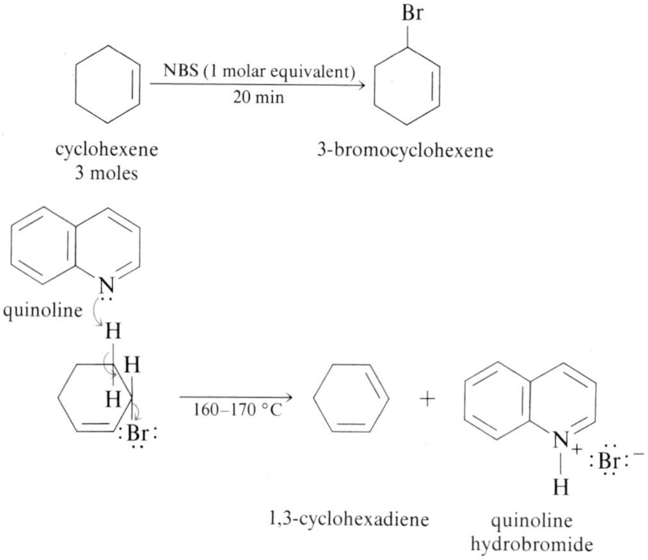

cyclohexene   3-bromocyclohexene
3 moles

quinoline

160–170 °C

1,3-cyclohexadiene   quinoline hydrobromide

Another preparation of 1,3-cyclohexadiene is the dehydrohalogenation of 1,2-dibromocyclohexane. The base used is the anion of isopropyl alcohol, prepared from the alcohol with sodium hydride in a high-boiling ether solvent.

$$\underset{\substack{\text{isopropyl alcohol}}}{\overset{\overset{\displaystyle CH_3}{|}}{CH_3CHOH}} + \underset{\substack{\text{sodium}\\\text{hydride}}}{NaH} \xrightarrow{\text{triglyme}} \underset{\substack{\text{sodium isopropoxide}}}{\overset{\overset{\displaystyle CH_3}{|}}{CH_3CHO^-\ Na^+}} + H_2\uparrow$$

$$\text{1,2-dibromocyclohexane} \quad \text{sodium isopropoxide} \qquad\qquad \text{1,3-cyclohexadiene} \qquad\quad \text{isopropyl}$$
$$\text{1 mole} \qquad\qquad\qquad \text{2.2 mole} \qquad\qquad\qquad\quad \text{bp 81 °C} \qquad\qquad \text{alcohol}$$
$$\text{70\%}$$

$$\text{triglyme} \equiv CH_3OCH_2CH_2OCH_2CH_2OCH_2CH_2OCH_3$$
$$\text{bp 222 °C}$$

1,3-Cyclohexadiene is distilled out of the reaction mixture as it is formed and is thus spared prolonged heating and contact with the other reagents.

Dienes are particularly sensitive to electrophiles, so acidic conditions are rarely used for their synthesis. 2,3-Dimethyl-1,3-butadiene is prepared, along with an interesting side-product, by the double dehydration of 2,3-dimethyl-2,3-butanediol.

$$\underset{\substack{\text{2,3-dimethyl-2,3-butanediol}\\\text{pinacol}}}{} \xrightarrow[95\,°C]{\text{HBr (48\%)}} \underset{\substack{\text{2,3-dimethyl-1,3-butadiene}\\57\%}}{} + \underset{\substack{\text{3,3-dimethyl-}\\\text{2-butanone}\\\text{pinacolone}\\\sim25\%}}{}$$

PROBLEM 11.7   2,3-Dimethyl-2,3-butanediol has the common name *pinacol*. The second product of the dehydration reaction, 2,3-dimethyl-2-butanone, commonly known as *pinacolone,* comes from the loss of one mole of water and a molecular rearrangement, known as the **pinacol-pinacolone rearrangement.** Write a detailed mechanism for the dehydration of pinacol in acid to 2,3-dimethyl-1,3-butadiene. Look closely at the structure of pinacolone and decide what type of rearrangement has occurred. Why does it occur? Write a mechanism for the formation of pinacolone. (Hint: You may find it helpful to review Section 5.6B.)

11.4

Organocopper Reagents Used to Form Carbon-Carbon Bonds

A. Lithium Organocuprates. Reactions with Halides

Grignard reagents and organolithium compounds are reactive towards carbonyl groups (Section 9.4B) but do not react readily with alkyl halides to give new carbon-carbon bonds. Copper, however, promotes reactions at electrophilic carbon atoms other than carbonyl groups. Organocopper reagents that are particularly useful are prepared by treating an organolithium compound with a copper(I) halide, usually copper(I) iodide.

$$2\ CH_3Li\ +\ CuI\ \xrightarrow[\substack{\text{diethyl ether}\\0\,°C}]{}\ (CH_3)_2CuLi\ +\ LiI$$

methyllithium   copper (I)   lithium   lithium
               iodide   dimethyl-   iodide
                              cuprate

Just as is the case with the Grignard reagent, the exact structure of the organocuprate reagent is not known, and the formula represents the stoichiometry observed for the reaction. Two organic groups appear to be associated with the copper atom in a negatively charged species that is a source of nucleophilic carbon atoms. The name *lithium dimethylcuprate* indicates that copper is associated with the anion in the compound. Primary alkyl halides give reasonably stable organocuprates. Vinyl and aryl organocuprates can also be prepared.

Organocuprates react with halides to give compounds with longer carbon chains. For example, lithium dibutylcuprate reacts with 1-bromopentane to give nonane.

$$CH_3CH_2CH_2CH_2CH_2Br\ \xrightarrow[\substack{\text{tetrahydrofuran}\\25\,°C,\ 1\ h}]{(CH_3CH_2CH_2CH_2)_2CuLi}\ CH_3CH_2CH_2CH_2CH_2CH_2CH_2CH_2CH_3$$

1-bromopentane   nonane
                                                                   98%

Organocuprates may be used to synthesize alkenes or polyenes. The unsaturation in the product may be derived from the organocuprate, from the halide with which it reacts, or from both. 3-Bromo-1-methylcyclohexene, for example, reacts with lithium diisopropenylcuprate to give 3-isopropenyl-1-methylcyclohexene.

3-bromo-1-methylcyclohexene   3-isopropenyl-1-methylcyclohexene
                                                            75%

The difference in reactivity between a lithium organocuprate and an organolithium reagent is illustrated by the reaction of 4-bromocyclohexanone with lithium diisopropenylcuprate.

4-bromocyclohexanone   4-isopropenylcyclohexanone
                                          65%

An organolithium reagent would have reacted with the carbonyl group (Section 9.4B) rather than with the alkyl halide function in 4-bromocyclohexanone.

PROBLEM 11.8 Complete the following equations.

(a) $CH_3CH_2CH_2CH_2Li\ +\ CuI\ \xrightarrow[\substack{\text{diethyl ether}\\0\,°C}]{}$ (b) $CH_3CH_2CH_2CH_2CH_2I\ \xrightarrow[\substack{\text{diethyl ether}\\25\,°C}]{(CH_3)_2CuLi}$

(c) 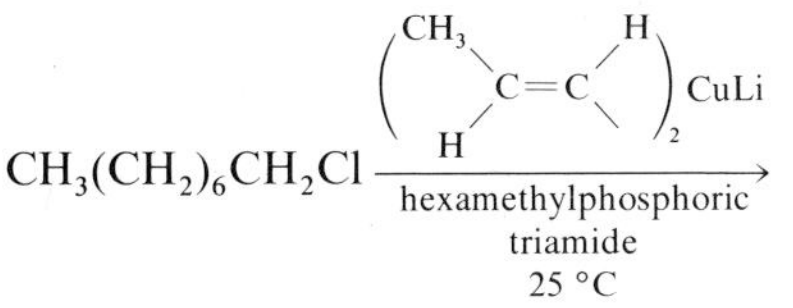 $CH_3(CH_2)_6CH_2Cl$ $\xrightarrow[\substack{\text{hexamethylphosphoric}\\\text{triamide}\\25\,°C}]{\left(\substack{CH_3\\\diagup\\C=C\\\diagdown\\H}\substack{H\\\diagdown\\\diagup}\right)_2 CuLi}$

(d) 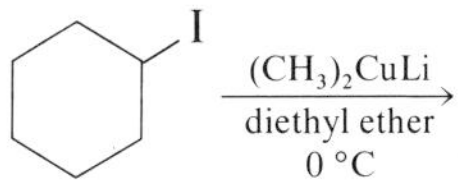 $\xrightarrow[\substack{\text{diethyl ether}\\0\,°C}]{(CH_3)_2CuLi}$

(e) 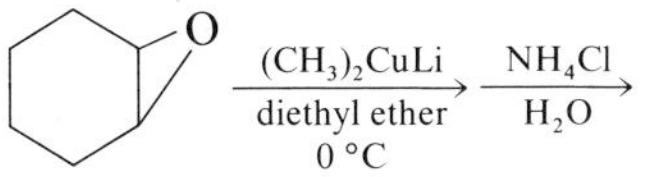 $\xrightarrow[\substack{\text{diethyl ether}\\0\,°C}]{(CH_3)_2CuLi}$ $\xrightarrow[H_2O]{NH_4Cl}$

PROBLEM 11.9   Juvenile hormones are substances that block the maturation of the pupae of insects. The juvenile hormone of the giant silkworm moth has the following structure.

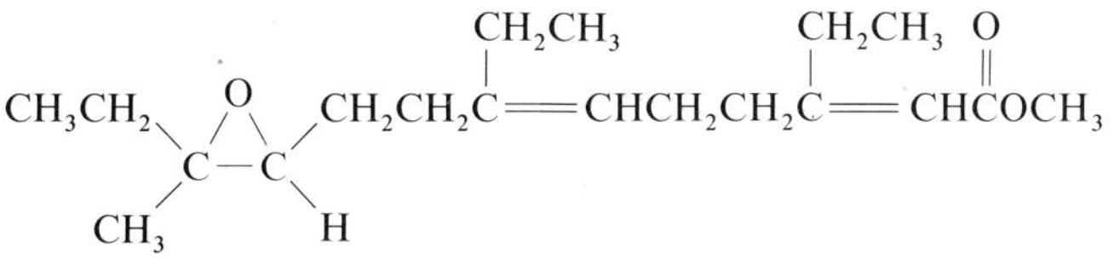

$$CH_3CH_2 \underset{\underset{CH_3}{|}}{\overset{O}{\underset{C-C}{\diagup\diagdown}}} \overset{H}{\underset{}{}}\,CH_2CH_2\overset{\overset{CH_2CH_3}{|}}{C}=CHCH_2CH_2\overset{\overset{CH_2CH_3}{|}}{C}=CH\overset{O}{\overset{||}{C}}OCH_3$$

An important step in the synthesis of the compound involves the following transformation. Suggest a reagent for the reaction.

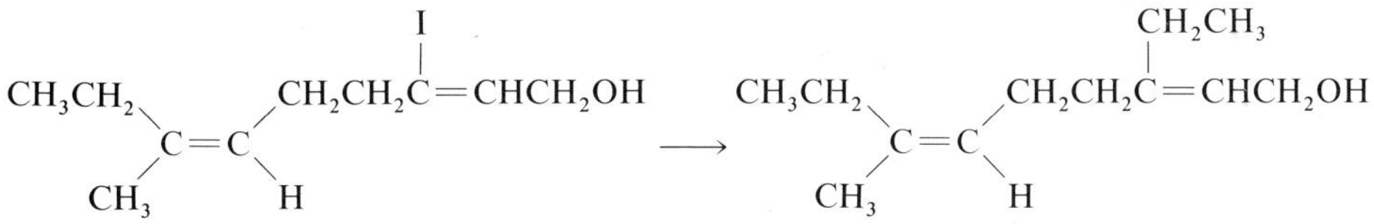

$$CH_3CH_2\overset{CH_3}{\underset{}{\diagup}}C=C\overset{H}{\underset{}{\diagdown}}\,CH_2CH_2\overset{\overset{I}{|}}{C}=CHCH_2OH \longrightarrow CH_3CH_2\overset{CH_3}{\underset{}{\diagup}}C=C\overset{H}{\underset{}{\diagdown}}\,CH_2CH_2\overset{\overset{CH_2CH_3}{|}}{C}=CHCH_2OH$$

B. Copper-Catalyzed Reactions in the Synthesis of Compounds with Multiple Triple Bonds. Intermediates in the Synthesis of Polyenes

Terminal alkynes show a special kind of reactivity because of the acidity of the hydrogen atom bonded directly to an *sp*-hybridized carbon atom. The anion derived from a terminal alkyne is prepared either by treating the alkyne with sodium amide in liquid ammonia or by preparing the Grignard reagent (Section 9.4A).

Organometallic reagents derived from terminal alkynes also form carbon-carbon bonds in reactions that are catalyzed by copper(I) salts. Two types of reactions are possible. In one, a terminal alkyne is treated with oxygen in the presence of copper(I) chloride and pyridine. A conjugated diyne in which a bond has been formed between two molecules of the alkyne used as a starting material is formed in high yield. This reaction is illustrated for 1-hexyne.

$$\underset{\substack{\text{1-hexyne}}}{CH_3CH_2CH_2CH_2C\equiv CH} \xrightarrow[\substack{CuCl\\20-40\,°C\\3\,h}]{O_2,\,\text{pyridine}} \underset{\substack{\text{5,7-dodecadiyne}\\97\%}}{CH_3CH_2CH_2CH_2C\equiv CC\equiv CCH_2CH_2CH_2CH_3}$$

The reaction is believed to proceed by the conversion of the alkyne, in the presence of the base pyridine, to a copper salt, which is oxidized by oxygen to the diyne. Copper(I) oxide is another product of the reaction. Although the detailed mechanism of the reaction is not known, it should be clear that a process that starts with two molecules of a terminal alkyne and ends with a carbon-carbon bond between the original terminal carbon atoms is an oxidation reaction (Section 7.6A). The terminal carbon atom of each 1-hexyne molecule loses a hydrogen atom and is bonded to another carbon atom in the product diyne. Hence, the reaction conditions call for base and an oxidizing agent, molecular oxygen in this case. The reaction is known as an **oxidative coupling reaction.** Note that the product of such a coupling reaction is a conjugated diyne.

Another type of coupling reaction catalyzed by copper(I) chloride occurs between an alkynylmagnesium bromide and a halide in which the halogen is on an sp^3-hybridized carbon atom adjacent to a triple bond. Such a sequence of reactions is shown for the preparation of 8-chloro-1,4-octadiyne from 5-chloro-1-pentyne and 3-bromo-1-propyne.

$$ClCH_2CH_2CH_2C \equiv CH + CH_3CH_2MgBr \xrightarrow{\text{tetrahydrofuran}}$$

5-chloro-1-pentyne   ethylmagnesium bromide

$$ClCH_2CH_2CH_2C \equiv CMgBr + CH_3CH_3\uparrow$$

5-chloro-1-pentynylmagnesium bromide

$$ClCH_2CH_2CH_2C \equiv CMgBr + BrCH_2C \equiv CH \xrightarrow[\text{tetrahydrofuran}]{\text{CuCl}}$$

5-chloro-1-pentynylmagnesium   3-bromo-1-propyne
bromide

$$ClCH_2CH_2CH_2C \equiv CCH_2C \equiv CH + MgBr_2$$

8-chloro-1,4-octadiyne
64%

The alkynyl Grignard reagent is prepared by an acid-base reaction between the acidic terminal alkyne 5-chloro-1-pentyne, and ethylmagnesium bromide. Note that the Grignard reagent formed does not react with the primary alkyl chloride function in 5-chloro-1-pentyne. It does displace the specially activated bromine atom in 3-bromo-1-propyne to form 8-chloro-1,4-octadiyne. The reaction proceeds best in the presence of copper(I) chloride in tetrahydrofuran. The two triple bonds are present in the product in a skipped arrangement.

As you recall from Section 6.8, halides in which the halogen atom is on a tetrahedral carbon atom adjacent to a double bond are known as *allylic halides*. Halides in which the halogen atom is on a tetrahedral carbon atom adjacent to a triple bond are called **propargylic halides.** Such halides have the same reactivity as allylic halides and for the same reason, namely they have the potential for delocalization of charge (or radical character) when an intermediate is formed.

Triple bonds are reduced to double bonds by a variety of methods. Catalytic hydrogenation over poisoned catalysts (Section 8.3A) and hydroboration followed by protonation (Section 8.3B) give cis alkenes; whereas dissolving metal reductions with sodium in liquid ammonia, for example, give trans alkenes (Section 8.3C).

The special reactivity of terminal alkynes is used in the two different coupling reactions described in this section to prepare compounds with more than one triple bond in them. The triple bonds may be *conjugated,* as is the result from the oxidative coupling of terminal alkynes, or they may be *skipped* from the reaction of an alkynyl Grignard with a propargylic halide. A triple bond is a reactive site of further transformation in a molecule. The reactions of greatest interest are those that allow for the precise placement of double bonds of known stereochemistry in a complex molecule.

PROBLEM 11.10   9,12-Octadecadiynal, an intermediate in the synthesis of one of the essential fatty acids, is synthesized in the laboratory by the following sequence of reactions.

(a)   Fill in structural formulas for all the products or intermediates designated by letter.

$$HC \equiv C(CH_2)_7 \overset{O}{\underset{O}{\diagdown}} \xrightarrow[\text{tetrahydrofuran}]{CH_3CH_2MgBr} A \xrightarrow[\Delta]{CH_3(CH_2)_4C \equiv CCH_2Br}{CuBr}$$

$$B \xrightarrow[\Delta]{H_3O^+} 9,12\text{-octadecadiynal}$$

(b)   To which fatty acid is 9,12-octadecadiynal related? How would you complete the synthesis? (Hint: Reviewing Sections 10.5A and 10.7C might be helpful.)

PROBLEM 11.11   Many compounds with conjugated triple bonds are found as constituents of plants. One of them, known as *artemisia ketone,* is isolated from the roots of *Artemisia vulgaris L.*, the common mugwort.

$$CH_3C \equiv CC \equiv CC \equiv CCH = CHCH_2CH_2\overset{\overset{\displaystyle O}{\|}}{C}CH_2CH_3$$

artemisia ketone

This ketone is synthesized by an oxidative coupling reaction of two terminal alkynes. What starting materials could be used for this synthesis? Write equations for the coupling reactions that could give rise to artemisia ketone. What complications do you expect to arise from using this reaction for the coupling of two different terminal alkynes?

11.5

The Diels-Alder Reaction

A. Introduction. Stereochemistry

One of the most important reactions of conjugated dienes is the 1,4-addition of another multiple bond to the conjugated system to give a six-membered ring. A classic example of this reaction is the formation of a substituted cyclohexene from 1,3-butadiene and maleic anhydride.

s-cis conformation of 1,3-butadiene	maleic anhydride	tetrahydrophthalic anhydride ~95%

With this equation we are extending to acyclic structures the convention that we adopted earlier for representing cyclic compounds (Sections 2.10 and 3.8). In this convention, each junction of two straight lines, or the end of a line, represents a carbon atom with the number of hydrogen atoms necessary to give the carbon four bonds.

Functional groups are represented by their usual symbols. A double carbon-carbon bond is shown by two parallel lines, a triple bond by three. This convention simplifies the drawing of complex molecules. A few examples of this convention applied to a variety of structures are shown below.

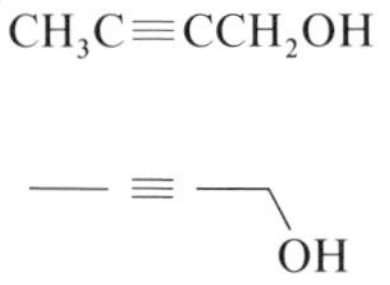

$$CH_3C{\equiv}CCH_2OH$$

$$CH_3CH_2\overset{\overset{\displaystyle O}{\|}}{C}CH_2CH_2Cl$$

2-butyn-1-ol

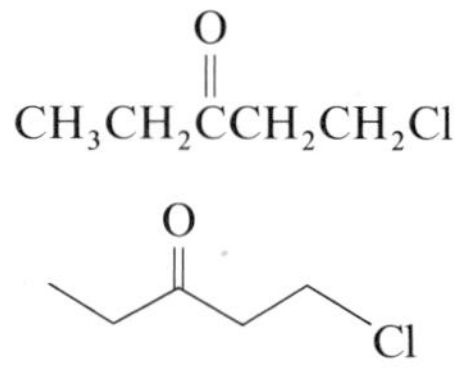

1-chloro-3-pentanone

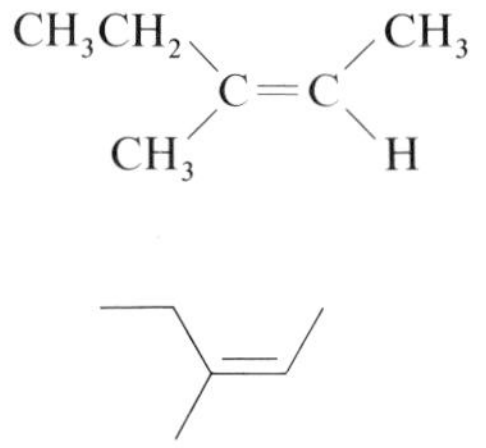

(Z)-3-methyl-2-pentene

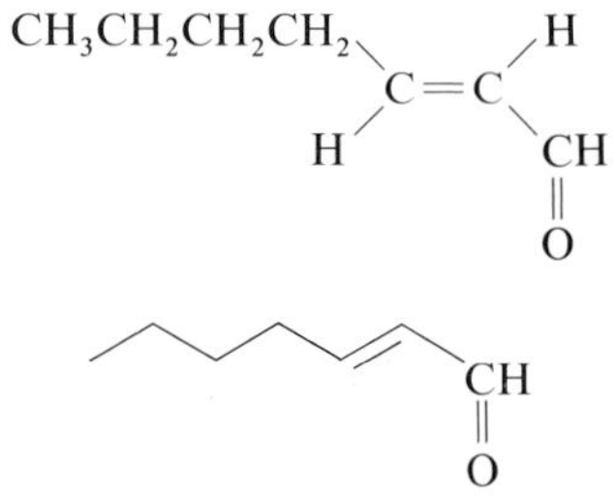

(E)-2-heptenal

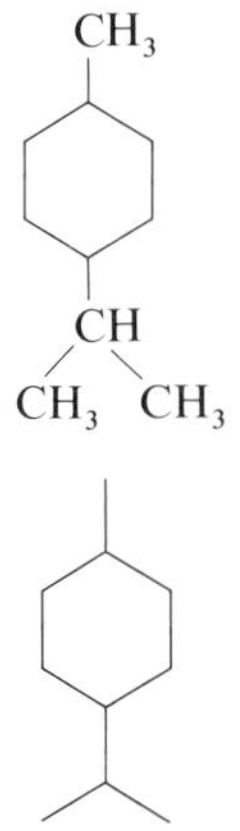

$$CH_3CH_2CH_2CH_2CH_2\overset{\overset{\displaystyle OH}{|}}{C}H\overset{\overset{\displaystyle O}{\|}}{C}H\overset{}{C}OH$$
$$\underset{\displaystyle CH_3}{}$$

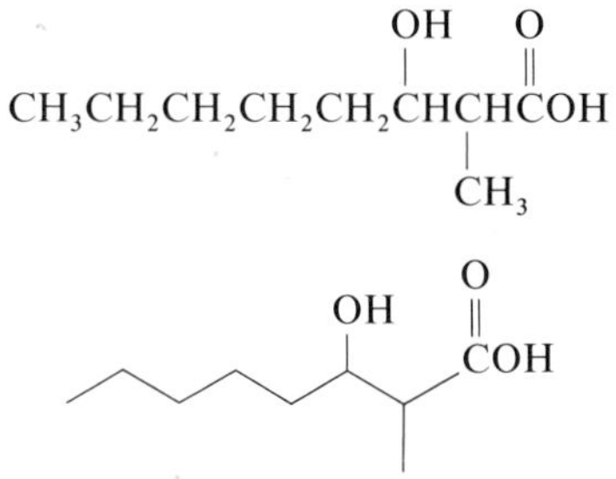

3-hydroxy-2-methyloctanoic acid

1-isopropyl-4-methylcyclohexane

The reaction of 1,3-butadiene with maleic anhydride is an example of the **Diels-Alder reaction,** named after the two German chemists, Otto Diels and Kurt Alder, who recognized the generality of the reaction and later jointly received the Nobel Prize in 1950 for their work. The addition reaction always has two components. One component is a **conjugated diene,** which may have many different types of substituents on it. The other component always has a double or a triple bond in it and is known as the **dienophile,** a compound that is attracted to and reacts with the diene. The dienophile may be a simple alkene or part of a diene system. The best and most reactive dienophiles usually have a carbonyl group or another electron-withdrawing group such as a cyano or nitro group conjugated with the double bond.

The Diels-Alder reaction involves a redistribution of electrons and bonds. Two double bonds disappear, two new single bonds are formed, and a double bond appears between two atoms that formerly shared a single bond. The reaction of 1,3-butadiene with maleic anhydride is rewritten showing the bonding changes that take place:

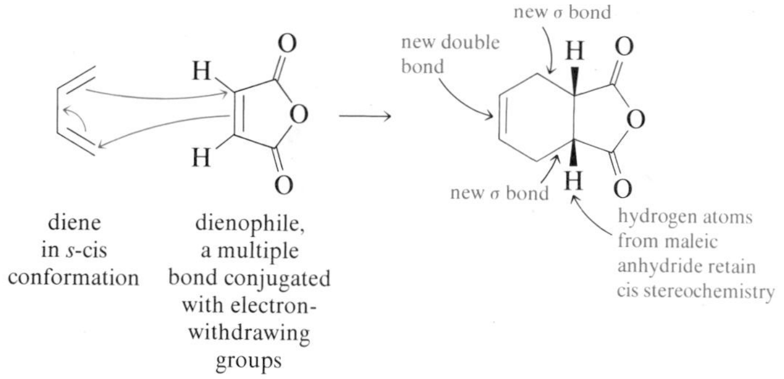

diene
in *s*-cis
conformation

dienophile,
a multiple
bond conjugated
with electron-
withdrawing
groups

The essential and important elements of a Diels-Alder reaction are outlined in the equation above. A conjugated diene reacts with a dienophile to give a six-membered ring with a double bond in it. The reaction is postulated to go by way of a cyclic transition state. The substituents on the dienophile retain their stereochemistry. The diene, in order to react, must be in the *s*-cis conformation.

The dienophile may contain a triple bond instead of a double bond. Esters of acetylenedicarboxylic acid are highly reactive dienophiles, as demonstrated by the reaction of diethyl acetylenedicarboxylate with (1*E*,3*E*)-1,4-diphenyl-1,3-butadiene.

(1*E*,3*E*)-1,4-diphenyl-
1,3-butadiene

diethyl
acetylenedicarboxylate

diethyl *cis*-3,6-diphenyl-
1,4-cyclohexadiene-
1,2-dicarboxylate
90%

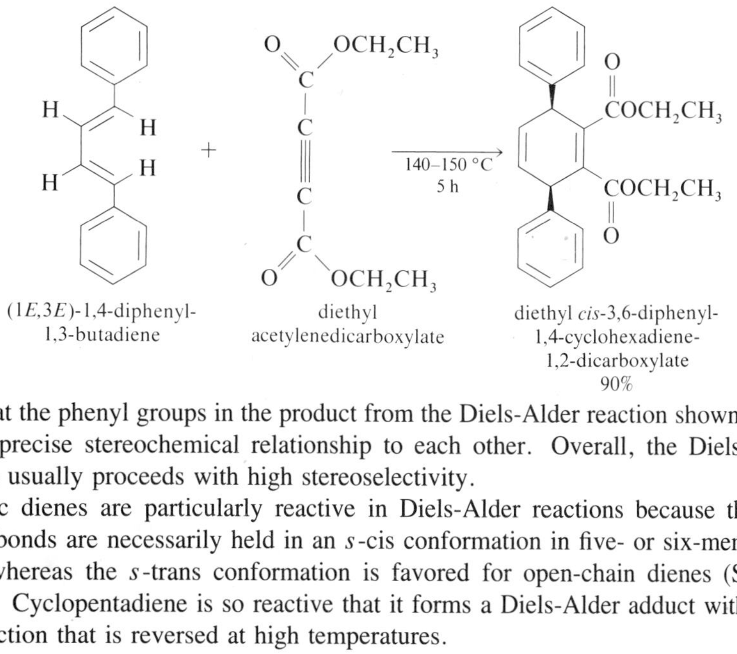

Note that the phenyl groups in the product from the Diels-Alder reaction shown above have a precise stereochemical relationship to each other. Overall, the Diels-Alder reaction usually proceeds with high stereoselectivity.

Cyclic dienes are particularly reactive in Diels-Alder reactions because the two double bonds are necessarily held in an *s*-cis conformation in five- or six-membered rings, whereas the *s*-trans conformation is favored for open-chain dienes (Section 11.1B). Cyclopentadiene is so reactive that it forms a Diels-Alder adduct with itself in a reaction that is reversed at high temperatures.

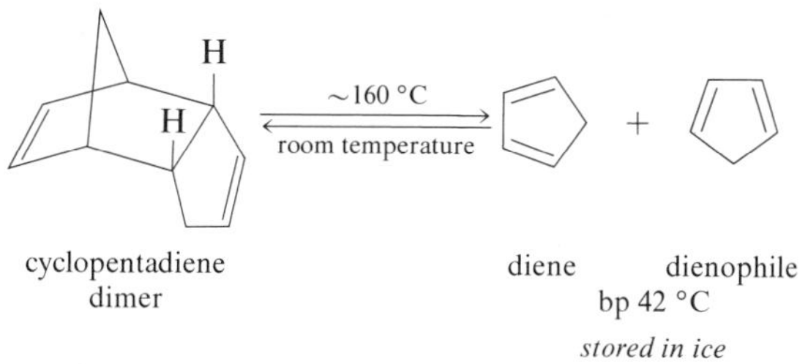

cyclopentadiene
dimer

diene   dienophile
bp 42 °C

stored in ice

In this reaction, one cyclopentadiene molecule acts as the diene, while one of the two double bonds in the other molecule is the dienophile. Cyclopentadiene is stored in the form of its dimer. Whenever some is needed, the dimer is heated to approximately 160 °C and the cyclopentadiene that distills out is collected in an ice-cooled container and used immediately for further reactions. On standing, cyclopentadiene reverts to its dimeric form. Freshly distilled cyclopentadiene will react with maleic anhydride or with propenal to give Diels-Alder adducts.

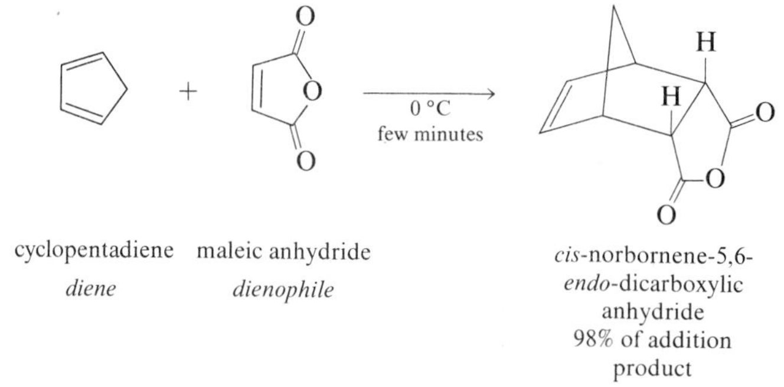

cyclopentadiene	maleic anhydride	*cis*-norbornene-5,6-
diene	*dienophile*	*endo*-dicarboxylic anhydride
		98% of addition product

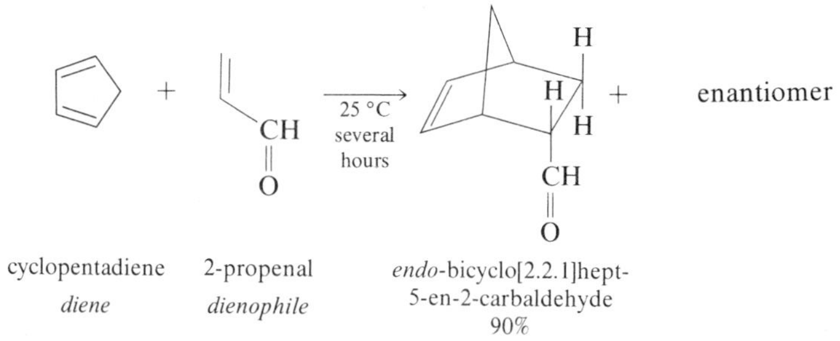

cyclopentadiene	2-propenal	*endo*-bicyclo[2.2.1]hept-
diene	*dienophile*	5-en-2-carbaldehyde
		90%

Reactions of cyclopentadiene with dienophiles give rise to a cyclohexene ring bridged by a single carbon atom. The common name of the saturated hydrocarbon having such a carbon skeleton is norbornane.

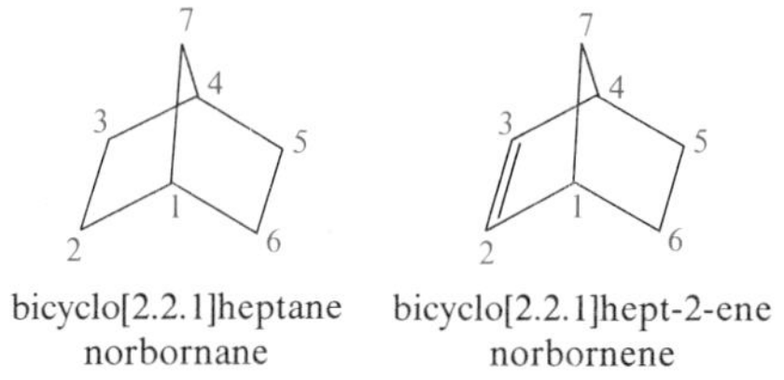

bicyclo[2.2.1]heptane	bicyclo[2.2.1]hept-2-ene
norbornane	norbornene

The systematic name for the compound is bicyclo[2.2.1]heptane, derived from the way seven atoms are held together in a structure with two rings, **a bicyclic structure.** The two rings have two carbon atoms in common, numbered 1 and 4 in the structural formula. These positions are called the **bridgehead positions.** Carbon atoms 1 and 4 are tied together by three bridges, two of which have two carbon atoms in them and one of which is a single carbon atom. The numbers in the brackets, [2.2.1], represent the numbers of atoms in the bridges. The presence of a substituent, or of a double bond in a bicyclic compound, is indicated by using prefixes and suffixes just as it is in other types of compounds. Thus the compound that would result if ethylene were to add to cyclopentadiene has the common name norbornene and the systematic name bicyclo-[2.2.1]hept-2-ene.

Reactions of cyclopentadiene with dienophiles bring up an interesting question of stereochemistry. The substituents on the bicyclic ring created by the Diels-Alder reaction are shown to be oriented away from the carbon bridge across the six-membered ring. Such an orientation is said to be **endo** or *into* the cavity on the bottom side of the bicyclic ring. Another orientation in which the substituent would be *out of* the cavity or on the **exo** side of the bicyclic ring is also possible, as shown below.

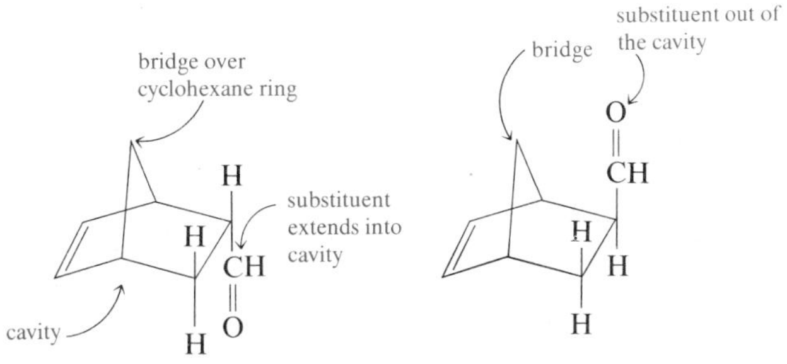

endo orientation
of aldehyde group
on bicyclic ring

exo orientation
of aldehyde group
on bicyclic ring

The reactions of cyclopentadiene with maleic anhydride, with propenal, and with itself give predominantly the endo orientation in the products, a stereochemistry that is quite general for many dienes and dienophiles. This phenomenon is discussed in much greater detail in Section 19.2B, where we examine the nature of the interaction between the π bonds of the diene and of the dienophile.

The significance of the electronic character of the diene and dienophile to the ease with which reaction takes place is evident if we compare the reaction of cyclo-pentadiene and maleic anhydride (p. 532) with the reaction of cyclopentadiene and ethylene. A mixture of maleic anhydride and cyclopentadiene must be cooled in ice to prevent the rapid reaction from becoming so vigorous that the low-boiling cy-clopentadiene (bp 42 °C) is lost. The reaction of cyclopentadiene with ethylene, on the other hand, requires the use of a steel reaction vessel called a bomb, in which the mixture of reagents can be maintained at 800–900 pounds per square inch (psi) at approximately 200 °C for seven hours.

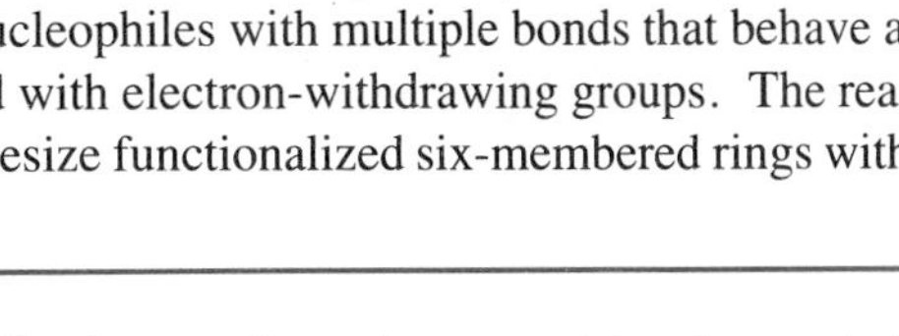

cyclopentadiene   ethylene

$$\xrightarrow[\substack{800-900\text{ psi} \\ 7\text{ h}}]{200\,°C}$$

norbornene
~60%

Classic Diels-Alder reactions may be regarded as reactions between electron-rich dienes, acting as nucleophiles with multiple bonds that behave as electrophiles because they are conjugated with electron-withdrawing groups. The reaction is a tremendously useful way to synthesize functionalized six-membered rings with high stereoselectivity.

PROBLEM 11.12 The following equations show a variety of reagents that undergo Diels-Alder reactions. Complete the equations, showing stereochemistry whenever you know what it will be.

(a)

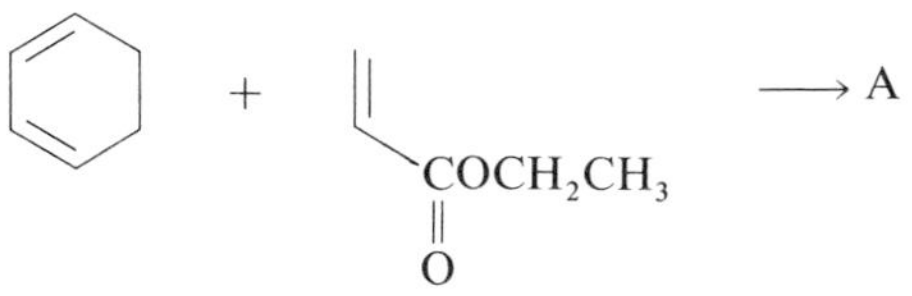

$\longrightarrow$ A

COCH$_2$CH$_3$
‖
O

(b) 1,3-butadiene + (E,E)-2-butenedinitrile (fumaronitrile) ⟶ B

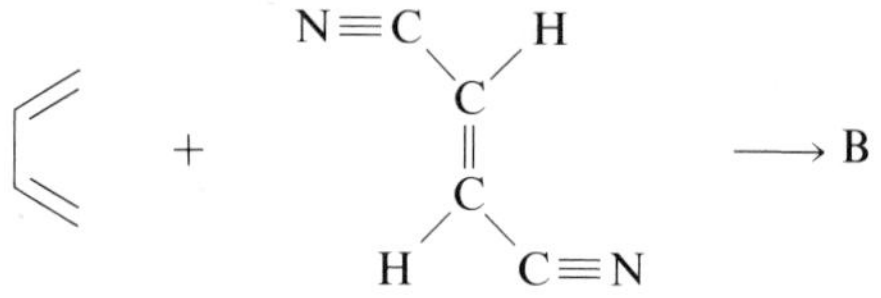

(c) 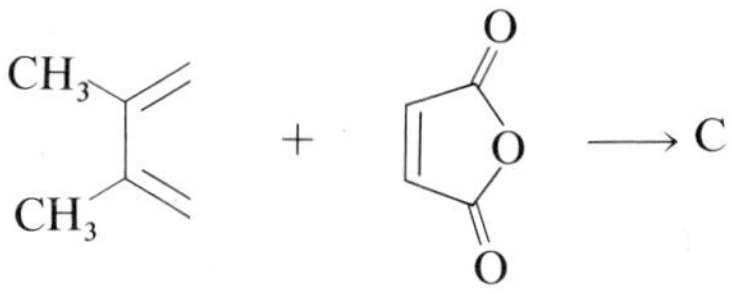 2,3-dimethyl-1,3-butadiene + maleic anhydride ⟶ C

(d) 1,3-butadiene + β-nitrostyrene (1-phenyl-2-nitroethylene) ⟶ D

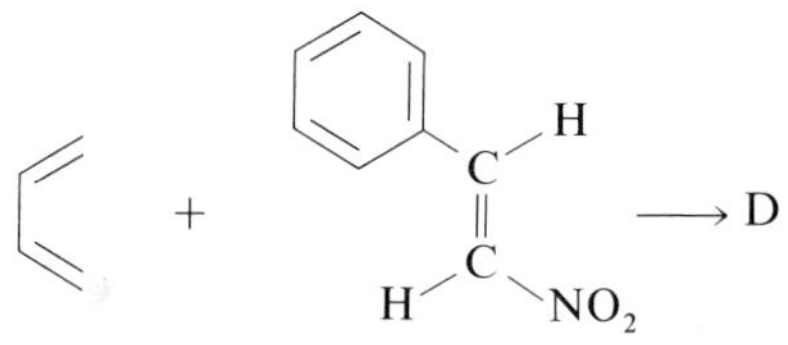

(e) 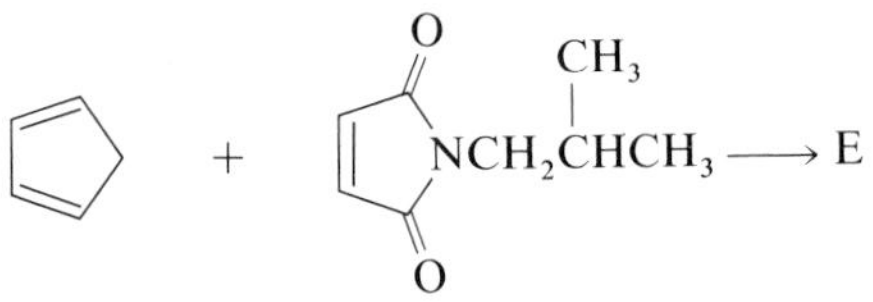 cyclopentadiene + N-isobutylmaleimide ⟶ E

(f) 1,3-cyclohexadiene + 2,3-dimethylmaleic anhydride ⟶ F

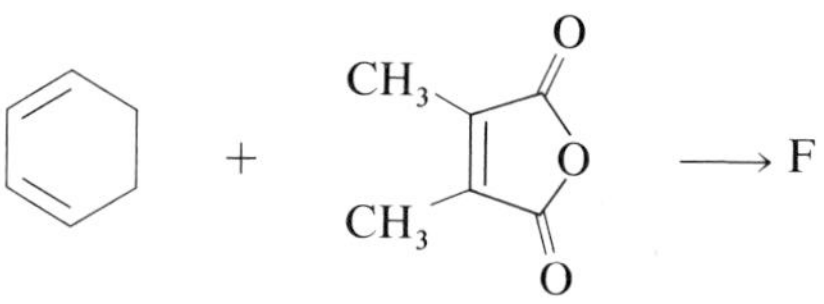

(g) 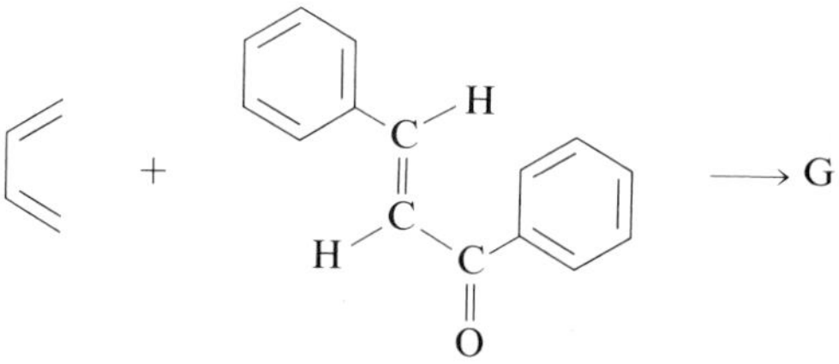 1,3-butadiene + chalcone (1,3-diphenyl-2-propen-1-one) ⟶ G

(h) 1,3-butadiene + acetylenedicarboxylic acid ⟶ H

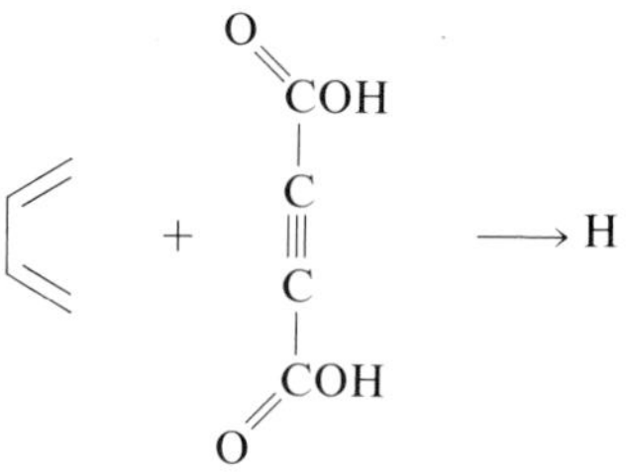

PROBLEM 11.13 The Diels-Alder reaction is reversible. Cyclohexene, for example, is broken down into a diene and a dienophile when it is exposed to a red-hot wire in the absence of air. Write an equation predicting the products of the reaction.

B. Diels-Alder Reactions of Unsymmetrical Dienes and Dienophiles

So far, we have discussed examples of Diels-Alder reactions for symmetrical dienes and dienophiles, so the question of regioselectivity has not come up. However, the reaction has high regioselectivity as well as high stereoselectivity. Some examples of reactions between unsymmetrically substituted dienes and dienophiles are given below.

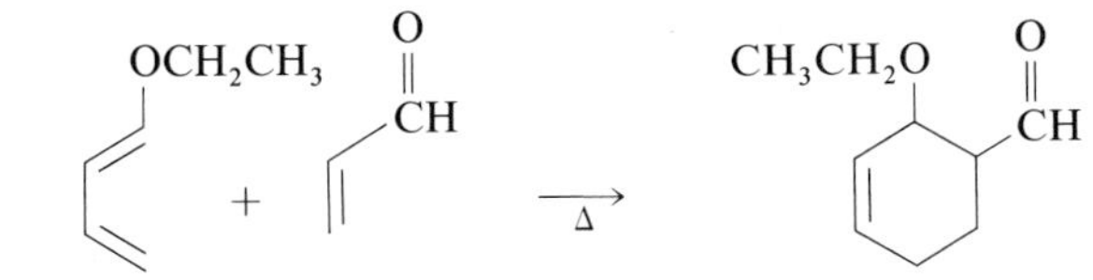

| 1-ethoxy-1,3-butadiene | propenal | 2-ethoxy-3-cyclohexenecarbaldehyde 58% |

| (E)-1,3-pentadiene | propenoic acid | 2-methyl-3-cyclohexenecarboxylic acid ~58% |

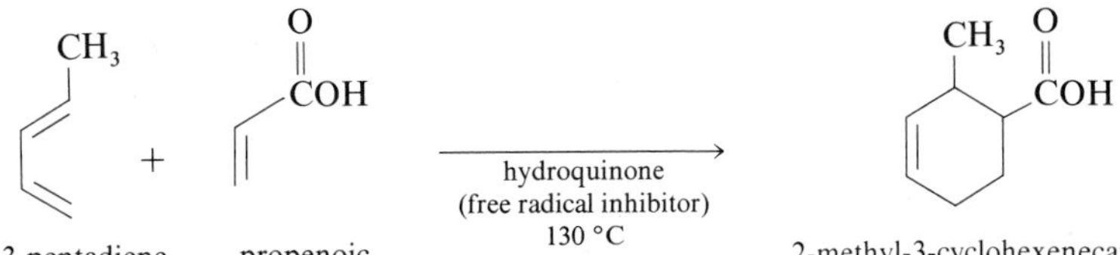

The experimental observations indicate that for most dienes substituted on the first carbon atom, reacting with most unsymmetrical dienophiles, the major product has the two substituents ortho to each other, to borrow a term from the nomenclature of aromatic compounds.

The regioselectivity observed in Diels-Alder reactions can be rationalized for the examples shown above by writing resonance contributors for the diene and dienophile and comparing different transition states. For example, the course of the reaction of 1-ethoxy-1,3-butadiene with propenal is quite easy to explain in this way.

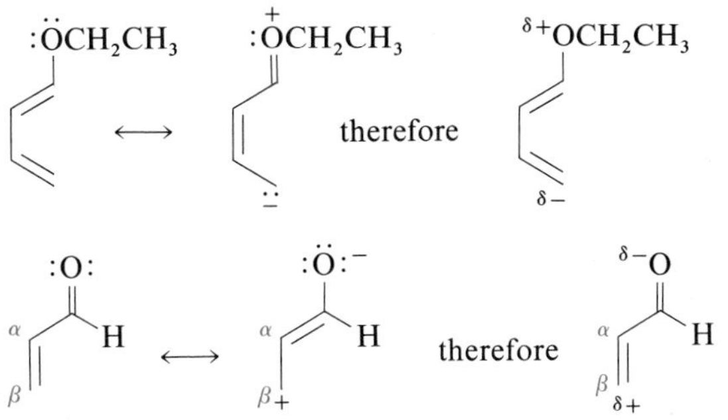

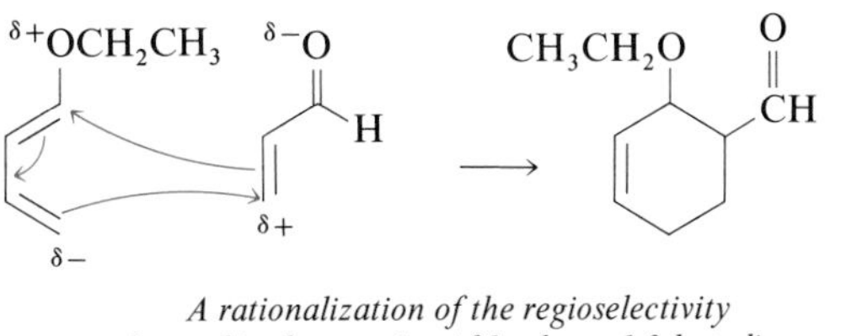

A rationalization of the regioselectivity
observed in the reaction of 1-ethoxy-1,3-butadiene
and propenal

Polarization of 1-ethoxy-1,3-butadiene is most likely to increase electron density at carbon 4 of the diene. The β-carbon atom of the α,β-unsaturated carbonyl compound, meanwhile, is deficient in electrons. In the transition state, the molecules line up so that the partial charges interact with each other, and the product in which the ethoxyl and aldehyde groups are ortho to each other is formed.

If the substituent is on the second carbon atom of the diene, we most frequently observe another type of product.

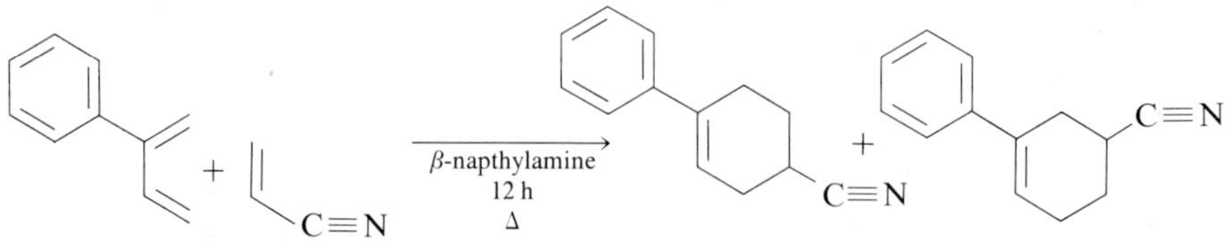

2-phenyl-1,3-butadiene	propenenitrile		5-cyano-2-phenyl-1-cyclohexene		4-cyano-2-phenyl-1-cyclohexene
			4	:	1
				32.8%	

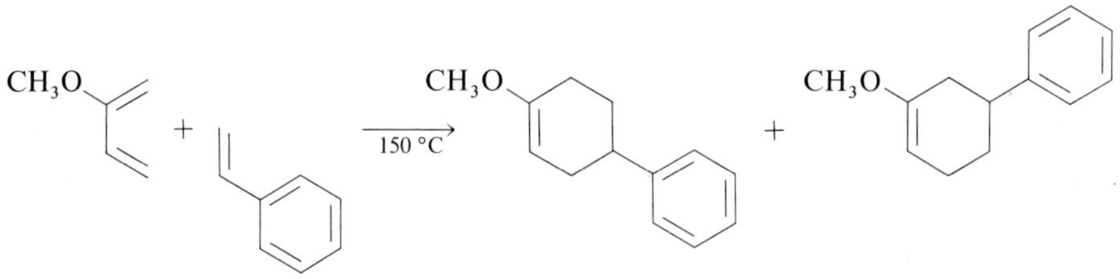

2-methoxy-1,3-butadiene	phenylethene	1-methoxy-4-phenylcyclohexene		2-methoxy-4-phenylcyclohexene
		12	:	1
			59%	

In these cases, the cyclohexene that is formed usually has the substituents para to each other. These products are more difficult to rationalize by simple electron pushing. Nevertheless, theoretical considerations outlined in Section 19.2B do make remarkably good predictions that the para orientation of the substituents would be favored. For us, it is sufficient to know that this is the experimental result in most cases.

The addition of 2-methoxy-1,3-butadiene to alkenes is an excellent synthetic route to substituted cyclohexanones. The Diels-Alder products in these cases are enol ethers and are easily hydrolyzed to the corresponding ketones (Problem 9.22).

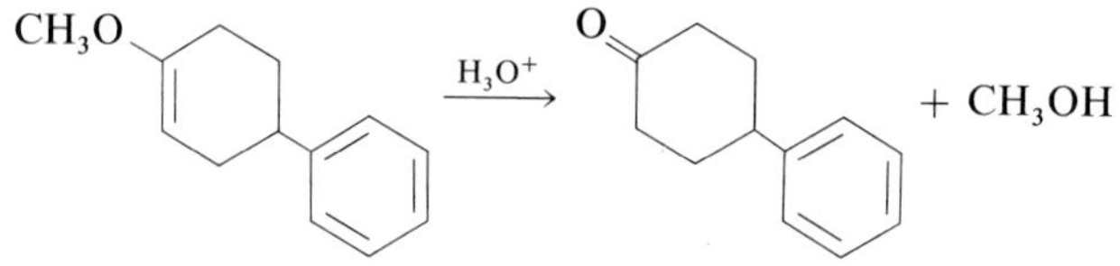

1-methoxy-4-phenylcyclohexene	4-phenylcyclohexanone	
	88%	

PROBLEM 11.14   The following reaction was carried out. Predict what the products will be.

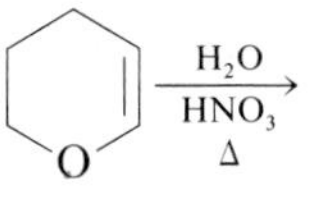

PROBLEM 11.15   The stereochemistry of the product of a Diels-Alder reaction is not always easy to determine. In one case, the decision about the stereochemistry of the reaction was made on the basis of the following set of reactions.

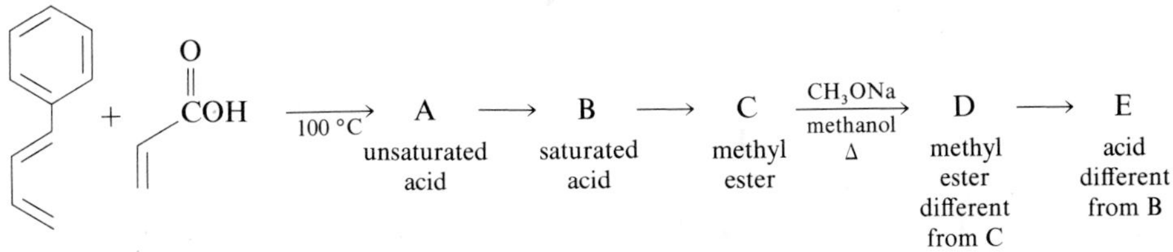

Supply reagents to carry out the transformations where they are missing, and write full structures, including stereochemistry, for Compounds A to E. What do these reactions prove about the stereochemistry of the original Diels-Alder adduct?

PROBLEM 11.16   Predict what the major product will be in each of the following reactions.

(a)

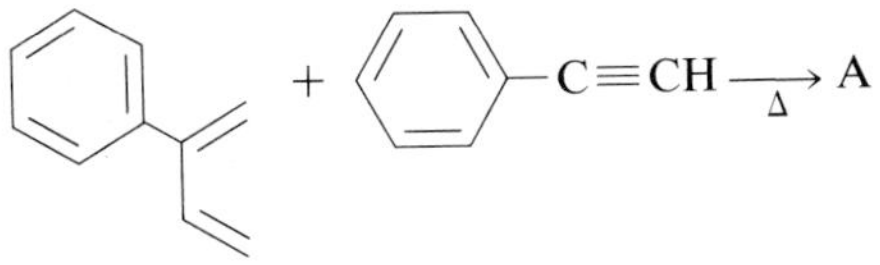

(b)

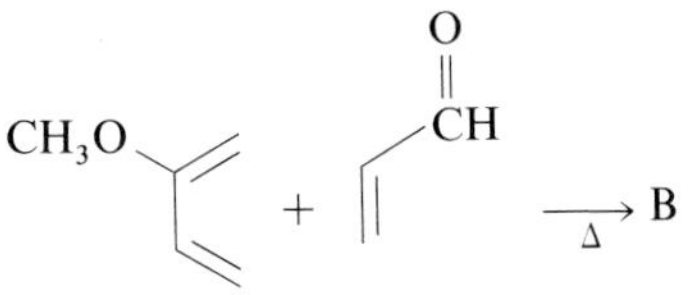

(c)

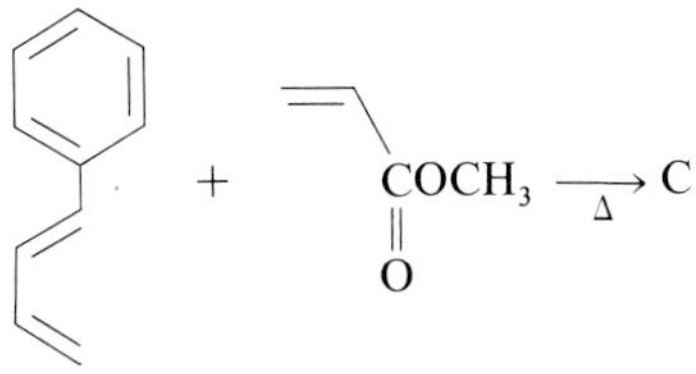

(d)

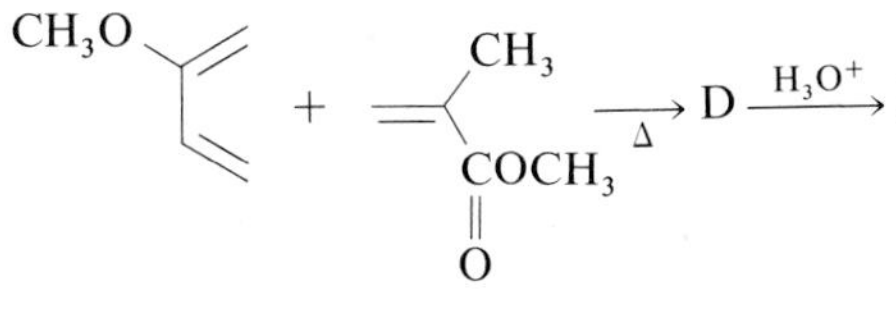

(e)

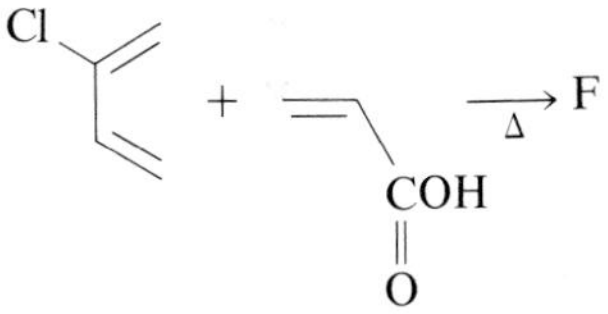

PROBLEM 11.17   In each case below supply the reagents that would be necessary to obtain the Diels-Alder adduct shown below.

(a)

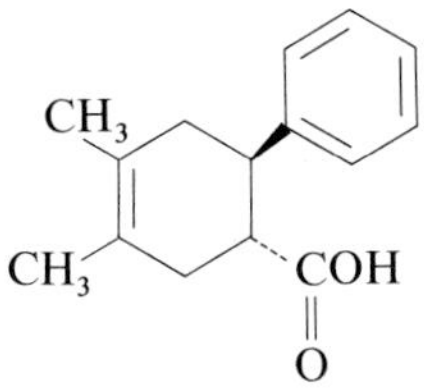

(b)

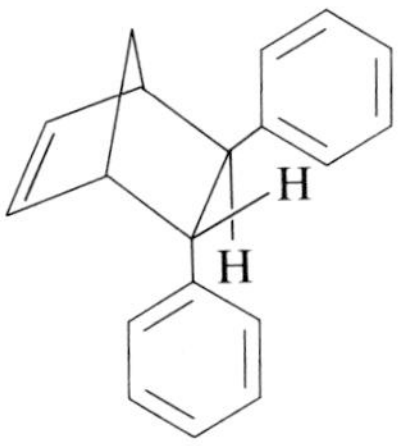

(c)

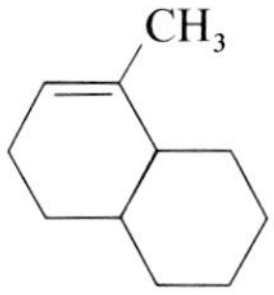

(d)

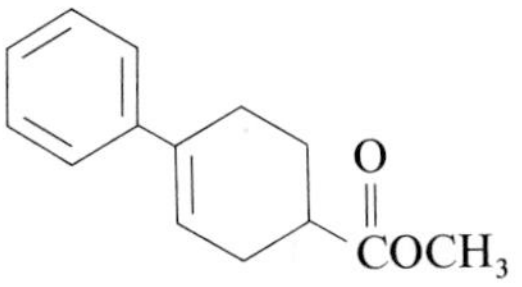

11.6

Terpenoids

A. Isoprene and the Isoprene Rule

For centuries, human beings have known that volatile oils with a variety of fragrances and flavors could be isolated from plants. These compounds occur in every part of the plant and are called **essential oils.** They are of great commercial importance in the perfume and flavoring industries. Many essential oils have also been used medicinally. The chemical constitutents of the essential oils have a large variety of structures, many of them containing rings, or one or more double bonds. Some of them are alcohols or ethers, others are ketones or aldehydes. After many years of carrying out structural determinations on these interesting and challenging compounds, chemists began to detect some patterns emerging. For example, most of these compounds are composed of even multiples of five carbon atoms. Whole families of compounds were discovered and classified according to their molecular formulas. Compounds with 10 carbon atoms, the **monoterpenes,** those with 15 carbon atoms, the **sesquiterpenes,** and those with 20 carbon atoms, **the diterpenes,** are the chief constituents of essential oils. Steroids are **triterpenes,** compounds with 30 carbon atoms, or are derived from them, while plant coloring materials such as carotene (from carrots) are **tetraterpenes** containing 40 carbon atoms. Rubber is a polymeric terpenoid (Section 20.5).

The structures of a few essential oils as well as some other plant constituents are shown below.

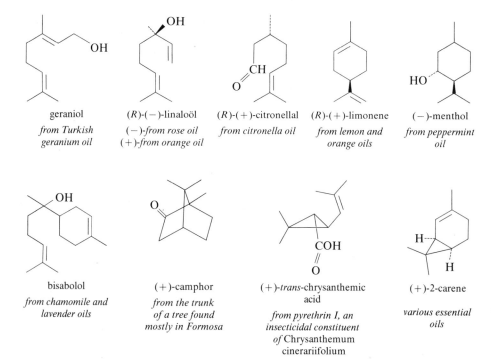

geraniol

from Turkish geranium oil

(*R*)-(−)-linaloöl

(−)-from rose oil
(+)-from orange oil

(*R*)-(+)-citronellal

from citronella oil

(*R*)-(+)-limonene

from lemon and orange oils

(−)-menthol

from peppermint oil

bisabolol

from chamomile and lavender oils

(+)-camphor

from the trunk of a tree found mostly in Formosa

(+)-*trans*-chrysanthemic acid

from pyrethrin I, an insecticidal constituent of Chrysanthemum cinerariifolium

(+)-2-carene

various essential oils

The last two compounds shown are interesting from a historical point of view. In 1920, the Yugoslav chemist Leopold Ruzicka, who did his scientific work in Switzerland and later won the Nobel Prize for discoveries in terpene chemistry, finished a proof of the structure of chrysanthemic acid. This compound is a constituent of a pyrethrin, a component of an insecticidal extract from a type of chrysanthemum that grows primarily in East Africa. (Pyrethrins have become important commercially as highly selective and biodegradable insecticides.) In the same year, the structure of 2-carene was determined by the British chemist John L. Simonsen. Ruzicka was struck by the structural resemblance between the two compounds and could see in both of them the elements of a five-carbon structural unit known as **isoprene.**

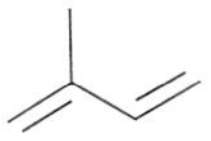

2-methyl-1,3-butadiene
isoprene

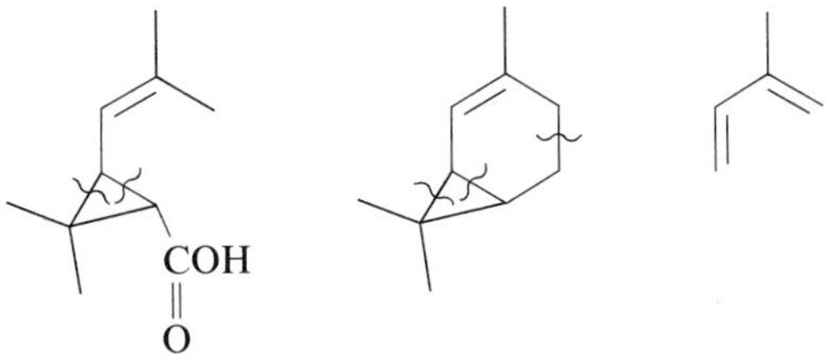

chrysanthemic   2-carene   isoprene in the
acid   *s*-cis conformation

In other compounds that were known at that time, especially the ones that did not contain a ring, another regularity could be seen. If the branched end of the isoprene molecule were to be called its head and the other end its tail, molecules that contained two or more isoprene units had the head of one isoprene unit attached to the tail of the next one in the chain.

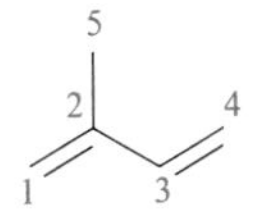

head of   tail of
isoprene   isoprene

four-carbon chain with
methyl branch on second carbon

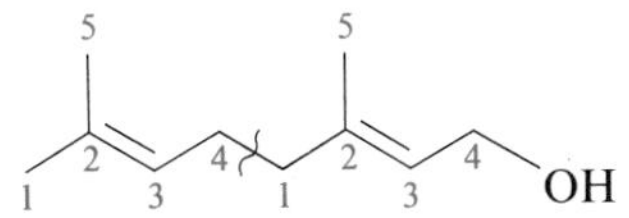

geraniol, a terpene made up of
two isoprene units attached
head-to-tail

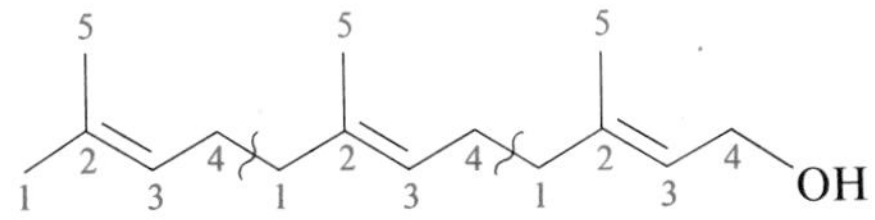

farnesol, scent of lily of the valley,
a sesquiterpene made up of three
isoprene units attached head-to-tail

Ruzicka hypothesized that these plant constituents were synthesized in nature by a head-to-tail connection of isoprene units. He examined a large number of terpenoid compounds over the years and found that the structures of most of them showed this regularity, which he called the isoprene rule.

The **isoprene rule** means that given a choice of possible structures for a compound that contains a multiple of five carbon atoms, chemists will favor one that appears to be made of isoprene units connected in a head-to-tail fashion. The isoprene rule says nothing about whether the compound contains double bonds or whether functional groups containing oxygen appear in the molecule. A wide variety of patterns of rings, unsaturation, and of alcohol, carbonyl, and carboxylic acid functional groups are present in these natural products. The isoprene rule applies only to the carbon skeleton of the molecule. When rings are present, they indicate that additional points of attachment between various isoprene units have also been formed. In those cases, it should still be possible to trace the main outline of the chain of isoprene units around the molecule. Some cyclic terpenoid compounds are dissected into isoprene units below.

| (−)-menthone | α-cadinene | β-selinene |
| *peppermint oils* | *oil of citronella* | *oil of celery* |

In some structures, it is possible to find the pattern in more than one way.

As molecules become larger, the final structure is often achieved only after some molecular rearrangements during the biosynthesis so that the isoprene rule does not apply to all parts of their structures, though it is still possible to find portions of the molecule that follow the rule.

PROBLEM 11.18 Find the isoprene units in the following compounds.

| menthofuran | bisabolol | camphor | linaloöl | camphene | α-pinene |

B. Biosynthesis of Terpenes

An unsaturated five-carbon alcohol, 3-methyl-3-buten-1-ol, present as its pyrophosphate ester, is the structural building block for terpenes in nature. Pyrophosphoric acid is an anhydride of phosphoric acid, and seems to be nature's way of creating good leaving groups. 3-Methyl-3-buten-1-yl pyrophosphate, known in the biochemical literature as isopentenyl pyrophosphate, is isomerized enzymatically to 3-methyl-2-buten-1-yl (dimethylallyl) pyrophosphate in a reaction that may be regarded as a protonation at one sp^2-hybridized carbon atom and a deprotonation of the incipient carbocation at another site to give the more highly substituted alkene. The participation of an enzyme, a highly specific biological catalyst (see Section 17.6 for an example of how an enzyme functions), ensures that no high-energy intermediate is ever formed at any point of the reaction.

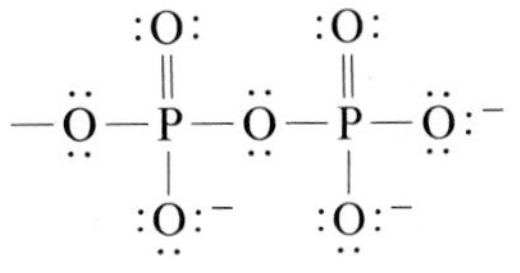

pyrophosphate group

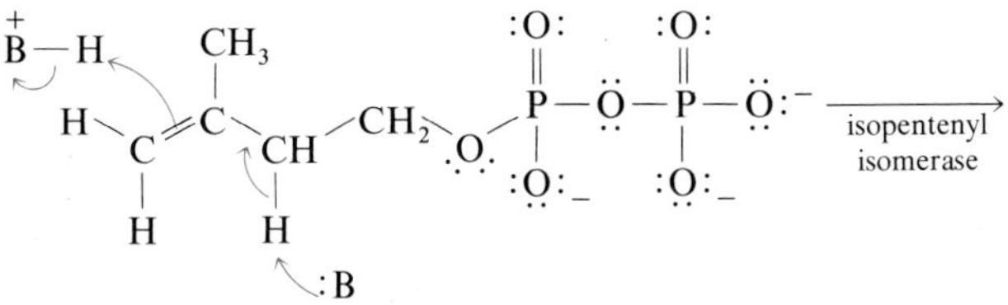

isopentenyl pyrophosphate
3-methyl-3-buten-1-yl
pyrophosphate

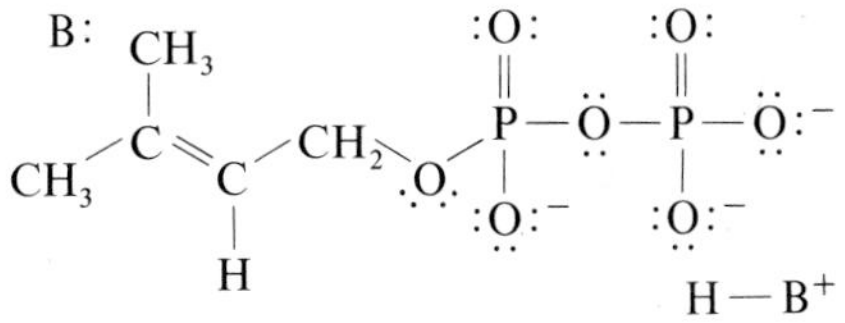

dimethylallyl pyrophosphate
3-methyl-2-buten-1-yl
pyrophosphate

The double bond in isopentenyl pyrophosphate serves as a nucleophile to displace pyrophosphate anion from dimethylallyl pyrophosphate, with the simultaneous loss of a proton to create a new double bond.

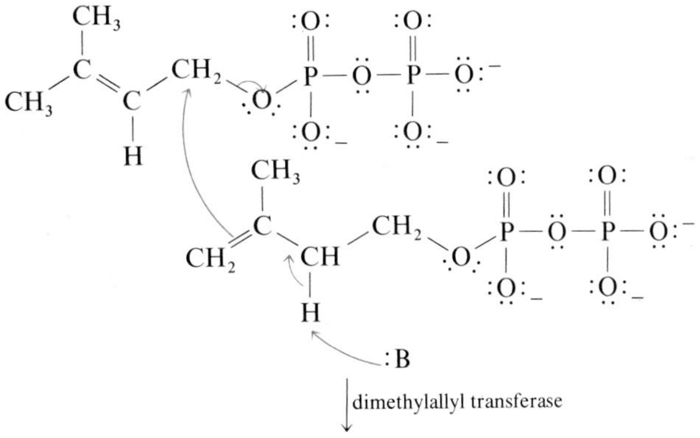

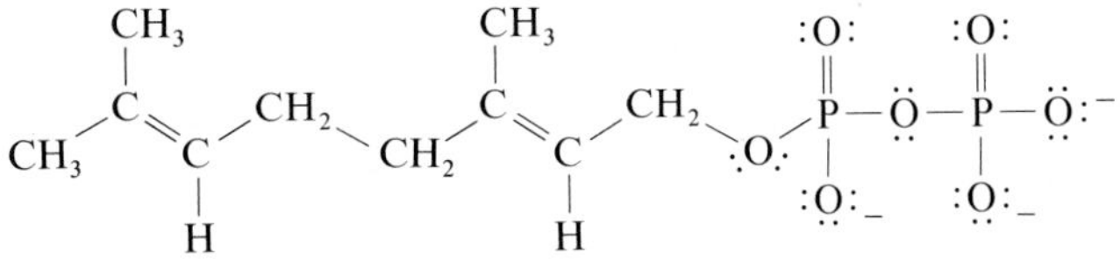

geranyl pyrophosphate

the pyrophosphate ester of geraniol

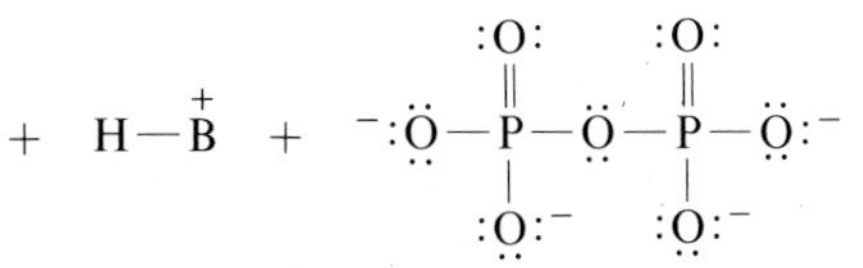

pyrophosphate ion

The product is the ten-carbon terpene, geraniol. Note that the mechanism proposed for terpene synthesis leads in each case to the head-to-tail connection of isoprene units. Five carbon atoms are added, and the end of the chain is left functionalized as a pyrophosphate ester that can serve as a leaving group for further reactions. Further protonations and deprotonations would lead to isomerizations of the positions and of the stereochemistry of double bonds. The oxygen-containing functional groups are modified by oxidation and reduction reactions, and new hydroxyl groups are created by the addition of water to double bonds or by enzymatic oxidation reactions at unactivated sites. Conversely, dehydration of alcohol functions introduces unsaturation into the systems. Biological reduction reactions convert alkenes into alkanes. Carbocations created by protonation of a double bond or loss of water from a protonated alcohol can attack a double bond in another part of the molecule to create rings. The next problem is a review of this familiar chemistry in the context of terpene reactions.

PROBLEM 11.19   In all of the following sections of this problem, you may use HB^+ as an acid and B: as a base as necessary. Water is always present as a nucleophile.

(a) Geraniol (p. 538) has a diastereomer called nerol. Nerol is also found in nature, but somewhat less abundantly than geraniol. Write a structural formula for nerol, and propose a mechanism for the isomerization of geraniol to nerol.

(b) Treatment of geraniol or nerol with aqueous acid gives rise to α-terpineol and terpin shown below. Nerol is converted into these compounds faster than geraniol is. Write mechanisms for the formation of α-terpineol and terpin from nerol.

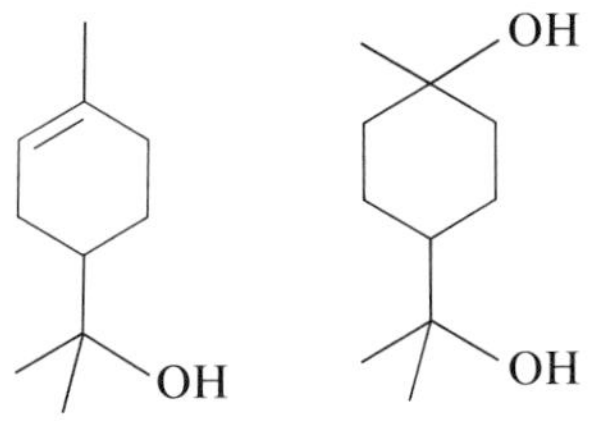

α-terpineol   terpin

(c) Limonene, a flavor constituent of citrus fruits (p. 538), is formed from α-terpineol and from terpin. Write equations showing how this would take place.

(d) Linaloöl (p. 538) may also be considered to arise from geraniol or nerol. Propose a mechanism for its formation.

PROBLEM 11.20   Isoprene undergoes a Diels-Alder reaction with itself to give rise to an optically inactive compound known for some time as dipentene. Dipentene was finally identified as the racemic form of a naturally occurring terpene. Write an equation for the reaction and identify the terpene, which appears on p. 538.

PROBLEM 11.21   Show how the sesquiterpene farnesol (p. 539) could be synthesized from geranyl pyrophosphate (p. 541).

PROBLEM 11.22   Diterpenes, compounds having 20 carbon atoms, are derived from a pyrophosphate with the following structure.

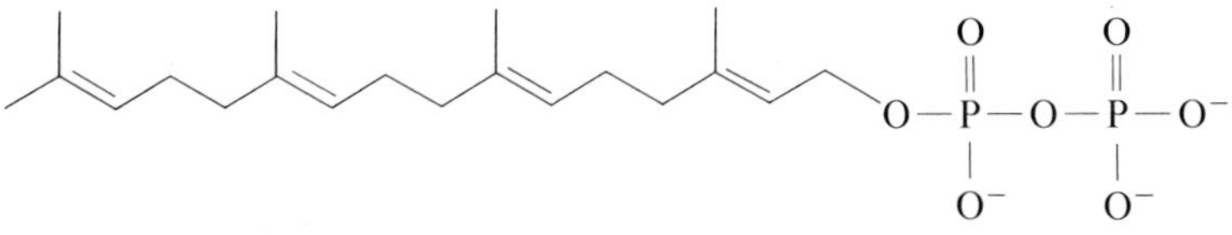

Propose a biosynthesis for this compound, sometimes called geranylgeranyl pyrophosphate.

C. Steroids Tracey

Steroids are triterpenoids, a group of compounds of plant or animal origin. They are widespread in nature and have important biological functions. You read about steroids that are hormones in Section 9.1B. Cholesterol (Sections 7.1B and 10.7C) is also a steroid.

Steroids are characterized by a tetracyclic structure. A typical steroid ring skeleton consists of three six-membered rings and a five-membered ring fused together. Depending upon their sources and their biological functions, steroids may also have a variety of functional groups substituted on them (Section 9.1B).

The rings in a steroid skeleton are designated as the A, B, C, and D rings.

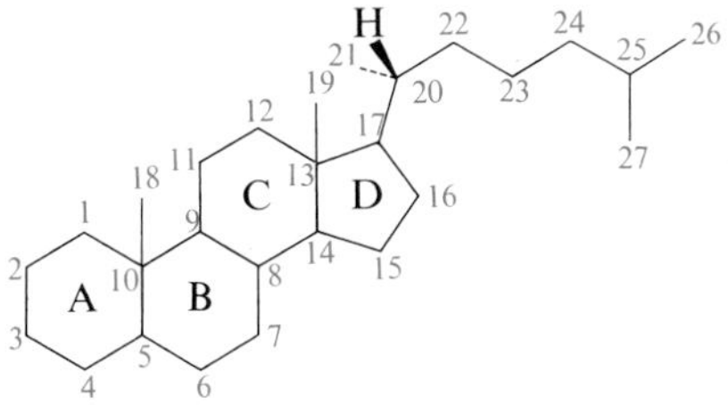

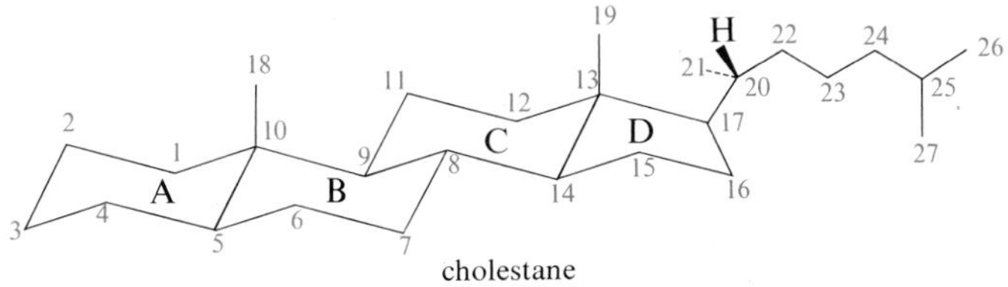

cholestane

The fusion of the rings in a steroid gives rigidity to the molecule. The conformational changes (Section 3.8C) that are easy for individual cyclohexane rings are not possible at room temperature for fused rings. Steroids have, therefore, been used for research into the influence of conformation on the course of organic reactions. Sir Derek H. R. Barton of Great Britain received the Nobel Prize in 1969 for recognizing that functional groups could vary in reactivity depending on whether they occupied an axial or an equatorial position on a ring. This way of thinking about stereochemistry, called **conformational analysis,** has greatly influenced the way chemists analyze reactivity.

Squalene, $C_{30}H_{50}$, was first isolated from shark liver oil in 1916. Since that time, research has established its importance in the formation of the steroid ring skeleton in living matter. Squalene is a triterpene, but one in which the isoprene rule is violated in one place. Rather than a head-to-tail arrangement of six units of isoprene, it appears to be made of two farnesyl units that have been connected tail-to-tail.

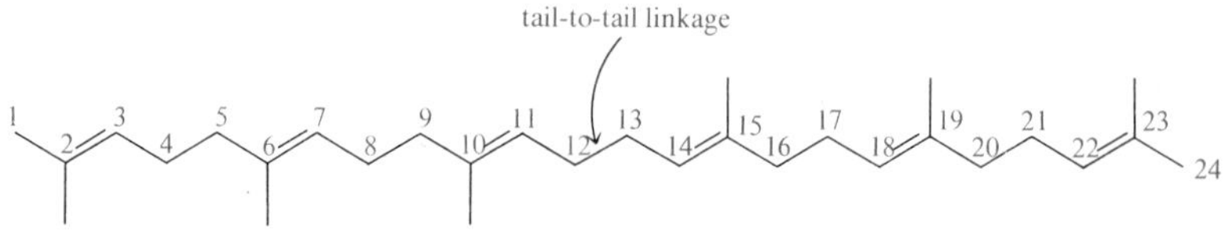

tail-to-tail linkage

squalene
(6*E*, 10*E*, 14*E*, 18*E*)-2,6,10,15,19,23-hexamethyl-
2,6,10,14,18,22-tetraicosahexaene

a triterpene

Much research has been carried out using selective labeling of carbon atoms and hydrogen atoms, and both types of experiments confirm the tail-to-tail connection. The latest evidence indicates that this process involves the formation and then the opening of a cyclopropane ring in the middle of the chain as two farnesyl pyrophosphate units interact.

The critical intermediate in the conversion of squalene to the steroid skeleton is an oxirane, squalene-2,3-oxide, which is transformed by enzymes into lanosterol, a steroid alcohol found in wool fat. Similar cyclization reactions take place in the laboratory with Lewis acid or Brønsted-Lowry acid catalysts. The squalene molecule is believed to be folded in a conformation that allows one double bond after another to react as a nucleophile with a cationic center as it develops nearby in the molecule. The initial leaving group that starts the process is the oxygen atom of the oxirane ring protonated or complexed in a way that makes it a better leaving group. Oxiranes, as strained small ring compounds, are reactive towards nucleophiles, especially in systems where the oxygen atom develops a positive charge by protonation (Section 8.5A).

Squalene oxide, the formation of the tetracyclic ring system, and the further 1,2-hydride and 1,2-methyl shifts that must take place to convert the skeleton to lanosterol are shown below.

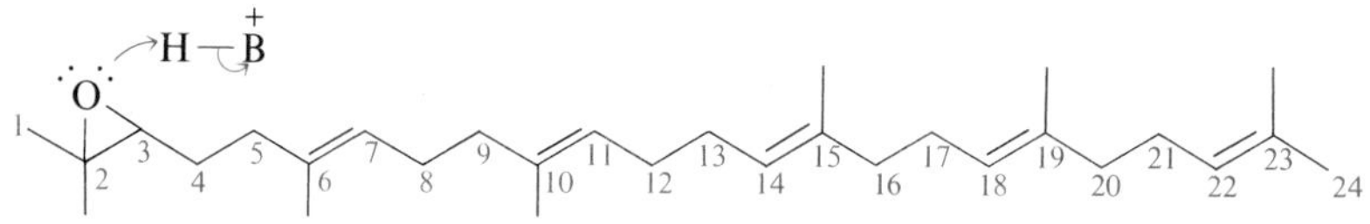

squalene oxide
protonation of oxide by enzyme

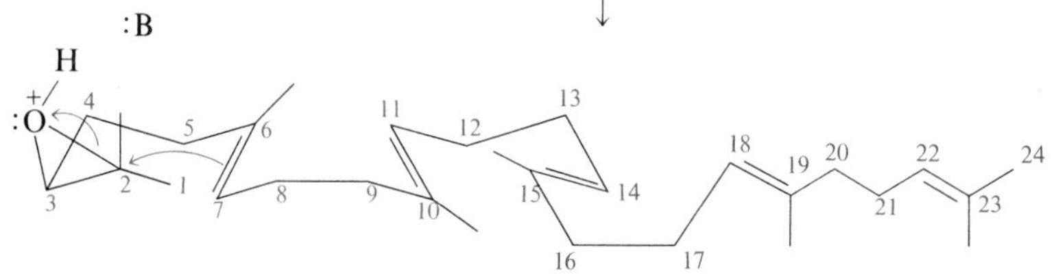

nucleophile opening protonated oxirane ring

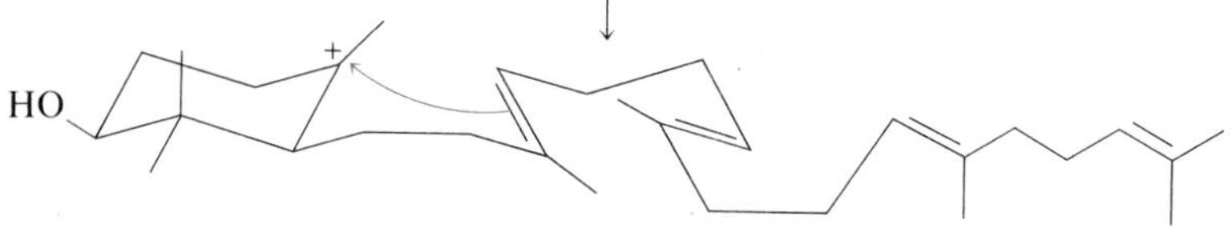

alkene reacting with carbocation

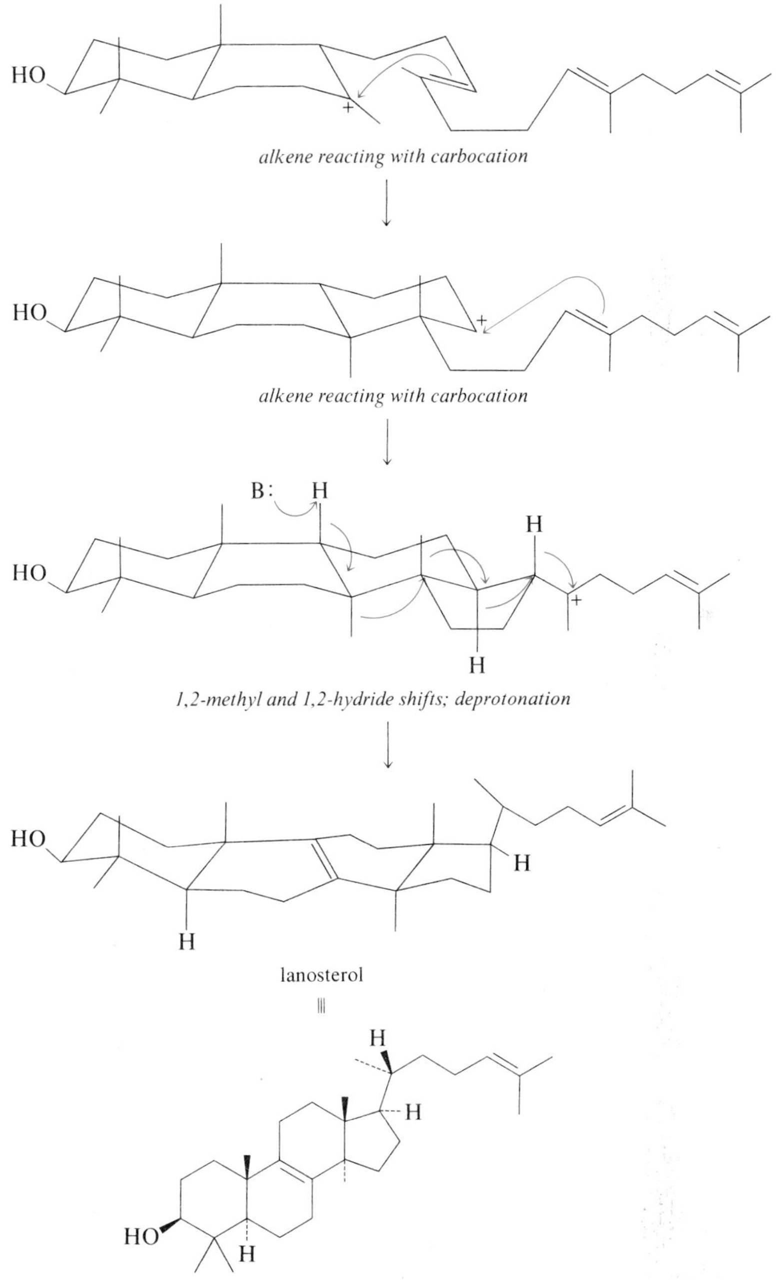

alkene reacting with carbocation

alkene reacting with carbocation

1,2-methyl and 1,2-hydride shifts; deprotonation

lanosterol

The whole process, catalyzed by an enzyme, proceeds with precise stereochemistry. The individual steps of the transformation, though, are all familiar to us as typical carbocation reactions (Section 5.6). A carbocationic center, as it is generated, behaves as a Lewis acid towards the π electrons of an adjacent double bond, which in turn creates a new cationic center, so the cyclization progresses. The initial cyclization product is a cation that differs from lanosterol in the placement of two methyl groups. The double bond is created in the B ring by a series of 1,2-shifts that result in a tertiary cation that undergoes the final deprotonation reaction.

Squalene is the biological precursor of many triterpenoids. Important among these are the steroids, one of which is cholesterol. During the biological conversion of lanosterol to cholesterol, three methyl groups are lost, the hydrocarbon chain is reduced to a saturated one, and the position of the double bond in ring B of the steroid nucleus is moved. The exact details of how these transformations are carried out are not known. More than one pathway may be involved. That lanosterol is converted to cholesterol by enzymes from the liver has been proved by a number of experiments. For example, when lanosterol, labeled with radioactive ^{14}C and carefully purified so that it contains no cholesterol, is incubated with cells from the liver of a rat, cholesterol containing radioactivity is recovered.

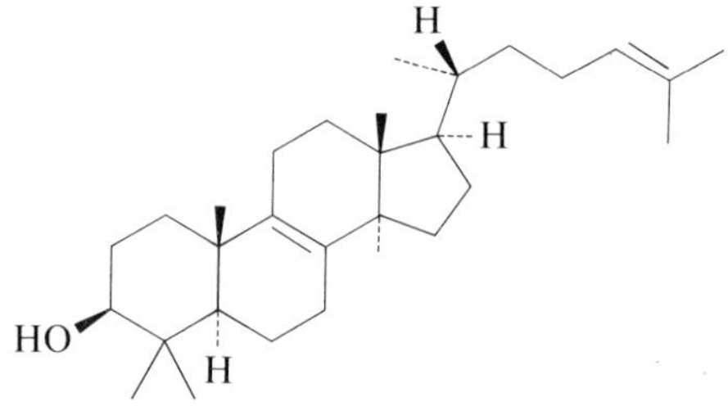

lanosterol

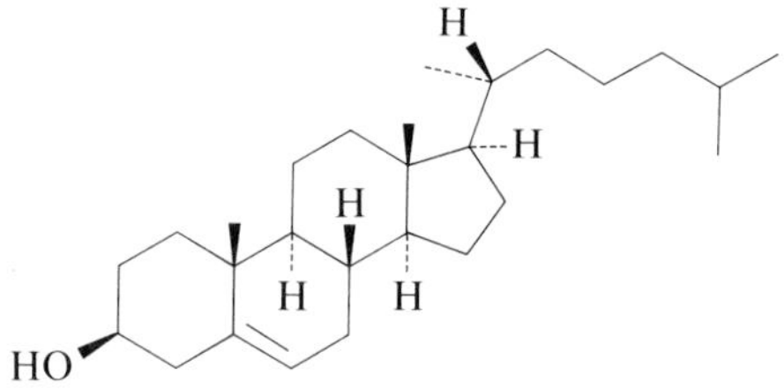

cholesterol

D. Carotenoids. Vitamin A

The **carotenoids** are compounds made up of eight isoprene units, usually containing some rings and a number of conjugated double bonds. The backbone of the chain appears to be constructed by a tail-to-tail union of two geranylgeranyl pyrophosphate units (Problem 11.22) in the same way that squalene is derived from a combination of farnesyl pyrophosphate units. The carotenoids are insoluble in water, but soluble in fat and in hydrocarbon solvents. All the carotenoids, which are generally yellow to red in color, are widely distributed in nature, in both plants and animals.

Lycopene is the parent carotenoid, related to all other known carotenoids by changes such as reduction of some of the double bonds, cyclization, isomerization of the position of a double bond, and introduction of functional groups containing oxygen.

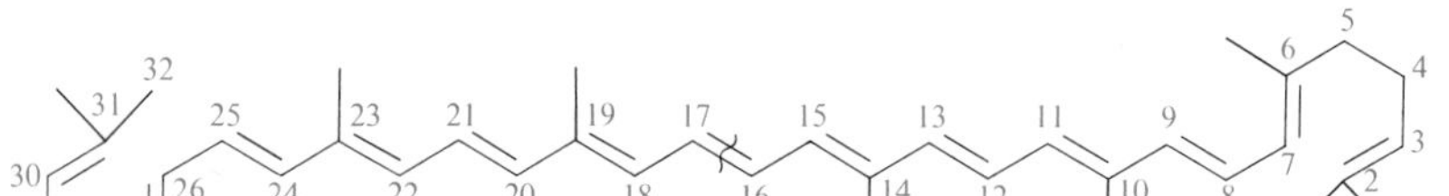

lycopene

dissected at the point of tail-to-tail
connection of two 20-carbon units

Note that the double bonds in lycopene are all trans in configuration and that except for the ends of the molecule, they are all in conjugation. As we see in Section 11.8A, this extended system of conjugation is responsible for the color of these compounds, which are the pigments in many plants. Lycopene is abundant in tomatoes, for example.

The plant pigment in carrots, carotene, is especially interesting because it is closely related to vitamin A, a diterpenoid vitamin essential to growth, to the health of membranes, and to vision. Two isomers of carotene, differing from each other in the position of one of the double bonds, are abundant in carrots. Both of them show strong vitamin A activity in biological tests.

β-carotene

α-carotene

CH_2OH

vitamin A
retinol

The derivation of the carotenes from lycopene by cyclization of the two ends of the molecule is evident. β-Carotene, in which all of the double bonds are in conjugation, is more abundant than α-carotene, in which one of the double bonds (the one in the ring on the right) has moved out of conjugation. Vitamin A contains the elements of half of a molecule of carotene and is an alcohol. It is pale yellow, in comparison with the deep red-orange color of crystalline carotene. Animal feeding experiments show that carotenes are transformed into vitamin A in the bodies of most mammals, but to differing extents.

Vitamin A, also called **retinol,** is involved in the process of vision as its aldehyde with the cis configuration at the double bond at the 11 position, numbered as is traditional for these systems. This compound, (11Z)-retinal, is bonded with an imine linkage (Section 9.6C) to a protein, opsin, in the visual pigment in our retinas. The complex molecule that results, called **rhodopsin,** absorbs light energy in the visible region of the spectrum. When light is absorbed, (11Z)-retinal undergoes isomerization to the all-trans configuration of retinal. In this form it can no longer stay bound to the protein and dissociates into (11E)-retinal and opsin, neither of which absorbs light in the visible region of the spectrum. This conversion is known as the **bleaching** of the visual pigment. (11E)-Retinal is reconverted to the cis form by an enzyme, binds once more to opsin, and the visual cycle begins again. Somehow the change in configuration of the double bond and the resulting dissociation of the visual pigment is translated by the retina into a message that travels through the optic nerve to the brain, to be interpreted there as sight and color vision. The processes described above in a highly simplified way are depicted on the next page.

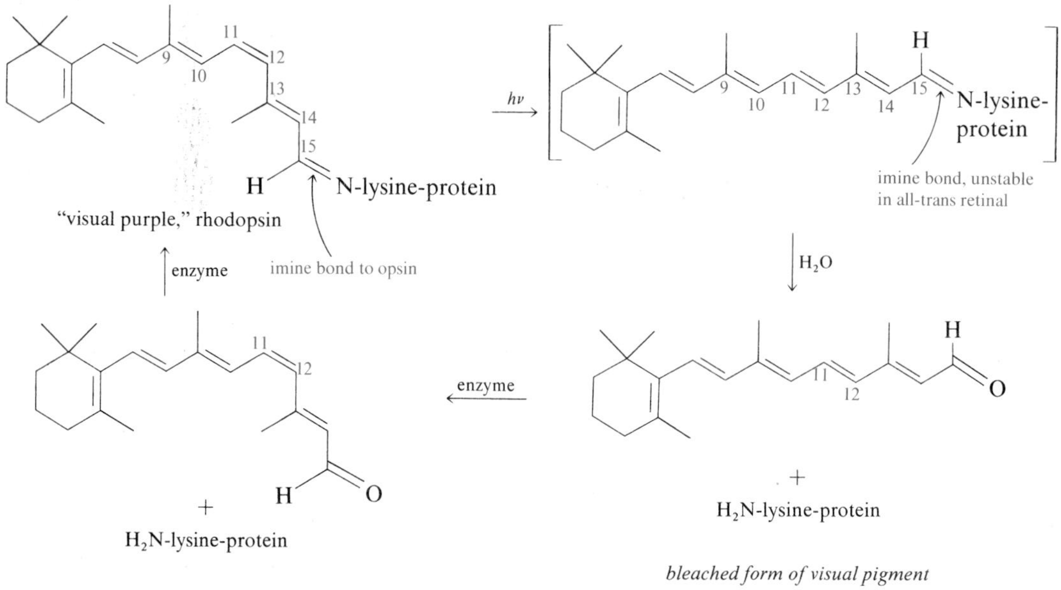

"visual purple," rhodopsin

imine bond to opsin

enzyme

hv

imine bond, unstable
in all-trans retinal

N-lysine-protein

H_2O

enzyme

+

H_2N-lysine-protein

+

H_2N-lysine-protein

bleached form of visual pigment

The absorption of light by conjugated systems such as vitamin A and how this can lead to isomerization of a double bond are discussed in Section 11.8A.

We do not understand all the details of the changes that take place between the moment when light falls on the retina and the creation of a visual image. Even if we did, we can still wonder at the way nature uses simple chemical reactions in systems of such precise form that one stereochemical change in a small part of a large molecule sets in motion events of such consquence.

PROBLEM 11.23

(a) Vitamin A, as the free alcohol, is sensitive to air. Its esters are more stable. It is found in fish oils as its hexadecanoate (palmitate) ester. What is the structure of this ester?

(b) Vitamin A is often sold as its acetate ester. Write an equation showing how vitamin A could be converted into its acetate ester in the laboratory.

PROBLEM 11.24 β-Carotene is converted by monoperoxyphthalic acid into a mixture of a mono-epoxide and a diepoxide. Which double bonds in β-carotene would be most vulnerable to oxidation by a peroxyacid? (Hint: You may wish to review Section 8.4D.) Write structural formulas for the mono- and the diepoxides of β-carotene.

11.7

Other Natural Products Containing Multiple Double Bonds

A. Reactions of Polyenes with Oxygen

The polyunsaturated fatty acids are converted by reactions with oxygen in air to compounds containing oxygen and conjugated double bonds. Molecular oxygen is a diradical, having two unpaired electrons, so the reaction of an unsaturated acid with oxygen has free radical intermediates. Oxidation of linoleic acid, catalyzed by an enzyme from soybeans, gives $(9Z,11E)$-13-hydroperoxy-9,11-octadecadienoic acid.

Labeling experiments have shown that the oxygen atoms in the product come from gaseous oxygen and not from water molecules.

$CH_3(CH_2)_4$ C=C C C=C $(CH_2)_7COH$

R—Ö—Ö·   linoleic acid

alkylperoxy radical

$CH_3(CH_2)_4$ C=C C C=C $(CH_2)_7COH$ + R—Ö—Ö—H

$CH_3(CH_2)_4CH$ C=C C=C $(CH_2)_7COH$

:Ö—Ö: $CH_3(CH_2)_4$ C—C C=C $(CH_2)_7COH$

:Ö
Ö·   H—R   (may be linoleic acid)

$CH_3(CH_2)_4CH$ C=C C=C $(CH_2)_7COH$

:Ö
:Ö—H   ·R

(9Z,11E)-13-hydroperoxy-9,11-octadecadienoic acid

We do not know the exact nature of the radical that does the first hydrogen abstraction from linoleic acid. Is is postulated to be some peroxy species. The hydrogen atoms that are allylic to two double bonds are particularly vulnerable to attack by a radical initiator. The carbon radical that forms is allylic and can be stabilized by delocalization of the radical center, giving rise to a conjugated double bond system. The new double bond formed between carbon atoms 11 and 12 is always trans. Combination of the carbon radical with molecular oxygen creates an alkylperoxy radical, which starts the chain reaction going again by abstracting an allylic hydrogen atom. The more polyunsaturated an acid is, the more sites there are for radical abstraction reactions, and the greater the complexity of the products. It is also possible for the alkylperoxy radicals to attack double bonds in other molecules of the unsaturated fatty acids, giving rise to large, complex molecules in which many acid units are held together by oxygen bridges. The characteristic drying of linseed oil on exposure to air is believed to consist of just such a linkage of linolenic acid (p. 470) molecules with oxygen to give a tough protective coating to paint. These reactions are also responsible for the development of rancidity in oils.

B. Prostaglandins

Prostaglandins and other related polyunsaturated acids containing 20 carbon atoms constitute a class of compounds with a wide range of important biological functions. The first prostaglandins were isolated from seminal fluid, hence their name. They have since been found to be widely distributed in all kinds of body tissues and have functions in reproduction, the nervous system, the intestinal system, blood clotting, and the production of allergic and inflammatory reactions.

The structural relationship between prostaglandin E_2 and arachidonic acid is shown below.

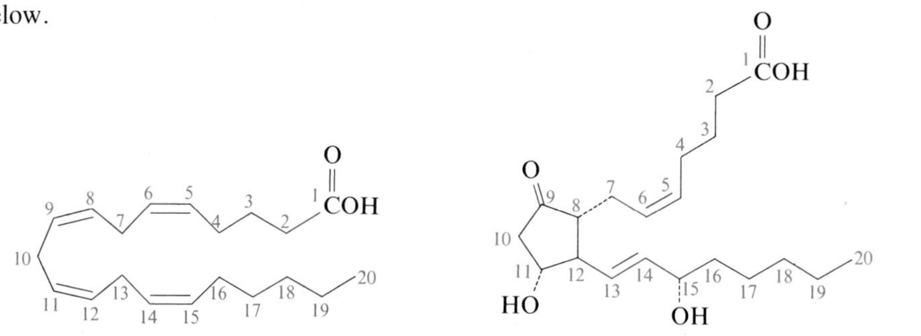

arachidonic acid

prostaglandin E_2

It has been established that two molecules of oxygen are used in the biological transformation of arachidonic acid to a prostaglandin, and that the oxygen atoms at carbon atoms 9 and 11 of a prostaglandin are derived from the same molecule of oxygen.

The enzymatic cyclization of arachidonic acid to prostaglandin is postulated to occur by way of a number of intermediates. To make the stages of the various free radical reactions clearer, transformations that probably take place more or less simultaneously in biological systems are shown here as separate steps.

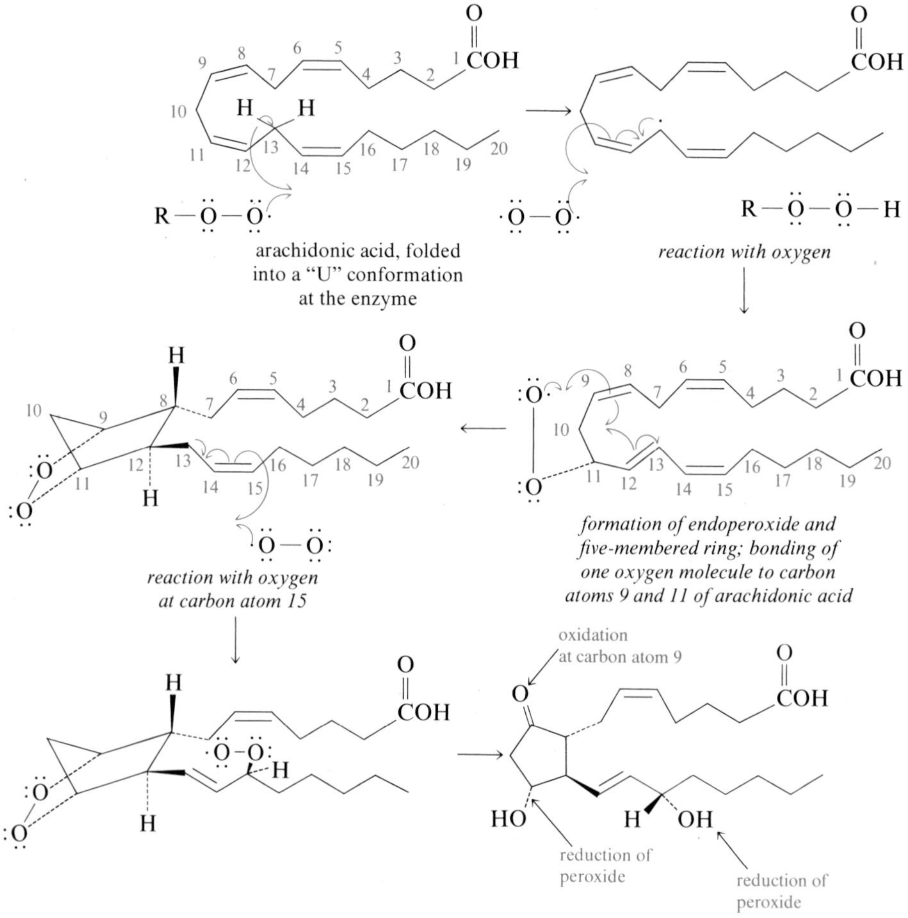

arachidonic acid, folded
into a "U" conformation
at the enzyme

reaction with oxygen

*reaction with oxygen
at carbon atom 15*

*formation of endoperoxide and
five-membered ring; bonding of
one oxygen molecule to carbon
atoms 9 and 11 of arachidonic acid*

oxidation
at carbon atom 9

reduction of
peroxide

reduction of
peroxide

prostaglandin E_2

An oxygen molecule binds to carbon 11 in arachidonic acid while one of the allylic hydrogen atoms is lost. In the process, the double bond moves from between carbon atoms 11 and 12 to between carbon atoms 12 and 13. The alkylperoxy radical that has formed attacks carbon atom 9 to give a cyclic peroxide. A radical intermediate forms at carbon 8, which is in position to attack carbon atom 12, forming a five-membered ring. A new allylic radical is formed at carbon atom 13, which becomes part of a double bond as the π bond between carbon atoms 14 and 15 is broken by reaction with another molecule of oxygen. The various prostaglandin structures shown below are related to this intermediate by reactions that you recognize, such as reduction of the double bond between carbon atoms 5 and 6, loss of water from the cyclopentane ring, and reduction of ketone functions to alcohols. Biologically, not all of these structures come from arachidonic acid.

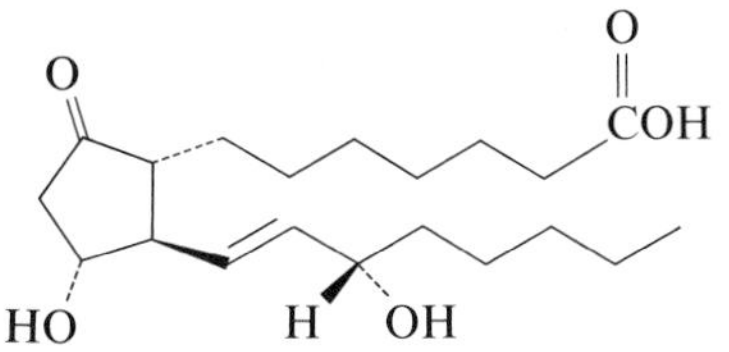

prostaglandin E_1

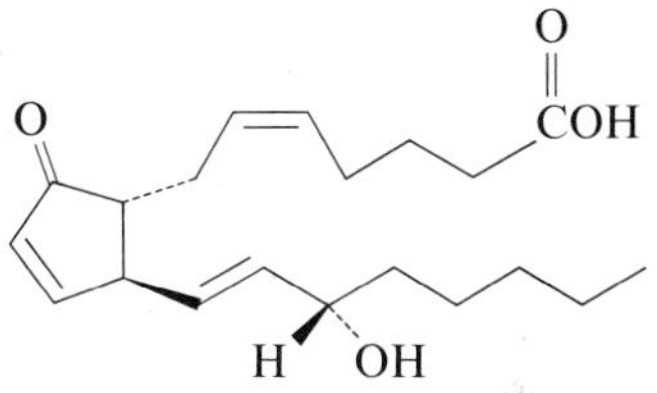

prostaglandin A_2

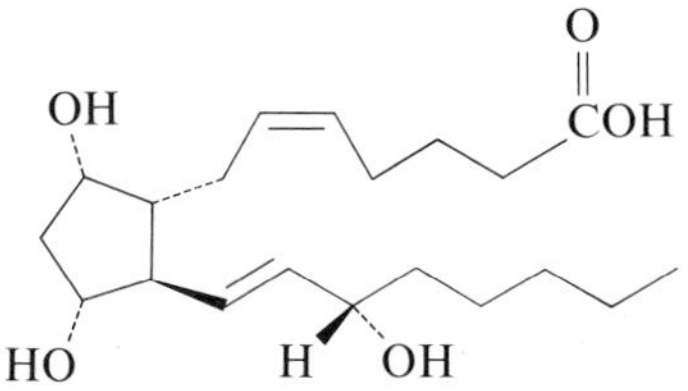

prostaglandin $F_{2\alpha}$

Another series of compounds with potent biological activity, also polyunsaturated 20-carbon acids, are the incongruously named leukotrienes, which have four double bonds. Of these, leukotriene C is the compound known as the slow-reacting substance of anaphylactic shock, the body's response to a foreign substance that provokes a severe allergic reaction, as in the case of asthma or insect stings. Exactly how leukotriene C is involved is not known. Apparently cells release it upon the arrival of antibodies formed in response to the presence of allergens, substances causing allergies.

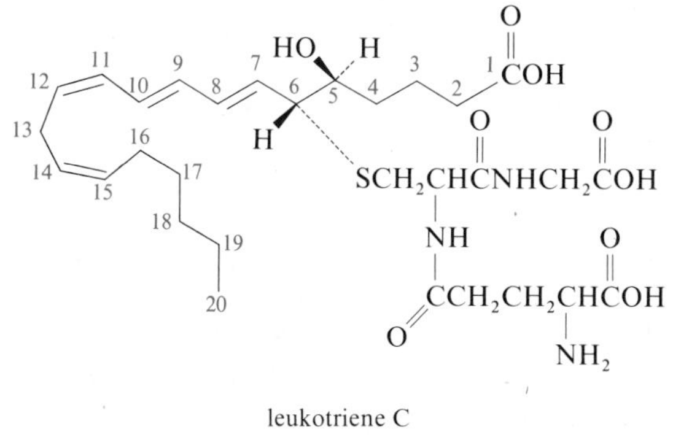

leukotriene C

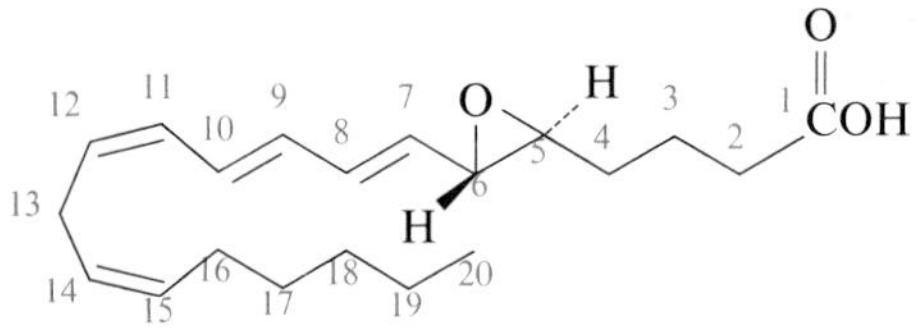

leukotriene A

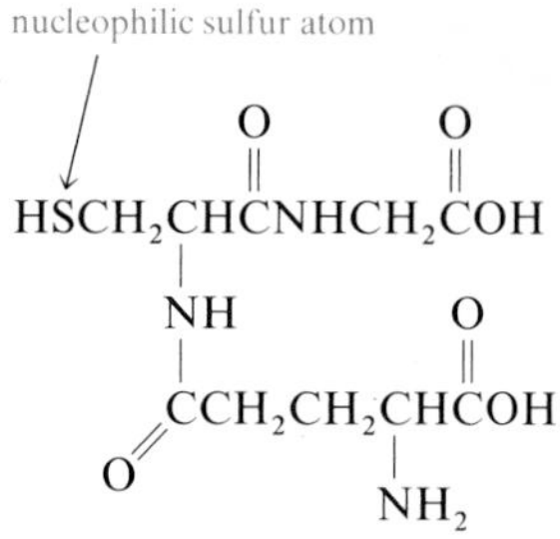

nucleophilic sulfur atom

glutathione

a peptide containing three amino acids and two peptide linkages

Both leukotriene A and leukotriene C are oxygenated at carbon atom 5 of the chain. Leukotriene A is an intermediate in the biosynthesis of leukotriene C. The transformation of A to C involves addition of glutathione (a peptide) to the oxirane. Glutathione includes the amino acid cysteine, which has a thiol group. The opening of the oxirane ring in leukotriene A results from attack by the nucleophilic sulfur atom of glutathione (Section 8.5C).

PROBLEM 11.25   (8Z,11Z,14Z)-8,11,14-Icosatrienoic acid is converted by an enzyme, soybean lipoxidase, and molecular oxygen to (8Z,11Z,13E)-15-hydroperoxy-8,11,13-icosatrienoic acid. Write out structural formulas for the starting material and product, and propose a mechanism for the conversion.

PROBLEM 11.26   Propose a mechanism for the reaction of the oxirane ring in leukotriene A with the peptide glutathione, paying special attention to stereochemistry. The abbreviated structures shown below represent the reactants. Remember that in an enzyme system, acids (HB$^+$) and bases (B:) are available as necessary.

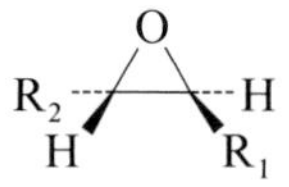

leukotriene A   glutathione

C. Pheromones

One of the most interesting discoveries in recent years is the fact that insects communicate with each other, and indeed with their environment, by means of organic chemicals secreted in minute amounts. These substances, known as **pheromones,** have a wide range of structures and include a variety of functional groups.

Because chemists have such small amounts of material to work with, the chemistry associated with the isolation of these compounds and the determination of their structures is particularly challenging. Many of the structural determinations were done in the laboratory on only a few milligrams of isolated material. Methods of separation and isolation, as well as spectroscopic techniques, were refined in order to deal with small quantities. For example, extraction of the abdominal tips of 500,000 virgin gypsy moth females yielded only 75 mg of the pheromone that attracts males. This quantity of active material had to be separated from approximately 250 g of fatty acids and their esters and 70 g of steroids, mostly cholesterol. The process of purification, which is very much like looking for the proverbial needle in a haystack, is aided by biological assay methods in which male gypsy moths are exposed to the fractions obtained in various separation steps. The fractions that do not attract the male moths are discarded while those that do are purified further until finally a single component that is highly attractive to the male is isolated.

The sex pheromone for the gypsy moth is a chiral oxirane. The dextrorotatory compound, known as disparlure, is a potent attractant to male gypsy moths while the enantiomer, the levorotatory compound, has no activity.

(+)-disparlure

the sex pheromone of the gypsy moth

The gypsy moth is a devastating pest of hardwood forests and fruit orchards because the larvae damage trees severely by eating their leaves. Female gypsy moths do not fly, so they must attract males if mating is to take place. The attractant, or pheromone, is volatile enough that it carries long distances in the air. Extremely small amounts of pheromone in a large volume of air are effective in attracting male moths.

A compound that serves as a sex pheromone is usually specific for one insect. Once we understand how pheromones work, we might be able to devise ways of controlling certain insect populations without harming the environment. At present, sex pheromones are used as lures in traps that allow foresters to estimate the population of a given insect in the region. For example, in 1980, a single Mediterranean fruit fly, caught in such a trap in a citrus grove in California, sounded the warning that a new infestation of the destructive pest was on the way and alerted the agricultural experts in that state to be prepared. The infestation became a disastrous reality in the summer of 1981.

The sex pheromones of many other insects are long-chain alcohols or esters, with one or more double bonds in them. In some cases, the double bonds are conjugated; in others they are not. They may have cis or trans stereochemistry. The sex pheromones of insects with colorful names such as the pink bollworm, the silkworm, and the red-banded leaf roller, have been isolated and their structures determined.

(*E*)-10-propyl-5,9-tridecadienyl acetate

from the pink bollworm

(10*E*,12*Z*)-10,12-hexadecadien-1-ol

from the silkworm

(*Z*)-11-tetradecenyl acetate

from the red-banded leaf roller

Terpenoid pheromones are believed to be related to the diet of the insect. The male boll weevil, the major cotton pest, secretes a mixture of terpenes that attracts females. Males that are fed on cotton buds are more attractive to the females than males fed an artificial diet. The terpene myrcene, found in the essential oils of the cotton bud, has been proposed as the biological precursor of the major sex pheromones of the boll weevil (Figure 11.4). There are two double bonds in the myrcene precursor, either of which can be protonated to give a tertiary carbocation. The other double bond in the molecule, acting as a nucleophile, attacks the cationic center to create a new cyclic cation. Loss of a proton from this cation gives the two major pheromones, both alcohols, one containing a six-membered ring and the other a four-membered one. Oxidation of the allylic alcohol function on the six-membered ring to an aldehyde and isomerization of the double bond by protonation and deprotonation give the other two components of the pheromone mixture.

A great deal of insect behavior is regulated by specific chemical substances that transmit one message or another. Interesting compounds are (*E*)-9-oxo-2-decenoic

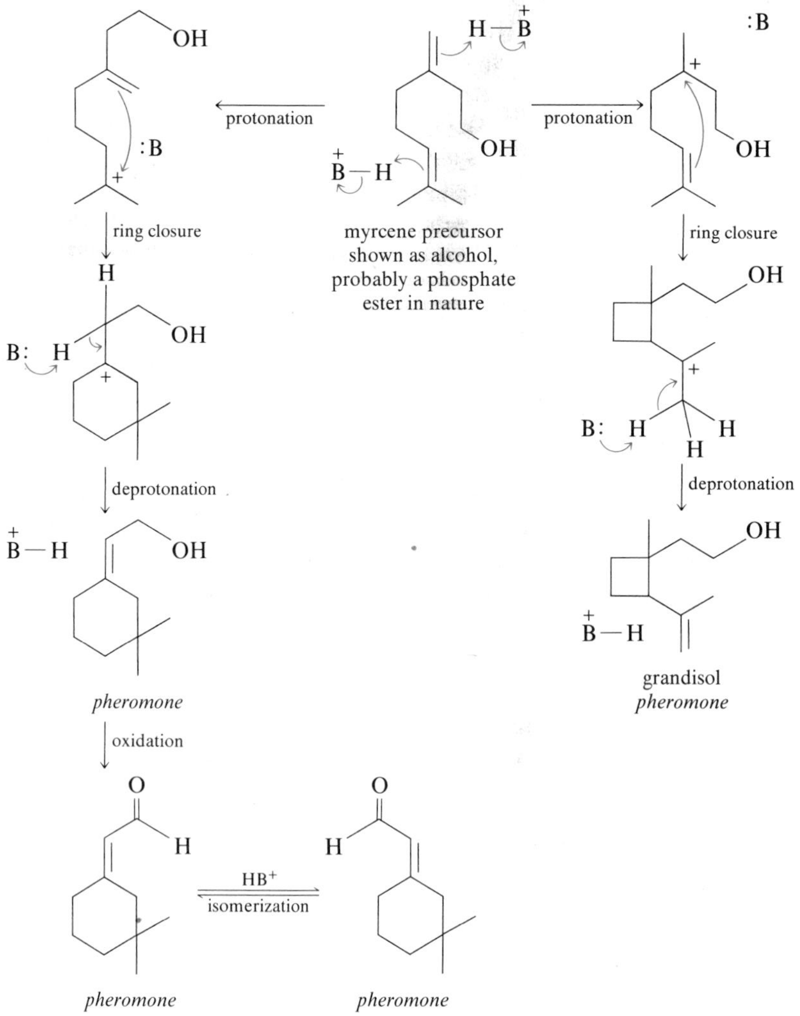

Figure 11.4 The transformation of the terpene myrcene into the sex pheromones of the boll weevil.

acid and its reduced form, (*E*)-9-hydroxy-2-decenoic acid, both secreted by the queen bee in a honey bee colony.

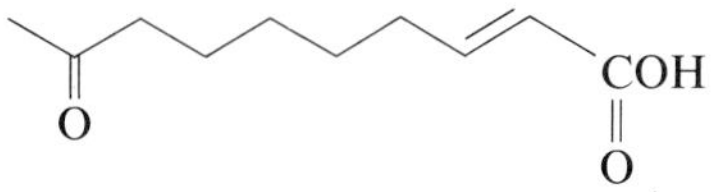

(E)-9-oxo-2-decenoic acid 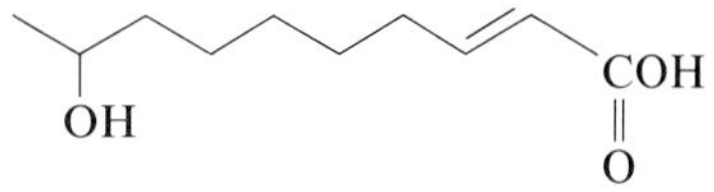   (E)-9-hydroxy-2-decenoic acid

The ketone is known as queen substance and is the compound that attracts drones to the queen in her mating flight. It also inhibits the development of ovaries in the other female bees in the colony so that they become worker bees. When a queen moves out of a hive and establishes a new colony, the same ketone attracts other bees around her to form a swarm, while the corresponding alcohol, (*E*)-9-hydroxy-2-decenoic acid, serves to calm and tranquilize the bees in the swarm so that they settle down and conserve energy in the part of their life cycle when they are exposed to danger. When a queen bee ages, she produces diminished amounts of queen substance. This is a signal to the workers that she must be replaced and a new queen must be nurtured and raised.

The most concentrated insect pheromone is the alarm pheromone, a chemical substance used to communicate disturbance in an insect colony. The simple ester 3-methylbutyl acetate is the alarm pheromone for the honey bee. Many other insect species use terpenes such as citronellal and limonene (p. 538) to signal alarm.

Even the fact of death is communicated by chemical signals in the insect world. Ants, for example, do not recognize a freshly frozen ant as dead, and treat it quite differently at first than they do after a while. Apparently, bacterial decomposition in the dead insect releases a number of fatty acids, such as oleic and linoleic acids. If mixtures of acids such as these are placed on a small piece of paper and put into the colony, the ants rapidly remove the paper and deposit it on a refuse pile. The same thing happens to a fellow ant once it has been dead for a while.

Organic compounds are used to regulate the relationship between an insect and its environment in many other ways. Plants synthesize compounds that prevent insects from eating them, and insects develop defensive secretions that keep birds and mammals away. Mammals, too, have chemical methods of communication that we are just beginning to recognize as important. The perfume industry, of course, has always understood the importance of scent as a means of communication.

PROBLEM 11.27   A chemical that protects a plant from being eaten by insects has been isolated and its structure determined in Japan. The compound, called shiromodiol diacetate, has the structure shown below.

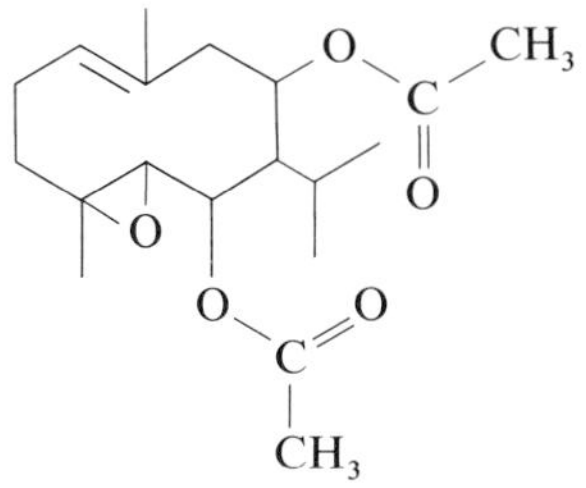

(a)   To what class of natural products does shiromodiol diacetate belong? Explain.

(b)   Shiromodiol diacetate has significant bands in its infrared spectrum at 1735 and 1240 cm^{-1}. Among the reactions used in the proof of the structure of the compound was hydrolysis with dilute aqueous base to give a Compound A, with absorption at 3400 cm^{-1} and no absorption between 1800 and 1700 cm^{-1} in the infrared. When Compound A is treated with chromium trioxide in pyridine, a reagent that behaves like pyridinium chlorochromate (Section 7.6C), Compound B, having absorption at 1720 and 1695 but not at 3400 cm^{-1}, is formed.

Write equations predicting what the structures of A and B are. Account for the changes observed in the infrared spectra for this series of compounds.

PROBLEM 11.28   The sex pheromone of the cabbage looper has the structure shown below.

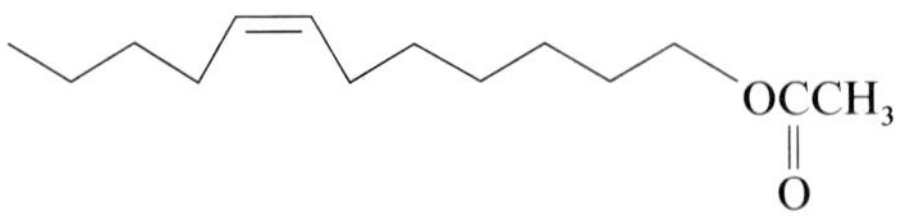

(Z)-7-dodecenyl acetate

(a)   Devise a synthesis for it. Organic reagents that are available to you are acetylene, butyl bromide, and 6-hydroxyhexanal. You may assume that the laboratory is well stocked with solvents, acids, bases, and catalysts. (Hint: You may find it helpful to review Sections 6.4B, 8.3, 9.4, and 9.7.)

(b)   The *(E)*-isomer of the pheromone has also been synthesized. How would you modify your synthesis to obtain that isomer?

11.8

Ultraviolet Spectroscopy

A. Transitions Between Electronic Energy Levels. The Chromophore

Absorption of energy corresponding to the ultraviolet and visible regions of the electromagnetic spectrum results in transitions between electronic energy levels in molecules. The π bond of an alkene, for example, is made up of a combination of two $2p$ orbitals on carbon, each containing one electron. The two atomic orbitals combine to give two molecular orbitals, a bonding π orbital that is lower in energy than the atomic orbitals, and an antibonding orbital that is higher in energy and is designated as π^* (Section 2.5A). The two electrons from the carbon atoms are usually found in the bonding π orbital, the state of lowest energy. If radiation corresponding to the difference in energy between the bonding orbital and the antibonding orbital is absorbed by a molecule, one of the electrons moves to the higher energy level. The molecule is said to undergo a $\pi \rightarrow \pi^*$ transition and to be raised from the **ground state** to an **excited state,** an unstable situation of high energy (Figure 11.5).

A bonding π orbital has two lobes, above and below the plane defined by the two carbon atoms of the alkene and the four atoms bonded to them. There is electron density between the two carbon atoms, and the necessity for parallel overlap of the p orbitals, if the π bond is to be maintained, prevents rotation of the two carbon atoms in relation to each other. These phenomena are responsible for the existence of cis and trans isomers of alkenes (Section 5.1A).

An antibonding π orbital not only has the node in the plane of the alkene, but also has a nodal plane bisecting the bond between the two carbon atoms (Figure 2.20, p.

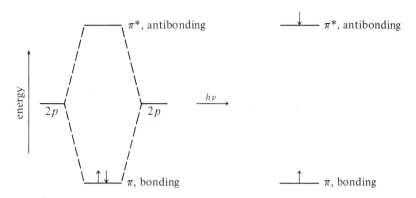

Figure 11.5   Schematic representation of the change in the electronic configuration of an alkene molecule upon absorption of electromagnetic radiation. A $\pi \rightarrow \pi^*$ transition.

54). If one of the electrons of a π orbital is promoted to a π^* orbital, the electron density between the two carbon atoms decreases. The bond between them becomes more like a single bond, and rotation of one carbon atom in relation to another one becomes easier. Thus, one of the reactions seen for alkenes when radiation is absorbed is cis-trans isomerization about the double bond. The isomerization of (11Z)-retinal to (11E)-retinal, upon the absorption of light, described in Section 11.6D as the primary event in vision, is based on this phenomenon.

A carbonyl group with its π bond also undergoes $\pi \rightarrow \pi^*$ transitions. In addition, the carbonyl group has two pairs of nonbonding electrons (represented by n) on the oxygen atom. The nonbonding electrons occupy orbitals designated as n orbitals, which are higher in energy than the bonding π orbital, but lower in energy than the antibonding π^* orbital. The energy levels for the π, n, and π^* orbitals for a carbonyl compound such as formaldehyde are shown in Figure 11.6.

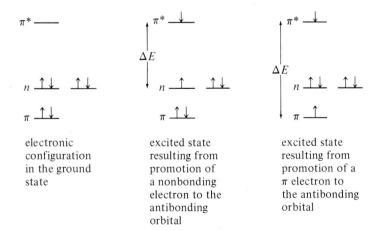

Figure 11.6   Electronic configuration and relative energy levels for the π and nonbonding (n) electrons in formaldehyde for the ground state and for the $n \rightarrow \pi^*$ and $\pi \rightarrow \pi^*$ excited state.

Just as nonbonding electrons are more readily available for protonation reactions than are π electrons, so they are also more easily promoted to the π^* orbital of the carbonyl group. Thus, the carbonyl group absorbs ultraviolet radiation in two regions. A lower energy transition corresponds to the promotion of one of the nonbonding electrons to the antibonding π orbital, an $n \rightarrow \pi^*$ transition. The transition corresponding to the promotion of an electron from a bonding π orbital to an antibonding π orbital, the $\pi \rightarrow \pi^*$ transition, requires more energy than the $n \rightarrow \pi^*$ transition.

A simple alkene such as ethylene has a $\pi \rightarrow \pi^*$ transition energy corresponding to radiation at wavelengths below 200 nm. A nanometer, nm, is 10^{-7} cm (10^{-9} m; 10 Å) and is the unit in which wavelengths in the ultraviolet and visible regions of the electromagnetic spectrum are expressed. Common ultraviolet and visible spectrophotometers do not record at wavelengths below 200 nm. The spectra of most organic compounds become interesting when there are two or more multiple bonds in conjugation. The extended interaction between the conjugated π orbitals lowers the amount of energy necessary for a $\pi \rightarrow \pi^*$ transition and increases the intensity of the absorption of radiation. If the conjugation becomes extended enough, absorption takes place in the visible region of the spectrum. For example, your eyes see carotene as red-orange because it is absorbing the violet-blue portion of the visible spectrum, leaving the more orange-red portion.

While the infrared spectrum of a compound is sensitive to exact structure and will distinguish similar compounds on the basis of differences in the fingerprint region (Section 7.7B), the ultraviolet spectrum of the compound shows only the presence or absence of certain portions of the molecule that undergo $\pi \rightarrow \pi^*$ or $n \rightarrow \pi^*$ transitions. *These distinctive groupings that absorb ultraviolet or visible radiation, usually conjugated double bonds, double bonds conjugated with carbonyl groups, or aromatic rings, are known as* **chromophores.**

Ultraviolet spectroscopy picks out the presence of such chromophores in otherwise complicated molecules, but says nothing about the large differences in structure that may exist. For example, cholesta-4-en-3-one and 4-methyl-3-penten-2-one have essentially the same ultraviolet spectrum, with a $\pi \rightarrow \pi^*$ transition around 240 nm, and an $n \rightarrow \pi^*$ transition around 310 nm.

4-methyl-3-penten-2-one   cholesta-4-en-3-one

the chromophore
common to both
compounds

Both compounds have a carbonyl group conjugated with a double bond. The carbonyl group is part of an alkyl ketone, and there are two alkyl groups substituted at the end carbon atom of the double bond. These things taken together contribute to the distinctive wavelength and intensity of the absorption of radiation, which is quite similar for both compounds. Only the chromophore in the large steroid molecule interacts with ultraviolet radiation in a way that results in absorption of energy in the region of the spectrum that our instruments scan. The rest of the molecule is invisible as far as this spectroscopic technique is concerned. This simplifying quality of ultraviolet spectroscopy has been extraordinarily useful in identifying compounds that have similar chromophores, regardless of the complexities and differences in other parts of their molecules. The same quality makes it useful as an analytical tool when the compound sought has an intense and distinctive absorption spectrum.

B. The Absorption Spectrum

An ultraviolet or visible spectrum of a compound is obtained by comparing the radiation absorbed by a solution of the compound with the radiation absorbed by a similar thickness of pure solvent. A sketch of an ultraviolet spectrophotometer is shown in Figure 11.7.

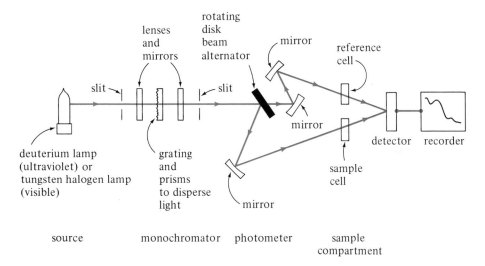

Figure 11.7   Schematic diagram of a typical ultraviolet-visible spectrophotometer.

In the last section, two aspects of an ultraviolet spectrum were mentioned: the wavelength at which absorption takes place, and the intensity of the absorption. The ultraviolet absorption spectrum of 1,3-pentadiene shown in Figure 11.8 is typical of a conjugated diene. First, note the simplicity of the spectrum in comparison with infrared spectra. One absorption appears as a broad band. Unless the radiation supplied to the compound is all of a single wavelength, molecular transitions will take place not only between one electronic energy level and a higher one, but between the closely spaced vibrational energy levels belonging to each electronic level (see Figure 7.1, p. 296). The infrared spectrum of the compound is a record of the transitions between those

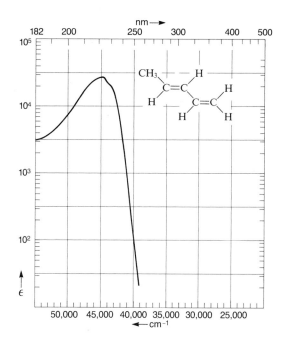

Figure 11.8   Ultraviolet spectrum of 1,3-pentadiene in heptane. (From the *UV Atlas of Organic Compounds*)

vibrational levels. The ultraviolet spectrum is really the envelope for a series of closely spaced transitions clustered around a major electronic transition.

The wavelength corresponding to the energy of the electronic transition is read off the spectrum, which plots **molar absorptivity, ϵ**, a measure of the intensity of the absorption of radiation characteristic of the compound, against wavelength (or wavenumber). The position of maximum absorption is recorded as λ_{max} 224 nm, for 1,3-pentadiene. The solvent can affect the spectrum (an alcohol interacts with the nonbonding electrons on a carbonyl group, for example) so the solvent is always specified, too. The complete designation of the wavelength of maximum absorption for the sample in Figure 11.8 would be:

$$\lambda_{max}^{heptane} \;\; 224 \text{ nm}$$

The molar absorptivity, ϵ, for 1,3-pentadiene is determined from the absorbance measured for the solution of the compound in heptane. The absorbance is related to the number of molecules of 1,3-pentadiene in the path of the light. This depends on the concentration of the solution and on the thickness of the cell through which the radiation passes. The thicker the cell, the longer the path length for the beam, and the more molecules will be encountered by the radiation on its way through, allowing for more absorption of energy. The molar absorptivity is related to the absorbance, A, by an equation

$$\epsilon = \frac{A}{bc}$$

in which A is the experimentally determined absorbance, b is the path length in centimeters, and c is the concentration in moles per liter. For 1,3-pentadiene ϵ is 26,000 L/mol·cm.

The information about the ultraviolet spectrum of a compound is thus given in two parts: the wavelength or wavelengths of maximum absorption, and the molar absorptivity for the compound at those wavelengths. For 1,3-pentadiene, the complete information would be that it has $\lambda_{max}^{heptane}$ 224 nm (ϵ 26,000). Note that the units of ϵ usually are not given.

C. The Relationship Between Structure and the Wavelength of Maximum Absorption

The wavelength of maximum absorption, λ_{max}, is dependent upon the exact structure of the chromophore. For example, 2,5-dimethyl-2,4-hexadiene has the ultraviolet spectrum shown in Figure 11.9. Note the general similarity of this spectrum to that of 1,3-pentadiene (Figure 11.8). The difference comes in the shift of the position of maximum absorption to a longer wavelength $\lambda_{max}^{heptane}$ 243 nm (ϵ 24,500). It takes less energy for the $\pi \rightarrow \pi^*$ transition to take place in the more highly substituted diene. This phenomenon is quite general with the position of λ_{max} being sensitive to extent of conjugation and to the degree of substitution on the conjugated system.

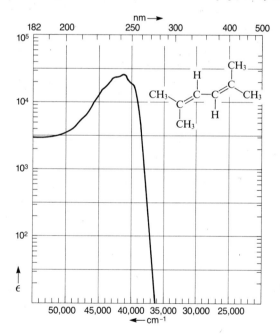

Figure 11.9   Ultraviolet spectrum of 2,5-dimethyl-2,4-hexa-diene. (From the *UV Atlas of Organic Compounds*)

The $\pi \rightarrow \pi^*$ transition gives rise to a more intense absorption band than the $n \rightarrow \pi^*$ transition for ketones. 4-Methylcyclohexanone has two absorption bands, $\lambda_{max}^{cyclohexane}$ 221 (ϵ 40.1) and 291 nm (ϵ 18.2). The lower wavelength, higher energy absorption is the $\pi \rightarrow \pi^*$ transition; the one at 291 nm corresponds to the $n \rightarrow \pi^*$ transition. Neither band is intense in a simple alkyl ketone.

The effect of conjugation on the spectrum of a ketone is shown in the spectra of 3-penten-2-one and 4-methyl-3-penten-2-one (Figure 11.10). For 3-penten-2-one, $\lambda_{max}^{ethanol}$ 220 (ϵ 13,000) and 311 nm (ϵ 35) are seen. The chromophore in this compound

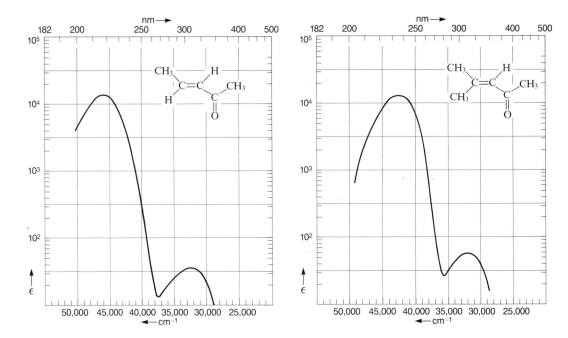

Figure 11.10   Ultraviolet spectra of 3-penten-2-one and
4-methyl-3-penten-2-one in ethanol. (From the
UV Atlas of Organic Compounds)

is different from the one in 4-methylcyclohexanone, and the $\pi \rightarrow \pi^*$ transition is related now to the entire conjugated system and not to any one π bond. While the λ_{max} values for 3-penten-2-one and 4-methylcyclohexanone may seem similar, the differences in the intensities of the absorption bands are so large that the spectrum of one compound cannot possibly be mistaken for that of the other.

When substitution on the double bond is increased by a methyl group on the conjugated double bond in 4-methyl-3-penten-2-one, the wavelength of the major absorption increases. For 4-methyl-3-penten-2-one, the absorption bands have $\lambda_{max}^{ethanol}$ 236 (ϵ 12,600) and 314 nm (ϵ 58).

When a system of really extended conjugation is present, such as in vitamin A or carotene, λ_{max} moves into the region of the spectrum from approximately 380 nm to 780 nm that is classified as the visible. The intensity of the absorption also increases. The major absorption bands in β-carotene, for example, occur at λ_{max}^{hexane} 425 (ϵ 103,000), 450 (ϵ 145,000), and 477 nm (ϵ 130,000). This absorption is in the blue-violet region of the visible spectrum. The color that we see for β-carotene is the portion of the visible spectrum left unabsorbed by the molecule.

Other systems that have useful features to their ultraviolet spectra are discussed in Section 12.8A. You should now be aware of the following important points:

1. Ultraviolet spectra point to the presence of distinctive groupings of atoms, usually involving conjugated systems called chromophores.
2. In general, the more extended the conjugation, the longer the wavelength at which absorption takes place and the greater the intensity of the absorption.
3. Added substitution, even of alkyl groups, on the conjugated system also increases the wavelength of absorption.

At a more advanced level, it is possible to be much more precise about how conjugation and substitution may affect absorption of energy in the ultraviolet region of the spectrum. Now, however, it is sufficient for you to appreciate and to be able to apply the principles described above.

PROBLEM 11.29   Which of the following compounds would you expect to absorb at the longest wavelength?

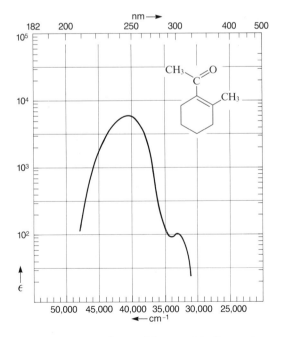

PROBLEM 11.30   The ultraviolet spectrum of 1-acetyl-2-methyl-1-cyclohexene in ethanol is shown in Figure 11.11. Identify the electronic transitions that are responsible for the different absorption bands in the spectrum. Determine λ_{max} and ϵ for the compound and present the information about the ultraviolet absorption of 1-acetyl-2-methyl-1-cyclohexene in the standard way.

Figure 11.11   Spectrum for Problem 11.30. (From the *UV Atlas of Organic Compounds*)

PROBLEM 11.31   Of the two isomeric hexadienols, 5-methyl-2,4-hexadien-1-ol and 2-methyl-3,5-hexadien-2-ol, one has $\lambda_{max}^{ethanol}$ 223 nm, and the other $\lambda_{max}^{ethanol}$ 236 nm. Draw structural formulas for the two compounds and assign each ultraviolet absorption band to the correct compound.

ADDITIONAL PROBLEMS

11.32   Complete the following equations, showing the product(s) you would expect. If the stereochemistry of a reaction is known, be sure to show it in your answer.

(a) 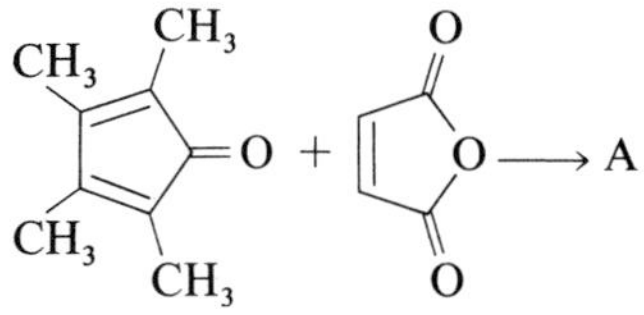 $\longrightarrow$ A

(b) $CH_3CH{=}CHCH{=}CHCH_3 \xrightarrow[\text{carbon tetrachloride}]{Br_2 \text{ (1 molar equivalent)}}$ B + C

(c) $CH_2{=}\overset{\underset{\displaystyle CH_3}{|}}{C}{-}\overset{\underset{\displaystyle CH_2CH_3}{|}}{C}{=}CH_2$ + NBS $\xrightarrow{\text{carbon tetrachloride}}$ D

(d) 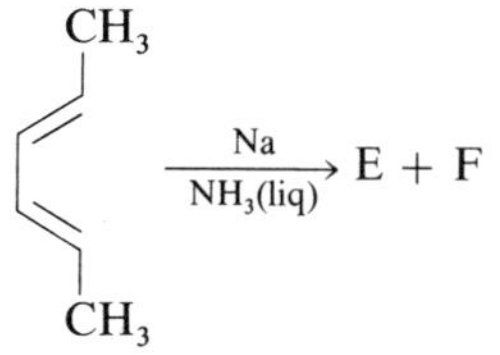 $\xrightarrow[\text{NH}_3\text{(liq)}]{\text{Na}}$ E + F   (e) $\longrightarrow$ G

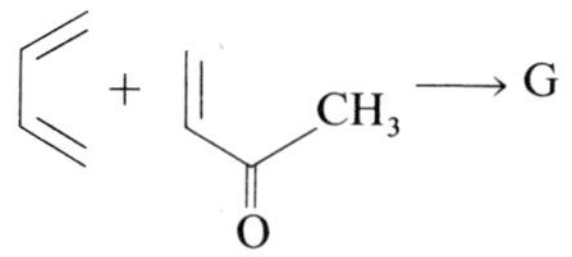

(f) $CH_3C{\equiv}CH \xrightarrow[\text{CuCl, pyridine}]{O_2}$ H   (g) 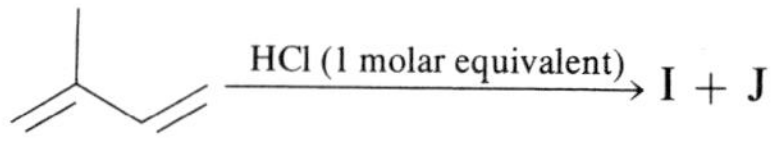 $\xrightarrow{\text{HCl (1 molar equivalent)}}$ I + J

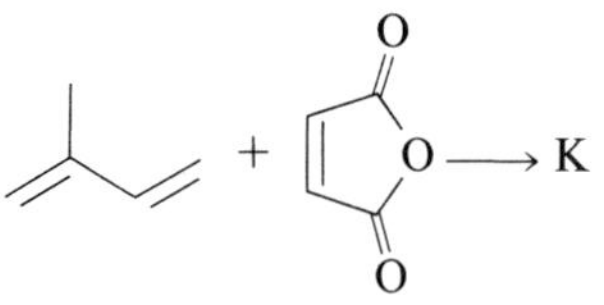

(h) $\longrightarrow$ K   (i) $\xrightarrow[\Delta]{H_2SO_4}$ L

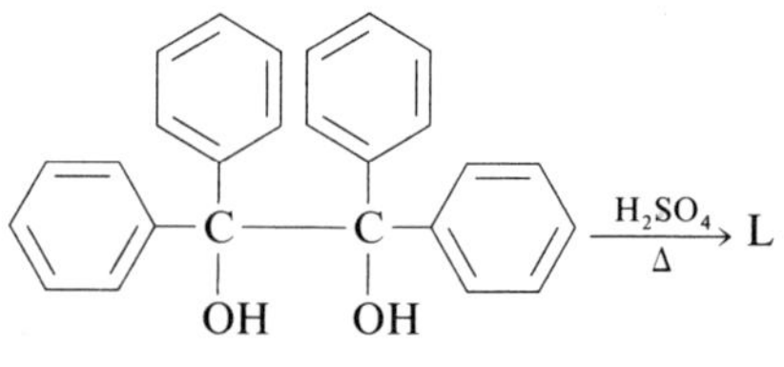

(j) $CH_3CH_2CH_2C{\equiv}CH \xrightarrow{CH_3CH_2MgBr}$ M $\xrightarrow[\substack{\text{CuCl} \\ \text{tetrahydrofuran}}]{BrCH_2C{\equiv}CCH_3}$ N

(k) 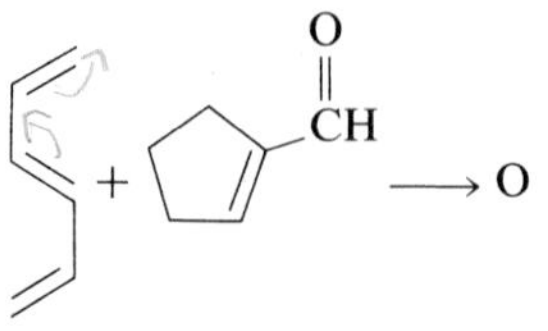 $\longrightarrow$ O

(l) 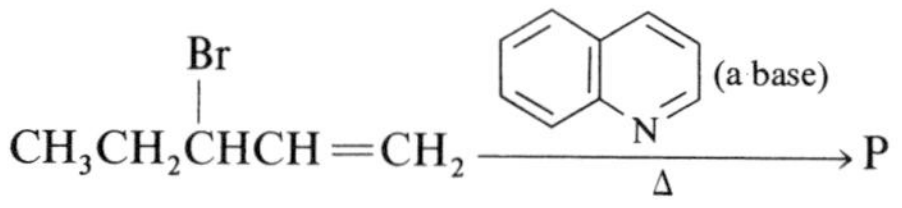 $CH_3CH_2\overset{\underset{\displaystyle Br}{|}}{C}HCH{=}CH_2 \xrightarrow[\Delta]{}$ P

(m) $CH_3C \equiv CCH_2C \equiv CCH_2CH_3 \xrightarrow[\substack{Pd/BaSO_4 \\ quinoline}]{H_2} Q$

(n)

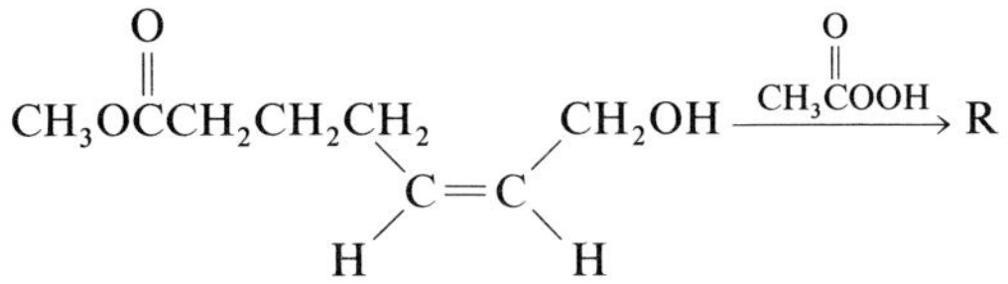

$$CH_3\overset{\overset{\displaystyle O}{\|}}{O}CCH_2CH_2CH_2 \underset{\underset{\displaystyle H}{\diagdown}}{} C = C \underset{\underset{\displaystyle H}{\diagdown}}{}CH_2OH \xrightarrow{\overset{\overset{\displaystyle O}{\|}}{CH_3COOH}} R$$

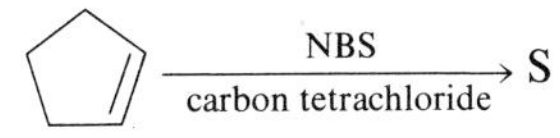

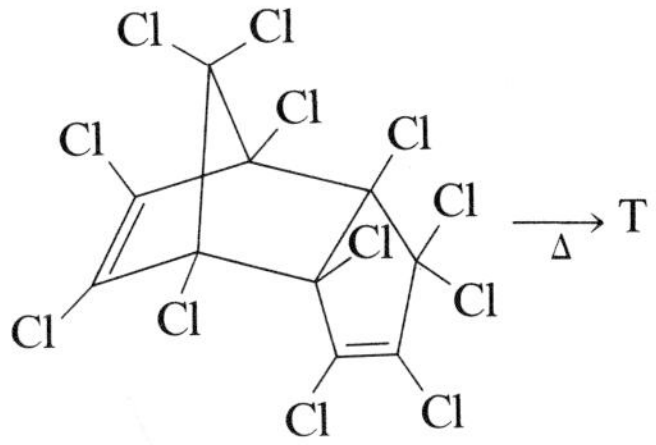

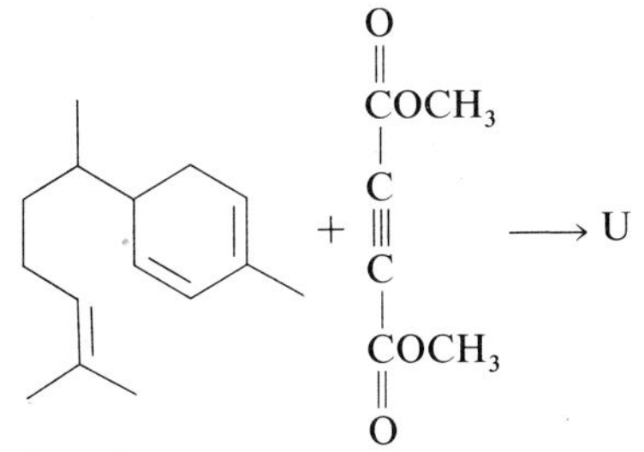

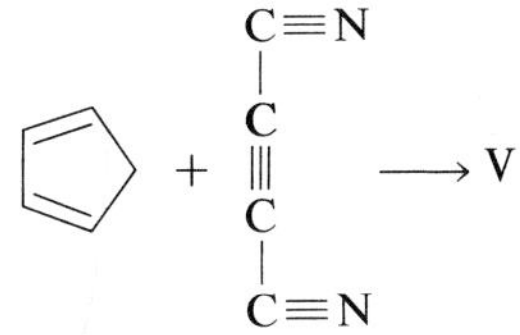

(o) $\xrightarrow[\text{carbon tetrachloride}]{\text{NBS}} S$ (p) $\xrightarrow{\Delta} T$

(q) $+$ $\longrightarrow U$ (r) $+$ $\longrightarrow V$

(s) $\xrightarrow[\substack{\text{diethylene glycol} \\ \Delta}]{H_2NNH_2, KOH} W \xrightarrow{\overset{\displaystyle H_2}{Pt}} X$

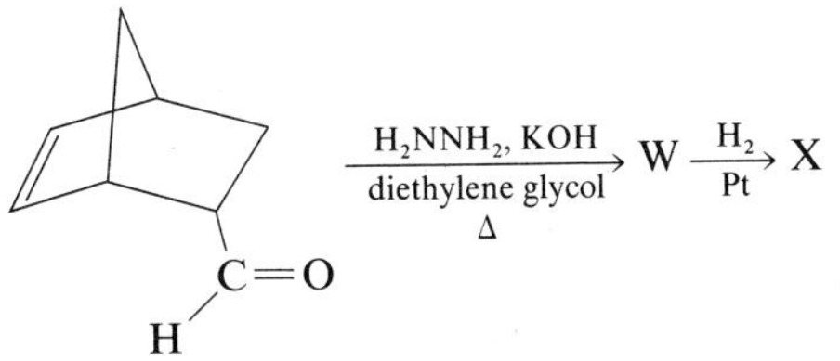

(t) $CH_2{=}CHCH_2CH_2CH{=}CH_2 \xrightarrow[\substack{\text{benzoyl peroxide} \\ \text{carbon tetrachloride}}]{\text{NBS}} Y$

(u) $CH_2{=}CHCH{=}CHCH_3 \xrightarrow[\text{tetrahydrofuran}]{\overset{\overset{\displaystyle O}{\|}}{Hg(OCCH_3)_2}, H_2O} Z \xrightarrow{NaBH_4, NaOH} AA$

(v) 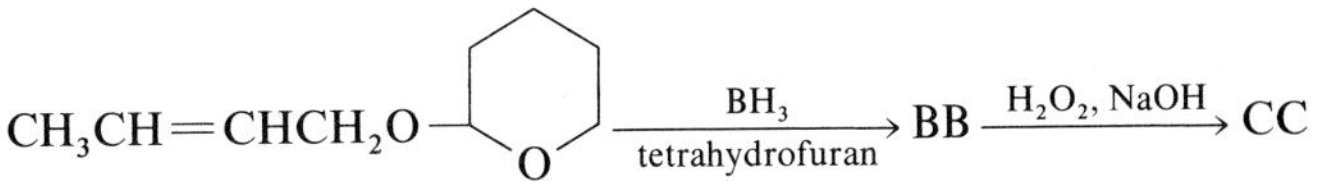 $CH_3CH{=}CHCH_2O{-}$ $\xrightarrow[\text{tetrahydrofuran}]{BH_3} BB \xrightarrow{H_2O_2, NaOH} CC$

11.33 Supply the reagents that would be necessary for the following transformations. More than one step may be needed in some cases.

(a)

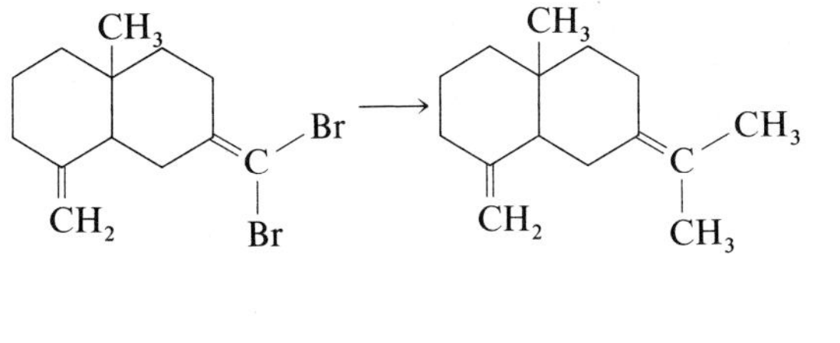

(b)

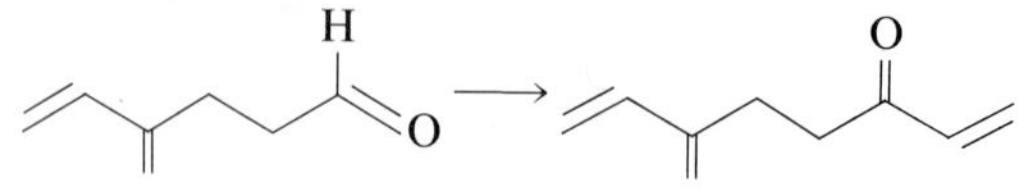

(c)

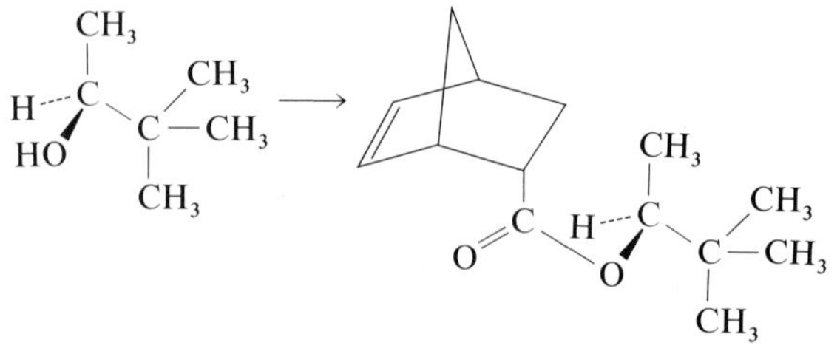

(d) $CH_3CH_2CH_2CH_2CH_2I \longrightarrow CH_3CH_2CH_2CH_2CH_2CH_2CH{=}CH_2$

(e)

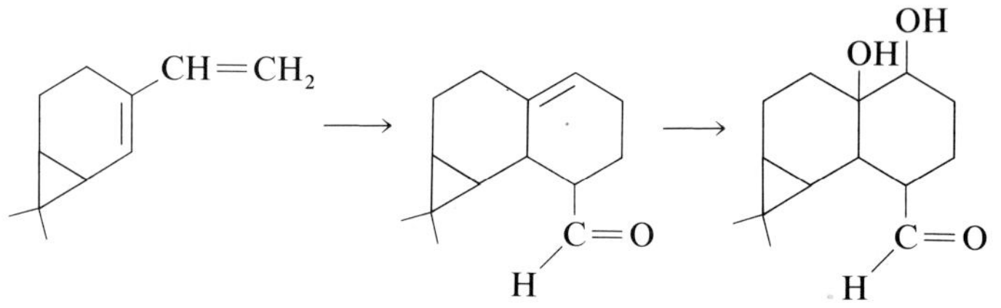

cis diol
formed

(f)

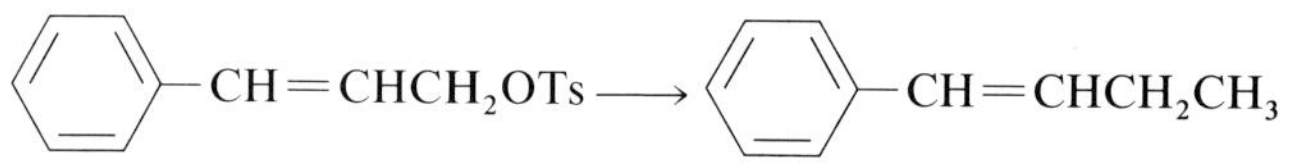

(g)

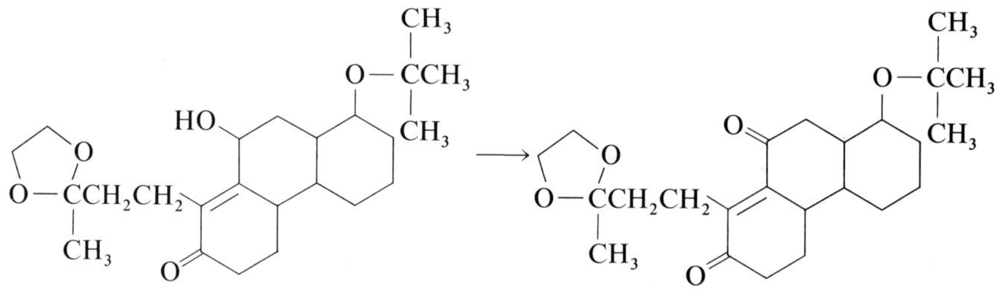

(h)

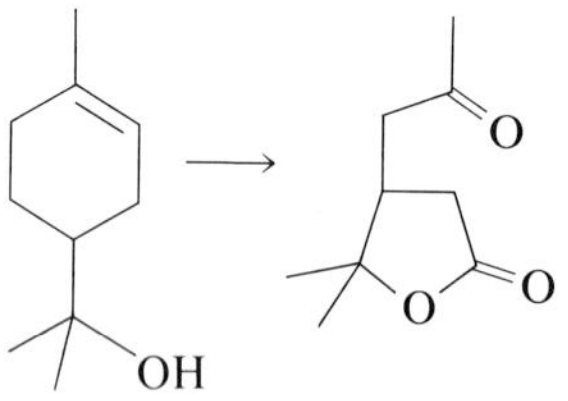

11.34   The following compounds are synthesized by Diels-Alder reactions. Write equations for their syntheses showing the diene and the dienophile that would have to be used in each case.

(a)

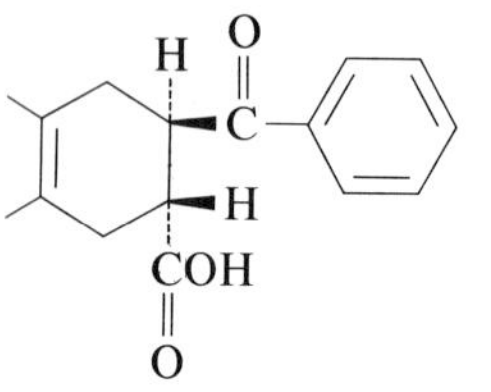

(b)

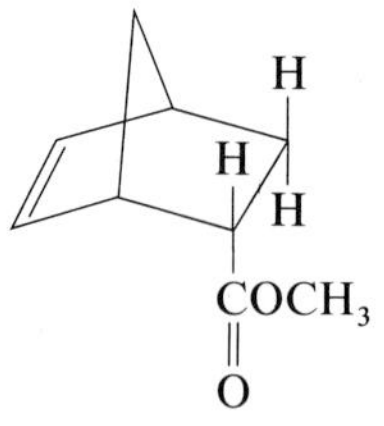

(c)

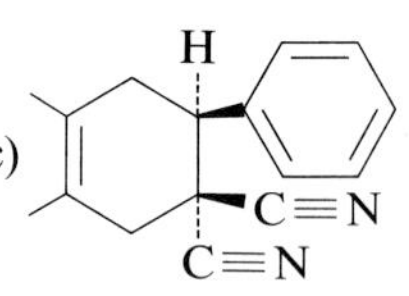

(d)

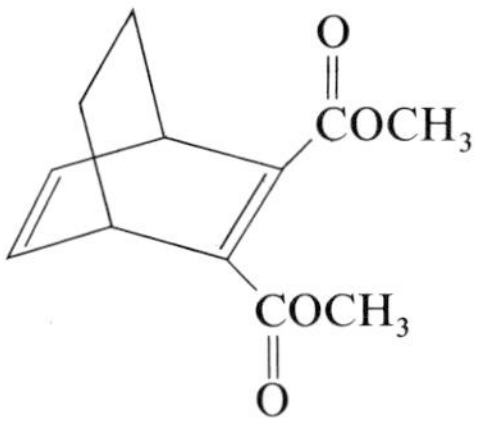

(e)

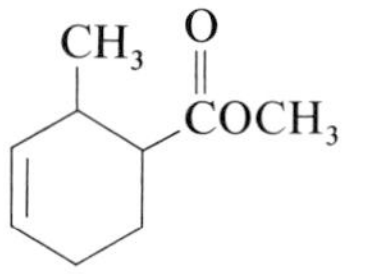

(f)

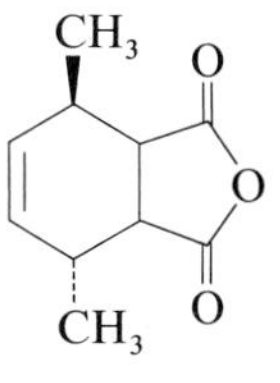

(g)

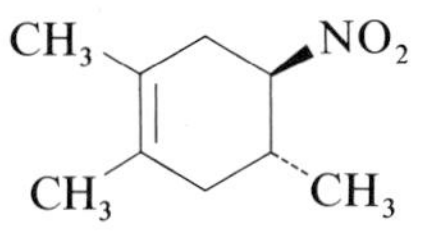

(h)

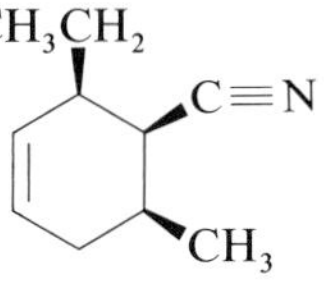

11.35 The following sequence of reactions was carried out. Supply structural formulas for the products or intermediates designated by letters.

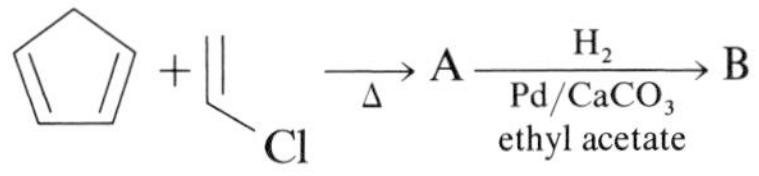

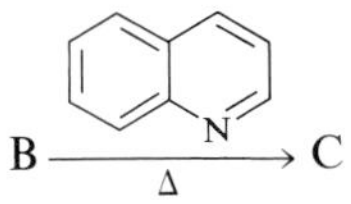

$$\text{cyclopentadiene} + \underset{Cl}{\text{acrylyl chloride}} \xrightarrow{\Delta} A \xrightarrow[\text{ethyl acetate}]{H_2 \; Pd/CaCO_3} B$$

$$B \xrightarrow[\Delta]{\text{quinoline}} C$$

$$B \xrightarrow[\text{diethyl ether}]{Mg} D \xrightarrow{CO_2} E \xrightarrow{H_3O^+} F$$

$$C \xrightarrow[\text{chloroform}]{Br_2} G$$

$$C \xrightarrow[\text{H}_2\text{O}]{KMnO_4, \; NaOH} H$$

$$C \xrightarrow{O_3} \xrightarrow{H_2O_2, \; NaOH} I$$

11.36 1,3-Butadiene adds to maleic anhydride at a reasonable rate at 100 °C and slowly at room temperature. 2,3-Di-*tert*-butyl-1,3-butadiene does not react with maleic anhydride at all under these conditions. Offer an explanation for these experimental observations.

11.37 1,3,5-Hexatriene reacts with 1 molar equivalent of bromine to give a mixture of 1,2- and 1,6-addition products. No 1,4-addition product is obtained. Explain this phenomenon by writing equations and showing the structures of the relevant intermediates.

11.38 Fill in the structural formulas, including stereochemistry, for the major products represented by letters in the following synthetic sequence.

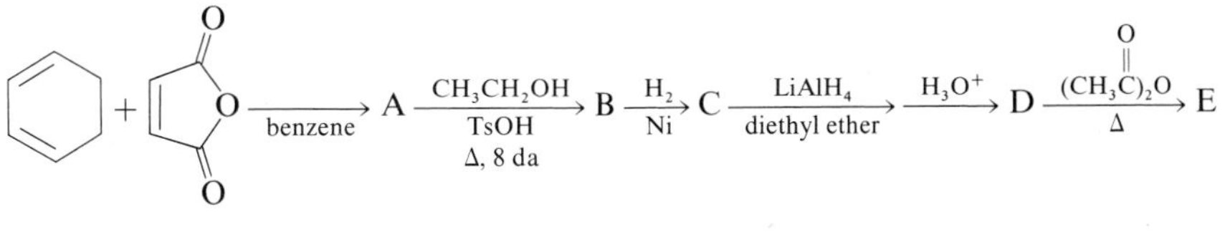

$$\text{1,3-cyclohexadiene} + \text{maleic anhydride} \xrightarrow[\text{benzene}]{} A \xrightarrow[\substack{\text{TsOH} \\ \Delta, \, 8 \, \text{da}}]{CH_3CH_2OH} B \xrightarrow[\text{Ni}]{H_2} C \xrightarrow[\text{diethyl ether}]{LiAlH_4} \xrightarrow{H_3O^+} D \xrightarrow[\Delta]{(CH_3C)_2O} E$$

11.39   Propose syntheses for the following compounds. In your laboratory, you have 1,3-butadiene and any organic reagents that contain three or fewer carbon atoms. There may be more than one way to carry out each synthesis.

(a)

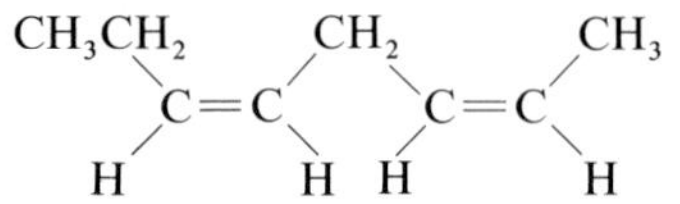

$$CH_3CH_2 \quad CH_2 \quad CH_3$$
$$C=C \qquad C=C$$
$$H \qquad H \ H \qquad H$$

(b)

$$\overset{O}{\overset{\|}{COCH_3}}$$
$$\overset{\|}{\underset{O}{COCH_3}}$$

(c)

$$\overset{Br}{\overset{|}{CH_3CH}} \qquad H$$
$$C=C$$
$$H \qquad CHCH_3$$
$$\overset{|}{Br}$$

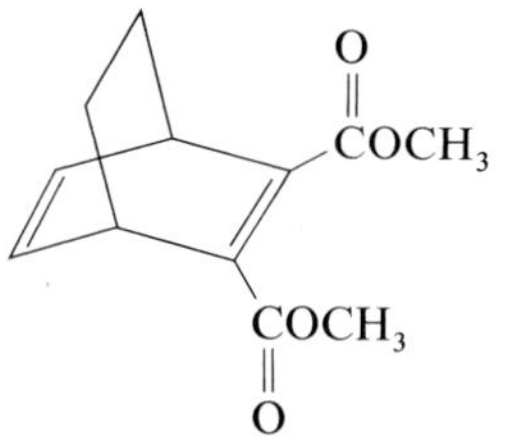

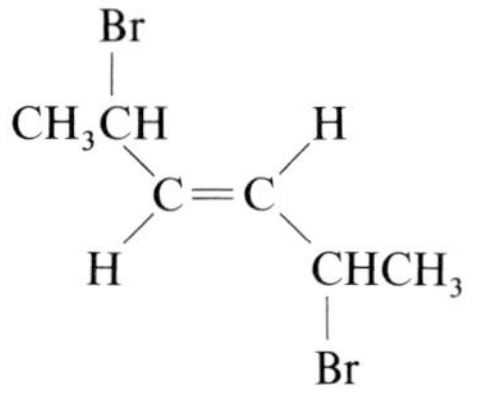

(d)

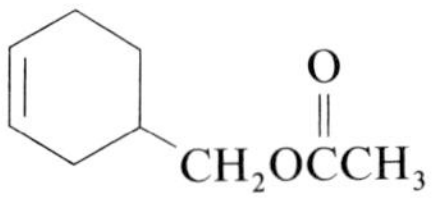

$$CH_2O\overset{O}{\overset{\|}{C}}CH_3$$

(e) 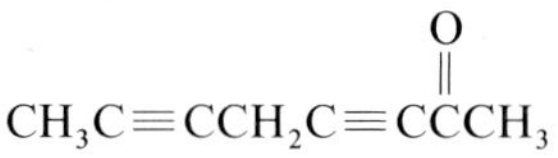 $CH_3C{\equiv}CCH_2C{\equiv}C\overset{O}{\overset{\|}{C}}CH_3$

(f) 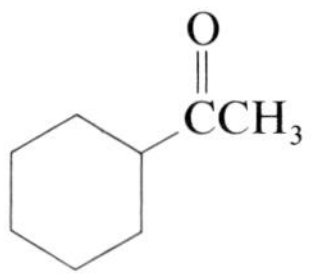 $\overset{O}{\overset{\|}{CCH_3}}$

(g) 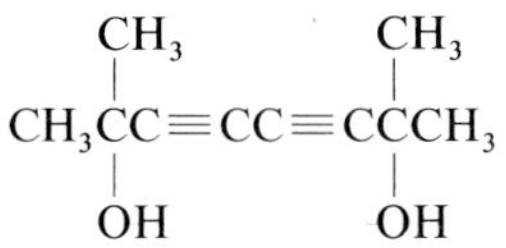 $\underset{\overset{|}{OH}}{CH_3\overset{\overset{CH_3}{|}}{C}C}{\equiv}CC{\equiv}C\underset{\overset{|}{OH}}{\overset{\overset{CH_3}{|}}{C}CH_3}$

11.40   A complete laboratory synthesis of leukotriene C was reported by E. J. Corey of Harvard University in 1980. Five of the steps in this synthesis are shown below. Complete each equation, showing what product you expect to get, or giving the mechanism for the transformation shown, as required.

1.

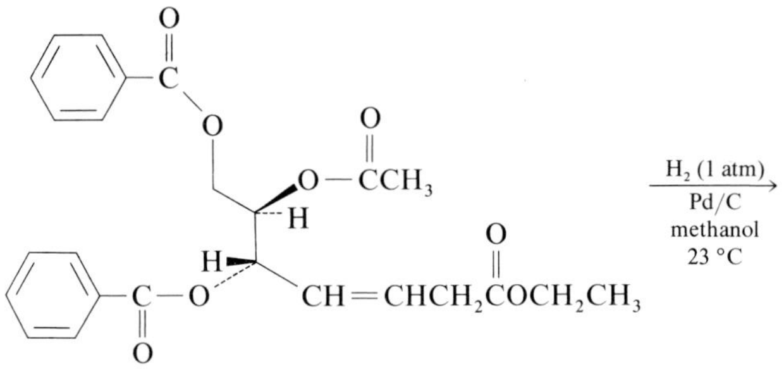

$$\xrightarrow[\substack{Pd/C \\ methanol \\ 23\,°C}]{H_2\ (1\ atm)}$$

2.   Product of 1 $\xrightarrow[\substack{HCl \\ 23\,°C \\ 72\ h}]{CH_3OH}$

Hint: Esters of aromatic carboxylic acids were not affected in this procedure.

3.

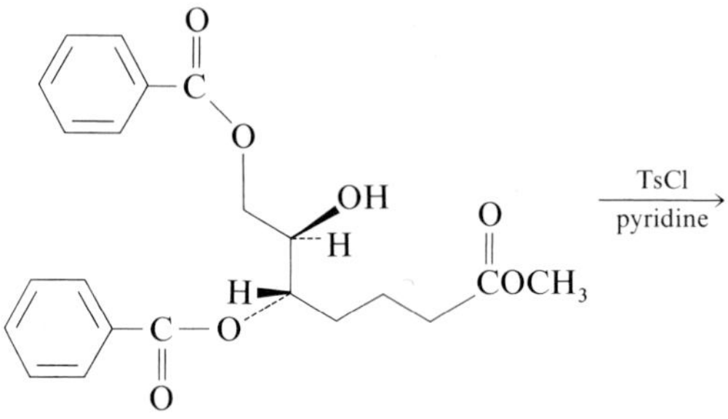

$$\xrightarrow[pyridine]{TsCl}$$

4. Product of 3 $\xrightarrow[\text{methanol}]{\text{K}_2\text{CO}_3 \text{ (excess)}}$ HOCH$_2$

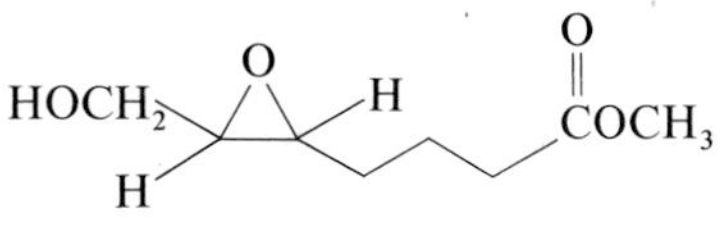

    What is the mechanism of this transformation?

5. Product of 4 $\xrightarrow[\substack{\text{pyridine} \\ \text{dichloromethane}}]{\text{CrO}_3}$

11.41 Predict the products of the following reactions of the terpene chrysanthemic acid and its esters by completing the equations.

(a) 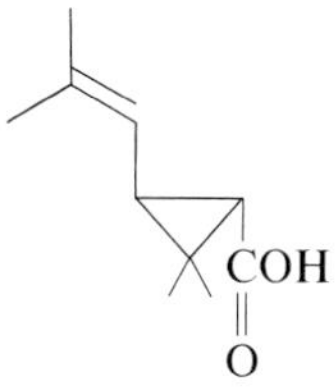 $\xrightarrow[\text{H}_2\text{O}]{\text{NaHCO}_3}$

(b) 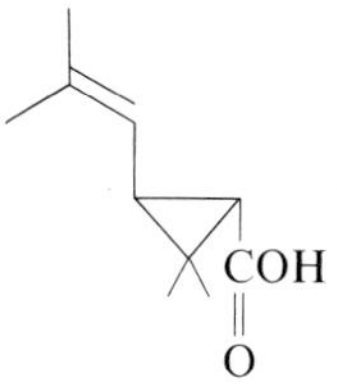 $\xrightarrow[\substack{\text{PtO}_2 \\ \text{acetic acid} \\ 20\,°\text{C}}]{\text{H}_2}$

(c) $\xrightarrow[\text{H}_2\text{SO}_4]{\text{CH}_3\text{OH}}$

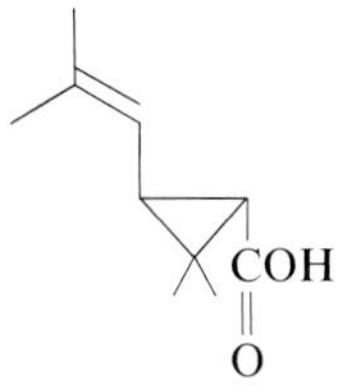

(d) Product of (c) $\xrightarrow[\substack{\text{H}_2\text{O} \\ \text{dioxane}}]{\text{OsO}_4}$

(e) Product of (c) $\xrightarrow{\text{O}_3}$ $\xrightarrow{\substack{\text{Zn} \\ \text{H}_2\text{O}}}$

(f) 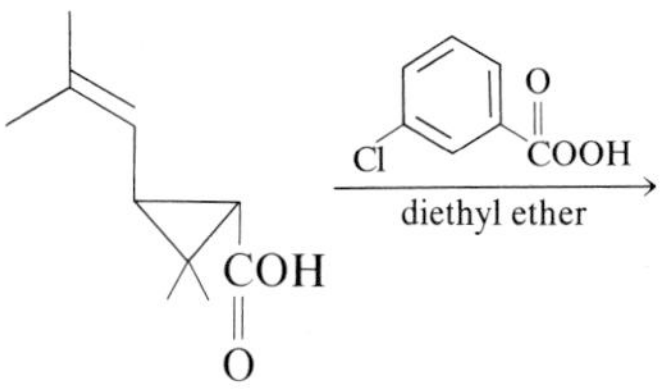 $\xrightarrow[\text{diethyl ether}]{}$

(g) Product of (f) $\xrightarrow[\substack{\text{H}_2\text{O} \\ \Delta}]{\text{CH}_3\text{NHCH}_3}$

11.42 Cholesteryl acetate is converted in the laboratory into the acetic ester of 7-dehydrocholesterol, a compound that is converted by sunlight to vitamin D$_3$ in the skin (Section 19.1B). 7-Dehydrocholesterol has a conjugated diene system in the B ring of the steroid nucleus, with double bonds between carbon atoms 5 and 6 and carbon atoms 7 and 8 (Section 11.6C). How would you convert cholesterol to 7-dehydrocholesteryl acetate?

11.43 A compound isolated from the roots of various *Coreopsis* species was thought to have the following structure.

$$\text{CH}_2{=}\text{CHC}{\equiv}\text{CC}{\equiv}\text{CC}{\equiv}\text{CC}{\equiv}\text{CCH}{=}\text{CH}_2$$

Synthesis of the compound above by an oxidative coupling reaction showed that the natural product did *not* have this structure. Which starting material would you use for the oxidative coupling reaction?

11.44   Two closely related terpene aldehydes, citral a and citral b, $C_{10}H_{16}O$, are isolated from lemon grass oil. Both citral a and citral b on ozonolysis give acetone, 4-oxopentanal, and ethanedial. Treatment of each citral with potassium permanganate gives a polyhydroxy compound, which on vigorous oxidation with chromium trioxide fragments to acetone, 4-oxopentanoic acid, and ethanedioic (oxalic) acid.

What structures are possible for the citrals? Which one is most likely to be the correct one? What must be the difference between the two compounds?

Write an equation showing the polyhydroxy compound that would be formed when citral is treated with potassium permanganate. Be sure that you understand why the polyhydroxy compound fragments the way it does.

11.45   (a)   Bisabolene, $C_{15}H_{24}$, is found widely distributed in nature, especially in myrrh and oil of bergamot. Hydrogenation over platinum in acetic acid gives Compound X, $C_{15}H_{30}$. How many units of unsaturation does bisabolene have? How many of these are rings?

   (b)   Bisabolene undergoes incomplete hydrogenation in cyclohexane to give Compound Y, $C_{15}H_{28}$. Ozonolysis of Y gives 6-methyl-2-heptanone and 4-methylcyclohexanone. What is the structure of Compound Y?

   (c)   Bisabolene itself when ozonized gives, among other products, acetone and 4-oxopentanoic acid. Assign a partial structure on the basis of these data. What uncertainty is there about the structure of bisabolene at this point?

   (d)   Bisabolene and an alcohol derived from bisabolene are among the products obtained from the acid-catalyzed cyclization of nerolidol, obtained from the flowers of bitter orange. The structural formula for nerolidol is shown below.

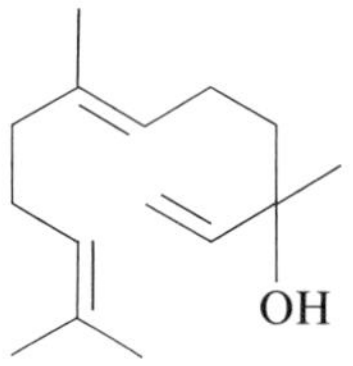

Propose a mechanism for an acid-catalyzed cyclization of nerolidol to bisabolene. What does this suggest about the structure of bisabolene when combined with the earlier data?

   (e)   What would be the structure of the bisabolol also formed in the acid-catalyzed cyclization of nerolidol?

   (f)   Bisabolene reacts easily with hydrogen chloride to give Compound Z, $C_{15}H_{27}Cl_3$. Compound Z is also formed when the mixture of bisabolene and bisabolol from nerolidol is treated with hydrogen chloride. What is the structure of Compound Z? Write equations showing its formation from bisabolene and bisabolol.

11.46   (a)   Compound A is a fragrant oil isolated from bay leaves, with the molecular formula $C_{10}H_{16}$. On prolonged hydrogenation in the presence of platinum, it is converted to Compound B, $C_{10}H_{22}$. How many units of unsaturation does Compound A contain? Does it have any rings?

   (b)   Compound A has λ_{max} 225 nm (ϵ 14,600) and reacts with maleic anhydride to give a Diels-Alder adduct. What do these facts tell us about the structure of Compound A?

   (c)   When Compound A is treated with sodium metal in alcohol, Compound C, $C_{10}H_{18}$, is formed. Compound C reacts with aqueous potassium permanganate to give a liquid polyhydroxy compound. Vigorous oxidation of this compound gives acetone, 4-oxopentanoic acid, and acetic acid. The results from the ozonolysis of Compound C, while less definitive, seem to support such a fragmentation. What structures are possible for Compound C? Which structure is likely to be the correct one?

   (d)   On the basis of the structure for Compound C, what structures are possible for Compound A? (Hint: You may want to review Section 11.1D.)

(e) When the primary product from the ozonolysis of Compound A is oxidized vigorously, succinic acid is obtained as the major fragment. What must the structure of Compound A be? Compound A reacts with only two molar equivalents of bromine. Suggest a structure for its tetrabromo product.

(f) An isomeric terpene, Compound D, has λ_{max} 237 nm. It is reduced by sodium and alcohol to the same Compound C as is Compound A. What is the structure of Compound D? How do you account for the difference in ultraviolet spectra between Compound A and Compound D?

11.47 Racemic disparlure, the sex pheromone of the gypsy moth (p. 553), was synthesized by the following sequence of reactions. Fill in the structures of the products or reagents designated by letters.

$$HC{\equiv}CH \xrightarrow[\text{tetrahydrofuran}]{\text{Na}} A \xrightarrow{B} CH_3(CH_2)_9C{\equiv}CH \xrightarrow[\text{diglyme}]{CH_3CH_2CH_2CH_2Li}$$

$$C \xrightarrow{D} CH_3(CH_2)_9C{\equiv}C(CH_2)_4\overset{\displaystyle CH_3}{\overset{|}{C}}HCH_3 \xrightarrow{E} F \xrightarrow{G} \text{racemic disparlure}$$

$$CH_3\overset{\displaystyle CH_3}{\overset{|}{C}}HCH_2CH_2CH_2CH_2OH \xrightarrow{H} D$$

11.48 (a) 1-Chloro-4,7,10,13-nonadecatetrayne, an intermediate in the synthesis of arachidonic acid (p. 550), has been synthesized by the following sequence of reactions. Fill in structural formulas for the products and intermediates designated by the letters.

$$CH_3(CH_2)_4C{\equiv}CH \xrightarrow[\text{diethyl ether}]{CH_3CH_2MgBr} A$$

$$A\ (2\text{ moles}) \xrightarrow[\text{CuCl, tetrahydrofuran}]{ClCH_2C{\equiv}CCH_2OH\ (1\text{ mole})} B \xrightarrow{H_3O^+} C$$

$$C \xrightarrow[\text{pyridine}]{PBr_3} D$$

$$ClCH_2CH_2CH_2C{\equiv}CCH_2C{\equiv}CH \xrightarrow[\text{diethyl ether}]{CH_3CH_2MgBr} E$$

$$D + E \xrightarrow[\text{CuCl, tetrahydrofuran}]{(\text{p. }528)} \text{1-chloro-4,7,10,13-nonadecatetrayne}$$

(b) What further steps are necessary to convert the tetrayne to arachidonic acid? Write equations showing how you would complete the synthesis.

11.49 The following reactions were carried out. Assign structures to the compounds designated by letters. Spectral data are given for the compounds. Assign as many of the infrared bands as you can.

$\xrightarrow[\text{6 h}]{100\ ^\circ\text{C}}$ A

ν_{max} (carbon tetrachloride) 2844, 1658, 1377, 1178, 1166, 1117 cm^{-1}

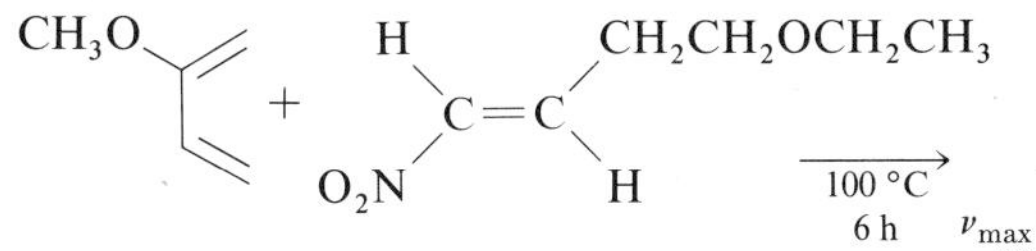

$$A \xrightarrow[\text{ethanol}]{H_3O^+} B$$

ν_{max} (carbon tetrachloride) 2844, 1722, 1350, 1117 cm^{-1}

11.50   Compound F, an intermediate for the synthesis of an antibiotic, was prepared by the following sequence of reactions. Fill in structural formulas for the reagents, intermediates, and products designated by letters.

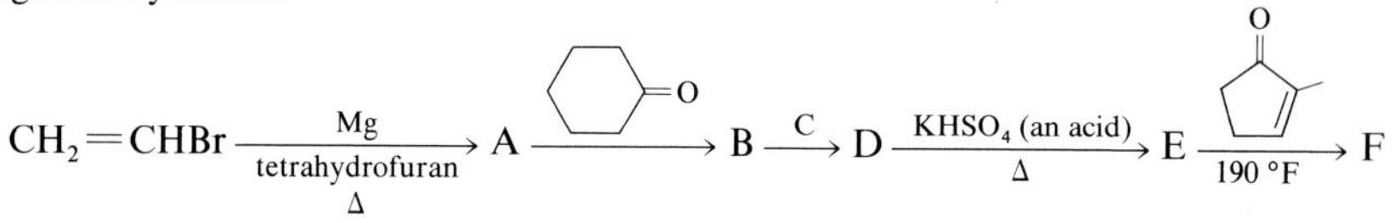

$$CH_2=CHBr \xrightarrow[\substack{\text{tetrahydrofuran} \\ \Delta}]{Mg} A \xrightarrow{\text{(cyclohexanone)}} B \xrightarrow{C} D \xrightarrow[\Delta]{KHSO_4\ (\text{an acid})} E \xrightarrow{190\ °F} F$$

11.51   The following reactions were carried out in the synthesis of the terpene neointermedeol. What are the structures of neointermedeol and of the intermediates in its synthesis?

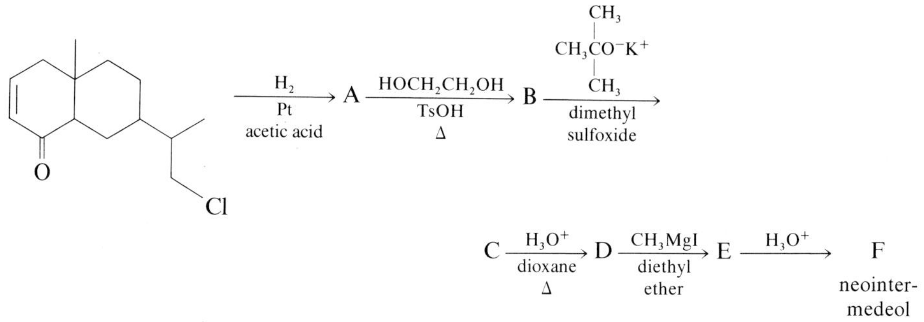

$$\xrightarrow[\substack{Pt \\ \text{acetic acid}}]{H_2} A \xrightarrow[\substack{TsOH \\ \Delta}]{HOCH_2CH_2OH} B \xrightarrow[\substack{CH_3CO^-K^+ \\ \text{dimethyl} \\ \text{sulfoxide}}]{} $$

$$C \xrightarrow[\substack{\text{dioxane} \\ \Delta}]{H_3O^+} D \xrightarrow[\substack{\text{diethyl} \\ \text{ether}}]{CH_3MgI} E \xrightarrow{H_3O^+} F\ \text{neointer-medeol}$$

11.52   Fulvoplumierin is a natural product with antibacterial activity. Several steps in the synthesis of the compound are shown below. Supply the reagents that would be necessary for each step.

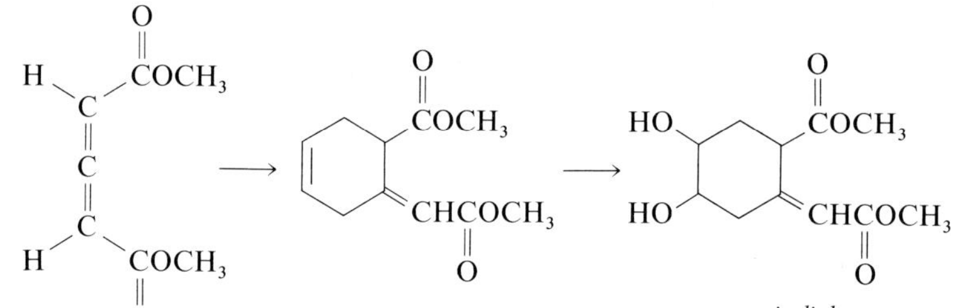

a cis diol

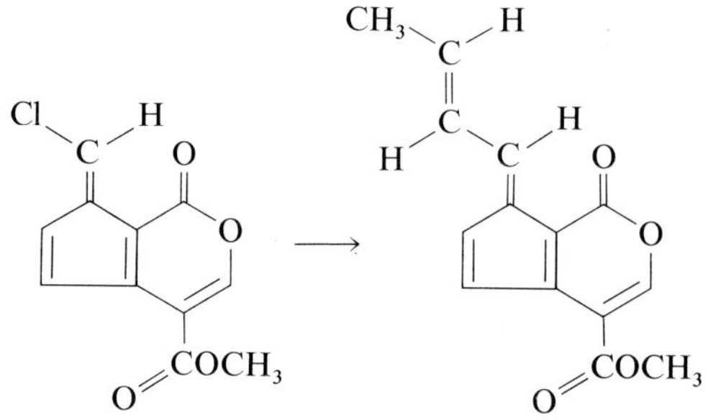

fulvoplumierin

11.53   The monoepoxide of β-carotene (Problem 11.24) rearranges in dilute hydrochloric acid to a compound having the structure shown below. Propose a mechanism for its formation from β-carotene monoepoxide.

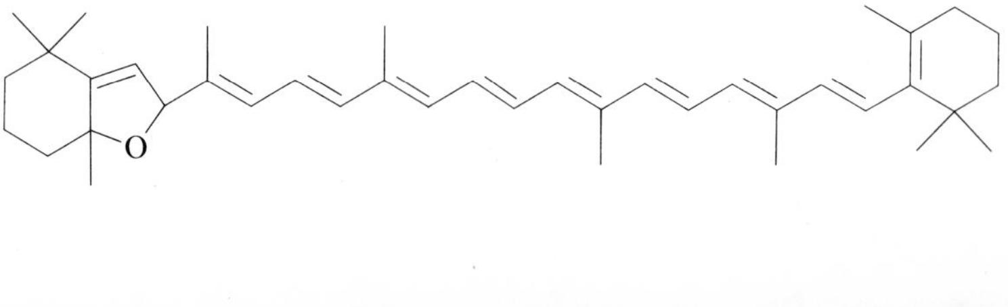

11.54   The following reactions were used in the synthesis of a prostaglandin. Supply the reagents that would be necessary for each transformation.

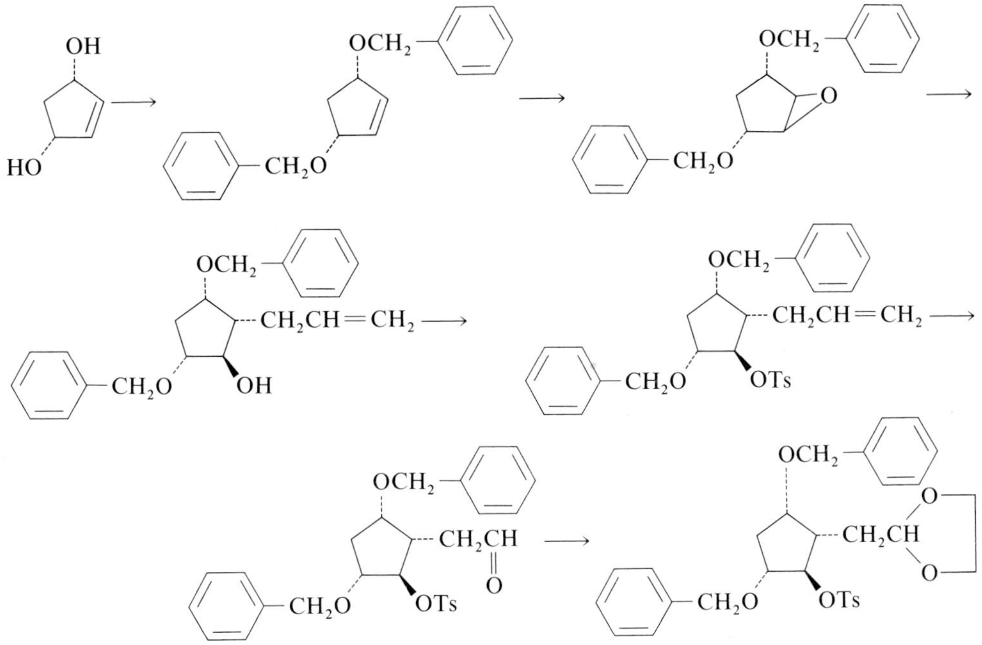

11.55   Steroids, because of the rigidity of their fused six-membered rings, have been used to study the stereochemistry of many reactions.

(a)   Predict which bromonium ion (Section 5.8B) is the intermediate in the addition of bromine to cholesterol (p. 546). What would be the stereochemistry of the dibromide that forms? (Hint: It may help you to review Section 8.5A.)

(b)   Two stereoisomeric bromohydrins derived from a steroid react with base at different rates to give oxiranes (Section 8.4E). Partial structures, A and B, for the two compounds are given below. Predict which compound produces a high yield of oxirane in 30 s and which one requires 24 h to give the same yield. How do you account for the difference in reactivity between the two isomers?

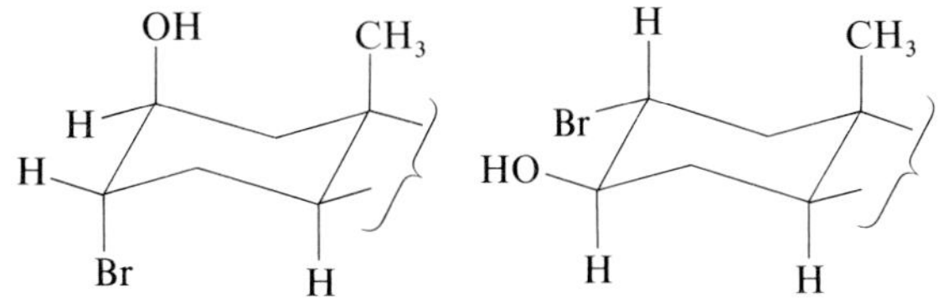

11.56   Two isomeric sesquiterpenes, zingiberene and β-curcumene, can be distinguished by ultraviolet spectroscopy. The first one has λ_{max} 260 nm; no ultraviolet absorption is reported for the other one. Structural formulas for the two compounds are shown below. Decide which one is zingiberene and which is β-curcumene.

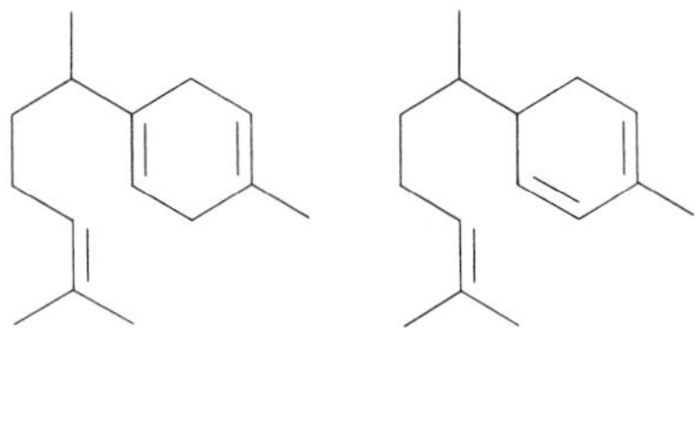

11.57   Three samples, labeled Compounds A, B, and C, are known to be unsaturated carboxylic acids. Ultraviolet spectra were taken of the three samples with the following results: Compound A, λ_{max}^{hexane} 302 nm (ϵ 36,500); Compound B, 208 nm (ϵ 12,000); Compound C, 261 nm (ϵ 25,000). Identify each structural formula given below as belonging to Compound A, B, or C.

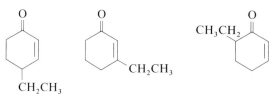

11.58   A compound with the molecular formula C_4H_6O has $\lambda_{max}^{ethanol}$ 219 nm (ϵ 16,600), 318 nm (ϵ 30).

(a)   How many units of unsaturation does the compound have?
(b)   Draw some structures compatible with its molecular formula and its ultraviolet spectrum.
(c)   What are some structures that would not be compatible with the ultraviolet spectrum, even though they fit the molecular formula?

11.59   Of the following three isomeric compounds, which one would you expect to absorb at the highest wavelength?

11.60   Irone, responsible for the fragrance of violets, is a mixture of isomers, for which structural formulas are given below. One of these compounds, β-irone, absorbs at λ_{max} 294.5 nm. The other two, α-irone and γ-irone, have absorption bands at λ_{max} 229 nm and 226.5 nm. Which of the structural formulas below belongs to β-irone?

The Chemistry of Aromatic Compounds

12

Aromaticity

Vanilla and oil of wintergreen are representatives of a class of flavoring materials isolated from plants and containing chemical substances called *aromatic* because of their characteristic fragrances.

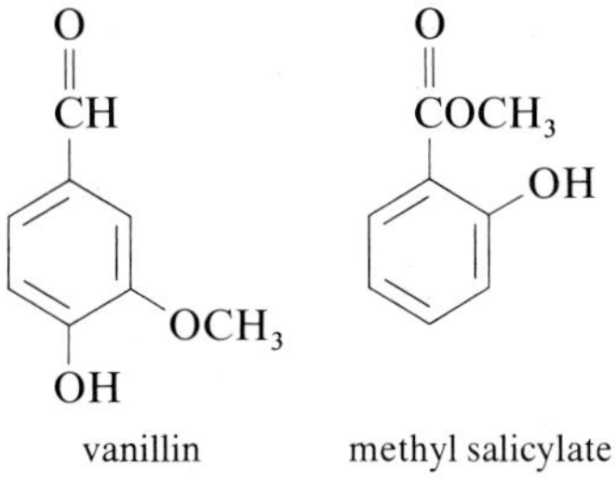

vanillin

constituent of vanilla

methyl salicylate

constituent of oil of wintergreen

When the structures of a number of these compounds were determined, they were all found to contain substituted benzene rings. In time, benzene and its derivatives were called **aromatic hydrocarbons** to distinguish them from saturated acyclic and cyclic hydrocarbons and simple unsaturated compounds. Still later, the term was redefined to include compounds that do not have a benzene ring but have chemical properties resembling those of benzene. Such compounds are said to have **aromaticity.**

Chemists are still arguing about what aromaticity means. At first the term was used to indicate that a compound had special stability, usually defined as *resonance energy,* and that it resisted the addition reactions expected of a molecule with that degree of unsaturation (Section 2.8). Benzene and substituted benzenes are unreactive to a large number of reagents that react with other functional groups. This unreactivity has been demonstrated many times in earlier chapters in reactions that transform substituents on aromatic rings but leave the rings untouched.

In 1926, Sir Robert Robinson, a British chemist who was among the first to suggest and use the ideas that form the basis for the mechanisms we now write for organic reactions, recognized that the compounds that had been classified as aromatic always had six electrons, either in π or nonbonding orbitals, in a planar ring. Benzene, discussed in Sections 2.8–2.10, is such a compound. Robinson suggested that there was a special stability associated with what he called an **aromatic sextet.** Some other compounds that exhibit the properties experimentally assigned to aromaticity, such as resistance to addition reactions, are shown below.

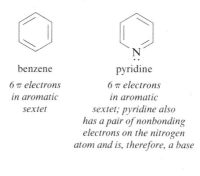

benzene

*6 π electrons
in aromatic
sextet*

pyridine

*6 π electrons
in aromatic
sextet; pyridine also
has a pair of nonbonding
electrons on the nitrogen
atom and is, therefore, a base*

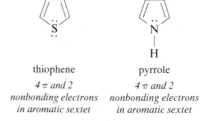

thiophene

*4 π and 2
nonbonding electrons
in aromatic sextet*

pyrrole

*4 π and 2
nonbonding electrons
in aromatic sextet*

When the theory of molecular bonding developed further, it became clear that a sextet was indeed a significant number, corresponding to the number of electrons that fill the three bonding molecular orbitals of benzene (Section 2.9). Any fewer electrons would leave unpaired electrons in the bonding orbitals and create a species with radical character. Any more would put electrons in antibonding molecular orbitals and create an unstable, high-energy species. Benzene has a filled shell of molecular orbitals, just as an element such as neon has a filled shell of atomic orbitals (Figure 12.1).

In 1931, the German chemist Erich Hückel pointed out that there were other numbers of electrons that corresponded to filled shells for smaller or larger rings. He postulated that *any conjugated monocyclic polyene that was planar and had $(4n + 2)\pi$ electrons, with n = 0, 1, 2, etc., would exhibit the special stability associated with aromaticity.* Since that time, researchers have attempted to synthesize compounds that would test the limits of Hückel's prediction.

Benzene is the ideal aromatic compound. It has $(4n + 2)$ or 6 π electrons with $n = 1$. The six carbon atoms in benzene are sp^2 hybridized so that their regular bond angles of 120° create a perfect hexagon. The molecule is planar and totally sym-

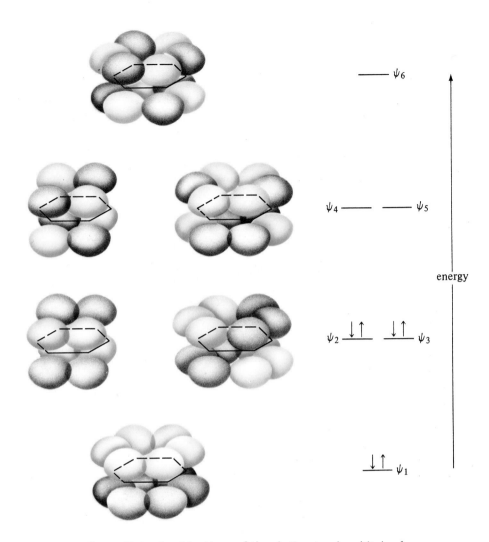

Figure 12.1   Combinations of the six 2*p* atomic orbitals of
benzene that will form the six molecular π
orbitals, and the relative energies of the three
bonding and three antibonding orbitals.

metrical. All the ring atoms are carbon so no differences in polarity are introduced by
the presence of other elements in the ring. The electrons are, therefore, delocalized
evenly over the entire system.

Cyclobutadiene, a conjugated cyclic polyene with four carbon atoms in the ring, is
unstable, and cyclooctatetraene, with an eight-membered ring, behaves as a polyene.

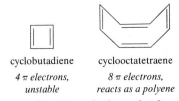

cyclobutadiene   cyclooctatetraene

4 π electrons,   *8 π electrons,*
unstable   *reacts as a polyene*

two monocyclic conjugated polyenes that do not have
(4n + 2) π electrons and are not aromatic

Cyclobutadiene has 4π electrons and thus does not meet the requirements of Hückel's Rule that an aromatic compound have $(4n + 2)$ π electrons. The compound has never been isolated, but has been glimpsed by infrared spectroscopy when it is generated at temperatures such as 4 K. Cyclooctatetraene is an ordinary polyene in its reactivity. It is not planar (a regular octagon would require bond angles of 135°), so there cannot be continuous conjugation of the double bonds around the ring. It, too, lacks the proper number of electrons to be aromatic, according to Hückel's Rule.

Larger rings with conjugated cis double bonds cannot be planar because the bond angles required for such geometrical figures are even larger than the 135° of a regular octagon. However, trans double bonds become possible for large rings. Such a configuration puts hydrogen atoms inside the ring where they interfere with each other and prevent planarity. Only when the ring contains 18 or more carbon atoms is this steric hindrance removed. None of these larger ring compounds, called **annulenes,** has the stability of benzene, although some of them have some of the properties associated with aromaticity, such as high resonance energy. [18]Annulene is shown below.

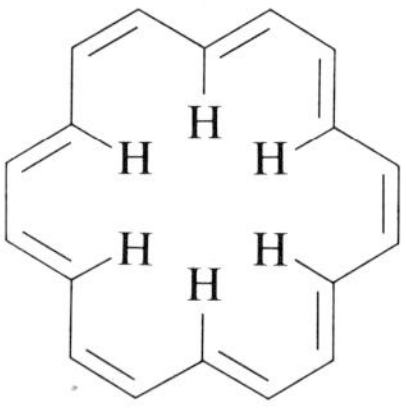

[18]annulene

planar

[18]annulene has $(4n + 2)$ π electrons with $n = 4$, but lacks the stability of benzene

Ions, as well as neutral molecules, can be said to have aromaticity. For example, cyclopentadiene (pK_a 15) is a strong acid when compared with other types of alkenes and can be deprotonated by *tert*-butoxide anion.

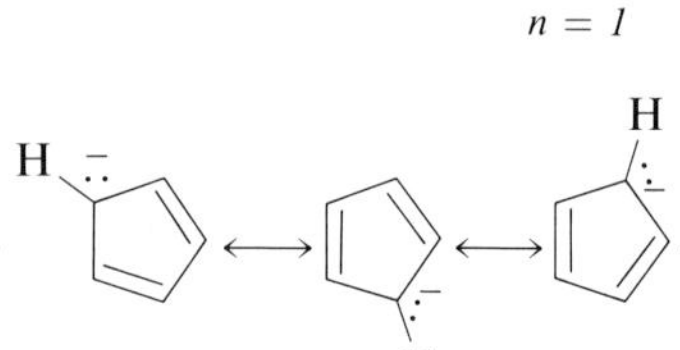

cyclopentadiene

pK_a *15*

cyclopentadienyl anion

$4n + 2 = 6$, $n = 1$

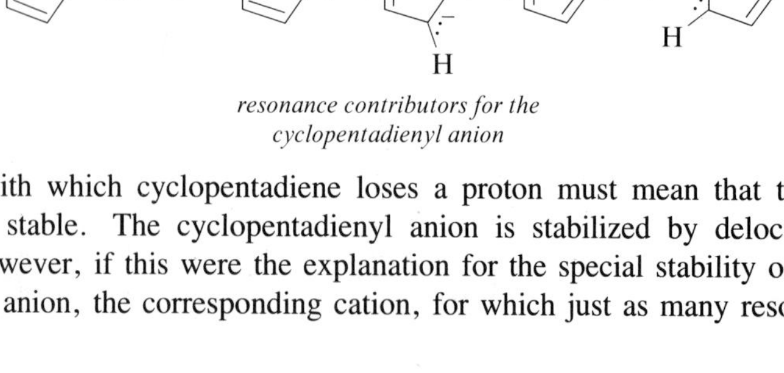

resonance contributors for the cyclopentadienyl anion

The ease with which cyclopentadiene loses a proton must mean that the anion is particularly stable. The cyclopentadienyl anion is stabilized by delocalization of charge. However, if this were the explanation for the special stability of the cyclopentadienyl anion, the corresponding cation, for which just as many resonance contributors can be written, should also be stable. But attempts to prepare such a cation have failed. The cyclopentadienyl anion has six electrons in a planar ring with the possibility of continuous delocalization of the electrons. It meets the criteria of Hückel's Rule. The cyclopentadienyl cation, on the other hand, is a 4π electron system and is expected to be unstable.

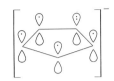
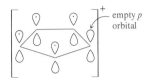

p orbitals of cyclopentadienyl anion
showing the presence
of 6 π electrons
(4n + 2, n = 1)

p orbitals of cyclopentadienyl cation
showing the presence
of only 4 π electrons

Cycloheptatriene is a compound with six π electrons in a ring. It is not aromatic because there cannot be continuous delocalization of the π electrons around the ring as long as one of the carbon atoms is an sp^3-hybridized carbon atom with no p orbital on it to overlap with the p orbitals on the other carbon atoms. Cycloheptatriene adds bromine, then easily loses hydrogen bromide to give a crystalline compound that is insoluble in nonpolar organic solvents such as ether but dissolves readily in cold water and gives an instant precipitate of silver bromide when silver nitrate is added to its solution. The compound melts at 203 °C with decomposition. All of these properties suggest that the monobromo compound ionizes as shown.

In other words, the compound is the salt of a stable carbocation, usually known as the **tropylium ion,** with bromide ion. The aromaticity possible for the cation so stabilizes it that the normal tendency to form covalent bonds between carbon and bromine is overcome.

PROBLEM 12.1   Write a mechanism showing the formation of tropylium bromide from the dibromide of 1,3,5-cycloheptatriene.

PROBLEM 12.2   What other dibromides of 1,3,5-cycloheptatriene besides the 1,6 addition product would be possible?

PROBLEM 12.3   Draw a representation of the tropylium ion showing the σ framework of the molecule and the *p* orbitals used for delocalization of charge. Include all of the π electrons available to the ion.

PROBLEM 12.4   Decide whether each of the following species would be aromatic or not.

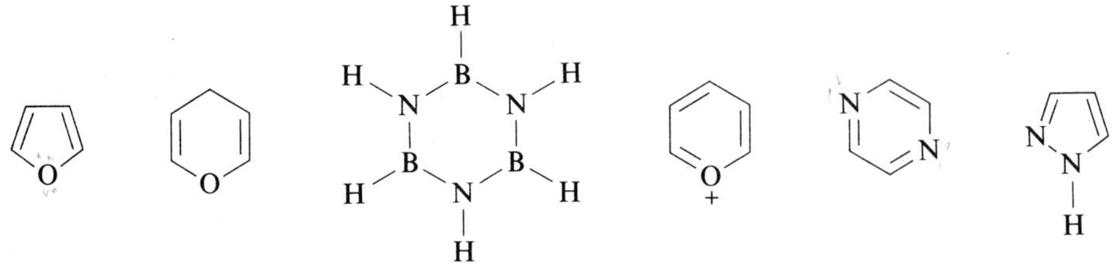

12.2

Kekulé Structures and Nomenclature for Aromatic Compounds

Michael Faraday isolated benzene in 1825 from the residues of gaseous fractions of coal. The structural formula that we now use for benzene was first written in 1865 by the German chemist August Kekulé after a dream in which he saw rows of atoms twisting with a snakelike motion until finally, to quote Kekulé, "One of the snakes had seized hold of its own tail, and the form whirled mockingly before my eyes. As if by a flash of lightning, I awoke, and this time also I spent the rest of the night in working out the consequences of the hypothesis." Kekulé also recognized that the chemical properties of benzene were not compatible with the alternating double and single bonds indicated by his structural formula and had by 1872 arrived at the idea that the double bonds were not localized. He wrote two structures, the ones that we now call resonance contributors for benzene (Section 2.9), to show the changing nature of the double bonds, a remarkable feat of chemical imagination for his time, when electrons had not yet been discovered.

Along with benzene, other hydrocarbons similarly unsaturated in their molecular formulas, but stable towards reagents that add to double bonds, were isolated from various sources, chiefly from fractions of coal tar. Some of these compounds are shown below with their names.

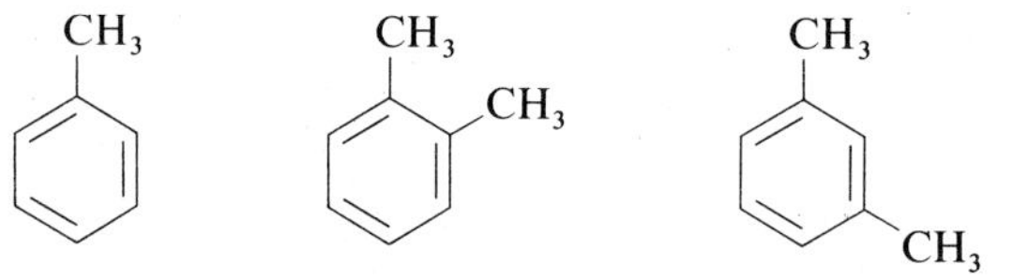

| methylbenzene | 1,2-dimethylbenzene | 1,3-dimethylbenzene |
| toluene | *o*-xylene | *m*-xylene |

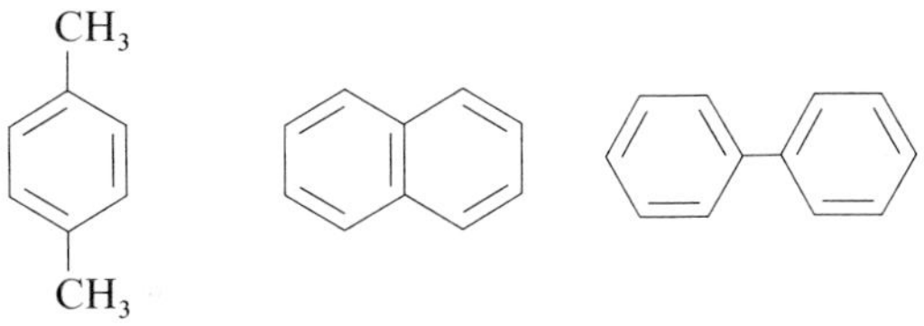

1,4-dimethylbenzene   naphthalene   biphenyl
p-xylene

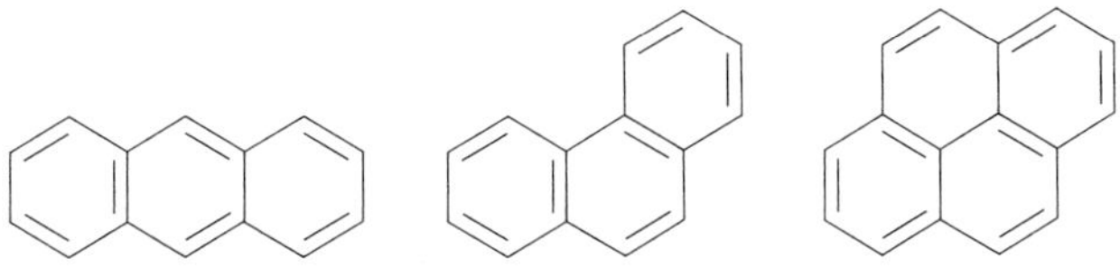

anthracene   phenanthrene   pyrene

Benzene, other aromatic hydrocarbons, and their alkyl derivatives as a class are known as **arenes.** Besides hydrocarbons, coal tar also contains aromatic compounds containing nitrogen, sulfur, and oxygen, such as the following. Those having elements other than carbon in the ring are also known as **heterocyclic compounds.** For example, pyridine, quinoline, indole, and thiophene, shown below, are heterocyclic.

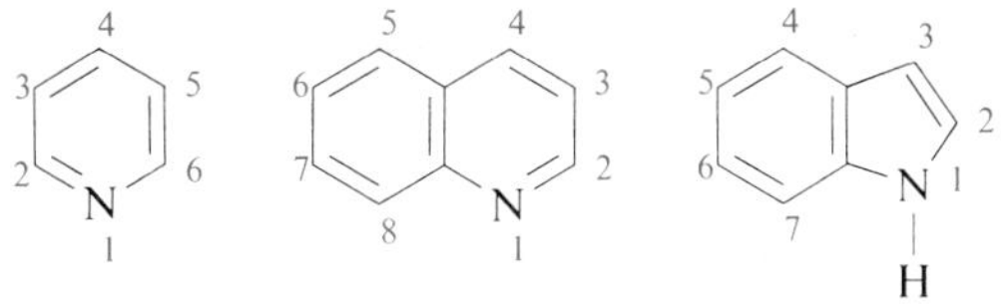

pyridine   quinoline   indole

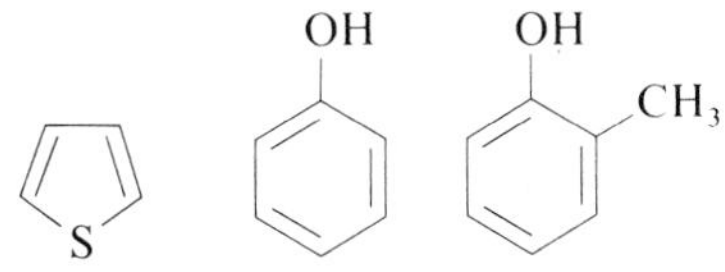

thiophene   phenol   *o*-cresol

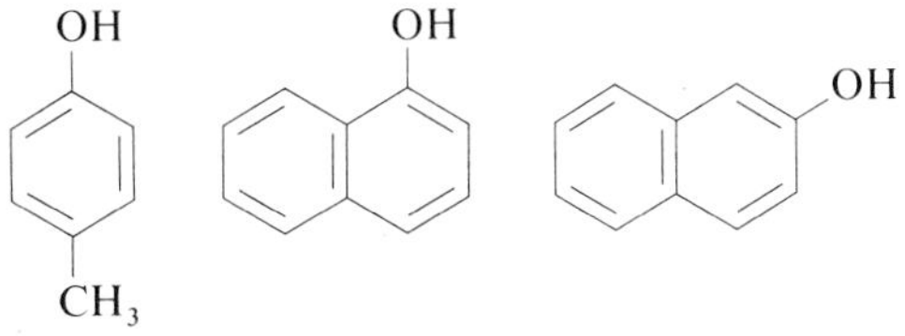

p-cresol   1-naphthol   2-naphthol
α-naphthol   β-naphthol

Other aromatic compounds are named as substitution products of the compounds shown above. The name of the parent compound takes on a prefix with the name of each substituent and a number indicating its position relative to the group that is already present. For benzene derivatives, the prefixes ortho, meta, and para are also acceptable to designate a 1,2-, 1,3-, or a 1,4- relationship between two substituents (Section 9.1C). Some examples of a variety of aromatic compounds will illustrate these rules.

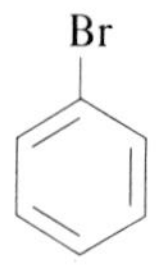

bromobenzene

1,4-dichlorobenzene
p-dichlorobenzene

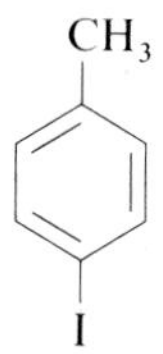

4-iodotoluene
p-iodotoluene

methyl group defines carbon 1

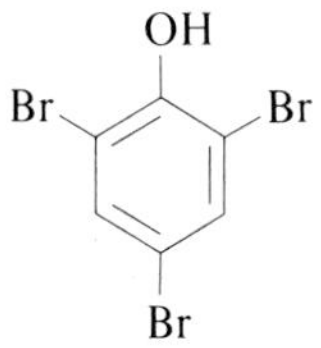

2,4,6-tribromophenol

hydroxyl group defines carbon 1

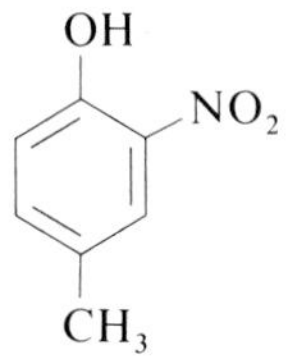

4-methyl-2-nitrophenol

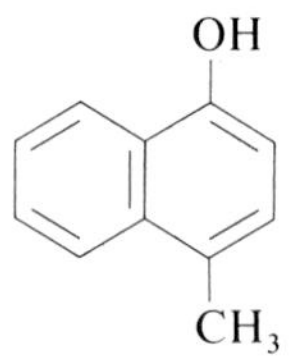

4-methyl-1-naphthol

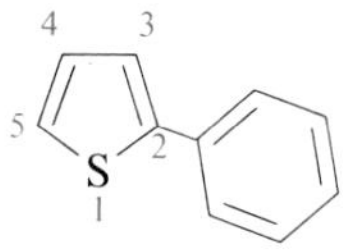

2-phenylthiophene

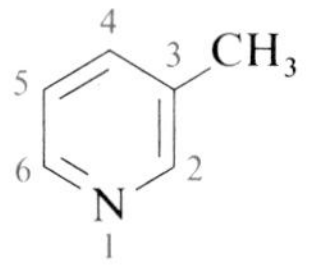

3-methylpyridine

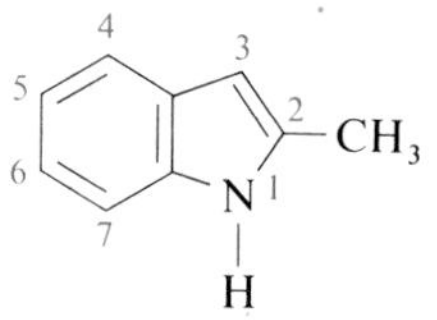

2-methylindole

As you can see from the examples above, some compounds contain more than one aromatic ring. In a case such as biphenyl, (p. 581) the two rings are joined by a single bond and are conjugated with each other. In other cases, such as naphthalene, anthracene, phenanthrene, and pyrene, the aromatic rings share sides and are said to contain **fused-ring systems,** or to be **polycyclic aromatic hydrocarbons.**

The fused-ring aromatic hydrocarbons no longer have completely identical resonance contributors, as benzene does. For example, the structural formulas with localized double and single bonds, the Kekulé structures, that can be written as resonance contributors for naphthalene are not equivalent.

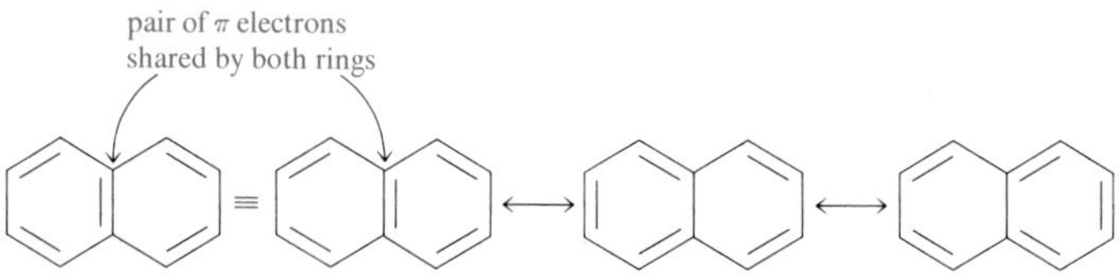

pair of π electrons shared by both rings

resonance contributors for naphthalene

The nonequivalence of resonance contributors is reflected in the chemical properties of polycyclic aromatic hydrocarbons. Anthracene and phenanthrene clearly illustrate the difference in reactivity between rings in such compounds. The center ring in anthracene is highly reactive. For example, anthracene reacts with maleic anhydride in a Diels-Alder reaction as if its central ring were a diene (Section 11.5).

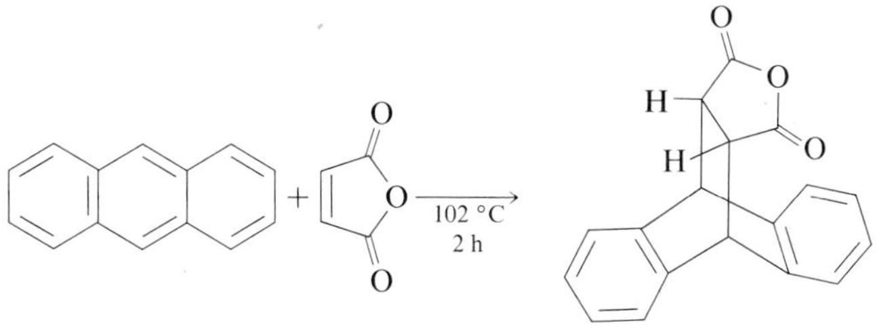

anthracene   maleic anhydride   Diels-Alder adduct of
anthracene and maleic anhydride
67%

Note that when maleic anhydride adds to the central ring, the two end rings remain aromatic. Similarly, when bromine is added to anthracene, it adds to the central ring in a reaction that resembles 1,4-addition to a diene (Section 11.1C).

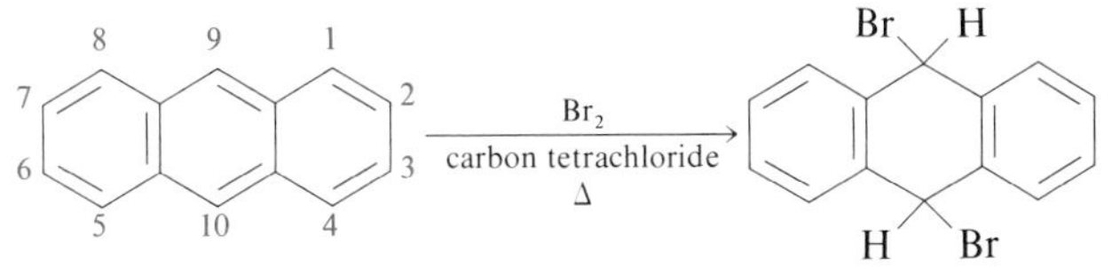

anthracene   9,10-dibromoanthracene

Phenanthrene also has a central ring that is more vulnerable to chemical reagents than the other two rings. Of the resonance contributors that can be written for phenanthrene, most have a double bond localized between the two carbon atoms usually numbered 9 and 10 for the phenanthrene system. Phenanthrene adds bromine to these positions on the ring as if it were an alkene (Section 5.8).

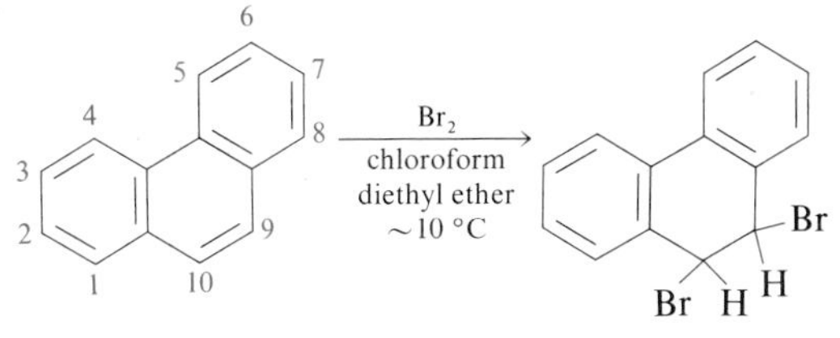

phenanthrene   9,10-dibromophenanthrene

Thus, while it is possible to talk in theoretical terms of molecular energy levels and the delocalization of electrons over entire molecules, Kekulé structures, with their reminder of alternating double and single bond character, are important when dealing with the experimental facts of the chemistry of aromatic compounds. Indeed, chemists have been unable to synthesize compounds for which electron delocalization over large molecules is possible, but for which Kekulé structures cannot be written. For these reasons, we retain the Kekulé structures for benzene and do not adopt the convention of a circle in a hexagon that is sometimes used to symbolize benzene.

PROBLEM 12.5 Name the following compounds.

(a) 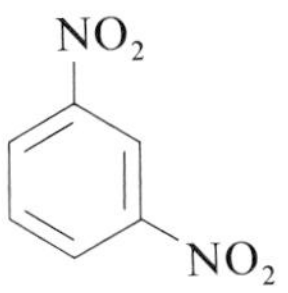   (b) 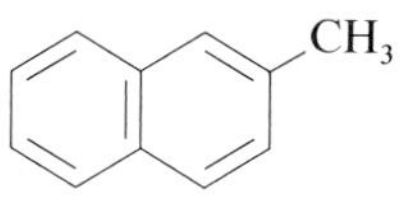   (c)

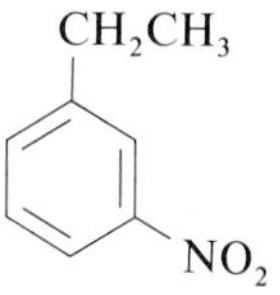

(d)    OH ... Cl

(e) 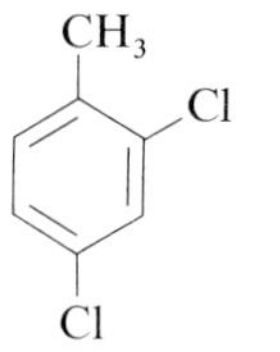   CH$_3$ ... Cl ... Cl

(f) 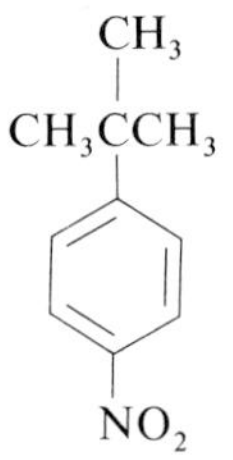   CH$_3$CCH$_3$ with CH$_3$ ... NO$_2$

PROBLEM 12.6   Draw structural formulas for the following compounds.

(a)   2,4-dinitrophenol   (b)   2,4,6-trimethylpyridine   (c)   8-hydroxyquinoline
(d)   2,4,6-trinitrotoluene (TNT)   (e)   4-bromo-2-methylnaphthalene
(f)   4,4'-dibromobiphenyl   (g)   8-methyl-1-naphthol   (h)   3-ethylthiophene
(i)   2,4-dibromo-1-pentylbenzene

PROBLEM 12.7   Draw resonance contributors for phenanthrene. Prove to yourself that the 9,10 bond in phenanthrene would be expected to behave more like a double bond than any of the other bonds in the molecule.

12.3

Electrophilic Aromatic Substitution Reactions

A. Substitution Reactions of Benzene. Experimental Observations

Benzene is rich in electrons, yet inert towards electrophilic addition reactions. Addition reactions would break up the aromatic sextet that gives benzene its special stability. *Benzene and other aromatic hydrocarbons will react,* however, *with electrophilic reagents in reactions in which an incoming group substitutes for one of the hydrogen atoms on the ring. Such reactions are called* **electrophilic aromatic substitution reactions.**

For example, benzene reacts with bromine in the presence of a catalyst such as iron or a Lewis acid such as aluminum chloride to give chiefly bromobenzene along with small amounts of *o*-dibromobenzene and *p*-dibromobenzene.

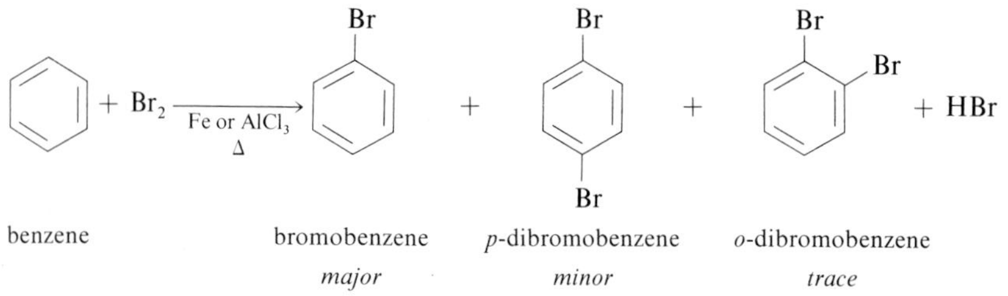

benzene   + Br$_2$ $\xrightarrow[\Delta]{\text{Fe or AlCl}_3}$ bromobenzene   +   *p*-dibromobenzene   +   *o*-dibromobenzene   + HBr

major   *minor*   *trace*

The other product of the reaction is hydrogen bromide.

In a similar fashion, benzene, if treated with nitric acid in concentrated sulfuric acid, is converted to nitrobenzene.

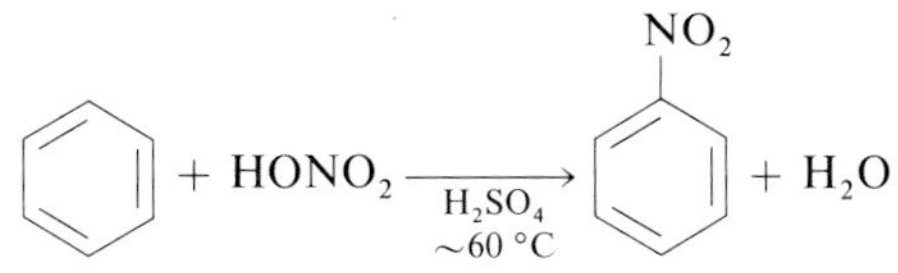

$$\text{benzene} + HONO_2 \xrightarrow[\sim 60\ ^\circ C]{H_2SO_4} \text{nitrobenzene} + H_2O$$

benzene   nitric acid   nitrobenzene

Different aromatic compounds vary in their susceptibility towards electrophilic substitution reactions. For example, benzene requires liquid bromine, a Lewis acid catalyst, and temperatures of approximately 80 °C to form bromobenzene. Phenol reacts with a dilute solution of bromine in acetic acid at 30 °C.

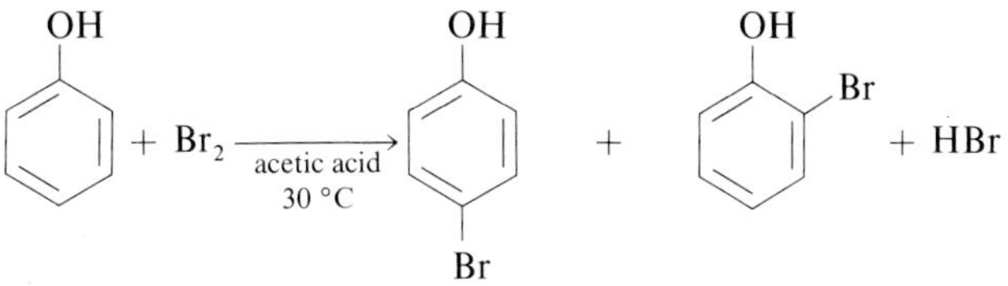

$$\text{phenol} + Br_2 \xrightarrow[30\ ^\circ C]{\text{acetic acid}} p\text{-bromophenol} + o\text{-bromophenol} + HBr$$

phenol   *p*-bromophenol   *o*-bromophenol   + HBr
                    88%   12%

p-Bromophenol is the major product of the reaction, along with some *o*-bromophenol. If phenol is treated with bromine in water, 2,4,6-tribromophenol is formed instantly.

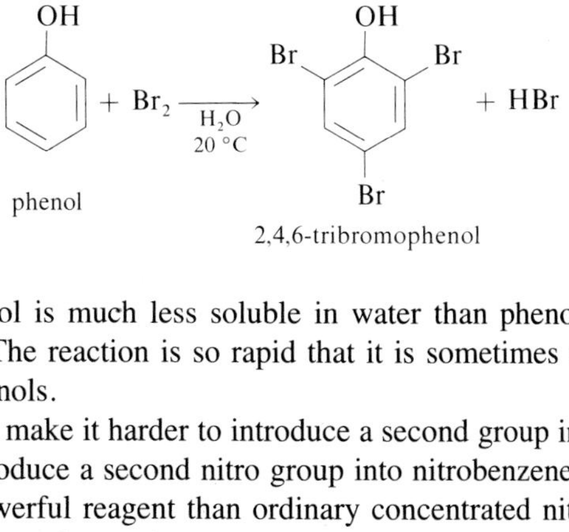

$$\text{phenol} + Br_2 \xrightarrow[20\ ^\circ C]{H_2O} \text{2,4,6-tribromophenol} + HBr$$

phenol   2,4,6-tribromophenol

2,4,6-Tribromophenol is much less soluble in water than phenol is and precipitates from the solution. The reaction is so rapid that it is sometimes used as a qualitative analysis test for phenols.

Some substituents make it harder to introduce a second group into an aromatic ring. For example, to introduce a second nitro group into nitrobenzene, fuming nitric acid, which is a more powerful reagent than ordinary concentrated nitric acid, and a temperature of 100 °C must be used.

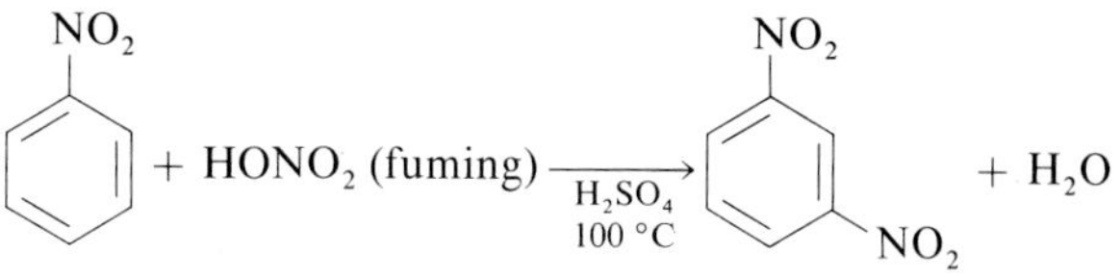

$$\text{nitrobenzene} + HONO_2 \text{ (fuming)} \xrightarrow[100\ ^\circ C]{H_2SO_4} m\text{-dinitrobenzene} + H_2O$$

nitrobenzene nitric acid   *m*-dinitrobenzene
                                 88%

These conditions are more severe than those used to introduce the first nitro group into benzene (above). The second nitro group takes a position meta to the first one.

On the other hand, phenol can be nitrated with dilute nitric acid at room temperature. The products are *o*-nitrophenol and *p*-nitrophenol.

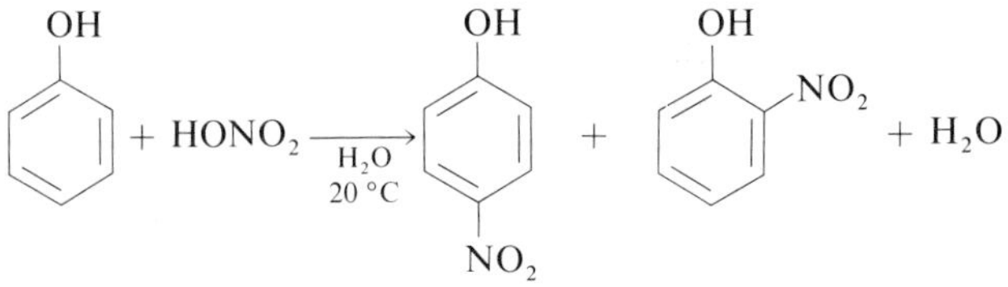

phenol	nitric acid	*p*-nitrophenol 60%	*o*-nitrophenol 40%

The experiments summarized in the equations above point to two important observations. First, *a group already present on the aromatic ring may make it easier or harder to introduce a second substitutent onto the ring. A group that makes it easier to introduce new substituents is said to* **activate** *the ring towards electrophilic substitution.* The hydroxyl group in phenol, for example, is a ring-activating substituent. *Groups that make the introduction of a second substituent harder are said to* **deactivate** *the ring towards electrophilic substitution.* The nitro group is such a ring-deactivating substituent.

The second observation is that *the position that a second substituent takes on the ring is influenced by the group that is already there.* When phenol undergoes further electrophilic substitution, the new group takes up a position ortho or para to the hydroxyl group. The hydroxyl group is said to **direct** the incoming substituent to those positions. The nitro group, on the other hand, directs the new substituent to the meta position.

Other substituents on the aromatic ring generally fall into one or the other of these categories. Some are ortho,para directing, others are meta directing. For example, an alkyl group on the aromatic ring is predominantly ortho,para directing, as shown by the nitration of ethylbenzene.

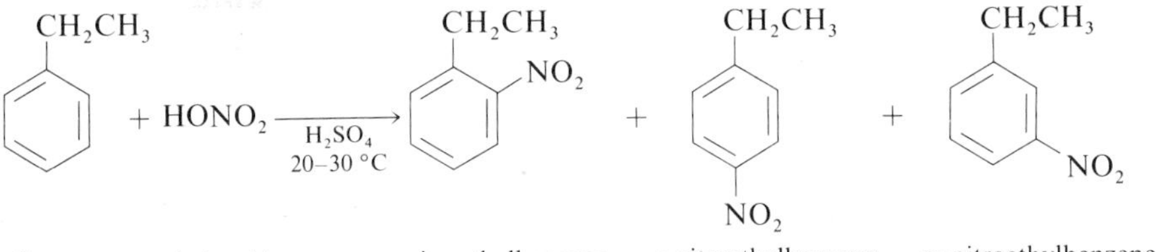

ethylbenzene	nitric acid	*o*-nitroethylbenzene 45%	*p*-nitroethylbenzene 48%	*m*-nitroethylbenzene 7%

The reaction gives a mixture, the chief components of which are *o*-nitroethylbenzene and *p*-nitroethylbenzene in approximately equal amounts. Very little meta substituted product is formed.

When methyl benzoate is nitrated, on the other hand, methyl *m*-nitrobenzoate is formed as the major product.

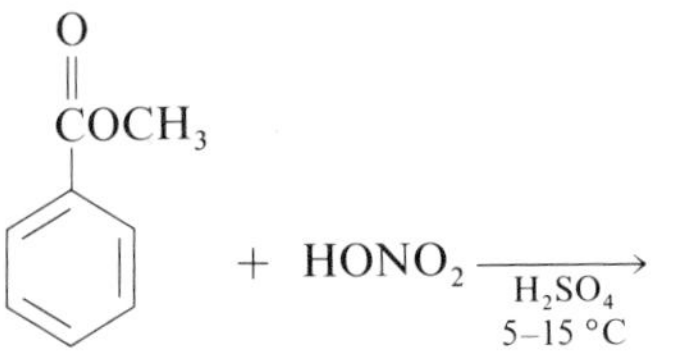

methyl benzoate	nitric acid	methyl *m*-nitrobenzoate ~85%

Thus, the ester function on the aromatic ring is meta directing. This reaction, however, does not require the harsh conditions necessary to introduce a second nitro group into nitrobenzene.

The results of many experiments like the ones shown in the equations in this section are summarized in Table 12.1 in which different substituents on the aromatic ring are classified as ortho,para directing and meta directing.

Most of the ortho,para-directing groups, except for the halogens, activate the ring towards further substitution. This effect is strong only for the hydroxyl group and for the unprotonated amino group. An examination of the structures of the groups that are ortho,para directing shows that except for alkyl groups, at least one pair of nonbonding electrons is available on the atom directly bonded to the aromatic ring. All the meta-directing groups either bear a positive charge on the atom directly bonded to the aromatic ring or have a structure that can be polarized to put partial positive charge there. The meta-directing groups at the top of the list are strongly ring deactivating. Experimentally, the others allow substitution reactions under relatively mild conditions even though they are meta directing. In other words, they do not deactivate the ring strongly.

In the next section, we examine the mechanism proposed to explain the trends summarized above.

TABLE 12.1 **Substituents on Aromatic Rings Classified as Ortho,Para- and Meta-Directing Groups**

Ortho,Para Directing		*Meta Directing*	
$-N(CH_3)_2$	very strongly activating	$-\overset{+}{N}(CH_3)_3$	very strongly deactivating
$-NH_2$		$-NO_2$	
$-OH$		$-C \equiv N$	
$-OCH_3$		$-SO_3H$	
$\underset{\displaystyle -NHCCH_3}{\overset{\displaystyle O \atop \|}{}}$		$\underset{\displaystyle -CH}{\overset{\displaystyle O \atop \|}{}}$	
$\underset{\displaystyle -OCCH_3}{\overset{\displaystyle O \atop \|}{}}$		$\underset{\displaystyle -CCH_3}{\overset{\displaystyle O \atop \|}{}}$	
$-R$		$\underset{\displaystyle -COH}{\overset{\displaystyle O \atop \|}{}}$	
$-Cl, Br, I$	mildly deactivating	$\underset{\displaystyle -COCH_3}{\overset{\displaystyle O \atop \|}{}}$	
		$\underset{\displaystyle -CNH_2}{\overset{\displaystyle O \atop \|}{}}$	
		$-NH_3^+$	

PROBLEM 12.8 Complete the following equations.

(a) 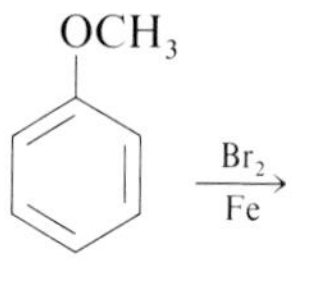 phenyl with OCH$_3$ substituent, $\xrightarrow[\text{Fe}]{\text{Br}_2}$

(b) 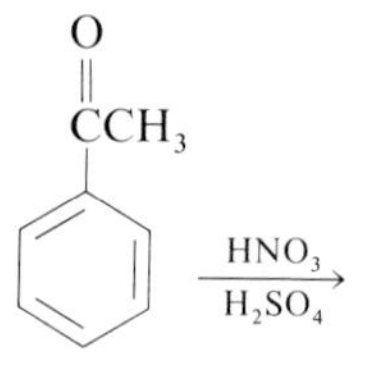 phenyl with $\overset{\displaystyle O}{\overset{\|}{C}}CH_3$ substituent, $\xrightarrow[\text{H}_2\text{SO}_4]{\text{HNO}_3}$

(c) phenyl with $\overset{\displaystyle O}{\overset{\|}{C}}OH$ substituent, $\xrightarrow[\text{Fe}]{\text{Br}_2}$

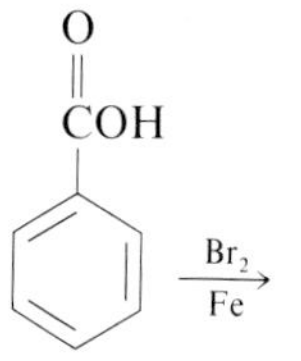

(d) 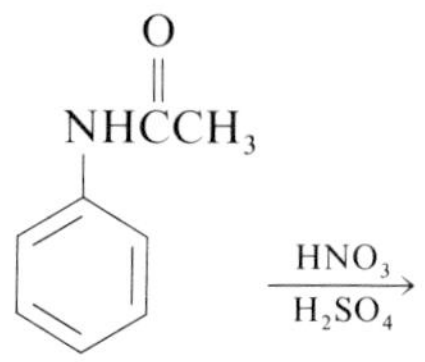 phenyl with $NH\overset{\displaystyle O}{\overset{\|}{C}}CH_3$ substituent, $\xrightarrow[\text{H}_2\text{SO}_4]{\text{HNO}_3}$

(e) 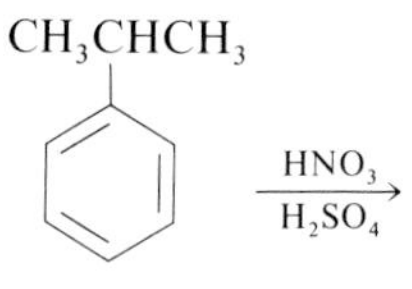 phenyl with CH_3CHCH_3 substituent, $\xrightarrow[\text{H}_2\text{SO}_4]{\text{HNO}_3}$

B. The Mechanism of Electrophilic Aromatic Substitution

Bromine and strong acids, as examples of reagents that introduce substituents on the aromatic ring, are electrophilic, electron-seeking reagents. Unless the ring is strongly activated, a Lewis acid must also be used with bromine. For example, a piece of iron is added to the reaction vessel in the bromination of benzene (p. 584). Iron reacts with bromine to give ferric bromide, which is a Lewis acid. The Lewis acid coordinates with the nonbonding electrons of bromine to polarize the bromine-bromine bond, making one of the bromine atoms more electrophilic and the other one a better leaving group.

$$3\,\text{Br}_2 + 2\,\text{Fe} \longrightarrow 2\,\text{FeBr}_3$$

bromine   iron   ferric bromide

a Lewis acid

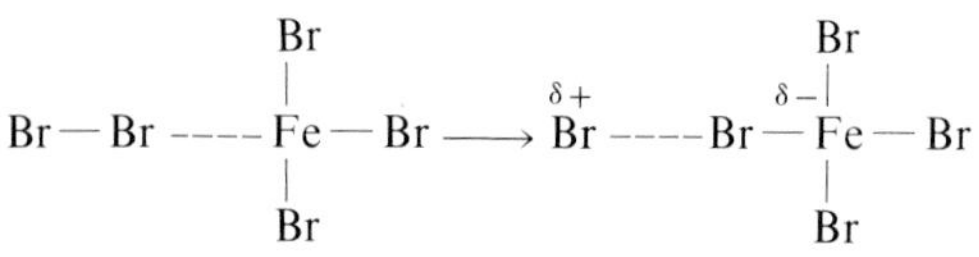

$$\text{Br}-\text{Br} ---- \overset{\displaystyle \text{Br}}{\underset{\displaystyle \text{Br}}{\text{Fe}}}-\text{Br} \longrightarrow \overset{\delta+}{\text{Br}} ---- \overset{\delta-}{\text{Br}} - \overset{\displaystyle \text{Br}}{\underset{\displaystyle \text{Br}}{\text{Fe}}}-\text{Br}$$

a complex of bromine
with a Lewis acid

The π electrons of benzene are attracted to the bromine atom at which positive charge is developing. A covalent bond forms between a carbon atom in benzene and a bromine atom as the other bromine atom leaves with the Lewis acid.

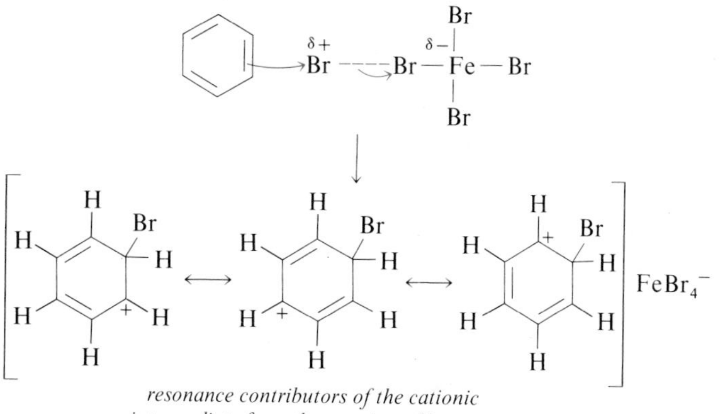

*resonance contributors of the cationic
intermediate from the reaction of benzene
with an electrophilic bromine atom*

The resulting species is a cation that loses a proton readily to restore the aromatic ring. Bromide ion (or $FeBr_4^-$) is a base strong enough to remove a proton from such an intermediate.

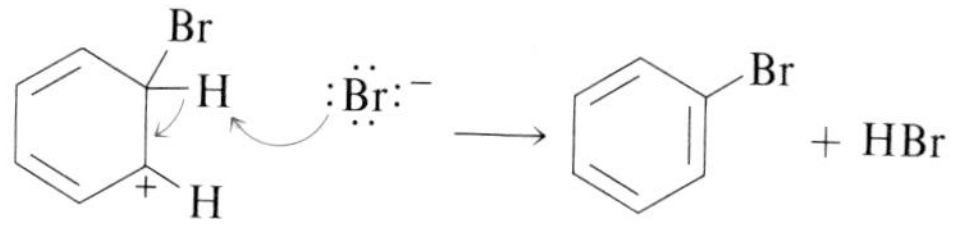

bromobenzene

The equations above represent *the essential steps in all electrophilic substitution reactions. The substituting reagent is polarized or ionized in such a way as to create an electron-deficient species. A cationic intermediate forms when bonding occurs between the electrons of the aromatic ring acting as the nucleophile and the electrophilic reagent. A proton is lost from the carbocation to restore the aromatic ring in which a substituent has replaced a hydrogen atom.*

In the nitration reaction, the electrophile is the nitronium ion, NO_2^+, created by protonation of nitric acid by the strong acid, sulfuric acid, and loss of water.

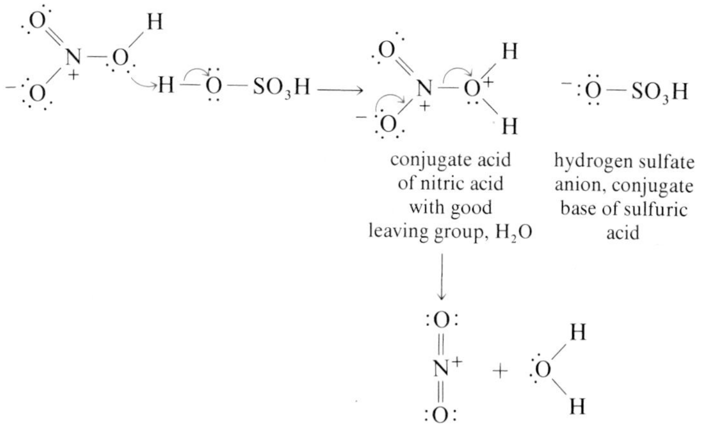

| conjugate acid of nitric acid with good leaving group, H_2O | hydrogen sulfate anion, conjugate base of sulfuric acid |

nitronium
ion

Nitronium ion reacts with the π electrons of the aromatic ring to give a cationic intermediate, which loses a proton to recreate the aromatic sextet.

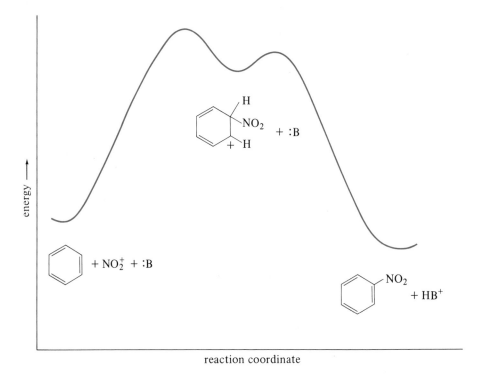

The rate-determining step in electrophilic aromatic substitution reactions is the formation of the carbocation intermediate. In going from benzene and nitronium ion to the transition state, we are not only breaking a bond but disrupting the aromaticity in the benzene ring as well. The energy barrier for this reaction is high (Figure 12.2).

The loss of a proton from the intermediate is a fast reaction. We expect this because the aromatic sextet is regenerated. It is possible to test whether the loss of the proton is the rate-determining step by substituting deuterium atoms for the hydrogen atoms on the aromatic ring. A deuterium atom is chemically similar to a hydrogen atom because it, too, has one proton and one electron. A deuterium atom, however, has twice the mass of a hydrogen atom. The breaking of a covalent bond results from an increase in the vibrational energy (Sections 7.7A and 7.7B) of the bond to a level represented by the bond dissociation energy (Section 5.4B). The vibrational energy, in turn, depends on the masses of the atoms involved in the bond. The vibrational energy of a bond is lower when the atoms in the bond have greater mass. Therefore, a $C-D$ bond has a lower ground state vibrational energy than a $C-H$ bond. For this reason, a reaction involving the breaking of a bond between carbon and deuterium has a higher

Figure 12.2 An energy diagram for the nitration of benzene.

energy of activation and is a slower reaction than the breaking of a carbon-hydrogen bond. If the removal of the proton from the cationic intermediate were the rate-determining step, we would expect to see a decrease in the rate if deuteriobenzene, C_6D_6, were used in the nitration reaction. The experiment has been tried, and no difference has been observed in the rate of nitration between C_6H_6 and C_6D_6 for a variety of reaction conditions. This experiment confirms the postulate that the formation of the cation should be the step with the highest energy of activation.

The reactivity of the aromatic ring towards acids is demonstrated when benzene, in which deuterium has been substituted for one of the hydrogen atoms, is put into aqueous sulfuric acid at room temperature.

deuteriobenzene   benzene

The deuterium atom is exchanged for a hydrogen atom. This observation means that in an acid medium the aromatic ring reacts constantly with protons. The protonation and deprotonation of the ring is usually not detected because the starting material and the product of the reaction are the same. The reaction can be detected, however, when the ring is labeled with isotopes of hydrogen. The reaction intermediate for this exchange reaction is a cation, arising from reaction of a proton with the π electrons of the aromatic ring.

C. Orientation in Electrophilic Aromatic Substitution

How does the mechanism proposed in Section 12.3B account for the different orientations seen for the incoming substituents when the aromatic ring already has a group on it? This is a question of the regioselectivity of a reaction, a question that has come up again and again in many different contexts. For example, we explored the addition of hydrogen bromide to propene in great detail (Section 5.3A); we studied which of all possible alkenes would form in elimination reactions (Section 6.9); and we wondered about how diborane added to double bonds (Section 7.3D). In each case, our answer came from a consideration of the transition states that would lead to the different products and a selection of the transition state of lowest energy, of greatest stability, as the one that would give rise to the major product. In considering the relative energies of different transition states, we assumed that they would be parallel to the relative energies of the possible reactive intermediates or of the products.

The rationalization of the regioselectivity seen in electrophilic aromatic substitution lies in a similar exercise. A consideration of the relative energies of transition states that would lead to the different products expected from a given substitution reaction should tell us which ones are the most stable and, therefore, most likely to yield products. In this discussion, the intermediates are assumed to resemble the transition states in structure and so are used as models.

When phenol reacts with bromine, the bromine atom can become attached to three possible positions: ortho, meta, and para to the hydroxyl group. The cationic intermediate that would result from bonding of the bromine atom to each of these positions, as well as resonance contributors for each one, are shown below.

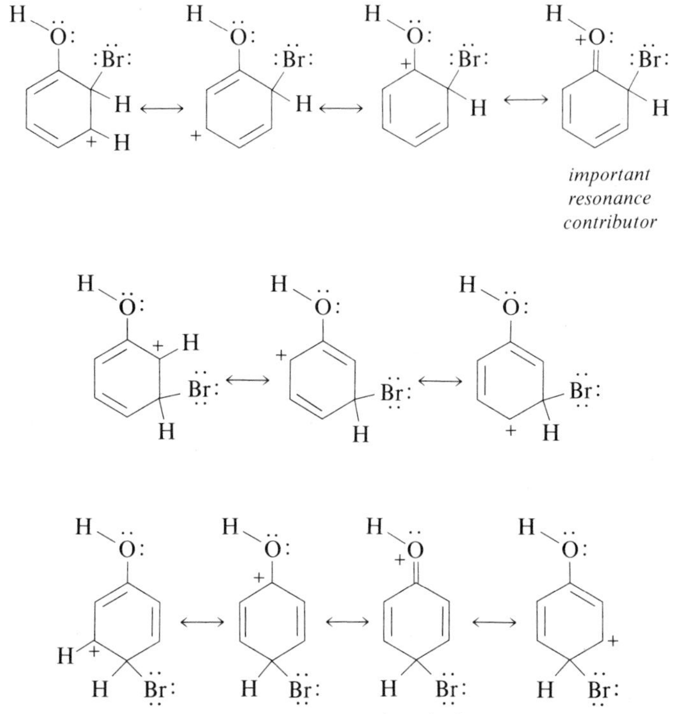

important resonance contributor

important resonance contributor

resonance contributors for the cationic intermediates that would result from bonding of a bromine atom to the ortho, meta, or para positions of phenol

The intermediates resulting from bonding of the bromine atom to the ortho and para positions of the ring have more resonance contributors than the intermediate that would give rise to *m*-bromophenol. In addition, for the intermediates arising from attack at the ortho or para positions, one of the resonance contributors has eight electrons around each atom, a particularly stable situation. For these reasons, we can conclude that the transition states for ortho and para substitution on a phenol ring are of lower energy than the one for meta substitution. The energies of activation for the formation of *o*-bromophenol and *p*-bromophenol are lower than for *m*-bromophenol. Most of the reaction will be channeled through the intermediates leading to those products.

The conclusions arrived at in the case of phenol will be true for any substituent in which there is a pair of nonbonding electrons on the atom bonded to the aromatic ring. In the case of alkyl groups, the electron-donating effect of the group stabilizes the intermediates resulting from ortho and para substitution more than it does the intermediate from meta substitution.

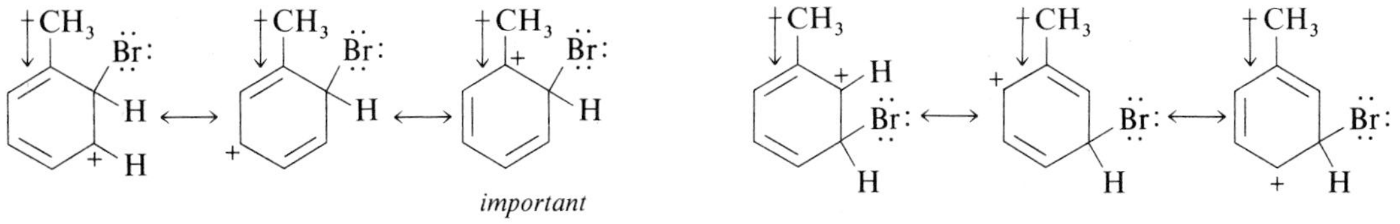

important resonance contributor

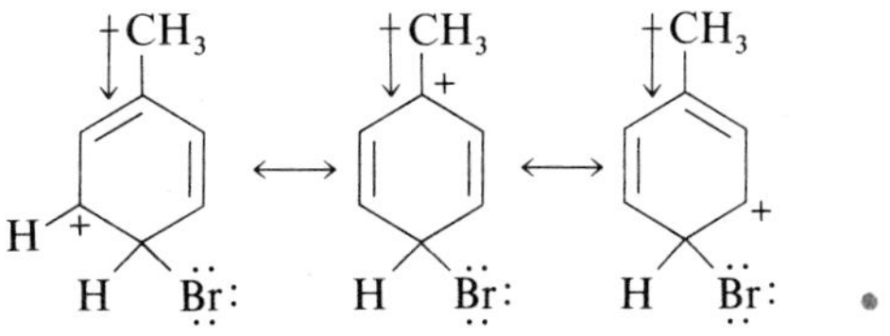

important resonance contributor

resonance contributors for the cationic intermediates that would
result from bonding of a bromine atom to the ortho, meta,
or para positions of toluene

When a strongly electron-withdrawing substituent is on the aromatic ring, it is harder to form the cationic intermediate. The cation is destabilized by the charge distribution in the polar electron-withdrawing group. The intermediate formed by reaction at the meta position is the most stable one because it is least destabilized. This concept is illustrated by the nitration of nitrobenzene.

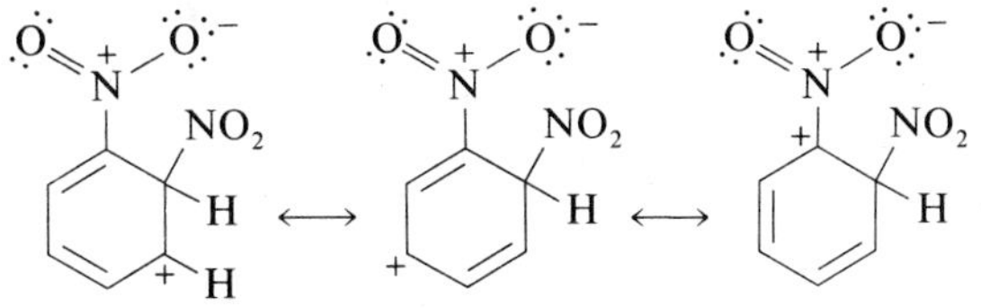

*unfavorable resonance contributor,
two adjacent positive charges*

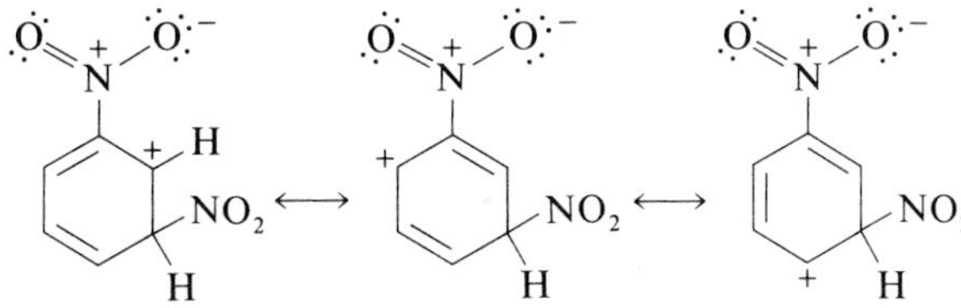

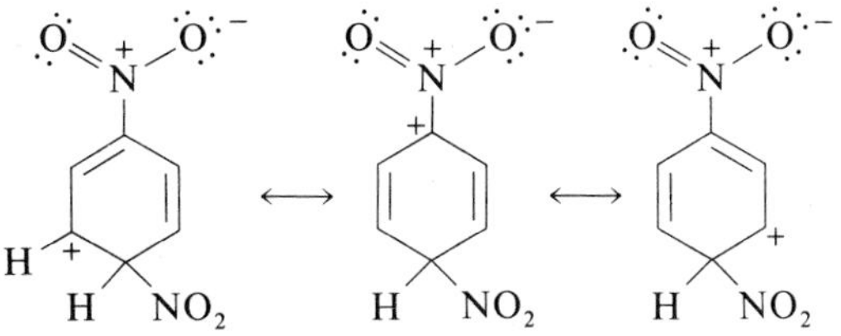

*unfavorable resonance contributor,
two adjacent positive charges*

resonance contributors for the cationic intermediates that would result from the bonding of
a nitronium ion to the ortho, meta, and para positions of nitrobenzene

None of the intermediates shown above has any special stabilization other than that which results from delocalization of charge in general. Among the intermediates resulting from bonding of a nitronium ion at the ortho and para positions of nitrobenzene, however, one resonance contributor is a particularly unfavorable one, resulting in the juxtaposition of two positive charges. As a result, the transition state arising from an initial approach of the nitronium ion to the meta position is the most stable among the three possible ones, and substitution on nitrobenzene takes place mostly at the meta position.

The second substitution of a halobenzene is an interesting case. The halogens are ring deactivating; their presence makes it harder for electrophilic aromatic substitution to take place. This would be expected from their electronegativities and thus their negative inductive effects. Yet they are ortho,para directing. As a positive charge develops on the aromatic ring during substitution, the nonbonding electrons of a halogen are drawn towards the charge and help to stabilize it. Thus, just as a halogen atom was found to stabilize a carbocation intermediate in the addition of hydrogen halides to an alkyne (Section 5.5A), so is it possible for halogen atoms to stabilize the carbocation intermediate in an electrophilic substitution reaction if the incoming substituent takes an ortho or para position. For example, the intermediate leading to *p*-dibromobenzene is favored over the one for *m*-dibromobenzene.

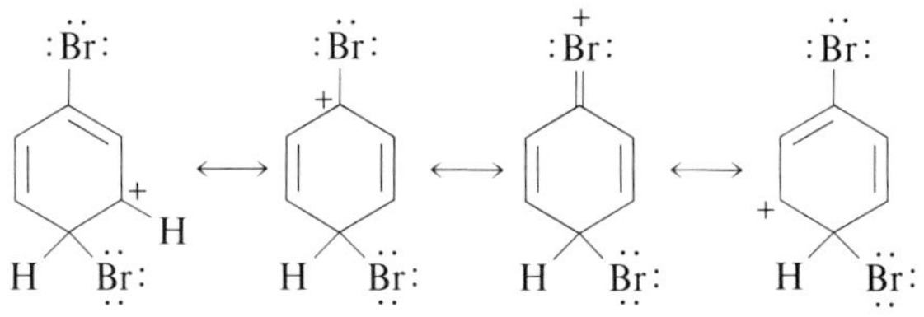

*delocalization of
charge to bromine
atom*

*resonance contributors for the intermediate for
p-dibromobenzene*

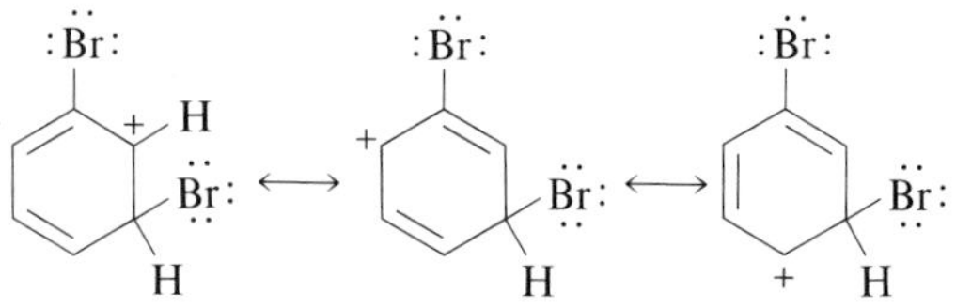

*no resonance contributor in which charge
can be delocalized to bromine atom*

*resonance contributors for the intermediate for
m-dibromobenzene*

The positive charge in the cationic intermediate for *p*-dibromobenzene can be delocalized to a bromine atom. Such a resonance contributor has eight electrons around each atom and is thus an important one. No such delocalization is possible for the intermediate that would give rise to *m*-dibromobenzene. Only a very small amount of the meta isomer (1.8%) is formed in the reaction of bromobenzene with bromine.

D. Steric Effects in Electrophilic Aromatic Substitution

The arguments presented in Section 12.3C for ortho,para direction by the hydroxyl group in phenol or the methyl group in toluene did not indicate any reason for preference between those positions on the basis of transition-state energies. If this were really so, twice as much ortho as para isomer should be formed because there are two ortho positions on the ring and only one para position. Statistically, it would be twice as likely that collision would occur with the reagents correctly oriented for reaction at the ortho position as at the para position. The experimental observations prove otherwise (p. 586). Actually, the ratios of ortho to para substitution can vary greatly, but substitution at the para position is generally favored over what would be expected statistically. The group already on the aromatic ring has some bulk and hinders the approach of the incoming reagent to the positions closest to it, the ortho positions. This idea has been tested by systematically increasing the bulk of a substituent and measuring the change in the composition of the product mixtures formed. Toluene, ethylbenzene, isopropylbenzene, and *tert*-butylbenzene are shown with numbers to indicate the percentage of substitution that takes place at each position on the ring in a nitration reaction.

percentage of ortho, meta, and para isomers
in the nitration of a series of alkylbenzenes

From the observations summarized above, it is quite clear that the proportion of ortho isomer in the product mixture becomes smaller as the bulk of the alkyl group on the aromatic ring becomes larger.

There is also evidence that the size of the incoming group affects the ratio of ortho to para isomers formed as products. For example, when bromobenzene reacts with chlorine, 42% of the product is *o*-bromochlorobenzene and 53% is the para isomer. If bromine, a larger atom, is the incoming group, only 13% *o*-dibromobenzene is formed, and 85% of the product mixture is *p*-dibromobenzene.

Both of these sets of data remind us that electronic factors are not the only ones to consider in thinking about a reaction. Steric interactions of the reagents are also important in determining relative energies of transition states and the resulting reaction pathways.

PROBLEM 12.9 The trimethylammonium ion, $(CH_3)_3\overset{+}{N}-$, is a strongly deactivating group that is also meta directing. Explain this fact. (Hint: A picture is worth a thousand words!)

PROBLEM 12.10 The following groups are not in Table 12.1. Predict whether they will be ortho,para or meta directing.

(a) CF_3- (b) $(CH_3)_3\overset{+}{P}-$ (c) O_2NCH_2- (d) $(CH_3)_2\overset{+}{S}-$

(e)   (f) $CH_3CH{=}CH-$

PROBLEM 12.11 The experimental observation outlined in the following equation has been made. How would you rationalize this observation? (Hint: What are the reacting species when *N*,*N*-dimethylaniline is dissolved in a mixture of concentrated nitric and sulfuric acids?)

$$CH_3NCH_3 \xrightarrow[\substack{H_2SO_4 \\ 5-10\,°C}]{HNO_3} \xrightarrow[H_2O]{NH_3} CH_3NCH_3 \quad NO_2$$

N,*N*-dimethylaniline   ~60%

PROBLEM 12.12 *p*-Nitrophenol is shaken with one equivalent of D_2O in the presence of the strong acid perchloric acid, $HClO_4$, at 100 °C for 100 hours. The original product has two deuterium atoms in it. One of these is lost instantly when the product is treated with ordinary water, H_2O. Propose a structure for the original product and explain the difference in the ease with which the two deuterium atoms can be removed from the molecule.

PROBLEM 12.13 How would you synthesize the following compounds from benzene or toluene and any other organic reagents containing not more than three carbon atoms? You may assume for the purposes of this exercise that ortho and para isomers are easily separable in the laboratory.

(a)
$$\overset{O}{\underset{\parallel}{C}}OH$$ with CH_3 para

(b) NO_2 ... Cl

(c) Br ... Cl

(d) CH_3, NO_2, NO_2

(e) CH_3CH_2CHOH

(f)
$$\overset{O}{\underset{\parallel}{C}}\!-\!NH_2$$ with NO_2

E. Alkylation of Arenes

In the reactions described in Section 12.3A, the electrophilic reagent was bromine or a nitronium ion. When it is cationic, carbon is also an electrophile; therefore, reactions that give rise to carbocations, when carried out in the presence of an aromatic ring, create carbon-carbon bonds. Three reactions give rise to carbocations: the protonation of an alkene, the protonation of an alcohol with the subsequent loss of water, and the ionization of an alkyl halide. All of these reactions are used to substitute carbon side chains on aromatic rings in a reaction known as the **Friedel-Crafts reaction,** named for Charles Friedel, a French chemist, and James Crafts, an American chemist, who developed these methods.

For example, sulfuric acid catalyzes the reaction of the alkene cyclohexene with benzene to give mostly cyclohexylbenzene and some *p*-dicyclohexylbenzene, even when an excess of benzene is used.

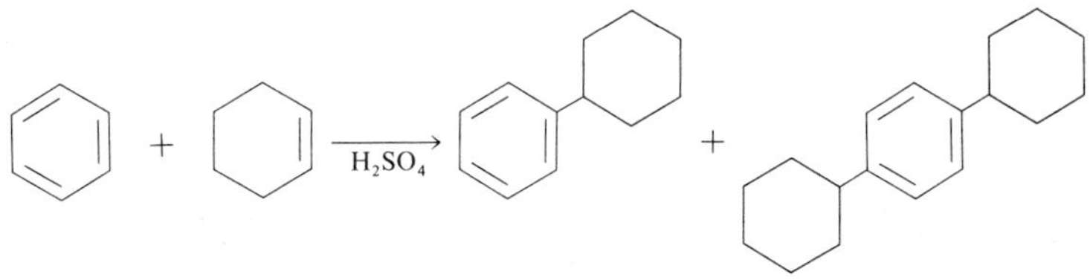

benzene   cyclohexene   cyclohexylbenzene   *p*-dicyclohexylbenzene
(used in excess)   *major*   *minor*

The reaction is an example of a **Friedel-Crafts alkylation,** and illustrates one of the difficulties of such reactions. The product of the reaction, an alkylbenzene, is more reactive towards electrophilic aromatic substitution than benzene itself. The alkyl group is ring activating, ortho,para directing. As a result, as soon as any quantity of mono-alkylated benzene accumulates in the reaction mixture, it starts to undergo further alkylation to give di- and even trisubstituted products.

The mechanism for the alkylation reaction with cyclohexene is straightforward, and follows the steps outlined in Section 12.3B. First, a good electrophile is created by protonation of cyclohexene by sulfuric acid, the catalyst in the reaction. A carbon-carbon bond forms when the electrophile reacts with the π electrons of the aromatic ring. A proton is lost from the cationic intermediate that is formed.

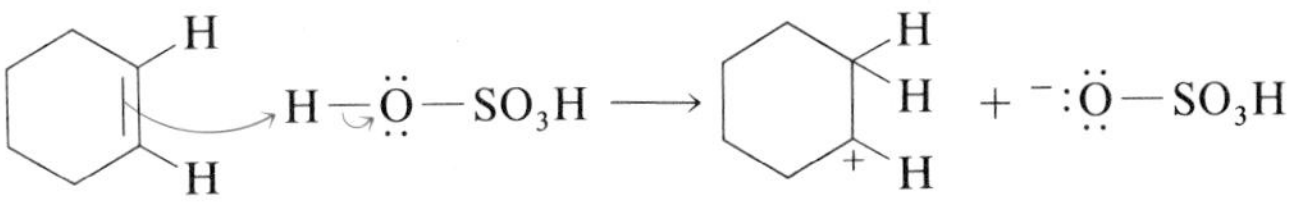

cyclohexyl cation
the electrophile

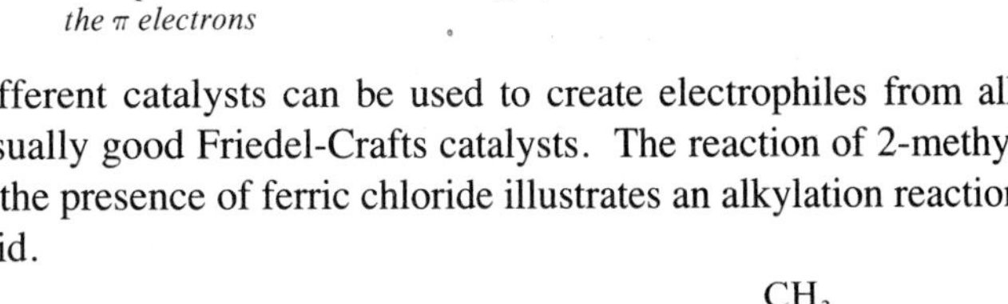

reaction of the   *the cationic intermediate*
electrophile with   *losing a proton.*
the π electrons

Many different catalysts can be used to create electrophiles from alkenes. Lewis acids are usually good Friedel-Crafts catalysts. The reaction of 2-methylpropene with benzene in the presence of ferric chloride illustrates an alkylation reaction catalyzed by a Lewis acid.

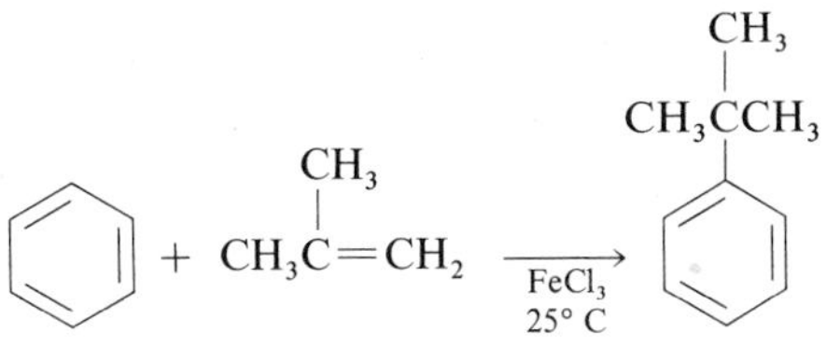

benzene   2-methylpropene   *tert*-butylbenzene
                                              89%

The electrophile is best represented as the most stable carbocation formed upon protonation of the alkene, whether the catalyst is a Brønsted-Lowry acid or a Lewis acid, because Lewis acids react with traces of water in the reaction mixture to form protic acids.

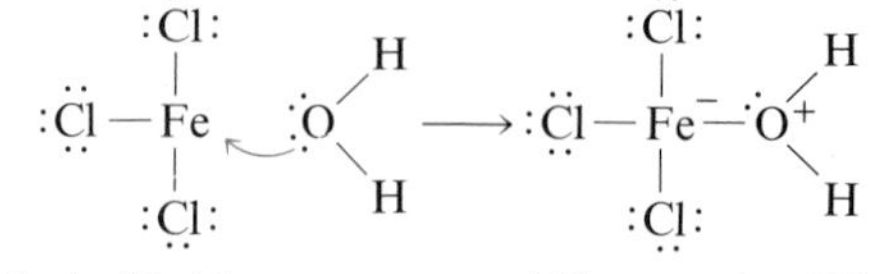

The structures of the reaction of ferric chloride with water:

ferric chloride   water   trichloroaquoiron(III) complex

a protic acid

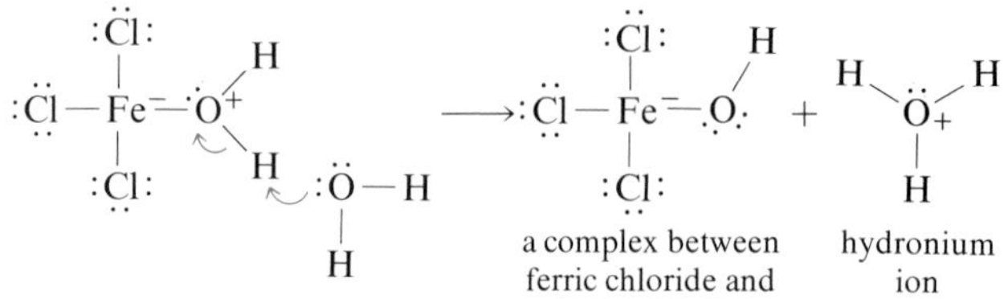

a complex between ferric chloride and hydroxide ion   hydronium ion

The protonation of 2-methylpropene gives the *tert-*butyl cation preferentially, and the carbon-carbon bond to the aromatic ring forms at the tertiary carbon atom.

Carbocations as electrophiles in the Friedel-Crafts alkylation reaction impose limits on the types of compounds that can be synthesized by this method. For example, it is difficult to put a primary alkyl chain on an aromatic ring. When benzene reacts with *n*-propyl chloride with aluminum chloride as a catalyst, about 40–50% of a mixture of two alkylbenzenes is formed. The composition of the mixture depends upon the temperature at which the reaction is run.

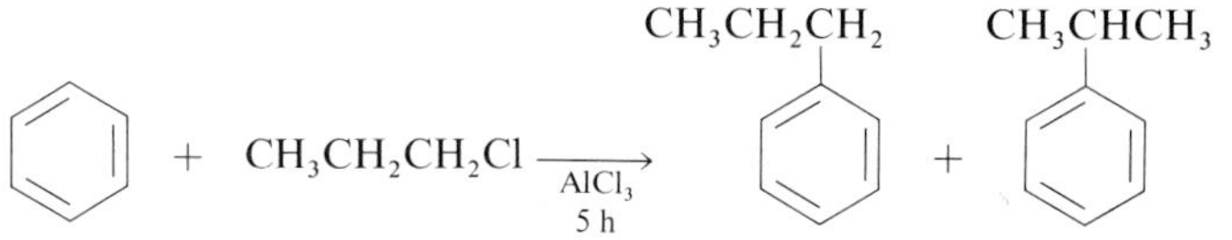

benzene   *n*-propyl chloride   propylbenzene   isopropylbenzene

			$CH_3CH_2CH_2$	CH_3CHCH_3
	at $-6\ ^\circ C$		60%	40%
	at $35\ ^\circ C$		40%	60%

benzene + $CH_3CH_2CH_2Cl$ $\xrightarrow[\text{5 h}]{AlCl_3}$

Aluminum chloride complexes with nonbonding electrons of the chlorine atom in *n*-propyl chloride, polarizing the carbon-chlorine bond. The intermediate formed must have enough carbocation character at the primary carbon atom so that some rearrangement to the more stable secondary carbocation takes place by a 1,2-hydride shift (Section 5.6B). The carbocation intermediates are seen as being tightly complexed with the aluminum halide in an ion pair.

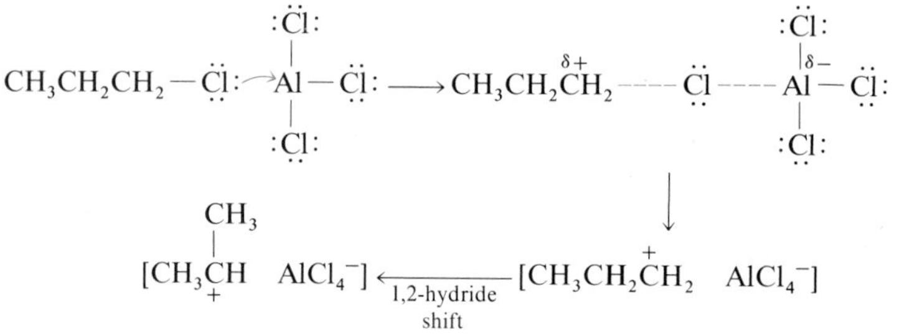

$$CH_3CH_2CH_2-\ddot{\underset{..}{Cl}}:\!\!\!-Al-\ddot{\underset{..}{Cl}}: \longrightarrow CH_3CH_2\overset{\delta+}{CH_2} ----\ddot{\underset{..}{Cl}}:----\overset{\delta-}{Al}-\ddot{\underset{..}{Cl}}:$$

$$[CH_3\underset{+}{CH}\ AlCl_4^-] \underset{\text{shift}}{\overset{}{\xleftarrow{\text{1,2-hydride}}}} [CH_3CH_2\overset{+}{CH_2}\ AlCl_4^-]$$

ion pair   ion pair
secondary carbocation   *primary carbocation*

We can see from the ratios of the two products from benzene and *n*-propyl chloride that free primary carbocations are not easily formed, even with a powerful Lewis acid such as aluminum chloride to pull off the chloride ion. At the lower temperature, the reaction is kinetically controlled (Section 5.3G) and most of the product comes from reaction at the first carbon of the propyl chain. Only at the higher temperature, when there would be a better chance of breaking the carbon-chlorine bond, does the rearranged product derived from the thermodynamically more stable cation become the major one. Clearly, the reaction is not a good one for the synthesis of either propylbenzene or isopropylbenzene.

There is no problem obtaining pure products if the alkyl halide used is one that gives rise to the most stable carbocation directly on ionization. Thus, *tert*-butyl chloride reacts with ethylbenzene to give *p-tert*-butylethylbenzene quantitatively.

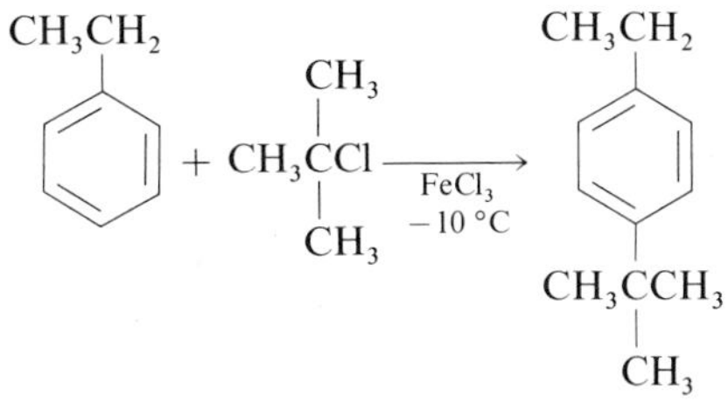

ethylbenzene   *tert*-butyl chloride   *p-tert*-butylethylbenzene

Carbocations are created by the reaction of alcohols with protic acids or with Lewis acids. Alcohols, therefore, are used as reagents in Friedel-Crafts alkylation reactions. In these reactions, also, the stability of the carbocation formed as the electrophile is important. Isopropyl alcohol reacts with benzene to give isopropylbenzene.

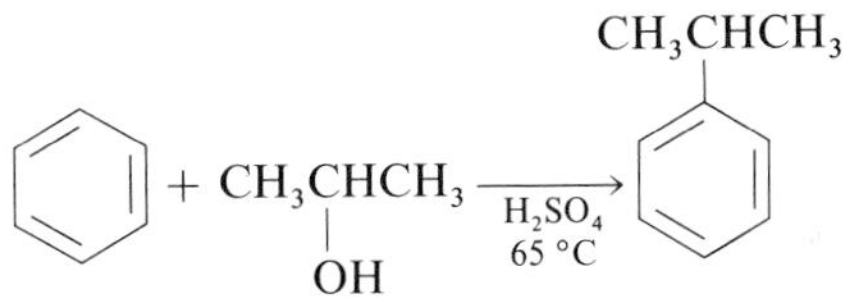

benzene   isopropyl alcohol   isopropylbenzene cumene 65%

When *n*-butyl alcohol reacts with benzene, with boron trifluoride as the catalyst, however, only *sec*-butylbenzene and the *p*-disubstituted compound are formed.

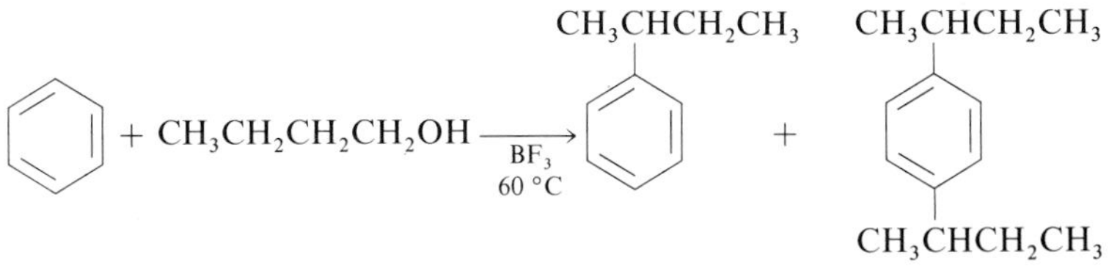

benzene   *n*-butyl alcohol   *sec*-butylbenzene 35%   *p*-di-*sec*-butylbenzene 25%

In this case, too, the formation of a primary cation and rearrangement to the secondary cation must be postulated to account for the results.

The Friedel-Crafts alkylation can be carried out with a variety of substituents on the aromatic ring. Nitrobenzene is so unreactive that it is used as a solvent for Friedel-Crafts reactions. If there is a ring-activating substituent also present, however, reaction becomes possible: 2-Nitromethoxybenzene reacts with isopropyl alcohol to give 4-isopropyl-2-nitromethoxybenzene. Hydrogen fluoride, a good Friedel-Crafts catalyst, is used to generate the electrophile.

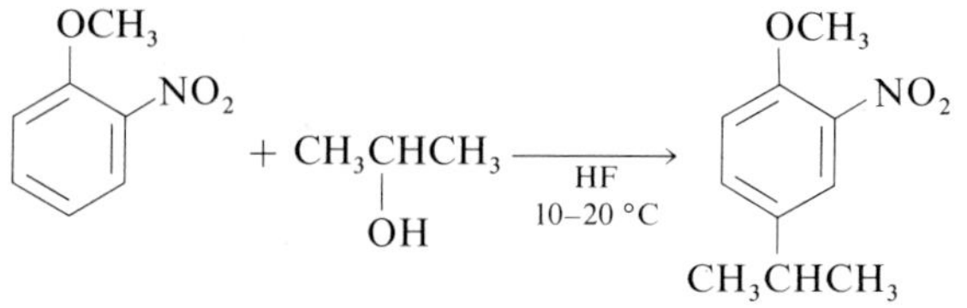

2-nitromethoxy-benzene + isopropyl alcohol $\xrightarrow[\text{10-20 °C}]{\text{HF}}$ 4-isopropyl-2-nitro-methoxybenzene

When there are two or more groups already on the aromatic ring, it is sometimes hard to predict where the next substituent will enter. In the case of 2-nitromethoxybenzene, the methoxyl group is ortho,para directing and the nitro group is meta directing. The isopropyl group could have been ortho to the methoxyl group and meta to the nitro group, or where it is, para to the methoxyl and meta to the nitro group. Because the isopropyl group is fairly large, the position para to the methoxyl group is preferred.

In 2-nitromethoxybenzene, both substituents directed the incoming substituent to the same positions on the aromatic ring. In cases where the directing influences of the groups already on the ring are in conflict, the group that is strongly activating and ortho, para directing determines the orientation of the incoming substituent. For example, 2-fluoromethoxybenzene undergoes nitration ortho and para to the methoxy group.

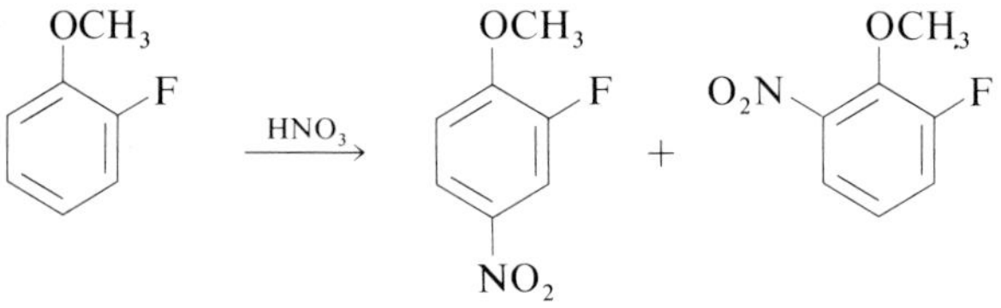

2-fluoromethoxybenzene $\xrightarrow{\text{HNO}_3}$ 2-fluoro-4-nitro-methoxybenzene + 2-methoxy-3-nitro-fluorobenzene

If the directing effects of the groups that are on the ring are more equal, then mixtures of products result.

F. Acylation of Arenes

Acid chlorides and acid anhydrides also serve as sources of electrophiles for Friedel-Crafts reactions. For example, benzene reacts with acetic anhydride in the presence of aluminum chloride to give acetophenone.

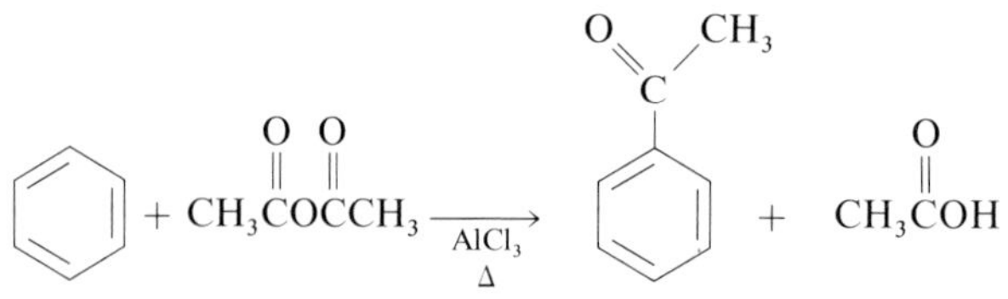

benzene + acetic anhydride $\xrightarrow[\Delta]{\text{AlCl}_3}$ acetophenone ~80% + acetic acid

The product is a ketone. In effect, an acyl group has been substituted on the aromatic ring, so the reaction is called the **Friedel-Crafts acylation** reaction.

An example in which an acid chloride is used is the acylation of bromobenzene with acetyl chloride.

Br — benzene ring + CH$_3$CCl (O double bond) $\xrightarrow[\Delta]{\text{AlCl}_3}$ p-bromoacetophenone + HCl

bromobenzene   acetyl chloride   *p*-bromoaceto-
                                                          phenone
                                                          70%

Both an acid chloride and an acid anhydride have good leaving groups (Section 10.4). Aluminum chloride can complex with either one of them to help remove the leaving group, creating an acylium cation.

$$CH_3\overset{:O:}{\overset{\|}{C}}-\overset{..}{\underset{..}{Cl}}:\rightarrow AlCl_3 \longrightarrow CH_3\overset{:O:}{\overset{\|}{C}}\overset{+}{\underset{..}{Cl}}-\overset{-}{AlCl_3} \longrightarrow [CH_3\overset{+}{C}=\overset{..}{\underset{..}{O}} \quad :\overset{..}{\underset{..}{Cl}}-\overset{-}{AlCl_3}]$$

complex   acylium ion
                                     ion pair

$$CH_3\overset{:O:}{\overset{\|}{C}}-\overset{:O:}{\underset{\underset{AlCl_3}{\nwarrow}}{\overset{..}{O}}}-\overset{:O:}{\overset{\|}{C}}CH_3 \longrightarrow CH_3\overset{:O:}{\overset{\|}{C}}\overset{+}{\underset{\overset{|}{-}AlCl_3}{\overset{..}{O}}}-\overset{:O:}{\overset{\|}{C}}CH_3 \longrightarrow [CH_3\overset{+}{C}=\overset{..}{\underset{..}{O}} \quad CH_3-\overset{:O:}{\overset{\|}{C}}-\overset{..}{\underset{..}{O}}-\overset{-}{AlCl_3}]$$

complex   acylium ion
                                     ion pair

The acylium ion is particularly stable because delocalization of its charge to the oxygen atom can occur.

$$CH_3\overset{+}{C}=\overset{..}{\underset{..}{O}} \longleftrightarrow CH_3C\equiv\overset{+}{O}:$$

particularly good
resonance contributor,
8 electrons around
each atom

resonance contributors of the acylium ion

No rearrangements of acylium ions are seen. Thus, the acid chloride from a long straight-chain acid can be used to put a straight carbon chain on an aromatic ring as an acyl group. The acid chloride of propanoic acid has been used to put a straight chain of three carbon atoms on the benzene ring.

benzene + CH$_3$CH$_2$CCl (O double bond) $\xrightarrow{\text{AlCl}_3}$ 1-phenyl-1-propanone (ring with CCH$_2$CH$_3$ and O double bond)

benzene   propanoyl chloride   1-phenyl-1-propanone
                                 65%

A Friedel-Crafts acylation reaction followed by reduction of the carbonyl group to a methylene group is the best way to introduce unbranched alkyl groups onto an aromatic ring. Thus Wolff-Kishner reduction (Section 9.8B) of 1-phenyl-1-propanone gives propylbenzene.

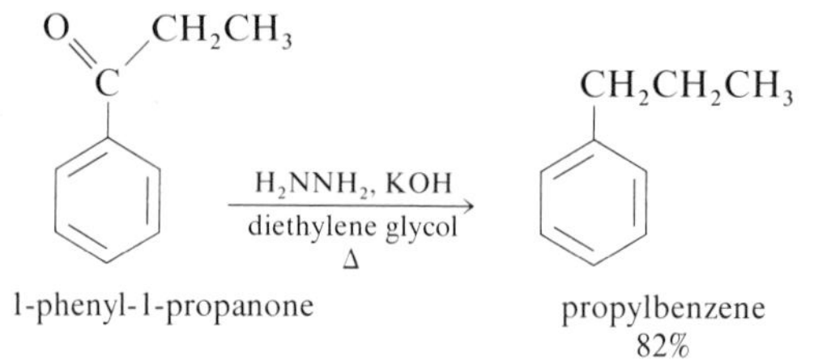

1-phenyl-1-propanone → propylbenzene
82%

(reagents: H₂NNH₂, KOH, diethylene glycol, Δ)

Another advantage of the acylation reaction is that the new carbonyl group on the ring deactivates it towards further substitution. Side products resulting from multiple substitution reactions are not a problem in Friedel-Crafts acylations.

Cyclic acid anhydrides are particularly useful because the leaving group remains bonded to the rest of the molecule and is present as a functional group that can undergo further reactions. Succinic anhydride reacts with benzene to give a ketone containing a carboxylic acid function.

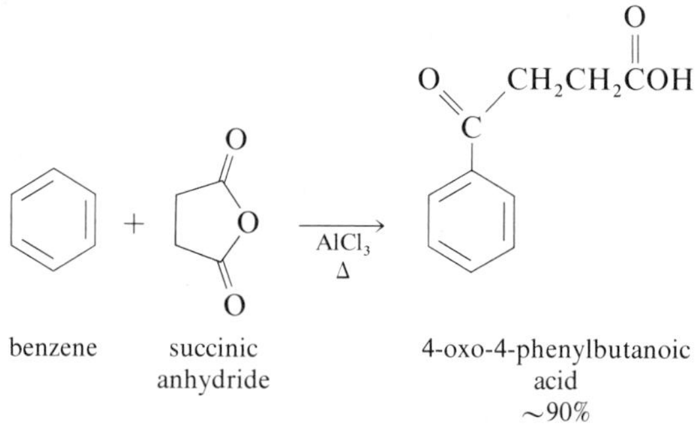

benzene   succinic anhydride $\xrightarrow{AlCl_3, \Delta}$ 4-oxo-4-phenylbutanoic acid
~90%

The electrophile that participates in this reaction has two functional groups on it. At one end is an acylium ion, at the other a carboxylate anion, complexed with aluminum chloride.

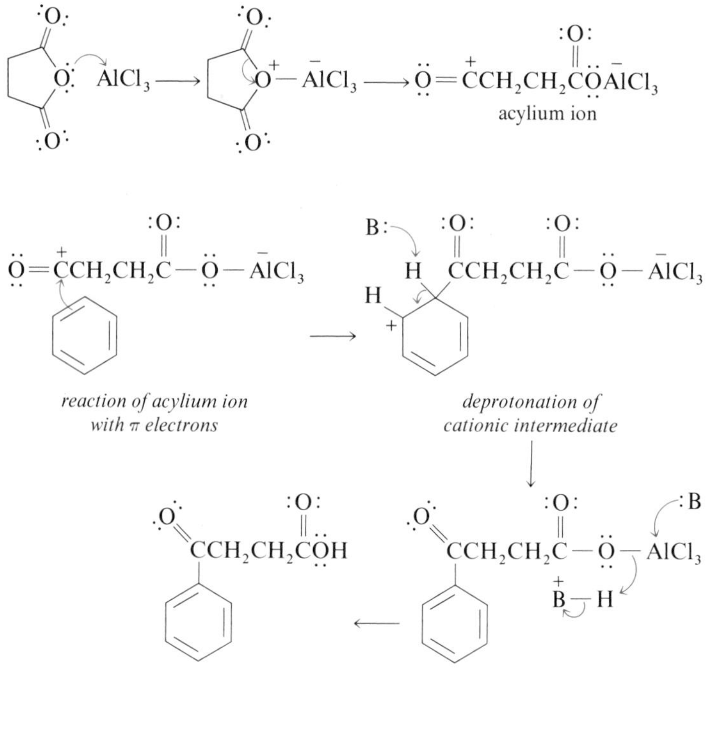

acylium ion

reaction of acylium ion with π electrons

deprotonation of cationic intermediate

Such a bifunctional molecule can undergo another Friedel-Crafts acylation reaction between the carboxylic acid group and the ring. Clemmensen reduction (Section 9.8A) of the ketone function to a methylene group gives 4-phenylbutanoic acid. This acid is converted into its acid chloride, which, on treatment with aluminum chloride, acylates the aromatic ring to give a cyclic ketone.

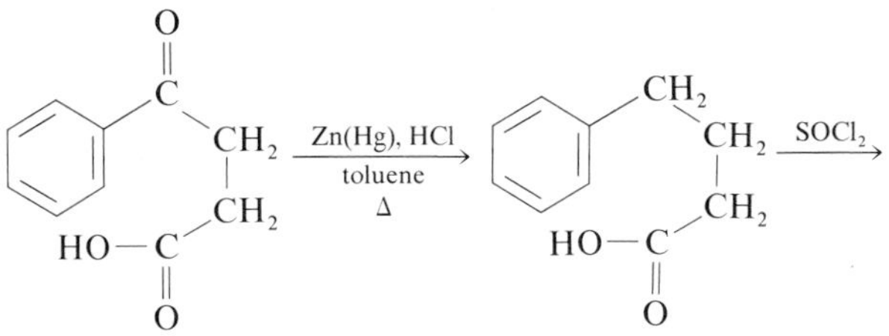

4-oxo-4-phenylbutanoic
acid

4-phenylbutanoic
acid
90%

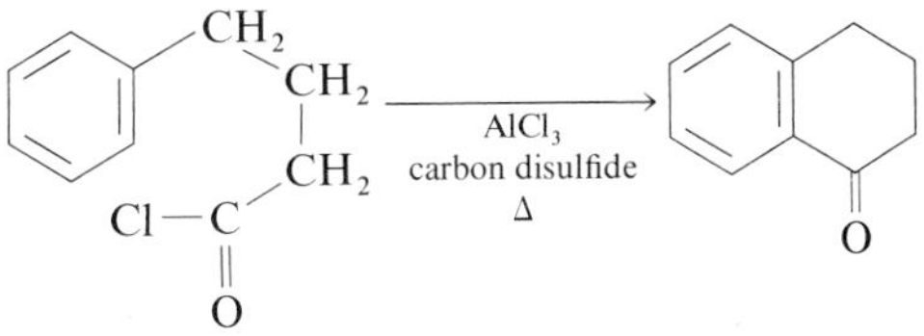

4-phenylbutanoyl
chloride

1-oxo-1,2,3,4-tetrahydro-
naphthalene
α-tetralone
∼80%

Steric considerations ensure that only the position ortho to the original point of attachment of the chain reacts, and a stable six-membered ring is formed. This sequence of reactions is a standard way of making fused-ring compounds. The presence of a reactive carbonyl group in the final product makes it possible to extend the synthesis.

PROBLEM 12.14   Predict the products, designated by letters, that would be formed when 1-oxo-1,2,3,4-tetrahydronaphthalene (above) reacts with the following reagents.

(a) $NaBH_4$, $CH_3CH_2OH \longrightarrow A$

(b) CH_3CH_2MgBr, diethyl ether $\longrightarrow B \xrightarrow[H_2O]{NH_4Cl} C$

(c) $NaC\equiv CCH_3$, $NH_3(liq) \longrightarrow D \xrightarrow[H_2O]{NH_4Cl} E$

(d) 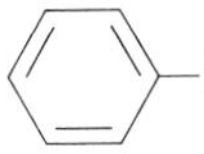 $-NHNH_2$, acetic acid $\longrightarrow F$

(e) $HOCH_2CH_2OH$, TsOH, toluene $\xrightarrow{\Delta} G$

Write equations showing the following further transformations.

(f) $A \xrightarrow[H_3PO_4, \Delta]{} H \xrightarrow[tetrahydrofuran]{BH_3} \xrightarrow{H_2O_2, OH^-} I$ (g) $C \xrightarrow[H_3PO_4, \Delta]{} J$

(h) $E \xrightarrow[\substack{Pd/BaSO_4 \\ quinoline}]{H_2} K \xrightarrow{9\text{-BBN}} L \xrightarrow{H_2O_2, OH^-} M$ (i) $E \xrightarrow{9\text{-BBN}} N \xrightarrow{H_2O_2, OH^-} O$

PROBLEM 12.15 Complete the following equations.

(a) 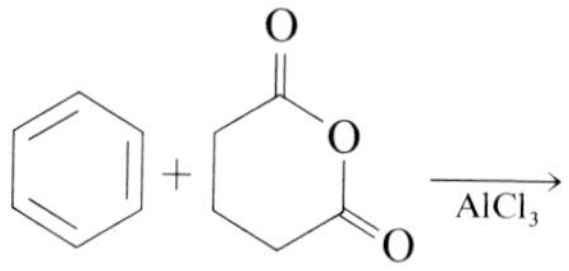 + $\xrightarrow{\text{AlCl}_3}$

(b) 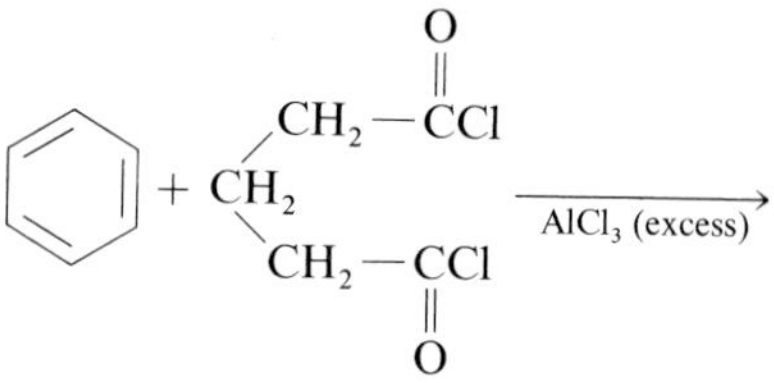 + $\underset{\substack{\text{CH}_2 \\ \downarrow}}{\overset{\text{CH}_2-\text{CCl}(=O)}{}}$ $\xrightarrow{\text{AlCl}_3 \text{ (excess)}}$

2 moles   1 mole

(c) 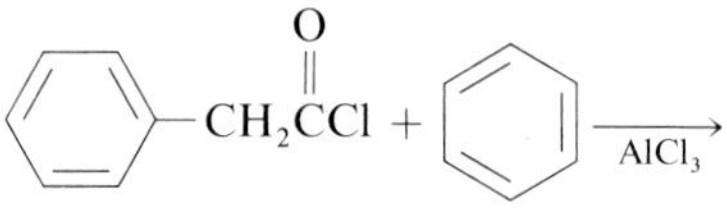 $-\text{CH}_2\overset{O}{\overset{||}{\text{C}}}\text{Cl}$ + $\xrightarrow{\text{AlCl}_3}$

(d) 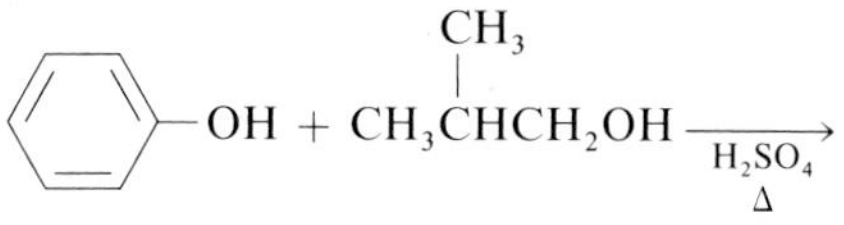 $-\text{OH}$ + $\text{CH}_3\overset{\text{CH}_3}{\underset{|}{\text{CH}}}\text{CH}_2\text{OH}$ $\xrightarrow[\Delta]{\text{H}_2\text{SO}_4}$

(e) $-\text{Cl}$ + $\text{CH}_3\overset{\text{CH}_3}{\underset{|}{\text{CH}}}\text{OH}$ $\xrightarrow[\Delta]{\text{H}_2\text{SO}_4}$

(f) + $\text{CH}_3\text{CH}_2\text{CH}=\text{CH}_2$ $\xrightarrow{\text{H}_2\text{SO}_4}$

(g) + $\text{CH}_3\overset{\text{CH}_3}{\underset{\underset{\text{Cl}}{|}}{\overset{|}{\text{C}}}}\text{CH}_2\text{CH}_3$ $\xrightarrow[\Delta]{\text{FeCl}_3}$

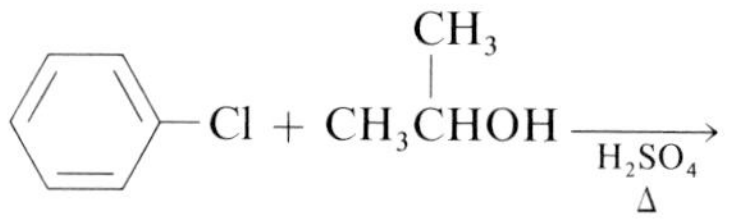

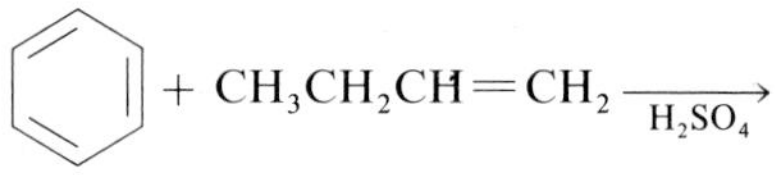

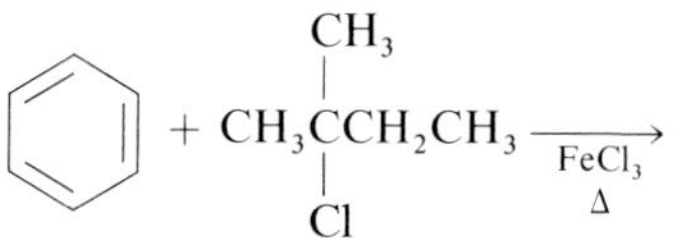

(h) + $\text{CH}_3\text{CH}_2\text{Br}$ $\xrightarrow{\text{AlCl}_3}$

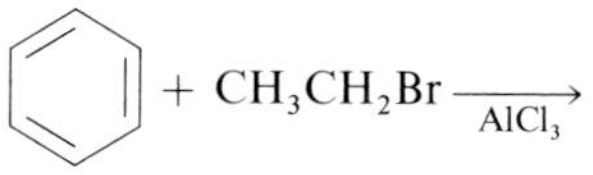

(i) 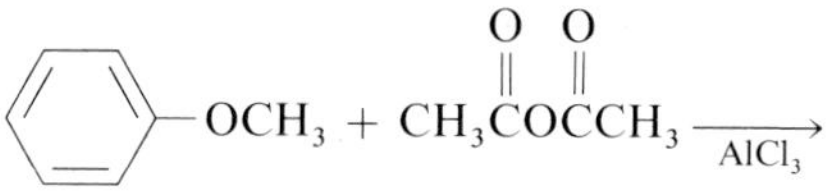 $-\text{OCH}_3$ + $\text{CH}_3\overset{O}{\overset{||}{\text{C}}}\text{O}\overset{O}{\overset{||}{\text{C}}}\text{CH}_3$ $\xrightarrow{\text{AlCl}_3}$

PROBLEM 12.16 Complete the following equations.

(a) 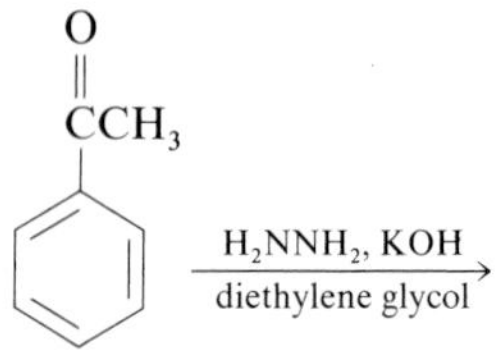 $\xrightarrow[\text{diethylene glycol}]{\text{H}_2\text{NNH}_2,\ \text{KOH}}$

(b) 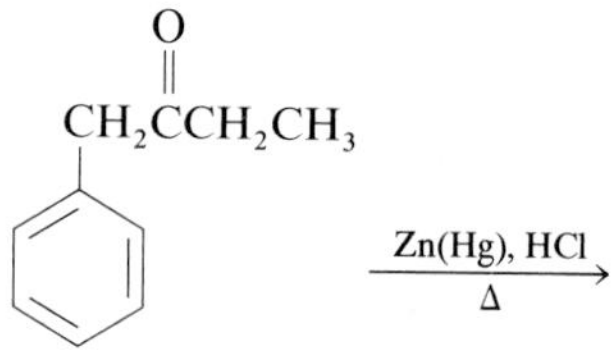 $\xrightarrow[\Delta]{\text{Zn(Hg), HCl}}$

(c) 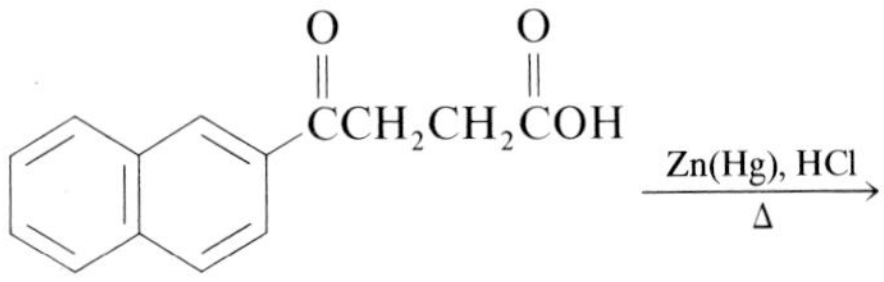 $\xrightarrow[\Delta]{\text{Zn(Hg), HCl}}$

(d)

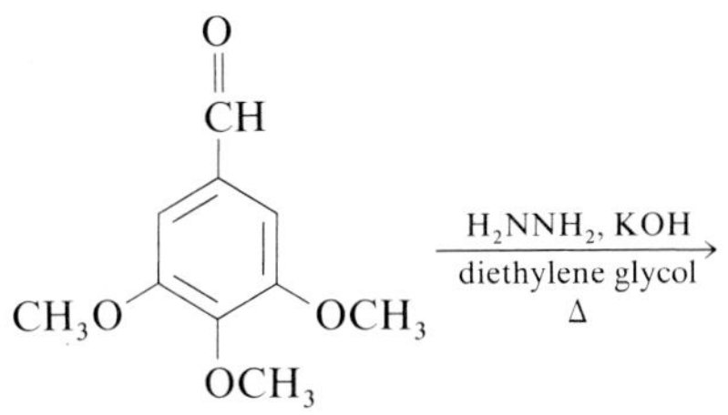

$$\xrightarrow[\substack{\text{diethylene glycol} \\ \Delta}]{H_2NNH_2,\ KOH}$$

(e)

$$\xrightarrow[\substack{BF_3 \\ \Delta}]{HSCH_2CH_2SH} \qquad \xrightarrow[\substack{\text{ethanol} \\ \Delta}]{\text{Raney Ni}}$$

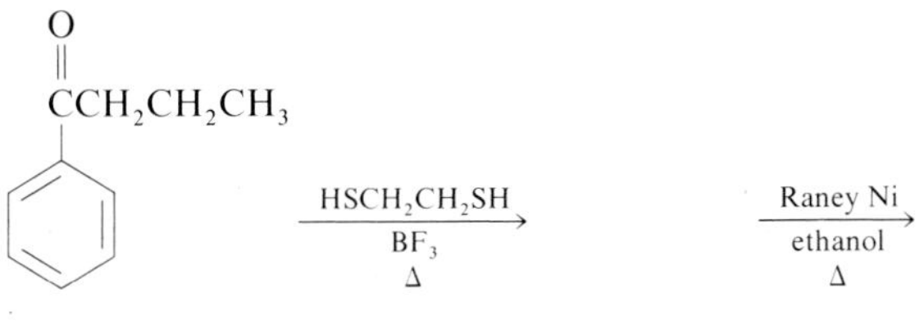

G. Sulfonation, Sulfonic Acids, and Their Derivatives

Benzene reacts with sulfuric acid in a reversible reaction to give benzenesulfonic acid. High yields of the sulfonic acid are obtained when water is removed from the reaction mixture by distillation with benzene or when sulfur trioxide is used as the electrophile.

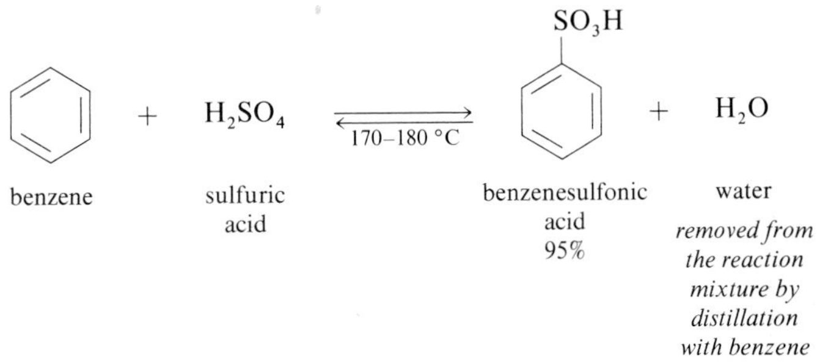

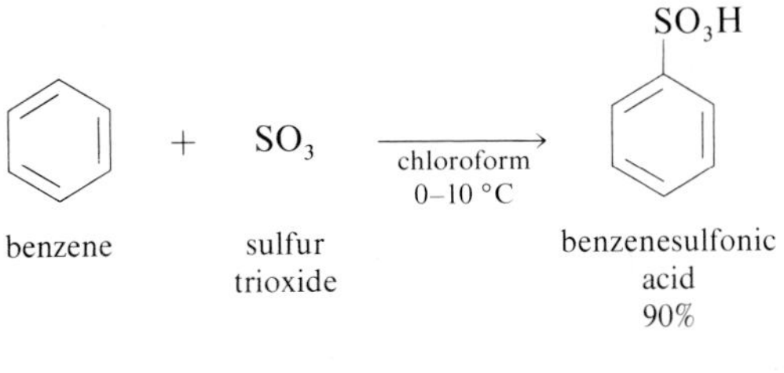

| benzene | sulfuric acid | | benzenesulfonic acid 95% | water *removed from the reaction mixture by distillation with benzene* |

| benzene | sulfur trioxide | | benzenesulfonic acid 90% |

The electrophile is believed to be sulfur trioxide, which is the anhydride of sulfuric acid, even when the acid is used as the reagent. For example, the rate of the reaction increases with increasing sulfur trioxide concentration in the sulfuric acid.

A mechanism for the reaction follows:

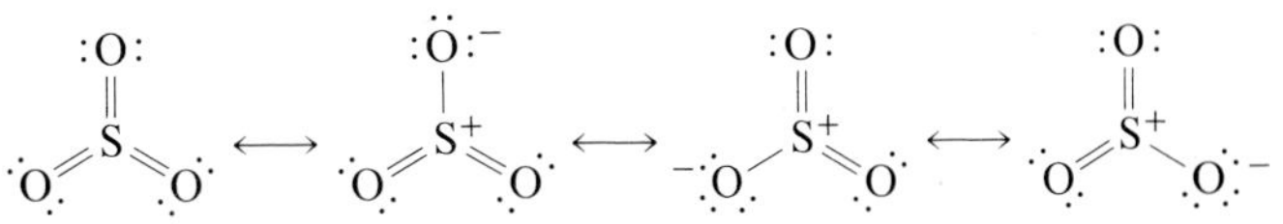

resonance contributors of sulfur trioxide

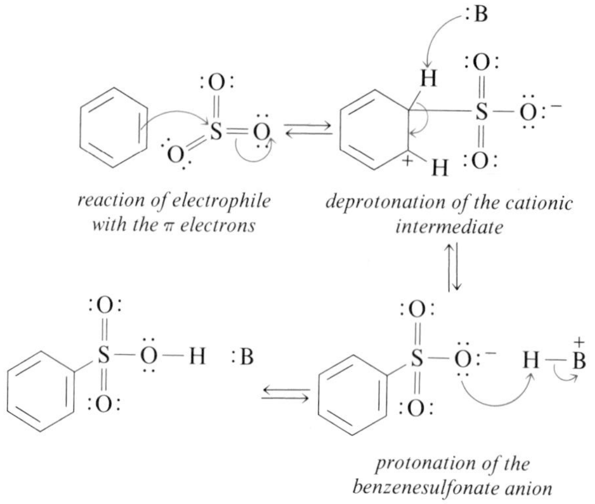

reaction of electrophile
with the π electrons

deprotonation of the cationic
intermediate

protonation of the
benzenesulfonate anion

The reverse reaction occurs when a sulfonic acid is heated in water.

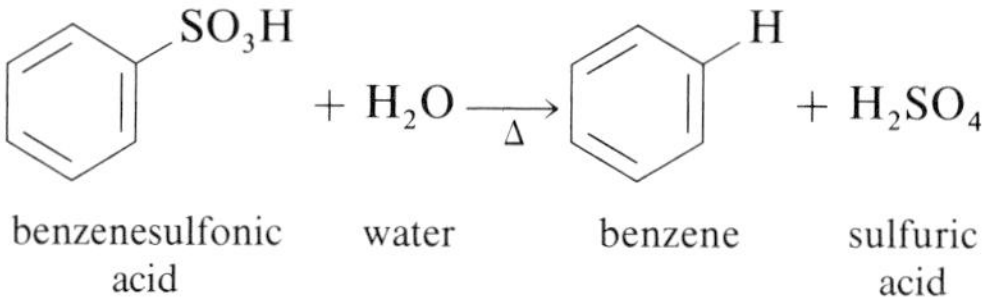

| benzenesulfonic acid | water | benzene | sulfuric acid |

The reaction is pictured as taking place by the loss of sulfur trioxide from the sulfonate anion and protonation of the developing anion.

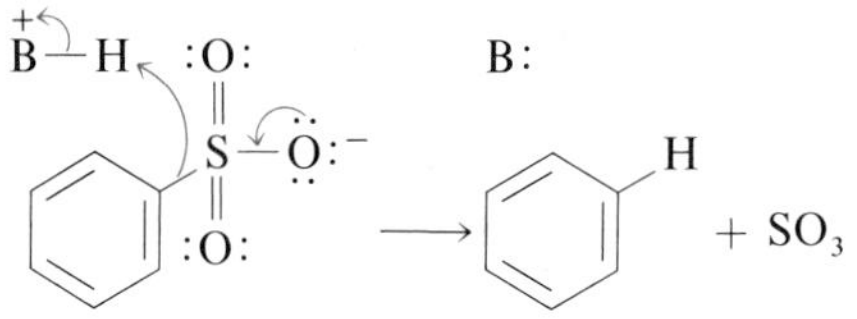

$$SO_3 + H_2O \rightleftharpoons H_2SO_4$$

PROBLEM 12.17   When toluene is sulfonated with concentrated sulfuric acid at 0 °C, the product mixture is 53% *p*-toluenesulfonic acid, 43% *o*-toluenesulfonic acid, and 4% *m*-toluenesulfonic acid. If the sulfonation is carried out at 100 °C, *p*-toluenesulfonic acid constitutes 79% of the product mixture. How do you explain the change in product composition associated with the temperature of the reaction? (Hint: You may find it helpful to review Section 5.3G.)

The products of sulfonation reactions, sulfonic acids, are organic acids related to sulfuric acid. Sulfonic acids, because of the resonance stabilization possible for the anion resulting from the loss of the proton on the hydroxyl group, are strong acids with pK_a values of approximately -0.5. The organic portion of the molecule ensures its solubility in organic solvents, so *p*-toluenesulfonic acid, for example, is often used when a strong acid that is soluble in an organic reaction mixture is required.

Sulfonic acids and carboxylic acids are closely related in their chemistry. For example, the acid chlorides of sulfonic acids are prepared by heating the sulfonic acids with thionyl chloride or phosphorus pentachloride (Section 10.6A). The conversions of methanesulfonic acid and sodium benzenesulfonate to the corresponding sulfonyl chlorides are examples of these reactions.

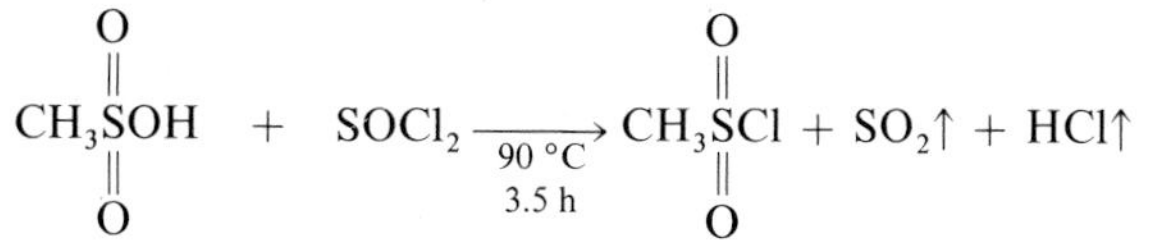

$$CH_3\overset{\displaystyle O}{\underset{\displaystyle O}{\overset{\|}{\underset{\|}{S}}}}OH + SOCl_2 \xrightarrow[\substack{90\,°C \\ 3.5\,h}]{} CH_3\overset{\displaystyle O}{\underset{\displaystyle O}{\overset{\|}{\underset{\|}{S}}}}Cl + SO_2\uparrow + HCl\uparrow$$

methanesulfonic   thionyl   methanesulfonyl
acid   chloride   chloride

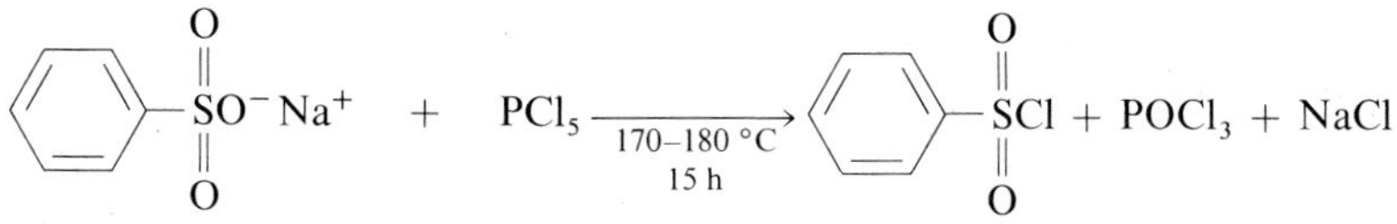

sodium benzenesulfonate   phosphorus   benzenesulfonyl
pentachloride   chloride

Sulfonyl chlorides react with nucleophiles the way that acyl chlorides do (Sections 10.4, 10.7A, and 10.8A). Thus, reactions with alcohols give sulfonic acid esters and reactions with ammonia or amines give amides of sulfonic acids called **sulfonamides.** In both cases, the reaction is carried out in the presence of base to neutralize the hydrochloric acid that is the other product of the reaction. In the reaction of *trans*-2-vinylcyclohexanol with *p*-toluenesulfonyl chloride, pyridine serves as solvent and base.

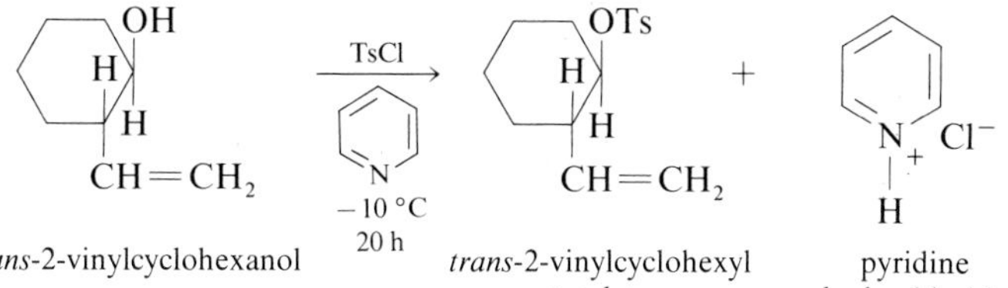

trans-2-vinylcyclohexanol   *trans*-2-vinylcyclohexyl
tosylate
88%
    pyridine
hydrochloride

As can be seen from the equation, the formation of the tosylate occurs with retention of configuration at the carbon atom to which the hydroxyl group is bonded. Many tosylates are used in studies of the mechanisms of organic reactions, so we are interested in the stereochemistry of their formation (Section 6.5D).

p-Toluenesulfonyl chloride heated with concentrated ammonium hydroxide solution gives *p*-toluenesulfonamide quantitatively.

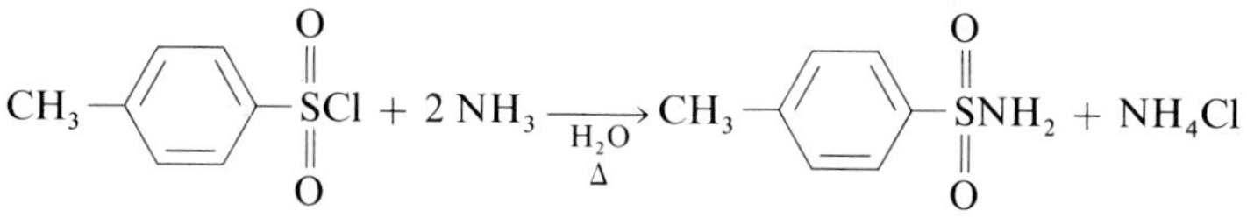

p-toluenesulfonyl   *p*-toluenesulfonamide
chloride   100%

The excess ammonia serves as the base for the reaction. When sulfonamides of other amines are prepared, another base is used to absorb the hydrochloric acid. For example, benzenesulfonyl chloride reacts with amines in the presence of aqueous sodium hydroxide.

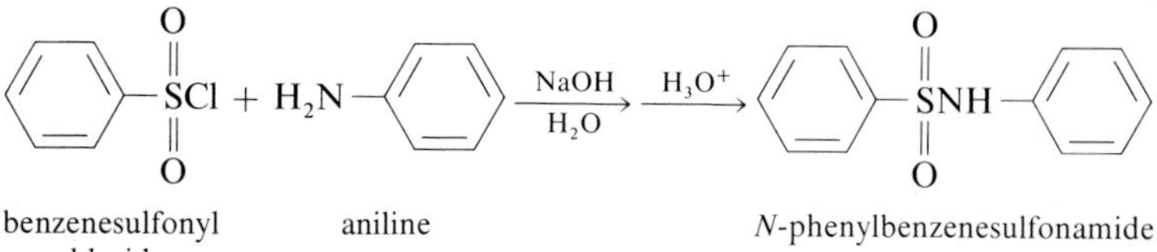

benzenesulfonyl   aniline   *N*-phenylbenzenesulfonamide
chloride

There is one important difference between a sulfonamide and the amide of a carboxylic acid. A sulfonamide is acidic if it has at least one hydrogen atom on the nitrogen atom of the amide group. Such a sulfonamide is a reasonably strong acid. Those in which the hydrocarbon portion of the amine is not large will dissolve in dilute sodium hydroxide. Thus, the product from the reaction of benzenesulfonyl chloride with aniline is soluble in aqueous sodium hydroxide and will precipitate when the solution is acidified. The acidity of a sulfonamide can be rationalized by pointing to the delocalization of electrons onto the two oxygen atoms of the sulfonamide function.

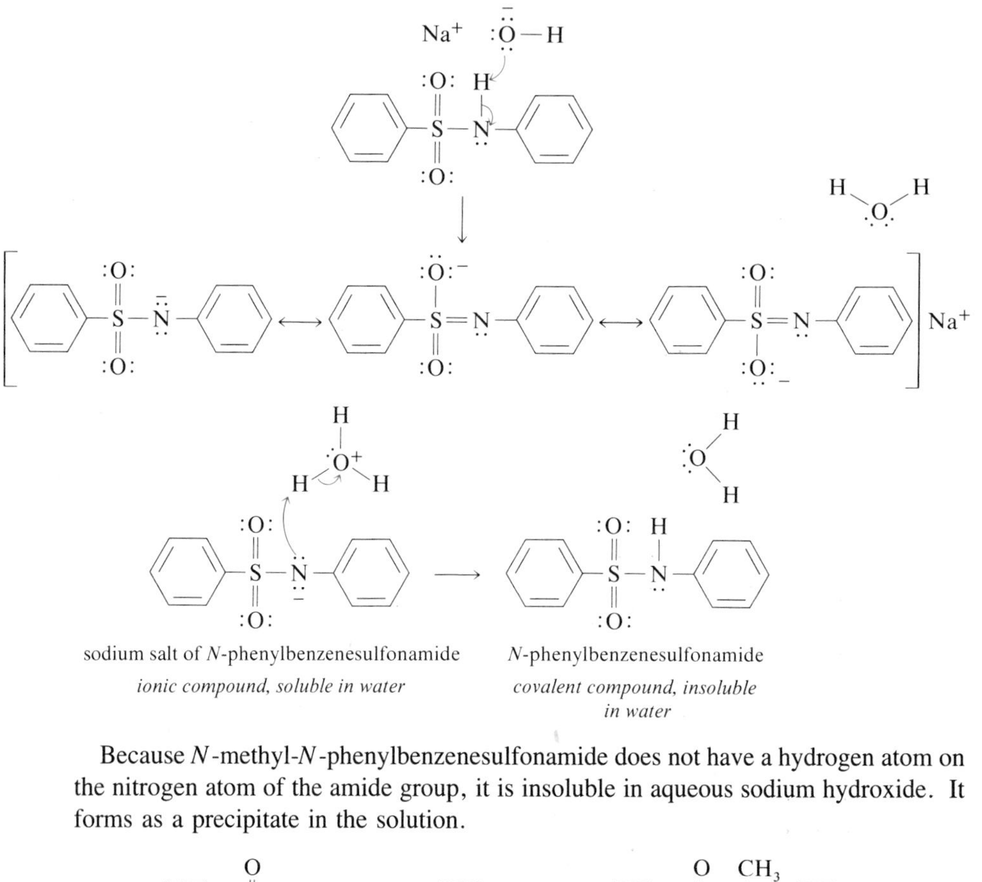

sodium salt of *N*-phenylbenzenesulfonamide

ionic compound, soluble in water

N-phenylbenzenesulfonamide

covalent compound, insoluble in water

Because *N*-methyl-*N*-phenylbenzenesulfonamide does not have a hydrogen atom on the nitrogen atom of the amide group, it is insoluble in aqueous sodium hydroxide. It forms as a precipitate in the solution.

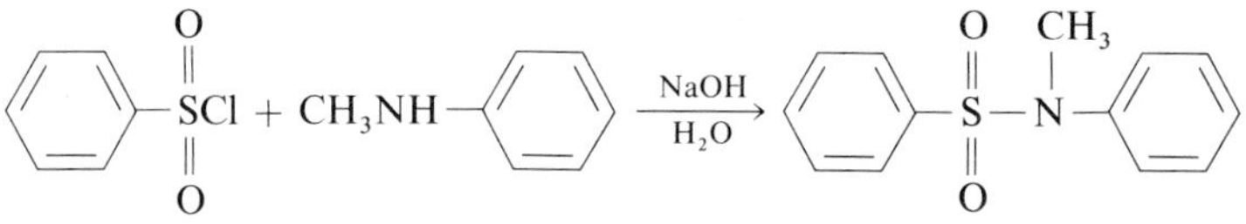

benzenesulfonyl chloride

N-methylaniline

N-methyl-*N*-phenylbenzenesulfonamide

not an acid, insoluble in base

All sulfonamides are solids with reasonably high melting points.

Sulfonyl chlorides, like acid chlorides and acid anhydrides, do not react to give stable compounds with amines that have no hydrogen atoms on the nitrogen atom.

PROBLEM 12.18   Write a detailed mechanism for the reaction of ammonia with *p*-toluenesulfonyl chloride. Analyze the reaction in the same way as we analyzed the reactions of acid chlorides and anhydrides in Section 10.4.

PROBLEM 12.19 Complete the following equations.

(a) CH$_3$—⬡—$\overset{\displaystyle O}{\underset{\displaystyle O}{\overset{\|}{\underset{\|}{S}}}}$OH + PCl$_5$ $\xrightarrow{\Delta}$

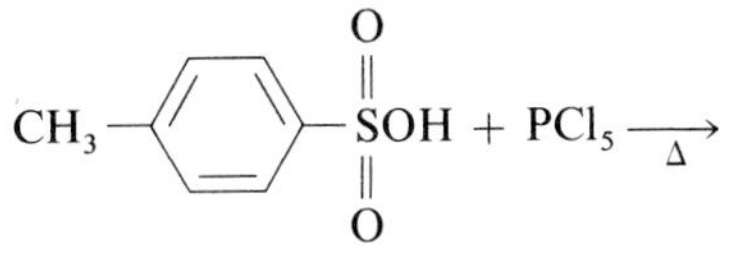

(b) CH$_3\overset{\displaystyle O}{\underset{\displaystyle O}{\overset{\|}{\underset{\|}{S}}}}$Cl + [cyclopentane with OH, H, H, CH$_3$ substituents] $\xrightarrow{\text{pyridine}}$

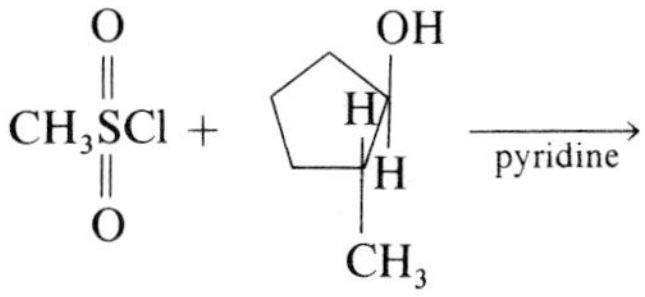

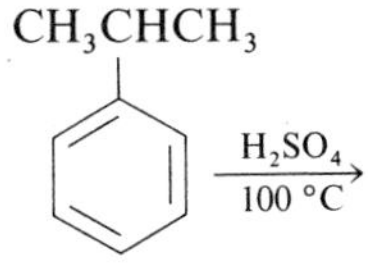

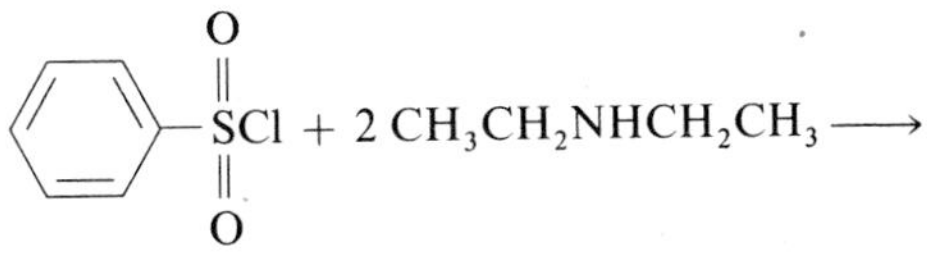

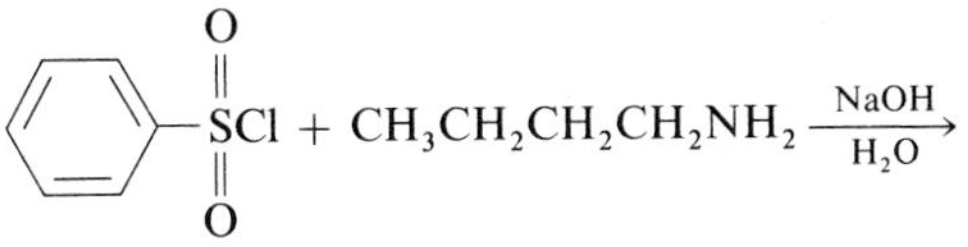

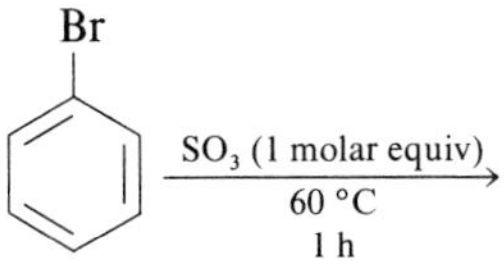

(c) [benzene with CH$_3$CHCH$_3$ group] $\xrightarrow[100\,°C]{H_2SO_4}$

(d) ⬡—$\overset{\displaystyle O}{\underset{\displaystyle O}{\overset{\|}{\underset{\|}{S}}}}$Cl + 2 CH$_3CH_2$NHCH$_2CH_3$ ⟶

(e) ⬡—$\overset{\displaystyle O}{\underset{\displaystyle O}{\overset{\|}{\underset{\|}{S}}}}$Cl + CH$_3CH_2CH_2CH_2NH_2$ $\xrightarrow[H_2O]{NaOH}$

(f) product of (e) $\xrightarrow{H_3O^+}$

(g) [benzene with Br] $\xrightarrow[\substack{60\,°C \\ 1\,h}]{SO_3\ (1\ \text{molar equiv})}$

(h) ⬡—$\overset{\displaystyle O}{\underset{\displaystyle O}{\overset{\|}{\underset{\|}{S}}}}$Cl + CH$_3CH_2\overset{\displaystyle CH_2CH_3}{\overset{|}{N}}CH_2CH_3$ $\xrightarrow[H_2O]{NaOH}$

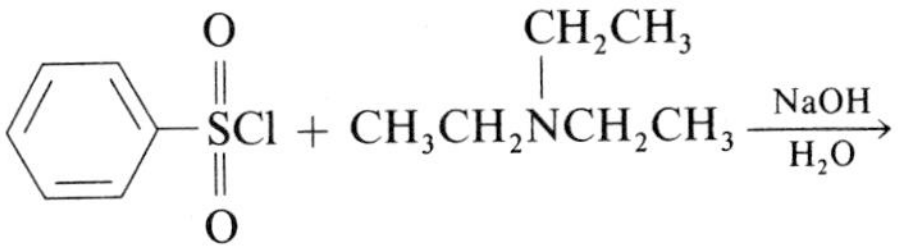

(i) [benzene with two CH$_3$ groups ortho] $\xrightarrow{\Delta}^{H_2SO_4}$

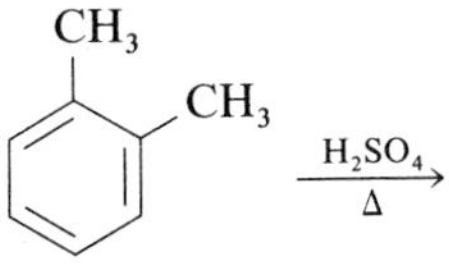

PROBLEM 12.20 Section 12.3 gives detailed discussions of several types of electrophilic aromatic substitution reactions. The examples used do not exhaust all the reagents that can act as electrophiles in aromatic substitution. In each of the following cases, predict the structures of the electrophile and of the product of the reaction.

(a) [benzene with CH$_3$NCH$_3$ group] + HONO ⟶

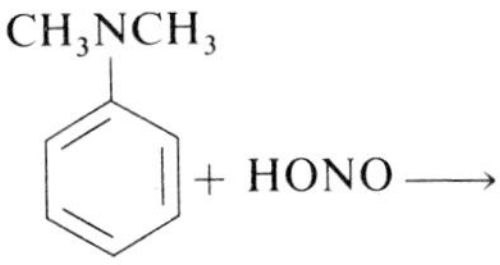

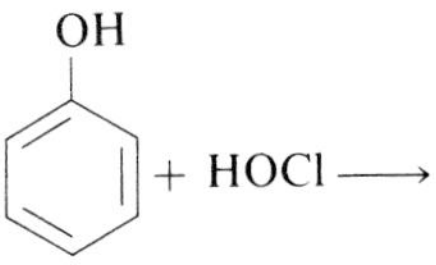

(b) [benzene with OH group] + HOCl ⟶

PROBLEM 12.21 Synthesize the following compounds from benzene and any other organic reagents containing three or fewer carbon atoms. There may be more than one way to carry out the synthesis. Aim for the fewest steps.

(a) ⬡—CH$_2$CH$_2$CH$_2$CH$_3$

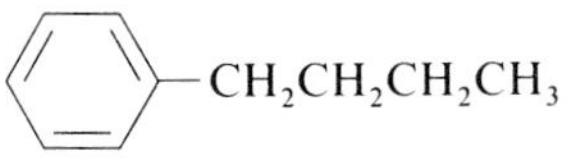

(b) ⬡—$\underset{\displaystyle H}{CH_2}$$\overset{}{\underset{}{C}}$=$\overset{}{\underset{\displaystyle H}{C}}$—$\overset{\displaystyle CH_2CH_3}{}$

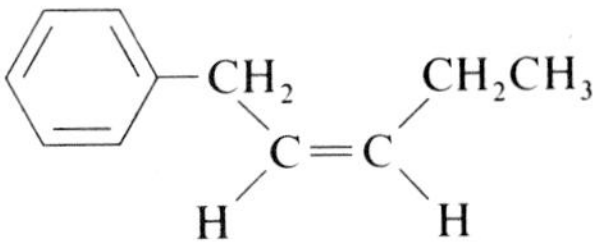

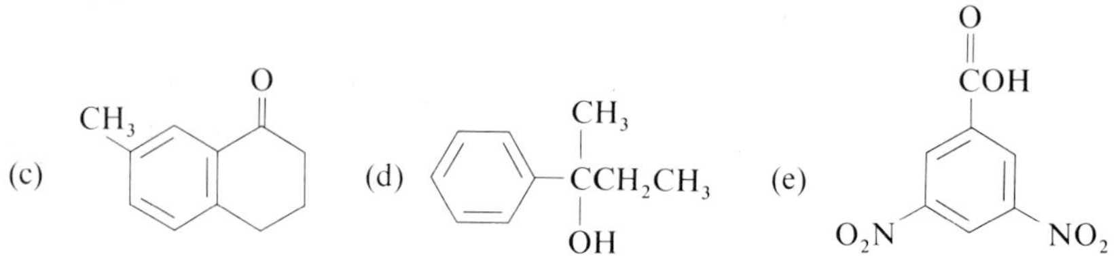

(c) [structure: 7-methyl-1-tetralone]   (d) [structure: 2-phenyl-2-butanol, with CH₃, CCH₂CH₃, OH] (e) [structure: 3,5-dinitrobenzoic acid with COH, O, O₂N, NO₂]

12.4

Reactions at Side Chains of Arenes

A. Substitution Reactions at the Benzylic Position. Halogenation Reactions

When toluene reacts with bromine in the presence of a Lewis acid, two products are formed: *o*-bromotoluene and *p*-bromotoluene.

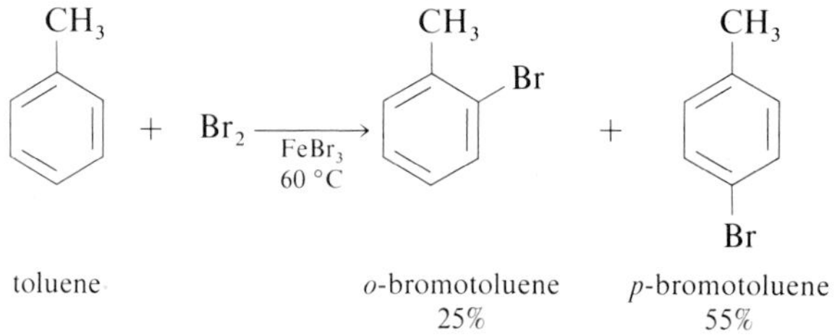

[reaction: toluene + Br_2 →(FeBr₃, 60 °C) *o*-bromotoluene + *p*-bromotoluene]

toluene   *o*-bromotoluene   *p*-bromotoluene
                                          25%   55%

Besides the hydrogen atoms on the aromatic ring, however, toluene has hydrogen atoms on the methyl group. If bromine and toluene are brought together in the absence of a catalyst, a third substitution product, in which one of the hydrogen atoms on the methyl group has been replaced, is also formed.

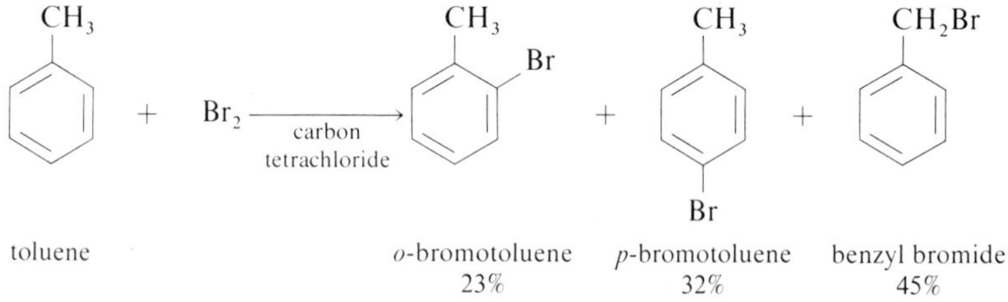

[reaction: toluene + Br_2 →(carbon tetrachloride) *o*-bromotoluene + *p*-bromotoluene + benzyl bromide]

toluene   *o*-bromotoluene   *p*-bromotoluene   benzyl bromide
                                    23%   32%   45%

The hydrogen atoms on the methyl group are on a benzylic carbon atom, one adjacent to an aromatic ring. Benzylic hydrogen atoms resemble allylic hydrogen atoms in their reactivity (Section 11.2). They are easily substituted in free-radical reactions because the radical that is formed when a benzylic hydrogen atom is abstracted is stabilized by resonance.

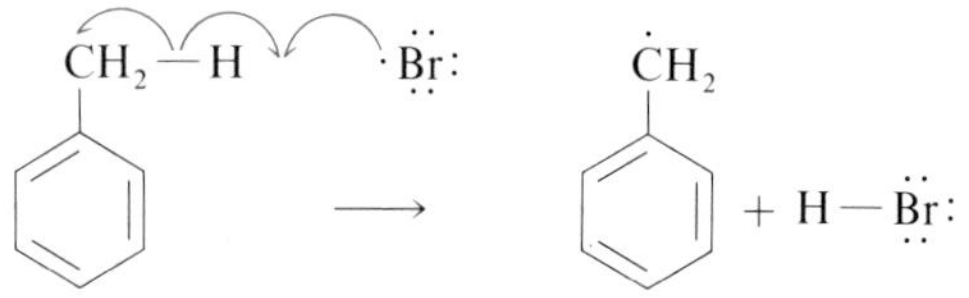

[reaction scheme showing formation of a benzyl radical: CH_2-H + $\cdot \ddot{Br}:$ → $\cdot CH_2$ + $H-\ddot{Br}:$]

formation of a benzyl radical

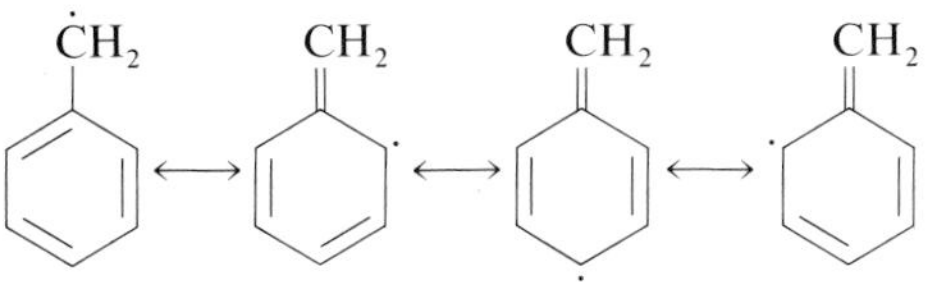

resonance stabilization of a benzyl radical

The deficiency of an electron, represented by the radical, is delocalized over the aromatic ring, giving rise to a relatively stable species. The radical abstracts a bromine atom from a bromine molecule, forming benzyl bromide and a new bromine atom to continue the reaction.

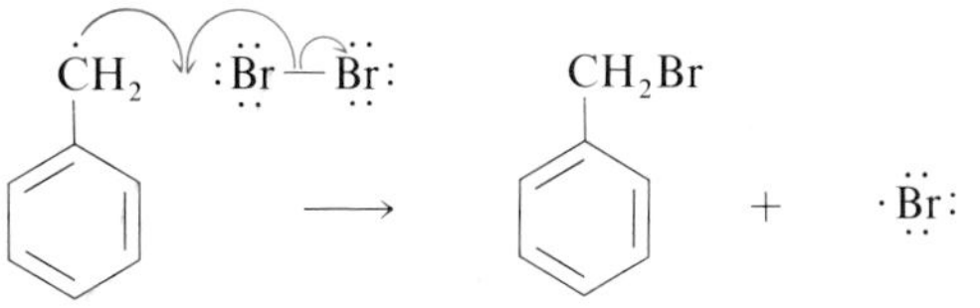

benzyl radical   bromine   benzyl bromide   bromine atom

Bromine can act as an electrophile towards the aromatic ring in toluene as well as a source of bromine atoms, which give rise to radical reactions at the side chain. Radical reactions with bromine are seen especially when heat or light is used to cause the dissociation of molecular bromine into two bromine atoms.

bromine   bromine
                           atoms

N-Bromosuccinimide, which was used to substitute bromine for allylic hydrogen (Section 11.2), has exactly the same kind of reactivity towards benzylic hydrogen atoms. It can be used to substitute selectively at that position, without any competing substitution on the aromatic ring. For example, a benzylic hydrogen atom on the carbon atom between two phenyl rings in diphenylmethane is selectively substituted by N-bromosuccinimide.

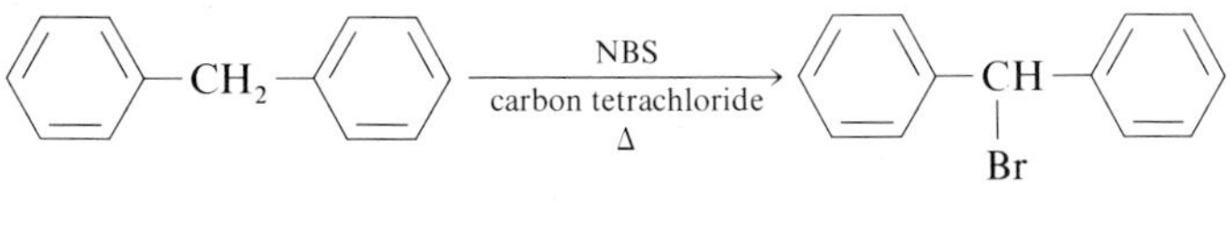

diphenylmethane   bromodiphenylmethane
                                                81%

The hydrogen atoms of a methyl group on naphthalene are also benzylic in nature and undergo the same substitution reaction.

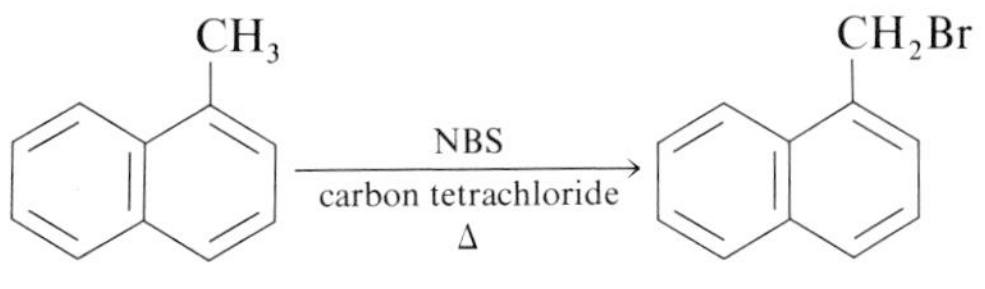

1-methylnaphthalene   1-(bromomethyl)naphthalene
                                            90%

Note that other types of hydrogen atoms are not affected by this substitution reaction. Benzylic radicals, along with allylic radicals, are much more stable than other types of radicals. Transition states leading to their formation are of lower energy than the transition states for competing reactions, so substitution reactions are channeled through such intermediates whenever possible.

A free radical that is stable enough to be observed as a discrete intermediate was first reported in 1900 by Moses Gomberg of the University of Michigan. He was trying to synthesize hexaphenylethane by the reaction of bromotriphenylmethane with silver. When he ran the reaction under an atmosphere of carbon dioxide, he obtained a white solid that was highly reactive towards oxygen and the halogens, including iodine. A solution of the solid in benzene prepared in the absence of air was yellow. The yellow color disappeared when a small amount of air was admitted into the reaction flask, but then developed again. Gomberg postulated that the yellow color was due to the formation of the triphenylmethyl radical in equilibrium with hexaphenylethane, and that the radical reacted with molecular oxygen, which is a diradical (Section 11.7A), to give a peroxide.

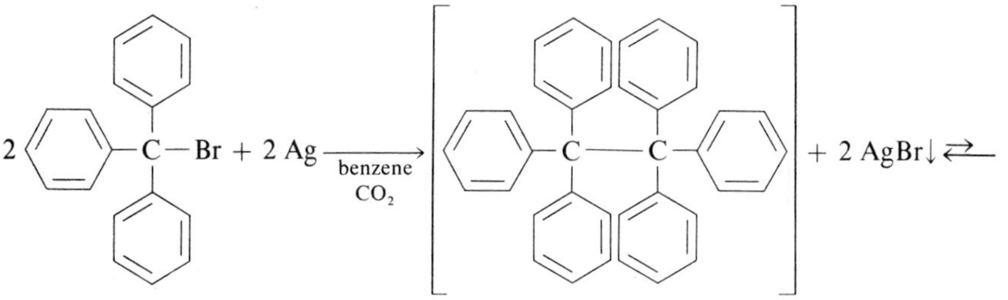

bromotriphenylmethane   hexaphenylethane
                                                         postulated as the
                                                         product

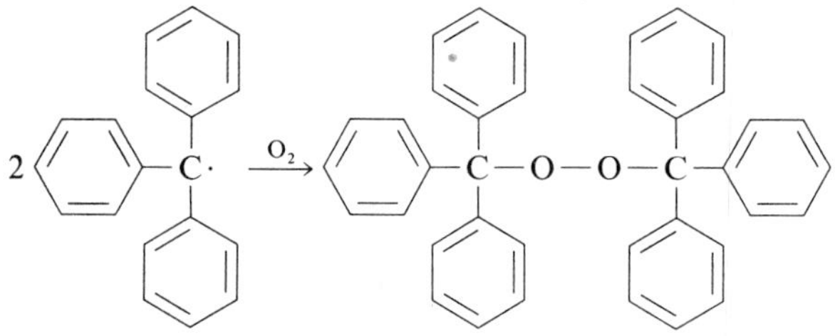

triphenylmethyl   peroxide, product
radical,   of the reaction
yellow   of the radical
                           with oxygen

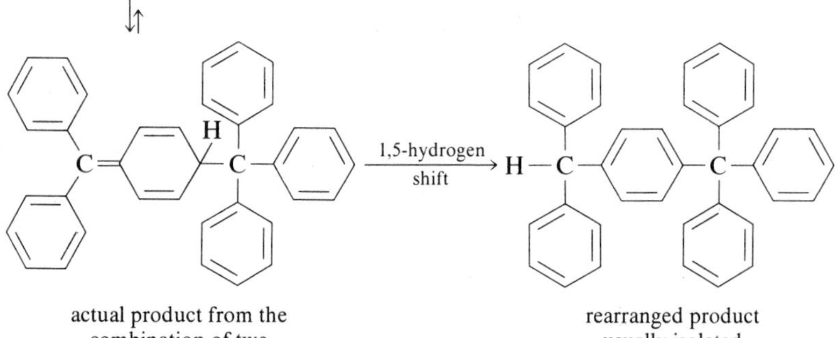

actual product from the   rearranged product
combination of two   usually isolated
triphenylmethyl radicals   from the reaction

Initially, many chemists opposed Gomberg's idea that a free radical could be stable enough to be observed as an intermediate in a reaction, but by 1911 a number of similar radicals had been prepared and characterized. Much later it was shown that a triphenylmethyl radical adds to the para postion of a phenyl ring in another radical to give the product observed and that hexaphenylethane, which would have a highly strained carbon-carbon single bond, does not form.

At present, radicals that are not only stable, but positively inert, have been synthesized. One of these is the perchlorotriphenylmethyl radical.

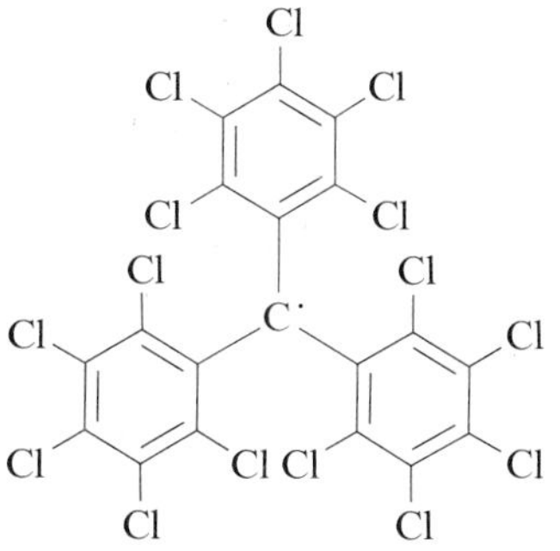

perchlorotriphenylmethyl radical
an inert free radical

This radical does not react with oxygen and is stable to 300 °C. The stability of the species is attributed to the presence of the large chlorine atoms that shield the radical center from contact with reagents.

PROBLEM 12.22   How do you explain the stability of the triphenylmethyl radical?

PROBLEM 12.23   Complete the following equations.

(a)

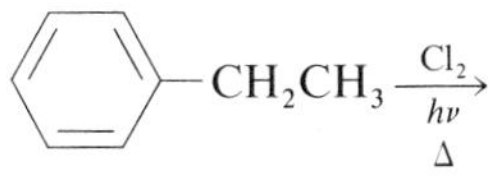

(b)

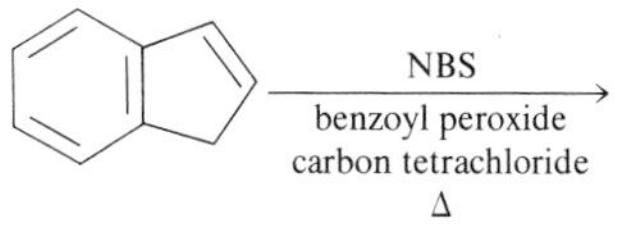

(c)

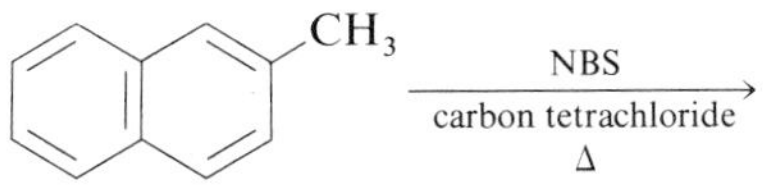

(d)

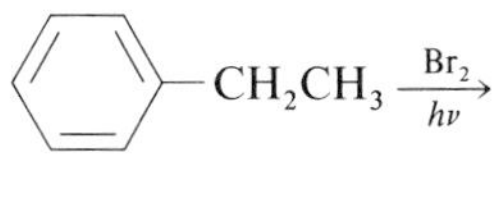

B. Oxidation of Side Chains in Arenes

The reactivity of the benzylic position of an alkylbenzene is reflected in the ease with which hydrocarbon side chains on aromatic rings undergo oxidation reactions. An alkane is normally inert to oxidation by reagents such as potassium permanganate (Sections 8.4A and 10.5A) or sodium dichromate (Section 7.6B).

Compounds having side chains on aromatic rings are oxidized to carboxylic acids by potassium permanganate or chromic acid. The aromatic ring itself is not affected. For example, *p*-nitrotoluene is converted to *p*-nitrobenzoic acid by heating for one hour with an acidic sodium dichromate solution.

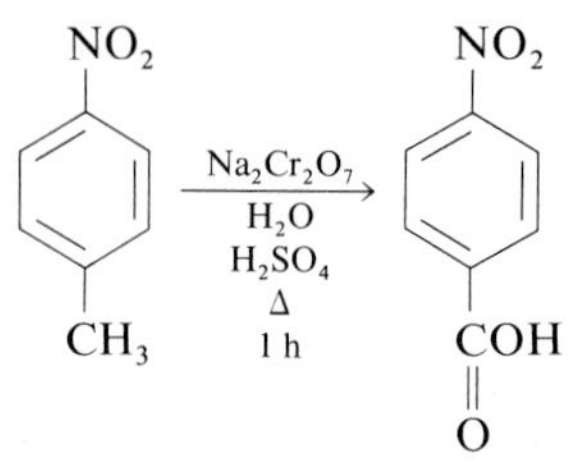

NO₂   NO₂

$$\xrightarrow[\substack{H_2O \\ H_2SO_4 \\ \Delta \\ 1\,h}]{Na_2Cr_2O_7}$$

CH₃   COH
‖
O

p-nitrotoluene   *p*-nitrobenzoic acid
86%

Under vigorous conditions, potassium permanganate oxidizes *p*-chlorotoluene to *p*-chlorobenzoic acid.

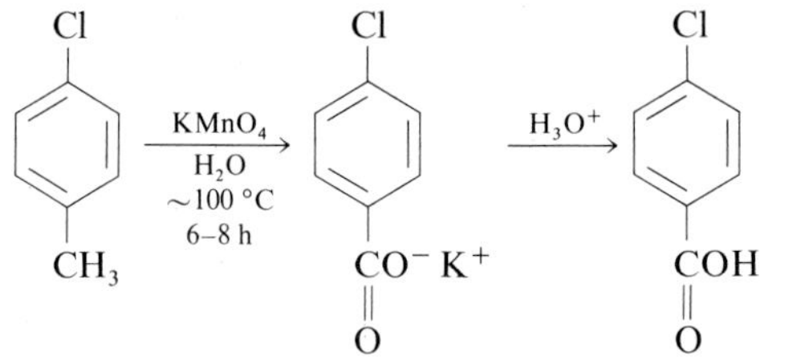

Cl   Cl   Cl

$$\xrightarrow[\substack{H_2O \\ \sim 100\,°C \\ 6{-}8\,h}]{KMnO_4}$$ $$\xrightarrow{H_3O^+}$$

CH₃   CO⁻ K⁺   COH
‖   ‖
O   O

p-chlorotoluene   potassium   *p*-chlorobenzoic
          p-chlorobenzoate   acid
                           89%

Reduction of potassium permanganate in water creates a basic solution even when no other base is used for catalysis (Section 8.4A). The acid is formed as its potassium salt and the reaction mixture must be acidified to recover the free carboxylic acid.

Even if the side chain on the aromatic ring has more than one carbon atom, the reaction produces an aromatic acid on vigorous oxidation. The other carbon atoms of the chain are lost, as is seen in the conversion of 1-phenylbutane to benzoic acid by aqueous sodium dichromate.

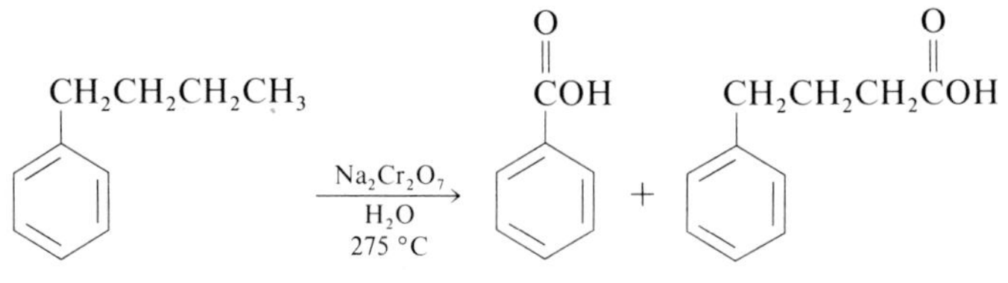

CH₂CH₂CH₂CH₃ $\overset{O}{\overset{\|}{C}}$OH   CH₂CH₂CH₂$\overset{O}{\overset{\|}{C}}$OH

$$\xrightarrow[\substack{H_2O \\ 275\,°C}]{Na_2Cr_2O_7}$$ +

1-phenylbutane   benzoic acid   4-phenylbutanoic
                   84%   acid
                                  14%

The major product of the reaction arises from oxidation at the benzylic carbon atom of the side chain. A side-product, 4-phenylbutanoic acid, comes from apparent oxidation of the methyl group at the end of the chain. Note that the reaction requires a temperature so high that the reaction must be carried out in a steel vessel, a ''bomb,'' under pressure.

All of these oxidation reactions are believed to proceed by the initial abstraction of a benzylic hydrogen atom by the oxidizing agent to give a stable benzylic radical. *tert*-Butyl groups are not oxidized, indicating the importance of this first step. The details of the mechanism from then on depend very much upon the oxidizing agent and the reaction conditions.

Before spectroscopic methods for determining the structures of compounds were developed, the oxidation of the side chains on aromatic rings was very useful. Note that the carboxylic acid group that is created always occupies the same position on the ring, relative to the other substituents, as the alkyl group from which it is derived. While many alkylbenzenes are liquids, the aromatic carboxylic acids are always solids. An oxidation reaction converts a substituted alkylbenzene to the corresponding substituted aromatic acid, and from the melting point of the acid, we can determine the structure of the original compound. For example, suppose we wish to distinguish three dimethylbenzenes, *o*-xylene, *m*-xylene, and *p*-xylene. They all have very similar boiling points, but are oxidized to dicarboxylic acids that are easy to distinguish.

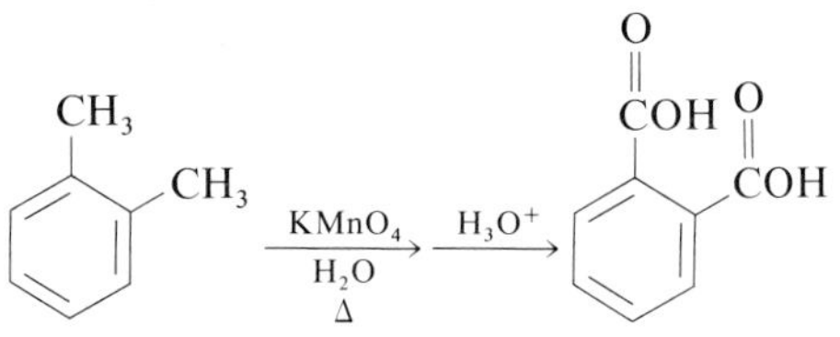

o-xylene
bp 144 °C

1,2-benzenedicarboxylic acid
phthalic acid
decomposes at 206 °C,
forms anhydride

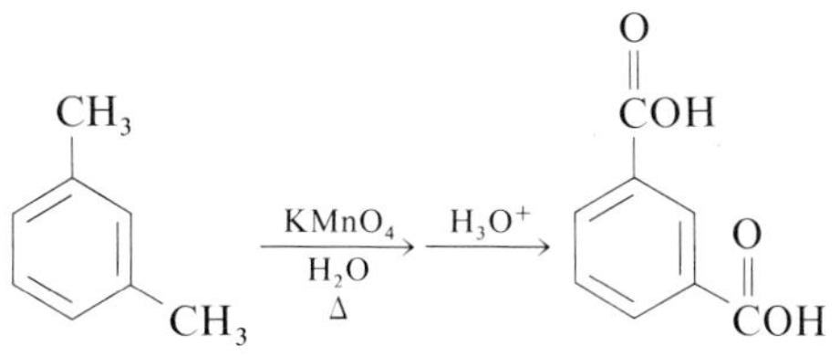

m-xylene
bp 139 °C

1,3-benzenedicarboxylic acid
isophthalic acid
mp 348 °C

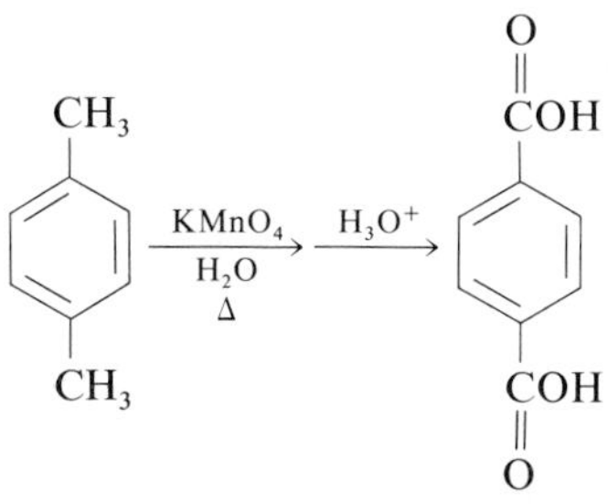

p-xylene
bp 138 °C

1,4-benzenedicarboxylic acid
terephthalic acid
sublimes at 300 °C

Note that the reactions that oxidize alkyl side chains require heating with the oxidizing agent for a period of time. If there are other functional groups on the side chain that can be oxidized, it is possible to select conditions so that they alone react and the side chain is not destroyed. For example, (*Z*)-1-phenylpropene is oxidized with cold dilute potassium permanganate to the diol.

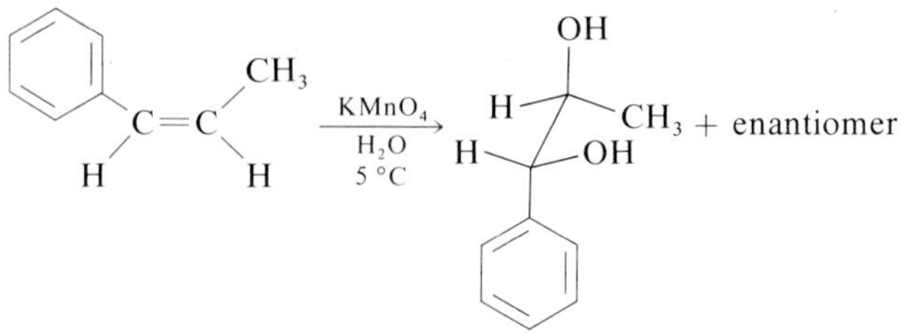

(Z)-1-phenylpropene   1-phenyl-1,2-propanediol
                                                41%

C. Oxidation of the Side Chain of an Arene by Oxygen

A tertiary benzylic hydrogen atom is particularly vulnerable to abstraction by free radicals, as we saw in Section 12.4A. This reaction allows for oxidation of alkylbenzenes with oxygen as the basis of an important commercial process that converts benzene and propene, both readily available hydrocarbons, into phenol and acetone.

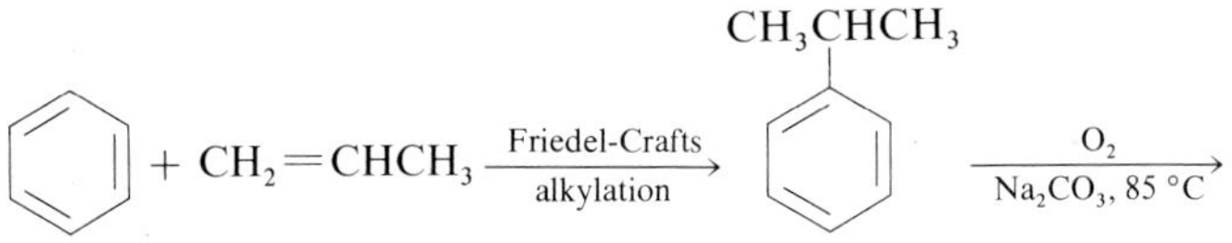

benzene   propene   isopropylbenzene
                                         cumene

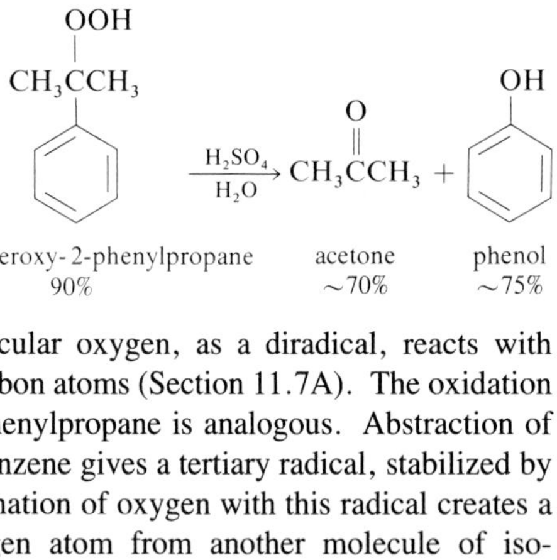

2-hydroperoxy-2-phenylpropane   acetone   phenol
            90%   ~70%   ~75%

We have seen reactions in which molecular oxygen, as a diradical, reacts with alkenes to give hydroperoxides at allylic carbon atoms (Section 11.7A). The oxidation of isopropylbenzene to 2-hydroperoxy-2-phenylpropane is analogous. Abstraction of a benzylic hydrogen atom from isopropylbenzene gives a tertiary radical, stabilized by delocalization to the aromatic ring. Combination of oxygen with this radical creates a peroxy radical, which abstracts a hydrogen atom from another molecule of isopropylbenzene to generate the reactive intermediate, the radical at the tertiary carbon atom, once more.

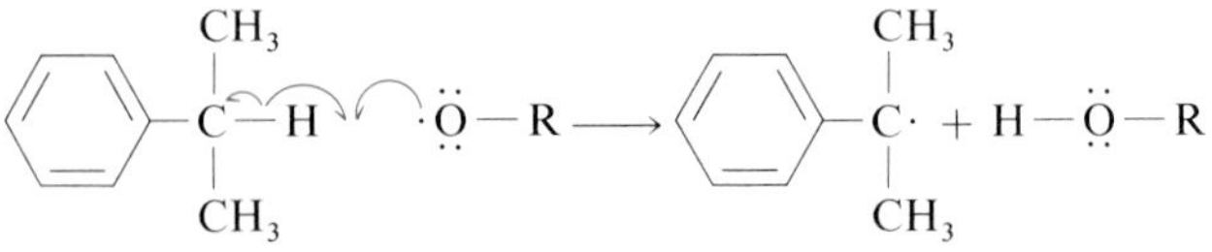

abstraction of hydrogen atom   stabilized radical
                                        benzylic and tertiary

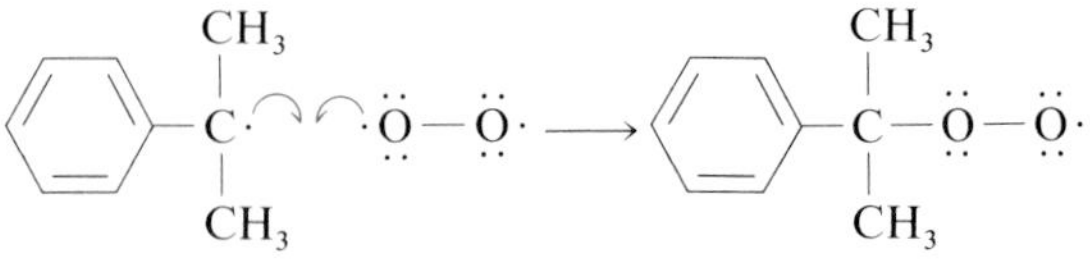

peroxy radical

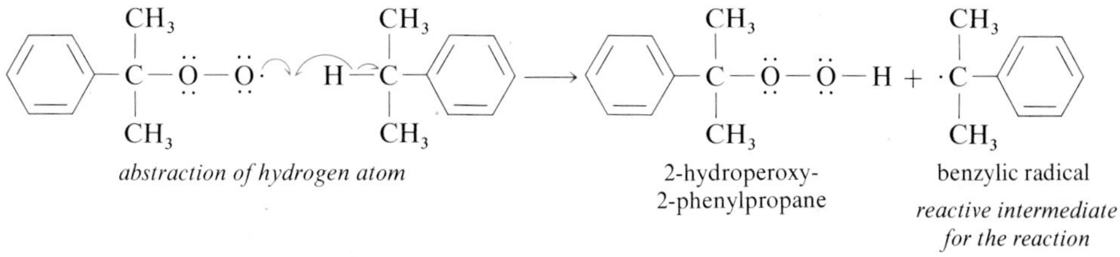

abstraction of hydrogen atom   2-hydroperoxy-   benzylic radical
                                        2-phenylpropane
                                                              reactive intermediate
                                                              for the reaction

2-Hydroperoxy-2-phenylpropane decomposes in dilute sulfuric acid to give phenol
and acetone. Protonation of the hydroperoxide in the acid solution would give a cation
that loses water as a leaving group as the phenyl group migrates to the oxygen atom that
is developing electron deficiency.

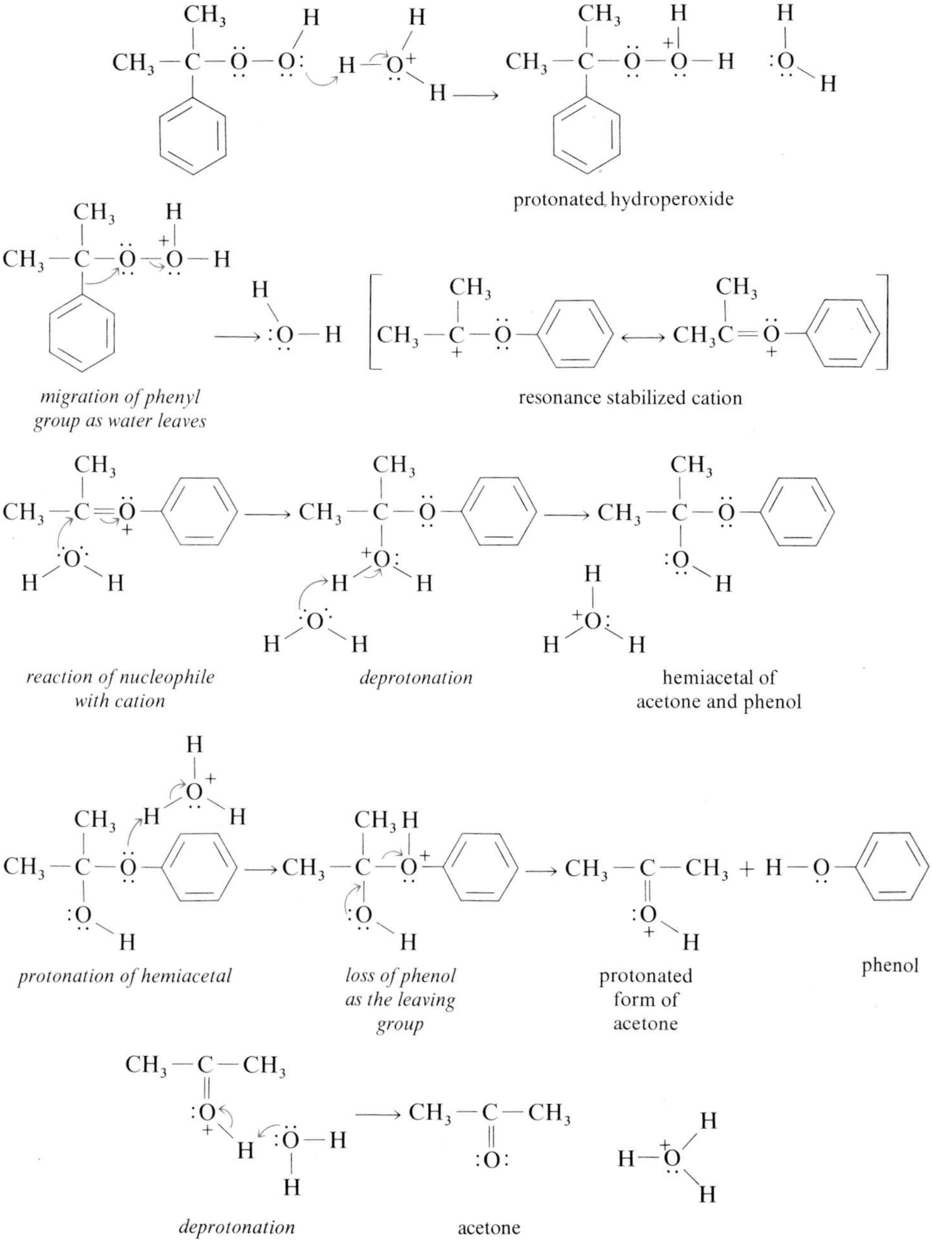

protonated hydroperoxide

migration of phenyl   resonance stabilized cation
group as water leaves

reaction of nucleophile   *deprotonation*   hemiacetal of
with cation   acetone and phenol

protonation of hemiacetal   *loss of phenol*   protonated   phenol
                               as the leaving   form of
                               group   acetone

deprotonation   acetone

The species that results reacts with water to give the hemiacetal of acetone with phenol. The hemiacetal breaks up in water, and acetone and phenol are produced.

PROBLEM 12.24 Complete the following equations.

(a) Ph—CH(OH)—Ph $\xrightarrow[\text{dichloromethane}]{\text{pyridinium } CrO_3Cl^-}$

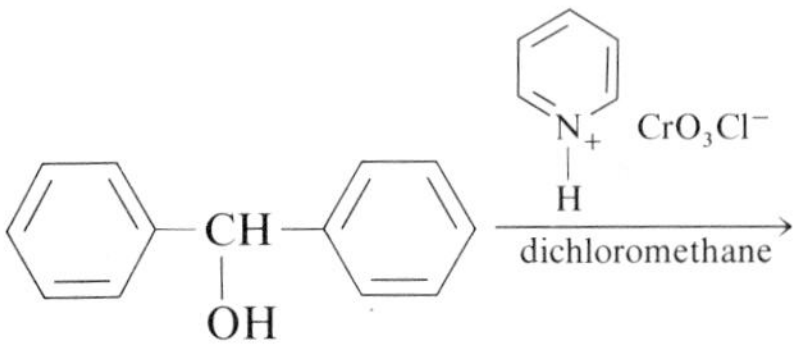

(b) Ph—CH_2CH_3 $\xrightarrow[\substack{\text{NaOH} \\ H_2O \\ \Delta}]{KMnO_4}$ $\xrightarrow{H_3O^+}$

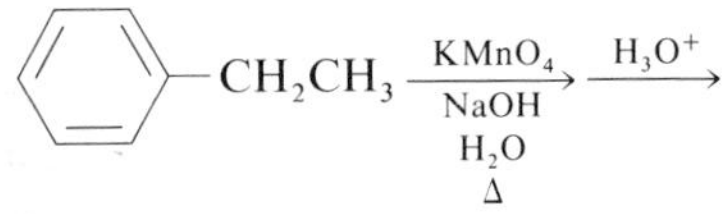

(c) indene $\xrightarrow[\substack{H_2O \\ 0\,°C \\ 30 \text{ min}}]{KMnO_4}$

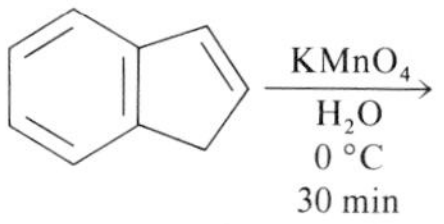

(d) (3-bromotoluene) $\xrightarrow[\substack{\text{NaOH} \\ H_2O \\ \Delta}]{KMnO_4}$ $\xrightarrow{H_3O^+}$

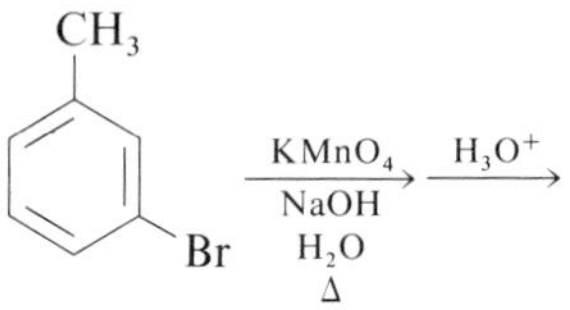

(e) (1,4-dimethylbenzene) $\xrightarrow[\substack{H_2O \\ H_2SO_4 \\ \Delta}]{Na_2Cr_2O_7}$

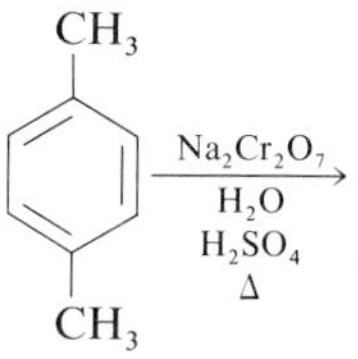

(f) $CH_3CHCH_2CH_3$-substituted benzene $\xrightarrow[\substack{Na_2CO_3 \\ H_2O \\ \Delta}]{O_2}$

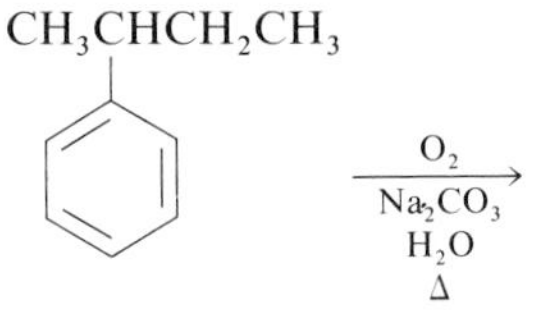

12.5

Nucleophilic Aromatic Substitution

An aryl halide that is not substituted by electron-withdrawing groups does not undergo S_N1 and S_N2 reactions. It is inert to the reagents that react with alkyl halides under the usual conditions. When there are electron-withdrawing groups, especially nitro groups ortho or para to the halogen, however, nucleophilic substitution takes place with relative ease. 2,4-Dinitrophenylhydrazine, the reagent used to characterize aldehydes and ketones, is synthesized by such a reaction.

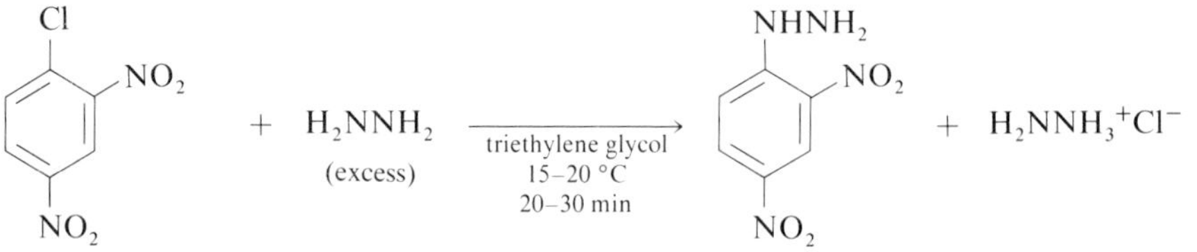

2,4-dinitrochlorobenzene (Cl, NO_2, NO_2) + H_2NNH_2 (excess) hydrazine $\xrightarrow[\substack{15-20\,°C \\ 20-30 \text{ min}}]{\text{triethylene glycol}}$ 2,4-dinitrophenylhydrazine ($NHNH_2$, NO_2, NO_2) + $H_2NNH_3^+Cl^-$ hydrazine hydrochloride

Hydrazine, a base and a nucleophile, displaces chloride ion from the aromatic ring. One of the products of the reaction is hydrogen chloride, which reacts with excess hydrazine to form a salt.

This same general reaction can occur with a variety of nucleophiles and leaving groups. For example, 2,4-dinitrochlorobenzene can be converted into 2,4-dinitrophenol by heating with sodium carbonate and water.

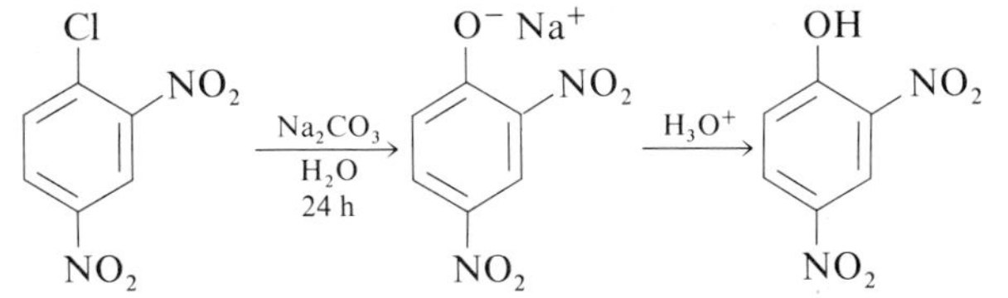

2,4-dinitro-
chlorobenzene

sodium
2,4-dinitrophenolate

2,4-dinitrophenol
91%

The nucleophile in this case is the hydroxide ion present in a solution of sodium carbonate in water.

A reagent that reacts very rapidly with amines is 2,4-dinitrofluorobenzene. It is used extensively in protein chemistry to label free amino groups on a protein or peptide chain (Section 17.3B). The typical reaction of the reagent with an amino acid, glycine, is illustrated below.

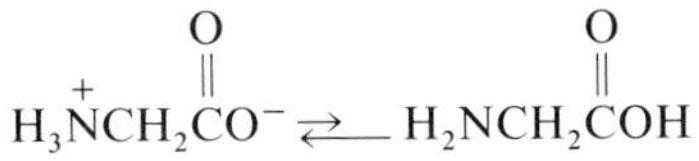

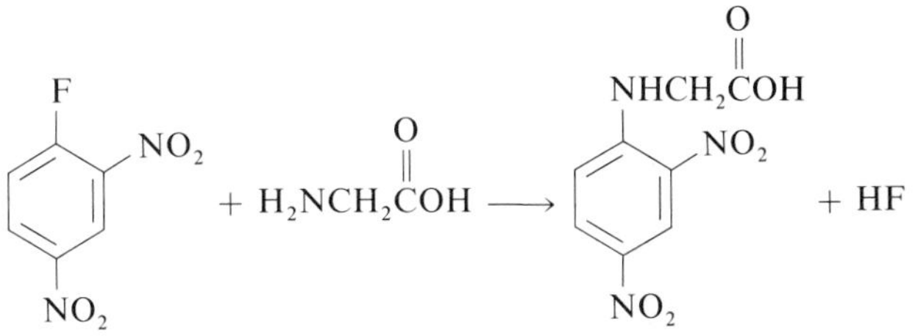

$$\overset{+}{H_3}NCH_2CO^- \rightleftharpoons H_2NCH_2COH$$

2,4-dinitrofluorobenzene   glycine   N-2,4-dinitrophenylglycine

Despite the fact that fluoride ion is a poor leaving group in nucleophilic substitution reactions of alkyl halides, it is an excellent leaving group in nucleophilic substitution reactions of aryl halides. In the following reaction in which the cyclic amine piperidine is used to displace halide ion from 2,4-dinitrohalobenzenes, the fluoro compound reacts 3300 times faster than the iodo compound.

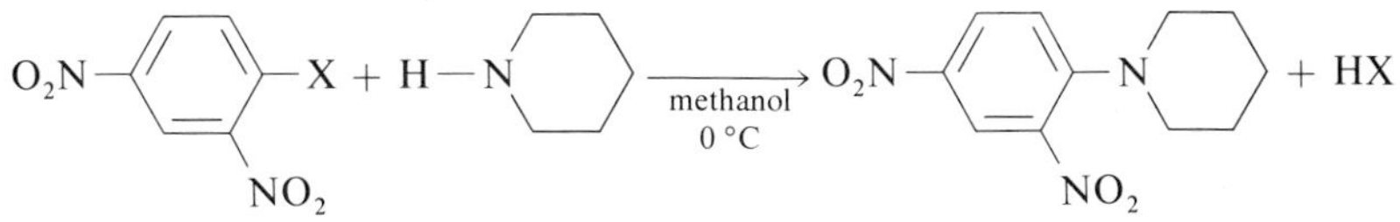

2,4-dinitrohalobenzene   piperidine   N-2,4-dinitrophenylpiperidine

X	k_{rel}
F	3300
I	1

Spectroscopic and kinetic evidence suggest that the reaction takes place by attack of the nucleophile on the aromatic ring at electron-deficient positions ortho and para to the electron-withdrawing nitro groups. An intermediate forms with a tetrahedral carbon atom, like the cation that is formed in electrophilic aromatic substitution. In this case, however, the intermediate is rich rather than poor in electrons and is stabilized by delocalization of negative charge to the nitro groups. A mechanism for the reaction of 2,4-dinitrofluorobenzene with dimethylamine is shown on the next page.

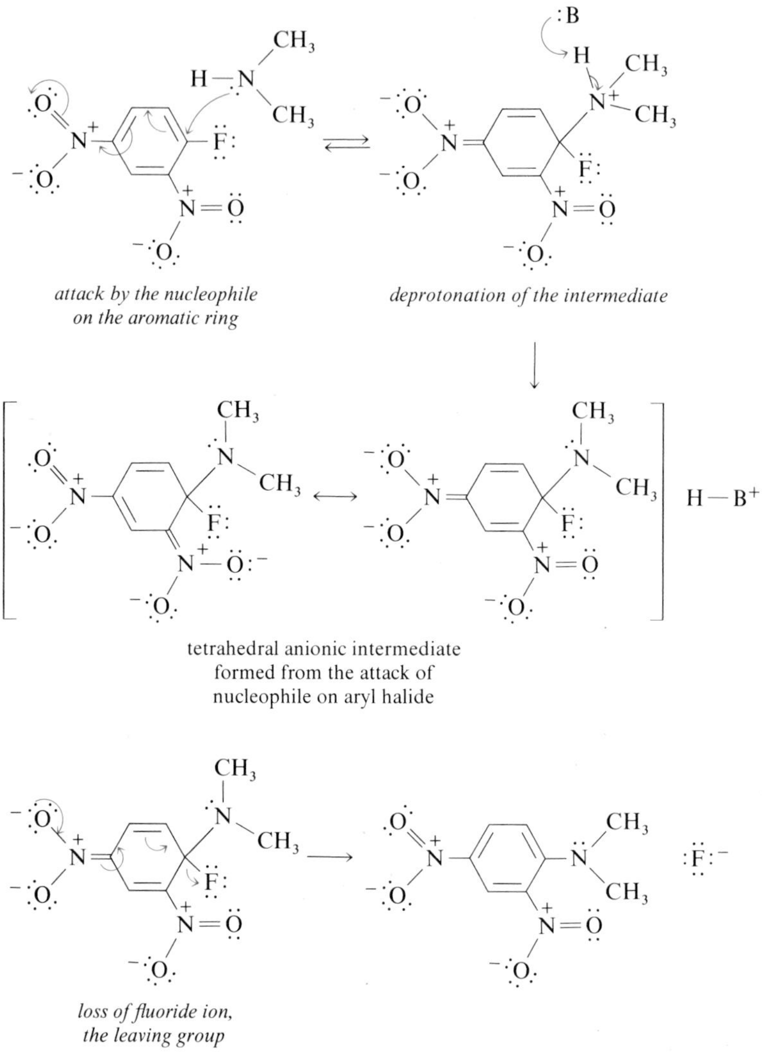

*attack by the nucleophile
on the aromatic ring*

deprotonation of the intermediate

tetrahedral anionic intermediate
formed from the attack of
nucleophile on aryl halide

*loss of fluoride ion,
the leaving group*

The rate-determining step is the formation of a tetrahedral intermediate by nucleophilic attack of the amine at the carbon atom bearing the halogen atom. A fluoro compound reacts much faster than an iodo compound (p. 619) for two reasons. First, fluorine is much more electronegative than iodine. The carbon atom to which the halogen atom is attached has a much larger positive charge in a fluorobenzene than in an iodobenzene. Second, a fluorine atom is much smaller than an iodine atom and, thus, allows the nucleophile to approach and bond to the carbon atom to give a tetrahedral intermediate much more easily.

Once the tetrahedral intermediate is formed, the amine loses a proton to a base and the resulting anion is stabilized by delocalization of charge to the nitro groups. Departure of a fluoride ion reestablishes aromaticity and gives the product. This step is fast relative to the first step of the reaction.

In some cases, especially when the incoming group and the leaving group are alkoxide ions, the intermediates are stable and can be isolated as salts.

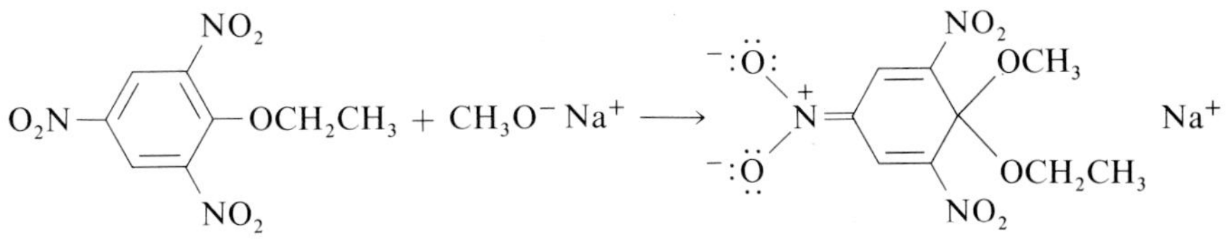

2,4,6-trinitroethoxybenzene   sodium methoxide   a Meisenheimer
                                                                    complex

These ionic compounds are called **Meisenheimer complexes** for the German chemist who first discovered them. The existence of such compounds is evidence that it is reasonable to postulate a tetrahedral anionic intermediate for nucleophilic substitution on aromatic rings.

Nucleophilic substitution on aryl halides, thus, does not go by an S_N2 mechanism, in which no intermediate is formed. Instead the reaction has two steps: addition of the nucleophile to give an intermediate, and elimination of the leaving group to restore the aromatic ring. The energy diagram for the nucleophilic aromatic substitution reaction resembles that for the electrophilic aromatic substitution reaction (Figure 12.2, p. 590).

PROBLEM 12.25   Complete the following equations.

(a)

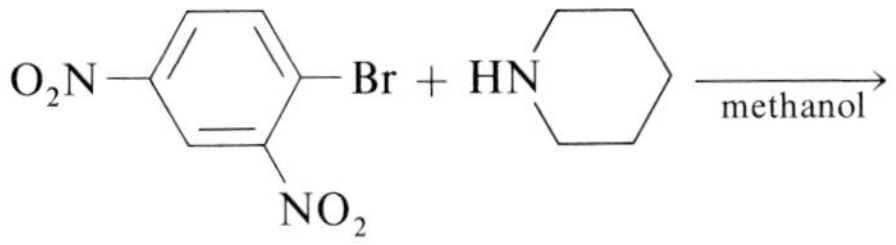

(b)

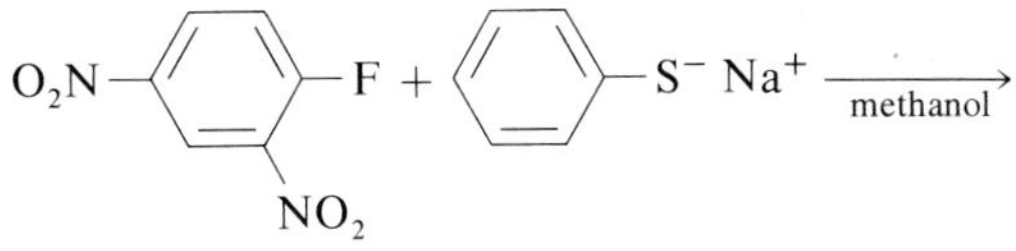

(c)

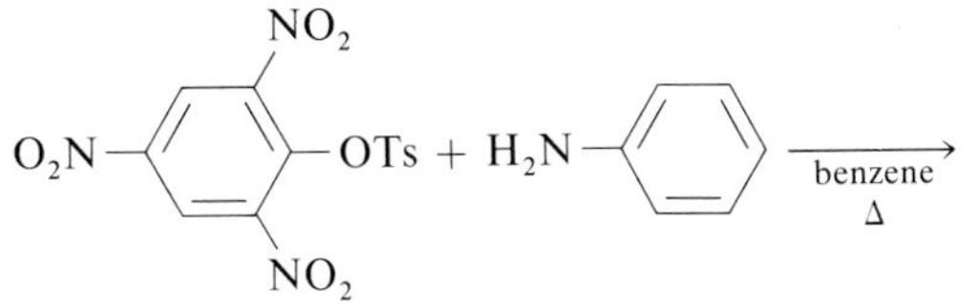

(d)

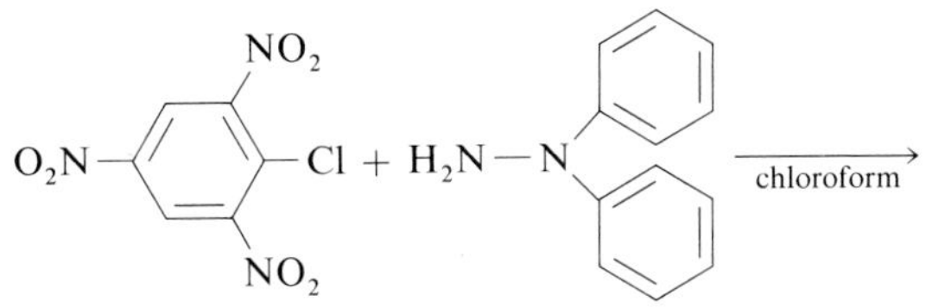

(e)

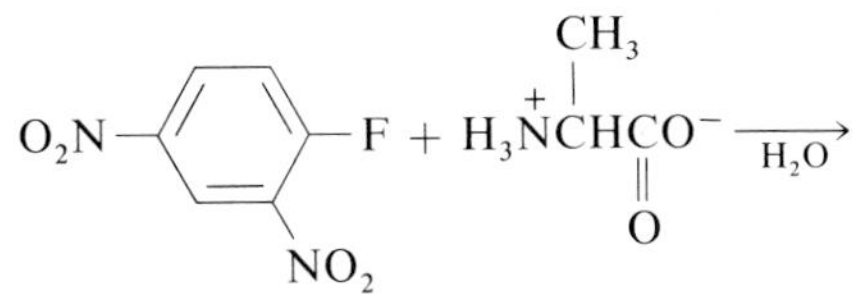

12.6

Phenols and Quinones

A. Acidity of Phenols

Phenols *are compounds in which a hydroxyl group is bonded directly to an sp^2-hybridized carbon atom of an aromatic ring. A typical phenol is much more acidic than an alcohol, but less so than a carboxylic acid.* Phenol itself has pK_a 10.0, while the pK_a of benzyl alcohol is approximately 16 and that of benzoic acid, 4.2. As was emphasized earlier (Sections 4.7 and 7.1A), the anion formed on the removal of a proton from an alcohol is strongly basic because the negative charge is localized on the oxygen atom. A carboxylate anion is much less basic than an alkoxide anion, because the negative charge is delocalized to two oxygen atoms. The anion of a phenol, a phenolate anion, also has delocalization of charge. Resonance contributors can be written showing negative charge at the ortho and para positions of the aromatic ring.

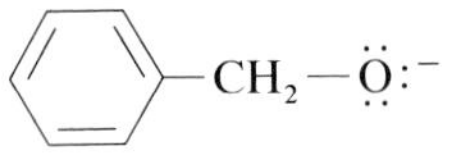

benzylate anion

no delocalization of charge
most basic

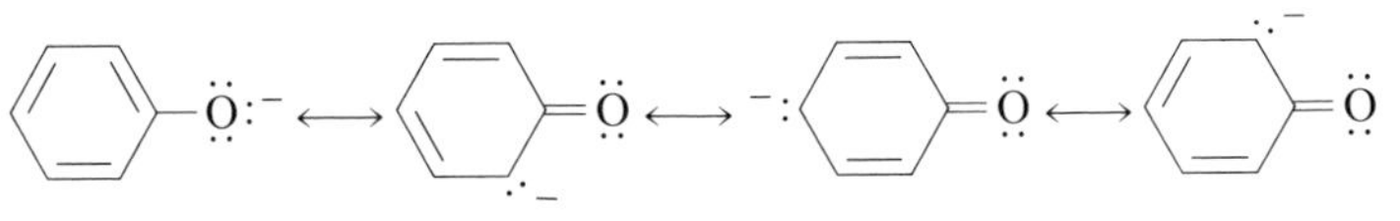

phenolate anion

delocalization of charge to aromatic ring

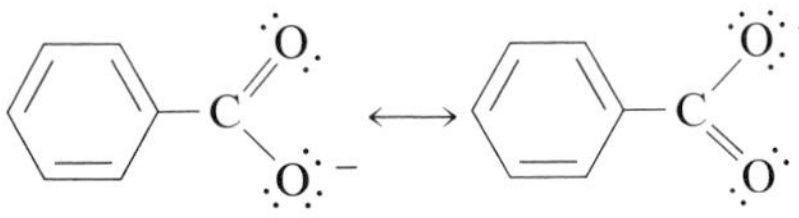

benzoate anion

delocalization of charge to 2 oxygen atoms
least basic

The benzoate anion has the highest stabilization and the lowest basicity of the three anions shown above. The two resonance contributors for the benzoate anion are equivalent, and both have the negative charge on the most electronegative of the atoms present. These factors stabilize the anion.

While there are more resonance contributors for the phenolate anion than there are for the benzoate anion, they are not all equivalent, and three of them put a negative charge on carbon instead of on the more electronegative oxygen atom. Thus, the phenolate anion is stabilized relative to benzylate anion, in which no delocalization is possible, but not as much as benzoate ion is. *Note that the presence of an aromatic ring in a compound bearing a hydroxyl group does not make it a phenol. The criterion is whether the hydroxyl group is on a tetrahedral carbon atom, in which case the compound is an alcohol, or directly on the aromatic ring, in which case it is a phenol.*

The acidity of a phenol has practical consequences in the laboratory. For example, a phenol can be separated from a carboxylic acid by taking advantage of the difference in acidity. Benzoic acid, a solid carboxylic acid, is insoluble in cold water but dissolves in a solution of sodium bicarbonate (Section 10.3A). The solution process is accompanied by the evolution of bubbles of carbon dioxide, indicating that carbonic acid is being formed (Section 4.7).

   A phenol, such as 2-naphthol, is also insoluble in water. It does not react with aqueous sodium bicarbonate either, because it is less acidic than carbonic acid. It is not a strong enough acid to protonate the bicarbonate anion. A phenol is a stronger acid than water, however, and easily protonates hydroxide ion. Thus, 2-naphthol will dissolve in a dilute solution of sodium hydroxide in water.

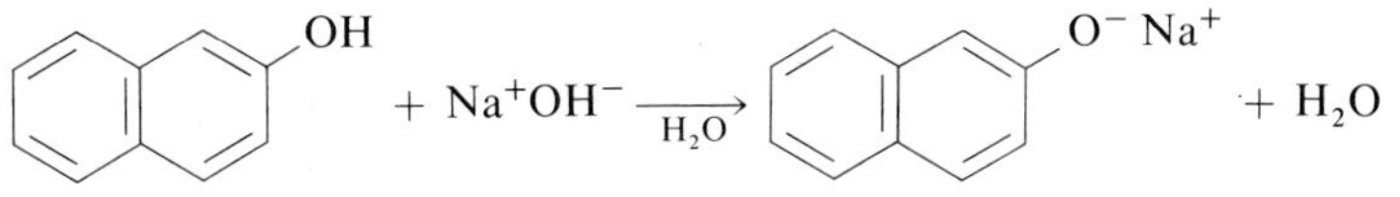

	2-naphthol		sodium 2-naphtholate	water
	$pK_a \sim 10$		*ionic compound*	pK_a 15.7
	covalent compound		*soluble in water*	
	insoluble in water			

The acidity of phenols is affected by substituents on the aromatic ring.

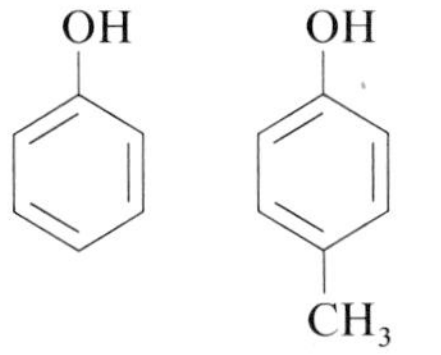

phenol   *p*-cresol
pK_a 10.0 pK_a 10.2

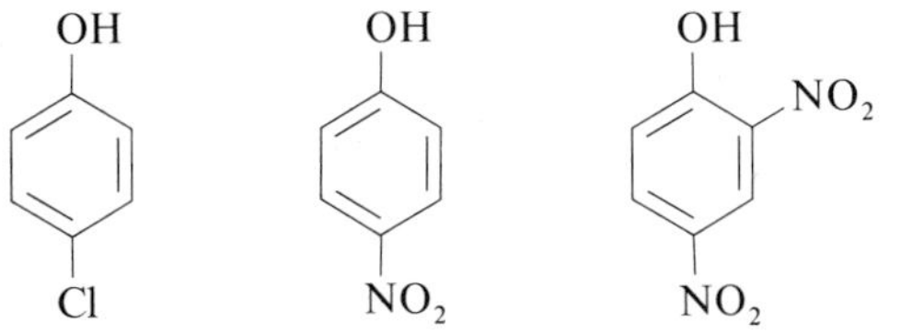

p-chlorophenol   *p*-nitrophenol   2,4-dinitrophenol
pK_a 9.38 pK_a 7.15 pK_a 4.02

Note that the effects of substituents on phenols parallel the effects that the same substituents have on the acidity of carboxylic acids (Section 10.1A).

PROBLEM 12.26

(a) The pK_a of *m*-nitrophenol is 8.39, while that of *p*-nitrophenol is 7.15. How would you rationalize these facts? The pK_a of phenol is 10.0.

(b) 2,4,6-Trinitrophenol, also called picric acid, has pK_a 0.25. Would you expect picric acid to be soluble in sodium bicarbonate solution? Explain.

PROBLEM 12.27   Assume that you are in a rather primitive laboratory and you need to make a distinction between a carboxylic acid, say *p*-chlorobenzoic acid, and a phenol, *p*-chlorophenol. You have no instruments, but you do have a supply of simple inorganic reagents such as sodium hydroxide, hydrochloric acid, and sodium bicarbonate. What would you do? How would you decide which compound was which?

B. Oxidation of Phenols. Quinones

The readiness of phenols to undergo electrophilic aromatic substitution reactions indicates an easy availability of electrons on the aromatic ring. *Oxidizing agents, which, like electrophiles, can be regarded as electron acceptors, also react with phenols as well as with aromatic amines. Two types of compounds that are oxidized with particular ease are aromatic compounds that have combinations of hydroxyl and/or amino groups ortho or para to each other. These compounds are oxidized to brightly colored carbonyl compounds known as* **quinones.**

An important example of this reactivity is the oxidation of 1,4-dihydroxybenzene, commonly known as hydroquinone, by silver ion. The reaction is used in developing photographic film, which contains finely divided grains of silver bromide in a gelatin layer. Light activates certain particles of silver bromide, so that they become especially susceptible to reduction to free silver. The development process converts these silver bromide particles to silver metal, which creates a darkening of the film at those points. A basic aqueous solution of hydroquinone is used as the reducing agent. The chemistry of the process in its bare essentials is shown below.

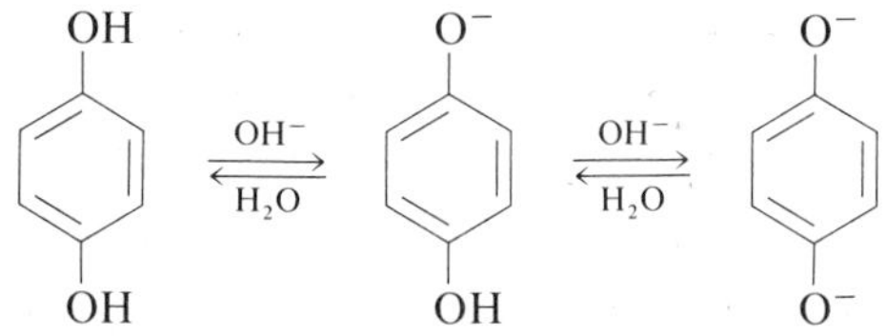

hydroquinone   anions of hydroquinone in
equilibrium with it in basic solution

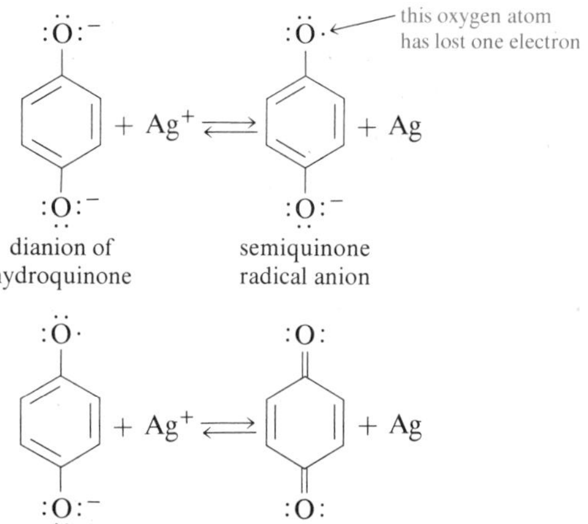

dianion of   semiquinone
hydroquinone   radical anion

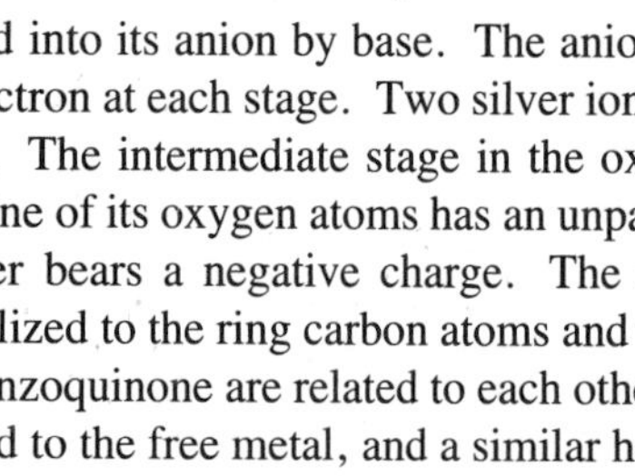

semiquinone   *p*-benzoquinone
radical anion   *yellow*

Hydroquinone is converted into its anion by base. The anion is oxidized in two steps by silver ion, losing an electron at each stage. Two silver ions gain one electron apiece to become metallic silver. The intermediate stage in the oxidation is a radical anion known as semiquinone. One of its oxygen atoms has an unpaired electron and thus has radical character; the other bears a negative charge. The radical character and the negative charge are delocalized to the ring carbon atoms and to the other oxygen atom.

Hydroquinone and *p*-benzoquinone are related to each other by oxidation-reduction, just as a metal ion is related to the free metal, and a similar half reaction can be written for them. The reactions are summarized below as two half reactions.

reduction half reaction: $Ag^+ + e^- \rightleftharpoons Ag$
                             silver ion   metallic silver

oxidation half reaction:

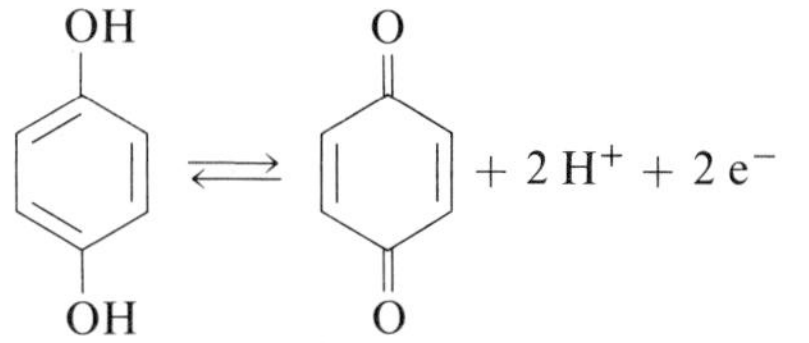

hydroquinone   *p*-benzoquinone

It is obvious from an examination of the two half reactions that it will take two silver ions to oxidize one molecule of hydroquinone to quinone.

Hydroquinone, which is colorless, forms a deeply colored complex with an equimolar amount of quinone. This complex, called **quinhydrone,** is much less soluble than either of its components. The two molecules of hydroquinone and quinone are held together by the attraction of the electron-deficient quinone ring for the electrons of the electron-rich hydroquinone ring. Such a complex is called a **charge-transfer complex,** indicating that some actual transfer of electronic charge takes place between the rings so that the bond between them has ionic character to a greater or lesser extent.

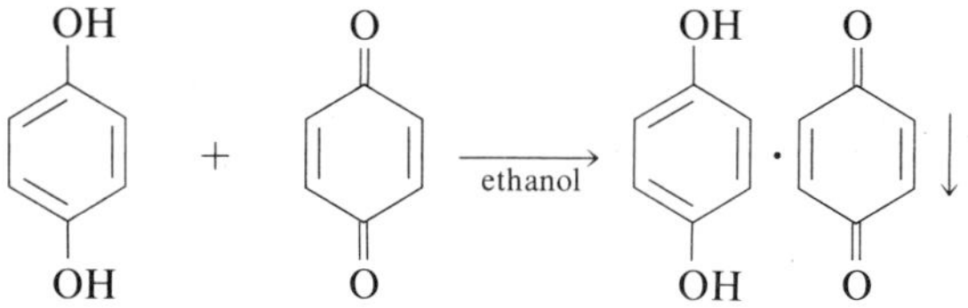

hydroquinone   *p*-benzoquinone   quinhydrone
  colorless   *yellow*   *deep green*

Quinones may have their two carbonyl groups para to each other, as in *p*-benzoquinone, or ortho to each other as in *o*-benzoquinone.

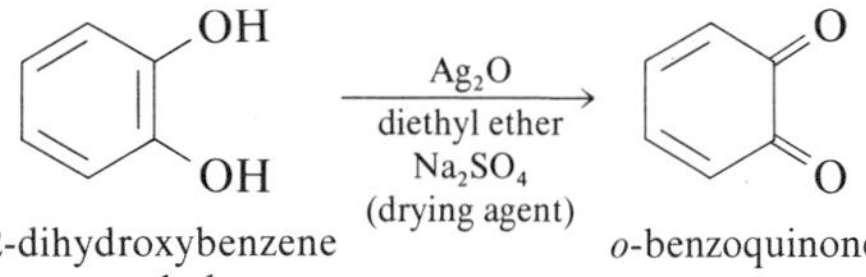

1,2-dihydroxybenzene   *o*-benzoquinone
      catechol

o-Benzoquinone is rather unstable, especially in the presence of moisture, and is prepared with very dry reagents. Some quinones are prepared by the direct oxidation of hydrocarbons. The vulnerability of the para positions in the central ring of anthracene, demonstrated by the addition of bromine and the Diels-Alder reaction in that ring (Section 12.2), allows the easy oxidation of anthracene to anthraquinone, a para quinone.

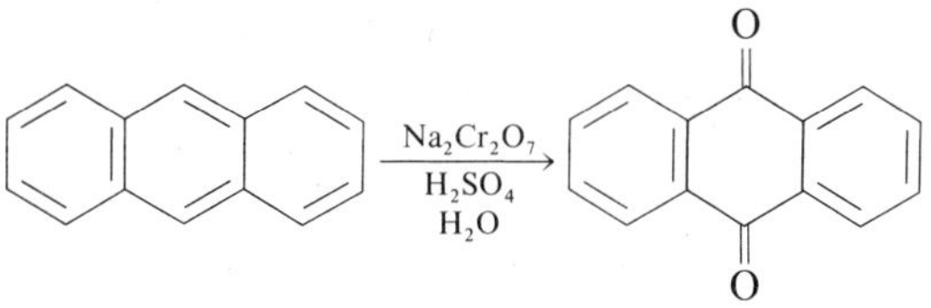

anthracene   anthraquinone
                                           ~90%
                                          yellow

Similarly, the central ring in phenanthrene undergoes oxidation to give an ortho quinone.

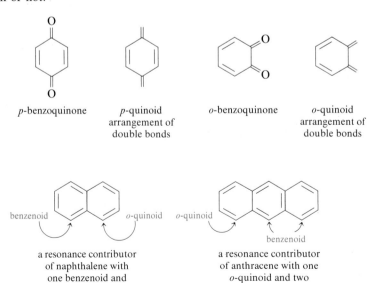

phenanthrene   phenanthraquinone
                                                         ~50%

                                                         orange

Note that a quinone no longer has aromatic character in the ring. It has a carbonyl group conjugated with a double bond, and many of its reactions are those of such compounds. Quinones will serve as dienophiles in Diels-Alder reactions (Section 11.5). *p*-Naphthoquinone, for example, adds 2,3-dimethyl-1,3-butadiene to give a reduced anthraquinone.

p-naphthoquinone   2,3-dimethyl-   *cis*-1,4,4a,9a-tetrahydro-
                            1,3-butadiene   2,3-dimethylanthraquinone
  dienophile   90%
                               diene

All quinones are intensely colored, with ortho quinones being more highly colored than para quinones. The chromophore responsible for their color is the particular arrangement of double bonds inside and outside the six-membered ring. The arrangements found in *p*-benzoquinone and *o*-benzoquinone are given the name of **quinoid** structures wherever they are found, whether there are oxygen atoms at the ends of the system or not.

p-benzoquinone   *p*-quinoid   *o*-benzoquinone   *o*-quinoid
                      arrangement of   arrangement of
                      double bonds   double bonds

benzenoid   *o*-quinoid   *o*-quinoid   benzenoid

a resonance contributor   a resonance contributor
of naphthalene with   of anthracene with one
one benzenoid and   *o*-quinoid and two
one *o*-quinoid ring   benzenoid rings

For example, the resonance contributors of polycyclic aromatic hydrocarbons can be analyzed in terms of whether the rings are benzenoid or quinoid in character. Resonance contributors that contain more benzenoid rings are the more important ones for polycyclic aromatic hydrocarbons.

Many important natural products are phenols or quinones. A number of the more complex structures, such as vitamin E and vitamin K, are discussed in Section 12.6D. Two simple plant products that are *p*-naphthoquinones are 2-hydroxy-*p*-naphthoquinone and 8-hydroxy-*p*-naphthoquinone.

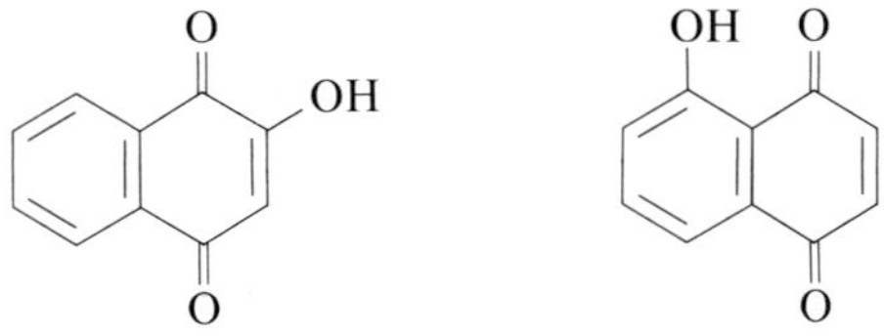

2-hydroxy-*p*-napthoquinone   8-hydroxy-*p*-naphthoquinone
Lawsone   Juglone

pigment in henna   *pigment from walnut shells*
orange dye   *brown dye*

The first compound is extracted from the leaves of the henna plant and is widely used by women in the Middle East to dye their hair, their palms, and their fingertips a reddish color. The second is the brown stain that is found in the outer soft shells of walnuts.

C. Oxidative Coupling Reactions of Phenols

Phenols with only one hydroxyl group undergo oxidation reactions with a wide variety of oxidizing agents to give compounds in which carbon-carbon bonds have been formed between aromatic rings. These reactions, like the ones in which the terminal carbon atoms of 1-alkynes are linked (Section 11.4B) *are called* **oxidative coupling reactions.** Unless the phenol is carefully chosen, a variety of products is obtained. 1-Naphthol, for example, gives three products when oxidized with ferric chloride.

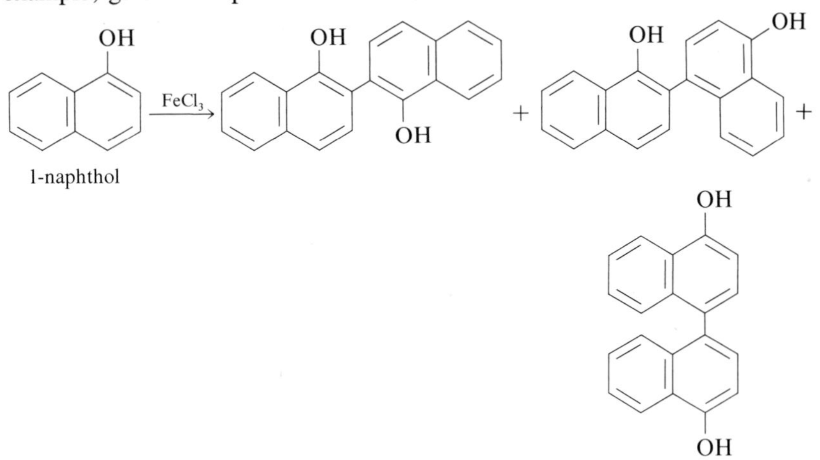

1-naphthol

The new carbon-carbon bond between aromatic rings is formed ortho or para to the hydroxyl group.

The mechanism suggested for the reaction starts with the removal of the equivalent of a hydrogen atom from the phenol to give a radical. Usually this happens by removal of a proton from the phenol by the basic reaction medium, and then the removal of an electron from the phenolate ion by the oxidizing agent. The oxygen radical is stabilized by delocalization of radical character to the ortho and para positions of the ring.

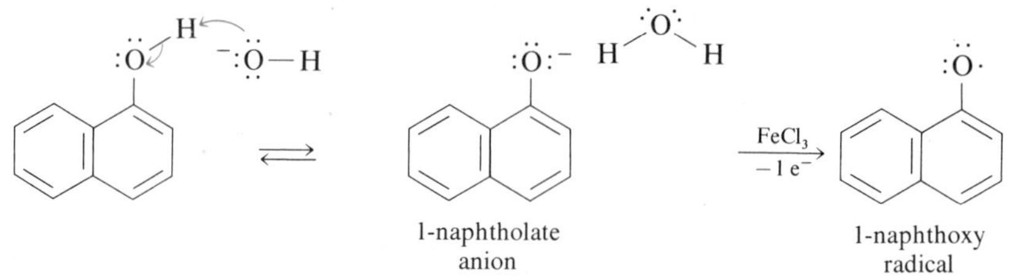

1-naphtholate
anion

1-naphthoxy
radical

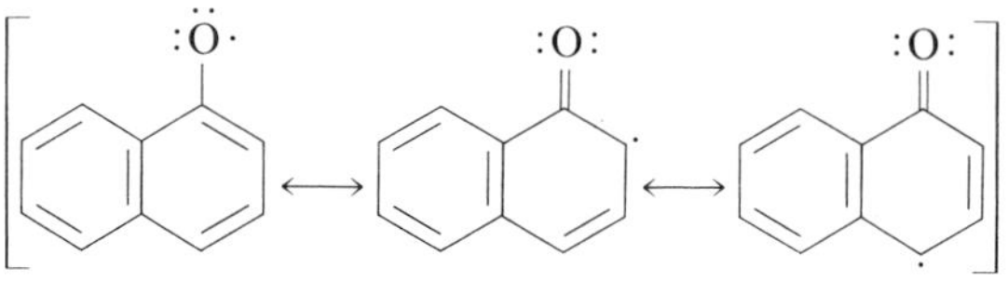

*resonance contributors of the
1-naphthoxy radical with radical
character ortho and para to
the oxygen atom*

Two radicals then combine to give an intermediate that tautomerizes to give the aromatic diphenol product.

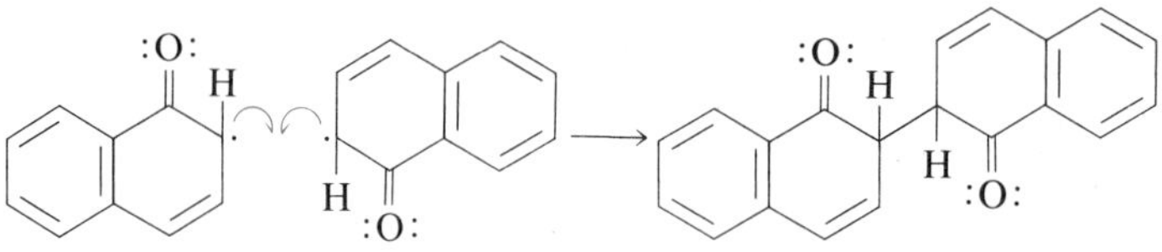

unstable intermediate from
the coupling of 2
radicals at the ortho position

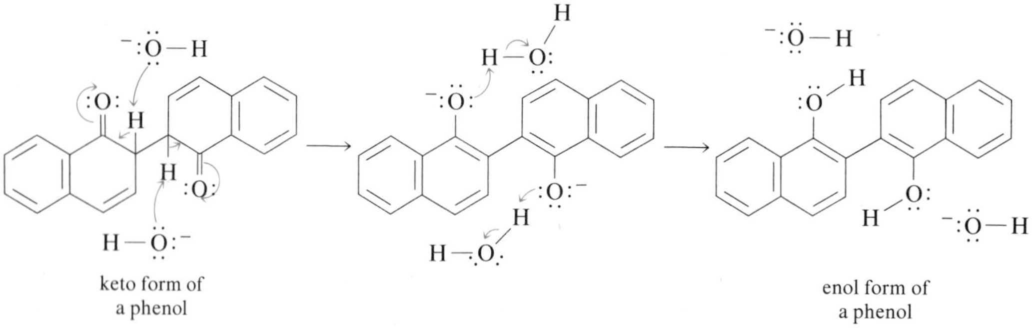

keto form of
a phenol

enol form of
a phenol

The oxidative coupling reactions of phenols are synthetically useful only when some of the positions on the ring are blocked by substituents so that a single product is formed in high yield. A good example of such a reaction is the oxidation of 2,6-di-*tert*-butylphenol by oxygen in a basic solution.

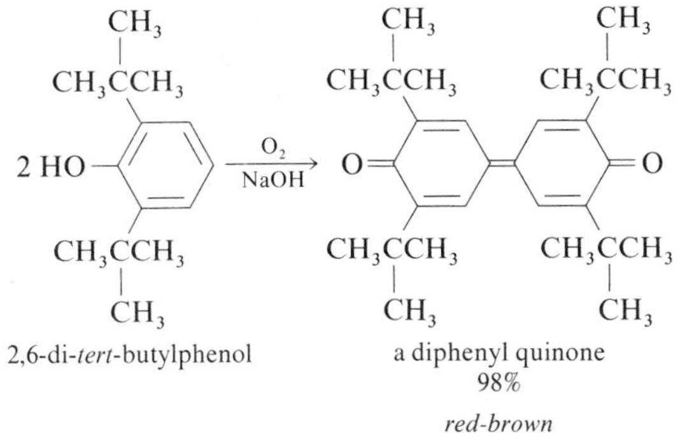

2,6-di-*tert*-butylphenol

a diphenyl quinone
98%

red-brown

The ortho positions of the phenol are blocked by groups so large that no coupling can take place there. Coupling takes place only at the para position, and the intermediate is further oxidized by oxygen to a quinone.

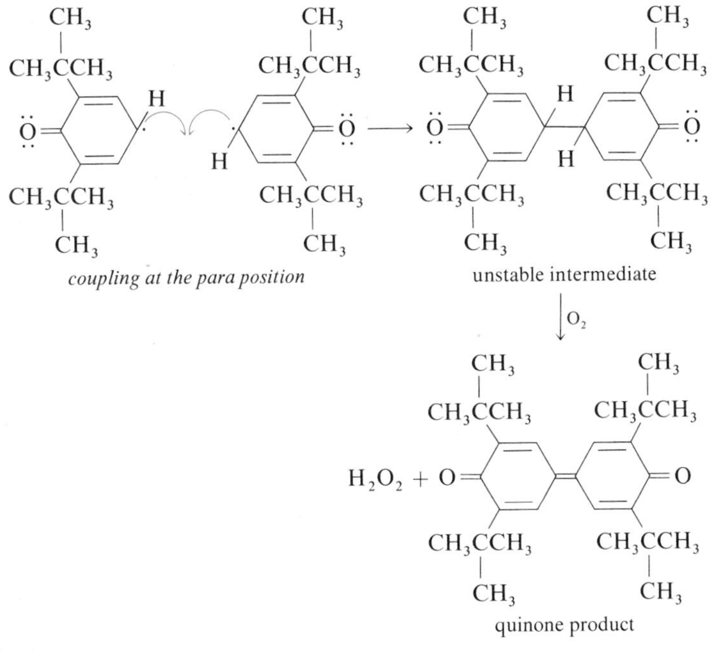

coupling at the para position

unstable intermediate

quinone product

Oxygen functions as an oxidizing agent at two stages: first to create the phenoxy radical, and second to remove hydrogen atoms from the intermediate formed in the coupling reaction and oxidize it to the quinone.

Another type of coupling reaction might be expected, that between a radical at the oxygen atom and another radical. Such products, which have ether bonds, become especially important when substituents prevent carbon-carbon coupling reactions. This kind of reaction is postulated to be the way the amino acid, diiodotyrosine is converted in biological systems into the thyroid hormone thyroxine.

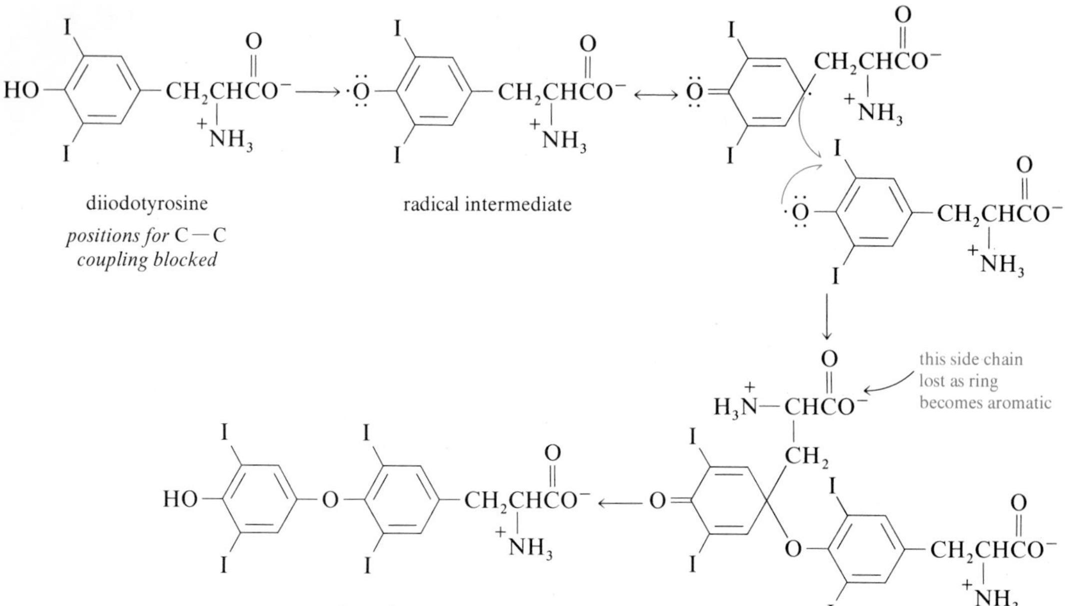

diiodotyrosine

positions for C—C
coupling blocked

radical intermediate

this side chain
lost as ring
becomes aromatic

thyroxine

Substituted phenols are used as **antioxidants** in processed foods. If you read the labels on cereals, cookies, rice products, and many other grocery items, you will almost certainly see BHA or BHT listed as an ingredient. The label will sometimes say that they were added to retard spoilage or rancidity. BHA stands for *butylated hydroxyanisole* and is a mixture of *tert*-butyl-methoxyphenols. BHT is *butylated hydroxytoluene*, 2,6-di-*tert*-butyl-4-methylphenol.

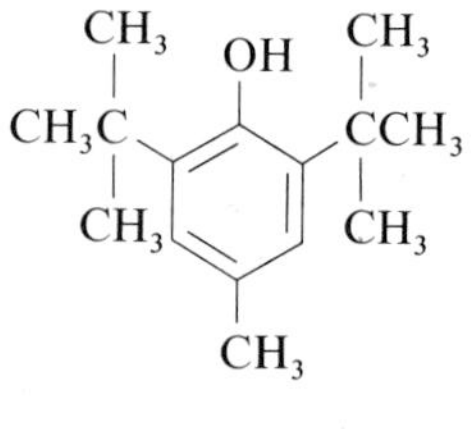

2,6-di-*tert*-butyl-4-methylphenol
butylated hydroxytoluene
BHT

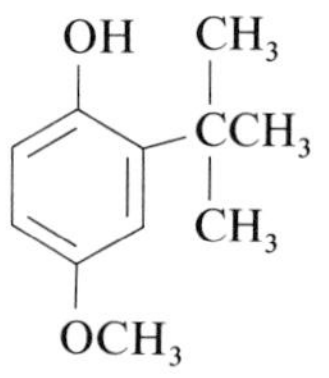

2-*tert*-butyl-4-methoxyphenol 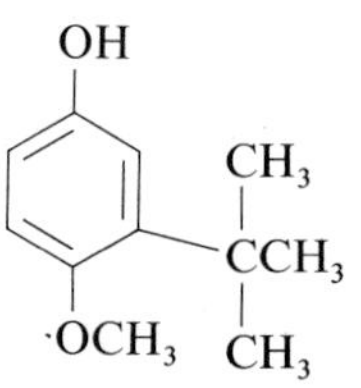   3-*tert*-butyl-4-methoxyphenol

butylated hydroxyanisole
BHA

two commercially important antioxidants

Oxygen from the air reacts with compounds containing double bonds by free-radical chain reactions (Section 11.7A). If these reactions are not stopped, the oils in food products, stored at room temperature in warehouses and on supermarket shelves for weeks, become oxidized and rancid. Not only do they taste bad, they are also toxic. As the food processing industry became more and more centralized, it became necessary for food to have a long shelf life. Therefore, chemists developed additives that break up the free-radical chain reactions by reacting themselves with any radicals that form. Phenols, as we have seen in this section, are reactive towards radicals. BHA and BHT give hindered phenoxy radicals that are unreactive and thus terminate free-radical chain reactions. As a result, they are used to retard radical-type oxidation reactions in food products, among other things. The natural antioxidant, found in foods that have high polyunsaturated fatty acid content such as vegetable oils, is vitamin E, discussed in the next section.

PROBLEM 12.28   Complete the following equations. To do so you must decide in each case whether the reagent over the arrow is an oxidizing agent or a reducing agent.

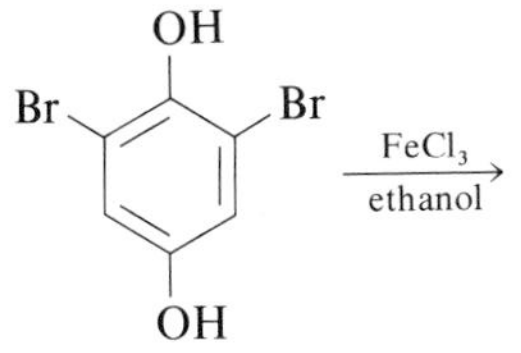

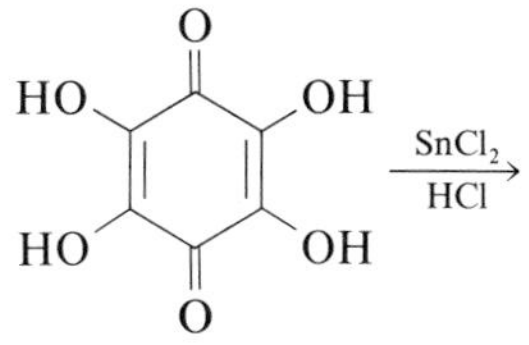

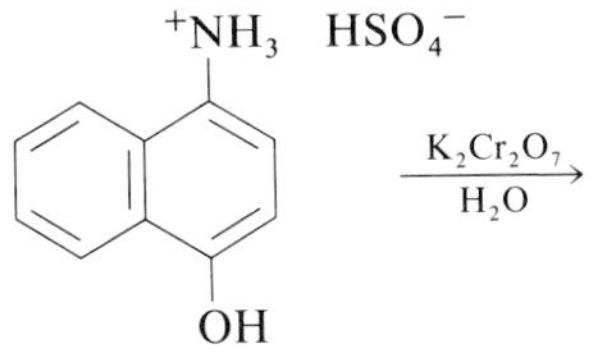

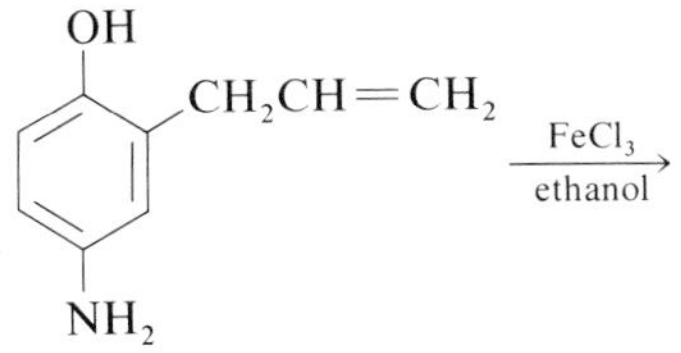

PROBLEM 12.29

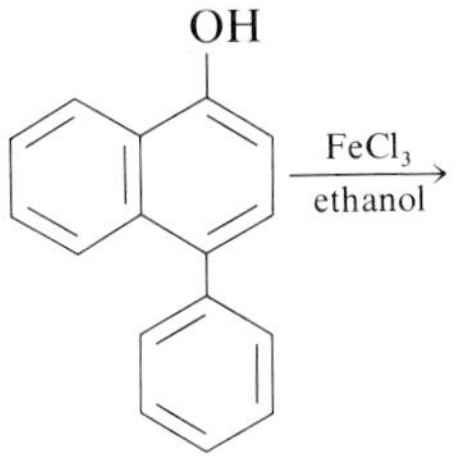

(a)   BHA (p. 630) is prepared commercially from p-methoxyphenol and 2-methylpropene. Suggest how this reaction could be carried out.

(b)   A similar reaction is used to make BHT (p. 630) from p-methylphenol and 2-methylpropene. Why does commercial BHA consist of a mixture of isomers, while the reaction to give BHT has greater regioselectivity?

D. Vitamin E and Vitamin K

Vitamin E, also called α-tocopherol, has phenolic and isoprenoid components. Its structure is derived from a trimethylhydroquinone and a chain containing four isoprene units, a phytyl group.

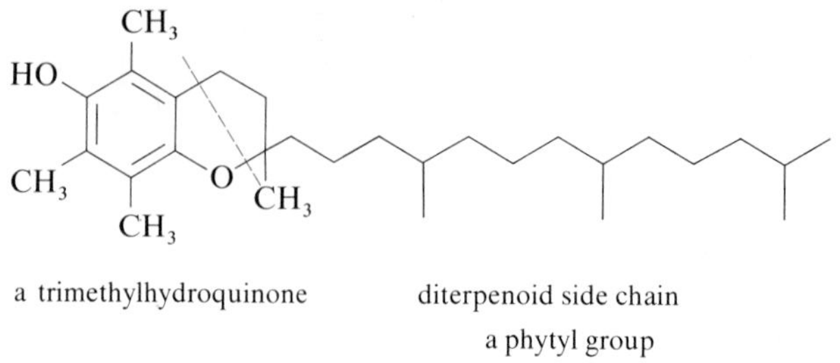

CH$_3$

HO

CH$_3$

CH$_3$

O CH$_3$

a trimethylhydroquinone

diterpenoid side chain

a phytyl group

α-tocopherol, vitamin E

Today, vitamin E is recognized for its biological role as an antioxidant. It is especially important in preventing the formation of hydroperoxides from polyunsaturated fatty acids (Section 11.7A) and is found most abundantly in seeds that are rich in such vegetable oils.

While a number of symptoms resulting from a deficiency of vitamin E are known, especially in premature babies, the total range of the activity of the vitamin has yet to be determined. Male rats fed a diet deficient in the vitamin are sterile; female rats conceive but cannot carry their babies to term. Vitamin E also affects blood clotting and is important in maintaining the integrity of cell membranes. Because human beings do not synthesize it in their bodies, they must obtain vitamin E from food. The best dietary sources are vegetable oils and wheat germ.

In 1930, researchers discovered that chicks developed hemorrhages when fed a highly artificial diet from which all fatty components had been removed. This observation led to the discovery that a fat-soluble vitamin, named vitamin K, was important in the clotting of blood. Later it was determined that a family of compounds has vitamin K activity. These compounds have a 2-methyl-1,4-naphthoquinone structure and differ in the isoprenoid side chains attached at carbon 3.

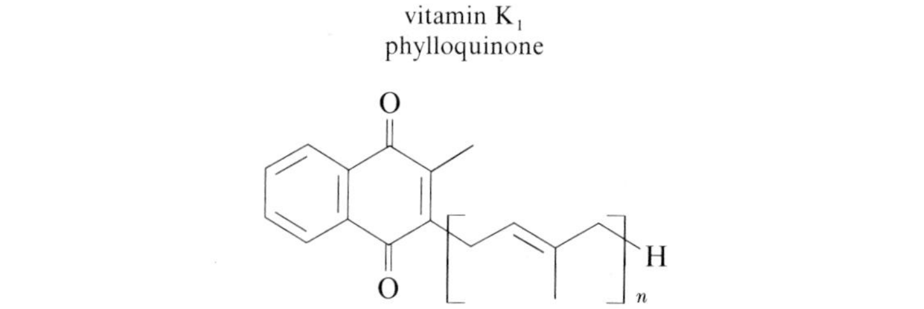

O

side chain derived from phytol

O

2-methyl-1,4-naphthoquinone

vitamin K$_1$
phylloquinone

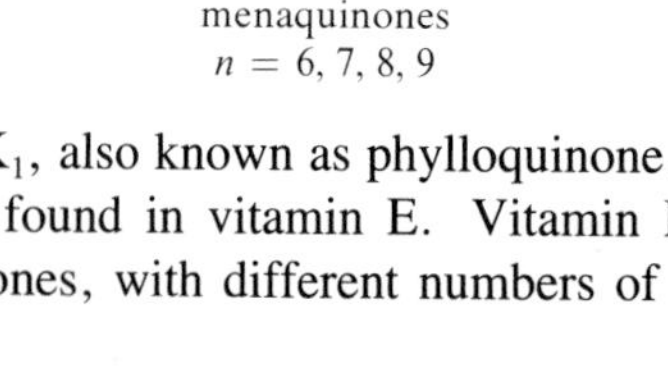

O

O

H

n

vitamin K$_2$
menaquinones
$n = 6, 7, 8, 9$

The side chain in vitamin K$_1$, also known as phylloquinone, is derived from the same diterpene alcohol, phytol, found in vitamin E. Vitamin K$_2$ is actually a group of compounds, the menaquinones, with different numbers of isoprene units in the side chain.

2-Methyl-1,4-naphthoquinone is a simple vitamin K analog that is practically as active in promoting blood clotting as vitamin K itself. The compound has been synthesized starting with a Diels-Alder reaction between 1,3-butadiene and 2-methyl-1,4-benzoquinone.

1,3-butadiene   2-methyl-1,4-
                 benzoquinone   84%

2-methyl-1,4-naphthoquinone

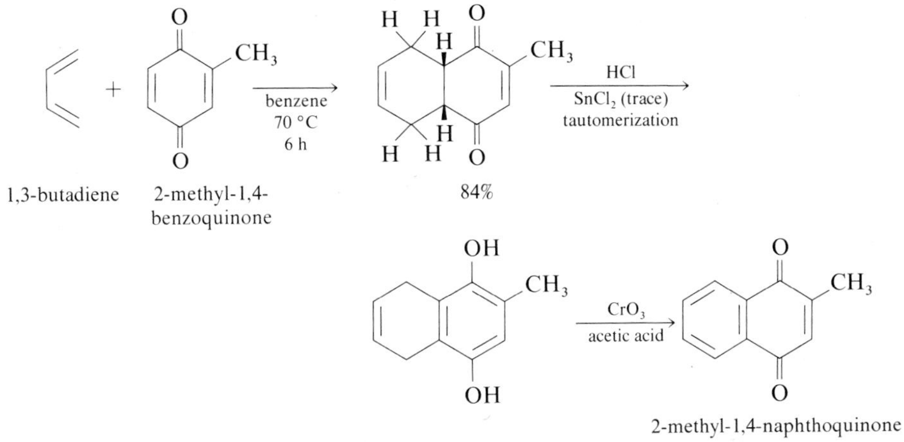

The product of the Diels-Alder reaction tautomerizes to a hydroquinone when treated with acid. Treatment of the hydroquinone with an oxidizing agent such as chromic acid leads to oxidation in both rings. The left-hand ring, which is at the oxidation level of a cyclohexadiene, becomes aromatic, and the hydroquinone becomes a quinone.

The K vitamins are found widely in the leaves of green plants. Vitamin K is also synthesized by bacteria in the lower intestines. Because the compounds are so widespread, it is unlikely that one would develop a deficiency unless one refused to eat any green vegetables at all.

A number of compounds are antagonists of vitamin K; they increase blood clotting times. Such compounds are known as **anticoagulants.** For example, salicylic acid, a product of the hydrolysis of aspirin, acetylsalicylic acid, has this effect. Two more powerful anticoagulants, dicoumarol and Warfarin, are also shown below.

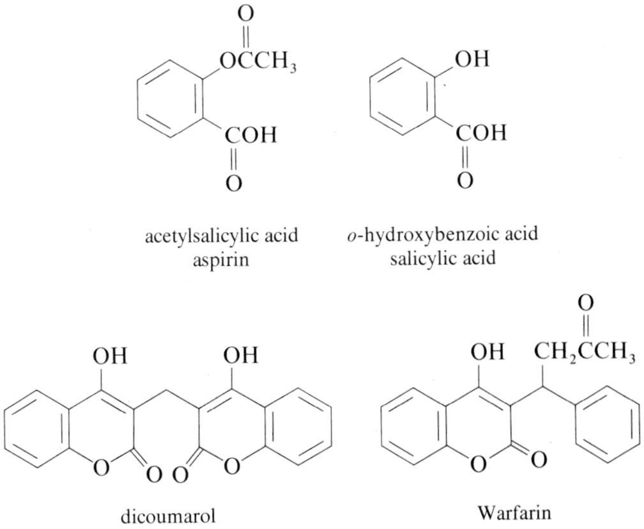

acetylsalicylic acid   *o*-hydroxybenzoic acid
aspirin   salicylic acid

dicoumarol   Warfarin

some compounds that increase blood clotting times
anticoagulants

Dicoumarol was discovered as a component of spoiled clover hay when animals that ate it started to bleed abnormally. It is now used medically to prevent blood clots. Warfarin is a synthetic compound, deliberately designed as an anticoagulant and used as a rat poison. Note that all these compounds have a phenolic or enolic hydroxyl group. Dicoumarol and Warfarin are also lactones of phenols. The phenol lactone system found in these two compounds is also found in coumarin, a fragrant constituent of clover.

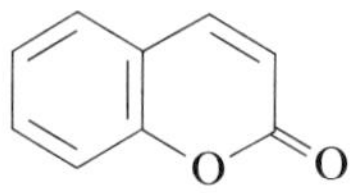

coumarin

a phenol lactone

Coumarin itself does not interfere with the clotting of blood and is used mostly as a flavoring agent.

PROBLEM 12.30   The anticoagulant Warfarin is used as its sodium salt. What would be the structure of this compound?

PROBLEM 12.31   2-Methyl-1,4-naphthoquinone labeled with radioactive ^{14}C at the methyl group was needed for tracer studies on the metabolism of the vitamin K analog. The synthesis used 2-bromonaphthalene as the starting material. $^{14}CO_2$ is also readily available. The last two steps of the synthesis are shown below. Show, with equations, how you would synthesize the labeled halide.

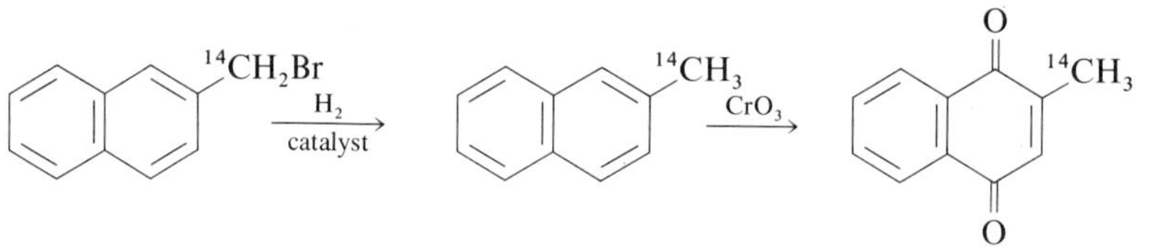

12.7

Arene Oxides

The body converts aromatic compounds into phenols in order to detoxify and excrete them. The totally nonpolar hydrocarbon is given polar functional groups that make it more soluble in the physiological solvent, water. Once a hydroxyl group is present on the hydrocarbon, it is used by the body to link the aromatic hydrocarbon residue to other functionalities that increase still further the solubility of the molecule in water. For example, sulfate esters of the phenol may be formed, or the hydroxyl group linked to glucuronic acid (p. 742) as an acetal. These functions serve to hold the hydrocarbon residue in solution and aid in its transport out of the body.

The liver is the body's major organ for the detoxification of foreign chemicals. Whether drugs, pollutants, food additives, or indeed some natural components of food, all chemical substances that cannot be usefully incorporated into the structure of our bodies must be excreted somehow. Many substances are transformed in oxidation reactions by enzymes found in abundance in the liver. Most important among these enzymes are a group of highly colored proteins known as **cytochromes.** These enzymes, which contain iron, are found in the microsomal fraction of liver cells. Their function is to convert the oxygen from the air into a highly reactive form that can attack systems normally resistant to oxidation by air.

In the case of aromatic hydrocarbons, the products of these oxidation reactions are unstable oxiranes known as **arene oxides.** Naphthalene, for example, is converted by microsomal enzymes from the liver of the rat into naphthalene-1,2-oxide.

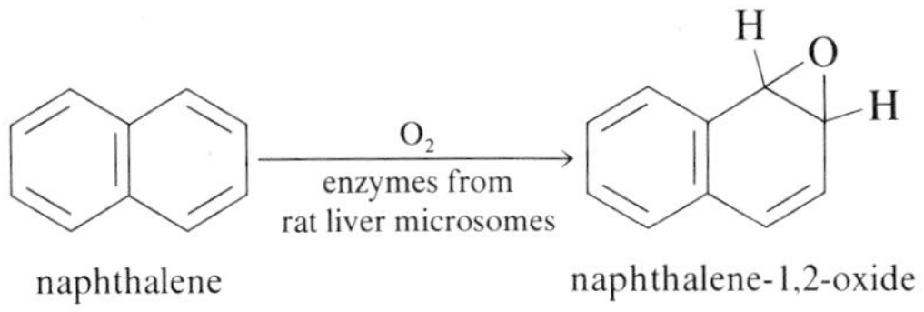

naphthalene   naphthalene-1,2-oxide

The reactive oxide is opened to a trans 1,2-diol by another enzyme. Experiments have shown that the oxygen atom in the benzylic hydroxyl group comes from molecular oxygen while the other one comes from water.

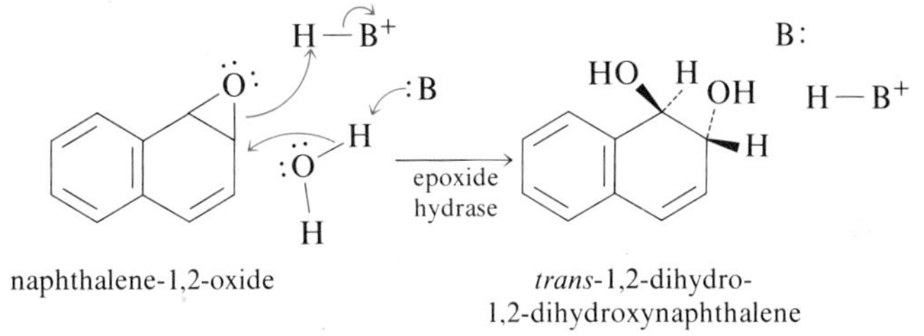

naphthalene-1,2-oxide   *trans*-1,2-dihydro-
                                             1,2-dihydroxynaphthalene

1-Naphthol is the other major metabolic product of naphthalene. It is formed in a spontaneous rearrangement of the unstable oxide and does not require an enzyme catalyst. The reaction takes place under basic, neutral, or acidic conditions. A carbocation intermediate is postulated. The mechanism for the isomerization was determined using naphthalene labeled with deuterium.

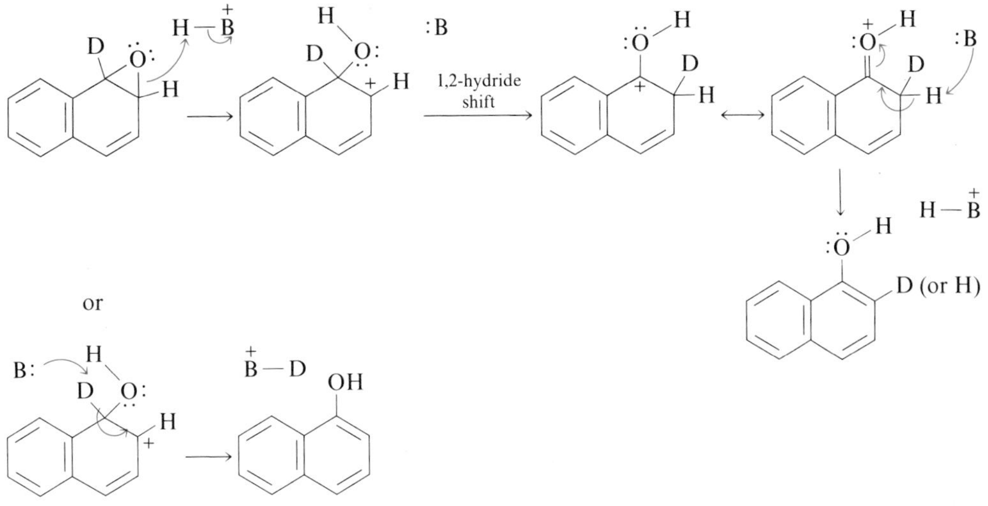

or

Two pathways are available to the cation that is initially formed. It may undergo a 1,2-hydride shift. The new cation that results from the shift is stabilized by delocalization of charge to an oxygen atom. Base removes a proton or a deuteron giving 1-naphthol. The original carbocation may also lose a proton without rearrangement. In the case of naphthalene-1,2-oxide labeled at carbon 1 by deuterium, such a pathway results in the loss of deuterium labeling. About 60% of the deuterium is retained when the labeled naphthalene-1,2-oxide rearranges in an acid-catalyzed reaction, indicating that the 1,2-hydride shift is an important reaction pathway.

The intense interest in the exact mechanism by which arene oxides react is the result of evidence that such oxides are responsible for the carcinogenicity of many aromatic hydrocarbons. It is ironic that the very same oxidation processes that detoxify harmful chemical substances and rid the body of them, may also create metabolites that are far more carcinogenic than the original hydrocarbons themselves.

Suspicion that substances in the residues from the burning of coal and wood cause cancer goes back to 1775 when a British doctor noticed that chimney sweeps were more likely to have cancer of the scrotum than men in other occupations. The harmful compounds, found widely in the environment in cigarette smoke, automobile exhausts, and industrial emissions, are polycyclic aromatic hydrocarbons. The one on which most interest has centered is benzo[a]pyrene. About 40 of its oxygenated metabolites have been isolated and characterized. One of them in particular has been shown to be a potent mutagen. Compounds that are mutagens are suspected of being carcinogens as well because they disrupt the genetic processes in cells. A cancer cell is a cell that has been transformed so that the processes that normally control growth and reproduction no longer operate.

The reactions that are postulated for the conversion of benzo[a]pyrene into its mutagenic metabolite are very much like those shown above for the biological conversion of naphthalene to *trans*-1,2-dihydro-1,2-dihydroxynaphthalene. A cytochrome catalyzes the conversion of benzo[a]pyrene into an arene oxide, which adds water enzymatically to give a trans diol. Another epoxidation takes place at the double bond in the same ring as the diol function, and two stereoisomeric diol-epoxides are formed (Figure 12.3, p. 637). The new oxirane ring can have two orientations with respect to the hydroxyl groups that are already present. The 7,8-diol-9,10-epoxide of benzo[a]pyrene with the oxirane ring trans to the benzylic hydroxyl group, called the anti isomer, is more potent as a mutagen than is the other isomer.

Benzo[a]pyrene-7,8-diol-9,10-epoxide disrupts the genetic mechanism of the cell in several ways. The chemistry that is relevant at this point involves the reactivity of oxiranes towards nucleophiles. There are many nucleophilic sites on a DNA molecule. One such site is an amino group on a heterocyclic ring in deoxyguanosine, which is part of the backbone of DNA (Section 18.6B).

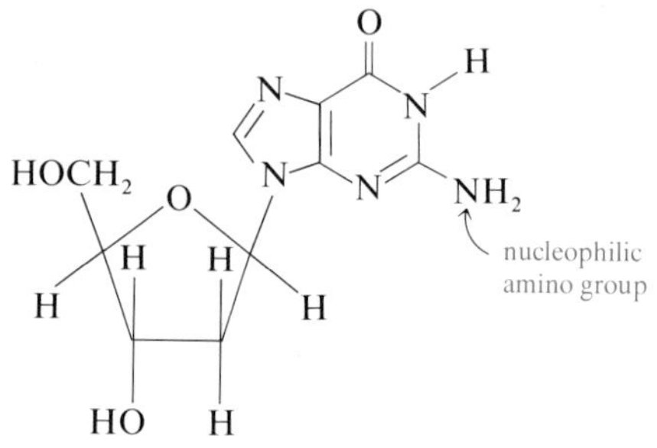

deoxyguanosine

a component of DNA

Reaction of the benzo[a]pyrene-7,8-diol-9,10-epoxide with DNA, degradation of the giant molecule, and purification of the fractions containing the hydrocarbon residue lead to the isolation of the compound shown on the next page.

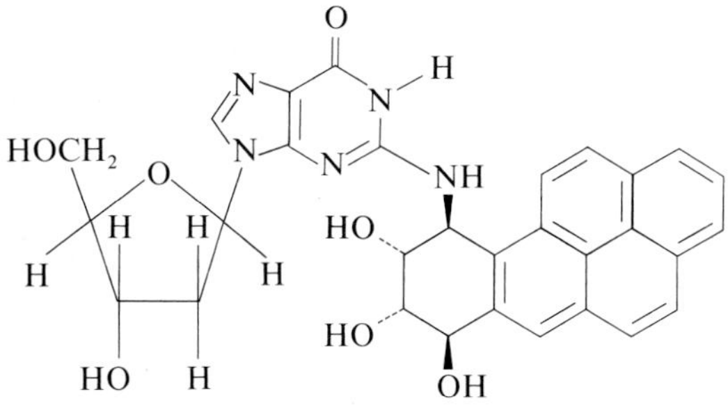
O
H
N
N
N
N
HOCH$_2$
O
NH
H H
HO
H
H
HO
HO H
OH

adduct of deoxyguanosine with benzo[a]pyrene-7,8-diol-9,10-epoxide

The amino group in deoxyguanosine reacts by a nucleophilic opening of the oxirane ring in benzo[a]pyrene-7,8-diol-9,10-epoxide. The same adduct is obtained when benzo[a]pyrene itself is incubated with mouse embryo cells, strongly indicating that the diol-epoxide is the metabolic intermediate responsible for reaction with DNA.

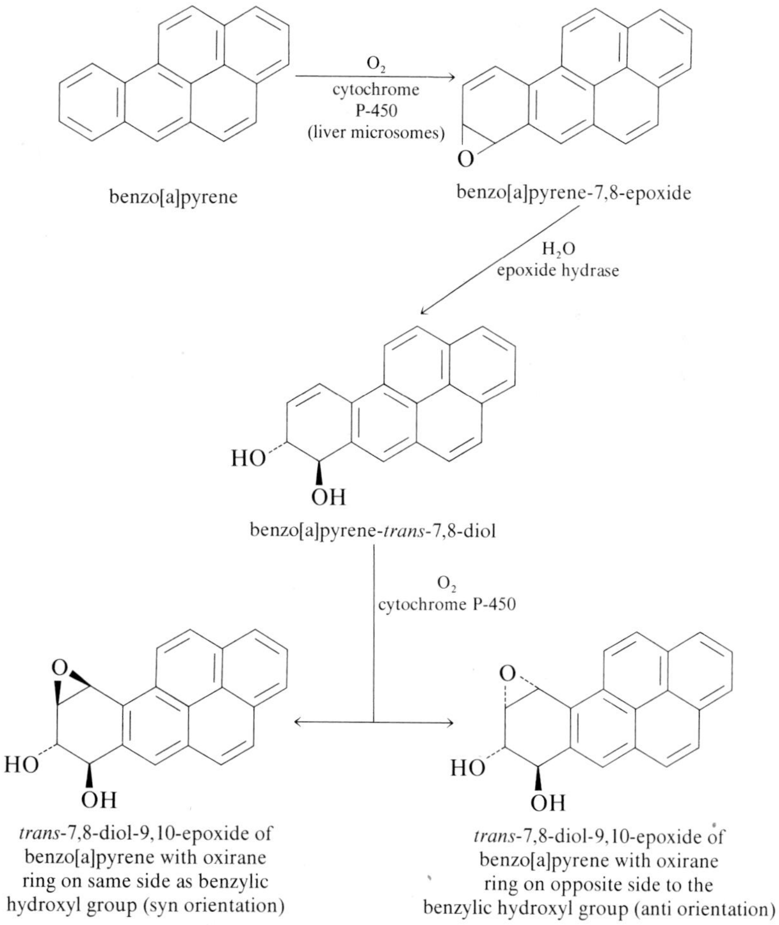
O_2
cytochrome
P-450
(liver microsomes)
O
benzo[a]pyrene
benzo[a]pyrene-7,8-epoxide
H_2O
epoxide hydrase
HO
OH
benzo[a]pyrene-*trans*-7,8-diol
O_2
cytochrome P-450
O
HO
OH
O
HO
OH
trans-7,8-diol-9,10-epoxide of benzo[a]pyrene with oxirane ring on same side as benzylic hydroxyl group (syn orientation)
trans-7,8-diol-9,10-epoxide of benzo[a]pyrene with oxirane ring on opposite side to the benzylic hydroxyl group (anti orientation)

Figure 12.3 The conversion of benzo[a]pyrene into its mutagenic metabolite.

If the aromatic hydrocarbon contains alkyl side chains, biological oxidations resembling the oxidation reaction discussed in Section 12.4C take place at the benzylic position. Benzene, with no alkyl side chain, is detoxified only slowly by the body and is a cumulative poison because it cannot be excreted rapidly. It affects bone marrow and causes aplastic anemia and leukemia. By contrast, toluene is much less toxic. The major pathway for its detoxification by the body is oxidation of the side chain to benzyl alcohol.

major oxidative route for the   *major oxidative route for the*
detoxification of benzene, slow   *detoxification of toluene, fast*

A realization of these facts has led to the substitution of toluene for benzene in the laboratory whenever possible.

PROBLEM 12.32   Two experiments provided part of the evidence for the formation of an arene oxide as an intermediate in the metabolism of naphthalene to *trans*-1,2-dihydro-1,2-dihydroxynaphthalene. In one, naphthalene was incubated with air containing $^{18}O_2$ in ordinary water, $H_2{}^{16}O$. In the other, naphthalene was incubated with ordinary air in $H_2{}^{18}O$. Go back to the equations on p. 635 and write mechanisms predicting the isotopic composition of the trans diol that was formed in each experiment described above.

12.8

The Spectroscopy of Aromatic Compounds

A. The Ultraviolet Spectroscopy of Aromatic Compounds

Ultraviolet and visible spectroscopy has played an important role in the determination of the structure of many vitamins, as we have already seen for the carotenes and vitamin A in Section 11.8C. In this section, some of the experimental observations that were used to determine the structure of vitamin E will be examined after we look at ultraviolet spectra of aromatic compounds, and especially of phenols and amines.

The aromatic ring itself is a chromophore. Benzene has major absorption bands at 180 (ϵ 60,000), 200 (ϵ 8000), and 254 nm (ϵ 212) with other bands at 234, 239, 243, 249, 261, and 268 nm. A series of such closely spaced bands is often seen in the ultraviolet spectrum of an aromatic compound and corresponds to the spacing of the vibrational energy levels in the molecule. The absorption band is said to have *structure*.

Substitution of an alkyl group on the aromatic ring slightly increases the wavelengths of maximum absorption. The ultraviolet spectrum of toluene, with λ_{max}^{hexane} 189 (ϵ 55,000) 208 (ϵ 7900) and 262 nm (ϵ 260) (Figure 12.4), shows the kind of structure typical of the spectrum of an aromatic hydrocarbon. The chromophore is an aromatic ring with one alkyl substituent on it.

For polycyclic hydrocarbons, the absorption bands in the ultraviolet spectrum shift to longer wavelengths and become more intense. For example, naphthalene has $\lambda_{max}^{methanol}$ 311 nm (ϵ 239) for the band corresponding to the absorption at 254 nm (ϵ 212) for benzene.

The interaction of the nonbonding electrons of an oxygen atom with the aromatic ring shifts the position of the absorption bands of benzene to higher values for phenols and aromatic ethers. For this effect to appear, the oxygen atom must, of course, be directly bonded to the aromatic ring. A comparison of the spectrum of methoxybenzene with that of toluene demonstrates this phenomenon (Figure 12.5). Methoxybenzene has $\lambda_{max}^{isooctane}$ 220 (ϵ 8100), 271 (ϵ 2200), and 278 nm (ϵ 2250).

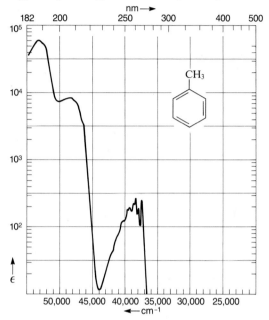

Figure 12.4   Ultraviolet spectrum of toluene in hexane. (Adapted from *UV Atlas of Organic Compounds*)

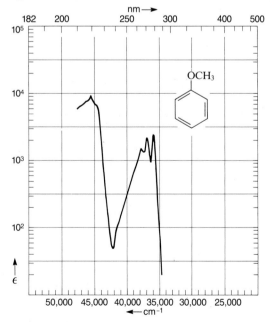

Figure 12.5   Ultraviolet spectrum of methoxybenzene in isooctane (Adapted from *UV Atlas of Organic Compounds*)

The conversion of a phenol into a phenolate anion results in a shift of the absorption maximum in its ultraviolet spectrum to even higher wavelengths. This phenomenon reflects the greater electron density in the aromatic ring that is possible when the phenol is deprotonated. The spectra of phenol taken in water and in aqueous sodium hydroxide show this effect (Figure 12.6). Both absorption bands for phenol (211 and 270 nm) move to longer wavelengths (235 and 287 nm) and increase in intensity for the phenolate anion. This change is typical of the spectra of phenols and can be used as a diagnostic test that the aromatic compound containing oxygen is indeed a phenol.

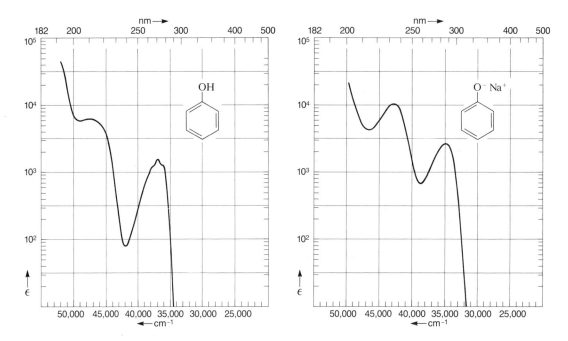

Figure 12.6   Ultraviolet spectra of phenol, taken in water and in aqueous sodium hydroxide (Adapted from *UV Atlas of Organic Compounds*)

For amines in which the nitrogen atom is bonded directly to an aromatic ring, interaction between the nonbonding electrons of the nitrogen atom and the ring is possible. Just as is the case for phenols and aryl ethers, the chromophore of the aromatic ring changes, and the absorption bands in the ultraviolet spectrum of such a compound appear at higher wavelength and become more intense. The spectrum of an aromatic amine when acid is added comes to resemble that of an alkylbenzene. Protonation of the nitrogen atom destroys the possibility of interaction between it and the ring. The spectra of aniline taken in a buffer at pH 8.0 and in aqueous sulfuric acid illustrate this phenomenon (Figure 12.7). Aniline has $\lambda_{max}^{H_2O}$ 196 (ϵ 34500), 230 (ϵ 8200), and 281 (ϵ 1400). The bands in the spectrum for the acidified solution at 200 (ϵ 7600) and 254 nm (ϵ 165) correspond to the bands at 230 and 281 nm for aniline in water. Thus, both the position of maximum absorption and the intensity of absorption decrease upon protonation of the nitrogen atom in an aromatic amine. The spectrum of the anilinium ion looks very much like that of toluene (Figure 12.4).

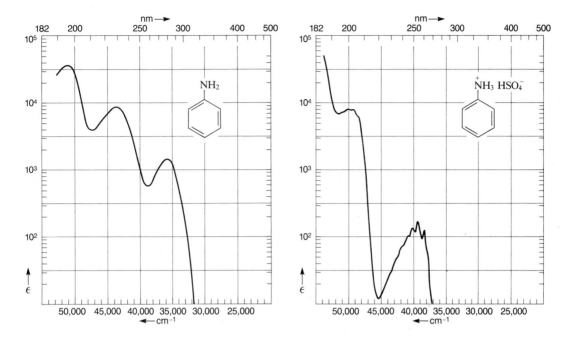

Figure 12.7   The ultraviolet spectra of aniline taken in a buffer at pH 8.0 and in aqueous sulfuric acid. (Adapted from *UV Atlas of Organic Compounds*)

PROBLEM 12.33   The ultraviolet spectrum shown in Figure 12.8 belongs to Compound X, $C_{10}H_{15}N$.

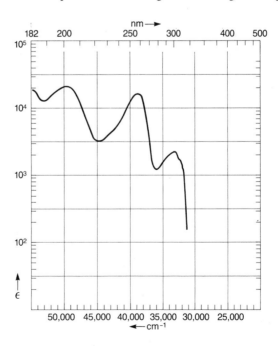

Figure 12.8   Spectrum for Problem 12.33. (Adapted from *UV . Atlas of Organic Compounds*)

Use the spectrum, taken in heptane, to distinguish between the following two possible structures for Compound X.

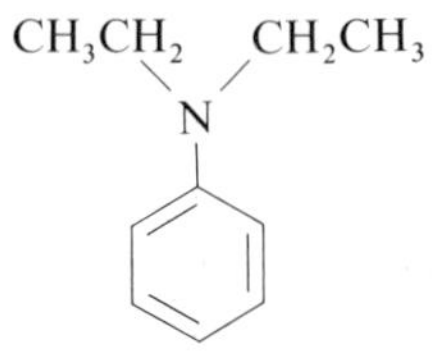

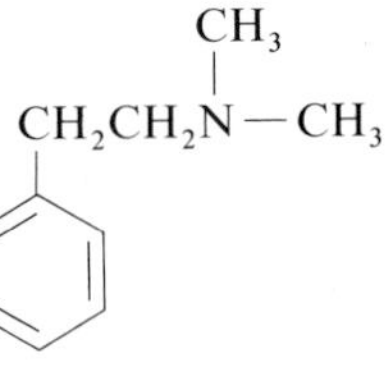

PROBLEM 12.34   Compound Y, C_7H_8O, has $\lambda_{max}^{methanol}$ 214 (ϵ 6030) and 273 nm (ϵ 1820). When Compound Y is put into methanol with some potassium hydroxide in it, it has bands at 238 (5510) and 282 nm (ϵ 1990). What is a possible structure for Compound Y?

The sensitivity of the ultraviolet spectrum of a compound to the particular chromophore that is present, and its insensitivity to any parts of the molecule that do not absorb radiation in the ultraviolet region make ultraviolet spectroscopy the ideal tool for determining the structure of vitamin E (Section 12.6D).

α-Tocopherol has many absorption bands in the ultraviolet. Its most intense band is at λ_{max} 297 nm (ϵ 3800). The position and intensity of this band change when base is added to the solution. A hydroquinone type structure was immediately suspected for the compound because hydroquinone has λ_{max} 294 nm (ϵ 3100). The chromophore in α-tocopherol was compared with those in some model compounds.

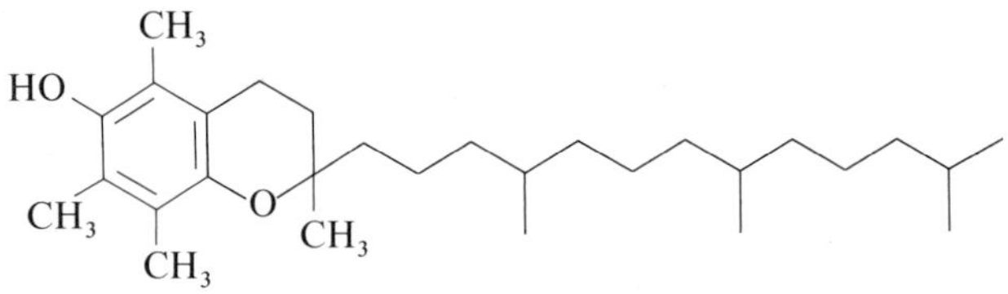

α-tocopherol
λ_{max} 297 nm (ϵ 3800)

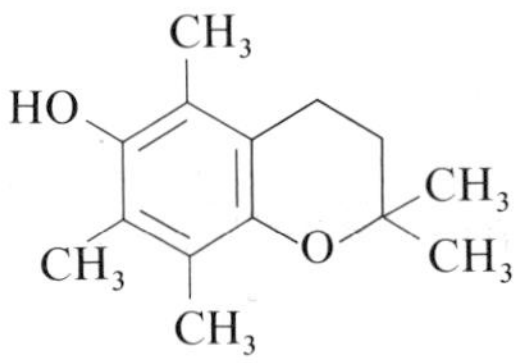

λ_{max} 297 nm (ϵ 3300) 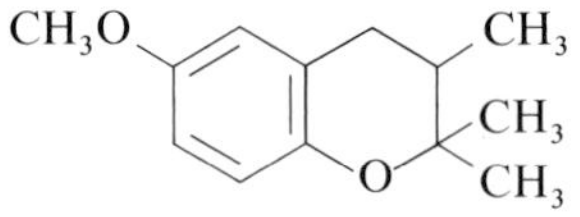 λ_{max} 280 nm (ϵ 3600)

the chromophore in α-tocopherol, compared with the chromophores in two model compounds

α-Tocopherol has the same chromophore as the model compound on the left, and their ultraviolet spectra are quite similar, even though the model compound has only a methyl group where α-tocopherol has the phytyl chain.

PROBLEM 12.35   Compound Z, C_7H_7BrO, has $\lambda_{max}^{methanol}$ 227 (ϵ 14,200), 281 (ϵ 1580), and 288 nm (ϵ 1280). The infrared absorption spectrum of Compound Z has no absorption bands of any intensity between 4000 to 3200 cm^{-1} and between 2000 to 1600 cm^{-1}. Assign a structure to Compound Z that fits these data.

B. Infrared Spectroscopy of Aromatic Compounds

The most typical absorption bands in the infrared spectrum of an aromatic hydrocarbon appear at 3100 to 3000 cm^{-1} for the stretching vibrations of the carbon-hydrogen bonds, and between 870 and 675 cm^{-1} for various bending vibrations of the carbon-hydrogen bonds and the ring. These bands are usually sharp strong bands in the spectra of aromatic compounds.

Benzene, which is a highly symmetrical compound, has a very simple infrared spectrum. Alkylbenzenes show features typical of the spectra of both alkanes (Section 7.7B) and arenes (Figure 12.9). It is possible to pick out of the spectrum of

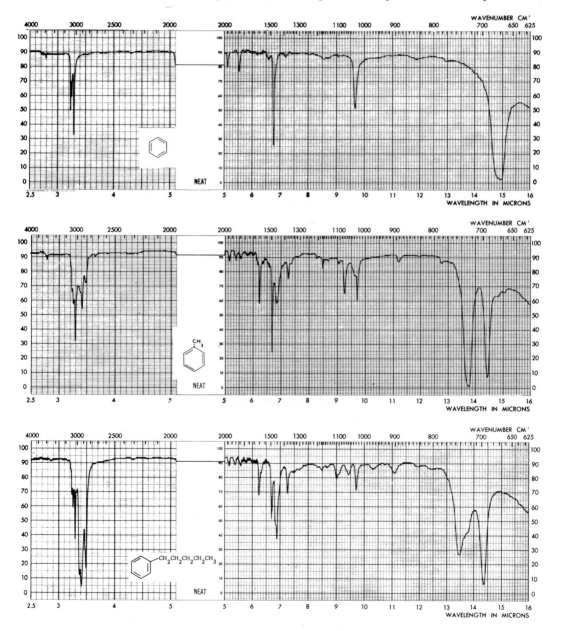

Figure 12.9   Infrared spectra of benzene, toluene, and 1-phenylpentane. (From *The Aldrich Library of Infrared Spectra*)

1-phenylpentane the stretching frequencies of the aromatic as well as the alkane carbon-hydrogen bonds (3050, 2900 cm^{-1}). In the lower frequency region of the spectrum, sharp bands at 1600, 1500, 750, and 700 cm^{-1} are also due to vibrations of the aromatic part of the molecule.

A comparison of the spectra of toluene and 1-phenylpentane reconfirms a point that was made in Section 7.7B. Infrared spectra of compounds of the same functional group type have similar spectra, but show distinct differences in the lower frequency regions of the spectra, known as the fingerprint region. As you become proficient in interpreting infrared spectra, you will learn that a consciousness of the general appearance of the spectrum, taking into account the relative intensities of bands and their appearance, whether sharp or broad, is as important as a determination of the frequencies of absorption. The general features typical of aromatic compounds containing other functional groups were previously discussed in connection with the spectra of conjugated carbonyl compounds (Section 9.10) and carboxylic acids and their derivatives (Section 10.11A).

PROBLEM 12.36   Infrared spectra are given in Figure 12.10 for the following compounds. Write out the structural formula of each compound and pick out the structural features that you expect to be able to identify by infrared spectroscopy. Match each compound with its spectrum.

1. 2-methylnaphthalene   2. 4-methyloctane   3. ethyl *p*-ethoxybenzoate
4. *m*-methoxyphenol   5. methyl pentanoate   6. *p*-chlorobenzoic acid
7. 3-chlorobutanoic acid   8. (2*E*,4*E*)-2,4-hexadiene

Compound A

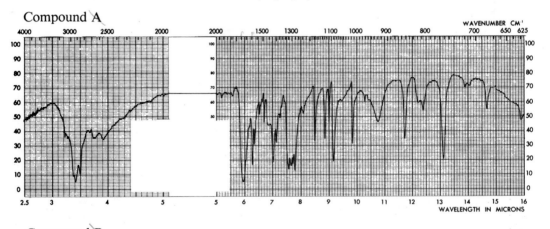

Compound B

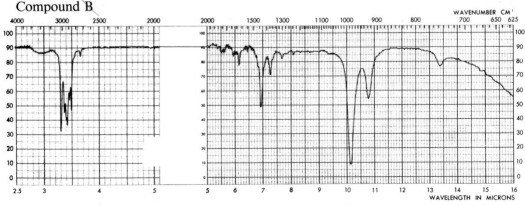

Figure 12.10   Spectra for Problem 12.36. (From *The Aldrich Library of Infrared Spectra*)

Compound C

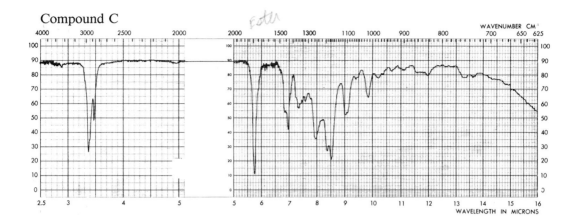

Compound D

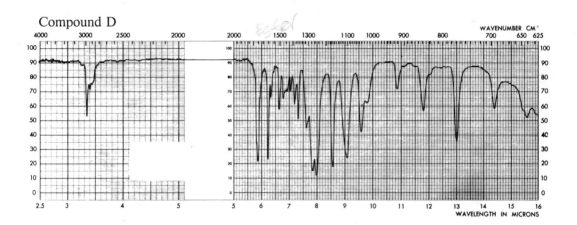

Compound E

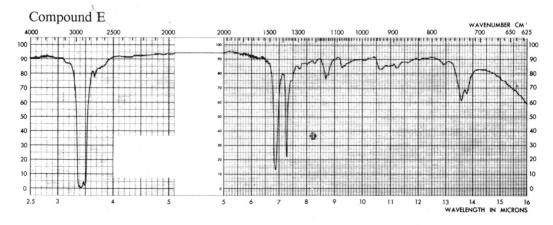

Figure 12.10   *(Continued)*

Compound F

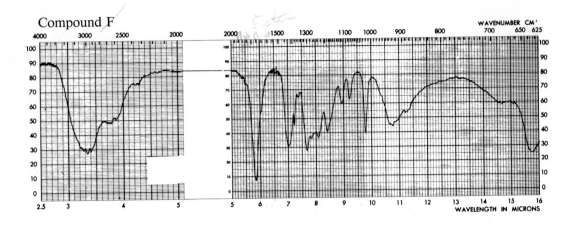

Compound G

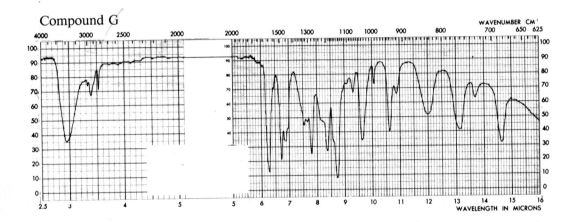

Compound H

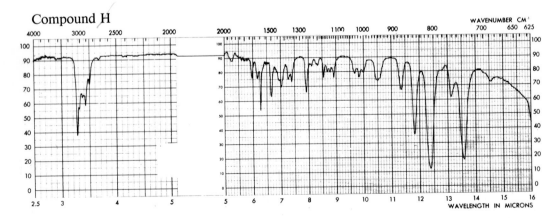

Figure 12.10   (*Continued*)

PROBLEM 12.37   We have three unknown hydrocarbons, Compounds A, B, and C. Only one of these compounds decolorizes bromine in carbon tetrachloride in the dark and reacts with dilute potassium permanganate. The infrared spectra of the compounds are given in Figure 12.11. What can you say about Compounds A, B, and C?

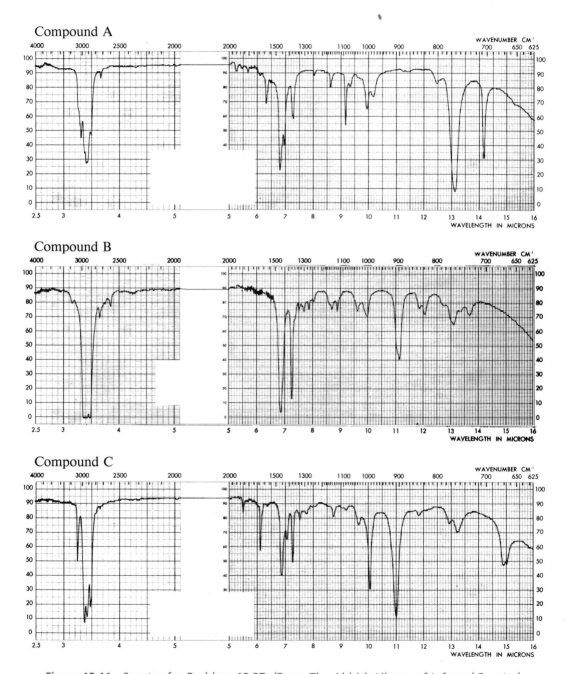

Figure 12.11   Spectra for Problem 12.37. (From *The Aldrich Library of Infrared Spectra*)

ADDITIONAL PROBLEMS

12.38   Name the following compounds.

(a) 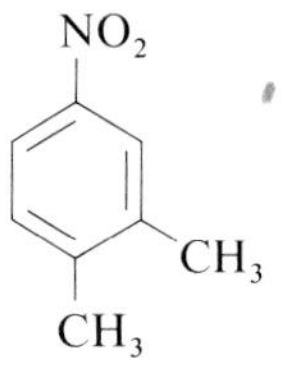 [structure: benzene ring with NO$_2$, CH$_3$, CH$_3$]

(b) 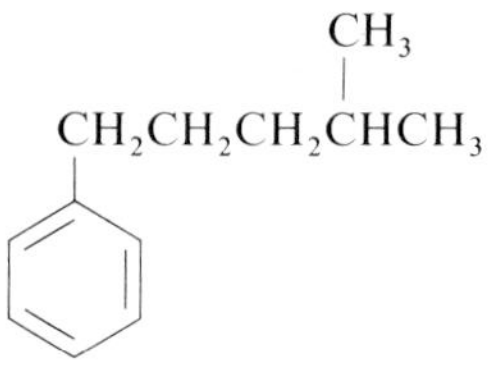 [structure: benzene with CH$_2$CH$_2$CH$_2$CHCH$_3$ with CH$_3$ branch]

(c) 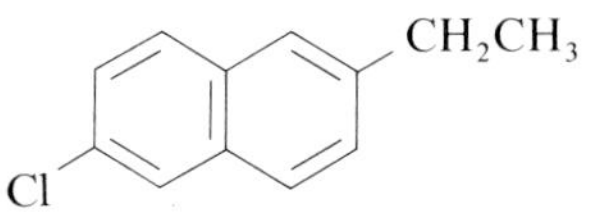 [structure: naphthalene with Cl and CH$_2$CH$_3$]

(d) 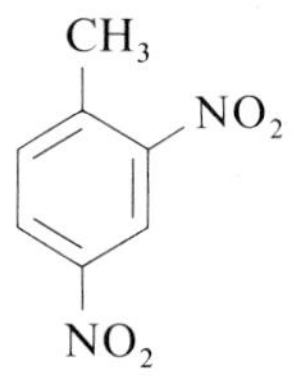 [structure: benzene with CH$_3$, NO$_2$, NO$_2$]

(e) 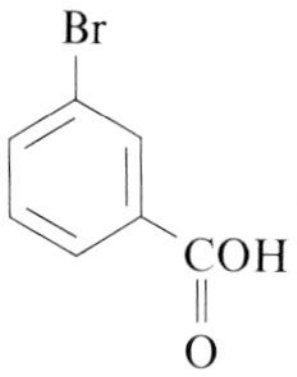 [structure: benzene with Br and COH with O]

(f) 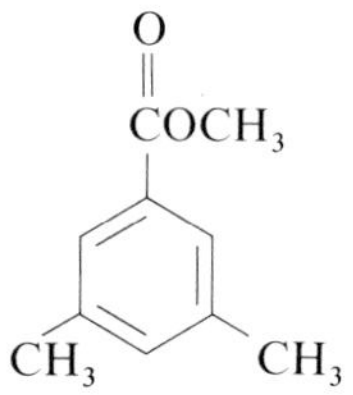 [structure: benzene with COCH$_3$ (with O), CH$_3$, CH$_3$]

(g) 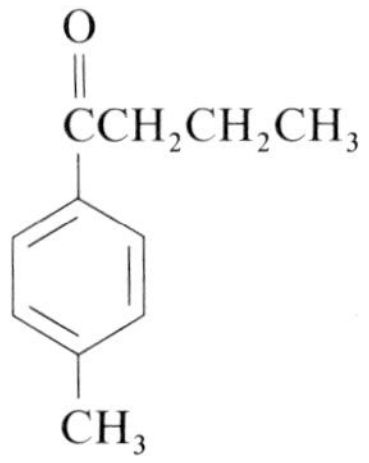 [structure: benzene with CCH$_2$CH$_2$CH$_3$ (O double bond) and CH$_3$]

(h) 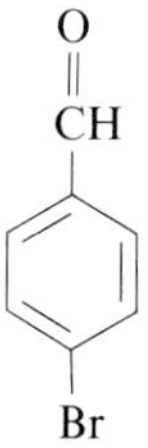 [structure: benzene with CH (O double bond) and Br]

12.39   Write structural formulas for the following compounds.

(a)   ethyl *m*-chlorobenzoate   (b)   1,4-dimethylnaphthalene
(c)   1-phenyl-3-methyl-1-hexanone   (d)   1-phenyl-1-chloropropane
(e)   *m*-bromonitrobenzene   (f)   *p*-chlorobromobenzene

12.40   The compound shown below has been isolated from human tubercle bacilli and given the common name of phthiocol. What would be a more systematic name for it?

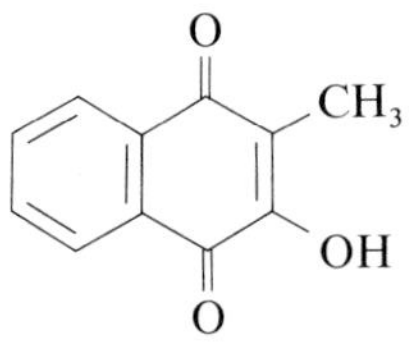

[structure: naphthoquinone with CH$_3$ and OH]

phthiocol

12.41   Complete the following equations.

(a) 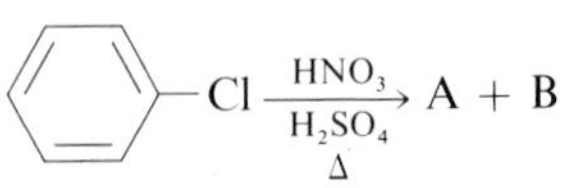 [benzene]—Cl $\xrightarrow[\Delta]{\text{HNO}_3 \atop \text{H}_2\text{SO}_4}$ A + B

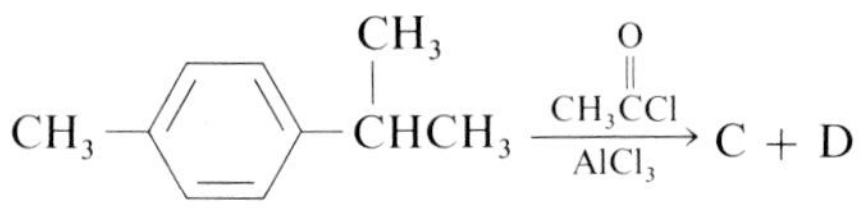

(b) CH$_3$—[benzene]—CHCH$_3$ (with CH$_3$ branch) $\xrightarrow[\text{AlCl}_3]{\text{CH}_3\text{CCl (O)}}$ C + D

(c) CH$_3$—[benzene]—CCH$_2$CH$_2$CH$_3$ (O double bond) $\xrightarrow[\text{ethylene glycol} \atop \Delta]{\text{H}_2\text{NNH}_2, \text{KOH}}$ E

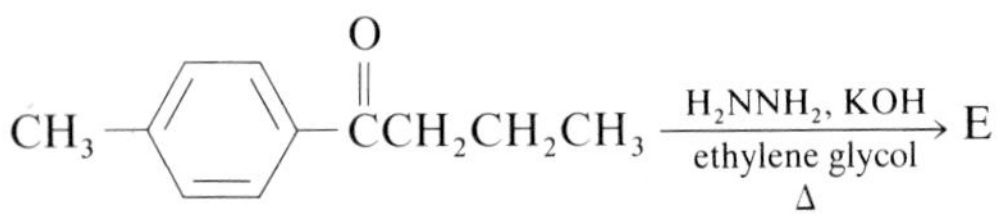

(d) CH_3—⬡—$\overset{\overset{O}{\|}}{C}CH_2CH_2CH_3 \xrightarrow[\Delta]{Br_2}$ F

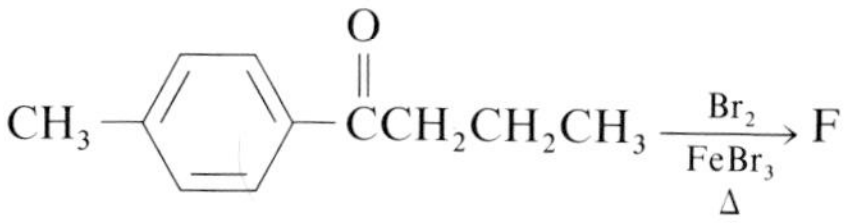

(e) ⬡⬡⬡ $\xrightarrow[\Delta]{\underset{\text{carbon tetrachloride}}{NBS}}$ G

(f) ⬡—$\overset{\overset{O}{\|}}{C}\underset{\underset{CH_3}{|}}{C}HCH_3 \xrightarrow[\Delta]{Zn(Hg),\ HCl}$ H

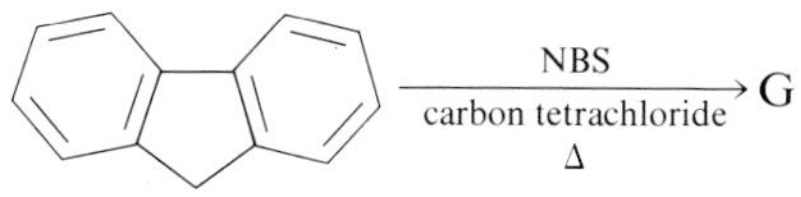

(g) ⬡—$\underset{\underset{CH_3}{|}}{C}HCH_3 \xrightarrow[AlCl_3]{\overset{\overset{CH_3CHCH_3}{|}}{\underset{}{Cl}}}$ I + J

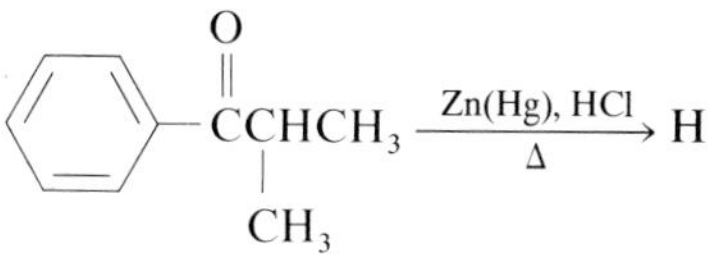

(h) HO—⬡—$\underset{\underset{CH_3}{|}}{\overset{\overset{CH_3}{|}}{C}}CH_3 \xrightarrow{O_2}{\Delta}$ K

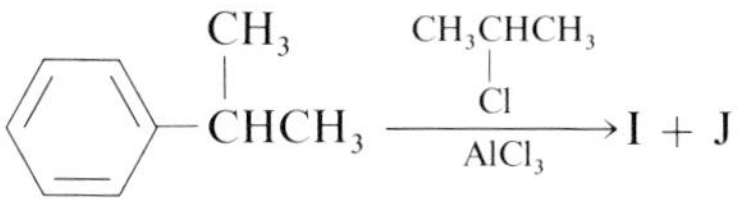

(i) structure of durohydroquinone $\xrightarrow{Ag_2O}$ L

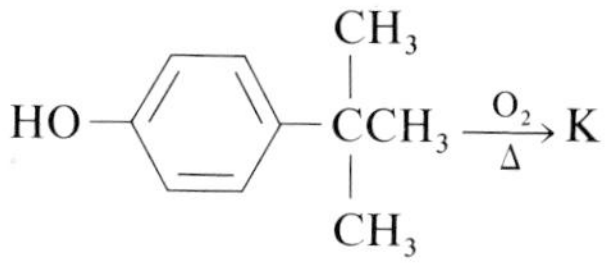

(j) methyl chrysene structure $\xrightarrow[\substack{H_2O \\ H_2SO_4}]{Na_2Cr_2O_7}$ M + N

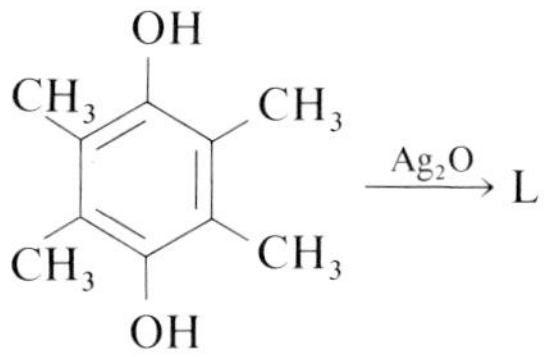

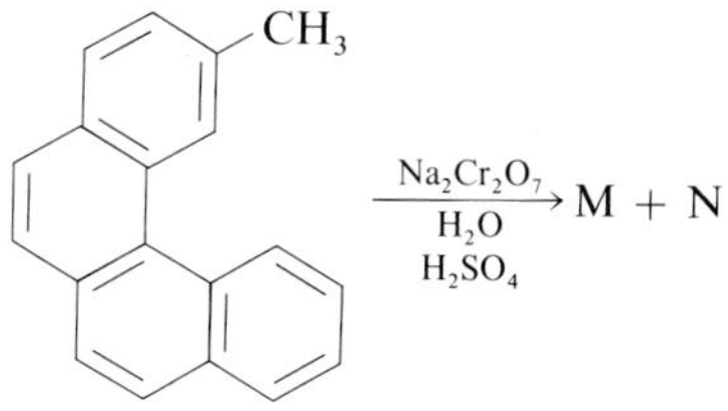

(k) ⬡—CH_3 + ⬠ $\xrightarrow{HF}$ O + P

(l) ⬡—CH_3 + (glutaric anhydride) $\xrightarrow{AlCl_3}$ Q + R

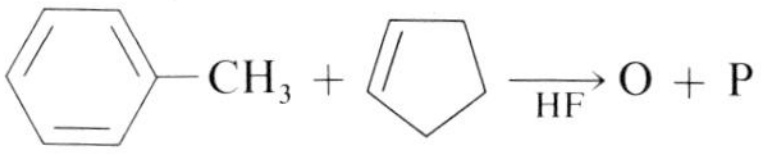

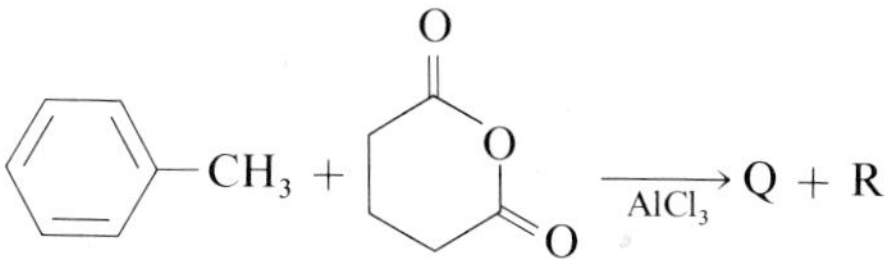

(m) (2-methyl-1,4-benzoquinone) + (2,3-dimethyl-1,3-butadiene) $\xrightarrow{\Delta}$ S

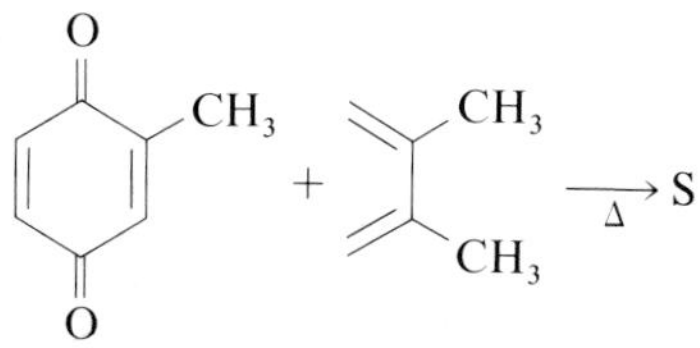

(n) O_2N—⬡(NO_2)—F + $CH_3ONa \xrightarrow{methanol}$ T

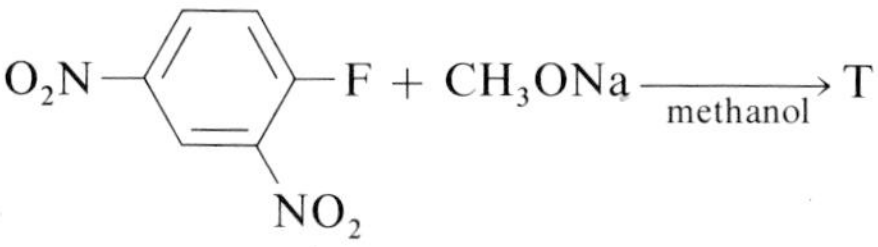

(o) Cl—⬡(D)—OH $\xrightarrow{H_3PO_4}$ U

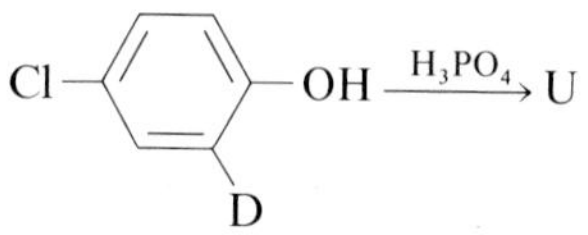

(p) (2-methylpyridine) $\xrightarrow[\Delta]{\underset{H_2O}{KMnO_4}}$ V

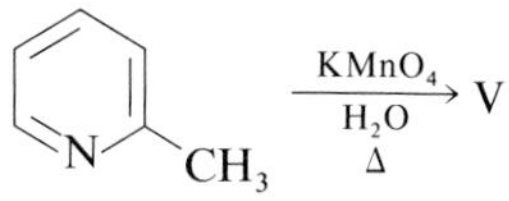

(q) ⬡—$NO_2 \xrightarrow[\Delta]{Br_2\ FeBr_3}$ W

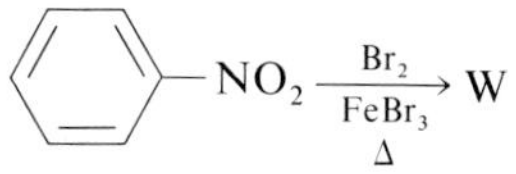

(r) 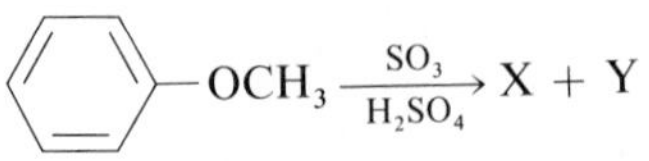   [benzene ring]—OCH_3 $\xrightarrow[\text{H}_2\text{SO}_4]{\text{SO}_3}$ X + Y

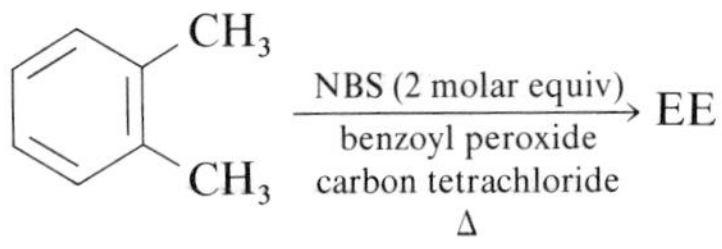

(s) 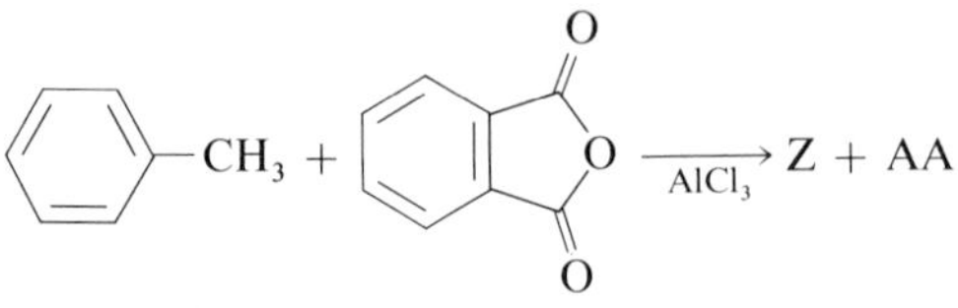   [benzene ring]—CH_3 + [phthalic anhydride] $\xrightarrow{\text{AlCl}_3}$ Z + AA

(t) 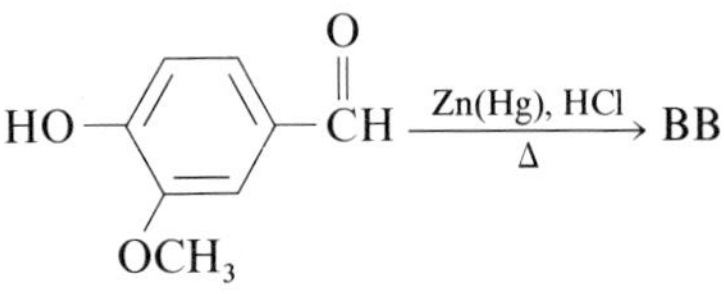   HO—[benzene ring with OCH₃]—$\overset{\overset{\textstyle O}{\|}}{C}H$ $\xrightarrow[\Delta]{\text{Zn(Hg), HCl}}$ BB

(u) 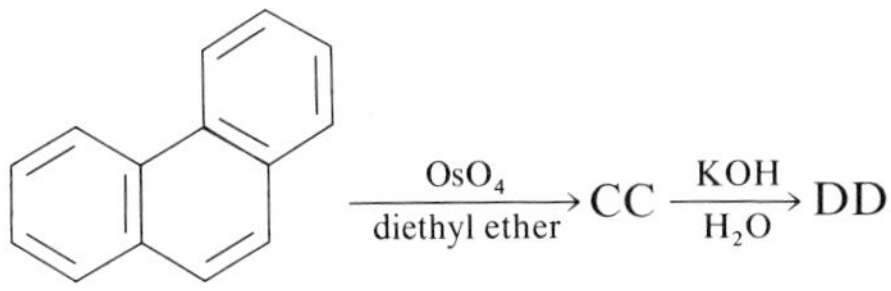   [phenanthrene] $\xrightarrow[\text{diethyl ether}]{\text{OsO}_4}$ CC $\xrightarrow{\text{KOH}}$ DD $\quad\text{H}_2\text{O}$

(v)   [benzene ring with two CH₃ groups] $\xrightarrow[\substack{\text{benzoyl peroxide} \\ \text{carbon tetrachloride} \\ \Delta}]{\text{NBS (2 molar equiv)}}$ EE

(w) 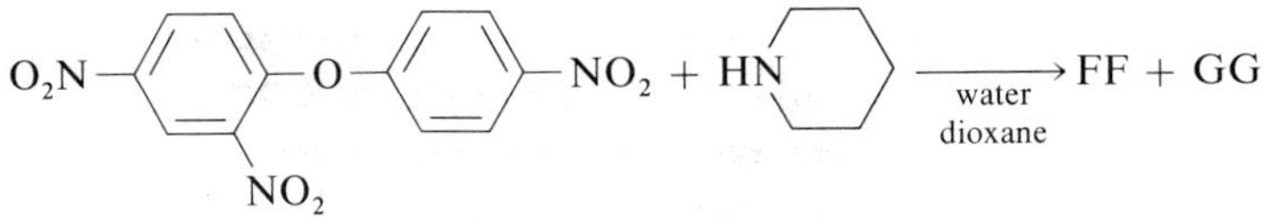 O_2N—[benzene ring with NO₂]—O—[benzene ring]—NO_2 + HN[piperidine] $\xrightarrow[\substack{\text{water} \\ \text{dioxane}}]{}$ FF + GG

(x) 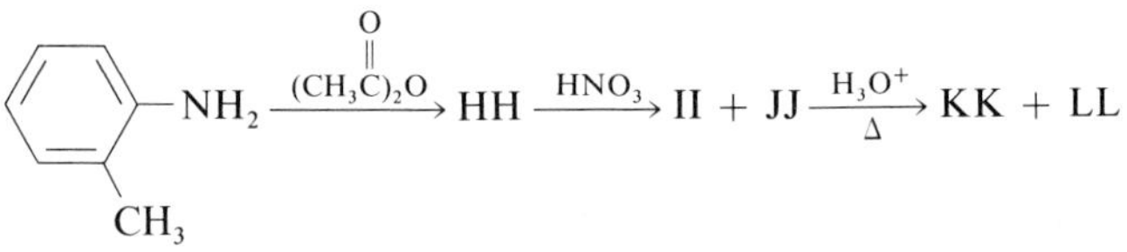   [benzene ring with CH₃]—NH_2 $\xrightarrow{(\text{CH}_3\overset{\overset{\textstyle O}{\|}}{C})_2\text{O}}$ HH $\xrightarrow{\text{HNO}_3}$ II + JJ $\xrightarrow[\Delta]{\text{H}_3\text{O}^+}$ KK + LL

(y) 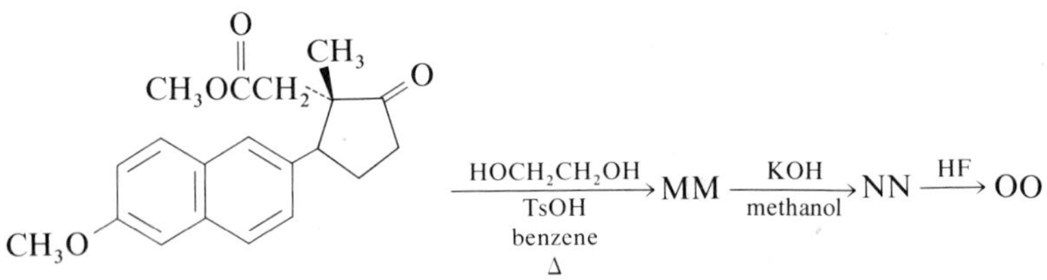   [substituted naphthalene–cyclopentanone compound with CH_3O and CH_3OCCH_2, CH_3 groups] $\xrightarrow[\substack{\text{TsOH} \\ \text{benzene} \\ \Delta}]{\text{HOCH}_2\text{CH}_2\text{OH}}$ MM $\xrightarrow[\text{methanol}]{\text{KOH}}$ NN $\xrightarrow{\text{HF}}$ OO

(z)   [benzene ring]—$\overset{\overset{\textstyle O}{\|}}{C}OH$ $\xrightarrow[\substack{\text{H}_2\text{SO}_4 \\ \Delta}]{\text{SO}_3 \text{ (excess)}}$ PP

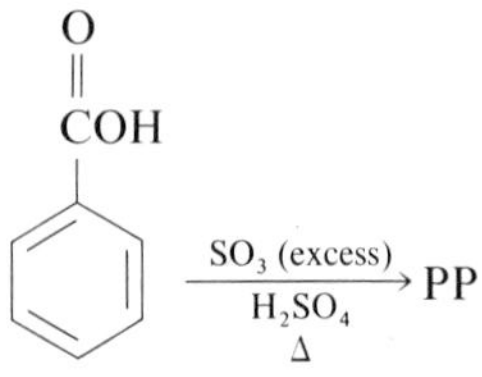

12.42   1,3,5-Trimethylbenzene is converted into 1,3,5-trimethylphenylacetic acid by the following sequence of reactions. Supply Reagents A and B, and propose a mechanism for the first step.

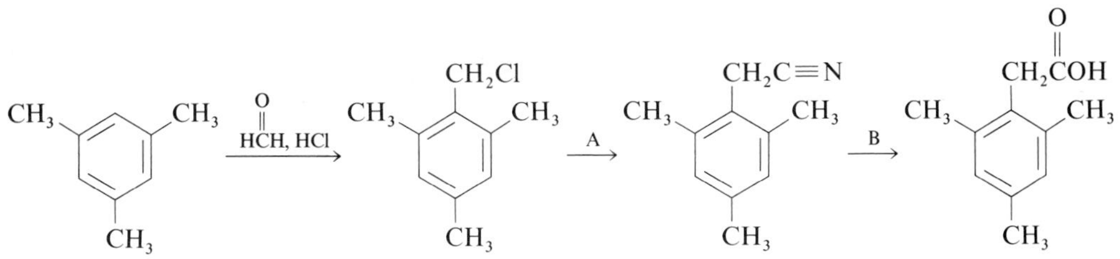

The following reaction scheme at the top of the page (compounds with CH₃ groups on benzene rings):

A series: mesitylene (1,3,5-trimethylbenzene) $\xrightarrow{\text{HCH, HCl}}$ ring with CH_2Cl and two CH_3 groups $\xrightarrow{A}$ ring with $CH_2C{\equiv}N$ and two CH_3 groups $\xrightarrow{B}$ ring with CH_2COH (with $=O$) and two CH_3 groups.

12.43 The following steps are used in the synthesis of the controversial sweetener saccharin. Fill in the structures of the intermediate products.

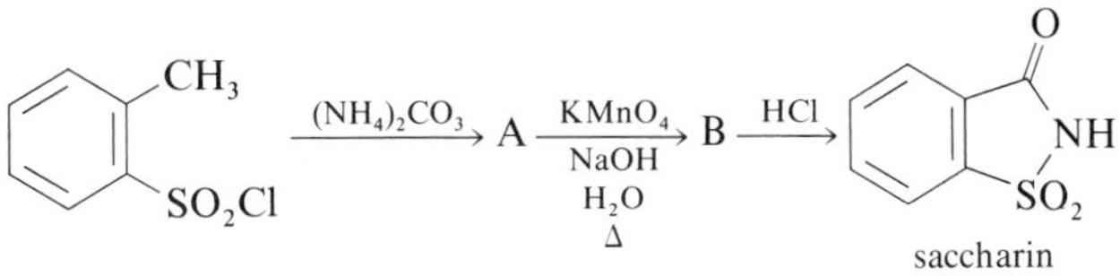

Scheme: *o*-toluenesulfonyl chloride (CH_3, SO_2Cl on benzene) $\xrightarrow{(NH_4)_2CO_3}$ A $\xrightarrow[\text{NaOH}\;H_2O\;\Delta]{\text{KMnO}_4}$ B $\xrightarrow{\text{HCl}}$ saccharin (benzene fused ring with C=O, NH, SO₂)

saccharin

12.44 The following reactions were carried out in research into the synthesis of terpenes. Give structural formulas for the intermediates and products indicated by the letters.

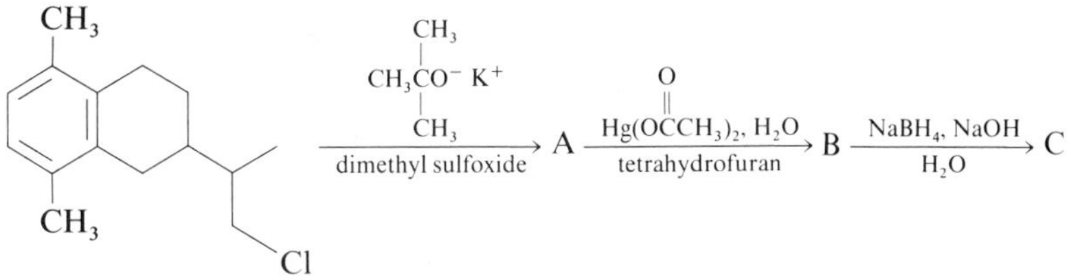

Scheme: bicyclic structure (dimethyl-substituted tetralin with a side chain ending in Cl) $\xrightarrow[\text{dimethyl sulfoxide}]{CH_3CO^- K^+ \text{ (with } CH_3,CH_3)}$ A $\xrightarrow[\text{tetrahydrofuran}]{\text{Hg(OCCH}_3)_2,\ H_2O}$ B $\xrightarrow[H_2O]{\text{NaBH}_4,\ \text{NaOH}}$ C

12.45 The following series of reactions was carried out to prepare compounds labeled with radioactive carbon, ^{14}C, in order to study the mechanism of solvolysis reactions.

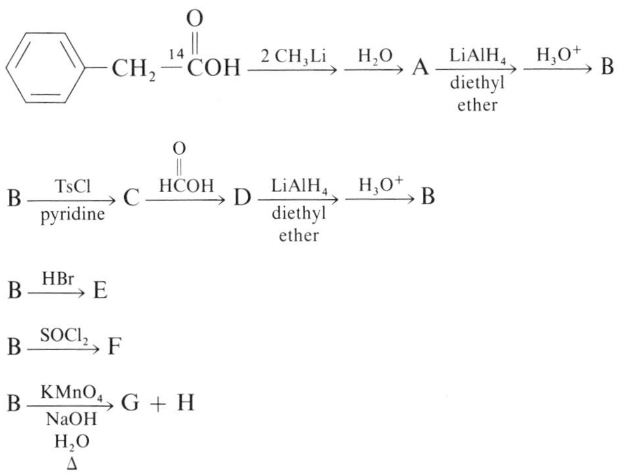

Scheme: $C_6H_5{-}CH_2{-}^{14}\overset{O}{\overset{\|}{C}}OH \xrightarrow{2\ CH_3Li} \xrightarrow{H_2O} A \xrightarrow[\text{diethyl ether}]{\text{LiAlH}_4} \xrightarrow{H_3O^+} B$

$B \xrightarrow[\text{pyridine}]{\text{TsCl}} C \xrightarrow{\text{H}\overset{O}{\overset{\|}{C}}\text{OH}} D \xrightarrow[\text{diethyl ether}]{\text{LiAlH}_4} \xrightarrow{H_3O^+} B$

$B \xrightarrow{\text{HBr}} E$

$B \xrightarrow{\text{SOCl}_2} F$

$B \xrightarrow[\substack{\text{NaOH}\\H_2O\\\Delta}]{\text{KMnO}_4} G + H$

(a) What are the structures of Compounds A–H?
(b) Two readily available sources of ^{14}C are $^{14}CO_2$ and $Na^{14}CN$. How would you synthesize from bromobenzene the radioactively labeled phenylacetic acid used as the starting material in the synthesis above?

12.46 (a) Dewar benzene (Problem 2.24) has actually been synthesized and shown to have reactivity that differs from benzene. The following reactions have been carried out with Dewar benzene. Predict what the products of these reactions are.

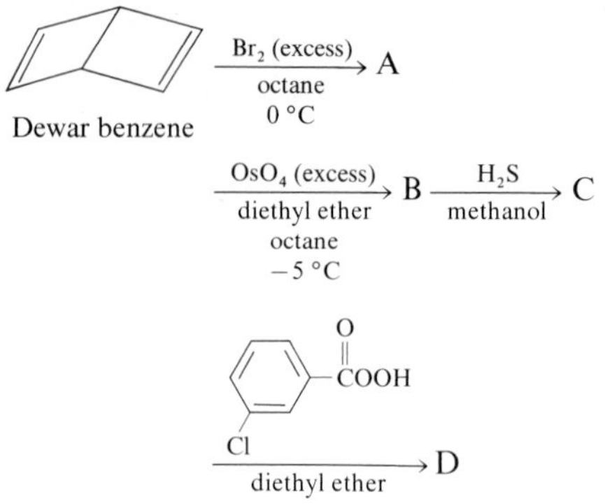

Dewar benzene

$\xrightarrow[\substack{\text{octane} \\ 0\,°C}]{\text{Br}_2\ (\text{excess})}$ A

$\xrightarrow[\substack{\text{diethyl ether} \\ \text{octane} \\ -5\,°C}]{\text{OsO}_4\ (\text{excess})}$ B $\xrightarrow[\text{methanol}]{\text{H}_2\text{S}}$ C

$\xrightarrow[\text{diethyl ether}]{}$ D

(b) When Dewar benzene is treated with either a Lewis acid, AlCl$_3$, or a protic acid, H$_2$SO$_4$, benzene is formed. Suggest a mechanism for the conversion of Dewar benzene to benzene in sulfuric acid.

12.47 The yellow pigment of cottonseed, gossypol, has the structure shown below. Recently there has been much interest in this compound because of reports from the People's Republic of China that it acts as a male contraceptive. 5-Isopropyl-6,7-dimethoxy-3-methyl-1-naphthol is available. Suggest how this compound could be used to create the backbone of gossypol.

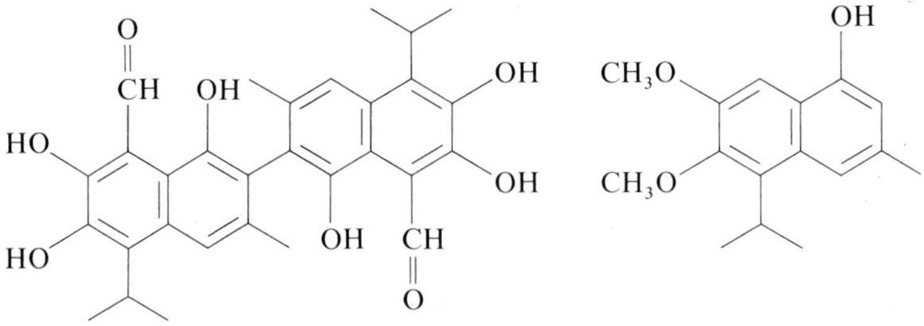

gossypol

5-isopropyl-6,7-dimethoxy-
3-methyl-1-naphthol

12.48 Propose syntheses for the following compounds starting with benzene, toluene, and any other organic compound having three or fewer carbon atoms. There may be more than one way to synthesize each compound. Try to find the shortest route.

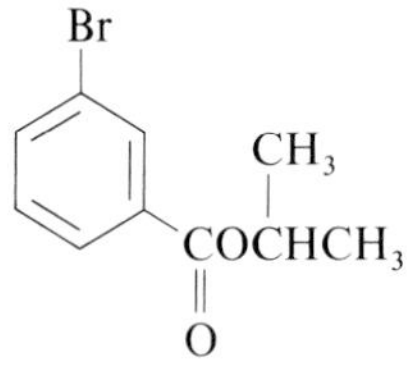

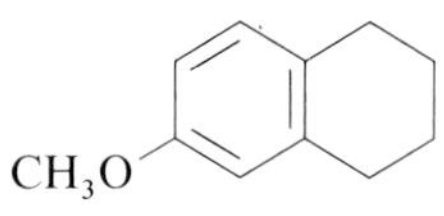

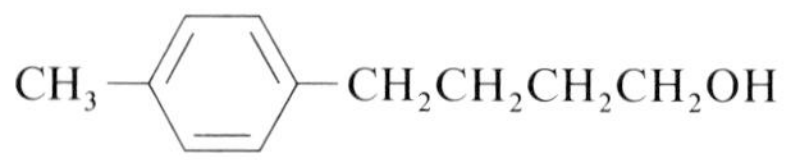

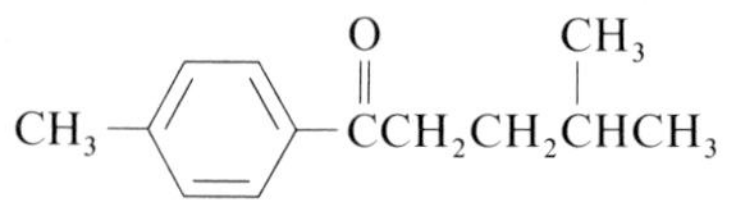

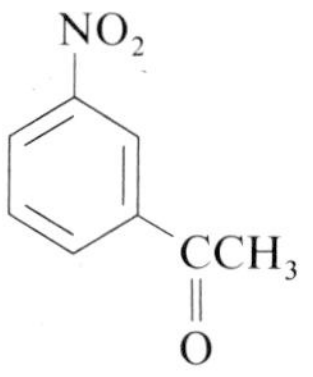

(g) H_2NC—⬡—CNH_2 (with O above each C) (h) Br—⬡—OCH_3 (i) 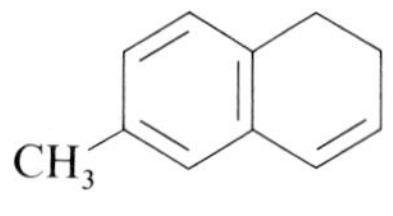 (naphthalene structure with CH_3)

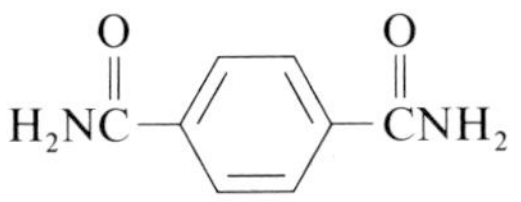

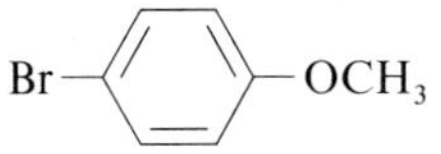

(j) 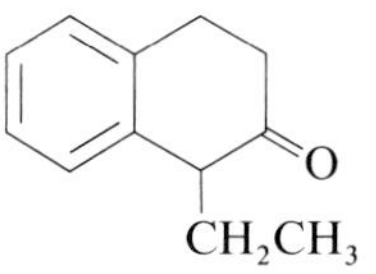 (tetralinone structure with CH_2CH_3 and O) (k) 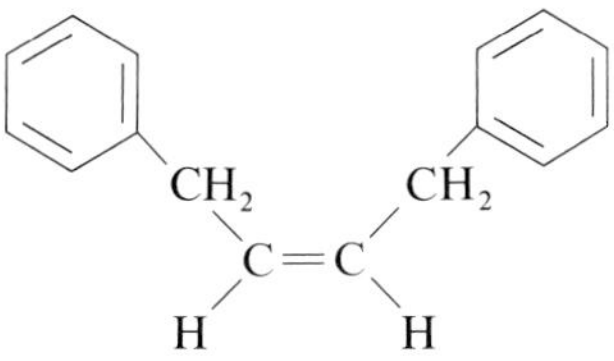 (diphenyl structure with CH_2, $C=C$, CH_2, and two H)

12.49 4-Deuteriotoluene was needed for an experiment examining the fate of an arene oxide derived from toluene. How would you synthesize 4-deuteriotoluene starting from toluene? Assume that heavy water, D_2O, is readily available.

12.50 Arrange the compounds in each of the following series in order of decreasing acidity.

(a) 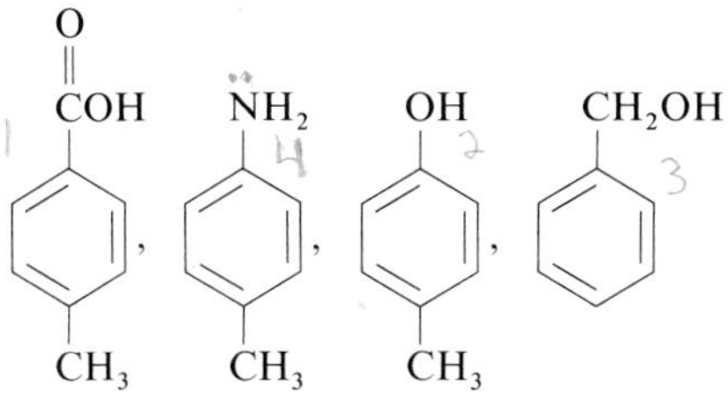 (four para-substituted benzenes with CH_3 groups: COH (with O), NH_2, OH, CH_2OH)

(b) 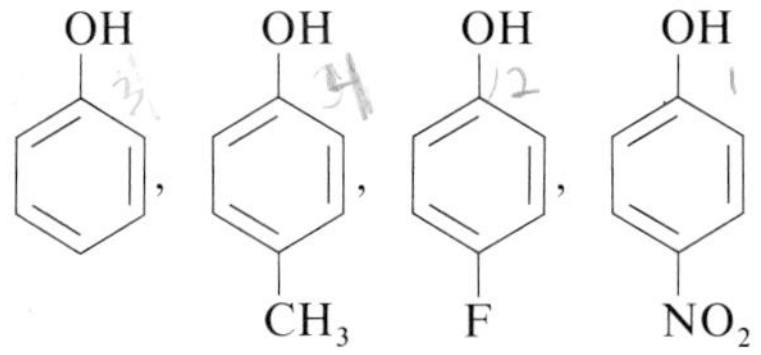 (four phenols: OH; OH with CH_3; OH with F; OH with NO_2)

(c) 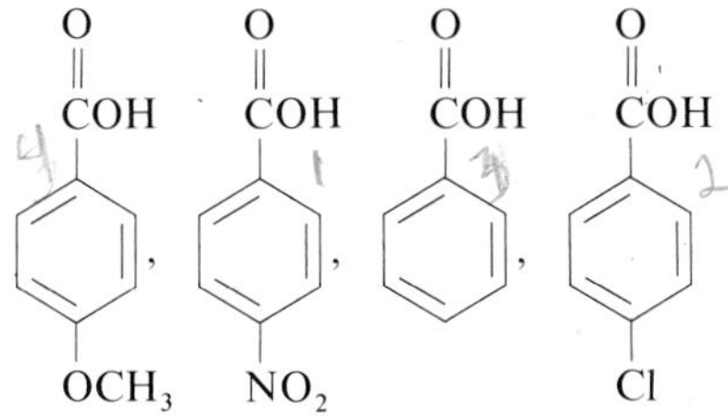 (four benzoic acids: COH with OCH_3; COH with NO_2; COH; COH with Cl, each with O)

12.51 1,3,5-Trimethylbenzene reacts by electrophilic aromatic substitution with iodine monochloride, ICl. Write an equation for this reaction showing the product you would expect. What is your reasoning in deciding what the product will be?

12.52 When allyl alcohol is treated with hydrogen fluoride in the presence of benzene, two products are formed. They are 3-phenyl-1-propene and 1,2-diphenylpropane. Write equations showing the mechanisms for the formation of these products.

12.53 When 2-hydroxybenzoic acid is heated with isobutyl alcohol in the presence of sulfuric acid, Compound A is formed. The same product is obtained if *tert*-butyl alcohol and sulfuric acid are used. What is the structure of Compound A? Write a mechanism that accounts for the experimental facts.

12.54   *p*-Nitrofluorobenzene reacts with a series of nucleophiles in methanol at 25 °C. The relative rates for its reactions are shown below. Explain the experimental observations. (Hint: Reviewing Section 6.2 may be helpful.)

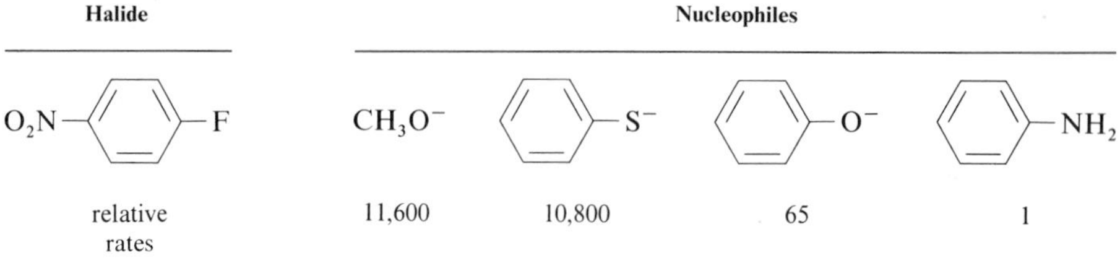

Halide	Nucleophiles			

O_2N—⟨ ⟩—F CH_3O^- ⟨ ⟩—S^- ⟨ ⟩—O^- ⟨ ⟩—NH_2

relative   11,600   10,800   65   1
rates

12.55   Several 2,4-dinitrochlorobenzenes were studied in their reaction with piperidine in ethanol as a solvent. Structural formulas for the compounds and their relative rates are shown below. What factor is influencing the rates in this series? How do you explain the effect? (Hint: A careful examination of the structural formula for the Meisenheimer complex on p. 621 will furnish a clue.)

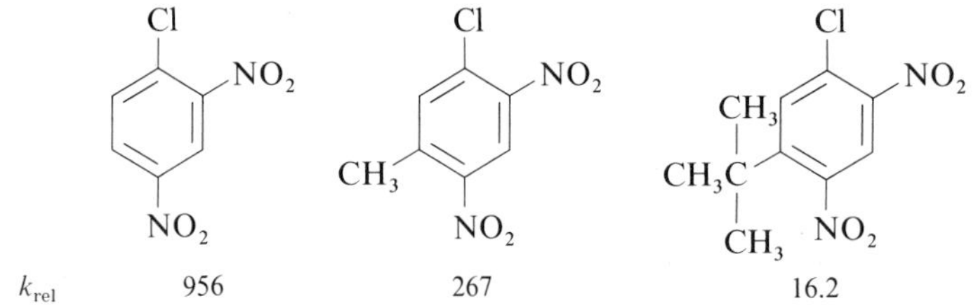

k_{rel} 956   267   16.2

12.56   An intensely blue hydrocarbon, called azulene, has the structure shown below. Predict whether it has aromaticity.

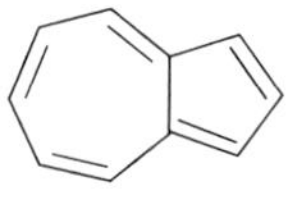

azulene

12.57   The heterocyclic compounds pyrrole and indole, shown below, are not very basic. Pyrrole has $pK_a \sim 15$, and thus has considerable acidity. Explain why a nitrogen compound such a pyrrole should be so much less basic and so much more acidic than ammonia ($pK_a \sim 36$).

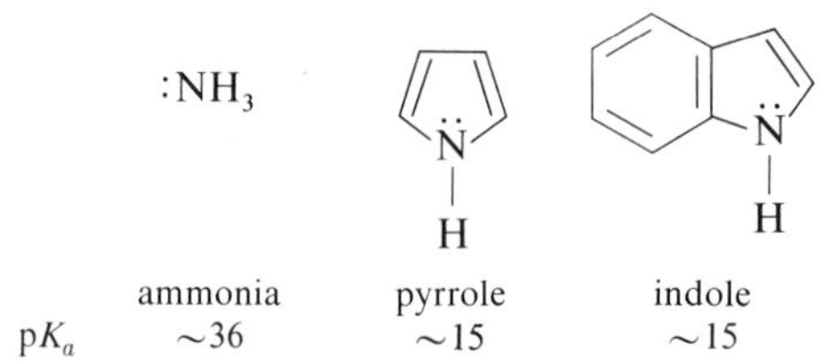

:NH_3

ammonia   pyrrole   indole
pK_a ~ 36 ~ 15 ~ 15

12.58   Experimentally, it has been found that the cyclononatetraenyl *anion* and the cyclooctatetraenyl *dianion* can be prepared and are reasonably stable species. The reaction of cyclononatetraene with the carbanion generated when dimethyl sulfoxide, CH_3SOCH_3, is treated with sodium hydride gives cyclononatetraenyl anion. Cyclooctatetraenyl dianion is formed when cyclooctatetraene reacts with two molar equivalents of potassium metal in tetrahydrofuran.

   Write equations for the reactions described above and explain the source of the stability of these hydrocarbon anions.

12.59 Heptafulvenes are compounds with a double bond attached to cycloheptatriene. The parent compound, shown below, is highly unstable, but substitution of two cyano groups for the hydrogen atoms on the double bond outside the ring gives a stable compound. How would you rationalize these experimental observations? (Hint: Writing down resonance contributors for both compounds will be helpful.)

heptafulvene   a dicyanoheptafulvene

unstable   *stable*

12.60 When the potassium salt of cyclopentadienyl anion is treated with iron(II) chloride, a stable compound in which an iron atom is sandwiched between two cyclopentadiene rings is formed. This compound, bis(cyclopentadienyl)iron, is also called ferrocene. Ferrocene undergoes aromatic substitution reactions. For example, heating ferrocene with acetic anhydride and phosphoric acid gives acetylferrocene. How do you explain the stability of ferrocene? (Hint: A similarly stable compound is not formed with iron(III). Looking at the periodic table may give you a clue.)

ferrocene   acetylferrocene

all C—H bonds on
ferrocene are equivalent

12.61 The infrared spectra of a series of compounds with their molecular formulas are given in Figure 12.12. Assign a structure that is compatible with all the data you have for each compound. Determining the units of unsaturation for each formula will be helpful.

Compound A, $C_{12}H_{12}$

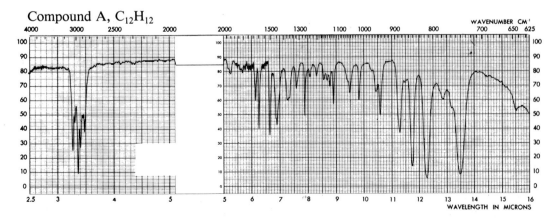

Figure 12.12 Spectra for Problem 12.61. (From *The Aldrich Library of Infrared Spectra*)

Compound B, C$_8$H$_8$O$_2$

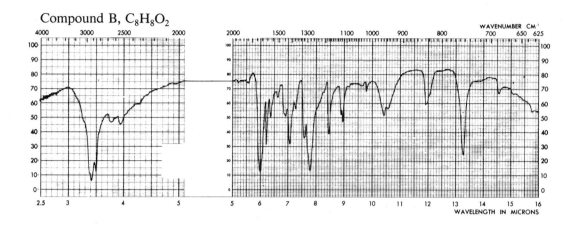

Compound C, C$_7$H$_8$O

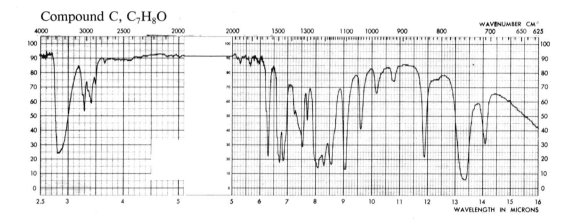

Compound D, C$_3$H$_5$ClO$_2$

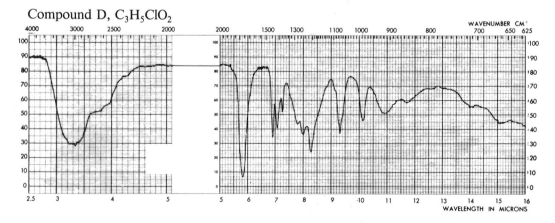

Figure 12.12 (*Continued*)

Compound E, C_7H_{12}

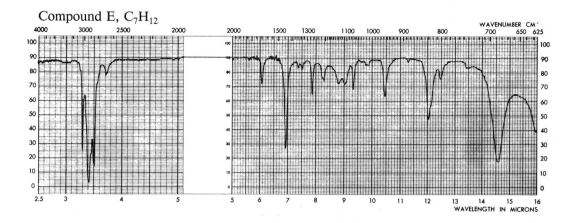

Compound F, $C_{14}H_{12}O_2$

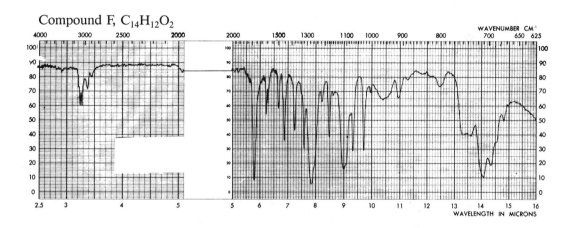

Compound G, C_9H_{12}

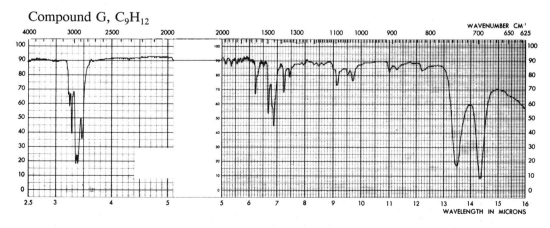

Figure 12.12 *(Continued)*

Compound H, $C_5H_{12}O_2$

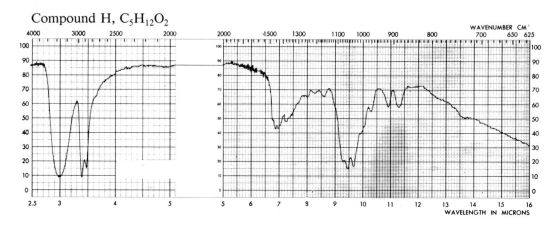

Figure 12.12   *(Continued)*

12.62   Two isomeric compounds have the molecular formula $C_8H_{11}N$. One of them, Compound A, has $\lambda_{max}^{methanol}$ 273 (ϵ 313), 264 (ϵ 374), and 259 nm (ϵ 326). The other, Compound B, has $\lambda_{max}^{methanol}$ 295 (ϵ 2040) and 246 nm (ϵ 11,000). Assign structures to the two compounds that fit these data.

12.63   Compound C, $C_8H_{10}O$, has $\lambda_{max}^{methanol}$ 267 (ϵ 106), 264 (ϵ 161), 258 (ϵ 219), 252 (ϵ 221), and 248 nm (ϵ 203). An isomer, Compound D, has $\lambda_{max}^{methanol}$ 289 (ϵ 2290) and 239 nm (ϵ 7050). What structures are possible for Compounds C and D? What further spectral information would be useful in assigning structures to these compounds?

12.64   Two isomeric compounds have the structural formulas shown below. One has $\lambda_{max}^{ethanol}$ 254 (log ϵ 2.34), 259 (log ϵ 2.38), 262 (log ϵ 2.37), 265 (log ϵ 2.27), and 268 (log ϵ 2.24) nm. The other has $\lambda_{max}^{ethanol}$ 246 (log ϵ 4.25), 284 (log ϵ 2.99), and 293 (log ϵ 2.84) nm. Which ultraviolet spectrum corresponds to which structure?

$$\langle\!\!\!\bigcirc\!\!\!\rangle\!-\!CH\!=\!CHCH_3 \qquad \langle\!\!\!\bigcirc\!\!\!\rangle\!-\!CH_2CH\!=\!CH_2$$

12.65   Vitamin B_6, pyridoxine, was isolated from rice bran and has the molecular formula $C_8H_{11}NO_3$. Its ultraviolet spectrum changes with pH. At pH 2.1, it has $\lambda_{max}^{H_2O}$ 292 (ϵ 6950) while at pH 10.2, $\lambda_{max}^{H_2O}$ 240 (ϵ 5500) and 315 nm (ϵ 5800) are observed.

(a)   When the acidic proton in vitamin B_6 is replaced by a methyl group, the spectrum has a single band at 280 nm (ϵ 5800) that does not change with pH. What conclusions can you reach about the structure of vitamin B_6 with this information?

(b)   Oxidation of the methyl ether of vitamin B_6 with barium permanganate in water at room temperature for 16 hours, filtration of the manganese dioxide that is formed, and acidification of the solution with sulfuric acid, gave two products, one a lactone and the other a dicarboxylic acid. Their structures are shown on the next page.

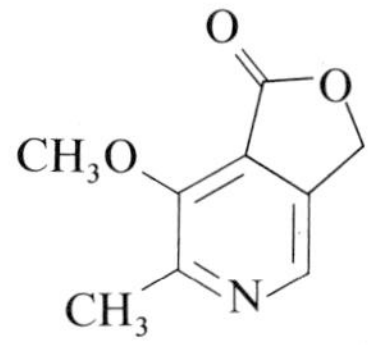

lactone
from the oxidation of
the methyl ether
of vitamin B_6

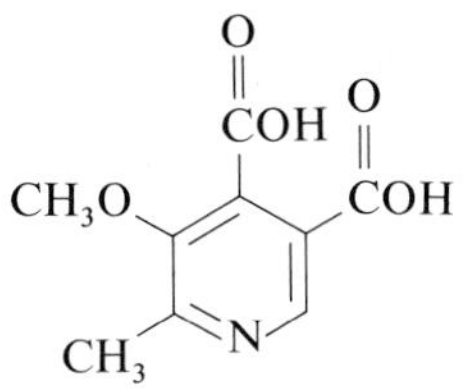

dicarboxylic acid
from the oxidation of
the methyl ether
of vitamin B_6

What must be the structure of vitamin B_6? Explain why two of the side chains on this aromatic ring were oxidized while the methyl group was untouched. Write equations showing a reaction pathway to the lactone.

(c) If you were trying to confirm the structure of vitamin B_6 by the use of ultraviolet spectroscopy, what simpler compounds would you use as models? Draw structural formulas for some that would be useful. Under what conditions would you want to look at the ultraviolet spectra of the models?

Nuclear Magnetic Resonance Spectroscopy and Mass Spectrometry

13

Introduction to Nuclear Magnetic Resonance Spectroscopy

Nuclear magnetic resonance spectra arise because certain atomic nuclei behave like small magnets. If a sample of a compound is placed in a strong external magnetic field, a certain number of the nuclei align themselves with the field. In this low-energy state, some of the nuclei absorb energy in the radiofrequency region of the electromagnetic spectrum and are raised to the higher-energy state in which the nuclei are aligned against the external field. The nuclear magnetic resonance spectrum of a compound is, like infrared (Section 7.7) and ultraviolet (Section 11.8) spectra, a record of transitions between different energy levels of the molecule.

Nuclei of hydrogen atoms are among those that absorb energy when placed in a magnetic field. The spectra that result are known as **proton magnetic resonance spectra.** Two examples of proton magnetic resonance spectra are shown in Figure 13.1.

Both spectra have a signal at the right-hand side of the spectrum, the absorption band for tetramethylsilane, $(CH_3)_4Si$. Tetramethylsilane, usually abbreviated TMS, is added to the solution of the compound being investigated to serve as a reference point in the spectrum. The divisions on the scale at the bottom of the spectra are called δ (delta) and are defined in Section 13.3. The signal for the hydrogen atoms in TMS appears at δ 0. The spectrum of iodomethane has one other signal at δ 2.10 while that of methyl acetate has two other signals, at δ 1.95 and δ 3.60.

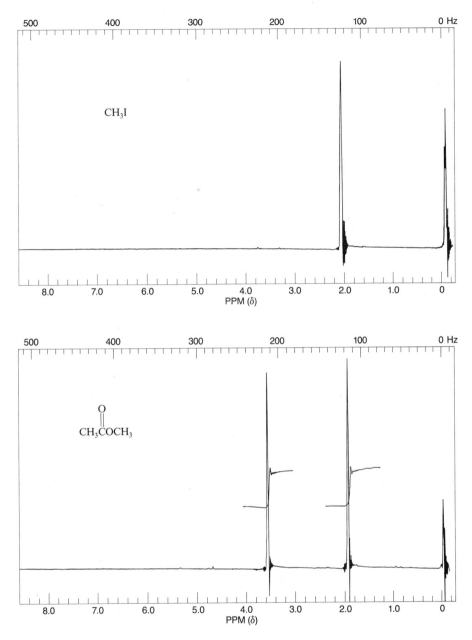

Figure 13.1 Proton magnetic resonance spectra of iodomethane and methyl acetate.

The three hydrogen atoms in iodomethane are said to be **chemically equivalent.** In methyl acetate, the hydrogen atoms on the methyl group adjacent to a carbonyl group are different from those on the methyl group adjacent to an oxygen atom. Whether hydrogen atoms are chemically equivalent or not can be determined by mentally substituting another group for each one of them in turn and seeing whether the same or a different compound results. For example, with iodomethane, substitution of a bromine atom for each of the three hydrogen atoms gives rise to the same compound.

I
|
C---H_b
H_c /   \ Br

substitution of H_a → bromoiodomethane

I
|
C---H_b
H_c /   \
     H_a

iodomethane

substitution of H_b →

I
|
C--- Br
H_c /   \
      H_a

bromoiodomethane

substitution of H_c ↘

I
|
C---H_b
Br /   \
     H_a

bromoiodomethane

The three hydrogen atoms are, therefore, chemically equivalent and give rise in the proton magnetic resonance spectrum to a single sharp peak, called a **singlet.**

The three hydrogen atoms of the acetyl group in methyl acetate are chemically equivalent but are different from the three hydrogen atoms of the methoxyl group. For example, substitution of a methyl group for a hydrogen atom on each part of the molecule leads to an entirely different compound in each case.

$$H_a \quad O \qquad H_b$$
$$| \quad\; \parallel \qquad\; |$$
$$H_a - C - C - O - C - H_b$$
$$| \qquad\qquad\quad |$$
$$H_a \qquad\qquad\; H_b$$

methyl acetate

substitution of any one of H_a by a methyl group →

$$\qquad H_a \quad O \qquad H_b$$
$$\qquad | \quad\; \parallel \qquad\; |$$
$$CH_3 - C - C - O - C - H_b$$
$$\qquad | \qquad\qquad\quad |$$
$$\qquad H_a \qquad\qquad\; H_b$$

methyl propanoate

substitution of any one of H_b by a methyl group →

$$H_a \quad O \qquad H_b$$
$$| \quad\; \parallel \qquad\; |$$
$$H_a - C - C - O - C - CH_3$$
$$| \qquad\qquad\quad |$$
$$H_a \qquad\qquad\; H_b$$

ethyl acetate

The hydrogen atoms of the acetyl group and of the methoxyl group give rise to different absorption bands, each one a singlet, in the proton magnetic resonance spectrum.

The spectrum of methyl acetate in Figure 13.1 shows another feature. *The stepwise tracing over the two bands is known as the* **integration of the spectrum.** The instrument is measuring the area under each absorption band and recording it as the height of a step in the tracing. An examination of the spectrum of methyl acetate shows that the two steps are each of the same height. In other words, the areas under the two peaks are equal. *In a proton magnetic resonance spectrum, the area under an absorption band is proportional to the number of hydrogen atoms giving rise to that signal.* Thus, the integration of this spectrum tells us that there are equal numbers of protons of two types in methyl acetate. Note that the integration does *not* tell us the absolute number of protons of each kind, but only the relative numbers.

Not all proton magnetic resonance spectra are as simple as the ones shown in Figure 13.1. In the compounds chosen here, even though there are a number of hydrogen atoms, either they are chemically equivalent or else they are isolated from each other

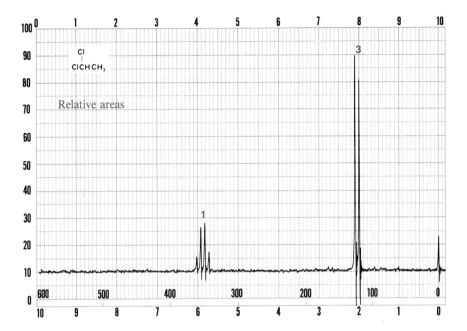

Figure 13.2   Proton magnetic resonance spectrum for 1,1-dichloroethane. (From *The Aldrich Library of NMR Spectra*)

by oxygen atoms or carbonyl groups. Figure 13.2 shows the proton magnetic resonance spectrum for 1,1-dichloroethane. In this compound, neighboring carbon atoms bear different types of hydrogen atoms.

The spectrum of 1,1-dichloroethane shows a profusion of peaks. Note, however, that there is a grouping of the bands; the peaks appear in two regions, one centered at $\delta\,2.05$ and the other at $\delta\,5.85$. The relative intensities of these bands (for all the small peaks in one group counted together) are about 3:1. (For the proton magnetic resonance spectra that do not show the actual integration curves, the *relative* areas under the bands will be given by numbers written above the peaks.) The band at $\delta\,2.05$ appears, therefore, to be the absorption band for the hydrogen atoms of the methyl group. The one at $\delta\,5.85$ belongs to the hydrogen atom on the carbon atom also bonded to two electronegative chlorine atoms.

The bands in the spectrum in Figure 13.2 are said to be split. This splitting arises from the interaction of the nuclei of the hydrogen atoms on neighboring carbon atoms and is known as **spin-spin coupling.** Section 13.4A examines how such interactions create the patterns we see.

Four essential pieces of information are thus obtained from a proton magnetic resonance spectrum. The number of different groupings of peaks tells us how many different kinds of hydrogen atoms are present in the compound. The δ values observed for the hydrogen atoms tell us something about the environment of the hydrogen atoms. The integration of the spectrum allows us to decide how many hydrogen atoms of each type are present in the molecule. Finally, the splitting patterns that appear in the spectrum alert us to hydrogen atoms whose nuclei are interacting with each other.

PROBLEM 13.1 For each of the following compounds, pick out the hydrogen atoms that are chemically equivalent to each other.

(a) $CH_3CH_2OCH_2CH_3$ (b) $CH_3CH_2CH_2Br$ (c)

$$\begin{array}{c} H \quad H \\ H - \bigcirc - OCH_2CH_3 \\ H \quad H \end{array}$$

(d) $CH_3\overset{\overset{\displaystyle CH_3}{|}}{C}HCH_2Cl$ (e) $CH_3CH_2\overset{\overset{\displaystyle O}{\|}}{C}CH_2CH_3$ (f) $CH_3CH_2CH_2CH_2CH_3$

PROBLEM 13.2 The spectra shown in Figure 13.3 belong to the following compounds. Assign each spectrum to the correct compound and discuss how you made your decision in each case. You can determine the structures by comparing the spectra with those already given in this section and reasoning by analogy.

(a)   acetone   (b)   1,2-dibromoethane   (c)   1,1,2-tribromoethane

Compound A

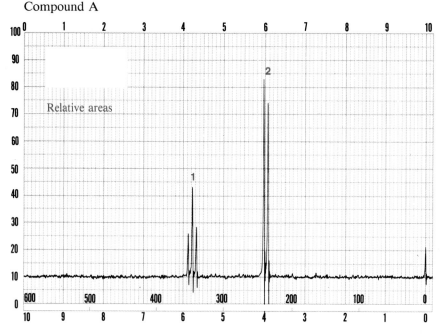

Figure 13.3 Spectra for Problem 13.2. (From *The Aldrich Library of NMR Spectra*)

Compound B

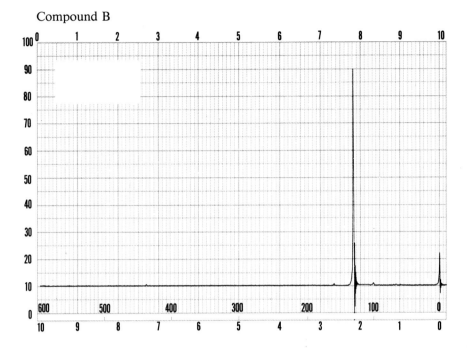

Compound C

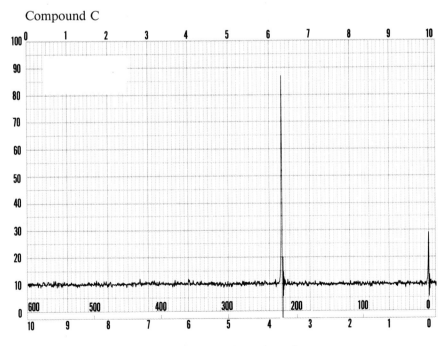

Figure 13.3   *(Continued)*

13.2

The Origin of Magnetic Resonance Spectra

Just as an electron in an atom is assigned a spin quantum number, the nuclei of certain isotopes also have spin, and are given a spin number. *Only atoms with nuclear spin give rise to nuclear magnetic resonance.*

Nuclei with an odd mass number have nuclear spin whether they have an odd or even atomic number. Thus the common isotope of hydrogen, ^1H, and the stable isotope of carbon, ^{13}C, which makes up 1.1% of natural carbon, have this property. Such a nucleus has a spin number, I, of $\frac{1}{2}$, and can have $(2I + 1)$ or two different energy states.

Nuclei of isotopes that have an even atomic number and an even mass number have no spin. The most abundant isotope of carbon, ^{12}C, and of oxygen, ^{16}O, fall in this category. They have a spin number, I, of 0.

Nuclei of isotopes that have an odd atomic number but an even mass number also possess nuclear spin. Deuterium, ^2H, and the common isotope of nitrogen, ^{14}N, belong to this category. These nuclei have a spin number, I, of 1.

The difference in nuclear spin energy levels for a hydrogen atom is very small at room temperature. For this reason the hydrogen nuclei in a given sample of a compound are almost evenly divided between the two energy levels. The nucleus is a charged particle, so a spinning nucleus gives rise to a magnetic field. Such a nucleus behaves like a small magnet and is affected by an external magnetic field. When a compound containing hydrogen atoms is placed in a strong magnetic field, 14,092 gauss, a small excess of the nuclei line up so that their magnetic moments are aligned with the external field. This situation is the lower energy state for the nuclei. Under these conditions, the difference between this lower energy level and the upper one corresponds to the energy of radiation in the radiofrequency range of the electromagnetic spectrum (Section 7.7A). If the sample is irradiated with radiation having wavelengths of approximately 5 meters or a frequency of 60,000,000 cycles per second or Hertz (Hz), energy is absorbed and nuclei are raised to the higher energy level in which the magnetic moments of the nuclei are opposed to the external magnetic field. The exact amount of energy required for the transition depends on the strength of the magnetic field experienced by the nuclei (Figure 13.4).

Once a nucleus is in the upper energy level, it returns to the lower level by various processes that involve losing energy to its surroundings. These processes are known as **relaxation** and are especially important in nuclear magnetic resonance spectroscopy because the difference in the number of nuclei in the two energy levels is so small. If the nuclei raised to the upper energy level do not return to the lower level, no more energy can be absorbed by the sample, and the spectrum ceases to be observed in a very short time. When this happens, the sample is said to be **saturated.**

Not all of the hydrogen nuclei in a molecule of an organic compound experience the same external magnetic field. *The electrons around the nucleus shield the nucleus from the effects of the external field. Thus, hydrogen atoms in different electronic environments in a molecule experience the external magnetic field to a different degree. Slightly different amounts of energy are, therefore, needed to promote the nuclei of different types of hydrogen atoms from the lower to the higher energy level.* The spectra in the last section demonstrated the practical consequences of this fact.

With most nuclear magnetic resonance spectrometers, the experiment is done by

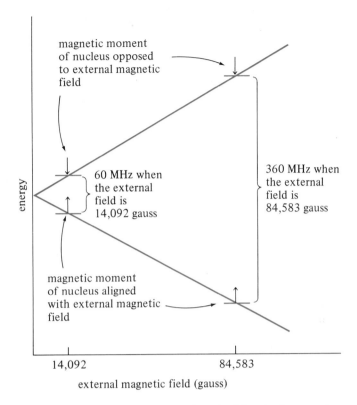

Figure 13.4   Energy levels associated with a hydrogen atom with nuclear spin number $I = 1/2$, in the presence of an external magnetic field.

irradiating the sample with a constant radiofrequency, usually 60,000,000 Hz (or 60 MHz) and varying the magnetic field slightly. Whenever the magnetic field actually felt at a nucleus reaches the value at which a transition between energy levels becomes possible with the absorption of energy at 60 MHz, absorption takes place. The instrument (Figure 13.5, p. 668) records the event as a spectrum.

The magnetic field used in nuclear magnetic resonance spectrometers must be highly homogeneous if small differences in energy are to be observed. One way to average out the field experienced by all of the molecules in a sample is to spin the tube containing the sample inside a cavity in the magnet. In practice, then, nuclear magnetic resonance spectra are usually taken of liquid samples or of solids in solution. The ideal solvent for proton magnetic resonance spectroscopy is carbon tetrachloride because it does not contain any hydrogen atoms. But carbon tetrachloride is not polar enough to dissolve many organic compounds, so solvents in which hydrogen atoms have been replaced by deuterium, which does not absorb in the same region of the spectrum as hydrogen does, are used. Among these, deuteriochloroform, $CDCl_3$, is the most generally useful.

Because it is difficult to measure the absolute value of an applied magnetic field, all nuclear magnetic resonance spectra are measured with reference to a standard compound, usually tetramethylsilane (Section 13.1).

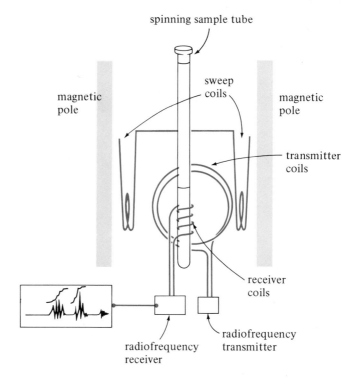

Figure 13.5   Schematic diagram of a typical nuclear magnetic resonance spectrometer.

PROBLEM 13.3   Predict whether the following isotopes will have a nuclear magnetic moment or not.

(a) ^{28}Si   (b) ^{15}N   (c) ^{19}F   (d) ^{31}P   (e) ^{11}B   (f) ^{32}S

13.3

The Chemical Shift

A. The Origin of the Chemical Shift. Chemical Shift Values for Hydrogen Atoms on Tetrahedral Carbon Atoms

The absorption of energy of a given radiofrequency by the nucleus of a hydrogen atom depends upon the effective magnetic field that it experiences. This, in turn, depends upon the environment of the hydrogen atom in the molecule as well as on the external magnetic field. Hydrogen atoms on carbon atoms bearing electronegative elements such as oxygen or the halogens have less electron density around them. They are said to be **deshielded** relative to the hydrogen atoms on the methyl groups in the reference compound, tetramethylsilane. The effect is cumulative; therefore, a hydrogen atom on a carbon atom bearing two halogen atoms is more deshielded than one on a carbon atom bearing only a single halogen atom. For example, the hydrogen atoms in iodomethane absorb at δ 2.10, those in diiodomethane at δ 3.88.

The effect of electronegativity can also be seen when the halogen substituents are varied. The hydrogen atoms on the carbon atom bearing the halogen atom become progressively more deshielded in the series iodomethane (δ 2.10), bromomethane (δ 2.70), chloromethane (δ 3.05), and fluoromethane (δ 4.30), as the electronegativity of the halogen increases. The deshielding of the hydrogen atoms is reflected in larger δ values.

CH_3F	CH_3Cl	CH_3Br	CH_3I
δ 4.30	δ 3.05	δ 2.70	δ 2.10
fluoromethane	chloromethane	bromomethane	iodomethane

←

*increasing deshielding of the
hydrogen atoms with increasing
electronegativity of the halogen*

For example, the spectrum of the dimethyl acetal of acetone (2,2-dimethoxypropane) (Figure 13.6) has two major peaks of equal intensity at δ 1.30 and 3.20. The signal at δ 1.30 is assigned to the hydrogen atoms of the methyl groups bonded to carbon, and that at δ 3.20 to the hydrogen atoms of the methyl groups bonded to oxygen. The hydrogen atoms of the methoxyl groups are *more highly deshielded* and *absorb at lower field* than do the other hydrogen atoms. These latter hydrogen atoms, at δ 1.30, are said to be *more* **shielded** and *to absorb at higher field*. Both groups of hydrogen atoms, however, are more deshielded, and absorb energy at lower field than the hydrogen atoms in TMS.

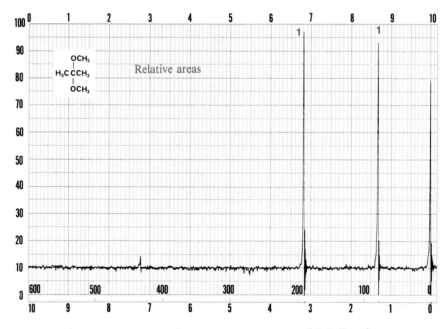

Figure 13.6   Proton magnetic resonance spectrum of 2,2-dimethoxypropane. (From *The Aldrich Library of NMR Spectra*)

The difference between the position of the absorption band for a given hydrogen atom and that of TMS is called the **chemical shift** *for that hydrogen.* The difference in energy between the absorption bands for TMS and for the hydrogen atoms in

2,2-dimethoxypropane can be read off the spectrum in two ways. The spectrum is recorded so that the space between the peak for TMS and the left-hand edge of the chart represents 600 Hz. Each small division from right to left, therefore, corresponds to 6 Hz. The positions of the absorption bands for the hydrogen atoms may be read directly off the chart. For example, the first absorption band occurs at a frequency, ν, of 78 Hz and the second one at ν 192 Hz downfield (to the left) from the signal for TMS. These chemical shift values depend upon the strength of the magnetic field of the instrument used, and therefore on the frequency, ν, of the radiofrequency radiation used to make the measurements. In most cases, chemical shifts measured in Hz are converted to δ, which is independent of instrumentation, and has units of ppm:

$$\delta = \frac{\nu \text{ sample } - \nu \text{ TMS}}{\nu \text{ applied of the instrument}} \times 10^6 \text{ ppm}$$

For a 60-MHz instrument,

$$\delta = \frac{\nu \text{ sample } - \nu \text{ TMS}}{60 \times 10^6} \times 10^6 \text{ ppm}$$

Thus, a band appearing 192 Hz downfield from TMS has

$$\delta = \frac{192 - 0}{60 \times 10^6} \times 10^6 \text{ ppm}, \quad \text{or} \quad \delta \text{ 3.20 ppm}$$

The differences in chemical shift that are being measured in a proton magnetic resonance spectrum are thus very small, on the order of parts per million in the frequency of the radiation being used to give rise to the nuclear transition. The entire range in chemical shift values for hydrogen atoms is approximately 20 ppm.

At the top of the chart, a third scale, the τ scale, is seen. On this scale TMS appears at 10 ppm and other hydrogen atoms usually have smaller values. We will not use this scale but mention it so that you will be careful not to confuse these numbers with chemical shift values expressed in δ units.

B. Special Effects Seen in Proton Magnetic Resonance Spectra of Compounds with π Bonds

The hydrogen atoms that are on sp^2-hybridized carbon atoms in double bonds and aromatic rings and the hydrogen atom on the carbonyl group of an aldehyde are much more deshielded than would be expected just from the effect of the electronegativity of the atoms to which they are bonded. For example, the hydrogen atoms on the double bond of cyclohexene absorb at δ 5.70; those on the aromatic ring in toluene, at δ 7.20 (Figure 13.7). In the spectrum for benzaldehyde, the signal for the hydrogen atom on the carbonyl group appears at δ 9.80 (Figure 13.8, p. 672). This peak appears far downfield from the range normally recorded for proton magnetic spectra, so the spectrum in Figure 13.8 has an expanded scale, with the left-hand edge of the spectrum being 1000 Hz downfield from TMS.

The hydrogen atom on the terminal carbon atom of a 1-alkyne, on the other hand, absorbs at δ 2.5, far upfield from a vinylic hydrogen atom. Such a hydrogen atom is bonded to an sp-hybridized carbon atom and should, in fact, be more deshielded than a hydrogen atom on an sp^2-hybridized carbon atom if electronegativity (Section 2.7) were the only factor to be considered.

These effects are explained by postulating that the motion of electrons in multiple bonds sets up induced magnetic fields around molecules containing such structural features. For certain orientations of the molecule, the induced magnetic field may

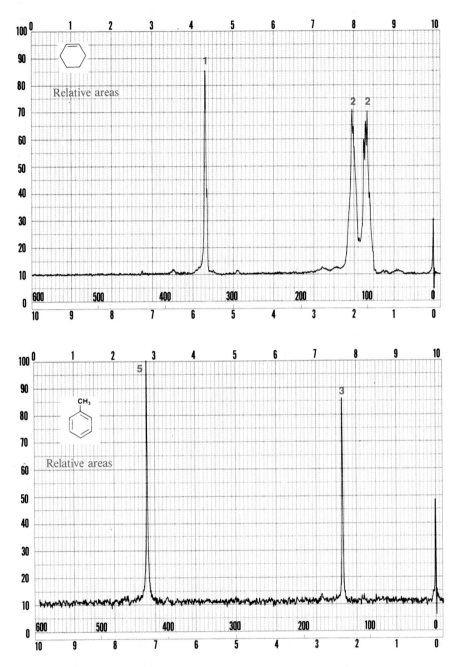

Figure 13.7   Proton magnetic resonance spectra for cyclohexene and toluene. (From *The Aldrich Library of NMR Spectra*)

either reinforce the external field or oppose it. Any hydrogen atoms that fall in the region of space around the molecule where the induced magnetic lines of force reinforce the external field appear to be deshielded in the spectrum. In other words, a lower external field is necessary for those nuclei to absorb energy and be promoted to the higher energy level. Hydrogen atoms that fall in the region of space in which the induced magnetic field opposes the external field are shielded. The external magnetic

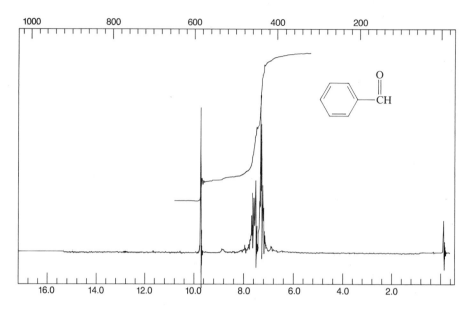

Figure 13.8 The proton magnetic resonance spectrum of benzaldehyde.

field must be increased before absorption of energy takes place. Thus, molecules with multiple bonds have deshielding and shielding regions about them. The origin of such regions is most easily seen for benzene. In Figure 13.9, the flow of electrons around the aromatic ring, the induced magnetic lines of force that result from this flow of charge, and the shielding and deshielding regions around the molecule are shown in a schematic way.

A difference in magnetic properties for different directions in space is called **magnetic anisotropy.** Molecules that have π bonds are magnetically anisotropic. Thus, benzene is a molecule that has magnetic anisotropy. In addition, an aromatic ring such as benzene is said to have a **ring current,** meaning a continuous movement of the π electrons around the ring. In the presence of the external magnetic field, this current gives rise to particularly strong induced lines of force, creating shielding and deshielding regions around the aromatic ring. The experimental detection of such a ring current in an unsaturated cyclic compound by nuclear magnetic resonance spectroscopy is a modern criterion of aromaticity. A comparison of the spectra of benzene and 1,3,5,7-cyclooctatetraene (Figure 13.10, p. 674) shows quite clearly the extra deshielding of the hydrogen atoms on the benzene ring. The chemical shift of the hydrogen atoms in 1,3,5,7-cyclooctatetraene is similar to that seen for the vinylic hydrogen atoms in cyclohexene (Figure 13.7). The spectrum of cyclooctatetraene supports the conclusion previously based on chemical evidence, and also on theory, that the double bonds in this cyclic polyene are localized and the compound does not have aromatic character (Section 12.1).

One of the most dramatic demonstrations of the effect of ring current in a conjugated cyclic polyene is seen in the spectrum of [18]annulene (Section 12.1). The spectrum of this compound has two bands, one at δ 8.9, and the other at δ -1.8 (1.8 ppm *to the right of, or upfield,* from TMS). These bands have relative intensities of 2:1. Thus, the low-field signal has been assigned to the hydrogen atoms around the outside of the ring, which lie in the deshielding region around the molecule. The high-field signal is

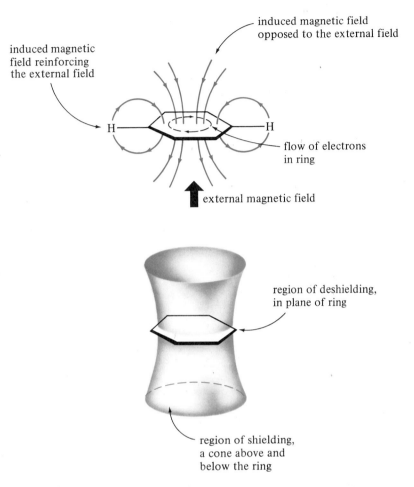

induced magnetic field reinforcing the external field

induced magnetic field opposed to the external field

H—

—H

flow of electrons in ring

external magnetic field

region of deshielding, in plane of ring

region of shielding, a cone above and below the ring

Figure 13.9   Ring current in benzene and induced magnetic lines of force. A simplified representation of shielding and deshielding regions around the aromatic ring.

assigned to the hydrogen atoms inside the ring. They lie in the shielding region created by the induced magnetic field.

Other functional groups that contain multiple bonds also have shielding and deshielding regions around them. These regions arise in the same general way as those around an aromatic ring, though it is a little more difficult to give a good simple picture of the circulation of electrons in the bonds and of the induced magnetic fields that result. The regions are shown as conical areas around the bonds for a carbonyl group, an alkene, and an alkyne in Figure 13.11 (p. 675).

The electronic currents induced in a double bond lie in planes above and below the double bond, just as is the case for an aromatic ring. In an alkyne, however, the absence of a nodal plane and the circular symmetry of the π electrons around the linear axis of the molecule allow the major electronic current to be around that axis. In this case, the shielding areas lie along the axis of the triple bond and the deshielding areas above and below it. Thus, the aldehyde hydrogen atom on a carbonyl group, or the hydrogen atoms on a double bond, lie in the plane of the molecule and in the de-

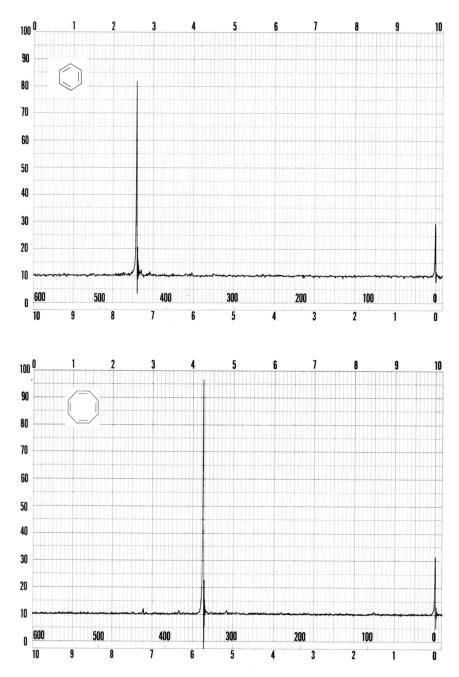

Figure 13.10 Proton magnetic resonance spectra for benzene and 1,3,5,7-cyclooctatetraene. (From *The Aldrich Library of NMR Spectra*)

shielding regions around those functional groups. The hydrogen atom on a triple bond, however, is shielded relative to one on a double bond but deshielded relative to one on a tetrahedral carbon atom.

Note that the hydrogen atoms on an aromatic ring are usually even more deshielded than those on an alkene double bond. This extra deshielding is attributed to the ring current, the possibility of the complete delocalization of electrons in the ring.

shielding regions

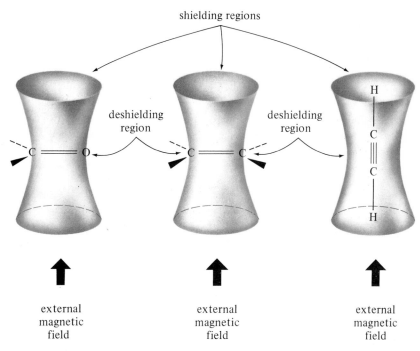

deshielding region

deshielding region

external magnetic field

external magnetic field

external magnetic field

Figure 13.11   Shielding and deshielding regions around functional groups containing multiple bonds.

PROBLEM 13.4   The proton magnetic resonance spectrum of thiophene (Section 12.2) is shown in Figure 13.12. The hydrogen atoms on the sp^2-hybridized carbon atoms of methyl vinyl sulfide, $CH_3SCH=CH_2$, for comparison, absorb at δ 4.95, 5.18, and 6.43. What conclusions can you draw from these data?

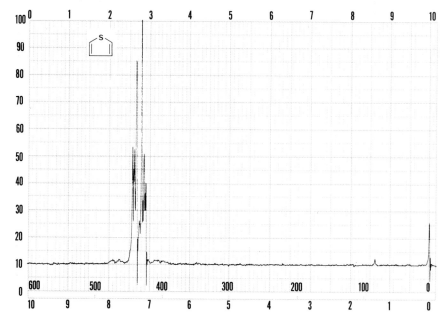

Figure 13.12   Spectrum for Problem 13.4. (From *The Aldrich Library of NMR Spectra*)

C. Chemical Shifts of Hydrogen Atoms of Hydroxyl Groups

The spectrum for methanol (Figure 13.13) shows the typical chemical shift for the hydrogen atom on a hydroxyl group. Methanol has two bands, at δ 3.40 for the three hydrogen atoms of the methyl group and at δ 4.50 for the hydrogen atom of the hydroxyl group.

The hydrogen atom of the hydroxyl group in a phenol (Figure 13.14) is usually more deshielded than that of an alcohol. The band for the hydrogen atom of the hydroxyl group in phenol appears at δ 5.80, while those for the aromatic hydrogen atoms, which

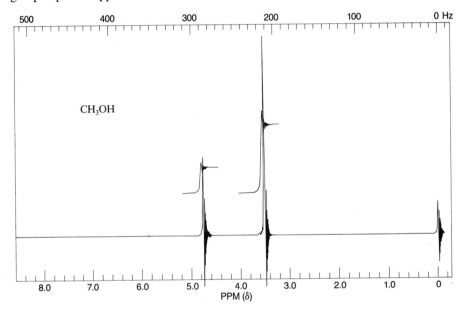

Figure 13.13 Proton magnetic resonance spectrum for methanol.

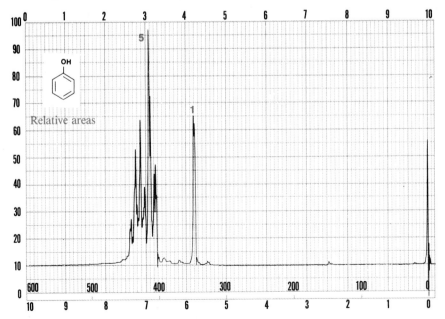

Figure 13.14 Proton magnetic resonance spectrum for phenol. (From *The Aldrich Library of NMR Spectra*)

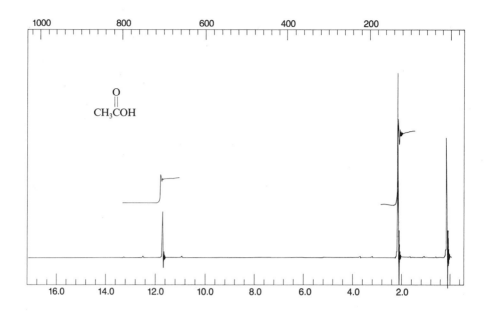

Figure 13.15   Proton magnetic resonance spectrum of acetic acid.

are no longer equivalent to each other and therefore split each other, appear at δ 6.7–7.4. In this case, as in the case of benzaldehyde (Figure 13.8), the presence of a substituent with considerable electronegativity differentiates the hydrogen atoms on the aromatic ring and gives rise to splitting patterns. Not all substitution on aromatic rings has this effect. The methyl group in toluene (Figure 13.7), for example, does not disturb the hydrogen atoms ortho to it sufficiently that splitting of the aryl hydrogen atoms is seen. In other words, the aryl hydrogen atoms in toluene, while not chemically equivalent, still have the same chemical shifts. *Hydrogen atoms having the same chemical shifts do not split each other.*

The proton on the carboxyl group of an acid is the most deshielded of all of these types of hydrogen atoms. The spectrum of acetic acid, taken like that for benzaldehyde, on an expanded scale, provides an example (Figure 13.15). The spectrum of acetic acid has two singlets, at δ 2.00 for the hydrogen atoms of the methyl group and at δ 11.57 for the acidic hydrogen atom.

In general, the chemical shifts of hydrogen atoms on oxygen are quite variable because they depend on whether hydrogen bonding is taking place, and this, in turn, depends on the concentration of the solution.

PROBLEM 13.5   If the solution of an alcohol is diluted with more $CDCl_3$, one of the bands in the proton magnetic resonance spectrum moves progressively upfield as dilution increases. Explain what is happening. Why does hydrogen bonding further deshield a hydrogen atom?

D. Typical Chemical Shifts in Proton Magnetic Resonance

A regularity in the chemical shifts shown by hydrogen atoms in different types of environments is apparent if the spectra given so far in this chapter are examined carefully. A summary of the chemical shifts usually observed for different kinds of hydrogen atoms is given in Table 13.1.

Note that the hydrogen atom on a tertiary carbon atom, a methine group, is more deshielded than those of methylene groups, which in turn are more deshielded than the

TABLE 13.1 Typical Chemical Shifts in Proton Magnetic Resonance

Type of Hydrogen Atom	δ, ppm*
RCH_3	0.9
RCH_2R acyclic	1.3
cyclic	1.5
R_3CH	1.5–2.0
$R_2C{=}CCH_3$ with R'	1.8
O ‖ $RCCH_3$	2.0–2.3
$ArCH_3$	2.3
$RC{\equiv}CH$	2.5
$RNHCH_3$	2–3
RCH_2X (X = Cl, Br, I)	3.5
O ‖ $ROCH_3, RCOCH_3$	3.8
$R_2C{=}CH_2$	5.0
$RCH{=}CR_2$	5.3
ArH	7.3
O ‖ RCH	9.7
RNH_2	1–3
$ArNH_2$	3–5
O ‖ $RCNHR$	5–9
ROH	5
$ArOH$	7
O ‖ $RCOH$	11

*The chemical shift values are given in ppm relative to tetramethylsilane at 0.00 ppm and are for the hydrogen atoms shown in color in the formulas. The values for hydrogen atoms on oxygen and nitrogen are highly dependent on solvent, concentration, and temperature.

hydrogen atoms of methyl groups. The effects of substituents are additive, so the chemical shift of a hydrogen atom may be different from the value given in the table if several factors influence it at once. Just as was the case with infrared spectra, it is important to pay attention to all the information available, including the molecular formula of the compound, the integration of the spectrum, and the appearance of spin-spin coupling, as well as to chemical shift values.

PROBLEM 13.6   Using the chemical shift values given in Table 13.1, assign the absorption bands between δ 1.0 to 3.0 in the spectrum of cyclohexene (Figure 13.7). Remember that values given in a table are general and approximate and that you have to reason by analogy in making assignments for the spectrum of a particular compound.

PROBLEM 13.7   Molecular formulas and proton magnetic resonance spectra for Compounds D through G are given in Figure 13.16. Assign structures to the compounds. [Hint: Calculating the units of unsaturation in a compound (Section 8.7) is a good way to start solving these kinds of problems.]

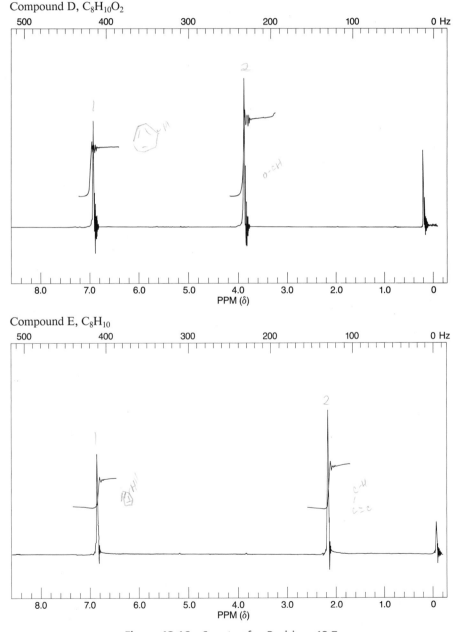

Compound D, $C_8H_{10}O_2$

Compound E, C_8H_{10}

Figure 13.16   Spectra for Problem 13.7

Compound F, C$_7$H$_7$Cl

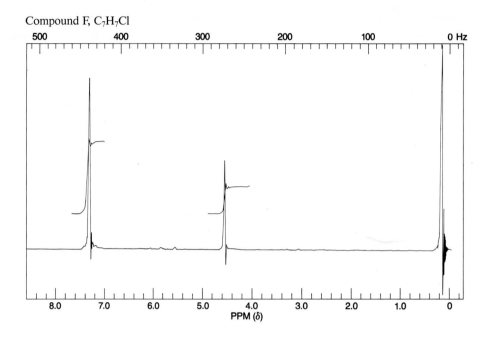

Compound G, C$_8$H$_8$O

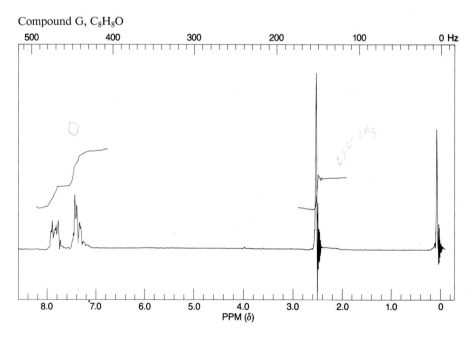

Figure 13.16 *(Continued)*

13.4

Spin-Spin Coupling

A. The Origin of Spin-Spin Coupling

Many of the spectra shown in the previous sections of this chapter have bands that are split into smaller peaks. These patterns arise from interactions between the nuclei of hydrogen atoms close enough that their nuclear spins affect each other.

In the spectrum of 1,1-dichloroethane (Figure 13.2) the hydrogen atoms of the methyl group appear as two peaks, called a **doublet,** separated from each other by 7 Hz. All of the methyl hydrogen atoms are, of course, equivalent to each other and have the same chemical shift, read off the spectrum as the central point of the doublet, δ 2.05. They appear as a doublet because of the influence of the single hydrogen atom on the adjacent carbon atom. In a given sample of the compound, about half of the hydrogen atoms at any one time have their nuclear spins aligned with the external field and half are aligned against it. The magnetic field experienced by the hydrogen atoms of the methyl group is influenced by the orientation of the nucleus of the hydrogen atom on the neighboring carbon atom. The hydrogen atoms on the methyl groups of half of the molecules in the sample experience a slightly decreased field, half a slightly increased field. This situation corresponds to two slightly different amounts of energy that are necessary for a transition between nuclear energy levels to take place for the hydrogen atoms of the methyl group. Two absorption bands of equal intensity appear in the spectrum.

If the hydrogen atoms on the methyl group in 1,1-dichloroethane are influenced by the hydrogen atom on the other carbon atom, the single hydrogen atom at δ 5.85 must also show the effect of the three hydrogen atoms of the methyl group. The presence of these three hydrogen atoms on the adjacent carbon atom splits the low-field absorption band into four peaks having relative intensities of 1:3:3:1. Such a pattern is called a **quartet.**

The magnetic moments of three hydrogen atoms can have four different types of interactions with the external magnetic field. These are shown in Figure 13.17 along

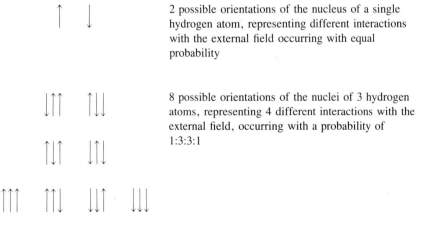

2 possible orientations of the nucleus of a single hydrogen atom, representing different interactions with the external field occurring with equal probability

8 possible orientations of the nuclei of 3 hydrogen atoms, representing 4 different interactions with the external field, occurring with a probability of 1:3:3:1

Figure 13.17 The interactions possible between the magnetic moments of the nuclei of the hydrogen atoms in 1,1-dichloroethane and an external magnetic field.

with the orientations of the nucleus of the single hydrogen atom as it was described above.

Three nuclei have four distinguishable ways in which they can be oriented with respect to an external magnetic field. Because individual nuclei cannot be labeled, all the cases in which one nucleus is oriented against the field and two with, are equal in energy, as are those in which two nuclei are against the field and one with. The low-field hydrogen atom may thus experience one of four slightly different magnetic fields, and the probability that any one of these situations will occur is $1:3:3:1$. Thus four peaks, a quartet, having those relative intensities arise. The distance between these peaks is the same as that seen for the methyl doublet.

The interaction between the nuclei of the different kinds of hydrogen atoms is called **spin-spin coupling.** *The interaction occurs through the intervening bonds. The strength of the interaction thus depends upon factors such as the number of bonds between the interacting nuclei, the types of bonds that intervene, and the stereochemical relationship between the atoms. The measure of the interaction is the* **coupling constant,** J**,** *given in Hz.* The value seen in the spectrum of 1,1-dichloroethane, approximately 7 Hz, is typical of coupling constants for alkyl hydrogen atoms on adjacent carbon atoms (Figure 13.18). Note that the bands arising from hydrogen atoms that are coupled to each other can be identified because the distances between the small peaks are always the same. The coupling between the different hydrogen atoms gives rise to one coupling constant, J.

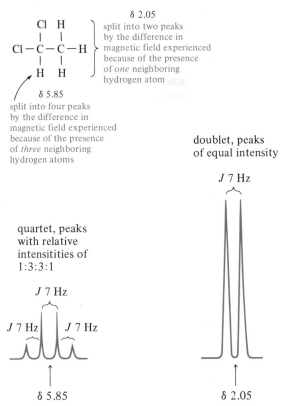

Figure 13.18   A schematic indication of the coupling between hydrogen atoms on neighboring carbon atoms in 1,1-dichloroethane.

For hydrogen atoms of widely differing chemical shifts on adjacent carbon atoms, the splitting pattern seen is thus determined by the number of hydrogen atoms on neighboring carbon atoms. *The number of peaks seen is N + 1 where N equals the number of neighboring hydrogen atoms that are chemically equivalent or that have equivalent coupling constants.* The patterns that are to be expected, and the relative intensities of the peaks within each pattern are given in Table 13.2.

TABLE 13.2 Splitting Patterns That Result from the Presence of N Neighboring Hydrogen Atoms

N	Multiplet $(N + 1)$	Relative Intensities of Peaks
0	singlet (1)	1
1	doublet (2)	$1:1$
2	triplet (3)	$1:2:1$
3	quartet (4)	$1:3:3:1$
4	quintet (5)	$1:4:6:4:1$
5	sextet (6)	$1:5:10:10:5:1$
6	septet (7)	$1:6:15:20:15:6:1$

PROBLEM 13.8 The spectrum of bromoethane is given in Figure 13.19. Assign chemical shifts to the different hydrogen atoms in bromoethane and analyze the spin-spin coupling. What is the coupling constant, J? Make sure that you understand why the methyl hydrogen atoms in bromoethane appear as a triplet by demonstrating to yourself the number of different orientations possible for the nuclei of the hydrogen atoms of the methylene group. What gives rise to the relative intensities seen for the individual peaks of the triplet?

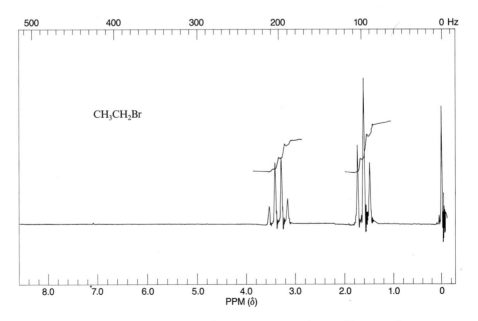

Figure 13.19 Proton magnetic resonance spectrum of bromoethane.

A look at other examples of spectra that show the effects of spin-spin coupling will be useful. In Figure 13.20 the spectra of 2-iodopropane and 1-iodopropane are shown. The spectrum of 2-iodopropane indicates the presence of two types of hydrogen atoms, in a ratio of 1:6. Those on the two methyl groups absorb at δ 1.86, and the hydrogen atom on the carbon atom bearing the iodine atom absorbs at δ 4.30. The signal for the methyl hydrogen atoms, which are all equivalent to each other, appears as a doublet

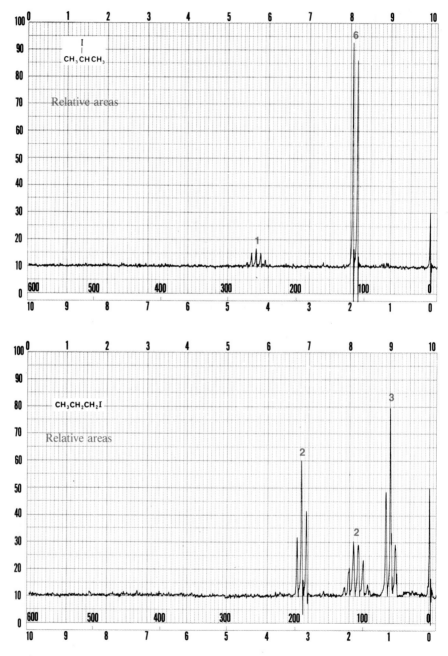

Figure 13.20   Proton magnetic resonance spectra of 2-iodopropane and
1-iodopropane. (From *The Aldrich Library of NMR Spectra*)

because of the presence of one hydrogen atom on the adjacent carbon atom. It is very important to understand that the methyl signal is a doublet *not* because there are two methyl groups, but because of spin-spin coupling. The influence of the six equivalent hydrogen atoms of the methyl groups likewise splits the absorption band for the hydrogen atom on the secondary carbon atom into (6 + 1) or 7 peaks. The intensities of the outer peaks of the septet are very weak, so they do not appear unless that portion of the spectrum is enhanced. The distance between each of the peaks of the septet is equal to the distance between the two peaks of the doublet. The value, read off the spectrum, is about 7 Hz, the usual coupling constant for hydrogen atoms on neighboring sp^3-hybridized carbon atoms.

doublet, split by
1 hydrogen atom on
neighboring carbon atom

δ 1.86

$$CH_3 - \underset{\underset{\underset{\delta\ 4.30}{H}}{|}}{\overset{\overset{CH_3}{|}}{C}} - I$$

septet, split by
6 hydrogen atoms on
neighboring carbon atoms

2-iodopropane

The spectrum of the isomeric 1-iodopropane has bands in three regions, a triplet centered at δ 1.00 (J 7 Hz), a sextet at δ 1.85 (J 7 Hz), and another triplet at δ 3.20 (J 7 Hz). These bands have relative intensities of 3:2:2. The band at δ 1.00 is assigned to the methyl group, while the one at δ 3.20 belongs to the hydrogen atoms on the carbon atom bearing the iodine atom. The hydrogen atoms on carbon 2 fall in between. The interactions that give rise to the splitting patterns are analyzed below.

triplet, split by   *triplet, split by*
2 hydrogen atoms on   *2 hydrogen atoms on*
neighboring carbon atom   *neighboring carbon atom*

$$I - CH_2 - CH_2 - CH_3$$
δ 3.20 δ 1.85 δ 1.00

sextet, split by
5 hydrogen atoms on
neighboring carbon atoms

1-iodopropane

The band for the hydrogen atoms on carbon 2 appears as a sextet because the coupling constants for the interactions of those hydrogen atoms with those on carbons 1 and 3 happen to be of the same magnitude. Thus, the hydrogen atoms on carbon 2 appear to be coupled with five equivalent hydrogen atoms. Different coupling constants for the hydrogen atoms on carbon 2 and those on carbon 1 and carbon 3 would have resulted in a more complex pattern.

PROBLEM 13.9   Compound H, C_9H_{12}, has the proton magnetic resonance spectrum shown in Figure 13.21. Assign a structure to the compound.

Compound H

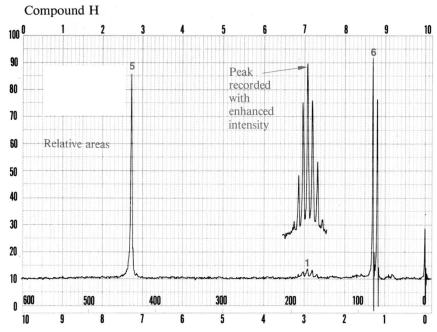

Figure 13.21   Spectrum for Problem 13.9 (From *The Aldrich Library of NMR Spectra*)

B. Spin-Spin Coupling in Compounds with Multiple Bonds

Not all hydrogen atoms couple with each other with *J* 7 Hz. It is easiest to see different types of coupling in the spectra of alkenes. The spectrum of vinyl acetate is shown in Figure 13.22.

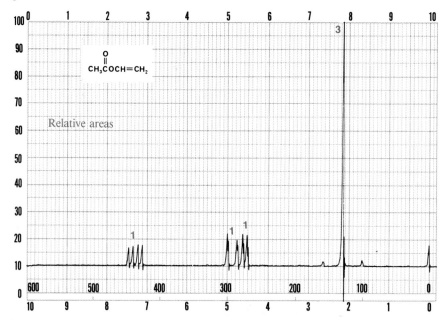

Figure 13.22   Proton magnetic resonance spectrum of vinyl acetate. (From *The Aldrich Library of NMR Spectra*)

Besides the singlet for the hydrogen atoms of the acetyl group, the spectrum of vinyl acetate has complex bands for three other hydrogen atoms. These bands are centered at δ 4.55, 4.86, and 7.28. Each of these bands is split into a doublet, and the peaks of each doublet are split again into another doublet. This pattern is called **a doublet of doublets.** Note that none of the groupings of four peaks is a quartet because the four peaks are not evenly spaced and do not have relative intensities of $1:3:3:1$.

The hydrogen atoms on the terminal carbon atom of the double bond are distinct from each other because one (δ 4.55) is trans to an oxygen atom and cis to a hydrogen atom on the other carbon atom of the double bond while the situation is reversed for the second hydrogen atom (δ 4.86). Three coupling constants are seen in the spectrum. The hydrogen atoms that are trans to each other on the double bond have the strongest interaction, J 14 Hz. The coupling constant for the hydrogen atoms that are cis to each other is 6.4 Hz, while the two hydrogen atoms on the same carbon atom have a small coupling constant, J 1.4 Hz.

The appearance of the splitting patterns seen in the spectrum of vinyl acetate can be duplicated by constructing a diagram. The diagram shows the lines that will result if the band for a hydrogen atom is split by interaction with two other hydrogen atoms with two different coupling constants. If a ruler is used so that the distances in the diagram are really proportional to the coupling constants, the patterns seen in the spectrum emerge (Figure 13.23).

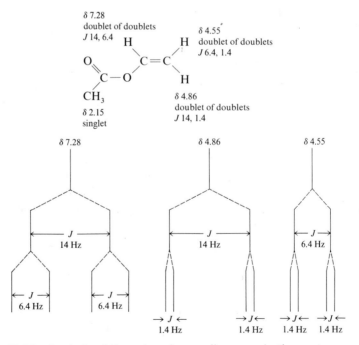

Figure 13.23 Analysis of the spin-spin coupling seen in the proton magnetic resonance spectrum of vinyl acetate.

Several different coupling constants are also seen in the spectra of aromatic compounds, represented here by 2-bromo-4-nitrotoluene (Figure 13.24). There are four different kinds of hydrogen atoms in 2-bromo-4-nitrotoluene. The methyl group at

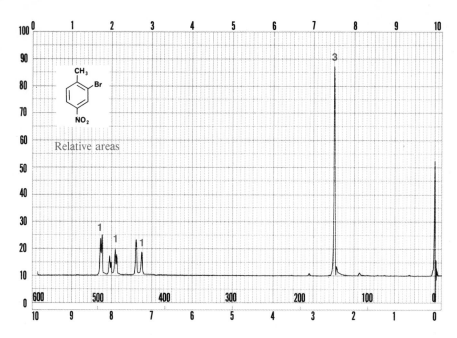

Figure 13.24   Proton magnetic resonance spectrum of 2-bromo-4-nitroluene. (From *The Aldrich Library of NMR Spectra*)

δ 2.50 is clearly separated from the aromatic hydrogen atoms, which absorb at δ 7.30, 7.98, and 8.25. In general, very little coupling is seen between hydrogen atoms on the side chains of aromatic rings and the hydrogen atoms of the ring. The signal at lowest field is assigned to the hydrogen atom that is between the nitro group and the bromine atom on the ring. It is split into a doublet by the hydrogen atom meta to it with a very small coupling constant, J 2 Hz. The signal at highest field is assigned to the hydrogen atom ortho to the methyl group. It is split into a doublet by the hydrogen atom ortho to it, J 8 Hz. Note that coupling between hydrogen atoms ortho to each other is stronger than coupling between the more distant meta hydrogens. Very little coupling is seen between hydrogen atoms that are para to each other, J about 0.7 Hz.

The hydrogen atom that is ortho to the nitro group alone gives rise to a doublet of doublets. This hydrogen atom interacts with the hydrogen atom that is ortho to it with one coupling constant, J 8 Hz, and with the hydrogen atom that is meta to it with another coupling constant, J 2 Hz.

δ 2.50
singlet

δ 7.30
doublet
J 8 Hz

δ 7.98
doublet of doublets
J 8, 2 Hz

δ 8.25
doublet
J 2 Hz

CH$_3$

H   Br

H   H

NO$_2$

2-bromo-4-nitrotoluene

Note that the proton magnetic resonance spectra of compounds that contain multiple bonds show coupling between hydrogen atoms that may have as many as four or five bonds between them. In general, coupling constants for these interactions are small but do give rise to complexity in the spectra observed.

A list of coupling constants for hydrogen atoms in various environments is given in

Table 13.3 (p. 690).

PROBLEM 13.10   Use the coupling constants shown for 2-bromo-4-nitrotoluene on p. 688 to construct a diagram like the one in Figure 13.23, showing the pattern of lines you would predict for the aryl hydrogen atoms in its spectrum.

PROBLEM 13.11   The signal for the hydrogen atoms on carbon 2 of 1-iodopropane (p. 684) may be regarded as being split into a quartet by the three hydrogen atoms on carbon 3 and each line of that quartet further split into a triplet by the hydrogen atoms on carbon 1. Assume that the coupling constant for both types of interactions is 7 Hz and construct a diagram like the one in Figure 13.23 to determine the number of lines you would expect to see for the pattern that results.

C. More Complex Splitting Patterns

The spin-spin splitting patterns seen in the spectra used as examples in the previous sections arise only when the difference in chemical shift between the hydrogen atoms that are interacting is large compared with their coupling constant, J. When hydrogen atoms of similar chemical shift interact, the patterns become much more complex. The spectrum for 1-chlorobutane (Figure 13.25) demonstrates this. In the spectrum of 1-chlorobutane, only the absorption band for the hydrogen atoms on carbon 1 (δ 3.55) is a recognizable triplet. These hydrogen atoms are deshielded by the chlorine atom and have a chemical shift quite different from those for the rest of the hydrogen atoms in the molecule. The hydrogen atoms on carbon atoms 2 and 3 are similar in chemical shift, but not identical. They couple with each other and also with either the hydrogen atoms on carbon 1 or those of the methyl group. Thus they appear as a complex pattern called a **multiplet.** The signal for the methyl group, however, which would be expected to appear as a triplet, is so distorted that it is hard to detect the pattern. This

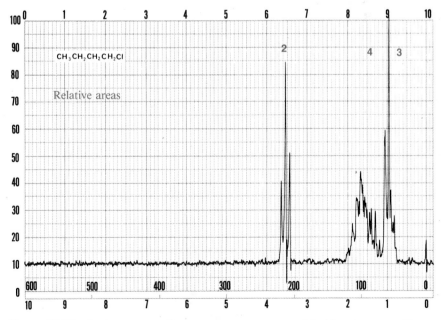

Figure 13.25   Proton magnetic resonance spectrum of 1-chlorobutane. (From *The Aldrich Library of NMR Spectra*)

TABLE 13.3   Coupling Constants Seen in Proton Magnetic Resonance Spectra

Type of Hydrogen Atoms	J, Hz
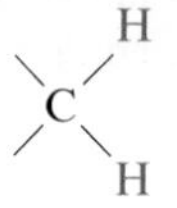	12–15
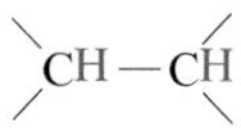	6–8
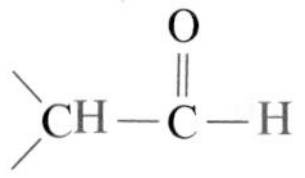	2–3
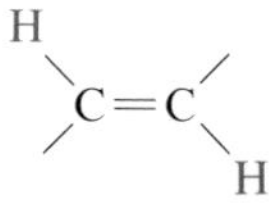	12–18
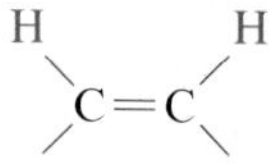	6–12
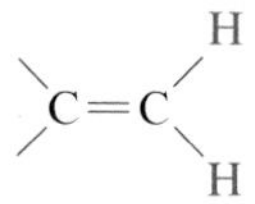	0–2
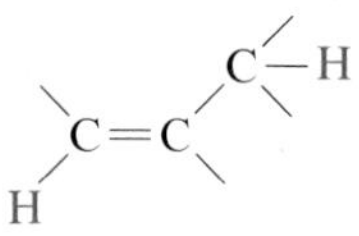	0–3
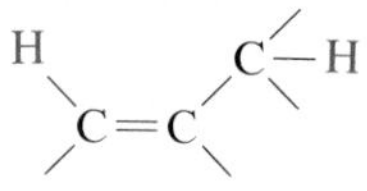	0–3
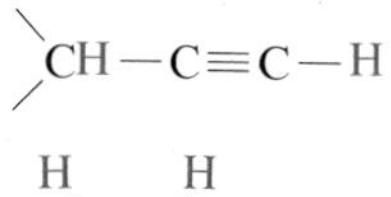	2–3
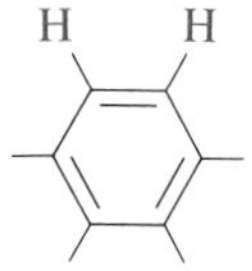	6–10
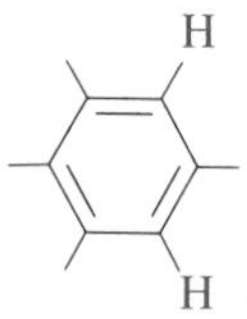	1–3
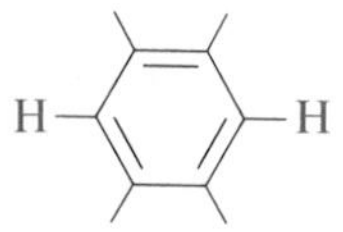	0–1

distortion of the triplet arises because the chemical shift for the hydrogen atoms of the methyl group is too close to that of methylene hydrogen atoms. The appearance of the spectrum of 1-chlorobutane is typical of that for substituted alkanes in which most of the hydrogen atoms have similar chemical shifts. The closer the chemical shifts of the interacting hydrogen atoms, the more distorted are the patterns that are seen.

Para-substituted aromatic compounds in which the two substituents differ from each other in electronic effects have typical symmetrical patterns in their proton magnetic resonance spectra (Figure 13.26). In the case of 1-bromo-4-methoxybenzene, because the chemical shifts of the hydrogen atoms ortho to the methoxy group (δ 6.69) and those

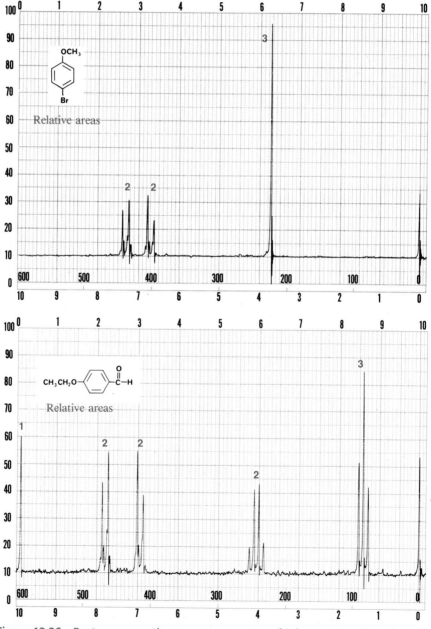

Figure 13.26 Proton magnetic resonance spectra of 1-bromo-4-methoxybenzene and *p*-ethoxybenzaldehyde. (From *The Aldrich Library of NMR Spectra*)

ortho to the bromine atom (δ 7.31) are quite different, the pattern is spread apart. Note the resemblance in the splitting pattern seen for the aromatic hydrogen atoms in the spectrum of *p*-ethoxybenzaldehyde to that for 1-bromo-4-methoxybenzene. In both cases we observe a complex pattern that is roughly a doublet of doublets and is symmetrical about a central point. The pattern typical of an ethyl group bonded to a deshielding atom such as oxygen or a halogen also shows clearly in the spectrum of *p*-ethoxybenzaldehyde.

PROBLEM 13.12   Proton magnetic resonance spectra for *o*-bromotoluene and *p*-bromotoluene are given in Figure 13.27. Assign each spectrum to the correct compound.

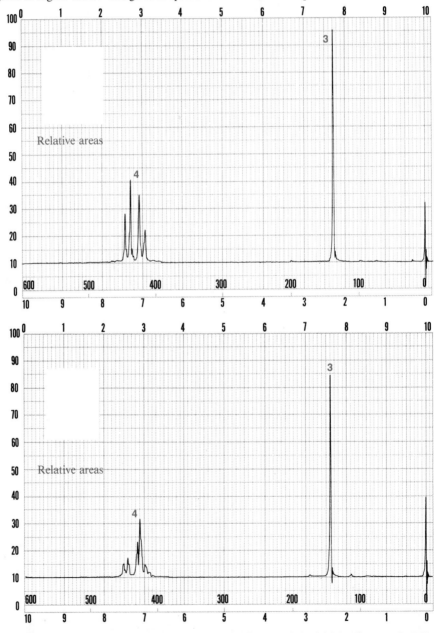

Figure 13.27   Spectra for Problem 13.12. (From *The Aldrich Library of NMR Spectra*)

PROBLEM 13.13   The proton magnetic resonance spectrum of 1-methoxy-2,4-dinitrobenzene is shown in Figure 13.28. Analyze the spectrum as completely as you can, deciding which signal belongs to which hydrogen atom and explaining the appearance of the spectrum.

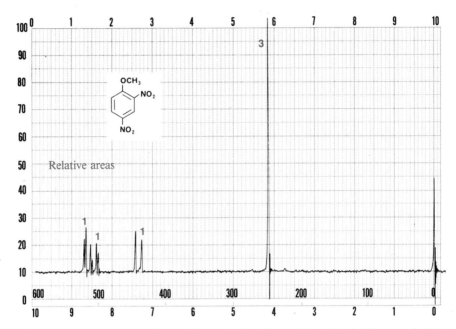

Figure 13.28   Spectrum for Problem 13.13. (From *The Aldrich Library of NMR Spectra*)

D. The Effect of Chemical Exchange on Spin-Spin Coupling

The spectrum of methanol, shown in Figure 13.13 in Section 13.3C, has two singlets, one for the hydrogen atoms of the methyl group and one for the hydrogen atom on oxygen. These two different types of hydrogen atoms are on adjacent atoms but do not appear to be coupled. This phenomenon is quite general for hydrogen atoms on oxygen atoms and on the nitrogen atoms of amino groups. Such hydrogen atoms hydrogen bond strongly and exchange with each other so rapidly that the effect of the nuclear spin of the exchanging hydrogen atom is averaged out. Thus the hydrogen atoms on an adjacent carbon atom experience only an average change of the external magnetic field and no spin-spin coupling occurs.

If a sample of an alcohol is carefully purified so that no trace of acid, which catalyzes the exchange reaction, remains, we can observe coupling between the hydrogen atom of the hydroxyl group with hydrogen atoms on an adjacent carbon atom. The spectrum of such a highly purified sample of methanol appears in Figure 13.29 (p. 694). In this spectrum of methanol, the band for the hydrogen atom of the hydroxyl group is a multiplet, while that for the methyl group is clearly a doublet. The coupling constant for the interaction between the two types of hydrogen atoms is approximately 5 Hz.

The spectrum of benzylamine (Figure 13.30, p. 694) has singlets for the hydrogen atoms on nitrogen (δ 1.48), on the methylene group deshielded by nitrogen and the aromatic ring (δ 3.89), and on the aromatic ring (δ 7.29). There is no coupling between the hydrogen atoms of the amino group and those of the methylene group, again because of rapid exchange of the hydrogen atoms on the nitrogen atom.

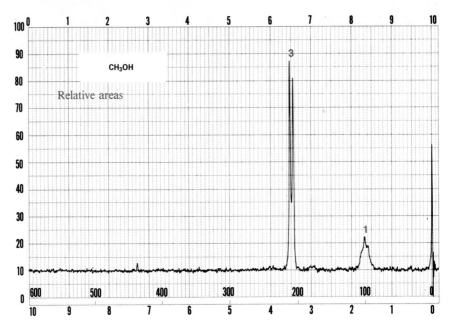

Figure 13.29   Proton magnetic resonance spectrum of highly purified methanol. (From *The Aldrich Library of NMR Spectra*)

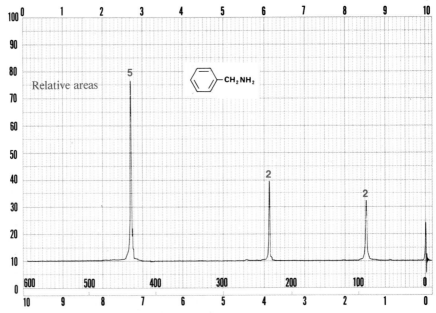

Figure 13.30   Proton magnetic resonance spectrum of benzylamine. (From *The Aldrich Library of NMR Spectra*)

Hydrogen atoms on the nitrogen atoms of amides are not so easily exchanged as those on amino groups. As a result, such hydrogen atoms do couple with hydrogen atoms on adjacent carbon atoms. The spectrum of *N*-methylacetamide, shown in Figure 13.31, has a doublet at δ 2.78, arising from coupling between the hydrogen atoms of the methyl group on the nitrogen with the amide hydrogen atom. The broad appearance of the band for the amide hydrogen atom (δ 6.70) (shown as an enhanced peak) arises

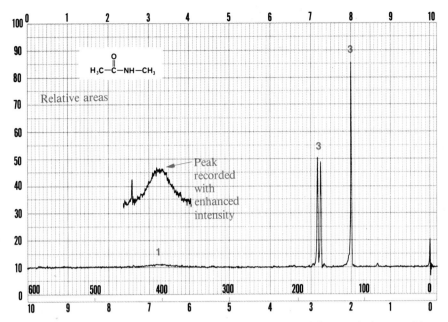

Figure 13.31   Proton magnetic resonance spectrum of *N*-methylacetamide.
(From *The Aldrich Library of NMR Spectra*)

not only from coupling but also from an interaction between the nucleus of nitrogen which has a magnetic moment (Section 13.2) and the nucleus of the hydrogen atom.

PROBLEM 13.14   Assign structures to the following compounds for which spectra and molecular formulas are given in Figure 13.32. Calculation of units of unsaturation and careful examination of chemical shifts, integration, and splitting patterns will be helpful in solving these problems.

Compound I, C_6H_{12}

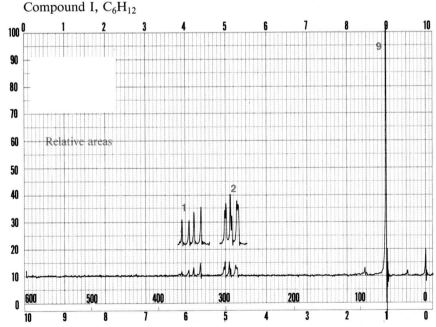

Figure 13.32   Spectra for Problem 13.14. (From *The Aldrich Library of NMR Spectra*)

Compound J, C_8H_8O

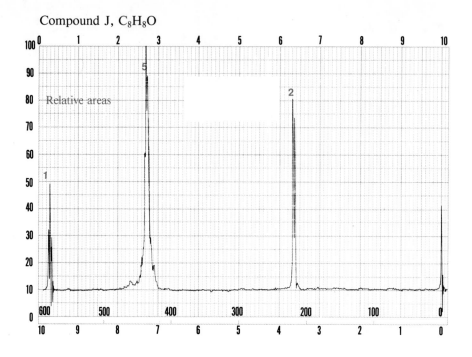

Compound K, C_3H_5Cl

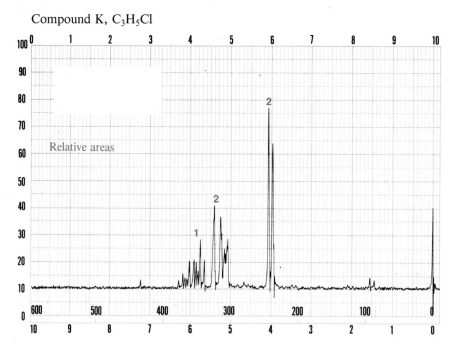

Figure 13.32   (*Continued*)

Compound L, C$_7$H$_7$ClO

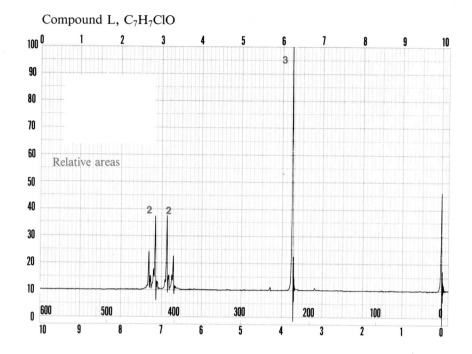

Compound M, C$_8$H$_{11}$N

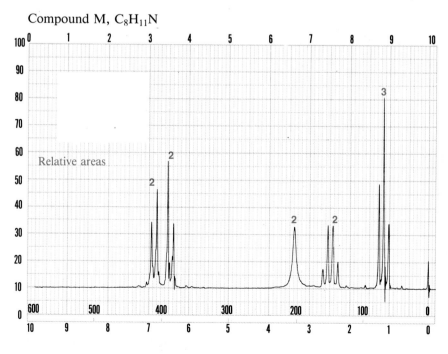

Figure 13.32   (*Continued*)

PROBLEM 13.15   Compound N has the molecular formula $C_6H_{14}O$ and the proton magnetic resonance spectrum shown in Figure 13.33. The important bands in its infrared spectrum are at 3000–2800, 1360 and 1100 cm^{-1}. Assign a structure to Compound N.

Compound N, $C_6H_{14}O$

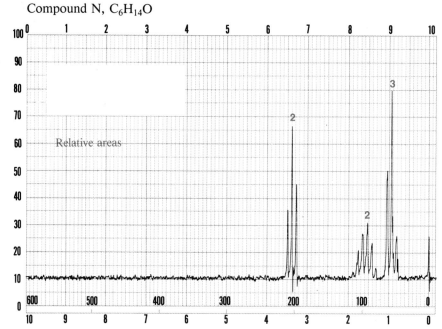

Figure 13.33   Spectrum for Problem 13.15. (From *The Aldrich Library of NMR Spectra*)

PROBLEM 13.16   The most important band in the infrared spectrum of Compound O is at 1710 cm^{-1}. Its proton magnetic resonance spectrum is shown in Figure 13.34. Assign a structure to it.

Compound O

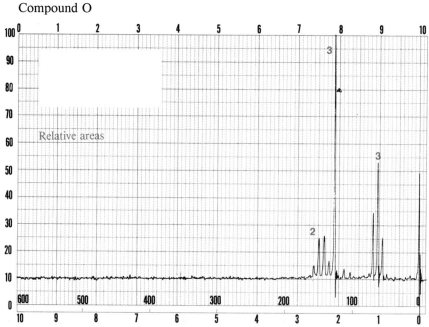

Figure 13.34   Spectrum for Problem 13.16. (From *The Aldrich Library of NMR Spectra*)

PROBLEM 13.17   The infrared spectrum of Compound P, C_3H_4O, has a broad strong band between 3500 and 3200 cm^{-1} and a sharp medium band at 2100 cm^{-1}. The proton magnetic resonance spectrum of Compound P is shown in Figure 13.35. Assign a structure to it and analyze the splitting patterns that are seen in the spectrum.

Compound P

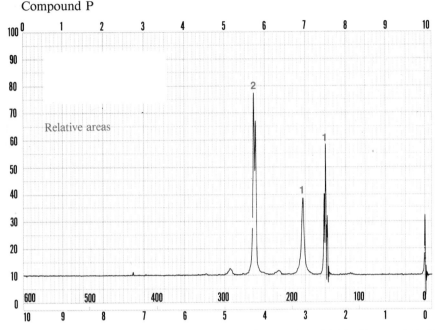

Figure 13.35   Spectrum for Problem 13.17. (From *The Aldrich Library of NMR Spectra*)

13.5

Carbon-13 Nuclear Magnetic Resonance Spectroscopy

The nucleus of the isotope of carbon with a mass of 13, ^{13}C, has a nuclear spin, just as the nucleus of the hydrogen atom does. This isotope of carbon has a natural abundance of about 1%, therefore approximately 1% of all the carbon atoms in any given sample of an organic compound are this isotope. Thus, it is possible to observe the nuclear magnetic resonance spectrum of the carbon atoms in an organic compound if a large enough sample is available, and computer techniques are used to average out the random background signals and collect and add up the weak positive signals that arise from the ^{13}C nuclei in the molecules.

The principles behind carbon-13 nuclear magnetic resonance spectroscopy are the same as those described for proton magnetic resonance. The nuclei, placed in an external magnetic field, absorb energy in the radiofrequency region in going from one energy level to another. Carbon nuclei may be shielded or deshielded depending upon their environment, just as the nuclei of hydrogen atoms are. Tetramethylsilane is used as a reference compound and the positions of the absorption bands for carbon atoms are reported in parts per million from TMS, just as for proton nuclear magnetic resonance.

In two respects, however, the carbon-13 nuclear magnetic resonance spectra that you will see differ from proton magnetic resonance spectra. First, the intensity of a band, in this case the height of the very narrow peak, is not directly related to the number of carbon atoms undergoing that energy transition. There is no integration on a carbon-13 nuclear magnetic resonance spectrum. The intensity of the signals in these spectra is related to the ways in which the nucleus of the carbon atom can return to the lower energy level after absorbing energy, in other words, to the relaxation of the nuclei (Section 13.2). In general, carbon atoms that bear hydrogen atoms give rise to more intense bands than those, such as the carbon atoms of carbonyl groups, that have no hydrogen atoms on them.

A second difference that you will notice is that the spectra do not show the effects of spin-spin coupling. The nuclei of the carbon atoms do couple with the nuclei of the hydrogen atoms that are on them. The coupling constants are so large, however, that spectra of great complexity result. It is possible to remove the coupling from a spectrum experimentally, and this is what has been done for the spectra you will see. Thus, the bands correspond to the different types of carbon atoms in the molecule but give no information about relative numbers of different kinds of carbon atoms, nor about the near neighbors of the atoms except by the chemical shift.

Spectra for 3-methyl-2-butanone and methyl benzoate are shown in Figure 13.36.

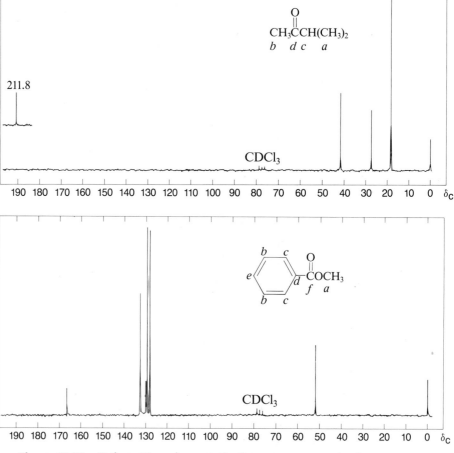

Figure 13.36   Carbon-13 nuclear magnetic resonance spectra for 3-methyl-2-butanone and methyl benzoate. (Adapted from Johnson and Jankowski. Reprinted by permission.)

The spectrum for 3-methyl-2-butanone shows four peaks. Assignment of the bands is not difficult to make. The carbon atom that is the most deshielded, with a signal at 211.8 ppm downfield from TMS (0 ppm), is the carbon atom of the carbonyl group. The spectrum had to be shifted upfield in order to bring the signal onto the chart. The carbon atoms that are most shielded (18.1 ppm) are the ones in the methyl groups that are farthest away from the carbonyl group. The carbon atom on the other methyl group absorbs at 27.3 ppm, and the carbon atom of the methine group next to the carbonyl group at 41.5 ppm. Note that methine carbon atoms are more deshielded than methyl carbon atoms in a similar environment. Note also that even the peaks that represent one carbon atom each are of unequal height.

The spectra show a set of three small peaks at 78 ppm. These peaks arise from the solvent, $CDCl_3$, and can be ignored in interpreting spectra.

The spectrum of methyl benzoate illustrates something that we already know from the chemistry of carbonyl compounds. The carbon atom of the carbonyl group of an ester is more shielded, because of the presence of nonbonding electrons on an adjacent oxygen atom, than the carbon atom of the carbonyl group of a ketone. The signal for the carbonyl group of an ester appears at 166.8 ppm downfield from TMS while the band for ketone appears 211.8 ppm downfield from TMS. The carbon atom of the methyl group on the oxygen atom in the ester is more highly deshielded than the carbon atoms of the methyl groups in 3-methyl-2-butanone and absorbs at 51.8 ppm. The four different types of carbon atoms of the aromatic ring are also highly deshielded and cluster around 128.5 to 132.9 ppm. Again, large differences can be seen in the intensities of the bands. The two bands of the lowest intensity at 130.3 and 166.8 ppm belong to carbon 1 in the aromatic ring and to the carbon atom of the carbonyl group. Neither of them bears a hydrogen atom.

Spectra for 2-octanol and methylenecyclobutane are given in Figure 13.37. The eight different carbon atoms of 2-octanol all have different absorption bands in its spectrum. The most highly shielded carbon atom (14.1 ppm) belongs to the methyl group farthest away from the hydroxyl group. The most highly deshielded carbon atom (67.8 ppm) is, of course, the one bonded to the oxygen atom. The others fall somewhere in between. The spectrum illustrates the power of the technique in giving structural information about an organic compound.

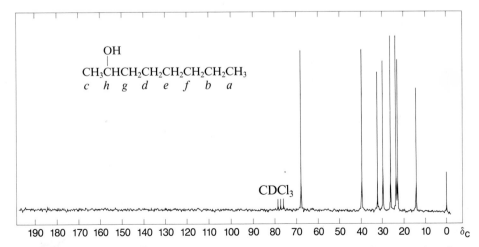

Figure 13.37   Carbon-13 nuclear magnetic resonance spectra of 2-octanol and methylenecyclobutane. (Adapted from Johnson and Jankowski. Reprinted by permission.)

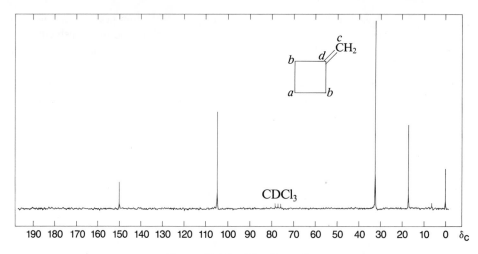

Figure 13.37 *(Continued)*

The spectrum of methylenecyclobutane has four bands. The two sp^2-hybridized carbon atoms of the double bond are much more deshielded than the sp^3-hybridized carbon atoms. The terminal carbon atom of the double bond absorbs at 104.8 ppm and the more highly substituted one at 150.4 ppm.

As a final example, the spectrum of cholesterol is shown in Figure 13.38. Twenty-six of the 27 carbon atoms in the molecule give rise to distinct absorption bands in the spectrum. The carbon atom to which the hydroxyl group is bonded (71.0 ppm) and the sp^2-hybridized carbon atoms of the double bond (120.9 and 141.7 ppm) are particularly easy to detect. As you can see, carbon-13 nuclear magnetic resonance spectra of highly complex molecules can be very informative about structural features.

Typical chemical shifts for different types of carbon atoms are given in Table 13.4.

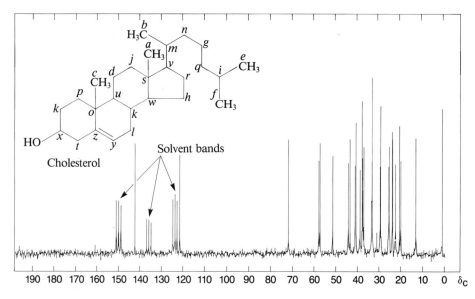

Figure 13.38   Carbon-13 nuclear magnetic resonance spectrum of cholesterol. (Adapted from Johnson and Jankowski. Reprinted by permission.)

TABLE 13.4   Chemical Shifts for Carbon Atoms in Carbon-13 Nuclear Magnetic Resonance Spectra

Type of Carbon Atom (in color)	Chemical Shift, ppm Downfield from TMS
RCH_2CH_3	$\sim13-16$
RCH_2CH_3	$\sim16-25$
R_3CH	$\sim25-38$
$\overset{\displaystyle O}{\overset{\|}{CH_3CR}}$	~30
$\overset{\displaystyle O}{\overset{\|}{CH_3COR}}$	~20
RCH_2Cl	$\sim40-45$
RCH_2Br	$\sim28-35$
RCH_2NH_2	$\sim37-45$
RCH_2OH	$\sim50-64$
$RC\equiv CH$	$\sim67-70$
$RC\equiv CH$	$\sim74-85$
$RCH=CH_2$	$\sim115-120$
$RCH=CH_2$	$\sim125-140$
$RC\equiv N$	$\sim117-125$
ArH	$\sim125-150$
$\overset{\displaystyle O}{\overset{\|}{RCOR'}}$	$\sim170-175$
$\overset{\displaystyle O}{\overset{\|}{RCOH}}$	$\sim177-185$
$\overset{\displaystyle O}{\overset{\|}{RCH}}$	$\sim190-200$
$\overset{\displaystyle O}{\overset{\|}{RCCH_3}}$	$\sim205-210$

As the first three entries in Table 13.4 show, a more highly substituted carbon atom usually has a higher chemical shift than a less substituted one with a similar environment. The numbers in the table are, as usual, to be applied with caution and only after taking into account all of the available information.

PROBLEM 13.18 Carbon-13 nuclear magnetic resonance spectra are given in Figure 13.39 for the following compounds. Decide which spectrum belongs to which compound.

(a) ⬡—NH₂   (b) $CH_3CH_2\overset{\overset{\displaystyle OH}{|}}{C}HCH_3$ (c) $CH_3CH_2\overset{\overset{\displaystyle O}{\|}}{C}O\overset{\overset{\displaystyle O}{\|}}{C}CH_2CH_3$

(d) ⬠O   (e) $CH_2{=}CHCH_2Br$ (f) ⬡—CH=CH₂

Spectrum I

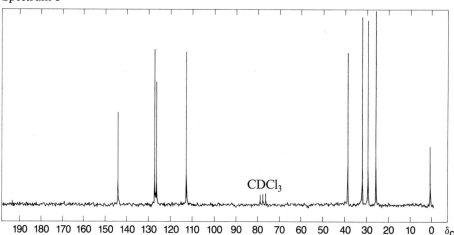

Spectrum II

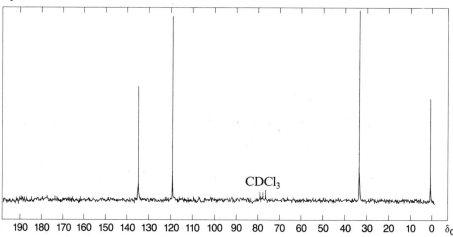

Figure 13.39 Spectra for Problem 13.18. (Adapted from Johnson and Jankowski. Reprinted by permission.)

Spectrum III

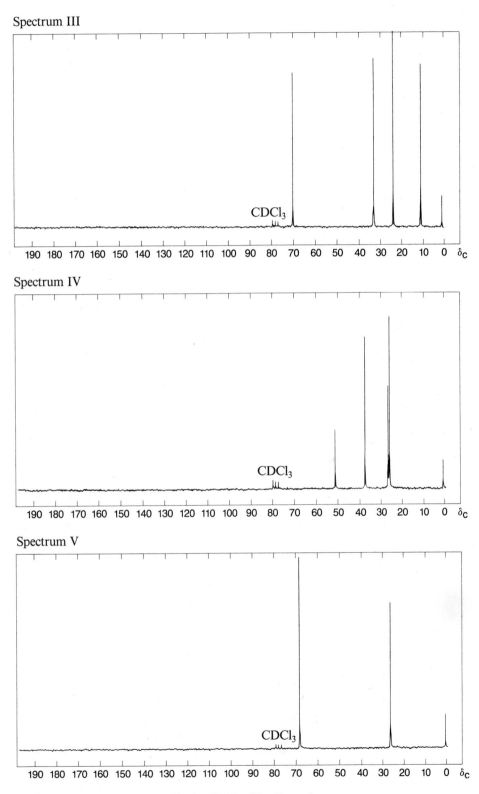

Spectrum IV

Spectrum V

Figure 13.39 *(Continued)*

Spectrum VI

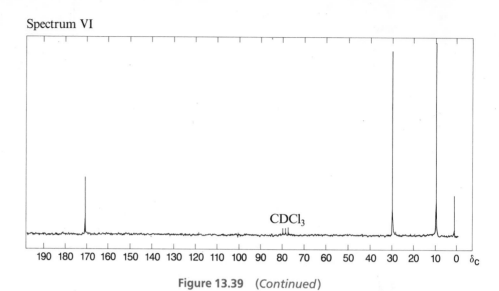

Figure 13.39 *(Continued)*

PROBLEM 13.19   Compound Q, C_5H_8O, has bands in the carbon-13 nuclear magnetic resonance spectrum at δ 23.4, 38.2, and 219.6. Assign a structure to it.

13.6

Mass Spectrometry

A. The Mass Spectrum

Mass spectrometry differs from the other forms of spectroscopy that have been presented so far in that a mass spectrum is not a record of the energy absorbed by a molecule in going from one energy level to another. Instead, *a **mass spectrum** is a record of the exact masses of a series of ions that form by fragmentation of a molecular species that is created in the collison of a molecule with a high-energy particle, usually an electron.* The collision knocks an electron out of the molecule, giving rise to a radical cation (a species with a positive charge and an unpaired electron) called the **molecular ion.** The molecular ion is unstable and fragments to give a series of other species such as smaller radical cations, carbocations, radicals, and neutral molecules. The positively charged species are separated according to their mass-to-charge ratios, m/e, by a variety of techniques. Most ions created in a mass spectrometer have a single positive charge; therefore, the separation is essentially according to their masses. For example, ions of different masses that are accelerated in an electric field are then deflected by different amounts when put through a magnetic field (Figure 13.40). Thus, ions of different masses impinge separately upon an electron multiplier and are recorded on a chart. A calibration of the instrument allows us to record not only the masses of the ions but the relative numbers of each kind of ion formed.

Actual spectra will help to make this clearer. The mass spectra of acetone and acetaldehyde, plotted from the original recordings as bar graphs, are shown in Figure 13.41. Lines appear in the spectra at numbers corresponding to the masses of the ions

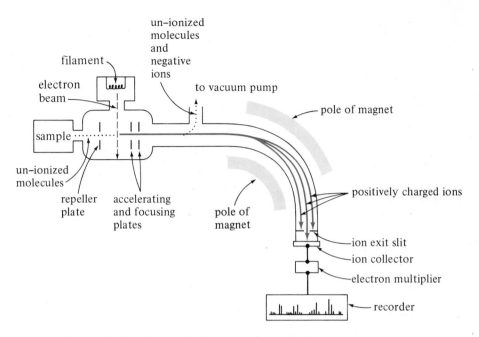

Figure 13.40   Schematic diagram of a typical mass spectrometer.

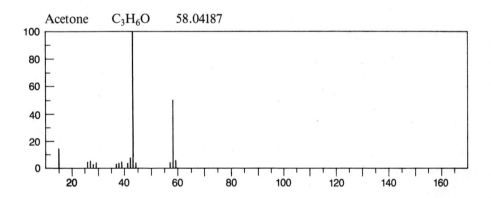

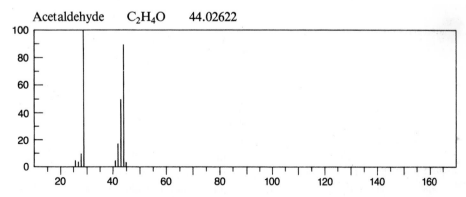

Figure 13.41   The mass spectra of acetone and acetaldehyde. (Adapted from Stenhagen et al. Reprinted by permission.)

that are formed when molecules of these compounds are bombarded with a high-energy beam of electrons. *The first reaction is the formation of the molecular ion by loss of an electron from the molecule.* In acetone and acetaldehyde, the nonbonding electrons on oxygen are the most easily lost.

$$\underset{\substack{\text{:O:} \\ CH_3 \quad CH_3}}{\overset{\text{radical cation}}{\xrightarrow{-e^-}}}$$

molecular ion of acetone
formed by loss of a
nonbonding electron
from oxygen

$C_3H_6O^{\ddot{+}}$, m/e 58

molecular ion of
acetaldehyde

$C_2H_4O^{\ddot{+}}$, m/e 44

The radical cation is shown localized on the oxygen atom of a carbonyl group. The *radical cations are created in the gas phase with a large amount of excess energy. They are unstable species and fragment in a series of unimolecular reactions.* An inspection of the mass spectrum of acetone indicates that the most abundant ion has m/e 43. *The most abundant ion in the mass spectrum of a compound is called the* **base peak.** The bar graphs are designed so that such an ion always has an intensity of 100%. The abundance of the other ions is shown relative to that of the base peak. Thus, the molecular ion of acetone, m/e 58, has an abundance of approximately 40% of the ion with m/e 43.

The base peak in the spectrum of acetone corresponds to the loss of a fragment with a mass of 15 units ($58 - 43 = 15$). A methyl group, CH_3, would have a mass of 15. Thus, it appears that the molecular ion of acetone fragments primarily by the loss of a methyl group.

$$\longrightarrow CH_3 - C \equiv \ddot{O}^+ \quad + \quad \cdot CH_3$$

molecular ion of acetone	acylium ion	methyl radical
m/e 58	$C_2H_3O^+$	*not charged,*
	m/e 43	*does not appear in the mass spectrum*

The reaction is shown as a homolytic cleavage (represented by fish hooks) of one of the bonds between the carbonyl group and a methyl group. As a consequence, an acylium ion, previously seen as an intermediate in Friedel-Crafts reactions of acid chlorides and anhydrides (Section 12.3F), and a methyl radical are formed. The methyl radical is not a charged particle and thus does not appear in the mass spectrum. There is, however, a small peak at m/e 15, which corresponds to the formation of a small amount of methyl cations by an alternate cleavage.

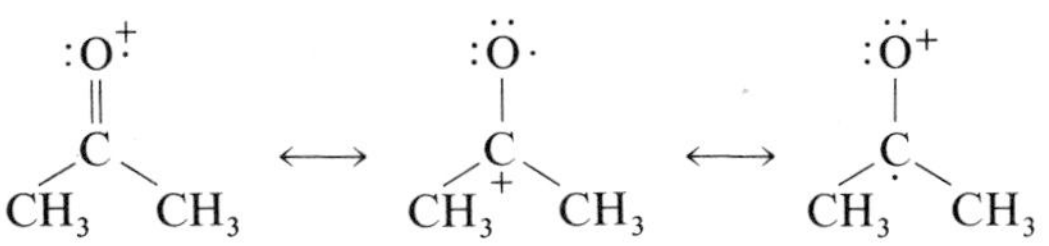

resonance contributors of the molecular ion of acetone

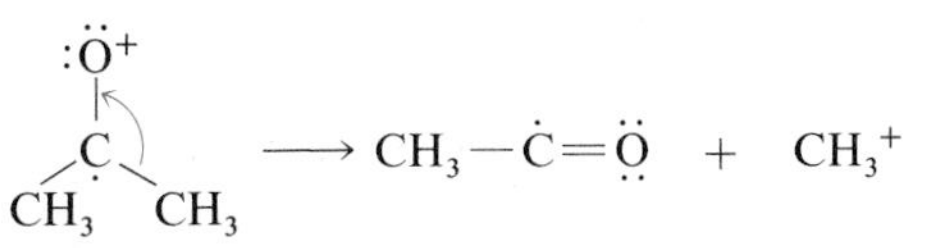

m/e 58   an acyl radical   methyl cation
m/e 15

The relative abundances of the acylium cation (m/e 43, 100%) and methyl cation (m/e 15, 15%) reflect the relative stabilities of the two species. In any case, it is important that the reactions are taking place in the gas phase in species that have much excess energy. Note, too, that any fragmentations written must preserve the numbers of unpaired electrons and charges present in the original species.

Besides the molecular ion, m/e 44, the mass spectrum of acetaldehyde shows important peaks at m/e 43 and m/e 29 (base peak). These peaks may be assigned to the ions that form on the loss of a hydrogen atom ($44 - 43 = 1$) and a methyl radical ($44 - 29 = 15$) from the molecular ion.

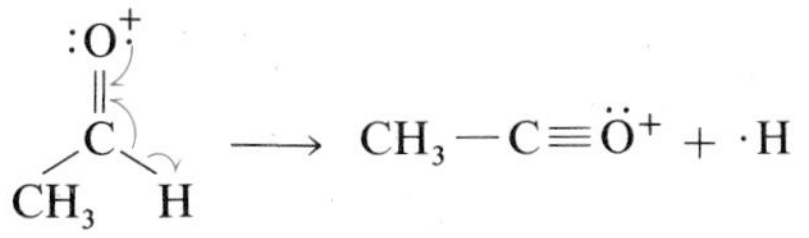

molecular ion   acylium ion   hydrogen
of acetaldehyde   atom
m/e 44 m/e 43

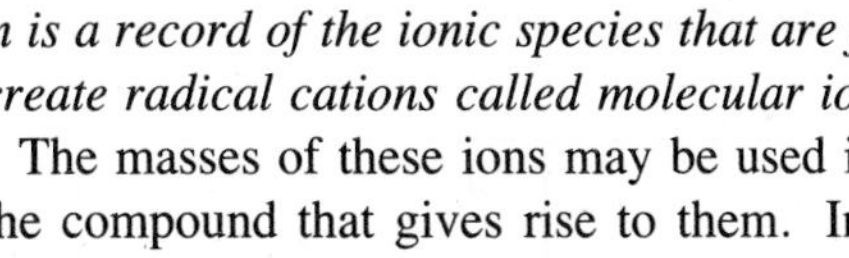

molecular ion   acylium ion   methyl
of acetaldehyde   radical
m/e 44 m/e 29

These spectra have served as an introduction to *the essential features of a mass spectrum. The spectrum is a record of the ionic species that are formed when a beam of electrons is used to create radical cations called molecular ions, which then fragment into smaller ions.* The masses of these ions may be used in making judgments about the structure of the compound that gives rise to them. In Section 13.6C, we discuss more fragmentation reactions after we examine more closely the information obtained from the molecular ion. Meanwhile, the following problems give you practice in writing molecular ions and fragmentation reactions for compounds related to acetone and acetaldehyde.

PROBLEM 13.20   The mass spectrum in Figure 13.42 is that of an aldehyde. Assign a structure for it, and write equations showing the formation of the molecular ion and the ion giving rise to the base peak.

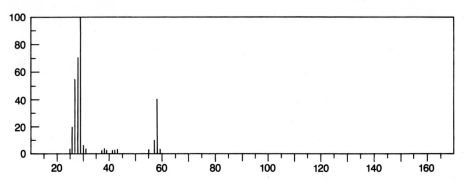

Figure 13.42 Spectrum for Problem 13.20. (Adapted from Stenhagen et al. Reprinted by permission.)

PROBLEM 13.21 The mass spectrum of a carboxylic acid is shown in Figure 13.43. Assign a structure to it, and write equations showing the formation of any ions having an abundance greater than 40%.

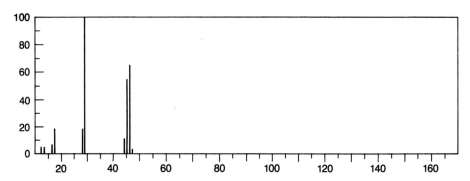

Figure 13.43 Spectrum for Problem 13.21. (Adapted from Stenhagen et al. Reprinted by permission.)

PROBLEM 13.22 Two mass spectra are shown in Figure 13.44. One belongs to 2-pentanone, the other to 3-pentanone. Which is which? Write equations showing how you made your decision.

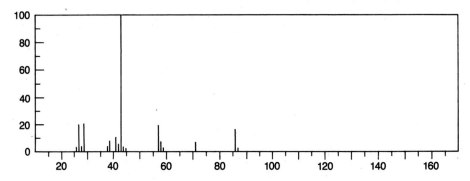

Figure 13.44 Spectra for Problem 13.22. (Adapted from Stenhagen et al. Reprinted by permission.)

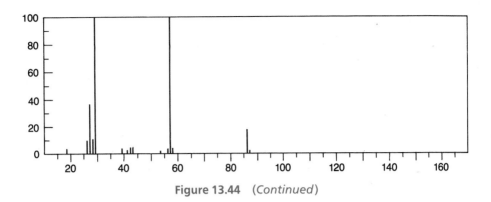

Figure 13.44   (*Continued*)

PROBLEM 13.23   The mass spectrum of 2,2-dimethylpropanal is given in Figure 13.45. Assign a structure to the base peak in the spectrum, and write an equation for its formation from the molecular ion. What would be the comparable species from the fragmentation for the molecular ion of acetaldehyde? Is such a species evident in the mass spectrum of acetaldehyde (Figure 13.41)? How do you explain the difference between the reactions of the molecular ions for these two compounds?

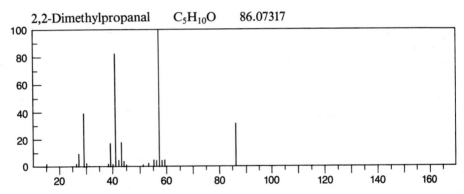

Figure 13.45   Spectrum for Problem 13.23. (Adapted from Stenhagen et al. Reprinted by permission.)

B. The Molecular Ion

In assigning the peaks at m/e 58 and m/e 44 to the molecular ions of acetone and acetaldehyde, respectively, we ignored the presence of small peaks at m/e 59 and m/e 45 in their spectra. These peaks arise from the presence of isotopes of carbon, hydrogen, and oxygen in these compounds. The common isotopes of elements that are of importance in organic chemistry, their atomic weights rounded off to the nearest mass unit, and their abundance relative to that of the most abundant isotope of that element (shown as 100%) are given in Table 13.5. Thus, any compound that contains a number of carbon, hydrogen, and oxygen atoms will have some peaks in its mass spectrum at mass values one or two (or more) units above the mass calculated from its molecular formula using common atomic weights. For example, a compound containing ten

TABLE 13.5   Natural Isotopic Abundances of Some Elements

	Most Common Isotopes			
Element	Mass	%	Mass	%
H	1	100	2	0.016
C	12	100	13	1.08
N	14	100	15	0.36
O	16	100	18	0.20
F	19	100	—	—
Cl	35	100	37	32.5
Br	79	100	81	98.0
I	127	100	—	—

carbon atoms would be expected to have an ion having an intensity of about 11% (10 × 1.08%) of the intensity of its molecular ion (M) at a mass of (M + 1). This peak would correspond to all of the species in which one of the carbon atoms in the molecule is ^{13}C instead of ^{12}C. The intensity of this peak, relative to the intensity of the molecular ion, would increase with increasing numbers of hydrogen and nitrogen atoms. Large numbers of oxygen atoms would increase the intensity of the (M + 2) peak, arising from molecular species in which ^{18}O instead of ^{16}O is found. The (M + 2) peak would also arise from species in which both ^{13}C and ^2H are found. The probability that there are such molecules increases with the total number of carbon and hydrogen atoms in the compound.

The most distinctive contributions to (M + 2) peaks in mass spectra come from the presence of the halogens bromine and chlorine in a molecule. Bromine has two isotopes, ^{79}Br and ^{81}Br, of nearly equal abundance. Thus, in a sample of a compound containing a bromine atom, half of the molecules contain ^{79}Br and half ^{81}Br. Every species containing bromine will appear as a pair of ions separated by two mass units. The spectrum of methyl bromide is shown in Figure 13.46. The molecular ion for $CH_3^{79}Br$ appears at m/e 94 and that for $CH_3^{81}Br$ at m/e 96. These two ions have roughly the same intensity, reflecting the almost equal abundance of the two isotopes of bromine in nature. Ions corresponding to $^{79}Br^+$ and $^{81}Br^+$ are seen at m/e 79 and m/e 81. The base peak in the spectrum is at m/e 15, corresponding to the methyl cation.

$$CH_3-{}^{79}\ddot{Br}: \xrightarrow{-e^-} CH_3-{}^{79}\ddot{Br}\!\cdot^+$$
$$m/e\ 94$$

$$CH_3-{}^{81}\ddot{Br}: \xrightarrow{-e^-} CH_3-{}^{81}\ddot{Br}\!\cdot^+$$
$$m/e\ 96$$

$$CH_3\!\!\frown\!\!{}^{79}\ddot{Br}\!\cdot^+ \longrightarrow \cdot CH_3 + {}^{79}\ddot{Br}\!:^+$$
$$m/e\ 94 \qquad\qquad m/e\ 79$$

$$CH_3\!\!\frown\!\!{}^{81}\ddot{Br}\!\cdot^+ \longrightarrow \cdot CH_3 + {}^{81}\ddot{Br}\!:^+$$
$$m/e\ 96 \qquad\qquad m/e\ 81$$

$$CH_3\!\!\frown\!\!\ddot{Br}\!\cdot^+ \longrightarrow CH_3^+ \ + \ :\ddot{Br}\cdot$$
$$m/e\ 94\ or\ m/e\ 96 \quad m/e\ 15 \quad \text{bromine atom}$$

not charged,
does not appear in
the mass spectrum

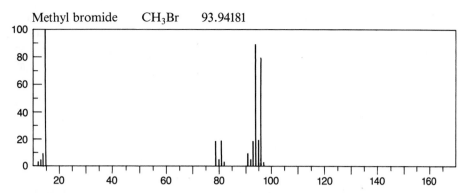

Methyl bromide CH_3Br 93.94181

Figure 13.46   Mass spectrum for methyl bromide. (Adapted from Stenhagen et al. Reprinted by permission.)

It is important to understand that the mass spectrum shows ions having different isotopic composition as individual peaks. The mass spectrum of a compound contains peaks corresponding to individual species containing all possible combinations of isotopes. The mass of a species as determined by mass spectroscopy is *not* the same as the molecular weight calculated by using the average atomic weights found in the periodic table.

Compounds containing a chlorine atom show an (M + 2) peak having an intensity of about a third that for the molecular ion. The spectrum of methyl chloride shows this feature (Figure 13.47). The molecular ion of methyl chloride for $CH_3{}^{35}Cl$ is at *m/e* 50. A peak having an intensity of about a third of the one at *m/e* 50 appears at *m/e* 52 and corresponds to the molecular ion for species having the composition $CH_3{}^{37}Cl$. The chief fragmentation seen for these molecular ions is the loss of a chlorine atom giving rise to a methyl cation, *m/e* 15.

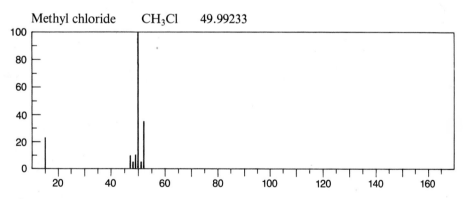

Methyl chloride CH_3Cl 49.99233

Figure 13.47   Mass spectrum of methyl chloride. (Adapted from Stenhagen et al. Reprinted by permission.)

PROBLEM 13.24   Write equations showing the formation of the molecular ions for methyl chloride and their fragmentation.

Not all compounds give molecular ions that are easily detectible in their mass spectra. The molecular ions formed by some compounds are so unstable that they fragment before they reach the detector. Alcohols and amines are some of the compounds that give very weak peaks for their molecular ions. The molecular ion for butylamine, for example, is so unstable that only a very small number survive to be recorded by the detector. You will have to inspect the spectrum very carefully to see it (*m/e* 73). The molecular ion for *tert*-butyl alcohol does not appear at all. The spectra for these compounds are shown in Figure 13.50 in the next section where we explore the fragmentations of such molecular ions.

The electron lost in the creation of a molecular ion may come from any bond in the molecule. *In general, a nonbonding electron is removed more easily than an electron in a π bond, which in turn is lost more easily than an electron from a σ bond.* The molecules for which mass spectra have been shown so far all had atoms bearing nonbonding electrons. The fragmentations seen were rationalized as reactions of a molecular ion arising from the loss of a nonbonding electron.

PROBLEM 13.25   The mass spectrum of a haloalkane is given in Figure 13.48. Assign a structure to the compound.

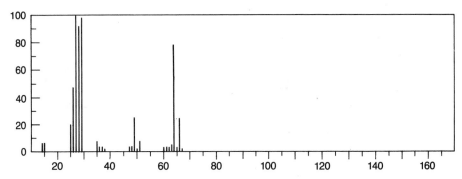

Figure 13.48   Spectrum for Problem 13.25. (Adapted from Stenhagen et al. Reprinted by permission.)

PROBLEM 13.26   The compound for which a mass spectrum appears in Figure 13.49 contains only one carbon atom and two types of halogen. After consulting Table 13.5, assign a structure to it and account for the peaks that appear in the spectrum.

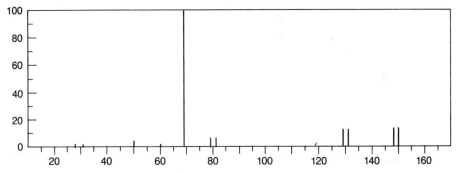

Figure 13.49   Spectrum for Problem 13.26. (Adapted from Stenhagen et al. Reprinted by permission.)

C. Important Fragmentation Pathways

Molecular ions may fragment unimolecularly by either heterolytic or homolytic cleavage of bonds adjacent to the site of the radical cation formed by the loss of an electron. Examples were shown in Sections 13.6A and 13.6B of how the loss of a nonbonding electron from an oxygen atom or a halogen atom created the radical cation. The molecular ion then loses different radical species and is converted to the fragment ions. Other types of reactions are also possible, but we will concentrate on the simpler fragmentation pathways.

The fragmentation of an amine or an alcohol is determined by the groups adjacent to the oxygen or the nitrogen atom. For example, the mass spectra of butylamine and *tert*-butyl alcohol each have one important peak (Figure 13.50).

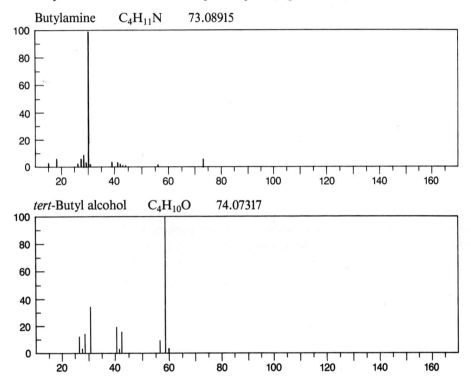

Butylamine $C_4H_{11}N$ 73.08915

tert-Butyl alcohol $C_4H_{10}O$ 74.07317

Figure 13.50 The mass spectra of butylamine and *tert*-butyl alcohol (Adapted from Stenhagen et al. Reprinted by permission.)

In each case, the base peak can be rationalized as arising from the homolytic cleavage of a bond one removed from the radical cation centered on the oxygen or the nitrogen atom.

$$CH_3CH_2CH_2-CH_2-\overset{..}{N}H_2 \xrightarrow{-e^-} CH_3CH_2\overset{\frown}{CH_2}-CH_2-\overset{.+}{N}H_2 \longrightarrow$$

butylamine

molecular ion
m/e 73

*very weak in
the spectrum*

$$CH_2=\overset{+}{N}H_2 + CH_3CH_2\overset{.}{C}H_2$$

an iminium
ion
m/e 30

1-propyl
radical

the base peak

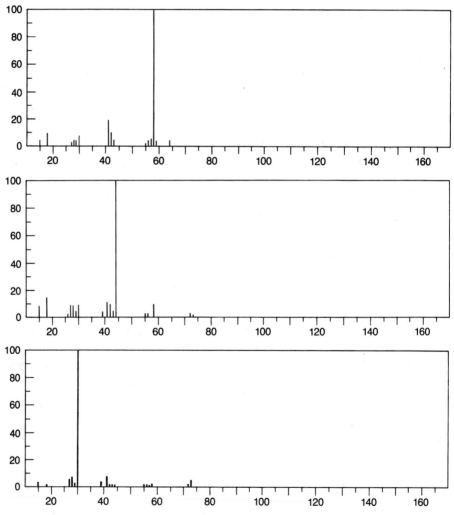

The stable cations, oxonium or iminium ions, are formed in these fragmentations. The molecular ion is usually weak or nonexistent in the spectra of alcohols and amines because of the ease with which the fragmentations occur.

PROBLEM 13.27 The mass spectra of three isomeric amines, *sec*-butylamine, isobutylamine, and *tert*-butylamine are given in Figure 13.51. Decide which spectrum belongs to which amine and write equations justifying your conclusions.

Figure 13.51   Spectra for Problem 13.27. (Adapted from Stenhagen et al. Reprinted by permission.)

The cleavage of bonds adjacent to carbonyl groups and the loss of halogen atoms or ions have already been discussed with examples in Sections 13.6A and 13.6B.

The mass spectra of alkanes and alkenes are more complicated because rearrangements of intermediate cations and radicals occur when the molecule is of any size. *The molecular ion of alkenes arises from the loss of an electron from the π bond*. Then an allylic bond in the molecular ion often cleaves to give an alkyl radical and an allylic cation. The ion arising from such processes is seen in the spectrum of 2-methyl-2-pentene (Figure 13.52).

2-Methyl-2-pentene C_6H_{12} 84.09390

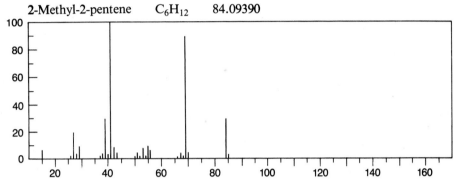

Figure 13.52 Mass spectrum for 2-methyl-2-pentene. (Adapted from Stenhagen et al. Reprinted by permission.)

The formation of the molecular ion and the cleavage of the allylic bond in 2-methyl-2-pentene is shown below.

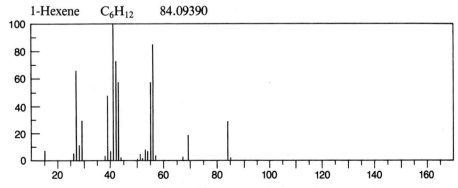

PROBLEM 13.28 The mass spectrum of 1-hexene is shown in Figure 13.53. Write equations showing how the ion giving rise to the base peak is formed.

1-Hexene C_6H_{12} 84.09390

Figure 13.53 Spectrum for Problem 13.28. (Adapted from Stenhagen et al. Reprinted by permission.)

Another category of fragmentation reaction that is important is cleavage at benzylic bonds. The mass spectra of toluene and of benzyl methyl ether illustrate such fragmentation (Figure 13.54).

Toluene C_7H_8 92.06260

Benzyl methyl ether $C_8H_{10}O$ 122.07317

Figure 13.54 Mass spectra of toluene and of benzyl methyl ether. (Adapted from Stenhagen et al. Reprinted by permission.)

The mass spectrum of toluene shows the molecular ion, *m/e* 92, and the base peak, *m/e* 91, which arises from the loss of a hydrogen atom from the molecular ion.

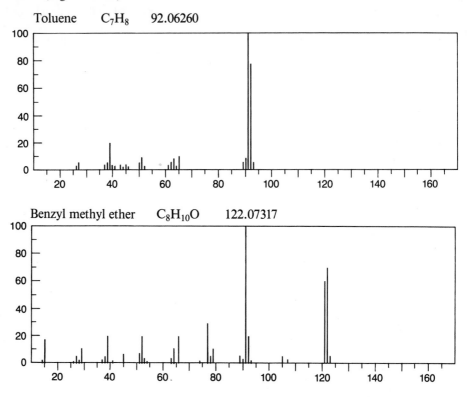

toluene

molecular ion
m/e 92

cycloheptatrienyl cation
tropylium ion
$C_7H_7^+$
m/e 91
a stable aromatic species

$+ H\cdot$

The cation formed on the loss of the hydrogen atom is believed to have the stable ring-expanded cycloheptatrienyl cation structure (Section 12.1) rather than that of a typical benzyl cation.

Such a cation also accounts for the base peak in the spectrum of benzyl methyl ether.

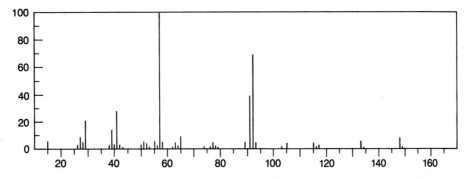

Another important peak, *m/e* 121, can be rationalized by the loss of a hydrogen atom to give the stable oxonium ion, a cleavage typical of compounds containing radical cations on oxygen atoms. Compounds containing benzene rings also show a peak at *m/e* 77, corresponding to $C_6H_5^+$, the phenyl cation. This peak is visible in the spectrum of benzyl methyl ether.

PROBLEM 13.29   Mass spectra for the isomeric compounds 2,2-dimethyl-1-phenylpropane and 2-methyl-3-phenylbutane are shown in Figure 13.55. Decide which spectrum belongs to which compound and write equations showing the formation for the ions giving rise to the base peaks.

Figure 13.55   Spectra for Problem 13.29. (Adapted from Stenhagen et al. Reprinted by permission.)

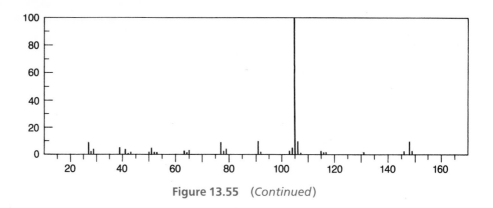

Figure 13.55   *(Continued)*

D. Rearrangements of Molecular Ions

One of the most common and most important rearrangements observed in the mass spectrometer is the transfer of a hydrogen atom from one part of a chain to a radical site at another part of the chain, usually through a six-membered cyclic transition state. The reaction is particularly easy to see in carbonyl compounds but is also one of the mechanisms by which rearrangements of alkenes occur. Such rearrangements are postulated to account for radical cations that give rise to peaks in the mass spectra of 2-methyl-1-pentene and 2-hexanone (Figure 13.56).

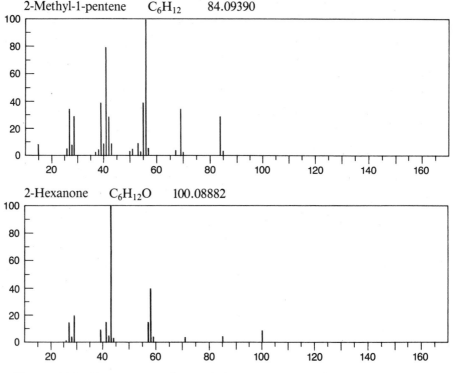

Figure 13.56   Mass spectra of 2-methyl-1-pentene and 2-hexanone. (Adapted from Stenhagen et al. Reprinted by permission.)

The base peak in the spectrum of 2-methyl-1-pentene has *m/e* 56, and results from the loss of a fragment having a mass of 28 units (84 − 56 = 28). Such a fragment could be ethylene, C_2H_4.

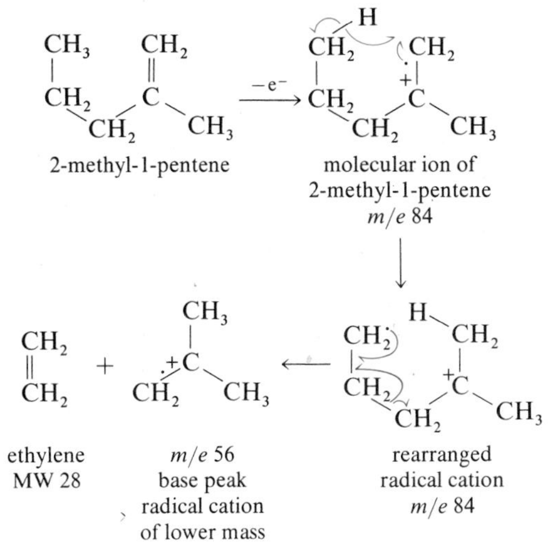

A hydrogen atom is shown being transferred from one carbon atom to another carbon atom, a radical site at another part of the chain, by a six-membered ring transition state. The new radical cation fragments to give a stable alkene and another radical cation of lower mass.

The base peak in the spectrum of 2-hexanone, *m/e* 43, arises from cleavage of a bond next to the carbonyl group. A significant peak, appearing at *m/e* 58 arises by way of a rearrangement.

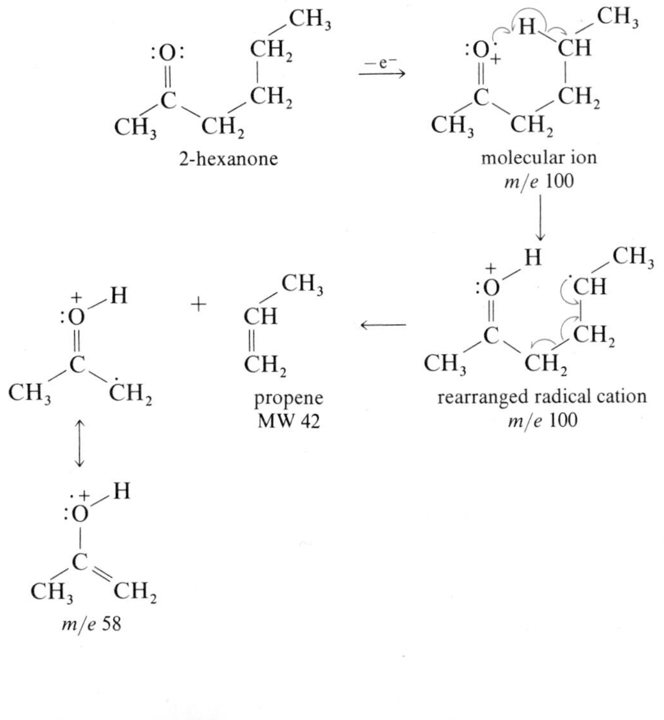

The radical cation with m/e 58 may be viewed as being tautomeric with the radical cation of acetone, and will undergo further fragmentations typical of acetone. This type of rearrangement is seen for a wide variety of compounds containing carbon-oxygen double bonds and is known as the **McLafferty rearrangement** for Fred W. McLafferty of Cornell University who first discovered how general it is.

PROBLEM 13.30   The mass spectrum of 3-methyl-2-pentanone is shown in Figure 13.57. Write equations accounting for formation of the ions with m/e 29, 43, 56, and 72. A review of the equations in Section 13.6A may be helpful.

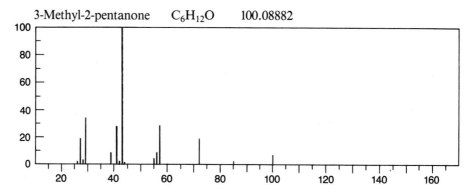

3-Methyl-2-pentanone $C_6H_{12}O$ 100.08882

Figure 13.57   Spectrum for Problem 13.30. (Adapted from Stenhagen et al. Reprinted by permission.)

PROBLEM 13.31   Two of the important peaks in the mass spectrum of 1-hexene (shown in Problem 13.28) at m/e 56 and 42 can be explained by writing mechanisms of the kind shown on p. 721 for 2-methyl-1-pentene. See whether you can use rearrangement and fragmentation reactions to arrive at structures for these radical cations. (Hint: Either carbon atom of the original double bond may have radical character or cation character.)

The mass spectrum of a compound of any complexity is unique in the pattern of ions and the relative intensities of these ions. As such, it can serve as a fingerprint for the compound. Only a very small sample of a compound, on the order of 10^{-6} g, is sufficient to obtain a mass spectrum. Thus this tool is a powerful one for identifying minute quantities of compounds. Mass spectrometers are now connected to instruments such as gas phase chromatographs, which separate mixtures efficiently, and computers, which automatically identify the components of the mixtures by comparing mass spectral information with stored patterns of known compounds. Such techniques make it possible to obtain information about compounds present in traces in complex mixtures, such as physiological fluids or discharges from industrial plants.

ADDITIONAL PROBLEMS

13.32   The proton magnetic resonance spectrum of Compound R, $C_3H_6Cl_2$, is shown in Figure 13.58. Assign a structure to the compound.

Compound R

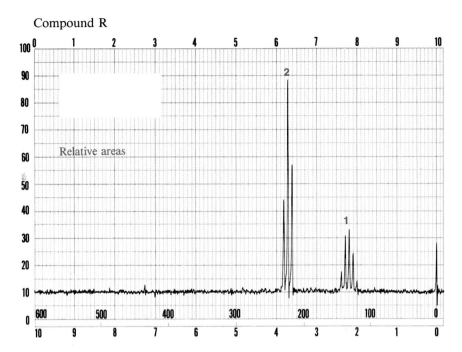

Figure 13.58 Spectrum for Problem 13.32. (From *The Aldrich Library of NMR Spectra*)

13.33 Compound S, $C_5H_{11}Br$, has the proton magnetic resonance spectrum shown in Figure 13.59. What is the structure of Compound S?

Compound S

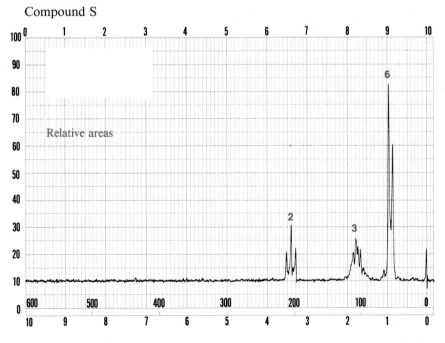

Figure 13.59 Spectrum for Problem 13.33. (From *The Aldrich Library of NMR Spectra*)

13.34   Compound T, C_3H_7Cl, has the proton magnetic resonance spectrum given in Figure 13.60. Assign a structure to T.

Compound T

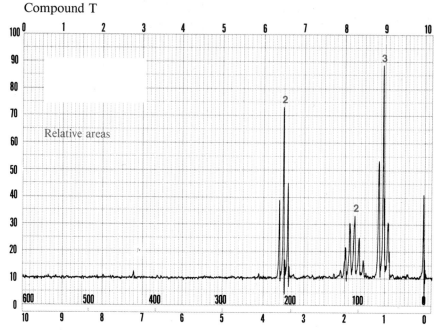

Figure 13.60   Spectrum for Problem 13.34. (From *The Aldrich Library of NMR Spectra*)

13.35   Proton magnetic resonance spectra for Compounds U–Z are given in Figure 13.61. Assign structures to the compounds.

Compound U, $C_8H_8O_2$

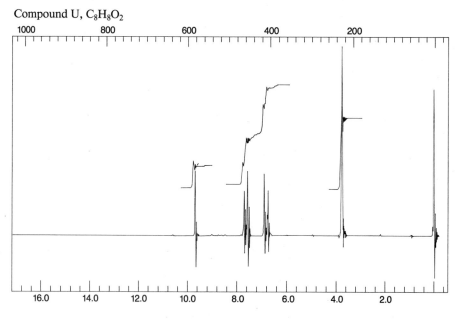

Figure 13.61   Spectra for Problem 13.35.

Compound V, C$_4$H$_8$O$_2$

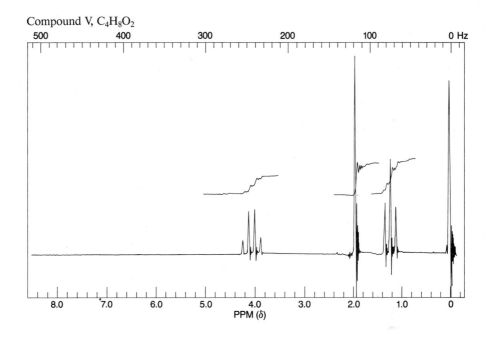

Compound W, C$_5$H$_8$O$_2$

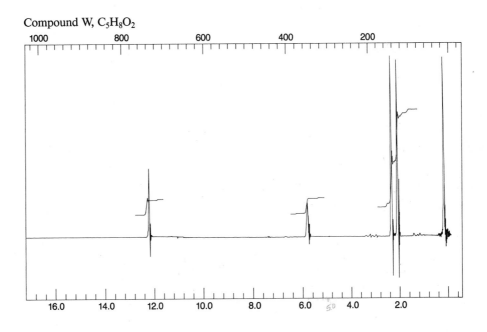

Figure 13.61 (*Continued*)

Compound X, C_7H_9N

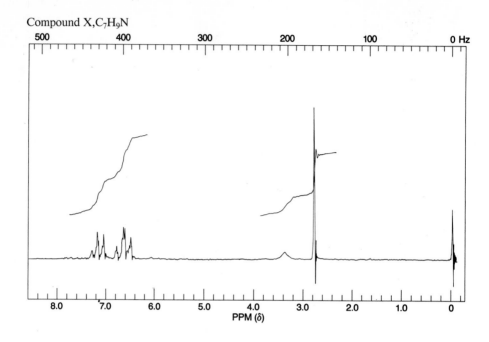

Compound Y, $C_7H_{12}O_3$

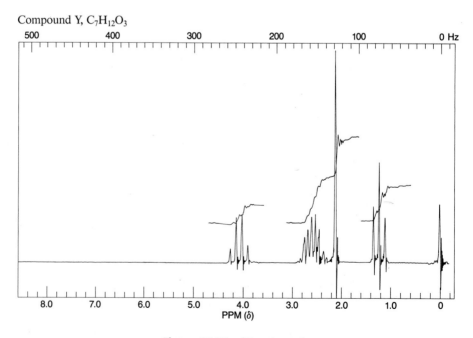

Figure 13.61 *(Continued)*

Compound Z, C₆H₁₅N

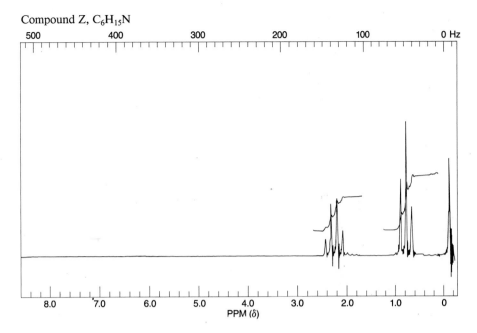

Figure 13.61 (*Continued*)

13.36 Trichloromethylbenzene has the proton magnetic resonance spectrum shown in Figure 13.62. Compare it with the spectrum of toluene (Figure 13.7) and explain the differences in the spectra.

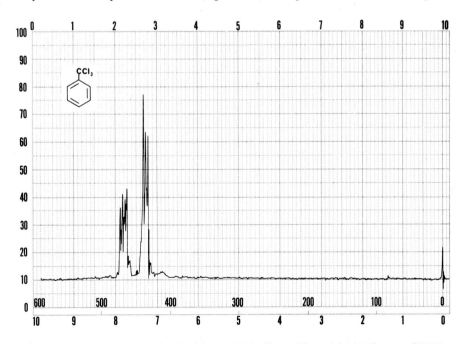

Figure 13.62 Spectrum for Problem 13.36. (From *The Aldrich Library of NMR Spectra*)

13.37   The proton magnetic resonance spectrum of 1-phenylpropane is given in Figure 13.63. Assign all of the bands in the spectrum and analyze the splitting patterns that are seen.

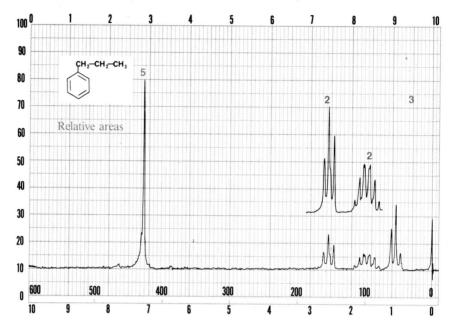

Figure 13.63   Spectrum for Problem 13.37. (From *The Aldrich Library of NMR Spectra*)

13.38   The proton magnetic resonance spectra of two isomeric esters, $C_9H_{10}O_2$, are shown in Figure 13.64. Assign structures to Compound AA and BB.

Compound AA

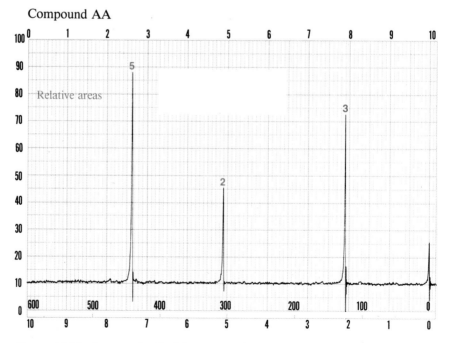

Figure 13.64   Spectra for Problem 13.38. (From *The Aldrich Library of NMR Spectra*)

Compound BB

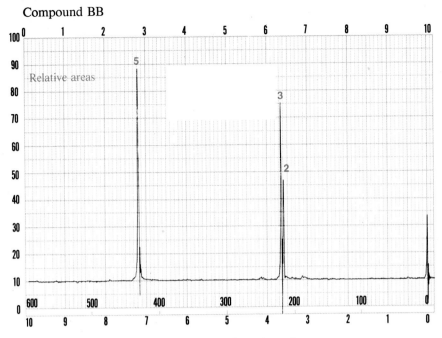

Figure 13.64 *(Continued)*

13.39   Compound CC, $C_8H_{10}O$, has the infrared, proton magnetic resonance, and mass spectra shown in Figure 13.65. Assign a structure to it and analyze the spectra.

Compound CC

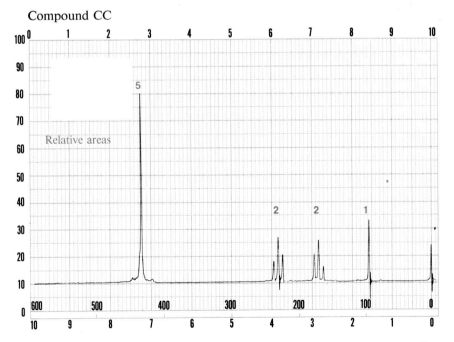

Figure 13.65   Spectra for Problem 13.39. (NMR spectrum from *The Aldrich Library of NMR Spectra*. Infrared spectrum from *The Aldrich Library of Infrared Spectra*. Mass spectrum adapted from Stenhagen et al. Reprinted by permission.)

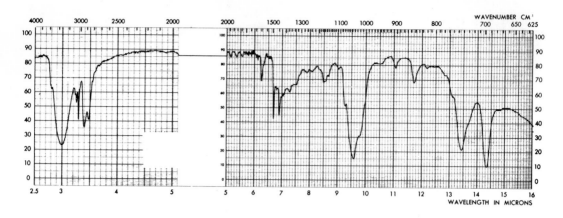

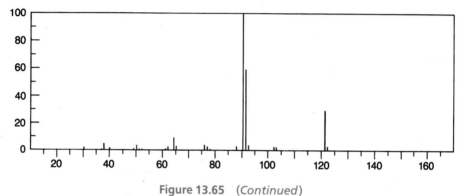

Figure 13.65 *(Continued)*

13.40 Compound DD has a strong band in its infrared spectrum at 1685 cm^{-1} and the proton magnetic resonance spectrum illustrated in Figure 13.66. Assign a structure to Compound DD.

Compound DD

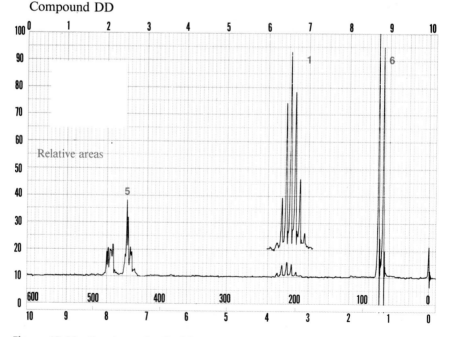

Figure 13.66 Spectrum for Problem 13.40. (From *The Aldrich Library of NMR Spectra*)

13.41 The proton magnetic resonance spectrum of 3,4-dichloroaniline appears in Figure 13.67. Assign all the bands in the spectrum and analyze the splitting patterns that are seen.

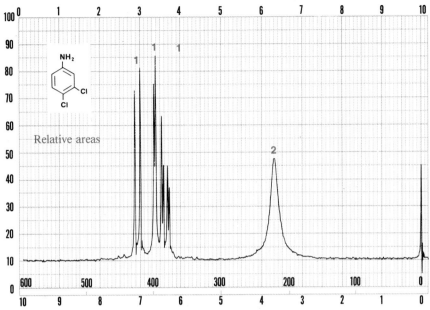

Figure 13.67 Spectrum for Problem 13.41. (From *The Aldrich Library of NMR Spectra*)

13.42 Compound EE, C_6H_5BrO, has $\lambda_{max}^{methanol}$ 226 (ϵ 12,100) and 283 nm (ϵ 1930). When potassium hydroxide is added to the solution, λ_{max} 245 (ϵ 15,600) and 300 nm (ϵ 2670) are observed. The infrared spectrum of Compound EE has a broad absorption band at 3400 cm^{-1}. The proton magnetic resonance spectrum of EE is shown in Figure 13.68. Assign a structure to it and account for the bands in its ultraviolet and infrared spectra.

Compound EE

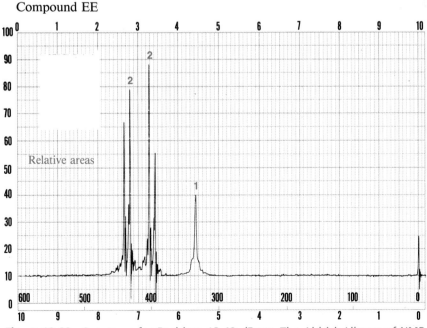

Figure 13.68 Spectrum for Problem 13.42. (From *The Aldrich Library of NMR Spectra*)

13.43 The proton magnetic resonance spectrum of *N*,*N*-dimethylformamide has two different chemical shifts for the hydrogen atoms on the methyl groups, as shown in Figure 13.69. The phenomenon observed is not due to spin-spin coupling. How do you know this? What does this observation tell us about the nature of the bond between the nitrogen atom and the carbonyl group? Draw structural formulas that account for the observation. (Hint: Writing resonance contributors for the compound may be helpful.)

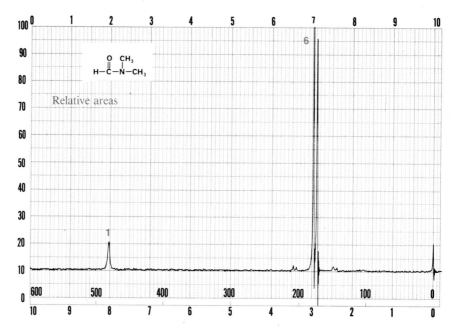

Figure 13.69   Spectrum for Problem 13.43. (From *The Aldrich Library of NMR Spectra*)

13.44 Carbon-13 NMR spectra for the following compounds are given in Figure 13.70. Decide which spectrum belongs to which compound. In each case, you should try to assign as many of the bands as you can.

(a) CH_3O—⟨benzene ring⟩—CH_2OH

(b) ⟨benzene ring⟩—$\overset{\overset{\displaystyle O}{\|}}{C}CH_2CH_2CH_3$

(c) $CH_3\overset{\overset{\displaystyle CH_3}{|}}{\underset{\underset{\displaystyle CH_3}{|}}{C}}OH$

(d) $CH_3CH_2CH_2CH_2CH_2\overset{\overset{\displaystyle O}{\|}}{C}OCH_3$

Spectrum I

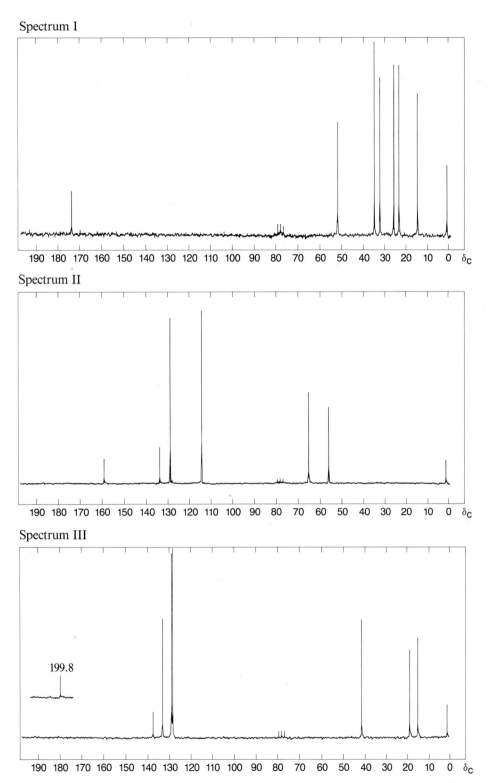

Spectrum II

Spectrum III

199.8

Figure 13.70 Spectra for Problem 13.45. (Adapted from Johnson and Jankowski. Reprinted by permission.)

Spectrum IV

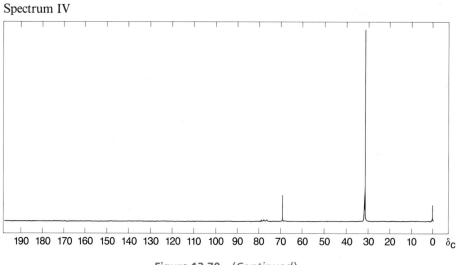

Figure 13.70   *(Continued)*

13.45   Two of the dimethylbenzenes have the bands given below in their carbon-13 nuclear magnetic resonance spectra. Assign structures to the compounds.

Compound FF: δ 21.3, 126.2, 128.2, 130.0, 137.6

Compound GG: δ 20.9, 129.0, 134.6

13.46   Compound HH, $C_4H_8Cl_2$, has carbon-13 nuclear magnetic resonance absorption bands at δ 30.0 and 44.1. Assign a structure to Compound HH.

13.47   Compound II contains only carbon, hydrogen, and a halogen. The mass spectrum is shown in Figure 13.71. Assign a structure to it.

Compound II

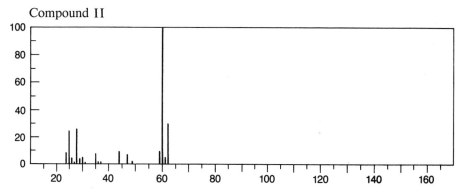

Figure 13.71   Spectrum for Problem 13.47. (Adapted from Stenhagen et al. Reprinted by permission.)

13.48   Compound JJ is an ester. It has a singlet in its proton magnetic resonance spectrum at δ 3.69. Identify the molecular ion in its mass spectrum shown in Figure 13.72. Assign a structure to it and to the ions having an intensity greater than 50%.

Compound JJ

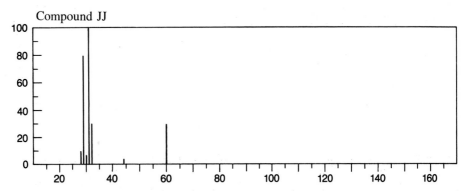

Figure 13.72   Spectrum for Problem 13.48. (Adapted from Stenhagen et al. Reprinted by permission.)

13.49   The molecular ion for Compound KK does *not* appear in its mass spectrum. The important fragments from the molecular ion, however, are quite clearly seen in Figure 13.73. Assign a structure to Compound KK.

Compound KK

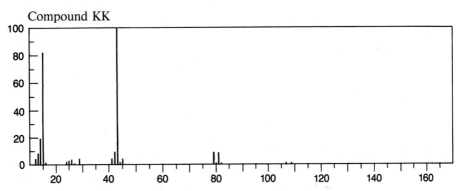

Figure 13.73   Spectrum for Problem 13.49. (Adapted from Stenhagen et al. Reprinted by permission.)

13.50   The mass spectra of two halogenated hydrocarbons appear in Figure 13.74. The molecular ion of Compound LL is also its base peak, but not so for Compound MM. Assign structures to the two compounds, and offer a rationalization for the stability of the molecular ion of Compound LL in comparison with that of Compound MM.

Compound LL

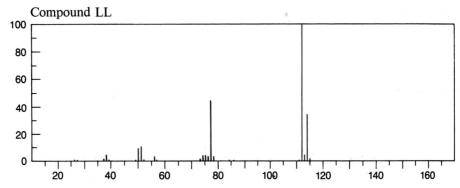

Figure 13.74   Spectra for Problem 13.50. (Adapted from Stenhagen et al. Reprinted by permission.)

Compound MM

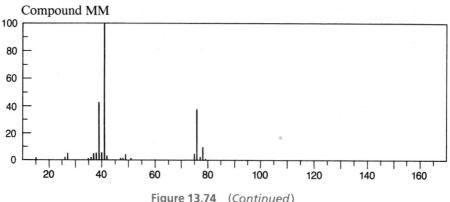

Figure 13.74   (*Continued*)

13.51 The mass spectra of two isomeric alcohols, Compounds NN and OO, are given in Figure 13.75. The molecular ion is visible in the spectrum of one, but not the other. Assign structures to the two alcohols, and write equations accounting for the formation of the base peak in each case.

Compound NN

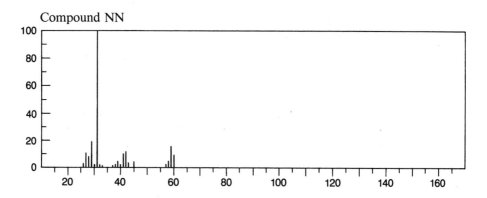

Compound OO

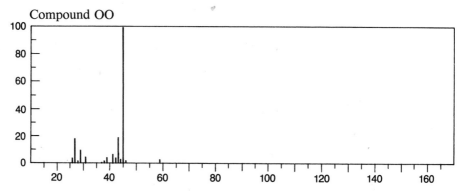

Figure 13.75   Spectra for Problem 13.51. (Adapted from Stenhagen et al. Reprinted by permission.)

13.52 The base peak in the mass spectrum of pentanal, shown in Figure 13.76, arises by way of a rearrangement. Write an equation that shows how the ion, represented by the base peak, comes into being.

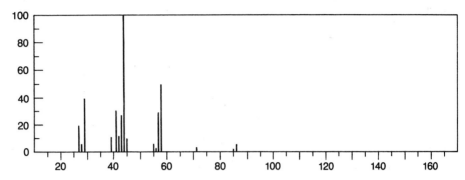

Figure 13.76 Spectrum for Problem 13.52. (Adapted from Stenhagen et al. Reprinted by permission.)

13.53 Assign a structure to the hydrocarbon Compound PP for which a mass spectrum is given in Figure 13.77, and write an equation showing how the ion giving rise to the base peak is formed. The molecular ion is seen in the spectrum.

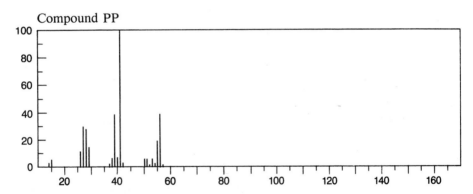

Figure 13.77 Spectrum for Problem 13.53. (Adapted from Stenhagen et al. Reprinted by permission.)

13.54   The proton magnetic resonance and the mass spectra of Compound QQ are given in Figure 13.78. Assign a structure to the compound.

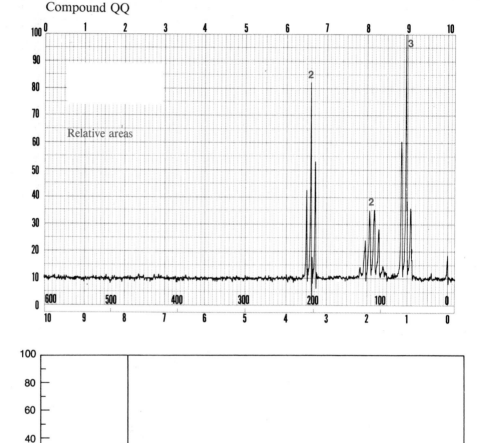

Figure 13.78   Spectra for Problem 13.54. (NMR spectrum from *The Aldrich Library of NMR Spectra*. Mass spectrum adapted from Stenhagen et al. Reprinted by permission.)

13.55 Compound RR has the mass and the proton magnetic resonance spectra shown in Figure 13.79. The infrared spectrum of RR has a major band at 1710 cm^{-1}. Assign a structure to the compound, and write equations showing how the major ions that appear in its mass spectrum are formed.

Compound RR

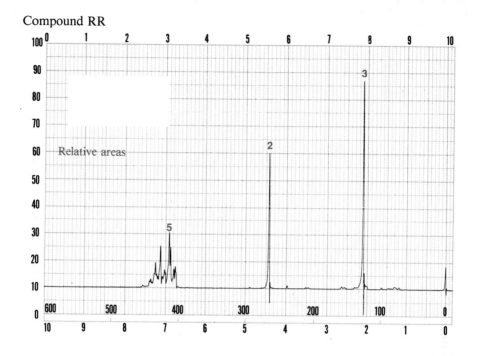

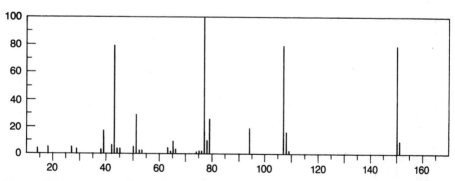

Figure 13.79 Spectra for Problem 13.55. (NMR spectrum from *The Aldrich Library of NMR Spectra*. Mass spectrum adapted from Stenhagen et al. Reprinted by permission.)

Carbohydrates

14

Introduction

Carbohydrates, *important constituents of both plants and animals, are polyhydroxy aldehydes or ketones.* Compounds classified as carbohydrates vary in structure from those consisting of a few carbon atoms to gigantic polymeric molecules having molecular weights in the millions.

Carbohydrates that cannot be broken down into simpler units by hydrolysis reactions are known as **monosaccharides.** The most common monosaccharide is glucose, $C_6H_{12}O_6$, the chief form in which carbohydrates are metabolized in our bodies. Fructose, $C_6H_{12}O_6$, is known as fruit sugar. Ribose, $C_5H_{10}O_5$, and 2-deoxyribose, $C_5H_{10}O_4$, are components of ribonucleic acids (RNA) and deoxyribonucleic acids (DNA), respectively, the giant molecules that play an important role in the storage and transmission of genetic information. These monosaccharides exist as their cyclic hemiacetals (Section 9.7C) in equilibrium with open-chain hydroxy aldehyde or ketone structures.

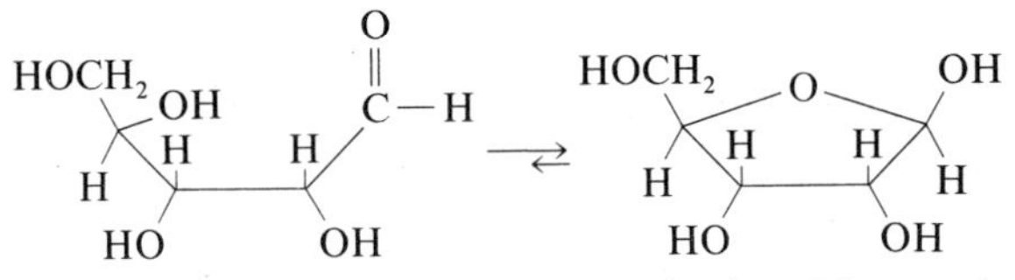

aldehyde form   hemiacetal form

ribose

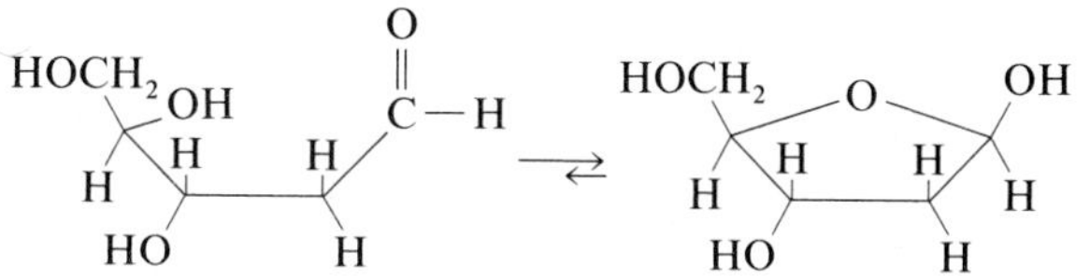

aldehyde form   hemiacetal form

2-deoxyribose

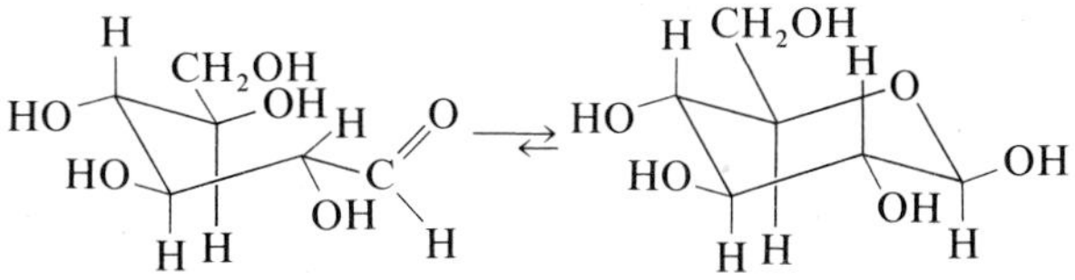

aldehyde form   hemiacetal form

glucose

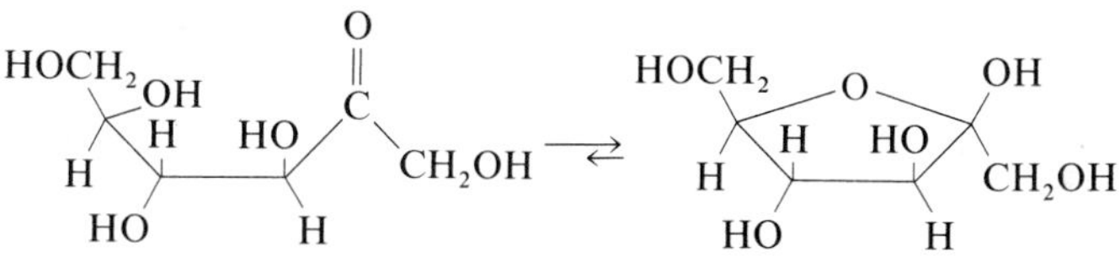

ketone form   hemiacetal form

fructose

Monosaccharides can be further subdivided according to the number of carbon atoms they contain and whether they are aldehydes (**aldoses**) or ketones (**ketoses**). Ribose, with five carbon atoms, is a pentose. It has an aldehyde function, so it is an aldo-pentose. Glucose is an aldohexose, while fructose is a ketohexose.

PROBLEM 14.1 Besides pentoses and hexoses, there are trioses with three carbon atoms, and tetroses with four carbon atoms. Classify the following monosaccharides according to the number of carbon atoms and the nature of the carbonyl function they contain.

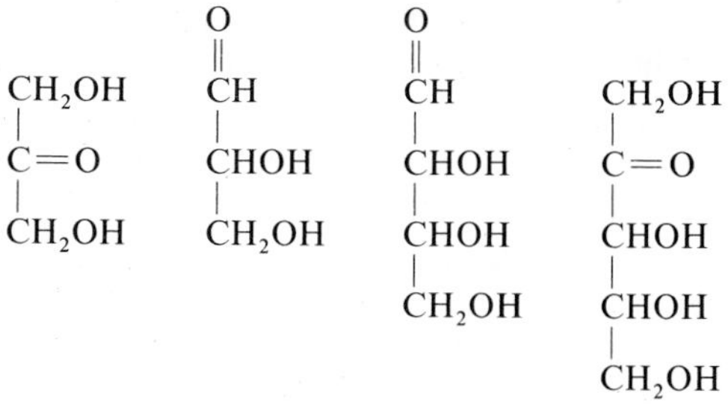

A   B   C   D

Hydroxyl and carbonyl functions are not the only ones to appear in monosaccharides. Sugars containing carboxylic acid groups and amino groups are common structural units in biologically important carbohydrates. Two such monosaccharides are 2-amino-2-deoxyglucose, $C_6H_{13}NO_5$, an amino sugar (also called glucosamine), and glucuronic acid, $C_6H_{10}O_7$, a sugar acid.

2-amino-2-deoxyglucose
glucosamine

glucuronic acid

Common table sugar, sucrose, has the molecular formula $C_{12}H_{22}O_{11}$. When it is boiled with water with a trace of acid, as in candy making, it is converted into a mixture of glucose and fructose. Therefore, it is classified as *a* **disaccharide,** *a compound made up of two monosaccharide units.*

$$C_{12}H_{22}O_{11} + H_2O \xrightarrow[\Delta]{H_3O^+} C_6H_{12}O_6 + C_6H_{12}O_6$$

sucrose   glucose   fructose

Maltose, another disaccharide, gives two molecules of glucose on hydrolysis.

$$C_{12}H_{22}O_{11} + H_2O \xrightarrow[\Delta]{H_3O^+} 2\, C_6H_{12}O_6$$

maltose   glucose

Similarly, *there are* **trisaccharides** *and* **tetrasaccharides** *that give three and four monosaccharide units, respectively, on hydrolysis. Compounds containing two to ten monosaccharide units are called* **oligosaccharides.** Saccharides in this molecular weight range are individual, identifiable compounds with definite structures and molecular weights. They are crystalline compounds that are water soluble and sweet to the taste. Oligosaccharides have many important physiological functions. For example, blood groups are determined by oligosaccharides combined with proteins on the surface of red blood cells. In fibrinogen, an important component in blood clotting, protein is also associated with oligosaccharides. Immune globulins, which are involved in the development of immunity to disease, too, are composed of carbohydrates and proteins. The chemistry of carbohydrates is, thus, directly relevant to the questions of cell recognition and regulation of cell growth, so important in the problem of cancer.

When there are more than ten monosaccharide units in the molecule, the compounds are defined as **polysaccharides.** *Cellulose* (Section 14.9B), the chief structural material of plants, is a high molecular weight polysaccharide made up of glucose units. Not all cellulose molecules have the same molecular weight. Instead, a given sample contains molecules in some range of molecular weights. *Starch* (Section 14.9A) is the polysaccharide form in which plants store glucose for their energy needs. Animals store glucose as another polysaccharide, *glycogen.*

Stereochemistry of Sugars

A. Relative and Absolute Configuration

The simplest optically active sugar is 2,3-dihydroxypropanal, glyceraldehyde. It has one asymmetric carbon atom and exists as two enantiomers, (+)-glyceraldehyde and (−)-glyceraldehyde. The structures of the two enantiomers and their configurations are shown below.

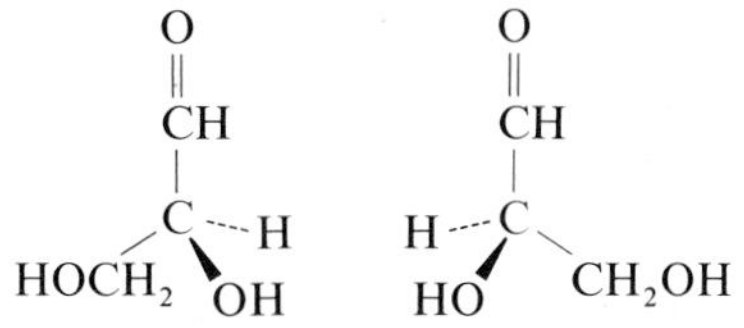

(*S*)-glyceraldehyde   (*R*)-glyceraldehyde

There is no way of knowing, just by looking at the structures drawn for the two enantiomers of glyceraldehyde, which one corresponds to (+)-glyceraldehyde. In other words, the actual arrangement of the atoms around the chiral center, the absolute configuration (Section 3.9F) of the compound, cannot be determined from the optical rotation of glyceraldehyde.

For about a century, the problem of absolute configuration plagued scientists working with optically active compounds. The phenomenon of the optical activity of sugars was known by J. B. Biot as early as 1815 (Section 3.9E). He was the first to recognize that optical activity must be a property of molecules and not of asymmetric crystals. Louis Pasteur, around 1850, carried out the first separation of optically active isomers of organic molecules with his research into salts of tartaric acid formed in the processes of wine making (Section 3.9E). The absolute configuration of (+)-tartaric acid was determined in 1951 by the Dutch chemist, J. M. Bijvoet, and his colleagues, using a sophisticated modification of x-ray diffraction. Their work was done in the laboratory named for van't Hoff, who had recognized a century earlier that the existence of optically active enantiomers could be explained by postulating a tetrahedral carbon atom as the chiral center (Section 3.9E).

Chemists recognized that they could relate the configurations of various optically active compounds to each other by chemical interconversions even when they did not know absolute configurations. The work on the stereochemistry of S_N1 and S_N2 reactions (Chapter 6), for example, was done in the 1930s and 1940s when the absolute configurations of the compounds were not known. The reasoning about inversion and retention of configuration remains valid because it refers only to the relative configurations of the molecules involved.

B. Glyceraldehyde as a Standard for the Assignment of Relative Configurations

In 1906, M. Rosanoff, an American carbohydrate chemist, suggested that (+)-glyceraldehyde be assigned a configuration and used as the standard to which the configurations of other sugars and, ultimately, other chiral compounds could be compared.

The configuration that was chosen for ($+$)-glyceraldehyde was the R configuration, and thus ($-$)-glyceraldehyde was assigned the S configuration.

$$
\begin{array}{cc}
\underset{\text{(R)-(+)-glyceraldehyde}}{\overset{\displaystyle \overset{O}{\underset{\displaystyle \parallel}{CH}}}{\underset{HO}{\overset{\displaystyle |}{\underset{}{}}}}} &
\underset{\text{(S)-(-)-glyceraldehyde}}{}
\end{array}
$$

(R)-(+)-glyceraldehyde   (S)-(−)-glyceraldehyde
configurations of (+)- and (−)-glyceraldehyde
as assigned in 1906

This assignment was chosen for ($+$)-glyceraldehyde because the hydroxyl group and the hydrogen atom at carbon 5 of ($+$)-glucose had been assigned the corresponding configuration in 1891 by Emil Fischer. ($+$)-Glyceraldehyde was the compound that could be related by chemical transformations to ($+$)-glucose, so the two compounds had to have the same stereochemistry at the asymmetric carbon atom closest to the primary alcohol group in each molecule. Rosanoff and Fischer, of course, had a 50% chance that their assignments were correct. ($+$)-Glyceraldehyde had to have either the R configuration or the enantiomeric S configuration.

Rosanoff's suggestion was widely accepted. Over the years, many compounds were synthesized from or degraded to ($+$)-glyceraldehyde so that other configurations could be established relative to the standard. Each compound that was related to ($+$)-glyceraldehyde could then serve as a standard for other compounds, until the relative configurations of many compounds were known. The following equation provides a simple example of the kinds of correlations that can be made. (R)-($+$)-Glyceraldehyde is oxidized to (R)-($-$)-glyceric acid (2,3-dihydroxypropanoic acid).

(R)-(+)-glyceraldehyde   oxidation →   (R)-(−)-glyceric acid

Because this reaction does not break bonds to the chiral center, the relative configurations of the two compounds must be the same.

(R)-($+$)-Isoserine (3-amino-2-hydroxypropanoic acid), an amino compound, serves to link ($-$)-lactic acid (2-hydroxypropanoic acid) from sour milk, with ($+$)-glyceraldehyde. Isoserine is converted to (R)-($-$)-glyceric acid by one reaction and to (S)-($-$)-3-bromo-2-hydroxypropanoic acid by another. The halogen-carbon bond in (S)-($-$)-3-bromo-2-hydroxypropanoic acid is reduced to give (R)-($-$)-lactic acid.

(R)-(−)-glyceric acid   ← HONO ←   (R)-(+)-isoserine   — NOBr →   (S)-(−)-3-bromo-2-hydroxypropanoic acid   — H₂ / catalyst →   (R)-(−)-lactic acid

Some of the reactions necessary for these conversions are discussed in Chapter 16. None of the transformations breaks bonds to the asymmetric carbon atom. Therefore, the configurations of all the compounds relative to (+)-glyceraldehyde are established.

PROBLEM 14.2   (−)-But-1-en-3-ol can be hydrogenated to (−)-2-butanol. Ozonolysis of the bute-nol, followed by mild oxidation, gives (−)-lactic acid. Write structures showing the relative configurations of all the compounds mentioned. Give the proper *R* or *S* designations to the compounds.

C. Fischer Projections as Two-Dimensional Representations of Chiral Compounds

Drawing one chiral center in a three-dimensional representation is not too difficult. When chemists had to draw molecules like glucose and compare stereochemistry at several chiral centers, they had to invent a convention for representing three-dimensional molecules in the plane of the paper in a consistent and simple way. Fischer introduced such a convention. He said that the sugars would be written with the backbone carbon chain in a vertical line with the aldehyde or ketone function at the top of the chain. The hydrogen atoms and hydroxyl groups would be placed to the left or right of this backbone. It would be understood that the hydrogens and hydroxyl groups were pointing up towards the viewer, while the vertical bonds were pointing away from the viewer. In his original paper, Fischer called this representation a *projection formula* because it is a projection of a particular three-dimensional conformation of the molecule onto a two-dimensional surface.

Any two bonds at an sp^3-hybridized carbon atom are in a plane at right angles to the plane defined by the other two bonds. The Fischer projection at a given carbon atom represents a projection in two dimensions of this view of the molecule. Lines intersecting at right angles represent the two planes (Figure 14.1).

The conformation that is drawn in a projection formula of any compound containing more than one asymmetric carbon atom happens to be the least stable, fully eclipsed

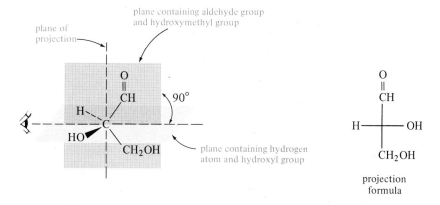

Figure 14.1   A representation of the planes that intersect at right angles at a tetrahedral carbon atom. An indication of the plane represented by the Fischer projection formula.

conformation of the molecule. Thus, Fischer projections do not represent the actual shape of the molecule as it would exist in the crystalline state or in solution. They are a convenient way of comparing the configurations of various chiral centers. Fischer's picture of the structure of glucose is shown below. The second representation uses wedges and dashes to remind us what the convention means. A sideways view of the molecule represented by the projection formula shows that the first and last carbons of the glucose molecule are in reality quite close to each other.

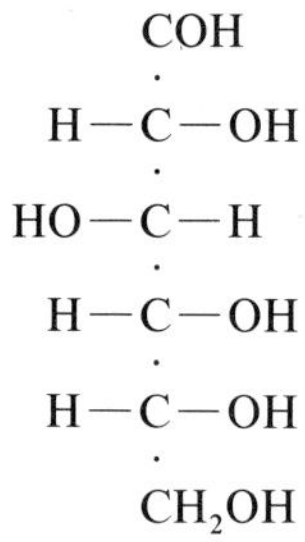

(+)-glucose as Fischer represented it in his original projection formula	*an indication of which bonds project up and which down at every chiral center viewed individually in (+)-glucose*	*sideways view*

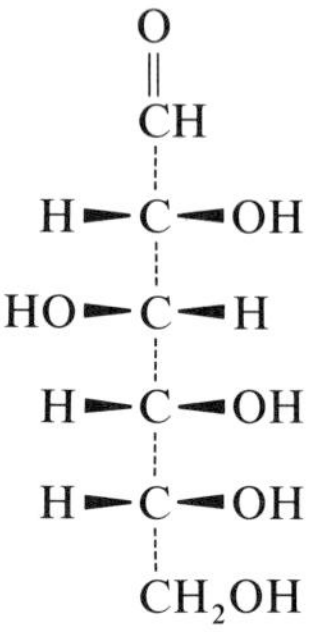

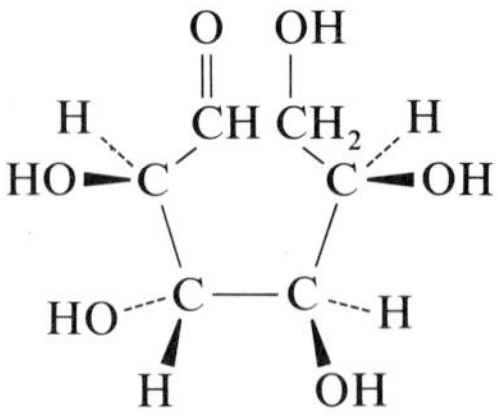

The original Fischer projections were later changed so that the dots, too, were replaced by lines that connected all the groups to the asymmetric carbon atoms. This is the form of Fischer projections now used in many books. In 1893, Victor Meyer introduced another convention that has been widely, but inconsistently, used. He suggested that the chiral centers in the Fischer projections be represented as the crossing point of the bonds and that no carbon be shown at that point. These conventions are illustrated with various representations of (R)-(+)-glyceraldehyde.

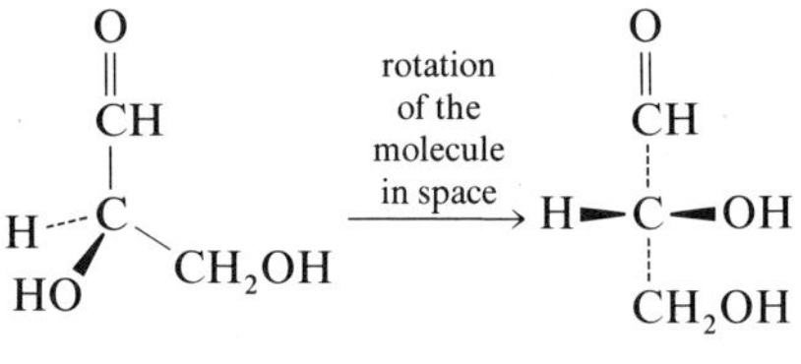

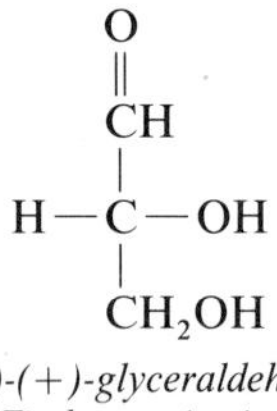

(R)-(+)-glyceraldehyde Fischer projection formula as often seen	*(R)-(+)-glyceraldehyde Victor Meyer's modification of the Fischer projection formula*

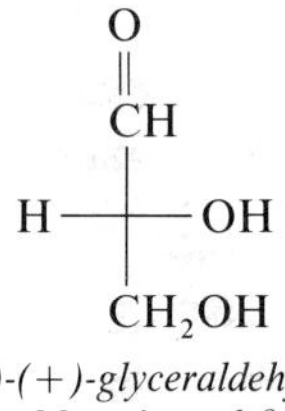

We will adopt Meyer's modification exclusively when we are using Fischer projections. The other form has no feature that clearly signals stereochemical intent in the way the formula is written. Many times confusion arises because Lewis structures, which look very similar to Fischer projections, are drawn for organic molecules. Lewis structures are not designed to give any stereochemical information. Unless a Fischer projection is identified as such every time it is used, there is nothing in its appearance to distinguish it from an ordinary representation of a tetrahedral carbon atom with no stereo-chemical information. The Meyer modification is clearly a different convention. Its appearance instantly alerts us to the specific directions in space that are intended for the four groups attached to the crossing point.

Projection formulas impose limitations upon us. For example, interchanging any two substituents at the chiral center converts the center into its enantiomeric configuration. The most obvious example of this is demonstrated below.

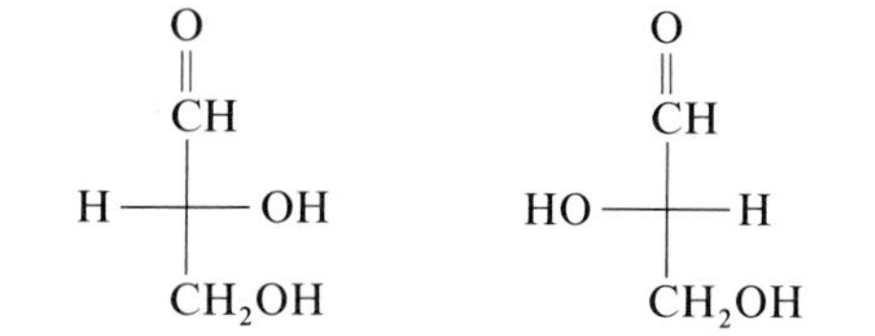

$$
\begin{array}{cc}
\underset{\text{CH}}{\overset{\overset{\displaystyle O}{\|}}{\text{}}} & \underset{\text{CH}}{\overset{\overset{\displaystyle O}{\|}}{\text{}}} \\
\text{H} \!-\!|\!-\! \text{OH} & \text{HO} \!-\!|\!-\! \text{H} \\
\text{CH}_2\text{OH} & \text{CH}_2\text{OH} \\
(R)\text{-}(+)\text{-glyceraldehyde} & (S)\text{-}(-)\text{-glyceraldehyde}
\end{array}
$$

exchange of two substituents at the asymmetric carbon atom converts the representation of one enantiomer into that of its mirror-image isomer

An interchange of the positions of the hydrogen atom and the hydroxyl group converts (R)-$(+)$-glyceraldehyde to (S)-$(-)$-glyceraldehyde. This is true of any other interchange of two substituents in compounds containing a single asymmetric carbon atom. The interchange of substituents on a single asymmetric carbon atom in the projection of a molecule containing several such centers will, of course, convert the projection into a representation of a diastereomeric molecule (Section 5.8B). All the structural formulas below are representations of (S)-$(-)$-glyceraldehyde arrived at by interchanging two substituents on the projection formula shown above for (R)-$(+)$-glyceraldehyde.

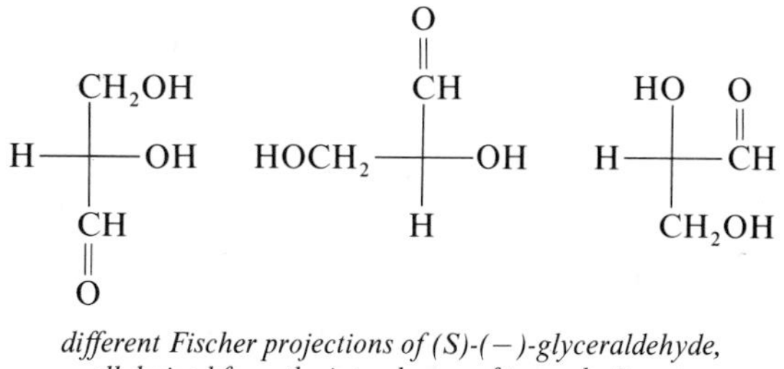

$$
\begin{array}{ccc}
\text{CH}_2\text{OH} & \underset{\text{CH}}{\overset{\overset{\displaystyle O}{\|}}{\text{}}} & \text{HO}\quad\text{O} \\
\text{H}\!-\!|\!-\!\text{OH} & \text{HOCH}_2\!-\!|\!-\!\text{OH} & \text{H}\!-\!|\!-\!\text{CH} \\
\underset{\text{O}}{\overset{\displaystyle \|}{\text{CH}}} & \text{H} & \text{CH}_2\text{OH}
\end{array}
$$

different Fischer projections of (S)-(−)-glyceraldehyde, all derived from the interchange of two substituents on the Fischer projection of (R)-(+)-glyceraldehyde

Two of the representations of (S)-$(-)$-glyceraldehyde also illustrate another point. The projection formula allows one to rotate the formula 180° *in* the plane of the paper and retain the same stereochemistry.

identical

$$
\begin{array}{ccc}
\underset{\text{CH}}{\overset{\overset{\displaystyle O}{\|}}{\text{}}} & \text{CH}_2\text{OH} & \underset{\text{CH}}{\overset{\overset{\displaystyle O}{\|}}{\text{}}} \\
\text{HO}\!-\!|\!-\!\text{H} & \text{H}\!-\!|\!-\!\text{OH} & \text{HO}\!-\!|\!-\!\text{H} \\
\text{CH}_2\text{OH} & \underset{\text{O}}{\overset{\displaystyle \|}{\text{CH}}} & \text{CH}_2\text{OH}
\end{array}
$$

rotate 180° in the plane of the paper

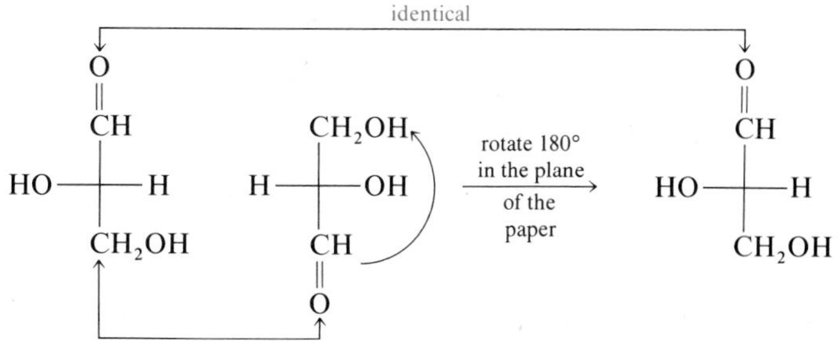

two pairs of substituents interchanged; molecule retains the same stereochemistry

This latest maneuver is the equivalent of interchanging two pairs of substituents at the chiral center. The aldehyde function was exchanged with the hydroxymethyl group, and the hydroxyl group with the hydrogen atom. Such a set of two transformations converts the projection formula into another projection of the *same* compound.

PROBLEM 14.3 Look at each pair of figures and decide whether they represent the same compound, enantiomers, or diastereomers. Check your conclusions by designating the configuration at each chiral center as R or S.

(a)

$$\begin{array}{c} CH_3 \\ H \!-\!\!\!-\!\!\!- Br \\ CH_2CH_3 \end{array} \qquad \begin{array}{c} Br \\ H \!-\!\!\!-\!\!\!- CH_3 \\ CH_2CH_3 \end{array}$$

(b)

$$\begin{array}{c} O \\ \parallel \\ COH \\ H \!-\!\!\!-\!\!\!- OH \\ CH_3 \end{array} \qquad \begin{array}{c} H \quad O \\ \parallel \\ HO \!-\!\!\!-\!\!\!- COH \\ CH_3 \end{array}$$

(c)

$$\begin{array}{c} CH_3 \\ H \!-\!\!\!-\!\!\!- Br \\ H \!-\!\!\!-\!\!\!- Br \\ CH_3 \end{array} \qquad \begin{array}{c} CH_3 \\ H \!-\!\!\!-\!\!\!- Br \\ H \!-\!\!\!-\!\!\!- CH_3 \\ Br \end{array}$$

(d)

$$\begin{array}{c} O \\ \parallel \\ COH \\ H \!-\!\!\!-\!\!\!- OH \\ HO \!-\!\!\!-\!\!\!- H \\ COH \\ \parallel \\ O \end{array} \qquad \begin{array}{c} H \quad O \\ \parallel \\ HO \!-\!\!\!-\!\!\!- COH \\ HO \!-\!\!\!-\!\!\!- COH \\ \parallel \\ H \quad O \end{array}$$

D. The Designation of Chiral Compounds as D and L

The R and S convention for designation of configuration was not introduced until the 1950s (Section 3.9F). Before that, chemists felt the need for some way of naming relative configurations. When Rosanoff suggested that a particular configuration be assigned to (+)-glyceraldehyde, he also suggested that the particular arrangement of atoms around the chiral center be called the D configuration. The capital letter D was chosen to distinguish it from *d*, which was being used ambiguously to indicate that the compound was dextrorotatory and also that it was related in structure to glucose. All compounds having an arrangement of atoms similar to that at the chiral center of (+)-glyceraldehyde at a comparable carbon atom would be members of a D family. Those with the opposite configuration at such a carbon atom would belong to an L family. This system is still used, especially in the biochemical literature, so you should recognize it. Except for a few cases where relationships are quite easy to see, the system led to many complications and inconsistencies and has been largely abandoned except for carbohydrates and amino acids. Some examples are given below. The chiral center that determines the family relationship in each compound is shaded in color.

$$\begin{array}{c} O \\ \parallel \\ CH \\ H \!-\!\!\!-\!\!\!- OH \\ CH_2OH \end{array} \qquad \begin{array}{c} O \\ \parallel \\ COH \\ H \!-\!\!\!-\!\!\!- OH \\ CH_3 \end{array} \qquad \begin{array}{c} O \\ \parallel \\ COH \\ HO \!-\!\!\!-\!\!\!- H \\ H \!-\!\!\!-\!\!\!- OH \\ COH \\ \parallel \\ O \end{array}$$

D-(+)-glyceraldehyde   D-(−)-lactic acid   D-(−)-tartaric acid

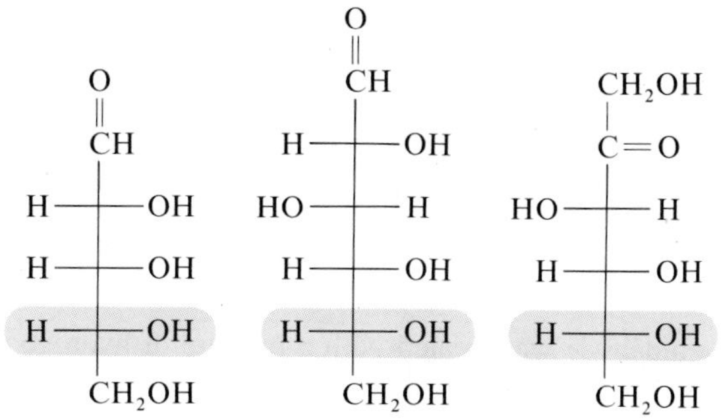

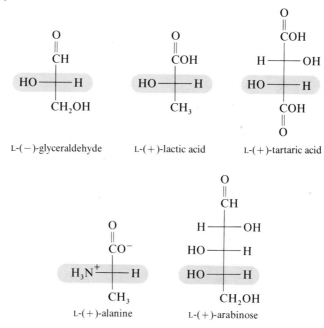

D-(−)-ribose   D-(+)-glucose   D-(−)-fructose

Notice that the symbol D has nothing to do with whether the compound is dextrorotatory or levorotatory, just as the designations *R* or *S* by themselves do not give that information.

Some compounds are members of the stereochemical family related to L-(−)-glyceraldehyde.

L-(−)-glyceraldehyde   L-(+)-lactic acid   L-(+)-tartaric acid

L-(+)-alanine   L-(+)-arabinose

The designation, once again, simply indicates that these compounds all have the same configuration at a certain chiral atom in the molecule.

PROBLEM 14.4   Designate the configuration as *R* or *S* at each chiral center in the compounds in the D and L series given above.

E. The Relative Configurations of (+)-Glyceraldehyde and (−)-Tartaric Acid

(+)-Glyceraldehyde is converted by a series of chemical reactions into tartaric acid. The experiment was done in 1917 by A. Wohl in Germany. He started with the dimethyl acetal of (+)-glyceraldehyde and hydrolyzed it to the free aldehyde in the reaction mixture.

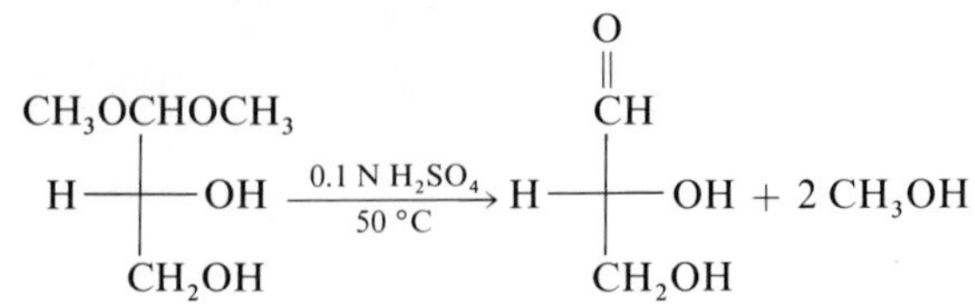

$$
\begin{array}{c}
\text{CH}_3\text{OCHOCH}_3 \\
\text{H}\!\!-\!\!\!\overset{\displaystyle |}{\underset{\displaystyle |}{}}\!\!-\!\!\text{OH} \\
\text{CH}_2\text{OH}
\end{array}
\xrightarrow[\text{50 °C}]{\text{0.1 N H}_2\text{SO}_4}
\begin{array}{c}
\overset{\text{O}}{\overset{\|}{\text{CH}}} \\
\text{H}\!\!-\!\!\!\overset{\displaystyle |}{\underset{\displaystyle |}{}}\!\!-\!\!\text{OH} \\
\text{CH}_2\text{OH}
\end{array}
+ \; 2\,\text{CH}_3\text{OH}
$$

dimethyl acetal of (+)-glyceraldehyde (+)-glyceraldehyde

The aldehyde was treated with a mixture of hydrogen cyanide and aqueous ammonia to give cyanohydrins, which were not isolated but hydrolyzed to hydroxyacids in the same reaction mixture.

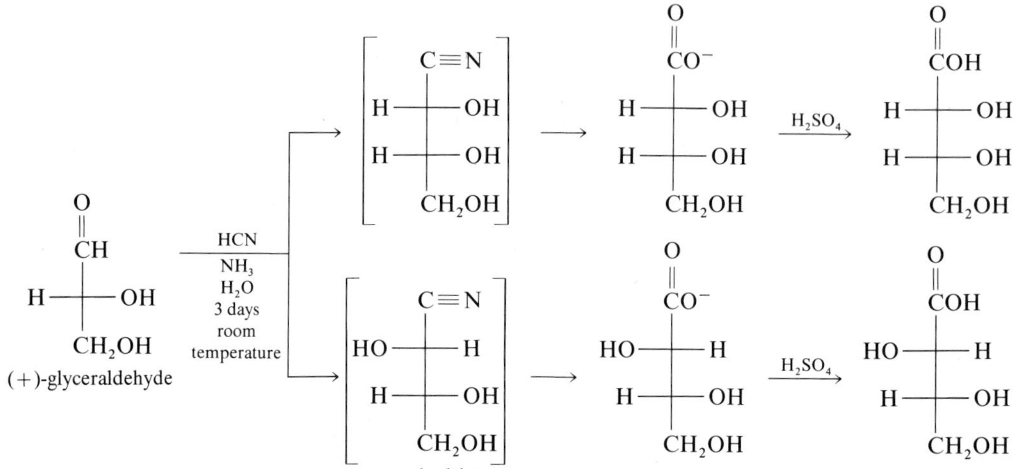

cyanohydrins from (+)-glyceraldehyde   2,3,4-trihydroxy-butanoic acids

In this process, a new chiral center is formed in the molecule. The products are diastereomers that are formed in unequal amounts and can be separated from each other by recrystallization, because they have different physical properties such as solubility. The trihydroxybutanoic acids were separated and then oxidized to tartaric acids.

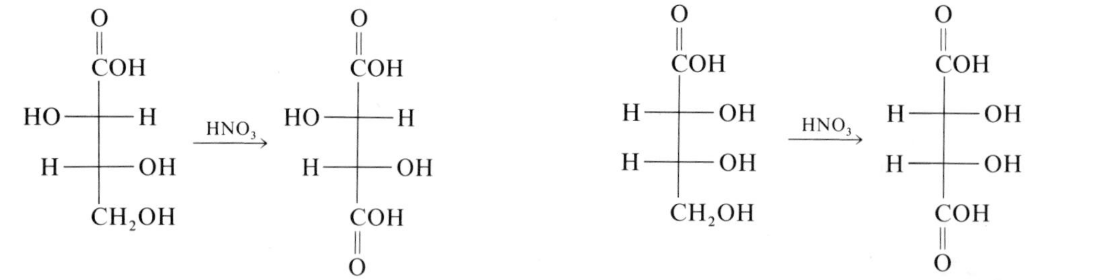

(2*S*,3*R*)-2,3,4-trihydroxybutanoic acid   (−)-tartaric acid (2*S*,3*S*)-2,3-dihydroxybutanedioic acid   (2*R*,3*R*)-2,3,4-trihydroxybutanoic acid   *meso*-tartaric acid (2*R*,3*S*)-2,3-dihydroxybutanedioic acid

The optically active tartaric acid formed in this synthesis is (−)-tartaric acid. The other tartaric acid is not optically active and cannot be resolved into optically active compounds. It is *meso*-tartaric acid, which is not optically active because the molecule has an internal plane of symmetry and is not chiral. With this experiment, the relative configurations of (+)-glyceraldehyde and (−)-tartaric acid were established.

PROBLEM 14.5   Draw the sawhorse formula for *meso*-tartaric acid that corresponds to the Fischer projection given above. Be sure you see the plane of symmetry in the molecule.

PROBLEM 14.6   Convert (−)-glyceraldehyde into tartaric acid. Show clearly the stereochemistry of the tartaric acids that would form, and indicate whether or not you would expect them to be optically active.

F. The Assignment of Absolute Configuration

(+)-Tartaric acid, the enantiomer of (−)-tartaric acid, is the isomer that occurs most abundantly in nature. Bijvoet used it in 1951 for the determination of absolute configuration. This experiment, mentioned in Section 14.2A, established the actual orientation in space of the various groups and atoms in (+)-tartaric acid.

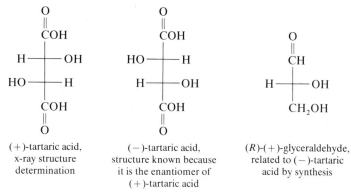

*various representations of the absolute
configuration of (+)-tartaric acid*

The determination of the actual structure of (+)-tartaric acid also established the absolute configuration of its enantiomer, (−)-tartaric acid, the structure of which had been related to that of (+)-glyceraldehyde by synthesis. (+)-Glyceraldehyde did, indeed, correspond to the R-configuration.

$\underset{\text{O}}{\overset{\text{O}}{\parallel}}$ COH	$\underset{\text{O}}{\overset{\text{O}}{\parallel}}$ COH	$\overset{\text{O}}{\overset{\parallel}{\text{CH}}}$
H——OH	HO——H	
HO——H	H——OH	H——OH
$\underset{\text{O}}{\overset{\parallel}{\text{COH}}}$	$\underset{\text{O}}{\overset{\parallel}{\text{COH}}}$	CH₂OH
(+)-tartaric acid, x-ray structure determination	(−)-tartaric acid, structure known because it is the enantiomer of (+)-tartaric acid	(R)-(+)-glyceraldehyde, related to (−)-tartaric acid by synthesis

*absolute configurations for (+)- and (−)-tartaric
acids and for (+)-glyceraldehyde*

With this one experiment, what had been known until that time as the relative configurations of hundreds of compounds suddenly became established as their absolute configurations. Assignments that had been made on the assumption that (+)-glyceraldehyde really did have the R-configuration were all correct. Ultimately, the original assignment went back to Fischer, who had decided arbitrarily that

(+)-glucose had a certain configuration at the chiral center farthest from the aldehyde group. He assigned that hydroxyl group as pointing to the right in his projection formula. He knew that all his information would also be valid if the structure of glucose was actually the mirror image of the one that he had assigned it, but he had to choose one structure in order to be able to make comparisons with those of other carbohydrates. His selection proved to be correct.

Glucose, in its open-chain form, has four different chiral centers and is, therefore, one of 2^4 or 16 different stereoisomers. The stereoisomers exist as eight pairs of enantiomers. Fischer's assignment of stereochemistry at each of the asymmetric carbon atoms in glucose was the result of painstaking experimental work and logical thought about the implications of those experimental observations. He was given the Nobel Prize for this work in 1902. His experiments and arguments will be retraced in Section 14.7B after we study the important reactions in the chemistry of carbohydrates.

PROBLEM 14.7   Write the structure of the enantiomer of D-(+)-glucose.

14.3

The Structure of Glucose

A. Glucose as a Pentahydroxyaldehyde

The open-chain form of glucose has been assigned the structure that we have been using for it on the basis of the following experimental data. Glucose has the molecular formula $C_6H_{12}O_6$. It reacts with mild oxidizing agents such as Tollens reagent (Section 9.9) and is therefore likely to be an aldehyde. When treated with acetic anhydride (Section 10.7A), it forms a pentaacetate and thus has five hydroxyl functions. Structures that have two hydroxyl functions on the same carbon atom do not represent stable species but are the hydrates of carbonyl compounds (Section 9.7A). Glucose must, therefore, have each hydroxyl group on a different carbon atom. The experimental observations described above are summarized below.

$$C_6H_7O(OCCH_3)_5 \xleftarrow{(CH_3C)_2O} C_6H_{12}O_6 \xrightarrow[\text{reagent}]{\text{Tollens}} \text{oxidation product,}$$

glucose pentaacetate, therefore 5 hydroxyl groups on separate carbon atoms in glucose

glucose

an acid, therefore aldehyde function in glucose

$$
\begin{array}{c}
O \\
\parallel \\
CH \\
| \\
CHOH \\
| \\
CHOH \\
| \\
CHOH \\
| \\
CHOH \\
| \\
CH_2OH
\end{array}
$$

structure of glucose, excluding stereochemistry, determined from chemical data

The addition of hydrogen cyanide to an aldehyde to give a cyanohydrin (Section 9.5), which was hydrolyzed to a carboxylic acid (Section 10.5C), was known in 1885 when the German chemist Heinrich Kiliani discovered that the reaction could be applied to sugars. He used the reaction to prove that glucose had a straight chain of six carbon atoms. Adding hydrogen cyanide to glucose and hydrolyzing the nitrile gave a hydroxycarboxylic acid with one more carbon atom than glucose. This was reduced with hydrogen iodide to an acid identified as heptanoic acid.

$$
\begin{array}{ccccc}
& & & \overset{O}{\overset{\|}{COH}} & \overset{O}{\overset{\|}{COH}} \\
\overset{O}{\overset{\|}{CH}} & \begin{bmatrix} C\equiv N \\ | \\ CHOH \end{bmatrix} & & | & | \\
| & CHOH & & CHOH & CH_2 \\
CHOH & | & & | & | \\
| & CHOH & & CHOH & CH_2 \\
CHOH \xrightarrow[H_2O]{HCN} & CHOH & \xrightarrow{H_2O} & CHOH \xrightarrow[\Delta]{\underset{P}{HI}} CH_2 \\
| & | & & | & | \\
CHOH & CHOH & & CHOH & CH_2 \\
| & | & & | & | \\
CHOH & CHOH & & CHOH & CH_2 \\
| & | & & | & | \\
CH_2OH & CH_2OH & & CH_2OH & CH_3 \\
\text{glucose} & & & & \text{heptanoic} \\
& & & & \text{acid}
\end{array}
$$

The conversion of glucose, a six-carbon sugar, into a known, straight-chain, seven-carbon acid showed that the carbonyl group to which the hydrogen cyanide added was at the end of a straight chain of six carbon atoms. Thus the assignment of the structure of glucose as 2,3,4,5,6-pentahydroxyhexanal was confirmed.

PROBLEM 14.8   When Kiliani tried the above sequence of reactions with fructose (p. 749), he obtained 2-methylhexanoic acid. Write equations showing the application of the same reactions to fructose. What do these reactions prove about the structure of fructose?

PROBLEM 14.9

(a) What do you think is the first step in the reduction of the polyhydroxyheptanoic acid to heptanoic acid with HI?

(b) Phosphorus is not necessary for the reaction. When phosphorus is not used, molecular iodine forms in the reaction mixture. The reaction works on ordinary alcohols and on alkyl iodides.

$$ROH + 2\ HI \longrightarrow RH + H_2O + I_2$$

$$RI + HI \longrightarrow RH + I_2$$

The table of bond dissociation energies (p. 185) tells us that the carbon-iodine bond is a relatively weak one. Use these facts to propose a mechanism for the conversion of an alcohol to an alkane by hydriodic acid.

B. Cyclic Structures of Monosaccharides

When a freshly prepared solution of a sample of glucose that has been recrystallized from methanol and has a melting point of 147 °C is put into the tube of a polarimeter, an initial specific rotation of $+113°$ is observed. As the solution stands in the polarimeter, the rotation falls until it reaches a value of $+52.5°$.

If glucose is recrystallized from water at high temperatures, another crystalline form is obtained. This form has a melting point of 150 °C. A freshly prepared solution of these crystals placed in the tube of a polarimeter has an initial specific rotation of $+19°$. On standing, the optical rotation of this solution rises to $+52.5°$. Either solution can be evaporated and recrystallized under one or the other of the conditions described above to give back the original two forms of glucose. Thus, the change in rotation is not a result of the decomposition of glucose in solution.

The change of optical rotation for a compound on standing in solution is called **mutarotation.** This phenomenon observed for the two forms of glucose with both of them arriving at the same final rotation for the solution suggests an equilibrium of two stereochemically different forms of glucose.

These facts have been interpreted to mean that glucose normally exists as a cyclic hemiacetal structure (Section 9.7C). The hydroxyl group on carbon 5 is close enough to the carbonyl group at carbon 1 that a six-membered ring can form.

α-D-glucose
mp 147 °C
[α] +113°

open-chain
form of
glucose

β-D-glucose
mp 150 °C
[α] +19°

When this cyclization occurs, a new asymmetric center is created. The orientation of the hydroxyl group on carbon 1 can be axial or equatorial. The form in which it is axial is called *α-glucose*, which crystallizes at ordinary temperatures and has an initial rotation of $+113°$. The form in which the hydroxyl group at carbon 1 is equatorial is called *β-glucose*. It crystallizes out of water at high temperatures and has an initial rotation of $+19°$. The two forms are in equilibrium with each other in aqueous solution through the open-chain form, which has the free aldehyde group.

The small concentration of this open-chain form is responsible for the reactions of glucose that are typical of aldehydes. This equilibrium is also responsible for the change in rotation from the initial values of $+113°$ for $α$-glucose and $+19°$ for $β$-glucose to the intermediate value of $+52.5°$ for the equilibrium mixture. This rotation corresponds to a mixture consisting of 36% $α$-glucose and 64% $β$-glucose. $β$-Glucose has all of the large substituents in the equatorial positions in the chair conformation of cyclohexane. It is more stable in solution than $α$-glucose where the hydroxyl group at carbon 1 is axial.

$α$-Glucose and $β$-glucose are stereoisomers that differ from each other at one chiral center, carbon 1. As such, they are diastereomers. *Diastereomers that differ from each other in stereochemistry at only one of many chiral centers are called* **epimers.** $α$-Glucose and $β$-glucose are epimers at carbon 1. Other sugars are epimers of glucose at other carbon atoms.

Epimers in which the difference in stereochemistry is at a potential carbonyl group in a cyclic hemiacetal are also called **anomers.** $α$-Glucose and $β$-glucose are, therefore, best defined as anomers of each other. *Carbon 1 is the anomeric carbon atom, bearing an anomeric hydroxyl group.*

A six-membered ring with an oxygen atom as part of the ring is related to the heterocyclic compound pyran.

pyran   α-D-glucopyranose   β-D-glucopyranose

A sugar in its six-membered cyclic form is called a **pyranose**. The names common in carbohydrate chemistry for the cyclic forms of glucose are α-D-glucopyranose and β-D-glucopyranose. The letter D defines the stereochemistry at carbon 5. The *gluco*-portion of the name defines the stereochemistry of the three asymmetric carbon atoms other than the anomeric one relative to carbon 5. Pyranose indicates that it is a six-membered cyclic structure, and α- and β- define the stereochemistry at the anomeric carbon atom in the cyclic form.

While we will use the chair form of the six-membered ring in talking about the stereochemistry and reactions of pyranoses, the six-membered ring is also widely represented in the planar form. This way of representing sugars is known as the **Haworth projection formula.** Such formulas are easily derived from the chair form.

α-D-glucopyranose   α-D-glucopyranose   β-D-glucopyranose
the chair form
                                      Haworth projection formulas for glucose

In the Haworth representations, the lower edge of the ring is defined as projecting out towards the viewer. The hydroxyl groups and other substituents are shown as being either above or below the plane of the ring.

The ring in cyclic hemiacetals does not have to be six membered. The hydroxyl group on carbon 4 in glucose is close enough to the carbonyl group to give a five-membered ring. Here the ring is related to the five-membered heterocycle furan. The sugar is called a glucofuranose in this form. The pyranose form is the most commonly observed in aldohexoses.

furan   α-D-glucofuranose   β-D-glucofuranose

Ketohexoses also give pyranose and furanose structures. The only crystalline form of fructose that is isolated is β-D-fructopyranose with an initial specific rotation of −133.5°. It undergoes rapid mutarotation to −92°. This mutarotation involves not only isomerization between the β- and α-pyranose forms, but conversion to the furanose forms as well (Figure 14.2). The fructofuranose structure is important because fructose is present in sucrose and in other oligosaccharides as the furanose form.

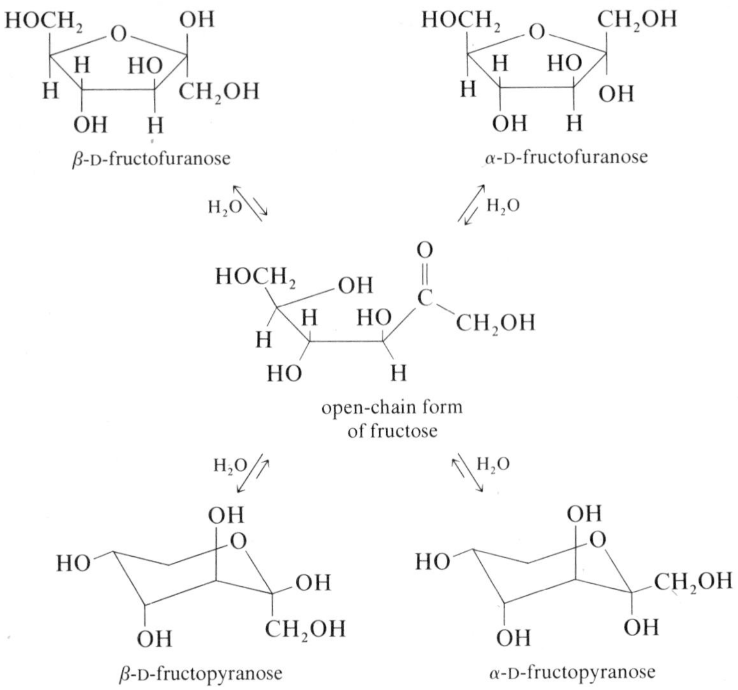

Figure 14.2 The mutarotation of fructose.

Ribose, an aldopentose, exists as the β-D-ribofuranose structure in compounds such as ribonucleic acids.

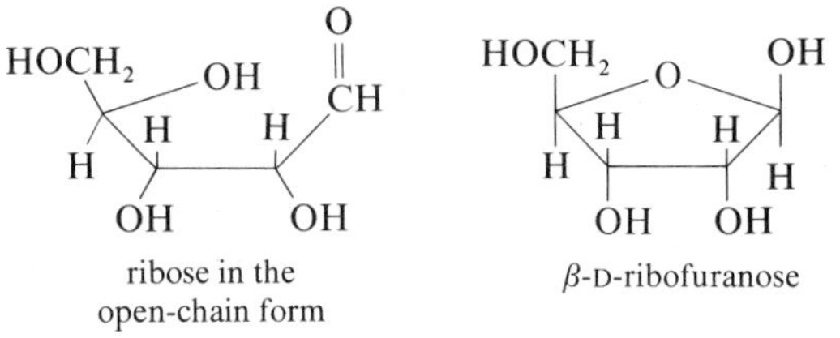

PROBLEM 14.10 In aqueous solution at equilibrium, 76% of ribose exists in a pyranose form. Write the structure of β-D-ribopyranose.

14.4

Reactions of Monosaccharides as Carbonyl Compounds

A. Glycoside Formation

When glucose is heated with methanol containing a little hydrogen chloride, two isomeric acetals are formed. In these compounds the hemiacetal function has been converted into the cyclic monomethyl acetal. Of the two isomers, the product with the methoxyl group axial, the α-isomer, is about 98% of the mixture. This is true whether one starts with α- or β-glucose.

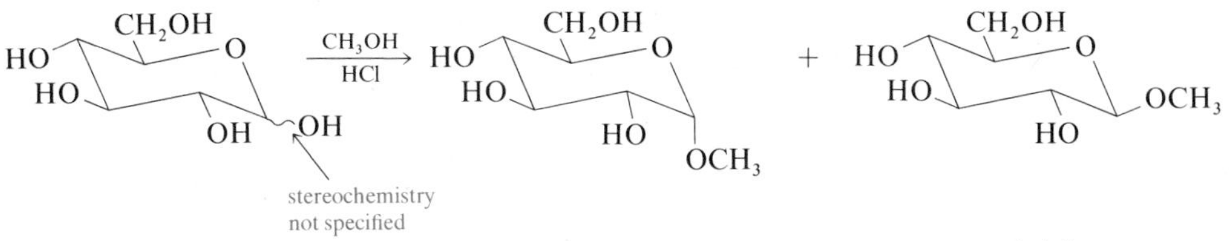

stereochemistry
not specified

α- or β-
D-glucopyranose

methyl α-D-
glucopyranoside
mp 166 °C
[α] +158°

major product

methyl β-D-
glucopyranoside
mp 105 °C
[α] −34°

minor product

As a class, carbohydrates with a full acetal linkage are known as **glycosides.** The compounds shown above are methyl glucosides, or methyl glucopyranosides to reflect their cyclic nature.

PROBLEM 14.11 Show the mechanism for the formation of methyl α-D-glucopyranoside from α-D-glucopyranose. How does the mechanism that you propose explain the fact that both α- and β-glucopyranoses give the same mixture of methyl pyranosides? (Hint: Reviewing Section 9.7B may help you to answer the question.)

Methyl glucopyranosides do not undergo mutarotation, nor do they show any of the aldehyde reactions that glucose exhibits. For example, they cannot be oxidized easily to a carboxylic acid. In the full acetal form, the carbonyl group is effectively protected.

The glucosides are stable in basic solutions. They are hydrolyzed easily in acidic solutions to give the equilibrium mixture of α-D-glucopyranose and β-D-glucopyranose.

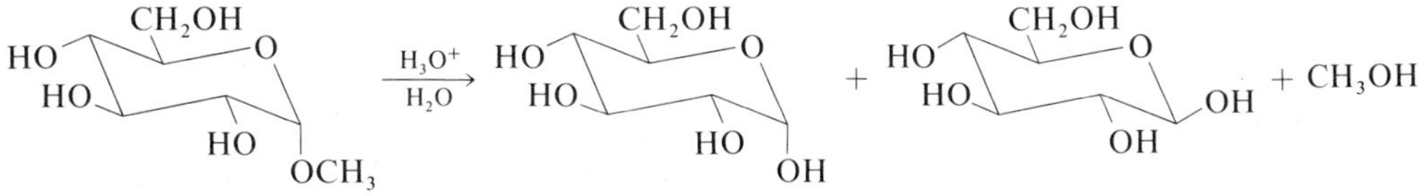

methyl α-D-
glucopyranoside

α-D-glucopyranose

β-D-glucopyranose

Enzymes, which are stereoselective biological catalysts, can be used to hydrolyze glucosides selectively. The enzyme maltase will cleave only α-glucosides, while the enzyme emulsin cleaves only β-glucosides.

PROBLEM 14.12 Give the detailed mechanism (Section 9.7B) for the hydrolysis of methyl α-D-glucopyranoside in dilute acid.

The glycosidic linkage occurs widely in nature. Such bonding between the anomeric carbon of one monosaccharide and a hydroxyl group of another is the way oligosaccharides and polysaccharides are formed. In plants, saccharides are also found bonded to a large number of alcohols and phenols in a variety of natural products of medicinal and other practical value. The hydroxyl compounds found bonded to sugars in glycosides are called **aglycons.**

An example of a glycoside with a long medicinal history is salicin, found in the bark of the willow tree. Salicin is the β-glycoside of o-(hydroxymethyl)phenol.

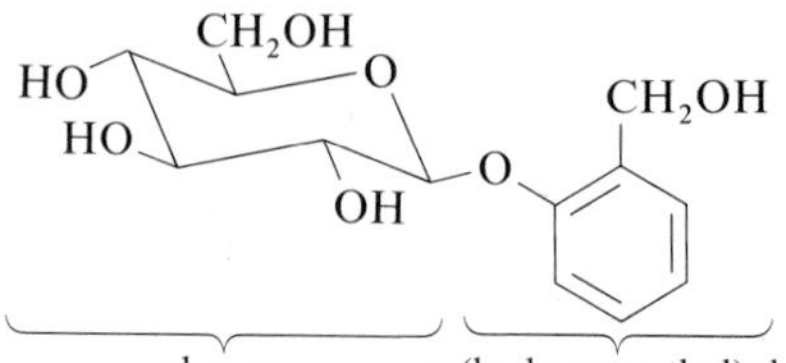

glucose o-(hydroxymethyl)phenol

a sugar   *an aglycon*

salicin, a naturally occurring glycoside

Willow bark preparations have been known since the time of the ancient Greeks as pain relievers. They were usually used externally because the willow juice is so bitter. Chemists sought to isolate the compound that gave willow its analgesic properties. The active component, salicin, was finally isolated from other plant sources and converted to salicylic acid. Salicylic acid itself has valuable medicinal properties, but it cannot be taken internally. Finally, in 1899, salicylic acid was converted into its acetyl derivative, acetylsalicylic acid, now commonly known as aspirin.

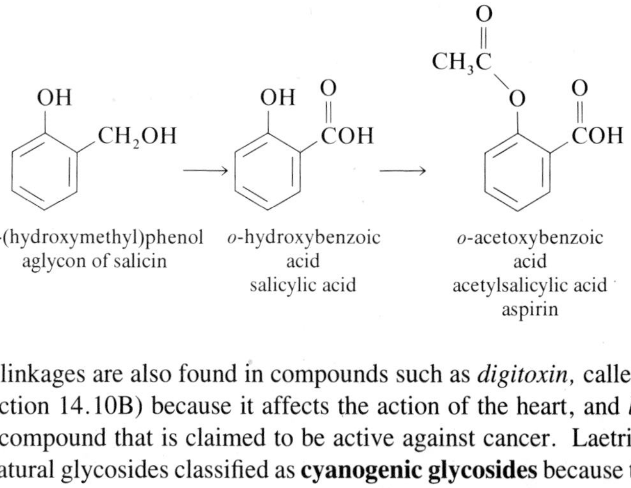

o-(hydroxymethyl)phenol o-hydroxybenzoic o-acetoxybenzoic
aglycon of salicin   acid   acid
                                    salicylic acid   acetylsalicylic acid
                                                                    aspirin

Glycosidic linkages are also found in compounds such as *digitoxin,* called a **cardiac glycoside** (Section 14.10B) because it affects the action of the heart, and *laetrile,* the controversial compound that is claimed to be active against cancer. Laetrile is one of a number of natural glycosides classified as **cyanogenic glycosides** because they release hydrogen cyanide when hydrolyzed either by acid or by enzymes and hence have considerable toxicity. The aglycon in laetrile is (R)-$(-)$-mandelonitrile.

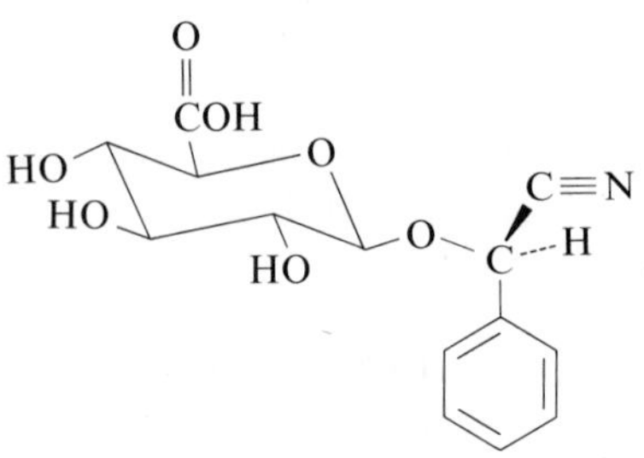

glucuronic (R)-$(-)$-mandelonitrile
acid

laetrile

PROBLEM 14.13   Outline the transformations necessary to convert salicin to salicylic acid and salicylic acid to aspirin.

PROBLEM 14.14   The compound arbutin is a glycoside isolated from the bearberry. It has found use as a diuretic (a medication that increases the production of urine) and antiseptic for the urinary tract. It can be hydrolyzed by emulsin (p. 757) to glucose and hydroquinone (p. 624). What is the structure of arbutin?

B. Glycosylamines

If a sugar is heated with an amine in the presence of a trace of acid, a *glycosidic linkage to nitrogen* is formed. *Such compounds are called* **glycosylamines.** The reaction works particularly well with aromatic amines.

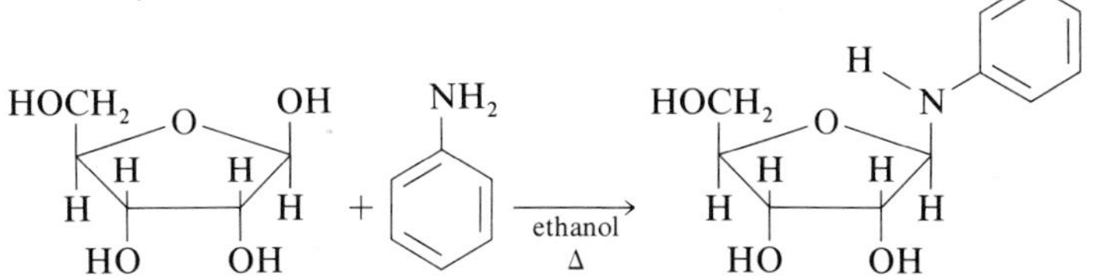

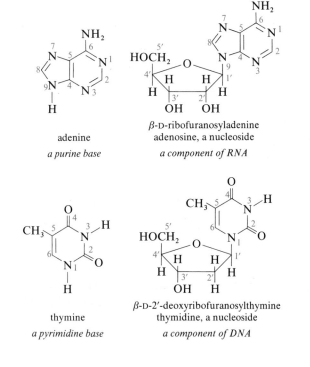

β-D-ribofuranose   aniline   *N*-phenyl-*β*-D-ribofuranosylamine

This kind of linkage between an amine and a sugar is biologically significant. *The structural units of ribonucleic acids (RNA) and deoxyribonucleic acids (DNA) are glycosylamines known as* **nucleosides.**

In nucleosides, the amine nitrogens are found in heterocyclic compounds known as pyrimidine and purine bases (Section 18.6A). The nitrogen is bonded to the anomeric carbon of D-ribofuranose or of 2-deoxy-D-ribofuranose.

adenine

a purine base

β-D-ribofuranosyladenine
adenosine, a nucleoside

a component of RNA

thymine

a pyrimidine base

β-D-2′-deoxyribofuranosylthymine
thymidine, a nucleoside

a component of DNA

These structural units are bonded to each other as the phosphoric acid esters of the sugar portions of the molecules. Such esters are discussed in Section 14.5B and the polymeric RNA and DNA molecules, in Section 18.6B.

C. Osazone Formation. Configurational Relationships Among Monosaccharides

Sugars have high solubility in water because they contain large numbers of hydroxyl groups that hydrogen bond strongly to water. Sugars tend to form syrups so that crystallization of the compounds is difficult. Emil Fischer discovered early in his research that sugars reacted with phenylhydrazine to give beautiful yellow crystalline compounds containing two phenylhydrazine residues. He named these compounds **osazones,** from *ose* for sugar and *azone* for phenylhydrazone.

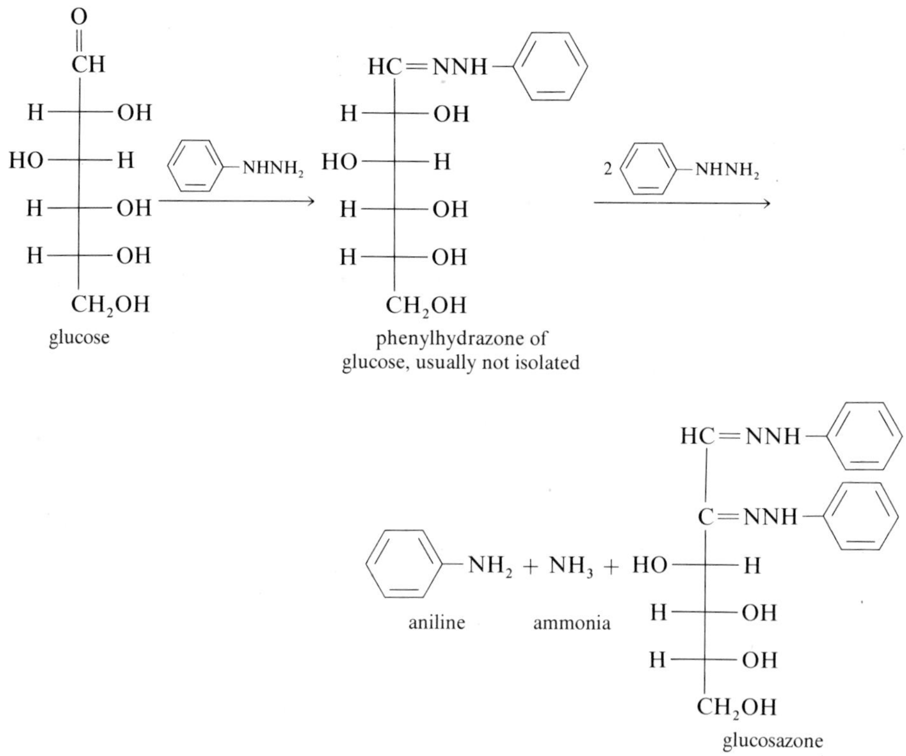

glucose

phenylhydrazone of
glucose, usually not isolated

glucosazone

aniline   ammonia

The phenylhydrazone of glucose does form as an intermediate in the production of osazones, but it is difficult to isolate unless one limits the quantity of phenylhydrazine very carefully. The formation of the osazone involves an oxidation-reduction reaction, with the second carbon of glucose being oxidized from a secondary alcohol to a carbonyl group and one molecule of phenylhydrazine being reduced to aniline and ammonia.

The forms in which different osazones crystallize are so distinctive that older biochemistry textbooks very often have colored photographs showing how they appear under the microscope. The preparation of an osazone and the examination of the crystals under a microscope was used in the laboratory to identify common sugars.

Fischer found that glucose, fructose, and mannose all gave the same osazone. He recognized that this meant that all three of these sugars had the same configuration at carbons 3, 4, and 5. Fructose, he knew, was a 2-ketohexose, so it did not have stereochemistry at carbon 2. This meant that glucose and mannose must be epimers at carbon 2, the carbon atom that loses its stereochemical identity when the osazone is formed.

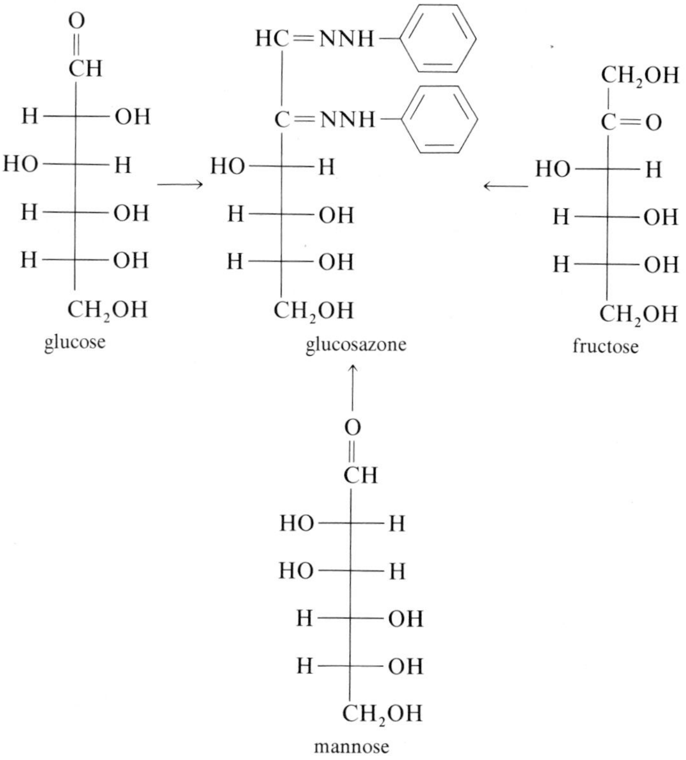

the configurational relationships of glucose,
fructose, and mannose

PROBLEM 14.15   Besides glucose and mannose, there is one other isomeric aldohexose commonly found in nature. It is D-(+)-galactose, which is an epimer of glucose at carbon 4. What is the structure of galactose? Would its osazone have the same structure as that of glucosazone?

PROBLEM 14.16   Using the information that mannose is epimeric with glucose at carbon 2 and galactose is epimeric at carbon 4, write structural formulas for α-D-mannopyranose and β-D-galactopyranose.

D. The Interconversion of Glucose, Mannose, and Fructose

Glucose, mannose, and fructose are converted to each other in the presence of bases. This transformation has the imposing name of the **Lobry de Bruyn–Alberda van Ekenstein rearrangement** for the two Dutch chemists who discovered the reaction at the end of the nineteenth century. If glucose is put into a solution of dilute sodium hydroxide, it is slowly transformed into fructose, mannose, and decomposition products.

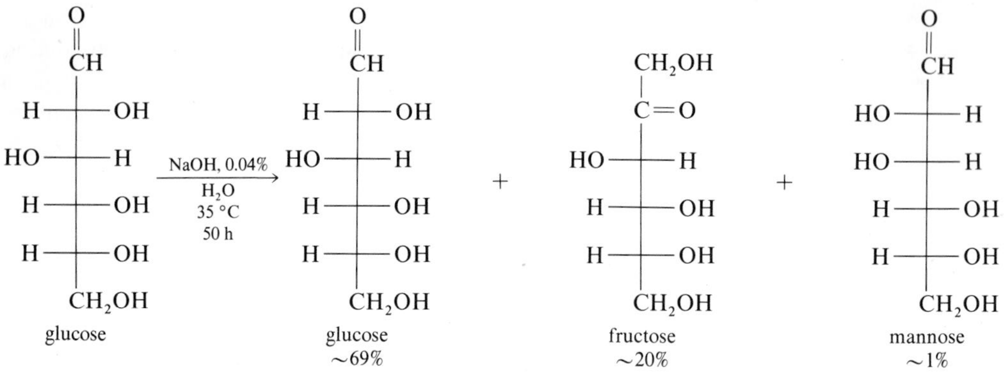

| glucose | glucose ~69% | fructose ~20% | mannose ~1% |

If the reaction is given more time, more fructose and mannose are formed, but the total recovery of hexoses falls to 70%.

The conversion of glucose to fructose is an important step in the process known as glycolysis by which glucose is metabolized in the body. This transformation is believed to proceed by an enolization reaction. The hydrogen atom on the carbon adjacent to the carbonyl group is acidic enough to be removed to give an enolate ion, which is protonated by water. The intermediate is a 1,2-enediol, where carbon atom 2 is now trigonal planar and has therefore lost its chirality.

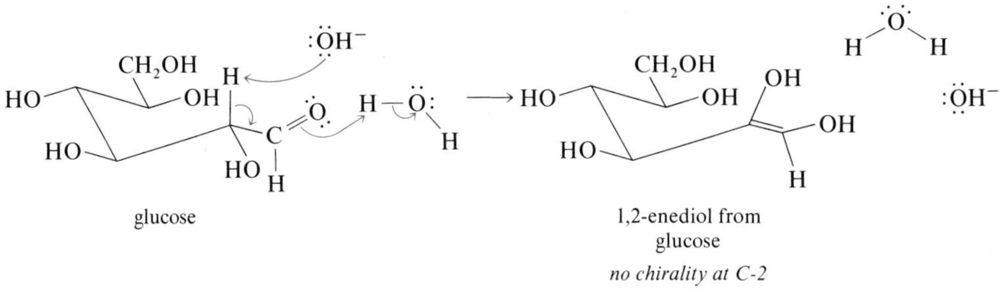

glucose

1,2-enediol from glucose

no chirality at C-2

The two protons on the enolic hydroxyl groups are acidic. Removal of the proton from the hydroxyl group on carbon 1 gives carbanion character at carbon 2. Protonation of the carbanion gives mannose or glucose, depending on the side of the molecule where the reaction takes place.

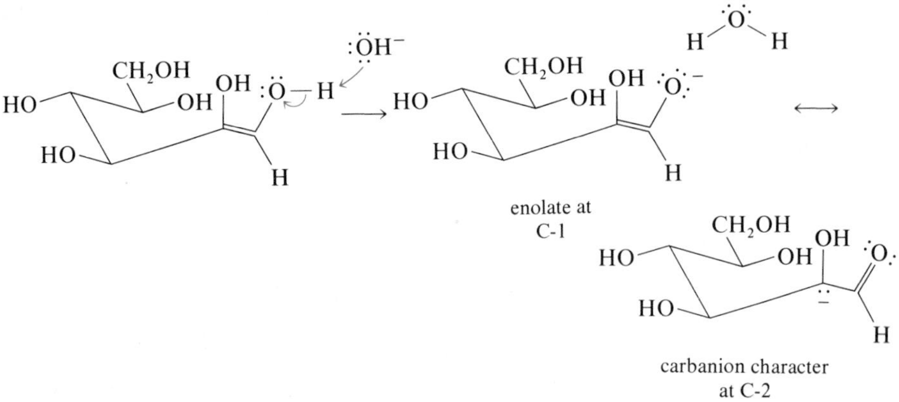

enolate at C-1

carbanion character at C-2

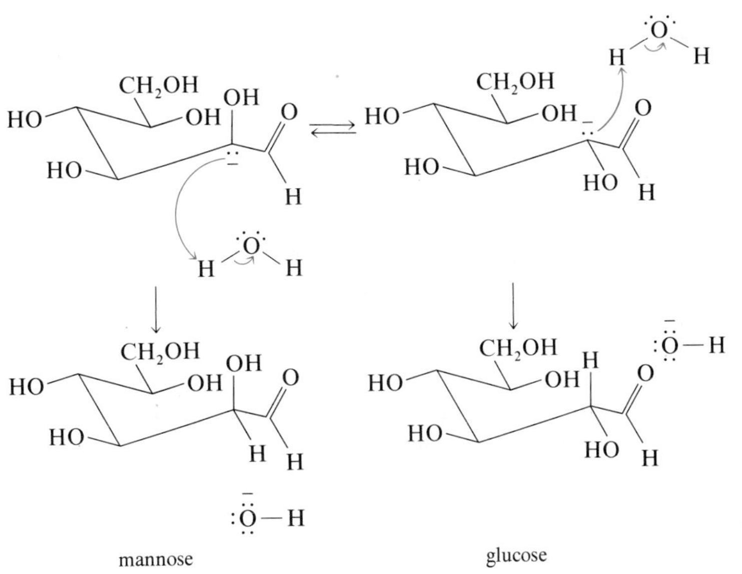

mannose   glucose

Removal of the proton from the hydroxyl group on carbon 2 yields the enolate ion that can be converted to fructose by protonation at carbon 1.

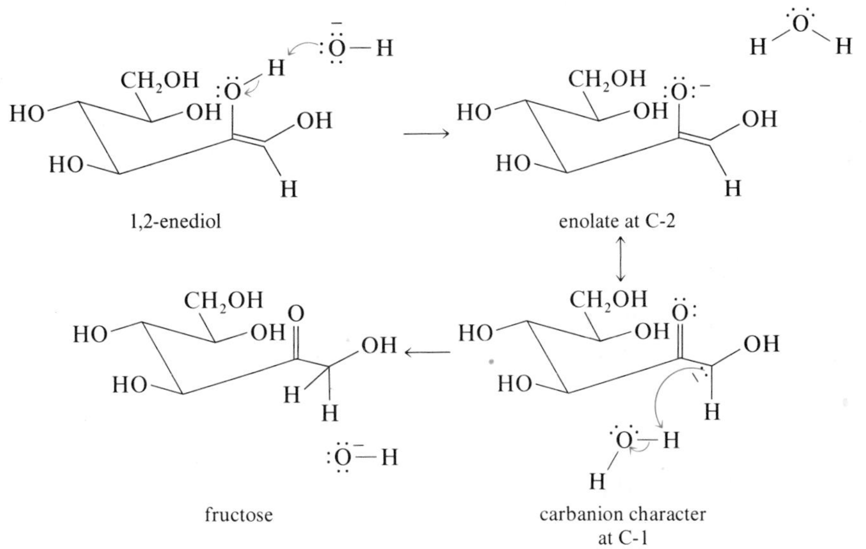

1,2-enediol   enolate at C-2

fructose   carbanion character
                                                                      at C-1

PROBLEM 14.17

(a) When the reaction described above was carried out in D_2O, it was found that mannose and glucose had deuterium attached to carbon 2, but in fructose deuterium was attached to carbon 1. Show how this evidence supports the mechanism proposed for the reaction.

(b) Mannose had 1.4 deuterium atoms attached to carbon per molecule of mannose. Fructose had an average of 1.7 deuterium atoms attached to carbon per molecule. Use the mechanism to explain how fructose could have more than one deuterium atom per molecule. What does the deuterium content of mannose suggest?

14.5

Reactions of Monosaccharides as Alcohols

A. Esterification

Carbohydrates contain many hydroxyl groups and thus will react with acid anhydrides (Section 10.7A) to give esters. The formation of a pentaacetate when glucose reacts with acetic anhydride has been given (p. 752) as evidence that glucose is an open-chain aldehyde with a hydroxyl group on each of the five other carbon atoms. At present, the pentaacetate of glucose is assigned a cyclic structure with an acetyl group at carbon 1 and not at carbon 5. This is an interesting example of how an experimental observation remains valid, while its interpretation may change with time and the accumulation of other experimental data.

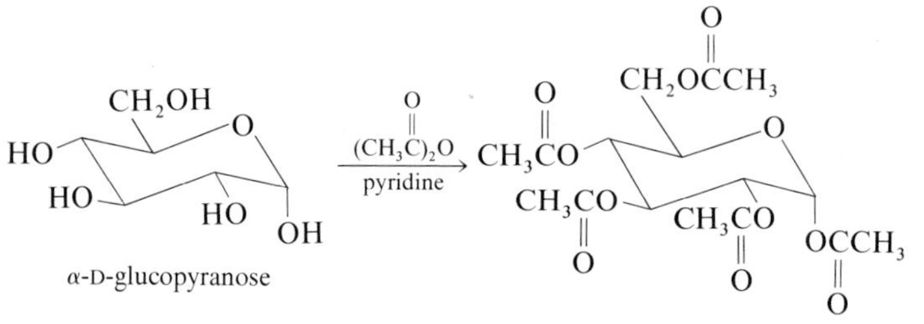

α-D-glucopyranose → α-D-glucopyranose pentaacetate

The ester functions in glucopyranose pentaacetate give normal ester reactions. For example, the acetyl groups can be hydrolyzed off with dilute base to regenerate glucose.

PROBLEM 14.18   Write the equation for the hydrolysis of glucose pentaacetate with dilute aqueous base. Write the mechanism for the reaction using

$$R-O-\overset{\overset{\displaystyle O}{\|}}{C}-CH_3$$

to represent glucose pentaacetate.

The acetoxy group at carbon 1 is more reactive than those at other carbons and can be replaced selectively. With hydrogen bromide in acetic acid, the α-bromide is formed.

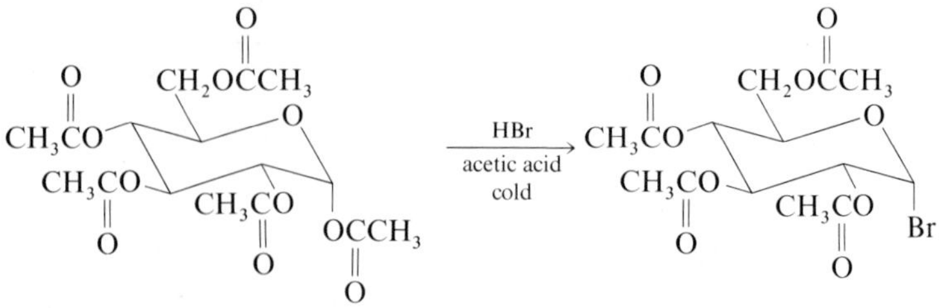

α-D-glucopyranose pentaacetate → α-D-glucopyranosyl bromide 2,3,4,6- tetraacetate

These bromides are important intermediates in the formation of glycosides. The synthesis of the β-glucoside, salicin (Section 14.4A), is an example of such a reaction.

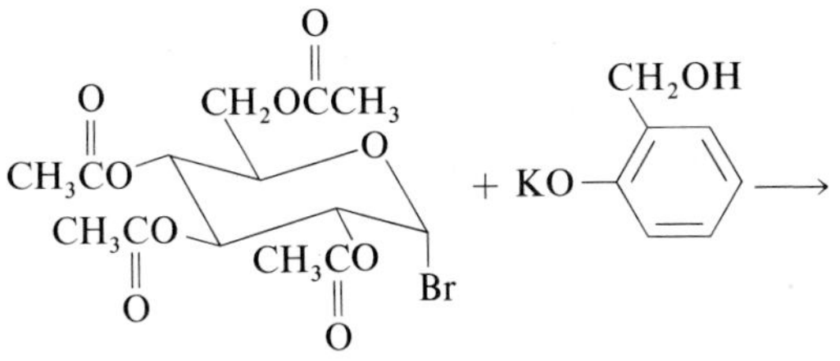

α-D-glucopyranosyl
bromide 2,3,4,6-tetraacetate

potassium
o-(hydroxymethyl)phenolate

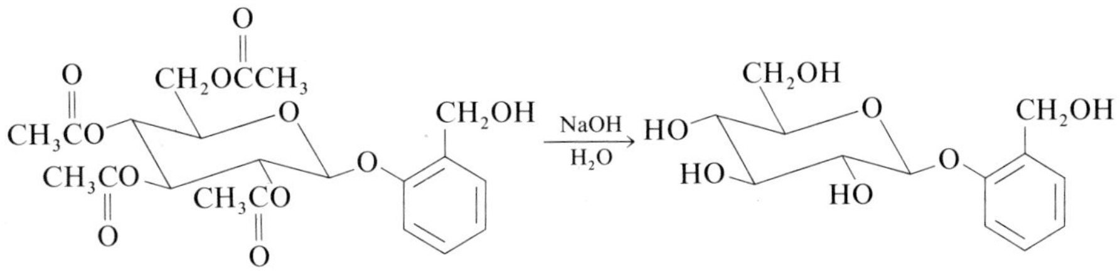

salicin tetraacetate   salicin

PROBLEM 14.19   Write the mechanism for the formation of the bromide at carbon 1 from glucose pentaacetate.

PROBLEM 14.20   Why is it necessary to use a basic hydrolysis to remove the acetate groups from salicin tetraacetate in the synthesis of salicin shown above?

PROBLEM 14.21   How would you classify the reaction in the first step of the preparation of salicin? What is the stereochemistry of the reaction?

B. Esters of Phosphoric Acid

In biological systems, the most important esters of sugars are the ones formed with phosphoric acid. There are three forms in which phosphoric acid appears in such systems. Two of these forms may be regarded as anhydrides of orthophosphoric acid, H_3PO_4.

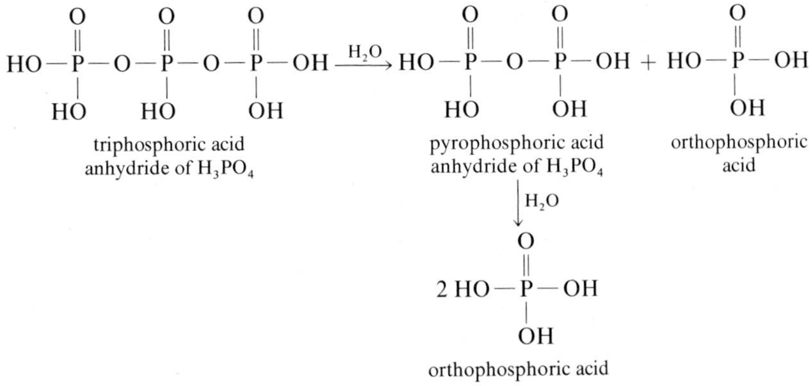

triphosphoric acid
anhydride of H_3PO_4

pyrophosphoric acid
anhydride of H_3PO_4

orthophosphoric
acid

orthophosphoric acid

All of these forms of phosphoric acid are seen in **nucleotides,** the phosphoric acid esters of nucleosides (Sections 14.4B and 18.6B). Adenosine triphosphate, the triphosphate ester at carbon 5 of ribose in adenosine (p. 759), is found extensively in living systems. Its presence is so synonymous with life as we know it that when experiments are designed to search for life in outer space, adenosine triphosphate is one of the key compounds sought. It stores chemical energy and makes it available to specific cell processes. It gives up energy by transferring a phosphate group to another molecule, which is then converted into a reactive form as its phosphate ester. A specific example that starts off the process of glycolysis, the metabolism of glucose in the body, is the conversion of glucose to glucose 6-phosphate.

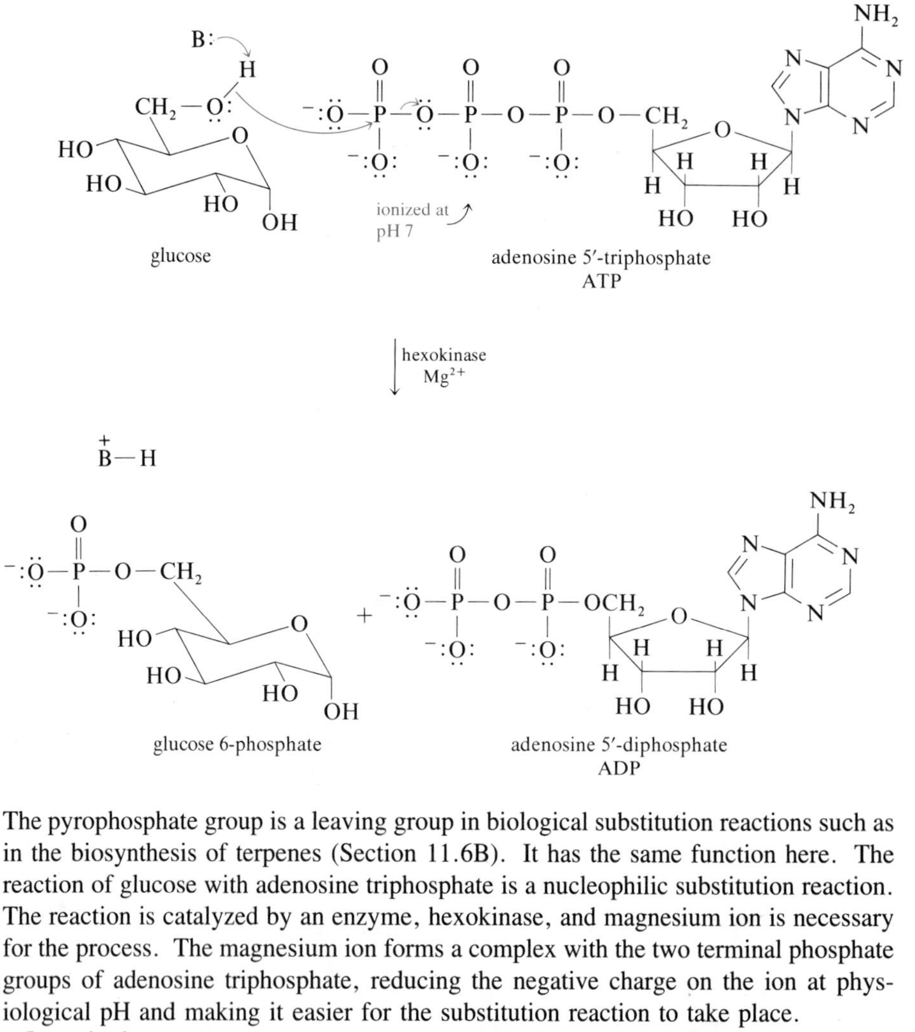

glucose

adenosine 5′-triphosphate
ATP

hexokinase
Mg^{2+}

glucose 6-phosphate

adenosine 5′-diphosphate
ADP

The pyrophosphate group is a leaving group in biological substitution reactions such as in the biosynthesis of terpenes (Section 11.6B). It has the same function here. The reaction of glucose with adenosine triphosphate is a nucleophilic substitution reaction. The reaction is catalyzed by an enzyme, hexokinase, and magnesium ion is necessary for the process. The magnesium ion forms a complex with the two terminal phosphate groups of adenosine triphosphate, reducing the negative charge on the ion at physiological pH and making it easier for the substitution reaction to take place.

Later in the process of glycolysis, a particularly reactive phosphate ester of an enol transfers a phosphate group back to the adenosine diphosphate and regenerates adenosine triphosphate. At this stage, some of the energy generated by the breakdown of glucose in the body is returned to storage in the triphosphate group. This reaction is also catalyzed by an enzyme in the body.

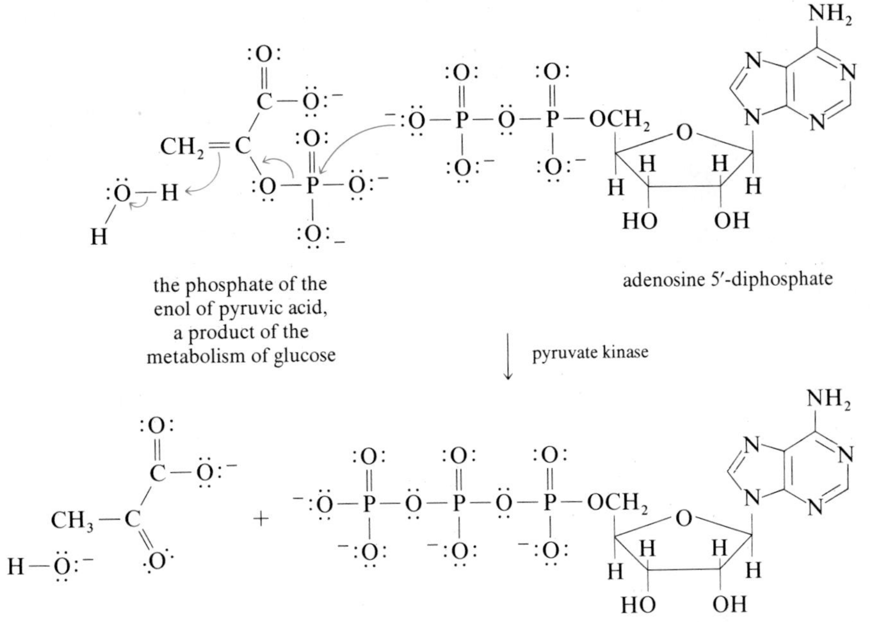

the phosphate of the
enol of pyruvic acid,
a product of the
metabolism of glucose

adenosine 5′-diphosphate

pyruvate kinase

pyruvate anion

adenosine 5′-triphosphate

The loss of the phosphate group from the enol oxygen takes place easily because the stable carbonyl group in pyruvic acid is generated by the reaction.

C. Ether Formation

If methyl α-D-glucopyranoside is treated with dimethyl sulfate in the presence of aqueous sodium hydroxide, the methyl ethers of the alcohol functions are formed. The base helps the ionization of the hydroxyl groups, creating good nucleophiles which then react by nucleophilic substitution with dimethyl sulfate.

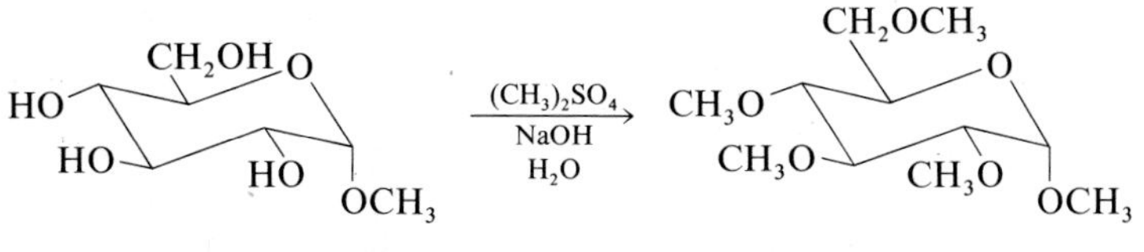

methyl α-D-glucopyranoside

methyl 2,3,4,6-*O*-
tetramethyl-α-D-glucopyranoside

The methyl sulfate anion is a good leaving group.

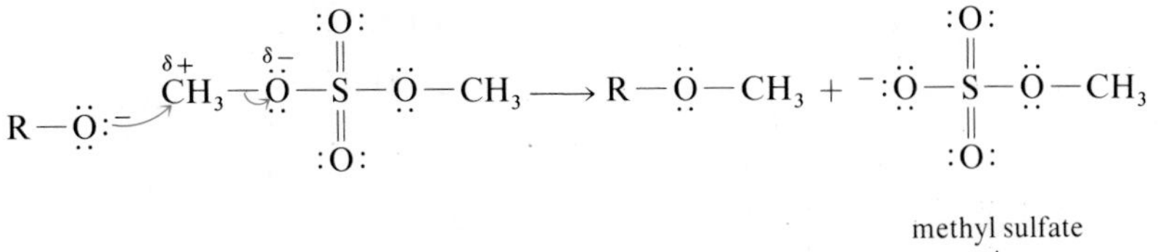

methyl sulfate
anion

The fully methylated sugar is named as a methyl glucopyranoside with four more methyl groups substituted on the oxygens at carbons 2, 3, 4, and 6. The name is therefore methyl 2,3,4,6-O-tetramethyl-α-D-glucopyranoside, with the capital O telling us that the substituents named are on oxygen, not carbon.

The methyl ethers formed are stable to bases and dilute acids and serve to protect the hydroxyl groups of the carbohydrate. The methoxyl group at the anomeric carbon is different in its reactivity from the other ether groups in the molecule. It is part of an acetal linkage and is therefore sensitive to acid hydrolysis. That one methyl group can be removed while leaving the others in place.

methyl 2,3,4,6-O-
tetramethyl-α-D-glucopyranoside

2,3,4,6-O-tetramethyl-
α-D-glucopyranose

This is possible because protonation of the oxygen atom of the methoxyl group and the loss of the methanol at the anomeric carbon results in a carbocation that is particularly well stabilized by delocalization of charge to the adjacent oxygen atom (Section 9.7B). Such stabilization is not possible for the cations that would result from the loss of the other methoxyl groups, so much more drastic conditions are necessary to remove them.

PROBLEM 14.22 Show the mechanism for the hydrolysis in dilute acid of methyl tetramethylglucopyranoside. Convince yourself that no special stabilization is possible for carbocations formed by the loss of a methoxyl group from carbon atoms other than the anomeric carbon atom.

PROBLEM 14.23 The methyl groups on the oxygens at carbons 2, 3, 4, and 6 of tetramethylglucopyranose can be removed by boiling with 57% hydriodic acid. One of the products of the reaction is iodomethane. Using ROCH$_3$ to represent the methyl ethers, write a mechanism for the removal of the methyl groups. Why is it possible to cleave an ether under these conditions, but not with dilute hydrochloric acid, for example?

The removal of the methyl group at the anomeric carbon regenerates the hemiacetal linkage. The ring is once more in equilibrium with the open-chain form and mutarotation takes place. Notice that the hydroxyl group on carbon 5 is not methylated and retains its properties as a secondary alcohol.

2,3,4,6-O-tetramethyl-
α-D-glucopyranose

open-chain
form

2,3,4,6-O-tetramethyl-
β-D-glucopyranose

In the next section, we will see how oxidation reactions involving that hydroxyl group can be used to confirm the size of the ring in pyranoses.

14.6

Oxidation Reactions of Sugars

A. Reducing Sugars. Aldonic Acids

Sugars are usefully classified as *reducing sugars* and *nonreducing sugars*. *Sugars such as glucose and fructose are* **reducing sugars.** *This means that they are easily oxidized by mild oxidizing agents.* The test for diabetes, in which urine is tested for glucose, is really a test for the presence of a reducing sugar.

Sugars in which the carbonyl group is tied up in an acetal linkage are **nonreducing sugars.** Methyl glucopyranoside is an example of a nonreducing sugar. Disaccharides, such as sucrose, in which the anomeric carbon atoms of both monosaccharide units are bonded to each other so that both carbonyl groups are protected, are also nonreducing sugars.

The aldehyde function in glucose, like most aldehydes, is oxidized by very mild oxidizing agents such as **Benedict's reagent,** which is a solution in aqueous base of copper(II) sulfate and sodium citrate (used to complex with the copper ion). This reaction is the traditional one used for the diabetes test.

$$
\begin{array}{c}
\overset{O}{\overset{\|}{C}}H \\
H\!-\!\!-\!OH \\
HO\!-\!\!-\!H \\
H\!-\!\!-\!OH \\
H\!-\!\!-\!OH \\
CH_2OH
\end{array}
\;\; + \; 2\,Cu^{2+} + 5\,OH^- \;\longrightarrow\;
\begin{array}{c}
\overset{O}{\overset{\|}{C}}\!-\!O^- \\
H\!-\!\!-\!OH \\
HO\!-\!\!-\!H \\
H\!-\!\!-\!OH \\
H\!-\!\!-\!OH \\
CH_2OH
\end{array}
\;\; + \; Cu_2O\!\downarrow + 3\,H_2O
$$

glucose   (blue)   anion of gluconic acid   (red)

Silver ion, in the presence of base, Tollens reagent (Section 9.9), will also serve as an oxidizing agent for reducing sugars. Glucose reduces silver ion to metallic silver.

$$
\begin{array}{c}
\overset{O}{\overset{\|}{C}}H \\
H\!-\!\!-\!OH \\
HO\!-\!\!-\!H \\
H\!-\!\!-\!OH \\
H\!-\!\!-\!OH \\
CH_2OH
\end{array}
\;\; + \; 2\,Ag(NH_3)_2{}^+ + 2\,OH^- \;\longrightarrow\;
\begin{array}{c}
\overset{O}{\overset{\|}{C}}O^-\,NH_4{}^+ \\
H\!-\!\!-\!OH \\
HO\!-\!\!-\!H \\
H\!-\!\!-\!OH \\
H\!-\!\!-\!OH \\
CH_2OH
\end{array}
\;\; + \; 2\,Ag\!\downarrow + NH_3 + H_2O
$$

glucose   ammonium gluconate

Both of these tests are also given by fructose. α-Hydroxyketones in general are oxidized very easily to diketones and react with Benedict's reagent or in the Tollens test.

$$R-\underset{\underset{OH}{|}}{C}H-\overset{\overset{O}{\|}}{C}-R' \xrightarrow[\substack{or \\ Ag(NH_3)_2{}^+}]{Cu^{2+}/OH^-} R-\overset{\overset{O}{\|}}{C}-\overset{\overset{O}{\|}}{C}-R'$$

Bromine in water is a mild oxidizing agent that will distinguish a ketose from an aldose. Only the aldehyde function is oxidized by this reagent. Thus, mannose is converted into the corresponding carboxylic acid, mannonic acid.

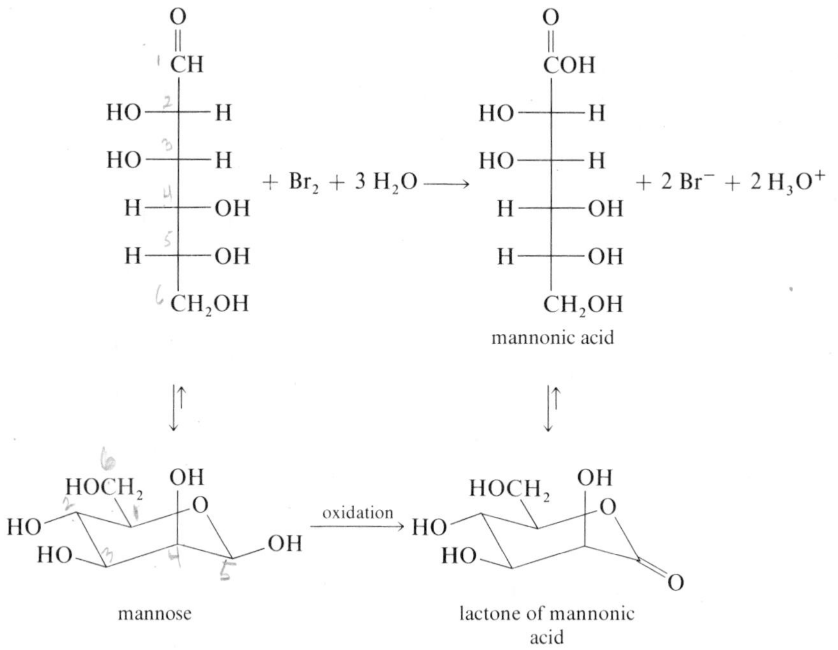

In acidic solutions, the hydroxyacid exists as a lactone (Section 10.7H).

The class of compounds in which carbon 1 of an aldose has been oxidized to a carboxylic acid is known as **aldonic acids.** In naming them, the **-ose** ending of the name of the sugar is converted into the **-onic acid** ending. Thus gluc*ose* gives gluc*onic acid* and mann*ose,* mann*onic acid.*

B. Nitric Acid as an Oxidizing Agent.
Aldaric Acids

If a more powerful oxidizing agent such as hot nitric acid is used, the sugar is oxidized at both ends of the chain to the dicarboxylic acid. Such acids are named as **aldaric acids.** Glucose is oxidized by nitric acid to glu**caric acid.**

$$
\begin{array}{ccc}
& \overset{\displaystyle O}{\overset{\displaystyle \|}{CH}} & \\
H\!\!-\!\!|\!\!-\!\!OH & & \\
HO\!\!-\!\!|\!\!-\!\!H & \xrightarrow[\Delta]{HNO_3} & \\
H\!\!-\!\!|\!\!-\!\!OH & & \\
H\!\!-\!\!|\!\!-\!\!OH & & \\
& CH_2OH &
\end{array}
\qquad
\begin{array}{c}
\overset{\displaystyle O}{\overset{\displaystyle \|}{COH}} \\
H\!\!-\!\!|\!\!-\!\!OH \\
HO\!\!-\!\!|\!\!-\!\!H \\
H\!\!-\!\!|\!\!-\!\!OH \\
H\!\!-\!\!|\!\!-\!\!OH \\
\underset{\underset{\displaystyle O}{\displaystyle \|}}{COH}
\end{array}
$$

glucose   glucaric acid

This reaction has played a large part in the determination of stereochemistry for sugars. In this oxidation reaction, carbon 1 and carbon 6 of the sugar are converted into the same functional group. Thus, any symmetry present in the rest of the molecule becomes apparent. The determination of optical activity for both glucaric acid and mannaric acid played an important part in the proof of the structure of glucose. Galactaric acid, derived from the other common natural aldohexose, has no optical activity.

$$
\begin{array}{ccc}
& \overset{\displaystyle O}{\overset{\displaystyle \|}{CH}} & \\
H\!\!-\!\!|\!\!-\!\!OH & & \\
HO\!\!-\!\!|\!\!-\!\!H & \xrightarrow[\Delta]{HNO_3} & \\
HO\!\!-\!\!|\!\!-\!\!H & & \\
H\!\!-\!\!|\!\!-\!\!OH & & \\
& CH_2OH &
\end{array}
\qquad
\begin{array}{c}
\overset{\displaystyle O}{\overset{\displaystyle \|}{COH}} \\
H\!\!-\!\!|\!\!-\!\!OH \\
HO\!\!-\!\!|\!\!-\!\!H \\
HO\!\!-\!\!|\!\!-\!\!H \\
H\!\!-\!\!|\!\!-\!\!OH \\
\underset{\underset{\displaystyle O}{\displaystyle \|}}{COH}
\end{array}
\qquad
\begin{array}{l}
\text{plane of} \\
\text{symmetry}
\end{array}
$$

galactose   galactaric acid

a meso compound

It is a meso form with a plane of symmetry between carbons 3 and 4. These experimental facts do not prove the structures of galactose, glucose, and mannose, of course. They simply limit the number of structures that are possible.

PROBLEM 14.24   Draw structural formulas for any other D-aldohexose(s) that would be oxidized to a meso aldaric acid.

C. Oxidations with Periodic Acid

An oxidizing agent ideally suited to carbohydrate chemistry is periodic acid, usually given the formula HIO_4. It exists in water solutions, as the dihydrated form, H_5IO_6. The reagent selectively cleaves the carbon-carbon bond in 1,2-diols, α-hydroxyaldehydes, and α-hydroxyethers. The reaction is carried out at or below room temperature in water solutions, an ideal solvent for carbohydrates, which are not very soluble in nonpolar organic solvents.

The reaction proceeds through a cyclic intermediate, as illustrated for *cis*-1,2-cyclopentanediol.

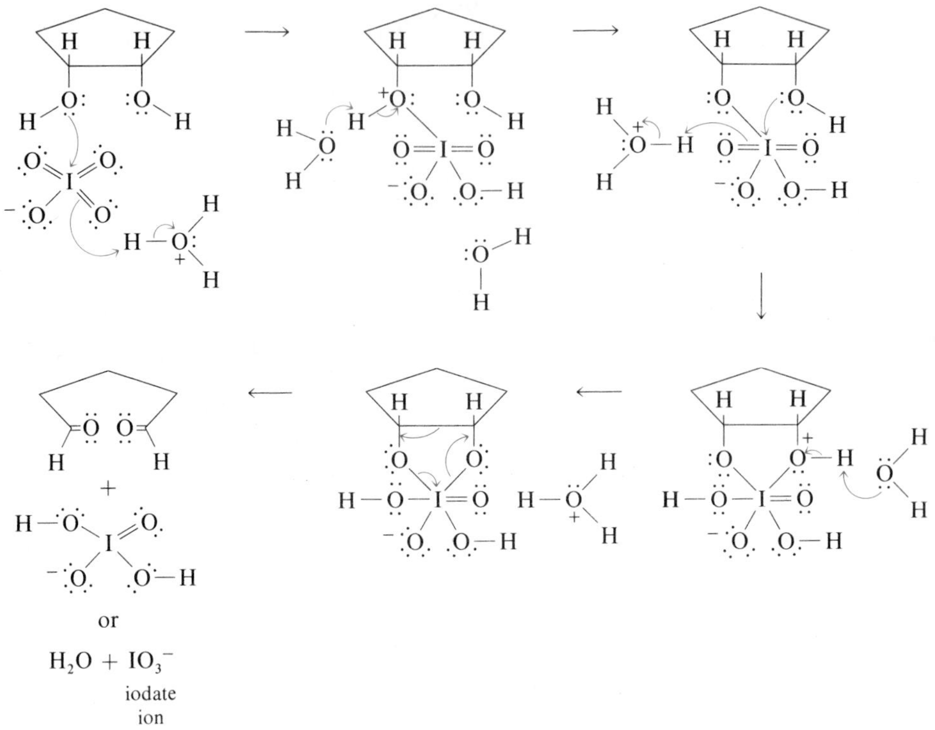

or

$$H_2O + IO_3^-$$

iodate
ion

As would be expected with such a cyclic intermediate, cis diols react faster than trans diols. The secondary alcohol groups in 1,2-cyclopentanediol end up as aldehyde functions in the product. Overall, the hydroxyl groups are converted to carbonyl groups as a result of cleavage of a carbon-carbon bond.

A polyhydroxy compound such a glucose reacts to give many fragments. Carbons 1 to 5 end up as formic acid and carbon 6 appears as formaldehyde when degradation is complete.

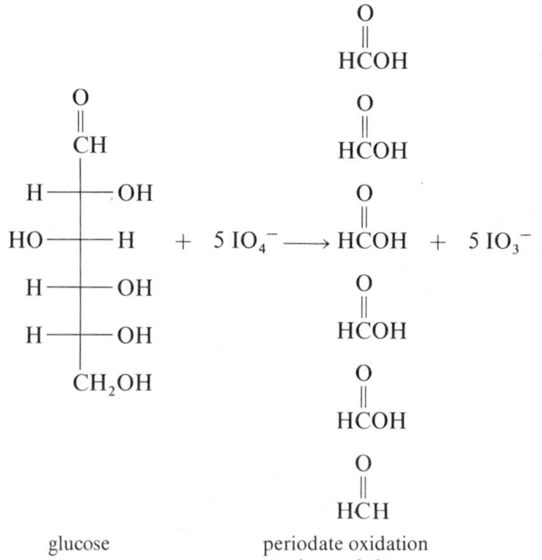

glucose

periodate oxidation
products of glucose

An interesting question is how the carbonyl group reacts with periodic acid. Isotopic labeling studies have shown that periodate ion attacks the carbonyl group directly as a nucleophile.

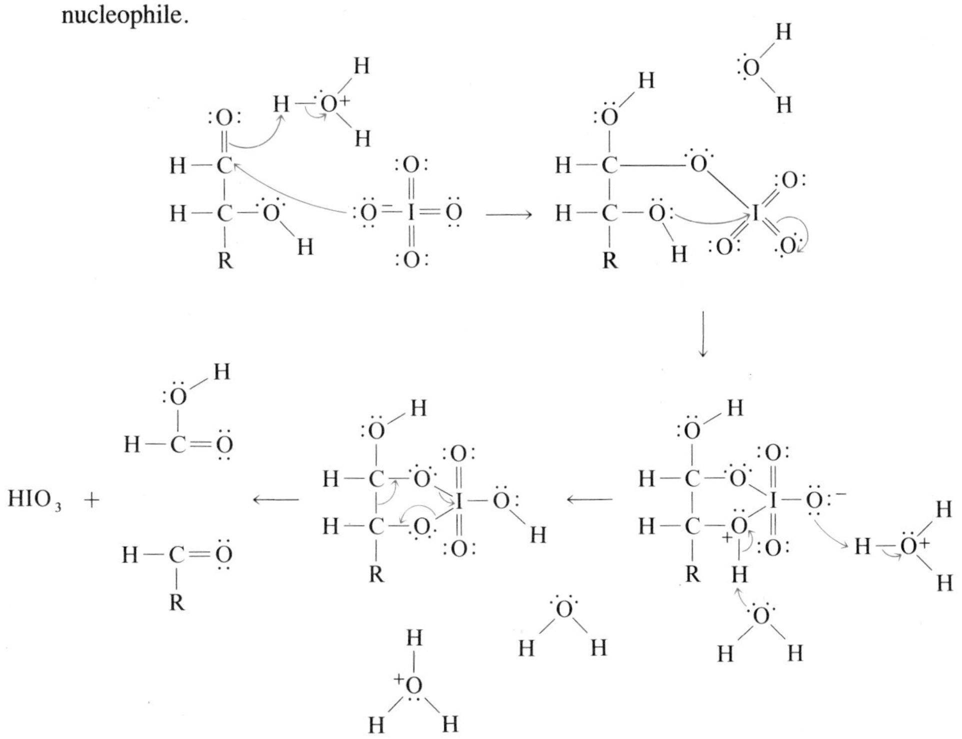

You should remember that the carbonyl group is oxidized to a carboxylic acid. Cleavage of a diol gives two carbonyl groups, which undergo further oxidation to acids if hydroxyl groups are adjacent to them. Only the last carbon atom of glucose, carbon 6, remains at the carbonyl stage as formaldehyde.

PROBLEM 14.25   Write out the mechanism for the periodate oxidation of a portion of the glucose molecule to be sure that you understand why the observed products are formed.

PROBLEM 14.26   Another possible mechanism for the periodate oxidation of α-hydroxyketones would be to postulate the formation of the hydrate of the carbonyl group to give a 1,2-diol that would then be oxidized in the usual way by periodate ion.

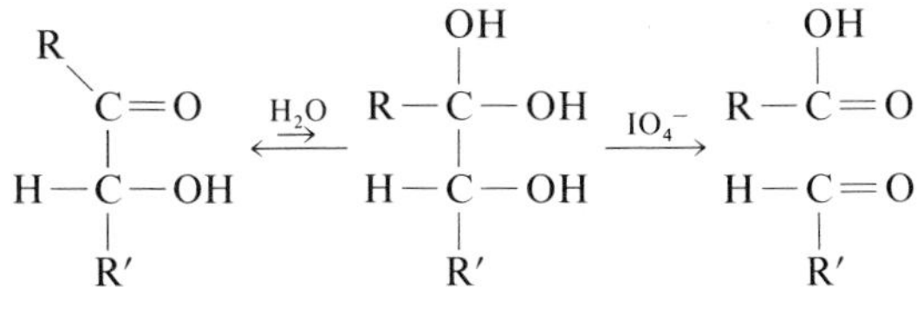

Two kinds of labeling experiments were done to determine the mechanism of the periodate oxidation of α-hydroxyketones. In one, water labeled at the oxygen, $H_2{}^{18}O$, was used as the solvent. In the other, labeled periodate ion, $I{}^{18}O_4{}^-$, was used. Look at the two possible mechanisms that have been described on this page and predict what observations in these experiments would support the mechanism that postulates direct nucleophilic attack by periodate ion on the carbonyl group.

Carboxylic acids do not participate in the oxidation. Glucaric acid, for example, reacts with only three moles of periodate ion.

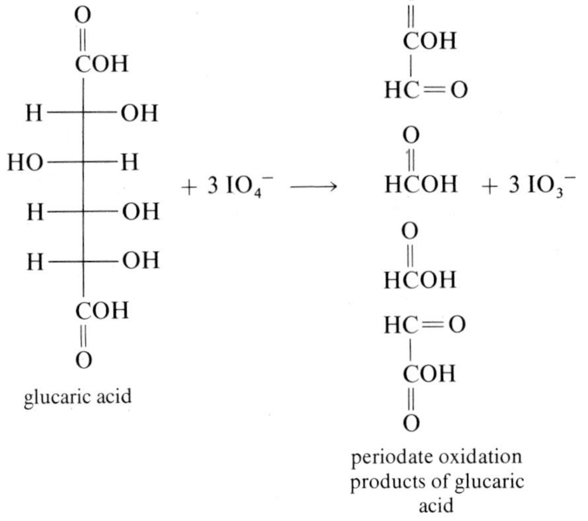

glucaric acid $+\ 3\ IO_4^-\ \longrightarrow$ periodate oxidation products of glucaric acid $+\ 3\ IO_3^-$

D. Oxidation Reactions Used to Establish Relative Configuration in Monosaccharides

The periodic acid oxidation of methyl glycosides was used to establish the size of the ring in the cyclic structure of a sugar and to correlate stereochemistry in different members of a series. Claude S. Hudson and his colleagues at the National Institutes of Health did elegant experimental work in this area in the 1930s and 1940s. They showed that methyl α-D-glucopyranoside and methyl α-D-mannopyranoside were both oxidized to the same dialdehyde. Note that the ring structure has only two adjacent diol units in the glucopyranoside and mannopyranoside. Formic acid is formed from carbon 3 in both cases.

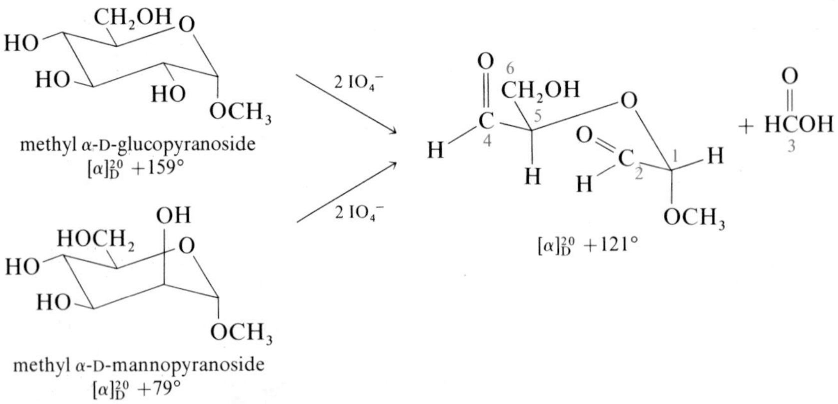

methyl α-D-glucopyranoside
$[\alpha]_D^{20}\ +159°$

methyl α-D-mannopyranoside
$[\alpha]_D^{20}\ +79°$

$2\ IO_4^-$

$[\alpha]_D^{20}\ +121°$ $+\ HCOH$

Stereochemistry at carbons 2, 3, and 4 is destroyed. The only asymmetric carbon atoms left in the dialdehyde are carbons 1 and 5. All the compounds assigned the methyl α-D-glycoside structure give the same dialdehyde, proving that they all have the same configuration at carbon 5, the chiral center that determines their membership in the D-family, and at carbon 1, which determines whether or not they are α-glycosides.

PROBLEM 14.27   Write out the equation necessary to prove to yourself that methyl α-D-galacto-pyranoside would give the same dialdehyde as the corresponding gluco- and mannopyranosides.

The dialdehydes formed from these oxidation reactions are actually not good com-pounds for structure determinations. They are unstable and form hydrates reversibly with water. In many cases, they give cyclic hydrates, creating new chiral centers, which complicate the assignment of stereochemistry.

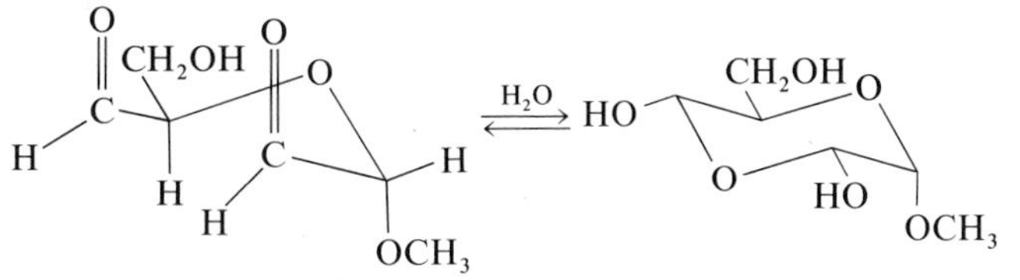

dialdehyde from IO_4^-
oxidation of α-D-glycosides

cyclic hydrate
of dialdehyde

For these reasons, Hudson made further structural comparisons by oxidizing the di-aldehydes to carboxylic acids with bromine water in the presence of strontium carbon-ate to keep the reaction mixture from getting acidic.

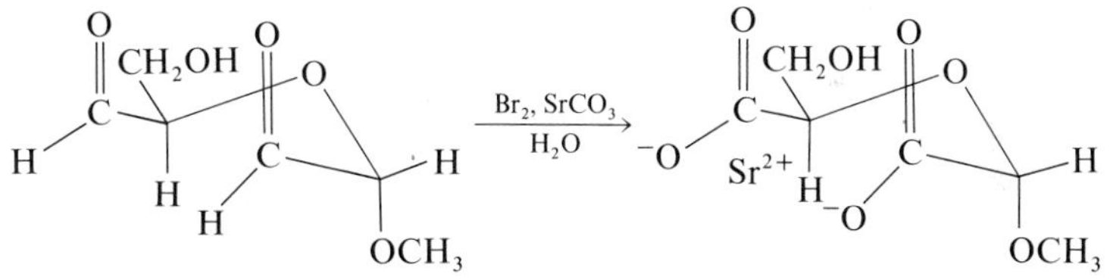

dialdehyde from periodate
oxidation of α-D-glycosides

dicarboxylic acid from
the oxidation of dialdehyde,
isolated as the strontium
salt

The same strontium salt is isolated from methyl α-D-glucopyranoside and from methyl α-D-mannopyranoside, once again proving the identity of stereochemistry at carbons 1 and 5 of the compounds.

PROBLEM 14.28   Strontium carbonate has a low solubility in water. How does it keep the solution from turning acidic? Why is it necessary to keep the system from getting acidic in this reaction?

Acid hydrolysis of the strontium salt of the dicarboxylic acid cleaves the acetal linkage. The fragments formed are oxidized with bromine water. ($-$)-Glyceric acid and oxalic acid are isolated and identified, usually as their calcium or barium salts.

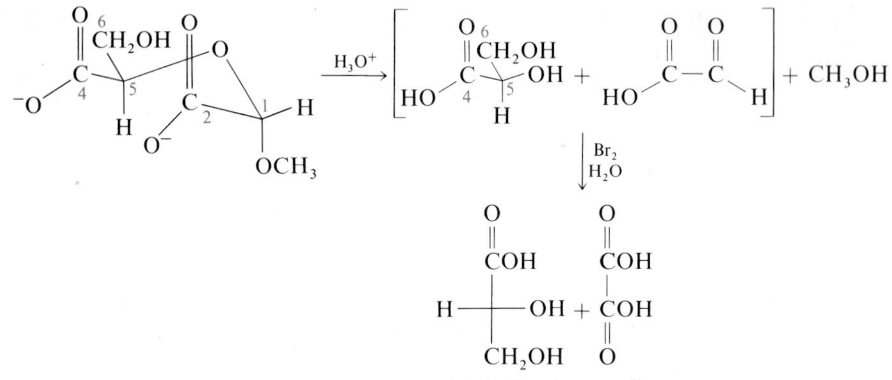

Carbon 1 loses its chirality when the acetal group is hydrolyzed. The only chiral center left untouched from the original sugar is carbon 5, which turns up in (−)-glyceric acid. In Section 14.2B, we learned that (−)-glyceric acid is related in configuration to (+)-glyceraldehyde and has the *R* or D configuration. These reactions, therefore, make a direct connection between the configuration at carbon 5 of hexoses and the configuration of (+)-glyceraldehyde.

PROBLEM 14.29 The aldopentoses also form pyranosides. When the degradative scheme outlined above was carried out for methyl β-D-arabinopyranoside, the strontium salt of the dicarboxylic acid had $[\alpha]_D^{20}$ +55.7°. The same series of reactions on methyl α-D-arabinopyranoside gave a salt with $[\alpha]_D^{20}$ of −55.5°. Methyl β-D-arabinopyranoside has the following structure.

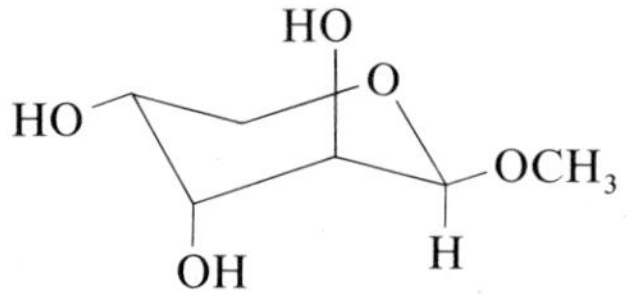

methyl β-D-arabinopyranoside

(a) Carry through the degradation on it, showing the intermediate dialdehyde and the salt of the diacid. Keep track of the asymmetric carbon atoms.

(b) What is the stereochemical relationship between the strontium salt obtained from the α-pyranoside and that from the β-pyranoside?

The dialdehydes formed by periodic acid oxidation of glycosides are reduced to primary alcohols with sodium borohydride. The reduction also solves the problem of handling the unstable dialdehydes while trying to determine structure. The reduction reaction removes the chirality at carbon 5 and allows a direct comparison of the stereochemistry at carbon 1 of the glycosides.

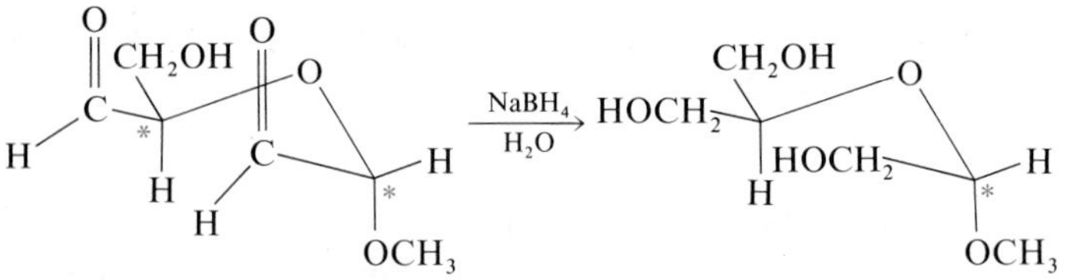

dialdehyde from periodate oxidation of α-D-glycosides, 2 asymmetric carbon atoms

reduction product of dialdehyde, 1 asymmetric carbon atom

PROBLEM 14.30   What is the stereochemical relationship between the dialdehyde derived from methyl α-D-glucopyranoside and the one obtained from methyl β-D-glucopyranoside? What is the stereochemical relationship between the two compounds that results from the $NaBH_4$ reduction of the two dialdehydes?

PROBLEM 14.31   What are the products you would expect from $NaBH_4$ reduction of glucose, mannose, galactose, and fructose? In each case, comment on: (a) whether you would expect a single product or a mixture to be formed; and (b) whether the products would be optically active. Is there a stereochemical relationship between the reduction products from fructose, glucose, and mannose?

E. The Determination of the Ring Structure in Sugars by Oxidation Reactions

The formation of formic acid and a dialdehyde in the periodate oxidation of glucopyranosides is a proof of the size of the ring. If, for example, the ring were five-membered, no formic acid would be expected, and the major fragment would be a trialdehyde.

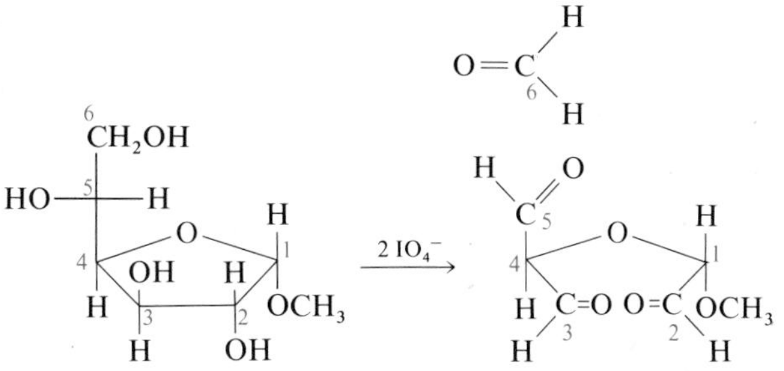

methyl α-D-glucofuranoside

periodate oxidation
products expected from
the methyl glucofuranoside

This is not what is observed for the major cyclic acetals formed by glucose under most reaction conditions.

The size of the ring in the cyclic forms of sugars can also be determined by oxidation of the methyl ethers of sugars. In the 1920s, chemists were keenly interested in discovering whether the aldopentoses and aldohexoses existed as six-membered rings. Sir Edmund L. Hirst showed that the oxidation of tetramethylglucose with concentrated nitric acid gave 2,3,4-trimethoxy-1,5-pentanedioic acid (trimethoxyglutaric acid) as well as 2,3-dimethoxy-1,4-butanedioic acid (dimethoxysuccinic acid).

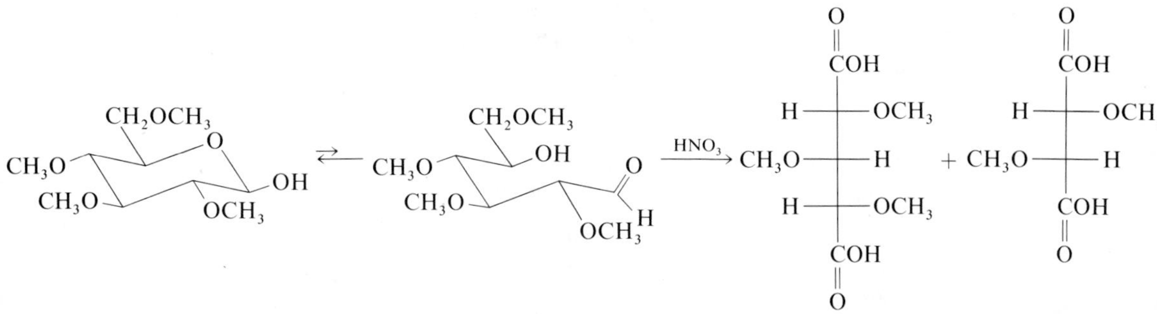

tetramethylglucose

trimethoxyglutaric
acid

dimethoxysuccinic
acid

Hirst's interpretation of these results was that the unmethylated hydroxyl group at carbon 5, a secondary alcohol group, is oxidized by hot nitric acid to a carbonyl group. The carbonyl group makes the carbon atoms adjacent to it vulnerable to further oxidation. Cleavage of the carbon-carbon bond at one side of the carbonyl group gives the glutaric acid and at the other, the succinic acid.

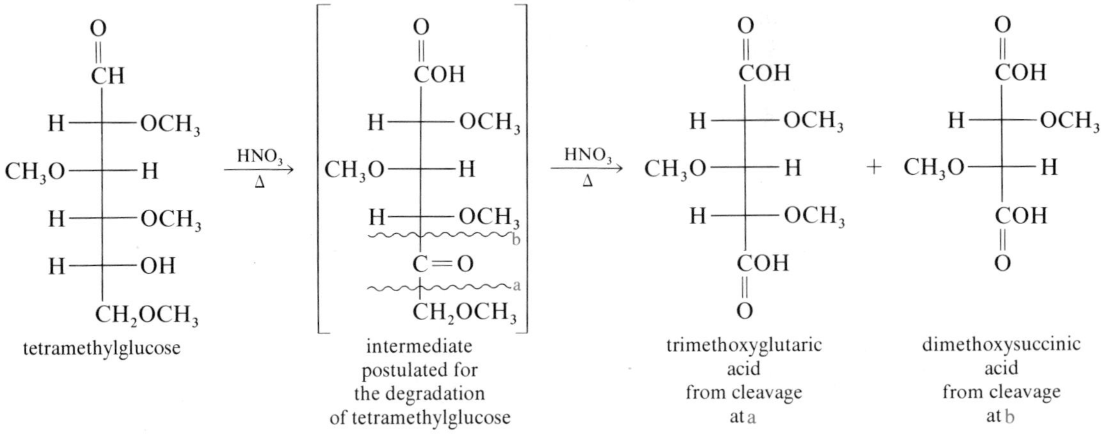

| tetramethylglucose | intermediate postulated for the degradation of tetramethylglucose | trimethoxyglutaric acid from cleavage at a | dimethoxysuccinic acid from cleavage at b |

If glucose had a furanose structure, the hydroxyl group at carbon 4 would be the one that was not methylated. Oxidation of this compound would give dimethoxysuccinic acid, but not the glutaric acid.

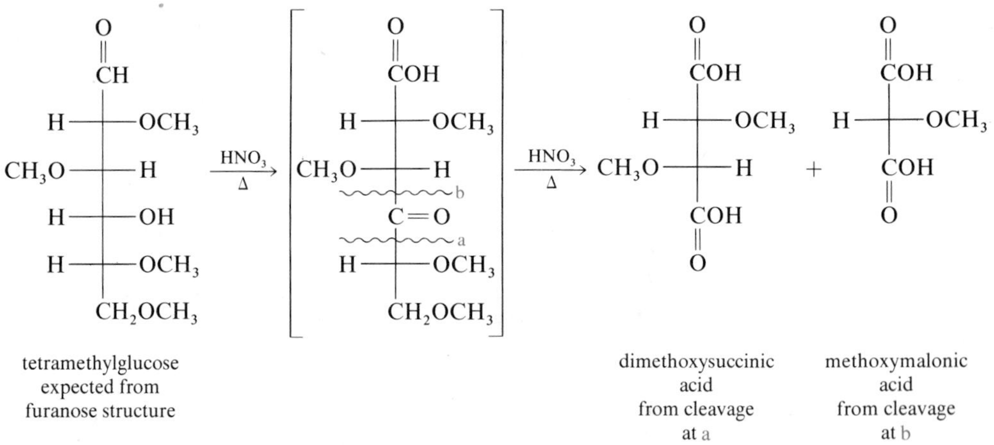

| tetramethylglucose expected from furanose structure | | dimethoxysuccinic acid from cleavage at a | methoxymalonic acid from cleavage at b |

The formation of trimethoxyglutaric acid is evidence that the original hemiacetal bond is to the hydroxyl group at carbon 5. The ring structure in the ordinary form of glucose is therefore a six-membered ring, a pyranose structure.

PROBLEM 14.32   The mixture of acids formed from the oxidation of glucose is actually converted into the dimethyl esters of the acids for isolation.

(a) What reagents would you use for this reaction? Write an equation for the conversion.

(b) Are both diacids optically active? Will the mixture of the dimethyl esters of the two acids show optical activity?

14.7

Synthetic Transformations of Monosaccharides

A. The Kiliani-Fischer Synthesis

The Kiliani reaction, the addition of hydrocyanic acid to a monosaccharide and the hydrolysis of the resulting cyanohydrin (Section 14.3A), gives a polyhydroxy-carboxylic acid. These carboxylic acids form lactones (Section 10.7H). γ-Lactones with a five-membered ring tend to be more stable than six-membered ring lactones, the δ-lactones. Lactones are found whenever chemists attempt to isolate the free sugar acids.

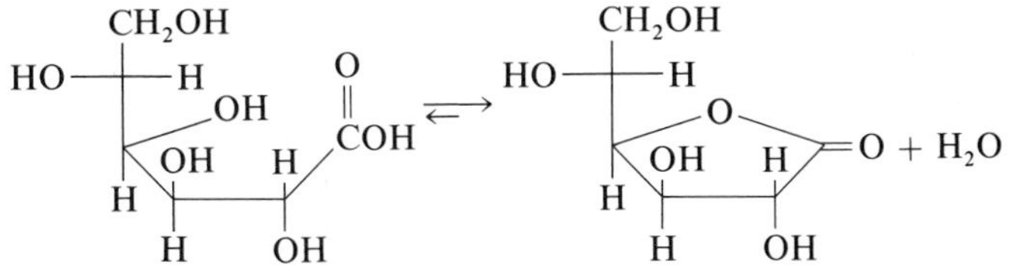

gluconic acid γ-gluconolactone

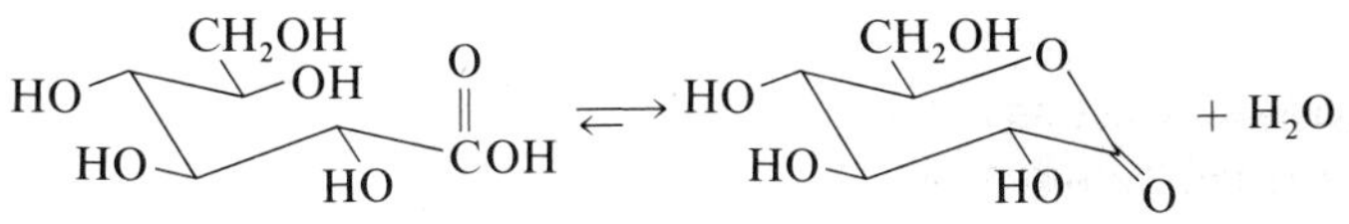

gluconic acid δ-gluconolactone

The sugar lactones were known in the early days of sugar chemistry. Fischer discovered in 1889 that the lactones could be reduced to aldoses. As his reducing agent he used sodium amalgam, a small amount of sodium metal dissolved in mercury in order to moderate the reactivity of sodium in water solutions. Sodium borohydride is now used for the same reduction. Sodium borohydride is added to a solution of δ-gluconolactone so that the borohydride reagent is never present in excess. The lactone is reduced to the hemiacetal form of glucose.

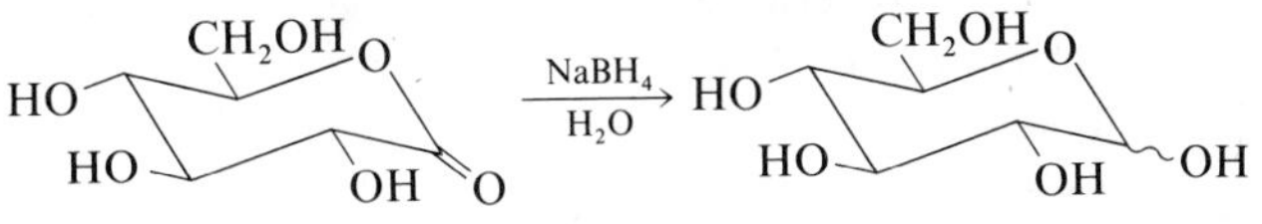

δ-gluconolactone   mixture of α- and β-
                                           D-glucopyranoses

PROBLEM 14.33   Why do we have to avoid an excess of sodium borohydride in the preceding reaction?

Fischer was able to reduce the lactone of the new carboxylic acid resulting from the Kiliani reaction to an aldose. The combination of the two methods, now known as the **Kiliani-Fischer synthesis,** enables us to synthesize a longer chain aldose from a shorter chain one.

The synthetic sequence was first applied to L-arabinose, a natural aldopentose, to give L-mannose and L-glucose. Fischer painstakingly proved in 1890 that the L-glucose he had obtained was the enantiomer of natural glucose. Since then, D-arabinose has been converted to D-glucose and D-mannose. The diastereomeric nitriles, which are epimers at carbon 2, are not produced in equal amounts in the addition reaction.

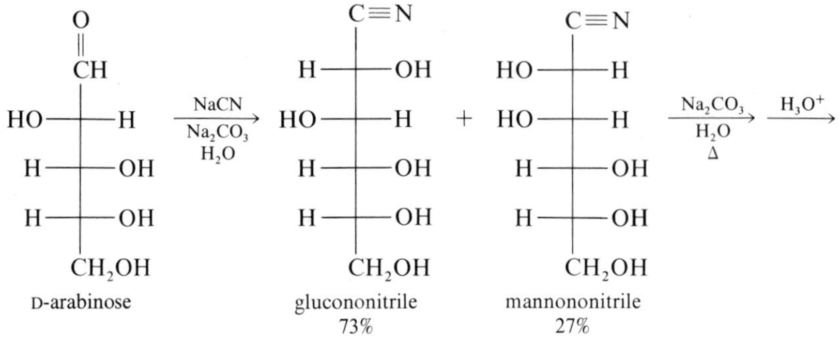

D-arabinose → (NaCN, Na$_2$CO$_3$, H$_2$O) gluconitrile 73% + mannonitrile 27% → (Na$_2$CO$_3$, H$_2$O, Δ; H$_3$O$^+$)

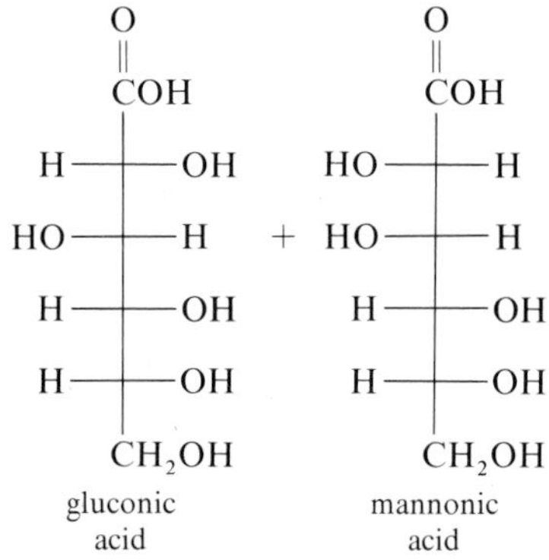

gluconic acid + mannonic acid

In a basic solution, the nitrile with the same stereochemistry as glucose is formed in larger amounts. The nitriles, which are easily hydrolyzed, are usually not isolated, but are converted directly to the carboxylic acids.

The diastereomeric carboxylic acids can be separated from each other. Their salts have different solubilities, and their lactones form with different degrees of ease. The separated acids, as their lactones, are reduced to the aldoses.

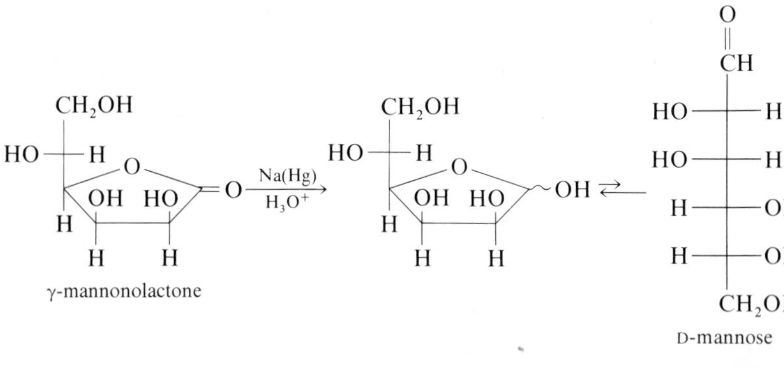

γ-mannonolactone → (Na(Hg), H$_3$O$^+$) → ⇌ D-mannose

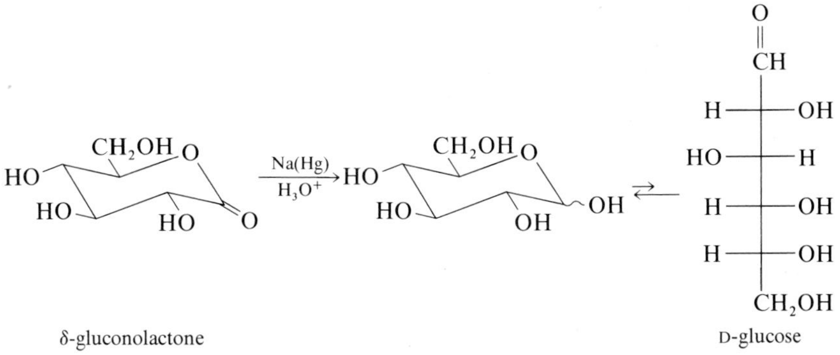

δ-gluconolactone   D-glucose

In this synthesis, mannonic acid gives the five-membered ring lactone, γ-mannono-lactone, while gluconic acid is isolated as the six-membered ring lactone, δ-gluconolactone. You should remember that when free carboxylic acids of sugars are isolated, they will form lactones, and that the lactones can be reduced to aldehydes.

PROBLEM 14.34   D-Glyceraldehyde can, in principle, be converted to two aldotetroses by the Kiliani-Fischer synthesis. One, D-erythrose, on oxidation with nitric acid gives *meso*-tartaric acid. The other, D-threose gives D-(−)-tartaric acid. Show the steps of the conversion and the oxidation to tartaric acids. Assign structures to D-erythrose and D-threose.

PROBLEM 14.35   In the same way, D-erythrose can be converted to D-arabinose and D-ribose. D-Arabinose, on oxidation with nitric acid, gives an optically active aldaric acid. The aldaric acid from D-ribose has no optical activity. Show how the reactions in Problem 14.34 and in this problem establish the stereochemistry of D-arabinose and D-ribose.

B. The Family of D-Aldoses

We have now seen all the different chemical transformations necessary to establish the structures of the 16 steroisomers of glucose, and of the pentoses and tetroses that connect them to D-(+)-glyceraldehyde. The compelling evidence for the stereochemistry of glucose came from many sources. We will retrace some of the reasoning that Fischer used to assign glucose its full stereochemical structure.

First of all, Fischer, in his projection formula of glucose, arbitrarily assigned the hydroxyl group to a position to the right of carbon 5.

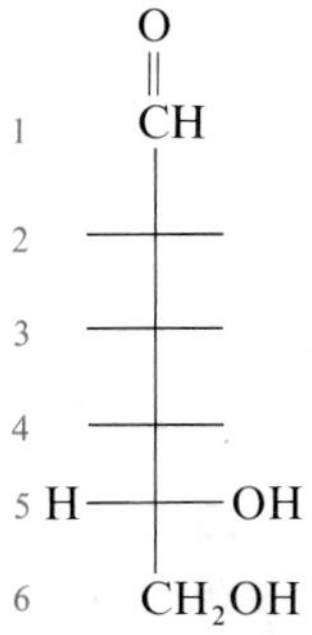

configuration at carbon 5 of glucose as
assigned by Fischer

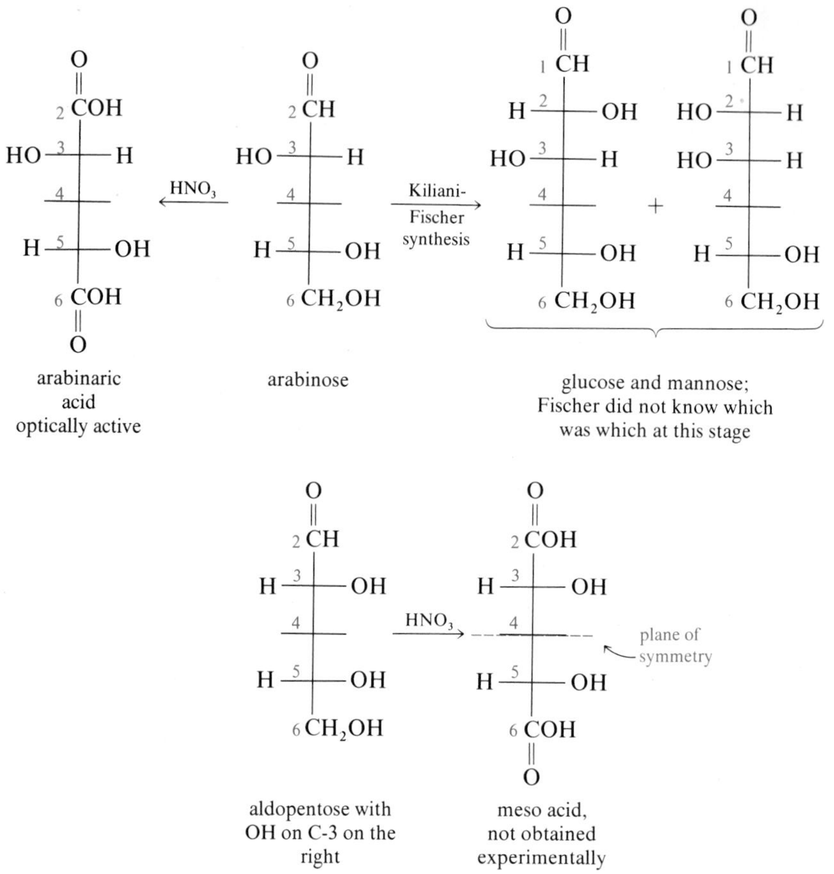

arabinaric
acid
optically active

arabinose

glucose and mannose;
Fischer did not know which
was which at this stage

aldopentose with
OH on C-3 on the
right

meso acid,
not obtained
experimentally

Figure 14.3 Reactions establishing the configuration of C-3 in glucose.

Glucose and mannose could both be prepared from arabinose and were epimers at carbon 2 (by synthesis and because they gave the same osazone). Arabinose gave an optically active aldaric acid and therefore had the hydroxyl group on the left on the carbon atom corresponding to carbon 3 in glucose (Figure 14.3). If that carbon atom had had opposite configuration in the aldopentose, the corresponding aldaric acid would have been a meso form with a plane of symmetry through the middle carbon atom.

Both glucose and mannose are oxidized to optically active aldaric acids (Figure 14.4). This means that the hydroxyl group on carbon 4 is on the right. Otherwise, one of the two would have given a meso aldaric acid. This meso aldaric acid, in fact, is galactaric acid, from galactose, the epimer of glucose at carbon 4.

Fischer now had a pair of compounds to which he had assigned relative stereochemistry at every asymmetric carbon atom. He knew that one was glucose and one was mannose, but he did not know which was which. He solved the problem by converting glucaric acid from glucose by a series of clever, selective reductions of a sugar lactone, of the aldehyde group, and of a lactone once more, into a new sugar, gulose (Figure 14.5, p. 784). This sugar belongs to the L-family and is glucose with aldehyde and hydroxymethyl groups interchanged. On oxidation with nitric acid, it gives the same dicarboxylic acid as glucose.

An examination of the structures that are possible for glucose and mannose shows that only one of them would be transformed into a new compound by the interchange

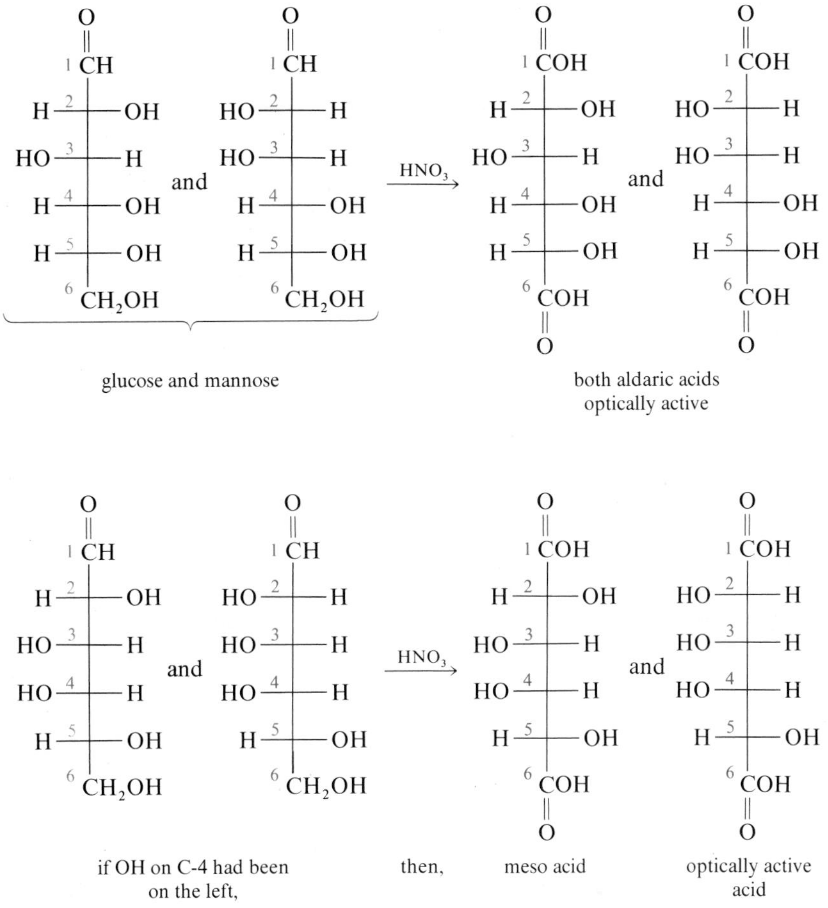

$$\underset{\text{glucose and mannose}}{\underbrace{
\begin{array}{c}
\overset{O}{\overset{\|}{\underset{1}{C}H}} \\
H\overset{2}{-}OH \\
HO\overset{3}{-}H \\
H\overset{4}{-}OH \\
H\overset{5}{-}OH \\
\underset{6}{C}H_2OH
\end{array}
\quad\text{and}\quad
\begin{array}{c}
\overset{O}{\overset{\|}{\underset{1}{C}H}} \\
HO\overset{2}{-}H \\
HO\overset{3}{-}H \\
H\overset{4}{-}OH \\
H\overset{5}{-}OH \\
\underset{6}{C}H_2OH
\end{array}
}}
\xrightarrow{HNO_3}
\begin{array}{c}
\overset{O}{\overset{\|}{\underset{1}{C}OH}} \\
H\overset{2}{-}OH \\
HO\overset{3}{-}H \\
H\overset{4}{-}OH \\
H\overset{5}{-}OH \\
\underset{6}{C}OH \\
\|\\ O
\end{array}
\quad\text{and}\quad
\begin{array}{c}
\overset{O}{\overset{\|}{\underset{1}{C}OH}} \\
HO\overset{2}{-}H \\
HO\overset{3}{-}H \\
H\overset{4}{-}OH \\
H\overset{5}{-}OH \\
\underset{6}{C}OH \\
\|\\ O
\end{array}$$

glucose and mannose   both aldaric acids
optically active

$$
\begin{array}{c}
\overset{O}{\overset{\|}{\underset{1}{C}H}} \\
H\overset{2}{-}OH \\
HO\overset{3}{-}H \\
HO\overset{4}{-}H \\
H\overset{5}{-}OH \\
\underset{6}{C}H_2OH
\end{array}
\quad\text{and}\quad
\begin{array}{c}
\overset{O}{\overset{\|}{\underset{1}{C}H}} \\
HO\overset{2}{-}H \\
HO\overset{3}{-}H \\
HO\overset{4}{-}H \\
H\overset{5}{-}OH \\
\underset{6}{C}H_2OH
\end{array}
\xrightarrow{HNO_3}
\begin{array}{c}
\overset{O}{\overset{\|}{\underset{1}{C}OH}} \\
H\overset{2}{-}OH \\
HO\overset{3}{-}H \\
HO\overset{4}{-}H \\
H\overset{5}{-}OH \\
\underset{6}{C}OH \\
\|\\ O
\end{array}
\quad\text{and}\quad
\begin{array}{c}
\overset{O}{\overset{\|}{\underset{1}{C}OH}} \\
HO\overset{2}{-}H \\
HO\overset{3}{-}H \\
HO\overset{4}{-}H \\
H\overset{5}{-}OH \\
\underset{6}{C}OH \\
\|\\ O
\end{array}
$$

if OH on C-4 had been   then, meso acid   optically active
on the left,   acid

Figure 14.4 Evidence for the configuration at C-4 in glucose.

of the aldehyde and hydroxymethyl groups. The other one is converted back into itself. This assertion can only be confirmed by a careful inspection of the structural formulas shown below and at the top of the next page.

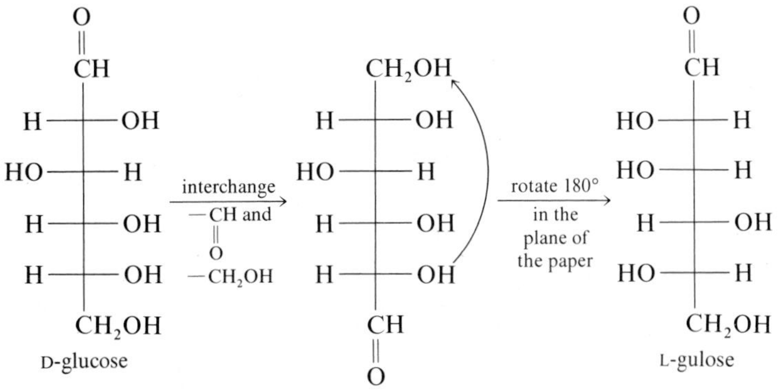

$$
\underset{\text{D-glucose}}{
\begin{array}{c}
\overset{O}{\overset{\|}{C}H} \\
H-OH \\
HO-H \\
H-OH \\
H-OH \\
CH_2OH
\end{array}
}
\xrightarrow[\substack{-CH\ and \\ \|\\ O \\ -CH_2OH}]{\text{interchange}}
\begin{array}{c}
CH_2OH \\
H-OH \\
HO-H \\
H-OH \\
H-OH \\
\overset{}{C}H \\
\|\\ O
\end{array}
\xrightarrow[\substack{\text{in the}\\ \text{plane of}\\ \text{the paper}}]{\text{rotate }180^\circ}
\underset{\text{L-gulose}}{
\begin{array}{c}
\overset{O}{\overset{\|}{C}H} \\
HO-H \\
HO-H \\
H-OH \\
HO-H \\
CH_2OH
\end{array}
}
$$

glucose gives a new sugar when CH *and* CH_2OH
groups are interchanged

$$
\begin{array}{c}
\overset{\displaystyle O}{\overset{\displaystyle \|}{\text{CH}}} \\
\text{HO}{-}\!\!-\!\!{-}\text{H} \\
\text{HO}{-}\!\!-\!\!{-}\text{H} \\
\text{H}{-}\!\!-\!\!{-}\text{OH} \\
\text{H}{-}\!\!-\!\!{-}\text{OH} \\
\text{CH}_2\text{OH}
\end{array}
\qquad
\begin{array}{c}
\text{CH}_2\text{OH} \\
\text{HO}{-}\!\!-\!\!{-}\text{H} \\
\text{HO}{-}\!\!-\!\!{-}\text{H} \\
\text{H}{-}\!\!-\!\!{-}\text{OH} \\
\text{H}{-}\!\!-\!\!{-}\text{OH} \\
\overset{\displaystyle CH}{\underset{\displaystyle \overset{\|}{O}}{}}
\end{array}
\qquad
\begin{array}{c}
\overset{\displaystyle O}{\overset{\displaystyle \|}{\text{CH}}} \\
\text{HO}{-}\!\!-\!\!{-}\text{H} \\
\text{HO}{-}\!\!-\!\!{-}\text{H} \\
\text{H}{-}\!\!-\!\!{-}\text{OH} \\
\text{H}{-}\!\!-\!\!{-}\text{OH} \\
\text{CH}_2\text{OH}
\end{array}
$$

interchange $-\overset{\|}{\underset{O}{\text{CH}}}$ and $-\text{CH}_2\text{OH}$ rotate 180° in the plane of the paper

D-mannose   D-mannose

*structure identical with
the one at
the far left*

mannose does not give a new sugar when $\overset{O}{\overset{\|}{\text{CH}}}$ *and
CH$_2$OH groups are interchanged*

Glucose was the compound that was converted experimentally into a new sugar by this maneuver; therefore, it was assigned the structure that it now has, and mannose was given the structure epimeric at carbon 2.

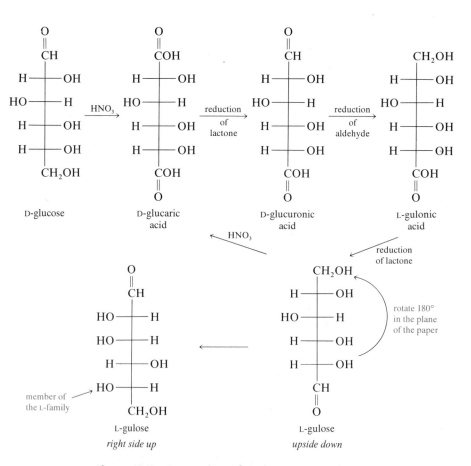

Figure 14.5 Conversion of D-glucose to L-gulose.

When Fischer started his work on carbohydrate chemistry in 1886, the only mono-saccharides that were known were L-arabinose, D-glucose, D-galactose, D-fructose, and L-sorbose, which is another 2-ketohexose. He and his coworkers were responsible for the synthesis and the investigation of the chemical properties of 12 of the 16 possible aldohexoses before his death in 1919. Two others were synthesized by that time in other laboratories. The last two were made in 1934. Stereochemical relationships in the entire family of monosaccharides from D-(+)-glyceraldehyde to the D-aldohexoses are shown in Figure 14.6 on p. 786.

PROBLEM 14.36   The reduction of L-sorbose, a 2-ketohexose, gives as one of the products a sugar alcohol, D-glucitol (also called sorbitol), which is the same alcohol obtained from the reduction of D-glucose. Assign a structure to L-sorbose.

14.8

Disaccharides

A. The Determination of the Structure of a Disaccharide. Lactose

Two disaccharides are commonly found in nature. Sucrose is the sugar derived from plants and is prepared commercially from sugar beets and sugar cane. Lactose is found in the milk of animals and was known as milk sugar when it was first isolated. Other common disaccharides are prepared by breaking down polysaccharides. Maltose is formed from the enzymatic hydrolysis of starch. The partial hydrolysis of cellulose gives cellobiose.

In disaccharides, two monosaccharide units are held together by a glycosidic linkage. The stereochemistry and the position of the linkage, and whether the monosaccharides are in the pyranose or the furanose form, must be determined in order to establish the structure of a disaccharide. The determination of the structure of lactose, begun in the 1880s by Fischer and completed by the English chemist Sir Walter N. Haworth in the 1920s, serves as an example of the kinds of chemical transformations and reasoning that led to the assignment of structure.

One of the most important pieces of information is whether a disaccharide is a reducing sugar or not. Lactose is a reducing sugar, which means that one of the monosaccharide units has a potential carbonyl group in the hemiacetal form. Lactose gives an osazone, which can be hydrolyzed in dilute acid to galactose and glucosazone. Lactose gives a positive Benedict's test (Section 14.6A) and is oxidized by bromine water to a carboxylic acid (usually present as a lactone). It undergoes mutarotation, which also indicates the possibility of an equilibrium between the open-chain aldehyde form and hemiacetal ring forms.

Lactose on hydrolysis with dilute acid gives galactose and glucose. If the hydrolysis is carried out on lactose after oxidation with bromine water, the formation of gluconic acid from the oxidation reaction indicates that the free aldehyde group is in the glucose residue. Carbon 1 of galactose must be tied up in a glycosidic linkage. The enzyme β-galactosidase cleaves lactose into galactose and glucose. This enzyme is specific for the β-stereochemistry of a glycosidic linkage.

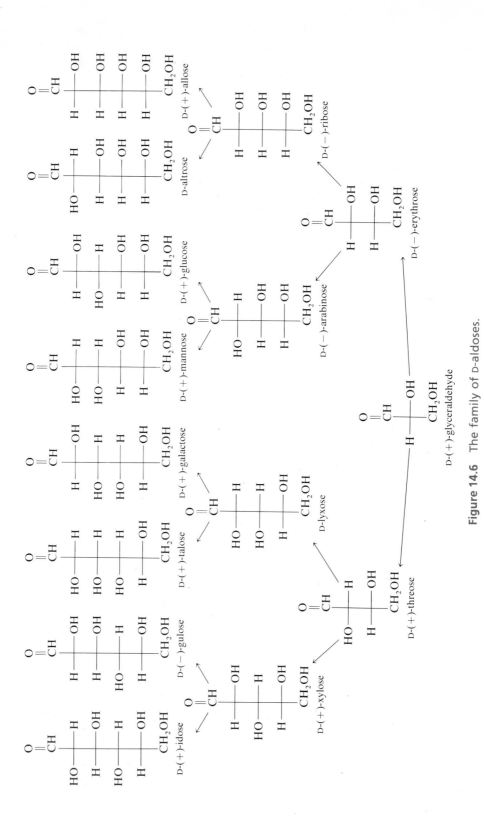

Figure 14.6 The family of D-aldoses.

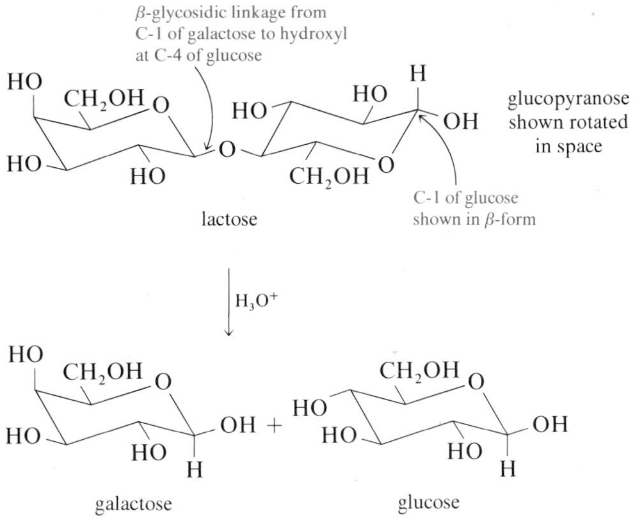

β-glycosidic linkage from
C-1 of galactose to hydroxyl
at C-4 of glucose

glucopyranose
shown rotated
in space

C-1 of glucose
shown in β-form

lactose

galactose   glucose

The scientific name for lactose describes it as a substituted glucopyranose. The glucose unit is substituted on the oxygen at carbon 4 by a galactopyranose substituent. This substituent is attached at carbon 1 with a β-configuration, so it is a β-D-galactopyranosyl group. One systematic name for lactose is therefore 4-*O*-(β-D-galactopyranosyl)-β-D-glucopyranose. Another way of indicating the linkage between the galactose and glucose units is to name the compound *O*-β-D-galactopyranosyl-(1→4)-β-D-glycopyranose. This nomenclature is useful with more complex saccharides.

The reactions described on p. 785 for lactose are shown below.

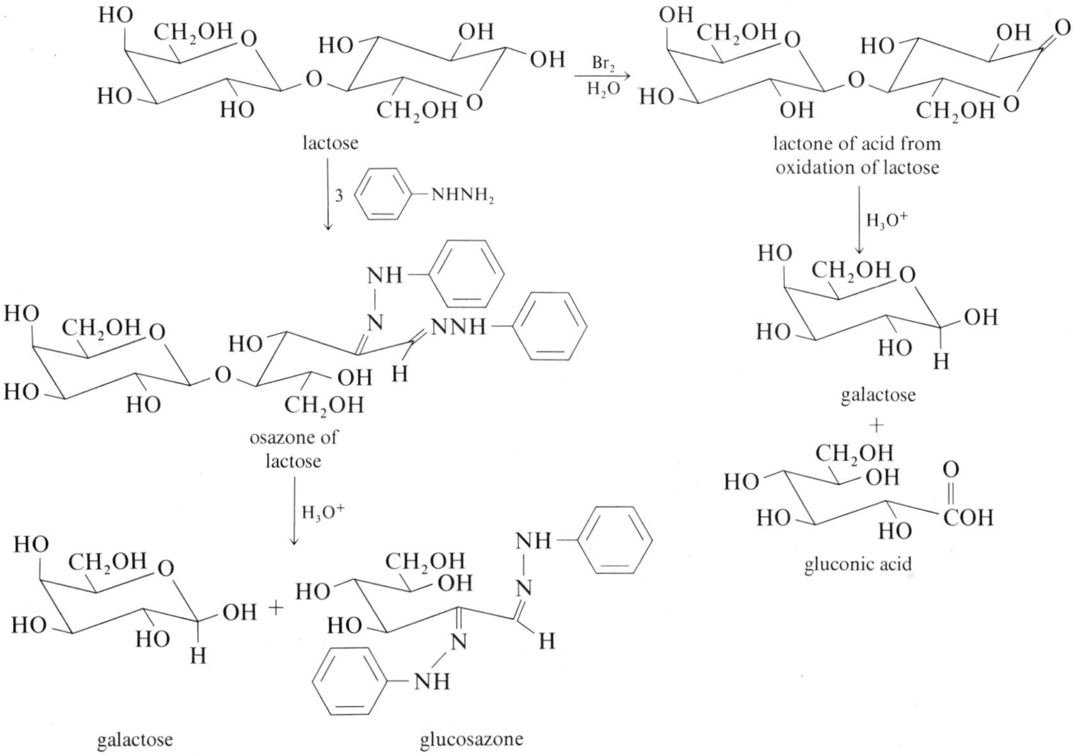

lactose

lactone of acid from
oxidation of lactose

osazone of
lactose

galactose

+

gluconic acid

galactose

glucosazone

The size of the rings in the galactose and glucose units and the point of attachment of the galactose unit to glucose were determined by a series of methylation experiments. When lactose is methylated and then hydrolyzed, 2,3,4,6-O-tetramethyl-D-galactose and 2,3,6-O-trimethyl-D-glucose are formed.

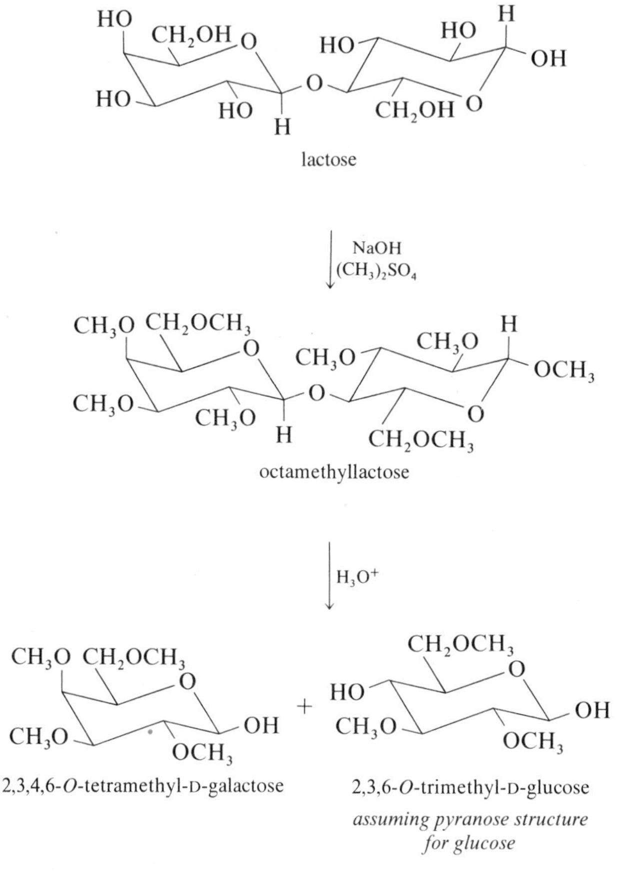

lactose

$$\downarrow \begin{array}{l} \text{NaOH} \\ \text{(CH}_3\text{)}_2\text{SO}_4 \end{array}$$

octamethyllactose

$$\downarrow \text{H}_3\text{O}^+$$

2,3,4,6-O-tetramethyl-D-galactose   2,3,6-O-trimethyl-D-glucose

*assuming pyranose structure
for glucose*

The presence of the hydroxyl group at carbon 5 as the only one that is not methylated in galactose indicates that galactose exists as a pyranose ring in lactose. The hydroxyl groups at carbons 4 and 5 are not methylated in the glucose unit. This could mean that the galactose is attached to the hydroxyl group at carbon 4 if glucose has the pyranose structure. On the other hand, if the glucose unit exists as a furanose, then the attachment between the two sugar units must be at carbon 5.

This question was resolved by fully methylating the acid from the oxidation of lactose, then hydrolyzing. Under the conditions of the hydrolysis, the methyl ester function, which had formed, is also hydrolyzed. The glucose portion of the molecule is isolated as 2,3,5,6-O-tetramethyl-γ-gluconolactone. The formation of the five-membered ring lactone shows clearly that the hydroxyl group at carbon 4 is part of the glycosidic linkage with galactose. That hydroxyl group is the only one left unmethylated on the glucose unit once the pyranose ring is opened by the oxidation reaction.

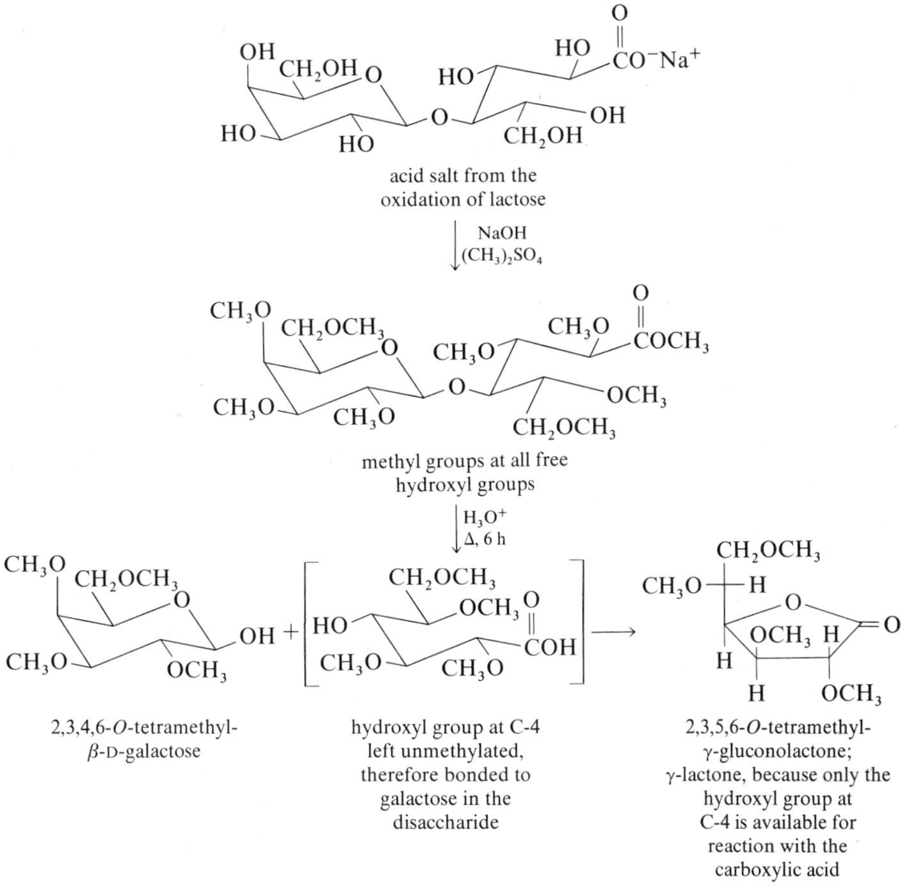

acid salt from the
oxidation of lactose

NaOH
$(CH_3)_2SO_4$

methyl groups at all free
hydroxyl groups

H_3O^+
Δ, 6 h

2,3,4,6-*O*-tetramethyl-
β-D-galactose

hydroxyl group at C-4
left unmethylated,
therefore bonded to
galactose in the
disaccharide

2,3,5,6-*O*-tetramethyl-
γ-gluconolactone;
γ-lactone, because only the
hydroxyl group at
C-4 is available for
reaction with the
carboxylic acid

B. The Structures of Maltose and Cellobiose

Similar types of experiments by Haworth in the 1920s established the structures of maltose and cellobiose. Maltose is 4-*O*-(α-D-glucopyranosyl)-D-glucopyranose, while cellobiose is 4-*O*-(β-D-glucopyranosyl)-D-glucopyranose. Both are reducing sugars that mutarotate. Both are hydrolyzed to two units of glucose. The difference between the two is in the stereochemistry of the glycosidic linkage. Maltose is hydrolyzed by the enzyme maltase, which is specific for α-glycosidic linkages, while emulsin, an enzyme specific for the cleavage of β-glycosidic linkages, is necessary for the hydrolysis of cellobiose (Section 14.4A).

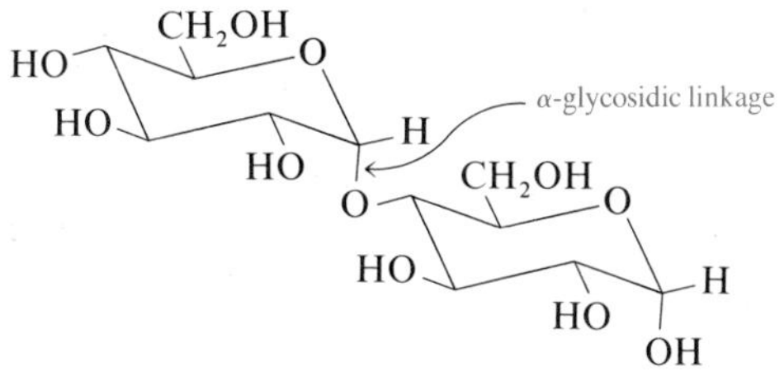

α-glycosidic linkage

maltose
product of the hydrolysis of starch

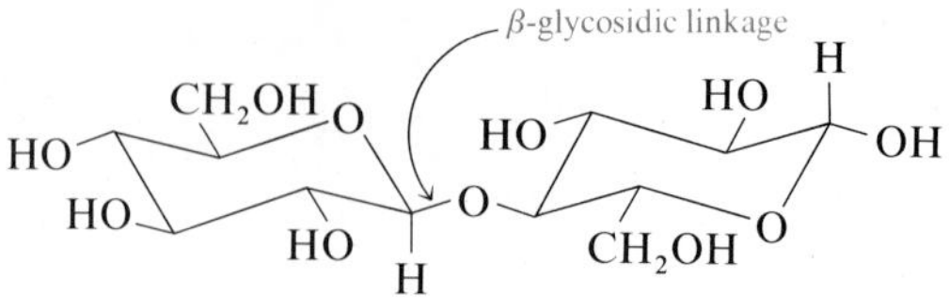

cellobiose

product of the hydrolysis of cellulose

These structural differences are important because they suggest similar structural differences that exist in the polysaccharides, starch and cellulose, from which these disaccharides are derived by hydrolysis reactions.

C. The Structure of Sucrose

Sucrose is not a reducing sugar, and it does not undergo mutarotation. It gives an osazone only after a long period of time. The osazone that is finally isolated is the osazone of glucose. Sucrose itself is dextrorotatory, $[\alpha]_D +66°$. Hydrolysis of sucrose converts it into glucose, $[\alpha]_D +52.5°$, and fructose, $[\alpha]_D -92°$. The resulting mixture is levorotatory. This phenomenon is known as the **inversion of sucrose.** The mixture of sugars that is formed is called **invert sugar.** Invert sugar is much sweeter than sucrose chiefly because of the great sweetness of fructose, eight times that of sucrose. Invert sugar is the chief component of honey. The preparation of all sugar syrups or candies in the kitchen involves boiling table sugar with water and a bit of acid such as vinegar, lemon juice, or cream of tartar (the monopotassium salt of (+)-tartaric acid). This process is nothing more than the preparation of invert sugar.

The fact that sucrose is not a reducing sugar means that there is no potential aldehyde or ketone function in the molecule in a hemiacetal form. Carbon 1 of glucose must be bonded to carbon 2 of fructose. Sucrose is hydrolyzed by maltase so the glucosidic linkage must be α in configuration.

The structure and stereochemistry of the fructose portion of the molecule was much more difficult to establish. Careful polarimetric studies by Hudson, started in 1908 and completed in 1934, and an x-ray structure determination in 1947 show that fructose is present as a β-fructofuranoside in sucrose.

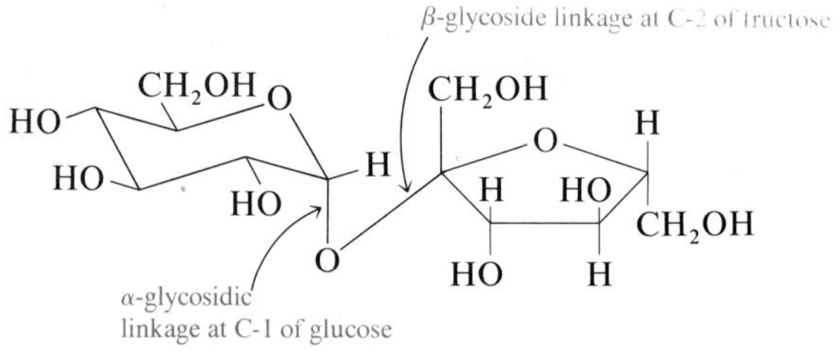

sucrose

(α-D-glucopyranosyl)-β-D-fructofuranoside

or

(β-D-fructofuranosyl)-α-D-glucopyranoside,

or

O-α-D-glucopyranosyl-(1→2)-β-D-fructofuranoside

PROBLEM 14.37   Write the equations that account for the formation of glucosazone from sucrose.

PROBLEM 14.38   Gentiobiose is a disaccharide found in a number of natural products. It has the formula $C_{12}H_{22}O_{11}$ and is a reducing sugar. It is hydrolyzed by emulsin to glucose. 2,3,4,6-O-Tetramethyl-D-glucopyranose and 2,3,4,-O-trimethyl-D-glucopyranose are obtained from the conversion of gentiobiose to its fully methylated form, followed by hydrolysis. What is the structure of gentiobiose? Give it a systematic name.

14.9

Polysaccharides

A. Starch

The two polysaccharides of overwhelming biological and economic importance are starch and cellulose. Plants store their reserve carbohydrates in the form of starch. We get the starch we need from roots, tubers, and seeds.

Starch is a high molecular weight polysaccharide that can be broken down completely to glucose by acid hydrolysis and by the enzyme maltase, which establishes the presence of α-linkages. Starch can be fractionated into two different kinds of molecules. One is amylose, which has an average molecular weight of approximately 10^6. It is primarily a polymeric chain of glucopyranose units attached to each other by α-glycosidic linkages between carbon 1 of one glucopyranose and the hydroxyl group on carbon 4 of the next one. The structure of amylose is proved by methylation and hydrolysis, which gives almost exclusively 2,3,6-O-trimethylglucose.

amylose
$n \sim 1000$–6000

Amylose and iodine form a complex that is blue, hence the familiar color reaction of iodine in starch.

The chief fraction of starch is amylopectin, which has a more complicated structure. It is highly polymeric, with as many as a million glucose units in a single molecule. Complete methylation and hydrolysis of amylopectin gives about 3% of 2,3-O-dimethylglucose, suggesting that some glucose units are connected to others through the hydroxyl group at carbon 6 as well as through the oxygens at carbons 1 and 4. Enzymatic studies and partial hydrolysis reactions have shown that amylopectin has a randomly branched structure in which the hydroxyl groups at carbon 6 of some glucose units are indeed involved.

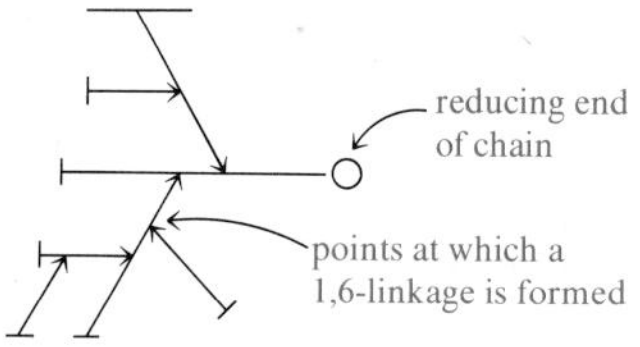

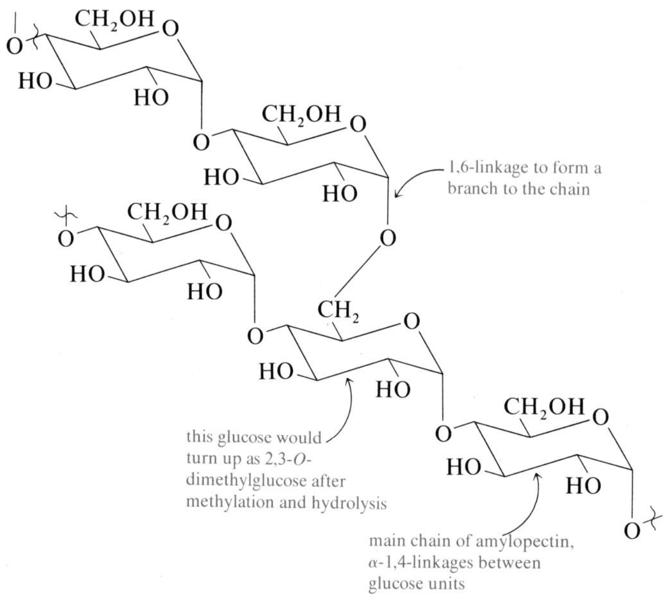

amylopectin, showing the branching of the chain

The branching seems to occur every 20 to 25 glucose units. The overall structure of amylopectin is like the branching of a tree.

Glycogen, the form in which animals store carbohydrates, is also composed of glucose units and is similar to amylopectin in structure except that the branching occurs at shorter intervals, with about 12 glucose units in each branch. It has been isolated from kidneys, brains, and skeletal and cardiac muscles of mammals and is especially abundant in the liver.

B. Cellulose

Cellulose is the organic substance that is most abundant in nature. It is the structural material of the higher plants and is found in all parts of plants. Wood is about 50% cellulose. Commercially important fibers such as cotton and flax consist almost completely of cellulose.

On hydrolysis, cellulose gives cellobiose and ultimately glucose. This establishes its structure as a linear chain of glucopyranose units attached to each other by β-glycosidic linkages from carbon 1 of one unit to the hydroxyl group on carbon 4 of another unit. The structure consists of long chains of six-membered rings in the most stable chair conformation with all the larger substituents in the equatorial positions.

glucopyranose structure
rotated in space

cellulose
$n \sim 5000\text{--}10,000$

The individual molecules of cellulose are associated with each other in regular structures that have crystalline properties. Approximately 100 to 200 cellulose molecules are held together in the larger unit. The exact nature of the interactions between the molecules has not been determined. Hydrogen bonding between adjacent strands of molecules does seem to be an important factor in determining the strength and rigidity of cellulose as a structural material.

Cellulose was the first polymeric material that was chemically modified to provide new materials useful to human beings. The polyhydroxyl functions in cellulose participate in the typical reactions of alcohols. Thus, cellulose with acetic anhydride in acetic acid with a little sulfuric acid as a catalyst is converted into its acetate. Exactly how many acetyl groups enter the molecule depends on the physical state of the cellulose at the beginning of the reaction, the exact reaction conditions, and how the material is treated afterwards. Some of the acetyl groups are hydrolyzed off by exposure to water during processing.

The most useful acetate is cellulose triacetate, which is soluble in mixtures of acetic anhydride and acetic acid.

fragment of
cellulose

fragment of
cellulose triacetate

All the free hydroxyl groups in cellulose are converted into ester functions. When a solution of cellulose triacetate is forced through small holes into a solution of dilute acetic acid, the water precipitates it in the form of a continuous thread that can be used to weave fabrics in the textile industry. This product is known commercially as Arnel.

The use of cellulose in a variety of other products is based on similar modifications of structure. The hydroxyl functions are converted to some other functional group. In the process, the cellulose molecule becomes somewhat degraded and much more soluble in organic solvents. In this soluble state, it is manipulated by extruding the solutions as sheets or fine streams into other solutions that precipitate the compounds or reverse the original chemical reactions, regenerating cellulose in a new, more useful form.

For example, the hydroxyl groups in cellulose are converted into alkoxide anions by base. The nucleophilic anions add to carbon disulfide to give compounds known as xanthate esters, which are stable as salts in basic solutions but lose carbon disulfide in acid.

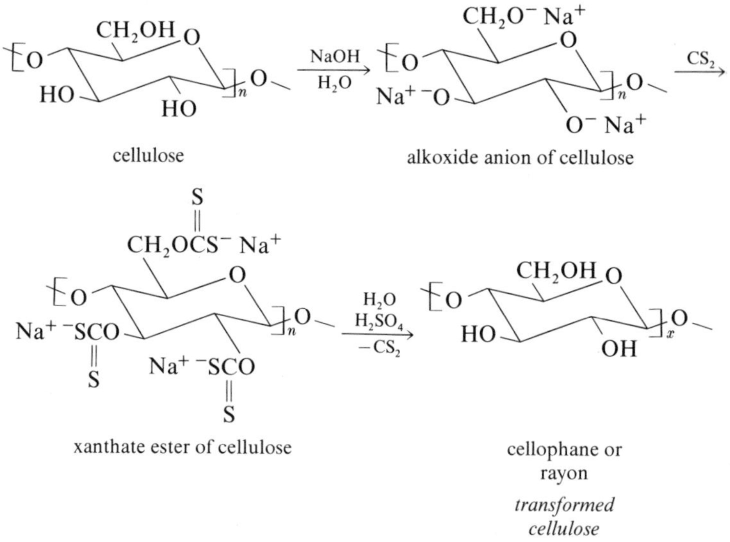

cellulose   alkoxide anion of cellulose

xanthate ester of cellulose   cellophane or rayon

transformed cellulose

The basic solution of cellulose xanthate salts can be forced out of spinners into dilute sulfuric acid to form rayon threads. If thin slits are used, sheets of cellophane are formed. Both rayon and cellophane are essentially cellulose in a transformed physical state. The hydrogen bonding between the molecules of cellulose in its original state has been disrupted by the chemical process. When the hydroxyl groups reform after the reactions are over, the cellulose molecules have been physically forced by the extrusion process into new conformations, which then have new physical properties.

PROBLEM 14.39 Cellulose is converted into a nitrate ester by treatment with a mixture of nitric and sulfuric acids. Cellulose trinitrate is called guncotton because of its explosive properties. Write an equation for the formation of cellulose trinitrate.

PROBLEM 14.40 Using ROH to represent cellulose, write the mechanism for the formation of cellulose xanthate.

PROBLEM 14.41 Among the hydrolysis products of chitin, the chief material in the shells of lobsters, is a disaccharide, chitobiose. It has the formula $C_{12}H_{24}O_9N_2$. Further hydrolysis gives N-acetylglucosamine (see p. 742 for glucosamine). Assign a structure to chitobiose, which seems to be an analog of cellobiose.

14.10

Other Natural Products Derived from Carbohydrates

A. Ascorbic Acid, Vitamin C

Vitamin C, ascorbic acid, is a sugar acid synthesized in plants and in the livers of most vertebrates, but not in human beings. Human beings must therefore constantly supplement their diets with the vitamin. The pathway by which the vitamin is synthesized in living organisms is reminiscent of the final steps in Fischer's proof of the structure of glucose, when he was converting D-glucose to L-gulose. The aldehyde group of D-glucuronic acid is reduced by an enzyme to give L-gulonic acid. L-Gulonic acid is converted to its γ-lactone by another enzyme and finally oxidized to L-ascorbic acid by yet another enzyme system.

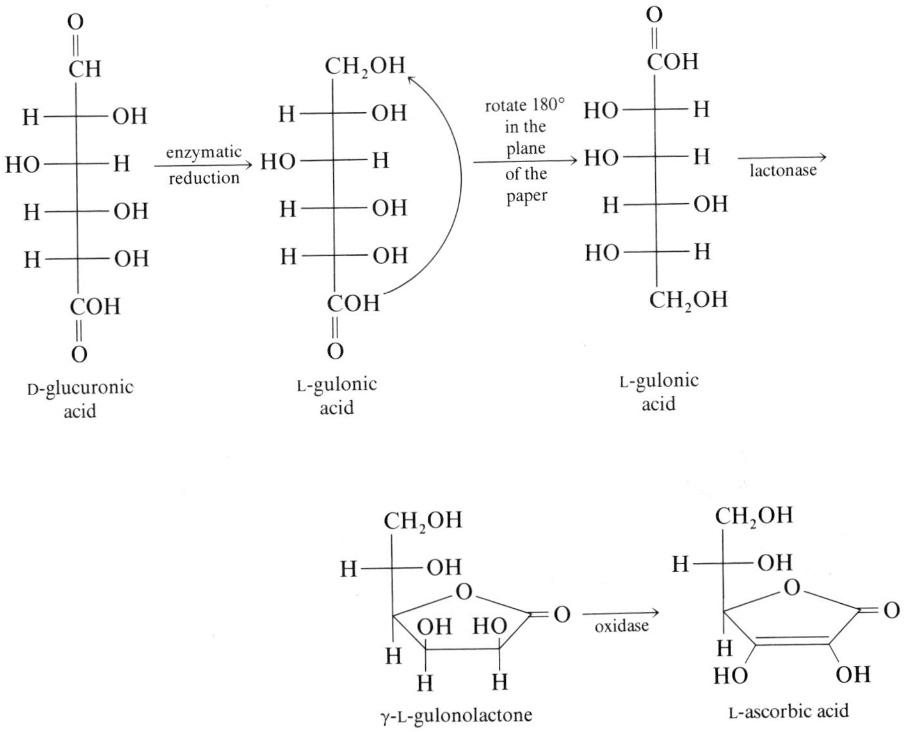

D-glucuronic acid → (enzymatic reduction) → L-gulonic acid → (rotate 180° in the plane of the paper) → L-gulonic acid → (lactonase)

γ-L-gulonolactone → (oxidase) → L-ascorbic acid

Fischer reduced the same γ-gulonolactone to L-gulose (p. 784).

Ascorbic acid is a lactone and an enediol. It is an unstable compound and a reducing agent. It easily undergoes further oxidation to L-dehydroascorbic acid, which also has some vitamin C activity. The activity is lost if the lactone ring is opened by hydrolysis.

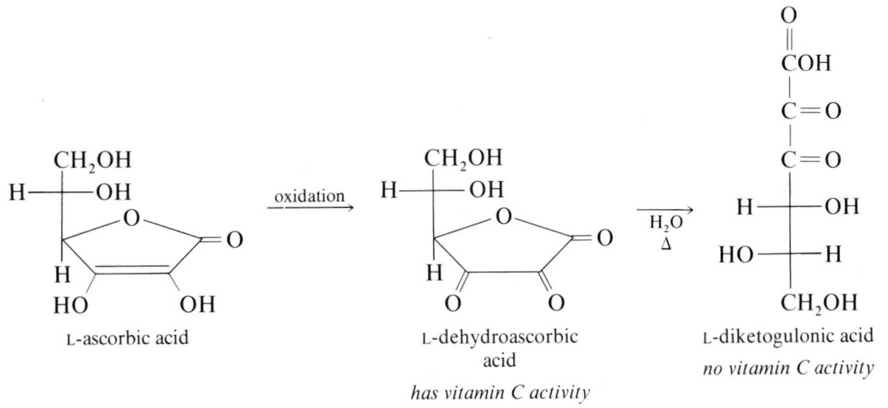

L-ascorbic acid → (oxidation) → L-dehydroascorbic acid *has vitamin C activity* → (H_2O, Δ) → L-diketogulonic acid *no vitamin C activity*

While it is known that ascorbic acid functions as a reducing agent in many biochemical reactions, the exact physiological function of the vitamin is unknown. It appears to be essential to the maintenance of the integrity of cells and tissues.

PROBLEM 14.42   The acidic properties of ascorbic acid come from the enol groups, not from the opening of the lactone ring to generate a carboxylic acid. Which proton is the most acidic one in ascorbic acid? (Hint: Write the anions that would result from the removal of each proton and see which anion would have the best delocalization of charge.)

B. Cardiac Glycosides

The cardiac glycosides are compounds some of which have an effect on the actions of the heart and many of which are highly toxic. A well-known example of a cardiac glycoside is digitoxin, isolated from the foxglove, *Digitalis purpurea*. This compound reduces the pulse rate, regularizes the rhythm of the heart, and strengthens the heart beat. It consists of a steroid aglycon attached to a trisaccharide consisting of three units of the sugar D-digitoxose.

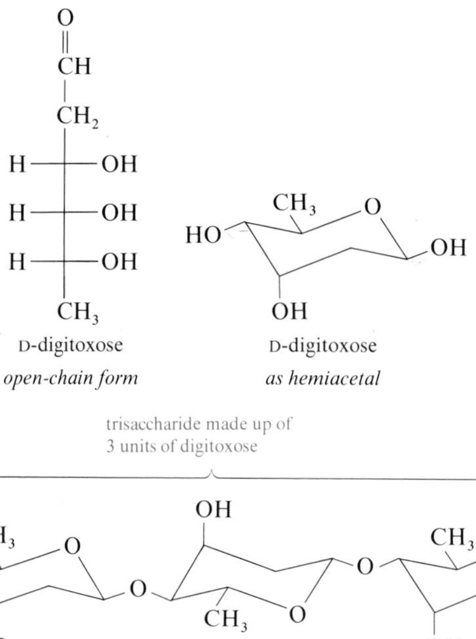

D-digitoxose
open-chain form

D-digitoxose
as hemiacetal

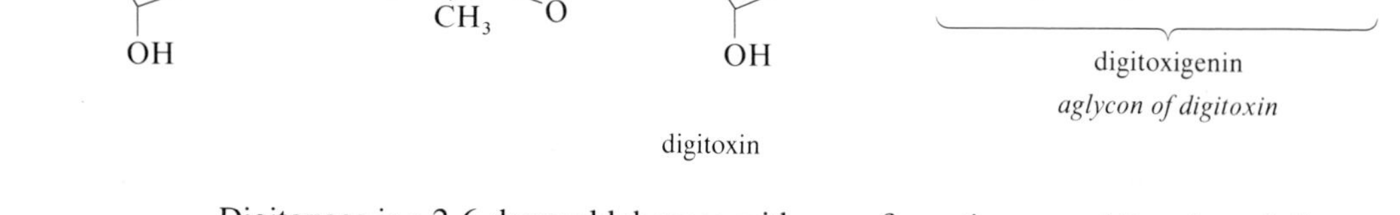

trisaccharide made up of
3 units of digitoxose

glycosidic linkage
to aglycon

digitoxigenin
aglycon of digitoxin

digitoxin

Digitoxose is a 2,6-deoxyaldohexose with a configuration resembling that of ribose. The trisaccharide unit has β-linkages from carbon 1 of one digitoxose unit to the hydroxyl group of carbon 4 of another one. The aglycon in digitoxin is a steroid called digitoxigenin. The five-membered ring α,β-unsaturated lactone unit and the hydroxyl group at the junction of the C and D rings of the steroid are found in other compounds having similar action on the heart, but the sugars of other cardiac glycosides vary.

PROBLEM 14.43   Uzarin is a cardiac glycoside with the same aglycon as digitoxin. It is attached by a β-glycosidic linkage to cellobiose. What is the structure of uzarin?

ADDITIONAL PROBLEMS

14.44 Write structures for the following compounds, showing stereochemistry when appropriate.

(a) *N*-(α-D-glucopyranosyl)methylamine
(b) methyl 2,3,4,6-tetra-*O*-methyl-β-D-mannopyranoside
(c) D-mannaric acid   (d) D-mannitol   (e) α-D-fructofuranose-6-phosphoric acid
(f) 2-amino-2-deoxy-D-galactose
(g) *O*-α-D-galactopyranosyl-(1→6)-*O*-α-D-glucopyranosyl-(1→2)-β-D-fructofuranoside
(h) D-galacturonic acid

14.45 Name the following compounds.

(a)

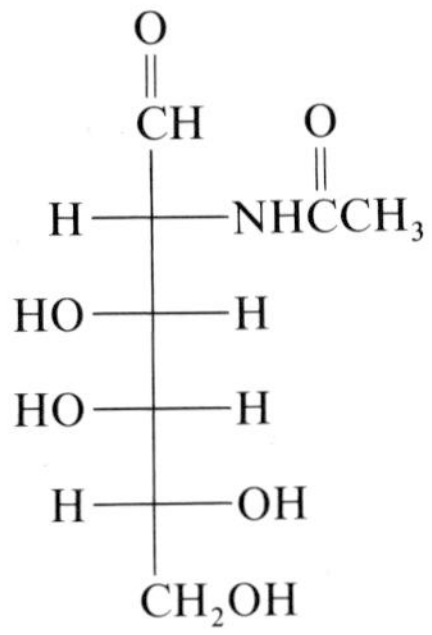

$$CHO$$
$$H \underline{\quad} NHCCH_3 \ (O)$$
$$HO \underline{\quad} H$$
$$HO \underline{\quad} H$$
$$H \underline{\quad} OH$$
$$CH_2OH$$

(b) *[cyclic structure: CH₂OH, HO, HO, HO, H, O, O—P—O⁻ phosphate]*

(c) *[furanose ring structure: HOCH₂, O, OCH₃, H, H, H, OH, H]*

(d) *[pyranose structure: HO, CH₂OH, O, HO, OH, O]*

(e)

$$CH (O)$$
$$HO \underline{\quad} H$$
$$H \underline{\quad} OH$$
$$H \underline{\quad} OH$$
$$HO \underline{\quad} H$$
$$CH_2OH$$

(f) *[furanose ring: HOCH₂, O, CH₃NCH₃, H, H, H, HO, OH]*

(g) *[pyranose structure with acetyl groups: CH₃CO (O), CH₂OCCH₃ (O), O, CH₃CO (O), CH₃CO (O), OCCH₃ (O)]*

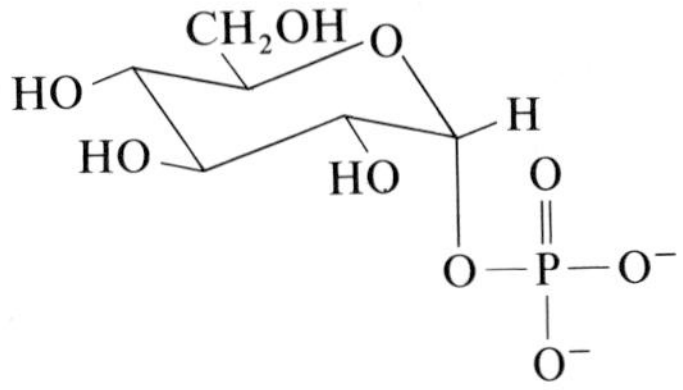

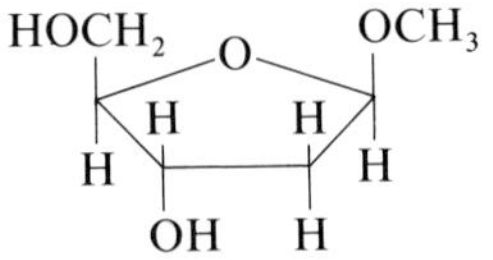

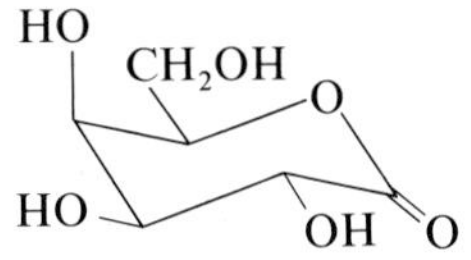

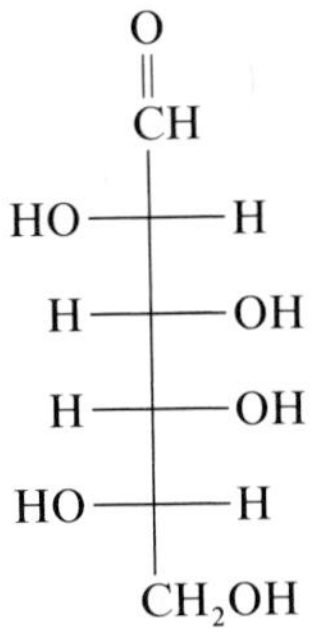

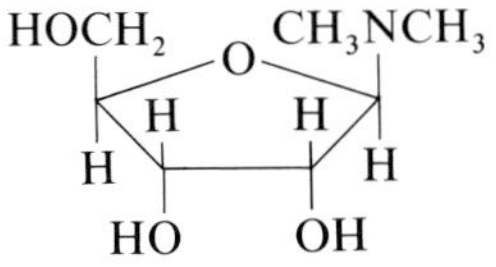

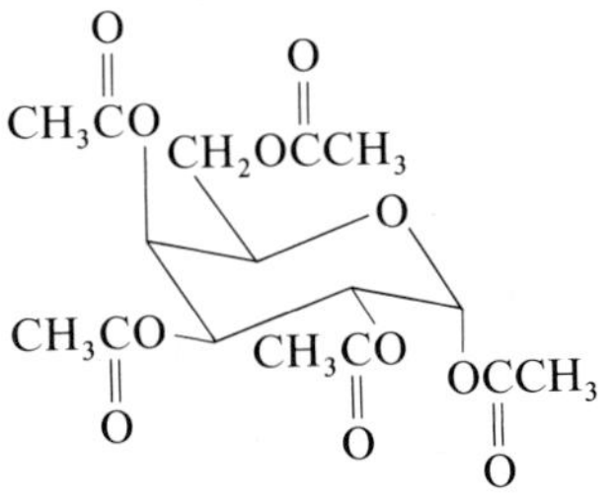

(h) *[furanose ring: CH₃OCH₂, O, OCH₃, H, H, H, CH₃O, OCH₃]*

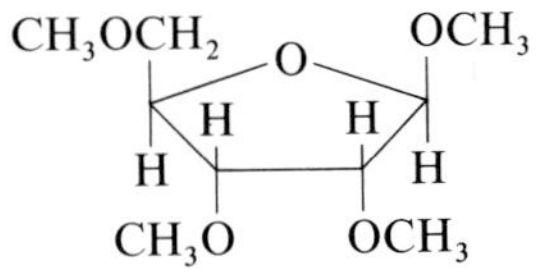

14.46 Using D-arabinose as a typical aldose, write equations for its reaction with the following reagents. Show stereochemistry where known.

(a) CH_3OH, HCl

(b) 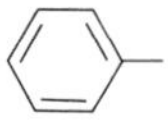   ⟨benzene⟩—$NHNH_2$ (excess), acetic acid   (c) HNO_3, Δ

(d) HIO_4 (excess), H_2O (e) Br_2, H_2O

(f)   product of (a) with NaOH, $(CH_3)_2SO_4$ (g)   product of (f) with dilute, aqueous HCl

(h) $CH_3\overset{O}{\overset{||}{C}}\overset{O}{\overset{||}{C}}CH_3$ (excess), pyridine   (i)   product of (h) with cold HBr

(j)   product of (i) with sodium phenolate   (k)   dilute, aqueous NaOH

(l) 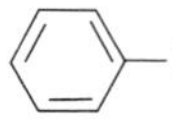   ⟨benzene⟩—NH_2, Δ   (m)   NaCN, Na_2CO_3, H_2O

14.47   Complete the following equations giving structural formulas for all the compounds and intermediates indicated by letters. Show stereochemistry whenever it is known.

(a)

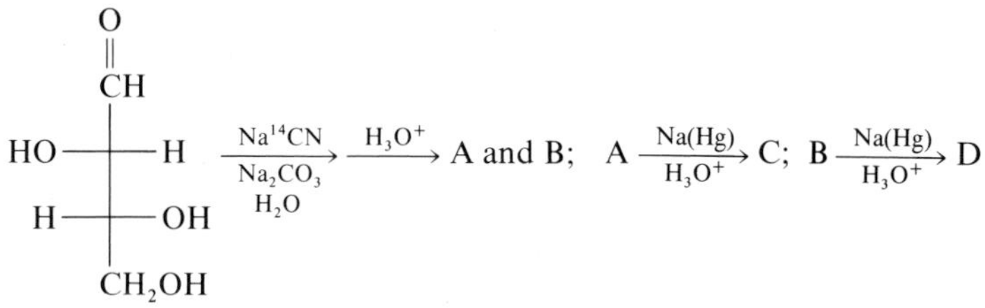

(b)

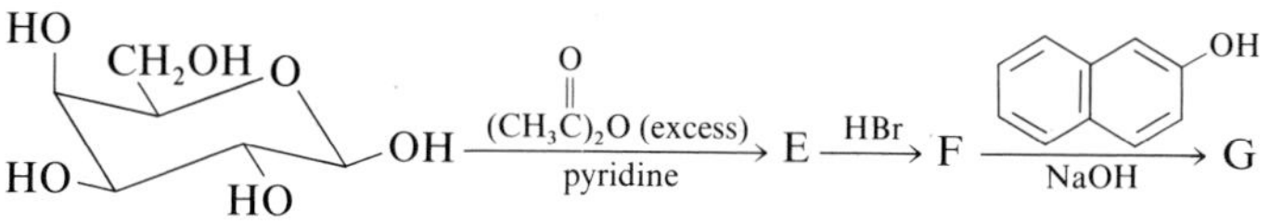

(c)

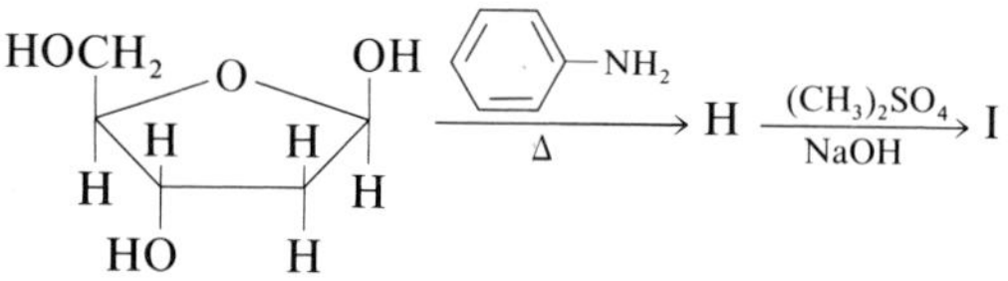

(d)

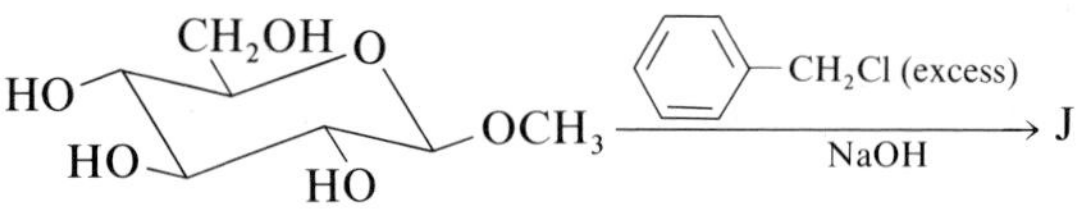

(e)

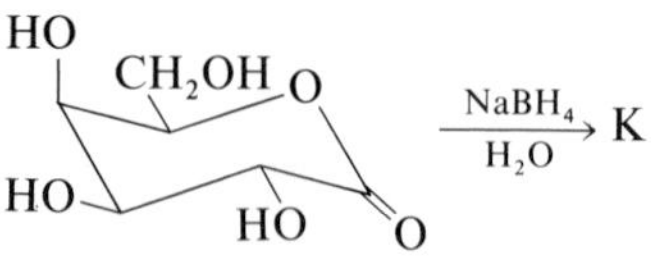

(f)

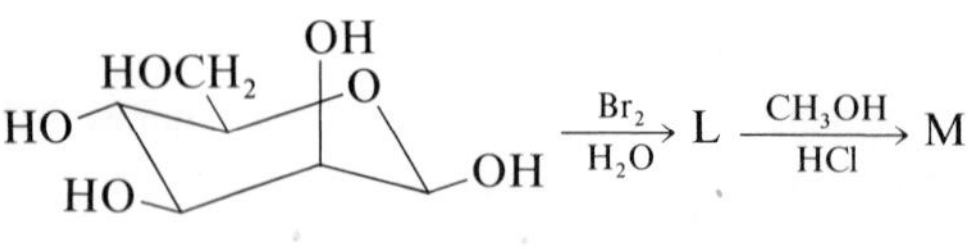

(g)

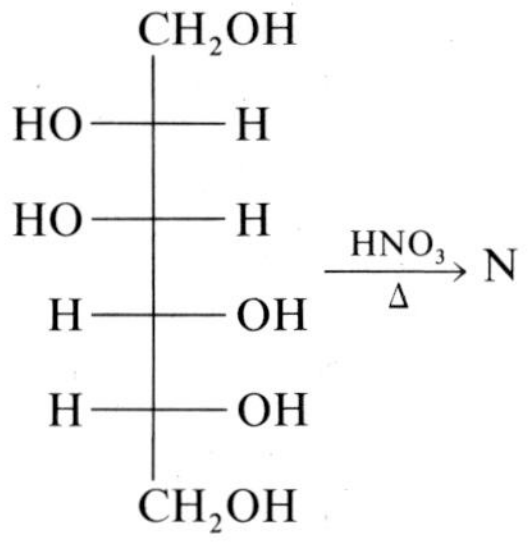

$$\text{CH}_2\text{OH}$$
$$\text{HO}\!-\!\!-\!\text{H}$$
$$\text{HO}\!-\!\!-\!\text{H}$$
$$\text{H}\!-\!\!-\!\text{OH}$$
$$\text{H}\!-\!\!-\!\text{OH}$$
$$\text{CH}_2\text{OH}$$

$\xrightarrow[\Delta]{\text{HNO}_3}$ N

(h)

$\xrightarrow[\Delta]{\text{NH}_3}$ O

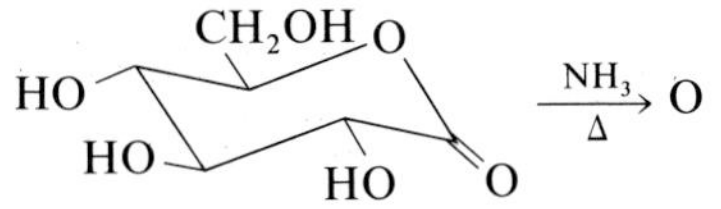

(i)

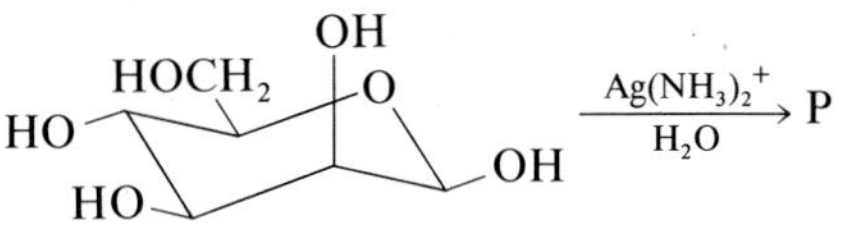

$\xrightarrow[\text{H}_2\text{O}]{\text{Ag(NH}_3)_2^+}$ P

(j)

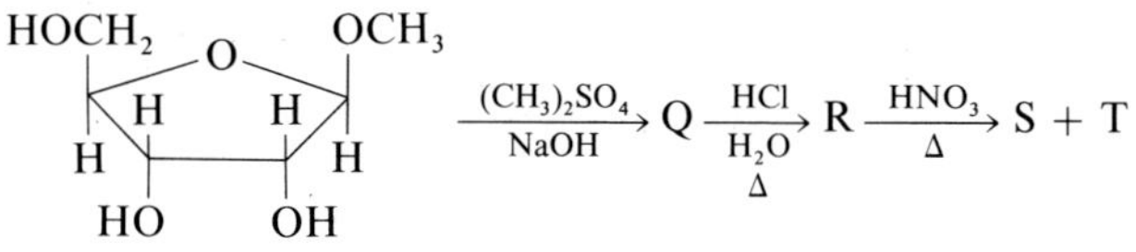

$\xrightarrow[\text{NaOH}]{\text{(CH}_3)_2\text{SO}_4}$ Q $\xrightarrow[\Delta]{\text{HCl}\atop\text{H}_2\text{O}}$ R $\xrightarrow[\Delta]{\text{HNO}_3}$ S + T

(k)

$\xrightarrow[\text{H}_2\text{O}]{\text{HCl}}$ U + V

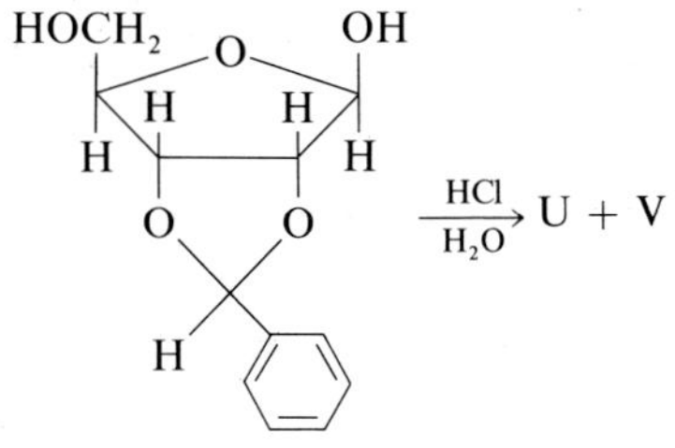

14.48 The absolute configuration of malic acid

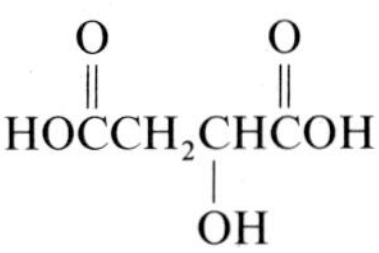

$$\underset{\underset{\text{OH}}{|}}{\text{HOCCH}_2\text{CHCOH}}$$

was determined by synthesis from (+)-tartaric acid (p. 751). (+)-Tartaric acid can be converted to (+)-malic acid by the following chemical reactions. Give the correct stereochemical structures for the compounds present at the different stages of the conversion.

(+)-tartaric acid $\xrightarrow[\text{HCl, }\Delta]{\text{CH}_3\text{CH}_2\text{OH (excess)}}$ A $\xrightarrow[\text{pyridine}]{\text{SOCl}_2}$ B (the reaction of thionyl chloride in pyridine goes with inversion of configuration)

B $\xrightarrow[\text{CH}_3\text{CH}_2\text{OH}]{\text{Zn (a reducing agent)}}$ C $\xrightarrow[\text{NaOH}\atop\Delta]{\text{H}_2\text{O}}$ $\xrightarrow{\text{HCl}\atop\text{H}_2\text{O}}$ (+)-malic acid

14.49 A new antiviral drug, given the trivial name Vira-A, has the scientific name 9-β-D-arabino-furanosyladenine. What is the structure of Vira-A? (The structure of adenine is found on p. 759.)

14.50 (a) When glucose is dissolved in D_2O, it is found that five deuterium atoms are rapidly incorporated into the molecule. Draw a cyclic structure of glucose and show which hydrogens have been rapidly replaced by deuterium atoms.

(b) When glucose is dissolved in $H_2{}^{18}O$, one oxygen atom in glucose slowly exchanges with the water so that glucose eventually contains one ^{18}O. Starting with a cyclic structure of glucose, write a mechanism that would explain this experimental observation.

14.51 Most methods for the preparation of glycosides lead to the β-anomers. It has been found that treatment of an alkyl tetraacetyl-β-D-glycoside with stannic chloride in dry chloroform converts the compound into the α-anomer, which is the more stable one for most glycosides. The reaction is shown below for glucose.

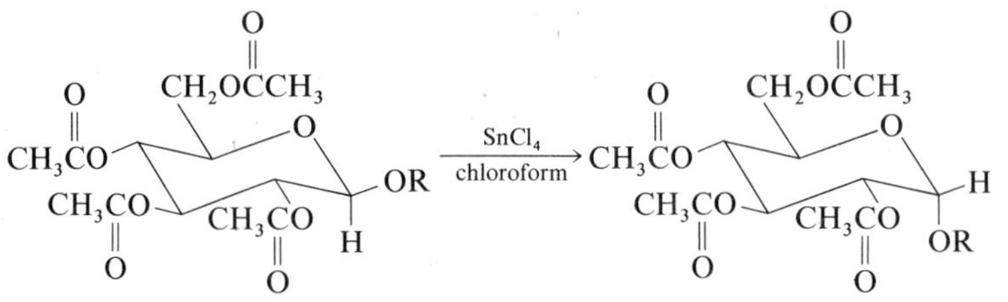

Postulate a possible mechanism for the reaction.

14.52 α-D-Glucose reacts in its furanose form with two molar equivalents of acetone to give cyclic acetals with the hydroxyl groups at carbons 1 and 2 and carbons 5 and 6. Such cyclic acetals are known as isopropylidene derivatives of sugars. Draw the structural formula for the 1,2,5,6-di-O-isopropylidene-D-glucofuranose, and propose a mechanism for its formation.

14.53 Isomaltose is a disaccharide formed by the enzymatic hydrolysis of amylopectin, a form of starch. Isomaltose is a reducing sugar and undergoes mutarotation. It can be hydrolyzed by maltase or by dilute acid to two units of glucose.

Isomaltose is oxidized by bromine water to isomaltobionic acid lactone. The lactone is converted by sodium hydroxide and dimethyl sulfate to the methyl ester of O-octamethyliso-maltobionic acid, which in turn is hydrolyzed to 2,3,4,6-tetra-O-methyl-D-glucopyranose and 2,3,4,5-tetra-O-methyl-D-gluconic acid. Write equations for all the reactions described, showing how they can be used to assign a structure to isomaltose. What is the systematic name for isomaltose?

14.54 A glycoside called gaultherin is isolated from the oil of wintergreen. Gaultherin is not a reducing sugar. It can be hydrolyzed by the enzyme primeverosidase into the disaccharide primeverose and the aglycon methyl salicylate. When hydrolyzed with dilute acid, primeverose, a reducing sugar, gives glucose and xylose.

(a) Xylose (p. 786) is present in primeverose in the pyranose form. Write the structure of β-D-xylopyranose.

(b) Primeverose is reduced by sodium borohydride. The reduction product is hydrolyzed by dilute acid to xylose and sorbitol (the reduction product of glucose). What does this prove about the structure of primeverose?

(c) Primeverose is synthesized from the reaction of α-D-xylopyranosyl bromide 2,3,4-tri-acetate with α-D-glucopyranosyl 1,2,3,4-tetraacetate in the presence of pyridine to give primeverose heptaacetate. Primeverose heptaacetate is then converted to primeverose by treatment with sodium methoxide in methanol. Write equations for these reactions showing the full structures of the reactants, including stereochemistry.

(d) The linkage between primeverose and the aglycon methyl salicylate is β. What is the full structure of gaultherin?

14.55 Cellobiose is prepared from cotton by the following sequence of reactions. Fill in structural formulas for the compounds. How would you classify the last reaction?

$$\text{cotton} \xrightarrow[\text{H}_2\text{SO}_4]{\overset{\overset{\displaystyle O}{\parallel}}{(\text{CH}_3\text{C})_2\text{O}}} \alpha\text{-cellobiose acetate} \xrightarrow[\text{methanol}]{\text{CH}_3\text{ONa}} \text{cellobiose and methyl acetate}$$

14.56 A trisaccharide A, $C_{18}H_{32}O_{16}$, was isolated from a culture of *Betacoccus arabinosaceous*. Total acidic hydrolysis of the carbohydrate gave two moles of glucose and one of galactose. Partial acidic hydrolysis of A gave glucose, galactose, lactose, and a disaccharide B. Trisaccharide A is cleaved by a β-glycosidase to galactose and disaccharide B. The other glycosidic linkage was determined to be α from the optical rotation of the trisaccharide.

Total methylation of A (eleven O-methyl groups) and then acid hydrolysis gave 2,3,4,6-tetra-O-methyl-D-glucopyranose; 2,3,4,6-tetra-O-methyl-D-galactopyranose; and 3,6-di-O-methyl-D-glucopyranose.

(a) Write structural formulas for trisaccharide A and disaccharide B.

(b) Trisaccharide A is a reducing sugar. Reduction of trisaccharide A with $NaBH_4$ gives an alcohol C, $C_{18}H_{34}O_{16}$. Alcohol C reacts with five moles of periodate with the production of two moles of formic acid and one of formaldehyde. Write the structural formula of alcohol C. On the structural formula of alcohol C, indicate clearly the points of cleavage by periodate ion, and which carbon atoms will appear as formic acid and which one as formaldehyde.

14.57 Patients with severe bleeding or burns, or who are undergoing surgery, need to have huge volumes of blood plasma replaced. It is not always possible to provide as much human blood plasma as necessary, so artificial mixtures called plasma volume extenders have been developed to be used as part of the replacement. For example, a chemical modification of the starch fraction amylopectin has been synthesized and used as a plasma extender.

The reaction used in the preparation of the plasma extender is shown below. Predict the structure of the modified polysaccharide formed, assuming that one hydroxyl group has been transformed for each glucose unit.

$$\text{amylopectin} + \text{NaOH} + \overset{\overset{\displaystyle O}{\diagup\;\diagdown}}{\text{CH}_2-\text{CH}_2} \xrightarrow[\text{H}_2\text{O}]{} \text{modified polysaccharide}$$

14.58 A cyanogenic glycoside by the name of amygdalin has been substituted on the market for the compound originally patented as laetrile. Amygdalin has the same aglycon, (R)-$(-)$-mandelonitrile, as laetrile. The aglycon is linked by a β-glycosidic linkage to gentiobiose (Problem 14.38). What is the structure of amygdalin?

14.59 The amygdalin on the market contains considerable amounts of amygdalin in which the asymmetric carbon in mandelonitrile has the S-configuration. Natural amygdalin, (R)-amygdalin, is easily epimerized to (S)-amygdalin by treatment with base. Write a mechanism for this epimerization. Why does it happen so easily?

The Chemistry of Carbanions; Enols and Enolate Anions

15

Enols

A. Carbanions as Reactive Intermediates. A Review

Carbanions *are reactive intermediates in which carbon has a pair of nonbonding electrons and bears a negative charge.* They are strong bases and nucleophiles (Sections 4.8, 6.11, and 9.4A).

Carbanions are generated in two main ways. One method is to remove a proton from an organic compound by using a base. Most carbon-hydrogen bonds are not acidic enough to be broken easily in this way unless there are electron-withdrawing groups on the carbon atom to help stabilize the negative charge on the carbanion. The most effective way to stabilize the negative charge on a carbanion is by delocalization of the charge to adjacent functional groups. For example, the anions derived from the deprotonation of nitromethane (pK_a 10.2, Section 4.8) and acetone (pK_a 20, Section 9.1A) are stabilized by resonance.

If a strong enough base is used, even the hydrogen atom on the triple bond of a terminal alkyne can be removed to give a carbanionic intermediate. Such synthetically useful intermediates are formed in the preparation of organosodium or organomagnesium derivatives of alkynes (Section 9.4A).

The other common method of generating carbanions is by the reaction of organic halides with metals to give organometallic reagents. Prepared in this way, Grignard reagents and organolithium reagents react as sources of groups containing strongly basic and nucleophilic carbon atoms. We explored their reactions both as bases and as nucleophiles in Section 9.4, and we examined the versatility of such reagents in forming new carbon-carbon bonds.

Carbanions generated by deprotonation of carbonyl compounds also react as nucleophiles with the electrophilic carbon atoms of carbonyl groups or alkyl halides to give new carbon-carbon bonds. The remainder of this chapter shows the synthetic and biological importance of these reactions.

B. Enols and Enolate Anions

The acidity of a hydrogen atom on the carbon atom adjacent to a carbonyl group varies with the structure of the carbonyl compound. For example, the following data are found in the table of pK_a values inside the front cover of the book.

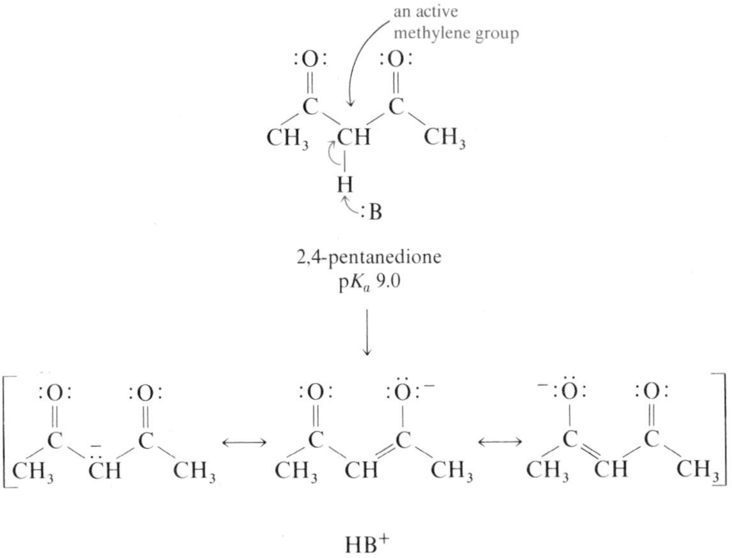
an active
methylene group
2,4-pentanedione
pK_a 9.0
HB$^+$

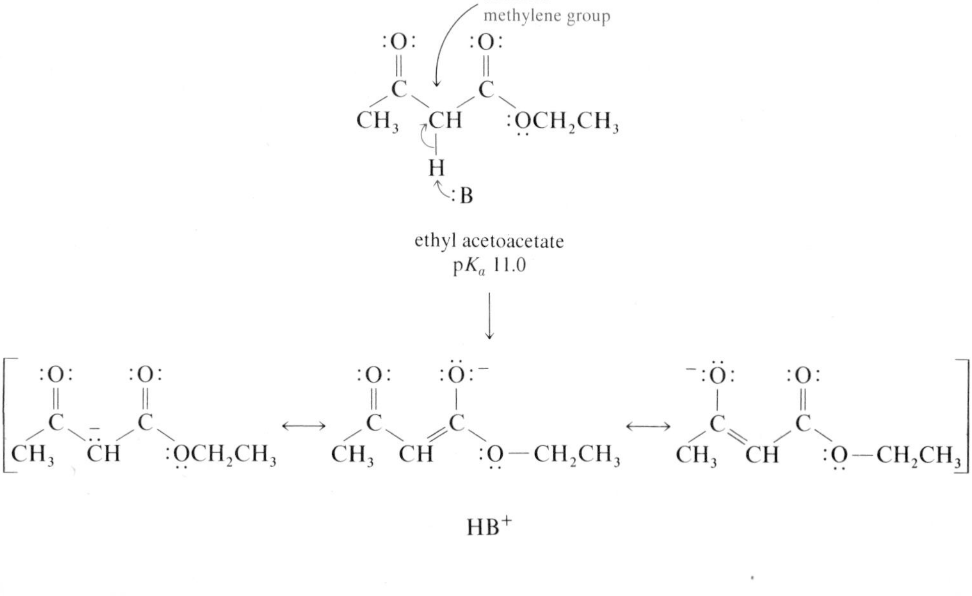
an active
methylene group
ethyl acetoacetate
pK_a 11.0
HB$^+$

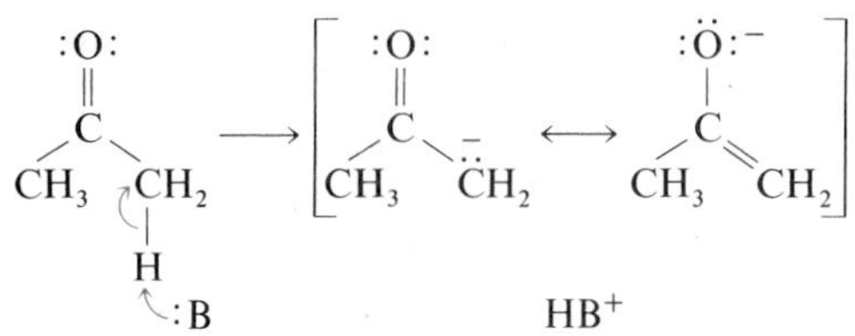

acetone
pK_a 20

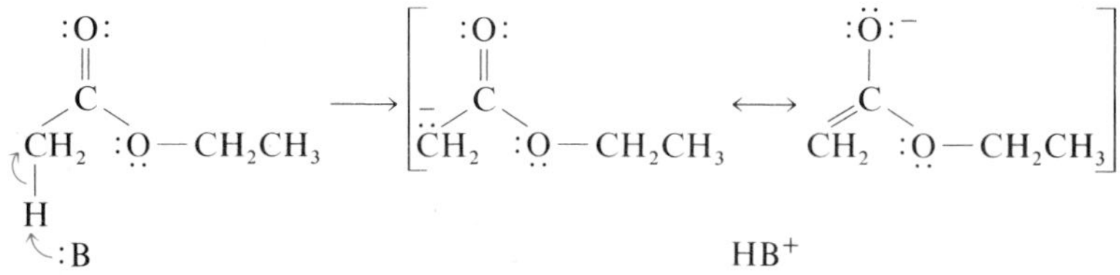

ethyl acetate
pK_a 23

Removal of a proton from an α-carbon atom in a carbonyl compound gives rise to an anion that is stabilized by delocalization of the negative charge to the electronegative oxygen atom of the carbonyl group. Such an anion is called an **enolate anion** (Section 9.1A). An examination of the structural formulas and the pK_a values shown above indicates that the compounds fall into two categories. 2,4-Pentanedione (pK_a 9.0) and ethyl 3-oxobutanoate (pK_a 11.0), which is commonly called ethyl acetoacetate, have reasonable acidity. In both of them the most acidic hydrogen atom is on the carbon atom that lies between two carbonyl groups, and resonance contributors can be drawn for their enolate anions showing the delocalization of the charge to two different oxygen atoms. *Compounds that contain a methylene group α to two carbonyl groups are said to contain an* **active methylene group.** The hydrogen atoms of an active methylene group are easily removed by bases because of the stability of the conjugate base, the enolate anion from such compounds.

Acetone (pK_a 20) and ethyl acetate (pK_a 23) are much less acidic than 2,4-pentanedione and ethyl acetoacetate. Removal of a proton from each of these compounds also creates an enolate anion, but in these cases the stabilization of the anion comes from delocalization of the charge to only one oxygen atom. The conjugate bases of acetone and ethyl acetate are not stabilized, relative to their acids, as much as the conjugate bases of compounds containing two carbonyl groups are.

The carbonyl group of an ester does not contribute as much to the stabilization of an enolate anion as the carbonyl group of a ketone does. Ethyl acetoacetate is a weaker acid than 2,4-pentanedione, and ethyl acetate is a weaker acid than acetone. The differences in polarity and electrophilicity between carbonyl groups in ketones (and aldehydes) and those in acid derivatives have been thoroughly discussed in several places, especially in connection with the reactivity of acid derivatives (Section 10.1B). The presence of unshared pairs of electrons on the oxygen atom of the alkoxyl group of the ester diminishes the effectiveness of the carbonyl group of the ester in delocalizing charge.

Besides carbonyl groups, there are other groups that increase the acidity of hydrogen atoms bonded to the carbon atoms adjacent to them. Two of the most important are the nitro group, which is even more effective than a carbonyl group in delocalizing charge (Section 4.8), and the cyano group, which is not quite as effective as a ketone in stabilizing charge but better than an ester group.

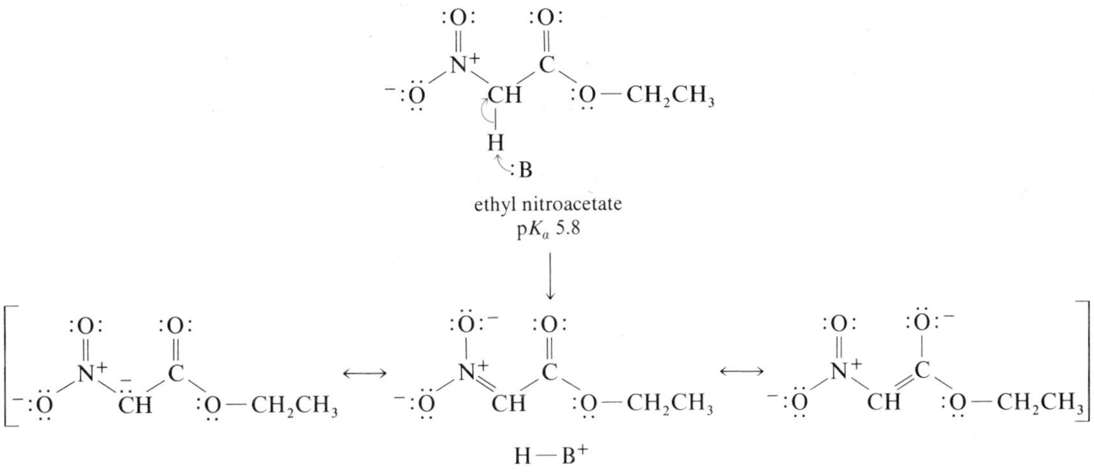

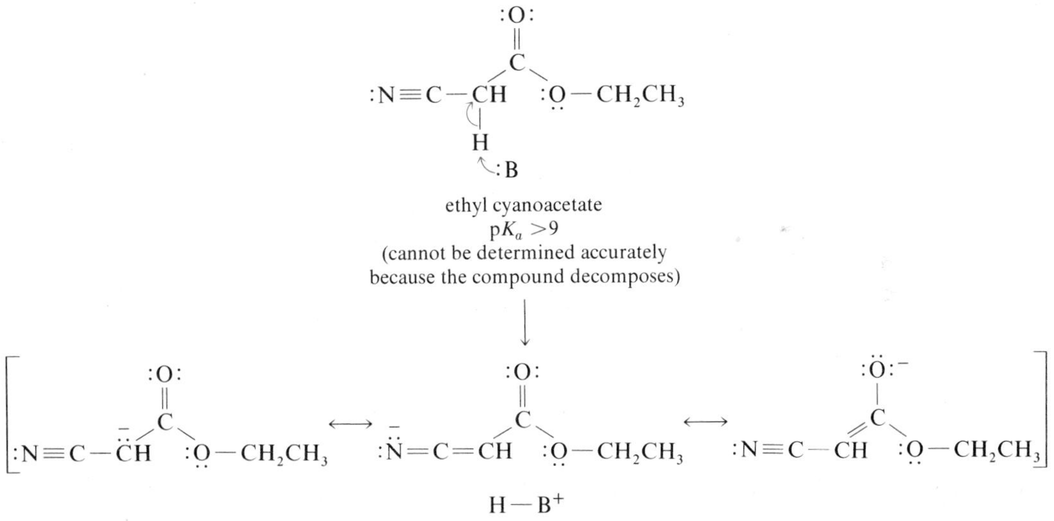

$$:\!O\!: \qquad :\!O\!:$$
$$\parallel \qquad\qquad \parallel$$
$$N^{+} \qquad\quad C$$
$$-:\!\ddot{O} \qquad CH \qquad :\!\underset{..}{O}-CH_2CH_3$$
$$H$$
$$\curvearrowleft :B$$

ethyl nitroacetate
pK_a 5.8

$$\Bigg[\;\; \begin{array}{ccc} :\!O\!: & :\!O\!: \\ \parallel & \parallel \\ N^{+} & C \\ -:\!\ddot{O} \quad \bar{C}H & :\!\underset{..}{O}-CH_2CH_3 \end{array} \;\longleftrightarrow\; \begin{array}{ccc} :\!\ddot{O}\!:^{-} & :\!O\!: \\ \parallel & \parallel \\ N^{+} & C \\ -:\!\ddot{O} \quad CH & :\!\underset{..}{O}-CH_2CH_3 \end{array} \;\longleftrightarrow\; \begin{array}{ccc} :\!O\!: & :\!\ddot{O}\!:^{-} \\ \parallel & \parallel \\ N^{+} & C \\ -:\!\ddot{O} \quad CH & :\!\underset{..}{O}-CH_2CH_3 \end{array} \;\Bigg]$$

H—B$^+$

$$:\!O\!:$$
$$\parallel$$
$$C$$
$$:N\!\equiv\!C-CH \qquad :\!\underset{..}{O}-CH_2CH_3$$
$$H$$
$$\curvearrowleft :B$$

ethyl cyanoacetate
pK_a >9
(cannot be determined accurately
because the compound decomposes)

$$\Bigg[\;\; \begin{array}{cc} :\!O\!: \\ \parallel \\ C \\ :N\!\equiv\!C-\bar{C}H \quad :\!\underset{..}{O}-CH_2CH_3 \end{array} \;\longleftrightarrow\; \begin{array}{cc} :\!O\!: \\ \parallel \\ C \\ :\!\ddot{N}\!=\!C\!=\!CH \quad :\!\underset{..}{O}-CH_2CH_3 \end{array} \;\longleftrightarrow\; \begin{array}{cc} :\!\ddot{O}\!:^{-} \\ \mid \\ C \\ :N\!\equiv\!C-CH \quad :\!\underset{..}{O}-CH_2CH_3 \end{array} \;\Bigg]$$

H—B$^+$

The process of converting a carbonyl compound into its enol is called **enolization.** The base-catalyzed enolization of carbonyl compounds is shown above. Acids also promote enolization as demonstrated by the acid-catalyzed enolization of acetone.

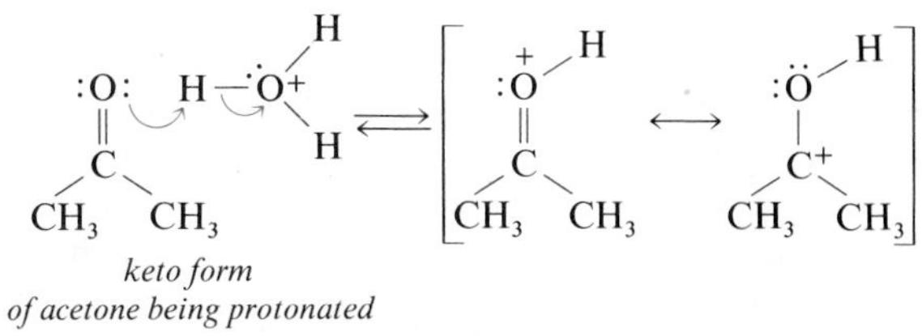

$$\begin{array}{cc} & H \\ :\!O\!: & H-\ddot{O}^{+} \\ \parallel & \\ C & H \\ CH_3 \quad CH_3 \end{array} \;\rightleftharpoons\; \Bigg[\;\; \begin{array}{cc} & H \\ :\!\overset{+}{O} & \\ \mid & \\ C & \\ CH_3 \quad CH_3 \end{array} \;\longleftrightarrow\; \begin{array}{cc} & H \\ :\!\ddot{O} & \\ \mid & \\ C^{+} & \\ CH_3 \quad CH_3 \end{array} \;\Bigg]$$

keto form
of acetone being protonated

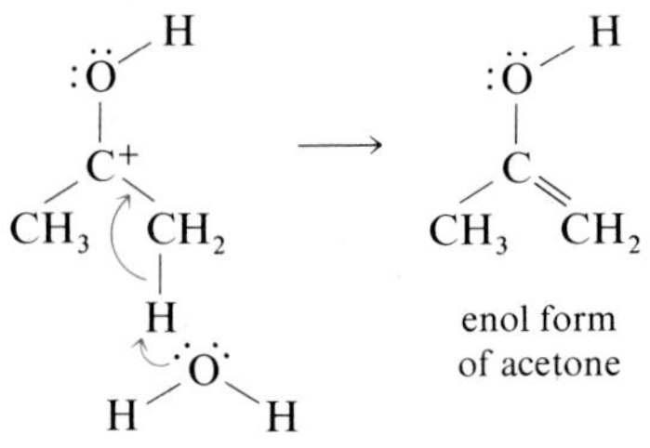

*loss of proton from
the α-carbon atom of the
protonated ketone*

enol form
of acetone

The ketone, protonated at the carbonyl group in acidic solution, loses a proton from the α-carbon atom to give an enol.

All carbonyl compounds are in equilibrium with their enol forms. The percentage of enol present at any given time depends upon the structure of the carbonyl compound and on other factors, such as the solvent. In the case of a compound with a single carbonyl group such as acetone, the equilibrium lies far on the side of the ketone. For 1,3-dicarbonyl compounds such as 2,4-pentanedione, the enol form is a significant part of the equilibrium mixture.

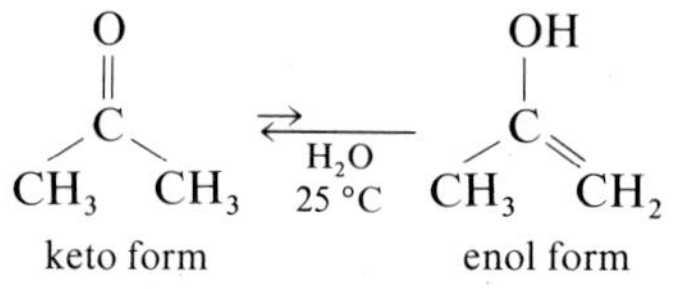

keto form

equilibrium lies
far on the side
of keto form

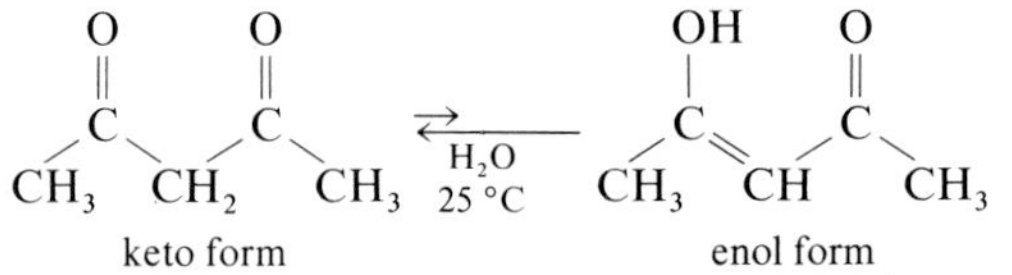

enol form

keto form
83% at equilibrium

enol form
17% at equilibrium

PROBLEM 15.1 At equilibrium, 34% of 1-phenyl-1,3-butanedione exists in its enol form, compared with 17% for 2,4-pentanedione under the same conditions (above). Write an equation for the enolization of 1-phenyl-1,3-butanedione and offer an explanation for its more extensive enolization.

C. Ambident Nucleophiles

An examination of the structure of an enolate anion shows us that it has electron density and negative charge at two sites. It is basic and nucleophilic at the α-carbon atom and at the oxygen atom of the carbonyl group; it can be protonated or can react with other electrophiles at each of those sites. An anion that has the capacity to react at either of two positions is called an **ambident anion** or **nucleophile.** Whether it reacts at the carbon atom or at the oxygen atom depends on the conditions, such as the positive ion

associated with the anion, the solvent, and the nature of the electrophile with which it
is reacting.

The enolate anion derived from diethyl malonate was used as a typical nucleophile
in nucleophilic substitution reactions in Section 6.11. *Enolate anions react with alkyl*
halides in nucleophilic substitution reactions known as **alkylation reactions.** The
enolate anion from ethyl acetoacetate also undergoes nucleophilic substitution reactions
with alkyl halides. When this reaction is carried out in ethanol, which hydrogen bonds
to the negatively charged oxygen atom of the enolate ion, the anion reacts as a carbon
nucleophile to form a new bond at the α-carbon atom of the ester.

$$CH_3CCH_2COCH_2CH_3 \xrightarrow[\text{ethanol}]{CH_3CH_2O^-Na^+} \left[\cdots \longleftrightarrow \cdots \right]$$

ethyl acetoacetate

$$\xrightarrow[\text{ethanol}]{CH_3Br} CH_3CCHCOCH_2CH_3 + NaBr$$
$$\quad\quad\quad\quad\quad\quad\quad\quad\quad\quad CH_3$$

ethyl 2-methylacetoacetate
83%

alkylation at the
carbon atom

The ambident nucleophile has undergone alkylation at the carbon atom. If the enolate
anion is generated in dimethylformamide, a polar solvent (Section 6.2B) that does not
hydrogen bond, reaction with *n*-butyl iodide gives 99% alkylation at the α-carbon
atom. If the alkyl halide used is changed to *n*-butyl bromide, 67% alkylation at carbon
and 33% alkylation at the oxygen atom results.

$$CH_3CCH_2COCH_2CH_3 \xrightarrow[\substack{\text{dimethylformamide} \\ 100\,°C}]{K_2CO_3} \left[\cdots \longleftrightarrow \cdots \right] K^+$$

$$\xrightarrow{CH_3CH_2CH_2CH_2X} CH_3CCHCOCH_2CH_3 \quad + \quad CH_3C=CHCOCH_2CH_3$$
$$\quad\quad\quad\quad\quad\quad\quad\quad\quad\quad CH_2CH_2CH_2CH_3 \quad\quad\quad\quad\quad\quad\quad\quad O$$

ethyl 2-butylacetoacetate	an enol ether
product from alkylation at the carbon atom	product from alkylation at the oxygen atom
X = I 99%	1%
X = Br 67%	33%

The experimental data indicate that the nature of the leaving group on the alkylating
agent and the nature of the solvent are important in determining whether the enolate
anion will react as a carbon nucleophile or an oxygen nucleophile. Protic solvents that
hydrogen bond with the oxygen anion increase alkylation at the carbon atom. Reagents
with large polarizable leaving groups such as the iodide ion react chiefly at the carbon

atom. Other experiments have shown that the nature of the metal ion associated with the enolate anion and whether or not it coordinates tightly with the oxygen atom also influence the ratio of carbon versus oxygen alkylation.

The reaction conditions for syntheses using enolate anions are usually chosen to give mostly alkylation at the carbon atom.

D. Relative Stabilities of Enolate Anions

An unsymmetrical carbonyl compound enolizes in two different ways. For example, when 2-methylcyclohexanone is heated with the base triethylamine in dimethylformamide in the presence of trimethylchlorosilane, a mixture of two trimethylsilyl enol ethers is formed.

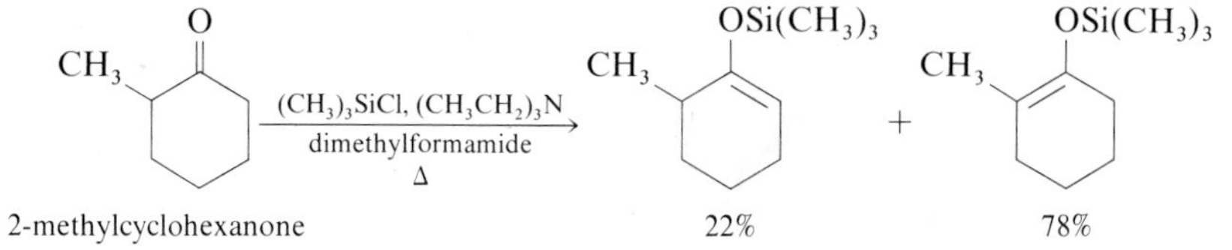

2-methylcyclohexanone   22%   78%

Trimethylchlorosilane is a reagent that reacts exclusively with the enolate anion at the oxygen atom because a strong silicon-oxygen bond is formed. Therefore, the reagent is used to trap the enolate anions as stable trimethylsilyl enol ethers. The composition of the mixture of enol ethers reflects the relative stabilities of the two enolate anions derived from 2-methylcyclohexanone.

more stable enolate anion, more highly substituted double bond

less stable enolate anion, less highly substituted double bond

major product of reaction

minor product of reaction

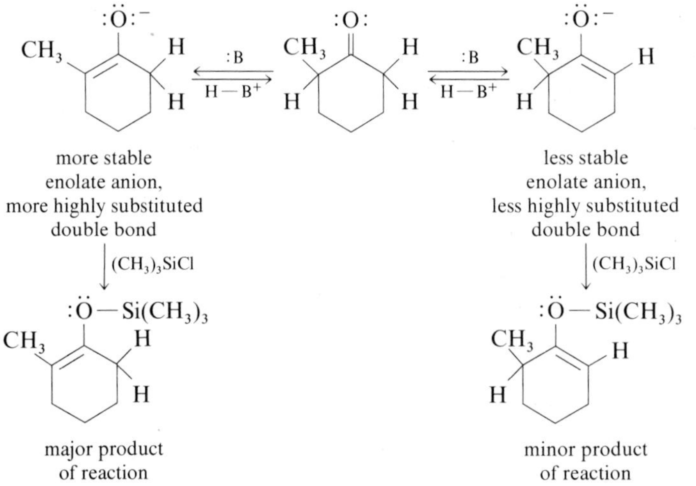

The enolate anion with the more highly substituted double bond is more stable than the other one and predominates under conditions that allow equilibrium to be established.

The conditions described for the reaction above allow for equilibration between the two enolate anions. The reaction is run at a high temperature. Triethylamine is not a strong base, therefore the conjugate acid of triethylamine as well as unreacted ketone molecules serve as acids to interconvert the enolate anions by way of the ketone. All these factors ensure that the mixture observed reflects the relative thermodynamic stabilities of the anions. *The more highly substituted enolate anion is said to be the* **thermodynamic enolate** of the ketone.

Enolate anions are versatile intermediates in reactions leading to new carbon-carbon bonds. Therefore, it is important that chemists know and be able to control the composition of mixtures obtained from unsymmetrical carbonyl compounds. Gilbert Stork at Columbia University and Herbert O. House at the Georgia Institute of Technology have made major contributions to methods for regioselective generation of enolate anions. For example, when 2-methylcyclohexanone is added slowly to a solution of the very strong base lithium diisopropylamide in 1,2-dimethoxyethane at 0 °C and the resulting mixture is then treated with trimethylchlorosilane in the presence of triethylamine, the trimethylsilyl ether of the less highly substituted enol is obtained almost exclusively.

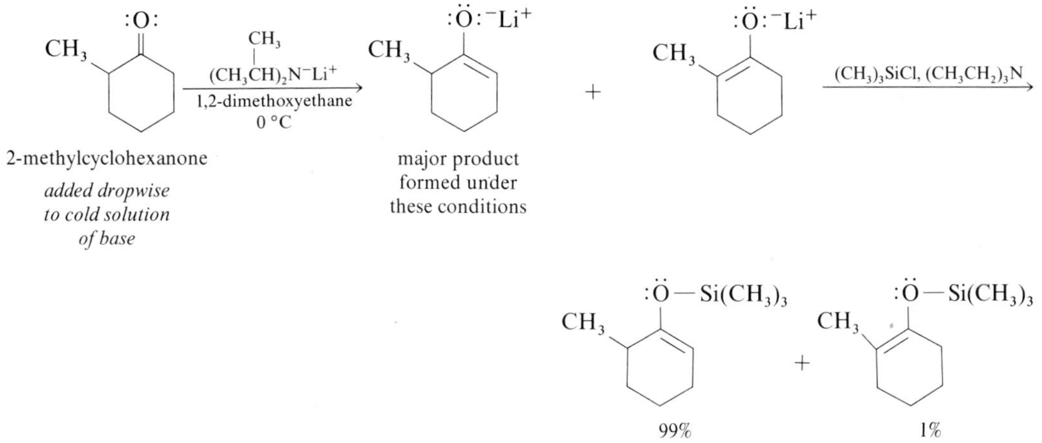

2-methylcyclohexanone

*added dropwise
to cold solution
of base*

major product
formed under
these conditions

99%   1%

The above reaction conditions do not allow for equilibration between the enolate anions. A very strong base, the diisopropylamide anion, is used to deprotonate the ketone. The conjugate acid of the anion, diisopropylamine ($pK_a \sim 36$), is not a strong enough acid to protonate the enolate anion once it is formed. The other acid in the system, the ketone itself ($pK_a \sim 16$), is never present at the same time as the enolate anion because it is added slowly to the solution of the base. An excess of base is always present and the ketone is deprotonated completely. The temperature is also kept low. Under these conditions *the major product is the* **kinetic enolate,** *the enolate that is formed the fastest,* not the one of greatest thermodynamic stability. Kinetic enolates tend to be the less highly substituted ones. *The rate-determining step in enolization is the breaking of the carbon-hydrogen bond at the α-carbon atom of the carbonyl compound.* The hydrogen atoms on the less highly substituted α-carbon atom are less sterically hindered and react fastest with the strong base.

The reasons for the choice of lithium diisopropylamide as the base are interesting. The reagent is easily prepared by treating diisopropylamine in an ether or hydrocarbon solvent with an alkyllithium. *n*-Butyllithium is usually used. The powerful base, *n*-butyl anion, deprotonates diisopropylamine completely, leaving a solution of lithium diisopropylamide.

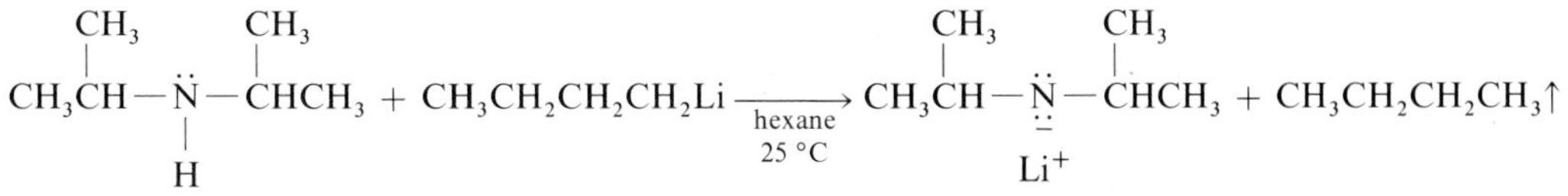

diisopropylamine   *n*-butyllithium   lithium diisopropylamide   butane

Solutions of the base in hydrocarbons are stable, but those in ether solvents are not. The base is strong enough to deprotonate ethers, so those solutions must be kept cold and used as soon as they are prepared.

The diisopropylamide anion is a strong base but is hindered enough that it is not a good nucleophile. Thus, it does not react with the alkyl halides and other reagents that react with the enolate anions in subsequent steps of the reaction. The conjugate acid of diisopropylamide anion, diisopropylamine, boils at 86 °C and is easily separated from the other products of the reactions. The small lithium ion, with its high concentration of positive charge, coordinates strongly with the oxygen atom and enhances the regioselectivity of the enolization.

PROBLEM 15.2   Predict what the major product of each of the following reactions would be.

$$
\text{(a)} \quad \text{CH}_3\text{CH}_2\overset{\displaystyle O}{\overset{\|}{\text{C}}}\text{CH}_3 \xrightarrow[\text{dimethylformamide} \atop \Delta]{(\text{CH}_3)_3\text{SiCl, (CH}_3\text{CH}_2)_3\text{N}}
$$

$$
\text{(b)} \quad \text{CH}_3\text{CH}_2\overset{\displaystyle O}{\overset{\|}{\text{C}}}\text{CH}_3 \xrightarrow[\text{1,2-dimethoxyethane}]{\overset{\displaystyle \text{CH}_3}{(\text{CH}_3\text{CH})_2\text{N}^-\text{Li}^+}} \xrightarrow{(\text{CH}_3)_3\text{SiCl, (CH}_3\text{CH}_2)_3\text{N}}
$$

E. Exchange of α-Hydrogen Atoms with Deuterium Atoms

One of the easiest ways to test for the presence of enolizable α-hydrogen atoms is to see whether any of the hydrogen atoms in a molecule can be exchanged for the deuterium in deuterium oxide. The hydrogen atom of an active methylene group exchanges with the solvent so readily that the reaction will take place in water with no added catalyst.

$$
\underset{\substack{\displaystyle | \\ \text{CH}_3 \\ \text{3-methyl-2,4-pentanedione}}}{\text{CH}_3\overset{O}{\overset{\|}{\text{C}}}\text{CH}\overset{O}{\overset{\|}{\text{C}}}\text{CH}_3} + \text{D}_2\text{O} \underset{\substack{\text{several} \\ \text{days}}}{\rightleftharpoons} \underset{\substack{\displaystyle | \\ \text{CH}_3 \\ \text{3-deuterio-3-methyl-} \\ \text{2,4-pentanedione}}}{\text{CH}_3\overset{O}{\overset{\|}{\text{C}}}\text{CD}\overset{O}{\overset{\|}{\text{C}}}\text{CH}_3} + \text{HOD}
$$

If 3-methyl-2,4-pentanedione is allowed to stand in deuterium oxide at room temperature for several days, one of its hydrogen atoms is exchanged for a deuterium atom. The reaction is an equilibrium reaction, and complete exchange requires that fresh deuterium oxide be added to the reaction mixture from time to time. Note that of all the α-hydrogen atoms in the molecule, the one alpha to two carbonyl groups exchanges under these conditions.

The α-hydrogen atoms of ordinary ketones can also be exchanged for deuterium atoms, but acid or base catalysis is usually required for the reaction to proceed at a reasonable rate. Acetone exchanges its hydrogens for deuterium when placed in deuterium oxide in the presence of a trace of sulfuric acid.

$$
\underset{\text{acetone}}{\text{CH}_3\overset{O}{\overset{\|}{\text{C}}}\text{CH}_3} + \text{D}_2\text{O} \xrightarrow{\text{H}_2\text{SO}_4} \text{CH}_3\overset{O}{\overset{\|}{\text{C}}}\text{CH}_2\text{D} \xrightarrow{\text{D}_2\text{O} \atop \text{H}_2\text{SO}_4} \xrightarrow{\text{D}_2\text{O} \atop \text{H}_2\text{SO}_4} \underset{\substack{>95\% \\ \text{deuterium}}}{\text{CD}_3\overset{O}{\overset{\|}{\text{C}}}\text{CD}_3}
$$

If the reaction time is extended and fresh deuterium oxide is added to the reaction mixture, eventually all of the α-hydrogen atoms in acetone are replaced by deuterium atoms.

PROBLEM 15.3 Complete the following equations.

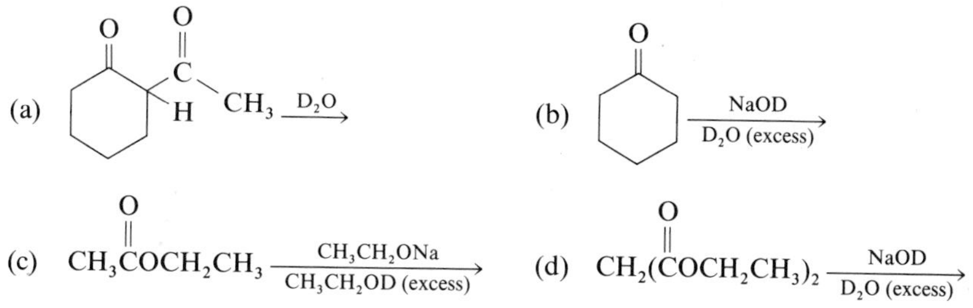

(a)
$\xrightarrow{\text{D}_2\text{O}}$

(b)
$\xrightarrow[\text{D}_2\text{O (excess)}]{\text{NaOD}}$

(c) $\text{CH}_3\overset{\text{O}}{\overset{\|}{\text{C}}}\text{OCH}_2\text{CH}_3 \xrightarrow[\text{CH}_3\text{CH}_2\text{OD (excess)}]{\text{CH}_3\text{CH}_2\text{ONa}}$

(d) $\text{CH}_2(\text{COCH}_2\text{CH}_3)_2 \xrightarrow[\text{D}_2\text{O (excess)}]{\text{NaOD}}$

15.2

Halogenation of Carbonyl Compounds

A. Halogenations in Acidic Media

Ketones react with halogens to give products in which a halogen atom is substituted for one of the α-hydrogen atoms. For example, acetone reacts with bromine to give chiefly 1-bromo-2-propanone.

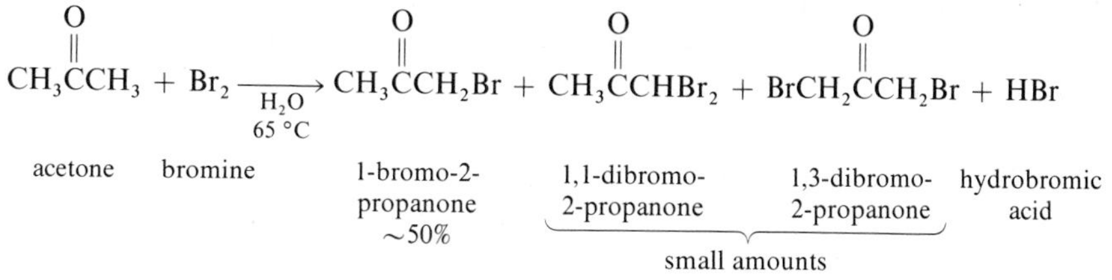

$$\text{CH}_3\overset{\text{O}}{\overset{\|}{\text{C}}}\text{CH}_3 + \text{Br}_2 \xrightarrow[\substack{\text{H}_2\text{O}\\65\,°\text{C}}]{} \text{CH}_3\overset{\text{O}}{\overset{\|}{\text{C}}}\text{CH}_2\text{Br} + \text{CH}_3\overset{\text{O}}{\overset{\|}{\text{C}}}\text{CHBr}_2 + \text{BrCH}_2\overset{\text{O}}{\overset{\|}{\text{C}}}\text{CH}_2\text{Br} + \text{HBr}$$

acetone   bromine   1-bromo-2-   1,1-dibromo-   1,3-dibromo- hydrobromic
                       propanone   2-propanone   2-propanone   acid
                         ~50%
                              small amounts

A solution of bromine in water is acidic, so the reaction is believed to go by an acid-catalyzed enolization of acetone (p. 805).

$$\text{Br}_2 + \text{H}_2\text{O} \rightleftharpoons \text{HBr} + \text{HOBr}$$

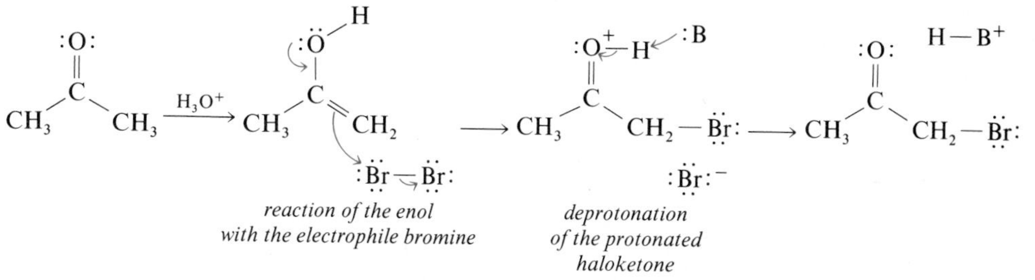

reaction of the enol   *deprotonation*
with the electrophile bromine   *of the protonated*
                                  haloketone

The enol reacts with the electrophilic halogen in a nucleophilic substitution reaction to give the protonated haloketone. Loss of the proton to one of the bases present in the mixture completes the reaction.

Chemists have studied the mechanism of such halogenation reactions in great detail. The rate of the halogenation is equal to the rate at which deuterium is introduced into acetone and depends only on the concentration of the ketone. These facts suggest that the formation of the enol from the ketone is the rate-determining step for both of these reactions and that once the enol is formed, it reacts quickly with any electrophiles that may be present.

The halogenation reaction is applicable to a wide range of compounds, as illustrated by another example, the chlorination of cyclohexanone.

| cyclohexanone | chlorine | | 2-chlorocyclo-hexanone ~60% | hydrochloric acid |

In this case too, one of the α-hydrogen atoms on the ketone is substituted by a halogen atom.

PROBLEM 15.4

(a) 3-Methyl-2,4-pentanedione reacts with bromine in water to give a product containing only one bromine atom. Write an equation for the reaction.

(b) If 3-methyl-2,4-pentanedione is dissolved in water and allowed to stand for a day, the initial rate at which this solution reacts with bromine is very fast. After a while, the bromine color disappears much more slowly but steadily. Explain these experimental observations.

PROBLEM 15.5   The α-hydrogen atom on a carboxylic acid can also be replaced by a halogen atom to give an α-haloacid. This reaction, which has the impressive name of the **Hell-Volhard-Zelinsky reaction,** works best under conditions in which the acid is first converted in catalytic amounts into its acid halide. Write a mechanism for the reaction and explain why the halogenation works better with the acid halide than with the acid itself.

B. Halogenations in Basic Solutions. The Haloform Reaction

In acidic solutions, only one of the α-hydrogen atoms in a ketone is replaced by a halogen atom in the major product. The reaction does give small amounts of di-halogenated products, as seen for acetone (p. 811), and the addition of more halogen and longer reaction times lead to further substitution. It is not difficult, however, to stop the halogenation when only one halogen atom has been introduced into the ketone. *If a carbonyl compound is treated with halogen in the presence of base, however, the situation is quite different,* as seen for the reaction of acetone with iodine in sodium hydroxide solution.

$$CH_3CCH_3 + I_2 \text{ (excess)} + NaOH \xrightarrow{H_2O} CHI_3\downarrow + CH_3CO^-Na^+$$

acetone   iodine   sodium   triiodomethane   sodium
                            hydroxide   iodoform   acetate

A basic solution of acetone gives a yellow precipitate almost as soon as iodine is added to the reaction mixture. The precipitate is triiodomethane, also known as iodoform. Many hospitals smell of iodoform, which is used as an antiseptic.

The reaction of methyl ketones with halogens in base is known as the **haloform reaction.** With chlorine, the product is trichloromethane (chloroform), and with bromine, it is tribromomethane (bromoform). The haloform reaction is typical of most compounds containing a methyl ketone unit as part of their structure, and of compounds that can be oxidized to a methyl ketone by the basic solution of halogen. For example, secondary alcohols such as 2-pentanol, in which the hydroxyl group is one carbon away from the end of the chain, also give a precipitate of iodoform when treated with iodine and sodium hydroxide.

$$CH_3CH_2CH_2CHCH_3 \xrightarrow[\substack{H_2O \\ oxidizing \\ agent}]{I_2,\ NaOH} CH_3CH_2CH_2CCH_3 \xrightarrow[H_2O]{I_2,\ NaOH}$$
$$\qquad\quad |$$
$$\qquad\quad OH$$

2-pentanol   2-pentanone

$$CHI_3\downarrow + CH_3CH_2CH_2CO^-Na^+ \xrightarrow{H_3O^+} CH_3CH_2CH_2COH$$

iodoform   sodium butanoate   butanoic acid

yellow   *carboxylic acid with one
                                             fewer carbon atom than
                                             the starting alcohol or
                                             ketone*

Removing the methyl group from a methyl ketone as iodoform leaves the remainder of the molecule in the form of a carboxylate anion, which gives a carboxylic acid on protonation. Thus the haloform reaction may also be used to synthesize a carboxylic acid having one fewer carbon atom than the starting ketone.

Iodoform is very insoluble in water; even small amounts of it precipitate and can be detected. This reaction, called the **iodoform reaction,** is useful as a qualitative analysis test for methyl ketones and compounds that can be oxidized to methyl ketones. Acetaldehyde is the only aldehyde that gives the test, and ethanol the only primary alcohol.

$$CH_3CH_2OH \xrightarrow[\substack{NaOH \\ H_2O}]{I_2} CH_3CH \xrightarrow[\substack{NaOH \\ H_2O}]{I_2} CHI_3\downarrow + HCO^-Na^+$$

ethanol   acetaldehyde   sodium
                                                         formate

The reaction takes place by a rapid replacement of the α-hydrogen atoms through the enolate ion present in the basic solution.

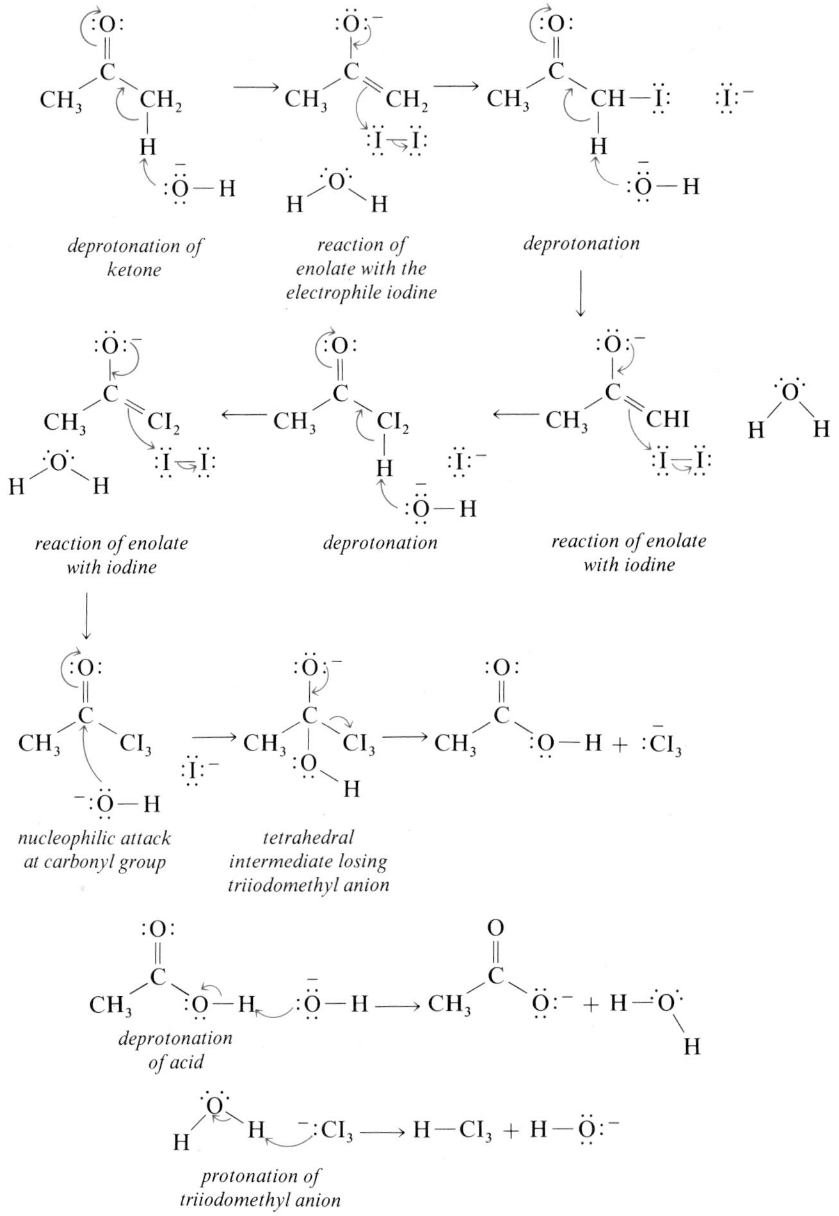

*deprotonation of
ketone*

*reaction of
enolate with the
electrophile iodine*

deprotonation

*reaction of enolate
with iodine*

deprotonation

*reaction of enolate
with iodine*

*nucleophilic attack
at carbonyl group*

*tetrahedral
intermediate losing
triiodomethyl anion*

*deprotonation
of acid*

*protonation of
triiodomethyl anion*

Once one of the α-hydrogen atoms is replaced by halogen, the remaining hydrogen atoms on the same carbon atom become more acidic because of the electron-withdrawing effect of the halogen. Each successive hydrogen atom is more easily substituted until the trihalogenated compound is formed. In some cases, such a product can be isolated, but most of the time it is so unstable that it falls apart when the carbonyl group is attacked by base. The methyl group adjacent to the carbonyl group is converted into a reasonably good leaving group by substitution of halogen atoms for hydrogen atoms. In the case of iodine, especially, the bulkiness of the iodine atoms results in a very crowded molecule. The electron-withdrawing effect of the three halogen atoms also increases the electrophilicity of the carbon atom of the carbonyl group. All these factors combine to make it possible to break a carbon-carbon bond in the ketone by nucleophilic attack of a hydroxyl ion on the carbonyl group, a reaction that we have not seen before. The triiodomethyl anion is a stabilized carbanion and picks up a proton from the solvent to give iodoform.

PROBLEM 15.6   The haloform reaction was used extensively early in this century in determining the structure of terpenes. In a typical sequence of reactions, γ-bisabolene, $C_{15}H_{24}$, was found to be hydrogenated over platinum in cyclohexane at room temperature to a tetrahydrobisabolene, $C_{15}H_{28}$. This compound absorbed one more mole of hydrogen when the solvent was changed to acetic acid to give hexahydrobisabolene, $C_{15}H_{30}$.

Tetrahydrobisabolene was subjected to ozonolysis and cleaved to form two fragments. One of the fragments was converted to 5-methylhexanoic acid when treated with bromine in sodium hydroxide solution. Reconstruct the carbon skeleton of γ-bisabolene from the information given here and your general knowledge about terpenes (Section 11.6). Locate the double bond that is resistant to hydrogenation.

PROBLEM 15.7   Complete the following equations.

(a) $CH_3CH_2OH + NaOH + Br_2 \xrightarrow[H_2O]{}$

(b) 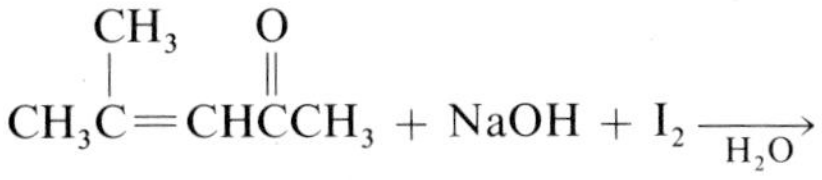 $CH_3C{=}CHCCH_3 + NaOH + I_2 \xrightarrow[H_2O]{}$

(c) 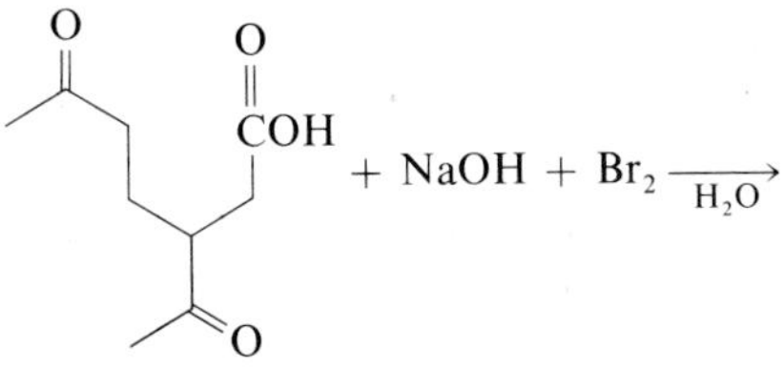 $+ NaOH + Br_2 \xrightarrow[H_2O]{}$

(d) 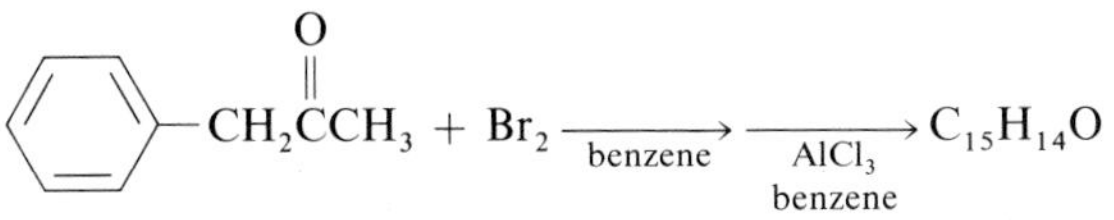 $-CH_2CCH_3 + Br_2 \xrightarrow[benzene]{} \xrightarrow[\substack{AlCl_3 \\ benzene}]{} C_{15}H_{14}O$

(Hint: Which would be the more stable enol form for 1-phenyl-2-propanone?)

(e) 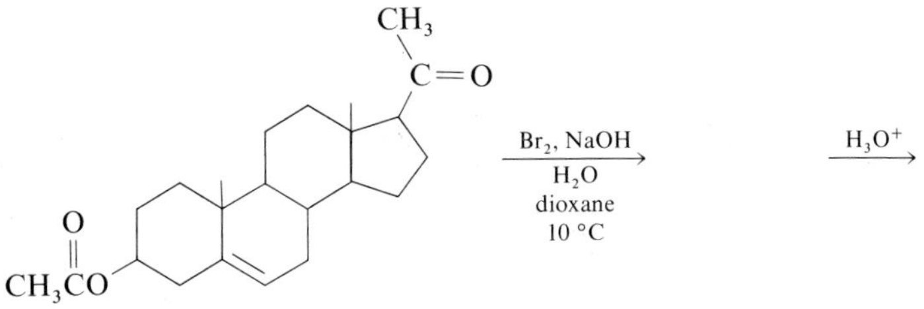 $\xrightarrow[\substack{H_2O \\ dioxane \\ 10\,°C}]{Br_2,\ NaOH} \xrightarrow{H_3O^+}$

(f) 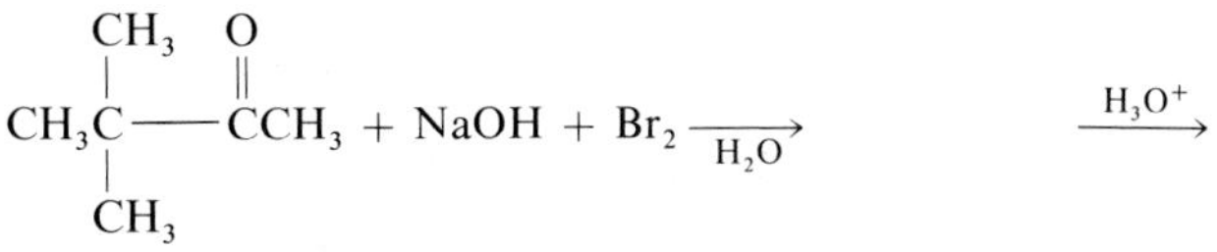 $CH_3C{-}CCH_3 + NaOH + Br_2 \xrightarrow[H_2O]{} \xrightarrow{H_3O^+}$

(g) 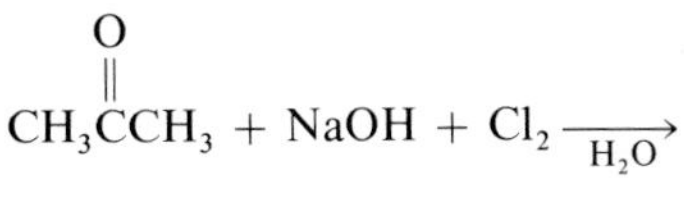 $CH_3CCH_3 + NaOH + Cl_2 \xrightarrow[H_2O]{}$

15.3

Alkylation of Ketones and Esters

A. Alkylation of Ketones

Alkylation of an enolate anion gives a compound with new substitution at the α-carbon atom (p. 807). For example, the enolate anion generated when 2-benzylcyclohexanone is deprotonated by lithium diisopropylamide reacts in a nucleophilic substitution reaction with methyl iodide.

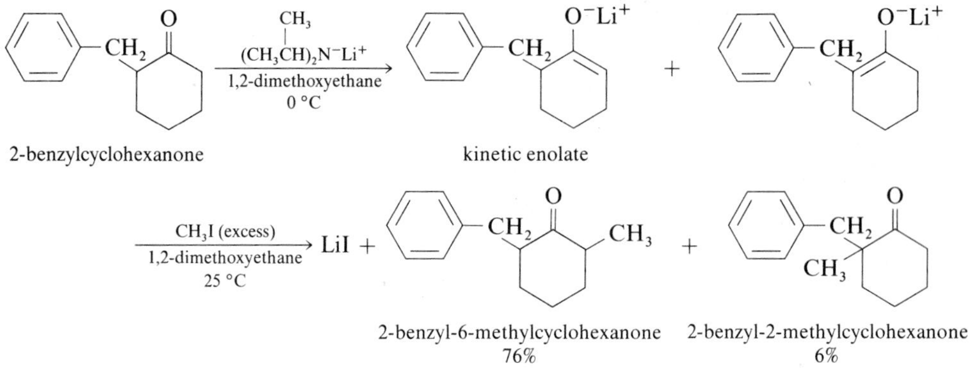

2-benzylcyclohexanone

kinetic enolate

2-benzyl-6-methylcyclohexanone 76%   2-benzyl-2-methylcyclohexanone 6%

The major product is derived from the kinetic enolate (Section 15.1D).

Intramolecular alkylation reactions of enolate anions have been used to make fused rings. The double bond in 2-allylcyclopentyl methyl ketone is converted into a primary alkyl halide by anti-Markovnikov addition of gaseous hydrogen bromide under free radical conditions (Section 5.4A). The product of this reaction cyclizes to give two different fused-ring compounds depending on whether conditions favoring the kinetic enolate or those for the thermodynamic enolate are used.

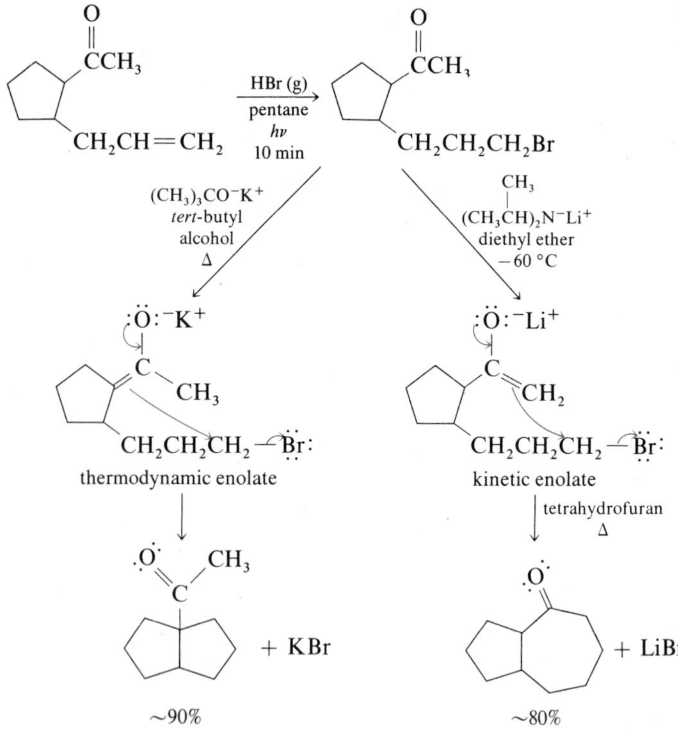

thermodynamic enolate

kinetic enolate

tetrahydrofuran
Δ

+ KBr   + LiBr

~90%   ~80%

B. Alkylation of Esters

Enolate anions of esters can also be prepared using lithium diisopropylamide. Methyl butanoate is converted into methyl 2-ethylbutanoate by alkylation of its enolate.

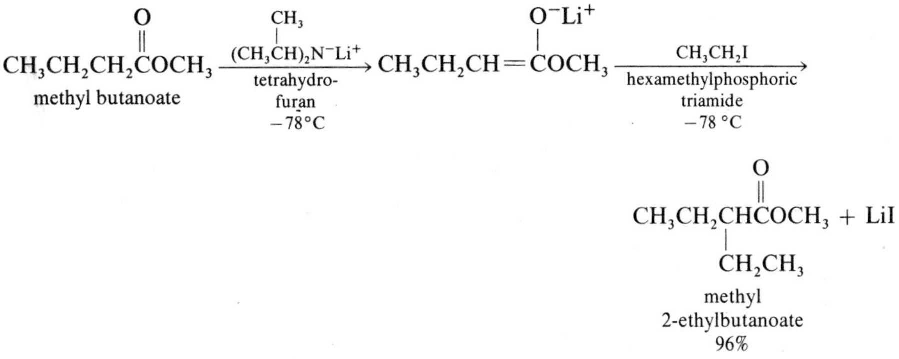

$$CH_3CH_2CH_2\overset{\overset{O}{\|}}{C}OCH_3 \xrightarrow[\substack{\text{tetrahydro-} \\ \text{furan} \\ -78°C}]{(CH_3CH)_2N^-Li^+} CH_3CH_2CH\overset{\overset{O^-Li^+}{|}}{=}COCH_3 \xrightarrow[\substack{\text{hexamethylphosphoric} \\ \text{triamide} \\ -78°C}]{CH_3CH_2I}$$

$$CH_3CH_2\overset{\overset{O}{\|}}{\underset{\underset{CH_2CH_3}{|}}{C}}HCOCH_3 + LiI$$

methyl
2-ethylbutanoate
96%

Lactones undergo similar reactions. For example, the lactone from 4-hydroxybutanoic acid is converted to the 2-allyl compound by deprotonation and alkylation.

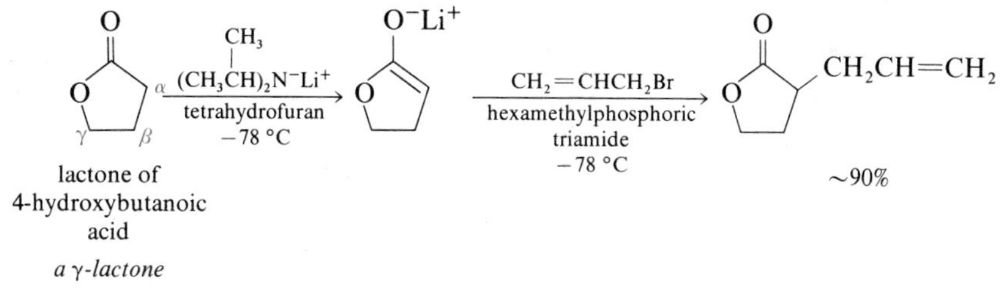

lactone of
4-hydroxybutanoic
acid

a γ-lactone

The reaction of lithium diisopropylamide with esters and lactones to give products of deprotonation reactions rather than nucleophilic substitution reactions at the carbonyl group (Section 10.4) demonstrates again how low the nucleophilicity of the di-isopropylamide anion is.

PROBLEM 15.8   Supply structural formulas for the reagents, intermediates, or products indicated by the letters.

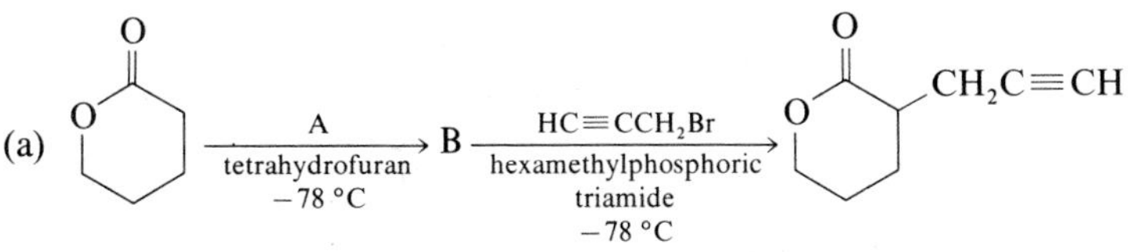

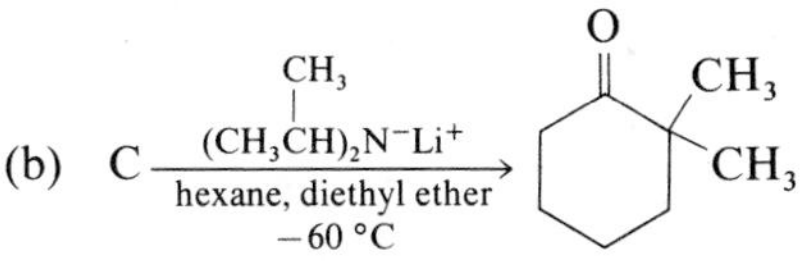

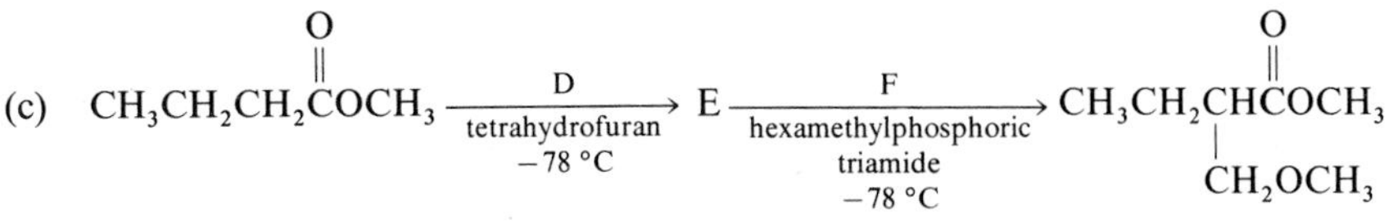

(a)

(b)

(c) $CH_3CH_2CH_2\overset{\overset{O}{\|}}{C}OCH_3 \xrightarrow[\substack{\text{tetrahydrofuran} \\ -78°C}]{D} E \xrightarrow[\substack{\text{hexamethylphosphoric} \\ \text{triamide} \\ -78°C}]{F} CH_3CH_2\overset{\overset{O}{\|}}{\underset{\underset{CH_2OCH_3}{|}}{C}}HCOCH_3$

(d)

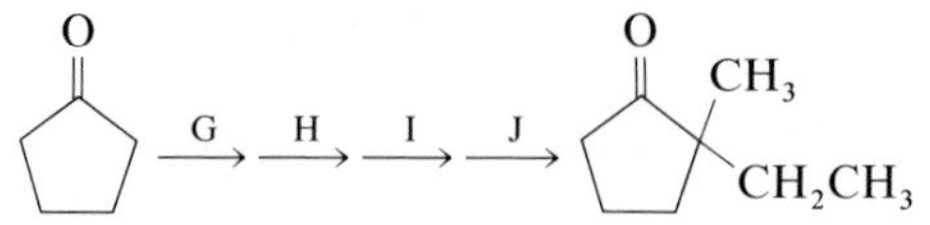

(e) $HOCH_2CH_2COCH_3$ $\xrightarrow[\substack{\text{tetrahydrofuran} \\ -78\ °C}]{(CH_3CH)_2N^-Li^+\ (2\ \text{molar equiv})}$ K $\xrightarrow{CH_3CH_2CH_2I\ (1\ \text{molar equiv})}$ $\xrightarrow{H_3O^+}$ L

15.4

Alkylation of Active Methylene Compounds

A. Alkylation of Ethyl Acetoacetate. Decarboxylation of β-Ketoacids

The enolate ions of compounds containing active methylene groups are formed easily and with high regioselectivity. For example, on p. 810 we saw that only one of the seven α-hydrogen atoms in 3-methyl-2,4-pentanedione was exchanged with deuterium oxide in an uncatalyzed reaction.

Ethyl acetoacetate enolizes regioselectively to one of the two possible enolate ions.

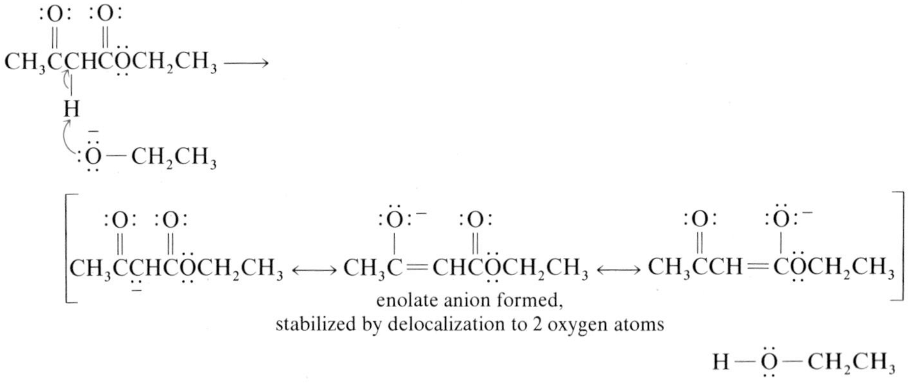

enolate anion formed,
stabilized by delocalization to 2 oxygen atoms

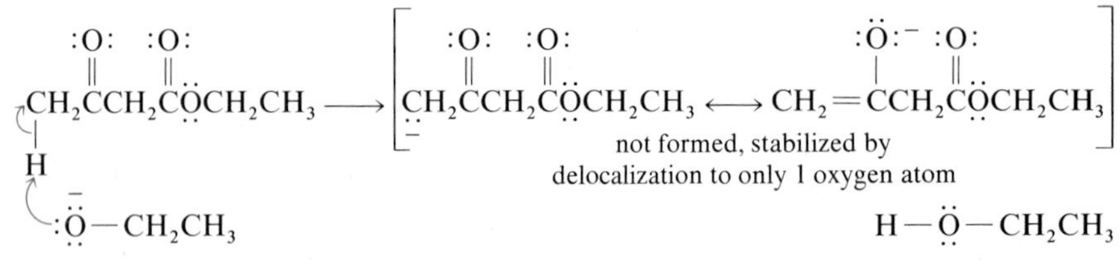

$H-\ddot{O}-CH_2CH_3$

not formed, stabilized by
delocalization to only 1 oxygen atom

$H-\ddot{O}-CH_2CH_3$

Ethyl 2-butylacetoacetate, the major product of the alkylation of ethyl acetoacetate with *n*-butyl iodide or *n*-butyl bromide (p. 807), is a β-ketoester. Saponification of the ester group with dilute base and careful acidification with cold dilute acid lead to the formation of an unstable β-ketocarboxylic acid. The acid loses carbon dioxide on heating.

$$\underset{\substack{\text{ethyl 2-butylacetoacetate}}}{\overset{\displaystyle \overset{O}{\overset{\|}{C}}}{\underset{\substack{|\\ CH_2CH_2CH_2CH_3}}{CH_3\overset{\displaystyle O}{\overset{\|}{C}}CHCOCH_2CH_3}}} \xrightarrow[\substack{H_2O\\25\,°C\\4\,h}]{5\%\ NaOH} \underset{\substack{\text{sodium 2-butylacetoacetate}}}{\overset{\displaystyle \overset{O\ \ \ O}{\overset{\|\ \ \ \|}{\ }}}{\underset{\substack{|\\ CH_2CH_2CH_2CH_3}}{CH_3CCHCO^-Na^+}}} \xrightarrow{H_2SO_4,\ cold,\ dilute}$$

$$\underset{\substack{\text{2-butylacetoacetic acid}}}{\underset{\substack{|\\ CH_2CH_2CH_2CH_3}}{CH_3\overset{O\ \ \ O}{\overset{\|\ \ \ \|}{C}}CHCOH}} \xrightarrow[\Delta]{} \underset{\substack{\text{2-heptanone}\\ \sim60\%}}{CH_3\overset{O}{\overset{\|}{C}}CH_2CH_2CH_2CH_2CH_3} + CO_2\uparrow$$

Thus, the ester group in ethyl acetoacetate serves to activate the α-hydrogen atoms, making their regioselective substitution possible. The ester group is removed easily once it has performed this function.

Carboxylic acids with carbonyl groups beta to the carboxyl function are unstable because they can react by means of a cyclic transition state, which allows transfer of a proton, loss of carbon dioxide, and the formation of an enol all to occur simultaneously.

cyclic transition state
for loss of carbon dioxide
by β-ketocarboxylic acid

enol form of
2-heptanone

| tautomerization

$$\underset{\substack{\text{2-heptanone}}}{\overset{:B\qquad\qquad O}{\overset{\qquad\qquad\quad\|}{CH_3CH_2CH_2CH_2CH_2CCH_3}}}\quad H-B^+$$

The species left when carbon dioxide is lost is the enol form of 2-heptanone. Deprotonation at the oxygen atom and protonation of the α-carbon atom converts it to the ketone.

The overall two-step sequence starting from a β-ketoester is a synthesis for ketones. Methyl ketones are formed when ethyl acetoacetate is used, but other β-ketoesters will also undergo the same reaction. For example, 2-ethylcyclopentanone is synthesized from ethyl 2-oxocyclopentanecarboxylate using similar reactions. The synthesis of the β-ketoesters that are the starting materials for these reactions is taken up in Section 15.5C.

ethyl 2-oxocyclo-
pentanecarboxylate

sodium salt of
the enolate of ethyl
2-oxocyclopentanecarboxylate

ethyl 1-ethyl-2-oxocyclo-
pentanecarboxylate

a β-ketoester

$$\text{ethyl 1-ethyl-2-oxocyclo-} \xrightarrow[\substack{H_2O \\ \Delta}]{H_2SO_4} \left[\text{1-ethyl-2-oxocyclo-} \right] \longrightarrow \text{CH}_2\text{CH}_3 + \text{CO}_2\uparrow$$

ethyl 1-ethyl-2-oxocyclo-
pentanecarboxylate

1-ethyl-2-oxocyclo-
pentanecarboxylic
acid

*a β-ketocarboxylic
acid*

2-ethylcyclopentanone

PROBLEM 15.9   Complete the following equations.

(a) $\overset{O}{\overset{\|}{\text{CH}_3\text{C}}}\text{CH}_2\overset{O}{\overset{\|}{\text{C}}}\text{OCH}_2\text{CH}_3 \xrightarrow[\text{ethanol}]{\text{CH}_3\text{CH}_2\text{ONa}} A \xrightarrow[\Delta]{\text{ClCH}_2\overset{O}{\overset{\|}{\text{C}}}\text{OCH}_2\text{CH}_3} B \xrightarrow[\Delta]{\text{H}_3\text{O}^+} C$

(b) $\text{C}_6\text{H}_5 - \overset{O}{\overset{\|}{\text{C}}}\text{CH}_2\overset{O}{\overset{\|}{\text{C}}}\text{OCH}_2\text{CH}_3 \xrightarrow[\text{ethanol}]{\text{CH}_3\text{CH}_2\text{ONa}} D \xrightarrow[\Delta]{\text{BrCH}_2\text{C}\equiv\text{CH}} E \xrightarrow[\Delta]{\text{H}_3\text{O}^+} F$

(c) $\overset{O}{\overset{\|}{\text{CH}_3\text{C}}}\text{CH}_2\overset{O}{\overset{\|}{\text{C}}}\text{OCH}_2\text{CH}_3 \xrightarrow[\text{ethanol}]{\text{CH}_3\text{CH}_2\text{ONa}} G \xrightarrow{\text{BrCH}_2\text{CH}_2\overset{O}{\overset{\|}{\text{C}}}\text{OCH}_2\text{CH}_3} H$

(d) $\overset{O}{\overset{\|}{\text{CH}_3\text{C}}}\text{CH}_2\overset{O}{\overset{\|}{\text{C}}}\text{OCH}_2\text{CH}_3 \xrightarrow[\text{ethanol}]{\text{CH}_3\text{CH}_2\text{ONa (1 molar equivalent)}} I \xrightarrow{\text{ClCH}_2\text{CH}_2\text{CH}_2\text{Br}} J$

(e) $\overset{O}{\overset{\|}{\text{CH}_3\text{C}}}\text{CH}_2\overset{O}{\overset{\|}{\text{C}}}\text{OCH}_2\text{CH}_3 \xrightarrow[\text{ethanol}]{\text{CH}_3\text{CH}_2\text{ONa}} K \xrightarrow{\text{CH}_2=\text{CHCH}_2\text{CH}_2\text{Br}} L \xrightarrow[\text{toluene}]{M} N \xrightarrow{\overset{I}{\overset{\|}{\text{CH}_3\text{CHCH}_3}}} O$

PROBLEM 15.10   Write a mechanism for the formation of 2-ethylcyclopentanone from 1-ethyl-2-oxocyclopentanecarboxylic acid (above).

B. Alkylation of Diethyl Malonate. Barbituric Acid Derivatives

Diethyl malonate also has an active methylene group and is a useful starting material for a number of syntheses. The enolate anion displaces bromide ion from *sec*-butyl bromide to give the substituted ester.

$$\overset{O}{\overset{\|}{\text{CH}_2(\text{COCH}_2\text{CH}_3)_2}} + \text{CH}_3\text{CH}_2\text{O}^-\text{Na}^+ \longrightarrow \text{Na}^+:\overset{O}{\overset{\|}{\text{CH}(\text{COCH}_2\text{CH}_3)_2}} + \text{CH}_3\text{CH}_2\text{OH}$$

diethyl malonate   sodium ethoxide

sodium salt of
the enolate of
diethyl malonate,
shown as carbanion

$$\underset{\text{Na}^+}{\overset{O \quad O}{\overset{\| \quad \|}{\text{CH}_3\text{CH}_2\text{OC}\overset{..}{\text{C}}\text{HCOCH}_2\text{CH}_3}}} + \underset{\textit{sec}\text{-butyl bromide}}{\overset{\text{CH}_3}{\overset{|}{\text{CH}_3\text{CH}_2\text{CHBr}}}} \longrightarrow \underset{\substack{\text{diethyl } \textit{sec}\text{-butylmalonate} \\ \sim 80\%}}{\overset{O \quad O}{\overset{\| \quad \|}{\underset{\overset{|}{\text{CH}_3\text{CH}_2\text{CHCH}_3}}{\text{CH}_3\text{CH}_2\text{OCCHCOCH}_2\text{CH}_3}}}} + \text{NaBr}$$

The substituted malonic ester is interesting for several reasons. When the diester is hydrolyzed, the resulting dicarboxylic acid is unstable and loses carbon dioxide on heating. This reaction is another example of the instability of carboxylic acids that have a carbonyl group beta to them (Section 15.4A). Thus, diethyl *sec*-butylmalonate can be saponified (Section 10.7D) and then heated with acid to remove one of the carboxylic acid groups.

$$
\underset{\text{diethyl } sec\text{-butylmalonate}}{\overset{\displaystyle CH_3 \quad\; O}{CH_3CH_2CHCH(\overset{\|}{C}OCH_2CH_3)_2}} \xrightarrow[\substack{H_2O \\ \Delta}]{KOH} \underset{\text{dipotassium } sec\text{-butylmalonate}}{\overset{\displaystyle CH_3 \quad\; O}{CH_3CH_2CHCH(\overset{\|}{C}O^-K^+)_2}} \xrightarrow[\substack{H_2O \\ \Delta,\, 3\,h}]{H_2SO_4}
$$

derived from the alkyl halide used in the substitution reaction

derived from diethyl malonate

$$
\longrightarrow \underset{\substack{\text{3-methylpentanoic acid} \\ \sim 60\%}}{\overset{\displaystyle CH_3 \quad\; O}{CH_3CH_2CHCH_2\overset{\|}{C}OH}} + CO_2\uparrow
$$

The loss of carbon dioxide does not start until the reaction mixture is acidified and heated; then the evolution of a gas is observed.

The sequence of reactions, alkylation of diethyl malonate, hydrolysis of the new diester, and then decarboxylation of the diacid, results in the conversion of *sec*-butyl bromide to a carboxylic acid having two more carbon atoms than the halide. This sequence is a general synthetic method for carboxylic acids. These acids may be regarded as substituted acetic acids. The carboxyl group and the α-carbon atom are derived from diethyl malonate, and the rest of the molecule from the alkyl halide used in the substitution reaction.

A monosubstituted malonic ester still has another hydrogen atom on the carbon atom between the two ester groups, so another enolization and alkylation are possible. The presence of one alkyl group on the active methylene group makes it a weaker acid, and sometimes a stronger base must be used to remove the second proton. For example, sodium *tert*-butoxide, which is a stronger base than sodium ethoxide, is used to form the enolate from diethyl isopropylmalonate in order to introduce a second alkyl group.

$$
\underset{\text{diethyl isopropylmalonate}}{\overset{\displaystyle CH_3 \quad\; O}{CH_3CHCH(\overset{\|}{C}OCH_2CH_3)_2}} + \underset{\substack{\text{sodium } tert\text{-} \\ \text{butoxide}}}{\overset{\displaystyle CH_3}{\underset{\displaystyle CH_3}{CH_3CO^-Na^+}}} \xrightarrow[\text{\textit{tert}-butyl alcohol}]{}
$$

$$
\underset{\displaystyle CH_3}{\overset{\displaystyle CH_3}{CH_3COH}} + \underset{Na^+}{\overset{\displaystyle CH_3 \quad O}{CH_3CHC(\overset{\|}{C}OCH_2CH_3)_2}} \xrightarrow[\Delta]{CH_3CH_2I} \underset{\substack{\text{diethyl ethylisopropylmalonate} \\ 65\%}}{\overset{\displaystyle CH_3 \quad O}{\underset{\displaystyle CH_2CH_3}{CH_3CHC(\overset{\|}{C}OCH_2CH_3)_2}}}
$$

Hydrolysis and decarboxylation of such a disubstituted malonic ester would lead to a carboxylic acid branched at the α-carbon atom.

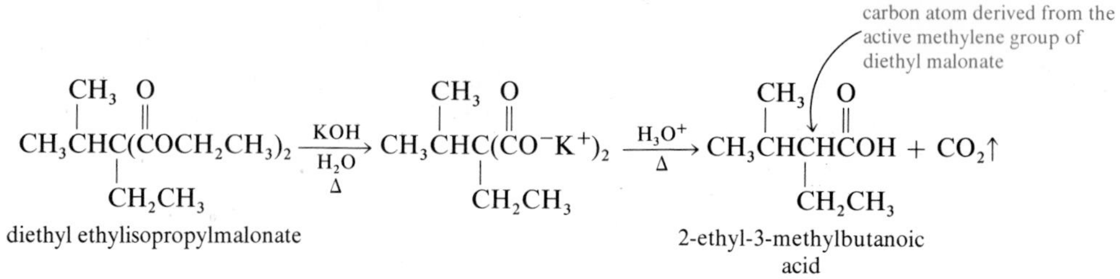

carbon atom derived from the active methylene group of diethyl malonate

$$CH_3CHC(COCH_2CH_3)_2 \xrightarrow[H_2O]{KOH} CH_3CHC(CO^-K^+)_2 \xrightarrow[\Delta]{H_3O^+} CH_3CHCHCOH + CO_2\uparrow$$

with CH_3, CH_2CH_3, CH_2CH_3 substituents

diethyl ethylisopropylmalonate

2-ethyl-3-methylbutanoic acid

Reactions that introduce substituents onto the active methylene group are interesting for yet another reason. Substituted malonic esters react with urea to give a variety of compounds of great medical interest, the barbituric acid derivatives.

derived from urea   derived from a substituted malonic ester

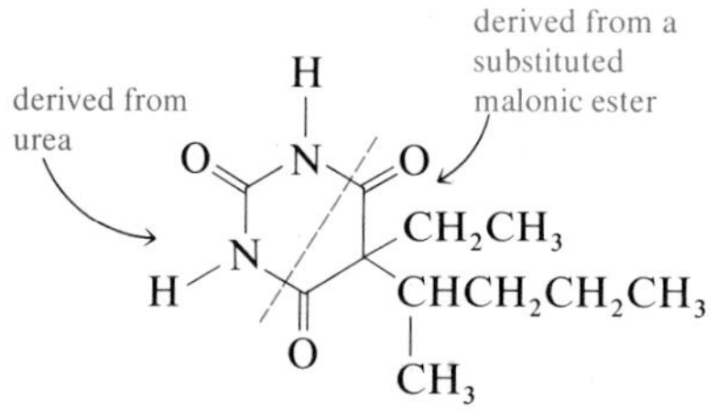

pentobarbital
Nembutal

*intermediate duration
of action*

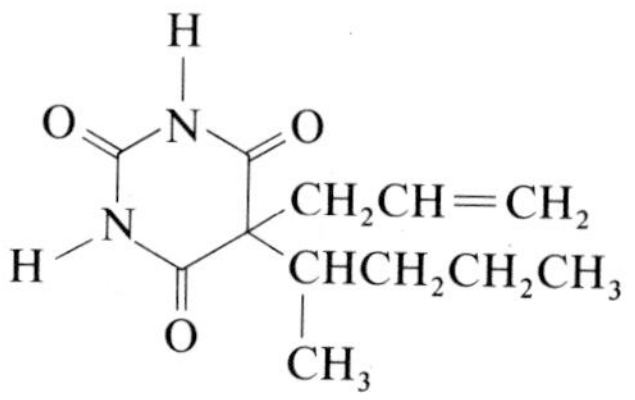

secobarbital
Seconal

*short duration
of action*

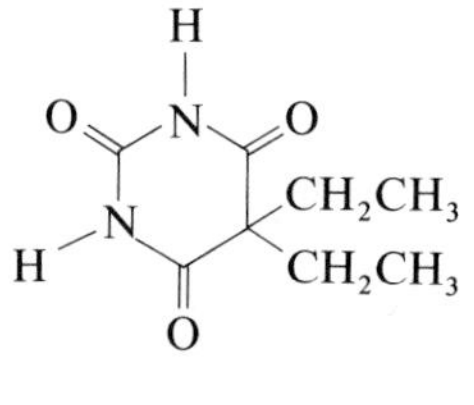

barbital
Veronal

*long duration
of action*

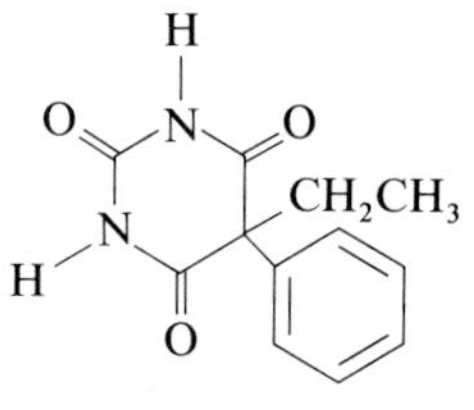

phenobarbital
Luminal

*long duration
of action*

The physiological properties of different barbituric acid derivatives vary, depending upon the substituents that are present on the carbon atom derived from the active methylene group of diethyl malonate. The drugs act as sedatives and hypnotics, or sleeping pills. Some of them act rapidly to induce sleep but do not have long-lasting effects. Others last longer, but are slower to establish their effect. Some, such as phenobarbital, also prevent convulsions. They act by depressing the activity of the central nervous system and, in high doses, also depress the respiratory system, which accounts for their toxicity. They are all addictive, meaning that a person who takes them with any regularity will suffer withdrawal symptoms when their use is discontinued.

The barbiturate Veronal is synthesized by the condensation of the diester of diethyl-malonic acid with urea in the presence of sodium methoxide.

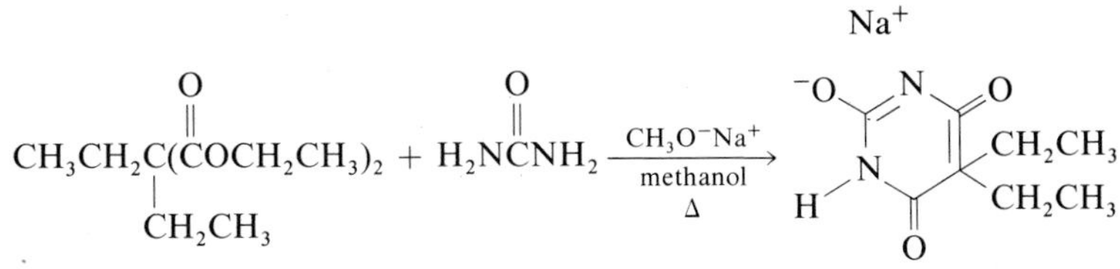

diethyl diethylmalonate   urea   sodium salt of
                                                     5,5-diethylbarbituric acid
                                                     Veronal

A six-membered ring containing two nitrogen atoms separated by one carbon atom is called a **pyrimidine ring.** This heterocyclic ring system is found in barbituric acid derivatives. Compounds such as thymine (p. 759) and uracil, bases found in DNA and RNA, are also pyrimidines and are discussed in greater detail in Section 18.6A.

The pyrimidines formed when malonic esters condense with urea are acidic, hence the name barbituric acid derivatives. If the carbon atom between the two carbonyl groups still has an enolizable hydrogen atom, an enol form having aromatic character can be written for the molecule. The relationship to pyrimidine shows up most clearly in this case. Even barbituric acid derivatives that have two substituents on the methylene group ionize readily because the conjugate base formed upon loss of a proton from a nitrogen atom is stabilized by delocalization of charge to two carbonyl groups.

pyrimidine

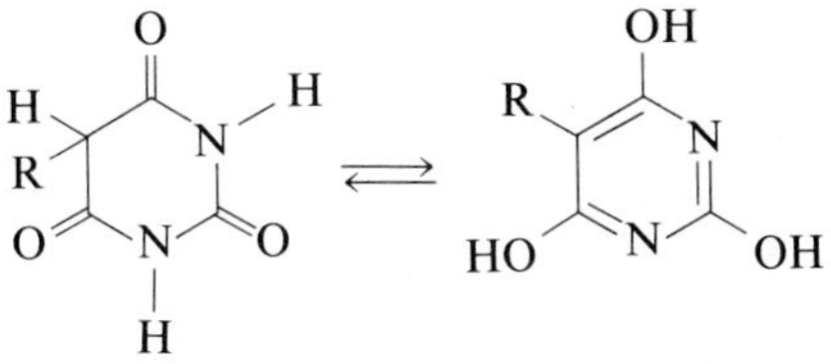

keto form of a   enol form of a
monosubstituted   monosubstituted
barbituric acid   barbituric acid

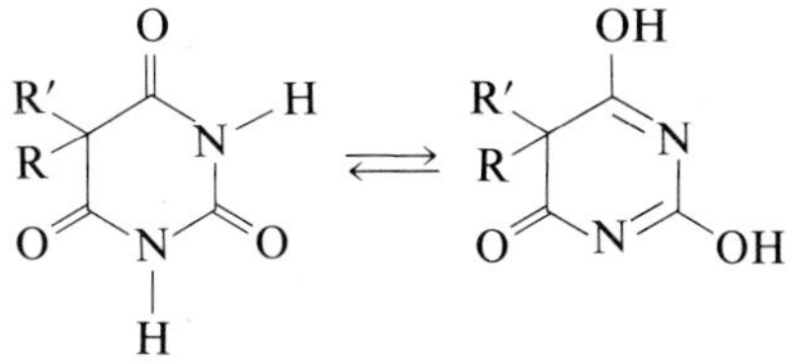

keto form of a   enol form of a
disubstituted   disubstituted
barbituric acid   barbituric acid

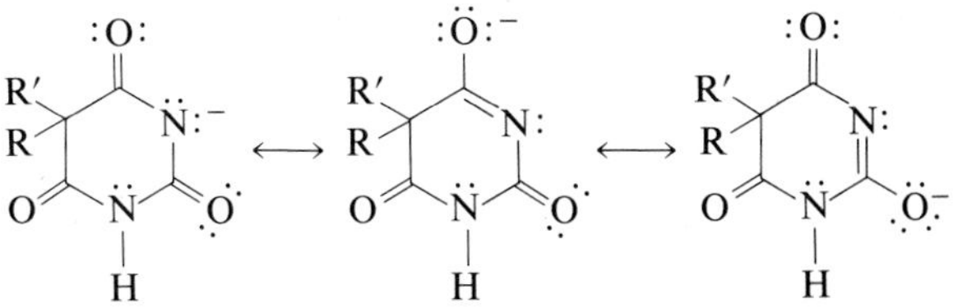

resonance contributors of the conjugate
base of a disubstituted barbituric acid

Most barbituric acid derivatives are prepared and sold as their sodium salts. For example, Veronal is isolated as a salt (p. 823).

PROBLEM 15.11　The barbiturate Amytal has the following structural formula. Write equations for its synthesis, starting from urea, diethyl malonate, and any other compounds containing not more than three carbon atoms.

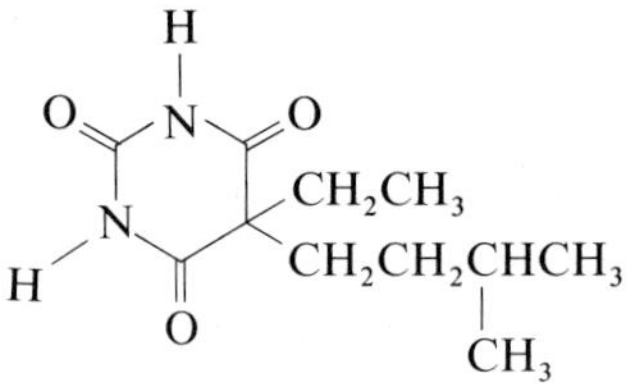

amobarbital
Amytal

PROBLEM 15.12　Complete the following equations.

(a)　$CH_2(COCH_2CH_3)_2 \xrightarrow[\text{ethanol}]{CH_3CH_2ONa} A \xrightarrow{CH_3CHCH_3 \text{ (Br)}} B$

(b)　$CH_3CH_2CH(COCH_2CH_3)_2 \xrightarrow[\text{tert-butyl alcohol}]{CH_3CONa \text{ (with } CH_3, CH_3)} C \xrightarrow{BrCH_2CH=CH_2} D$

(c)　$CH_2(COCH_2CH_3)_2 \xrightarrow[\text{ethanol}]{CH_3CH_2ONa \text{ (1 molar equivalent)}} E \xrightarrow{BrCH_2CH_2CH_2Cl}$

$F \xrightarrow[\text{ethanol}]{CH_3CH_2ONa \text{ (1 molar equivalent)}} G \xrightarrow[\Delta]{H_3O^+} H$

(d)　$CH_2CH_2(O) + NaCH(COCH_2CH_3)_2 \longrightarrow I$

(Hint: A review of Section 8.5C may be helpful.)

(e)　$CH_2(COCH_2CH_3)_2 \xrightarrow[\text{ethanol}]{CH_3CH_2ONa} J \xrightarrow{\text{(benzyl)}-CH_2Cl} K$

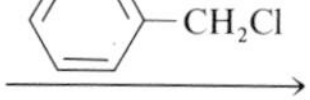

PROBLEM 15.13　Write a mechanism for the decarboxylation of malonic acid.

15.5

Reactions of Enolate Anions with Carbonyl Compounds

A. The Aldol Condensation

The reaction of nucleophilic carbon atoms with electrophilic carbon atoms to form carbon-carbon bonds is quite general. Such reactions form the basis of many synthetic transformations including the ones described in the last two sections. Reactions of carbanions with carbonyl groups are particularly important in the creation of complex structures. Such reactions have already been discussed in connection with the addition of organometallic reagents to the carbonyl groups of aldehydes and ketones (Section 9.4B), of acids (Section 10.10B), and of acid derivatives (Section 10.10A).

Enolate ions also react with carbonyl groups to form new carbon-carbon bonds. For example, acetaldehyde reacts with itself in acid or base to give a β-hydroxyaldehyde called **aldol** because it is both an **ald**ehyde and an alcoh**ol**.

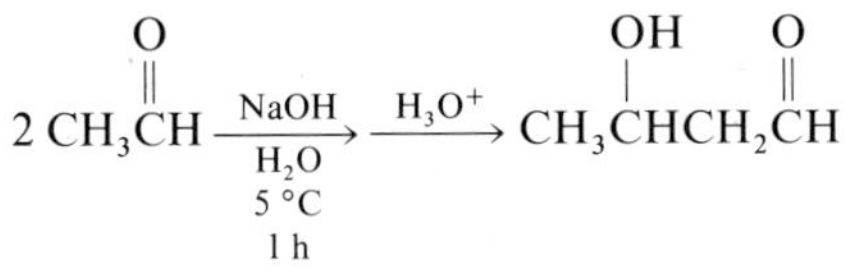

$$2\ CH_3CH \xrightarrow[\substack{H_2O \\ 5\ ^\circ C \\ 1\ h}]{NaOH} \xrightarrow{H_3O^+} CH_3CHCH_2CH$$

acetaldehyde   3-hydroxybutanal
aldol
50%

The reaction proceeds by enolization of acetaldehyde and attack by the enolate ion from one molecule of acetaldehyde on the carbonyl group of another aldehyde molecule.

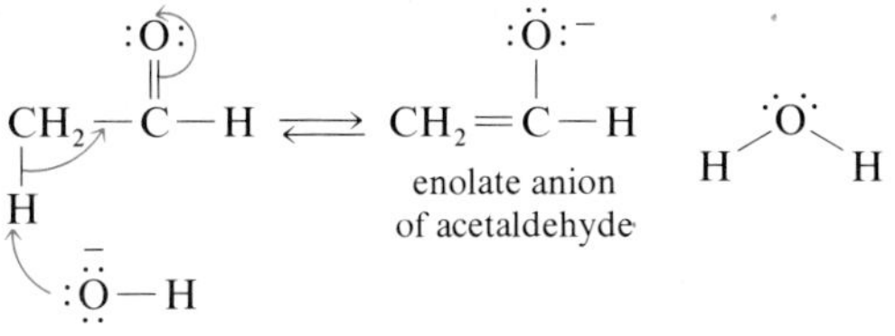

enolate anion
of acetaldehyde

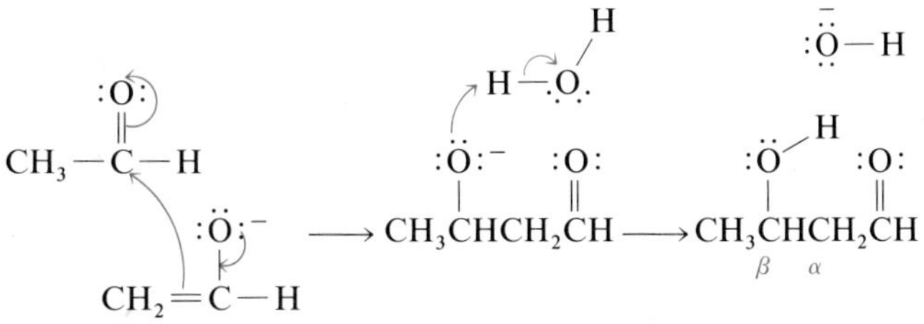

reaction of enolate anion from
one molecule of aldehyde with
carbonyl group of
another one   protonation
of the alkoxide
ion

The resulting alkoxide ion is protonated by the solvent, and the catalyst, hydroxide ion, is regenerated. The product of the reaction is a β-hydroxyaldehyde.

Such reactions in which the enolate ion of one carbonyl compound reacts with the carbonyl group of another one are called **aldol condensations.** The reaction is called a *condensation* because one larger molecule is created from the union of two smaller ones.

Aldol condensations are reversible. The reverse reaction, illustrated below with 3-hydroxybutanal, is called the **retroaldol reaction.**

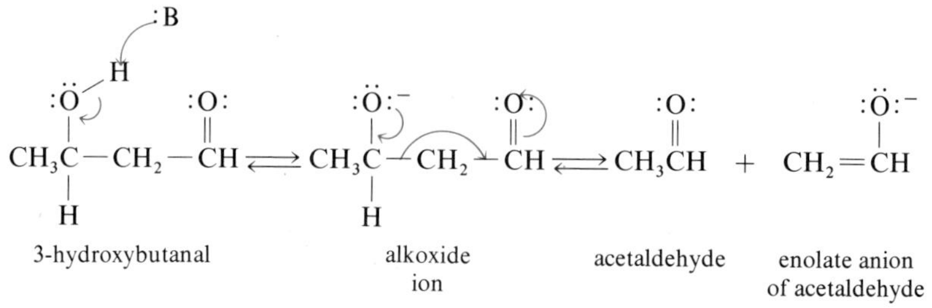

| 3-hydroxybutanal | alkoxide ion | acetaldehyde | enolate anion of acetaldehyde |

In basic solution, the hydroxyl group in the product of an aldol reaction loses a proton to give an alkoxide ion. Once the alkoxide ion is there, the formation of a carbonyl group provides the driving force for cleaving the molecule into the two fragments from which it was originally formed, an aldehyde and an enolate ion. The combination of the enolate ion with the aldehyde is the key step in the aldol condensation.

The aldol reaction has several complications. First, β-hydroxyaldehydes are rather unstable and dehydrate easily to give compounds in which the double bond is in conjugation with the carbonyl group. If the reaction is carried out at higher temperatures, the unsaturated compound is the product.

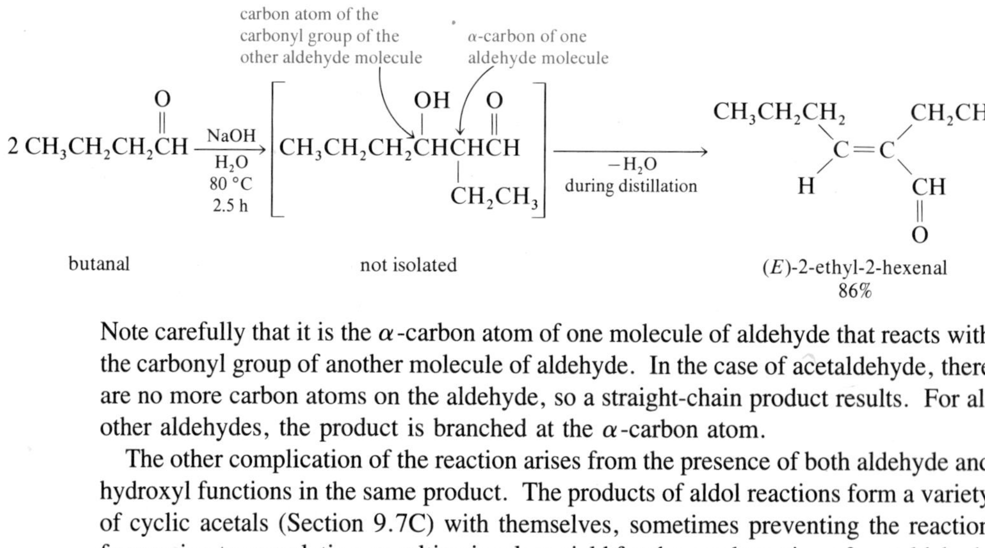

| butanal | not isolated | (E)-2-ethyl-2-hexenal 86% |

Note carefully that it is the α-carbon atom of one molecule of aldehyde that reacts with the carbonyl group of another molecule of aldehyde. In the case of acetaldehyde, there are no more carbon atoms on the aldehyde, so a straight-chain product results. For all other aldehydes, the product is branched at the α-carbon atom.

The other complication of the reaction arises from the presence of both aldehyde and hydroxyl functions in the same product. The products of aldol reactions form a variety of cyclic acetals (Section 9.7C) with themselves, sometimes preventing the reaction from going to completion, resulting in a low yield for the condensation of acetaldehyde with itself (p. 825).

Ketones do not condense with themselves in aldol reactions as readily as aldehydes do. The carbonyl group of a ketone is more hindered and less electrophilic than the carbonyl group of an aldehyde. In the equilibrium between a ketone and its aldol condensation product, the ketone is favored unless the reaction conditions are such that the product is removed so that it cannot undergo the base-catalyzed retroaldol reaction.

The simplest ketone, acetone, has been condensed with itself in acid to give an α,β-unsaturated ketone.

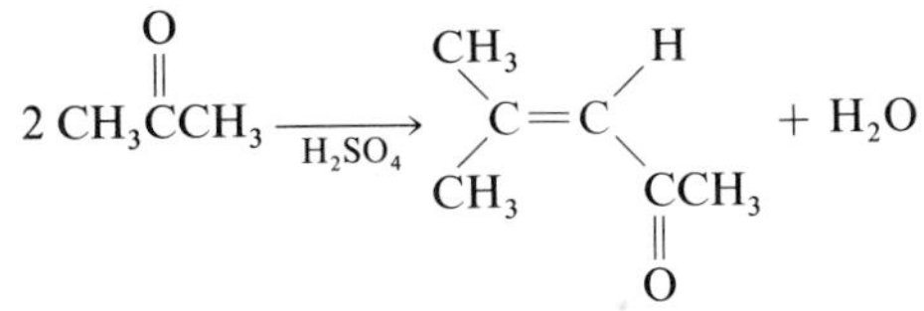

$$2\ CH_3\overset{\displaystyle O}{\overset{\displaystyle \|}{C}}CH_3 \xrightarrow[\ H_2SO_4\ ]{} \begin{array}{c} CH_3 \qquad H \\ \diagdown \quad \diagup \\ C = C \\ \diagup \qquad \diagdown \\ CH_3 \qquad CCH_3 \\ \qquad\quad \| \\ \qquad\quad O \end{array} +\ H_2O$$

acetone   4-methyl-3-penten-2-one

Under acidic conditions, the enol derived from one molecule of acetone (p. 806) reacts with the protonated carbonyl group of another one.

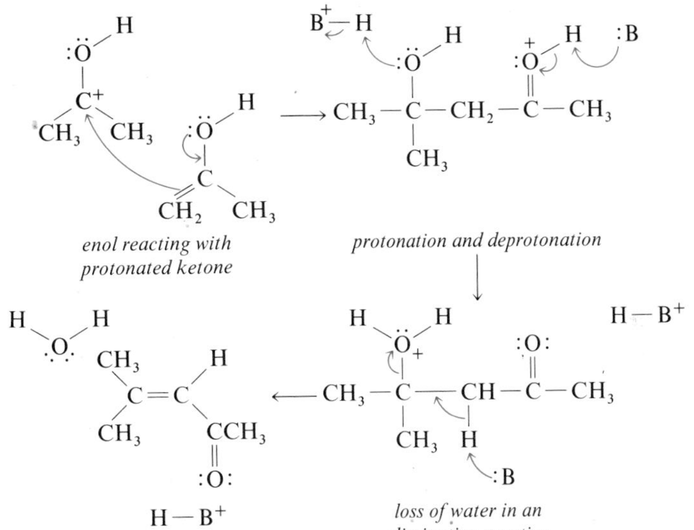

enol reacting with   *protonation and deprotonation*
protonated ketone

loss of water in an
elimination reaction

The intermediate β-hydroxyketone is dehydrated instantly in acid, and the reaction is thus drawn to completion.

Aldol condensations of ketones are favored if five- or six-membered rings form in the reaction. 2,8-Nonadione, for example, cyclizes to 1-acetyl-2-methylcyclohexene.

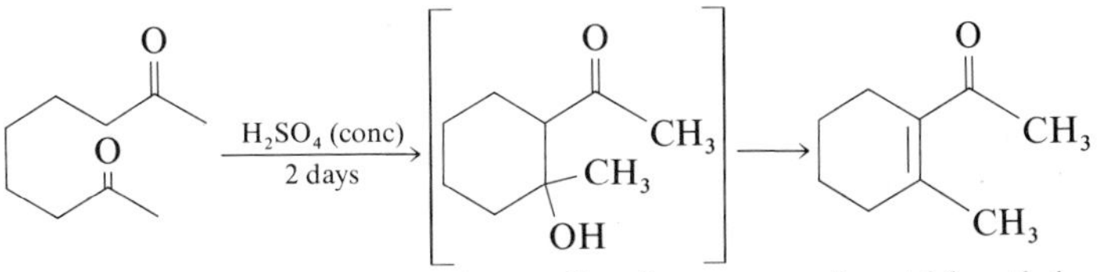

2,8-nonadione   intermediate from   1-acetyl-2-methyl-
                           the aldol condensation   cyclohexene

---

PROBLEM 15.14   Suggest a mechanism for the conversion of 2,8-nonadione to 1-acetyl-2-methyl-cyclohexene.

The two carbonyl compounds that participate in an aldol condensation do not have to be the same. As long as one can provide enolate anions and the other has a reactive carbonyl group for them to attack, aldol reactions between different compounds are possible. For example, 2-pentanone is converted into the kinetic enolate by lithium diisopropylamide at low temperatures. Under these conditions, self-condensation of the ketone, which is an unfavorable reaction anyway, does not occur. The enolate anion reacts cleanly with the carbonyl group of butanal in the second step of the reaction.

$$CH_3CH_2CH_2\overset{\overset{\displaystyle O}{\|}}{C}CH_3 \xrightarrow[\substack{\text{tetrahydrofuran}\\ -78\,°C}]{(CH_3CH)_2N^-Li^+} CH_3CH_2CH_2\overset{\overset{\displaystyle O^-Li^+}{|}}{C}=CH_2 \xrightarrow[\text{butanal}]{CH_3CH_2CH_2\overset{\overset{\displaystyle O}{\|}}{C}H}$$

2-pentanone   kinetic enolate

$$CH_3CH_2CH_2\overset{\overset{\displaystyle O}{\|}}{C}CH_2\overset{\overset{\displaystyle OH}{|}}{C}HCH_2CH_2CH_3 \xrightarrow[\substack{\text{benzene}\\ \Delta}]{TsOH} CH_3CH_2CH_2\overset{\overset{\displaystyle O}{\|}}{C}CH=CHCH_2CH_2CH_3$$

originally carbonyl group
of aldehyde

6-hydroxy-4-nonanone   5-nonen-4-one
65%   72%

If one of the participants in the aldol condensation does not have any α-hydrogen atoms, the possibility for competing reactions is reduced. One reactant supplies enolate ions and the other one, the carbonyl group that reacts with them. Benzaldehyde is an aldehyde that reacts with enolate ions but cannot form them itself. It reacts with the enolate of acetone, for example, to give an α,β-unsaturated ketone.

$$\text{C}_6\text{H}_5-\overset{\overset{\displaystyle O}{\|}}{C}H + CH_3\overset{\overset{\displaystyle O}{\|}}{C}CH_3 \xrightarrow[H_2O]{NaOH} \xrightarrow{H_3O^+} \underset{H}{\overset{C_6H_5}{}}C=C\underset{\overset{\overset{\displaystyle \|}{C}CH_3}{\overset{\displaystyle \|}{O}}}{\overset{H}{}}$$

benzaldehyde   acetone   (E)-4-phenyl-3-buten-2-one
~70%

In condensation reactions involving aryl aldehydes, the intermediate loses water with great ease because the product of the dehydration reaction has the double bond in conjugation not only with the carbonyl group but with the aromatic ring as well. In most cases, the unsaturated compounds obtained from aldol condensations have the E-configuration.

Whenever you see a compound with an α,β-unsaturated carbonyl function in it, you should suspect that the compound was synthesized by an aldol condensation. You can dissect each such product shown in equations above to discover the carbonyl compounds from which it was created.

α-carbon atom of the
other reactant

$$\underset{\substack{\text{carbonyl carbon}\\ \text{of one reactant}}}{\overset{CH_3CH_2CH_2}{\underset{H}{\diagup}}}C \overset{\curvearrowright}{=} C\underset{\overset{\overset{\displaystyle \|}{CH}}{\overset{\displaystyle O}{}}}{\overset{CH_2CH_3}{\diagdown}} \longleftarrow \underset{H}{\overset{CH_3CH_2CH_2}{\diagup}}C=O \qquad \underset{H}{\overset{H}{\diagup}}C\underset{\overset{\overset{\displaystyle \|}{CH}}{\overset{\displaystyle O}{}}}{\overset{CH_2CH_3}{\diagdown}}$$

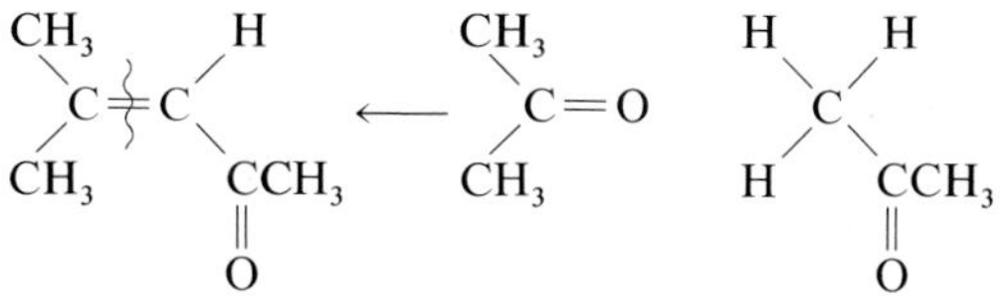

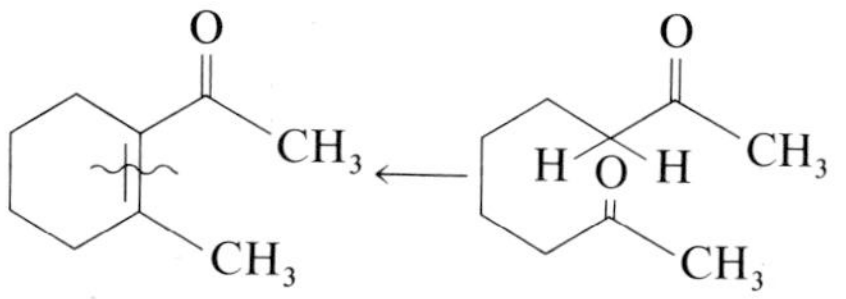

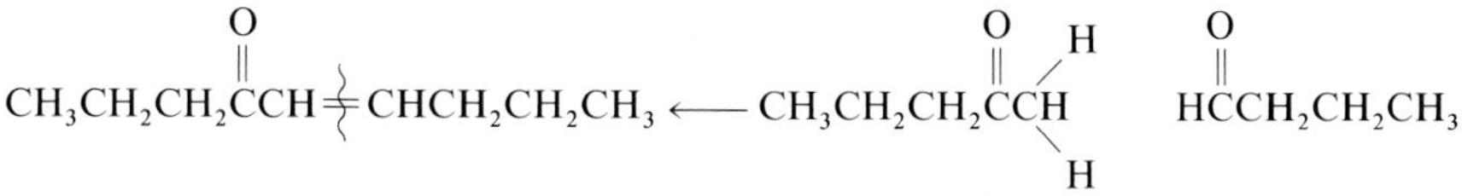

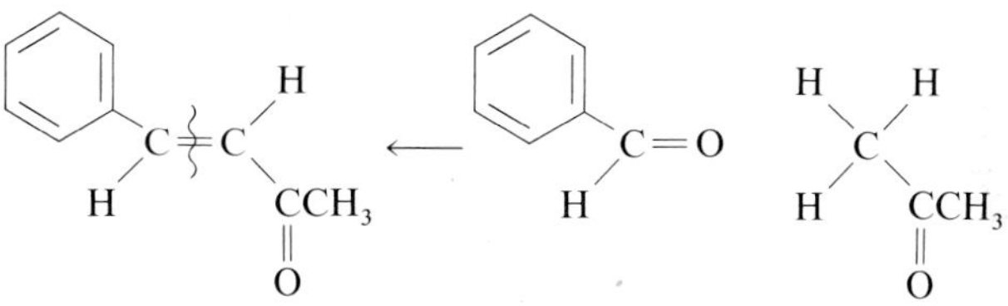

Note that in dissecting the structure, we reconstruct the carbonyl group of one reagent by mentally replacing a double bond to carbon with a double bond to oxygen. We also mentally restore two hydrogen atoms to the α-carbon atom of the other reagent.

PROBLEM 15.15   Complete the following equations.

(a) 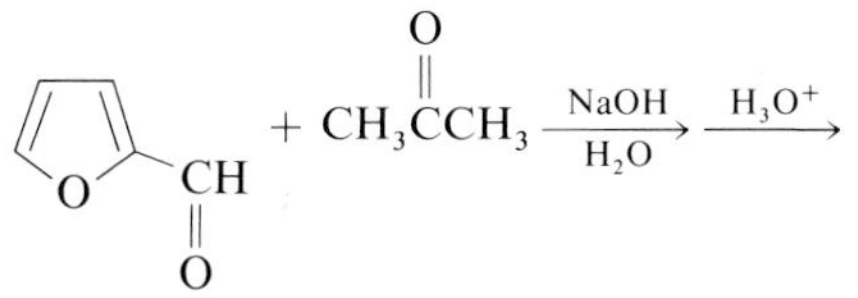
$$\text{furfural} + CH_3CCH_3 \xrightarrow[H_2O]{NaOH} \xrightarrow{H_3O^+}$$

(b) 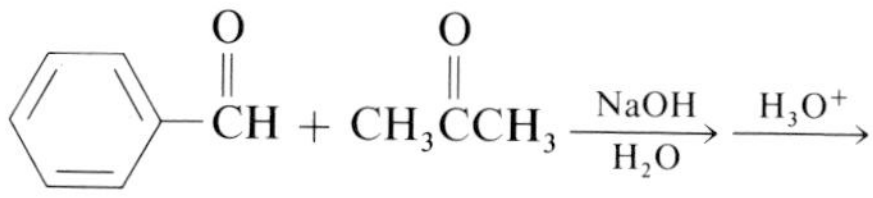
$$\phi\text{—CH} + CH_3CCH_3 \xrightarrow[H_2O]{NaOH} \xrightarrow{H_3O^+}$$

2 moles   1 mole

(c) 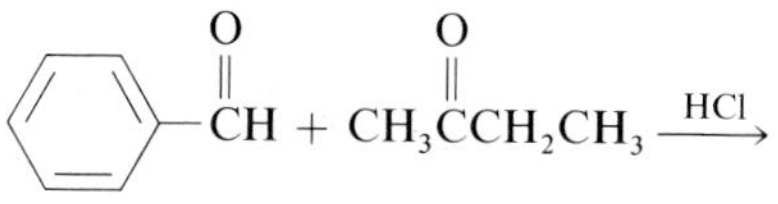
$$\phi\text{—CH} + CH_3CCH_2CH_3 \xrightarrow{HCl}$$

(Hint: Which is the more stable enol?)

(d) 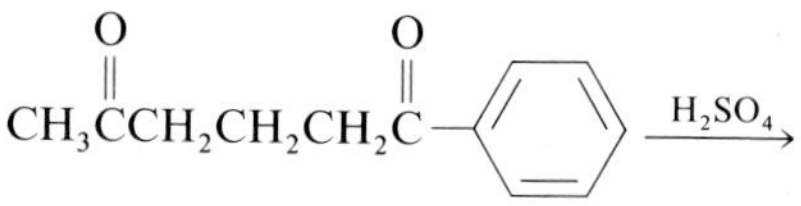
$$CH_3CCH_2CH_2CH_2C\text{—}\phi \xrightarrow{H_2SO_4}$$

(e) 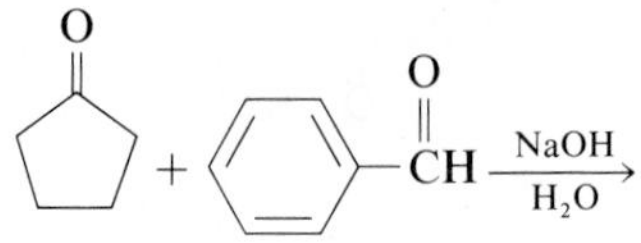 + $\xrightarrow[\text{H}_2\text{O}]{\text{NaOH}}$

(f) 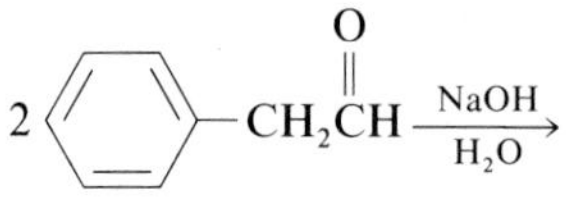 2 $\xrightarrow[\text{H}_2\text{O}]{\text{NaOH}}$

(g) 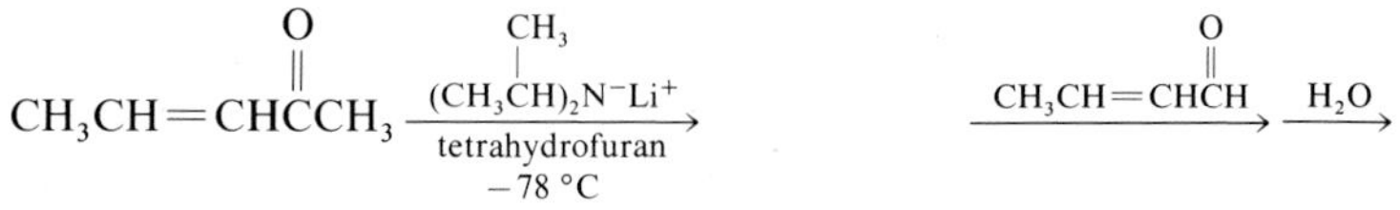 $CH_3CH{=}CHCCH_3$ $\xrightarrow[\substack{\text{tetrahydrofuran} \\ -78\ ^\circ\text{C}}]{(CH_3CH)_2N^-Li^+}$ $CH_3CH{=}CHCH$ $\xrightarrow{\text{H}_2\text{O}}$

PROBLEM 15.16   Write out the structures of the starting materials that you would need to synthesize the following compounds.

(a)

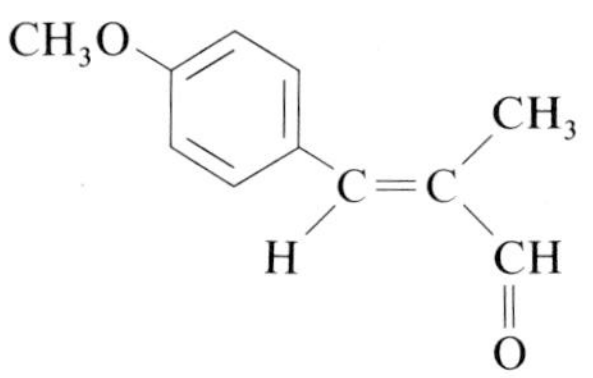

(b) 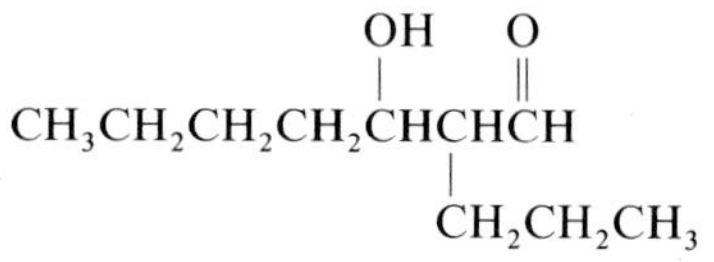 $CH_3CH_2CH_2CH_2\overset{\text{OH}}{\underset{\underset{CH_2CH_2CH_3}{|}}{\text{CH}}}\overset{\text{O}}{\text{CHCH}}$

(c)

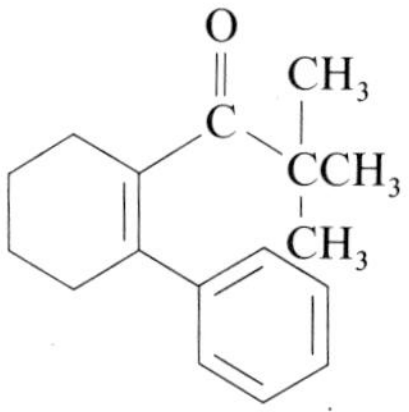

(d)

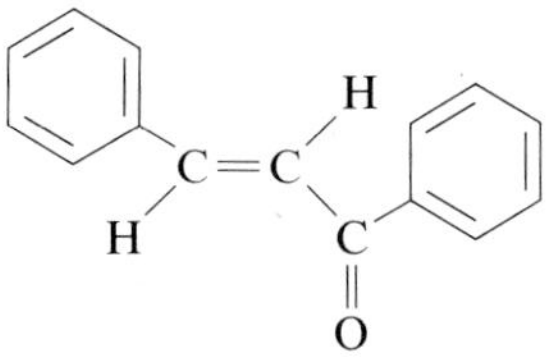

PROBLEM 15.17   The following experimental observations were made.

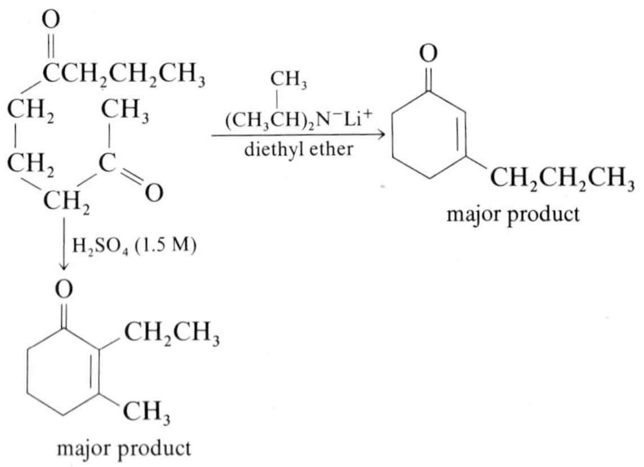

major product

Write mechanisms to rationalize these observations.

B. The Biological Importance of the Retroaldol Reaction.

The reversibility of the aldol reaction (p. 826) is of great biological importance because it is a key step in the metabolism of glucose. In the body, glucose is broken down into lactic acid by means of a process known as *glycolysis*. The metabolism of glucose starts with its conversion to glucose 6-phosphate by adenosine triphosphate (Section 14.5B). Glucose 6-phosphate is then converted into fructose 6-phosphate in a reversible process that is pictured as going through an enediol intermediate (Section 14.4D). In the next step of glycolysis, fructose 6-phosphate is further activated by the formation of a second phosphate ester linkage, this time to the hydroxyl group at carbon 1 of the molecule.

Fructose 1,6-diphosphate is cleaved into two triose phosphates by a retroaldol reaction.

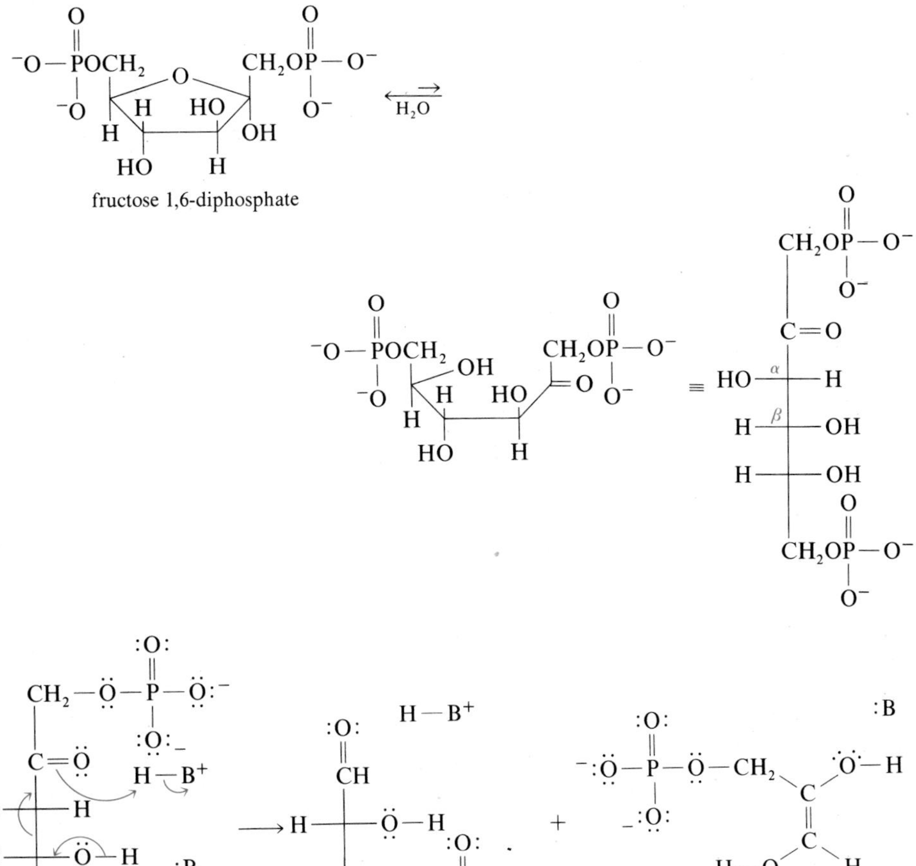

fructose 1,6-diphosphate

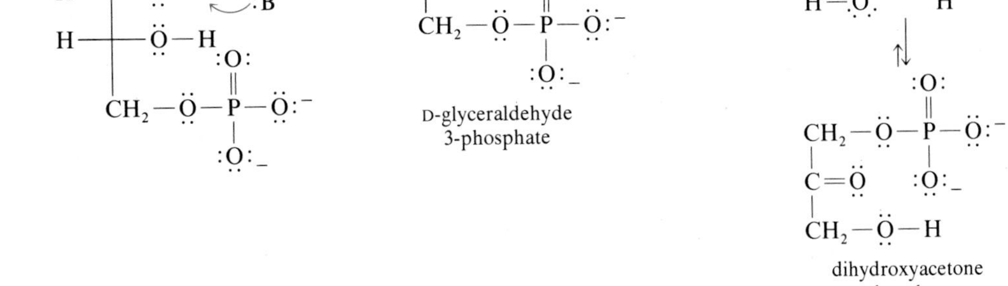

D-glyceraldehyde
3-phosphate

dihydroxyacetone
phosphate

The furanose form of fructose 1,6-diphosphate is in equilibrium with the ring-opened form, which is a β-hydroxyketone. Removing a proton from the hydroxyl group beta to the carbonyl group starts the reverse aldol reaction, which cleaves the molecule into two smaller carbohydrates, each one containing one of the phosphate groups. One of them is the ester of the smallest chiral aldose possible, D-glyceraldehyde 3-phosphate. It retains the stereochemistry at the carbon atom that was originally carbon 5 of glucose. The other fragment is dihydroxyacetone phosphate, the ester of the smallest ketose. The later stages of glycolysis transform dihydroxyacetone phosphate into D-glyceraldehyde 3-phosphate, and D-glyceraldehyde 3-phosphate into the anion of lactic acid. The soreness experienced in muscles during vigorous exercise is the result of the accumulation of lactic acid.

dihydroxyacetone phosphate   D-glyceraldehyde 3-phosphate   lactate ion   phosphate ion

---

PROBLEM 15.18   Write a detailed mechanism for the conversion of dihydroxyacetone phosphate into D-glyceraldehyde 3-phosphate. (Hint: Reviewing Section 14.4D may be useful.)

C. The Formation of β-Ketoesters. Claisen and Dieckmann Condensations

The aldol condensation involves the reaction of enolate ions with the carbonyl groups of aldehydes or ketones. In a similar reaction, *enolate ions from esters react with the carbonyl groups of acid derivatives to form β-ketoesters.* Such reactions are **acylation reactions** of enolate anions. In the reaction, the α-carbon atom of the compound giving rise to the enolate ion ends up bonded to an acyl group. For example, the enolate anion of *tert*-butyl 2-methylpropanoate reacts with benzoyl chloride to give *tert*-butyl 2,2-dimethyl-3-oxo-3-phenylpropanoate.

tert-butyl 2-methylpropanoate   *tert*-butyl 2,2-dimethyl-3-oxo-3-phenylpropanoate
78%

The ester is acylated at the α-carbon by this sequence of reactions.

When the enolate ion from an ester reacts with the carbonyl group of another ester, the reaction is known as the **Claisen condensation.** The classic example of this acylation reaction is the condensation of two molecules of ethyl acetate to give the β-ketoester ethyl acetoacetate.

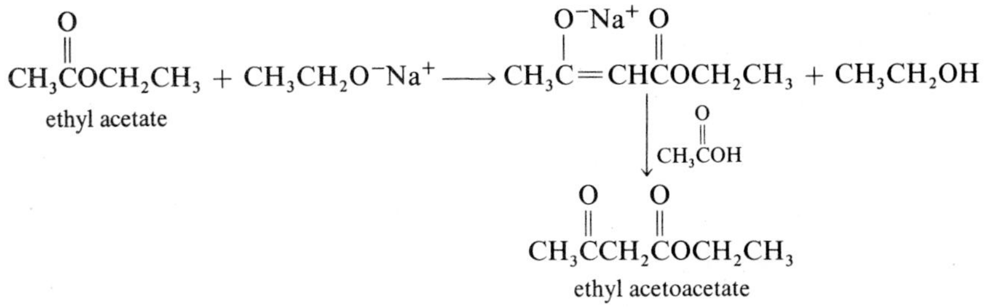

$$\underset{\text{ethyl acetate}}{CH_3\overset{\displaystyle O}{\overset{\|}{C}}OCH_2CH_3} + CH_3CH_2O^-Na^+ \longrightarrow CH_3\overset{\displaystyle O^-Na^+}{\overset{|}{C}}=CH\overset{\displaystyle O}{\overset{\|}{C}}OCH_2CH_3 + CH_3CH_2OH$$

$$\Big\downarrow\; CH_3\overset{\displaystyle O}{\overset{\|}{C}}OH$$

$$\underset{\text{ethyl acetoacetate}}{CH_3\overset{\displaystyle O}{\overset{\|}{C}}CH_2\overset{\displaystyle O}{\overset{\|}{C}}OCH_2CH_3}$$

The formation of the enolate ion of the active methylene group of the product provides the driving force for the completion of the reaction. The alkoxyl group on the ester function serves as a leaving group from the tetrahedral intermediate formed when the carbonyl group is attacked by an enolate ion.

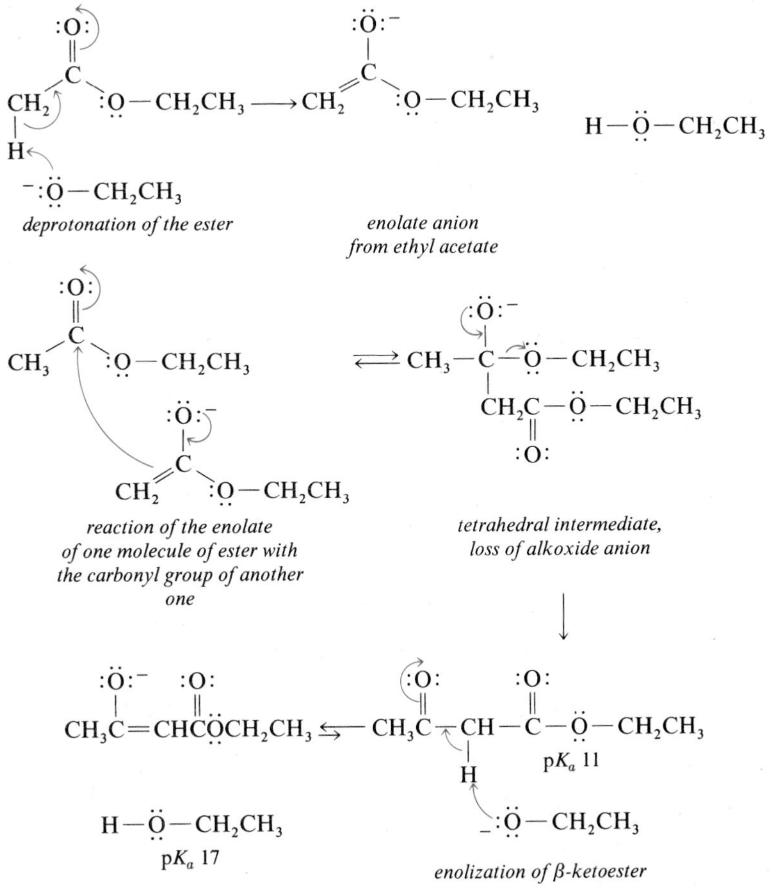

deprotonation of the ester   *enolate anion from ethyl acetate*

reaction of the enolate of one molecule of ester with the carbonyl group of another one   *tetrahedral intermediate, loss of alkoxide anion*

enolization of β-ketoester

The last stage of the reaction is written as an equilibrium reaction, showing the reversible protonation and deprotonation of the active methylene group in the β-ketoester. The equilibrium lies far towards the enolate anion of the β-ketoester because the active methylene compound is much more acidic (pK_a 11) than ethanol (pK_a 17). This step is an important one in determining the yield of the product. The β-keto-ester is unstable in base and undergoes a reverse reaction that is similar to the reversal of the aldol condensation. The ketone function in the β-ketoester is vulnerable to attack by the strongly nucleophilic alkoxide ion, giving the same tetrahedral intermediate as the condensation reaction does.

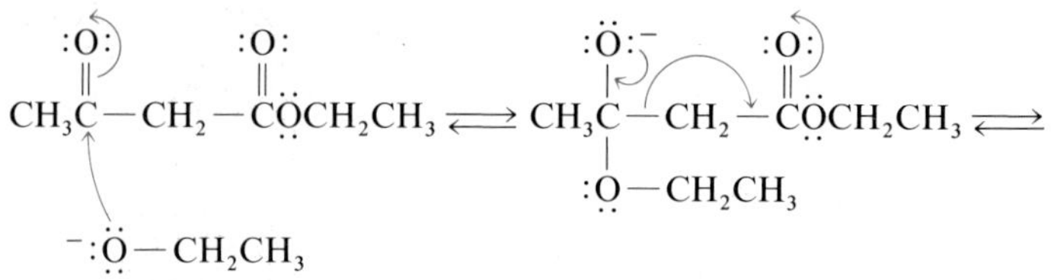

alkoxide anion
attacking the
carbonyl group

tetrahedral
intermediate losing
the enolate anion
of ethyl acetate

ethyl acetate   enolate anion

If the enolate ion serves as the leaving group, the condensation reaction is reversed. The best yields for the forward reaction are obtained when ethanol is distilled out and more ester is added to the reaction mixture. The equilibrium then lies towards complete conversion of the ketoester to its enolate anion, in which form it is stable. The enolate ion is converted to the free ketoester with an organic acid, such as acetic acid.

Just as aldol reactions are possible between different aldehydes, or between aldehydes and ketones, condensations between different esters have also been tried. If one of the esters does not have enolizable α-hydrogen atoms, the reaction works quite well. For example, a mixed condensation between ethyl benzoate and ethyl acetate is used in industry to synthesize ethyl 3-phenyl-3-oxopropanoate. Ethyl acetate and sodium are added slowly to ethyl benzoate. Under these conditions, as soon as the enolate ion of ethyl acetate is formed, it is more likely to encounter a molecule of ethyl benzoate than another molecule of ethyl acetate.

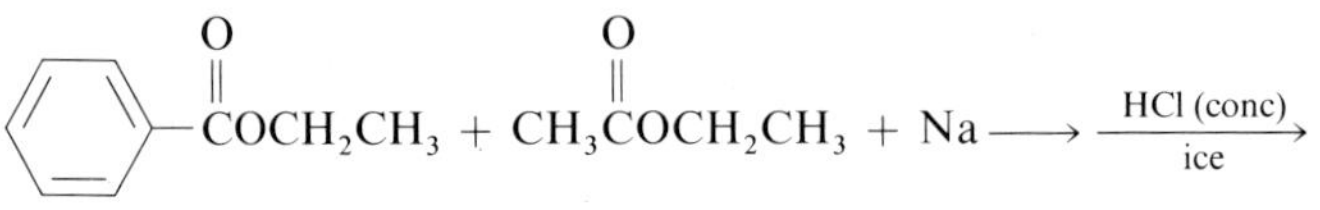

ethyl benzoate   ethyl acetate

ethyl 3-phenyl-
3-oxopropanoate
68%

ethanol

Sodium metal is used to generate the enolate ion in order to avoid the problem of the reversibility of the reaction that is seen when alkoxide ions are used as the catalysts. Some ethyl acetoacetate is formed as a side-product.

If two ester functions are separated by four or five carbon atoms in the same molecule, then the carbonyl group of one ester is in an ideal position to accept the enolate anion created alpha to the other ester group, giving rise to a cyclic β-ketoester. This variation of a Claisen condensation is called the **Dieckmann condensation** and is best illustrated by the formation of ethyl 2-oxocyclopentanecarboxylate.

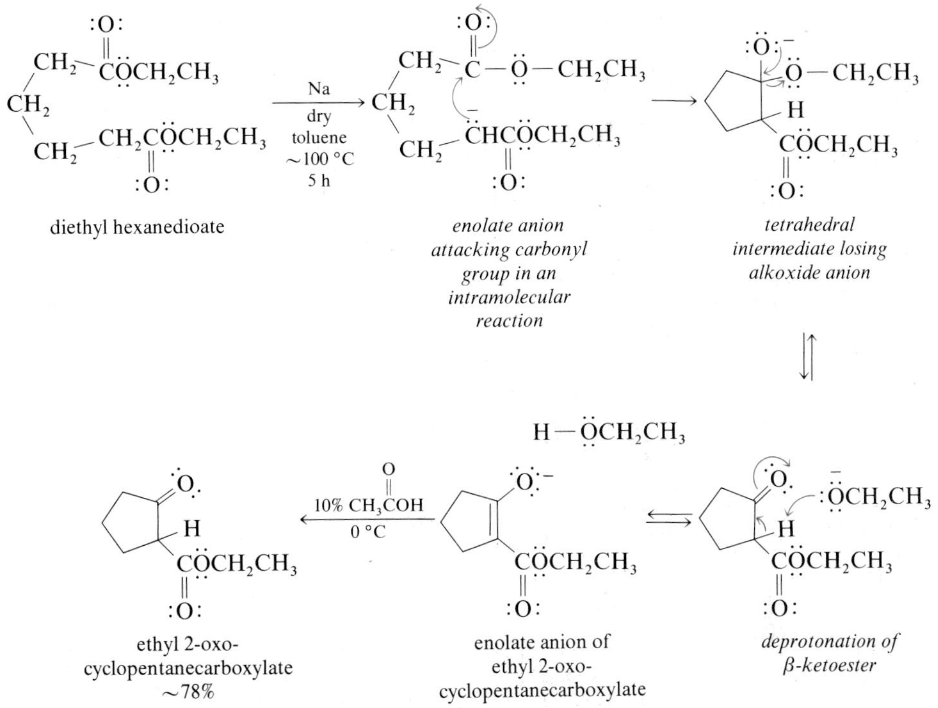
Na
dry
toluene
~100 °C
5 h
diethyl hexanedioate
enolate anion attacking carbonyl group in an intramolecular reaction
tetrahedral intermediate losing alkoxide anion
10% CH3COH
0 °C
ethyl 2-oxo-cyclopentanecarboxylate
~78%
enolate anion of ethyl 2-oxo-cyclopentanecarboxylate
deprotonation of β-ketoester

All the factors that affect the Claisen condensation also apply to the Dieckmann condensation. The reaction is reversible, and the β-ketoester is stable in basic solution only as its enolate anion.

D. A Unified Look at Condensation Reactions

Anions from esters can be used for condensations with aldehydes or ketones as well as with other esters. For example, ethyl acetate will condense with benzaldehyde to give an unsaturated carboxylic acid derivative.

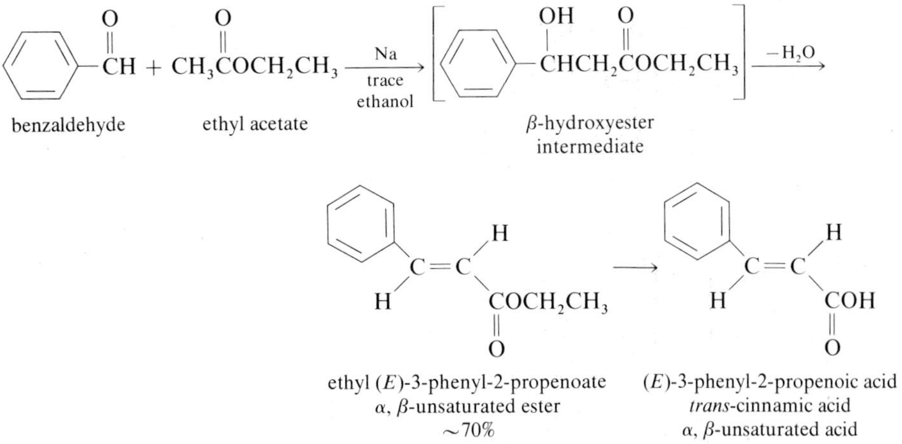
Na
trace
ethanol
−H2O
benzaldehyde
ethyl acetate
β-hydroxyester intermediate
ethyl (E)-3-phenyl-2-propenoate
α, β-unsaturated ester
~70%
(E)-3-phenyl-2-propenoic acid
trans-cinnamic acid
α, β-unsaturated acid

In another example, the active methylene group of diethyl malonate condenses with formaldehyde under mild conditions.

$$CH_2(COCH_2CH_3)_2 + 2\ HCH \xrightarrow[H_2O]{K_2CO_3} \begin{array}{c} HOCH_2 \quad COCH_2CH_3 \\ \diagdown \diagup \\ C \\ \diagup \diagdown \\ HOCH_2 \quad COCH_2CH_3 \end{array}$$

diethyl malonate   formaldehyde   diethyl bis(hydroxymethyl)-propanedioate

In both of these reactions, only one of the compounds used in the reaction has an enolizable hydrogen atom, so the reaction goes one way in high yield. You will also notice that the enolate ion reacts much more readily with the carbonyl group of the aldehyde that is present than with the carbonyl group of an ester. In the case of benzaldehyde, the intermediate hydroxy compound dehydrates easily to give a conjugated alkene. In the case of formaldehyde, the product retains the β-hydroxy groups.

It must be clear by now that a large variety of reactions are possible between enolate ions and compounds containing carbonyl groups. Many of them are known individually by the names of their discoverers or developers. But such an array of different names and structures tends to hide the underlying unity and simplicity of this type of reaction. In deciding what kinds of reactions are possible, we have to ask three questions:

1. Is there an enolizable hydrogen atom (recognizing that those alpha to nitro and cyano groups also count as acidic hydrogen atoms)?
2. Is there a carbonyl group that can be attacked by the enolate anion? Is it in the same molecule? If a five- or six-membered ring can form, it usually will.
3. Does the carbonyl group that is being attacked have a leaving group, usually the alkoxyl group of an ester? If there is one present, it is lost in going to the product. These types of reactions may be classified as Claisen condensations. If there is no good leaving group, the product is a β-hydroxycarbonyl compound or its dehydration product. Such reactions are aldol-type reactions.

All you need to bring order to what appears to be a bewildering array of reactions is a sharp eye for these features and a willingness to practice making predictions on a wide variety of examples. Some of them are provided for you in the next problem.

PROBLEM 15.19 Complete the following equations, giving structures for the intermediates and products designated by the letters.

(a)
$$\text{C}_6\text{H}_5\text{—CCH}_3 + N\equiv CCH_2COCH_2CH_3 \xrightarrow[\substack{\text{acetic acid}\\ \text{benzene}\\ \Delta}]{CH_3CO^-NH_4^+} A$$

(b)
$$\text{C}_6\text{H}_5\text{—C—C}_6\text{H}_5 + \begin{array}{c} CH_2COCH_2CH_3 \\ | \\ CH_2COCH_2CH_3 \end{array} \xrightarrow[\substack{\text{tert-butyl}\\ \text{alcohol}}]{CH_3COK} B$$

(c) C_6H_5—$\overset{O}{\overset{\|}{C}}H$ + $CH_2(COCH_2CH_3)_2$ $\xrightarrow[\text{benzene}\; \Delta]{\text{piperidinium benzoate}}$ C

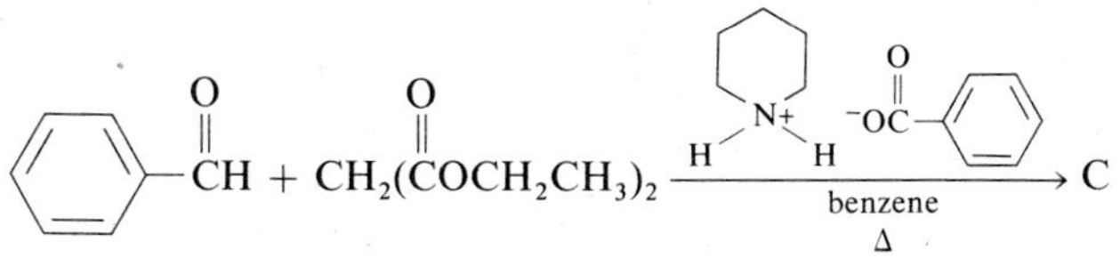

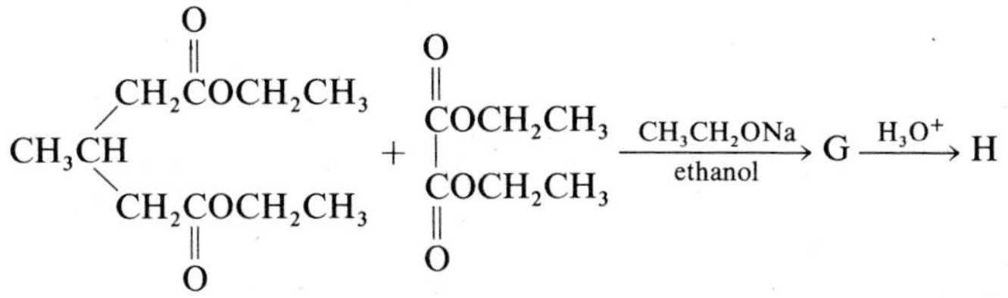

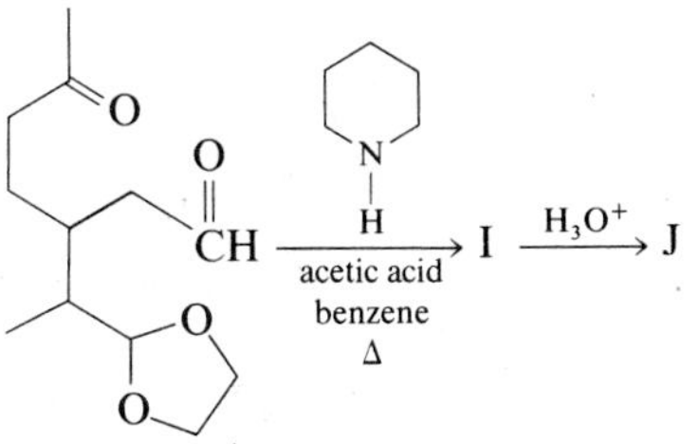

(d) 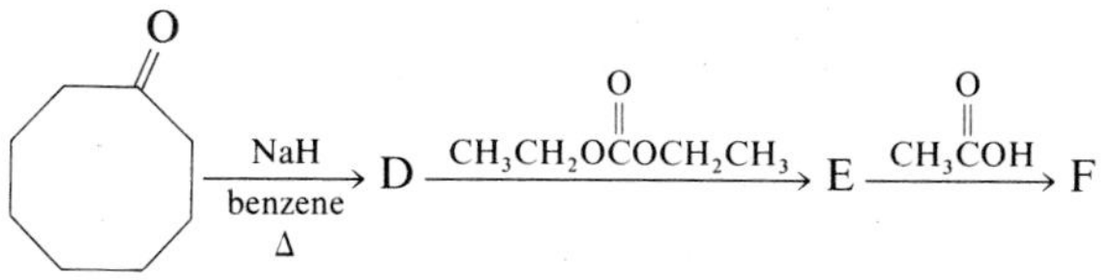 (cyclooctanone) $\xrightarrow[\text{benzene}\; \Delta]{\text{NaH}}$ D $\xrightarrow{CH_3CH_2OCOCH_2CH_3}$ E $\xrightarrow{CH_3COH}$ F

(e) $CH_3CH\big\langle{\overset{O}{\overset{\|}{CH_2COCH_2CH_3}}}\atop{\overset{\|}{CH_2COCH_2CH_3}\atop O}$ + $\overset{O}{\overset{\|}{COCH_2CH_3}}\atop{\overset{\|}{COCH_2CH_3}\atop O}$ $\xrightarrow[\text{ethanol}]{CH_3CH_2ONa}$ G $\xrightarrow{H_3O^+}$ H

(f) (structure) $\xrightarrow[\substack{\text{acetic acid} \\ \text{benzene} \\ \Delta}]{\text{piperidine}}$ I $\xrightarrow{H_3O^+}$ J

(g) $CH_3CH_2\big\langle{\overset{O}{\overset{\|}{C(COCH_2CH_3)_2}}}\atop{CH_2\atop{CH_2-CH(COCH_2CH_3)_2}}$ $\xrightarrow[\substack{\text{diethyl ether} \\ \Delta}]{CH_3CH_2ONa}$ K $\xrightarrow[\Delta]{H_3O^+}$ L

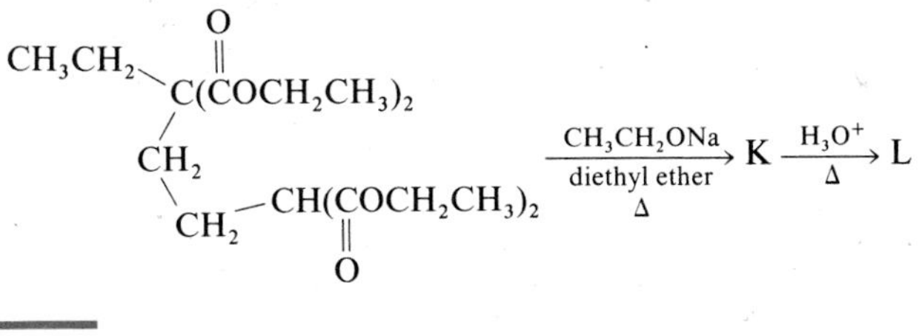

PROBLEM 15.20 Suggest syntheses for the following compounds. Assume that ethyl acetoacetate, diethyl malonate, and ethyl cyanoacetate are readily available and that you have a good supply of aromatic aldehydes and any other compounds containing not more than four carbon atoms.

(a) 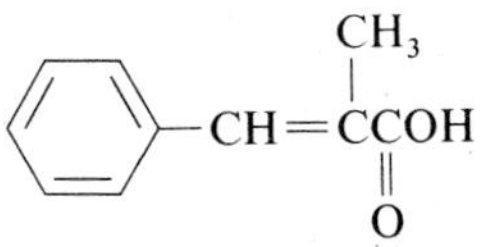 C_6H_5—$CH=\overset{CH_3}{\underset{\overset{\|}{O}}{C}}COH$

(b) (2,6-dimethoxyphenyl)—$CH=CHCOH$, $\overset{O}{\overset{\|}{}}$ with CH_3O and OCH_3

(c) $CH_3CH=CHCH=CH\overset{O}{\overset{\|}{C}}OH$

(d) 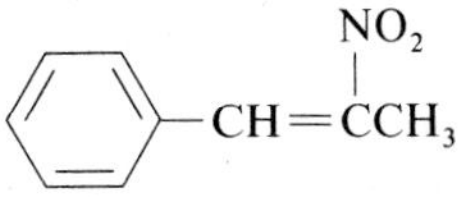 C_6H_5—$CH=\overset{NO_2}{\underset{}{C}}CH_3$

(e) 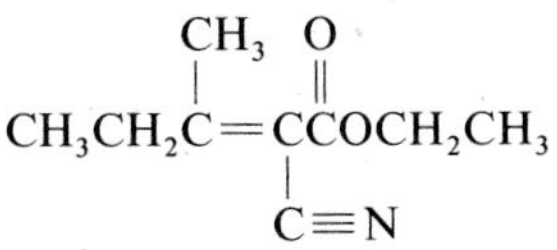 $CH_3CH_2\overset{CH_3}{\underset{\overset{|}{C}\equiv N}{C}}=\overset{O}{\overset{\|}{C}}COCH_2CH_3$

(f) 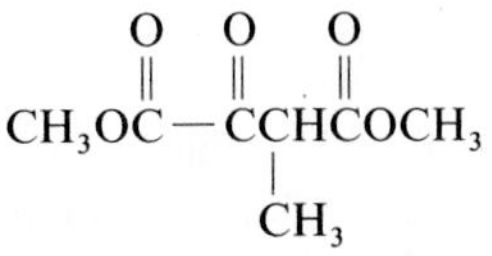 $CH_3O\overset{O}{\overset{\|}{C}}-\overset{O}{\overset{\|}{C}}CH\overset{O}{\overset{\|}{C}}OCH_3$, with CH_3 branch

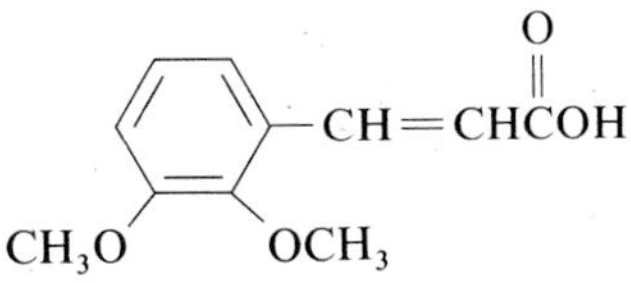

(g)

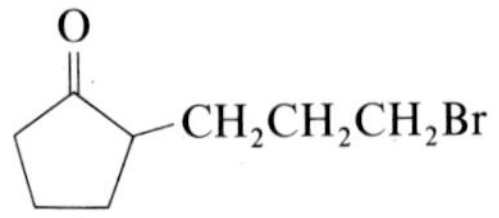

O
‖
⬡—CH₂CH₂CH₂Br

PROBLEM 15.21   The following reaction takes place.

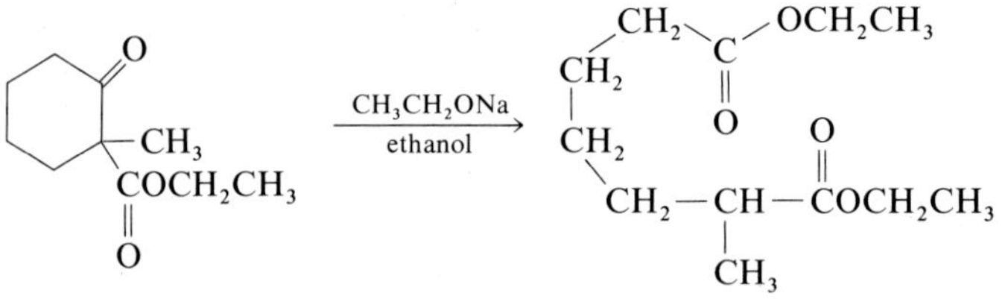

⬡(O) —CH₃ / COCH₂CH₃ (with ‖O) → [CH₃CH₂ONa / ethanol] →

CH₂—C—OCH₂CH₃ (‖O) / CH₂ / CH₂ / CH₂—CH—COCH₂CH₃ (‖O) / CH₃

(a)   Write a mechanism for the reaction.

(b)   When the diester produced in the reaction above is treated with one molar equivalent of sodium hydride with the careful exclusion of any alcohol, ethyl 3-methyl-2-oxocyclohexane-carboxylate is formed in 90% yield. Explain why that compound, and not the original ethyl 1-methyl-2-oxocyclohexanecarboxylate, is formed under these conditions.

PROBLEM 15.22   The disubstituted malonic ester required for the synthesis of phenobarbital (p. 822) cannot be prepared by the usual route in which the enolate anion from diethyl malonate is used in nucleophilic substitution reactions, because ordinary aryl halides do not undergo such substitution reactions. Work out a synthesis for diethyl ethylphenylmalonate starting from phenylacetonitrile using condensation reactions and alkylation reactions of active methylene compounds.

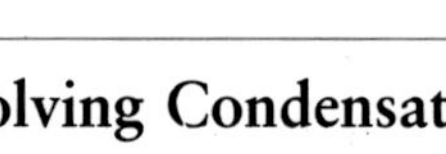

⬡—CH₂C≡N

phenylacetonitrile

15.6

Biological Transformations Involving Condensation Reactions. The Synthesis of Fatty Acids

Syntheses of many complex molecules in biological systems start with the two carbon atoms supplied by acetic acid, present in the body as a degradation product of glucose and of fatty acids. The reactions by which acetic acid is converted to longer-chain molecules are condensation reactions, completely analogous to the reactions that have been described as aldol condensations or Claisen condensations earlier in this chapter. They differ from the reactions carried out in the laboratory in that they are catalyzed by enzymes and therefore do not require strong bases to generate reactive intermediates such as carbanions. Many of them require several steps to bring about a single transformation. The overall conversions, however, can be analyzed in terms of the reactions and intermediates that are already familiar.

Long-chain fatty acids are synthesized in the body starting from acetyl coenzyme A, the thioester of acetic acid with the coenzyme derived from the vitamin pantothenic acid (Section 10.7G). The first step in the synthesis converts acetyl coenzyme A to an active methylene compound, malonyl coenzyme A.

$$\underset{\substack{\uparrow \\ \alpha\text{-carbon atom} \\ \text{of acetyl CoA}}}{\overset{\displaystyle O \atop \displaystyle \|}{CH_3CSCoA}} + \underset{\substack{\text{biological} \\ \text{source of} \\ CO_2}}{HCO_3^-} + ATP \xrightarrow[\substack{\text{acetyl CoA} \\ \text{carboxylase}}]{} \underset{\substack{\uparrow \\ \text{derived from the} \\ \alpha\text{-carbon atom} \\ \text{of acetyl CoA}}}{\overset{\substack{\text{derived from} \\ CO_2}}{\overset{\displaystyle O \quad O \atop \displaystyle \| \quad \|}{HOCCH_2CSCoA}}} + ADP + PO_4^{3-}$$

malonyl CoA

The reaction, described in chemical terms, involves generation of a carbanion at the α-carbon atom of acetyl CoA and addition of that carbanion to carbon dioxide. Carbon dioxide is present in the cell as bicarbonate ion. Energy for the reaction is derived from the energy-rich bonds of adenosine triphosphate, ATP, which is converted to adenosine diphosphate, ADP, and phosphate ion in the process. The reaction is assisted by an enzyme that catalyzes the introduction of carboxyl groups, acetyl CoA carboxylase.

Malonyl coenzyme A has an active methylene group. It undergoes condensation and substitution reactions selectively at the carbon atom between the two carbonyl groups. Before malonyl CoA participates in the biosynthesis of fatty acids, it is transferred in a transesterification reaction from coenzyme A to a thiol group on a protein known as acyl carrier protein.

$$\underset{\substack{\text{acyl carrier} \\ \text{protein}}}{\overset{:O: \quad :O: \atop \| \quad \|}{HOCCH_2C - \overset{\cdot\cdot}{S} - CoA}} \atop H - \overset{\cdot\cdot}{S} - ACP \longrightarrow \overset{:O: \quad :O: \atop \| \quad \|}{HOCCH_2C - \overset{\cdot\cdot}{\underset{+}{S}} - ACP} \atop \underset{-:\overset{\cdot\cdot}{S} - CoA}{H} \longrightarrow \underset{\substack{\text{malonyl acyl} \\ \text{carrier protein}}}{\overset{:O: \quad :O: \atop \| \quad \|}{HOCCH_2CSACP}} + \underset{\substack{\text{coenzyme A}}}{H\overset{\cdot\cdot}{S}CoA}$$

As this new thioester, it reacts with an acetyl group, also in the form of a thioester, on the enzyme that catalyzes this condensation reaction.

$$\underset{\substack{\text{malonyl acyl} \\ \text{carrier protein}}}{\overset{O \quad O \atop \| \quad \|}{{}^-OCCH_2CSACP}} + \underset{}{\overset{O \atop \|}{CH_3CS}} - enzyme \xrightarrow[\text{synthase}]{} \underset{\substack{\text{acetoacetyl} \\ \text{acyl carrier protein}}}{\overset{O \quad O \atop \| \quad \|}{CH_3CCH_2CSACP}} + HS - enzyme + CO_2$$

In the process, the malonyl group is converted to an acetoacetyl group. The molecule of carbon dioxide that was used earlier to convert acetyl coenzyme A into malonyl coenzyme A is released. The condensation of malonyl acyl carrier protein with the acetyl group can be rationalized using the symbolism of organic chemistry.

$$\overset{:O: \quad :O: \atop \| \quad \|}{{}^-:\overset{\cdot\cdot}{O} - C - CH_2CSACP} \atop CH_3C - \overset{\cdot\cdot}{S} - enzyme \atop \underset{:\overset{\cdot\cdot}{O}:}{\|} \longrightarrow \overset{\cdot\cdot}{O} = C = \overset{\cdot\cdot}{O} \quad + \quad \overset{:O: \atop \|}{CH_2 - CSACP} \atop CH_3 - C - \overset{\cdot\cdot}{S} - enzyme \atop \underset{-:\overset{\cdot\cdot}{O}:}{} \quad H - B^+$$

$$\downarrow$$

$$\overset{:O: \quad :O: \atop \| \quad \|}{CH_3CCH_2\overset{\cdot\cdot}{C}SACP}$$

$$H - \overset{\cdot\cdot}{S} - enzyme \quad :B$$

The sequence of steps shown above represents the reactions that take place in the cell, but does not necessarily correspond to them in detail. The mysteries that still surround the steps of various biological transformations are minimized by invoking enzymes as catalysts. In Chapter 17, it becomes apparent that the catalytic powers of enzymes are the result of the precise forms in which these large and complex molecules exist. In them, reactive groups are positioned so that they interact with great specificity.

A comparison of acetoacetyl acyl carrier protein with acetyl coenzyme A shows us that the overall transformation is the equivalent of a Claisen condensation between two molecules of an ester of acetic acid. The reactions that convert the β-ketoester into a saturated fatty acid are familiar ones.

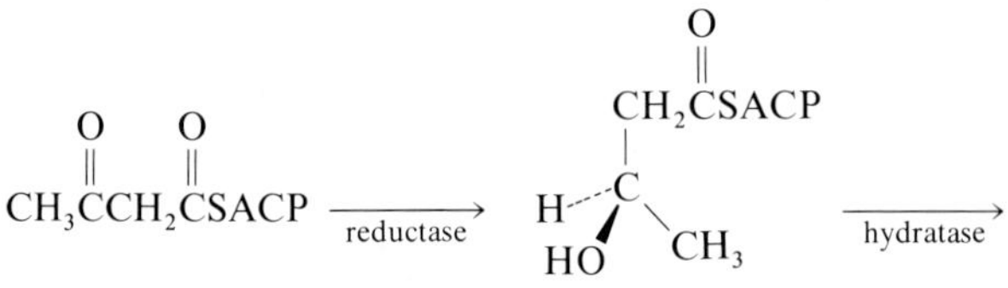

$$\underset{\substack{\text{acetoacetyl acyl} \\ \text{carrier protein}}}{\text{CH}_3\text{CCH}_2\text{CSACP}} \xrightarrow{\text{reductase}} \underset{\substack{(R)\text{-3-hydroxybutanoyl} \\ \text{acyl carrier protein}}}{\text{(structure)}} \xrightarrow{\text{hydratase}}$$

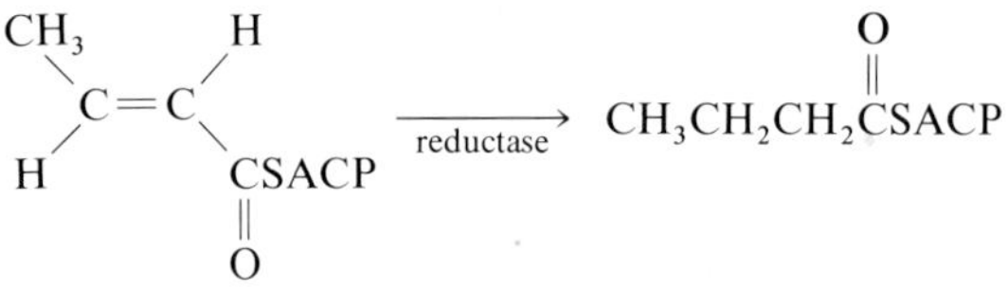

$$\underset{\substack{(E)\text{-2-butenoyl acyl} \\ \text{carrier protein}}}{\text{(structure)}} \xrightarrow{\text{reductase}} \underset{\substack{\text{butanoyl acyl} \\ \text{carrier protein}}}{\text{CH}_3\text{CH}_2\text{CH}_2\text{CSACP}}$$

The transformation involves reduction of the ketone group to an alcohol, dehydration of the alcohol to an α,β-unsaturated ester, and reduction of the double bond to give a saturated ester. In Section 18.6C, the structural properties of some of the coenzymes that function as biological oxidizing and reducing agents are examined. Note especially that the reduction of the ketone function in acetoacetyl acyl carrier protein gives rise to only one of the two enantiomers possible when an asymmetric carbon atom is created in the product. This stereoselectivity is characteristic of biological reducing agents and differs from the stereochemistry seen when sodium borohydride, for example, is used to carry out such a reduction (Section 9.3).

The four-carbon acyl group is transferred from acyl carrier protein to the same enzyme that brought the acetyl group into condensation with malonyl acyl carrier protein, and the cycle starts again. At the end of this process, a six-carbon straight-chain acyl group is present on the acyl carrier protein and is cycled through the synthesis once more. After seven cycles, the long-chain fatty acid hexadecanoic acid (palmitic acid) is released into body fluids from the acyl carrier protein or transferred to coenzyme A for other reactions. Thus, the fatty acid is systematically built up by the introduction of two-carbon units, activated first by conversion to malonyl coenzyme A.

PROBLEM 15.23 Write out equations for the conversion of butanoyl-*S*-enzyme to hexanoyl acyl carrier protein.

PROBLEM 15.24 The body degrades a saturated fatty acid using reactions that may be regarded as the reverse of those used in the synthesis. The fatty acid is dehydrogenated to an unsaturated acid, hydrated to a β-hydroxy acid, and then oxidized to a β-keto acid. A crucial step in the degradation is the reaction of a β-keto acid (present as the thioester of coenzyme A) with coenzyme A itself to give acetyl coenzyme A and the thioester of a new fatty acid that contains two carbon atoms fewer than the starting compound. Show the mechanism of this degradation using 3-oxohexanoyl coenzyme A as the starting material. You may use HB$^+$ and B: as general acids and bases in the reaction.

15.7

Reactions of Nucleophiles with α,β-Unsaturated Carbonyl Compounds

A. Reactions of Nitrogen, Oxygen, and Sulfur Nucleophiles with α,β-Unsaturated Carbonyl Compounds

A double bond that is conjugated with a carbonyl group reacts with nucleophiles. The behavior of compounds with this type of structure contrasts with that of ordinary alkenes, which usually react with electrophiles but not with nucleophiles. The reactions of alkenes with electrophiles such as acids, halogens, and oxidizing agents were treated in Chapters 5 and 8. But we did not observe any reactions that were started by the attack of a nucleophile on the double bond.

In contrast, diethylamine adds to the double bond that is in conjugation with a carbonyl group in methyl vinyl ketone.

$$\underset{\substack{\text{methyl vinyl}\\\text{ketone}}}{\text{CH}_2{=}\text{CHCCH}_3} + \underset{\substack{\text{diethylamine}}}{\underset{\overset{|}{\text{H}}}{\text{CH}_3\text{CH}_2\text{NCH}_2\text{CH}_3}} \xrightarrow[\substack{0\ °\text{C}}]{\text{acetic acid}} \underset{\substack{4\text{-}(N,N\text{-diethylamino})\text{-}2\text{-}\\\text{butanone}\\68\%}}{\overset{\text{CH}_2\text{CH}_3}{\underset{}{\text{CH}_3\text{CH}_2\text{NCH}_2\text{CH}_2\text{CCH}_3}}}$$

The reaction proceeds under mild acid catalysis and the nucleophilic nitrogen atom binds to the carbon atom beta to the carbonyl group.

An examination of the polarity of an α,β-unsaturated carbonyl compound shows us why this reaction takes place.

$$\overset{:\text{O}:}{\underset{}{\text{CH}_2{=}\text{CH}-\text{CCH}_3}} \longleftrightarrow \overset{:\ddot{\text{O}}:^-}{\underset{+}{\text{CH}_2{=}\text{CH}-\text{CCH}_3}} \longleftrightarrow \overset{:\ddot{\text{O}}:^-}{\underset{+}{\text{CH}_2-\text{CH}{=}\text{CCH}_3}}$$

resonance contributors for methyl vinyl ketone

$$\overset{\text{O}^{\delta-}}{\underset{}{\underset{\delta+}{\text{CH}_2}{=}\text{CH}-\underset{\delta+}{\text{CCH}_3}}}$$

*the polarization of an
α,β-unsaturated
carbonyl compound*

Resonance contributors can be written for methyl vinyl ketone showing that the presence of the carbonyl group in conjugation with the double bond leads to a polarization of the double bond so that the β-carbon atom has a partial positive charge. The electrophilicity of the carbon atom of the carbonyl group is transferred to the β-carbon atom of the double bond by conjugation. Methyl vinyl ketone is the simplest α,β-unsaturated ketone possible, but the electronic properties that we describe for it can be extended to all other α,β-unsaturated carbonyl compounds. The same electronic properties are also found in other compounds that have functional groups analogous in their polarization to the carbonyl group, such as the nitro and cyano groups. In all these compounds, we expect the β-carbon atom to be vulnerable to attack by nucleophiles. Such alkenes are also called **electrophilic alkenes.**

The mechanism for the reaction of methyl vinyl ketone and diethylamine is shown in detail below.

protonation of the carbonyl group   *nucleophilic attack at the β-carbon atom*   *deprotonation*

intermediate enol undergoing tautomerization

Because the experimental directions require acetic acid as a catalyst, the reaction is shown as starting with an initial protonation of the carbonyl group. This step increases the positive character at the carbon atom of the carbonyl group and also at the β-carbon atom. Bonding of the nucleophilic amine with the β-carbon atom of methyl vinyl ketone leads to an intermediate that is an enol. The enol tautomerizes back to the more stable keto form by losing a proton at the oxygen atom and gaining one at the α-carbon atom.

The double bond in α,β-unsaturated carbonyl compounds is oxidized only slowly by peroxyacids (Section 8.4D) but undergoes rapid oxidation when a nucelophilic oxidizing agent, a basic solution of hydrogen peroxide, is used. Propenal, for example, is successfully oxidized to an oxirane in this way.

propenal   oxide of propenal ~80%

a substituted oxirane

This reaction is another example of the vulnerability to nucleophilic attack of a double bond conjugated with a carbonyl group. Hydrogen peroxide is converted into its anion in basic solution. The anion reacts at the β-carbon atom of the unsaturated system to give an enolate ion as an intermediate.

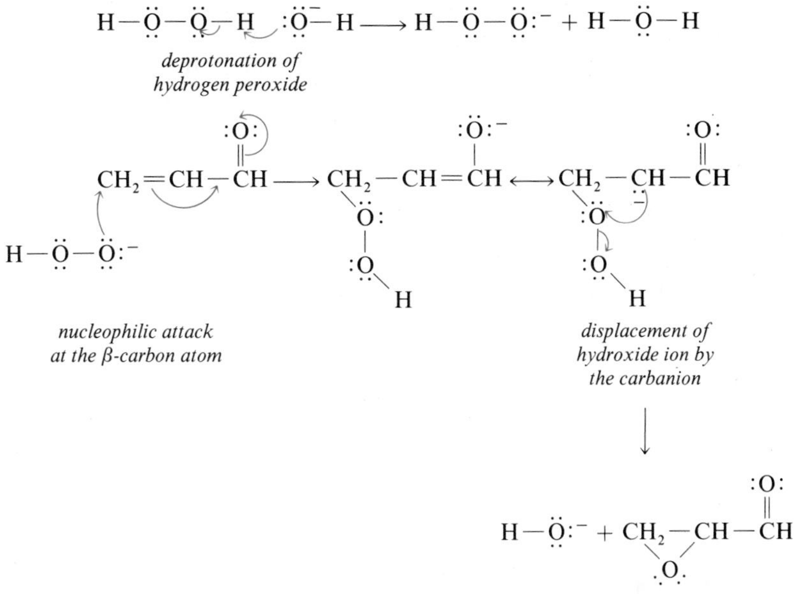

$$H-\ddot{O}-\ddot{O}-H \quad :\ddot{O}-H \longrightarrow H-\ddot{O}-\ddot{O}:^- + H-\ddot{O}-H$$

*deprotonation of
hydrogen peroxide*

nucleophilic attack
at the β-carbon atom

displacement of
hydroxide ion by
the carbanion

$$H-\ddot{O}:^- + \underset{\underset{\displaystyle \ddot{O}.}{\diagdown\diagup}}{CH_2-CH-\overset{\displaystyle :O:}{\overset{\displaystyle \|}{CH}}}$$

The carbanionic α-carbon atom of the enolate ion displaces hydroxide ion, breaking the weak oxygen-oxygen bond in the hydroperoxy group to give the oxirane.

Sulfur compounds also serve as nucleophiles towards α,β-unsaturated carbonyl compounds. For example, hydrogen sulfide reacts with two molecules of methyl propenoate to give an organic sulfide.

$$2\ \underset{\text{methyl propenoate}}{CH_2=CH\overset{\displaystyle O}{\overset{\displaystyle \|}{C}}OCH_3} + H_2S \xrightarrow[\substack{\text{ethanol}\\ \Delta}]{CH_3CO^-Na^+} \underset{\substack{\text{bis(2-carbomethoxyethyl)}\\ \text{sulfide}\\ \sim 75\%}}{CH_3O\overset{\displaystyle O}{\overset{\displaystyle \|}{C}}CH_2CH_2SCH_2CH_2\overset{\displaystyle O}{\overset{\displaystyle \|}{C}}OCH_3}$$

Again, we can write a mechanism showing nucleophilic attack at the β-carbon atom of the unsaturated system.

nucleophilic attack
at the β-carbon
atom of
methyl propenoate

deprotonation

protonation of the
enolate anion

$$\underset{}{CH_3O\overset{\displaystyle O}{\overset{\displaystyle \|}{C}}CH_2CH_2SCH_2CH_2\overset{\displaystyle O}{\overset{\displaystyle \|}{C}}OCH_3} \quad \longleftarrow \longleftarrow \quad \underset{\text{methyl 3-mercaptopropanoate}}{H-\ddot{S}-CH_2CH_2\overset{\displaystyle :O:}{\overset{\displaystyle \|}{C}}OCH_3}$$

*new nucleophile that reacts
with another molecule of
methyl propenoate*

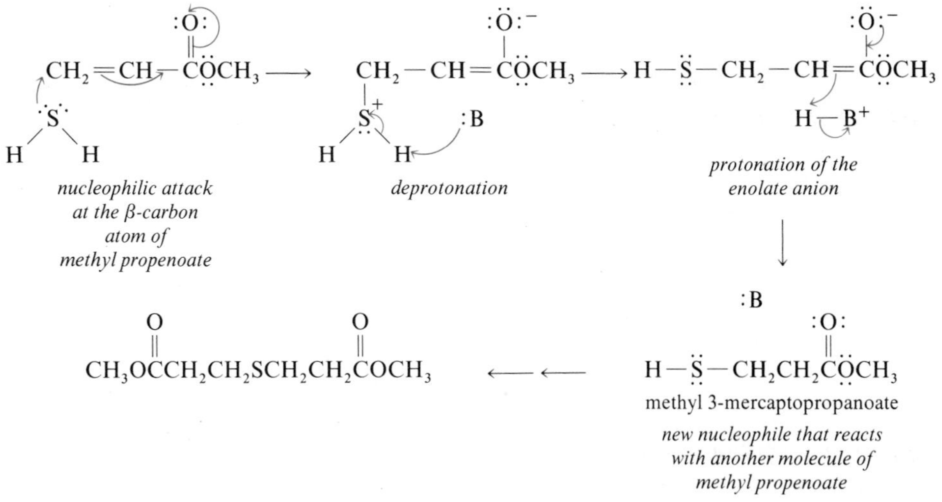

The first step of the reaction gives a thiol, which then adds to another molecule of methyl propenoate to give the final product.

Why does a nucleophile react with the β-carbon atom of the α,β-unsaturated carbonyl system and not with the electrophilic carbon atom of the carbonyl group? In some cases, of course, it does react at the carbonyl group. For example, in many instances sodium borohydride and lithium aluminum hydride can reduce carbonyl groups in α,β-unsaturated carbonyl compounds without touching the double bond, especially if the reaction is carried out quickly at low temperatures (Section 9.3). Many α,β-unsaturated carbonyl compounds add Grignard reagents at the carbonyl group (Section 9.4B) or react with reagents such as hydroxylamine to give carbonyl derivatives (Section 9.6A).

Whether a reaction takes place at the carbonyl group or at the β-carbon atom depends on the nature of the nucleophile and on the structure of the carbonyl compound. For example, the hydride ion, a small species with a high concentration of charge, reacts primarily at the carbonyl group. A sulfur atom, large and polarizable, is more likely to react at the β-carbon atom. In the next section we will explore how the change in the nature of an organometallic reagent changes the site at which it reacts with an α,β-unsaturated carbonyl compound.

PROBLEM 15.25   Complete the equation on p. 843 by showing the details of the mechanism for the reaction of methyl 3-mercaptopropanoate with methyl propenoate.

PROBLEM 15.26   Complete the following equations, giving structural formulas for reagents or products.

(a) $+ \; H_2O_2 \; \xrightarrow[\text{methanol}]{\text{NaOH}} \; A$

(b) $B + C \longrightarrow$

(c) $CH_2=CHC\equiv N$ (2 molar equiv) $+ \; NH_3 \; \xrightarrow{H_2O} \; D$

(d) $E + F \xrightarrow{CH_3ONa}$

(e) $+ \; CH_3CH_2NHCH_2CH_3$ (1 molar equiv) $\xrightarrow[\text{ether}]{\text{diethyl}} \; G$

B. Reaction of Organocuprate Reagents with α,β-Unsaturated Carbonyl Compounds. A Comparison of 1,2- and 1,4-Addition Reactions

An α,β-unsaturated carbonyl compound can react with a nucleophile at the carbonyl group to give a product in which the carbonyl function has been modified. This reaction is called a **1,2-addition** to the conjugated system in analogy with the reactions of conjugated dienes (Section 11.1C). Alternatively, as was seen in the last section (15.7A), the nucleophile may attack the β-carbon atom, giving a product in which the carbonyl group is regenerated from an enolate intermediate. The case in which the nucleophile is bonded to the β-carbon atom is known as **1,4-addition** or **conjugate addition.**

A Grignard reagent reacts with α,β-unsaturated ketones by both 1,2- and 1,4-addition. In the presence of catalytic amounts of copper(I) salts, however, 1,4-addition becomes the major reaction. The reactions of 3,5,5-trimethyl-2-cyclohexenone with methylmagnesium bromide with and without added copper(I) chloride illustrate this phenomenon.

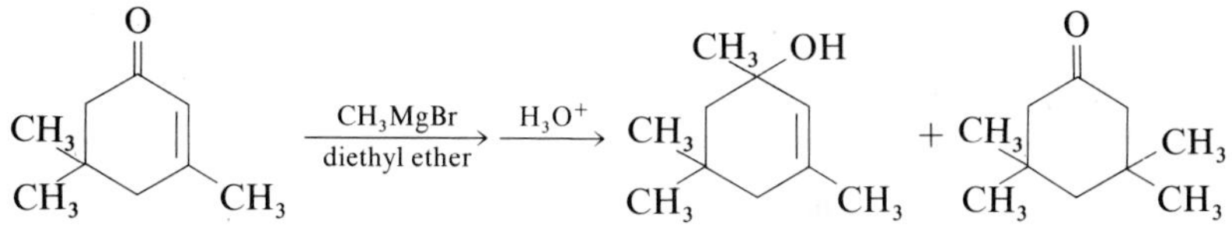

3,5,5-trimethyl-2-
cyclohexenone

1,3,5,5-tetramethyl-
2-cyclohexen-1-ol
major product
1,2-addition product
91%

3,3,5,5-tetramethyl-
cyclohexanone
minor product
1,4-addition product
1.5%

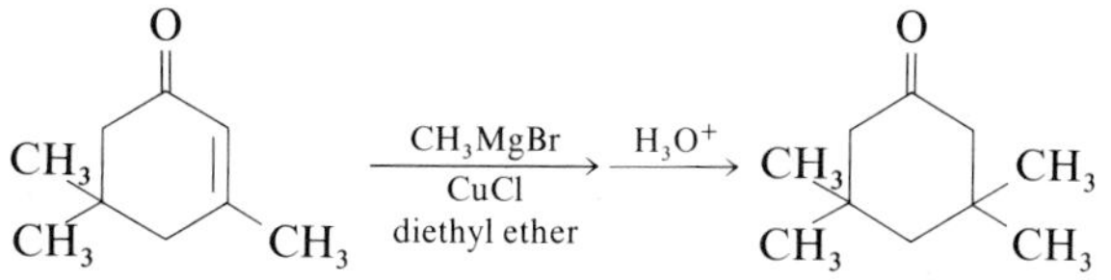

3,3,5,5-tetramethyl-
cyclohexanone
1,4-addition product
83%

The use of copper(I) halides in catalytic amounts thus changes the mode of reaction of the Grignard reagent and promotes reaction at the β-carbon atom of an α,β-unsaturated system. A similar catalytic effect of copper(I) halides on the formation of carbon-carbon bonds was seen in the reactions of alkynyl Grignard reagents with propargylic halides (Section 11.4B).

Organocuprates (Section 11.4A) also react with α,β-unsaturated carbonyl compounds at the β-carbon atom to give high yields of conjugate addition products. For example, 2-cyclohexenone is converted almost quantitatively to 3-methylcyclohexanone by lithium dimethylcuprate in a reaction run at low temperatures to improve selectivity.

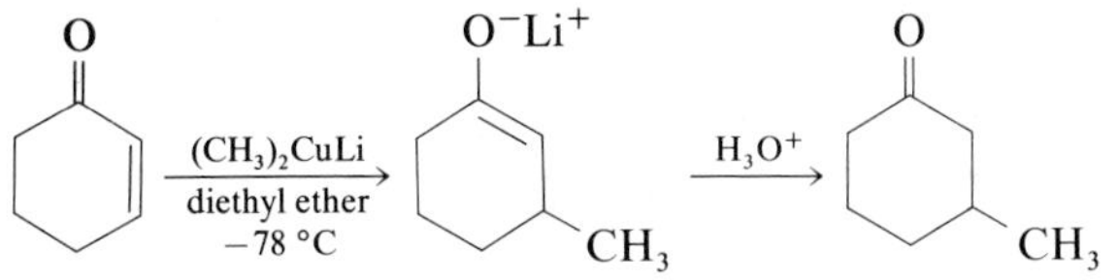

2-cyclohexenone

enolate anion
from conjugate
addition

3-methylcyclohexanone
97%

We do not know the exact mechanism of the reaction. An excess of the organocuprate is required and chemists have proposed mechanisms postulating complexation between the organometallic reagent and the carbonyl compound. An enolate anion is formed as an intermediate and can be used in further reactions forming new carbon-carbon bonds.

For example,

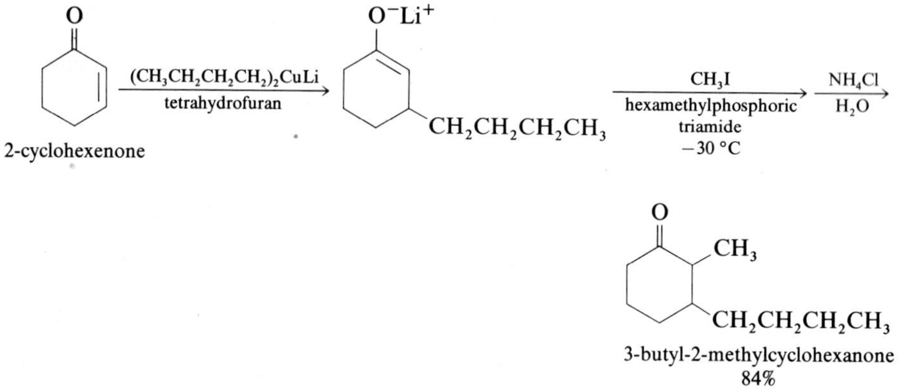

2-cyclohexenone

3-butyl-2-methylcyclohexanone
84%

The product obtained can be rationalized by postulating nucleophilic attack at the β-carbon atom giving an enolate anion, which is then alkylated by methyl iodide.

Conjugate addition of organocuprate reagents or of Grignard reagents in the presence of copper salts is important in creating new carbon skeletons in organic syntheses.

PROBLEM 15.27 Complete the following equations, giving structural formulas for reagents, intermediates, or products.

(a)

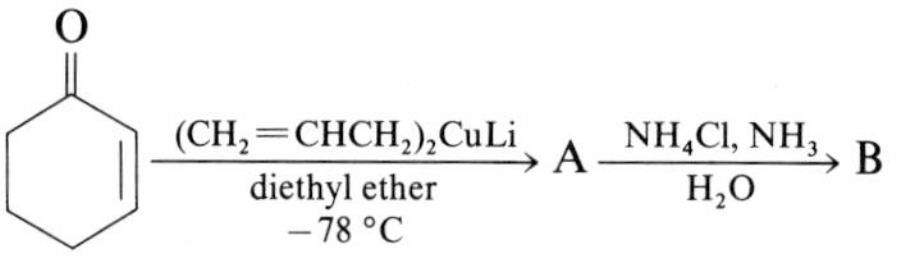

(b)

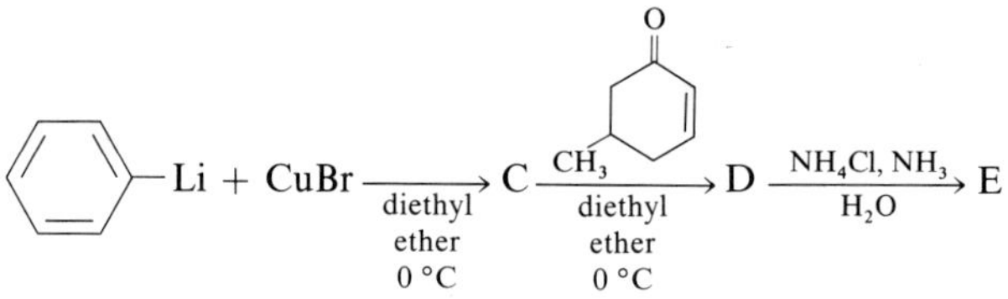

(c)

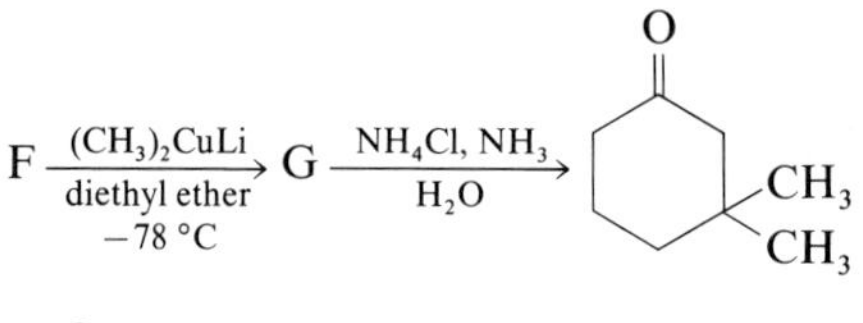

$$F \xrightarrow[\substack{\text{diethyl ether} \\ -78\,°C}]{(CH_3)_2CuLi} G \xrightarrow[H_2O]{NH_4Cl,\ NH_3}$$

(d)

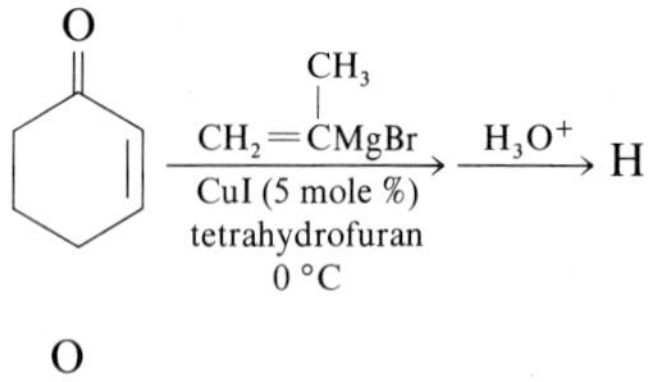

(e)

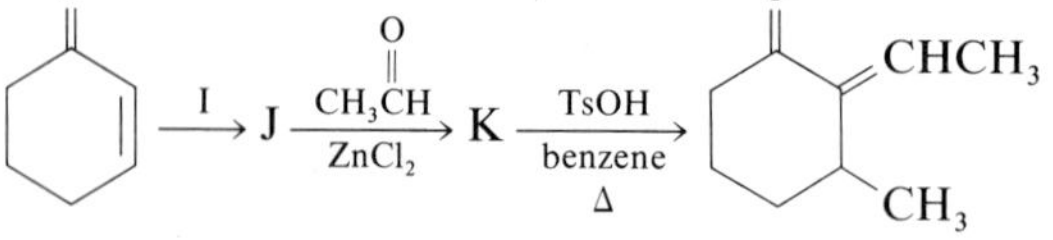

PROBLEM 15.28 Part of a synthesis of pentalenene, a hydrocarbon related to pentalenolactone, an antibiotic that inhibits the synthesis of nucleic acids in bacterial cells, uses the following reaction. Write structural formulas for reagents, intermediates, and products indicated by letters.

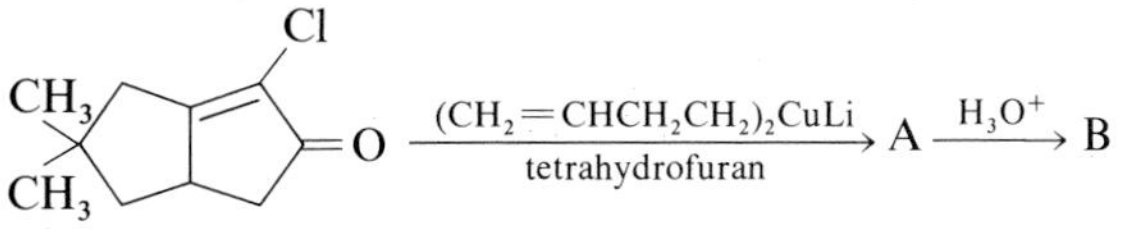

$$\text{[structure]} \xrightarrow[\text{tetrahydrofuran}]{(CH_2=CHCH_2CH_2)_2CuLi} A \xrightarrow{H_3O^+} B$$

C. Reactions of Enolate Anions with α,β-Unsaturated Carbonyl Compounds. The Michael Reaction

Enolate anions react with α,β-unsaturated carbonyl compounds at the β-carbon atom. For example, the anion generated by treating diethyl malonate with sodium metal adds to methyl propenoate.

$$\underset{\text{diethyl malonate}}{CH_2(COCH_2CH_3)_2} + Na \longrightarrow Na^+ : \bar{C}H(COCH_2CH_3)_2 \xrightarrow{\underset{\substack{\text{methyl}\\\text{propenoate}}}{\overset{\beta}{CH_2}=\overset{\alpha}{CH}\overset{O}{\overset{\|}{C}}OCH_3}}$$

$$\underset{\substack{\text{ethyl methyl 2-carboethoxy-}\\\text{pentanedioate}\\76\%}}{(CH_3CH_2OC)_2CH\underset{\beta}{CH_2}\underset{\alpha}{CH_2}COCH_3}$$

In a similar way, both the hydrogen atoms from the active methylene group in ethyl acetoacetate are replaced by alkyl groups through the addition of anions to methyl vinyl ketone.

$$\underset{\text{ethyl acetoacetate}}{CH_3\overset{O}{\overset{\|}{C}}CH_2\overset{O}{\overset{\|}{C}}OCH_2CH_3} + 2\ \underset{\substack{\text{methyl vinyl ketone}}}{\underset{\beta}{CH_2}=\underset{\alpha}{CH}\overset{O}{\overset{\|}{C}}CH_3} \xrightarrow[\substack{\text{ethanol}\\35\text{–}40\ ^\circ C}]{CH_3CH_2O^-\ Na^+}$$

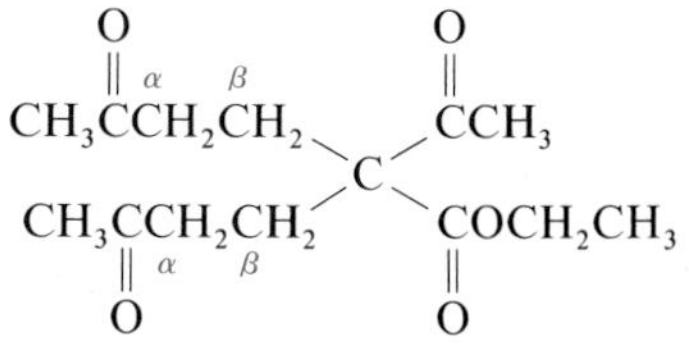

$$\begin{array}{c} CH_3\overset{O}{\overset{\|}{C}}\underset{\alpha}{CH_2}\underset{\beta}{CH_2}\diagdown \quad \overset{O}{\overset{\|}{C}}CH_3 \\ \qquad\qquad C \\ CH_3\overset{O}{\overset{\|}{C}}\underset{\alpha}{CH_2}\underset{\beta}{CH_2}\diagup \quad COCH_2CH_3 \\ 92\% \end{array}$$

The carbanion does not have to be derived from a carbonyl compound. Nitro groups also stabilize carbanions (Section 15.1B). For example, 2-nitropropane will add to methyl vinyl ketone in the presence of sodium methoxide.

$$\underset{\text{2-nitropropane}}{CH_3\overset{\displaystyle CH_3}{\underset{\displaystyle |}{C}}HNO_2} + \underset{\text{methyl vinyl ketone}}{CH_2{=}CH\overset{\displaystyle O}{\overset{\displaystyle \|}{C}}CH_3} \xrightarrow[\text{diethyl ether}]{CH_3O^-Na^+} \underset{\substack{\text{5-methyl-5-nitro-2-hexanone}\\69\%}}{CH_3\overset{\displaystyle CH_3}{\underset{\displaystyle \underset{\displaystyle NO_2}{|}}{C}}CH_2CH_2\overset{\displaystyle O}{\overset{\displaystyle \|}{C}}CH_3}$$

A nitro group can also polarize a double bond and make it susceptible to attack by an anion. The enolate ion from dimethyl malonate adds to 1-nitro-2-phenylethene in high yield.

$$\underset{\text{dimethyl malonate}}{CH_2(\overset{\displaystyle O}{\overset{\displaystyle \|}{C}}OCH_3)_2} + \underset{\text{1-nitro-2-phenylethene}}{\underset{\beta \quad\quad \alpha}{CH{=}CHNO_2}} \xrightarrow[\text{methanol}]{CH_3O^-Na^+} \underset{\substack{\text{methyl 2-carbomethoxy-}\\\text{3-phenyl-4-nitrobutanoate}\\92\%}}{(CH_3O\overset{\displaystyle O}{\overset{\displaystyle \|}{C}})_2CH\underset{\beta \quad \alpha}{CHCH_2NO_2}}$$

Note that in all these examples, a stabilized carbanion adds to the β-carbon atom of a double bond that is made electrophilic by an electron-withdrawing substituent.

The reactions described above are known as **Michael reactions.** *The compound that furnishes the nucleophilic carbon atom is called a* **Michael donor,** *and the compound containing the polarized double bond is called a* **Michael acceptor.** Though the reactions of Michael acceptors with the oxygen, nitrogen, and sulfur nucleophiles used as examples in Section 15.7A are not strictly Michael reactions, they are frequently called *Michael-type reactions.* Such additions of nucleophilic atoms other than carbon to Michael acceptors are biologically important reactions. Proteins and components of RNA and DNA contain nucleophilic amino, thiol, or hydroxyl groups. Michael acceptors are toxic to living organisms because of their ability to react with these groups. For example, one of the defensive secretions of the East African termite is dodeca-1-en-3-one. The termite uses it to kill other insects. The molecule acts as a solvent that penetrates the cuticle of the victim. Chemists believe that the substance is toxic because it reacts with biological nucleophiles. It might, for example, deactivate an enzyme by reacting with an amino or thiol group in a critical part of the molecule.

Michael acceptors are one example of a class of toxic compounds known as **alkylating agents** because they introduce alkyl groups onto nucleophilic sites in biologically important molecules. Other compounds such as methyl iodide and dimethyl sulfate also have considerable toxicity for the same reason.

PROBLEM 15.29   Write an equation showing how a thiol, RSH, might react with dodeca-1-en-3-one.

PROBLEM 15.30   Show the structure and the stabilization of the carbanion derived from 2-nitropropane.

PROBLEM 15.31   Write structural formulas that rationalize the polarity of the double bond in 1-nitro-2-phenylethene.

PROBLEM 15.32   Complete the following equations.

(a) $CH_2(C\equiv N)_2 + 2\ CH_2{=}CHCCH_3 \xrightarrow[\text{benzene}]{\text{Na}}$
with the ketone O above.

(b) $CH_3CHCCH_2COCH_2CH_3 + CH_3C{-}C{=}CH_2 \xrightarrow[\text{ethanol}]{\text{KOH}}$
with two O's, a CH$_3$ substituent, and a lower CH$_3$.

(c) 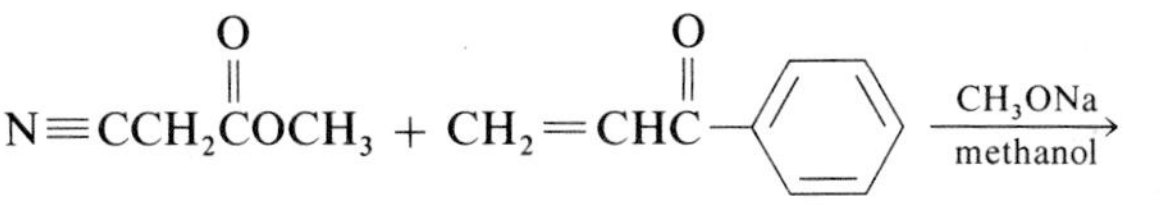 $N{\equiv}CCH_2COCH_3 + CH_2{=}CHC{-} \text{(phenyl ring)} \xrightarrow[\text{methanol}]{CH_3ONa}$

(d) 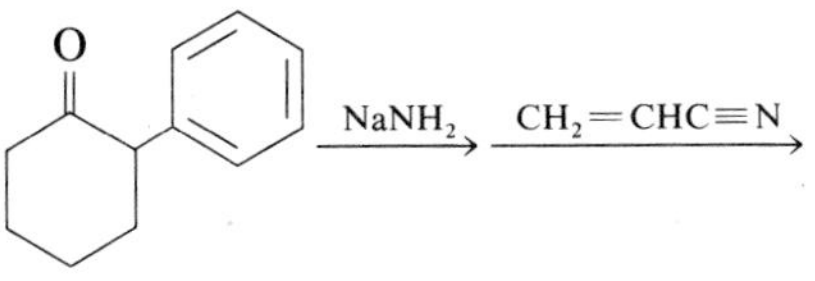 (cyclohexanone with phenyl substituent) $\xrightarrow{NaNH_2}$ $CH_2{=}CHC{\equiv}N \longrightarrow$

(Hint: Which enolate anion would you expect to be more stable?)

(e) 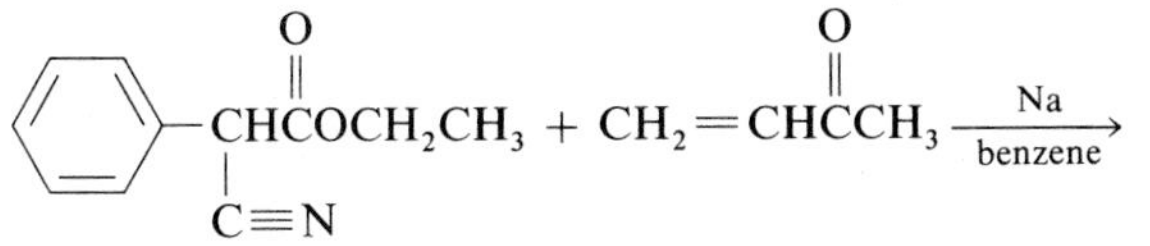 (phenyl)$-CHCOCH_2CH_3 + CH_2{=}CHCCH_3 \xrightarrow[\text{benzene}]{\text{Na}}$
with $C{\equiv}N$ substituent.

(f) $CH_3NO_2 + CH_3C{=}CHCCH_3 \xrightarrow{CH_3CH_2NHCH_2CH_3}$
with CH$_3$ and O above.

(g) 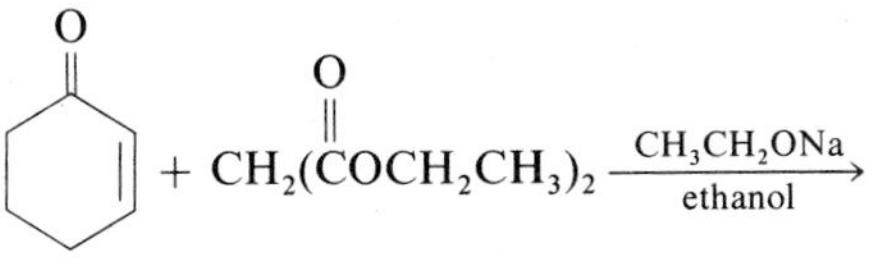 (cyclohexenone) $+ CH_2(COCH_2CH_3)_2 \xrightarrow[\text{ethanol}]{CH_3CH_2ONa}$

PROBLEM 15.33 The following sequence of reactions was used in a synthesis of pentalenolactone (Problem 15.28). Supply structural formulas for reagents, intermediates, and products designated by the letters.

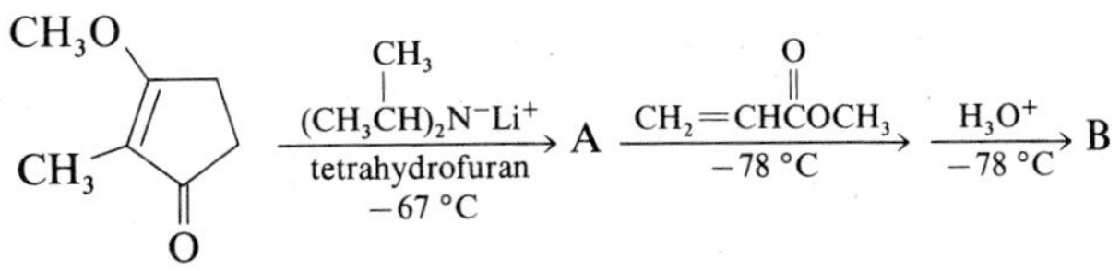

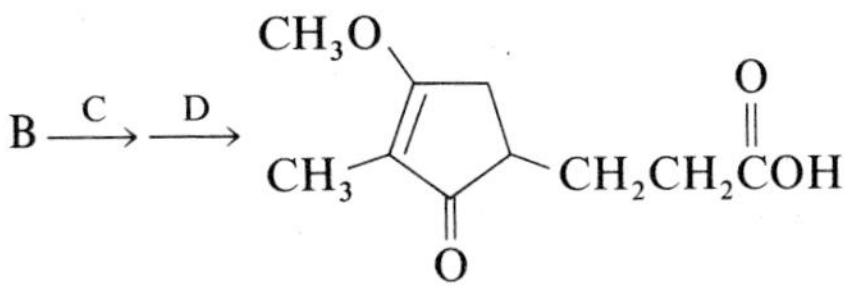

(cyclopentenone with CH$_3$O and CH$_3$ substituents) $\xrightarrow[\substack{\text{tetrahydrofuran} \\ -67\,°C}]{(CH_3CH)_2N^-Li^+}$ **A** $\xrightarrow[-78\,°C]{CH_2{=}CHCOCH_3}$ $\xrightarrow[-78\,°C]{H_3O^+}$ **B**

B $\xrightarrow{C}$ $\xrightarrow{D}$ (cyclopentenone with CH$_3$O, CH$_3$, and CH_2CH_2COH substituents)

PROBLEM 15.34 The ketoacid that is the product of the reaction sequence in Problem 15.33 is converted into another ketoacid when it is treated successively with methyllithium and dilute hydrochloric acid. Write equations that show how the conversion occurs.

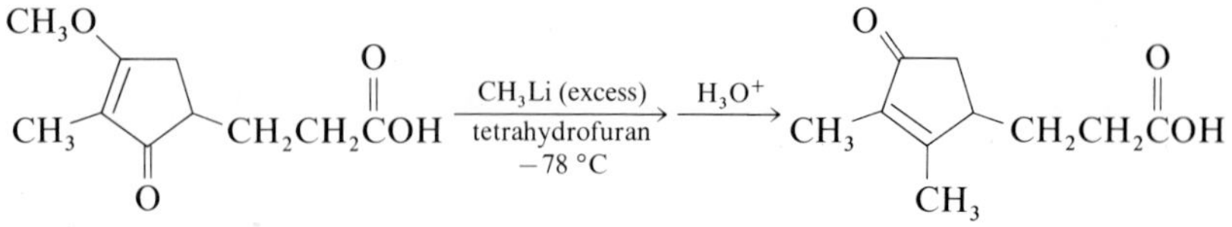

How do you explain the selectivity of the reaction with methyllithium?

PROBLEM 15.35   The synthesis of pentalenolactone can be continued by the conversion of the new ketoacid from Problem 15.34 to a bicyclic diketone. How would you carry out this conversion, which has two steps?

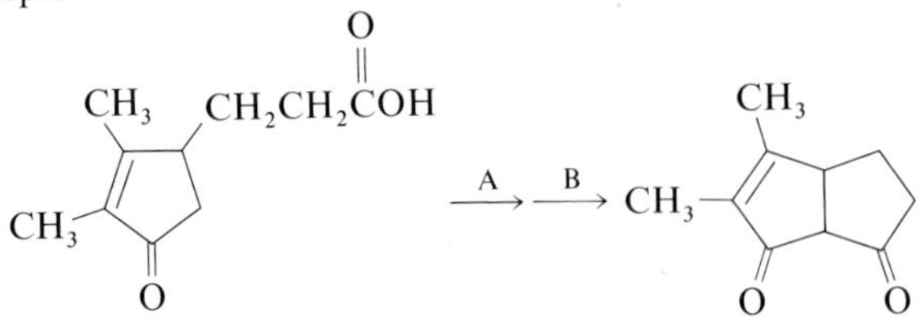

D. Cyclization Reactions of Products of Michael Reactions. The Robinson Annulation Reaction

The product of a Michael reaction is often a carbonyl compound that will enolize further under the conditions of the reaction, or in the presence of added base. If the structure of the molecule permits, the new enolate ion reacts with other functional groups within the molecule to form rings. For example, 2-methyl-1,3-cyclohexadione adds to methyl vinyl ketone to give a methyl ketone that is in equilibrium with several enolate ions.

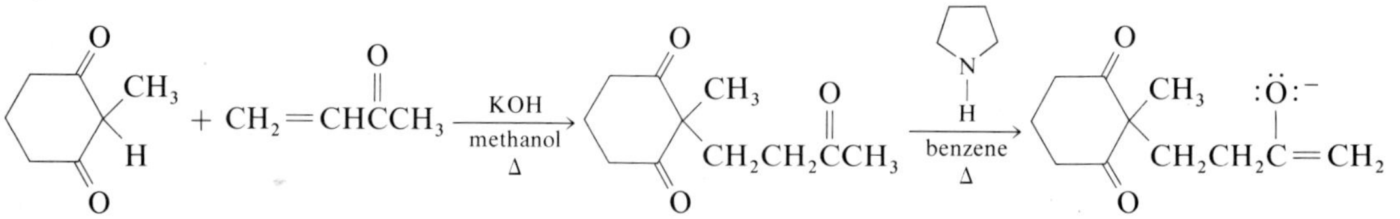

2-methyl-1,3-
cyclohexadione

methyl vinyl
ketone

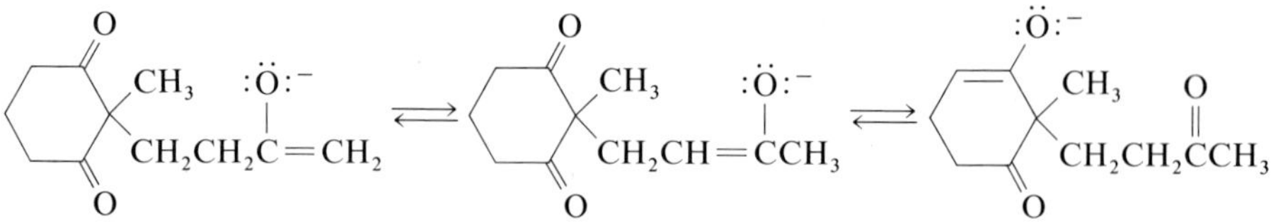

equilibrating enolate anions

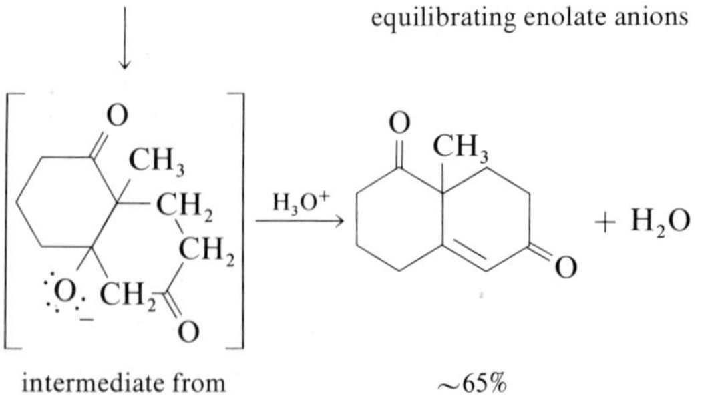

intermediate from
aldol condensation

~65%

In one of the ions, the carbanionic center is in a position to form a six-membered ring by an aldol condensation with one of the two equivalent carbonyl groups of the cyclohexadione. Protonation and loss of water gives the cyclic α,β-unsaturated ketone as the final product of the reaction.

This method of attaching new rings to functionalized cyclopentane or cyclohexane rings is known as the **Robinson annulation reaction,** after Sir Robert Robinson who developed it and used it extensively in his research. In recognition of his many contributions to organic chemistry, Robinson was awarded a Nobel Prize in 1947.

In the Robinson annulation reaction, a single six-membered ring is converted into two fused six-membered rings in essentially one step. The new ring contains an α,β-unsaturated ketone function, and may be used in further transformations. The starting material may also contain other functional groups that are retained in the product. Such is the case with the 2-methylcyclohexadione. Thus, the product is reactive at several points. For example, reaction of sodium acetylide (Section 9.4B) with the ketone group in the product above gives an alcohol with an alkyne function that is converted in acid to an α,β-unsaturated methyl ketone.

The final stage of the transformation may be visualized as an acid-catalyzed hydration of the alkyne to a methyl ketone (Section 5.5C) and dehydration of the tertiary alcohol. The new α,β-unsaturated methyl ketone can be used as a Michael acceptor, so it is possible to build still another ring onto the system.

Even the small portion of the synthesis outlined above shows how structural elements that are important in the synthesis of steroids can be introduced. For example, the final product, as it is written, shows rings A and B of a steroid system with the beginning of the construction of ring C. The methyl group in the angle between rings A and B, and the α,β-unsaturated ketone function in ring A are features that are found in naturally occurring steroids (p. 372).

In many Robinson annulation reactions, a compound that will give rise to methyl vinyl ketone in basic solution is used instead of the α,β-unsaturated ketone itself, which is unstable. The Michael acceptor is generated easily from compounds that have good leaving groups on the β-carbon atom. In one example, methyl vinyl ketone is prepared right in the reaction mixture from 4-(N,N-diethylamino)-2-butanone. The Michael donor is the enolate ion from ethyl 2-oxocyclohexanecarboxylate.

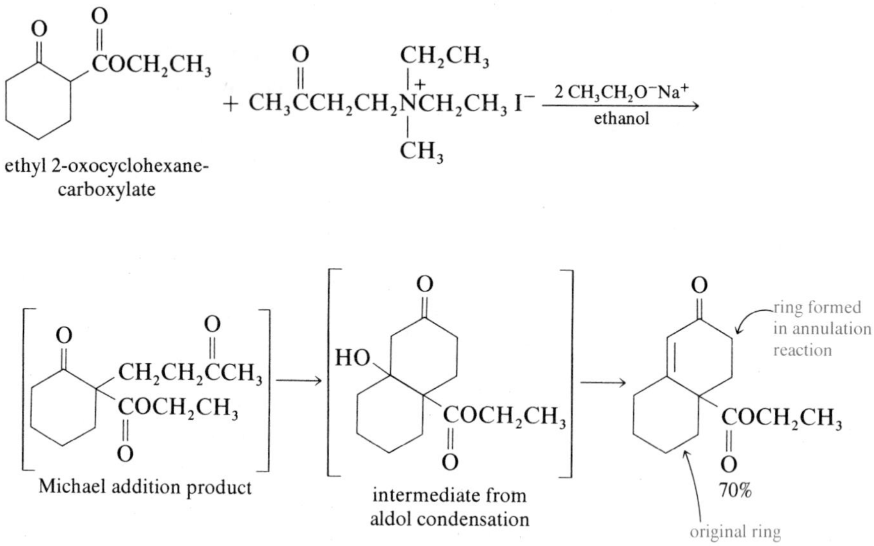

ethyl 2-oxocyclohexane-
carboxylate

Michael addition product

intermediate from
aldol condensation

ring formed
in annulation
reaction

70%

original ring

4-(N,N-Diethylamino)-2-butanone reacts with methyl iodide to give a compound with a positively charged nitrogen atom bonded to four organic groups. Such compounds, called quaternary ammonium compounds, contain a good leaving group, an amine, and undergo elimination reactions with ease. These reactions are discussed in detail in Section 16.7A. The anion from ethyl 2-oxocyclohexanecarboxylate adds to the α,β-unsaturated ketone that is generated in the reaction mixture, and the intermediate ionizes under the reaction conditions so that the final product is the result of the aldol cyclization reaction.

The Robinson annulation reaction builds a six-membered ring containing an α,β-*unsaturated carbonyl function by combining a Michael reaction on methyl vinyl ketone, or an equivalent compound,* with an internal condensation reaction in the product of the first step. It creates fused-ring systems containing carbonyl functions that allow further synthetic transformations of the molecule.

PROBLEM 15.36   Complete the following equations showing first the Michael addition product, then the cyclization product that you would expect in each case.

(a)

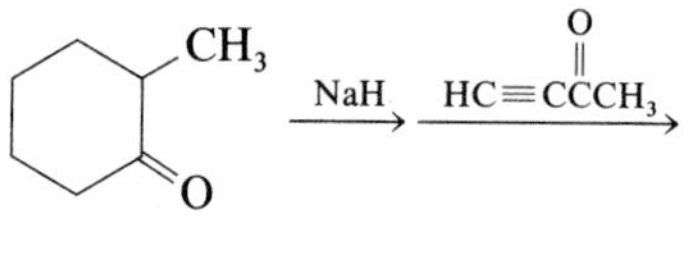

(b)

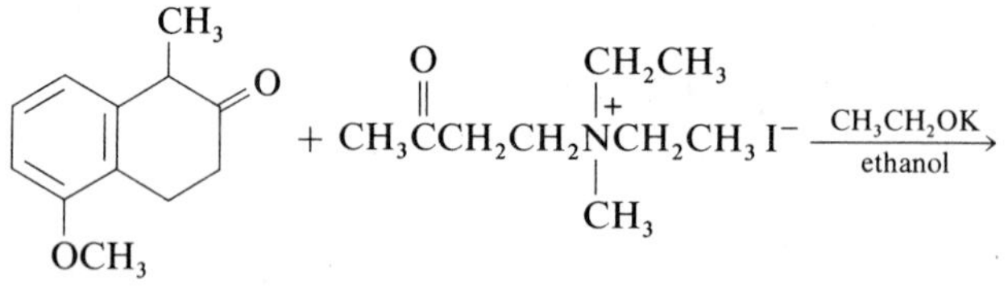

(c)

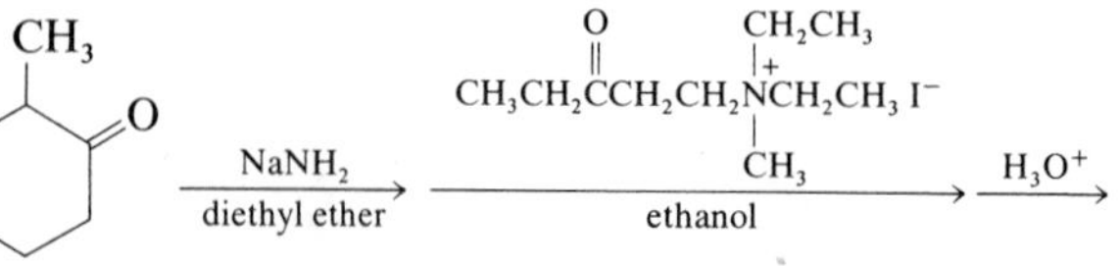

(d)

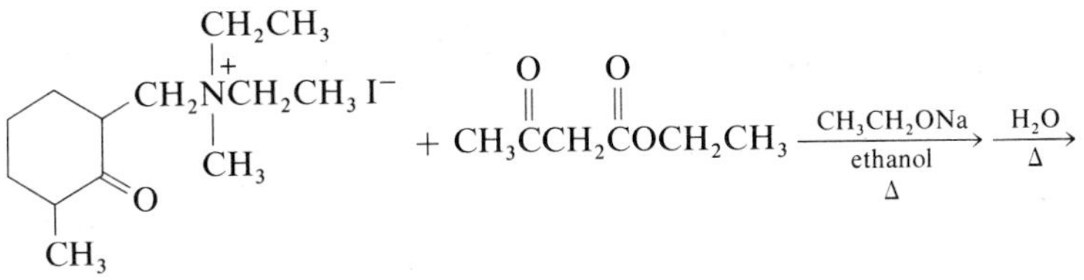

(e)

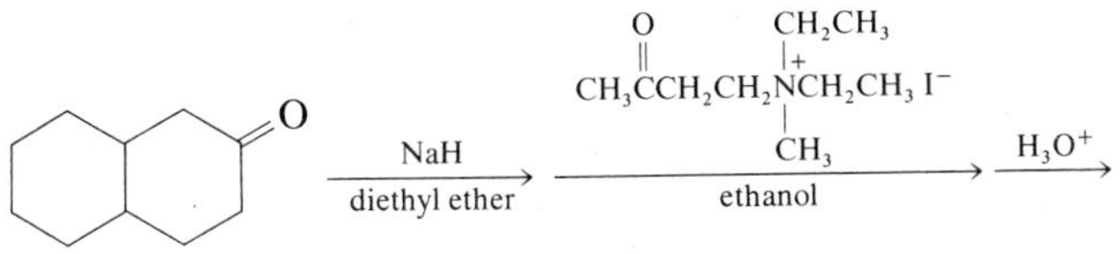

Carbanions Stabilized by Phosphorus. The Wittig Reaction

Chemists continue to search for new ways of stabilizing carbanions so that they can be generated under conditions that allow their use in different types of syntheses. Researchers have found that elements from the third row of the periodic table, such as sulfur and phosphorus, stabilize carbanions adjacent to them in a variety of structures.

A positively charged phosphorus atom is formed when a phosphine, the phosphorus analog of an amine, reacts with an alkyl halide. The phosphonium compound can be deprotonated at the carbon atom adjacent to the phosphorus atom to give a carbanion stabilized by the positively charged phosphorus atom. This sequence of reactions is shown for triphenylphosphine.

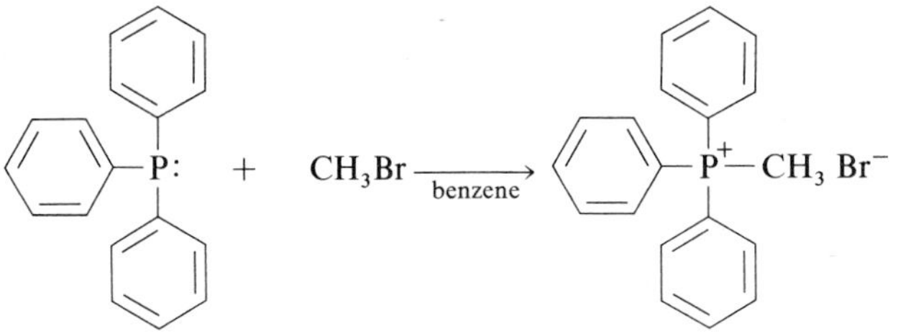

triphenylphosphine   methyl bromide   methyltriphenylphosphonium
                                                     bromide

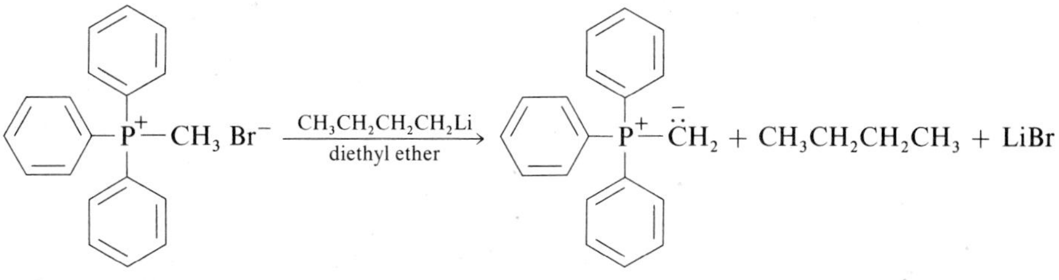

methyltriphenylphosphonium   a phosphonium ylide   butane
          bromide

Such a stabilized carbanion is known as a *phosphonium ylide. An* **ylide** (rhymes with "ill lid") *is a reactive species having a positively charged atom next to one that is negatively charged.* The phosphonium ylide can be written with a double bond symbolizing a donation of electrons from the $2p$ orbital on the carbon atom to the empty $3d$ orbitals of the phosphorus atom.

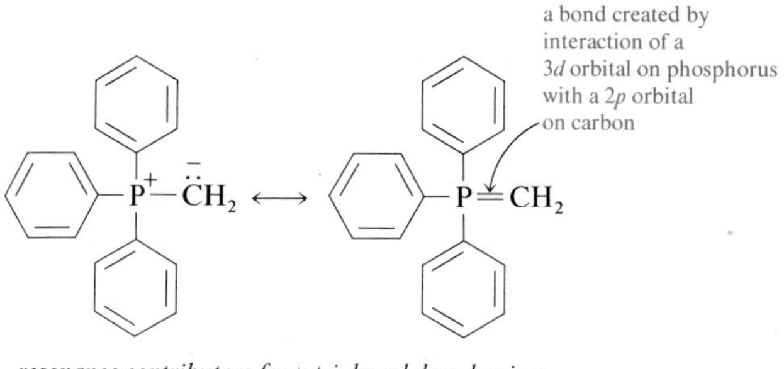

a bond created by interaction of a $3d$ orbital on phosphorus with a $2p$ orbital on carbon

resonance contributors for a triphenylphosphonium ylide

In the resonance contributor with a double bond between the phosphorus and the carbon atoms, neither atom has a formal charge. There are ten electrons around phosphorus in this resonance contributor, but this is permissible for an element in the third row.

Phosphorus ylides may be prepared with a wide range of structures. A particularly interesting class of ylides comes from α-haloesters such as ethyl bromoacetate.

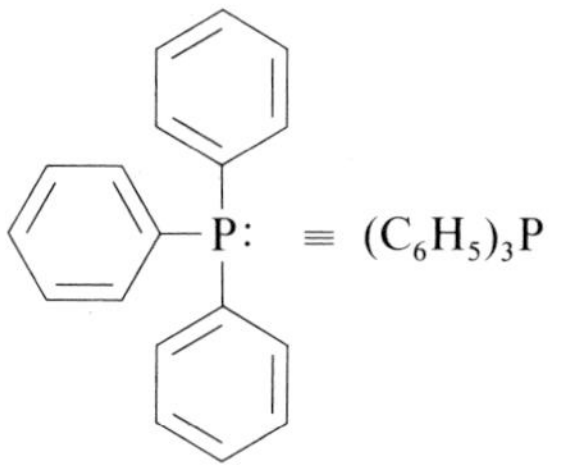

$$\text{(C}_6\text{H}_5\text{)}_3\text{P} + \text{BrCH}_2\overset{\overset{\textstyle O}{\|}}{\text{C}}\text{OCH}_2\text{CH}_3 \longrightarrow \overset{\overset{\textstyle Br^-}{\;}}{\text{(C}_6\text{H}_5\text{)}_3\overset{+}{\text{P}}\text{CH}_2}\overset{\overset{\textstyle O}{\|}}{\text{C}}\text{OCH}_2\text{CH}_3 \xrightarrow[\text{H}_2\text{O}]{\text{NaOH}}$$

triphenylphosphine   ethyl bromoacetate   phosphonium salt

$$\text{(C}_6\text{H}_5\text{)}_3\overset{+}{\text{P}}-\overset{..}{\overset{-}{\text{C}}}\text{H}\overset{\overset{\textstyle O}{\|}}{\text{C}}\text{OCH}_2\text{CH}_3$$

phosphonium ylide

A comparison of the equations above with the one on p. 853 indicates that it is much easier to remove a proton from the phosphonium salt derived from ethyl bromoacetate than from methyltriphenylphosphonium bromide. The base used to deprotonate methyltriphenylphosphonium bromide is butyllithium, a strong base. The reaction is carried out in a nonprotic solvent, dry ether. The methylene group between the positively charged phosphorus atom and the carbonyl group of the ester, however, is much more acidic. Aqueous sodium hydroxide is a strong enough base to remove a proton, and the resulting ylide is stable in water, a medium that must be avoided with most other types of ylides. Phosphorus ylides with carbonyl groups on the carbanionic center are generally more stable and less reactive than other types of phosphorus ylides.

Phosphorus ylides react with aldehydes or ketones to introduce a carbon-carbon double bond selectively in place of the carbonyl group. For example, the ylide from methyltriphenylphosphonium bromide reacts with cyclohexanone to give methylene-cyclohexane.

$$(C_6H_5)_3\overset{+}{P}-CH_3 \;+\; CH_3CH_2CH_2CH_2Li \;\xrightarrow[\substack{-78\,°C}]{\text{tetrahydrofuran}}\; (C_6H_5)_3P=CH_2 \;+\; CH_3CH_2CH_2CH_3\uparrow$$

methyltriphenylphosphonium   *n*-butyllithium
bromide

$+$ LiBr

tetrahydrofuran

(C_6H_5)_3P=O + =CH_2

triphenylphosphine   methylenecyclohexane
oxide

The base used to deprotonate the phosphonium salt in this reaction is the anion from *n*-butyllithium. The ylide reacts quickly with the ketone to give an alkene and tri-phenylphosphine oxide. The reaction is believed to start by nucleophilic attack of the ylide on the carbonyl group. The nature of the intermediate in such reactions depends on reaction conditions. At low temperatures, an unstable four-membered ring inter-mediate, an oxaphosphetane, has been observed by nuclear magnetic resonance spec-troscopy. In the presence of lithium bromide, an ionic intermediate stabilized by interaction with lithium and bromide ions also forms. Bond reorganization in either intermediate leads to the products.

$$(C_6H_5)_3P=CH_2 \longleftrightarrow (C_6H_5)_3\overset{+}{P}-\overset{-}{C}H_2$$

=Ö:

$\longrightarrow$

Ö
P(C_6H_5)_3

$^-$CH_2$-\overset{+}{P}$(C_6H_5)_3

*nucleophilic attack at
the carbonyl group*

oxaphosphetane
intermediate

LiBr

Ö:$^-$Li$^+$

$\longrightarrow$

=CH_2 + Ö=P(C_6H_5)_3 $\longleftrightarrow$ $^-$:Ö$-\overset{+}{P}$(C_6H_5)_3

CH_2$-$P(C_6H_5)_3

Br$^-$

ionic intermediate

Triphenylphosphine oxide is a stable compound and the formation of the phosphorus-oxygen bond is part of the driving force of the reaction. In any event, the product always has the group that can be written as doubly bonded to the phosphorus atom in the ylide, now doubly bonded to the carbon atom of the carbonyl group in the aldehyde or ketone that is used in the reaction.

For example, benzaldehyde is converted into an α,β-unsaturated ester with the phosphonium ylide from ethyl bromoacetate.

$$(C_6H_5)_3\overset{+}{P}\!-\!CH_2COCH_2CH_3 \xrightarrow[\text{ethanol}]{CH_3CH_2O^-Na^+} (C_6H_5)_3P\!=\!CHCOCH_2CH_3 \longrightarrow$$

phosphonium salt from
ethyl bromoacetate

phosphonium ylide

ethyl (*E*)-3-phenyl-2-propenoate
100%.

triphenylphosphine
oxide

$+ (C_6H_5)_3P\!=\!O$

The product above is formed exclusively with the trans configuration around the double bond. But this is not always so. In most cases where stereoisomerism is possible, some of each isomer forms. The composition of the mixture depends on factors such as the solvent used, the presence of inorganic salts in the reaction mixture, and the temperature at which the reaction is carried out.

In both cases discussed above, the carbonyl group has been replaced by a double bond to a carbon atom that was the anionic center of a phosphonium ylide. This method of forming double bonds quite selectively is called the **Wittig reaction** after the German chemist Georg Wittig, who developed it. The reaction is a useful one because it introduces double bonds into molecules with precision. Other methods such as the dehydration of alcohols (Sections 5.5B and 5.6B) or elimination reactions (Section 6.9) often give mixtures of products. Wittig reactions usually take place at low temperatures and under basic conditions so that rearrangements of complex molecules do not occur. Note that in its essentials the Wittig reaction resembles the aldol condensation (Section 15.5A).

PROBLEM 15.37   Review in your mind the difficulties involved in preparing methylenecyclohexane selectively by other methods starting with cyclohexanone. How would you introduce first a carbon atom, then a double bond?

Phosphorus ylides may be prepared from halides containing double bonds, so the Wittig reaction can be used to synthesize polyenes. It has been used extensively in the synthesis of natural products that contain multiple double bonds, such as pheromones (Section 11.7C), β-carotene, and vitamin A and its derivatives (Section 11.6D). An example of the application of a Wittig reaction to the synthesis of an alkene of biological interest is the preparation of the ester of the acid derived from vitamin A (p. 547). The phosphonium ylide incorporating the ring structure of vitamin A and of β-carotene is added to a polyene aldehyde ester that completes the rest of the chain.

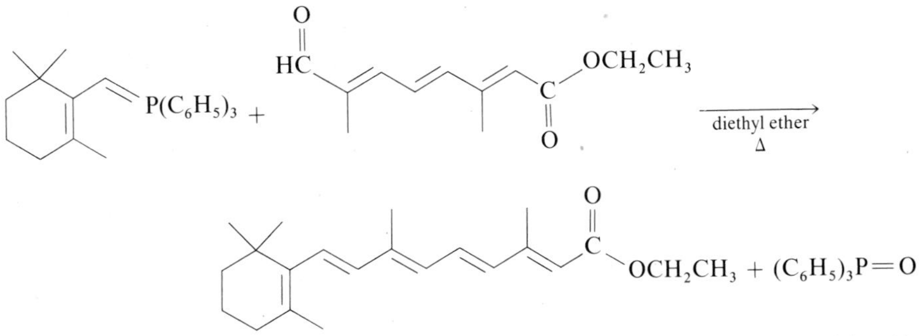

ethyl ester of vitamin A acid
65%

triphenylphosphine
oxide

In this case, researchers report the formation of a small amount of the compound with a cis double bond. Note that the carbonyl group of the aldehyde, rather than that of the ester, reacts.

The Wittig reaction has also been used to make modifications on ketone steroids. 4-Cholesten-3-one, for example, is converted to the corresponding methylene compound.

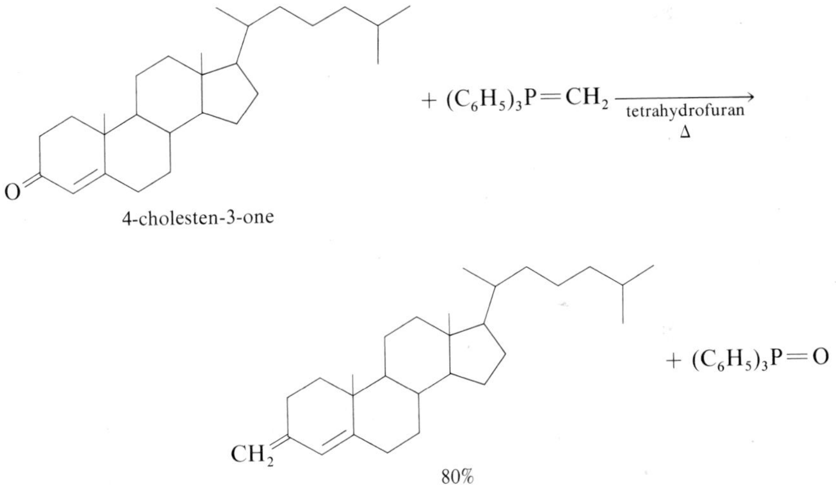

4-cholesten-3-one

$+ (C_6H_5)_3P=CH_2 \xrightarrow[\Delta]{\text{tetrahydrofuran}}$

$+ (C_6H_5)_3P=O$

80%

Note that the phosphorus ylide does not add to the α,β-unsaturated carbonyl system at the β-carbon atom, as do other carbanions in Michael reactions, but reacts directly with the carbonyl group.

The Wittig reaction can be used to synthesize many interesting compounds. Some of them are included in other syntheses later in this book.

PROBLEM 15.38   Complete the following equations.

(a)

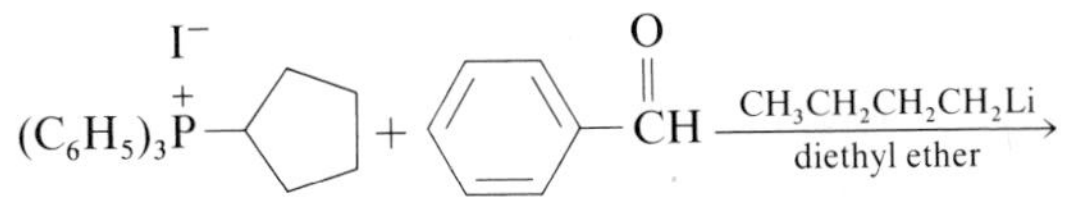

(b) $(C_6H_5)_3\overset{+}{P}CH_2\overset{O}{\overset{\|}{C}}OCH_2CH_3 \;\; Br^-\;\; + \;\; CH_3CH{=}CH\overset{O}{\overset{\|}{C}}H \;\; \xrightarrow[\text{ethanol}]{CH_3CH_2ONa}$

(c) 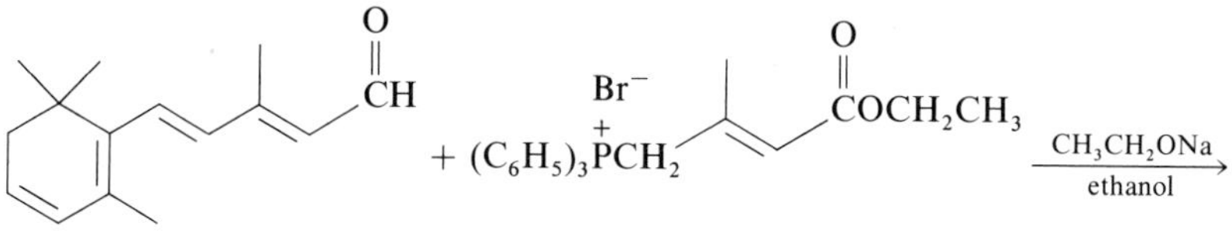   [cyclohexadiene ring with dimethyl and methyl substituents and a diene chain ending in CHO] $+\;(C_6H_5)_3\overset{+}{P}CH_2$ [chain ending in $\overset{O}{\overset{\|}{C}}OCH_2CH_3$] $\;Br^-\;\xrightarrow[\text{ethanol}]{CH_3CH_2ONa}$

(d) 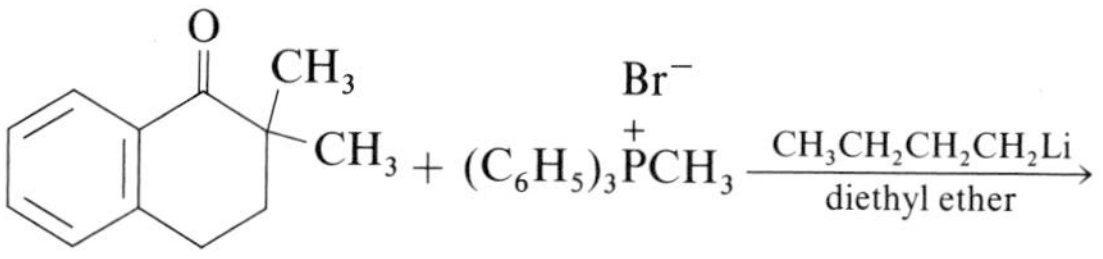   [bicyclic tetralone structure with C=O and two CH_3 groups] $+\;(C_6H_5)_3\overset{+}{P}CH_3 \;\; Br^-\; \xrightarrow[\text{diethyl ether}]{CH_3CH_2CH_2CH_2Li}$

(e) 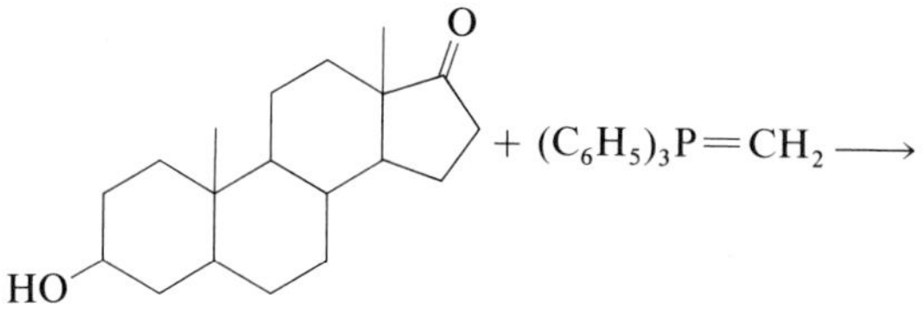   [steroid structure with HO at one end and C=O (ketone) on the five-membered ring] $+\;(C_6H_5)_3P{=}CH_2 \longrightarrow$

15.9

Carbanions Stabilized by Sulfur

A. Sulfonium Ylides

A sulfur atom, like a phosphorus atom, can stabilize an adjacent carbanion by accepting electrons into its $3d$ orbitals. Several types of anions containing sulfur have been developed as synthetic intermediates by Elias J. Corey of Harvard University. Among these, sulfonium ylides most resemble the Wittig reagent. For example, a sulfonium ylide forms on removal of a proton from trimethylsulfonium iodide (prepared from dimethyl sulfide and methyl iodide) by means of a strong base such as sodium hydride.

$$CH_3SCH_3 \;+\; CH_3I \;\longrightarrow\; CH_3\overset{\overset{\displaystyle CH_3}{|}}{\underset{+}{S}}CH_3\;I^-$$

dimethyl sulfide   methyl iodide   trimethylsulfonium
iodide

$$CH_3\overset{\overset{\displaystyle CH_3}{|}}{\underset{+}{\ddot{S}}}{-}CH_3\;I^- + NaH \xrightarrow[\substack{\text{dimethyl}\\ \text{sulfoxide}}]{} CH_3\overset{\overset{\displaystyle CH_3}{|}}{\underset{+}{\ddot{S}}}{-}\ddot{C}H_2 + H_2\uparrow + NaI$$

In a sulfonium ylide, the sulfur atom is positively charged and the carbon atom is negatively charged.

a bond created by
interaction of a 3*d*
orbital on sulfur
with a 2*p* orbital
on carbon

$$CH_3 \quad \quad \quad \quad CH_3$$
$$>\overset{+}{\underset{..}{S}}-\overset{..}{C}H_2 \longleftrightarrow \quad >\underset{..}{S}=CH_2$$
$$CH_3 \quad \quad \quad \quad CH_3$$

resonance contributors for the sulfonium ylide
generated from trimethylsulfonium iodide

Sulfonium ylides react with carbonyl compounds to give oxiranes. For example, 2-phenyloxirane is formed when the sulfonium ylide from trimethylsulfonium iodide reacts with benzaldehyde.

a sulfonium   benzaldehyde   2-phenyloxirane   dimethyl sulfide
ylide   75%

The products obtained can be rationalized by recognizing that sulfonium ylides are nucleophilic at the carbanionic carbon atom and also contain a good leaving group, dimethyl sulfide in this case. The carbanion attacks the electrophilic carbon atom of the carbonyl compound to give an alkoxide ion intermediate.

nucleophilic attack   *alkoxide ion*
at the carbonyl group   *intermediate losing*
                               dimethyl sulfide

An intramolecular nucleophilic substitution reaction by the alkoxide ion on the carbon atom bearing the good leaving group completes the reaction. The overall reaction can be described as an addition of a methylene group to the carbon-oxygen double bond of the aldehyde.

It is interesting to compare the reactions of phosphonium and sulfonium ylides with carbonyl compounds. The phosphonium ylide replaces the carbon-oxygen double bond with a carbon-carbon double bond. The sulfonium ylide gives an oxirane. The difference in the reactivity of these two ylides derives from the relative energies of the phosphorus-oxygen and sulfur-oxygen bonds. The formation of a phosphine oxide with a strong phosphorus-oxygen bond favors creation of the carbon-carbon double bond in the Wittig reaction.

B. Dithiane Anions

Corey developed another type of carbanion stabilized by sulfur, the 1,3-dithiane anion, in a search for the synthetic equivalent of a carbonyl group that would serve as a nucleophile in a nucleophilic substitution reaction. In the reactions we have studied so far, the carbon atom of a carbonyl group is an electrophilic center and is attacked by nucleophiles. Corey reasoned that a method to introduce carbonyl groups by a nucleophilic substitution reaction initiated by a potential carbonyl group would be useful in syntheses. He chose the cyclic thioacetal of aldehydes, the 1,3-dithiane system, as the source of a carbanionic equivalent of a carbonyl group.

The conversion of formaldehyde to 6-undecanone illustrates the chief features of this method.

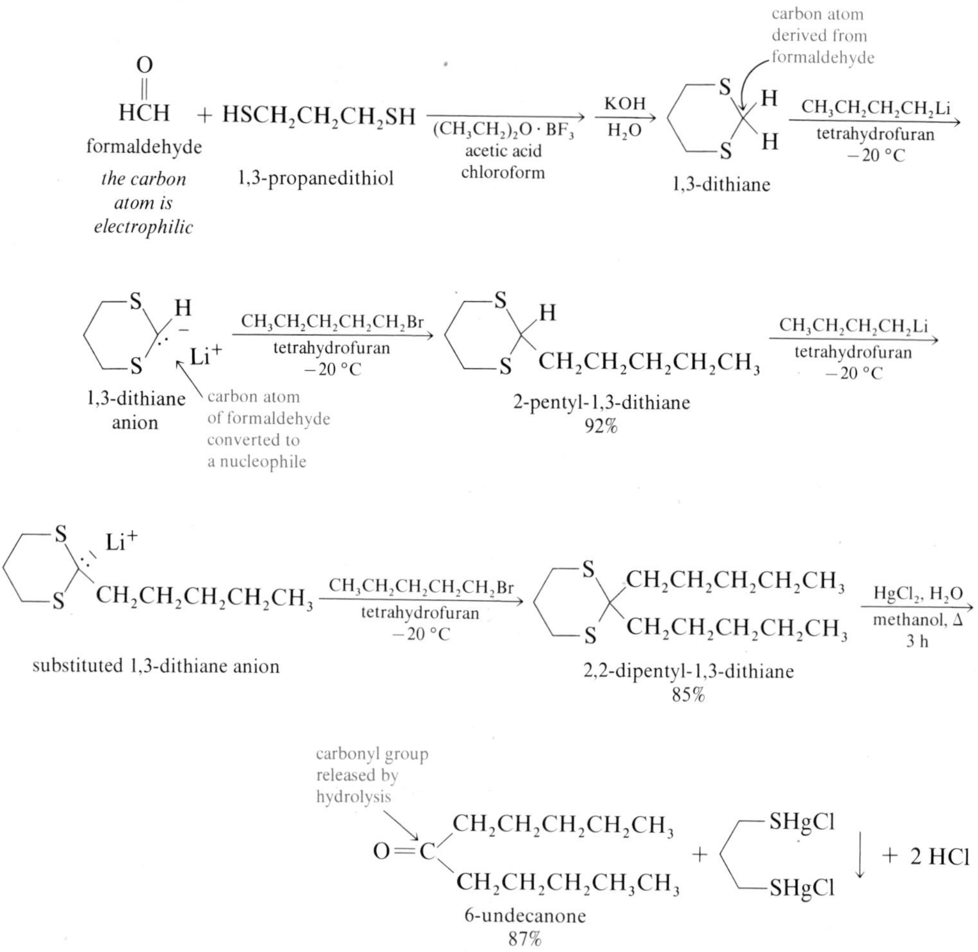

This sequence of reactions allows for the reversal of polarity of the carbonyl group, resulting in greater versatility in the design of syntheses.

The thioacetal (Section 9.7D) is prepared from the aldehyde by treatment with 1,3-propanedithiol in the presence of an acid catalyst. A 1,3-dithiane is deprotonated at the carbon atom between the two sulfur atoms by a strong base, such as *n*-butyllithium, to give a carbanion stabilized by delocalization of charge to the sulfur atoms. The carbanion is a strong nucleophile and reacts with alkyl halides in nucleophilic substitution reactions. In the synthesis shown above, the deprotonation and alkylation of 1,3-dithiane occurs twice. The product of these reactions, 2,2-dipentyl-1,3-dithiane, contains a carbonyl group hidden in a thioacetal function. The carbonyl group is released by hydrolysis in the presence of mercury(II) salts, which serve as Lewis acids to coordinate with the sulfur atoms and help cleave the sulfur-carbon bonds. The mechanism proposed for the reaction starts with reaction between a sulfur atom, a Lewis base, and mercury(II), a Lewis acid, creating a good leaving group at sulfur.

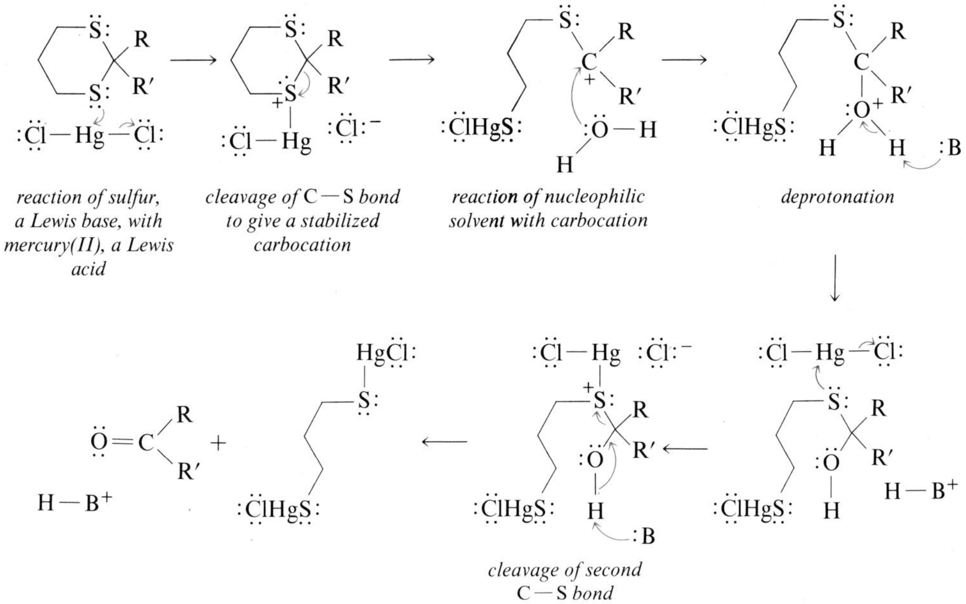

reaction of sulfur,   *cleavage of* C — S *bond*   *reaction of nucleophilic*   *deprotonation*
a Lewis base, with   *to give a stabilized*   *solvent with carbocation*
mercury(II), a Lewis   *carbocation*
acid

cleavage of second
C — S *bond*

The carbocationic intermediate reacts with water, and the second carbon-sulfur bond is broken with further assistance from mercury(II).

    1,3-Dithiane anions react also with other types of electrophilic carbon atoms such as carbonyl groups. The addition of the carbanion to a carbonyl compound must be carried out at very low temperatures if the carbonyl compound has enolizable hydrogen atoms. The reaction of the anion from 1,3-dithiane with benzaldehyde, which has no α-hydrogen atoms, goes well.

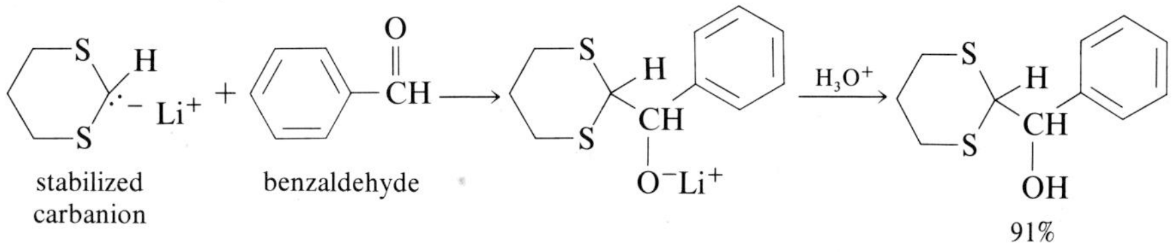

stabilized   benzaldehyde   91%
carbanion

The carbanion also reacts with oxiranes, attacking the less highly substituted carbon atom of unsymmetrical oxiranes (Section 8.5C).

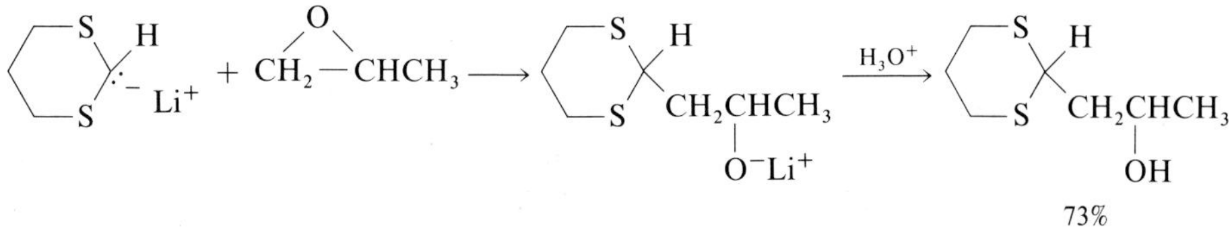

                                                  73%

The reactions of dithiane anions with carbonyl compounds and with oxiranes are routes to potential hydroxycarbonyl compounds.

PROBLEM 15.39   Show how the following conversions can be carried out.

(a) 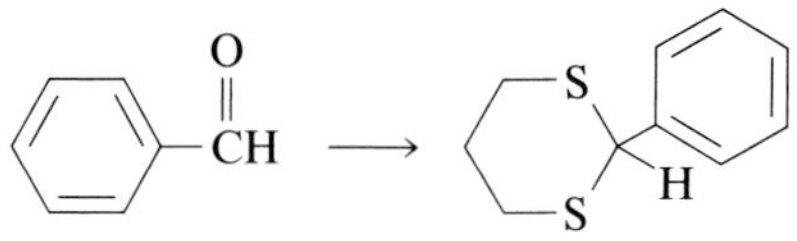   C₆H₅–CH(=O)   ⟶   2-phenyl-1,3-dithiane (with H)

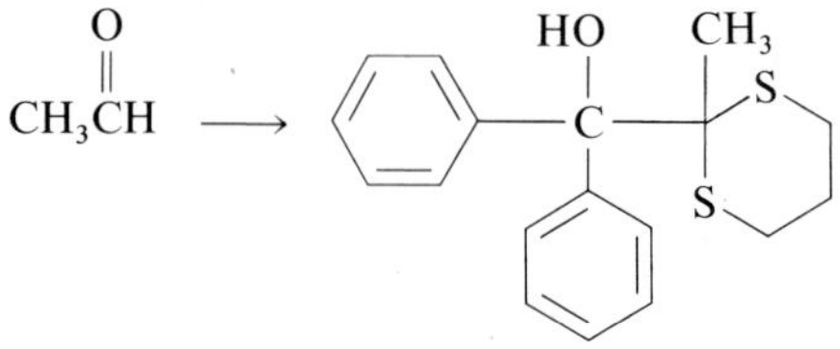

(b)   2-phenyl-1,3-dithiane (with H)   ⟶   2-phenyl-2-(1-methylpropyl)-1,3-dithiane $CHCH_2CH_3$ / CH_3

(c)   1,3-dithiane (with two H)   ⟶   bicyclic product with OH, H and S–S dithiane

(d) $HCH(=O)$ ⟶ $C_6H_5-CH_2CH_2\overset{O}{\overset{\|}{C}}CH_2CH_2-C_6H_5$

(e) $C_6H_5-\overset{O}{\overset{\|}{C}}-C_6H_5$ ⟶   epoxide $C_6H_5-\overset{O}{C}-CH_2$ (with C_6H_5)

(f) $CH_3CH(=O)$ ⟶   cyclohexyl with OH, S, CH_3 dithiane

(g) $HCH(=O)$ ⟶   1,3-dithiane with H and CH_2CH_2OH

(h) $CH_3CH(=O)$ ⟶ $C_6H_5-\overset{HO}{\underset{C_6H_5}{C}}-$ 2-methyl-1,3-dithiane with CH_3, S, S

(i) 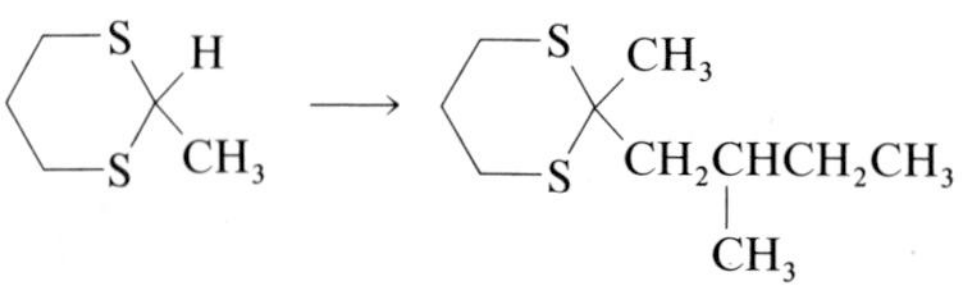   1,3-dithiane with H and CH_3 ⟶   1,3-dithiane with CH_3 and $CH_2CHCH_2CH_3$ / CH_3

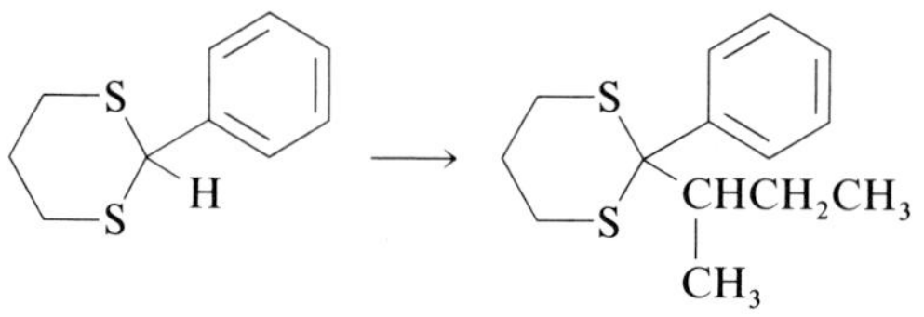

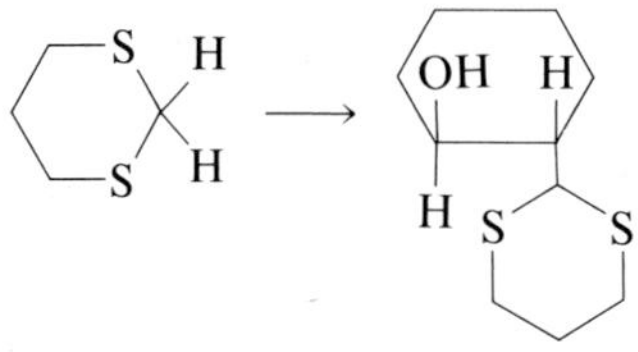

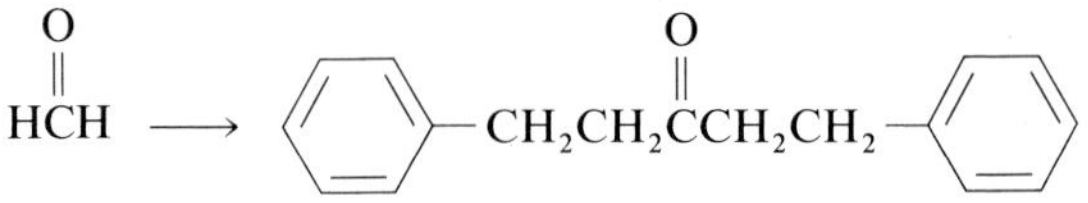

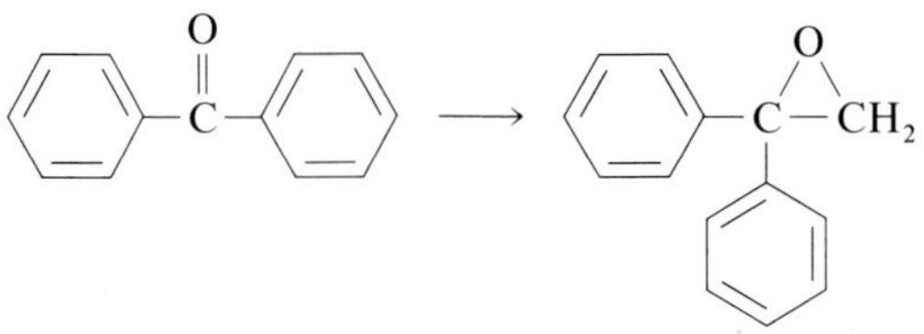

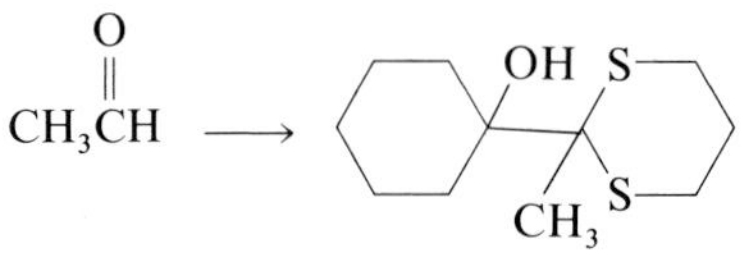

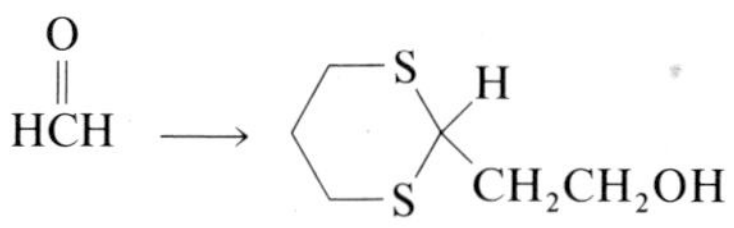

PROBLEM 15.40 Dimethyl sulfoxide, CH_3SOCH_3, is deprotonated by sodium hydride to give the methylsufinyl carbanion. Write an equation for the formation of this anion. How do you account for the acidity of the hydrogen atoms in dimethyl sulfoxide? What other species does this anion resemble? Predict how it will react with benzophenone and with ethyl cyclohexanecarboxylate.

15.10

Organopalladium Reagents as Electrophiles in Reactions with Carbanions

An area of research that is growing rapidly is the use of transition metals to complex with and stabilize reactive intermediates in synthetic reactions. A transition metal may function as either an acceptor (Section 8.2C) or a donor of electrons. As an electron donor, a transition metal stabilizes carbocations and thus makes them more available as reactive intermediates. In addition, the transition metal, by its presence, often directs the regiochemistry and stereochemistry of new bonds that are formed.

In this section we explore synthetic methods developed by Barry M. Trost at the University of Wisconsin, using organopalladium reagents to activate allylic acetates towards nucleophilic substitution reactions. First, let us consider an example of how the presence of palladium changes the selectivity observed in a nucleophilic substitution reaction. When the anion of methyl benzenesulfonylacetate is added to 3-acetoxy-8-bromo-1-octene in dimethylformamide, nucleophilic substitution at the primary alkyl halide function occurs.

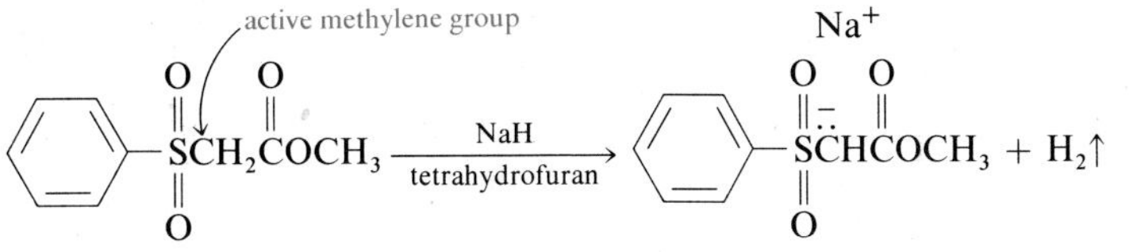

methyl benzenesulfonylacetate

anion of methyl benzenesulfonylacetate

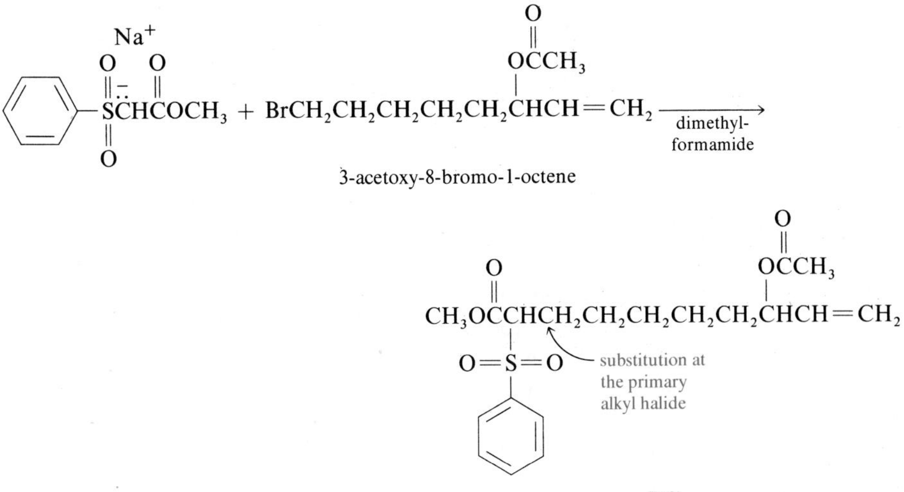

3-acetoxy-8-bromo-1-octene

substitution at the primary alkyl halide

75%

If the same reagents are mixed in the presence of tetrakis(triphenylphosphine)-palladium(0), however, displacement of the allylic acetate group at the other end of the molecule occurs.

$$BrCH_2CH_2CH_2CH_2CH_2\overset{\overset{\displaystyle O}{\overset{\displaystyle \|}{OCCH_3}}}{CH}CH=CH_2 \; + \; \text{(anion of methyl benzenesulfonylacetate)} \xrightarrow[\substack{\text{tetrahydrofuran} \\ \Delta}]{[(C_6H_5)_3P]_4\ Pd}$$

3-acetoxy-8-bromo-1-octene   anion of methyl
                                             benzenesulfonylacetate

$$BrCH_2CH_2CH_2CH_2CH_2CH=CHCH_2\overset{\overset{\displaystyle O}{\displaystyle \|}}{C}HCOCH_3 \; + \; BrCH_2CH_2CH_2CH_2CH_2CHCH=CH_2$$

E and *Z* isomers

*major component of the
mixture*

77%

*minor component of
the mixture*

A mixture of products is obtained from the reaction catalyzed by palladium. The incoming nucleophile reacts at both ends of the allylic system. The major product is derived from reaction at the less hindered, terminal carbon atom of the alkene.

Organopalladium compounds also affect the stereochemistry of substitution reactions. For example, methyl *cis*-3-acetoxy-4-cyclohexenecarboxylate undergoes substitution of the acetoxy group by the enolate anion of dimethyl malonate in the presence of palladium with retention of configuration.

methyl *cis*-3-acetoxy-
4-cyclohexenecarboxylate

$$+ \; CH_3O\overset{\overset{\displaystyle O}{\displaystyle \|}}{C}\overset{..}{C}H\overset{\overset{\displaystyle O}{\displaystyle \|}}{C}OCH_3 \xrightarrow[\substack{(C_6H_5)_3P \\ \text{tetrahydrofuran} \\ \Delta}]{[(C_6H_5)_3P]_4Pd}$$

Na$^+$

substitution with
retention of
configuration

The following mechanism has been proposed to account for the regioselectivity and stereoselectivity seen for reactions of allyl acetates in the presence of palladium. The organopalladium compound, which has 18 electrons (Section 8.2C) around it when palladium is bound to four triphenylphosphines, is in equilibrium with forms in which one or more of the ligands has dissociated. When a ligand is lost from the palladium, the palladium metal reacts with the alkene, accepting the π electrons of the double bond.

$$[(C_6H_5)_3P]_4Pd \rightleftharpoons [(C_6H_5)_3P]_3Pd \; + \; (C_6H_5)_3P$$

$$\updownarrow$$

$$[(C_6H_5)_3P]_2Pd \; + \; (C_6H_5)_3P$$

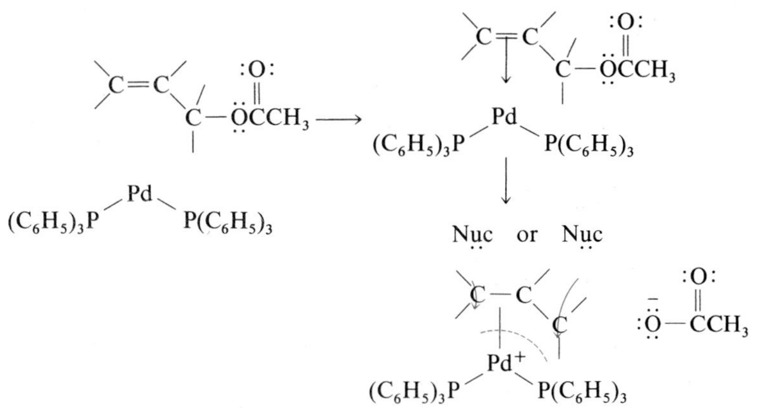

a π-allyl complex of the
allyl acetate with palladium,
a stabilized allylic cation;
reacts with nucleophiles
at either end of the allylic system

The palladium-alkene complex loses acetate ion with delocalization of the positive charge to the palladium atom. The complex that results, known as a π-allyl complex of palladium, is an electrophile and will react at either end with nucleophiles. Usually the least hindered carbon atom will form the new bond to the incoming nucleophile. The nucleophile will approach from the side of the molecule away from the palladium, which functions as the new leaving group in a nucleophilic substitution reaction. Thus, the retention of configuration seen in the reaction of the enolate anion of dimethyl malonate with methyl *cis*-3-acetoxy-4-cyclohexenecarboxylate is the result of two consecutive inversions of configuration, the first one occurring when palladium displaces acetate ion, and the second one when the nucleophile displaces palladium.

The synthesis of a functionalized allyl acetate and its transformation in a palladium-catalyzed carbon-carbon bond-forming reaction is illustrated below.

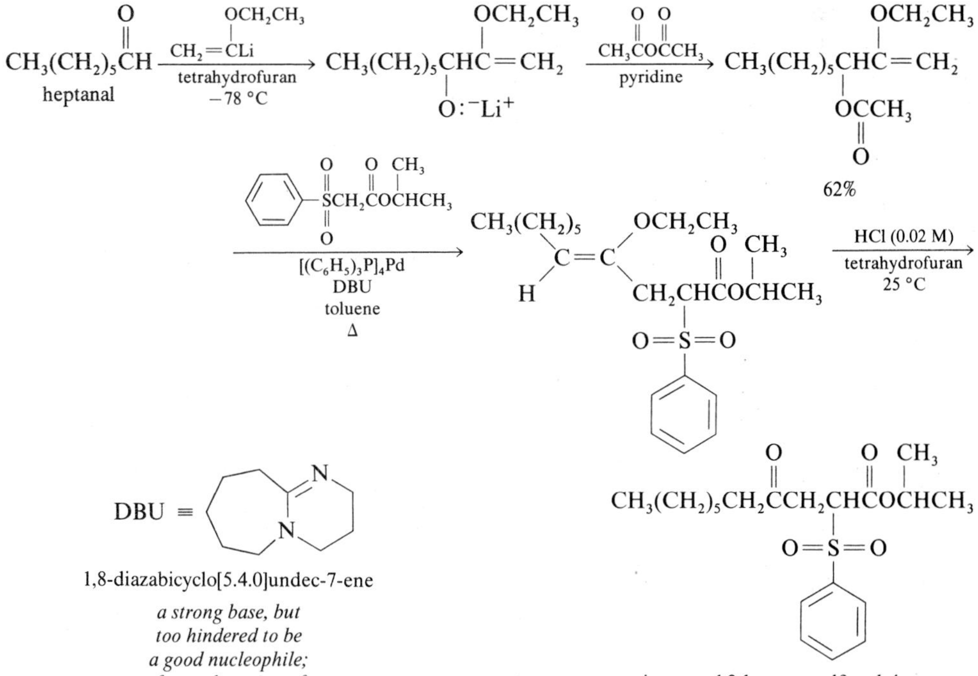

DBU $\equiv$

1,8-diazabicyclo[5.4.0]undec-7-ene

*a strong base, but
too hindered to be
a good nucleophile;
forms the anion of
isopropyl benzenesulfonylacetate*

isopropyl 2-benzenesulfonyl-4-oxo-
undecanoate

The alkoxide ion produced when a functionalized organolithium reagent reacts with heptanal is trapped with acetic anhydride. The resulting allylic acetate reacts with the anion of isopropyl benzenesulfonylacetate to give a chain-lengthened enol ether. The enol ether function can be hydrolyzed to give a ketone under very mild conditions (Section 11.5B), leaving the ester group untouched.

Carbon-sulfur bonds are cleaved easily by reduction (Section 9.7D). Removal of the benzenesulfonyl group with reagents such as sodium amalgam, Na(Hg), in the presence of a mild acid is possible. The anion of a benzenesulfonylacetic ester is, therefore, the synthetic equivalent of a two-carbon ester fragment. The reactions described above are an introduction to the use of transition metals to promote greater selectivity in synthetic organic chemistry.

PROBLEM 15.41   What would be the product of the reduction of the carbon-sulfur bond in isopropyl 2-benzenesulfonyl-4-oxoundecanoate with sodium amalgam?

PROBLEM 15.42   Predict what the products of the following reactions will be. Ignore stereo-isomerism in your answers.

(a) 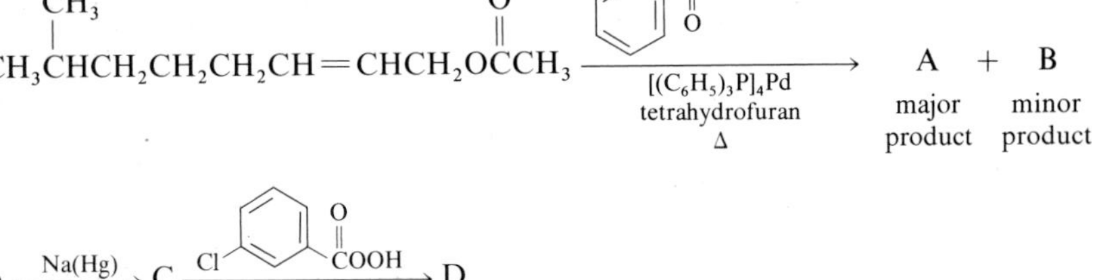   A + B
major product minor product

(b)

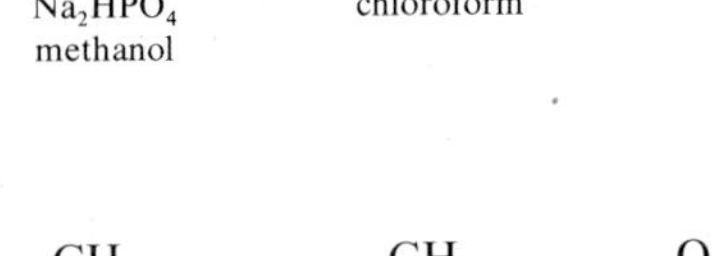

(c) 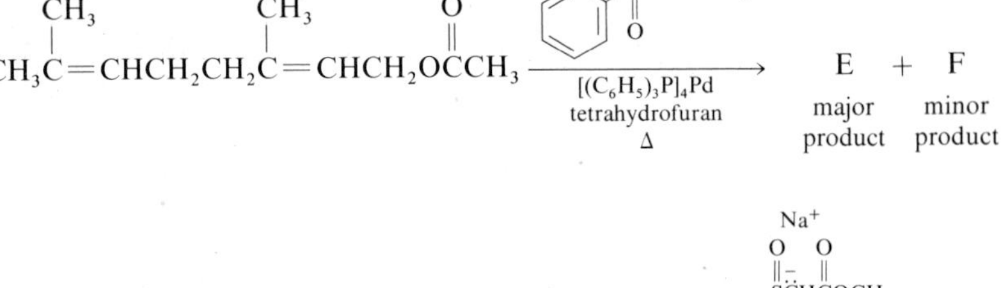   E + F
major product minor product

(d)

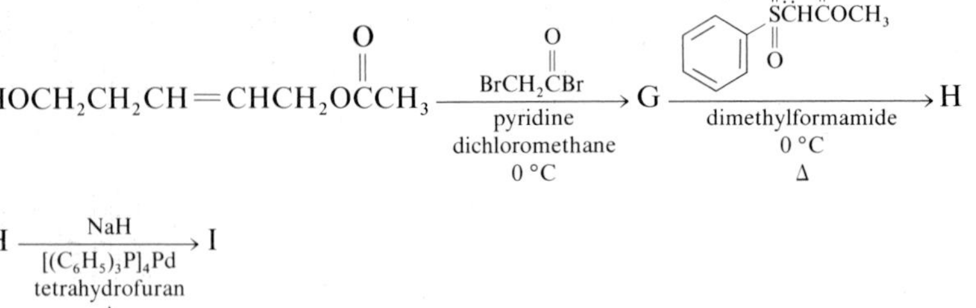

(e)

ADDITIONAL PROBLEMS

15.43 Complete the following equations, giving structural formulas for all intermediates or products designated by letters.

(a) $CH_3\overset{O}{\overset{||}{C}}CH=\overset{CH_3}{\overset{|}{C}}CH_3 + CH_2(\overset{O}{\overset{||}{C}}OCH_3)_2 \xrightarrow[\text{methanol}]{CH_3ONa} A \xrightarrow[\text{methanol}]{CH_3ONa} B$

(b) 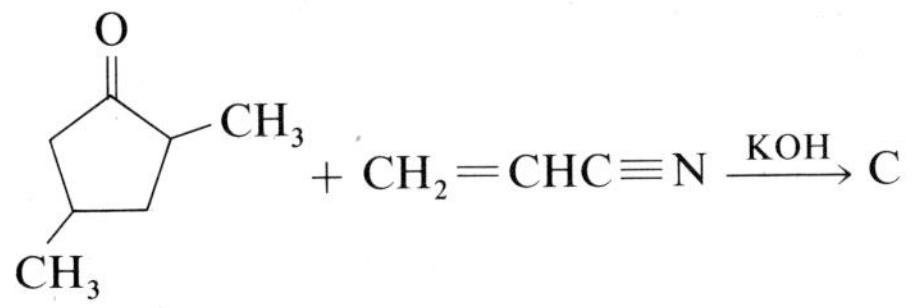 $+ CH_2=CHC\equiv N \xrightarrow{KOH} C$

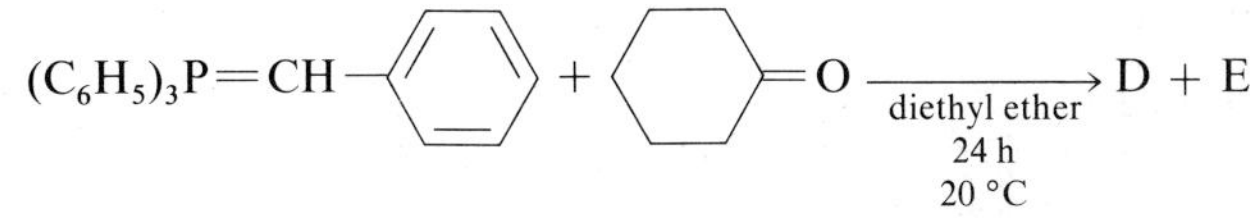

(c) $(C_6H_5)_3P=CH-$ $+$ $=O \xrightarrow[\substack{\text{diethyl ether}\\24\ h\\20\ °C}]{} D + E$

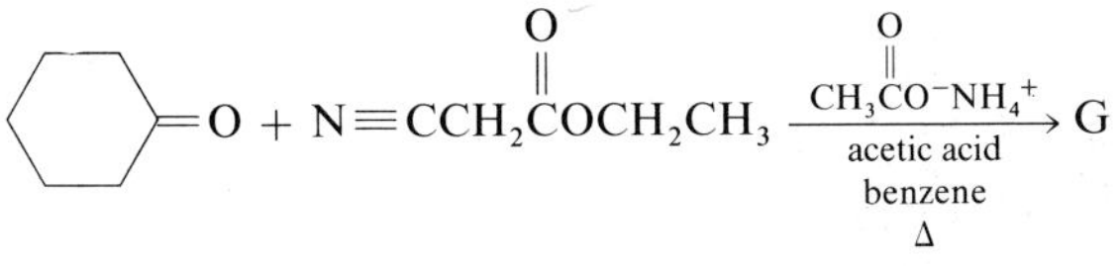

(d) $-CH=CH-\overset{O}{\overset{||}{C}}-$ $+$ $N-H \xrightarrow[\Delta]{\text{ethanol}} F$

(e) $=O + N\equiv CCH_2\overset{O}{\overset{||}{C}}OCH_2CH_3 \xrightarrow[\substack{\text{acetic acid}\\\text{benzene}\\\Delta}]{CH_3CO^-NH_4^+} G$

(f) $+ CH_3\overset{CH_3}{\underset{+}{\overset{|}{S}}}-\overset{-}{CH_2} \xrightarrow[\substack{\text{dimethyl}\\\text{sulfoxide}}]{} H + I$

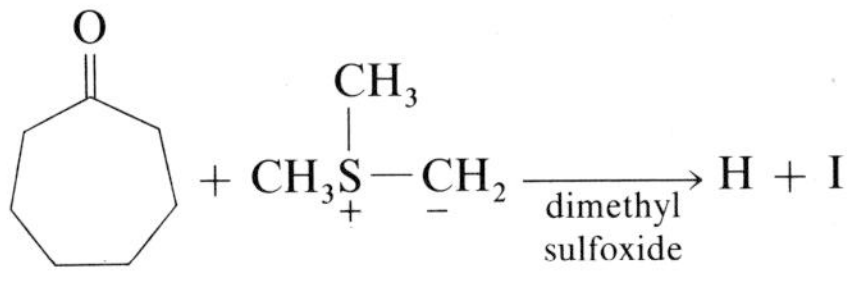

(g) $CH_2(\overset{O}{\overset{||}{C}}OCH_2CH_3)_2 \xrightarrow[\text{ethanol}]{CH_3CH_2ONa} J \xrightarrow{CH_3\overset{CH_3}{\overset{|}{C}}HCH_2Br} K \xrightarrow[\Delta]{H_3O^+} L$

(h) 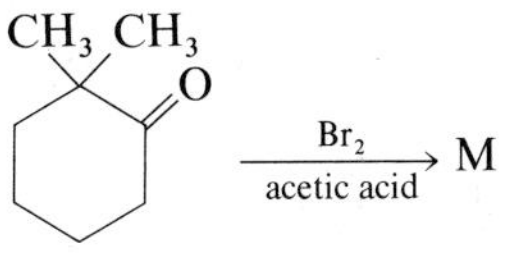 $\xrightarrow[\text{acetic acid}]{Br_2} M$

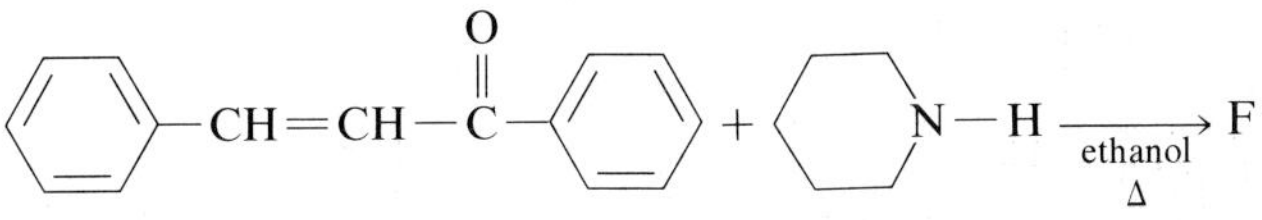

(i) 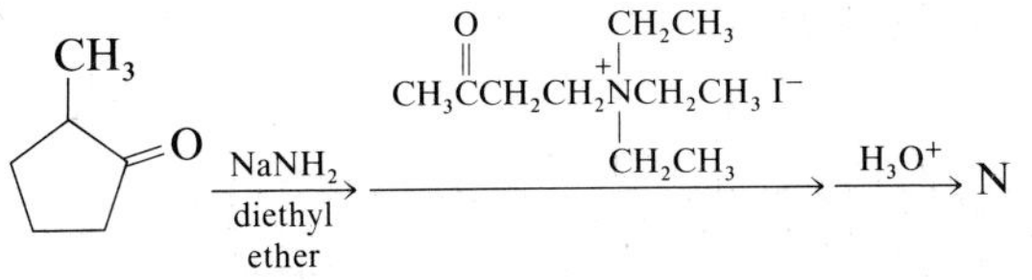 $\xrightarrow[\substack{\text{diethyl}\\\text{ether}}]{NaNH_2} \xrightarrow{CH_3\overset{O}{\overset{||}{C}}CH_2CH_2\overset{CH_2CH_3}{\underset{CH_2CH_3}{\overset{+|}{N}}}CH_2CH_3\ I^-} \xrightarrow{H_3O^+} N$

(j) 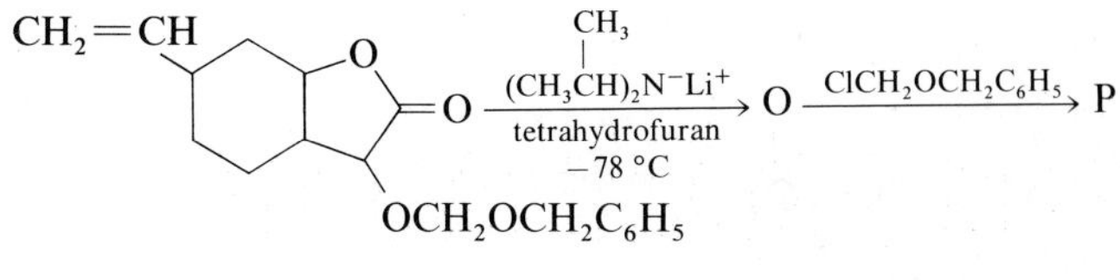 $\xrightarrow[\substack{\text{tetrahydrofuran}\\-78\ °C}]{(CH_3CH)_2N^-Li^+} O \xrightarrow{ClCH_2OCH_2C_6H_5} P$

(k)

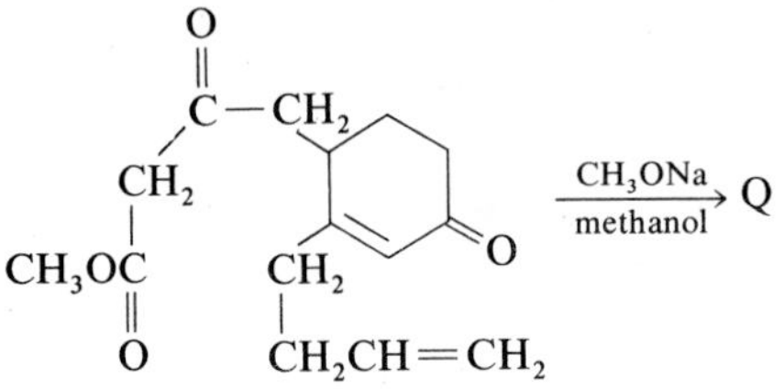

$$\xrightarrow[\text{methanol}]{\text{CH}_3\text{ONa}} Q$$

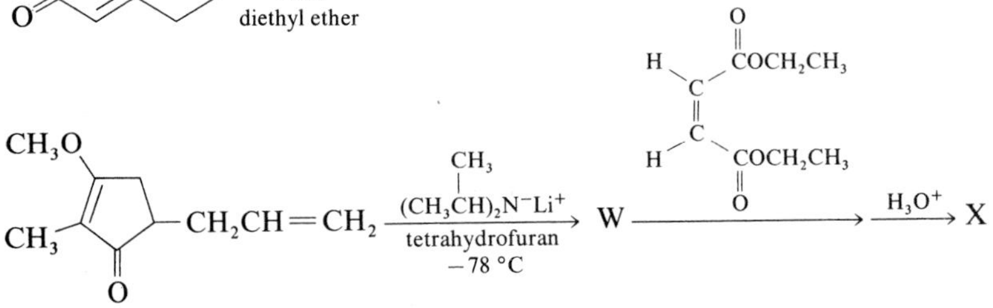

(l)

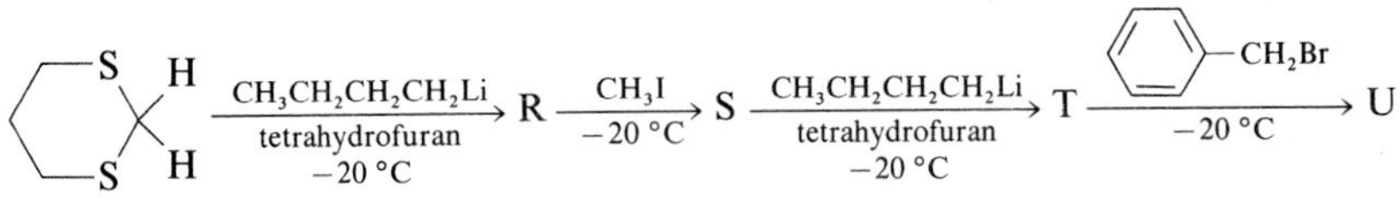

$$\xrightarrow[\substack{\text{tetrahydrofuran} \\ -20\ ^\circ\text{C}}]{\text{CH}_3\text{CH}_2\text{CH}_2\text{CH}_2\text{Li}} R \xrightarrow[-20\ ^\circ\text{C}]{\text{CH}_3\text{I}} S \xrightarrow[\substack{\text{tetrahydrofuran} \\ -20\ ^\circ\text{C}}]{\text{CH}_3\text{CH}_2\text{CH}_2\text{CH}_2\text{Li}} T \xrightarrow[-20\ ^\circ\text{C}]{} U$$

(m)

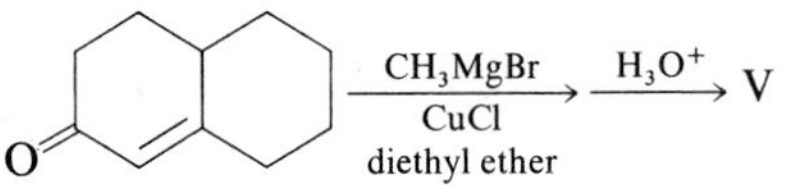

$$\xrightarrow[\substack{\text{CuCl} \\ \text{diethyl ether}}]{\text{CH}_3\text{MgBr}} \xrightarrow{\text{H}_3\text{O}^+} V$$

(n)

$$\xrightarrow[\substack{\text{(CH}_3\text{CH})_2\text{N}^-\text{Li}^+ \\ \text{tetrahydrofuran} \\ -78\ ^\circ\text{C}}]{} W \xrightarrow{} \xrightarrow{\text{H}_3\text{O}^+} X$$

(o)

$$\xrightarrow[\substack{\text{diethyl ether} \\ -78\ ^\circ\text{C}}]{} Y \xrightarrow[\text{H}_2\text{O}]{\text{NH}_4\text{Cl, NH}_3} Z$$

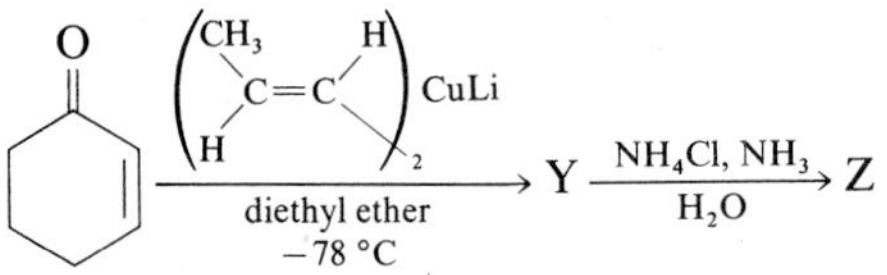

15.44 The following equations provide more practice in recognizing reactions. Give structural formulas for all compounds designated by letters.

(a) $(\text{C}_6\text{H}_5)_3\text{P} + \text{BrCH}_2\text{CH}=\text{CHCOCH}_2\text{CH}_3 \xrightarrow[\text{benzene}]{} A \xrightarrow[\text{H}_2\text{O}]{\text{NaOH}} B$

(b) 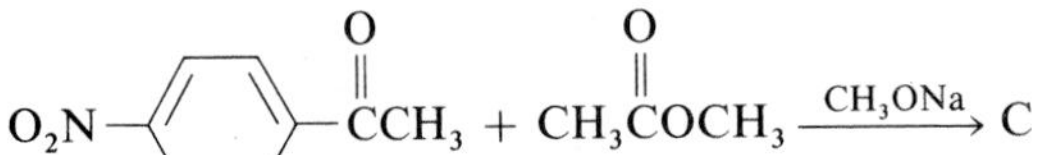 $\text{O}_2\text{N}-\!\!\bigcirc\!\!-\overset{\text{O}}{\overset{\|}{\text{C}}}\text{CH}_3 + \text{CH}_3\overset{\text{O}}{\overset{\|}{\text{C}}}\text{OCH}_3 \xrightarrow{\text{CH}_3\text{ONa}} C$

(c)

$$\xrightarrow{\text{CH}_3\text{CH}_2\text{ONa}} D$$

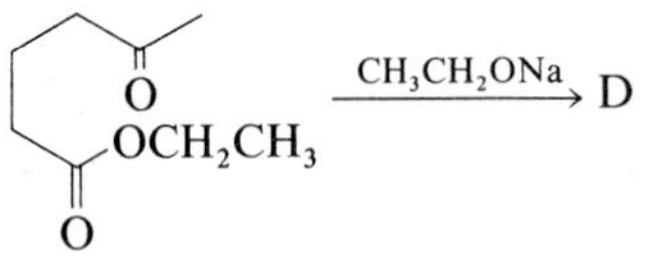

(d) $\text{CH}_2(\overset{\text{O}}{\overset{\|}{\text{C}}}\text{OCH}_2\text{CH}_3)_2 \xrightarrow[\text{ethanol}]{\text{CH}_3\text{CH}_2\text{ONa}} E \xrightarrow{} F$

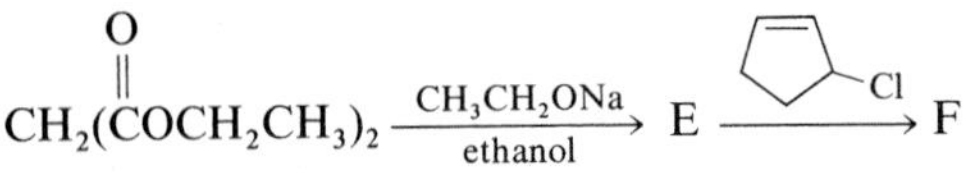

(e) $\text{CH}_3\overset{\text{O}}{\overset{\|}{\text{C}}}\text{H} + \text{CH}_2(\overset{\text{O}}{\overset{\|}{\text{C}}}\text{OCH}_2\text{CH}_3)_2 \xrightarrow[\Delta]{(\text{CH}_3\text{C})_2\text{O}} G$

(f) $CH_3CCH_2COCH_2CH_3 \xrightarrow[H_2O]{NaOH} H \xrightarrow[NaOH]{C_6H_5COCl} I$

(with two C=O groups on the starting diketoester, and the benzoyl chloride reagent shown as a benzene ring with –CCl and =O)

(g) $CH_3CCH_2COCH_2CH_3$ (with two C=O) + (2,3-dimethoxybenzaldehyde: benzene ring with –CH=O, OCH_3, and CH_3O substituents) $\xrightarrow[\text{acetic acid}]{\text{piperidine}} J$

(h) $CH_3O\text{–}C_6H_4\text{–}CCH_3$ (para-methoxyacetophenone, C=O) $+ I_2 \xrightarrow[\substack{H_2O \\ \text{dioxane}}]{NaOH} K + L$

(i) (1,4-naphthoquinone, with two O) $\xrightarrow[\text{methanol}]{H_2O_2,\ NaOH} M$

(j) (2-phenyl-1,3-dithiane: ring with two S and H) $\xrightarrow[\substack{\text{tetrahydrofuran} \\ -20\,°C}]{CH_3CH_2CH_2CH_2Li} N \xrightarrow[-20\,°C]{CH_3CHI\ (CH_3)} O$

(k) (cyclohexane with exocyclic =CH_2, a –CH_2CH_2Br chain, an isopropyl group, and a –CH_2CH_2OH chain) $\xrightarrow[\text{dichloromethane}]{\text{pyridinium chlorochromate}} P \xrightarrow[\substack{\text{tert-butyl} \\ \text{alcohol}}]{CH_3C(CH_3)_2COK\ ...} \xrightarrow{H_3O^+} Q$

(l) $C_6H_5\text{–}CH_2C\equiv N \xrightarrow[\text{ethanol}]{CH_3CH_2ONa} R \xrightarrow[\substack{\text{toluene} \\ \Delta}]{CH_3CH_2OCOCH_2CH_3} S$

(m) (2-methoxybenzaldehyde: benzene ring with –CH=O and OCH_3) $+ CH_3CH_2NO_2 \xrightarrow[\substack{\text{toluene} \\ \Delta}]{CH_3CH_2CH_2CH_2NH_2} T$

(n) $CH_3(CH_2)_4CH_2CCH_2CH_2CH_2OH$ (C=O) $\xrightarrow[\text{dichloromethane}]{\text{pyridinium chlorochromate}} U \xrightarrow[\substack{H_2O \\ \text{tetrahydro-furan}}]{KOH} V$

(o) (1,3-dioxolane with H)$\text{–}CH_2CH_2CHCH=CHCH=CHCH$ (with CH_2CH_3 substituent, and terminal C=O) $\xrightarrow{(C_6H_5)_3P=CHCOCH_3} W \xrightarrow[\text{cold}]{H_3O^+} X$

(p)

$$\xrightarrow[\substack{\text{ethyl acetate}\\\text{trimethylamine}}]{\substack{H_2\\Pt/C}} Y \xrightarrow{\text{NaOH}} Z$$

15.45 More reactions are given below. Give structural formulas for all compounds designated by letters.

(a) $CH_2(COCH_2CH_3)_2 \xrightarrow[\substack{\text{diethyl}\\\text{ether}}]{Na} A \xrightarrow{\quad\quad} B$

(b) $\xrightarrow[\text{methanol}]{H_2O_2,\ NaOH} C$

(c) $\xrightarrow[\substack{RhCl[P(C_6H_5)_3]_3}]{H_2} D$

(d) $CH_3CH_2C \xrightarrow[\substack{\text{tetrahydrofuran}\\-72\ ^\circ C}]{(CH_3CH)_2N^-Li^+} E \xrightarrow{\quad\quad} \xrightarrow[H_2O]{NH_4Cl} F$

(e) $\xrightarrow[\substack{\text{tetrahydrofuran}\\0\ ^\circ C}]{(CH_3CH)_2N^-Li^+} G \xrightarrow{(CH_3)_3SiCl,\ (CH_3CH_2)_3N} H$

(f) $\xrightarrow[\substack{\text{tetrahydrofuran}\\0\ ^\circ C}]{(CH_3CH)_2N^-Li^+} I \xrightarrow{CH_3I} J$

(g) $\xrightarrow[\substack{\text{tetrahydrofuran}\\-20\ ^\circ C}]{CH_2-CHCH(OCH_2CH_3)_2} \xrightarrow{H_3O^+} K \xrightarrow[\text{pyridine}]{(CH_3C)_2O} L \xrightarrow[\substack{\text{BF}_3\\\text{tetrahydrofuran}}]{HgO,\ H_2O} M + N$

(h) 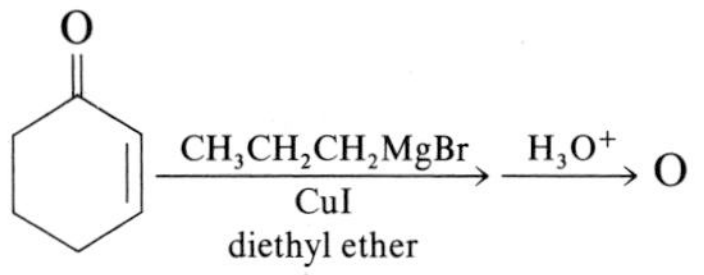
$$\xrightarrow[\substack{\text{CuI} \\ \text{diethyl ether}}]{\text{CH}_3\text{CH}_2\text{CH}_2\text{MgBr}} \xrightarrow{\text{H}_3\text{O}^+} \text{O}$$

(i) 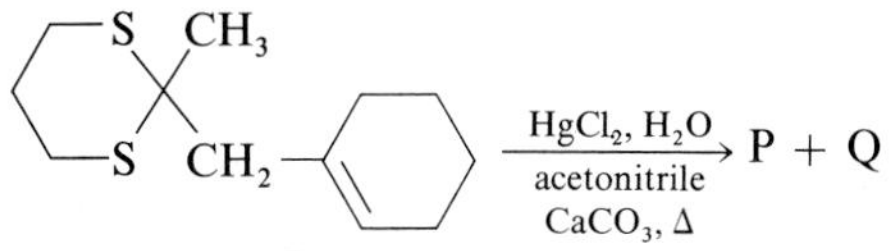
$$\xrightarrow[\substack{\text{acetonitrile} \\ \text{CaCO}_3, \Delta}]{\text{HgCl}_2, \text{H}_2\text{O}} \text{P} + \text{Q}$$

(j) 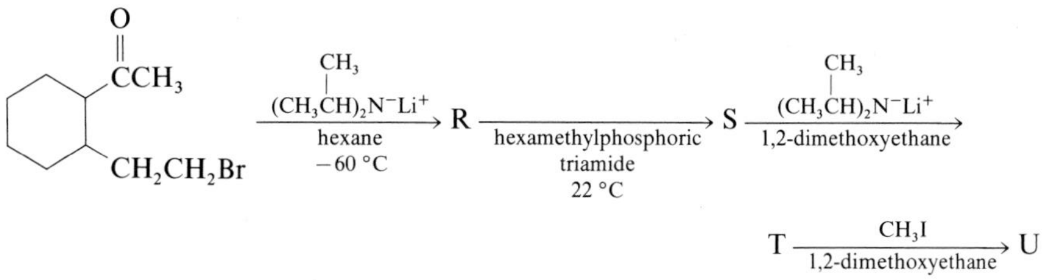
$$\xrightarrow[\substack{\text{hexane} \\ -60\,°\text{C}}]{(\text{CH}_3\text{CH})_2\text{N}^-\text{Li}^+} \text{R} \xrightarrow[\substack{\text{hexamethylphosphoric} \\ \text{triamide} \\ 22\,°\text{C}}]{} \text{S} \xrightarrow[\substack{1,2\text{-dimethoxyethane}}]{(\text{CH}_3\text{CH})_2\text{N}^-\text{Li}^+}$$

$$\text{T} \xrightarrow[1,2\text{-dimethoxyethane}]{\text{CH}_3\text{I}} \text{U}$$

(k) 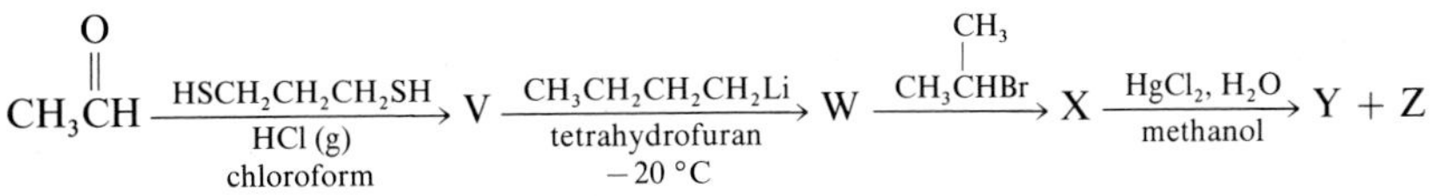 $\text{CH}_3\overset{\text{O}}{\overset{\|}{\text{CH}}} \xrightarrow[\substack{\text{HCl (g)} \\ \text{chloroform}}]{\text{HSCH}_2\text{CH}_2\text{CH}_2\text{SH}} \text{V} \xrightarrow[\substack{\text{tetrahydrofuran} \\ -20\,°\text{C}}]{\text{CH}_3\text{CH}_2\text{CH}_2\text{CH}_2\text{Li}} \text{W} \xrightarrow[]{\text{CH}_3\text{CHBr}} \text{X} \xrightarrow[\text{methanol}]{\text{HgCl}_2, \text{H}_2\text{O}} \text{Y} + \text{Z}$

(l) 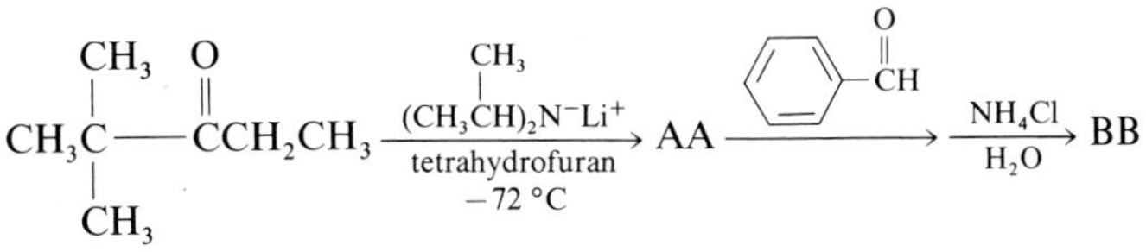
$$\xrightarrow[\substack{\text{tetrahydrofuran} \\ -72\,°\text{C}}]{(\text{CH}_3\text{CH})_2\text{N}^-\text{Li}^+} \text{AA} \xrightarrow[]{} \xrightarrow[\text{H}_2\text{O}]{\text{NH}_4\text{Cl}} \text{BB}$$

15.46 Supply reagents for the following transformations. More than one step may be required.

(a) 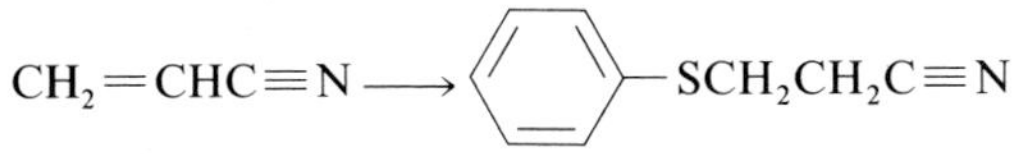 $\text{CH}_2{=}\text{CHC}{\equiv}\text{N} \longrightarrow$

(b)

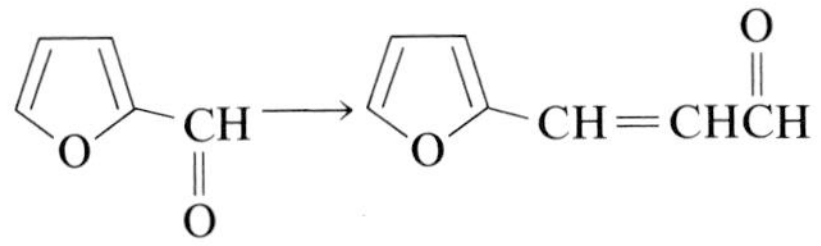

(c)

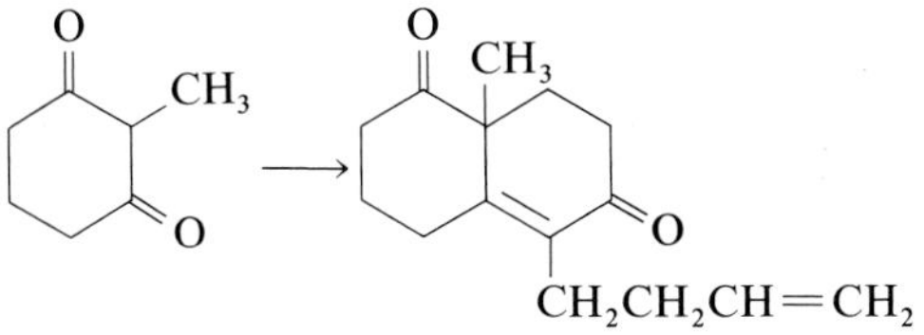

(d)

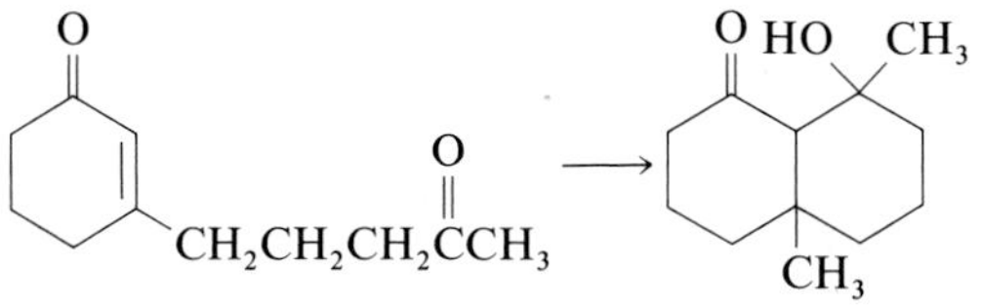

(e)

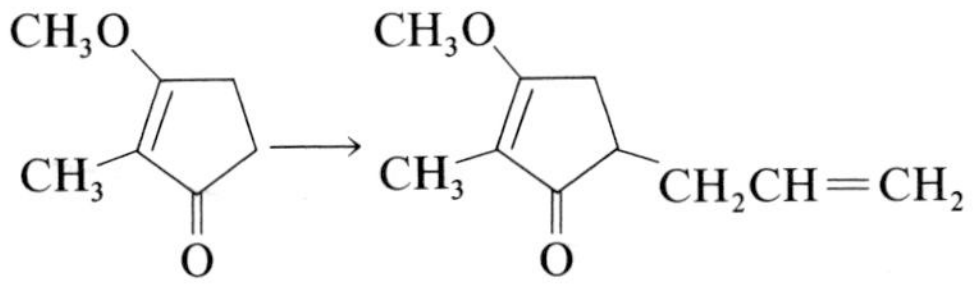

(f)

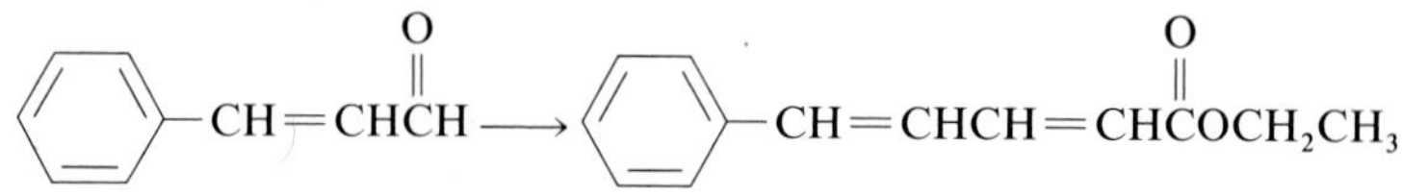

(g)

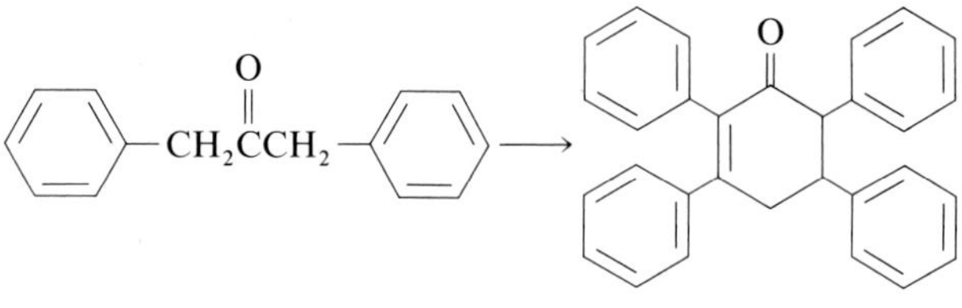

(Hint: A review of Sections 15.5A and 15.7C will be useful.)

(h)

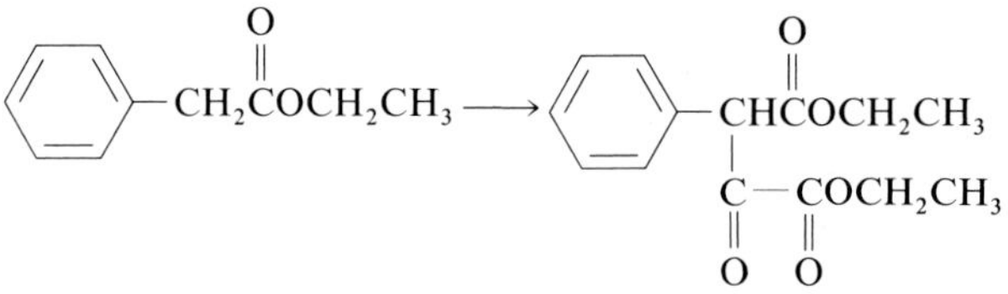

(i) 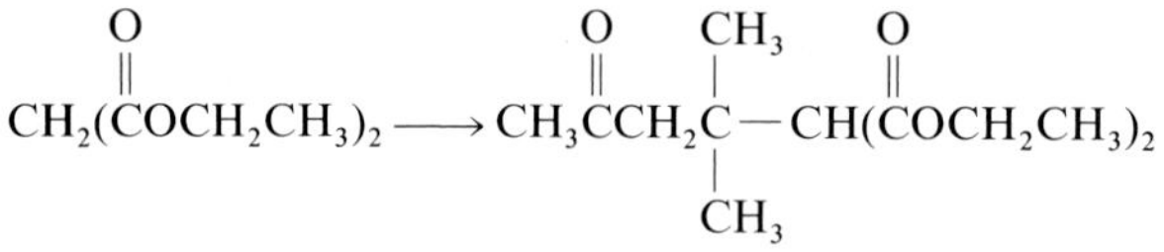 $CH_2(COCH_2CH_3)_2 \longrightarrow CH_3CCH_2C-CH(COCH_2CH_3)_2$

(j)

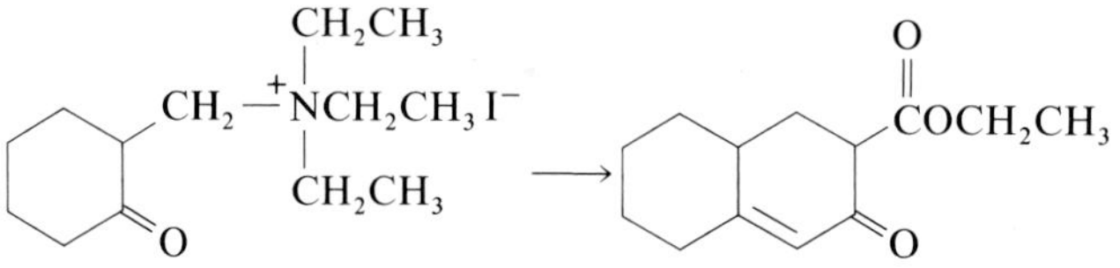

(k) 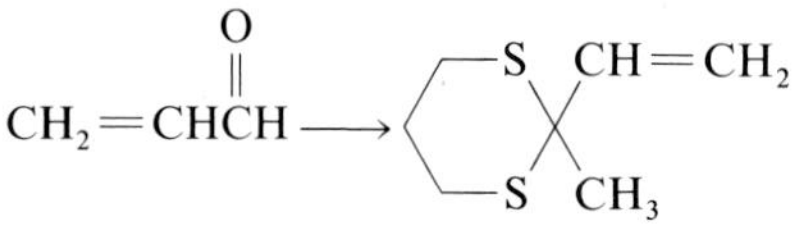 $CH_2{=}CHCH \longrightarrow$

(l)

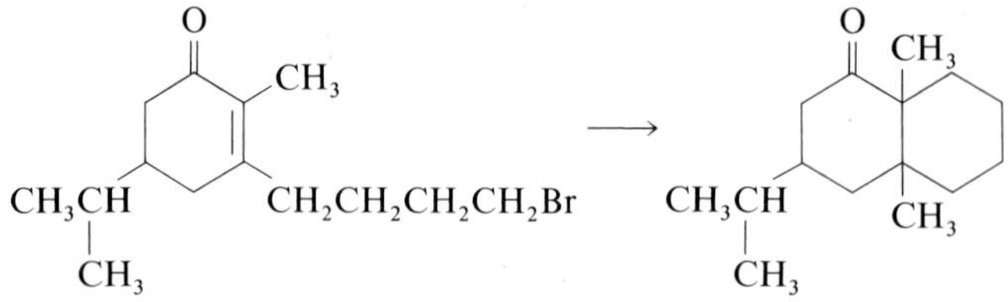

(m)

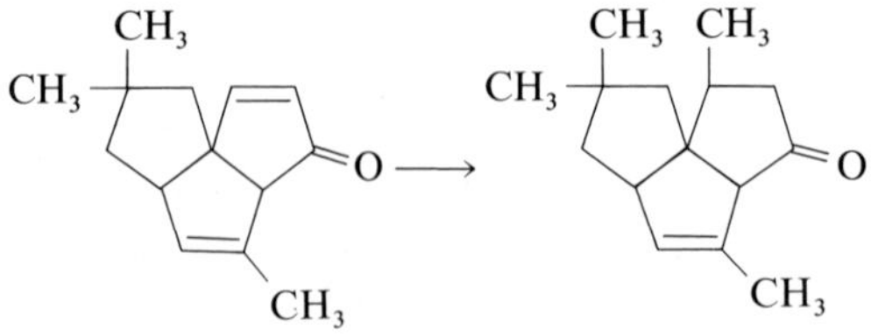

(n) 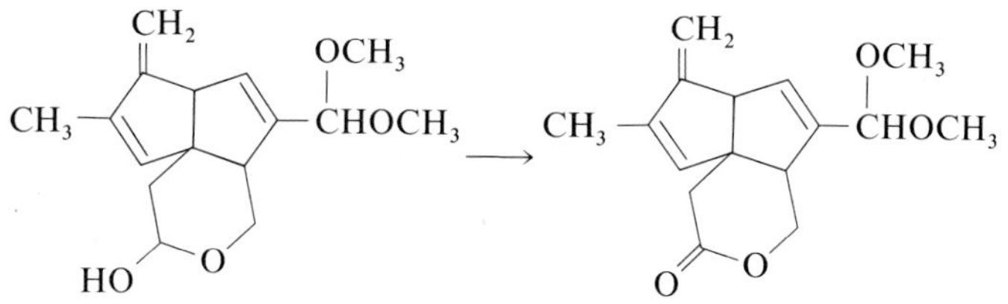
CH$_2$
OCH$_3$
CH$_3$
CHOCH$_3$
HO
O
CH$_2$
OCH$_3$
CH$_3$
CHOCH$_3$
O
O

(Hint: A review of Section 7.6C will be helpful.)

(o) 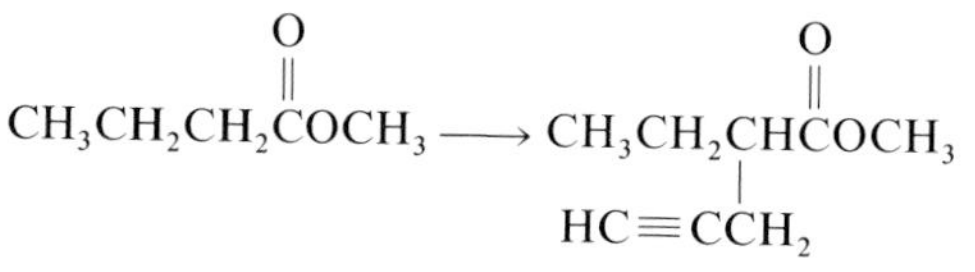
O
CH$_3$CH$_2$CH$_2$COCH$_3$
O
CH$_3$CH$_2$CHCOCH$_3$
HC≡CCH$_2$

(Hint: A review of Section 15.3B will be helpful.)

(p) 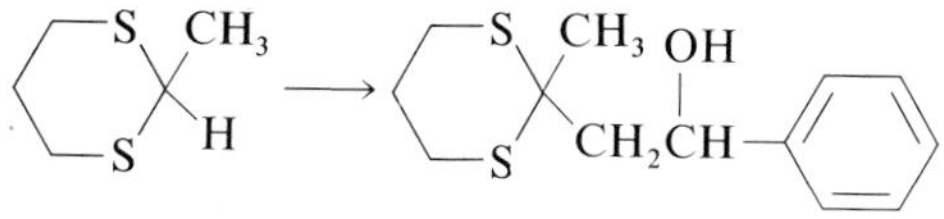
S
CH$_3$
S
H
S
CH$_3$
OH
S
CH$_2$CH

15.47 Gilbert Stork of Columbia University discovered that **enamines,** nitrogen analogs of enolate anions, can be used as sources of nucleophilic carbon atoms in syntheses. The following equations show the preparation and reactions of an enamine. Provide a mechanism for each of the steps.

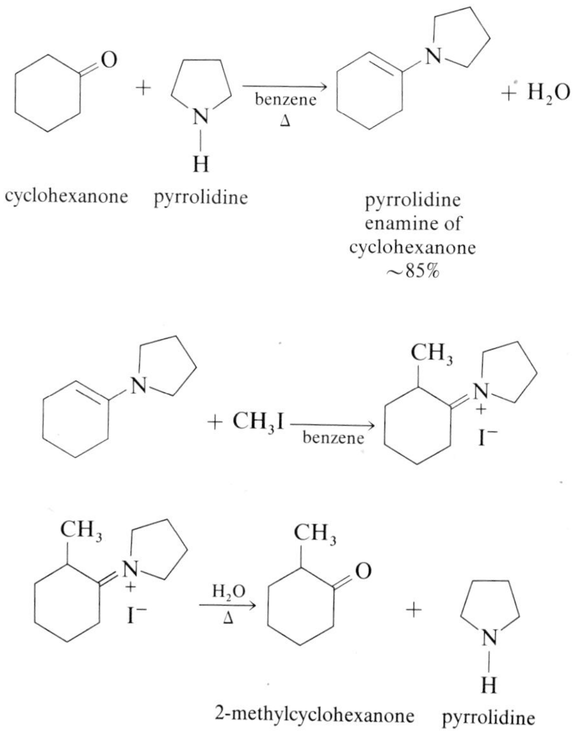
O
+
N
H
benzene
Δ
N
+ H$_2$O
cyclohexanone
pyrrolidine
pyrrolidine enamine of cyclohexanone ~85%
N
+ CH$_3$I
benzene
CH$_3$
N
+
I$^-$
CH$_3$
N
+
I$^-$
H$_2$O
Δ
CH$_3$
O
+
N
H
2-methylcyclohexanone
pyrrolidine

15.48 A synthesis of disparlure, the sex pheromone of the gypsy moth (p. 553), was carried out in the following way. Provide structural formulas for all the compounds or intermediates that are designated by letters.

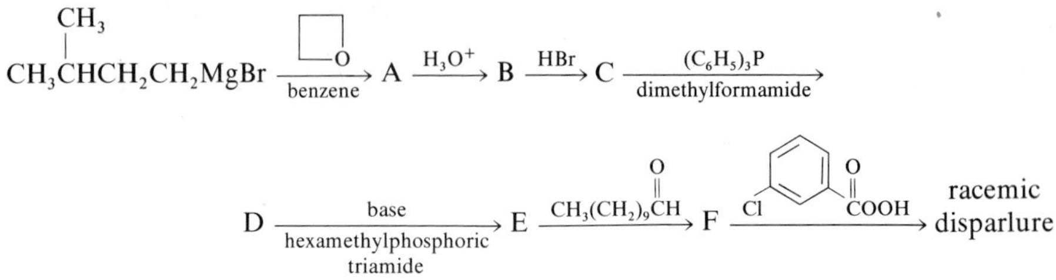

$$\underset{\text{benzene}}{\text{CH}_3\text{CHCH}_2\text{CH}_2\text{MgBr} \xrightarrow{\text{O}}} A \xrightarrow{\text{H}_3\text{O}^+} B \xrightarrow{\text{HBr}} C \xrightarrow[\text{dimethylformamide}]{\text{(C}_6\text{H}_5)_3\text{P}}$$

$$D \xrightarrow[\substack{\text{hexamethylphosphoric} \\ \text{triamide}}]{\text{base}} E \xrightarrow{\text{CH}_3(\text{CH}_2)_9\overset{\text{O}}{\overset{\|}{\text{CH}}}} F \xrightarrow{\text{Cl} \quad \text{COOH}} \substack{\text{racemic} \\ \text{disparlure}}$$

The Wittig reaction in hexamethylphosphoric triamide (Section 6.2B) gives mostly the (*Z*)-alkene.

15.49 The following sequence of reactions was carried out. Fill in structural formulas for the intermediates and products indicated by the letters.

$$\underset{\text{BF}_4^-}{\overset{+}{\text{S}}-\text{CH}_2\text{CH}_3} \xrightarrow[\text{tetrahydrofuran}]{\underset{\text{CH}_3}{\overset{\text{CH}_3}{\text{CH}_3-\text{C}-\text{Li}}}} A \xrightarrow{\quad} B + C$$

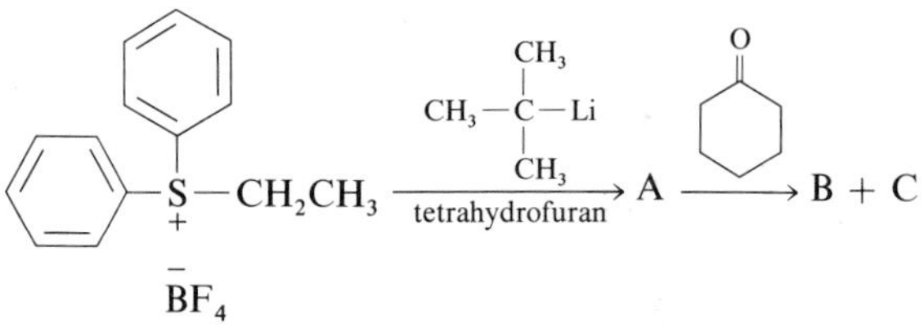

15.50 The following sequence of reactions was carried out. Fill in structural formulas for the intermediates and products indicated by the letters.

$$\xrightarrow[\text{ether}]{\substack{\text{LiAlH}_4 \\ \text{diethyl}}} \xrightarrow{\text{H}_3\text{O}^+} \underset{\substack{\text{(see Problem} \\ \text{10.64)}}}{A} \xrightarrow{\text{PBr}_3} B \xrightarrow[\text{CH}_3\text{CH}_2\text{ONa}]{\text{CH}_2(\overset{\text{O}}{\overset{\|}{\text{COCH}_2\text{CH}_3})_2}}$$

$$C \xrightarrow[\substack{\text{H}_2\text{O} \\ \text{methanol}}]{\text{KOH}} D \xrightarrow[\Delta]{\text{H}_3\text{O}^+} E \xrightarrow[\substack{\text{H}_2\text{SO}_4 \\ \Delta}]{\text{CH}_3\text{OH}} F \xrightarrow[\text{benzene}]{\text{NaH}} \xrightarrow[\Delta]{\text{H}_3\text{O}^+} G\ (\text{C}_{11}\text{H}_{18}\text{O})$$

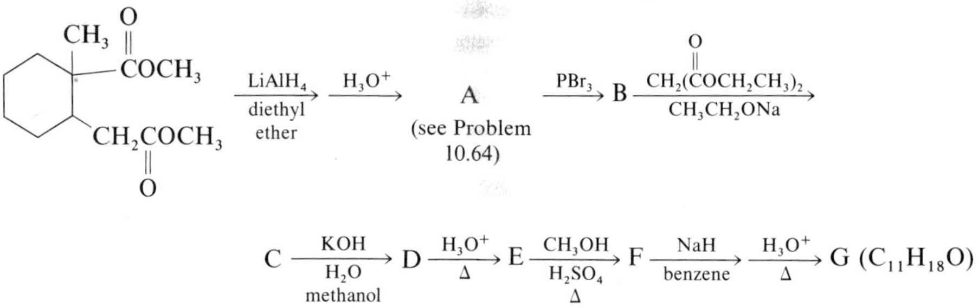

15.51 The following reaction is observed. Suggest a mechanism that accounts for the product.

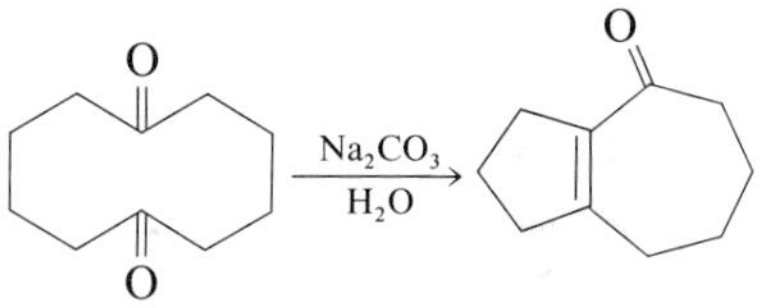

$$\xrightarrow[\text{H}_2\text{O}]{\text{Na}_2\text{CO}_3}$$

96%

15.52 The following sequence of reactions was used to determine that the sesquiterpene *β*-curcumene is a mixture of the two isomeric compounds shown below.

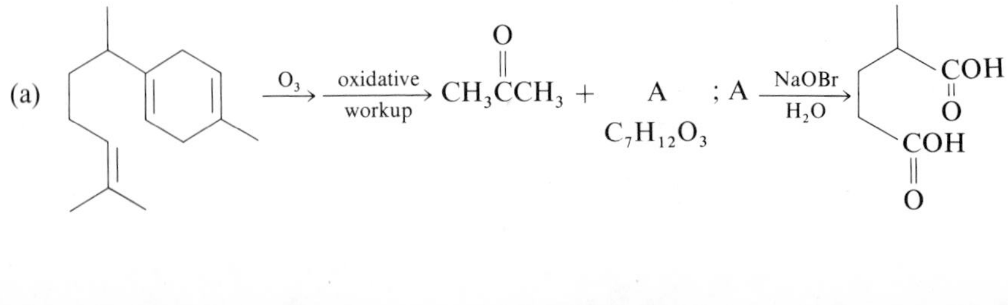

(a) $\xrightarrow[\text{workup}]{\overset{\text{O}_3}{\text{oxidative}}} \underset{\text{C}_7\text{H}_{12}\text{O}_3}{\text{CH}_3\overset{\text{O}}{\overset{\|}{\text{CCH}_3}} + \quad A} \quad ; A \xrightarrow[\text{H}_2\text{O}]{\text{NaOBr}}$

(b)

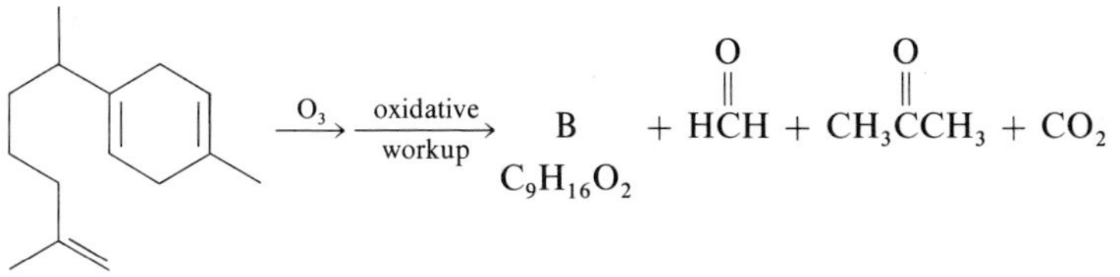

$$\xrightarrow[\text{workup}]{O_3 \quad \text{oxidative}} \mathbf{B} \underset{C_9H_{16}O_2}{} + HCH + CH_3CCH_3 + CO_2$$

(with carbonyl groups O on HCH and CH₃CCH₃)

Assign structures to Compounds A and B and explain how they were formed in the process of ozonolysis and decomposition of the ozonides in the absence of a reducing agent.

15.53 The following reaction is observed. Write equations showing a mechanism that would account for the experimental observations.

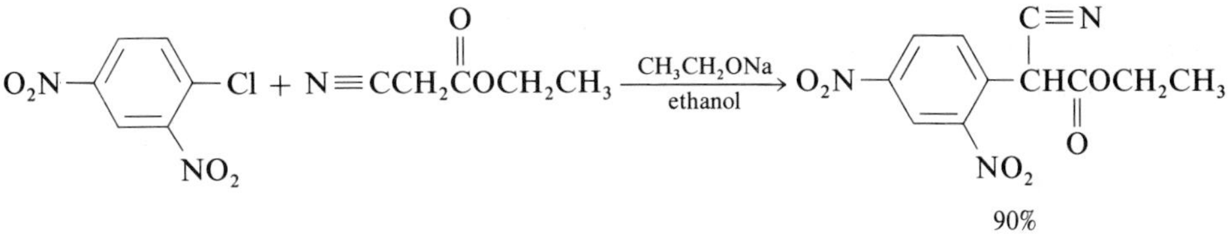

$$O_2N\text{—}\underset{NO_2}{\bigcirc}\text{—}Cl + N\equiv CCH_2\overset{O}{\underset{\|}{C}}OCH_2CH_3 \xrightarrow[\text{ethanol}]{CH_3CH_2ONa} O_2N\text{—}\underset{NO_2}{\bigcirc}\text{—}\underset{\underset{O}{\|}}{\underset{|}{C}}\overset{C\equiv N}{\underset{H}{C}}OCH_2CH_3$$

90%

15.54 Predict what will happen in the following sequence of reactions.

(a) $CH_3\overset{O}{\underset{\|}{S}}\text{—}\underset{\|}{N}CH_3 + CH_3CH_2CH_2CH_2Li \xrightarrow{\text{tetrahydrofuran}} A \xrightarrow{\bigcirc\text{—}\overset{O}{\underset{\|}{C}}\text{—}\bigcirc} B \xrightarrow{H_2O} C$

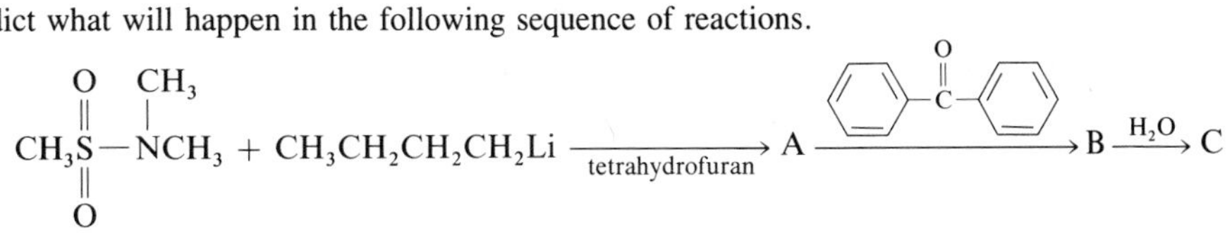

(with O and CH₃ on the sulfur/nitrogen)

(b) $A + \bigcirc\text{—}\overset{O}{\underset{\|}{C}}OCH_2CH_3 \xrightarrow{\text{tetrahydrofuran}} D \xrightarrow{H_3O^+} E$

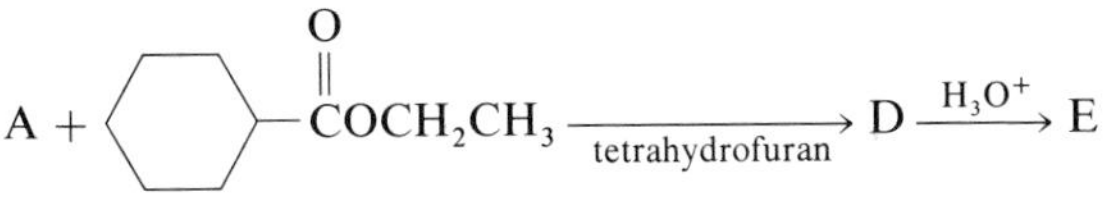

15.55 A carbanion adjacent to a sulfone, in which the sulfur atom bears two oxygen atoms, is a weaker base than a carbanion adjacent to a sulfoxide, in which there is only one oxygen atom on the sulfur atom. How would you rationalize this fact?

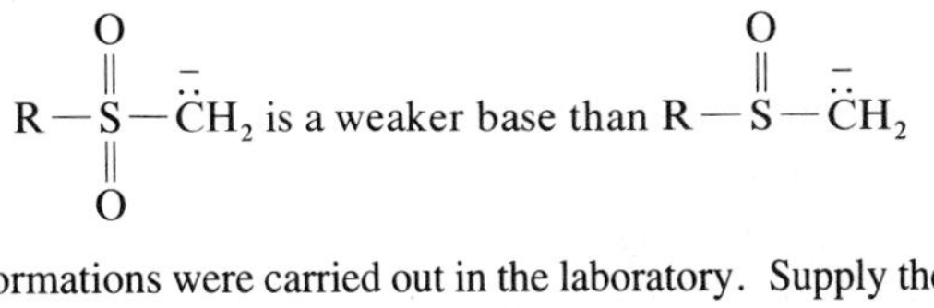

$$R\text{—}\overset{O}{\underset{\underset{O}{\|}}{\underset{\|}{S}}}\text{—}\overset{..}{C}H_2 \text{ is a weaker base than } R\text{—}\overset{O}{\underset{\|}{S}}\text{—}\overset{..}{C}H_2$$

15.56 The following transformations were carried out in the laboratory. Supply the reagents that would be necessary for each step.

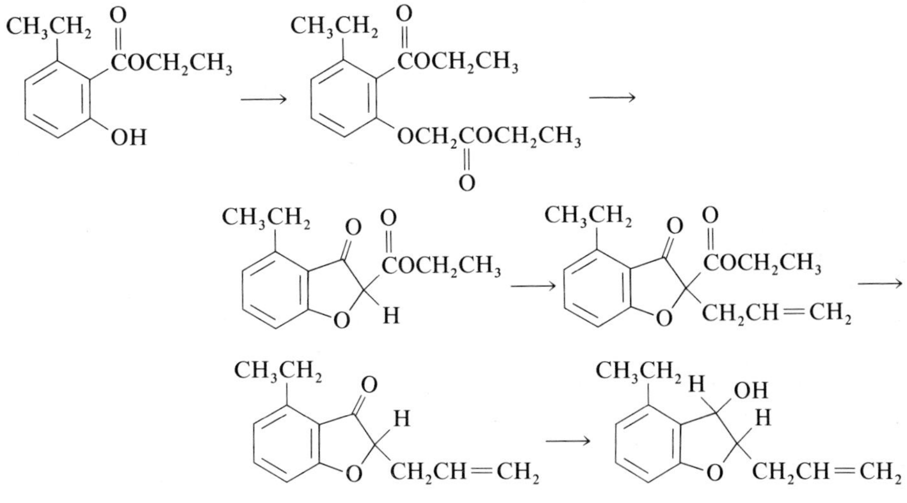

(Series of structural transformations of ethyl-substituted aromatic/benzofuranone compounds)

15.57   The following reaction is observed. Write a mechanism for the reaction that accounts for the product obtained.

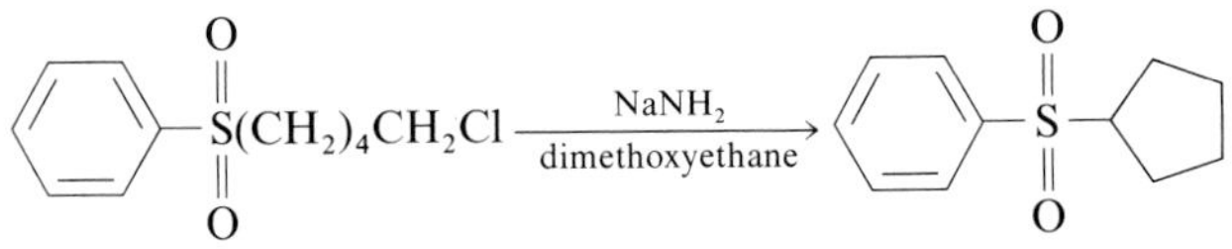

15.58   For each of the following sets of compounds, decide which is the most acidic.

(a) 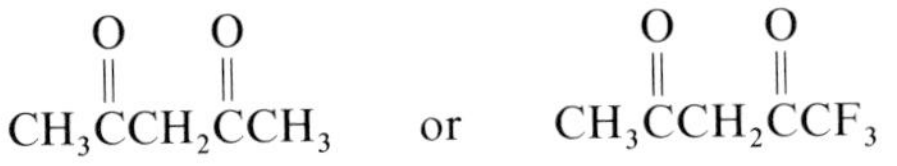 $CH_3CCH_2CCH_3$ or $CH_3CCH_2CCF_3$

(b) 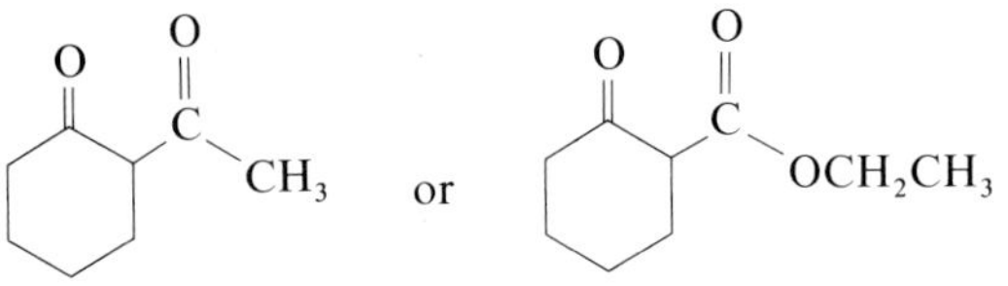   or

(c) 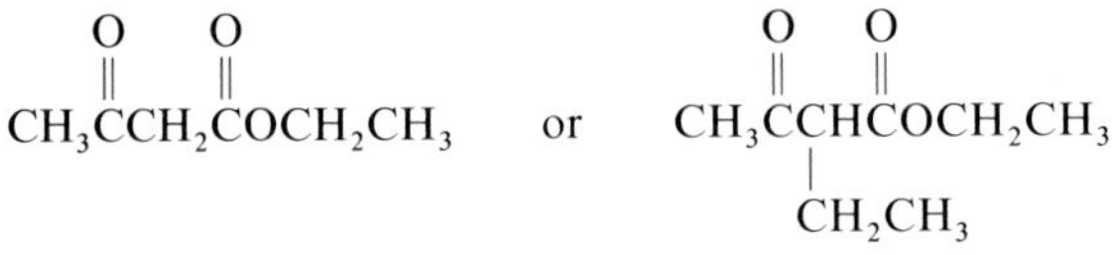 $CH_3CCH_2COCH_2CH_3$ or $CH_3CCHCOCH_2CH_3$
                                                            $|$
                                                        CH_2CH_3

(d)  $O_2NCH_2COCH_2CH_3$, $O_2NCH_2NO_2$,   or $O_2NCH_2CCH_3$

(e) 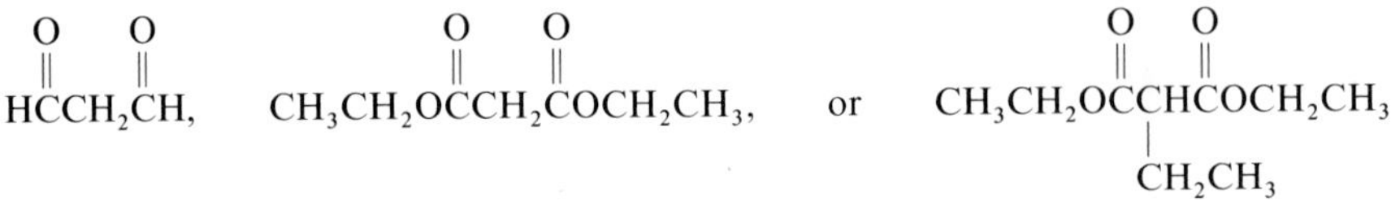 $HCCH_2CH$, $CH_3CH_2OCCH_2COCH_2CH_3$,   or $CH_3CH_2OCCHCOCH_2CH_3$
                                                                                            $|$
                                                                                        CH_2CH_3

15.59   Shikimic acid is an important intermediate in the biosynthesis of amino acids containing aromatic rings. Shikimic acid itself is synthesized in the body from phosphoenolpyruvic acid, the phosphate ester of the enol of pyruvic acid. Various steps of this synthesis are given below. Show the mechanisms for the different transformations, supplying general acid or base (HB^+ and $B:$) catalysis as necessary. The named intermediates are known; the unnamed one is postulated.

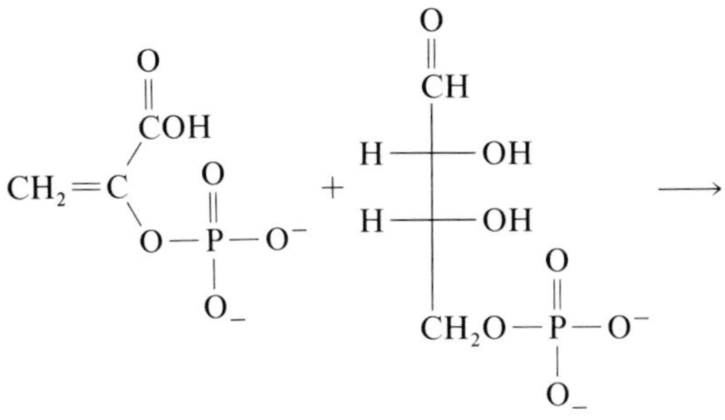

phosphoenolpyruvic
acid

erythrose
4-phosphate

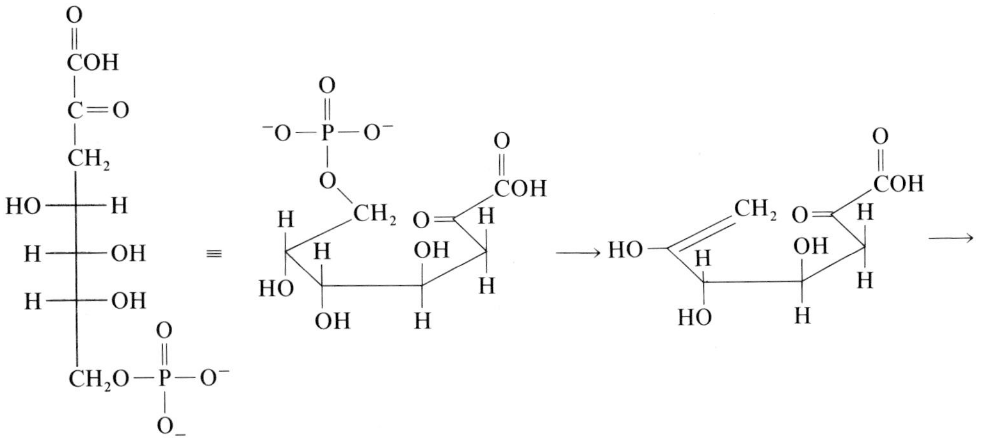

3-deoxy-α-arabinoheptulosonic
acid 7-phosphate

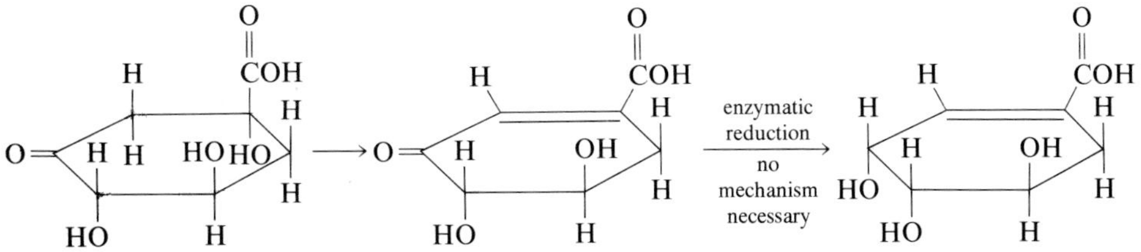

5-dehydroquinic
acid

5-dehydroshikimic
acid

enzymatic
reduction
no
mechanism
necessary

shikimic
acid

15.60 The following synthesis was used to test the feasibility of using organopalladium reagents to catalyze intramolecular cyclizations. Supply structural formulas for the starting materials, A and B, and for the reagents necessary for each step. (Hint: Reviewing Sections 11.5, 15.8, and 15.10 will be helpful.)

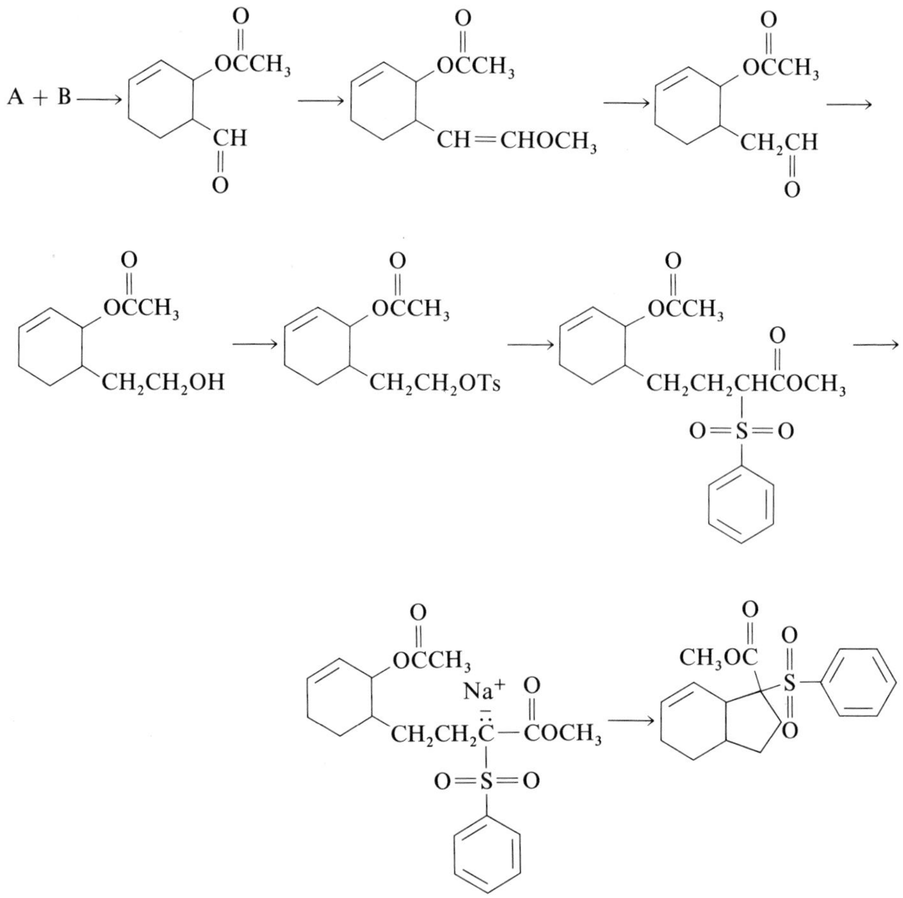

A + B ⟶ [structure: cyclohexene ring with OCCH₃ (O double bond) ester and CH=O substituent] ⟶ [structure with OCCH₃ ester and CH=CHOCH₃] ⟶ [structure with OCCH₃ ester and CH₂CH=O] ⟶

[structure with OCCH₃ ester and CH₂CH₂OH] ⟶ [structure with OCCH₃ ester and CH₂CH₂OTs] ⟶ [structure with OCCH₃ ester and CH₂CH₂CHCOCH₃ with O=S=O phenyl group] ⟶

[structure with OCCH₃ ester and CH₂CH₂C̈⁻—COCH₃ with Na⁺ and O=S=O phenyl group] ⟶ [bicyclic structure with CH₃OC, O=S= phenyl and O]

15.61   (a)   The following sequence of reactions was used to convert the isobutyl enol ether of 1,3-cyclohexadione to 4-propylcyclohexanone. Fill in structural formulas for the intermediates, products, and the reagent indicated by the letters.

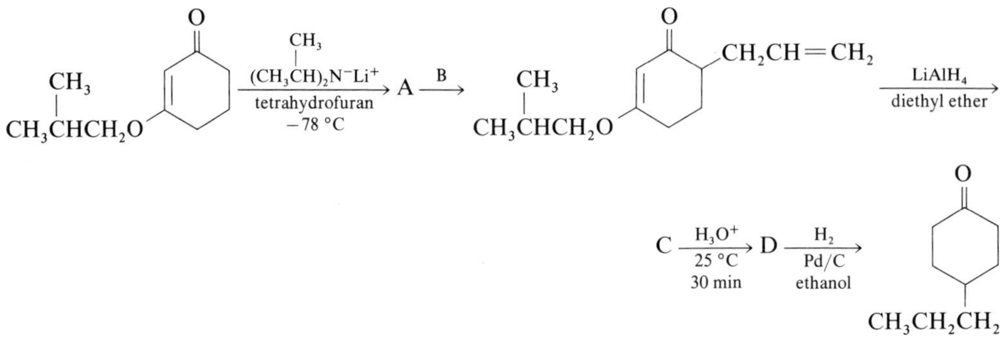

[structure: CH₃CHCH₂O (isobutyl) enol ether of cyclohexenone]
$\xrightarrow[\substack{\text{tetrahydrofuran} \\ -78\,°C}]{(CH_3CH)_2N^-Li^+}$ A $\xrightarrow{B}$ [structure: CH₃CHCH₂O enol ether of cyclohexenone with CH₂CH=CH₂ group] $\xrightarrow[\text{diethyl ether}]{LiAlH_4}$

C $\xrightarrow[\substack{25\,°C \\ 30\ min}]{H_3O^+}$ D $\xrightarrow[\substack{Pd/C \\ ethanol}]{H_2}$ [structure: cyclohexanone with CH₃CH₂CH₂ substituent]

(b)   The isobutyl enol ether of 1,3-cyclohexadione is prepared in 90% yield by heating the dione with isobutyl alcohol and *p*-toluenesulfonic acid in benzene. Write a mechanism for the reaction that shows why it is so easy to make this enol ether.

15.62 Helenalin, a natural product isolated from *Helenium autumnale*, is of interest because it is cytotoxic. Some steps in a recent synthesis of the compound are shown below. Fill in the reagents that could be used for these transformations.

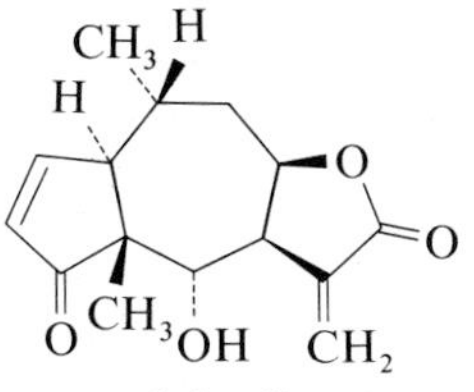

helenalin

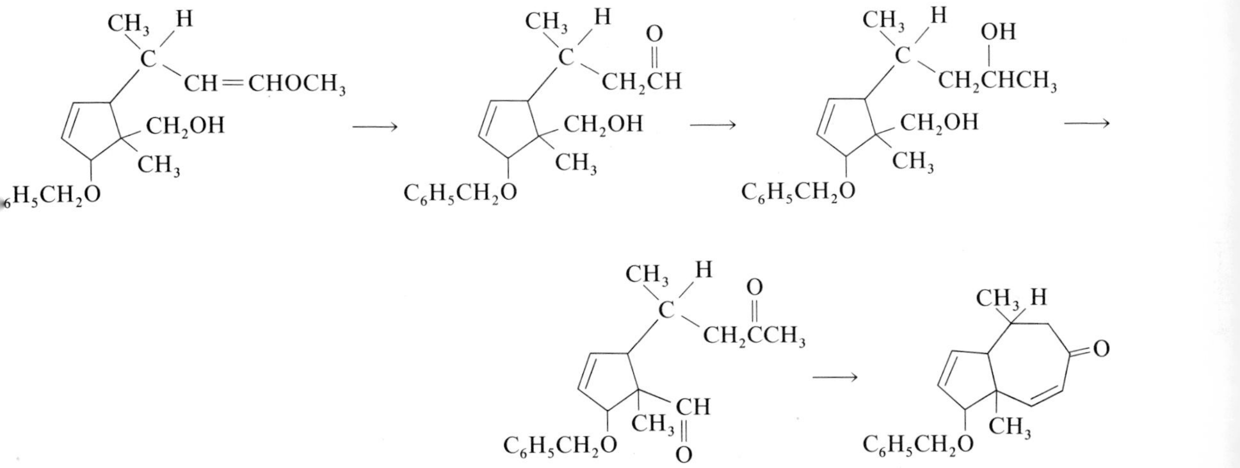

15.63 1,3-Diketones decompose when heated with base. For example, the following reaction takes place:

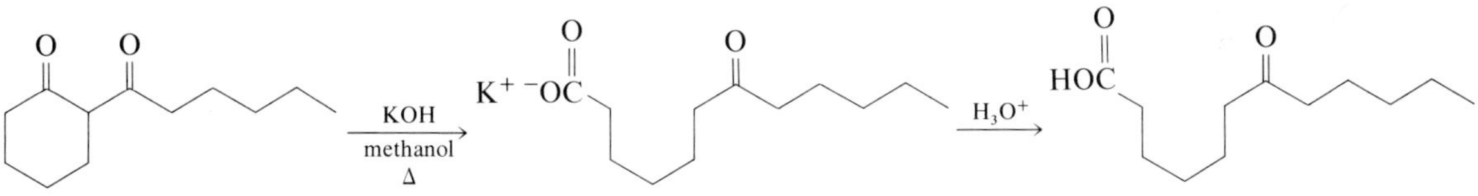

15.64 A synthesis of β-carotene (p. 547) uses the following two reagents.

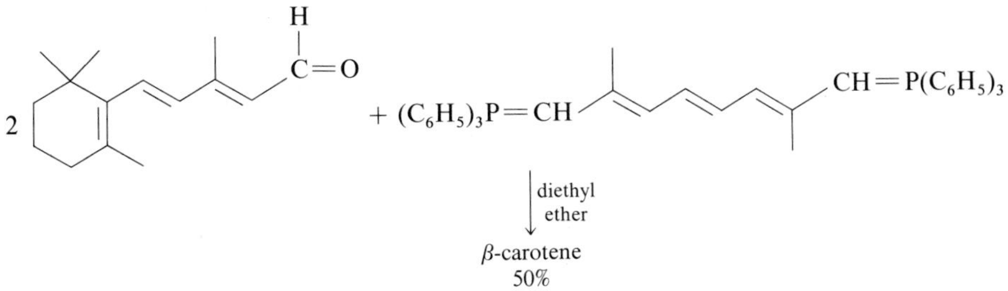

diethyl
ether

β-carotene
50%

The double Wittig reagent is synthesized using the following sequence of reactions. Fill in the structural formulas for the compounds or intermediates indicated by the letters.

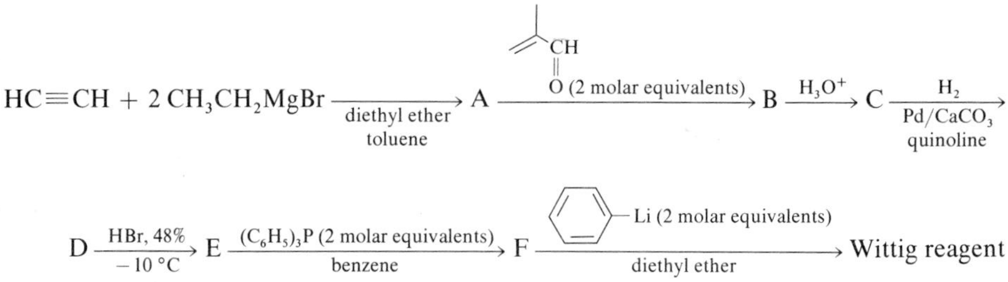

$$HC\equiv CH + 2\ CH_3CH_2MgBr \xrightarrow[\substack{\text{diethyl ether}\\\text{toluene}}]{} A \xrightarrow[]{\substack{\text{(2 molar equivalents)}}} B \xrightarrow{H_3O^+} C \xrightarrow[\substack{Pd/CaCO_3\\\text{quinoline}}]{H_2}$$

$$D \xrightarrow[-10\,°C]{\text{HBr, 48\%}} E \xrightarrow[\text{benzene}]{(C_6H_5)_3P\ \text{(2 molar equivalents)}} F \xrightarrow[\text{diethyl ether}]{\text{Li (2 molar equivalents)}} \text{Wittig reagent}$$

15.65 The following cyclization reaction is observed. Propose a mechanism that accounts for the product obtained.

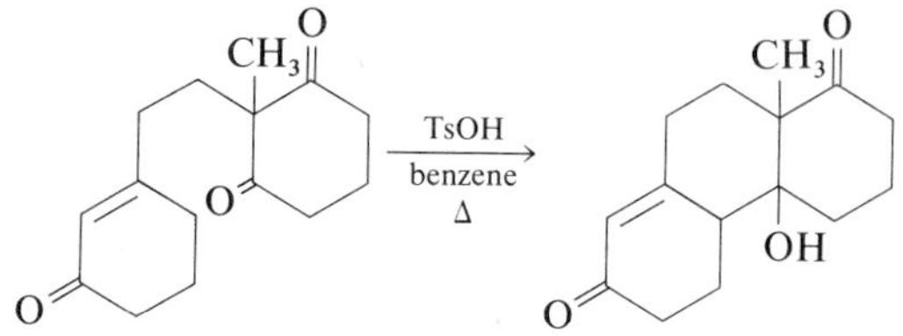

15.66 Organic chemists have tried to discover whether the metal ion in an enolate salt is primarily associated with a negatively charged oxygen atom or with a negatively charged carbon atom in the ambident nucleophile (Section 15.1C). One of the research tools applied to this question is nuclear magnetic resonance spectroscopy. The proton magnetic resonance spectra of the lithium salt of the enolate ion from 1-phenyl-2-methyl-1-propanone in a variety of solvents such as benzene or tetrahydrofuran have two singlets in the region δ 1.0–2.0.

Compare structural formulas for the two resonance contributors for the enolate ion from 1-phenyl-2-methyl-1-propanone and decide what conclusion can be drawn from proton magnetic resonance data.

15.67 D-(+)-Glyceraldehyde (Section 14.2D) is converted in basic or acidic solution to a mixture of D-fructose and D-sorbose, both shown in their open-chain forms below. The conversion proceeds faster if dihydroxyacetone is added to the solution. Write equations showing how D-(+)-glyceraldehyde is transformed into the mixture of hexoses.

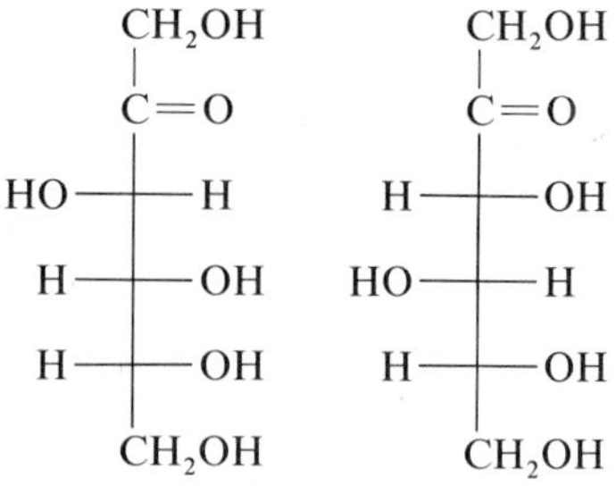

15.68 The sex attractant of a female beetle that bores into the ponderosa pine has an unusual bicyclic structure, which is an internal cyclic acetal between a methyl ketone and a diol. The compound, brevicomin, exists in its bicyclic form as the exo and endo isomers, each of which is a racemic mixture. Both diastereomeric forms are found in nature but only the exo isomer has biological activity.

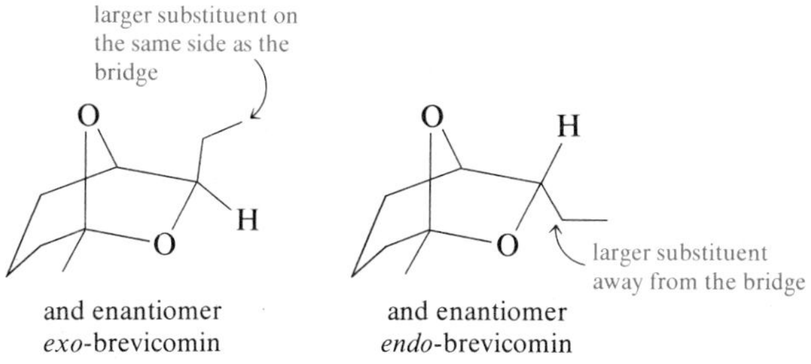

larger substituent on
the same side as the
bridge

and enantiomer
exo-brevicomin

larger substituent
away from the bridge

and enantiomer
endo-brevicomin

Both diastereomers of brevicomin were synthesized by the following sequence of reactions. Fill in the structures of the intermediates and products indicated by letters. Be as specific as you can about stereochemistry at each stage.

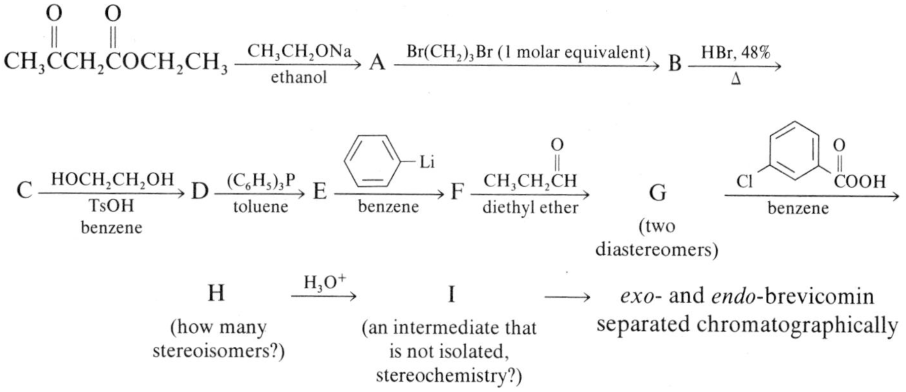

$$CH_3CCH_2COCH_2CH_3 \xrightarrow[\text{ethanol}]{CH_3CH_2ONa} A \xrightarrow{Br(CH_2)_3Br \text{ (1 molar equivalent)}} B \xrightarrow[\Delta]{HBr, 48\%}$$

$$C \xrightarrow[\substack{TsOH \\ benzene}]{HOCH_2CH_2OH} D \xrightarrow[toluene]{(C_6H_5)_3P} E \xrightarrow[benzene]{\text{---Li}} F \xrightarrow[\substack{diethyl\ ether}]{CH_3CH_2CH} G \xrightarrow{\underset{benzene}{Cl\text{---}COOH}}$$

(two
diastereomers)

$$H \xrightarrow{H_3O^+} I \longrightarrow \text{ } exo\text{- and } endo\text{-brevicomin}$$

(how many
stereoisomers?)

(an intermediate that
is not isolated,
stereochemistry?)

separated chromatographically

Another synthesis of *exo*-brevicomin is given below. Complete the steps for this synthesis also.

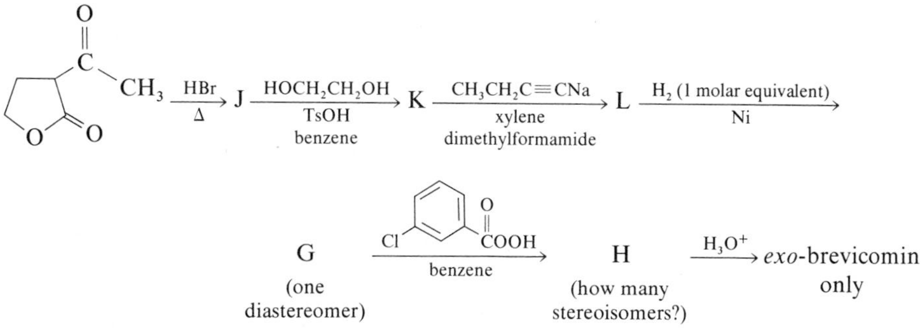

$$\xrightarrow[\Delta]{HBr} J \xrightarrow[\substack{TsOH \\ benzene}]{HOCH_2CH_2OH} K \xrightarrow[\substack{xylene \\ dimethylformamide}]{CH_3CH_2C\equiv CNa} L \xrightarrow[Ni]{H_2 \text{ (1 molar equivalent)}}$$

$$G \xrightarrow[benzene]{Cl\text{---}COOH} H \xrightarrow{H_3O^+} exo\text{-brevicomin}$$

(one
diastereomer)

(how many
stereoisomers?)

only

15.69 Work out syntheses for the following compounds. You have triphenylphosphine, benzene, and any other organic reagents containing not more than four carbon atoms available to you.

(a) 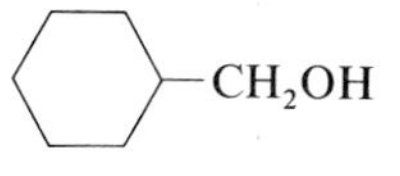 [cyclohexyl]—CH$_2$OH (b) 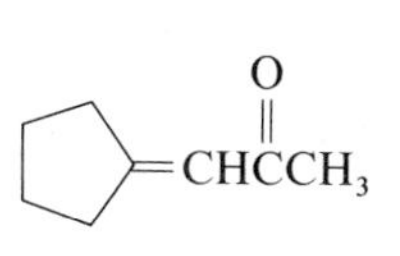 [cyclopentylidene]=CHCCH$_3$ with O

(c) 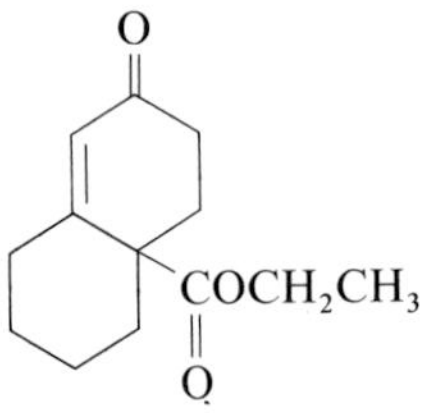   (d)

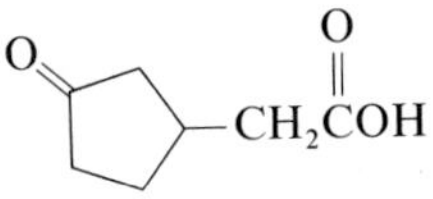

(e) 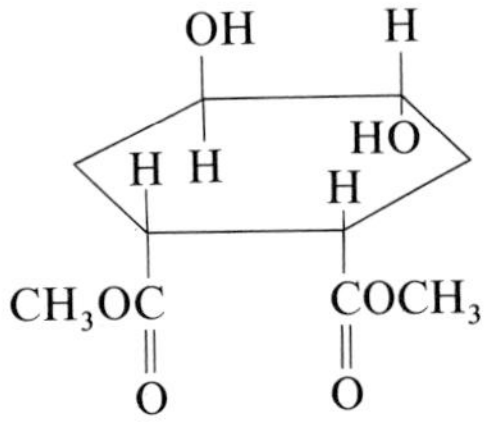 and enantiomer   (f)

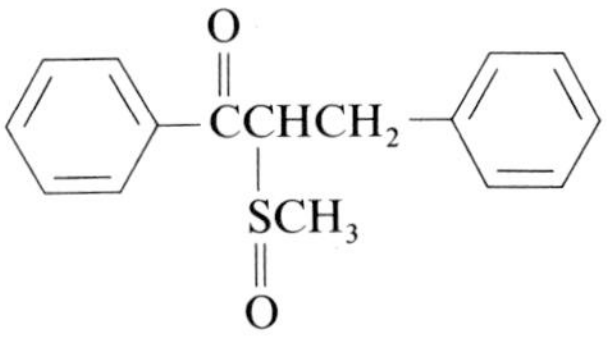

(g) 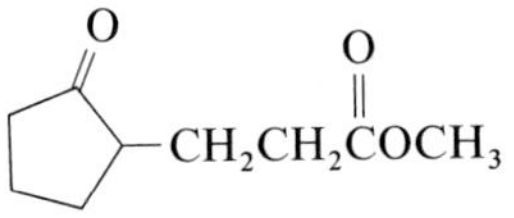   (h)

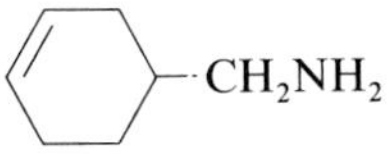

15.70 An intermediate for the synthesis of quadrone, a sesquiterpene with possible antitumor activity, was prepared by the following reactions. Fill in structural formulas for the compounds represented by letters.

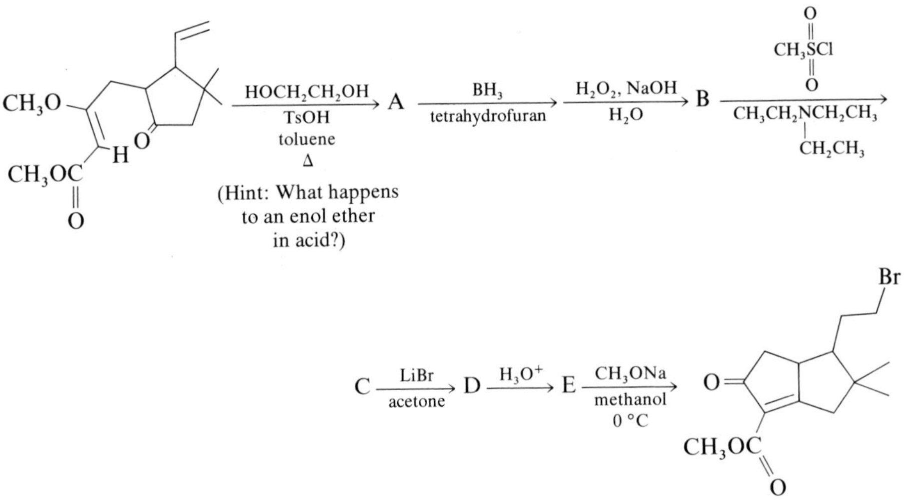

(Hint: What happens to an enol ether in acid?)

15.71 The preparation of intermediates in the synthesis of coriolin, a natural product with antibacterial and antitumor activity, involves the epoxidation of double bonds. Predict what the products in the two steps of the synthesis shown below will be.

(a)

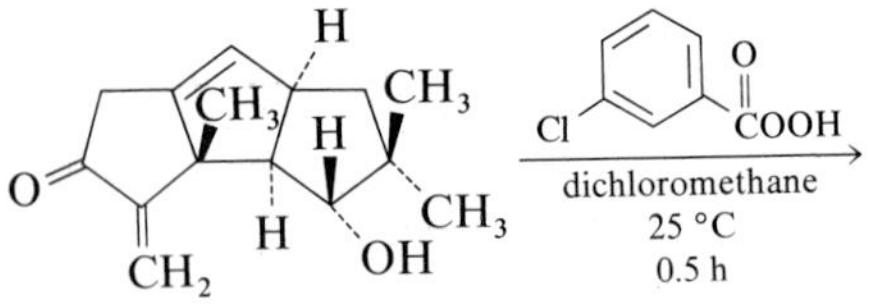

(b)

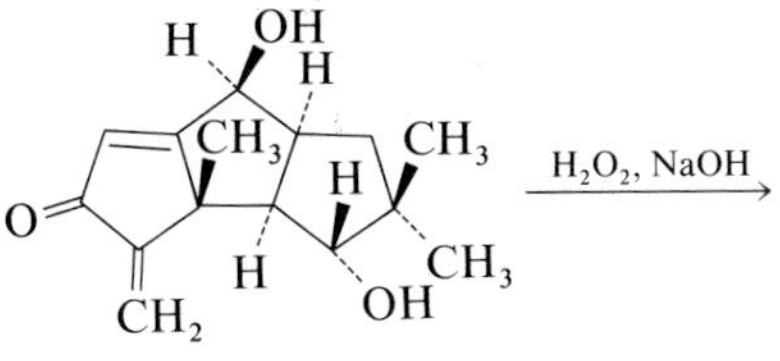

$$\xrightarrow{\text{H}_2\text{O}_2,\ \text{NaOH}}$$

15.72 The following experimental observation has been made. Propose a detailed mechanism for the reaction.

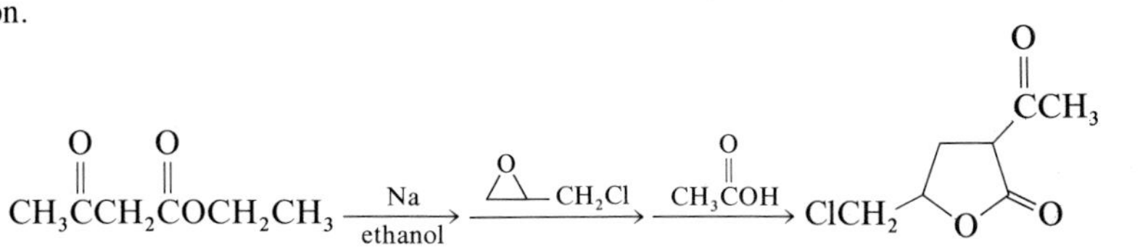

$$\text{CH}_3\text{CCH}_2\text{COCH}_2\text{CH}_3 \xrightarrow[\text{ethanol}]{\text{Na}} \xrightarrow{\triangle-\text{CH}_2\text{Cl}} \xrightarrow{\text{CH}_3\overset{\text{O}}{\text{COH}}} \text{ClCH}_2$$

15.73 When the ketolactone shown below is heated with concentrated hydrochloric acid, the product is 5-chloro-2-pentanone. Treatment of 5-chloro-2-pentanone with aqueous sodium hydroxide gives cyclopropyl methyl ketone in about 80% yield. How would you rationalize these observations?

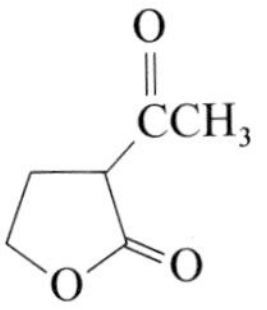

ketolactone used as
starting material

15.74 Recently it has been discovered that it is possible to prepare dianions of carbonyl compounds if very strong bases are used. One such dianion derived from a substituted acetoacetic ester was used in the synthesis of dihydrojasmone, a compound valuable in perfumery. The synthesis is outlined below with some hints. Fill in structural formulas for the intermediates and products designated by letters.

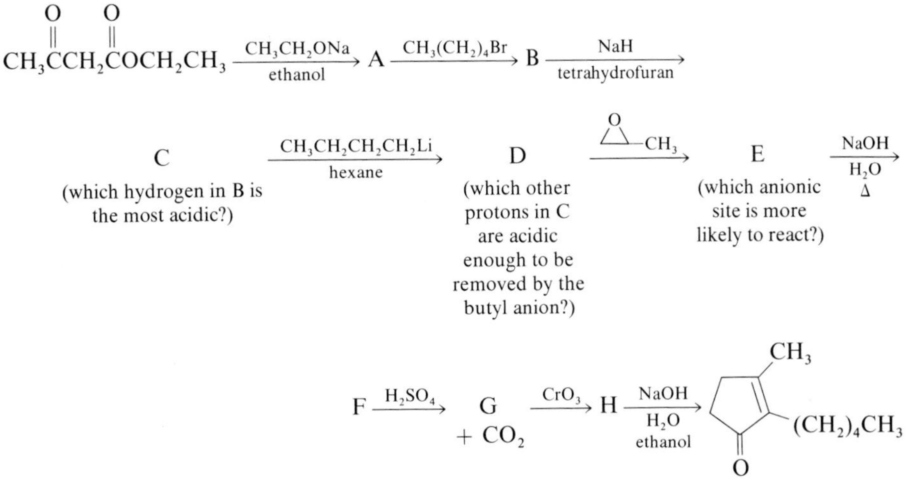

$$\text{CH}_3\text{CCH}_2\text{COCH}_2\text{CH}_3 \xrightarrow[\text{ethanol}]{\text{CH}_3\text{CH}_2\text{ONa}} \text{A} \xrightarrow{\text{CH}_3(\text{CH}_2)_4\text{Br}} \text{B} \xrightarrow[\text{tetrahydrofuran}]{\text{NaH}}$$

C $\xrightarrow[\text{hexane}]{\text{CH}_3\text{CH}_2\text{CH}_2\text{CH}_2\text{Li}}$ D $\xrightarrow{\triangle-\text{CH}_3}$ E $\xrightarrow[\underset{\Delta}{\text{H}_2\text{O}}]{\text{NaOH}}$

(which hydrogen in B is
the most acidic?)

(which other
protons in C
are acidic
enough to be
removed by the
butyl anion?)

(which anionic
site is more
likely to react?)

F $\xrightarrow{\text{H}_2\text{SO}_4}$ G $\xrightarrow{\text{CrO}_3}$ H $\xrightarrow[\text{ethanol}]{\underset{\text{H}_2\text{O}}{\text{NaOH}}}$

+ CO₂

dihydrojasmone

What other product is possible in the last step of the synthesis? How do you explain the regioselectivity of the reaction?

15.75 Compound A, having the formula $C_{12}H_{20}O_2$, is isolated from a fungus. The compound has bands at 2978(s), 2935(s), 2850(s), 1735(s), 1450(m), 1365(m), 1225(s), and 975(s) cm^{-1} in its infrared spectrum. Its proton magnetic resonance spectrum shows a doublet corresponding to three hydrogens at δ 1.2 and a band corresponding to two hydrogens at δ 5.3 ppm.

Compound A was treated with ozone in ethanol and ethyl acetate; the reaction mixture was then heated with formic acid and hydrogen peroxide and finally heated under reflux with potassium hydroxide in alcohol. Acidification of the reaction mixture gave octanedioic acid and 3-hydroxybutanoic acid.

Compound A reacts with hydrogen in the presence of platinum to give Compound B, $C_{12}H_{22}O_2$. When Compound B is heated with aqueous base, then acidified, Compound C, $C_{12}H_{24}O_3$, is obtained. Treatment of Compound C with chromium trioxide in acetic acid gives Compound D, $C_{12}H_{22}O_3$. Compound D reacts with iodine in base to give iodoform and undecanedioic acid.

Assign structures to Compounds A, B, C, and D that are compatible with these facts.

15.76 The proton magnetic resonance spectrum of ethyl (E)-3-phenylpropenoate is shown in Figure 15.1. Assign the bands in the spectrum, and analyze any splitting that is observed. How do you account for the large difference in chemical shift for the two vinylic hydrogen atoms?

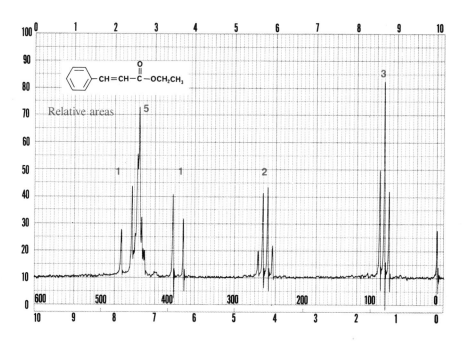

Figure 15.1 Spectrum for Problem 15.77. (From *The Aldrich Library of NMR Spectra*)

15.77 The carbon-13 nuclear magnetic resonance spectrum of (E)-2-butenoic acid has bands at 18.0, 122.6, 147.5, and 172.3 ppm downfield from TMS. Assign these bands to the carbon atoms in (E)-2-butenoic acid, and explain your reasoning in making these assignments. Table 13.4, Section 13.5, may be useful.

The Chemistry of Amines

16

16.1

Introduction and Nomenclature

A. Introduction

Amines, *compounds in which one or more of the hydrogen atoms of an ammonia molecule have been substituted by an organic group,* are found widely in nature. Many plants synthesize complex amines called **alkaloids,** some of which have medicinal and poisonous properties. Morphine, one of the most effective painkillers known; quinine, an important medication for malaria; and nicotine, the toxic component in tobacco, are among some of the alkaloids that have been isolated and used by mankind.

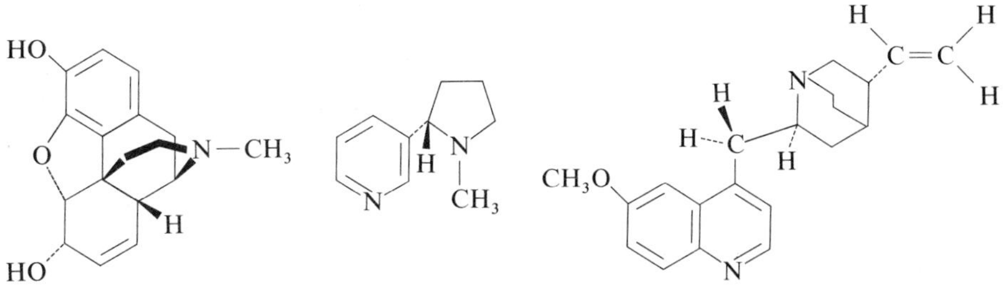

(−)-morphine   (−)-nicotine   (−)-quinine

some naturally occurring amines present in plants

Proteins, *the complex molecules that constitute large portions of our bodies and of the enzyme systems that catalyze chemical processes in our bodies, are made up of* **amino acids.** *In such amino acids, an amino group is substituted on the second carbon atom of a carboxylic acid.* Proteins are formed when amino acids are bonded together by amide bonds, called **peptide linkages** (Section 10.8B), between the amino group of one amino acid and the carboxylic acid group of another one.

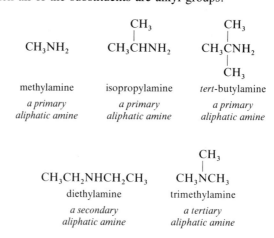

an amino acid

a fragment of a
protein showing 2
amino acid units

*amino groups, in a free amino acid, and in an
amide bond in part of a protein chain*

The synthetic methods for preparing amino acids resemble those used for simple alkylamines, and that part of their chemistry is included in this chapter. The structural features of amino acids that are significant in the chemistry of peptides and proteins are discussed in the next chapter.

The decomposition of amino acids and proteins in decaying animal matter gives rise to many simple amines such as methylamine, 1,4-butanediamine, and 1,5-pentanediamine. The odor of fish comes from low molecular weight amines, such as methylamine, and the common names for 1,4-butanediamine and 1,5-pentanediamine are putrescine and cadaverine, respectively.

$$CH_3NH_2 \quad NH_2CH_2CH_2CH_2CH_2NH_2 \quad NH_2CH_2CH_2CH_2CH_2CH_2NH_2$$

methylamine   1,4-butanediamine   1,5-pentanediamine
putrescine   cadaverine

some amines found in animal tissues

Amines are classified as **primary, secondary,** *and* **tertiary** *depending on the number of organic groups on the nitrogen atom.* Amines are also divided into **aromatic amines,** in which at least one of the organic substituents is an aryl group, and **aliphatic amines,** in which all of the substituents are alkyl groups.

$$CH_3NH_2 \qquad \overset{\displaystyle CH_3}{\underset{}{\overset{|}{CH_3CHNH_2}}} \qquad \overset{\displaystyle CH_3}{\underset{\underset{CH_3}{|}}{\overset{|}{CH_3CNH_2}}}$$

methylamine   isopropylamine   *tert*-butylamine

*a primary
aliphatic amine*   *a primary
aliphatic amine*   *a primary
aliphatic amine*

$$CH_3CH_2NHCH_2CH_3 \qquad \overset{\displaystyle CH_3}{\underset{}{\overset{|}{CH_3NCH_3}}}$$

diethylamine   trimethylamine

*a secondary
aliphatic amine*   *a tertiary
aliphatic amine*

NH$_2$ NHCH$_3$ CH$_3$CH$_2$NCH$_2$CH$_3$

aniline   *N*-methylaniline   *N,N*-diethylaniline

a primary   *a secondary*   *a tertiary*
aromatic amine   *aromatic amine*   *aromatic amine*

Note that the designations *primary, secondary,* and *tertiary* refer to the substitution on the nitrogen atom and not to the carbon atom to which the nitrogen is attached, as in the case of alcohols and alkyl halides.

Primary and secondary amines participate in hydrogen bonding, both as donors and as acceptors (Section 1.8B). Nitrogen is not as electronegative as oxygen; therefore, the nitrogen-hydrogen bond is less polar than the oxygen-hydrogen bond, and hydrogen bonding for amines is weaker than it is for alcohols. The boiling points of amines are lower than those of alcohols of similar molecular weight, but higher than the boiling points of hydrocarbons and other compounds where no hydrogen bonding is possible. The trend in boiling points is illustrated with pentane, butylamine, and *n*-butyl alcohol.

CH$_3$CH$_2$CH$_2$CH$_2$CH$_3$ CH$_3$CH$_2$CH$_2$CH$_2$NH$_2$ CH$_3$CH$_2$CH$_2$CH$_2$OH

pentane   butylamine   *n*-butyl alcohol
MW 72   MW 73   MW 74
bp 36 °C   bp 78 °C   bp 118 °C
no hydrogen bonding   strong hydrogen bonding

the effect of hydrogen bonding on the boiling
points of compounds of comparable molecular weight

Primary, secondary, and tertiary amines in which the organic groups are not too large (Section 1.8B) are soluble in water. The nonbonding electrons on a nitrogen atom are more available to the hydrogen atom of water in a hydrogen bond than are the electrons on an oxygen atom. The basicity of amines is related to this phenomenon. The reactions of amines as bases were introduced in Section 4.3 and are discussed in greater detail in Section 16.2.

B. Nomenclature of Amines

Simple amines are named by combining the name of the alkyl group that is present with the ending **-amine.** If there are several alkyl groups on the nitrogen atom, they are named in order of increasing complexity, and the prefixes **di-** and **tri-** are used to indicate the presence of two or three alkyl groups of the same kind. *In IUPAC nomenclature, the suffix* **-amine** *is substituted for the final - e in the name of the alkane and a number is used to indicate the position of the amino group on the chain or ring.*

For more complex amines, the prefix **amino-** is used with a number to indicate the position of the amino group on the hydrocarbon chain. Other substituents on the nitrogen atom are indicated by an *N* before the name of the substituent.

CH$_2$CH$_2$CH$_3$
|
CH$_3$CH$_2$NH$_2$ CH$_3$NHCH$_3$ CH$_3$CH$_2$CH$_2$NCH$_2$CH$_2$CH$_3$

ethylamine   dimethylamine   tri-*N*-propylamine

CH$_3$
|
CH$_3$CHCH$_2$CH$_2$NH$_2$ HOCH$_2$CH$_2$CH$_2$CH$_2$NH$_2$

3-methyl-1-butylamine   4-amino-1-butanol

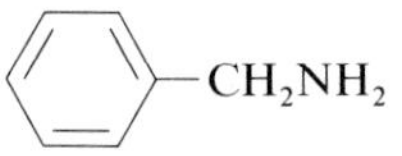

benzylamine 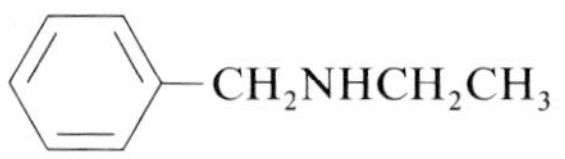   *N*-ethylbenzylamine

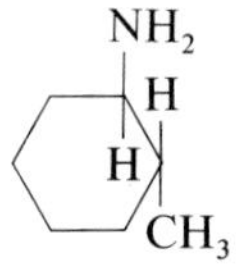

trans-2-methyl-1-cyclohexanamine   *N,N*-dimethyl-3-methyl-2-pentanamine
trans-(2-methylcyclohexyl)amine   2-(*N,N*-dimethylamino)-3-methylpentane

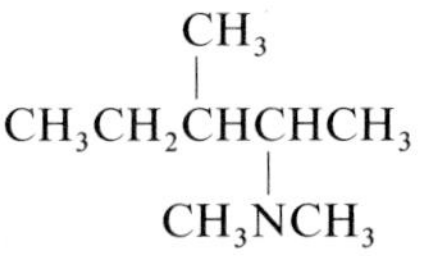

The simplest aromatic amine is **aniline,** and many aromatic amines are named as substituted anilines. Exceptions are the *amino derivatives of toluene, which are usually called* **toluidines.** When there are a large number of substituents on the aromatic ring, the prefix **amino-** is used to locate the amino group. In some amines, nitrogen is part of a ring. Such compounds are called **heterocycles,** meaning that they have atoms other than carbon in a ring. Many of them have common names, and some that you should recognize are shown below.

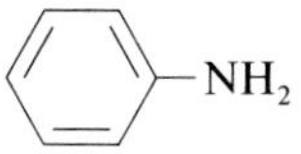

aniline 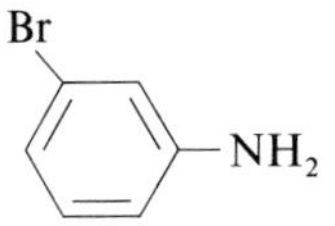   *m*-bromoaniline 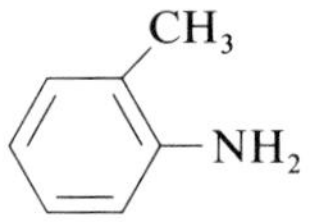   *o*-toluidine    *p*-toluidine

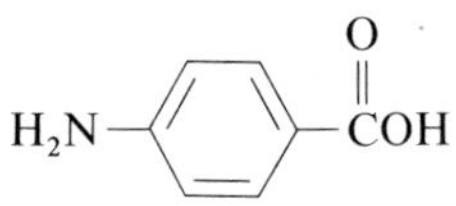

p-aminobenzoic acid 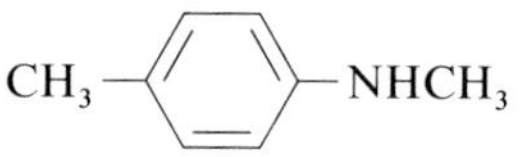   *N*-methyl-*p*-toluidine 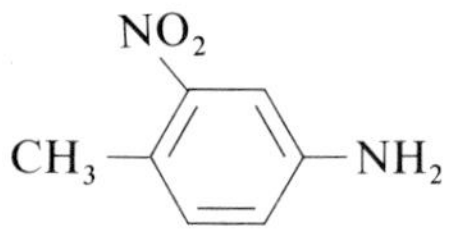   4-methyl-3-nitroaniline

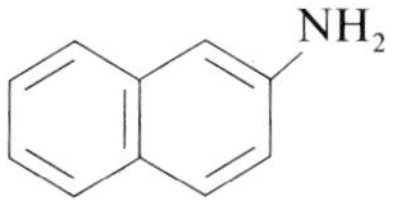

2-naphthylamine
2-aminonaphthalene
β-naphthylamine

pyridine 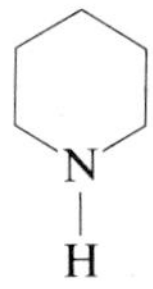   piperidine 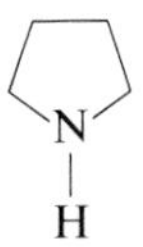   pyrrolidine

The chemistry of heterocyclic amines is discussed in Chapter 18.

PROBLEM 16.1   Name the following compounds.

(a) 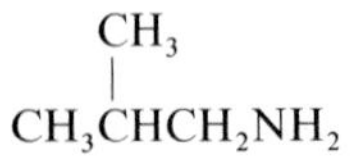 $CH_3CHCH_2NH_2$ (b)

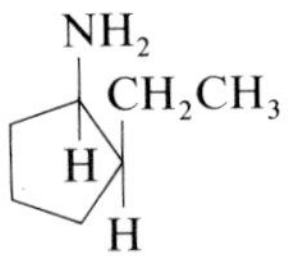

(c) $CH_3CH_2\overset{\underset{\textstyle CH_2CH_3}{|}}{N}CH_2CH_2CH_2CH_3$ (d) $\triangleright\!\!-CH_2CH_2CH_2NH_2$

(e)   (aromatic ring with NO_2)$-NH_2$ (f) $Br-$(aromatic ring with Br)$-NH_2$

16.2

The Basicity of Amines

A. The Basicity of Alkylamines

Amines are the most important organic bases. The relative basicities of amines are determined by the availability of the nonbonding electrons on the nitrogen atom to a proton or a Lewis acid, and by the stabilization of the positively charged nitrogen atom by solvation, or in some special cases, by resonance. In Chapter 4, the basicities of various species were related to the pK_a values of their conjugate acids. For an amine too, the larger the pK_a of its conjugate acid, the weaker the acid, and the stronger the corresponding base.

Amines in which alkyl groups are substituted on the nitrogen atom have basicities similar to ammonia. The conjugate acids are alkylammonium ions with pK_a values of approximately 10.8 to 9.5. Methylamine, dimethylamine, and trimethylamine are protonated by water just as ammonia is. All low molecular weight amines in which the nitrogen atom is a significant portion of the molecule dissolve in water to give solutions that turn red litmus paper blue.

$$NH_3 \ + \ H_2O \ \rightleftharpoons \ NH_4^+ \ + \ OH^-$$

ammonia pK_a 15.7   ammonium
ion
pK_a 9.2

$$CH_3NH_2 \ + \ H_2O \ \rightleftharpoons \ CH_3NH_3^+ \ + \ OH^-$$

methylamine pK_a 15.7   methylammonium
ion
pK_a 10.6

$$(CH_3)_2NH \ + \ H_2O \ \rightleftharpoons \ (CH_3)_2NH_2^+ \ + \ OH^-$$

dimethylamine pK_a 15.7   dimethylammonium
ion
pK_a 10.7

$$(CH_3)_3N \ + \ H_2O \ \rightleftharpoons \ (CH_3)_3NH^+ \ + \ OH^-$$

trimethylamine pK_a 15.7   trimethylammonium
ion
pK_a 9.8

Substitution on the nitrogen atom of one methyl group in methylamine and two methyl groups in dimethylamine increases basicity. Trimethylamine is slightly more basic than ammonia but is a weaker base than dimethylamine. The three alkyl groups around the nitrogen atom interfere with protonation and with stabilization of the cation by solvation, making the amine less basic. The reduced basicity of tertiary amines is

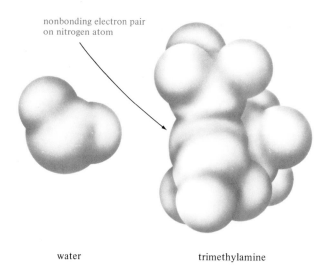

water   trimethylamine

Figure 16.1   Steric hindrance of a tertiary amine towards
protonation.

an example of a steric effect that becomes more important if the alkyl groups on the
nitrogen atom are larger, or if the acid is bulkier than a proton, as would be the case
with a Lewis acid with large groups on it (Figure 16.1).

PROBLEM 16.2   For the reaction

$$R_3\overset{+}{N}:\overset{-}{B}R_3' \rightleftharpoons R_3N: + BR_3'$$

the following dissociation constants were measured:

$H_3\overset{+}{N}:\overset{-}{B}(CH_3)_3$ $CH_3\overset{+}{N}H_2:\overset{-}{B}(CH_3)_3$ $(CH_3)_2\overset{+}{N}H:\overset{-}{B}(CH_3)_3$ $(CH_3)_3\overset{+}{N}:\overset{-}{B}(CH_3)_3$

K_{diss} 4.6 K_{diss} 0.0350 K_{diss} 0.0214 K_{diss} 0.477

(a)   Write the expression for the dissociation constant for the reaction.
(b)   Give a brief rationalization of the experimental facts that were observed.

B. The Basicity of Aromatic Amines

*An amine with a nitrogen atom bonded directly to an aromatic ring is a much weaker
base than ammonia.* For example, the conjugate acid of aniline is slightly more acidic
(pK_a 4.6) than acetic acid (pK_a 4.8).

$$\langle\!\!\!\bigcirc\!\!\!\rangle\!\!-NH_2 + H_2O \rightleftharpoons \langle\!\!\!\bigcirc\!\!\!\rangle\!\!-\overset{+}{N}H_3 + OH^-$$

aniline   anilinium ion
base pK_a 4.6
                                     conjugate acid

The nonbonding electrons on the nitrogen atom in aniline are said to be less available
for reactions with acids because they can be delocalized to the aromatic ring.
Resonance contributors for aniline indicate that it would have decreased electron

density at the nitrogen atom and increased electron density at the ortho and para positions of the ring.

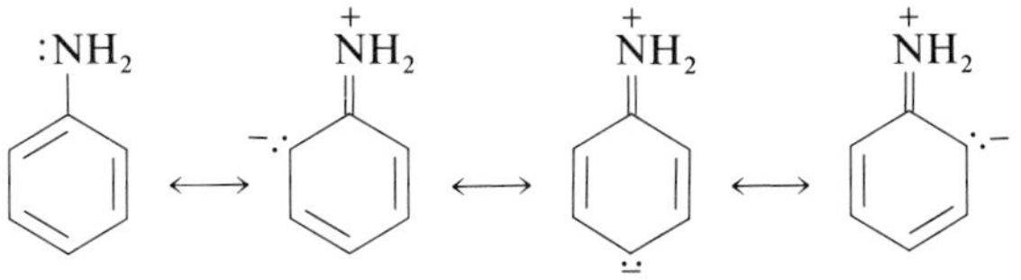

resonance contributors for aniline

The most important resonance contributor is the first one, in which the aromatic sextet is intact and there is no separation of charge. The other resonance contributors are also significant, as demonstrated by the great ease with which aromatic amines react with electrophilic reagents at the ortho and para positions. The amino group on an aromatic ring is a ring-activating and ortho, para-directing substituent (Section 12.3A).

Amines in which the hydrocarbon portion of the molecule has more than five carbon atoms are not soluble in water. Most amines that are not soluble in water will dissolve in dilute hydrochloric acid because they are protonated by the acid to give water-soluble ions. The solid amine, *p*-toluidine, for example, is water insoluble in its unprotonated, covalent form, but is converted into a water-soluble ionic salt by dilute hydrochloric acid. A strong base such as hydroxide ion removes a proton from the conjugate acid of *p*-toluidine, which reverts to its nonpolar, water-insoluble form as *p*-toluidine and precipitates from solution.

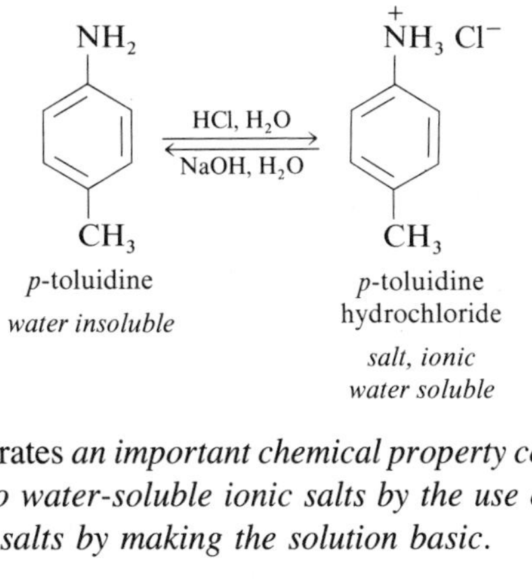

p-toluidine   *p*-toluidine
water insoluble   hydrochloride
                                             salt, ionic
                                             water soluble

The equation above illustrates *an important chemical property common to most amines. They can be converted to water-soluble ionic salts by the use of dilute acids and can be recovered from their salts by making the solution basic.*

PROBLEM 16.3   Complete the following equations.

(a) $CH_3 \overset{+}{N}H_3 + Cl^- + Na^+ + OH^- \xrightarrow[H_2O]{}$

(b) 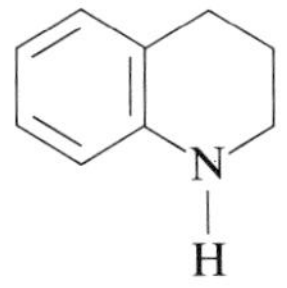 $+ H_3O^+ + Cl^- \xrightarrow[H_2O]{}$ (c) $(CH_3CH_2CH_2)_3N \xrightarrow[\text{diethyl ether}]{HCl}$

(d) 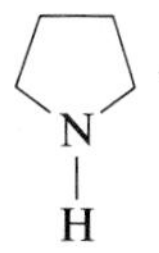 $+ BF_3 \xrightarrow[\text{diethyl ether}]{}$ (e) 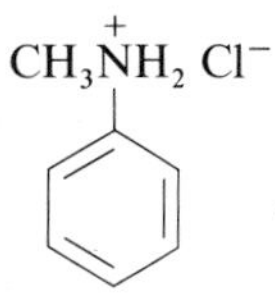 $+ Na^+ + OH^- \xrightarrow[H_2O]{}$

PROBLEM 16.4   The pK_a of the conjugate acid of diphenylamine is 0.8. Is diphenylamine a weak or strong base? How would you rationalize this experimental observation?

16.3

Preparation of Amines

A. A Review of Reactions That Create Carbon-Nitrogen Bonds

Carbon-nitrogen bonds are formed by nucleophilic substitution reactions of alkyl halides (Section 6.11) and carboxylic acid derivatives (Section 10.8A), by electrophilic substitution reactions of aromatic compounds (Section 12.3A), and by nucleophilic addition reactions of carbonyl compounds (Sections 9.6 and 15.7A). You may find it useful to review these reactions as they apply to the synthesis of amines.

Ammonia, as a nucleophile, reacts with alkyl halides to give amines. The reaction of methyl iodide with ammonia illustrates the course of this reaction.

$$CH_3 - \ddot{I}: \longrightarrow CH_3 - \overset{H}{\underset{H}{\overset{|}{\overset{+}{N}}}} - H \quad :\ddot{I}:^- \longrightarrow CH_3 - \ddot{N}H_2 + \overset{+}{N}H_4 \quad :\ddot{I}:^-$$

$$H - \underset{H}{\overset{\cdot\cdot}{N}} - H \qquad\qquad H - \underset{H}{\overset{\cdot\cdot}{N}} - H$$

methyl iodide and ammonia	methylammonium iodide and excess ammonia	methylamine	ammonium iodide

Ammonia displaces iodide ion to give methylammonium iodide. The reaction stops at this point if there is no excess ammonia present. Ammonia deprotonates methylammonium ion and frees methylamine, which is also a nucleophile and reacts further with methyl iodide. Dimethylamine and trimethylamine are also products of this reaction.

$$CH_3NH_2 \xrightarrow[NH_3]{CH_3I} CH_3NHCH_3 \xrightarrow[NH_3]{CH_3I} CH_3\overset{CH_3}{\overset{|}{N}}CH_3$$

methylamine	dimethylamine	trimethylamine

It is difficult to prepare pure primary amines by this method.

When this reaction is used for synthetic purposes, a large excess of ammonia is used, as illustrated in the preparation of the amino acid glycine from chloroacetic acid.

$$ClCH_2\overset{O}{\overset{||}{C}}OH + 2\,NH_3 \xrightarrow[25\,°C]{H_2O} H_3\overset{+}{N}CH_2\overset{O}{\overset{||}{C}}O^- + \overset{+}{N}H_4\,Cl^-$$

chloroacetic acid 1 mole	ammonia 60 moles	glycine 65%	ammonium chloride

The same precautions must be used when preparing a secondary amine to avoid getting a tertiary amine. For example, an excess of aniline is used in its reaction with benzyl chloride to prevent the formation of *N*,*N*-dibenzylaniline.

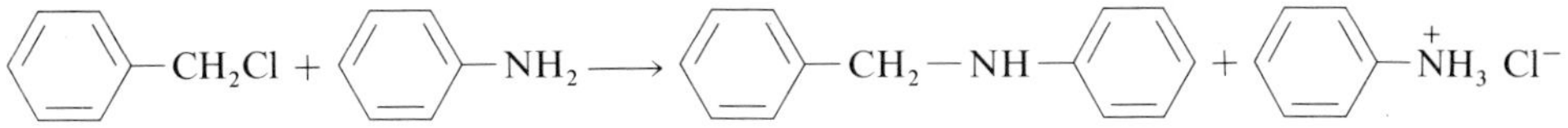

benzyl chloride   aniline   *N*-benzylaniline   aniline
1 mole   4 moles   85%   hydrochloride

Carbon-nitrogen bonds are also formed when nitro compounds are made. The nitrite anion can be used as a nucleophile to make aliphatic nitro compounds from primary and secondary alkyl halides. Bromocyclopentane is converted to nitrocyclopentane in this way.

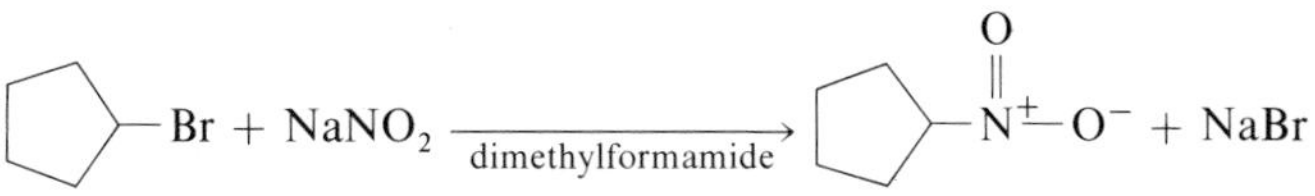

bromocyclopentane   sodium   nitrocyclopentane   sodium
    nitrite   57%   bromide

Aromatic nitro compounds are prepared by electrophilic aromatic substitution reactions, as illustrated for the preparation of 2,4,6-trimethylnitrobenzene.

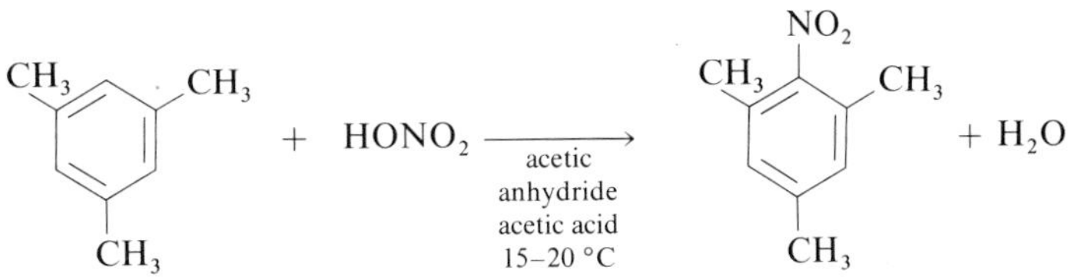

1,3,5-trimethylbenzene   nitric acid   2,4,6-trimethylnitrobenzene

Both aliphatic and aromatic nitro compounds are reduced to amines by a variety of reagents (Section 16.3B).

The reactions of amines or amine derivatives, such as hydroxylamine, with aldehydes and ketones create carbon-nitrogen double bonds. Benzaldehyde reacts with hydroxylamine to give an oxime and with aniline to give an imine (p. 453).

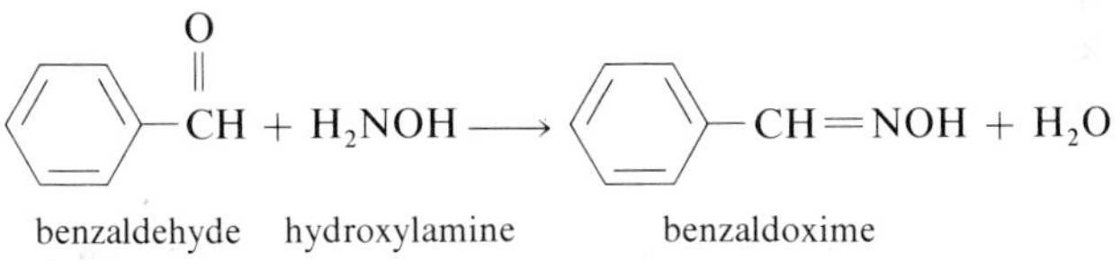

benzaldehyde   hydroxylamine   benzaldoxime

In Section 16.3D, the reduction of the carbon-nitrogen double bond to give amino compounds is explored in detail.

Finally, acid derivatives such as acid chlorides and acid anhydrides also react with amines to give amides (Section 10.8A), which can also be reduced to amines (Section 10.9). Thus benzoyl chloride reacts with 1,5-pentanediamine (cadaverine) in the presence of base to give the diamide.

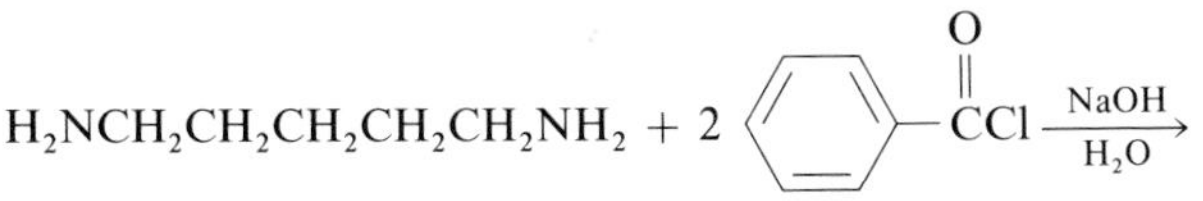

$H_2NCH_2CH_2CH_2CH_2CH_2NH_2$ + 2

1,5-pentanediamine   benzoyl chloride

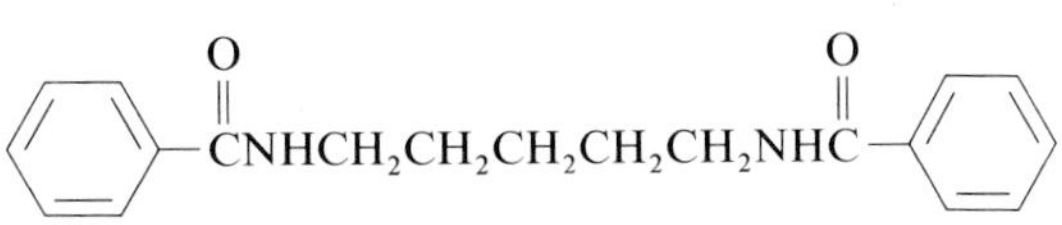

dibenzamide of
1,5-pentanediamine

The reaction of benzoyl chloride with amines under the Schotten-Baumann conditions (Section 10.4) was developed at the end of the last century as a method for isolating low molecular weight, water-soluble amines from biological samples.

The conditions necessary to convert some of these compounds containing nitrogen to amines and the useful applications of such syntheses of amino compounds are explored in the following sections.

PROBLEM 16.5 Write a Lewis structure for nitrite ion and a mechanism for the substitution reaction of cyclopentyl bromide with sodium nitrite.

PROBLEM 16.6 Supply the starting materials that would be necessary to make the following compounds. Assume that aniline, benzene, and any other organic compounds containing not more than three carbon atoms are available.

(a) a phenyl–C(CH$_3$)=NOH

(b) CH_3CHCO^- with $\overset{+}{N}H_3$

(c) phenyl–NHCH$_3$

(d) $CH_3CH_2CH=N$–phenyl

(e) $CH_3\overset{O}{\overset{\|}{C}}NHCH_2$–phenyl

B. Reduction of Nitro Compounds

Aromatic amines are most conveniently synthesized by nitration of an aromatic ring and reduction of the nitro group to an amino group. The reduction may be carried out using tin or iron and hydrochloric acid, in which case the amine is obtained as the salt. Catalytic reduction of a nitro group with hydrogen gas and a catalyst, usually nickel in some form, gives the free amine directly. The reduction of nitrobenzene to aniline is a classic example of the reduction of a nitro group with a metal and an acid.

nitrogen bonded to 2 oxygen atoms, oxidized state

$$2\ C_6H_5-NO_2 + 3\ Sn + 14\ HCl \xrightarrow{\Delta} 2\ C_6H_5-\overset{+}{N}H_3\ Cl^- + 3\ SnCl_4 + 4\ H_2O$$

nitrobenzene — aniline hydrochloride

nitrogen bonded to 2 hydrogen atoms, reduced state

$$C_6H_5-\overset{+}{N}H_3\ Cl^- + NaOH \longrightarrow C_6H_5-NH_2 + NaCl + H_2O$$

aniline hydrochloride — aniline

Tin(0) is oxidized to tin(IV) while the nitrogen atom is reduced to the amine. The amine is in the form of its hydrochloride salt at the end of the reduction, and must be treated with a strong base, such as sodium hydroxide, to be set free as aniline.

Catalytic reduction is part of a synthesis of *p*-aminobenzoic acid from *p*-nitrobenzoic acid.

$$\text{O}_2\text{N}-\underset{}{\bigcirc}-\overset{\displaystyle\text{O}}{\overset{\|}{\text{C}}}\text{OH} \xrightarrow[\substack{\text{Pt/C}\\\text{methanol}\\\Delta}]{\text{H}_2} \text{H}_2\text{N}-\underset{}{\bigcirc}-\overset{\displaystyle\text{O}}{\overset{\|}{\text{C}}}\text{OH}$$

p-nitrobenzoic acid   *p*-aminobenzoic acid

PROBLEM 16.7   How would you synthesize *p*-nitrobenzoic acid from toluene?

---

    p-Aminobenzoic acid is a compound with many interesting physiological properties. It is incorporated by bacteria into the vitamin folic acid, which is essential to bacterial growth. The ability of sulfa drugs to inhibit bacterial growth results from the structural resemblance between a sulfa drug, such as sulfanilamide, and *p*-aminobenzoic acid.

fragment derived from
p-aminobenzoic acid

$$\text{H}_2\text{N}\underset{\text{HO}}{\overset{\text{N}\quad\text{N}}{\bigcirc\hspace{-2mm}\bigcirc}}\text{N}-\text{CH}_2\text{NH}-\bigcirc-\overset{\displaystyle\text{O}}{\overset{\|}{\text{C}}}\text{NHCHCH}_2\text{CH}_2\overset{\displaystyle\text{O}}{\overset{\|}{\text{C}}}\text{OH}$$

folic acid

aromatic
amino
group

NH$_2$ NH$_2$

planar
aromatic
ring

ĊOH   O$=$S$=$O
‖
O   NH$_2$

acidic hydrogen
atom

    Both *p*-aminobenzoic acid and sulfanilamide have the weakly basic amino group on an aromatic ring para to a functional group with a central atom doubly bonded to oxygen and an acidic hydrogen atom bonded to an electronegative atom such as oxygen or nitrogen. The sites of basicity and acidity, and the spatial arrangement of these functional groups, are similar enough in the two compounds that the bacterial enzyme that synthesizes folic acid mistakes the sulfa drug for *p*-aminobenzoic acid. The sulfa drug takes the place of *p*-aminobenzoic acid at the catalytic surface of the enzyme, preventing the synthesis of the essential vitamin and disrupting metabolic processes in the bacteria. Because human beings do not synthesize folic acid, sulfa drugs do not disrupt human metabolism as they do bacterial metabolism.

    Esters of *p*-aminobenzoic acid are local anesthetics. Ethyl *p*-aminobenzoate is commonly known as Benzocaine, and 2-(*N*,*N*-dimethylamino)ethyl *p*-aminobenzoate is the local anesthetic procaine widely used in dentistry as its hydrochloride salt

$$\underset{\substack{\text{ethyl } p\text{-aminobenzoate} \\ \text{Benzocaine}}}{\overset{\overset{\displaystyle O}{\underset{\displaystyle \|}{}} \overset{\displaystyle \overset{\displaystyle \ddot{C}OCH_2CH_3}{\big|}}{}}{}}$$

ethyl *p*-aminobenzoate
Benzocaine

2-(*N*,*N*-dimethylamino)ethyl
p-aminobenzoate
procaine

Novocain. *p*-Aminobenzoic acid derivatives are also used widely in suntan lotions to absorb ultraviolet radiation and to screen the skin from the more harmful wavelengths of the light.

PROBLEM 16.8

(a) A laboratory synthesis of *p*-aminobenzoic acid starts with *p*-toluidine. The first step of the synthesis is protection of the amino group as its acetamide. Show how you would complete the synthesis.

(b) A careful adjustment of the acidity of the system is necessary to isolate *p*-aminobenzoic acid. Why is the pH of the system so important?

PROBLEM 16.9   Write the structural formula for the local anesthetic Novocain, procaine hydrochloride, $C_{11}H_{16}N_2O_2 \cdot HCl$. (Hint: Which is the most basic site in procaine, above)?

Dinitration of benzene to *m*-dinitrobenzene results in an aromatic ring with two nitrogen substituents meta to each other (Section 12.3A). It is possible to reduce one of the nitro groups selectively to an amine with ammonium sulfide.

benzene   *m*-dinitrobenzene   *m*-nitroaniline

$$+ \; 6\,NH_3 + 3\,S + 2\,H_2O$$

This reaction is useful because it allows one amino group to be transformed in a variety of ways, while retaining the option of reducing the nitro group to create a new amino group at a later stage of synthesis. In Section 16.6D, we will see examples of how such manipulations are employed.

Simple aliphatic nitro compounds can also be reduced in high yield with metal and acid or by catalytic hydrogenation to alkylamines. Butylamine is prepared by these two methods.

$$\underset{\text{1-nitrobutane}}{CH_3CH_2CH_2CH_2NO_2} + 2\,Fe + 7\,HCl \longrightarrow \underset{\text{butylamine hydrochloride}}{CH_3CH_2CH_2CH_2\overset{+}{N}H_3 \; Cl^-} + 2\,FeCl_3 + 2\,H_2O$$

$$\underset{\text{1-nitrobutane}}{CH_3CH_2CH_2CH_2NO_2} \xrightarrow[\substack{Ni \\ \text{high pressure}}]{H_2} \underset{\text{butylamine}}{CH_3CH_2CH_2CH_2NH_2} + 2\,H_2O$$

The reactions are entirely analogous to those seen for aromatic nitro compounds.

Nitroalkenes are easily prepared from the condensation of aldehydes with nitro-methane. This reaction is important in the synthesis of β-phenylethylamines, a class of amines of great importance in biological systems. *Adrenalin,* the hormone secreted by the adrenal glands to mobilize the body to energetic action when either danger or pleasure is anticipated, is a β-phenylethylamine as is *norepinephrine,* another amine involved in the transmission of nerve impulses, and *mescaline,* the hallucinogenic alkaloid of the peyote cactus.

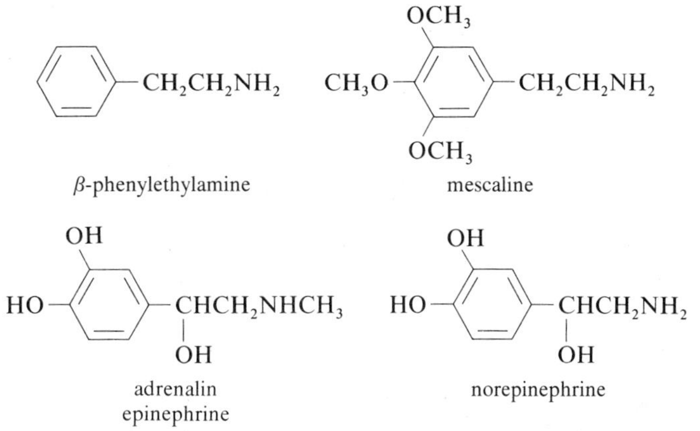

β-phenylethylamine   mescaline

adrenalin
epinephrine

norepinephrine

biologically important β-phenylethylamines

Scientists have devoted considerable study to the ways in which compounds such as adrenalin are generated in the body and how they are deactivated and removed from the body when the momentary need is over. A large number of compounds that are structurally similar to these hormones have been synthesized so that the relationships between structure and biological activity can be studied.

A synthesis of a biologically active β-phenylethylamine starts with the condensation of nitromethane with 3,4-methylenedioxybenzaldehyde. The reaction is catalyzed by base and involves the addition of the carbanion from nitromethane to the carbonyl group of the aldehyde (Section 15.5D). The intermediate alcohol loses water easily to give a double bond conjugated with the aromatic ring and with the nitro group.

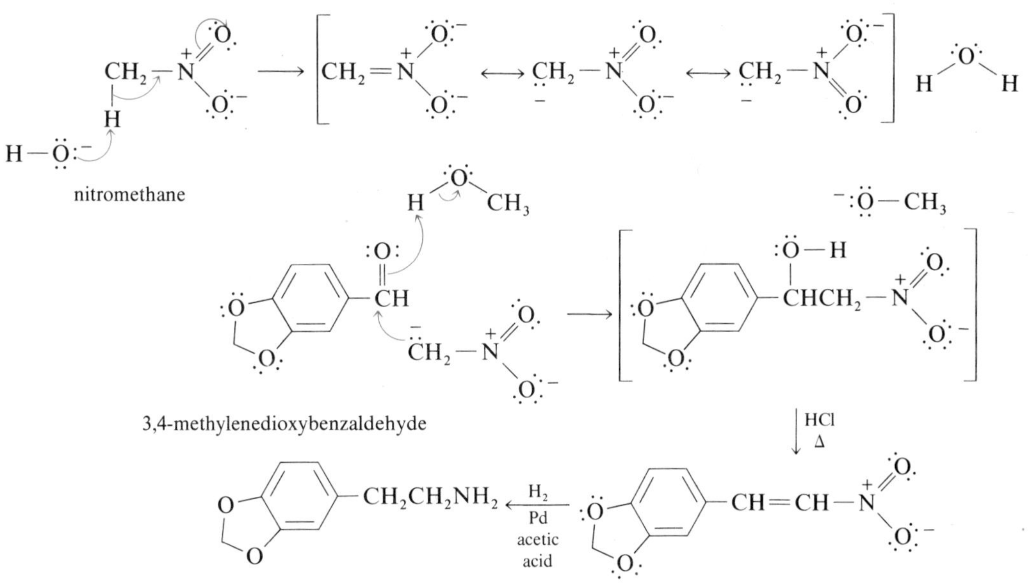

nitromethane

3,4-methylenedioxybenzaldehyde

β-(3,4-methylenedioxyphenyl)ethylamine   2-(3,4-methylenedioxyphenyl)-1-nitroethene

Catalytic hydrogenation of the double bond and the nitro group at the same time produces β-(3,4-methylenedioxyphenyl)ethylamine.

PROBLEM 16.10   Starting with benzene, toluene, and any other organic compound of three carbon atoms or less, synthesize the following amines using the reduction of a nitro compound at some stage of the synthesis.

(a) 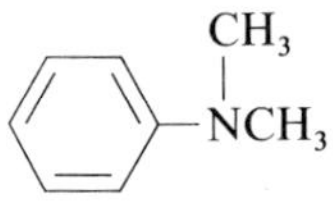 [structure: benzene ring with —N(CH₃)CH₃ substituent]   (b) 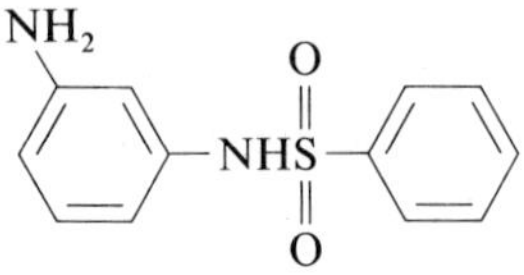 [structure: benzene ring with NH₂ substituent and —NHSO₂— connecting to another benzene ring]

(c) 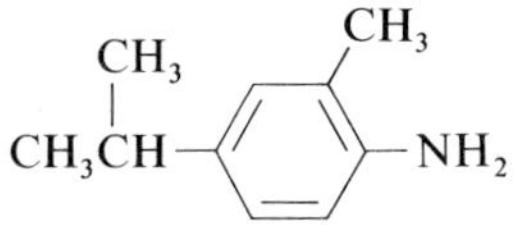 CH₃CH(CH₃)—[benzene ring with CH₃ and —NH₂ substituents]   (d)  O₂N—[benzene ring]—NH₂

(e) 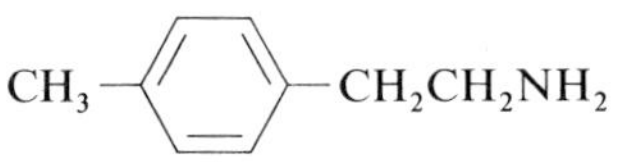 CH₃—[benzene ring]—CH₂CH₂NH₂   (f) 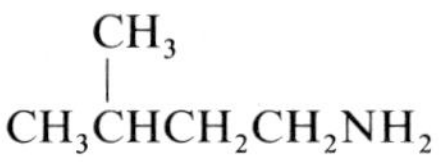 CH₃CHCH₂CH₂NH₂ with CH₃ branch

(g) 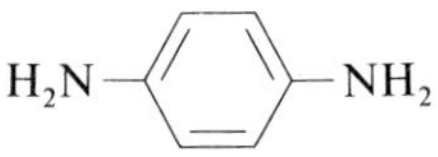 H₂N—[benzene ring]—NH₂

C. Gabriel-Type Syntheses of Primary Amines

Nucleophilic substitution reactions of ammonia with alkyl halides lead to mixtures of primary, secondary, and tertiary amines unless a large excess of ammonia is used (Section 16.3A). Pure primary amines are prepared more conveniently if the nitrogen atom is protected so that alkylation can take place only once. Such a protected nitrogen atom is present in phthalimide, which is acidic enough (pK_a 7.4) to be deprotonated easily to a nitrogen anion in the salt of phthalimide.

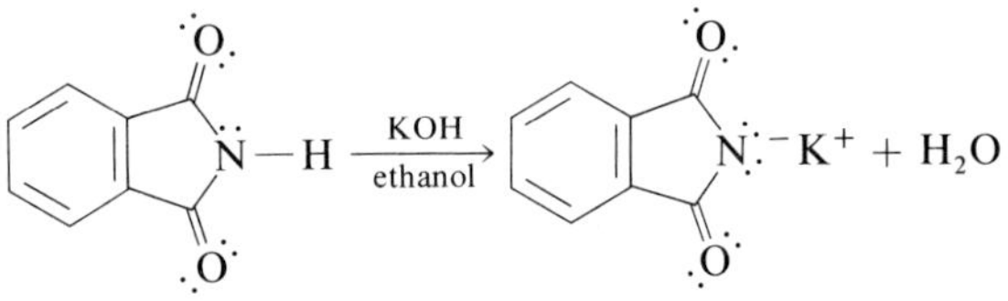

[structure: phthalimide with N—H, reacting with KOH/ethanol to give potassium phthalimidate N:⁻ K⁺ + H₂O]

phthalimide
pK_a 7.4

potassium phthalimidate
85%

Amines are normally weak acids. Ammonia has pK_a 36, for example. The acidity of phthalimide can be rationalized by the possibilities for delocalization of the charge on the nitrogen anion in its conjugate base.

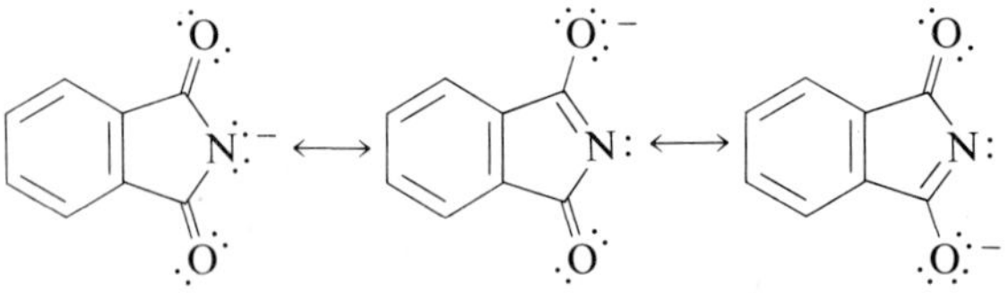

[structure: three resonance contributors for the anion of phthalimide]

resonance contributors for the anion of phthalimide

The phthalimidate anion is a good nucleophile that participates in a wide variety of nucleophilic substitution reactions. It can be used to prepare primary amines that are difficult to make in the pure state by means of the simple alkylation of ammonia. For example, benzyl chloride reacts with potassium phthalimidate to give *N*-benzylphthalimide, which is hydrolyzed in acid to the primary amine and phthalic acid.

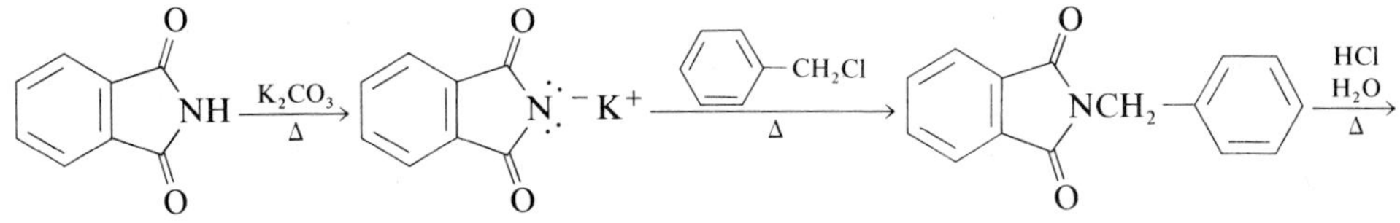

phthalimide   potassium phthalimidate   *N*-benzylphthalimide

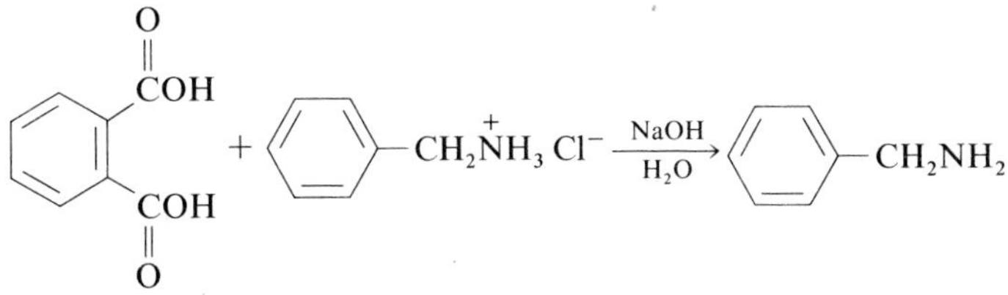

phthalic acid   benzylamine   benzylamine
                        hydrochloride

Such a sequence of reactions is called the **Gabriel phthalimide synthesis of primary amines.**

The phthalimide synthesis is most useful for amines containing other functional groups. 4-Aminobutanoic acid, which functions as a chemical agent in the transmission of nerve impulses, is synthesized by the alkylation of potassium phthalimidate with 4-chlorobutanenitrile.

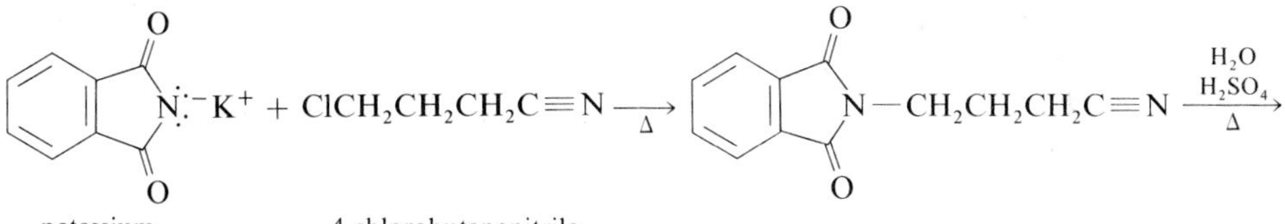

potassium   4-chlorobutanenitrile
phthalimidate

phthalic acid   4-aminobutanoic acid
                                                          γ-aminobutyric acid
                                                          GABA

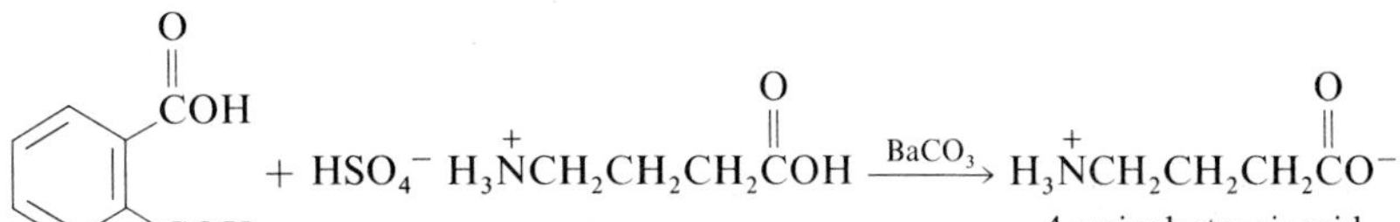

The carboxylic acid group of the amino acid is also protected as a nitrile until hydrolysis, the last step of the synthesis.

PROBLEM 16.11   4-Chlorobutanoic acid was not used as the reagent to alkylate potassium phthalimidate in the synthesis shown above. Why not?

PROBLEM 16.12   4-Chlorobutanenitrile is synthesized from 1,3-propanediol. How would you carry out the synthesis?

The phthalimide synthesis of the amino acid methionine brings together reactions met in other contexts and serves as a review of some important principles. Methionine, p. 963, is one of the amino acids essential for human beings. An **essential amino acid** is one that must be supplied by the diet because the organism does not synthesize it. In this synthesis, the potential amino group of the amino acid is protected as the nitrogen atom in phthalimide. The carboxylic acid group and the α-carbon atom of the amino acid are introduced by way of diethyl malonate (Section 15.4B). The first step of the synthesis is the conversion of diethyl malonate into diethyl bromomalonate by bromination of the enolate (Section 15.2A) so that it can undergo nucleophilic substitution.

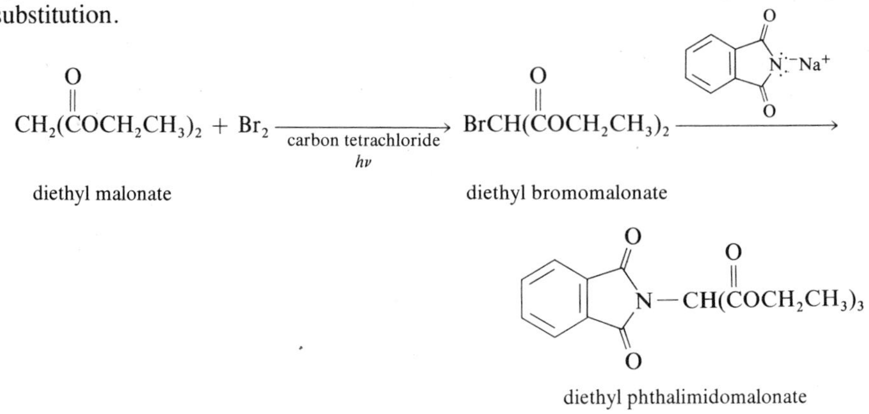

diethyl malonate   diethyl bromomalonate

diethyl phthalimidomalonate

The reaction of diethyl bromomalonate with sodium phthalimidate gives a compound that has an active hydrogen on the carbon atom between the two ester groups (Section 15.1B). Conversion of this compound to its enolate with sodium ethoxide and reaction of the enolate with 2-chloro-1-methylthioethane introduces the rest of the methionine molecule.

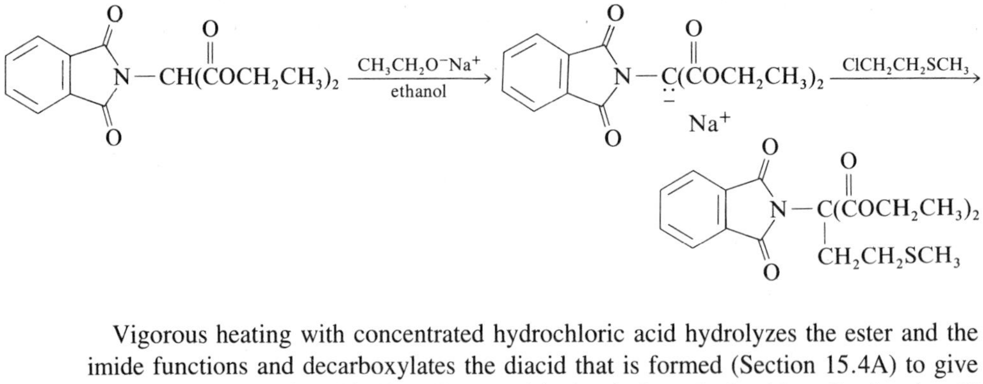

Vigorous heating with concentrated hydrochloric acid hydrolyzes the ester and the imide functions and decarboxylates the diacid that is formed (Section 15.4A) to give the hydrochloride of methionine. Free methionine is then obtained by adjusting the pH of the solution with pyridine.

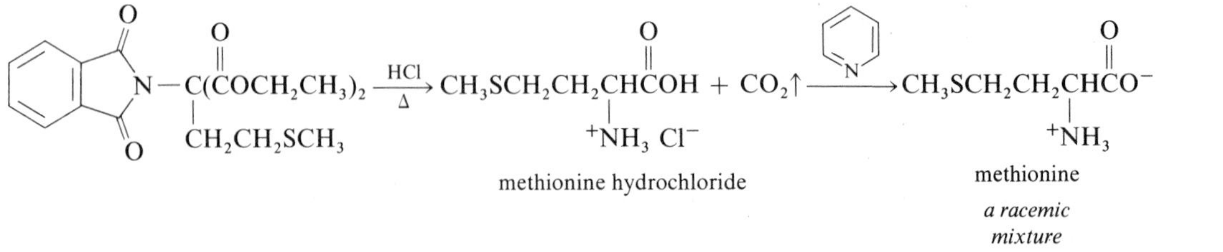

methionine hydrochloride   methionine

a racemic mixture

The methionine synthesized in this way is racemic. The synthesis starts with diethyl malonate, which is achiral, and no chirality is introduced at any point until the decarboxylation step, in which there is an equal probability that either carboxylic acid group will be lost.

PROBLEM 16.13 2-Chloro-1-methylthioethane can be synthesized from 2-mercaptoethanol, $HSCH_2CH_2OH$. Outline a synthesis for it, taking advantage of the different acidities of the thiol and the hydroxyl groups (Section 4.5) and the difference in nucleophilicity between sulfur and oxygen anions (Section 6.2).

PROBLEM 16.14 Use the phthalimide synthesis to synthesize phenylalanine and aspartic acid.

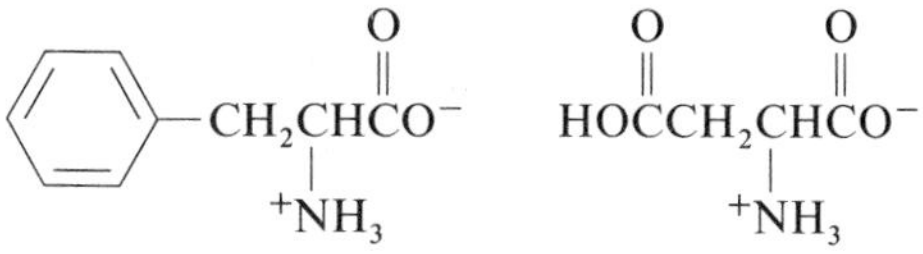

phenylalanine   aspartic acid

D. Reductions of Carbon-Nitrogen Multiple Bonds

Imines and oximes, with carbon-nitrogen double bonds, are formed when amines or hydroxylamines react with aldehydes and ketones (Sections 9.6A and 16.3A). Nitriles, with carbon-nitrogen triple bonds, are made easily in nucleophilic substitution reactions of primary, allyl, or benzyl halides with cyanide ion (Section 6.11). These compounds can all be reduced to amines.

The phenylimine of benzaldehyde, which is formed when benzaldehyde and aniline condense (p. 453), is reduced by sodium borohydride to a secondary amine. The imine is like a carbonyl group in its polarization and undergoes hydride reduction in the same way.

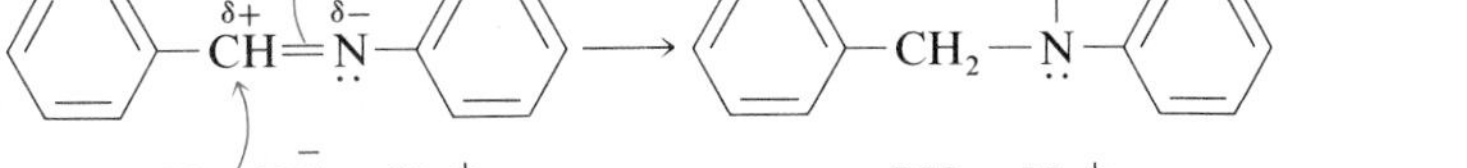

phenylimine of benzaldehyde   N-benzylaniline
97%

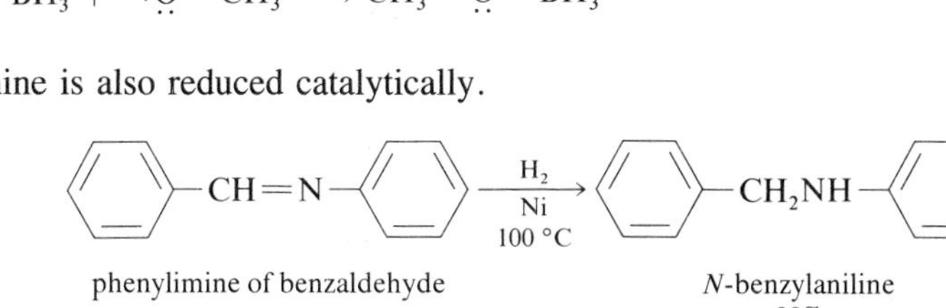

$$BH_3 + {}^-\!:\!\ddot{O}\!-\!CH_3 \longrightarrow CH_3\!-\!\ddot{O}\!-\!\bar{B}H_3$$

The imine is also reduced catalytically.

phenylimine of benzaldehyde   N-benzylaniline
99%

In fact, it is not necessary to isolate an imine in order to prepare the corresponding amine. The imines generated from aldehydes and ketones with ammonia are highly unstable. In a process known as **reductive amination,** it is possible to form the imine and reduce it in the same step by mixing the carbonyl compound and ammonia in the presence of hydrogen gas and a catalyst. Thus acetophenone is converted into α-phenylethylamine.

$$\text{acetophenone} + NH_3 \longrightarrow \left[\text{imine} \right] \xrightarrow[\text{Ni}]{H_2} \text{α-phenylethylamine} + H_2O$$

acetophenone   imine α-phenylethylamine
                    postulated as intermediate   64%

Oximes, as well as imines, are reduced by hydrogen to amines. The oxime of pentanal, for example, is converted to pentylamine in this way.

$$CH_3CH_2CH_2CH_2CH=NOH \xrightarrow[\substack{Ni \\ 100\,°C}]{H_2} CH_3CH_2CH_2CH_2CH_2NH_2 + H_2O$$

oxime of pentanal   pentylamine
                                        62%

Another reagent for reducing oximes is sodium in ethanol. Heptylamine is prepared by this method.

$$CH_3(CH_2)_5CH=NOH \xrightarrow[\text{ethanol}]{Na} CH_3(CH_2)_5CH_2NH_2$$

oxime of heptanal   heptylamine
                                        ~65%

Nitriles require the more powerful hydride reagent, lithium aluminum hydride, for their reduction (Section 10.9). For example, octanenitrile is reduced in this way to octylamine.

$$CH_3(CH_2)_6C\equiv N \xrightarrow[\text{diethyl ether}]{LiAlH_4} \xrightarrow{H_2O} CH_3(CH_2)_6CH_2NH_2$$

octanenitrile   octylamine
                                        90%

Catalytic reduction of nitriles is also possible. Thus, butyl bromide may be converted to a primary amine containing one more carbon atom by a sequence of nucleophilic substitution and catalytic reduction reactions.

$$CH_3CH_2CH_2CH_2Br + K^+C\equiv N^- \longrightarrow CH_3CH_2CH_2CH_2C\equiv N + K^+Br^- \xrightarrow[\text{Ni}]{H_2}$$

butyl bromide   pentanenitrile

$$CH_3CH_2CH_2CH_2CH_2NH_2$$
                                        pentylamine
                                        90%

Amides, formed from the reaction of ammonia or amines with acid derivatives (Section 10.8A), are also reduced to amines by lithium aluminum hydride. The kind of amine formed depends on the structure of the amide undergoing reduction. The details of these reactions were explored in Section 10.9, so the reduction of α-phenoxyacetamide serves as a reminder of this route to amines.

$$\text{α-phenoxyacetamide} \xrightarrow[\text{diethyl ether}]{LiAlH_4} \xrightarrow{H_2O} \text{β-phenoxyethylamine}$$

α-phenoxyacetamide β-phenoxyethylamine
                                        80%

PROBLEM 16.15 Show how the following transformations could be carried out. Several steps may be required.

(a) 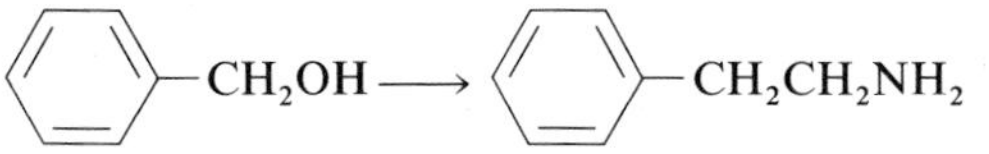 ⟨benzene ring⟩—CH$_2$OH $\longrightarrow$ ⟨benzene ring⟩—CH$_2$CH$_2$NH$_2$

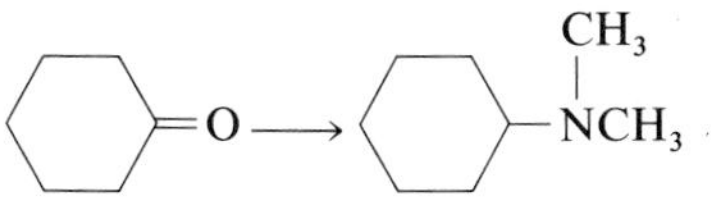

(b) ⟨cyclohexane⟩=O $\longrightarrow$ ⟨cyclohexane⟩—N(CH$_3$)CH$_3$

(c) CH$_3$O—⟨benzene ring⟩ $\longrightarrow$ CH$_3$O—⟨benzene ring⟩—NH$_2$

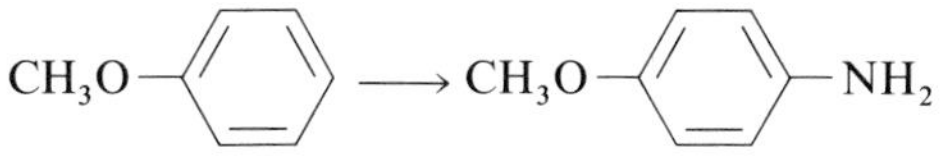

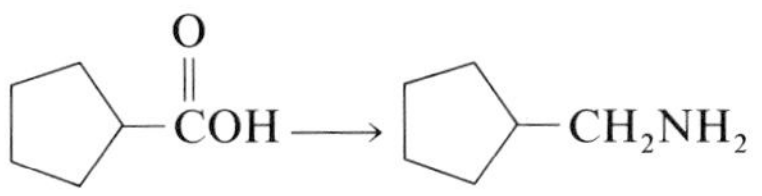

(d) CH$_3$CH$_2$CH$_2$CH$_2$OH $\longrightarrow$ CH$_3$CH$_2$CH$_2$CHCH$_2$CH$_3$ with NH$_2$ substituent

(e) ⟨cyclopentane⟩—C(=O)OH $\longrightarrow$ ⟨cyclopentane⟩—CH$_2$NH$_2$

(f) CH$_3$CH$_2$CH$_2$CH(=O) $\longrightarrow$ ⟨benzene ring⟩—N(CH$_2$CH$_2$CH$_2$CH$_3$)CH$_2$CH$_2$CH$_2$CH$_3$

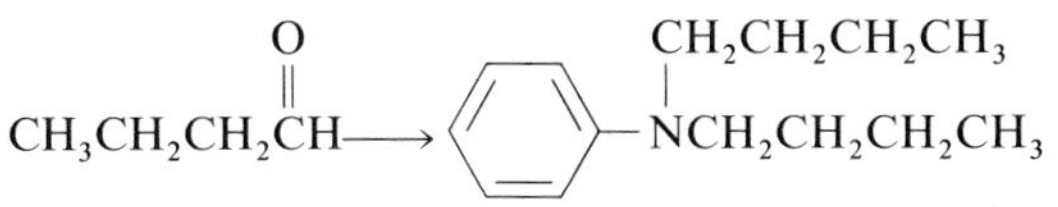

(g) CH$_3$CH$_2$CH$_2$CH$_2$Br $\longrightarrow$ CH$_3$(CH$_2$)$_6$CH$_2$NH$_2$ (three different syntheses required)

(h) 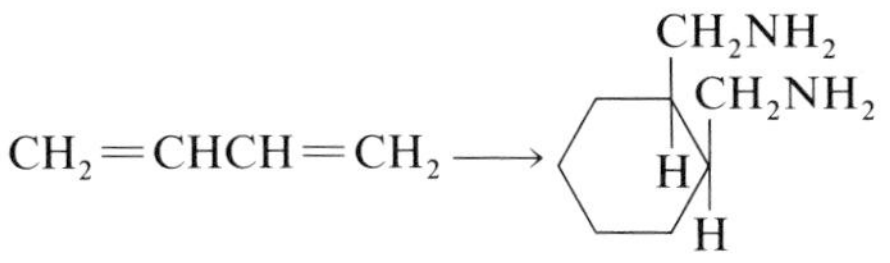 CH$_2$=CHCH=CH$_2$ $\longrightarrow$ ⟨cyclohexane with CH$_2$NH$_2$, CH$_2$NH$_2$, H, H substituents⟩

16.4

Separations of Mixtures of Organic Compounds

A. Isolation of Pure Compounds from Reaction Mixtures. A Brief Introduction

The equations for the reactions presented in this book indicate, whenever good experimental evidence is available, how much of the desired product is formed. You have probably noticed how seldom it is that an organic reaction gives a quantitative amount, 100% yield, of the compound being synthesized. In most cases, a mixture is present at the end of the reaction. The mixture may contain unreacted starting material, especially if one of the reagents was used in excess, and products of side reactions. In the laboratory, the reaction mixture often has to be subjected to what is called a *work-up* before the product can be isolated.

Some separations of mixtures are now done by **chromatographic methods.** A mixture is put through a column from which the different components emerge at different rates, depending upon their structures and hence their polarities. The columns are made up in a variety of ways, but they always involve one phase that remains in place, a **stationary phase,** and another one that flows past it, the **mobile phase.** The components of the mixture are held back by the stationary phase or moved forward by the mobile phase in different ways depending upon the structure of each compound and the composition of the two phases. Ultimately, it is possible to separate even compounds with relatively similar structures by choosing the right conditions. The technique is enormously powerful, and the experimental details are highly specialized for each area of interest.

In most cases, however, a cruder separation based on the acid-base properties of the components is made first. Sometimes this, along with other bulk separation procedures such as distillation of a liquid product or recrystallization of a solid, is sufficient to isolate the desired product. If not, then chromatography is used on the part of the mixture that is still unseparated.

The reaction of benzoyl chloride with aniline to give benzanilide is a relatively simple one. Let us examine it closely to see all the species that may be present in the reaction mixture at the end of the reaction.

benzoyl chloride
bp 197 °C
water insoluble

aniline
bp 184 °C
water insoluble

benzanilide
mp 163 °C
water insoluble

sodium benzoate
water soluble

aniline
water insoluble

sodium chloride
water soluble

For the purposes of the separation, we have to be conscious of the physical properties of the reagents and the products. Are they liquids or solids? Are they water soluble or insoluble? If they are liquids, are they relatively low boiling, that is, below 80°C and, therefore, easily removed by evaporation?

When the reaction of benzoyl chloride with aniline is over, a solid product, benzanilide, is present in the reaction mixture. It is contaminated with some unreacted aniline. The water in which the reaction was run has sodium benzoate dissolved in it (from the reaction of benzoyl chloride with sodium hydroxide, an unavoidable side-reaction), along with sodium chloride (from the neutralization of hydrochloric acid by sodium hydroxide).

The separation of this mixture is straightforward. A filtration will separate the solid benzanilide from the water-soluble salts. The solid is washed with dilute hydrochloric acid to convert the aniline clinging to its surface into a water-soluble salt and wash it away from the benzanilide. An amide is *not* basic and does not react with the cold dilute acid to a significant extent.

benzanilide
neutral
water insoluble
acid insoluble

aniline
basic
water insoluble
soluble in dilute acid

benzanilide
unchanged

aniline
hydrochloride
ionic compound
water soluble

If the benzanilide is then rinsed with water, the product is essentially pure but wet. It can be dried and used this way, or recrystallized from some organic solvent such as ethanol, if a very pure product is desired.

B. The Separation of Enantiomers. Resolution

The separation of mixtures according to acidity or basicity works with compounds that belong to different functional group classes and have different properties. Chromatographic techniques will separate not only compounds belonging to different functional group classes, but members of the same functional group class with differing molecular structures. Diastereomers, such as cis and trans isomers of alkenes or cyclic compounds, can also be separated chromatographically.

What about compounds that differ from each other only in the configuration, *R* or *S*, of asymmetric carbon atoms? Compounds that have only one asymmetric carbon atom exist as two mirror-image isomers, enantiomers. Enantiomers have identical physical properties except for the direction in which they rotate the plane of polarized light (Sections 3.9A and 3.9D). Enantiomers are indistinguishable in the polarity of their molecules and have identical solubilities in ordinary solvents, so they cannot be separated by recrystallization or chromatography in the usual way. They have identical boiling points, so distillation does not separate them.

When chiral compounds containing a single asymmetric carbon atom are prepared in the laboratory, starting with achiral reagents, both enantiomers are formed in equal amounts, and the product is a racemic mixture (Section 3.9D). We have seen many examples of such reactions, including the reduction of ketones with metal hydride reagents (Section 9.3) and addition of organometallic reagents and hydrogen cyanide to carbonyl compounds (Sections 9.4B and 9.5, Problem 9.15). Once a racemic mixture is formed, the two enantiomers in the mixture are inseparable by the techniques normally used to separate structural isomers and diastereomers.

As discussed earlier (Section 5.8B), compounds that have two different asymmetric carbon atoms have four stereoisomers, or two sets of enantiomers. The members of one pair of enantiomers are diastereomers of the compounds in the other set of enantiomers and have physical properties sufficiently different from them that in principle, at least, they can be separated. These stereochemical relationships were explored thoroughly with examples from the addition of bromine to alkenes (Section 5.8B) and from oxidation reactions of alkenes with potassium permanganate (Section 8.4) and with peroxyacids (Sections 8.4D and 8.5). Diastereomers, whether cis-trans isomers or arising from differences in configuration at asymmetric carbon atoms, have different physical properties and can be separated by ordinary techniques such as recrystallization, distillation, and chromatography.

The separation of a mixture of enantiomers is called the **resolution of a racemic mixture.** Louis Pasteur performed a very unusual resolution using a pair of tweezers on the crystals of sodium ammonium tartrate and picking out left-handed and right-handed crystals under a microscope (Section 3.9E). Living organisms, with their enzyme systems that are highly sensitive to the stereochemistry of the compounds with which they interact, perform resolutions all the time when they metabolize one enantiomer and reject another one. Sometimes, this process can be put to good use. For example, if a rabbit is fed racemic 2-methylcyclohexanone, it excretes only one metabolite, $(1S,2S)$-$(+)$-2-methylcyclohexanol, a reduction product of the ketone.

Other biological resolutions are performed by molds and bacteria. Yeast cells, if allowed to ferment in a medium containing sugar and a racemic amino acid, consume the enantiomer of the amino acid that is found naturally in proteins and leave the other.

Most vertebrates, including human beings, also separate enantiomers biologically. For example, if racemic alanine (p. 963) is fed to a human being, the (*S*)-isomer is metabolized and the (*R*)-isomer is excreted in the urine. Researchers have observed that in most biological resolutions, only one enantiomer, the one that cannot be used by the living organism, is recovered.

A more useful way to separate enantiomers from each other is to convert them into compounds that are diastereomers, which then have different physical properties and, as a result, are separable. Reactions of enantiomers with achiral reagents cause no change in their relationship as mirror-image isomers of each other. Only when the reagent itself is chiral is new stereochemistry introduced into the molecule and with it, the possibility of a physical separation. The most widely used technique for the resolution of mixtures of enantiomers depends on the reaction of a chiral acid with a racemic base (or a chiral base with a racemic acid) to give a mixture of diastereomeric salts. The salts are separated by recrystallization, and the individual enantiomers, as well as the chiral reagent, are recovered by further acid-base reactions. These ideas will become clearer as we work our way through an actual resolution.

In Section 8.5, we saw how the opening of an oxirane ring could give rise to stereoisomeric compounds. The oxirane from (*Z*)-2-butene reacts slowly with ammonia at room temperature to give a racemic mixture of 3-amino-2-butanols.

cis-2,3-dimethyloxirane

a meso compound

(2*S*,3*S*)-(+)-3-amino-2-butanol 50%

(2*R*,3*R*)-(−)-3-amino-2-butanol 50%

racemic mixture
$[\alpha]_D^{25} = 0.00°$

The products are drawn using sawhorse formulas in a way that indicates the origins of the two isomers. They arise from the equal probability that ammonia will attack either one of the carbon atoms of the oxirane ring. As ammonia attacks, a carbon-oxygen bond breaks, ultimately giving a hydroxyl group in an anti orientation to the amino group.

The product as it is isolated from the reaction mixture has no optical activity. If the racemic mixture is treated with acetic anhydride in pyridine, both the amino group and the hydroxyl group react.

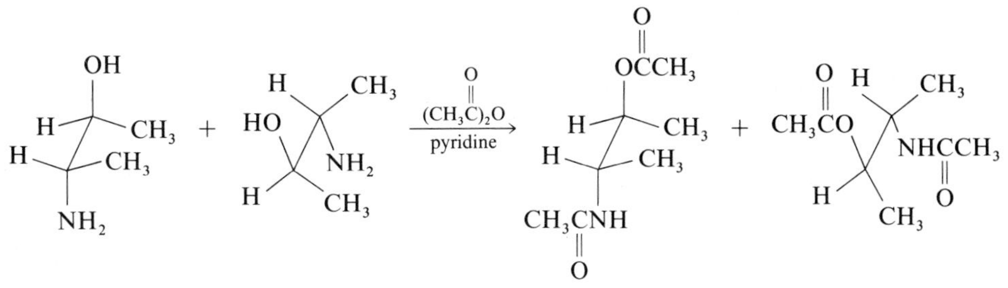

racemic mixture of
3-amino-2-butanols
$[\alpha]_D^{25} = 0.00°$

racemic mixture of the
diacetyl derivatives of
3-amino-2-butanols
$[\alpha]_D^{25} = 0.00°$

Acetic anhydride is not a chiral reagent. The formation of acetyl derivatives at the amino and hydroxyl groups in the 3-amino-2-butanols does not change the stereochemical relationship of the two amino alcohols to each other. It is clear that their diacetyl derivatives are also mirror-image isomers of each other. One racemic mixture is converted to another one, and still no optical activity is observed.

If, however, (+)-tartaric acid, a naturally occurring chiral acid that has played a very important role in the study of stereochemistry (Section 14.2F), is used as the reagent, the picture changes.

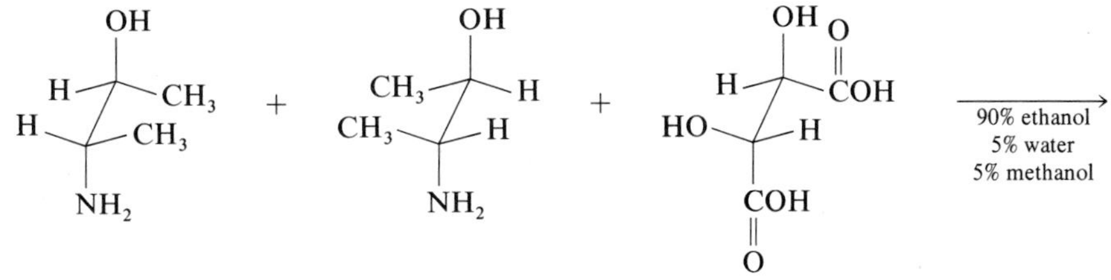

(2S,3S)-(+)-3-amino-
2-butanol

(2R,3R)-(−)-3-amino-
2-butanol

(2R,3R)-2,3-dihydroxybutanedioic
acid
(+)-tartaric acid
1 mole

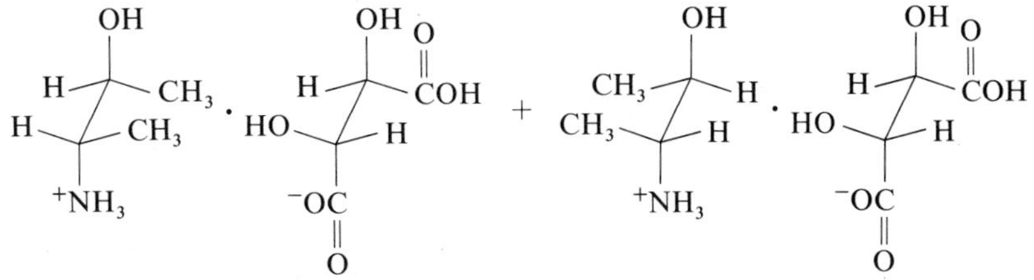

90% ethanol
5% water
5% methanol

1 mole

salt of (2S,3S)-(+)-amine
with (2R,3R)-(+)-acid

*more soluble in the
reaction mixture,
remains in solution*

salt of (2R,3R)-(−)-amine
with (2R,3R)-(+)-acid

*less soluble in the
reaction mixture,
crystallizes out*

diastereomers of each other

A proton is transferred from one of the carboxylic acid groups of (+)-tartaric acid to the amino group of 3-amino-2-butanol. Two salts are formed, one from each enantiomer of the amine. The new compounds are now diastereomers of each other. If you examine closely the structural formulas written for the salts, you will see that the two compounds are *not* mirror-image isomers of each other. The amine portions of the molecules remain enantiomeric, but attached to each one is the ion of an acid that is identical in both cases. If all four asymmetric carbon atoms in each salt are taken into account, they do not have the opposite configuration at each point. One salt could be called the (S,S,R,R)-salt and the other one the (R,R,R,R)-salt. The two compounds are thus stereoisomers, but not enantiomers. By definition (Section 5.1A), all stereo-isomers that are not enantiomers are diastereomers of each other.

The diastereomeric salts have different physical properties. The salt derived from the $(2R,3R)$-amine is less soluble in the reaction mixture than the other one and crystallizes out of solution. Though in practice in the laboratory, several further recrystallization steps are necessary for a complete separation of the diastereomers, in principle the separation is made as soon as two diastereomers with different solubilities in the reaction mixture are formed.

Once the diastereomeric salts are separated, the amino alcohols, now as the individual enantiomers, can be recovered by treatment with aqueous sodium hydroxide.

salt of $(2S,3S)$-$(+)$-amine with $(2R,3R)$-$(+)$-acid $\xrightarrow[\text{H}_2\text{O}]{\text{NaOH}}$ $(2S,3S)$-$(+)$-amine $[\alpha]_D^{25}\ +16.91°$ *covalent compound recovered by distillation* + disodium $(+)$-tartrate *ionic compound does not distill*

salt of $(2R,3R)$-$(-)$-amine with $(2R,3R)$-$(+)$-acid $\xrightarrow[\text{H}_2\text{O}]{\text{NaOH}}$ $(2R,3R)$-$(-)$-amine $[\alpha]_D^{25}\ -17.05°$ *covalent compound recovered by distillation* + disodium $(+)$-tartrate *ionic compound does not distill*

The strong base, hydroxide ion, removes a proton from the organic ammonium ion and regenerates the free amine, which is separated from the water-soluble, nonvolatile, disodium salt of tartaric acid by distillation. Each enantiomeric amine is recovered separately and has optical activity, the specific rotation of the $(2S,3S)$-isomer being $+16.91°$, and that of the $(2R,3R)$- isomer $-17.05°$. Within the limits of experimental error, arising chiefly from the possibility that the separation was not absolutely complete, the two enantiomers are seen to have optical rotations that are equal in magnitude but opposite in sign.

The process described above is the resolution of a racemic mixture. The chiral reagent used, $(+)$-tartaric acid in this case, *is called the* **resolving agent.** A large number of resolving agents are available, many of them acids and bases from natural sources, as well as new ones created by modifications of naturally chiral compounds. Some synthetic compounds that are easily resolved are then used as resolving agents themselves.

Resolutions of racemic mixtures are not easy procedures. They require a great deal of patience and skill, and some good luck, too. In many cases, they do depend on taking advantage of the properties of organic compounds as acids and bases. *The most important step in any resolution, however, is the conversion of the mixture of enantiomers into a mixture of diastereomers with physical properties that allow for their separation.*

The formation of diastereomers need not result in isolable compounds in different flasks. For example, if a chromatography column were to contain a chiral adsorbent, one enantiomer would interact more strongly than the other with the material in the column and a separation would take place. Such methods are already being used, and will certainly be used more and more in the future as new chiral adsorbents and complexing agents are developed.

PROBLEM 16.16   Write structural formulas showing the details for the resolution of a racemic mixture of α-phenylethylamine with ($+$)-tartaric acid. The separation takes place in methanol as the solvent. The salt of the (S)-($-$)-amine with ($+$)-acid is the one with the lower solubility.

PROBLEM 16.17   Racemic 2-hydroxy-2-phenylacetic acid, prepared by the hydrolysis of the cyanohydrin of benzaldehyde (p. 462) is resolved with (S)-($-$)-α-phenylethylamine. The (S,S)-salt is less soluble in water than the (R,S)-salt. Outline a resolution of the acid.

PROBLEM 16.18   What would have happened if racemic 2-hydroxy-2-phenylacetic acid had been treated with racemic α-phenylethylamine instead of one of its enantiomers? Explore the stereochemical consequences of such a reaction. How many separable fractions would result? Would they have optical activity?

16.5

Rearrangements to Nitrogen Atoms

A. The Beckmann Rearrangement

Oximes, when treated with a strong protic acid or a Lewis acid, are converted to amides in a reaction known as the **Beckmann rearrangement.** *For example, acet-*anilide is formed when the oxime of acetophenone is treated with trifluoroacetic acid.

acetophenone oxime   acetanilide
                                               91%

An inspection of the equation suggests that the phenyl group migrates from a carbon atom to a nitrogen atom during the reaction. Such a migration is reminiscent of the 1,2-shifts of alkyl groups to electron-deficient carbon atoms in rearrangements of carbocations (Section 5.6B).

The mechanism of the Beckmann rearrangement is shown below.

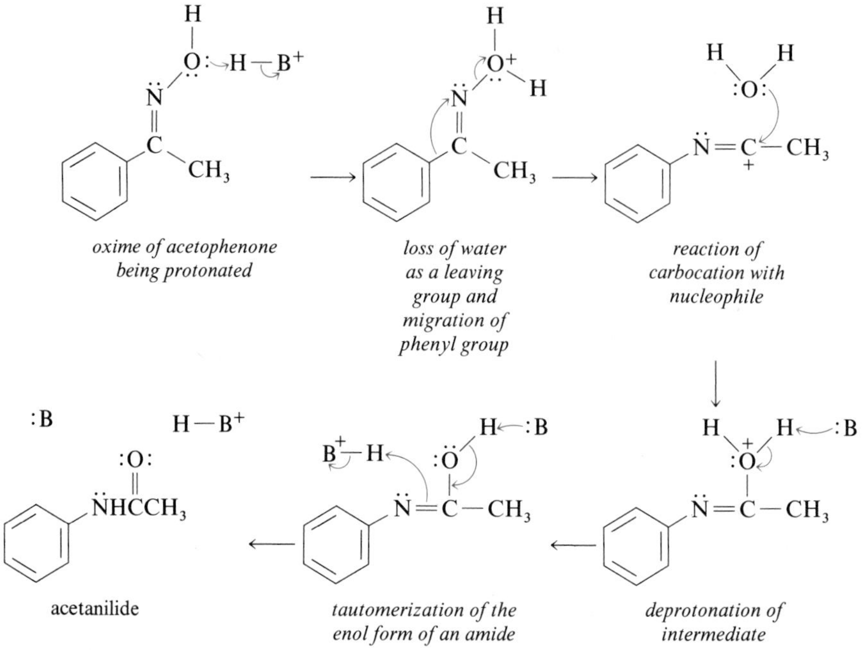

oxime of acetophenone being protonated	*loss of water as a leaving group and migration of phenyl group*	*reaction of carbocation with nucleophile*

acetanilide	*tautomerization of the enol form of an amide*	*deprotonation of intermediate*

The first step of the reaction is a familiar one. The hydroxyl group of the oxime is converted into a good leaving group. As the leaving group moves away, the substituent that can approach the nitrogen atom from the rear and assist in the departure of the leaving group migrates to the nitrogen atom, leaving behind a carbocation. The carbocation reacts with water to give an amide. Hydrolysis of the amide would give an amine derived from one portion of the oxime molecule and a carboxylic acid derived from the other portion.

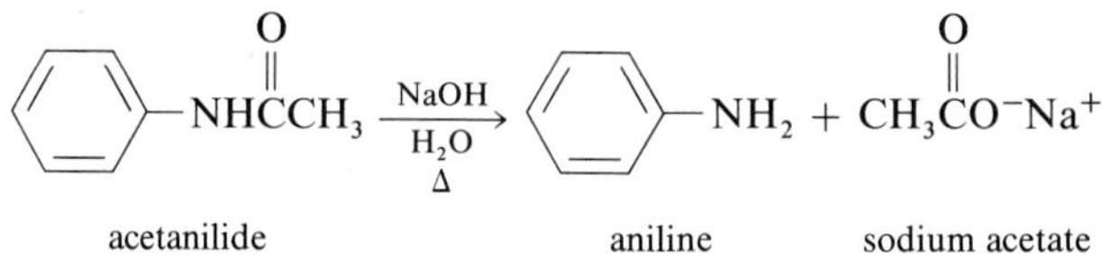

acetanilide	aniline	sodium acetate

In a reaction of great commercial importance, the amide that is formed is of interest in its own right. ε-Caprolactam, the starting material for a polymer, *nylon 6,* which is used extensively in tire cords, is made from cyclohexanone oxime.

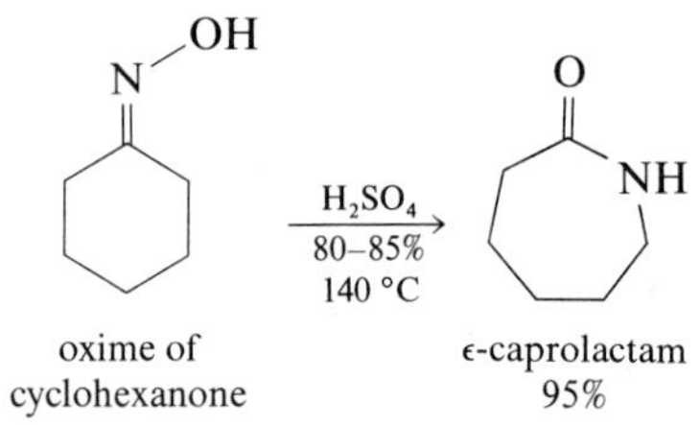

oxime of cyclohexanone	ε-caprolactam 95%

The Beckmann rearrangement of the oxime of a cyclic ketone gives rise to a **lactam,** a cyclic amide, with one more atom in the ring than the original ketone. ϵ-Caprolactam is polymerized (Section 20.2B) to give a linear polyamide.

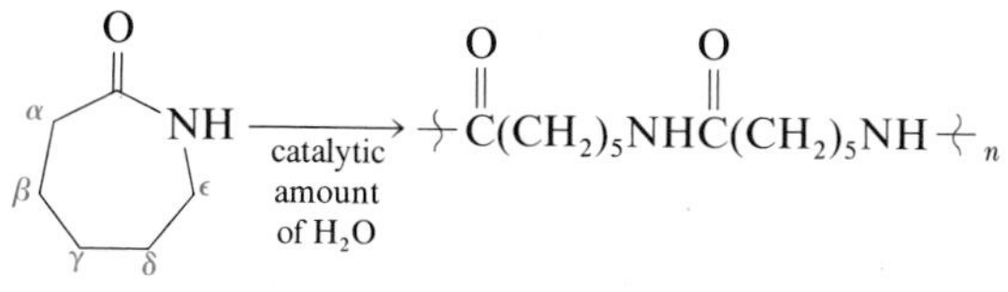

ϵ-caprolactam   nylon 6

Many analogs of cholesterol have been synthesized for animal experiments designed to find out more about the genesis and metabolism of cholesterol in the body. Another example of an application of the Beckmann rearrangement is such a synthesis of a compound related to cholesterol but having a nitrogen atom in the D-ring (Section 11.6C).

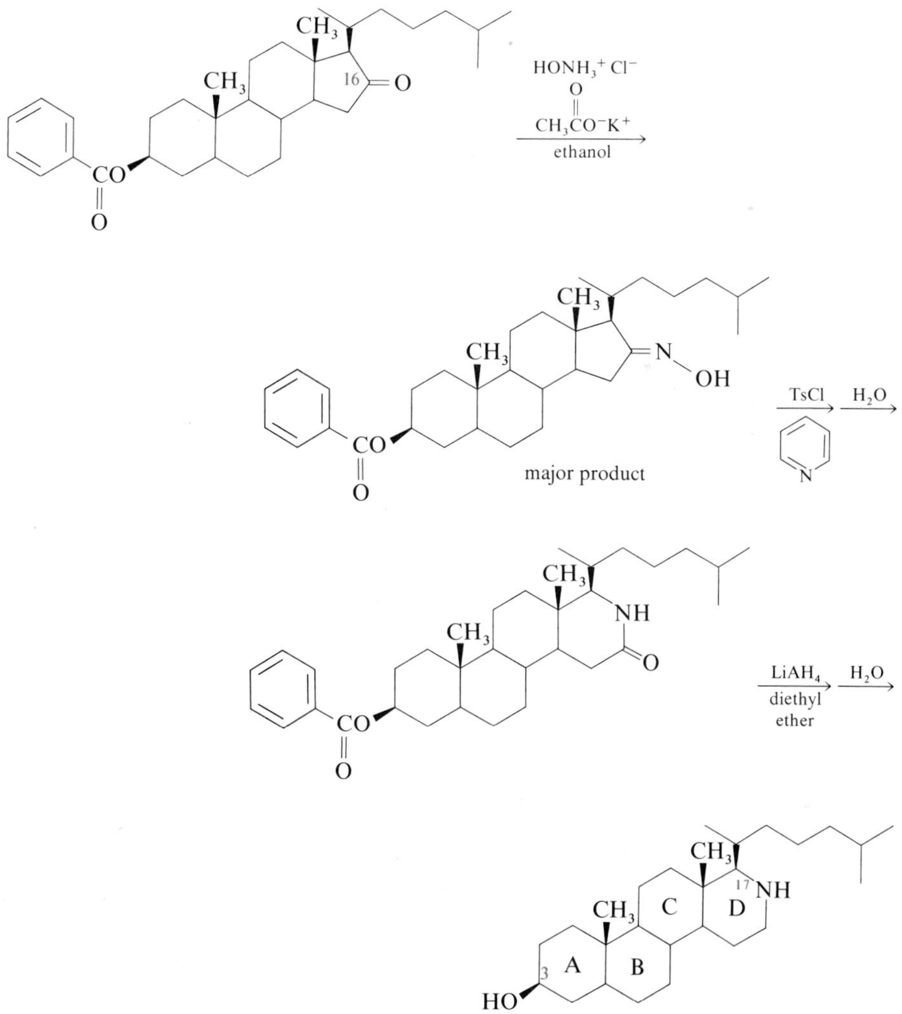

major product

17-aza-D-homocholestanol

Several features of the sequence of reactions shown above are interesting. First, the 16-oxo steroid gives two oximes that are stereoisomers. The one shown has the hydroxyl group pointing away from the bulky side-chain. The stereochemistry of the oxime determines the structure of the six-membered ring lactam formed in the next step of the reaction. The reagent for the rearrangement is tosyl chloride. The oxime is converted to its tosylate, which has a good leaving group, the tosylate anion. The side of the five-membered ring that is away from the departing tosylate anion migrates. Finally, lithium aluminum hydride reduces both the cyclic amide and the benzoic ester that protects the 3-hydroxyl group during the reaction. The name of the final product, 17-aza-D-homocholestanol, indicates that a nitrogen atom is at position 17 in a molecule that is like cholestanol (cholesterol in which the double bond in ring B has been hydrogenated) in all respects except that ring D is homologous (Section 3.1) to ring D in cholestanol (that is, has one more atom in that ring than cholestanol does).

PROBLEM 16.19

(a) Show the formation of the tosylate from the oxime and the mechanism for the rearrangement of the tosylate in the second step of the reaction sequence above.
(b) Show the structure of the minor oxime that would be formed in the first step of the reaction sequence.
(c) Show the structure of the lactam and the aza steroid that would be formed from the Beckmann rearrangement of the minor oxime [from part (b) of this problem].
(d) Why did the 3-hydroxyl group have to be protected as the benzoic ester before the reactions shown above could be carried out on the steroid?

B. The Hofmann Rearrangement

In the **Hofmann rearrangement,** *a primary amide is transformed into a primary amine containing one less carbon atom.* The preparation of 3,4-dimethoxyaniline from 3,4-dimethoxybenzamide illustrates this reaction.

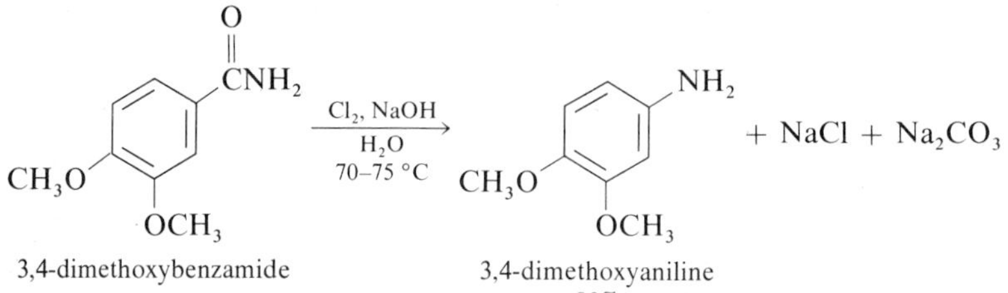

Like the Beckmann rearrangement, the Hofmann rearrangement involves the migration of an organic group from a carbon atom to a nitrogen atom. The leaving group during the migration is halide ion. The mechanism is shown below.

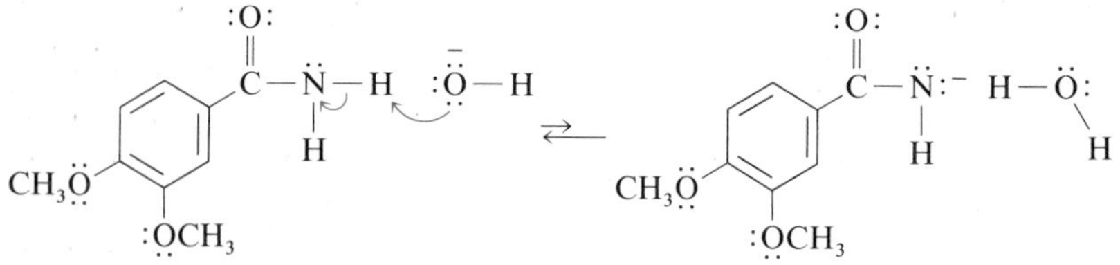

deprotonation of the amide

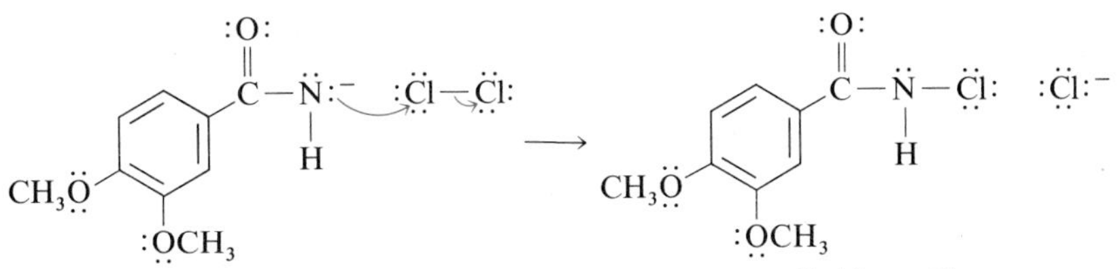

reaction of anion of the amide
with the electrophilic halogen

an N-chloroamide

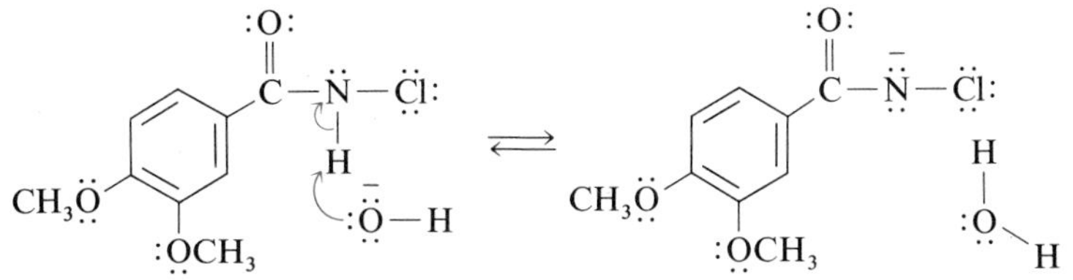

deprotonation of the N-chloroamide

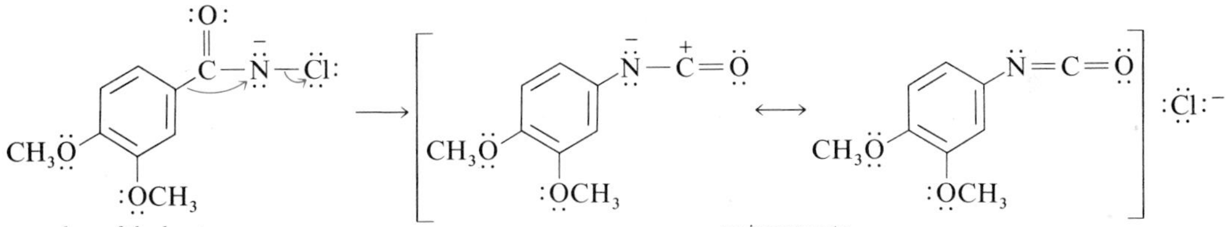

loss of the leaving group,
chloride ion, with
migration of the
aryl group

an isocyanate

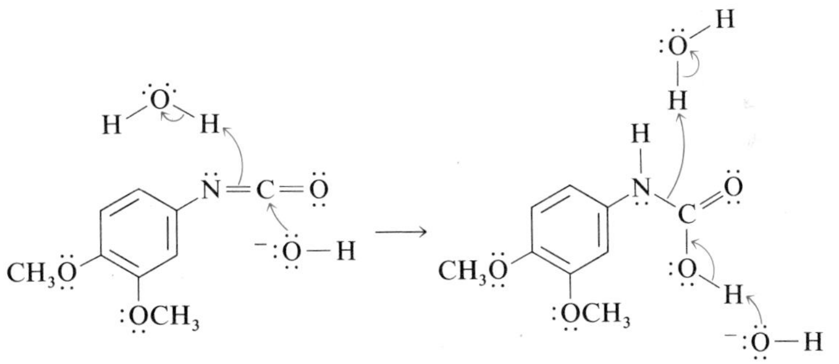

nucleophilic attack
at the carbonyl group
and protonation

loss of carbon dioxide from
the unstable intermediate

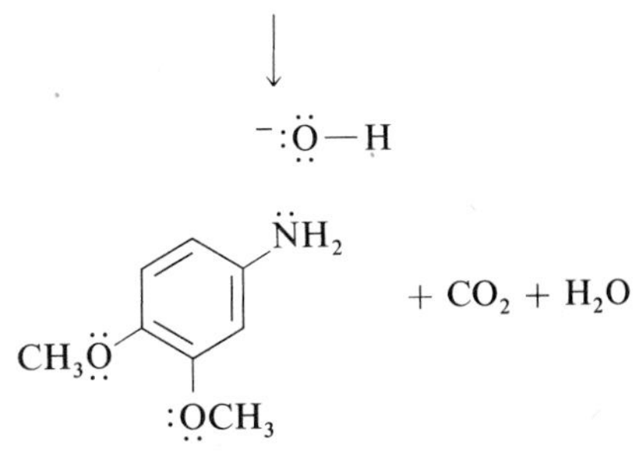

$+ CO_2 + H_2O$

The hydrogen atoms on the nitrogen atom of an amide are weakly acidic and can be removed by a base to give a nitrogen anion, which reacts with chlorine. The N-chloroamide is much more acidic than the original amide so the loss of the second proton from the nitrogen atom is easier than the loss of the first one. The anion of the *N*-chloroamide is unstable and loses chloride ion with the simultaneous migration of the aryl group from the carbon atom to the nitrogen atom. The intermediate formed at this stage is an organic isocyanate, a compound that reacts readily with base and water to give the free amine and carbon dioxide. The reaction described works with other halogens, but the amide must be unsubstituted on the nitrogen atom. The carbon atom of the carbonyl group of the amide is lost as carbon dioxide.

PROBLEM 16.20   Why is the *N*-chloroamide more acidic than the original unsubstituted amide?

Amines that are difficult to prepare by other methods can be made by the Hofmann rearrangement. An example is the synthesis of neopentylamine. Direct nucleophilic substitution of neopentyl bromide with ammonia is not possible because of steric hindrance to an S_N2 reaction by the methyl groups on the β-carbon atom (Section 6.5C). *tert*-Butyl cyanide, which would be a precursor to the amine by a reductive pathway, cannot be prepared easily either because the tertiary halide, *tert*-butyl bromide, undergoes elimination rather than substitution reactions with cyanide ion (Section 6.9B). Treatment of 3,3-dimethylbutanamide with bromine and sodium hydroxide converts it to neopentylamine.

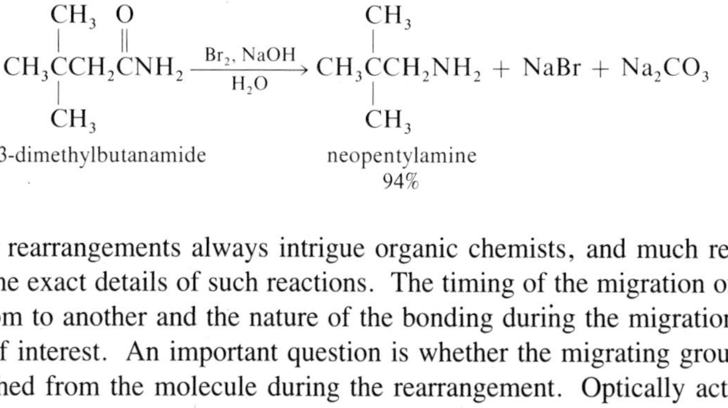

$$\underset{\substack{\text{3,3-dimethylbutanamide}}}{\overset{\displaystyle CH_3 \quad O}{\underset{\displaystyle CH_3}{CH_3\!\overset{|}{\underset{|}{C}}CH_2\overset{\parallel}{C}NH_2}}} \xrightarrow[\text{H}_2\text{O}]{\text{Br}_2,\ \text{NaOH}} \underset{\substack{\text{neopentylamine}\\94\%}}{\overset{\displaystyle CH_3}{\underset{\displaystyle CH_3}{CH_3\!\overset{|}{\underset{|}{C}}CH_2NH_2}}} + \text{NaBr} + \text{Na}_2\text{CO}_3$$

Molecular rearrangements always intrigue organic chemists, and much research is focused on the exact details of such reactions. The timing of the migration of a group from one atom to another and the nature of the bonding during the migration process are always of interest. An important question is whether the migrating group is ever totally detached from the molecule during the rearrangement. Optically active compounds in which an asymmetric carbon atom is the point of attachment of the migrating group have been used to investigate this question in both the Beckmann and the Hofmann rearrangements. The results of such experiments are illustrated below.

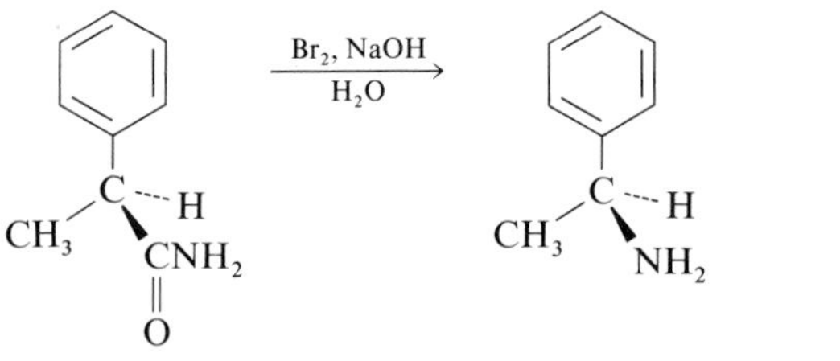

(*S*)-(+)-2-phenylpropanamide   (*S*)-(−)-α-phenylethylamine
96% optically pure; characterized
as its acetamide

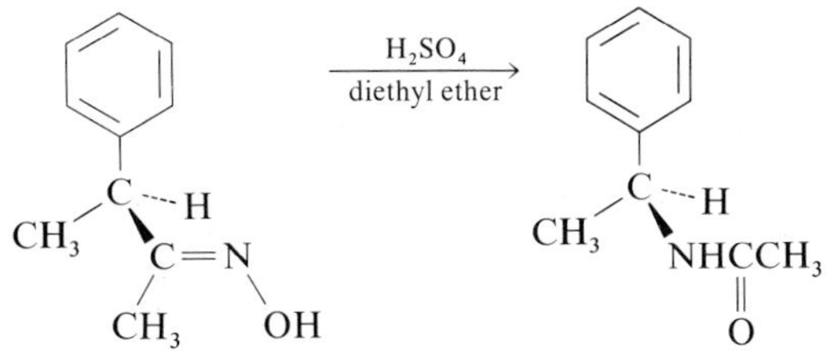

(S)-$(+)$-3-phenyl-2-butanone
oxime

(S)-$(-)$-N-acetyl-α-
phenylethylamine
99% optically pure

In each reaction, an α-phenylethyl group migrates from a carbon atom to a nitrogen atom with a pair of bonding electrons, and with almost complete retention of configuration. The alkyl group must be partially bonded first to the carbon atom and then to the nitrogen atom at the same face of the asymmetric carbon atom at all times during the rearrangement. A detached intermediate would quickly lose its chirality. That migration takes place without a loss of the original point of attachment is also indicated by the formation of 3,4-dimethoxyaniline from 3,4-dimethoxybenzamide (p. 912).

PROBLEM 16.21   In the Beckmann rearrangement, a water molecule is the leaving group on the nitrogen atom. A water molecule also reacts with the carbocation intermediate that forms to give the final amide product (p. 910). Another mechanistic question that chemists have studied is whether the water was ever free of the molecule during the migration. The following experimental observation was made when water containing ^{18}O was added to the reaction mixture.

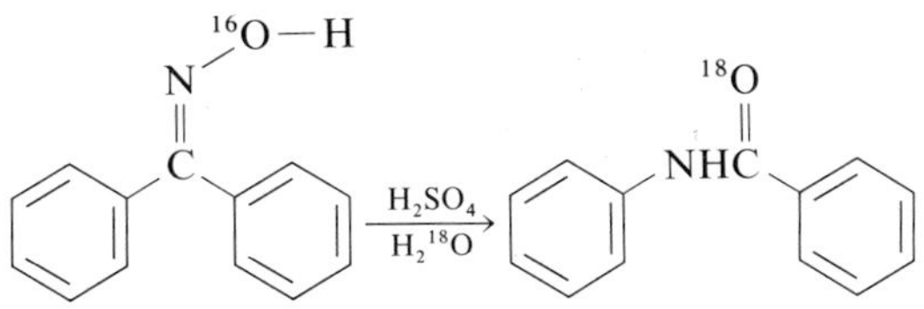

Write a mechanism that accounts for this observation, and explain how it clarifies whether water reacts intramolecularly or intermolecularly in the rearrangement.

C. Nitrogen Derivatives of Carbonic Acid

The aryl isocyanate that is the reaction intermediate shown on p. 913 for the Hofmann rearrangement is related to compounds that are nitrogen-containing derivatives of carbonic acid.

carbonic
acid

phosgene

*acid chloride of
carbonic acid*

urea

*amide of
carbonic acid*

$$\underset{\text{diethyl carbonate}}{\underset{\textit{ester of carbonic acid}}{CH_3CH_2O\overset{\overset{\displaystyle O}{\|}}{C}OCH_2CH_3}} \qquad \underset{\substack{\text{carbamic acid} \\ \textit{half amide of} \\ \textit{carbonic acid}}}{H_2N\overset{\overset{\displaystyle O}{\|}}{C}OH} \qquad \underset{\substack{\textit{tert}\text{-butyl chlorocarbonate} \\ \textit{half ester, half acid chloride} \\ \textit{of carbonic acid}}}{CH_3\overset{\overset{\displaystyle CH_3}{|}}{\underset{\underset{\displaystyle CH_3}{|}}{C}}-O\overset{\overset{\displaystyle O}{\|}}{C}Cl}$$

$$\underset{\substack{\text{ethyl carbamate} \\ \text{urethane} \\ \textit{half amide, half ester} \\ \textit{of carbonic acid}}}{H_2N\overset{\overset{\displaystyle O}{\|}}{C}OCH_2CH_3} \qquad \underset{\substack{\text{phenyl isocyanate} \\ \textit{an isocyanate}}}{\langle\bigcirc\rangle - N=C=O} \qquad \underset{\text{dicyclohexylcarbodiimide}}{\langle\bigcirc\rangle - N=C=N - \langle\bigcirc\rangle}$$

some derivatives of carbonic acid

Carbonic acid itself is unstable and decomposes into carbon dioxide and water. Carbamic acid is also unstable, giving carbon dioxide and ammonia. In fact, derivatives of carbonic acid in which only one of the two acid groups has been substituted all lose carbon dioxide with ease. The ones in which both sides of carbonic acid are substituted, such as phosgene, urea, and ethyl carbamate, are stable. Phosgene was used in World War I as a poison gas. It reacts with water to give carbon dioxide and hydrogen chloride, which causes fluid to accumulate rapidly in the lungs. Urea is a tremendously important compound in mammalian metabolism. It is the chief form in which materials from the breakdown of proteins are excreted from the body. An average man excretes about 30 g of urea a day.

The unsymmetrical derivatives of carbonic acid retain the reactivity of each type of functional group present. For example, *tert*-butyl chlorocarbonate is an ester and an acid chloride. It reacts readily as an acid chloride with amines to give carbamates, which in turn are easily hydrolyzed to regenerate the amino group, again demonstrating the instability of derivatives of carbonic acid from which carbon dioxide can be lost.

$$\underset{\substack{\textit{tert}\text{-butyl} \\ \text{chlorocarbonate}}}{CH_3\overset{\overset{\displaystyle CH_3}{|}}{\underset{\underset{\displaystyle CH_3}{|}}{C}}-O\overset{\overset{\displaystyle O}{\|}}{C}Cl} + \underset{\text{amine}}{2RNH_2} \longrightarrow \underset{\substack{\text{a } \textit{tert}\text{-butyl} \\ \text{carbamate}}}{CH_3\overset{\overset{\displaystyle CH_3}{|}}{\underset{\underset{\displaystyle CH_3}{|}}{C}}-O\overset{\overset{\displaystyle O}{\|}}{C}NHR} + RNH_3^+Cl^-$$

$$\downarrow H_3O^+$$

$$\underset{\substack{\text{amine} \\ \text{regenerated}}}{CO_2 + RNH_2} \longleftarrow \underset{\substack{\text{an } N\text{-substituted} \\ \text{carbamic acid}}}{\left[HO\overset{\overset{\displaystyle O}{\|}}{C}NHR \right]} + \underset{\substack{\text{methylpropene} \\ \text{from } \textit{tert}\text{-butyl} \\ \text{cation}}}{CH_3\overset{\overset{\displaystyle CH_3}{|}}{C}=CH_2}$$

This sequence of reactions is important in syntheses with amino acids in which it is necessary to protect the amino group while reactions are taking place at the carboxylic acid group (Section 17.4A). The protecting group must be removable under mild conditions that do not destroy amide bonds. Compounds related to *tert*-butyl chlorocarbonate serve to introduce a useful protecting group.

Isocyanates and carbodiimides are reactive compounds that add nucleophiles such as water, alcohols, and amines, with ease. 1-Naphthylisocyanate is used, for example, to convert liquid alcohols into solid derivatives useful for their identification.

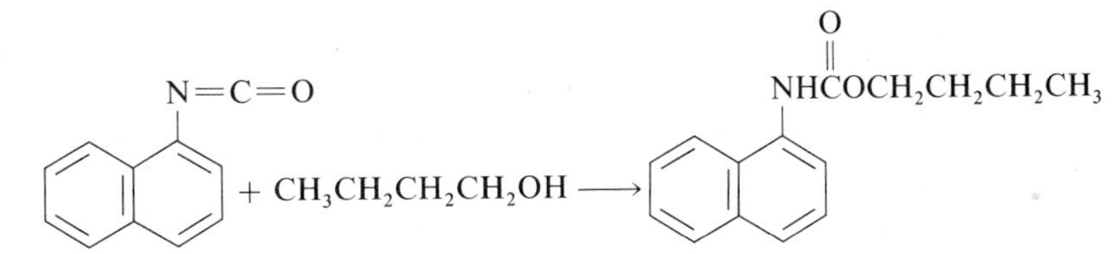

1-naphthylisocyanate	*n*-butyl alcohol	butyl *N*-(1-naphthyl)carbamate
	bp 118 °C	butyl naphthylurethane
		mp 71 °C

The reaction of polyfunctional isocyanates with polyfunctional alcohols is also of considerable industrial importance. Such reactions give polyurethane polymers, which are used in foam cushions, fibers with elastic qualities, tire treads, and coatings for floors, among many other applications (Section 20.4B).

Dicyclohexylcarbodiimide is most useful in promoting the formation of amide bonds. It reacts with carboxylic acids to give intermediates that then react with amines to give amides. The reaction is used in the synthesis of peptides (Section 17.4B).

PROBLEM 16.22   Complete the following equations.

(a) $\overset{\displaystyle O}{\overset{\|}{C}}$lCCl + NH$_3$ (excess) $\longrightarrow$

(b) $\overset{\displaystyle O}{\overset{\|}{C}}$lCCl + ⬡—CH$_2$OH (1 molar equiv) $\longrightarrow$

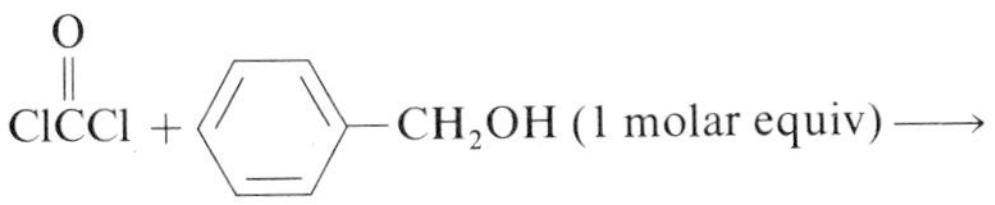

(c)   ⬡—N=C=O + CH$_3$CH$_2$CH$_2$NH$_2$ $\longrightarrow$

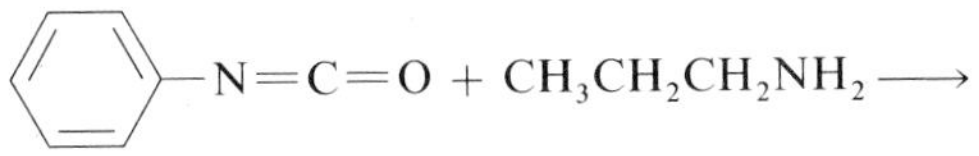

(d)   CH$_3$CH$_2$O$\overset{\displaystyle O}{\overset{\|}{C}}NH_2$ + H$_2$O $\xrightarrow[\Delta]{H_3O^+}$

(e)   ⬡—N=C=S + ⬡—NH$_2$ $\longrightarrow$
       phenyl isothiocyanate

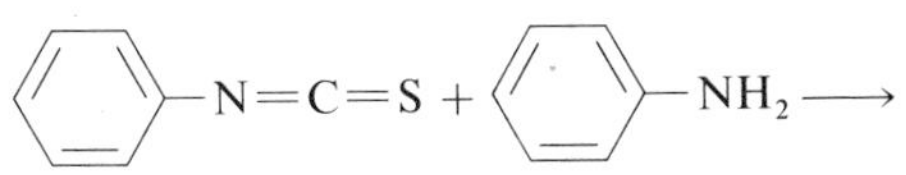

PROBLEM 16.23   One of the pesticides developed in response to the discovery that chlorinated hydrocarbons such as DDT accumulate in the environment is carbaryl. It is a carbamate with the structure shown on the next page.

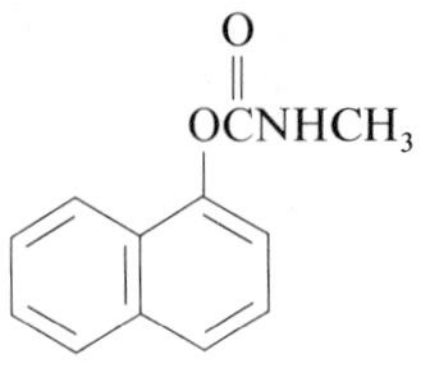

carbaryl

an insecticide that is
biodegradable

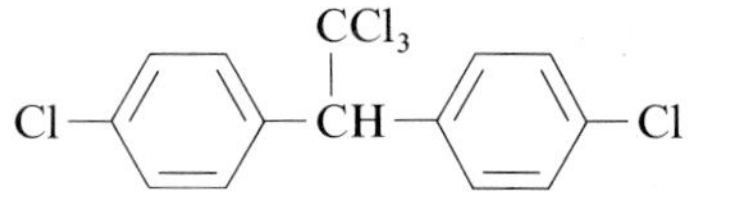

1,1,1-trichloro-2,2-bis(*p*-chlorophenyl)ethane
DDT

an insecticide that is
not biodegradable

Write equations showing a mechanism for its degradation by water in the environment. You may assume that the pH of the soil is either below or above 7.

16.6

Nitrosation Reactions

A. Nitrous Acid

Nitrous acid, HNO_2, is an unstable species that exists only as its salts, or in solution in equilibrium with a number of other species depending upon the acidity of the solution and the other ions present. In the laboratory, it is generated as needed by treating sodium nitrite with a strong mineral acid, usually hydrochloric acid, at 0–5 °C.

$$Na^+ + NO_2^- + H_3O^+ + Cl^- \xrightarrow[0\ °C]{H_2O} HO-N=O + Na^+ + Cl^-$$

nitrous acid
pK_a 3.23

It is difficult to pin down the exact nature of the reacting species in a solution of nitrous acid. The acid is in equilibrium with its anhydride, dinitrogen trioxide, which, in turn, is in equilibrium with nitric oxide and nitrogen dioxide.

$$2\ HO-N=O \rightleftharpoons H_2O + O=N-O-N=O \rightleftharpoons N=O + NO_2$$

nitrous
acid

dinitrogen trioxide

nitric
oxide

nitrogen
dioxide

a brown gas

In strongly acidic solutions, nitrous acid is protonated to its conjugate acid, which loses water to give the nitrosonium ion. In the presence of halide ions, nitrosyl halides also form.

$$HO-N=O + H_3O^+ \rightleftharpoons \overset{H}{\underset{+}{HO}}-N=O \longrightarrow H_2O + \overset{+}{N}=O$$

nitrous acid

conjugate acid
of nitrous acid

nitrosonium ion

$$\overset{+}{N}=O + Cl^- \rightleftharpoons Cl-N=O$$

nitrosonium
ion

nitrosyl
chloride

All of these species will be treated as sources of nitrosonium ion in the reactions to be explored. The nitrosonium ion is an electrophilic species, reminiscent of the nitronium ion postulated as the reacting species in electrophilic aromatic substitution reactions (Section 12.3B).

$$:N\equiv\overset{+}{O}: \longleftrightarrow :\overset{+}{N}=\overset{..}{\underset{..}{O}}$$

resonance contributors of the nitrosonium ion

Nitrosonium ion is a weaker electrophile towards the aromatic ring than nitronium ion, but reacts readily with the nonbonding electrons on nitrogen atoms to initiate reactions that vary with the structure of the amine. *Reactions of amines with nitrosonium ions are known as* **nitrosation reactions.**

B. Nitrosation of Alkyl Amines

The reaction of a primary alkyl amine, represented below by butylamine, with nitrous acid, generated from sodium nitrite and hydrochloric acid, yields a mixture of products including alcohols, alkenes, and alkyl halides.

$$CH_3CH_2CH_2CH_2NH_2 \xrightarrow[0\,°C]{\underset{HCl}{NaNO_2}} N_2\uparrow + CH_3CH_2CH_2CH_2OH + CH_3CH_2\underset{\underset{OH}{|}}{C}HCH_3 +$$

butylamine	*n*-butyl alcohol 25%	*sec*-butyl alcohol 13%

$$CH_3CH_2CH_2CH_2Cl + CH_3CH_2\underset{\underset{Cl}{|}}{C}HCH_3 + CH_3CH_2CH=CH_2 + CH_3CH=CHCH_3$$

n-butyl chloride 5%	*sec*-butyl chloride 3%	1-butene 26%	2-butene 10%

The reaction is initiated by attack of the nonbonding electrons on the nitrogen atom of the amino group on the electrophilic nitrosonium ion. Deprotonation and protonation reactions lead to the formation of an alkyl diazonium ion, which is an unstable species containing a very good leaving group, a nitrogen molecule.

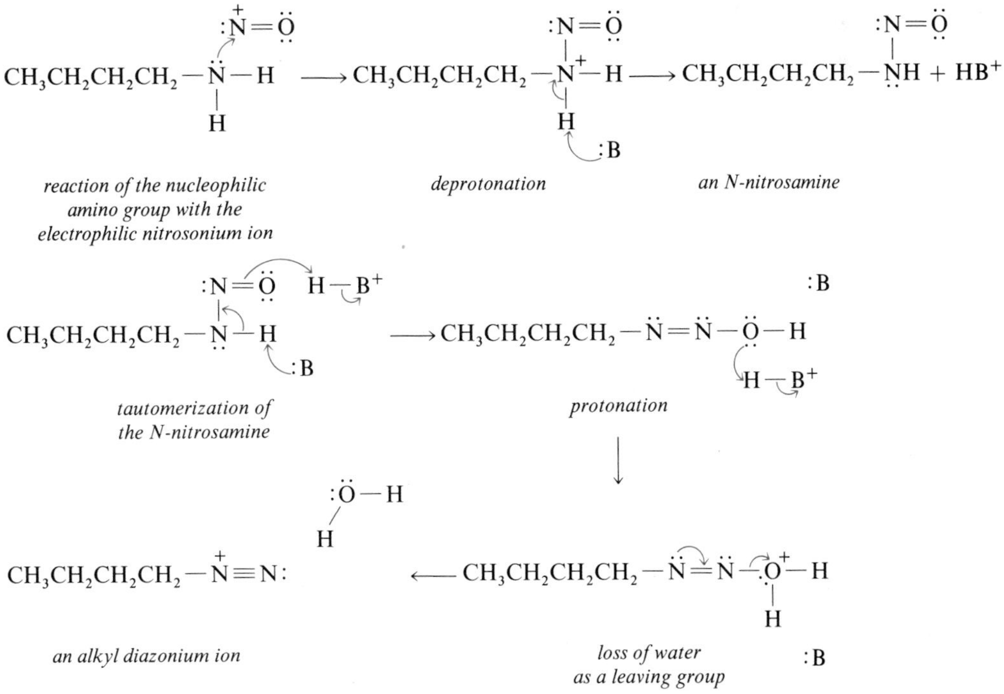

reaction of the nucleophilic amino group with the electrophilic nitrosonium ion   *deprotonation*   *an N-nitrosamine*

tautomerization of the N-nitrosamine   *protonation*

an alkyl diazonium ion   *loss of water as a leaving group*   :B

One of the important stages in the reaction is the formation of an *N*-nitrosamine. If there is no other hydrogen atom on the amino group, as is the case with secondary amines, the reaction stops at this stage.

Alkyl diazonium ions lose nitrogen spontaneously even at low temperatures. The nitrogen molecule may be displaced in S_N2 reactions by nucleophiles present in the reaction mixture, or lost as the leaving group in an E_2 elimination reaction.

$$CH_3CH_2CH_2CH_2 \overset{+}{-} N\equiv N: \xrightarrow{S_N2} :N\equiv N: + CH_3CH_2CH_2CH_2 \overset{+}{-}\overset{H}{O}-H \longrightarrow CH_3CH_2CH_2CH_2OH$$

n-butyl alcohol

$$CH_3CH_2CH_2CH_2 \overset{+}{-} N\equiv N: \xrightarrow{S_N2} CH_3CH_2CH_2CH_2Cl + :N\equiv N:$$

n-butyl chloride

$$CH_3CH_2CH{-}CH_2 \overset{+}{-} N\equiv N: \xrightarrow{E_2} CH_3CH_2CH=CH_2 + :N\equiv N:$$

1-butene

Alternatively, the nitrogen molecule may be lost first to give a carbocation intermediate, which rearranges to a more stable carbocation that, in turn, reacts with nucleophiles in S_N1 reactions or loses a proton in an E_1 reaction.

$$CH_3CH_2CH_2CH_2 \overset{+}{-} N\equiv N: \xrightarrow{ionization} CH_3CH_2CH_2\overset{+}{C}H_2 + :N\equiv N:$$

a primary carbocation

$$CH_3CH_2CH{-}\overset{+}{C}H_2 \xrightarrow{1,2\text{-hydride shift}} CH_3CH_2\overset{+}{C}HCH_3$$

a secondary carbocation

E_1 loss of a proton H_2O Cl^-

$$CH_3CH=CHCH_3 \qquad CH_3CH_2CHCH_3 \qquad CH_3CH_2CHCH_3$$
$$\qquad\qquad\qquad\qquad OH \qquad\qquad\qquad Cl$$

2-butene   *sec*-butyl   *sec*-butyl
                   alcohol   chloride

In the butyldiazonium ion, the leaving group is on a primary carbon atom, and most of the products of the reaction are formed by S_N2 and E_2 reactions.

The formation of nitrogen gas from the reaction of an amine with nitrous acid can be used as a qualitative analysis test for primary amines. Historically, the *van Slyke method,* which is the measurement of the volume of nitrogen evolved from amino acids treated with nitrous acid, was used for the quantitative analysis of amino acids.

$$\overset{\displaystyle O}{\overset{\displaystyle \|}{RCHCO^-}} \xrightarrow[HCl]{NaNO_2} N_2\uparrow + \text{organic products}$$
$$|$$
$$^+NH_3$$

amino acid   1 mole
1 mole

Secondary alkyl amines react with nitrous acid to give *N*-nitrosamines, compounds that are of great biological interest because they are known to be potent mutagens and carcinogens. One of the most potent is *N*-nitrosodimethylamine, formed when dimethylamine reacts with nitrous acid.

$$CH_3-\overset{\overset{\textstyle CH_3}{|}}{\underset{\underset{\textstyle \overset{+}{:}N=\ddot{O}}{\searrow}}{N}}-H \longrightarrow CH_3-\overset{\overset{\textstyle CH_3}{|}}{\underset{\underset{\textstyle :N=\ddot{O}}{|}}{\overset{+}{N}}}-H \overset{\frown}{\quad}:B \longrightarrow CH_3-\overset{\overset{\textstyle CH_3}{|}}{\underset{}{\ddot{N}}}-N=\ddot{O} \qquad H-B^+$$

N-nitrosodimethylamine

Since the discovery of the carcinogenicity of nitrosamines, extensive research has been conducted on the presence in biological systems of secondary amines capable of forming nitrosamines and of sources of nitrite in food. Combinations of the two seem to be found with disturbing frequency. Dimethylamine itself is found in a number of fish and meat products. Dimethylamine, methylethylamine, and the cyclic secondary amine pyrrolidine (p. 888) are found in tobacco smoke. Sodium nitrite is used as a preservative for meats such as bacon, cold cuts, and frankfurters. Nitrates, which are used widely as fertilizers, are also reduced to nitrites by the body and by some plants, so that residues of nitrates may also contribute to nitrite intake. Hydrochloric acid in gastric juice generates nitrous acid from nitrites that are eaten. This combination of factors has aroused serious concern and has led to attempts to find ways of preserving foods without using nitrites.

Tertiary amines also react with nitrous acid, though not as readily as primary and secondary amines do at low pH and low temperatures. At higher temperatures and higher pH, a tertiary amine is cleaved to give an *N*-nitroso secondary amine. The third group of the amine is converted into a carbonyl compound if there is a hydrogen atom on the carbon atom bonded to the nitrogen. The reaction of tribenzylamine with nitrous acid is typical.

$$\text{tribenzylamine} \xrightarrow[\substack{\text{sodium acetate}\\ \text{pH 4-5}\\ 90\ °C}]{NaNO_2,\ CH_3COH} \text{N-nitrosodibenzylamine} + \text{benzaldehyde}$$

| tribenzylamine | | *N*-nitrosodibenzylamine | benzaldehyde |

PROBLEM 16.24   Complete the following equations showing the product(s) that you would expect in each case.

(a) $CH_3CH_2CH_2\underset{\underset{\textstyle NH_2}{|}}{C}HCH_3 \xrightarrow[\substack{H_2O\\ 0\ °C}]{NaNO_2,\ HCl}$

(b) piperidine $\xrightarrow[\substack{H_2O\\ 0\ °C}]{NaNO_2,\ HCl}$

(c) $CH_3CH_2CH_2\underset{\underset{\textstyle CH_3}{|}}{N}CH_3 \xrightarrow[\substack{O\\ \|\\ CH_3CO^-Na^+\\ \Delta}]{NaNO_2,\ CH_3\overset{O}{\overset{\|}{C}}OH}$

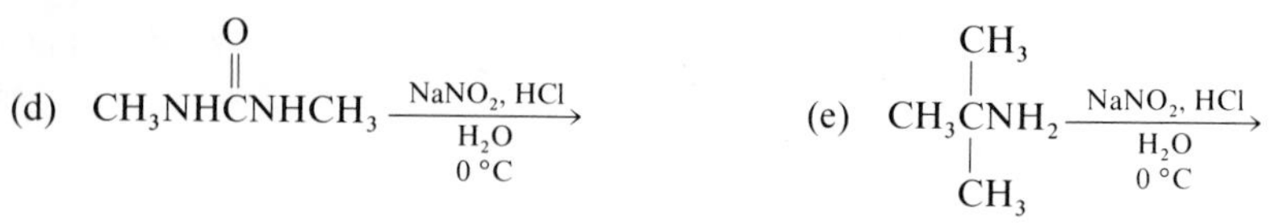

(d) $\underset{\text{O}}{\overset{\overset{\displaystyle \text{O}}{\|}}{\text{CH}_3\text{NHCNHCH}_3}} \xrightarrow[\substack{\text{H}_2\text{O} \\ 0\,°\text{C}}]{\text{NaNO}_2,\ \text{HCl}}$ (e) $\underset{\text{CH}_3}{\overset{\text{CH}_3}{\text{CH}_3\text{CNH}_2}} \xrightarrow[\substack{\text{H}_2\text{O} \\ 0\,°\text{C}}]{\text{NaNO}_2,\ \text{HCl}}$

PROBLEM 16.25   A secondary amine is soluble in the acidic solution used for a nitrosation reaction. The *N*-nitroso secondary amine that is the product of the reaction usually separates out of the solution as an insoluble oil or precipitate. Why?

C. Nitrosation of Aromatic Amines

Primary, secondary, and tertiary aromatic amines undergo the same kinds of reactions with nitrous acid as their aliphatic counterparts do. Aryl diazonium ions, which are formed from primary aromatic amines, in contrast to alkyl diazonium ions, however, are stable enough in solution at low temperatures that they can serve as reagents in reactions in which the nitrogen molecule is replaced in a controlled way. These reactions are of major synthetic importance and are discussed in detail in Section 16.6D.

The mechanism described on p. 919 for the formation of an alkyl diazonium ion also applies to the reaction of an aromatic amine with nitrous acid. Aniline, for example, gives benzenediazonium chloride when treated with sodium nitrite and hydrochloric acid at 0 °C. This is called a **diazotization reaction.**

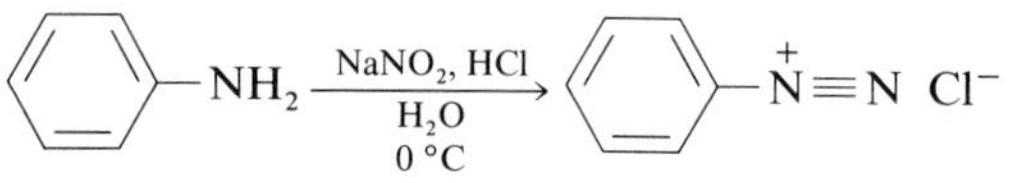

aniline

benzenediazonium chloride

soluble in water
stable at 0 °C

N-Methylaniline, on the other hand, is a secondary aromatic amine and gives an *N*-nitrosamine under the same conditions.

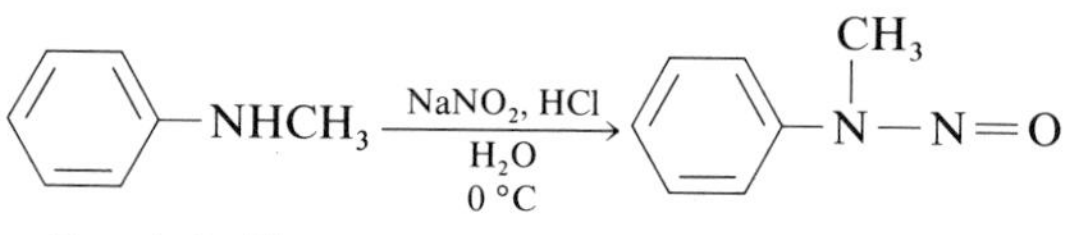

N-methylaniline

N-methyl-*N*-nitrosoaniline

insoluble in water

Tertiary aromatic amines react in two ways. If the position of the aromatic ring para to the amino group is unsubstituted, an electrophilic substitution reaction takes place with the nitrosonium ion acting as the electrophile.

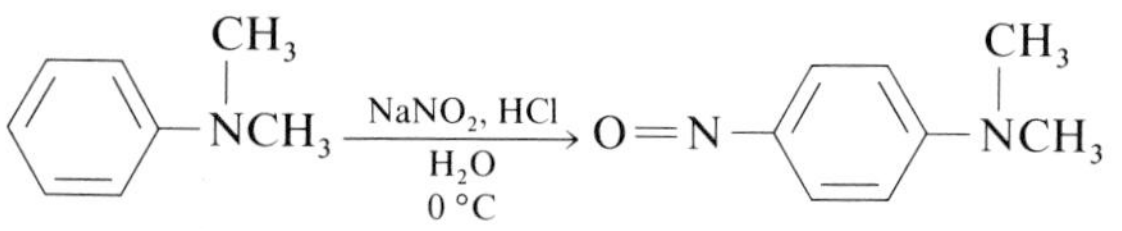

N,N-dimethylaniline

N,N-dimethyl-*p*-nitrosoaniline

If there is a substituent in the para position, reaction at the amino group (p. 921) takes place. For example, *N,N*-dimethyl-*p*-nitroaniline gives formaldehyde and *N*-methyl-*N*-nitroso-*p*-nitroaniline.

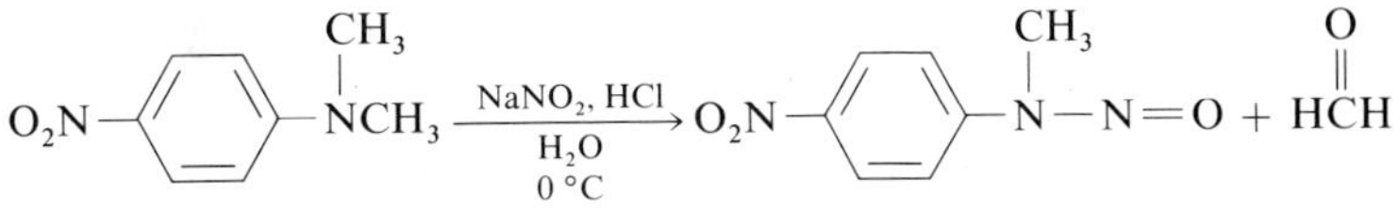

N,N,-dimethyl-*p*-nitroaniline   *N*-methyl-*N*-nitroso-   formaldehyde
                                      p-nitroaniline
                                            90%

PROBLEM 16.26   Complete the following equations.

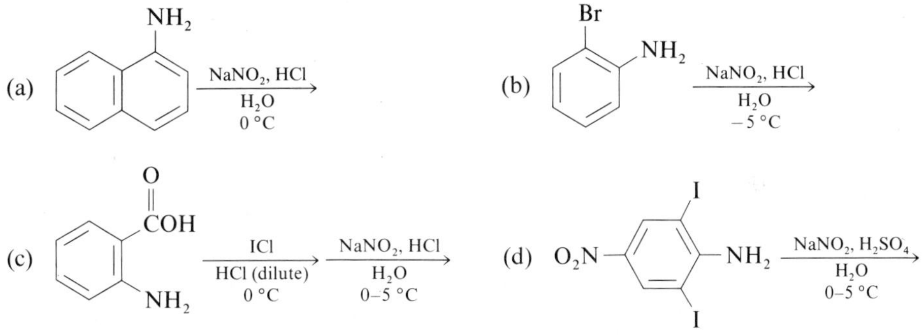

D. Substitution of Nitrogen in Aryl Diazonium Ions

Nitrogen, as a leaving group, can be replaced in a variety of substitution reactions on aryl diazonium salts. Some of the reactions require catalysis by copper metal or copper(I) salts and are known as **Sandmeyer reactions.** The exact mechanism of the substitution reactions is difficult to establish, and probably involves radical inter-mediates in some cases and ionic substitution reactions in others. The important fact to remember in each case is that a primary amine group on an aromatic ring is replaced selectively by halogen, cyano, hydroxyl, or hydrogen. Some examples will illustrate the usefulness of the reaction.

When toluene is brominated, a mixture of isomers is formed, including *o*-bromotoluene, *p*-bromotoluene, and even benzyl bromide when the reaction is carried out in the light (Section 12.4A). If pure *o*-bromotoluene is needed, the best way to make it is by diazotization of *o*-toluidine and treatment of the resulting *o*-toluenediazonium bromide with copper metal or copper(I) bromide.

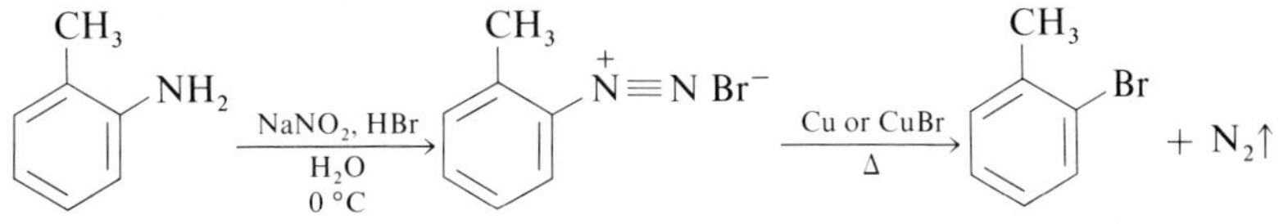

o-toluidine   *o*-toluenediazonium   *o*-bromotoluene
                            bromide

Entirely analogous reactions are used to introduce chlorine regioselectively.

Elemental fluorine and iodine cannot be used for electrophilic aromatic substitution reactions. Aromatic fluoro and iodo compounds are prepared from the corresponding amines. Aniline, for example, is converted into fluorobenzene and iodobenzene.

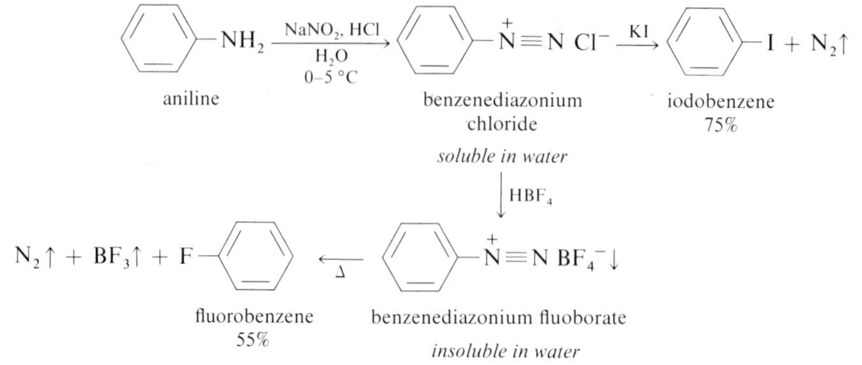

aniline   benzenediazonium chloride   iodobenzene 75%

soluble in water

fluorobenzene 55%   benzenediazonium fluoborate

insoluble in water

Iodide ion replaces the nitrogen in benzenediazonium chloride directly to give iodobenzene. To make fluorobenzene, it is necessary to replace the chloride ion in benzenediazonium chloride with fluoborate anion by using fluoboric acid. The diazonium fluoborate is much less soluble in water than the chloride is, and it precipitates. It is isolated, dried, and decomposed by heating to give fluorobenzene, boron trifluoride, and nitrogen.

A cyano group is another group that cannot be introduced directly onto an aromatic ring because aryl halides do not undergo nucleophilic substitution reactions easily (Section 12.5). Copper(I) cyanide is used to replace nitrogen in an aryl diazonium salt by a cyano group. Again, the reaction proceeds with complete regioselectivity as shown by the conversion of *p*-toluidine to *p*-toluonitrile.

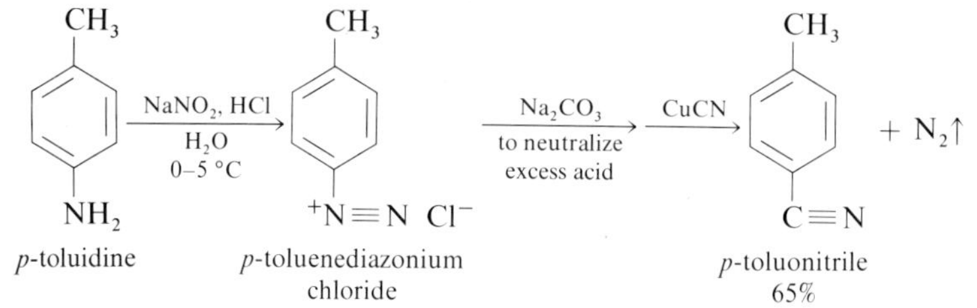

p-toluidine   *p*-toluenediazonium chloride   *p*-toluonitrile 65%

In all of the reactions described above, phenols are produced as side products when the diazonium ion reacts with water. The replacement of nitrogen by a hydroxyl group can be made the chief reaction by preparing the diazonium salt in an acid with a conjugate base that is a particularly poor nucleophile, and then heating the reaction mixture to give the phenol. For example, *m*-nitrophenol, which cannot be prepared by the direct nitration of phenol (Section 12.3C), is made in this way. Note the use of sulfuric acid in the diazotization reaction.

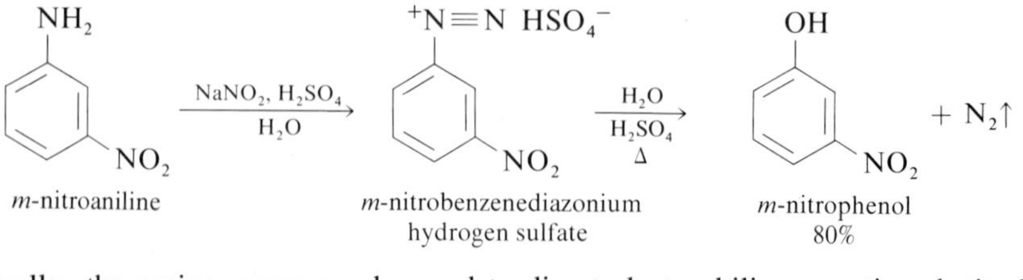

m-nitroaniline   *m*-nitrobenzenediazonium hydrogen sulfate   *m*-nitrophenol 80%

Finally, the amino group can be used to direct electrophilic aromatic substitution reactions on the ring, and can then be removed by diazotization and replacement by hydrogen in a reduction reaction. The preparation of 2,4,6-tribromobenzoic acid illustrates this sequence of reactions. Bromination of benzoic acid would be expected to give *m*-bromobenzoic acid as the chief product because the carboxyl group is ring deactivating and meta directing (Section 12.3A). The amino group in *m*-aminobenzoic acid is used to activate the ring and direct bromine atoms ortho and para to itself. The amino group is then removed by diazotization and reduction of the diazonium ion with hypophosphorous acid, H_3PO_2, a reducing agent.

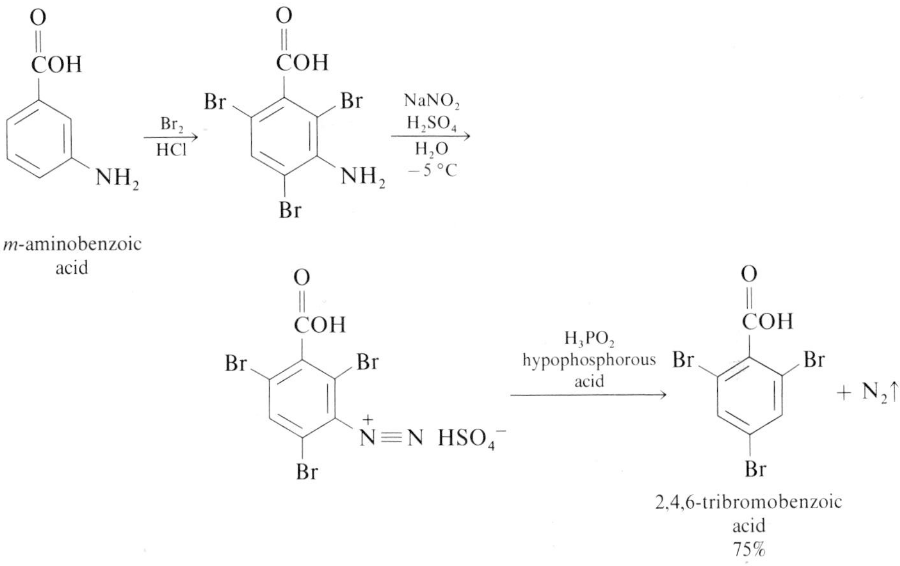

The reactions outlined in the equations in this section represent ways in which functional groups can be positioned quite selectively on aromatic rings.

PROBLEM 16.27   The starting materials for the syntheses in Section 16.6D can be prepared by standard aromatic substitution reactions and further manipulations of the substituents. How would you prepare each of the following starting materials from benzene or toluene?

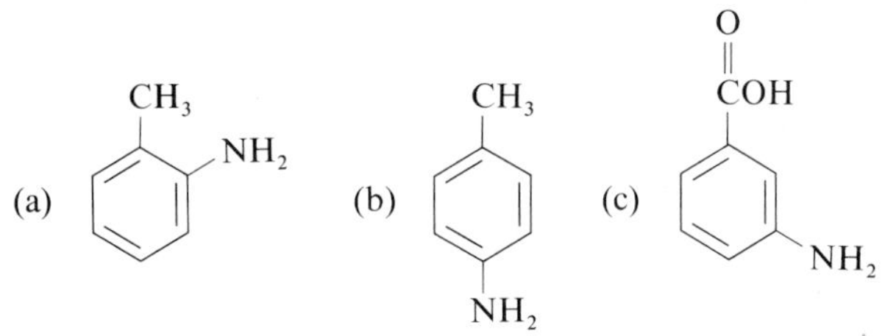

PROBLEM 16.28   Devise syntheses for the following compounds starting from benzene or toluene.

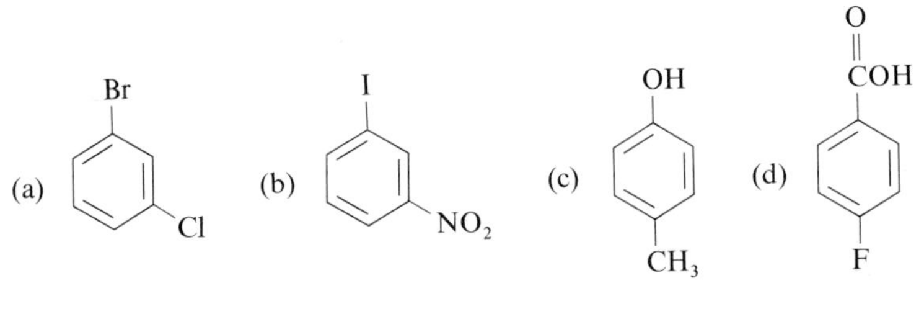

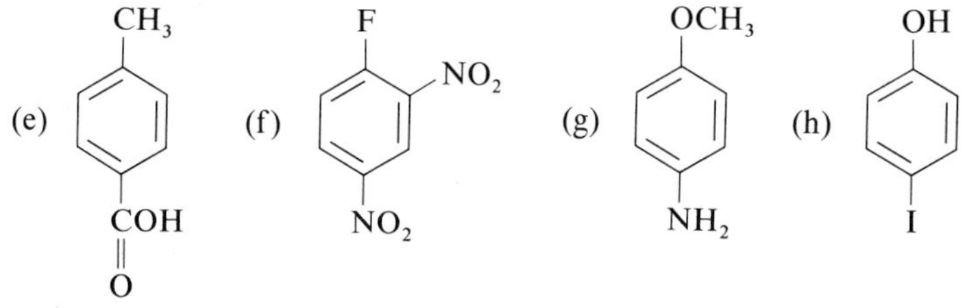

(e) CH$_3$ / COH || O

(f) F / NO$_2$ / NO$_2$

(g) OCH$_3$ / NH$_2$

(h) OH / I

E. Benzyne and Its Reactions

An interesting phenomenon is observed when *o*-aminobenzoic acid, also known as anthranilic acid, is diazotized. Both nitrogen and carbon dioxide are lost from the molecule to give a very reactive intermediate. This intermediate, called **benzyne,** is a reactive dienophile (Section 11.5) and also adds nucleophiles. The generation of benzyne by way of the benzenediazonium-2-carboxylate is shown below. An alkyl nitrite is used with acid to make the diazonium salt of anthranilic acid. The diazonium chloride is then converted into the internal salt by removing the elements of hydrogen chloride with silver oxide.

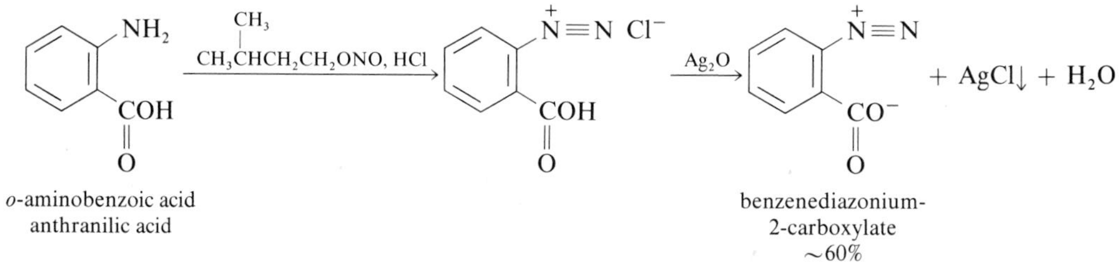

o-aminobenzoic acid
anthranilic acid

benzenediazonium-
2-carboxylate
∼60%

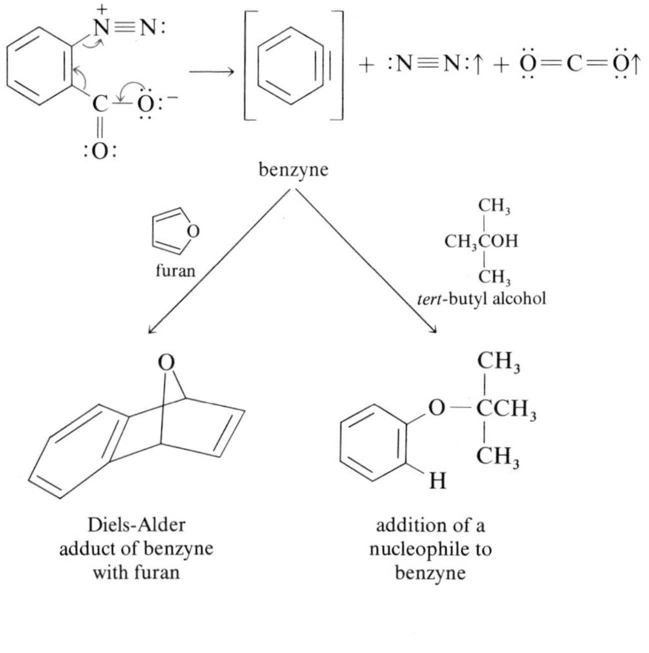

benzyne

furan

$$CH_3CH(CH_3)COH(CH_3)$$
tert-butyl alcohol

Diels-Alder
adduct of benzyne
with furan

addition of a
nucleophile to
benzyne

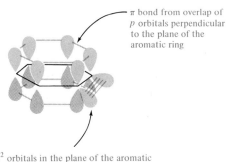

π bond from overlap of
p orbitals perpendicular
to the plane of the
aromatic ring

*sp*² orbitals in the plane of the aromatic
ring, overlapping to create a strained
bond between two of the carbon atoms

Figure 16.2   A representation of the orbitals involved in π
bonding in benzyne.

Benzyne is postulated to have a triple bond in the benzene ring and is still an aromatic system. The aromatic sextet within the six-membered ring is undisturbed. Between two of the carbon atoms there is an additional highly strained and, therefore, very reactive π bond, which can be represented as shown in Figure 16.2.

Other systems provide evidence for benzynes as reaction intermediates. One of the most convincing experiments indicating such an intermediate uses iodobenzene, labeled with the radioactive isotope of carbon, ^{14}C, at the carbon atom bearing the iodine atom. The iodobenzene that is used is synthesized from aniline specifically labeled at the carbon atom bearing the amino group by diazotization and replacement of nitrogen by iodide ion. This sequence of reactions is a good example of an application of the syntheses discussed in Section 16.6D. The labeled iodobenzene is then subjected to a nucleophilic substitution reaction using the powerful nucleophile and strong base amide anion in liquid ammonia. The aniline formed in the reaction has the radioactivity distributed approximately equally at the carbon atom originally bearing the amino group and the carbon atom adjacent to it. In the equation below, an asterisk is used to indicate the radioactive carbon atom.

The experimental observation that radioactivity is distributed to two adjacent positions is explained by postulating a benzyne intermediate for the reaction. The strongly basic amide anion is thought to carry out an E_2 elimination reaction on iodobenzene to give benzyne, which then adds ammonia with roughly equal probability at the two adjacent carbon atoms.

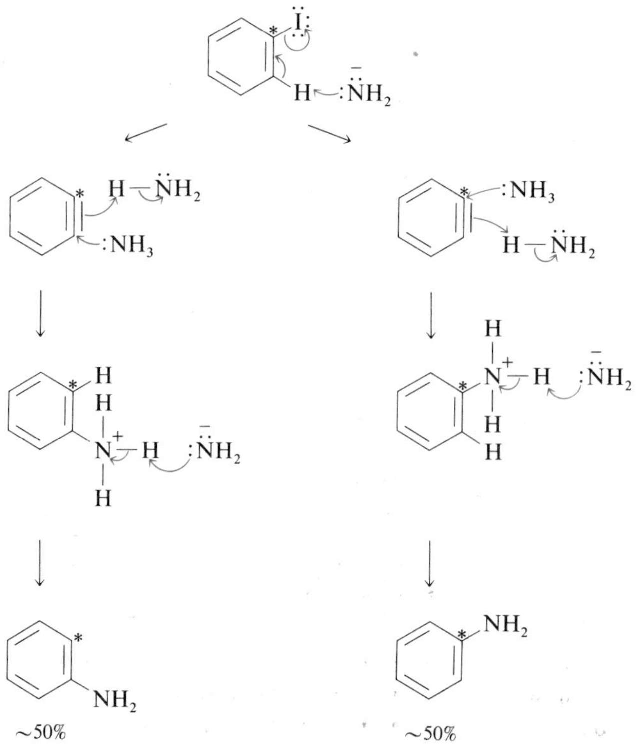

The reactions of aryl halides with potassium amide, which appear to be simple nucleophilic substitution reactions, are in fact **elimination-addition reactions.**

PROBLEM 16.29   When *o*-bromofluorobenzene is treated with magnesium metal in the presence of cyclopentadiene (Section 11.5), the reaction shown below is observed.

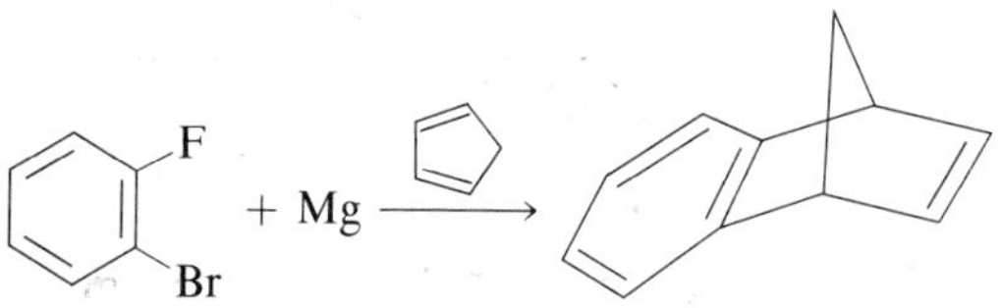

Postulate a mechanism for the reaction.

F. The Reaction of Diazonium Ions with Phenols and Amines

A reaction of diazonium ions that is of immense practical importance to the dye industry is the coupling reaction in which the diazonium ion acts as an electrophile and substitutes the activated aromatic ring of either a phenol or an aromatic amine. A typical reaction is that of benzenediazonium chloride and phenol.

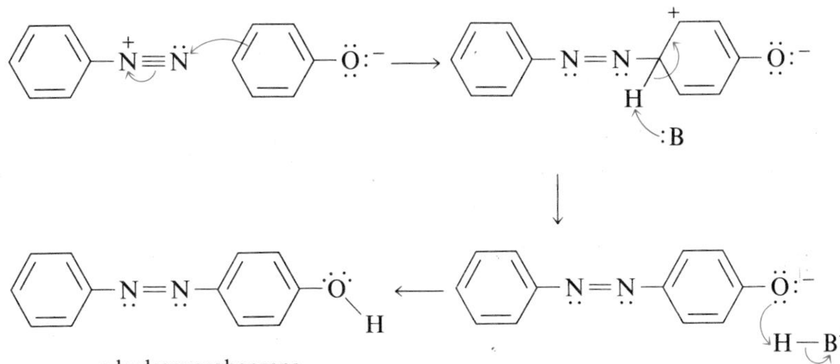

p-hydroxyazobenzene
$\lambda_{max}^{ethanol}$ 349 nm (ϵ 26,300)

The reaction with a phenol is carried out in a weakly basic solution. Electrophilic attack by the benzenediazonium ion takes place at one of the activated positions of the ring, ortho or para to the hydroxyl group. The product is an **azo compound,** containing a nitrogen-nitrogen double bond, which can have cis and trans isomers just as the carbon-carbon double bond does. The compound with two phenyl groups at the ends of the azo linkage is azobenzene, of which both isomers are known.

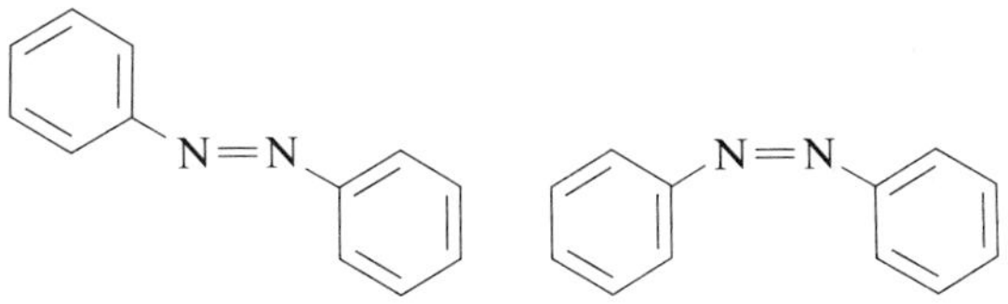

(*E*)-azobenzene   (*Z*)-azobenzene
geometrical isomers of azobenzene

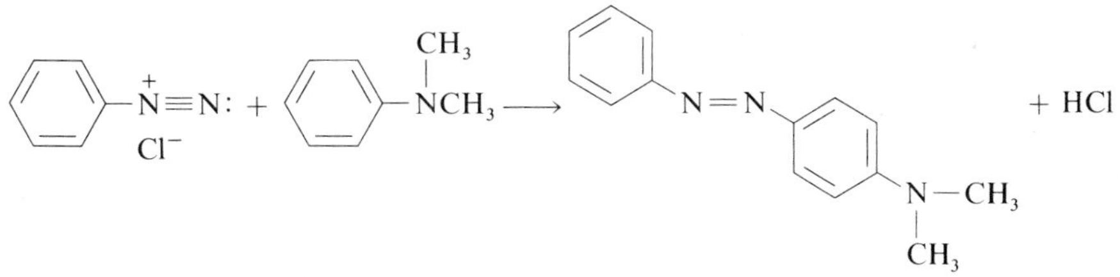

Benzenediazonium chloride also couples with tertiary aromatic amines such as *N,N*-dimethylaniline, for example, to give *p*-dimethylaminoazobenzene, a dye known as Butter Yellow that was used to color margarine until it was discovered to be carcinogenic.

benzenediazonium   *N,N*-dimethylaniline   *p*-dimethylaminoazobenzene
    chloride   Butter Yellow
                                                        $\gamma_{max}^{ethanol}$ 408 nm (ϵ 27,540)

Compounds in which the azo linkage is bound to two aromatic rings are highly colored. These compounds have extended systems of conjugation that push their absorption maxima into the visible range of the spectrum (Section 11.8C). (*E*)-Azobenzene, with $\lambda_{max}^{ethanol}$ 318 nm (ϵ 21,380), absorbs in the near ultraviolet region and is orange in color. This phenomenon is the basis for a qualitative analysis test for primary aromatic amines. The formation of a red color when diazotized amine solution is added to 2-naphthol indicates the presence of an aromatic diazonium ion, and thus an aromatic primary amine.

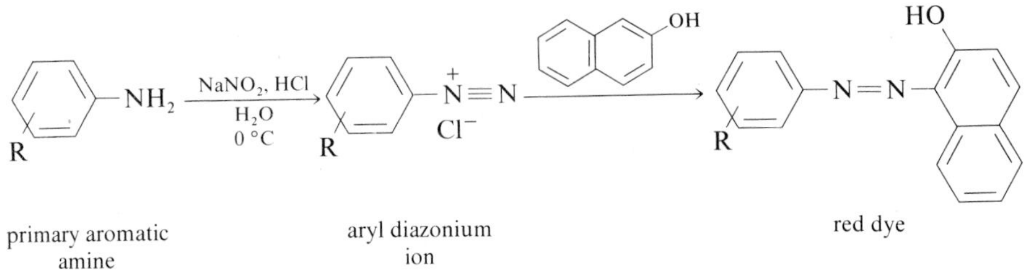

primary aromatic
amine

aryl diazonium
ion

red dye

Hydroxyl and amino groups, especially when they are placed ortho or para to the azo bond, intensify the color of the compounds. For example, p-hydroxyazobenzene has $\lambda_{max}^{ethanol}$ 349 nm (ϵ 26,300) while p-dimethylaminoazobenzene has $\lambda_{max}^{ethanol}$ 408 nm (ϵ 27,540). Thus, each of these compounds absorbs at a longer wavelength than azobenzene itself does. In each case, the molar absorptivity, ϵ, is also higher. Compounds in which an electron-donating group on one ring is conjugated with an electron-withdrawing group on another ring have especially deep colors. The azobenzene substituted with a p-nitro group on one ring and a p-dimethylamino group on the other illustrates this phenomenon.

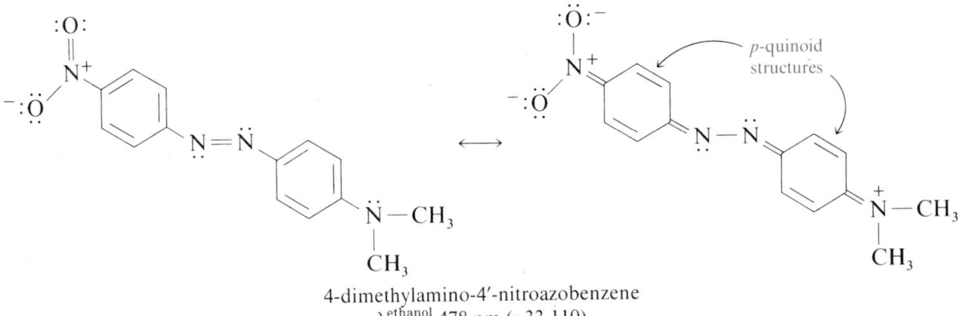

4-dimethylamino-4'-nitroazobenzene
$\lambda_{max}^{ethanol}$ 478 nm (ϵ 33,110)

The deep color of the compound is attributed to one of its resonance contributors that is quinoid in structure, a feature that often leads to deep color in compounds (Section 12.6B).

The aromatic rings also allow the introduction of a variety of other functional groups that can interact chemically with the acidic, basic, or simply polar sites in the fibers used in making cloth or paper. A large variety of dyes, tailored to fit the chemical nature of the material to be dyed, have been created by extensions of the reactions shown above.

Many dyes have sulfonic acid groups on them, which enable them to adhere firmly to fibers such as silk and wool, which are composed chiefly of proteins and have basic functional groups on them. Methyl orange, synthesized from sulfanilic acid and *N,N*-dimethylaniline, is such a dye.

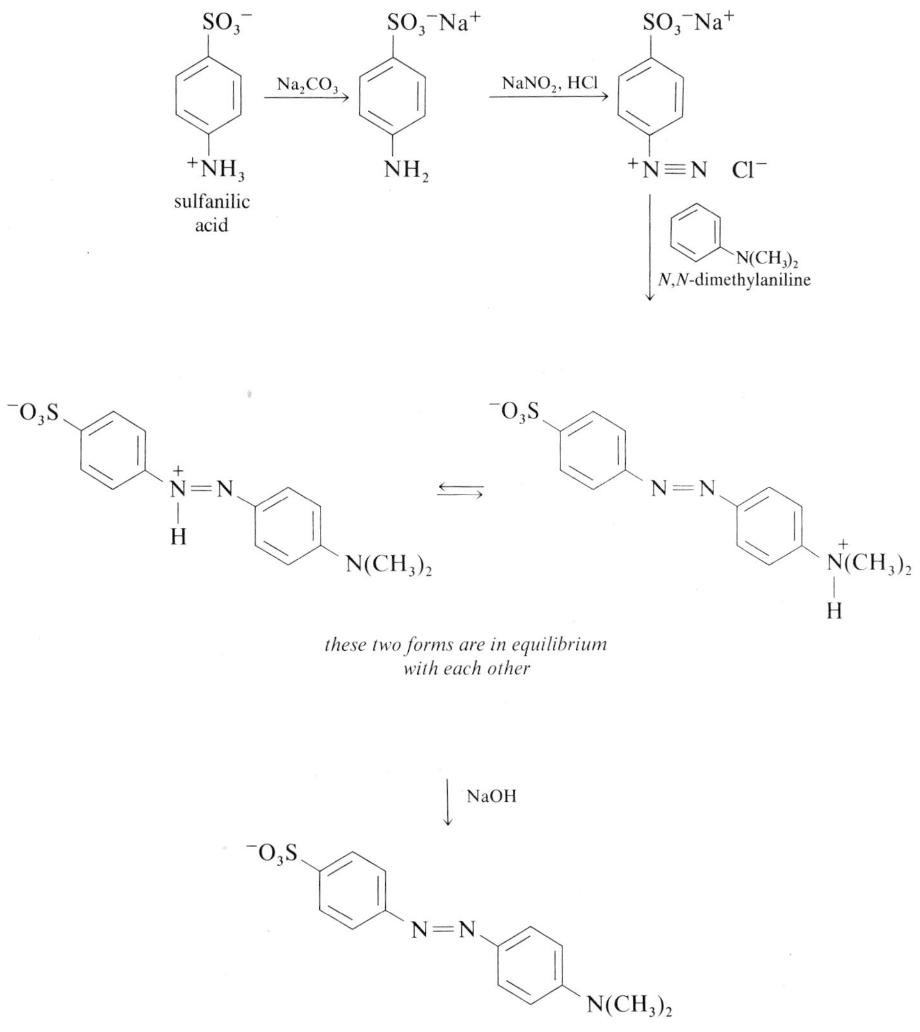

these two forms are in equilibrium with each other

methyl orange

Methyl orange is also an acid-base indicator. In dilute solutions at pH greater than 4.4, it is yellow, λ_{max} 460 nm. When acid is added to the system, methyl orange is protonated to form a dipolar ion that predominates at pH below 3.2. The protonated form has λ_{max} at 520 nm and appears red to the eye.

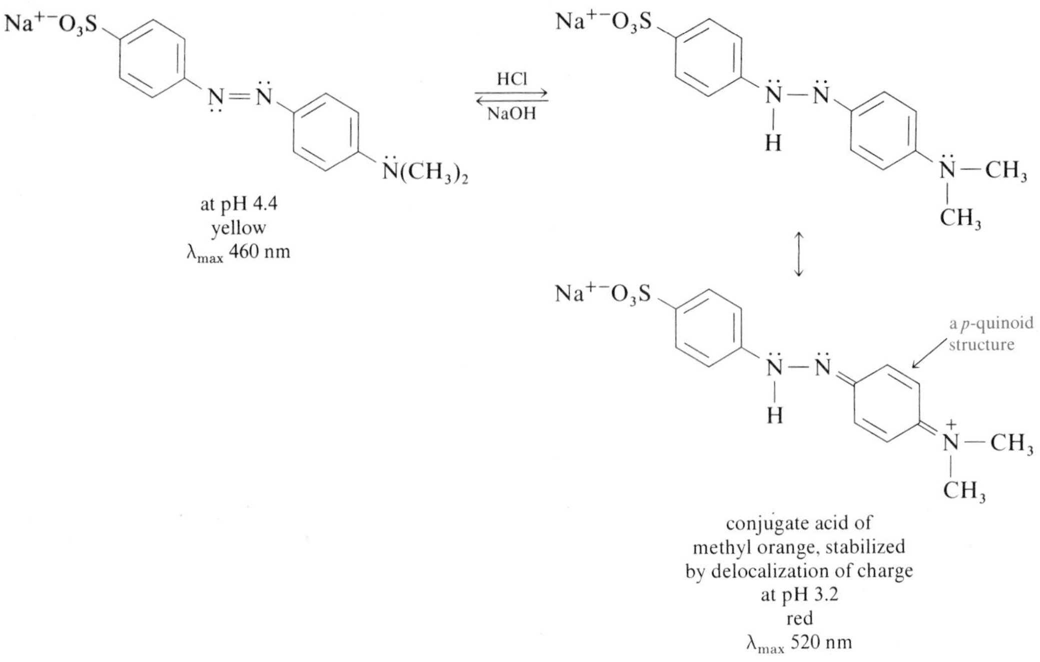

$Na^{+-}O_3S$ —[ring]— $N=N$ —[ring]— $\ddot{N}(CH_3)_2$

at pH 4.4
yellow
λ_{max} 460 nm

$\xrightarrow[\text{NaOH}]{\text{HCl}}$

$Na^{+-}O_3S$ —[ring]— $\underset{H}{\overset{}{\ddot{N}}}-\ddot{N}$ —[ring]— $\underset{CH_3}{\overset{}{\ddot{N}}}-CH_3$

$\updownarrow$

$Na^{+-}O_3S$ —[ring]— $\underset{H}{\overset{}{\ddot{N}}}-\ddot{N}$ =[p-quinoid ring]= $\underset{CH_3}{\overset{+}{N}}-CH_3$

a *p*-quinoid structure

conjugate acid of
methyl orange, stabilized
by delocalization of charge
at pH 3.2
red
λ_{max} 520 nm

Methyl orange and its conjugate acid have two different types of chromophores (Section 11.8A) and hence two different absorption spectra in the range of visible light. The different colors that result allow us to use the compound to detect a change in the acidity of a system around the range of pH at which methyl orange is protonated and deprotonated.

PROBLEM 16.30   There are three nitrogen atoms in methyl orange. Why is it protonated on the particular nitrogen atom shown in the equation above?

PROBLEM 16.31   Para red is a dye much used for cotton. The cotton fabric is soaked in one of the components of the dye and then the diazonium salt derived from the other component is added to the system so that the azo dye forms directly inside the fibers and is trapped there. Para red has the structure shown below. Show by an equation the components that you would choose for its synthesis.

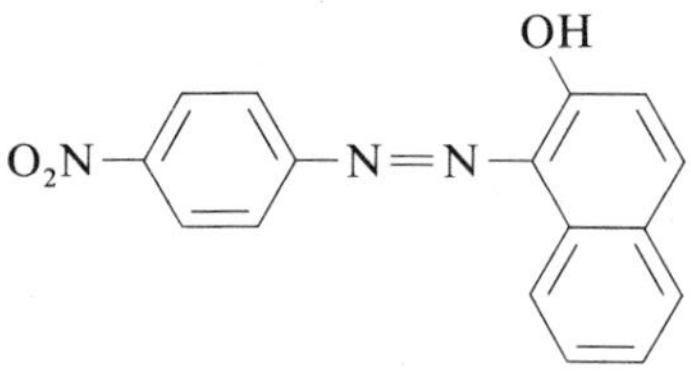

O_2N —[ring]— $N=N$ —[naphthalene with OH]

para red

PROBLEM 16.32   Extensive conjugation must be present in an azo dye before it has a blue color. Such a blue dye is synthesized by the following sequence of reactions. Supply structural formulas for the intermediates or products designated by letters.

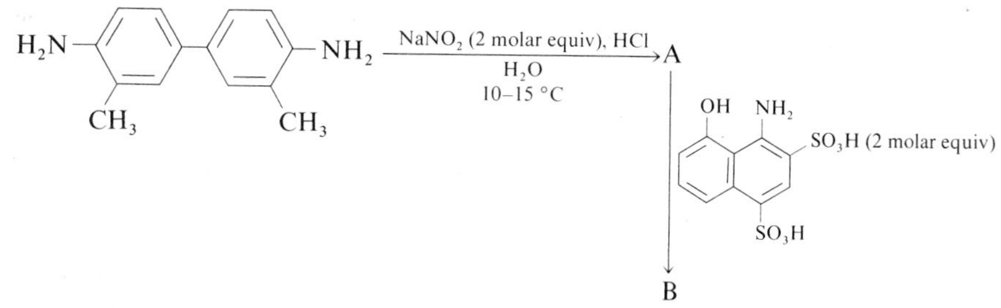

$$H_2N-\text{[biphenyl with two CH}_3\text{]}-NH_2 \xrightarrow[\substack{H_2O \\ 10-15\,°C}]{\text{NaNO}_2\ (2\ \text{molar equiv}),\ \text{HCl}} A$$

OH NH$_2$

SO$_3$H (2 molar equiv)

SO$_3$H

B

a blue dye

16.7

The Chemistry of Quaternary Nitrogen Compounds

A. The Hofmann Elimination Reaction

In Section 16.3A, we saw that ammonia can react with an alkyl halide such as methyl iodide to form a mixture of primary, secondary, and tertiary amines. If an excess of methyl iodide is used, a tertiary amine will react still further with methyl iodide to give *an ammonium salt with four organic groups on the nitrogen atom. The resulting compounds are known as* **quaternary ammonium salts.**

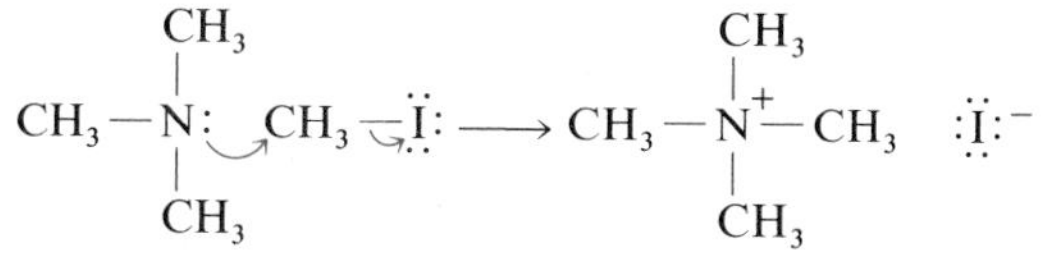

trimethylamine methyl
iodide

tetramethylammonium
iodide

*a quaternary ammonium
salt*

Quaternary ammonium salts are of interest for two reasons. First, the tertiary amine group that is part of their structure is a good leaving group, so quaternary ammonium salts undergo substitution and elimination reactions (Section 15.7D, for example). Second, the quaternary ammonium group is incorporated into systems responsible for the transmission of nerve impulses in our bodies, and many poisonous compounds and drugs that affect the nervous system also contain that functional group. Some of these compounds of biological interest are examined in Section 16.7B.

Tetramethylammonium iodide is converted into the corresponding hydroxide when it is treated with moist silver oxide. When tetramethylammonium hydroxide is heated, methanol and trimethylamine are formed in a typical S_N2 reaction.

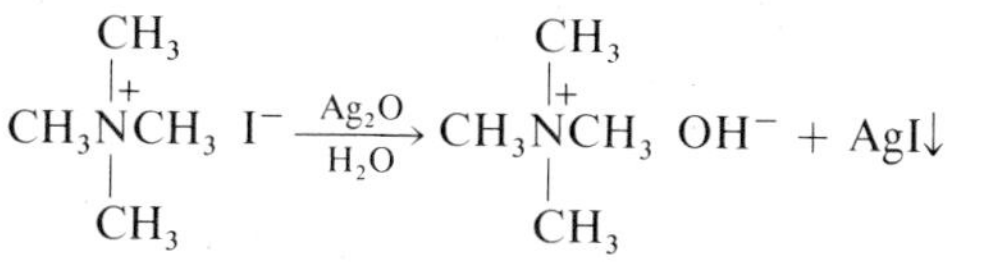

$$CH_3\overset{+}{N}CH_3\ I^- \xrightarrow[H_2O]{Ag_2O} CH_3\overset{+}{N}CH_3\ OH^- + AgI\downarrow$$

tetramethylammonium
iodide

tetramethylammonium
hydroxide

$$H-\ddot{O}:^- \quad CH_3 - N^+ \underset{CH_3}{\overset{CH_3}{\langle}} CH_3 \xrightarrow{\Delta} CH_3\ddot{O}H \; + \; :N \underset{CH_3}{\overset{CH_3}{\langle}} CH_3$$

methanol   trimethylamine

A quaternary ammonium hydroxide derived from an amine with alkyl groups larger than the methyl group will also show some S_N2 reactions, but the chief reaction for such a compound becomes an elimination reaction. *sec*-Butylamine, for example, can be converted into *N,N,N*-trimethyl-*sec*-butylammonium hydroxide and heated to give mostly 1-butene, with a very small amount of 2-butene in the product mixture.

$$CH_3CH_2CHCH_3 \xrightarrow{CH_3I \;(excess)} CH_3CH_2CHCH_3 \xrightarrow[H_2O]{Ag_2O} CH_3CH_2CHCH_3 \xrightarrow{150\,°C}$$

$$\underset{NH_2}{|} \qquad\qquad \underset{\overset{|+}{CH_3NCH_3}\; I^-}{} \qquad \underset{\overset{|+}{CH_3NCH_3}\; OH^-}{}$$

$$\underset{CH_3}{|} \qquad\qquad \underset{CH_3}{|}$$

sec-butylamine   *N,N,N*-trimethyl-   *N,N,N*-trimethyl-
                       sec-butylammonium   *sec*-butylammonium
                       iodide   hydroxide

$$\overset{CH_3}{\underset{|}{}}$$
$$CH_3CH_2CH{=}CH_2 \;+\; CH_3CH{=}CHCH_3 \;+\; CH_3NCH_3 \;+\; H_2O$$

1-butene   2-butene   trimethylamine
95%   5%

This elimination reaction is known as the **Hofmann elimination.** Experimental evidence, including the kind of stereochemical evidence developed in Section 6.9C for other elimination reactions, suggests that the reaction follows mainly the E_2 mechanism. Yet, in this elimination reaction, the formation of the least substituted alkene, instead of the thermodynamically more stable internal alkene, is favored. The reason for this orientation in the reaction has been a subject of much dispute, with steric arguments and electronic factors being used to rationalize the observations. The leaving group is a trialkylamine, in contrast to a halide ion in the dehydrohalogenation reactions seen in Chapter 6. An important factor seems to be the degree to which the transition state for the elimination reaction resembles the product alkene. A poorer leaving group, such as the amine, gives a transition state in which the carbon-hydrogen bond has been broken more than the bond to the leaving group.

*transition state in which
considerable double-bond
character has developed*

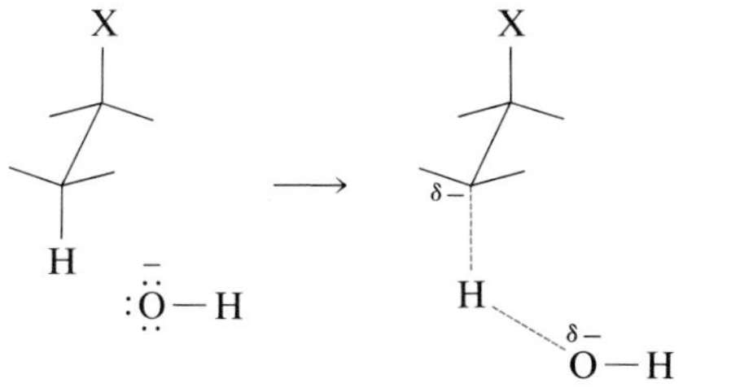

*transition state in which
carbon-hydrogen bond has
started to break without
the development of much
double-bond character*

Under such conditions, the acidity of the hydrogen atom being removed becomes more important in determining the relative energies of transition states than the stability of the alkenes that may eventually result. Thus, the product resulting from the removal of a hydrogen atom from the less highly substituted β-carbon atom is the major one.

The Hofmann elimination is useful because it allows for the preparation of less highly substituted alkenes. For example, we expect treatment of 1-bromo-1-methyl-cyclopentane with alcoholic potassium hydroxide (Section 6.9B) to give chiefly 1-methylcyclopentene.

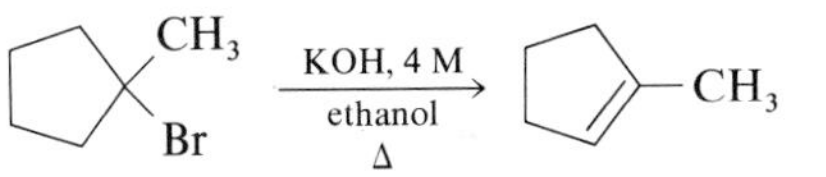

1-bromo-1-methylcyclopentane   1-methylcyclopentene

When 1-amino-1-methylcyclopentane is converted into the quaternary ammonium hydroxide, and the base heated, methylenecyclopentane is the major product.

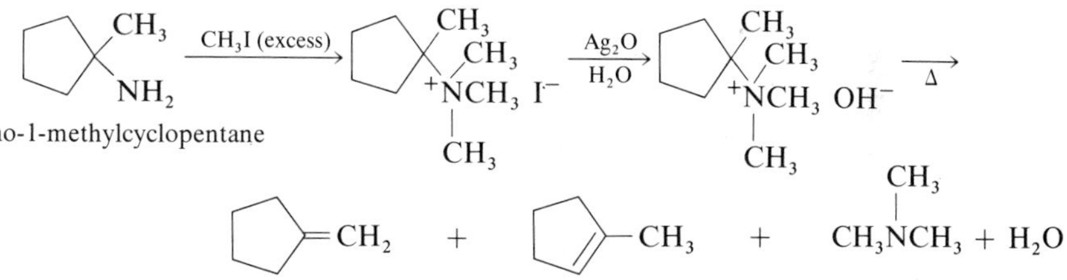

1-amino-1-methylcyclopentane

methylenecyclopentane   1-methylcyclopentene
        91%   9%

Historically, the reaction has also been useful in determining the structure of alkaloids because it gives information about the position of the nitrogen atom in cyclic compounds. A simple example will demonstrate. The structure of piperidine, a cyclic amine found as part of many alkaloids, was proved in 1881 by August Wilhelm von Hofmann, for whom the elimination reaction is named. Piperidine is converted to the quaternary ammonium hydroxide by treatment with excess methyl iodide and then with moist silver oxide. Heating the hydroxide gives the Hofmann elimination.

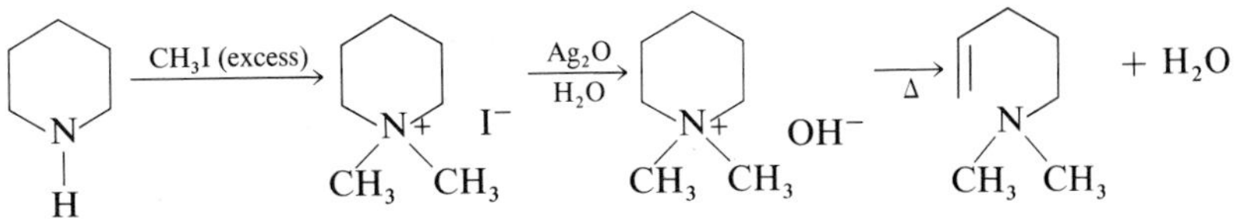

After the first Hofmann elimination reaction, the trialkylamine leaving group is still attached to the molecule through the other side of the ring. A second methylation and decomposition of the quaternary ammonium hydroxide is necessary to remove the nitrogen atom from the molecule as a tertiary amine.

The structures of many alkaloids were investigated using these reactions. The number of steps necessary to eliminate the nitrogen atom from the molecule provided information about its position in the molecule and especially about the number of rings in which it was a member. The location of the nitrogen atom was also narrowed down by the determination of the structure of the alkene that was formed.

PROBLEM 16.33   Predict what the chief products will be when N-ethyl-N-methylpiperidinium iodide is treated with moist silver oxide and then heated.

PROBLEM 16.34   The quinuclidine ring system is part of the structure of quinine (p. 885). How many times would the Hofmann elimination have to be repeated to free the nitrogen atom from the compound? What would be the structure of the alkene that results from the sequence of reactions? Write equations illustrating your answer.

quinuclidine

PROBLEM 16.35   Hofmann applied his elimination method to an investigation of the structure of coniine, the poisonous alkaloid in hemlock. The compound causes a paralysis of the motor nerves; death results from the victim's inability to breathe. Hemlock was the potion used in the execution of the Greek philosopher Socrates.

(a)   Coniine is 2-propylpiperidine. Write out equations for its degradation by the Hofmann method. Coniine is known to give a mixture of alkenes by this method. What structures are possible for the alkenes?

(b)   Coniine is present in nature in both the levorotatory and dextrorotatory forms. Levorotatory coniine has the R configuration. Draw a correct three-dimensional structural formula for coniine.

B. Biologically Active Quaternary Nitrogen Compounds

Choline, 2-(trimethylammonium)ethanol, is an important participant in the transmission of nerve impulses. It is also a component of lecithin, a major constituent of cell membranes. Choline is converted in nerve endings to its acetic ester, acetylcholine, by an enzyme with the participation of the coenzyme, acetyl coenzyme A (Section 10.7G).

$$\underset{\substack{| \\ CH_3}}{\overset{\substack{CH_3 \\ +|}}{CH_3NCH_2CH_2OH}} + \underset{\substack{acetyl \\ coenzyme\ A}}{CoA\overset{O}{-}\overset{\|}{S}CCH_3} \xrightarrow[\substack{choline \\ acetylase}]{} \underset{\substack{| \\ CH_3 \\ acetylcholine}}{\overset{\substack{CH_3 \\ +|}}{CH_3NCH_2CH_2O}}\overset{O}{\overset{\|}{C}}CH_3 + \underset{coenzyme\ A}{CoA-SH}$$

At the moment of the transmission of the nerve impulse, acetylcholine moves across the synapse, the point at which a nervous impulse passes from one nerve cell to the other. In motor nerves, this triggers the contraction of the muscle. An enzyme at the receptor site called acetylcholine esterase catalyzes the hydrolysis of acetylcholine back to choline and acetic acid; the nerve impulse stops and the muscle relaxes.

$$\xrightarrow[\substack{acetylcholine \\ esterase}]{} \underset{\substack{| \\ CH_3 \\ choline}}{\overset{\substack{CH_3 \\ +|}}{CH_3NCH_2CH_2OH}} + \underset{acylated \ enzyme}{CH_3COR}$$

hydroxyl function
at enzyme

Anything that interferes with this sequence of synthesis and destruction of acetylcholine at the nerve junction creates paralysis. The enzyme acetylcholine esterase is particularly vulnerable. Though the bond that is hydrolyzed at the enzyme active site is an ester linkage, the positively charged quaternary ammonium center is necessary to position acetylcholine so that hydrolysis can take place. When we explore the structure of proteins in the next chapter, we will see how the exact locations of acidic, basic, charged, or even hydrophobic portions of molecules are crucial to interactions between enzymes and the molecules on which they act. The importance of the quaternary ammonium group to the action of acetylcholine esterase is made apparent by the paralyzing effect of a number of compounds that are quaternary ammonium salts. (+)-Tubocurarine, the active component of the curare poison used by the Indians of South America on the tips of their arrows to paralyze their prey, is an example of such a naturally occurring quaternary ammonium compound. A synthetic drug, decamethonium bromide, mimics the presence of the two quaternary ammonium salt centers and the distance between them in tubocurarine.

tubocurarine

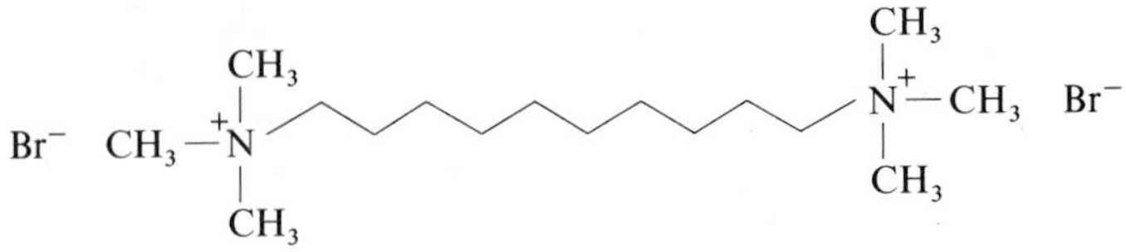

decamethonium bromide

Both tubocurarine and decamethonium bromide are used in surgery as muscle relax-ants. They act as inhibitors of acetylcholine esterase, meaning that they occupy the active site of the enzyme and interact with it because of the presence of the quaternary ammonium groups. They prevent the enzyme from carrying out its proper function of destroying acetylcholine, and thus prevent the proper transmission of nerve impulses. An excess of tubocurarine kills by paralysis of the respiratory muscles.

Another potent poison is found in the fungus *Amanita muscaria,* commonly called the fly agaric because it was used as a fly poison at one time. The alkaloid muscarine is found in the fungus along with choline and acetylcholine.

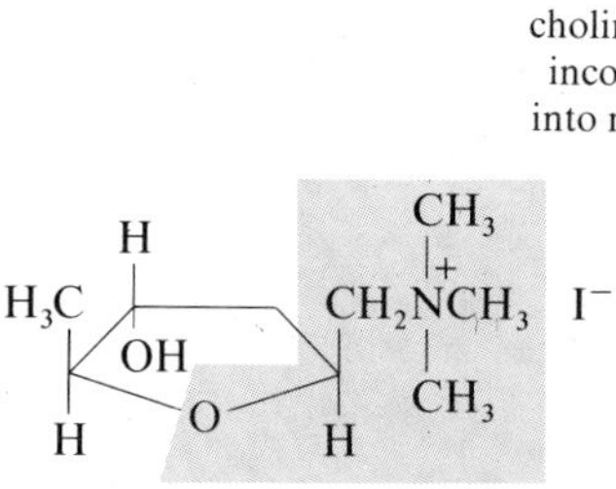

choline residue
incorporated
into muscarine

(+)-muscarine

Choline is incorporated into the structure of muscarine as shown by the colored portion of the formula. (+)-Muscarine inhibits acetylcholine esterase, but (−)-muscarine has very little toxicity, a good example of the high stereoselectivity of enzymatic reactions.

The function of choline in lecithin is quite different from its function in the trans-mission of nerve impulses. Lecithin is a triglyceride (Section 10.7C) in which one of the primary alcohol groups of glycerol is esterified by phosphoric acid. The phosphoric acid also forms an ester link to choline. The other two alcohol functions in glycerol are esterified by fatty acids, usually stearic acid and oleic acid.

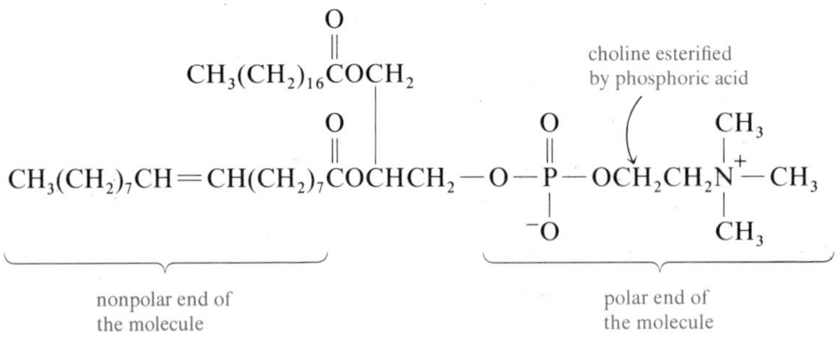

nonpolar end of
the molecule

polar end of
the molecule

choline phosphoglyceride
lecithin

Lecithin is one of a series of compounds known as phospholipids. At the pH prevalent in the body, the phosphoric acid group is negatively charged, while the quaternary ammonium group in the choline residue is positively charged. The mole-cule as a whole is neutral, but quite highly charged at one end, and hydrocarbon-like and nonpolar at the other. The hydrocarbon portions of phospholipid molecules are

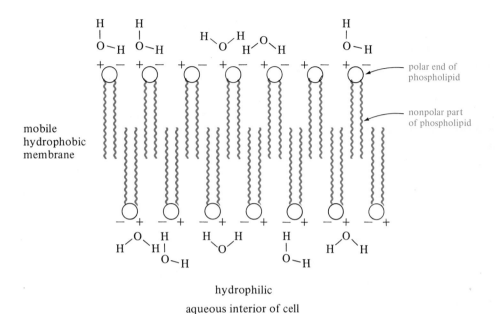

aqueous exterior of cell

hydrophilic

Figure 16.3 Model of a cell membrane composed of a phospholipid bilayer.

believed to interact with each other to give cell membranes consisting of two layers of phospholipids with the polar phosphoric acid-choline ends facing out and in towards the cell interior. The layers have a high electrical resistance and are not easily penetrated by highly polar molecules. The polar parts of the membrane are placed where they interact with water in the cell and surrounding fluids. The two layers of lipid molecules can move sideways past each other, giving the membranes flexibility and mobility (Figure 16.3).

16.8

The Infrared Spectroscopy of Amines

The infrared spectra of primary amines show two stretching frequencies for the nitrogen-hydrogen single bonds around 3500 and 3400 cm^{-1} as the most distinctive bands. In addition, bending frequencies for the nitrogen-hydrogen bond appear around 1650 to 1590 cm^{-1} and 900 to 650 cm^{-1}. The carbon-nitrogen single bond stretching frequency, around 1220 to 1020 cm^{-1}, is not as strong and as characteristic as the similar band for the carbon-oxygen single bond (Sections 7.7C and 10.11A). These features of the infrared spectra of primary amines are seen in the spectra for propylamine and aniline (Figure 16.4).

Only a single band of variable intensity appears in the region for the nitrogen-hydrogen single bond stretching frequency, at about 3300 to 3400 cm^{-1} in the spec-

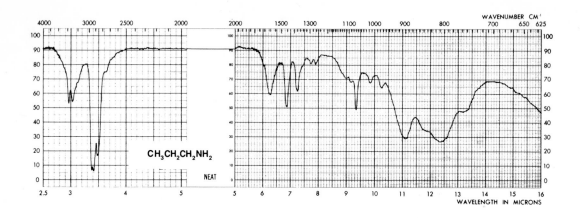

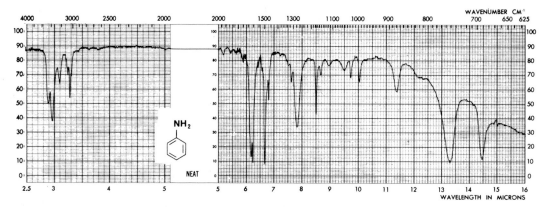

Figure 16.4 Infrared spectra for propylamine and aniline. (From *The Aldrich Library of Infrared Spectra*)

tra of secondary amines. The band is more intense for aromatic amines than for alkylamines. There is, of course, no band in this region for tertiary amines. The spectra of *N*-methylcyclohexylamine, *N*-methylaniline, and *N,N*-diethylcyclohexylamine illustrate these points (Figure 16.5). All of these spectra show a band around 2800 cm^{-1} for the stretching frequency for a carbon-hydrogen single bond on a carbon atom adjacent to a nitrogen atom.

The infrared spectroscopy of amides was discussed in Section 10.11B.

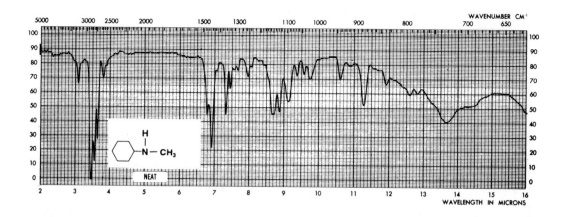

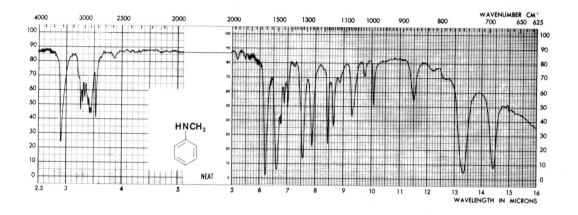

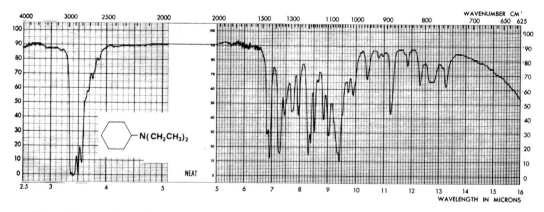

Figure 16.5   Infrared spectra for *N*-methylcyclohexylamine, *N*-methylaniline, and *N,N*-diethylcyclohexylamine. (From *The Aldrich Library of Infrared Spectra*)

PROBLEM 16.36 Infrared spectra are given in Figure 16.6 for triethylamine, *N*-ethylaniline, and butylamine. Which spectrum belongs to which compound?

Compound A

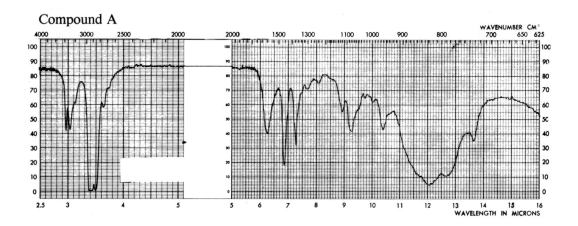

Compound B

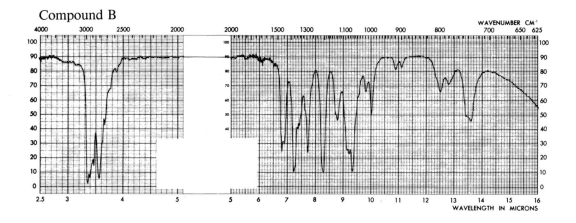

Compound C

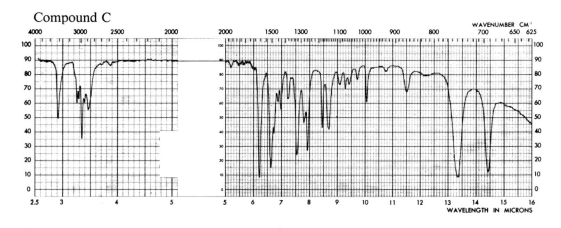

Figure 16.6 Spectra for Problem 16.36. (From *The Aldrich Library of Infrared Spectra*)

PROBLEM 16.37 The infrared spectra of butanamide, 6-bromohexanoic acid, and heptylamine appear in Figure 16.7. Assign each spectrum to the correct compound.

Compound D

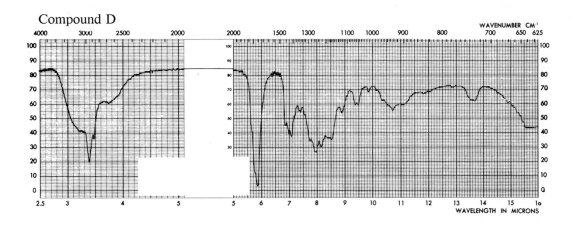

Compound E

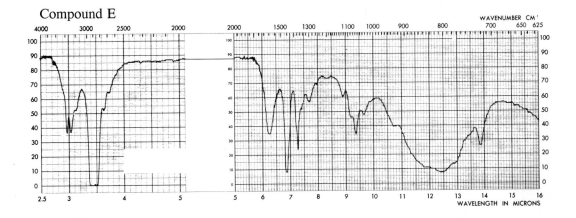

Compound F

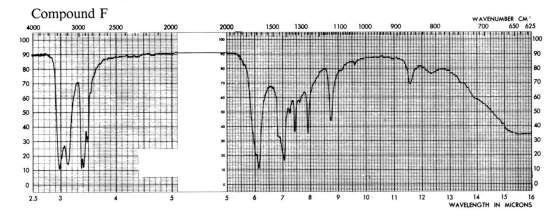

Figure 16.7 Spectra for Problem 16.37. (From *The Aldrich Library of Infrared Spectra*)

ADDITIONAL PROBLEMS

16.38 Name the following compounds.

(a) 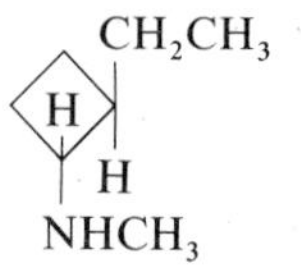 [structure with CH$_2$CH$_3$, H, H, NHCH$_3$]

(b) [pyridine ring with N and CH$_3$]

(c) CH$_3$CH$_2$NCH$_2$CH$_3$ (with CH$_2$CH$_3$ substituent on N)

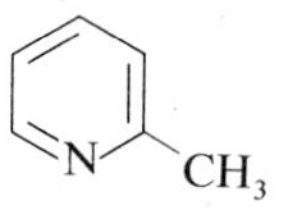

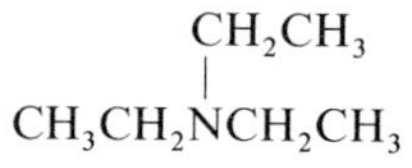

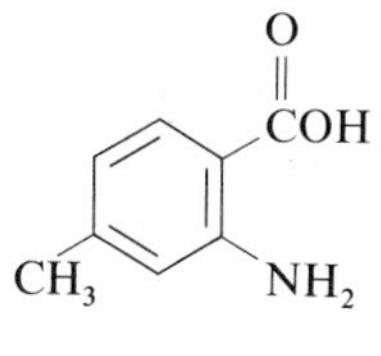

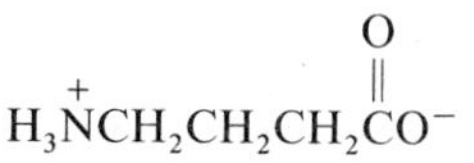

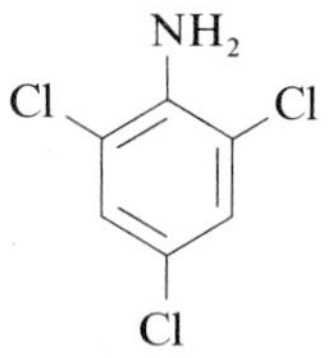

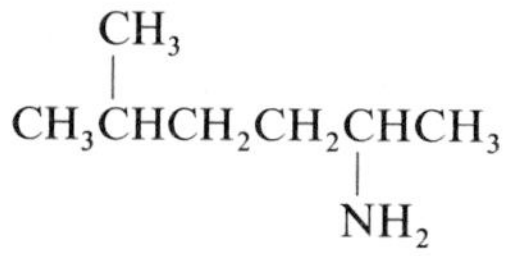

(d) [benzene ring with COH (C=O), CH$_3$, NH$_2$ substituents]

(e) H$_3$NCH$_2$CH$_2$CH$_2$CO$^-$ (with C=O)

(f) [benzene ring with NH$_2$, Cl, Cl, Cl substituents]

(g) CH$_3$CHCH$_2$CH$_2$CHCH$_3$ (with CH$_3$ and NH$_2$ substituents)

(h) [cyclohexane with H, H, NH$_2$, HO substituents]

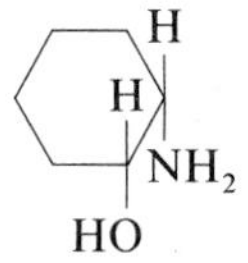

(i) 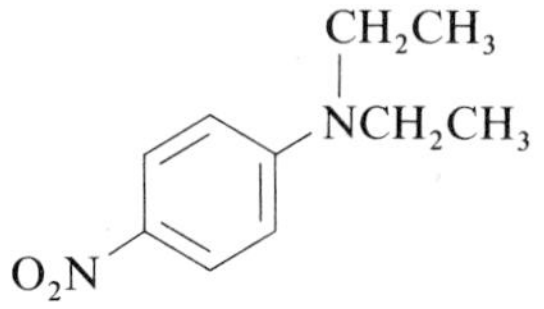 [benzene ring with N(CH$_2$CH$_3$)(CH$_2$CH$_3$) and O$_2$N substituents]

16.39 Give structural formulas for the following compounds.

(a) *trans*-2-aminocyclopentanol   (b) *N*-methyl-*N*-propylcyclohexylamine
(c) 3,5-dinitroaniline   (d) *N*,*N*-dimethyl-*p*-methoxyaniline   (e) *m*-toluidine
(f) 2,5-diaminooctane   (g) 3-hexanamine
(h) 4-amino-2,2-dimethylpentanoic acid

16.40 Complete the following equations giving structural formulas for all intermediates and products designated by letters.

(a) 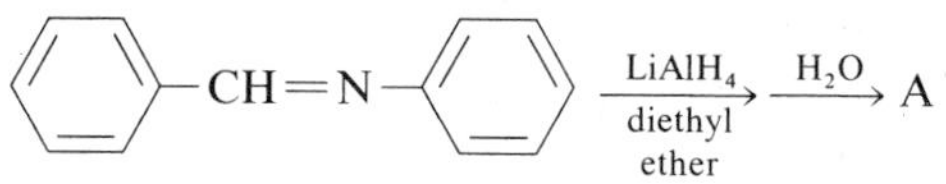 [benzene]—CH=N—[benzene] $\xrightarrow[\text{diethyl ether}]{\text{LiAlH}_4}$ $\xrightarrow{\text{H}_2\text{O}}$ A

(b) HOC—[benzene]—CNH$_2$ (both C=O) $\xrightarrow[\text{H}_2\text{O}]{\text{Cl}_2,\ \text{NaOH}}$ $\xrightarrow{\text{NH}_3,\ \text{pH 8}}$ B

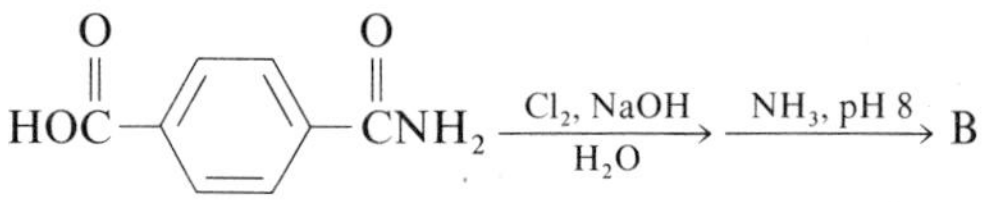

(c) 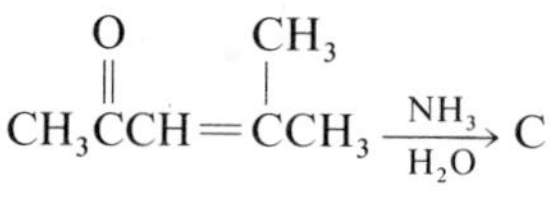 CH$_3$CCH=CCH$_3$ (C=O, CH$_3$) $\xrightarrow[\text{H}_2\text{O}]{\text{NH}_3}$ C

(d) 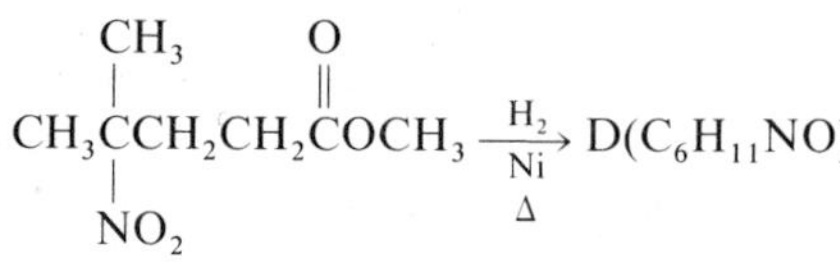 CH$_3$CCH$_2$CH$_2$COCH$_3$ (with CH$_3$, O, NO$_2$) $\xrightarrow[\Delta]{\text{H}_2 \ \text{Ni}}$ D(C$_6$H$_{11}$NO)

(e) H$_2$N—[benzene]—SO$_2$NH$_2$ $\xrightarrow[\text{Fe}]{\text{Br}_2}$ E

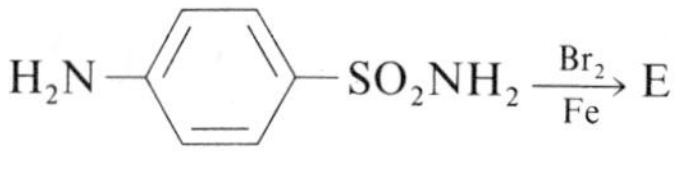

(f) 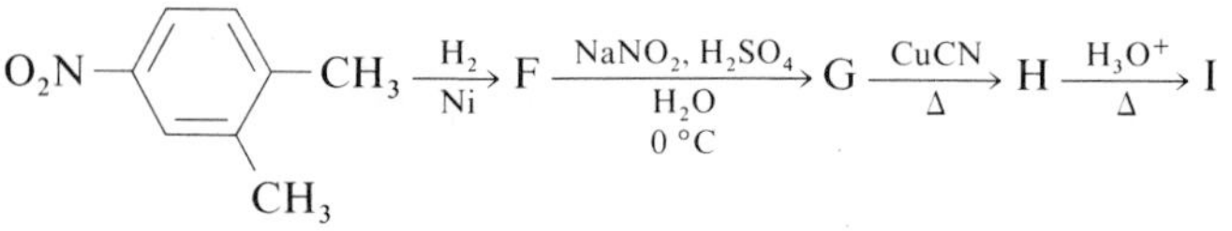 O$_2$N—[benzene with CH$_3$, CH$_3$]—CH$_3$ $\xrightarrow[\text{Ni}]{\text{H}_2}$ F $\xrightarrow[\substack{\text{H}_2\text{O} \\ 0\,°\text{C}}]{\text{NaNO}_2,\ \text{H}_2\text{SO}_4}$ G $\xrightarrow[\Delta]{\text{CuCN}}$ H $\xrightarrow[\Delta]{\text{H}_3\text{O}^+}$ I

(g) CH$_3$CH$_2$CCH$_2$CH$_2$CH$_3$ (with C=O) $\xrightarrow[\text{Na}_2\text{CO}_3]{\text{HONH}_3\ \text{HSO}_4^-}$ J + J′ $\xrightarrow{\text{H}_2\text{SO}_4}$ K + L

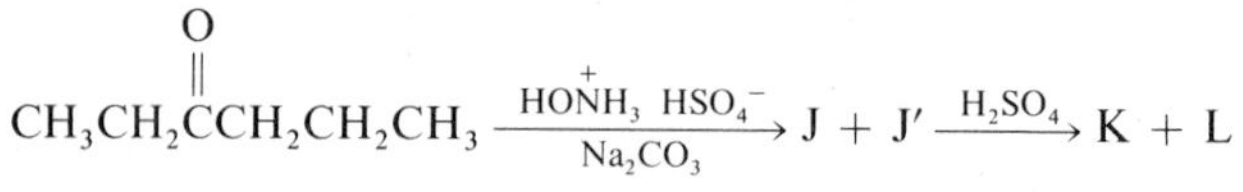

(h) $CH_3\overset{\overset{\displaystyle O}{\|}}{C}CH_2CH_2CH_2CH_2CH_3 \xrightarrow[\substack{Ni \\ \Delta}]{NH_3,\ H_2} M$

(i) $CH_3CH_2CH_2NO_2 \xrightarrow[NaOH]{\overset{\overset{\displaystyle O}{\|}}{H}CH} N \xrightarrow{H_2SO_4,\ Fe} O \xrightarrow{Ca(OH)_2} P$

(j) $Br\!-\!\!\bigcirc\!\!-\!Br \xrightarrow{CH_3NHCH_3\ (2\ molar\ equiv)} Q \xrightarrow{BaO} R \xrightarrow{CH_3I\ (2\ molar\ equiv)} S \xrightarrow[\substack{H_2O \\ \Delta}]{Ag_2O} T + U$

(k) [cyclohexanone with C_6H_5 and $\overset{+}{C_6H_5}$ substituents] $\xrightarrow[\substack{pyridine \\ \Delta}]{HO\overset{+}{N}H_3Cl^-} V \xrightarrow[\substack{diethyl \\ ether}]{LiAlH_4} \xrightarrow{H_3O^+} W \xrightarrow[H_2O]{NaOH} X$

(l) $^-O_3S\!-\!\!\bigcirc\!\!-\!\overset{+}{N}H_3 \xrightarrow[H_2O]{Na_2CO_3} Y \xrightarrow[\substack{H_2O \\ 0-5\ ^\circ C}]{NaNO_2,\ HCl} Z$

16.41   The following problems offer some more practice in recognizing reactions.

(a) $CH_3CH_2O\!-\!\!\bigcirc\!\!-\!NH_2 \xrightarrow[\substack{acetic \\ acid}]{Br_2} A \xrightarrow[\substack{H_2O \\ 0\ ^\circ C}]{NaNO_2,\ HCl} B \xrightarrow{H_3PO_2} C$

(b) $HO\!-\!\!\bigcirc\!\!-\!NH_2 \xrightarrow[\substack{H_2O \\ 0\ ^\circ C}]{NaNO_2,\ H_2SO_4} D \xrightarrow[\substack{Cu \\ \Delta}]{KI} E$

(c) [3,5,5-trimethylcyclohexanone $=O$] $\xrightarrow[\substack{Ni \\ \Delta,\ pressure}]{NH_3,\ H_2} F$

(d) $CH_3\overset{\overset{\displaystyle O}{\|}}{C}CH_3 \xrightarrow[HCl\ (catalyst)]{\bigcirc\!\!-\!NH_2} G \xrightarrow{NaOH} H$

(e) [3-bromobenzamide] $\overset{\overset{\displaystyle O}{\|}}{C}NH_2 \xrightarrow[H_2O]{Br_2,\ NaOH} I$

(f) [2-aminonaphthalene, NH_2] $\xrightarrow[\substack{H_2O \\ 0-5\ ^\circ C}]{NaNO_2,\ HCl} J \xrightarrow{HBF_4} K \xrightarrow{\Delta} L$

(g) $CH_3\underset{\underset{\displaystyle Cl}{|}}{CH}\overset{\overset{\displaystyle O}{\|}}{C}OCH_2CH_3 \xrightarrow[\substack{dimethyl \\ sulfoxide}]{NaNO_2} M$

(h) [benzene with $C\!\equiv\!N$ and CH_3 groups] $\xrightarrow[\substack{diethyl \\ ether}]{LiAlH_4} \xrightarrow{H_2O} N$

(i) $CH_3\overset{\overset{\displaystyle O}{\|}}{C}NH_2 \xrightarrow[H_2O]{Br_2,\ KOH} O$

(j) [3-methylaniline $-NH_2$] $+$ [benzaldehyde $-\overset{\overset{\displaystyle O}{\|}}{C}H \longrightarrow P \xrightarrow[Ni]{H_2} Q$

(k) 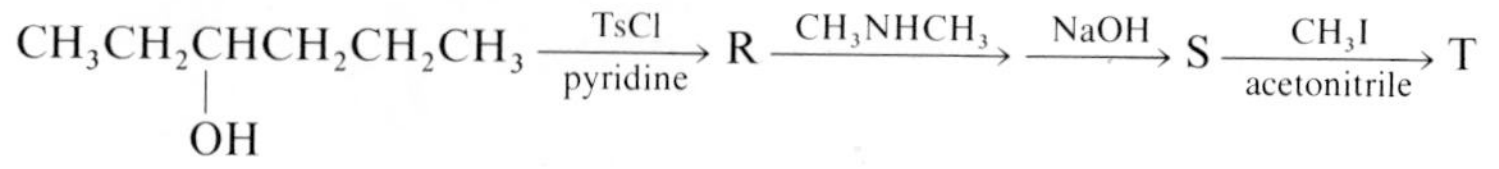 $CH_3CH_2CHCH_2CH_2CH_3$ $\xrightarrow[\text{pyridine}]{\text{TsCl}}$ R $\xrightarrow{CH_3NHCH_3}$ $\xrightarrow{\text{NaOH}}$ S $\xrightarrow[\text{acetonitrile}]{CH_3I}$ T

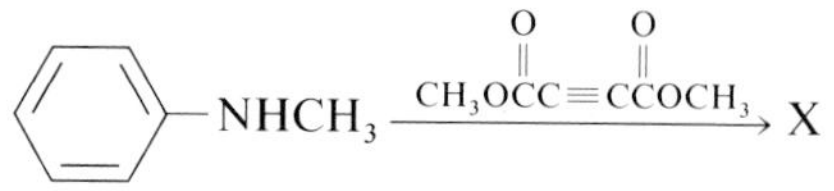

         | $\overset{|}{O}H$

(l) 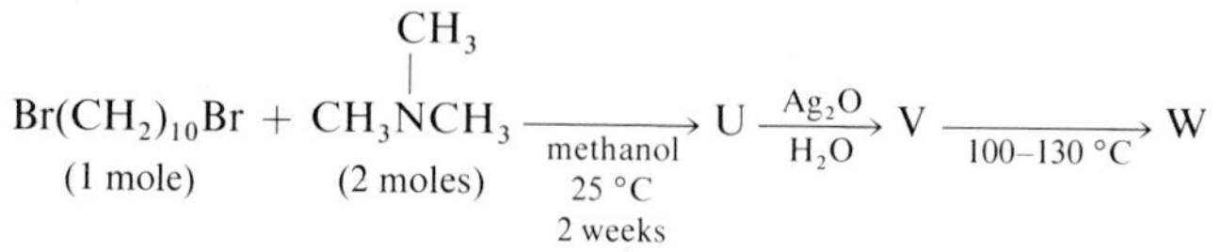 $Br(CH_2)_{10}Br$ + $CH_3\overset{\overset{\displaystyle CH_3}{|}}{N}CH_3$ $\xrightarrow[\substack{\text{methanol}\\25\,°C\\2\text{ weeks}}]{}$ U $\xrightarrow[H_2O]{Ag_2O}$ V $\xrightarrow{100\text{–}130\,°C}$ W

     (1 mole)   (2 moles)

(m)   ⟨benzene⟩—$NHCH_3$ $\xrightarrow{CH_3O\overset{\overset{\displaystyle O}{||}}{C}C\equiv C\overset{\overset{\displaystyle O}{||}}{C}OCH_3}$ X

(n) O_2N—⟨benzene⟩—NH_2 $\xrightarrow[\text{acetic acid}]{\text{ICl (excess)}}$ Y $\xrightarrow[\substack{H_2O\\5\,°C}]{\text{NaNO}_2,\ H_2SO_4}$ $\xrightarrow{KI}$ Z

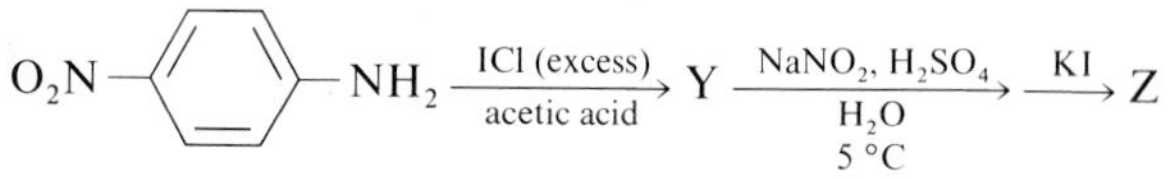

16.42   More practice problems follow. Give structural formulas for all the products and intermediates designated by letters.

(a) 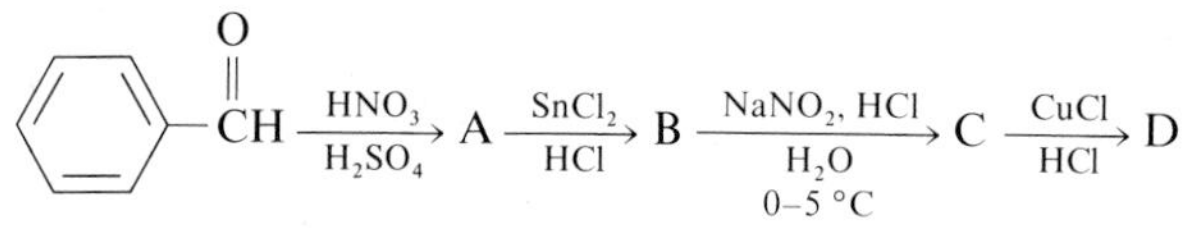   ⟨benzene⟩—$\overset{\overset{\displaystyle O}{||}}{C}H$ $\xrightarrow[H_2SO_4]{HNO_3}$ A $\xrightarrow[\text{HCl}]{SnCl_2}$ B $\xrightarrow[\substack{H_2O\\0\text{–}5\,°C}]{\text{NaNO}_2,\ \text{HCl}}$ C $\xrightarrow[\text{HCl}]{\text{CuCl}}$ D

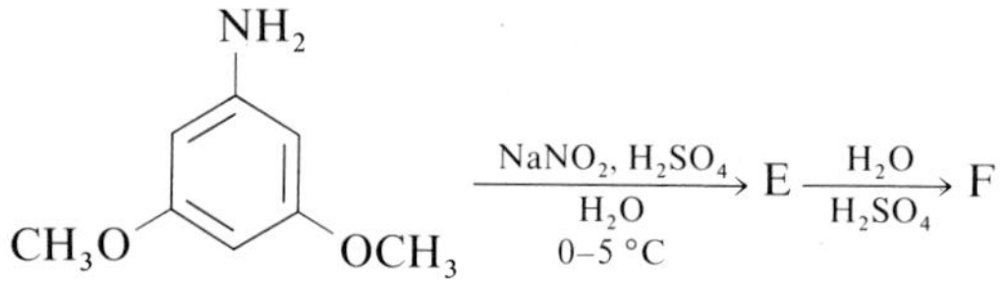

(b) CH_3O—⟨benzene with NH_2⟩—OCH_3 $\xrightarrow[\substack{H_2O\\0\text{–}5\,°C}]{\text{NaNO}_2,\ H_2SO_4}$ E $\xrightarrow[H_2SO_4]{H_2O}$ F

(c)   Br—⟨benzene⟩—OCH_3 $\xrightarrow[\text{NH}_3\ \text{(liq)}]{\text{NaNH}_2}$ G + H

(d) O_2N—⟨benzene with CH_3, NO_2⟩ $\xrightarrow[\text{ethanol}]{(NH_4)_2S}$ I $\xrightarrow[\substack{H_2O\\0\text{–}5\,°C}]{\text{NaNO}_2,\ H_2SO_4}$ J $\xrightarrow[H_2SO_4]{H_2O}$ K

(e)   ⟨benzene⟩—$CH_2C\equiv N$ $\xrightarrow[\substack{\text{Ni}\\NH_3}]{H_2}$ L

(f)   ⟨cyclohexane⟩=O $\xrightarrow[\substack{\text{sodium acetate}\\H_2O}]{\overset{+}{HONH_3Cl^-}}$ M $\xrightarrow[\text{ethanol}]{\text{Na}}$ N

(g)   ⟨benzene with $\overset{\overset{\displaystyle O}{||}}{C}OH$ and NH_2⟩ $\xrightarrow[\substack{H_2O\\3\text{–}5\,°C}]{\text{NaNO}_2,\ \text{HCl}}$ O $\xrightarrow{\text{⟨benzene⟩—}\overset{\overset{\displaystyle CH_3}{|}}{N}CH_3}$ P (methyl red)

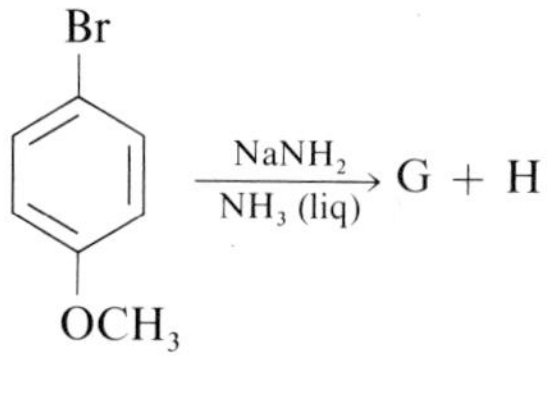

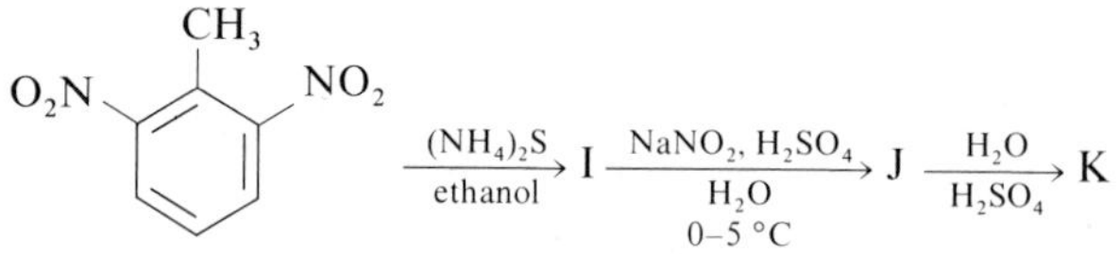

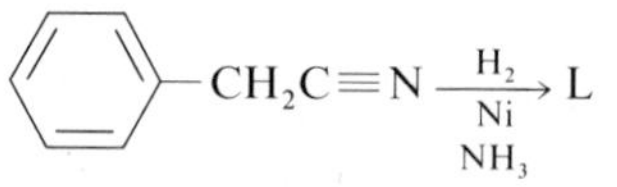

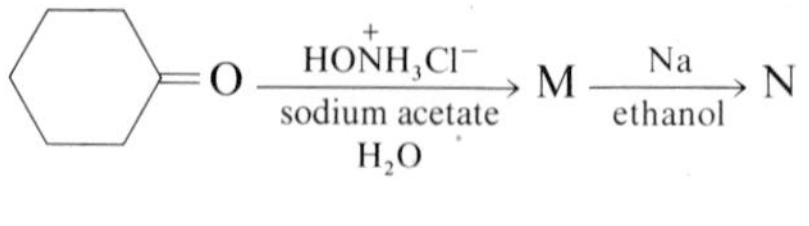

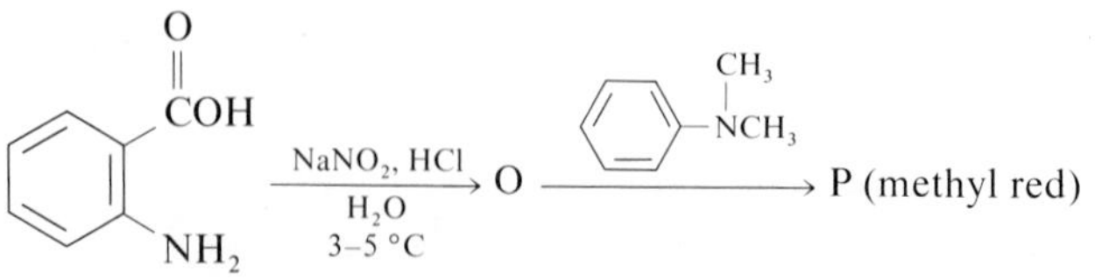

(h)

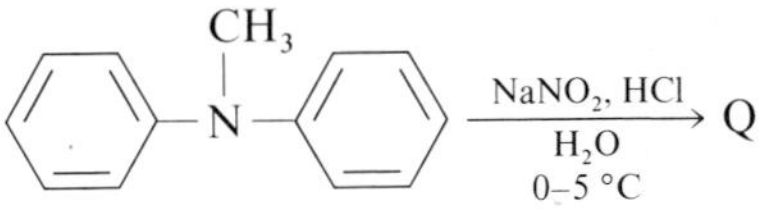

$$\xrightarrow[\substack{H_2O \\ 0-5\,°C}]{NaNO_2,\ HCl} Q$$

(i)

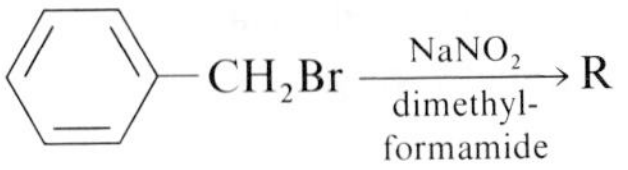

$-CH_2Br \xrightarrow[\substack{\text{dimethyl-} \\ \text{formamide}}]{NaNO_2} R$

(j) 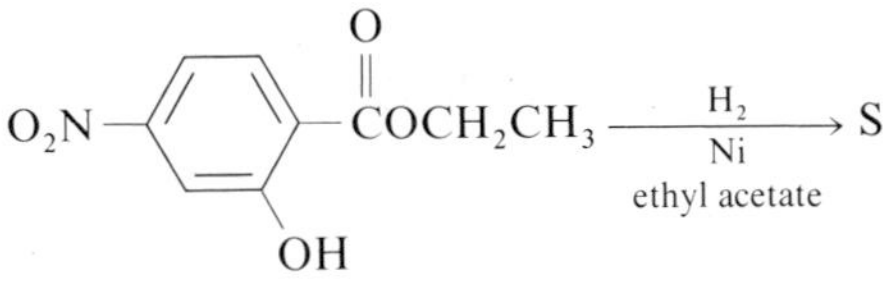 O_2N-

$\overset{\overset{\displaystyle O}{\|}}{-COCH_2CH_3} \xrightarrow[\substack{Ni \\ \text{ethyl acetate}}]{H_2} S$

with OH substituent

(k)

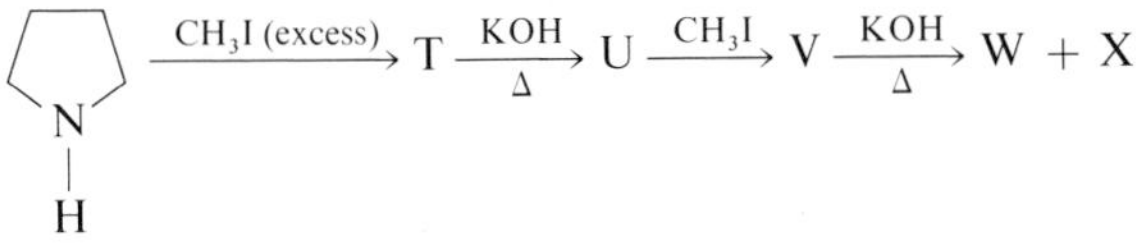

$\xrightarrow{CH_3I\ (excess)} T \xrightarrow[\Delta]{KOH} U \xrightarrow{CH_3I} V \xrightarrow[\Delta]{KOH} W + X$

(l) 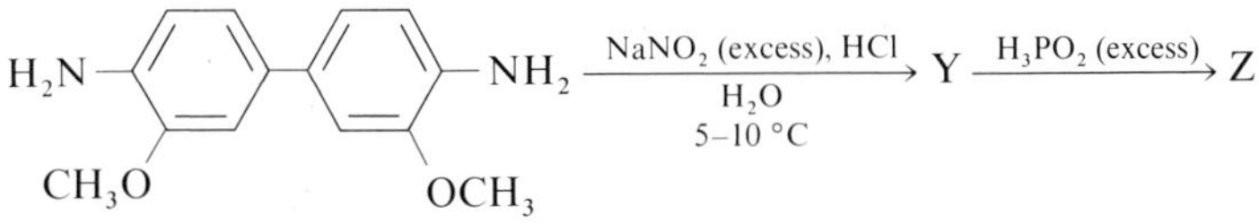 H_2N-

$-NH_2 \xrightarrow[\substack{H_2O \\ 5-10\,°C}]{NaNO_2\ (excess),\ HCl} Y \xrightarrow{H_3PO_2\ (excess)} Z$

with CH_3O and OCH_3 substituents

16.43 Show how you would carry out the following transformations. Some of them may require more than one step.

(a)

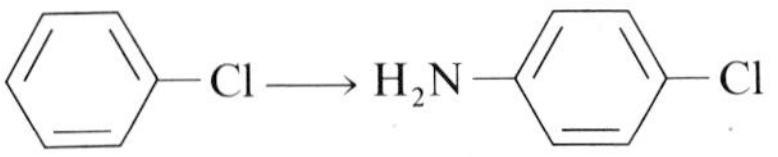

$-Cl \longrightarrow H_2N-$

$-Cl$

(b)

$-CH_3 \longrightarrow CH_3 \overset{\overset{\displaystyle CH_3}{|}}{N}-$

$\overset{\overset{\displaystyle O}{\|}}{-COH}$

(c)

$-CH_2Br \longrightarrow$

$-CH_2\overset{\overset{\displaystyle CH_2CH_3}{|}}{CH}CH_2NH_2$

(d)

$\overset{\overset{\displaystyle O}{\|}}{-CH} \longrightarrow$

$-CH_2CH_2\underset{\underset{\displaystyle NH_2}{|}}{CH}CH_3$

(e)

$-Br \longrightarrow I-$

$-Br$

(f)

$-CH_3 \longrightarrow H_2NCH_2-$

$-CH_2NH_2$

(g) $HO\overset{\overset{\displaystyle O}{\|}}{C}CH_2\overset{\overset{\displaystyle CH_3}{|}}{C}HCH_2CH_2\overset{\overset{\displaystyle O}{\|}}{C}OH \longrightarrow H_2NCH_2\overset{\overset{\displaystyle CH_3}{|}}{C}HCH_2CH_2NH_2$

(h) $CH_3CH_2\overset{\overset{\displaystyle O}{\|}}{C}CH_2CH_3 \longrightarrow CH_3CH_2\overset{\overset{\displaystyle O}{\|}}{C}NHCH_2CH_3$

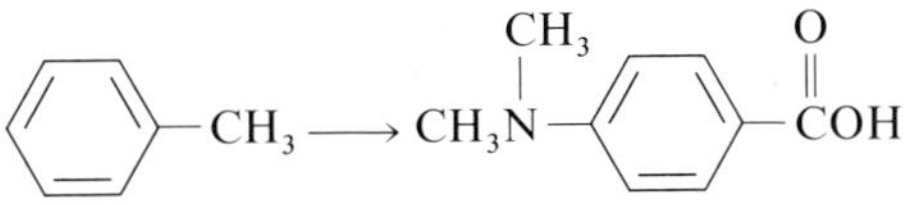

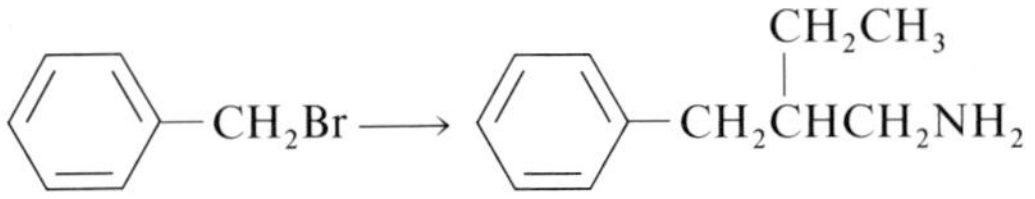

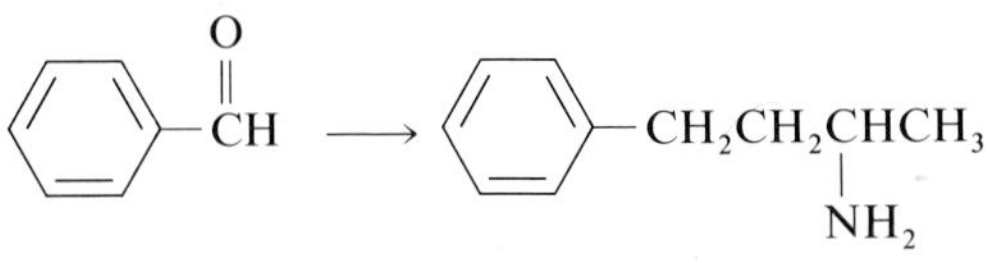

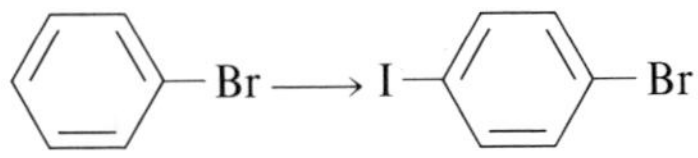

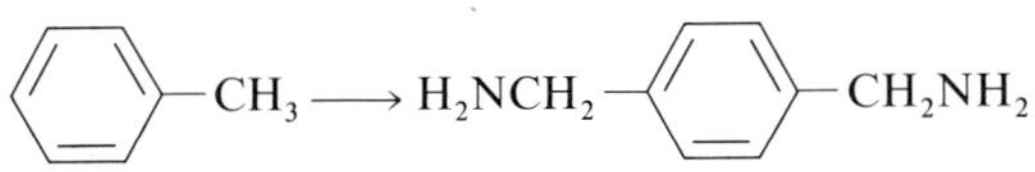

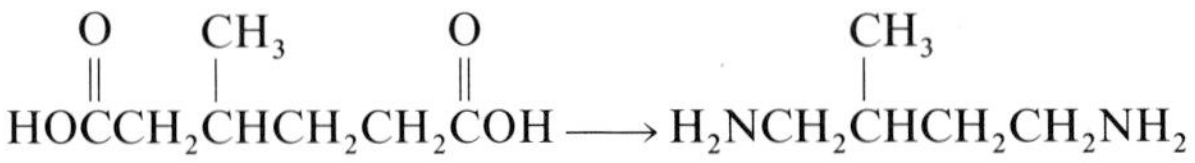

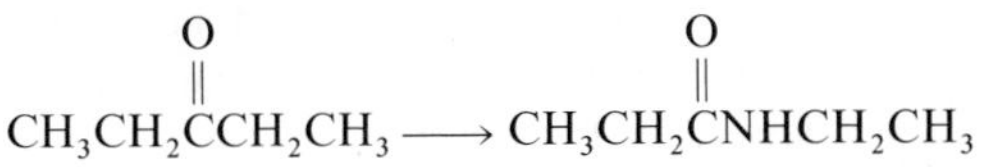

(i) $CH_3CH_2CH_2CH_2CH_2CH_2OH \longrightarrow CH_3CHCH_2CH_2CH_2CH_2CH_2CH_3$
$\qquad\qquad\qquad\qquad\qquad\qquad\qquad\qquad\qquad\qquad\qquad\quad |$
$\qquad\qquad\qquad\qquad\qquad\qquad\qquad\qquad\qquad\qquad\quad NO_2$

16.44   The following sequence of reactions was used to synthesize an amino acid. Fill in structural formulas for intermediates or products represented by letters.

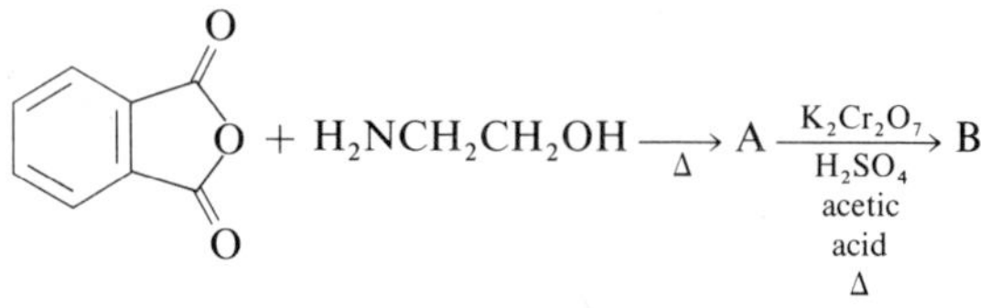

$B \xrightarrow[\substack{HCl \\ \Delta}]{H_2O} C + D$

$D \xrightarrow[pyridine]{} \text{an amino acid}$

16.45   An isomer of alanine, which is 2-aminopropanoic acid, is 3-aminopropanoic acid, also called β-alanine. It is prepared by a Gabriel synthesis in which the phthalimidate anion is added to propenenitrile in a Michael-type reaction (Section 15.7A). Write equations showing the mechanism of this step of the reaction. What other reactions must be carried out to get 3-aminopropanoic acid by this route?

$$CH_2{=}CHC{\equiv}N$$
propenenitrile

16.46   The nucleophilic substitution reaction of 2,4-dinitrochlorobenzene (Section 12.5) with a series of substituted anilines was studied. The rate of the reaction was found to depend on the substituent at the para position of the aniline. Write a general equation for the reaction described, and predict the order of reactivity for the following anilines, showing clearly which one would be expected to react the fastest and which the slowest.

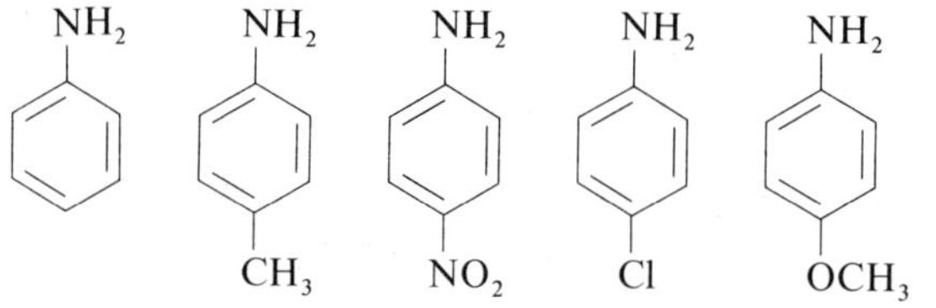

16.47   When (aminomethyl)cyclopentane is treated with sodium nitrite in water with sodium dihydrogen phosphate as the acid, the following alcohol products are observed.

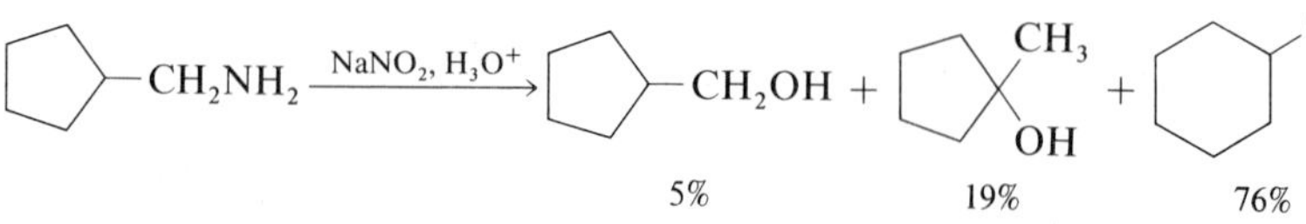

Write a detailed mechanism for the reaction showing how each of the products could have been formed.

16.48   The five-membered D-ring of a steroid, estrone, is converted into a six-membered ring by the following sequence of reactions. Fill in the intermediate structures, and propose a mechanism for the step in which expansion of the D-ring occurs.

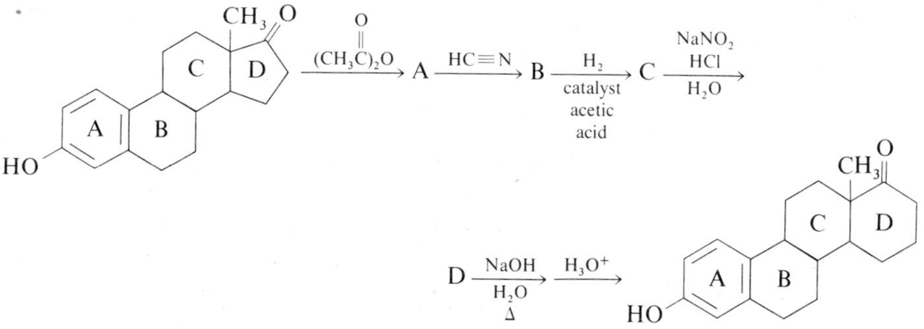

16.49 In Section 12.3G, the reactions of benzenesulfonyl chloride with aniline and *N*-methylaniline and its lack of reactivity with a tertiary amine were described. The acid-base properties of the sulfonamides were emphasized. The reaction of benzenesulfonyl chloride with amines in the presence of aqueous sodium hydroxide is the basis of a qualitative analysis test, known as the **Hinsberg test,** to distinguish low molecular weight primary, secondary, and tertiary amines. Using hexylamine, dipropylamine, and triethylamine to represent those classes of amines, write equations showing what you would expect to happen with each one under the following conditions. At each point, be sure to describe what you expect to see in the test tube. All of the above amines are liquids; all sulfonamides are solids.

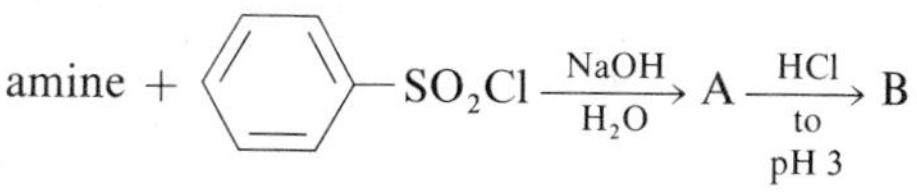

How could these reactions be used to distinguish primary, secondary, and tertiary amines?

16.50 Propose syntheses for the following compounds starting with benzene, toluene, and any other organic compounds containing three or fewer carbon atoms.

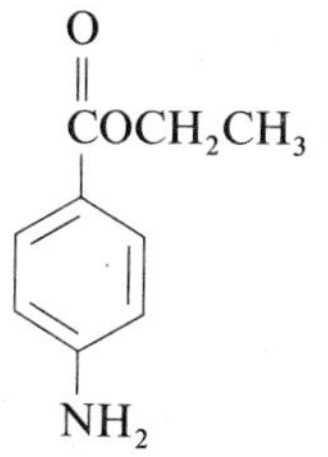

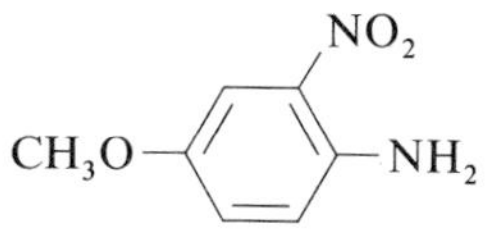

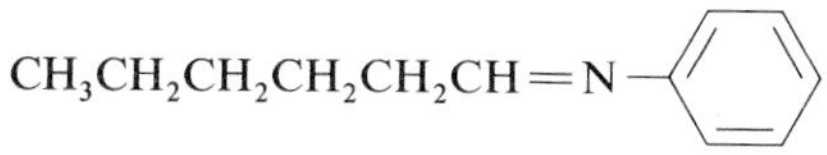

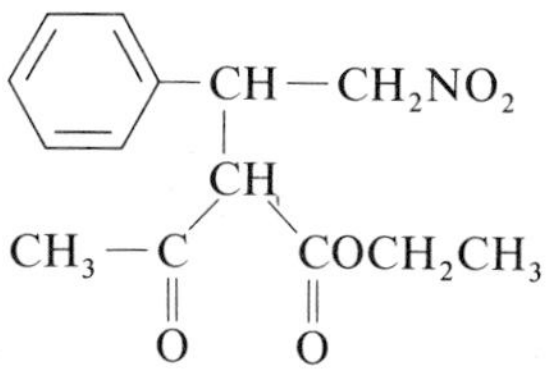

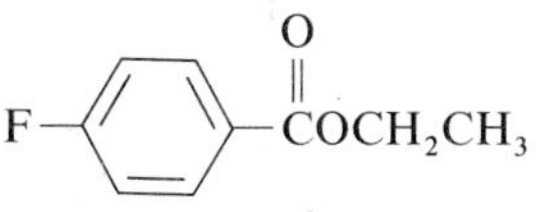

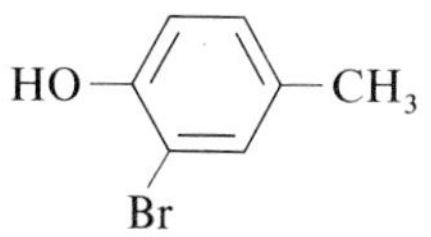

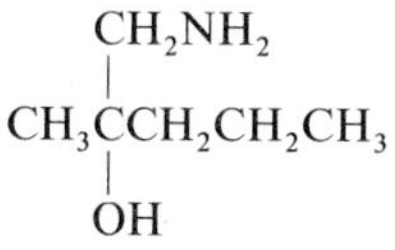

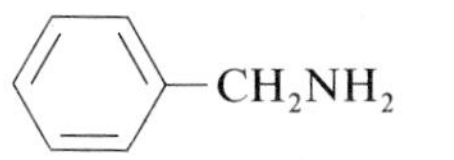

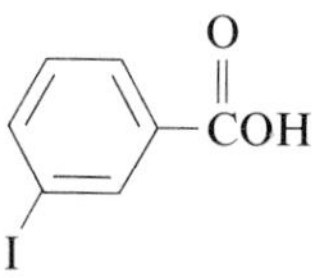

(j) 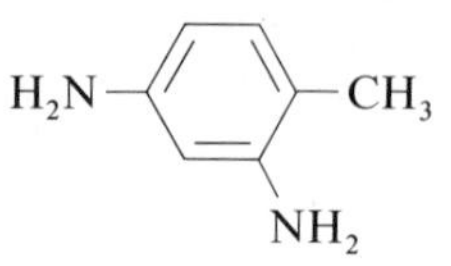 H$_2$N—⟨benzene ring⟩—CH$_3$ with NH$_2$

(k) 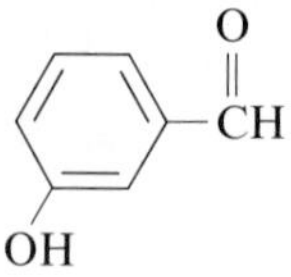 ⟨benzene ring⟩—CH with O (double bond) and OH

(l) 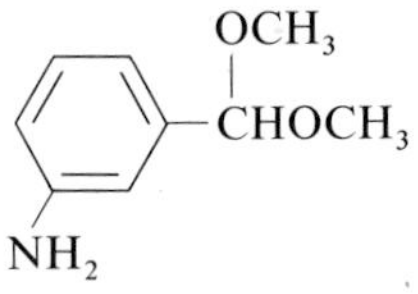 ⟨benzene ring⟩—CHOCH$_3$ with OCH$_3$ and NH$_2$

16.51 2-Amino-1-(4-hydroxyphenyl)propane was prepared in a racemic form in experiments on the synthesis of analogs of epinephrine. It was resolved by the use of ($+$)-tartaric acid (Section 16.4B). The levorotatory enantiomer of 2-amino-1-(4-hydroxyphenyl)propane was synthesized from (S)-($-$)-tyrosine by reactions that preserve the configuration of the asymmetric carbon atom.

Write structural formulas that show the absolute configuration of ($-$)- and ($+$)-2-amino-1-(4-hydroxyphenyl)propane, and outline a resolution of the racemic mixture.

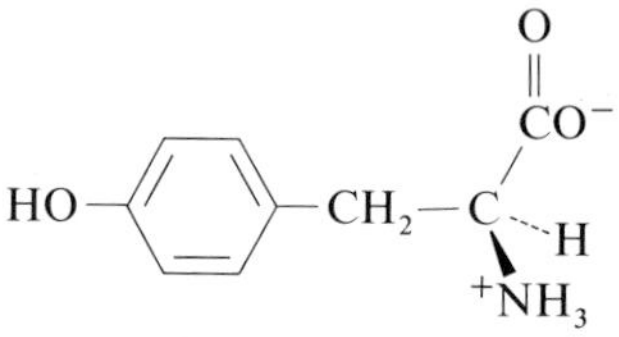

HO—⟨benzene ring⟩—CH$_2$—C with CO$^-$ (C=O), H, $^+$NH$_3$

(S)-($-$)-tyrosine

16.52 The following pK_a values were measured for the conjugate acids of aniline and substituted anilines. Rationalize the trends that are apparent in the data.

(a)

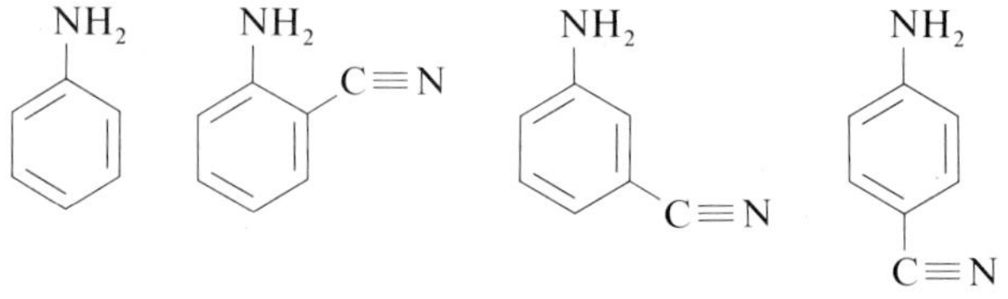

aniline	2-cyanoaniline	3-cyanoaniline	4-cyanoaniline
pK_a 4.60	0.95	2.75	1.74

(b)

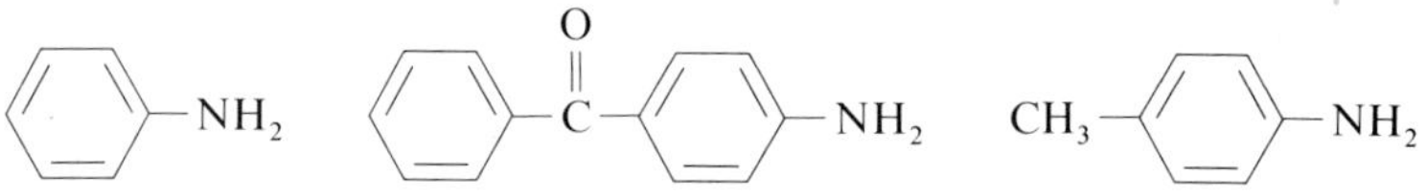

benzylamine	4-aminophenyl ketone	4-methylaniline
pK_a 4.60	2.17	5.10

16.53 A bicyclic amine has been synthesized by the following reactions. Provide structural formulas for the compounds designated by letters.

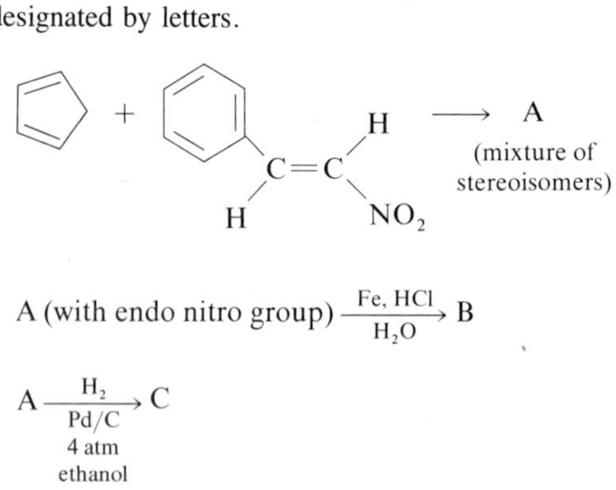

⟨cyclopentadiene⟩ + ⟨phenyl⟩—C=C with H and NO$_2$ ⟶ A (mixture of stereoisomers)

A (with endo nitro group) $\xrightarrow[\text{H}_2\text{O}]{\text{Fe, HCl}}$ B

A $\xrightarrow[\substack{\text{Pd/C} \\ \text{4 atm} \\ \text{ethanol}}]{\text{H}_2}$ C

The structure of Compound C was confirmed by the reactions that follow. Give structural formulas for Compounds A to H.

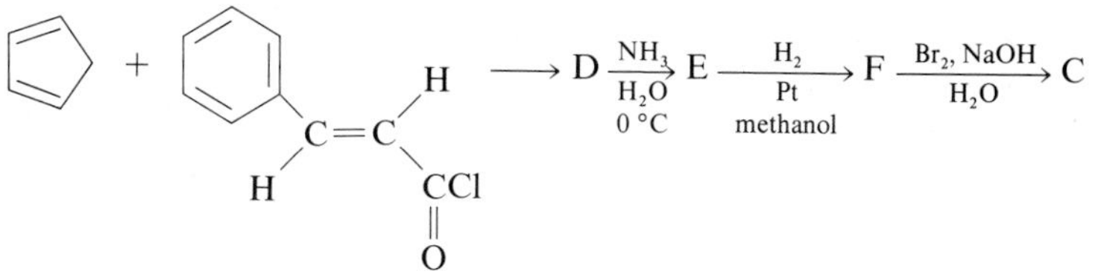

$$D \xrightarrow[\text{H}_2\text{O}]{\text{NaOH}} G \xrightarrow{\text{H}_3\text{O}^+} H$$

16.54 Trimethylenecyclopropane, C_6H_6, is isomeric with benzene. A sample of the compound was synthesized in order to study its physical and chemical properties.

(a) The synthesis is outlined below. Fill in the structural formulas for the compounds designated by letters.

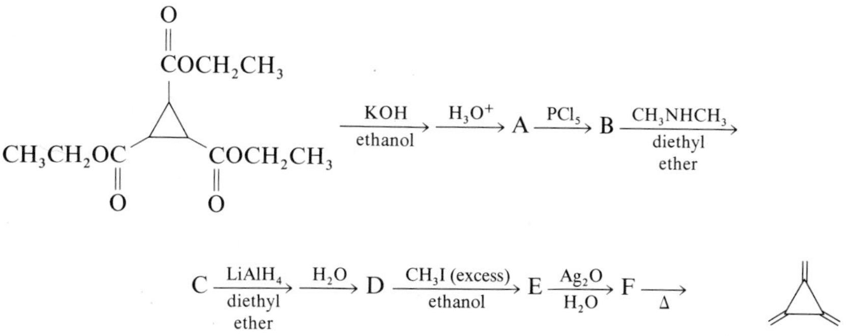

(b) Compound E was also prepared by the following route.

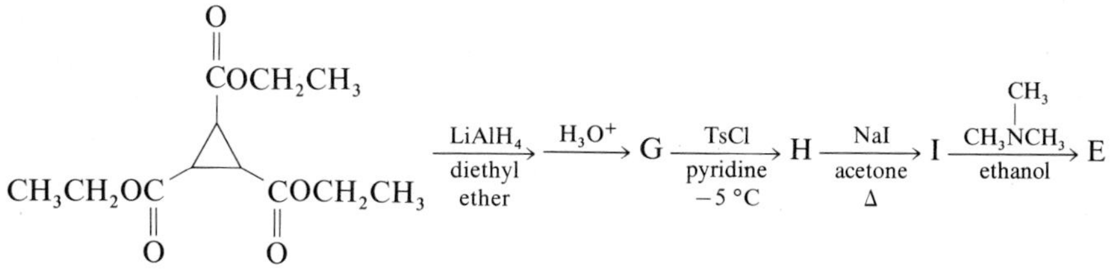

(c) The conversion of Compound G to Compound I could in theory also be carried out with hydriodic acid. The acid was not used because molecular rearrangement was feared. Write equations showing what rearrangement is possible and explain why special care was necessary in this synthesis.

(d) What stereoisomerism is possible for the triethyl cyclopropanetricarboxylate used as the starting material for these syntheses? Draw structural formulas that clearly illustrate your answers.

(e) Trimethylenecyclopropane can also be prepared from Compound I. Write equation(s) showing how you would carry out this transformation.

16.55 The following compounds were prepared as intermediates in the synthesis of compounds analogous to alkaloids. Supply reagents for the transformations shown below.

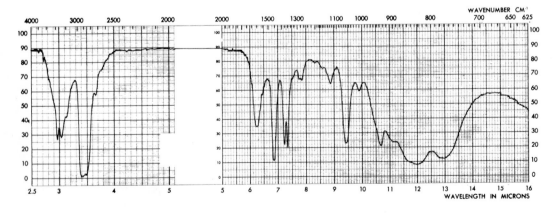

16.56 The dye Butter Yellow, *p*-dimethylaminoazobenzene (p. 929), has λ_{max} 408 nm in neutral solution. When acid is added to the solution, two different species are detected spectroscopically. One has λ_{max} 320 nm, the other, λ_{max} 510 nm. Write structural formulas for the species that have chromophores giving rise to these two different absorption bands.

16.57 The infrared and proton magnetic resonance spectra for Compound G are given in Figure 16.8. Assign a structure to the compound.

Figure 16.8 Spectra for Problem 16.57. (From *The Aldrich Library of Infrared Spectra* and *The Aldrich Library of NMR Spectra*)

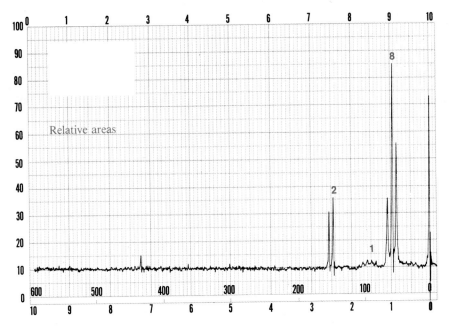

Figure 16.8 *(Continued)*

16.58   Infrared spectra for the following compounds are given in Figure 16.9.   Which spectrum belongs
to which compound?

(a)   octadecanamide   (b)   isopropylamine

(c)   *N*-methylbenzylamine   (d)   *N*-benzylformamide

(e)   *N,N*-dimethylbenzylamine   (f)   *N,N*-dibutylformamide

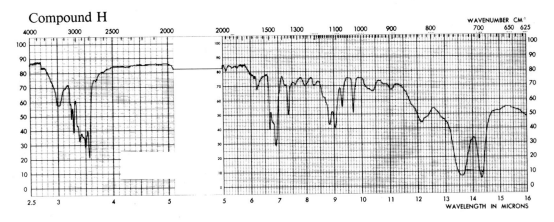

Figure 16.9   Spectra for Problem 16.58. (From *The Aldrich Library of Infrared Spectra*)

Compound I

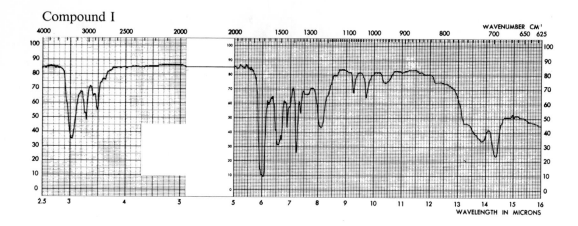

Compound J

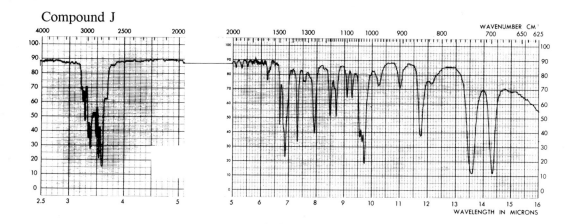

Compound K

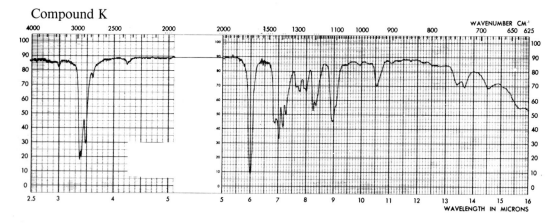

Figure 16.9 *(Continued)*

Compound L

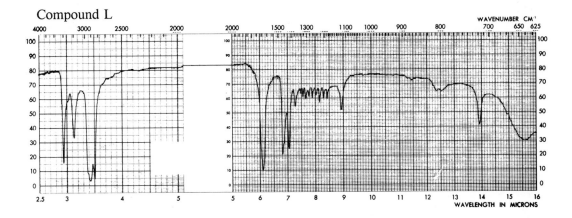

Compound M

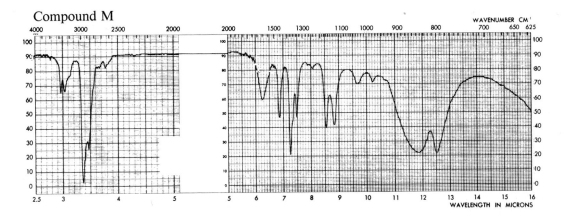

Figure 16.9 (*Continued*)

16.59   Compound N, $C_7H_7NO_3$, is converted into Compound O, C_7H_9NO, when treated with reducing agents such as tin in hydrochloric acid. Proton magnetic resonance spectra for N and O are shown in Figure 16.10. Assign structures to these compounds. How do you account for the difference that is apparent in these two spectra in the region where the hydrogen atoms on aromatic rings absorb.

Compound N

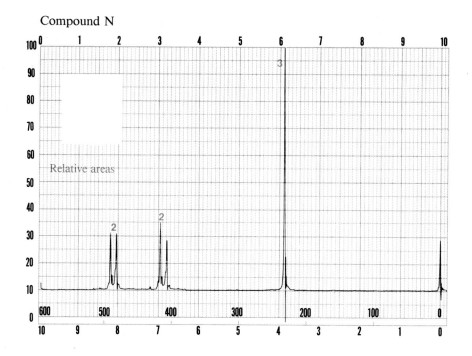

Compound O

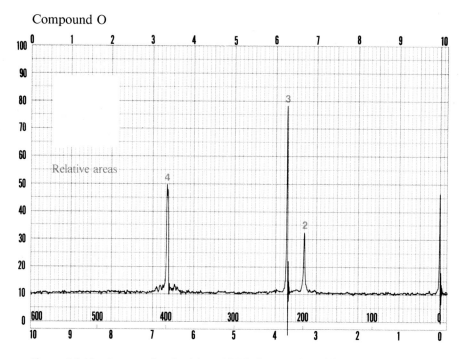

Figure 16.10   Spectra for Problem 16.59. (From *The Aldrich Library of NMR Spectra*)

16.60   Compound P, $C_{10}H_{15}N$, has the proton magnetic resonance spectrum shown in Figure 16.11. Assign a structure to the compound.

Compound P

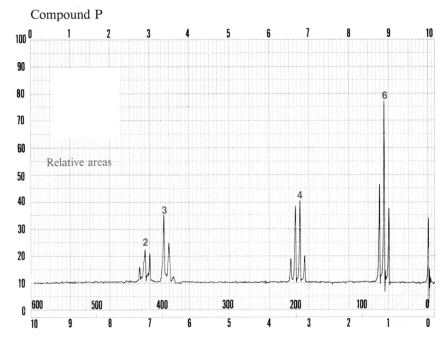

Figure 16.11 Spectrum for Problem 16.60. (From *The Aldrich Library of NMR Spectra*)

16.61 Compound Q, $C_8H_{11}NO$, has the proton magnetic resonance spectrum shown in Figure 16.12. Assign a structure to the compound that is consistent with its formula and its spectrum.

Compound Q

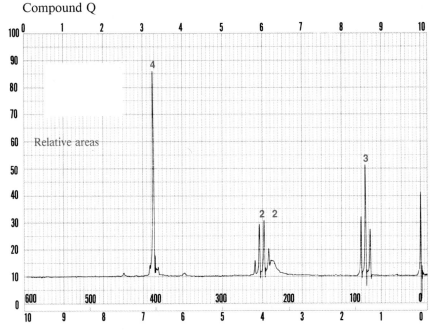

Figure 16.12 Spectrum for Problem 16.61. (From *The Aldrich Library of NMR Spectra*)

16.62 The carbon-13 nuclear magnetic resonance spectrum of *N*-ethylacetamide has bands at 14.6, 22.8, 34.4, and 171.0 ppm downfield from TMS. Assign these bands to the various carbon atoms in the compound. Table 13.4, Section 13.5 may be useful.

16.63   Compound R, C_3H_9N, has two bands in its carbon-13 nuclear magnetic resonance spectrum, at 26.2 and 42.8 ppm downfield from TMS. Assign a structure to Compound R.

16.64   Compound S, $C_8H_{11}N$, has bands in its carbon-13 nuclear magnetic resonance spectrum at 40.2, 112.7, 116.6, 129.0, and 150.7 ppm downfield from TMS. Compound S is not soluble in water, but does dissolve in dilute hydrochloric acid. It does not react with benzenesulfonyl chloride in the presence of sodium hydroxide. Assign a structure to Compound S and assign the bands that appear in its spectrum.

16.65   Compound T, $C_9H_{13}N$, has the infrared and proton magnetic resonance spectra shown in Figure 16.13. It has $\lambda_{max}^{CH_3OH}$ 286 (ϵ 1530) and 232.5 (ϵ 9920), and $\lambda_{max}^{CH_3OH,HCl}$ 250 (ϵ 314) and 206 nm (ϵ 7120). Assign a structure to Compound T and discuss the important features of its spectra.

Compound T

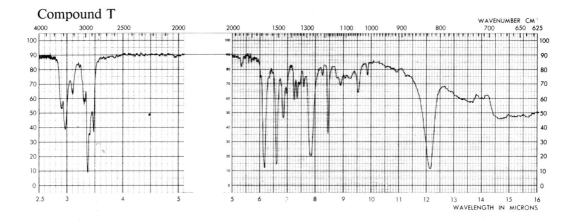

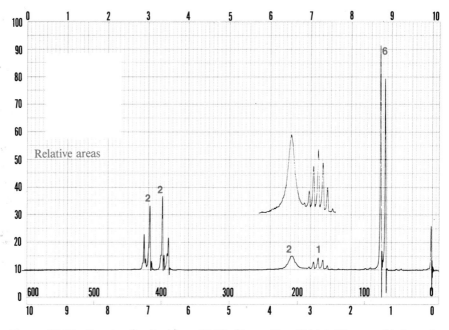

Figure 16.13   Spectra for Problem 16.65. (From *The Aldrich Library of Infrared Spectra* and *The Aldrich Library of NMR Spectra*)

Amino Acids, Peptides, Proteins

17

Introduction

A protein *is a high molecular weight compound composed of amino acids bonded to each other by* **peptide linkages.** Proteins have long been recognized as one of the primary constituents of living matter. Even in plants, where carbohydrates are more abundant as structural material, proteins are present in the parts of the plant that are responsible for its growth and reproduction. Ribonucleic acids and deoxyribonucleic acids, which transmit hereditary information by directing the synthesis of proteins, are found associated with proteins in the nuclei and ribosomes of all cells.

Most of the chemical reactions in cells are catalyzed by **enzymes,** which are complex and highly specialized proteins. About two thousand different enzymes, each of which catalyzes a different reaction, are known. In contrast to chemical reactions carried out in the laboratory, which often require high temperatures and strong acid or base for catalysis and give many side-products as well as the desired one, enzymes function in our bodies at 37 °C and at a nearly neutral pH to give the single desired product in 100% yield.

The fundamental structure of proteins is in principle very simple. Protein molecules consist of long chains of amino acids bonded to each other by amide linkages between the carboxylic acid group of one amino acid and the amino group in another (Sections 10.8B and 16.1A). *These chains are known as* **polypeptides** and may contain 50 to 300 amino acid residues. A protein may consist of a single polypeptide chain or of several associated with each other. Molecular weights for proteins range from around 5000 to 1,000,000.

About 20 different amino acids are the units that recur in proteins. The number of ways in which these amino acids may be combined in sequence to give a protein is staggeringly large, even for relatively small proteins. The amino acids may be regarded as the letters of an alphabet from which an infinite variety of words can be created.

Although the amino acid units of a polypeptide are linked to each other in a linear fashion, the polypeptide itself assumes a folded conformation that is typical of that particular molecule and is to a great extent responsible for its biological activity. In many cases, the polypeptide chain forms a right-handed coil around a central axis. *This conformation of the polypeptide is known as the **α-helix**, and several such helices may be wound around each other to give strands of **fibrous** proteins* such as those found in hair and wool and in the tough proteins of horns and nails. *The α-helices may also fold back upon themselves to form parts of the spherical structures typical of **globular proteins.*** All of the enzymes are globular proteins, as are proteins such as hemoglobin in blood cells and myoglobin, which performs a similar oxygen-transporting function in the muscles. The specific conformation of each protein is necessary to bring distant portions of the polypeptide chain together to create the folds and pockets lined with the particular amino acids that function in the catalytic processes of the enzyme.

*The order of amino acids in the backbone of a polypeptide is known as the **primary structure** of the peptide. The conformation of the peptide chain gives it a **secondary structure**. Further bending and folding of the peptide chain creates the **tertiary structure** of proteins. Finally, several polypeptide units may be associated with each other and with other simpler molecules such as sugars, inorganic residues, or co-enzymes to give what is known as the **quaternary structure** of proteins.*

In the next sections of the chapter, we explore in greater detail the chemistry of amino acids and peptides in order to understand better the complex proteins and their functions as structural units and catalysts in living organisms.

17.2

The Structure and Properties of Amino Acids

A. Structure and Stereochemistry of Amino Acids. A Review

The amino acids found in proteins derived from animals and higher plants are all α-aminocarboxylic acids of a particular stereochemical configuration (Sections 3.9B and 16.1A). The older literature refers to them as L-amino acids, meaning that their configuration is related to that of L-($-$)-glyceraldehyde (Section 14.2D).

an L-amino acid   L-($-$)-glyceraldehyde

configurational relationship between L-amino acids and L-($-$)-glyceraldehyde

The acidity and basicity of amino acids, derived from the presence of both a carboxyl and an amino group, are important properties. Additional basic or acidic functional groups in the R group of amino acids also influence their chemistry. It is most useful to classify amino acids according to their ionic state at the pH that exists in the cells and fluids of living organisms. The amino acids that are commonly found in proteins will be listed in Section 17.2C after we discuss the reactions of amino acids as acids and bases in the next section.

B. Amino Acids as Acids and Bases

Amino acids have relatively high melting points, usually decomposing above 200 °C. They have good solubility in water and low solubility in nonpolar solvents. They have large dipole moments, and their solutions in water have high dielectric constants. All of these facts suggest that the units in the crystal lattice and in solution are charged species.

An examination of the relative basicities of a carboxylate anion and of an amino group indicate that an amino acid should exist as a dipolar ion, also called a **zwitterion,** in which the amino group is protonated and the carboxyl group exists as a carboxylate anion (Problem 4.33). In other words, the acidic group in an amino acid is a substituted ammonium ion and the basic group is the carboxylate anion. The acid-base reactions of an amino acid are shown below.

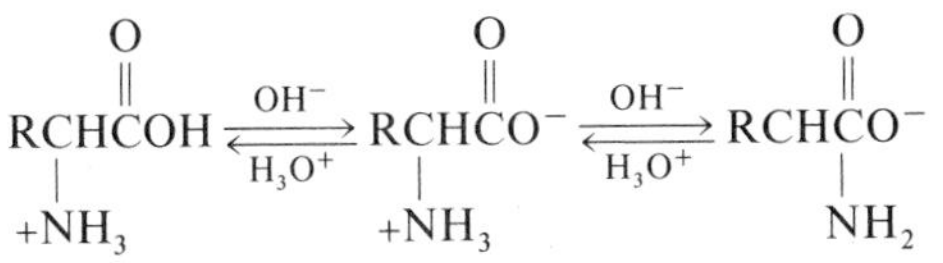

$$\underset{\underset{1}{\underset{+\text{NH}_3}{|}}}{\text{RCHCOH}} \underset{\text{H}_3\text{O}^+}{\overset{\text{OH}^-}{\rightleftharpoons}} \underset{\underset{2}{\underset{+\text{NH}_3}{|}}}{\text{RCHCO}^-} \underset{\text{H}_3\text{O}^+}{\overset{\text{OH}^-}{\rightleftharpoons}} \underset{\underset{3}{\underset{\text{NH}_2}{|}}}{\text{RCHCO}^-}$$

From this equation it is clear that at very low pH, the amino acid exists in a form in which it is protonated at both the amino group and the carboxyl group (species 1). At such a pH, it bears a net positive charge and is a diprotic acid. At high pH, the amino acid bears a net negative charge and has two basic sites at which it can be protonated (species 3). Note that the equation suggests that it will be protonated first at the amino group, then at the carboxylate anion. This is in accord with the relative basicities of the two groups.

At some intermediate pH, the amino acid exists as a zwitterion with no net charge (species 2). The pH at which this is true is known as the pH_{I}, or the **isoelectric point,** for the amino acid. At this pH, the amino acid is stationary in an electric field and migrates neither to the negative pole nor to the positive pole because the charges on it are balanced. In contrast, at low pH the amino acid, bearing a positive charge, migrates to the negative pole, while at high pH, it migrates to the positive pole of an electric field.

The behavior of the fully protonated amino acid as a diprotic acid can be seen clearly if a titration is carried out. For the typical amino acid, two pK_a values are determined. When half of a molar equivalent of base is added to the amino acid alanine ($R = CH_3$) in its fully protonated form, the pH of the solution is 2.34. At this point, the concentration of the zwitterionic form 2 equals the concentration of the diprotic acid 1. According to the Henderson-Hasselbach equation (Problem 4.35), the pK_a is equal to pH when the concentration of an acid and its conjugate base are equal. Therefore, the pK_a of the carboxylic acid group in alanine must be 2.34.

$$pK_a = \text{pH} + \log\frac{[\text{HA}]}{[\text{A}^-]}$$

when $[\text{HA}] = [\text{A}^-]$, $\log\dfrac{[\text{HA}]}{[\text{A}^-]} = 0$

and $pK_a = \text{pH}$

$$\text{At pH 2.34,} \quad \left[\begin{matrix} & \overset{\displaystyle O}{\underset{\displaystyle \|}{}} \\ \text{CH}_3\text{CHCOH} \\ | \\ \text{+NH}_3 \end{matrix} \right] = \left[\begin{matrix} & \overset{\displaystyle O}{\underset{\displaystyle \|}{}} \\ \text{CH}_3\text{CHCO}^- \\ | \\ \text{+NH}_3 \end{matrix} \right]$$

$$\text{therefore, } pK_a \text{ of } \underset{\displaystyle \underset{\displaystyle \text{+NH}_3}{|}}{\overset{\displaystyle \overset{\displaystyle O}{\|}}{\text{CH}_3\text{CHCOH}}} = 2.34$$

When a molar equivalent of base has been added, the zwitterionic form 2 predominates. The pH at this point is 6.02, the isoelectric point of alanine. Finally, addition of another half of a molar equivalent of base gives a pH of 9.69, which corresponds to the pK_a of the ammonium ion derived from the amino acid. At this point, equal concentration of zwitterion 2 and its conjugate base 3 exist.

The pH_I is the arithmetic mean of the pK_a values for the two acid groups in the compound.

$$pH_I = \frac{1}{2}(pK_{a_1} + pK_{a_2})$$

An isoelectric point of approximately 6 is typical for the amino acids that do not have additional acidic or basic groups on their side chains.

A pK_a of 2.34 for the carboxylic acid group of alanine indicates that it is a stronger acid than a typical aliphatic carboxylic acid such as acetic acid, which has pK_a 4.76. The protonated amino group with its positive charge has an electron-withdrawing effect that strengthens the acid (Section 4.7B). The amino group is also affected by the presence of the adjacent carboxylic acid group. It is less basic and its conjugate acid more acidic than is the case for an ordinary aliphatic amine. The conjugate acid of methyl amine, for example, has pK_a 10.6, in contrast to pK_a 9.69 for the protonated amino group in alanine.

Amino acids that have additional amino or carboxyl groups in them have different isoelectric points that reflect their tendency either to be further protonated or to lose an additional proton. The amino acid lysine, for example, has three pK_a values: 2.18 for the carboxylic acid group, 8.95 for the α-amino group, and 10.53 for the amino group on carbon 6, the ϵ-amino group.

$$\underset{\displaystyle \underset{\displaystyle \text{+NH}_3}{|}}{\overset{\displaystyle \overset{\displaystyle O}{\|}}{\underset{+}{\text{H}_3\text{N}}\overset{\epsilon}{\text{C}}\text{H}_2\overset{\delta}{\text{C}}\text{H}_2\overset{\gamma}{\text{C}}\text{H}_2\overset{\beta}{\text{C}}\text{H}_2\overset{\alpha}{\text{C}}\text{HCO}^-}} \underset{\text{H}_3\text{O}^+}{\overset{\text{OH}^-}{\rightleftharpoons}} \underset{\displaystyle \underset{\displaystyle \text{NH}_2}{|}}{\overset{\displaystyle \overset{\displaystyle O}{\|}}{\overset{+}{\text{H}_3\text{N}}\text{CH}_2\text{CH}_2\text{CH}_2\text{CH}_2\text{CHCO}^-}}$$

form that predominates at physiological pH   form that predominates at pH 9.74

$$\text{H}_3\text{O}^+ \updownarrow \text{OH}^- \qquad\qquad\qquad \text{H}_3\text{O}^+ \updownarrow \text{OH}^-$$

$$\underset{\displaystyle \underset{\displaystyle \text{+NH}_3}{|}}{\overset{\displaystyle \overset{\displaystyle O}{\|}}{\overset{+}{\text{H}_3\text{N}}\text{CH}_2\text{CH}_2\text{CH}_2\text{CH}_2\text{CHCOH}}} \qquad \underset{\displaystyle \underset{\displaystyle \text{NH}_2}{|}}{\overset{\displaystyle \overset{\displaystyle O}{\|}}{\text{H}_2\text{N}\text{CH}_2\text{CH}_2\text{CH}_2\text{CH}_2\text{CHCO}^-}}$$

lysine at low pH   lysine at high pH

The isoelectric point for lysine is 9.74. Note that the ϵ-amino group is more basic than the α-amino group in lysine, and hence in lysine the zwitterionic form is shown with the protonated amino group at carbon 6. The isoelectric point of lysine indicates that lysine would bear a net positive charge at pH 7, roughly the pH of cells. Only if the pH of the solution were raised to 9.74 would most of the lysine be converted to the zwitterionic form with no net charge.

Glutamic acid, on the other hand, has an isoelectric point of 3.2. The carboxylic acid group on carbon 4 is like a normal aliphatic carboxylic acid group and has pK_a 4.25. The other carboxylic acid group has pK_a 2.19 and the protonated amino group, 9.67.

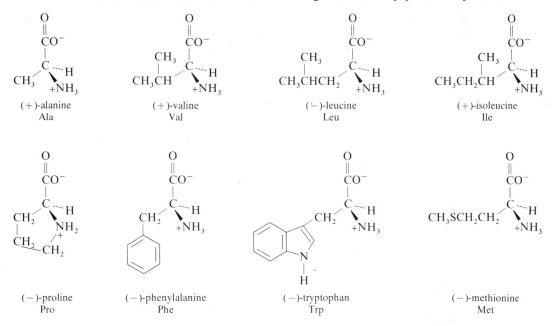

$$\underset{\underset{+NH_3}{|}}{HOCCH_2CH_2CHCO^-} \quad \overset{OH^-}{\underset{H_3O^+}{\rightleftarrows}} \quad \underset{\underset{+NH_3}{|}}{^-OCCH_2CH_2CHCO^-}$$

form that predominates at pH 3.2   form that predominates at physiological pH

$$H_3O^+ \updownarrow OH^- \qquad\qquad H_3O^+ \updownarrow OH^-$$

$$\underset{\underset{+NH_3}{|}}{HOCCH_2CH_2CHCOH} \qquad \underset{\underset{NH_2}{|}}{^-OCCH_2CH_2CHCO^-}$$

glutamic acid at low pH   glutamic acid at high pH

At physiological pH glutamic acid bears a net negative charge. Only when the pH is lowered to 3.2 does it exist as the zwitterion with no net charge.

C. A Classification of the Amino Acids

The common amino acids can be classified into four groups on the basis of their properties at physiological pH. The first group contains the amino acids that have no polar substituents on their side chains. These all have isoelectric points close to 6 and have no net charge at pH 6–7. They are shown in Figure 17.1, along with the abbreviations of their names that will be used in writing formulas for peptides and proteins.

(+)-alanine
Ala

(+)-valine
Val

(−)-leucine
Leu

(+)-isoleucine
Ile

(−)-proline
Pro

(−)-phenylalanine
Phe

(−)-tryptophan
Trp

(−)-methionine
Met

Figure 17.1   Amino acids with nonpolar R groups, with no net charge at physiological pH.

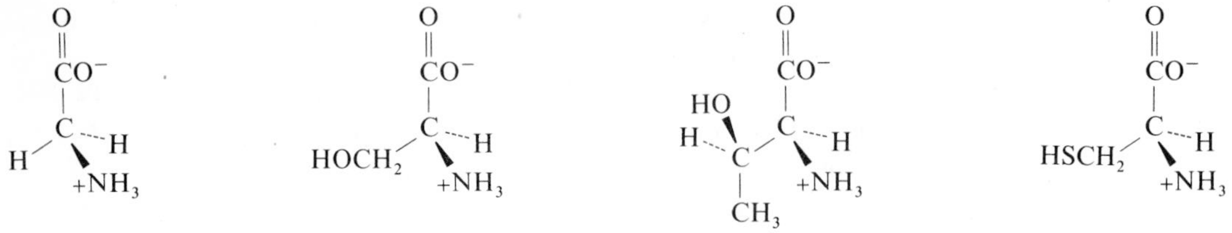

glycine Gly; (−)-serine Ser; (−)-threonine Thr; (−)-cysteine Cys

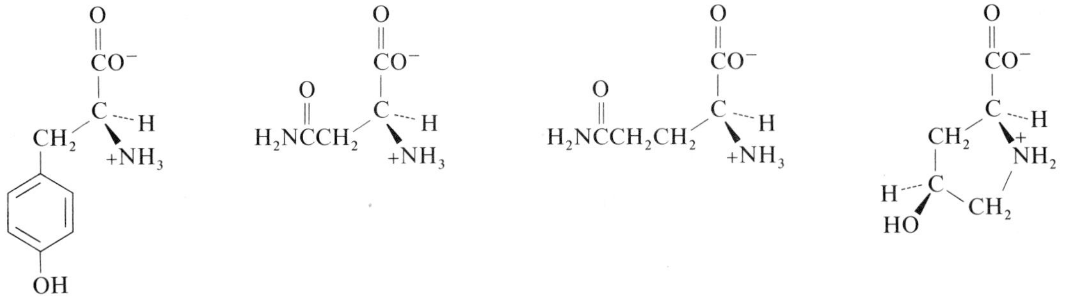

(−)-tyrosine Tyr; (−)-asparagine Asn; (+)-glutamine Gln; (−)-4-hydroxyproline Hypro

Figure 17.2 Amino acids with polar R groups, with no net charge at physiological pH.

Other amino acids have polar functional groups that can participate in hydrogen bonding or function as nucleophiles in reactions, but are not ionized at physiological pH (Figure 17.2). Glycine is included among these because it does not have a large, hydrophobic side chain.

Two amino acids have extra carboxylic acid groups on them and therefore have a net negative charge at physiological pH (Figure 17.3).

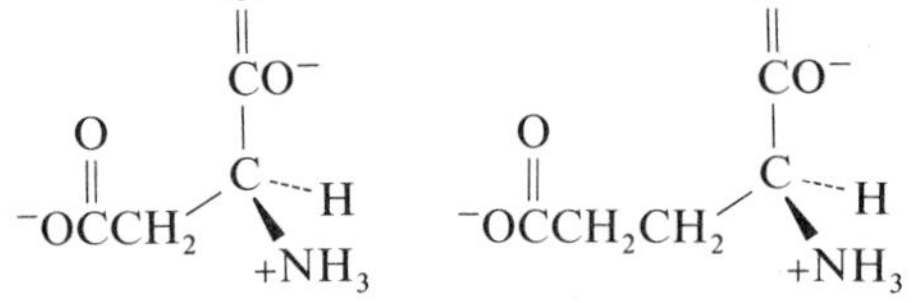

(+)-aspartic acid Asp; (+)-glutamic acid Glu

Figure 17.3 Amino acids that are negatively charged at physiological pH.

Finally, three amino acids have basic functional groups and are, therefore, positively charged at the pH of cellular fluids (Figure 17.4).

Many of the amino acids are synthesized in the body from ammonia and the degradation products of carbohydrates such as α-ketocarboxylic acids in the presence of reducing enzymes. The reaction, in fact, greatly resembles the reductive amination reactions we studied in Section 16.3D. The synthesis of glutamic acid illustrates this reaction.

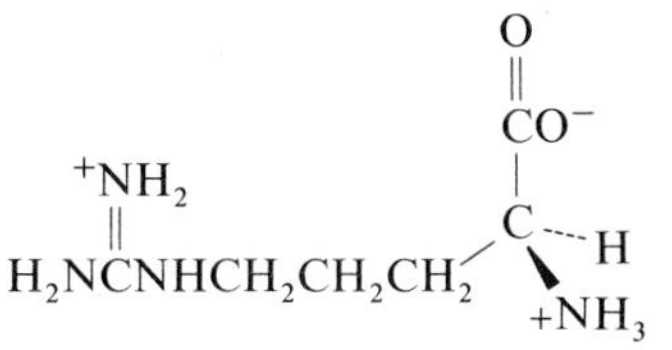

(+)-arginine
Arg

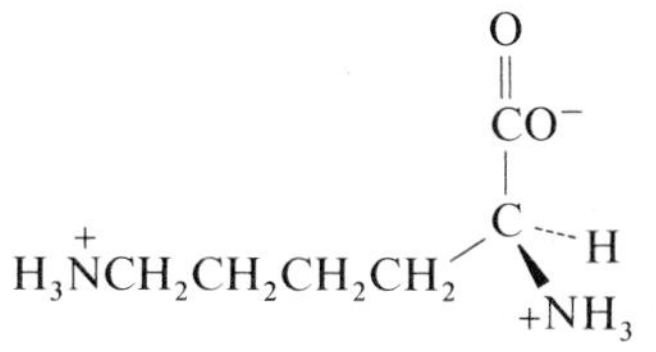

(+)-lysine
Lys

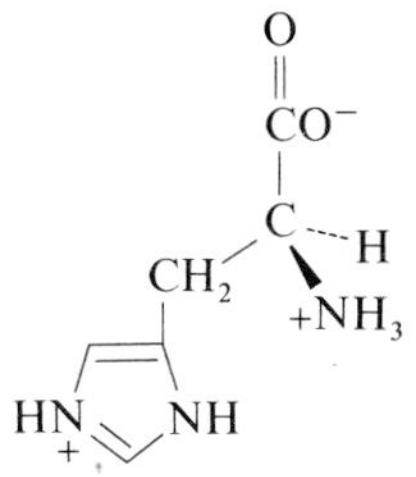

(−)-histidine
His

Figure 17.4 Amino acids that are positively charged at physiological pH.

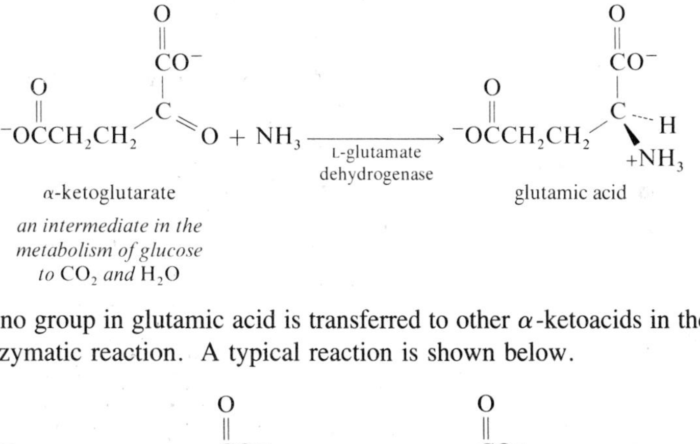

α-ketoglutarate

an intermediate in the metabolism of glucose to CO_2 *and* H_2O

glutamic acid

The amino group in glutamic acid is transferred to other α-ketoacids in the body in another enzymatic reaction. A typical reaction is shown below.

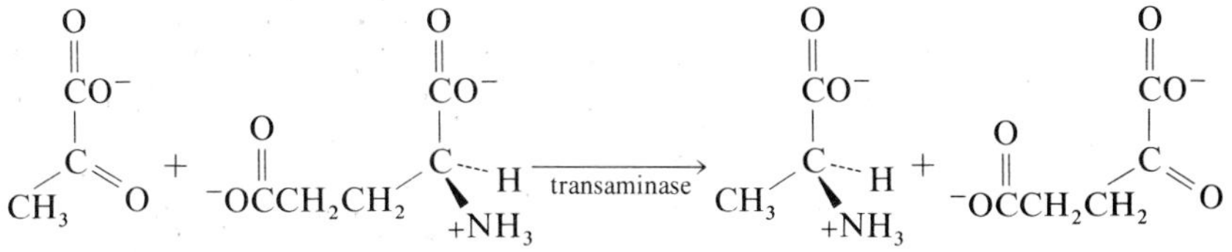

pyruvate anion glutamate anion alanine α-ketoglutarate anion

Such a transformation is called a **transamination** because an amino group is transferred from one species to another. A ketoacid is converted into the corresponding amino acid while glutamic acid becomes α-ketoglutaric acid, which then reacts with ammonia to regenerate glutamic acid.

Because the human body cannot synthesize some of the amino acids necessary for protein synthesis, they must be obtained from food. These **essential amino acids** are arginine, histidine, isoleucine, leucine, lysine, methionine, phenylalanine, threonine, tryptophan, and valine.

PROBLEM 17.1

(a)   Would you expect the carboxylic acid group on carbon 3 of aspartic acid to be more or less acidic than the carboxylic acid group on carbon 4 of glutamic acid (p. 963)? Explain your answer.

(b)   Would the isoelectric point for aspartic acid be higher or lower than the one for glutamic acid?

PROBLEM 17.2   The amino acids are assigned to the L-family according to their stereochemistry at carbon 2. Can they all be assigned the same configuration, R or S, at that carbon atom?

PROBLEM 17.3   Some of the amino acids have more than one asymmetric carbon atom. Pick them out. Assign configuration at these other asymmetric carbon atoms where possible.

PROBLEM 17.4

(a)   Arginine (p. 965) is the most basic of the amino acids. The pK_a of the conjugate acid of the guanidinyl group on its side chain is 13.2. Why is the guanidinyl group so much more basic than the amino acid group in lysine (p. 962)?

(b)   Tryptophan is classified as a nonpolar amino acid even though it has a nitrogen atom in an indole ring (Section 18.1) as part of its side chain. Why is tryptophan not basic? (Hint: A review of Section 12.1 and of Problem 12.57 may be helpful.)

PROBLEM 17.5   Amino acids have the reactions that are typical of primary alkyl amines and of carboxylic acids. In addition, other functional groups on the side chains also have their typical reactions. Complete the following equations to show the products that would be expected from the reagents shown. For some of the reactions, it may be necessary to remember that a small amount of the uncharged form of the amino acid is always in equilibrium with the zwitterionic form.

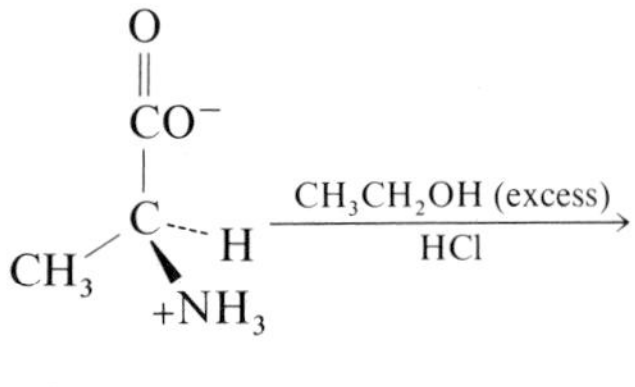

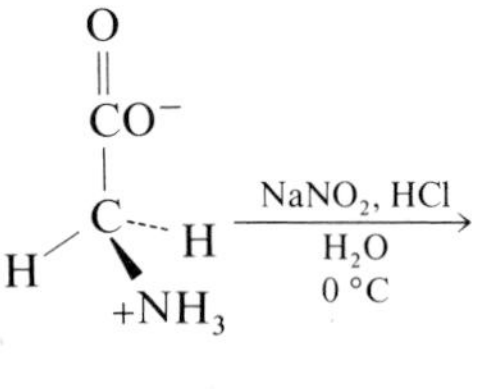

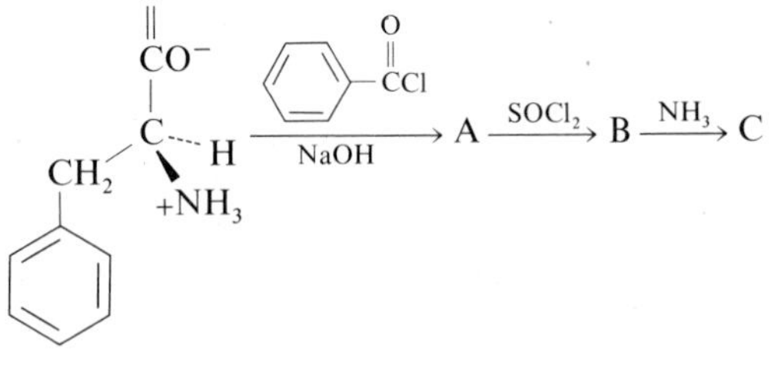

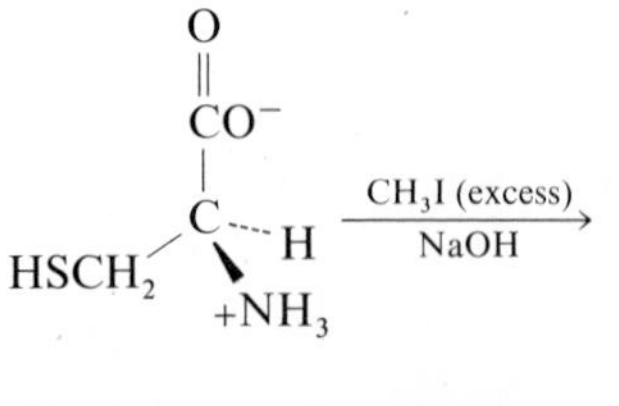

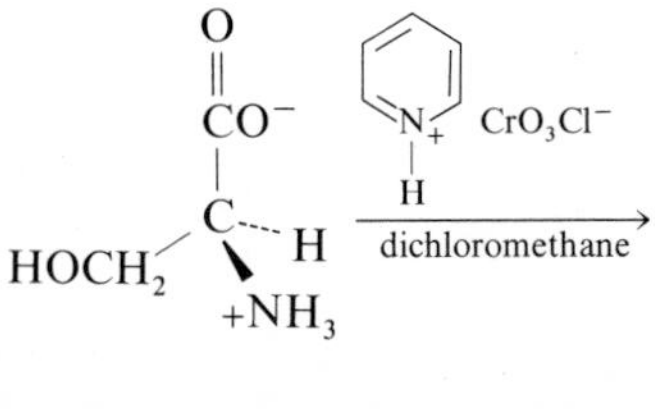

(f)

$$
\begin{array}{c}
\text{O} \\
\parallel \\
\text{CO}^- \\
\mid \\
\text{C}\cdots\text{H} \\
\text{CH}_2 \quad {}^+\text{NH}_3
\end{array}
\xrightarrow[\text{H}_2\text{O}]{\text{NaOH}}
$$

(with a *p*-hydroxyphenyl (—C$_6$H$_4$—OH) group on CH$_2$)

(g)

$$
\begin{array}{c}
\text{O} \\
\parallel \\
\text{CH}_3 \quad \text{CO}^- \\
\mid \qquad \mid \\
\text{CH}_3\text{CHCH}_2 \;\; \text{C}\cdots\text{H} \\
\qquad\qquad {}^+\text{NH}_3
\end{array}
\;\; \text{(+ C}_6\text{H}_5\text{CHO)} \longrightarrow
$$

(h)

$$
\begin{array}{c}
\text{O} \\
\parallel \\
\text{CO}^- \\
\mid \\
{}^+\text{H}_3\text{NCH}_2\text{CH}_2\text{CH}_2\text{CH}_2 \;\; \text{C}\cdots\text{H} \\
\qquad\qquad\qquad {}^+\text{NH}_3
\end{array}
\xrightarrow{(\text{CH}_3\text{C})_2\text{O (excess)}}
$$

(i)

$$
\begin{array}{c}
\text{O} \\
\parallel \\
\text{CO}^- \\
\mid \\
\text{C}\cdots\text{H} \\
\text{CH}_2 \quad {}^+\text{NH}_3
\end{array}
\xrightarrow{\text{HNO}_3}
$$

(with a *p*-hydroxyphenyl (—C$_6$H$_4$—OH) group on CH$_2$)

17.3

The Determination of the Structure of Peptides and Proteins

A. The Degradation of Proteins into Peptides and Amino Acids. Hydrolysis with Acids

Amino acids are found in nature linked together by amide bonds, not only in the large molecules known as proteins, but in smaller molecules, many of which have hormonal activity. Such small molecules, containing only a few amino acid residues, are known as **peptides.** Proteins are cleaved into peptides in the process of digestion by special- ized enzymes called **proteases.**

It is also possible to degrade proteins into peptides or amino acids by acidic hydrol- ysis of the amide bonds (Section 10.8B). If the protein is subjected to concentrated hydrochloric acid at 37 °C for several hours, it is degraded into smaller peptide residues. If, on the other hand, it is heated with 20% hydrochloric acid for a period of days, all the amide bonds are cleaved and the protein is degraded completely into the individual amino acids, some of which are destroyed by this treatment. Tryptophan does not survive vigorous acid hydrolysis, and serine and threonine, which contain hydroxyl groups, are partially destroyed. Asparagine and glutamine, which have amide groups on their side chains, are converted to aspartic acid and glutamic acid by this treatment, and ammonium chloride is formed as another product of the hydrolysis.

Rapid progress in the determination of the structure of proteins followed the devel- opment of chromatographic methods (Section 16.4A) for the analysis of mixtures of peptides and of amino acids. Many different techniques are now used, and automated equipment can deliver a complete analysis of the amino acid content of a sample in a few hours. Some methods depend on an electric field to separate the amino acids according to the charge they bear at a given pH. In other chromatographic methods such as paper chromatography, thin-layer chromatography, or column chromatography, amino acids of differing polarities are distributed differently between two phases, one of which is stationary and the other, mobile. The two phases may be two liquid phases of different polarities, or a solid phase and a liquid phase. The more polar amino acids are held back by the more polar phase so that separations take place. At the end of the process, the amino acids are usually treated with ninhydrin, which reacts with amino acids other than proline and 4-hydroxyproline to give a purple dye.

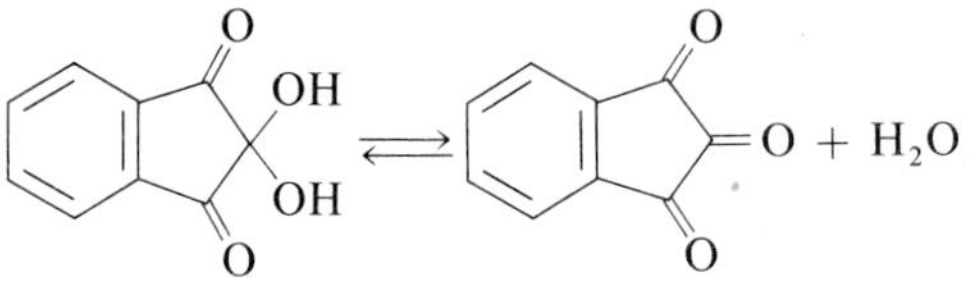

ninhydrin

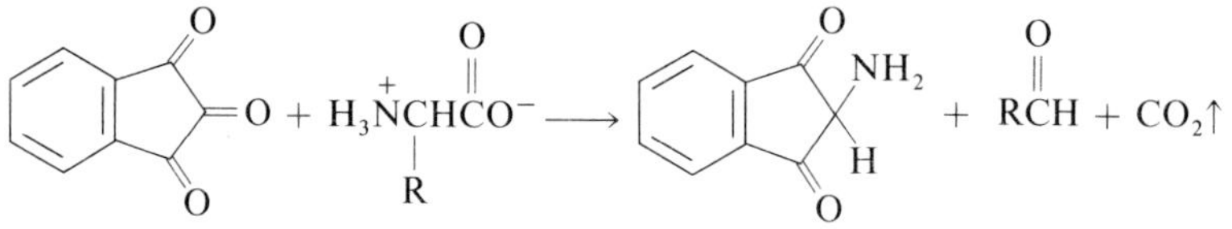

same product
from all amino acids
except proline
and 4-hydroxyproline

different
aldehyde
from each
amino acid

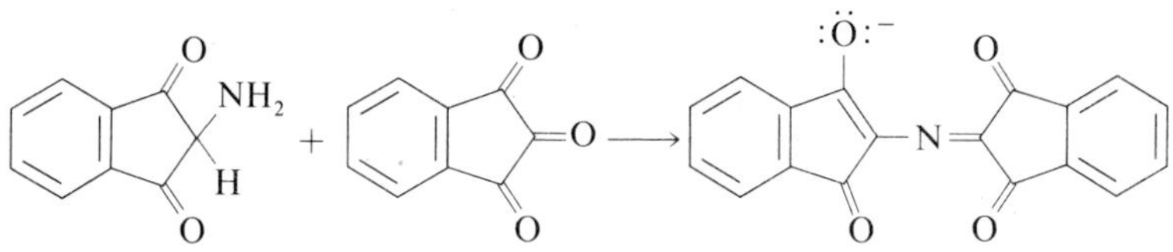

same product from all
amino acids except proline
and 4-hydroxyproline

from ninhydrin

purple color

If the experimental details are carefully standardized, the exact positions at which the dye spots appear on the paper sheet or thin-layer plate can be used to identify the amino acids present. The patterns of spots are sometimes called *maps* of amino acids. The same kinds of techniques can also be used to separate and identify peptides. Similarly, the absorption of the dye from ninhydrin is detected spectroscopically and used to plot the appearance of the separated amino acids as they leave the column in column chromatography. Once again, careful standardization of conditions allows chemists to compare the experimental pattern of absorption peaks with known patterns to identify the amino acids present.

Separations may also be carried out on the derivatives of amino acids instead of the amino acids themselves. Automated systems that depend on chromatographic separation of the phenylthiohydantoins (Section 17.3B) of the amino acids in a protein have been developed as the latest and most rapid method of analysis.

PROBLEM 17.6 Show the mechanism for the formation of the purple dye from the reaction of excess ninhydrin with the amino derivative of ninhydrin shown in the last line of the equation above.

Peptides are named for the amino acids that constitute them, starting with the amino acid that has the free amino group. That end of the peptide chain is known as the **N-terminal amino acid** *while the amino acid that has the free carboxylic acid group is called the* **C-terminal amino acid.** In naming peptides, starting with the N-terminal amino acid, each amino acid except the C-terminal one is treated as if it were an alkyl substituent on the next unit in the chain. This nomenclature is illustrated for a tetrapeptide.

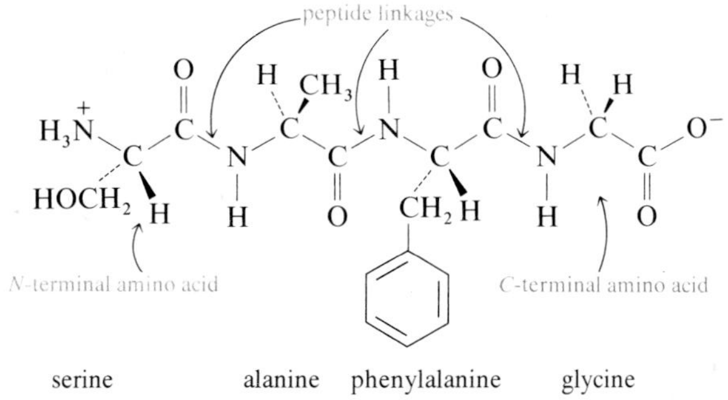

serine   alanine   phenylalanine   glycine

serylalanylphenylalanylglycine

Ser-Ala-Phe-Gly

N-terminal   C-terminal
amino acid   amino acid

When written out in full, the name of a peptide becomes cumbersome if it is of any size at all. Abbreviations for the names of the amino acids are used to make the name of a peptide more manageable and also to substitute for the full structural formula. Separating the abbreviations by hyphens indicates that the amino acids are bonded to each other by peptide linkages in the order shown. The first amino acid listed is the N-terminal amino acid, and the last one shown is the C-terminal amino acid. Separating the symbols for the amino acids by commas indicates that the amino acids are present in the peptide but the order of their bonding is not known. For example, complete hydrolysis of the tetrapeptide shown above would yield the information that it contained the amino acids alanine, glycine, phenylalanine, and serine, listed in alphabetical order, but no information on how they were bonded to each other.

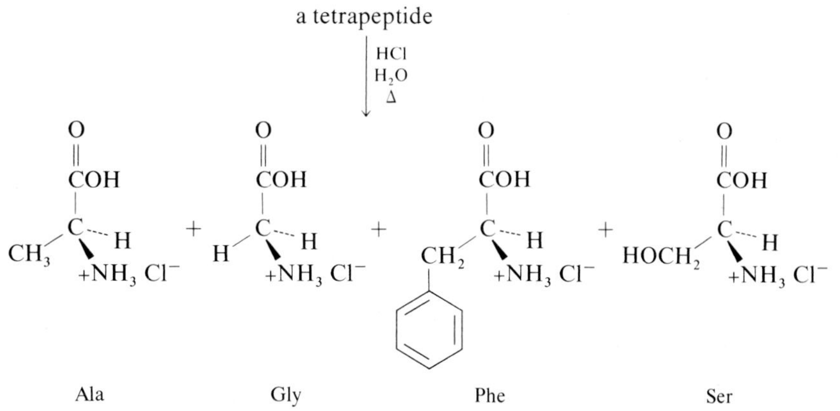

Ala   Gly   Phe   Ser

a tetrapeptide $\xrightarrow[\substack{H_2O \\ \Delta}]{HCl}$ Ala, Gly, Phe, Ser as their hydrochloride salts

*no information about the order in which
the amino acids are bonded to each other*

PROBLEM 17.7   Four amino acids can be combined to give 24 different peptides. Write complete structures for four peptides containing alanine, glycine, phenylalanine, and serine that are different from the one shown on p. 969. Show stereochemistry, and label the *N*-terminal and *C*-terminal amino acids in each one.

PROBLEM 17.8   A peptide has the structure Leu-Asn-Lys-Tyr. Write out its structure, and then show the fragments that you would expect to get from its complete hydrolysis with hydrochloric acid.

A peptide, as well as an amino acid, exists as a dipolar ion at physiological pH. Peptides also have typical isoelectric points depending on the amino acids present in them. The carboxylic acid groups and amino groups that are part of the peptide linkages no longer have acidic or basic properties, so only the *N*-terminal amino group, the *C*-terminal carboxylic acid group, and any acidic or basic groups that are present in the side chains of the peptide will be ionized. The tetrapeptide Ser-Ala-Phe-Gly (p. 969) would have an isoelectric point around 6.0.

PROBLEM 17.9   For each of the following peptides, predict whether the isoelectric point would be below, above or approximately 6. Predict whether each peptide would move to the negative or positive pole of an electric field, or do neither, if the pH of the solution were kept at 6.0.

(a) Gly-Phe-Thr-Lys   (b) Tyr-Ala-Val-Asn   (c) Trp-Glu-Leu

(d) Pro-Hypro-Gly

B. End-Group Analysis of Peptides and Proteins

Chemists have developed a series of reagents that react selectively with the *N*-terminal amino acid of a polypeptide, transforming it into some derivative that can be separated from all of the other amino acids in the chain and identified. One of the most useful of these is 2,4-dinitrofluorobenzene, **Sanger's reagent,** developed by Frederick Sanger during his determination of the structure of insulin. Sanger was given the Nobel Prize in 1958 for being the first to establish the sequence of amino acids in a protein.

2,4-Dinitrofluorobenzene, abbreviated as DNFB, is an aryl halide with nitro groups on the aromatic ring in positions to activate the halogen towards nucleophilic substitution (Section 12.5). The free amino group of the *N*-terminal amino acid displaces fluoride ion from the reagent, giving rise to a dinitroarylamino group. Complete hydrolysis of the peptide gives free amino acids from all of the residues except the *N*-terminal one, which turns up labeled with the aryl group. The process is illustrated for the tripeptide Ser-Gly-Val, written as the form in which the amino group is not protonated.

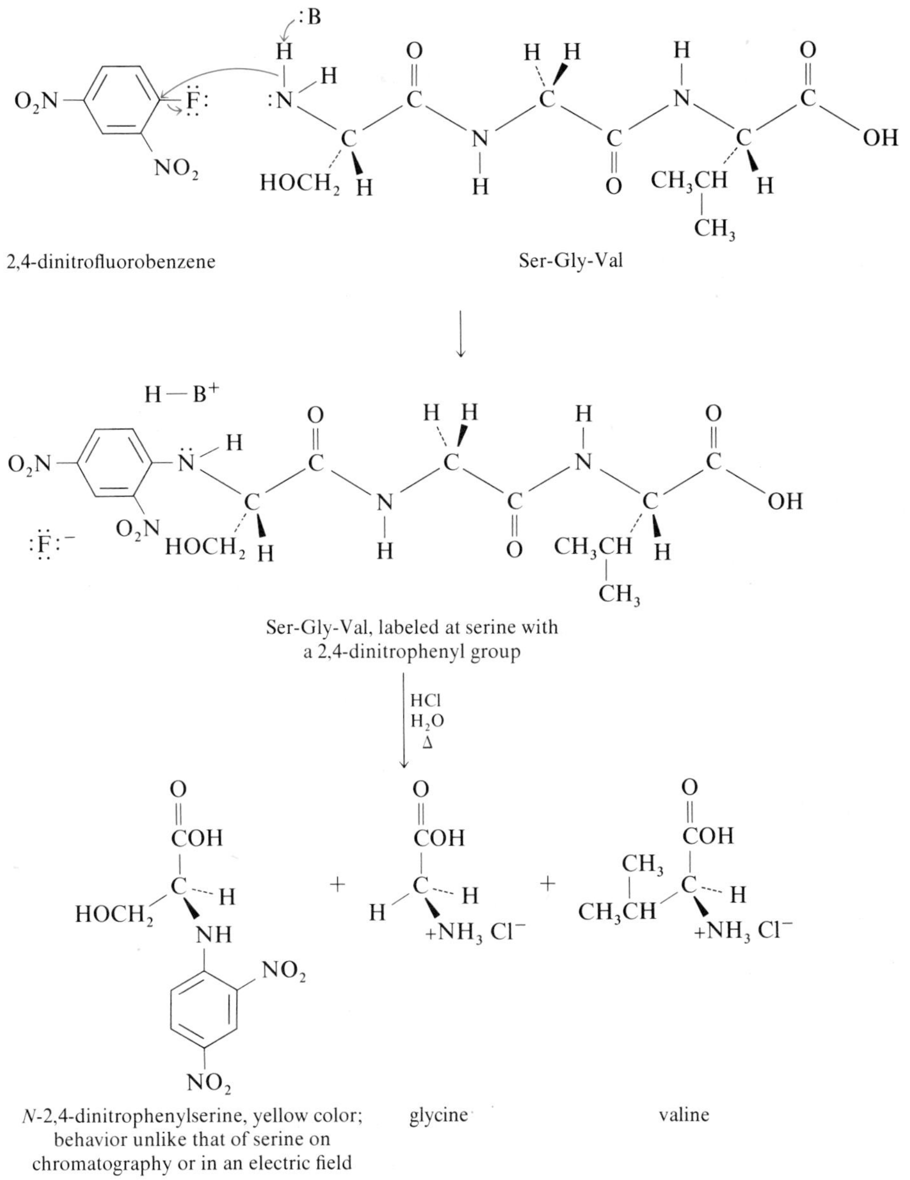

:B

O_2N— —F: :N—H   C=O   H H   H   O

2,4-dinitrofluorobenzene   Ser-Gly-Val

$H—B^+$

O_2N—   N—H

O_2N

:F:⁻   HOCH₂ H

Ser-Gly-Val, labeled at serine with
a 2,4-dinitrophenyl group

HCl
H₂O
Δ

O
‖
COH   +   H—C---H   +   CH₃CH   C---H
HOCH₂—C---H   +NH₃ Cl⁻   +NH₃ Cl⁻
         NH
         NO₂

         NO₂

N-2,4-dinitrophenylserine, yellow color;   glycine   valine
behavior unlike that of serine on
chromatography or in an electric field

The presence of the dinitrophenyl group on the nitrogen atom of serine converts it from an alkyl amine to an aryl amine and changes its acid-base properties drastically. The labeled serine behaves differently from ordinary serine in all the techniques used to make separations of amino acids and is easily identified by comparison with an authentic sample of *N*-2,4-dinitrophenylserine.

PROBLEM 17.10   2,4-Dinitrofluorobenzene can be used to determine whether lysine is the *N*-terminal amino acid of a peptide or whether it is in the middle of the peptide chain. Draw segments of hypothetical peptide chains with lysine in either position and show with equations how the determination of the position of the lysine molecule could be carried out.

Another reagent used to analyze for *N*-terminal amino acids is phenyl isothiocyanate. This reagent, known as **Edman's reagent,** reacts with the free amino group to give a thiourea that is then cleaved off in anhydrous acid. The peptide fragment has a new *N*-terminal amino acid that can be subjected to the degradation reaction again, while the amino acid that is lost appears as a substituted phenylthiohydantoin.

phenyl isothiocyanate

Ala-Leu-Gly

phenylthiohydantoin
derived from alanine, racemized

Leu-Gly

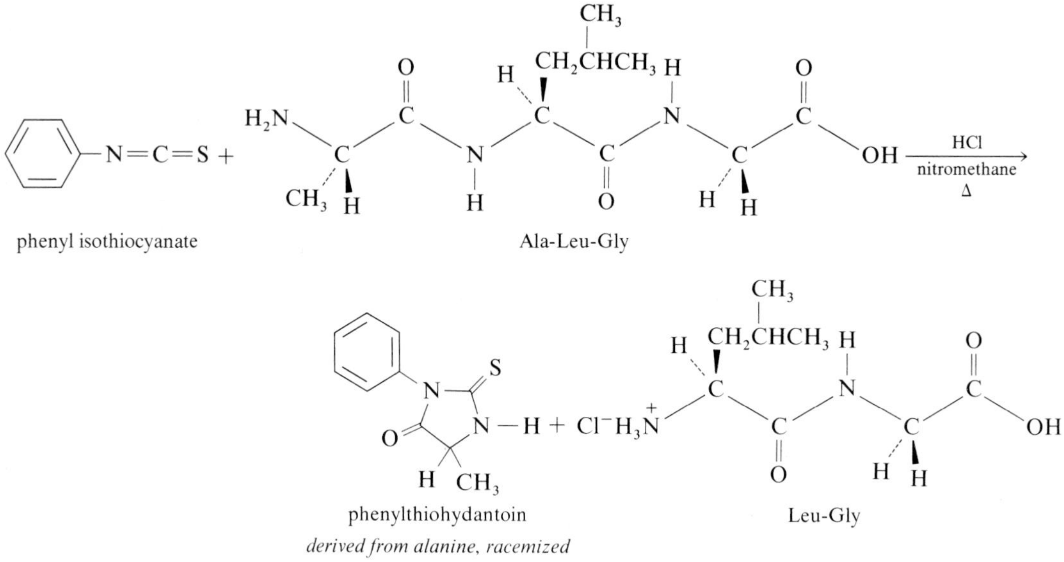

The substituted phenylthiohydantoin can be identified by comparing it with phenyl-thiohydantoins prepared from known amino acids. The reaction proceeds by way of an unstable intermediate that rearranges to give the phenylthiohydantoin that is isolated.

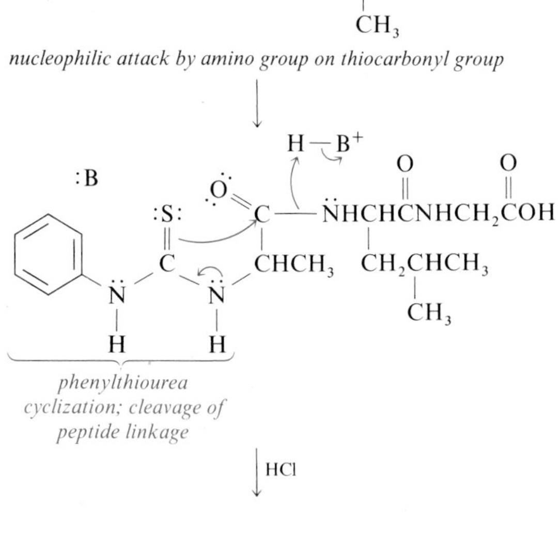

nucleophilic attack by amino group on thiocarbonyl group

phenylthiourea
cyclization; cleavage of
peptide linkage

HCl

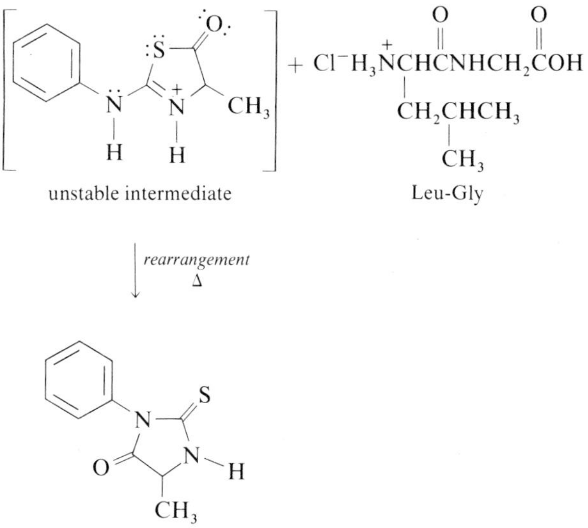

unstable intermediate

Leu-Gly

rearrangement
Δ

phenylthiohydantoin
derived from alanine

In principle, a large polypeptide could be degraded one amino acid at a time and each new *N*-terminal amino acid identified as its phenylthiohydantoin (see p. 976). In general, though, the practical problems involved in handling small quantities of material and side reactions that take place with some of the amino acids limit the number of times the reaction can be used.

The enzyme carboxypeptidase selectively removes the *C*-terminal acid of a peptide. This process creates a new *C*-terminal residue, which is then also attacked and removed by the enzyme, so this method must be used in connection with studies of the rate at which different free amino acids appear in the solution.

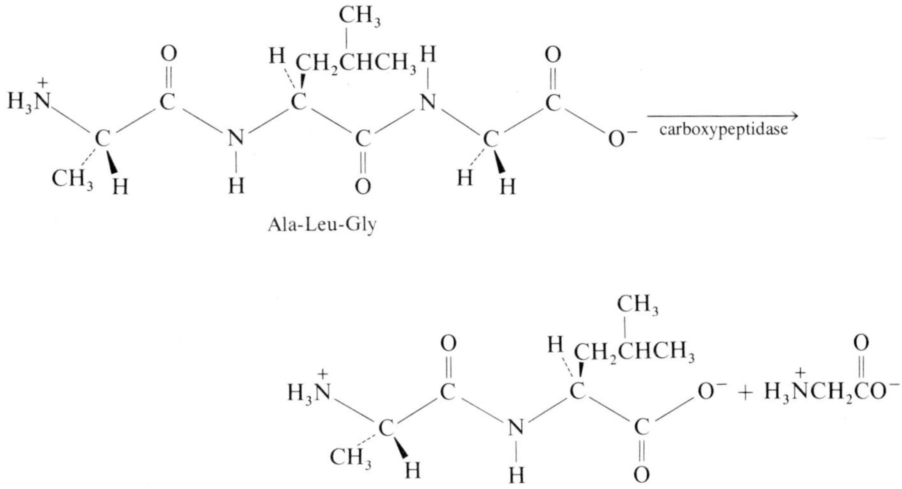

Ala-Leu-Gly

Ala-Leu   Gly

If the tripeptide Ala-Leu-Gly were treated with carboxypeptidase, the amino acid glycine would be liberated into the solution first. After a while, leucine and alanine would also appear.

PROBLEM 17.11 When the hydrolysis of a peptide by carboxypeptidase is carried out in water containing ^{18}O, the isotopic ^{18}O appears in the carboxylic acid groups of all of the amino acids except the *C*-terminal one, thus allowing the *C*-terminal amino acid to be identified. Using Gly-Ala as an example, write a mechanism employing general acid and base catalysis and $H_2^{18}O$ to show why this is so.

End-group analysis of many biologically interesting peptides is complicated by the presence of the *C*-terminal amino acid as an amide. When glutamic acid is the *N*-terminal amino acid, it often forms a cyclic amide known as pyroglutamic acid, symbolized by a suggestion of a ring in its three-letter abbreviation. An important hormone secreted by the hypothalamus gland, the thyrotropin-releasing factor, has both of these features.

γ-lactam structure;
pyroglutamic acid,
cyclic amide of
N-terminal glutamic acid

C-terminal acid,
proline, as amide

Glu-His-Pro-NH$_2$
thyrotropin-releasing factor
a tripeptide hormone

Such peptides do not react with reagents such as Sanger's reagent or Edman's reagent at the *N*-terminal amino acid, nor is the *C*-terminal amino acid cleaved off by carboxypeptidase. Hydrolysis of the peptide would give a mole of ammonia, as well as the three amino acids, glutamic acid, histidine, and proline.

C. The Degradation of Proteins and Peptides by Enzymes. The Disulfide Bridge

If hydrolysis of a protein is carried out by cold, concentrated hydrochloric acid for relatively short periods of time, a variety of peptide fragments are obtained. The protein may also be more selectively cleaved at certain points of the peptide chain by enzymes. Trypsin, for example, attacks only the peptide bonds formed by the carboxylic acid groups of the basic amino acids lysine and arginine. The newly formed peptide fragments have lysine or arginine as their *C*-terminal amino acids, except for the fragment resulting from the original *C*-terminal end of the protein.

Another enzyme, chymotrypsin, is less selective. It hydrolyzes the bonds formed by the carboxyl groups of tyrosine, phenylalanine, tryptophan, and methionine. The usefulness of these enzymes in selective cleavages of peptide bonds is illustrated by the determination of the structure of somatostatin, a tetradecapeptide that inhibits the secretion of pituitary growth hormone.

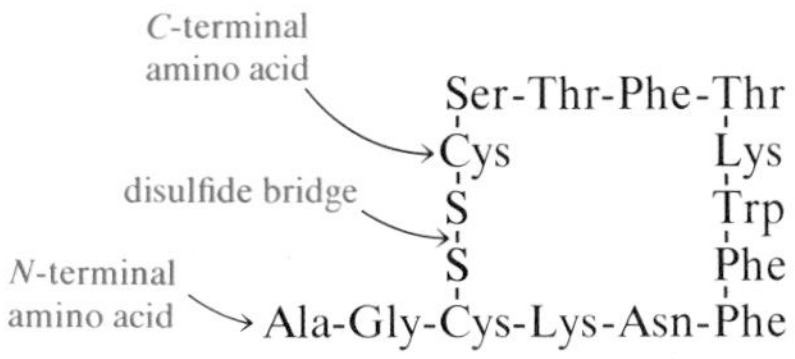
C-terminal amino acid
disulfide bridge
N-terminal amino acid
Ser-Thr-Phe-Thr
Cys
Lys
S
Trp
S
Phe
Ala-Gly-Cys-Lys-Asn-Phe

the structure of somatostatin, showing the disulfide bridge between two cysteine residues

The *C*-terminal amino acid of somatostatin and the amino acid third from the *N*-terminal end are cysteine. They are found bound to each other in their oxidized form to give a disulfide bridge between the two ends of the peptide. Such bridges are important in stabilizing the tertiary structure of many proteins.

The oxidized form of cysteine is known as cystine, a diamino acid with a disulfide bond. Cystine is reduced to give two molecules of cysteine.

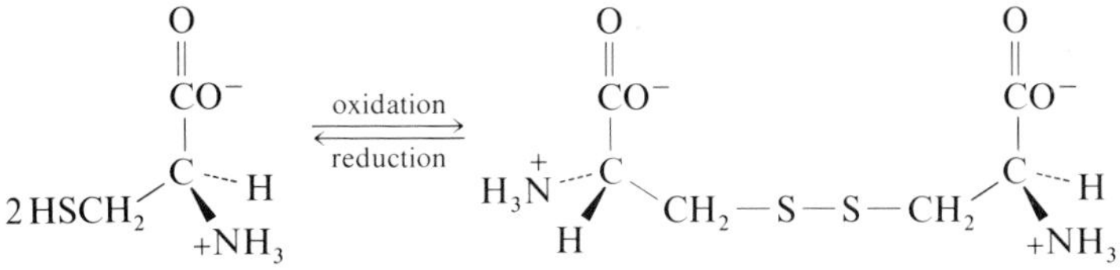
oxidation
reduction

cysteine cystine

The proof of the structure of somatostatin starts with the reduction of the disulfide linkage with 2-mercaptoethanol, $HSCH_2CH_2OH$, so that cystine is converted to cysteine. The thiol groups on the cysteine residues are then converted to inert groups by alkylation with iodoacetic acid.

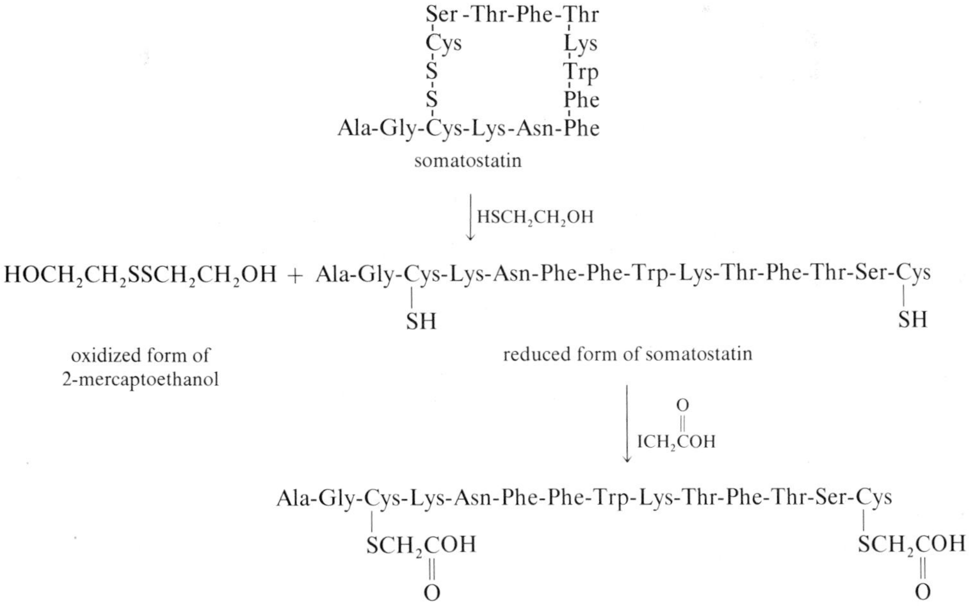
somatostatin
HSCH2CH2OH
HOCH2CH2SSCH2CH2OH + Ala-Gly-Cys-Lys-Asn-Phe-Phe-Trp-Lys-Thr-Phe-Thr-Ser-Cys
oxidized form of 2-mercaptoethanol
reduced form of somatostatin
ICH2COH
Ala-Gly-Cys-Lys-Asn-Phe-Phe-Trp-Lys-Thr-Phe-Thr-Ser-Cys

the reduced form of somatostatin with the thiol groups of cysteine residues converted to acid residues

This step is necessary to prevent the cysteines from being oxidized back to a disulfide bridge by air.

The peptide is cleaved at two points by trypsin.

somatostatin
(reduced and alkylated)

$\downarrow$ trypsin digestion

Ala-Gly-Cys-Lys   +   Asn-Phe-Phe-Trp-Lys + Thr-Phe-Thr-Ser-Cys
            | |
        SCH₂COH   SCH₂COH
            ‖   ‖
            O   O

Digestion of somatostatin with chymotrypsin gives more fragments than digestion with trypsin.

somatostatin
(reduced and alkylated)

$\downarrow$ chymotrypsin digestion

Ala-Gly-Cys-Lys-Asn-Phe + Phe + Trp + Lys-Thr-Phe + Thr-Ser-Cys
            | |
        SCH₂COH   SCH₂COH
            ‖   ‖
            O   O

If the structures of the individual smaller peptides that result from acid or enzymatic hydrolyses could be determined, it would be possible to reconstruct the structure of somatostatin by piecing together the different overlapping sequences of amino acids found in the fragments. The structures of small peptides are determined by reactions that selectively identify their *N*-terminal and *C*-terminal amino acids by end-group analysis (Section 17.3B). Thus, for example, if a determination were made that the *N*-terminal amino acid in somatostatin is alanine and the *C*-terminal acid is cysteine, the fragments obtained from the enzymatic cleavages could be combined to give the full structure of the molecule. The fragments from trypsin and those from chymotrypsin are lined up below showing how the different points of cleavage provide overlapping information about the order of the amino acids in the chain.

peptides from trypsin cleavage of somatostatin

Ala-Gly-Cys-Lys/Asn-Phe-Phe-Trp-Lys/Thr-Phe-Thr-Ser-Cys

Ala-Gly-Cys-Lys-Asn-Phe/Phe/Trp/Lys-Thr-Phe/Thr-Ser-Cys

peptides from chymotripsin cleavage of somatostatin

The total structure of somatostatin has also been determined by repeated use of Edman's reagent. The chain was degraded one amino acid at a time starting from the *N*-terminal end, and each amino acid was identified as its phenylthiohydantoin (Section 17.3B).

PROBLEM 17.12   A heptapeptide is isolated from milk protein when it is degraded by a protease from the bacterium *Bacillus subtilis*. It contains the amino acids arginine, glycine, isoleucine, phenylalanine, proline (2 residues), and valine.

(a)   Treatment with trypsin gives arginine and a hexapeptide. What do you know about the structure of the original peptide now?

(b) Some of the fragments that could be obtained by partial hydrolyses are Pro-Phe-Ile, Arg-Gly-Pro, Pro-Phe-Ile-Val, and Pro-Pro. Assign a structure to the heptapeptide.

PROBLEM 17.13   The hormone arginine vasopressin, which is important in the regulation of water balance in the body, is a nonapeptide with one disulfide bridge. In the proof of its structure, the disulfide bridge was cleaved by oxidation with peroxyformic acid, which oxidizes cysteine residues to cysteic acid. During degradation, the cysteines appear as cysteic acid residues, also symbolized below by Cys.

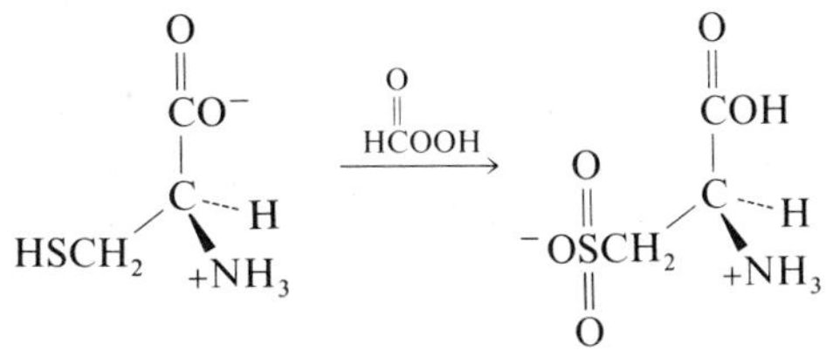

cysteine   cysteic acid

Complete hydrolysis of the oxidized peptide with hydrochloric acid gave *three moles of ammonia* and the amino acids Arg, Asp, Cys(2), Glu, Gly, Phe, Pro, and Tyr. 2,4-Dinitrofluorobenzene reacted with one of the cysteic acid residues. Trypsin gave an octapeptide with arginine as its *C*-terminal residue, and glycinamide, Gly-NH$_2$, as the fragments. Partial acid hydrolysis gave the following fragments.

Asp-Cys, Phe-Glu, Cys-Tyr-Phe, Glu-Asp-Cys, Phe-Glu-Asp

In addition, a fragment in which cysteine was the *N*-terminal amino acid, and which contained arginine, glycine, and proline in an undetermined order, was obtained.

(a) What is the order of amino acids derived from the degradation experiments?

(b) What is the full correct structure for arginine vasopressin?

17.4

The Synthesis of Peptides and Proteins in the Laboratory

A. Protection of Reactive Functional Groups

The synthesis of a peptide requires reactions to form amide bonds between the amino group of one amino acid and the carboxylic acid group of another one. If a peptide of a given structure is to be prepared, the amino acids must be added on to the chain in precise order. The reaction involves a nucleophilic substitution by an amino group at a carbonyl group of a carboxylic acid, and yet there are nucleophilic groups such as other amino groups, thiols, and alcohols on the side chains of amino acids that would also be expected to react with carboxylic acids. The reactions to form the backbone of the peptide chain must take place with the amino group at carbon 2 and not with any of the other nucleophilic functional groups.

The free carboxylic acid group does not react readily with amines to give amides. Usually an acid chloride or an acid anhydride is used, so that a better leaving group than a hydroxyl will be present, when an acid is converted into an amide (Section 10.8A).

Briefly then, *there are two major aspects to peptide synthesis. Functional groups must be selectively protected so that reaction can be directed to the part of the molecule that we wish to use in the synthesis. The carboxylic acid group of an amino acid must be made reactive enough to form an amide bond under conditions that do not destroy peptide bonds and other functional groups that may already be present in the molecule.*

In order to simplify the problem that faces peptide chemists, we will limit ourselves to reactions for the protection of the α-amino group in amino acids. The art of protecting different types of functional groups in a protein synthesis is the subject of lively research now, and many different and highly selective protecting groups are being developed. In principle, the arguments that we will make about the α-amino group extend to the problems of protecting all functional groups.

The most important factor in the choice of a protecting group for the amino group is the ease with which the protecting group can be removed later in the synthesis. Amines have been protected as their amides, for example, when electrophilic aromatic substitution reactions were to be performed on aromatic amines (Problem 16.8). An ordinary amide would not be a suitable protecting group in a peptide synthesis because there would be no way to remove the protecting group except by hydrolysis, which would also cleave peptide bonds.

Groups that can be cleaved under anhydrous acidic conditions that do not affect peptide bonds are benzyl or *tert*-butyl carbamates (Section 16.5C). The carbamates are prepared by reaction of the amino acid with a half ester of carbonic acid, the other half of which has a good leaving group on it. Benzyl chloroformate, for example, gives benzyl carbamates with amino acids.

alanine   benzyl chloroformate   *N*-benzyloxycarbonylalanine
                                                 N-carbobenzoxyalanine
                                                 N-Cbz-Ala

This protecting group is given the symbol *N*-Cbz in abbreviations of amino acids or peptides.

tert-Butyl chloroformate is not very stable, so *tert*-butyl azidoformate, in which the azide ion is the leaving group in the substitution reaction, is more commonly used as a reagent for protecting amino groups.

phenylalanine   *tert*-butyl azidoformate   *N-tert*-butoxycarbonylphenylalanine
                                                          Boc-Phe

The presence of the *tert*-butoxycarbonyl protecting group is indicated by the abbreviation Boc.

The benzyl and *tert*-butyl groups were chosen because they both yield stable carbocations or radicals. Under acidic conditions, in the absence of water, they are both cleaved to give carbamic acids, which lose carbon dioxide and regenerate the amino group. The *tert*-butoxycarbonyl group comes off with acids such as dry trifluoroacetic acid in dichloromethane or dry hydrogen chloride in ethers.

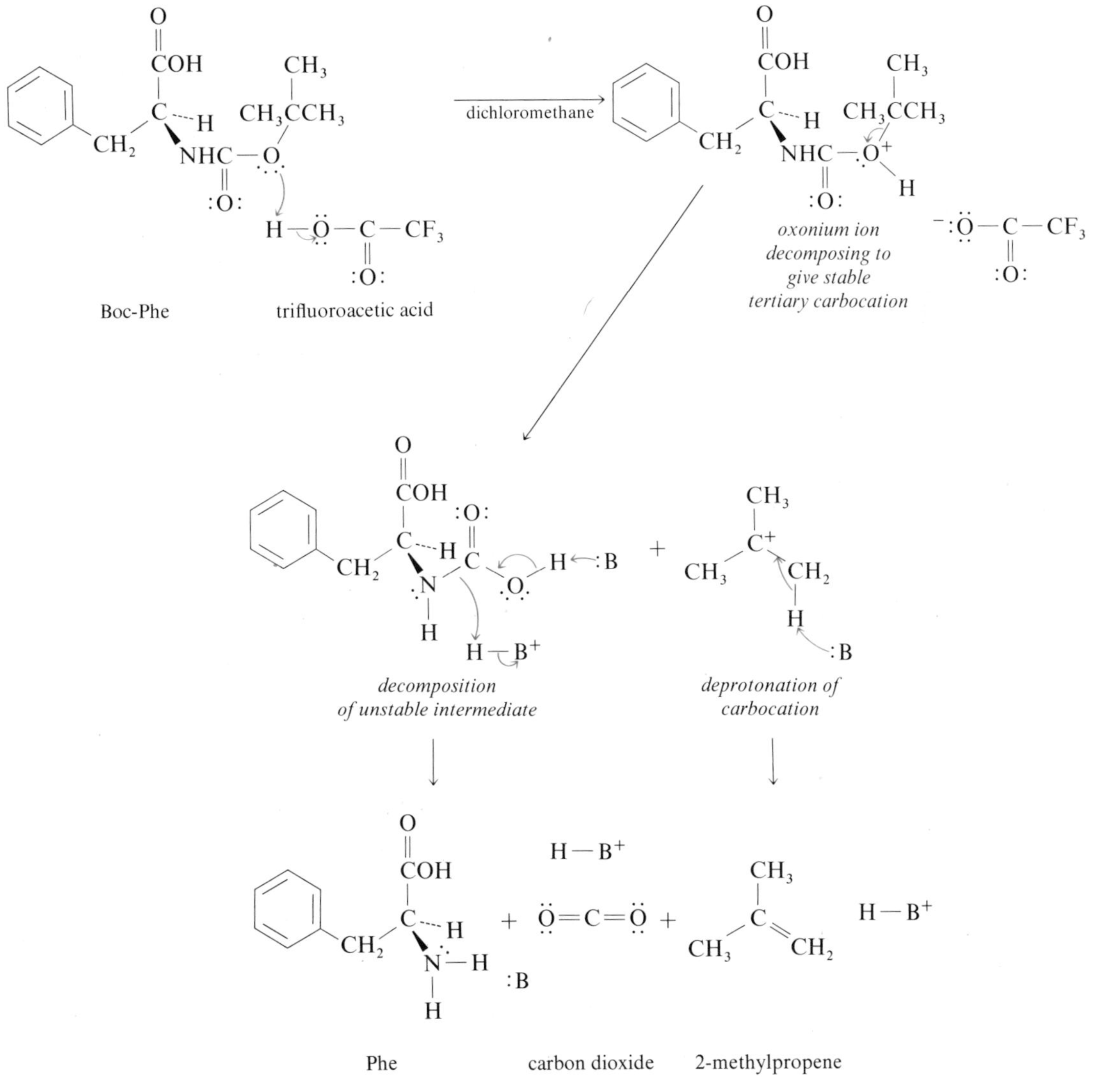

Boc-Phe   trifluoroacetic acid $\xrightarrow{\text{dichloromethane}}$ *oxonium ion decomposing to give stable tertiary carbocation*

decomposition of unstable intermediate   *deprotonation of carbocation*

Phe   carbon dioxide   2-methylpropene

The products, other than the amino acid, are the gases, carbon dioxide and 2-methylpropene, which comes from the *tert*-butyl cation.

The benzyloxycarbonyl group is removed by a stronger acid, such as dry hydrogen bromide dissolved in trifluoroacetic or acetic acid.

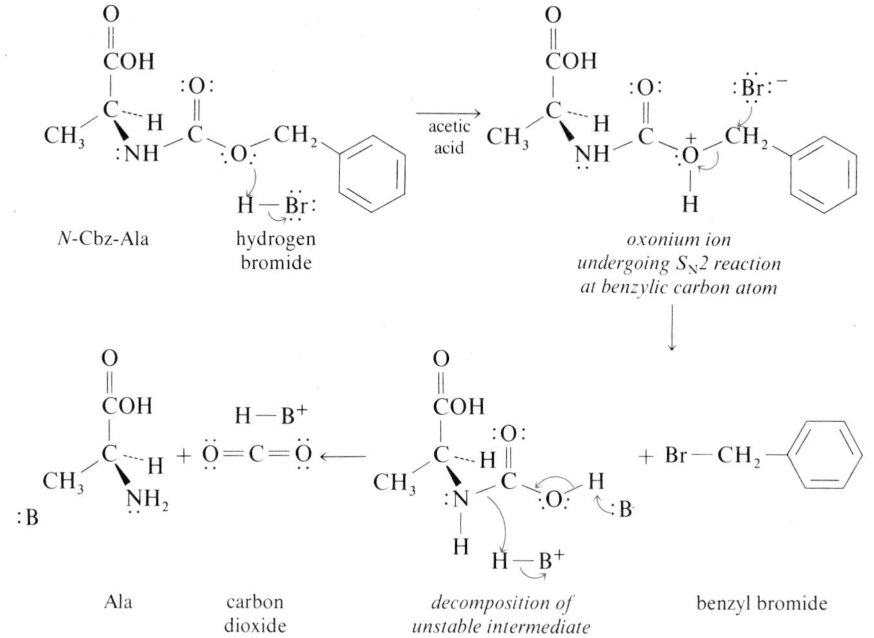
acetic acid
N-Cbz-Ala
hydrogen bromide
oxonium ion undergoing S_N2 reaction at benzylic carbon atom
Ala
carbon dioxide
decomposition of unstable intermediate
benzyl bromide

Alternatively the benzyl-oxygen bond is cleaved by hydrogenation.

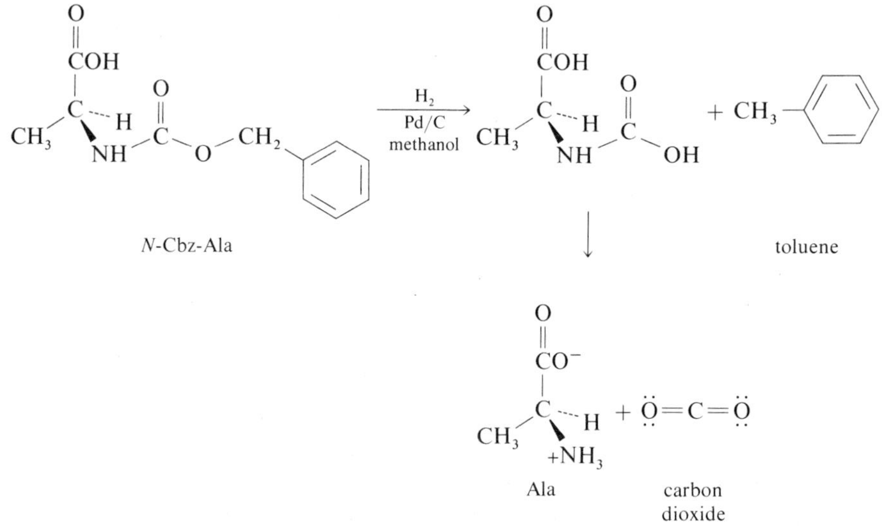
H_2
Pd/C
methanol
N-Cbz-Ala
toluene
Ala
carbon dioxide

The peptide bonds are stable to both of these sets of conditions. As carbamates, the amino groups are no longer basic and nucleophilic, and thus they are protected from reaction until the appropriate time in the synthesis when the protecting group is removed.

The carboxylic acid group of the amino acid is usually protected as a relatively unreactive ester such as the methyl or ethyl ester.

B. Activation of the Carboxyl Group and the Formation of Peptide Bonds

For the peptide bond to be formed, it is necessary that the hydroxyl group present in the carboxylic acid function be converted into a good leaving group. This is done in two general ways. One of them was introduced in Section 16.5C. In the presence of dicyclohexylcarbodiimide, an acid reacts with an amine to give an amide.

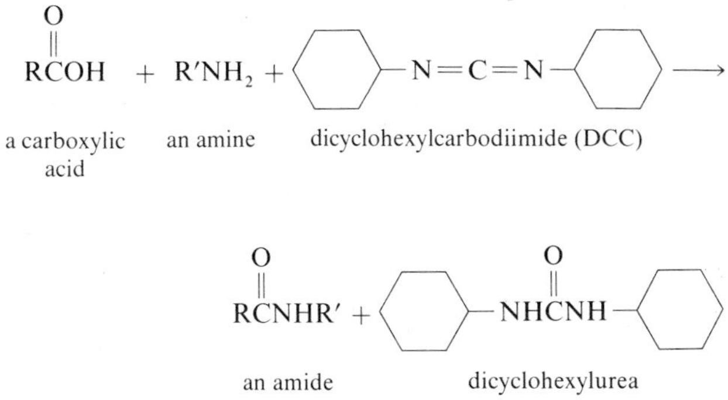

a carboxylic   an amine   dicyclohexylcarbodiimide (DCC)
acid

an amide   dicyclohexylurea

Dicyclohexylcarbodiimide (abbreviated as DCC in equations) converts the carboxylic acid into an intermediate that has the properties of an acid anhydride. An acid anhydride may in fact be formed by the further reaction of the acid with the intermediate.

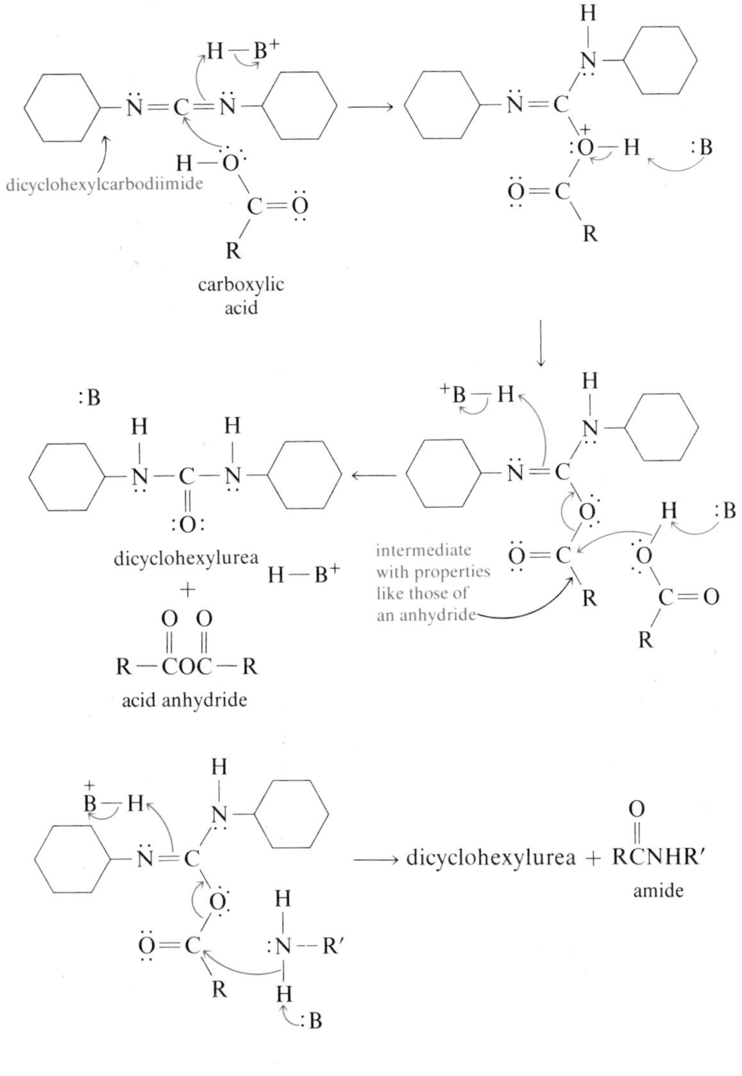

In either case, *the hydroxyl group of the carboxylic acid is converted into a good leaving group, so that reaction with an amine to give an amide takes place.*

The hydroxyl group is also converted into good leaving groups by the preparation of active esters of the amino acid. The *p*-nitrophenyl ester is often used. For example, the *p*-nitrophenyl ester of glycine, protected at the amino group by the benzyloxycarbonyl group, is prepared. The activated carboxyl group then reacts with the amino acid valine.

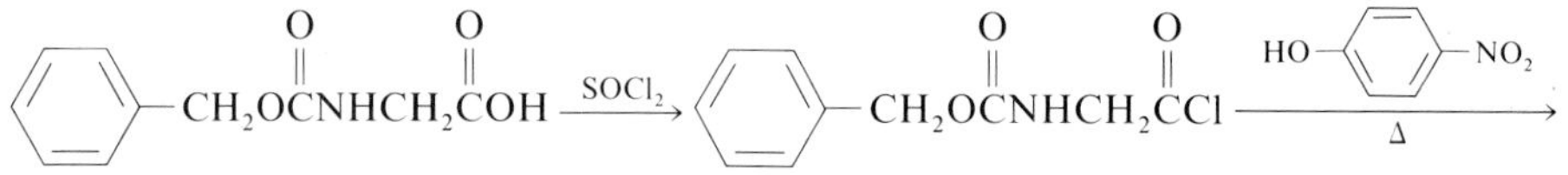

N-Cbz-Gly

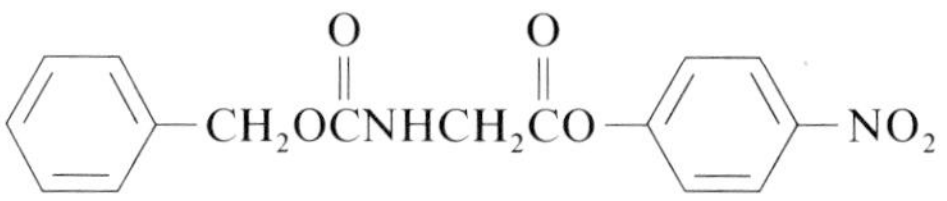

p-nitrophenyl ester of
N-benzyloxycarbonylglycine

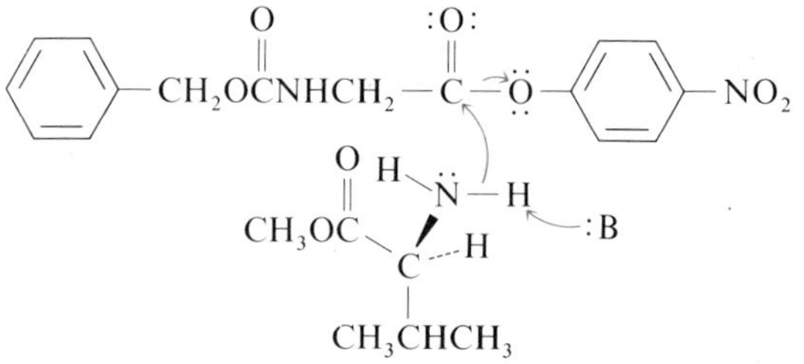

valine
*protected at
its carboxyl group
as the methyl ester*

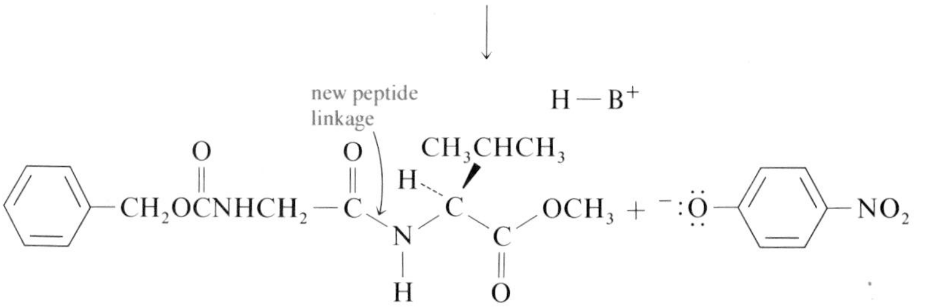

N-Cbz-Gly-Val-OMe
75%

p-nitrophenolate anion

*good leaving group,
stabilized by resonance
and electron-withdrawing
effect of nitro group*

Removal of the protecting group on the glycyl residue, the *N*-terminal amino acid of the dipeptide synthesized above, allows for the extension of the chain by adding another protected and activated amino acid residue, this time using alanine.

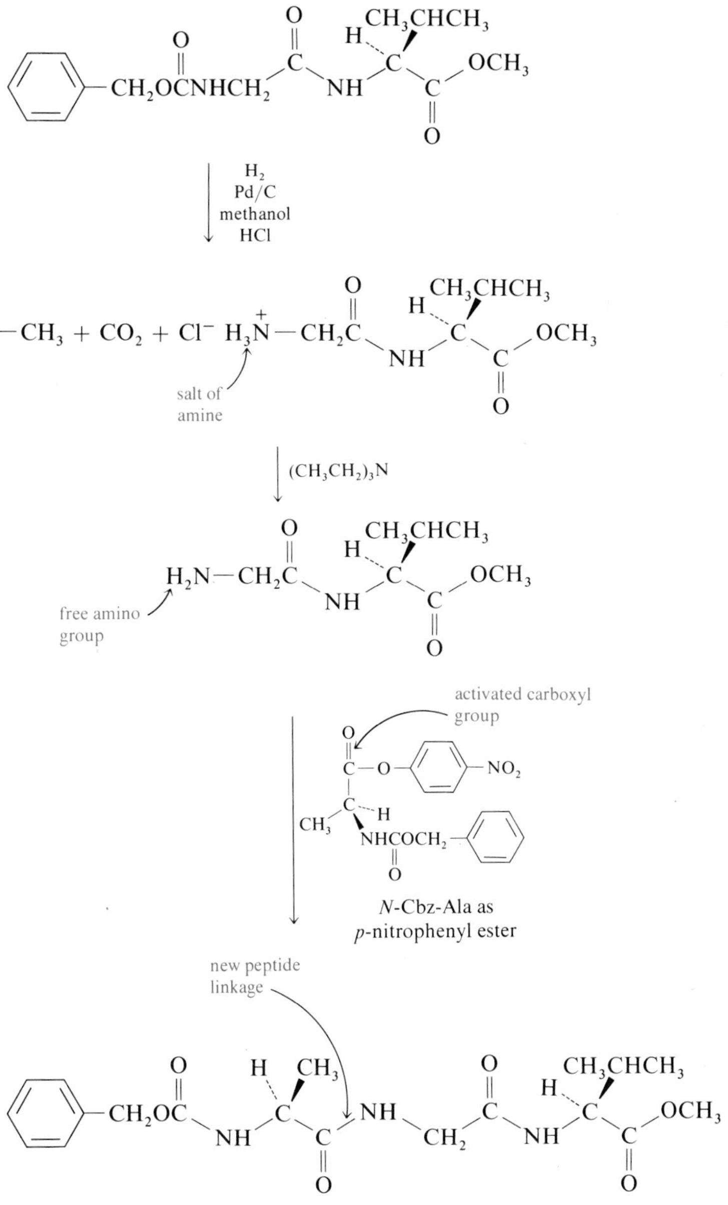

$$\text{C}_6\text{H}_5\text{—CH}_2\text{OCNHCH}_2\text{—C(=O)—NH—CH(CH}_2\text{CHCH}_3\text{)—C(=O)—OCH}_3$$

H₂
Pd/C
methanol
HCl

$$\text{C}_6\text{H}_5\text{—CH}_3 + \text{CO}_2 + \text{Cl}^- \ \text{H}_3\overset{+}{\text{N}}\text{—CH}_2\text{C(=O)—NH—CH(CH}_2\text{CHCH}_3\text{)—C(=O)—OCH}_3$$

salt of amine

$(\text{CH}_3\text{CH}_2)_3\text{N}$

$$\text{H}_2\text{N—CH}_2\text{C(=O)—NH—CH(CH}_2\text{CHCH}_3\text{)—C(=O)—OCH}_3$$

free amino group

activated carboxyl group

N-Cbz-Ala as
p-nitrophenyl ester

new peptide linkage

N-Cbz-Ala-Gly-Val-OMe
55%

Some of the practical problems plaguing the synthesis of large peptides and proteins become apparent even with the very small synthesis represented by the two equations above. Many manipulations are involved, and unless the yields are very high in each step, overall yields tend to be miniscule. In the example above, without even taking into account any losses that were connected with the conversion of amino acids into their N-protected p-nitrophenyl esters, the formation of the two peptide bonds alone gave an overall yield of $0.75 \times 0.55 \times 100$ or 41%. The peptide at that stage still must have the protecting groups removed to resemble a natural peptide.

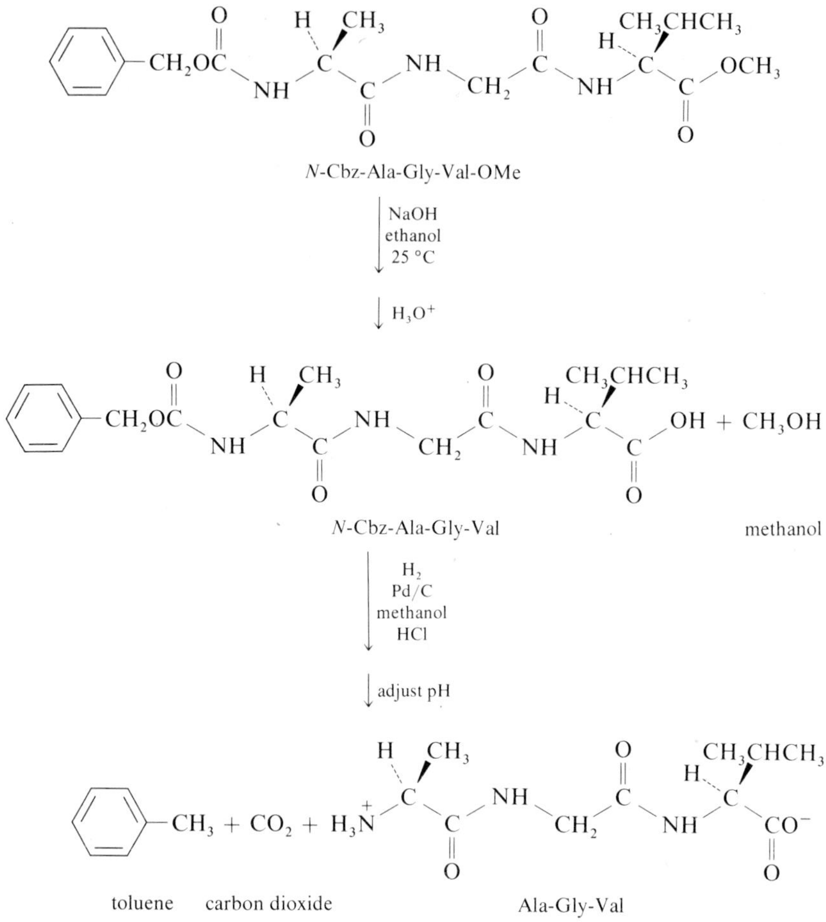

N-Cbz-Ala-Gly-Val-OMe

NaOH
ethanol
25 °C

H_3O^+

N-Cbz-Ala-Gly-Val + CH$_3$OH

methanol

H_2
Pd/C
methanol
HCl

adjust pH

toluene   carbon dioxide   Ala-Gly-Val

The conditions for the formation of the *p*-nitrophenyl ester are rather harsh, and might not be suitable for activating the carboxylic acid group of a larger peptide. Other types of esters, formed in the presence of dicyclohexylcarbodiimide, are being used in many peptide syntheses. In esters of *N*-hydroxysuccinimide, resonance contributors of the succinimide structure stabilize the leaving group.

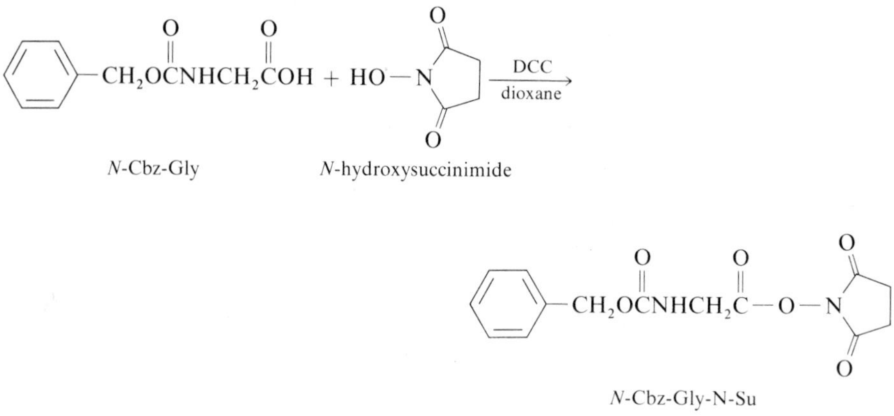

N-Cbz-Gly   N-hydroxysuccinimide

DCC
dioxane

N-Cbz-Gly-N-Su

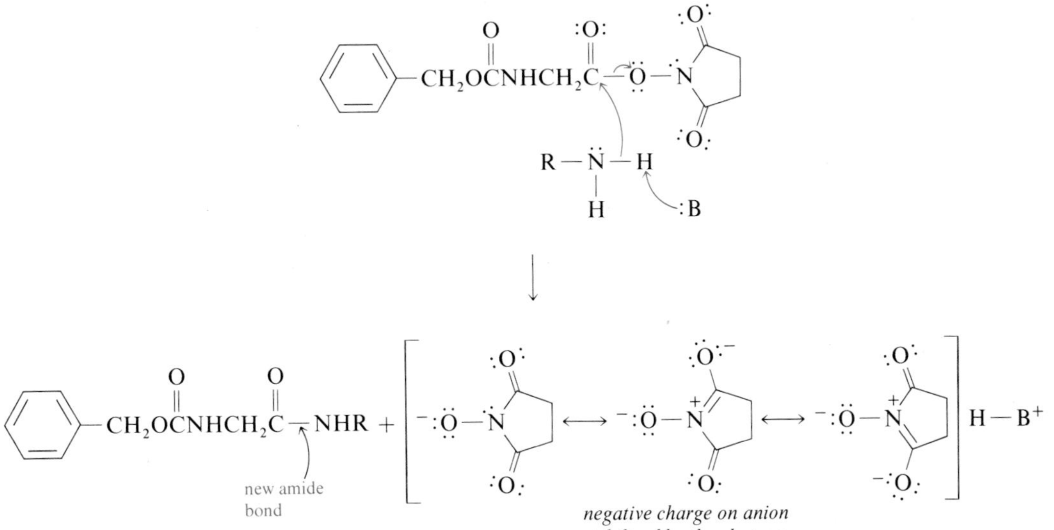

The top reaction scheme shows the mechanism of peptide bond formation:

$$\text{C}_6\text{H}_5\text{-CH}_2\text{OCNHCH}_2\overset{:\ddot{O}:}{\underset{}{C}}\text{-}\overset{..}{O}\text{-N(succinimide)} \quad + \quad R\text{-}\ddot{N}\text{-H} \cdots :B$$

$$\downarrow$$

$$\text{C}_6\text{H}_5\text{-CH}_2\text{OCNHCH}_2\overset{O}{C}\text{-NHR} \quad + \quad \left[\text{succinimide anion resonance structures} \right] \text{H-B}^+$$

new amide bond

negative charge on anion stabilized by the electron-withdrawing effect of the adjacent nitrogen atom

---

PROBLEM 17.14 Complete the following equations.

(a)
$$\text{CH}_3\text{CO}-\underset{|}{\overset{|}{\text{C}}}\text{...} \quad \text{CH}_3\overset{\text{CH}_3}{\underset{\text{CH}_3}{\text{CO}}}\text{-CNHCH}_2\text{CNHCH}_2\text{COH} \xrightarrow[\text{dichloromethane}]{\text{CF}_3\text{COH}} \text{A}$$

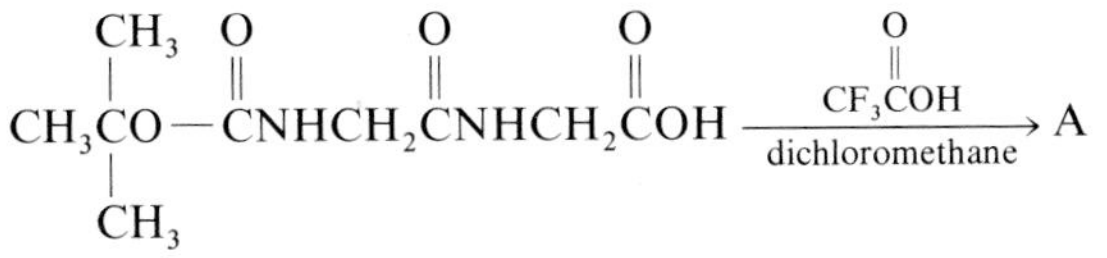

(b)
$$\text{CH}_3\text{CHCH}_2\underset{\text{CH}_3}{\overset{\text{CO}^-}{\text{C}}}\text{...H} \; {}^+\text{NH}_3 \quad + \quad \text{C}_6\text{H}_5\text{-CH}_2\text{OCCl} \longrightarrow \text{B} \xrightarrow[\text{DCC}]{\text{HO-N(succinimide)}} \text{C}$$

(c)
$$\text{CH}_3\text{CH}\underset{\text{CH}_3}{\overset{\text{CO}^-}{\text{C}}}\text{...H} \; {}^+\text{NH}_3 \quad + \quad \text{CH}_3\underset{\text{CH}_3}{\overset{\text{CH}_3}{\text{C}}}\text{-O-CN}_3 \longrightarrow \text{D}$$

(d)
$$\text{CH}_3\underset{\text{NHCOCH}_2\text{C}_6\text{H}_5}{\overset{\text{COH}}{\text{C}}}\text{...H} \xrightarrow{\text{SOCl}_2} \text{E} \xrightarrow{\text{HO-C}_6\text{H}_4\text{-NO}_2} \text{F} \xrightarrow{\text{H}_2\text{NCH}_2\text{COCH}_3} \text{G} \xrightarrow[\text{ethanol}]{\text{NaOH}}$$

$$\text{H} \xrightarrow[\substack{\text{Pt/C}\\\text{methanol}\\\text{HCl}}]{\text{H}_2} \text{I} \xrightarrow[\text{pH 7}]{\text{pyridine, CH}_3\text{COH}} \text{J}$$

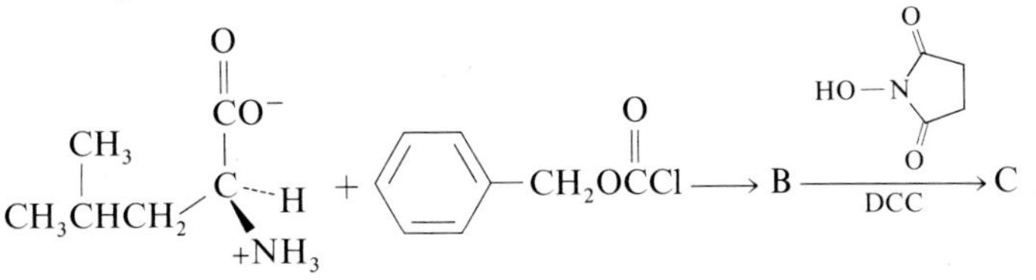

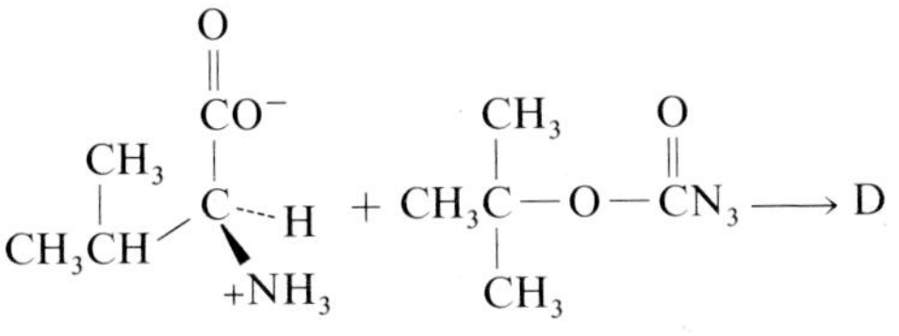

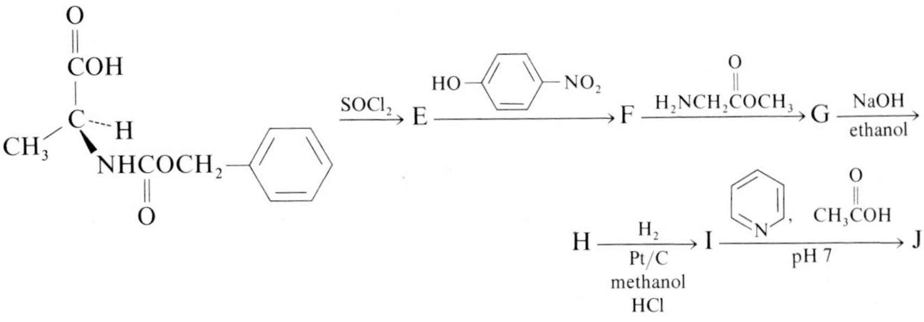

C. Synthesis of a Peptide

Methionine enkephalin is a pentapeptide found in the brain and believed to function as a natural painkiller by interacting with the same sites in the brain with which potent opiates such as morphine (Section 16.1A) react. Its synthesis illustrates concretely some of the general points about the protection of functional groups and the activation of carboxyl groups that we have discussed so far. Methionine enkephalin has the structure Tyr-Gly-Gly-Phe-Met. It was synthesized in two fragments that were then joined. The synthesis of the smaller fragment Phe-Met is shown below.

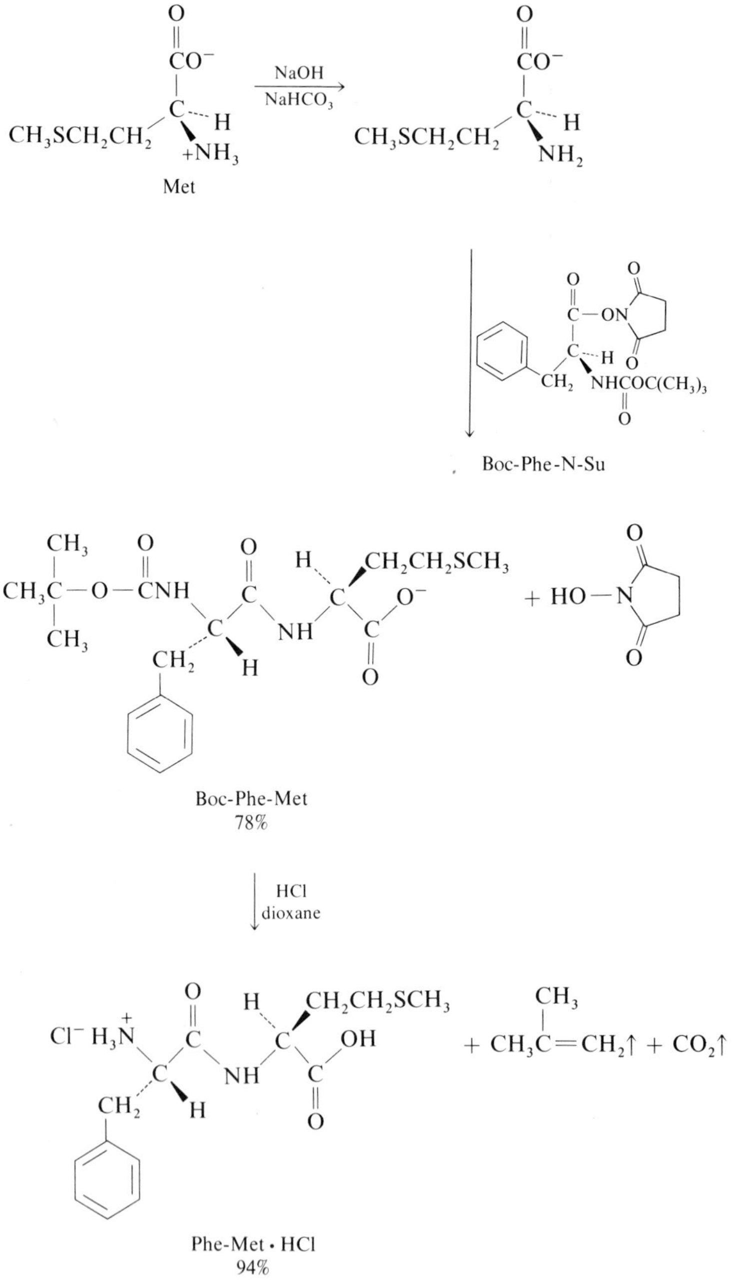

Boc-Phe-N-Su

Boc-Phe-Met
78%

HCl
dioxane

Phe-Met · HCl
94%

Phenylalanine, with its amino group protected as the *tert*-butoxycarbonyl group and its carboxyl group activated as the *N*-hydroxysuccinimide ester, reacts with methionine to give the protected dipeptide *N*-*tert*-butoxycarbonylphenylalanylmethionine, Boc-Phe-Met. The protecting group is removed with dry hydrogen chloride in dioxane, and the amino group of the dipeptide is protected as its hydrochloride salt until it is used in the next step of the synthesis.

Glycylglycine is available from chemical supply houses. Tyrosine, protected as the *tert*-butyl ether at the phenolic oxygen, and with a *tert*-butoxycarbonyl group at the amino group, reacts as its *N*-hydroxysuccinimide ester with glycylglycine to give a protected tripeptide. The carboxyl group of the tripeptide is activated by conversion to the *N*-hydroxysuccinimide ester with dicyclohexylcarbodiimide. The tripeptide and the dipeptide fragments are combined in the presence of base. Removal of the protecting groups and adjustment of pH gives the pentapeptide methionine enkephalin.

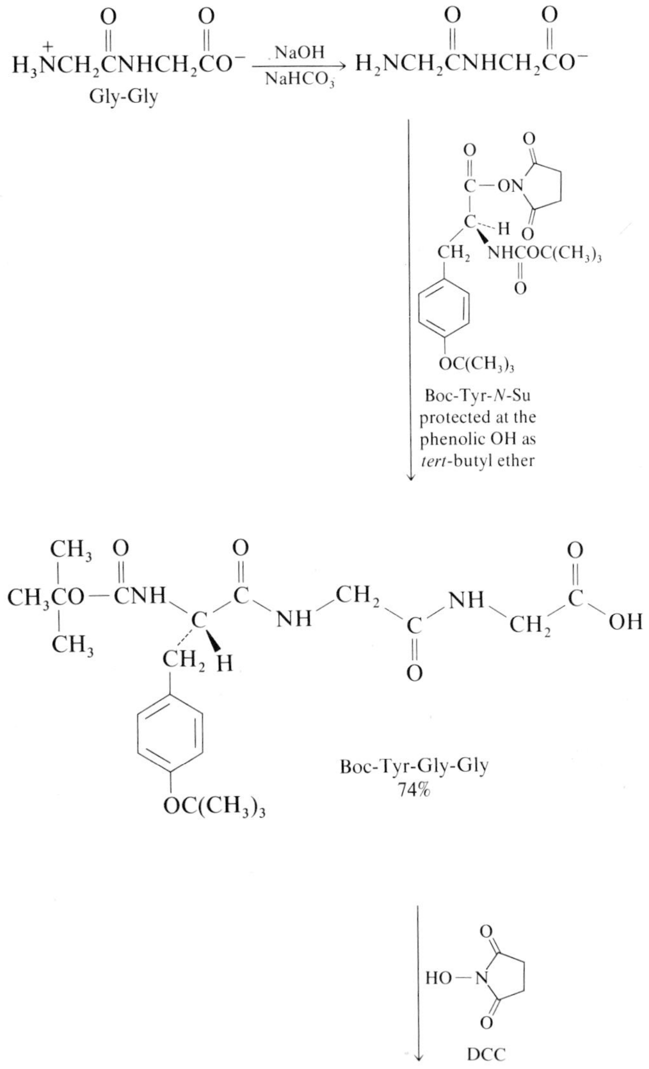

Gly-Gly

Boc-Tyr-*N*-Su
protected at the
phenolic OH as
tert-butyl ether

Boc-Tyr-Gly-Gly
74%

DCC

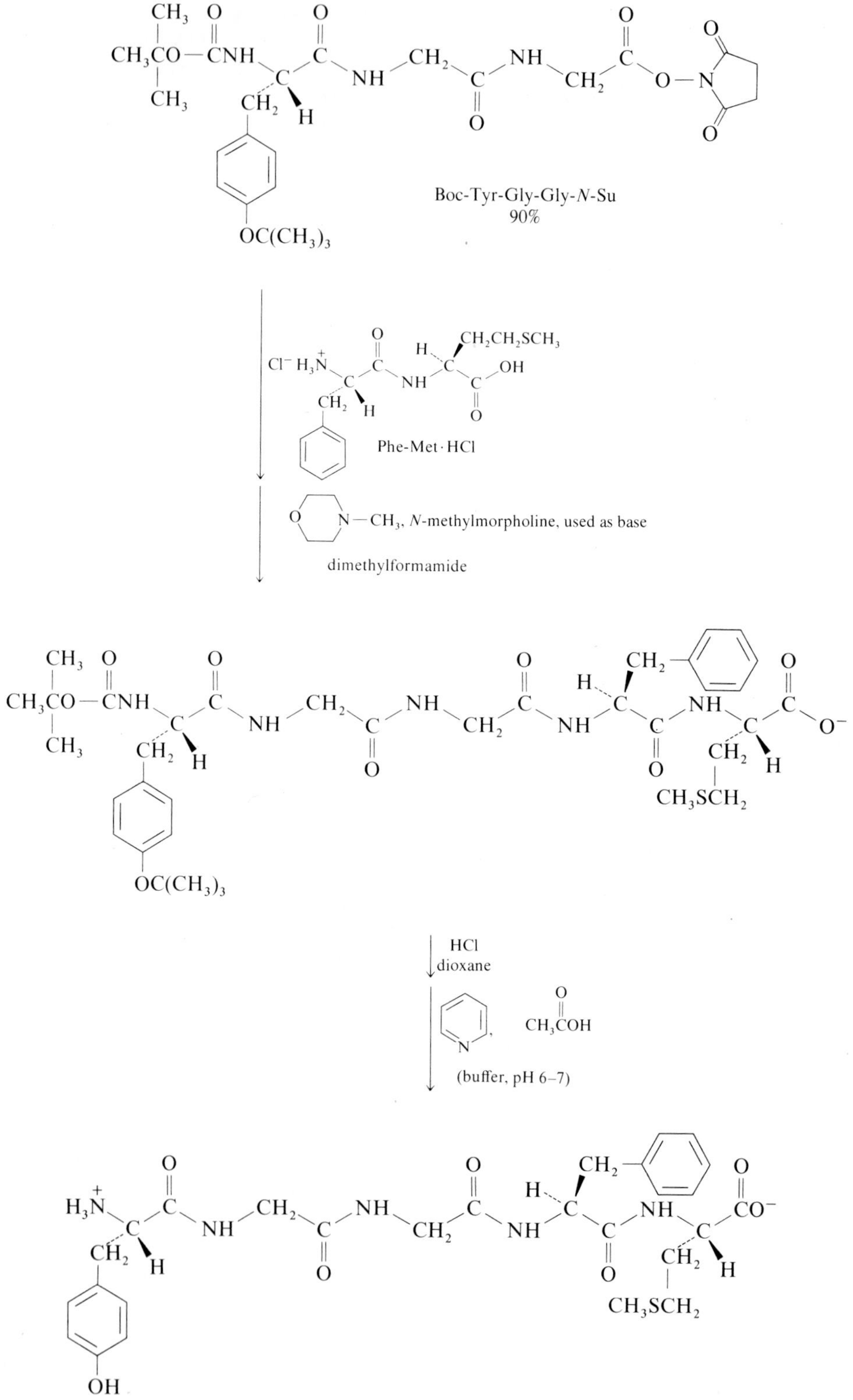

Boc-Tyr-Gly-Gly-*N*-Su
90%

Phe-Met·HCl

N-methylmorpholine, used as base

dimethylformamide

HCl
dioxane

(buffer, pH 6–7)

Tyr-Gly-Gly-Phe-Met
methionine enkephalin
73%

Methionine enkephalin is not a large peptide, yet its synthesis demonstrates the problems that become more important as the peptide to be synthesized becomes larger and the number of times that intermediates have to be isolated and purified increases. Each manipulation represents losses of material and time. Though some fairly large peptides were synthesized by methods similar to those shown for the enkephalin, real progress in the synthesis of larger peptides and proteins awaited the development of the technique of solid-phase synthesis by Robert B. Merrifield in 1963, discussed in the next section.

PROBLEM 17.15   Why was the phenolic hydroxyl group of tyrosine protected in the synthesis of methionine enkephalin? Write an equation predicting what would happen if the tyrosine were not protected.

PROBLEM 17.16   Along with methionine enkephalin, another pentapeptide with similar physiological properties, leucine enkephalin, is found in the brain. It has the same structure as methionine enkephalin, except that leucine is substituted for methionine as the *C*-terminal amino acid. Leucine enkephalin has also been synthesized by a method similar to that used for methionine enkephalin. Propose a detailed synthesis for leucine enkephalin.

D. Solid-Phase Peptide Synthesis

Solid-phase peptide synthesis was designed by Robert B. Merrifield of Rockefeller University to simplify the problems of isolation and purification of peptides when many steps are necessary for a synthesis. He reasoned that if the growing peptide could be incorporated into a solid polymer, it could be handled as a solid and purified thoroughly at each stage of the synthesis by repeated washing and filtration steps, minimizing losses of peptides. The rates of the reactions for forming peptide bonds and the degree of completion of formation could be increased greatly by using large excesses of protected amino acids, which could then be washed away from the extended peptide chain on its solid support. Thus, it would be possible to add one amino acid at a time in precise order to build up large molecules of high purity in good yield. Ideally, the whole process would be automated so that the multiple cycles necessary to form a peptide bond, purify the extended protected peptide, remove the protecting group, and add a different amino acid to the newly exposed *N*-terminal amino acid could be repeated over and over again with minimal attention.

It was necessary to establish a solid support system that could be covalently bonded to the *C*-terminal amino acid of the peptide with a link that would be stable to the reagents used in making the peptide bonds and removing protecting groups. In addition, the support had to be insoluble in the solvents used in the reactions. It had to have structural stability and maintain its particle size so that it would survive the physical manipulations of repeated washings and filtrations. It had to be chemically inert except at the site where the original attachment of the peptide chain was made. Its structure had to allow the solutions of reagents to flow through it so that reactions could take place on a large portion of its surface area.

Much research has been done in searching for a polymeric material that would meet all these requirements. The material now most widely used is a polystyrene that is cross-linked with divinylbenzene (Section 20.2A). The polystyrene is formed into beads approximately 5×10^{-3} cm in diameter. The cross-linking gives the polymer a high molecular weight and very low solubility in organic solvents. The amount of cross-linking is adjusted so that the polymer has some flexibility and swells to double the size of the dry beads when it is placed in an organic solvent for the reactions. This allows reactions to take place inside the beads as well as on the surface. Even though the beads are so small that they can hardly be seen by the naked eye, each one has room for the growth of a trillion (10^{12}) peptide chains.

The polystyrene resin is converted by a Friedel-Crafts reaction with chloromethyl methyl ether into a polymer with chloromethyl groups on approximately 22% of the aromatic rings.

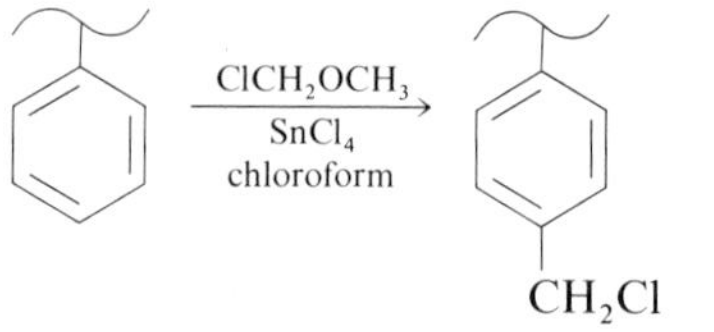

$$\text{(fragment of polystyrene)} \xrightarrow[\substack{\text{SnCl}_4 \\ \text{chloroform}}]{\text{ClCH}_2\text{OCH}_3} \text{(polystyrene, converted to a benzyl chloride, CH}_2\text{Cl)}$$

fragment of polystyrene

polystyrene, converted to a benzyl chloride

PROBLEM 17.17   Propose a mechanism for the reaction shown above.

The *C*-terminal amino acid of the peptide is attached to the polystyrene as a benzyl ester. The amino groups of the amino acids to be added to the chain are protected by *tert*-butoxycarbonyl groups, which can be removed under conditions that do not affect the benzylic ester bond (Section 17.4A). Dicyclohexylcarbodiimide (DCC) is used to activate the carboxylic acid group of the amino acid to be added to the chain (Section 17.4B). The different steps in a solid-phase synthesis of a pentapeptide, Gly-Val-Gly-Ala-Pro, using abbreviations for the amino acids and protecting groups, are shown below.

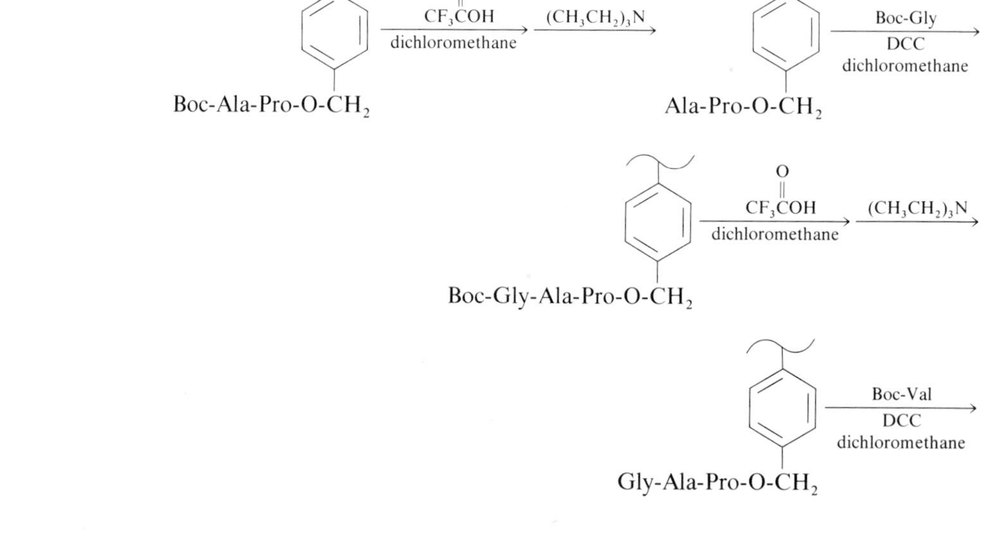

$$\text{CH}_2\text{Cl} \xrightarrow[\substack{(\text{CH}_3\text{CH}_2)_3\text{N} \\ \text{dichloromethane}}]{\text{Boc-Pro}} \text{Boc-Pro-O-CH}_2 \xrightarrow[\substack{\text{dichloromethane}}]{\overset{\text{O}}{\overset{\|}{\text{CF}_3\text{COH}}}} \xrightarrow{(\text{CH}_3\text{CH}_2)_3\text{N}} \text{Pro-O-CH}_2 \xrightarrow[\substack{\text{DCC} \\ \text{dichloromethane}}]{\text{Boc-Ala}}$$

$$\text{Boc-Ala-Pro-O-CH}_2 \xrightarrow[\substack{\text{dichloromethane}}]{\overset{\text{O}}{\overset{\|}{\text{CF}_3\text{COH}}}} \xrightarrow{(\text{CH}_3\text{CH}_2)_3\text{N}} \text{Ala-Pro-O-CH}_2 \xrightarrow[\substack{\text{DCC} \\ \text{dichloromethane}}]{\text{Boc-Gly}}$$

$$\text{Boc-Gly-Ala-Pro-O-CH}_2 \xrightarrow[\substack{\text{dichloromethane}}]{\overset{\text{O}}{\overset{\|}{\text{CF}_3\text{COH}}}} \xrightarrow{(\text{CH}_3\text{CH}_2)_3\text{N}}$$

$$\text{Gly-Ala-Pro-O-CH}_2 \xrightarrow[\substack{\text{DCC} \\ \text{dichloromethane}}]{\text{Boc-Val}}$$

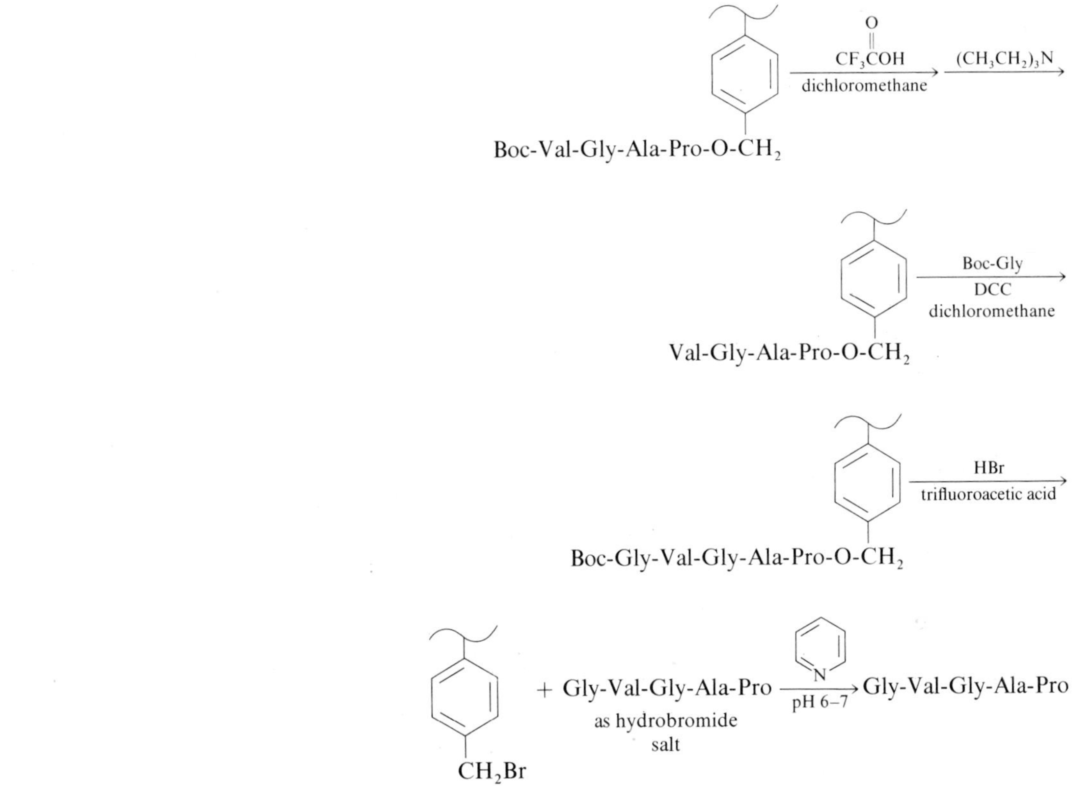

Boc-Val-Gly-Ala-Pro-O-CH$_2$ $\xrightarrow[\text{dichloromethane}]{\text{CF}_3\text{COH}}$ $\xrightarrow{(\text{CH}_3\text{CH}_2)_3\text{N}}$

Val-Gly-Ala-Pro-O-CH$_2$ $\xrightarrow[\text{dichloromethane}]{\substack{\text{Boc-Gly} \\ \text{DCC}}}$

Boc-Gly-Val-Gly-Ala-Pro-O-CH$_2$ $\xrightarrow[\text{trifluoroacetic acid}]{\text{HBr}}$

CH$_2$Br   + Gly-Val-Gly-Ala-Pro $\xrightarrow[\text{pH 6-7}]{}$ Gly-Val-Gly-Ala-Pro

as hydrobromide salt

The reagents for the peptide synthesis have all been chosen so that the reaction sequences are repetitious and, therefore, can be automated. The *tert*-butoxycarbonyl group is removed with trifluoroacetic acid, which leaves behind a protonated amino group. Triethylamine removes the proton and gives the free amino group, which reacts with the activated carboxylic acid group from a new protected amino acid. Finally, the strong acid, hydrogen bromide in trifluoroacetic acid, cleaves both the protecting group and the benzylic ester bond. Note that aqueous acids, which would hydrolyze peptide bonds, are avoided at all stages. Besides the reagents that are shown, there are a number of intermediate stages in which the polymer, with the attached peptide chains, is washed with solvents and filtered to purify it. In all, 11 separate steps are used for each amino acid added to the chain.

PROBLEM 17.18   Write equations showing how you would carry out a solid-phase synthesis of Phe-Leu-Gly. Show full structures of the benzylic ester, any protecting groups, and any reagents used to promote the formation of the peptide bond at least once. Show the full structure of the peptide at each stage of the synthesis.

The solid-phase peptide synthesis has been used to prepare a number of naturally occurring proteins, including insulin and the enzyme ribonuclease, which was synthesized in 1969 by Merrifield. Ribonuclease has 124 amino acid residues. Its synthesis required 369 chemical reactions and 11,931 steps in the automated process. The introduction of one amino acid took approximately four hours. This is a much more rapid and efficient process than is possible in classic solution-phase chemistry, but it does not begin to approach the efficiency with which proteins are synthesized in the living cell. A bacterial cell completes protein synthesis in seconds and manages to keep about three thousand different protein syntheses going simultaneously with no confusion of reagents and products.

17.5

Conformation of Peptides and the Structures of Proteins

Peptides consisting of a number of amino acids assume conformations that are typical of their particular structure. The carbonyl group and the nitrogen and hydrogen atoms of the peptide bond, as well as the two carbon atoms to which the carbonyl and amino groups are bonded, lie in a plane. Resonance interaction between the nitrogen atom and the carbonyl group gives a partial double-bond character to the carbon-nitrogen bond and prevents free rotation.

The oxygen atom of the carbonyl group and the hydrogen atom on the nitrogen atom are found trans to each other. The bond lengths and bond angles typical of the peptide group are illustrated in Figure 17.5.

All the amino acids found in natural peptides and proteins have the same configuration, and this imparts a stereochemical regularity to peptide chains. They assume two major conformations depending mainly on the types of amino acids that make up the primary structure. One common conformation that we have already mentioned in Section 17.1 is the α-helix. This structure was proposed in 1951 on the basis of x-ray diffraction experiments by L. Pauling and R. Corey. The peptide chain

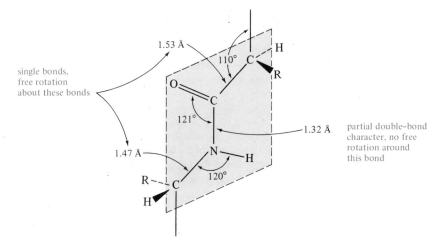

shaded area represents the plane
of the peptide linkage and the
two adjacent carbon atoms

Figure 17.5   Bond lengths and bond angles typical of the peptide group.

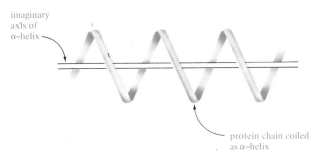

α–helix showing hydrogen bonding
between peptide groups

imaginary
axis of
α–helix

protein chain coiled
as α–helix

Figure 17.6   The α-helix, showing hydrogen bonding between peptide groups
and the coiling conformation of the peptide chain.

forms a right-handed coil containing 3.7 amino acids per turn of the coil. The structure is given rigidity and stability by hydrogen bonding between the hydrogen atom on a nitrogen atom and the carbonyl group of the amino acid four units down the chain at the next turn of the coil. The spacing of the amino acids is such that every peptide linkage is involved in hydrogen bonding (Figure 17.6).

The α-helix is particularly important in the structural proteins such as **α-keratins,** which make up the protein of skin, nails, horns, hair, and feathers. Proteins that form structures such as hair and wool are rich in the oxidized form of cysteine, cystine (Section 17.3C). In them, several α-helices are further coiled around each other to give multiple strands that are held together by disulfide bridges. A permanent wave, in fact, involves reduction of some of the disulfide linkages in hair so that it loses its natural shape, which is determined genetically. Then the hair is held in a new conformation on curlers and reoxidized to form new disulfide linkages.

Not all proteins have α-helical structures. Silk is an example of a protein, also called **fibroin,** that has a **β-keratin** or **pleated sheet** conformation. The peptide chains are more extended in this conformation and lie side-by-side, with hydrogen bonding between the chains. In this structure, maximum hydrogen bonding is possible when one polypeptide chain runs in the opposite direction from the adjacent one (Figure 17.7, p. 994). Such structures are most stable when the side chains of the amino acid residues are small and not charged. Silk fibroin consists mostly of glycine and alanine, for example.

Another important structural protein is **collagen,** which makes up connective tissues. When collagen is boiled in water, gelatin is formed. Collagens are rich in glycine, alanine, and proline, and contain 4-hydroxyproline, an amino acid that is seldom found

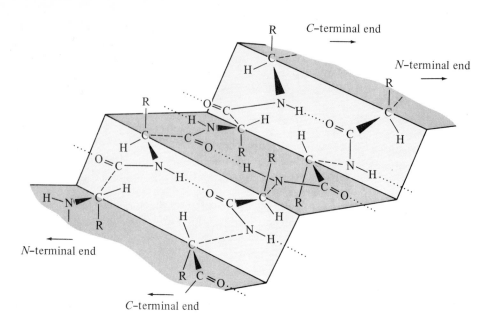

Figure 17.7   The pleated sheet structure of proteins.

in other proteins. The structure of collagens consists of a triple helix of polypeptide chains, each of which is a left-handed helix.

Keratins and collagens are known as *fibrous proteins*. The other large class of proteins is the *globular proteins* in which the tertiary structure of the protein tends to be spherical or ellipsoidal, with much coiling of the polypeptide chain. Many proteins and all enzymes have this type of structure.

Parts of globular proteins have the α-helix conformation. Other parts are more randomly coiled. Not all side chains in amino acids fit well into an α-helix. Proline, in particular, causes a bend in the α-helix structure because the nitrogen of the amino acid is part of a five-membered ring. Furthermore, proline has no amide hydrogen atom to participate in hydrogen bonding. If a polypeptide is rich in lysine and arginine, the side chains are positively charged at physiological pH and repel each other, disrupting the helical structure. Conversely, strong ionic interactions between carboxylate anions on the side chains of glutamic and aspartic acid residues and cationic centers on lysine and arginine residues may be more important than hydrogen bonding in determining conformation. Disulfide bridges between different parts of the peptide chain also stabilize particular conformations of the protein. The whole molecule is arranged so that the hydrophobic side chains are protected on the inside of a globular protein where they are stabilized by van der Waals interactions. The more polar and charged amino acid residues are on the surface of the protein in contact with the polar solvent water.

Thus, *the overall structure of a protein in its natural state is determined by a multitude of interactions made possible by the specific order of amino acids in the chain. Among these, we have seen attractions between positively and negatively charged sites on the chain or repulsions between sites of similar charge, hydrogen bonding, van der Waals forces acting between hydrocarbon-like side chains on amino acids, and disulfide bridges.* In the next section, we look more closely at the structure of an enzyme, chymotrypsin, to see how its function is determined by these complex structural factors.

A protein retains its natural structure within a narrow range of pH, temperature, and ionic strength of the solution. Addition of acid, base, metal ions, or urea (which disrupts hydrogen bonding in a protein) to a solution of a protein changes its structure to a *denatured* form in which biological activity is lost. An increase in temperature does the same thing—that is what happens when an egg is cooked. Sometimes a denatured protein will return to the natural form and recover most of its biological activity, if, for example, the optimum pH is restored. In other cases, the damage is irreversible. The particular structure that is native to a given protein is the favored one, and the protein will return to it if possible. The enzyme ribonuclease, for example, has four disulfide bridges between eight cysteine residues, widely spaced in the molecule. If the disulfide bridges are cleaved by reduction, the protein is completely denatured by urea and loses all of its enzymatic activity. Removal of the urea allows the molecule to regain its activity, indicating that it folds back into the same tertiary structure, and even reforms the same disulfide links by oxidation in air. A random reconnecting of eight cysteine residues into four disulfide bridges would give more than one hundred different possibilities, yet only one set of four disulfide bridges is finally formed.

Some proteins such as hemoglobin have quaternary structure. In hemoglobin, four polypeptide chains, each of which has tertiary structure, are associated with each other and with four molecules of heme. Heme is a heterocyclic ring (Section 18.7B) that coordinates with iron(II) ions to serve the oxygen-carrying function of the blood (Figure

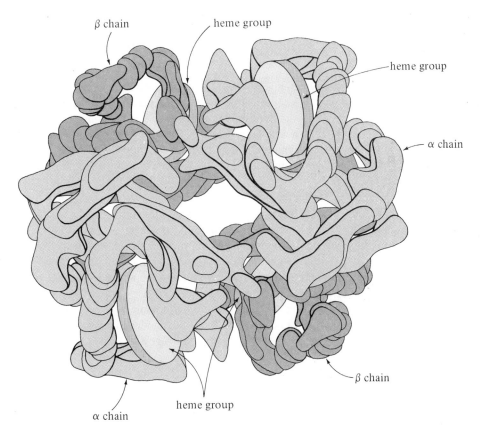

Figure 17.8   Schematic representation of the structure of hemoglobin, showing four polypeptide chains and four heme molecules. (Adapted from "The Hemoglobin Molecule," by M. F. Pertuz. Copyright © 1964 by Scientific American, Inc. All rights reserved.)

17.8). In hemoglobin, the placement of the heme units in pockets of the protein that are lined with hydrophobic amino acid side chains protects iron(II) against being oxidized irreversibly by oxygen to iron(III), something that happens easily in ordinary solutions of iron(II) compounds. One ligand to the iron is a histidine residue, which also enables the iron to bind loosely and reversibly with oxygen. Carbon monoxide binds more strongly and irreversibly to hemoglobin than oxygen does. This can lead to carbon monoxide poisoning. Heavy smokers, who inhale large doses of carbon monoxide in tobacco smoke, have about 20% of their hemoglobin bound up in this nonfunctional form.

A protein such as hemoglobin, in which a small molecule is associated with the protein, is called a **conjugated protein.** The small molecule, such as heme, when associated with the protein, is known as a **prosthetic group.** Many enzymes are proteins conjugated with coenzymes that include compounds known to be vitamins.

17.6

Chymotrypsin. A Look at the Functioning of an Enzyme

Chymotrypsin, one of the digestive enzymes secreted by the pancreas, belongs to a family of enzymes that cleave proteins into smaller peptides. The group, which includes trypsin, is known as the **serine proteases,** because the side chain of serine plays an important part in their catalytic activity. Chymotrypsin hydrolyzes the peptide bond at the carboxylic acid group of amino acids having large hydrophobic side chains such as phenylalanine, tryptophan, or tyrosine (Section 17.3C).

Chymotrypsin is formed in the body when two dipeptide units, amino acids 14 and 15, and 147 and 148, are removed from a precursor molecule, chymotrypsinogen, which has 245 amino acid residues. What is left consists of three polypeptide chains held together by two disulfide bridges. There are three other disulfide bridges in the molecule, stabilizing its conformation. The enzyme has two important regions. The folding of the molecule brings histidine at position 57, aspartic acid at position 102, and serine at position 195 close together in what is known as the **active site** of the enzyme. Close by is a region lined with hydrophobic groups that accepts and positions the proper portion of the peptide chain for cleavage. This region is known as the **binding site.** Figure 17.9 shows a drawing of chymotrypsin in which each amino acid residue, except for those involved in the active site and in the disulfide bridges, is represented only by carbon 2.

The way in which chymotrypsin acts has been determined by many experiments with different types of compounds that are hydrolyzed by the enzyme. Some reagents deactivate the enzyme, and a degradation of the protein shows which amino acids have been affected. Spectroscopic methods have also been used to follow the course of protonation and deprotonation reactions. The mechanism for the hydrolysis of the peptide bond by the enzyme is one that is entirely familiar to you. The only new facts will be the specific reagents that appear in the role of acid, base, and nucleophile. The steps of the hydrolysis reaction are represented schematically on the next three pages.

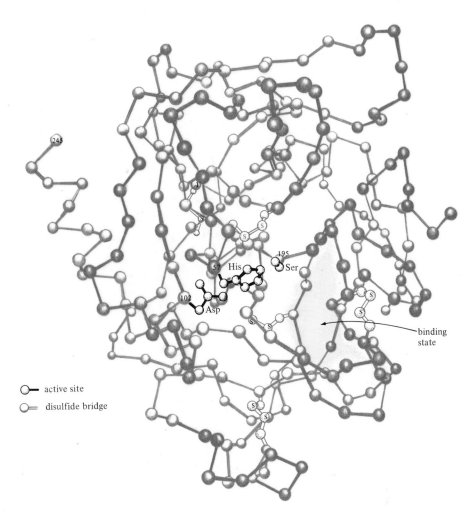

Figure 17.9   A representation of chymotrypsin showing the placement of histidine, aspartic acid, and serine at the active site. The hydrophobic pocket that binds the peptide residue during cleavage is below and to the right of serine-195. (Adapted from "A Family of Protein-Cutting Proteins," by Robert M. Stroud. Copyright © 1974 by Scientific American, Inc. All rights reserved.)

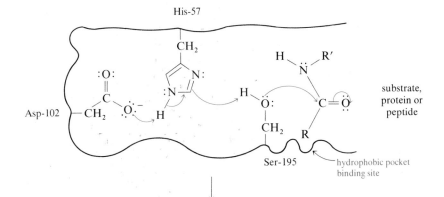

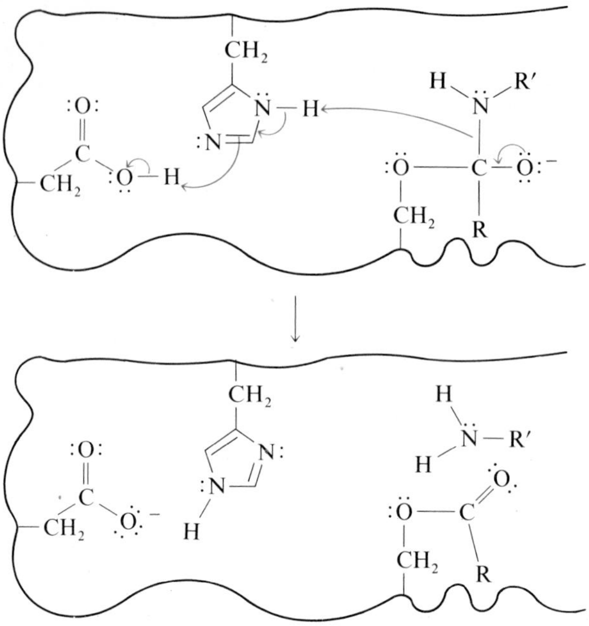

tetrahedral
intermediate

The nucleophile, the hydroxyl group on the serine, attacks the carbonyl group of the peptide bond to give a tetrahedral intermediate (Section 10.4). Serine is made more nucleophilic by transferring its proton to histidine, which in turn is able to accept that proton because it can transfer a proton to a carboxylate anion on a conveniently placed aspartic acid residue. The tetrahedral intermediate at the carbonyl group breaks up following transfer of a proton from histidine to the amide nitrogen atom, creating an amine as a leaving group. The histidine reclaims its proton from aspartic acid. At this stage, the enzyme is acylated at the serine residue. A transfer of an acyl group from the peptide to the enzyme has taken place. Serine is regenerated by a similar sequence of steps, now with water as a participant.

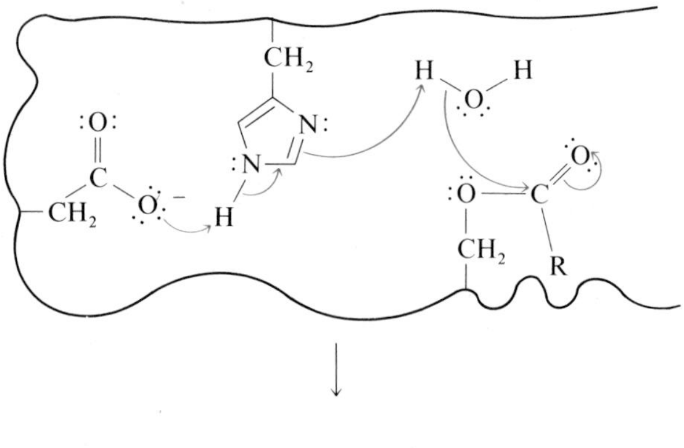

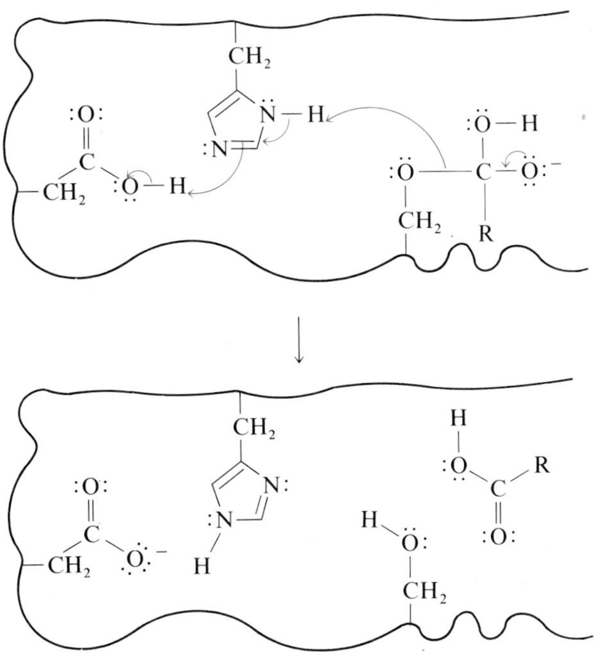

The ester bond at serine-195 is hydrolyzed and the peptide fragments diffuse away from the active site, which is then available for cleavage of another peptide bond. A process that takes strong aqueous acid or base and several hours of heating to 100 °C in the laboratory (Section 10.8B) is achieved at pH 6–7, at 37 °C, in a fraction of a second in the body.

The activity of an enzyme provides a powerful demonstration of the effect that lowering the energy of the transition state can have on the overall rate of the reaction. The form of an enzyme molecule creates the maximum degree of coordination between the different steps necessary for the cleavage of the peptide bond, ensuring that the transition state for the reaction is achieved with a minimum energy of activation.

Chymotrypsin is only one of the thousands of enzymes that catalyze the chemical processes that support life. The structures of only a few of these enzymes are known well enough that we can begin to understand how they function. Proteins that have similar functions in different organisms have remarkably similar, if not identical, structures. It is as if nature finds certain solutions to chemical problems and uses them over and over again with minor modifications to accommodate the unique characteristics of different species.

ADDITIONAL PROBLEMS

17.19 Write full structures for the following compounds:

(a) glycylalanyllysine   (b) ethyl alaninate   (c) *N*-benzyloxycarbonylmethionine
(d) *N-tert*-butoxycarbonylvaline   (e) *N*-2,4-dinitrophenylleucine
(f) *p*-nitrophenyl tyrosinate

17.20   Name the following compounds.

(a)

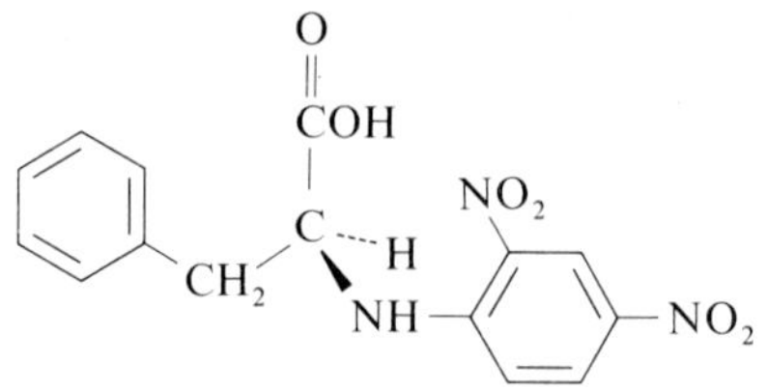

O
‖
COH
|
CH₂—C⋯H
NH—(NO₂)(NO₂ ring)

(b)

SH
|
CH₂ H
O | O
‖   C   ‖
H₃N⁺—C—C—NH—C—C—NH—C—C—O⁻
| H   ‖   ‖ |
H   H   O   O   CH₂ H

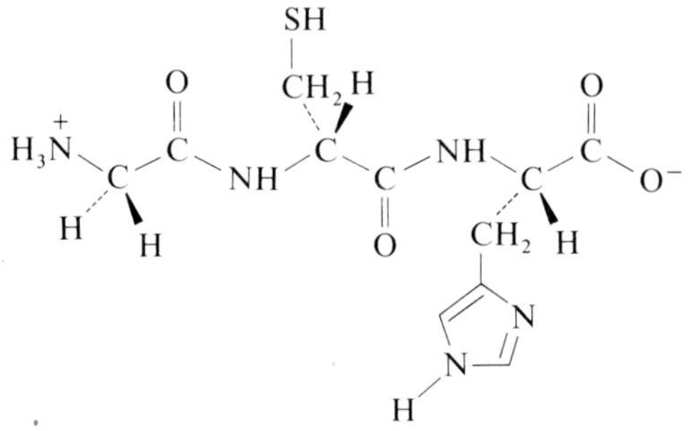

(imidazole ring with N, N—H)

(c)

O
‖
COH
|
C⋯H
H   NH—C(=O)—(phenyl)

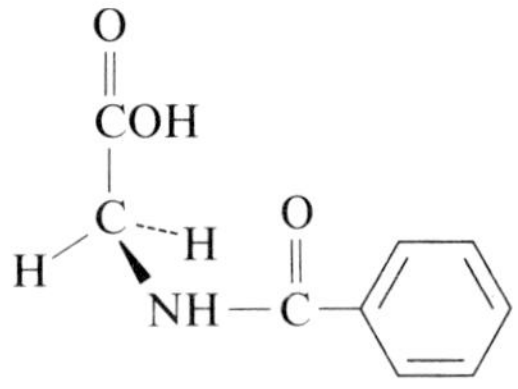

(d)

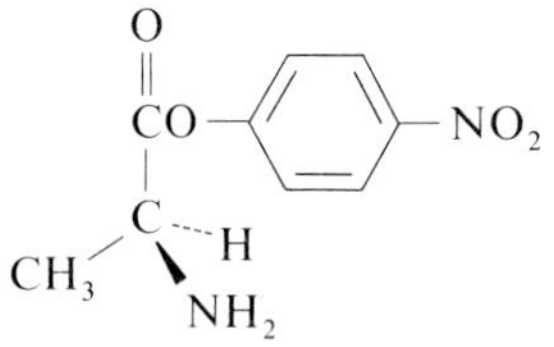

O
‖
CO—(ring)—NO₂
|
CH₃—C⋯H
NH₂

(e)

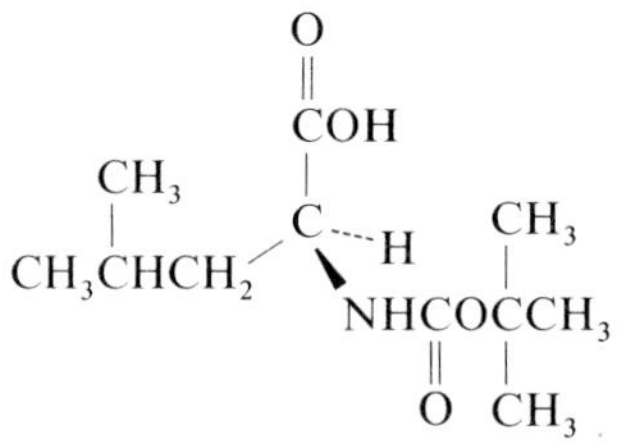

O
‖
COH
CH₃ |
| C⋯H   CH₃
CH₃CHCH₂   NHCOCCH₃
             ‖ |
             O CH₃

(f)

O
‖
COCH₃
|
HSCH₂—C⋯H
NH₂

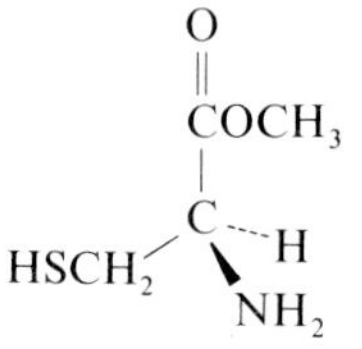

17.21 Using valine as a typical amino acid, write equations showing the reactions that you would predict with the following reagents.

(a) 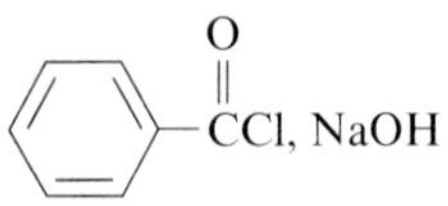 (phenyl)—C(=O)Cl, NaOH   (b) CH₃OH, HCl   (c) NaNO₂, HCl, H₂O, 0°C

(d) 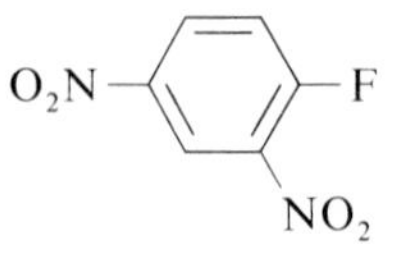 O₂N—(ring)—F (with NO₂)   (e) 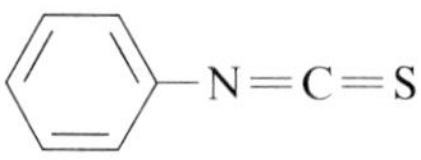 (phenyl)—N=C=S   (f) product of (e), acid, Δ

(g) 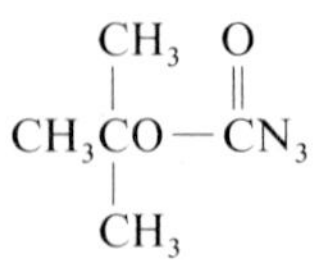 CH₃   O
     | ‖
     CH₃CO—CN₃
     |
     CH₃

(h) product of (g) + (cyclohexyl)—N=C=N—(cyclohexyl)

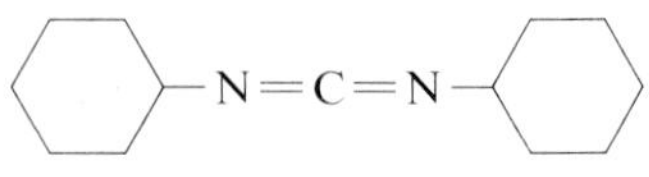

(i)   product of (h) +

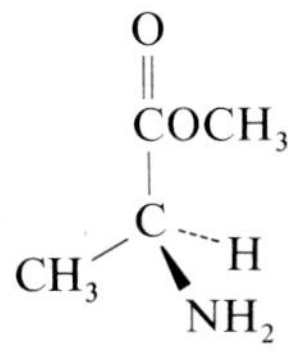

(j)

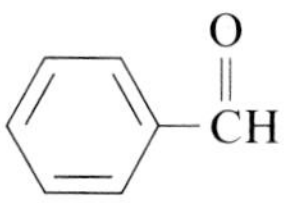

17.22   Complete the following equations.

(a)

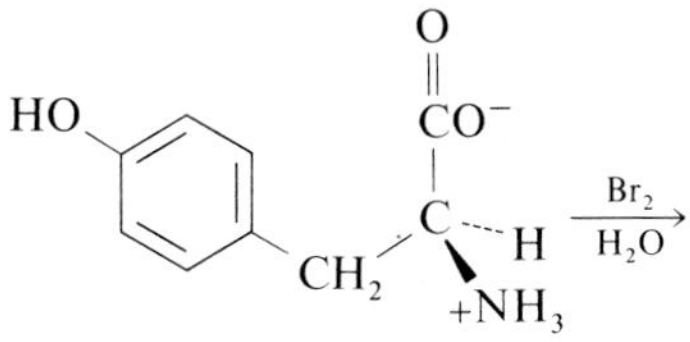

$\xrightarrow[H_2O]{Br_2}$

(b) 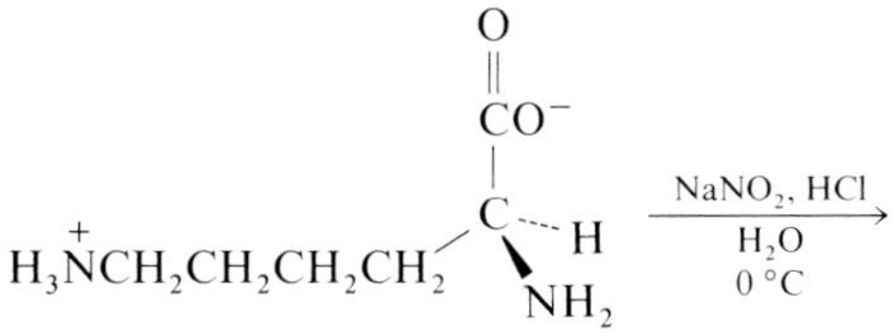 $H_3\overset{+}{N}CH_2CH_2CH_2CH_2$

$\xrightarrow[\substack{H_2O \\ 0\,^{\circ}C}]{NaNO_2,\ HCl}$

(c)

$\xrightarrow{CH_3COCCH_3}$

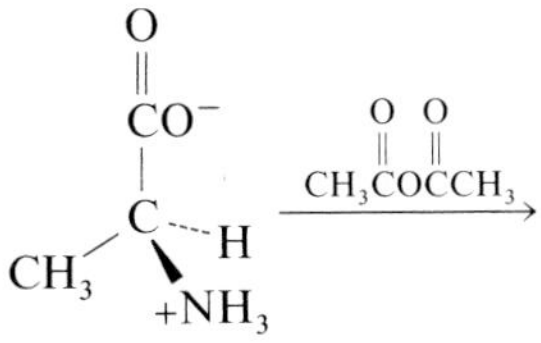

(d) 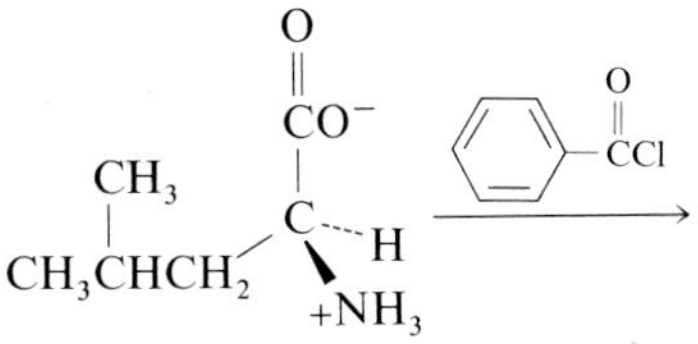 CH_3CHCH_2

(e)

$\xrightarrow[HCl]{CH_3CH_2OH\ (excess)}$

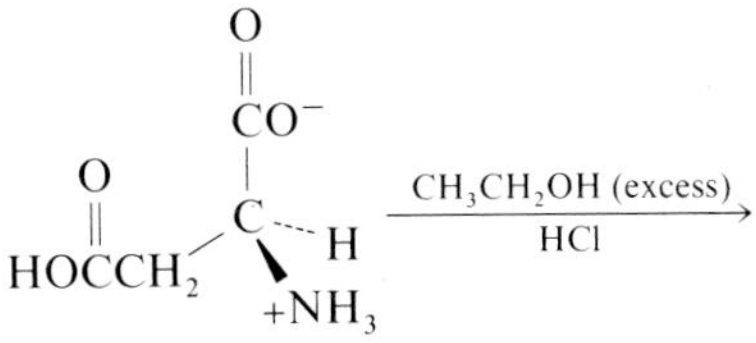

(f)

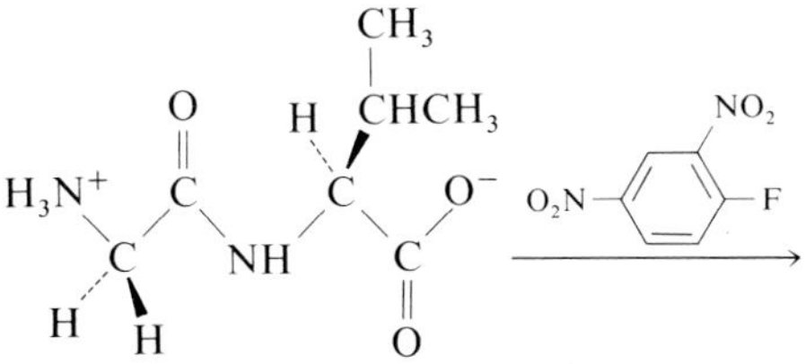

(g)   product of (f) $\xrightarrow[\Delta]{H_2O,\ HCl}$

(h)

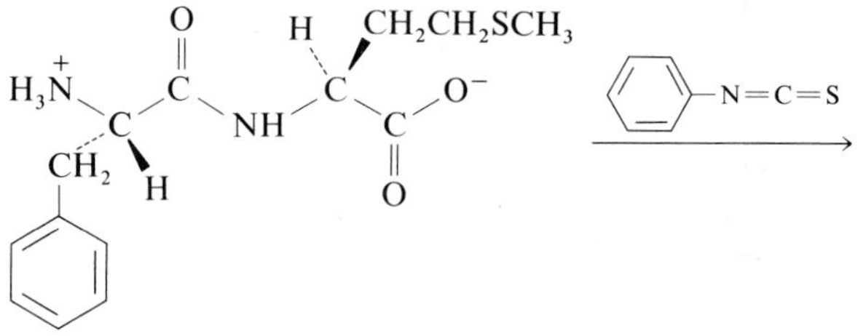

(i) product of (h) $\xrightarrow[\text{nitromethane}]{\text{HCl}}$
Δ

(j)

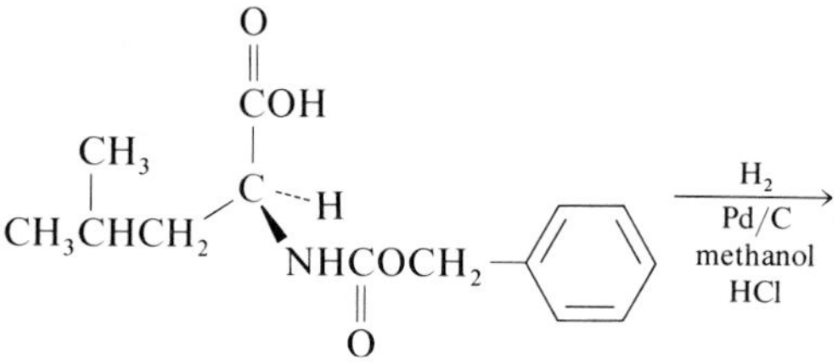

(k)

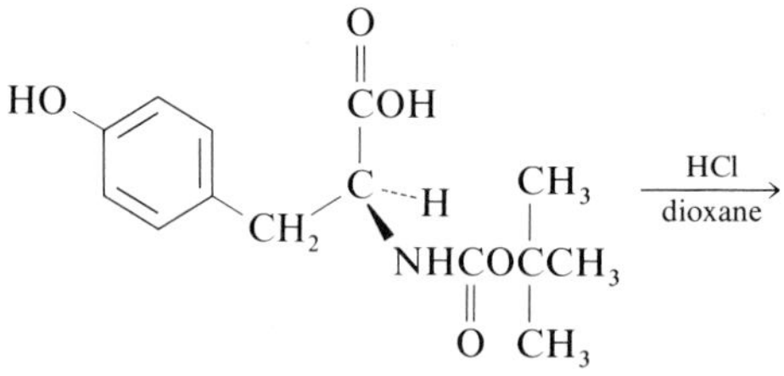

(l)

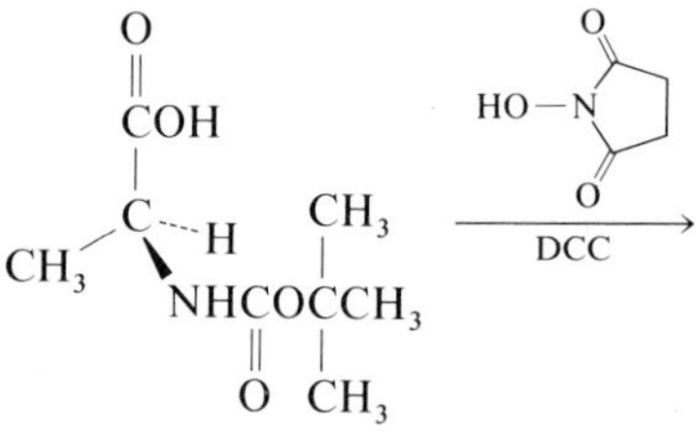

17.23 For each of the following peptides, show the full structure and the ionic state in which the compound would exist at pH 1, 6, and 11.

(a) Phe-Val-Asp   (b) Lys-Ala-Gly   (c) Leu-Tyr-Gly-NH$_2$ (d) Met-Gln-Ala

17.24 Since the discovery in 1975 that there are small peptides in the brain that have analgesic properties, chemists have studied the synthesis of analogs of these compounds. One such synthetic peptide that is 1500 times as active an analgesic agent as the naturally occurring methionine enkephalin (Section 17.4C) has the following structure.

Tyr-D-Ala-Gly-Phe-Pro-NH$_2$

(a) Write the full structure for this peptide, showing correct stereochemistry at every chiral center in the molecule.

(b) Synthesize a three-amino acid portion of the peptide, starting from either end of the chain, showing necessary protecting groups and reagents. Assume that all the necessary amino acids are available to you with the proper stereochemistry. Also assume that any group you use to protect an amino group can also be used to protect a hydroxyl function and can be removed under the same experimental conditions from both groups.

17.25 Lysine vasopressin, also a naturally occurring hormone, has the same structure as arginine vasopressin (Problem 17.13) except that lysine is substituted for arginine. Lysine vasopressin was synthesized as outlined below. Fill in the synthesis by writing structures for the intermediates and products indicated by letters.

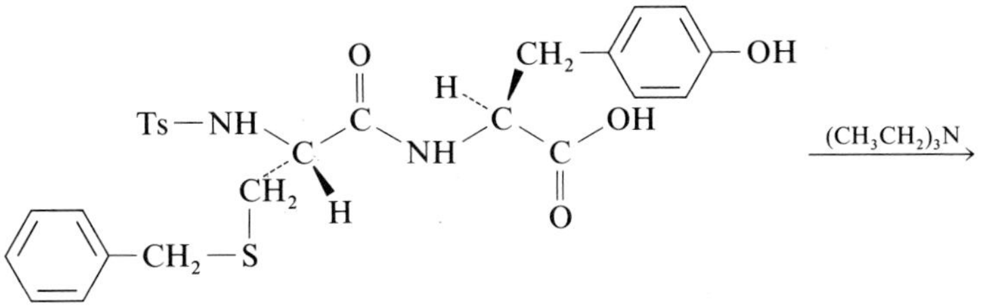

$$\text{A} \xrightarrow[\text{CH}_3\text{CHCH}_2\text{OCCl}]{} \text{B} \xrightarrow{\text{Phe-Gln-Asn}} \text{C}$$
a protected pentapeptide

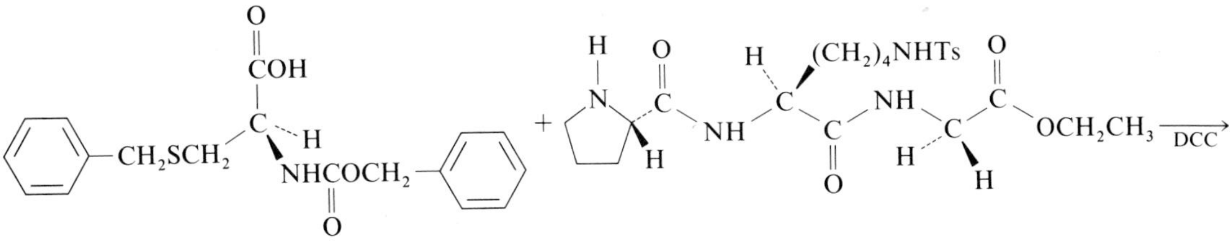

$$\text{D} \xrightarrow[\text{ethanol}]{\text{NH}_3} \text{E} \xrightarrow[\text{acetic acid}]{\text{HBr}} \text{F}$$
a protected tetrapeptide,   *a protected tetrapeptide,*
as the glycyl ester   *as glycyl amide*

$$\text{F} + \text{C} \xrightarrow{\text{DCC}} \text{G} \xrightarrow[\text{NH}_3 \text{ (liq)}]{\text{Na}} \text{H} \xrightarrow{\substack{\text{air} \\ \text{oxidation}}} \text{I}$$
a protected   lysine vasopressin
vasopressin

17.26 Synthetic, racemic D,L-leucine is resolved using an enzyme, hog renal acylase, isolated from hog kidneys. *Hog renal acylase catalyzes the hydrolysis of the amides of L-amino acids only.* The reactions are

$$\text{D,L-leucine} \xrightarrow{(\text{CH}_3\text{C})_2\text{O}} N\text{-acetyl-D,L-leucine} \xrightarrow[\substack{\text{H}_2\text{O, pH 7 maintained} \\ \text{by buffer}}]{\text{hog renal acylase}} \text{mixture}$$

(a) Write equations for the reactions described, being sure to show the correct stereochemistry for the species present at each step. What are the components of the mixture formed at the end of the enzymatic hydrolysis?

(b)   How would you separate the components of the mixture that results from the enzymatic treatment? Write equations showing what you would have to do to recover the individual compounds. (Hint: What ionic species are present in the solution when the reaction is over?)

(c)   Write equations to describe what you would have to do to recover pure D-leucine.

Some useful information: acetic acid, $pK_a = 4.8$; leucine, $pH_I = 6.0$. Leucine and N-acetylleucine are crystalline solids when neutral.

17.27   In the body, an amino acid is prepared for incorporation into a peptide chain by activation of its carboxyl group through the formation of an acylphosphate linkage. This reaction takes place with ATP (Section 14.5B) to give an acylphosphate group at carbon 5 of the ribose unit, and a pyrophosphate anion. Outline a possible mechanism for the activation reaction using any amino acid of your choice.

17.28   The octapeptide xenopsin from the frog *Xenopus laevis* has a powerful contractile effect on muscles. The N-terminal amino acid of xenopsin is glutamic acid, which exists in the natural peptide as the cyclic amide pyroglutamic acid (Section 17.3B). Carboxypeptidase releases leucine first. Trypsin digestion yields ⌐Glu-Gly-Lys-Arg and ⌐Glu-Gly-Lys-Arg-Pro-Trp. Chymotrypsin gives the same hexapeptide as trypsin and Ile-Leu. What is the structure of xenopsin?

17.29   The extract of the European mistletoe *Viscum album* contains a series of pharmacologically active peptides known as viscotoxins. The amino acid sequence of one of them, viscotoxin A$_2$, was determined by using oxidation to break disulfide linkages, followed by digestion of the peptide with trypsin and chymotrypsin. The structures of the peptide fragments that could be isolated from the two enzymatic cleavages are given below.

Trypsin peptides: Asn-Ile-Tyr-Asn-Thr-Cys-Arg,
Lys-Ser-Cys-Cys-Pro-Asn-Thr-Thr-Gly-Arg,
Ile-Ile-Ser-Ala-Ser-Thr-Cys-Pro-Ser-Tyr-Pro-Asp-Lys,
Phe-Gly-Gly-Gly-Ser-Arg,
Ser-Cys-Cys-Pro-Asn-Thr-Thr-Gly-Arg.

Chymotrypsin peptides: Asn-Thr-Cys-Arg-Phe,
Gly-Gly-Gly-Ser-Arg-Glu-Val-Cys-Ala-Ser-Leu,
Lys-Ser-Cys-Cys-Pro-Asn-Thr-Thr-Gly-Arg-Asn-Ile-Tyr,
Ser-Gly-Cys-Lys-Ile-Ile-Ser-Ala-Ser-Thr-Cys-Pro-Ser-Tyr-Pro-Asp-Lys.

Viscotoxin A$_2$ has 46 amino acid residues and a molecular weight of 4833. The N-terminal and C-terminal amino acids are both lysine. What is the order of the amino acids in viscotoxin A$_2$?

17.30   Fructose 1,6-diphosphate is converted into D-glyceraldehyde 3-phosphate and dihydroxyacetone phosphate (1,3-dihydroxy-2-propanone phosphate) by an enzyme known as aldolase. The enzymatic reaction proceeds through the formation of a Schiff base (an imine) between the carbonyl group of fructose and the ϵ-amino group of lysine.

(a)   Write the mechanism for the formation of the Schiff base at the active site of the enzyme, using HB$^+$ and B: to represent any general acids and bases that you may need for the reaction.

(b)   Write the mechanism for the conversion of this enzyme-substrate complex to D-glyceraldehyde 3-phosphate and the Schiff base of dihydroxyacetone phosphate.

(c)   If the imine function in the enzyme-substrate complex is reduced with sodium borohydride, and the enzyme is hydrolyzed, a lysine derivative of dihydroxyacetone is isolated. Propose a structure for this compound.

17.31 Peptide P is found widely distributed in the body, especially in the nervous system. It is believed to mediate the body's response to pain. The following data were obtained in a determination of its structure:

1. Vigorous acidic hydrolysis of Peptide P gives Arg, Glu (2), Gly, Leu, Lys, Met, Phe (2), Pro (2). Enzymatic hydrolysis gives Arg, Gln (2), Gly, Leu, Lys, Met, Phe (2), Pro (2). Peptide P has 11 amino acid units in all.
2. When Peptide P is treated with phenyl isothiocyanate, the phenylthiohydantoins derived from arginine, proline, lysine, and proline can be degraded off in that order.
3. Incubation of Peptide P with chymotrypsin gives Peptide A and Peptide B.
4. Peptide A contains Arg, Gln (2), Lys, Phe, Pro (2). Degradation of Peptide A with Edman's reagent gives the same phenylthiohydantoins derived from the intact Peptide P. Carboxypeptidase releases first Phe, then Gln to the solution.
5. Peptide B reacts with phenyl isothiocyanate to give the phenylthiohydantoins derived from Phe, Gly, Leu in that order.
6. Peptide P is a strongly basic peptide with an isoelectric point above 8.9. No amino acid is released to the solution when the native peptide is incubated with carboxypeptidase.
7. If Peptide P is incubated with 0.03 M HCl at 110 °C for 8–12 h (a procedure developed to hydrolyze carboxylic acid amide bonds, but to leave most peptide bonds untouched), the resulting Peptide C now reacts with carboxypeptidase. Gly, Met, Leu, Phe appear in the solution. The order in which these amino acids were released was not determined.

(a) What is the *N*-terminal amino acid of Peptide P?

(b) What is the *C*-terminal amino acid of Peptide P?

(c) What is the sequence of amino acids in Peptide A?

(d) What is the sequence of amino acids in Peptide B?

(e) What is the complete structure of Peptide B?

(f) What is the sequence of amino acids in Peptide C?

(g) What is the structure of Peptide P?

The solid state synthesis of Peptide P was carried out.

1. Glutamine residues were protected at the α-amino group with *tert*-butoxycarbonyl groups and activated at the carboxylic acid as the *p*-nitrophenyl esters. Write equations, starting with glutamine and showing its conversion to Boc-Gln-*O*-*p*-nitrophenyl, giving the full structure of the protected, activated amino acid.
2. Lysine was protected at the α-amino group by a *tert*-butoxycarbonyl group and at the ϵ-amino group by a benzyloxycarbonyl group. Write the full structure of lysine, as it would be used in the peptide synthesis.
3. The peptide linkages other than those to glutamine were created by activating with dicyclohexylcarbodiimide the carboxylic acid group of the amino acid being added. Show the mechanism for the reaction of Boc-Phe (written out in full) with dicyclohexylcarbodiimide, and the addition of the *N*-terminal end of a peptide fragment (of your choice) to the activated amino acid to form a new peptide bond.
4. The Boc groups in this synthesis were removed by the addition of 4 M HCl dissolved in dioxane. Write the equation for removing the protecting group from the peptide fragment you synthesized in part 3.
5. It was discovered that methionine esters could not be formed efficiently with the benzylic chloride functions on the polymeric support because the side chain of methionine reacted with the benzylic chloride. Indicate the reaction that you would expect between the side chain of Boc-Met and benzyl chloride.

17.32 The amino acid proline is synthesized in the body from glutamic acid. This synthesis takes place in the following steps. Show a mechanism for the transformation in each step.

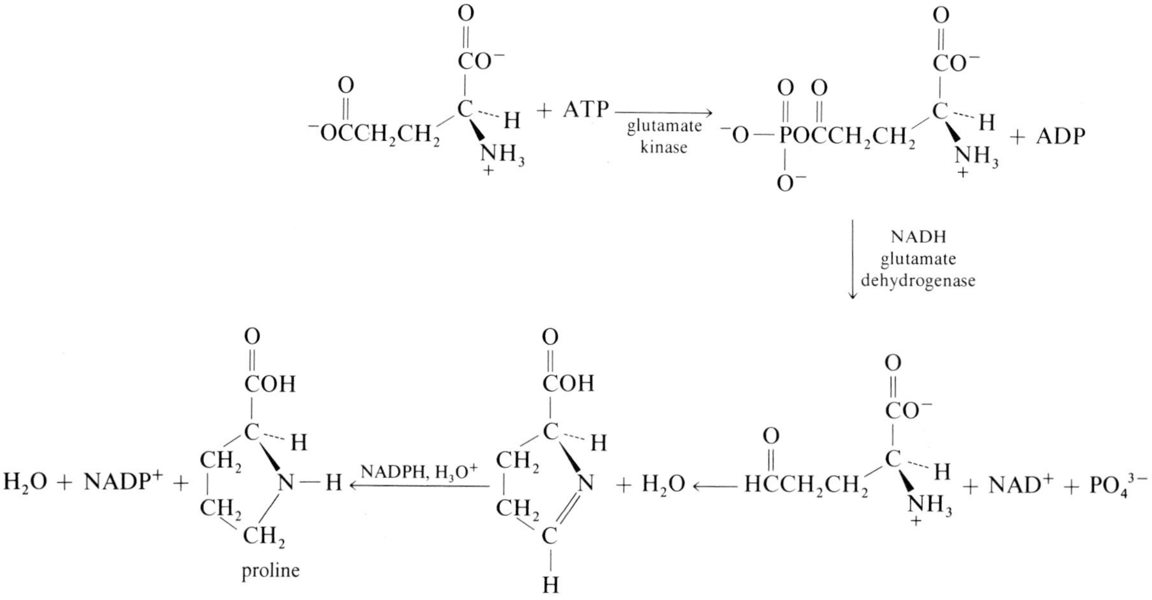

ATP = adenosine triphosphate (Section 14.58)

NADH, NADPH = 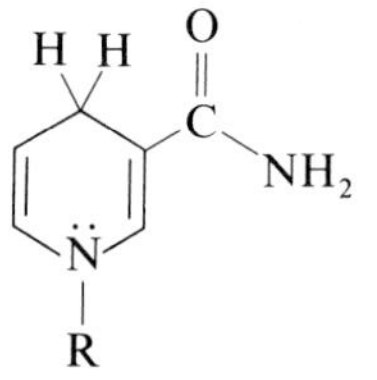 (Section 18.6C)

NAD⁺, NADP⁺ 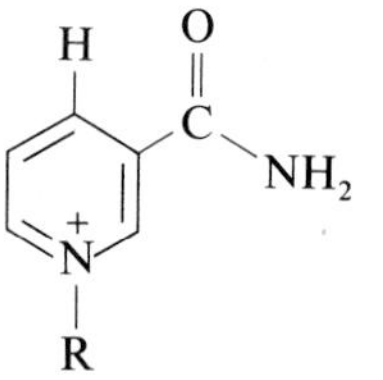 = (Section 18.6C)

17.33 The human growth hormone isolated from the pituitary gland has 188 amino acid residues and has been synthesized by a solid-state technique similar to that of Merrifield (Section 17.4D). The first five amino acids in the chain, starting from the C-terminal unit, are phenylalanine, glycine, cysteine, serine, and glycine, in that order. Devise a solid-state synthesis of this portion of the chain. Assume that the amino acids cysteine and serine are available with benzyl groups protecting the thiol and the hydroxyl group on their side chains. The benzyl group can be removed by reduction by sodium in ammonia.

17.34 A new amino acid, β-carboxyaspartic acid, was discovered in 1981 when researchers noticed that the acid hydrolysis of proteins associated with the ribosomes of the bacteria *Escherichia coli* gave more aspartic acid than hydrolyses catalyzed by base. The amino acid has the structure shown on the next page.

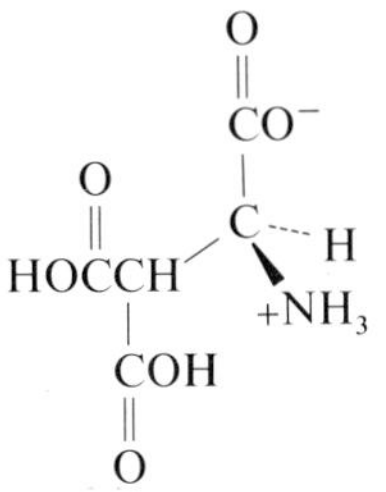

β-carboxyaspartic acid

(a) Why is there a difference in the products obtained from acid-catalyzed and base-catalyzed hydrolyses of proteins containing this amino acid?

(b) The new amino acid was synthesized in order to compare its spectral properties with those of products from the hydrolysis of the ribosomal proteins. The reactions used in the synthesis are shown below. Propose structures for the compounds designated by letters.

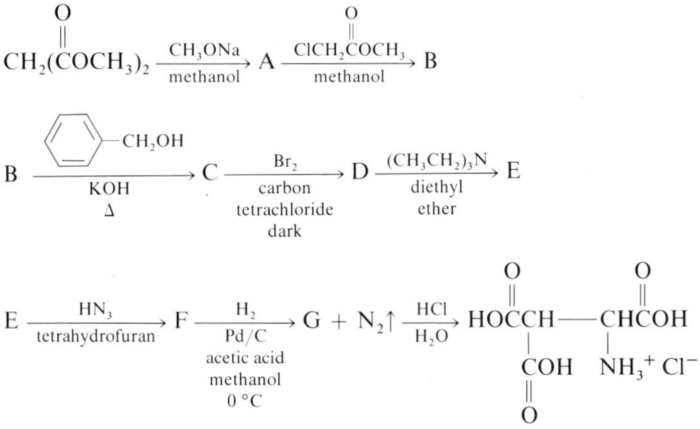

HN₃ = hydrazoic acid

(c) The experimental directions specify that it is important to exclude light completely in the reaction of Compound C with bromine in carbon tetrachloride. What side reactions would be possible if the reaction mixture were exposed to light? (Hint: A review of Section 12.4A may be helpful.)

(d) Why does the reaction of Compound E with hydrazoic acid (HN₃) to give Compound F go with the orientation observed?

17.35 The amino acid proline has been used as the starting material for a number of chiral amines. One such synthesis is outlined below. Supply structural formulas for the compounds designated by letters.

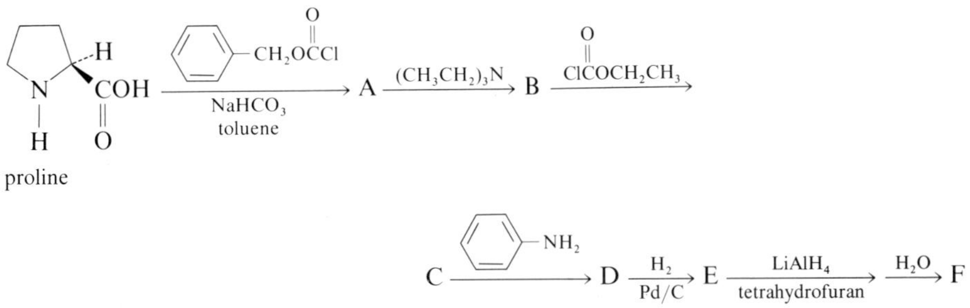

17.36 Distamycin A, a compound isolated from the fermentation broth of the microorganism *Streptomyces distallicus* has antiviral and antitumor activity. The compound has the following structure.

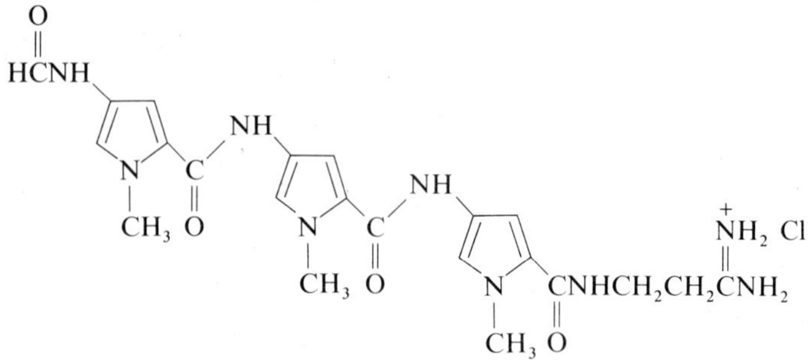

distamycin A

A synthesis of the compound took advantage of the techniques developed for the synthesis of peptides. The synthesis is outlined below. Supply structural formulas for the compounds and intermediates indicated by the letters.

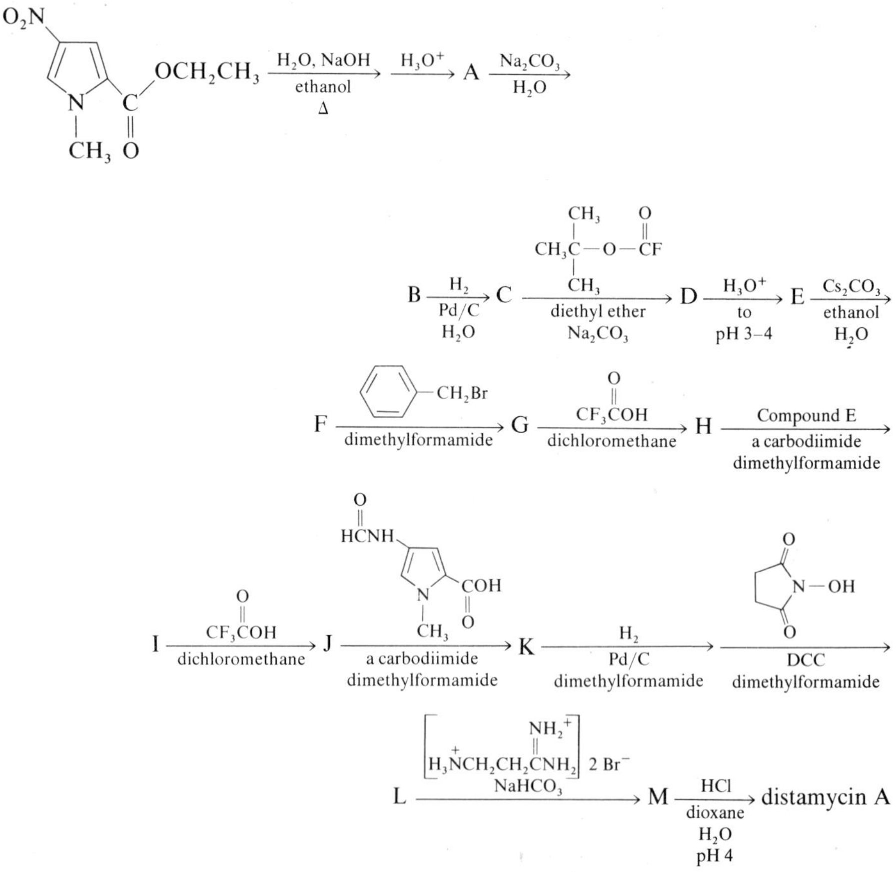

The Chemistry of Heterocyclic Compounds

18

Nomenclature

Organic compounds containing rings are divided into two large classes. Those that have only carbon atoms in the ring are known as **carbocyclic compounds,** *while those that have atoms of elements other than carbon as part of the ring are called* **heterocyclic compounds.** Heterocyclic compounds are so widespread in nature and of such importance chemically that we have not been able to go very far in our discussion of organic chemistry without encountering a large number of them.

The chemistry of heterocyclic compounds is so vast that this chapter will by necessity be a highly selective look at some of the chemistry of such systems. We emphasize reactions that expand and reinforce your understanding of the basic chemical principles that have been the theme of the rest of the book.

Heterocyclic compounds may be classified in many ways: by the size of the ring, by the nature and number of the atoms other than carbon (**heteroatoms**) in the ring, by the degree of unsaturation in the ring, and by whether or not the compound has aromatic character.

Aromatic heterocyclic compounds were included in Section 12.1 to show that nonbonding electrons on heteroatoms could be considered part of an aromatic sextet. The important aromatic heterocyclic systems that contain a single heteroatom are shown on p. 1010. Nonbonding electrons shown inside a ring are part of the aromatic sextet.

Note that some heterocyclic compounds have additional nonbonding electrons on nitrogen, oxygen, or sulfur atoms that do not contribute to the aromaticity of the compound. Some of the chemical consequences of these structural properties are explored in Section 18.2.

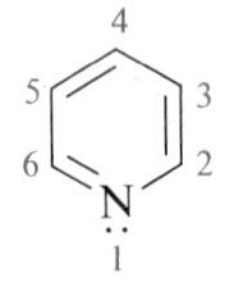

pyridine 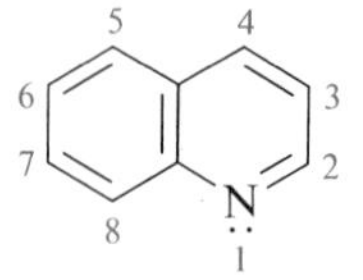   quinoline 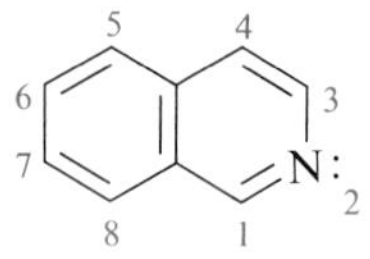   isoquinoline

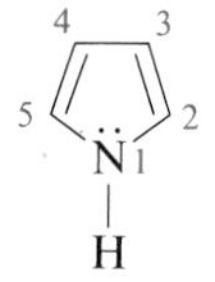

pyrrole 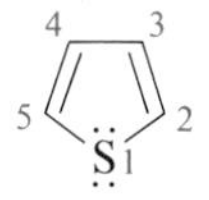   thiophene 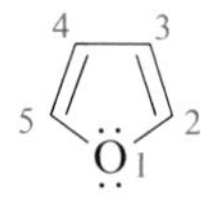   furan 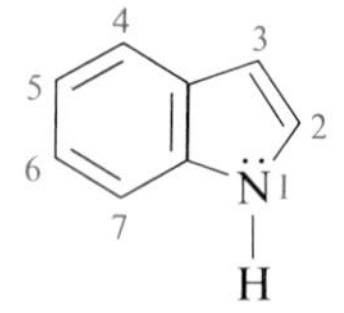   indole

The heteroatom in a heterocycle is given the lowest possible number consistent with an orderly progression around the ring system. In isoquinoline, a carbon atom has a lower number than the nitrogen atom to preserve an orderly sequence around the periphery of the two rings. Substituted heterocyclic compounds are named in the same way as other compounds, by giving the name of the substituent and a number to indicate its position on the parent ring system. The name of the substituent may appear as either a prefix or a suffix to the name of the heterocycle, depending upon the rules for the nomenclature of such functional groups that you learned in earlier chapters.

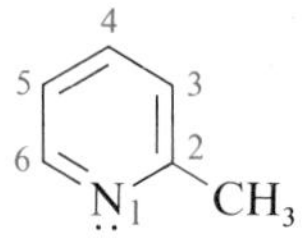

2-methylpyridine 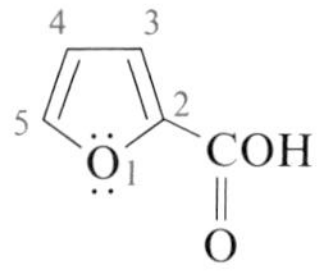   2-furancarboxylic   3-ethylpyrrole
                          acid
                          furoic acid

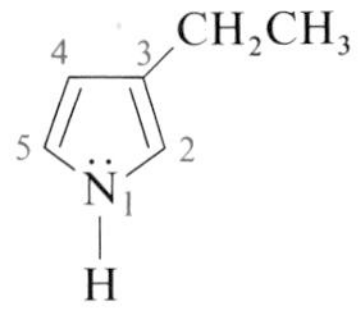

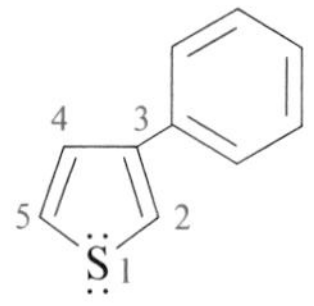

3-phenylthiophene 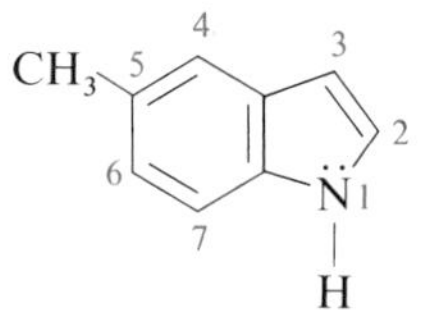   5-methylindole 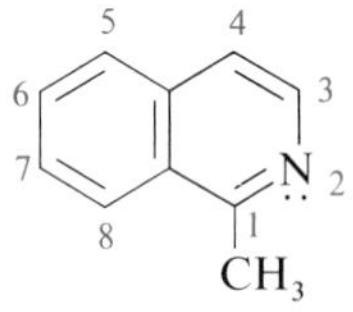   1-methylisoquinoline

The compounds shown above have only a single heteroatom in their rings. There is another important series of heterocyclic compounds with some aromatic character that have two or more heteroatoms, one of which is nitrogen. In the five-membered ring systems, these compounds all have names ending in **-azole.** The rest of the name indicates what other heteroatoms are present.

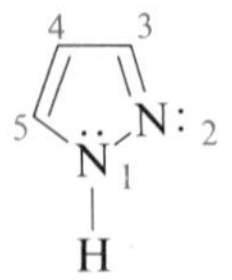

pyrazole 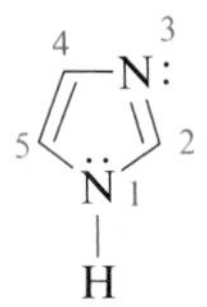   imidazole

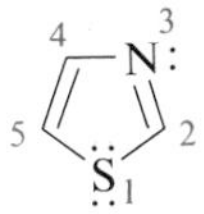

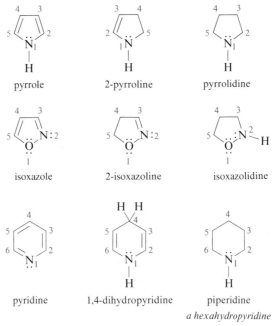

thiazole

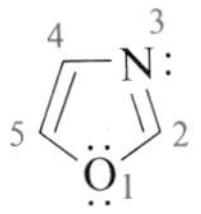

oxazole 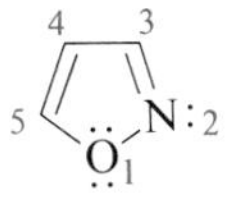   isoxazole

The names *pyrazole* and *imidazole* are given to the two isomeric compounds containing two nitrogen atoms in the ring. The name *thiazole* indicates that the ring has a sulfur atom and a nitrogen atom in it, while *oxazole* points to the presence of oxygen and nitrogen. These names are reserved for the systems in which the two heteroatoms are separated by a carbon atom. A compound in which oxygen and nitrogen are adjacent to each other is known as *isoxazole*.

The introduction of another nitrogen atom into a pyridine ring gives three isomeric compounds, the most important of which is pyrimidine.

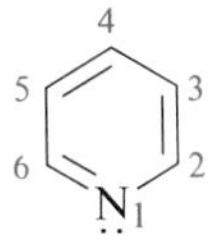

pyridine 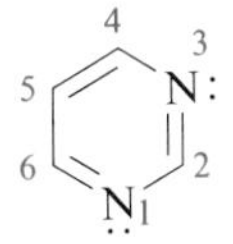   pyrimidine

The compounds discussed so far in this section have the maximum degree of unsaturation. More saturated heterocyclic compounds are also important. For example, you are already familiar with tetrahydrofuran as the fully saturated form of furan. Some other examples of heterocyclic compounds with varying degrees of unsaturation are shown below.

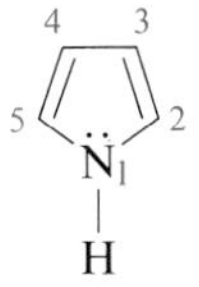

pyrrole   2-pyrroline   pyrrolidine

isoxazole   2-isoxazoline   isoxazolidine

pyridine   1,4-dihydropyridine   piperidine
                                          a hexahydropyridine

⟵ increasing unsaturation

Smaller ring heterocycles are also important. For example, substituted *azetidinones* are components of a number of important antibiotics such as penicillin and cephalosporin.

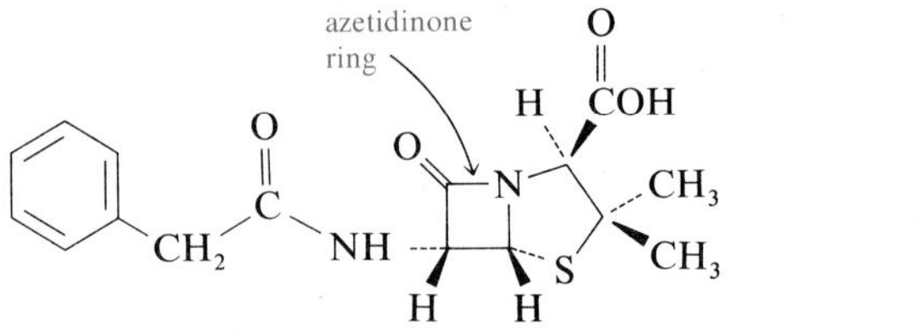

penicillin G

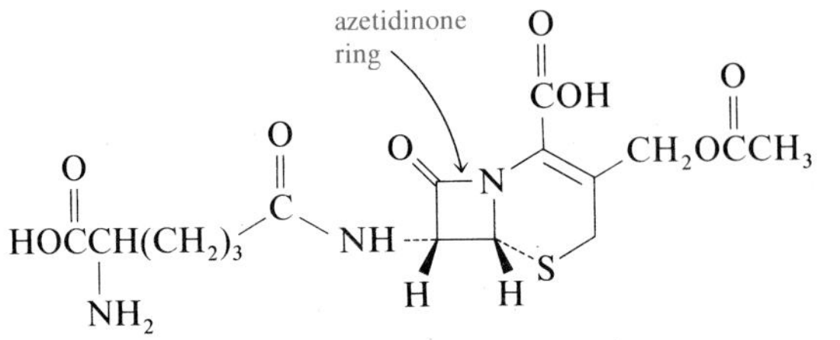

cephalosporin C

The reactivity of the carbonyl group in the strained four-membered amide ring (a β-lactam, Section 16.5A) present in these antibiotics is responsible for their biological activity. The compounds act as acylating agents and disrupt the synthesis of bacterial membranes.

PROBLEM 18.1 Name the following compounds.

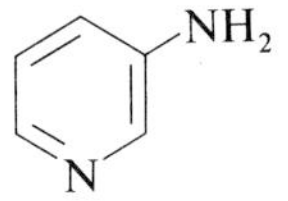

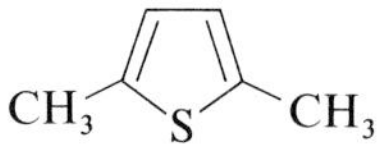

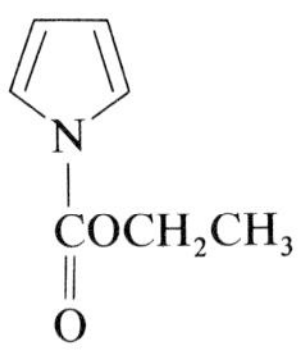

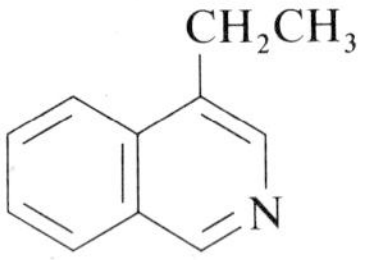

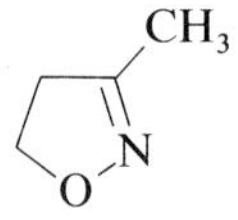

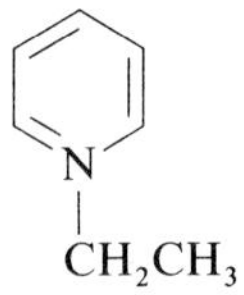

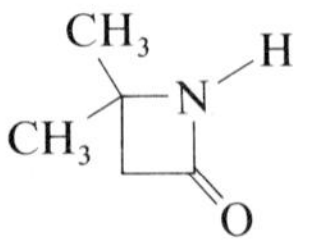

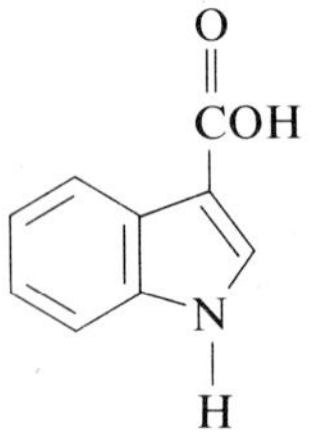

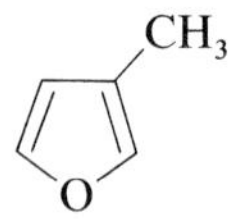

(j) 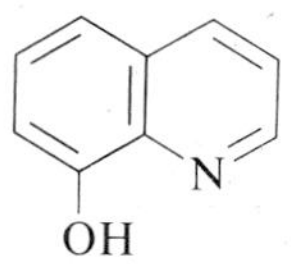   (k) 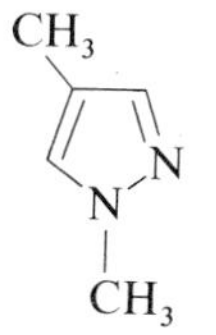   (l) 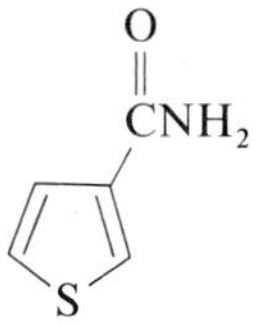   (m)

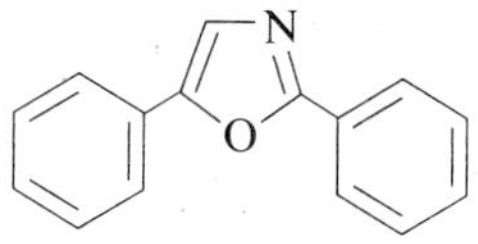

PROBLEM 18.2 Write structural formulas for the following compounds.

(a) 4-ethylindole   (b) 6-aminoquinoline   (c) 2-methyl-5-phenylthiophene
(d) 1,4-dimethylisoquinoline   (e) 5-phenylisothiazole   (f) 2,5-dimethylfuran
(g) 3-ethylisoxazole   (h) 5-chloropyrimidine   (i) 1,2-dihydropyridine
(j) 2,4-dimethyloxazole   (k) 3-methyltetrahydrofuran   (l) 3,5-dimethylpyrazole
(m) 3-ethyl-2-pyrrolecarboxylic acid

18.2

Aromaticity of Heterocycles

Aromatic heterocyclic compounds resemble benzene in that each one can be shown experimentally to have resonance energy. In other words, they are each more stable than would be expected of a similar compound with localized double bonds. Resonance energies, determined by heats of combustion for different heterocycles, are shown in Table 18.1. Benzene and cyclopentadiene are included for comparison.

None of the heterocyclic compounds appears to have as much resonance stabilization as benzene. All of the five-membered ring heterocycles, however, are much more stable than would be expected of a cyclic diene, represented in Table 18.1 by cyclopentadiene. Thiophene, pyrazole, pyridine, pyrrole, and imidazole would be expected to behave in their chemical reactivity as aromatic compounds. In Section 18.5

TABLE 18.1	Resonance Energies for Some Cyclic Compounds (Determined from Heats of Combustion)	
Compound	*Resonance Energy (kcal/mol)*
benzene	36
thiophene	29
pyrazole	29
pyridine	28
pyrrole	22
imidazole	22
furan	16
cyclopentadiene	3

we examine electrophilic aromatic substitution reactions of some heterocycles to see to what extent this prediction is correct.

Based on the numbers in the table, furan stands out as the compound with the least resonance stabilization of the simple heterocycles. The loss of aromaticity is not a large barrier to the reactions of furan. In general, it undergoes addition rather than substitution reactions with much greater ease than the other heterocyclic compounds. Examples of this type of reactivity are given in the next section.

The small degree of aromaticity of furan is also illustrated by the relative ease with which it serves as a diene in the Diels-Alder reaction. It is much less reactive than cyclopentadiene, which as a cyclic cisoid diene reacts readily with all kinds of dienophiles (Sections 11.5 and 19.2B), but more so than thiophene or pyrrole. Furan reacts only with highly reactive dienophiles such as dimethyl acetylenedicarboxylate (Section 11.5) and benzyne (Section 16.6E).

The Diels-Alder reactions of furan have been investigated extensively in attempts to synthesize cantharidin, the active ingredient in Spanish fly, which is an extract of the beetle *Cantharis vesicatoria*. Cantharidin is a vessicant, which means that it raises blisters. It has an unwarranted reputation as an aphrodisiac, which may be the reason for the interest in developing a synthetic source. An analysis of the structure of cantharidin shown below suggests that it could be synthesized from furan and dimethylmaleic anhydride.

cantharidin   furan   dimethylmaleic
                                                        anhydride

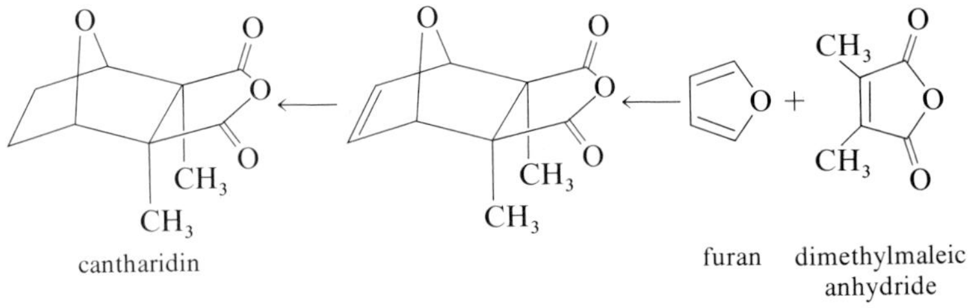

Furan adds to maleic anhydride to give an exo adduct that is unstable and dissociates easily back to the reactants.

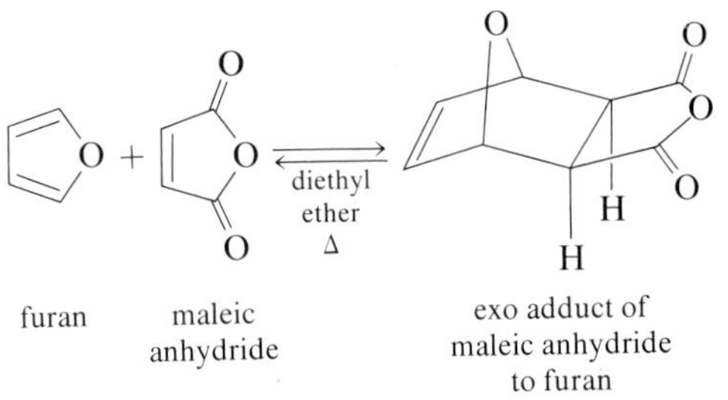

furan   maleic   exo adduct of
             anhydride   maleic anhydride
                                  to furan

Dimethylmaleic anhydride does not add to furan at all. The two methyl groups on the double bond of the dienophile affect its reactivity in two ways. First, the methyl groups, as electron-donating groups, increase electron density in the double bond and make the compound a poorer dienophile. Second, the methyl groups also increase steric crowding at the reaction site and thus hinder the reaction.

William Dauben of the University of California at Berkeley has devised a direct synthesis of cantharidin using a maleic anhydride derivative in which the two methyl groups are hidden as part of a five-membered ring containing sulfur. The steric effect of the methyl groups is diminished because only methylene groups appear on the double bond, and even they are held back as part of a ring. The presence of a sulfur atom is also expected to reduce the electron-donating effect of the alkyl groups. The Diels-Alder reaction occurs under high pressure.

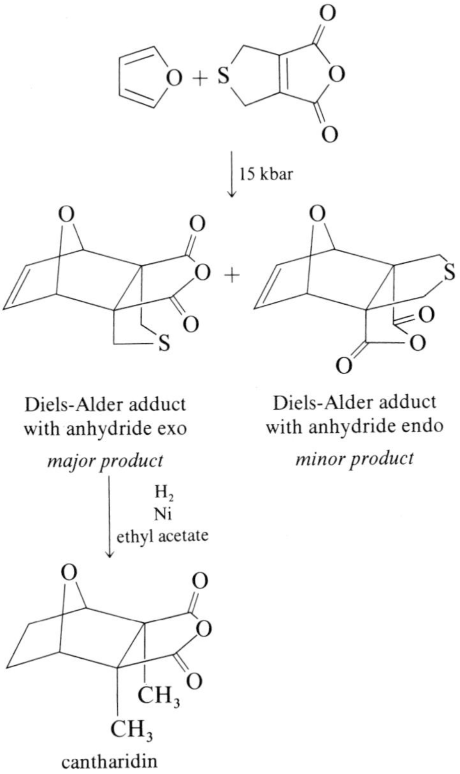

Diels-Alder adduct   Diels-Alder adduct
with anhydride exo   with anhydride endo

major product   *minor product*

cantharidin

The reason that high pressure is used in this reaction is interesting. The transition state for the Diels-Alder reaction occupies a smaller volume than do the individual reagents in position to react. Therefore, increasing the pressure on the reaction mixture should favor the transition state and speed the progress of the reaction. The combination of high pressure and a carefully designed dienophile led to a successful reaction. The major product of the reaction has the stereochemistry necessary for the synthesis of cantharidin. Reduction of the double bond and reductive cleavage of the carbon-sulfur bonds by hydrogen on Raney nickel (Section 9.7D) complete the synthesis.

18.3

Reactions of Heterocycles as Acids and Bases

An examination of the structural formulas for aromatic heterocycles raises interesting questions about the acidity and basicity of these compounds. For example, pyrrole, in which the nonbonding electrons on the nitrogen atom are part of the aromatic sextet, has low basicity, and when it accepts a proton, it does so on the carbon atom adjacent to the nitrogen atom. On the other hand, the proton on the nitrogen atom is removed by hydroxide ion to give the conjugate base of pyrrole. Salts containing the pyrrole anion are easily prepared in this way. The pK_a values of the different species related to pyrrole are compared with the corresponding values for ammonium ion and ammonia, taken from the table of pK_a values in the front cover of the book.

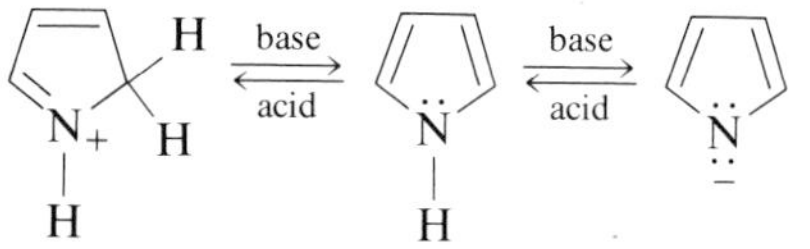

conjugate acid of pyrrole pK_a -3.80 pyrrole pK_a ~15 conjugate base of pyrrole

$$^{+}NH_4 \underset{acid}{\overset{base}{\rightleftharpoons}} :NH_3 \underset{acid}{\overset{base}{\rightleftharpoons}} :\ddot{N}H_2$$

ammonium ion conjugate acid of ammonia pK_a 9.4   ammonia pK_a 36   amide anion conjugate base of ammonia

The nonbonding electrons on the nitrogen atom in pyrrole are much less available for protonation than the electron pair on ammonia. The conjugate acid of pyrrole, when it is formed, is a strong acid with pK_a -3.80. Pyrrole itself, with pK_a ~15, is a much stronger acid than ammonia. The conjugate base of pyrrole is still an aromatic species but now has an extra pair of nonbonding electrons on the nitrogen atom. The anion is stabilized by delocalization of the negative charge. The negative charge on the amide, NH_2^- ion, by contrast, is localized on a single atom.

PROBLEM 18.3 Write resonance contributors for furan and thiophene, pyrrole, and the conjugate base of pyrrole. Which resonance contributors for the neutral molecules are the important ones?

The reactions of imidazole as an acid or a base are important in many biological systems, as we saw in the discussion of the enzyme chymotrypsin (Section 17.6). Imidazole is a stronger base than pyrrole, because on the second nitrogen atom it has a pair of nonbonding electrons that are not part of the aromatic sextet. It also has a proton that is lost in base and thus is an acid. The conjugate acid of imidazole is stabilized by delocalization of the charge to both of the nitrogen atoms, giving two equivalent resonance contributors. The various species involved in the protonation and deprotonation reactions of imidazole are shown below.

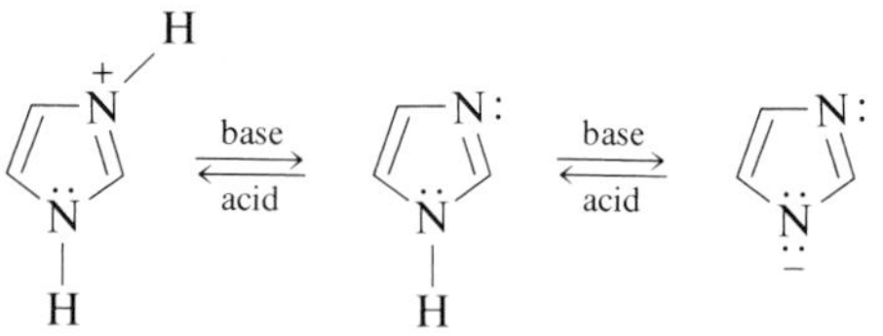

conjugate acid of imidazole pK_a 6.95   imidazole pK_a 14.5   conjugate base of imidazole

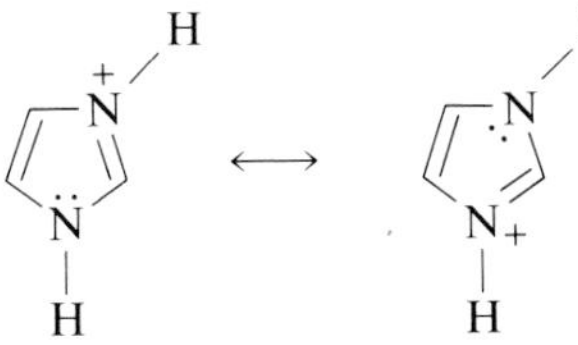

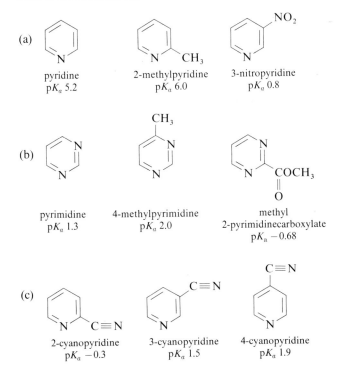

resonance contributors for the
conjugate acid of imidazole

A comparison of the basicities of the six-membered ring heterocycles pyridine and pyrimidine is also interesting. Pyridine is a weak base. The conjugate acid has pK_a 5.2. The nonbonding electrons on the nitrogen atom are not part of the aromatic sextet but are present on a nitrogen atom that is sp^2 hybridized. It is postulated that the electrons are in an sp^2 hybrid orbital, rather than an sp^3 hybrid orbital as is the case for ammonia. Thus, they are held more closely to the nucleus of the nitrogen atom than is usual for amines because of the greater s character of the orbital (Section 2.7B).

Introducing a second nitrogen atom into the pyridine ring lowers the basicity of the molecule still further. The conjugate acid of pyrimidine has pK_a 1.3. Nitrogen is more electronegative than carbon. The inductive effect of the second nitrogen atom makes the electrons on the first nitrogen less available for protonation.

Pyridine is the heterocycle that most resembles benzene in its structure and stability. Electrophilic substitution reactions of pyridine are discussed in Section 18.5B, where a closer comparison of the two aromatic compounds is made.

PROBLEM 18.4   Discuss the trends that you see in the following sets of pK_a values for the conjugate acids of the bases shown.

(a)

pyridine
pK_a 5.2

2-methylpyridine
pK_a 6.0

3-nitropyridine
pK_a 0.8

(b)

pyrimidine
pK_a 1.3

4-methylpyrimidine
pK_a 2.0

methyl
2-pyrimidinecarboxylate
pK_a −0.68

(c)

2-cyanopyridine
pK_a −0.3

3-cyanopyridine
pK_a 1.5

4-cyanopyridine
pK_a 1.9

PROBLEM 18.5   The conjugate acid of quinoline (p. 1010) has pK_a 4.9, while the pK_a of the conjugate acid of isoquinoline is 5.4. Offer a rationalization for these facts. What pK_a value would you expect to observe for the conjugate acid of 2-phenylpyridine?

18.4

Synthesis of Some Aromatic Heterocyclic Compounds

A general way to synthesize heterocyclic compounds is by cyclization of a dicarbonyl compound with a nucleophilic reagent that introduces the desired heteroatom or atoms. An example of such a synthesis is the preparation of 2-methyl-5-phenylpyrrole from 1-phenyl-1,4-pentanedione and ammonia.

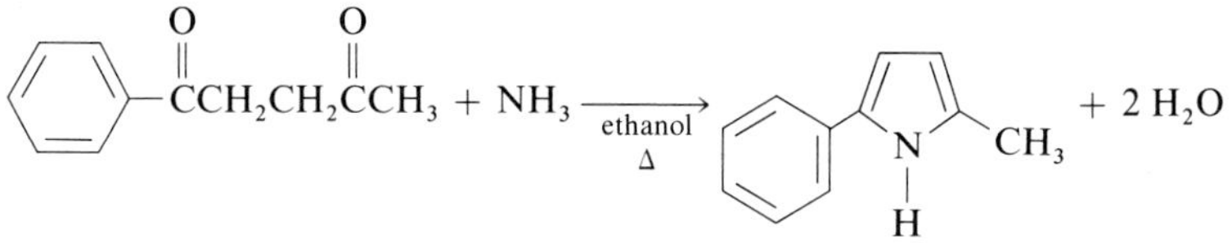

1-phenyl-1,4-pentanedione   ammonia   2-methyl-5-phenylpyrrole
                                                        70%

The reaction may be envisioned as occurring by way of aminoacetal intermediates resulting from nucleophilic attack of ammonia on a carbonyl group with subsequent cyclization to a stable five-membered ring and dehydration. The last stage of the mechanism, shown below, represents two sequences of protonation and elimination reactions.

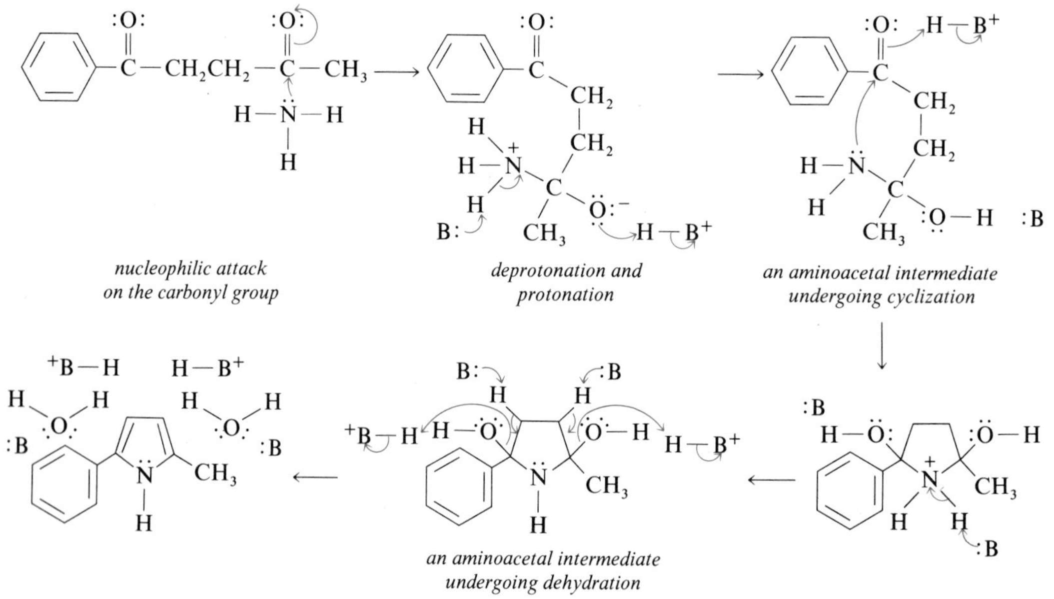

nucleophilic attack on the carbonyl group   *deprotonation and protonation*   *an aminoacetal intermediate undergoing cyclization*

an aminoacetal intermediate undergoing dehydration

If 1-phenyl-1,4-pentanedione is heated with phosphorous pentasulfide, which actually has the molecular formula P_4S_{10}, 2-methyl-5-phenylthiophene is formed.

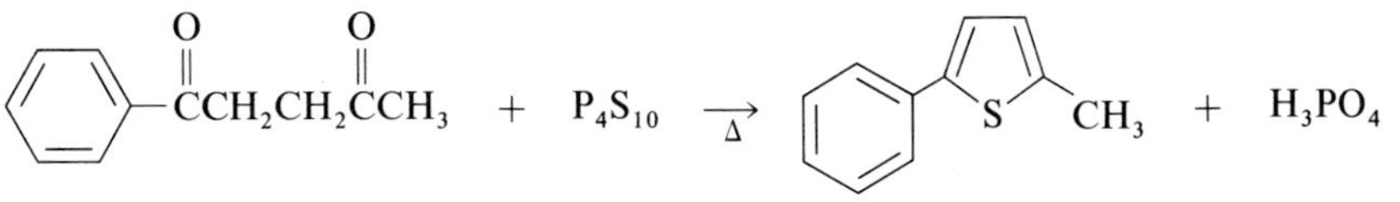

| 1-phenyl-1,4-pentanedione | phosphorus pentasulfide | 2-methyl-5-phenylthiophene ~65% | phosphoric acid |

The structure of the phosphorus sulfide is too complex to allow a simple mechanism to be written, but the reagent acts by replacing carbonyl groups with thiocarbonyl functions, perhaps by way of the enol of the ketone. Phosphorus, which is pentavalent in the sulfide, is converted to phosphoric acid. The cyclization reaction may be represented as an attack by a nucleophilic sulfur atom on the other carbonyl group in the molecule.

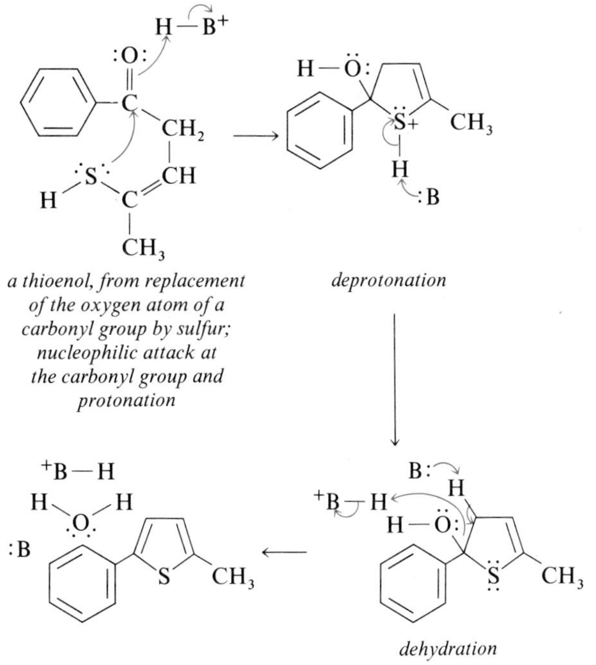

a thioenol, from replacement of the oxygen atom of a carbonyl group by sulfur; nucleophilic attack at the carbonyl group and protonation

deprotonation

dehydration

In an analogous reaction, 1-phenyl-1,4-pentanedione cyclizes to 2-methyl-5-phenylfuran when it is heated in acid. In the absence of a nucleophilic reagent that supplies a heteroatom other than oxygen, the oxygen heterocycle is formed.

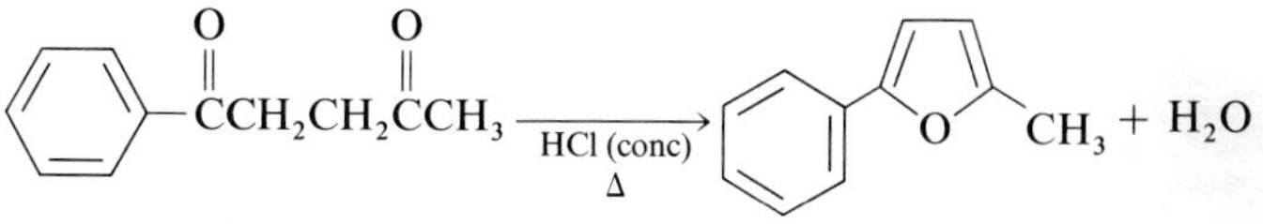

| 1-phenyl-1,4-pentanedione | 2-methyl-5-phenylfuran |

In this case the enol form of one of the ketone functions serves as the nucleophile to form a cyclic hemiacetal that dehydrates.

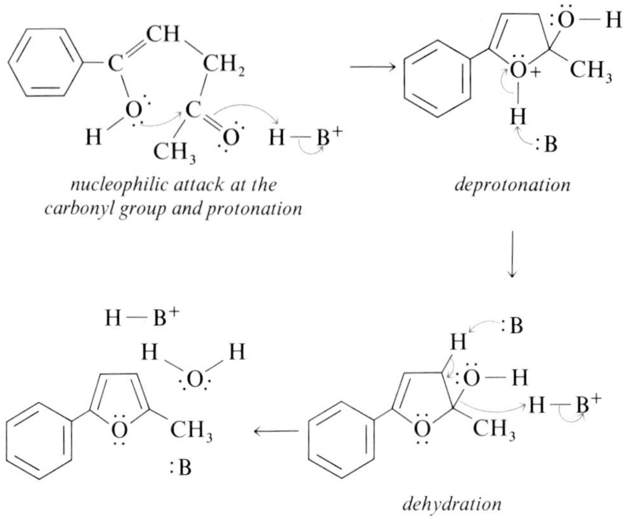

nucleophilic attack at the carbonyl group and protonation

deprotonation

dehydration

In all of these reactions, the driving force is the formation of a stable five-membered ring with aromaticity. Thiophene and pyrrole are quite stable to reactions that lead to the opening of the ring. Furan, however, may be regarded as a cyclic hemiacetal that has been dehydrated, and it is hydrolyzed back to a dicarbonyl compound easily when heated with dilute acid.

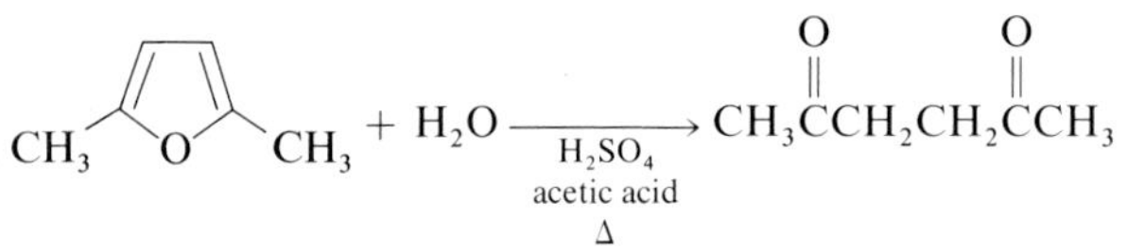

2,5-dimethylfuran

2,5-hexanedione
86%

Reagents with two adjacent heteroatoms such as hydrazine and hydroxylamine react with 1,3-dicarbonyl compounds to give pyrazoles and isoxazoles. For example, 2,4-pentanedione reacts with hydrazine to form 3,5-dimethylpyrazole and with hydroxylamine to give 3,5-dimethylisoxazole.

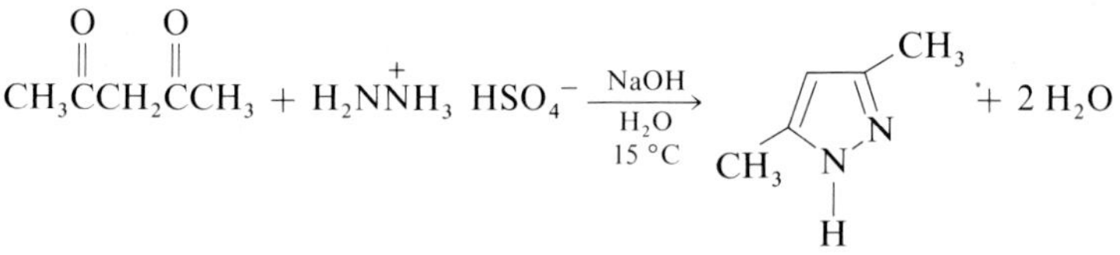

2,4-pentanedione   hydrazine sulfate

3,5-dimethylpyrazole
~80%

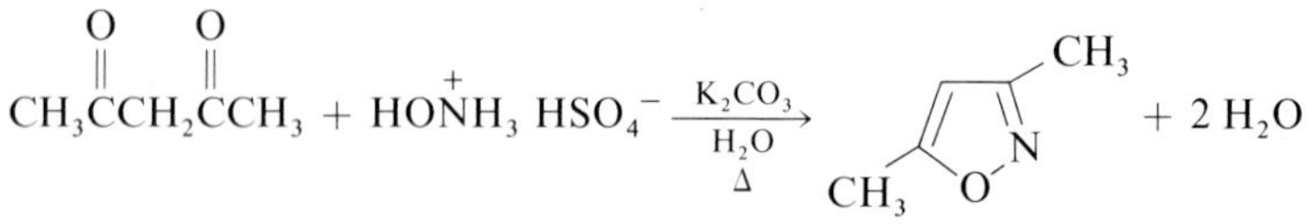

2,4-pentanedione   hydroxylamine sulfate

3,5-dimethylisoxazole
84%

Hydrazine and hydroxylamine are both basic reagents that are most easily stored and handled as their salts. In the presence of bases such as hydroxide or carbonate ions, the free nucleophiles are generated and react with the carbonyl compounds. A mechanism for the formation of the isoxazole is shown below.

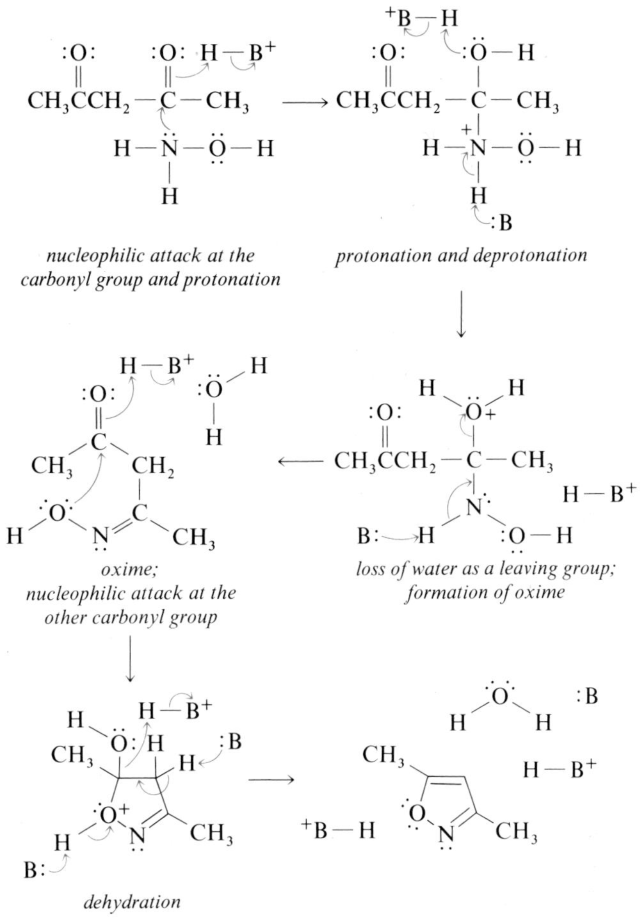

$$CH_3CCH_2-C-CH_3 \quad \text{H}-\text{B}^+ \quad \longrightarrow \quad CH_3CCH_2-C-CH_3$$

*nucleophilic attack at the
carbonyl group and protonation*

protonation and deprotonation

$$CH_3CCH_2-C-CH_3$$

*oxime;
nucleophilic attack at the
other carbonyl group*

*loss of water as a leaving group;
formation of oxime*

dehydration

Chemists have postulated the formation of an oxime at one of the carbonyl groups and the cyclization of the compound by attack of the oxygen atom of the oxime, serving as a nucleophile, on the other carbonyl group. The resulting intermediate dehydrates easily to give the aromatic ring.

PROBLEM 18.6 Write a mechanism for the formation of 3,5-dimethylpyrazole from 2,4-pentanedione and hydrazine.

Imidazoles are synthesized from two carbonyl compounds joined together with nitrogen atoms derived from ammonia. For example, when a mixture of 1,2-diphenyl-1,2-ethanedione (benzil) and benzaldehyde is heated with ammonium acetate in glacial acetic acid, 2,4,5-triphenylimidazole is obtained.

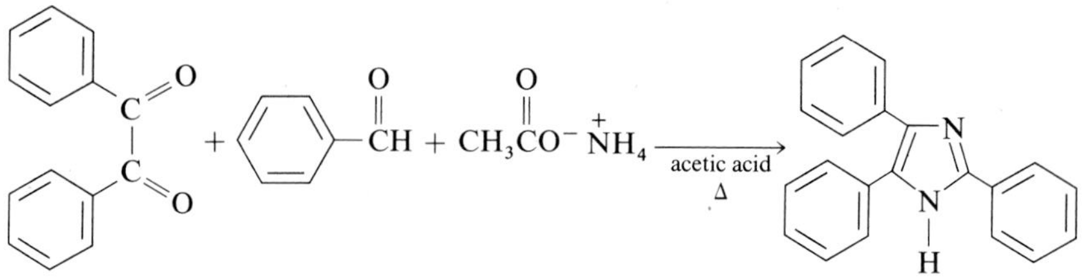

1,2-diphenyl-1,2-   benzaldehyde   ammonium   2,4,5-triphenylimidazole
ethanedione   acetate   90%
benzil

The three carbon atoms in the ring are derived from the carbon atoms of the carbonyl groups in the organic reagents, while the nitrogen atoms come from ammonia, which is in equilibrium with ammonium ions in its salt with the weak acid, acetic acid.

Of the six-membered ring heterocycles, pyridine and various simple substituted pyridines can be obtained conveniently from natural sources. A great deal of work has been done, however, on the synthesis of pyrimidines because of their importance as drugs, such as barbiturates, and as bases found in nucleic acids. The synthesis of barbiturates from derivatives of diethyl malonate and urea (Section 15.3A) is a good illustration of the most general way to create the pyrimidine ring. A 1,3-dicarbonyl compound is condensed with a reagent that is structurally related to urea. The products formed depend on the exact substitution present on each fragment. Two examples are given below.

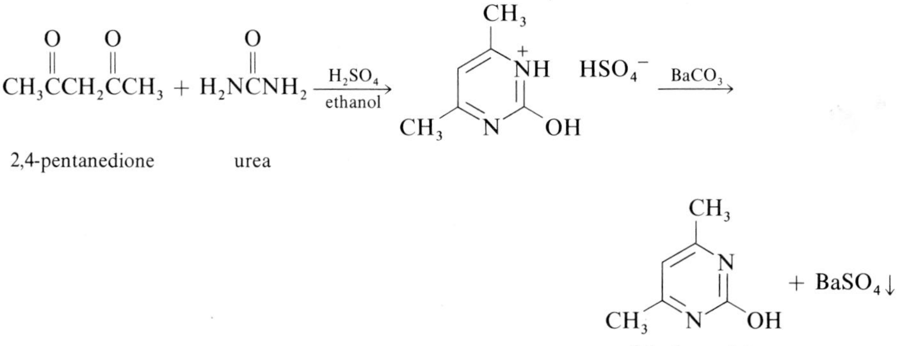

2,4-pentanedione   urea

2-hydroxy-4,6-
dimethylpyrimidine

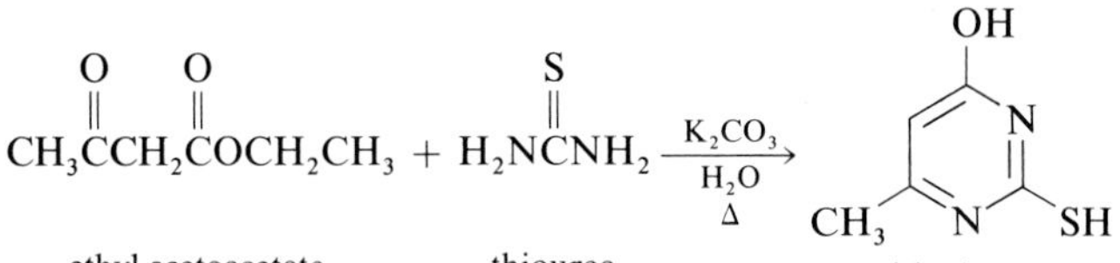

ethyl acetoacetate   thiourea   4-hydroxy-
2-mercapto-
6-methylpyrimidine
95%

In the reaction of 2,4-pentanedione with urea, the ketone carbonyl groups do not have good leaving groups on them. The reaction proceeds by condensation of the amino groups of urea with the carbonyl groups of the ketone and tautomerization to the aromatic system.

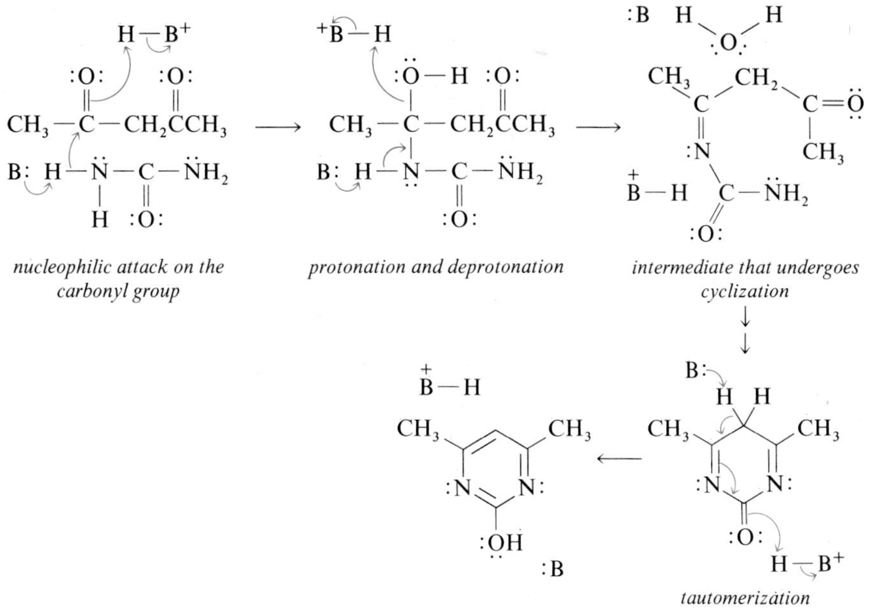

nucleophilic attack on the carbonyl group

protonation and deprotonation

intermediate that undergoes cyclization

tautomerization

When an ester group is one of the carbonyl functions, as it is in diethyl malonate or ethyl acetoacetate, an alkoxide group serves as a leaving group, providing a different pathway for the condensation reaction.

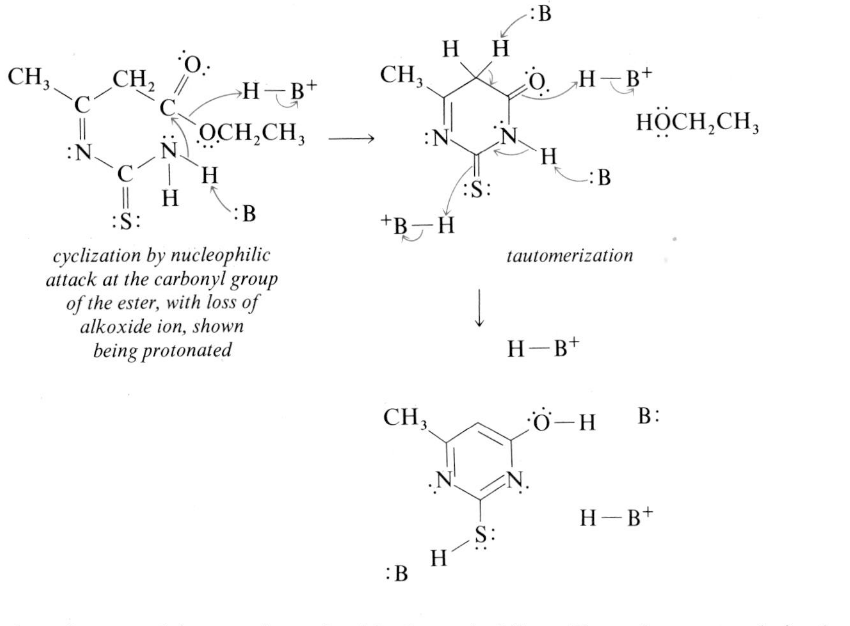

cyclization by nucleophilic attack at the carbonyl group of the ester, with loss of alkoxide ion, shown being protonated

tautomerization

The carbonyl group of the ester is retained in the pyrimidine. The carbon atoms derived from carbonyl groups of esters, or of urea (and thiourea) are marked by hydroxyl (or thiol) substituents on the fully aromatic form of the ring after tautomerization.

PROBLEM 18.7   Complete the following equations.

(a) 
$$\text{Ph}-\overset{\overset{\displaystyle O}{\|}}{C}CH_2\overset{\overset{\displaystyle O}{\|}}{C}-\text{Ph} \xrightarrow[\Delta]{HONH_3^+ \ Cl^-, \ NaOH}$$

(b) $CH_3\overset{\overset{\displaystyle O}{\|}}{C}CH_2CH_2\overset{\overset{\displaystyle O}{\|}}{C}CH_3 \xrightarrow[\Delta]{P_4S_{10}}$

(c) $H\overset{\overset{\displaystyle O}{\|}}{C}CH_2CH_2\overset{\overset{\displaystyle O}{\|}}{C}H \xrightarrow{HCl}$

(d) $CH_3\overset{\overset{\displaystyle O}{\|}}{C}CH_2CH_2\overset{\overset{\displaystyle O}{\|}}{C}CH_3 \xrightarrow[\Delta]{P_4Se_{10}}$

(Hint: Where is selenium
in the periodic table?)

(e) 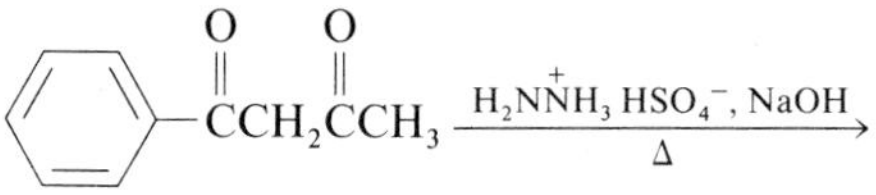
$$\text{Ph}-\overset{\overset{\displaystyle O}{\|}}{C}CH_2\overset{\overset{\displaystyle O}{\|}}{C}CH_3 \xrightarrow[\Delta]{H_2N\overset{+}{N}H_3 \ HSO_4^-, \ NaOH}$$

(f) $CH_3\overset{\overset{\displaystyle O}{\|}}{C}CH_2CH_2\overset{\overset{\displaystyle O}{\|}}{C}CH_3 \xrightarrow[\Delta]{CH_3NH_2}$

18.5

Substitution Reactions of Heterocyclic Compounds

A. Aromatic Substitution Reactions of Five-Membered Ring Heterocycles

The five-membered aromatic heterocycles are all more reactive towards electrophiles than benzene is. In many ways, the reactivity of the ring resembles that of phenol in the ease with which substitution takes place. For example, thiophene reacts with bromine to give a mixture of bromothiophenes.

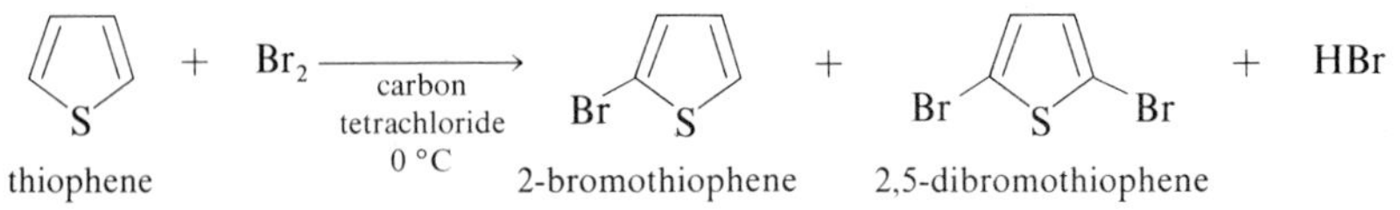

thiophene   2-bromothiophene   2,5-dibromothiophene

Furan, on the other hand, reacts with bromine by a 1,4-addition reaction, another indication of the relatively low aromaticity of this heterocycle. When the reaction is carried out in methanol, the product that is isolated is formed by solvolysis of the intermediate dibromide.

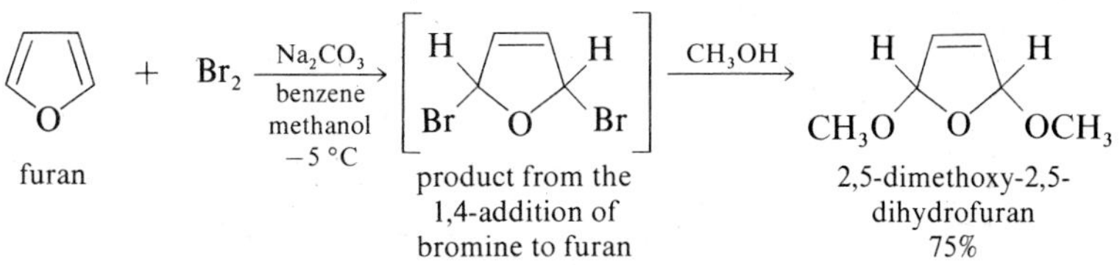

furan   product from the   2,5-dimethoxy-2,5-
                         1,4-addition of   dihydrofuran
                         bromine to furan   75%

Aromatic heterocycles also undergo nitration reactions. But the mixture of nitric acid and sulfuric acid used for the nitration of benzene (Section 12.3A) destroys the heterocycles, so a milder nitrating agent prepared by dissolving nitric acid in acetic anhydride is used. Acetic anhydride acts as a dehydrating agent to create nitronium ions from nitric acid. Substitution in thiophene takes place chiefly at the carbon atom adjacent to the heteroatom.

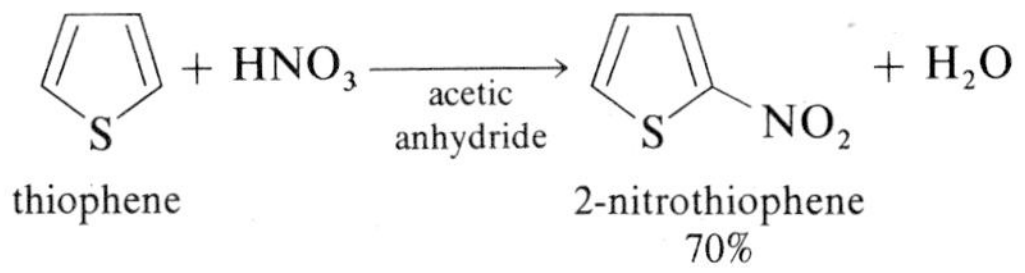

thiophene   2-nitrothiophene
                                   70%

The regioselectivity of the substitution reaction in these heterocycles can be rationalized by the same kind of reasoning that was used to explain the directing effects of substituents on benzene (Section 12.3C). The resonance contributors for the intermediates that would result from attack of the nitronium ion at carbon 2 and at carbon 3 of the thiophene ring are compared below.

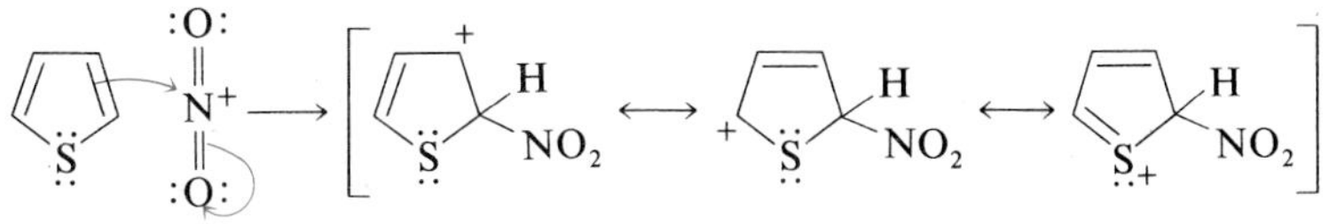

resonance contributors for the intermediate
formed by attack of nitronium ion at carbon 2

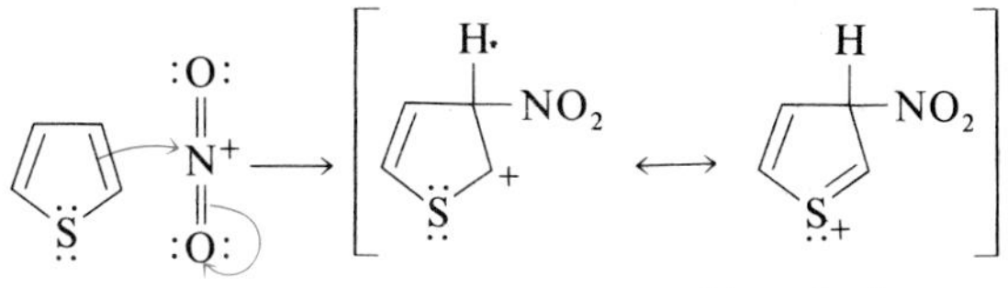

resonance contributors for the
intermediate formed by attack of
nitronium ion at carbon 3

The carbocation formed when nitronium ion attacks carbon 2 of thiophene is more stable than the other intermediate because greater delocalization of charge is possible for it. The reaction thus goes by way of the path leading through that intermediate and the lower energy transition state corresponding to its formation. The intermediate carbocation loses a proton easily and the product has the stable aromatic ring.

Friedel-Crafts acylation reactions are another type of aromatic substitution reaction (Section 12.3F) that can be carried out with such heterocycles. Thiophene, for example, reacts with benzoyl chloride in the presence of aluminum chloride to give phenyl 2-thienyl ketone.

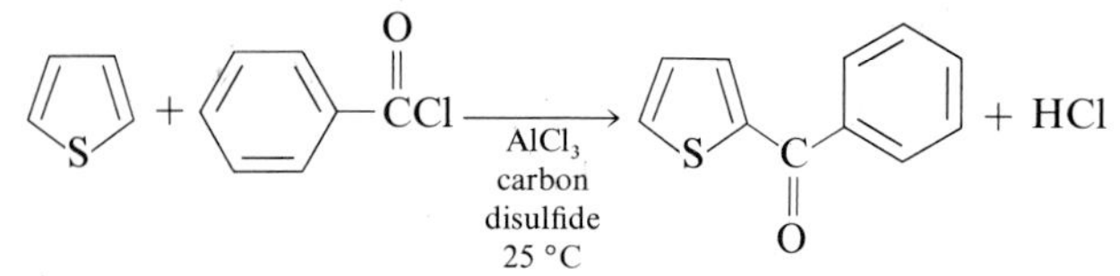

thiophene   benzoyl chloride   2-benzoylthiophene
                                             phenyl 2-thienyl ketone
                                                       90%

Substituted aromatic heterocycles usually undergo the reactions typical of the functional groups that are present. For example, a hydrogen atom on the methyl group of 3-methylthiophene is substituted by bromine when N-bromosuccinimide is used in a typical free-radical substitution reaction (Sections 11.2 and 12.4A).

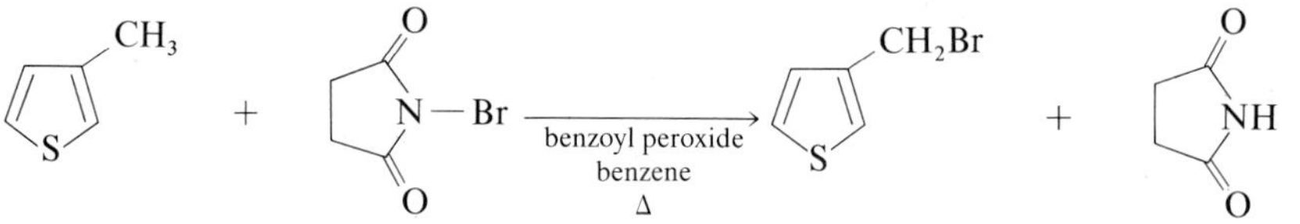

3-methylthiophene N-bromosuccinimide   3-(bromomethyl)thiophene   succinimide
                                                          75%

From the reactions shown above, you can see that the five-membered ring aromatic heterocycles behave much like benzene and its derivatives in electrophilic substitution reactions. Substitution occurs preferentially at carbon 2. The presence of the heteroatom makes the ring more reactive than benzene, so some of the reaction conditions must be modified. Substituents on a heterocyclic ring retain many of the reactions typical of their functional groups.

PROBLEM 18.8   Complete the following equations, giving structural formulas for all the intermediates and products designated by letters.

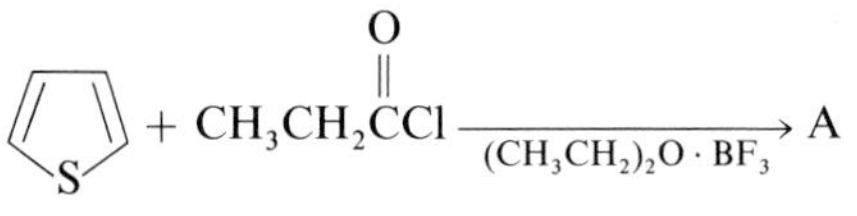

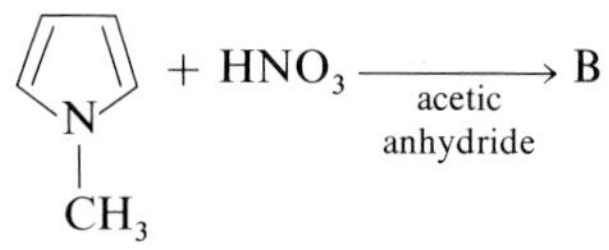

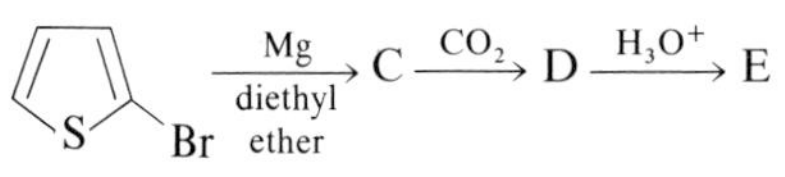

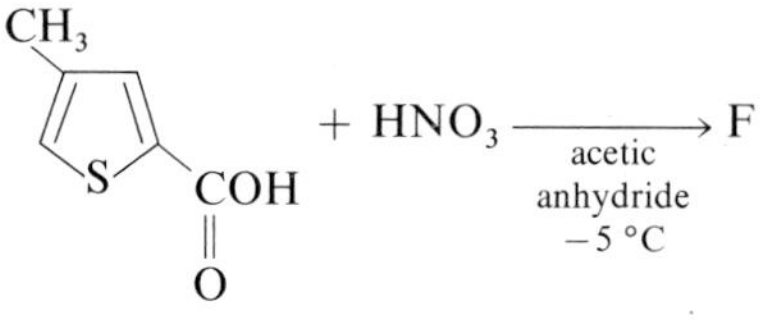

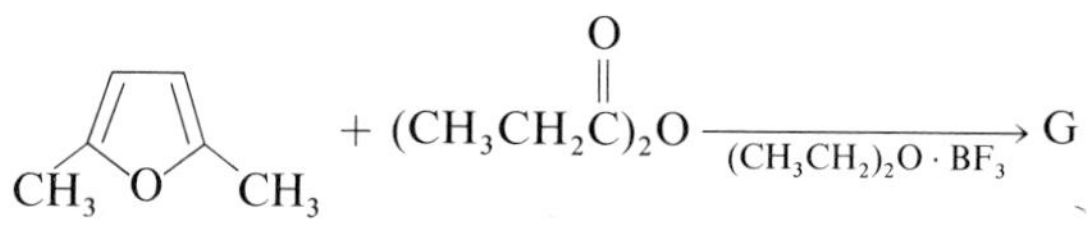

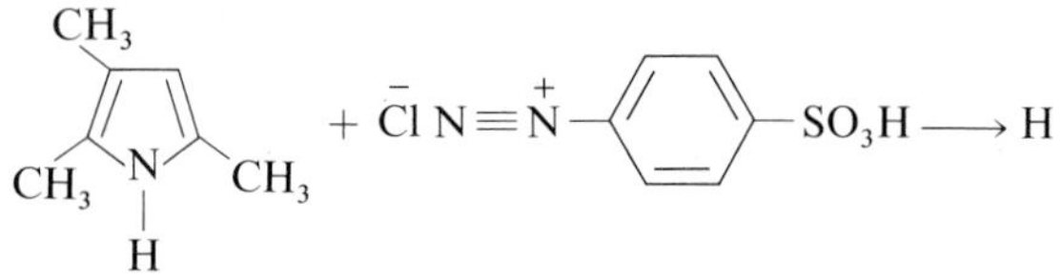

(g)
NO2
N
CH3
+ (CH3C)2O
O
(CH3CH2)2O · BF3
100 °C
I

(h)
N
H
KOH
J
CH3I
K

(i)
S
+
O
O
O
AlCl3
nitrobenzene
0 °C
L

(j)
S
CCH3
O
NaNH2
NH3 (liq)
M
O
CH3CH2OCOCH2CH3
N
H3O+
O

(k)
N
N
H
CH3
+ HNO3
H2SO4
Δ
P + Q

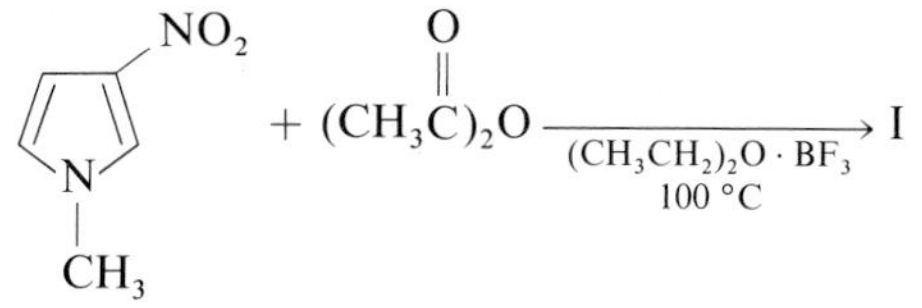

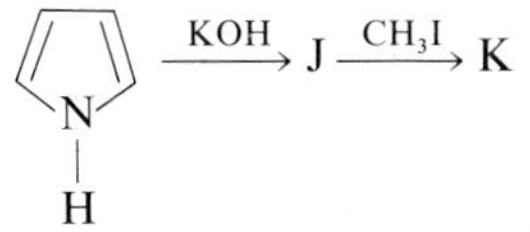

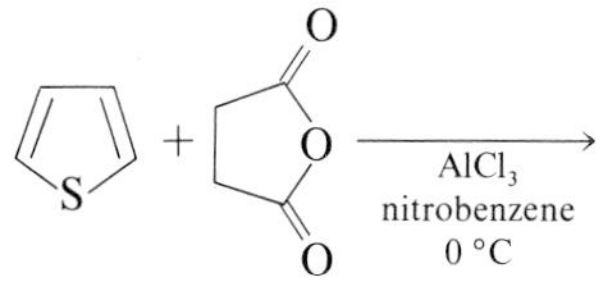

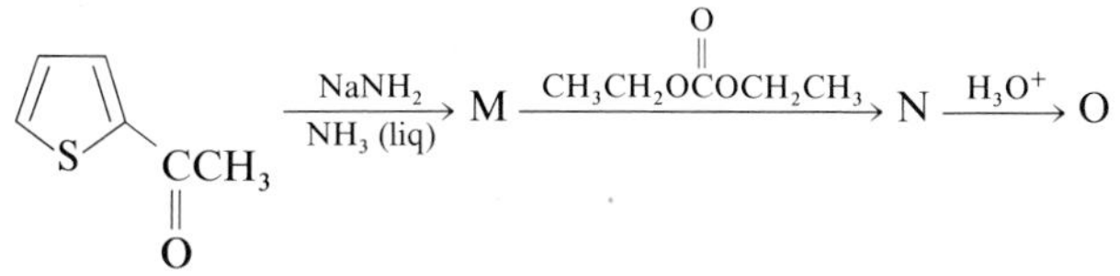

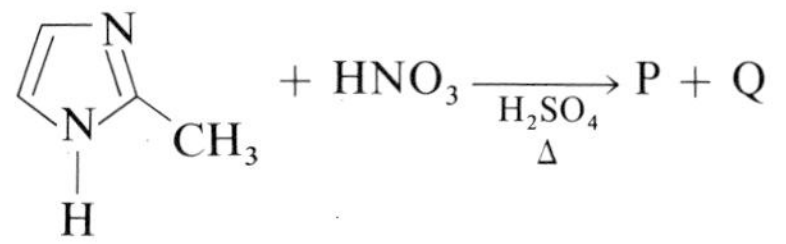

PROBLEM 18.9 2-Acetyl-1-methylpyrrole, when it is treated with nitric acid in acetic anhydride at 0 °C, gives two nitration products. Predict what their structures are by reasoning about the relative stabilities of the intermediates formed. Predict which isomer is the major one.

B. Aromatic Substitution Reactions of Pyridine

While the five-membered ring heterocycles are much more reactive towards electrophilic substitution than benzene is, pyridine is much less reactive. The substitution of the more electronegative nitrogen atom for one of the carbon atoms in the ring decreases the availability of electrons and makes it harder for an electrophilic attack to take place. The conditions that are necessary to bring about substitution on the pyridine ring are often more severe than those required to carry out multiple substitutions on nitrobenzene. An example is the nitration of pyridine, which takes place at 330 °C.

N
+ HNO3
KNO3
330 °C
NO2
N

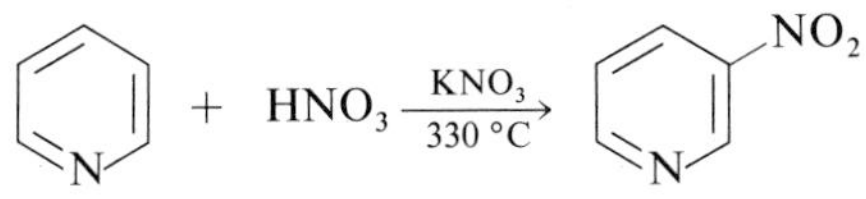

pyridine 3-nitropyridine 15%

Even at this temperature only a small amount of the pyridine is nitrated.

Substitution takes place preferentially at carbon 3 of the pyridine ring. The nitrogen atom in the ring deactivates the positions that are ortho and para to it more than it deactivates the meta positions. In this, it has the same effect as a nitro group on benzene. Part of the effect arises because pyridine is protonated or coordinates with Lewis acids under the conditions necessary for most substitution reactions. For example, in the nitration reaction, the species undergoing substitution is the pyridinium ion and not pyridine itself.

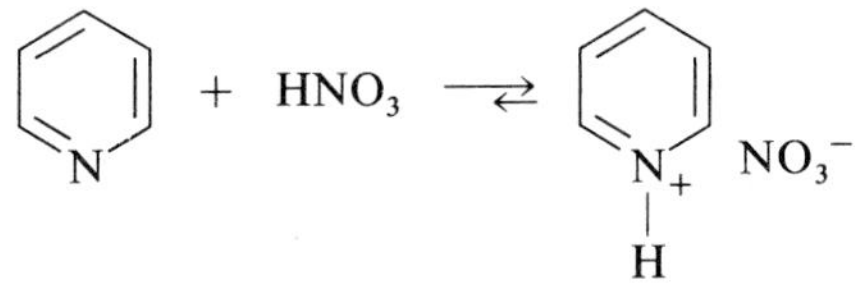

pyridine   nitric acid   pyridinium
                                          nitrate

The resonance contributors for the intermediates that arise when an electrophile attacks at carbons 2 and 3 of the pyridine ring are shown below.

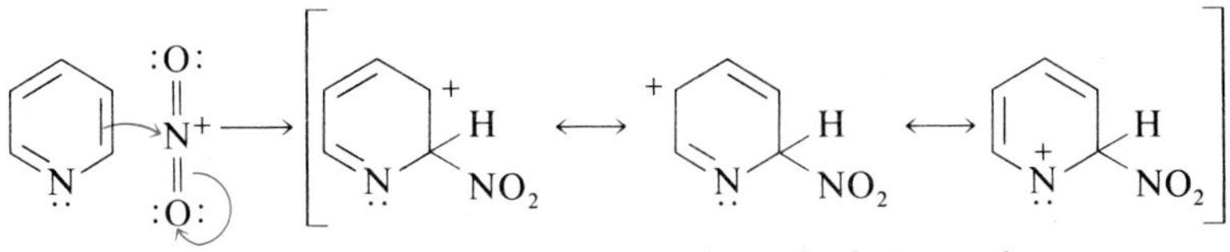

resonance contributors for the intermediate
formed by electrophilic attack at carbon 2

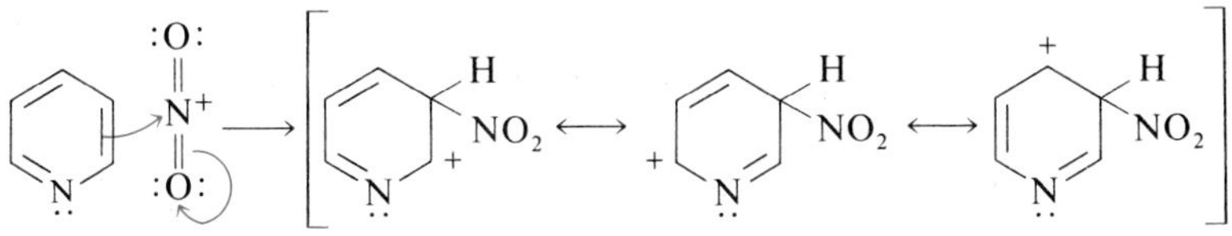

resonance contributors for the intermediate
formed by electrophilic attack at carbon 3

Comparing the formulas shows that the intermediate formed by electrophilic attack at carbon 3 is more stable than the other one. Delocalization of the positive charge in the intermediate from reaction at carbon 2 puts the charge on the nitrogen atom that would already be bearing a positive charge because of prior protonation or coordination with a Lewis acid. The last resonance contributor in that set also shows nitrogen, an element that is more electronegative than carbon, with only a sextet of electrons. The delocalization of charge in the intermediate from reaction at carbon 3 does not require such a high-energy situation.

While pyridine itself is resistant to electrophilic attack, pyridines substituted by electron-donating groups react more easily. The amino group is particularly effective in activating the ring to substitution. 2-Aminopyridine, which is readily prepared from pyridine (p. 1030), is brominated easily.

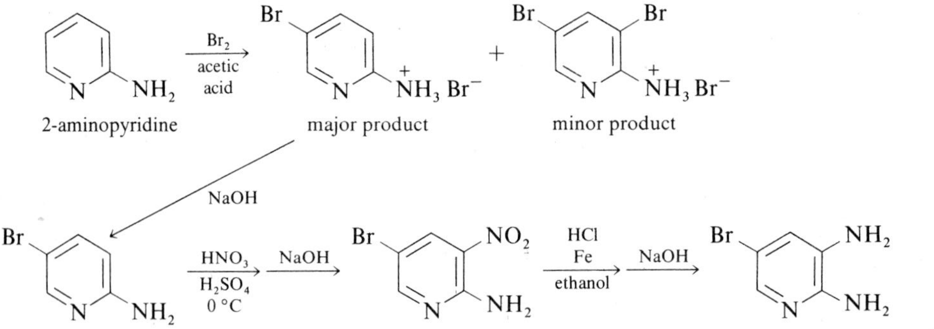

2-aminopyridine   major product   minor product

2-amino-5-
bromopyridine

2-amino-5-bromo-
3-nitropyridine
~80%

2,3-diamino-5-bromopyridine
~70%

The new substituent enters the position para to the activating group and meta to the deactivating ring nitrogen atom. The second substitution reaction also occurs easily. (Compare these reaction conditions with those for the nitration of pyridine, p. 1027.) The last two reactions shown above demonstrate that substituents on the pyridine ring can be transformed into other functional groups by the same reactions used to make benzene derivatives. The aromatic nitro compound is reduced to an aromatic amine with a metal and hydrochloric acid (Section 16.3B).

Pyridine is highly resistant to oxidation, unlike the five-membered ring heterocycles, which are destroyed by strong oxidizing agents. For example, a complex of chromium trioxide and hydrochloric acid with pyridine, pyridinium chlorochromate, is used as an oxidizing agent for alcohols that may be sensitive to acid, or when a primary alcohol is oxidized to an aldehyde (Section 7.6C). The ease with which an aromatic ring is oxidized is related to the electron density in the ring. The same factor influences the ease with which electrophilic aromatic substitution occurs. Thus, the five-membered ring heterocycles undergo aromatic substitution reactions rapidly and are also susceptible to oxidation. Pyridine is not substituted easily and resists oxidation.

Methylpyridines, for example, are oxidized by hot potassium permanganate to the corresponding carboxylic acids (Section 12.4B). The reaction is illustrated by the conversion of 4-methylpyridine to 4-pyridinecarboxylic acid.

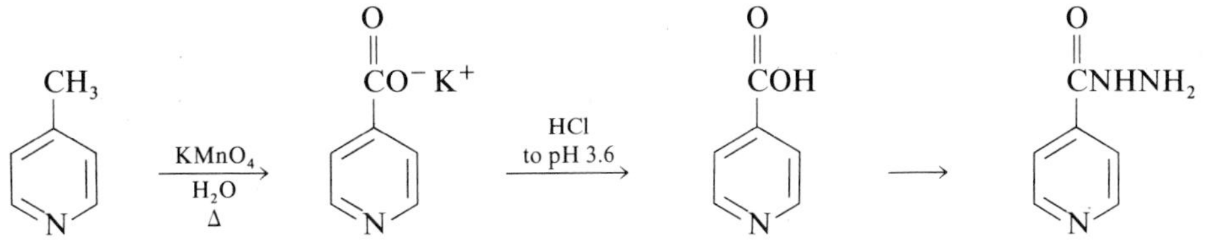

| 4-methylpyridine | potassium 4-pyridinecarboxylate | 4-pyridinecarboxylic acid isonicotinic acid 76% | isonicotinic acid hydrazide isoniazid *a drug used in the treatment of tuberculosis* |

4-Pyridinecarboxylic acid, also known as isonicotinic acid, is of medical interest. Isonicotinic acid hydrazide is a potent drug in the treatment of tuberculosis.

Pyridine derivatives differ most from benzene derivatives in the ease with which they react in nucleophilic aromatic substitution reactions. Pyridines with halogen at the 2 or 4 position have the kind of reactivity seen for 2,4-dinitrohalobenzenes (Section 12.5). For example, 2-bromopyridine reacts with aniline to give 2-(N-phenylamino)-pyridine.

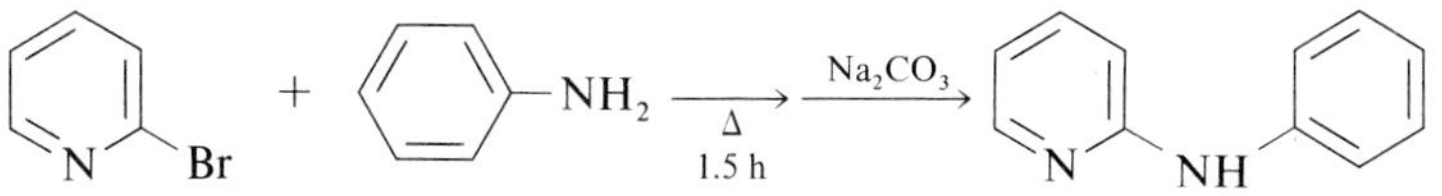

| 2-bromopyridine | aniline | 2-(N-phenylamino)pyridine |

PROBLEM 18.10   Propose a detailed mechanism for the reaction of 2-bromopyridine with aniline.

Pyridine itself is vulnerable to nucleophilic attack. The most useful reaction of this type is the formation of 2-aminopyridine when pyridine is treated with sodium amide.

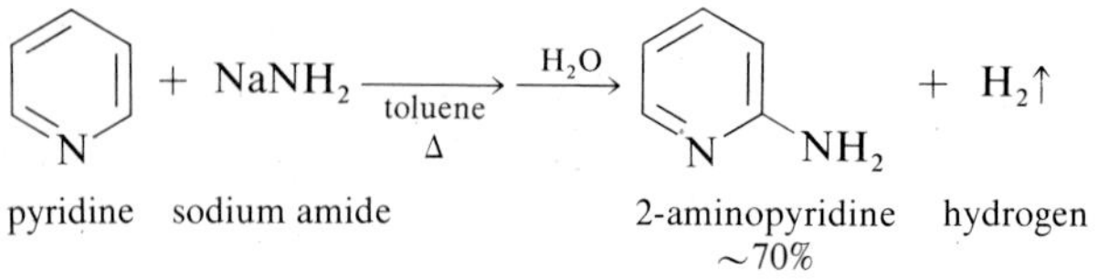

pyridine   sodium amide   2-aminopyridine   hydrogen
                                                   ~70%

The reaction starts with an attack of amide anion on carbon 2 of the ring and loss of hydride ion, which deprotonates the amino group. The product, 2-aminopyridine, is formed when water is added to the reaction mixture.

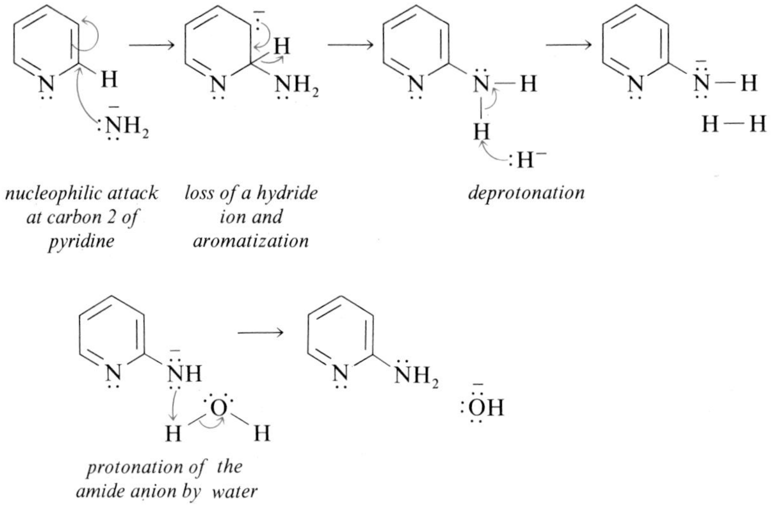

nucleophilic attack   loss of a hydride   deprotonation
at carbon 2 of   ion and
pyridine   aromatization

protonation of the
amide anion by water

Aminopyridines prepared in this way are valuable reagents for preparing more highly substituted pyridines, as we saw earlier in this section.

The reactions of pyridine in electrophilic and nucleophilic aromatic substitution reactions are predictable on the basis of its electronic character. Electrophilic substitution occurs predominantly at carbon 3 and nucleophilic substitution at carbon 2. In most cases, the reactions of substituents on pyridine resemble those of similar substituents on benzene.

PROBLEM 18.11   Pyridine reacts with phenyllithium to give 2-phenylpyridine and lithium hydride in a reaction analogous to the reaction with amide anion.

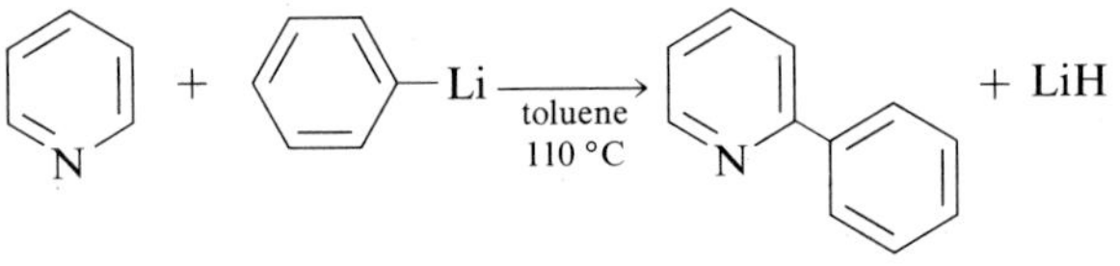

Suggest a mechanism for this reaction. How does the reaction start? What is the leaving group? Why is such a leaving group feasible in this reaction and in the reaction of amide anion with pyridine?

PROBLEM 18.12   Complete the following equations.

(a)

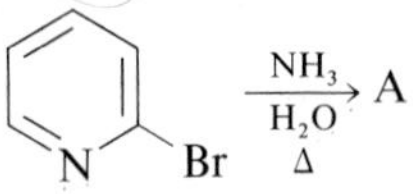

(b)   [2,4,6-trimethylpyridine] $\xrightarrow[\text{H}_2\text{SO}_4 \atop 100\,°\text{C}]{\text{KNO}_3}$ B

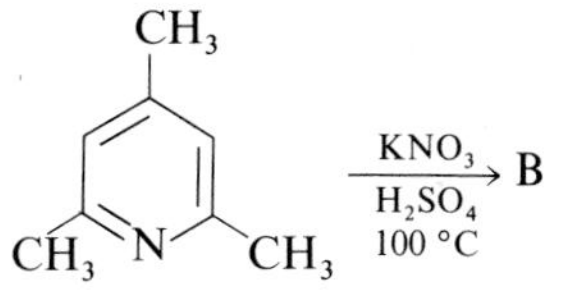

(c)   [2-aminopyridine] + [benzoyl chloride, C$_6$H$_5$—CCl with O] $\xrightarrow{\text{pyridine}}$ C

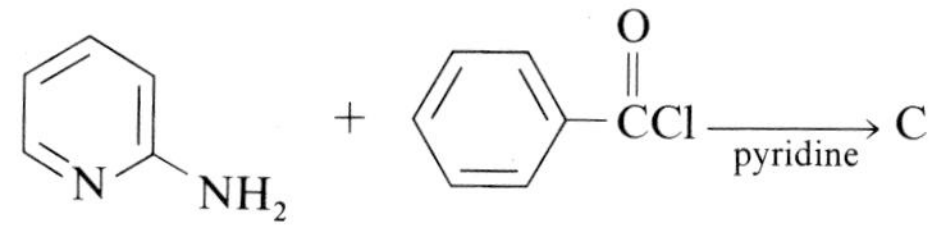

(d)   [6-methyl-2-aminopyridine, H$_2$N and CH$_3$] $\xrightarrow[\text{ethanol}]{\text{Br}_2\ (2\ \text{molar equiv})}$ D

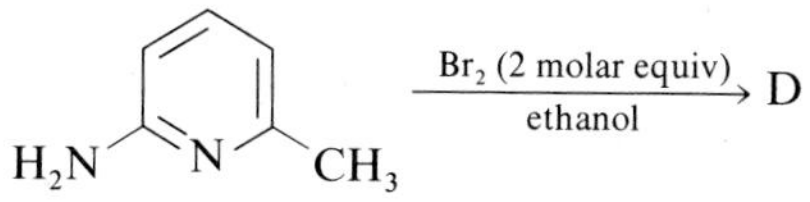

(e)   [2-aminopyridine] $\xrightarrow[\text{H}_2\text{O} \atop 0\,°\text{C}]{\text{NaNO}_2,\ \text{HI}}$ E

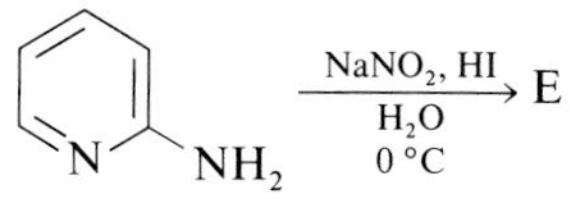

(f)   [nicotinamide, pyridine-3-carboxamide, CNH$_2$ with O] $\xrightarrow{\text{Br}_2,\ \text{NaOH}}$ F

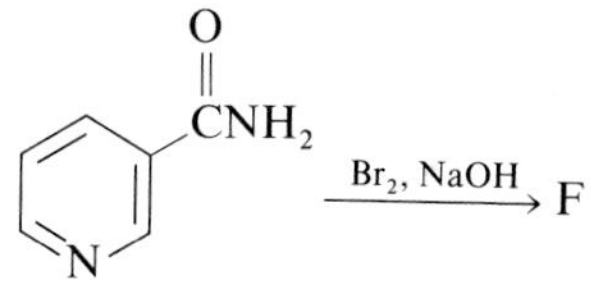

(g)   [4-nitro-2-methylpyridine, NO$_2$ and CH$_3$] $\xrightarrow[\text{catalyst}]{\text{H}_2}$ G

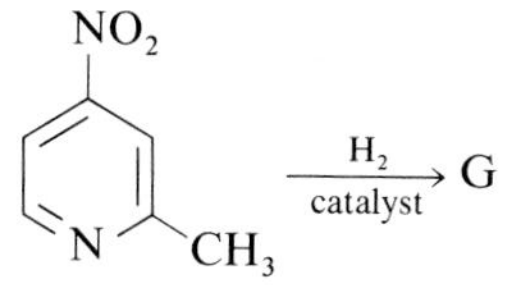

(h)   [nicotinic acid, pyridine-3-carboxylic acid, COH with O] $\xrightarrow{\text{SOCl}_2}$ H [anisole, C$_6$H$_5$—OCH$_3$] $\xrightarrow{\text{AlCl}_3}$ I

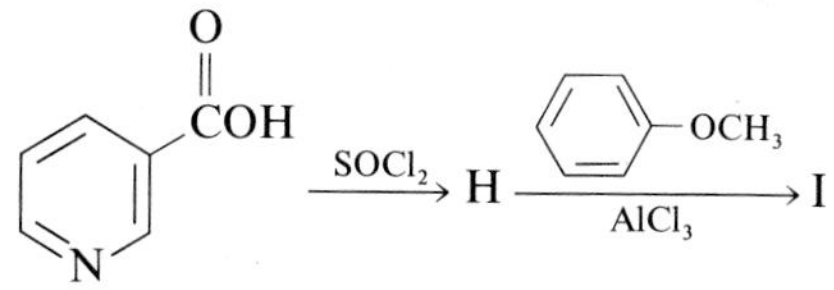

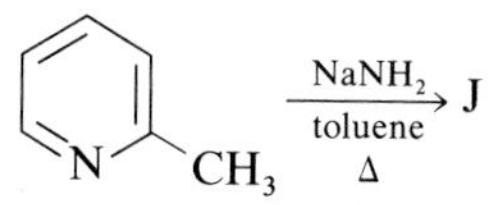

(i)   [2-methylpyridine, CH$_3$] $\xrightarrow[\text{toluene} \atop \Delta]{\text{NaNH}_2}$ J

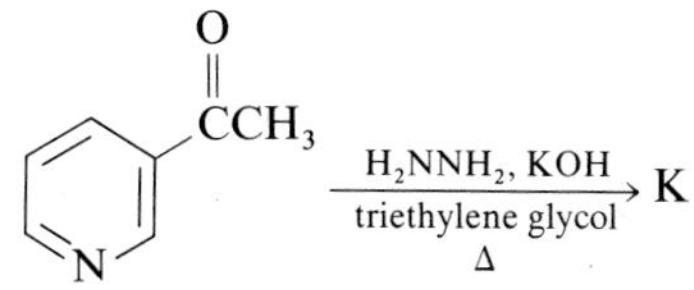

(j)   [3-acetylpyridine, CCH$_3$ with O] $\xrightarrow[\text{triethylene glycol} \atop \Delta]{\text{H}_2\text{NNH}_2,\ \text{KOH}}$ K

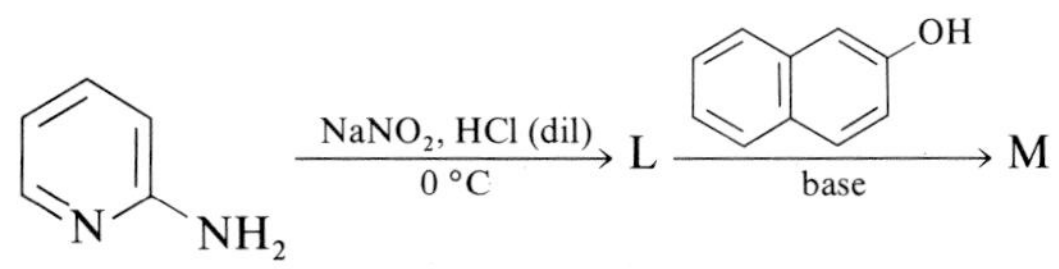

(k)   [2-aminopyridine, NH$_2$] $\xrightarrow[0\,°\text{C}]{\text{NaNO}_2,\ \text{HCl (dil)}}$ L [2-naphthol, OH] $\xrightarrow{\text{base}}$ M

Pyridines and Pyrimidines of Biological Interest

A. Pyrimidines and Purines

The chemistry of pyrimidines and purines is of interest because these heterocyclic rings are found in deoxyribonucleic acids (DNA) and ribonucleic acids (RNA), the complex molecules that transmit genetic information and mediate the synthesis of proteins in the cell. The particular structures and chemical properties of several pyrimidines and purines determine the interactions possible between different strands of DNA and between molecules of DNA and RNA. These interactions are believed to be largely responsible for the storage and transmission of genetic information in cells. The ways in which these heterocycles influence the structure and function of DNA and RNA are discussed in greater detail in the next section. In this section, the chemistry of the heterocyclic ring systems is examined.

The pyrimidine and purine bases that are found in DNA and RNA are shown below.

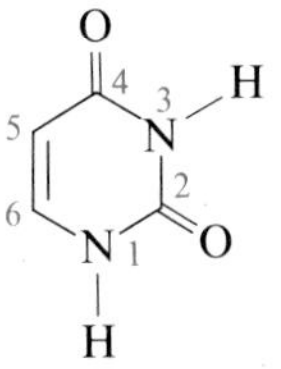

uracil

found in RNA

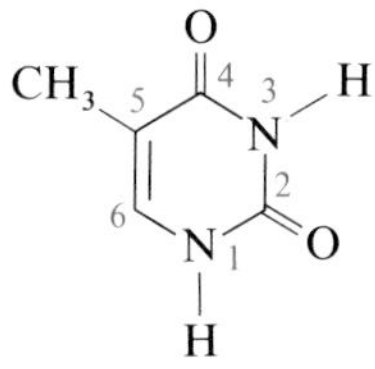

thymine

found in DNA

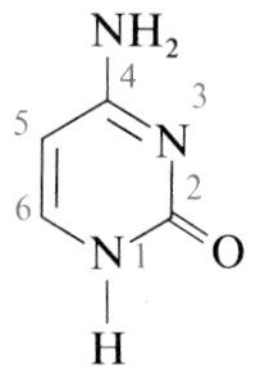

cytosine

*found in both
RNA and DNA*

pyrimidine bases

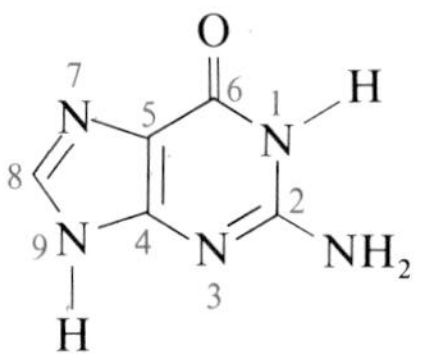

guanine

*found in both
RNA and DNA*

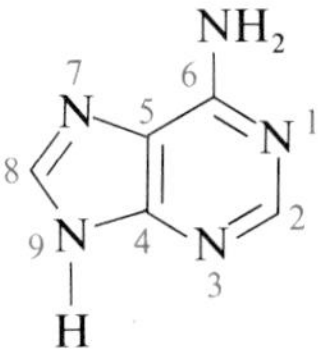

adenine

*found in both
RNA and DNA*

purine bases

An examination of the structures of the purines shows that they contain an imidazole ring fused to a pyrimidine ring. The numbering system shown is the one commonly used for these compounds in biological systems.

The tautomeric forms of the bases shown above are the ones that are important in water at pH 7, the conditions under which these bases are found in nucleic acids. The question of tautomerism is important because the exact location of hydrogen atoms on oxygen and nitrogen atoms determines the way the bases interact with each other by hydrogen bonding (Section 18.6B).

A number of other purines and pyrimidines are of interest. Among them are purines such as xanthine, hypoxanthine, and uric acid that occur in nature . Caffeine, found in coffee, tea, and cocoa, theobromine from cocoa, and theophylline from tea, are methylated xanthines.

hypoxanthine

xanthine

uric acid

caffeine

found in tea, coffee, maté leaves, guarana paste, cola nuts

theobromine

principal alkaloid of cacao bean, also in cola nuts and tea

theophylline

small amounts in tea

Caffeine is a powerful stimulant of the central nervous system. Theophylline is a milder stimulant of the central nervous system and, at the same time, is a relaxant of smooth muscles. Theobromine does not have much activity as a stimulant. Hypoxanthine, xanthine, and uric acid are products of the metabolism of the purine bases adenine and guanine. The disease, gout, results from the faulty metabolism and excretion of uric acid.

The purine and pyrimidine bases occupy a central place in the metabolic processes of a cell because they are involved in regulating protein synthesis. Their importance has led chemists to design medications that incorporate these ring systems and mimic the structures of the bases found in RNA and DNA. Researchers hope that such compounds will be useful in disrupting metabolic processes in cancer cells. Many compounds of this type have been synthesized and tested. Two of the ones that have been used in the treatment of cancer are 5-fluorouracil and 6-mercaptopurine.

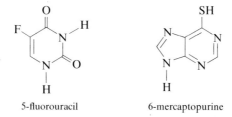

5-fluorouracil

6-mercaptopurine

5-Fluorouracil was designed to resemble thymine in its structural dimensions. It interacts with enzymes that function in the synthesis of nucleotides of uracil and thymine and prevents normal metabolic processes from taking place. 6-Mercaptopurine resembles adenine in structure, except that the nucleophilic sulfur atom replaces a nucleophilic nitrogen atom. It acts in cells by blocking the synthesis of adenine and guanine nucleotides. Cancer cells grow in an uncontrolled way compared with normal cells. The designers of these antitumor compounds hoped that the metabolic processes in cancer cells would therefore be more vulnerable to disruptive drugs than those of normal cells. To some extent this is true, but all the medications that are used in the treatment of cancer are also extremely toxic to the rest of the human body.

The roles that purine and pyrimidine bases play in biological systems give importance to their chemical reactions. The chemistry of pyrimidines resembles that of pyridines, except that the extra nitrogen atom in the heterocycle makes the ring even more resistant to electrophilic attack. For nitration of the ring to occur, for example, a pyrimidine must have two ring-activating substituents. Uracil, which is 2,4-dihydroxypyrimidine in one of its tautomeric forms, can be nitrated.

uracil in   uracil in   2,4-dihydroxy-5-nitropyrimidine
its keto form   its enol form   5-nitrouracil
                                            99%

$$\xrightarrow[50-60\,°C]{HNO_3}$$

Note that the substituent enters the 5 position of the pyrimidine ring in an electrophilic substitution reaction. Arguments such as the ones given in the case of substitution reactions on pyridine can be used to show that attack at the position meta to the two nitrogen atoms of the ring and ortho and para to the hydroxyl groups gives the most stable intermediates (Section 18.5B).

Aminopyrimidines and aminopurines react with dilute nitrous acid to give the corresponding hydroxy compounds (Section 16.6D). This reaction is believed to be responsible for the damage that nitrous acid (and thus nitrites) does to DNA, giving rise to mutations in organisms such as yeast. For example, cytosine is converted to uracil by the action of nitrous acid, as illustrated by the conversion of the nucleoside cytidine to uridine by this reagent.

$$\xrightarrow[\substack{acetic\ acid,\ H_2O\\sodium\ acetate\\0\,°C}]{NaNO_2}$$

cytidine   uridine

Cytidine and uridine are ribonucleosides, but similar reactions occur with aminopyrimidines and purines in deoxyribonucleic acids. In the reactions that are involved in the transmission of genetic information, cytosine is paired with guanine by precise hydrogen bonding interactions. Uracil interacts with adenine in a similar way. The conversion of cytosine to uracil by a chemical reaction changes the way the bases pair and thus changes the genetic code (Section 18.6B).

Functional groups on pyrimidines and purines have the same general kinds of reactivity that they do on a benzene ring. The next problem is an exercise on the transformations of functional groups on purines and pyrimidines.

PROBLEM 18.13 Complete the equations below giving structural formulas for all the intermediates and products designated by letters.

(a)

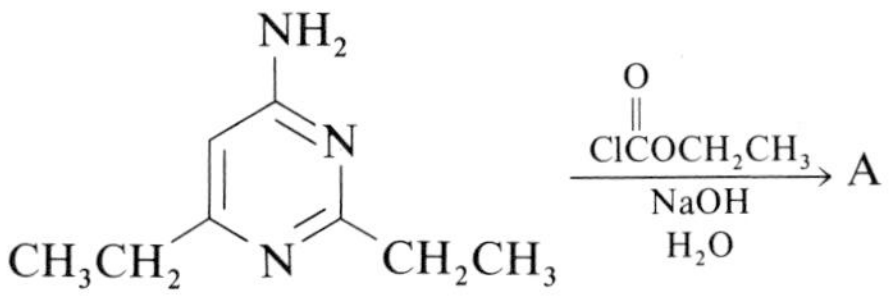

(b)

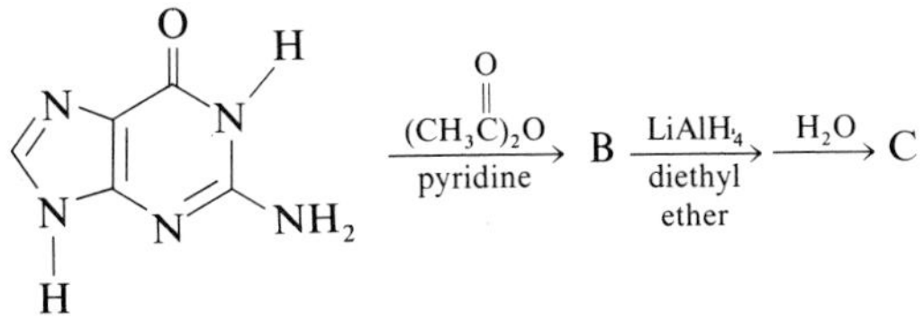

(c)

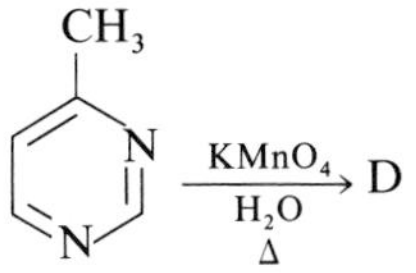

(d)

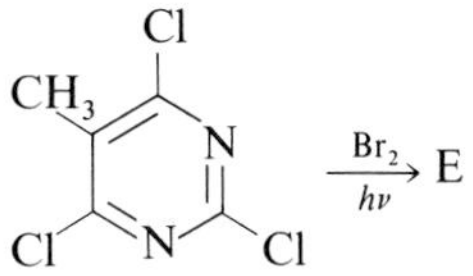

(e)

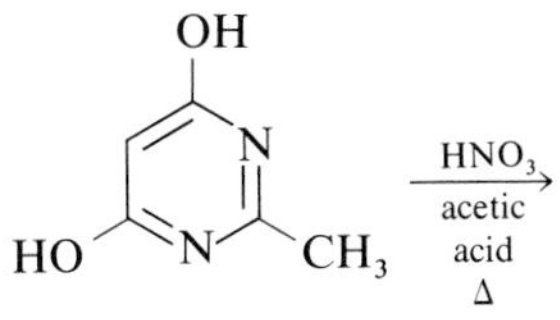

(f)

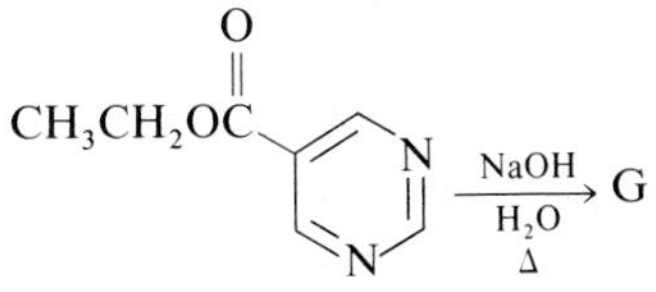

(g)

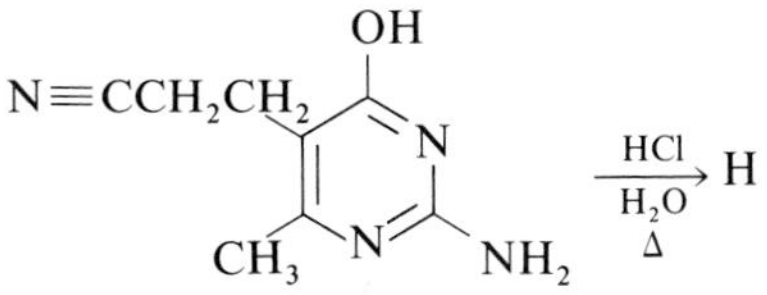

(h)

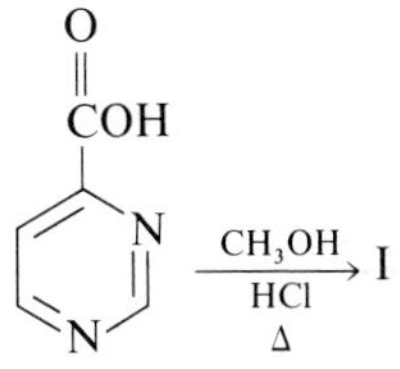

(i)

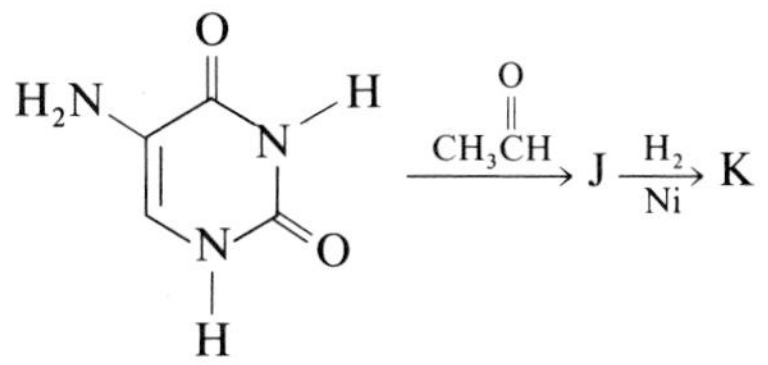

(j)

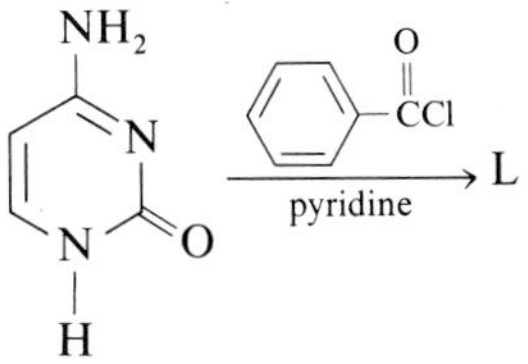

B. Ribonucleic Acids and Deoxyribonucleic Acids

Some of the largest molecules known, often with molecular weights over a million, are the nucleic acids, especially the deoxyribonucleic acids, found in the nuclei of cells. The nucleic acids have backbones made up of aldopentoses in their furanose form, held together as the phosphate esters of the hydroxyl groups at carbons 3 and 5. In ribonucleic acids, RNA, the sugar is D-ribose. In deoxyribonucleic acids, DNA, it is D-2-deoxyribose. Each sugar unit is also bonded at carbon 1 as its β-N-glycoside to a nitrogen heterocycle (p. 759). The N-glycosides of ribose and deoxyribose with these bases are called **nucleosides.** The phosphate esters of nucleosides are known as

nucleotides (Section 14.5B). RNA and DNA are polymers of nucleotides, and as such are also called **polynucleotides.** The structural formulas and names for the nucleotides commonly found in RNA and DNA are shown below.

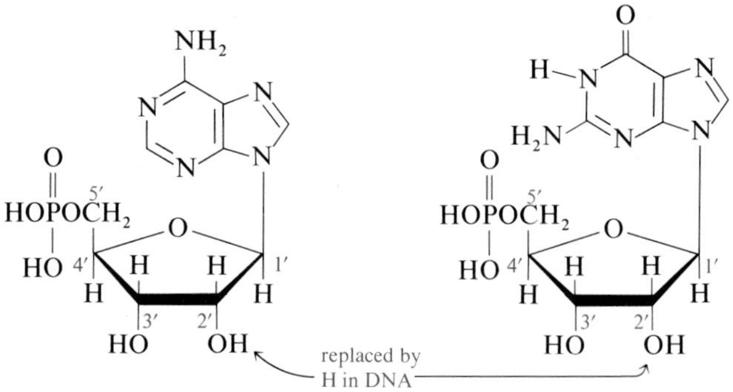

5'-adenylic acid
adenosine 5'-phosphate
5'-AMP

5'-guanylic acid
guanosine 5'-phosphate
5'-GMP

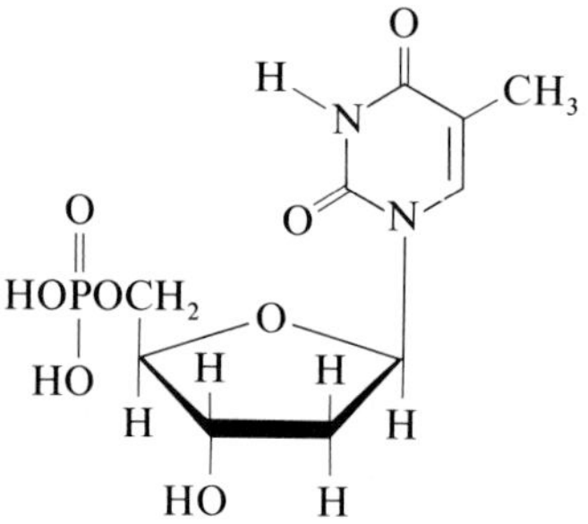

replaced by
H in DNA

5'-thymidylic acid (DNA)
thymidine 5'-phosphate
5'-TMP

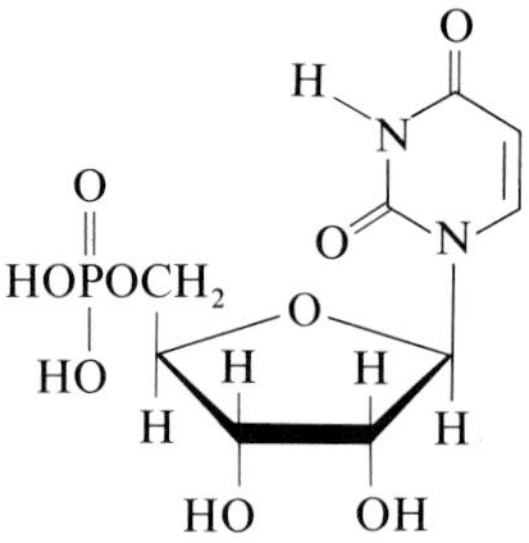

5'-uridylic acid (RNA)
uridine 5'-phosphate
5'-UMP

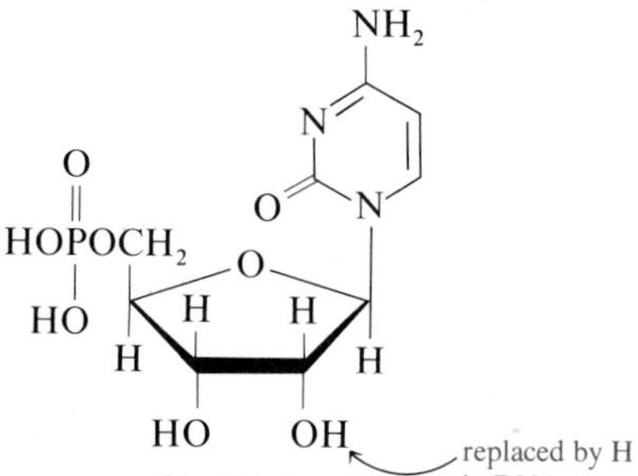

5'-cytidylic acid
cytidine 5'-phosphate
5'-CMP

replaced by H
in DNA

In this section, we examine how the properties of the parts of a nucleic acid contribute to the overall structure of these giant molecules and to their biological functions as they are understood at present.

The history of the determination of the structures and functions of RNA and DNA is a long one, starting with the isolation of a material rich in phosphorus from the nuclei of pus cells and from the sperm of salmon by Friedrich Miescher in Germany in 1868. This material was first called *nuclein* and then *nucleic acid.* Early in this century, the components of nucleic acids, the heterocyclic bases, sugars, and phosphoric acids, were identified. In the 1920s, the structures of the individual nucleotides were determined. Beginning in 1939, the British chemist Sir Alexander Todd investigated the structures of the nucleotides and showed that they are linked in polynucleotides as phosphoric acid esters at the hydroxyl groups of carbons 3 and 5 of the sugar units. For his contribution to the determination of the structure of RNA and DNA, he received the Nobel Prize in 1957.

Another large step was taken in the 1950s, when Erwin Chargaff of Columbia University investigated deoxyribonucleic acids from a large number of sources such as as viruses, bacteria, molds, yeast, insects, plants, and mammals. He found that while nucleic acids from different organisms contain differing amounts of the purine and pyrimidine bases, in each case the molar ratio of adenine to thymine and of guanine to cytosine is close to 1.00. He reasoned that the constancy of this relationship in such a variety of organisms could not be accidental; it suggested some association of adenine with thymine and of guanine with cytosine in deoxyribonucleic acid.

At about the same time, researchers were making progress in developing techniques for determining the structures of complex organic molecules by x-ray diffraction. The patterns that develop on a photographic plate when a crystal is exposed to x rays can be interpreted in terms of the locations of the atoms that diffract the x rays, and hence in terms of molecular structure. Dorothy Crowfoot Hodgkins of the United Kingdom is one of the pioneers in this area of research. She received the Nobel Prize in 1964 for her work on the structure of vitamin B_{12}. By the 1950s, the technique was advanced enough that it was being used to investigate the structures of proteins and nucleic acids. The best x-ray crystallographic data were obtained by Rosalind Franklin in the laboratories of Maurice Wilkins at King's College, London. She worked with the crystalline sodium salt of deoxyribonucleic acid from the thymus gland of the calf. The patterns that are seen in the photograph of the diffraction pattern are best interpreted if it is assumed that DNA has a helical structure with a distance of 3.4 Å between the different nucleotide units and a diameter of 20 Å.

James Watson and Francis Crick, working in the Cavendish Laboratory at Cambridge University, recognized that a DNA molecule consisting of a single helical strand having these dimensions would not have the density that had already been determined for it. Very shortly after they saw the x-ray data obtained by Franklin, they worked out the idea that in the DNA molecule two strands are twisted around each other in a double helix, arranged so that the more hydrophobic nitrogen bases are inside the helix and the hydrophilic sugar and phosphate groups are on the outside. They proposed, in a paper published in 1953 along with papers by Wilkins and by Franklin describing the x-ray crystallographic data, that the two strands of the double helix were held together by hydrogen bonding between the bases, adenine with thymine and guanine with cytosine. Thus, while there is no regularity to the order in which the bases appear on the backbone of either chain, there is a one-to-one correspondence in the number of adenine units to thymine units, and of guanine to cytosine for the two strands taken together. Adenine and thymine are said to constitute one **base pair** and guanine and cytosine, another one. The hydrogen bonding responsible for the pairing of nucleotides containing these bases is shown schematically on the next page.

the hydrogen bonding between
nucleotides that is responsible
for the pairing of the bases

The diameter of the helix is such that a purine base on one strand must be matched with a pyrimidine base on the other. The flat rings of the bases lie parallel to each other in a stack up the axis of the double helix. The sugar units to which they are attached and the phosphate linkages between the sugars together form a twisting ribbon on the outside of the helix.

Just as a polypeptide chain has direction in the sense that it has an *N*-terminal amino acid and a *C*-terminal amino acid (Section 17.3A), so also does a polynucleotide have direction. An unbonded phosphate group is present at carbon 5 of the final sugar unit at one end of the chain, and another is observed at carbon 3 of the analogous unit at the other end of the molecule. The two strands of the double helix are polynucleotide chains, headed in opposite directions. A fragment of a double helix and a view of the overall shape of the helix are shown in Figure 18.1 to illustrate some of these relationships.

Even before a structure had been assigned to DNA, experimental evidence had accumulated indicating that DNA is involved in the storage and transmission of genetic information. Watson and Crick recognized that the pairing of the bases on one strand of the double helix with those on the other strand provided a mechanism by which genetic information could be transmitted. Either strand of a double helix could in principle, if separated from its partner, direct the synthesis of a new strand containing purine and pyrimidine bases exactly the same as those of the missing strand, as shown in Figure 18.2 (p. 1040).

Watson and Crick proposed as the "central dogma" of the new science of molecular genetics that the bases are like the letters of an alphabet, that genes are in essence molecules of DNA, and that the order of the bases on the backbone of the poly-nucleotide chain constitutes a genetic code. The code is reproduced when DNA fosters the synthesis of molecules of ribonucleic acid. The order of bases on the RNA chains is determined when their bases pair by means of hydrogen bonding with those on a strand of DNA. Ribonucleic acids, in turn, direct the synthesis of proteins.

Ribonucleic acid molecules are generally smaller than deoxyribonucleic acids. Their

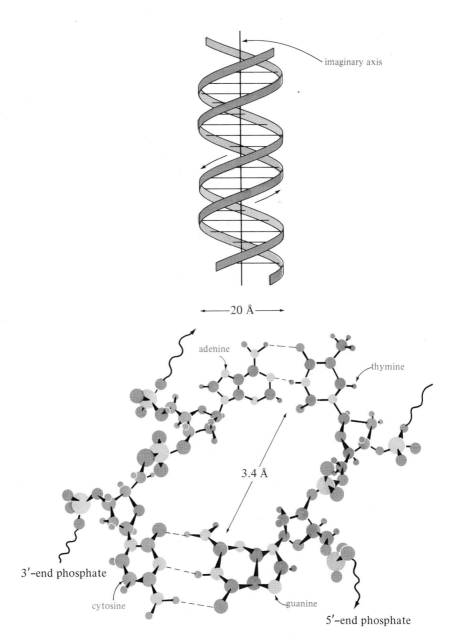

imaginary axis

←——— 20 Å ———→

adenine

thymine

3.4 Å

3′-end phosphate

cytosine

guanine

5′-end phosphate

Figure 18.1   The double helix.

molecular weights range from 30,000 to 2,500,000. A ribonucleic acid does not form a double helix because the hydroxyl group present at carbon 2 of the sugar units (unlike the hydrogen atom occupying that position in 2-deoxyribose) is too bulky to allow two strands of RNA to come close enough to interact in that way. The RNA molecules exist as single strands that fold and loop within themselves with base pairing taking place at some points. In RNA, adenosine hydrogen bonds with uracil and guanine bonds with cytosine. Because RNA does not have a double helix with every base on one strand paired with the corresponding base on an opposite strand, analyses of samples of RNA do not show the regularity in the molar ratios of purine bases to pyrimidine bases that is characteristic of DNA. This is an important and striking experimental difference between the two types of polynucleotides.

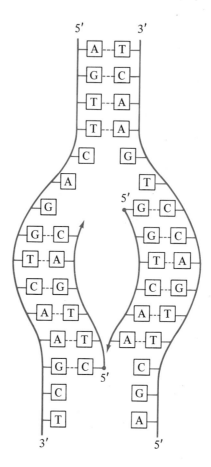

Figure 18.2   A schematic representation of the conversion of one double helix into two identical double helices. Replication of DNA.

Experimentally, it has been found that three adjacent bases on a kind of ribonucleic acid known as **messenger RNA** are the code for a given amino acid. Historically, the first such three-nucleotide code was identified when it was found that polyuridylic acid directed the synthesis of the peptide polyphenylalanine. Thus, three uracil bases in a row are the code for the amino acid phenylalanine. Most amino acids have more than one code of three nucleotides. There are also portions of RNA molecules that do not code for anything, as far as we know. Other three-nucleotide sequences serve as punctuation marks, such as a signal to stop the synthesis of a polypeptide. Scientists have also found that some nucleotides are part of a code for two different amino acids. The code sequences for them overlap. The amino acids to be incorporated in the protein chain are transported to the messenger RNA by another type of RNA known as **transfer RNA.** Each amino acid has at least one transfer RNA that is specific for it.

Methods have been invented to determine the sequence of the bases in very small samples of DNA. It is now possible, in principle, to compare the sequence of bases on a segment of DNA with the structure of the polypeptide that is synthesized from that gene. Complete analyses have been carried out for a few reasonably complex proteins such as the β-chain of mouse hemoglobin.

Watson, Crick, and Wilkins received the Nobel Prize in 1962 for their work on the structure of DNA. The model they developed for the genetic code has led to a tremendous amount of research into the chemistry of polynucleotides and the ways they interact with other constituents of cells. The deoxyribonucleic acid on which the x-ray crystallographic work was done was a highly purified and crystallized sample. Even such a sample was observed to have a different structure depending upon the humidity of its surroundings. The picture of the molecular structure of DNA, and of the way it may function, that emerges from the x-ray data is a simplified one. It must be modified when the activity of deoxyribonucleic acid in the living cell is considered.

Basic proteins rich in lysine and arginine, known as **histones,** are found with DNA in the nucleus of cells from the thymus gland of the calf. Histones have isoelectric points, pH_I, above 10 (Section 17.2B). They are positively charged at pH 7 and are held by ionic and hydrogen bonds to the phosphate groups on the double helix, stabilizing the molecule. A polypeptide chain lies along the groove in the helix structure and is bonded at several points to the polynucleotide strand. The combined polypeptide-polynucleotide structure is known as a **nucleoprotein.**

In the nuclei of cells of the calf thymus gland there is also a smaller fraction of acidic proteins in which glutamic acid and aspartic acid units are particularly abundant. These are also found associated with DNA and with histones. The proteins constitute about 30% of the dry weight of the nucleus of a cell, suggesting that they have an important function there.

Just as proteins have primary, secondary, and tertiary structure, so does DNA. The primary structure of DNA is the order of the nucleotides in the backbone of a single strand. The synthesis of a polynucleotide involves the orderly placement of one nucleotide unit after another in a chain held together by phosphate ester linkages. In the body, this process is catalyzed by an enzyme, DNA polymerase.

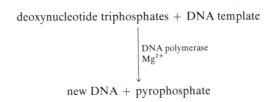

deoxynucleotide triphosphates + DNA template

DNA polymerase
Mg^{2+}

new DNA + pyrophosphate

The secondary structure of DNA is the right-handed helical shape that characterizes a strand of DNA, and the double helix that forms when the complementary strand is synthesized. In Figure 18.1 (p. 1039), the double helix was shown as if its central axis were a straight line. It is seldom so. The double helix itself curves and twists in a variety of ways that are termed supercoiling, which constitutes the tertiary structure of DNA. Some forms of supercoiling are shown in Figure 18.3 (p. 1042).

Proteins maintain the tertiary structures that are responsible for their biological activity only within a narrow range of pH, temperature, and solvent composition. When conditions are changed too drastically, the protein is denatured (Section 17.5). The same is true of DNA molecules. They, too, are denatured by changes in pH and temperature and by the addition of a variety of chemical substances to their solutions. The most important cause of denaturation is the disruption of the ionic and hydrogen bonds that give the double helix its stability. For example, an increase in pH would result in the loss of protons from the basic proteins associated with the phosphate groups on the double helix. The ionic bonds that hold them together would be disrupted, and the excess of negative charge on the helix would destabilize it.

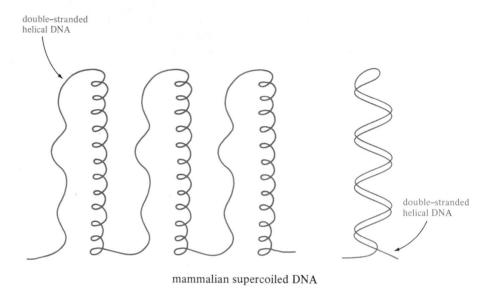

double–stranded helical DNA

double–stranded helical DNA

mammalian supercoiled DNA

Figure 18.3   Supercoiling of DNA molecules.

PROBLEM 18.14   The portions of a DNA strand in which adenine is paired with thymine are the first to pull apart when DNA is denatured. In fact, the stability of a particular form of DNA, as indicated by the temperature to which it can be heated before it melts is directly related to the relative amount of guanine (and cytosine) that it contains. How do you explain this experimental observation? (Hint: Another look at the diagram of the pairing of bases, p. 1038, may be helpful.)

Any reaction that interferes with a correct pairing of bases by hydrogen bonding changes the way in which the genetic code is transmitted. An incorrect transmission of genetic information by DNA may lead to a mutation in the species. Mutations take place all the time in nature, but some compounds, known as **mutagens,** greatly increase the rate at which mutations occur. Many mutagens are also **carcinogens,** compounds that increase the probability that a malignant tumor will develop in a living organism.

For a DNA molecule to serve as a template, either for the synthesis of other DNA molecules or for the synthesis of RNA molecules, which then direct protein synthesis, the double helix must unwind to a greater or lesser extent. In other words, some disruption of the most stable arrangement of the giant molecule must take place. The basic protein molecules associated with DNA in multicelled organisms play an important part in the regulation of the process. They must somehow be detached to expose part of the polynucleotide before it can serve as a template. Exactly how this happens and what starts the process in living organisms is not yet understood.

The story of how genetic information is stored and transmitted is still unraveling. Recent advances in recombinant DNA technology allow DNA molecules to be cleaved selectively, then combined with fragments of DNA molecules from other organisms. These experiments have greatly increased public awareness of the potential for useful, and perhaps some harmful, manipulations of genes. In 1975, scientists who were working in this research area stopped to consider the possible consequences of what they were doing. As a result of their deliberations, controls have been put on the types of experiments that can be done and on the laboratory procedures and precautions that must be followed to minimize the risk. This is the first time that scientists have ever joined together to impose formal controls on their own work.

At the time of this writing, corporations are being formed to develop the potential for using transformed biological systems to synthesize commercially useful products such as human insulin for diabetics, growth hormones, and various immunological factors that may be useful in boosting resistance to disease, including cancer.

PROBLEM 18.15   Write structural formulas for the following compounds. A review of the structures on p. 1036 may be helpful.

(a)   adenosine 3′,5′-diphosphate   (b) deoxyguanosine 5′-phosphate   (c) cytidine

(d)   thymidine 3′-phosphate   (e) guanosine   (f) uridine 2′-phosphate

---

PROBLEM 18.16   Some of the most potent mutagens have three-membered heterocyclic rings such as oxiranes (Section 12.7) or aziridines (three-membered rings containing nitrogen). Write an equation suggesting why such compounds are so destructive to DNA.

C. Biological Oxidation-Reduction Reactions

A large number of biological oxidation-reduction reactions are catalyzed by enzymes that are associated with coenzymes that have nicotinamide as part of their structures. In these coenzymes, nicotinamide, one of the B vitamins, is bonded to ribose at the nitrogen atom of the pyridine ring as an N-glycoside. The molecule also contains the nucleotide adenosine 5′-phosphate.

Reactions of interest occur in the nicotinamide portion of the molecule, so in equations the structural formula of nicotinamide adenine dinucleotide is abbreviated by using R to symbolize everything that is bonded to the nitrogen atom of the pyridine ring.

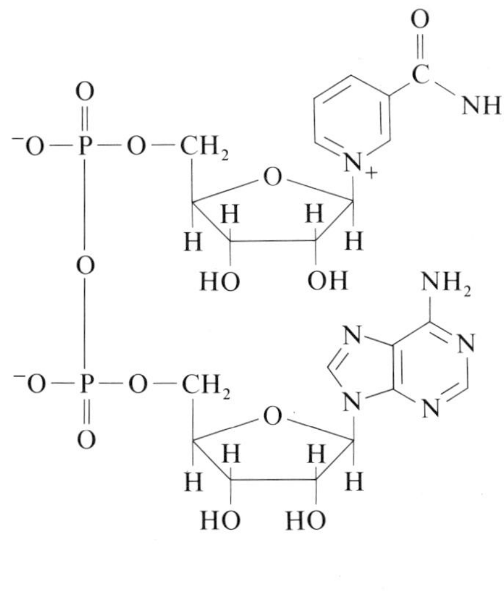

nicotinamide adenine dinucleotide
NAD⁺

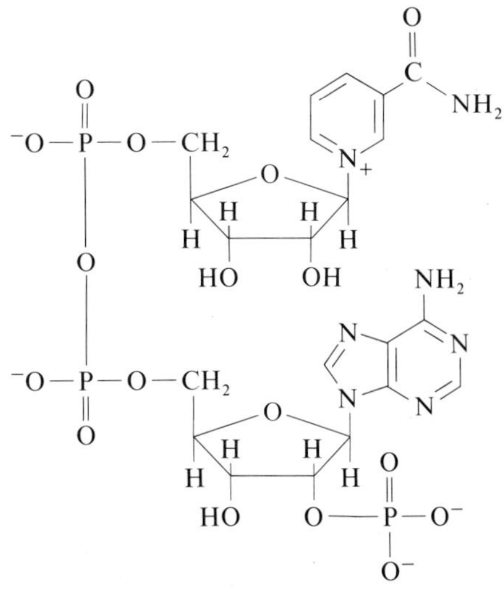

nicotinamide adenine dinucleotide phosphate
NADP⁺

The presence of a positive charge on the pyridine ring is important. The nitrogen atom of the ring is in the form of a quaternary ammonium ion, and the ring is therefore highly activated towards nucleophilic attack, even more so than is pyridine itself (Section 18.5B).

The coenzymes NAD^+ and $NADP^+$ are associated with a large number of enzymes known as dehydrogenases. A typical reaction catalyzed by a dehydrogenase found in the liver is the oxidation of ethanol to acetaldehyde.

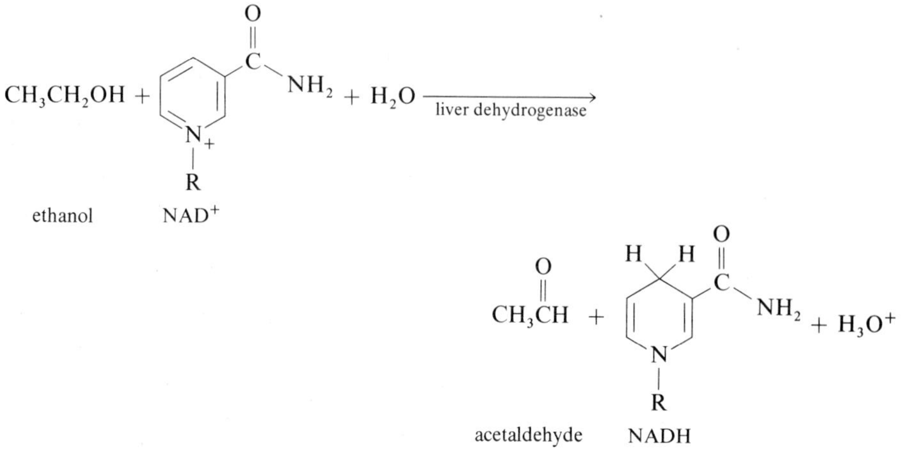

In the process, NAD^+ is reduced to dihydronicotinamide adenine dinucleotide, abbreviated NADH. A proton is also generated in the reaction.

The reaction proceeds by the transfer of a hydride ion from the alcohol to NAD^+. For example, if ethanol labeled with deuterium at carbon 1 is used, deuterium appears in the reduction product of NAD^+.

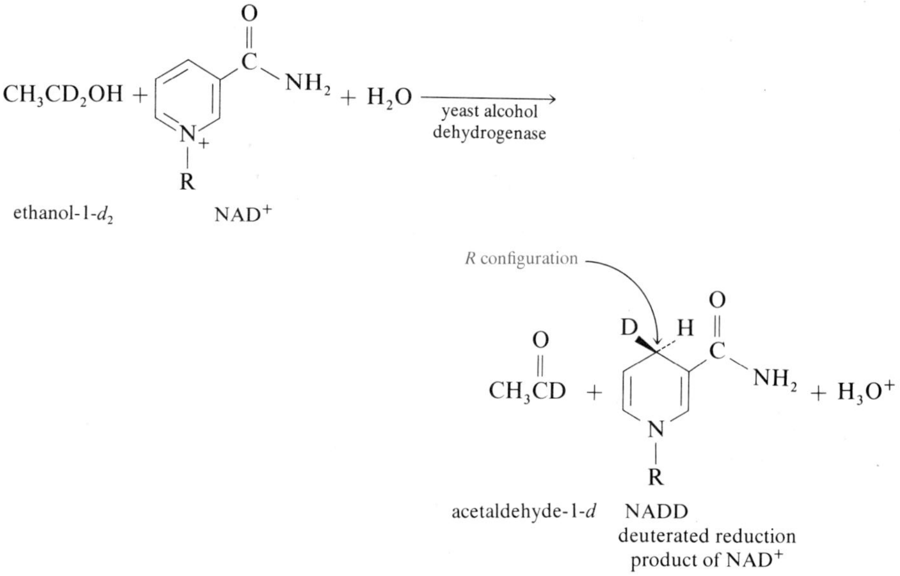

The deuterium atom is transferred with high stereoselectivity to one face of the pyridine ring so that the reduction product always has the R configuration at carbon 4 with this enzyme and many others. Not all enzymes have the same stereoselectivity. A number of enzymes catalyze the transfer of hydride ion to the other face of the nicotinamide residue.

Dehydrogenases catalyze reduction as well as oxidation reactions. The reduced nicotinamide serves as the reducing agent also with high stereoselectivity. For example, if acetaldehyde-1-*d*, the product formed in the reaction shown above, is reduced with NADH, ethanol-1-*d* having the *S*-configuration is produced.

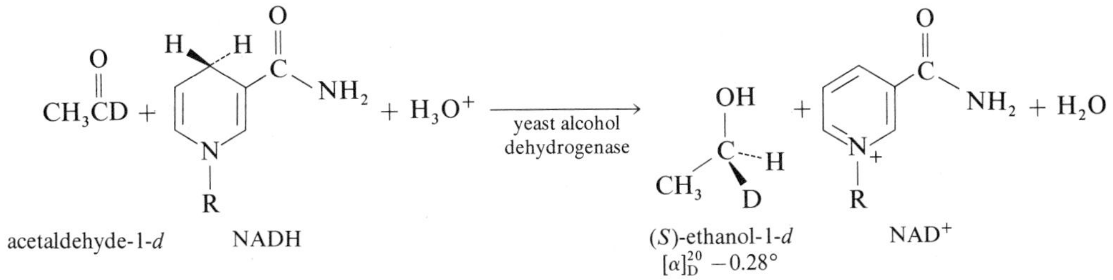

acetaldehyde-1-*d*   NADH   (*S*)-ethanol-1-*d*   NAD$^+$
                                                                                                        $[\alpha]_D^{20}$ $-0.28°$

If, on the other hand, NADD is used to reduce unlabeled acetaldehyde, the product is (*R*)-ethanol-1-*d*.

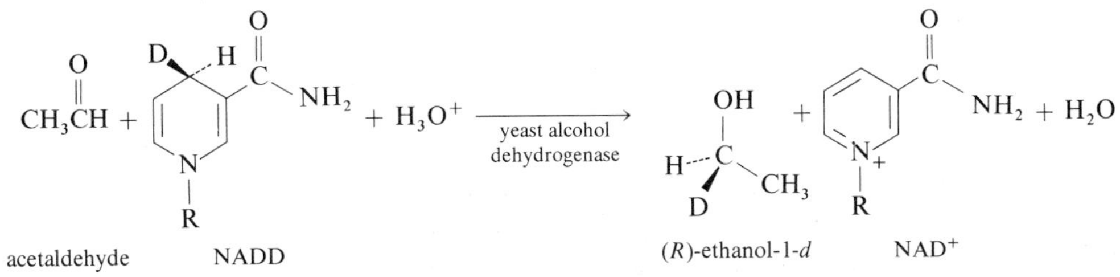

acetaldehyde   NADD   (*R*)-ethanol-1-*d*   NAD$^+$

In other words, the enzyme distinguishes between the two faces of the planar carbonyl group. The incoming hydride ion is always attached to the same side of the plane of the carbonyl group and is detached from only one face of the nicotinamide ring.

The reduction of a carbonyl compound by NADH is entirely analogous to reduction with a metal hydride such as sodium borohydride except for the stereoselectivity of the NADH reaction.

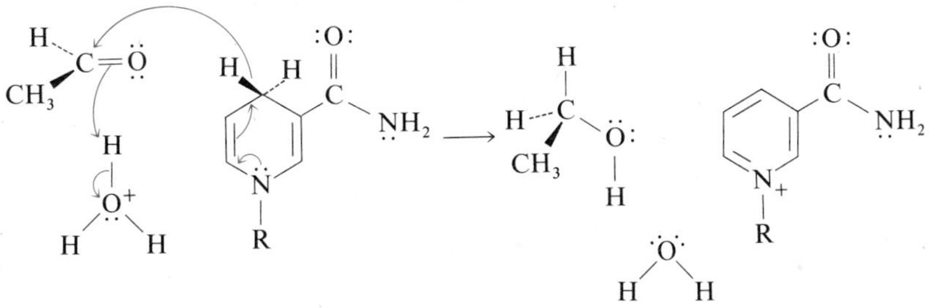

Ethanol that is not labeled with deuterium has no chirality. However, if one of the hydrogen atoms of the methylene group is replaced by another group, such as deuterium, one enantiomer of a pair is formed. Replacement of the other hydrogen atom by deuterium gives the mirror-image isomer of the first compound. *The two hydrogen atoms on carbon 1 of ethanol are called* **enantiotopic** *because enantiomeric compounds are formed by replacing one or the other by another group.* For example, the two secondary hydrogen atoms on carbon 2 of butane are also enantiotopic. Replacement of one of them by a chlorine atom gives (R)-2-chlorobutane, while replacement of the other forms the enantiomeric (S)-2-chlorobutane (Section 3.9A). An achiral reagent does not distinguish between enantiotopic hydrogen atoms, but a chiral one such as an enzyme does. Thus, when butane reacts with the achiral reagent, chlorine, there is an equal probability that each one of the two hydrogen atoms on carbon 2 will be replaced, and equal numbers of molecules of (R)- and (S)-2-chloro-butane are formed. When ethanol reacts with NAD$^+$ at the active site of the enzyme yeast alcohol dehydrogenase, however, only the hydrogen atom that occupies a certain position in space is transferred to NAD$^+$.

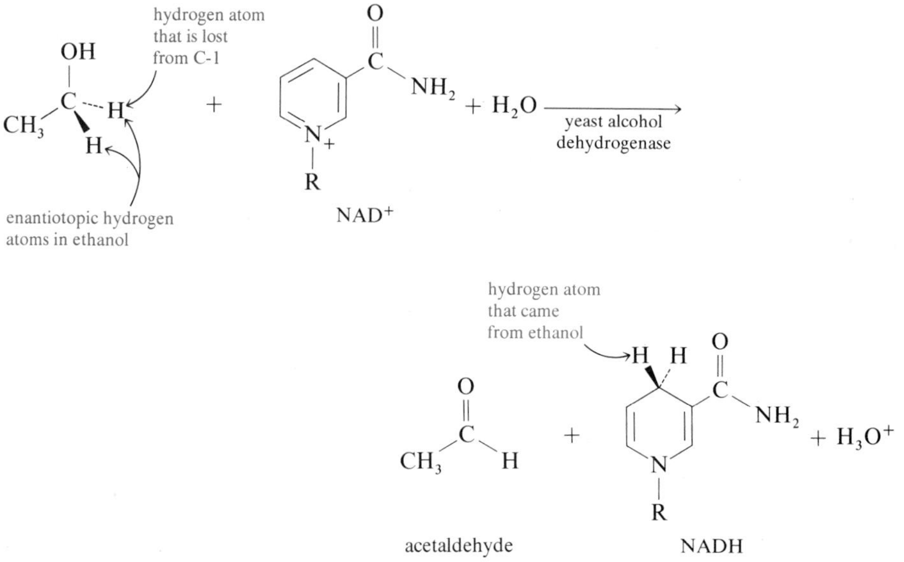

This phenomenon is only detectible when the ethanol is labeled with deuterium, but it occurs whether the label is there or not.

PROBLEM 18.17   Write equations predicting what the products of the reaction of (R)-ethanol-1-d and of (S)-ethanol-1-d with NAD$^+$ will be.

An enzyme can distinguish between the two enantiotopic hydrogen atoms because the ethanol molecule fits the active site of the enzyme better in one orientation than in the mirror image of that orientation. Ethanol will fit the active site well in only one of two enantiomeric positions if the active site binds three parts of the molecule. There is no stereoselectivity if the active site interacts with only two parts of the molecule (Figure 18.4).

In a similar way, the two faces of nicotinamide in NAD$^+$ may be said to be enantiotopic. Depending upon the nature of the enzyme, one face or the other receives the incoming hydride ion stereoselectively. While nicotinamide itself is achiral because it has a plane of symmetry that coincides with the plane of the ring, the presence of ribose units in the coenzyme makes the molecule as a whole chiral. But even if it were not, the arguments that were made about ethanol could be used to show that the active site of an enzyme would interact differently with two enantiomeric orientations of a nicotinamide ring.

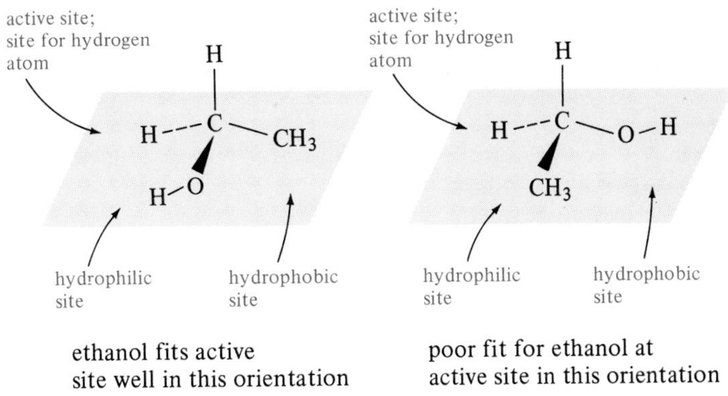

Figure 18.4 Two possible ways in which ethanol can interact with the active site of an enzyme.

PROBLEM 18.18 For the sake of an exercise, assume that the active site of an enzyme interacts with the nitrogen atom in the pyridine ring of nicotinamide, with the carbonyl group of the amide function, and with carbon 5 of the ring as a hydrophobic site. Prove to yourself that the two faces of an achiral nicotinamide molecule are enantiotopic and thus distinguishable by such an enzyme.

PROBLEM 18.19 Inspect the following structural formulas and decide which of these compounds contain enantiotopic hydrogen atoms.

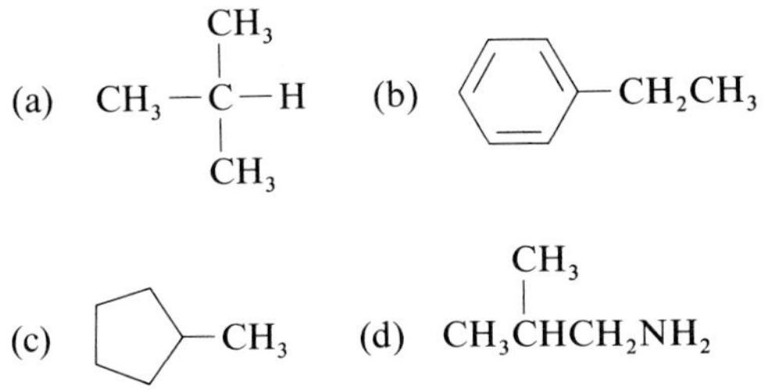

PROBLEM 18.20 **Diastereotopic** hydrogen atoms are defined as those hydrogen atoms which, when replaced by another group, give rise to diastereomers. An example would be the hydrogen atoms that are shown in color in the Fischer projection shown below.

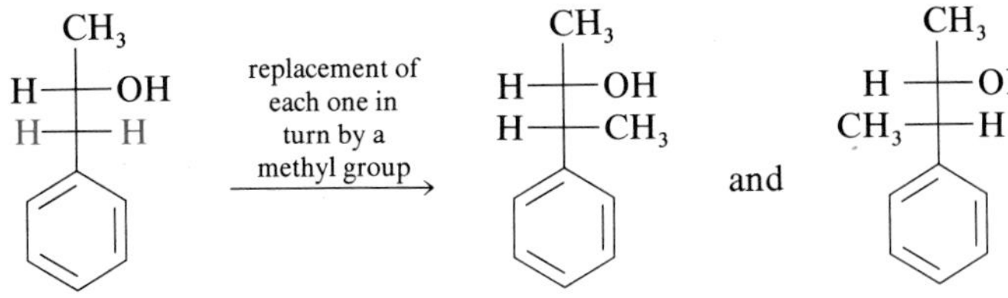

Are there any diastereotopic hydrogen atoms in any of the compounds shown in Problem 18.19? How about the compounds shown below?

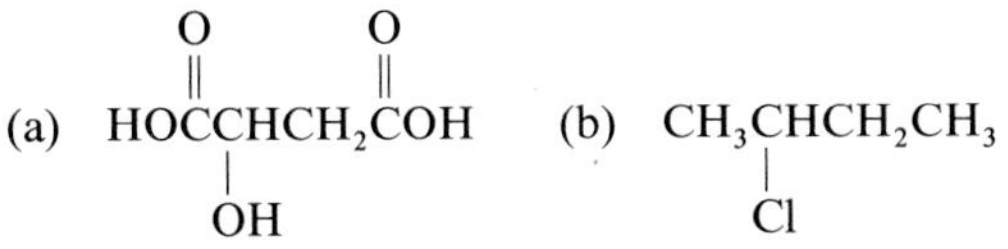

$$\text{(c)} \quad CH_3\overset{\overset{\displaystyle CH_3}{|}}{C}HCH_2CH_3 \qquad \text{(d)} \quad \text{C}_6\text{H}_5\overset{\overset{\displaystyle }{}}{C}H\overset{\underset{\displaystyle NH_2}{|}}{C}H_3$$

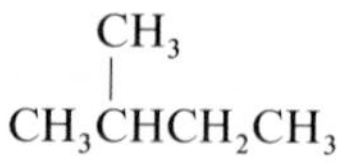

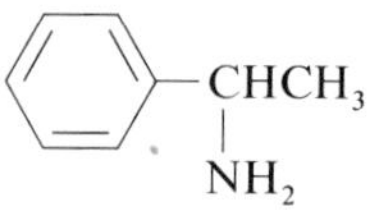

PROBLEM 18.21 NAD^+ participates in the oxidation of testosterone. The enzyme that catalyzes this reaction, testosterone dehydrogenase, has a stereoselectivity opposite to that of yeast alcohol dehydrogenase. Write an equation showing the products that you expect from the reaction. Trace the fate of the hydrogen atoms removed from testosterone (p. 372).

PROBLEM 18.22   Nicotinamide is given that name because it is the amide of an acid derived from nicotine (p. 885). Write equations showing how you would synthesize nicotinamide from nicotine.

18.7

Five-Membered Heterocycles of Biological Interest

A. Sulfur Heterocycles

One of the first vitamins to be discovered was thiamine, vitamin B_1. Thiamine pyrophosphate is found in every cell of the body and functions as a coenzyme in reactions that convert pyruvate ion, one of the products of glycolysis, to acetaldehyde and acetyl coenzyme A (Section 10.7G). Thiamine is thus involved in essential metabolic processes. A deficiency of vitamin B_1 in human beings leads to a disease of the nervous system known as beriberi.

In thiamine, a thiazole ring is bonded to an aminopyrimidine ring by way of a methylene group.

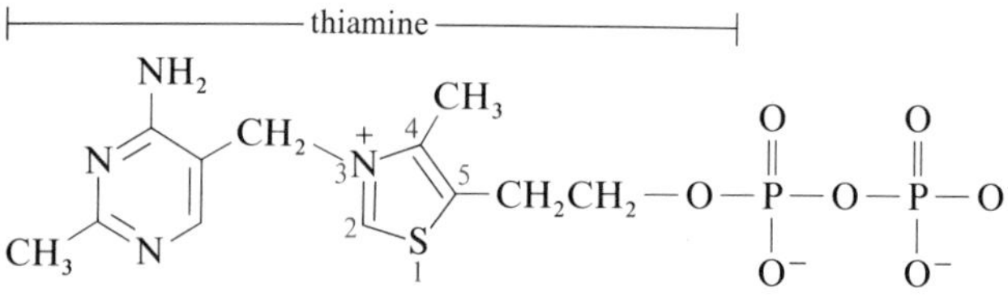

thiamine pyrophosphate

An interesting and important structural feature of the vitamin is the presence of a quaternary nitrogen atom in the thiazole ring. The hydrogen atom on carbon 2 of the thiazolinium ion, between the positively charged nitrogen atom and the sulfur atom, is extraordinarily acidic. In the 1950s, Ronald Breslow of Columbia University found that such a hydrogen atom in a simple analog of thiamine exchanges with deuterium in deuterium oxide in the absence of either acid or base.

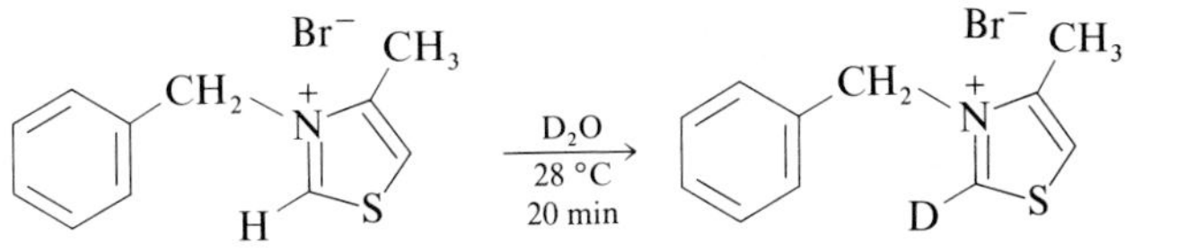

3-benzyl-4-methylthiazolinium
bromide

3-benzyl-4-methylthiazolinium-2-*d*
bromide

If the course of the reaction is followed by nuclear magnetic resonance spectroscopy, it is found that half of the hydrogen atoms at carbon 2 in a sample of the thiazolinium compound are replaced by deuterium atoms in 20 minutes at 28 °C. The rate of the reaction is extraordinarily fast for the breaking of a carbon-hydrogen bond, especially as no strong base is present.

The deuterium exchange reaction proceeds by removal of a proton from the thiazolinium ion to give a carbanionic species that is an ylide.

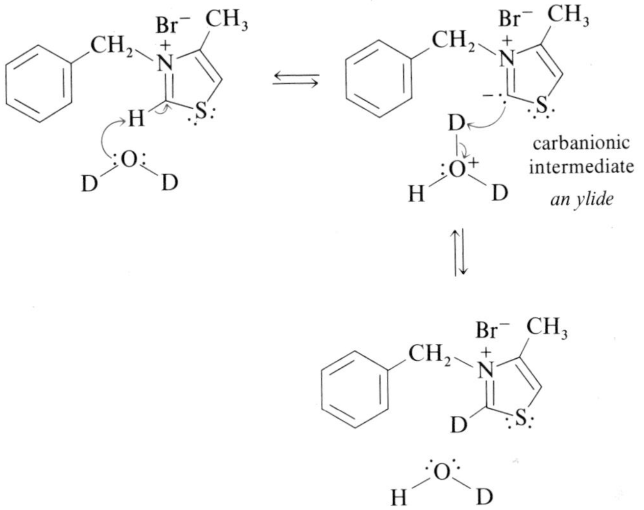

carbanionic
intermediate
an ylide

The negative charge is on a carbon atom adjacent to a positively charged atom (Sections 15.8 and 15.9A), which stabilizes the carbanion. The sulfur atom of the thiazolinium ion can also stabilize the carbanion by accepting some electron density into its empty *d* orbitals. The combination of these two factors makes it possible for the thiazolinium ring of thiamine to ionize at neutral pH, in other words, under the conditions that exist in human cells.

An important reaction that is catalyzed by thiamine pyrophosphate, in the presence of magnesium ions and an enzyme from brewer's yeast known as pyruvate decarboxylase, is the decarboxylation of an α-ketoacid to an aldehyde.

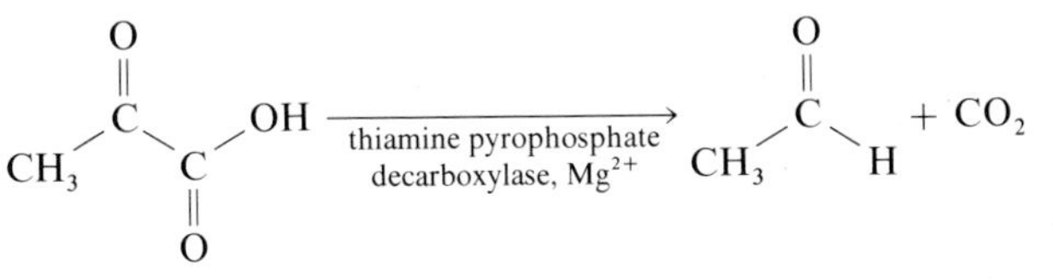

pyruvic acid

acetaldehyde   carbon
dioxide

The reaction is postulated to occur through the nucleophilic addition of the ylide from thiamine to the ketone function in pyruvic acid.

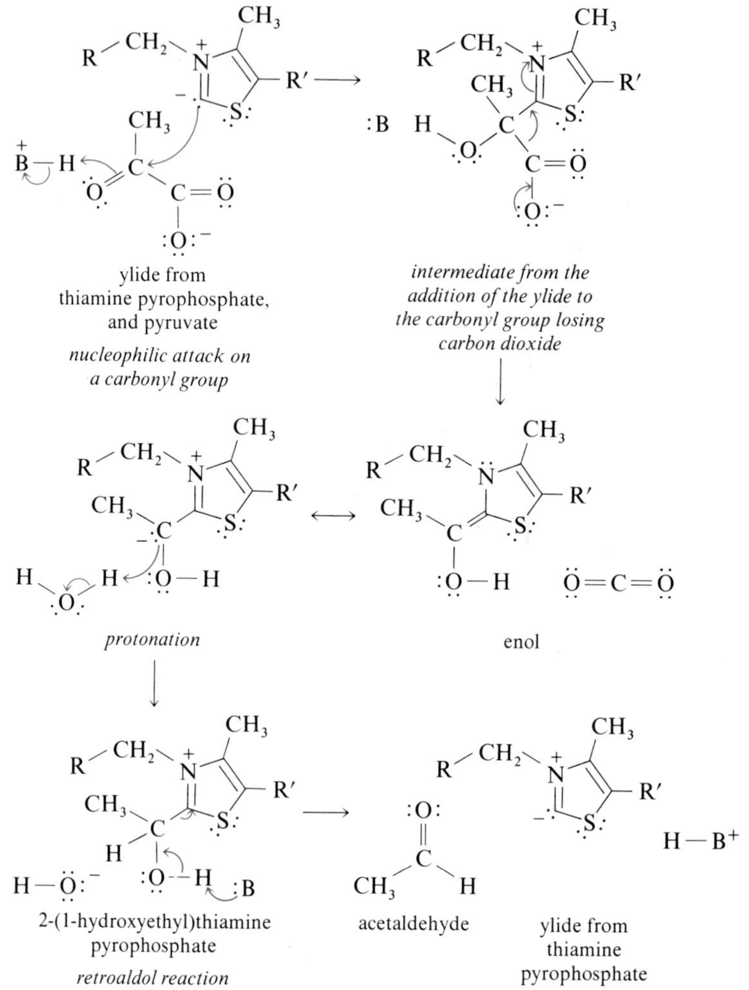

ylide from
thiamine pyrophosphate,
and pyruvate

*nucleophilic attack on
a carbonyl group*

intermediate from the
addition of the ylide to
the carbonyl group losing
carbon dioxide

protonation

enol

2-(1-hydroxyethyl)thiamine
pyrophosphate

retroaldol reaction

acetaldehyde

ylide from
thiamine
pyrophosphate

The intermediate formed by the addition reaction decarboxylates easily because the carbanion that is formed in the process is stabilized by delocalization of the charge to the nitrogen atom. Protonation of the carbanion gives a second intermediate, the 2-(1-hydroxyethyl)thiamine pyrophosphate. This compound has been isolated and has vitamin B$_1$ activity. A retroaldol reaction (Section 15.5A) of this intermediate gives acetaldehyde and the ylide, which is a good leaving group because of its stability.

PROBLEM 18.23 Thiamine is reasonably stable in acid. In pure water it falls apart into a thiazole and a pyrimidine. In strong base the thiazole ring opens. Suggest mechanisms for the two reactions shown below.

(a)

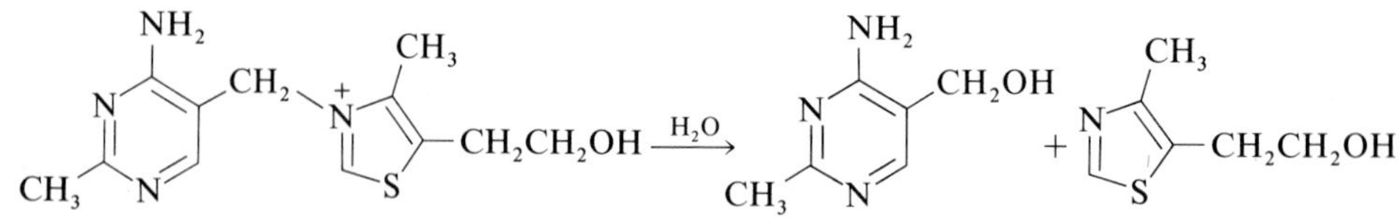

(b)

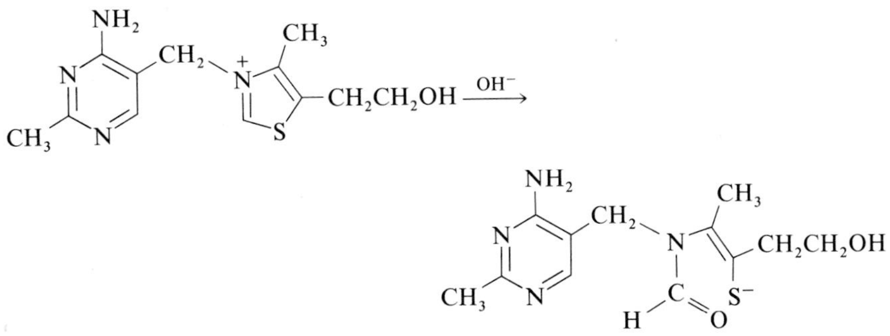

NH$_2$

CH$_3$

CH$_2$

N$^+$

CH$_3$

N

S

CH$_3$

—CH$_2$CH$_2$OH $\xrightarrow{\text{OH}^-}$

NH$_2$

CH$_3$

CH$_2$

N

CH$_3$

N

CH$_2$CH$_2$OH

C

S$^-$

H   O

B. Nitrogen Heterocycles. Pyrrole

Pyrrole has a special role in the chemistry of living organisms. The conversion of light energy from the sun to the energy stored in the chemical bonds of carbohydrates synthesized by green plants is mediated by compounds known as **chlorophylls.** The esssential structural feature of chlorophylls is a system of four pyrrole rings held together by bridges, each containing a single carbon atom. This ring system, known as **porphin,** also appears in heme, the prosthetic group of the proteins hemoglobin (Section 17.5) and myoglobin, which are responsible for the transport and storage of oxygen in the body tissues of warm-blooded animals.

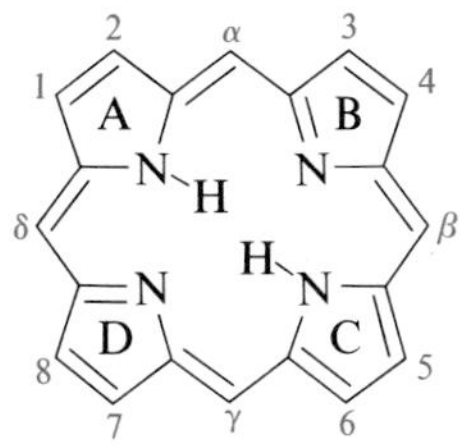

porphin

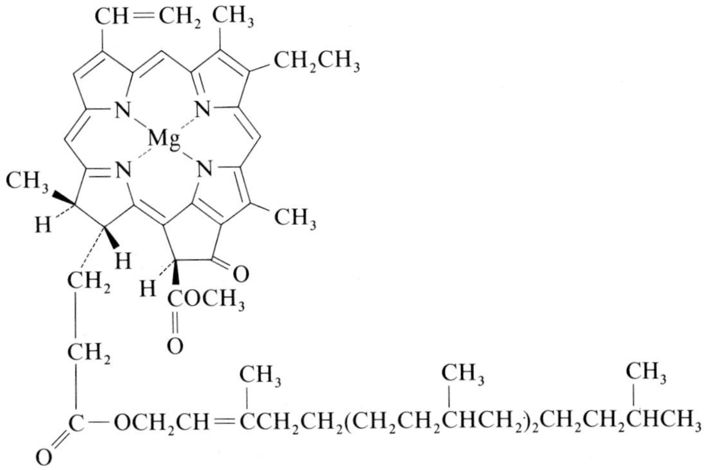

CH=CH$_2$ CH$_3$

CH$_2$CH$_3$

N   N

Mg

CH$_3$

N   N

H

CH$_3$

CH$_2$ H   COCH$_3$

CH$_2$ O   O

CH$_3$ CH$_3$ CH$_3$

C—OCH$_2$CH=CCH$_2$CH$_2$(CH$_2$CH$_2$CHCH$_2$)$_2$CH$_2$CH$_2$CHCH$_3$

O

chlorophyll a

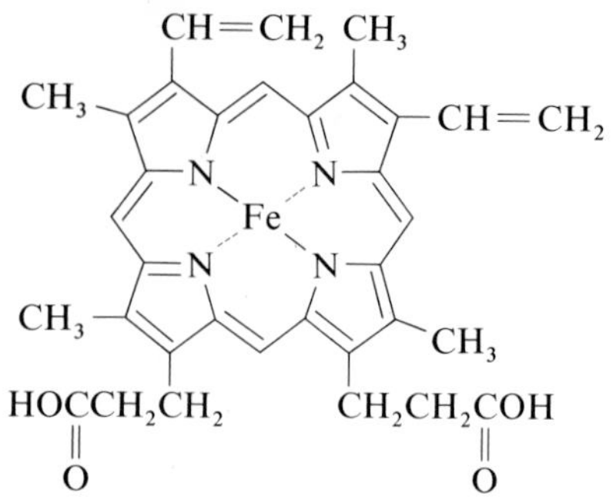

heme

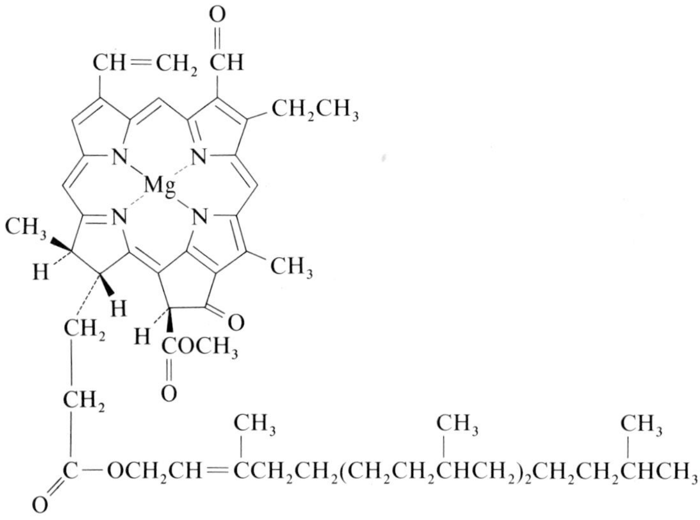

chlorophyll b

Porphin, with the four nitrogen atoms of the pyrrole rings pointing towards the center of a larger ring system, complexes efficiently with metal ions. In heme, the ion is iron(II); in chlorophylls it is magnesium(II). The metal ions may also complex with additional ligands above and below the plane of the heterocycle (Section 17.5).

The ring systems present in heme and chlorophylls have various substituents on the periphery of the porphin ring. Such variously substituted porphins are given the general name of **porphyrins.** The porphyrin in heme has the same degree of unsaturation in the heterocycle as porphin. Those in the chlorophylls are dihydroporphyrins, with one of the double bonds in the D-ring of porphin being reduced, and are given the special name of **chlorins.** The compounds shown above are a few of many known porphyrins. The structural details described show the complexity of structure and the possibilities for isomerism in these systems.

The porphyrin ring system is very stable and has aromatic character. Porphin and heme have 22 π electrons, but only 18 of these are part of a cyclic array for which resonance contributors can be written. In this respect, a porphyrin resembles an [18]annulene (Section 12.1) with $(4n + 2)$ π electrons, $n = 4$.

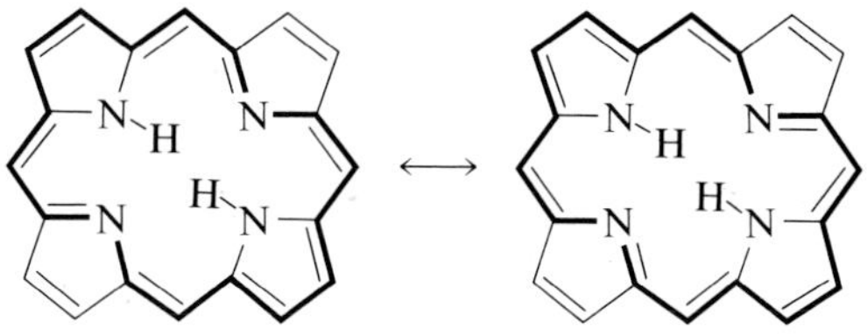

resonance contributors for porphin drawn in analogy to [18]annulene

Any attempt to include in the resonance contributors the four π electrons that are outside the thick black line gives structures that require that the protons on two of the nitrogen atoms also be moved. Such a tautomeric transformation does occur in porphyrins. However, two structures that are related to each other by a change in the location of atoms as well as electrons are not resonance contributors. The extended conjugation present in porphyrins is also responsible for the deep colors of these compounds (Sections 11.8C and 16.6F).

Porphyrins are synthesized in nature with remarkable ease. The basic unit that combines with itself to give porphyrins substituted in a variety of patterns is a trisubstituted pyrrole called porphobilinogen. This compound is synthesized in living organisms by the condensation of two molecules of 5-amino-4-oxopentanoic acid (δ-aminolevulinic acid).

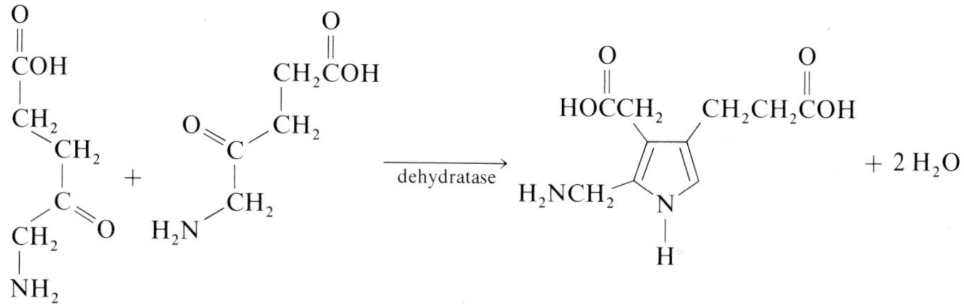

5-amino-4-oxopentanoic
acid
δ-aminolevulinic acid

porphobilinogen

In the presence of the enzyme deaminase (and by simply heating with acid, for that matter) porphobilinogen is converted into uroporphyrinogen I in high yield. A porphyrinogen is oxidized by air to the more stable aromatic porphyrin, unless special precautions are taken to exclude air.

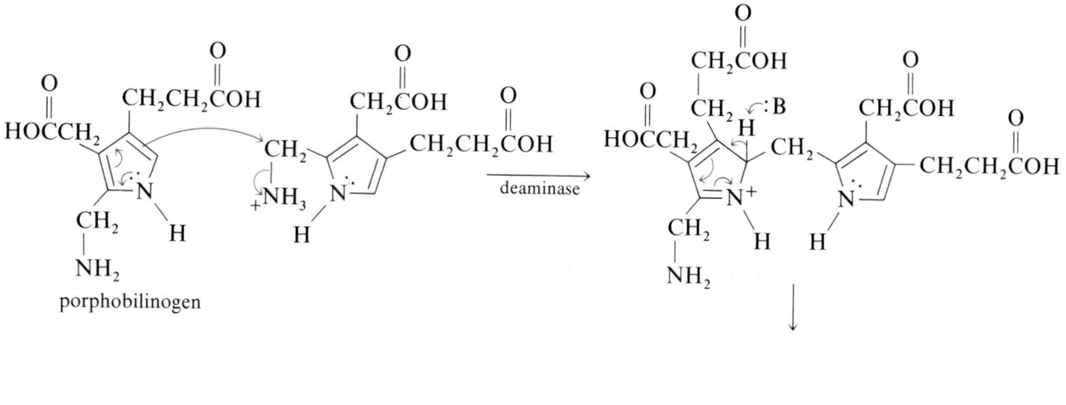

porphobilinogen

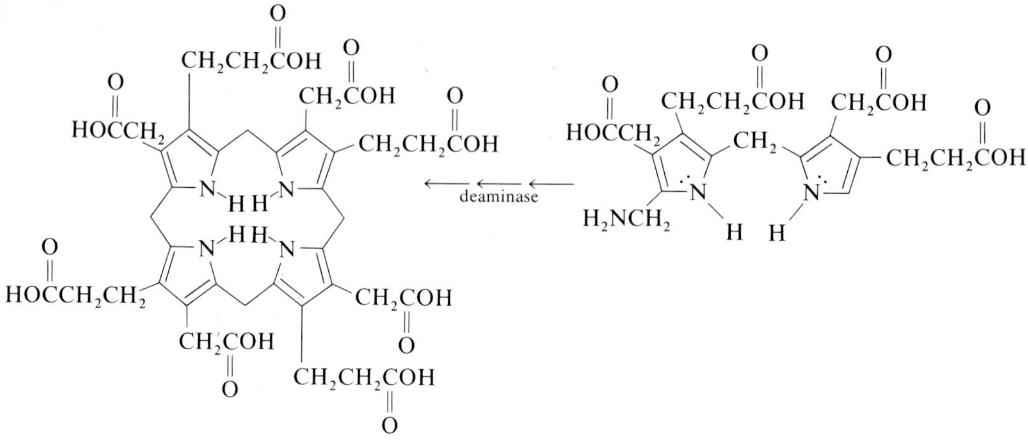

uroporphyrinogen I, ~90%

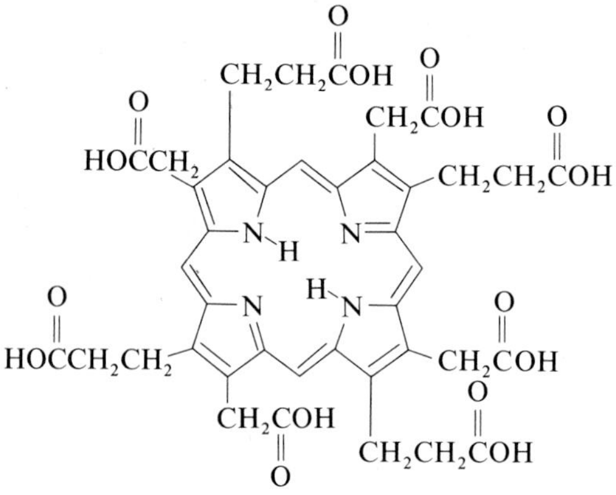

uroporphyrin I

When two enzymes are present, the unsymmetrical porphyrin that is the structural precursor of heme and chlorophyll is formed.

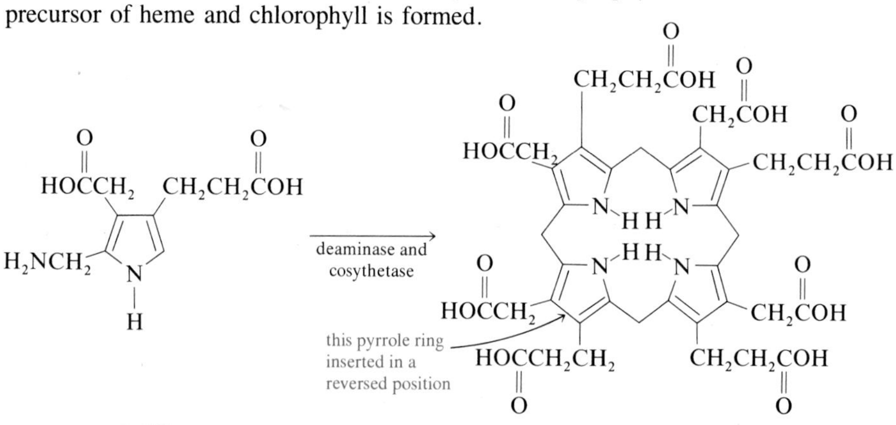

porphobilinogen

uroporphyrinogen III

PROBLEM 18.24   In one of the experiments designed to uncover the mechanism for the formation of uroporphyrinogen III, 5-amino-4-oxopentanoic acid labeled with ^{13}C at carbon 5 was synthesized and converted to porphobilinogen. Predict where the labeling appears in the pyrrole.

18.8

Alkaloids

A. Tropane Alkaloids

Alkaloids, heterocyclic compounds containing nitrogen, are found in nature, most often in the seeds, the leaves, and the bark of plants. Alkaloids are bases, many of which have a bitter taste and profound physiological effects.

A group of alkaloids containing a pyrrolidine ring that is bridged by three carbon atoms between the second and fifth carbon atoms of the heterocyclic ring are known as **tropane alkaloids.** To this family belong cocaine, from the leaves of the coca shrub, and atropine, which is the racemic form of (−)-hyoscyamine, obtained from henbane and the deadly nightshades.

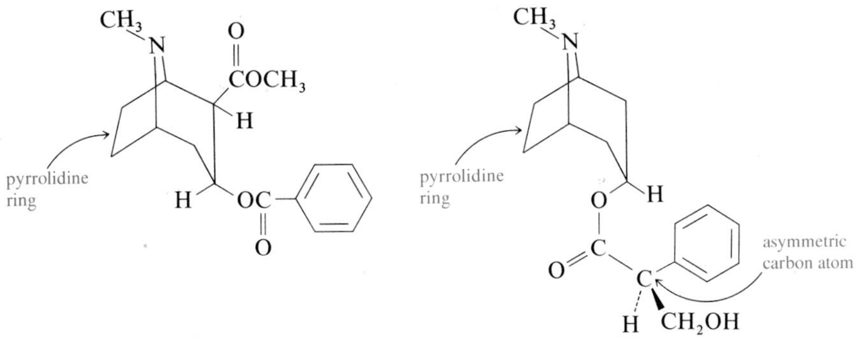

(−)-cocaine

(−)-hyoscyamine
atropine = (±)-hyoscyamine

Cocaine is a stimulant of the central nervous system and a local anesthetic because it blocks the transmission of nerve impulses. The drug is toxic and addictive. For this reason a series of compounds that mimic the action of cocaine as a local anesthetic but lack its more harmful properties has been synthesized. Among them is Novocain, p. 896.

Atropine acts to relax the smooth muscles and thereby ease intestinal and bronchial spasms. Among other medical applications, atropine is used to dilate the pupil of the eye to allow examination of the retina. (−)-Hyoscyamine is the ester of a heterocyclic amino alcohol known as tropine, and of an acid, (S)-(−)-tropic acid, to which it can be hydrolyzed in cold water.

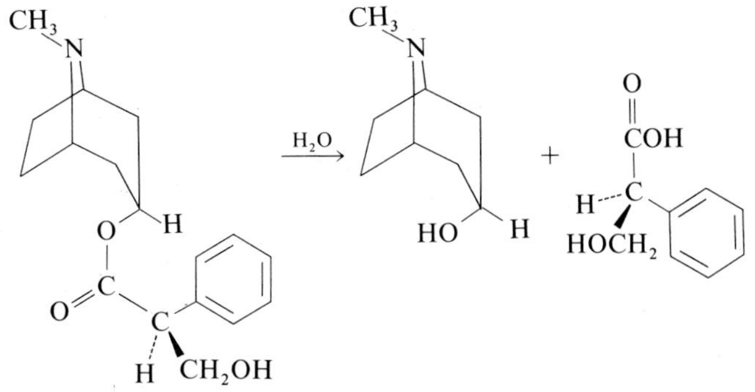

(−)-hyoscyamine   tropine   (S)-(−)-tropic acid

Tropic acid is easily racemized, so (−)-hyoscyamine is converted into optically inactive atropine by warming with base, and is hydrolyzed by basic solutions into tropine and racemic tropic acid.

PROBLEM 18.25   Write an equation outlining a mechanism for converting (−)-hyoscyamine to atropine by aqueous base.

PROBLEM 18.26   Tropine is oxidized by chromic acid to a ketone, tropinone. When tropinone is reduced by reagents such as sodium amalgam in alcohol, tropine is not formed. Instead another alcohol, also $C_8H_{15}NO$, called ψ-tropine, is obtained. ψ-Tropine can be oxidized back to tropinone. Write equations showing what is happening.

Cocaine (p. 1055) has the same bicyclic ring structure as atropine, except that the pattern of substitution is different. Hydrolysis of cocaine gives (−)-ecgonine, benzoic acid, and methanol.

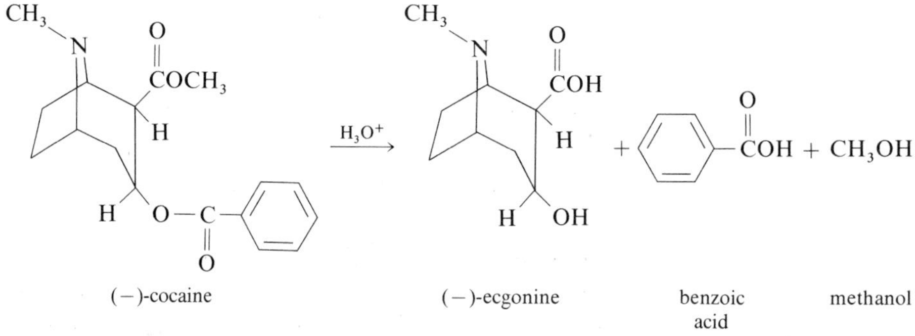

The relationship between cocaine and atropine becomes clear when (−)-ecgonine is oxidized with chromic acid. Tropinone (Problem 18.26) is one of the products obtained along with other compounds from oxidative cleavage of the ring.

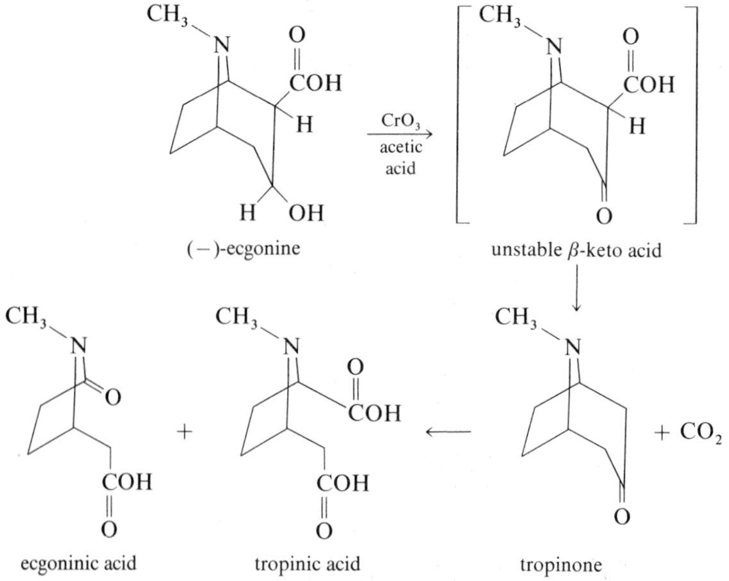

PROBLEM 18.27   A base isomeric with ecgonine has been synthesized from tropinone by the addition of hydrocyanic acid and the hydrolysis of the resulting compound. Propose a structure for this base, known as α-ecgonine, by writing equations for the reactions described. Is there any ambiguity about the structure you have obtained?

B. Indole Alkaloids

A large and important class of alkaloids is structurally related to the amino acid tryptophan and contains the aromatic indole ring system (p. 1010). Among the indole alkaloids, one appears to be of central importance in physiology. This compound, 5-hydroxytryptamine, also known as serotonin, is widely distributed in nature and stimulates a variety of smooth muscles and nerves. It has an essential function in the central nervous system as a neurotransmitter. Several drugs that interfere with the metabolism of serotonin in the brain because of their structural similarity to the compound are known to induce mental changes, including symptoms that resemble those seen in schizophrenia. Structural formulas for serotonin and for three indole alkaloids that cause hallucinations are given below.

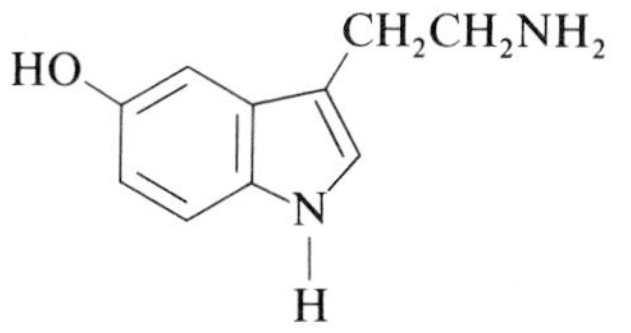

5-hydroxytryptamine
serotonin

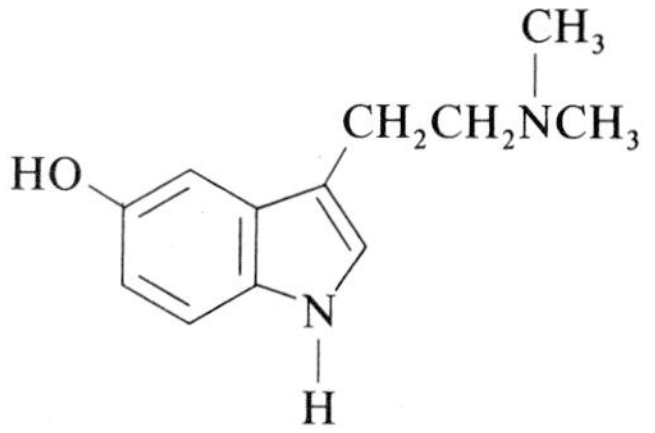

N,N-dimethyl-5-hydroxytryptamine
bufotenin

*a psychoactive drug from
the cahobe bean*

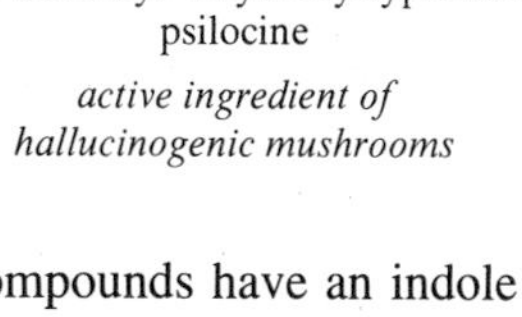

N,N-dimethyl-4-hydroxytryptamine
psilocine

*active ingredient of
hallucinogenic mushrooms*

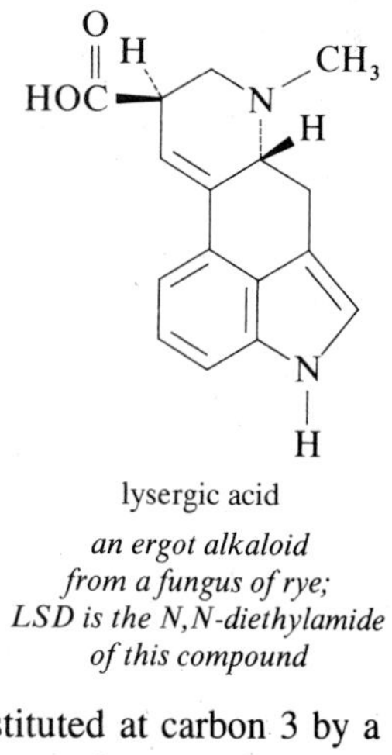

lysergic acid

*an ergot alkaloid
from a fungus of rye;
LSD is the N,N-diethylamide
of this compound*

All of the compounds have an indole ring substituted at carbon 3 by a two-carbon chain ending in an amino group. Serotonin contains a primary amine, while bufotenin, psilocine, and lysergic acid are tertiary amines. In lysergic acid, the side chain is incorporated into two other rings. Serotonin, bufotenin, and psilocine also have phenolic hydroxyl groups on the indole ring. Serotonin is synthesized in mammals from the amino acid tryptophan, first by hydroxylation of the aromatic ring (Section 12.7) and then by decarboxylation.

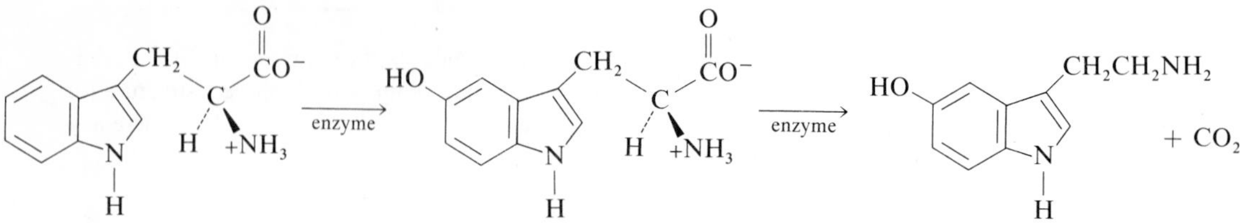

tryptophan → 5-hydroxytryptophan → serotonin + CO_2

PROBLEM 18.28 A synthesis of racemic lysergic acid (p. 1057) is outlined below. Fill in the structural formulas for the compounds designated by letters.

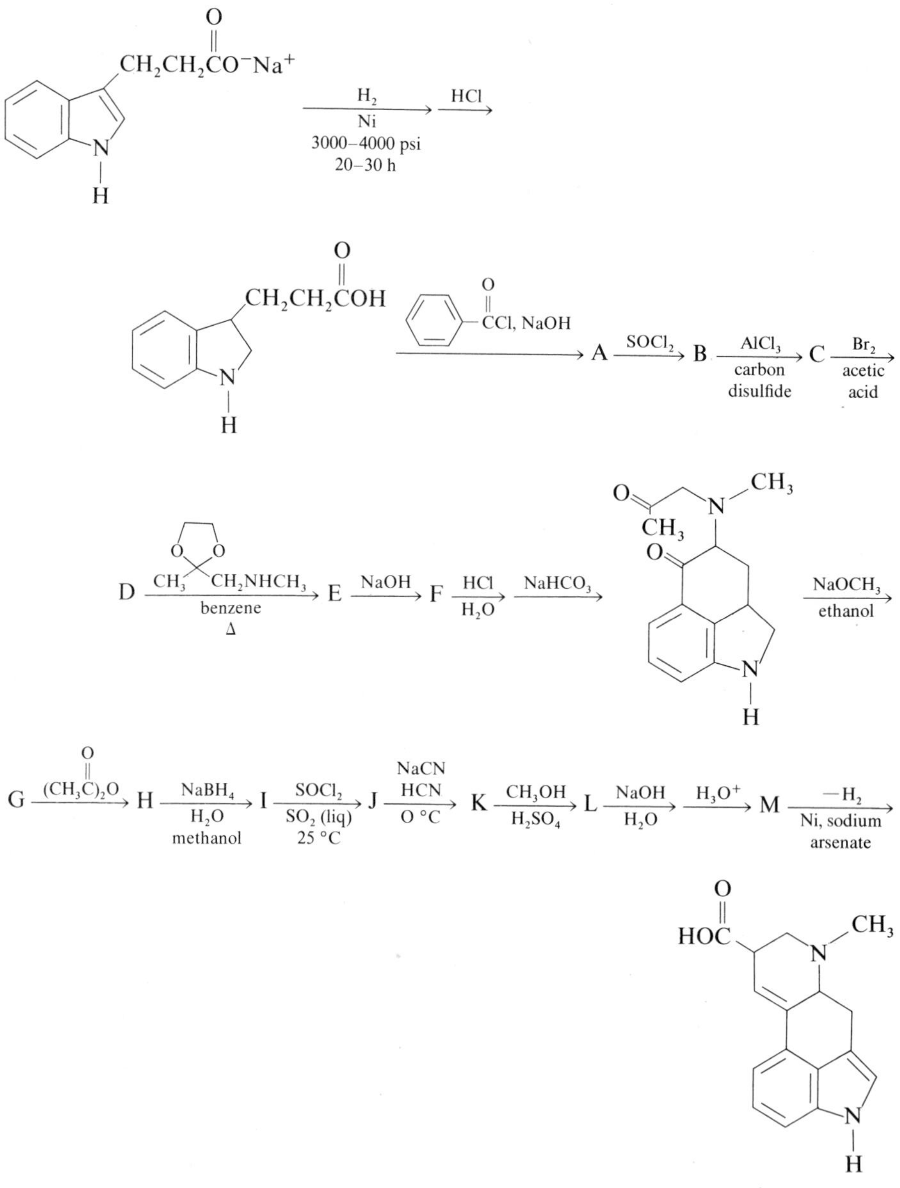

$(\pm)$-lysergic acid

C. Isoquinoline Alkaloids

The most effective painkiller known to medicine is the alkaloid morphine, isolated from the juice obtained from unripe seed pods of the opium poppy, *Papaver somniferum.* Apparently morphine changes the perception of pain even when the pain itself is not much diminished. For this reason, the drug is valuable in medical practice. Unfortunately, morphine also has two severe disadvantages as a medication. It is addictive and the body builds up a tolerance to it so that larger and larger doses may be necessary to provide the same relief from pain. The drug also depresses the function of the brain center that controls respiration, so that large doses of morphine (or of heroin, its synthetic diacetyl derivative) can kill by causing respiratory arrest. Another opium alkaloid that occurs in nature is codeine, a monomethyl ether of morphine. Codeine is also a painkiller and is especially useful as a cough suppressant.

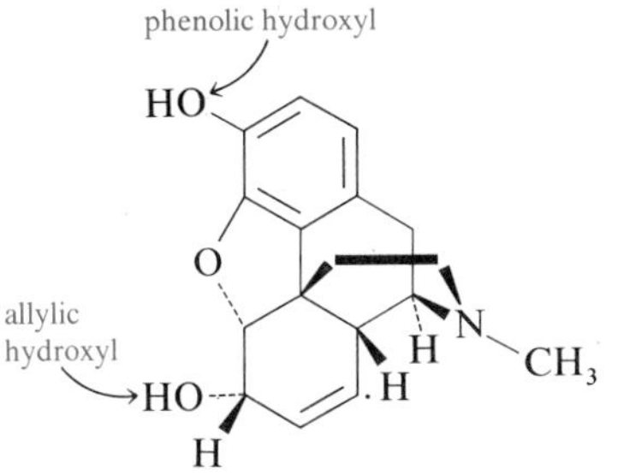

morphine 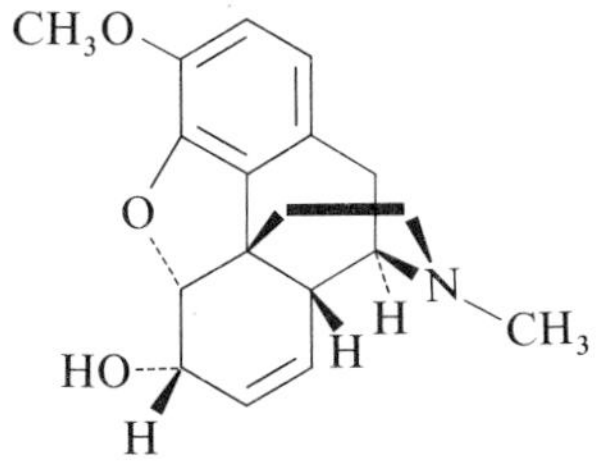   codeine

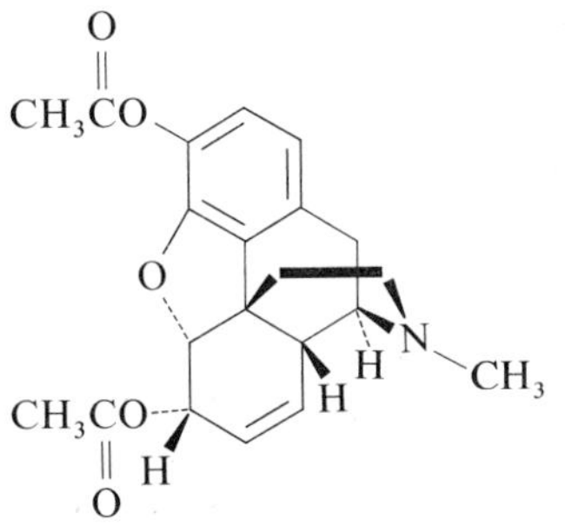

heroin

*the opium alkaloids morphine and codeine, and the
synthetic derivative heroin*

All of these alkaloids have as part of their structures the benzylisoquinoline unit, seen more easily in the structural formula of papaverine, another opium alkaloid.

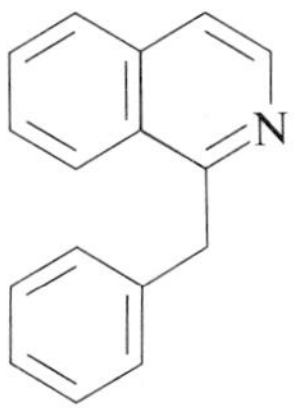

1-benzylisoquinone 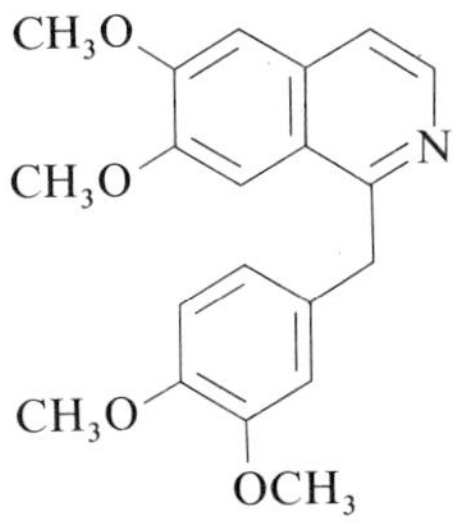   papaverine

These alkaloids, as well as some simpler ones, are synthesized in plants from the amino acid tyrosine. Tyrosine is converted first by oxidation of the aromatic ring

(Section 12.7) into (3,4-dihydroxyphenyl)alanine (DOPA), a compound that is valuable in treating Parkinson's disease. (3,4-Dihydroxyphenyl)alanine is converted to β-(3,4-dihydroxyphenyl)ethylamine by decarboxylation and to (3,4-dihydroxyphenyl)pyruvic acid by transamination (Section 17.2C).

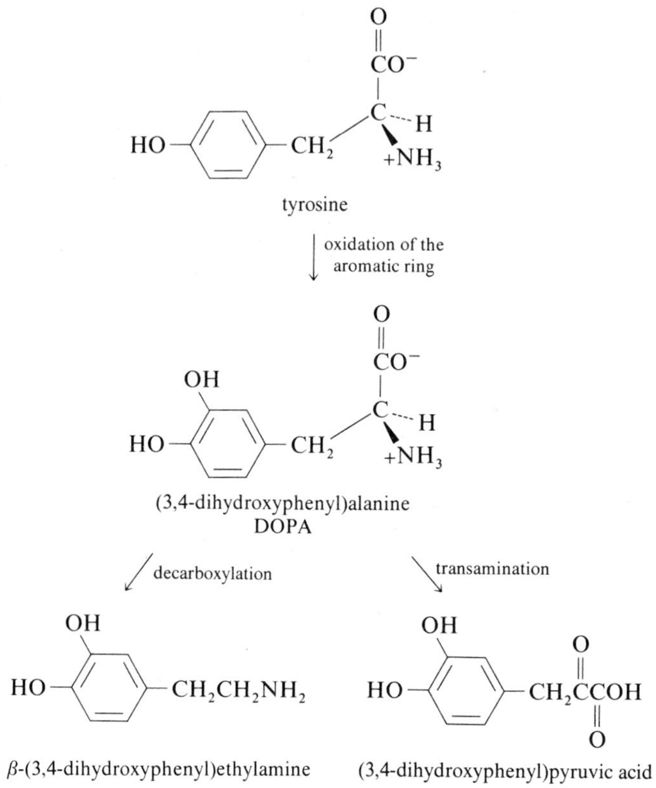

tyrosine

oxidation of the aromatic ring

(3,4-dihydroxyphenyl)alanine
DOPA

decarboxylation

transamination

β-(3,4-dihydroxyphenyl)ethylamine

(3,4-dihydroxyphenyl)pyruvic acid

The benzylisoquinoline skeleton of the opium alkaloids is created by condensation of β-(3,4-dihydroxyphenyl)ethylamine with (3,4-dihydroxyphenyl)pyruvic acid followed by an electrophilic aromatic substitution reaction that forms the heterocyclic ring. The aromatic ring, with the two hydroxyl groups on it, is highly activated so a strong electrophile is not necessary for the reaction.

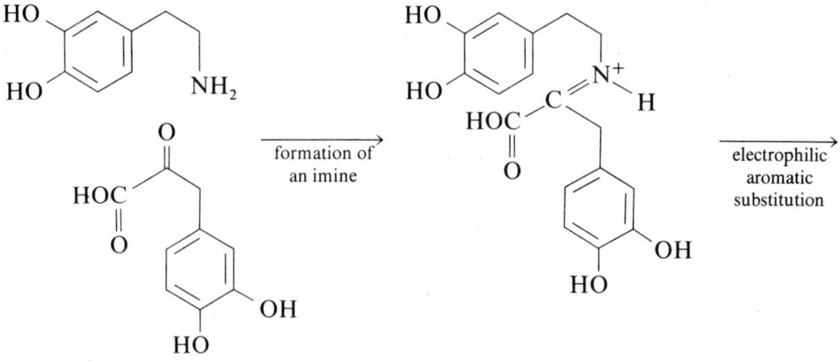

formation of an imine

electrophilic aromatic substitution

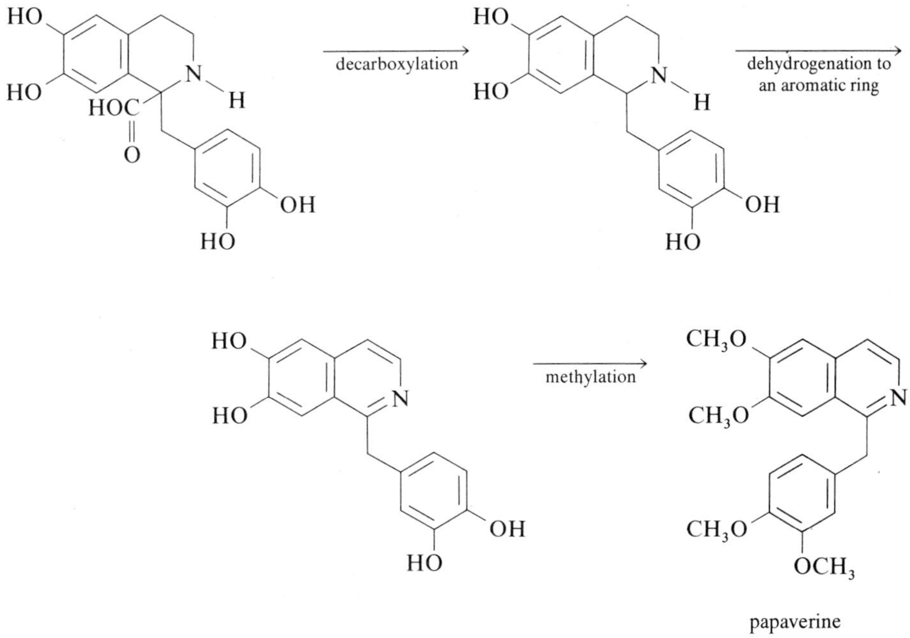

papaverine

The intermediate formed by closure of the heterocyclic ring loses carbon dioxide and then hydrogen to form the stable aromatic isoquinoline. Methylation of the hydroxyl groups gives papaverine.

PROBLEM 18.29   Write equations suggesting mechanisms for the first three reactions in the reaction scheme above. Use HB^+ and $B:$ as the general acids and bases necessary in the reactions.

A key step in the biosynthesis of morphine is the oxidative coupling of the two phenolic rings (Section 12.6C) in a benzylisoquinoline. Coupling occurs between the positions ortho to one hydroxyl group and para to the other one.

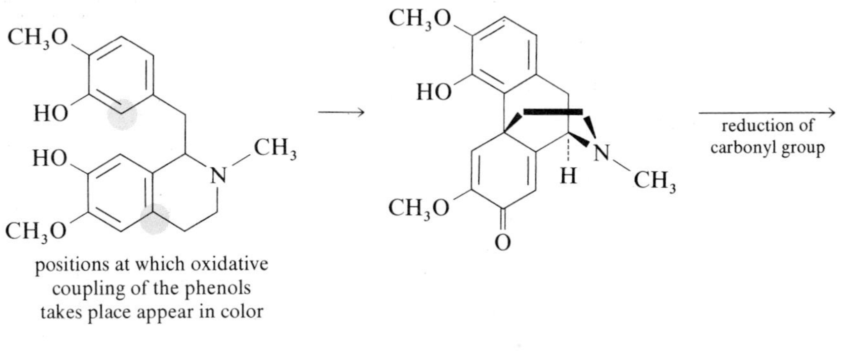

positions at which oxidative
coupling of the phenols
takes place appear in color

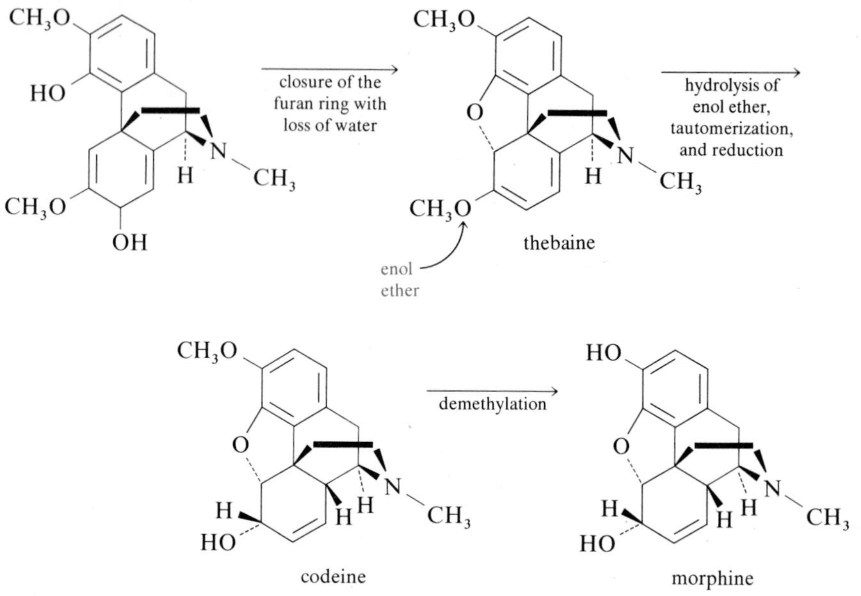

Thebaine, an intermediate in the synthesis of codeine and morphine, is also an opium alkaloid.

PROBLEM 18.30   Using a 2-methoxy-5-methylphenol as a model for each of the two phenolic rings that undergo oxidative coupling in the precursor to morphine, write a mechanism for the coupling reaction. A review of Section 12.6C may be helpful.

PROBLEM 18.31   Suggest a mechanism for the formation of the furan ring in thebaine above and for the hydrolysis of the enol ether in thebaine and the tautomerization of the product to give the α,β-unsaturated ketone precursor to codeine.

PROBLEM 18.32   Tubocurarine, the potent curare alkaloid poison used by South American Indians on the tips of their hunting arrows, is a bis(benzylisoquinoline) alkaloid. Dissect its structural formula shown on p. 937 and outline the benzylisoquinoline units present.

PROBLEM 18.33   Complete the following equations, supplying structural formulas for any intermediates and products indicated by letters.

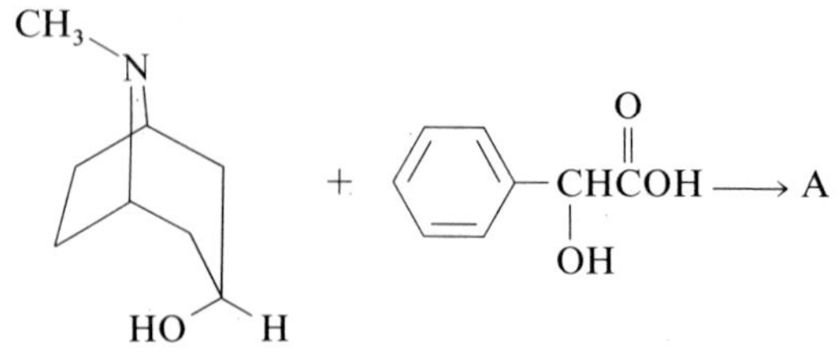

(b)

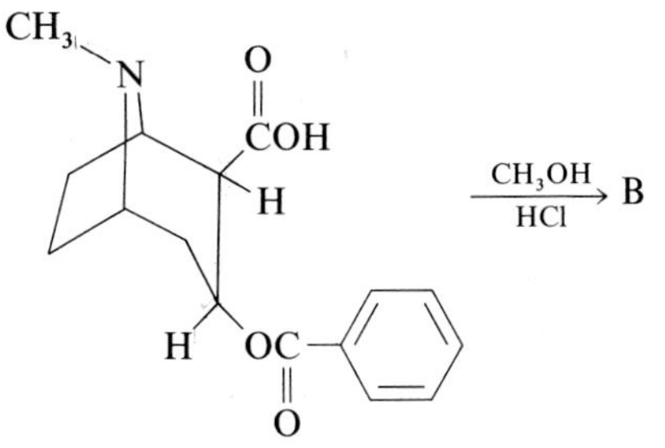

$$\xrightarrow[\text{HCl}]{\text{CH}_3\text{OH}} \text{B}$$

(c)

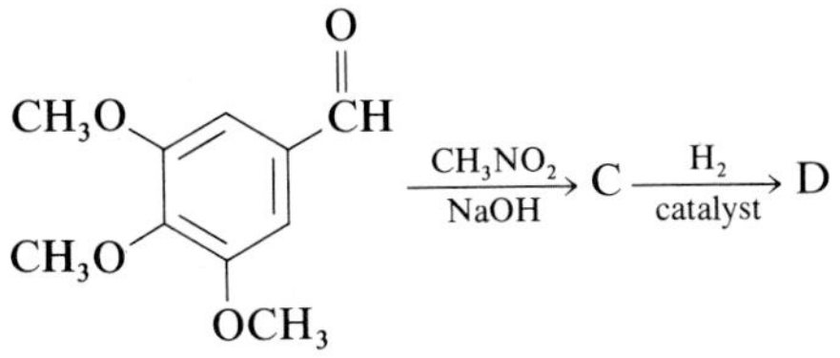

$$\xrightarrow[\text{NaOH}]{\text{CH}_3\text{NO}_2} \text{C} \xrightarrow[\text{catalyst}]{\text{H}_2} \text{D}$$

(d)

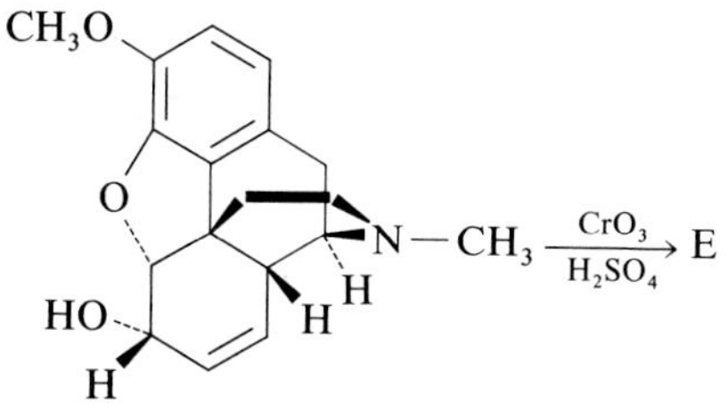

codeine

$$\xrightarrow[\text{H}_2\text{SO}_4]{\text{CrO}_3} \text{E}$$

(e)

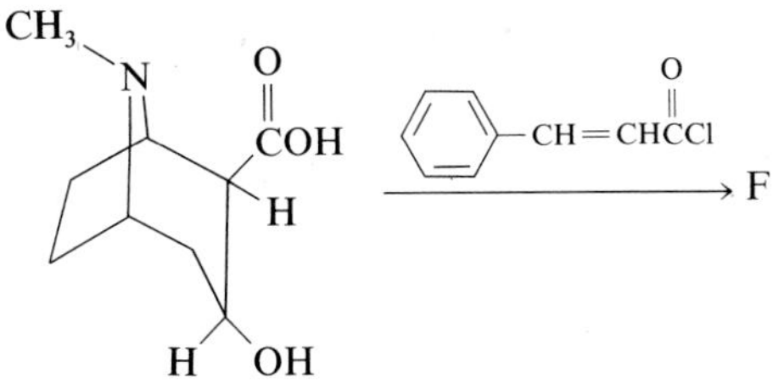

$$\xrightarrow{\quad\quad} \text{F}$$

(f)

$$\xrightarrow[\substack{\text{tetrahydrofuran} \\ -78\,°\text{C}}]{(\text{CH}_3\text{CH})_2\text{N}^- \text{Li}^+ \ (2 \text{ molar equiv})} \text{G}$$

$$\text{G} \xrightarrow[-78\,°\text{C}]{\text{BrCH}_2\overset{\text{O}}{\overset{\|}{\text{C}}}\text{OCH}_3 \ (1 \text{ molar equiv})} \xrightarrow{\text{H}_3\text{O}^+} \text{H}$$

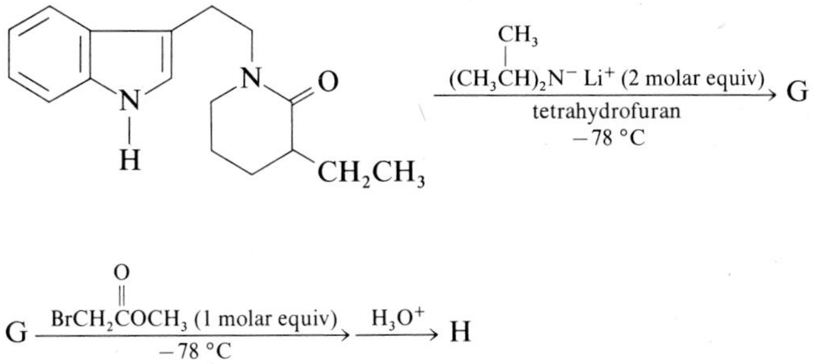

ADDITIONAL PROBLEMS

18.34 Name the following compounds.

(a)

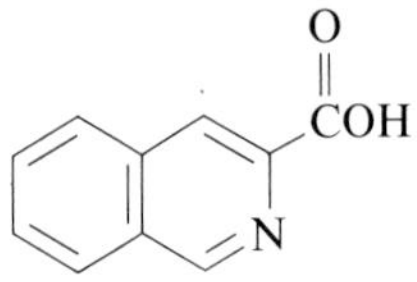

(b)

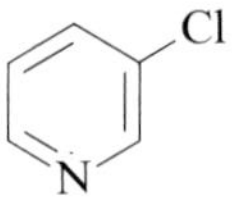

(c)

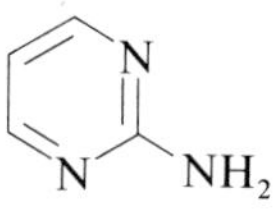

(d)

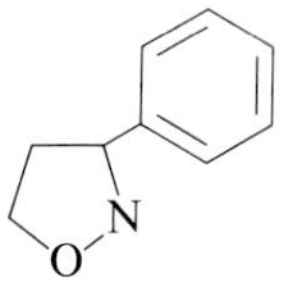

(e)

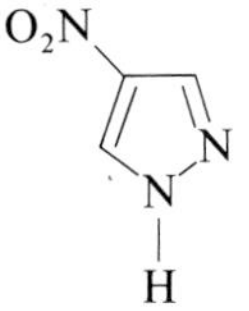

(f)

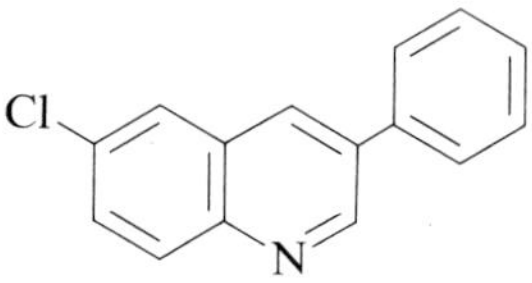

(g)

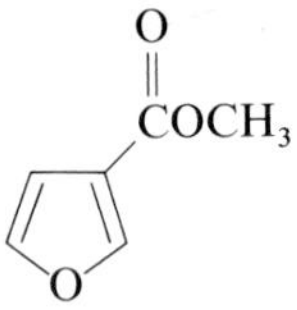

(h)

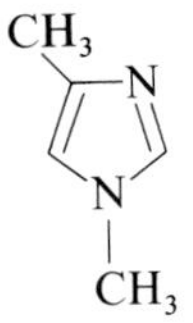

(i)

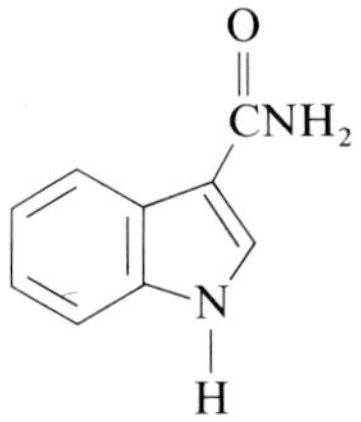

(j)

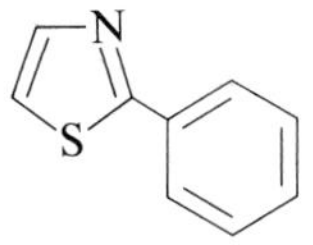

(k)

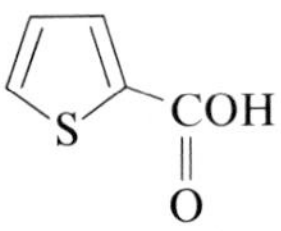

18.35 Complete the following equations giving structural formulas for all the intermediates and products designated by letters.

(a) 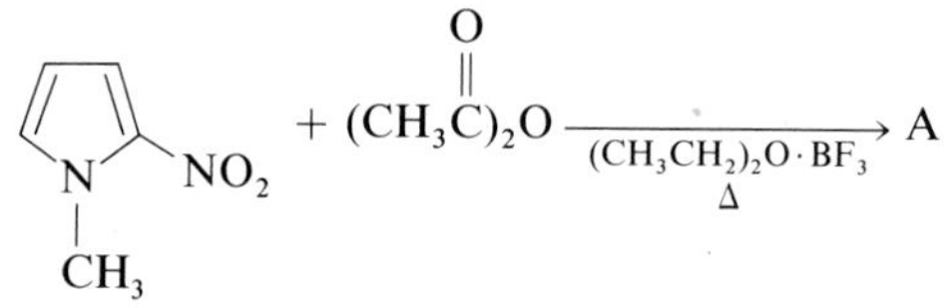 $+ (CH_3\overset{O}{\overset{\|}{C}})_2O \xrightarrow[\Delta]{(CH_3CH_2)_2O \cdot BF_3}$ A

(b) $+ H_2SO_4 \longrightarrow$ B

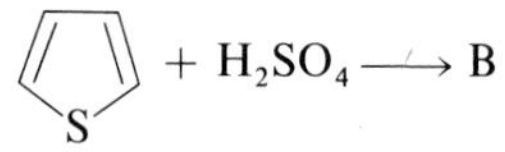

(c) 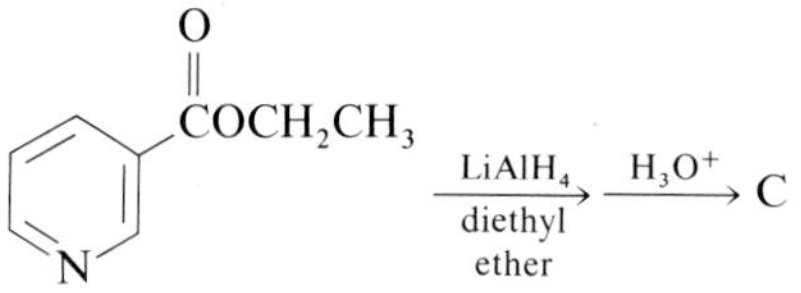 $\xrightarrow[\substack{\text{diethyl} \\ \text{ether}}]{LiAlH_4} \xrightarrow{H_3O^+}$ C

(d) $+ H_2 \xrightarrow{Ni}$ D

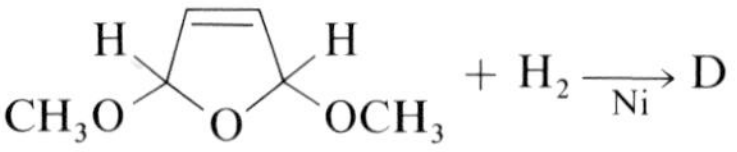

(e) 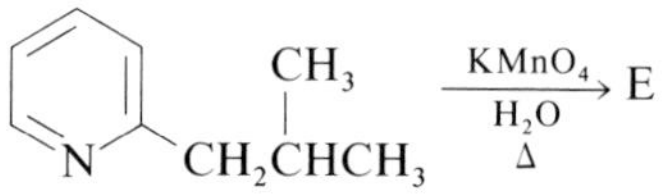 $\xrightarrow[\Delta]{KMnO_4 \atop H_2O}$ E

(f) $+ (CH_3CH_2CH_2\overset{O}{\overset{\|}{C}})_2O \xrightarrow[\substack{\text{acetic} \\ \text{acid}}]{BF_3}$ F

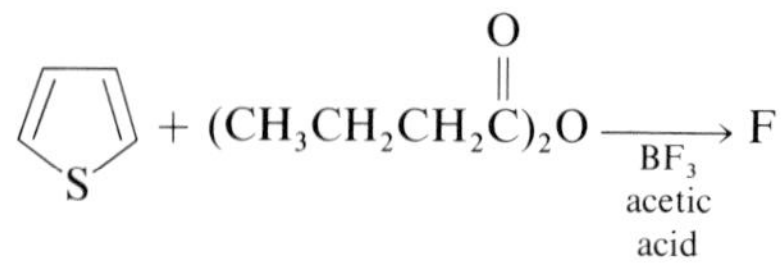

(g) 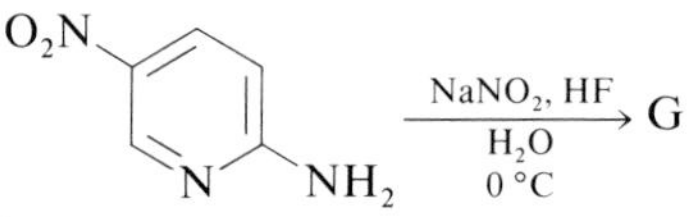 $\xrightarrow[\substack{H_2O \\ 0\,°C}]{NaNO_2, \; HF}$ G

(h) $+$ $-NH_2 \xrightarrow{\Delta}$ H

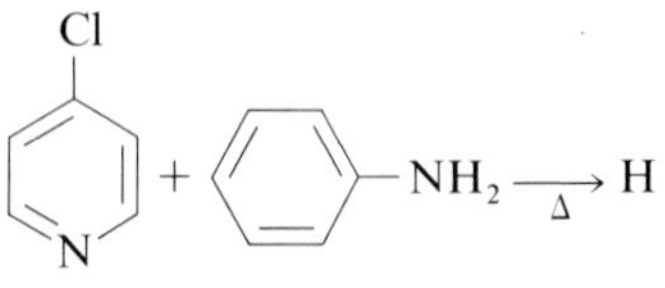

(i) 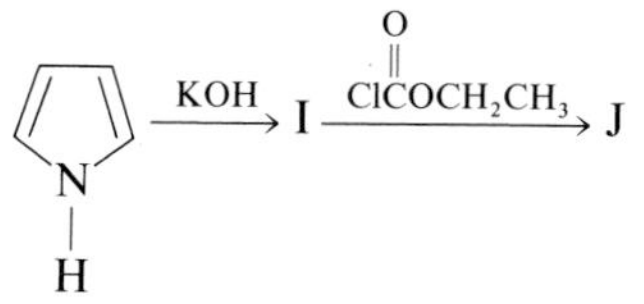 [pyrrole] $\xrightarrow{\text{KOH}}$ I $\xrightarrow{\text{ClCOCH}_2\text{CH}_3}$ J

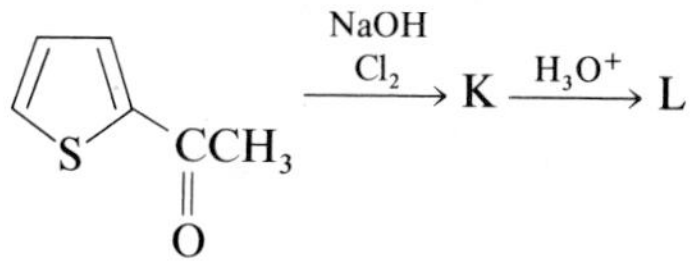

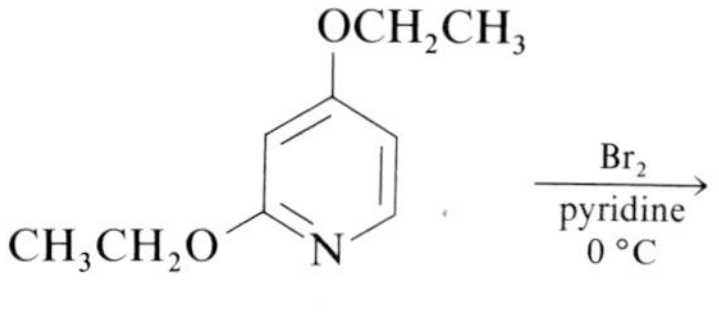

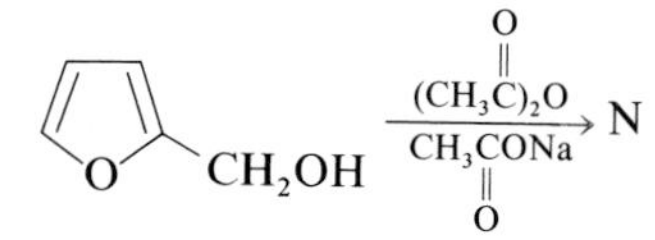

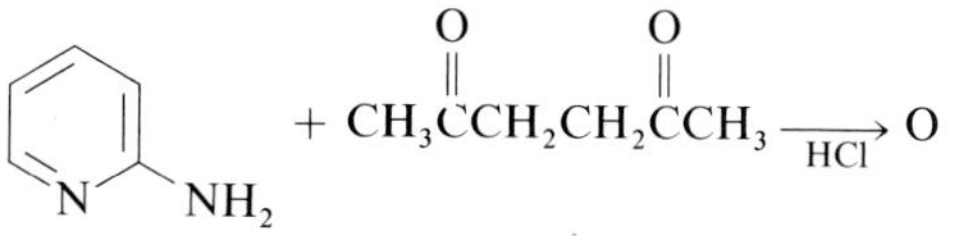

(j) [2-acetylthiophene] $\xrightarrow[\text{Cl}_2]{\text{NaOH}}$ K $\xrightarrow{\text{H}_3\text{O}^+}$ L

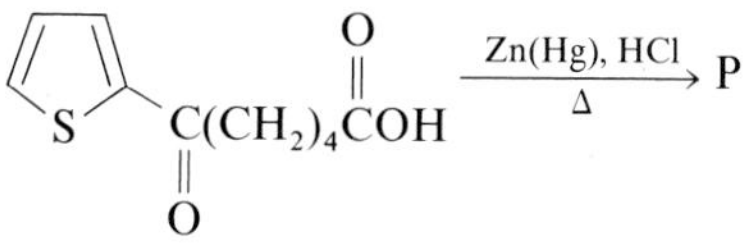

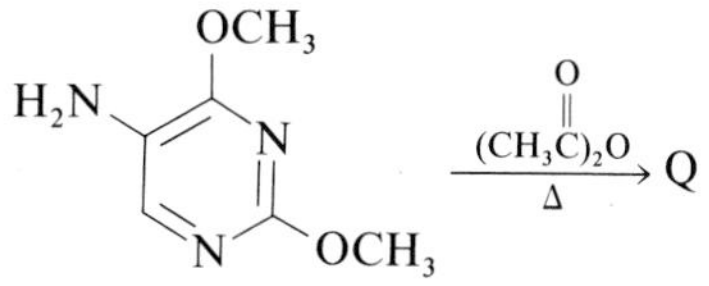

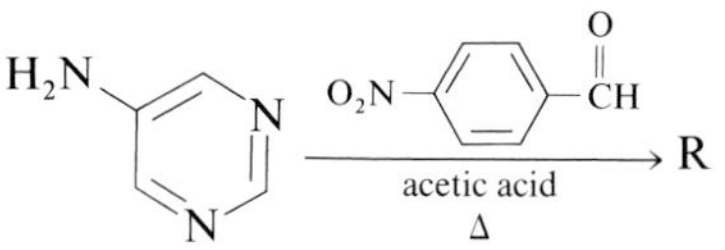

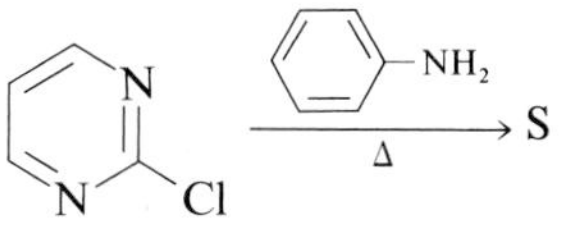

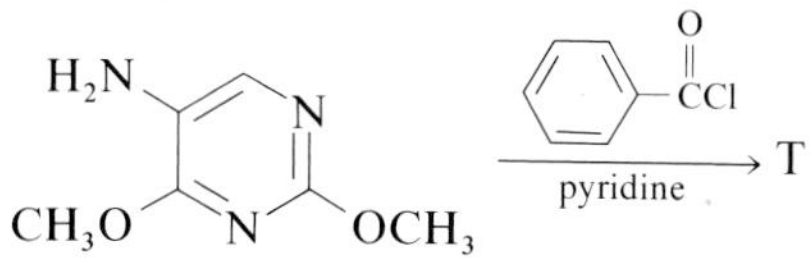

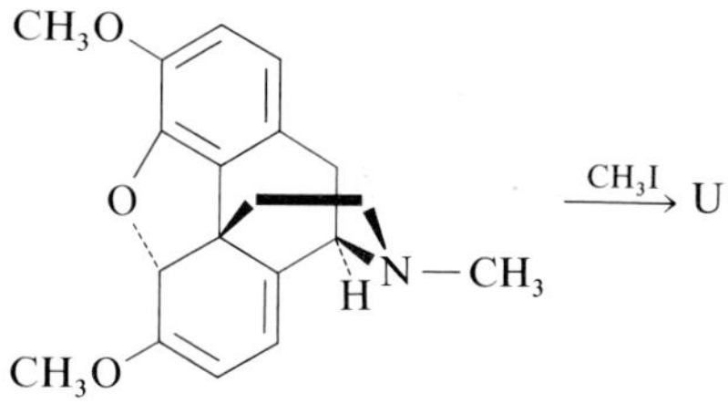

(k) [pyridine with OCH$_2$CH$_3$ and CH$_3$CH$_2$O] $\xrightarrow[\substack{\text{pyridine} \\ 0\,°\text{C}}]{\text{Br}_2}$ M

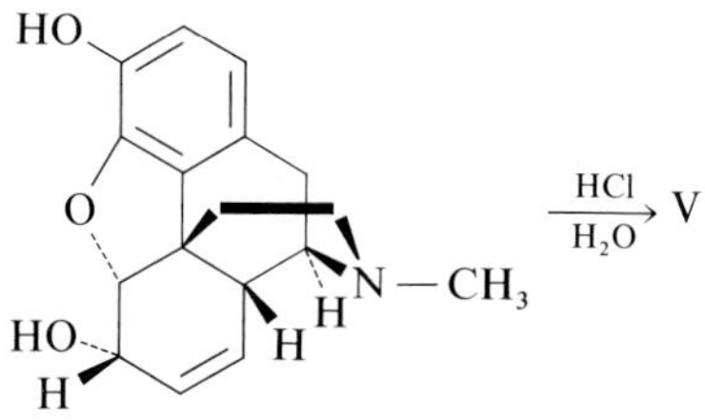

(l) [furfuryl alcohol, furan-CH$_2$OH] $\xrightarrow[\text{CH}_3\text{CONa}]{(\text{CH}_3\text{C})_2\text{O}}$ N

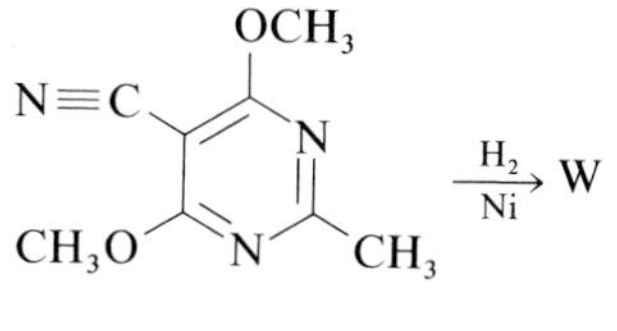

(m) [2-aminopyridine] $+ \text{CH}_3\text{CCH}_2\text{CH}_2\text{CCH}_3$ $\xrightarrow{\text{HCl}}$ O

(n) [thiophene-C(CH$_2$)$_4$COH] $\xrightarrow[\Delta]{\text{Zn(Hg), HCl}}$ P

(o) [pyrimidine with H$_2$N, OCH$_3$, OCH$_3$] $\xrightarrow[\Delta]{(\text{CH}_3\text{C})_2\text{O}}$ Q

(p) [5-aminopyrimidine] $\xrightarrow[\substack{\text{acetic acid} \\ \Delta}]{\text{O}_2\text{N}-\text{C}_6\text{H}_4-\text{CHO}}$ R

(q) [2-chloropyrimidine] $\xrightarrow[\Delta]{\text{C}_6\text{H}_5-\text{NH}_2}$ S

(r) [pyrimidine with H$_2$N, CH$_3$O, OCH$_3$] $\xrightarrow[\text{pyridine}]{\text{C}_6\text{H}_5-\text{CCl}}$ T

(s) thebaine $\xrightarrow{\text{CH}_3\text{I}}$ U

(t) morphine $\xrightarrow[\text{H}_2\text{O}]{\text{HCl}}$ V

(u) [pyrimidine with N≡C, OCH$_3$, CH$_3$O, CH$_3$] $\xrightarrow[\text{Ni}]{\text{H}_2}$ W

(v) [2-amino-5-chloropyrimidine] $\xrightarrow[\substack{\text{H}_2\text{O} \\ 0\,°\text{C}}]{\text{NaNO}_2,\ \text{HCl}}$ X

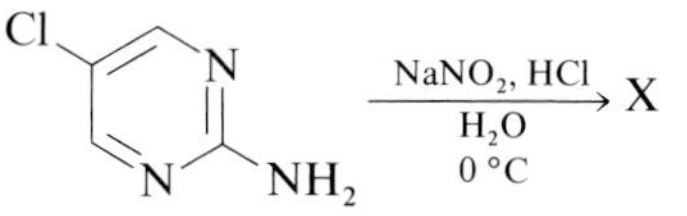

18.36 Complete the following equations giving structural formulas for all the intermediates and products designated by letters.

(a) $\xrightarrow[\text{diethyl ether}]{\text{LiAlH}_4}$ $\xrightarrow{\text{H}_2\text{O}}$ A

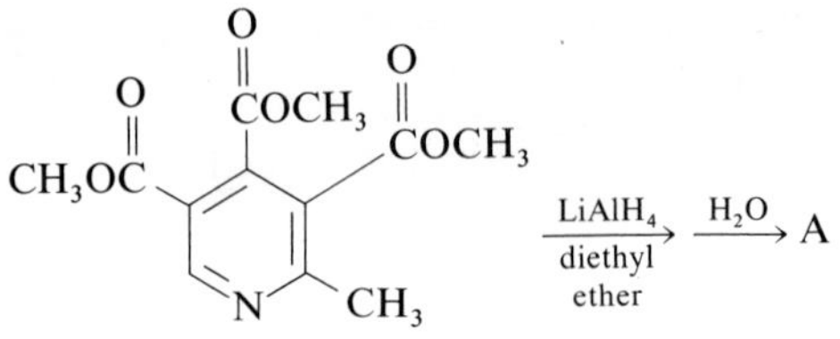

(b) $\xrightarrow[\text{methanol}]{\text{—CH}_2\text{NH}_2}$ B $\xrightarrow[\text{ethanol}]{\text{H}_2\ \text{PtO}_2}$ C

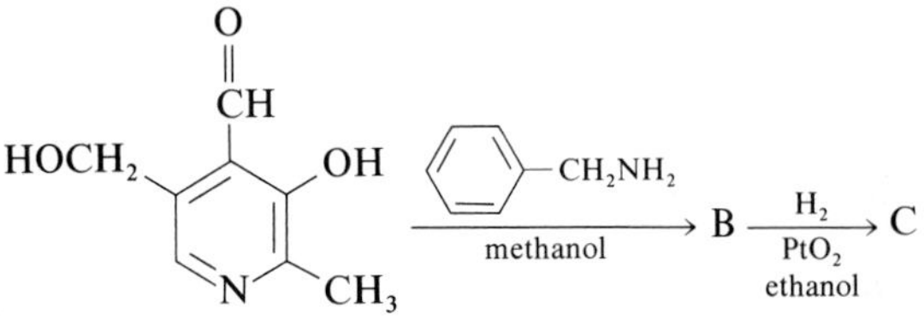

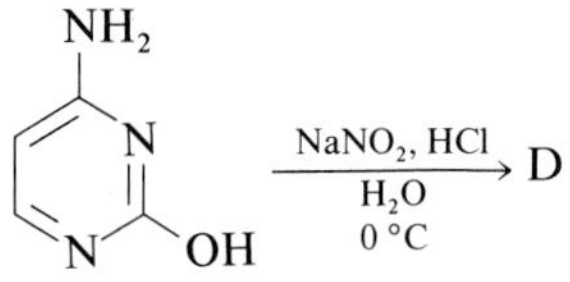

(c) $\xrightarrow[\substack{\text{H}_2\text{O} \\ 0\,°\text{C}}]{\text{NaNO}_2,\ \text{HCl}}$ D

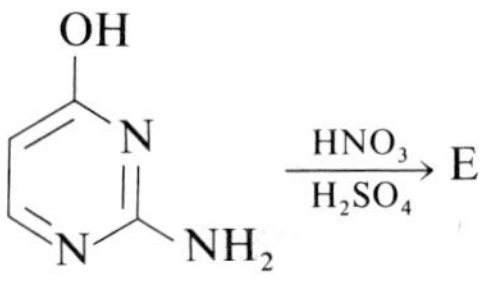

(d) $\xrightarrow[\text{H}_2\text{SO}_4]{\text{HNO}_3}$ E

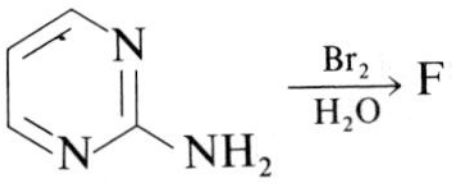

(e) $\xrightarrow[\text{H}_2\text{O}]{\text{Br}_2}$ F

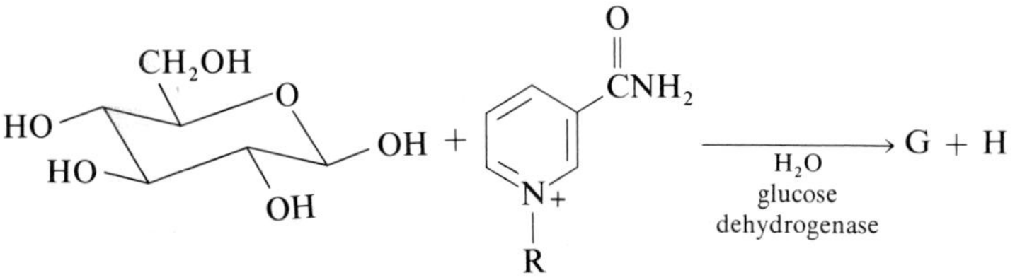

(f) $\xrightarrow[\substack{\text{H}_2\text{O} \\ \text{glucose} \\ \text{dehydrogenase}}]{}$ G + H

(g) $\xrightarrow[\substack{\text{HCl} \\ \Delta}]{}$ I

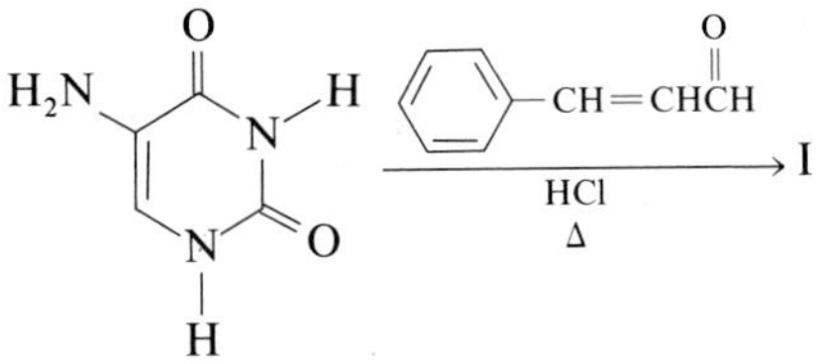

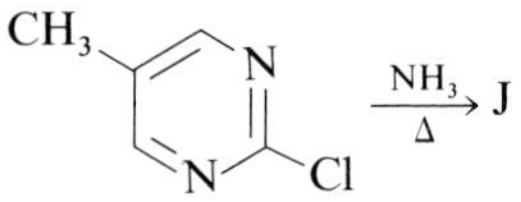

(h) $\xrightarrow[\Delta]{\text{NH}_3}$ J

(i) $\xrightarrow[\substack{\text{ethanol} \\ \text{H}_3\text{O}^+}]{\text{H}_2\ \text{PtO}_2}$ K

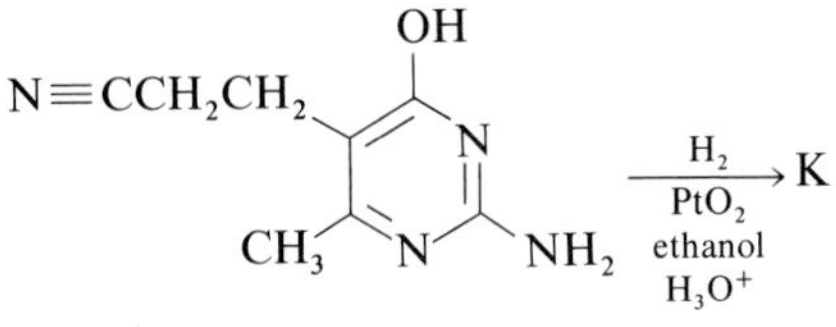

(j) $\xrightarrow[\Delta]{(\text{CH}_3\text{C})_2\text{O}}$ L

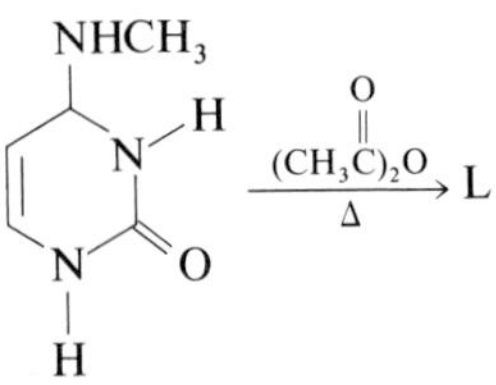

(k) $\xrightarrow{\text{Cl}_2 \atop 200\,°\text{C}}$ M

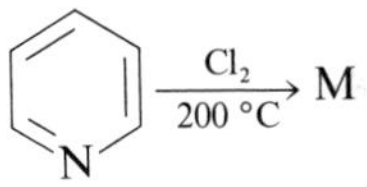

(l) $\xrightarrow[\substack{\text{toluene} \\ \Delta}]{\text{NaNH}_2}$ N

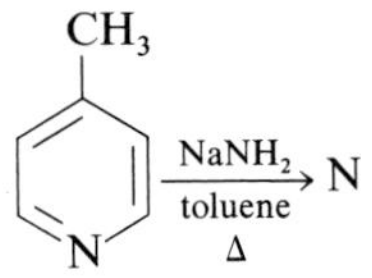

(m)

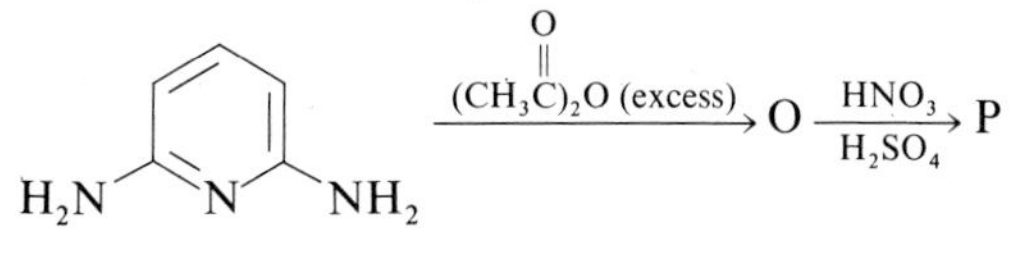

(n)

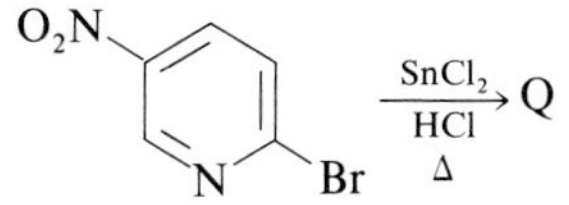

(o)

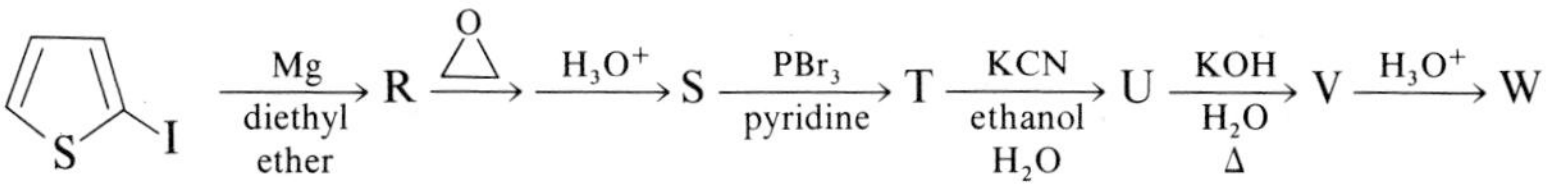

(p)

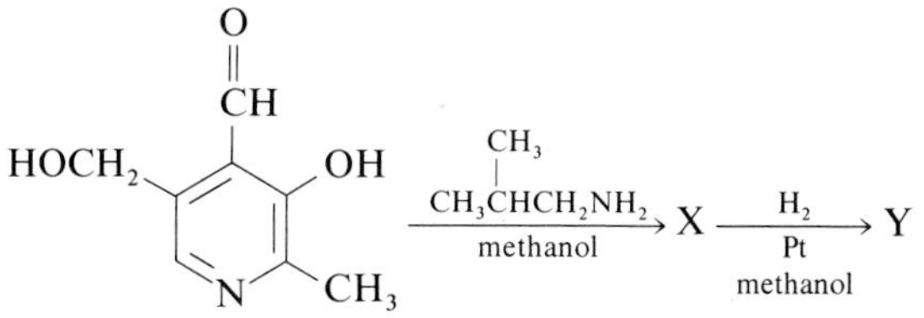

(q)

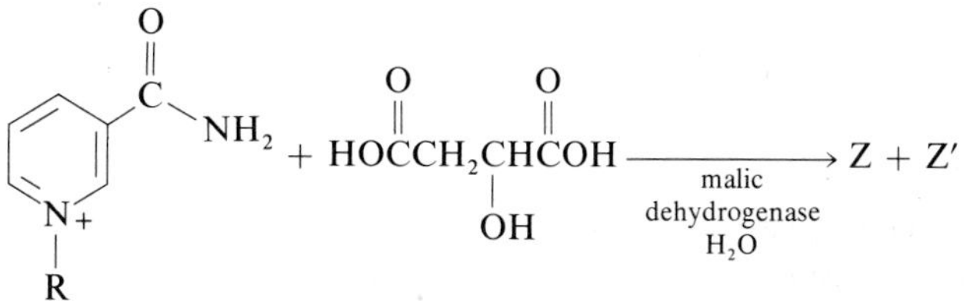

malic acid

18.37 The following reactions are observed. Write detailed mechanisms showing how the transformations could have taken place.

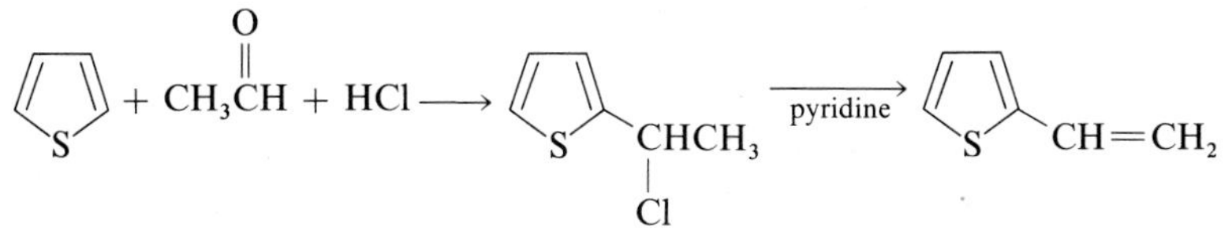

18.38 The two enantiomeric ethanol-1-*d* species obtained in the reactions shown on p. 1045 were collected at first in quantities too small to allow measurement of their optical rotations. The fact that they were enantiomers was proved by the following sequence of reactions.

$$CH_3CD + NADH \longrightarrow \text{alcohol A} \xrightarrow[\text{pyridine}]{\text{TsCl}} B \xrightarrow[\substack{H_2O \\ \Delta}]{\text{NaOH}} C$$

$$C + NAD^+ \longrightarrow \underset{\substack{\text{containing} \\ \text{no deuterium}}}{\text{acetaldehyde}} + \underset{\substack{\text{containing 1} \\ \text{molar equivalent} \\ \text{of deuterium}}}{\text{NADD}}$$

$$A + NAD^+ \longrightarrow \text{acetaldehyde} + NADH \quad \text{containing}$$

A + NAD⁺ → acetaldehyde + NADH

containing
deuterium

containing

no deuterium

Write equations showing what is happening in this series of experiments. Be sure to use correct stereochemical representations of all the compounds involved. How do these experiments prove that the alcohols obtained in the original experiments are enantiomeric?

18.39 Arrange the following species in order of increasing acidity and give your reasons for making these decisions.

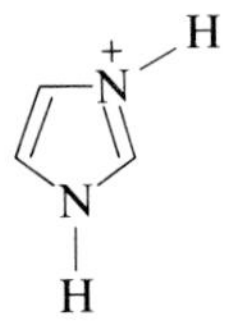

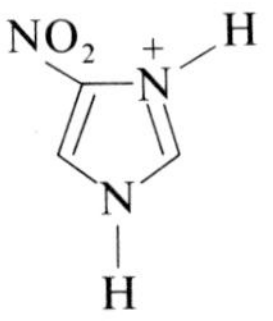

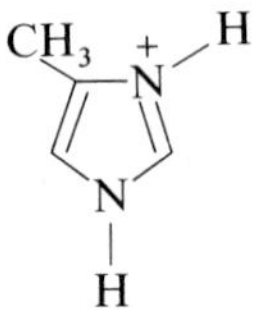

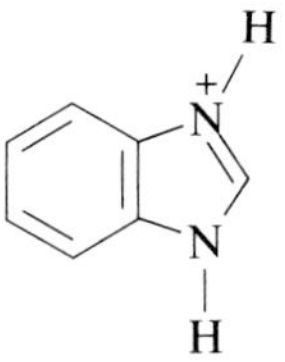

18.40 Propose syntheses for the following compounds. You must start with thiophene, furan, pyrrole, or a methylpyridine as one of the reagents for each synthesis. The other organic reagents you use may not contain more than seven carbon atoms.

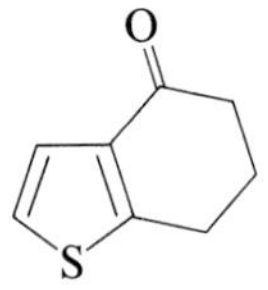

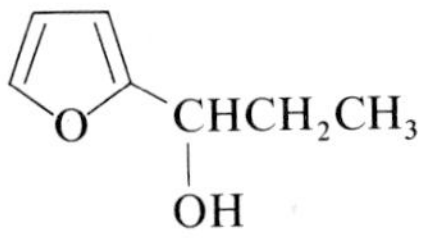

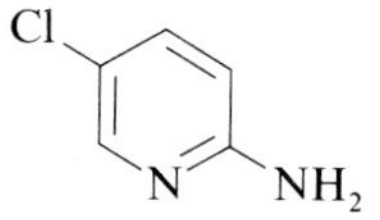

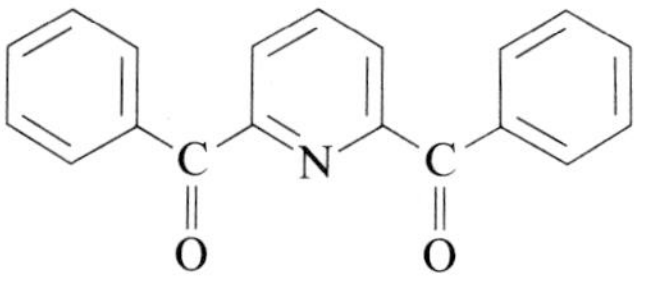

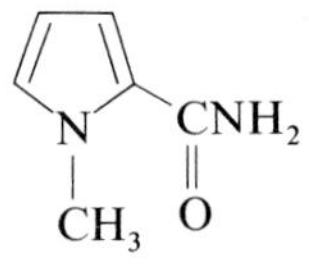

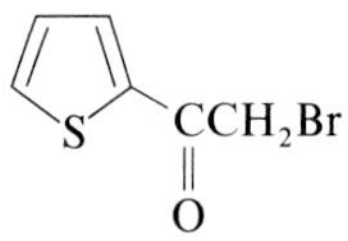

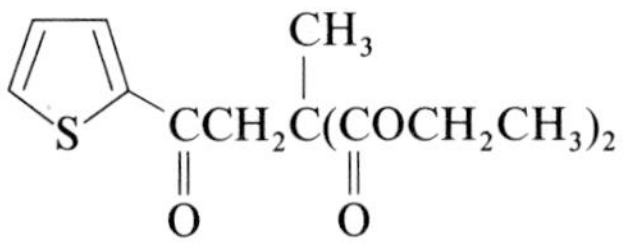

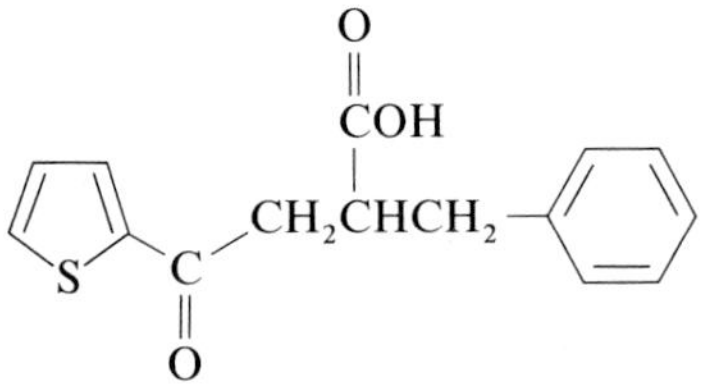

18.41 When phenylhydrazine is heated with 2-butenoic acid, water is evolved and a cyclic compound having the molecular formula $C_{10}H_{12}N_2O$ is formed. Propose a structure for the product of this reaction. (Hint: A review of Section 15.7A may be helpful.)

Concerted Reactions

19

19.1

Introduction

A. Some Examples of Concerted Reactions

Organic reactions can be classified into two types. Some proceed through the formation of an intermediate reactive species, such as a carbocation, a carbanion, or a radical. The addition of hydrogen bromide to propene is one of many examples of such a reaction (Section 5.3). Nucleophilic substitution of an alkyl halide by an S_N1 mechanism is another example of a reaction that proceeds by way of a reactive intermediate, a carbocation (Sections 6.3 and 6.6).

Other reactions go from starting materials directly to products by way of a single transition state, involving no reactive intermediate. Bimolecular nucleophilic substitution reactions (Section 6.5) and Diels-Alder reactions (Section 11.5) occur in this way. The energy diagram shown in Figure 6.1 (Section 6.5B) is typical of such reactions.

Both the S_N2 reaction and the Diels-Alder reaction are highly stereoselective. Bimolecular nucleophilic substitution at a tetrahedral carbon atom takes place with inversion of configuration at the carbon atom in all cases in which stereochemistry has been observed (Section 6.5D). The Diels-Alder reaction also has definite stereochemical results. The stereochemistry of the dienophile is retained in the product, as is the relative stereochemistry of the groups at the ends of the diene system (Section 11.5).

Reactions that take place in one step with high stereoselectivity without the formation of a reactive intermediate are known as **concerted reactions.** Thus, bimolecular nucleophilic substitution reactions and additions of dienes to dienophiles, the Diels-Alder reaction, are examples of concerted reactions. In this chapter, we explore a series of chemical transformations that have been classified as concerted reactions. In all of these reactions, rearrangements of the positions of σ and π bonds take place by way of cyclic transition states. Most of the reactions are insensitive to conditions such as solvent polarity and catalysis. They are believed to proceed in a single step in which the reorganization of the bonds takes place simultaneously.

Experimental observations for three types of concerted reactions—cycloaddition reactions, electrocyclic reactions, and sigmatropic rearrangements—are given below.

The Diels-Alder reaction (Section 11.5) is an example of a **cycloaddition reaction,** defined as a reaction that forms a ring by addition to an alkene or an alkyne. Cycloaddition reactions are further classified according to the number of electrons involved in the transition state. *A Diels-Alder reaction is called a [4 + 2] cycloaddition reaction because one component of the reaction, the diene, contributes four π electrons while two π electrons come from the dienophile. A total of six electrons is involved in the transition state.*

diene	dienophile	cyclic transition state	product
4 π electrons	*2 π electrons*	*6 π electrons*	

The Diels-Alder reaction is an example of a **thermal cycloaddition reaction.** In contrast, some cycloaddition reactions occur only when one of the reactants absorbs light energy. For example, 2-butene does not add to itself to give a four-membered ring unless it is irradiated with ultraviolet radiation at 214 nm. The products obtained depend upon the stereochemistry of the starting material.

(Z)-2-butene → 214 nm liquid phase → (E)-2-butene *major product* + 1-butene + *methyl groups retain the cis orientation in each alkene unit*

(E)-2-butene → 214 nm liquid phase → (Z)-2-butene + 1-butene + *methyl groups retain the trans orientation in each alkene unit*

Reactions that take place upon the absorption of ultraviolet or visible radiation by molecules are called **photochemical reactions.** The addition of 2-butene to itself upon the absorption of ultraviolet radiation is a stereoselective photochemical [2 + 2] cycloaddition reaction, which is explored in greater detail in Section 19.2A.

The major photochemical reaction for both (*E*)- and (*Z*)-2-butene is the familiar stereochemical isomerization of the alkene (Sections 5.1A and 11.8A). A small amount of isomerization of the position of the double bond also takes place to give 1-butene.

Throughout the rest of the chapter, we will contrast *thermal* reactions with *photochemical* reactions. The word thermal is used in this sense to mean that the reaction proceeds in the ground state of the reactants in contrast to the excited state achieved upon absorption of ultraviolet or visible radiation (Section 11.8A). A thermal reaction may proceed at room temperature or even at very low temperatures. For example, the Diels-Alder reaction of cyclopentadiene with maleic anhydride, a thermal reaction, is carried out at 0 °C (Section 11.5A).

Electrocyclic reactions take place when polyenes react intramolecularly to give cyclic alkenes, or when cyclic alkenes undergo an opening of the ring to give acyclic polyenes. These reactions, which are the reverse of each other, occur with high stereoselectivity. For example, dimethyl *cis*-3-cyclobutene-1,2-dicarboxylate, when heated to 140 °C, gives a dimethyl 2,4-hexadienedioate, which is almost entirely the (*Z*,*E*)-isomer.

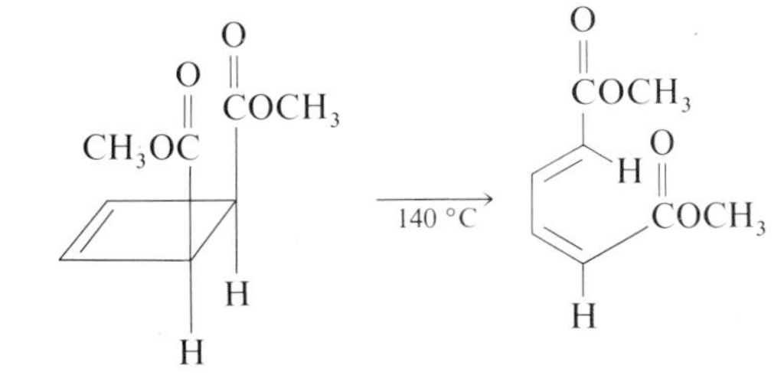

dimethyl *cis*-3-
cyclobutene-1,2-dicarboxylate

dimethyl (2*Z*,4*E*)-2,4-
hexadienedioate

In principle, the reaction is reversible, but the strain of a double bond in a small ring such as cyclobutene and the conjugation of the double bonds with the carbonyl groups stabilize the acyclic diene with respect to the starting material, and the reaction goes to completion at relatively low temperatures.

A conjugated triene, on the other hand, undergoes easy ring closure to give a stable six-membered ring. (2*E*,4*Z*,6*E*)-2,4,6-Octatriene gives *cis*-5,6-dimethyl-1,3-cyclohexadiene, but the (*Z*,*Z*,*E*) isomer closes to the corresponding trans compound.

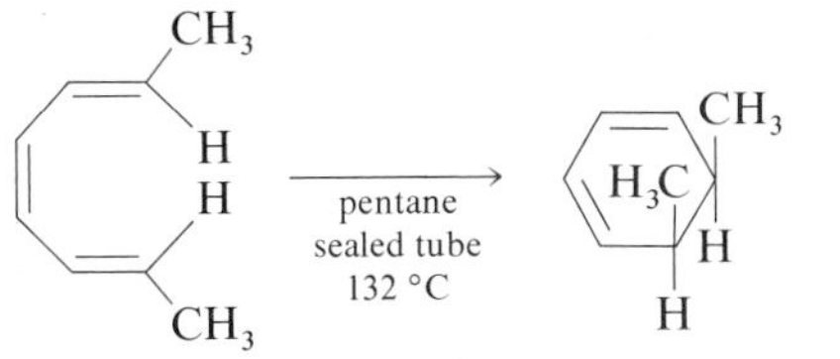

(2*E*,4*Z*,6*E*)-2,4-6-
octatriene

cis-5,6-dimethyl-
1,3-cyclohexadiene

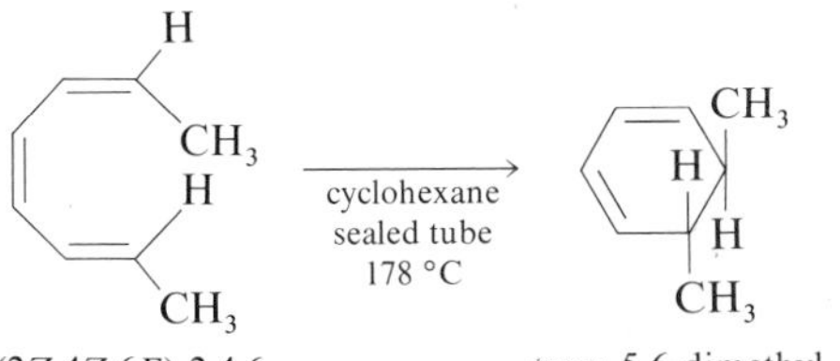

(2*Z*,4*Z*,6*E*)-2,4,6-
octatriene

trans-5,6-dimethyl-
1,3-cyclohexadiene

Thus these reactions too are highly stereoselective.

In **sigmatropic rearrangements,** a substituent and the pair of electrons that bond it to the rest of the molecule move along a conjugated system to a new position. The substituent that moves may be a hydrogen atom or an organic group. It may move from carbon to carbon, or from some other element, such as oxygen, to carbon. In the simplest case, (Z)-1,3-pentadiene rearranges into itself, a reaction that can be detected only when the molecule is labeled with deuterium.

(Z)-1,3-pentadiene   (Z)-1,3-pentadiene

(Z)-1,3-pentadiene-1,1-d_2 (Z)-1,3-pentadiene-5,5-d_2

(Z)-1,3-pentadiene-5,5-d_2 (1E,3Z)-1,3-pentadiene-1,5-d_2

In the rearrangement, a hydrogen (or deuterium) atom, along with its σ bond, moves five atoms away along a conjugated diene. Such a movement of a σ bond is called a sigmatropic rearrangement. The equation above illustrates [1,5] sigmatropic shifts of hydrogen and deuterium atoms. A σ bond is broken between the hydrogen atom and a carbon atom assigned the number 1 as the starting point of the rearrangement. A new σ bond is formed between the hydrogen atom, the first atom of the migrating group, and carbon 5 of the polyene systems, hence the designation [1,5] for the shift. We will examine other types of sigmatropic rearrangements in Sections 19.1B and 19.5, and the nomenclature will become clearer with practice. The important idea is that there has been a rearrangement involving the breaking of a σ bond in one portion of the molecule, the forming of one in another part of the molecule, and a new arrangement of the π bonds between these two points.

Sigmatropic rearrangements of hydrogen atoms are intramolecular reactions. They have been shown to be unimolecular and to be unaffected by solvent polarity. The reaction above is stereoselective with the hydrogen (or deuterium) atom leaving the top side of the methyl group on one end of the pentadiene and arriving at the top side of the methylene group of the terminal double bond. According to all these experimental observations, sigmatropic rearrangements are also concerted reactions.

This section provides an introduction to the experimental facts about a variety of reactions that can be classified together. They are all concerted reactions involving the stereoselective reorganization of the electrons in π and σ bonds, usually through cyclic transition states. In the next section, we will explore some of the attempts to correlate these experimental observations by means of theoretical explanations. The concepts and the language that have been developed during the evolution of these attempts have transformed the way chemists think and talk about a wide range of reactions.

PROBLEM 19.1   To prove that the photochemical cycloaddition reactions of the 2-butenes shown on p. 1070 were indeed stereoselective, it was necessary to examine the products formed when a mixture of (*Z*)- and (*E*)-2-butenes were irradiated with ultraviolet radiation at 214 nm. Predict what the products observed in such experiments would be.

PROBLEM 19.2   The following sequence of reactions was used to synthesize (*Z*)-1,3-pentadiene-1,1-d_2. Fill in structures for the compounds represented by letters.

$$CH_3CH\!=\!CHCH_2\overset{\overset{\displaystyle O}{\|}}{C}OCH_3 \xrightarrow[\text{diethyl ether}]{LiAlD_4} \xrightarrow{H_3O^+} A \xrightarrow[\text{pyridine}]{TsCl} B \xrightarrow[\text{acetone}]{NaI}$$

(*Z*)-isomer

$$C \xrightarrow[\text{methanol}]{\overset{\displaystyle CH_3}{\overset{\displaystyle |}{CH_3NCH_3}}} D \xrightarrow[\text{H}_2O]{Ag_2O} E \xrightarrow{\Delta} \underset{H}{\overset{CH_3}{\diagup}} C\!=\!C \underset{H}{\overset{CH=CD_2}{\diagdown}}$$

B. The Evolution of the Theory of Concerted Reactions

In the 1960s, chemists began to develop theoretical explanations for the experimental facts described in the previous section and many others observed in similar reactions. The need not only for explanations, but also for theories that would suggest new experiments, arose mostly out of greatly expanded research into the chemistry and synthesis of natural products. An illustration of the close relationship between such practical research and the development of theory in organic chemistry is the work done on vitamin D by Egbert Havinga of the Netherlands. A form of vitamin D is synthesized by the body in the skin on exposure to sunlight.

7-dehydrocholesterol

$$\xrightarrow[\substack{\text{an electrocyclic}\\ \text{opening of the}\\ \text{B-ring}}]{h\nu}$$

$$\xrightarrow[\substack{\text{a [1,7]}\\ \text{sigmatropic}\\ \text{rearrangement}}]{37\ ^\circ C,}$$

vitamin D$_3$
cholecalciferol

The overall conversion may be classified as an electrocyclic opening of the B-ring of 7-dehydrocholesterol to a conjugated triene and then a sigmatropic rearrangement of a hydrogen atom from the methyl group to the other end of the conjugated system, a [1,7] sigmatropic shift. Recent research has shown that the sigmatropic rearrangement takes place slowly at body temperature, so vitamin D_3 continues to be synthesized in the skin for up to three days after exposure to the sun.

The reactions shown above are only two of many of this kind observed for the vitamin D system. The reactions have different stereochemistry depending on whether they occur photochemically or thermally, as will be seen in Section 19.3B. Havinga's very precise determination of the course of these reactions provided information that led his colleague Luitzen J. Oosterhoff to suggest that the stereochemistry characteristic of these reactions is determined by the symmetry of the molecular orbitals involved in the transformations.

Meanwhile, in the 1960s, Robert B. Woodward at Harvard University was doing research on the synthesis of vitamin B_{12}. During the course of the synthesis, he observed the precise stereochemistry with which cyclohexadienes were formed from conjugated trienes and then converted back into trienes in thermal and photochemical reactions. At that time, the reactions in which cyclobutenes open to butadienes, such as the example given on p. 1071, Section 19.1A, were known. Woodward and his colleague Roald Hoffmann, in thinking about these observations, also arrived at the idea that the symmetry of the molecular orbitals that participate in the chemical reaction determines the course of the reaction. In a series of papers published in 1965, they applied their ideas systematically to electrocyclic, cycloaddition, and sigmatropic reactions. They proposed what they called the principle of the **conservation of orbital symmetry** in concerted reactions. The recognition of the generality and the significance of this way of examining organic reactions was only one of the many contributions made by Woodward, who received the Nobel Prize in 1965 for his work in organic chemistry.

In the most general terms, the principle means that in a concerted reaction, the molecular orbitals of the starting materials must be transformed into the molecular orbitals of the products in a smooth, continuous way. This is possible only if the orbitals that are transformed into each other have similar symmetry. To see what is meant by the symmetry of a molecular orbital, we will review molecular orbitals.

C. A Review of π Molecular Orbitals

The bonding and antibonding π orbitals of ethylene (Figure 2.20, Section 2.5A) and the electronic configuration of ethylene in the ground and the excited states (Section 11.8A) are shown in Figure 19.1. The π molecular orbital has a nodal plane containing the two carbon atoms and the four hydrogen atoms. The π^* orbital has a second nodal plane besides the one that includes the σ-bonded backbone of the molecule, and is of higher energy than the bonding π orbital. The more nodes there are in a molecular orbital (and in an atomic orbital), the higher the energy level of the orbital. Ethylene is a stable molecule in the ground state because the two electrons from the two p atomic orbitals are in the bonding π orbital. Absorption of ultraviolet radiation by ethylene raises the molecule to the excited state in which one electron has been promoted to the π^* orbital (Section 11.8A).

The nodal characteristics and the relative energies of the π and π^* orbitals of ethylene are important because they represent the properties of all other alkenes with

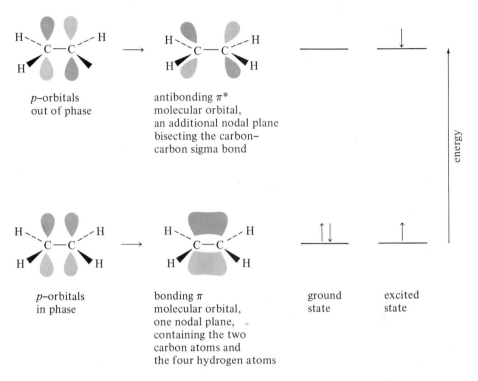

p–orbitals
out of phase

antibonding π^*
molecular orbital,
an additional nodal plane
bisecting the carbon–
carbon sigma bond

p–orbitals
in phase

bonding π
molecular orbital,
one nodal plane,
containing the two
carbon atoms and
the four hydrogen atoms

ground
state

excited
state

energy

Figure 19.1   The π and π^* orbitals of ethylene. The electronic configuration of ethylene in the ground state and after the absorption of a photon.

one double bond and two π electrons. We assume that substituents on the double bond do not change these properties of the molecular orbitals sufficiently to change the conclusions that can be drawn by looking at the simplified picture.

Just as the molecular orbital picture that was developed for ethylene can represent all compounds with a single π orbital and two electrons in it, a similar molecular orbital picture for 1,3-butadiene (Figure 11.2, Section 11.1B) will be used whenever we talk about a conjugated diene. The four molecular orbitals, designated as ψ_1, ψ_2, ψ_3, and ψ_4 are shown again in Figure 19.2 (p. 1076). They are constructed by successively introducing additional nodes into the array of p atomic orbitals, while retaining the overall symmetry of the system. Note how the energy level of the orbital is raised as the number of nodes in the orbital increases. In 1,3-butadiene, the symmetry of the molecule dictates that carbon 1 is equivalent to carbon 4 and that carbon 2 is equivalent to carbon 3 (and that carbon 1 is *not* equivalent to carbon 2, and carbon 3 *not* to carbon 4). In constructing molecular orbitals, therefore, if we place a node between carbon 1 and carbon 2, a node must also appear between carbon 4 and carbon 3.

In the ground state of 1,3-butadiene, the bonding molecular orbitals, ψ_1 and ψ_2, are occupied by two electrons each. The antibonding molecular orbitals, ψ_3 and ψ_4, are empty.

The molecular orbitals shown in Figure 19.2 will be used whenever we talk about a conjugated diene with four π electrons in the system. Once again, we will make the assumption that the substituents on the diene system will not change the nodal properties of the molecular orbitals sufficiently to affect our conclusions.

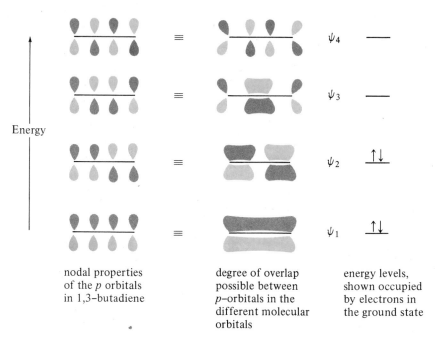

Energy

| nodal properties of the *p* orbitals in 1,3–butadiene | degree of overlap possible between *p*–orbitals in the different molecular orbitals | energy levels, shown occupied by electrons in the ground state |

Figure 19.2   The four molecular orbitals of butadiene, created from four *p* atomic orbitals.

PROBLEM 19.3   How many nodes are there in each of the molecular orbitals ψ_1, ψ_2, ψ_3, and ψ_4 of 1,3-butadiene?

What is the symmetry of a molecular orbital to which Woodward and Hoffmann refer in their principle? Each molecular orbital can be examined for symmetry in ways that are analogous to the method we use when we decide whether molecules are chiral or not. An examination of all the ways this can be done is beyond the scope of this book, but one example will be given to show that in principle each molecular orbital can be classified as symmetrical or antisymmetrical in some way.

The two π molecular orbitals shown in Figure 19.1 can be described as **symmetrical** or **antisymmetrical** with respect to a plane bisecting the carbon-carbon bond. If we imagine such a plane, we see that it cuts the π orbital into two halves. The left-hand side of the orbital is an exact mirror image of the right-hand side. We can thus say that the π orbital is symmetrical with respect to this plane. The π^* orbital is not. For example, to the left of the plane on the top there is a grey lobe, while to the right of the plane at the top and at the same distance away from the plane, there is a colored lobe. In other words, the sign of the wave function is reversed as it goes through the plane. The π^* orbital is thus antisymmetrical with respect to a plane bisecting the carbon-carbon bond.

PROBLEM 19.4   Decide which of the molecular orbitals of 1,3-butadiene (Figure 19.2) are symmetrical and which antisymmetrical with respect to a plane bisecting the bond between carbon 2 and carbon 3.

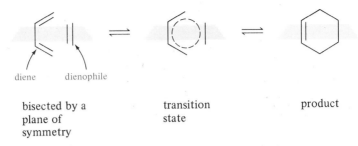

diene   dienophile

bisected by a   transition   product
plane of   state
symmetry

Figure 19.3   A demonstration of the conservation of symmetry in the Diels-Alder
reaction.

In a concerted reaction, the symmetry that is present in the reactants is maintained
during the course of the reaction and is also present in the product of the reaction. The
Diels-Alder reaction provides a simple demonstration of this principle (Figure 19.3).
The reactants, the diene and the dienophile, each have a plane of symmetry. They must
approach each other in the transition state in such a way that the same plane of
symmetry is maintained. The product, cyclohexene, also retains the same plane of
symmetry.

D. Interactions Between Molecular Orbitals

Interactions between the molecular orbitals of reagents and their transformations into
the molecular orbitals of the products have been described in many different ways. One
of the simplest and most useful is to postulate that a reaction takes place when the
highest energy molecular orbital that contains electrons of one reagent interacts with the
lowest energy molecular orbital that does not contain electrons of the other reagent.
This is just like saying that a Lewis base, with electrons to donate, reacts with a Lewis
acid, with an empty orbital ready to receive them, or that a nucleophile reacts with an
electrophile. The only difference is that the orbitals performing these roles are now
examined very precisely in order to explain the highly specific chemistry and stereochemistry observed for concerted reactions.

The orbitals that interact have been called the **frontier orbitals** for the reaction by
Kenichi Fukui of Japan. His contributions to the development of this way of looking
at chemical reactions were recognized in 1981 by the award of the Nobel Prize in
Chemistry jointly to him and to Roald Hoffmann. The orbitals are given the names
highest occupied molecular orbital, abbreviated **HOMO,** and **lowest unoccupied
molecular orbital,** or **LUMO.** For ethylene, for example, the π orbital is the HOMO
and the π^* orbital the LUMO in the ground state. When ethylene is raised to its excited
state in a photochemical reaction, the π^* orbital, with one electron in it, becomes its
HOMO (Figure 19.1). In the ground state, ψ_2 is the HOMO and ψ_3 the LUMO of
1,3-butadiene (Figure 19.2). These relationships will be taken up again in the next
section when we consider how to apply them to the reactions actually observed for
compounds with molecular orbitals like those of ethylene and butadiene.

Why are these particular orbitals so important in determining the course of a reaction? The electrons in the HOMO of a molecule are like the outer-shell electrons of an
atom. They are the electrons that can be removed with the least expenditure of energy
because they are already in a higher energy level than any of the other electrons in the
molecule. For example, they are the electrons that are lost when the molecule is
ionized in the gas phase, and spectroscopic techniques that cause this ionization can be

used to determine the energy level of the HOMO for a molecule. The LUMO of a molecule is the orbital to which the electrons can be transferred with the least expenditure of energy.

The higher the HOMO of a molecule, the more easily electrons can be removed from it. The lower the LUMO of a molecule, the more easily electrons can be transferred into it. Therefore, the interaction between a molecule with a high HOMO and one with a low LUMO is particularly strong. In general, the smaller the difference in energy between the HOMO of one molecule and the LUMO of the other species with which it is reacting, the stronger is the interaction between the two molecules.

PROBLEM 19.5   Assume that an electron is raised from the HOMO to the LUMO of a molecule upon absorption of energy. Show the electronic configuration of 1,3-butadiene after it absorbs ultraviolet radiation, and decide which molecular orbital will be the HOMO and which the LUMO of the diene in the excited state.

19.2

Cycloaddition Reactions of Carbon Compounds

A. The Photochemical Dimerization of Alkenes

The usefulness of the theoretical picture of the interactions between molecular orbitals that was developed in the last section is demonstrated most simply in an examination of the cycloaddition reactions of alkenes. In Figure 19.4, the HOMO of one molecule of (E)-2-butene is shown approaching the LUMO of another molecule of (E)-2-butene in the ground state, if ultraviolet radiation is not used. An examination of the two orbitals shows that if they approach each other in the manner shown, it will not be possible for a cyclobutane ring to form in a concerted manner. The bottom right-hand lobe of the HOMO is in phase with the top right-hand lobe of the LUMO and could interact to give a σ bond, but the lobes that are facing each other on the left-hand side are out of phase and would only interact in an antibonding way. The second bond necessary to complete the ring would not form.

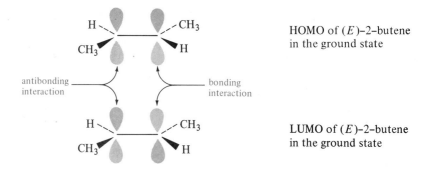

HOMO of (E)–2–butene in the ground state

antibonding interaction — bonding interaction

LUMO of (E)–2–butene in the ground state

Figure 19.4   A schematic representation of the interaction of the HOMO and LUMO of (E)-2-butene in the ground state.

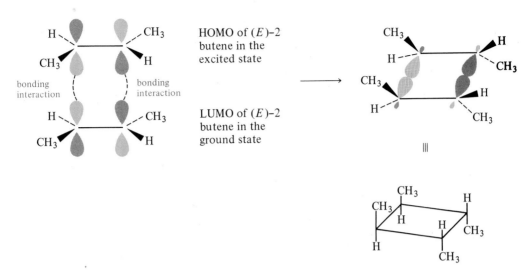

HOMO of (*E*)-2
butene in the
excited state

LUMO of (*E*)-2
butene in the
ground state

Figure 19.5   A schematic representation of the interaction of the HOMO of
(*E*)-2-butene in the excited state with the LUMO of (*E*)-2-butene in
the ground state.

If one of the molecules of (*E*)-2-butene is raised to the excited state, however, its
HOMO will have a different symmetry, as shown in the energy diagram in Figure 19.1.
The concentration of (*E*)-2-butene molecules in the excited state is low, therefore a
molecule in the excited state is most likely to encounter and react with a molecule of
(*E*)-2-butene in the ground state. The molecular orbital picture for such an interaction
is shown in Figure 19.5. In this case, bonding interactions are possible at both ends
of the alkene bonds as the two molecules approach each other face-to-face. Only one
of the two possible stereochemical outcomes for the two (*E*)-2-butene molecules
reacting in this way has been drawn. The methyl groups on the left-hand carbon atom
of each 2-butene are on the same side of the double bond and are cis to each other in
the cyclobutane, and the same is true of the methyl groups on the other carbon atom
of the reagent. The stereochemistry of the starting materials has been preserved in the
product.

The molecular orbital picture thus provides an explanation for the lack of reactivity
of two molecules of 2-butene in the ground state and for the stereoselective dimerization that occurs in the excited state.

PROBLEM 19.6   Another stereoisomer is also formed when (*E*)-2-butene is irradiated. Draw the
molecular orbital picture showing how it arises.

PROBLEM 19.7   Draw molecular orbital pictures showing how the two cyclobutanes that are formed
when (*Z*)-2-butene is irradiated at 214 nm come into being.

While the nature of the molecular orbitals involved when two alkene molecules react
photochemically may seem a rather exotic and abstract topic to worry about, exactly
such a photochemical reaction occurs when DNA is damaged by ultraviolet radiation.

This reaction is believed to be responsible for most of the harmful effects of such radiation on living organisms. For example, ultraviolet radiation kills bacteria quite efficiently and is used to sterilize equipment. Excessive ultraviolet radiation from sunlight (or tanning lamps) is also responsible for causing skin cancer. Both of these effects arise from damage done to DNA so that it can no longer function properly to direct the synthesis of proteins in the cell.

When thymine residues are adjacent to each other on a strand of DNA they can undergo a photochemical [2 + 2] cycloaddition reaction to give a cyclobutane compound known as a *thymine dimer*.

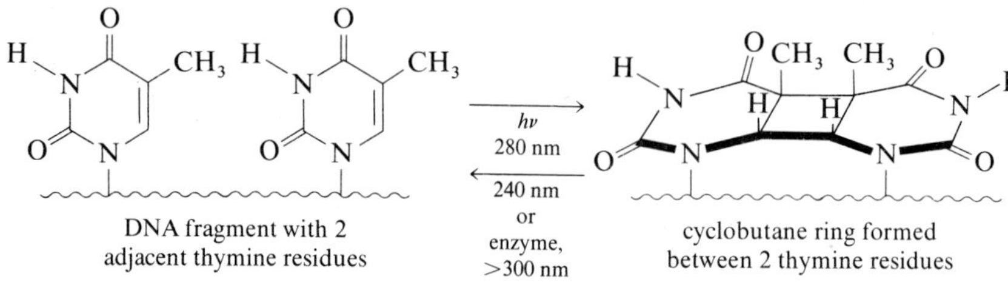

DNA fragment with 2 adjacent thymine residues

$h\nu$ 280 nm
240 nm or enzyme, >300 nm

cyclobutane ring formed between 2 thymine residues

The photochemical dimerization of thymine disrupts the normal functioning of DNA (Section 18.6B).

Fortunately for creatures that must live under constant exposure to ultraviolet radiation, DNA that is damaged by the formation of dimers can be repaired. The cell has enzymes that repair DNA by cutting out the damaged parts and filling in the missing units. The dimer of thymine also absorbs ultraviolet radiation and undergoes an opening of the cyclobutane ring back to thymine. These pathways provide a mechanism for DNA repair, but some damage always remains.

PROBLEM 19.8 The chromophore (Section 11.8A) in thymine is an α,β-unsaturated carbonyl group. Such a functional group in the excited state adds to alkenes or alkynes in a [2 + 2] cycloaddition reaction. Given this information, predict the products of the following reactions.

(a) 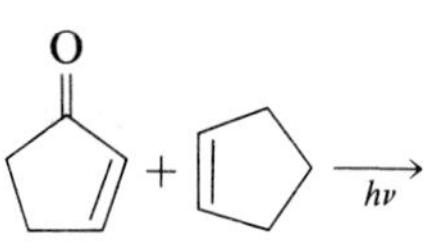 [structure] + [structure] $\xrightarrow{h\nu}$

(b) 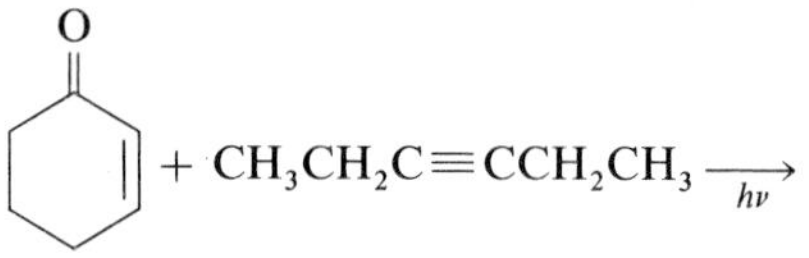 [structure] + $CH_3CH_2C{\equiv}CCH_2CH_3$ $\xrightarrow{h\nu}$

(c) 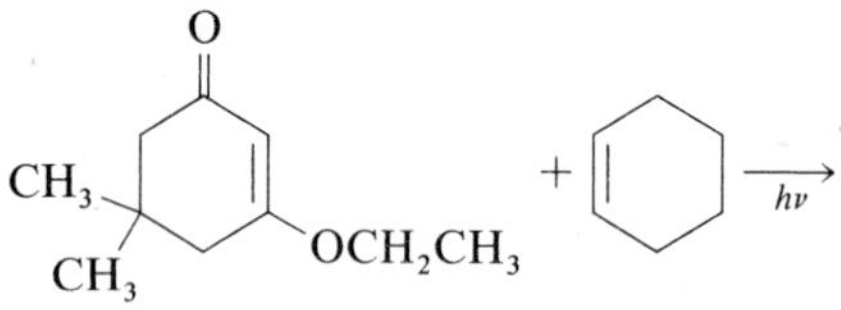 [structure] + [structure] $\xrightarrow{h\nu}$

PROBLEM 19.9 The photochemical [2 + 2] cycloaddition reaction of alkenes has been used in several syntheses of grandisol, a sex pheromone of the boll weevil (p. 554, Section 11.7C), as the step creating the cyclobutane ring. The cycloaddition products obtained in each case are shown below. Complete the equations showing what two reagents must have been used in each synthesis.

(a) 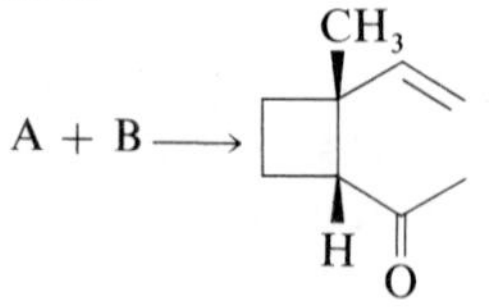 A + B $\longrightarrow$ [structure]

(Hint: One of them is *not* ethylene.)

(b)   C + D ⟶

(c)   E + F ⟶

(d)   G + H ⟶

PROBLEM 19.10   When thymine is irradiated in solution, dimers other than the one shown on p. 1080 are produced. What other structures are possible for thymine dimers?

B. The Diels-Alder Reaction

A molecular orbital picture of the Diels-Alder reaction shows the interaction of the HOMO of the diene with the LUMO of the dienophile (Figure 19.6). An examination of the HOMO of the diene shows that the lower lobes of the orbital at carbons 1 and 4 of the conjugated system have the same color as the corresponding upper lobes of the LUMO of the dienophile. Bonding interactions can develop between them if the dienophile approaches the ends of the diene at one face of the molecule. Note that it does not make any difference whether the dienophile is above or below the diene. Direct overlap of the lobes of the orbitals that are in phase gives rise to two bonding σ orbitals and the formation of a six-membered ring.

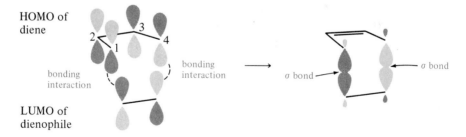

Figure 19.6   A schematic representation of the favorable interaction between the HOMO of a diene and the LUMO of a dienophile.

The substituents at the ends of the diene system and on the double bond of the dienophile retain the stereochemistry they had in the starting reagents because there has been no twisting around single bonds during the reaction. The only part of the picture that remains puzzling is how the orbitals between carbons 2 and 3 of the diene system form the π bond in the product. At first glance, it looks as if the lobes of the p orbitals that are left over at carbons 2 and 3 will be antibonding. This is so because we are looking at only two of the orbitals that participate in the reaction, and, for the sake of simplicity, are ignoring the others. As the σ bonds form, the interactions between the lobes of the p orbitals at carbons 1 and 2 and those at carbons 3 and 4, which are shown as bonding in the HOMO of the diene, change. The static picture that is concentrating on only one part of the process does not tell the whole story.

PROBLEM 19.11   Draw a molecular orbital picture of the Diels-Alder reaction using the LUMO of the diene and the HOMO of the dienophile to prove to yourself that predictions about the stereochemistry of the reaction and the nature of the transition state of the reaction remain as described above.

How can this picture be used to account for the predominance of endo stereochemistry (p. 533) in Diels-Alder reactions? The molecular orbital picture of the cycloaddition of cyclopentadiene with maleic anhydride is shown in Figure 19.7. The LUMO for maleic anhydride is shown to include the π bonds of the carbonyl groups that are conjugated with the double bond of the diene. The picture drawn is that of the LUMO of a 1,3,5-hexatriene system.

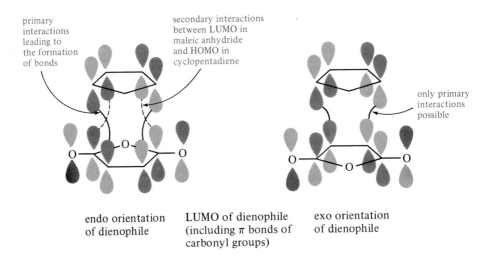

Figure 19.7   Schematic representation of the interactions between the HOMO of cyclopentadiene and the LUMO of maleic anhydride when the anhydride is endo to the cyclopentadiene ring and when it is exo.

PROBLEM 19.12   Very soon we will need to look at the HOMO of a 1,3,5-hexatriene also. Starting with six *p* orbitals in a row, showing the different signs for their lobes by different kinds of shading, construct the six π-type molecular orbitals of 1,3,5-hexatriene. Remember to increase the number of nodes in the orbitals systematically and to maintain the overall symmetry of the system. Show the relative energies for the molecular orbitals, decide which ones are bonding, which antibonding, where the six π electrons will be, and which is the HOMO for the system. The LUMO is already shown in Figure 19.7.

If maleic anhydride approaches the diene so that the major part of the molecule lies underneath the cyclopentadiene ring, two kinds of interactions are possible. First, bonding interactions occur between the lobes at the ends of the diene system and the lobes at the double bond in the dienophile. These interactions give rise to the stable six-membered ring in the product of the reaction. Second, additional weaker interactions are possible between the other lobes of the diene system and those on the carbon atoms of the carbonyl groups. These interactions do not lead to bonding because such a reaction would destroy the stable carbonyl groups and also lead to highly strained systems with multiple rings. However, they do stabilize the transition state and lower the energy of activation for the reaction. Thus, reaction with the endo orientation is faster than reaction with the exo orientation of the dienophile in which such secondary interactions are not possible (Figure 19.7). As a result, more of the product has the endo stereochemistry.

In Section 11.5B, the regioselectivity observed in reaction of unsymmetrically substituted dienes and dienophiles was discussed in terms of resonance contributors. Computer programs for calculating the molecular orbitals and other electronic properties of a wide range of molecules are now available. The qualitative results that organic chemists get by pushing electrons are confirmed in many cases (and contradicted in a few) by theoretical calculations that describe molecular orbitals more precisely than we have been doing. For example, we have shown all the lobes of the p orbitals making up the molecular orbitals to be the same size. Actually, the lobes are predicted by theory to be of different sizes, reflecting the different probabilities of finding an electron at any point along a conjugated system for each molecular orbital. Calculations like these for compounds such as 1-ethoxybutadiene and propenal (p. 535) confirm that the most important interaction for the HOMO of 1-ethoxybutadiene and the LUMO of propenal is between carbon 4 of the diene system and carbon 3 of propenal, thus determining the orientation found in the product.

19.3

Electrocyclic Reactions

A. Interconversion of Cyclobutenes and Conjugated Dienes

Cyclobutene rings open to conjugated dienes when heated (Section 19.1A). For example, the ring in *cis*-3,4-dimethylcyclobutene opens smoothly to form (2Z,4E)-2,4-hexadiene.

cis-3,4-dimethyl-
cyclobutene

(2Z,4E)-2,4-
hexadiene
>99% of the
dienes formed

On the other hand, *trans*-3,4-dimethylcyclobutene gives (2*E*,4*E*)-2,4-hexadiene.

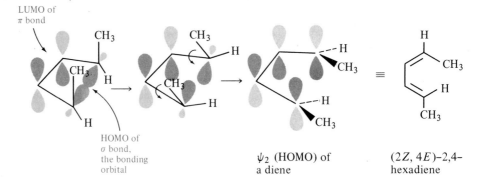

trans-3,4-dimethyl-   (2E,4E)-2,4-
cyclobutene   hexadiene

The reaction consists of the breaking of a σ bond and the overlap of the lobes of that orbital with those of the adjacent π bond so that two new π orbitals are formed to accept the four bonding electrons of the diene system. Using the language of frontier molecular orbitals (p. 1077), the reaction may be described as the addition of the HOMO of the σ orbital to the LUMO of the π bond in the cyclobutene (Figure 19.8).

LUMO of
π bond

CH$_3$ CH$_3$ H

CH$_3$ H   CH$_3$

H   CH$_3$ H   ---H

H   CH$_3$ CH$_3$ $\equiv$ CH$_3$

H   H   ---H   H

CH$_3$ CH$_3$

HOMO of
σ bond,
the bonding ψ_2 (HOMO) of   (2Z, 4E)–2,4–
orbital   a diene   hexadiene

Figure 19.8   A schematic representation of the interactions of molecular orbitals that give rise to the stereoselective ring opening of *cis*-3,4-dimethylcyclobutene in a thermal reaction.

The HOMO of the σ bond is the bonding orbital, shown with two lobes of the sp^3 hybrid orbitals on the carbon atoms overlapping. In order for the new π bonds to develop, bonding interactions must be possible between the lobes of the orbitals shown for the π bond in cyclobutene and the lobes that are freed from the σ bond as it breaks. For this to happen, the carbon atoms that are part of the breaking bond must rotate in order to bring the lobes of the σ orbital into line with those lobes of similar sign on the other two carbon atoms. As indicated in Figure 19.8, both carbon atoms must rotate in the same direction, either clockwise or counterclockwise, for the new π bonds to form smoothly. If you follow the motions of the two methyl groups during the reaction, you will see that they both move in the same direction as the carbon atoms to which they are bonded, and the product is (2*Z*,4*E*)-2,4-hexadiene. *This type of motion in which the carbon atoms at the ends of the developing polyene system rotate in the same direction is said to be* **conrotatory.** Thus, *cis*-3,4-dimethylcyclobutene, when heated, undergoes conrotatory ring opening to (2*Z*,4*E*)-2,4-hexadiene.

PROBLEM 19.13   Draw the molecular orbital picture for the thermal opening of the ring in *trans*-3,4-dimethylcyclobutene.

PROBLEM 19.14   The following equilibrium has been observed. How do you account for this isomerization? Note that none of the other possible stereoisomers (what are they?) is formed.

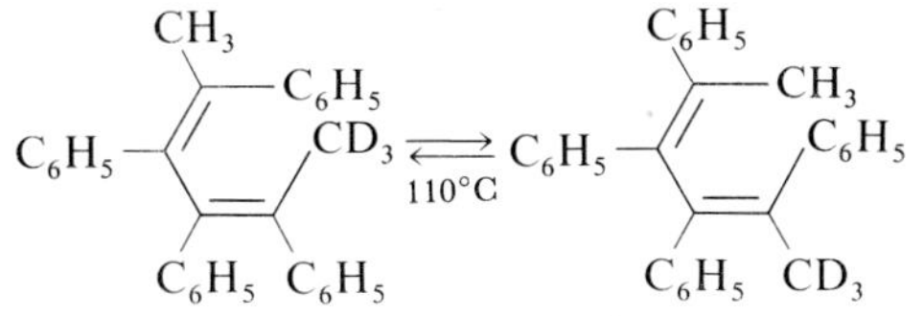

When *cis*-3,4-dimethylcyclobutene undergoes a photochemical ring opening, a mixture of products is obtained, among them (2*E*,4*E*)-2,4-hexadiene.

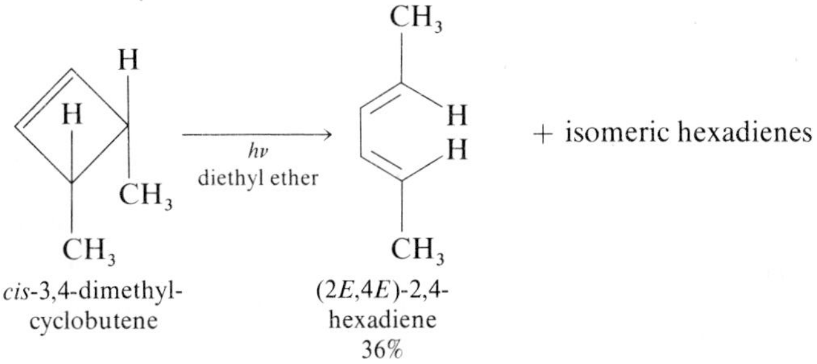

cis-3,4-dimethyl-
cyclobutene

(2*E*,4*E*)-2,4-
hexadiene
36%

+ isomeric hexadienes

Practically none of this product was seen in the thermal ring opening of *cis*-3,4-dimethylcyclobutene (p. 1083). (2*E*,4*E*)-2,4-Hexadiene is believed to arise from the excited state of the cyclobutene. One molecular orbital picture that can be used to rationalize this experimental observation is shown in Figure 19.9.

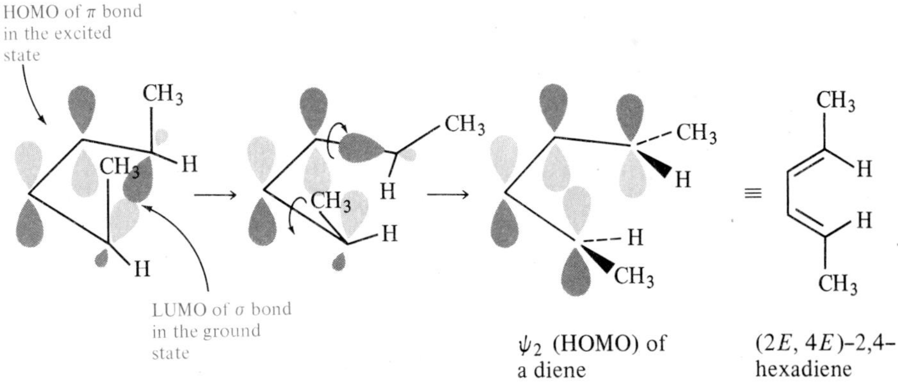

HOMO of π bond
in the excited
state

LUMO of σ bond
in the ground
state

ψ_2 (HOMO) of
a diene

(2*E*, 4*E*)-2,4-
hexadiene

Figure 19.9   A schematic representation of the interactions of molecular orbitals that give rise to the stereoselective ring opening of *cis*-3,4-dimethylcyclobutene in a photochemical reaction.

In the photochemical reaction, the π bond of the cyclobutene is assumed to have absorbed ultraviolet energy so that an electron has been promoted to the π^* antibonding orbital. This antibonding orbital now plays the role of the HOMO for that portion of the molecule, while the antibonding σ orbital is used as the LUMO. As is seen in Figure 19.9, the new π bonds can form smoothly only if the two carbon atoms of the σ bond rotate in opposite directions. One carbon atom must rotate clockwise and the other one counterclockwise so that the lobes being freed from the σ bond can overlap with lobes of the same sign from the original π bond. *When the carbon atoms at the* *ends of the developing polyene rotate in opposite directions with respect to each other, their motion is said to be* **disrotatory.** The methyl groups move with the carbon atoms that are rotating, and (2E,4E)-2,4-hexadiene is formed.

   (2E,4E)-2,4-Hexadiene undergoes a photochemical ring-closure reaction to give cis-3,4-dimethylcyclobutene.

(2E,4E)-2,4-   cis-3,4-dimethyl-
hexadiene   cyclobutene

minor product

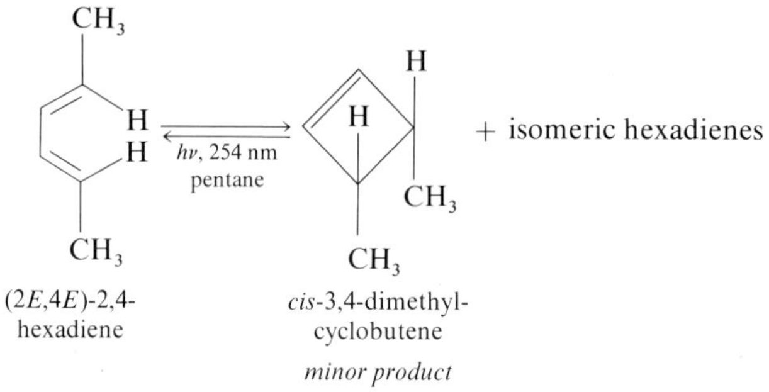

   The photochemical ring-closure reaction is also a disrotatory process. *The stereochemistry of the ring-closure reaction is determined by the symmetry of the HOMO of the polyene system undergoing the reaction.* The HOMO for the diene in the excited state is ψ_3 (Figure 19.2, Problem 19.5). If the end carbon atoms of the diene system are to form a σ bond, they must rotate towards each other, one of them clockwise and the other counterclockwise, to bring the lobes of the same sign into contact so that overlap and bonding can take place (Figure 19.10).

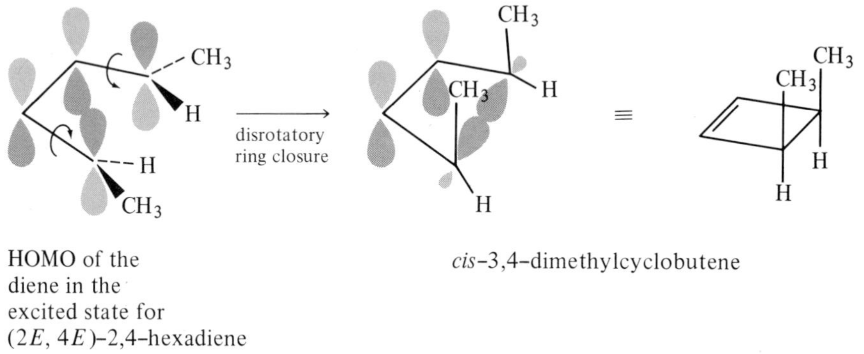

HOMO of the   cis-3,4-dimethylcyclobutene
diene in the
excited state for
(2E, 4E)-2,4-hexadiene

Figure 19.10   A schematic representation of the stereochemistry of
                   photochemical ring closure of a diene.

   In summary, cyclobutenes and conjugated dienes are interconverted in a conrotatory way in the ground state. Photochemical interconversions of these systems occur in a disrotatory fashion. Note that in all these reactions, we are dealing with systems that either start or end with four π electrons. The difference in the stereochemical course of the thermal and photochemical reactions can be rationalized by considering the difference in the symmetry of the molecular orbitals of the reacting molecules in the ground state and in the excited state.

PROBLEM 19.15   Predict what the product of this reaction will be.

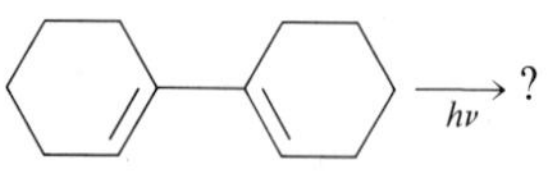

PROBLEM 19.16 1,3-Butadiene selectively labeled with deuterium at carbons 1 and 4 of the diene system has been prepared. For the isomer shown below, predict the structures of the products you would get in a thermal and in a photochemical cyclization reaction. Use molecular orbital pictures to explain your conclusions.

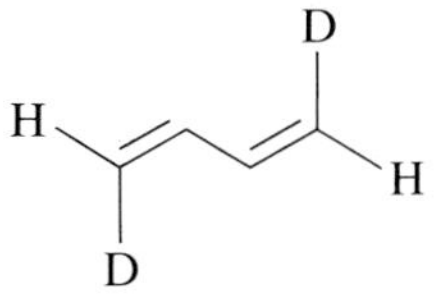

B. Interconversion of Cyclohexadienes and Trienes

The stereoselective ring-closure reactions of (2*E*,4*Z*,6*E*)- and (2*Z*,4*Z*,6*E*)-2,4,6-octatriene to *cis*- and *trans*-5,6-dimethyl-1,3-cyclohexadiene, respectively, have already been shown (p. 1071, Section 19.1A). An examination of the stereochemistry of the reactions indicates that the ring is closing in a disrotatory fashion. Thus, the thermal ring-closure reaction for a conjugated triene takes place with stereochemistry opposite to that observed for a diene. This experimental observation can be understood by looking at the nature of the molecular orbital undergoing the transformation. The HOMO of (2*E*,4*Z*,6*E*)-2,4,6-octatriene is shown in Figure 19.11.

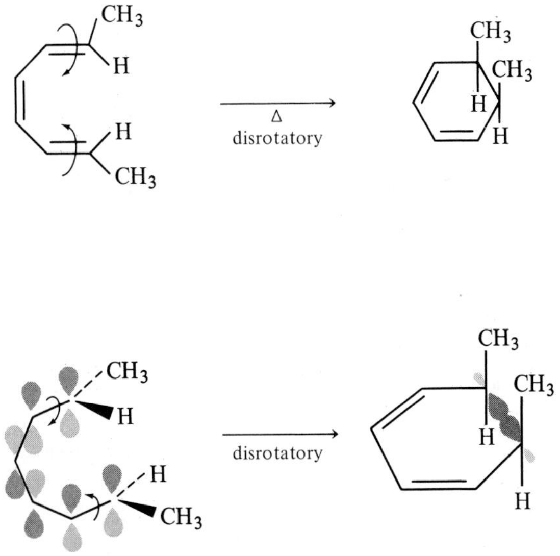

HOMO of the triene
in the ground state

(2*E*, 4*Z*, 6*E*)–2,4,6–
octatriene

cis–5,6–dimethyl–
1,3–cyclohexadiene

Figure 19.11 Schematic representation of the molecular orbital governing the stereochemistry of the thermal ring closure of a triene to a cyclohexadiene.

The HOMO of a triene system has two nodes besides the node in the plane of the molecule (Problem 19.12). The two end carbon atoms of the triene system must rotate towards each other if lobes having the same sign are to overlap to form a bond. The thermal ring-closure reaction is thus disrotatory.

PROBLEM 19.17   Go through the exercise of drawing the molecular orbitals for (2Z,4Z,6E)-2,4,6-octatriene and demonstrate to yourself that *trans*-5,6-dimethyl-1,3-cyclohexadiene is the product of a disrotatory ring closure.

When (2E,4Z,6E)-2,4,6-octatriene is irradiated, *trans*-5,6-dimethyl-1,3-cyclo-hexadiene is formed, along with stereoisomers of the octatriene. Under the conditions of the reaction, an equilibrium is set up between the cyclohexadiene and the triene.

(2E,4Z,6E)-2,4-6-   *trans*-5,6-dimethyl-
octatriene   1,3-cyclohexadiene

(2E,4E,6E)-2,4,6-octatriene
(mostly)

The photochemical reaction is conrotatory (Figure 19.12).

When the triene absorbs ultraviolet radiation, an electron is promoted from the HOMO to the LUMO, which then becomes the HOMO of the excited state for the triene. It is the symmetry of this orbital that governs the stereochemistry of the photochemical ring-closure reaction of the triene. In order to cause the lobes of the same sign on the end carbon atoms of the polyene to overlap, the carbon atoms, and the methyl groups on them, must rotate in the same direction, either clockwise or counterclockwise. The photochemical conrotatory ring closure of (2E,4Z,6E)-2,4,6-octatriene gives *trans*-5,6-dimethyl-1,3-cyclohexadiene, in contrast to the thermal disrotatory ring closure of the triene, which gives rise to the cis isomer. In the case of the triene-cyclohexadiene transformations, just as for the diene-cyclobutene inter-conversions, thermal and photochemical reactions proceed with opposite stereo-chemistry.

PROBLEM 19.18   In Figure 19.11, only one of two possible disrotatory ring-closure reactions for the (2E,4Z,6E)-2,4,6-octatriene was shown. Similarly, in Figure 19.12 only one kind of con-rotatory motion was indicated. Draw molecular orbital pictures to explore the consequences of the other possible disrotatory and conrotatory motions of the end carbon atoms of this polyene system. What is the stereochemical relationship of the new products to those shown in Figures 19.11 and 19.12?

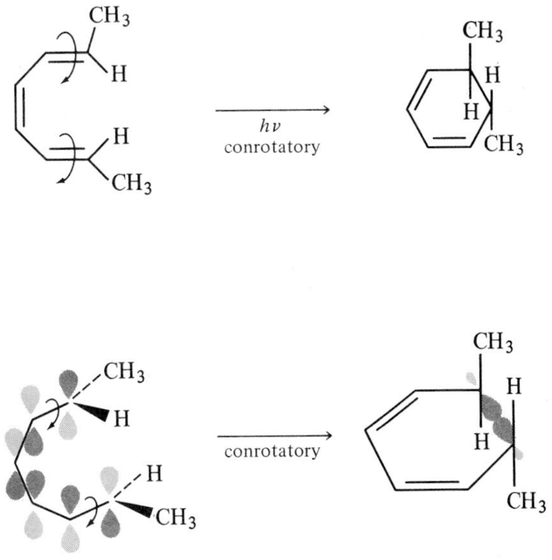

HOMO of the triene
in the excited state

$(2E, 4Z, 6E)$–2,4,6–octatriene

trans–5,6–dimethyl–
1,3–cyclohexadiene

Figure 19.12 Schematic representation of the molecular
orbitals governing the stereochemistry of the
photochemical ring closure of a triene to a
cyclohexadiene.

Photochemical ring-opening reactions of cyclohexadienes are well known and have
been most extensively investigated in connection with studies on vitamin D. Vitamin
D is a mixture of compounds that are effective in preventing and curing rickets, a
disease in which bones become soft and easily deformed. The disease and its relationship to a dietary deficiency that can be remedied by cod liver oil has been
recognized for over a hundred years. It was later recognized that certain foods contain
compounds that can be transformed into vitamin D by irradiation. These compounds,
called *provitamins,* occur widely in both plants and animals. They are all steroids that
have a diene system in the B-ring. They have no vitamin D activity until that ring is
opened upon irradiation (Section 19.1B).

The most widely investigated provitamin D is *ergosterol.* Ergosterol undergoes a
photochemical ring opening to give a compound called *precalciferol.* All the compounds that are provitamins for vitamin D have the same structure except for variations
in the side chain at carbon 17 of the steroid ring. The transformation of ergosterol to
precalciferol is a conrotatory opening of the B-ring. The stereochemical relationship of
the angular methyl group at carbon 10 and the hydrogen atom at carbon 9 is particularly
important, especially for the transformation of precalciferol into vitamin D_2 in the next
step of the reaction. The photochemical reaction is reversible. The conrotatory opening of the cyclohexadiene ring gives precalciferol in which the methyl group on carbon
10 lies toward carbon 9 of the steroid system.

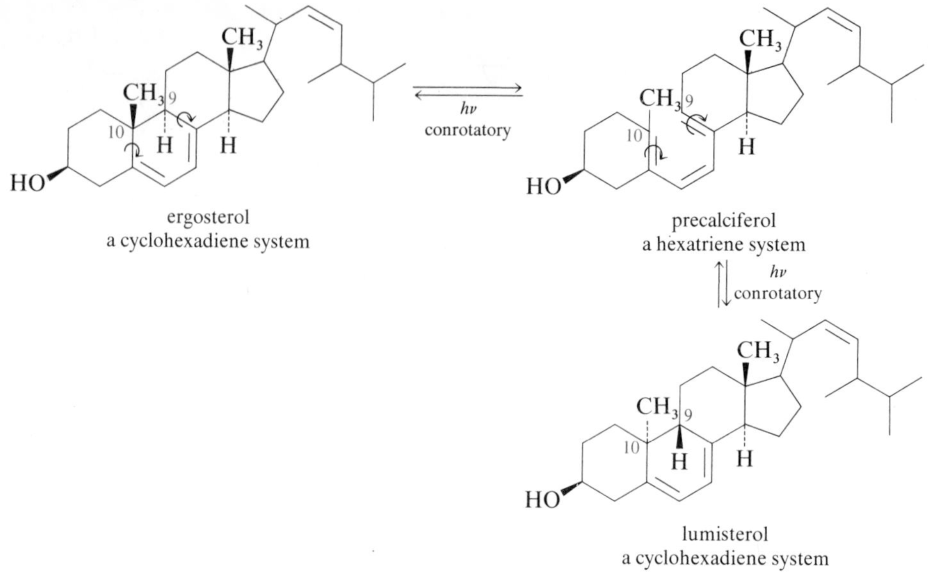

ergosterol
a cyclohexadiene system

precalciferol
a hexatriene system

lumisterol
a cyclohexadiene system

Precalciferol is converted back to ergosterol by a reversal of this motion, but a new cyclohexadiene is formed if ring closure takes place with further rotation of the carbon atoms of the triene in the same direction. This cyclohexadiene is a stereoisomer of ergosterol and is called *lumisterol*. In it, the methyl group at carbon 10 lies below, and the hydrogen atom at carbon 9 above, the plane of the steroid ring.

The most important reaction of precalciferol, from a biological viewpoint, is its transformation into calciferol by means of a [1,7] sigmatropic rearrangement (p. 1102).

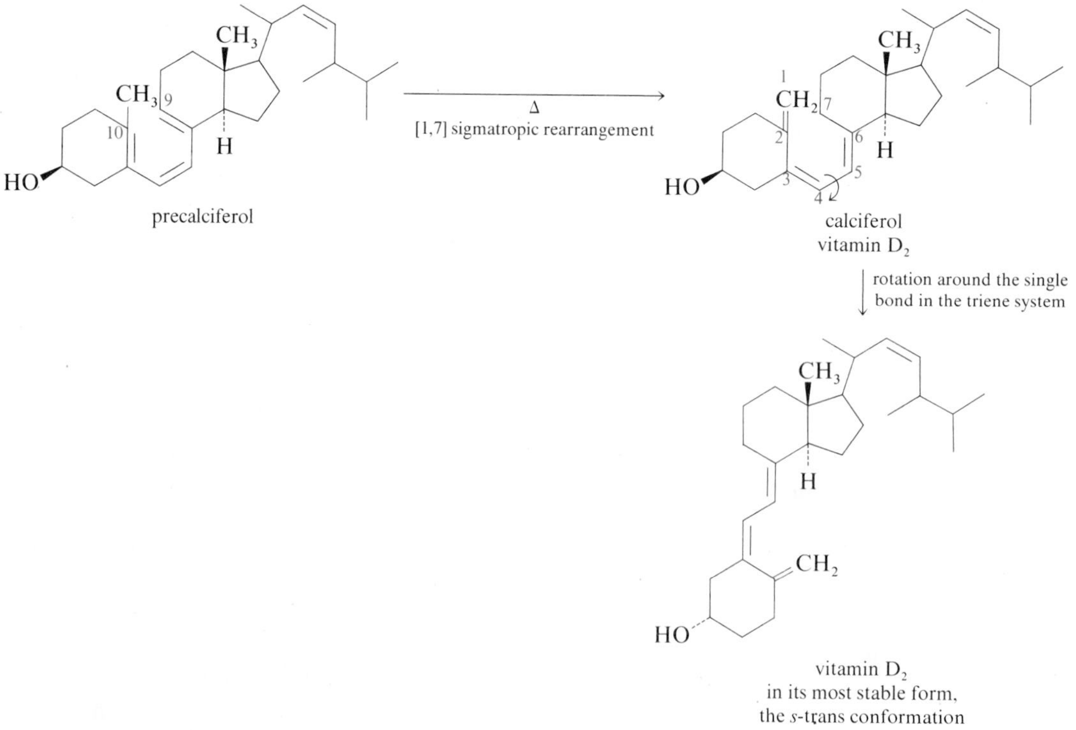

precalciferol

[1,7] sigmatropic rearrangement

calciferol
vitamin D$_2$

rotation around the single bond in the triene system

vitamin D$_2$
in its most stable form,
the *s*-trans conformation

The shift of the hydrogen atom results in a change in the position of the double bonds in the triene, allowing the molecule to assume a more stable conformation through rotation around the central single bond of the polyene system. Vitamin D_2 is usually shown in the *s*-trans conformation at that bond.

When precalciferol is heated, two disrotatory ring-closure reactions occur. The photochemical and thermal electrocyclic reactions that have been observed starting with the irradiation of ergosterol are summarized below.

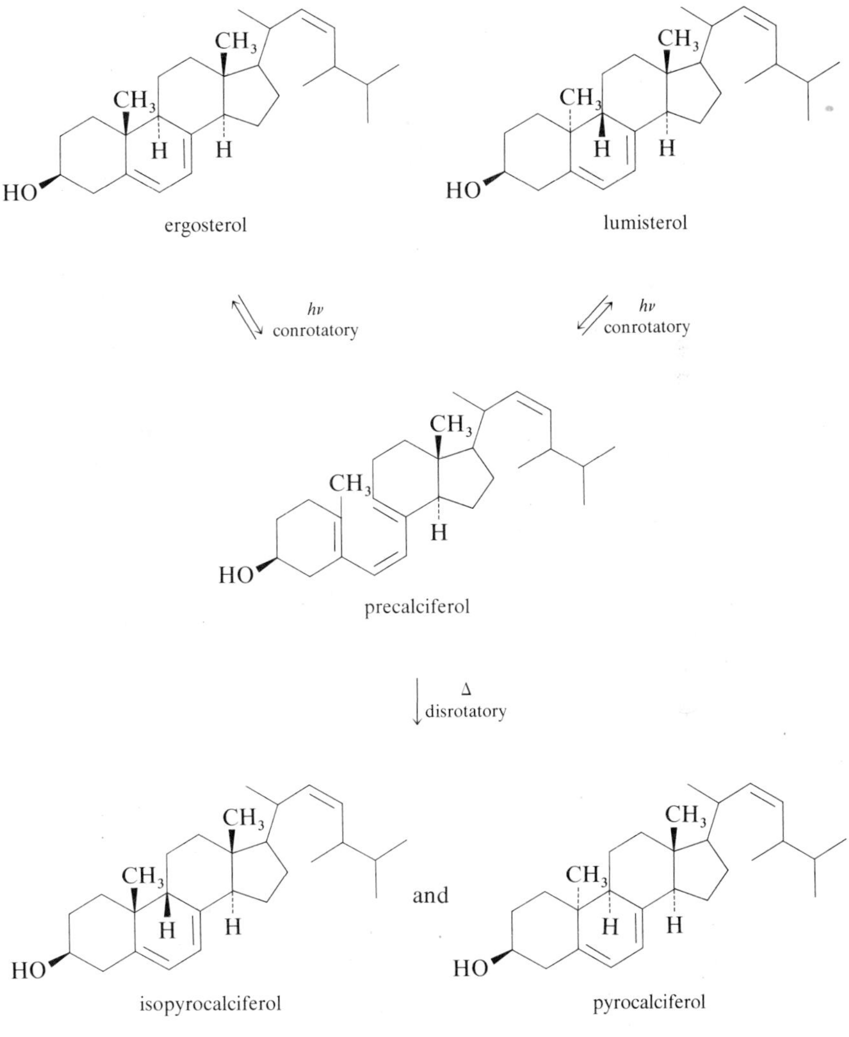

ergosterol

lumisterol

hv
conrotatory

hv
conrotatory

precalciferol

Δ
disrotatory

isopyrocalciferol

and

pyrocalciferol

It is easy to understand why chemists were intrigued when they observed reactions in which the stereochemistry depended so precisely on reaction conditions. To explain these observations, they developed for the interacting systems the theoretical models discussed in this section.

PROBLEM 19.19 Using molecular orbital pictures, show how precalciferol is converted thermally into pyrocalciferol and isopyrocalciferol.

PROBLEM 19.20   Predict what the products will be when the terpene allo-ocimene is irradiated.

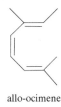

allo-ocimene

C. Some Generalizations about Cycloaddition and Electrocyclic Reactions

A cycloaddition reaction takes place in a concerted manner in the ground state only if $4n + 2$ electrons, where $n = 1, 2, \ldots$, are involved in the transition state. A cycloaddition reaction involving reagents with a total of $4n$ electrons participating in the transition state is concerted only in the excited state. These rules apply when the stereochemistry of the reaction has two unsaturated components approaching each other face to face, and when bonding occurs on the same side of the molecule at each end of the diene or alkene system. This is the stereochemistry that is shown for the Diels-Alder reaction in Figures 19.6 and 19.7, for example.

These rules, known as the **Woodward-Hoffmann rules,** indicate how the total number of electrons involved in the transition state of a cycloaddition reaction governs the manner of the cycloaddition (Table 19.1). The table simply restates the generalizations formulated in the first paragraph. A concerted cycloaddition reaction is said to be *allowed* for a system with $4n$ electrons if one of the components of the reaction is raised to the excited state, but is *forbidden* if both of the components are in the ground state. The word *forbidden* does not mean that a reaction cannot ever take place. It simply means that if a reaction is observed to occur under these forbidden conditions, it is not a concerted reaction.

Note that the generalization refers to the number of electrons and not to the number of orbitals. It is quite easy to remember the rules if you can keep firmly in mind that the Diels-Alder reaction is a $4n + 2$ system in which $n = 1$, and that it is allowed in the ground state. What is allowed in the ground state is forbidden in the excited state, and any change in the number of electrons by two, either up or down, changes the nature of the allowed reaction.

The experimental data that have been given for electrocyclic reactions also allow us to make generalizations in terms of the number of electrons involved in the transition state. These Woodward-Hoffmann rules are summarized in Table 19.2. Once again,

---

TABLE 19.1   Rules for Concerted Cycloaddition Reactions in Which Both Components Retain Stereochemistry

Total Number of Electrons in the Transition State	*Concerted Cycloaddition Reaction*	
$4n$	allowed in the excited state	forbidden in the ground state
$4n + 2$	forbidden in the excited state	allowed in the ground state

TABLE 19.2 **Rules for the Stereochemistry of Concerted Electrocyclic Reactions**

Total Number of Electrons in the Transition State	*Stereochemistry of the Allowed Concerted Reaction in the Ground State*	*Stereochemistry of the Allowed Concerted Reaction in the Excited State*
$4n$	conrotatory	disrotatory
$4n + 2$	disrotatory	conrotatory

we should stress that these rules do not mean that other types of reactions do not take place, but only that if a reaction is observed to disobey these rules, it is taking place by a different mechanism and is not a concerted reaction.

The two sets of rules given in Tables 19.1 and 19.2 sum up the experimental observations that have been described in this chapter. The basis for explaining these observations is molecular orbital theory. The theory allows us to make predictions about systems for which experimental data are not yet available, and about systems for which we do not happen to know the data. One of the predictions based on the rules is that systems containing four electrons on three atoms will add to alkenes in a concerted manner to give five-membered rings. The most fruitful experimental application of this prediction is described in the next section.

PROBLEM 19.21 A useful synthesis of substituted phenanthrenes is the photochemical cyclization of 1,2-diarylethenes. The reaction gives a dihydrophenanthrene, which is then oxidized easily by oxygen or iodine to the phenanthrene. The reaction is shown below for 1,2-diphenylethene.

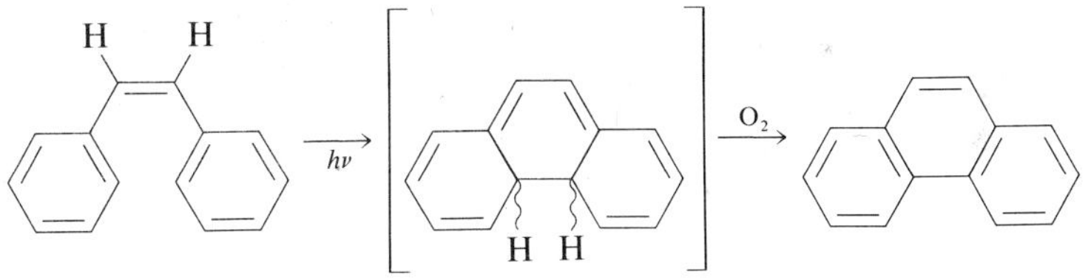

a dihydrophenanthrene intermediate

If all oxygen is removed from the reaction mixture before the irradiation, the dihydrophenanthrene can be isolated. How would you classify this reaction according to the Woodward-Hoffmann rules? What kind of stereochemistry would you expect the dihydrophenanthrene to have?

PROBLEM 19.22 The Woodward-Hoffmann rules predict that ionic species should also undergo electrocyclic reactions. For example, the following reaction has been observed.

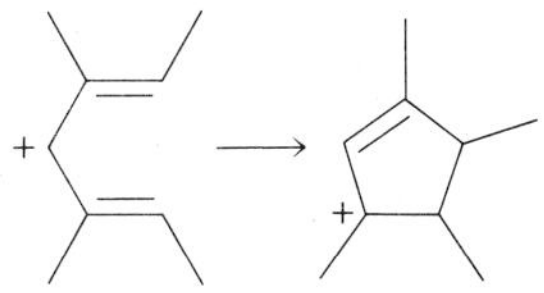

How would you classify this reaction of a pentadienyl cation? What stereochemistry do you expect for the product cation? (Hint: How many electrons are involved in the transition state?)

PROBLEM 19.23 When cyclopentadiene and 3-iodo-2-methyl-1-propene are allowed to react in liquid sulfur dioxide in the presence of silver trifluoroacetate, the products that are formed can be rationalized as arising from the bicyclic tertiary cation shown below. How would you classify the reaction? What is a likely intermediate? How many electrons are involved in the transition state?

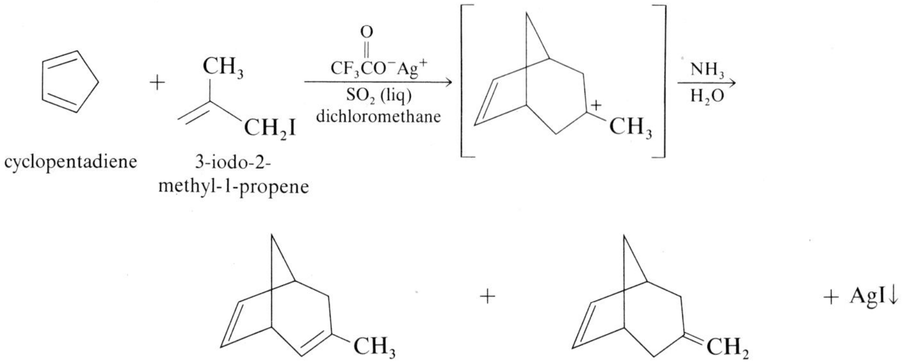

cyclopentadiene   3-iodo-2-methyl-1-propene

3-methylbicyclo[3.2.1]octa-2,6-diene
40%

3-methylenebicyclo[3.2.1]oct-6-ene
16%

+ AgI↓

1,3-Dipolar Additions. Cycloadditions of Compounds Containing Nitrogen and Oxygen

Molecular orbital theory is by no means limited to simple cycloaddition and electrocyclic reactions; it is a powerful tool that can be applied to many different types of reactions. One of the most significant extensions of the theory was made by Rolf Huisgen of West Germany, who recognized that certain species containing oxygen or nitrogen could serve as sources of four electrons in [4 + 2] cycloaddition reactions and could thus be used to create a variety of heterocyclic compounds. These reactions have all the characteristics of concerted reactions, including insensitivity to reaction conditions such as solvent polarity and high stereoselectivity in cases where this has been investigated. Some of these reactions are explored below.

Some species, while reasonably stable and isolable, have structures that can best be represented as a series of resonance contributors, one of which has a formal positive charge on one atom and a formal negative charge two atoms away. Structures characterized by this distribution of formal charge are called **1,3-dipolar species.** Ozone is such a 1,3-dipolar species (Section 8.6A), and so is phenyl azide. All the resonance contributors of phenyl azide have formal charges on various atoms. The 1,3-dipolar character of the compound is represented by the last structural formula shown below.

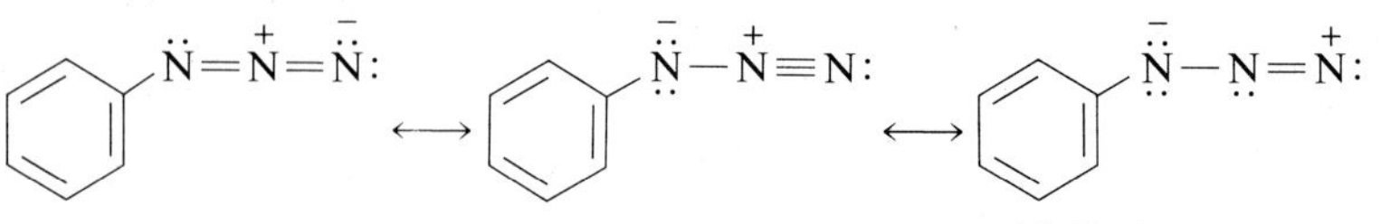

1,3-dipolar character

resonance contributors of phenyl azide

1,3-Dipolar species add to alkenes or alkynes (called **dipolarophiles**) in cycloaddition reactions. The addition of ozone or phenyl azide to an alkene involves the participation of a pair of π electrons and a pair of nonbonding electrons from the dipolar species, and a pair of π electrons from the alkene.

2,3-dimethyl-
2-butene

*a 2-electron
system*

ozone

*a 4-electron
system*

molozonide of
2,3-dimethyl-2-butene

*a [4 + 2] cycloaddition
product*

cyclopentene

*a 2-electron
system*

phenyl azide

*a 4-electron
system*

a triazoline

*a [4 + 2] cycloaddition
product*

and enantiomer

A description of each reaction from an electronic point of view would classify it as a [4 + 2] cycloaddition reaction, which is allowed in the ground state. The four electrons in the dipolar species are distributed over three p orbitals, and all three atoms in the dipolar species participate in the bonding. The product is, therefore, a five-membered ring, containing two atoms from the double bond of the alkene and three atoms from the dipolar species.

An important 1,3-dipolar species is diazomethane, CH_2N_2, a yellow gas that is explosive and highly toxic. Diazomethane is always handled in ether solution and protected from heat and light in order to decrease the danger of an explosive decomposition.

The structure of diazomethane, like those of ozone and phenyl azide, can only be described by writing a series of resonance contributors.

$$:\overset{-}{C}H_2 - \overset{+}{N} \equiv N: \longleftrightarrow CH_2 = \overset{+}{N} = \overset{..}{\overset{-}{N}}: \longleftrightarrow :\overset{-}{C}H_2 - N = \overset{+}{N}: \longleftrightarrow \overset{+}{C}H_2 - \overset{..}{N} = \overset{..}{\overset{-}{N}}:$$

most important resonance
contributors

1,3-dipolar character

resonance contributors for diazomethane

The first two are the most important resonance contributors for diazomethane because in them each atom is surrounded by an octet of electrons. The other two contributors show the 1,3-dipolar character of the molecule.

Diazomethane reacts with alkenes as a 1,3-dipolar species to give five-membered heterocyclic rings. For example, it reacts with stereoisomeric esters to give diastereomeric products, as shown on the next page.

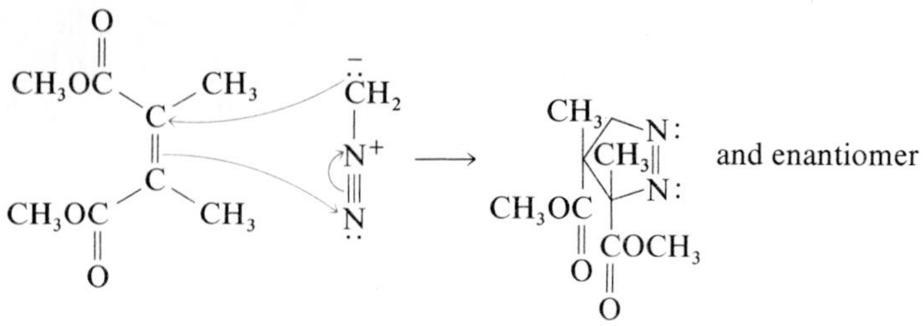

dimethyl
dimethylmaleate

diazomethane

a pyrazoline

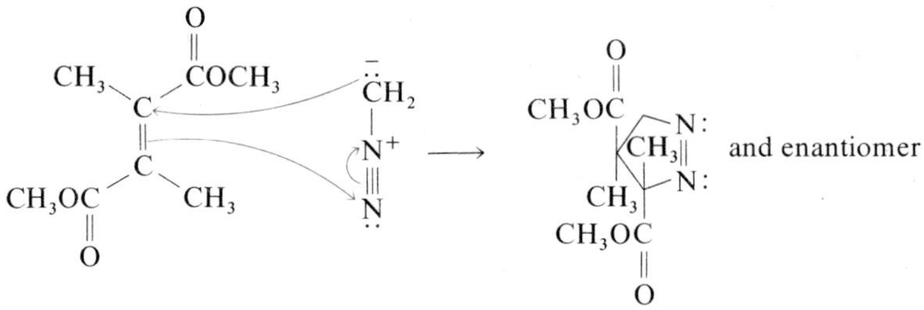

dimethyl dimethylfumarate

diazomethane

a pyrazoline

a 2-electron system

a 4-electron system

*a [4 + 2] cycloaddition
product*

The retention of the stereochemical relationships of the substituents in the isomeric alkenes in the product pyrazolines clearly indicates that the reaction is stereoselective. Diazomethane adds to one face or the other of the alkene molecule in a concerted fashion.

Diazomethane adds to other alkenes as well. For example, it reacts with styrene to give a pyrazoline.

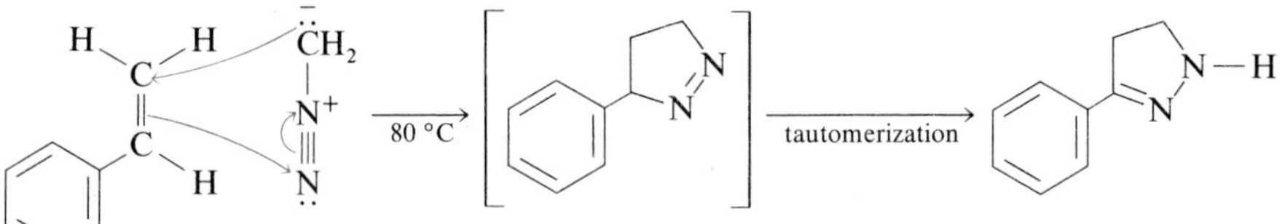

styrene

diazomethane

a 1-pyrazoline

a 2-pyrazoline

*a 2-electron
system*

*a 4-electron
system*

*a [4 + 2] cycloaddition
product*

The original products from the addition of diazomethane to alkenes are 1-pyrazolines, in which the double bond in the heterocyclic ring is between the two nitrogen atoms. Such compounds usually tautomerize easily to a more stable form, the 2-pyrazolines, which are the isolated products.

Another compound with 1,3-dipolar character is formed when an *N*-substituted hydroxylamine reacts with a carbonyl compound. Such compounds are called *nitrones*. For example, benzaldehyde reacts with hydroxylamine to give an oxime (p. 893) and with *N*-methylhydroxylamine to give a nitrone.

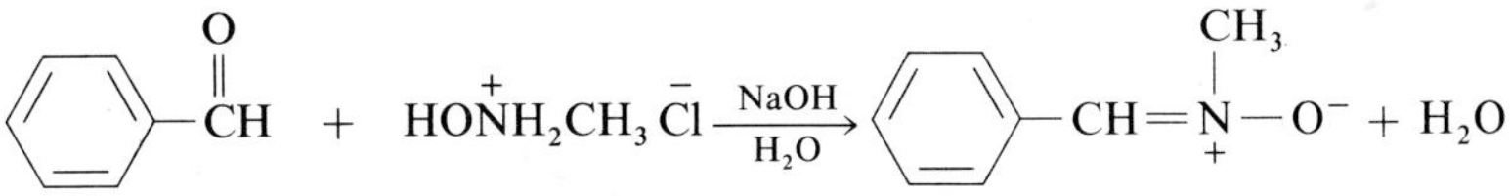

benzaldehyde   *N*-methylhydroxylamine   *N*-methyl-*C*-phenylnitrone
             hydrochloride   100%

The mechanism for the reaction of a substituted hydroxylamine with a carbonyl compound is similar to that for the formation of all the ammonia derivatives of aldehydes and ketones (Section 9.6B). The difference in products arises from the lack of a second hydrogen atom on the nitrogen atom of the substituted hydroxylamine.

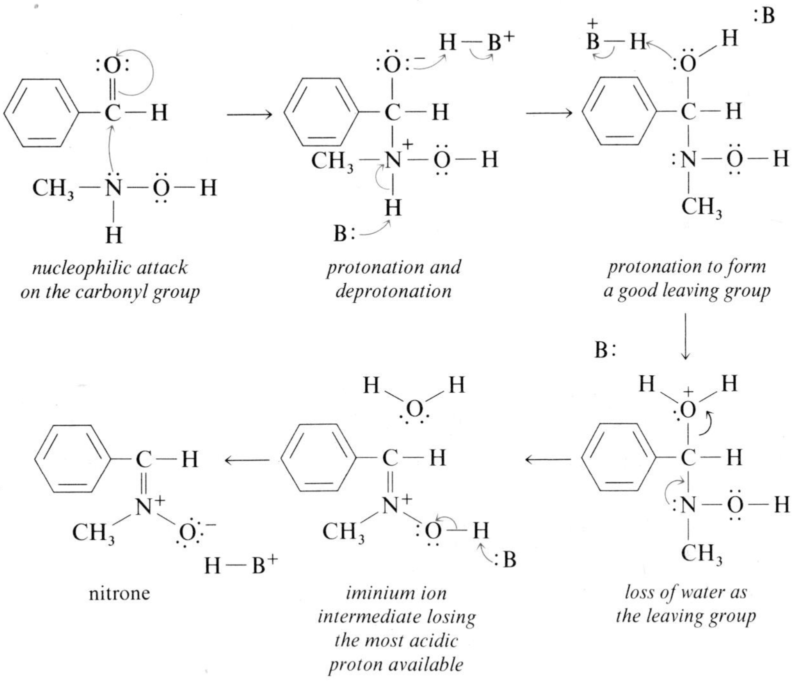

nucleophilic attack   *protonation and*   *protonation to form*
on the carbonyl group   *deprotonation*   *a good leaving group*

nitrone   *iminium ion*   *loss of water as*
             intermediate losing   *the leaving group*
             the most acidic
             proton available

The iminium ion intermediate formed in the synthesis of an oxime has a hydrogen atom on the positively charged nitrogen atom. That hydrogen atom is the most acidic one in the molecule and is lost in the last step of the reaction. In the reaction of an *N*-substituted hydroxylamine, the iminium ion intermediate has no hydrogen atom on the quaternary nitrogen atom. In this case, the most acidic hydrogen atom is the one on the hydroxyl group. Removal of that proton by a base gives a nitrone.

The nitrone is a 1,3-dipolar compound and reacts as such with alkenes. For example, it adds to styrene to give a five-membered ring heterocycle.

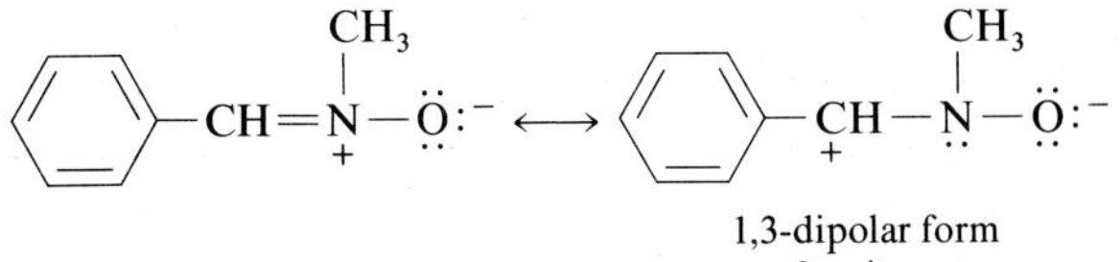

1,3-dipolar form
of a nitrone

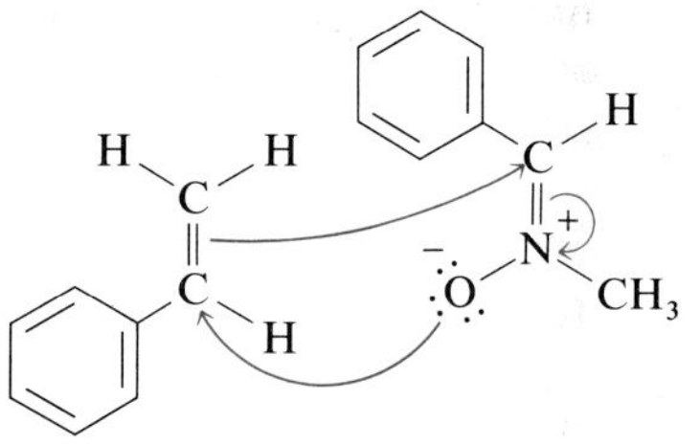

styrene   *N*-methyl-*C*-phenylnitrone

a 2-electron   *a 4-electron*
system   *system*

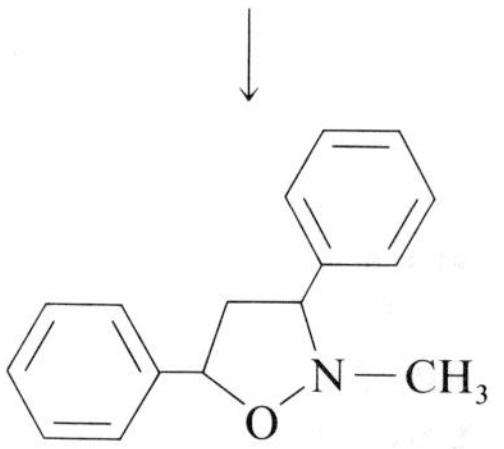

cis and *trans*-2-methyl-
3,5-diphenylisoxazolidine
95%

racemic
a [4 + 2] cycloaddition product

The saturated five-membered ring heterocycle containing adjacent oxygen and nitrogen atoms is known as an isoxazolidine (p. 1011). The formation of an isoxazolidine from a nitrone and an alkene is pictured as a [4 + 2] cycloaddition reaction. The nitrone contributes four electrons, two from a π bond and two nonbonding electrons from the oxygen atom, to the transition state of the reaction.

The regioselectivity shown for the addition of nitrones to alkenes is quite general. Terminal alkenes give products in which the substituent is at carbon 5 of the isoxazolidine.

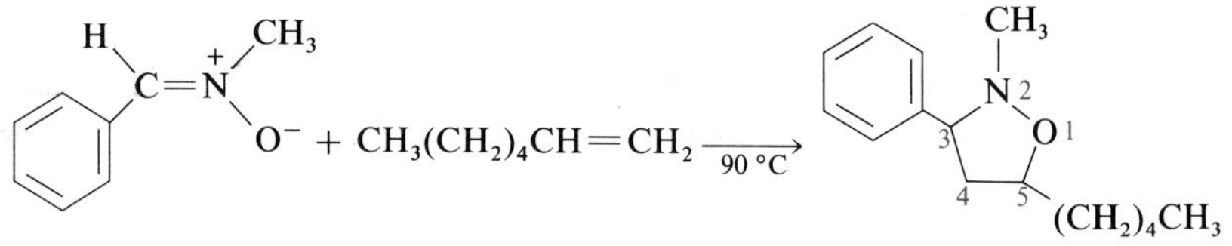

N-methyl-*C*-phenylnitrone   1-heptene   2-methyl-5-pentyl-3-
phenylisoxazolidine
76%

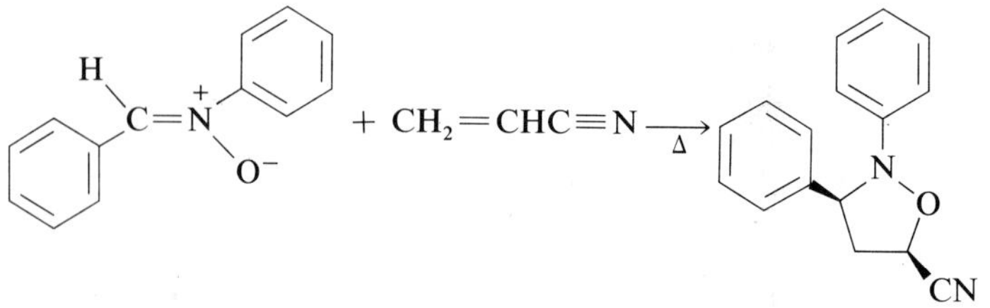

C,N-diphenylnitrone   propenenitrile   *cis*-5-cyano-2,3-
diphenylisoxazolidine
100%

It is not necessary to synthesize and isolate a nitrone in order to take advantage of the reaction leading to isoxazolidines. Mixtures of a carbonyl compound, an *N*-

substituted hydroxylamine, and an alkene heated together give high yields of isoxazolidines. For example, butanal, *N*-phenylhydroxylamine, and styrene react to give 2,5-diphenyl-3-propylisoxazolidine.

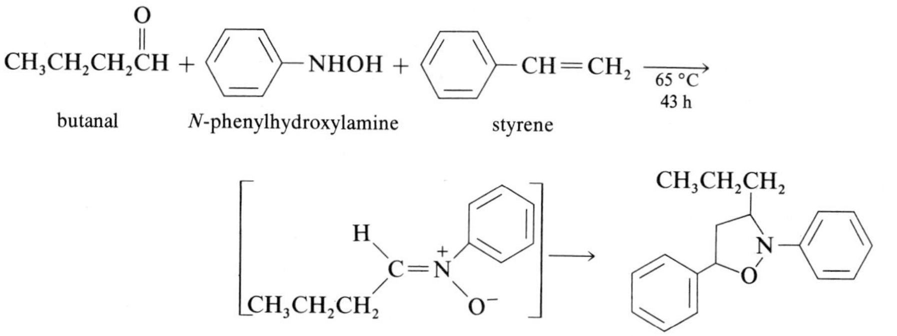

$$CH_3CH_2CH_2\overset{\overset{\displaystyle O}{\|}}{C}H + \langle\!\!\!\bigcirc\!\!\!\rangle\!\!-\!NHOH + \langle\!\!\!\bigcirc\!\!\!\rangle\!\!-\!CH\!=\!CH_2 \xrightarrow[43\,h]{65\,°C}$$

butanal   N-phenylhydroxylamine   styrene

$$\left[\begin{array}{c} \overset{H}{\underset{CH_3CH_2CH_2}{\diagdown}}C\!=\!\overset{+}{N}\!\!-\!\!\overset{\langle\bigcirc\rangle}{\underset{O^-}{}} \end{array} \right] \longrightarrow$$

N-phenyl-C-propylnitrone   2,5-diphenyl-3-propylisoxazolidine
(stereochemistry not investigated)
99%

The reactions described above are only a few examples of the kinds of heterocyclic systems that can be created by the addition of 1,3-dipoles to alkenes or alkynes.

PROBLEM 19.24   Write equations predicting what the products of the following reactions will be.

(a) 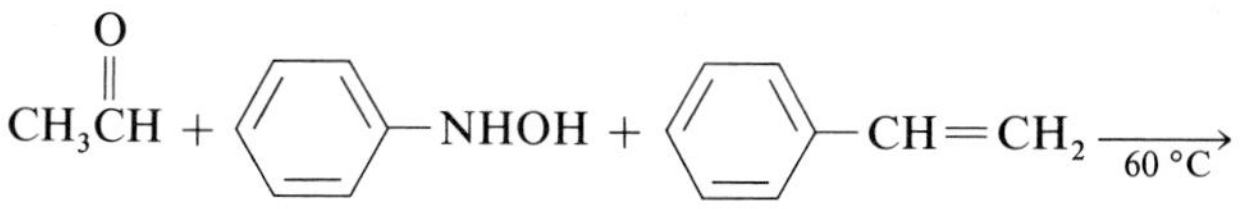 $CH_3\overset{\overset{\displaystyle O}{\|}}{C}H + \langle\!\!\!\bigcirc\!\!\!\rangle\!\!-\!NHOH + \langle\!\!\!\bigcirc\!\!\!\rangle\!\!-\!CH\!=\!CH_2 \xrightarrow{60\,°C}$

(b) $CH_2\!=\!\overset{+}{N}\!=\!N^- (1\ molar\ equiv) + CH_2\!=\!CHCH\!=\!CH_2 \xrightarrow[diethyl\ ether]{}$

(c) 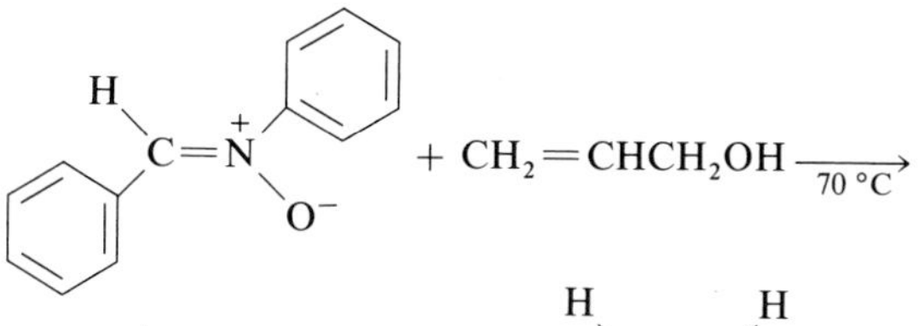 $\overset{H}{\underset{}{\diagdown}}C\!=\!\overset{+}{N}\!\!-\!\!O^- + CH_2\!=\!CHCH_2OH \xrightarrow{70\,°C}$

(d) 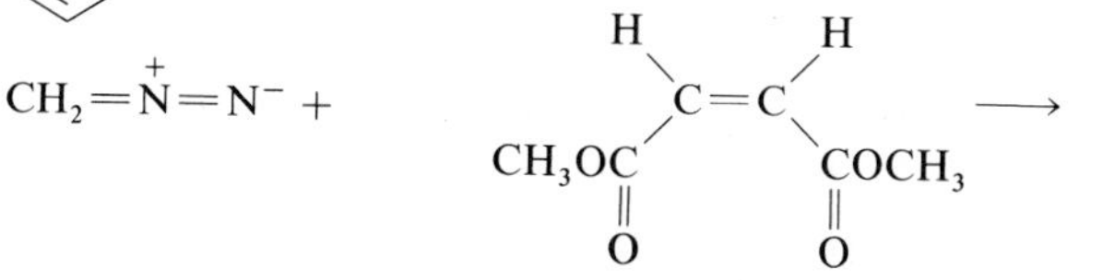 $CH_2\!=\!\overset{+}{N}\!=\!N^- + \underset{CH_3O\overset{}{\underset{\|}{C}}}{\overset{H}{\diagdown}}C\!=\!C\underset{\overset{\|}{C}OCH_3}{\overset{H}{\diagup}} \longrightarrow$

(e) 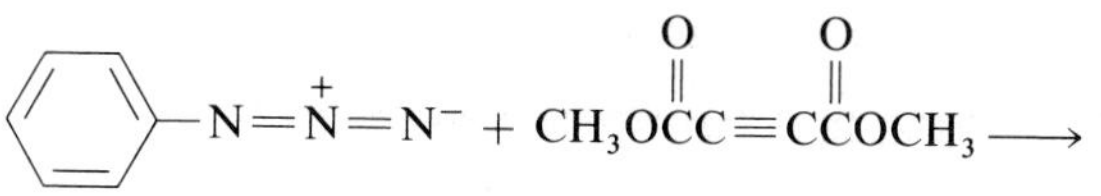 $\langle\!\!\!\bigcirc\!\!\!\rangle\!\!-\!N\!=\!\overset{+}{N}\!=\!N^- + CH_3O\overset{\overset{\displaystyle O}{\|}}{C}C\!\equiv\!C\overset{\overset{\displaystyle O}{\|}}{C}OCH_3 \longrightarrow$

(f) $CH_2\!=\!\overset{+}{N}\!=\!N^- + HC\!\equiv\!CH \xrightarrow{20\,°C}$

(g) 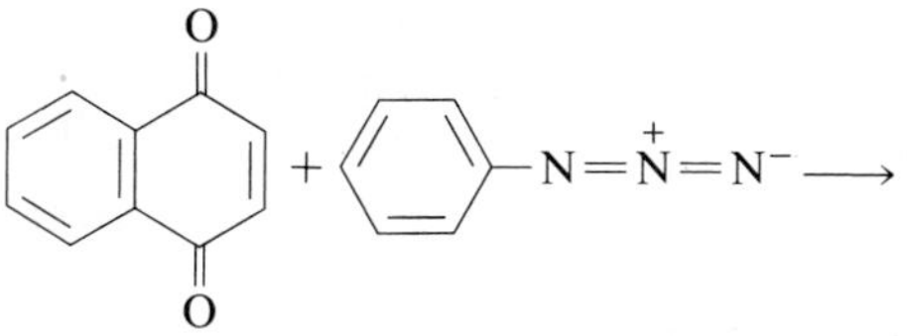   (naphthoquinone) $+ \langle\!\!\!\bigcirc\!\!\!\rangle\!\!-\!N\!=\!\overset{+}{N}\!=\!N^- \longrightarrow$

(h)

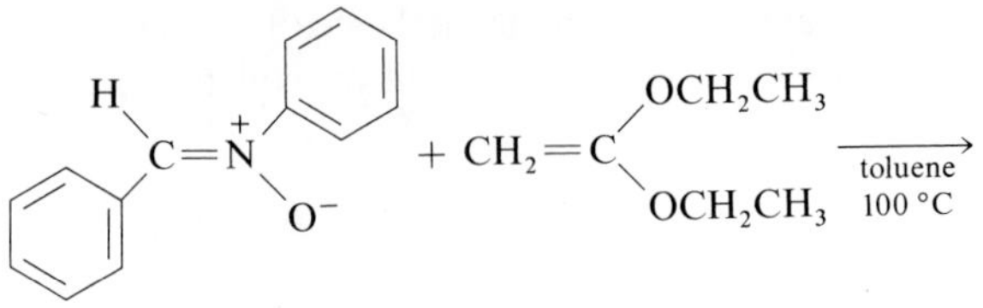

(i)  Cl—⬡—CH + HONH₂CH₃ Cl⁻ $\xrightarrow[\Delta]{\text{Na}}$ ethanol

$$ \text{Cl}-\bigcirc-\overset{\overset{\displaystyle O}{\|}}{\text{CH}} + \text{HO}\overset{+}{\text{N}}\text{H}_2\text{CH}_3\ \text{Cl}^- $$

PROBLEM 19.25 The following heterocyclic compounds are obtained by 1,3-dipolar cycloaddition reactions. For each compound, draw the structure of the alkene (or alkyne) and the 1,3-dipolar species that would be required for the synthesis.

(a)

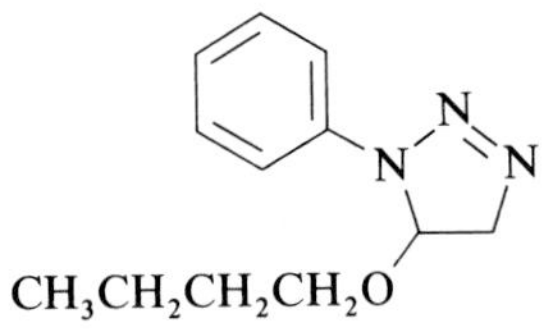

CH₃CH₂CH₂CH₂O

(b)

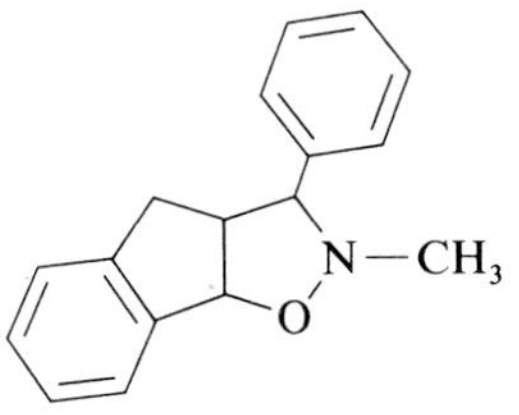

(c)

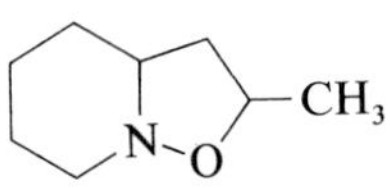

—CH₃

(d)

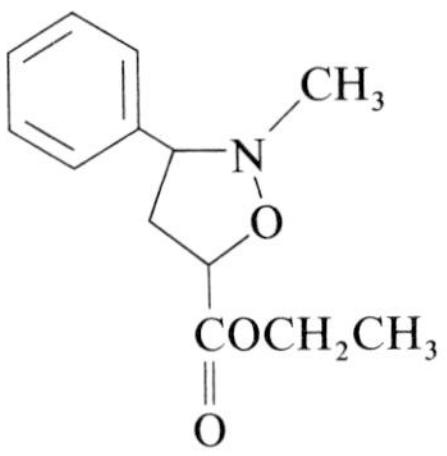

COCH₂CH₃

(e)

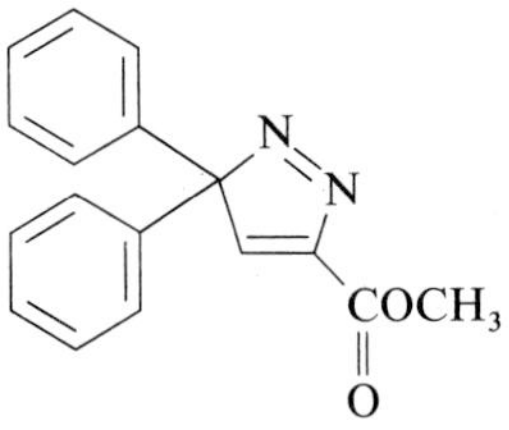

COCH₃

(f)

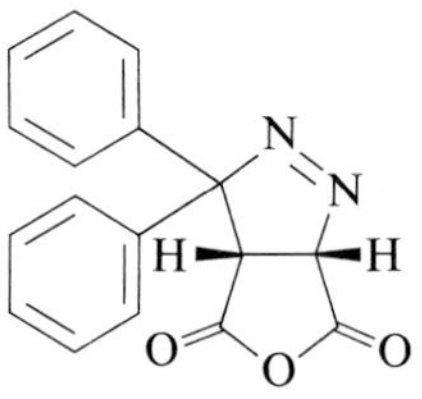

(g)

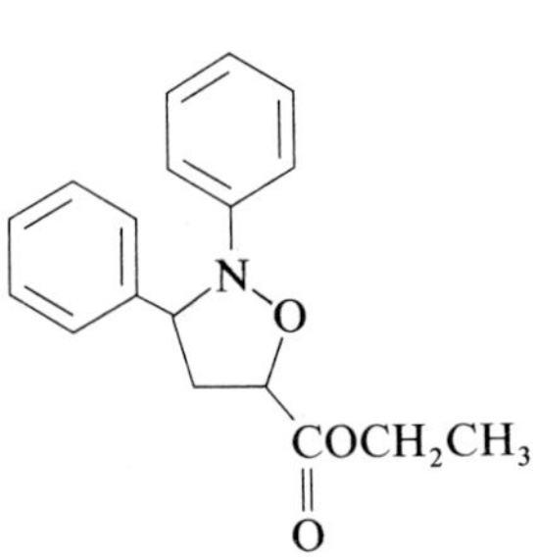

COCH₂CH₃

(h)

CH₃CH₂CH₂CH₂O

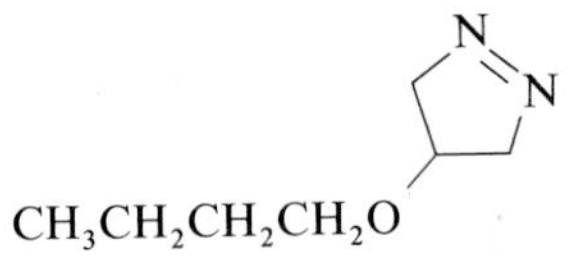

19.5

Sigmatropic Rearrangements

A. Hydrogen Shifts

A sigmatropic rearrangement involves the migration of a σ bond from one position to another one along a polyene chain while there is a simultaneous shift of π bonds. In Section 19.1A, such a migration of a hydrogen atom and its σ bond in (Z)-1,3-pentadiene was described (p. 1072). The migration occurs in a highly stereoselective way with the hydrogen atom moving along one surface of the polyene system.

The stereoselectivity of the reaction can be explained if a molecular orbital picture is drawn. In Figure 19.13, the process is depicted as the interaction of the HOMO of the carbon-hydrogen σ bond at the tetrahedral carbon atom with the LUMO of the diene that is the remainder of the molecule. A bonding interaction can take place between the *s* orbital of the hydrogen atom and the molecular orbital of the diene system at carbon 5 only if the hydrogen atom moves from the top of carbon 1 to the top of carbon 5. A new diene system with double bonds between carbons 1 and 2 and carbons 3 and 4 forms as the hydrogen atom migrates.

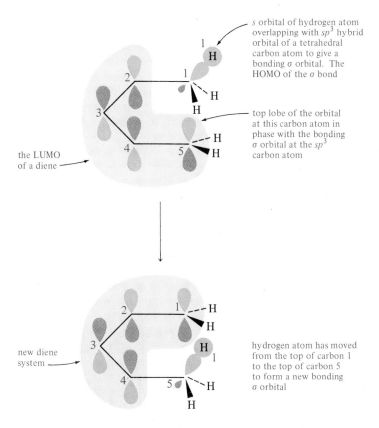

Figure 19.13   A molecular orbital picture of the stereochemistry of a [1,5] sigmatropic hydrogen migration.

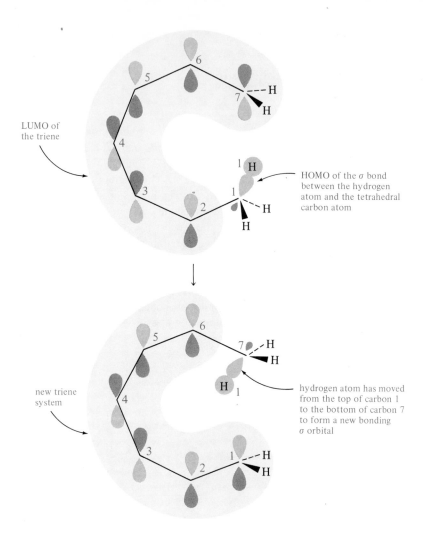

Figure 19.14   A molecular orbital picture of the stereochemistry of a [1,7] sigmatropic hydrogen migration.

A similar look at the [1,7] sigmatropic shift that occurs when precalciferol is converted to calciferol (p. 1090, Section 19.3B) suggests an interesting prediction (Figure 19.14). The reaction is pictured as an interaction between the HOMO of the σ bond that is migrating and the LUMO of the triene system. If you examine the symmetry of these orbitals, you see that it is the bottom lobe of the orbital at carbon 7 that is in phase with the σ bond at carbon 1. The hydrogen atom must move from the top of carbon 1 to the bottom face of carbon 7 for a bonding interaction to occur between the hydrogen atom and the newly forming tetrahedral carbon atom. Such a migration is stereochemically possible in this kind of an open-chain triene because one end of the molecule lies over the top of the other end in one conformation of the system.

A more general way of looking at these reactions would be to consider the total number of electrons involved in the transition state. The [1,5] hydrogen shift occurs with a reorganization of six electrons, two from the σ bond and four in the diene system. In the drawing of the molecular orbital interactions for the reaction shown in

Figure 19.13, the reaction has been treated as such a [4 + 2] interaction. A further generalization is that all sigmatropic rearrangements involving $4n + 2$ electrons in the transition state take place in a concerted fashion in the ground state with the stereochemistry shown in Figure 19.13. The hydrogen and alkyl shifts seen in the rearrangements of carbocations (Section 5.6B) are concerted [1,2] sigmatropic shifts that take place along one face of the molecule. In these cases, only the two electrons of the migrating σ bond are involved in the transition state, as the bond shifts to a cationic carbon atom with an empty orbital. Such a case would be a rearrangement of the system with $4n + 2$ electrons where $n = 0$.

Using the same language, a [1,7] hydrogen shift would be described as having six electrons from the triene system and two electrons from the σ bond involved in the transition state, for a total of eight electrons. Thus, such a reaction is a $4n$ system, with $n = 2$. Rearrangements involving the $4n$ electrons are concerted in the ground state when the stereochemistry is that shown in Figure 19.14, with migration taking place from the top face of one part of the molecule to the bottom face of the other end.

Just as for electrocyclic reactions and cycloaddition reactions, in sigmatropic rearrangements, the symmetry of the orbitals involved and the resulting stereochemistry of the reactions change in the excited state. The problem below provides an opportunity for you to prove this to yourself.

PROBLEM 19.26   When a hexatriene system undergoes a photochemical [1,7] hydrogen shift, the hydrogen atom migrates from the top of carbon 1 to the top of carbon 7. This observation can be rationalized by drawing a molecular orbital picture similar to the one shown in Figure 19.14, but using the HOMO that is appropriate for the excited state of the triene. Prove to yourself that this is so. You may want to refer back to Problem 19.12 to see what the HOMO of the excited state of a triene looks like. You must now, of course, use the LUMO of the σ bond to the hydrogen atom.

B. The Cope Rearrangement

Not all sigmatropic rearrangements involve migration of hydrogen atoms. A large and important group of reactions takes place with the migration of carbon σ bonds. A 1,2-alkyl shift in a carbocation is an example that you have already learned. A group of such reactions that do not have ionic intermediates is known as the **Cope rearrangement,** for Arthur Cope of the Massachusetts Institute of Technology who discovered and studied them.

A typical Cope rearrangement is the interconversion of 3-methyl-1,5-hexadiene and (*E*)-1,5-heptadiene.

3-methyl-1,5-hexadiene   (*E*)-1,5-heptadiene

Either compound when heated to approximately 300 °C is converted into a mixture of the two. The two alkenes are similar to each other in energy, so an equilibrium is established and the reaction does not go to completion in one direction or the other.

If the diene is designed so that at least one of the double bonds in the rearranged diene is conjugated with an aryl group or a carbonyl group, then the reaction takes place at lower temperatures to give good yields of the new diene. An example of such a reaction is the rearrangement of 3-methyl-4-phenyl-1,5-hexadiene.

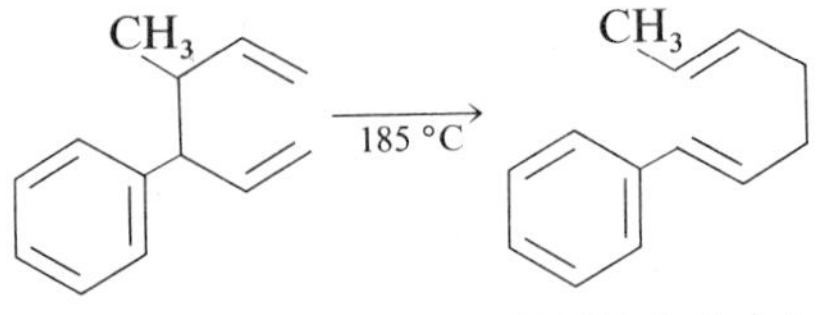

3-methyl-4-phenyl-
1,5-hexadiene

(1E,5E)-1-phenyl-
1,5-heptadiene
90%

In the product, (1E,5E)-1-phenyl-1,5-heptadiene, one of the double bonds is conjugated with the aromatic ring, and the other one is an internal double bond. The product diene is more stable than 3-methyl-4-phenyl-1,5-hexadiene, in which both of the double bonds are in terminal positions.

Another example of a Cope rearrangement in which one of the double bonds is conjugated in the product diene is shown below.

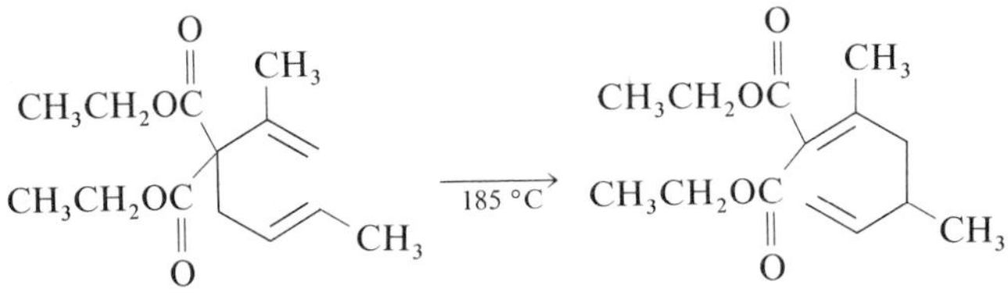

(E)-3,3-dicarboethoxy-
2-methyl-1,5-heptadiene

ethyl 2-carboethoxy-3,5-dimethyl-
2,6-heptadienoate

(E)-3,3-Dicarboethoxy-2-methyl-1,5-heptadiene is a derivative of diethyl malonate. After rearrangement, one of the double bonds is conjugated with both of the ester groups. The product diene is therefore more stable than the starting diene, so that compound predominates at equilibrium.

A reexamination of the rearrangement of 3-methyl-1,5-hexadiene to (E)-1,5-heptadiene (p. 1103) shows that the reaction involves breaking a σ bond at one part of the diene chain and forming a σ bond at another, while the π bonds in the molecule shift position.

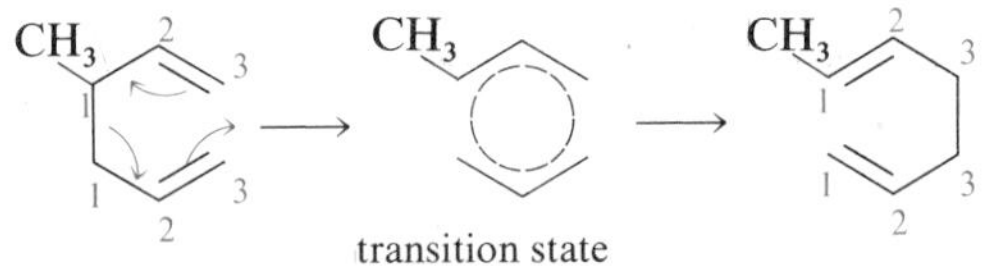

transition state

a [3,3] sigmatropic rearrangement

The arrows indicate that the Cope rearrangement has a transition state involving six electrons, two from the σ bond that is breaking and four from the two π bonds. The hexadiene is numbered with the carbon atoms of the breaking σ bond as the first carbon atoms of both halves of the chain. In this way of looking at the molecule, the new σ bond forms at the third atom of each half of the chain. In other words, the σ bond has migrated three atoms away for the top half of the molecule and also three atoms away for the bottom half. For this reason the reaction is called a [3,3] sigmatropic rearrangement.

PROBLEM 19.27   Complete the following equations.

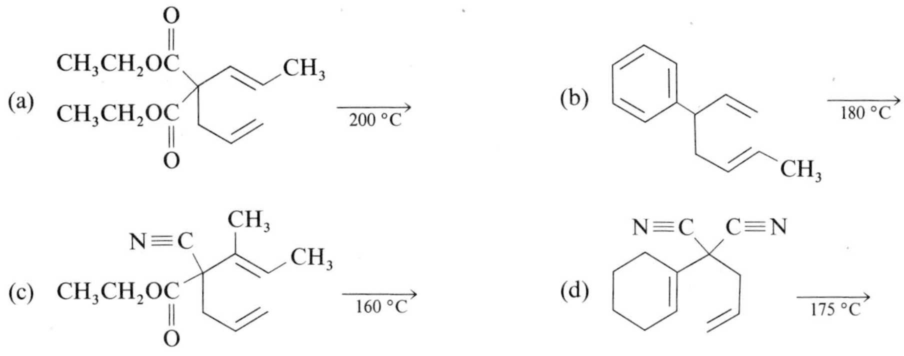

(a)
$$\xrightarrow{200\ °C}$$

(b)
$$\xrightarrow{180\ °C}$$

(c)
$$\xrightarrow{160\ °C}$$

(d)
$$\xrightarrow{175\ °C}$$

PROBLEM 19.28   A synthesis of grandisol, a sex pheromone of the male boll weevil (p. 554), started with a metal-catalyzed dimerization of isoprene to form what the researchers hoped would be a cyclobutane with the following structure. The product that is isolated when the reaction mixture is allowed to exceed room temperature is 1,5-dimethyl-1,5-cyclooctadiene. How would you rationalize this observation?

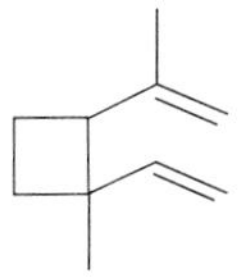

C. The Claisen Rearrangement

Sigmatropic rearrangements involving the cleavage of a σ bond at an oxygen atom are called **Claisen rearrangements.** The same Ludwig Claisen who discovered the condensation reaction of esters to form β-ketoesters (Section 15.5C) also discovered that the allyl enol ether of ethyl acetoacetate undergoes a rearrangement to an acetoacetic ester substituted at the α-carbon atom by an allyl group.

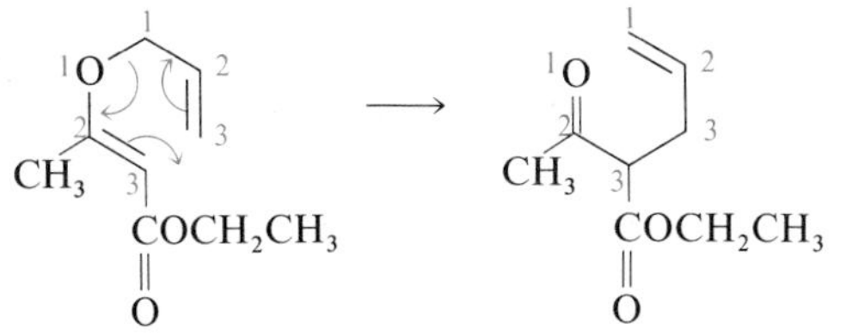

O-allyl ether of   ethyl 2-allylacetoacetate
the enol of ethyl   85%
acetoacetate

the Claisen rearrangement, a [3,3] sigmatropic
rearrangement

A σ bond is broken between two atoms, an oxygen atom and a carbon atom in this case, and another one is formed three atoms away on each portion of the chain. The σ bond migrates three atoms away from its starting position for both portions of the chain that participate in the rearrangement; thus, this reaction is also called a [3,3] sigmatropic

rearrangement. The driving force for the reaction is the formation of the stable carbonyl group in the product.

The Claisen rearrangement is quite general. The rearrangement of the allyl ethers of
phenols is another example of the reaction. For example, allyl phenyl ether, easily
prepared from phenol, rearranges to give *o*-allylphenol.

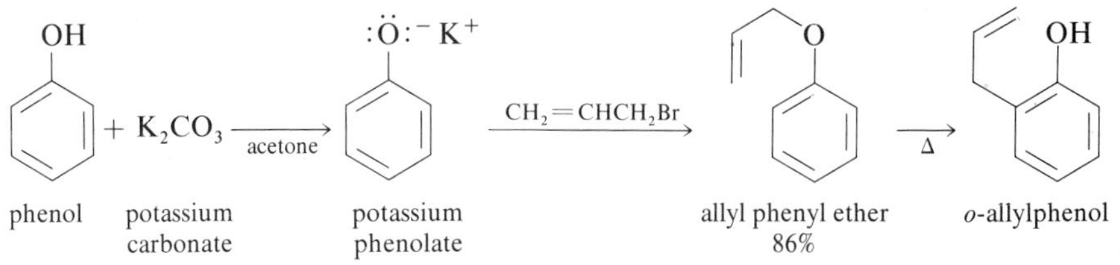

The rearrangement is postulated to go through a cyclic transition state.

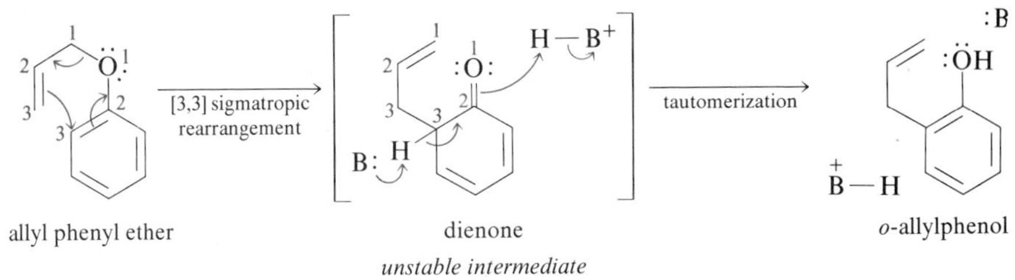

Two of the π electrons of the aromatic ring participate in the reaction, which involves
a total of six electrons in the transition state. The mechanism shown above requires that
an unstable dienone be an intermediate in the reaction. Such a dienone is expected to
tautomerize rapidly to the stable phenol by loss of a proton at the aromatic ring and
protonation of the oxygen atom.

A consequence of the mechanism is that the carbon atom attached to the aromatic
ring is not the carbon atom that was bonded to the oxygen atom in the ether. This
postulate can be tested. If a 2-butenyl instead of an allyl group is present in a phenyl
ether, the product phenol has a methyl branch on the side chain.

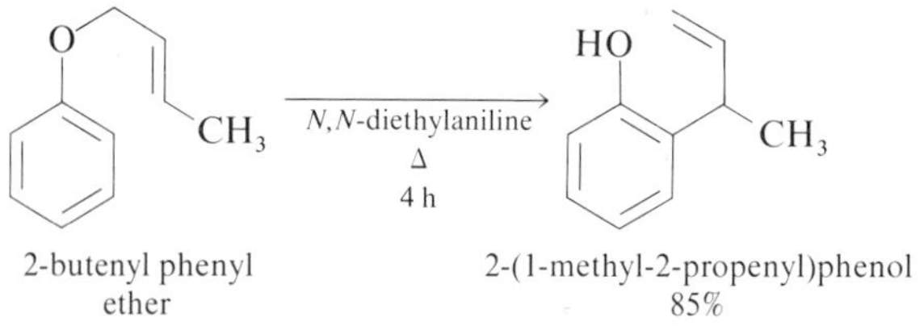

This experiment, and others like it, show that attachment between the aromatic ring
and the allyl side chain takes place three carbon atoms away from the oxygen atom.
The allyl group is not detached from the oxygen atom and reattached to the aromatic
ring at the original position of bonding.

An allyl phenyl ether in which the ortho positions are blocked rearranges to put the
allyl group on the para position of the ring. Thus, allyl 2,6-dimethylphenyl ether
rearranges to 4-allyl-2,6-dimethylphenol.

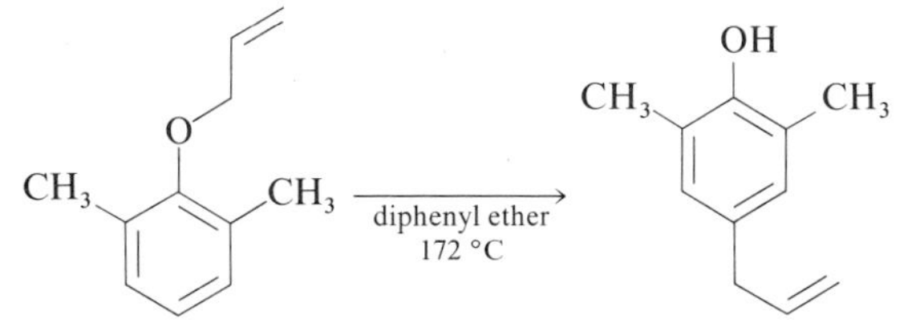

allyl 2,6-dimethylphenyl
ether

4-allyl-2,6-dimethylphenol

The reaction is postulated to go through a dienone intermediate in which rearrangement to the ortho position has occurred, followed by a second rearrangement to the para position.

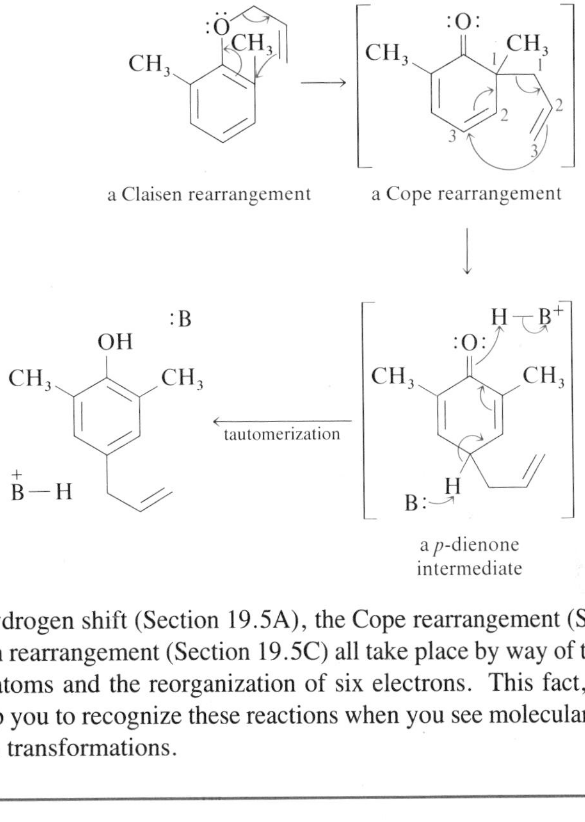

a Claisen rearrangement

a Cope rearrangement

tautomerization

a *p*-dienone
intermediate

The [1,5] hydrogen shift (Section 19.5A), the Cope rearrangement (Section 19.5B), and the Claisen rearrangement (Section 19.5C) all take place by way of transition states involving six atoms and the reorganization of six electrons. This fact, kept firmly in mind, will help you to recognize these reactions when you see molecular structures that allow for such transformations.

PROBLEM 19.29 Complete the following equations.

(a) $CH_2=CHOCH_2CH=CH_2 \xrightarrow{255\ °C}$

(b)

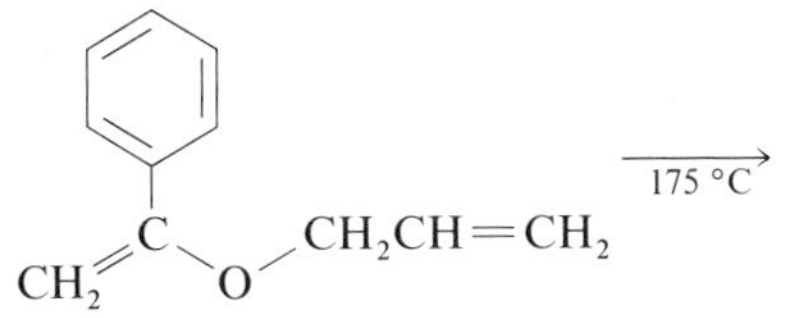

(c)

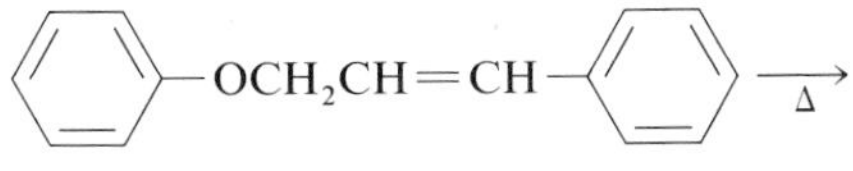

(d) 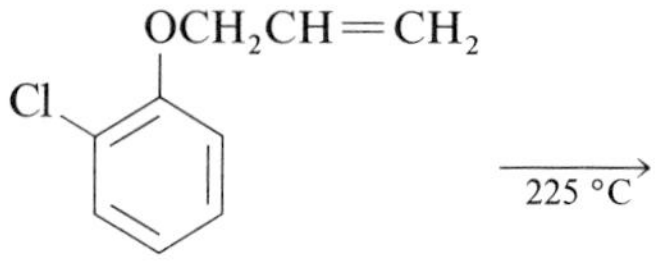 [structure: 2-chlorophenyl allyl ether] $OCH_2CH{=}CH_2$, Cl — $\xrightarrow{225\ °C}$

(e) 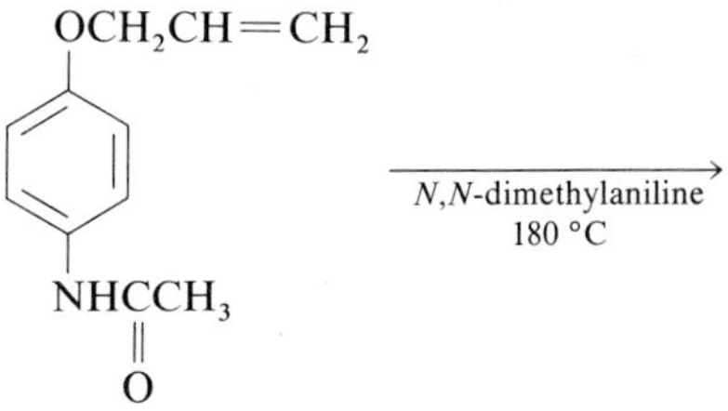 [structure: 4-allyloxy acetanilide] $OCH_2CH{=}CH_2$ / $NHCCH_3$ / $\overset{\|}{O}$ — $\xrightarrow[180\ °C]{N,N\text{-dimethylaniline}}$

(f) 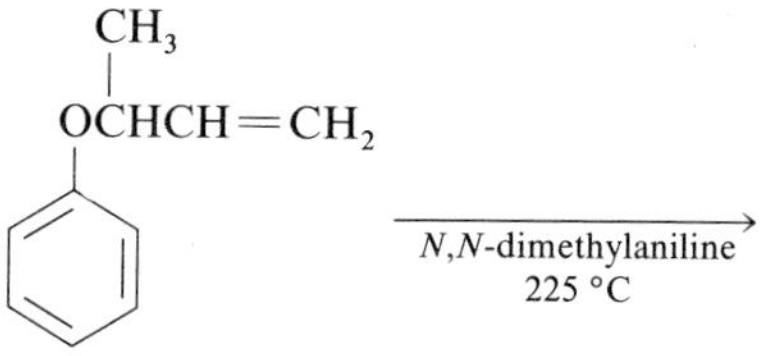 [structure: phenyl ether] $\underset{|}{\overset{CH_3}{}}$ $OCHCH{=}CH_2$ — $\xrightarrow[225\ °C]{N,N\text{-dimethylaniline}}$

(g) 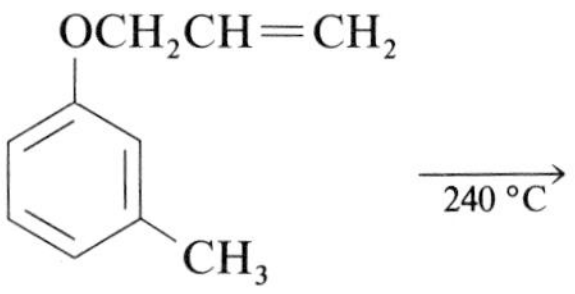 [structure: 3-methylphenyl allyl ether] $OCH_2CH{=}CH_2$ / CH_3 — $\xrightarrow{240\ °C}$

19.6

Reactions of Carbenes

Carbenes are reactive intermediates that have two single bonds and a pair of non-bonding electrons around a carbon atom. Thus, they are neutral species that must acquire another pair of electrons to complete a stable octet. Carbenes, and their addition to alkenes to give cyclopropanes, were introduced briefly in Section 6.10. In this section, we reexamine this transformation as a concerted reaction.

The reaction of a carbene such as dibromomethylene, which is generated by the action of a strong base on tribromomethane, with alkenes is stereoselective. The species reacts with (Z)-2-butene to give *cis*-1,1-dibromo-2,3-dimethylcyclopropane and with (E)-2-butene to give the trans isomer.

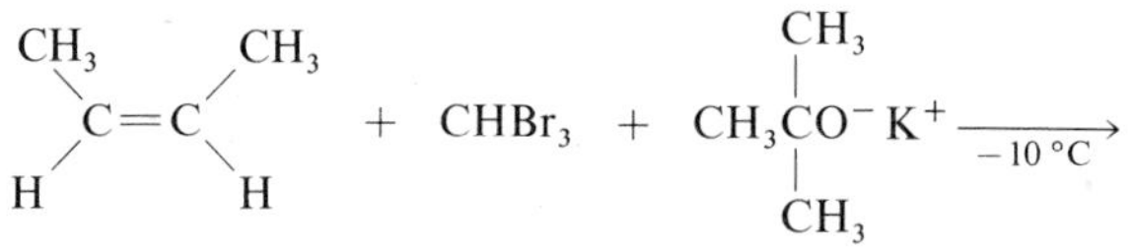

$$\underset{H}{\overset{CH_3}{}}C{=}C\underset{H}{\overset{CH_3}{}} \ +\ CHBr_3\ +\ CH_3\underset{CH_3}{\overset{CH_3}{C}}O^-K^+ \xrightarrow{-10\ °C}$$

(Z)-2-butene | tribromomethane bromoform | potassium *tert*-butoxide

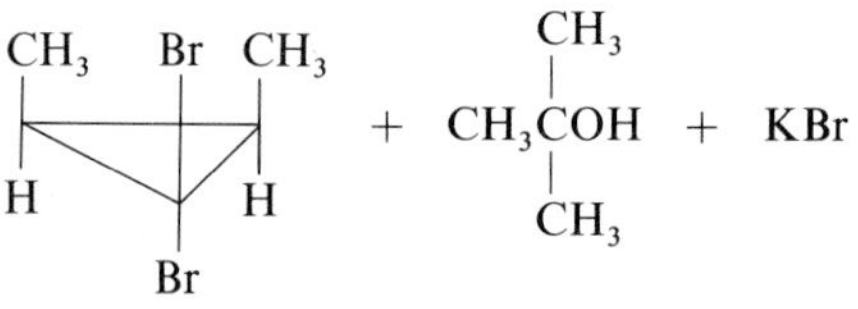

$$\underset{H}{\overset{CH_3}{\bigtriangleup}}\ \underset{}{\overset{Br}{}}\ \underset{H}{\overset{CH_3}{}} \ +\ CH_3\underset{CH_3}{\overset{CH_3}{C}}OH\ +\ KBr$$
Br

cis-1,1-dibromo-2,3-dimethylcyclopropane 80% | *tert*-butyl alcohol | potassium bromide

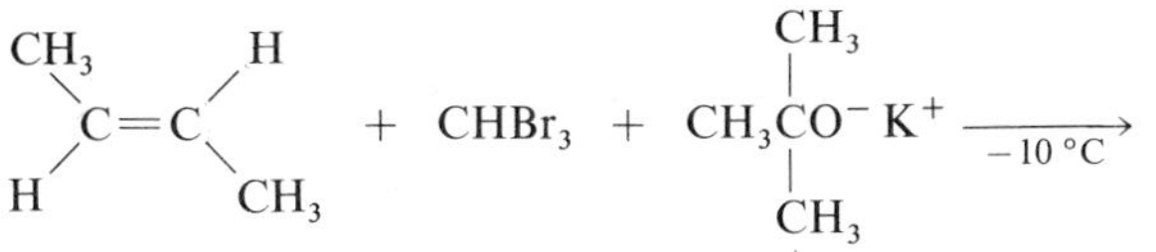

(E)-2-butene   tribromomethane   potassium
                    bromoform   *tert*-butoxide

trans-1,1-dibromo-2,3-dimethylcyclopropane   *tert*-butyl   potassium
68%   alcohol   bromide

racemic

The strong base *tert*-butoxide anion removes the proton from tribromomethane giving the tribromomethyl anion, which loses bromide ion in an α-elimination to form dibromomethylene. Chemists postulate that the carbene adds to the double bond in a single step in which both bonds to the carbon atoms of the alkene are formed simultaneously.

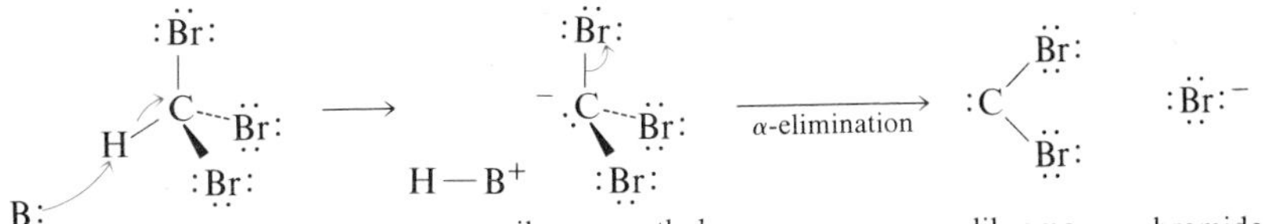

tribromomethyl   dibromo-   bromide
anion   methylene   ion

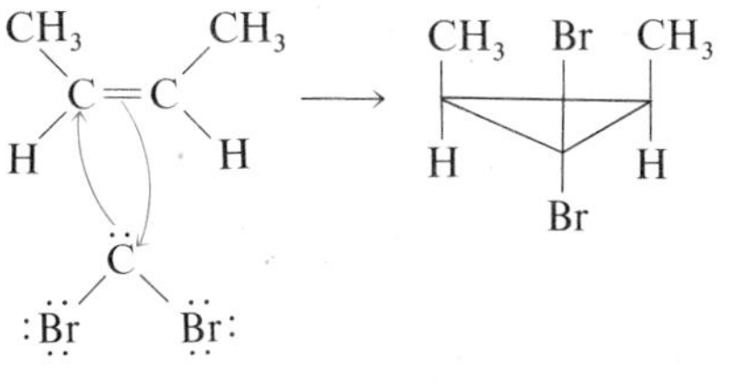

Because the reaction takes place in a single step, the stereochemistry of the alkene is retained in the product cyclopropane. In this reaction, two σ bonds to the same atom, in this case a carbon atom, are formed simultaneously in a concerted manner.

The number of electrons participating in the reaction, however, presents a problem at first glance. Four electrons are involved in the transition state, two from the π bond of the alkene, and two from the carbene. This reaction resembles the concerted cyclo-addition of two ethylene units, which is forbidden in the ground state (Table 19.1, p. 1092). The difference between the thermal cycloaddition of two alkene units to give cyclobutane and the cycloaddition of a carbene to an alkene lies in the stereochemistry that is possible for the two types of reactions. The four-carbon atom framework of a cyclobutane molecule requires that the two double bonds approach each other and react in a face-to-face manner (Figure 19.5, Section 19.2A). Otherwise, an impossibly twisted and strained four-membered ring would form. An examination of the orbitals of a carbene shows that a bonding overlap between the HOMO of a carbene and the LUMO of an alkene (or the other way around) is possible (Figure 19.15).

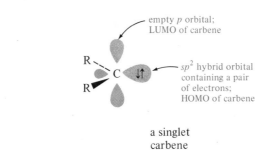

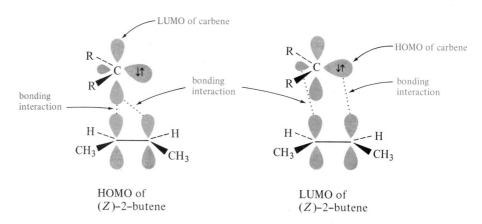

Figure 19.15   A representation of the interactions possible between the molecular orbitals of a singlet carbene and an alkene.

A carbene has three pairs of electrons around the central carbon atom. It is a trigonal carbon atom with the three pairs of electrons each occupying an sp^2 hybrid orbital, and an empty *p* orbital perpendicular to the plane defined by the carbon atom and the two substituents on it. The form of the carbene described above is called its **singlet state,** in which both nonbonding electrons occupy a single orbital and have opposing spins. For some carbenes, another electronic state in which the nonbonding electrons become unpaired and each one occupies a different orbital is more stable. Such carbenes are said to be in the **triplet state.** Triplet carbenes, because of their unpaired electrons, behave more like radicals than singlet carbenes do, and do not add to double bonds in concerted reactions.

The cycloaddition reactions of dibromomethylene are reactions of the singlet state of the carbene. Note that reaction is possible only if the carbene approaches the alkene sideways so that the plane defined by the carbon atom and its two substituents parallels the plane of the alkene. In this orientation, the empty *p* orbital of the carbene is pointing towards the electrons of the π bond. The carbene differs from an alkene in having this empty orbital, so a sideways approach of the carbene to the π bond and reaction to give a three-membered ring is possible without the creation of an impossible amount of strain in the transition state.

PROBLEM 19.30   Prove to yourself that an approach of the carbene in which its nonbonding electrons are pointing towards the π orbital of the alkene cannot lead to a concerted cycloaddition reaction.

Adding carbenes to double bonds is a useful way to synthesize cyclopropanes. One method often used for this reaction involves an organozinc reagent known as the **Simmons-Smith reagent,** so called for the two American chemists, Howard E. Simmons and Ronald D. Smith, who developed it. The reagent is prepared by treating metallic zinc with a salt of copper to make a zinc-copper couple, which reacts with diiodomethane to give the active species, usually formulated as $CH_2(ZnI_2)$. The exact structure of the reagent is not known, but it adds a methylene group to a wide variety of alkenes. The preparation of the Simmons-Smith reagent and its reaction with 2-cyclohexenone is shown below.

$$CH_2I_2 \quad + \quad Zn(Cu) \quad \xrightarrow{\text{diethyl ether}} \quad CH_2(ZnI_2)$$

diiodomethane	zinc-copper couple	organozinc reagent Simmons-Smith reagent

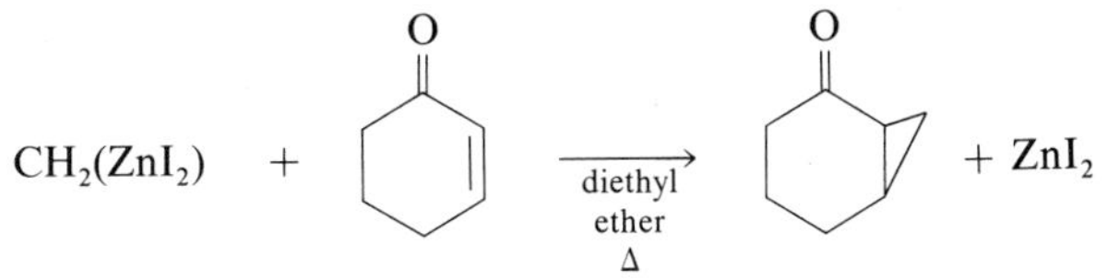

Simmons-Smith reagent   2-cyclohexenone   bicyclo[4.1.0]heptan-2-one 90%

PROBLEM 19.31   Complete the following equations.

(a) 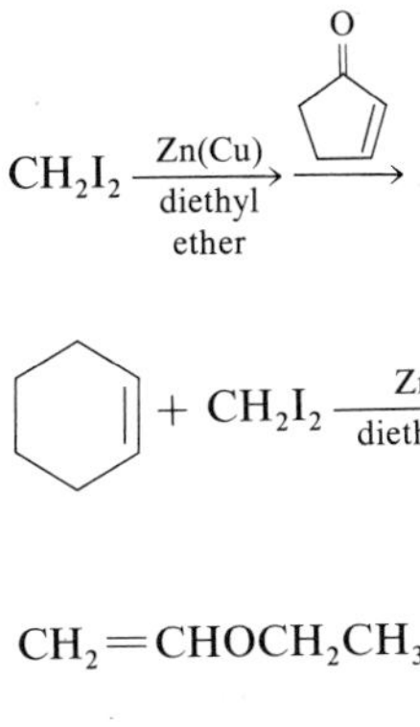 $CH_2I_2 \xrightarrow[\text{diethyl ether}]{Zn(Cu)}$

(b) 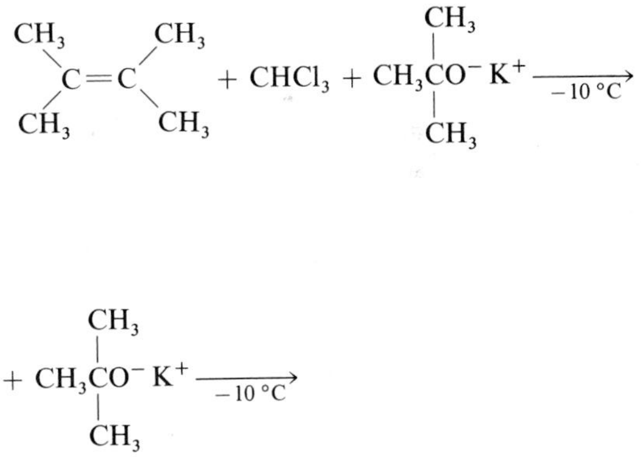 $+ \; CHCl_3 + CH_3CO^- K^+ \xrightarrow{-10\,°C}$

(c) $+ \; CH_2I_2 \xrightarrow[\Delta]{Zn(Cu) \atop \text{diethyl ether}}$

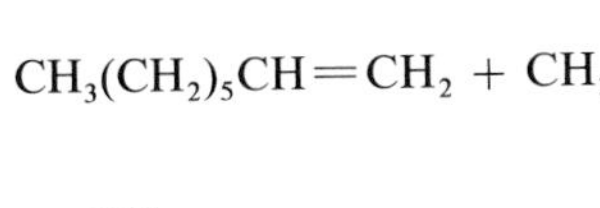

(d) $CH_2{=}CHOCH_2CH_3 + CHCl_3 + CH_3CO^- K^+ \xrightarrow{-10\,°C}$

(e) $CH_3(CH_2)_5CH{=}CH_2 + CH_2I_2 \xrightarrow[\Delta]{Zn(Cu) \atop \text{diethyl ether}}$

(f) 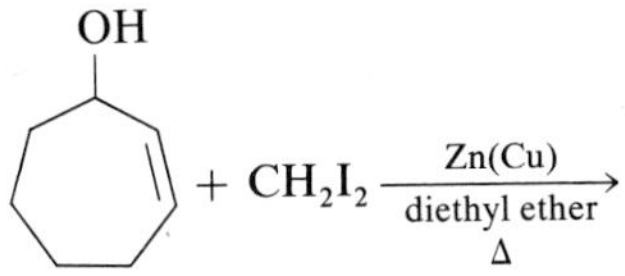 $+ \; CH_2I_2 \xrightarrow[\Delta]{Zn(Cu) \atop \text{diethyl ether}}$

(g) 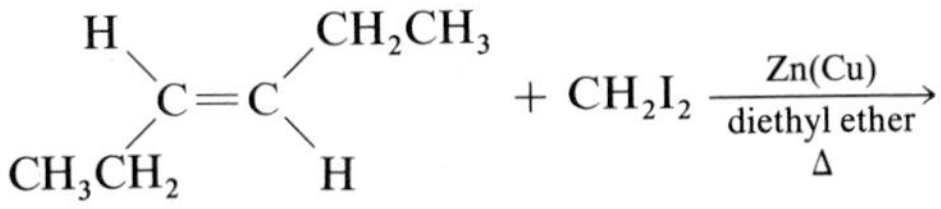 $+ \; CH_2I_2 \xrightarrow[\Delta]{Zn(Cu) \atop \text{diethyl ether}}$

(h) 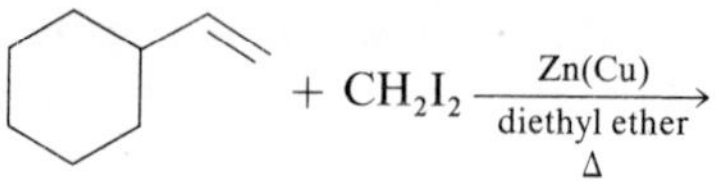 + CH_2I_2 $\xrightarrow[\substack{\text{diethyl ether} \\ \Delta}]{\text{Zn(Cu)}}$

(i)

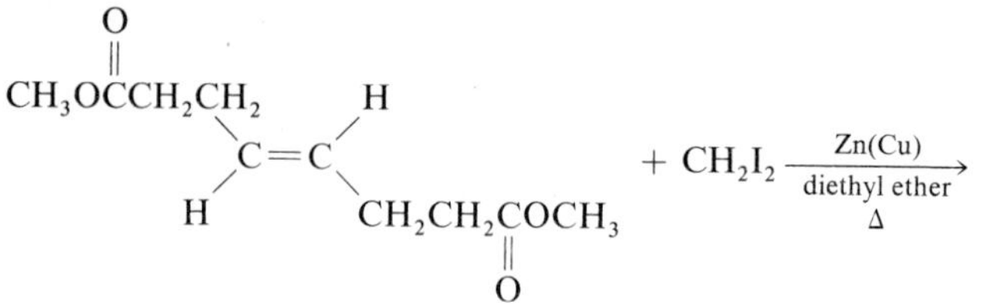

$$CH_3OCCH_2CH_2 \quad \overset{O}{\underset{\|}{}}$$

$$\underset{H}{\overset{H}{\underset{\diagdown}{}}} C = C \underset{CH_2CH_2COCH_3}{\overset{H}{\diagup}}$$

+ CH_2I_2 $\xrightarrow[\substack{\text{diethyl ether} \\ \Delta}]{\text{Zn(Cu)}}$

with the $CH_2CH_2COCH_3$ group having $\overset{\|}{O}$ below.

(j)

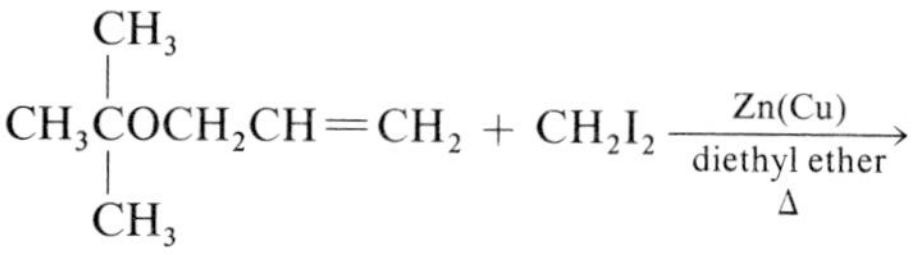

$$\underset{\underset{CH_3}{|}}{\overset{\overset{CH_3}{|}}{CH_3COCH_2CH=CH_2}} + CH_2I_2 \xrightarrow[\substack{\text{diethyl ether} \\ \Delta}]{\text{Zn(Cu)}}$$

(k)

$$\underset{\underset{CH_3}{|}}{\overset{\overset{CH_3}{|}}{CH_3C=CH_2}} + CHBr_3 + CH_3CO^- K^+ \xrightarrow[-10\,°C]{}$$

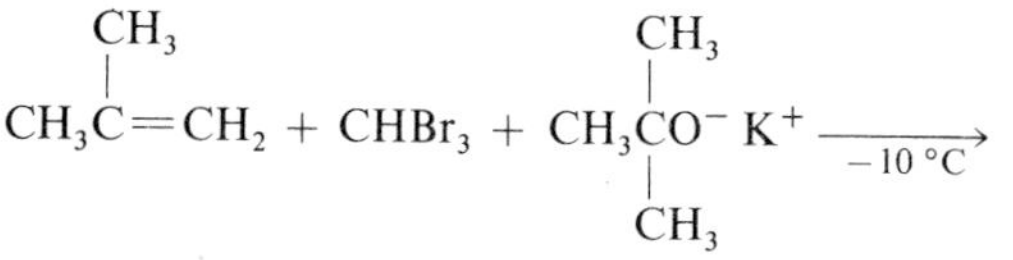

In summary, a consideration of the symmetry of molecular orbitals and a classification of a large number of organic reactions according to the number of electrons participating in the transition state have provided chemists with the tools to explain experimental phenomena that look unconnected at first glance. It is important to keep in mind, however, that in each case the theory is a rationalization of the experimental facts and an indication of new directions for experimentation. The concepts described in this chapter make it easier to correlate and remember facts about known reactions and to predict new reactions. The mere fact that we can describe reactions in a certain way is no guarantee that they actually occur by those pathways. Experimental and theoretical work is still being done to extend the limits of our understanding of reactivity, regioselectivity, and stereoselectivity in the transformations of organic compounds.

ADDITIONAL PROBLEMS

19.32   Complete the following equations. Be sure to show stereochemistry when it can be predicted.

(a) 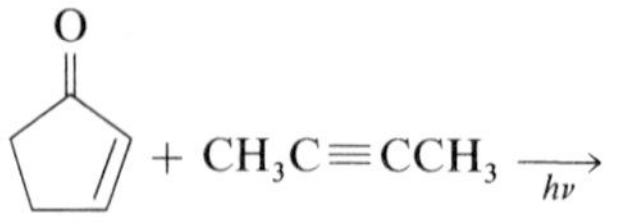 + $CH_3C\equiv CCH_3$ $\xrightarrow[h\nu]{}$

(b) 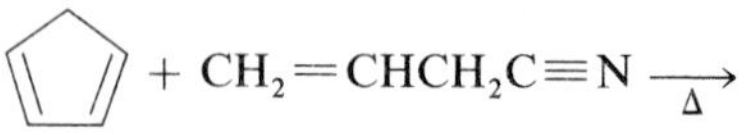 + $CH_2=CHCH_2C\equiv N$ $\xrightarrow[\Delta]{}$

(c) 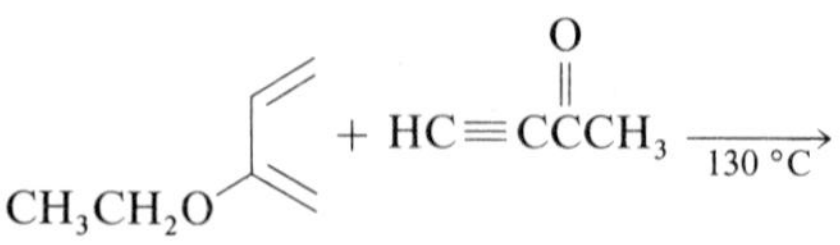 + $HC\equiv CCCH_3$ $\xrightarrow[130\,°C]{}$ (with CH_3CH_2O and $\overset{O}{\underset{\|}{}}$)

(d) 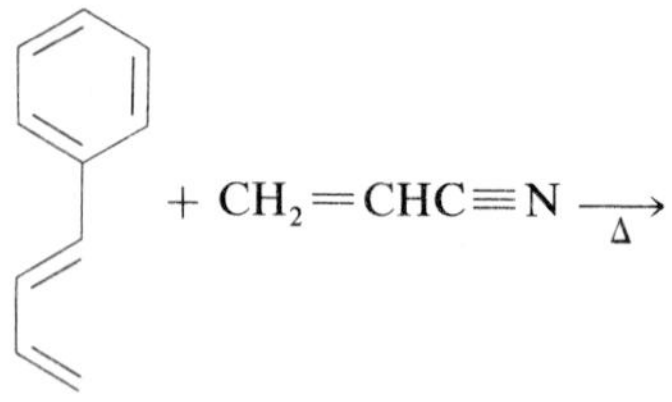 + $CH_2=CHC\equiv N$ $\xrightarrow[\Delta]{}$

(e)

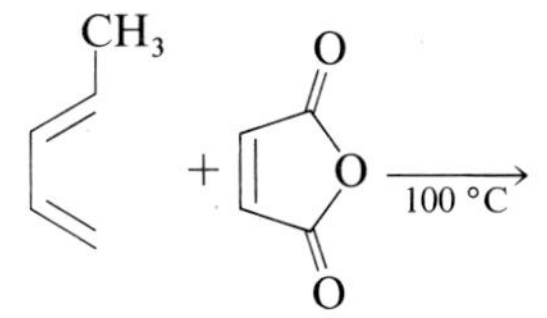

(f)

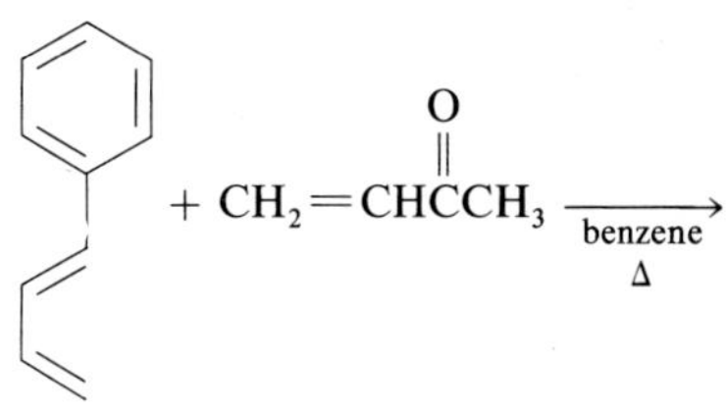

(g)

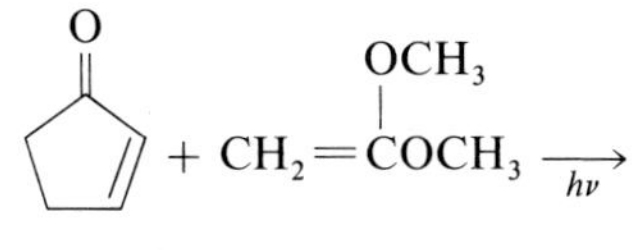

(h)

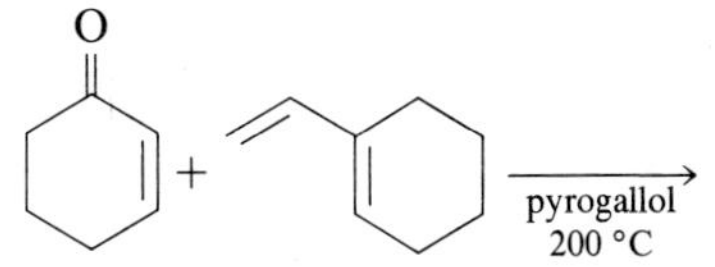

(i)

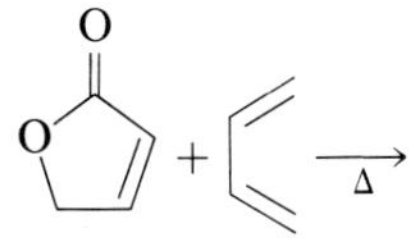

(j)

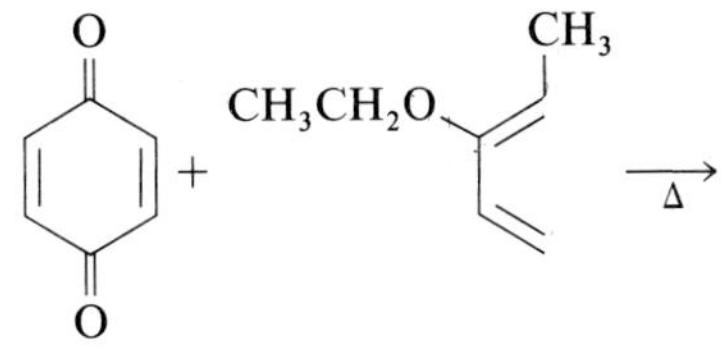

(k)

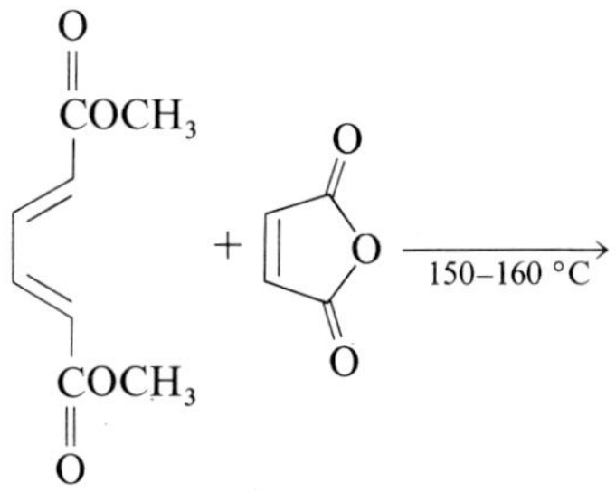

(l)

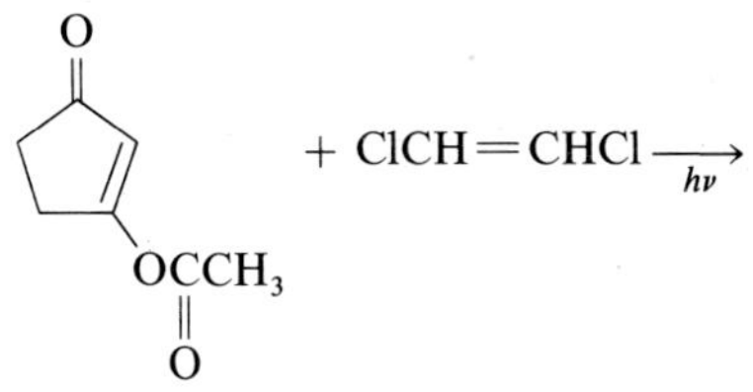

19.33 Complete the following equations. Show stereochemistry where it is known.

(a)

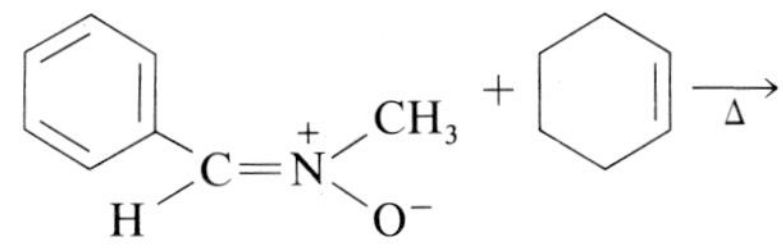

(b)

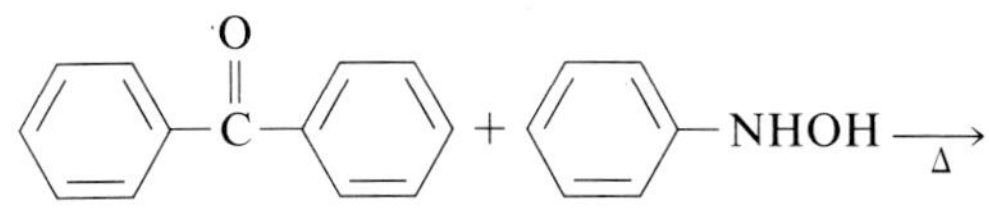

(c)

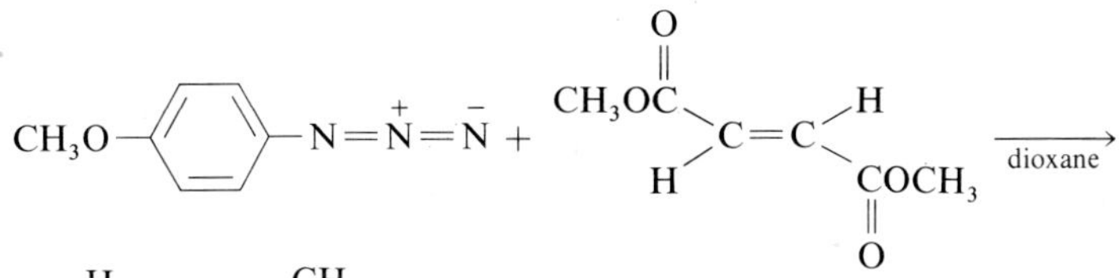

(d)

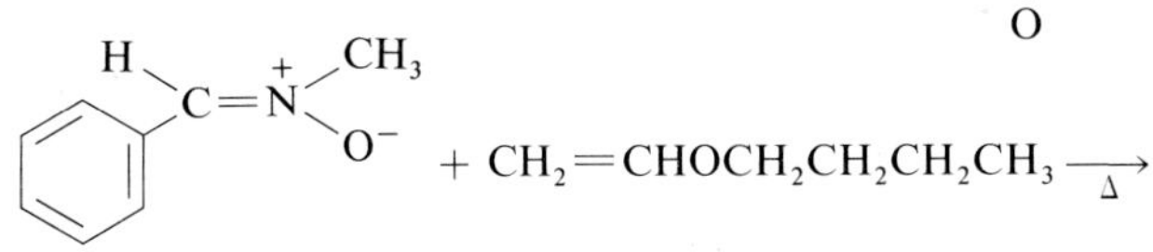

(e)

(f) 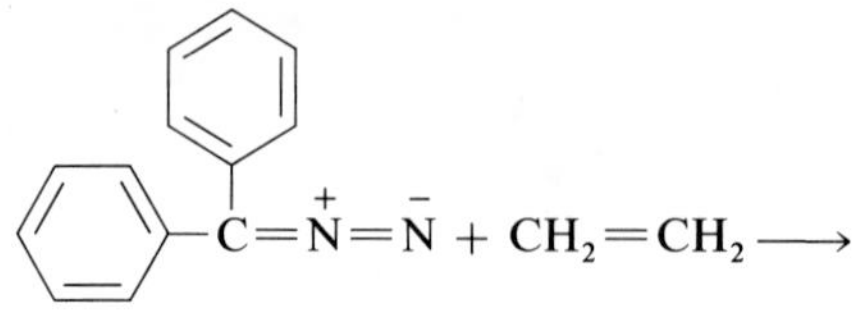 [structure: diphenyl C=N⁺=N⁻] + CH$_2$=CH$_2$ ⟶

(g) 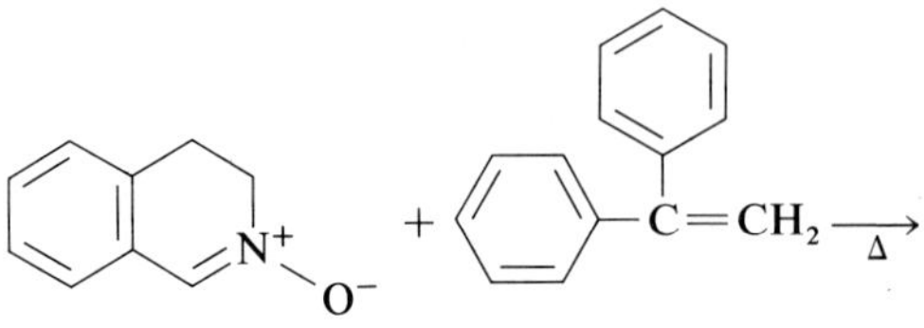 [isoquinoline N-oxide structure, N⁺–O⁻] + [Ph$_2$C=CH$_2$ structure] $\xrightarrow{\Delta}$

(h) 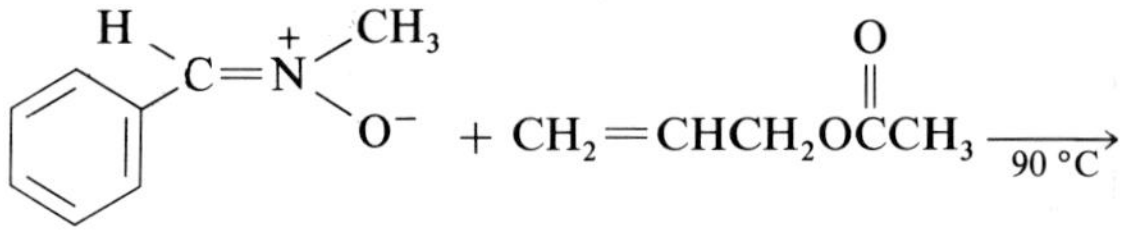 [structure: H, C₆H₅ C=N⁺(CH$_3$) O⁻] + CH$_2$=CHCH$_2$OCCH$_3$ (with C=O) $\xrightarrow{90\ °C}$

(i) 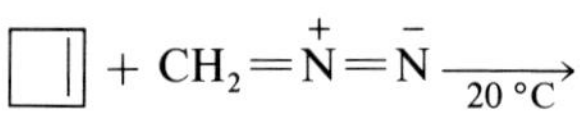 [cyclobutene square] + CH$_2$=N⁺=N⁻ $\xrightarrow{20\ °C}$

19.34   Complete the following equations.

(a) 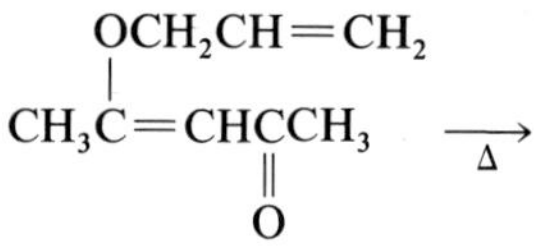   [structure: OCH$_2$CH=CH$_2$ on CH$_3$C=CHCCH$_3$ with C=O] $\xrightarrow{\Delta}$

(b) 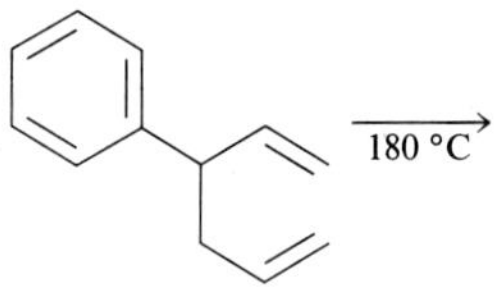   [phenyl-substituted 1,4-pentadiene structure] $\xrightarrow{180\ °C}$

(c)   [structure: N≡C, N≡C on carbon bearing CH$_2$CH$_3$ and a CH=CH–CH$_3$ chain and CH$_2$CH=CH$_2$] $\xrightarrow{150\ °C}$

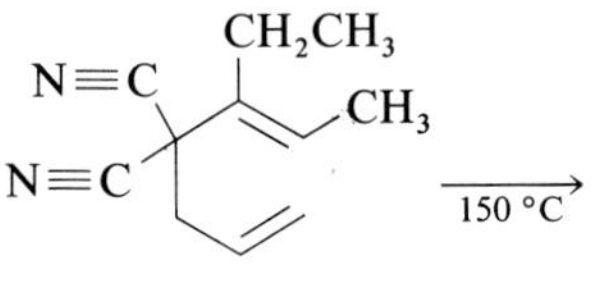

(d)   [phenyl–OCH$_2$C(CH$_3$)=CH$_2$ structure] $\xrightarrow[200\ °C]{N,N\text{-diethylaniline}}$

(e) 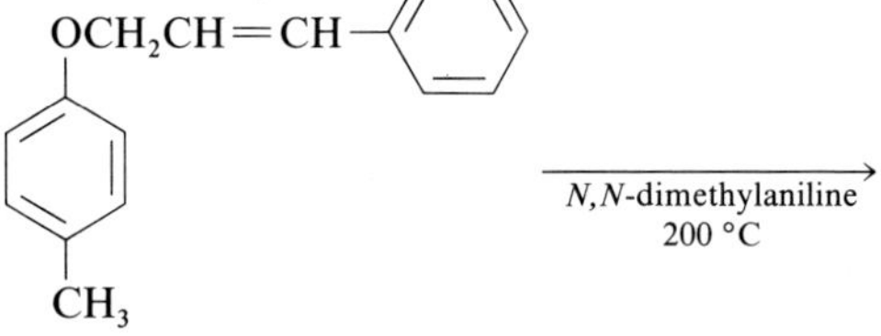   [p-methylphenyl O–CH$_2$CH=CH–C$_6$H$_5$ structure with CH$_3$] $\xrightarrow[200\ °C]{N,N\text{-dimethylaniline}}$

(f) 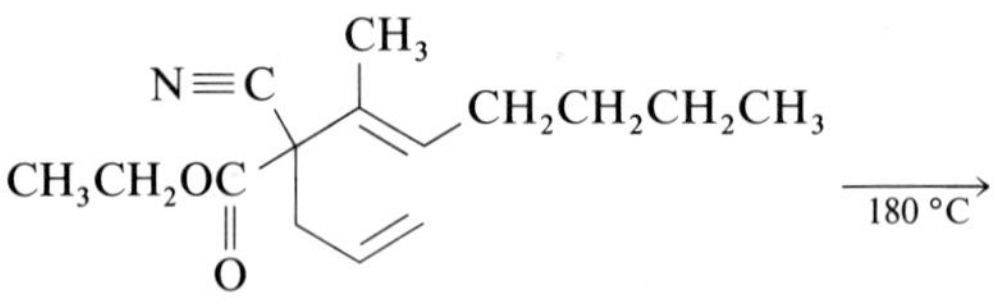   [structure: CH$_3$CH$_2$OC(=O) on carbon with N≡C, CH$_3$, CH=CH–CH$_2$CH$_2$CH$_2$CH$_3$, and CH$_2$CH=CH$_2$] $\xrightarrow{180\ °C}$

(g) 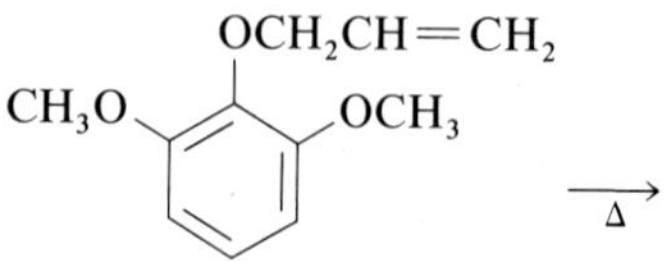   [benzene ring with OCH$_2$CH=CH$_2$, CH$_3$O, and OCH$_3$ substituents] $\xrightarrow{\Delta}$

19.35 Predict what the products of the following reactions will be.

(a)

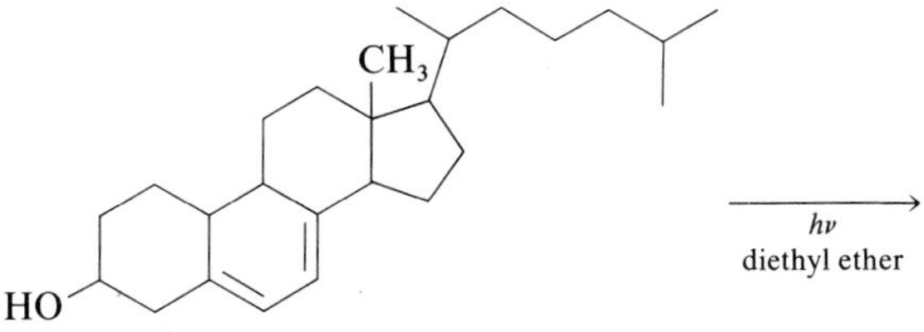

7-dehydro-19-norcholesterol
(7-dehydrocholesterol lacking the
methyl group numbered carbon 19)

$$\xrightarrow[\text{diethyl ether}]{h\nu}$$

(b)

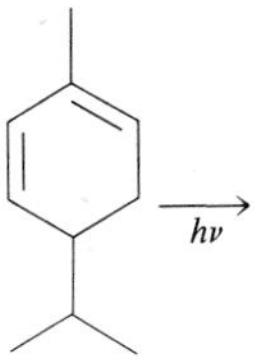

$$\xrightarrow{h\nu}$$

(c)

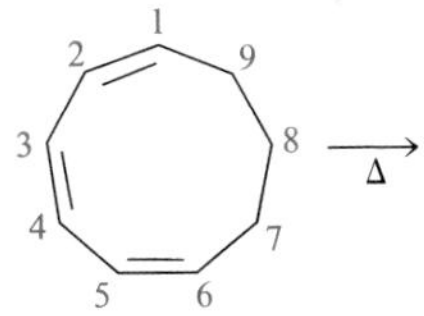

$$\xrightarrow{\Delta}$$

(d)

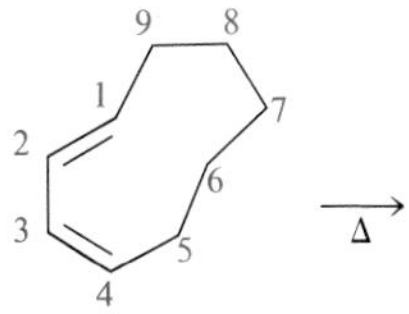

$$\xrightarrow{\Delta}$$

(e)

$$\xrightarrow{h\nu}$$

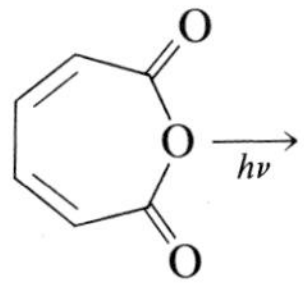

(f)

$$\xrightarrow{\Delta}$$

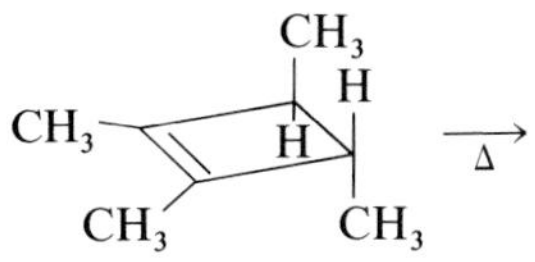

19.36 Complete the following equations.

(a)

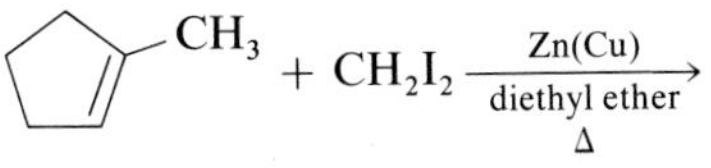

$+ CH_2I_2 \xrightarrow[\Delta]{\text{Zn(Cu)}}$

(b)

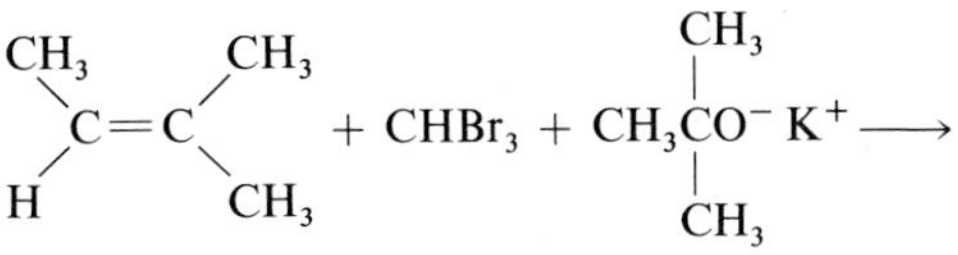

$+ CHBr_3 + CH_3CO^- K^+ \longrightarrow$

(c) 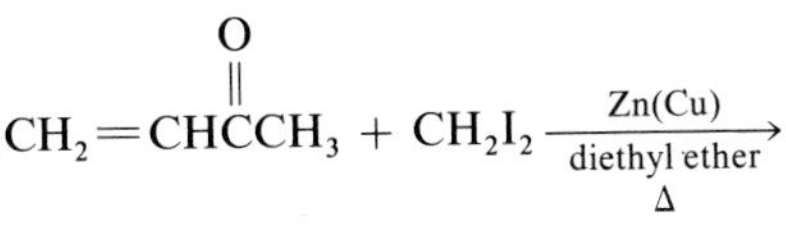 $CH_2{=}CHCCH_3 + CH_2I_2 \xrightarrow[\Delta]{\text{Zn(Cu)}}$

(d)

$+ CH_2I_2 \xrightarrow[\Delta]{\text{Zn(Cu)}}$

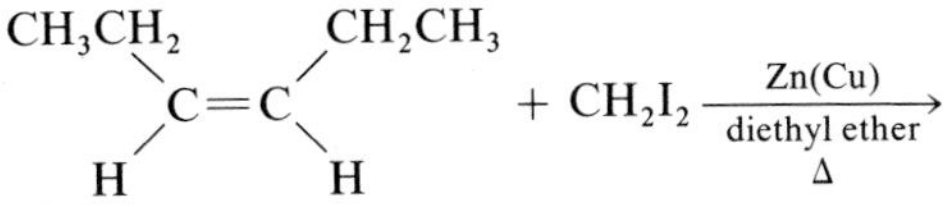

(e) 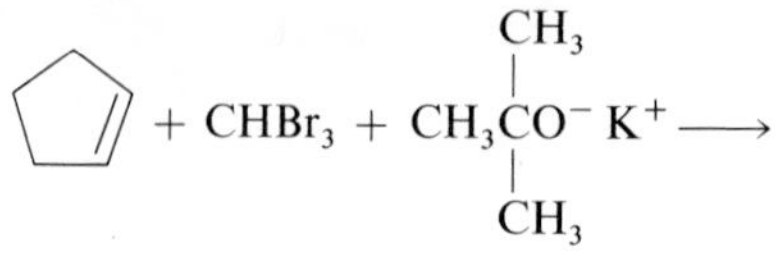 $\bigcirc\!\!\!\!\!/\,$ + CHBr$_3$ + CH$_3\overset{\underset{\displaystyle CH_3}{|}}{\underset{\underset{\displaystyle CH_3}{|}}{C}}O^-$ K$^+$ $\longrightarrow$

(f) 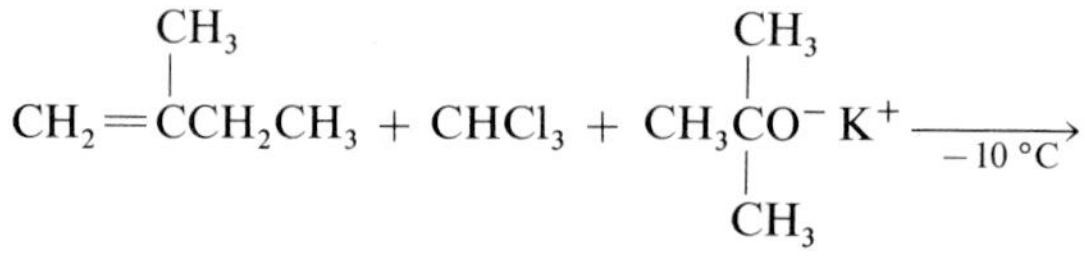 CH$_2$=$\overset{\underset{\displaystyle CH_3}{|}}{C}CH_2CH_3$ + CHCl$_3$ + CH$_3\overset{\underset{\displaystyle CH_3}{|}}{\underset{\underset{\displaystyle CH_3}{|}}{C}}O^-$ K$^+$ $\xrightarrow{-10\,°C}$

19.37   Allyl phenyl ether, specifically labeled with radioactive ^{14}C at the terminal alkene carbon atom was synthesized by the following sequence of reactions.

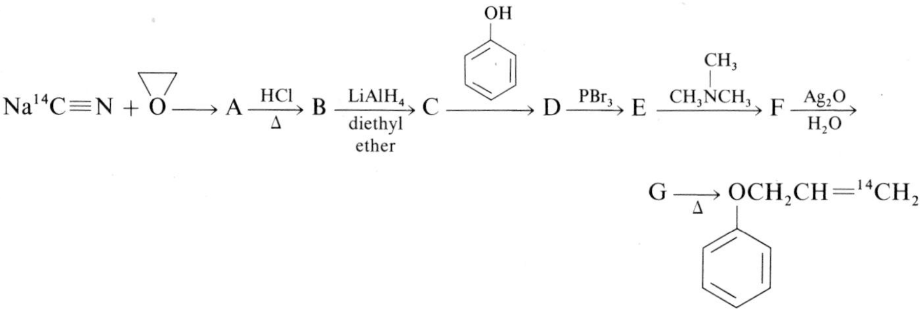

Na^{14}C≡N + $\triangle\!\!\!O$ $\longrightarrow$ A $\xrightarrow[\Delta]{HCl}$ B $\xrightarrow[\substack{diethyl \\ ether}]{LiAlH_4}$ C $\xrightarrow{}$ D $\xrightarrow{PBr_3}$ E $\xrightarrow{CH_3\overset{\underset{CH_3}{|}}{N}CH_3}$ F $\xrightarrow{\substack{Ag_2O \\ H_2O}}$

G $\xrightarrow{\Delta}$ OCH$_2$CH=^{14}CH$_2$ (phenyl)

(a) Fill in the structures of the compounds represented by the letters.
(b) Why was this particular way of introducing the double bond used rather than another method such as the dehydration of an alcohol?
(c) When the labeled allyl ether is heated, 2-allylphenol is formed. This phenol is subjected to the following reactions. Fill in the structures of the compounds formed in this sequence of reactions.

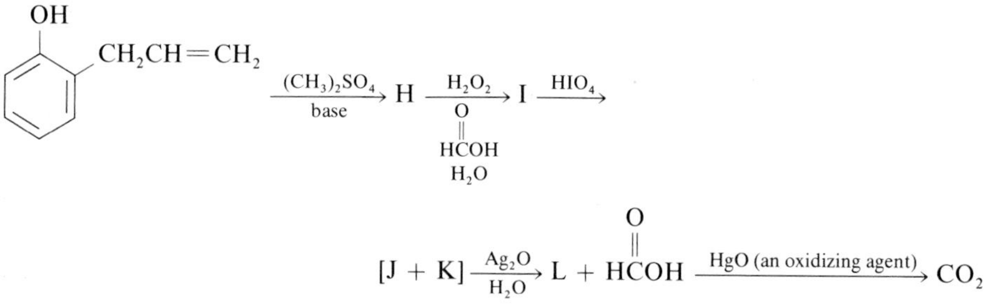

OH CH$_2$CH=CH$_2$ (phenol) $\xrightarrow[base]{(CH_3)_2SO_4}$ H $\xrightarrow[\substack{O \\ \| \\ HCOH \\ H_2O}]{H_2O_2}$ I $\xrightarrow{HIO_4}$

$[J + K]\xrightarrow[H_2O]{Ag_2O}$ L + H$\overset{\overset{\displaystyle O}{\|}}{C}$OH $\xrightarrow{HgO\ (an\ oxidizing\ agent)}$ CO$_2$

(d) Will the CO$_2$ that is obtained as a product of the sequence of reactions shown above have the radioactivity introduced into the allyl ether in the first synthesis?
(e) 2,6-Dimethylphenyl allyl ether, similarly labeled with ^{14}C, was synthesized, rearranged, and degraded by the reactions shown in parts (a) and (c). Will the CO$_2$ obtained from that experiment be radioactive?

19.38   The following experimental observations were made:

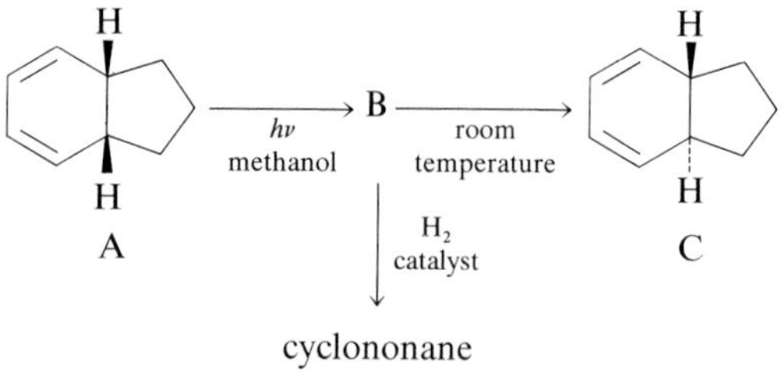

A $\xrightarrow[methanol]{h\nu}$ B $\xrightarrow{room\ temperature}$ C

B $\xrightarrow{\substack{H_2 \\ catalyst}}$ cyclononane

When *cis*-bicyclo[4.3.0]nona-2,4-diene (Compound A) is irradiated at $-20\ °C$ in methanol, Compound B is obtained. Compound B has λ_{max} 290 nm (ϵ 2050) and bands in the infrared at 1645, 1621, 1598, 975, 960, and 670 cm^{-1}. It has two broad multiplets in its nuclear magnetic resonance spectrum, six protons in the region of 6.2–4.7 ppm and six protons at 2.6 to 0.6 ppm. If the temperature is kept below 0 °C, Compound B can be hydrogenated to cyclononane. At room temperature, it is converted to *trans*-bicyclo[4.3.0]nona-2,4-diene (Compound C). How would you explain these observations? What is the structure of Compound B?

19.39 (1Z,3E)-1,3-Cyclooctadiene when heated to 80 °C gives bicyclo[4.2.0]oct-7-ene (shown below). (1Z,3Z)-1,3-Cyclooctadiene undergoes a photochemical ring closure to give the same compound, including stereochemistry, but does not give a cyclobutene product when heated. How do you explain these observations?

When bicyclo[4.2.0]oct-7-ene is heated above 300 °C, the cyclobutene ring opens and (1Z,3Z)-1,3-cyclooctadiene is formed. Is this opening of the ring a concerted reaction?

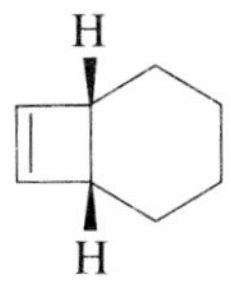

bicyclo[4.2.0]oct-7-ene

19.40 Cyclopentadiene reacts with (E)-1,2-dichloro-1,2-difluoroethylene to give two different products, shown below.

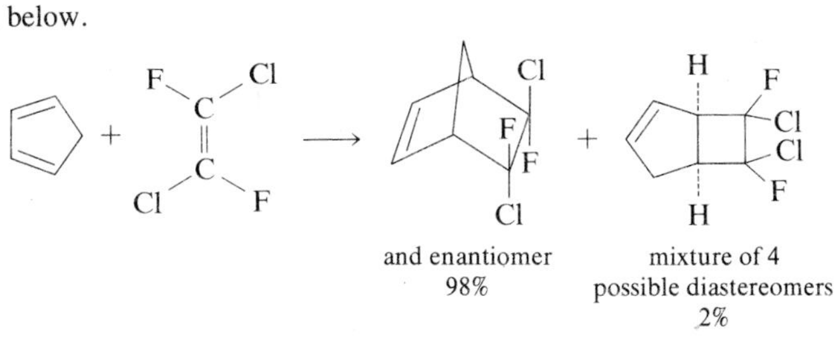

and enantiomer
98%

mixture of 4
possible diastereomers
2%

How do you account for the difference in stereoselectivity in the reactions leading to the two types of products?

19.41 The following experimental observation was made.

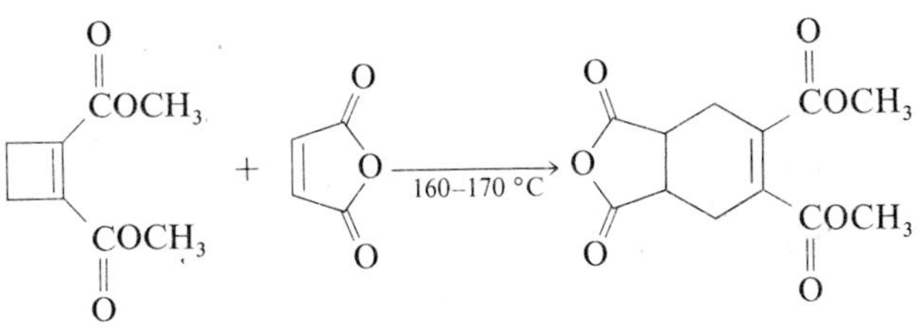

Write equations showing a mechanism for the transformation.

19.42 Of the two compounds shown below, one undergoes opening of the cyclobutene ring easily, whereas the other requires high temperatures. Predict which one reacts with ease, and explain why this is so.

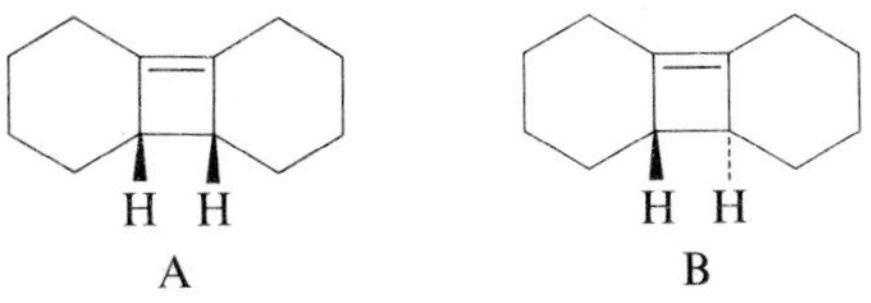

H H
A

H H
B

19.43   The cyclobutene diol shown below gives 3,4-dimethyl-2,5-dione, among other products, on heating. How would you rationalize the formation of this product?

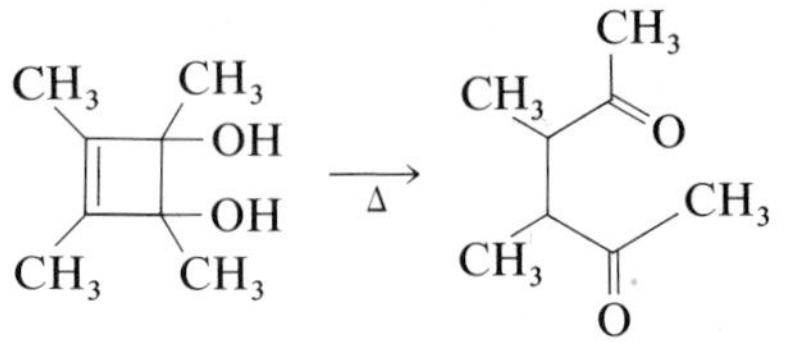

19.44   The following sequence of reactions was carried out.

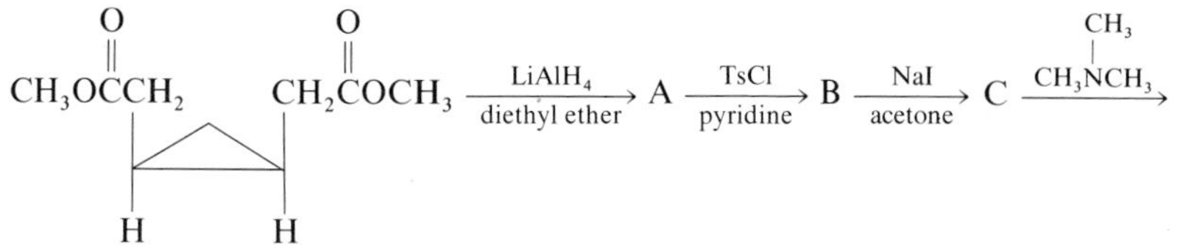

$$CH_3OCCH_2 \qquad CH_2COCH_3 \xrightarrow[\text{diethyl ether}]{LiAlH_4} A \xrightarrow[\text{pyridine}]{TsCl} B \xrightarrow[\text{acetone}]{NaI} C \xrightarrow{CH_3NCH_3}$$

$$D \xrightarrow[H_2O]{Ag_2O} E \xrightarrow{\Delta} \underset{C_7H_{10}}{F} + \underset{C_7H_2}{G}; F \xrightarrow[PtO_2]{H_2} \text{cycloheptane}$$

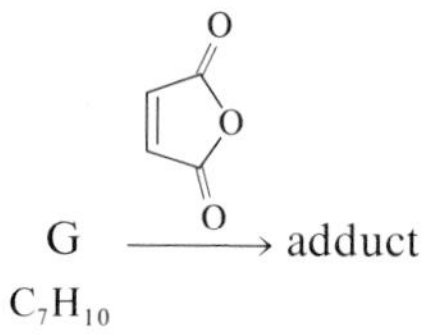

$$\underset{C_7H_{10}}{G} \xrightarrow{} \text{adduct}$$

Compound F has no absorption in the ultraviolet region of the spectrum above 215 nm. It has no bands in the infrared at 990 and 909 cm^{-1}, regions typical of C—H bending frequencies of terminal alkenes. Compound F has three multiplets in its nuclear magnetic resonance spectrum, with relative intensities of 4:4:2. Four of the protons lie in the region for vinylic protons, the other six are in the aliphatic region. Compound G has λ_{max}^{hexane} 248 nm. Fill in the structures for Compounds A to G. How do you account for the formation of Compounds F and G?

19.45   The Hoffmann degradation of tropine, obtained from the hydrolysis of atropine (Section 18.8A) gives 5-(N,N-dimethylamino)-1,3-cycloheptadiene, which isomerizes at 150 °C to 1-(N,N-dimethylamino)-1,3-cycloheptadiene. What kind of a reaction is occurring?

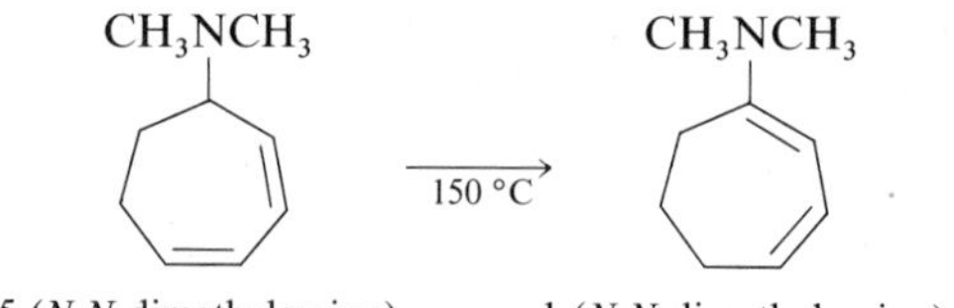

5-(N,N-dimethylamino)-   1-(N,N-dimethylamino)-
1,3-cycloheptadiene   1,3-cycloheptadiene

19.46 The following sequence of reactions was carried out. Assign structures to the compounds designed by the letters.

The following sequence of reactions was carried out. Assign structures to the compounds designated by the letters.

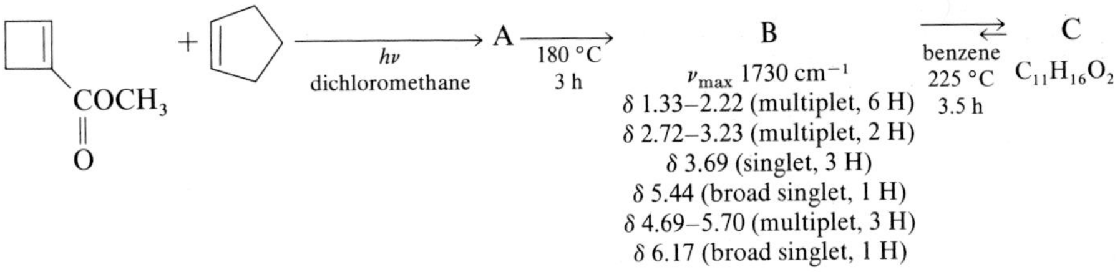

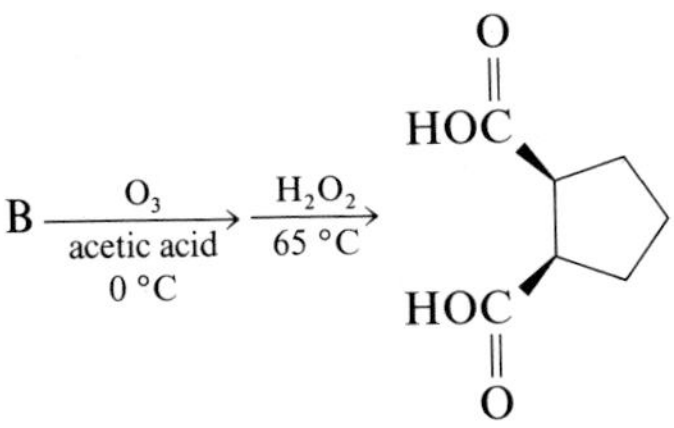

$B \xrightarrow[\text{acetic acid} \atop 0\,°C]{O_3} \xrightarrow[65\,°C]{H_2O_2}$

Reagent structures and NMR data for B:

ν_{max} 1730 cm^{-1}
δ 1.33–2.22 (multiplet, 6 H)
δ 2.72–3.23 (multiplet, 2 H)
δ 3.69 (singlet, 3 H)
δ 5.44 (broad singlet, 1 H)
δ 4.69–5.70 (multiplet, 3 H)
δ 6.17 (broad singlet, 1 H)

C$_{11}$H$_{16}$O$_2$

Macromolecular Chemistry

20

20.1

Introduction to Macromolecules

A. Macromolecules of Biological Importance

The interactions that take place between different parts of large molecules are important in determining their form and biological function. The stereoselectivity and the chemical specificity of reactions that take place at the active sites of enzymes, for example, have been emphasized many times in this book. In Section 17.6, we traced the specificity of an enzyme to the precise form that the large protein molecule has in its natural state.

Proteins are only one category of large molecules that have biological functions. Others are the nucleic acids, DNA and RNA (Section 18.6B). The structural materials of plants, cellulose (Section 14.9B) and lignin, are also giant molecules. The spaces within the long fibers formed by cellulose are filled by lignins, which are complex molecules containing carbon-oxygen and carbon-carbon bonds between adjoining units of phenylpropanes with varying patterns of hydroxyl and methoxyl substitution on the aromatic rings.

a fragment of lignin

The points at which the phenylpropane units may be linked are varied. The molecule is not linear, but has many points of linkage so that it forms a three-dimensional network. The combination of the long crystalline fibers of cellulose with the network of lignin molecules penetrating them creates the strong, rigid structure of woody plants.

Rubber is another large molecule that owes its useful properties to its molecular size. When rubber is decomposed by heat, isoprene is formed (Section 11.6A). The structure and chemistry of rubber is discussed in Section 20.5.

Proteins, cellulose, lignins, rubber, and the nucleic acids are naturally occurring representatives of a class of organic compounds of biological and, more recently, industrial importance. All these compounds have large molecular size in common. For this reason, they are known as **macromolecules,** or giant molecules. Their molecular structures can be dissected into smaller units. For example, proteins have units of amino acids, cellulose has units of glucose, and rubber has units of isoprene. *The small unit that appears again and again in the macromolecule is called a* **monomer** *and the large molecule that is composed of these units is known as a* **polymer.** These terms were defined earlier in connection with the reaction of 2-methylpropene with itself in acid (Section 5.6C). For example, isoprene is a monomer, and rubber is a polymer made up of many units of isoprene bonded together. *The process by which a monomer is converted into a polymer is called* **polymerization.**

In some macromolecules, all the monomeric units are the same. Rubber contains only isoprene units. Cellulose is made up only of glucose units. In proteins or nucleic acids, on the other hand, the repeating monomeric units are not all identical. In Chapter 17, we saw that a large number of different protein molecules are possible because of the different order in which the 20 amino acids are incorporated into peptide chains. DNA and RNA are also made up of nucleotides with different structures, arranged in a chain with a varying order. *Polymers in which there may be more than one repeating unit are called* **copolymers.** *Thus, rubber and cellulose are simple polymers, while proteins and nucleic acids are complex copolymers.*

Polymers may also be classified according to the structure of the giant molecule. Rubber, cellulose, proteins, and nucleic acids are all **linear polymers.** *The backbone of the macromolecule* in each case *consists of a long chain of monomer units held together by covalent bonds.* The principal linkage between monomers in proteins is the peptide bond; in cellulose, it is the glycosidic linkage. Isoprene units are held together by covalent bonds between carbon atoms in rubber. In linear polymers, once the chain is formed, other types of interactions, such as hydrogen bonding in proteins and van der Waals interactions between different parts of the hydrocarbon chain in rubber, contribute to the overall shapes of the macromolecules.

Lignins represent another class of polymers, those in which there is additional covalent bonding between the monomer units. Not only do the monomers form long chains, but there are also covalent links between adjacent chains. *Such polymers are said to be* **cross-linked.** These links may be close together, as is the case in lignin, or they may be relatively far apart, as with some synthetic polymers. Such bonding is important in determining the properties of a polymer and its potential uses. In later sections of this chapter, we will examine the relationship between structure and properties and see how the introduction of cross-linking affects these characteristics and thus changes the uses of a polymer.

Human beings have made use of macromolecular substances for a long time. The structural rigidity of wood, which results from the properties of cellulose and lignin, is useful in construction. Many macromolecules have fibrous structures that can be spun into threads and then used to weave fabrics. Cellulose fibers in cotton and linen and protein fibers in silk and wool are used in clothing and household furnishings. Modern methods of transportation would not be possible without rubber.

Human beings have also discovered ways to modify natural macromolecules to make them more useful. For example, a modification of cellulose led to the invention of paper, an important technological advance. Because only limited modifications could be made on natural molecules without destroying their essential structures, chemists started to think about creating entirely new polymeric materials starting from small reactive molecules. The remainder of this chapter is the story of their successes in this endeavor.

B. Macromolecules of Industrial Importance

By the middle of the nineteenth century, organic chemists were discovering that some of their experiments gave rise to high molecular weight substances that were usually considered to be evidence of failed reactions. It was not until the early part of this century that chemists started to create polymers deliberately. To do this, they designed reactions that allowed them to control the average molecular weight, and hence the properties, of the large molecules that were being formed. Hermann Staudinger of Germany, a pioneer in this field, was one of the first to recognize that control of the conditions of polymerization was essential to the synthesis of useful substances. He received the Nobel Prize in 1953 for his highly original work in this area of chemistry.

At first, chemists in this field tried to imitate nature. For example, the first really successful synthetic fiber was nylon, a polyamide created by the American chemist Wallace Carothers in the 1930s. The structure of nylon resembles a protein with its multitude of amide bonds, but it is much more regular in its repeating units. Several different kinds of nylons are shown on the next page, along with the structural units that give rise to such polyamides.

$$\underset{\substack{\text{nylon 66, a polyamide of 1,6-hexanediamine and} \\ \text{hexanedioic acid (adipic acid)}}}{+\overset{\text{O}}{\overset{\|}{\text{C}}}(\text{CH}_2)_4\overset{\text{O}}{\overset{\|}{\text{C}}}\text{NH}(\text{CH}_2)_6\text{NH}\underset{n}{+}} \qquad \text{H}_2\text{N}(\text{CH}_2)_6\text{NH}_2 \qquad \text{HO}\overset{\text{O}}{\overset{\|}{\text{C}}}(\text{CH}_2)_4\overset{\text{O}}{\overset{\|}{\text{C}}}\text{OH}$$

nylon 66, a polyamide of 1,6-hexanediamine and
hexanedioic acid (adipic acid)

$$+\overset{\text{O}}{\overset{\|}{\text{C}}}(\text{CH}_2)_{10}\overset{\text{O}}{\overset{\|}{\text{C}}}\text{NH}(\text{CH}_2)_6\text{NH}\underset{n}{+} \qquad \text{H}_2\text{N}(\text{CH}_2)_6\text{NH}_2 \qquad \text{HO}\overset{\text{O}}{\overset{\|}{\text{C}}}(\text{CH}_2)_{10}\overset{\text{O}}{\overset{\|}{\text{C}}}\text{OH}$$

nylon 612, a polyamide of 1,6-hexanediamine and
dodecanedioic acid

$$+\text{NH}(\text{CH}_2)_5\overset{\text{O}}{\overset{\|}{\text{C}}}\underset{n}{+} \qquad \text{H}_2\text{N}(\text{CH}_2)_5\overset{\text{O}}{\overset{\|}{\text{C}}}\text{OH}$$

nylon 6, a polyamide of 6-aminohexanoic acid

n = number of repeating units in the polymer chain;
for nylons that are useful as fibers, n = 50 to 120

The numbers after the word *nylon* indicate the structure of the polyamide. Nylon 66 is made from an amine and an acid, each containing six carbon atoms. Nylon 612, on the other hand, contains amine units with six carbon atoms and acid units with 12. Nylons of the appropriate molecular weight, where n is 50 to 120, can be formed into threads and used to make fabrics. In an early practical application, nylon was used to replace silk in women's stockings. It has since been used in a wide variety of fabrics, from carpets to parachutes to clothing of every description. The chemistry of the polymerization processes that give rise to nylons is discussed in Sections 20.2B and 20.4A. First, however, in the next section we will examine how the properties of macromolecules are determined by molecular size and shape as well as by the nature of the repeating unit.

Since World War II, with the discovery that useful macromolecular compounds could be synthesized in the laboratory and then produced on a large scale in factories, much research has been concentrated on creating and developing new polymers. More organic chemists now work in industry in the area of polymer chemistry than in any other field. Objects made of polymers are so pervasive in everyday life that their names, and especially their trademarks, are part of our vocabulary. News stories about the health-related and environmental effects of polymers and the monomers that go into their manufacture are common. Structural formulas for some familiar polymers, along with the monomers from which they are made, are given below and on the next page.

$$+\text{CH}_2\text{CH}_2\underset{n}{+} \qquad \text{CH}_2{=}\text{CH}_2$$

polyethylene   ethylene

$$\overset{\text{Cl}}{\underset{\substack{|}}{+\text{CH}_2\text{CH}\underset{n}{+}}} \qquad \text{CH}_2{=}\text{CHCl}$$

poly(vinyl chloride)   vinyl chloride
PVC

$$+\text{CF}_2\text{CF}_2\underset{n}{+} \qquad \text{CF}_2{=}\text{CF}_2$$

polytetrafluoroethylene   tetrafluoroethylene
Teflon

$$\underset{\substack{\text{a polyacrylonitrile}\\\text{Orlon, Acrilan}}}{\underset{|}{\overset{\overset{\displaystyle C\equiv N}{|}}{+CH_2CH+_n}}} \qquad \underset{\text{acrylonitrile}}{CH_2=CHC\equiv N}$$

$$\underset{\substack{\text{poly(ethylene terephthalate)}\\\text{a polyester}\\\text{Dacron, Terylene}}}{+OCH_2CH_2O\overset{\displaystyle O}{\overset{||}{C}}-\!\!\left\langle\!\!\bigcirc\!\!\right\rangle\!\!-\overset{\displaystyle O}{\overset{||}{C}}+_n} \qquad \underset{\text{ethylene glycol}}{HOCH_2CH_2OH} \qquad \underset{\text{terephthalic acid}}{HO\overset{\displaystyle O}{\overset{||}{C}}-\!\!\left\langle\!\!\bigcirc\!\!\right\rangle\!\!-\overset{\displaystyle O}{\overset{||}{C}}OH}$$

$$\underset{\text{a polyurethane}}{+\overset{\displaystyle O}{\overset{||}{C}}NH-\!\!\left\langle\!\!\bigcirc\!\!\right\rangle\!\!-NH\overset{\displaystyle O}{\overset{||}{C}}OCH_2CH_2O+_n} \qquad \underset{\text{ethylene glycol}}{HOCH_2CH_2OH} \qquad \underset{\substack{m\text{-phenylenediisocyanate}\\\text{a diisocyanate}}}{O=C=N-\!\!\left\langle\!\!\bigcirc\!\!\right\rangle\!\!-N=C=O}$$

Polymers contain a wide variety of functional groups. These large molecules are made using essentially the same reactions we learned for low molecular weight compounds. Later in this chapter, we concentrate on how these reactions generate the large molecules with physical properties that make them useful in practical applications.

C. The Special Properties of Large Molecules

If you read the literature of polymer chemistry, you will find that researchers are intensely interested in the molecular weight of macromolecules. Every synthetic process is judged primarily by the range of molecular weights of the polymers formed. A synthesis of a polymer differs in this way from that of a low molecular weight organic compound. All the preparations in the book up to now have given compounds with well defined structures. The synthesis of a polymer results in a mixture of compounds having a range of molecular weights. For example, the equation for the preparation of nylon 66 from hexanedioic acid and 1,6-hexanediamine can only indicate an approximate structure for the polymer that is formed.

$$n\ \underset{\text{hexanedioic acid}}{HO\overset{\displaystyle O}{\overset{||}{C}}(CH_2)_4\overset{\displaystyle O}{\overset{||}{C}}OH} + n\ \underset{\substack{\text{1,6-hexanediamine}\\\text{hexamethylenediamine}}}{H_2N(CH_2)_6NH_2}$$

$$\downarrow$$

$$n\ \underset{\text{salt of amine and acid}}{{}^-O\overset{\displaystyle O}{\overset{||}{C}}(CH_2)_4\overset{\displaystyle O}{\overset{||}{C}}O^-} \qquad n\ H_3\overset{+}{N}(CH_2)_6\overset{+}{N}H_3$$

$$\left|\ \begin{array}{l}220\ °C\\\sim200\text{--}250\ \text{psi}\\1\text{--}2\ \text{h}\end{array}\right.$$

$$\left|\ 270\text{--}280\ °C\right.$$

$$\underset{\text{nylon 66}}{+NH(CH_2)_6NH\overset{\displaystyle O}{\overset{||}{C}}(CH_2)_4\overset{\displaystyle O}{\overset{||}{C}}+_n} + 2n\ H_2O\uparrow$$

The exact structure of the polymer depends upon how many of the repeating units become bonded together before the polymer chain stops growing. The repetitive structure of the polymer is indicated by the subscript n outside the brackets that enclose the repeating unit. The larger n is, the higher the molecular weight of the polymer. The molecules of a polymer formed in any one experiment do not have identical numbers of repeating units and thus differ in molecular weight. The product is a mixture of similar molecules that differ somewhat in size.

The concern with molecular weights of polymers is well justified because the physical properties, and hence the practical uses, of a polymer depend heavily upon its molecular weight. For example, nylon with a low molecular weight has no useful properties. It is a brittle solid. Only when the molecular weight reaches 10,000 does nylon begin to show properties that make it useful as a fiber. Nylons that have molecular weights above 100,000 no longer give good fibers, but have high resistance to heat and high mechanical strength, so they are used in other industrial applications. One such nylon, reinforced with glass fibers, is used instead of steel to make valve covers for automobile engines.

Molecular weight is, of course, a reflection of molecular size. The molecules of a polymer must have a minimum size before the interactions that are important in determining the properties of the substance can take place. These interactions may be between different parts of the same molecule or between adjacent molecules. All the factors that we have discussed in connection with the tertiary structure of proteins, such as hydrogen bonding, dipole-dipole interactions, and van der Waals interactions are also important in determining the shapes of other macromolecules. The nature of the repeating unit in a polymer is, of course, of primary importance in determining which types of interactions are possible. For a nylon, which has a large number of amide linkages, hydrogen bonding may be important. In polyethylene, which is like a gigantic alkane molecule, van der Waals interactions between different parts of the molecule and between adjacent chains are most likely.

The shape of a polymer molecule and the types of interactions that it can have with neighboring molecules are determined by the regularity with which repeating units appear in the chain and by the stereochemistry at points where stereoisomerism is possible. For example, just like the low molecular weight alkenes, a polymer may contain cis or trans double bonds. Rubber is an example of a polymer with cis double bonds. The trans isomer is a natural product known as *gutta percha,* and has some properties that are quite different from those of rubber (Section 20.5A).

If the repeating units in a polymer are chiral, as is the case for amino acids in proteins and glucose in cellulose, the macromolecule as a whole will retain chirality. Even a polymer formed from an achiral monomer, such as propene, has the possibility of stereoisomerism at the newly created tetrahedral carbon atoms along the backbone of the chain (Section 20.3A). In summary, polymers can exist as stereoisomers with differing stereochemistry at double bonds or at tetrahedral carbon atoms.

A given polymer can also exist in a large number of conformations arising from free rotations of the atoms that make up the backbone of the chain. If you think back to the conformational isomerism that is possible for butane with four carbon atoms in its chain (Section 3.5B), you will realize that molecules in a given sample of a polymer must exist in many different and constantly changing conformations. For polymers, too, anti and gauche arrangements of groups on adjacent carbon atoms are preferred over eclipsed conformations. The conformations that are favored and the range of motions possible for different parts of polymer chains are important in determining the properties of a polymer and how it will behave under different conditions.

One of the properties of a polymer that is very important for practical applications is how it behaves at different temperatures. The interactions between large molecules of a polymer create solids that have a high degree of structural regularity. Linear polyethylene (p. 1123) is a highly crystalline substance with a melting point of approximately 135 °C. The polyethylene produced by some manufacturing processes has branching on the chain and does not form as highly crystalline a solid. Branched polyethylene has a lower melting point, about 120 °C, than linear polyethylene.

Some polymers, such as rubber, do not pack together well in regular structures and exist mostly as amorphous solids. Many polymers are partly crystalline, meaning that when the liquid polymer is cooled, part of it arranges itself into regions with a high degree of order and other parts solidify before such arrangement takes place. Nylon is such a partly crystalline polymer. These different kinds of interactions between polymer chains are represented in Figure 20.1.

Whether a polymer has crystalline regions or not determines properties such as its flexibility and mechanical strength. For example, the amorphous and coiled structure of rubber is responsible for its elasticity. *Polymers with such elastic qualities are known as* **elastomers.** Semicrystalline polymers are rather hard and can be drawn out into strong fibers. The process of drawing them out increases the alignment of the polymer chains and thus their crystallinity. *Unless a polymer has some crystalline regions, it cannot be formed into* **fibers;** thus, an amorphous polymer does not make good fibers. *Not all amorphous polymers have elasticity; those that do not are called* **plastics** and are used in applications in which they can be molded under heat and

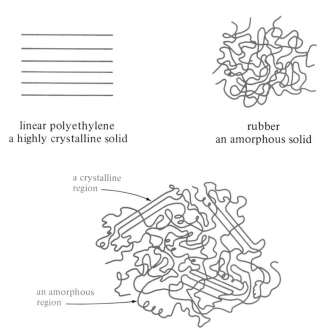

linear polyethylene
a highly crystalline solid

rubber
an amorphous solid

a crystalline region

an amorphous region

nylon

a solid that has crystalline
and amorphous regions

Figure 20.1   Schematic representations of crystalline,
amorphous, and semicrystalline polymers.

pressure into objects such as toys and household goods. *Polymers are thus classified as elastomers, fibers, and plastics* according to their properties and their uses.

A polymer can exist in three physical forms at differing temperatures. At low temperatures, it exists as a solid. As we saw above, in the solid state it may be partly or highly crystalline, or it may be amorphous. A solid that is not crystalline is called a **glass.** The quality of transparency, for example, is typical of a solid that is a glass, while a crystalline solid is opaque.

Upon warming to a given temperature, each polymer softens to a more pliable state but does not actually melt. This temperature varies a bit with the method used to make the determination, but is typical for each compound. For rubber, this temperature is $-70\ ^{\circ}C$; for nylon 66 it is $50\ ^{\circ}C$. Thus, rubber, as we know it, is in an intermediate state in which it is not a rigid solid but is not a liquid either. If a rubber ball is made very cold by putting it into liquid nitrogen ($-196\ ^{\circ}C$), it loses all of its bounce and shatters if thrown against a hard surface. This is typical of the brittle character of a glass.

At an even higher temperature, the polymer melts and becomes a liquid. This temperature is the melting point of the polymer, comparable to the melting point of a lower molecular weight organic compound. The melting point of raw rubber is $30\ ^{\circ}C$; for nylon 66 it is $265\ ^{\circ}C$. Knowledge of these properties of a polymer is clearly important. For example, raw rubber is useful at temperatures between $-70\ ^{\circ}C$ and $+30\ ^{\circ}C$. Other elastomers also show these properties in the range of temperature between the point at which they solidify and their melting point. Similarly, a nylon to be used as a valve cover on an automobile engine must retain its shape and must therefore remain a solid at high temperatures. The nylon actually used for this purpose withstands exposure to temperatures of approximately $120\ ^{\circ}C$ for long periods of time.

D. Types of Polymerization Reactions

Some of the polymers shown on pp. 1123–1124 contain all of the atoms of the monomeric unit within the polymer. For example, poly(vinyl chloride) consists of a large number of molecules of vinyl chloride bonded to each other by carbon-carbon bonds created at the carbon atoms of the double bond in the monomer.

$$n\ CH_2{=}CHCl \xrightarrow[\substack{(CH_3(CH_2)_{10}CO)_2 \\ water \\ gelatin \\ 50\ ^{\circ}C \\ 14\text{–}18\ h}]{} {+}CH_2\overset{\displaystyle Cl}{\underset{\displaystyle |}{CH}}{\xrightarrow{\ }}_n$$

vinyl chloride
monomer

poly(vinyl chloride)
polymer
$\sim 90\%$

Such *polymers are called* **addition polymers** *because they are created by the addition of one molecule of monomer to another one.*

A polyamide, on the other hand, is formed when a diamine reacts with a diacid with the loss of molecules of water (p. 1124, Section 20.1C). Such polymers are called **condensation polymers,** *formed by the condensing together of two different types of functional groups with the elimination of a small stable molecule such as water.* The water that is formed is driven off at high temperatures as steam to bring the reaction to completion.

A much more useful way to classify polymerization reactions is by their mechanisms. The types of functional groups present and their reactivity under different conditions determines the extent of polymerization and therefore the molecular weights of the products. Polymerization reactions, like other organic reactions, can have cationic, anionic, or radical intermediates. Polymers formed in reactions that proceed through these different intermediates often have different properties.

How a polymer chain grows is of great importance in determining how long the chain will be and therefore the molecular weight of the polymer. *The reactions that give rise to polymers have been classified into two main types* with this question in mind. *One type is called* **chain-growth polymerization.** *In this type of polymerization, a reactive intermediate is formed and reacts rapidly with a monomer molecule to give a new reactive intermediate, which in turn reacts with yet another monomer.* The monomer is consumed rapidly and always adds onto a chain that gets longer and longer. A new reactive center is created at the end of the polymer chain, so the reaction perpetuates itself until the system runs out of monomer or the reactive intermediate is destroyed by some end reaction. You probably recognize that this description corresponds to the steps of a free-radical chain reaction (Sections 5.4A and 12.4A). There is an initiation step, many propagation steps in which each reactive intermediate generates a new reactive intermediate, and finally a termination step. Though so far we have seen such a reaction only for free radicals, under the proper conditions chain reactions with cationic or anionic intermediates are also possible. Some of them will be examined in Sections 20.2B and 20.2C. The important factor in all of these reactions is the presence of a reactive intermediate that directs the course of the reaction, making it more probable at one molecular site than at others. In fact, in such a reaction the monomer units are usually incapable of reacting with each other until some reagent is added to create the reactive intermediate. The polymerization of an alkene, such as the reaction of vinyl chloride (p. 1127), is usually a chain-growth reaction.

The other type of polymerization is called a **step-growth reaction.** *In such a reaction, the monomer units contain functional groups that are capable of reacting with each other without the formation of a reactive intermediate.* The reactions are generally slower than those in chain-growth polymerizations, and the reaction sites are more random. The formation of a polyamide from a diamine and a diacid is an example of such a reaction. A closer look at the individual steps of the reaction demonstrates the problems associated with such a process.

First step of polymerization

$$H_2N(CH_2)_6NH_2 + HOC(CH_2)_4COH \longrightarrow H_2N(CH_2)_6NHC(CH_2)_4COH + H_2O$$

1,6-hexanediamine   hexanedioic acid

monomers

reactivity of amine function and carboxyl function not much different from reactivity of these functional groups in monomers

Second step of polymerization

$$2\ H_2N(CH_2)_6NH_2 + 2\ HOC(CH_2)_4COH + 2\ H_2N(CH_2)_6NHC(CH_2)_4COH$$

$$\downarrow$$

$$\underset{\text{monoamide}}{H_2N(CH_2)_6\overset{\overset{\displaystyle O}{\|}}{N}HC(CH_2)_4\overset{\overset{\displaystyle O}{\|}}{C}OH} \ + \ \underset{\substack{\text{diamide from 1 diamine}\\\text{and 2 diacid units}}}{HO\overset{\overset{\displaystyle O}{\|}}{C}(CH_2)_4\overset{\overset{\displaystyle O}{\|}}{C}NH(CH_2)_6NH\overset{\overset{\displaystyle O}{\|}}{C}(CH_2)_4\overset{\overset{\displaystyle O}{\|}}{C}OH}$$

$$+ \ \underset{\substack{\text{diamide from 1 diacid}\\\text{and 2 diamine units}}}{H_2N(CH_2)_6\overset{\overset{\displaystyle O}{\|}}{N}HC(CH_2)_4\overset{\overset{\displaystyle O}{\|}}{C}NH(CH_2)_6NH_2} \ + \ 3\ H_2O$$

The first step of the polymerization gives an amide formed from the diamine and the diacid. The amide still retains amino and carboxylic acid groups, and the reactivities of these functional groups are not much different from the reactivities of the same functional groups in the monomers. In the next stage, we get a random reaction of the carboxylic function of the monoamide with the diamine, and of the amine function of the monoamide with another molecule of diacid. Instead of a steady and directed growth of a chain, a mixture of molecules results. Large increases in chain length will take place only when monomer units are used up and amides containing several units combine with each other. The polymerization process takes place in jumps rather than by a unit-by-unit addition to a continuously growing chain.

PROBLEM 20.1   Write equations showing the intermediate steps in the formation of poly(vinyl chloride), which is formed by a free-radical chain reaction (p. 1127). The chain initiator is the dodecanoyloxy radical,

$$CH_3(CH_2)_{10}\overset{\overset{\displaystyle O}{\|}}{C}-O\cdot$$

PROBLEM 20.2   Draw structural formulas for the monomeric units in the following polymers.

(a) $\left[\!\!\left[SCH_2\overset{\overset{\displaystyle O}{\|}}{C}\right]\!\!\right]_n$ (b) $\left[\!\!\left[CHCH_2\right]\!\!\right]_n$ with phenyl group

(c) $\left[\!\!\left[CH_2\underset{\underset{\displaystyle CH_3}{|}}{C}H\right]\!\!\right]_n$ (d) $\left[\!\!\left[\overset{\overset{\displaystyle O}{\|}}{C}NH-\!\!\bigcirc\!\!-NH\overset{\overset{\displaystyle O}{\|}}{C}OCH_2CH_2O\right]\!\!\right]_n$

(e) $\left[\!\!\left[CH_2\underset{\underset{\underset{\underset{\displaystyle O}{\|}}{\displaystyle COCH_3}}{|}}{\overset{\overset{\displaystyle CH_3}{|}}{C}}\right]\!\!\right]_n$

Mechanisms of Chain-Growth Polymerization Reactions

A. Free-Radical Reactions

The polymerization of styrene in the presence of a small amount of added benzoyl peroxide is a typical example of a free-radical polymerization reaction.

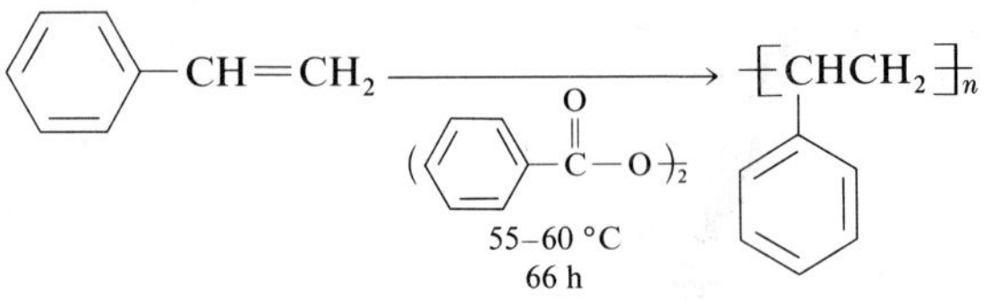

styrene

polystyrene
~100%

The reaction takes place in three stages. First, benzoyl peroxide dissociates into two benzoyloxy radicals (Section 5.4A). The benzoyloxy radical, an electrophile, reacts with the π electrons of the double bond in styrene to create a new radical, the stable benzylic radical, in an *initiation step*.

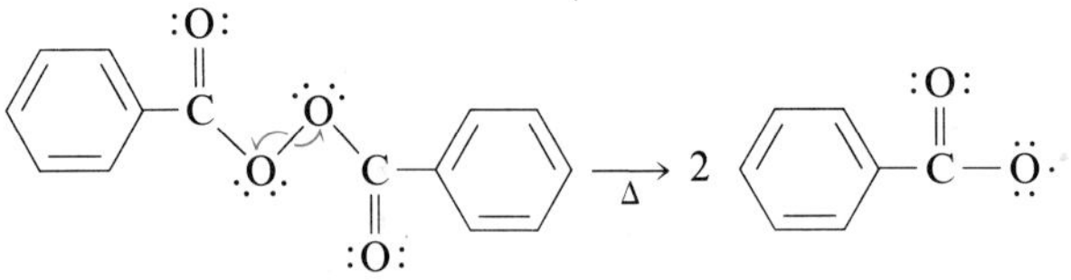

benzoyl peroxide

benzoyloxy radical
an oxy radical

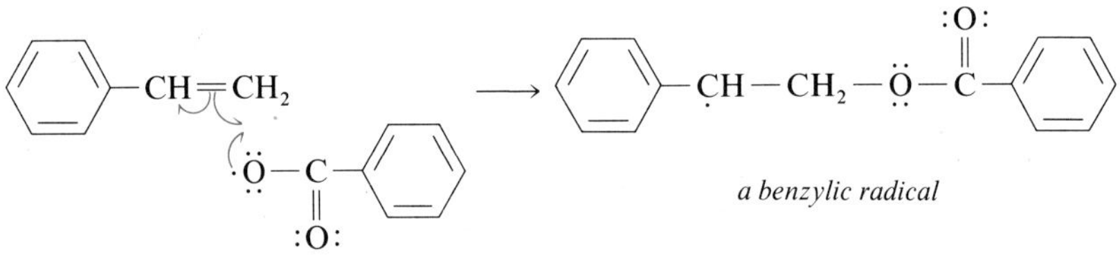

a benzylic radical

initiation step

The benzylic radical attacks another molecule of styrene, creating yet another radical intermediate.

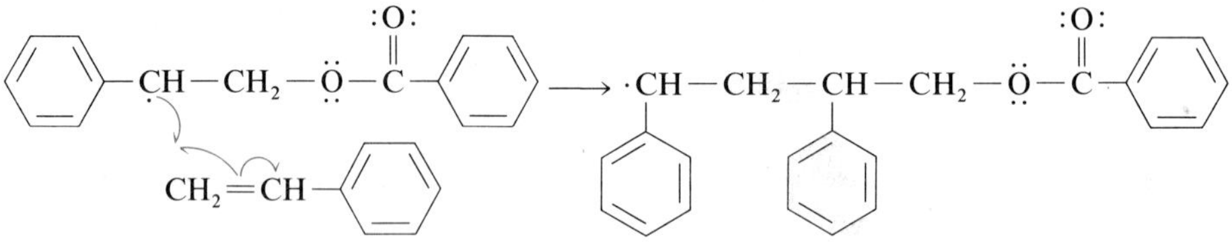

chain-propagation step

Each new radical continues to react with monomer in a series of *chain-propagating steps*. The growth of the chain of polystyrene is fast; it has been calculated that approximately 1500 monomer units are added to the chain each second.

The reaction continues until all the monomer is consumed or until the radical intermediate is destroyed by one of a number of *termination reactions*. For example, a combination of two radicals will stop the polymerization process. Most polystyrene chains stop growing as a result of a combination of two polystyryl radicals.

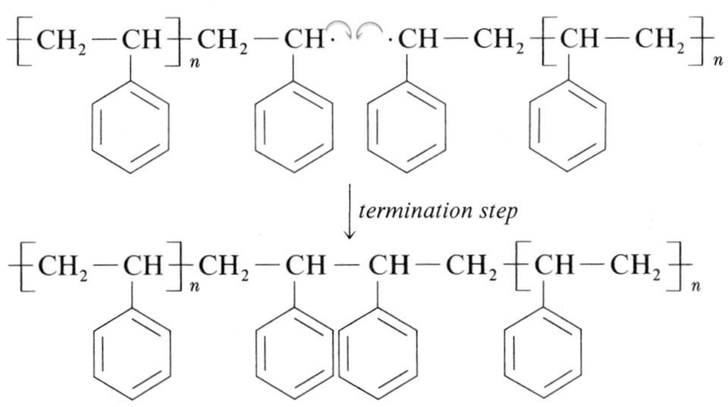

$$\left[CH_2-CH\right]_n CH_2-CH\cdot \quad \cdot CH-CH_2\left[CH-CH_2\right]_n$$

(each CH bears a phenyl group)

termination step →

$$\left[CH_2-CH\right]_n CH_2-CH-CH-CH_2\left[CH-CH_2\right]_n$$

(each CH bears a phenyl group)

"dead" polymer

The polymer radical may also abstract a hydrogen atom from another chain, creating a new radical in what is known as a *chain-transfer reaction*.

$$\left[CH_2-CH\right]_n CH_2-CH\cdot \; + \; \left[CH_2-CH-CH_2-CH-CH_2-CH\right]_x \longrightarrow$$

$$\left[CH_2-CH\right]_n CH_2-CH_2 \; + \; \left[CH_2-\dot{C}-CH_2-CH-CH_2-CH\right]_x$$

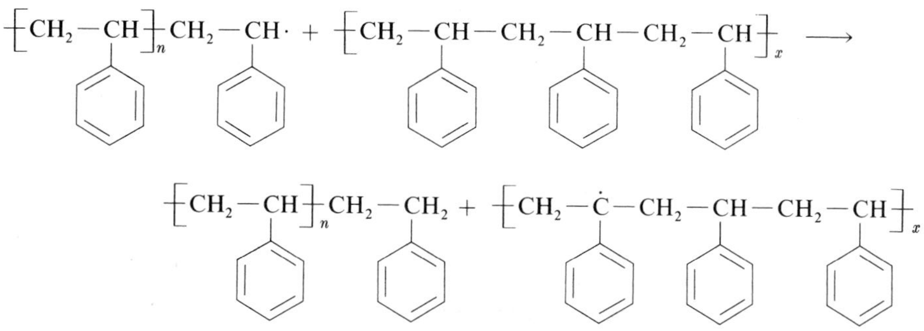

"dead" polymer   a new radical created by abstraction of a
                        benzylic hydrogen atom from the
                        middle of a polymer chain;
                        a chain-transfer reaction

The new radical is in the middle of a chain, not at the end as it was in the growing polymer. It too can react with monomer, giving rise to branching of the chain.

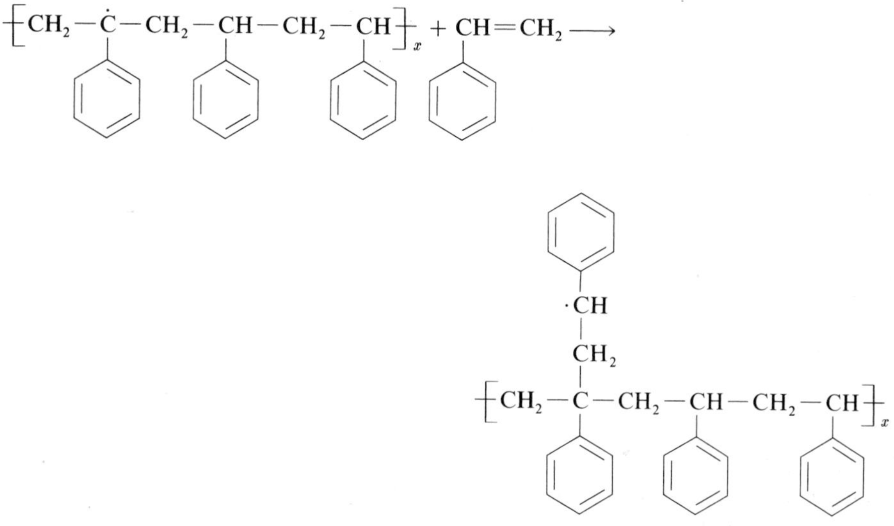

$$\left[CH_2-\dot{C}-CH_2-CH-CH_2-CH\right]_x \; + \; CH=CH_2 \longrightarrow$$

$$\cdot CH$$
$$|$$
$$CH_2$$
$$|$$
$$\left[CH_2-C-CH_2-CH-CH_2-CH\right]_x$$

branched chain

Note that in writing the structure of the polymer, we do not worry about the ends of the molecule. The end groups are an insignificant portion of the total chain and are not necessarily identical for every molecule in the sample. The properties of the polymer are determined largely by the main body of the chain. Polystyrene, polymerized simply by adding a small amount of benzoyl peroxide to the monomer, reaches molecular weights of about 2,500,000. The polymer produced in this way is amorphous.

Polystyrene has many uses. The polymer may be molded into cases for radios and batteries. Toys and all kinds of containers are made from it. If a low-boiling hydrocarbon such as pentane or hexane is included in the polystyrene during processing, when the polymer softens as it is heated, the alkane vaporizes and creates bubbles that expand the polystyrene into a rigid but lightweight foam. This foam has very good insulating properties and is used as insulation in construction. Pellets of the foam are supplied to manufacturers, who use pressure and some heat to mold them into ice chests and disposable plastic cups for hot drinks. Egg cartons, which have to be rigid but cushiony at the same time, are made of polystyrene foam. The foam can be molded to the exact shape of an object that needs to be protected during transportation, thus many delicate instruments, or bottles of chemicals, now travel in their own polystyrene foam sheaths.

Useful as polystyrene is, a multitude of copolymers designed to have particular properties have been invented. For example, a copolymer of styrene with acrylonitrile is superior to polystyrene alone in toughness, stability to light, and resistance to chemicals. The polymer, produced by polymerization of a mixture of the two monomers, is believed to have mostly an alternating arrangement of the two units in the chain.

styrene	acrylonitrile	copolymer of styrene
~1 mole	1 mole	and acrylonitrile

The exact arrangement of monomer units in a copolymer depends on the relative reactivities and the relative concentrations of the two monomers in the reaction mixture. The styrene-acrylonitrile copolymers have exceptional clarity and strength and have been made into a variety of objects such as lenses for automobile headlights, household containers, and many devices for medical use, such as disposable syringes and parts for artificial kidneys.

Another important copolymer of styrene is the cross-linked polymer it forms with divinylbenzenes.

copolymer of styrene and p-divinylbenzene
showing how the p-divinylbenzene forms a
cross-link between two polymer chains

Because *p*-divinylbenzene has two alkene functions, it becomes part of two polymer chains, forming a link between them. Depending upon the relative amounts of styrene and *p*-divinylbenzene used, the links may be close together or widely separated.

In one practical application, the cross-linked copolymer of styrene and *p*-divinylbenzene, modified by incorporating polar functional groups such as a sulfonic acid onto the aromatic rings, is used as an ion-exchange resin. The whole polymer molecule acts as a gigantic insoluble anion, which can exchange its cations with those in the solution that flows through it. Such resins are used to soften water by exchanging sodium ions for calcium and magnesium ions in hard water (Section 10.3B), both in water-treatment plants and water-softening units in homes.

A styrene-divinylbenzene copolymer is also the support system for an automated solid-phase synthesis of peptides and proteins, described in Section 17.4D.

PROBLEM 20.3   For each of the following polymers, pick out the monomers that would be used in a polymerization reaction. Say whether each one is a copolymer or a simple polymer.

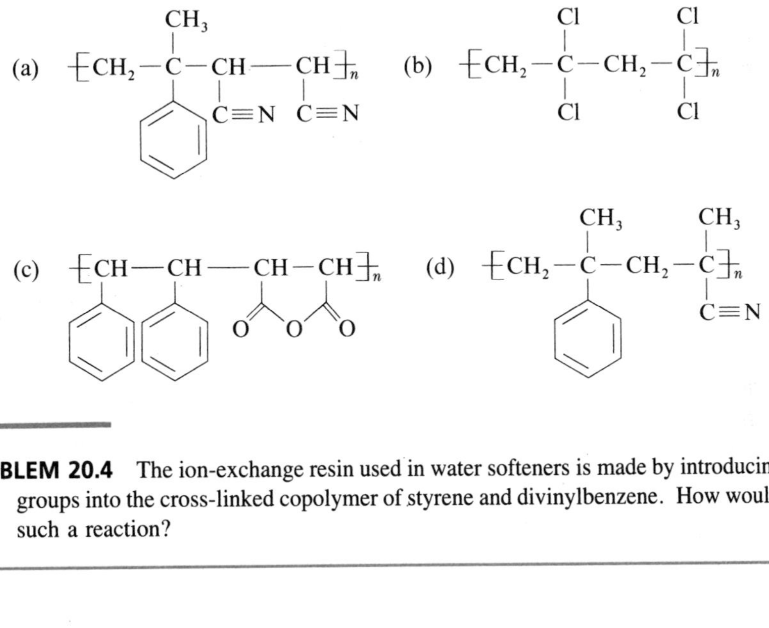

PROBLEM 20.4   The ion-exchange resin used in water softeners is made by introducing sulfonic acid groups into the cross-linked copolymer of styrene and divinylbenzene. How would you carry out such a reaction?

B. Anionic Reactions

Some polymerization reactions are catalyzed by alkali metals or by organometallic compounds. The reactive species that participates in the growth of the chain in these cases is a carbanion. 2-Phenylpropene (α-methylstyrene) polymerizes by such a reaction.

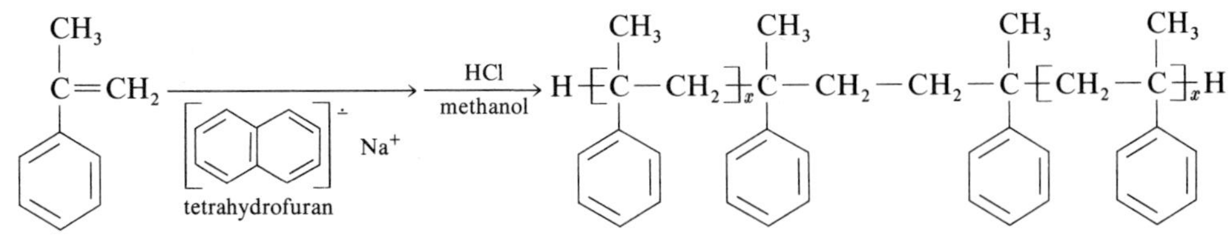

The reaction is catalyzed by the radical anion (Section 8.3C) of naphthalene in tetrahydrofuran. The radical anion of naphthalene transfers an electron (and the accompanying negative charge) to α-methylstyrene.

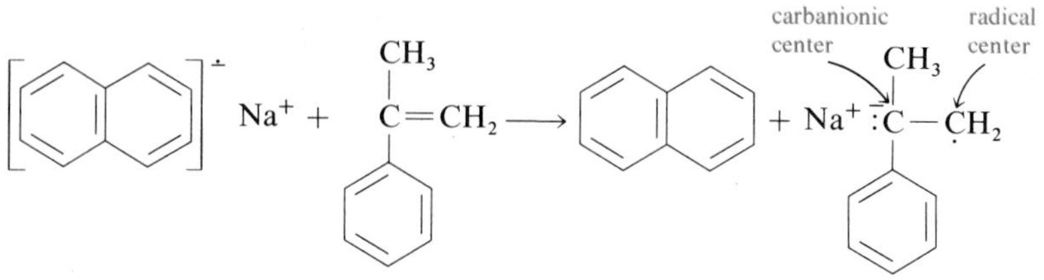

sodium naphthalide; α-methylstyrene   naphthalene   sodium salt of
radical anion of   the radical anion
naphthalene   of α-methylstyrene

Note that one electron has been added to the alkene. For the sake of clarity, this electron and the two electrons of the π bond are shown as if they were localized to give a radical center and a carbanionic center in the radical anion. This is, of course, a highly simplified way of representing the reactive species but one that is useful in rationalizing the course of the reaction, which proceeds by a combination of two radicals to give a dianion.

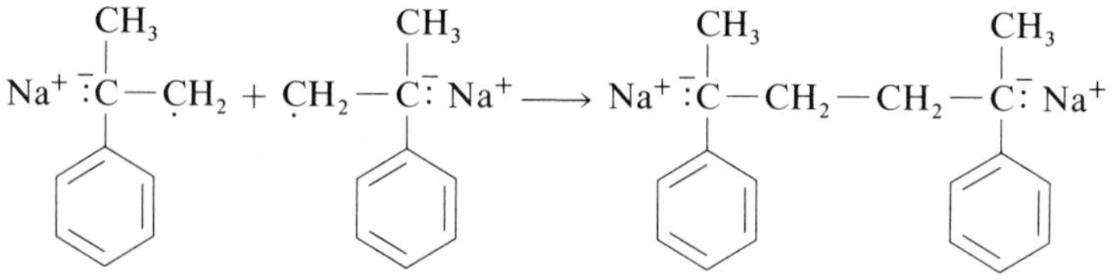

sodium salt of the radical   dianion
anion of α-methylstyrene   product of the reaction
                                                of 2 radicals

The dianion is coordinated with sodium ions. It grows by adding monomer units at either end, creating new anions, which are stabilized by their association with cations.

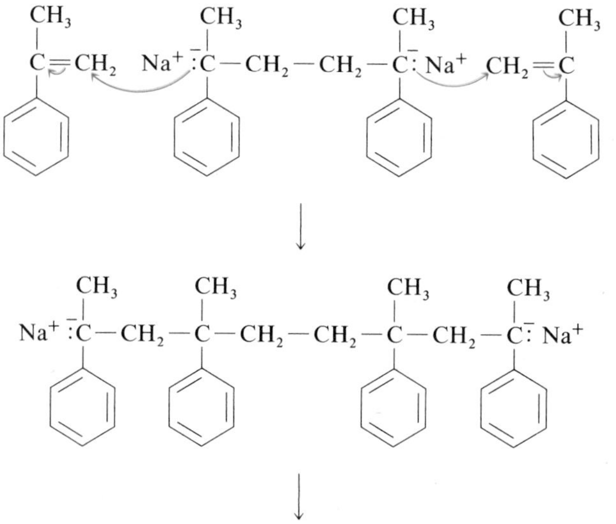

high molecular weight dianion ($\sim$50,000)

HCl
methanol

poly(α-methylstyrene)

The chain grows at both ends until it reaches an equilibrium with unreacted monomer, or until some reagent that reacts with a carbanion is added to the reaction mixture. For example, the anions may be protonated by a dilute solution of hydrochloric acid in methanol.

This type of polymerization reaction was investigated extensively by Michael Szwarc at the State University of New York at Syracuse who called the systems *living polymers,* because the chains remain reactive until some reagent is deliberately added to stop the reaction. Note that two anionic centers have no tendency to react with each other. Thus, *termination reactions resulting from the combination of two growing chains, an important reaction for free radicals, do not occur in anionic polymerizations.* If the reagents are pure and the reaction mixture is protected from moisture, the polymer chains remain reactive. An anionic polymerization differs from a free-radical reaction in another way. All chains start at the same time, at the moment the anionic initiator is added to the reaction mixture, and the intermediates react with monomer at the same rate so polymers of remarkably uniform chain length and molecular weight are formed. There is a much larger range of molecular weight for a given sample of a polymer prepared by free-radical reactions.

The living polymer can be used for copolymerization of two monomers in such a way that a large portion of the polymer chain consists of units of one monomer and another portion contains only units of the second one. Such a polymer is called a **block copolymer** in contrast to a **random or alternating copolymer.** For example, the free-radical polymerization of a mixture of styrene and acrylonitrile gives an alternating copolymer (Section 20.2A). It is also possible to create a block copolymer of these two monomers. (See Figure 20.2, p. 1136.)

By controlling the concentration of the initiator and the amount of the first monomer used, the length of the chain in the first dianion can be determined. Addition of a known amount of a second monomer will result in the growth of the chain to create a new dianion. In principle, alternating additions of styrene and acrylonitrile will give a long chain in which there are sections consisting of polystyrene attached to sections that are polyacrylonitrile.

Nucleophilic opening of a ring to give an intermediate that will propagate the chain reaction is an important method of anionic polymerization. Nylon 6, for example, is

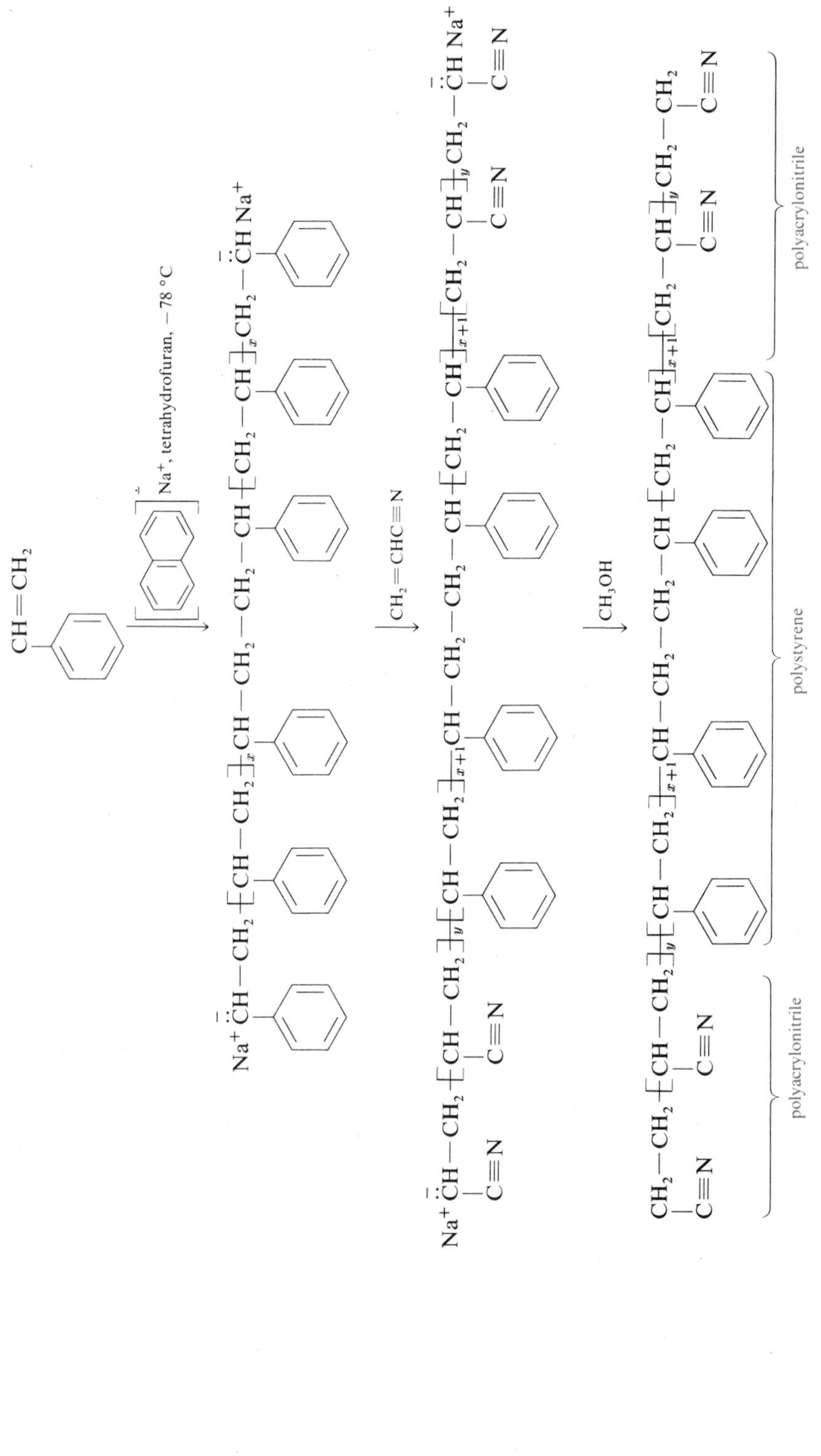

Figure 20.2 Formation of a block copolymer by the anionic polymerization of styrene and acrylonitrile.

prepared by the polymerization of the seven-membered ring lactam, ϵ-caprolactam (Section 16.5A). The reaction is catalyzed by any one of a number of amines added to the reaction mixture as the salt of an organic acid.

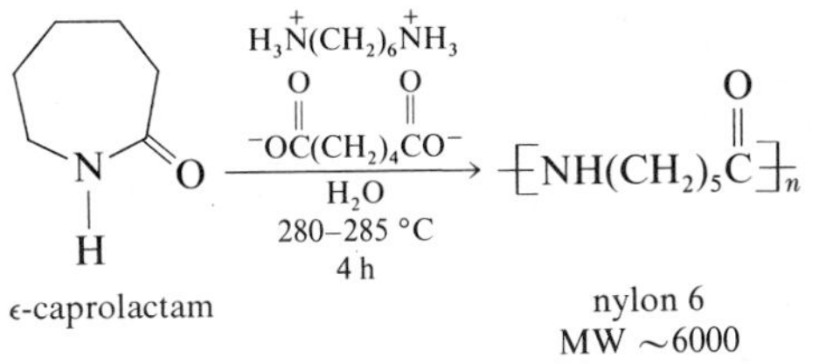

ϵ-caprolactam

nylon 6
MW ~ 6000

Caprolactam polymerizes so easily that water alone in catalytic amounts at high temperatures is enough to start the reaction. The reaction proceeds by attack of a nucleophile on the carbonyl group of the cyclic amide, giving rise to an amine, which then reacts with the carbonyl group of another molecule of lactam. The reaction perpetuates itself because a new nucleophilic center is created every time a lactam ring is opened.

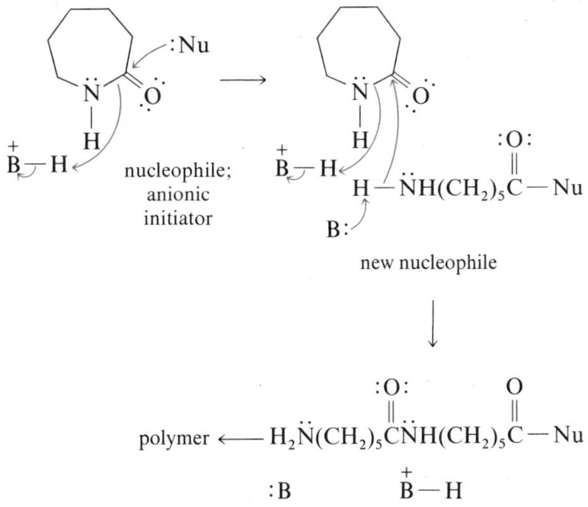

nucleophile; anionic initiator

new nucleophile

polymer

The polymerization of caprolactam can be classified as a chain-growth reaction with anionic intermediates.

PROBLEM 20.5 The following polymerization reactions have been observed. Complete the equations showing the intermediates that would be responsible for the polymerization.

(a)

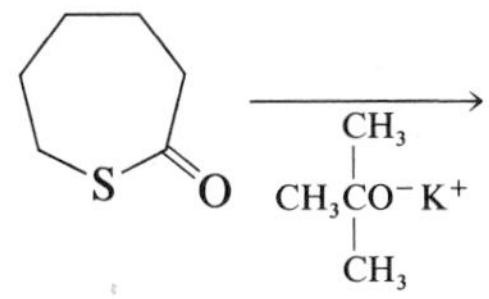

(b)

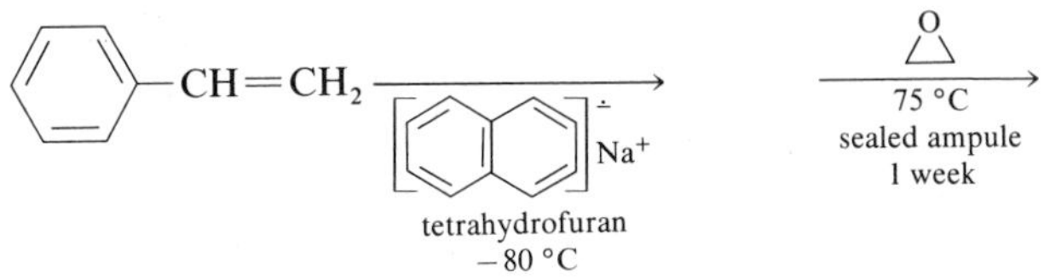

tetrahydrofuran
$-80\ °C$

$75\ °C$
sealed ampule
1 week

$$
\text{(c)} \quad CH_2 = CH\overset{\displaystyle O}{\overset{\displaystyle \|}{C}}OCH_3 + CH_2 = CHCl \xrightarrow[\substack{\text{acetone} \\ 50\,°C}]{\text{benzoyl peroxide}}
$$

(d) [benzene ring]—CH=CH$_2$ + [maleic anhydride] $\xrightarrow[\substack{\text{benzene} \\ 80\,°C}]{\text{benzoyl peroxide}}$

(e) CH$_3$CH—CH$_2$ [epoxide] $\xrightarrow[\Delta]{\text{KOH}}$

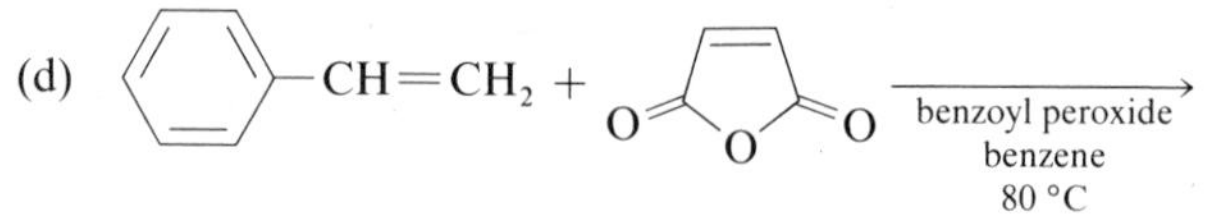

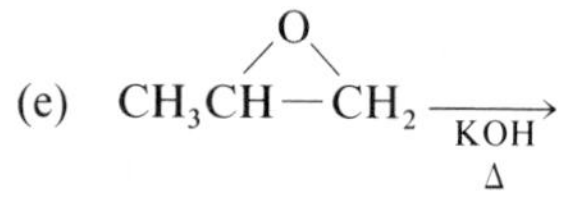

(Hint: At which carbon atom of the ring is the attack of the nucleophile more likely?)

PROBLEM 20.6   A block copolymer of methyl 2-methylpropenoate (methyl methacrylate) and iso-propyl propenoate (isopropyl acrylate) has been prepared. Write equations showing how you would proceed to carry out such a preparation.

C. Cationic Reactions

Acid-catalyzed polymerization of an alkene was one of the earliest reactions that we explored (Section 5.6C). Sulfuric acid, for example, will cause the polymerization of styrene, as will Lewis acids.

When Lewis acids such as stannic chloride are used as polymerization catalysts, traces of water or hydrogen halides are necessary for the reaction to take place, suggesting that some protic acid intermediate is involved. Carbocations are the chain-propagating species.

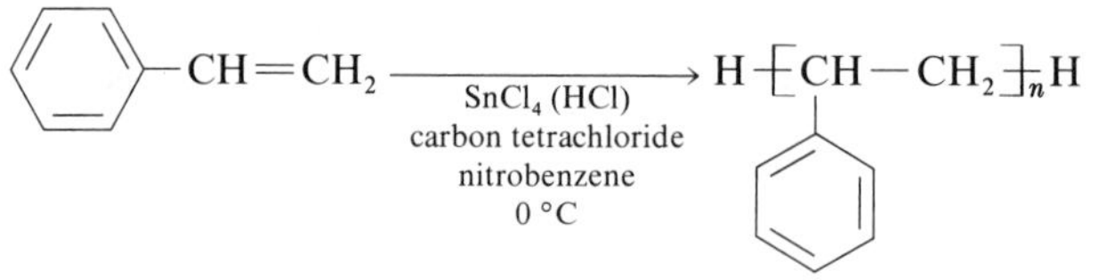

[benzene ring]—CH=CH$_2$ $\xrightarrow[\substack{\text{SnCl}_4\ (\text{HCl}) \\ \text{carbon tetrachloride} \\ \text{nitrobenzene} \\ 0\,°C}]{}$ H$\dashv$CH—CH$_2\dashv_n$H [benzene ring]

styrene   amorphous polystyrene

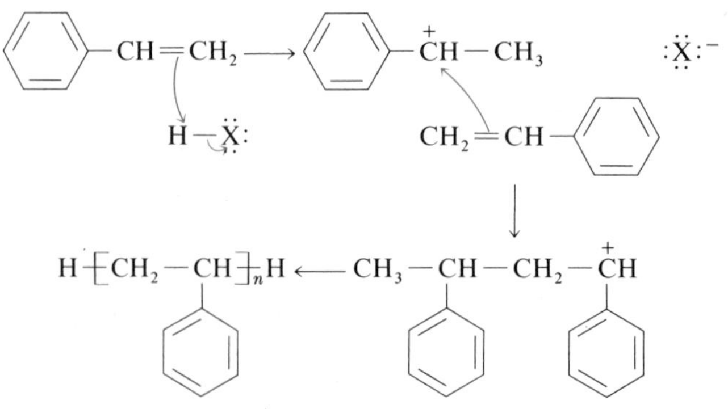

[benzene ring]—CH=CH$_2$ $\longrightarrow$ [benzene ring]—$\overset{+}{C}$H—CH$_3$ $:\ddot{X}:^-$

H—$\ddot{X}:$ CH$_2$=CH—[benzene ring]

$\downarrow$

H$\dashv$CH$_2$—CH$\dashv_n$H [benzene ring] $\longleftarrow$ CH$_3$—CH—CH$_2$—$\overset{+}{C}$H [benzene rings]

As with anionic polymerization, chain-termination reactions from the combination of two reactive intermediates are not possible in cationic polymerization. The carbocation intermediates, however, are much more likely to undergo other types of termination reactions than anions are. The reactions that terminate chains in cationic polymerizations are the familiar reactions of carbocations. Among them are reaction with a nucleophile, loss of a proton to give an alkene, and abstraction of a hydride ion from another molecule.

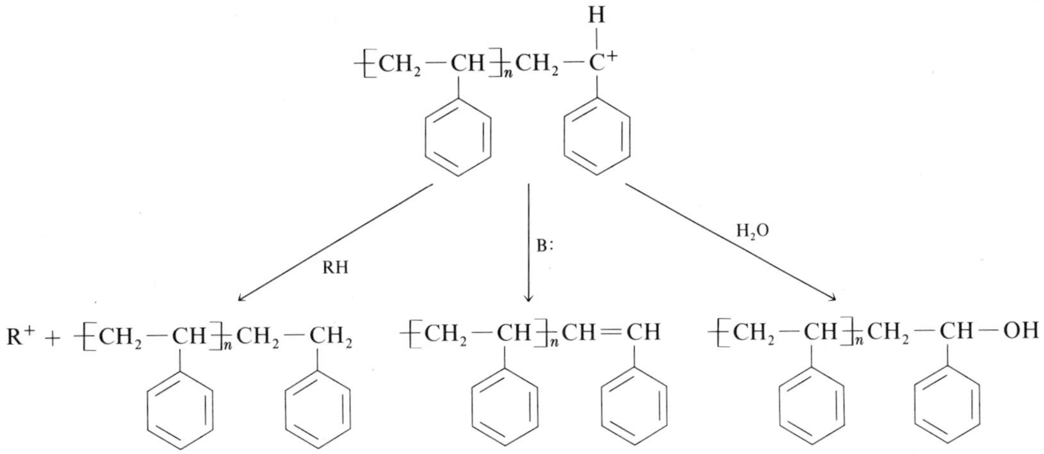

termination of the chain reaction by abstraction of hydride from another molecule

termination of the chain reaction by deprotonation of the cation

termination of the chain reaction by a combination with a nucleophile

Cyclic ethers also undergo polymerization reactions by cationic mechanisms. The most interesting of these are the reactions of oxetane and tetrahydrofuran.

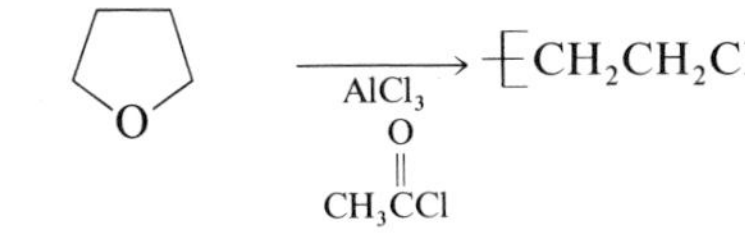

oxetane

a polyether
95%

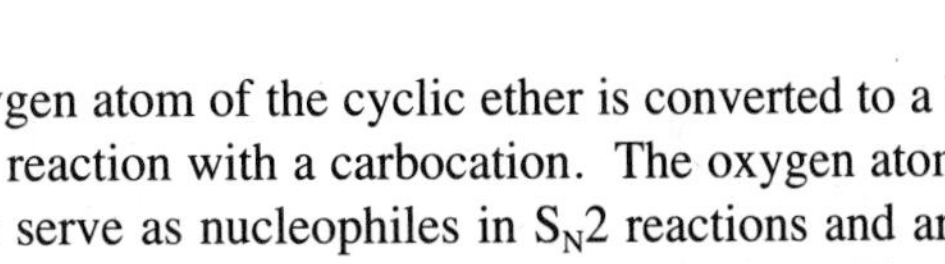

tetrahydrofuran

a polyether

In each case, the oxygen atom of the cyclic ether is converted to a good leaving group by protonation or by reaction with a carbocation. The oxygen atoms of the remaining ether molecules then serve as nucleophiles in S_N2 reactions and are converted in turn to reactive intermediates that propagate the chain reaction. The sequence is shown below for oxetane.

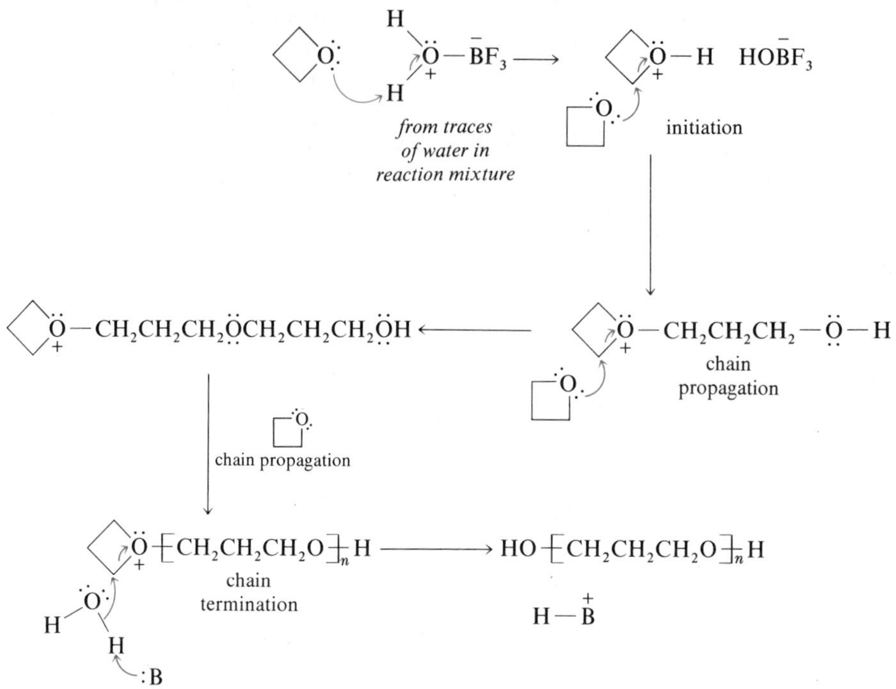

initiation

chain propagation

chain propagation

chain termination

Termination of the reaction occurs when the growing chain reacts with some nucleo-phile such as water.

PROBLEM 20.7   What will be the reactive species that starts off the polymerization of tetrahydro-furan in the presence of aluminum chloride and acetyl chloride? Write equations showing the intiation and chain-propagating steps of the reaction.

PROBLEM 20.8   The following compounds polymerize under the conditions shown. Complete the equations showing the intermediates that must be involved in the reactions.

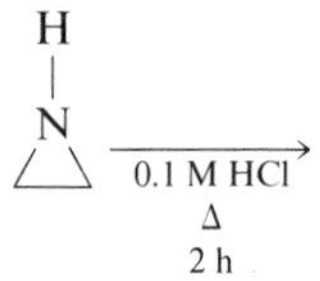

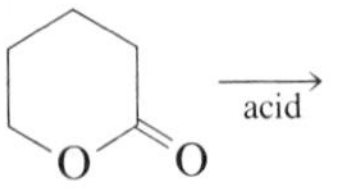

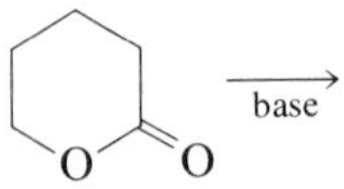

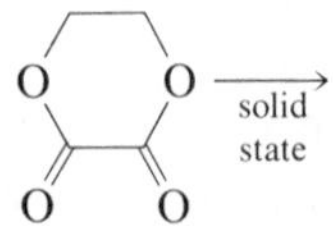

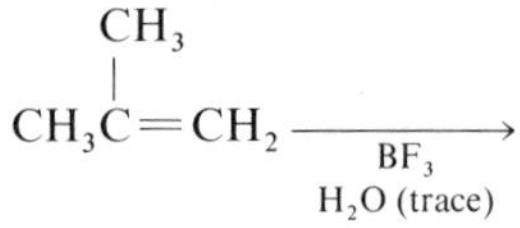

Polymerization Reactions with Controlled Stereochemistry

A. Stereochemical Regularity in Polymer Structures

The repeating nature of the structure of a polymer chain allows for three possible kinds of stereochemical relationships among the substituents on the chain. These are illustrated for polypropylene, formed when propene (propylene) is polymerized.

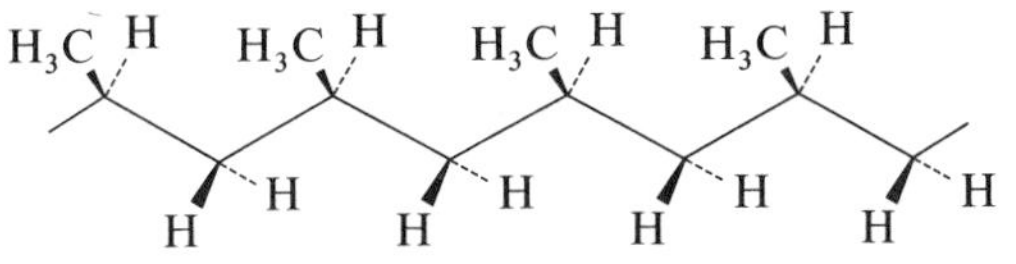

isotactic polypropylene

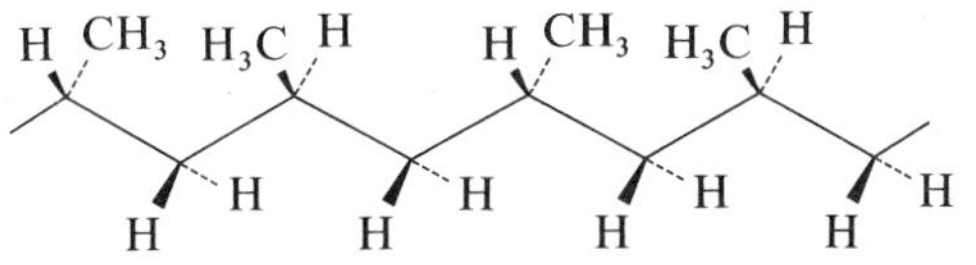

syndiotactic polypropylene

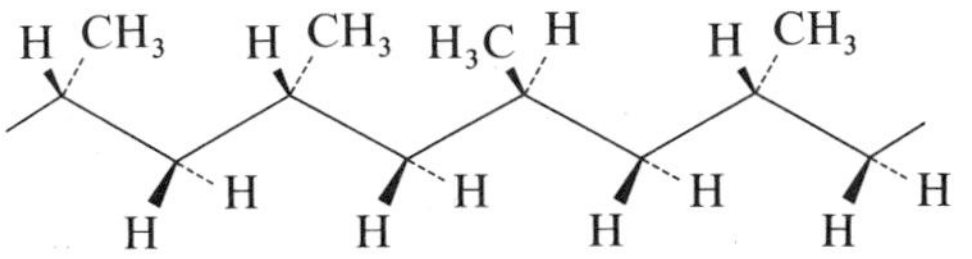

atactic polypropylene

In one of these examples, if the chain is stretched out into its most extended form, all the methyl groups are on the same side of the chain. *The configuration at each center is the same. Such a polymer is said to be an* **isotactic polymer.** In a **syndiotactic polymer,** there is a *regular alternation in the configuration at the tetrahedral carbon atoms* bearing the methyl groups. One methyl group is behind the plane of the paper, the next one in front, and the third one behind again. *The third form* of polypropylene, *an* **atactic polymer,** *has no stereochemical regularity.* The arrangement of the methyl groups is completely random.

Whether a polymer is atactic, isotactic, or syndiotactic is important in determining its properties. Only isotactic and syndiotactic hydrocarbon polymers have the regularity of structure that makes it possible for them to be crystalline, and thus have the physical properties necessary for many applications. Indeed, early attempts to polymerize propylene using radical and cationic initiators resulted in soft polymers that were of no practical use. It was not until it became possible to make isotactic polypropylene with the development of new polymerization catalysts in the 1950s (Section 20.3B) that the polymer was used industrially. Many household objects such as pails, dishpans, and plastic containers are made of polyethylene or polypropylene. Crystalline polypropylene, for example, has a melting point around 170 °C, and can therefore be used in objects that will be exposed to boiling water. It can also be made into tough films such as those used in packaging food.

The polymerization reactions described in the previous sections do not give rise to polymers with a high degree of stereochemical regularity. With a careful choice of catalysts and solvent systems, however, it is possible to control the stereochemistry of polymerization for some monomers. In the next section, we will explore these reactions.

PROBLEM 20.9   Write structural formulas showing fragments of isotactic, syndiotactic, and atactic polystyrene.

PROBLEM 20.10   Why is isotactic or syndiotactic polypropylene more highly crystalline than the atactic polymer?

PROBLEM 20.11   A free-radical polymerization reaction of methyl methacrylate (Problem 20.6) produces atactic poly(methyl methacrylate). Give the structural formula for a fragment of such a polymer chain.

B. Heterogeneous Catalysis. Ziegler-Natta Catalysts

In the 1950s, Karl Ziegler of West Germany and Giulio Natta of Italy developed a catalyst system that polymerizes alkenes to give linear polymers with high stereoselectivity. These catalysts, which consist of mixtures of transition metal halides with organometallic compounds, mostly trialkylaluminums, are not soluble in the alkane solvents that are used for these reactions. The reaction mixture is, therefore, a heterogeneous one, and polymerization takes place at the surface of the catalyst. The use of these catalysts has revolutionized the production of alkene polymers. Polyethylene prepared in this way, for example, is almost completely linear and highly crystalline, whereas other methods of polymerization result in a branched product of much lower strength and chemical resistance. We have already described the usefulness of crystalline polypropylene prepared with such catalysts. Ziegler and Natta were jointly awarded the Nobel Prize in 1963 for the work that led to the discovery of these catalysts.

A typical reaction using a Ziegler-Natta catalyst is the preparation of crystalline polystyrene.

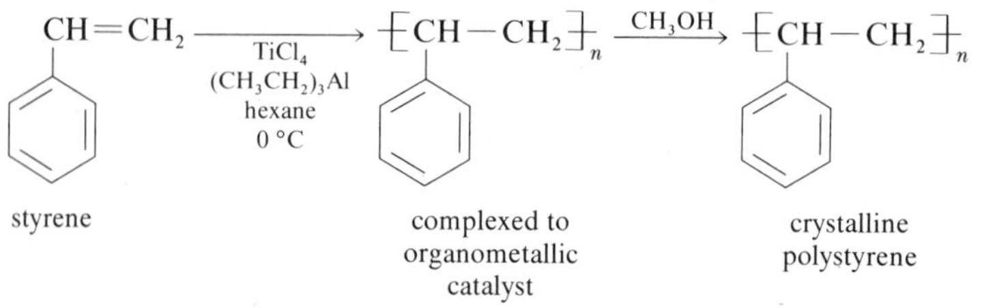

In a similar reaction, ethylene gives a polymer described as high density polyethylene to distinguish it from the softer, more highly branched polymer obtained by other methods.

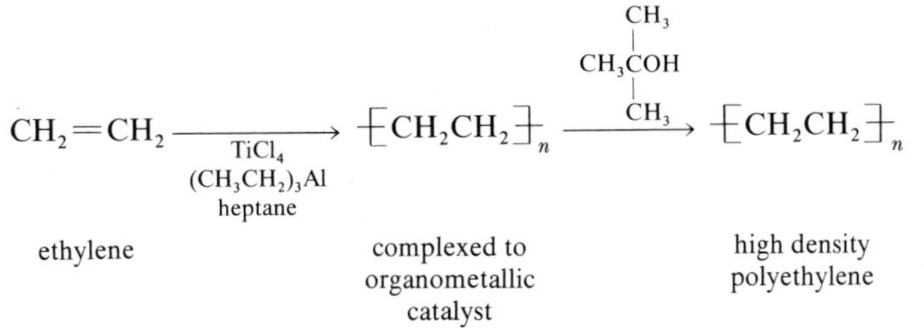

$$CH_2{=}CH_2 \xrightarrow[\substack{TiCl_4 \\ (CH_3CH_2)_3Al \\ heptane}]{} \overbrace{}^{} \left[CH_2CH_2\right]_n \xrightarrow[\substack{CH_3 \\ CH_3COH \\ CH_3}]{} \left[CH_2CH_2\right]_n$$

ethylene   complexed to   high density
organometallic   polyethylene
catalyst

Titanium tetrachloride is a typical transition metal halide, and triethylaluminum is the kind of organometallic compound used in the catalyst. They react with each other to give an organotitanium compound, which is believed to play the most important part in the catalysis.

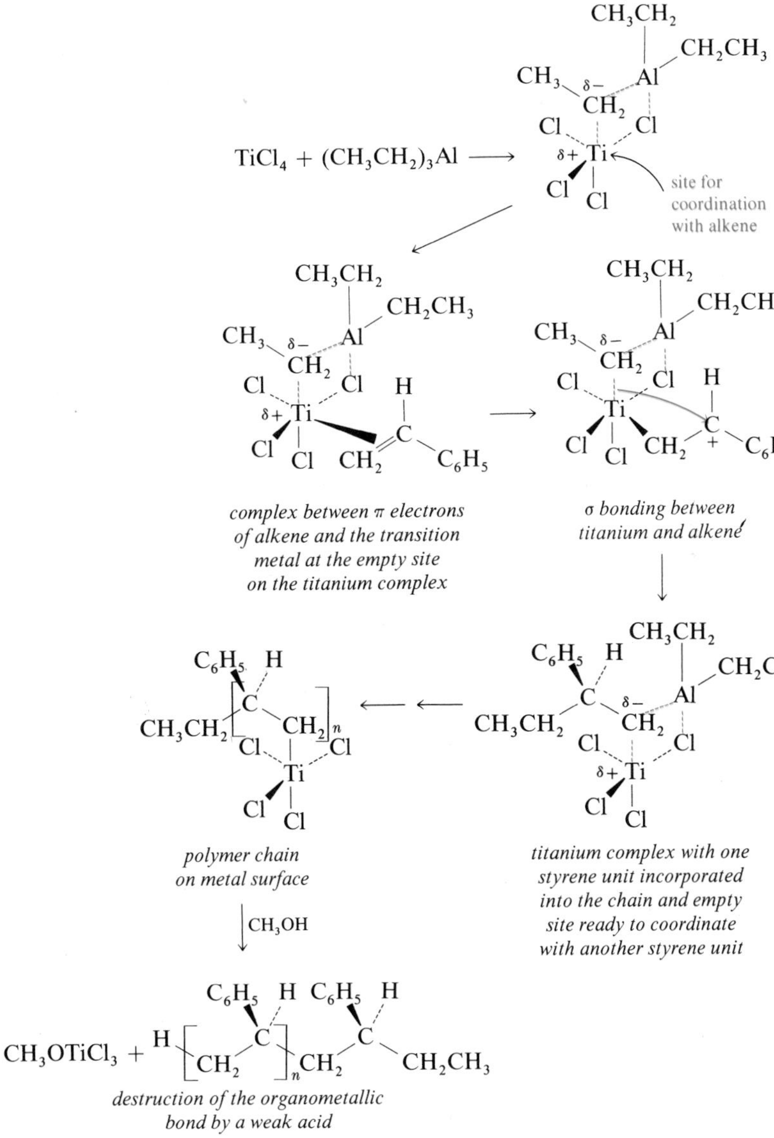

complex between π electrons of alkene and the transition metal at the empty site on the titanium complex

σ bonding between titanium and alkene

polymer chain on metal surface

titanium complex with one styrene unit incorporated into the chain and empty site ready to coordinate with another styrene unit

destruction of the organometallic bond by a weak acid

The reaction is pictured as proceeding by a π complex between the transition metal and the monomer. As bonding develops between the metal and the alkene, the metal complex acquires a partial negative charge while the benzylic carbon atom acquires a partial positive charge. At some point, the alkyl group already bonded to the metal moves to the developing carbocation with a pair of electrons. The new organotitanium compound that is formed is able to repeat this process, successively inserting monomer units between the metal atom and the rest of the growing polymer chain. The reaction takes place in a highly stereoselective way for reasons that are not well understood, partly because the exact nature of the catalyst is not known. A highly simplified representation was used above. Exactly what the organoaluminum compounds are doing during the reaction and whether it takes place at a single metal atom or with the cooperation of several is not known. The stereoselectivity is believed to arise from the stereochemistry imposed on the monomer as it is adsorbed onto the catalyst. There is no doubt that the nature of the metal catalyst is important in determining the stereochemistry of the reaction. For example, isoprene can be polymerized to give all *cis*- or all *trans*-1,4-polyisoprene stereoselectively, depending on the choice of transition metal in the catalyst (Section 20.5B).

PROBLEM 20.12 Complete the following equations showing the structure of the polymers that you would expect to get. Where it is known with certainty, the tacticity of the polymer is indicated. Show it in the structural formula for the polymer. Assume that all copolymers have regular alternating structures.

(a) $CH_3CH_2\overset{\overset{\displaystyle CH_3}{|}}{C}HCH{=}CH_2 \xrightarrow[\substack{TiCl_3 \\ (CH_3CH_2)_2AlCl}]{} \xrightarrow{CH_3CH_2OH}$ isotactic

(b) $CH_2{=}CHOCH_3 \xrightarrow[\substack{VCl_3 \\ (CH_3\overset{\overset{\displaystyle CH_3}{|}}{C}HCH_2)_3Al \\ heptane}]{} \xrightarrow{CH_3CH_2OH}$ isotactic

(c) $CH_2{=}CH_2 + \underset{\underset{\displaystyle H}{\diagup}}{\overset{\overset{\displaystyle CH_3}{\diagdown}}{C}}{=}\underset{\underset{\displaystyle H}{\diagdown}}{\overset{\overset{\displaystyle CH_3}{\diagup}}{C}} \xrightarrow[\substack{VCl_4 \\ [CH_3(CH_2)_4CH_2]_3Al \\ heptane \\ -30\,°C}]{} \xrightarrow{CH_3CH_2OH}$ syndiotactic

(d) $CH_2{=}\overset{\overset{\displaystyle CH_3}{|}}{\underset{\underset{\displaystyle O}{\|}}{C}}COCH_3 \xrightarrow[\substack{VCl_3 \\ (CH_3\overset{\overset{\displaystyle CH_3}{|}}{C}HCH_2)_3Al \\ heptane}]{} \xrightarrow{CH_3CH_2OH}$ highly crystalline stereochemistry not known

(e) $CH_3CH_2CH{=}CH_2 \xrightarrow[\substack{TiCl_3 \\ (CH_3CH_2)_3Al}]{} \xrightarrow{CH_3CH_2OH}$

20.4

Step-Growth Polymerization

A. Polyamides and Polyesters

The preparation of nylon 66 from a mixture of 1,6-hexanediamine and adipic acid (pp. 1124 and 1128) is an example of step-growth polymerization reactions. Nylon 66 is a typical polyamide. Polyamides are also formed in chain-growth processes, as was demonstrated for the anionic polymerization of ε-caprolactam to nylon 6 (p. 1137, Section 20.2B).

Other polyamides are prepared by the reaction of acid chlorides with amines in the presence of sodium hydroxide, conditions of the Schotten-Baumann reaction for the preparation of simple amides (Sections 10.4 and 16.3A). Nylon 610 is synthesized in this way in a two-phase system. The reaction is run in a blender to mix the water and organic layers thoroughly.

$$H_2N(CH_2)_6NH_2 \ + \ \underset{\substack{\text{decanedioyl chloride}\\\text{sebacyl chloride}}}{Cl\overset{O}{\overset{\|}{C}}(CH_2)_8\overset{O}{\overset{\|}{C}}Cl} \ \xrightarrow[\substack{H_2O\\ \text{tetrachloroethylene}\\ \text{blend}\\ \text{2 min}}]{NaOH} \ \underset{\substack{\text{poly(hexamethylenesebacamide)}\\ \text{nylon 610}\\ 85\%\\ MW \sim 20,000}}{\left[NH(CH_2)_6NH\overset{O}{\overset{\|}{C}}(CH_2)_8\overset{O}{\overset{\|}{C}}\right]_n}$$

1,6-hexanediamine

Poly(ethylene terephthalate), a typical polyester, on the other hand, is prepared by two transesterification reactions (Section 10.7F) from dimethyl terephthalate and ethylene glycol. If the lower-boiling alcohol component that is formed at each stage is continuously removed from the reaction mixture, the polymerization is driven to completion. The first stage of the polymerization reaction is shown below.

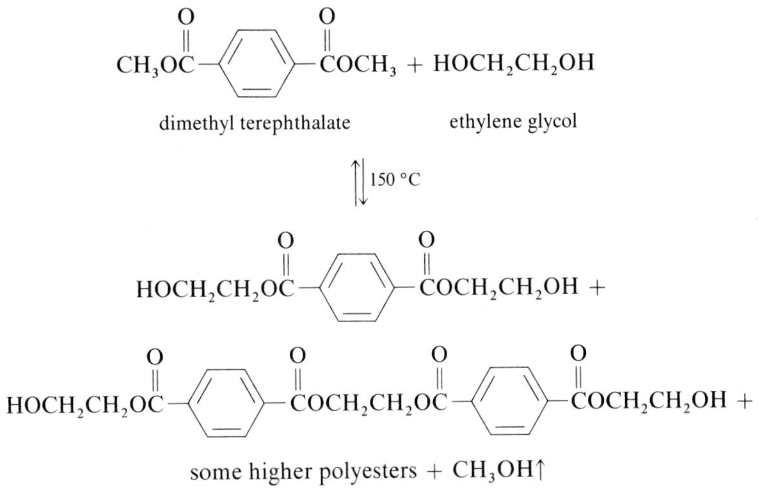

$$CH_3O\overset{O}{\overset{\|}{C}}-\!\!\!\bigcirc\!\!\!-\overset{O}{\overset{\|}{C}}OCH_3 + HOCH_2CH_2OH$$

dimethyl terephthalate   ethylene glycol

$$\Updownarrow 150\ °C$$

$$HOCH_2CH_2O\overset{O}{\overset{\|}{C}}-\!\!\!\bigcirc\!\!\!-\overset{O}{\overset{\|}{C}}OCH_2CH_2OH \ +$$

$$HOCH_2CH_2O\overset{O}{\overset{\|}{C}}-\!\!\!\bigcirc\!\!\!-\overset{O}{\overset{\|}{C}}OCH_2CH_2O\overset{O}{\overset{\|}{C}}-\!\!\!\bigcirc\!\!\!-\overset{O}{\overset{\|}{C}}OCH_2CH_2OH \ +$$

some higher polyesters + $CH_3OH\uparrow$

Polyamides and polyesters are put to many uses. Both of them are important in the synthetic fibers industry. It is difficult, for example, to find any cotton clothing that does not contain some polyester fiber. Strong, relatively inflexible, but very light polyester films are being used in sails for racing yachts and to build the wings of human-powered planes.

Polyamides interact with water by hydrogen bonding at the amide bonds. The properties of a polyamide change somewhat, therefore, with changes in humidity. The chains of nylon 66, for example, become more mobile with increasing humidity as the hydrogen bonds between chains are replaced by hydrogen bonding to water molecules. It is possible to design nylons that are particularly resistant to the absorption of moisture by making the hydrocarbon-like portions of the molecule larger or giving the structures more rigidity by using components containing rings. The polyamides containing aromatic rings form highly crystalline structures of exceptional strength. Fibers made from these polyamides are stiffer than steel at a much lower density. Such polymers are being used to reinforce tires, to make bullet-proof vests, and in lightweight but very strong cords and cables.

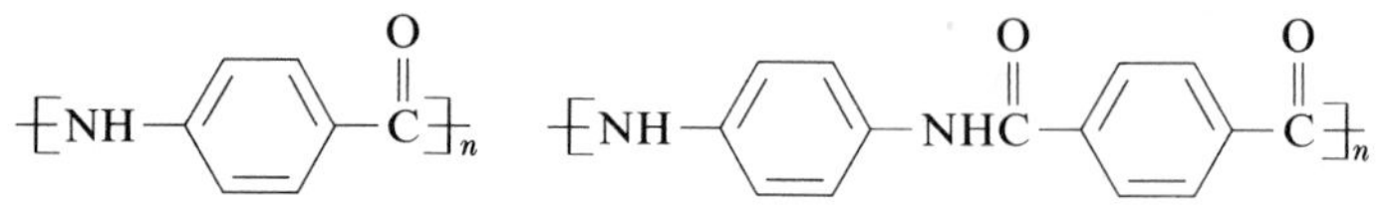

Kevlar

two polyamides of particular molecular rigidity

Polymers of every description are finding use in health care. Not only are all kinds of disposable bottles, syringes, and laboratory ware being made from different polymers, but actual replacements for parts of human bodies are being created. Every year, several million artificial parts are implanted into individuals who have suffered some loss either as a result of accident or disease. The development of biomaterials, artificial substances that are compatible with human tissues, is an important area of research. In some cases, the implant must become a permanent part of the body and not be rejected by it. Poly(ethylene terephthalate) mesh tubes, for example, are used to replace blood vessels, in hopes that the human tissue will grow into and around the mesh to make the implant part of the body's structure. In other cases, an implant may be necessary for a period of time, but should eventually be replaced by the body's own tissues. In these cases, a polyester such as poly(lactic acid) can be used. This polymer is gradually hydrolyzed to lactic acid by the body, then metabolized to carbon dioxide and water, just as natural lactic acid is.

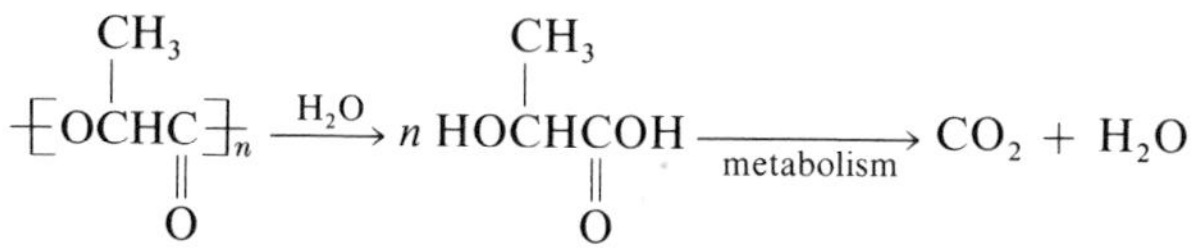

poly(lactic acid)

a polyester

lactic acid

The implant is therefore absorbed by the body and leaves no permanent residue. This is only a small sample of the applications of the thousands of synthetic materials being created.

PROBLEM 20.13 Two important commercial products are a polycarbonate and a polyester having the following structures.

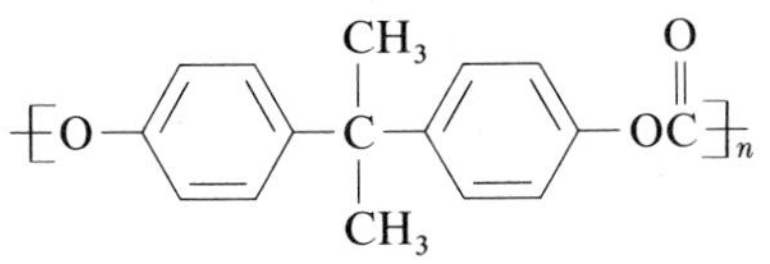

a polycarbonate 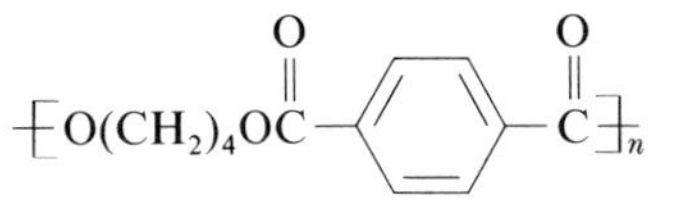   a polyester

Both are prepared by a transesterification reaction (Section 10.7F). Dissect the structure of each polymer, decide what types of reagents you would need, and write equations showing in general terms how you would prepare these polymers.

PROBLEM 20.14   Nylon 11 has the structure shown below. How would you synthesize it? Write an equation showing the monomer that you would need and the general conditions for the polymerization reaction. Is this monomer likely to be available as a lactam?

$$\left[NH(CH_2)_{10}\overset{\overset{\displaystyle O}{\|}}{C}\right]_n$$

nylon 11

B. Polyurethanes

The reaction that forms the backbone of a polyurethane is the addition of an alcohol to an isocyanate (Section 16.5C). For example, ethylene glycol reacts with 4,4′-diphenylmethane diisocyanate to give such a polymer.

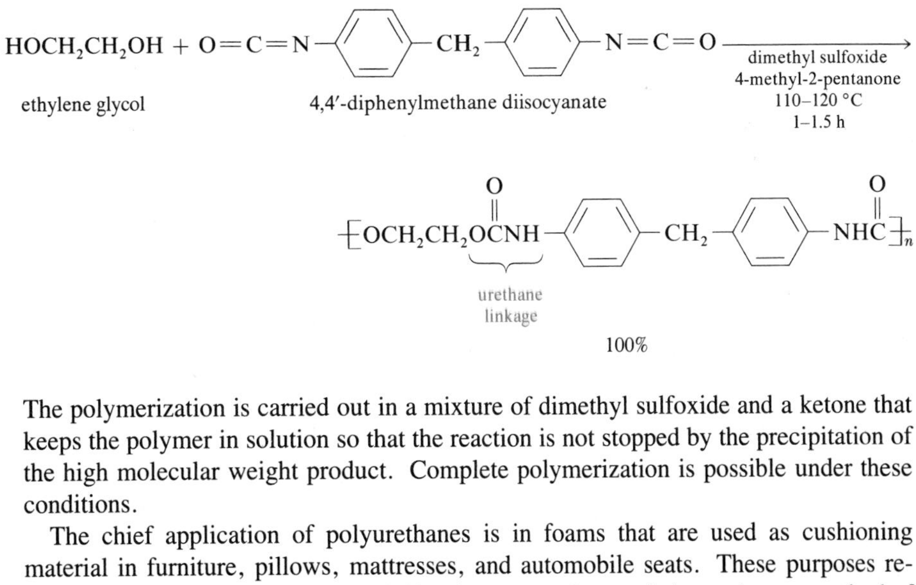

$$HOCH_2CH_2OH + O=C=N-\!\!\!\!\bigcirc\!\!\!\!-CH_2-\!\!\!\!\bigcirc\!\!\!\!-N=C=O \xrightarrow[\substack{\text{dimethyl sulfoxide}\\ \text{4-methyl-2-pentanone}\\ 110–120\,°C\\ 1–1.5\,h}]{}$$

ethylene glycol   4,4′-diphenylmethane diisocyanate

urethane linkage

100%

The polymerization is carried out in a mixture of dimethyl sulfoxide and a ketone that keeps the polymer in solution so that the reaction is not stopped by the precipitation of the high molecular weight product. Complete polymerization is possible under these conditions.

The chief application of polyurethanes is in foams that are used as cushioning material in furniture, pillows, mattresses, and automobile seats. These purposes require a moderate amount of cross-linking between polymer chains and some method of creating bubbles in the melted polyurethane. The compounds that have rubbery properties usually have long, flexible polymer chains made from diol units that are themselves polymers. When cross-linking is desired, an excess of the diisocyanate is used so that some of the polymer chains end in unreacted isocyanate functions. This type of a molecule is represented by a wavy line joining the two isocyanate groups. The polymeric diisocyanate reacts with the urethane linkages in other polymer chains to link them.

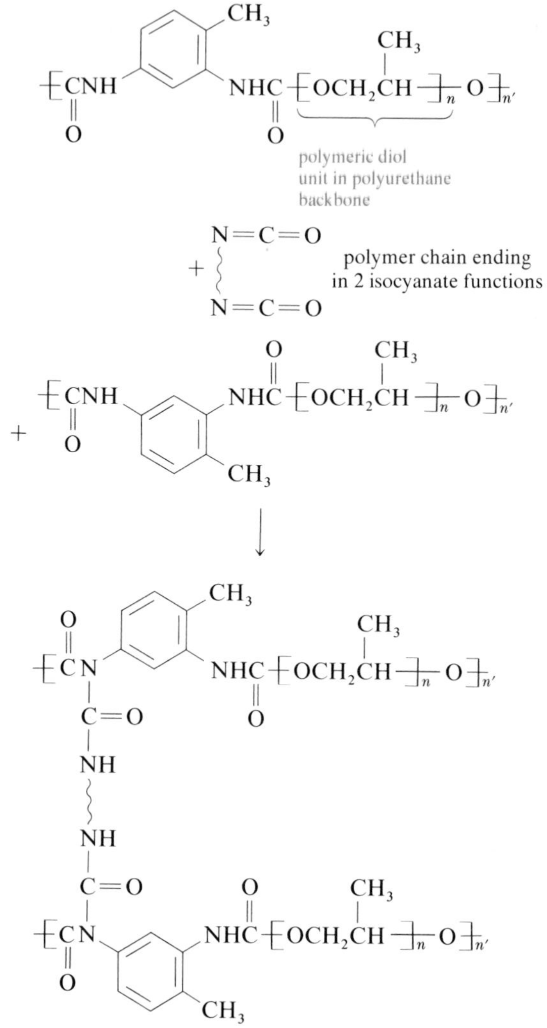

polymeric diol
unit in polyurethane
backbone

polymer chain ending
in 2 isocyanate functions

cross-linked polyurethane

If the cross-linking chains are long and flexible, the polyurethane network will also be flexible and, hence, an elastomer. If a large number of cross-links are formed, and the cross-linking chains are short and rigid, the whole polymer will also be hard and inflexible.

An isocyanate reacts with water to give carbon dioxide and an amine (Section 16.5C). This reaction is one method used to create polyurethane foam (Figure 20.3). A small amount of water is added to the hot polymer at a stage when an excess of isocyanate end groups is present. The carbon dioxide forms bubbles that expand in the hot polyurethane, giving it a foamy texture that it retains when it cools. Meanwhile, the new amine groups that are created in the process react with the remaining excess isocyanate groups to give urea bonds. At the end of the polymerization process, no reactive end groups are left, and the structure of the foam is reinforced by new cross-linking reactions.

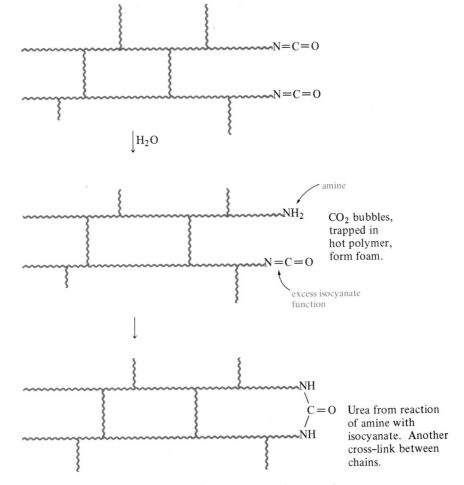

Figure 20.3   Reaction of isocyanate end groups in polyurethane to give carbon dioxide and further cross-linking of the polymer.

PROBLEM 20.15   Complete the following equations.

(a) $\overset{O}{\overset{\|}{Cl}}C O(CH_2)_4 O \overset{O}{\overset{\|}{C}} Cl + H_2N(CH_2)_6NH_2 \xrightarrow[\substack{NaOH \\ H_2O}]{}$

(b)   [structure: benzene ring with CH$_3$ and two N=C=O groups] $+ HOCH_2CH_2OH \xrightarrow[\substack{\text{dimethyl sulfoxide} \\ \text{4-methyl-2-pentanone} \\ 115\,°C}]{}$

(c) $O=C=N(CH_2)_6N=C=O + HO(CH_2)_4OH \xrightarrow[185–195\,°C]{}$

C. Polymers from Condensation Reactions of Formaldehyde

Polymers from the reactions of formaldehyde with phenols have been known for over a hundred years. Leo Hendrik Baekeland of the United States was the first chemist to patent such materials at the beginning of this century. The polymer, Bakelite, is a stiff, three-dimensional network with very little solubility in organic solvents and a high resistance to electricity and to heat. It is used in a wide variety of household objects and electrical fixtures.

Phenols condense with formaldehyde under either acidic or basic conditions. Under acidic conditions, polymerization proceeds to give a network of phenol rings held together at the ortho and para positions by methylene groups. These polymers have a broad range of molecular weights in any one sample.

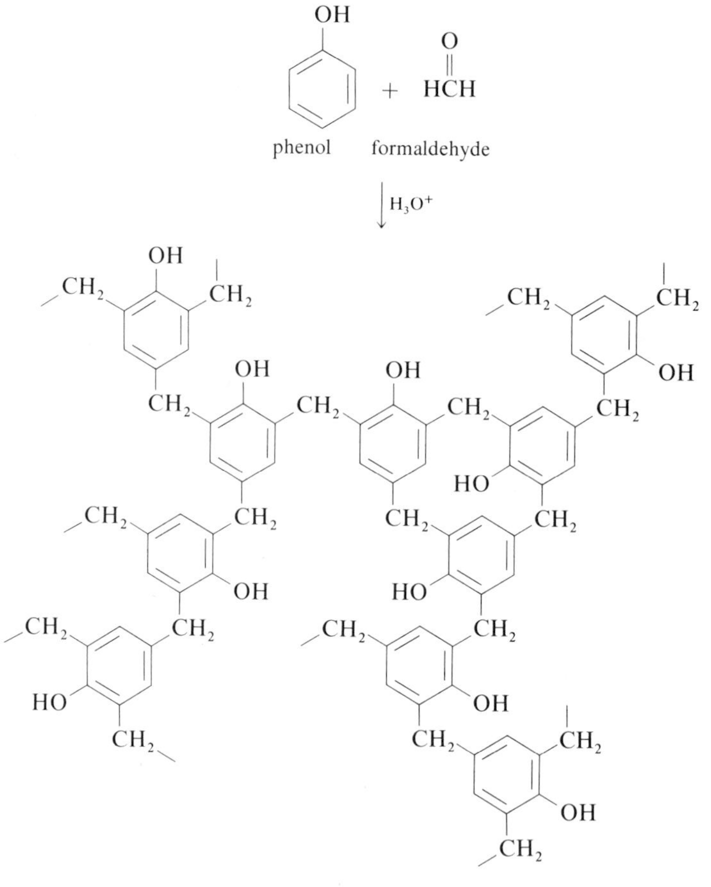

The reaction is postulated to proceed by protonation of the carbonyl group of the aldehyde (or its hydrate) by the acid catalyst to give an electrophile that reacts with the phenol ring, which is highly activated towards electrophilic aromatic substitution at the ortho and para positions (Section 12.3A). The resulting benzylic alcohols are also converted into electrophilic cations by protonation and loss of water, and further electrophilic substitution of phenols takes place to create the network. The first stages of the process are shown below.

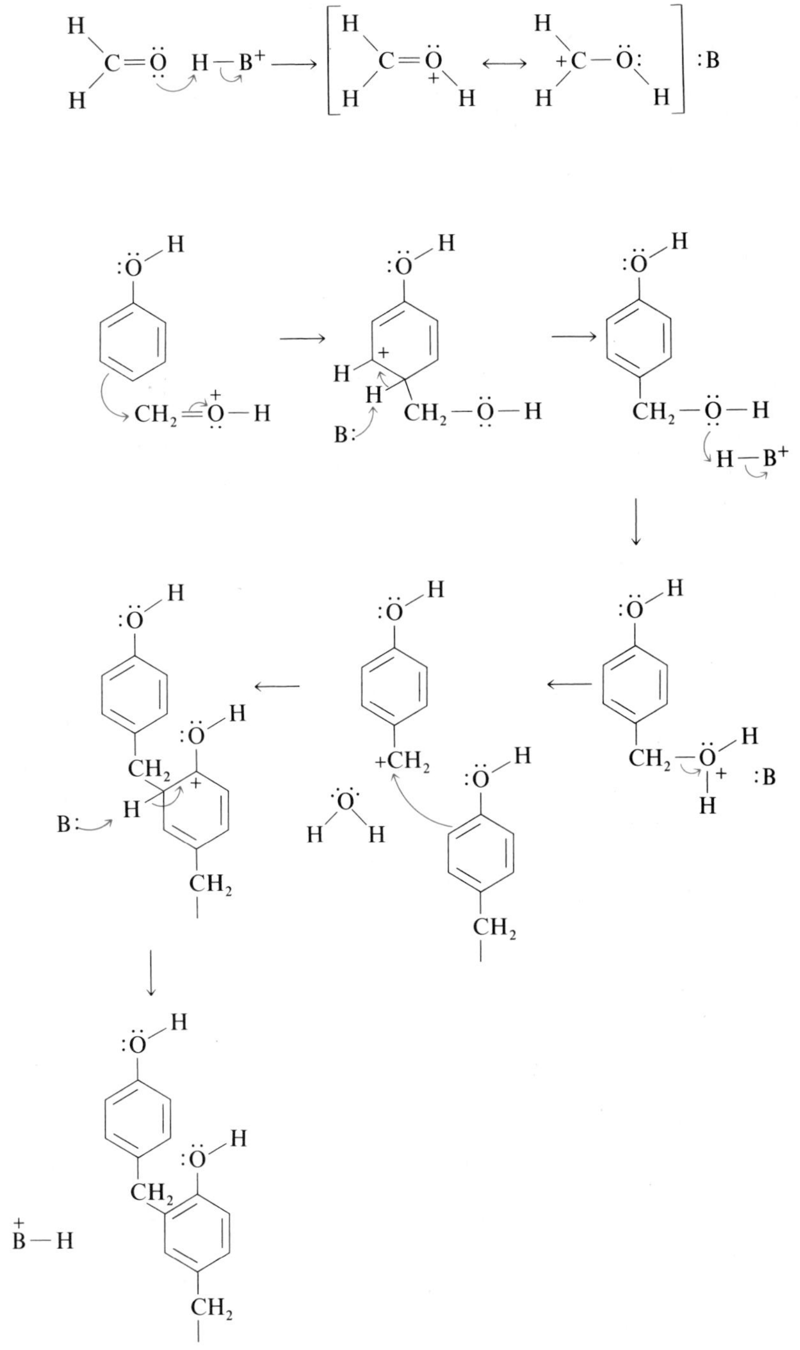

beginning of polymer network

Commercially important plastics are also obtained when urea and the heterocyclic triamine melamine react with formaldehyde. In both of these reactions, amino groups add to formaldehyde to give intermediates with hydroxymethyl groups.

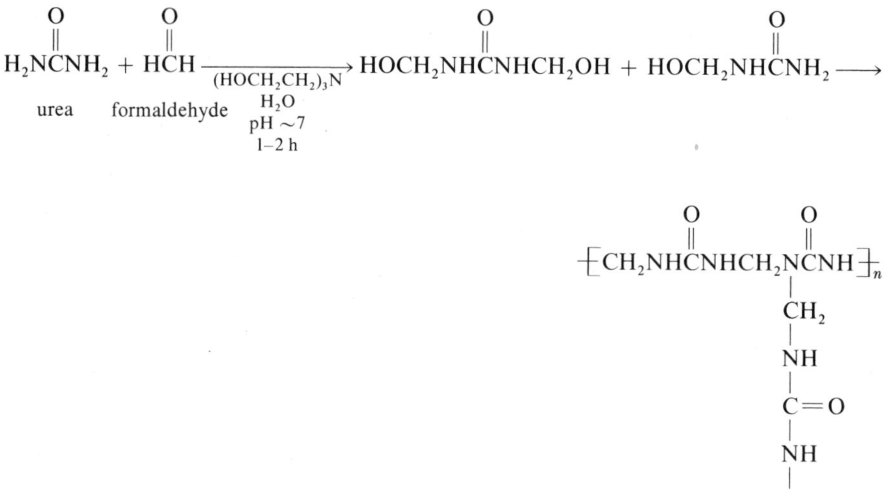
urea
formaldehyde
$(HOCH_2CH_2)_3N$
H_2O
pH ~7
1–2 h

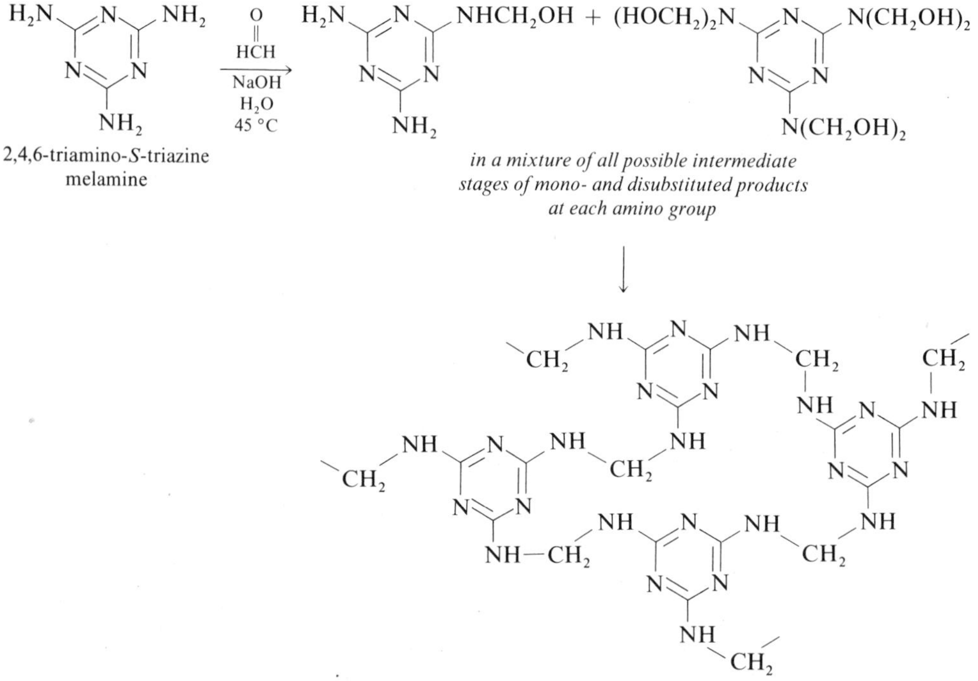
NaOH
H_2O
45 °C
2,4,6-triamino-*S*-triazine
melamine
in a mixture of all possible intermediate stages of mono- and disubstituted products at each amino group

The resins obtained from these reactions have many applications. A small amount of urea-formaldehyde resin applied to a cotton fabric gives it resistance to creasing. In fact, the predominant odor in a large fabric shop is very often that of formaldehyde. Waste wood products are converted into boards by blending sawdust or wood chips with the resin and curing the mixture in a hot mold. Packaging paper is made more resistant to moisture by treating the paper pulp with urea-formaldehyde resin. Melamine resins are used in very much the same ways.

PROBLEM 20.16 Show the details of the reaction of urea with formaldehyde to produce a cross-linked polymer.

D. Epoxy Resins

Epoxy resins are most familiar in everyday use as adhesives. The resin is formed upon mixing two solutions, one of a polymer that contains oxirane rings that will form cross-links between polymer chains when treated with a nucleophile, and the other containing a polyamino compound, which is the nucleophilic reagent starting the process and forming some of the cross-links itself. The polymer itself contains hydroxyl groups, ether linkages, and oxirane rings. All these functional groups, along with the amino groups that are added during the cross-linking process, hydrogen bond and coordinate strongly with surfaces such as glass, ceramics, and metal. The result is a resin that glues such surfaces together with tremendous force. The most common epoxy resin is made from the oxirane derived from 3-chloro-1-propene and 2,2-bis(4'-hydroxyphenyl)propane.

$$\text{ClCH}_2\overset{\displaystyle O}{\overset{\displaystyle \diagup\!\!\!\diagdown}{\text{CH}-\text{CH}_2}} \;+\; \text{HO}-\!\!\langle\!\!\bigcirc\!\!\rangle\!\!-\!\!\overset{\displaystyle \text{CH}_3}{\underset{\displaystyle \text{CH}_3}{\text{C}}}\!\!-\!\!\langle\!\!\bigcirc\!\!\rangle\!\!-\text{OH} \;\xrightarrow[\Delta]{\text{NaOH}}$$

oxirane from   2,2-bis(4'-hydroxyphenyl)propane
3-chloro-1-propene
epichlorohydrin

$$\overset{\displaystyle O}{\overset{\diagup\!\!\!\diagdown}{\text{CH}_2\text{—CHCH}_2}}\!\!\left[\text{O}-\!\langle\!\bigcirc\!\rangle\!-\overset{\text{CH}_3}{\underset{\text{CH}_3}{\text{C}}}\!-\!\langle\!\bigcirc\!\rangle\!-\text{O}-\text{CH}_2\overset{\text{OH}}{\underset{}{\text{CHCH}_2}}\right]_n\!\!\text{O}-\!\langle\!\bigcirc\!\rangle\!-\overset{\text{CH}_3}{\underset{\text{CH}_3}{\text{C}}}\!-\!\langle\!\bigcirc\!\rangle\!-\text{O}-\text{CH}_2\overset{\displaystyle O}{\overset{\diagup\!\!\!\diagdown}{\text{CH}-\text{CH}_2}}$$

The oxirane is used in excess so that the end groups of the polymer chains are oxiranes.

The oxirane derived from 3-chloro-1-propene has the common name epichlorohydrin, and is a toxic compound suspected of being a carcinogen. The reason for its toxicity is the ease with which it reacts with nucleophiles to link them together, a property that it must retain with biological nucleophiles. When the oxirane ring is opened by the attack of one nucleophile, the alkoxide ion that is formed displaces chloride ion intramolecularly to give a new oxirane. This ring, in turn, opens in another nucleophilic substitution reaction.

$$\text{H}-\ddot{\text{O}}-\!\langle\!\bigcirc\!\rangle\!-\overset{\text{CH}_3}{\underset{\text{CH}_3}{\text{C}}}\!-\!\langle\!\bigcirc\!\rangle\!-\ddot{\text{O}}\overset{\curvearrowright}{=}\text{H} \quad\longrightarrow\quad \text{H}-\ddot{\text{O}}-\!\langle\!\bigcirc\!\rangle\!-\overset{\text{CH}_3}{\underset{\text{CH}_3}{\text{C}}}\!-\!\langle\!\bigcirc\!\rangle\!-\ddot{\text{O}}{:}^{-}\;\; {:}\overset{\text{H}}{\underset{\text{H}}{\text{O}}}$$

$$^{-}{:}\ddot{\text{O}}-\text{H}$$

nucleophile

$$\left(\text{represented by R}-\!\langle\!\bigcirc\!\rangle\!-\ddot{\text{O}}{:}^{-}\right)$$

$$:\ddot{C}l-CH_2-\overset{\overset{\overset{\textstyle \cdot\ddot{O}\cdot}{\displaystyle |}}{\displaystyle }}{CH}-CH_2 \qquad \longrightarrow \qquad :\ddot{C}l-CH_2-CH-CH_2-\ddot{O}-\langle\!\!\langle\bigcirc\rangle\!\!\rangle-R$$

$$-:\ddot{O}-\langle\!\!\langle\bigcirc\rangle\!\!\rangle-R$$

intermediate alkoxide ion

| intramolecular
$S_N 2$

:B

OH

$$R-\langle\!\!\langle\bigcirc\rangle\!\!\rangle-O-CH_2\overset{}{C}HCH_2-O-\langle\!\!\langle\bigcirc\rangle\!\!\rangle-R \longleftarrow$$

2 phenol units, linked by
reaction with 1 oxirane molecule

$$:\ddot{C}l:^- \qquad H\frown B^+$$

$$\overset{\overset{\textstyle \cdot\ddot{O}\cdot}{}}{CH_2}-CHCH_2-\ddot{O}-\langle\!\!\langle\bigcirc\rangle\!\!\rangle-R$$

$$R-\langle\!\!\langle\bigcirc\rangle\!\!\rangle-\ddot{O}:^-$$

Cross-linking takes place when an amine reacts with the oxirane end groups of the polymer. The linking groups may be the amine or the alkoxide ion that is created when the oxirane ring is opened (Figure 20.4). The cross-linking takes place very fast so the amine and the linear polymer are mixed just before they are applied to the surfaces that they are to bond together. The bonding is so strong that mended objects frequently break elsewhere before they fall apart at the site of repair.

PROBLEM 20.17 The diphenol used in the preparation of polycarbonates and epoxy resin has the trivial name bisphenol A because it is synthesized from phenol and acetone. Propose a mechanism for its formation from these reagents.

$$HO-\langle\!\!\langle\bigcirc\rangle\!\!\rangle-\overset{\overset{\textstyle CH_3}{\displaystyle |}}{\underset{\underset{\textstyle CH_3}{\displaystyle |}}{C}}-\langle\!\!\langle\bigcirc\rangle\!\!\rangle-OH$$

2,2-bis(4′-hydroxyphenyl)propane
bisphenol A

PROBLEM 20.18 The commercial synthesis of the oxirane derived from 3-chloropropene starts with propene and involves the use of gaseous chlorine at high temperatures, then chlorine with water at low temperatures, and finally calcium hydroxide. Write equations for these reactions showing how they lead to the formation of 2-(chloromethyl)oxirane. Classify each reaction according to its mechanistic type.

20.5

A Natural Macromolecule. Rubber

A. The Structure of Rubber and Gutta Percha

Rubber is produced by a number of plants that grow in tropical regions. The most important of these is a tree, *Hevea brasiliensis,* which was originally found in Brazil but now grows mostly in southeast Asia. The polymer is contained in a fluid known

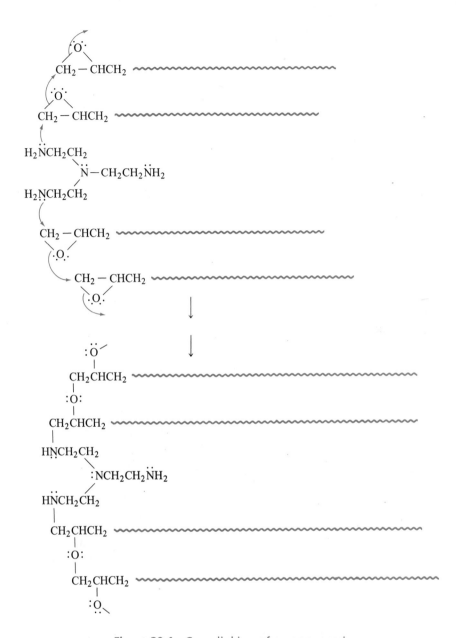

Figure 20.4   Cross-linking of an epoxy resin.

as *latex,* which is synthesized by cells under the bark of the tree. Latex is obtained by making cuts in the bark and collecting the liquid that flows out. *Hevea brasiliensis* synthesizes latex with great speed, and a tree may be tapped every other day to harvest rubber.

Rubber is a polymer made of isoprene units and contains cis double bonds. An isomeric polymer in which the double bonds are trans is obtained from trees of the *Dichopsis* genus, also found in southeast Asia. This polymeric material, which has properties quite different from those of rubber, is called *gutta percha.*

isoprene unit

CH$_3$ H

C = C

$-$CH$_2$ CH$_2$ $-$[CH$_2$ CH$_2$$]_nCH_2$ CH$_2$$-$

C = C   C = C

CH$_3$ H   CH$_3$ H

rubber
cis double bonds
$n \sim 1500{-}15{,}000$
MW $\sim 100{,}000{-}1{,}000{,}000$

isoprene unit

CH$_2$ H   [CH$_2$ H]   CH$_2$ H

C = C   C = C   C = C

CH$_3$ CH$_2$ CH$_3$ CH$_2$$]_n$ CH$_3$ CH$_2$

gutta percha
trans double bonds
$n \sim 100$
MW ~ 7000

Both rubber and gutta percha are synthesized in plants from isopentenyl pyrophosphate. The reactions that give rise to gutta percha are like the ones in the biosynthesis of terpenes such as geraniol and farnesol, which also have trans double bonds (Section 11.6B). The molecular weight of gutta percha is approximately 7000, while that of rubber usually ranges from 100,000 to the millions. The biosynthesis of rubber is obviously directed by a different type of enzyme than that of gutta percha because of the stereoselective formation of cis double bonds in rubber.

The physical properties of rubber and gutta percha are quite different, as would be expected from the different shapes of their molecules. Rubber has a much more folded structure than gutta percha. The cis arrangement of the double bonds makes it harder for adjacent rubber molecules to fit close to each other in the ordered way that results in a crystalline nature. Thus, rubber is highly amorphous. Because of the random coiling of its large molecules, rubber can be stretched easily to many times its natural length. When it is stretched, the molecules are forced into a more orderly arrangement, which is unstable for the polymer, so it snaps back into the more random coiled structure when the tension is released.

The structure of gutta percha allows for a closer packing of the polymer molecules next to each other so that the polymer is much more crystalline in its natural state than rubber. In general, gutta percha is harder and less flexible than rubber. For example, the relatively hard, inflexible covers of golf balls are made mostly of gutta percha. It is also a good electrical insulator and is used in many electrical applications such as casings for cables.

Raw rubber is affected by a number of environmental factors such as temperature (Section 20.1C) and oxygen. The allylic positions next to the double bonds, for example, are vulnerable to reaction with oxygen as a free radical (Section 11.7A). The double bonds themselves react with ozone, which is always present but is especially abundant in air polluted by exhaust fumes from automobiles. Rubber is affected by organic solvents such as gasoline and oil, and tends to swell and dissolve in such hydrocarbons. Light is also destructive to rubber. For these reasons, almost all rubber used industrially is treated in a process known as vulcanization to increase its stability. Vulcanized rubber has greater strength, less stickiness, and greater elasticity than raw rubber. It is less soluble in most solvents, and tends to retain its flexibility at lower temperatures.

Charles Goodyear of the United States discovered vulcanization in 1839 when he found that heating rubber with sulfur improved the properties of the rubber. The sulfur reacts irreversibly with rubber to link different polymer chains and parts of the same chain together to give greater stability to the molecule. The physical properties of rubber are changed even when only a few percent by weight of sulfur is used. The chemistry is complicated, and cyclic structures containing sulfur as well as sulfur bridges between chains are formed.

B. Synthetic Rubbers

Rubber is important to transportation as well as to many other aspects of modern technology. The world wars made it clear that industrial states could easily lose the supply of natural rubber that came to them from tropical countries, especially from southeast Asia, and research was started into ways of making synthetic rubber. One of the earliest replacements for natural rubber was neoprene, developed by DuPont in the United States. It is the polymer of 2-chloro-1,3-butadiene, a molecule that resembles isoprene, except that a chlorine atom replaces the methyl group on the backbone of the chain, hence the common name of chloroprene for the compound.

A diene may polymerize by 1,2- or by 1,4-addition to the conjugated double bonds (Section 11.1C). 2-Chloro-1,3-butadiene polymerizes in a free-radical reaction by both types of addition.

$$CH_2\!=\!\overset{\overset{\displaystyle Cl}{|}}{C}CH\!=\!CH_2 \xrightarrow[\substack{\text{reaction carried out}\\ \text{in an emulsion}}]{\text{radical initiator}}$$

chloroprene

1,4-addition   1,4-addition   1,2-addition   1,4-addition

a fragment of poly(chloroprene)
neoprene

The portions of the chain that result from 1,2-addition to the butadiene contain chlorine atoms that are on a tertiary and allylic carbon atom. Heating causes isomerization of the allylic halide to the isomer in which the more stable internal double bond is present. Further heating with a metal oxide gives cross-linking of different polymer chains.

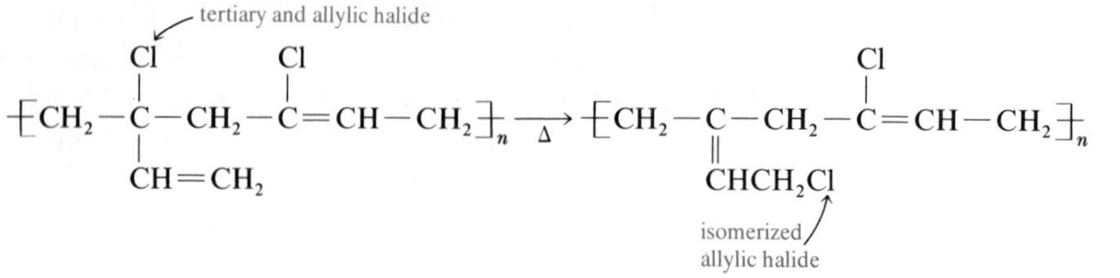

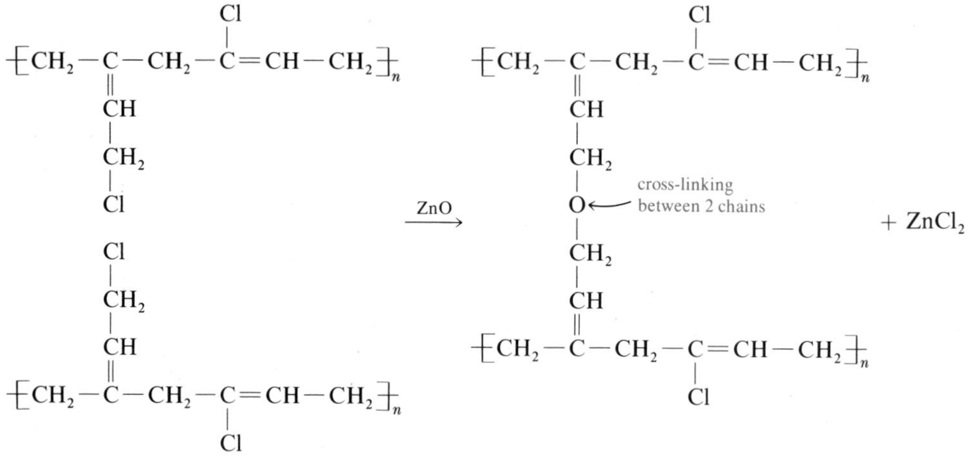

The cross-link is an ether group formed by nucleophilic displacement of the reactive allylic halides on adjacent chains. Note that the unreactive vinylic chlorine atoms are not affected either in the isomerization step or in the nucleophilic substitution reaction.

The stereochemistry of the double bond created between the second and third carbon atoms of a butadiene by 1,4-addition depends upon the reaction conditions used. The polychloroprene, neoprene, formed in emulsion by a free-radical polymerization, has mostly trans double bonds in the chain.

Neoprene is more resistant than natural rubber to oils and solvents and to reaction with ozone. It is tougher and resists wear better than rubber. It is used mostly in applications where its toughness and resistance to oil and grease are important, such as in gaskets, sealing rings, and engine mountings. It is also used in making protective gloves and aprons.

Many attempts have been made to polymerize isoprene to create a synthetic rubber that matched the properties of natural rubber. It was not until the Ziegler-Natta catalysts became available in the 1950s that isoprene was polymerized stereoselectively to give 1,4-polyisoprene with cis double bonds using one catalyst and 1,4-polyisoprene with trans double bonds using another one.

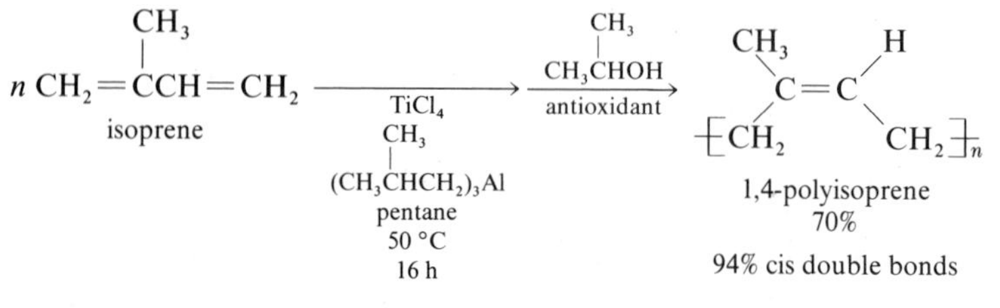

$$n\ CH_2\!=\!\underset{\underset{CH_3}{|}}{C}CH\!=\!CH_2 \xrightarrow[\text{VCl}_3\text{ on clay}]{} \xrightarrow[\text{antioxidant}]{CH_3OH} \left[\begin{array}{c} \underset{\underset{CH_2}{|}}{C}H_3\quad CH_2 \\ C\!=\!C \\ \underset{\underset{}{}}{CH_2}\qquad H \end{array}\right]_n$$

isoprene

$$\underset{(CH_3CHCH_2)_3Al}{\underset{|}{CH_3}}$$

$$\underset{(CH_3CH_2CHCH_2O)_2Ti}{\underset{|}{CH_3CH_2}}$$

benzene

1,4-polyisoprene
100%

all trans double bonds

A change of the transition metal used as the primary catalyst from titanium to vanadium changes the stereoselectivity of the polymerization reaction.

In this chapter, we have explored the chemical reactions used to transform low molecular weight organic compounds into giant molecules that imitate natural molecules in their properties and usefulness. Most of all, however, we have observed the inventiveness and ingenuity of the human mind when faced with practical problems. Human beings have a great deal of curiosity about the limits of their understanding of natural phenomena. The experimental observations that are accessible to the scientist spawn theories, which are followed by new experiments to test the theories. Polymer chemistry is an area where a questioning attitude, a constant asking of "what if?" has been particularly fruitful.

On the other hand, it must also be recognized that the age of polymers has been a mixed blessing. Plastics are useful because they have good mechanical strength and are chemically unreactive. These same properties keep plastics from degrading naturally when they are discarded. Many of them, when they burn, depolymerize to give toxic degradation products, creating especially dangerous situations when they are used in carpeting and furniture for homes and institutions. The raw materials for polymers come from petroleum, so a nonrenewable resource is being converted into plastics, many of which find only temporary use and are then discarded to create problems for the solid-waste disposal systems of towns and cities. Each one of us contributes to this waste. Perhaps the time has come to ask ourselves which applications of polymers are a wise use of our resources, and which are, at best, a minor convenience that we could do without.

Human ingenuity now has new tasks. The materials of the future must come increasingly from renewable sources and be less wasteful both of substance and of energy. Plants are especially versatile in converting the energy of the sun into a wide variety of natural products that can be transformed into other useful substances. Microorganisms, now being modified by genetic engineering, are already used in fermentation processes that convert readily available plant products such as glucose to both simpler and more complex chemicals. Organic chemists in the future will be increasingly challenged to perfect methods for the transformation of common substances isolated from natural sources into other useful products. The stimulating interaction between experimental observation and careful thinking about the implications of the observed phenomenon that is the basis of the science of organic chemistry must continue if human beings are to use the resources of the earth wisely in meeting their needs.

PROBLEM 20.19   What are the products that you would expect to get from the reaction of rubber with ozone? Does the process that takes place in the environment end up with a reductive or an oxidative decomposition of the ozonide?

PROBLEM 20.20   1,3-Butadiene, when polymerized in the presence of titanium tetrabromide and triethylaluminum, gives a 1,4-polybutadiene with mostly cis double bonds. Write an equation for this reaction.

PROBLEM 20.21   When 1-methoxy-1,3-butadiene is polymerized with a free-radical initiator, a low molecular weight amorphous polymer is obtained. The same diene, polymerized in the presence of vanadium trichloride and triisobutylaluminum, gives a highly crystalline, high molecular weight polymer with trans double bonds. Write equations showing the differences that may be present in the structures of the two polymers.

PROBLEM 20.22   A copolymer made of acrylonitrile, butadiene, and styrene is finding wide application because it combines some of the properties of rubber with the toughness of an acrylonitrile-styrene copolymer (Section 20.2A). Write an equation for the preparation of such a polymer.

ADDITIONAL PROBLEMS

20.23   Write equations predicting what the structures of the polymers resulting from the following reactions would be.

(a) 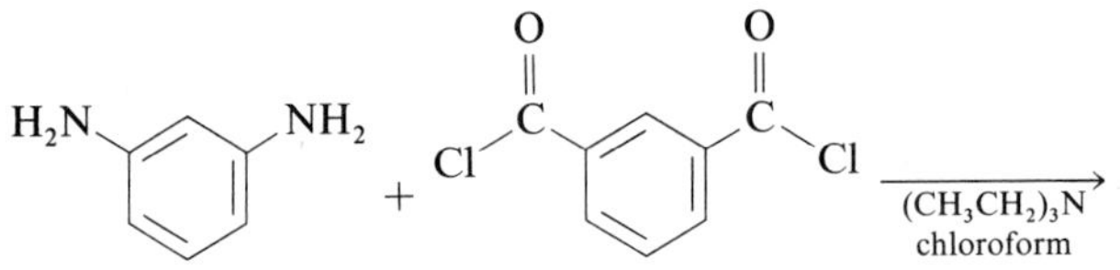 H_2N—⟨benzene ring⟩—NH_2 + Cl—C(=O)—⟨benzene ring⟩—C(=O)—Cl $\xrightarrow[\text{chloroform}]{(CH_3CH_2)_3N}$

(b) $CH_2{=}CH_2$ + $CH_2{=}\overset{\overset{\displaystyle CH_3}{|}}{\underset{\underset{\displaystyle O}{\|}}{C}}COCH_3 \xrightarrow{\text{peroxide}}$

(c) ⟨dianhydride structure⟩ + H_2N—⟨biphenyl⟩—NH_2 $\xrightarrow{\text{dimethylformamide}}$ $\xrightarrow{\Delta}$

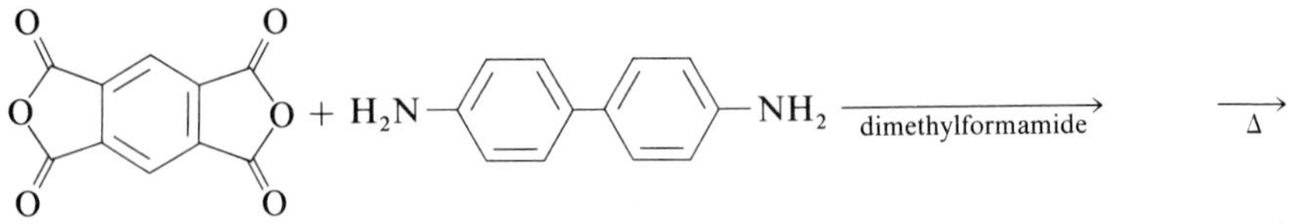

(d) 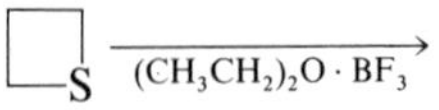 ⟨thietane ring⟩$_S$ $\xrightarrow{(CH_3CH_2)_2O \cdot BF_3}$

(e) 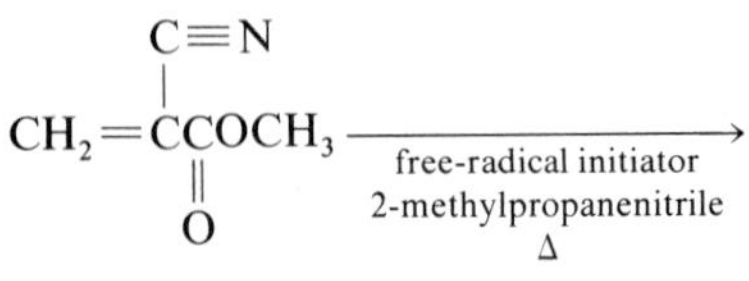 $CH_2{=}\overset{\overset{\displaystyle C{\equiv}N}{|}}{\underset{\underset{\displaystyle O}{\|}}{C}}COCH_3 \xrightarrow[\substack{\text{2-methylpropanenitrile} \\ \Delta}]{\text{free-radical initiator}}$

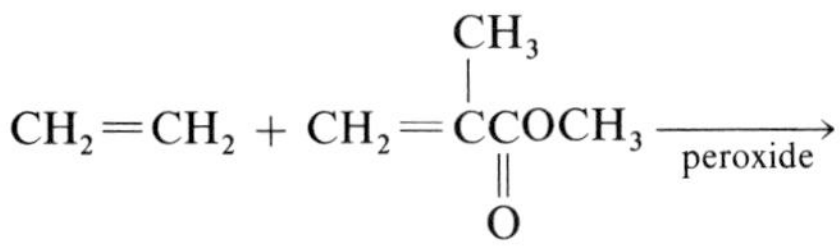

(f) 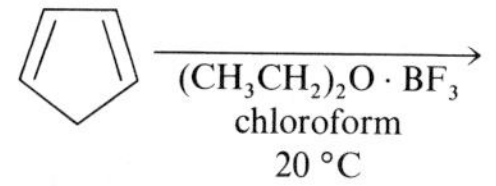 cyclopentadiene $\xrightarrow[\substack{\text{chloroform} \\ 20\,°C}]{(CH_3CH_2)_2O \cdot BF_3}$

(Hint: How does 1,3-butadiene polymerize?)

(g) 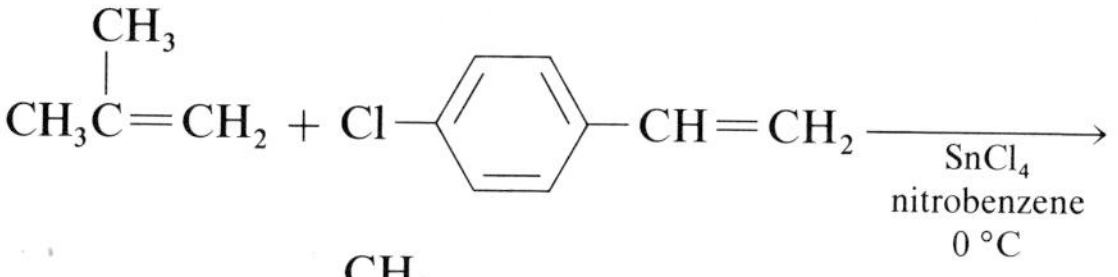 $CH_3\overset{\overset{\displaystyle CH_3}{|}}{C}=CH_2 + Cl\text{—}\underset{}{\bigcirc}\text{—}CH=CH_2 \xrightarrow[\substack{\text{nitrobenzene} \\ 0\,°C}]{SnCl_4}$

(h) 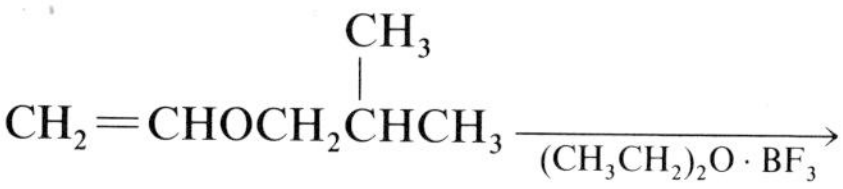 $CH_2=CHOCH_2\overset{\overset{\displaystyle CH_3}{|}}{C}HCH_3 \xrightarrow{(CH_3CH_2)_2O \cdot BF_3}$

(i) 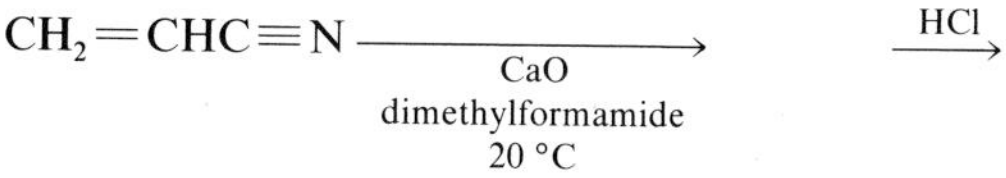 $CH_2=CHC\equiv N \xrightarrow[\substack{\text{dimethylformamide} \\ 20\,°C}]{CaO} \xrightarrow{HCl}$

(j) 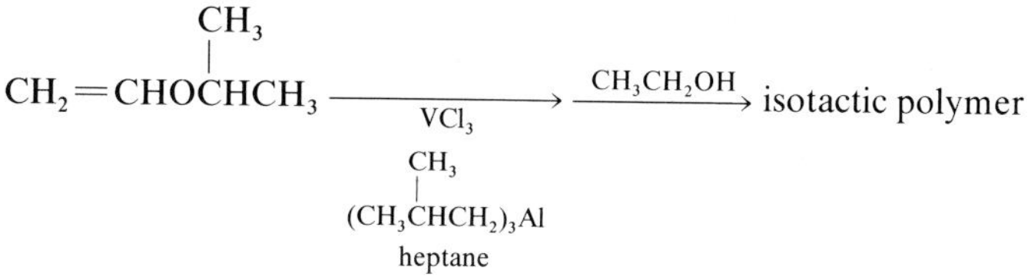 $CH_2=CHO\overset{\overset{\displaystyle CH_3}{|}}{C}HCH_3 \xrightarrow[\substack{CH_3 \\ | \\ (CH_3CHCH_2)_3Al \\ \text{heptane}}]{VCl_3} \xrightarrow{CH_3CH_2OH}$ isotactic polymer

(k) 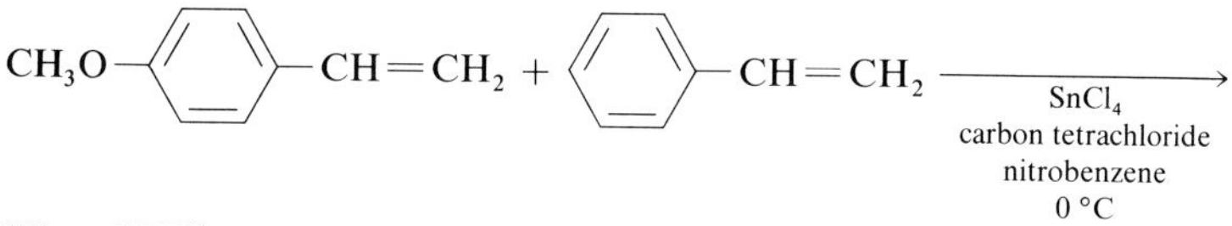 $CH_3O\text{—}\bigcirc\text{—}CH=CH_2 + \bigcirc\text{—}CH=CH_2 \xrightarrow[\substack{\text{carbon tetrachloride} \\ \text{nitrobenzene} \\ 0\,°C}]{SnCl_4}$

(l) 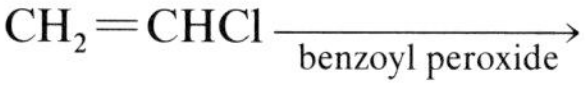 $CH_2=CHCl \xrightarrow{\text{benzoyl peroxide}}$

(m) 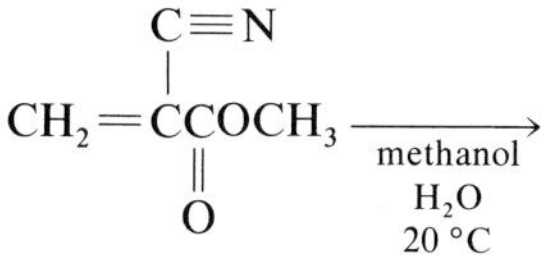 $CH_2=\overset{\overset{\displaystyle C\equiv N}{|}}{\underset{\underset{\displaystyle O}{\|}}{C}}COCH_3 \xrightarrow[\substack{\text{methanol} \\ H_2O \\ 20\,°C}]{}$

(n) $HO\text{—}\bigcirc\text{—}\overset{\overset{\displaystyle CH_3}{|}}{\underset{\underset{\displaystyle CH_3}{|}}{C}}\text{—}\bigcirc\text{—}OH + Cl\overset{\overset{\displaystyle O}{\|}}{C}Cl \xrightarrow[H_2O]{NaOH}$

(o) $CH_2=CH_2 + \bigcirc \xrightarrow[\substack{(hexyl)_3Al \\ \text{heptane}}]{VCl_4} \xrightarrow{CH_3CH_2OH}$

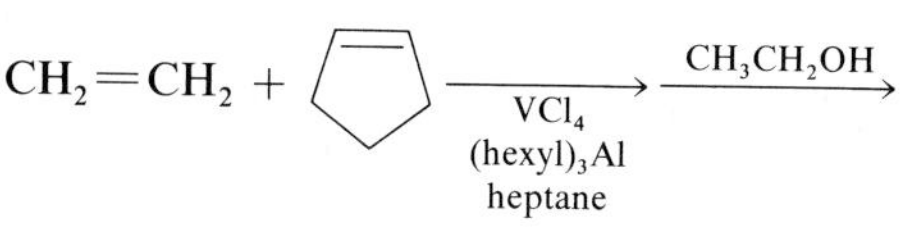

(p) 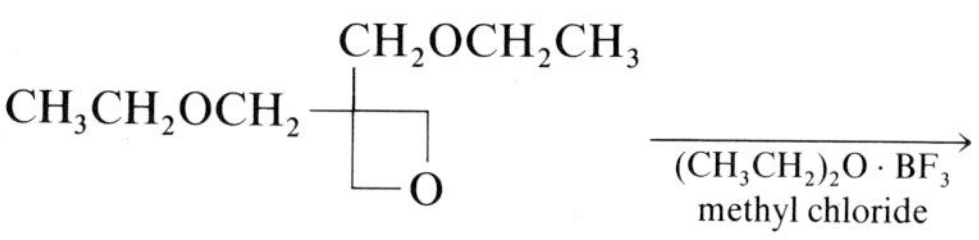 $CH_3CH_2OCH_2\text{—}\square\text{—}CH_2OCH_2CH_3$ (oxetane with O at bottom) $\xrightarrow[\text{methyl chloride}]{(CH_3CH_2)_2O \cdot BF_3}$

(q) 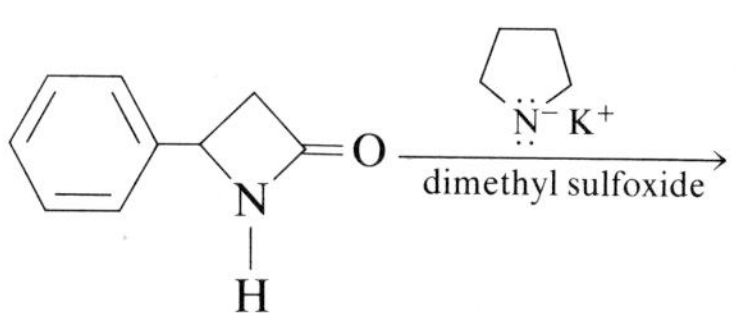 phenyl-β-lactam (4-phenylazetidin-2-one) $\xrightarrow[\text{dimethyl sulfoxide}]{\overset{\displaystyle \ddot{N}^-K^+}{\text{pyrrolidide}}}$

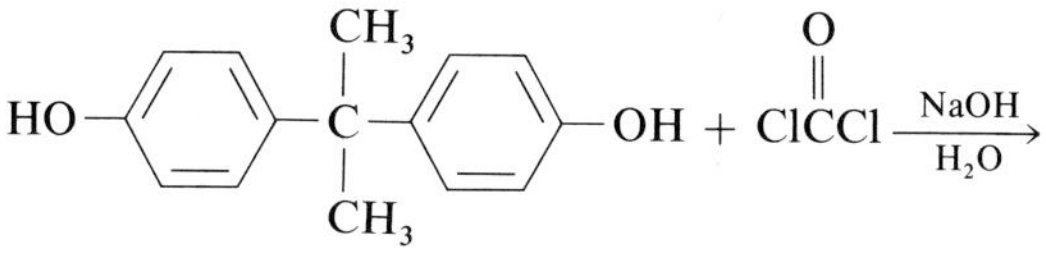

20.24　Unsaturated esters having the general formula $CH_2{=}CHCOR$ and acrylonitrile, $CH_2{=}CHC{\equiv}N$, are successfully polymerized in chain reactions using free-radical or anionic initiators. Polymerization reactions do not take place when cationic initiators are used. How would you rationalize these experimental observations?

20.25　The reactivity of an alkene monomer in a free-radical polymerization reaction as measured by the rate of the propagation step depends upon the substituent that is on the double bond. By this measure, the following order of reactivity has been observed.

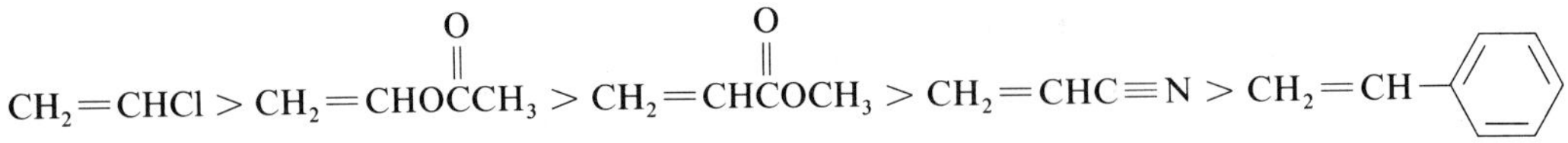

$$CH_2{=}CHCl > CH_2{=}CHO\overset{O}{\overset{\|}{C}}CH_3 > CH_2{=}CH\overset{O}{\overset{\|}{C}}OCH_3 > CH_2{=}CHC{\equiv}N > CH_2{=}CH{-}\langle\text{phenyl}\rangle$$

How would you rationalize this order of reactivity?

20.26　(a)　The following polyimide was developed to create a plastic film that retains its usefulness over a very wide range of temperatures. The repeating unit of the polymer is shown below. What monomeric units would you need to synthesize this polymer?

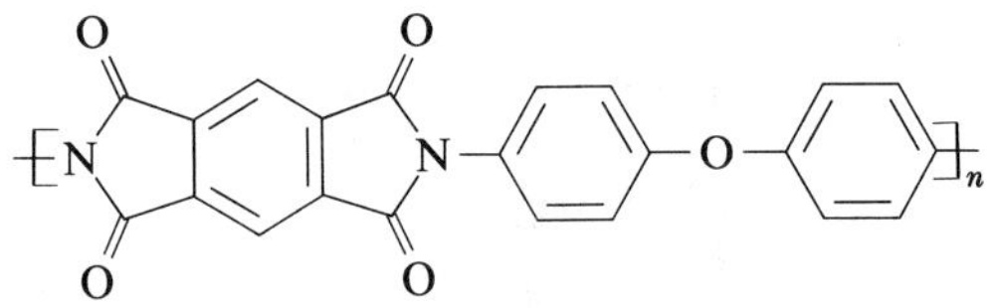

(b)　The polymer above can be transformed into another more soluble polymer by reaction with amines. Write an equation predicting what would happen to the polymer if it were treated with 2-(N-methylamino)ethanol, $CH_3NHCH_2CH_2OH$.

20.27　Ion-exchange resins containing sulfonic acid groups are strongly acidic. Weakly acidic ion-exchange resins have also been made. One of them has the following partial structure. How would you make such a resin?

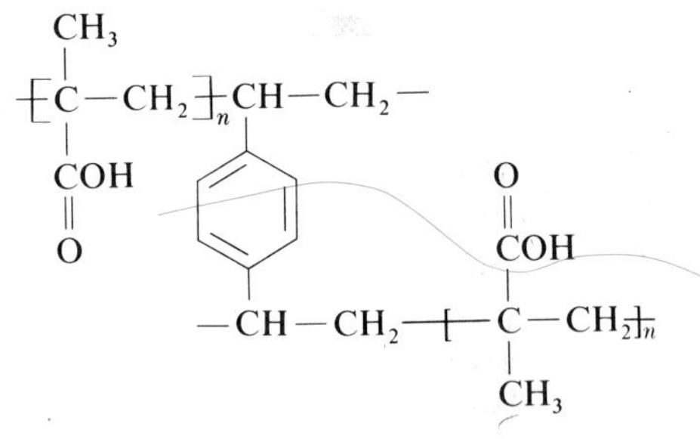

20.28　In the presence of a free-radical initiator, such as high-energy γ-radiation, acrylamide polymerizes at the double bond to give a hydrocarbon chain with amide substituents on it. When a strong base is used to catalyze the polymerization of acrylamide, however, a polyamide that can be hydrolyzed to 3-aminopropanoic acid (β-alanine) is obtained. Write mechanisms that explain the difference in products for these two reactions, shown below.

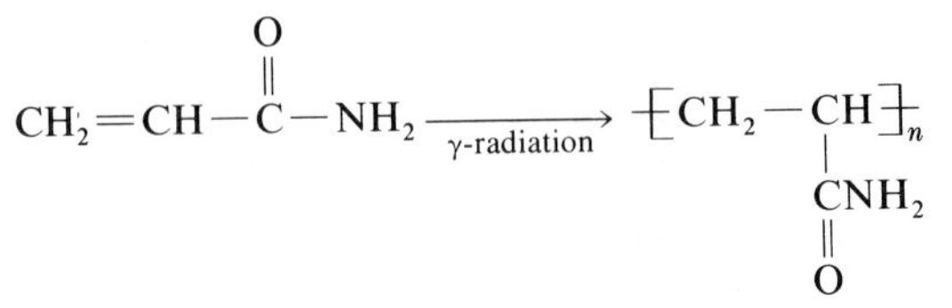

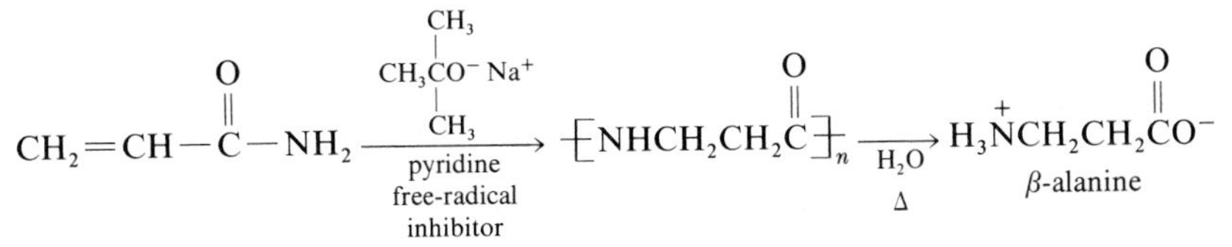

$$CH_2{=}CH{-}\underset{\underset{\text{O}}{\|}}{C}{-}NH_2 \xrightarrow[\substack{\text{pyridine}\\\text{free-radical}\\\text{inhibitor}}]{\substack{CH_3\\|\\CH_3CO^-\ Na^+\\|\\CH_3}} \underset{n}{\left[\text{NHCH}_2\text{CH}_2\underset{\underset{\text{O}}{\|}}{C}\right]} \xrightarrow[\Delta]{H_2O} \underset{\beta\text{-alanine}}{H_3\overset{+}{N}CH_2CH_2\underset{\underset{\text{O}}{\|}}{C}O^-}$$

20.29   A polyester, polypivalolactone, has a particularly highly oriented fiber, which recovers very quickly from deformation. This is a desirable property in fibers used in carpets, otherwise the carpet shows footprints. The polyester is synthesized from either one of the following monomers. Suggest a method for synthesizing the polymer from each of the monomers. For example, what type of catalyst would you use, and what reaction conditions would be necessary for each process?

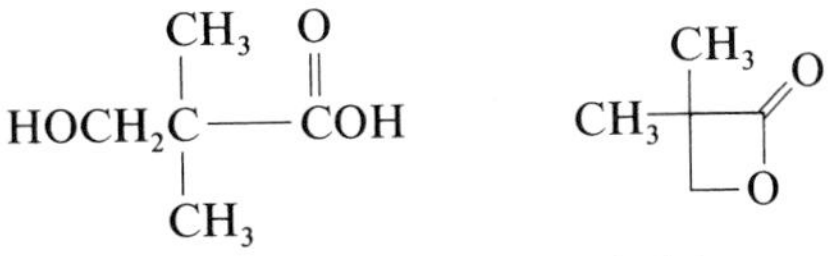

$$\underset{\underset{\text{3-hydroxy-2,2-dimethylpropanoic}}{\text{acid}}}{HOCH_2\overset{\overset{CH_3}{|}}{\underset{\underset{CH_3}{|}}{C}}{-}\overset{\overset{\text{O}}{\|}}{C}OH} \qquad \underset{\text{pivalolactone}}{CH_3\text{—}}$$

20.30   The polymerization of (S)-$(-)$-2-methyloxirane, when catalyzed by solid potassium hydroxide, gives a crystalline, optically active polymer. If the polymerization of the chiral oxirane is catalyzed by iron(III) chloride in ether, however, the resulting polymer is not optically active. Propose mechanisms for the two different types of polymerizations that account for the different stereochemistry observed.

20.31   One of the new polyamides that has high molecular rigidity has the following repeating unit.

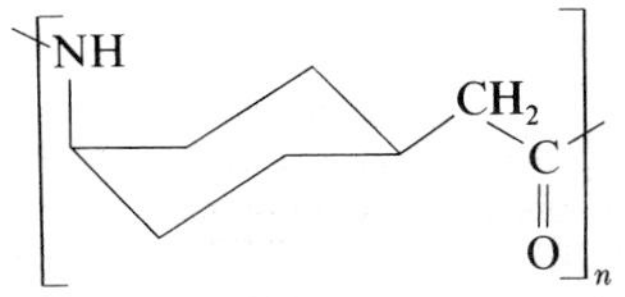

$$\left[\underset{\text{C}}{\overset{\text{NH}}{}}\cdots\underset{\underset{\underset{\text{O}}{\|}}{C}}{CH_2}\right]_n$$

It is prepared by the polymerization of a bicyclic lactam.

(a)   What is the structure of a lactam that would give rise to the polymer shown above?

(b)   This lactam is synthesized in a series of steps starting with 1,3-cyclohexadiene and vinyl acetate and ending with a Beckmann rearrangement. Propose a synthesis for the lactam.

20.32   Under the proper conditions, the carbon-oxygen double bond of a carbonyl group takes part in polymerization reactions. The two reactions shown below have been observed. Write mechanisms for each one showing why the polymerization proceeds as it does.

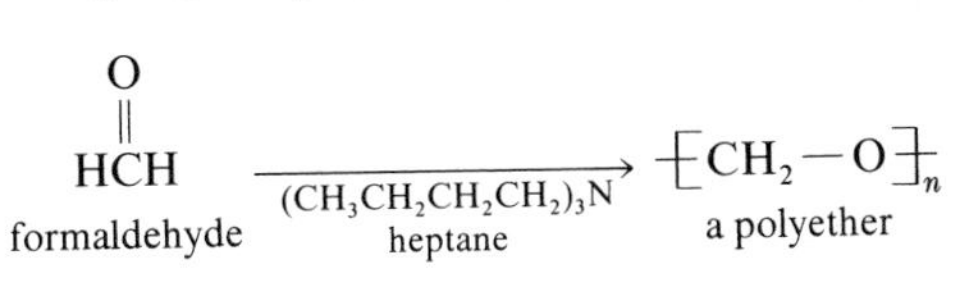

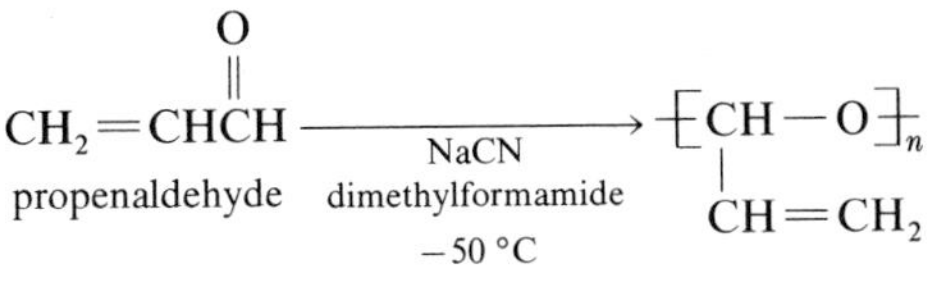

$$\underset{\text{formaldehyde}}{\overset{\overset{\text{O}}{\|}}{H}CH} \xrightarrow[\text{heptane}]{(CH_3CH_2CH_2CH_2)_3N} \underset{n}{\underset{\text{a polyether}}{\left[CH_2{-}O\right]}}$$

$$\underset{\text{propenaldehyde}}{CH_2{=}CH\overset{\overset{\text{O}}{\|}}{C}H} \xrightarrow[\substack{\text{dimethylformamide}\\-50\ °C}]{NaCN} \underset{n}{\left[\underset{\underset{CH{=}CH_2}{|}}{CH}{-}O\right]}$$

Index

Boldface page number indicates text page on which indexed term is defined.

Characteristic Infrared Absorption Frequencies

Bond Type	Stretching, cm^{-1}	Bending, cm^{-1}
C—H alkanes	2960–2850 (s)	1470–1350 (s)
C—H alkenes	3080–3020 (m)	1000–675 (s)
C—H aromatic	3100–3000 (v)	870–675 (v)
C—H aldehyde	2900, 2700 (m)	
C—H alkyne	3300 (s)	
C≡C alkyne	2260–2100 (v)	
C≡N nitrile	2260–2220 (v)	
C=C alkene	1680–1620 (v)	
C=C aromatic	1600–1450 (v)	
C=O ketone	1725–1705 (s)	
C=O aldehyde	1740–1720 (s)	
C=O α, β-unsaturated ketone	1685–1665 (s)	
C=O aryl ketone	1700–1680 (s)	
C=O ester	1750–1735 (s)	
C=O acid	1725–1700 (s)	
C=O amide	1690–1650 (s)	
O—H alcohols (not hydrogen bonded)	3650–3590 (v)	
O—H hydrogen bonded	3600–3200 (s, broad)	
O—H acids	3000–2500 (s, broad)	
N—H amines	3500–3300 (m)	1620–1590 (v)
N—H amides	3500–3350 (m)	1655–1510 (s)
C—O alcohols, ethers, esters	1300–1000 (s)	
C—N amines, aliphatic	1220–1020 (w)	
C—N amines, aromatic	1360–1250 (s)	
NO_2 nitro	1560–1515 (s)	
	1385–1345 (s)	

s = strong absorption w = weak absorption
m = medium absorption v = variable absorption

Typical Chemical Shifts in Proton Magnetic Resonance

Type of Hydrogen Atom	δ, ppm*	Type of Hydrogen Atom	δ, ppm*
RCH_3	0.9	$R_2C{=}CH_2$	5.0
RCH_2R acyclic	1.3	$RCH{=}CR_2$	5.3
cyclic	1.5	ArH	7.3
R_3CH	1.5–2.0	$\overset{O}{\overset{\|}{RCH}}$	9.7
$R_2C{=}\underset{\underset{R'}{\|}}{C}CH_3$	1.8	RNH_2	1–3
		$ArNH_2$	3–5
$\overset{O}{\overset{\|}{RCCH_3}}$	2.0–2.3	$\overset{O}{\overset{\|}{RCNHR}}$	5–9
$ArCH_3$	2.3	ROH	5
$RC{\equiv}CH$	2.5	ArOH	7
$RNHCH_3$	2–3	$\overset{O}{\overset{\|}{RCOH}}$	11
RCH_2X (X = Cl, Br, I)	3.5		
$ROCH_3$, $\overset{O}{\overset{\|}{RCOCH_3}}$	3.8		

*The chemical shift values are given in ppm relative to tetramethylsilane at 0.00 ppm and are for the hydrogen atoms shown in color in the formulas. The values for hydrogen atoms on oxygen and nitrogen are highly dependent on solvent, concentration, and temperature.